Log on.

Tune in.

Succeed.

Your steps to success.

STEP 1: Register

All you need to get started is a valid email address and the access code below. To register, simply:

1. Go to www.mybiology.com
2. Click the appropriate book cover.
   Cover must match the textbook edition being used for your class.
3. Click **"Register"** under **"First-Time User?"**
4. Leave **"No, I Am a New User"** selected.
5. Using a coin, scratch off the silver coating below to reveal your access code.
   Do not use a knife or other sharp object, which can damage the code.
6. Follow the on-screen instructions to complete registration. You may want to click Log In Now on your registration confirmation screen and bookmark that page.

   During registration, you will establish a personal login name and password to use for logging into the website. You will also be sent a registration confirmation email that contains your login name and password.

Your Access Code is:

Note: If there is no silver foil covering the access code, it may already have been redeemed, and therefore may no longer be valid. In that case, you can purchase access online using a major credit card. To do so, go to www.mybiology.com and click "Buy Access" and follow the on-screen instructions.

STEP 2: Log in

1. Go to www.mybiology.com and **"Log In."**
2. Click your book cover and under enter the login name and password that you created during registration. Cover must match the textbook edition being used for your class. If unsure of this information, refer to your registration confirmation email.
3. Click **"Log In."**

STEP 3: (Optional) Join a class

Instructors have the option of creating an online class for you to use with this website. If your instructor decides to do this, you'll need to complete the following steps using the Class ID your instructor provides you. By "joining a class," you enable your instructor to view the scored results of your work on the website in his or her online gradebook.

To join a class:

1. Log into the website. For instructions, see "STEP 2: Log in."
2. Click **"Join a Class"** near the top right.
3. Enter your instructor's **"Class ID"** and then click **"Next."**
4. At the Confirm Class page you will see your instructor's name and class information. If this information is correct, click Next.
5. Click **"Enter Class Now"** from the Class Confirmation page.
- *To confirm your enrollment in the class, check for your instructor and class name at the top right of the page. You will be sent a class enrollment confirmation email.*
- *As you complete quizzes on the website from now through the class end date, your results will post to your instructor's gradebook, in addition to appearing in your personal view of the Results Reporter.*

To log into the class later, follow the instructions under "STEP 2: Log in."

Got technical questions?

SITE REQUIREMENTS

For the latest updates on Site Requirements, go to www.mybiology.com, choose your text cover, and click Site Reqs.

WINDOWS
OS: Windows XP
Resolution: 1024 x 768
Plugins: Latest version of Flash/QuickTime/Shockwave (as needed)
Browsers: Internet Explorer 6.0 (XP only); Internet Explorer 7.0; Firefox 2.0.
Internet connection: 56K minimum

MACINTOSH
OS: 10.3.x; 10.4
Resolution: 1024 x 768
Plugins: Latest version of Flash/QuickTime/Shockwave (as needed)
Browsers: Safari 1.3 (10.3.x only); Safari 2.0 (10.4.x only); Firefox 2.0
Internet connection: 56K minimum

Register and log in

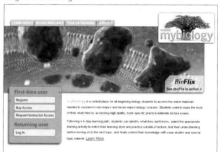

Join a class

Features of the Student Website for *Biology: Concepts & Connections,* **Sixth Edition**

Chapter Guide

The Chapter Guide organizes media under central concepts that correlate directly to the text.

Bioflix™

These 3-D animations and interactive, self-paced tutorials bring difficult biological concepts to life. Topics include cellular respiration, mitosis, and meiosis. Prepare for exams by completing a tutorial, printing a personal review sheet, and taking a quiz.

MP3 Tutors

Study anytime, anywhere, with these downloadable audio tutorials narrated by the author, Eric Simon. Nineteen new MP3 Tutors focus on tough terms and provide a unique alternative to flashcards.

The New York Times Current Event Articles

Current articles are posted monthly along with quiz questions to help you link the topics you study in the course to the latest news coverage.

Discovery Channel Video Clips

These brief two-five minutes video clips from the Discovery Channel cover topics from fighting cancer to antibiotic resistance to introduced species.

Activities

Explore approximately 200 activities, including animations, interactive review exercises, and videos. Test your knowledge with Activities Quizzes.

Process of Science Investigations

Perform fifty-six virtual investigations that develop data collection and analysis skills.

Quizzes

Assess your understanding with over 3,600 multiple-choice questions. Each chapter includes a Pre-Test to diagnose current knowledge, an Activities Quiz with graphics that tests understanding of the media activities in the chapter, and a comprehensive Post-Test (50 questions per chapter on average). Each quiz has hints, immediate feedback, grading, and e-mailable results.

Cumulative Quizzes

Create customized tests on multiple chapters at once by choosing the chapters and number of questions.

E-book

Navigate the on-line e-Book to electronically view and search the textbook and related images.

Art and Videos

View all the art from the textbook and a collection of 85 videos.

Word Study Tools

Use electronic Flash Cards to test your knowledge of the key terms and definitions for each chapter or multiple chapters. Includes audio pronunciations for selected terms. Key Terms allow you to study the list of key terms for each chapter, looking up definitions and hearing selected audio pronunciations. Word roots are also provided to improve vocabulary skills. The Glossary provides definitions of boldface terms and selected audio pronunciations.

Web Links and References

Organized by chapter, Web Links and References allow you to extend your knowledge through links to relevant websites, access news links on recent developments in biology, and consult an archive of relevant biology news articles and lists of further readings for each chapter.

BIOLOGY

CONCEPTS & CONNECTIONS

SIXTH EDITION

NEIL A. CAMPBELL

JANE B. REECE | *Berkeley, California*

MARTHA R. TAYLOR | *Ithaca, New York*

ERIC J. SIMON | *New England College*

JEAN L. DICKEY | *Clemson University*

PEARSON

Benjamin
Cummings

San Francisco Boston New York Cape Town Hong Kong London Madrid
Mexico City Montreal Munich Paris Singapore Sydney Tokyo Toronto

Editor in Chief:
Beth Wilbur

Senior Acquisitions Editor:
Chalon Bridges

**Executive Director
of Development:**
Deborah Gale

Senior Editorial Manager:
Ginnie Simione Jutson

Senior Development Editors:
*Evelyn Dahlgren and
Susan Teahan*

Development Editor:
Debbie Hardin

Project Manager:
Mary Douglas

Art Development:
Precision Graphics

**Senior Supplements
Project Editor:**
Susan Berge

Assistant Editor:
Rebecca Johnson

Executive Marketing Manager:
Lauren Harp

Senior Marketing Manager:
Jay Jenkins

Managing Editor:
Mike Early

Production Supervisor:
Lori Newman

Production Management:
*Progressive Publishing
Alternatives*

Compositor:
*Progressive Information
Technologies*

Art Coordinator:
*Progressive Publishing
Alternatives*

Interior Designer:
Jana Anderson

Cover Designer:
*Riezebos Holzbaur Design
Group*

Art and Design Director:
Mark Ong

Design Manager:
Marilyn Perry

Senior Photo Editor:
Donna Kalal

Illustrators:
*Greg Williams and
Precision Graphics*

Photo Researcher:
Kristin Piljay

**Director, Image Resource
Center:**
Melinda Patelli

**Image Rights and Permissions
Manager:**
Zina Arabia

**Image Permissions
Coordinator:**
Michelina Viscusi

Manufacturing Buyer:
Michael Penne

Text printer:
*RR Donnelley and Sons,
Willard*

Cover printer:
Phoenix Color Corp.

Cover Photo Credit:
Leopard (*Panthera pardus*) in grass, looking up, in
Sabi Sands Conservancy, Mpumalanga Province,
South Africa; Heinrich van den Berg, Getty
Images

Library of Congress Cataloging-in-Publication Data

Biology : concepts and connections / Neil A. Campbell . . . [et al.].—6th ed.
   p. cm.

Includes index.

ISBN 978-0-321-48984-5

1. Biology—Textbooks. I. Campbell, Neil A., 1946–2004

QH308.2.B56448 2008
570—dc22   2007052356

ISBN 0321489845 / 9780321489845 (Student edition)
ISBN 0321547918 / 9780321547910 (Professional copy)
ISBN 013135566X / 9780131355668 (NASTA)
ISBN 0321547896 / 9780321547897 (Books a la Carte)

1 2 3 4 5 6 7 8 9 10—**DOW**—12 11 10 09 08
www.pearsonhighered.com

Benjamin Cummings
1301 Sansome St.
San Francisco, CA 94111
www.pearsonhighered.com

About the Authors

Neil A. Campbell (1946–2004) taught general biology for over 30 years and with Dr. Reece coauthored *Biology,* now in its eighth edition and the most widely used text for biology majors. His enthusiasm for sharing the fun of science with students stemmed from his own undergraduate experience. He began at Long Beach State College as a history major, but switched to zoology after general education requirements "forced" him to take a science course. Following a B.S. from Long Beach, he earned an M.A. in zoology from UCLA and a Ph.D. in plant biology from the University of California, Riverside. He published numerous articles on how certain desert plants thrive in salty soil and how the sensitive plant (*Mimosa*) and other legumes move their leaves. His diverse teaching experiences included courses for nonbiology majors at Cornell University, Pomona College, and San Bernardino Valley College, where he received the first Outstanding Professor Award in 1986. For many years, Dr. Campbell was a visiting scholar in the Department of Botany and Plant Sciences at UC Riverside, which recognized him as the university's Distinguished Alumnus for 2001. In addition to *Biology,* Dr. Campbell coauthored *Essential Biology* and *Essential Biology with Physiology.* Neil Campbell died in October of 2004. While he is greatly missed by his many friends throughout the biology community, his coauthors remain inspired by his visionary dedication to education and committed to searching for ever better ways to engage students in the wonders of biology.

Jane B. Reece has worked in biology publishing since 1978, when she joined the editorial staff of Benjamin Cummings. Her education includes an A.B. in biology from Harvard University (where she was initially a philosophy major), an M.S. in microbiology from Rutgers University, and a Ph.D. in bacteriology from the University of California, Berkeley. At UC Berkeley, and later as a postdoctoral fellow in genetics at Stanford University, her research focused on genetic recombination in bacteria. Dr. Reece taught biology at Middlesex County College (New Jersey) and Queensborough Community College (New York). During her 12 years as an editor at Benjamin Cummings, she played a major role in a number of successful textbooks. She is a coauthor of *Biology, Essential Biology,* and *Essential Biology with Physiology.*

Martha R. Taylor has been teaching biology for over 30 years. She earned her B.A. in biology from Gettysburg College. After teaching biology in high school and community college, she went on to earn her M.S. and Ph.D. in science education from Cornell University. She was assistant director of the Office of Instructional Support at Cornell for seven years. She has taught introductory biology for both majors and nonmajors at Cornell University for many years. Based on her experiences working with students in classrooms, laboratories, and tutorials, Dr. Taylor is committed to helping students create their own knowledge of and appreciation for biology. She has been the author of the *Student Study Guide* for all eight editions of *Biology* by Drs. Campbell and Reece.

Eric J. Simon is an associate professor of biology at New England College in Henniker, New Hampshire. He teaches introductory biology to both science majors and nonscience majors, as well as upper-level courses in genetics, microbiology, and molecular biology. Dr. Simon received a B.A. in biology and computer science and an M.A. in biology from Wesleyan University and a Ph.D. in biochemistry from Harvard University. His research focuses on innovative ways to use technology to improve teaching and learning in the science classroom, particularly for nonscience majors. Dr. Simon is also a coauthor of *Essential Biology* and *Essential Biology with Physiology.*

Jean L. Dickey is a professor of biology at Clemson University. She had no idea that science was interesting until her senior year in high school, when a scheduling problem landed her in advanced biology. Abandoning plans to study English or foreign languages, she enrolled in Kent State University as a biology major. After receiving her B.S. in biology, she went on to earn a Ph.D. in ecology and evolution from Purdue University. Since joining the faculty at Clemson in 1984, Dr. Dickey has specialized in teaching nonscience majors, including a course designed for preservice elementary teachers and workshops for in-service teachers. She also developed an investigative laboratory curriculum for general biology and is the author of *Laboratory Investigations for Biology.*

To the student: *How to use this book*

Introduce yourself to the chapter.

Find out where you're going. Use the ***chapter outline*** to preview the chapter.

Focus on what's most important.

Get the big picture. Look for the ***main headings*** (orange bars) that organize the chapter into major sections.

Understand biology one concept at a time. Each module features a ***central concept***, announced in its heading.

Use both text and figures as you study. The ***figures*** illuminate the text and vice versa. Text and figures are always together—so you'll never have to turn a page to find what you need.

---

chapter 1

Biology: *Exploring Life*

Themes in the Study of Biology

1.1 In life's hierarchy of organization, new properties emerge at each level

1.2 Living organisms interact with their environments, exchanging matter and energy

1.3 Cells are the structural and functional units of life

Evolution, the Core Theme of Biology

1.4 The unity of life: All forms of life have common features

1.5 The diversity of life can be arranged into three domains

1.6 Evolution explains the unity and diversity of life

The Process of Science

1.7 Scientists use two main approaches to learn about nature

1.8 With hypothesis-based science, we pose and test hypotheses

Biology and Everyday Life

1.9 Biology, technology, and society are connected in important ways

1.10 Evolution is connected to our everyday lives

Cloning of Plants and Animals

11.14 Plant cloning shows that differentiated cells may retain all of their genetic potential

One of the most important "take home lessons" from this chapter is that differentiated cells express only a small percentage of their genes. So then how do we know that all the genes are still present? And if all the genes are still there, do differentiated cells retain the potential to express them?

One way to approach these questions is to see if a differentiated cell can generate a whole new organism. In plants, this ability is common, as was first demonstrated during the 1950s by F. C. Steward and his students at Cornell University. As shown in Figure 11.14, they found that when they transferred cells from a carrot to a culture medium, a single cell could begin dividing and eventually grow into an adult plant, a genetic replica of the parent plant. The fact that a mature plant cell can dedifferentiate (reverse its differentiation) and then give rise to all the different kinds of specialized cells of a new plant shows that differentiation does not necessarily involve irreversible changes in the plant's DNA.

Plant cloning is now used extensively in agriculture. For some plants, such as orchids, cloning is the only commercially practical means of reproducing plants. In other cases, cloning has been used to reproduce a plant with valuable characteristics, such as the ability to resist a plant pathogen. You yourself may be a plant cloner if you have ever grown a new plant from a cutting.

But is this sort of cloning possible in animals? An indication that differentiation need not impair an animal cell's genetic potential is the natural process of **regeneration**, the regrowth of lost body parts. When a salamander loses a leg, for example, certain cells in the leg stump dedifferentiate, divide, and then

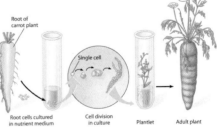

Figure 11.14 Growth of a carrot plant from a differentiated root cell

redifferentiate, giving rise to a new leg. Many animals can regenerate lost parts, especially among the invertebrates, and in a few relatively simple animals, isolated differentiated cells can dedifferentiate and then develop into an organism. Further evidence for the complete genetic potential of animal cells comes from cloning experiments, our next topic.

? How does the cloning of plants from differentiated cells support the view that differentiation is based on the control of gene expression rather than on irreversible changes in the genome?

■ Cloning shows that all the genes of a fully differentiated plant are still present, but some may be turned off.

Dining in the Trees

female leopard (*Panthera pardus*) and her young daughter have
st killed a duiker, a small antelope and a preferred leopard meal.
he pair could start their feast on the spot. But the mother instinc-
vely knows that eating in the open might allow other predators,
ich as lions or hyenas, to steal their kill. The duiker is best con-
med up in a tree, out of reach of thieves. The older leopard has
auled the family's prey up into the branches many times before.
ut this afternoon, the cub follows her mother's example and car-
es the meal herself. The mother, featured on the cover of this
ook, looks up from below. She'll wait to eat, watching first to see
ow well her daughter has learned the lessons she'll need to survive.

Leopards are the strongest climbers among the big cats that
ake up the genus *Panthera*. Scientific evidence indicates that
embers of *Panthera*, which include leopards, lions, tigers, and
guars, last shared a common ancestor about 6 million years
go. This linkage still shows in these cats' large size, hunting
rowess, and unique ability to roar. The leopard species, which
found across Africa and Asia, has the widest distribution of
ese big cats and displays many adaptations that enhance its
urvival.

African leopards, for example, have features that help them
rive alongside Africa's regal lion. Leopards tend to be nocturnal
unters, whereas lions hunt during the day. And leopards take a
uch larger variety of prey ite
about 58 kilograms (128 p
4 pounds) for females, Afri
e size of a typical lion. But
unning speeds of 64 kilomet
perior jumping skills that p
rough the air horizontally an
exceptional climber. Power
ve the leopard the ability to p
d the additional strength n
mes dragging a carcass as hig

Other characteristics of th
nerally loners and maintair
es. Thus, they do not compet
hiskers on the face and head
getation while hunting at ni
at, often an array of black

ou can review metric measuremen

ochre fur, provides excellent camouflage in the dry grasses of the
African plains. Leopards also thrive in desert areas, hilly country,
or jungle terrain, and their coat color can vary from pale cream to
deep gold, enabling them to blend in with their surroundings.

This habitat flexibility of leopards is linked to their amazing
diet flexibility. Leopards consume everything from beetles and
birds to baboons, hares, and smaller cats. Antelopes such as the
duiker or the larger impala are a frequent favorite among leopards
living in the African savannah. Observers have noted leopards
killing impala larger than themselves and still managing to drag
the carcass high into tree branches.

So far, the African leopard has fared better in the modern era

Discover. The *opening essays* introduce the
chapter topic through stories that will pique
your curiosity.

Never get lost. *Figures* describing a process
take you through a series of numbered steps keyed
to explanations in the text.

Interact. *Media references* direct
you to related media resources
including BioFlix™ 3-D Animations,
MP3 Tutors, and Web Activities.

Test yourself. Get immediate
feedback with a *checkpoint question*
at the end of each module.

11.7 Small RNAs play multiple roles in controlling gene expression

Recall that only 1.5% of the human genome—and a similarly
small percentage of the genomes of many other multicellular eukaryotes—codes for proteins. Another very small fraction of
DNA consists of genes for ribosomal RNA and transfer RNA.
Until recently, most of the remaining DNA was considered to be
"noncoding," meaning that it neither coded for proteins nor was
transcribed into functional RNA of the few known types. In
other words, it was thought not to contain meaningful genetic
information. However, a flood of recent data has contradicted
this view. Biologists currently think that a significant amount of
the genome may be transcribed into non-protein-coding RNAs,
including a variety of small RNAs. While many questions about
the functions of these RNAs remain unanswered, researchers are
uncovering more evidence of their biological roles every day.

In 1993, researchers discovered small single-stranded RNA
molecules, called microRNAs (miRNAs), that can bind to complementary sequences on mRNA molecules (Figure 11.7).
Each miRNA, typically about 20 nucleotides long, ❶ associates
with a large protein complex. The complex can ❷ bind to any
mRNA molecule with the complementary sequence. Then the
miRNA-protein complex either ❸ degrades the target mRNA
or ❹ blocks its translation. It has been estimated that miRNAs
may regulate the expression of up to one-third of all human
genes, a striking figure given that miRNAs were unknown a
mere 20 years ago.

Researchers can take advantage of the miRNA mechanism
to artificially control gene expression. For example, injecting
miRNAs into a cell can turn off expression of a gene with a
sequence that matches the miRNA. This procedure is called
RNA interference (RNAi). As you learned in Chapter 10, some
viruses that infect cells have double-stranded RNA genomes.
The RNAi pathway may have evolved as a natural defense
against infection by such viruses.

Biologists are excited about these recent discoveries, which
hint at a large, diverse population of RNA molecules in the cell

that play crucial roles in regulating gene expression—and have
gone largely unnoticed until now. Clearly, we must revise the
long-standing view that because they code for proteins,
mRNAs are the most important RNAs in terms of cellular
function.

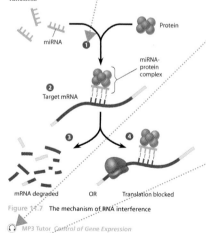

Figure 11.7   The mechanism of RNA interference

🎧 MP3 Tutor *Control of Gene Expression*

❓ If a gene has the sequence AATTCGCG, what would be
the sequence of an miRNA that turns off the gene?

■ The gene will be transcribed as the mRNA sequence UUAAGCGC, an
miRNA of sequence AAUUCGCG would bind to and disable this mRNA.

To the student: *How to use this book*
(CONTINUED)

Learn about biology in your world.

Make a connection. *Connection modules* relate biology to your life and interests—from global warming to blood doping to sperm motility to sustainable agriculture.

Understand how evolution works.
NEW! *Evolution Connection modules in every chapter* relate evolution to a wide spectrum of biology topics and help explain the mechanisms underlying evolution and the evidence for it.

Meet the people behind the science. Scientists discuss their careers, their research, and their thoughts about the future in *Talking About Science modules.*

7.13 Photosynthesis moderates global warming

The greenhouse in Figure 7.13A is used to grow plants where the weather outside is too cold. The glass or plastic walls of a greenhouse allow solar radiation to pass through. The sunlight heats the soil, which in turn warms the air. The walls trap the warm air, raising the temperature inside.

An analogous process, called the **greenhouse effect,** operates on a global scale (Figure 7.13B). Solar radiation reaching Earth's atmosphere includes ultraviolet radiation and visible light. As we discuss in the next module, the ozone layer filters out most of the damaging UV radiation. Visible light passes through and is absorbed by the planet's surface, warming it. Heat is radiated by the warmed planet, and these longer, infrared wavelengths are absorbed by gases in the atmosphere, which then reflect some of the heat back to Earth. This natural heating effect is highly beneficial. Without it, Earth would be much colder and much less hospitable to life.

The gases in the atmosphere that absorb heat radiation are called greenhouse gases. Some occur naturally, such as water vapor, carbon dioxide (CO_2), and methane (CH_4), while others are synthetic, such as chlorofluorocarbons (CFCs). Human activities are adding to the levels of these greenhouse gases.

Carbon dioxide is one of the most important greenhouse gases. You have just learned that CO_2 is a raw material for photosynthesis and a waste product of cellular respiration. These two processes, taking place in microscopic chloroplasts and mitochondria, keep carbon cycling between inorganic and organic compounds on a global scale. Photosynthetic organisms absorb billions of tons of CO_2 each year. Most of that fixed carbon returns to the atmosphere via cellular respiration, the action of decomposers, and fires. But much of it remains locked in large tracts of forests and undecomposed organisms. And large amounts of carbon are in long-term storage in fossil fuels buried deep under Earth's surface.

Before 1850, carbon dioxide was estimated to make up less than 0.03% of the air we breathe. Since the start of the Industrial Revolution, the atmospheric concentration of CO_2 has increased about 40%, mostly due to the combustion of carbon-based fossil fuels, such as coal, oil, and gasoline.

Increasing concentrations of greenhouse gases have been linked to **global warming,** a slow but steady rise in Earth's surface temperature. Predicted changes of just a few degrees over the next 50 years may have dramatic and wide-ranging consequences. These would likely include melting of polar ice, rising [...] weather patterns, droughts, and the spread [...]

[...] he rise in atmospheric CO_2 levels during the [...] ded with widespread deforestation, which [...] bal warming problem by reducing an effec [...] forests are cleared for lumber or agriculture, [...] growth increases the demand for fossil fuels,

Figure 7.13A Plants growing in a greenhouse

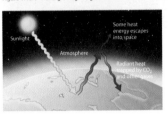
Figure 7.13B CO_2 in the atmosphere and global warming

CO_2 levels will continue to rise. We discuss global warming in more detail in Module 38.5.

Can photosynthesis help mitigate this increase in atmospheric CO_2? As you read in the chapter introduction, "energy plantations" hold out the promise of a cleaner, renewable fuel source. Recent studies show that increasing CO_2 levels will increase plant production somewhat, but far less than predicted even a decade ago. Slowing the destruction of forests will sustain their photosynthetic and carbon-storing contributions. Taking a lesson from plants, we can explore technologies that utilize solar energy for some of our energy needs. Almost all life on Earth depends on the ability of plants and other photosynthetic organisms to convert light energy to the chemical energy of food molecules. Their contribution to life on Earth may also come to include increased removal of CO_2 from the atmosphere.

? Explain the greenhouse effect.

■ Sunlight warms Earth's surface, which radiates heat to the atmosphere. CO_2 and other greenhouse gases absorb and radiate some heat back to Earth.

Chapter 7 *Photosynthesis: Using Light to Make Food* **119**

22.5 The evolution of lungs facilitated the movement of tetrapods onto land

The colonization of land by vertebrates was one of the pivotal milestones in the history of life. The evolution of legs from fins may be the most obvious change in body design, but the refinement of lung breathing was just as important. And while skeletal changes were undoubtedly required in the transition from fins to legs, the evolution of lungs for breathing on land also required skeletal changes. Interestingly, current fossil evidence supports the hypothesis that the beginning changes in the front fins and shoulder girdle of tetrapod ancestors may actually have been breathing adaptations that enabled the kind of push up needed in shallow water for a fish to gulp air.

Paleontologists have uncovered numerous transitional forms in tetrapod evolution (see Module 19.4). It now seems clear that tetrapods first evolved in shallow water from what some researchers jokingly call "fishapods." These ancient forms had both gills and lungs. The adaptations for air-breathing evident in their fossils include a stronger and elongated snout and a muscular neck that enabled the animal to lift the head clear of water and into the unsupportive air. Strengthening of the lower jaw may have facilitated the pumping motion presumed to be used by early air-breathing tetrapods and still employed by frogs to inflate their lungs. The recently discovered 375 million year old fossil of *Tiktaalik* (Figure 22.5) illustrates some of these air-breathing adaptations.

The first tetrapods on land diverged into three major lineages: amphibians, reptiles (including birds), and mammals. Most amphibians have small lungs and rely heavily on the diffusion of gases across body surfaces. Reptiles (including all

Figure 22.5 A cast of a fossil of *Tiktaalik*. Note the elongated snout and strong shoulder support.

birds) and mammals rely o[...] general, the size and complex [...] an animal's metabolic rate (an [...] ple, the lungs of birds and [...] temperatures are maintained b [...] greater area of exchange surf [...] sized amphibians and nonbir [...] lower metabolic rate.

We explore the mammalian [...]

? How might adaptations [...] the evolution of tetrapo [...]

■ [...] in the shoulder girdle and limb [...] ancestors lift their heads above water

---

18.16 Sean Carroll talks about the evolution of animal diversity

> . . . from so simple a beginning endless forms most beautiful and most wonderful have been, and are being evolved.
>
> —Charles Darwin, *The Origin of Species*

This chapter, a brief tour of invertebrate life, introduced you to some of the immense variety of animal bodies—flower-like sea anemones, echinoderms in the shapes of stars and cucumbers, and arthropods with countless configurations of appendages and color patterns, to name just a few. Data from molecular biology are providing new insights into the evolutionary relationships among these animal groups (see Module 18.15). But how can we explain the evolution of such strikingly different forms from a common protistan ancestor? Dr. Sean Carroll (Figure 18.16) of the Howard Hughes Medical Institute and the University of Wisconsin-Madison is a pioneer in the field of evolutionary developmental biology ("evo devo"). His book *Endless Forms Most Beautiful: The New Science of Evo Devo,* details the new understanding of the evolution of diversity that emerged when exciting new discoveries in the genetics and evolution of development were integrated with existing lines of evidence from the fossil record, comparative anatomy, and molecular biology. In a recent interview, Dr. Carroll described how surprising revelations forced scientists to think differently about the evolution of animal diversity.

The conventional wisdom was that humans were very complex and really—in terms of genetic material—fundamentally different from things like fruit flies, butterflies, or worms. Maybe we'd have something more in common with a mouse or an ape, but not necessarily with the rest of the animal kingdom. But the huge discovery of evo devo was that the genes that build the bodies and body parts [...] organs of f[...] and with virt[...]

Figure 18.16 Sean Carroll, talking with Jane Reece

control the events that lead to the different body forms of different animals. How is that possible? Dr. Carroll explains.

It turns out that these body-building genes have very sophisticated and extensive sets of regulatory instructions [control sequences in the DNA that regulate gene expression; see Module 11.5] that surround them or are embedded within them.

The key genetic differences are not in the segments of DNA that are transcribed and translated into proteins, but in the segments of DNA that control when and where genes are transcribed and translated into proteins, or expressed (see Module 11.5). Given different sets of instructions, the same gene can direct the formation of a fly eye or a human eye. The [...]

Chapter Review

Reviewing the Concepts

An Overview of Photosynthesis (Introduction–7.5)

Photosynthesis is summarized by the following equation:

$$6\,CO_2 + 6\,H_2O + Light\ energy \rightarrow C_6H_{12}O_6 + 6\,O_2$$

Plants, algae, and some bacteria are photoautotrophs, the producers of food consumed by virtually all organisms. Chloroplasts appear to have evolved from an endosymbiotic photosynthetic prokaryote (**Introduction–7.1**). In plants, photosynthesis occurs primarily in the leaves, within chloroplasts, which contain stroma and stacks of thylakoids called grana (**7.2**).

The Two Stages of Photosynthesis. The O_2 liberated by photosynthesis comes from H_2O (**7.3**). In photosynthesis, H_2O is oxidized and CO_2 is reduced (**7.4**). The light reactions convert light energy to chemical energy and produce O_2. The Calvin cycle assembles sugar molecules from CO_2 using ATP and NADPH from the light reactions (**7.5**).

The Light Reactions: Converting Solar Energy to Chemical Energy (7.6–7.9)

Photosystems. Certain wavelengths of visible light, absorbed by pigments such as chlorophyll and carotenoids, drive photosynthesis (**7.6**). Thylakoid membranes contain multiple photosystems, each consisting of light-harvesting complexes of pigments and a reaction center complex with a primary electron acceptor that receives excited electrons from a reaction center chlorophyll *a* (**7.7**).

The Light Reactions. Two photosystems absorb photons and transfer energy to chlorophyll P680 and P700. Their excited electrons are passed from primary electron acceptors to electron transport chains. Electrons shuttle from photosystem II to I, providing energy to make ATP. Electrons from photosystem I reduce NADP⁺ to NADPH. Photosystem II regains electrons by splitting water, releasing O_2 (**7.8**).

Photophosphorylation. ATP is synthesized by chemiosmosis. The electron transport chain pumps H⁺ into the thylakoid space. The concentration gradient drives H⁺ back across the membrane through ATP synthase and powers the phosphorylation of ADP to produce ATP (**7.9**).

The Calvin Cycle: Converting CO₂ to Sugars (7.10)

The Calvin cycle, occurring in the stroma, consists of carbon fixation, reduction, release of G3P, and regeneration of RuBP. Using carbon from CO_2, electrons from NADPH, and energy from ATP, the cycle constructs G3P, which is used to build glucose and other organic molecules (**7.10**).

Photosynthesis Reviewed and Extended (7.11–7.12)

A summary diagram of the two stages of photosynthesis is shown at the top of the next column (**7.11**).

C₃, C₄, and CAM plants. In C₃ plants, a drop in CO_2 and rise in O_2 when stomata close on hot, dry days divert the Calvin cycle to photorespiration. C₄ plants first fix CO_2 into a four-carbon compound that provides CO_2 to the Calvin cycle. CAM plants open stomata at night, making a four-carbon compound used as a CO_2 source during the day (**7.12**).

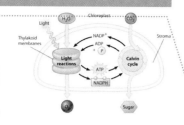

Photosynthesis, Solar Radiation, and Earth's Atmosphere (7.13–7.14)

Excess CO_2 is contributing to global warming. Photosynthesis, which removes CO_2 from the atmosphere, moderates this warming (**7.13**). Solar radiation converts O_2 high in the atmosphere to ozone (O_3), which shields organisms from damaging UV radiation. Industrial chemicals called CFCs have caused dangerous thinning of the ozone layer, but international restrictions on CFC use is allowing recovery (**7.14**).

Connecting the Concepts

1. This diagram compares the chemiosmotic synthesis of ATP in mitochondria and chloroplasts. In both cases, label the structures involved and indicate which side of the membrane has the higher H⁺ concentration. Then label on the right the locations within the chloroplast.

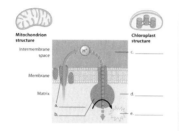

2. Continue your comparison of chemiosmosis and electron transport in mitochondria and chloroplasts. In each c...
   a. where do the electrons come from?
   b. where do the electrons get their energy?
   c. what picks up the electrons at the end of the chain?
   d. how is the energy given up by the electrons used?

Chapter 7 *Photosynthesis: Using Light to Make Food*

3. Complete this summary map of photosynthesis.

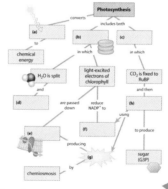

Testing Your Knowledge

Multiple Choice

4. Photosynthesis consumes _____ and produces _____.
   a. O_2 ... H_2O
   b. H_2O ... CO_2
   c. CO_2 ... chlorophyll
   d. H_2O ... O_2
   e. glucose ... O_2

5. Which of the following are produced by reactions that take place in the thylakoids and consumed by reactions in the stroma?
   a. CO_2 and H_2O
   b. NADP⁺ and ADP
   c. ATP and NADPH
   d. ATP, NADPH, and O_2
   e. CO_2 and ATP

6. In photosynthesis, _____ is oxidized and _____ is reduced.
   a. glucose ... oxygen
   b. carbon dioxide ... water
   c. water ... carbon dioxide
   d. glucose ... carbon dioxide
   e. water ... oxygen

7. Why is it difficult for most plants to carry out photosynthesis in very hot, dry environments such as deserts?
   a. The light is too intense and overpowers the pigment molecules.
   b. The closing of stomata keeps CO_2 from entering and O_2 from leaving the plant.
   c. They must rely on photorespiration to make ATP.
   d. Global warming is intensified in a desert environment.
   e. CO_2 builds up in the leaves, blocking carbon fixation.

8. When light strikes chlorophyll molecules, they lose electrons, which are ultimately replaced by
   a. splitting water.
   b. breaking down ATP.
   c. oxidizing NADPH.
   d. fixing carbon.
   e. oxidizing glucose.

9. What is the role of NADP⁺ in photosynthesis?
   a. It assists chlorophyll in capturing light.
   b. It acts as the primary electron acceptor for the photosystems.
   c. As part of the electron transport chain, it helps to synthesize ATP.
   d. It assists photosystem II in the splitting of water.
   e. It accepts electrons and carries them to the Calvin cycle.

10. The reactions of the Calvin cycle are not directly dependent on light, but they usually do not occur at night. Why? (*Explain your answer.*)
   a. It is often too cold at night for these reactions to take place.
   b. Carbon dioxide concentrations decrease at night.
   c. The Calvin cycle depends on products of the light reactions.
   d. Plants usually close their stomata at night.
   e. Most plants do not make four-carbon compounds, which they would need for the Calvin cycle at night.

11. How many "turns" of the Calvin cycle are required to produce one molecule of glucose? (Assume one CO_2 is fixed in each turn of the cycle.)
   a. 1   b. 2   c. 3   d. 6   e. 12

12. Which of the following does *not* occur during the Calvin cycle?
   a. carbon fixation
   b. oxidation of NADPH
   c. consumption of ATP
   d. regeneration of RuBP, the CO_2 acceptor
   e. release of oxygen

Describing, Comparing, and Explaining

13. What are the major inputs and outputs of the two stages of photosynthesis?

14. What do plants do with the sugar they produce in photosynthesis?

Applying the Concepts

15. Most experts now agree that global warming is occurring, and in response to its potential threat, a number of countries have made a commitment to reduce CO_2 emissions significantly. However, some countries oppose taking strong action at this time. Several reasons are cited: First, a few experts think that the apparent warming trend may be just a random fluctuation in temperature. Second, if the temperature increase is real, it has yet to be shown that it is caused by increased CO_2. Some people also believe that it would be difficult to cut CO_2 emissions without sacrificing economic growth. Do you think we should have more evidence before taking action? Or is it better to play it safe and act now to reduce CO_2 emissions? What are the possible costs and benefits of each of these two strategies?

16. The use of biomass energy (see chapter introduction) avoids many of the problems associated with gathering, refining, transporting, and burning fossil fuels. Yet biomass energy is not without its own set of problems. What challenges do you think would arise from a large-scale conversion to biomass energy? How do these challenges compare with those encountered with fossil fuels? Which set of challenges do you think is more likely to be overcome eventually? Do you think any one type of energy has more benefits and fewer costs than the others? Which one, and why?

Answers to all questions can be found in Appendix 4.

For Practice Quizzes, BioFlix, MP3 Tutors, and Activities, go to www.mybiology.com.

Feel confident going into the test.

Review the main points. *Reviewing the Concepts* helps you do just that, with helpful summary diagrams and references back to the text.

Connect the chapter's key concepts. *Connecting the Concepts* activities test your ability to link topics from different modules and include concept mapping, labeling, and categorizing exercises.

Prepare for the test using the multiple-choice questions that appear in the *Testing Your Knowledge* section.

Test your understanding with *questions that ask you to describe, compare, and explain*. If you can restate a concept in your own words, you've probably learned it.

Get involved. *Applying the Concepts* questions help connect biology to your life and society.

Check your knowledge by comparing your answers to those found in *Appendix 4*.

Get a personal study plan by doing the *chapter pre-test or post-test*.

Make the most of your study time with mybiology.com

mybiology is a comprehensive resource that helps you study more efficiently and provides all of the study tools you need, including quizzes, an integrated eBook, MP3 Tutors, and BioFlix™—movie-quality animations that tackle the toughest topics in biology.

The 4-step structure of **mybiology** allows you to identify what you don't know, select the appropriate learning activity to match your learning style, and practice outside of lecture.

① *Focus Your Effort* uses a pre-test as a roadmap to identify your study needs.

② *Direct Your Learning* utilizes the resources that support the various learning styles, from MP3 Tutors for auditory learners to BioFlix for visual learners and the eBook for those who prefer to read.

③ *Test Yourself* offers a post-test to check your understanding and prepare for exams.

④ *Extend Your Knowledge* allows you to go beyond the chapter material by exploring case studies in the process of science and viewing Discovery Channel videos, etc.

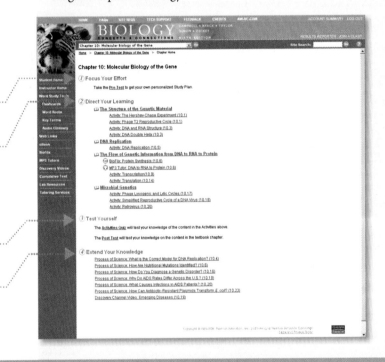

Also located on mybiology.com

*Bio*Flix

NEW! BioFlix are interactive, self-paced tutorials that help you master the toughest topics in biology. You can prepare for exams by viewing a 3-D animation, completing the tutorial, printing a personal review sheet, and taking a quiz.

BioFlix Tutorial Topics
- Tour of an Animal Cell
- Tour of a Plant Cell
- Cellular Respiration
- Photosynthesis
- Mitosis
- Meiosis
- Protein Synthesis
- How Neurons Work

MP3 Tutors
Expanded! *MP3 Tutor Sessions* are downloadable audio tutorials, narrated by author Eric Simon, allowing you to study anytime, anywhere. **NEW!** *19 MP3 Tutors* focus on tough terms and provide a unique alternative to flashcards.

Preface

Inspired by the enthusiastic responses of the many users of the fifth edition, we authors have worked closely together to create a new, sixth edition that reflects current progress in biology and the evolving needs of today's courses and students. For this edition, the author team is fortunate to include Jean Dickey of Clemson University. Jean brings a wealth of experience teaching introductory biology, superb communication skills, and special expertise in ecology. She is also the author of the popular lab manual *Laboratory Investigations for Biology*, which has accompanied our book since its first edition.

The aim of *Biology: Concepts & Connections* has always been to engage students from a wide variety of majors in the wonders of the living world. Most of these students will not become biologists themselves, but their lives will be touched by biology every day. Today, understanding the concepts of biology and their connections to our lives is more important than ever. Whether we're concerned with our own health or the health of our planet, a familiarity with biology is essential. This basic knowledge and an appreciation for how science works have become elements of good citizenship in an era when informed evaluations of health issues, environmental problems, and applications of new technology are critical.

The "connections" to which the title of this book refers include many such practical applications of biology—and go beyond them. Biology has important connections with the other natural sciences and with the humanities and social sciences as well. And the study of life has no coherence without an understanding of the connections between the different areas of biology and an appreciation of the grand unifying theme of evolution. From its first edition, the hallmarks of this book have included an emphasis on connections within biology and between biology and other fields. In this sixth edition, we have increased the emphasis on connections to our everyday world while also doing more to help students connect the concepts of biology to evolution.

We could not have hoped to meet our ambitious goals for this book without extensive discussions with teaching colleagues around the world and feedback from many of the hundreds of instructors and hundreds of thousands of students who have used our earlier editions. We have been gratified by their enthusiasm for the book and have paid close attention to their thoughtful suggestions for improvement. For this edition, we set out to create a book that would be more effective than ever as a learning tool and would also integrate smoothly with the rich program of supporting materials at **mybiology**.

How can we help students learn—and enjoy—biology? How can we help instructors teach biology? Our responses to these questions are reflected in the teaching strategies we bring to the book. Below we describe our main strategies as they are embodied in *Biology: Concepts & Connections*, Sixth Edition.

Focus Students on the Main Ideas of Biology

Biology is a vast subject that gets bigger every year, but an introductory biology course is still only one or two terms long. In that brief time, we explore all of life, from molecules to ecosystems, while also trying to share the excitement of research breakthroughs. For beginning students confronting this avalanche of information, it can seem as important to memorize all the scientific terms and facts as it is to master and apply the major ideas. This situation changes, however, when students acquire a framework of key biological concepts into which they can fit the many new things they learn. It is this framework of concepts that will serve them long after they have forgotten specific facts and terms.

Concept Modules *Biology: Concepts & Connections* was the first introductory biology textbook to use concept modules to help students recognize and focus on the main ideas of each chapter. The heading of each module is not simply a topic title but a carefully crafted statement of a biological concept. Each concept heading serves as a focal point for a module, and the module's text and illustrations converge on that concept with explanation and, often, evidence. For example, "Sensory receptors convert stimulus energy to action potentials" announces a key concept in Chapter 29 (Module 29.2). The text and illustrations introduce the general principles of sensory reception and transduction, using the human sense of taste as an example. In this and other modules, we integrate the words and pictures to an unprecedented degree: The text walks the student through the illustrations, just as an instructor would do in class. In teaching a sequential process, such as the functioning of the taste-bud receptor cells (Figure 29.2), we number the steps in the text to correspond to numbered steps in the figure. The synergy between a module's verbal and graphic components transforms the concept heading into an idea with meaning to the student, serving both visual and nonvisual learners.

Integrated Media Relevant modules feature references to the BioFlix™ 3-D animations with tutorials, MP3 Tutors, Web Activities, and Web Process of Science Investigations available at **mybiology**. You can read about these valuable learning tools on page xx.

Checkpoint Questions The checkpoint question at the end of each module encourages students to test themselves as they proceed through a chapter. Some questions simply ask the student to restate the main concept or a corollary; others test understanding of the supporting evidence or ask the student to connect the concept to another concept in the book; still others require the student to carry out a calculation using information in the module. Feedback is provided on the spot: The answer is printed upside down beneath the question.

Help Students Connect the Concepts

How do we help students see the connections between concepts? First, we group the modules under prominent main headings, printed on orange bars, which form an overarching framework for the chapter. Students first see these main headings and their subordinate concepts and connections in the outline at the beginning of the chapter. Within the chapter, we often make use of overviews and reviews, and explanatory transitions at the beginnings or ends of modules help the modules flow as a continuous story. The chapter summary at the end of the chapter is organized using the main chapter headings. This Reviewing the Concepts section includes helpful visuals. Following the summary, a section called Connecting the Concepts helps students tie the chapter's concepts together for themselves. This section includes concept maps for students to complete and other graphic activities.

Relate Biological Concepts to Everyday Life

Connection Modules Students are more motivated to study biology when they can connect it to their own lives and interests—for example, to health issues, economic problems, environmental quality, ethical controversies, and social responsibility. In this edition, gold Connection tabs mark the numerous application modules that go beyond the core biological concepts. You can preview the Connection module headings on the front endpapers of the book.

Chapter-Opening Essays Most of the illustrated essays that introduce the chapters discuss topics that relate to readers' lives and interests. For example, Chapter 3, "The Molecules of Cells," opens with an essay on lactose intolerance. And a new essay for Chapter 26, "Hormones and the Endocrine System," discusses endocrine disruption caused by pollutants in the environment. Other introductory essays continue our earlier tradition of featuring a nonhuman organism and describing how it is adapted to its environment. One example is our new essay for Chapter 1, which focuses on African leopards, our cover organism. Our hope is that essays like this one will nurture students' appreciation for biological diversity.

Applying the Concepts Many of the questions in Applying the Concepts, found at the end of every chapter, encourage students to use the concepts they have learned in thinking about various social and environmental issues.

Adapt This Book to Fit Your Course

Though a biology textbook's table of contents must be linear, biology itself is more like a web of related concepts without a single starting point or prescribed path. Courses can navigate this network starting with molecules, with ecology, or somewhere in between, and most courses omit some topics. *Biology: Concepts & Connections* is uniquely suited to serve this variety of courses. The seven units of the book are largely self-contained, and in a number of the units, chapters can be assigned in a different order without much loss of coherence.

Moreover, the modular format of the chapters makes it easy to omit or to relocate modules within a syllabus.

Relate Biological Concepts to Evolution

The history of life on Earth goes back more than 3.5 billion years, and this past is the key to the present diversity of organisms. As the unifying theme of this textbook, evolution elevates biology from a collection of facts to a coherent study of changing life on a changing planet. In *Biology: Concepts & Connections*, students study the structure, function, and behavior of organisms in an evolutionary context. And throughout the book, students learn to view the unity and diversity of life—the similarities and differences among organisms—as the dual consequences of descent with modification. Our reorganized and enhanced coverage of evolution in Unit Three of this edition strengthens this unifying theme.

In addition, this edition has a new feature called Evolution Connections. In every chapter, one or more modules, identified by rust-colored tabs, highlight the relevance of evolution to some aspect of the chapter content. Examples include Module 12.21 on what genome sequences reveal about evolution and Module 20.2 on how the body forms of aquatic animals reflect natural selection.

Help Students Learn the Process of Science

A biology course should familiarize students with the scientific process—in particular, with the posing and testing of hypotheses. With a solid introduction to the process of science in Chapter 1, students will be equipped to appreciate the many examples throughout the book of how scientific concepts emerge from observations and experimental evidence. The book also puts human faces on science with Talking About Science modules, marked with teal tabs.

Many of the questions in Applying the Concepts, at the end of each chapter, give students personal practice with science as a process. At **mybiology**, students can access a wealth of interactive activities, including Process of Science Investigations, YouDecide, and GraphIt! In fact, this book will work best for students who participate actively in learning about biological concepts and their applications.

■ ■ ■

Introductory biology is the only science course that many students will take during their college years. Long after today's students have forgotten most of the specific content of their biology course, they will be left with general impressions and attitudes about science and scientists. We hope that this new edition of *Biology: Concepts & Connections* helps make those impressions positive and supports instructors' goals for sharing the fun of biology. To help us produce an even better text in the next edition, please send your comments and suggestions to us in care of our editor, Chalon Bridges, at Pearson Benjamin Cummings, 1301 Sansome Street, San Francisco, CA 94111. You can also send e-mail to chalon.bridges@pearson.com.

Jane Reece, Martha Taylor, Eric Simon, and Jean Dickey

New to This Edition

In preparing this new edition, we had ambitious goals for improvement in four main areas. The first major goal was to strengthen our coverage of evolution, the all-important core theme of biology. To more clearly relate evolution to *all* of biology and to address common misconceptions about its scope and relevance, we decided to include Evolution Connection modules in each and every chapter. We also worked to restructure Unit Three, "Concepts of Evolution," to better emphasize Darwin's theory, microevolution, speciation, and macroevolution. In Unit Four, "The Evolution of Biological Diversity," we continued our quest to tell the story of diversity in an evolutionary context by, among other changes, presenting the evolution of vertebrates and invertebrates in separate chapters. Our second major goal was to improve our presentation of ecology, an area of increasing practical importance and interest to students; we wanted to infuse Unit Seven with current content and interesting new examples and applications. Our third major goal was, as always, to improve the book's pedagogy in every chapter to make learning biology more fun as well as more successful. Finally, we sought to update the science in a way that would give students an understanding of the exciting field that biology is today.

Below are just a few highlights of new content and organizational improvements in *Biology: Concepts & Connections,* Sixth Edition.

Chapter 1, "Biology: Exploring Life," has a new opening essay on the African leopard, pictured on the cover of this book. In addition to emphasizing the core biological theme of evolution, the chapter now highlights three other themes: emergent properties (Module 1.1), interactions between organisms and their environments (Module 1.2), and cells as a prime example of the correlation of biological structure and function (Module 1.3).

Unit One, The Life of the Cell,

the book's introduction to the basic chemistry, structure, and energetics of cells, has benefited from a number of improvements. Chapter 2 now includes an abbreviated periodic table to help illustrate atomic structure and chemical behavior. (Appendix 2 contains a complete periodic table.) Chapters 4 and 5 have been revised to more clearly introduce the important topic of cellular membranes: New Module 4.5, in the "Tour of the Cell" chapter, introduces the structure and function of membranes, formerly in Chapter 5, "The Working Cell." Chapter 5 now begins with a more detailed exploration of membranes and then proceeds to major sections on energy and enzymes. This new organization better fits the location of Chapter 5 between Chapter 4 (on cells) and Chapter 6 (on cellular respiration). Chapter 5 also features an Evolution Connection on the evolution of membranes.

Unit Two, Cellular Reproduction and Genetics,

incorporates important recent advances in the field. Chapter 10, "Molecular Biology of the Gene," has a new module (10.21) on prions and viroids. Chapter 11, now entitled "How Genes Are Controlled," has been reorganized to improve flow and enhance understanding; it includes a streamlined discussion of development and new coverage of microRNAs and RNA interference (Module 11.7). Chapter 12, "DNA Technology and Genomics," has been extensively revised. For example, it has a new module (12.14) on STR (short tandem repeat) analysis and its use in DNA profiling. Within the genomics section of the chapter, new Module 12.18 explains the main findings of the Human Genome Project and discusses how that information is being used. New Figure 12.12 shows how primers are used to amplify specific DNA sequences during the polymerase chain reaction (PCR). Chapter 12 also includes a new Module 12.20 on proteomics and an Evolution Connection (Module 12.21) on what genome studies reveal about evolution.

Unit Three, Concepts of Evolution,

has been restructured, as mentioned in the first paragraph. In Chapter 13, "How Populations Evolve," changes include a new Figure 13.4H showing gradual transition of land-dwelling ancestors of whales to life in the sea—an example of fossil evidence for evolution. Also, the new Module 13.6 introduces evolutionary (phylogenetic) trees and shows how biologists use them to show patterns of evolution. Common misconceptions about evolution are explicitly addressed. Chapter 14, "The Origin of Species," now focuses on speciation (material on macroevolution has been moved to Chapter 15). A new opening essay tells the story of the dramatic speciation of cichlids in Lake Victoria only 100,000 years ago and the dramatic decline in species in the past 30 years, and the cichlid example is used throughout the chapter. Chapter 15, "Tracing Evolutionary History," has been revised to provide a cohesive presentation of all the material on macroevolution (some moved from Chapter 14 and some from 16). The chapter retains the modules on phylogeny. New Module 15.4 gives an overview of the major events in life's history. Module 15.11 has new material on the important role of developmental genes in evolution; a new figure on gene expression in threespine stickleback fish is used to illustrate differences in gene regulation. Module 15.16 has interesting new material on using phylogenetic trees to make inferences about a common ancestor, pointing out that dinosaurs may have brooded their eggs and "sung." Module 15.18 has been dedicated to the discussion of molecular clocks, which can help track evolutionary time; it includes a figure showing how scientists have dated the origin of HIV infection in humans.

Unit Four, The Evolution of Biological Diversity,

has been significantly restructured. Modules about the origin of life have been moved into Unit Three, where they are integrated with other material on the history of life. Chapter 18, which included all of animal diversity in previous editions, now includes only invertebrates and features a new opening essay on octopuses. Chapter 19 now covers vertebrate diversity as well as human evolution (formerly the sole topic of Chapter 19). Throughout the unit, the presentation of key concepts has been improved, and new and revised modules provide compelling evolutionary explanations for some of the most striking organismal diversity. Chapter 16, "The Origin and Evolution of Microbial Life: Prokaryotes and Protists," is now more focused on diversity. New Module 16.12 "Secondary endosymbiosis is the key to protist diversity" builds on new Module 4.16, "Mitochondria and chloroplasts evolved by endosymbiosis," to explain the current hypothesis for protist evolution. Revised Module 16.5 includes increased coverage of biofilms, aggregates of bacteria whose significance in infectious disease and other fields is just beginning to be recognized. In Chapter 17, "Plants, Fungi, and the Colonization of Land," Module 17.12 now describes adaptations that attract and reward pollinators, and considers the effect of these adaptations on the reproductive success of flowering plants. Chapter 18, "The Evolution of Invertebrate Diversity," is now more limited in scope. New Module 18.12 highlights features that have made insects the most numerous and diverse group of animals, and the role of homeotic genes in the evolution of insect diversity. In new Talking About Science Module 18.16, leading researcher Sean Carroll reflects on how recent discoveries in evolutionary developmental biology (evo-devo) have led to astonishing insights into the process of animal evolution. Chapter 19, "The Evolution of Vertebrate Diversity," now includes the vertebrate diversity from Chapter 18 in the previous edition as well as our coverage of human evolution. New Evolution Connection Module 19.4, on tetrapod evolution, tracks transitional fossil forms related to the ancient tetrapod Tiktaalik—the "missing link" between water and land animals. Modules 19.12 and 19.13, on the evolution of our upright posture and enlarged brain, have been updated to include recent fossil evidence. A heavily revised Module 19.14 on the divergence of Neanderthals and *Homo sapiens* includes discussion of Neanderthals in popular culture, analysis of DNA from Neanderthal bones, and recent sequencing of Neanderthal DNA. New Figure 19.15 illustrates the global spread of *Homo sapiens*. New Table 19.17 helps students see the relationships between UV radiation and vitamin deficiencies that probably influenced the evolution of differences in human skin pigmentation. Throughout Unit IV, phylogenetic trees are rendered in a graceful new style that is more attractive and easier for students to interpret.

Unit Five, Animals: Form and Function,

has been updated throughout. In addition, Chapter 20, "Unifying Concepts of Animal Structure and Function," includes a new Module 20.12 on the integumentary system. Chapter 21, "Nutrition and Digestion," has a new Module 21.11 on the liver, emphasizing the hepatic portal vein and liver functions related to digestion (including material that was formerly in Chapter 25). In Chapter 27, "Reproduction and Embryonic Development," our discussion of sexual and asexual reproduction is split into separate modules (27.1 and 27.2) to help students better compare and contrast the two modes of reproduction. In the same chapter, the coverage of STDs and contraception has been expanded and updated, and "Principles of

Embryonic Development" has been streamlined. Chapter 28, "Nervous Systems," includes a new Module 28.18 on fMRI (functional magnetic resonance imaging). Chapter 30, "How Animals Move," has a new opening essay on the "Man vs. Horse" race (which compares human and horse anatomy and function), improved coverage of the vertebrate skeleton and its evolution (Module 30.3), and a new Connection module (30.12) on how muscle fiber characteristics affect athletic performance.

Unit Six, Plants: Form and Function,

has been refined and revised to make this subject more interesting to students. The first chapter, Chapter 31, opens with a new essay on the use of "extreme tree climbing" in studying the world's tallest plants, and Module 31.1 is a new Connection on the link between the cultivation of wheat and the development of civilization. Other additions include mention of how tree rings can provide data on global climate change (Module 31.8). Chapter 32, "Plant Nutrition and Transport," has a new opening essay connecting to environmental action in the wake of Hurricane Katrina, and new coverage of composting and the use of fertilizer.

Unit Seven, Ecology,

has been extensively revised. The opening to the first chapter, Chapter 34, introduces the biosphere with examples of life in two extreme environments: the snow leopard in a Himalayan alpine meadow and the Yeti crab, which lives around hydrothermal vents at the bottom of the ocean near Easter Island. Heavily revised Module 34.1, with new figures, explains what ecologists study and how organisms interact with their environment at several levels. Module 34.6, also heavily revised, emphasizes the roles of sunlight and substrate in marine ecosystems; new Figure 34.6A, showing ocean zones, includes examples of organisms that live in each zone. The water cycle diagram formerly in Chapter 37 is now Figure 34.17. The revision of Module 34.8, on terrestrial biomes, includes mention of global warming effects. Chapter 35, "Behavioral Adaptations to the Environment," has a new opener on the monogamous prairie vole, and modules within the chapter weave in examples relating to this animal. For example, the heavily revised Module 35.3, "Behavior is the result of both genetic and environmental factors," explains the genetic basis of vole mating, as well as cross-fostering experiments using rats. New Module 35.23 discusses how studies of animal behavior (including vole mating) can help explain some aspects of human behavior. Chapter 36, "Population Ecology" (formerly "Population Dynamics"), has a new opening essay on the invasive Nile perch in Lake Victoria and the extinction of cichlid species. In Module 36.4, new data tables help students understand models of population growth. Module 36.9, on human population growth, has been reorganized and revised to better emphasize major concepts and to help students apply the lessons of Module 36.4. New Connection Module 36.10 uses age structure diagrams to show the effect of the post–World War II baby boom on the future of programs such as Medicare and Social Security; the accompanying figure shows age structures for the United States in 1974, 2000, and 2025. New Connection

Module 36.11, on the ecological footprint concept, emphasizes in words and pictures that overconsumption is potentially as damaging to the Earth as overpopulation. In Chapter 37, the new opener discusses the Mzima Springs ecosystem, where hippo dung plays a central role. The chapter's major section on community ecology has been reorganized for greater effectiveness. New Module 37.2, "Interspecific interactions are fundamental to community structure," briefly introduces categories of interspecific interactions and includes a table showing the effect of each category on the interacting individuals. The five modules that follow (all either new or revised) explain and give examples of competition, mutualism, predation, herbivory, and parasitism and disease. New

Module 37.7, "Parasites and pathogens can affect community composition," uses chestnut blight—one of the few examples where disease impact has been studied in a natural system—to describe the consequences of a disease epidemic on a forest community. Sudden oak death, an epidemic that is still in its early stages, brings the topic into the present day. Revised Module 37.11, on keystone species, uses the high-interest topic of coral reefs to show the effect of a keystone species on its community. Chapter 38, "Conservation Biology," provides new information on global climate change from the 2007 report by the IPCC (Intergovernmental Panel on Climate Change). Our chapter now has four new modules that cover the latest findings in this crucial area: Modules 38.5–38.8.

Acknowledgments

Biology: Concepts & Connections, Sixth Edition, results from the combined efforts of many people, and the authors wish to extend heartfelt thanks to all those who contributed to this and previous editions. Our work on this edition was shaped by input from the biologists acknowledged in the Sixth Edition reviewer list on p. xv, who shared with us their experiences teaching introductory biology and provided specific suggestions for improving the book. In particular, we would like to thank Ed Zalisko, Blackburn College, and Jon Hoekstra, Gainesville College. The unsolicited comments and suggestions we received from other biologists and from biology students were also extremely helpful. In addition, this book has benefited in countless ways from the stimulating contacts we had with numerous biologists during the recent preparation of the larger text, Biology, Eighth Edition.

Eric Simon would like to thank Mark J. Daly, Ph.D. (Center for Human Genetic Research, Harvard Medical School), Kevin McMahon (New Hampshire State Police Forensics Laboratory), Marshall Simon, Dr. Michael C. Ain (Johns Hopkins Children's Center), Jamey Barone, Amanda Marsh, the Cystic Fibrosis Foundation, and Emily Posner of the Meg Perry Healthy Soil Project of the Common Ground Collective. Dr. James Newcomb, New England College, provided expert content and pedagogical advice for a number of chapters. EJS would also like to thank his colleagues at New England College for their support: Lori Bergeron, Amanda and Giulia Bussone, Sachie Howard, Mark Mitch, Maria Colby, Ed Cooper, Shannan Hudgins, and Linden Jackett. Jean Dickey thanks her colleagues at Clemson University for their support and assistance.

The superb publishing team for this edition was headed up by editor-in-chief Beth Wilbur and senior acquisitions editor Chalon Bridges. We cannot thank them enough for their unstinting efforts on behalf of the book and for their commitment to excellence in biology education. We are fortunate to have had the contributions of executive director of development Deborah Gale and senior editorial manager Ginnie Simione Jutson. Directing the project on a daily basis was our beloved editorial project manager Mary Douglas. We enormously appreciate her dedication to this project, her firm yet gentle guidance, and her positive approach to this work and to life. We are similarly grateful to our exceptional senior developmental editor Evelyn Dahlgren for her thoughtful and steadfast commitment to quality and her leadership of the talented developmental editorial team of Susan Teahan and Debbie Hardin. We thank them for their thoroughness and hard work; the book is far better than it would have been without their efforts. Thanks also to senior supplements project editor Susan Berge for her oversight of the supplements program and to assistant editor Rebecca Johnson for her efficient and calm support of the editorial team and her coordination of various supplements. We wish to express our appreciation to Linda Davis, president of Pearson Math, Economics, and Science, and Paul Corey, president of Pearson Science, for their ongoing support.

This book and all the other components of the teaching package are both attractive and pedagogically effective in large part because of the hard work and creativity of the production professionals on our team. We wish to thank the managing editor for production, Michael Early, and our very talented and hard-working production supervisor, Lori Newman. We thank Heather Meledin and Crystal Clifton of Progressive Publishing Alternatives for overseeing production, copyediting, and indexing and Progressive Information Technologies for composition. We are also grateful to permissions editor Sue Ewing.

For users of this book, the illustrations and photos are as important as the prose. We thank developmental artist Andrew Recher and his team at Precision Graphics for illustrations. We are indebted to senior photo editor Donna Kalal and photo researcher Kristin Piljay. In addition, we thank design director Mark Ong and designer Jana Anderson for the text design. And many thanks go to Yvo Riezebos for his patient and creative work on our beautiful cover.

The value of Concepts & Connections as a learning tool is greatly enhanced by the talents of outside contributors. Our thanks to April Lynch for writing the chapter opening essay for Chapter 1 and the back cover story. We much appreciate the hard work and creativity of the following supplements contributors: Ed Zalisko (Instructor's Guide to Text and Media); Richard Liebaert (Student Study Guide); David Reid, Jon Hoekstra, Richard Myers, Linda Brooke Stabler, Mimi Bres, Arnold Weisshaar, and David Mirman (Test Bank contributors); and Sarah Alvanipour, Richard Myers, Chris Romero, and Jennifer Katcher (Test Bank accuracy checkers). We thank the many contributors to the wonderful package of electronic media that accompanies the book: James Newcomb (author of new MP3 Tutor scripts and questions); Dawn Keller (selector of monthly New York Times articles and author of accompanying quiz questions); Martin Zahn, Helen Walter, Sarah Alvanipour, Richard Myers, Jon Hoekstra, and Dawn Keller (authors of website questions); Michael Wenzel (website accuracy checker); Jan McDearmon (website copyeditor); and Nina Lewallen Hufford (correlator of all Concepts media for text and website). Playing key roles in the development and production of the electronic supplements were media producer Ericka O'Benar and media project editor Brienn Buchanan. Thank you, one and all!

For their important roles in marketing the book, we are very grateful to Lauren Harp, executive marketing manager, and Jay Jenkins, senior marketing manager. The members of the Pearson Science sales team have continued to help us connect with biology instructors and their teaching needs. We thank them for all their hard work and enthusiastic support.

Finally, we are deeply grateful to our families and friends for their support, encouragement, and patience throughout this project. Our special thanks to: Paul, Dan, Maria, Armelle, and Sean (J.R.); Josie, Alice, Jack, David, Paul, and Ava (M.T.); Amanda, Reed, and Forest (E.S.); and Katherine and Jessie (J.D.).

Jane Reece, Martha Taylor, Eric Simon, and Jean Dickey

Reviewers

Reviewers of the Sixth Edition

Tanveer Abidi, *Kean University*
John Aliff, *Georgia Perimeter College*
Yael Avissar, *Rhode Island College*
Andrei Barkovskii, *Georgia College
    and State University*
Michael Battaglia, *Greenville Technical College*
Dennis Bogyo, *Valdosta State University*
Robert Boyd, *Auburn University*
Chad Brommer, *Emory University*
Ray Burton, *Germanna Community College*
Beth Campbell, *Itawamba Community College*
John Campbell, *University of Central Oklahoma*
John Capeheart, *University of Houston–
    Downtown*
David Chambers, *Northeastern University*
Craig Clifford, *Northeastern State University,
    Tahlequah*
Richard Cobb, *South Maine Community College*
Mary Colavito, *Santa Monica College*
Jennifer Cooper, *Itawamba Community College*
Jessica Crowe, *South Georgia College*
Jean DeSaix, *University of North Carolina
    at Chapel Hill*
Mary Dettman, *Seminole Community College*
Kathy Diamond, *College of San Mateo*
Lee Edwards, *Greenville Technical College*
William Ezell, *University of North Carolina
    at Pembroke*
Edward Fliss, *St. Louis Community College,
    Florissant Valley*
Linda Flora, *Montgomery County Community
    College*
Linda Gardner, *San Diego Mesa College*
Janet Gaston, *Troy University*
Bagie George, *Georgia Gwinnett College*
Laura Grayson-Roselli, *Burlington County College*
Reba Harrell, *Hinds Community College*
Jon Hoekstra, *Gainesville State College*
Kelly Hogan, *University of North Carolina
    at Chapel Hill*
Barbara Hunnicutt, *Seminole Community College*
Robert Iwan, *Inver Hills Community College*
Charles Jacobs, *Henry Ford Community College*
Roishene Johnson, *Bossier Parish Community
    College*
Jennifer Katcher, *Pima Community College*
Cindy Klevickis, *James Madison University*
MaryLynne LaMantia, *Golden West College*
Thomas Lammers, *University of
    Wisconsin–Oshkosh*
Peggy Lepley, *Cincinnati State University*
Eric Lovely, *Arkansas Tech University*
Paul Lurquin, *Washington State University*
Juan Morata, *Miami Dade College*
James Newcomb, *New England College*
Mark Paulissen, *Northeastern State University,
    Tahlequah*
John Peters, *College of Charleston*
Michael Read, *Germanna Community College*
Erin Rempala, *San Diego Mesa College*
Tim Revell, *Mt. San Antonio College*
Laura Ritt, *Burlington County College*

Lynn Rivers, *Henry Ford Community College*
Lynette Rushton, *South Puget Sound
    Community College*
Connie Rye, *East Mississippi Community College*
Beverly Schieltz, *Wright State University*
Daniela Shebitz, *Kean University*
Cara Shillington, *Eastern Michigan University*
Marc Smith, *Sinclair Community College*
Michael Smith, *Western Kentucky University*
Ralph Sorensen, *Gettysburg College*
Ruth Sporer, *Rutgers University*
Linda Brooke Stabler, *University of Central
    Oklahoma*
Christopher Tabit, *University of West Georgia*
Franklin Te, *Miami Dade College*
Gene Thomas, *Solano Community College*
Ken Thomas, *Northern Essex Community College*
Virginia Turner, *Harper College*
Cinnamon VanPutte, *Southwestern Illinois College*
Rani Vajravelu, *University of Central Florida*
Martin Vaughan, *Indiana University–Purdue
    University*
Mark Venable, *Appalachian State University*
Lura Williamson, *University of New Orleans*
Robert Wise, *University of Wisconsin-Oshkosh*
Mary Wisgirda, *San Jacinto College*
Michael Womack, *Macon State University*
Maury Wrightson, *Germanna Community College*
Tumen Wuliji, *Langston University*
Mark Wygoda, *McNeese State University*
Gregory Zagursky, *Radford University*
Martin Zahn, *Thomas Nelson Community College*
Edward Zalisko, *Blackburn College*
David Zeigler, *University of North Carolina
    at Pembroke*

Reviewers of Previous Editions

Michael Abbott, *Westminster College*
Daryl Adams, *Mankato State University*
Dawn Adrian Adams, *Baylor University*
Olushola Adeyeye, *Duquesne University*
Shylaja Akkaraju, *Bronx Community College*
Felix Akojie, *Paducah Community College*
Dan Alex, *Chabot College*
Sylvester Allred, *Northern Arizona University*
Jane Aloi-Horlings, *Saddleback College*
Loren Ammerman, *University of Texas at Arlington*
Dennis Anderson, *Oklahoma City Community
    College*
Marjay Anderson, *Howard University*
Bert Atsma, *Union County College*
Gail Baker, *LaGuardia Community College*
Mark Barnby, *Ohlone College*
Chris Barnhart, *University of San Diego*
Stephen Barnhart, *Santa Rosa Junior College*
William Barstow, *University of Georgia*
Kirk A. Bartholomew, *Central Connecticut State
    University*
Gail Baughman, *Mira Costa College*
Jane Beiswenger, *University of Wyoming*
Tania Beliz, *College of San Mateo*
Lisa Bellows, *North Central Texas College*
Ernest Benfield, *Virginia Polytechnic Institute*

Rudi Berkelhamer, *University of California, Irvine*
Harry Bernheim, *Tufts University*
Richard Bliss, *Yuba College*
Lawrence Blumer, *Morehouse College*
Mehdi Borhan, *Johnson County Community College*
Kathleen Bossy, *Bryant College*
William Bowen, *University of Arkansas at Little Rock*
Robert Boyd, *Auburn University*
Bradford Boyer, *State University of New York,
    Suffolk County Community College*
Paul Boyer, *University of Wisconsin–Parkside*
William Bradshaw, *Brigham Young University*
Agnello Braganza, *Chabot College*
James Bray, *Blackburn College*
Peggy Brickman, *University of Georgia*
Chris Brinegar, *San Jose State University*
Becky Brown-Watson, *Santa Rosa Junior College*
Charles Brown, *Santa Rosa Junior College*
Carole Browne, *Wake Forest University*
Virginia Buckner, *Johnson County Community
    College*
Joseph C. Bundy, Jr., *University of North Carolina
    at Greensboro*
Warren Buss, *University of Northern Colorado*
Michael Bucher, *College of San Mateo*
Linda Butler, *University of Texas at Austin*
Jerry Button, *Portland Community College*
Carolee Caffrey, *University of California, Los Angeles*
George Cain, *University of Iowa*
James Cappuccino, *Rockland Community College*
M. Carabelli, *Broward Community College*
Cathryn Cates, *Tyler Junior College*
Russell Centanni, *Boise State University*
Van Christman, *Ricks College*
Ruth Chesnut, *Eastern Illinois University*
Vic Chow, *San Francisco City College*
Mary Colavito-Shepanski, *Santa Monica College*
Bob Cowling, *Ouachita Technical College*
Don Cox, *Miami University*
Robert Creek, *Western Kentucky University*
Hillary Cressey, *George Mason University*
Norma Criley, *Illinois Wesleyan University*
Judy Daniels, *Monroe Community College*
Michael Davis, *Central Connecticut State University*
Lewis Deaton, *University of Louisiana-Lafayette*
Lawrence DeFilippi, *Lurleen B. Wallace College*
James Dekloe, *Solano Community College*
Loren Denney, *Southwest Missouri State University*
Jean DeSaix, *University of North Carolina
    at Chapel Hill*
Veronique Delesalle, *Gettysburg College*
Mary Dettman, *Seminole Community College*
Kathleen Diamond, *College of San Mateo*
Alfred Diboll, *Macon College*
Jean Dickey, *Clemson University*
Stephen Dina, *St. Louis University*
Robert P. Donaldson, *George Washington University*
Gary Donnermeyer, *Iowa Central Community
    College*
Charles Duggins, *University of South Carolina*
Susan Dunford, *University of Cincinnati*
Betty Eidemiller, *Lamar University*
Jamin Eisenbach, *Eastern Michigan University*

Norman Ellstrand, *University of California, Riverside*
Thomas Emmel, *University of Florida*
Cindy Erwin, *City College of San Francisco*
David Essar, *Winona State University*
Gerald Esch, *Wake Forest University*
Cory Etchberger, *Longview Community College*
Nancy Eyster-Smith, *Bentley College*
Laurie Faber, *Grand Rapids Community College*
Terence Farrell, *Stetson University*
Shannon Kuchel Fehlberg, *Colorado Christian University*
Jerry Feldman, *University of California, Santa Cruz*
Eugene Fenster, *Longview Community College*
Dino Fiabane, *Community College of Philadelphia*
Kathleen Fisher, *San Diego State University*
Dennis Forsythe, *The Citadel Military College of South Carolina*
Robert Frankis, *College of Charleston*
James French, *Rutgers University*
Bernard Frye, *University of Texas at Arlington*
Anne Galbraith, *University of Wisconsin–LaCrosse*
Robert Galbraith, *Crafton Hills College*
Rosa Gambier, *State University of New York, Suffolk County Community College*
George Garcia, *University of Texas at Austin*
Sandi Gardner, *Triton College*
Gail Gasparich, *Towson University*
Shelley Gaudia, *Lane Community College*
Douglas Gayou, *University of Missouri–Columbia*
Robert Gendron, *Indiana University of Pennsylvania*
Rebecca German, *University of Cincinnati*
Grant Gerrish, *University of Hawaii*
Frank Gilliam, *Marshall University*
Patricia Glas, *The Citadel Military College of South Carolina*
David Glenn-Lewin, *Wichita State University*
Robert Grammer, *Belmont University*
Peggy Green, *Broward Community College*
Miriam L. Greenberg, *Wayne State University*
Sylvia Greer, *City University of New York*
Dana Griffin, *University of Florida*
Richard Groover, *J. Sargeant Reynolds Community College*
Peggy Guthrie, *University of Central Oklahoma*
Maggie Haag, *University of Alberta*
Richard Haas, *California State University, Fresno*
Martin Hahn, *William Paterson College*
Leah Haimo, *University of California, Riverside*
James Hampton, *Salt Lake Community College*
Blanche Haning, *North Carolina State University*
Richard Hanke, *Rose State College*
Laszlo Hanzely, *Northern Illinois University*
David Harbster, *Paradise Valley Community College*
Jim Harris, *Utah Valley Community College*
Mary Harris, *Louisiana State University*
Chris Haynes, *Shelton State Community College*
Janet Haynes, *Long Island University*
Jean Helgeson, *Collin County Community College*
Ira Herskowitz, *University of California, San Francisco*
Paul Hertz, *Barnard College*
Margeret Hicks, *David Lipscomb University*
Jean Higgins-Fonda, *Prince George's Community College*
Phyllis Hirsch, *East Los Angeles College*
William Hixon, *St. Ambrose University*
Carl Hoagstrom, *Ohio Northern University*
Kim Hodgson, *Longwood College*
John Holt, *Michigan State University*

Laura Hoopes, *Occidental College*
Lauren Howard, *Norwich University*
Robert Howe, *Suffolk University*
George Hudock, *Indiana University*
Michael Hudecki, *State University of New York, Buffalo*
Kris Hueftle, *Pensacola Junior College*
Catherine Hurlbut, *Florida Community College*
Charles Ide, *Tulane University*
Mark Ikeda, *San Bernardino Valley College*
Georgia Ineichen, *Hinds Community College*
Charles Jacobs, *Henry Ford Community College*
Fred James, *Presbyterian College*
Ursula Jander, *Washburn University*
Alan Jaworski, *University of Georgia*
R. Jensen, *Saint Mary's College*
Russell Johnson, *Ricks College*
Florence Juillerat, *Indiana University–Purdue University at Indianapolis*
Tracy Kahn, *University of California, Riverside*
Hinrich Kaiser, *Victor Valley College*
Klaus Kalthoff, *University of Texas at Austin*
Tom Kantz, *California State University, Sacramento*
Jennifer Katcher, *Pima Community College*
Marlene Kayne, *The College of New Jersey*
Judy Kaufman, *Monroe Community College*
Mahlon Kelly, *University of Virginia*
Kenneth Kerrick, *University of Pittsburgh at Johnstown*
Joyce Kille-Marino, *College of Charleston*
Joanne Kilpatrick, *Auburn University, Montgomery*
Stephen Kilpatrick, *University of Pittsburgh at Johnstown*
Lee Kirkpatrick, *Glendale Community College*
Peter Kish, *Southwestern Oklahoma State University*
Robert Koch, *California State University, Fullerton*
Eliot Krause, *Seton Hall University*
Kevin Krown, *San Diego State University*
Mary Rose Lamb, *University of Puget Sound*
Carmine Lanciani, *University of Florida*
Vic Landrum, *Washburn University*
Deborah Langsam, *University of North Carolina at Charlotte*
Geneen Lannom, *University of Central Oklahoma*
Brenda Latham, *Merced College*
Steven Lebsack, *Linn-Benton Community College*
Karen Lee, *University of Pittsburgh at Johnstown*
Tom Lehman, *Morgan Community College*
Richard Liebaert, *Linn-Benton Community College*
Kevin Lien, *Portland Community College*
Harvey Liftin, *Broward Community College*
Ivo Lindauer, *University of Northern Colorado*
William Lindsay, *Monterey Peninsula College*
Kirsten Lindstrom, *Santa Rosa Junior College*
Melanie Loo, *California State University, Sacramento*
Dave Loring, *Johnson County Community College*
James Mack, *Monmouth University*
David Magrane, *Morehead State University*
Joan Maloof, *Salisbury State University*
Joseph Marshall, *West Virginia University*
Presley Martin, *Drexel University*
William McComas, *University of Iowa*
Steven McCullagh, *Kennesaw State College*
Mitchell McGinnis, *North Seattle Community College*
James McGivern, *Gannon University*
Colleen McNamara, *Albuquerque TVI Community College*
Scott Meissner, *Cornell University*

Joseph Mendelson, *Utah State University*
Timothy Metz, *Campbell University*
Iain Miller, *University of Cincinnati*
Robert Miller, *University of Dubuque*
V. Christine Minor, *Clemson University*
Brad Mogen, *University of Wisconsin–River Falls*
James Moné, *Millersville University*
Richard Mortensen, *Albion College*
Henry Mulcahy, *Suffolk University*
Christopher Murphy, *James Madison University*
James Nivison, *Mid Michigan Community College*
Peter Nordloh, *Southeastern Community College*
Stephen Novak, *Boise State University*
Bette Nybakken, *Hartnell College*
Michael O'Donnell, *Trinity College*
Karen Olmstead, *University of South Dakota*
Steven Oliver, *Worcester State College*
Steven O'Neal, *Southwestern Oklahoma State University*
Lowell Orr, *Kent State University*
William Outlaw, *Florida State University*
Kevin Padian, *University of California at Berkeley*
Kay Pauling, *Foothill College*
Debra Pearce, *Northern Kentucky University*
David Pearson, *Bucknell University*
Patricia Pearson, *Western Kentucky University*
Kathleen Pelkki, *Saginaw Valley State University*
Andrew Penniman, *Georgia Perimeter College*
Gary Peterson, *South Dakota State University*
Margaret Peterson, *Concordia Lutheran College*
Russell L. Peterson, *Indiana University of Pennsylvania*
Paula Piehl, *Potomac State College*
Ben Pierce, *Baylor University*
Barbara Pleasants, *Iowa State University*
Kathryn Podwall, *Nassau Community College*
Judith Pottmeyer, *Columbia Basin College*
Donald Potts, *University of California, Santa Cruz*
Nirmala Prabhu, *Edison Community College*
James Pru, *Belleville Area College*
Rongsun Pu, *Kean University*
Charles Pumpuni, *Northern Virginia Community College*
Rebecca Pyles, *East Tennessee State University*
Bob Ratterman, *Jamestown Community College*
Jill Raymond, *Rock Valley College*
Brian Reeder, *Morehead State University*
Bruce Reid, *Kean College*
David Reid, *Blackburn College*
Stephen Reinbold, *Longview Community College*
Michael Renfroe, *James Madison University*
Douglas Reynolds, *Central Washington University*
Fred Rhoades, *Western Washington University*
John Rinehart, *Eastern Oregon University*
Laura Ritt, *Burlington County College*
Lynn Rivers, *Henry Ford Community College*
Bruce Robart, *University of Pittsburgh at Johnstown*
Jennifer Roberts, *Lewis University*
Laurel Roberts, *University of Pittsburgh*
Duane Rohlfing, *University of South Carolina*
Jeanette Rollinger, *College of the Sequoias*
Steven Roof, *Fairmont State College*
Jim Rosowski, *University of Nebraska*
Stephen Rothstein, *University of California, Santa Barbara*
Donald Roush, *University of North Alabama*
Lynette Rushton, *South Puget Sound Community College*

Linda Sabatino, *State University of New York, Suffolk County Community College*
Douglas Schamel, *University of Alaska, Fairbanks*
Douglas Schelhaas, *University of Mary*
Fred Schindler, *Indian Hills Community College*
Robert Schoch, *Boston University*
Brian Scholtens, *College of Charleston*
Julie Schroer, *Bismarck State College*
Fayla Schwartz, *Everett Community College*
Judy Shea, *Kutztown University of Pennsylvania*
Thomas Shellberg, *Henry Ford Community College*
Lisa Shimeld, *Crafton Hills College*
Brian Shmaefsky, *Kingwood College*
Mark Shotwell, *Slippery Rock University*
Jane Shoup, *Purdue University*
Michele Shuster, *New Mexico State University*
Linda Simpson, *University of North Carolina at Charlotte*
Gary Smith, *Tarrant County Junior College*
Phil Snider, *University of Houston*
Gary Sojka, *Bucknell University*
Ralph Sorensen, *Gettysburg College*
David Stanton, *Saginaw Valley State University*

Amanda Starnes, *Emory University*
John Stolz, *Duquesne University*
Ross Strayer, *Washtenaw Community College*
Donald Streuble, *Idaho State University*
Mark Sugalski, *New England College*
Gerald Summers, *University of Missouri–Columbia*
Marshall Sundberg, *Louisiana State University*
Christopher Tabit, *State University of West Georgia*
David Tauck, *Santa Clara University*
Hilda Taylor, *Acadia University*
Kenneth Thomas, *Hillsborough Community College*
Kathy Thompson, *Louisiana State University*
Laura Thurlow, *Jackson Community College*
Anne Tokazewski, *Burlington County College*
Bruce Tomlinson, *State University of New York, Fredonia*
Nancy Tress, *University of Pittsburgh at Titusville*
Donald Trisel, *Fairmont State College*
Kimberly Turk, *Mitchell Community College*
Michael Twaddle, *University of Toledo*
Leslie VanderMolen, *Humboldt State University*
John Vaughan, *Georgetown College*

Ann Vernon, *St. Charles County Community College*
Rukmani Viswanath, *Laredo Community College*
Mary Beth Voltura, *State University of New York, Cortland*
Jerry Waldvogel, *Clemson University*
Robert Wallace, *Ripon College*
Patricia Walsh, *University of Delaware*
James Wee, *Loyola University*
Harrington Wells, *University of Tulsa*
Larry Williams, *University of Houston*
Mary Jo Witz, *Monroe Community College*
Neil Woffinden, *University of Pittsburgh at Johnstown*
Mark Wygoda, *McNeese State University*
William Yurkiewicz, *Millersville University of Pennsylvania*
Ray S. Williams, *Appalachian State University*
Sandra Winicur, *Indiana University, South Bend*
Patrick Woolley, *East Central College*
Tony Yates, *Seminole State College*
William Yurkiewicz, *Millersville University*
Edward J. Zalisko, *Blackburn College*
Uko Zylstra, *Calvin College*

Supplements for the Instructor

Lectures

Instructor Resource CD/DVD Set for *Biology: Concepts & Connections,* Sixth Edition

(0-321-54824-8/978-0-321-54824-5)
The instructor media for *Biology: Concepts & Connections,* Sixth Edition, is combined into one chapter-by-chapter resource along with a Quick Reference Guide. CD/DVDs provide convenient one-stop access to all the visual media for each chapter. Assets are now organized by chapter folders, making it easier to access files. The CD/DVDs include the following:

Prepared PowerPoint™ Tools

The following lecture tools are provided in Powerpoint™ format.

- **Lecture Presentations for every chapter** allow instructors to lecture immediately and with confidence, supported by a rich array of content, including additional notes and examples not found in the text, plus links to videos and animations, including BioFlix™ 3-D animations.

- **Art and Photos** includes all the art, photos, and tables from the book embedded in PowerPoint™. All figure labels are editable and selected figures are layered for step-by-step presentation.

- More than 100 **Extra Animations** on topics across all units provide additional lecture resources.

- **Introductory Biology Lecture Launcher (Discovery Channel™) Videos** bring biology to life and show scientists in action. Questions before and after each video elicit student input and teaching tips in the Notes field provide alternative questions to spark discussions.

- **Active Lecture Questions** for use with and without clickers help prompt classroom discussions and elicit student input.

- **Lectures accompanying *Current Issues in Biology* (volumes 1–5) articles from *Scientific American*** features concept summaries, class response questions, and art and images from every article to enhance class discussions on issues of interest to students.

- **Quiz Show** offers unit-level quizzes in an interactive game show format.

JPEGs of Art and Photos

- All the art, photos, and tables from the book with and without labels are provided.

- More than 270 extra photos provide additional resources for your lectures.

Animation and Video Files

- **BioFlix™ 3-D Animations** invigorate lectures with movie-quality 3-D animations. Topics include: Tour of an Animal Cell, Tour of a Plant Cell, Cellular Respiration, Photosynthesis, Mitosis, Meiosis, Protein Synthesis, and How Neurons Work.

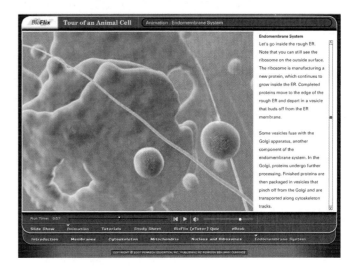

- Over 200 animations in Flash format

- 90 videos in Flash and MPEG formats

Also available in the Instructor Resource CD/DVD set is the *Instructor Guide* in Word format. The **Test Bank CD-ROM** includes test bank questions in Pearson's exclusive TestGen software, which allows instructors to make use of the same high-quality questions on Course Compass™, Blackboard, and WebCT, with results that flow instantly into their online GradeBooks. The Test Bank is also available in Word format.

Instructor's Guide to Text and Media, Sixth Edition

(0-321-54783-7/978-0-321-54783-5)
Edward Zalisko, *Blackburn College*
This newly streamlined guide provides an outstanding set of resources, including objectives, a guide to teaching resources, teaching tips and activities, and a section devoted to student misconceptions about subject matter, for each chapter. References to all media resources available to instructors and students, plus lists of key terms and relevant word roots, are also provided.

Transparency Acetates for *Biology: Concepts & Connections,* Sixth Edition

(0-321-54731-4/978-0-321-54731-6)
More than 650 full-color acetates include all illustrations and tables from the text, many of which incorporate photographs. Selected figures illustrating key concepts are broken down into layers for step-by-step lecture presentation.

Assessment

Printed Test Bank for *Biology: Concepts & Connections,* Sixth Edition

(0-3215-4792-6/978-0-321-54792-7)

Thoroughly revised and updated, and carefully reviewed for accuracy, the test bank features 40–70 multiple-choice questions per chapter, authored by instructors who are experts in each subject area. Chapters now also feature brand-new, art-based scenario questions to stimulate visual and critical thinking. The Test Bank is also available as part of the Instructor Resource CD/DVD in Pearson's exclusive TestGen software.

Course Management

CourseCompass™ with eBook for *Biology: Concepts & Connections,* Sixth Edition

(0-321-54785-3/978-0-321-54785-9)

URL: http://cmc.aw.com/coursecompass

CourseCompass™ combines the strength of Benjamin Cummings content with state-of-the-art eLearning tools, including a complete electronic version of the text. CourseCompass™ is a nationally hosted, dynamic, interactive online course management system powered by Blackboard, leaders in the development of Internet-based learning tools. This easy-to-use and customizable program enables instructors to tailor content and functionality to meet individual course needs. Every CourseCompass™ course includes a range of preloaded content, such as testing and assessment question pools, chapter-level objectives, chapter summaries, photos, illustrations, videos, animations, and web activities—all designed to help students master core course objectives.

CourseCompass™ for *Biology: Concepts & Connections,* Sixth Edition

(0-321-54779-9/, 978-0-321-54779-8)

CourseCompass™ Student Access Kit for *Biology: Concepts & Connections,* Sixth Edition

(0-321-54784-5/978-0-321-54784-2)

WebCT Premium for *Biology: Concepts & Connections,* Sixth Edition

(0-321-54787-X/978-0-321-54787-3)

WebCT Student Access Kit for *Biology: Concepts & Connections,* Sixth Edition

(0-321-54823-X/978-0-321-54823-8)

WebCT Open Access for *Biology: Concepts & Connections,* Sixth Edition

(0-321-54780-2/978-0-321-54780-4)

Blackboard Premium for *Biology: Concepts & Connections,* Sixth Edition

(0-321-54781-0/978-0-321-54781-1)

Blackboard Open Access for *Biology: Concepts & Connections,* Sixth Edition

(0-321-54782-9/978-0-321-54782-8)

Blackboard Student Access Kit for *Biology: Concepts & Connections,* Sixth Edition

(0-321-54823-X/978-0-321·54823-8)

Biology: Concepts & Connections, Sixth Edition, CourseSmart eTextbook

(0-321-54447-1/978-0-321-54447-6)

CourseSmart is an exciting new *choice* for students looking to save money. As an alternative to purchasing the print textbook, students can purchase an electronic version of the same content and save up to 50% off the suggested list price of the print text. With a CourseSmart eTextbook, students can search the text, make notes online, print out reading assignments that incorporate lecture notes, and bookmark important passages for later review. For more information, or to purchase access to the CourseSmart eTextbook, visit www.coursesmart.com.

Labs

Annotated Instructor's Edition for Laboratory Investigations for Biology, Second Edition

(0-8053-6792-6/978-0-805-36792-8)

Jean L. Dickey, *Clemson University*

The *Annotated Instructor's Edition* includes margin notes with teacher overviews, time requirements, helpful hints, and suggestions for extending or supplementing labs; answers to questions in the Student Edition; and suggestions for adapting the labs to a 2-hour period.

Preparation Guide for Laboratory Investigations for Biology, Second Edition

(0-8053-6771-3/978-0-805-36771-3)

Jean L. Dickey, *Clemson University*

This guide provides complete instructions on how to order materials and plan, set up, and run labs smoothly.

Symbiosis Lab Authoring Kit

Build a customized lab manual, choosing the labs you want, importing artwork from our graphics library, and even adding your own notes, syllabi, or other material.

Instructor's Lab Manual for BiologyLabs On-Line

(0-8053-7018-8/978-0-805-37018-8)

URL: http://www.biologylabsonline.com

This printed manual provides the individual and group assignments from the Student Lab Manual for BiologyLabs On-Line. Includes a subscription to BiologyLabs On-Line.

Explorations in Basic Biology, Eleventh Edition

(0-132-22913-7/978-0-132-22913-5)

Stanley E. Gunstream

Thinking About Biology: An Introductory Laboratory Manual, Third Edition

(0-132-30736-7/978-0-132-30736-9)

Kathleen T. McWhorter, *Niagara County Community College;* Mimi Bres, *Prince George's Community College;* Arnold Weisshaar, *Prince George's Community College*

Supplements for the Student

Books a la Carte for *Biology: Concepts & Connections*, Sixth Edition

(0-321-54789-6/978-0-321-54789-7)

Study Tools

 mybiology for *Biology: Concepts & Connections*, Sixth Edition

(www.mybiology.com)

mybiology is a comprehensive study resource that provides all of the study tools students need, including quizzes, an integrated eBook, MP3 Tutors, and BioFlix™ movie-quality animations that tackle the toughest topics in biology. The four-step structure of **mybiology** allows students to identify what they don't know, select the appropriate learning activity to match their learning style, and practice outside of lecture. This edition includes 19 additional MP3 Tutors on tough terms, narrated by co-author Eric Simon, providing a unique alternative to flashcards.

BioFlix

NEW! BioFlix are interactive, self-paced tutorials that help you master the toughest topics in biology. You can prepare for exams by viewing a 3-D animation, completing the tutorial, printing a personal review sheet, and taking a quiz.

BioFlix Tutorial Topics
- Tour of an Animal Cell
- Tour of a Plant Cell
- Cellular Respiration
- Photosynthesis
- Mitosis
- Meiosis
- Protein Synthesis

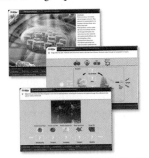

mybiology.com Premium Website

(0-321-54728-4/978-0-321-54728-6)

mybiology.com Website with eBook

(0-321-54729-2/978-0-321-54729-3)

Study Guide for *Biology: Concepts & Connections*, Sixth Edition

(0-321-54825-6/978-0-321-54825-2)

Richard Liebaert, *Linn-Benton Community College*

Students can master key concepts and earn a better grade with the thought-provoking exercises found in this study guide. A wide range of questions and activities help students test their understanding of biology. The study guide also includes key terms, word roots, and a listing of student media activities at **mybiology**.

Study Card for *Biology: Concepts & Connections*, Sixth Edition

(0-321-54777-2/978-0-321-54777-4)

This fold-out quick-reference card provides students with an overview of the entire book, helping them see the connections between topics and understand the big picture.

Current Topics in Biology

Exclusive to Benjamin Cummings and free when bundled with Benjamin Cummings' nonmajors biology texts, each volume of *Current Issues* contains a set of timely and accessible articles from *Scientific American* that are sure to stimulate thought among students. Volume 5, the most recent edition, includes articles on deciphering diet advice, the genetics of alcoholism, and the role of plants in global warming, among others. Each article ends with a set of questions that allow students to test their comprehension, practice their writing skills, and apply their imagination to the topic.

Current Issues in Biology, Vol 1

(0-805-37507-4/978-0-805-37507-7)

Current Issues in Biology, Vol 2

(0-805-37108-7/978-0-805-37108-6)

Current Issues in Biology, Vol 3

(0-805-37527-9/978-0-805-37527-5)

Current Issues in Biology, Vol 4

(0-805-33566-8/978-0-805-33566-8)

Current Issues in Biology, Vol 5

(0-321-54187-1/978-0-321-54187-1)

Special Topics in Biology (booklets)

- *Alzheimer's Disease*
  (978-0-1318-3834-5/0-1318-3834-2)

- *Biological Terrorism*
  (978-0-8053-4868-2/0-8053-4868-9)

- *Biology of Cancer*
  (978-0-8053-4867-5/0-8053-4867-0)

- *Emerging Infectious Diseases*
  (978-0-8053-3955-0/0-8053-3955-8)

- *Gene Therapy*
  (978-0-8053-3819-5/0-8053-3819-5)

- *Genetic Testimony: A Guide to Forensic DNA Profiling*
  (978-0-1314-2338-1/0-1314-2338-X)

- *HIV and AIDS*
  (978-0-8053-3956-7/0-8053-3956-6)

- *Mad Cows and Cannibals: A Guide to the Transmissible Spongiform Encephalopathies*
  (978-0-1314-2339-8/0-1314-2339-8)

- *Stem Cells and Cloning*
  (978-0-8053-4864-4/0-8053-4864-6)

- *Understanding the Human Genome Project*
  (978-0-8053-4877-4/0-8053-4877-8)

Labs

Laboratory Investigations for Biology, Second Edition

(0-8053-6789-6/978-0-805-36789-8)
Jean L. Dickey, *Clemson University*
An investigative approach actively involves students in the process of scientific discovery by allowing them to make observations, devise techniques, and draw conclusions. Twenty carefully chosen laboratory topics encourage students to use their critical thinking skills to solve problems using the scientific method.

Student Lab Manual for BiologyLabs On-Line, First Edition

(0-8053-7017-X/978-0-805-37017-1)
This printed student manual provides background information, instructions, assignments, and group assignments for Biology-Labs On-Line. Includes a subscription to BiologyLabs On-Line.

Explorations in Basic Biology, Eleventh Edition

(0-132-22913-7/978-0-132-22913-5)
Stanley E. Gunstream

Thinking About Biology: An Introductory Laboratory Manual, Third Edition

(0-132-30736-7/978-0-132-30736-9)
Kathleen T. McWhorter, Niagara County Community College; Mimi Bres, Prince George's Community College; Arnold Weisshaar, Prince George's Community College

Detailed Contents

UNIT TWO Cellular Reproduction and Genetics

9 Patterns of Inheritance 152

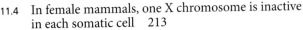

17 Plants, Fungi, and the Colonization of Land 340

18 The Evolution of Invertebrate Diversity 364

19   The Evolution of Vertebrate Diversity   388

UNIT FIVE   Animals: Form and Function

20   Unifying Concepts of Animal Structure and Function   412

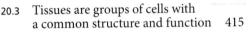

21   Nutrition and Digestion   428

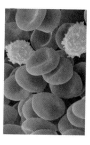

25 Control of Body Temperature and Water Balance 504

26 Hormones and the Endocrine System 516

27 Reproduction and Embryonic Development 532

UNIT SEVEN Ecology

38 Conservation Biology   762

Biology:
Exploring Life

Dining in the Trees

A female leopard (*Panthera pardus*) and her young daughter have just killed a duiker, a small antelope and a preferred leopard meal. The pair could start their feast on the spot. But the mother instinctively knows that eating in the open might allow other predators, such as lions or hyenas, to steal their kill. The duiker is best consumed up in a tree, out of reach of thieves. The older leopard has hauled the family's prey up into the branches many times before. But this afternoon, the cub follows her mother's example and carries the meal herself. The mother, featured on the cover of this book, looks up from below. She'll wait to eat, watching first to see how well her daughter has learned the lessons she'll need to survive.

Leopards are the strongest climbers among the big cats that make up the genus *Panthera*. Scientific evidence indicates that members of *Panthera,* which include leopards, lions, tigers, and jaguars, last shared a common ancestor about 6 million years ago. This linkage still shows in these cats' large size, hunting prowess, and unique ability to roar. The leopard species, which is found across Africa and Asia, has the widest distribution of these big cats and displays many adaptations that enhance its survival.

African leopards, for example, have features that help them thrive alongside Africa's regal lion. Leopards tend to be nocturnal hunters, whereas lions hunt during the day. And leopards take a much larger variety of prey items than lions. With average weights of about 58 kilograms (128 pounds) for males and 38 kilograms (84 pounds) for females, African leopards are only about a third the size of a typical lion. But leopards are slightly faster, with top running speeds of 64 kilometers (40 miles) per hour.[1] And with superior jumping skills that propel leopards to 6 meters (20 feet) through the air horizontally and about 3 meters vertically, the cat is an exceptional climber. Powerful jaw, neck, and shoulder muscles give the leopard the ability to pounce and kill prey with a quick bite and the additional strength needed to haul prey upward, sometimes dragging a carcass as high as 15 meters into a tree.

Other characteristics of the leopard contribute to their survival. With the exception of mothers rearing cubs, leopards are generally loners and maintain separate, nonoverlapping territories. Thus, they do not compete with each other for prey. Sensitive whiskers on the face and head help the cat navigate through dense vegetation while hunting at night. The leopard's signature spotted coat, often an array of black patches dappled across tawny or

ochre fur, provides excellent camouflage in the dry grasses of the African plains. Leopards also thrive in desert areas, hilly country, or jungle terrain, and their coat color can vary from pale cream to deep gold, enabling them to blend in with their surroundings.

This habitat flexibility of leopards is linked to their amazing diet flexibility. Leopards consume everything from beetles and birds to baboons, hares, and smaller cats. Antelopes such as the duiker or the larger impala are a frequent favorite among leopards living in the African savannah. Observers have noted leopards killing impala larger than themselves and still managing to drag the carcass high into tree branches.

So far, the African leopard has fared better in the modern era than its leopard cousins farther north. In Asia, the pressures of human population growth mean less land for leopards to roam, less prey, and more hunters after the cats' beautiful fur. In southern Africa, vast protected refuges, such as the Sabi Sands Reserve of South Africa, where the leopard on the cover was photographed, help maintain the cats' numbers. But the African leopard still faces threats from illegal hunting. Farmers and ranchers, enraged when leopards snatch their livestock, also target the cats.

With unique characteristics that let them thrive across a diversity of habitats, leopards offer wonderful examples of the adaptations of organisms that fit them to their environment. Traits such as coat camouflage, strong jaws, and climbing prowess make leopards successful hunters and enhance their ability to survive and reproduce. These adaptations are the result of evolution, the process of change that has transformed life on Earth from its earliest beginnings to the diversity of organisms living today. In this chapter, we begin our exploration of **biology**, the scientific study of life. We start by introducing some common themes in biology, which will help guide our study of life, its evolution, and all its diversity. ■ ■ ■

---

[1]You can review metric measurements in Appendix 1.

1.1 In life's hierarchy of organization, new properties emerge at each level

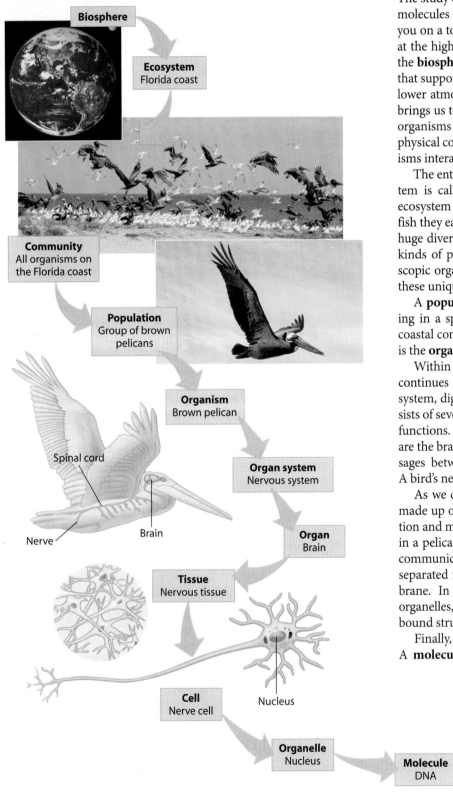

Biosphere

Ecosystem
Florida coast

Community
All organisms on
the Florida coast

Population
Group of brown
pelicans

Organism
Brown pelican

Organ system
Nervous system

Spinal cord

Brain

Nerve

Organ
Brain

Tissue
Nervous tissue

Cell
Nerve cell

Nucleus

Organelle
Nucleus

Molecule
DNA

Atom

Figure 1.1 Life's hierarchy of organization

The study of life extends from the microscopic scale of cells and molecules to the global scale of the biosphere. Figure 1.1 takes you on a tour of these levels of biological organization, starting at the highest level. At the upper left we take a distant view of the **biosphere,** which consists of all the environments on Earth that support life—most regions of land, bodies of water, and the lower atmosphere. A closer look at one of these environments brings us to the level of an **ecosystem,** which consists of all the organisms living in a particular area, as well as all the nonliving, physical components of the environment with which the organisms interact, such as air, soil, water, and sunlight.

The entire array of organisms inhabiting a particular ecosystem is called a **community.** The community in the coastal ecosystem shown in Figure 1.1 includes the pelicans and the fish they eat, as well as seagulls; raccoons that eat pelican eggs; a huge diversity of insects, molluscs, and worms; many different kinds of plants and fungi; and enormous numbers of microscopic organisms, such as protists, algae, and bacteria. Each of these unique forms of life is called a species.

A **population** consists of all the individuals of a species living in a specified area, such as all the brown pelicans in the coastal community. Below the population level in the hierarchy is the **organism,** an individual living thing.

Within a complex organism such as a pelican, life's hierarchy continues to unfold. An **organ system,** such as a circulatory system, digestive system, or nervous system (shown here) consists of several **organs** that work together in performing specific functions. For instance, the main organs of the nervous system are the brain, the spinal cord, and the nerves that transmit messages between the spinal cord and other parts of the body. A bird's nervous system controls its ability to fly.

As we continue down through the hierarchy, each organ is made up of several different **tissues,** each with a specific function and made up of a group of similar cells. The nervous tissue in a pelican's brain has millions of nerve cells organized into a communication network of spectacular complexity. A **cell** is separated from its environment by a boundary called a membrane. In the nerve cell shown here, you can see several organelles, such as the nucleus. An **organelle** is a membrane-bound structure that performs a specific function in a cell.

Finally, we reach the level of molecules in the hierarchy. A **molecule** is a cluster of atoms held together by chemical bonds. We show as our example DNA (deoxyribonucleic acid). DNA molecules provide the blueprint for constructing the organism's other molecules and transmit this information from parents to offspring. In the computer graphic of a small section of DNA at the bottom of Figure 1.1, each of the spheres represents an atom, the smallest particle of ordinary matter.

Now work your way in the opposite direction in Figure 1.1, building life's hierarchy from molecules to the biosphere. It takes many molecules to build organelles and make a cell, many cells to make a tissue, several kinds of tissues to make an organ, and so on. At each new level, notice that there are novel properties that emerge, properties that were not part of the components of the preceding level. For example, life emerges at the level of the cell—a test tube full of organelles is not alive. Such properties illustrate an important theme of biology, called **emergent properties.** The familiar saying that "the whole is greater than the sum of its parts" captures this idea. The emergent properties of the whole result from the specific arrangement and interactions of the component parts.

In the next two modules, we look at more themes in the study of life: the interaction of organisms with the environment and the cell as the structural and functional unit of life.

Web Activity *The Levels of Life Card Game*

? Which of the following levels of biological organization includes all others in the list: cell, molecule, organ, tissue?

■ Organ

1.2 Living organisms interact with their environments, exchanging matter and energy

Organisms interact with both the living and nonliving components of their environment. Figure 1.2 is a simplified diagram of these interactions in an African savannah. Plants and other photosynthetic organisms are the **producers** that provide the food for a typical ecosystem. A tree, for example, absorbs water (H_2O) and minerals from the soil, and its leaves take in carbon dioxide (CO_2) from the air. In photosynthesis, a tree's leaves use energy from sunlight to convert CO_2 and H_2O to sugar and oxygen (O_2). The tree releases O_2 to the air, and its roots help form soil by breaking up rocks. Thus, both organism and environment are affected by the interactions between them.

The **consumers** of the ecosystem eat plants and other animals. The giraffe in Figure 1.2 eats leaves; the leopard described in the chapter introduction eats meat. All animals take in oxygen from the air and release carbon dioxide. Their wastes return other chemicals to the environment.

Another vital part of the ecosystem includes the bacteria, fungi, and small animals in the soil that decompose wastes and the remains of dead organisms. These decomposers act as recyclers, changing complex matter into simpler mineral nutrients that plants can use.

The dynamics of ecosystems include two major processes—the recycling of chemical nutrients and the flow of energy. These processes are illustrated in Figure 1.2. The most basic chemicals necessary for life—carbon dioxide, oxygen, water, and various minerals—flow from the air and soil to plants, to animals and decomposers, and back to the air and soil. As shown by the blue arrows in the figure, chemical nutrients cycle within an ecosystem.

By contrast, an ecosystem gains and loses energy constantly. Energy flows into the ecosystem when plants and other photosynthesizers absorb light energy from the sun (yellow arrow) and convert it to the chemical energy of sugars and other complex molecules. Chemical energy (green arrow) is then passed through a series of consumers and, eventually, decomposers, powering each organism in turn. In the process of these energy conversions between and within organisms, some

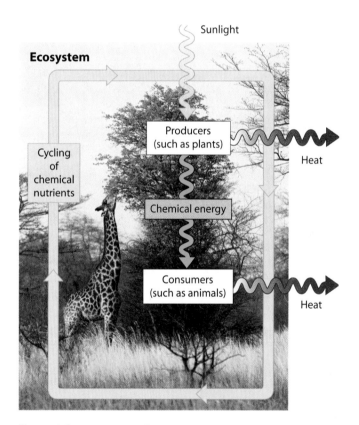

Figure 1.2 The cycling of nutrients and flow of energy in an ecosystem

energy is converted to heat, which is then lost from the system (red arrows). In contrast to chemical nutrients, which recycle *within* an ecosystem, energy flows *through* an ecosystem, entering as light and exiting as heat.

? Explain how the photosynthesis of plants functions in both the cycling of chemical nutrients and the flow of energy in an ecosystem.

■ Photosynthesis uses light to convert carbon dioxide and water to energy-rich food, making it the pathway by which both chemical nutrients and energy become available to most organisms.

1.3 Cells are the structural and functional units of life

The cell has a special place in the hierarchy of biological organization. It is the level at which the properties of life emerge—the lowest level of structure that can perform all activities required for life. A cell can regulate its internal environment, take in and use energy, respond to its environment, and develop and maintain its complex organization. The ability of cells to give rise to new cells is the basis for all reproduction and for the growth and repair of multicellular organisms. Your every movement and thought are based on the activities of muscle cells and nerve cells. Even a global process such as the cycling of carbon is the result of cellular activities, including the photosynthesis of plant cells and the cellular respiration of nearly all cells, a process that breaks down sugar for energy and releases carbon dioxide.

Cells illustrate another theme of biology: *the correlation of structure and function.* Experience shows you that form generally fits function. A screwdriver tightens or loosens screws, a

hammer pounds nails. Because of their form, these tools can't do each other's jobs. Applied to biology, this theme of form fitting function is a guide to the structure of life at all its organizational levels. As you can see in Figure 1.1, the long extension of the nerve cell enables it to transmit impulses throughout the body. Often, analyzing a biological structure gives us clues about what it does and how it works.

The properties of life emerge from the ordered interactions of the structures of a cell. Such a combination of components forms a more complex organization called a *system.* Cells are examples of biological systems, as are organisms and ecosystems. Systems and their emergent properties are not unique to life. Consider a box of bicycle parts. When all of the individual parts are properly assembled, the result is a mechanical system you can use for exercise or transportation.

The emergent properties of life, however, are particularly challenging to study because of the unrivaled complexity of biological systems. At the cutting edge of large-scale research today is an approach called **systems biology.** The goal of systems biology is to construct models for the dynamic behavior of whole biological systems, ranging from the functioning of the biosphere to the complex molecular machinery of a cell.

All cells share many characteristics. For example, every cell is enclosed by a membrane that regulates the passage of materials between the cell and its surroundings. And every cell uses DNA as its genetic information. There are two basic kinds of cells. Figure 1.3 shows artificially colored photographs of these cells taken with an electron microscope. A **prokaryotic cell** is much simpler and usually much smaller than a eukaryotic cell. The cells of the microorganisms we commonly call bacteria are prokaryotic. Forms of life such as plants, animals, and fungi are composed of eukaryotic cells. As you can see in Figure 1.3, a **eukaryotic cell** is subdivided by internal membranes into many different functional compartments, or organelles, including the nucleus that houses the cell's DNA.

In the next module, we see how DNA unifies all of life, as we begin to explore the core theme of life—evolution.

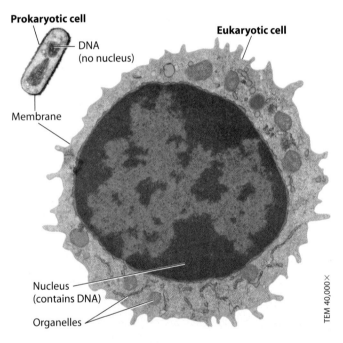

Figure 1.3 Contrasting the size and complexity of prokaryotic and eukaryotic cells. (Cells are shown approximately 40,000 times their real size.)

? Explain why cells are considered the basic units of life.

■ They are the lowest level in the hierarchy of biological organization at which the properties of life emerge.

1.4 The unity of life: All forms of life have common features

All cells have DNA, and the continuity of life is based on this genetic material. DNA is the substance of **genes,** the units of inheritance that transmit information from parents to offspring. The molecular structure of DNA accounts for this function. Let us explain: Each DNA molecule is made up of two long chains coiled together into what is called a double helix (see Figure 1.1). The chains are made up of four kinds of chemical building blocks.

Figure 1.4A on the next page diagrams a short section of one of the chains and indicates the four different building blocks, called bases, with different colors and letter abbreviations of their names.

The way DNA encodes a cell's information is analogous to the way we arrange letters of the alphabet into precise sequences with specific meanings. The word *rat,* for example, conjures up an image of a rodent; *tar* and *art,* which contain the same letters,

A
C
T
A
T
A
C
C
G
T
A
G
T
A

Figure 1.4A One chain of a DNA molecule, its message written in the order of its four bases

mean very different things. We can think of the four building blocks as the alphabet of inheritance. Specific sequential arrangements of these four chemical letters encode precise information in genes, which are typically hundreds or thousands of "letters" long. A bacterial gene may direct the cell to "Build a purple pigment." A particular human gene may mean "Make the hormone insulin." All forms of life use essentially the same genetic code, which has made it possible to engineer cells that produce proteins normally found only in some other organism. Thus, bacteria can be used to produce insulin by inserting a gene for human insulin into bacterial cells.

The diversity of life arises from differences in DNA sequences—in other words, from variations on the common theme of storing genetic information in DNA. Bacteria and humans are different because they have different genes. But both sets of instructions are written in the same language.

Figure 1.4B illustrates some of the other properties that are common to all organisms.

(1) *Order.* All living things exhibit complex organization, as seen in the highly ordered structure of this sunflower.

(2) *Regulation.* The environment outside an organism may change markedly, but mechanisms maintain an organism's internal environment within limits that sustain life. Regulation of the amount of blood flowing through the blood vessels in the large ears of this jackrabbit helps maintain a constant body temperature by adjusting heat exchange with the air.

(3) *Growth and development.* Inherited information carried by genes controls the pattern of growth and development of organisms such as this Nile crocodile.

(4) *Energy processing.* Organisms take in energy and transform it to perform all of life's activities. When this bear eats the fish, it will use the chemical energy stored in the fish to power its own activities and chemical reactions (metabolism).

(5) *Response to the environment.* All organisms respond to environmental stimuli. This Venus flytrap closed its trap rapidly in response to the stimulus of a damselfly landing on it.

(6) *Reproduction.* Organisms reproduce their own kind. This emperor penguin is protecting its baby.

(7) *Evolutionary adaptation.* The appearance of this pygmy seahorse camouflages it in its environment. Such adaptations evolve over many generations as individuals with traits best suited to their environment pass them to offspring.

The organisms in Figure 1.4B illustrate some of the amazing diversity of life. In the next module, we see how biologists attempt to organize this diversity.

? What is the chemical basis for all of life's kinship?

■ DNA as the genetic material

(1) Order   **(2) Regulation**   **(3) Growth and development**   **(4) Energy processing**

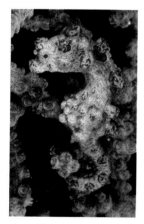

(5) Response to the environment   **(6) Reproduction**   **(7) Evolutionary adaptation**

Figure 1.4B Some important properties of life

The diversity of life can be arranged into three domains

We can think of biology's enormous scope as having two dimensions. The "vertical" dimension, which we examined in Module 1.1, is the size scale that stretches from molecules to the biosphere. But biology's scope also has a "horizontal" dimension, spanning across the great diversity of organisms existing now and over life's long history.

Grouping Species Diversity is a hallmark of life. Biologists have so far identified and named about 1.8 million **species,** the term used for a particular type of organism, such as *Panthera pardus,* the leopard. Researchers identify thousands of additional species each year. Estimates of the total number of species range from 10 million to over 400 million. As an example, Figure 1.5A shows a small sample of the diversity of moths and butterflies. How do we make sense of so much diversity?

There seems to be a human tendency to group diverse items according to similarities. We may speak of bears or butterflies, though we recognize that each group includes many different species. We may even sort groups into broader categories, such as mammals and insects. Taxonomy, the branch of biology that names and classifies species, arranges them into a hierarchy of broader and broader groups, from genus, family, order, class, and phylum, to kingdom.

The Three Domains of Life Until the last few decades, most biologists adopted a taxonomic scheme that divided all of the diversity of life into five kingdoms. But new methods, such as comparison of DNA sequences, have led to an ongoing reevaluation of the number and boundaries of kingdoms. New classification schemes have been proposed that range from six to dozens of kingdoms. As that debate continues, however, most biologists now agree that the kingdoms of life can be organized into three overarching groups called **domains.** The organisms in Figure 1.5B on the facing page are representatives of the three domains.

Domain **Bacteria** and domain **Archaea** both consist of prokaryotes, organisms with prokaryotic cells. Most prokaryotes are unicellular and microscopic. The photos of these prokaryotes in Figure 1.5B were made with a scanning electron microscope, and the number along the side indicates how many times these images were magnified. (We will discuss microscopy in Chapter 4.) In the five-kingdom system, bacteria and archaea were combined in a single kingdom. But newer evidence from comparisons of DNA and other molecules suggests that they represent two very distinct branches of life, as you'll learn in Chapters 15 and 16.

Bacteria are the most diverse and widespread prokaryotes and are now divided among several kingdoms. In the photo of bacteria in Figure 1.5B, each of the rod-shaped structures is a bacterial cell.

Many of the prokaryotes known as archaea live in Earth's extreme environments, such as salty lakes and boiling hot springs. Domain Archaea also includes multiple kingdoms. The photo of archaea in Figure 1.5B shows a colony composed of many cells.

Figure 1.5A   Drawers of diversity: some of the tens of thousands of species in the moth and butterfly collection at the National Museum of Natural History in Washington, D.C.

All the eukaryotes, organisms with eukaryotic cells, are now grouped into the various kingdoms of domain **Eukarya.** As you learned in Module 1.3, eukaryotic cells have a nucleus and other internal structures called organelles.

Most aquatic or moist habitats support members of domain Eukarya called protists. Protists are a diverse collection of mostly single-celled organisms. Some protists, including those commonly called algae, make their own food by photosynthesis. Another assortment of protists, commonly called protozoans, are animal-like in that they eat other organisms. The top center photo in Figure 1.5B shows a number of different protists in a drop of water viewed with a microscope (magnified 275 times). The large, irregular, bluish cell in the center is an amoeba, and the smaller cells are mostly single-celled algae. Also present are multicellular algae (the rodlike organisms). In the five-kingdom scheme, protists were classified in a single kingdom, but evidence from molecular studies now indicates that they include several different evolutionary lineages. The recent trend has been to split the protists into several kingdoms to accurately reflect their evolutionary relationships.

The three remaining kingdoms within Eukarya contain multicellular eukaryotes. These three kingdoms are distinguished partly by their modes of nutrition. Kingdom Plantae consists of plants, which produce their own food by photosynthesis and have cells with rigid walls made of cellulose. The photo representing kingdom Plantae in Figure 1.5B is a tropical bromeliad, a plant native to the Americas.

Kingdom Fungi, represented by the mushrooms in Figure 1.5B, is a diverse group that includes molds, yeasts, and mushrooms. Fungi are mostly decomposers. They break down the remains of dead organisms and organic wastes, such as leaf litter and animal feces, and absorb the nutrients into their cells.

Animals obtain food by ingestion, which means they eat other organisms. Most animals are motile and are made of cells that lack rigid walls. Representing the kingdom Animalia (animals), the sloth in Figure 1.5B resides in the trees of American rain forests. There are actually members of three groups in the sloth photo. The sloth is clinging to a tree (kingdom Plantae), and the greenish tinge in the animal's hair is a luxuriant growth of photosynthetic prokaryotes (domain Bacteria). This photograph exemplifies a theme reflected in our book's title: connections between living things. The sloth depends on trees for food and shelter; the tree uses nutrients from the decomposition of the sloth's feces; the prokaryotes gain access to the sunlight necessary for photosynthesis by living on the sloth; and the sloth is camouflaged from predators by its green coat.

Life's diversity and its interconnectedness are evident almost everywhere. We have looked at life's unity and surveyed its diversity. In the next module, we describe how evolution explains both the unity and the diversity of life.

Web Activity *Classification Schemes*

? To which of the three domains of life do we belong?

■ Eukarya

Domain Bacteria

Bacteria (multiple kingdoms)

SEM 3,250×

Domain Archaea

Archaea (multiple kingdoms)

SEM 25,000×

Domain Eukarya

Protists (multiple kingdoms)

LM 275×

Kingdom Plantae

Kingdom Fungi

Kingdom Animalia

Figure 1.5B The three domains of life

1.6 Evolution explains the unity and diversity of life

Figure 1.6A
Charles Darwin
in 1859

In November 1859, Charles Darwin (Figure 1.6A) published one of the most important and influential books ever written. Entitled *On the Origin of Species by Means of Natural Selection,* Darwin's book was an immediate bestseller and soon made his name almost synonymous with the concept of evolution. Darwin stands out in history with people like Newton and Einstein, scientists who synthesized ideas with great explanatory power. Such comprehensive ideas are called theories.

The Origin of Species articulated two main points. First, Darwin presented evidence to support the idea of **evolution**—that species living today are descendants of ancestral species. Darwin called his evolutionary theory "descent with modification." It was an insightful phrase, as it captured both the unity of life (descent from a common ancestor) and the diversity of life (modification as species diverged from their ancestors).

Darwin's second point was to propose a mechanism for evolution, a mechanism he called **natural selection.** Darwin synthesized this idea from observations that by themselves were neither profound nor original. Others had the pieces of the puzzle, but Darwin saw how they fit together. He inferred natural selection by connecting two readily observable features of life:

OBSERVATION #1: **Individual variation.** Individuals in a population vary in many heritable traits.

OBSERVATION #2: **Overproduction of offspring.** A population of any species has the potential to produce far more offspring than will survive to produce offspring of their own.

INFERENCE #1: **Unequal reproductive success.** From these two observations, Darwin inferred that individuals are unequal in their likelihood of surviving and reproducing. Those individuals with heritable traits best suited to the environment will leave the greatest number of healthy, fertile offspring.

INFERENCE #2: **Over time, favorable traits accumulate in a population.** Over many generations, a higher and higher proportion of individuals will have the advantageous traits.

Unequal reproductive success is what Darwin called natural selection, because the natural environment "selects" for individuals with certain traits. And the product of natural selection is evolutionary adaptation, the accumulation of favorable traits in a population over time.

Figure 1.6B uses a simple example to show how natural selection works. ❶ An imaginary beetle population has colonized an area where the soil has been blackened by a recent brush fire. Initially, the population varies extensively in the

inherited coloration of individuals, from very light gray to charcoal. ❷ A predatory bird eats the beetles it sees most easily, the light-colored ones. This selective predation favors the survival and reproductive success of the darker beetles compared to the lighter ones. ❸ The surviving beetles have reproduced. The population is now quite different from the original one, as natural selection has produced a change in the frequencies of the different-colored beetles.

Our example also illustrates that natural selection is not a creative process but an editing mechanism. It can only select among the variations that are present in the population.

We can summarize natural selection as follows: Natural selection occurs as heritable variations are exposed to environmental factors that favor the reproductive success of some individuals over others. In our example, birds are the "selecting" environmental factor.

Darwin realized that numerous small changes in populations caused by natural selection could eventually lead to major alterations of species. He proposed that new species could evolve as a result of the gradual accumulation of changes over long periods of time. We will explore evolution and natural selection in more detail in Chapters 13 and 14.

We see the exquisite results of natural selection in every kind of organism. Each species has its own set of evolutionary adaptations that evolved by means of natural selection. In Figure 1.6C, on the facing page, we see the pangolin, a mammal that lives in

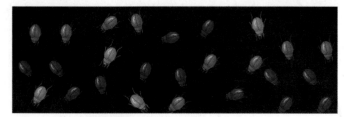

❶ Population with varied inherited traits

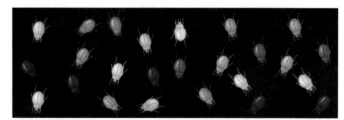

❷ Elimination of individuals with certain traits

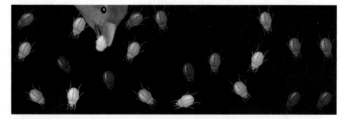

❸ Reproduction of survivors

Figure 1.6B An example of natural selection in action

East African rain forests. One of its main adaptations, its tough body armor of overlapping scales, protects it from most predators. The pangolin also has an unusually long tongue, which it uses to prod termites and ants out of their nests. Another mammal shown in Figure 1.6C, the killer whale (orca), is adapted for life at sea. It breathes air through nostrils on the top of its head and communicates with its companions by emitting clicking sounds. Orcas use sound echoes to detect obstacles and schools of fish or other prey. The pangolin's armor and the orca's echolocating ability arose over many, many generations as individuals with heritable traits that made them better adapted to their environment had the greatest reproductive success.

Evolution is biology's core theme—the one idea that makes sense of all we know about life.

Web Process of Science *How Do Environmental Changes Affect a Population?*

? How does natural selection adapt a population of organisms to its environment?

■ On average, those individuals with heritable traits best suited to the local environment produce the greatest number of offspring that survive and reproduce. This increases the frequency of those traits in the population.

Killer whale

Pangolin

Figure 1.6C Examples of adaptations to different environments

The Process of Science

1.7 Scientists use two main approaches to learn about nature

The word *science* is derived from a Latin verb meaning "to know." Science is a way of knowing. It stems from our curiosity about ourselves and the world around us. Science seeks natural causes for natural phenomena. Thus the scope of science is limited to the study of structures and processes that we can observe and measure, either directly or with the help of tools such as microscopes that extend our senses. This dependence on direct observations that other people can confirm distinguishes science from belief in the supernatural. Science can neither prove nor disprove the existence of a God or supernatural power—for such questions are outside the bounds of science.

Biology blends two main scientific approaches: discovery science, which is mostly about *describing* nature, and hypothesis-based science, which is mostly about *explaining* nature. Most research combines these two forms of inquiry.

Discovery Science Verifiable observations and measurements are the data of discovery science. In biology, discovery science describes life at its many levels, from the biosphere down to cells and molecules. An example is the sequencing of the human genome. While this research involves complicated methods and instruments, it is essentially just a detailed dissection and description of human DNA.

Discovery science can lead to important conclusions based on a type of logic called *inductive reasoning*. This kind of reasoning derives general principles from a large number of specific observations. "All organisms are made of cells" is an inductive conclusion based on the discovery of cells in every microscopic biological specimen observed by biologists over two centuries of time. The careful observations of discovery science and the inductive conclusions they sometimes produce are fundamental to our understanding of nature.

Hypothesis-Based Science The observations of discovery science stimulate us to seek natural causes and explanations for those observations. Such inquiry usually involves the proposing and testing of hypotheses. A **hypothesis** is a proposed explanation for a set of observations. A good hypothesis leads to predictions that scientists can test by recording additional observations or by designing experiments.

Deductive reasoning is the logic used in hypothesis-based science to come up with ways to test hypotheses. In deduction, the reasoning flows from the general to the specific. From general premises, we extrapolate to the specific results we should expect if the premises are true. If all organisms are made of cells (premise 1), and humans are organisms (premise 2), then humans are composed of cells (deduction). This deduction is a prediction that can be tested by examining human tissues.

Theories in Science How is a theory different from a hypothesis? A **theory** in science is much broader in scope. It explains a great diversity of observations and is supported by a large and usually growing body of evidence. And theories continue to generate new hypotheses, which can be tested.

🎧 MP3 Tutor *The Process of Science*

? What is the difference between discovery science and hypothesis-based science?

■ In the first, scientists observe and describe objects and phenomena; in the second, they propose hypotheses, make deductions, and test predictions.

1.8 With hypothesis-based science, we pose and test hypotheses

We will explore the elements of hypothesis-based science with the help of two scientific investigations, one from everyday life and one from a research project on snakes.

A Case Study from Everyday Life We all use hypotheses in solving everyday problems. Let's say, for example, that your flashlight fails during a campout. That's an observation. The question is obvious: Why doesn't the flashlight work? Two reasonable hypotheses based on past experience are that either the batteries in the flashlight are dead or the bulb is burned out. Figure 1.8A diagrams this campground inquiry.

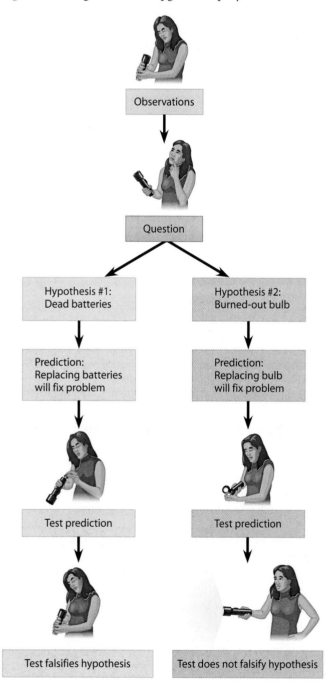

Figure 1.8A An example of hypothesis-based science

Now let's use deductive reasoning. In the process of science, deduction usually takes the form of predictions of experimental results or observations we should expect *if* a particular hypothesis is correct. We then test our predictions by carrying out the appropriate experiments or observations to see whether or not the results are expected. This deductive testing uses *"If . . . then"* logic. For example, *if* the dead-batteries hypothesis is correct and I replace the batteries with new ones, *then* the flashlight should work. If the flashlight still does not work, we can test our other hypothesis by replacing the flashlight bulb.

The flashlight example illustrates two important points. First, a hypothesis must be *testable*—there must be some way to check its validity. Second, a hypothesis must be *falsifiable*—there must be some observation or experiment that could show that it is not true. As shown on the left in Figure 1.8A, the hypothesis that dead batteries are the sole cause of the problem was falsified by replacing the batteries with new ones. As shown on the right, the burned-out-bulb hypothesis is the more likely explanation. Notice that testing supports a hypothesis not by proving that it is correct but by not eliminating it through falsification. Perhaps the bulb was simply loose and the new bulb was inserted correctly. No amount of experimental testing can *prove* a hypothesis beyond a shadow of doubt, because it is impossible to exhaust all alternative hypotheses. A hypothesis gains credibility by surviving various attempts to falsify it.

A Case Study of Hypothesis-Based Science One way to learn more about how hypothesis-based science works is to examine a case study from actual scientific research.

The story begins with a set of observations and generalizations from discovery science. Many poisonous animals are brightly colored, often with distinctive patterns. This so-called warning coloration apparently says "dangerous species" to potential predators. But there are also mimics. These imposters look like poisonous species but are actually harmless. A question that follows from these observations is: What is the function of such mimicry? A reasonable hypothesis is that such deception is an evolutionary adaptation that reduces the harmless animal's risk of being eaten.

In 2001, biologists David and Karin Pfennig, along with William Harcombe, one of their undergraduate students, designed an elegant set of field experiments to test the hypothesis that mimics benefit because predators confuse them with the harmful species. A poisonous snake called the eastern coral snake has warning coloration: bold, alternating rings of red, yellow, and black (Figure 1.8B). Predators rarely attack these snakes. The predators do not learn this avoidance behavior by trial and error; a first encounter with a coral snake would usually be deadly. A predator making that mistake will not pass its genes on to any more offspring. Natural selection has apparently increased the frequency of predators that inherit an instinctive avoidance of the coral snake's coloration.

The nonpoisonous scarlet king snake mimics the ringed coloration of the coral snake (Figure 1.8C). Both types of snakes live in North and South Carolina, but the king snakes'

Figure 1.8B   Eastern coral snake (poisonous)

Figure 1.8C   Scarlet king snake (nonpoisonous)

geographic range also extends into regions where no coral snakes are found.

The geographic distribution of these snakes made it possible for the researchers to test a key prediction of the mimicry hypothesis: Mimicry should help protect king snakes from predators, but only in regions where coral snakes also live. Avoiding snakes with warning coloration is an adaptation of predator populations that evolved in areas where the poisonous coral snakes are present. The mimicry hypothesis predicts that predators adapted to the warning coloration of coral snakes will attack king snakes less frequently than will predators in areas where coral snakes are absent.

To test this prediction, Harcombe made hundreds of artificial snakes out of wire covered with a claylike substance called plasticine. He made two versions of fake snakes: an *experimental group* with the red, black, and yellow ring pattern of king snakes and a *control group* of plain brown artificial snakes as a basis of comparison.

The researchers placed equal numbers of the two types of artificial snakes in field sites throughout North and South Carolina, including the region where coral snakes are absent. After four weeks, they retrieved the snakes and recorded how many had been attacked by looking for bite or claw marks. The most common predators were foxes, coyotes, and raccoons, but black bears also attacked some of the artificial snakes (Figure 1.8D).

The data fit the key prediction of the mimicry hypothesis. Compared to the brown artificial snakes, the ringed snakes were attacked by predators less frequently only in field sites within the geographic range of the poisonous coral snakes. The bar graph in Figure 1.8E summarizes the results.

This case study is an example of a **controlled experiment,** one that is designed to compare an experimental group (the artificial king snakes, in this case) with a control group (the artificial brown snakes). Ideally, the experimental and control groups differ only in the one factor the experiment is designed to test—in our example, the effect of the snakes' coloration on the behavior of predators. Without the control group, the researchers would not have been able to rule out the number of predators in the different test areas as the cause of the different number of attacks on the artificial king snakes. The experimental design left coloration as the only factor that could account for the low predation rate on the artificial king snakes placed within the range of coral snakes.

Both of these case studies illustrate the "scientific method," a process of inquiry that involves observations, questions, hypotheses, predictions, and tests of predictions. But there is no rule book that says science must follow rigid steps. As in all quests, science includes elements of challenge, adventure, and luck, along with careful planning, reasoning, creativity, cooperation, competition, patience, and persistence.

Web Process of Science *How Does Acid Precipitation Affect Trees?*

? Why is it difficult to draw a conclusion from an experiment that does not include a control group?

■ Without a control group, you don't know if the experimental outcome is due to the variable you are trying to test or to some other variable.

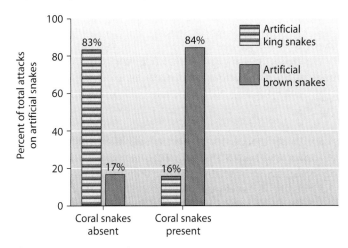

Figure 1.8E   Results of mimicry experiment

Figure 1.8D   Artificial king snake that was not attacked (left); artificial brown snake that was attacked by a bear (right)

1.9 Biology, technology, and society are connected in important ways

Figure 1.9 Biology and technology in the news

Many issues facing society are related to biology (Figure 1.9). Most of these issues also involve our expanding technology. Science and technology are interdependent, but their basic goals differ. The goal of science is to understand natural phenomena. In contrast, the goal of **technology** is to apply scientific knowledge for some specific purpose. Scientists often speak of "discoveries," while engineers more often speak of "inventions." The beneficiaries of those inventions also include scientists, who use new technology in their research. And scientific discoveries often lead to the development of new technologies.

Technology depends less on the curiosity that drives basic science than it does on the needs and wants of people and on the social environment of the times. Debates about technology center more on "should we do it" than "can we do it." Should insurance companies have access to individuals' DNA information? Should we permit research with embryonic stem cells?

Technology has improved our standard of living in many ways, but not without adverse consequences. Technology that keeps people healthier has enabled Earth's population to grow more than tenfold in the past three centuries, to more than 6 billion. The environmental effects of this growth can be devastating. Global warming, toxic wastes, acid rain, deforestation, nuclear accidents, and extinction of species are just some of the repercussions of more and more people wielding more and more technology. Science can help us identify such problems and provide insight into what course of action may prevent further damage. But solutions to these problems have as much to do with politics, economics, and cultural values as with science and technology. Now that science and technology have become such powerful aspects of society, every citizen has a responsibility to develop a reasonable amount of scientific literacy. The crucial science-technology-society relationship is a theme that adds to the significance of any biology course.

? How do science and technology interact?

■ New scientific discoveries may lead to new technologies; new technologies may increase the ability of scientists to search for new knowledge.

1.10 Evolution is connected to our everyday lives

Evolution is the core theme of biology. To emphasize the centrality of evolution to biology, we have added an Evolution Connection module to each chapter in this book. But is evolution connected to our everyday lives? And if so, in what ways?

Evolution tells us that all living species are descendants of ancestral species that have become modified as natural selection adapts populations to their environments. As environments change, populations change. What accounts for these changes?

Biologists now recognize that differences in DNA among individuals, populations, and species reflect the pattern of evolutionary change. Scientists have determined the order of billions of DNA bases in thousands of different species. Comparisons of those sequences allow us to identify genes shared across many species, study their actions in other species, and, in some cases, search for new medical treatments. Identifying beneficial genes in relatives of our crop plants has permitted the breeding or genetic engineering of enhanced crops.

The recognition that DNA differs between people has led to the use of DNA tests to identify individuals. DNA profiling is now used to help convict or exonerate the accused, determine paternity, and identify remains.

Evolution teaches us that the environment matters because it is a powerful selective force for traits that best adapt populations to their environment. We are major agents of environmental change when we take drugs to combat infection or grow crops in pesticide-dependent monocultures or alter most of Earth's habitats. We have seen the effects of such environmental changes in antibiotic-resistant bacteria, pesticide-resistant pests, endangered species, and increasing rates of extinction.

How can evolutionary theory help? It can help us be more judicious in our use of antibiotics and pesticides, and develop strategies for conservation efforts. It can help us create flu vaccines and HIV drugs by tracing the evolution of these viruses.

We hope this book will help you develop an appreciation for evolution and for biology and help you apply that understanding to evaluating issues ranging from your personal health to the well-being of the whole world. Biology offers us a deeper understanding of ourselves and our planet and a chance to more fully appreciate life in all its diversity.

? How might an understanding of evolution contribute to the development of new drugs?

■ As one example, we can find organisms that share our genes and similar cellular processes and test the actions of potential drugs in these organisms.

Reviewing the Concepts

Biology is the scientific study of life (**Introduction**).

Themes in the Study of Biology (1.1–1.3)

Life's hierarchy of organization unfolds as follows: biosphere > ecosystem > community > population > organism > organ system > organ > tissue > cell > organelle > molecule. Starting at the bottom of the hierarchy, with each step "upward," new properties emerge as a result of interactions among components of the lower levels (**1.1**).

Ecosystems are characterized by the cycling of chemical nutrients from the atmosphere and soil to producers to consumers to decomposers and back to the environment. Energy flows one way through an ecosystem from the sun to producers to consumers and exits as heat (**1.2**).

A **cell** is the basic unit of life, the lowest level of organization that can perform all activities required for life. Eukaryotic cells contain membrane-enclosed organelles, including a DNA-containing nucleus. Prokaryotic cells are smaller and lack such organelles (**1.3**).

Systems biology models the complex interactions of biological systems, such as the molecular interactions within a cell (**1.3**).

Evolution, the Core Theme of Biology (1.4–1.6)

DNA is the genetic information responsible for heredity and for programming the production of an organism's molecules. Each species' genes are coded in the sequences of the four building blocks (bases) making up DNA's two helically coiled chains. All organisms share a set of common features: ordered structures, regulation of internal conditions, growth and development, energy processing, response to environmental stimuli, the ability to reproduce, and evolutionary adaptations (**1.4**).

Three domains. Taxonomy names species and classifies them into a system of broader groups. Domain Bacteria and Archaea consist of prokaryotes. The eukaryotic domain Eukarya includes various protist kingdoms and the kingdoms Fungi, Plantae, and Animalia (**1.5**).

Evolution and natural selection. Evolution explains the unity and diversity of life. Charles Darwin synthesized the theory of evolution by natural selection. Natural selection is an editing mechanism that occurs when populations of organisms, having inherited variations, are exposed to environmental factors that favor the reproductive success of some individuals over others. All organisms have adaptations that have evolved by means of natural selection (**1.6**).

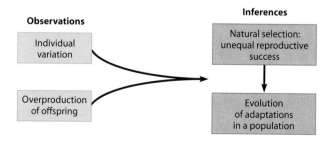

The Process of Science (1.7–1.8)

Scientific inquiry. In discovery science, scientists carefully observe and describe some aspect of the world and use inductive reasoning to draw general conclusions. In hypothesis-based science, they attempt to explain observations by testing hypotheses (**1.7**).

Hypothesis-based science involves observations, questions, hypotheses as tentative explanations, deductions leading to predictions, and then tests of the predictions to see if the hypotheses are falsifiable. Deductive reasoning is used: *If* a hypothesis is correct *and* we test one of its predictions (by performing an experiment or making observations), *then* a particular outcome will occur. In experiments designed to test hypotheses, the use of control groups and experimental groups helps to control variables (**1.8**).

Biology and Everyday Life (1.9–1.10)

Biology, technology, and society. Technological advances stem from scientific research, and research benefits from technology. The science-technology-society relationship is an important aspect of a biology course (**1.9**).

Evolution affects everyday life in medicine, agriculture, forensics, and conservation. Environmental changes are powerful selective pressures on the adaptive traits of many populations (**1.10**).

Connecting the Concepts

1. Biology can be described as having both a vertical scale and a horizontal scale. Explain what that means.

2. Complete the following map organizing some of biology's major concepts.

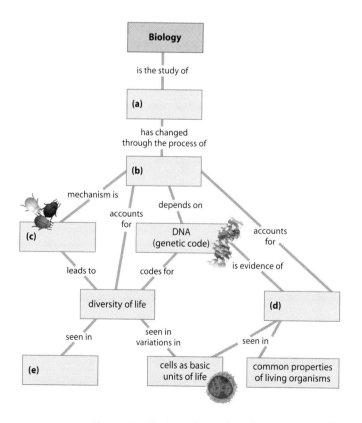

Testing Your Knowledge

Multiple Choice

3. Which of the following best describes the logic of the scientific process?
   a. If I generate a testable hypothesis, tests and observations will support it.
   b. If my prediction is correct, it will lead to a testable hypothesis.
   c. If my observations are accurate, they will support my hypothesis.
   d. If my hypothesis is correct, I can expect certain test results.
   e. If my tests are right, they will prove my hypothesis.

4. Amoebas and bacteria are grouped into different domains because
   a. amoebas eat bacteria.
   b. bacteria are not made of cells.
   c. bacterial cells lack a membrane-enclosed nucleus.
   d. bacteria decompose amoebas.
   e. amoebas are photosynthetic.

5. A biologist studying interactions among the protists in an ecosystem could *not* be working at which level in life's hierarchy? (*Choose carefully and explain your answer.*)
   a. the population level   d. the organism level
   b. the molecular level   e. the organ level
   c. the community level

6. Which of the following questions is outside the realm of science?
   a. Which organisms play the most important role in energy input to a savannah?
   b. What percentage of music majors take a biology course?
   c. What is the physical nature of the universe?
   d. What is the nature of the supernatural?
   e. What is the historical basis for the division of Earth's human population into ethnic groups?

7. Which of the following statements best distinguishes hypotheses from theories in science?
   a. Theories are hypotheses that have been proved.
   b. Hypotheses are tentative guesses; theories are correct answers to questions about nature.
   c. Hypotheses usually are narrow in scope; theories have broad explanatory power.
   d. Hypotheses and theories are different terms for essentially the same thing in science.
   e. Theories cannot be falsified; hypotheses are expected to be falsified.

8. Which of the following best demonstrates the unity among all living organisms?
   a. descent with modification   d. natural selection
   b. the structure and function of DNA   e. the three domains
   c. emergent properties

9. The core idea that makes sense of all of biology is
   a. the process of science.
   b. the correlation of function with structure.
   c. systems biology.
   d. evolution.
   e. unity in diversity.

Describing, Comparing, and Explaining

10. In an ecosystem, how is the movement of energy similar to that of chemical nutrients, and how is it different?

11. Explain the role of heritable variations in Darwin's theory of natural selection.

12. Explain what is meant by this statement: The scientific process is not a rigid method.

13. Contrast technology with science. Give an example of each to illustrate the difference.

14. Explain what is meant by this statement: Natural selection is an editing mechanism rather than a creative process.

Applying the Concepts

15. The graph below shows the results of an experiment in which mice learned to run through a maze.

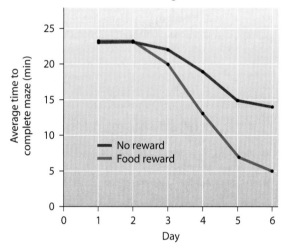

   a. State the hypothesis and prediction that you think this experiment tested.
   b. Which was the control group and which the experimental? Why was a control group needed?
   c. List some variables that must have been controlled so as not to affect the results.
   d. Do the data support your hypothesis? Explain.

16. In an experiment similar to the mimicry experiment described in Module 1.8, a researcher found that there were more predator attacks on artificial king snakes in areas with coral snakes than in areas outside the range of coral snakes. From this the researcher concluded that the mimicry hypothesis is false. Do you think this conclusion is justified? Why or why not?

17. The fruits of wild species of tomato are tiny compared to the giant beefsteak tomatoes available today. This difference in fruit size is almost entirely due to the larger number of cells in the domesticated fruits. Plant biologists have recently discovered genes that are responsible for controlling cell division in tomatoes. Why would such a discovery be important to producers of other kinds of fruits and vegetables? To the study of human development and disease? To our basic understanding of biology?

18. The news media and popular magazines frequently report stories that are connected to biology. In the next 24 hours, record all the ones you hear or read about in three different sources and briefly describe the biological connections you perceive in each story.

Answers to all questions can be found in Appendix 4.

For Practice Quizzes, BioFlix, MP3 Tutors, and Activities, go to www.mybiology.com.

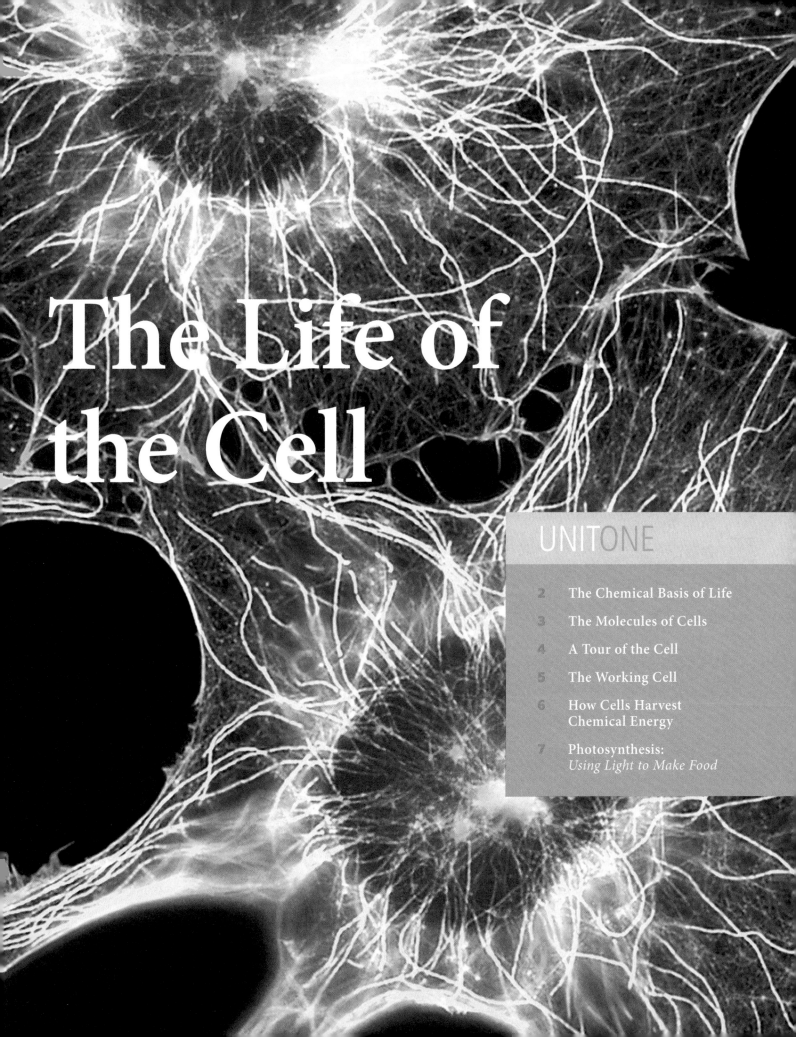

The Life of
the Cell

The Chemical Basis of Life

A researcher chiseling into a tree in Peru to find lemon ants (opposite)

A "Devil's Garden" in the Peruvian Amazon (left)

Who Tends This Garden?

The Amazonian rain forest in South America is a showcase for the diversity of life on Earth. Colorful birds, insects, and other animals live among a myriad of trees, shrubs, vines, and wild-flowers, and an excursion along a waterway or a forest path typi-cally reveals a lush variety of plant life. Visitors traveling near the Amazon's headwaters in Peru are therefore surprised to come across tracts of forest like that seen in the foreground of the photo at the top of this page. This patch is almost completely dominated by a single plant species—a willowy flowering tree called the lemon ant tree (*Duroia hirsute*). At first, it may appear that the garden is planted and maintained by local people. However, the indigenous people are as mystified as any visitors. They call these stands of trees "devil's gardens," from a legend attributing them to an evil forest spirit. Local legend is that the devil's gar-dens are home to a mythical dwarf who is half monkey and half human, with one hoof and one human foot. This creature is said to change his appearance into that of a friend or relative and then lead lone travelers in circles until they are lost.

Seeking a scientific explanation for the "devil's garden," a re-search team recently solved the mystery of these unusual forests. Graduate student Megan Frederickson, working with Deborah Gordon, a professor at Stanford University, showed that the "farmers" who create and maintain these gardens are ac-tually ants that live in the hollow stems of the trees. Indeed, the common name for the trees comes from the lemon ant (*Myrmelachista schumanni*) usually found living there. The ants do not plant the trees, of course, but they prevent other kinds of plants from growing in the garden by injecting intruders with a poisonous chemical.

How was the mystery solved? To test the hypothesis that the ants were killing competing trees, Frederickson did field experi-ments in Peru. She planted saplings of another tree species in-side and outside ten devil's gardens. At the base of some saplings, a sticky insect barrier was applied. The other saplings had no barrier to ant traffic. The researchers observed ants making in-jections from the tips of their abdomens into leaves of the un-protected saplings in their gardens (see photo). Within one day, these leaves developed dead areas. But the ants did not harm the saplings outside their gardens. These results and observations indicate that it is the ants who tend the gardens.

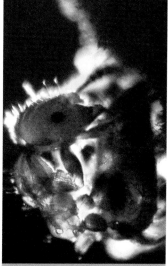

An ant injecting formic acid into a competitor's leaves

The chemical the ants use to weed their garden turns out to be formic acid. You may have experienced formic acid if you have ever been stung by a bee or wan-dered into a patch of sting-ing nettles. This substance is produced by many species of ants and, in fact, got its name from the Latin word for ant, *formica*. In many species, the formic acid probably serves as a disin-fectant that protects the ants against microbial parasites.

The lemon ant is the first ant species found to use formic acid as a herbicide. This use of a chemical is an important addition to the list of functions mediated by chemicals in the insect world. Scientists already know that chemicals play important roles in insect communication, attraction of mates, and de-fense against predators.

Research on devil's gardens is only one example of the rele-vance of chemistry to the study of life. Chemicals play many more roles in life than herbicides produced by ants, of course: They are the very stuff making up our bodies, those of other organisms, and the physical environment.

In this chapter, we make connections to the themes introduced in Chapter 1. One of these themes is the organization of life into a hierarchy of structural levels, with additional properties emerging at each successive level. The community formed of lemon ants and the garden of lemon ant trees they weed is one example. The interaction of these two species provides a home for colonies of ants and a habitat free of plant competitors for the trees.

Emergent properties are also apparent at the lowest levels of biological organization—such as the ordering of atoms into mol-ecules and the interactions of those molecules. We begin our study of biology with the basic concepts of chemistry that will apply throughout our study of life. ■ ■ ■

2.1 Living organisms are composed of about 25 chemical elements

Why does a biology textbook begin with a chapter on chemistry? To learn about life, you will study the structures and functions of living organisms. In the process, you will travel through all the levels of the hierarchy of biological organization described in Module 1.1, from molecules to ecosystems. At the base of this hierarchy are atoms and molecules. You will see that the properties of life emerge from the arrangement of these chemical parts into higher and higher levels of structural organization.

Living organisms are composed of **matter,** which is anything that occupies space and has mass. (In everyday language, we can think of mass as an object's weight.) Matter, in forms as diverse as rock, water, gases, and living organisms, is composed of chemical elements. An **element** is a substance that cannot be broken down to other substances by ordinary chemical means. Today, chemists recognize 92 elements occurring in nature; gold, copper, carbon, and oxygen are some examples. (Chemists have also made dozens more.) Each element has a symbol, the first letter or two of its English, Latin, or German name. For instance, the symbol for sodium, Na, is from the Latin word *natrium;* the symbol O stands for the English word *oxygen.*

About 25 chemical elements are essential to life. As you can see in Table 2.1, four of these—oxygen (O), carbon (C), hydrogen (H), and nitrogen (N)—make up about 96% of the weight of the human body, as well as that of most other living organisms. These four elements are the main ingredients of biological molecules such as proteins, sugars, and fats. Calcium (Ca), phosphorus (P), potassium (K), sulfur (S), sodium (Na), chlorine (Cl), and magnesium (Mg) account for most of the remaining 4% of the human body. These elements are involved in such important functions as bone formation, nerve signaling, and DNA synthesis.

TABLE 2.1	ELEMENTS IN THE HUMAN BODY	
Element	**Symbol**	**Percentage of Human Body Weight**
Oxygen	O	65.0 ⎫
Carbon	C	18.5 ⎪
Hydrogen	H	9.56 ⎬ 96.3
Nitrogen	N	3.3 ⎭
Calcium	Ca	1.5
Phosphorus	P	1.0
Potassium	K	0.4
Sulfur	S	0.3
Sodium	Na	0.2
Chlorine	Cl	0.2
Magnesium	Mg	0.1

Trace elements (less than 0.01%): boron (B), chromium (Cr), cobalt (Co), copper (Cu), fluorine (F), iodine (I), iron (Fe), manganese (Mn), molybdenum (Mo), selenium (Se), silicon (Si), tin (Sn), vanadium (V), and zinc (Zn).

The **trace elements** listed at the bottom of the table are essential, but only in minute quantities. We explore the importance of trace elements to your health next.

Web Process of Science *Connection: How Are Space Rocks Analyzed for Signs of Life?*

? Which four chemical elements are most abundant in living matter?

Oxygen, carbon, hydrogen, and nitrogen ∎

2.2 Trace elements are common additives to food and water

Some trace elements, such as iron (Fe), are needed by all forms of life. Iron makes up only 0.004% of your body mass but is vital for energy processing and for transporting oxygen in your blood. Other trace elements, such as iodine (I), are required only by certain species. The average human needs about 0.15 milligram (mg) of iodine each day. Iodine is an essential ingredient of a hormone produced by the thyroid gland, which is located in the neck. An iodine deficiency in the diet causes the thyroid gland to grow to abnormal size, a condition called goiter. The goiter of the woman shown in Figure 2.2A can probably be reversed by iodine supplements. Adding iodine to table salt has reduced the incidence of goiter in many countries. Indeed, the salt on your table is probably iodized; check

the label on the box. Unfortunately, iodized salt is not available or affordable everywhere, and goiter still affects many thousands of people in developing nations.

Iodine is just one example of a trace element added to food or water to improve health. For

Figure 2.2A Goiter in a Malaysian woman, a symptom of iodine deficiency

more than 50 years, the American Dental Association has supported fluoridation of community drinking water supplies as a public health measure. Fluoride is a form of fluorine (F), an element in Earth's crust that is found in small amounts in all water sources. In many areas, fluoride is added as part of the municipal water treatment process to raise levels to a concentration that can reduce tooth decay. If you mostly drink bottled water, your fluoride intake may be reduced, although some bottled water now contains added fluoride.

Chemicals are added to food to help preserve it, make it more nutritious, or simply make it look better. Look at the nutrition facts label from the side of the cereal box in Figure 2.2B to see a familiar example of how foods are fortified with mineral elements. Iron, for example, is a trace element commonly added to foods. (You can actually see the iron that has been added to a fortified cereal by crushing the cereal and then stirring a magnet through it.) Also note that the nutrition facts label lists numerous vitamins that are added to improve the nutritional value of the cereal. For instance, the cereal in this example supplies 10% of the recommended daily value for vitamin A. Vitamins consist of more than one element and are examples of compounds, which we consider next.

? In addition to iron, what other trace elements have been added to the cereal in Figure 2.2B? Does *Total* provide the "total" amount needed of these elements?

■ Zinc and copper: Total provides 100% of the zinc but only 4% of the copper needed in a day.

Figure 2.2B Nutrition facts from a fortified cereal

2.3 Elements can combine to form compounds

A **compound** is a substance consisting of two or more different elements combined in a fixed ratio. Compounds are much more common than pure elements. In fact, few elements exist in a pure state in nature.

Many compounds consist of only two elements; for instance, table salt (sodium chloride, NaCl) has equal parts of the elements sodium and chlorine. Pure sodium is a metal and pure chlorine is a poisonous gas. Chemically combined, however, they form an edible compound (Figure 2.3). The elements hydrogen (H) and oxygen (O) exist as gases. Combined in a ratio of 2 : 1, they form the most abundant compound on Earth, water (H_2O). These are simple examples of organized matter having emergent properties: A compound has characteristics different from those of its elements.

Most of the compounds in living organisms contain at least three or four elements, mainly carbon, hydrogen, oxygen, and nitrogen. Sugar, for example, is formed of just carbon, hydrogen, and oxygen. The formic acid produced by ants (see chapter introduction) is also formed from those three elements. Proteins are compounds containing carbon, hydrogen, oxygen, nitrogen, and a small amount of sulfur. Different arrangements of the atoms of these elements determine unique properties for each compound.

Each element has its own kind of atom. In the next module we explore the structure of an atom and then see how this structure determines the chemical properties of elements.

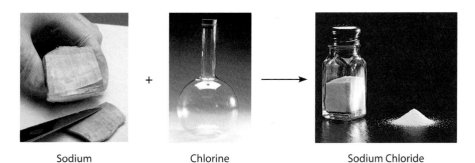

Sodium   Chlorine   Sodium Chloride

Figure 2.3 The emergent properties of the edible compound sodium chloride

? Explain how table salt illustrates the theme of emergent properties.

■ We are able to eat table salt, showing that it has properties different from those of a metal and a poisonous gas.

Atoms consist of protons, neutrons, and electrons

An **atom,** named from a Greek word meaning "indivisible," is the smallest unit of matter that still retains the properties of an element. Atoms are so small that it would take about a million of them to stretch across the period printed at the end of this sentence.

Subatomic Particles Physicists have split the atom into more than a hundred types of subatomic particles. However, only three kinds of particles are relevant to the chemistry of life. A **proton** is a subatomic particle with a single positive electrical charge (+). An **electron** is a subatomic particle with a single negative electrical charge (−). A third type of particle, the **neutron,** is electrically neutral (has no electrical charge).

Figure 2.4A shows two very simple models of an atom of the element helium (He), the "lighter-than-air" gas that makes balloons rise. Notice that two neutrons and two protons are tightly packed in the atom's central core, or **nucleus.** Two electrons orbit the nucleus at nearly the speed of light. The attraction between the negatively charged electrons and the positively charged protons keeps the electrons near the nucleus. The left-hand model shows the number of electrons in the atom. The right-hand model, slightly more realistic, shows a spherical cloud of negative charge created by the rapidly orbiting electrons around the nucleus. Neither model is drawn to scale. In real atoms, the electrons are very much smaller than the protons and neutrons, and the electron cloud is much bigger compared to the nucleus. Imagine that this atom was the size of the baseball stadium of the New York Yankees: The nucleus would be the size of a fly in center field, and the electrons would be like two tiny gnats buzzing around the stadium.

Atomic Number and Atomic Mass Elements differ in the number of subatomic particles in their atoms. All atoms of a particular element have the same unique number of protons. This number is the element's **atomic number.** Thus, an atom of helium, with 2 protons, has an atomic number of 2. Carbon, with 6 protons, has an atomic number of 6 (Figure 2.4B). Note that in these atoms, the atomic number is also the number of electrons. Unless otherwise indicated, an atom has an equal number of protons and electrons and thus its net electrical charge is 0 (zero).

An atom's **mass number** is the sum of the protons and neutrons in its nucleus. For helium, the mass number is 4; for carbon, it is 12 (Figures 2.4A and 2.4B). The mass of a proton and the mass of a neutron are almost identical and are expressed in a unit of measurement called the *dalton.* Protons and neutrons each have masses close to 1 dalton. An electron has only about $\frac{1}{2000}$ the mass of a proton, so it contributes little to an atom's mass. Thus an atom's **atomic mass** (or weight) is approximately equal to its mass number, the sum of its protons and neutrons.

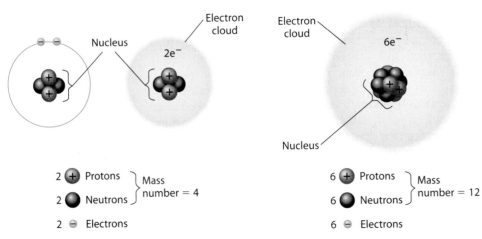

Figure 2.4A   Two models of a helium atom

Figure 2.4B   Model of a carbon atom

Isotopes All atoms of an element have the same atomic number, but some atoms of an element may differ in mass number. The different **isotopes** of an element have the same numbers of protons and electrons and behave identically in chemical reactions, but they have different numbers of neutrons. Table 2.4 shows the numbers of subatomic particles in the three isotopes of carbon. Carbon-12 (also written ^{12}C), with 6 neutrons, accounts for about 99% of the carbon in nature. Most of the remaining 1% consists of carbon-13 (^{13}C), with 7 neutrons. A third isotope, carbon-14 (^{14}C), with 8 neutrons, occurs in minute quantities. Notice that all three isotopes have 6 protons—otherwise, they would not be carbon.

Both ^{12}C and ^{13}C are stable isotopes, meaning their nuclei do not have a tendency to lose particles. The isotope ^{14}C, on the other hand, is unstable, or radioactive. A **radioactive isotope** is one in which the nucleus decays spontaneously, giving off particles and energy. Radiation from decaying isotopes can damage cellular molecules and thus can pose serious risks to living organisms. But radioactive isotopes can be helpful, as in their use in dating fossils (see Module 15.5). They are also used in biological research and medicine, as we see next.

TABLE 2.4	ISOTOPES OF CARBON		
	Carbon-12	Carbon-13	Carbon-14
Protons	6	6	6
Neutrons	6	7	8
Electrons	6	6	6

Web Activity *Structure of the Atomic Nucleus*

? A nitrogen atom has 7 protons, and its most common isotope has 7 neutrons. A radioactive isotope of nitrogen has 9 neutrons. What is the atomic number and mass number of this radioactive nitrogen?

■ Atomic number = 7; mass number = 16

2.5   Radioactive isotopes can help or harm us

Living cells cannot distinguish between isotopes of the same element. Consequently, organisms take up and use compounds containing radioactive isotopes in the usual way. Because radioactivity is easily detected by instruments, radioactive isotopes are useful as tracers—biological spies, in effect—for monitoring the fate of atoms in living organisms.

Basic Research   Biologists often use radioactive tracers to follow molecules as they undergo chemical changes in an organism. For example, researchers have used carbon dioxide (CO_2) containing the radioactive isotope ^{14}C to study photosynthesis. Using sunlight to power the conversion, plants take in CO_2 from the air and use it to make sugar molecules. Radioactively labeled CO_2 has enabled researchers to trace the sequence of molecules plants make in the chemical route from CO_2 to sugar.

Medical Diagnosis   Radioactive tracers are also used in medicine. Certain kidney disorders, for example, are diagnosed by injecting a radioactive chemical into a patient's blood and then measuring the amount of radioactive material passed in the urine. In most diagnostic uses of radioactive tracers, the patient receives only a tiny amount of an isotope that decays completely in minutes or hours.

Radioactive tracers are often used for diagnosis in combination with sophisticated imaging instruments. Figure 2.5A shows a patient being examined by a technique called PET (positron-emission tomography), which can detect locations of radioactive materials that have been introduced into the body. PET is useful for diagnosing certain heart disorders and cancers and for basic research on the brain (see Module 28.17).

The early detection of Alzheimer's disease may be a new use for such techniques. This devastating illness gradually destroys a person's memory and ability to think. As the disease progresses, the brain becomes riddled with deposits (plaques) of a protein called beta-amyloid. Researchers have identified a protein

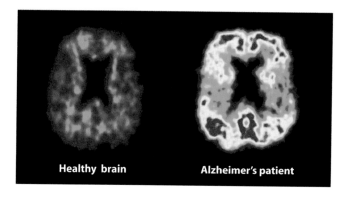

Figure 2.5B   PET images of brains of a healthy person (left) and a person with Alzheimer's disease (right). Red and yellow colors indicate high levels of PIB bound to beta-amyloid plaques.

molecule called PIB that binds to beta-amyloid. PIB contains a radioactive isotope that can be detected on a PET scan. Figure 2.5B shows PET images of the brains of a healthy person (left) and a person with Alzheimer's (right) injected with PIB. Notice that the brain of the Alzheimer's patient has high levels of PIB (red and yellow areas), whereas the unaffected person's brain has lower levels (blue). New therapies are focused on limiting the production of beta-amyloid or clearing it from the brain. A diagnostic test using PIB would allow researchers to monitor the effectiveness of new drugs in people living with the disease.

Dangers   Though radioactive isotopes have many beneficial uses, uncontrolled exposure to them can harm living organisms by damaging molecules, especially DNA. The particles and energy thrown off by radioactive atoms can break chemical bonds and also cause abnormal bonds to form. The explosion of a nuclear reactor at Chernobyl, Ukraine, in 1986 released large amounts of radioactive isotopes into the environment, killing 30 people within a few weeks. The survivors have suffered increased rates of thyroid cancer and increased rates of birth defects in their children, and thousands may be at increased risk of future cancers. A 2005 United Nations study predicted that 4,000 people will eventually die as a result of radiation exposure from the Chernobyl disaster (although some experts think the number of casualties might be much higher).

Natural sources of radiation can also pose a threat. Radon, a radioactive gas, may be a cause of lung cancer. Radon can contaminate buildings in regions where underlying rocks naturally contain uranium, a radioactive element. Homeowners can buy a radon detector or hire a company to test their home to ensure that radon levels are safe. If levels are found to be unsafe, technology exists to remove radon from homes.

?   Why are radioactive isotopes useful as tracers in research on the chemistry of life?

■ Organisms incorporate radioactive isotopes of an element into their molecules just as they do nonradioactive isotopes, and researchers can use special scanning devices to detect the presence of the radioactive isotopes.

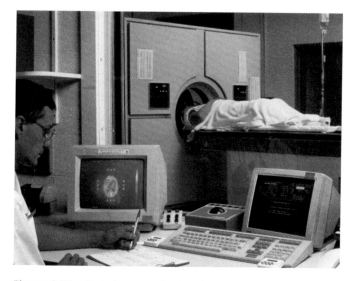

Figure 2.5A   Technician monitoring the output of a PET scanner

2.6 Electron arrangement determines the chemical properties of an atom

Only electrons are directly involved in the chemical activity of an atom. Electrons vary in the amount of energy they possess. The farther an electron is from the nucleus, the greater its energy. Electrons occur only at certain energy levels, called **electron shells.** Depending on its atomic number, an atom may have one, two, or more electron shells surrounding the nucleus.

Figure 2.6 is an abbreviated version of the *periodic table of the elements* (see also Appendix 2). It shows the distribution of electrons for the first 18 elements, arranged in rows according to the number of electron shells (one, two, or three) in their atoms. Note that the number of electrons (and protons) increases by one as you read from left to right to the next element in each row.

It is the number of electrons in the outermost shell that determines the chemical properties of an atom. The first electron shell is full with 2 electrons, so hydrogen and helium are the only elements in the first row. For the second and third rows, the outer shell can hold up to 8 electrons (four pairs). Atoms whose outer shells are not full (have unpaired electrons) tend to interact with other atoms—that is, to participate in chemical reactions.

Look at the electron shells of the atoms of the four elements that are the main components of biological molecules (highlighted in blue in Figure 2.6). Because their outer shells are incomplete, all these atoms react readily with other atoms. The hydrogen atom has only 1 electron in its single electron shell, which can accommodate 2 electrons. Atoms of carbon, nitrogen, and oxygen also are highly reactive because their outer shells, which can hold 8 electrons, are incomplete. In contrast, the helium atom has a first-level shell that is full with 2 electrons. Neon and argon also have full outer electron shells. As a result, these elements are chemically inert (unreactive).

How do chemical interactions between atoms enable them to fill their outer electron shells? When two atoms with incomplete outer shells react, each atom either shares, donates, or receives outer electrons, so that both partners end up with completed outer shells. These interactions usually result in atoms staying close together, held by attractions known as **chemical bonds.** In the next two modules, we look at two important types of chemical bonds.

Web Activity *Build an Atom*

Web Activity *Electron Arrangement*

? How many electrons and electron shells does a sodium atom have? How many electrons are in its outer shell?

◼ 11 electrons; three electron shells; one electron in the outermost shell

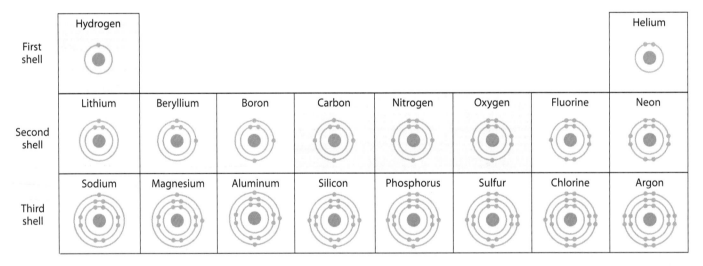

Figure 2.6   The electron shell diagrams of the first 18 elements in the periodic table

2.7 Ionic bonds are attractions between ions of opposite charge

Figure 2.7A at the top of the next page shows how a sodium atom and a chlorine atom form the compound sodium chloride (NaCl). Notice that sodium has only 1 electron in its outer shell, whereas chlorine has 7. When these atoms interact, the sodium atom donates its single outer electron to chlorine. Sodium now has only two shells, the outer shell having a full set of 8 electrons. When chlorine accepts sodium's electron, its own outer shell is now full with 8 electrons.

Because electrons are negatively charged particles, the electron transfer between the two atoms moves one unit of negative charge from sodium to chlorine. Sodium, with 11 protons but now only 10 electrons, has a net electrical charge of $+1$. Chlorine, having gained an extra electron, now has a net electrical charge of -1. In each case, an atom has become what is called an ion. An **ion** is an atom or molecule with an electrical charge resulting from a gain or loss of one

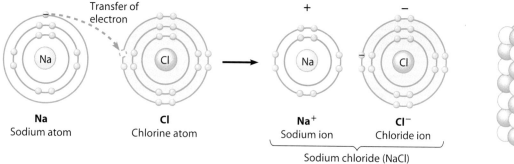

Figure 2.7A Formation of an ionic bond, producing sodium chloride

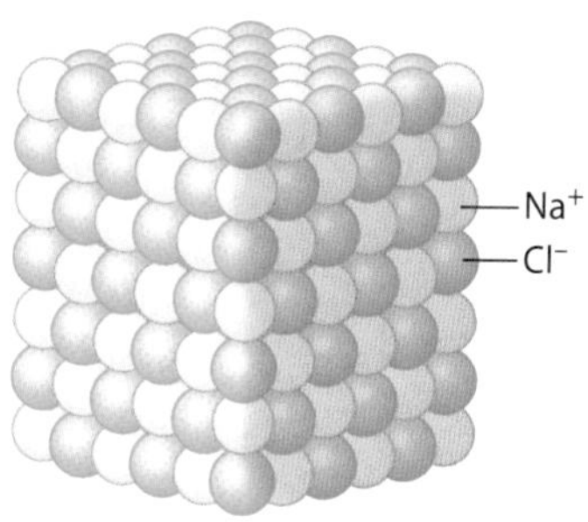

Figure 2.7B A crystal of sodium chloride

or more electrons. As you can see in Figure 2.7A, the ion formed from chlorine is called a chloride ion. Two ions with opposite charges attract each other; when the attraction holds them together, it is called an **ionic bond.** The resulting compound, in this case NaCl, is electrically neutral.

Sodium chloride is a familiar type of **salt,** a synonym for an ionic compound. Salts often exist as crystals in nature. Figure 2.7B shows the atoms in a crystal of sodium chloride. An NaCl crystal can be of any size (there is no fixed number of

ions), but sodium and chloride ions are always present in a 1 : 1 ratio. The ratio of ions differs with the kind of salt.

Web Activity *Ionic Bonds*

? Explain what holds together the atoms in a crystal of table salt (NaCl).

■ Opposite charges attract. The positively charged sodium ions (Na⁺) and the negatively charged chloride ions (Cl⁻) are held together by ionic bonds, the attractions between oppositely charged ions.

2.8 Covalent bonds join atoms into molecules through electron sharing

The second kind of strong chemical bond is the **covalent bond,** in which two atoms *share* one or more pairs of outer-shell electrons. Two or more atoms held together by covalent bonds form a **molecule.** For example, a covalent bond connects two hydrogen atoms in a molecule of the gas H_2.

Table 2.8 shows four ways to represent this molecule. The symbol H_2, called the molecular formula, tells you that the molecule consists of two atoms of hydrogen. The electron configuration diagram shows that the atoms share two electrons; as a result, both atoms fill their outer (and only) shells. The third column shows a structural formula. The line between the hydrogen atoms represents the single covalent bond formed by the sharing of a pair of electrons. In an O_2 molecule, for example, the two oxygen atoms share two pairs of electrons, forming a **double bond.** A double bond is indicated by a pair of lines. Space-filling models, shown in the fourth column, use color-coded balls to symbolize atoms and show a molecule's shape.

The number of covalent bonds an atom can form is equal to the number of additional electrons needed to fill its outer shell. This number is called the valence, or bonding capacity, of an atom. Looking back at Figure 2.6, we see that H can form one bond; O can form two; N, three; and C, four.

H_2 and O_2 are molecules composed of only one element. An example of a molecule that is a compound is methane (CH_4), a major component of natural gas. Water is also a compound, and we look more closely at water molecules next.

Web Activity *Covalent Bonds*

? What is chemically nonsensical about this structure?

H—C＝C—H

■ Each carbon atom has only three covalent bonds instead of the required four.

TABLE 2.8	ALTERNATIVE WAYS TO REPRESENT FOUR COMMON MOLECULES		
Molecular Formula	Electron-Distribution Diagram	Structural Formula	Space-Filling Model
H_2	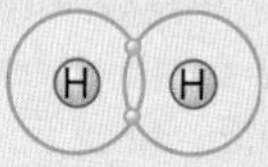	H—H Single bond	
O_2	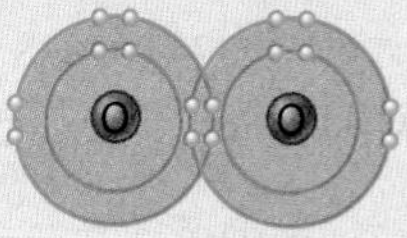	O＝O Double bond	
CH_4 Methane	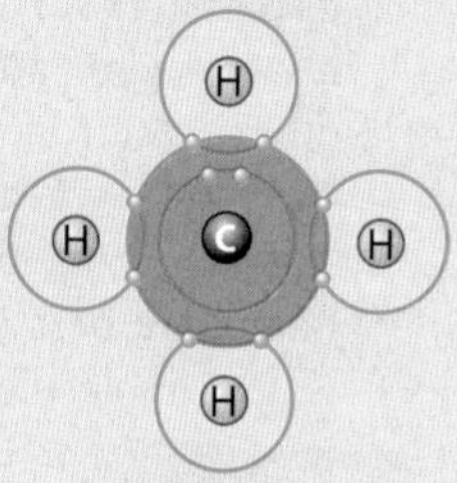	H—C—H (with H above and below)	
H_2O Water	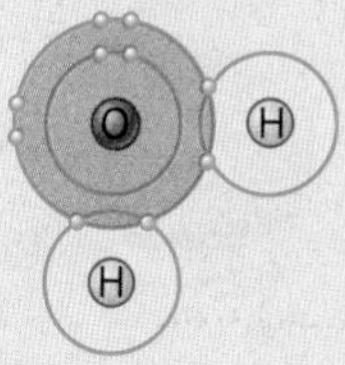	O—H with H	

2.9 Unequal electron sharing creates polar molecules

A water molecule (H₂O), as shown in Figure 2.9, consists of two hydrogen atoms covalently bonded to a single oxygen atom. Atoms in a covalently bonded molecule are in a constant tug-of-war for the shared electrons of their covalent bonds. An atom's attraction for shared electrons is called its **electronegativity.** The more electronegative an atom, the more strongly it pulls shared electrons toward its nucleus. In molecules of only one element, such as O_2 and H_2, the two identical atoms exert an equal pull on the electrons. The bonds in such molecules are said to be **nonpolar covalent bonds** because the electrons are shared equally between the atoms. Some compounds—for instance, methane (CH_4), shown in Table 2.8—also have nonpolar bonds. The atoms carbon and hydrogen are not substantially different in electronegativity.

In contrast to O_2, H_2, and CH_4, water is composed of atoms with different electronegativities. Oxygen is one of the most electronegative of the elements. (Nitrogen is also highly electronegative.) As indicated by the arrows in Figure 2.9, oxygen attracts the shared electrons in H_2O much more strongly than does hydrogen, so that the shared electrons spend more time

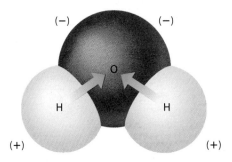

Figure 2.9   A water molecule

near the oxygen atom than near the hydrogen atoms. This unequal sharing of electrons produces a **polar covalent bond.** In a polar covalent bond, the pulling of shared, negatively charged electrons closer to the more electronegative atom makes that atom partially negative and the other atom partially positive. Thus, in H_2O, the oxygen atom actually has a slight negative charge and each hydrogen atom a slight positive charge. Because of its V shape and its polar covalent bonds, water is a **polar molecule**—that is, it has an unequal distribution of charges. It is slightly negative at the oxygen end of the molecule (apex of the V) and slightly positive at each of the two hydrogen ends.

Web Activity *Nonpolar and Polar Molecules*

? Why is it unlikely that two neighboring water molecules would be arranged like this?

■ The hydrogen atoms of one molecule, with their partial positive charge, would repel the hydrogen atoms of the adjacent molecule.

2.10 Hydrogen bonds are weak bonds important in the chemistry of life

In living organisms, most of the strongest chemical bonds are covalent, linking atoms to form a cell's molecules. But crucial to the functioning of a cell are weaker bonds within and between molecules. Most important large molecules are held in their three-dimensional shape by weak bonds. In addition, molecules in a cell may adhere briefly by weak bonds, respond to one another in some way, and then separate.

When a hydrogen atom is part of a polar covalent bond, its partial positive charge allows it to share attractions with other electronegative atoms, such as oxygen or nitrogen. These weak but important bonds are best illustrated with water molecules, as shown in Figure 2.10. The charged

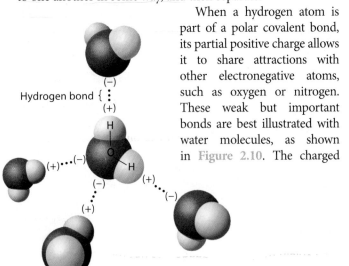

Hydrogen bond {

Figure 2.10   Hydrogen bonds between water molecules

regions on each water molecule are electrically attracted to oppositely charged regions on neighboring molecules. Because the positively charged region in this special type of bond is always a hydrogen atom, the bond is called a **hydrogen bond.** As Figure 2.10 shows, the negative (oxygen) pole on each water molecule can form hydrogen bonds (dotted lines) to two hydrogen atoms. And each hydrogen atom can form a hydrogen bond with a negatively charged oxygen atom. Thus, each H_2O molecule can hydrogen-bond to as many as four partners.

You will learn in Chapter 3 how hydrogen bonds help to create a protein's shape (and thus its function) and hold the two strands of a DNA molecule together. Indeed, these weak bonds are involved in many of the activities of a cell.

In the next few modules, we explore how water's polarity and hydrogen bonds give it unique, life-supporting properties. The abundance of water on Earth and its emergent properties make it possible for life to thrive on Earth.

Web Activity *Water's Polarity and Hydrogen Bonding*

? What enables neighboring water molecules to hydrogen-bond to one another?

■ The molecules are polar, with the negative end (oxygen end) of one molecule attracted to the positive end (hydrogen end) of its neighbor.

2.11 Hydrogen bonds make liquid water cohesive

Hydrogen bonds between molecules of liquid water last for only a few trillionths of a second, yet at any instant, many of the molecules are hydrogen-bonded to others. This tendency of molecules to stick together, called **cohesion,** is much stronger for water than for most other liquids. The cohesion of water is important in the living world. Trees, for example, depend on cohesion to help transport water and nutrients from their roots to their leaves. The evaporation of water from a leaf exerts a pulling force on water within the veins of the leaf. Because of cohesion, the force is relayed through the veins all the way down to the roots. **Adhesion,** the clinging of one substance to another, also plays a role. The adhesion of water to the cell walls of a plant's thin veins also helps counter the downward pull of gravity.

Related to cohesion is **surface tension,** a measure of how difficult it is to stretch or break the surface of a liquid. Hydrogen bonds give water unusually high surface tension, making it behave as though it were coated with an invisible film. You can observe the surface tension of water by slightly overfilling a glass; the water will stand above the rim. The water strider in Figure 2.11 takes advantage of the high surface tension of water to "stride" across ponds without breaking the surface.

○ MP3 Tutor *The Properties of Water*
Web Activity *Cohesion of Water*

? When you look at the "beads" of sweat on your face following a hard workout, can you explain what holds those drops together?

Figure 2.11 Surface tension allows a water strider to walk on water

■ The cohesion of water molecules and its high surface tension hold droplets of water together. The adhesion of water to your skin helps hold the beads in place.

2.12 Water's hydrogen bonds moderate temperature

If you have ever burned your finger on a metal pot while waiting for the water in it to boil, you know that water heats up much more slowly than metal. In fact, because of hydrogen bonding, water has a better ability to resist temperature change than most other substances. Because of this property, Earth's giant water supply moderates temperatures, helping to keep them within limits that permit life.

Temperature and heat are related but different. A swimmer crossing San Francisco Bay has a higher temperature than the water, but the bay contains far more heat because of its immense volume. **Heat** is the amount of energy associated with the movement of atoms and molecules in a body of matter. **Temperature** measures the intensity of heat—that is, the *average* speed of molecules rather than the *total* amount of heat energy in a body of matter.

Heat must be absorbed in order to break hydrogen bonds, and heat is released when hydrogen bonds form. To raise the temperature of water, heat energy must first disrupt hydrogen bonds before water molecules can move faster. Thus, water absorbs a large amount of heat while warming up only a few degrees. Conversely, when water is cooled, more hydrogen bonds form, releasing heat and slowing the cooling process.

A large body of water can store a huge amount of heat from the sun during warm periods. At cooler times, heat given off from the gradually cooling water can warm the air. That's why coastal areas generally have milder climates than inland regions. Water's resistance to temperature change also stabilizes ocean temperatures, creating a favorable environment for marine life. And at approximately 66% of your body weight, water helps moderate your internal temperature.

Another way water moderates temperatures is by evaporative cooling. When a substance evaporates, the surface of the liquid remaining behind cools down as the molecules with the greatest energy (the "hottest" ones)

Figure 2.12 Evaporative cooling

leave. It's as if the ten fastest runners on the track team left school, lowering the average speed of the remaining team. Evaporative cooling helps prevent some land-dwelling organisms from overheating. Evaporation from a plant's leaves keeps them from becoming too warm in the sun, just as sweating helps dissipate our excess body heat (Figure 2.12). On a much larger scale, the evaporation of surface waters cools tropical seas.

? Explain the popular adage "It's not the heat, it's the humidity."

■ High humidity hampers cooling by slowing the evaporation of sweat.

2.13 Ice is less dense than liquid water

Water exists in nature as a gas (water vapor), liquid, and solid. Unlike most substances, water is less dense as a solid than as a liquid. This unusual property is due to hydrogen bonds.

As water freezes, each molecule forms stable hydrogen bonds with four neighbors, holding them at "arm's length" and creating a three-dimensional crystal. In Figure 2.13A, compare the spaciously arranged molecules in the ice crystal with the more tightly packed molecules in the liquid water. The ice crystal has fewer molecules than an equal volume of liquid water. Therefore, ice is less dense and floats on top of liquid water.

If ice sank, then eventually ponds, lakes, and even oceans in cooler regions would freeze solid. Instead, when a deep body of water cools, the floating ice insulates the water below from colder air above. This "blanket" prevents the water below from freezing and allows fish and many other aquatic forms of life to survive under the frozen surface.

In the Arctic, this frozen surface serves as the winter hunting ground for polar bears (Figure 2.13B). The shrinking of this ice cover as a result of global warming may doom these bears.

> **?** Explain how the freezing of water can crack boulders.

> ■ Water expands as it freezes because the water molecules become spaced farther apart in forming ice crystals. When there is water in a crevice of a boulder, expansion of the water due to freezing may crack the rock.

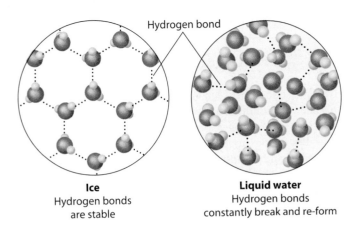

Hydrogen bond

Ice
Hydrogen bonds are stable

Liquid water
Hydrogen bonds constantly break and re-form

Figure 2.13A Hydrogen bonds between water molecules in ice and water

Figure 2.13B Polar bear on shrinking ice

2.14 Water is the solvent of life

If you add a teaspoon of salt to a glass of water, the salt will dissolve and eventually become evenly mixed with the water, forming a solution. A **solution** is a liquid consisting of a uniform mixture of two or more substances. The dissolving agent (in this case, water) is the **solvent,** and a substance that is dissolved (salt) is a **solute.** An **aqueous solution** (from the Latin *aqua,* water) is one in which water is the solvent.

Water's versatility as a solvent results from the polarity of its molecules. Figure 2.14 shows how a teaspoon of table salt dissolves in water. At the surface of each grain, or crystal, of salt, the sodium and chloride ions are exposed to water. These ions and the water molecules are attracted to each other due to their opposite charges. The oxygen ends (red) of the water molecules have a partial negative charge and cling to the positive sodium ions (Na^+). Hydrogens (○), with their partial positive charge, are attracted to the negative chloride ions (Cl^-). Working inward from the surface of each salt crystal, water molecules eventually surround and separate all the ions. Water dissolves other ionic compounds as well. Seawater, for instance, contains a great variety of dissolved ions, as do living cells.

A compound doesn't need to be ionic to dissolve in water. A spoonful of sugar will also dissolve in a glass of water. Polar molecules such as sugar dissolve as they become surrounded by polar water molecules, which form hydrogen bonds with them. Even large molecules, such as proteins, can dissolve if they have ionic and polar regions on their surface. As the solvent inside all cells, in blood, and in plant sap, water dissolves an enormous variety of solutes necessary for life.

> **?** Why are blood and most other biological fluids classified as aqueous solutions?

> ■ The solvent is water.

Figure 2.14 A crystal of salt (NaCl) dissolving in water

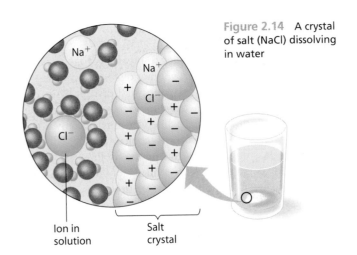

Ion in solution

Salt crystal

2.15 The chemistry of life is sensitive to acidic and basic conditions

In aqueous solutions, a very small percentage of the water molecules actually break apart (dissociate) into ions. The ions formed are called hydrogen ions (H^+) and hydroxide ions (OH^-). Hydrogen and hydroxide ions are very reactive. The proper balance of these ions is critical for the chemical processes within an organism.

Some chemical compounds contribute additional H^+ to an aqueous solution, whereas others remove H^+ from it. A compound that donates hydrogen ions to solutions is called an **acid.** One example of a strong acid is hydrochloric acid (HCl), the acid in your stomach. In solution, HCl dissociates completely into H^+ and Cl^-. An acidic solution has a higher concentration of H^+ than OH^-.

A **base** is a compound that accepts hydrogen ions and removes them from solution. Some bases, such as sodium hydroxide (NaOH), do this by donating OH^-; the OH^- combines with H^+ to form H_2O, thus reducing the H^+ concentration. The more basic a solution, the higher its OH^- concentration and the lower its H^+ concentration. Basic solutions are also called alkaline.

We use the **pH scale** to describe how acidic or basic a solution is (pH stands for potential of hydrogen). As shown in Figure 2.15, the scale ranges from 0 (most acidic) to 14 (most basic). Each pH unit represents a tenfold change in the concentration of H^+ in a solution. For example, lemon juice at pH 2 has 10 times more H^+ than an equal amount of grapefruit juice at pH 3 and 100 times more H^+ than tomato juice at pH 4.

Pure water and aqueous solutions that are neither acidic nor basic are said to be neutral; they have a pH of 7. They contain some hydrogen and hydroxide ions, but the concentrations of the two kinds of ions are equal. The pH of the solution inside most living cells is close to 7. Even a slight change in pH can be harmful because the proteins and complex molecules in cells are extremely sensitive to the concentrations of H^+ and OH^-.

The pH of human blood is very close to 7.4. A person cannot survive for more than a few minutes if the blood pH drops to 7 or rises to 7.8. If you add a small amount of a strong acid to pure water, the pH drops from 7.0 to 2.0. If the same amount of acid is added to blood, however, the pH decrease is only from 7.4 to 7.3. How can the acid have so much less effect on the pH of blood? Biological fluids contain

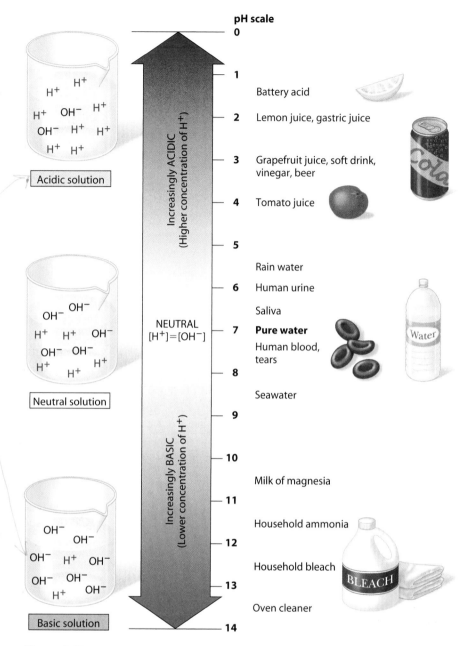

Figure 2.15   The pH scale reflects the relative concentrations of H^+ and OH^-

buffers, substances that minimize changes in pH. They do so by accepting H^+ when it is in excess and donating H^+ when it is depleted. There are several types of buffers that contribute to the pH stability in blood and many other internal solutions.

In the next module we explore how changes in acidity can have environmental consequences.

Web Activity Acids, Bases, and pH

?   Compared to a basic solution at pH 9, the same volume of an acidic solution at pH 4 has __ times more hydrogen ions (H^+).

100,000

2.16 Acid precipitation and ocean acidification threaten the environment

Considering the dependence of all life on water, contamination of rivers, lakes, seas, and precipitation poses serious environmental problems. In Chapter 38, you will learn about threats to water quality that result from human activity. Here we consider two threats that relate to changes in pH. The burning of fossil fuels (coal, oil, and gas) releases air-polluting compounds and large amounts of CO_2 into the atmosphere. The chemical reactions of these compounds with water alter the delicate balance of conditions for life by affecting water pH.

Sulfur oxides and nitrous oxides released by burning fossil fuels react with water in the air to form strong acids, which fall to Earth with rain or snow. **Acid precipitation** refers to rain, snow, or fog with a pH lower than 5.6 (the pH of uncontaminated rain). Rain with a pH between 2 and 3—more acidic than vinegar—has been recorded in the eastern United States. Acid precipitation has killed fish and damaged other forms of life in lakes and streams. It also changes soil chemistry and has negatively affected some North American and European forests. In the United States, amendments made in 1990 to the Clean Air Act have resulted in a decrease in acid precipitation.

Carbon dioxide is a major product of fossil fuel combustion, and its steadily increasing release into the atmosphere is predicted to double relative to 1880 levels by the year 2065. About half of the CO_2 stays in the atmosphere, acting like a reflective blanket over the planet that prevents heat from radiating into outer space. We will discuss the problems of global warming associated with this "greenhouse" effect later. Some CO_2 is taken up by plants for their photosynthesis. But about 30% is absorbed by the oceans. An increase in CO_2 absorption is expected to change ocean chemistry and harm marine life and ecosystems.

One effect of CO_2 absorption is to lower the pH of seawater. This change then decreases the concentration of carbonate

Figure 2.16 Coral reef that could be threatened by ocean acidification

ions, which are required by corals and other organisms to produce the calcium carbonate that makes up their skeletons or shells. This process is called calcification. Studies have shown that coral reef calcification decreases in response to a decrease in the concentration of carbonate ions. Coral reef ecosystems act as havens for a great diversity of organisms (Figure 2.16). There are other calcifying organisms that would also be affected by changes in ocean chemistry. Many of these are important food sources for salmon, herring, and other ocean fishes. Decreased calcification is likely to affect marine food webs and may substantially alter the productivity and biodiversity of the Earth's oceans.

Web Process of Science *How Does Acid Precipitation Affect Trees?*

? What is the relationship between fossil fuel consumption and coral reefs?

◼ Some of the increased CO_2 released by burning fossil fuels dissolves in and lowers the pH of the oceans. A lower pH reduces levels of carbonate ions, which then lowers the rate of calcification by coral animals.

2.17 The search for extraterrestrial life centers on the search for water

When astronomers search for signs of extraterrestrial life on distant planets, why do they look for evidence of water? Water is the substance that makes life as we know it possible. The chemical reactions of life take place in the watery environment of cells, and many of those reactions involve water as a reactant or product. Most organisms consist of more than 65% water by weight. The emergent properties of water molecules contribute to Earth's fitness for life. These properties include the cohesion of water molecules, water's ability to moderate temperature, the insulation of bodies of water by floating ice, and water's versatility as a solvent. Life on Earth began in water and evolved there for 3 billion years before spreading onto land. Is it possible that some form of life has evolved other places?

Researchers with the National Aeronautics and Space Administration (NASA) have enthusiastically disclosed evidence that water was once abundant on Mars. In January 2004, NASA succeeded in landing two golf-cart-sized rovers, named *Spirit* and *Opportunity*, on Mars. These robotic geologists used sophisticated instruments to determine the composition of rocks and send back images of rock formations. *Opportunity* landed within easy driving distance of a geologic formation that could answer the question of whether water was once plentiful on Mars. Its instruments detected jarosite, a mineral that is formed only in the presence of water. Other chemical evidence indicated that water once permeated Martian rocks. And pictures sent back revealed physical evidence of past water.

New images taken by NASA's Mars *Global Surveyor* spacecraft indicate that not only did the red planet have a water-filled ancient past but liquid water may have flowed recently on its surface. Two gullies that were photographed in 1999 and 2001 were imaged again in 2004 and 2005, revealing new light-colored deposits that are consistent with water flowing through the gullies (Figure 2.17). Scientists have proposed that reservoirs of water beneath Mars's surface could harbor microbial life. Water that may have burst through the surface and created such gullies provides possible evidence of underground liquid water.

Is it possible that life is not unique to planet Earth? Chapter 15 presents evidence that life could originate in the chemical and physical environment of early Earth. Finding evidence of life elsewhere would support the hypothesis that the chemical evolution of life is possible.

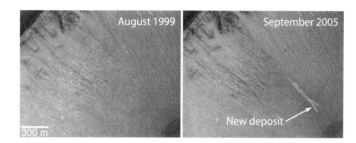

Figure 2.17 Gullies on Mars that might indicate recent water flow

? Why is the existence of water important in the search for extraterrestrial life?

■ Water plays important roles in life as we know it, from moderating temperatures on the planet to functioning as the solvent of life.

Chemical Reactions

2.18 Chemical reactions make and break bonds, changing the composition of matter

The basic chemistry of life has an overriding theme: The structure of atoms and molecules determines the way they behave. As we have seen, the chemical properties of an atom are determined by the number and arrangement of its subatomic particles, particularly its electrons. Other properties emerge when atoms combine to form molecules and when molecules interact, as in liquid water and ice. Water is a good example, because its emergent properties sustain all life on Earth.

Hydrogen and oxygen can react to form water:

$$2\,H_2 + O_2 \rightarrow 2\,H_2O$$

This is a **chemical reaction,** the making and breaking of chemical bonds, leading to changes in the composition of matter. In this case, two molecules of hydrogen (2 H_2) react with one molecule of oxygen (O_2) to produce two molecules of water (2 H_2O). The arrow indicates the conversion of the starting materials, called the **reactants** (H_2 and O_2), to the resulting **product** (H_2O). Notice that the same *numbers* of hydrogen and oxygen atoms appear on the left and right sides of the arrow, although they are grouped differently. Chemical reactions do not create or destroy matter; they only rearrange it in various ways. As shown in Figure 2.18, the covalent bonds holding hydrogen atoms together in H_2 and those holding oxygen atoms

together in O_2 are broken, and new bonds are formed to yield the H_2O product molecules.

Organisms cannot make water from H_2 and O_2, but they do carry out a great number of chemical reactions that rearrange matter in significant ways. Let's examine one that is essential to life on Earth: photosynthesis. The raw materials of photosynthesis are carbon dioxide (CO_2), which is taken from the air, and water (H_2O), which is absorbed from the soil. Within plant cells, sunlight powers the conversion of these reactants to a sugar product called glucose ($C_6H_{12}O_6$) and oxygen (O_2), a by-product that the plant releases to the air. The following chemical shorthand summarizes the process:

$$6\,CO_2 + 6\,H_2O \rightarrow C_6H_{12}O_6 + 6\,O_2$$

Although photosynthesis is actually a sequence of many chemical reactions, we see that we end up with the same number and kinds of atoms we started with. Matter has simply been rearranged, with an input of energy provided by sunlight.

Living cells routinely carry out thousands of chemical reactions. Most of these reactions involve compounds of the element carbon. We look at the carbon compounds of cells in more detail in Chapter 3.

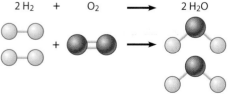

Figure 2.18 Breaking and making of bonds in a chemical reaction

? Fill in the blanks with the correct numbers in the following chemical process:

$$C_6H_{12}O_6 + \_\_O_2 \rightarrow \_\_CO_2 + \_\_H_2O$$

What process do you think this reaction represents? (Hint: think about an energy-producing reaction in your cells.)

■ $C_6H_{12}O_6 + 6\,O_2 \rightarrow 6\,CO_2 + 6\,H_2O$; The breakdown of sugar in the presence of oxygen to carbon dioxide and water, with the release of energy for cells.

Chapter Review

Reviewing the Concepts

Elements, Atoms, and Molecules (2.1–2.10)

Elements. Of the 25 chemical elements essential to life, carbon, hydrogen, oxygen, and nitrogen make up the bulk of living matter (**2.1**). Trace elements are essential to human health and may be added to food or water (**2.2**). Chemical elements combine in fixed ratios to form compounds (**2.3**).

Atoms. Atoms, the smallest unit of an element, consist of protons and neutrons (found in the nucleus) and electrons (arranged in electron shells). Atoms of an element with varying numbers of neutrons are called isotopes (**2.4**). Though potentially harmful, radioactive isotopes can be useful in medicine and science (**2.5**).

Chemical bonds. An atom whose outermost electron shell is not full tends to interact with other atoms and gain, lose, or share electrons, resulting in attractions called chemical bonds (**2.6**). Electron gain and loss create charged atoms, called ions. Oppositely charged ions attract one another in ionic bonds (**2.7**). In covalent bonds, atoms complete their outer electron shells by sharing electrons (**2.8**).

Polarity. A molecule is nonpolar when its covalently bonded atoms share electrons equally. In a polar molecule, such as water, electrons are not shared equally due to differences in electronegativity (**2.9**). The slightly positively charged H atoms in one water molecule may be attracted to the partial negative charge of neighboring O or N atoms, forming weak but important hydrogen bonds (**2.10**).

Water's Life-Supporting Properties (2.11–2.17)

Hydrogen bonds make water molecules cohesive, creating surface tension and allowing water to move from plant roots to leaves (**2.11**). Water's ability to store heat moderates body temperature and climate. It takes a lot of energy to disrupt hydrogen bonds, so water is able to absorb a great deal of heat energy without a large increase in temperature. As water cools, heat is released as hydrogen bonds form. A water molecule takes energy with it when it evaporates, leading to evaporative cooling (**2.12**). Hydrogen bonds hold molecules in ice farther apart than in liquid water. Floating ice protects lakes and oceans from freezing solid (**2.13**).

Solutions. Polar or charged solutes dissolve when water molecules surround them, forming aqueous solutions (**2.14**). A compound that releases H^+ in solution is an acid, and one that accepts H^+ is a base. Acidity is measured on the pH scale, from 0 (most acidic) to 14 (most basic). The pH of most cells is close to 7 (neutral) and kept that way by buffers (**2.15**).

pH and the environment. Some ecosystems are damaged by acid precipitation. The acidification of the ocean may threaten coral reefs and other marine organisms (**2.16**).

Extraterrestrial life. The search for life on other planets focuses on finding evidence of water (**2.17**).

Chemical Reactions (2.18)

Matter is rearranged in chemical reactions as bonds are broken and formed to convert reactants to products (**2.18**).

Connecting the Concepts

1. Fill in the blanks in this concept map to help you tie together the key concepts concerning elements, atoms, and molecules.

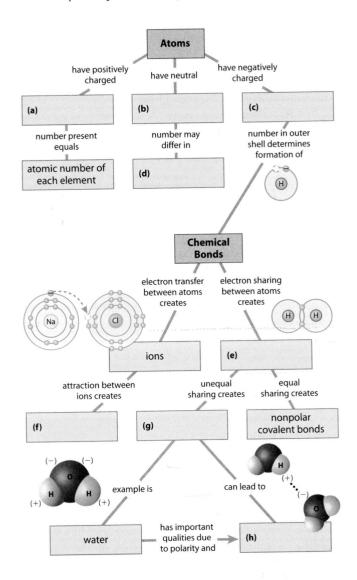

2. Create a concept map to organize your understanding of the life-supporting properties of water. A sample map is in the answer section, but the value of this exercise is in the thinking and integrating you must do to create your own map.

Testing Your Knowledge

Multiple Choice

3. Changing the _____ would change it into an atom of a different element.
   a. number of electrons surrounding the nucleus of an atom
   b. number of bonds formed by an atom
   c. number of protons in the nucleus of an atom
   d. electrical charge of an atom
   e. number of neutrons in the nucleus of an atom

4. Your body contains the smallest amount of which of the following?
   a. nitrogen
   b. phosphorus
   c. carbon
   d. oxygen
   e. hydrogen

5. A solution at pH 6 contains _____ than the same amount of a solution at pH 8.
   a. 2 times more H^+
   b. 4 times more H^+
   c. 100 times more H^+
   d. 4 times less H^+
   e. 100 times less H^+

6. Most of the unique properties of water result from the fact that water molecules
   a. are very small.
   b. are held together by covalent bonds.
   c. easily separate from one another.
   d. are constantly in motion.
   e. are polar and form hydrogen bonds.

7. A sulfur atom has 6 electrons in its outer shell. As a result, it forms _____ covalent bonds with other atoms. (*Explain your answer.*)
   a. 2   c. 4   e. 8
   b. 3   d. 6

8. What does the word *trace* mean when you're talking about a *trace element*?
   a. The element is required in very small amounts.
   b. The element can be used as a label to trace atoms through an organism's body.
   c. The element is very rare on Earth.
   d. The element enhances health but is not essential for the organism's long-term survival.
   e. The element passes rapidly through the organism.

9. A can of cola consists mostly of sugar dissolved in water, with some carbon dioxide gas that makes it fizzy and makes the pH less than 7. In chemical terms, you could say that cola is an aqueous solution where water is the _____, sugar is a _____, and carbon dioxide makes the solution _____.
   a. solvent . . . solute . . . basic
   b. solute . . . solvent . . . basic
   c. solvent . . . solute . . . acidic
   d. solute . . . solvent . . . acidic
   e. not enough information to say

10. Radioactive isotopes can be used in medical studies because
    a. they allow researchers to time how long processes take.
    b. they are more reactive than nonradioactive isotopes.
    c. the cell does not recognized the extra protons in the nucleus, so isotopes are readily used in cellular processes.
    d. their location or quantity can be determined because of their radioactivity.
    e. their extra neutrons produce different colors that can be traced through the body.

True/False (*Change false statements to make them true.*)

11. Table salt, water, and carbon are compounds.

12. The smallest particle of an element is a molecule.

13. A bathtub full of lukewarm water may hold more heat than a teakettle full of boiling water.

14. If the atoms in a molecule share electrons equally, the molecule is said to be nonpolar.

15. Ice floats because water molecules in ice are more tightly packed than in liquid water.

16. Atoms in a water molecule are held together by the sharing of electrons.

17. Most acid precipitation results from the presence of pollutants from aerosol cans and air conditioners.

18. An atom that has gained or lost electrons is called an ion.

Describing, Comparing, and Explaining

19. Make a sketch that shows how water molecules hydrogen-bond with one another. Why do water molecules form hydrogen bonds? What unique properties of water result from water's tendency to form hydrogen bonds?

20. Describe two ways in which the water in your body helps stabilize your body temperature.

21. Compare covalent and ionic bonds.

22. What is an acid? A base? How is the acidity of a solution described?

Applying the Concepts

23. The diagram below shows the arrangement of electrons around the nucleus of a fluorine atom (left) and a potassium atom (right). What kind of bond do you think would form between these two atoms?

Fluorine atom   Potassium atom

24. Animals obtain energy through a series of chemical reactions in which sugar ($C_6H_{12}O_6$) and oxygen gas (O_2) are reactants. This process produces water (H_2O) and carbon dioxide (CO_2) as waste products. How might you use a radioactive isotope to find out whether the oxygen in CO_2 comes from sugar or oxygen gas?

25. Look back at the abbreviated periodic table of the elements in Figure 2.6. If two or more elements are in the same row, what do they have in common? If two elements are in the same column, what do they have in common?

26. One solution to the problem of acid precipitation is to use nuclear energy to produce electricity. Development of nuclear power in the United States virtually stopped after the accident at the Three Mile Island power plant in Pennsylvania in 1979. But proponents of nuclear power contend that accelerated development of nuclear energy is the answer to U.S. energy needs. Besides reducing acid precipitation, what are other potential benefits of nuclear power? What are its possible costs and dangers? Do you think we ought to pursue the development of nuclear power? Why or why not? If a power plant were to be built near your home, would you prefer it to be a nuclear or coal-fired plant? Why?

Answers to all questions can be found in Appendix 4.

For Practice Quizzes, BioFlix, MP3 Tutors, and Activities, go to www.mybiology.com.

The Molecules of Cells

Model of a milk-digesting enzyme

Got Lactose?

Is a big glass of milk a way to a healthy diet—or an upset stomach? For much of the world's population, the answer is often the latter. Most of the world's people cannot easily digest milk-based foods. Those who can happily devour ice cream and cheese, scientists have found, are in the minority, benefiting from a lucky connection between human evolution and the major molecules that shape life.

Milk and other dairy products have long been recognized as highly nutritious foods, rich in protein and minerals necessary for healthy teeth and strong bones. But for millions of people, those health benefits come with a heavy dose of digestive discomfort. Such people suffer from lactose intolerance, or the inability to properly break down lactose, the main sugar found in milk.

For those with lactose intolerance, the problem starts once lactose passes through the stomach and enters the small intestine. There, to absorb the sugar, digestive cells need to secrete an enzyme called lactase, which is necessary for the breakdown of lactose. (An enzyme, such as the one shown above, is a protein that speeds up a specific chemical reaction.) Those with lactose intolerance, however, produce insufficient amounts of the enzyme. Without enough lactase, lactose cannot be digested or absorbed properly, leading to uncomfortable symptoms that include nausea, cramps, and bloating.

Scientists who study the problem have known for many years that the condition is closely tied to age and heritage. When it comes to milk, most people are born equal. Nearly everyone produces enough lactase in infancy to digest the lactose in breast milk and dairy products. Thus, milk—high in protein, fats, and sugars—provides excellent nourishment for infants. But after the age of 2, lactase levels start to decline in most of the world's populations, as if the lactase-producing mechanism in the body is being slowly shut down. In the United States, researchers have noted, as many as 75% of African Americans and Native Americans and 90% of Asian Americans are lactase-deficient once they reach their teenage years. People of European descent make up one of the few groups with relatively low levels of lactose intolerance; only about 15% are believed to suffer from the problem.

Why would lactose intolerance be linked with ancestry? It appears that the majority of humans do not produce lactase as adults. Indeed, lactose-intolerance is the normal human condition. The ability to digest milk after childhood is due to a mutation that allows lactase production to continue into adulthood. About 9,000 years ago, this beneficial mutation spread among cultures in northern Europe as they began to raise dairy herds and consume greater quantities of dairy products. And the mutation was passed down to their descendants.

What works for this lactose-tolerant minority, however, can mean more complicated eating for the rest of the world. There is no treatment for the underlying genetic cause of lactase deficiency, and lactose is now used widely in everything from bottled salad dressings and lunch meat to prescription drugs. But the condition's uncomfortable symptoms can be controlled through diet. People with lactose intolerance must watch what they eat and how much milk-based food they consume. Some avoid foods that contain lactose altogether, and doctors and nutritionists advise such people to take calcium supplements and vitamins. Others turn to substitutes: In many Asian cultures, for example, beverages have long been made from soy or rice instead of milk. Other foods are prepared from milk that has been pretreated with lactase. Lactase, in pill form, can also be taken along with food to ease digestion.

Lactose intolerance, with its interplay between genes and milk sugar, illustrates the importance of biological molecules to the functioning of living cells and to human health. In people who easily digest milk, lactose, a sugar, is broken down by lactase, a protein, which is produced by a gene, made of DNA, a nucleic acid. If the gene for lactase production is not active, lactase is not present. And the presence of lactase can mean the difference between delight and discomfort when someone contemplates an ice cream sundae. Such molecular interactions, repeated in countless variations, drive all biological processes. In this chapter, we explore the structure and function of sugars, proteins, fats, and nucleic acids—the biological molecules that are essential to life. We begin with a look at carbon, the versatile atom at the center of life's molecules. ■ ■ ■

3.1 Life's molecular diversity is based on the properties of carbon

Almost all the molecules a cell makes are composed of carbon atoms bonded to one another and to atoms of other elements. Carbon is unparalleled in its ability to form large, diverse molecules. The diversity of carbon compounds is the foundation for the myriad of molecules and chemical processes required for life. This diversity has also made possible the great diversity of organisms that have evolved on Earth.

Carbon-based molecules are called **organic compounds**. As we discussed in Chapter 2, an element's chemical properties are determined by the electrons in the outermost shell of its atoms. A carbon atom has four outer electrons in a shell that holds eight. Carbon completes its outer shell by sharing electrons with other atoms in four covalent bonds (see Module 2.8). Thus, each carbon atom is a connecting point from which a molecule can branch in up to four directions.

Figure 3.1A illustrates three representations of methane (CH_4), one of the simplest organic molecules. The structural formula shows that covalent bonds link four hydrogen atoms to the carbon atom. Each of the four lines in the formula represents a pair of shared electrons. The two models show that methane is three-dimensional, with the space-filling version portraying its overall shape more accurately. The ball-and-stick model shows that carbon's four bonds (the gray "sticks") angle out toward the corners of an imaginary tetrahedron (an object with four triangular sides). The red lines trace this shape, with the four hydrogen atoms of methane at the corners of the tetrahedron. This tetrahedral shape occurs wherever a carbon atom participates in four single bonds. Different bond angles occur where carbon atoms form double bonds. Large organic molecules can have elaborate shapes. And, as we will see many times, the shape of a molecule often determines its function.

Methane and other compounds composed of only carbon and hydrogen are called **hydrocarbons**. Figure 3.1B illustrates some variations in hydrocarbon structure. Carbon atoms, with attached hydrogens, can bond together in chains of various lengths to form compounds such as ethane or propane, a gas used as household fuel. The chain of carbon atoms in an organic molecule is called a **carbon skeleton** (shaded in gray in Figure 3.1B). Carbon skeletons can be unbranched, as in butane, or

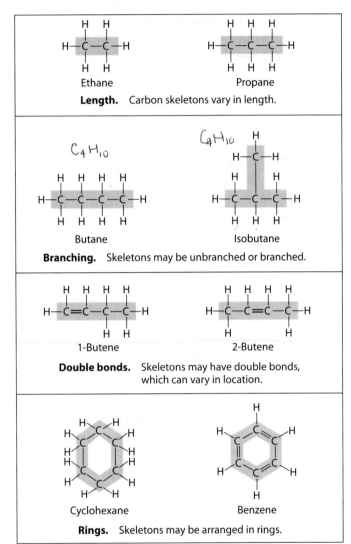

Figure 3.1B Variations in carbon skeletons

branched, as in isobutane. Carbon skeletons may also include double bonds, as in 1-butene and 2-butene. These two compounds have the same molecular formula, C_4H_8, but differ in the position of their double bond. Compounds with the same formula but different structures are called **isomers**. Their different shapes result in unique properties. Cyclohexane and benzene, the only hydrocarbons in the figure that are liquids rather than gases, are examples of carbon skeletons arranged in rings.

A general characteristic of all hydrocarbons is that they are nonpolar molecules due to their nonpolar C—H bonds.

Web Activity *Diversity of Carbon-Based Molecules*

? Why do isomers, which have the same formula (same number of atoms), have different properties?

■ Isomers have different structures, or shapes, and the shape of a molecule usually helps determine the way it functions—how it interacts with other molecules.

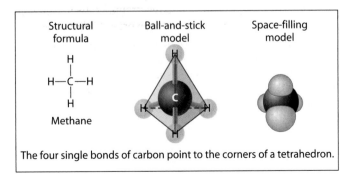

Figure 3.1A Three representations of methane (CH_4)

3.2 Characteristic chemical groups help determine the properties of organic compounds

The unique properties of an organic compound depend on the size and shape of its carbon skeleton and on the groups of atoms that are attached to that skeleton.

Table 3.2 illustrates six chemical groups important in the chemistry of life. The first five chemical groups shown are **functional groups.** They affect a molecule's function by participating in chemical reactions in characteristic ways. These groups are polar, because their oxygen or nitrogen atoms exert a strong pull on shared electrons. This polarity tends to make the compounds containing these groups **hydrophilic** (water-loving) and therefore soluble in water—a necessary condition for their roles in water-based life. The methyl group is nonpolar and not reactive, but it affects molecular shape and thus function.

A **hydroxyl group** consists of a hydrogen atom bonded to an oxygen atom, which in turn is bonded to the carbon skeleton. Ethanol, shown in the table, and other organic compounds containing hydroxyl groups are called alcohols.

In a **carbonyl group** a carbon atom is linked by a double bond to an oxygen atom. If the carbon of the carbonyl group is at the end of a carbon skeleton, the compound is called an aldehyde; if it is within the chain, the compound is called a ketone. Sugars contain a carbonyl group and several hydroxyl groups.

A **carboxyl group** consists of a carbon double-bonded to an oxygen and also bonded to a hydroxyl group. The carboxyl group acts as an acid by contributing an H^+ to a solution (see Module 2.15) and becoming ionized. Compounds with carboxyl groups are called **carboxylic acids.** Acetic acid, shown in the table, gives vinegar its sour taste.

An **amino group** is composed of a nitrogen bonded to two hydrogen atoms and the carbon skeleton. It acts as a base by picking up an H^+ from a solution. Organic compounds with an amino group are called **amines.** Amino acids, the building blocks of proteins, contain a carboxyl and an amino group.

A **phosphate group** consists of a phophorus atom bonded to four oxygen atoms. It is usually ionized and attached to the carbon skeleton by one of its oxygen atoms. This structure is

abbreviated as Ⓟ in this text. Compounds with phosphate groups are called organic phosphates and are often involved in energy transfers, as is the energy compound ATP.

A **methyl group** consists of a carbon bonded to three hydrogens. Compounds with methyl groups are called methylated compounds. The addition of a methyl group to the component of DNA shown in the table affects the expression of genes.

Figure 3.2 compares female and male sex hormones. They differ only in the chemical groups, highlighted with colored boxes. These subtle differences result in the different actions of these molecules, which help produce the features of females and males in humans and other vertebrates.

Web Activity *Functional Groups*

? Identify the chemical groups that do *not* contain carbon.

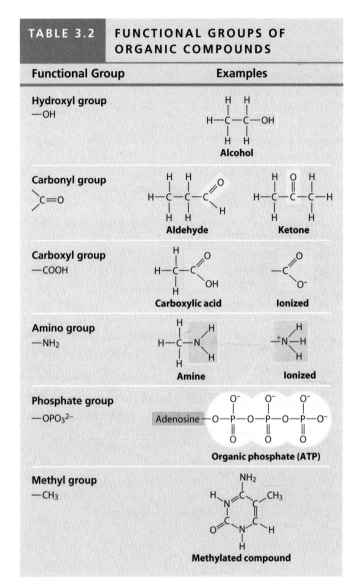

TABLE 3.2 FUNCTIONAL GROUPS OF ORGANIC COMPOUNDS

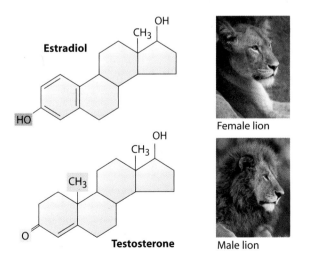

Figure 3.2   Differences in the chemical groups of sex hormones

■ The amino, hydroxyl, and phosphate groups

3.3 Cells make a huge number of large molecules from a small set of small molecules

The four main classes of large biological molecules are carbohydrates, lipids, proteins, and nucleic acids. On a molecular scale, many of these molecules are gigantic; in fact, biologists call them **macromolecules.** Proteins, for example, may consist of thousands of covalently connected atoms.

Cells make most of their large molecules by joining smaller molecules into chains called **polymers** (from the Greek *polys,* many, and *meros,* part). A polymer is a long molecule consisting of many identical or similar building blocks strung together, much as a train consists of many individual cars. The building blocks of polymers are called **monomers.**

The Diversity of Polymers Living cells make a vast number of different polymers. For proteins alone, there are about a trillion different kinds in nature, and the variety is potentially endless. Remarkably, a cell makes all of its diverse macromolecules from a small list of ingredients—about 40 to 50 common components and a few others that are rare. Proteins, for example, are built from only 20 kinds of amino acids. DNA is built from just four kinds of monomers called nucleotides. As we will see, the key to the great diversity of protein and DNA molecules is arrangement—variation in the sequence in which their monomers are strung together.

The variety in polymers accounts for the uniqueness of each organism. The monomers used to make polymers, however, are essentially universal. Your proteins and those of a tree or ant are assembled from the same 20 amino acids; the amino acids are just arranged in different sequences. Life has a simple yet elegant molecular logic: Small molecules common to all organisms are ordered into large molecules, which vary from species to species and even from individual to individual.

Making Polymers Cells link monomers together to form polymers by a **dehydration reaction,** a reaction that removes a molecule of water. As you can see in Figure 3.3A, an unlinked monomer has a hydroxyl group (—OH) at one end and a hydrogen atom (—H) at the other. For each monomer added to a chain, a water molecule (H_2O) is removed. Notice that two monomers contribute to the H_2O molecule, one monomer (the one at the right end of the short polymer in this example) losing a hydroxyl group and the other monomer losing a hydrogen atom. As this occurs, a new covalent bond forms, linking the two monomers. Dehydration reactions are the same regardless of the type of polymer the cell is producing.

Breaking Polymers Cells not only make macromolecules but also have to break them down. For example, food that an organism ingests is often in the form of polymers that are much too large to enter its cells. To digest these polymers, a cell carries out **hydrolysis.** Essentially the reverse of a dehydration reaction, hydrolysis means to break (*lyse*) with water (*hydro-*), and cells break bonds between monomers by adding water to them, as Figure 3.3B shows. In the process, a hydroxyl group

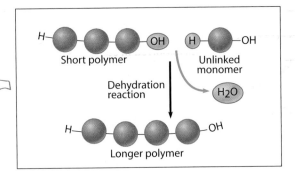

Figure 3.3A Dehydration reactions build a polymer chain

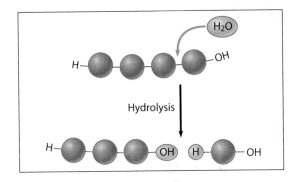

Figure 3.3B Hydrolysis breaks a polymer chain

joins to one monomer, and a hydrogen joins to the adjacent monomer.

The lactose-intolerant individuals you learned about in the chapter introduction are unable to hydrolyze such a bond in lactose because they lack the enzyme lactase. Both dehydration reactions and hydrolysis require the help of enzymes to make and break bonds. **Enzymes** are specialized macromolecules that speed up chemical reactions in cells.

In the remainder of the chapter, we explore each of the four classes of large biological molecules. We will see that polysaccharides, proteins, and nucleic acids are polymers assembled by dehydration reactions from their monomers. Lipids are not true polymers. However, some lipids are formed by dehydration reactions from several smaller molecules.

Like water and simple organic molecules, large biological molecules have unique emergent properties arising from the orderly arrangement of their atoms. As you will see, for these molecules of life, structure and function are inseparable.

Web Activity *Making and Breaking Polymers*

? Suppose you eat some cheese. What reactions must occur for the amino acid monomers in the protein of the cheese to be converted to proteins in your body?

■ In digestion, the amino acids are released by hydrolysis. New proteins are formed in your body from these monomers in dehydration reactions.

3.4 Monosaccharides are the simplest carbohydrates

The name **carbohydrate** refers to a class of molecules ranging from the small sugar molecules dissolved in soft drinks to large polysaccharides, such as the starch molecules we consume in pasta and potatoes.

The carbohydrate monomers (single-unit sugars) are **monosaccharides** (from the Greek *monos-*, single, and *sacchar,* sugar). The honey shown in Figure 3.4A consists mainly of monosaccharides called glucose and fructose. These and other single-unit sugars can be hooked together by a dehydration reaction to form more complex sugars and polysaccharides.

Monosaccharides generally have molecular formulas that are some multiple of CH_2O. For example, the formula for glucose, a common monosaccharide of central importance in the chemistry of life, is $C_6H_{12}O_6$. Figure 3.4B illustrates the molecular structure of glucose; its carbons are numbered 1 to 6. This structure also shows the two trademarks of a sugar: a number of hydroxyl groups ($-OH$) and a carbonyl group ($>C=O$). The hydroxyl groups make a sugar an alcohol, and the carbonyl group, depending on its location, makes it either an aldose (an aldehyde sugar) or a ketose (a ketone sugar). As you see in Figure 3.4B, glucose is an aldose and fructose is a ketose. (Note that most names for sugars end in *-ose.*)

If you count the numbers of different atoms in the fructose molecule in Figure 3.4B, you will find that its molecular formula is $C_6H_{12}O_6$, identical to that of glucose. Thus, glucose and fructose are isomers; they differ only in the arrangement of their atoms (in this case, the positions of the carbonyl groups, highlighted in blue). Seemingly minor differences like this give isomers different properties, such as how they react with other molecules. In this case, the differences also make fructose taste considerably sweeter than glucose.

The carbon skeletons of both glucose and fructose are six carbon atoms long, whereas other monosaccharides have three to seven carbon atoms. The five-carbon sugars, called pentoses, and the six-carbon sugars, called hexoses, are among the most common.

It is convenient to draw sugars as if their carbon skeletons were linear, but in aqueous solutions, most monosaccharides form rings, as shown for glucose in Figure 3.4C. To form the glucose ring, carbon 1 bonds to the oxygen attached to carbon 5. As shown in the middle representation, the ring diagram of glucose and other sugars may be abbreviated by not showing the carbon atoms at the corners of the ring. Also, the bonds in the ring are often drawn with varied thickness, indicating that the ring is a relatively flat structure with atoms and functional groups, such as $-OH$, extending above and below it. The simplified ring symbol on the right is often used in this book to represent glucose.

Monosaccharides, particularly glucose, are the main fuel molecules for cellular work. Cells also use the carbon skeletons of monosaccharides as raw material for manufacturing other kinds of organic molecules, including amino acids. Monosaccharides that cells do not use immediately are usually incorporated into disaccharides and polysaccharides, as described next.

Web Activity *Models of Glucose*

? Write the formula for a monosaccharide that has three carbons.

■ $C_3H_6O_3$

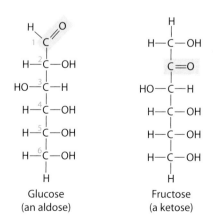

Figure 3.4A   Bees with honey, a mixture of two monosaccharides

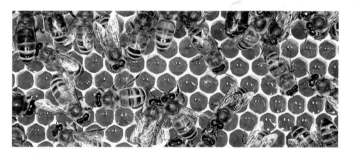

Glucose
(an aldose)

Fructose
(a ketose)

Figure 3.4B   Structures of glucose and fructose

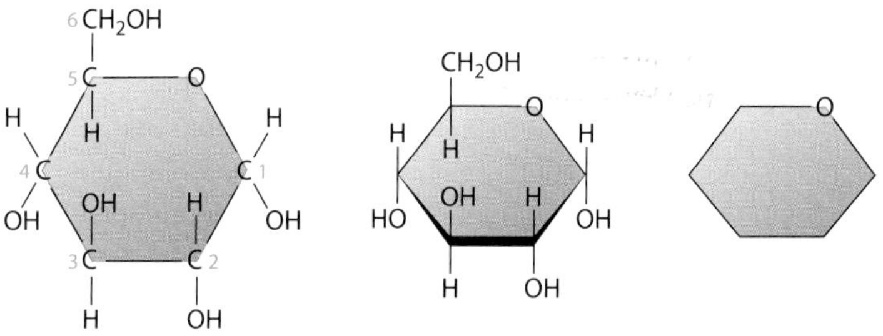

Structural
formula

Abbreviated
structure

Simplified
structure

Figure 3.4C   Three representations of the ring form of glucose

3.5 Cells link two single sugars to form disaccharides

Cells construct a **disaccharide** from two monosaccharides by a dehydration reaction. Figure 3.5 shows how maltose, also called malt sugar, forms from two glucose monomers. One monomer gives up a hydroxyl group and the other gives up a hydrogen atom from a hydroxyl group. As H_2O forms, an oxygen atom is left, linking the two monomers. Maltose, which is common in germinating seeds, is used in making beer, malted milk shakes, and malted milk candy.

The most common disaccharide is sucrose, which is made of a glucose monomer linked to a fructose monomer. The main carbohydrate in plant sap, sucrose nourishes all the parts of the plant. We extract it from the stems of sugarcane or the roots of sugar beets to use as table sugar.

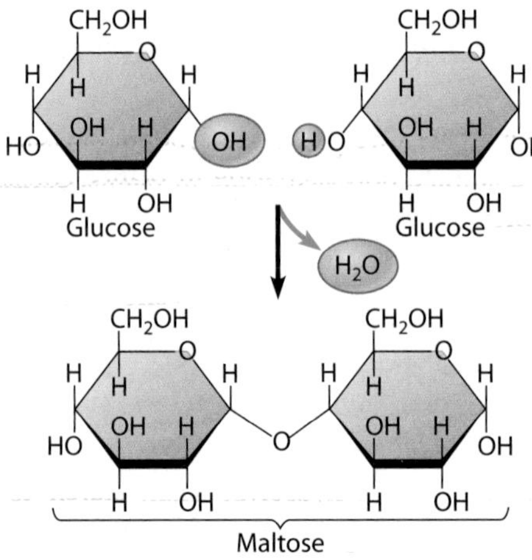

? Lactose, as you read in the chapter introduction, is the disaccharide sugar in milk. It is formed from glucose and galactose. The formula for both these monosaccharides is $C_6H_{12}O_6$. What is the formula for lactose?

■ $C_{12}H_{22}O_{11}$

Figure 3.5 Disaccharide formation by a dehydration reaction

Connection

3.6 What is high-fructose corn syrup and is it to blame for obesity?

If you want to sweeten your coffee or tea, you probably reach for sugar—the disaccharide sucrose. But if you drink sodas or fruit drinks, you're probably consuming the monosaccharides of sucrose in the form of high-fructose corn syrup. In fact, if you look at the label of almost any processed food, you will see high-fructose corn syrup listed as one of the ingredients (Figure 3.6). And if you listen to news reports or read health-related articles, you have probably heard high-fructose corn syrup named as a likely culprit in the "obesity epidemic."

What is high-fructose corn syrup (HFCS)? Let's start with the corn syrup part. The main carbohydrate in corn is starch, a polysaccharide. Industrial processing hydrolyzes starch with enzymes, breaking it into its component monomers, glucose. But glucose does not taste as sweet to us as sucrose. Fructose, on the other hand, tastes much sweeter than both glucose and sucrose. When a new process was developed in the 1970s that used an enzyme to rearrange the atoms of glucose into the sweeter isomer, fructose (see Figure 3.4B), the high-fructose corn syrup industry was born. (High-fructose corn syrup is a bit of a misnomer, however, because the fructose syrup produced is combined with regular corn syrup to produce a mixture of about 55% fructose and 45% glucose, not much different from the proportions in sucrose.)

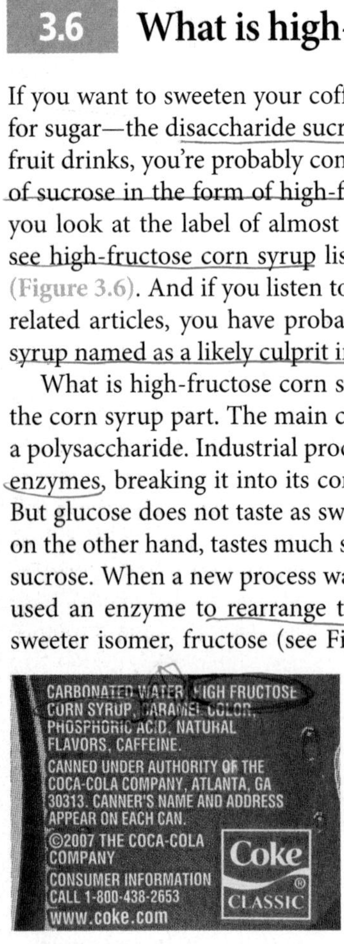

Figure 3.6 HFCS, a main ingredient of soft drinks and processed foods

This clear, goopy liquid is cheaper than sugar and easier to mix into drinks and processed food. And it contains basically the same monosaccharides as sucrose, the disaccharide it is replacing. So is there a problem with HFCS? Some point to circumstantial evidence. From 1980 to 2000, the incidence of obesity doubled in the United States. In that same time period, the consumption of HFCS more than tripled, whereas the consumption of refined cane and beet sugar decreased 21%. The combined per capita consumption of HFCS and refined sugars was 33 kilograms (73 lb) in 1980 and 41.4 kg (91.3 lb) in 2000, a 25% increase. In 2005, that consumption was down somewhat to 39.7 kg: 19.2 kg of HFCS and 20.5 kg of sucrose. That still translates to about 26 teaspoons of added sweeteners a day.

So, is HFCS to blame for increases in obesity, type 2 diabetes, and other chronic diseases associated with increased weight? Scientific studies are ongoing, and the jury is still out. There is consensus, however, that overconsumption of either sweetener along with dietary fat and decreased physical activity contribute to weight gain. In addition, high sugar consumption also tends to replace eating more varied and nutritious foods. Sugars have been described as "empty calories" because they contain only neglibile amounts of other nutrients. For good health, we require proteins, fats, vitamins, and minerals, as well as complex carbohydrates, the topic of the next module.

? How is high-fructose corn syrup made from corn?

■ Corn starch is hydrolyzed to glucose, then enzymes convert glucose to fructose. This fructose is combined with corn syrup to produce HFCS.

Polysaccharides are polymers of monosaccharides linked together by dehydration reactions. Polysaccharides may function as storage molecules or as structural compounds. Figure 3.7 illustrates three common types of polysaccharides.

Starch, a storage polysaccharide in plants, consists entirely of glucose monomers. Starch molecules coil into a helical shape because of the angles of the bonds joining their glucose units. A starch helix may be unbranched (as shown in the figure) or branched. Plant cells, like animal cells, need sugar for energy and as raw material for building other molecules. Plant cells often contain starch granules from which they can withdraw glucose by hydrolysis. Humans and most other animals also have enzymes that can hydrolyze plant starch to provide a source of glucose. Potatoes and grains, such as wheat, corn, and rice, are the major sources of starch in the human diet.

Animals store excess sugar in the form of another glucose polysaccharide, called **glycogen.** Glycogen is more highly branched than starch, as shown in the figure. Most of our glycogen is stored as granules in our liver and muscle cells, which hydrolyze the glycogen to release glucose when it is needed.

Cellulose, the most abundant organic compound on Earth, forms cablelike fibrils in the tough walls that enclose plant cells. Cellulose is also a polymer of glucose, but its glucose monomers are linked together in a different orientation. (Carefully compare the oxygen "bridges" highlighted in yellow between glucose monomers in starch, glycogen, and cellulose in the figure.) Arranged parallel to each other, cellulose molecules are joined by hydrogen bonds, forming part of a fibril. In wood, layers of cellulose fibrils combine with other polymers, making a material strong enough to support trees hundreds of feet high.

Most animals do not have enzymes that can hydrolyze the glucose linkages in cellulose. Therefore, cellulose is not a nutrient for humans, although it does contribute to digestive system health. The cellulose that passes unchanged through our digestive tract is commonly known as "insoluble fiber." Fresh fruits, vegetables, and grains are rich in fiber. Animals that do derive nutrition from cellulose, such as cows and termites, have cellulose-hydrolyzing prokaryotes inhabiting their digestive tracts.

Another structural polysaccharide, **chitin,** is used by insects and crustaceans to build their exoskeleton, the hard case enclosing the animal. Chitin is also found in the cell walls of fungi. Humans use chitin to make a strong and flexible surgical thread that decomposes after a wound or incision heals.

Almost all carbohydrates are hydrophilic owing to the many hydroxyl groups attached to their sugar monomers (see Figure 3.4B). Thus, cotton bath towels, which are mostly cellulose, are quite water absorbent. Next we look at a class of macromolecules that are not soluble in water.

Web Activity *Carbohydrates*

❓ Compare and contrast starch and cellulose, two plant polysaccharides.

■ Both are polymers of glucose, but the bonds between glucose monomers have different shapes. Starch functions mainly for sugar storage. Cellulose is a structural polysaccharide that is the main material of cell walls.

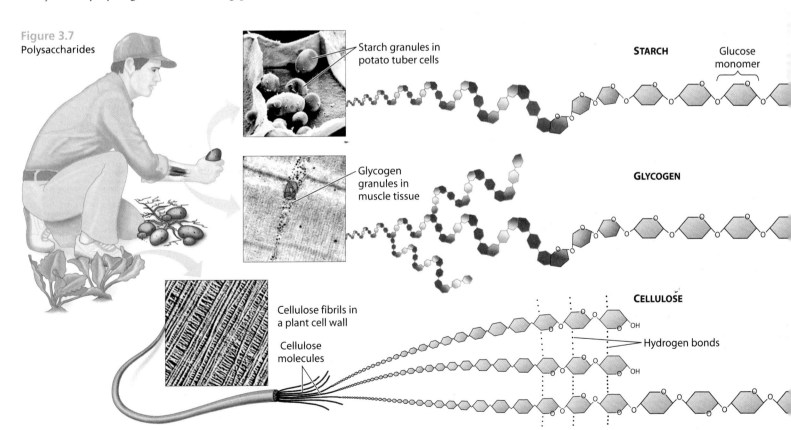

Figure 3.7
Polysaccharides

Starch granules in potato tuber cells

Glycogen granules in muscle tissue

Cellulose fibrils in a plant cell wall

Cellulose molecules

STARCH — Glucose monomer

GLYCOGEN

CELLULOSE

Hydrogen bonds

3.8   Fats are lipids that are mostly energy-storage molecules

Lipids are diverse compounds that are grouped together because they share one trait: They mix poorly, if at all, with water. Lipids consist mainly of carbon and hydrogen atoms linked by nonpolar covalent bonds. In contrast to carbohydrates and most other biological molecules, lipids are **hydrophobic** (water-fearing). You can see the effect of this chemical difference in a bottle of salad dressing: The oil (a type of lipid) separates from the vinegar (which is mostly water). Other oils make the feathers in Figure 3.8A repel water, which you see as beads on the surface. By keeping feathers from absorbing water, oils help waterfowl such as ducks stay afloat.

Oils are a liquid fat. A **fat** is a large lipid made from two kinds of smaller molecules: glycerol and fatty acids. Shown at the top in Figure 3.8B, glycerol is an alcohol with three carbons, each bearing a hydroxyl group. A fatty acid consists of a carboxyl group (the functional group that gives these molecules the name fatty *acid*) and a hydrocarbon chain, usually 16 or 18 carbon atoms in length. The carbons in the chain are linked to each other and to hydrogen atoms by nonpolar covalent bonds, making the hydrocarbon chain hydrophobic.

The main function of fats is energy storage. A gram of fat stores more than twice as much energy as a gram of a polysaccharide such as starch. This compact energy storage enables a mobile animal, such as a duck or a human, to get around much better than if the animal had to lug its stored energy around in the bulkier form of a carbohydrate. In addition to storing energy, fatty tissue cushions vital organs and insulates the body.

Figure 3.8B shows how one fatty acid molecule can link to a glycerol molecule by a dehydration reaction. Linking three fatty acids to glycerol produces a fat, as illustrated in Figure 3.8C. A synonym for fat is *triglyceride*, a term you may see on food labels or on medical tests for fat in the blood. The three fatty acids in a fat are often of different kinds, as in our example.

As shown in the third fatty acid in Figure 3.8C, some fatty acids contain double bonds, which cause kinks (or bends) in the carbon chain. Double bonds prevent the maximum number

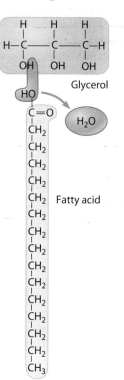

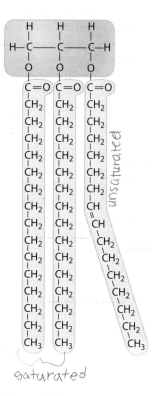

Figure 3.8B   A dehydration reaction linking a fatty acid to glycerol

Figure 3.8C   A fat molecule made from glycerol and three fatty acids

of hydrogen atoms from bonding to a carbon skeleton. Fatty acids and fats with double bonds in the carbon chain are said to be **unsaturated**—that is, having less than the maximum number of hydrogens. Fats with the maximum number of hydrogens are said to be **saturated.** The kinks in unsaturated fats prevent the molecules from packing tightly together and solidifying at room temperature. Corn oil, olive oil, and other vegetable oils are unsaturated fats. When you see "hydrogenated vegetable oils" on a margarine label, it means that unsaturated fats have been converted to saturated fats by adding hydrogen. Unfortunately, hydrogenation also creates *trans* fats, a form of fat that recent research associates with heart disease.

Most plant fats are unsaturated oils, whereas most animal fats are saturated. Their fatty acid chains pack closely together, which is why butter and beef fat are solid at room temperature. Diets rich in saturated fats and *trans* fats may contribute to cardiovascular disease by promoting atherosclerosis. In this condition, lipid-containing deposits called plaques build up within the walls of blood vessels, reducing blood flow.

?   On a food package, what does "unsaturated fats" mean?

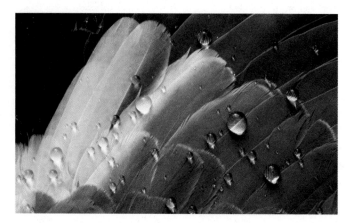

Figure 3.8A   Water beading on the oily coating of feathers

■ Unsaturated fats are fats with some double bonds in the carbon chain of their fatty acids. These fats have fewer hydrogen atoms than they would without the double bonds.

3.9 Phospholipids and steroids are important lipids with a variety of functions

Cells could not exist without **phospholipids,** the major component of cell membranes. Phospholipids are structurally similar to fats, but they contain only two fatty acids attached to glycerol instead of three. In place of the third fatty acid is a negatively charged phosphate group. The structure of phospholipids provides a classic example of how form fits function, as we will explore further in Module 4.5. The hydrophilic and hydrophobic ends of multiple molecules assemble in a bilayer of phospholipids to form membranes (Figure 3.9A). The hydrophobic tails cluster in the center, and the hydrophilic phosphate heads face the watery environment on both sides of the membrane.

Steroids are lipids whose carbon skeleton contains four fused rings, as shown in the structural diagram of cholesterol in Figure 3.9B. (The diagram omits the carbons making up the rings and most of the chain and also the hydrogens attached to these carbons.) **Cholesterol** is a common component in animal cell membranes, and animal cells also use it as a starting material for making other steroids, including sex hormones. Different steroids vary in the chemical groups attached to the rings, as you saw in Figure 3.2. Too much cholesterol in the blood may contribute to atherosclerosis.

Web Activity *Lipids*

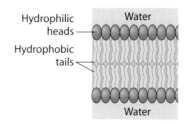

Figure 3.9A   Section of a phospholipid membrane

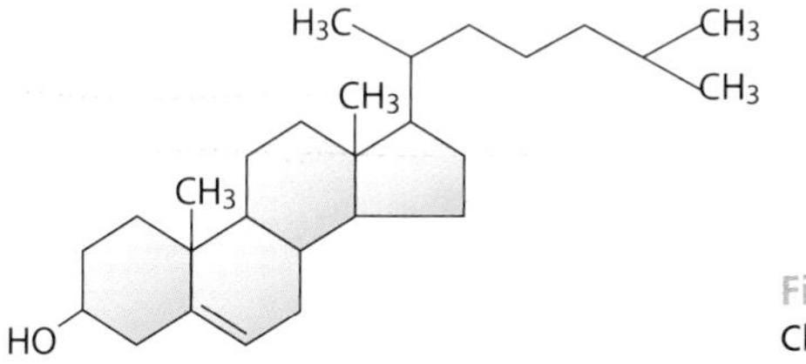

Figure 3.9B Cholesterol, a steroid

HO

? Why are human sex hormones considered lipids?

■ They are steroids, one of the hydrophobic compounds grouped as lipids.

Connection

3.10 Anabolic steroids pose health risks

Anabolic steroids are synthetic variants of the male hormone testosterone. Testosterone causes a general buildup of muscle and bone mass in males during puberty and maintains masculine traits throughout life. Because anabolic steroids structurally resemble testosterone, they also mimic some of its effects. (Anabolic comes from *anabolism,* the building of substances by the body.)

As prescription drugs, anabolic steroids are used to treat general anemia and diseases that destroy body muscle. However, some individuals abuse these drugs, with serious consequences. Overdosing may cause violent mood swings ("steroid rage") and deep depression. The liver may be damaged, leading to cancer. The use of anabolic steroids can also alter cholesterol levels and lead to high blood pressure, increasing the risk of cardiovascular problems. Use of these drugs often makes the body reduce its output of natural male sex hormones, which can cause shrunken testicles, reduced sex drive, infertility, and breast enlargement in men. Use in women has been linked to menstrual cycle disruption and development of masculine characteristics. A serious effect in teens is that bones may stop growing, stunting growth.

Despite the risks, some athletes use steroids to gain a competitive edge. Meanwhile, sports organizations ban their use, implement drug testing, and penalize violators in an effort to counter an environment in which athletes feel they have to take performance-enhancing drugs just to stay competitive.

In 2003, the discovery of a new designer steroid rocked the sports world. THG, short for tetrahydrogestrinone, is a chemically modified ("designer") steroid intended to avoid detection in ordinary drug testing. Once the U.S. Anti-Doping Agency received a sample of the drug, the agency was able to develop a test for it. This test revealed the substance's use among some high-profile track and field athletes as well as professional football players. The U.S. Food and Drug Administration declared THG an illegal steroid, and athletic governing bodies from the Olympics to swimming, soccer, football, and baseball organizations now prohibit its use.

Unscrupulous chemists, as well as dishonest trainers, coaches, and athletes, will continue to try to find ways to best the system, while the U.S. Congress, professional sports authorities, and most high school and college athletic programs struggle to keep the competition fair and protect the health of athletes.

? What kind of carbon ring structure would you find in an anabolic steroid?

■ Four fused carbon rings: three six-sided and one five-sided, as in all steroids

3.11 Proteins are essential to the structures and functions of life

The name *protein*, from the Greek word *proteios*, meaning "first place," suggests the importance of this class of macromolecules. A **protein** is a polymer constructed from amino acid monomers. Each of our many thousands of different proteins has a unique three-dimensional structure that corresponds to a specific function. Proteins are important to the structures of cells and organisms and participate in everything they do.

Probably the most important role for proteins is as *enzymes,* the chemical catalysts that speed and regulate virtually all chemical reactions in cells. Lactase, which you read about in the chapter introduction, is just one of the thousands of different enzymes produced by body cells.

In Figure 3.11, you can see examples of two other types of proteins. *Structural proteins* are found in hair and the fibers that make up connective tissues such as tendons and ligaments. Muscles contain *contractile proteins*.

Other types of proteins include *defensive proteins,* such as the antibodies of the immune system, and *signal proteins,* such as many of the hormones and other messengers that help coordinate body activities by communicating between cells. *Receptor proteins* may be built into cell membranes and transmit signals into cells. Hemoglobin in red blood cells is a *transport protein* that delivers O_2 to working muscles. Different transport proteins move sugar molecules into cells for energy. Other proteins are *storage proteins,* such as ovalbumin, the protein of egg white, which serves as a source of amino acids for developing embryos. Milk proteins provide amino acids for baby mammals, and plants have storage proteins for the developing embryos in seeds.

As we see next, this huge diversity of proteins is based on just 20 building blocks.

Web Activity *Protein Functions*

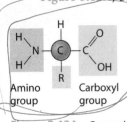

Figure 3.11 Structural proteins of hair, tendons, and ligaments; and contractile proteins of muscles

? Which of the following is not a protein: hemoglobin, cholesterol, ovalbumin, an enzyme, an antibody?

■ Cholesterol

3.12 Proteins are made from amino acids linked by peptide bonds

Of all of life's molecules, proteins are the most elaborate and the most diverse in structure and function. Protein diversity is based on differing arrangements of a common set of just 20 amino acids. **Amino acids** all have an amino group and a carboxyl group (which makes it an acid, hence the name *amino acid*). As you can see in the general structure shown in Figure 3.12A, both of these functional groups are covalently bonded to a central carbon atom, called the alpha carbon. Also bonded to the alpha carbon is a hydrogen atom and a chemical group symbolized by the letter R. The R group, also called the side chain, differs with each amino acid. In the simplest amino acid (glycine), the R group is just a hydrogen atom. In others, such as those in Figure 3.12B, the R group consists of one or more carbon atoms with various chemical groups attached. The structure of the R group determines the specific properties of each of the 20 amino acids that are found in proteins (see Appendix 3).

The amino acids in Figure 3.12B represent two main types, hydrophobic and hydrophilic. Leucine (abbreviated Leu) is an example of an amino acid whose R group is nonpolar and hydrophobic. Serine (Ser), with a hydroxyl group in its R group,

Figure 3.12A General structure of an amino acid

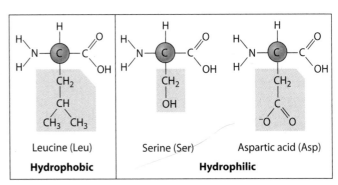

Figure 3.12B Examples of amino acids with hydrophobic and hydrophilic R groups.

is an example of an amino acid with a polar, hydrophilic R group. Aspartic acid (Asp) is acidic and negatively charged at the pH of a cell. (Indeed, the amino and carboxyl groups of amino acids are usually ionized at cellular pH, as shown in Table 3.2.) Other amino acids have basic R groups and are positively charged. Amino acids with polar and charged R groups help proteins dissolve in the aqueous solutions inside cells.

Now that we have examined amino acids, let's see how they are linked to form polymers. Cells join amino acids together in a dehydration reaction that links the carboxyl group of one amino acid to the amino group of the next amino acid as a water molecule is removed (Figure 3.12C). The resulting covalent linkage is called a **peptide bond.** The product of the reaction shown in the figure is called a *di*peptide, because it was made from *two* amino acids. Additional amino acids can be added by the same process to form a chain of amino acids, a **polypeptide.** To release amino acids from the polypeptide by hydrolysis, a molecule of H_2O must be added back to each peptide bond.

How is it possible to make thousands of different kinds of proteins from just 20 amino acids? The answer has to do with sequence. You know that thousands of English words can be made by varying the sequence of 26 letters. Although the protein alpha-bet is slightly smaller (just 20 "letters"), the "words" are much longer. Most polypeptides are at least 100 amino acids in length; some are a thousand or more. Each polypeptide has a unique sequence of amino acids. But a long polypeptide chain of specific sequence is not the same as a protein, any more than a long strand of yarn is the same as a sweater. A functioning protein is one or more polypeptide chains precisely coiled, twisted, and folded into a unique three-dimensional shape.

? In what way is the production of a dipeptide similar to the production of a disaccharide?

■ In both cases, the monomers are joined by a dehydration reaction.

Figure 3.12C
Peptide bond formation

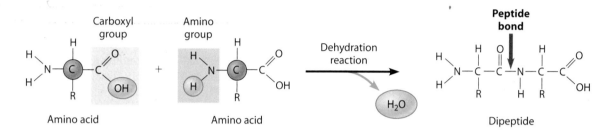

Amino acid   Amino acid   Dipeptide

3.13   A protein's specific shape determines its function

Figure 3.13A shows a ribbon model of lysozyme, an enzyme found in our tears and white blood cells. Lysozyme consists of one long polypeptide, represented by the purple ribbon. Lysozyme's general shape is called globular. This overall shape is more apparent in Figure 3.13B, a space-filling model of lysozyme. (The colors represent different atoms: gray for carbon, red for oxygen, blue for nitrogen. The barely visible yellow balls are sulfur atoms that form the stabilizing bonds shown as yellow lines in the ribbon model.) Most enzymes and other proteins are globular. Structural proteins, such as those making up hair and tendons, are typically long and thin—fibrous.

General shape is one thing; specific shape is another. The coils and twists of lysozyme's polypeptide ribbon appear haphazard, but they represent the molecule's specific, three-dimensional shape, and this shape is what determines its specific function. Nearly all proteins must recognize and bind to some other molecule to function. Lysozyme, for example, can destroy bacterial cells, but first it must bind to specific molecules on the bacterial cell surface. Lysozyme's specific shape enables it to recognize and attach to its molecular target, which fits into the groove you see on the right in the figures.

The dependence of protein function on a protein's specific shape becomes clear when proteins are altered. In a process called **denaturation,** polypeptide chains unravel, losing their specific shape and, as a result, their function. Changes in salt concentration and pH can denature many proteins, as can excessive heat. For example, visualize what happens when you fry an egg. Heat quickly denatures the clear proteins surrounding the yolk, making them solid, white, and opaque. One of the reasons why extremely high fevers are so dangerous is that some proteins in the body become denatured and cannot function.

Given the proper cellular environment, a newly synthesized polypeptide chain spontaneously folds into its functional shape. We examine the four levels of a protein's structure next.

🎧 **MP3 Tutor** *Protein Structures and Function*

? Why does a denatured protein no longer function normally?

■ The function of each protein is a consequence of its specific shape, which is lost when a protein denatures.

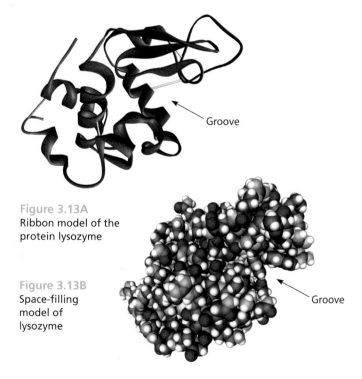

Figure 3.13A
Ribbon model of the protein lysozyme

Groove

Figure 3.13B
Space-filling model of lysozyme

Groove

3.14 A protein's shape depends on four levels of structure

Primary Structure The **primary structure** of a protein is its unique sequence of amino acids. As an example, let's consider transthyretin, an important transport protein found in our blood. It is a globular molecule whose specific shape enables it to transport two key chemicals throughout the body, vitamin A and a hormone from the thyroid gland. A complete molecule of transthyretin has four polypeptide chains, each made up of 127 amino acids. Figure 3.14A on the next page shows part of one of these chains partially unraveled for a closer look at its primary structure. The three-letter abbreviations represent amino acids. One of the 20 amino acids occupies each of the 127 positions along the chain.

For this or any other protein to perform its specific function, it must have the correct amino acids arranged in a precise order. The primary structure of a protein is determined by inherited genetic information. Even a slight change in a protein's primary structure may affect its overall shape and its ability to function. For instance, a single amino acid change in hemoglobin, the oxygen-carrying blood protein, causes sickle-cell disease, a serious blood disorder.

Secondary Structure In the second level of protein structure, parts of the polypeptide coil or fold into local patterns called **secondary structure.** Coiling of a polypeptide chain results in a secondary structure called an **alpha helix;** a certain kind of folding leads to a **pleated sheet.** Both of these patterns are maintained by regularly spaced hydrogen bonds between the hydrogens of the $>$N—H groups and the oxygens of the $>$C$=$O groups of neighboring peptide bonds along the polypeptide chain. The hydrogen bonds are represented in Figure 3.14B by a row of dots. Because the R groups of the amino acids are not involved in the hydrogen bonds forming these secondary structures, they are omitted from the diagrams.

Transthyretin has only one alpha-helix region (see Figure 3.14C). In contrast, many fibrous proteins, such as the structural protein of hair, have the alpha-helix structure over most of their length. Pleated sheets make up the core of many globular proteins, as is the case for transthyretin. Pleated sheets also dominate some fibrous proteins, including the silk protein of a spider's web, shown to the left. The teamwork of so many hydrogen bonds makes each silk fiber of a web as strong as steel. Potential uses of spider silk proteins include surgical thread, fishing line, and bulletproof vests.

Tertiary Structure The term **tertiary structure** refers to the overall three-dimensional shape of a polypeptide. As already mentioned, most tertiary structures can be roughly described as either globular or fibrous. As you can see in Figure 3.14C, a transthyretin polypeptide has a generally globular shape, which results from the compact combination of an alpha helix and several pleated-sheet regions. The indentations and bulges arising from its particular arrangement of coils and folds give the polypeptide the specific shape that fits it to its function.

Tertiary structure generally results from interactions among the R groups of the amino acids making up the polypeptide. For example, globular proteins found in aqueous solutions such as transthyretin are folded so that the hydrophobic R groups are on the inside of the molecule and the hydrophilic groups on the outside, exposed to water. In addition to the clustering of hydrophobic groups, hydrogen bonding between polar side chains and ionic bonding of some of the charged R groups help maintain the tertiary structure. A protein's shape may be reinforced further by covalent bonds called disulfide bridges. The yellow lines in Figure 3.13A represent disulfide bridges.

Quaternary Structure Many proteins consist of two or more polypeptide chains, or subunits. Such proteins have a **quaternary structure,** resulting from the association of the subunits. Figure 3.14D shows a complete transthyretin molecule with its four identical globular subunits.

Another example of a protein with quaternary structure is collagen, shown to the right. Collagen is a fibrous protein with helical subunits intertwined into a larger triple helix. This arrangement gives the long fibers great strength, suited to their function as the girders of connective tissue in skin, bone, tendons, and ligaments. (Collagen accounts for 40% of the protein in a human body.)

Many other proteins have subunits that are different from one another. For example, the oxygen-transporting molecule hemoglobin has four subunits of two distinct types (see Figure 22.11). Each subunit also has a nonprotein component, called a heme, with an iron atom that binds oxygen.

Polypeptide chain

Collagen

What happens if a protein folds incorrectly? Many diseases, such as Alzheimer's and Parkinson's, involve an accumulation of misfolded proteins. Prions are infectious mishapened proteins that are associated with serious degenerative brain diseases such as mad cow disease (see Module 10.21). Such diseases reinforce the theme that structure fits function: A protein's unique three-dimensional shape determines its proper functioning.

Web Activity *Protein Structure*

? If a genetic mutation changes the primary structure of a protein, how might this destroy the protein's function?

■ Primary structure, the amino acid sequence, affects the secondary structure, which affects the tertiary structure, which affects the quaternary structure (if any). Thus primary structure affects the shape of the protein, and the function of a protein depends on its shape.

Four Levels of Protein Structure

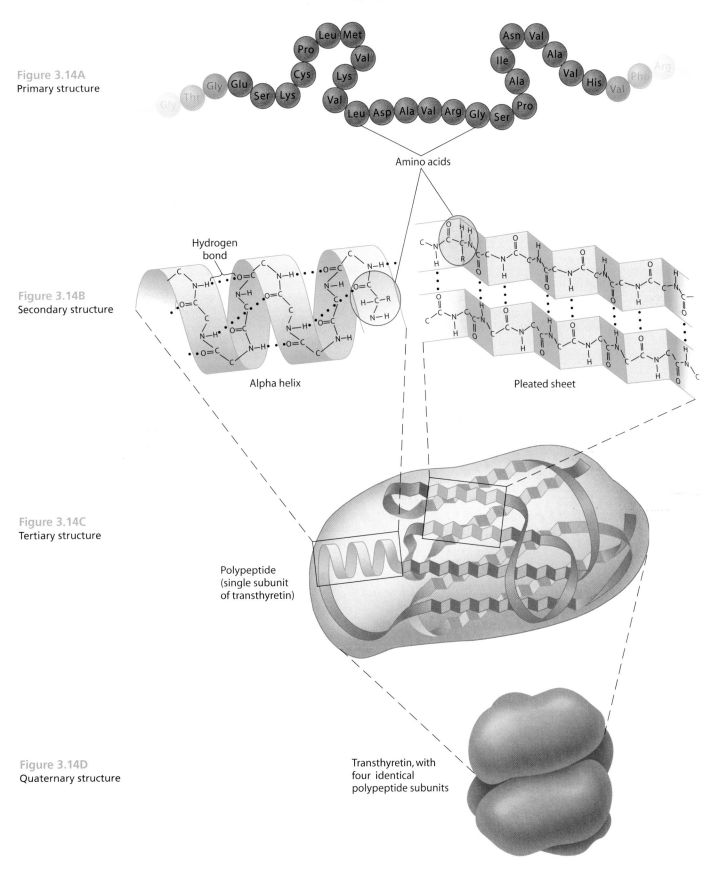

Figure 3.14A
Primary structure

Amino acids

Figure 3.14B
Secondary structure

Hydrogen bond

Alpha helix

Pleated sheet

Figure 3.14C
Tertiary structure

Polypeptide
(single subunit
of transthyretin)

Figure 3.14D
Quaternary structure

Transthyretin, with
four identical
polypeptide subunits

3.15 Linus Pauling contributed to our understanding of the chemistry of life

Linus Pauling, one of the giants of 20th-century science, made extraordinary contributions to both science and human affairs. Driven by the desire "to understand the world," Pauling started out in chemistry and physics. He published a series of papers on chemical bonding that eventually led to a Nobel Prize in Chemistry in 1954. By that time, Pauling was also studying biological molecules. In 1935, he and a colleague discovered how oxygen molecules attach to the iron atoms of hemoglobin. He also discovered how an abnormal hemoglobin molecule causes sickle-cell disease. And it was Pauling who first described the two fundamental secondary structures of proteins, the alpha helix and the pleated sheet (Figure 3.15). In an interview a number of years ago, he recounted that work:

> I began trying to find the structure of proteins in 1937, and didn't succeed. So I began working with my collaborators to determine the three-dimensional structures of amino acids and simple peptides. In 1937, no one had yet determined such structures. . . . And then in 1948 I found the alpha-helix and pleated-sheet structures in proteins. . . . I'm surprised that nobody else had done this job in the 11 years that intervened—in a sense, surprised that I hadn't done it in '37, when my ideas were all the right ones. I just hadn't worked hard enough.

Pauling's efforts were not limited to science. He also became the scientific community's leading advocate for halting the testing of nuclear weapons. As a result of his politics, the U.S. State Department viewed him as a threat and revoked his passport. Nevertheless, in 1963 he received the Nobel Peace Prize for helping produce a ban on nuclear testing. Pauling is the only person to receive two unshared Nobel Prizes.

The cancellation of Pauling's passport in 1952 kept him from working directly with scientists in Europe. This may have

Figure 3.15
Linus Pauling with a model of the alpha helix in 1948

prevented Pauling from earning a third Nobel Prize, for at that time he was hot on the trail of the structure of DNA. But his triple-helix model was wrong, and it was James Watson and Francis Crick, working in England, who came up with the correct solution—a double helix, which we describe in the next module.

? What contribution did Pauling make to our understanding of protein structure?

■ He identified the alpha helix and pleated sheet of secondary structure.

Nucleic Acids

3.16 Nucleic acids are information-rich polymers of nucleotides

The primary structure of polypeptides determines the shape of a protein. But what determines the primary structure? The amino acid sequence of a polypeptide is programmed by a discrete unit of inheritance known as a **gene.** Genes consist of **DNA (deoxyribonucleic acid),** one of the two types of polymers known as **nucleic acids.**

A gene does not put its genetic DNA information to work directly. It works through an intermediary—the second type of nucleic acid known as **ribonucleic acid (RNA).** DNA is transcribed into RNA, which is then translated into the primary structure of polypeptides. We return to this chain of command and the functions of DNA and RNA later in the book.

The monomers that make up nucleic acids are **nucleotides.** As indicated in Figure 3.16A, each nucleotide has three parts.

One part is a five-carbon sugar (blue); DNA has deoxyribose (shown in Figure 3.16A), whereas RNA has a slightly different sugar called ribose. Linked to one side of the sugar in both types of nucleic acid is a phosphate group (yellow). Linked to the sugar's other side is a nitrogenous base (green), a molecular structure containing carbon and nitrogen. (The nitrogen atoms tend to take up H^+ from solution, which is why it is called a nitrogenous *base*.) DNA has the nitrogenous bases adenine (A), thymine (T), cytosine (C), and guanine (G). RNA also has A, C, and G, but instead of thymine, it has uracil (U).

Like polysaccharides and polypeptides, a nucleic acid polymer—a polynucleotide—forms from its monomers by dehydration reactions. In this process, the phosphate group of one nucleotide bonds to the sugar of the next monomer. The result

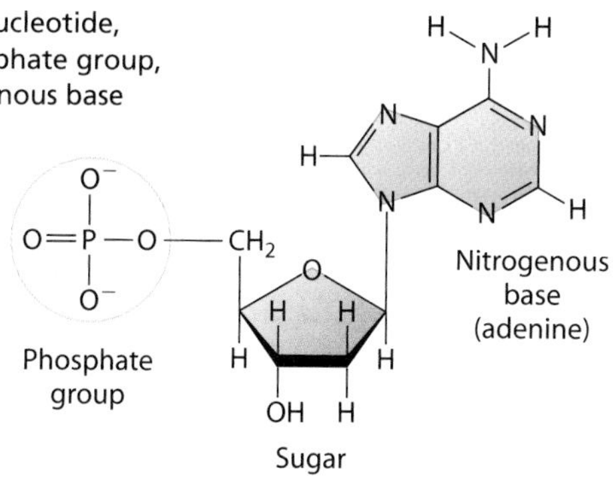

Figure 3.16A A nucleotide, consisting of a phosphate group, sugar, and a nitrogenous base

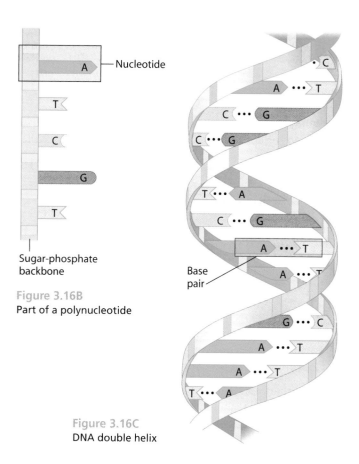

Phosphate group

Sugar

Nitrogenous base (adenine)

Nucleotide

Sugar-phosphate backbone

Figure 3.16B
Part of a polynucleotide

Base pair

Figure 3.16C
DNA double helix

is a repeating sugar-phosphate backbone in the polymer, as represented by the blue and yellow ribbon in Figure 3.16B.

RNA usually consists of a single polynucleotide strand, but DNA is a **double helix,** in which two polynucleotides wrap around each other (Figure 3.16C). The nitrogenous bases protrude from the two sugar-phosphate backbones into the center of the helix. There they always pair up as shown: A pairs with T, and C pairs with G. The two DNA chains are held in a double helix by hydrogen bonds (dotted lines) between their paired bases. Most DNA molecules have thousands or even millions of base pairs.

An organism's genes determine the proteins and thus the structures and functions of its body. Let's return to the subject of the chapter introduction—lactose intolerance—to conclude our study of biological molecules. In the next chapter, we move up in the biological hierarchy to the level of the cell.

🎧 MP3 Tutor *DNA Structure*

Web Activity *Nucleic Acid Structure*

Web Process of Science *Connection: What Factors Determine the Effectiveness of Drugs?*

? How are the two types of nucleic acids functionally related?

■ The hereditary material of DNA contains the instructions for the primary structure of polypeptides. RNA is the intermediary that translates those instructions into the order of amino acids.

3.17 Lactose tolerance is a recent event in human evolution

As you'll recall from the chapter introduction, the majority of people stop producing the enzyme lactase in early childhood and thus do not easily digest the milk sugar lactose. Researchers suspected that there must be a genetic link to explain the regional distribution of lactose tolerance and intolerance. In 2002, American and Finnish scientists completed a study of the genes of 196 lactose-intolerant adults of African, Asian, and European descent. They determined that lactose intolerance is actually the human norm. It is "lactose tolerance" that represents a relatively recent mutation in the human genome.

The ability to make lactase into adulthood is concentrated in people of northern European descent, and the researchers speculated that lactose tolerance became widespread among this group because it offered a survival advantage. In northern Europe's relatively cold climate, only one harvest a year is possible. For earlier humans, that meant nonplant sources of food often had to be found. Cattle were first domesticated in northern Europe about 9,000 years ago. With milk and other dairy products at hand year-round, natural selection would have favored anyone with a mutation that kept the lactase gene switched on.

Researchers wondered whether the lactose tolerance mutation found in Europeans might be present in other cultures who kept dairy herds. Indeed, a 2006 study compared the genetic makeup and lactose tolerance of 43 ethnic groups in East Africa. The researchers identified three mutations, all different from each other and from the European mutation, that keep the lactase gene permanently turned on. The mutations appear to have occurred beginning around 7,000 years ago, around the time that archaeological evidence shows the domestication of cattle in these African regions.

Mutations that confer a selective advantage, such as surviving cold winters or withstanding drought by drinking milk, spread rapidly in these early pastoral peoples. Their evolutionary and cultural history is thus recorded in their genes and in their continuing ability to digest milk.

? Explain how lactose intolerance involves three of the four major classes of biological macromolecules.

■ Lactose, milk sugar, is a carbohydrate that is hydrolyzed by the enzyme lactase, a protein. The ability to make this enzyme and the regulation of when it is made is coded for in DNA, a nucleic acid.

Chapter Review

Reviewing the Concepts

Introduction to Organic Compounds (3.1–3.3)

Carbon.   Carbon's ability to bond with four other atoms is the basis for building large and diverse organic compounds. Hydrocarbons are composed of only carbon and hydrogen. Isomers have the same molecular formula but different structures (**3.1**). Characteristic chemical groups, such as the hydrophilic functional groups, give organic molecules specific properties (**3.2**).

Polymers.   A huge number of different polymers are built from a small number of monomers (**3.3**).

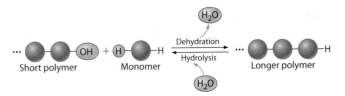

Short polymer | Monomer | Dehydration / Hydrolysis | Longer polymer

Carbohydrates (3.4–3.7)

Monosaccharides.   A monosaccharide has a formula that is a multiple of (CH_2O) and contains hydroxyl groups and a carbonyl group. Two monosaccharides join to form a disaccharide (**3.4–3.5**). High-fructose corn syrup is commonly added to drinks and processed foods (**3.6**).

Polysaccharides are polymers of monosaccharides, such as glucose. Starch and glycogen are storage polysaccharides; cellulose is structural, found in plant cell walls. Chitin is a component of insect exoskeletons and fungal cell walls (**3.7**).

Lipids (3.8–3.10)

Lipids, diverse compounds composed largely of carbon and hydrogen, are grouped together because they are hydrophobic. Fats, also called triglycerides, are energy-storage molecules consisting of glycerol linked to three fatty acids (**3.8**). Other lipids include phospholipids (found in cell membranes), and steroids such as cholesterol (**3.9**). Use of anabolic steroids by athletes can cause serious health problems (**3.10**).

Proteins (3.11–3.15)

Proteins are involved in almost all of a cell's activities; as enzymes, they regulate chemical reactions (**3.11**).

Amino acids.   Protein diversity is based on different sequences of amino acids, monomers that contain an amino group, a hydrogen atom, and an R group, all attached to a central carbon. The R groups distinguish 20 different amino acids, each with specific properties. Amino acids are linked together by peptide bonds to form polypeptides (**3.12**).

Protein structure.   A protein consists of one or more polypeptide chains folded into a unique shape that determines the protein's function (**3.13**). A protein's primary structure is the sequence of amino acids forming its polypeptide chains. Its secondary structure is the coiling or folding of the chain, stabilized by hydrogen bonds. Tertiary structure is the overall three-dimensional shape of a polypeptide, resulting from interactions among R groups. Proteins made of more than one polypeptide have quaternary structure (**3.14**). Linus Pauling made important contributions to our understanding of proteins (**3.15**)

Nucleic Acids (3.16–3.17)

Nucleic acids.   DNA and RNA serve as the blueprints for proteins and thus control the life of a cell. The monomers of nucleic acids are nucleotides, composed of a sugar, a phosphate group, and a nitrogenous base. DNA is a double helix; RNA is a single-stranded polynucleotide (**3.16**). Mutations in DNA have led to lactose tolerance in several human groups whose ancestors raised dairy cattle (**3.17**).

Connecting the Concepts

1. The diversity of life is staggering. Yet the molecular logic of life is simple and elegant: Small molecules common to all organisms are ordered into unique macromolecules. Explain why carbon is central to this diversity of organic molecules. How do carbon skeletons, chemical groups, monomers, and polymers relate to this molecular logic of life?

2. There are four classes of organic molecules that are essential to all living organisms. Complete the following table to help review the structures and functions of these macromolecules.

Classes of macromolecules and their components	Functions	Examples
Carbohydrates CH$_2$OH structure Monosaccharides	Energy for cell, raw material	a. _____
	b. _____	Starch, glycogen
	Plant cell support	c. _____
Lipids (don't form polymers) Glycerol · Fatty acid Components of a fat molecule	Energy storage	d. _____
	e. _____	Phospholipids
	Hormones	f. _____
Proteins g. _____ h. _____ i. _____ Amino acid	j. _____	Lactase
	k. _____	Hair, tendons
	l. _____	Muscles
	Transport	m. _____
	Communication	Signal proteins
	n. _____	Antibodies
	Storage	Egg albumin
	Receive signals	Receptor protein
Nucleic Acids o. _____ p. _____ q. _____ Nucleotide	Heredity	r. _____
	s. _____	DNA and RNA

Testing Your Knowledge

Multiple Choice

3. A glucose molecule is to starch as (*Explain your answer.*)
   a. a steroid is to a lipid.
   b. a protein is to an amino acid.
   c. a nucleic acid is to a polypeptide.
   d. a nucleotide is to a nucleic acid.
   e. an amino acid is to a nucleic acid.

4. What makes a fatty acid an acid?
   a. It does not dissolve in water.
   b. It is capable of bonding with other molecules to form a fat.
   c. It has a carboxyl group that donates an H^+ ion to a solution.
   d. It contains only two oxygen atoms.
   e. It is a polymer made of many smaller subunits.

5. Where in the tertiary structure of a water-soluble protein would you most likely find an amino acid with a hydrophobic R group?
   a. at both ends of the polypeptide chain
   b. on the outside, next to the water
   c. covalently bonded to another R group
   d. on the inside, away from water
   e. hydrogen-bonded to nearby amino acids

6. Cows can derive nutrients from cellulose because
   a. they produce enzymes that recognize the shape of the glucose-glucose bonds and hydrolyze them.
   b. they rechew their cud to break down cellulose fibers.
   c. one of their stomachs contains prokaryotes that can hydrolyze the bonds of cellulose.
   d. their intestinal tract contains termites, which produce enzymes to hydrolyze cellulose.
   e. they convert cellulose to starch and can digest starch.

7. A shortage of phosphorus in the soil would make it especially difficult for a plant to manufacture
   a. DNA.   d. fatty acids.
   b. proteins.   e. sucrose.
   c. cellulose.

8. Lipids differ from other macromolecules in that they
   a. are much larger.
   b. are not truly polymers.
   c. do not have specific shapes.
   d. are nonpolar and therefore hydrophilic.
   e. contain nitrogen atoms.

9. Which functional group (or groups) is polar and tends to make organic compounds hydrophilic?
   a. carbonyl   d. carboxyl
   b. amino   e. all of the above
   c. hydroxyl

10. Unsaturated fats
    a. are more common in animals than in plants.
    b. have fewer fatty acid molecules per fat molecule.
    c. are associated with greater health risks than are saturated fats.
    d. have double bonds in their fatty acid chains.
    e. are usually solid at room temperature.

Describing, Comparing, and Explaining

11. List three different kinds of lipids and describe their functions.

12. Explain why heat, pH changes, and other environmental changes can interfere with a protein's function.

13. How can a cell make many different kinds of protein out of only 20 amino acids? Of the myriad possibilities, how does the cell "know" which proteins to make?

14. Briefly describe the various functions performed by proteins in a cell.

15. Explain how DNA controls the functions of a cell.

16. Sucrose is broken down in your intestine to the monosaccharides glucose and fructose, which are then absorbed into your blood. What is the name of this type of reaction? Using this diagram of sucrose, show how this would occur.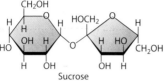

17. Circle and name the functional groups in this organic molecule. What type of compound is this? For which class of macromolecules is it a monomer?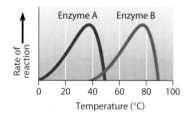

Applying the Concepts

18. Toward the end of his 93 years. Pauling was most often associated with his controversial belief that large doses of vitamin C can help prevent the common cold, cancer, and other diseases. Imagine you are applying for a research grant from the National Institutes of Health to evaluate Pauling's claims. How would you go about setting up an experimental study to determine whether vitamin C can prevent colds? How would you evaluate your results?

19. Enzymes usually function best at an optimal pH and temperature. The following graph shows the effectiveness of two enzymes at various temperatures.

    a. At which temperature does enzyme A perform best? Enzyme B?
    b. One of these enzymes is found in humans and the other in thermophilic (heat-loving) bacteria. Which enzyme would you predict comes from which organism?
    c. From what you know about enzyme structure, explain why the rate of the reaction catalyzed by enzyme A slows down at temperatures above 40 °C.

20. Some scientists hypothesize that life elsewhere in the universe might be based on the element silicon rather than on carbon. Look at the electron shell diagrams in Figure 2.6. What properties does silicon share with carbon that would make silicon-based life more likely than, for example, neon-based or sulfur-based life?

Answers to all questions can be found in Appendix 4.

For Practice Quizzes, BioFlix, MP3 Tutors, and Activities, go to www.mybiology.com.

A Tour of the Cell

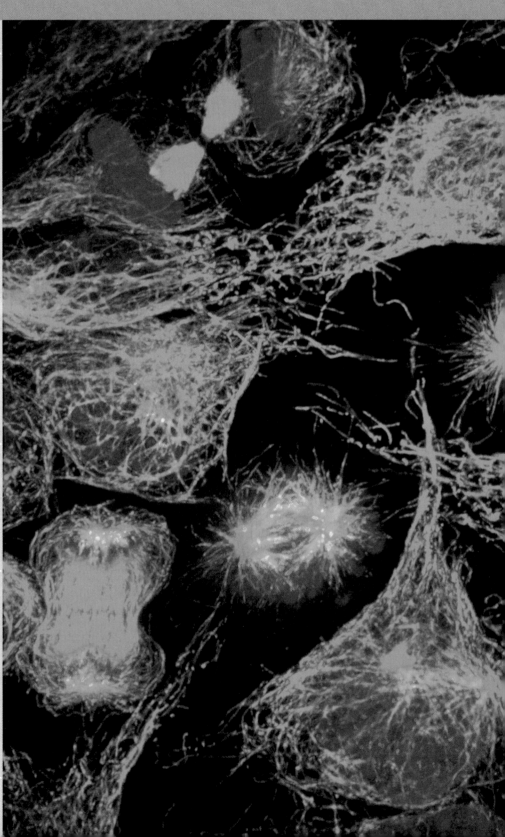

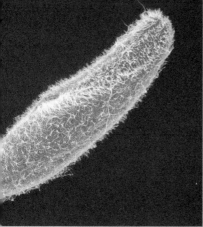

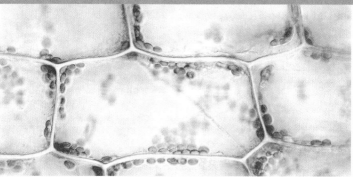

Cells on the Move

Movement might seem like a characteristic of life—at least animal life and the lives of some single-celled organisms. For example, any biology student who has tried to chase a *Paramecium* under a microscope knows that some protists move quite fast. When you think about cells of the human body, you probably know that sperm cells swim and white blood cells creep about, engulfing bacteria and debris from injury sites. You may also realize that, unfortunately, cancer cells move, letting go of connections with neighboring cells and migrating to other tissues, where their rapid cell divisions establish secondary tumors. The **micrograph** (photo taken through a microscope) on the facing page shows beautiful but deadly cancer cells in the midst of dividing. (Part of their beauty comes from the fluorescently colored stains that selectively attach to certain components of the cells.)

But what about plant cells? Or bacteria? Do they move? In 1665 Robert Hooke used a crude microscope to examine a piece of bark (cork). Hooke compared these structures to "little rooms"—*cellulae* in Latin—and the term *cells* stuck. These cells certainly did not move, because they were merely the outer cell walls of former plant cells.

But some of the cells viewed by Hooke's contemporary, Antoni van Leeuwenhoek, did move. Working in the late 1600s with refined lenses that he crafted, Leeuwenhoek examined numerous subjects, from human blood and sperm to pond water. His reports to the Royal Society of London included drawings and descriptions of his discoveries. About what he found in the scrapings from his teeth he wrote, "I then most always saw, with great wonder, that in the said matter there were many very little living animalcules, very prettily a-moving. The biggest sort . . . shot through the spittle like a pike does through the water. The second sort . . . oft-times spun round like a top." Indeed, Leeuwenhoek was the first to observe and describe bacteria.

Leeuwenhoek established that some bacterial cells move. In fact, we now know that some move at incredible speeds. For example, *Bdellovibrio bacteriophorus* is a bacterium that attacks other bacteria. It can charge its prey at up to 100 μm/sec,* comparable to a human running 600 km/hr (about 375 mi/hr). It

*Metric measurements are described in Table 4.1 on the following page.

then bores into its hapless victim by spinning at 100 revolutions per second. Most bacteria aren't quite that fast, but they can locomote with whipping flagella or by spiraling through their environment.

But can plant cells move as well? If you look through a microscope at a thin piece of the common aquarium plant *Elodea*, you may see green chloroplasts marching in an orderly procession around the perimeter of the cells. The micrograph pictured above shows what such plant cells would look like, minus the movement. The fact is that all living cells have movement inside, even if they are anchored within a leaf or attached to other cells in your body. Whereas some cells are motile and move from place to place, all cells have internal structures that move about and interact.

Since the days of Leeuwenhoek and Hooke, improved microscopes have vastly expanded our view of the intricate structures within a cell. Photography and electronic imaging now enable biologists to capture microscope images directly. But just as Hooke and Leeuwenhoek illustrated what they saw, drawings and diagrams are still important in helping us understand the microscopic world. You will see in this chapter, for example, that the use of different colors in drawings makes it easier to distinguish between the various cellular structures, most of which are actually colorless. You will also notice that drawings are often paired with micrographs of cell structures. This pairing provides the best of both worlds: The micrograph shows a structure as a biologist sees it, and the drawing helps you understand the microgaph by emphasizing specific details.

But neither drawings nor micrographs allow you to see the movement that is characteristic of all living cells. For that you need to look for yourself through a microscope or view videos in lecture or on websites, such as the one associated with this book. As you study the still images in this textbook, remind yourself that these cellular parts aren't static—they move.

This chapter focuses on the cellular structures that microscopes have revealed and describes their functions in a living cell. Cells are dynamic, moving, living systems. As you learn about their parts, remember that the phenomenon we call life emerges from the arrangement and interactions of the components of a cell. ■ ■ ■

4.1 Microscopes reveal the world of the cell

Our understanding of nature often parallels the invention and refinement of instruments that extend human senses to new limits. Before microscopes were first used in the 17th century, no one knew that living organisms were composed of cells. The first microscopes were light microscopes, like the ones you may use in a biology laboratory. A **light microscope (LM),** such as the one shown in Figure 4.1A, works by passing visible light through a specimen, such as a microorganism or a thin slice of animal or plant tissue. Glass lenses in the microscope bend the light to magnify the image of the specimen and project the image into the viewer's eye or onto photographic film or a video screen.

Magnification is the increase in the apparent size of an object. Figure 4.1B shows a protist called *Paramecium*. The notation "LM 230×" printed along the right edge of this micrograph is one we use throughout this book. It tells you that the photograph was taken through a light microscope and that this image is about 112 times the actual size of the organism. This *Paramecium* is about 0.67 mm in length. Table 4.1 gives the most common units of length that biologists use. (As you can see, the units are metric, so conversions are easy.) Although this image could be magnified further, additional magnification would not allow us to see more detail. Indeed, at higher magnification, the image would become blurrier. Light microscopes can effectively magnify objects only about 1,000 times.

Thus, another important factor in microscopy (the use of a microscope) is *resolution,* a measure of the clarity of an image. Resolution is the ability of an optical instrument to show two close objects as separate. For example, what looks to your unaided eye like a single star in the sky may be resolved as two separate stars with the help of a telescope.

Any optical device—be it an eye, a telescope, or a microscope—has a limit to its resolution. The human eye cannot resolve details finer than 0.1 mm. The light microscope can resolve objects as small as 0.2 micrometer, about the size of the smallest bacterium. (As you can see in the table, a micrometer (μm) is one-thousandth of a millimeter. The μ in the abbreviation for micrometer is the Greek letter mu.) But no matter how many times its image of such a bacterium is magnified, the light microscope cannot show the details of this small cell's structure.

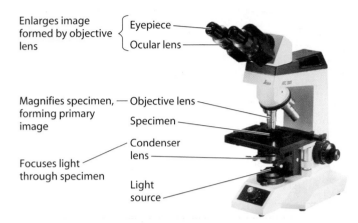

Enlarges image formed by objective lens
— Eyepiece
— Ocular lens

Magnifies specimen, forming primary image
— Objective lens
— Specimen

Focuses light through specimen
— Condenser lens
— Light source

Figure 4.1A Light microscope (LM)

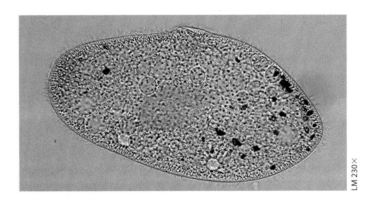

LM 230×

Figure 4.1B Light micrograph of a protist, *Paramecium*

From the year 1665, when Robert Hooke discovered cells, until the middle of the 20th century, biologists had only light microscopes for viewing cells. With these microscopes, they discovered a great deal, including microorganisms, the cells composing animal and plant tissues, and also some of the structures within cells. By the mid-1800s, these discoveries led to the **cell theory,** which states that all living things are composed of cells and that all cells come from other cells.

Our knowledge of cell structure took a giant leap forward as scientists began using the electron microscope in the 1950s. Instead of light, the **electron microscope (EM)** uses a beam of electrons. The EM has a much greater resolution than the light microscope. It can distinguish biological structures as small as about 2 nanometers (nm), a hundredfold improvement over the light microscope. This high resolution has allowed biologists to explore cellular *ultrastructure,* the complex internal anatomy of a cell. The highest power electron micrographs you will see in this book have magnifications of about 100,000 times.

Figures 4.1C and 4.1D on the next page show images produced by two kinds of electron microscopes. Biologists use the **scanning electron microscope (SEM)** to study the detailed architecture of cell surfaces. The SEM uses an electron beam to

TABLE 4.1	MEASUREMENT EQUIVALENTS
1 meter (m) = 10^0 m = 39.37 inches	
1 millimeter (mm) = 10^{-3} m (1/1,000 m) = 0.04 inch	
1 micrometer (μm) = 10^{-3} mm = 10^{-6} m (1/1,000,000 m)	
1 nanometer (nm) = 10^{-3} μm = 10^{-9} m (1/1,000,000,000 m)	
1 meter = 10^3 mm = 10^6 μm = 10^9 nm	

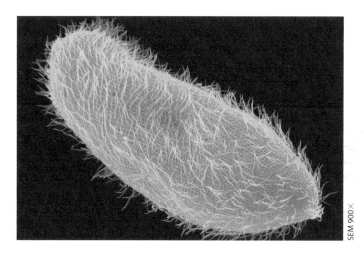

Figure 4.1C   Scanning electron micrograph of *Paramecium*

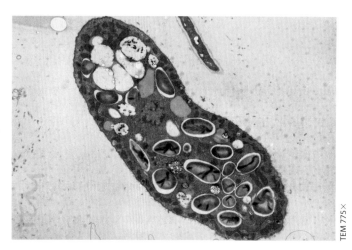

Figure 4.1D   Transmission electron micrograph of *Paramecium*

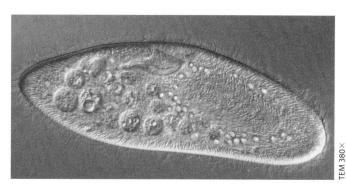

Figure 4.1E   Micrograph produced by differential interference-contrast microscopy of *Paramecium*

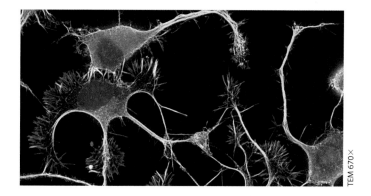

Figure 4.1F   Micrograph produced by fluorescence confocal microscopy, showing nerve cells growing in tissue culture

scan the surface of a cell or group of cells that has been coated with a thin film of metal. When the surface is hit by the beam, it emits electrons. The electrons are detected by a device that then translates their pattern into an image projected onto a video screen. The scanning electron micrograph in Figure 4.1C highlights the numerous cilia on a *Paramecium,* projections it uses for movement. Many structural details of cell surfaces have been discovered using the SEM. As you can see, the SEM produces images that look three-dimensional.

The **transmission electron microscope (TEM)** (Figure 4.1D) is used to study the details of internal cell structure. Specimens are cut into extremely thin sections and stained with atoms of heavy metals, which attach to certain cellular structures more than others. The TEM aims an electron beam through the thin section, just as a light microscope aims a beam of light through a specimen. The image is created by the pattern of transmitted electrons. Instead of using glass lenses, the TEM uses electromagnets as lenses to bend the paths of the electrons, magnifying and focusing an image onto a viewing screen or photographic film. The micrograph in Figure 4.1D shows internal details of *Paramecium* as seen with the TEM. SEMs and TEMs are initially black and white but are often artificially colorized to highlight or clarify structural features.

Electron microscopes have truly revolutionized the study of cells and their structures. Nonetheless, they have not replaced the light microscope. One problem is that electron microscopes cannot be used to study living specimens because the methods used to prepare the specimen kill the cells. For a biologist studying a living process, such as the movement of *Paramecium,* a light microscope equipped with a video camera is more suitable than either an SEM or a TEM.

There are different types of light microscopy, each using different techniques to enhance contrast and selectively highlight cellular components. Figure 4.1E shows a *Paramecium,* as seen using differential interference-contrast microscopy. This optical technique amplifies differences in density while still allowing living cells to be examined. Figure 4.1F is a striking example of fluorescence and confocal microscopy. In this technique, specific molecules are tagged with fluorescent dyes, which selectively bind to various cellular molecules. A type of "optical sectioning" technique then brings a very thin section of the cell into focus. By capturing sharp, in-focus images at many different planes, a three-dimensional reconstruction of the sample can be created. You will see many beautiful examples of fluorescence microscopy in this textbook.

Web Activity *Metric System Review*

? Which type of microscope would you use to study (a) the changes in shape of a living human white blood cell; (b) the finest details of surface texture of a human hair; (c) the detailed structure of an organelle in a human liver cell?

■ (a) light microscope; (b) scanning electron microscope; (c) transmission electron microscope

Most cells are microscopic

There is a reason that our knowledge of cells depended on the development of microscopes: Most cells cannot be seen with the unaided eye. Figure 4.2A shows the size range of cells compared with objects both larger and smaller. The smallest cells are bacteria called mycoplasmas, with diameters as small as 0.2 μm (see Table 4.1). Some of the bulkiest cells are bird eggs, and the longest human cells are certain muscle and nerve cells. Most cells lie between these extremes, in the range indicated by the yellow area in the figure below. The scale is logarithmic, with the length labels on the left ascending in powers of ten, to accommodate the range of sizes shown. Thus, most plant and

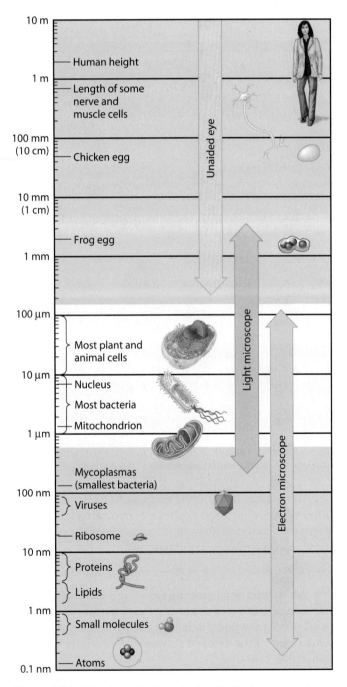

Figure 4.2A   The sizes of cells and related objects

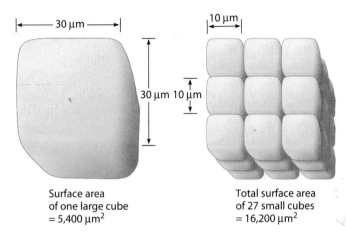

Surface area
of one large cube
= 5,400 μm²

Total surface area
of 27 small cubes
= 16,200 μm²

Figure 4.2B   Effect of cell size on surface area

animal cells, with diameters ranging from 10 to 100 μm, are ten times larger than most bacteria.

The logistics of carrying out a cell's functions sets limits on cell size. At minimum, a cell must be able to house enough DNA, protein molecules, and internal structures to survive and reproduce. The maximum size of a cell is influenced by its requirement for enough surface area to obtain adequate nutrients and oxygen from the environment and dispose of wastes. Size is also limited by the distance these materials must diffuse within a cell.

Large cells have more surface area than small cells, but large cells have much less surface area *relative to their volume* than small cells of the same shape. Figure 4.2B illustrates the relationship of surface area to volume using cubes to represent cells. The figure shows one large cube and 27 small ones. The total volume is the same in both cases:

Volume = 30 μm * 30 μm * 30 μm = 27,000 μm³

In contrast to the total volume, the total surface areas are very different. A cube has six sides, thus its surface area is six times the area of one side. The surface areas are as follows:

Area of large cube = 6 * (30 μm * 30 μm) = 5,400 μm²
Area of one small cube = 6 * (10 μm * 10 μm) = 600 μm²

Combining all 27 of the small cubes, the total surface area is 27 × 600 μm², which equals 16,200 μm², three times the surface area of the large cube. Thus, we see that a large cell has a much smaller surface area relative to its volume than smaller cells have. The need for a surface area large enough to service a cell's volume helps explain the microscopic size of most cells.

In the next module, we look briefly at the smallest cells, the prokaryotic cells of domains Bacteria and Archaea.

Web Process of Science *Connection: What Is the Size and Scale of Our World?*

? To convince yourself of the larger surface area to volume of small cells, compare the ratio of surface area to volume of the large cube and one small cube.

■ large cube: 5,400/27,000 = 1/5; small cube: 600/1,000 = 3/5 (volume is 10 μm × 10 μm × 10 μm = 1,000 μm³)

Prokaryotic cells are structurally simpler than eukaryotic cells

Two kinds of structurally different cells have evolved over time. Bacteria and archaea consist of **prokaryotic cells,** whereas all other forms of life (protists, fungi, plants, and animals) are composed of **eukaryotic cells.**

All cells have several basic features in common. They are all bounded by a membrane, called a **plasma membrane.** All cells have **chromosomes** carrying genes made of DNA. And all cells contain **ribosomes,** tiny structures that make proteins according to instructions from the genes. Eukaryotic cells are distinguished by having a membrane-enclosed nucleus, which houses most of their DNA. The entire region betweeen the nucleus and the plasma membrane is called the **cytoplasm.** The term *cytoplasm* is also used for the interior of a prokaryotic cell. Eukaryotic cells contain many membrane-enclosed organelles, another key difference from prokaryotic cells.

It takes an electron microscope to see clearly the structural details of any cell, and this is especially true of prokaryotic cells (Figure 4.3). Most prokaryotic cells range from 1 to 10 μm in length, about one-tenth the size of a typical eukaryotic cell. A prokaryotic cell lacks a nucleus (its name comes from the Greek *pro,* before, and *karyon,* kernel, referring to the nucleus).

The cutaway diagram in Figure 4.3 reveals the structure of a generalized prokaryotic cell. The DNA of a prokaryotic cell is coiled into a region called the **nucleoid** (nucleus-like), but in contrast to the nucleus of eukaryotic cells, no membrane surrounds the DNA. The ribosomes of prokaryotes are smaller and differ somewhat from those of eukaryotes.

These molecular differences are the basis for the action of some antibiotics, which specifically target prokaryotic ribosomes, interrupting protein synthesis for the bacterium but not for the eukaryote taking the drug.

Outside the plasma membrane (shown here in gray) of most prokaryotes is a fairly rigid, chemically complex cell wall (orange). The wall protects the cell and helps maintain its shape. In some prokaryotes, another layer, a sticky outer coat called a capsule (shown here in yellow), surrounds the cell wall and further protects the cell surface. Capsules also help glue prokaryotes to surfaces, such as sticks and rocks in fast-flowing streams or tissues within the human body. In addition to capsules, some prokaryotes have surface projections. Short projections called pili (singular, *pilus*) help attach prokaryotes to surfaces. Longer projections called **flagella** (singular, *flagellum*) may propel the prokaryotic cell through its liquid environment.

Prokaryotes will be described in more detail in Chapter 16. Eukaryotic cells are the main focus of this chapter.

Web Activity *Prokaryotic Cell Structure and Function*

? List three features that are common to prokaryotic and eukaryotic cells. List three features that differ.

■ Both types of cells have plasma membranes, chromosomes containing DNA, and ribosomes. Prokaryotic cells are smaller, do not have a nucleus that houses their DNA or other membrane-enclosed organelles, and have smaller, somewhat different ribosomes.

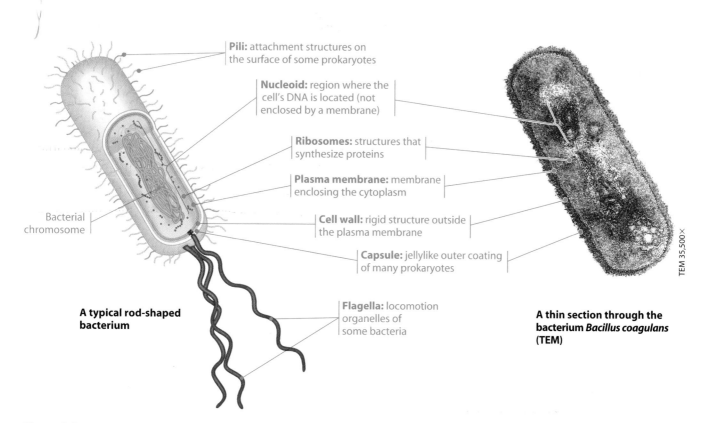

Pili: attachment structures on the surface of some prokaryotes

Nucleoid: region where the cell's DNA is located (not enclosed by a membrane)

Ribosomes: structures that synthesize proteins

Plasma membrane: membrane enclosing the cytoplasm

Cell wall: rigid structure outside the plasma membrane

Capsule: jellylike outer coating of many prokaryotes

Bacterial chromosome

A typical rod-shaped bacterium

Flagella: locomotion organelles of some bacteria

TEM 35,500×

A thin section through the bacterium *Bacillus coagulans* (TEM)

Figure 4.3 A structural diagram (left) and electron micrograph (right) of a typical prokaryotic cell

4.4 Eukaryotic cells are partitioned into functional compartments

All eukaryotic cells (Greek *eu*, true, and *karyon*, kernel)—whether from animals, plants, protists, or fungi—are fundamentally similar to one another and profoundly different from prokaryotic cells. Let's look at an animal cell and a plant cell as representatives of the eukaryotes.

Figure 4.4A is a diagram of an idealized animal cell. No cell would look exactly like what this diagram depicts. In this text, we color-code the various organelles and other structures for easier identification. And recall from the chapter introduction that in living cells these structures are moving and interacting.

The nucleus, with its nuclear membrane, multiple chromosomes, and nucleolus, is the most obvious difference between a prokaryotic and eukaryotic cell. A eukaryotic cell also has various **organelles** ("little organs") in the cytoplasm. These membrane-bounded structures perform specific functions in the cell.

The structures and organelles of eukaryotic cells can be organized into four basic functional groups as follows. ① The nucleus, ribosomes, endoplasmic reticulum, and Golgi apparatus function in manufacturing. ② Organelles involved in breakdown or hydrolysis of molecules include lysosomes, vacuoles, and peroxisomes. ③ Mitochondria in all cells and chloroplasts in plant cells are involved in energy processing. ④ Structural support, movement, and communication among cells are the functions of components of the cytoskeleton, plasma membrane, and cell wall. These cellular components are identified in the figures that follow and will be examined in detail in the chapter.

In essence, the internal membranes of a eukaryotic cell partition it into compartments. Many of the chemical activities of cells—activities known collectively as **cellular metabolism**—occur within organelles. In fact, many enzymatic proteins essential for metabolic processes are built into the membranes of organelles. The fluid-filled spaces within organelles are important as sites where specific chemical conditions are maintained, conditions that vary from one organelle to another and that favor the metabolic processes occurring in each kind of organelle. For example, while a part of the endoplasmic reticulum is engaged in making steroid hormones, neighboring peroxisomes may be detoxifying harmful compounds and making

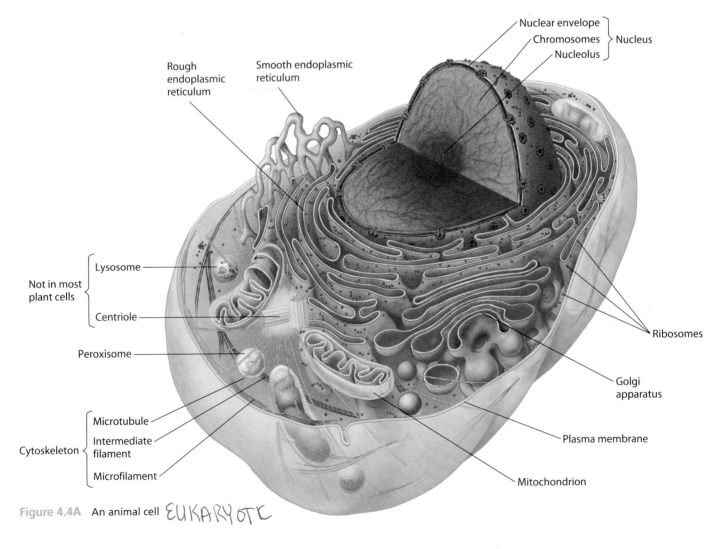

Figure 4.4A   An animal cell EUKARYOTC

hydrogen peroxide (H_2O_2) as a poisonous by-product of their activities. But because the H_2O_2 is confined within peroxisomes, where it is quickly converted to H_2O by resident enzymes, the rest of the cell is protected from destruction.

Almost all of the organelles and other structures appearing in Figure 4.4A are also present in plant cells. As you can see in Figure 4.4A, there are a few exceptions: Lysosomes and centrioles are not found in plant cells. Some animal cells have flagella or cilia (not shown in Figure 4.4A). Among plants, only the sperm cells of a few plant species have flagella. (The eukaryotic and prokaryotic flagella differ in both structure and function.)

A plant cell (Figure 4.4B) has some structures that an animal cell lacks. For example, a plant cell has a rigid, rather thick cell wall (as do the cells of fungi and many protists). Cell walls protect cells and help maintain their shape. Chemically different from prokaryotic cell walls, plant cell walls contain the polysaccharide cellulose. Plasmodesmata are channels through cell walls that connect adjacent cells. An important organelle found in plant cells is the chloroplast, where photosynthesis occurs. (Chloroplasts are also found in algae and some other protists.) Unique to plant cells is a large central vacuole, a compartment that stores water and a variety of chemicals.

Although we have emphasized organelles, eukaryotic cells contain nonmembranous structures as well. Among them are the cytoskeleton, which consists of protein tubes called microtubules and other protein filaments, and ribosomes. You can see by the many brown dots in both figures that ribosomes, the sites of protein synthesis, occur throughout the cytoplasm, as they do in prokaryotic cells. In addition to ribosomes in the cytoplasmic fluid, eukaryotic cells have many ribosomes attached to parts of the endoplasmic reticulum (making it "rough") and to the outer membrane of the nucleus.

Before we begin our in-depth tour of the cell, let's explore the structure of membranes—those all-important components that enclose and partition cells into functional compartments.

(Bio Flix) **BioFlix** *Tour of an Animal Cell*

(Bio Flix) **BioFlix** *Tour of a Plant Cell*

(🎧) **MP3 Tutor** *Cell Organelles*

? Which of the following organelles does not belong in the list: mitochondrion, chloroplast, ribosome, lysosome, peroxisome? Why?

■ Ribosome, because it is the only organelle in the list that is not bounded by a membrane

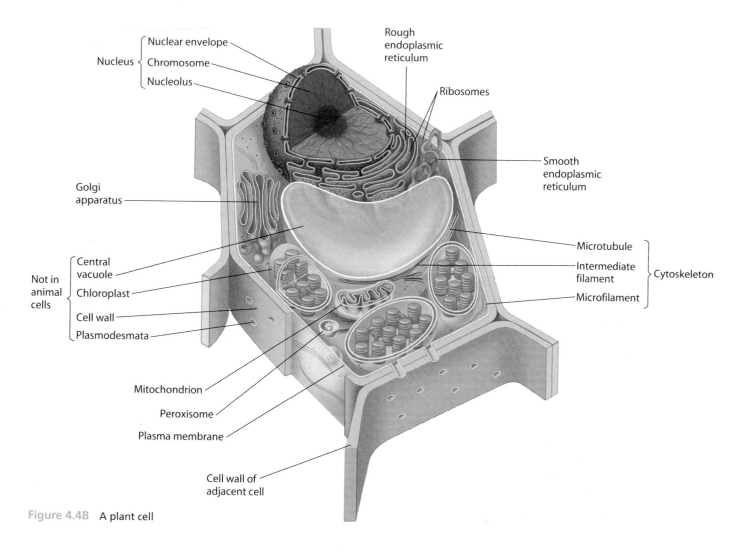

Figure 4.4B A plant cell

The structure of membranes correlates with their functions

For all cells, the plasma membrane forms a boundary between the living cell and its surroundings and controls the traffic of materials into and out of the cell. For a structure that separates life from nonlife, the plasma membrane is amazingly thin. It would take a stack of more than 8,000 membranes to equal the thickness of this page. The plasma membrane and the internal membranes of a eukaryotic cell perform diverse functions—functions that depend on the structure of these membranes.

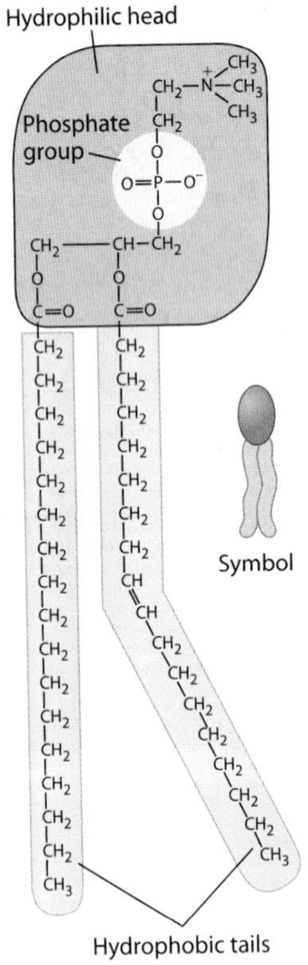

Phospholipids are the main components of biological membranes. As you learned in Module 3.9, a phospholipid has two distinct regions: a negatively charged and thus hydrophilic phosphate group (head) and two nonpolar, hydrophobic fatty acid tails. The structural formula of a phospholipid in Figure 4.5A highlights these two parts. The phospholipid symbol shown in the figure is used in this book in depicting membranes.

The structure of phospholipid molecules is well suited to their role in membranes. Phospholipids form a two-layer sheet called a phospholipid bilayer. Their hydrophilic heads face outward, exposed to the aqueous solution on both sides of a membrane, and their hydrophobic tails point inward, mingling together and shielded from water (Figure 4.5B). Embedded in this lipid bilayer or attached to its

Figure 4.5A
Phospholipid molecule

Hydrophilic head
Phosphate group
CH₃
Symbol
Hydrophobic tails

HYDROPHOBIC ≥ HYDROPHILIC

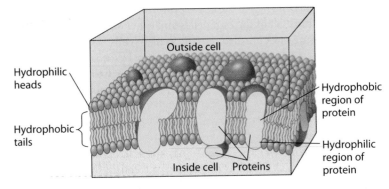

Hydrophilic heads

Hydrophobic tails

Outside cell

Hydrophobic region of protein

Hydrophilic region of protein

Inside cell   Proteins

Figure 4.5B   Phospholipid bilayer with associated proteins

surfaces are diverse proteins. The interior portions of membrane proteins are hydrophobic. The exterior parts of these proteins and the proteins attached to the membrane are hydrophilic.

The permeability of a membrane to various substances relates to the properties of the lipid bilayer. Nonpolar molecules such as O_2 and CO_2 can easily pass through its hydrophobic interior. Other functions of a membrane depend largely on the kinds of proteins associated with the membrane. Some of these proteins form channels that allow specific ions and other hydrophilic molecules to cross the membrane.

We will return to a more detailed look at the structure and function of biological membranes in Chapter 5. Now we begin a more detailed tour of the cell.

? Describe the structure of a biological membrane.

A membrane is a phospholipid bilayer with associated proteins. The hydrophilic heads of the phospholipids face the water on both sides of the membrane; the hydrophobic tails mingle in the center.

Cell Structures Involved in Manufacturing and Breakdown

The nucleus is the cell's genetic control center

The **nucleus** contains most of the cell's DNA and controls the cell's activities by directing protein synthesis. Eukaryotic chromosomes are made up of a material called **chromatin**, which is a complex of proteins and DNA. Chromatin appears as a diffuse mass, as shown in the TEM (left) and diagram (right) of a nucleus in Figure 4.6. As a cell prepares to divide, the DNA is copied and the thin chromatin fibers coil up, becoming thick enough to be visible with a light microscope as the familiar separate structures we know as chromosomes.

Enclosing the nucleus is a **nuclear envelope,** a double membrane perforated with protein-lined

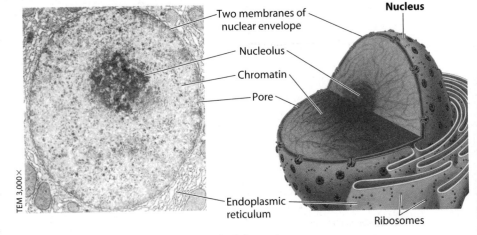

TEM 3,000×

Two membranes of nuclear envelope

Nucleolus

Chromatin

Pore

Endoplasmic reticulum

Nucleus

Ribosomes

Figure 4.6   TEM (left) and diagram (right) of the nucleus

pores that control the flow of materials into and out of the nucleus. As you can see in the diagram in Figure 4.6, the nuclear envelope connects with the cell's network of membranes called the endoplasmic reticulum.

The **nucleolus,** a prominent structure in the nucleus, is the site where a special type of RNA called ribosomal RNA (rRNA) is synthesized according to instructions in the DNA. Proteins brought in through the nuclear pores from the cytoplasm are assembled with this rRNA to form the subunits of ribosomes. These subunits then exit through the pores to the cytoplasm, where they will join to form functional ribosomes.

The nucleus directs protein synthesis by making another type of RNA, messenger RNA (mRNA) according to instructions in the DNA. The messenger RNA moves through the pores to the cytoplasm and is translated there by ribosomes into the amino acid sequences of proteins. Protein synthesis will be explored in more detail in Chapter 10.

? What are the main functions of the nucleus?

■ To contain DNA and pass it on to daughter cells in cell division, to build ribosomes, to copy DNA instructions into RNA

4.7 Ribosomes make proteins for use in the cell and export

Ribosomes are the cellular components that carry out protein synthesis. Cells that have high rates of protein synthesis have a particularly large number of ribosomes. For example, a human pancreas cell producing digestive enzymes may contain a few million ribosomes. Not surprisingly, cells active in protein synthesis also have prominent nucleoli, which, as you just learned, function in ribosome synthesis.

Ribosomes are found in two locations in the cell (Figure 4.7). *Free ribosomes* are suspended in the fluid of the cytoplasm, while *bound ribosomes* are attached to the outside of the endoplasmic reticulum or nuclear envelope. Free and bound ribosomes are structurally identical, and ribosomes can alternate between the two locations (free or bound to the ER). As shown in the diagram, ribosomes are composed of a large and small subunit.

Most of the proteins made on free ribosomes function within the cytoplasm; examples are enzymes that catalyze the first steps of sugar breakdown. In Module 4.9, you will see how bound ribosomes make proteins that will be inserted into membranes, packaged in certain organelles, or exported from the cell. Cells that specialize in protein secretion—for instance, the cells of the pancreas that secrete digestive enzymes—have a

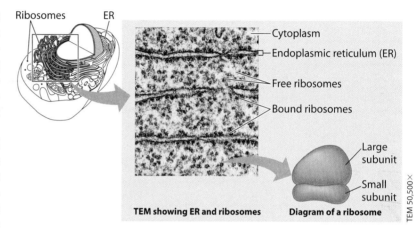

Figure 4.7   Ribosomes

high proportion of bound ribosomes that produce proteins for export.

? What role do ribosomes play in carrying out the genetic instructions of a cell?

■ Ribosomes synthesize proteins according to the instructions carried by messenger RNA from the DNA in the nucleus.

4.8 Overview: Many cell organelles are connected through the endomembrane system

Many of the membranes of the eukaryotic cell are part of an **endomembrane system.** Some of these membranes are physically connected and some are related by the transfer of membrane segments by tiny **vesicles** (sacs made of membrane). The endomembrane system includes the nuclear envelope, endoplasmic reticulum, Golgi apparatus, lysosomes, vacuoles, and the plasma membrane. Many of these organelles work together in the synthesis, storage, and export of molecules. We focus on these interrelated membranes in Modules 4.9–4.12.

An extensive network of flattened sacs and tubules called the **endoplasmic reticulum (ER)** is the prime example of the direct interrelatedness of parts of the endomembrane system. (The term *endoplasmic* means "within the cytoplasm," and

reticulum is Latin for "little net.") As illustrated in Figure 4.6, membranes of the ER are continuous with the nuclear envelope. As we discuss next, there are two regions of ER, smooth ER and rough ER, which differ in structure and function. The membranes that form them, however, are connected.

The tubules and sacs of the ER enclose an interior space that is separate from the cytoplasmic fluid. Dividing the cell into separate compartments is a major function of the endomembrane system.

? Which structure includes all others in the list: rough ER, smooth ER, endomembrane system, nuclear envelope?

■ Endomembrane system

4.9 The endoplasmic reticulum is a biosynthetic factory

The diagram in Figure 4.9A shows a cutaway view of the interconnecting membranes of the smooth and rough ER. These two types of ER can be distinguished in the electron micrograph. **Smooth endoplasmic reticulum** is called *smooth* because it lacks attached ribosomes. **Rough endoplasmic reticulum** has ribosomes that stud the outer surface of the membrane, and thus it appears *rough* in the electron micrograph.

Smooth ER The smooth ER of various cell types functions in diverse metabolic processes. Enzymes of the smooth ER are important in the synthesis of lipids, including oils, phospholipids, and steroids. In vertebrates, for example, cells of the ovaries and testes synthesize the steroid sex hormones. These cells are rich in smooth ER, a structural feature that fits their function by providing ample space for steroid synthesis.

Our liver cells also have large amounts of smooth ER, with additional kinds of functions. Certain enzymes in the smooth ER of liver cells help process drugs and other potentially harmful substances. The sedative phenobarbital and other barbiturates are examples of the drugs detoxified by these enzymes. Undesirable complications can result when liver cells respond to drugs. As the cells are exposed to such chemicals, the amount of smooth ER and its detoxifying enzymes increase, thereby increasing the rate of detoxification and thus the body's tolerance to the drugs. This means that higher and higher doses of a drug are required to achieve a particular effect, such as sedation. Another complication is that detoxifying enzymes often cannot distinguish among related chemicals. As a result, the growth of smooth ER in response to one drug can increase tolerance to other drugs, including important medicines. Barbiturate use, for example, can decrease the effectiveness of certain antibiotics.

Smooth ER has yet another function, the storage of calcium ions. In muscle cells, for example, a specialized smooth ER membrane pumps calcium ions into the interior of the ER. When a nerve signal stimulates a muscle cell, calcium ions rush from the smooth ER into the cytoplasmic fluid and trigger contraction of the cell.

Rough ER One of the functions of rough ER is to make more membrane. Phospholipids made by enzymes of the ER are inserted into the membrane. As a result, the rough ER membrane enlarges, and some of this membrane is transferred to other components of the endomembrane system in vesicles.

The bound ribosomes that attach to rough ER produce proteins that will be inserted into the ER membrane, transported to other organelles, or secreted by the cell. An example of a secretory protein is insulin, a hormone secreted by certain cells in the pancreas.

Figure 4.9B shows the synthesis, modification, and packaging of a secretory protein. ❶ As the polypeptide is synthesized by a bound ribosome, it is threaded into the cavity of

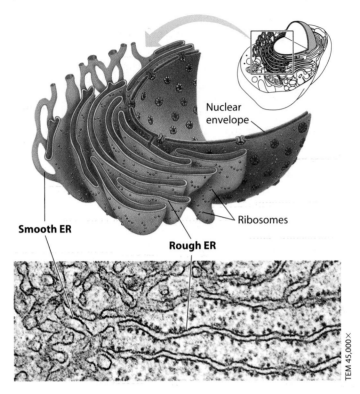

Smooth ER
Rough ER
Nuclear envelope
Ribosomes
TEM 45,000×

Figure 4.9A Smooth and rough endoplasmic reticulum

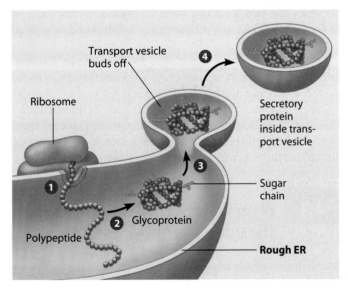

Transport vesicle buds off
Ribosome
Polypeptide
Glycoprotein
Secretory protein inside transport vesicle
Sugar chain
Rough ER

Figure 4.9B Synthesis and packaging of a secretory protein by the rough ER

the rough ER through a pore formed by a protein complex. As it enters, the new protein folds into its three-dimensional shape. ❷ Short chains of sugars are often linked to the polypeptide, making the molecule a **glycoprotein** (*glyco-* means "sugar"). ❸ When the molecule is ready for export from the ER, it is packaged in a **transport vesicle,** a vesicle that is in transit from one part of the cell to another. ❹ This vesicle buds off from the ER membrane. The protein now travels to the Golgi apparatus (described in the next module) for further processing. From there, a transport vesicle containing the finished molecule will make its way to the plasma membrane and release its contents from the cell.

Web Activity *Overview of Protein Synthesis*

? Contrast the form and functions of smooth ER and rough ER.

■ *Smooth ER:* no bound ribosomes. Functions include lipid synthesis; destruction of toxic substances; regulation of muscle contraction by release of calcium. *Rough ER:* ribosomes attach to surface; functions include production of membrane by adding proteins and phospholipids; bound ribosomes produce proteins that are modified and packaged in transport vesicles

4.10 The Golgi apparatus finishes, sorts, and ships cell products

After leaving the ER, many transport vesicles travel to the **Golgi apparatus.** The Golgi apparatus was named after Italian biologist and physician Camillo Golgi, whose career spanned the turn of the 20th century. Using the light microscope, Golgi and his contemporaries discovered this membranous organelle and a number of others in animal and plant cells. The electron microscope confirmed Golgi's discovery. Electron micrographs reveal that the Golgi apparatus consists of flattened sacs stacked on top of each other. As you can see in Figure 4.10, the sacs are not interconnected like ER sacs. A cell may contain only a few Golgi stacks or hundreds. The number of Golgi stacks correlates with how active the cell is in secreting proteins—a multistep process that, as we have just seen, is initiated in the rough ER.

The Golgi apparatus performs several functions in close partnership with the ER. Serving as a molecular warehouse and finishing factory, a Golgi apparatus receives and modifies products manufactured by the ER. One side of a Golgi stack serves as a receiving dock for transport vesicles produced by the ER. These vesicles join and form a new Golgi sac. The other side of the Golgi, the shipping side, gives rise to vesicles, which bud off and travel to other sites.

Products of the ER are usually modified during their transit from the receiving to the shipping side of the Golgi. For example, various Golgi enzymes modify the carbohydrate portions of glycoproteins. Molecular identification tags, such as phosphate groups, may be added that serve to mark and sort molecules into different batches for different destinations.

Until recently, the Golgi was viewed as a static structure, with products in various stages of processing moved from sac to sac by transport vesicles. Recent research has given rise to a new *maturation model* in which entire sacs "mature" as they move from the receiving to the shipping side, carrying and modifying their cargo as they go. The shipping side of the Golgi stack serves as a depot from which finished secretory products, packaged in transport vesicles, move to the plasma membrane for export from the cell. Alternatively, finished products may become part of the plasma membrane itself or part of another organelle, such as a lysosome, which we discuss next.

? What is the relationship of the Golgi apparatus to the ER in a protein-secreting cell?

■ The Golgi receives transport vesicles that bud from the ER and that contain proteins synthesized by ribosomes attached to the ER. The Golgi finishes processing the proteins and then dispatches transport vesicles that secrete the proteins to the outside of the cell.

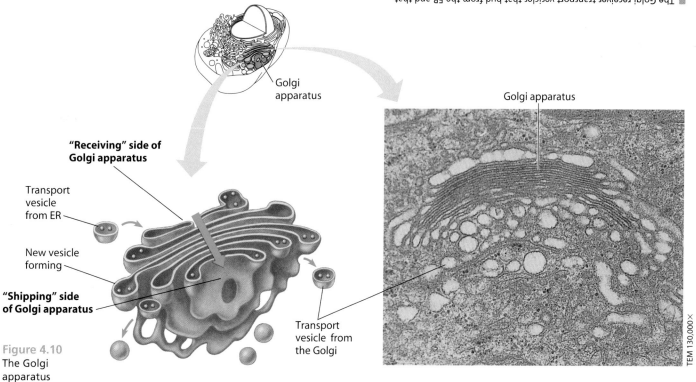

Golgi apparatus

Golgi apparatus

"Receiving" side of Golgi apparatus

Transport vesicle from ER

New vesicle forming

"Shipping" side of Golgi apparatus

Transport vesicle from the Golgi

TEM 130,000×

Figure 4.10
The Golgi apparatus

4.11 Lysosomes are digestive compartments within a cell

A **lysosome** consists of digestive enzymes enclosed in a membranous sac. The name *lysosome* is derived from two Greek words meaning "breakdown body." The enzymes and membranes of lysosomes are made by rough ER and then transferred to the Golgi apparatus for further processing. Lysosomes illustrate the main theme of eukaryotic cell structure—compartmentalization. The lysosomal membrane encloses a compartment in which digestive enzymes are provided with an acidic environment and are safely isolated from the rest of the cell.

Lysosomes have several types of digestive functions. Many protists engulf food particles into cytoplasmic sacs called food vacuoles. As Figure 4.11A shows, lysosomes fuse with the food vacuoles and then break down the food, releasing nutrients to the cell. Our white blood cells ingest bacteria into vacuoles, and lysosomal enzymes emptied into these vacuoles rupture the bacterial cell walls. Lysosomes also serve as recycling centers for animal cells. Damaged organelles or small amounts of cell fluid become enclosed in a membrane vesicle. A lysosome fuses with such a vesicle (Figure 4.11B) and dismantles its contents, making organic molecules available for reuse. With the help of lysosomes, a cell continually renews itself.

The cells of people with inherited lysosomal storage diseases lack one or more lysosomal enzymes. The lysosomes become engorged with undigested material, eventually interfering with cellular function. In Tay-Sachs disease, for example, a lipid-digesting enzyme is missing, and brain cells become impaired by an accumulation of lipids. A child with Tay-Sachs disease will die within a few years.

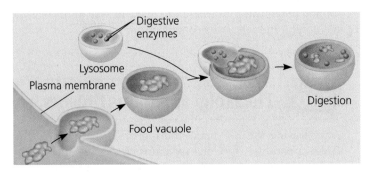

Figure 4.11A Lysosome fusing with a food vacuole and digesting food

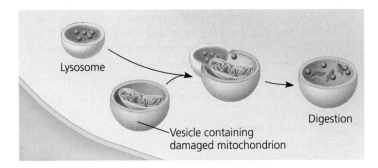

Figure 4.11B Lysosome fusing with vesicle containing damaged organelle and digesting and recyling its contents

? How is a lysosome like a recycling center?

■ It breaks down damaged organelles and recycles their molecules.

4.12 Vacuoles function in the general maintenance of the cell

Vacuoles are membranous sacs that have a variety of functions. In Module 4.11, we saw that the food vacuole functions in collaboration with a lysosome. Figure 4.12A shows a plant cell's **central vacuole,** which has hydrolytic functions like a lysosome. The central vacuole also helps the cell grow in size by absorbing water and enlarging, and it can store vital chemicals or waste products. Vacuoles in flower petals contain pigments that attract pollinating insects. Central vacuoles may also contain poisons that protect the plant against predators.

Figure 4.12B shows a very different kind of vacuole in the protist *Paramecium*. Notice the two contractile vacuoles, looking somewhat like wheel hubs with radiating spokes. The "spokes" collect excess water from the cell, and the hub expels it to the outside. Freshwater protists constantly take in water from their environment. Without a way to get rid of the excess water, the cell fluid would become too dilute to support life, and eventually the cell would swell and burst.

? Is a food vacuole part of the endomembrane system?

■ Yes, it forms by pinching in from the plasma membrane, which is part of the endomembrane system.

Figure 4.12A Central vacuole in a plant cell

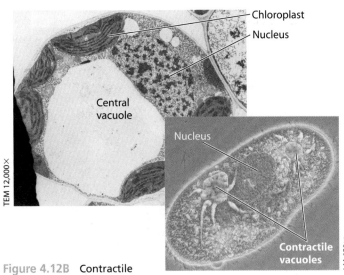

Figure 4.12B Contractile vacuoles in *Paramecium*, a single-celled organism

4.13 A review of the structures involved in manufacturing and breakdown

Figure 4.13 summarizes the relationships among the major organelles of the endomembrane system. You can see the direct *structural* connections between the nuclear envelope, rough ER, and smooth ER. The red arrows show their *functional* connections, as membranes and proteins produced by the ER travel in transport vesicles to the Golgi and from there to other destinations. Some vesicles develop into lysosomes or vacuoles.

Transport vesicles carry secretory proteins or other products to the plasma membrane. When vesicles fuse with the membrane, the products are secreted from the cell and the vesicle membrane is added to the plasma membrane.

A **peroxisome** is an organelle that is not part of the endomembrane system but is involved in various metabolic functions, including the breakdown of fatty acids to be used as fuel and the detoxification of alcohol and other harmful substances.

Next we consider two organelles that are also not part of the endomembrane system—mitochondria and chloroplasts. Both are involved in energy processing.

Web Activity *The Endomembrane System*

? How do transport vesicles help tie together the endomembrane system?

ʇuǝɯǝldǝ ƃuᴉꞁꞁoɟ ʇɥǝ ǝupɯǝɯqɯǝ ɟo sʇuǝuodɯoɔ
ɹǝʌo ǝsoꞁɔuǝ ʎǝɥʇ sǝɔuɐʇsqns puɐ sǝuɐɹqɯǝɯ ǝʌoɯ sǝꞁɔᴉsǝʌ ʇɹodsuɐɹꞁ ■

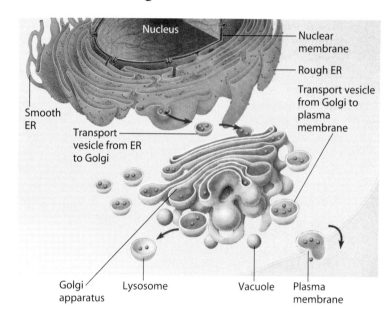

Figure 4.13 Connections among the organelles of the endomembrane system

---

Energy-Converting Organelles

4.14 Mitochondria harvest chemical energy from food

Mitochondria (singular, *mitochondrion*) are organelles that carry out cellular respiration in nearly all eukaryotic cells, converting the chemical energy of foods such as sugars to the chemical energy of a molecule called ATP (adenosine triphosphate). ATP is the main energy source for cellular work.

The mitochondrion's structure suits its function. It is enclosed by two membranes, each a phospholipid bilayer with a unique collection of embedded proteins (Figure 4.14). The mitochondrion has two internal compartments. The first is the **intermembrane space,** the narrow region between the inner and outer membranes. The inner membrane encloses the second compartment, the **mitochondrial matrix,** which contains the mitochondrial DNA and ribosomes, as well as many enzymes that catalyze some of the reactions of cellular respiration. The inner membrane is highly folded, with protein molecules that make ATP embedded in it. The folds, called cristae, increase the membrane's surface area, enhancing the mitochondrion's ability to produce ATP. We discuss the role of mitochondria in cellular respiration in more detail in Chapter 6.

? What is cellular respiration?

ꓒꓔ∀ ɟo ɯɹoɟ ǝɥʇ uᴉ ʎƃɹǝuǝ ꞁɐɔᴉɯǝɥɔ oʇ
sǝꞁnɔǝꞁoɯ ɹǝɥʇo puɐ sɹɐƃns ɟo ʎƃɹǝuǝ ꞁɐɔᴉɯǝɥɔ ǝɥʇ sʇɹǝʌuoɔ ʇɐɥʇ ssǝɔoɹd ∀ ■

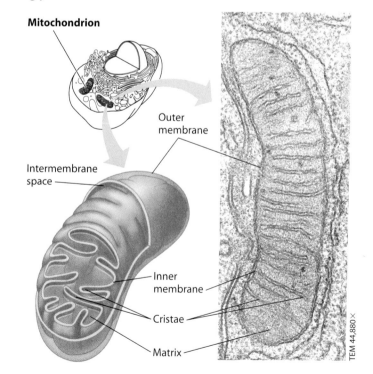

Figure 4.14 The mitochondrion

Chloroplasts convert solar energy to chemical energy

Most of the living world runs on the energy provided by photosynthesis, the conversion of light energy from the sun to the chemical energy of sugar molecules. **Chloroplasts** are the photosynthesizing organelles of all photosynthetic eukaryotes. The chloroplast's solar power system is much more successful than anything yet produced by human ingenuity.

Befitting an organelle that carries out complex, multistep processes, internal membranes partition the chloroplast into compartments (Figure 4.15). The chloroplast is enclosed by an inner and outer membrane separated by a thin intermembrane space. The compartment inside the inner membrane holds a thick fluid called **stroma,** which contains the chloroplast DNA and ribosomes as well as many enzymes. A network of inter-

connected sacs called **thylakoids** is inside the chloroplast. The compartment inside these sacs is called the thylakoid space. In some regions, thylakoids are stacked like poker chips; each stack is called a **granum** (plural, *grana*). The grana are the chloroplast's solar power packs—the sites where the green chlorophyll molecules embedded in their membranes trap solar energy. In Chapter 7, we discuss the role of each part of a chloroplast in converting solar energy to chemical energy.

Web Activity *Build a Chloroplast and a Mitochondrion*

? Which membrane in a chloroplast appears to be the most extensive? Why might this be so?

■ The thylakoid membranes are most extensive, providing a large area of membrane that contains chlorophyll for photosynthesis.

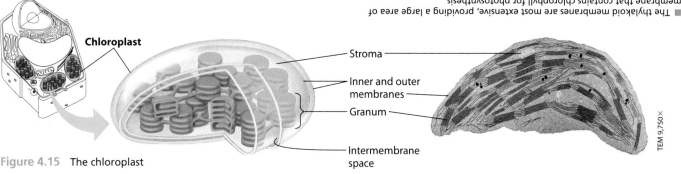

Figure 4.15   The chloroplast

Labels: Chloroplast; Stroma; Inner and outer membranes; Granum; Intermembrane space

TEM 9,750×

Mitochondria and chloroplasts evolved by endosymbiosis

Did it seem unusual to you that mitochondria and chloroplasts contain DNA and ribosomes? They have a single circular DNA molecule, similar in structure to the chromosome of prokaryotes. Their ribosomes are more similar to prokaryotic ribosomes than to eukaryotic ribosomes. Mitochondria and chloroplasts reproduce by a splitting process that is similar to that of certain prokaryotes. Both organelles are surrounded by a

double membrane, and the inner membrane has many similarities to the plasma membranes of living prokaryotes.

The hypothesis of **endosymbiosis** proposes that mitochondria and chloroplasts were formerly small prokaryotes that began living within larger cells. The term *endosymbiont* refers to a cell that lives within another cell, called the host cell. These small prokaryotes may have gained entry to the larger cell as undigested prey or internal parasites (Figure 4.16).

By whatever means the relationships began, we can hypothesize how the symbiosis could have become beneficial. A host could use nutrients released from photosynthetic endosymbionts. And in a world that was becoming increasingly aerobic from the photosynthesis of prokaryotes, a host would have benefited from endosymbionts that were able to use oxygen to release large amounts of energy in cellular respiration. Over time, the host and endosymbionts would have become increasingly interdependent, eventually becoming a single organism.

All eukaryotes have mitochondria, but not all eukaryotes have chloroplasts. The hypothesis of serial endosymbiosis proposes that mitochondria evolved before chloroplasts.

? Can you see in Figure 4.16 how the outer membrane of mitochondria and chloroplasts may have originated?

■ The outer membrane would have been part of the host plasma membrane that enclosed the prokaryote as it was engulfed by the host cell.

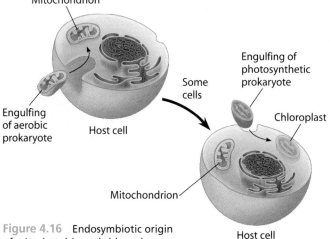

Figure 4.16   Endosymbiotic origin of mitochondria and chloroplasts

Labels: Mitochondrion; Engulfing of aerobic prokaryote; Host cell; Some cells; Engulfing of photosynthetic prokaryote; Chloroplast; Mitochondrion; Host cell

4.17 The cell's internal skeleton helps organize its structure and activities

Biologists once thought that organelles floated freely in the cell. But improvements in both light microscopy and electron microscopy have allowed scientists to uncover a network of protein fibers, collectively called the **cytoskeleton,** extending throughout the cytoplasm of a cell. These fibers function like a skeleton in providing for both structural support and cell motility. As discussed in the chapter introduction, cell movement includes both internal movement and the locomotion of a cell. These movements generally require the interaction of the cytoskeleton with proteins called motor proteins. Mounting evidence also suggests that the cytoskeleton helps regulate activities by transmitting signals from the cell's surface to its interior.

Three main kinds of fibers make up the cytoskeleton: microfilaments, the thinnest fiber; microtubules, the thickest; and intermediate filaments, in between in thickness. Figure 4.17 shows micrographs of three cells of the same type, each stained with a different fluorescent dye that selectively highlights one of these types of fibers.

Microfilaments, also called actin filaments, are solid rods composed mainly of globular proteins called actin, arranged in a twisted double chain (bottom left of Figure 4.17). Microfilaments form a three-dimensional network just inside the plasma membrane that helps support the cell's shape.

Microfilaments are also involved in cell movements. As we will see in Chapter 30, actin filaments and thicker filaments made of the motor protein myosin interact to cause contraction of muscle cells. Localized contractions brought about by actin and myosin are involved in the amoeboid (crawling) movement of the protist *Amoeba* and certain of our white blood cells.

Intermediate filaments are made of various fibrous proteins and have a ropelike structure. Intermediate filaments serve mainly to reinforce cell shape and to anchor certain organelles. The nucleus is held in place by a cage of intermediate filaments. While microfilaments may be disassembled and reassembled elsewhere, intermediate filaments are often more permanent fixtures in the cell. The outer layer of our skin consists of dead skin cells full of intermediate filaments made of keratin proteins.

Microtubules are straight, hollow tubes composed of globular proteins called tubulins. As shown in the bottom right of Figure 4.17, microtubules elongate by adding subunits consisting of tubulin pairs. They are readily disassembled in a reverse manner, and the tubulin subunits can then be reused elsewhere in the cell. In many animal cells, microtubules grow out from a "microtubule-organizing center" called a centrosome. Within the centrosome is a pair of **centrioles** (see Figure 4.4A). We will return to centrioles when we discuss cell division in Chapter 8.

Microtubules shape and support the cell and also act as tracks along which organelles equipped with motor proteins can move. For example, a lysosome might "walk" along a microtubule to reach a food vacuole. Microtubules also guide the movement of chromosomes when cells divide, and as we see next, they are the main components of cilia and flagella.

? Which component of the cytoskeleton is most important in (a) holding the nucleus in place within the cell; (b) guiding transport vesicles from the Golgi to the plasma membrane; (c) contracting muscle cells?

■ (a) intermediate filaments; (b) microtubules; (c) microfilaments

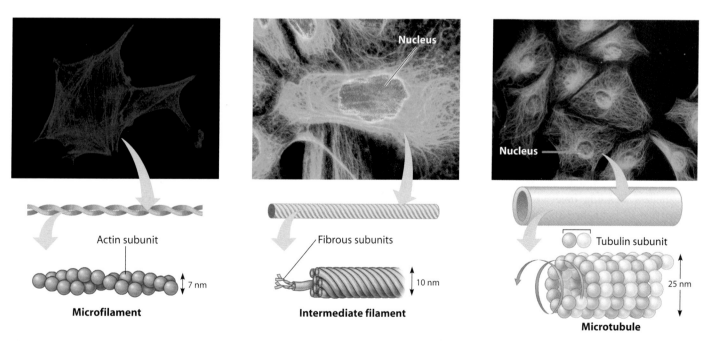

Figure 4.17 Fibers of the cytoskeleton: microfilaments are stained red (left), intermediate filaments are stained yellow-green (center), and microtubules are stained green (right)

4.18 Cilia and flagella move when microtubules bend

The role of the cytoskeleton in movement is clearly seen in eukaryotic flagella and cilia, the locomotor appendages that protrude from certain cells. The short, numerous appendages that propel protists such as *Paramecium* (see Figure 4.1C) are called **cilia** (singular, *cilium*). Longer than cilia and usually limited to one or a few per cell, flagella are found on many protists (see Module 16.15).

Some cells of multicellular organisms also have cilia or flagella. For example, Figure 4.18A shows cilia on cells lining the human windpipe. In this case, the cilia sweep mucus containing trapped debris out of our lungs. Most animals and some plants have flagellated sperm. A flagellum, shown in Figure 4.18B, propels the cell by an undulating whiplike motion. In contrast, cilia work more like the coordinated oars of a rowing team.

Though different in length and beating pattern, flagella and cilia have a common structure and mechanism of movement (Figure 4.18C). Both flagella and cilia are composed of microtubules wrapped in an extension of the plasma membrane. A ring of nine microtubule doublets surrounds a central pair of microtubules. This arrangement, found in nearly all eukaryotic flagella and cilia, is called the 9 + 2 pattern.

The microtubule assembly extends into an anchoring structure called a **basal body,** which has a pattern of nine microtubule triplets arranged in a ring. When a cilium or flagellum begins to grow, the basal body may act as a foundation for microtubule assembly from tubulin subunits. Basal bodies are very similar in structure to centrioles, which are found in the centrosome (microtubule-organizing center) of animal cells.

How does this microtubule assembly produce the bending movement of these organelles? Bending involves motor proteins called dynein arms that are attached to each outer microtubule doublet. Using energy from ATP, the dynein arms grab an adjacent doublet and exert a sliding force as they start to "walk" along it. The arms then release and reattach a little farther along the doublet. The doublets are held together by cross-linking proteins (not illustrated) and radial spokes extending to the central microtubules. If the doublets were not held in place, the walking action would make them slide past each other.

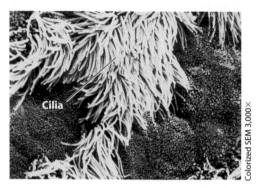

Figure 4.18A   Cilia on cells lining the respiratory tract

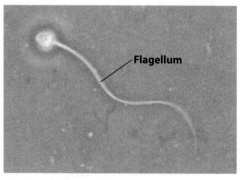

Figure 4.18B   Undulating flagellum on a sperm cell

Electron micrographs of cross sections:

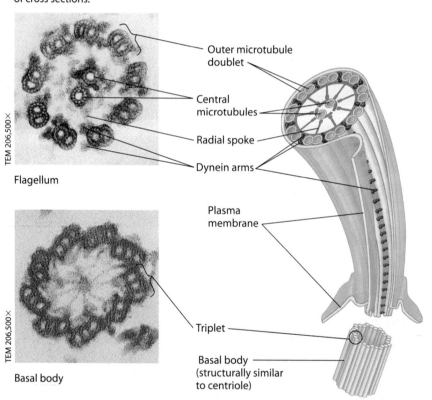

Flagellum

Basal body

Outer microtubule doublet
Central microtubules
Radial spoke
Dynein arms
Plasma membrane
Triplet
Basal body (structurally similar to centriole)

Figure 4.18C   Structure of a eukaryotic flagellum or cilium

Instead, the movements of the dynein arms cause the microtubules (and consequently the flagellum or cilium) to bend.

Web Activity *Cilia and Flagella*

?   Compare and contrast cilia and flagella.

■ Both cilia and flagella share the same 9 + 2 pattern of microtubules and mechanism for bending. Cilia are shorter, more numerous, and beat in a coordinated oar-like pattern. The longer flagella, which are limited to one or more per cell, undulate like a whip.

4.19 Problems with sperm motility may be environmental or genetic

Human sperm quality varies among men and in different areas. In the last 50 years, there has been an apparent decline in developed countries in sperm quality—lower sperm counts, higher proportions of malformed sperm, and reduced motility. Various environmental factors are being studied as possible causes. One hypothesis links this trend to an increase in hormonally active chemicals in the environment and in people's bodies.

Research has indicated that one common group of chemicals, collectively called phthalates, interferes with sex hormones and adversely affects sperm quality in rodents. Phthalates are used in cosmetics and deodorants and are found in the plastics of food packaging, medical tubing, and children's toys. Critics of these animal studies have claimed that the environmental exposures of people are beneath the levels causing effects in rodents. Several new studies, however, have indicated that normal levels of human exposure to these chemicals may result in impaired sperm quality.

Results of a five-year study of 463 men who had come to a hospital for infertility treatment found that the men with higher concentrations of breakdown products of phthalates in their urine had lower sperm counts and motility. The range in levels of phthalates measured in these men were typical of the general population. But does a statistical correlation indicate cause and effect? Humans are not laboratory animals, and it is difficult to design a study that controls for all variables. Research continues on the potential reproductive health risks of hormone-disrupting chemicals in the environment.

Other problems with sperm motility are clearly genetic. Primary ciliary dyskinesia (PCD), also known as immotile cilia syndrome, is a fairly rare disease charactized by recurrent infections of the respiratory tract and immotile sperm. Compare the cross section of the flagellum of a sperm of a male with PCD in Figure 4.19 with the upper TEM in Figure 4.18C. Can you notice the absense of the dynein arms? How does that explain the seemingly unrelated symptoms of PCD? If microtubules cannot bend (see Module 4.18), then cilia cannot help cleanse the respiratory tract and sperm cannot swim.

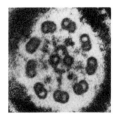

Figure 4.19 Cross section of immotile sperm flagellum

? Why does a lack of dynein arms affect the action of both cilia and flagella?

■ The arrangement of microtubule doublets with attached dynein motor proteins that cause them to bend is the same in cilia and flagella.

4.20 The extracellular matrix of animal cells functions in support, movement, and regulation

The plasma membrane is usually regarded as the boundary of the cell, but most cells synthesize and secrete materials that are external to the plasma membrane. These extracellular structures are essential to many cell functions.

Animal cells produce an elaborate **extracellular matrix (ECM)** (Figure 4.20). This layer helps hold cells together in tissues and protects and supports the plasma membrane. The main components of the ECM are glycoproteins (proteins bonded with carbohydrates). The most abundant glycoprotein is collagen, which forms strong fibers outside the cell. The collagen fibers are embedded in a network woven from other types of glycoproteins. Large complexes form when hundreds of these glycoproteins connect to a central long polysaccharide molecule (green in the figure). The ECM may attach to the cell through other glycoproteins that bind to membrane proteins called integrins. **Integrins** span the membrane, attaching on the other side to proteins connected to microfilaments of the cytoskeleton.

As their name implies, integrins have the function of integration: They transmit information between the ECM and the cytoskeleton, thus integrating changes occurring outside and inside the cell. Research shows that the ECM can regulate a cell's behavior, directing the path along which embryonic cells move and even influencing the activity of genes. The ECM of a particular tissue may help coordinate the behavior

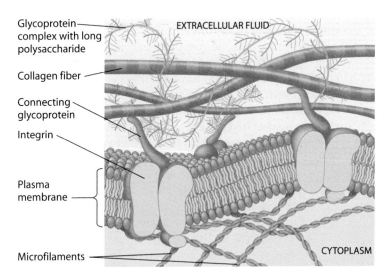

Glycoprotein complex with long polysaccharide

Collagen fiber

Connecting glycoprotein

Integrin

Plasma membrane

Microfilaments

EXTRACELLULAR FLUID

CYTOPLASM

Figure 4.20 The extracellular matrix (ECM) of an animal cell

of all the cells in that tissue. Direct connections between cells also function in this coordination, as we discuss next.

? Refer to Figure 4.20 and describe the three structures that provide support to the plasma membrane.

■ The membrane is attached through membrane proteins (integrins) to the microfilaments of the cytoskeleton and the glycoproteins and collagen fibers of the ECM.

4.21 Three types of cell junctions are found in animal tissues

Neighboring cells in animal tissues often adhere, interact, and communicate through specialized junctions between them. Figure 4.21 illustrates the three types of intercellular junctions.

At *tight junctions*, the membranes of neighboring cells are very tightly pressed against each other, knit together by proteins. Forming continuous seals around cells, tight junctions prevent leakage of extracellular fluid across a layer of epithelial cells. Such a sheet of tissue lines the digestive tract, preventing the contents from leaking into surrounding tissues.

Anchoring junctions function like rivets, fastening cells together into strong sheets. Intermediate filaments made of sturdy keratin proteins anchor these junctions in the cytoplasm. Anchoring junctions are common in tissues subject to stretching or mechanical stress, such as skin and heart muscle.

Gap junctions (also called communicating junctions) are channels that allow small molecules to flow through protein-lined pores between neighboring cells. For example, the flow of ions through gap junctions in the cells of heart muscle coordinates their contraction. These junctions are especially common in animal embryos, where chemical communication between cells is essential for development.

Web Activity *Cell Junctions*

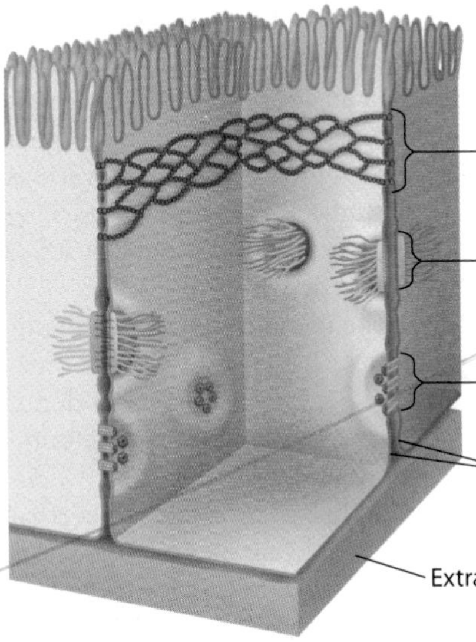

Figure 4.21 Three types of cell junctions in animal tissues

? A muscle tear injury would probably involve the rupture of which type of cell junction?

anchoring junction ▪

4.22 Cell walls enclose and support plant cells

The **cell wall** is one feature that distinguishes plant cells from animal cells. This rigid extracellular structure not only protects the cells but provides the skeletal support that keeps plants upright on land. Typically 10 to 100 times thicker than the plasma membrane, plant cell walls consist of fibers of cellulose embedded in a matrix of other polysaccharides and proteins. This tough, fibers-in-a-matrix construction resembles that of fiberglass, a manufactured product also noted for its strength.

Figure 4.22 shows the layered arrangement of plant cell walls. Cells initially lay down a relatively thin and flexible primary wall, which allows the growing cell to continue to enlarge. Some cells then add a secondary wall deposited in laminated layers. Rigid molecules called lignin strengthen the cell walls of many plants. These strong cell walls are the main component of wood. Between adjacent cells is a layer of sticky polysaccharides (shown here in dark brown) that glues the cells together.

Despite their thickness, plant cell walls do not totally isolate the cells from each other. To function in a coordinated way as part of a tissue, the cells must have cell junctions, structures that connect them to one another. As Figure 4.22 shows, numerous **plasmodesmata** (singular, *plasmodesma*), channels between adjacent plant cells, form a circulatory and communication system connecting the cells in plant tissues. Notice that the plasma membrane and the cytoplasm of the cells extend through the plasmodesmata, so that water and other small molecules can readily pass from cell to cell. Through plasmo-

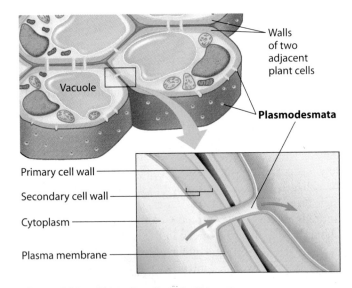

Figure 4.22 Plant cell walls and cell junctions

desmata, the cells of a plant tissue share water, nourishment, and chemical messages.

? Which animal cell junction is analagous to a plasmodesma?

a gap junction ▪

4.23 Review: Eukaryotic cell structures can be grouped on the basis of four basic functions

We have introduced many important cell structures in this chapter. To provide a framework for this information and reinforce the theme that structure is correlated with function, we have grouped the eukaryotic cell organelles into four categories by general function, as reviewed in Table 4.23.

The first category is manufacturing. Here we include not only the synthesis of molecules but also their transport within the cell. The second category includes three organelles that break down and recycle materials that are harmful or no longer needed. (Vacuoles are included, although, being multifunctional, they do not fit neatly into any one of our categories.) The third category includes the two energy-processing organelles. The fourth category is support, movement, and intercellular communication. These three functions are related because for movement to occur, there must be some sort of rigid support against which force can be applied. And when a supporting structure forms the cell's outer boundary, it is necessarily involved in the cell's communication with its neighbors.

Within each of the four categories, a structural similarity underlies the general function of each component. In the first category, manufacturing depends heavily on a network of metabolically active membranes. In the second category, all the organelles listed are composed of membranous sacs, inside of which materials can be safely broken down. In the third category, expanses of metabolically active membranes and intermembrane compartments within the organelles make it possible for chloroplasts and mitochondria to perform the complex energy conversions that power the cell. Even in the diverse fourth category, there is a common structural theme in the various fibers involved in the functioning of these cellular systems.

We can summarize further by emphasizing that these cellular structures form an integrated team and that the properties of life emerge at the level of the cell from the coordinated functions of the team members. We will see many examples of the integrated actions of cellular structures in later chapters.

TABLE 4.23 EUKARYOTIC CELL STRUCTURES AND FUNCTIONS

1. Manufacturing

Nucleus	DNA synthesis; RNA synthesis; assembly of ribosomal subunits (in nucleoli)
Ribosomes	Polypeptide (protein) synthesis
Rough ER	Synthesis of membrane lipids and proteins, secretory proteins, and hydrolytic enzymes; formation of transport vesicles
Smooth ER	Lipid synthesis; detoxification in liver cells; calcium ion storage
Golgi apparatus	Modification and transport of macromolecules; formation of lysosomes and transport vesicles

2. Breakdown

Lysosomes (in animal cells and some protists)	Digestion of ingested food, bacteria, and a cell's damaged organelles and macromolecules for recycling
Vacuoles	Digestion (like lysosomes); storage of chemicals; cell enlargement; water balance
Peroxisomes (not part of endomembrane system)	Diverse metabolic processes, with breakdown of H_2O_2 by-product

3. Energy Processing

Mitochondria	Conversion of chemical energy of food to chemical energy of ATP
Chloroplasts (in plants and some protists)	Conversion of light energy to chemical energy of sugars

4. Support, Movement, and Communication Between Cells

Cytoskeleton (including cilia, flagella, and centrioles in animal cells)	Maintenance of cell shape; anchorage for organelles; movement of organelles within cells; cell movement; mechanical transmission of signals from exterior of cell to interior
Extracellular matrix (in animals)	Binding of cells in tissues; surface protection; regulation of cellular activities
Cell junctions	Communication between cells; binding of cells in tissues
Cell walls (in plants, fungi, and some protists)	Maintenance of cell shape and skeletal support; surface protection; binding of cells in tissues

All organisms share the fundamental features of (1) consisting of cells, each enclosed by a membrane that maintains internal conditions very different from the surroundings; (2) having DNA as the genetic material; and (3) carrying out metabolism, which involves the interconversion of different forms of energy and of chemical materials. We expand on the subjects of membranes and metabolism in Chapter 5.

Web Activity *Review: Animal Cell Structure and Function*

Web Activity *Review: Plant Cell Structure and Function*

? How do mitochondria, smooth ER, and the cytoskeleton all contribute to the contraction of a muscle cell?

■ Mitochondria supply energy in the form of ATP. The smooth ER helps regulate contraction by the uptake and release of calcium. Microfilaments function as the actual contractile apparatus.

Chapter Review

Reviewing the Concepts

Introduction to the Cell (4.1–4.5)

Microscopy. The light microscope can magnify cells, both living and preserved, up to 1,000 times. The greater magnification and resolution of the scanning and transmission electron microscopes reveal the ultrastructure of cells (**4.1**). The microscopic size of most cells ensures a sufficient surface area across which nutrients and wastes can move to service the cell volume (**4.2**).

Prokaryotic and eukaryotic cells. All cells have a plasma

membrane, DNA, ribosomes, and cytoplasm. Prokaryotic cells are small, relatively simple cells that do not have a membrane-bounded nucleus (**4.3**). All other forms of life are composed of more complex eukaryotic cells, distinguished by the presence of a true nucleus. Membranes form the boundaries of the many eukaryotic organelles, compartmentalizing the interior of the cell and facilitating a variety of metabolic activities (**4.4**).

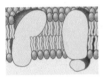
Biological membranes are composed of a phospholipid bilayer with proteins embedded in and attached to it. The functions of membranes relate to the properties of phospholipids and the types of associated proteins (**4.5**).

Cell Structures Involved in Manufacturing and Breakdown (4.6–4.13)

Nucleus. Surrounded by a porous nuclear envelope, the nucleus contains the DNA that carries the cell's hereditary blueprint and directs its activities. The nucleolus assembles ribosomes (**4.6**).

Ribosomes, composed of ribosomal RNA and proteins, synthesize proteins according to directions from DNA (**4.7**).

Endomembrane system is a collection of membranous organelles that manufacture, modify, store, and distribute cell products (**4.8**). The various organelles of the endomembrane system are interconnected structurally and functionally (**4.13**).

Endoplasmic reticulum (ER) is a membranous network of tubes and sacs. Smooth ER synthesizes lipids, processes toxins and drugs in liver cells, and stores and releases calcium ions in muscle cells. Rough ER manufactures membranes, and ribosomes on its surface produce proteins that are secreted, inserted into membranes, or transported in vesicles to other organelles (**4.9**).

Golgi apparatus consists of stacks of sacs that modify ER products, then ship them to other organelles or to the cell surface (**4.10**).

Lysosomes are membrane compartments with enzymes that function in digestion and recycling within the cell (**4.11**).

Vacuoles. Plant cells contain a large central vacuole, with lysosomal, storage, and growth functions. Some protists have contractile vacuoles (**4.12**).

Peroxisomes are organelles that produce and break down H_2O_2 as a by-product of their various metabolic functions (**4.13**).

Energy-Converting Organelles (4.14–4.16)

Mitochondria carry out cellular respiration, using the energy in food to make ATP for cellular work (**4.14**).

Chloroplasts are present in plants and some protists, converting solar energy to chemical energy in sugars (**4.15**). These organelles may have evolved by endosymbiosis of prokaryote cells in a host cell (**4.16**).

Internal and External Support: The Cytoskeleton and Cell Surfaces (4.17–4.22)

Cytoskeleton is a structural network of protein fibers. Microfilaments of actin enable cells to change shape and move. Intermediate filaments reinforce the cell and anchor certain organelles. Microtubules give the cell rigidity and act as tracks for organelle movement (**4.17**). Eukaryotic cilia and flagella are locomotor appendages made of microtubules in a 9 + 2 arrangement (**4.18**). Environmental chemicals or genetic disorders may interfere with sperm motility (**4.19**).

Animal cells. The extracellular matrix of animal cells consists mainly of glycoproteins. Tight junctions bind cells to form leakproof sheets. Anchoring junctions rivet cells into strong tissues. Gap junctions allow substances to flow from cell to cell (**4.20–4.21**).

Plant cells are supported by rigid cell walls made largely of cellulose. Plasmodesmata are connecting channels between plant cells (**4.22**).

Functional Categories of Cell Structures (4.23)

Eukaryotic organelles and structures fall into four functional groups: (1) manufacturing; (2) breakdown; (3) energy processing; and (4) support, movement, and communication between cells. These functions correlate with the structures of these cellular components (**4.23**).

Connecting the Concepts

1. Label the structures in this diagram of an animal cell. Review the functions of each of these organelles.

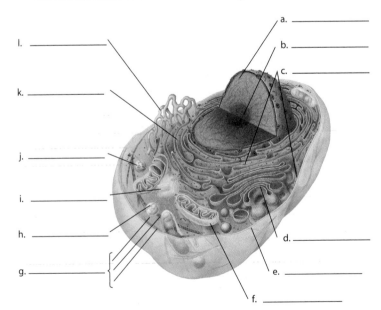

2. List some structures found in animal cells but not in plant cells.

3. List some structures found in plant cells but not in animal cells.

Testing Your Knowledge

Multiple Choice

4. The ultrastructure of a chloroplast is best studied using a
   a. light microscope.
   b. telescope.
   c. scanning electron microscope.
   d. transmission electron microscope.
   e. light microscope and fluorescent dyes.

5. The cells of an ant and a horse are, on average, the same small size; a horse just has more of them. What is the main advantage of small cell size?
   a. Small cells are less likely to burst than large cells.
   b. A small cell has a larger plasma membrane surface area than does a large cell.
   c. Small cells can better take up sufficient nutrients and oxygen to provide for their needs.
   d. It takes less energy to make an organism out of small cells.
   e. Small cells require less oxygen than do large cells.

6. Which of the following clues would tell you whether a cell is prokaryotic or eukaryotic?
   a. the presence or absence of a rigid cell wall
   b. whether or not the cell is partitioned by internal membranes
   c. the presence or absence of ribosomes
   d. whether or not the cell carries out cellular metabolism
   e. whether or not the cell contains DNA

7. Which state correctly describes bound ribosomes?
   a. Bound ribosomes are enclosed in a membrane.
   b. Bound and free ribosomes are structurally different.
   c. Bound ribosomes are most commonly found on the surface of the plasma membrane.
   d. Bound ribosomes generally synthesize membrane proteins and secretory proteins.
   e. Bound ribosomes produce the subunits of microtubules, microfilaments, and intermediate filaments.

Choose from the following cells for questions 8–12:
   a. muscle cell in thigh of long-distance runner
   b. pancreatic cell that secretes digestive enzymes
   c. ovarian cell that produces the steroid hormone estrogen
   d. cell in tissue layer lining digestive tract
   e. white blood cell that engulfs bacteria

8. In which cell would you find the most lysosomes?

9. In which cell would you find the most mitochondria?

10. In which cell would you find the most smooth ER?

11. In which cell would you find the most rough ER?

12. In which cell would you find the most tight junctions?

13. A type of cell called a lymphocyte makes proteins that are exported from the cell. Which of the following is the path of a protein from the site where its polypeptides are made to its export?
   a. chloroplast . . . Golgi . . . lysosomes . . . plasma membrane
   b. Golgi . . . rough ER . . . smooth ER . . . transport vesicle
   c. rough ER . . . Golgi . . . transport vesicle . . . plasma membrane
   d. smooth ER . . . Golgi . . . lysosome . . . plasma membrane
   e. nucleus . . . rough ER . . . Golgi . . . plasma membrane

14. Which of the following structures is *not* directly involved in cell support or movement?
   a. microfilament
   b. flagellum
   c. microtubule
   d. lysosome
   e. cell wall

Describing, Comparing, and Explaining

15. What three cellular components are shared by prokaryotic and eukaryotic cells?

16. Briefly describe the three kinds of junctions that can connect animal cells, and compare their functions.

17. What general function do the chloroplast and mitochondrion have in common? How are their functions different?

18. How does a eukaryotic cell benefit from its internal membranes?

19. Describe two different ways in which the motion of cilia can function in organisms.

20. Explain how a protein inside the ER can be exported from the cell without ever crossing a membrane.

21. Is this statement true or false? "Animal cells have mitochondria; plant cells have chloroplasts." Explain.

22. Describe the structure of the plasma membrane of an animal cell. What would be found directly inside and outside the membrane?

Applying the Concepts

23. Imagine a spherical cell with a radius of 10 μm. What is the cell's surface area in μm^2? Its volume, in μm^3? What is the ratio of surface area to volume for this cell? Now do the same calculations for a second cell, this one with a radius of 20 μm. Compare the surface-to-volume ratios of the two cells. How is this comparison significant to the functioning of cells? (*Note:* For a sphere of radius r, surface area $= 4\pi r^2$ and volume $= \frac{4}{3}\pi r^3$. Remember that the value of π is 3.14.)

24. Cilia are found on cells in almost every organ of the human body, and the malfunction of cilia is involved in several human disorders. During embryological development, for example, cilia generate a leftward flow of fluid that initiates the left-right organization of the body organs. Some individuals with primary ciliary dyskinesia (see Module 4.19) exhibit *situs inversus,* in which internal organs such as the heart are on the wrong side of the body. Explain why this reversed arrangement may be a symptom of PCD.

25. The cells of plant seeds store oils in the form of droplets enclosed by membranes. Unlike typical biological membranes, this oil droplet membrane consists of a single layer of phospholipids rather than a bilayer. Draw a model for a membrane around such an oil droplet. Explain why this arrangement is more stable than a bilayer of phospholipids.

26. Doctors at a California university removed a man's spleen, standard treatment for a type of leukemia, and the disease did not recur. Researchers kept the spleen cells alive in a nutrient medium. They found that some of the cells produced a blood protein that showed promise as a treatment for cancer and AIDS. The researchers patented the cells. The patient sued, claiming a share in profits from any products derived from his cells. The California Supreme Court ruled against the patient, stating that his suit "threatens to destroy the economic incentive to conduct important medical research." The U.S. Supreme Court agreed. Do you think the patient was treated fairly? Is there anything else you would like to know about this case that might help you decide?

Answers to all questions can be found in Appendix 4.

For Practice Quizzes, BioFlix, MP3 Tutors, and Activities, go to www.mybiology.com.

Turning on the Lights to Be Invisible

How does one hide in the open ocean? Animals on land can crouch behind trees or rocks, and many use amazing camouflage disguises. There are insects that look like sticks, toads that resemble rocks, rabbits that blend in with the snow. But would camouflage work when there is nothing around but water?

Ocean predators often hunt by looking up, searching for a silhouette of their prey above them. Can an animal hide its silhouette? The answer is yes, if it turns on the lights. The deep-sea firefly squid (*Watasenia scintillans*) shown here has light-producing organs called photophores dotting its ventral (lower) surface. These organs emit a soft glow that matches the light filtering down from above. During the day, the squid remain in deep, cold water, and their photophores glow blue, matching the blue wavelengths of light that penetrate into deeper water. As night approaches, the squid come closer to the warmer surface, and their glow turns to green, again matching the ambient light.

These squid aren't the only ones to use this mechanism, called counter-illumination. Many marine invertebrates and fishes avoid their predators by producing light to hide their silhouettes. The cookiecutter shark (*Isistius brasiliensis*), shown to the right, uses a variation on this theme to help it hunt. This small shark has densely packed photophores on its underside, except on a small patch behind its head. This dark band may appear as a small fish to large predators like tuna and mackerel. When they approach from below to prey on this "small fish," the cookiecutter shark snags a cookie-sized bite from the potential predator's side and darts off.

The light these animals produce, referred to as bioluminescence, comes from a chemical reaction that converts chemical energy into visible light. These reactions involve enzymes located within membranes in special light-producing cells. In the presence of oxygen (O_2), an enzyme called luciferase catalyzes the conversion of a molecule called luciferin to a molecule that emits light. (Both are named for the Latin word *lucifer*, meaning light bearer.) You may be familiar with bioluminescence in the familiar fireflies that use particular patterns of flashes of yellow or green light to attract mates. Such light production is fairly rare on land, but it is actually the norm in the ocean, where an estimated 90% of deep-sea marine life produce bioluminescence.

Hiding behind light is obviously not effective in depths where there is no light. Emitting light serves different purposes in the dark. Animals can use it to attract mates or attract prey. Sometimes they do both at once. The anglerfish shown above is a deep-sea dweller with a huge head, enormous mouth, and a bioluminescent fishing lure hanging above its mouth. Actually only the female "fishes." The tiny male is attracted to a female in the dark, probably by the unique shape of her glowing lure as well as her chemical smell. He latches onto the female with his sharp teeth and transforms into a permanently attached, parasitic mate. He loses his eyes and all internal organs except his testes. The female must attract prey to feed both herself and her attached sperm-producing "mates." Unlike the squid and most other bioluminescent animals, the anglerfish doesn't produce her own light but uses a glowing glob of bacteria housed in her lure to attract prey. Some fishes eat fecal pellets that are colonized by bioluminescent bacteria. As they approach these glowing tidbits, the anglerfish's huge jaws snatch them up.

Life and death stories involving bioluminescence abound in the ocean. Some microorganisms light up when attacked, drawing the attention of larger predators that may come to feed on their attackers. Certain squid expel a cloud of bioluminescence instead of ink to escape. Whether for camouflage, attraction of mates or prey, repulsion of predators, or communication, the production of light by living organisms is a widespread phenomenon. It has evolved in certain bacteria, algae, mushrooms, many marine invertebrates, and fishes. Different versions of light-emitting luciferin and catalyzing luciferase have evolved in these various groups. Bioluminescence demonstrates how living cells put energy to work by means of enzyme-controlled chemical reactions. Light is a form of energy, and the ability of bioluminescent organisms to convert chemical energy to light energy is just one example of the myriad of energy conversions central to life.

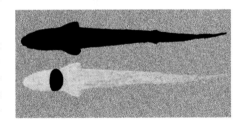

Silhouette of the cookiecutter shark. Photophores mask the silhouette except for a small dark patch.

Many of these energy-converting reactions of cells take place in organelles, such as those in the light-producing cells in a squid. And the enzymes that control these reactions are often embedded in membranes of the organelle. Indeed, everything that happens when a squid turns on the lights to hide has some relation to the topics of this chapter: how working cells use membranes, energy, and enzymes. ▪ ▪ ▪

5.1 Membranes are a fluid mosaic of phospholipids and proteins

Recall from Chapter 4 that biological membranes are composed of phospholipids and proteins. Figure 5.1A reviews the structure of the plasma membrane of an animal cell. Like other membranes, it is commonly described as a **fluid mosaic.** The word *mosaic* denotes a surface made of small pieces. A membrane is a "mosaic" in having diverse protein molecules embedded in a framework of phospholipids. The membrane is "fluid" in that most of these molecules can drift about in the membrane.

Double bonds in the unsaturated fatty acid tails of many phospholipids produce kinks that prevent phospholipids from packing tightly together, keeping the membrane fluid. In animal cells, the steroid cholesterol (see Module 3.9), wedged into the bilayer, helps stabilize the membrane at warm temperatures but also helps keep the membrane fluid at lower temperatures. In a cell, the phospholipid bilayer remains about as fluid as salad oil.

The word *mosaic* can refer not only to the position of proteins in the phospholipid bilayer but also to the varied functions of these proteins. More than 50 kinds of proteins have been found in the plasma membrane of red blood cells. Different types of cells contain different membrane proteins, and the various membranes within a cell each have unique proteins.

Two of the functions of membrane proteins are illustrated in Figure 5.1A. Some proteins give the membrane a stronger *framework.* These proteins called integrins, span the membrane and attach to the cytoskeleton on the inside and the extracellular matrix (ECM) on the outside (see Module 4.20). Glycoproteins are involved in *cell–cell recognition,* a second function of plasma membrane proteins. The outside surface of the membrane has carbohydrates (chains of sugars, blue in Figure 5.1A) bonded to proteins or lipids in the membrane. The carbohydrates vary among species, among individuals, and even among cell types. Many function as identification tags that are recognized by other cells. This cell–cell recognition allows cells in an embryo to sort themselves into tissues and organs. It also enables cells of the immune system to recognize and reject foreign cells, such as infectious bacteria. A third function specific to plasma membrane proteins is forming *junctions* between cells (see Module 4.21).

Many membrane proteins are *enzymes.* Figure 5.1B shows how enzymes may work as a team to carry out sequential steps in a pathway. Other proteins function as *receptors* for chemical messengers from other cells (Figure 5.1C). A receptor protein has a shape that fits a specific messenger, such as a hormone. Often the binding of the messenger to the receptor triggers a chain reaction involving other proteins, which relay the message to molecules that perform specific functions inside the cell. This message-transfer process is called *signal transduction* and will be described in more detail in Chapter 11.

A final important function of membrane proteins is in *transport* (Figure 5.1D). Membranes exhibit **selective permeability;** that is, they allow some substances to cross more easily than others. Their hydrophobic interior is one reason for this selective permeability. Nonpolar, hydrophobic molecules can easily pass through membranes. In contrast, polar molecules and ions are not soluble in lipids. Many essential molecules, such as glucose and ions, require transport proteins to enter or leave the cell.

Web Activity *Membrane Structure*

Web Activity *Selective Permeability of Membranes*

Web Activity *Signal Transduction*

Web Process of Science *How Do Cells Communicate with Each Other?*

? Review the six different types of functions that membrane proteins can perform.

■ Support by attaching to the cytoskeleton and ECM, cell–cell recognition, intercellular junctions , enzymes, signal transduction, and transport.

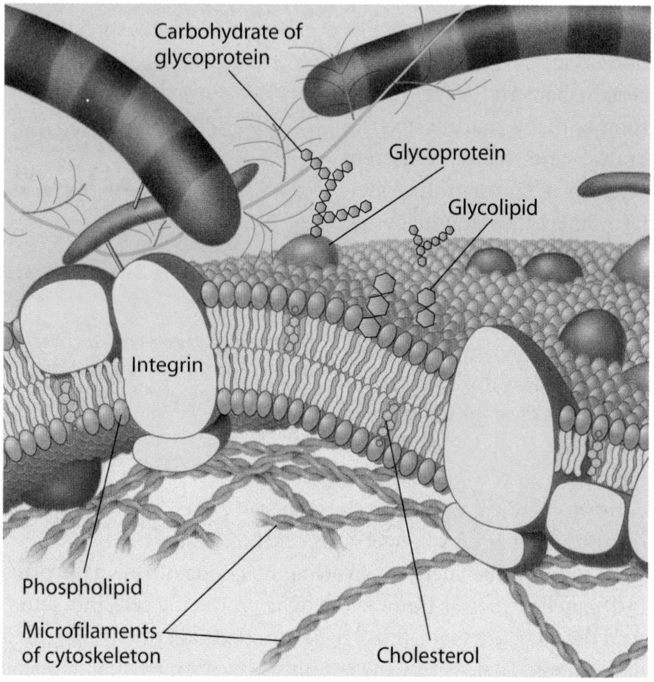

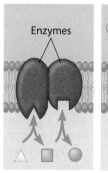
Figure 5.1B
Enzyme activity

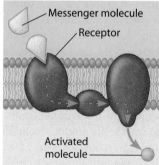
Figure 5.1C
Signal transduction

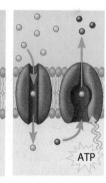
Figure 5.1D
Transport

Figure 5.1A The plasma membrane and extracellular matrix of an animal cell

5.2 Membranes form spontaneously, a critical step in the origin of life

Phospholipids, the key ingredients of biological membranes, were probably among the first organic molecules that formed from chemical reactions on early Earth (see Module 15.2). These lipids could spontaneously self-assemble into simple membranes, as we can demonstrate in a test tube. When a mixture of phospholipids and water is shaken, the phospholipids organize into bilayers surrounding water-filled bubbles (Figure 5.2). This assembly requires neither genes nor other information beyond the properties of the phospholipids themselves.

The formation of membrane-enclosed collections of molecules was a critical step in the evolution of the first cells. A membrane can enclose a solution that is different in composition from its surroundings. A plasma membrane that allows cells to regulate their chemical exchanges with the environment is a basic requirement for life. Indeed, all cells are enclosed by a plasma membrane that is similar in structure and function—one of the features that illustrates the evolutionary unity of life.

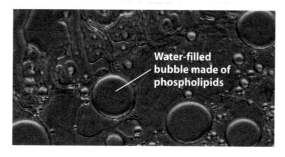

Figure 5.2 Artificial membrane-bound sacs

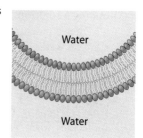

? To the right is a diagram of a section of one of the membrane sacs shown in the photo above. Describe its structure.

■ The phospholipids form a bilayer. The hydrophobic fatty acid tails cluster in the center and the hydrophilic phosphate heads face the water on both sides.

5.3 Passive transport is diffusion across a membrane with no energy investment

Diffusion is the tendency for particles of any kind to spread out evenly in an available space, moving from where they are more concentrated to regions where they are less concentrated. Molecules vibrate and move randomly as a result of a type of energy called thermal motion (heat). Diffusion requires no work; it results from the thermal motion of atoms and molecules.

The figures to the right will help you to visualize diffusion across a membrane. Figure 5.3A shows a solution of green dye separated from pure water by a membrane. Assume that this membrane has microscopic pores through which dye molecules can move. Thus, it is permeable to the dye. Although each molecule moves randomly, there will be a net movement from the side of the membrane where dye molecules are more concentrated to the side where they are less concentrated. Put another way, the dye diffuses down its **concentration gradient.** Eventually, the solutions on both sides have equal concentrations of dye. At this dynamic equilibrium, molecules still move back and forth, but there is no *net* change in concentration on either side of the membrane. Figure 5.3B illustrates the important point that two or more substances diffuse independently of each other; that is, each diffuses down its own concentration gradient.

Because a cell does not perform work when molecules diffuse across its membrane, the diffusion of a substance across a biological membrane is called **passive transport.** Much of the traffic across cell membranes occurs by diffusion. In our lungs, diffusion down concentration gradients is the sole means by which oxygen (O_2), essential for metabolism, enters red blood cells and carbon dioxide (CO_2), a metabolic waste, passes out of them.

Both O_2 and CO_2 are small, nonpolar molecules that diffuse easily across the phospholipid bilayer of a membrane. But can ions and polar molecules also move by passive transport into or

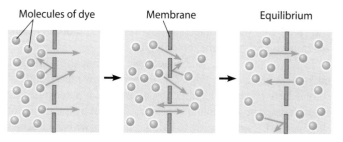

Figure 5.3A Passive transport of one type of molecule

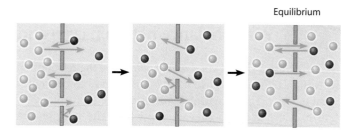

Figure 5.3B Passive transport of two types of molecules

out of a cell? They can if they are moving down their concentration gradients and if they have transport proteins to provide a pathway across the membrane.

Web Activity Diffusion

? Why is diffusion across a membrane called passive transport?

■ The cell does not expend energy to transport substances that are diffusing down their concentration gradients.

5.4 Osmosis is the diffusion of water across a membrane

One of the most important substances that crosses membranes by passive transport is water. In the next module, we consider the critical balance of water between a cell and its environment. But first let's explore a physical model of the diffusion of water molecules across a selectively permeable membrane, a process called **osmosis.** Remember that a selectively permeable membrane allows some substances to cross more easily than others.

The top of Figure 5.4 shows what happens if a membrane permeable to water but not to a solute (such as glucose) separates two solutions with different concentrations of solute. (A solute is a substance that dissolves in a liquid solvent, producing a solution.) The solution on the right side initially has a higher concentration of solute than that on the left. As you can see, water crosses the membrane until the solute concentrations (molecules per milliliter of solution) are equal on both sides.

In the close-up view at the bottom of Figure 5.4, you can see what happens at the molecular level. Clusters of polar water molecules form weak bonds with solute molecules, so that fewer water molecules are free to diffuse across the membrane. The less concentrated solution on the left, with fewer solute molecules, has more free water molecules. There is a net movement of water down its own concentration gradient, from the solution with the lower solute concentration and more free water molecules to that with the higher solute concentration and fewer free water molecules. The result is the difference in water levels you see at the top of Figure 5.4.

Here we show only one type of solute, but the same net movement of water would occur no matter how many kinds of solutes were present. The direction of osmosis is determined by the difference in *total* solute concentration, not by the nature of the solutes. Let's now apply to living cells what we have learned about osmosis in artificial systems.

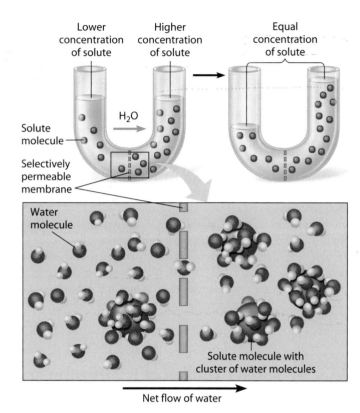

Figure 5.4   Osmosis, the diffusion of water across a membrane

? Indicate the direction of net water movement between two solutions—a 0.5% sucrose solution and a 2% sucrose solution—separated by a membrane not permeable to sucrose.

■ From the 0.5% sucrose solution (lower solute concentration) to the 2% sucrose solution (higher solute concentration)

5.5 Water balance between cells and their surroundings is crucial to organisms

Biologists use a special vocabulary to describe the relationship between a cell and its surroundings with regard to the movement of water. The term **tonicity** describes the ability of a solution to cause a cell to gain or lose water. The tonicity of a solution mainly depends on its concentration of solutes that cannot cross the plasma membrane relative to the concentrations of solutes in the cell.

Figure 5.5 on the facing page illustrates how the principles of osmosis and tonicity apply to cells. The effects of solutions of various tonicities on an animal cell are shown at the top of the illustration; the effect of the same solutions on a plant cell are shown at the bottom of the figure. When an animal cell, such as the red blood cell shown in part A of the figure, is immersed in a solution that is **isotonic** to the cell (*iso,* same, and *tonos,* tension), the cell's volume remains constant. The solute concentration of a cell and its isotonic environment are essentially equal, and the cell gains water at the same rate that it loses it. Red blood cells are transported in the isotonic plasma of the blood, and intravenous (IV) fluids administered in hospitals must also be isotonic to blood cells. The cells of most animals are bathed in an extracellular fluid that is isotonic to the cells. Many marine animals, such as sea stars and crabs, are isotonic to seawater.

Part B of the figure shows what can happen to an animal cell in a **hypotonic** solution (*hypo,* below), a solution with a solute concentration lower than that of the cell. The cell gains water, swells, and may burst (lyse) like an overfilled balloon. Part C shows the opposite case—an animal cell in a **hypertonic** solution (*hyper,* above), a solution with a higher solute concentration. The cell shrivels and can die from water loss.

For an animal to survive in a hypotonic or hypertonic environment, it must have a way to prevent excessive uptake or excessive loss of water. The control of water balance is called **osmoregulation.** For example, a freshwater fish, which lives in a hypotonic environment, has kidneys and gills that work constantly to prevent an excessive buildup of water in the body. (We discuss osmoregulation further in Chapter 25.)

Water balance issues are somewhat different for the cells of plants, prokaryotes, and fungi because of their cell walls. As shown in part D, a plant cell immersed in an isotonic solution is flaccid (limp). In contrast, a plant cell is turgid (very firm), which is the healthy state for most plant cells, in a hypotonic environment (part E). To become turgid, a plant cell needs a net inflow of water. Although the somewhat elastic cell wall expands a bit, the pressure it exerts prevents the cell from taking in too much water and bursting, as an animal cell would in a hypotonic environment. Some plants, including most houseplants, depend on their turgid cells for mechanical suppport.

Part F shows that in a hypertonic environment, a plant cell is no better off than an animal cell. As a plant cell loses water, it shrivels, and its plasma membrane pulls away from the cell wall. This process, called plasmolysis, causes the plant to wilt and can be lethal to the cell and the plant.

Web Activity *Osmosis and Water Balance in Cells*

Web Process of Science *How Does Osmosis Affect Cells?*

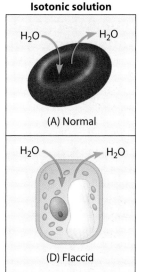 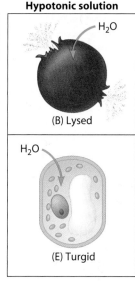 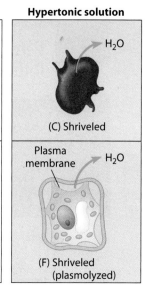

	Isotonic solution	Hypotonic solution	Hypertonic solution
Animal cell	(A) Normal	(B) Lysed	(C) Shriveled
Plant cell	(D) Flaccid	(E) Turgid	(F) Shriveled (plasmolyzed)

Figure 5.5 How animal and plant cells behave in different solutions

? Explain the function of the contractile vacuoles in the freshwater *Paramecium* shown in Figure 4.12B in terms of what you have just learned about water balance in cells.

■ The pond water in which *Paramecium* lives is hypotonic to the cell, and thus there is a constant net osmosis of water into the cell. The contractile vacuoles expel this excess water, preventing the cell from bursting.

5.6 Transport proteins may facilitate diffusion across membranes

Recall that nonpolar, hydrophobic molecules can dissolve in the lipid bilayer of a membrane and cross it with ease. Numerous substances that do not diffuse freely across membranes because of their polarity or charge can move across a membrane with the help of specific transport proteins. When one of these proteins makes it possible for a substance to move down its concentration gradient, the process is called **facilitated diffusion.** Without the protein, the substance does not cross the membrane or it diffuses across it too slowly to be useful to the cell. Facilitated diffusion is a type of passive transport because it does not require energy. As in all passive transport, the driving force is the concentration gradient.

Figure 5.6 shows a common type of transport protein, which provides a hydrophilic channel that some molecules or ions use as a tunnel through the membrane. Another type of transport protein binds its passenger, changes shape, and releases its passenger on the other side. In both cases, the transport protein is specific for the substance it helps move across the membrane. The greater the number of transport proteins for a particular solute present in a membrane, the faster the solute's rate of diffusion across the membrane.

Substances that use facilitated diffusion for crossing cell membranes include a number of sugars, amino acids, ions—and even water. The water molecule is very small, but because it

is polar (see Module 2.9), its diffusion through a membrane's hydrophobic interior is relatively slow. The very rapid diffusion of water into and out of certain cells, such as plant cells, kidney cells, and red blood cells, is made possible by transport proteins called **aquaporins.** In the next module, we meet the scientist who discovered these water channels and received the 2003 Nobel Prize in Chemistry for his work.

Web Activity *Facilitated Diffusion*

? How do transport proteins contribute to a membrane's selective permeability?

■ Transport proteins are specific for the solutes they transport. Thus, the numbers and kinds of different transport protein molecules embedded in the membrane affect its permeability to various solutes.

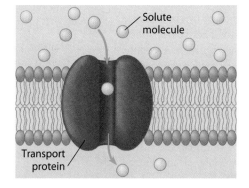

Figure 5.6
Transport protein providing a channel for the diffusion of a specific solute across a membrane

5.7 Peter Agre talks about aquaporins, water-channel proteins found in some cells

Figure 5.7   Peter Agre

An aquaporin is an hourglass-shaped protein that spans the membrane and allows entry or exit of up to 3 billion water molecules per second. The presence of aquaporins in cells that require the rapid diffusion of water illustrates the correlation between structure and function in biological membranes.

In a recent interview, Dr. Peter Agre (Figure 5.7), who is a medical doctor and a professor at the Johns Hopkins University School of Medicine, described his research that led to the discovery of aquaporins:

I'm a blood specialist (hematologist), and my particular interest has been proteins found in the plasma membrane of red blood cells. When I joined the faculty at Hopkins, I began to study the Rh blood antigens. Rh is of medical importance because of Rh incompatibility, which occurs when Rh-negative mothers have Rh-positive babies. Membrane-spanning proteins like the Rh-protein are really messy to work with. But we worked out a method to isolate and partially purify the Rh protein. Our sample seemed to consist of two proteins, but we were sure that the smaller one was just a breakdown product of the larger one. We were completely wrong.

Using antibodies we made to the protein, we showed it to be one of the most abundant proteins in red cell membranes (200,000 copies per cell!) and even more abundant in certain kidney cells.

We asked Dr. Agre why cells have water-channel proteins.

Not all cells do. Before our discovery, however, many physiologists thought that diffusion through the phospholipid bilayer was enough for getting water into and out of *all* cells. Others said this couldn't be enough, especially for cells whose water permeability needs to be very high or regulated. For example, our kidneys must filter and reabsorb many liters of water every day. . . . People whose kidney cells have defective aquaporin molecules need to drink 20 liters of water a day to prevent dehydration. In addition, some patients make too much aquaporin, causing them to retain too much fluid. Fluid retention in pregnant women is caused by the synthesis of too much aquaporin. Knowledge of aquaporins may in the future contribute to the solution of medical problems.

? Why are aquaporins important in kidney cells?

■ Kidney cells must reabsorb a large amount of water from the urine before it is excreted.

5.8 Cells expend energy in the active transport of a solute against its concentration gradient

In **active transport** a cell must expend energy to move a solute *against* its concentration gradient—that is, across a membrane toward the side where the solute is more concentrated. The cell's energy molecule ATP (described in more detail in Module 5.13) supplies the energy for most active transport.

Figure 5.8 shows a simple model of an active transport system that moves a solute out of the cell against its concentration gradient. ❶ The process begins when solute on the cytoplasmic side of the plasma membrane attaches to a specific binding site on the transport protein. ❷ ATP then transfers one of its phosphate groups to the transport protein, ❸ causing it to change shape in such a way that the solute is released on the other side of the membrane. ❹ Then, the phosphate group detaches, and the transport protein returns to its original shape, ready for a new round of active transport. (As indicated in the figure, two solute ions or molecules are often transported together.)

Active transport allows a cell to maintain concentrations of small molecules that are different from concentrations in its surroundings. For example, the inside of an animal cell has a higher concentration of potassium ions (K^+) and a lower concentration of sodium ions (Na^+) than the solution outside the cell. The generation of nerve signals depends on these concentration differences. A transport protein called the sodium-potassium pump helps cells maintain these steep gradients by shuttling Na^+ and K^+ across the membrane against their concentration gradients.

Web Activity *Active Transport*

? Cells actively transport Ca^{2+} out of the cell. Is calcium more concentrated inside or outside the cell? Explain.

■ Outside the cell. If the cell uses active transport, it is moving calcium against its concentration gradient.

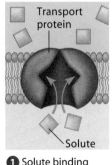

❶ Solute binding

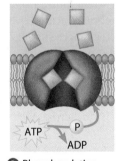

❷ Phosphorylation

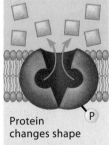

❸ Transport

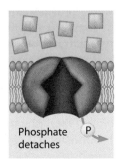

❹ Protein reversion

Figure 5.8   Active transport of a solute across a membrane

Exocytosis and endocytosis transport large molecules across membranes

So far, we've focused on how water and small solutes enter and leave cells. The story is different for large molecules.

A cell uses the process of **exocytosis** (from the Greek *exo,* outside, and *kytos,* cell) to export bulky materials such as proteins or polysaccharides. As shown in Figure 4.13, a transport vesicle filled with macromolecules buds from the Gogli apparatus and moves to the plasma membrane. Once there, the vesicle fuses with the plasma membrane, and the vesicle's contents spill out of the cell while the vesicle membrane becomes part of the plasma membrane. When we weep, for instance, cells in our tear glands use exocytosis to export a salty solution containing proteins. In another example, certain cells in the pancreas manufacture the hormone insulin and secrete it into the bloodstream by exocytosis.

Endocytosis (*endo,* inside) is a transport process that is the opposite of exocytosis. In endocytosis, a cell takes in substances. A depression in the plasma membrane pinches in and forms a vesicle enclosing material that had been outside the cell.

Figure 5.9 shows three kinds of endocytosis. The top diagram illustrates **phagocytosis,** or "cellular eating." A cell engulfs a particle by wrapping extensions called pseudopodia around it and packaging it within a membrane-enclosed sac large enough to be called a vacuole. As Module 4.11 described, the vacuole then fuses with a lysosome, whose hydrolytic enzymes digest the contents of the vacuole. The micrograph on the top right shows an amoeba taking in a food particle via phagocytosis.

The center diagram shows **pinocytosis,** or "cellular drinking." The cell "gulps" droplets of fluid into tiny vesicles. Pinocytosis is not specific; it takes in any and all solutes dissolved in the droplets. The micrograph in the middle shows pinocytosis vesicles forming (arrows) in a cell lining a small blood vessel.

Receptor-mediated endocytosis is highly specific, in contrast to pinocytosis. Receptor proteins for specific molecules are embedded in regions of the membrane that are lined by a layer of coat proteins. The bottom diagram shows that the plasma membrane has indented to form a coated pit, whose receptor proteins have picked up particular molecules from the surroundings. The coated pit then pinches closed to form a vesicle that carries the molecules into the cytoplasm. The micrograph shows material bound to receptor proteins inside a coated pit.

Our cells use receptor-mediated endocytosis to take in cholesterol from the blood for synthesis of membranes and as a pre-

cursor for other steroids. Cholesterol circulates in the blood in particles called low-density lipoproteins (LDLs). LDLs bind to receptor proteins and then enter cells by endocytosis. In humans with the inherited disease hypercholesterolemia, LDL receptor proteins are defective and cholesterol accumulates to high levels in the blood, leading to atherosclerosis (see Modules 9.12 and 23.6).

Web Activity *Exocytosis and Endocytosis*

? As a cell grows, its plasma membrane expands. Does this involve endocytosis or exocytosis? Explain.

■ Exocytosis. When a transport vesicle fuses with the plasma membrane, the vesicle membrane becomes part of the plasma membrane.

Phagocytosis

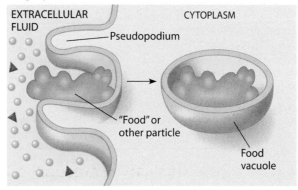

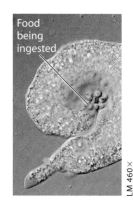

Pinocytosis

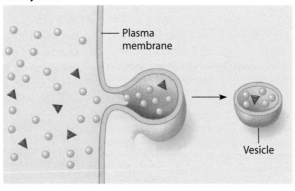

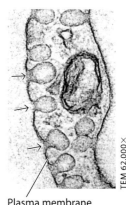

Receptor-mediated endocytosis

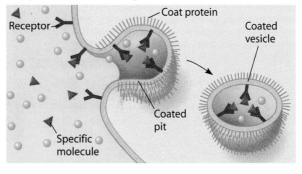

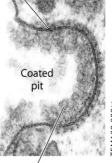

Figure 5.9   Three kinds of endocytosis

5.10 Cells transform energy as they perform work

The title of this chapter is "The Working Cell." But just what type of work does a cell do? You just learned that a cell actively transports substances across membranes. The cell also builds those membranes and the transport proteins embedded in them. A cell is a miniature chemical factory in which thousands of reactions occur within a microscopic space. Some of these reactions break down sugars and other fuels to release energy. The cell uses this energy in other reactions to manufacture cellular parts and products and to move cellular structures or even the whole cell. Bioluminescent organisms use energy in chemical reactions to produce light (see chapter introduction). Life requires energy, and cells are energy transformers. To understand how the cell works, you must have a basic knowledge of energy.

Energy is defined as the capacity to perform work. Work is performed when an object is moved against an opposing force, such as gravity or friction. Put another way, energy is the capacity to rearrange matter. There are two basic forms of energy: kinetic energy and potential energy.

Kinetic energy is the energy of motion. Moving objects can perform work by transferring motion to other matter. A pool player uses the motion of the cue stick to push the cue ball, which in turn moves the other balls. The contraction of leg muscles pushes bicycle pedals, turning the wheels and moving the bike and its rider up a hill (Figure 5.10A). **Heat,** or thermal energy, is a form of kinetic energy associated with the random movement of atoms or molecules. Light is another kind of kinetic energy that can be harnessed to perform work, such as powering photosynthesis in green plants.

Potential energy, the second form of energy, is stored energy that an object possesses as a result of its location or structure. Water behind a dam, for instance, possesses energy because of its altitude above sea level. A cyclist at the top of a hill has potential energy as a result of elevation (Figure 5.10B).

Likewise, molecules possess potential energy because of the arrangement of their atoms. **Chemical energy** is a term that refers to the potential energy available for release in a chemical reaction. Chemical energy is the most important type of energy for living organisms—it is the energy that is available to do the work of the cell. And just as the potential energy of the cyclist can be converted to kinetic energy in the speedy ride downhill (Figure 5.10C), the chemical energy of molecules can be released to power the work of the cell. Life depends on the fact that energy can be converted from one form to another. Next we consider some basic principles of energy transformation.

🎧 MP3 Tutor *Basic Energy Concepts*

Web Activity *Energy Transformations*

❓ How can an object at rest have energy?

■ It can have potential energy as a result of its location.

Figure 5.10A Kinetic energy, the energy of motion

Figure 5.10B Potential energy, stored energy as a result of location or structure

Figure 5.10C Potential energy being converted to kinetic energy

Thermodynamics is the study of energy transformations that occur in a collection of matter. Scientists use the word *system* for the matter under study and refer to the rest of the universe—everything outside the system—as the *surroundings.* A system can be an electric power plant, a single cell, or the entire planet. An organism is an open system; that is, it exchanges both energy and matter with its surroundings.

The First Law of Thermodynamics According to the **first law of thermodynamics,** also known as the law of energy conservation, the energy in the universe is constant. Energy can be transferred and transformed, but it cannot be created or destroyed. A power plant does not create energy; it merely converts it from one form (such as the energy stored in coal) to the more convenient form of electricity. A light bulb does not make light; it converts electricity to light energy. By converting sunlight to chemical energy, a green plant acts as an energy transformer, not an energy producer.

The Second Law of Thermodynamics If energy cannot be destroyed, then why can't organisms simply recycle their energy? It turns out that during every energy transfer or transformation, some energy becomes unusable—unavailable to do work. In most energy transformations, some energy is converted to heat, the energy associated with random molecular motion. Heat is a disordered form of energy, and its release makes the universe more random. Scientists use a quantity called **entropy** as a measure of disorder, or randomness. The more randomly arranged a collection of matter is, the greater its entropy. According to the **second law of thermodynamics,** energy conversions increase the entropy (disorder) of the universe.

Figure 5.11 compares a car and a cell to show how energy is transformed and transferred and how entropy increases as a result. Automobile engines and living cells use the same basic process to make the chemical energy stored in their fuel available for work. The engine of an automobile mixes oxygen with gasoline in an explosive chemical reaction that pushes the pistons, which eventually move the wheels. The waste products emitted from the exhaust pipe are mostly carbon dioxide and water, energy-poor, simple molecules. Only about 25% of the chemical energy stored in gasoline is converted to the kinetic energy of the car's movement; the rest is lost as heat energy.

Cells also use oxygen in the process of cellular respiration, which harvests chemical energy from food molecules. Just like the car, the waste products are mostly carbon dioxide and water. Cells are more efficient than car engines, however, converting about 40% of the chemical energy stored in foods to energy for cellular work. The other 60% generates heat, which explains why vigorous exercise makes you hot. For the bicyclist pumping his way up the hill in Figure 5.10A, sweating and evaporative cooling help release this excess heat to the surroundings.

According to the second law of thermodynamics, energy transformations result in the universe becoming more disordered. How, then, can we account for biological order? A cell creates its structures from less organized materials. For example, amino acids are ordered into the specific sequences of polypeptides. Although this increase in order corresponds to a decrease in entropy, it is accomplished at the expense of ordered forms of energy taken in from the surroundings. As shown in Figure 5.11, cells extract the chemical energy of glucose and return disordered heat and the lower-energy molecules of carbon dioxide and water to their surroundings. In a thermodynamic sense, a cell or an organism is an island of low entropy in an increasingly random universe.

? How does the second law of thermodynamics help explain the diffusion of a solute across a membrane?

■ Diffusion across a membrane results in equal concentrations of solute molecules, which is a more disordered arrangement (higher entropy) than a high concentration on one side and a low concentration on the other.

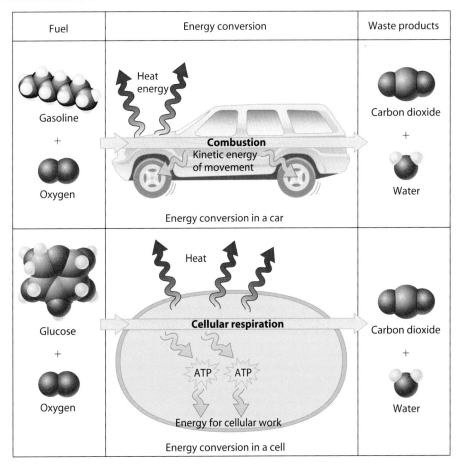

Figure 5.11 Energy transformations (with an increase in entropy) in a car and a cell

5.12 Chemical reactions either release or store energy

Chemical reactions are of two types: They either release energy, or they require an input of energy and store energy.

An **exergonic reaction** is a chemical reaction that releases energy (exergonic means "energy outward"). As shown in Figure 5.12A, an exergonic reaction begins with reactants whose covalent bonds contain more energy than those in the products. The reaction releases to the surroundings an amount of energy equal to the difference in potential energy between the reactants and the products.

As an example of an exergonic reaction, consider what happens when wood burns. One of the major components of wood is cellulose, a large carbohydrate composed of many glucose monomers. Each glucose is rich in potential energy. The burning of wood releases the energy of glucose as heat and light. Carbon dioxide and water are the products of the reaction.

Burning is one way to release energy from chemicals. Cells release energy by means of a different exergonic process, called cellular respiration. **Cellular respiration** is a chemical process that uses oxygen to convert the chemical energy stored in fuel molecules to a form of chemical energy that the cell can use to perform work. Burning and cellular respiration are alike in being exergonic. They differ in that burning is essentially a one-step process that releases all of a substance's energy at once. Cellular respiration, on the other hand, involves many steps, each a separate chemical reaction; you can think of it as a "slow burn." As you saw in Figure 5.11, some of the energy released from glucose by cellular respiration escapes as heat, but a substantial amount of released energy is converted to the chemical energy of ATP. Cells use ATP as an immediate source of energy. We explore ATP in more detail in the next module.

The other type of chemical reaction requires a net input of energy. **Endergonic reactions** yield products that are rich in potential energy (endergonic means "energy inward"). As you can see in Figure 5.12B, on the bottom right of this page, an endergonic reaction starts out with reactant molecules that contain relatively little potential energy. Energy is absorbed from the surroundings as the reaction occurs, so the products of an endergonic reaction store more energy than the reactants did. The energy is potential energy stored in the covalent bonds of the product molecules. And as the graph shows, the amount of additional energy stored in the products equals the difference in potential energy between the reactants and the products.

Photosynthesis, the process by which plant cells make sugar, is an example of an endergonic process. Photosynthesis starts with energy-poor reactants (carbon dioxide and water molecules) and, using energy absorbed from sunlight, produces energy-rich sugar molecules.

All cells use endergonic reactions in the process of building the macromolecules that form the structure and perform the functions of the cell.

Every working cell in every organism carries out thousands of exergonic and endergonic reactions. The total of an organism's chemical reactions is called **metabolism** (from the Greek *metabole*, change). We can picture a cell's metabolism as a road map of thousands of chemical reactions, arranged as intersecting metabolic pathways. A **metabolic pathway** is a series of chemical reactions that either builds a complex molecule or breaks down a complex molecule into simpler compounds. The "slow burn" of cellular respiration is an example of a metabolic pathway in which a sequence of reactions slowly releases the potential energy stored in sugar.

All of an organism's activities require energy, which is obtained from sugar and other molecules by the exergonic reactions of cellular respiration. Cells then use that energy in endergonic reactions to make molecules and do the work of the cell. **Energy coupling**—the use of energy released from exergonic reactions to drive essential endergonic reactions—is a crucial ability of all cells. ATP molecules are the key to energy coupling. In the next module, we explore the structure and function of ATP.

Web Activity *Chemical Reactions and ATP*

? Cellular respiration is an exergonic process. Remembering that energy must be conserved, what becomes of the energy extracted from food during cellular respiration?

■ Some of it is stored in ATP molecules; the rest is released as heat.

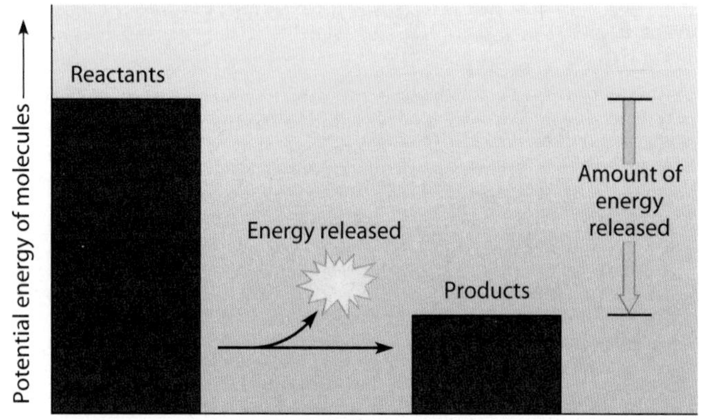

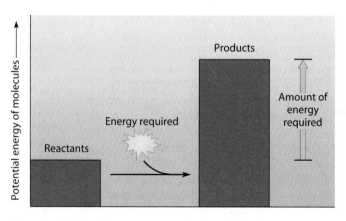

Figure 5.12A   Exergonic reaction, energy released

Figure 5.12B   Endergonic reaction, energy required

ATP powers nearly all forms of cellular work. The structure of ATP, or adenosine triphosphate, is shown in Figure 5.13A. The adenosine part of ATP consists of adenine, a nitrogenous base (see Module 3.16), and ribose, a five-carbon sugar. The triphosphate part is a chain of three phosphate groups (each symbolized by $\textcircled{P}$). All three phosphate groups are negatively charged (see Table 3.2). These like charges are crowded together, and their mutual repulsion makes the triphosphate chain of ATP the chemical equivalent of a compressed spring.

As a result, the bonds connecting the phosphate groups are unstable and can readily be broken by hydrolysis (the addition of water). Notice in Figure 5.13A, below, that when the bond to the third group breaks, a phosphate group leaves ATP—which becomes ADP (adenosine diphosphate)—and energy is released.

Thus, the hydrolysis of ATP is an exergonic reaction—it releases energy. How does the cell couple this reaction to an endergonic one? It usually does so by transferring the third phosphate group from ATP to some other molecule. This transfer is called **phosphorylation,** and most cellular work depends on ATP energizing molecules by phosphorylating them.

There are three main types of cellular work: chemical, mechanical, and transport. As shown in Figure 5.13B, to the right, ATP drives all three types of work. In chemical work, the phosphorylation of reactants provides energy to drive the endergonic synthesis of products. In an example of mechanical work, the transfer of phosphate groups to special motor proteins in muscle cells causes the proteins to change shape and pull on actin filaments, in turn causing the cells to contract. In transport work, as discussed in Module 5.8, ATP drives the active transport of solutes across a membrane against their concentration gradient by phosphorylating certain membrane proteins.

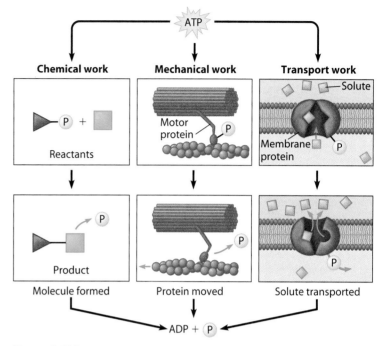

Figure 5.13B How ATP powers cellular work

Work can be sustained because ATP is a renewable resource that cells regenerate. Figure 5.13C, below, shows the ATP cycle. Energy released in exergonic reactions, such as the breakdown of glucose during cellular respiration, is used to regenerate ATP from ADP. In this endergonic (energy-storing) process, a phosphate group is bonded to ADP. A cell at work uses ATP continuously, and the ATP cycle runs at an astonishing pace. In fact, a working muscle cell may consume and regenerate 10 million ATP molecules each second.

But even with a constant supply of energy, few metabolic reactions would occur without the assistance of enzymes. We explore these biological catalysts next.

Web Activity *The Structure of ATP*

? Explain how ATP transfers energy from exergonic to endergonic processes in the cell.

■ Exergonic processes phosphorylate ADP to form ATP. ATP transfers energy to endergonic processes by phosphorylating other molecules.

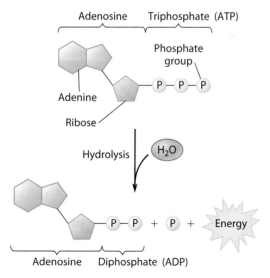

Figure 5.13A The structure and hydrolysis of ATP. The reaction of ATP and water yields ADP, a phosphate group, and energy.

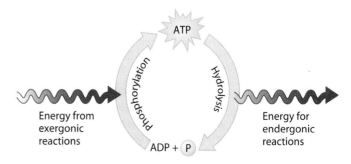

Figure 5.13C The ATP cycle

5.14 Enzymes speed up the cell's chemical reactions by lowering energy barriers

Proteins, DNA, carbohydrates, phospholipids—most of the cell's molecules are rich in potential energy. Why don't these molecules spontaneously break down into simpler, less energetic molecules? For the same reason that wood doesn't normally burst into flames. There is an energy barrier that must be overcome before a chemical reaction can begin. Energy must be absorbed to contort or weaken bonds in reactant molecules so that they can break and new bonds can form. We call this the **energy of activation (E_A),** and we can think of it as the amount of energy needed to push the reactants over an energy barrier, or "hill," so that the "downhill" part of the reaction can begin.

Now we have a dilemma. Most of the essential reactions of metabolism must occur quickly and precisely for a cell to survive. How can the specific reactions that a cell requires get over that energy barrier? One way to speed reactions is to add heat. Certainly, adding a match to kindling will start a fire. But heating a cell would speed up all reactions, not just the necessary ones, and too much heat would obviously kill a cell.

The solution lies in **enzymes**—proteins that function as biological catalysts, increasing the rate of a reaction without being consumed by the reaction. (Some RNA molecules also function as enzymes.) An enzyme speeds up a reaction by lowering the E_A barrier. In Figure 5.14, a graph shows the effect of an enzyme on the reaction it catalyzes. The black curve represents the course of the reaction without an enzyme; the E_A barrier is higher than in the reaction with an enzyme (red curve). Notice that the net change in energy from reactants to products, however, is the same for both curves.

? Explain why an enzyme cannot change an endergonic reaction into an exergonic one.

■ Although an enzyme speeds a reaction by lowering the energy of activation, it does not change the difference in energy between reactants and products. An endergonic reaction would still require an input of energy.

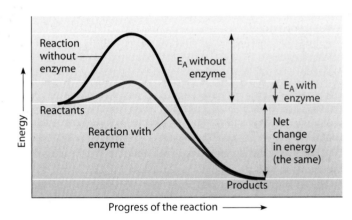

Figure 5.14 The effect of an enzyme is to lower E_A

5.15 A specific enzyme catalyzes each cellular reaction

As a protein, an enzyme has a unique three-dimensional shape, and that shape determines which chemical reaction the enzyme catalyzes. A specific reactant that an enzyme acts on is called the enzyme's **substrate.** A substrate fits into a region of the enzyme called an **active site.** An active site is typically a pocket or groove on the surface of the enzyme formed by only a few of the enzyme's amino acids. The rest of the protein maintains the shape of the active site. Enzymes are specific because their active sites fit only specific substrate molecules.

Figure 5.15 illustrates the catalytic cycle of an enzyme. Our example is the enzyme sucrase, which catalyzes the hydrolysis of sucrose (table sugar) to glucose and fructose. (Most enzymes have names that end in -ase, and many are named for their substrate.) ❶ Sucrase starts with an empty active site. ❷ Sucrose enters the active site, attaching by weak bonds. The active site changes shape slightly so that it embraces the substrate more snugly, like a firm handshake. This **induced fit** may strain substrate bonds or place chemical groups of the amino acids making up the active site in position to catalyze the reaction. (In reactions involving two or more reactants, the active site may hold the substrates in the proper orientation for a reaction to occur.) ❸ The strained bond reacts with water, and the

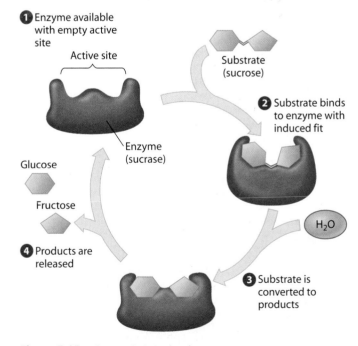

Figure 5.15 The catalytic cycle of an enzyme

substrate is converted (hydrolyzed) to the products glucose and fructose. ❹ The enzyme releases the products and emerges unchanged from the reaction. Its active site is now available for another substrate molecule, and another round of the cycle can begin. A single enzyme molecule may act on thousands or even millions of substrate molecules per second.

Optimal Conditions for Enzymes As with all proteins, an enzyme's shape is central to its function, and this three-dimensional shape is affected by the environment. For every enzyme, there are conditions under which it is most effective. Temperature, for instance, affects molecular motion, and an enzyme's optimal temperature produces the highest rate of contact between reactant molecules and the enzyme's active site. Higher temperatures denature the enzyme, altering its specific shape and destroying its function. Most human enzymes work best at 35–40°C (95–104°F), close to our normal body temperature of 37°C. Prokaryotes that live in hot springs, however, contain enzymes with optimal temperatures of 70°C (158°F) or higher.

The optimal pH for most enzymes is near neutrality, in the range of 6–8. Outside this range, enzyme action may be impaired. Buffers (see Module 2.15) regulate cellular pH, helping to maintain the normal chemical functioning of cells.

Cofactors Many enzymes require nonprotein helpers called **cofactors.** The cofactors of some enzymes are inorganic, such as the ions of zinc, iron, or copper. If the cofactor is an organic molecule, it is called a **coenzyme.** Most vitamins are important in nutrition because they function as coenzymes.

Web Activity *How Enzymes Function*

Web Process of Science *How Is the Rate of Enzyme Catalysis Measured?*

? Explain how an enzyme speeds up a specific reaction.

■ An enzyme lowers the energy of activation needed for a reaction to proceed by the specific binding of its substrate in its active site. With an induced fit, the enzyme strains bonds that need to break or orients substrates so that reactants are converted to products.

5.16 Enzyme inhibitors block enzyme action and can regulate enzyme activity in a cell

A chemical that interferes with an enzyme's activity is called an inhibitor. If an inhibitor attaches to the enzyme by covalent bonds, the inhibition is usually irreversible. Inhibition is reversible when weak interactions bind inhibitor and enzyme.

Some inhibitors resemble the enzyme's normal substrate and compete for the active site. As shown in the lower left of Figure 5.16, such a **competitive inhibitor** reduces the enzyme's productivity by blocking substrates from entering the active site. Competitive inhibition can be overcome by increasing the concentration of substrate, making it more likely that a substrate molecule will be nearby when an enzyme becomes available.

In contrast, a **noncompetitive inhibitor** does not enter the active site. Instead, it binds to the enzyme somewhere else, and its binding changes the shape of the enzyme so that the active site no longer fits the substrate.

Humans have developed many uses for enzyme inhibitors as beneficial drugs and pesticides, as well as poisons intended for use in warfare. Many antibiotics inhibit specific enzymes in disease-causing bacteria. Penicillin, for instance, blocks an enzyme that bacteria use in making cell walls. Other enzyme inhibitors are HIV drugs and some medicines for depression, and many cancer drugs are inhibitors that disrupt enzymes involved in cell division.

Toxins and poisons are often irreversible enzyme inhibitors. Nerve gases bind in the active site of an enzyme vital to the transmission of nerve impulses. The inhibition of this enzyme leads to rapid paralysis of vital functions and death. Pesticides such as malathion and parathion are toxic to insects because they also irreversibly inhibit this enzyme.

Although such inhibitors are deadly and enzyme inhibition itself sounds harmful, in fact, cells use inhibitors as important regulators of cell metabolism. As discussed in Module 5.12, most chemical reactions are organized into metabolic pathways

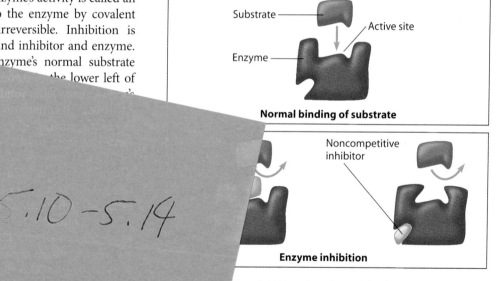

Substrate — Active site — Enzyme

Normal binding of substrate

Noncompetitive inhibitor

Enzyme inhibition

How inhibitors interfere with substrate binding

...molecule is altered in a series of steps, each catalyzed by a specific enzyme, to form a final product. If a cell is producing more of that product than it needs, the product may act as an inhibitor of one of the enzymes early in the pathway. This type of inhibition, whereby a metabolic reaction is blocked by its products, is called **feedback inhibition** and is one of the most important mechanisms that regulate metabolism. In the next chapter, we will see how feedback inhibition helps regulate ATP production in cellular respiration.

? What is the advantage of feedback inhibition to a cell?

■ It prevents the cell from wasting valuable resources by synthesizing more of a particular product than is necessary.

Chapter Review

Reviewing the Concepts

Membrane Structure and Function (5.1–5.9)

Membranes are a fluid mosaic, with protein molecules embedded in a phospholipid bilayer. Functions of membrane proteins include attachment to the cytoskeleton and extracellular matrix, cell-cell recognition, intercellular junctions, enzymatic activity, signal transduction, and transport. Phospholipids spontaneously form membranes, a necessary step in the evolution of life (**5.1–5.2**).

Diffusion, Passive Transport, Active Transport (5.3–5.8).

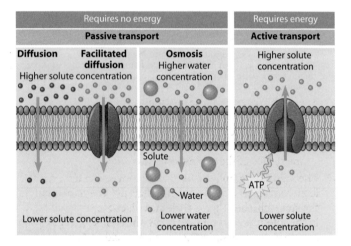

Requires no energy			Requires energy
Passive transport			**Active transport**
Diffusion	**Facilitated diffusion**	**Osmosis** Higher water concentration	Higher solute concentration
Higher solute concentration			
Lower solute concentration	Solute Water Lower water concentration		ATP Lower solute concentration

Tonicity. Cells shrink in a hypertonic solution and swell in a hypotonic solution. In isotonic solutions, animal cells are normal, but plant cells are limp. The control of water balance is called osmoregulation (**5.5**). Aquaporins are water channels in cells with high water-transport needs (**5.7**).

Exocytosis and Endocytosis. In moving large molecules across a membrane, a vesicle may fuse with the membrane and expel its contents (exocytosis), or the membrane may fold inward, enclosing material from the outside (**5.9**).

Energy and the Cell (5.10–5.13)

Energy is the capacity to perform work. Kinetic energy is the energy of motion. Potential energy is energy stored in the location or structure of matter. Chemical energy is potential energy stored in chemical bonds (**5.10**). According to the laws of thermodynamics, energy can change form but cannot be created or destroyed. Energy transformations increase disorder, or entropy, and some energy is lost as heat (**5.11**).

Chemical Reactions. Exergonic reactions release energy. Endergonic reactions require energy and yield products rich in potential energy. Metabolism includes all of a cell's chemical reactions. Energy coupling uses exergonic reactions to drive endergonic ones (**5.12**). ATP performs cellular work by phosphorylation, transferring a phosphate group to make molecules more reactive (**5.13**).

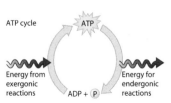

ATP cycle
ATP
Energy from exergonic reactions
ADP + P
Energy for endergonic reactions

How Enzymes Function (5.14–5.16)

Enzymes are protein catalysts that decrease the energy of activation (E_A) needed to begin a reaction (**5.14**). Each type of enzyme has a unique active site that binds specifically with its substrate. Temperature and pH influence enzyme activity. Some enzymes require cofactors, such as metal ions or organic coenzymes (**5.15**).

Inhibitors. A competitive inhibitor competes with the substrate for the active site. A noncompetitive inhibitor alters an enzyme's function by changing its shape. Poisons, pesticides, and drugs may inhibit enzymes. Feedback inhibition helps regulate metabolism (**5.16**).

Connecting the Concepts

1. Fill in the following concept map to review the processes by which molecules move across membranes.

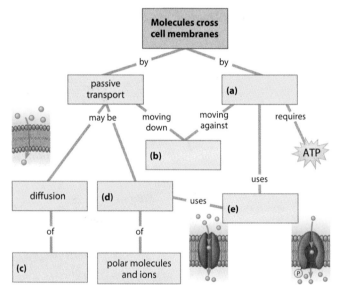

Molecules cross cell membranes
by — passive transport — may be — diffusion
by — (a) — requires — ATP
moving down — moving against — (b)
uses — (e) — uses — (d)
(c) — of
polar molecules and ions — of

2. Label the parts of this diagram illustrating the catalytic cycle of an enzyme.

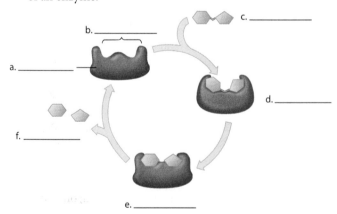

a. _____
b. _____
c. _____
d. _____
e. _____
f. _____

Testing Your Knowledge

Multiple Choice

3. Which best describes the structure of a cell membrane?
   a. proteins between two bilayers of phospholipids
   b. proteins embedded in a bilayer of phospholipids
   c. a bilayer of protein coating a layer of phospholipids

d. phospholipids between two layers of protein
e. cholesterol embedded in a bilayer of phospholipids

4. Consider the following: chemical bonds in the gasoline in a car's gas tank and the movement of the car along the road; a climber at the top of a hill and the hike she took to get there. The first parts of these situations illustrate _____, and the second parts illustrate _____.
   a. the first law of thermodynamics . . . the second law
   b. kinetic energy . . . potential energy
   c. an exergonic reaction . . . an endergonic reaction
   d. potential energy . . . kinetic energy
   e. the second law of thermodynamics . . . the first law

5. A plant cell placed in distilled water will _____; an animal cell placed in distilled water will _____.
   a. burst . . . burst
   b. become flaccid . . . shrivel
   c. become flaccid . . . be normal in shape
   d. become turgid . . . be normal in shape
   e. become turgid . . . burst

6. The sodium concentration in a cell is ten times less than the concentration in the surrounding fluid. How can the cell move sodium out of the cell? (*Explain.*)
   a. passive transport
   b. diffusion
   c. active transport
   d. osmosis
   e. any of these processes

7. The synthesis of ATP from ADP and Ⓟ
   a. is an exergonic process.
   b. involves the hydrolysis of a phosphate bond.
   c. transfers a phosphate, priming a protein to do work.
   d. stores energy in a form that can drive cellular work.
   e. releases energy.

8. Facilitated diffusion across a membrane requires _____ and moves a solute _____ its concentration gradient.
   a. transport proteins . . . up (against)
   b. transport proteins . . . down
   c. energy . . . up
   d. energy and transport proteins . . . up
   e. energy and transport proteins . . . down

Describing, Comparing, and Explaining

9. What are aquaporins? Where would you expect to find these structures?

10. How do the two laws of thermodynamics apply to living organisms?

11. What are the main types of cellular work? How does ATP provide the energy for this work?

12. Why is the barrier of the energy of activation beneficial for organic molecules? Explain how enzymes lower E_A.

13. How do the components and structure of cell membranes relate to the functions of membranes?

14. Sometimes inhibitors can be harmful to a cell; often they are beneficial. Explain.

Applying the Concepts

15. Explain how each of the following food preservation methods would interfere with a microbe's enzyme activity and ability to break down food: canning (heating), freezing, pickling (soaking in acetic acid), salting.

16. A biologist performed two series of experiments on lactase, the enzyme that hydrolyzes lactose to glucose and galactose. First, she made up 10% lactose solutions containing different concentrations of enzyme and measured the rate at which galactose was produced (grams of galactose per minute). Results of these experiments are shown in Table A below. In the second series of experiments (Table B), she prepared 2% enzyme solutions containing different concentrations of lactose and again measured the rate of galactose production.

Table A: Rate and Enzyme Concentration

Lactose concentration	10%	10%	10%	10%	10%
Enzyme concentration	0%	1%	2%	4%	8%
Reaction rate	0	25	50	100	200

Table B: Rate and Substrate Concentration

Lactose concentration	0%	5%	10%	20%	30%
Enzyme concentration	2%	2%	2%	2%	2%
Reaction rate	0	25	50	65	65

   a. Graph and explain the relationship between the reaction rate and the enzyme concentration.
   b. Graph and explain the relationship between the reaction rate and the substrate concentration. How and why did the results of the two experiments differ?

17. The following graph shows the rate of reaction for two different enzymes: One is pepsin, a digestive enzyme found in the stomach; the other is trypsin, a digestive enzyme found in the intestine. As you may know, the stomach secretes digestive juices containing hydrochloric acid. Which curve belongs to which enzyme?

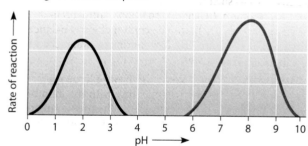

A lysosome, a digestive organelle in a cell, has an internal pH of around 4.5. Draw a curve on the graph that you would predict for a lysosomal enzyme, labeling its optimal pH.

18. Lead acts as an enzyme inhibitor and can interfere with the development of the nervous system. One manufacturer of lead-acid batteries instituted a "fetal protection policy" that banned female employees of childbearing age from working in areas where they might be exposed to high levels of lead. Women were involuntarily transferred to lower-paying jobs in lower-risk areas. A group of employees challenged the policy, claiming that it deprived women of job opportunities available to men. The U.S. Supreme Court ruled the policy illegal. What rights and responsibilities of employers, employees, and government agencies are in conflict in this situation? Who should determine what makes a safe environment?

Answers to all questions can be found in Appendix 4.

For Practice Quizzes, BioFlix, MP3 Tutors, and Activities, go to www.mybiology.com.

How Is a Marathoner Different from a Sprinter?

Athletes who participate in track competitions usually have a favorite event in which they excel. For some runners, this event may be a sprint, a short race of only 100 or 200 m. For others, it may be a race of 1,500, 5,000, or even 10,000 m. It is unusual to find a runner who competes equally well in both 100-m and 10,000-m races; runners just seem to feel more comfortable running races of particular lengths. But why?

Could it be that runners' bodies "tell" them which races are best for them? There are indications that this is indeed the case. The muscles that move our legs contain two main types of muscle fibers, called slow and fast muscle fibers. Slow muscle fibers (also called "slow-twitch" fibers) are muscle cells that can sustain repeated contractions but don't generate a lot of quick power for the body. They perform better in endurance exercises, like long-distance running, which require slow, steady muscle activity. Fast muscle fibers ("fast-twitch" fibers) are cells that can contract more quickly and powerfully than slow fibers but fatigue much more easily; they function best for short bursts of intense activity, like weight lifting or sprinting.

All human muscles contain both slow and fast fibers, but muscles differ in the percentage of each. The percentage of each fiber type in a particular muscle also varies from person to person. For example, in the quadriceps muscles of the thigh, most marathon runners have about 80% slow fibers, whereas sprinters have half as many slow fibers and about 60% fast fibers. These differences, which are genetically determined, undoubtedly help account for our differing athletic capabilities. No amount of training will turn a marathoner into a sprinter or vice versa!

What makes these two types of muscle fibers perform so differently? An important part of the answer is that they use different processes for making ATP (see Module 5.13), the molecule that supplies the energy for muscle contraction. While both types of muscle cells break down sugar (chiefly glucose) to make chemical energy available for ATP production, slow fibers do it *aerobically,* using oxygen (O_2), while fast fibers can work *anaerobically,* without oxygen.

The structures of these two kinds of muscle cells correlate with their differing functions. Slow fibers have many mitochondria, the organelles where aerobic ATP production occurs. And slow fibers contain many molecules of myoglobin, a red protein related to hemoglobin that, like hemoglobin, is a carrier of O_2 molecules. Myoglobin helps supply mitochondria with oxygen. The aerobic harvesting of energy from sugar by muscle cells (or other cells) is called **cellular respiration.** This process yields carbon dioxide (CO_2), water (H_2O), and a large amount of ATP—perfect for sustaining muscle contractions over a long period of time.

Fast muscle fibers, on the other hand, are thicker than slow ones, have fewer mitochondria, and have much less myoglobin. The thickness of fast fibers enhances their power. But they rapidly deplete their oxygen supply and switch to an energy-harvesting process that produces much less ATP per glucose molecule. Thus, they can't completely break down glucose to CO_2. Instead, they produce lactic acid, which lowers the pH in muscles, making them ache and fatigue. Fast muscle fibers are best at supplying short bursts of power. Anaerobic ATP production in our muscles is only effective for a minute or so.

Muscle cells are not the only cells in our body that break down sugars and other food molecules for ATP production. All our cells harvest chemical energy from food, as do the cells of all other organisms, eukaryotic and prokaryotic alike. Cells of most eukaryotic organisms—protists, fungi, plants, and animals— function like slow muscle fibers in that they carry out the aerobic process of cellular respiration. Before learning more about how cellular respiration powers the work of cells, let's take a step back and consider the bigger picture of energy flow through an ecosystem. In other words, where does the energy for cellular work originate? ■ ■ ■

6.1   Photosynthesis and cellular respiration provide energy for life

Life requires energy. A cell uses energy to build and maintain its structure, transport materials, manufacture products, move, grow, and reproduce. This energy ultimately comes from the sun. Through the process of photosynthesis, green plants, algae, and photosynthetic protists and bacteria convert light energy into chemical energy. Photosynthesis is the topic of Chapter 7. But a brief overview here of the relationship between cellular respiration and photosynthesis will illustrate how these two processes provide energy for life. As Figure 6.1 shows, in photosynthesis the energy of sunlight is used to rearrange the atoms of CO_2 and H_2O to produce glucose and O_2. In cellular respiration, O_2 is consumed as glucose is broken down to CO_2 and H_2O; the cell captures the energy released in ATP.

This figure also shows that in these energy conversions some energy is lost as heat. Life on Earth is solar powered, and energy makes a one-way trip through an ecosystem. Chemicals, however, are recycled. For example, the CO_2 and H_2O released by cellular respiration are converted through photosynthesis to glucose and O_2, which are then used in respiration.

Before we take a closer look at how our cells harvest energy through cellular respiration, let's consider the connection between breathing and respiration.

Web Activity *Build a Chemical Recycling System*

Figure 6.1 The connection between photosynthesis and cellular respiration

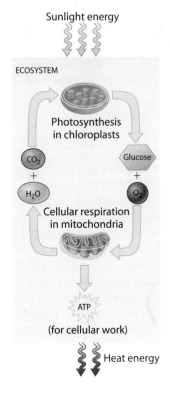

? What is misleading about the following statement? "Plants perform photosynthesis and animals perform cellular respiration."

■ The statement implies that cellular respiration does not occur in plants. In fact, almost all eukaryotic cells use cellular respiration to obtain energy.

6.2   Breathing supplies oxygen to our cells for use in cellular respiration and removes carbon dioxide

We often use the word *respiration* as a synonym for "breathing," the meaning of its Latin root. In this sense, respiration refers to an exchange of gases: An organism obtains O_2 from its environment and releases CO_2 as a waste product. Biologists also define respiration as the aerobic harvesting of energy from food molecules by cells. This process is called cellular respiration to distinguish it from breathing.

Breathing and cellular respiration are closely related. As the runner in Figure 6.2 breathes in air, her lungs take up O_2 and pass it to her bloodstream. The bloodstream carries the O_2 to her muscle cells. Mitochondria in the muscle cells use the O_2 in cellular respiration, harvesting energy from glucose and other organic molecules to generate ATP, which the cells then use to contract. The runner's bloodstream and lungs also perform the vital function of disposing of the CO_2 waste produced by cellular respiration.

? How is your breathing related to your cellular respiration?

■ In breathing, CO_2 and O_2 are exchanged between your lungs and the air. In cellular respiration, cells use O_2 to break down fuel, releasing CO_2 as a waste product.

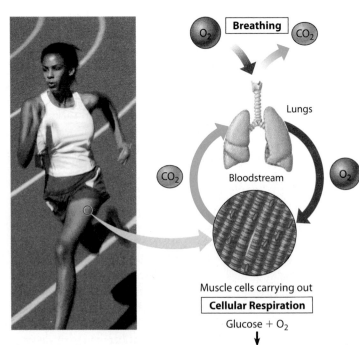

Figure 6.2 The connection between breathing and cellular respiration

6.3 Cellular respiration banks energy in ATP molecules

As the runner example in Figure 6.2 implies, oxygen usage is only a means to an end. Generating ATP for cellular work is the fundamental function of cellular respiration.

The balanced chemical equation in Figure 6.3 summarizes cellular respiration as carried out by cells that use O_2 in harvesting energy from glucose. Throughout this chapter, we use glucose as a representative food molecule, although cells also "burn" many other organic molecules in cellular respiration. The equation tells us that the atoms of the starting (reactant) molecules glucose and O_2 regroup to form the products CO_2 and H_2O. In this exergonic process, the chemical energy of the bonds in glucose is transferred and stored (or "banked") in the chemical bonds of ATP (see Modules 5.11–5.13). The series of arrows in Figure 6.3 indicates that cellular respiration consists of many steps, not just a single reaction.

Cellular respiration can produce up to 38 ATP molecules for each glucose molecule, representing about 40% of the energy in glucose. The rest of the energy is released as heat (see Module 5.11). This may seem inefficient, but it compares

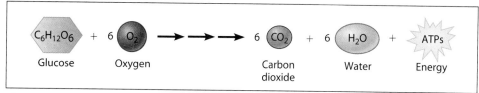

Figure 6.3 Summary equation for cellular respiration:
$C_6H_{12}O_6 + 6\ O_2 \rightarrow 6\ CO_2 + 6\ H_2O + energy$

very well with the efficiency of most energy-conversion systems. The average automobile engine is able to convert only about 25% of the energy in gasoline to the kinetic energy of movement.

How great are the energy needs of a cell? If ATP could not be regenerated through cellular respiration, humans would use up nearly their body weight in ATP each day. Let's consider the energy requirements for various human activities next.

? Why are sweating and other body-cooling mechanisms necessary during vigorous exercise?

■ The demand for ATP is supported by an increased rate of cellular respiration, but about 60% of the energy from food produces body heat instead of ATP.

Connection

6.4 The human body uses energy from ATP for all its activities

Your body requires a continuous supply of energy just to stay alive—to keep the heart pumping, to breathe, and to maintain body temperature. Your brain requires a huge amount of energy—its cells burn about 120 g (1/4 lb) of glucose a day and account for about 15% of total oxygen consumption. Maintaining brain cells and other life-sustaining activities uses as much as 75% of the energy a person takes in as food during a typical day.

Above and beyond the energy we need for body maintenance, cellular respiration provides energy for voluntary activities. Table 6.4 shows the amount of energy it takes to perform some of these activities. The energy units are **kilocalories (kcal)**, the

TABLE 6.4	ENERGY CONSUMED BY VARIOUS ACTIVITIES (IN KCAL)
Activity	**Kcal Consumed per Hour by a 67.5-kg (150-lb) Person***
Running (7 min/mi)	979
Dancing (fast)	510
Bicycling (10 mph)	490
Swimming (2 mph)	408
Walking (3 mph)	245
Dancing (slow)	204
Sitting (writing)	28

*Not including kcal needed for body maintenance

quantity of heat required to raise the temperature of 1 kilogram (kg) of water by 1°C. (The "calories" listed on food packages are actually kilocalories.) The values shown do not include the energy the body consumes for its basic life-sustaining activities. Thus, sleeping or lying quietly does not consume any energy above the energy used in maintenance.

The U.S. National Academy of Sciences estimates that the average adult human needs to take in food that provides about 2,200 kcal of energy per day. This includes the energy expended in both maintenance and voluntary activity.

Obviously, the human diet includes more than just glucose. But taking glucose as our example, how much would it take to provide that 2,200 kcal? Burning a mole* of glucose (about 180 g) releases 686 kcal of heat. Cellular respiration captures 40% of that energy in ATP, about 275 kcal. Thus, 8 moles of glucose (about 1.44 kg or 3.1 lb) is needed to fuel the production of ATP equal to 2,200 kcal.

? Walking at 3 mph, how far would you have to travel to "burn off" the equivalent of an extra slice of pizza, which has about 475 kcal? How long would that take?

■ About 6 miles; about 2 hours (Now you understand why it is said that the most effective exercise for losing weight is pushing away from the table!)

*A mole is the molecular mass of a molecule (sum of the masses of all its atoms) in grams and represents 6.02×10^{23} molecules (Avogadro's number). The molecular mass of glucose ($C_6H_{12}O_6$) is 180; thus, a mole of glucose = 180 g.

Just how do our cells extract energy from organic molecules? The energy available to a cell is contained in the arrangement of electrons in the chemical bonds that hold an organic molecule like glucose together. During cellular respiration, electrons are transferred to oxygen as the carbon-hydrogen bonds of glucose are broken and the hydrogen-oxygen bonds of water form. Oxygen very strongly attracts electrons, and an electron loses potential energy when it "falls" to oxygen. If you burn a cube of sugar, this electron fall happens very rapidly, releasing energy in the form of heat and light. Cellular respiration is a more controlled descent of electrons—more like stepping down an energy staircase, with energy released in small amounts that can be stored in the chemical bonds of ATP.

In the cellular respiration equation in Figure 6.5A, you cannot see any electron transfers. What you do see are changes in hydrogen atom distribution. Glucose loses hydrogen atoms as it is converted to carbon dioxide; simultaneously, oxygen (O_2) gains hydrogen atoms in being converted to water. These hydrogen movements represent electron transfers because, as we saw in Chapter 2, each hydrogen atom consists of an electron and a proton.

The movement of electrons from one molecule to another is an oxidation-reduction reaction, or **redox reaction** for short. In a redox reaction, the loss of electrons from one substance is called **oxidation,** and the addition of electrons to another substance is called **reduction.** A molecule is said to become oxidized when it loses one or more electrons and reduced when it gains one or more electrons. Because an electron transfer requires both a donor and an acceptor, oxidation and reduction always go together. You can see the overall results of the redox reactions of cellular respiration in Figure 6.5A: Glucose loses electrons (in H atoms) and becomes oxidized, while O_2 gains electrons (in H atoms) and becomes reduced. The electrons lose potential energy along the way, and energy is released.

Two key players in the process of oxidizing glucose are an enzyme called **dehydrogenase** and a coenzyme called NAD^+. **NAD^+** (nicotinamide adenine dinucleotide) is an organic molecule that cells make from the vitamin niacin and use to shuttle electrons in redox reactions. The top equation in Figure 6.5B at the top of the next column depicts the oxidation of an organic molecule. We show only its three carbons (⬤) and a few of its other atoms. Dehydrogenase strips two hydrogen atoms from this molecule. Simultaneously, as shown in the lower equation, NAD^+ picks up the two electrons (one of which neutralizes its positive charge) and becomes reduced to NADH. One proton

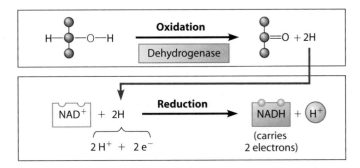

Figure 6.5B   A pair of redox reactions, occuring simultaneously

(H^+) is released. (Reduced NADH is represented throughout this chapter as a light brown box carrying two blue electrons.)

Using the energy staircase analogy of electrons falling from glucose to oxygen, the transfer of electrons from an organic molecule to NADH represents the first step. Figure 6.5C shows NADH delivering these electrons to the rest of the staircase—an **electron transport chain.** The steps in the chain are electron carrier molecules, shown here as purple ovals, built into the inner membrane of a mitochondrion. At the bottom of the staircase is O_2, the final electron acceptor.

The electron transport chain involves a series of redox reactions in which electrons pass from carrier to carrier down to oxygen. The redox steps in the staircase release energy in amounts small enough to be used by the cell to make ATP.

With an understanding of this basic mechanism of electron transfer and energy release, we can now explore cellular respiration in more detail.

?   What chemical characteristic of the element oxygen accounts for its function in cellular respiration?

■ Oxygen is very electronegative (see Module 2.9), meaning that it is very powerful in pulling electrons from other elements.

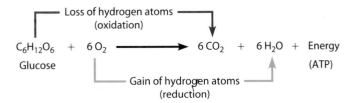

Figure 6.5A   Rearrangement of hydrogen atoms (with their electrons) in the redox reactions of cellular respiration

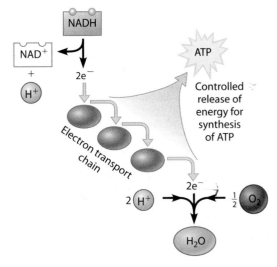

Figure 6.5C   In cellular respiration, electrons fall down an energy staircase and finally reduce O_2.

6.6 Overview: Cellular respiration occurs in three main stages

Cellular respiration consists of a sequence of steps that can be divided into three main stages. Figure 6.6 below gives an overview of the three stages and shows where they occur in a eukaryotic cell. (In prokaryotic cells that use aerobic respiration, these steps occur in the cytoplasm, and the electron transport chain is built into the plasma membrane.)

Stage 1: Glycolysis (shown with an aqua background throughout) occurs in the cytoplasmic fluid of the cell, that is, outside the organelles. Glycolysis begins respiration by breaking glucose into two molecules of a three-carbon compound called pyruvate.

Stage 2: The **citric acid cycle** (shown in this chapter with a light salmon color) takes place within the mitochondria. It completes the breakdown of glucose by decomposing a derivative of pyruvate to carbon dioxide. As suggested by the smaller ATP symbols in the diagram, the cell makes a small amount of ATP during these first two stages. The main function of glycolysis and the citric acid cycle, however, is to supply the third stage of respiration with electrons (shown with gold arrows).

Stage 3: Oxidative phosphorylation (a purple background) involves the electron transport chain (see Module 6.5) and a process known as chemiosmosis. NADH and a related electron carrier, $FADH_2$ (flavin adenine dinucleotide), shuttle electrons to the electron transport chain embedded in the inner mitochondrion membrane. Most of the ATP produced by cellular respiration is generated by oxidative phosphorylation, which uses the energy released by the downhill fall of electrons from NADH and $FADH_2$ to O_2 to phosphorylate ADP. (Recall from

Module 5.13 that cells generate ATP by adding a phosphate group to ADP.)

What couples the electron transport chain to ATP synthesis? As the electron transport chain passes electrons down the energy staircase, it also pumps hydrogen ions (H^+) across the inner membrane into the narrow intermembrane space. The result is a concentration gradient of H^+ across the membrane. In **chemiosmosis,** the potential energy of this concentration gradient is used to make ATP. The concentration gradient drives the diffusion of H^+ through **ATP synthases,** protein complexes built into the inner membrane that synthesize ATP. The details of this process are explored in Module 6.10. In 1978, British biochemist Peter Mitchell was awarded the Nobel Prize for developing the theory of chemiosmosis.

The small amount of ATP produced in glycolysis and the citric acid cycle is made by substrate-level phosphorylation, a process we discuss in more detail in the next module. In the next several modules, we look more closely at the three stages of cellular respiration and the two mechanisms of ATP synthesis.

(Bio Flix) **BioFlix** *Cellular Respiration*

Web Activity *Overview of Cellular Respiration*

? Of the three main stages of cellular respiration represented in Figure 6.6, which one uses oxygen to extract chemical energy from organic compounds?

■ Oxidative phosphorylation, using the electron transport chain, which eventually transfers electrons to oxygen

Figure 6.6
An overview of cellular respiration

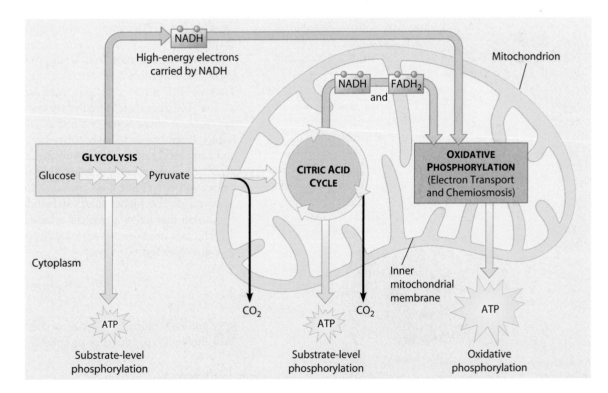

6.7 Glycolysis harvests chemical energy by oxidizing glucose to pyruvate

Now that we have introduced the major players and processes, it's time to focus on the individual stages of cellular respiration. The term for the first stage, *glycolysis*, means "splitting of sugar" (*glyco,* sweet, and *lysis,* split), and that's exactly what happens during this phase.

Figure 6.7A below gives an overview of glycolysis in terms of input and output. Glycolysis begins with a single molecule of glucose and concludes with two molecules of pyruvate. (Pyruvate is the ionized form of pyruvic acid.) The gray balls represent the carbon atoms in each molecule; glucose has six, and these same six end up in the two molecules of pyruvate (three in each). The straight arrow from glucose to pyruvate represents nine chemical steps, each catalyzed by its own enzyme. As these reactions occur, the cell reduces two molecules of NAD^+, forming two molecules of NADH, and produces two molecules of ATP by substrate-level phosphorylation.

In **substrate-level phosphorylation** (see Figure 6.7B, next column) an enzyme transfers a phosphate group from a substrate molecule directly to ADP, forming ATP. This process produces a small amount of ATP in both glycolysis and the citric acid cycle.

Thus, the energy extracted from glucose during glycolysis is banked in a combination of ATP and NADH. The cell can use the energy in ATP immediately, but for it to use the energy in NADH, electrons from NADH must pass down the electron transport chain located in the inner mitochondrial membrane. And the pyruvate still holds most of the energy of glucose; these molecules will be oxidized in the citric acid cycle.

Let's take a closer look at glycolysis. Figure 6.7C (next page) shows all the organic compounds that form in the nine chemical reactions of glycolysis. Commentary on the left highlights the main features of the reactions. The gray balls represent the carbon atoms in each of the compounds named on the right.

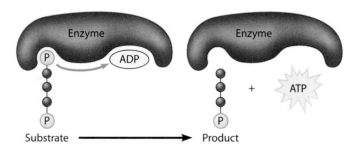

Figure 6.7B Substrate-level phosphorylation: transfer of a phosphate group Ⓟ from a substrate to ADP, producing ATP

The compounds that form between the initial reactant, glucose, and the final product, pyruvate, are known as **intermediates.** Glycolysis is an example of a metabolic pathway in which each chemical step leads to the next one. For instance, the intermediate glucose-6-phosphate is the product of step 1 and the reactant for step 2. Similarly, fructose-6-phosphate is the product of step 2 and the reactant for step 3. Also essential are the specific enzymes that catalyze each of the chemical steps; however, to keep the figures simple, we have not included enzymes in the diagrams.

As indicated in Figure 6.7C, the steps of glycolysis can be grouped into two main phases. Steps ❶–❹, the energy investment phase, actually *consume* energy. In this phase, ATP is used to energize a glucose molecule, which is then split into two small sugars that are now primed to release energy. The figure follows both of these three-carbon sugars through the second phase.

Steps ❺–❾, the energy payoff phase, *yield* energy for the cell. In this phase, two NADH molecules are produced for each initial glucose molecule, and four ATP molecules are generated. Since the first phase uses two molecules of ATP, *the net gain to the cell is two ATP molecules for each glucose molecule that enters glycolysis.*

These two ATP molecules from glycolysis account for only 5% of the energy that a cell can harvest from a glucose molecule. The two NADH molecules generated during step 5 account for another 16%, but their stored energy is not available for use in the sabsence of O_2. Some organisms—yeasts and certain bacteria, for instance—can satisfy their energy needs with the ATP produced by glycolysis alone. And the fast-twitch muscle fibers described in the chapter introduction may use this anaerobic production of ATP for short periods. Most cells and organisms, however, have far higher energy demands. The stages of cellular respiration that follow glycolysis release much more energy. In the next two modules, we see what happens in most organisms after glycolysis forms pyruvate.

🎧 MP3 Tutor *Cellular Respiration Part 1: Glycolysis*

Web Activity *Glycolysis*

? For each glucose molecule processed, what are the net molecular products of glycolysis?

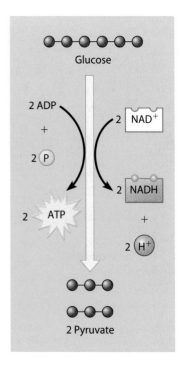

Figure 6.7A
An overview
of glycolysis

Glucose

2 ADP
+
2 Ⓟ

2 NAD^+

2 ATP

2 NADH
+
2 H^+

2 Pyruvate

■ Two molecules of pyruvate, two molecules of ATP, and two molecules of NADH

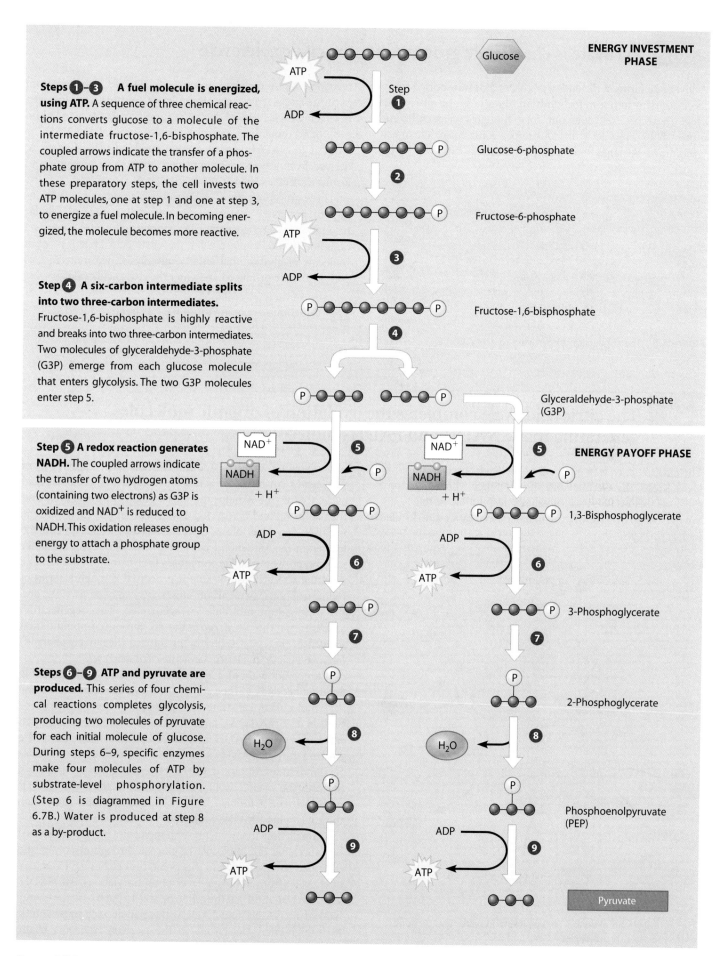

Steps ①–③ A fuel molecule is energized, using ATP. A sequence of three chemical reactions converts glucose to a molecule of the intermediate fructose-1,6-bisphosphate. The coupled arrows indicate the transfer of a phosphate group from ATP to another molecule. In these preparatory steps, the cell invests two ATP molecules, one at step 1 and one at step 3, to energize a fuel molecule. In becoming energized, the molecule becomes more reactive.

Step ④ A six-carbon intermediate splits into two three-carbon intermediates. Fructose-1,6-bisphosphate is highly reactive and breaks into two three-carbon intermediates. Two molecules of glyceraldehyde-3-phosphate (G3P) emerge from each glucose molecule that enters glycolysis. The two G3P molecules enter step 5.

Step ⑤ A redox reaction generates NADH. The coupled arrows indicate the transfer of two hydrogen atoms (containing two electrons) as G3P is oxidized and NAD^+ is reduced to NADH. This oxidation releases enough energy to attach a phosphate group to the substrate.

Steps ⑥–⑨ ATP and pyruvate are produced. This series of four chemical reactions completes glycolysis, producing two molecules of pyruvate for each initial molecule of glucose. During steps 6–9, specific enzymes make four molecules of ATP by substrate-level phosphorylation. (Step 6 is diagrammed in Figure 6.7B.) Water is produced at step 8 as a by-product.

ENERGY INVESTMENT PHASE

Glucose

Step ①

Glucose-6-phosphate

②

Fructose-6-phosphate

③

Fructose-1,6-bisphosphate

④

Glyceraldehyde-3-phosphate (G3P)

ENERGY PAYOFF PHASE

⑤ NAD^+ NADH + H^+

1,3-Bisphosphoglycerate

⑥   ADP   ATP

3-Phosphoglycerate

⑦

2-Phosphoglycerate

⑧ H_2O

Phosphoenolpyruvate (PEP)

⑨   ADP   ATP

Pyruvate

Figure 6.7C Details of glycolysis

Chapter 6 How Cells Harvest Chemical Energy

Pyruvate is chemically groomed for the citric acid cycle

As pyruvate forms at the end of glycolysis, it is transported from the cytoplasm into a mitochondrion, the site of the citric acid cycle. Pyruvate itself does not enter the citric acid cycle. As shown in Figure 6.8, it first undergoes some major chemical "grooming." A large, multienzyme complex catalyzes three

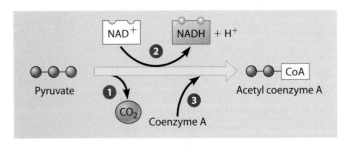

Figure 6.8   The conversion of pyruvate to acetyl CoA

reactions: ❶ A carboxyl group ($-COO^-$) is removed from pyruvate and given off as a molecule of CO_2 (this is the first step in which CO_2 is released during respiration); ❷ the two-carbon compound remaining is oxidized while a molecule of NAD^+ is reduced to NADH; and ❸ a compound called coenzyme A, derived from a B vitamin, joins with the two-carbon group to form a molecule called acetyl coenzyme A.

These grooming steps—a chemical "haircut and conditioning" of pyruvate—set up the second major stage of cellular respiration. Acetyl coenzyme A, abbreviated **acetyl CoA,** is a high-energy fuel molecule for the citric acid cycle. For each molecule of glucose that entered glycolysis, two molecules of acetyl CoA are produced and enter the citric acid cycle.

? Which molecule in Figure 6.8 has been reduced?

■ NAD^+ is reduced to NADH.

The citric acid cycle completes the oxidation of organic molecules, generating many NADH and $FADH_2$ molecules

The citric acid cycle is often called the Krebs cycle in honor of Hans Krebs, the German-British researcher who worked out much of this cyclic phase of cellular respiration in the 1930s. We present an overview figure first, followed by a more detailed look at this cycle.

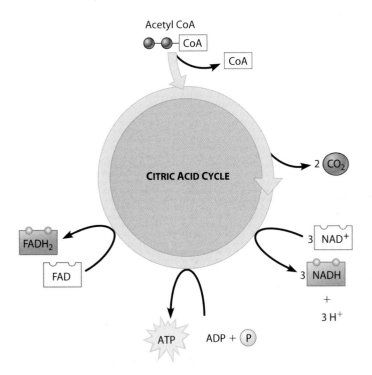

Figure 6.9A   An overview of the citric acid cycle: Two carbons enter the cycle through acetyl CoA, and 2 CO_2, 3 NADH, 1 $FADH_2$, and 1 ATP exit the cycle.

As shown in Figure 6.9A, only the two-carbon acetyl part of the acetyl CoA molecule actually participates in the citric acid cycle. Coenzyme A helps the acetyl group enter the cycle and then splits off and is recycled. Not shown in this figure are the multiple steps that follow, each catalyzed by a specific enzyme located in the mitochondrial matrix or embedded in the inner membrane. The acetyl group joins a four-carbon molecule. The resulting six-carbon molecule is processed through a series of redox reactions, two carbon atoms are removed as CO_2, and the four-carbon molecule is regenerated. This regeneration accounts for the word *cycle*; the six-carbon compound first formed in the cycle is citrate, the ionized (negatively charged) form of citric acid. Hence the name *citric acid cycle*.

Compared with glycolysis, the citric acid cycle pays big energy dividends to the cell. Each turn of the cycle makes one ATP molecule by substrate-level phosphorylation (shown at the bottom of Figure 6.9A). It also produces four other energy-rich molecules: three NADH molecules and one molecule of the electron carrier, $FADH_2$. Since the citric acid cycle processes two molecules of acetyl CoA for each initial molecule of glucose, the overall yield per molecule of glucose is 2 ATP, 6 NADH, and 2 $FADH_2$. This yield is considerably more than the 2 ATP plus 2 NADH produced by glycolysis alone.

Overall, how many energy-rich molecules has the cell gained by processing one molecule of glucose through glycolysis and the citric acid cycle? Up to this point, the cell has gained a total of 4 ATP (from substrate-level phosphorylation), 10 NADH, and 2 $FADH_2$. Still, for the cell to be able to put to use the energy banked in NADH and $FADH_2$, these molecules must shuttle their high-energy electrons to the electron transport chain. There the energy from the oxidation of organic molecules can be used for the oxidative phosphorylation of ADP to ATP. Before we look at how oxidative phosphorylation works, you may want to examine the inner workings of the citric acid cycle in Figure 6.9B, below.

MP3 Tutor *Cellular Respiration Part 2: Citric Acid Cycle and Electron Transport Chain*

Web Activity *The Citric Acid (Krebs) Cycle*

? What is the total number of NADH molecules generated during the complete breakdown of one glucose molecule to six carbon dioxide molecules? (*Hint:* Combine the outputs of Modules 6.7–6.9.)

■ 10 NADH: 2 from glycolysis; 2 from the grooming of pyruvate; 6 from the citric acid cycle. (Did you remember to double the output due to the sugar-splitting step of glycolysis?)

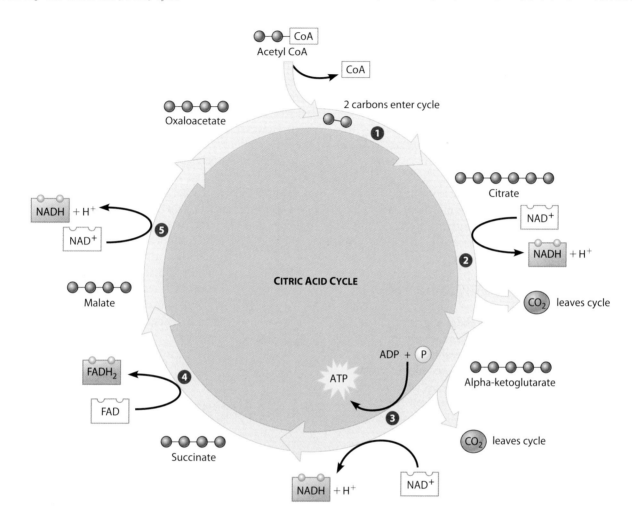

Step ❶
Acetyl CoA stokes the furnace.

A turn of the citric acid cycle begins (top center) as enzymes strip the CoA portion from acetyl CoA and combine the remaining two-carbon acetyl group with the four-carbon molecule oxaloacetate (top left) already present in the mitochondrion. The product of this reaction is the six-carbon molecule citrate. Citrate is the ionized form of citric acid. All the acid compounds in this cycle exist in the cell in their ionized form, hence the suffix -ate.

Steps ❷ – ❸
NADH, ATP, and CO_2 are generated during redox reactions.

Successive redox reactions harvest some of the energy of the acetyl group by stripping hydrogen atoms from organic acid intermediates (such as alpha-ketoglutarate) and producing energy-laden NADH molecules. In two places, an intermediate compound loses a CO_2 molecule. Energy is harvested by substrate-level phosphorylation of ADP to produce ATP. A four-carbon compound called succinate emerges at the end of step 3.

Steps ❹ – ❺
Redox reactions generate $FADH_2$ and NADH.

Enzymes rearrange chemical bonds, eventually completing the citric acid cycle by regenerating oxaloacetate. Redox reactions reduce the electron carriers FAD and NAD^+ to $FADH_2$ and NADH, respectively. One turn of the citric acid cycle is completed with the conversion of a molecule of malate to oxaloacetate. This compound is then ready to start the next turn of the cycle by accepting another acetyl group from acetyl CoA.

Figure 6.9B   Details of the citric acid cycle

6.10 Most ATP production occurs by oxidative phosphorylation

The final stage of cellular respiration is oxidative phosphorylation, which involves the electron transport chain and chemiosmosis. Oxidative phosphorylation is a clear illustration of structure fitting function: The spatial arrangement of electron carriers built into a membrane makes it possible for the mitochondrion to use the chemical energy released by redox reactions to create a hydrogen ion (H^+) gradient and then, through the process called chemiosmosis, use the energy stored in that gradient to drive ATP synthesis.

Figure 6.10 expands on our earlier discussion of oxidative phosphorylation and chemiosmosis in Module 6.6. It shows that the electron transport chain is built into the inner membrane of the mitochondrion. The folds (cristae) of this membrane enlarge its surface area, providing space for thousands of copies of the electron transport chain and many ATP synthase complexes.

Starting on the left in Figure 6.10, the gold arrow traces the path of electron flow from the shuttle molecules NADH and $FADH_2$ through the electron transport chain to oxygen, the final electron acceptor. This is the critical role that oxygen plays in cellular respiration. Each oxygen atom ($\frac{1}{2}O_2$) accepts two electrons from the chain and picks up two hydrogen ions from the surrounding solution to form H_2O, one of the final products of cellular respiration.

Most of the carrier molecules reside in four main protein complexes, while two mobile carriers transport electrons between the complexes. All of the carriers bind and release electrons in redox reactions, passing electrons down the "energy staircase." Three of the protein complexes use the energy released from these electron transfers to actively transport H^+ across the membrane, from where H^+ is less concentrated to

where it is more concentrated. The green vertical arrows indicate that the hydrogen ions are transported from the matrix of the mitochondrion (its innermost compartment) into the mitochondrion's narrow intermembrane space.

The resulting H^+ gradient stores potential energy, much the way a dam stores energy by holding back the elevated water behind it. The energy stored by a dam can be harnessed to do work (such as generating electricity) when the water is allowed to rush downhill, turning giant wheels called turbines. Similarly, the ATP synthases built into the inner mitochondrial membrane act like miniature turbines. The hydrogen ions tend to be driven across the membrane by the energy of their concentration gradient. However, the membrane is not permeable to hydrogen ions, and they can only cross through a channel in the ATP synthase, as shown on the far right of the figure. Hydrogen ions rush back "downhill" through an ATP synthase, spinning a component of the complex, just as water turns the turbines in a dam. The rotation activates catalytic sites in the synthase that attach phosphate groups to ADP molecules to generate ATP.

So why is this process called oxidative phosphorylation? The energy derived from the *oxidation*-reduction reactions of the electron transport chain is used to *phosphorylate* ADP. The exergonic reactions of electron transport produce an H^+ gradient. Through chemiosmosis, the energy stored in this H^+ gradient drives the endergonic synthesis of ATP.

Web Activity *Electron Transport and Chemiosmosis*

? What effect would an absence of oxygen (O_2) have on the process illustrated in Figure 6.10?

■ There would be no ATP produced. Without oxygen to "pull" electrons down the electron transport chain, the energy stored in NADH cannot be harnessed for ATP synthesis.

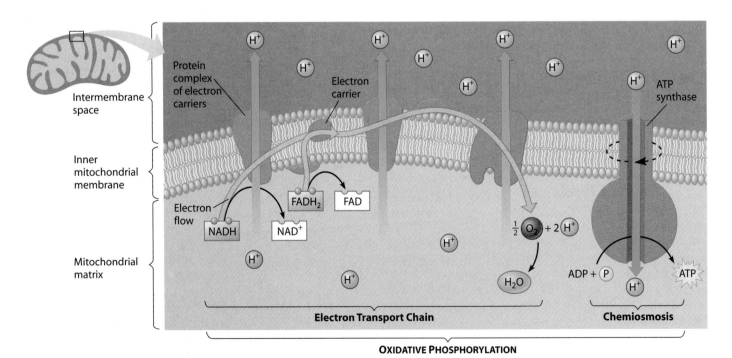

Figure 6.10 Oxidative phosphorylation, using electron transport and chemiosmosis in the mitochondrion

6.11 Certain poisons interrupt critical events in cellular respiration

A number of poisons produce their deadly effects by interfering with some of the events we have just discussed. Figure 6.11 shows the places where three different categories of poisons obstruct cellular respiration.

Poisons in one category block the electron transport chain. A substance called rotenone, for instance, binds tightly with one of the electron carrier molecules in the first protein complex, preventing electrons from passing to the next carrier molecule. Rotenone is often used to kill pest insects and fish. By blocking the electron transport chain near its start and thus preventing ATP synthesis, rotenone literally starves an organism's cells of energy.

Two other electron transport blockers, cyanide and carbon monoxide, bind with an electron carrier in the fourth protein complex. Here they block the passage of electrons to oxygen. This blockage is like turning off a faucet; electrons cease to flow through the "pipe." The result is the same as that of rotenone: No H$^+$ gradient is generated, and no ATP is made. Cyanide was the lethal agent in the Tylenol murders of 1982. Seven people in the Chicago area died after ingesting Tylenol capsules that had been laced with cyanide. The perpetrator of that crime was never caught.

A second kind of respiratory poison inhibits ATP synthase. On the right side of the figure, the antibiotic oligomycin blocks the passage of H$^+$ through the channel in ATP synthase. Oligomycin is used on the skin to combat fungal infections. It kills fungal cells by preventing them from using the potential energy of the H$^+$ gradient to make ATP. (Because the drug cannot get past the outer layer of dead cells into the living skin cells, they are protected from its effects.)

A third kind of poison, collectively called uncouplers, makes the membrane of the mitochondrion leaky to hydrogen ions. Electron transport continues, but ATP cannot be made because leakage of H$^+$ through the membrane abolishes the H$^+$ gradient. Cells continue to consume oxygen, often at a higher than normal rate, but to no avail, for they cannot make any ATP through chemiosmosis because no H$^+$ gradient exists.

A highly toxic uncoupler called dinitrophenol (DNP) is shown in the figure. DNP poisoning produces an enormous increase in metabolic rate, profuse sweating as the body attempts to dissipate excess heat, collapse, and then death. For a short time in the 1940s, some physicians prescribed DNP in

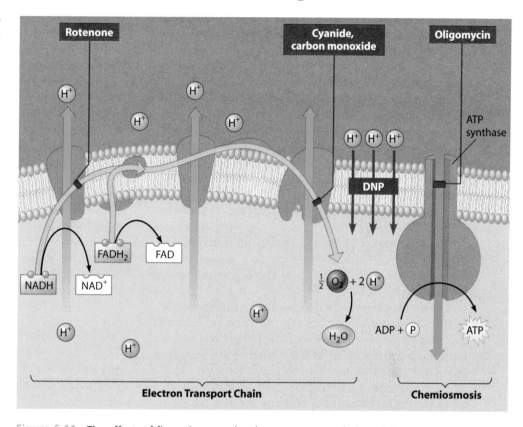

Figure 6.11 The effects of five poisons on the electron transport chain and chemiosmosis

low doses as weight loss pills, but fatalities soon made it clear that there were far safer ways to lose weight. When DNP is present, all steps of cellular respiration except chemiosmosis continue to run, consuming fuel molecules, even though almost all the energy is lost as heat.

Poisons do serve some productive purposes. In addition to being useful as pesticides or antibiotics, toxins may be used in biochemical research. Discovering exactly what these substances do to the cell's respiratory machinery has, in many cases, helped biochemists understand how that machinery works. The effects of uncouplers, for example, made it clear that ATP synthesis is a complicated activity involving the distinct, but related, processes of electron transport and generation of a membrane H$^+$ gradient.

The function of cellular respiration is to generate ATP for cellular work. Without energy, a cell cannot live. We review cellular respiration and the essential production of ATP in the next module.

? The poison DNP causes what one biochemist calls "mitochondrial wheel-spinning." Explain this metaphor.

■ Like an automobile wasting fuel by spinning its wheels and going nowhere, a mitochondrion poisoned with DNP consumes fuel and powers electron transport but makes no ATP. The DNP destroys the H$^+$ gradient required for the chemiosmotic synthesis of ATP.

6.12 Review: Each molecule of glucose yields many molecules of ATP

Now that we have looked at all stages of cellular respiration, let's review what the cell accomplishes by oxidizing a molecule of glucose. Figure 6.12 shows all the stages and indicates where they occur in a eukaryotic cell. At the bottom of the figure is a tally of ATP molecules, showing the potential energy payoff for a typical cell. If you wish to refer back to earlier modules, this diagram summarizes glycolysis (Module 6.7), the chemical grooming of pyruvate (Module 6.8), the citric acid cycle (Module 6.9), and oxidative phosphorylation (Module 6.10).

Starting on the left, glycolysis, occurring in the cytoplasmic fluid, and the citric acid cycle, occurring in the mitochondrial matrix, contribute a net total of 4 ATP per glucose molecule by substrate-level phosphorylation. The cell harvests much more energy than this from the carrier molecules NADH and $FADH_2$, which are produced by glycolysis, the grooming of pyruvate, and the citric acid cycle. The energy of the electrons they carry is used to make an estimated 34 molecules of ATP using the electron transport chain and chemiosmosis in oxidative phosphorylation.

Let's see where the numbers in the diagram come from. Our model assumes that each NADH that transfers a pair of high-energy electrons from a food molecule to the electron transport chain contributes enough to the mitochondrion's H^+ gradient to generate 3 ATP. Another assumption is that each $FADH_2$ molecule yields only 2 ATP because it contributes its electrons later in the electron transport chain (see Figure 6.10). These numbers are maximums; the actual amounts may vary. For instance, some of the energy of the H^+ gradient may be used for transport work instead of ATP production. Also, as shown in the diagram, a shuttle mechanism passes the electrons from NADH produced in glycolysis into the mitochondrion. Depending on the type of shuttle, either NAD^+ or FAD picks up the electrons (so the yield of oxidative phosphorylation may be 32 or 34 ATP). The total net yield of ATP molecules per glucose molecule has a theoretical maximum of about 38.

More important than the actual numbers of ATP molecules is the point that a cell can harvest a great deal of energy from glucose—up to about 40% of the molecule's potential energy. Because most of the ATP generated by cellular respiration results from oxidative phosphorylation, the ATP yield depends heavily on an adequate supply of oxygen to the cell. Without oxygen to function as the final electron acceptor in the electron transport chain, chemiosmosis ceases, and cells die from energy starvation. However, as we discussed in the chapter introduction, muscle cells may continue to function for a while without oxygen. We look at how some cells and organisms can oxidize organic fuel and generate ATP *without* oxygen next.

Web Process of Science *How Is the Rate of Cellular Respiration Measured?*

? What would a cell's net ATP yield per glucose be in the presence of the poison DNP? (See Module 6.11.)

■ 4 ATP, all from substrate-level phosphorylation. The uncoupler would destroy the H^+ concentration gradient necessary for chemiosmosis.

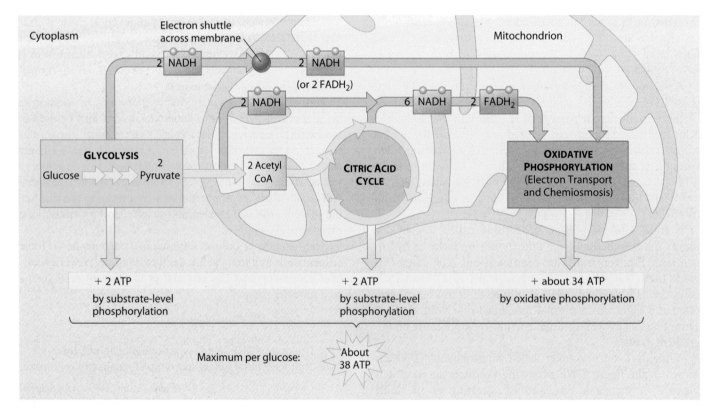

Figure 6.12 An estimated tally of the ATP produced by substrate-level and oxidative phosphorylation in cellular respiration

6.13 Fermentation enables cells to produce ATP without oxygen

The metabolic pathway that generates ATP during fermentation is glycolysis, the same pathway that functions in the first stage of cellular respiration. Remember that glycolysis uses no O_2; it simply generates a net gain of 2 ATP while oxidizing glucose to two molecules of pyruvate and reducing NAD^+ to NADH. The yield of 2 ATP is certainly a lot less than the possible 38 ATP per glucose generated during aerobic respiration, but it is enough to keep your muscles contracting for a short while when the need for ATP outpaces the delivery of O_2 via the bloodstream. And many microorganisms supply all their energy needs with the 2 ATP/glucose yield of glycolysis.

There is more to fermentation, however, than just glycolysis. To oxidize glucose in glycolysis, NAD^+ must be present as an electron acceptor. This is no problem under aerobic conditions, because the cell regenerates its pool of NAD^+ when NADH passes its electrons into the mitochondrion to be transported to the electron transport chain. Fermentation provides an anaerobic path for recycling NADH back to NAD^+.

Lactic Acid Fermentation Your muscle cells, a few other cell types, and certain bacteria can regenerate NAD^+ by a process called **lactic acid fermentation,** illustrated in Figure 6.13A. You can see that NADH is oxidized to NAD^+ as pyruvate is reduced to lactate (the ionized form of lactic acid). The lactate that builds up in muscle cells during strenuous exercise is carried in the blood to the liver, where it is converted back to pyruvate. (Panting after heavy exercise provides extra O_2 that repays the "oxygen debt," returning muscles to aerobic respiration and converting lactate back to pyruvate.)

The dairy industry uses lactic acid fermentation by bacteria to make cheese and yogurt. Other types of microbial fermentation turn soybeans into soy sauce and cabbage into sauerkraut.

Alcohol Fermentation For thousands of years people have used **alcohol fermentation** in brewing, winemaking, and baking. Yeasts are single-celled fungi that normally use aerobic respiration to process their food. But they are also able to survive in anaerobic environments. Yeasts and certain bacteria recycle their NADH back to NAD^+ while converting pyruvate to CO_2 and ethanol (Figure 6.13B). The CO_2 provides the bubbles in beer and champagne. Bubbles of CO_2 generated by baker's yeast cause bread dough to rise. Ethanol (ethyl alcohol), the two-carbon product, is toxic to the organisms that produce it. Yeasts release their alcohol wastes to their surroundings, where it usually diffuses away. When yeasts are confined in a wine vat, they die when the alcohol concentration reaches 14%.

Unlike muscle cells and yeasts, many prokaryotes that live in stagnant ponds and deep in the soil are **obligate anaerobes,** meaning they require anaerobic conditions and are poisoned by oxygen. Yeasts and many other bacteria are facultative anaerobes. A **facultative anaerobe** can make ATP either by fermentation or by oxidative phosphorylation, depending on whether O_2 is available. On the cellular level, our muscle cells behave as facultative anaerobes.

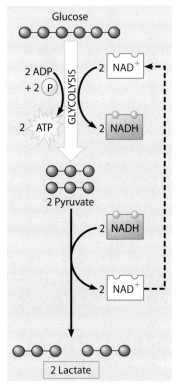

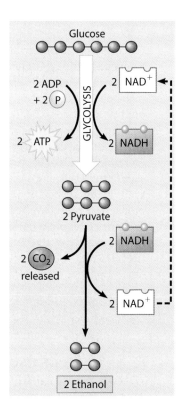

Figure 6.13A Lactic acid fermentation oxidizes NADH to NAD^+ and produces lactate.

Figure 6.13B Alcohol fermentation oxidizes NADH to NAD^+ and produces ethanol and CO_2.

Figure 6.13C Fermentation vats for wine

For a facultative anaerobe, pyruvate is a fork in the metabolic road. If oxygen is available, the organism will always use the more productive aerobic respiration. Thus, to make wine and beer, yeasts must be grown anaerobically so that they will ferment sugars and produce ethanol. For this reason, the large fermentation vats in Figure 6.13C are equipped with one-way gas valves that vent off excess CO_2 but keep air out.

Web Activity *Fermentation*

? A glucose-fed yeast cell is moved from an aerobic environment to an anaerobic one. For the cell to continue generating ATP at the same rate, how would its rate of glucose consumption need to change?

■ The cell must consume glucose at a rate about 19 times the consumption rate in the aerobic environment (2 ATP by fermentation versus 38 ATP by cellular respiration).

6.14   Glycolysis evolved early in the history of life on Earth

Glycolysis is the universal energy-harvesting process of life. If we looked inside a bacterial cell, inside one of our own body cells, or inside virtually any other living cell, we would find the metabolic machinery of glycolysis. The role of glycolysis in both fermentation and respiration has an evolutionary basis. Ancient prokaryotes probably used glycolysis to make ATP long before oxygen was present in Earth's atmosphere. The oldest known fossils of bacteria date back over 3.5 billion years, but significant levels of O_2 did not accumulate in the atmosphere until about 2.7 billion years ago. For almost a billion years, prokaryotes must have generated ATP exclusively from glycolysis, which does not require oxygen.

The fact that glycolysis occurs in almost all organisms suggests that it evolved very early in ancestors common to all the domains of life. The location of glycolysis within the cell also implies great antiquity; the pathway does not require any of the membrane-bounded organelles of the eukaryotic cell, which evolved more than a billion years after the prokaryotic cell. Glycolysis is a metabolic heirloom from early cells that continues to function in fermentation and as the first stage in the breakdown of organic molecules by cellular respiration.

? List some of the characteristics of glycolysis that indicate that it is an ancient metabolic system.

■ Glycolysis occurs universally (functioning in both fermentation and respiration), does not require oxygen, and does not occur in a membrane-bounded organelle.

Interconnections Between Molecular Breakdown and Synthesis

6.15   Cells use many kinds of organic molecules as fuel for cellular respiration

Throughout this chapter, we have spoken of glucose as the fuel for cellular respiration. But free glucose molecules are not common in our diet. We obtain most of our calories as fats, proteins, sucrose and other disaccharide sugars, and starch (a polysaccharide). You consume all these types of molecules when you eat a bag of peanuts, for instance.

Figure 6.15 illustrates how a cell uses three kinds of molecules to make ATP. A wide range of carbohydrates (polysaccharides and sugars) can be funneled into glycolysis, as shown by the blue arrows in the diagram. For example, enzymes in our digestive tract hydrolyze starch to glucose, which is then broken down by glycolysis and the citric acid cycle. Similarly, glycogen, the polysaccharide stored in our liver and muscle cells, can be hydrolyzed to glucose to serve as fuel between meals.

Proteins (purple arrows) can be used for fuel, but first they must be digested to their constituent amino acids. Typically, a cell will use most of the amino acids to make its own proteins. Enzymes will convert excess amino acids to intermediates of glycolysis or the citric acid cycle, and their energy is then harvested by cellular respiration. During the conversion, the amino groups are stripped off and later disposed of in urine.

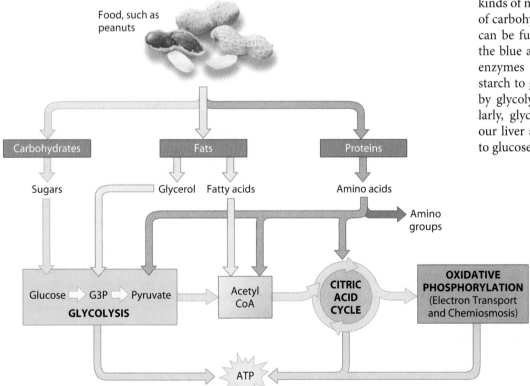

Figure 6.15   Pathways that break down various food molecules

Fats make excellent cellular fuel because they contain many hydrogen atoms and thus many energy-rich electrons. As the diagram on the previous page shows (tan arrows), the cell first hydrolyzes fats to glycerol and fatty acids. It then converts the glycerol to glyceraldehyde-3-phosphate (G3P), one of the intermediates in glycolysis. The fatty acids are broken into two-carbon fragments that enter the citric acid cycle as acetyl CoA. Processed this way, a gram of fat yields more than twice as much ATP as a gram of starch. Because so many calories are stockpiled in each gram of fat, a person must expend a large amount of energy to burn fat stored in the body. This explains why it is so difficult for a dieter to lose excess fat.

? Animals store most of their energy reserves as fats, not as polysaccharides. What is the advantage of this mode of storage for an animal?

■ Most animals are mobile and benefit from a compact and concentrated form of energy storage. Because fats are hydrophobic, they can be stored without extra water associated with them, as is the case with hydrophilic carbohydrates (see Module 3.8).

6.16 Food molecules provide raw materials for biosynthesis

Not all food molecules are destined to be oxidized as fuel for making ATP. Food also provides the raw materials a cell uses for biosynthesis—the production of organic molecules using energy-requiring metabolic pathways. A cell must be able to make its own molecules to build its structures and perform its functions. Some raw materials, such as amino acids, can be incorporated directly into an organism's macromolecules. However, cells also make molecules that are not present in food by using some of the intermediate compounds of glycolysis and the citric acid cycle.

Figure 6.16 outlines the pathways by which cells make three classes of organic molecules using some of the intermediate molecules of glycolysis and the citric acid cycle. By comparing Figures 6.15 and 6.16, we see clear connections between the energy-harvesting process of cellular respiration and the biosynthetic pathways used to construct all parts of the cell.

Basic principles of supply and demand regulate these pathways. If there is an excess of a certain amino acid, for example, the pathway that synthesizes it is switched off. The most common mechanism for this control is feedback inhibition: The end product inhibits an enzyme that catalyzes an early step in the pathway (see Module 5.16). Feedback inhibition also controls cellular respiration. If ATP accumulates in a cell, it inhibits an early enzyme in glycolysis, slowing down respiration and conserving resources. On the other hand, the same enzyme is activated by a buildup of ADP in the cell, signaling the need for more energy.

The cells of all living organisms—including those of the panda shown in Figure 6.16 and the bamboo plants it eats—have the ability to harvest energy from the breakdown of organic molecules. When the process is cellular respiration, the atoms of the starting materials end up in CO_2 and H_2O. In contrast, the ability to make organic molecules from CO_2 and H_2O is not universal. Animal cells lack this ability, but plant cells can actually produce organic molecules from inorganic ones using the energy of sunlight. This process, photosynthesis, is the subject of Chapter 7.

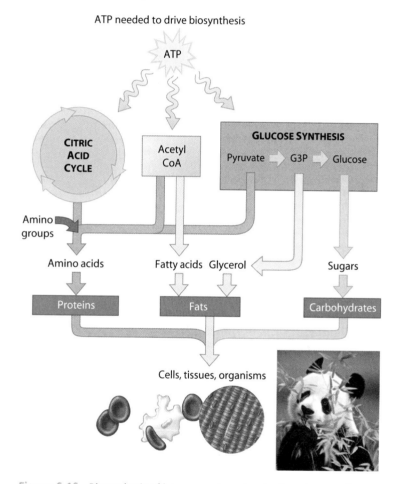

Figure 6.16 Biosynthesis of large organic molecules from intermediates of cellular respiration

? Explain how one can gain weight and store fat even when on a low-fat diet. (*Hint:* Look for G3P and acetyl CoA in Figures 6.15 and 6.16.)

■ If caloric intake is excessive, body cells use metabolic pathways to convert the excess to fat. The glycerol and fatty acids of fats are made from G3P and acetyl CoA, respectively, both produced from the oxidation of carbohydrates.

Chapter Review

Reviewing the Concepts

Introduction to Cellular Respiration (Introduction–6.5)

Photosynthesis and cellular respiration provide energy for life. Photosynthesis uses solar energy to produce glucose and O_2 from CO_2 and H_2O. O_2 is consumed during the oxidation of glucose to CO_2 and H_2O (**Introduction–6.1**). Breathing provides for the exchange of O_2 and CO_2 between an organism and its environment (**6.2**). Cellular respiration banks energy in ATP, which powers cellular work (**6.3–6.4**).

Oxidation/reduction. Electrons lose potential energy during their transfer from organic compounds to oxygen. Dehydrogenase removes electrons (in hydrogen atoms) from fuel molecules (oxidation) and transfers them to NAD^+ (reduction). NADH passes electrons to an electron transport chain. As electrons "fall" from carrier to carrier and finally to O_2, energy is released in small quantities (**6.5**).

Stages of Cellular Respiration and Fermentation (6.6–6.14)

Cellular respiration produces ATP. A small amount of ATP is made in glycolysis and the citric acid cycle. In oxidative phosphorylation, cells use the energy released by "falling" electrons to pump H^+ across a membrane. In a mechanism called chemiosmosis, the energy of the H^+ gradient is harnessed to make ATP (**6.6**). Cellular respiration can be divided into three stages:

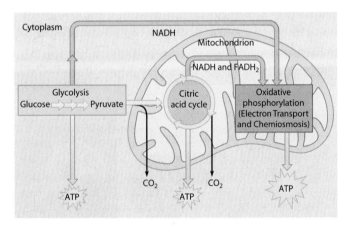

Glycolysis. ATP is used to prime a glucose molecule, which is split in two. These three-carbon intermediates are oxidized and converted to two molecules of pyruvate, yielding a net of 2 ATP and 2 NADH. ATP is formed by substrate-level phosphorylation, in which a phosphate group is transferred from an organic molecule to ADP (**6.7**).

Citric acid cycle. First, enzymes process pyruvate, releasing CO_2 and producing NADH and acetyl CoA (**6.8**). In the citric acid cycle, the two-carbon acetyl group is added to a four-carbon compound, forming citrate, which is degraded back to the starting four-carbon compound. For each turn of the cycle, 2 CO_2 are released; the energy yield is 1 ATP, 3 NADH, and 1 $FADH_2$ (**6.9**).

Oxidative phosphorylation. Electrons from NADH and $FADH_2$ travel down the electron transport chain to oxygen, which picks up H^+ to form water. Energy released by these redox reactions is used to pump H^+ into the space between the membranes of the mitochondrion. In chemiosmosis, the H^+ diffuses back across the inner membrane (down its concentration gradient) through ATP synthase complexes, driving the synthesis of ATP (**6.10**). Various poisons can block the movement of electrons, block the flow of H^+ through ATP synthase, or allow H^+ to leak through the membrane (**6.11**).

Energy tally. Substrate-level phosphorylation and oxidative phosphorylation produce up to 38 ATP molecules for every glucose molecule that enters cellular respiration (**6.12**).

Fermentation. Under anaerobic conditions, muscle cells, yeasts, and certain bacteria produce small amounts of ATP by glycolysis. NAD^+ is recycled from NADH as pyruvate is converted to lactate (lactic acid fermentation) or alcohol and CO_2 (alcohol fermentation) (**6.13**).

Glycolysis occurs in nearly all organisms and probably evolved in ancient prokaryotes before there was O_2 in the atmosphere (**6.14**).

Interconnections Between Molecular Breakdown and Synthesis (6.15–6.16)

Food as fuel. Carbohydrates, fats, and proteins can all fuel cellular respiration when they are converted to molecules that enter glycolysis or the citric acid cycle (**6.15**).

Food as building blocks. Cells use some food molecules and intermediates from glycolysis and the citric acid cycle as raw materials. Biosynthesis consumes ATP. Metabolic pathways are often regulated by feedback inhibiton. All organisms can harvest energy from organic molecules. Plants can also make organic molecules from inorganic sources by photosynthesis (**6.16**).

Connecting the Concepts

1. Fill in the blanks in this summary map to help you review the key concepts of cellular respiration.

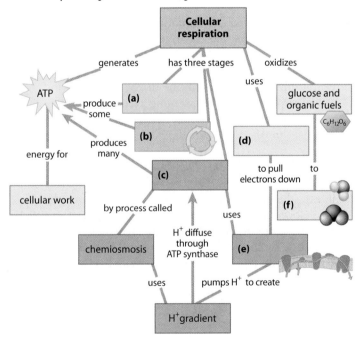

Testing Your Knowledge

Multiple Choice

2. What is the role of oxygen in cellular respiration?
   a. It is reduced in glycolysis as glucose is oxidized.
   b. It provides electrons to the electron transport chain.
   c. It combines with the carbon removed during the citric acid cycle to form CO_2.
   d. It is required for the production of heat and light.
   e. It accepts electrons from the electron transport chain.

3. When the poison cyanide blocks the electron transport chain, glycolysis and the citric acid cycle soon grind to a halt as well. Why do you think they stop?
   a. They both run out of ATP.
   b. Unused O_2 interferes with cellular respiration
   c. They run out of NAD^+ and FAD.
   d. Electrons are no longer available.
   e. They run out of ADP.

4. A biochemist wanted to study how various substances were used in cellular respiration. In one experiment, he allowed a mouse to breathe air containing O_2 "labeled" by a particular isotope. In the mouse, the labeled oxygen first showed up in
   a. ATP.   d. CO_2.
   b. glucose ($C_6H_{12}O_6$).   e. H_2O.
   c. NADH.

5. In glycolysis, _____ is oxidized and _____ is reduced.
   a. NAD^+ ... glucose   d. glucose ... NAD^+
   b. glucose ... oxygen   e. ADP ... ATP
   c. ATP ... ADP

6. Which of the following is the most immediate source of energy for making most of the ATP in your cells?
   a. the reduction of oxygen
   b. the transfer of ℗ from intermediate substrates to ADP
   c. the movement of H^+ across a membrane down its concentration gradient
   d. the splitting of glucose into two molecules of pyruvate
   e. electrons moving through the electron transport chain

7. In which of the following is the first molecule becoming reduced to the second molecule?
   a. pyruvate → acetyl CoA
   b. pyruvate → lactate
   c. glucose → pyruvate
   d. $NADH + H^+ → NAD^+ + 2 H$
   e. $C_6H_{12}O_6 → 6 CO_2$

8. Which of the following is a true distinction between cellular respiration and fermentation?
   a. NADH is oxidized by the electron transport chain in respiration only.
   b. Only respiration oxidizes glucose.
   c. Fermentation is an example of an endergonic reaction; cellular respiration is an exergonic reaction.
   d. Substrate-level phosphorylation is unique to fermentation; cellular respiration uses oxidative phosphorylation.
   e. Fermentation is the metabolic pathway found in prokaryotes; cellular respiration is unique to eukaryotes.

Describing, Comparing, and Explaining

9. Which of the three stages of cellular respiration is considered the most ancient? Explain your answer.

10. Explain in terms of cellular respiration why we need oxygen and why we exhale carbon dioxide.

11. Compare and contrast fermentation as it occurs in human muscle cells and as it occurs in yeast cells.

12. Explain how your body can convert excess carbohydrates in the diet to fats. Can excess carbohydrates be converted to protein? What else must be supplied?

Applying the Concepts

13. An average adult human requires 2,200 kcal of energy per day. Suppose your diet provides an average of 2,300 kcal per day. How many hours per week would you have to walk to burn off the extra calories? Swim? Run? (See Table 6.4.)

14. Your body makes NAD^+ and FAD from two B vitamins, niacin and riboflavin. The recommended dietary allowance for niacin is 20 mg daily and for riboflavin, 1.7 mg. These amounts are thousands of times less than the amount of glucose your body needs each day to fuel its energy needs. Why is the daily requirement for these vitamins so small?

15. In a detail of the citric acid cycle not shown in Figure 6.9B, succinate is converted to a compound called fumarate, with the release of H^+. You are studying this reaction using a suspension of bean cell mitochondria and a blue dye that loses its color as it takes up H^+. You know that the higher the concentration of succinate, the more rapid the decolorization of the dye. You set up reaction mixtures with mitochondria, dye, and three different concentrations of succinate (0.1 mg/L, 0.2 mg/L, and 0.3 mg/L). Which of the following graphs represents the results you would expect, and why?

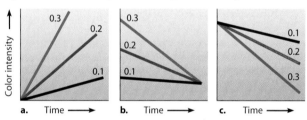

16. ATP synthase enzymes are found in the prokaryotic plasma membrane and in mitochondria and chloroplasts. What does this suggest about the evolutionary relationship of these eukaryotic organelles to prokaryotes?

17. Excess consumption of alcohol by a pregnant woman can cause a complex of birth defects called fetal alcohol syndrome (FAS). Symptoms of FAS include head and facial irregularities, heart defects, mental retardation, and behavioral problems. The U.S. Surgeon General's Office recommends that pregnant women abstain from drinking alcohol, and the government has mandated a warning label on liquor bottles: "Women should not drink alcoholic beverages during pregnancy because of the risk of birth defects." Imagine you are a server in a restaurant. An obviously pregnant woman orders an alcoholic drink. How would you respond? Is it the woman's right to make those decisions about her unborn child's health? Do you bear any responsibility in the matter? Is a restaurant responsible for monitoring the health habits of its customers?

Answers to all questions can be found in Appendix 4.

For Practice Quizzes, BioFlix, MP3 Tutors, and Activities, go to www.mybiology.com.

7

Photosynthesis:
Using Light to Make Food

Plant Power

With every new car, MP3 player, and cell phone, the world's demand for energy grows. In an era when most energy is generated from pollution-laden fossil fuels, more power often means more air pollution, more acid precipitation, and more of the "greenhouse gases" that contribute to global warming. Is there a better energy solution, one that could somehow produce power without harming our atmosphere? Some scientists believe they've found such a solution—and if you can see green through a nearby window, their answer might literally be in sight.

Energy researchers from Brazil to upstate New York have seized on the potential of plant power, turning to trees and other green vegetation to create "energy plantations" as a promising fuel source. In many ways, this solution draws from an ancient one, relying on one of the oldest energy pathways on the planet—**photosynthesis.** In this process, green plants, algae, and certain bacteria transform light energy to chemical energy stored in the bonds of the sugar they make from carbon dioxide and water. During the process, they also produce the oxygen we and all other aerobic organisms (including the plants themselves) need for cellular respiration. And they store an energy source that could help solve the human power problem.

As you learned in Chapter 6, both carbon dioxide and water are waste products of cellular respiration. Plants take in this cellular "exhaust," bringing in CO_2 through their leaves and pulling up water through their roots. Their leaves also absorb the solar energy of sunlight. In a sort of molecular shuffling driven by this energy, carbon dioxide and water are converted to a simple sugar and oxygen gas is released. One way to summarize this process is shown in the following equation.

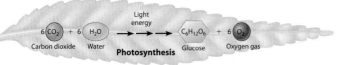

Photosynthesis happens on a microscopic level, but carried out repeatedly in plants around the world, it is responsible for an enormous amount of product. Earth's plants and other photosynthetic organisms make about 160 billion metric tons (176 billion tons) of carbohydrate each year. Almost all food can be traced back to plants—either eaten directly or fed to animals that are then consumed. And for most of human history, burning plant material has been a major source of heat, light, and cooking fuel. Its replacement by fossil fuels is relatively recent and centered in the developed world.

Now, as fossil fuel supplies dwindle and pollution accumulates, scientists are once again looking to plants and researching ways to make them a more efficient energy source. Certain types of fast-growing trees have shown promise as fuel to grow and burn in power generation facilities. Willows, for example, grow quickly and resprout once they are cut, limiting the amount of new planting, herbicides, and pesticides required to produce each generation of trees. They also grow much faster than most trees: Willow crops should be able to produce about 5–8 dry tons of wood per acre each year, according to researchers in New York. In comparison, natural forests produce only about 0.5–1.0 ton of wood per acre. Other fast-growing tree species, including sycamore, poplar, eucalyptus, and black locust, are also being tested as fuel sources.

Fuel from energy plantations is not only a renewable energy source; it can also be environmentally safer than fossil fuels. The products of burned wood do not include the sulfur contaminants released by fossil fuels, which contribute to acid rain. To be sure, burning wood fuel releases carbon dioxide into the air. But as energy plantation crops are quickly regrown, their photosynthesizing removes carbon dioxide from the atmosphere even as recently harvested trees release the gas through a power plant's smokestack. In contrast, the burning of fossil fuels, which come from the remains of ancient organisms, releases back carbon dioxide that was removed from the atmosphere through photosynthesis hundreds of millions of years ago.

Today, this "biomass energy" accounts for only about 4% of all energy consumed in the United States. But the idea has drawn enough support to be studied or tried in many countries and may play a role in meeting future energy needs.

In this chapter, you will learn how photosynthesis works. Because photosynthesis can seem like a complex process, we begin with some basic concepts. Then we look at the two stages of photosynthesis: the light reactions, in which solar energy is captured and transformed into chemical energy, and the Calvin cycle, in which the chemical energy is used to make organic molecules. Finally, we explore ways in which photosynthesis affects our global environment. ■ ■ ■

7.1 Autotrophs are the producers of the biosphere

Figure 7.1A   Forest plants

Figure 7.1B   Wheat field

Figure 7.1C   Kelp, a large alga

Figure 7.1D   Micrograph of cyanobacteria (photosynthetic bacteria)

Plants are **autotrophs** (meaning "self-feeders" in Greek) in that they make their own food and thus sustain themselves without consuming organic molecules derived from any other organisms. Plant cells capture light energy that has traveled 150 million kilometers from the sun and convert it to chemical energy. Using this energy, plants make their own organic molecules and are the ultimate source of organic molecules for almost all other organisms. They are often referred to as the **producers** of the biosphere because they produce its food supply. Actually, plants are not the only producers; algae, certain other protists, and some prokaryotes also make food molecules from carbon dioxide, water, and other inorganic materials. All organisms that produce organic molecules from inorganic molecules using the energy of light are called **photoautotrophs.**

The photographs on this page illustrate some of the diversity among photoautotrophs. On land, plants, such as those in the forest scene in Figure 7.1A and the wheat field in Figure 7.1B, are the predominant producers. In aquatic environments, algae and photosynthetic bacteria are the main food producers. Figure 7.1C shows kelp, a large alga that forms extensive underwater "forests" off the coast of California. Figure 7.1D is a micrograph of prokaryotic cyanobacteria, abundant and important producers in freshwater and marine ecosystems.

In this chapter, we focus on photosynthesis in plants, which takes place in chloroplasts. The remarkable ability to harness light energy and use it to drive the synthesis of organic compounds emerges from the structural organization of these organelles: Photosynthetic pigments, enzymes, and other molecules are grouped together in membranes, enabling the sequences of reactions to be carried out efficiently. The process of photosynthesis most likely originated in a group of bacteria that had infolded regions of the plasma membrane containing such clusters of enzymes and molecules. In fact, chloroplasts appear to have originated from a photosynthetic prokaryote that lived inside a eukaryotic cell (see Module 4.16). Let's begin with an overview of the location and structure of plant chloroplasts.

? What do "self-feeding" autotrophs require from the environment in order to make their own food?

■ Light, carbon dioxide, and water. (Plants also require inorganic minerals, as you'll learn in Chapter 32.)

Photosynthesis occurs in chloroplasts in plant cells

All green parts of a plant have chloroplasts in their cells and can carry out photosynthesis. In most plants, however, the leaves have the most chloroplasts (about half a million per square millimeter of leaf surface) and are the major sites of photosynthesis. Their green color is from **chlorophyll,** a light-absorbing pigment in the chloroplasts that plays a central role in converting solar energy to chemical energy.

Figure 7.2 zooms in on a leaf to show the actual sites of photosynthesis. At the top is a cross section (slice) of a leaf as it would look under a light microscope. Chloroplasts are concentrated in the cells of the **mesophyll,** the green tissue in the interior of the leaf. Carbon dioxide enters the leaf, and oxygen exits, by way of tiny pores called **stomata** (singular, *stoma,* meaning "mouth"). Water absorbed by the roots is delivered to the leaves in veins.

As you can see in the light micrograph of a single cell (second panel down), each mesophyll cell has numerous chloroplasts. The bottom drawing and electron micrograph show the structures in a single chloroplast. Membranes in the chloroplast form the framework where many of the reactions of photosynthesis occur, just as mitochondrial membranes do for the energy-harvesting machinery we discussed in Chapter 6. An envelope of two membranes encloses an inner compartment in the chloroplast, which is filled with a thick fluid called **stroma.** Suspended in the stroma is a system of interconnected membranous sacs, called **thylakoids,** which enclose another compartment, called the thylakoid space. (As you will see later, this thylakoid space plays a role analogous to the intermembrane space of a mitochondrion.) In some places, thylakoids are concentrated in stacks called **grana** (singular, *granum*). Built into the thylakoid membranes are the chlorophyll molecules that capture light energy. The thylakoid membranes also house much of the machinery that converts light energy to chemical energy, which is used in the stroma to make sugar.

Later in the chapter, we examine the function of these structures in more detail. But first, let's look more closely at the general process of photosynthesis.

BioFlix *Photosynthesis*

Web Activity *The Sites of Photosynthesis*

? How do the reactant molecules of photosynthesis reach the chloroplasts in leaves?

■ CO_2 enters leaves through stomata, and H_2O enters the roots and is carried to leaves through veins.

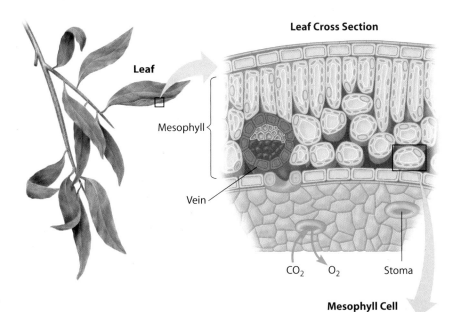

Leaf

Leaf Cross Section

Mesophyll

Vein

CO_2 O_2 Stoma

Mesophyll Cell

LM 2,600×

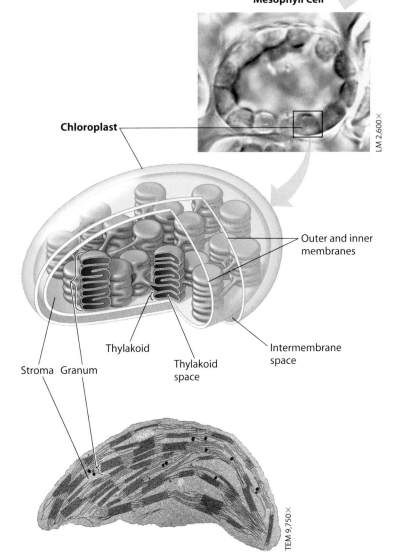

Chloroplast

Outer and inner membranes

Intermembrane space

Stroma Granum

Thylakoid

Thylakoid space

TEM 9,750×

Figure 7.2 The location and structure of chloroplasts

7.3 Plants produce O_2 gas by splitting water

The leaves of plants that live in lakes and ponds are often covered with bubbles like the ones shown in Figure 7.3A. The bubbles are oxygen gas (O_2) produced during photosynthesis.

The basic equation for photosynthesis was determined in the 1800s, and most scientists assumed that plants produce O_2 by extracting it from CO_2. In the 1950s, scientists tested this hypothesis by using a heavy isotope of oxygen, ^{18}O, to follow the fate of oxygen atoms during photosynthesis. (This was one of the first uses of isotopes as tracers in biological research. To review isotopes and their use as tracers, see Module 2.5.) In the photosynthesis equations in Figure 7.3B, the orange type denotes ^{18}O. The equations here are written in a slightly more detailed form than the summary photosynthesis equation you often see. These show that water is actually both a reactant and a product in the reaction. (As is often done, glucose is shown as a product, although the direct product of photosynthesis is a three-carbon sugar, which can be used to make glucose.)

In experiment 1, a plant given carbon dioxide containing ^{18}O gave off no labeled (^{18}O-containing) oxygen gas. But in experiment 2, a plant given water containing ^{18}O did produce labeled O_2. These experiments showed that the O_2 produced during photosynthesis comes from water and not from CO_2. It takes two water (H_2O) molecules to make each molecule of O_2.

Knowing where the O_2 comes from gives us a hint of what else happens during photosynthesis. Additional experiments have revealed that the oxygen atoms in CO_2 and the hydrogens in the reactant H_2O molecules end up in the sugar molecule and in water that is formed anew. Figure 7.3C summarizes the fates of all the atoms that start out in the reactant molecules of photosynthesis.

? Photosynthesis produces 160 billion metric tons of carbohydrate a year. Where does most of the mass of this huge amount of organic matter come from?

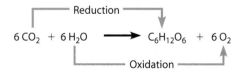

■ Mostly from CO_2 in the air, which provides both the carbon and oxygen in carbohydrate. Water supplies only the hydrogen.

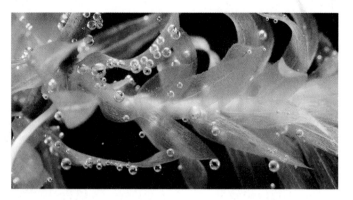

Figure 7.3A Oxygen bubbles on the leaves of an aquatic plant

Experiment 1

$$6\,CO_2 + 12\,H_2O \longrightarrow C_6H_{12}O_6 + 6\,H_2O + 6\,O_2$$

Not labeled

Experiment 2

$$6\,CO_2 + 12\,H_2O \longrightarrow C_6H_{12}O_6 + 6\,H_2O + 6\,O_2$$

Labeled

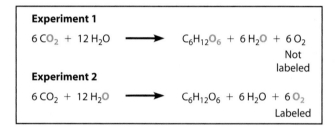

Figure 7.3B Experiments tracking the oxygen atoms in photosynthesis

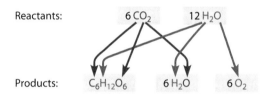

Figure 7.3C Fates of all the atoms in photosynthesis

7.4 Photosynthesis is a redox process, as is cellular respiration

What actually happens when photosynthesis converts CO_2 and water into sugar and O_2? Photosynthesis is a redox (oxidation-reduction) process, just as cellular respiration is (see Module 6.5). As indicated in the summary photosynthesis equation (Figure 7.4A), when water molecules are split apart, yielding O_2, they are actually oxidized; that is, they lose electrons, along with hydrogen ions (H^+). Meanwhile, CO_2 is reduced to sugar as electrons and hydrogen ions are added to it. Oxidation and reduction go hand in hand.

Now move from the food-producing equation for photosynthesis to the energy-releasing equation for cellular respiration (Figure 7.4B). Overall, cellular respiration harvests energy

Reduction

$$6\,CO_2 + 6\,H_2O \longrightarrow C_6H_{12}O_6 + 6\,O_2$$

Oxidation

Figure 7.4A Photosynthesis (uses light energy)

Oxidation

$$C_6H_{12}O_6 + 6\,O_2 \longrightarrow 6\,CO_2 + 6\,H_2O$$

Reduction

Figure 7.4B Cellular respiration (releases chemical energy)

stored in a glucose molecule by oxidizing the sugar and reducing O_2 to H_2O. This process involves a number of energy-releasing redox reactions, with electrons losing potential energy as they travel down an energy "hill" from sugar to O_2. Along the way, the mitochondrion uses some of the energy to synthesize ATP, as we saw in Chapter 6.

In contrast, the food-producing redox reactions of photosynthesis involve an uphill climb. As water is oxidized and CO_2 is reduced during photosynthesis, electrons gain energy by being boosted up an energy hill. The light energy captured by chlorophyll molecules in the chloroplast provides the boost for the electrons. Photosynthesis converts light energy to chemical energy and stores it in the chemical bonds of sugar molecules, which can provide energy for later use or raw materials for biosynthesis.

> **?** Which redox process, photosynthesis or cellular respiration, is endergonic? (*Hint:* See Module 5.12.)

■ Photosynthesis

7.5 Overview: The two stages of photosynthesis are linked by ATP and NADPH

The equation for photosynthesis is a simple summary of a very complex process. Actually, photosynthesis occurs in two stages, each with multiple steps. Figure 7.5 shows the inputs and outputs of the two stages and how the stages are related.

The **light reactions** (left in figure) include the steps that convert light energy to chemical energy and produce O_2. The light reactions occur in the thylakoid membranes. Water is split, providing a source of electrons and giving off O_2 gas as a by-product. Light energy absorbed by chlorophyll molecules built into the membranes is used to drive the transfer of electrons and H^+ from water to $NADP^+$, reducing it to NADPH. NADPH is an electron carrier similar to the NADH that transports electrons in cellular respiration. NADPH temporarily stores the electrons and provides "reducing power" to the Calvin cycle. The light reactions also generate ATP from ADP and a phosphate group.

In summary, the light reactions of photosynthesis are the steps that absorb solar energy and convert it to chemical energy stored in ATP and NADPH. Notice that these reactions produce no sugar; sugar is not made until the Calvin cycle, the second stage of photosynthesis.

The **Calvin cycle** occurs in the stroma of the chloroplast (Figure 7.5). It is a cyclic series of reactions that assembles sugar molecules using CO_2 and the energy-containing products of the light reactions. This second stage of photosynthesis is named for American biochemist and Nobel laureate Melvin Calvin. In the 1940s, Calvin and his colleagues traced the path of carbon in the cycle, using the radioactive isotope ^{14}C to label the carbon in CO_2. The incorporation of carbon from CO_2 into organic compounds, shown in the figure as CO_2 entering the Calvin cycle, is called **carbon fixation.** After carbon fixation, enzymes of the cycle make sugars by further reducing the carbon compounds.

As the figure suggests, it is NADPH produced by the light reactions that provides the electrons for reducing carbon in the Calvin cycle. And ATP from the light reactions provides chemical energy that powers several of the steps of the Calvin cycle. The Calvin cycle is sometimes referred to as the dark reactions, or light-independent reactions, because none of the steps

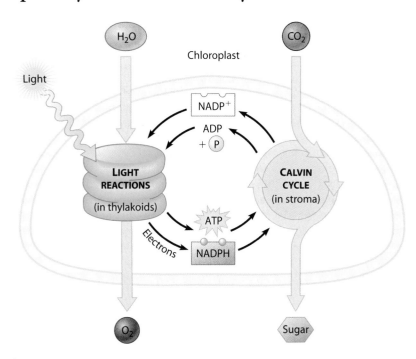

Figure 7.5 An overview of the two stages of photosynthesis that take place in a chloroplast

requires light directly. However, in most plants, the Calvin cycle runs during daytime, when the light reactions power the cycle's sugar assembly line by supplying it with NADPH and ATP.

The word *photosynthesis* encapsulates the two stages. *Photo-*, from the Greek word for light, refers to the light reactions; *synthesis*, meaning "putting together," refers to sugar construction by the Calvin cycle. In the next several modules, we look at these two stages in more detail. But first, let's consider some of the properties of light, the energy source that powers photosynthesis.

🎧 **MP3 Tutor** *Photosynthesis*

Web Activity *Overview of Photosynthesis*

> **?** For chloroplasts to produce sugar from carbon dioxide in the dark, they would need to be supplied with _____ and _____.

■ ATP . . . NADPH

7.6   Visible radiation drives the light reactions

What exactly do we mean when we say that photosynthesis is powered by light energy from the sun? Sunlight is a type of energy called electromagnetic energy or radiation.

Electromagnetic energy travels in space as rhythmic waves analogous to those made by a pebble dropped in a puddle of water. Figure 7.6A shows the **electromagnetic spectrum,** the full range of electromagnetic wavelengths from the very short gamma rays to the very long-wavelength radio waves. As you can see in the center of the figure, visible light—the radiation your eyes see as different colors—is only a small fraction of the spectrum. It consists of wavelengths from about 380 nm to about 750 nm. The distance between the crests of two adjacent waves is called a **wavelength** (illustrated at the bottom of the figure). Shorter wavelengths have more energy than longer ones. In fact, wavelengths that are shorter than those of visible light have enough energy to damage organic molecules such as proteins and nucleic acids. This is why ultraviolet (UV) radiation in sunlight can cause sunburns and skin cancer.

Figure 7.6B shows what happens to visible light in the chloroplast. Light-absorbing molecules called pigments, built into the thylakoid membranes, absorb some wavelengths of light and reflect or transmit other wavelengths. We do not see the absorbed wavelengths; their energy has been absorbed by pigment molecules. What we see when we look at a leaf are the green wavelengths that the pigment transmits and reflects.

Different pigments absorb light of different wavelengths, and chloroplasts contain several kinds of pigments. Chlorophyll *a*, which participates directly in the light reactions, absorbs mainly blue-violet and red light. It looks grass-green because it reflects mainly green light. A very similar molecule, chlorophyll *b*, absorbs mainly blue and orange light and reflects (appears) yellow-green. Chlorophyll *b* broadens the range of light that a

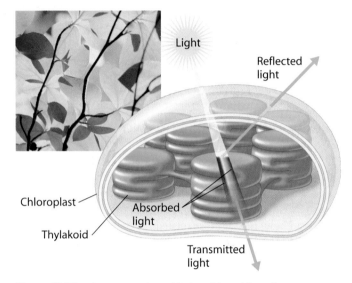

Figure 7.6B   The interaction of light with a chloroplast

plant can use by conveying absorbed energy to chlorophyll *a*, which then puts the energy to work in the light reactions.

Chloroplasts also contain a family of pigments called carotenoids, which are various shades of yellow and orange. The spectacular colors of fall foliage in certain parts of the world are due partly to the yellow-orange hues of longer lasting carotenoids that show through once the green chlorophyll breaks down. Carotenoids broaden the spectrum of colors that can drive photosynthesis by passing energy to chlorophyll *a*, as chlorophyll *b* does. However, a more important function of at least some carotenoids seems to be *photoprotection:* They absorb and dissipate excessive light energy that would otherwise damage chlorophyll or interact with oxygen to form reactive oxidative molecules that can damage cell molecules. Similar carotenoids, which we obtain from carrots and some other plants, have a photoprotective role in our eyes.

The theory of light as waves explains most of light's properties. However, light also behaves as discrete packets of energy called photons. A **photon** is a fixed quantity of light energy, and, as you have just learned, the shorter the wavelength, the greater the energy. Each type of pigment absorbs certain wavelengths of light because it is able to absorb the specific amounts of energy in those photons. Next we see what happens when a pigment molecule such as chlorophyll absorbs a photon of light.

Web Activity *Light Energy and Pigments*

Web Process of Science *How Does Paper Chromatography Separate Plant Pigments?*

? You may hear about the proposed health benefits of "phytochemicals" found in deep orange or red fruits and vegetables. How might such chemicals benefit a cell?

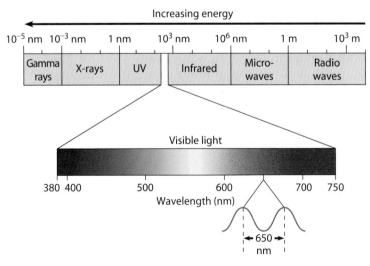

Figure 7.6A   The electromagnetic spectrum and the wavelengths of visible light. (A wavelength of 650 nm is illustrated.)

■ As antioxidants that protect from reactive forms of oxidative molecules

7.7 Photosystems capture solar power

When a pigment molecule absorbs a photon, one of the pigment's electrons jumps to an energy level farther from the nucleus. In this location, the electron has more potential energy, and we say that the electron has been raised from a ground state to an excited state. The excited state is very unstable. Generally, when isolated pigment molecules absorb light, their excited electrons drop back down to the ground state in a billionth of a second, releasing their excess energy as heat. This conversion of light energy to heat is what makes a black car so hot on a sunny day (black pigments absorb all wavelengths of light).

Some isolated pigments, including chlorophyll, emit light as well as heat after absorbing photons. We can demonstrate this phenomenon in the laboratory with a chlorophyll solution, as shown on the left in Figure 7.7A. When illuminated, the chlorophyll emits photons that produce a reddish afterglow called fluorescence. The right side of Figure 7.7A illustrates what happens in fluorescence: An absorbed photon boosts an electron to an excited state, from which it falls back to the ground state, emitting its energy as heat and light.

But chlorophyll behaves very differently in isolation than it does in an intact chloroplast. In its native habitat of the thylakoid membrane, chlorophyll passes off its excited electron to a neighboring molecule before it has a chance to drop back to the ground state.

In the thylakoid membrane, chlorophyll molecules are organized along with other pigments and proteins into clusters called photosystems (Figure 7.7B). A **photosystem** consists of a number of light-harvesting complexes surrounding a reaction center complex. The light-harvesting complexes consist of pigment molecules (which may include chlorophyll *a*,

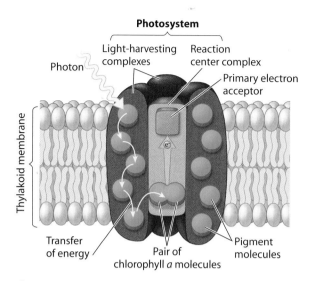

Photosystem — Light-harvesting complexes — Reaction center complex — Photon — Primary electron acceptor — Thylakoid membrane — Transfer of energy — Pair of chlorophyll *a* molecules — Pigment molecules

Figure 7.7B Light-excited chlorophyll embedded in a photosystem: Its electron is transferred to a primary electron acceptor before it returns to ground state

chlorophyll *b*, and carotenoids) bound to proteins. Collectively, the light-harvesting complexes function as a light-gathering antenna. The pigments absorb photons and pass the energy from molecule to molecule (thin yellow arrows) until it reaches the reaction center. The **reaction center complex** contains a pair of chlorophyll *a* molecules and a molecule called the primary electron acceptor, which is capable of accepting electrons and becoming reduced. The solar-powered transfer of an electron from the reaction center chlorophyll *a* to the primary electron acceptor is the first step of the light reactions.

Two types of photosystems have been identified, and they cooperate in the light reactions. They are referred to as photosystem I and photosystem II, in order of their discovery, although photosystem II actually functions first in the sequence of steps that make up the light reactions. Each photosystem has a characteristic reaction center. In photosystem II, the chlorophyll *a* of the reaction center is called P680 because the light it absorbs best is red light with a wavelength of 680 nm. The reaction center chlorophyll of photosystem I is called P700 because the wavelength of light it absorbs best is 700 nm (in the far-red part of the spectrum). Let's now see how the two photosystems work together in the light reactions to generate ATP and NADPH.

? Compared to a solution of isolated chlorophyll, why do intact chloroplasts release less heat and fluorescence when illuminated?

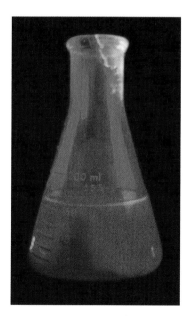

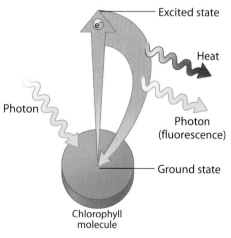

Excited state — Heat — Photon — Photon (fluorescence) — Ground state — Chlorophyll molecule

Figure 7.7A A solution of chlorophyll glowing red when illuminated (left); a diagram of an isolated, light-excited chlorophyll molecule that releases a photon of red light (right)

■ In the chloroplasts, the light-excited electrons are trapped by a primary electron acceptor rather than immediately giving up all their energy as heat and light.

7.8 Two photosystems connected by an electron transport chain generate ATP and NADPH

In the light reactions, light energy is transformed into the chemical energy of ATP and NADPH. In this process, electrons removed from water molecules pass from photosystem II to photosystem I to NADP$^+$. Between the two photosystems, the electrons move down an electron transport chain (similar to the one in cellular respiration) and provide energy for the synthesis of ATP.

Let's follow the flow of electrons (represented by gold arrows) in Figure 7.8A below, which shows the two photosystems embedded in a thylakoid membrane. ❶ A pigment molecule in a light-harvesting complex absorbs a photon of light. The energy is passed to other pigment molecules and finally to the reaction center of photosystem II, where it excites an electron of chlorophyll P680 to a higher energy state. ❷ This electron is captured by the primary electron acceptor. ❸ Water is split, and its electrons are supplied one by one to P680, each replacing an electron lost to the primary electron acceptor. The oxygen atom combines with an oxygen from another split water molecule to form a molecule of O$_2$.

❹ Each photoexcited electron passes from photosystem II to photosystem I via an electron transport chain. The exergonic "fall" of electrons provides energy for the synthesis of ATP by pumping H$^+$ across the membrane. ❺ Meanwhile, light energy excites an electron of chlorophyll P700 in the reaction center of photosystem I. The primary electron acceptor captures the electron, and an electron from the bottom of the electron transport chain replaces the lost electron in P700. ❻ The excited electron of photosystem I is passed through a short electron transport chain to NADP$^+$, reducing it to NADPH.

Figure 7.8B, right, provides a mechanical analogy to help you review how the two photosystems cooperate in generating ATP and NADPH. The input of light energy, represented by the large

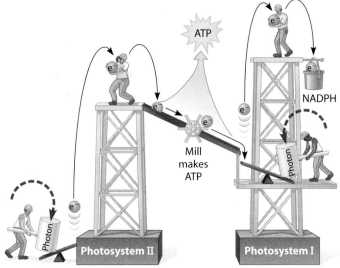

Figure 7.8B   A mechanical analogy of the light reactions

yellow mallets, boosts electrons in both photosystems up to the excited state. The electrons are caught by the primary electron acceptor on top of the platform in each photosystem. Photosystem II passes the electrons through an ATP mill. Photosystem I hands its electrons off to reduce NADP$^+$ to NADPH.

NADPH, ATP, and O$_2$ are the products of the light reactions. Next we look in more detail at how ATP is formed.

? Tracing the light reactions in Figure 7.8A, there is a flow of electrons from _____ molecules to _____, which is reduced to _____, the source of electrons for sugar synthesis in the _____ cycle.

■ water . . . NADP$^+$ . . . NADPH . . . Calvin

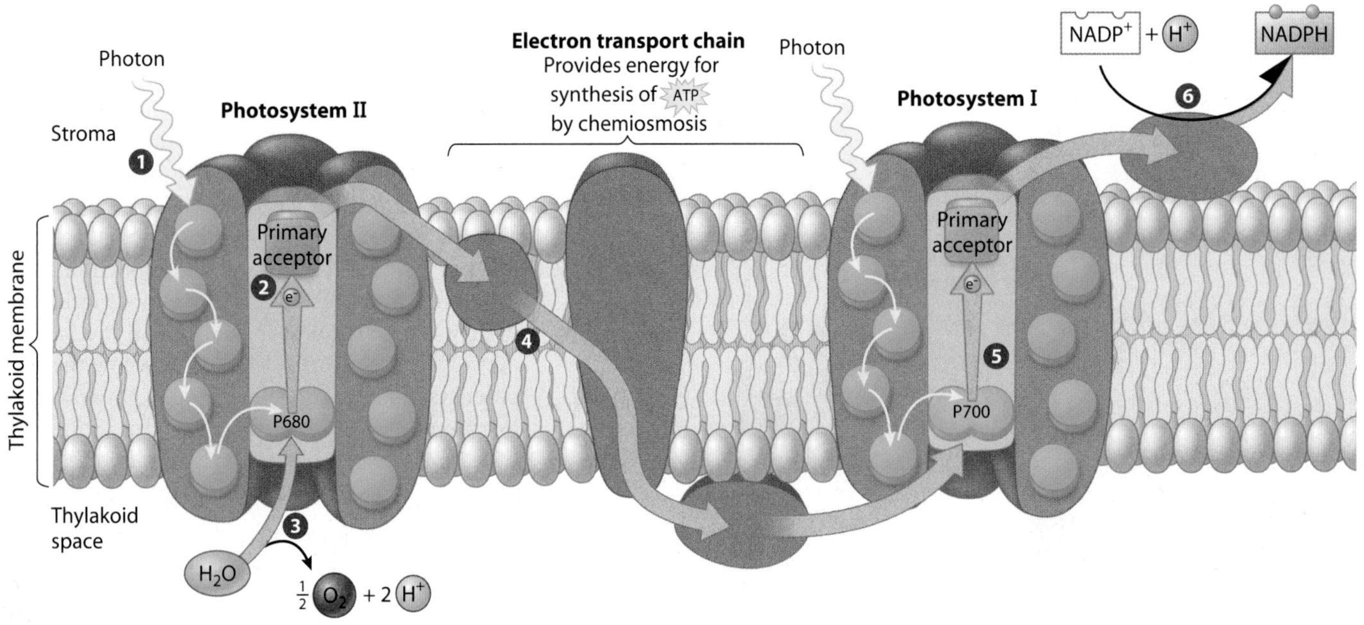

Figure 7.8A   Electron flow in the light reactions of photosynthesis: Both photosystems and the electron transport chain that connects them are located in the thylakoid membrane. The energy from light drives electrons from water to NADPH.

7.9 Chemiosmosis powers ATP synthesis in the light reactions

You first encountered chemiosmosis in Modules 6.6 and 6.10 as the mechanism of oxidative phosphorylation (ATP formation) in a mitochondrion. Chemiosmosis is also the mechanism that generates ATP in a chloroplast. Recall that the process of chemiosmosis drives ATP synthesis using the potential energy of a concentration gradient of hydrogen ions (H^+) across a membrane. The gradient is created when an electron transport chain uses the energy released as it passes electrons down the chain to pump hydrogen ions across a membrane.

Figure 7.9 illustrates the relationship between chloroplast structure and function in the light reactions of photosynthesis. As in Figure 7.8A, we show the two photosystems and electron transport chains, all located within the thylakoid membrane of a chloroplast. Here you can see that as photoexcited electrons are passed down the electron transport chain connecting the two photosystems, hydrogen ions are pumped across the membrane from the stroma into the thylakoid space (inside the thylakoid sacs). This generates a concentration gradient across the membrane.

The flask-shaped structure in the figure represents an ATP synthase complex, which is like the one we saw in the mitochon-drion. The energy of the concentration gradient drives H^+ across the membrane through ATP synthase. As you learned in Module 6.10, ATP synthase couples the flow of H^+ to the phosphorylation of ADP. In photosynthesis, this chemiosmotic production of ATP is called **photophosphorylation** because the initial energy input is light energy.

How does photophosphorylation compare with oxidative phosphorylation? In cellular respiration, the high-energy electrons passed down the electron transport chain come from the oxidation of food molecules. In photosynthesis, light energy is used to drive electrons to the top of the transport chain. Mitochondria transfer chemical energy from food to ATP; chloroplasts transform light energy into the chemical energy of ATP.

Notice that in the light-driven flow of electrons through the two photosystems, the final electron acceptor is $NADP^+$, not O_2 as in cellular respiration. Electrons do not end up at a low energy level in H_2O, as they do in respiration. Instead, they are stored at a high state of potential energy in NADPH.

In summary, the light reactions provide the chemical energy (ATP) and reducing power (NADPH) for the next stage of photosynthesis, the Calvin cycle. Module 7.10 describes how that cycle makes sugar.

Web Activity *The Light Reactions*

? What is the advantage of the light reactions producing NADPH and ATP on the stroma side of the thylakoid membrane?

■ The Calvin cycle, which consumes the NADPH and ATP, occurs in the stroma.

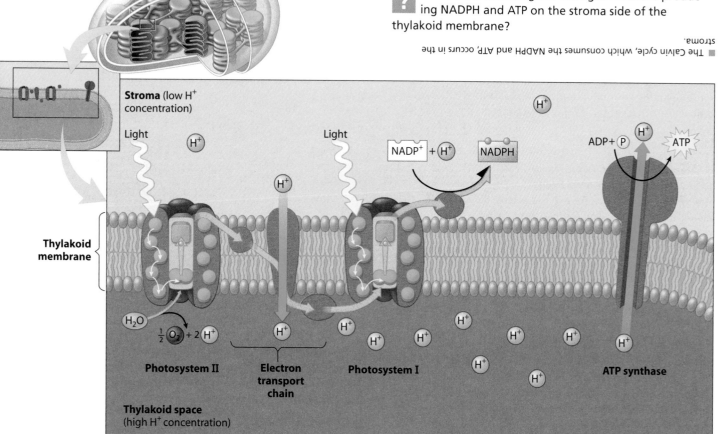

Figure 7.9 The production of ATP by chemiosmosis in photosynthesis: The small diagram on the upper left illustrates the location of the components of the light reactions in a thylakoid membrane. Numerous copies of these components are present in each thylakoid.

7.10 ATP and NADPH power sugar synthesis in the Calvin cycle

Input $\begin{cases} CO_2 \\ ATP \\ NADPH \end{cases}$

CALVIN CYCLE

Output: G3P

Figure 7.10A An overview of the Calvin cycle

The Calvin cycle functions like a sugar factory within a chloroplast. As **Figure 7.10A** shows, inputs to this all-important food-making process are CO_2 (from the air) and ATP and NADPH (both generated by the light reactions). Using carbon from CO_2, energy from ATP, and high-energy electrons from NADPH, the Calvin cycle constructs an energy-rich, three-carbon sugar, glyceraldehyde-3-phosphate (G3P). A plant cell can use G3P to make glucose and other organic molecules as needed. (You already met G3P in glycolysis: It is the three-carbon sugar formed by the splitting of glucose.)

Figure 7.10B presents the details of the Calvin cycle. It is called a cycle because, like the citric acid cycle in cellular respiration, the starting material is regenerated as the process occurs. In this case, the starting material is a five-carbon sugar named ribulose bisphosphate (RuBP). ❶ In the carbon fixation step, the enzyme rubisco attaches CO_2 to RuBP. ❷ In the next

step, a reduction, NADPH reduces the organic acid 3-PGA to G3P with the assistance of ATP. To make a molecule of G3P, the cycle must incorporate the carbon atoms from three molecules of CO_2. The cycle actually incorporates one carbon at a time, but we show it starting with three CO_2 molecules so that we end up with a complete G3P molecule.

For this to be a cycle, RuBP must be regenerated. ❸ For every three CO_2 molecules fixed, one G3P molecule leaves the cycle as product, and the remaining five G3P molecules are rearranged, ❹ using energy from ATP to regenerate three molecules of RuBP. Review the steps of the Calvin cycle below.

Note that for the net synthesis of one G3P molecule, the Calvin cycle consumes nine ATP and six NADPH molecules, which were provided by the light reactions.

Web Activity *The Calvin Cycle*

? To synthesize one glucose molecule, the Calvin cycle must turn six times, using 6 CO_2, 18 ATP, and 12 NADPH. Explain why this high number of ATP and NADPH molecules is consistent with the value of glucose as an energy source.

■ Glucose is a valuable energy source because it is highly reduced, storing lots of potential energy in its electrons. The more energy a molecule stores, the more energy and reducing power required to produce that molecule.

Step ❶ Carbon fixation. An enzyme called rubisco combines CO_2 with a five-carbon sugar called ribulose bisphosphate (abbreviated RuBP). The unstable product splits into two molecules of the three-carbon organic acid, 3-phosphoglyceric acid (3-PGA). For three CO_2 entering, six 3-PGA result.

Step ❷ Reduction. Two chemical reactions (indicated by the two blue arrows) consume energy from six molecules of ATP and oxidize six molecules of NADPH. Six molecules of 3-PGA are reduced, producing six molecules of the energy-rich three-carbon sugar, G3P.

Step ❸ Release of one molecule of G3P. Five of the G3Ps from step 2 remain in the cycle. The single molecule of G3P you see leaving the cycle is the net product of photosynthesis. A plant cell uses G3P to make glucose and other organic compounds.

Step ❹ Regeneration of RuBP. A series of chemical reactions uses energy from ATP to rearrange the atoms in the five G3P molecules (15 carbons total), forming three RuBP molecules (15 carbons). These can start another turn of the cycle.

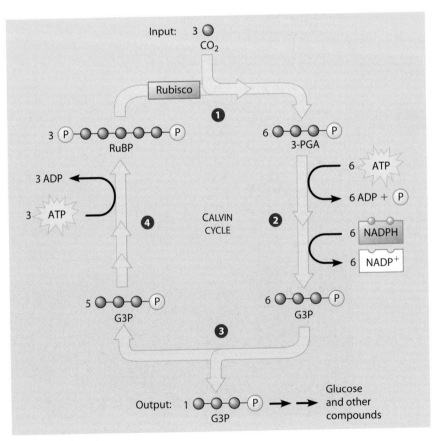

Figure 7.10B Details of the Calvin cycle, which takes place in the stroma of a chloroplast

7.11 Review: Photosynthesis uses light energy, CO_2, and H_2O to make food molecules

Life on Earth is solar powered. As we have discussed, most of the living world depends on the food-making machinery of photosynthesis. Figure 7.11 summarizes the two stages of this vital process and reviews where they occur in the chloroplast. The production of sugar from CO_2 is an emergent property of the structural arrangement of a chloroplast, which integrates the two stages of photosynthesis.

Starting on the left in the diagram, you see a summary of the light reactions, which occur in the thylakoid membranes (green). Two photosystems in the membranes capture solar energy, using it to energize electrons. Simultaneously, water is split, and O_2 is released. The photosystems transfer photo-excited electrons through electron transport chains, where energy is harvested to make ATP, and to $NADP^+$, where electrons are stored in the high-energy compound NADPH.

The chloroplast's sugar factory is the Calvin cycle, the second stage of photosynthesis. In the stroma, the enzyme rubisco combines CO_2 with RuBP, and ATP and NADPH are used to reduce 3-PGA to G3P. Sugar molecules made from G3P serve as a plant's own food supply. Plants use about 50% of the carbohydrate they make as fuel for cellular respiration. Sugars also serve as starting material for making other organic molecules, such as the structural molecule cellulose. Most plants make considerably more food each day than they need. They stockpile the excess sugar as starch, storing it in roots, tubers, and fruits.

Plants (and other photosynthesizers) not only feed themselves but are also the ultimate source of food for virtually all other organisms. Humans and other animals make none of

their own food and are totally dependent on the organic matter made by photosynthesizers. Even the energy we acquire when we eat meat was originally captured by photosynthesis. The energy in a hamburger, for instance, came from sunlight that was originally converted to a chemical form in chloroplasts in the cells of the grasses eaten by cattle. Photosynthesis is the ultimate source of the food we eat and the oxygen we breathe.

This review of photosynthesis is an appropriate place to reflect on the metabolic ground we have covered in this chapter and the previous one. In Chapter 6, we saw that virtually all organisms, plants included, use cellular respiration to obtain the energy they need from fuel molecules such as glucose. We followed the chemical pathways of glycolysis and the citric acid cycle, which break glucose down and release energy from it. We have now come full circle, seeing how plants trap sunlight energy and use it to make glucose from the raw materials carbon dioxide and water.

In tracing glucose synthesis and its breakdown, we have also seen that cells use several of the same mechanisms—electron transport, redox reactions, and chemiosmosis—in energy storage (photosynthesis) and energy harvest (cellular respiration).

Web Process of Science *How Is the Rate of Photosynthesis Measured?*

? Explain why a poison that inhibits an enzyme of the Calvin cycle will also inhibit the light reactions.

■ The light reactions require ADP and $NADP^+$, which are not recycled from ATP and NADPH when the Calvin cycle stops.

Figure 7.11
A summary of the chemical processes of photosynthesis

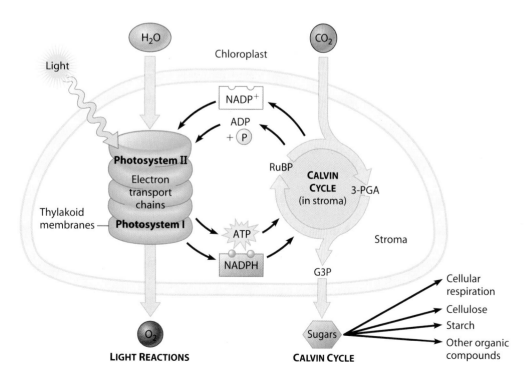

7.12 Adaptations that save water in hot, dry climates evolved in C₄ and CAM plants

The algal ancestors of plants lived in water. Since plants first moved onto land about 475 million years ago, they have been adapting to the challenges of terrestrial life, particularly the problem of dehydration. Some solutions involve trade-offs, such as the compromise between the requirements for photosynthesis and preventing excessive water loss. Closing stomata on a hot, dry day is an adaptation that reduces water loss, but it also prevents CO_2 from entering the leaf and O_2 from leaving. As a result, CO_2 levels get very low in the leaf, while O_2 from the light reactions builds up. These conditions reduce photosynthesis and favor an apparently wasteful process called photorespiration.

In most plants, initial fixation of carbon occurs when the enzyme rubisco adds CO_2 to RuBP (see Figure 7.10B). Such plants are called **C₃ plants** because the first organic compound produced is the three-carbon compound 3-PGA. C₃ plants are common and widely distributed; they include soybeans, oats, wheat, and rice. One of the problems that farmers face in growing C₃ plants is that hot, dry weather can decrease crop yield. When stomata close to reduce water loss and O_2 builds up in a leaf, rubisco adds O_2 instead of CO_2 to RuBP. A two-carbon product of this reaction is then broken down by the cell to CO_2 and H_2O. This process is called **photorespiration.** Unlike photosynthesis, photorespiration yields no sugar, and unlike cellular respiration, it produces no ATP. It can, however, drain away as much as 50% of the carbon fixed by the Calvin cycle.

According to one hypothesis, photorespiration is an evolutionary relic from when the atmosphere had less O_2 than it does today. In the ancient atmosphere that prevailed when rubisco first evolved, the inability of the enzyme's active site to exclude O_2 would have made little difference. It is only after O_2 became so concentrated in the atmosphere that the "sloppiness" of rubisco presented a problem. There are also alternative hypotheses for photorespiration, including that it plays a protective role in plants when the products of the light reactions build up in a cell (as occurs when the Calvin cycle has slowed due to a lack of CO_2).

C₄ Plants In certain plant species, alternate modes of carbon fixation have evolved that save water without shutting down photosynthesis. **C₄ plants** are so named because they precede the Calvin cycle by first fixing CO_2 into a four-carbon compound. When the weather is hot and dry, a C₄ plant keeps its stomata mostly closed, thus conserving water. At the same time, it continues making sugars by photosynthesis, using the pathway and the two types of cells shown in the left side of Figure 7.12. An enzyme in the mesophyll cells has a high affinity for CO_2 and can continue to fix carbon even when the CO_2 concentration in the leaf is low. The resulting four-carbon compound then acts as a carbon shuttle; it moves into bundle-sheath cells, which are packed around the veins of the leaf, and releases CO_2. Thus, the CO_2 concentration in these cells remains high enough for the Calvin cycle to make sugars and avoid photorespiration. Corn and sugarcane are examples of agriculturally important C₄ plants. C₄ photosynthesis is advantageous in hot, dry climates, which is where C₄ plants evolved and thrive today.

CAM Plants A second photosynthetic adaptation to arid conditions has evolved in pineapples, many cacti, and succulent (water-storing) plants, such as aloe and jade plants. Collectively called **CAM plants,** these species are adapted to very dry climates. A CAM plant (right side of Figure 7.12) conserves water by opening its stomata and admitting CO_2 only at night. When CO_2 enters the leaves, it is fixed into a four-carbon compound, as in C₄ plants. The four-carbon compound in a CAM plant banks CO_2 at night and releases it to the Calvin cycle during the day. This keeps photosynthesis operating during the day, even though the leaf's stomata remain closed. CAM stands for crassulacean acid metabolism, after the plant family Crassulaceae, in which this water-saving adaptation was first discovered.

In C₄ plants, carbon fixation and the Calvin cycle occur in different types of cells. In CAM plants, these processes occur in the same cells, but at different times. Keep in mind that CAM, C₄, and C₃ plants all eventually use the Calvin cycle to make sugar from CO_2. The C₄ and CAM pathways are two evolutionary solutions to the problem of maintaining photosynthesis with stomata partially or completely closed on hot, dry days.

Web Activity *Photosynthesis in Dry Climates*

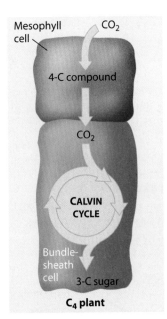
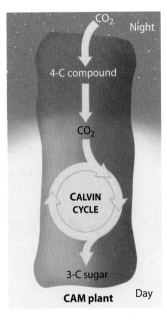

Figure 7.12 Comparison of photosynthesis in C₄ and CAM plants: In both pathways, CO_2 is first incorporated into a four-carbon compound, which then provides CO_2 to the Calvin cycle.

? How would you expect the relative abundance of C₃ versus C₄ and CAM species to change in a geographic region where the climate becomes much hotter and drier?

■ C₄ and CAM species would replace many of the C₃ species.

7.13 Photosynthesis moderates global warming

The greenhouse in Figure 7.13A is used to grow plants where the weather outside is too cold. The glass or plastic walls of a greenhouse allow solar radiation to pass through. The sunlight heats the soil, which in turn warms the air. The walls trap the warm air, raising the temperature inside.

An analogous process, called the **greenhouse effect,** operates on a global scale (Figure 7.13B). Solar radiation reaching Earth's atmosphere includes ultraviolet radiation and visible light. As we discuss in the next module, the ozone layer filters out most of the damaging UV radiation. Visible light passes through and is absorbed by the planet's surface, warming it. Heat is radiated by the warmed planet, and these longer, infrared wavelengths are absorbed by gases in the atmosphere, which then reflect some of the heat back to Earth. This natural heating effect is highly beneficial. Without it, Earth would be much colder and much less hospitable to life.

The gases in the atmosphere that absorb heat radiation are called greenhouse gases. Some occur naturally, such as water vapor, carbon dioxide (CO_2), and methane (CH_4), while others are synthetic, such as chlorofluorocarbons (CFCs). Human activities are adding to the levels of these greenhouse gases.

Carbon dioxide is one of the most important greenhouse gases. You have just learned that CO_2 is a raw material for photosynthesis and a waste product of cellular respiration. These two processes, taking place in microscopic chloroplasts and mitochondria, keep carbon cycling between inorganic and organic compounds on a global scale. Photosynthetic organisms absorb billions of tons of CO_2 each year. Most of that fixed carbon returns to the atmosphere via cellular respiration, the action of decomposers, and fires. But much of it remains locked in large tracts of forests and undecomposed organisms. And large amounts of carbon are in long-term storage in fossil fuels buried deep under Earth's surface.

Before 1850, carbon dioxide was estimated to make up less than 0.03% of the air we breathe. Since the start of the Industrial Revolution, the atmospheric concentration of CO_2 has increased about 40%, mostly due to the combustion of carbon-based fossil fuels, such as coal, oil, and gasoline.

Increasing concentrations of greenhouse gases have been linked to **global warming,** a slow but steady rise in Earth's surface temperature. Predicted changes of just a few degrees over the next 50 years may have dramatic and wide-ranging consequences. These would likely include melting of polar ice, rising sea levels, extreme weather patterns, droughts, and the spread of tropical diseases.

Unfortunately, the rise in atmospheric CO_2 levels during the last century coincided with widespread deforestation, which aggravated the global warming problem by reducing an effective CO_2 sink. As forests are cleared for lumber or agriculture, and as population growth increases the demand for fossil fuels,

Figure 7.13A Plants growing in a greenhouse

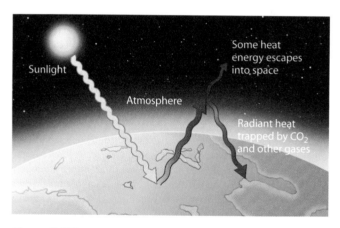

Figure 7.13B CO_2 in the atmosphere and global warming

Some heat energy escapes into space

Sunlight

Atmosphere

Radiant heat trapped by CO_2 and other gases

CO_2 levels will continue to rise. We discuss global warming in more detail in Module 38.5.

Can photosynthesis help mitigate this increase in atmospheric CO_2? As you read in the chapter introduction, "energy plantations" hold out the promise of a cleaner, renewable fuel source. Recent studies show that increasing CO_2 levels will increase plant production somewhat, but far less than predicted even a decade ago. Slowing the destruction of forests will sustain their photosynthetic and carbon-storing contributions. Taking a lesson from plants, we can explore technologies that utilize solar energy for some of our energy needs. Almost all life on Earth depends on the ability of plants and other photosynthetic organisms to convert light energy to the chemical energy of food molecules. Their contribution to life on Earth may also come to include increased removal of CO_2 from the atmosphere.

? Explain the greenhouse effect.

Sunlight warms Earth's surface, which radiates heat to the atmosphere. CO_2 and other greenhouse gases absorb and radiate some heat back to Earth.

7.14 Mario Molina talks about Earth's protective ozone layer

Figure 7.14A   Mario Molina

As the process of photosynthesis consumes CO_2, it produces the O_2 on which plants, animals, and most other organisms depend for cellular respiration. This O_2 has another benefit: High in the atmosphere it is converted to ozone (O_3), which plays a protective role for life on Earth. Among the scientists who study the ozone layer is Mario Molina of the University of California, San Diego (Figure 7.14A), who in 1995 shared a Nobel Prize for his research on how certain pollutants are damaging that layer. In an interview, Dr. Molina explained why the ozone layer is important:

The ozone layer shields the Earth's surface from powerful ultraviolet radiation that comes from the sun. This UV radiation is harmful to organisms, including humans. For example, UV radiation causes sunburn, and skin cancer can be a cumulative result of exposure. There is also evidence that UV can damage crops. Certain developing animals, such as the larvae of fish, seem to be particularly sensitive.

He described how ozone forms and how it is destroyed:

The ozone forms when high-energy solar radiation breaks apart O_2 molecules and frees oxygen atoms. These then react with unbroken O_2 molecules. The result is ozone, which has three oxygen atoms (O_3). So ozone is continuously forming by the action of sunlight on the atmosphere. This is balanced by continuous destruction of the ozone molecules when they react with other chemical compounds that are naturally present in the atmosphere. Humans have disrupted that balance by releasing certain industrial chemicals that hasten this destruction.

The research that won Dr. Molina the Nobel Prize dealt with the destruction of ozone by one particular class of industrial chemicals, called chlorofluorocarbons, or CFCs (mentioned as greenhouse gases in Module 7.13). In 1974, when Dr. Molina and his colleagues first published their CFC-ozone depletion hypothesis, CFCs were used in large amounts as refrigerants, as propellants in spray cans, as solvents, and in the process for making plastic foams:

We predicted that the continuous release of CFCs would damage the protective ozone layer. CFCs are very stable compounds, and this stability allows them to make it up to the ozone layer, which is about 15 miles above the surface of the Earth. There solar radiation converts them to very reactive chemicals called free radicals, which then destroy the ozone.

How did scientists and the general public react to the alarming possibility that the ozone layer was thinning?

At first, there was very little reaction because not many people were aware of the importance of invisible things like the ozone layer and

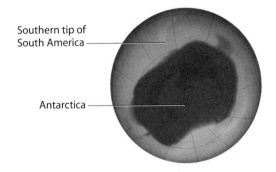

Figure 7.14B   The ozone hole in the Southern Hemisphere, spring 2006

UV radiation. Experts in our field quickly realized that the prediction of the CFC-ozone depletion theory was something to worry about, but many other scientists were skeptical, which is natural for scientists. Then, as we and others began doing experiments to test the idea, the evidence became strong, and more and more people, including politicians, became concerned about ozone depletion. Then, in 1985, scientists documented a drastic depletion of the ozone layer over Antarctica—an ozone hole.

Figure 7.14B shows an image produced from atmospheric data from 2006 of the "ozone hole" over Antarctica. Greens and yellows are where there is more ozone; blue and purple colors are where there is the least. This thinning of the ozone layer appears every spring. What is being done to correct this problem?

The Montreal Protocol was an international agreement that required a complete phasing out of CFC production by developed countries by 1996. So that's already happened. These countries now use other refrigerants, which are destroyed in the atmosphere before they reach the ozone layer. Because CFCs are so stable, however, we predict that the ozone layer won't recover until the middle of the twenty-first century.

We asked Dr. Molina what he has learned since 1974 about the interface between science and politics:

One lesson is that science is not always something that politicians care very much about. Often, supporting science is considered a luxury, though actually it is a good investment. Another challenge is that many issues related to science and technology are long-term issues that require patience and long-term commitment.

Whether an environmental problem involves CFCs or carbon dioxide emissions, the scientific research is often complicated, and solutions are often complex and expensive. The connections between science, technology, and society, so clearly exemplified by the work of Mario Molina, are a major theme of this book. This theme will come up again in the next unit, on cellular reproduction and genetics.

?   Where does the ozone layer in Earth's atmosphere come from?

■ Photosynthesis releases O_2. High in the atmosphere, radiation from the sun converts some of the O_2 to ozone, O_3.

Chapter Review

Reviewing the Concepts

An Overview of Photosynthesis (Introduction–7.5)

Photosynthesis is summarized by the following equation:

$$6\,CO_2 + 6\,H_2O + \text{Light energy} \rightarrow C_6H_{12}O_6 + 6\,O_2$$

Plants, algae, and some bacteria are photoautotrophs, the producers of food consumed by virtually all organisms. Chloroplasts appear to have evolved from an endosymbiotic photosynthetic prokaryote (**Introduction–7.1**). In plants, photosynthesis occurs primarily in the leaves, within chloroplasts, which contain stroma and stacks of thylakoids called grana (**7.2**).

The Two Stages of Photosynthesis. The O_2 liberated by photosynthesis comes from H_2O (**7.3**). In photosynthesis, H_2O is oxidized and CO_2 is reduced (**7.4**). The light reactions convert light energy to chemical energy and produce O_2. The Calvin cycle assembles sugar molecules from CO_2 using ATP and NADPH from the light reactions (**7.5**).

The Light Reactions: Converting Solar Energy to Chemical Energy (7.6–7.9)

Photosystems. Certain wavelengths of visible light, absorbed by pigments such as chlorophyll and carotenoids, drive photosynthesis (**7.6**). Thylakoid membranes contain multiple photosystems, each consisting of light-harvesting complexes of pigments and a reaction center complex with a primary electron acceptor that receives excited electrons from a reaction center chlorophyll *a* (**7.7**).

The Light Reactions. Two photosystems absorb photons and transfer energy to chlorophyll P680 and P700. Their excited electrons are passed from primary electron acceptors to electron transport chains. Electrons shuttle from photosystem II to I, providing energy to make ATP. Electrons from photosystem I reduce $NADP^+$ to NADPH. Photosystem II regains electrons by splitting water, releasing O_2 (**7.8**).

Photophosphorylation. ATP is synthesized by chemiosmosis. The electron transport chain pumps H^+ into the thylakoid space. The concentration gradient drives H^+ back across the membrane through ATP synthase and powers the phosphorylation of ADP to produce ATP (**7.9**).

The Calvin Cycle: Converting CO₂ to Sugars (7.10)

The Calvin cycle, occurring in the stroma, consists of carbon fixation, reduction, release of G3P, and regeneration of RuBP. Using carbon from CO_2, electrons from NADPH, and energy from ATP, the cycle constructs G3P, which is used to build glucose and other organic molecules (**7.10**).

Photosynthesis Reviewed and Extended (7.11–7.12)

A summary diagram of the two stages of photosynthesis is shown at the top of the next column (**7.11**).

C₃, C₄, and CAM plants. In C_3 plants, a drop in CO_2 and rise in O_2 when stomata close on hot, dry days divert the Calvin cycle to photorespiration. C_4 plants first fix CO_2 into a four-carbon compound that provides CO_2 to the Calvin cycle. CAM plants open stomata at night, making a four-carbon compound used as a CO_2 source during the day (**7.12**).

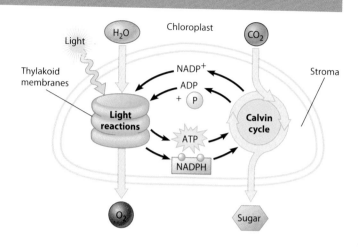

Photosynthesis, Solar Radiation, and Earth's Atmosphere (7.13–7.14)

Excess CO_2 is contributing to global warming. Photosynthesis, which removes CO_2 from the atmosphere, moderates this warming (**7.13**). Solar radiation converts O_2 high in the atmosphere to ozone (O_3), which shields organisms from damaging UV radiation. Industrial chemicals called CFCs have caused dangerous thinning of the ozone layer, but international restrictions on CFC use is allowing recovery (**7.14**).

Connecting the Concepts

1. This diagram compares the chemiosmotic synthesis of ATP in mitochondria and chloroplasts. In both cases, label the structures involved and indicate which side of the membrane has the higher H^+ concentration. Then label on the right the locations within the chloroplast.

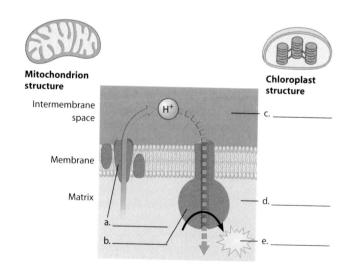

2. Continue your comparison of chemiosmosis and electron transport in mitochondria and chloroplasts. In each case,
   a. where do the electrons come from?
   b. where do the electrons get their energy?
   c. what picks up the electrons at the end of the chain?
   d. how is the energy given up by the electrons used?

3. Complete this summary map of photosynthesis.

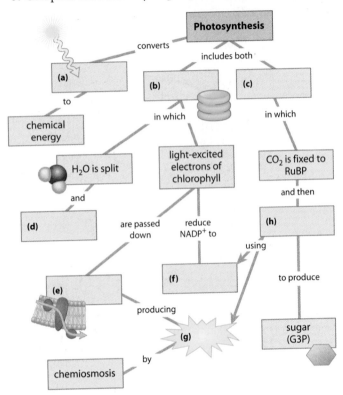

Testing Your Knowledge

Multiple Choice

4. Photosynthesis consumes _____ and produces _____.
   a. O_2 . . . H_2O
   b. H_2O . . . CO_2
   c. CO_2 . . . chlorophyll
   d. H_2O . . . O_2
   e. glucose . . . O_2

5. Which of the following are produced by reactions that take place in the thylakoids and consumed by reactions in the stroma?
   a. CO_2 and H_2O
   b. $NADP^+$ and ADP
   c. ATP and NADPH
   d. ATP, NADPH, and O_2
   e. CO_2 and ATP

6. In photosynthesis, _____ is oxidized and _____ is reduced.
   a. glucose . . . oxygen
   b. carbon dioxide . . . water
   c. water . . . carbon dioxide
   d. glucose . . . carbon dioxide
   e. water . . . oxygen

7. Why is it difficult for most plants to carry out photosynthesis in very hot, dry environments such as deserts?
   a. The light is too intense and overpowers the pigment molecules.
   b. The closing of stomata keeps CO_2 from entering and O_2 from leaving the plant.
   c. They must rely on photorespiration to make ATP.
   d. Global warming is intensified in a desert environment.
   e. CO_2 builds up in the leaves, blocking carbon fixation.

8. When light strikes chlorophyll molecules, they lose electrons, which are ultimately replaced by
   a. splitting water.
   b. breaking down ATP.
   c. oxidizing NADPH.
   d. fixing carbon.
   e. oxidizing glucose.

9. What is the role of $NADP^+$ in photosynthesis?
   a. It assists chlorophyll in capturing light.
   b. It acts as the primary electron acceptor for the photosystems.
   c. As part of the electron transport chain, it helps to synthesize ATP.
   d. It assists photosystem II in the splitting of water.
   e. It accepts electrons and carries them to the Calvin cycle.

10. The reactions of the Calvin cycle are not directly dependent on light, but they usually do not occur at night. Why? (*Explain your answer.*)
    a. It is often too cold at night for these reactions to take place.
    b. Carbon dioxide concentrations decrease at night.
    c. The Calvin cycle depends on products of the light reactions.
    d. Plants usually close their stomata at night.
    e. Most plants do not make four-carbon compounds, which they would need for the Calvin cycle at night.

11. How many "turns" of the Calvin cycle are required to produce one molecule of glucose? (Assume one CO_2 is fixed in each turn of the cycle.)
    a. 1   b. 2   c. 3   d. 6   e. 12

12. Which of the following does *not* occur during the Calvin cycle?
    a. carbon fixation
    b. oxidation of NADPH
    c. consumption of ATP
    d. regeneration of RuBP, the CO_2 acceptor
    e. release of oxygen

Describing, Comparing, and Explaining

13. What are the major inputs and outputs of the two stages of photosynthesis?

14. What do plants do with the sugar they produce in photosynthesis?

Applying the Concepts

15. Most experts now agree that global warming is occurring, and in response to its potential threat, a number of countries have made a commitment to reduce CO_2 emissions significantly. However, some countries oppose taking strong action at this time. Several reasons are cited: First, a few experts think that the apparent warming trend may be just a random fluctuation in temperature. Second, if the temperature increase is real, it has yet to be shown that it is caused by increased CO_2. Some people also believe that it would be difficult to cut CO_2 emissions without sacrificing economic growth. Do you think we should have more evidence before taking action? Or is it better to play it safe and act now to reduce CO_2 emissions? What are the possible costs and benefits of each of these two strategies?

16. The use of biomass energy (see chapter introduction) avoids many of the problems associated with gathering, refining, transporting, and burning fossil fuels. Yet biomass energy is not without its own set of problems. What challenges do you think would arise from a large-scale conversion to biomass energy? How do these challenges compare with those encountered with fossil fuels? Which set of challenges do you think is more likely to be overcome eventually? Do you think any one type of energy has more benefits and fewer costs than the others? Which one, and why?

Answers to all questions can be found in Appendix 4.

For Practice Quizzes, BioFlix, MP3 Tutors, and Activities, go to www.mybiology.com.

Cellular Reproduction and Genetics

The Cellular Basis of Reproduction and Inheritance

Rain Forest Rescue

Deep in a Hawaiian rain forest, the thrumming of a helicopter cuts through the dense tropical air. After touching down in a clearing, a pair of rescuers jump out, one of them carrying a precious bundle. A 20-minute hike through dense foliage brings them to their "patient": the last known wild *Cyanea kuhihewa* plant, a member of the bellflower family (pictured at right). The rescuers now stand in the middle of a life-or-death battle, armed with the pollen of a species on the very brink of extinction.

Despite its tiny size, Hawaii is home to over 300 endangered or threatened species, including more endangered plants than any other state in the United States. Of these, 11 plant species have fewer than five representatives growing anywhere in the wild. At the time of this rescue attempt in 2002, the small bellflower was among the rarest of the rare, with only a single surviving member in the forest.

So when this lone wild bellflower bloomed, scientists from the National Tropical Botanical Garden on Kauai saw an opportunity. Their goal was to promote sexual reproduction, the reproductive process that involves fertilization, the union of a sperm and an egg. Using a fine brush, the botanical rescuers transferred sperm-carrying pollen from a garden-grown plant onto the egg-containing wild bloom. If sperm and egg join, the fertilized egg may divide and eventually form an embryo and then a seed. Upon germination of the seed, the embryo may further develop into a juvenile and later into an adult plant.

Development from a fertilized egg to a new adult organism is one phase of a multicellular organism's **life cycle,** the sequence of stages leading from the adults of one generation to the adults of the next. The other phase of the life cycle is reproduction, the formation of new individuals from preexisting ones. The reproductive phase of the life cycle entails the creation of offspring carrying genetic information, as DNA, from their parents. Sperm and egg each carry one set of genetic information—one copy of the organism's **genome.** Thus, the offspring of sexual reproduction inherit traits from two parents. In the case of *C. kuhihewa,* sexual reproduction could provide desperately needed genetic diversity.

Unfortunately, the bellflower fertilization attempt failed. Even worse, the last remaining wild specimen died in 2003. But hope remains for *C. kuhihewa*. The garden's botanists are now trying asexual reproduction, the production of offspring by a single parent. Many plants that normally reproduce sexually can be induced to reproduce asexually in the laboratory. Indeed, work at Kauai's botanical garden has produced several new bellflower plants through asexual reproduction, using stem tissue snipped from wild plants in the 1990s.

These examples of sexual and asexual reproduction illustrate the main point of this chapter: Cell division is at the heart of organismal reproduction. Plants created from cuttings result from repeated cell divisions, as does the more familiar development of a multicellular organism from a fertilized egg. And eggs and sperm themselves result from cell division of a special kind. In this chapter, we discuss the two main types of cell division and their functions in organisms. ■ ■ ■

8.1 Like begets like, more or less

The ability of organisms to reproduce their own kind is the one characteristic that best distinguishes living things from nonliving matter (see Module 1.4 to review the characteristics of life). Only amoebas produce more amoebas, only people make more people, and only maple trees produce more maple trees. These simple facts of life have been recognized for thousands of years and are summarized by the age-old saying "Like begets like."

In a strict sense, the adage "Like begets like" applies only to **asexual reproduction,** the creation of genetically identical offspring by a single parent, without the participation of sperm and egg. For example, the amoeba in Figure 8.1A has duplicated its **chromosomes,** the structures that contain most of the organism's DNA. After doubling, identical chromosomes were allocated to opposite sides of the parent cell. When the parent cell divides, the resulting two daughter amoebas will be genetically identical to each other and to the original parent. (Biologists traditionally use the word "daughter" in this context only to indicate offspring, not to imply gender.) In asexual reproduction, there is one simple principle of inheritance: The lone parent and each of its offspring have identical genes.

The photograph of the family in Figure 8.1B makes the point that in a *sexually* reproducing species, like does not precisely beget like. Offspring produced by **sexual reproduction** generally resemble their parents more closely than they resemble unrelated individuals of the same species, but they are not identical to their parents or to each other. Each offspring inherits a unique combination of genes from its two parents, and this one-and-only set of genes programs a unique combination of traits. As a result, sexual reproduction can produce great variation among offspring. Notice in the photo that despite the family resemblances, each family member has a unique appearance. You probably resemble your parents more closely than a

Figure 8.1B Sexual reproduction produces offspring with unique combinations of genes.

stranger, but you do not look exactly like either parent or any of your siblings (unless you are an identical twin).

Long before anyone knew about genes, chromosomes, or the underlying principles of inheritance, people recognized that individuals of sexually reproducing species are highly varied. Furthermore, people learned to develop domestic breeds of plants and animals by controlling sexual reproduction. A domestic breed displays particular traits from among the great variety of traits found in the species as a whole. For example, though all domestic dogs belong to a single species, each breed of dog exhibits much less variability than the species as a whole. In producing each kind of domestic dog, breeders selected from a varied population of dogs certain individuals with specific traits and allowed these individuals to mate and produce offspring. The ancestry of a dog breed such as the German shepherd can be traced back many generations, during which breeders reduced variability in the breed by rigidly selecting for mating only those dogs with specific traits. In a sense, selective breeding is an attempt to make like beget like more than it does in nature.

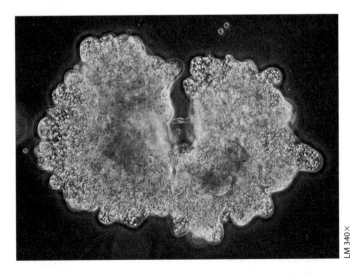

LM 340×

Figure 8.1A A single-celled amoeba producing a genetically identical offspring cell through asexual reproduction

? In terms of the "like begets like" adage, contrast asexual with sexual reproduction.

■ Asexual reproduction produces genetically identical offspring that inherit all their DNA from a single parent. The offspring of sexual reproduction show a family resemblance, but siblings vary because they inherit different combinations of genes from the two parents.

8.2 Cells arise only from preexisting cells

In 1858, German physician Rudolf Virchow stated an important biological principle: "Where a cell exists, there must have been a preexisting cell." Like many significant ideas now taken for granted, Virchow's principle (which he summarized as "Every cell from a cell") is both simple and profound. It tells us that the perpetuation of life, including all aspects of reproduction and inheritance, is based on the reproduction of cells, or **cell division.**

Cell division plays several important roles in the lives of organisms. In unicellular organisms, cell division can reproduce an entire organism. Cell division on a larger scale allows some multicellular organisms to reproduce asexually (such as plants that grow from cuttings). Among sexually reproducing organisms, cell division is the basis of sperm and egg formation. Cell division also enables sexually reproducing organisms to develop from a single cell—the fertilized egg, or zygote—into an adult organism. After an organism is fully grown, cell division continues to function in renewal and repair, replacing cells that die from normal wear and tear or accidents.

So far, we have discussed only the division of eukaryotic cells, and we will emphasize them in this chapter. Prokaryotes, however, also illustrate Virchow's principle, and in the next module, we look briefly at prokaryotic cell division.

> **?** What function does cell division play in an amoeba? What functions does it play in your body?
>
> ■ Reproduction; development, growth, and repair

8.3 Prokaryotes reproduce by binary fission

Prokaryotes (bacteria and archaea) reproduce by a type of cell division called **binary fission** ("dividing in half"). In typical prokaryotes, most genes are carried on a circular DNA molecule that, with associated proteins, constitutes the organism's single chromosome. Although prokaryotic chromosomes are

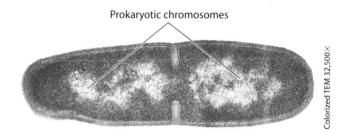

Prokaryotic chromosomes

Colorized TEM 32,500×

Figure 8.3B   Electron micrograph of a dividing bacterium

much smaller than those of eukaryotes, duplicating them in an orderly fashion and distributing the copies equally to two daughter bacteria is still a formidable task. Consider, for example, that when stretched out, the chromosome of the bacterium *Escherichia coli* is about 500 times longer than the cell itself. Accurately replicating this molecule when it is coiled and packed inside the cell is no small achievement.

Figure 8.3A illustrates binary fission in a prokaryote. ❶ As the chromosome is duplicating, one copy moves toward the opposite end of the cell. ❷ Meanwhile, as chromosome duplication progresses, the cell elongates. ❸ When chromosome duplication is complete and the bacterium has reached about twice its initial size, the plasma membrane grows inward, dividing the parent cell into two daughter cells. Figure 8.3B is an electron micrograph of a dividing bacterium at a stage similar to step 3 in Figure 8.3A.

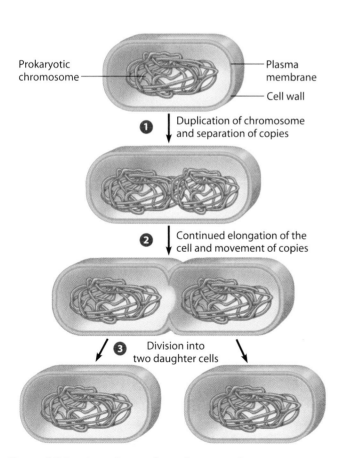

Prokaryotic chromosome — Plasma membrane — Cell wall

❶ Duplication of chromosome and separation of copies

❷ Continued elongation of the cell and movement of copies

❸ Division into two daughter cells

Figure 8.3A   Binary fission of a prokaryotic cell

> **?** Why is binary fission classified as asexual reproduction?
>
> ■ Because the genetically identical offspring inherit their DNA from a single parent.

8.4   The large, complex chromosomes of eukaryotes duplicate with each cell division

Eukaryotic cells are more complex and generally much larger than prokaryotic cells, and they have many more genes. Human cells, for example, carry about 25,000 genes, versus about 3,000 for a typical bacterium. Almost all the genes in the cells of humans, and in all other eukaryotes, are found in the cell nucleus, grouped into multiple chromosomes. (The exceptions include genes on the small DNA molecules of mitochondria and, in plants, of chloroplasts.)

Most of the time, chromosomes exist as a diffuse mass of long, thin fibers. This material, called **chromatin,** is a combination of DNA and protein molecules. As a cell prepares to divide, its chromatin coils up, forming compact, distinct chromosomes. In this state, the chromosomes become visible under the light microscope. Figure 8.4A is a micrograph of a plant cell that is about to divide; each dark thread is an individual chromosome. In fact, chromosomes (from the Greek *chromo,* colored, and *somes,* bodies) get their name from the fact that they are colored by certain stains used in microscopy.

Like a prokaryotic chromosome, each eukaryotic chromosome contains one long DNA molecule bearing hundreds or thousands of genes and, attached to the DNA, a number of protein molecules. However, the eukaryotic chromosome has a much more complex structure than the prokaryotic chromosome. The eukaryotic chromosome includes many more protein molecules, which help maintain the chromosome structure and control the activity of its genes. The number of chromosomes in a eukaryotic cell depends on the species. For example, human body cells generally have 46 chromosomes, while the body cells of a dog have 78.

Well before a eukaryotic cell begins to divide, it duplicates all of its chromosomes. The DNA molecule of each chromosome

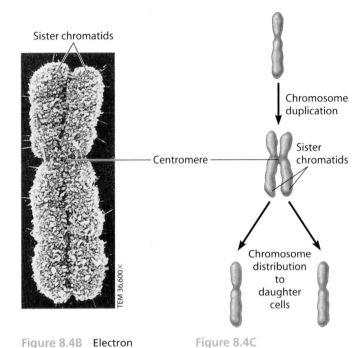

Figure 8.4B   Electron micrograph of a duplicated chromosome

Figure 8.4C Chromosome duplication and distribution

is copied, and new protein molecules attach as needed. The result is that each chromosome now consists of two copies called **sister chromatids,** which contain identical copies of the DNA molecule. Figure 8.4B is an electron micrograph of a human chromosome that has duplicated. The two chromatids are joined together especially tightly at a narrow "waist" called the **centromere.** The fuzzy appearance of the chromosome comes from the intricate twists and folds of its two chromatin fibers.

Figure 8.4C is a simple diagram making the point that when the cell divides, the sister chromatids of a duplicated chromosome separate from each other. Once separated from its sister, each chromatid is called a chromosome, and it is identical to the chromosome the cell started with. One of the new chromosomes goes to one daughter cell, and the other goes to the other daughter cell. In this way, each daughter cell receives a complete and identical set of chromosomes. In humans, for example, a typical dividing cell has 46 duplicated chromosomes (or 92 chromatids), and each of the two daughter cells that results from it has 46 single chromosomes.

?   When does a chromosome consist of two identical chromatids?

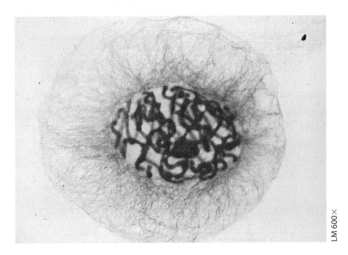

Figure 8.4A   A plant cell (from an African blood lily) just before division

■ When the cell is preparing to divide and has duplicated its chromosomes, but before the duplicates actually separate

The cell cycle multiplies cells

How do chromosome duplication and cell division fit into the life of a cell—and the life of an organism? As you already know, cell division is essential to life. Cell division is the basis of reproduction for every organism. It enables a multicellular organism to grow to adult size. It also replaces worn-out or damaged cells, keeping the total cell number in a mature individual relatively constant. In your own body, for example, millions of cells must divide every second to maintain the total number of about 100 trillion cells. Some cells divide once a day, others less often, and highly specialized cells, such as our mature muscle cells, not at all.

The process of cell division is a key component of the **cell cycle,** an ordered sequence of events that extends from the time a cell is first formed from a dividing parent cell until its own division into two cells. The cell cycle consists of two broad stages: a growing stage (called interphase), during which the cell roughly doubles everything in its cytoplasm and precisely replicates its chromosomal DNA, and the actual cell division (called the mitotic phase).

As Figure 8.5 indicates, most of the cell cycle is spent in **interphase.** This is a time when a cell's metabolic activity is very high and the cell performs its various functions within the organism. Moreover, a cell in interphase increases its supply of proteins, creates more cytoplasmic organelles (such as mitochondria and ribosomes), and grows in size. Additionally, the chromosomes duplicate during this period. Typically, interphase lasts for at least 90% of the total time required for the cell cycle.

Interphase can be divided into three subphases: the G_1 phase ("first gap"), the S phase, and the G_2 phase ("second gap"). During all three subphases, the cell grows. However, chromosomes are duplicated only during the S phase. S stands for synthesis of DNA—also known as DNA replication—a process that will be discussed in Chapter 10. At the beginning of the S phase, each chromosome is single. At the end of this phase, after DNA replication, the chromosomes are double, each consisting of two sister chromatids. To summarize interphase, a cell grows (G_1), continues to grow as it copies its chromosomes (S), and then grows more as it completes preparations for cell division (G_2).

The **mitotic phase (M phase),** the part of the cell cycle when the cell actually divides, accounts for only about 10% of the total time required for the cell cycle. The mitotic phase is divided into two stages, called mitosis and cytokinesis, although the second stage begins before the first one ends. In **mitosis** (light yellow area in the figure), the nucleus and its contents, including the duplicated chromosomes, divide and are evenly distributed to form two daughter nuclei. During **cytokinesis,** the cytoplasm is divided in two. The combination of mitosis and cytokinesis produces two genetically identical daughter cells, each with a single nucleus, surrounding cytoplasm, and plasma membrane. Each newly produced daughter cell may then proceed through G_1 and repeat the cycle.

Mitosis is unique to eukaryotes and may be an evolutionary solution to the problem of allocating identical copies of a large amount of genetic material in a number of separate chromosomes to two daughter cells. Mitosis is a remarkably accurate mechanism. Experiments with yeast, for example, indicate that an error in chromosome distribution occurs only once in about 100,000 cell divisions.

A living cell viewed through a light microscope undergoes dramatic changes in appearance during the mitotic phase. During interphase, the cell's individual chromosomes are not distinguishable because they are in the form of loosely packed chromatin fibers. With the onset of mitosis, however, striking changes are visible in the chromosomes and other structures, as we see in the next module.

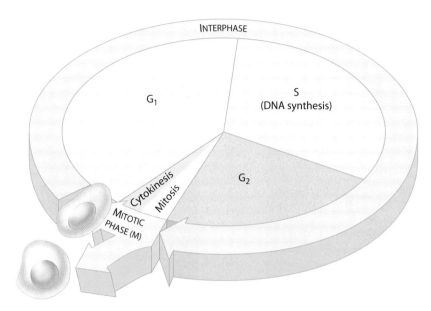

Figure 8.5 The eukaryotic cell cycle

(Bio Flix) **BioFlix** *Mitosis*

(headphones) **MP3 Tutor** *Mitosis*

Web Activity *The Cell Cycle*

? A researcher treats cells with a chemical that prevents DNA synthesis from starting. This treatment would trap the cells in which part of the cell cycle?

■ G_1

8.6 Cell division is a continuum of dynamic changes

The light micrographs in Figure 8.6 show the cell cycle for an animal cell—in this case, from a newt. Interphase is included, but the emphasis is on the dramatic changes that occur during cell division, the mitotic phase. Mitosis is a continuum of changes, but biologists distinguish five main stages: **prophase, prometaphase,** **metaphase, anaphase,** and **telophase.** The drawings in Figure 8.6 show details not visible in the micrographs. For simplicity, only four chromosomes are drawn.

The chromosomes are the stars of the mitotic drama, and their movements depend on the **mitotic spindle,** a

Figure 8.6 The stages of cell division

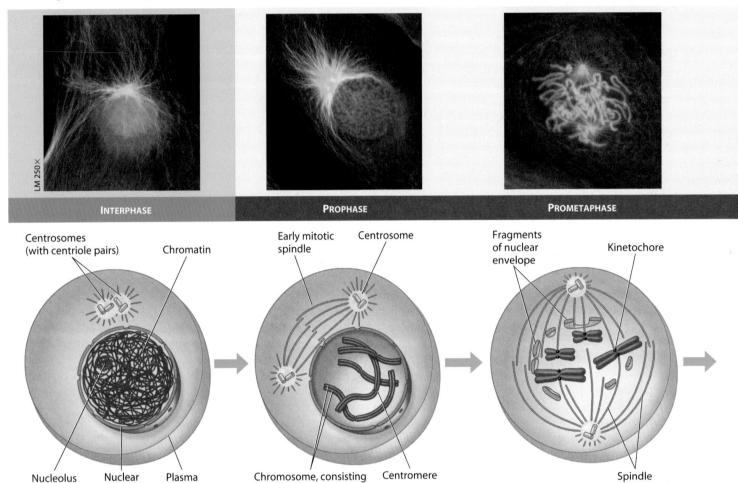

Interphase Interphase is the period of cell growth when the cell synthesizes new molecules and organelles. At the point shown here, late interphase (G_2), the cell looks much the same as it does throughout interphase. Nonetheless, by the G_2 stage the cell has doubled much of its earlier contents and the cytoplasm contains two centrosomes. Within the nucleus, the chromosomes are duplicated, but they cannot be distinguished individually because they are still in the form of loosely packed chromatin. The nucleus also contains one or more nucleoli, an indication that the cell is actively making proteins. Nucleoli are where the parts of ribosomes are assembled before export to the cytoplasm (see Chapter 10).

Prophase During prophase, changes occur in both the nucleus and the cytoplasm. Within the nucleus, the chromatin fibers become more tightly coiled and folded, forming discrete chromosomes that can be seen with the light microscope. The nucleoli disappear. Each duplicated chromosome appears as two identical sister chromatids joined together, with a narrow "waist" at the centromere. In the cytoplasm, the mitotic spindle begins to form as microtubules rapidly grow out from the centrosomes, which begin to move away from each other.

Prometaphase The nuclear envelope breaks into fragments and disappears. Microtubules emerging from the centrosomes at the poles (ends) of the spindle reach the chromosomes, now highly condensed. At the centromere region, each sister chromatid has a protein structure called a kinetochore (shown as a black dot). Some of the spindle microtubules attach to the kinetochores, throwing the chromosomes into agitated motion. Other spindle microtubules make contact with microtubules coming from the opposite pole. Forces exerted by protein "motors" associated with spindle microtubules move the chromosomes toward the center of the cell.

football-shaped structure of microtubules that guides the separation of the two sets of daughter chromosomes. The spindle microtubules emerge from two **centrosomes,** clouds of cytoplasmic material that in animal cells contain centrioles (see Figure 4.4A). (Centrosomes are also known as *microtubule-organizing centers,* a term describing their function.) The role of centrioles in cell division is a mystery; destroying them experimentally does not interfere with normal spindle formation, and plant cells lack them entirely.

Web Process of Science *How Much Time Do Cells Spend in Each Phase of Mitosis?*

? An organism called a plasmodial slime mold is one large cytoplasmic mass with many nuclei. Explain how such a "megacell" could form.

■ Mitosis occurs repeatedly without cytokinesis.

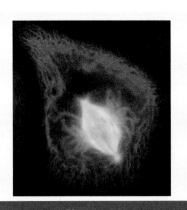 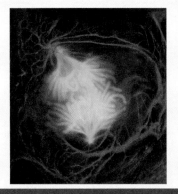 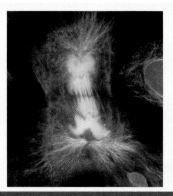

| METAPHASE | ANAPHASE | TELOPHASE AND CYTOKINESIS |

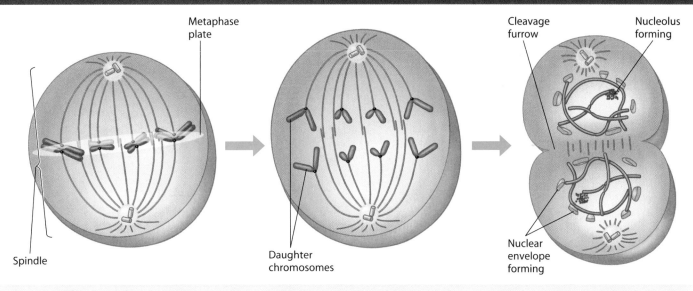

Metaphase plate

Spindle

Daughter chromosomes

Cleavage furrow

Nucleolus forming

Nuclear envelope forming

Metaphase At metaphase, the mitotic spindle is fully formed, with its poles at opposite ends of the cell. The chromosomes convene on the metaphase plate, an imaginary plane equidistant between the two poles of the spindle. The centromeres of all the chromosomes are lined up on the metaphase plate. For each chromosome, the kinetochores of the two sister chromatids face opposite poles of the spindle. The microtubules attached to a particular chromatid all come from one pole of the spindle, and those attached to its sister chromatid come from the opposite pole.

Anaphase Anaphase begins when the two centromeres of each chromosome come apart, separating the sister chromatids. Once separate, each sister chromatid is considered a full-fledged (daughter) chromosome. Motor proteins of the kinetochores, powered by ATP, "walk" the daughter chromosomes centromere-first along the microtubules toward opposite poles of the cell. As this happens, the spindle microtubules attached to the kinetochores shorten. However, the spindle microtubules not attached to chromosomes lengthen. The poles are moved farther apart, elongating the cell. Anaphase is over when equivalent—and complete—collections of chromosomes have reached the two poles of the cell.

Telophase Telophase is roughly the reverse of prophase. The cell elongation that started in anaphase continues. Daughter nuclei appear at the two poles of the cell as nuclear envelopes form around the chromosomes. Meanwhile, the chromatin fiber of each chromosome uncoils, and nucleoli reappear. At the end of telophase, the mitotic spindle disappears. Mitosis, the equal division of one nucleus into two genetically identical daughter nuclei, is now finished.

Cytokinesis Cytokinesis, the division of the cytoplasm, usually occurs along with telophase, with two daughter cells completely separating soon after the end of mitosis. In animal cells, cytokinesis involves a cleavage furrow, which pinches the cell in two.

8.7 Cytokinesis differs for plant and animal cells

Cytokinesis, or division of one cell into two, typically begins during telophase, although it may begin in late anaphase. In animal cells, cytokinesis occurs by a process known as cleavage. As shown in Figure 8.7A, the first sign of cleavage is the appearance of a **cleavage furrow,** a shallow groove in the cell surface. At the site of the furrow, the cytoplasm has a ring of microfilaments made of actin, associated with molecules of the protein myosin. (Actin and myosin are the same proteins responsible for muscle contraction—see Module 30.8.) When the actin microfilaments interact with the myosin, the ring contracts, like the pulling of drawstrings. The cleavage furrow deepens and eventually pinches the parent cell in two, producing two completely separate cells, each with its own nucleus and share of cytoplasm.

Cytokinesis is markedly different in plant cells, which possess cell walls (Figure 8.7B). During telophase, vesicles containing cell wall material (tan in figure) collect at the middle

of the parent cell. The vesicles fuse, forming a membranous **cell plate.** The cell plate grows outward, accumulating more cell wall materials as more vesicles fuse with it. Eventually, the outer edges of the cell plate fuse with the plasma membrane, and the cell plate's contents join the parental cell wall. The result is two daughter cells, each bounded by its own plasma membrane and cell wall.

Web Activity *Mitosis and Cytokinesis Video*

? Contrast cytokinesis in animals with cytokinesis in plants.

In animals, cytokinesis involves a cleavage furrow in which contracting microfilaments pinch the cell in two. In plants, it involves formation of a cell plate, a fusion of vesicles that forms new membrane and walls between the cells.

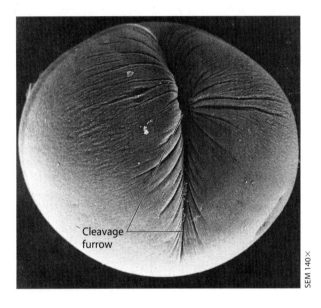

Cleavage furrow

SEM 140×

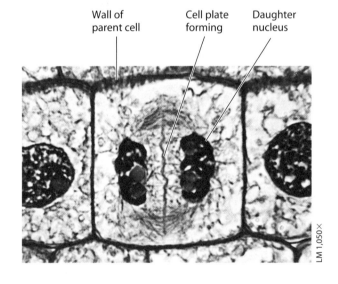

Wall of parent cell   Cell plate forming   Daughter nucleus

LM 1,050×

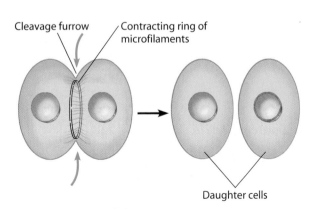

Cleavage furrow   Contracting ring of microfilaments

Daughter cells

Figure 8.7A   Cleavage of an animal cell

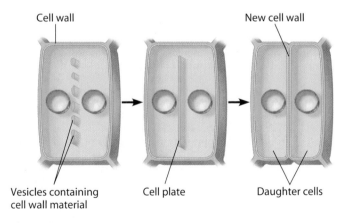

Cell wall   New cell wall

Vesicles containing cell wall material   Cell plate   Daughter cells

Figure 8.7B   Cell plate formation in a plant cell

8.8 Anchorage, cell density, and chemical growth factors affect cell division

For a plant or an animal to grow and develop normally and to maintain its tissues once full grown, it must be able to control the timing of cell division in different parts of its body. For example, in the adult human, skin cells and the cells lining the digestive tract divide frequently throughout life, replacing cells that are constantly being abraded and sloughed off. In contrast, cells in the human liver usually do not divide unless the liver is damaged. Cell division in this case repairs wounds.

By growing animal cells in culture, researchers have been able to identify many factors, both physical and chemical, that can influence cell division. For example, cells fail to divide if an essential nutrient is left out of the culture medium. And most types of mammalian cells divide in culture only if certain specific growth factors are included. A **growth factor** is a protein secreted by certain body cells that stimulates other cells to divide (Figure 8.8A). Researchers have discovered at least 50 different growth factors that can trigger cell division. Different cell types respond specifically to certain growth factors or a combination of growth factors. For example, injury to the skin causes blood platelets to release a protein called platelet-derived growth factor. This protein promotes the rapid growth of connective tissue cells that help seal the wound.

The effect of a physical factor on cell division is clearly seen in **density-dependent inhibition**, a phenomenon in which crowded cells stop dividing. For example, animal cells growing on the surface of a dish multiply to form a single layer and usually stop dividing when they touch one another (Figure 8.8B). If some cells are removed, those bordering the open space begin dividing again and continue until the vacancy is filled. What actually causes the cessation of growth? Studies of cultured cells suggest that physical contact of cell-surface proteins between adjacent cells is responsible for inhibiting cell division.

Most animal cells also exhibit **anchorage dependence;** they must be in contact with a solid surface—such as the inside of a culture dish or the extracellular matrix of a tissue—to divide. But density-dependent inhibition mediated by the availability of growth factors is probably the most important regulatory mechanism controlling the division of the body's cells. How exactly do growth factors work? We pursue answers to this question in the next module.

? Compared to a control culture, the cells in an experimental culture are fewer but much larger when they cover the dish surface and stop growing. What is a reasonable hypothesis for this difference?

■ The experimental culture is deficient in one or more growth factors.

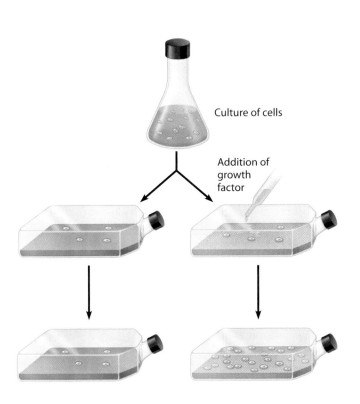

Figure 8.8A An experiment demonstrating the effect of growth factors on the division of cultured animal cells

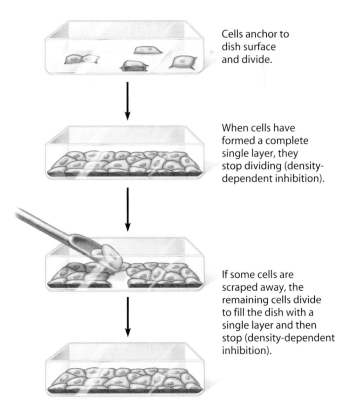

Figure 8.8B An experiment demonstrating density-dependent inhibition, using animal cells grown in culture

8.9 Growth factors signal the cell cycle control system

In a living animal, most cells are anchored in a fixed position and bathed in a solution of nutrients supplied by the blood, yet they usually do not divide unless they are signaled by other cells to do so. Growth factors are the main signals, and their role in promoting cell division leads us back to our earlier discussion of the cell cycle.

The sequential events of the cell cycle, represented by the circle of flat blocks in Figure 8.9A, are directed by a distinct cell cycle control system, represented by the knob in the center. Analogous to the control device of an automatic washing machine, the **cell cycle control system** is a cyclically operating set of molecules in the cell that both triggers and coordinates key events in the cell cycle. The cell cycle is *not* like a row of falling dominoes, with each event causing the next one in line. Within the M phase, for example, metaphase does not automatically lead to anaphase. Instead, proteins of the cell cycle control system must trigger the separation of sister chromatids that marks the start of anaphase.

A checkpoint in the cell cycle is a critical control point where stop and go-ahead signals can regulate the cycle. (The signals are transmitted within the cell by signal transduction pathways—see Figure 5.1C.) Animal cells generally have built-in stop signals that halt the cell cycle at checkpoints until overridden by go-ahead signals.

The red barriers in Figure 8.9A represent three major checkpoints in the cell cycle: during the G_1 and G_2 subphases of interphase and in the M phase. Intracellular signals detected by the control system tell it whether key cellular processes up to each point have been completed and thus whether or not the cell cycle should proceed past that point. The control system also receives messages from outside the cell, indicating both general environmental conditions and the presence of specific signal molecules from other cells. When the cell cycle control system gets a go-ahead signal at the G_1 checkpoint, for example, the cell soon enters the S phase of the cell cycle.

For many cells, the G_1 checkpoint seems to be the most important. If a cell receives a go-ahead signal—for example, from a growth factor—at the G_1 checkpoint, it will usually complete its cycle and divide. But if it does not receive a go-ahead signal at G_1, it will exit the cell cycle, switching into a nondividing state called the G_0 phase. For instance, nondividing nerve cells and muscle cells in your body are in the G_0 phase.

Figure 8.9B shows a simplified model for how a growth factor might affect the cell cycle control system at the G_1 checkpoint. A cell that responds to a growth factor (▼) has molecules of a specific receptor protein in its plasma membrane. Binding of the growth factor to the receptor triggers a signal transduction pathway in the cell, a pathway that in this case leads to cell division. The "signals" are changes that a molecule induces in the next molecule in the pathway. Via a series of relay molecules, a signal finally reaches the cell cycle control system and overrides the brakes that otherwise prevent progress of the cell cycle. In Figure 8.9B, the cell cycle is set off from the cell in a separate diagram because, unlike the components of the timer in a washing machine, the proteins making up the control system in the cell are not actually located together in one place.

Research on the control of the cell cycle is one of the hottest areas in biology today. This research is leading to a better understanding of cancer, which we discuss next.

? At which of the three checkpoints described in this module do the chromosomes exist as duplicated sister chromatids?

■ G_2 and M

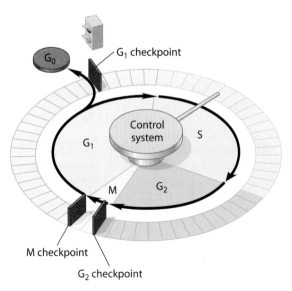

Figure 8.9A Mechanical model for the cell cycle control system

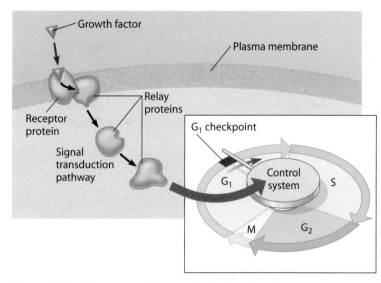

Figure 8.9B How a growth factor signals the cell cycle control system

8.10 Growing out of control, cancer cells produce malignant tumors

Cancer, which currently claims the lives of one out of every five people in the United States and other developed nations, is a disease of the cell cycle. Cancer cells do not respond normally to the cell cycle control system; they divide excessively and can invade other tissues of the body. If unchecked, cancer cells may continue to grow until they kill the organism.

The abnormal behavior of cancer cells begins when a single cell undergoes transformation, a process that converts a normal cell to a cancer cell. The body's immune system normally recognizes a transformed cell as abnormal and destroys it. However, if the cell evades destruction, it may proliferate to form a **tumor,** an abnormally growing mass of body cells. If the abnormal cells remain at the original site, the lump is called a **benign tumor.** Benign tumors can cause problems if they grow in and disrupt certain organs, such as the brain, but often they can be completely removed by surgery.

In contrast, a **malignant tumor** can spread into neighboring tissues and other parts of the body, displacing normal tissue and interrupting organ function as it goes (Figure 8.10). Cancer cells may separate from the original tumor or secrete signal molecules that cause blood vessels to grow toward the tumor. A few tumor cells may then enter the blood and lymph vessels of the circulatory system and move to other parts of the body, where they may proliferate and form new tumors. This spread of cancer cells via the circulatory system beyond their original site is called **metastasis.** An individual with a malignant tumor is said to have cancer.

Cancers are named according to the organ or tissue in which they originate. Liver cancer, for example, starts in liver tissue and may or may not spread from there. Based on their site of origin, cancers are grouped into four categories. **Carcinomas** are cancers that originate in the external or internal coverings of the body, such as the skin or the lining of the intestine. **Sarcomas** arise in tissues that support the body, such as bone and muscle. Cancers of blood-forming tissues, such as bone marrow, spleen, and lymph nodes, are called **leukemias** and **lymphomas.**

From studying cancer cells in culture, researchers have learned that these cells do not heed the normal signals that regulate the cell cycle. For example, cancer cells do not exhibit density-dependent inhibition; they continue to divide even at high densities, piling up on one another. Many cancer cells have defective cell cycle control systems that proceed past checkpoints even in the absence of growth factors. Other cancer cells themselves synthesize growth factors that make them divide continuously. If cancer cells do stop dividing, they seem to do so at random points in the cell cycle, rather than at the normal cell cycle checkpoints. Moreover, in culture, cancer

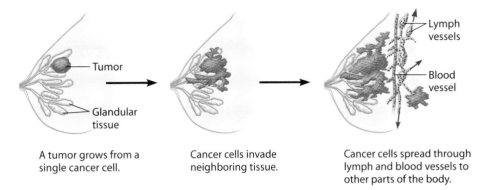

A tumor grows from a single cancer cell.

Cancer cells invade neighboring tissue.

Cancer cells spread through lymph and blood vessels to other parts of the body.

Figure 8.10 Growth and metastasis of a malignant (cancerous) tumor of the breast

cells are "immortal"; they can go on dividing indefinitely, as long as they have a supply of nutrients (whereas normal mammalian cells divide only about 20 to 50 times before they stop).

Luckily, many tumors can be successfully treated. A tumor that appears to be localized may be removed surgically. Alternatively, it can be treated with high-energy radiation, which damages DNA in cancer cells much more than it does in normal cells, apparently because cancer cells have lost the ability to repair such damage. However, there is sometimes enough damage to normal body cells to produce bad side effects. For example, radiation damage to cells of the ovaries or testes can lead to sterility.

To treat widespread or metastatic tumors, chemotherapy is used. During periodic chemotherapy treatments, drugs are administered that disrupt specific steps in the cell cycle. For example, the drug paclitaxel (trade name Taxol) "freezes" the mitotic spindle, which stops actively dividing cells from proceeding past metaphase. (Interestingly, Taxol was originally discovered in the bark of the Pacific yew tree, found mainly in the northwestern United States.) Vinblastin, a chemotherapeutic drug first obtained from the periwinkle plant (found in the rain forests of Madagascar), prevents the spindle from forming in the first place.

The side effects of chemotherapy are due to the drugs' effects on normal cells that rapidly divide. For example, nausea results from chemotherapy's effects on intestinal cells, hair loss from effects on hair follicle cells, and susceptibility to infection from effects on immune cell production.

We will return to the topic of cancer in Chapter 11, after studying the structure and function of genes. We will see that cancer results from changes in genes for proteins that control cell division.

 What is metastasis?

■ Metastasis is the spread of cancer cells via the circulatory system from their original site of formation to sites in the body.

8.11 Review: Mitosis provides for growth, cell replacement, and asexual reproduction

The three micrographs in Figures 8.11A–8.11C summarize the roles that mitotic cell division plays in the lives of multicellular organisms. Figure 8.11A shows some of the cells from the tip of a rapidly growing onion plant root. Notice the several cells whose nuclei are in various stages of mitosis (as evidenced by the visibly compact chromosomes). Cell division in the root tip produces new cells, which elongate and bring about growth of the root.

Figure 8.11B shows a dividing bone marrow cell. Mitotic cell division within the red marrow of your body's bones (particularly within your ribs, vertebrae, breastbone, and pelvis) continuously creates new blood cells that replace older ones. Similar processes replace cells throughout your body. For example, dividing cells within your epidermis continuously replace dead cells that slough off the surface of your skin.

Figure 8.11C is a micrograph of a hydra, a common inhabitant of freshwater lakes. A hydra is a tiny multicellular animal that reproduces by either sexual or asexual means. This individual is reproducing asexually by budding. A bud starts out as a mass of mitotically dividing cells growing on the side of the parent. The bud develops into a small hydra like the one in the photo here. Eventually, the offspring detaches from the parent and takes up life on its own. The offspring is literally a "chip off the old block," being genetically identical to (a clone of) its parent.

In all three of these cases—the growing onion root, replacement of blood cells, and budding hydra—the new cells have exactly the same number and types of chromosomes as the parent cells because of the way duplicated chromosomes divide during mitosis. Mitosis makes it possible for organisms to grow, regenerate and repair tissues, and reproduce asexually by producing cells that carry the same genes as the parent cells.

If we examine the cells of any individual organism, we see that almost all of them contain the same number and types of chromosomes. Likewise, if we examine cells from different individuals of any one species, we see that they have the same number and types of chromosomes. In the next module, we take a closer look at how chromosomes are organized in cells.

> **?** If a human skin cell with 46 chromosomes divides by mitosis, each daughter cell will have ___ chromosomes.
>
> 46 ■

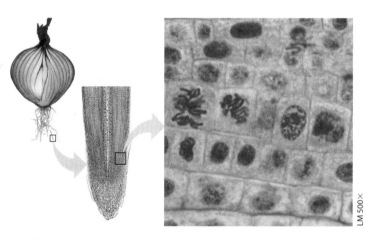

Figure 8.11A   Growth (in an onion root)

LM 500×

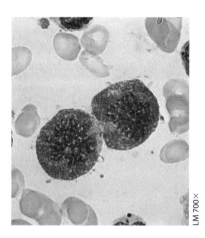

Figure 8.11B   Cell replacement (in bone marrow)

LM 700×

Figure 8.11C   Asexual reproduction (of a hydra)

LM 20×

Meiosis and Crossing Over

8.12 Chromosomes are matched in homologous pairs

In humans, a typical body cell, called a **somatic cell,** has 46 chromosomes. If we use a microscope to examine human chromosomes in metaphase of mitosis, we see that each chromosome (consisting of two sister chromatids) has a twin chromosome that is identical in length and centromere position. Figure 8.12 (at the top of the next page) illustrates one pair of metaphase chromosomes. Altogether, the cell contains 23 such matched pairs of duplicated chromosomes. Other species have different numbers of chromosomes, but these too usually occur in matched pairs. Moreover, when treated with special dyes, the chromosomes of a pair display matching staining patterns (represented by colored stripes in Figure 8.12). Notice that each chromosome consists of two sister chromatids joined at the centromere. The two chromosomes composing a pair are called **homologous chromosomes** (or homologs) because they both carry genes controlling the same inherited characteristics. For

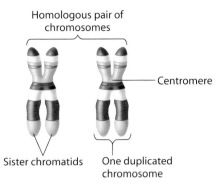

Figure 8.12
A homologous pair of chromosomes

Homologous pair of chromosomes

Centromere

Sister chromatids

One duplicated chromosome

example, if a gene that determines whether a person has freckles is located at a particular place, or **locus** (plural, *loci*), on one chromosome—within the narrow orange band in our drawing, for instance—then the other chromosome of the homologous pair also has a gene for freckles at that locus. (However, the two homologs may have different versions of the freckles gene, perhaps one that promotes freckles and one that does not.)

The two distinct chromosomes called X and Y are an important exception to the general pattern of homologous chromosomes. Human females have a homologous pair of X chromosomes (XX), but males have one X and one Y chromosome (XY). Only small parts of the X and Y are homologous; most of the genes carried on the X chromosome do not have counterparts on the tiny Y, and the Y chromosome has genes lacking on the X. Because they determine an individual's sex, the X and Y chromosomes are called **sex chromosomes** (although they carry genes that perform other functions as well). The other 22 pairs of chromosomes are called **autosomes.**

For both autosomes and sex chromosomes, we inherit one chromosome of each pair from our mother and the other from our father, as discussed in the next module.

? **What is the explanation for our having two of each type of chromosome?**

■ We inherit one of each kind from each parent.

8.13 Gametes have a single set of chromosomes

Having two sets of chromosomes, one inherited from each parent, is a key factor in the human life cycle (outlined in Figure 8.13) and in the life cycles of all other species that reproduce sexually. (We examine other life cycles in Unit IV.)

Any cell with two homologous sets of chromosomes is called a **diploid cell,** and the total number of chromosomes is called the diploid number (abbreviated $2n$). For humans, the diploid number is 46; that is, $2n = 46$. Humans are said to be diploid

organisms because almost all our cells are diploid. The exceptions are the egg and sperm cells, collectively known as **gametes.** Each gamete has a single set of chromosomes: 22 autosomes plus a single sex chromosome, either X or Y. A cell with a single chromosome set is called a **haploid cell.** For humans, the haploid number (abbreviated n) is 23; that is, $n = 23$.

In humans, sexual intercourse allows a haploid sperm cell from the father to reach and fuse with a haploid egg cell of the mother in the process of **fertilization.** The resulting fertilized egg, called a **zygote,** is diploid. It has two haploid sets of chromosomes: one set from the mother and a homologous set from the father. The life cycle is completed as a sexually mature adult develops from the zygote. Mitotic cell division ensures that all somatic cells of the human body receive copies of all of the zygote's 46 chromosomes.

All sexual life cycles, including our own, involve an alternation of diploid and haploid stages. Having haploid gametes keeps the chromosome number from doubling in each generation. Gametes are made by a special sort of cell division called meiosis, which occurs only in reproductive organs (ovaries and testes in animals). Whereas mitosis produces daughter cells with the same numbers of chromosomes as the parent cell, meiosis reduces the chromosome number by half. We turn to meiosis next.

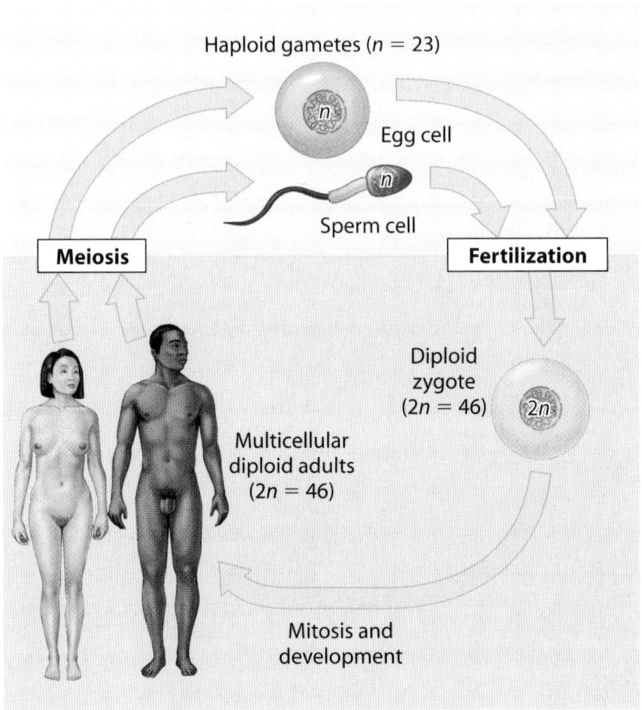

Web Activity *Asexual and Sexual Life Cycles*

? Imagine you stain a human cell and view it under a microscope. You observe 23 chromosomes, including a Y chromosome. You could conclude that this must be a _____ cell taken from the organ called the _____.

Haploid gametes ($n = 23$)

n

Egg cell

n

Sperm cell

Meiosis

Fertilization

Diploid zygote ($2n = 46$)

$2n$

Multicellular diploid adults ($2n = 46$)

Mitosis and development

Figure 8.13 The human life cycle

■ sperm . . . testes

8.14 Meiosis reduces the chromosome number from diploid to haploid

Meiosis is a type of cell division that produces haploid gametes in diploid organisms. Many of the stages of meiosis closely resemble corresponding stages in mitosis. Meiosis, like mitosis, is preceded by the duplication of chromosomes. However, this single duplication is followed by two consecutive cell divisions, called meiosis I and meiosis II. These

Figure 8.14 The stages of meiosis

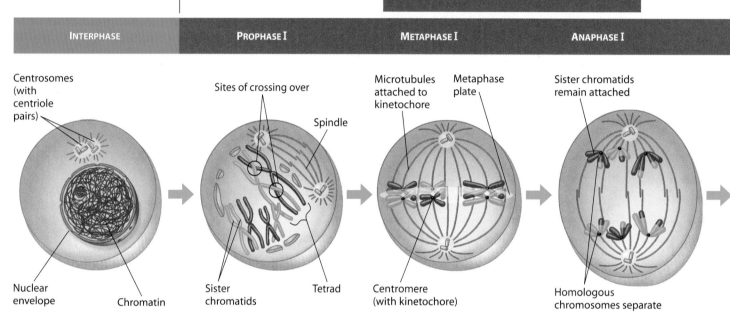

MEIOSIS I: Homologous chromosomes separate

| INTERPHASE | PROPHASE I | METAPHASE I | ANAPHASE I |

Centrosomes (with centriole pairs)

Sites of crossing over

Spindle

Microtubules attached to kinetochore

Metaphase plate

Sister chromatids remain attached

Nuclear envelope

Chromatin

Sister chromatids

Tetrad

Centromere (with kinetochore)

Homologous chromosomes separate

Interphase
Like mitosis, meiosis is preceded by an interphase, during which the chromosomes duplicate. At the end of this interphase, each chromosome consists of two genetically identical sister chromatids attached together. But at this stage the chromosomes are not yet visible under the microscope except as a mass of chromatin. The cell's centrosome has also duplicated by the end of this interphase.

Prophase I
Prophase I is the most complex phase of meiosis and typically occupies over 90% of the time required for meiotic cell division. Early in this phase, the chromatin coils up, so that individual chromosomes become visible with the microscope. In a process called **synapsis**, homologous chromosomes, each composed of two sister chromatids, come together as pairs. The resulting structure, consisting of four chromatids, is called a tetrad. During synapsis, chromatids of homologous chromosomes exchange segments in a process called crossing over. Because the

versions of the genes on a chromosome (or one of its chromatids) may be different from those on its homologue, crossing over rearranges genetic information. As we will discuss in Module 8.18, the genetic shuffling produced by crossing over can make an important contribution to the genetic variability resulting from sexual reproduction.

As prophase I continues, the chromosomes condense further as the nucleoli disappear. Now the centrosomes move away from each other, and a spindle starts to form between them. The nuclear envelope breaks into fragments, and the chromosome tetrads, captured by spindle microtubules, are moved toward the center of the cell.

Metaphase I
At metaphase I, the chromosome tetrads are aligned on the metaphase plate, midway between the two poles of the spindle. Each chromosome is condensed and thick, with its sister chromatids still attached at their centromeres. Spindle microtubules are attached to kinetochores at the centromeres. In each tetrad, the homologous chromosomes are

held together at sites of crossing over. Notice that, for each tetrad, the spindle microtubules attached to one of the homologous chromosomes come from one pole of the cell, and the microtubules attached to the other homologous chromosome come from the opposite pole. With this arrangement, the homologous chromosomes of each tetrad are poised to move toward opposite poles of the cell.

Anaphase I
Like anaphase of mitosis, anaphase I of meiosis is marked by the migration of chromosomes toward the two poles of the cell. In contrast to mitosis, however, the sister chromatids making up each doubled chromosome remain attached at their centromeres. Only the tetrads (pairs of homologous chromosomes) split up. Thus, in the drawing you see three still-doubled chromosomes (the haploid number) moving toward each spindle pole. If this were anaphase of mitosis, you would see six daughter chromosomes moving toward each pole.

divisions result in four daughter cells (rather than the two daughter cells of mitosis), each with a single haploid set of chromosomes. Thus, meiosis produces daughter cells with only half as many chromosomes as the parent cell. The drawings in Figure 8.14 show the two meiotic divisions for an animal cell with a diploid number of 6.

BioFlix *Meiosis*

MP3 Tutor *Meiosis*

? A cell has the haploid number of chromosomes, but each chromosome has two chromatids. The chromosomes are arranged at the center of the spindle. What is the meiotic stage?

■ Metaphase II

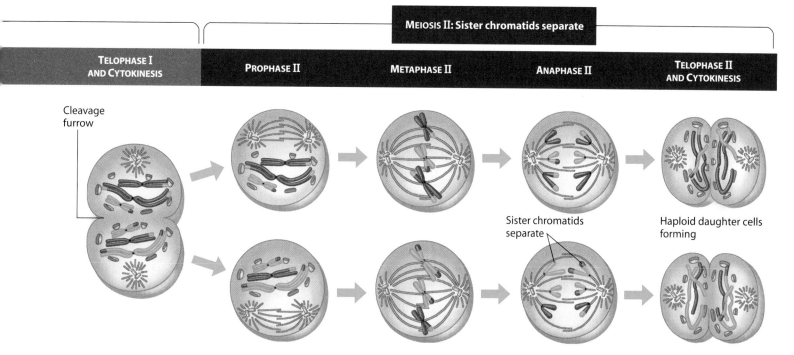

| MEIOSIS II: Sister chromatids separate |
| TELOPHASE I AND CYTOKINESIS | PROPHASE II | METAPHASE II | ANAPHASE II | TELOPHASE II AND CYTOKINESIS |

Cleavage furrow

Sister chromatids separate

Haploid daughter cells forming

Telophase I and Cytokinesis

In telophase I, the chromosomes arrive at the poles of the cell. When the chromosomes finish their journey, each pole of the cell has a haploid chromosome set, although each chromosome is still in duplicate form at this point. In other words, each chromosome still consists of two sister chromatids. Usually, cytokinesis occurs along with telophase I, and two haploid daughter cells are formed.

Following telophase I in some organisms, the chromosomes uncoil and the nuclear envelope re-forms, and there is an interphase before meiosis II begins. In other species, daughter cells produced in the first meiotic division immediately begin preparation for the second meiotic division. In either case, no chromosome duplication occurs between telophase I and the onset of meiosis II.

Meiosis II

In organisms having an interphase after meiosis I, the chromosomes condense again and the nuclear envelope breaks down during prophase II. In any case, meiosis II is essentially the same as mitosis. The important difference is that meiosis II starts with a haploid cell.

During prophase II, a spindle forms and moves the chromosomes toward the middle of the cell. During metaphase II, the chromosomes are aligned on the metaphase plate as they are in mitosis, with the kinetochores of the sister chromatids of each chromosome pointing toward opposite poles. In anaphase II, the centromeres of sister chromatids finally separate, and the sister chromatids of each pair, now individual daughter chromosomes, move toward opposite poles of the cell. In telophase II, nuclei form at the cell poles, and cytokinesis occurs at the same time. There are now four daughter cells, each with the haploid number of (single) chromosomes.

8.15 Mitosis and meiosis have important similarities and differences

We have now described the two ways that cells of eukaryotic organisms divide. Mitosis, which provides for growth, tissue repair, and asexual reproduction, produces daughter cells genetically identical to the parent cell. Meiosis, needed for sexual reproduction, yields haploid daughter cells—cells with one member of each homologous chromosome pair.

For both mitosis and meiosis, the chromosomes duplicate only once, in the preceding interphase. Mitosis involves one division of the nucleus, and it is usually accompanied by cytokinesis, producing two identical cells. Meiosis entails two nuclear and cytoplasmic divisions, yielding four haploid cells.

Figure 8.15 compares mitosis and meiosis, tracing these two processes for a diploid parent cell with four chromosomes. Homologous chromosomes are those matching in size.

All the events unique to meiosis occur during meiosis I. In prophase I, duplicated homologous chromosomes pair to form **tetrads,** four chromatids, with each pair of sister chromatids joined at their centromeres. Crossing over occurs between homologous (nonsister) chromatids. In metaphase I, tetrads (not individual chromosomes) are aligned at the metaphase plate. During anaphase I, homologous pairs of chromosomes separate, but the sister chromatids of each chromosome stay together. At the end of meiosis I, there are two haploid cells, but each chromosome still has two sister chromatids.

Meiosis II is virtually identical to mitosis and separates sister chromatids. But unlike mitosis, each daughter cell produced by meiosis II has only a *haploid* set of chromosomes.

🎧 **MP3 Tutor** *Mitosis–Meiosis Comparison*

❓ Explain how mitosis conserves chromosome number while meiosis reduces the number from diploid to haploid.

In mitosis, the duplication of chromosomes is followed by one division of the cell. In meiosis, homologous chromosomes separate in the first of two cell divisions; after the second division, each new cell ends up with just a single haploid set.

Figure 8.15 Comparison of mitosis and meiosis

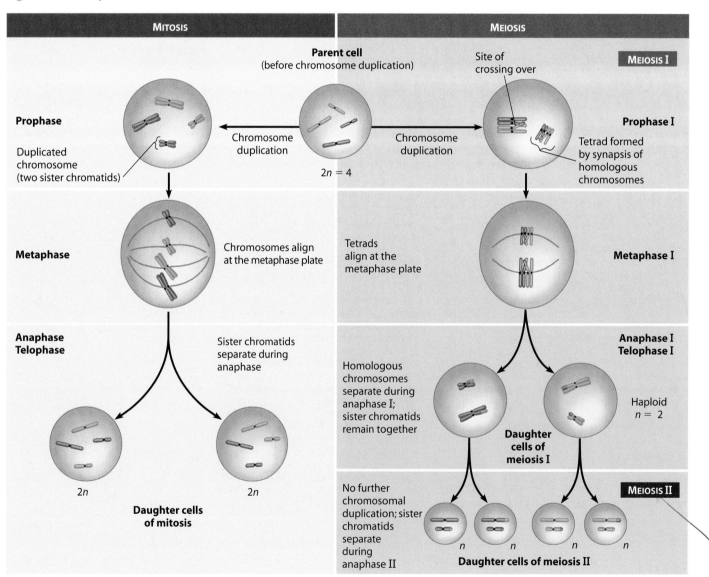

Independent orientation of chromosomes in meiosis and random fertilization lead to varied offspring

As we discussed in Module 8.1, offspring that result from sexual reproduction are highly varied; they are genetically different from their parents and from one another. How do we account for this genetic variation? As you will learn in later chapters, mutations are the original source of genetic diversity: Changes in an organism's DNA create different versions of genes. Once these differences arise, reshuffling of the different versions during sexual reproduction produces genetic variation. When we discuss natural selection and evolution in Unit III, we will see that this genetic variety in offspring is the raw material for natural selection.

Figure 8.16 illustrates one way in which the process of meiosis contributes to genetic differences in gametes. The

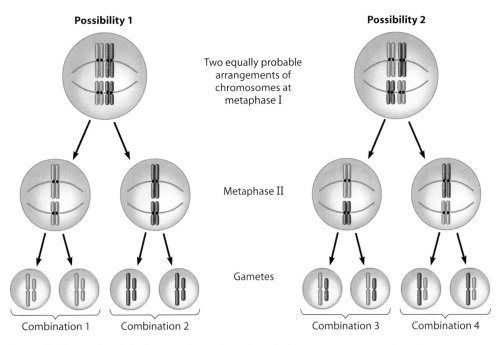

Possibility 1 · **Possibility 2**

Two equally probable arrangements of chromosomes at metaphase I

Metaphase II

Gametes

Combination 1 · Combination 2 · Combination 3 · Combination 4

Figure 8.16 Results of the independent orientation of chromosomes at metaphase I

figure shows how the arrangement of homologous chromosome pairs at metaphase of meiosis I affects the resulting gametes. Here we use color to distinguish chromosomes that originated from the organism's two parents (red for maternal and blue for paternal). The colors highlight the important fact that each maternal chromosome differs genetically from its paternal homolog. In fact, each chromosome you received from your mother carries many genes that are different versions from those on the homologous chromosome you received from your father.

The orientation of the homologous pairs of chromosomes (tetrads) at metaphase I—whether the maternal or paternal chromosome is closer to a given pole—is as random as the flip of a coin. Thus, there is a 50% chance that a particular daughter cell will get the maternal (red) chromosome of a certain homologous pair and a 50% chance that it will receive the paternal (blue) chromosome. In this example, there are two possible ways that the two tetrads can align during metaphase I. In possibility 1, the tetrads are oriented with both red chromosomes on the same side of the metaphase plate. Therefore, the gametes produced in possibility 1 can each have only red *or* blue chromosomes (bottom row, combinations 1 and 2).

In possibility 2, the tetrads are oriented differently, with one red and one blue chromosome on each side of the metaphase I plate. This arrangement produces gametes that each have one red and one blue chromosome. Furthermore, half the gametes have a big blue chromosome and a small red one (combination 3), and half have a big red one and a small blue one (combination 4).

So we see that for this example, a total of four chromosome combinations is possible in the gametes, and in fact the organism will produce gametes of all four types in roughly equal quantities. For a species with more than two pairs of chromosomes, such as the human, *all* the chromosome pairs orient independently at metaphase I. (Chromosomes X and Y behave as a homologous pair in meiosis.)

For any species, the total number of combinations of chromosomes that meiosis can package into gametes is 2^n, where n is the haploid number. For the organism in this figure, $n = 2$, so the number of chromosome combinations is 2^2, or 4. For a human ($n = 23$), there are 2^{23}, or about 8 million possible chromosome combinations! This means that each gamete you produce contains one of roughly 8 million possible combinations of chromosomes inherited from your mother and father.

How many possibilities are there when a gamete from one individual unites with a gamete from another individual in fertilization? In humans, the random fusion of a single sperm with a single ovum during fertilization will produce a zygote with any of about 64 trillion (8 million × 8 million) combinations of chromosomes! While the random nature of fertilization adds a huge amount of potential variability to the offspring of sexual reproduction, there is in fact even more variety created during meiosis, as we see in the next two modules.

? A particular species of worm has a diploid number of 10. How many chromosomal combinations are possible for gametes formed by meiosis?

■ 32; 2n = 10, so n = 5 and 2ⁿ = 32

8.17 Homologous chromosomes can carry different versions of genes

So far, we have focused on genetic variability in gametes and zygotes at the whole-chromosome level. We have yet to discuss the actual genetic information—the genes—contained in the chromosomes of gametes and zygotes. The question we need to answer now is this: What is the significance of the independent orientation of metaphase chromosomes at the level of genes?

Let's take a simple example, the single tetrad in Figure 8.17A. The letters on the homologous chromosomes represent genes. Recall that homologous chromosomes have genes for the same characteristic at corresponding loci. Our example involves hypothetical genes controlling the appearance of mice. *C* and *c* are different versions of a gene for one characteristic, coat color; *E* and *e* are different versions of a gene for another characteristic, eye color. (As you'll learn in later chapters, different versions of a gene contain slightly different nucleotide sequences in the chromosomal DNA.)

Let's say that *C* represents the gene for the brownish color of the mouse on the left in Figure 8.17B and that *c* represents the gene for white coat color (mouse on right). In the chromosome

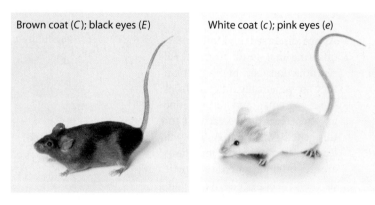

Brown coat (*C*); black eyes (*E*) | White coat (*c*); pink eyes (*e*)

Figure 8.17B   Coat-color and eye-color traits in mice

diagram, notice that *C* is at the same locus on the red homolog as *c* is on the blue one. Likewise, gene *E* (for black eyes) is at the same locus as *e* (pink eyes).

The fact that homologous chromosomes can bear two different kinds of genetic information for the same characteristic (for instance, coat color) is what really makes gametes—and therefore offspring—different from one another. In our example, a gamete carrying a red chromosome would have genes specifying brownish coat color and black eye color, while a gamete with the homologous blue chromosome would have genes for white coat and pink eyes. Thus, we see how a tetrad with genes shown for only two characteristics can yield two genetically different kinds of gametes. In the next module, we go a step further and see how this same tetrad can actually yield *four* different kinds of gametes.

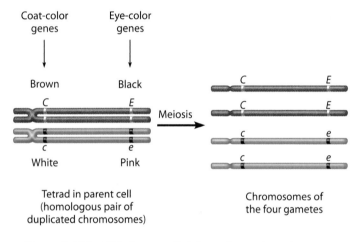

Figure 8.17A   Differing genetic information on homologous chromosomes

? In the tetrad of Figure 8.17A, use labels to distinguish the homologous pair of chromosomes from sister chromatids.

8.18 Crossing over further increases genetic variability

Crossing over is an exchange of corresponding segments between two homologous chromosomes. The micrograph and drawing in Figure 8.18A show the results of crossing over between two homologous chromosomes during prophase I of meiosis. The chromosomes are a tetrad—four chromatids, with each pair of sister chromatids joined at their centromeres. The sites of crossing over appear as X-shaped regions; each is called a **chiasma** (Greek for "cross"). A chiasma (plural, *chiasmata*) is a place where two homologous (nonsister) chromatids are attached to each other. Figure 8.18B (on the facing page) illustrates how crossing over can produce new combinations of genes, using as examples the hypothetical mouse genes mentioned in the previous module.

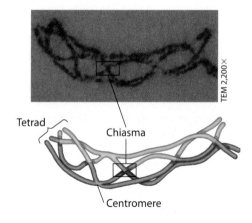

Figure 8.18A
Chiasmata

Crossing over begins very early in prophase I of meiosis. At that time, homologous chromosomes are paired along their lengths. Each gene on one homolog is aligned precisely with the corresponding gene on the other homolog.

At the top of the figure is a tetrad with coat-color (*C, c*) and eye-color (*E, e*) genes labeled. ❶ The DNA molecules of two nonsister chromatids—one maternal (red) and one paternal (blue)—break at the same place. ❷ Immediately, the two broken chromatids join together in a new way (red to blue and blue to red). In effect, the two homologous segments trade places, or cross over, producing hybrid (red/blue and blue/red) chromosomes with new combinations of maternal and paternal genes. ❸ When the homologous chromosomes separate in anaphase I, each contains a new segment originating from its homolog. ❹ Finally, in meiosis II, the sister chromatids separate, each going to a different gamete.

In this example, if there were no crossing over, meiosis could produce only two genetic types of gametes. These would be the ones ending up with the "parental" types of chromosomes, carrying either genes *C* and *E* or genes *c* and *e*. These are the same two kinds of gametes we saw in Figure 8.17A. With crossing over, two other types of gametes can result. One of these carries genes *C* and *e,* and the other carries genes *c* and *E.* Chromosomes with these combinations of genes would not exist if not for crossing over; thus, they are called "recombinant" because they result from **genetic recombination,** the production of gene combinations different from those carried by the original chromosomes.

In meiosis in humans, an average of one to three crossover events occur per chromosome pair. Thus, if you were to examine a chromosome from one of your gametes, you would most likely find that it is not exactly like any one of your own chromosomes. Rather, it is probably a patchwork of segments derived from a pair of homologous chromosomes, cut and pasted together to form a chromosome with a unique combination of genes.

We have now examined three sources of genetic variability in sexually reproducing organisms: independent orientation of chromosomes at metaphase I, random fertilization, and crossing over during prophase I of meiosis. When we take up molecular genetics in Chapter 10, we will see yet another source of variability—mutations, or rare changes in the DNA of genes. The different versions of genes that homologous chromosomes may have at each locus originally arise from mutations, so mutations are ultimately responsible for genetic diversity in living organisms.

Our discussion of meiosis to this point has focused on the process as it normally and "correctly" occurs. In the next, and last, major section of the chapter, we consider some of the consequences of errors in the process.

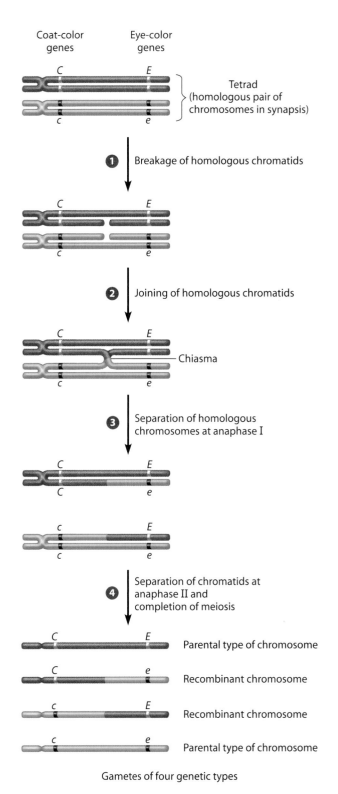

Figure 8.18B How crossing over leads to genetic recombination

❓ Describe how crossing over and the random alignment of homologous chromosomes account for the genetic variation among gametes formed by meiosis.

■ (1) Crossing over creates recombinant chromosomes having a combination of genes that were originally on different, though homologous, chromosomes. (2) Homologous chromosome pairs are oriented randomly at metaphase of meiosis I (see Figure 8.16).

8.19 A karyotype is a photographic inventory of an individual's chromosomes

Errors in meiosis can lead to gametes containing chromosomes in abnormal numbers or with major alterations in their structures. Fertilization of these abnormal gametes results in offspring with chromosomal abnormalities. Such conditions can be readily detected in a **karyotype,** an ordered display of magnified images of an individual's chromosomes arranged in pairs, starting with the longest. A karyotype shows the chromosomes condensed and doubled, as they appear in metaphase of mitosis.

To prepare a karyotype, medical scientists often use lymphocytes, a type of white blood cell. A blood sample is treated with a chemical that stimulates mitosis. After growing in culture for several days, the cells are treated with another chemical to arrest mitosis at metaphase, when the chromosomes, each consisting of two joined sister chromatids, are most highly condensed. Figure 8.19 outlines the steps in one method for the preparation of a karyotype from a blood sample.

The photograph in step 5 shows the karyotype of a normal human male. The 46 chromosomes of a single diploid cell are arranged in 23 homologous pairs: autosomes numbered from 1 to 22 and one pair of sex chromosomes (X and Y in this case). The chromosomes have been stained to reveal band patterns, which are helpful in differentiating the chromosomes and in detecting structural abnormalities. Among the alterations detected by karyotyping is trisomy 21, the basis of Down syndrome, which we discuss next.

? How would the karyotype of a human female differ from the male karyotype in Figure 8.19?

■ Instead of an XY combination for the sex chromosomes, there would be a homologous pair of X chromosomes (XX).

Figure 8.19 Preparation of a karyotype from a blood sample

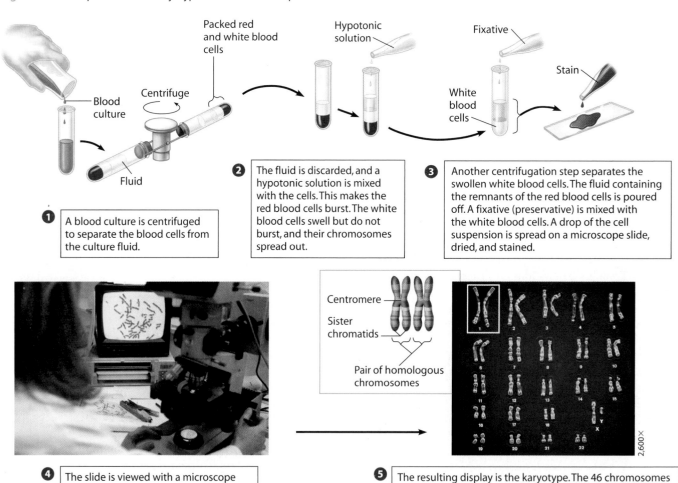

① A blood culture is centrifuged to separate the blood cells from the culture fluid.

② The fluid is discarded, and a hypotonic solution is mixed with the cells. This makes the red blood cells burst. The white blood cells swell but do not burst, and their chromosomes spread out.

③ Another centrifugation step separates the swollen white blood cells. The fluid containing the remnants of the red blood cells is poured off. A fixative (preservative) is mixed with the white blood cells. A drop of the cell suspension is spread on a microscope slide, dried, and stained.

④ The slide is viewed with a microscope equipped with a digital camera. A photograph of the chromosomes is entered into a computer, which electronically arranges them by size and shape.

⑤ The resulting display is the karyotype. The 46 chromosomes here include 22 pairs of autosomes and 2 sex chromosomes, X and Y. Although difficult to discern in the karyotype, each of the chromosomes consists of two sister chromatids lying very close together (see diagram).

8.20 An extra copy of chromosome 21 causes Down syndrome

The karyotype in Figure 8.19 shows the normal human complement of 23 pairs of chromosomes. However, the karyotype in Figure 8.20A is different; notice that there are three number 21 chromosomes, making 47 chromosomes in total. This condition is called **trisomy 21.**

In most cases, a human embryo with an abnormal number of chromosomes is spontaneously aborted (miscarried) long before birth. But some aberrations in chromosome number, including trisomy 21, appear to upset the genetic balance less drastically, and individuals carrying them can survive. These individuals have a characteristic set of symptoms, called a syndrome. A person with an extra copy of chromosome 21, for instance, has a condition called **Down syndrome** (named after John Langdon Down, who characterized the syndrome in 1866).

Trisomy 21 is the most common chromosome number abnormality. Affecting about one out of every 700 children born, it is also the most common serious birth defect in the United States. Chromosome 21 is one of our smallest chromosomes, but an extra copy produces a number of effects. Down syndrome (Figure 8.20B) includes characteristic facial features, notably a round face, a skin fold at the inner corner of the eye, a flattened nose bridge, and small, irregular teeth, as well as short stature, heart defects, and susceptibility to respiratory infections, leukemia, and Alzheimer's disease.

People with Down syndrome usually have a life span shorter than normal. They also exhibit varying degrees of mental retardation. However, some individuals with the syndrome live to middle age or beyond, and many are socially adept and able to hold jobs. A few women with Down syndrome have had children, though most people with the syndrome are sexually underdeveloped and sterile. Half the eggs produced by a woman with Down syndrome will have the extra chromosome 21, so there is a 50% chance that she will transmit the syndrome to her child.

Figure 8.20B   A child with Down syndrome

As indicated in Figure 8.20C, the incidence of Down syndrome in the offspring of normal parents increases markedly with the age of the mother. Down syndrome strikes less than 0.05% of children (fewer than one in 2,000) born to women under age 30. The risk climbs to just under 1% (ten in 1,000) for mothers at age 40 and is even higher for older mothers. Because of this relatively high risk, pregnant women over 35 are candidates for fetal testing for trisomy 21 and other chromosomal abnormalities (see Module 9.10).

What causes trisomy 21? We address that question in the next module.

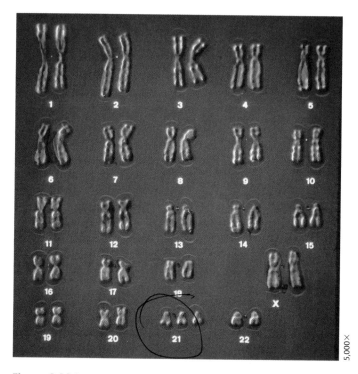

Figure 8.20A   A karyotype for trisomy 21 (Down syndrome)

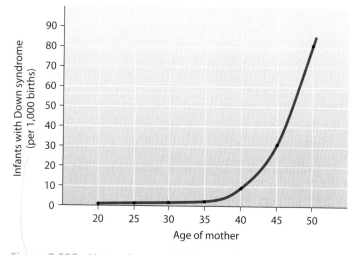

Figure 8.20C   Maternal age and incidence of Down syndrome

? For mothers of age 47, the risk of having a baby with Down syndrome is about _____ per thousand births, or _____%.

■ 40 . . . 4

8.21 Accidents during meiosis can alter chromosome number

Meiosis occurs repeatedly in a human lifetime as the testes or ovaries produce gametes. Almost always, the meiotic spindle distributes chromosomes to daughter cells without error. But there is an occasional mishap, called a **nondisjunction,** in which the members of a chromosome pair fail to separate. Figures 8.21A and 8.21B illustrate two ways that nondisjunction can occur. For simplicity, we use a hypothetical organism whose diploid chromosome number is 4. In both figures, the cell at the top is diploid ($2n = 4$), with two pairs of homologous chromosomes undergoing anaphase of meiosis I.

Sometimes, as in Figure 8.21A, a pair of homologous chromosomes does not separate during meiosis I. In this case, even though the rest of meiosis occurs normally, all the resulting gametes end up with abnormal numbers of chromosomes. Two of the gametes have three chromosomes; the other two gametes have only one chromosome each. In Figure 8.21B, meiosis I is normal, but one pair of sister chromatids fails to move apart during meiosis II. In this case, two of the resulting gametes are abnormal; the other two gametes are normal.

If an abnormal gamete produced by nondisjunction unites with a normal gamete in fertilization, the result is a zygote with an abnormal number of chromosomes. Mitosis will then transmit the anomaly to all embryonic cells. If this were a real organism and it survived, it would have an abnormal karyotype and probably a syndrome of disorders caused by the abnormal number of genes. For example, if there is nondisjunction affecting human chromosome 21, some resulting gametes will carry an extra chromosome 21. If one of these gametes unites with a normal gamete, trisomy 21 (Down syndrome) will result.

Nondisjunction explains how abnormal chromosome numbers come about, but what causes nondisjunction in the first place? We do not yet know the answer. We do know, however, that meiosis begins in a woman's ovaries before she is born but is not completed until years later, at the time of an ovulation (see Module 27.5). Because only one egg cell usually matures each month, a cell might remain arrested in the mid-meiosis state for decades. Some research points to an age-dependent error in one of the checkpoints that coordinates the process of meiosis. Nondisjunction can also affect chromosomes other than 21, as we see next.

? Explain how nondisjunction could result in a diploid gamete.

■ A diploid gamete would result if the nondisjunction affected all the chromosomes during one of the meiotic divisions.

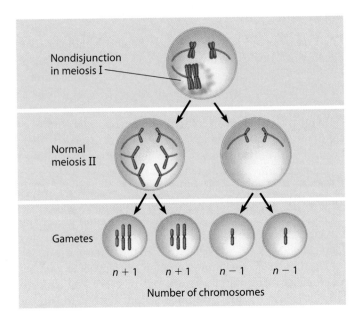

Figure 8.21A   Nondisjunction in meiosis I

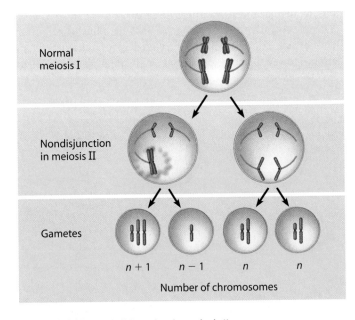

Figure 8.21B   Nondisjunction in meiosis II

8.22 Abnormal numbers of sex chromosomes do not usually affect survival

Nondisjunction in meiosis does not affect just autosomes, such as chromosome 21. It can also lead to abnormal numbers of sex chromosomes, X and Y. Unusual numbers of sex chromosomes

seem to upset the genetic balance less than unusual numbers of autosomes. This may be because the Y chromosome is very small and carries relatively few genes. Furthermore, mammalian cells usually operate with only one functioning X chromosome because other copies of the chromosome become inactivated in each cell (see Module 11.4).

Table 8.22 lists the most common human sex chromosome abnormalities. An extra X chromosome in a male, making him XXY, occurs approximately once in every 2,000 live births (once in every 1,000 male births). Men with this disorder, called *Klinefelter syndrome,* have male sex organs, but the testes are abnormally small and the individual is sterile. The syndrome often includes breast enlargement and other female body characteristics. The person is usually of normal intelligence. Klinefelter syndrome is also found in individuals with more than three sex chromosomes, such as XXYY, XXXY, or XXXXY. These abnormal numbers of sex chromosomes probably result from multiple nondisjunctions; such men are more likely to have developmental disabilities than XY or XXY individuals.

Human males with an extra Y chromosome (XYY) do not have any well-defined syndrome, although they tend to be taller than average. Females with an extra X chromosome (XXX) cannot be distinguished from XX females except by karyotype.

Females who are lacking an X chromosome are designated XO; the O indicates the absence of a second sex chromosome. These women have *Turner syndrome.* They have a characteristic appearance, including short stature and often a web of skin extending between the neck and the shoulders. Women with Turner syndrome are sterile because their sex organs do not

TABLE 8.22	ABNORMALITIES OF SEX CHROMOSOME NUMBER IN HUMANS		
Sex Chromosomes	Syndrome	Origin of Nondisjunction	Frequency in Population
XXY	Klinefelter syndrome (male)	Meiosis in egg or sperm formation	$\frac{1}{2,000}$
XYY	None (normal male)	Meiosis in sperm formation	$\frac{1}{2,000}$
XXX	None (normal female)	Meiosis in egg or sperm formation	$\frac{1}{1,000}$
XO	Turner syndrome (female)	Meiosis in egg or sperm formation	$\frac{1}{5,000}$

fully mature at adolescence. If left untreated, girls with Turner syndrome have poor development of breasts and other secondary sexual characteristics. Artificial administration of estrogen can alleviate these symptoms. Girls with Turner syndrome have normal intelligence. The XO condition is the sole known case where having only 45 chromosomes is not fatal in humans.

The sex chromosome abnormalities described here illustrate the crucial role of the Y chromosome in determining sex. In general, a single Y chromosome is enough to produce "maleness," even in combination with several X chromosomes. The absence of a Y chromosome yields "femaleness."

? What is the *total* number of chromosomes you would expect to find in the karyotype of a female with Turner syndrome?

■ 45

Evolution Connection

8.23 New species can arise from errors in cell division

Errors in meiosis or mitosis do not always lead to problems. In fact, biologists hypothesize that such errors have been instrumental in the evolution of many species. Numerous plant species, in particular, seem to have originated from accidents during cell division that resulted in extra sets of chromosomes. The new species is polyploid, meaning that it has more than two sets of homologous chromosomes in each somatic cell. At least half of all species of flowering plants are polyploid, including such useful ones as wheat, potatoes, apples, and cotton.

Let's consider one scenario by which a diploid (2n) plant species might generate a tetraploid (4n) plant. Imagine that, like many plants, our diploid plant produces both sperm and egg cells and can self-fertilize. If meiosis fails to occur in the plant's reproductive organs and gametes are instead produced by mitosis, the gametes will be diploid. The union of a diploid (2n) sperm with a diploid (2n) egg during self-fertilization will produce a tetraploid (4n) zygote, which may develop into a mature tetraploid plant that can itself reproduce by self-fertilization. The tetraploid plants will constitute a new species, one that has evolved in just one generation.

Although polyploid animal species are less common than polyploid plants, they are known to occur among the fishes and amphibians. Moreover, researchers in Chile have identified the first candidate for polyploidy among the mammals, a rat whose cells seem to be tetraploid. Tetraploid organisms are sometimes strikingly different from their recent diploid ancestors—larger, for example. Scientists don't yet understand exactly how polyploidy brings about such differences.

? What is a polyploid organism?

■ One with more than two sets of homologous chromosomes in its genome

8.24 Alterations of chromosome structure can cause birth defects and cancer

Even if all chromosomes are present in normal numbers, abnormalities in chromosome structure may cause disorders. Breakage of a chromosome can lead to a variety of rearrangements affecting the genes of that chromosome. Figure 8.24A shows three types of rearrangement. (The pink arrows indicate chromosome breaks.) If a fragment of a chromosome is lost, the remaining chromosome will then have a **deletion.** If a fragment from one chromosome joins to a sister chromatid or homologous chromosome, it will produce a **duplication.** If a fragment reattaches to the original chromosome but in the reverse direction, an **inversion** results.

Inversions are less likely than deletions or duplications to produce harmful effects, because in inversions all genes are still present in their normal number. Many deletions in human chromosomes, however, cause serious physical and mental problems. One example is a specific deletion in chromosome 5 that causes *cri du chat* ("cry of the cat") syndrome. A child born with this syndrome is mentally retarded, has a small head with unusual facial features, and has a cry that sounds like the mewing of a distressed cat. Such individuals usually die in infancy or early childhood.

Another type of chromosomal change is chromosomal **translocation,** the attachment of a chromosomal fragment to a nonhomologous chromosome. Figure 8.24B shows a translocation that is reciprocal; that is, two nonhomologous chromosomes exchange segments. Like inversions, translocations may or may not be harmful. Some people with Down syndrome have only part of a third chromosome 21; as the result of a translocation, this partial chromosome is attached to another (nonhomologous) chromosome.

Whereas chromosomal changes present in sperm or egg can cause congenital disorders, such changes in a somatic cell may contribute to the development of cancer. For example, a chromosomal translocation in somatic cells in the bone marrow is associated with *chronic myelogenous leukemia (CML).* CML is one of the most common of the leukemias, the cancers affecting cells that give rise to white blood cells (leukocytes). In the cancerous cells of most CML patients, a part of chromosome 22 has switched places with a small fragment from a tip of chromosome 9 (Figure 8.24C). This reciprocal translocation activates a gene that leads to leukemia. The chromosome ending up with the activated cancer-causing gene is called the "Philadelphia chromosome," after the city where it was discovered.

Because the chromosomal changes in cancer are usually confined to somatic cells, cancer is not usually inherited. We'll return to cancer in Chapter 11. In Chapter 9, we continue our study of genetic principles, looking first at the historical development of the science of genetics and then at the rules governing the way traits are passed from parents to offspring.

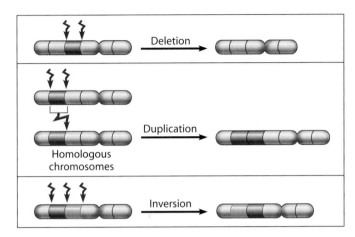

Figure 8.24A Alterations of chromosome structure involving one chromosome or a homologous pair

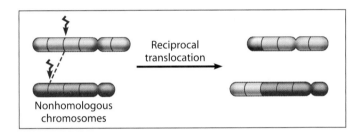

Figure 8.24B Chromosomal translocation between nonhomologous chromosomes

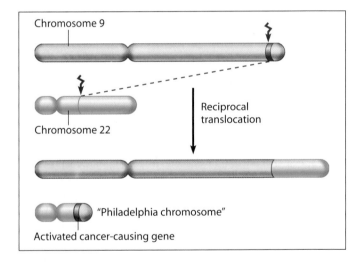

Figure 8.24C The translocation associated with chronic myelogenous leukemia

? How is reciprocal translocation different from normal crossing over?

■ Reciprocal translocation swaps chromosome segments between nonhomologous chromosomes. Crossing over normally exchanges corresponding segments between homologous chromosomes.

Reviewing the Concepts

Connections Between Cell Division and Reproduction (Introduction–8.3)

Cell division is at the heart of the reproduction of cells and organisms, because cells come only from preexisting cells. Some organisms reproduce asexually, and their offspring are all genetic copies of the parent and of each other. Others reproduce sexually, creating a variety of offspring, each with a unique combination of traits (**Introduction–8.2**). Prokaryotic cells reproduce asexually by cell division. As the cell replicates its single chromosome, the copies move apart; the growing membrane then divides the cell (**8.3**).

The Eukaryotic Cell Cycle and Mitosis (8.4–8.11)

Chromosomes. A eukaryotic cell has many more genes than a prokaryotic cell, and they are grouped into multiple chromosomes in the nucleus. Each chromosome contains a very long DNA molecule associated with proteins. Individual chromosomes are visible only when the cell is in the process of dividing; otherwise, they are in the form of thin, loosely packed chromatin fibers. Before a cell starts dividing, the chromosomes duplicate, producing sister chromatids (containing identical DNA) joined together at the centromere. Cell division involves the separation of sister chromatids and results in two daughter cells, each containing a complete and identical set of chromosomes (**8.4**).

The eukaryotic cell cycle (8.5):

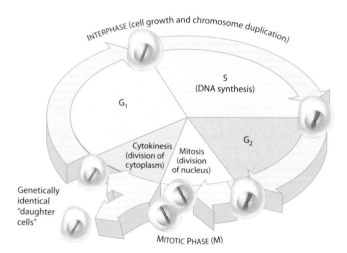

Mitosis distributes duplicated chromosomes into two daughter nuclei. After the chromosomes coil up, a mitotic spindle made of microtubules moves them to the middle of the cell. The sister chromatids then separate and move to opposite poles of the cell, where two new nuclei form. Cytokinesis, in which the cell divides in two, overlaps the end of mitosis (**8.6**). In animals, cytokinesis occurs by a constriction of the cell (cleavage). In plants, a membranous cell plate splits the cell in two (**8.7**).

Cell cycle control. Most animal cells divide only when stimulated, and some not at all. In laboratory cultures, most normal cells divide only when attached to a surface. They continue dividing until they touch one another. Growth factors are proteins secreted by cells that stimulate other cells to divide (**8.8**).

A set of proteins within the cell controls the cell cycle. Signals affecting critical checkpoints in the cell cycle determine whether a cell will go through the complete cycle and divide. The binding of growth factors to specific receptors on the plasma membrane is usually necessary for cell division (**8.9**).

Cancer cells divide excessively to form masses called tumors. Malignant tumors can invade other tissues. Radiation and chemotherapy are effective as cancer treatments because they interfere with cell division (**8.10**). When the cell cycle operates normally, mitotic cell division functions in growth, replacement of damaged and lost cells, and asexual reproduction (**8.11**).

Meiosis and Crossing Over (8.12–8.18)

Homologous chromosomes. The somatic (body) cells of each species contain a specific number of chromosomes; for example, human cells have 46, making up 23 pairs (two sets) of homologous chromosomes. The chromosomes of a homologous pair carry genes for the same characteristics at the same place, or locus (**8.12**). Cells with two sets of homologous chromosomes are said to be diploid. Gametes—eggs and sperm—are haploid cells with a single set of chromosomes. Sexual life cycles involve the alternation of haploid and diploid stages (**8.13**):

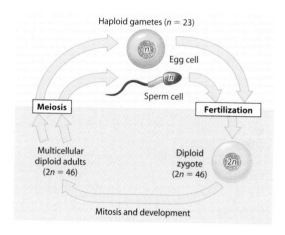

Meiosis, like mitosis, is preceded by chromosome duplication, but in meiosis the cell divides twice to form four daughter cells. The first division, meiosis I, starts with synapsis, the pairing of homologous chromosomes. In crossing over, homologous chromosomes exchange corresponding segments. Meiosis I separates each homologous pair and produces two daughter cells, each with one set of chromosomes. Meiosis II is essentially the same as mitosis: In each of the cells, the sister chromatids of each chromosome separate; the result is a total of four haploid cells (**8.14**). Figure 8.15 reviews and compares mitosis and meiosis (**8.15**).

Genetic variation. Each chromosome of a homologous pair differs at many points from the other member of the pair. Random arrangements of chromosome pairs at metaphase I of meiosis leads to many different combinations of chromosomes in eggs and sperm. Random fertilization of eggs by sperm greatly increases this variation (**8.16**). The differences between homologous chromosomes are based on the fact that they can bear different versions of a gene at corresponding loci (**8.17**). Genetic recombination, which results from crossing over during prophase I of meiosis, increases variation still further (**8.18**).

Alterations of Chromosome Number and Structure (8.19–8.24)

Chromosome number. A karyotype is an ordered arrangement of images of a cell's chromosomes (**8.19**). A person may have an abnormal number of chromosomes, which causes problems. Down syndrome is caused by trisomy 21, an extra copy of chromosome 21 (**8.20**). An abnormal chromosome count is the result of nondisjunction, the failure of a homologous pair of chromosomes to separate during meiosis I or of sister chromatids to separate during meiosis II (**8.21**). Nondisjunction can also produce gametes with extra or missing sex chromosomes, leading to varying degrees of malfunction in humans but not usually affecting survival (**8.22**). Nondisjunction can produce polyploid organisms, ones with extra sets of chromosomes. Such errors in mitosis are important in the evolution of new species (**8.23**). Chromosome breakage can lead to rearrangements—deletions, duplications, inversions, and translocations—that can produce genetic disorders or, if the changes occur in somatic cells, cancer (**8.24**).

Connecting the Concepts

1. Complete the following table to compare mitosis and meiosis.

	Mitosis	Meiosis
Number of chromosomal duplications		
Number of cell divisions		
Number of daughter cells produced		
Number of chromosomes in daughter cells		
How chromosomes line up during metaphase		
Genetic relationship of daughter cells to parent cell		
Functions performed in the human body		

Testing Your Knowledge

Multiple Choice

2. If an intestinal cell in a grasshopper contains 24 chromosomes, a grasshopper sperm cell contains _____ chromosomes.
   a. 3
   b. 6
   c. 12
   d. 24
   e. 48

3. Which of the following phases of mitosis is essentially the opposite of prophase in terms of nuclear changes?
   a. telophase
   b. metaphase
   c. S phase
   d. interphase
   e. anaphase

4. A biochemist measured the amount of DNA in cells growing in the laboratory and found that the quantity of DNA in a cell doubled
   a. between prophase and anaphase of mitosis.
   b. between the G_1 and G_2 phases of the cell cycle.
   c. during the M phase of the cell cycle.
   d. between prophase I and prophase II of meiosis.
   e. between anaphase and telophase of mitosis.

5. Which of the following is *not* a function of mitosis in humans?
   a. repair of wounds
   b. growth
   c. production of gametes from diploid cells
   d. replacement of lost or damaged cells
   e. multiplication of somatic cells

6. A micrograph of a dividing cell from a mouse showed 19 chromosomes, each consisting of two sister chromatids. During which of the following stages of cell division could this picture have been taken? (*Explain your answer.*)
   a. prophase of mitosis
   b. telophase II of meiosis
   c. prophase I of meiosis
   d. anaphase of mitosis
   e. prophase II of meiosis

7. Cytochalasin B is a chemical that disrupts microfilament formation. This chemical would interfere with
   a. DNA replication.
   b. formation of the mitotic spindle.
   c. cleavage.
   d. formation of the cell plate.
   e. crossing over.

8. It is difficult to observe individual chromosomes during interphase because
   a. the DNA has not been replicated yet.
   b. they are in the form of long, thin strands.
   c. they leave the nucleus and are dispersed to other parts of the cell.
   d. homologous chromosomes do not pair up until division starts.
   e. the spindle must move them to the metaphase plate before they become visible.

9. A fruit fly somatic cell contains 8 chromosomes. This means that _____ different combinations of chromosomes are possible in its gametes.
   a. 4
   b. 8
   c. 16
   d. 32
   e. 64

10. If a fragment of a chromosome breaks off and then reattaches to the original chromosome but in the reverse direction, the resulting chromosomal abnormality is called
    a. a deletion.
    b. an inversion.
    c. a translocation.
    d. a nondisjunction.
    e. a reciprocal translocation.

11. Why are individuals with an extra chromosome 21, which causes Down syndrome, more numerous than individuals with an extra chromosome 3 or chromosome 16?
    a. There are probably more genes on chromosome 21 than on the others.
    b. Chromosome 21 is a sex chromosome and chromosomes 3 and 16 are not.
    c. Down syndrome is not more common, just more serious.
    d. Extra copies of the other chromosomes are probably fatal.
    e. Chromosome 21 is more likely to produce a nondisjunction error than other chromosomes.

Describing, Comparing, and Explaining

12. Briefly describe how three different processes that occur during a sexual life cycle increase the genetic diversity of offspring.

13. In the light micrograph below of dividing cells near the tip of an onion root, identify a cell in interphase, prophase, metaphase, anaphase, and telophase. Describe the major events occurring at each stage.

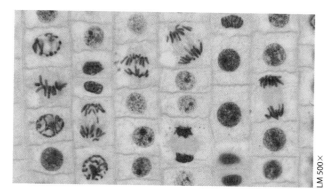

LM 500×

14. Discuss the factors that control the division of eukaryotic cells grown in the laboratory. Cancer cells are easier to grow in the lab than other cells. Why do you suppose this is the case?

15. Compare cytokinesis in plant and animal cells.

16. Sketch a cell with three pairs of chromosomes undergoing meiosis, and show how nondisjunction can result in the production of gametes with extra or missing chromosomes.

Applying the Concepts

17. Suppose you read in the newspaper that a genetic engineering laboratory has developed a procedure for fusing two gametes from the same person (two eggs or two sperm) to form a zygote. The article mentions that an early step in the procedure prevents crossing over from occurring during the formation of the gametes in the donor's body. The researchers are in the process of determining the genetic makeup of one of their new zygotes. Which of the following predictions do you think they would make? Justify your choice, and explain why you rejected each of the other choices.
    a. The zygote would have 46 chromosomes, all of which came from the gamete donor (its one parent), so the zygote would be genetically identical to the gamete donor.
    b. The zygote *could* be genetically identical to the gamete donor, but it is much more likely that it would have an unpredictable mixture of chromosomes from the gamete donor's parents.
    c. The zygote would not be genetically identical to the gamete donor, but it would be genetically identical to one of the donor's parents.
    d. The zygote would not be genetically identical to the gamete donor, but it would be genetically identical to one of the donor's grandparents.

18. Bacteria are able to divide on a much faster schedule than eukaryotic cells. Some bacteria can divide every 20 minutes, while the minimum time required by eukaryotic cells in a rapidly developing embryo is about once per hour, and most cells divide much less often than that. State several testable hypotheses explaining why bacteria can divide at a faster rate than eukaryotic cells.

19. Red blood cells, which carry oxygen to body tissues, live for only about 120 days. Replacement cells are produced by cell division in bone marrow. How many cell divisions must occur each second in your bone marrow just to replace red blood cells? Here is some information to use in calculating your answer: There are about 5 million red blood cells per cubic millimeter (mm^3) of blood. An average adult has about 5 L (5,000 cm^3) of blood. (*Hint:* What is the total number of red blood cells in the body? What fraction of them must be replaced each day if all are replaced in 120 days?)

20. A mule is the offspring of a horse and a donkey. A donkey sperm contains 31 chromosomes and a horse egg cell 32 chromosomes, so the zygote contains a total of 63 chromosomes. The zygote develops normally. The combined set of chromosomes is not a problem in mitosis, and the mule combines some of the best characteristics of horses and donkeys. However, a mule is sterile; meiosis cannot occur normally in its testes (or ovaries). Explain why mitosis is normal in cells containing both horse and donkey chromosomes but the mixed set of chromosomes interferes with meiosis.

21. Every year about a million Americans are diagnosed as having cancer. This means that about 75 million Americans now living will eventually have cancer, and one in five will die of the disease. There are many kinds of cancers and many causes of the disease. For example, smoking causes most lung cancers. Overexposure to ultraviolet rays in sunlight causes most skin cancers. There is evidence that a high-fat, low-fiber diet is a factor in breast, colon, and prostate cancers. And agents in the workplace such as asbestos and vinyl chloride are also implicated as causes of cancer. Hundreds of millions of dollars are spent each year in the search for effective treatments for cancer; far less money is spent preventing cancer. Why might this be the case? What kinds of lifestyle changes could we make to help prevent cancer? What kinds of prevention programs could be initiated or strengthened to encourage these changes? What factors might impede such changes and programs? Should we devote more of our resources to treating cancer or preventing it? Why?

Answers to all questions can be found in Appendix 4.

For Practice Quizzes, BioFlix, MP3 Tutors, and Activities, go to www.mybiology.com.

Patterns of Inheritance

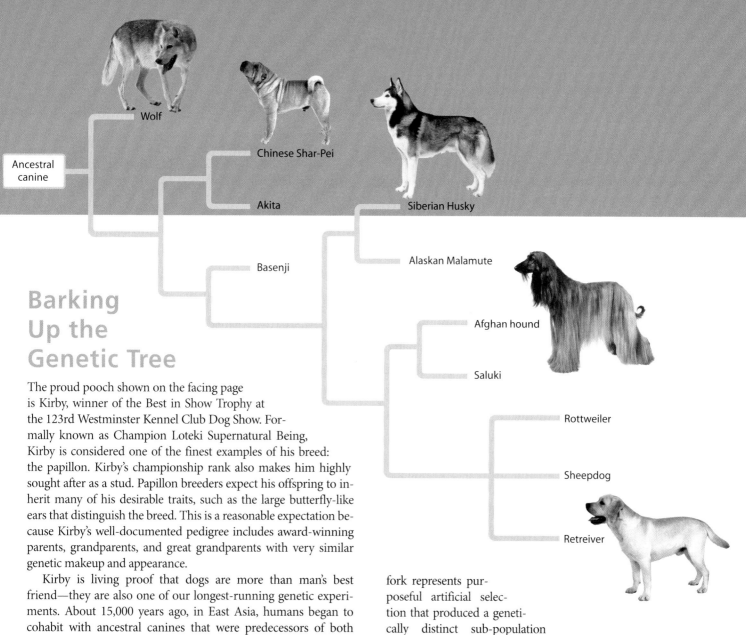

Barking Up the Genetic Tree

The proud pooch shown on the facing page is Kirby, winner of the Best in Show Trophy at the 123rd Westminster Kennel Club Dog Show. Formally known as Champion Loteki Supernatural Being, Kirby is considered one of the finest examples of his breed: the papillon. Kirby's championship rank also makes him highly sought after as a stud. Papillon breeders expect his offspring to inherit many of his desirable traits, such as the large butterfly-like ears that distinguish the breed. This is a reasonable expectation because Kirby's well-documented pedigree includes award-winning parents, grandparents, and great grandparents with very similar genetic makeup and appearance.

Kirby is living proof that dogs are more than man's best friend—they are also one of our longest-running genetic experiments. About 15,000 years ago, in East Asia, humans began to cohabit with ancestral canines that were predecessors of both modern wolves and dogs. As people settled into permanent, geographically isolated settlements, different populations of canines were separated from one another and eventually became inbred. At the same time, different groups of people chose dogs with different traits, depending on their needs. Herders selected dogs that were good at controlling flocks of animals. Hunters chose dogs that were good at retrieving wounded prey. Continued over millennia, the result of such genetic tinkering is an incredibly diverse array of dog body types and behaviors, from tiny, feisty Chihuahuas to huge, docile St. Bernards. But no matter what appearance a dog sports, we can ultimately explain their inborn traits through **genetics,** the science of heredity. In each breed, a distinct genetic makeup results in a distinct set of physical and behavioral characteristics.

Our understanding of canine genetics took a big leap forward in 2003 when researchers sequenced the complete genome of a dog. Using the genome sequence and a wealth of other data, canine geneticists produced the family tree of dog breeds shown above. Published in 2006, the tree is based on genetic analysis of 414 dogs from 85 recognized breeds. The analysis shows that the canine family tree includes a series of well-defined branch points. Each fork represents purposeful artificial selection that produced a genetically distinct sub-population with specific desired traits. The genetic tree shows that the most ancient breeds, those with genetic makeups most closely resembling the wolf, are Asian species such as the shar-pei and Akita. Subsequent genetic splits created distinct breeds in Africa (basenji), the Arctic (Alaskan malamute and Siberian husky), and the Middle East (Afghan hound and saluki). The remaining breeds, primarily of European ancestry, can be grouped by genetic makeup into those bred for guarding (for example, the rottweiler), herding (such as various sheepdogs), and hunting (including the golden retriever and beagle).

Although humans have been applying genetics for thousands of years—by breeding food crops as well as domesticated animals—the biological basis of selective breeding has only recently been understood. In this chapter, we examine the rules that govern how inherited traits are passed from parents to offspring. We look at several kinds of inheritance patterns and how they allow us to predict the ratios of offspring with particular traits. Most importantly, we uncover a basic biological concept: how the behavior of chromosomes during gamete formation and fertilization (discussed in Chapter 8) accounts for the patterns of inheritance we observe. ■ ■ ■

9.1 The science of genetics has ancient roots

Attempts to explain inheritance date back at least to ancient Greece. The physician Hippocrates (approximately 460–370 B.C.) suggested an explanation called pangenesis. According to this idea, particles called pangenes travel from each part of an organism's body to the eggs or sperm and are then passed to the next generation; moreover, changes that occur in the body during an organism's life are passed on in this way. The Greek philosopher Aristotle (384–322 B.C.) rejected this idea as simplistic, saying that what is inherited is the potential to produce body features rather than particles of the features themselves.

Actually, pangenesis proves incorrect on several counts. The reproductive cells are not composed of particles from somatic (body) cells, and changes in somatic cells do not influence eggs and sperm. For instance, no matter how much you enlarge your biceps by lifting weights, muscle cells in your arms do not transmit genetic information to your gametes, and your offspring will not be changed by your weight-lifting efforts. This may seem like common sense today, but the pangenesis hypothesis and the idea that traits acquired during an individual's lifetime are passed on to offspring prevailed well into the 19th century.

By observing inheritance patterns in ornamental plants, biologists of the early 19th century established that offspring inherit traits from both parents. The favored explanation of inheritance then became the "blending" hypothesis, the idea that the hereditary materials contributed by the male and female parents mix in forming the offspring in the way that blue and yellow paints blend to make green. For example, according to this hypothesis, after the genetic information for the colors of black and chocolate brown Labrador retrievers is blended, the colors should be as inseparable as paint pigments. But this is not what happens: Instead, the offspring of a purebred black Lab and a purebred brown one will all be black, but some of the dogs in the next generation will be brown. The blending hypothesis was finally rejected because it does not explain how traits that disappear in one generation can reappear in later ones.

? Horse breeders sometimes speak of "mixing the bloodlines" of two pedigrees. How does this language relate back to the "blending" model of inheritance?

■ It implies that offspring are a blend of two parents, as in a liquid mixture.

9.2 Experimental genetics began in an abbey garden

The modern science of genetics began in the 1860s, when an Augustinian monk named Gregor Mendel deduced the fundamental principles of genetics by breeding garden peas (Figure 9.2A). Mendel lived and worked in an abbey in Brunn, Austria (now Brno, in the Czech Republic). Strongly influenced by his study of physics, mathematics, and chemistry at the University of Vienna, his research was both experimentally and mathematically rigorous, and these qualities were largely responsible for his success.

In a paper published in 1866, Mendel correctly argued that parents pass on to their offspring discrete heritable factors. (It is interesting to note that Mendel's publication came just seven years after Darwin's 1859 publication of *The Origin of Species,* making the 1860s a banner decade in the development of modern biology.) In his paper, Mendel stressed that the heritable factors (today called genes) retain their individuality generation after generation; that is, genes are like marbles of different colors. Just as marbles retain their colors permanently and do not blend, no matter how they are mixed, genes permanently retain their identities.

Mendel probably chose to study garden peas because they had short generation times, they produced large numbers of offspring from each mating, and they came in many readily distinguishable varieties. For example, one variety has purple flowers, and another variety has white flowers. A heritable feature that varies among individuals, such as flower color, is called a **character.** Each variant for a character, such as purple or white flowers, is called a **trait.**

Perhaps the most important advantage of pea plants as an experimental model was that Mendel could strictly control matings. As Figure 9.2B shows, the petals of the pea flower almost completely enclose the reproductive organs: the stamens and carpel. Consequently, pea plants usually **self-fertilize** in nature. That is, sperm-carrying pollen grains released from the stamens land on the egg-containing carpel of the same flower. Mendel could ensure self-fertilization by covering a flower with a small bag so that no pollen from another plant could reach the carpel. When he wanted **cross-fertilization** (fertilization of one

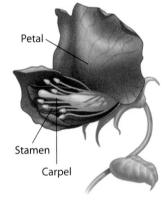

Figure 9.2B Anatomy of a garden pea flower (with one petal removed to improve visibility)

Figure 9.2A Gregor Mendel

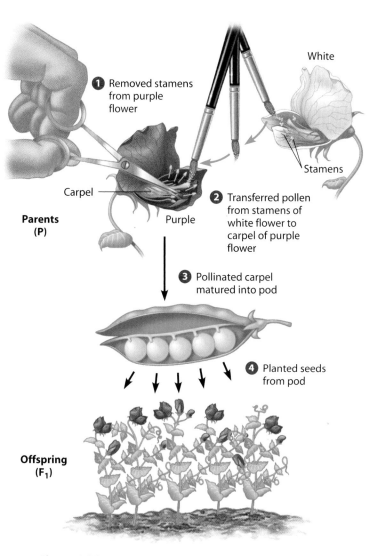

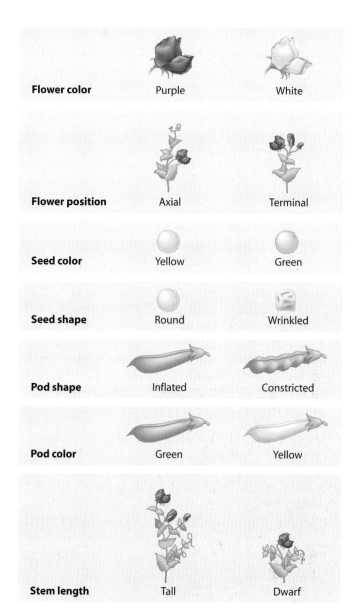

Parents (P)

1. Removed stamens from purple flower

Carpel

Purple

White

Stamens

2. Transferred pollen from stamens of white flower to carpel of purple flower

3. Pollinated carpel matured into pod

4. Planted seeds from pod

Offspring (F₁)

Figure 9.2C Mendel's technique for cross-fertilization of pea plants

Flower color	Purple	White
Flower position	Axial	Terminal
Seed color	Yellow	Green
Seed shape	Round	Wrinkled
Pod shape	Inflated	Constricted
Pod color	Green	Yellow
Stem length	Tall	Dwarf

Figure 9.2D The seven pea characters studied by Mendel

plant by pollen from a different plant), he used the method shown in Figure 9.2C. ❶ He prevented self-fertilization by cutting off the immature stamens of a plant before they produced pollen. ❷ To cross-fertilize the stamenless flower, he dusted its carpel with pollen from another plant. After pollination, ❸ the carpel developed into a pod, containing seeds (peas) that ❹ he planted. The seeds grew into offspring plants. Through these methods, Mendel could always be sure of the parentage of new plants.

Mendel's success was due not only to his experimental approach and choice of organism but also to his selection of characteristics to study. He chose to observe seven characters, each of which occurred in two distinct forms (Figure 9.2D). Mendel worked with his plants until he was sure he had **true-breeding** varieties—that is, varieties for which self-fertilization produced offspring all identical to the parent. For instance, he identified a purple-flowered variety that, when self-fertilized, produced offspring plants that all had purple flowers.

Now Mendel was ready to ask what would happen when he crossed his different true-breeding varieties with each other.

For example, what offspring would result if plants with purple flowers and plants with white flowers were cross-fertilized? In the language of plant and animal breeders and geneticists, the offspring of two different varieties are called **hybrids,** and the cross-fertilization itself is referred to as a hybridization, or simply a **cross.** The true-breeding parental plants are called the **P generation** (P for parental), and their hybrid offspring are the **F₁ generation** (F for filial, from the Latin word for "son"). When F₁ plants self-fertilize or fertilize each other, their offspring are the **F₂ generation.** We turn to Mendel's results next.

? Why was the development of true-breeding varieties critical to the success of Mendel's experiments?

■ True-breeding varieties allowed Mendel to predict the outcome of specific crosses and therefore run controlled experiments.

Mendel's law of segregation describes the inheritance of a single character

Mendel performed many experiments in which he tracked the inheritance of characters that occur in two forms, such as flower color. The results led him to formulate several ideas about inheritance. Let's look at some of his experiments and follow the reasoning that led to his hypotheses.

Figure 9.3A starts with a cross between a pea plant with purple flowers and one with white flowers. This is called a **monohybrid cross** because the parent plants differ in only one character. Mendel observed that F_1 plants produced by crossing these two true-breeding parents all had purple flowers; they were not a lighter purple, as predicted by the blending hypothesis. Was the white-flowered plant's genetic contribution to the hybrids lost? By mating the F_1 plants, Mendel found the answer to be no. Out of 929 F_2 plants, Mendel found that 705 (about $\frac{3}{4}$) had purple flowers and 224 (about $\frac{1}{4}$) had white flowers, a ratio of about three plants with purple flowers to every one with white flowers in the F_2 generation. Mendel reasoned that the heritable factor for white flowers did not disappear in the F_1 plants, but that only the purple-flower factor was affecting F_1 flower color. He also deduced that the F_1 plants must have carried two factors for the flower-color character, one for purple and one for white.

Mendel observed these same patterns of inheritance for six other pea plant characters (see Figure 9.2D). From these results, Mendel developed four hypotheses, which we describe here using modern terminology (such as "gene" instead of "heritable factor"):

1. *There are alternative versions of genes that account for variations in inherited characters.* For example, the gene for flower color in pea plants exists in two versions, one for purple and the other for white. The alternative versions of a gene are now called **alleles.**

2. *For each character, an organism inherits two alleles, one from each parent.* These alleles may be the same or different. An organism that has two identical alleles for a gene is said to be **homozygous** for that gene (and is called a homozygote). An organism that has two different alleles for a gene is said to be **heterozygous** for that gene (and is called a heterozygote).

3. *If the two alleles of an inherited pair differ, then one determines the organism's appearance and is called the **dominant allele**; the other has no noticeable effect on the organism's appearance and is called the **recessive allele.*** We use uppercase letters to represent dominant alleles and lowercase letters to represent recessive alleles.

4. *A sperm or egg carries only one allele for each inherited character because allele pairs separate (segregate) from each other during the production of gametes.* This statement is now known as the **law of segregation.** When sperm and egg unite at fertilization, each contributes its allele, restoring the paired condition in the offspring.

Figure 9.3B shows how Mendel explained the results given in Figure 9.3A. In this example, the italic letter *P* represents the

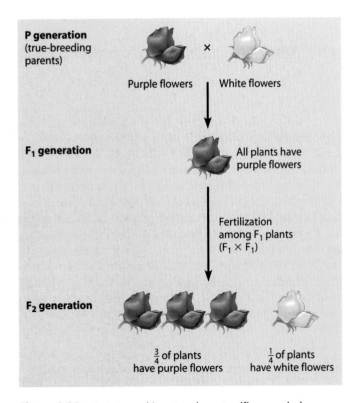

Figure 9.3A Crosses tracking one character (flower color)

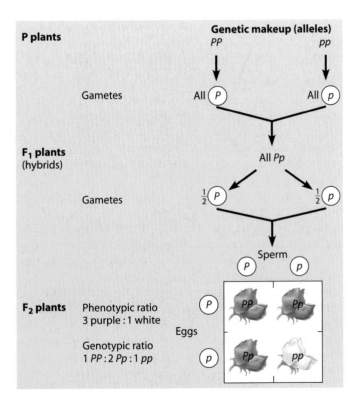

Figure 9.3B Explanation of the crosses in Figure 9.3A

dominant allele (for purple flowers) and *p* stands for the recessive allele (for white flowers). At the top in Figure 9.3B, you see the alleles carried by the parental plants. Both plants were true-breeding, and Mendel's first two hypotheses propose that one parental variety had two alleles for purple flowers (*PP*) and the other had two alleles for white flowers (*pp*).

Consistent with hypothesis 4, the gametes of Mendel's parental plants each carried one allele; thus, the parental gametes in Figure 9.3B are either *P* or *p*. As a result of fertilization, the F₁ hybrids each inherited one allele for purple flowers and one for white. Hypothesis 3 explains why all of the F₁ hybrids (*Pp*) had purple flowers: The dominant *P* allele has its full effect in the heterozygote, while the recessive *p* allele has no effect on flower color.

Mendel's hypotheses also explain the 3 : 1 ratio ($\frac{3}{4}$ purple flowers to $\frac{1}{4}$ white flowers) in the F₂ generation. Because the F₁ hybrids are *Pp*, they make gametes *P* and *p* in equal numbers. The bottom diagram in Figure 9.3B, called a **Punnett square,** shows the four possible combinations of gametes.

The Punnett square shows the proportions of F₂ plants predicted by Mendel's hypotheses. If a sperm carrying allele *P* fertilizes an egg carrying allele *P*, the offspring (*PP*) will produce purple flowers. Mendel's hypotheses predict that this combination will occur in $\frac{1}{4}$ of the offspring. As shown in the Punnett square, the hypotheses also predict that $\frac{2}{4}$ of the offspring will inherit one *P* allele and one *p* allele. These offspring (*Pp*) will all have purple flowers because *P* is dominant. The remaining $\frac{1}{4}$ of the F₂ plants will inherit two *p* alleles and will have white flowers.

Because an organism's appearance does not always reveal its genetic composition, geneticists distinguish between an organism's expressed, or physical, traits, called its **phenotype** (such as purple or white flowers), and its genetic makeup, its **genotype** (in this example, *PP*, *Pp*, or *pp*). So now we see that Figure 9.3A shows the phenotypes and Figure 9.3B the genotypes in our sample crosses. For the F₂ plants, the ratio of plants with purple flowers to those with white flowers (3 : 1) is called the phenotypic ratio. The genotypic ratio, as shown by the Punnett square, is 1 *PP* : 2 *Pp* : 1 *pp*.

Mendel found that each of the seven characteristics he studied exhibited the same inheritance pattern: One parental trait disappeared in the F₁ generation, only to reappear in one-fourth of the F₂ offspring. The mechanism underlying this inheritance pattern is stated by Mendel's law of segregation: *Pairs of alleles segregate (separate) during gamete formation.* The fusion of gametes at fertilization creates allele pairs once again. Research since Mendel's time has established that the law of segregation applies to all sexually reproducing organisms, including humans. We'll return to Mendel and his experiments with pea plants in Module 9.5. But first let's see how some of the concepts we discussed in Chapter 8 fit with what we have said about genetics so far.

Web Activity *Monohybrid Cross*

? How can two plants with different genotypes for a particular inherited character be identical in phenotype?

■ One could be homozygous for the dominant allele, whereas the other is heterozygous.

9.4 Homologous chromosomes bear the alleles for each character

Figure 9.4 shows three gene loci on a pair of homologous chromosomes (homologues)—chromosomes that carry genes controlling the same inherited characteristics. Recall from Module 8.12 that every diploid cell, whether from pea plant or human, has two sets of homologous chromosomes. One set comes from the organism's female parent, the other from the male parent.

The labeled bands on the chromosomes in the figure represent three gene loci, specific locations of genes along the chromosome. The matching colors of corresponding loci on the two homologues highlight the fact that homologous chromosomes have genes for the same characters located at the same positions along their lengths. However, as the uppercase and lowercase letters next to the loci indicate, two homologous chromosomes may bear either the same alleles or different ones. Thus, we see the connection between Mendel's laws and homologous chromosomes: *Alleles (alternative versions) of a gene reside at the same locus on homologous chromosomes.*

The diagram here also serves as a review of some of the genetic terms we have encountered to this point. We will return to the chromosomal basis of inheritance in more detail beginning with Module 9.16.

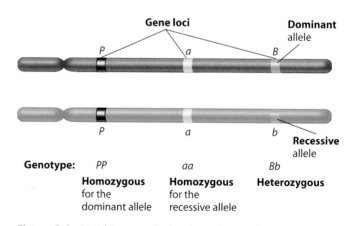

Figure 9.4   Matching gene loci on homologous chromosomes

? An individual is heterozygous, *Gg*, for a gene. According to the law of segregation, each gamete formed by this individual will have *either* the *G* allele *or* the *g* allele. Recalling what you learned about meiosis in Chapter 8, explain the physical basis for this segregation of alleles.

■ The *G* and *g* alleles are located at the same gene locus on homologous chromosomes, which separate during meiosis and are packaged in separate gametes.

The law of independent assortment is revealed by tracking two characters at once

Two of the seven characters Mendel studied were seed shape and seed color. Mendel's seeds were either round or wrinkled in shape, and they were either yellow or green in color. From monohybrid crosses, Mendel knew that the allele for round shape (designated *R*) was dominant to the allele for wrinkled shape (*r*) and that the allele for yellow seed color (*Y*) was dominant to the allele for green seed color (*y*). What would result from a mating of parental varieties differing in two characters—a **dihybrid cross?** Mendel crossed homozygous plants having round yellow seeds (genotype *RRYY*) with plants having wrinkled green seeds (*rryy*). As shown in Figure 9.5A, the union of *RY* and *ry* gametes yielded hybrids heterozygous for both characteristics (*RrYy*)—that is, *dihybrids*. As we would expect, all of these offspring, the F₁ generation, had round yellow seeds. But were the two characters transmitted from parents to offspring as a package, or was each character inherited independently of the other?

The question was answered when Mendel allowed fertilization to occur among the F₁ plants. If the genes for the two characters were inherited together, as shown on the left in the figure, then the F₁ hybrids would produce only the same two kinds (genotypes) of gametes—*RY* and *ry*—that they received from their parents. This hypothesis predicts that the phenotypic ratio of the F₂ generation will be 3 : 1 (three plants with round yellow seeds for every one with wrinkled green seeds),

as in the left Punnett square. If, however, the two seed characters segregated independently, then the F₁ generation would produce four gamete genotypes—*RY, rY, Ry,* and *ry*—in equal quantities. The Punnett square on the right shows all possible combinations of alleles that can result in the F₂ generation from the union of four kinds of sperm with four kinds of eggs. This Punnett square shows that there are nine different genotypes in the F₂. However, there are only four phenotypes, with a ratio of 9 : 3 : 3 : 1.

The Punnett square on the right also reveals that a dihybrid cross is equivalent to two monohybrid crosses occurring simultaneously. From the 9 : 3 : 3 : 1 ratio, we can see that there are 12 plants with round seeds to 4 with wrinkled seeds and 12 yellow-seeded plants to 4 green-seeded ones. These 12 : 4 ratios each reduce to 3 : 1, which is the F₂ ratio for a monohybrid cross. Mendel tried his seven pea characters in various dihybrid combinations and always obtained data close to the predicted 9 : 3 : 3 : 1 ratio. These results supported the hypothesis that *each pair of alleles segregates independently of other pairs of alleles during gamete formation.* This is called Mendel's **law of independent assortment.**

Figure 9.5B shows how this law applies to the inheritance of two characters in Labrador retrievers: black versus chocolate coat color and normal vision versus the eye disorder called progressive retinal atrophy (PRA). As you would

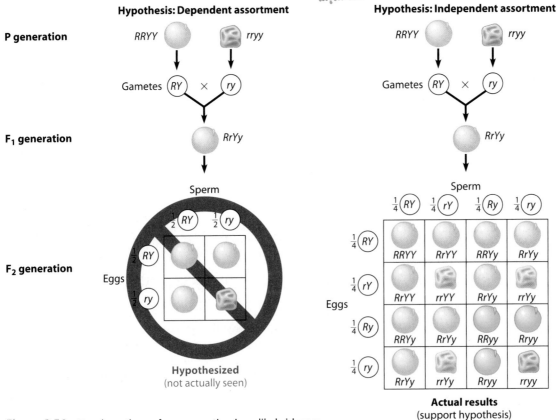

Figure 9.5A Two hypotheses for segregation in a dihybrid cross

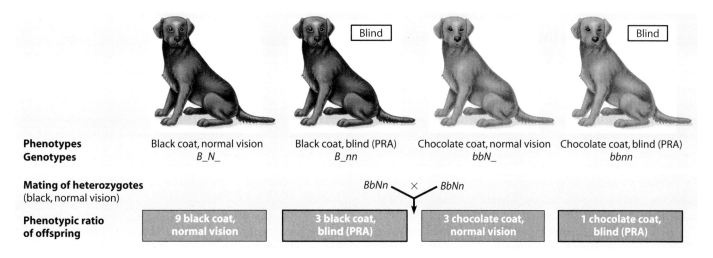

| Phenotypes
Genotypes | Black coat, normal vision
B_N_ | Black coat, blind (PRA)
B_nn | Chocolate coat, normal vision
bbN_ | Chocolate coat, blind (PRA)
bbnn |

Mating of heterozygotes
(black, normal vision)

BbNn × BbNn

| Phenotypic ratio
of offspring | 9 black coat,
normal vision | 3 black coat,
blind (PRA) | 3 chocolate coat,
normal vision | 1 chocolate coat,
blind (PRA) |

Figure 9.5B   Independent assortment of two genes in the Labrador retriever

expect, these characters are controlled by separate genes. Black Labs have at least one copy of an allele called *B*, which gives their hairs densely packed granules of a dark pigment. The *B* allele is dominant to *b*, which leads to a less tightly packed distribution of pigment granules. As a result, the coats of dogs with genotype *bb* are chocolate in color. The allele that causes PRA, called *n*, is recessive to allele *N*, which is necessary for normal vision. Thus, only dogs of genotype *nn* become blind from PRA. (In the figure, blanks in the genotypes indicate a second allele of either sort. And if you're wondering about yellow Labs, their color is controlled by a different gene altogether.)

The lower part of Figure 9.5B shows what happens when we mate two heterozygous Labs, both of genotype *BbNn*. The F$_2$ phenotypic ratio will be nine black dogs with normal eyes to three black with PRA to three chocolate with normal eyes to one chocolate with PRA. These results are analogous to the results in Figure 9.5A and demonstrate that the *B* and *N* genes are inherited independently.

? Predict the phenotypes of offspring obtained by mating a black Lab homozygous for both coat color and normal eyes with a chocolate Lab that is blind from PRA.

■ All offspring would be black with normal eyes (BBNN × bbnn → BbNn).

9.6   Geneticists use the testcross to determine unknown genotypes

Suppose you have a Labrador retriever with a chocolate coat. Referring to Figure 9.5B, you can tell that its genotype must be *bb*, the only combination of alleles that produces the chocolate-coat phenotype. But what if you had a black Lab? It could have one of two possible genotypes—*BB* or *Bb*—and there is no way

Testcross:

Genotypes   B_   ×   bb

Two possibilities for the black dog:

BB   or   Bb

Gametes   B   B   b

Offspring   Bb   Bb   bb
               All black   1 black : 1 chocolate

Figure 9.6   Using a testcross to determine genotype

to tell which is correct simply by looking at the dog. To determine your dog's genotype, you need to perform a **testcross,** a mating between an individual of unknown genotype (your dog) and a homozygous recessive (*bb*) individual—in this case, a chocolate Lab.

Figure 9.6 shows the offspring that could result from such a mating. If, as shown on the left, the black parent's genotype is *BB*, we would expect all the offspring to be black because a cross between genotypes *BB* and *bb* can produce only *Bb* offspring. On the other hand, if the black parent is *Bb*, we would expect both black (*Bb*) and chocolate (*bb*) offspring. Thus, the appearance of the offspring reveals the original black dog's genotype.

Mendel used testcrosses to verify that he had true-breeding varieties of plants. The testcross continues to be an important tool of geneticists for determining genotypes.

Web Activity *Dihybrid Cross*

? You use a testcross to determine the genotype of a Lab with normal eyes. Half of the offspring of the testcross are normal and half develop PRA. What is the genotype of the normal parent?

uN ■

9.7 Mendel's laws reflect the rules of probability

Mendel's strong background in mathematics served him well in his studies of inheritance. He understood, for instance, that the segregation of allele pairs during gamete formation and the re-forming of pairs at fertilization obey the rules of probability—the same rules that apply to the tossing of coins, the rolling of dice, and the drawing of cards. Mendel also appreciated the statistical nature of inheritance. He knew that he needed to obtain large samples—count many offspring from his crosses—before he could begin to interpret inheritance patterns.

Let's see how the rules of probability apply to inheritance. The probability scale ranges from 0 to 1. An event that is certain to occur has a probability of 1, whereas an event that is certain not to occur has a probability of 0. The probabilities of all possible outcomes for an event must add up to 1. With a coin, the chance of tossing heads is $\frac{1}{2}$, and the chance of tossing tails is $\frac{1}{2}$. In a standard deck of 52 playing cards, the chance of drawing a jack of diamonds is $\frac{1}{52}$, and the chance of drawing any card other than the jack of diamonds is $\frac{51}{52}$.

An important lesson we can learn from coin tossing is that for each and every toss of the coin, the probability of heads is $\frac{1}{2}$. In other words, the outcome of any particular toss is unaffected by what has happened on previous attempts. Each toss is an *independent event.*

If two coins are tossed simultaneously, the outcome for each coin is an independent event, unaffected by the other coin. What is the chance that both coins will land heads up? The probability of such a *compound event* is the product of the probabilities of each independent event—for the

coins, $\frac{1}{2} \times \frac{1}{2} = \frac{1}{4}$. This is called the **rule of multiplication,** and it holds true for genetics as well as coin tosses. Figure 9.7 represents a cross between F$_1$ Labrador retrievers that have the *Bb* genotype for coat color. The genetic cross is represented by the tossing of two coins that represent the two gametes (a dime for the egg and a penny for the sperm); the heads side of each coin stands for the dominant *B* allele and the tails side of each coin the recessive *b* allele. What is the probability that a particular F$_2$ dog will have the *bb* genotype? To produce a *bb* offspring, both egg and sperm must carry the *b* allele. The probability that an egg will have the *b* allele is $\frac{1}{2}$, and the probability that a sperm will have the *b* allele is also $\frac{1}{2}$. By the rule of multiplication, the probability that two *b* alleles will come together at fertilization is $\frac{1}{2} \times \frac{1}{2} = \frac{1}{4}$. This is exactly the answer given by the Punnett square in Figure 9.7. If we know the genotypes of the parents, we can predict the probability for any genotype among the offspring.

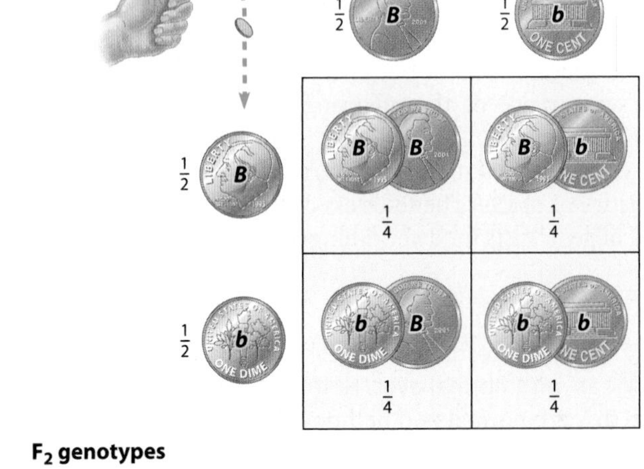

Now consider the probability that an F$_2$ Lab will be heterozygous for the coat-color gene. As Figure 9.7 shows, there are two ways in which F$_1$ gametes can combine to produce a heterozygous offspring. The dominant (*B*) allele can come from the egg and the recessive (*b*) allele from the sperm, or vice versa. The probability that an event can occur in two or more alternative ways is the *sum* of the separate probabilities of the different ways; this is known as the **rule of addition.** Using this rule, we can calculate the probability of an F$_2$ heterozygote as $\frac{1}{4} + \frac{1}{4} = \frac{1}{2}$.

By applying the rules of probability to segregation and independent assortment, we can solve some rather complex genetics problems. For instance, we can predict the results of trihybrid crosses, in which three different characters are involved. Consider a cross between two organisms that both have the genotype *AaBbCc*. What is the probability that an offspring from this cross will be a recessive homozygote for all three genes (*aabbcc*)? Since each allele pair assorts independently, we can treat this trihybrid cross as three separate monohybrid crosses:

Aa × *Aa*:   Probability of *aa* offspring = $\frac{1}{4}$

Bb × *Bb*:   Probability of *bb* offspring = $\frac{1}{4}$

Cc × *Cc*:   Probability of *cc* offspring = $\frac{1}{4}$

Because the segregation of each allele pair is an independent event, we use the rule of multiplication to calculate the probability that the offspring will be *aabbcc*:

$$\frac{1}{4} aa \times \frac{1}{4} bb \times \frac{1}{4} cc = \frac{1}{64}$$

We could reach the same conclusion by constructing a 64-section Punnett square, but that would take a lot of space!

Web Activity *Gregor's Garden*

? A plant of genotype *AABbCC* is crossed with an *AaBbCc* plant. What is the probability of an offspring having the genotype *AABBCC*?

F₁ genotypes

Bb female
Formation of eggs

Bb male
Formation of sperm

F₂ genotypes

Figure 9.7   Segregation and fertilization as chance events

$\frac{1}{16}$ (that is, $\frac{1}{2} \times \frac{1}{4} \times \frac{1}{2}$) ■

9.8 Genetic traits in humans can be tracked through family pedigrees

Mendel's laws apply to the inheritance of many human traits. Figure 9.8A illustrates alternative forms of three human characters that are each thought to be determined by simple dominant-recessive inheritance at one gene locus. (The genetic basis of many other human characters—such as eye and hair color—involves several genes and is not well understood.) If we call the dominant allele of any such gene *A*, the dominant phenotype results from either the homozygous genotype *AA* or the heterozygous genotype *Aa*. Recessive phenotypes always result from the homozygous genotype *aa*. In genetics, the word *dominant* does not imply that a phenotype is either normal or more common than a recessive phenotype; wild-type traits (those prevailing in nature) are not necessarily specified by dominant alleles. In genetics, dominance means that a heterozygote (*Aa*), carrying only a single copy of a dominant allele, displays the dominant phenotype. By contrast, the phenotype of the corresponding recessive allele is seen only in a homozygote (*aa*). Recessive traits are often more common in the population than dominant ones. For example, the absence of freckles is more common than their presence.

How can we determine the inheritance pattern of a particular human trait? Since we obviously cannot control human matings, geneticists must analyze the results of matings that have already occurred. First, the geneticist collects as much information as possible about a family's history for the trait of interest. The next step is to assemble this information into a family tree—the family **pedigree.** (You may associate pedigrees with purebred animals—such as racehorses or championship dogs—but they can represent human matings just as well.) To analyze the pedigree, the geneticist applies logic and the Mendelian laws you have learned.

Let's apply this approach to the example in Figure 9.8B, which shows a pedigree that traces the incidence of free versus attached earlobes. The letter *F* stands for the dominant allele for free earlobes, and *f* symbolizes the recessive allele for attached earlobes. In the pedigree, squares represent males, circles represent females, and colored symbols indicate that the person is affected by the trait under investigation (in this case, attached earlobes). The earliest generation studied is at the top of the pedigree.

By applying Mendel's laws, we can deduce that the attached allele is recessive because that is the only way that

Figure 9.8A Examples of single-gene inherited traits in humans

one of the third-generation sisters (at the bottom of the pedigree) could have attached earlobes when both her parents did not. We can therefore label all the individuals with attached earlobes in the pedigree (that is, all those with colored circles or squares) as homozygous recessive (*ff*). Mendel's laws also enable us to deduce the genotypes for most of the people in the pedigree. For example, both of the second generation parents must have carried the *f* allele (which they passed on to the affected daughter) along with the *F* allele that gave them free earlobes. The same must be true of the first set of grandparents because they both had free earlobes but their two sons had attached earlobes.

Notice that we cannot deduce the genotype of every member of the pedigree. For example, the sister with the free earlobe must have at least one *F* allele, but it is possible that she could be *FF* or *Ff*. We cannot distinguish between these two possibilities using the available data.

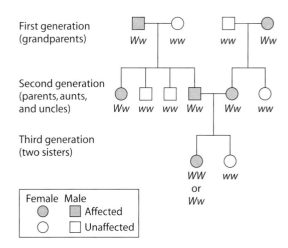

Figure 9.8B Pedigree showing inheritance of attached versus free earlobe in a hypothetical family

? Imagine the second sister (who has attached earlobes) married a man with free earlobes and they had a daughter. What phenotype of this daughter would allow us to deduce all three genotypes in that family?

■ If the daughter had free earlobes, then she and her father must be *ff* and her mother must be *Ff*.

9.9 Many inherited disorders in humans are controlled by a single gene

The genetic disorders listed in Table 9.9 at the bottom of the next page are known to be inherited as dominant or recessive traits controlled by a single gene locus. These disorders show simple inheritance patterns like the traits Mendel studied in pea plants. The genes discussed in this module are all located on autosomes (chromosomes other than the sex chromosomes X and Y; see Module 8.12).

Recessive Disorders Most human genetic disorders are recessive. They range in severity from relatively mild, such as albinism (lack of pigmentation), to life-threatening, such as cystic fibrosis. Most people who have recessive disorders are born to normal parents who are both heterozygotes—that is, who are **carriers** of the recessive allele for the disorder but are phenotypically normal.

Using Mendel's laws, we can predict the fract ion of affected offspring likely to result from a mating between two carriers. Consider a form of inherited deafness caused by a recessive allele (Figure 9.9A). Suppose two heterozygous carriers (*Dd*) had a child. What is the probability that this child would be deaf? As the Punnett square in Figure 9.9A shows, each child of two carriers has a $\frac{1}{4}$ chance of inheriting two recessive alleles. Thus, we can say that about one-fourth of the children of this marriage are likely to be deaf. We can also say that a hearing ("normal") child from such a marriage has a $\frac{2}{3}$ chance of being a carrier (that is, two out of three of the offspring with the hearing phenotype are likely to be carriers). We can apply this same method of pedigree analysis and prediction to any genetic trait controlled by a single gene locus.

The most common fatal genetic disease in the United States is **cystic fibrosis (CF)**. Affecting about 30,000 children and adults in the United States and 70,000 worldwide, the CF allele is carried by about one in 25 people of European ancestry. A person with two copies of this allele has cystic fibrosis, which is characterized by an excessive secretion of very thick mucus from the lungs, pancreas, and other organs. This mucus can interfere with breathing, digestion, and liver function and makes the person vulnerable to recurrent bacterial infections. Although there is no cure for this fatal disease, strict adherence to a daily health regimen—including gentle pounding on the chest and back to clear the airway, inhaled antibiotics, and a high-calorie diet—can have a profound impact on the health of the affected person. Once invariably fatal in childhood, tremendous advances in treatment have raised the median survival age of Americans with CF to 37.

Like cystic fibrosis, most genetic disorders are not evenly distributed across all ethnic groups. Such uneven distribution is the result of prolonged geographic isolation of certain populations. For example, the isolated lives of the early inhabitants of Martha's Vineyard (an island off the coast of Massachusetts) led to frequent marriages between close relatives. Consequently, the frequency of an allele that causes deafness was high, and the deafness allele was rarely transmitted to outsiders.

With the increased mobility in most societies today, it is relatively unlikely that two carriers of a rare, harmful allele will meet and mate. However, the probability increases greatly if close relatives marry and have children. People with recent common ancestors are more likely to carry the same recessive alleles than are unrelated people. Therefore, a mating of close relatives, called **inbreeding,** is more likely to produce offspring homozygous for recessive traits. Such effects can be observed in many types of inbred animals. For example, dogs that have been inbred for appearance may have serious genetic disorders, such as weak hip joints, eye problems, or undesirable behaviors. The detrimental effects of inbreeding are also seen in some endangered species, such as cheetahs.

Geneticists debate the extent to which human inbreeding increases the risk of inherited diseases. Many harmful mutations have such severe effects that a homozygous embryo spontaneously aborts long before birth. Still, most societies have taboos and laws forbidding marriages between close relatives. These rules may have arisen out of the observation that stillbirths and birth defects are more common when parents are closely related.

Dominant Disorders Although many harmful alleles are recessive, a number of human disorders are caused by dominant alleles. Some are nonlethal conditions, such as extra fingers and toes (called polydactyly) or webbed fingers and toes.

One serious dominant disorder is **achondroplasia,** a form of dwarfism. In people with this disorder, the head and torso of the body develop normally, but the arms and legs are short (Figure 9.9B, top of next page). About one out of every 25,000 people has achondroplasia. The homozygous dominant genotype causes death of the embryo, and therefore only heterozygotes, individuals with a single copy of the defective allele, have this disorder. (This also means that a person with achondroplasia

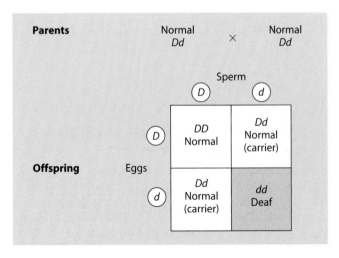

Figure 9.9A Offspring produced by parents who are both carriers for a recessive disorder

has a 50% chance of passing the condition on to any children.) Therefore, all those who do not have achondroplasia, more than 99.99% of the population, are homozygous for the recessive allele. This example makes it clear that a dominant allele is not necessarily more common in a population than the corresponding recessive allele.

Dominant alleles that cause lethal diseases are, in fact, much less common than lethal recessives. One reason for this difference is that the dominant lethal allele cannot be carried by heterozygotes without affecting them. Many lethal dominant alleles result from mutations in a sperm or egg that subsequently kill the embryo. And if the afflicted individual is born but does not survive long enough to reproduce, he or she will not pass on the lethal allele to future generations. This is in contrast to lethal recessive mutations, which are perpetuated from generation to generation by healthy (unaffected) heterozygous carriers.

A lethal dominant allele can escape elimination, however, if it does not cause death until a relatively advanced age. One such example is the allele that causes **Huntington's disease,** a degenerative disorder of the nervous system that usually does not appear until 35 to 45 years of age. Once the deterioration of the nervous system begins, it is irreversible and inevitably

Figure 9.9B Dr. Michael C. Ain, a specialist in the repair of bone defects caused by achondroplasia and related disorders

fatal. Because the allele for Huntington's disease is dominant, any child born to a parent with the allele has a 50% chance of inheriting the allele and the disorder. This example makes it clear that the dominant allele is not necessarily "better" than the corresponding recessive allele.

Until relatively recently, the onset of symptoms was the only way to know if a person had inherited the Huntington's allele. This is no longer the case. By analyzing DNA samples from a large family with a high incidence of the disorder, geneticists tracked the Huntington's allele to a locus near the tip of chromosome 4, and the gene has now been sequenced. This information led to development of a test that can detect the presence of the Huntington's allele in an individual's genome. This is one of several genetic tests currently available. We'll discuss this topic in greater depth in the next module.

? Peter is a 30-year-old man whose father died of Huntington's disease. Neither Peter's mother nor a much older sister show any signs of Huntington's. What is the probability that Peter has inherited Huntington's disease?

$\frac{1}{2}$ ■

TABLE 9.9	SOME AUTOSOMAL DISORDERS IN HUMANS		
Disorder	**Major Symptoms**	**Incidence**	**Comments**
Recessive disorders			
Albinism	Lack of pigment in skin, hair, and eyes	$\frac{1}{22,000}$	Prone to skin cancer
Cystic fibrosis	Excess mucus in lungs, digestive tract, liver; increased susceptibility to infections; death in early childhood unless treated	$\frac{1}{2,500}$ Caucasians	See Module 9.9
Galactosemia	Accumulation of galactose in tissues; mental retardation; eye and liver damage	$\frac{1}{100,000}$	Treated by eliminating galactose from diet
Phenylketonuria (PKU)	Accumulation of phenylalanine in blood; lack of normal skin pigment; mental retardation	$\frac{1}{10,000}$ in U.S. and Europe	See Module 9.10
Sickle-cell disease	Sickled red blood cells; damage to many tissues	$\frac{1}{400}$ African-Americans	See Module 9.13
Tay-Sachs disease	Lipid accumulation in brain cells; mental deficiency; blindness; death in childhood	$\frac{1}{3,500}$ Jews from central Europe	See Module 4.11
Dominant disorders			
Achondroplasia	Dwarfism	$\frac{1}{25,000}$	See Module 9.9
Alzheimer's disease (one type)	Mental deterioration; usually strikes late in life	Not known	
Huntington's disease	Mental deterioration and uncontrollable movements; strikes in middle age	$\frac{1}{25,000}$	See Module 9.9
Hypercholesterolemia	Excess cholesterol in blood; heart disease	$\frac{1}{500}$ are heterozygous	See Module 9.11

9.10 New technologies can provide insight into one's genetic legacy

Some prospective parents are aware that they have an increased risk of having a baby with a genetic disease. For example, many pregnant women over age 35 know that they have a heightened risk of bearing children with Down syndrome (see Module 8.20), and some couples are aware that a certain genetic disease runs in their families. These potential parents may want to learn more about their own and their baby's genetic makeup. Modern technologies offer ways to obtain such information before conception, during pregnancy, and after birth.

Genetic Testing Because most children with recessive disorders are born to healthy parents, the genetic risk for many diseases is determined by whether the prospective parents are carriers of the recessive allele. For an increasing number of genetic disorders, including Tay-Sachs disease, sickle-cell disease, and one form of cystic fibrosis (see Table 9.9), tests are available that can distinguish between individuals who have no disease-causing alleles and those who are heterozygous carriers. Other parents may know that a dominant but late-appearing disease, such as Huntington's disease, runs in their family. Such

people may benefit from genetic tests for dominant alleles. Information from genetic testing (also called genetic screening) can inform decisions about whether to have a child.

Fetal Testing Several technologies are available for detecting harmful genetic conditions in a fetus. Genetic testing before birth requires the collection of fetal cells. In a procedure called **amniocentesis,** performed between weeks 14 and 20 of pregnancy, a physician carefully inserts a needle into the mother's uterus, avoiding the fetus (Figure 9.10A, left). The physician extracts about 10 milliliters (2 teaspoonsful) of the amniotic fluid that bathes the developing fetus. Tests for genetic disorders can be performed on fetal cells (mostly shed skin) present in the fluid. Before testing, these cells are usually cultured in the laboratory for several weeks. By then, enough dividing cells can be harvested to allow karyotyping and the detection of chromosomal abnormalities such as Down syndrome. Biochemical tests can also be performed on the cultured cells, revealing conditions such as Tay-Sachs disease.

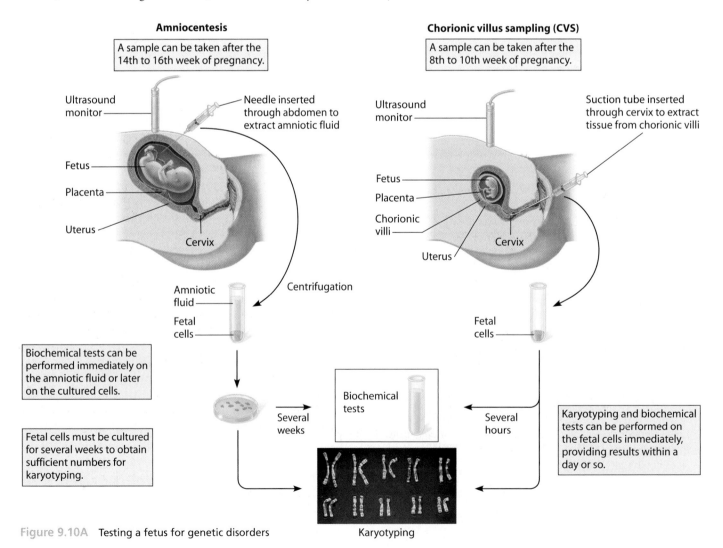

Figure 9.10A Testing a fetus for genetic disorders

In another procedure, **chorionic villus sampling (CVS),** a physician extracts a tiny sample of chorionic villus tissue from the placenta, the organ that carries nourishment and wastes between the fetus and the mother. The tissue can be obtained using a narrow, flexible tube inserted through the mother's vagina and cervix into the uterus (see Figure 9.10A, right). Results of karyotyping and some biochemical tests can be available within 24 hours. The speed of CVS is an advantage over amniocentesis. Another advantage is that CVS can be performed as early as the 8th to 10th week of pregnancy.

Unfortunately, both amniocentesis and CVS pose some risk of complications, such as maternal bleeding, miscarriage, or premature birth. Complication rates for amniocentesis and CVS are about 1% and 2%, respectively. Because of the risks, these procedures are usually reserved for situations in which the chance of a genetic disorder is reasonably high. Recently, medical scientists have developed methods for isolating fetal cells from the mother's blood or cervical mucus. Although very few in number, these cells can be cultured and tested with little risk to the mother or child.

Blood tests on the mother at 15 to 20 weeks of pregnancy can help identify fetuses at risk for certain birth defects—and thus candidates for further testing that may require more invasive procedures (such as amniocentesis). The most widely used blood test measures the mother's blood level of alpha-fetoprotein (AFP), a protein produced by the fetus. High levels of AFP may result from Down syndrome or neural tube defects in the fetus. (The neural tube is an embryonic structure that develops into the brain and spinal cord.) For a more complete risk profile, a woman's doctor may order a "triple screen test," which measures AFP as well as two other hormones produced by the placenta. Abnormal levels of these substances in the maternal blood may also point to a risk of Down syndrome.

Fetal Imaging Other techniques enable a physician to examine a fetus directly for anatomical deformities. The most widely used procedure is **ultrasound imaging,** which uses sound waves to produce a picture of the fetus. Figure 9.10B shows an ultrasound scanner, which emits high-frequency sounds, beyond the range of hearing. When the sound waves bounce off the fetus, the echoes produce an image on the monitor. The inset image in Figure 9.10B shows a fetus at about 18 weeks. Ultrasound imaging is noninvasive (no foreign objects are inserted into the mother's body) and has no known risk. In another imaging method, fetoscopy, a needle-thin tube containing a fiber optic viewing scope is inserted into the uterus. Fetoscopy can provide highly detailed images of the fetus but, unlike ultrasound, carries risk of complications.

Newborn Screening Some genetic disorders can be detected at birth by simple tests that are now routinely performed in most hospitals in the United States. One common screening program is for phenylketonuria (PKU), a recessively inherited disorder that occurs in about one out of every 10,000 births in the United States. Children with this disease cannot properly break down the naturally occurring amino acid phenylalanine; an accumulation of phenylalanine may lead to mental retardation. However, if the deficiency is

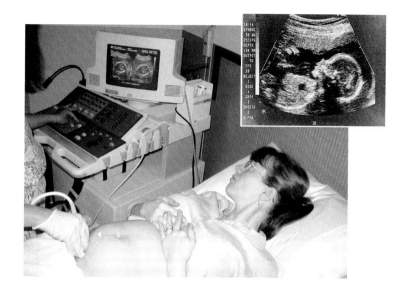

Figure 9.10B Ultrasound scanning of a fetus

detected in the newborn, a special diet low in phenylalanine can usually prevent retardation. Unfortunately, few other genetic disorders are currently treatable.

Ethical Considerations As new technologies such as fetal imaging and testing become more widespread, geneticists are working to make sure that they do not cause more problems than they solve. Consider the tests for identifying carriers of recessive diseases. Such information may enable people with family histories of genetic disorders to make informed decisions about having children. But these new methods for genetic screening pose problems, too. If confidentiality is breached, will carriers be stigmatized? For example, will they be denied health or life insurance, even though they themselves are healthy? Will misinformed employers equate "carrier" with disease? Geneticists stress that patients seeking genetic testing should receive counseling both before and after to clarify their family history, to explain the test, and to help them cope with the results. But will sufficient numbers of genetic counselors be available to help individuals understand their test results?

Couples at risk for conceiving children with genetic disorders may now learn a great deal about their unborn children. In particular, CVS gives parents a chance to become informed very early in pregnancy. What is to be done with such information? If fetal tests reveal a serious disorder, the parents must choose between terminating the pregnancy and preparing themselves for a baby with severe problems. Identifying a genetic disease early can give families time to prepare—emotionally, medically, and financially.

Advances in biotechnology offer possibilities for reducing human suffering, but not before key ethical issues are resolved. The dilemmas posed by human genetics reinforce one of this book's themes: the immense social implications of biology.

? What is the primary benefit of genetic screening by CVS? What is the primary risk?

■ CVS allows genetic screening to be performed very early in pregnancy, but it carries a risk of miscarriage.

9.11 Incomplete dominance results in intermediate phenotypes

Mendel's laws explain inheritance in terms of discrete factors—genes—that are passed along from generation to generation according to simple rules of probability. Mendel's laws are valid for all sexually reproducing organisms, including garden peas, show dogs, and human beings. But just as the basic rules of musical harmony cannot account for all the rich sounds of a symphony, Mendel's laws stop short of explaining some patterns of genetic inheritance. In fact, for most sexually reproducing organisms, cases where Mendel's laws can strictly account for the patterns of inheritance are relatively rare. More often, the inheritance patterns are more complex, as we will see in this and the next four modules.

The F₁ offspring of Mendel's pea crosses always looked like one of the two parental varieties. This situation is called **complete dominance;** the dominant allele had the same phenotypic effect whether present in one or two copies. But for some characters, the appearance of F₁ hybrids falls *between* the phenotypes of the two parental varieties, an effect called **incomplete dominance.** For instance, as Figure 9.11A illustrates, when red snapdragons are crossed with white

snapdragons, all the F₁ hybrids have pink flowers. This third phenotype results from flowers of the heterozygote having less red pigment than the red homozygotes.

Incomplete dominance does *not* support the blending hypothesis, which would predict that the red and white traits could never be retrieved from the pink hybrids. As the Punnett square at the bottom of Figure 9.11A shows, the F₂ offspring appear in a phenotypic ratio of one red to two pink to one white, because the red and white alleles segregate during gamete formation in the pink F₁ hybrids. In incomplete dominance, the phenotypes of heterozygotes differ from the two homozygous varieties, and the genotypic ratio and the phenotypic ratio are both 1 : 2 : 1 in the F₂ generation.

We also see examples of incomplete dominance in humans. One case involves a recessive allele (*h*) that can cause *hypercholesterolemia,* dangerously high levels of cholesterol in the blood. Normal individuals are *HH*. Heterozygotes (*Hh;* about one in 500 people) have blood cholesterol levels about twice normal. They are unusually prone to atherosclerosis, the blockage of arteries by cholesterol buildup in artery walls, and they may have heart attacks from blocked heart arteries by their mid-30s. Hypercholesterolemia is even more serious in homozygous individuals (*hh;* about one in a million people). Homozygotes have about five times the normal amount of blood cholesterol and may have heart attacks as early as age 2.

Figure 9.11B illustrates the molecular basis for hypercholesterolemia. The dominant allele, which normal individuals carry in duplicate (*HH*), specifies a cell-surface protein called an LDL receptor. LDLs, or low-density lipoproteins, are cholesterol-containing particles in the blood. The LDL

P generation

Red
RR   ×   White
rr

Gametes (R)   (r)

F₁ generation

Pink
Rr

Gametes ½(R) ½(r)

F₂ generation

Sperm
½(R) ½(r)

Eggs
½(R)   RR | rR
½(r)   Rr | rr

Figure 9.11A Incomplete dominance in snapdragon color

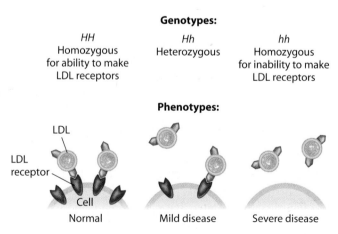

Genotypes:

| *HH* | *Hh* | *hh* |
| Homozygous for ability to make LDL receptors | Heterozygous | Homozygous for inability to make LDL receptors |

Phenotypes:

LDL

LDL receptor

Cell

Normal   Mild disease   Severe disease

Figure 9.11B Incomplete dominance in human hypercholesterolemia

receptors pick up LDL particles from the blood and promote their uptake by cells that break down the cholesterol. This process helps prevent the accumulation of cholesterol in arteries. Without the receptors, lethal levels of LDL build up in the blood. Heterozygotes (*Hh*) have only half the normal number of LDL receptors, and homozygous recessives (*hh*) have none.

Web Activity *Incomplete Dominance*

? Why is a testcross unnecessary to determine whether a snapdragon with red flowers is homozygous or heterozygous?

■ Because the homozygotes and heterozygotes differ in phenotype: red flowers for the dominant homozygote and pink flowers for the heterozygote

9.12 Many genes have more than two alleles in the population

So far, we have discussed inheritance patterns involving only two alleles per gene. But most genes can be found in populations in more than two versions, known as multiple alleles. Although any particular individual carries, at most, two different alleles for a particular gene, in cases of multiple alleles, more than two possible alleles exist in the wider population.

For instance, the **ABO blood group** phenotype in humans involves three alleles of a single gene. These three alleles, in various combinations, produce four phenotypes: A person's blood group may be either O, A, B, or AB. These letters refer to two carbohydrates, designated A and B, that may be found on the surface of red blood cells. A person's red blood cells may have carbohydrate A (type A blood), carbohydrate B (type B), both (type AB), or neither (type O). Matching compatible blood groups is critical for safe blood transfusions. If a donor's blood cells have a carbohydrate (A or B) that is foreign to the recipient, then the recipient's immune system produces blood proteins called antibodies (see Module 24.8) that bind specifically to the foreign carbohydrates and cause the donor blood cells to clump together, potentially killing the recipient. The clumping reaction is also the basis of a blood-typing test performed in the laboratory. Figure 9.12 shows which combinations of blood groups result in clumping.

The four blood groups result from various combinations of the three different alleles, symbolized as I^A (for the enzyme that adds carbohydrate A to red blood cells), I^B (for B), and i (for neither A nor B). Each person inherits one of these alleles from each parent. Because there are three alleles, there are six possible genotypes, as listed in the figure. Both the I^A and I^B alleles are dominant to the i allele. Thus, $I^A I^A$ and $I^A i$ people have type A blood, and $I^B I^B$ and $I^B i$ people have type B. Recessive homozygotes, *ii*, have type O blood because their blood cells have neither the A nor the B carbohydrate. The I^A and I^B alleles are **codominant;** both alleles are expressed in heterozygous individuals ($I^A I^B$), who have type AB blood. Note that codominance (the expression of both alleles) is different from incomplete dominance (the expression of one intermediate trait).

? Maria has type O blood and her sister has type AB blood. The girls know that both of their maternal grandparents are type A. What are the genotypes of the girls' parents?

■ Their mother is $I^A i$; their father is $I^B i$.

Blood Group (Phenotype)	Genotypes	Red Blood Cells	Antibodies Present in Blood	Reaction When Blood from Groups Below Is Mixed with Antibodies from Groups at Left			
				O	A	B	AB
O	*ii*		Anti-A Anti-B				
A	$I^A I^A$ or $I^A i$	Carbohydrate A	Anti-B				
B	$I^B I^B$ or $I^B i$	Carbohydrate B	Anti-A				
AB	$I^A I^B$		—				

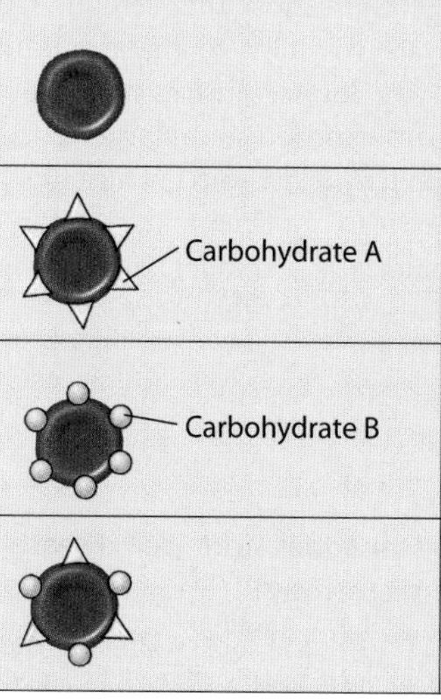

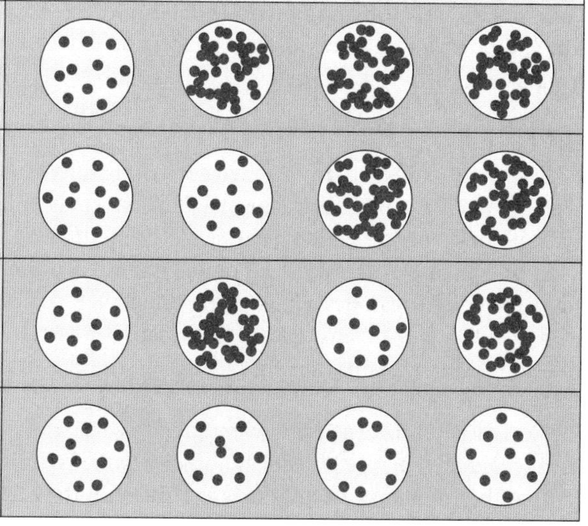

Figure 9.12 Multiple alleles for the ABO blood groups

9.13 A single gene may affect many phenotypic characters

All of our genetic examples to this point have been cases in which each gene specifies only one hereditary character. Most genes, however, influence multiple characters, a property called **pleiotropy** (from the Greek *pleion,* more).

An example of pleiotropy in humans is sickle-cell disease, a disorder characterized by the diverse symptoms shown in Figure 9.13. The direct effect of the sickle-cell allele is to make red blood cells produce abnormal hemoglobin molecules. These molecules tend to link together and crystallize, especially when the oxygen content of the blood is lower than usual because of high altitude, overexertion, or respiratory ailments. As the hemoglobin crystallizes, the normally disk-shaped red blood cells deform to a sickle shape with jagged edges, as shown in the micrograph. Sickled cells are destroyed rapidly by the body, and the destruction of these cells may seriously lower the individual's red cell count, causing anemia and general weakening of the body. Also, because of their angular shape, sickled cells do not flow smoothly in the blood and tend to accumulate and clog tiny blood vessels. Blood flow to body parts is reduced, resulting in periodic fever, severe pain, and damage to various organs, including the heart, brain, and kidneys. Sickled cells also accumulate in the spleen, damaging it. Blood transfusions and certain drugs may relieve some of the symptoms, but there is no cure, and sickle-cell disease kills about 100,000 people in the world annually.

In most cases, only people who are homozygous for the sickle-cell allele suffer from the disease. Heterozygotes, who have one sickle-cell allele and one nonsickle allele, are usually healthy, although in rare cases they may experience some pleiotropic effects when oxygen in the blood is severely reduced, such as at very high altitudes. These effects may occur because the nonsickle and sickle-cell alleles are codominant: Both alleles are expressed in heterozygous individuals, and their red blood cells contain both normal and abnormal hemoglobin. Heterozygotes are said to have "sickle-cell trait."

Sickle-cell disease is the most common inherited disorder among people of African descent, striking one in 400 African-Americans. About one in ten African-Americans is a carrier—a heterozygote. Among Americans of other ancestry, the sickle-cell allele is extremely rare.

One in ten is an unusually high frequency of carriers for an allele with such harmful effects in homozygotes. We might expect that the frequency of the sickle-cell allele in the population would be much lower because many homozygotes die before passing their genes to the next generation. The high frequency appears to be a vestige of the roots of African-Americans. Sickle-cell disease is most common in tropical Africa, where the deadly disease malaria is also prevalent. The protistan parasite that causes malaria spends part of its life cycle inside red blood cells. When it enters those of a person with the sickle-cell allele, it triggers sickling. The body destroys most of the sickled cells, and the parasite does not grow well in those that remain. Consequently, sickle-cell carriers are resistant to malaria, and in many parts of Africa, they live longer and have more offspring than noncarriers who are exposed to malaria. In this way, malaria has kept the frequency of the sickle-cell allele relatively high in much of the African continent. To put it in evolutionary terms, as long as malaria is a danger, individuals with the sickle-cell allele have a selective advantage.

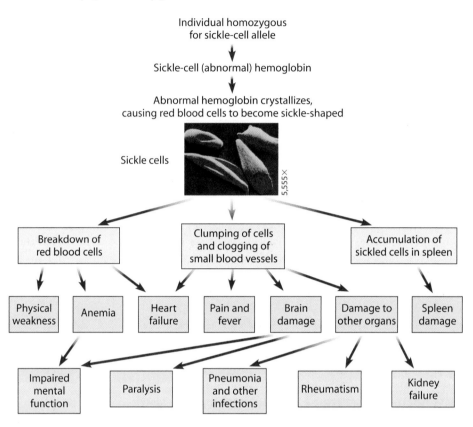

Individual homozygous
for sickle-cell allele

↓

Sickle-cell (abnormal) hemoglobin

↓

Abnormal hemoglobin crystallizes,
causing red blood cells to become sickle-shaped

Sickle cells

5,555×

Breakdown of red blood cells

Clumping of cells and clogging of small blood vessels

Accumulation of sickled cells in spleen

Physical weakness | Anemia | Heart failure | Pain and fever | Brain damage | Damage to other organs | Spleen damage

Impaired mental function | Paralysis | Pneumonia and other infections | Rheumatism | Kidney failure

Figure 9.13 Sickle-cell disease, multiple effects of a single human gene

? How does sickle-cell disease exemplify the concept of pleiotropy?

Homozygosity for the sickle-cell allele causes abnormal hemoglobin, and the impact of the abnormal hemoglobin on the shape of red blood cells leads to a cascade of symptoms in multiple organs of the body. ■

9.14 A single character may be influenced by many genes

Mendel studied genetic characters that could be classified on an either-or basis, such as purple or white flower color. However, many characteristics, such as human skin color and height, vary in a population along a continuum. Many such features result from **polygenic inheritance,** the additive effects of two or more genes on a single phenotypic character. (This is the converse of pleiotropy, in which a single gene affects several characters.)

Let's consider a hypothetical example. Assume that the continuous variation in human skin color is controlled by three genes that are inherited separately, like Mendel's pea genes. (Actually, genetic evidence indicates that *at least* three genes control this character.) The "dark-skin" allele for each gene (*A, B,* or *C*) contributes one "unit" of darkness to the phenotype and is incompletely dominant to the other allele (*a, b,* or *c*). A person who is *AABBCC* would be very dark, whereas an *aabbcc* individual would be very light. An *AaBbCc* person (resulting, for example, from a mating between an *AABBCC* person and an *aabbcc* person) would have skin of an intermediate shade. Because the alleles have an additive effect, the genotype *AaBbCc* would produce the same skin color as any other genotype with just three dark-skin alleles, such as *AABbcc.*

The Punnett square in Figure 9.14 shows all possible genotypes of offspring from a mating of two triple heterozygotes (the F₁ generation here). The row of squares below the Punnett square shows the seven skin-color phenotypes that would theoretically result from this mating. The seven bars in the graph at the bottom of the figure depict the relative numbers of each of the phenotypes in the F₂ generation. In real human populations, skin color has even more variations than shown in the figure, in part for reasons we discuss in the next module.

Up to this point in the chapter, we have presented four types of inheritance patterns that are extensions of Mendel's laws of inheritance: incomplete dominance, codominance, pleiotropy, and polygenic inheritance. It is important to realize that these patterns are extensions of Mendel's model, rather than exceptions to it. From Mendel's garden pea experiments came data supporting a particulate theory of inheritance, with the particles (genes) being transmitted according to the same rules of chance that govern the tossing of coins. The particulate theory holds true for all inheritance patterns. In the next module, we consider another important source of deviation from Mendel's standard model: the effect of the environment.

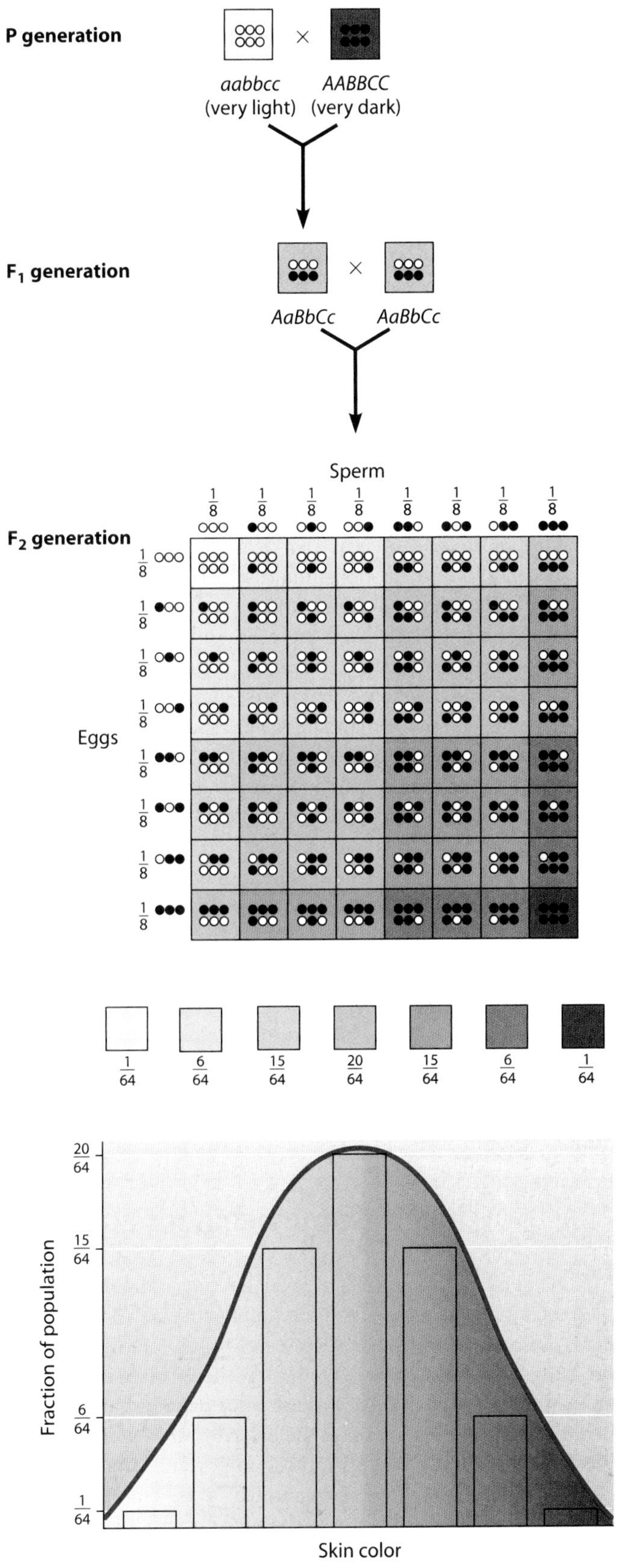

? Based on the skin-color model in Figure 9.14, an *AaBbcc* individual would be indistinguishable in phenotype from which of the following individuals: *AAbbcc, aaBBcc, AabbCc, Aabbcc,* or *aaBbCc?*

All except *Aabbcc*

Figure 9.14 A model for polygenic inheritance of skin color

9.15 The environment affects many characters

In the previous module, we saw how a set of three hypothetical human skin-color genes could produce seven different skin-color phenotypes. If we examine a real human population for the skin-color phenotype, we would see more shades than just seven. The true range might be similar to the entire spectrum of color under the bell-shaped curve in Figure 9.14. In fact, no matter how carefully we characterize the genes for skin color, a purely genetic description will always be incomplete. This is because some intermediate shades of skin color result from the effects of environmental factors, such as exposure to the sun.

Many characters result from a combination of heredity and environment. For example, although a single tree is locked into its inherited genotype, its leaves vary in size, shape, and color, depending on exposure to wind and sun and the tree's nutritional state. For humans, nutrition influences height; exercise alters build; sun-tanning darkens the skin; and experience improves performance on intelligence tests. As geneticists learn more and more about our genes, it is becoming clear that many human phenotypes—such as risk of heart disease and cancer and susceptibility to alcoholism and schizophrenia—are influenced by both genes and environment.

Whether human characters are more influenced by genes or by the environment—nature or nurture—is a very old and hotly contested debate. For some characters, such as the ABO blood group, a given genotype mandates a very specific phenotype. In contrast, a person's blood count of red and white cells varies quite a bit, depending on such factors as the altitude, the customary level of physical activity, and the presence of infectious agents.

It is important to realize that the individual features of any organism arise from a combination of genetic and environmental factors. Simply spending time with identical twins will convince anyone that environment, and not just genes, affects a

Figure 9.15 Varying phenotypes due to environmental factors in genetically identical twins

person's traits (Figure 9.15). However, there is an important difference between these two sources of variation: Only genetic influences are inherited. Any effects of the environment are not passed on to the next generation.

? Why was Mendel able to ignore environmental influences on his pea plants ?

■ The characters he chose for study were all entirely genetically determined.

The Chromosomal Basis of Inheritance

9.16 Chromosome behavior accounts for Mendel's laws

Mendel published his results in 1866, but not until long after he died did biologists understand the significance of his work. Cell biologists worked out the processes of mitosis and meiosis in the late 1800s (see Chapter 8 to review these processes). Then, around 1900, researchers began to notice parallels between the behavior of chromosomes and the behavior of Mendel's heritable factors. One of biology's most important concepts—the **chromosome theory of inheritance**—was emerging. The chromosome theory states that genes occupy specific loci (positions) on chromosomes and it is the chromosomes that undergo segregation and independent assortment during

meiosis. Thus, it is the behavior of chromosomes during meiosis and fertilization that accounts for inheritance patterns.

We can see the chromosomal basis of Mendel's laws by following the fates of two genes during meiosis and fertilization in pea plants. In Figure 9.16, we picture the genes for seed shape (alleles R and r) and seed color (Y and y) as black bars on different chromosomes. We start with the F_1 generation, in which all plants have the $RrYy$ genotype. To simplify the diagram, we show only two of the seven pairs of pea chromosomes and three of the stages of meiosis: metaphase I, anaphase I, and metaphase II.

To see the chromosomal basis of the law of segregation, let's follow just the homologous pair of long chromosomes, the ones carrying R and r, taking either the left or the right branch from the F_1 cell. Whichever arrangement the chromosomes assume at metaphase I, the two alleles *segregate* as the homologous chromosomes separate in anaphase I. And at the end of meiosis II, a single long chromosome ends up in each of the gametes. Fertilization then recombines the two alleles at random, resulting in F_2 offspring that are $\frac{1}{4}$ RR, $\frac{1}{2}$ Rr, and $\frac{1}{4}$ rr. The ratio of round to wrinkled phenotypes is thus $3:1$ (12 round to 4 wrinkled), the ratio Mendel observed (see Figure 9.3A).

To see the chromosomal basis of the law of independent assortment, follow both the long and short (nonhomologous) chromosomes through the figure below. Two alternative, equally likely arrangements of tetrads can occur at metaphase I. The nonhomologous chromosomes (and their genes) assort independently, leading to four gamete genotypes. Random fertilization leads to the $9:3:3:1$ phenotypic ratio in the F_2 generation, as you saw in Figure 9.5A.

🎧 **MP3 Tutor** *Chromosomal Basis of Inheritance*

? Which of Mendel's laws have their physical basis in the following phases of meiosis: (a) the orientation of homologous chromosome pairs in metaphase I; (b) the separation of homologues in anaphase I?

■ (a) the law of independent assortment; (b) the law of segregation

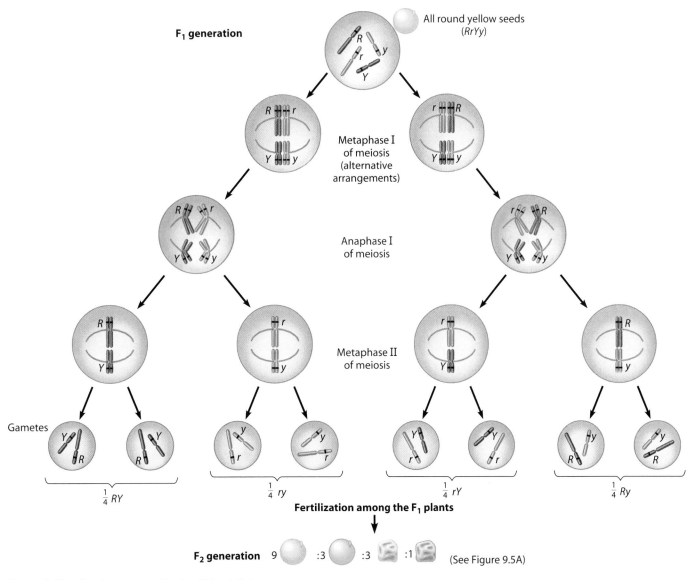

Figure 9.16 The chromosomal basis of Mendel's laws

9.17 Genes on the same chromosome tend to be inherited together

In 1908, British biologists William Bateson and Reginald Punnett (originator of the Punnett square) observed an inheritance pattern that seemed inconsistent with Mendelian laws. Bateson and Punnett were working with two characters in sweet peas, flower color and pollen shape. They crossed doubly heterozygous plants (*PpLl*) that exhibited the dominant traits: purple flowers (expression of the *P* allele) and long pollen grains (expression of the *L* allele). The corresponding recessive traits are red flowers (in *pp* plants) and round pollen (in *ll* plants).

The top part of Figure 9.17 illustrates Bateson and Punnett's experiment. When they looked at just one of the two characters (that is, either cross *Pp* × *Pp* or cross *Ll* × *Ll*), they found that the dominant and recessive alleles segregated, producing a phenotypic ratio of approximately 3 : 1 for the offspring, in agreement with Mendel's law of segregation. However, when the biologists combined their data for the two characters, they did not see the predicted 9 : 3 : 3 : 1 ratio. Instead, as shown in the table, they found a disproportionately large number of plants with just two of the predicted phenotypes: purple long (almost 75% of the total) and red round (about 14%). The other two phenotypes (purple round and red long) were found in far fewer numbers than expected. What can account for these results?

The number of genes in a cell is far greater than the number of chromosomes; in fact, each chromosome has hundreds or thousands of genes. Genes located close together on the same chromosome tend to be inherited together and are called **linked genes.** Linked genes generally do not follow Mendel's law of independent assortment.

As shown in the "Explanation" in the figure, the sweet-pea genes for flower color and pollen shape are located on the same chromosome. Thus, meiosis in the heterozygous (*PpLl*) sweet-pea plant yields mostly two genotypes of gametes (*PL* and *pl*) rather than equal numbers of the four types of gametes that would result if the flower-color and pollen-shape genes were not linked. The large numbers of plants with purple long and red round traits in the Bateson-Punnett experiment resulted from fertilization among the *PL* and *pl* gametes. But what about the smaller numbers of plants with purple round and red long traits? As you will see in the next module, the phenomenon of crossing over accounts for these offspring.

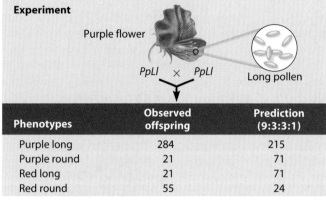

Experiment

Purple flower — *PpLl* × *PpLl* — Long pollen

Phenotypes	Observed offspring	Prediction (9:3:3:1)
Purple long	284	215
Purple round	21	71
Red long	21	71
Red round	55	24

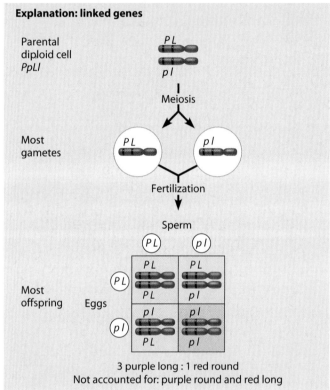

Explanation: linked genes

Parental diploid cell *PpLl*

Meiosis

Most gametes

Fertilization

Sperm

Most offspring   Eggs

3 purple long : 1 red round
Not accounted for: purple round and red long

Figure 9.17 Experiment involving linked genes in the sweet pea

? Why do linked genes tend to be inherited together?

■ Because their loci are close together on the same chromosome

9.18 Crossing over produces new combinations of alleles

In Module 8.18, we saw that during meiosis, crossing over between homologous chromosomes produces new combinations of alleles in gametes. Figure 9.18A (top of next page) reviews this process, showing that two linked genes can give rise to four different gamete genotypes. Gametes with genotypes *AB* and *ab* carry parental-type chromosomes that have not been altered by crossing over. In contrast, gametes with

genotypes *Ab* and *aB* are recombinant gametes. The exchange of chromosome segments during crossing over has produced new combinations of alleles. We can now understand the results of the Bateson-Punnett experiment presented in the previous module: The small fraction of offspring with the recombinant phenotype must have resulted from fertilization involving recombinant gametes.

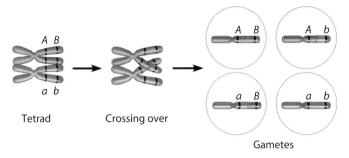

Tetrad   Crossing over

Gametes

Figure 9.18A   Review: Production of recombinant gametes

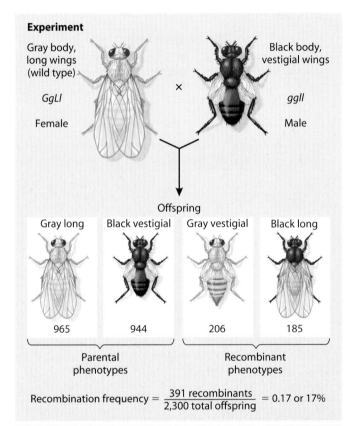

Experiment

Gray body, long wings (wild type)

GgLl

Female

×

Black body, vestigial wings

ggll

Male

Offspring

Gray long	Black vestigial	Gray vestigial	Black long
965	944	206	185

Parental phenotypes   Recombinant phenotypes

$$\text{Recombination frequency} = \frac{391 \text{ recombinants}}{2{,}300 \text{ total offspring}} = 0.17 \text{ or } 17\%$$

Explanation

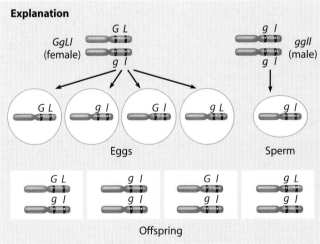

GgLl (female)   G L / g l

ggll (male)   g l / g l

G L   g l   G l   g L   g l

Eggs   Sperm

Offspring

Figure 9.18C   Fruit fly experiment demonstrating the role of crossing over in inheritance

The discovery of how crossing over creates gamete diversity confirmed the relationship between chromosome behavior and heredity. Some of the most important early studies of crossing over were performed in the laboratory of American embryologist Thomas Hunt Morgan in the early 1900s. Morgan and his colleagues used the fruit fly *Drosophila melanogaster* in many of their experiments (Figure 9.18B). Often seen flying around ripe fruit, *Drosophila* is a good research animal for genetic studies because it can be easily and inexpensively grown and can produce several generations in a matter of weeks or months.

Figure 9.18B
Drosophila melanogaster

Figure 9.18C shows one of Morgan's experiments, a cross between a wild-type fruit fly (gray body and long wings) and a fly with a black body and undeveloped, or vestigial, wings. Morgan knew the genotypes of these flies from previous studies. Here we use the following gene symbols:

G = gray body (dominant)

g = black body (recessive)

L = long wings (dominant)

l = vestigial wings (recessive)

In mating a gray fly with long wings (genotype *GgLl*) with a black fly with vestigial wings (genotype *ggll*), Morgan performed a testcross (see Module 9.6). If the genes were not linked, then independent assortment would produce offspring in a phenotypic ratio of $1:1:1:1$ ($\frac{1}{4}$ gray body, long wings; $\frac{1}{4}$ black body, vestigial wings; $\frac{1}{4}$ gray body, vestigial wings; and $\frac{1}{4}$ black body, long wings). But because these genes were linked, Morgan obtained the results shown in the top part of Figure 9.18C: Most of the offspring had parental phenotypes, but 17% of the offspring flies were recombinants. The percentage of recombinants is called the **recombination frequency.**

When Morgan first obtained these results, he did not know about crossing over. To explain the ratio of offspring, he hypothesized that the genes were linked and that some mechanism occasionally broke the linkage. Tests of the hypothesis proved him correct, establishing that crossing over was the mechanism that "breaks linkages" between genes.

The lower part of Figure 9.18C explains Morgan's results in terms of crossing over. A crossover between chromatids of homologous chromosomes in parent *GgLl* broke linkages between the *G* and *L* alleles and between the *g* and *l* alleles, forming the recombinant chromosomes *Gl* and *gL*. Later steps in meiosis distributed the recombinant chromosomes to gametes, and random fertilization produced the four kinds of offspring Morgan observed.

? Return to the data in Figure 9.17. What is the recombination frequency for the flower-color and pollen-length genes?

■ 11%, or $\frac{42}{381}$

9.19 Geneticists use crossover data to map genes

Working with *Drosophila*, T. H. Morgan and his students (Alfred H. Sturtevant in particular) greatly advanced our understanding of genetics. One of Sturtevant's major contributions to genetics was an approach for using crossover data to map gene loci. Sturtevant started by assuming that the chance of crossing over is approximately equal at all points along a chromosome. He then hypothesized that the farther apart two genes are on a chromosome, the higher the probability that a crossover will occur between them. His reasoning was elegantly simple: The greater the distance between two genes, the more points there are between them where crossing over can occur. With this principle in mind, Sturtevant began using recombination data from fruit fly crosses to assign to genes relative positions on chromosomes—that is, to map genes.

Figure 9.19A represents a part of the chromosome that carries the linked genes for black body (*g*) and vestigial wings (*l*) that we described in Module 9.18. This same chromosome also carries a gene that has a recessive allele (we'll call it *c*) determining cinnabar eye color, a brighter red than the wild-type color. Figure 9.19B shows the actual crossover (recombination) frequencies between these alleles, taken two at a time: 17% between the *g* and *l* alleles, 9% between *g* and *c*, and 9.5% between *l* and *c*. Sturtevant reasoned that these values represent the relative distances between the genes. Because the crossover frequencies between *g* and *c* and between *l* and *c* are approximately half that between *g* and *l*, gene *c* must lie roughly midway between *g* and *l*. Thus, the sequence of these genes on one of the fruit fly chromosomes must be *g-c-l* (or the equivalent *l-c-g*).

Years later it was learned that Sturtevant's assumption that crossovers are equally likely at all points on a chromosome was not exactly correct. Still, his method of mapping genes worked, and it proved extremely valuable in establishing the relative positions of many other fruit fly genes. Eventually, enough data were accumulated to reveal that *Drosophila* has four groups of genes, corresponding to its four pairs of chromosomes. Figure 9.19B is a genetic map showing just five of the gene loci on part of one chromosome: the loci labeled *g, c,* and *l* and two others. Notice that eye color is a characteristic affected by more than one gene. Here we see the cinnabar-eye and brown-eye genes; still other eye-color genes are found elsewhere (see Module 9.21). For all these genes, however, the wild-type allele specifies red eyes.

Today, with DNA technology, geneticists can determine the actual distances in nucleotides between linked genes. The new

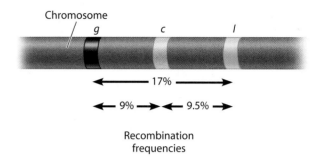

Figure 9.19A Mapping genes from crossover data

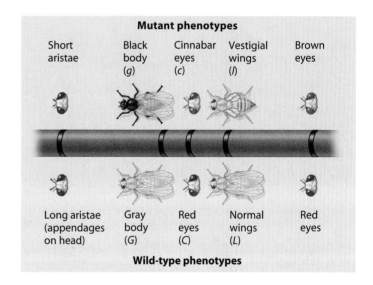

Figure 9.19B A partial genetic map of a fruit fly chromosome

genetic maps confirm the relative positions established by Sturtevant's mapping method.

Web Activity *Linked Genes and Crossing Over*

? You design *Drosophila* crosses to provide recombination data for a gene not included in Figure 9.19B. The gene has recombination frequencies of 5% with the vestigial-wing (*l*) locus and 5% with the cinnabar-eye (*c*) locus. Where is it located on the chromosome?

■ About halfway between the vestigial and cinnabar loci

Sex Chromosomes and Sex-Linked Genes

9.20 Chromosomes determine sex in many species

Many animals, including fruit flies and all mammals, have a pair of **sex chromosomes,** designated X and Y, that determine an individual's sex (Figure 9.20A, top of next page). Figure 9.20B

reviews what you learned in Chapter 8 about sex determination in humans. Individuals with one X chromosome and one Y chromosome are males; XX individuals are females. Human

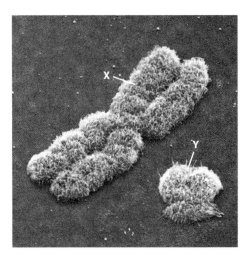

Figure 9.20A The human sex chromosomes

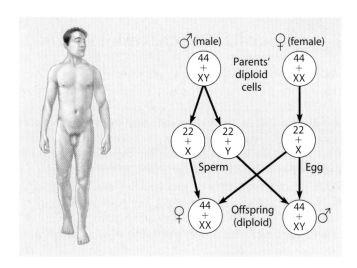

Figure 9.20B The X-Y system

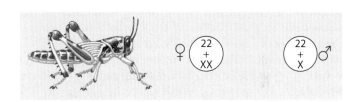

Figure 9.20C The X-O system

Figure 9.20D The Z-W system

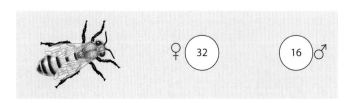

Figure 9.20E Sex determination by chromosome number

males and females both have 44 autosomes (nonsex chromosomes). As a result of chromosome segregation during meiosis, each gamete contains one sex chromosome and a haploid set of autosomes (22 in humans). All eggs contain a single X chromosome. Of the sperm cells, half contain an X chromosome and half contain a Y chromosome. An offspring's sex depends on whether the sperm cell that fertilizes the egg bears an X or a Y. In the fruit fly's X-Y system, sex is determined primarily by the number of X chromosomes, although the Y chromosome is essential for sperm formation.

The genetic basis of sex determination in humans is not yet completely understood, but one gene on the Y chromosome plays a crucial role. This gene, discovered by a British research team in 1990, is called *SRY* (for sex-determining region of Y) and triggers testis development. In the absence of *SRY,* an individual develops ovaries rather than testes. *SRY* codes for proteins that regulate other genes on the Y chromosome. These genes in turn produce proteins necessary for normal testis development.

The X-Y system is only one of several sex-determining systems. For example, grasshoppers, roaches, and some other insects have an X-O system, in which O stands for the absence of a sex chromosome (Figure 9.20C). Females have two X chromosomes (XX); males have only one sex chromosome (XO). Males produce two classes of sperm (X-bearing and lacking any sex chromosome), and sperm cells determine the sex of the offspring at fertilization.

In contrast to the X-Y and X-O systems, *eggs* determine sex in certain fishes, butterflies, and birds (Figure 9.20D). The sex chromosomes in these animals are designated Z and W. Males have the genotype ZZ; females are ZW. In this system, sex is determined by whether the egg carries a Z or a W.

Some organisms lack sex chromosomes altogether. In most ants and bees, sex is determined by chromosome *number,* rather than by sex chromosomes (Figure 9.20E). Females develop from fertilized eggs and thus are diploid. Males develop from unfertilized eggs—they are fatherless—and are haploid.

Most animals have two separate sexes; that is, individuals are either male or female. Many plant species have sperm-bearing and egg-bearing flowers borne on different individuals. Some plant species, such as date palms and marijuana, have the X-Y system of sex determination; others, such as the wild strawberry, have the Z-W system. However, most plant species and some animal species have individuals that produce both sperm and eggs. In such species, all individuals have the same complement of chromosomes.

? During fertilization in humans, what determines the sex of the offspring?

■ Whether the egg is fertilized by a sperm bearing an X chromosome (producing a female offspring) or by a sperm with a Y chromosome (producing a male)

9.21 Sex-linked genes exhibit a unique pattern of inheritance

Besides bearing genes that determine sex, the so-called sex chromosomes also contain genes for characters unrelated to femaleness or maleness. A gene located on either sex chromosome is called a **sex-linked gene,** although in humans the term has historically referred specifically to a gene on the X chromosome. (Be careful not to confuse the term *sex-linked gene,* which refers to a gene on a sex chromosome, with the term *linked genes,* which refers to genes on the same chromosome that tend to be inherited together.)

The figures here illustrate inheritance patterns for white eye color in the fruit fly, an X-linked recessive trait. Wild-type fruit flies have red eyes; white eyes are very rare (Figure 9.21A). We use the uppercase letter R for the dominant, wild-type, red-eye allele and r for the recessive, white-eye allele. Because these alleles are carried on the X chromosome, we show them as superscripts to the letter X. Thus, red-eyed male fruit flies have the genotype $X^R Y$; white-eyed males are $X^r Y$. The Y chromosome does not have a gene locus for eye color; therefore, the male's phenotype results entirely from his single X-linked gene. In the female, $X^R X^R$ and $X^R X^r$ flies have red eyes, and $X^r X^r$ flies have white eyes.

A white-eyed male ($X^r Y$) will transmit his X^r to all of his female offspring, but to none of his male offspring. This is because his daughters, in order to be female, must inherit his X chromosome, but his sons must inherit his Y chromosome. As shown in Figure 9.21B, when the female parent is a dominant homozygote ($X^R X^R$) and the male parent is $X^r Y$, all the offspring have red eyes, but the female offspring are all carriers of the allele for white eyes ($X^R X^r$). When those offspring are bred to each other, the classical 3 : 1 phenotypic ratio of red eyes to white eyes appears among the offspring (Figure 9.21C). However, there is a twist: The white-eyed trait shows up only in males. All the females have red eyes, whereas half the males have red eyes and half have white eyes. All females inherit at least one dominant allele (from their father); half of them are homozygous dominant, whereas the other half are heterozygous carriers, like their mother. Among the males, half of them inherit the recessive allele their mother was carrying, producing the white-eye phenotype.

Because the white-eye allele is recessive, a female will have white eyes only if she receives that allele on both X chromosomes. For example, if a heterozygous female mates with a white-eyed male, there is a 50% chance that each offspring will have white eyes (resulting from genotype $X^r X^r$ or $X^r Y$), regardless of sex (Figure 9.21D). Daughters with red eyes are heterozygotes, whereas red-eyed male offspring completely lack the recessive allele.

Web Activity *Sex-Linked Genes*

Web Process of Science *How Is the Chi-Square Test Used in Genetic Analysis?*

? A white-eyed female *Drosophila* is mated with a red-eyed (wild-type) male. What result do you predict for the numerous offspring?

■ All female offspring will be red-eyed but heterozygous ($X^R X^r$); all male offspring will be white-eyed ($X^r Y$).

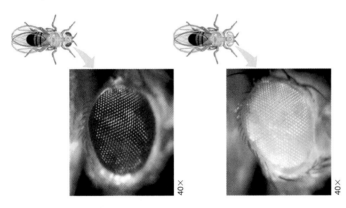

Figure 9.21A   Fruit fly eye color, a sex-linked characteristic

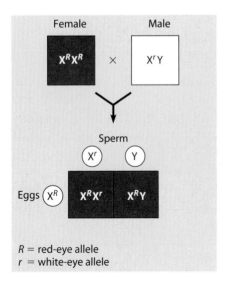

Figure 9.21B   Homozygous, red-eyed female × white-eyed male

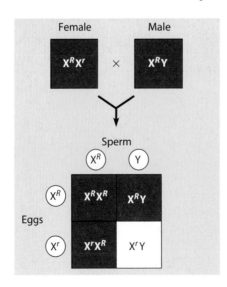

Figure 9.21C   Heterozygous female × red-eyed male

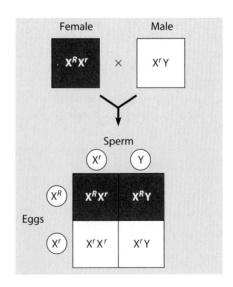

Figure 9.21D   Heterozygous female × white-eyed male

9.22 Sex-linked disorders affect mostly males

As in fruit flies, a number of human conditions result from sex-linked (X-linked) recessive alleles. Like a male fruit fly, if a man inherits only one sex-linked recessive allele—from his mother—the allele will be expressed. In contrast, a woman has to inherit two such alleles—one from each parent—in order to exhibit the trait. Thus, recessive sex-linked traits are expressed much more frequently in men than in women.

Hemophilia is a sex-linked recessive trait with a long, well-documented history. Hemophiliacs bleed excessively when injured because they lack one or more of the proteins required for blood clotting. A high incidence of hemophilia plagued the royal families of Europe. Queen Victoria (1819–1901) of England was a carrier of the hemophilia allele. She passed it on to one of her sons and two of her daughters. Through marriage, her daughters then introduced the disease into the royal families of Prussia, Russia, and Spain. Thus, the age-old practice of strengthening international alliances by marriage effectively spread hemophilia through the royal families of several nations. The pedigree in Figure 9.22 traces the disease through one branch of the family. As you can see in the pedigree, Alexandra, like her mother and grandmother, was a carrier, and Alexis had the disease.

Another recessive disorder in humans that is sex-linked is **red-green color blindness,** a malfunction of light-sensitive cells in the eyes. Color blindness is actually a class of disorders that involves several X-linked genes. A person with normal color vision can see more than 150 colors. In contrast, someone with red-green color blindness can see fewer than 25. Mostly males are affected, but heterozygous females have some defects.

Figure 9.22 Hemophilia in the royal family of Russia (half-filled symbols represent heterozygous carriers)

Duchenne muscular dystrophy, a condition characterized by a progressive weakening of the muscles and loss of coordination, is another human sex-linked recessive disorder. The first symptoms appear in early childhood, when the child begins to have difficulty standing up. He is inevitably wheelchair-bound by age 12. Eventually, he becomes severely wasted, and normal breathing becomes difficult. Affected individuals rarely live past their early 20s.

? Neither Tom nor Sue has hemophilia, but their first son does. If the couple has a second child, what is the probability that he or she will also have the disease?

■ $\frac{1}{4}$ ($\frac{1}{2}$ chance of a male child × $\frac{1}{2}$ chance that he will inherit the mutant X)

9.23 The Y chromosome provides clues about human male evolution

Barring mutations, the human Y chromosome passes essentially intact from father to son. By analyzing Y DNA, researchers can learn about the ancestry of human males.

In 2003, geneticists discovered that about 8% of males currently living in central Asia have Y chromosomes of striking genetic similarity. Further analysis traced their common genetic heritage to a single man living about 1,000 years ago. In combination with historical records, the data led to the speculation that the Mongolian ruler Genghis Kahn may be responsible for the spread of the unusual chromosome to nearly 16 million men living today. A similar study of Irish men in 2006 suggested that nearly 10% of them were descendants of Niall of the Nine Hostages, a warlord who lived during the 5th century.

Another study of Y DNA seemed to confirm the claim by the Lemba people of southern Africa that they are descended from ancient Jews. Sequences of Y DNA distinctive of the Jewish priestly caste called Cohanim (descendants of Moses'

brother Aaron, according to the Bible) are found at high frequencies among the Lemba.

The discovery of the sex chromosomes and their pattern of inheritance was one of many breakthroughs in understanding how genes are passed from one generation to the next. During the first half of the 20th century, geneticists rediscovered Mendel's work, reinterpreted his laws in light of chromosomal behavior during meiosis, and firmly established the chromosome theory of inheritance. The chromosome theory set the stage for another explosion of experimental work in the second half of the 20th century. This work was in molecular genetics, an area we explore in the next three chapters.

? Why is the Y chromosome particularly useful in tracing recent human heritage?

■ Because it is passed directly from father to son, forming an unbroken chain of male lineage

Reviewing the Concepts

Mendel's Laws (Introduction–9.10)

Mendelian genetics. The historical roots of genetics, the science of heredity, date back to ancient attempts at selective breeding **(Introduction–9.1).** Modern genetics began with Gregor Mendel's quantitative experiments. Mendel crossed pea plants and traced traits from generation to generation. He hypothesized that there are alternative versions of genes, the units that determine heritable traits **(9.2).** Mendel's law of segregation predicts that alleles separate in gametes **(9.3–9.4):**

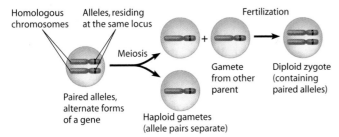

Homologous chromosomes
Alleles, residing at the same locus
Meiosis
Fertilization
Gamete from other parent
Diploid zygote (containing paired alleles)
Paired alleles, alternate forms of a gene
Haploid gametes (allele pairs separate)

When the two alleles of a gene in a diploid individual are different, the dominant allele determines the inherited trait, whereas the recessive allele has no effect. Mendel's law of independent assortment states that the alleles of a pair segregate independently of other allele pairs during gamete formation **(9.5).**

Mendelian crosses. The offspring of a testcross, a mating between an individual of unknown genotype and a homozygous recessive individual, can reveal the unknown's genotype **(9.6).** Inheritance follows the rules of probability. The rule of multiplication calculates the probability of two independent events. The rule of addition calculates the probability of an event that can occur in alternative ways **(9.7).**

Human genetics. The inheritance of many human traits follows Mendel's laws. Family pedigrees can be used to determine individual genotypes **(9.8).** Many inherited disorders in humans are controlled by a single gene **(9.9).** New technologies—including carrier screening, fetal testing, fetal imaging, and newborn screening—can provide insight for reproductive decisions but create new ethical dilemmas **(9.10).**

Variations on Mendel's Laws (9.11–9.15)

Mendel's laws extended. Mendel's laws are valid for all sexually reproducing species, but genotype often does not dictate phenotype in the simple way his laws describe:

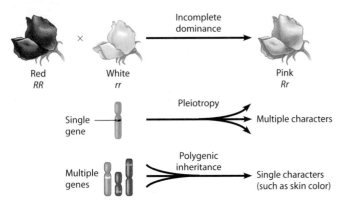

Red RR × White rr
Incomplete dominance
→ Pink Rr

Single gene
Pleiotropy
→ Multiple characters

Multiple genes
Polygenic inheritance
→ Single characters (such as skin color)

In a population, there may be multiple alleles for a character, such as the three alleles for the ABO blood group. The alleles determining the A and B blood factors are codominant; that is, both are expressed in a heterozygote **(9.11–9.14).** Many traits are affected, in varying degrees, by both genetic and environmental factors **(9.15).**

The Chromosomal Basis of Inheritance (9.16–9.19)

Genes and chromosomes. Genes are located on chromosomes, whose behavior during meiosis and fertilization accounts for inheritance patterns **(9.16).** Certain genes are linked; they tend to be inherited together because they reside close together on the same chromosome **(9.17).** Crossing over can separate linked alleles, producing gametes with recombinant chromosomes **(9.18).** Recombination frequencies can be used to map the relative positions of genes on chromosomes **(9.19).**

Sex Chromosomes and Sex-Linked Genes (9.20–9.23)

Sex chromosomes determine sex in many species. In mammals, a male is XY, and a female is XX. The Y chromosome has genes for the development of testes, whereas an absence of the Y allows ovaries to develop. Other systems of sex determination exist in other animals and plants **(9.20).** All genes on the sex chromosomes are said to be sex-linked. However, the X chromosome carries many genes unrelated to sex **(9.21).** Most sex-linked human disorders are due to recessive alleles and are seen mostly in males. A male receiving a single X-linked recessive allele from his mother will have the disorder; a female must receive the allele from both parents to be affected **(9.22).** Because they are passed on intact from father to son, Y chromosomes can provide data about recent human evolutionary history **(9.23).**

Connecting the Concepts

1. Complete this concept map to help you review some key concepts of genetics.

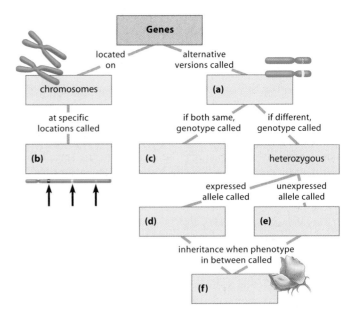

Genes
located on
alternative versions called
chromosomes
(a)
at specific locations called
if both same, genotype called
if different, genotype called
(b)
(c)
heterozygous
expressed allele called
unexpressed allele called
(d)
(e)
inheritance when phenotype in between called
(f)

Testing Your Knowledge

Multiple Choice

2. Edward was found to be heterozygous (*Ss*) for sickle-cell trait. The alleles represented by the letters *S* and *s* are
   a. on the X and Y chromosomes.
   b. linked.
   c. on homologous chromosomes.
   d. both present in each of Edward's sperm cells.
   e. on the same chromosome but far apart.

3. Whether an allele is dominant or recessive depends on
   a. how common the allele is, relative to other alleles.
   b. whether it is inherited from the mother or the father.
   c. which chromosome it is on.
   d. whether it or another allele determines the phenotype when both are present.
   e. whether or not it is linked to other genes.

4. Two fruit flies with eyes of the usual red color are crossed, and their offspring are as follows: 77 red-eyed males, 71 ruby-eyed males, 152 red-eyed females. The allele for ruby eyes is
   a. autosomal (carried on an autosome) and dominant.
   b. autosomal and recessive.
   c. sex-linked and dominant.
   d. sex-linked and recessive.
   e. impossible to determine without more information.

5. In some of his experiments, Mendel studied the inheritance patterns of two characters at once—flower color and pod color, for example. He did this to find out
   a. whether genes for the two characters are inherited together or separately.
   b. how many genes are responsible for determining a character.
   c. whether genes are on chromosomes.
   d. the distance between genes on a chromosome.
   e. how many different genes a pea plant has.

6. A man who has type B blood and a woman who has type A blood could have children of which of the following phenotypes?
   a. A or B only   d. A, B, or O
   b. AB only   e. A, B, AB, or O
   c. AB or O

Additional Genetics Problems

7. Why do more men than women have color blindness?

8. In fruit flies, the genes for wing shape and body stripes are linked. In a fly whose genotype is *WwSs*, *W* is linked to *S*, and *w* is linked to *s*. Show how this fly can produce gametes containing four different combinations of alleles. Which are parental-type gametes? Which are recombinant gametes? How are the recombinants produced?

9. Adult height in humans is at least partially hereditary; tall parents tend to have tall children. But humans come in a range of sizes, not just tall and short. Explain an extension of Mendel's model that could produce the hereditary variation in human height.

10. Tim and Jan both have freckles (see Module 9.8), but their son Mike does not. Show with a Punnett square how this is possible. If Tim and Jan have two more children, what is the probability that both will have freckles?

11. Both Tim and Jan (problem 10) have a widow's peak (see Module 9.8), but Mike has a straight hairline. What are their genotypes? What is the probability that Tim and Jan's next child will have freckles and a straight hairline?

12. In rabbits, black hair depends on a dominant allele, *B*, and brown hair on a recessive allele, *b*. Short hair is due to a dominant allele, *S*, and long hair to a recessive allele, *s*. If a true-breeding black, short-haired male is mated with a brown, long-haired female, describe their offspring. What will be the genotypes of the offspring? If two of these F_1 rabbits are mated, what phenotypes would you expect among their offspring? In what proportions?

13. A fruit fly with a gray body and red eyes (genotype *BbPp*) is mated with a fly having a black body and purple eyes (genotype *bbpp*). What offspring, in what proportions, would you expect if the body-color and eye-color genes are on different chromosomes (unlinked)? When this mating is actually carried out, most of the offspring look like the parents, but 3% have a gray body and purple eyes, and 3% have a black body and red eyes. Are these genes linked or unlinked? What is the recombination frequency?

14. A series of matings shows that the recombination frequency between the black-body gene (problem 13) and the gene for dumpy (shortened) wings is 36%. The recombination frequency between purple eyes and dumpy wings is 41%. What is the sequence of these three genes on the chromosome?

15. A couple are both phenotypically normal, but their son suffers from hemophilia, a sex-linked recessive disorder. What fraction of their children are likely to suffer from hemophilia? What fraction are likely to be carriers?

16. Heather was surprised to discover she suffered from red-green color blindness. She told her biology professor, who said, "Your father is color-blind too, right?" How did her professor know this? Why did her professor not say the same thing to the color-blind males in the class?

Applying the Concepts

17. In 1981, a stray black cat with unusual rounded, curled-back ears was adopted by a family in Lakewood, California. Hundreds of descendants of this cat have since been born, and cat fanciers hope to develop the "curl" cat into a show breed. The curl allele is apparently dominant and autosomal (carried on an autosome). Suppose you owned the first curl cat and wanted to breed it to develop a true-breeding variety. Describe tests that would determine whether the curl gene is dominant or recessive and whether it is autosomal or sex-linked. Explain why you think your tests would be conclusive. Describe a test to determine that a cat is true-breeding.

Answers to all questions can be found in Appendix 4.

For Practice Quizzes, BioFlix, MP3 Tutors, and Activities, go to www.mybiology.com.

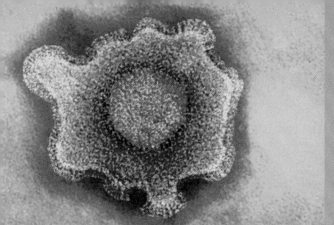

Sabotage Inside Our Cells

A saboteur drifts stealthily toward its target: a vital factory. Stopped at the perimeter by a guard, the intruder presents counterfeit identification to gain entry. Once inside, it surveys the scene and makes a quick decision: The time is not yet ripe for sabotage. So it lies low and waits silently, undetected, until it receives a go signal. The intruder now acts quickly, hijacking the factory machinery and diverting production to its own diabolical ends. Controlled by the intruder, the factory now manufactures replicas of the saboteur! When these duplicates are ready, they break their way out, destroying the factory as they exit. With the ruins behind them, they move off silently in search of new targets.

The scenario just described is played out millions of times each year. What is it? Industrial sabotage? Military espionage? In fact, the saboteur in this story is a herpesvirus, the type of virus that causes cold sores, genital herpes, chicken pox, and a number of other diseases. The colorized transmission electron microscope image above (at a magnification of 250,000×) and the computer model to the left show the structure of the herpesvirus.

Viruses share some of the characteristics of living organisms, such as genetic material in the form of nucleic acid packaged within a highly organized structure. A virus is generally not considered alive, however, because it is not cellular and cannot reproduce on its own. A **virus** is simply nucleic acid wrapped in a coat of protein and, for herpesviruses and some other animal viruses, a membranous envelope. Although a herpesvirus is fairly large as viruses go—about 200 nm across—its diameter is less than $\frac{1}{100}$ that of a typical human cell. Just about all a herpesvirus or any other virus can do is infect a host. It is the host that provides most of the tools and raw materials needed to duplicate the virus.

Once in the body, a herpesvirus tumbles along until it finds a suitable target cell, recognized when protein molecules on the outside of the virus fit into protein receptor molecules on the surface of the cell. Not recognizing the threat, the cell takes in the virus. Once inside the cell, the DNA of the herpesvirus enters the nucleus. In the nuclei of certain nerve cells, the viral DNA can remain dormant for long periods of time, until activated by a signal such as cellular stress. When activated, the viral DNA hijacks the cell's own molecules and organelles to produce new copies of the virus. Virus production eventually results in destruction of host cells—causing the sores that are characteristic of herpes diseases. The released viruses can then infect other cells.

Once a person is infected with a herpesvirus, the virus remains permanently latent (dormant) in the body, its DNA integrated into the chromosomes of nerve cells. Although many people never develop symptoms, over 75% of American adults are thought to carry herpes simplex 1 (which causes cold sores), and over 20%, herpes simplex 2 (which causes genital herpes). Herpesviruses are somewhat unusual in being able to remain latent inside our cells. Another virus with this ability is HIV, the virus that causes AIDS.

Because viruses have much less complex structures than cells, they are relatively easy to study on the molecular level, far easier than Mendel's peas or Morgan's fruit flies. For this reason, we owe our first glimpses of the functions of DNA, the molecule that controls hereditary traits, to the study of viruses.

This chapter is about **molecular biology,** the study of DNA and how it serves as the chemical basis of heredity. Here we explore the structure of DNA, how it replicates (the molecular basis of why offspring resemble their parents), and how it controls the cell by directing RNA and protein synthesis. We also look at viruses that infect bacteria, animals, and plants. We end with an examination of bacterial genetics. To start the chapter, we recount the story of how we know that DNA is the genetic material, a story in which a virus played a major role. ■ ■ ■

10.1 Experiments showed that DNA is the genetic material

Today, even schoolchildren have heard of DNA, and scientists routinely manipulate DNA in the laboratory and use it to change the heritable characteristics of cells. Early in the 20th century, however, the precise identity of the molecules of inheritance was unknown. Biologists knew that genes are located on chromosomes. Therefore, the two chemical components of chromosomes—DNA and protein—were the candidates for the genetic material. Until the 1940s, the case for proteins seemed stronger because proteins appeared to be more structurally complex and functionally specific. Biologists finally established the role of DNA in heredity through studies involving bacteria and the viruses that infect them.

We can trace the discovery of the genetic role of DNA back to 1928. British medical officer Frederick Griffith was studying two strains of a bacterium: a harmless strain and a pathogenic (disease-causing) strain that causes pneumonia. Griffith was surprised to find that when he killed the pathogenic bacteria and then mixed the bacterial remains with living harmless bacteria, some living bacterial cells were converted to the disease-causing form. Furthermore, all of the descendants of the transformed bacteria inherited the newly acquired ability to cause disease. Clearly, some chemical component of the dead bacteria could act as a "transforming factor" that brought about a heritable change.

Most biologists doubted that DNA could be Griffith's transforming factor, primarily because so little was known about DNA. However, in 1952, American biologists Alfred Hershey and Martha Chase performed a very convincing set of experiments. They showed that DNA is the genetic material of a virus called T2, which infects the bacterium *Escherichia coli* (*E. coli*). Bacterial viruses are called **bacteriophages** ("bacteria-eaters"), or **phages** for short. Figure 10.1A shows the structure of phage T2, which consists solely of DNA (blue) and protein (yellow). Resembling a lunar landing craft, T2 has a DNA-containing head and a hollow tail with six jointed fibers extending from it. The fibers attach to the surface of a susceptible bacterium. Hershey and Chase knew that T2 could reprogram its host cell to produce new phages, but they did not know which component—DNA or protein—was responsible for this ability.

Hershey and Chase found the answer by devising an experiment to determine what kinds of molecules the phage transferred to *E. coli* during infection. Their experiment used only a few relatively simple tools: chemicals containing radioactive isotopes (see Module 2.5); a radioactivity detector; a kitchen blender; and a centrifuge, a device that spins test tubes to separate particles of different weights. (These are still basic tools of molecular biology.)

Hershey and Chase used different radioactive isotopes to label the DNA and protein in T2. First, they grew T2 with *E. coli* in a solution containing radioactive sulfur (yellow in Figure 10.1B). Protein contains sulfur but DNA does not, so as new phages were made, the radioactive sulfur atoms were

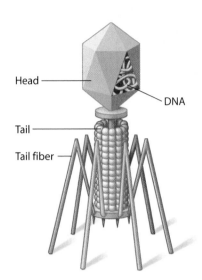

Figure 10.1A   Phage T2

incorporated only into their proteins. The researchers grew a separate batch of phages in a solution containing radioactive phosphorus (green). Because nearly all the phage's phosphorus is in DNA, this labeled only the phage DNA.

Armed with the two batches of labeled T2, Hershey and Chase were ready to perform the experiment outlined in Figure 10.1B. ❶ They allowed the two batches of T2 to infect separate samples of nonradioactive bacteria. ❷ Shortly after the onset of infection, they agitated the cultures in a blender to shake loose any parts of the phages that remained outside the bacterial cells. ❸ They then spun the mixtures in a centrifuge. The cells were deposited as a pellet at the bottom of the centrifuge tubes, but phages and parts of phages, being lighter, remained suspended in the liquid. ❹ The researchers then measured the radioactivity in the pellet and the liquid.

Hershey and Chase found that when the bacteria had been infected with T2 phages containing labeled protein, the radioactivity ended up mainly in the liquid, which contained phages but not bacteria. This result suggested that the phage protein did not enter the cells. But when the bacteria had been infected with phages whose DNA was tagged, then most of the radioactivity was in the bacteria pellet. When these bacteria were returned to liquid growth medium, the bacterial cells were soon destroyed, lysing (breaking open) and releasing new phages that contained radioactive phosphorus in their DNA but no radioactive sulfur in their proteins.

Hershey and Chase concluded that T2 injects its DNA into the host cell, leaving virtually all its protein outside (as shown in Figure 10.1B). More importantly, they demonstrated that it is the injected DNA molecules that cause the cells to produce additional phage DNA and proteins—indeed, new complete phages. This indicated that the DNA contained the instructions for making phages. Figure 10.1C outlines the reproductive cycle for phage T2 as we now understand it.

The Hershey-Chase results, added to earlier evidence, convinced most scientists that DNA is the hereditary material. What happened next was one of the most celebrated quests in the history of science: the effort to figure out the structure of DNA and how this structure enables the molecule to store genetic information and transmit it from parents to offspring.

Web Activity *The Hershey-Chase Experiment*

Web Activity *Phage T2 Reproductive Cycle*

? What convinced Hershey and Chase that DNA, rather than protein, is the genetic material of phage T2?

■ Radioactively labeled phage DNA, but not labeled protein, entered the host cell during infection and directed the synthesis of new viruses.

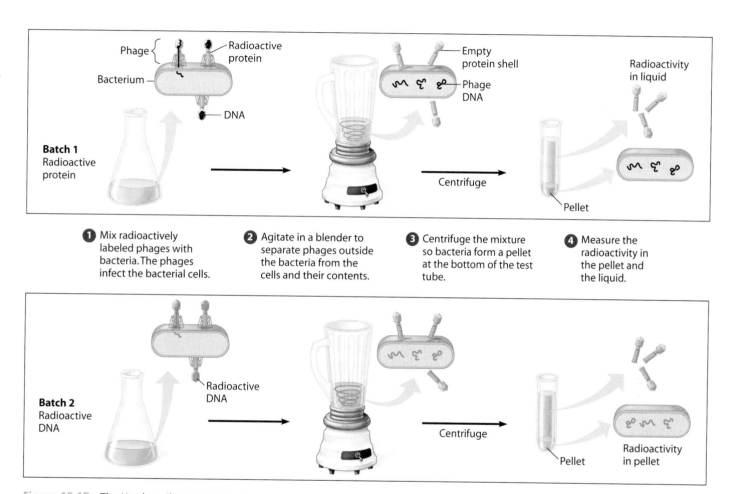

Batch 1
Radioactive protein

❶ Mix radioactively labeled phages with bacteria. The phages infect the bacterial cells.

❷ Agitate in a blender to separate phages outside the bacteria from the cells and their contents.

❸ Centrifuge the mixture so bacteria form a pellet at the bottom of the test tube.

❹ Measure the radioactivity in the pellet and the liquid.

Batch 2
Radioactive DNA

Figure 10.1B The Hershey-Chase experiment

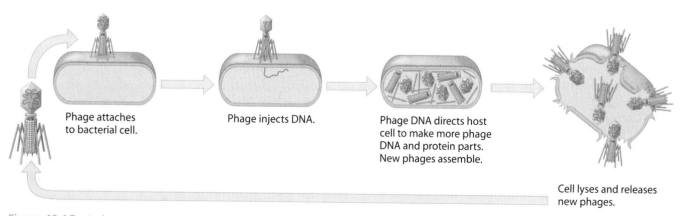

Phage attaches to bacterial cell.

Phage injects DNA.

Phage DNA directs host cell to make more phage DNA and protein parts. New phages assemble.

Cell lyses and releases new phages.

Figure 10.1C A phage reproductive cycle

10.2 DNA and RNA are polymers of nucleotides

By the time Hershey and Chase performed their experiments, much was already known about DNA. Scientists had identified all its atoms and knew how they were covalently bonded to one another. What was not understood was the specific arrangement of atoms that gave DNA its unique properties—the capacity to store genetic information, copy it, and pass it from generation to generation. However, only one year after Hershey and Chase published their results, scientists figured out the three-dimensional structure of DNA and the basic strategy of how it works. We will examine that momentous discovery in Module 10.3. First, let's look at the underlying chemical structure of DNA and its chemical cousin RNA.

Recall from Module 3.16 that DNA and RNA are nucleic acids consisting of long chains (polymers) of chemical units (monomers) called **nucleotides.** A very simple diagram of such a polymer, or **polynucleotide,** is shown on the far left in Figure 10.2A. This chain shows one arrangement of the four types of nucleotides that make up DNA. Each type of DNA nucleotide has a different nitrogenous base: adenine (A), cytosine (C), thymine (T), or guanine (G). Because nucleotides can occur in a polynucleotide in any sequence and polynucleotides

vary in length from long to very long, the number of possible polynucleotides is enormous.

Looking more closely at our polynucleotide, we see in the center of Figure 10.2A that each nucleotide consists of three components: a nitrogenous base (in DNA, A, C, T, or G), a sugar (blue), and a phosphate group (yellow). The nucleotides are joined to one another by covalent bonds between the sugar of one nucleotide and the phosphate of the next. This results in a **sugar-phosphate backbone,** a repeating pattern of sugar-phosphate-sugar-phosphate. The nitrogenous bases are arranged as appendages all along this backbone.

Examining a single nucleotide in even more detail (on the right in Figure 10.2A), we note the chemical structure of its three components. The phosphate group has a phosphorus atom (P) at its center and is the source of the *acid* in nucleic acid. The sugar has five carbon atoms (shown in red here for emphasis)—four in its ring and one extending above the ring. The ring also includes an oxygen atom. The sugar is called deoxyribose because, compared with the sugar ribose, it is missing an oxygen atom. (Notice that the C atom in the lower right corner of the ring is bonded to an H atom instead of to an

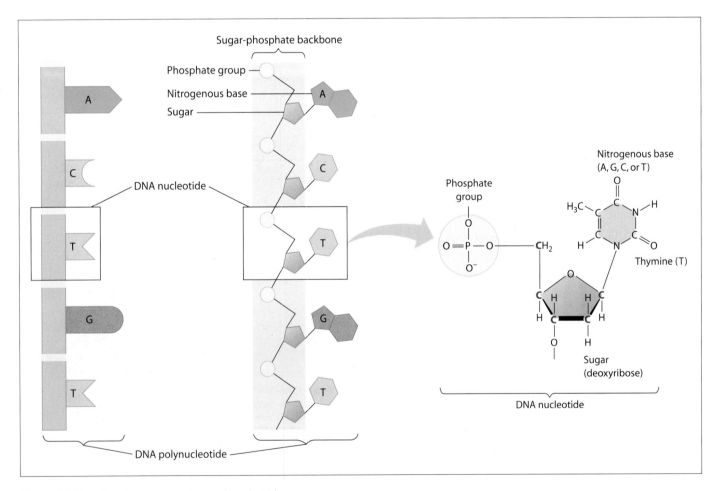

Figure 10.2A The structure of a DNA polynucleotide

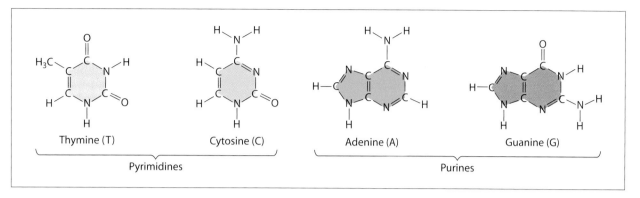

Figure 10.2B   Nitrogenous bases of DNA

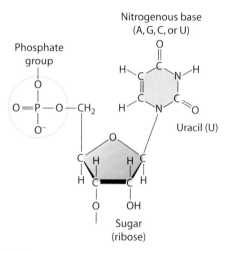

Figure 10.2C   An RNA nucleotide

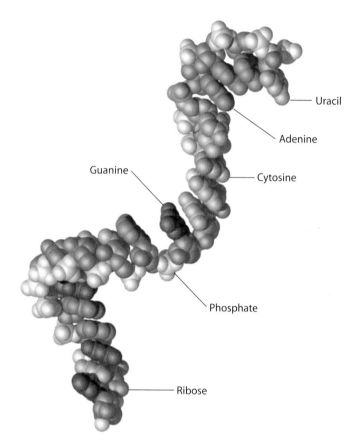

Figure 10.2D   Part of an RNA polynucleotide

—OH group, as it is in ribose; see Figure 10.2C.) The full name for DNA is deoxyribonucleic acid, with the "nucleic" portion of the word coming from DNA's location in the nuclei of eukaryotic cells. The nitrogenous base (thymine, in our example) has a ring consisting of nitrogen and carbon atoms with various functional groups attached. In contrast to the acidic phosphate group, nitrogenous bases are basic (hence their name).

The four nucleotides found in DNA differ only in their nitrogenous bases. Figure 10.2B shows the structures of the four nitrogenous bases in DNA. At this point, the structural details are not as important as the fact that the bases are of two types. **Thymine (T)** and **cytosine (C)** are single-ring structures called *pyrimidines*. **Adenine (A)** and **guanine (G)** are larger, double-ring structures called *purines*. The one-letter abbreviations can be used for either the bases alone or for the nucleotides containing them.

What about RNA? As its name—ribonucleic acid—implies, its sugar is ribose rather than deoxyribose. Notice the ribose in the RNA nucleotide in Figure 10.2C; unlike deoxyribose, the sugar ring has an —OH group attached to the C atom at its lower-right corner. Another difference between RNA and DNA is that instead of thymine, RNA has a nitrogenous base called **uracil (U)**. (You can see the structure of uracil in Figure 10.2C; it is very similar to thymine.) Except for the presence of ribose and

uracil, an RNA polynucleotide chain is identical to a DNA polynucleotide chain. Figure 10.2D is a computer graphic of a piece of RNA polynucleotide about 20 nucleotides long. The yellow phosphorus atoms at the center of the phosphate groups makes it easy to spot the sugar-phosphate backbone.

In this module, we reviewed the structure of a polynucleotide. In the next module, we'll see how two polynucleotides join together in a molecule of DNA.

 Compare and contrast DNA and RNA polynucleotides.

Both are polymers of nucleotides consisting of a sugar + a nitrogenous base + a phosphate; in RNA, the sugar is ribose; in DNA, it is deoxyribose. Both RNA and DNA have the bases A, G, and C, but DNA has a T and RNA has a U.

10.3 DNA is a double-stranded helix

After the 1952 Hershey-Chase experiment convinced most biologists that DNA was the material that stored genetic information, a race was on to determine how the structure of this molecule could account for its role in heredity. By that time, the arrangement of covalent bonds in a nucleic acid polymer was well established, and researchers focused on discovering the three-dimensional structure of DNA. First to the finish line were two scientists who were relatively unknown at the time—American James D. Watson and Englishman Francis Crick.

The brief but celebrated partnership that solved the puzzle of DNA structure began soon after the 23-year-old Watson journeyed to Cambridge University, where Crick was studying protein structure with a technique called X-ray crystallography. While visiting the laboratory of Maurice Wilkins at King's College in London, Watson saw an X-ray crystallographic image of DNA produced by Wilkins's colleague Rosalind Franklin (Figure 10.3A). A careful study of the image enabled Watson to deduce the basic shape of DNA to be a helix with a uniform diameter of 2 nanometers (nm), with its nitrogenous bases stacked about one-third of a nanometer apart. (For comparison, the plasma membrane of a cell is about 8 nm thick.) The diameter of the helix suggested that it was made up of two polynucleotide strands. The presence of two strands accounts for the now-familiar term **double helix.**

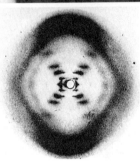

Figure 10.3A   Rosalind Franklin and her X-ray image

Watson and Crick began trying to construct a double helix that would conform both to Franklin's data and to what was then known about the chemistry of DNA (Figure 10.3B). Franklin had concluded that the sugar-phosphate backbones must be on the outside of the double helix, forcing the nitrogenous bases to swivel to the interior of the molecule. But how were the bases arranged in the interior of the double helix?

At first, Watson and Crick imagined that the bases paired like with like—for example, A with A and C with C. But that kind of pairing did not fit the X-ray data, which suggested that the DNA molecule has a *uniform* diameter. An AA pair would be almost twice as wide as a CC pair, causing bulges in the molecule. It soon became apparent that a double-ringed base (purine) must always be paired with a single-ringed base (pyrimidine) on the opposite strand. Moreover, Watson and Crick realized that the individual structures of the bases dictated the pairings even more specifically. Each base has chemical side groups that can best form hydrogen bonds with one appropriate partner (to review the hydrogen bond, see Module 2.10). Adenine can best form hydrogen bonds with thymine, and guanine with cytosine. In the biologist's shorthand, A pairs with T, and G pairs with C. A is also said to be "complementary" to T, and G to C.

Watson and Crick's pairing scheme not only fit what was known about the physical attributes and chemical bonding of DNA, but also explained some data obtained several years earlier by American biochemist Erwin Chargaff. Chargaff had discovered that the amount of adenine in the DNA of any one species was equal to the amount of thymine and that the amount of guanine was equal to that of cytosine. Chargaff's rules, as they are called, are explained by the fact that A on one of DNA's polynucleotide chains always pairs with T on the other polynucleotide chain, and G on one chain pairs only with C on the other chain.

You can picture the model of the DNA double helix proposed by Watson and Crick as a twisted rope ladder with wooden rungs (Figure 10.3C). The side ropes are the equivalent of the sugar-phosphate backbones, and the rungs represent pairs of nitrogenous bases joined by hydrogen bonds.

Figure 10.3D shows three representations of the double helix. The shapes of the base symbols in the ribbonlike diagram on the left indicate the bases' complementarity. In the center is an atomic-level version showing four base pairs, with the helix untwisted and the hydrogen bonds specified by dotted lines. You can see that the two sugar-phosphate backbones of the double helix are oriented in opposite directions. (Notice that the sugars on the two strands are upside down

Figure 10.3B   Watson and Crick in 1953 with their model of the DNA double helix

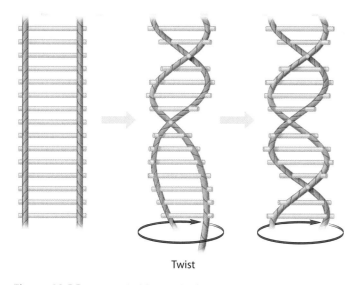

Twist

Figure 10.3C A rope-ladder model for the double helix

with respect to each other.) On the right is a computer graphic showing every atom of part of a double helix. The atoms that compose the deoxyribose sugars are shown as blue, phosphate groups as yellow, and nitrogenous bases as shades of green and orange.

Although the Watson-Crick base-pairing rules dictate the side-by-side combinations of nitrogenous bases that form the rungs of the double helix, they place no restrictions on the *sequence* of nucleotides along the length of a DNA strand. In fact, the sequence of bases can vary in countless ways, and each gene has a unique order of nucleotides, or base sequence.

In April 1953, Watson and Crick shook the scientific world with a succinct paper explaining their molecular model for DNA in the journal *Nature*. In 1962, Watson, Crick, and Wilkins received the Nobel Prize for their work. (Rosalind Franklin probably would have received the prize as well but for her death from cancer in 1958; Nobel Prizes are never awarded posthumously.) Few milestones in the history of biology have had as broad an impact as the discovery of the double helix, with its AT and CG base pairing.

The Watson-Crick model gave new meaning to the words *genes* and *chromosomes*—and to the chromosome theory of inheritance (see Module 9.16). With a complete picture of DNA, we can see that the genetic information in a chromosome must be encoded in the nucleotide sequence of the molecule. One powerful aspect of the Watson-Crick model is that the structure of DNA suggests a molecular explanation for genetic inheritance, as we see in the next module.

Web Activity *DNA and RNA Structure*

Web Activity *DNA Double Helix*

? Along one strand of a double helix is the nucleotide sequence GGCATAGGT. What is the complementary sequence for the other DNA strand?

■ CCGTATCCA

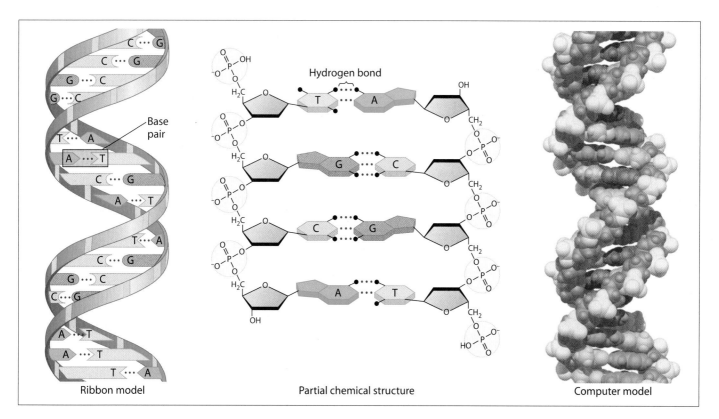

Ribbon model   Partial chemical structure   Computer model

Figure 10.3D Three representations of DNA

10.4 DNA replication depends on specific base pairing

One of biology's overarching themes—the relationship between structure and function—is evident in the double helix. The idea that there is specific pairing of bases in DNA was the flash of inspiration that led Watson and Crick to the correct structure of the double helix. At the same time, they saw the functional significance of the base-pairing rules. They ended their classic 1953 paper with this statement: "It has not escaped our notice that the specific pairing we have postulated immediately suggests a possible copying mechanism for the genetic material."

The logic behind the Watson-Crick proposal for how DNA is copied—by specific pairing of complementary bases—is quite simple. You can see this by covering one of the strands in the parental DNA molecule in Figure 10.4A. You can determine the sequence of bases in the covered strand by applying the base-pairing rules to the unmasked strand: A pairs with T, G with C.

Watson and Crick predicted that a cell applies the same rules when copying its genes. As shown in Figure 10.4A, the two strands of parental DNA (blue) separate. Each then becomes a template for the assembly of a complementary strand from a supply of free nucleotides (gray). The nucleotides line up one at a time along the template strand in accordance with the base-pairing rules. Enzymes link the nucleotides to form the new DNA strands. The completed new molecules, identical to the parental molecule, are known as daughter DNA. The copying mechanism is analogous to using a photographic negative to make a positive image, which can in turn be used to make another negative, and so on.

Watson and Crick's model predicts that when a double helix replicates, each of the two daughter molecules will have one old strand, which was part of the parental molecule, and one newly created strand. This model for DNA replication is known as the **semiconservative model** because half of the parental molecule is maintained (conserved) in each daughter molecule. The semiconservative model of replication was confirmed by experiments performed in the 1950s.

Although the general mechanism of DNA replication is conceptually simple, the actual process involves complex biochemical gymnastics. Some of the complexity arises from the fact that the helical DNA molecule must untwist as it replicates and must

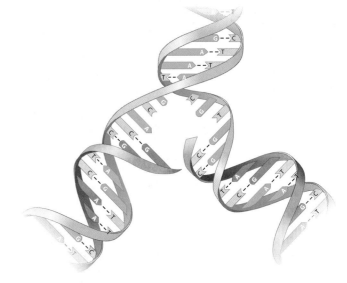

Figure 10.4B Untwisting and replication of DNA

copy its two strands roughly simultaneously (Figure 10.4B). Another challenge is the speed of the process. *E.coli*, with about 4.6 million DNA base pairs, can copy its entire genome in less than an hour. Humans, with over 6 billion base pairs in 46 diploid chromosomes, require only a few hours. And yet the process is amazingly accurate; typically, only about one DNA nucleotide per several billion is incorrectly paired. In the next module, we take a closer look at the mechanisms of DNA replication that allow it to proceed with such speed and accuracy.

Web Process of Science *What Is the Correct Model for DNA Replication?*

? How does complementary base pairing make possible the replication of DNA?

■ When the two strands of the double helix separate, each serves as a "mold" for the base-pairing of the new complementary strands.

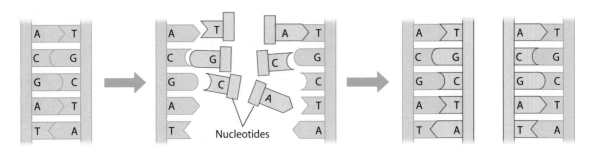

Figure 10.4A
A template model for DNA replication   Parental molecule of DNA   Both parental strands serve as templates   Nucleotides   Two identical daughter molecules of DNA

DNA replication proceeds in two directions at many sites simultaneously

Altogether, DNA replication requires the cooperation of more than a dozen enzymes and other proteins. Replication of a DNA molecule begins at special sites called *origins of replication,* stretches of DNA having a specific sequence of nucleotides where proteins attach to the DNA and separate the strands. As shown in Figure 10.5A, replication then proceeds in both directions, creating what are called replication "bubbles." The parental DNA strands (blue) open up as daughter strands (gray) elongate on both sides of each bubble. The DNA molecule of a eukaryotic chromosome has many origins where replication can start simultaneously, shortening the total time needed for the process. Thus, thousands of bubbles can be present at once. Eventually, all the bubbles merge, yielding two completed daughter DNA molecules.

Figure 10.5B shows the molecular building blocks of a tiny segment of DNA, reminding us that the DNA's sugar-phosphate backbones run in opposite directions. Notice that each strand has a 3′ ("three-prime") end and a 5′ end. The primed numbers refer to the carbon atoms of the nucleotide sugars. At one end of each DNA strand, the sugar's 3′ carbon atom is attached to an —OH group; at the other end, the sugar's 5′ carbon has a phosphate group.

The opposite orientation of the strands is important in DNA replication. The enzymes that link DNA nucleotides to a growing daughter strand, called **DNA polymerases,** add nucleotides only to the 3′ end of the strand, never to the 5′ end. Thus, a daughter DNA strand can only grow in the 5′ → 3′ direction. You see the consequences of this enzyme specificity in Figure 10.5C. The forked structure represents one side of a replication bubble. One of the daughter strands (shown in gray) can be synthesized in one continuous piece by a DNA polymerase working toward the forking point of the parental DNA. However, to make the other daughter strand, polymerase molecules must work outward from the forking point. This new strand is synthesized in short pieces as the fork opens up. Another enzyme, called **DNA ligase,** then links (ligates) the pieces together into a single DNA strand.

In addition to their roles in linking nucleotides together, DNA polymerases carry out a proofreading step that quickly removes nucleotides that have base-paired incorrectly during replication. DNA polymerases and DNA ligase are also involved in repairing DNA damaged by harmful radiation (such as ultraviolet light and X-rays) or toxic chemicals in the environment, such as those found in tobacco smoke.

DNA replication ensures that all the somatic cells in a multicellular organism carry the same genetic information. It is also the means by which genetic instructions are copied for the next generation of the organism. In the next module, we begin to pursue the connection between DNA instructions and an organism's phenotypic traits.

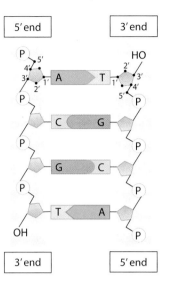

Figure 10.5B The opposite orientations of DNA strands

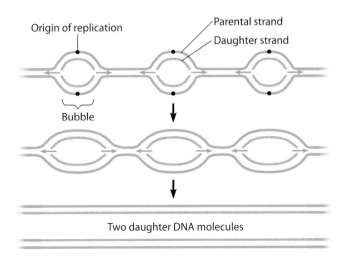

Figure 10.5A Multiple "bubbles" in replicating DNA

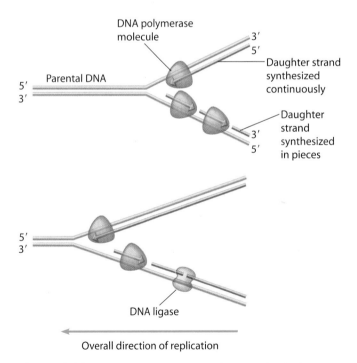

Figure 10.5C How daughter DNA strands are synthesized

Web Activity *DNA Replication*

? What is the function of DNA polymerase in DNA replication?

■ This enzyme lines up new nucleotides along an existing strand according to the base-pairing rules and then covalently connects the nucleotides into the new strand.

10.6 The DNA genotype is expressed as proteins, which provide the molecular basis for phenotypic traits

With our knowledge of DNA, we can now define genotype and phenotype more precisely than we did in Chapter 9. An organism's genotype, its genetic makeup, is the heritable information contained in its DNA. The phenotype is the organism's specific traits. So what is the molecular connection between genotype and phenotype?

The answer is that the DNA inherited by an organism specifies traits by dictating the synthesis of proteins. In other words, proteins are the links between genotype and phenotype. However, a gene does not build a protein directly. Rather, a gene dispatches instructions in the form of RNA, which in turn programs protein synthesis. This central concept in biology (termed the "central dogma" by Francis Crick in 1956) is summarized in Figure 10.6A. The chain of command is from DNA in the nucleus of the cell (purple area) to RNA to protein synthesis in the cytoplasm (tan area). The two main stages are **transcription,** the transfer of genetic information from DNA into an RNA molecule, and **translation,** the transfer of the information in the RNA into a protein. In the next nine modules, we will explore the steps in this flow of molecular information from gene to protein.

The relationship between genes and proteins was first proposed in 1909, when English physician Archibald Garrod suggested that genes dictate phenotypes through enzymes, the proteins that catalyze chemical processes in the cell. Garrod's idea came from his observations of inherited diseases. He hypothesized that an inherited disease reflects a person's inability to make a particular enzyme, and he referred to such diseases as "inborn errors of metabolism." He gave as one example the hereditary condition called alkaptonuria, in which the urine appears black because it contains a chemical called alkapton. Garrod reasoned that normal individuals have an enzyme that breaks down alkapton, whereas alkaptonuric individuals cannot make the enzyme. Garrod's hypothesis was ahead of its time, but research conducted decades later proved him right. In the intervening years, biochemists accumulated evidence that cells make and break down biologically important molecules via metabolic pathways, as in the synthesis of an amino acid or the breakdown of a sugar. As we described in Unit I, each step in a metabolic pathway is catalyzed by a specific enzyme. Therefore, individuals lacking one of the enzymes for a pathway are unable to complete it.

The major breakthrough in demonstrating the relationship between genes and enzymes came in the 1940s from the work of American geneticists George Beadle and Edward Tatum with the bread mold *Neuro-spora crassa* (Figure 10.6B). Beadle and Tatum studied strains of the mold that were unable to grow on a simple growth medium. Each of these so-called nutritional mutants turned out to lack an enzyme in a metabolic pathway that produced some molecule the mold needed, such as an amino acid. Beadle and Tatum also showed that each mutant was defective in a single gene. This result suggested the one gene–one enzyme hypothesis: the function of a gene is to dictate the production of a specific enzyme.

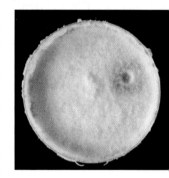

Figure 10.6B *Neurospora crassa* growing in a culture dish

The one gene–one enzyme hypothesis has been amply confirmed, but with some important modifications. First it was extended beyond enzymes to include *all* types of proteins. For example, keratin—the structural protein of hair—and the hormone insulin are two examples of proteins that are not enzymes. So biologists began to think in terms of one gene–one *protein*. However, many proteins are made from two or more polypeptide chains (see Module 3.14), with each polypeptide specified by its own gene. For example, hemoglobin, the oxygen-transporting protein in your blood, is built from two kinds of polypeptides, encoded by two different genes. Thus, Beadle and Tatum's hypothesis has come to be restated as the one gene–one *polypeptide* hypothesis.

(Bio Flix) Bioflix *Protein Synthesis*

(🎧) MP3 Tutor *DNA to RNA to Protein*

Web Process of Science *How Are Nutritional Mutations Identified?*

? What are the functions of transcription and translation?

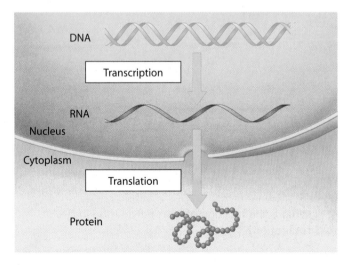

Figure 10.6A Flow of genetic information in a eukaryotic cell

■ Transcription is the transfer of information from DNA to RNA. Translation is the use of the RNA as information for making a protein.

10.7 Genetic information written in codons is translated into amino acid sequences

Genes provide the instructions for making specific proteins. But a gene does not build a protein directly. The bridge between DNA and protein synthesis is the nucleic acid RNA: DNA is transcribed into RNA, which is then translated into protein. Put another way, cells are governed by a molecular chain of command: DNA → RNA → protein.

Transcription and translation are linguistic terms, and it is useful to think of nucleic acids and proteins as having languages. To understand how genetic information passes from genotype to phenotype, we need to see how the chemical language of DNA is translated into the different chemical language of proteins.

What, exactly, is the language of nucleic acids? Both DNA and RNA are polymers made of nucleotide monomers. In DNA, there are four types of nucleotides, which differ in their nitrogenous bases (A, T, C, and G). The same is true for RNA, although it has the base U instead of T.

Figure 10.7 focuses on a small region of one of the genes (gene 3, shown in light blue) carried by a DNA molecule. DNA's language is written as a linear sequence of nucleotide bases on a polynucleotide, a sequence such as the one you see on the enlarged DNA strand in the figure. Specific sequences of bases, each with a beginning and an end, make up the genes on a DNA strand. A typical gene consists of hundreds or thousands of nucleotides in a specific sequence.

The pink strand underneath the enlarged DNA region represents the results of transcription: an RNA molecule. The process is called transcription because the nucleic acid language of DNA has been rewritten (transcribed) as a sequence of bases on RNA; the language is still that of nucleic acids.

Notice that the nucleotide bases on the RNA molecule are complementary to those on the DNA strand. As we will see in Module 10.9, this is because the RNA was synthesized using the DNA as a template.

The purple chain represents the results of translation, the conversion of the nucleic acid language into the polypeptide language (recall that proteins consist of one or more polypeptides). Like nucleic acids, polypeptides are polymers, but the monomers that compose them are the 20 amino acids common to all organisms. Again, the language is written in a linear sequence, and the sequence of nucleotides of the RNA molecule dictates the sequence of amino acids of the polypeptide. The RNA acts as a messenger carrying genetic information from DNA.

During translation, there is a change in language from the nucleotide sequence of the RNA into the amino acid sequence of the polypeptide. The brackets below the RNA indicate how genetic information is coded in nucleic acids. Notice that each bracket encloses *three* nucleotides on RNA. Recall that there are only four different kinds of nucleotides in DNA (A, G, C, T) and RNA (A, G, C, U). In translation, these four must somehow specify 20 amino acids. If each nucleotide base specified one amino acid, only 4 of the 20 amino acids could be accounted for. What if the language consisted of two-letter code words? If we read the bases of a gene two at a time, AG, for example, could specify one amino acid, whereas AT could designate a different amino acid. However, when the 4 bases are taken in doublets, there are only 16 (that is, 4^2) possible arrangements—still not enough to specify all 20 amino acids.

Triplets of bases are the smallest "words" of uniform length that can specify all the amino acids. Suppose each code word in DNA consists of a triplet, with each arrangement of three consecutive bases specifying an amino acid. Then there can be 64 (that is, 4^3) possible code words—more than enough to specify the 20 amino acids. Indeed, there are enough triplets to allow more than one coding for each amino acid. For example, the base triplets AAT and AAC both code for the same amino acid (leucine).

Experiments have verified that the flow of information from gene to protein is based on a **triplet code:** The genetic instructions for the amino acid sequence of a polypeptide chain are written in DNA and RNA as a series of three-base words, called **codons.** Notice in the figure that three-base codons in the DNA are transcribed into complementary three-base codons in the RNA, and then the RNA codons are translated into amino acids that form a polypeptide. We turn to the codons themselves in the next module.

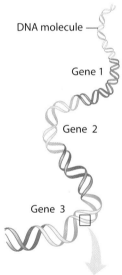

DNA molecule

Gene 1

Gene 2

Gene 3

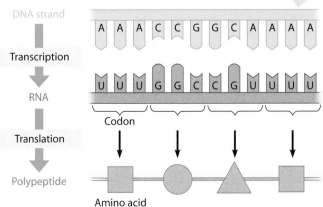

DNA strand

| A | A | A | C | C | G | G | C | A | A | A | A |

Transcription

RNA

| U | U | U | G | G | C | C | G | U | U | U | U |

Codon

Translation

Polypeptide

Amino acid

? A particular protein is 100 amino acids long. How many nucleotides are necessary to code for this protein?

Figure 10.7 Transcription and translation of codons

The genetic code is the Rosetta stone of life

In 1799, a large stone tablet was found in Rosetta, Egypt, carrying the same lengthy inscription in three ancient languages: Egyptian hieroglyphics, Egyptian script, and Greek. This stone provided the key that enabled scholars to crack the previously indecipherable hieroglyphic code.

To crack the genetic code, scientists wrote their own Rosetta stone. It was based on information gathered from a series of elegant experiments that disclosed the amino acid translations of each of the nucleotide-triplet code words. The first codon was deciphered in 1961 by American biochemist Marshall Nirenberg. He synthesized an artificial RNA molecule by linking together identical RNA nucleotides having uracil as their base. No matter where this message started or stopped, it could contain only one type of triplet codon: UUU. Nirenberg added this "poly U" to a test-tube mixture containing ribosomes and the other ingredients required for polypeptide synthesis. This mixture translated the poly U into a polypeptide containing a single kind of amino acid, phenylalanine. Thus, Nirenberg learned that the RNA codon UUU specifies the amino acid phenylalanine (Phe). By variations on this method, the amino acids specified by all the codons were soon determined.

The **genetic code** is the set of rules giving the correspondence between codons in RNA and amino acids in proteins. As Figure 10.8A shows, 61 of the 64 codons code for amino acids. The triplet AUG has a dual function: It codes for the amino acid methionine (Met) and also can provide a signal for the start of a polypeptide chain. Three of the other codons (in red boxes in the figure) do not designate amino acids. They are the stop codons that mark the end of translation.

Notice in Figure 10.8A that there is redundancy in the code but no ambiguity. For example, although codons UUU and UUC both specify phenylalanine (redundancy), neither of them ever represents any other amino acid (no ambiguity). The codons in the figure are the triplets found in RNA. They have a straightforward, complementary relationship to the codons in DNA. The nucleotides making up the codons occur in a linear order along the DNA and RNA, with no gaps or "punctuation" separating the codons.

As an exercise in translating the genetic code, consider the 12-nucleotide segment of DNA in Figure 10.8B. Let's read this as a series of triplets. Using the base-pairing rules (with U in RNA instead of T), we see that the RNA codon corresponding to the first transcribed DNA triplet, TAC, is AUG. As you can see in Figure 10.8A, AUG indicates, "place Met as the first amino acid in the polypeptide." The second DNA triplet, TTC, dictates RNA codon AAG, which designates lysine (Lys) as the second amino acid. We continue until we reach a stop codon.

The genetic code is nearly universal, shared by organisms from the simplest bacteria to the most complex plants and animals. In experiments, bacteria can translate human genetic messages, and human cells can translate bacterial RNA. A language shared by all living things must have evolved early enough in the history of life to be present in the common ancestors of all modern organisms. A shared genetic vocabulary is a reminder of the kinship that connects all life on Earth.

? **Translate the RNA sequence CCAUUUACG into the corresponding amino acid sequence.**

■ Pro-Phe-Thr

	Second base				
	U	**C**	**A**	**G**	
U	UUU ⎤ Phe / UUC ⎦ / UUA ⎤ Leu / UUG ⎦	UCU ⎤ / UCC / Ser / UCA / UCG ⎦	UAU ⎤ Tyr / UAC ⎦ / UAA Stop / UAG Stop	UGU ⎤ Cys / UGC ⎦ / UGA Stop / UGG Trp	U C A G
C	CUU ⎤ / CUC / Leu / CUA / CUG ⎦	CCU ⎤ / CCC / Pro / CCA / CCG ⎦	CAU ⎤ His / CAC ⎦ / CAA ⎤ Gln / CAG ⎦	CGU ⎤ / CGC / Arg / CGA / CGG ⎦	U C A G
A	AUU ⎤ / AUC ⎥ Ile / AUA ⎦ / AUG Met or start	ACU ⎤ / ACC / Thr / ACA / ACG ⎦	AAU ⎤ Asn / AAC ⎦ / AAA ⎤ Lys / AAG ⎦	AGU ⎤ Ser / AGC ⎦ / AGA ⎤ Arg / AGG ⎦	U C A G
G	GUU ⎤ / GUC / Val / GUA / GUG ⎦	GCU ⎤ / GCC / Ala / GCA / GCG ⎦	GAU ⎤ Asp / GAC ⎦ / GAA ⎤ Glu / GAG ⎦	GGU ⎤ / GGC / Gly / GGA / GGG ⎦	U C A G

First base (left axis), Third base (right axis)

Figure 10.8A Dictionary of the genetic code (RNA codons)

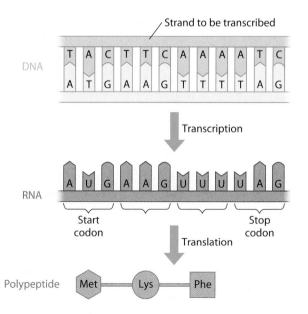

Figure 10.8B Deciphering the genetic information in DNA

Transcription produces genetic messages in the form of RNA

In eukaryotic cells, transcription, the transfer of genetic information from DNA to RNA, occurs in the nucleus. (The nucleus, after all, contains the DNA; see Figure 10.6A for a review.) An RNA molecule is transcribed from a DNA template by a process that resembles the synthesis of a DNA strand during DNA replication. Figure 10.9A is a close-up view of this process. As with replication, the two DNA strands must first separate at the place where the process will start. In transcription, however, only one of the DNA strands serves as a template for the newly forming molecule. The nucleotides that make up the new RNA molecule take their places one at a time along the DNA template strand by forming hydrogen bonds with the nucleotide bases there. Notice that the RNA nucleotides follow the same base-pairing rules that govern DNA replication, except that U, rather than T, pairs with A. The RNA nucleotides are linked by the transcription enzyme **RNA polymerase,** symbolized in the figure by the large gray shape in the background.

Figure 10.9B is an overview of the transcription of an entire prokaryotic gene. (We focus on prokaryotes here; eukaryotic transcription is a similar process but more complex.) Specific sequences of nucleotides along the DNA mark where transcription of a gene begins and ends. The "start transcribing" signal is a nucleotide sequence called a **promoter.** A promoter is a specific binding site for RNA polymerase and determines which of the two strands of the DNA double helix is used as the template in transcription.

❶ The first phase of transcription, called initiation, is the attachment of RNA polymerase to the promoter and the start of RNA synthesis. ❷ During a second phase of transcription, the RNA elongates. As RNA synthesis continues, the RNA strand peels away from its DNA template, allowing the two separated DNA strands to come back together in the region already transcribed. ❸ Finally, in the third phase, termination, the RNA polymerase reaches a sequence of bases in the DNA template

called a **terminator.** This sequence signals the end of the gene; at that point, the polymerase molecule detaches from the RNA molecule and the gene.

In addition to producing RNA that encodes amino acid sequences, transcription makes two other kinds of RNA that are involved in building polypeptides. We discuss these three kinds of RNA in the next three modules.

🎧 MP3 Tutor *DNA Transcription*

Web Activity *Transcription*

? What is a promoter? What molecule binds to it?

■ A promoter is a specific nucleotide sequence at the start of a gene where RNA polymerase attaches and begins transcription.

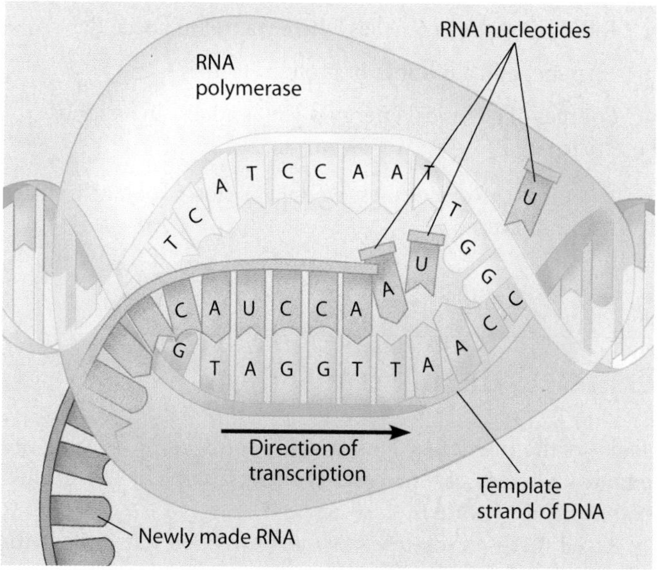

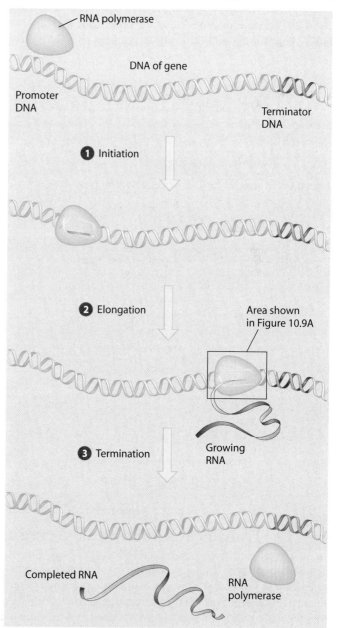

Figure 10.9B   Transcription of a gene

Figure 10.9A   A close-up view of transcription

10.10 Eukaryotic RNA is processed before leaving the nucleus

The kind of RNA that encodes amino acid sequences is called **messenger RNA (mRNA)** because it conveys genetic information from DNA to the translation machinery of the cell. Messenger RNA is transcribed from DNA, and the message in the mRNA is then translated into polypeptides. In prokaryotic cells, which lack a nucleus, transcription and translation occur in the same place (the cytoplasm). In eukaryotic cells, however, mRNA molecules and other RNA molecules required for translation must exit the nucleus via the nuclear pores and enter the cytoplasm, where the machinery for polypeptide synthesis is located.

Before leaving the nucleus as mRNA, eukaryotic transcripts are modified, or processed, in several ways. One kind of RNA processing is the addition of extra nucleotides to the ends of the RNA transcript (Figure 10.10). These additions include a small cap (a single G nucleotide) at one end and a long tail (a chain of 50 to 250 adenine nucleotides) at the other end. The cap and tail facilitate the export of the mRNA from the nucleus, protect

the mRNA from attack by cellular enzymes, and help ribosomes bind to the mRNA. The cap and tail themselves are not translated into protein.

Eukaryotes require an additional type of RNA processing because, in most protein-coding genes, the DNA sequence that codes for the polypeptides is not continuous. Most genes of plants and animals include internal noncoding regions called **introns** (for "intervening sequences"). The coding regions—the parts of a gene that are expressed as amino acids—are called **exons.** As Figure 10.10 shows, both exons (darker color) and introns (lighter color) are transcribed from DNA into RNA. However, before the RNA leaves the nucleus, the introns are removed, and the exons are joined to produce an mRNA molecule with a continuous coding sequence. (The short noncoding regions just inside the cap and tail are considered parts of the first and last exons.) This cutting-and-pasting process is called **RNA splicing.** In most cases, RNA splicing is catalyzed by a complex of proteins and small RNA molecules, but sometimes the RNA transcript itself catalyzes the process. In other words, RNA can sometimes act as an enzyme that removes its own introns! As we will see in the next chapter (Module 11.6), RNA splicing also provides a means to produce multiple polypeptides from a single gene.

> **?** Explain why many eukaryotic genes are longer than the mRNA that leaves the nucleus.
>
> ◼ These genes have introns, noncoding sequences of nucleotides that are spliced out of the final RNA transcripts.

We are now ready to see how the translation process works. Translation of mRNA into protein involves more complicated machinery than transcription, including:

◼ Transfer RNA, another kind of RNA molecule

◼ Ribosomes, the organelles where translation occurs

◼ Enzymes and a number of protein "factors"

◼ Sources of chemical energy, such as adenosine triphosphate (ATP)

In the next two modules, we take a closer look at transfer RNA and ribosomes.

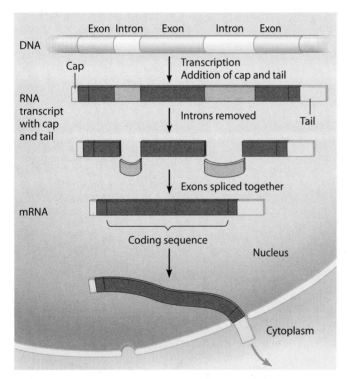

Figure 10.10   The production of eukaryotic mRNA

10.11 Transfer RNA molecules serve as interpreters during translation

Translation of any language requires an interpreter, someone who can recognize the words of one language and convert them to another. Translation of a message carried in mRNA into the amino acid language of proteins also requires an interpreter. To convert the three-letter words (codons) of nucleic

acids to the one-letter, amino acid words of proteins, a cell employs a molecular interpreter, a special type of RNA called **transfer RNA (tRNA).**

A cell that is ready to carry out translation has in its cytoplasm a supply of amino acids, either obtained from food or made from other chemicals. The amino acids themselves cannot recognize the codons in the mRNA. The amino acid tryptophan, for example, is no more attracted by codons for tryptophan than by any other codons. It is up to the cell's molecular interpreters, tRNA molecules, to match amino acids to the appropriate codons to form the new polypeptide. To perform this task, tRNA molecules must carry out two functions: (1) picking up the appropriate amino acids and (2) recognizing the appropriate codons in the mRNA. The unique structure of tRNA molecules enables them to perform both tasks.

As shown in Figure 10.11A, a tRNA molecule is made of a single strand of RNA—one polynucleotide chain—consisting of about 80 nucleotides. By twisting and folding upon itself, tRNA forms several double-stranded regions in which short stretches of RNA base-pair with other stretches. A single-stranded loop at one end of the folded molecule contains a special triplet of bases called an **anticodon.** The anticodon triplet is complementary to a codon triplet on mRNA. During translation, the anticodon on tRNA recognizes a particular codon on mRNA by using base-pairing rules. At the other end of the tRNA molecule is a site where an amino acid can attach.

In the modules that follow, in which we trace the process of translation, we represent tRNA with the simplified shape that is shown on the right in Figure 10.11A. This symbol emphasizes two parts of the molecule—the anticodon and the amino acid attachment site—that give tRNA its ability to match a particular nucleic acid word (codon) with its corresponding protein word (amino acid). Although all tRNA molecules are similar, there is a slightly different variety of tRNA for each amino acid.

Each amino acid is joined to the correct tRNA by a specific enzyme. There is a family of 20 versions of these enzymes, one enzyme for each amino acid. Each enzyme specifically binds one type of amino acid to all tRNA molecules that code for that amino acid, using a molecule of ATP as energy to drive the reaction. The resulting amino acid–tRNA complex can then furnish its amino acid to a growing polypeptide chain, a process that we describe in Module 10.12.

The computer graphic in Figure 10.11B shows a tRNA molecule (red and yellow) and an ATP molecule (purple) bound to the enzyme molecule (blue). In this picture, you can see the proportional sizes of these three molecules. The amino acid that would attach to the tRNA is not shown; it would be less than half the size of the ATP.

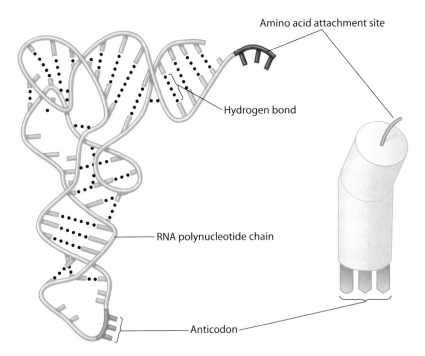

Figure 10.11A The structure of tRNA

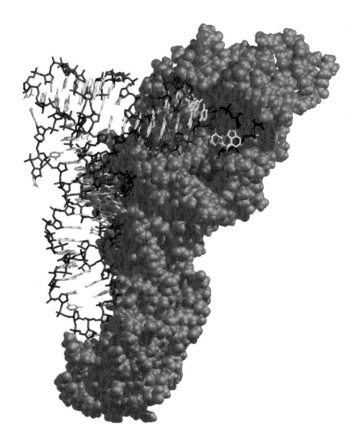

Figure 10.11B A molecule of tRNA binding to an enzyme molecule (blue)

? What is an anticodon, and what is its function?

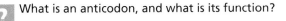

■ It is the base triplet of a tRNA molecule that couples the tRNA to a complementary codon in the mRNA. This is a key step in translating mRNA to polypeptide.

10.12 Ribosomes build polypeptides

We have now looked at many of the components a cell needs to carry out translation: instructions in the form of mRNA molecules, tRNA to interpret the instructions, a supply of amino acids, enzymes for attaching amino acids to tRNA, and ATP for energy. The final components needed are the ribosomes, organelles in the cytoplasm that coordinate the functioning of the mRNA and tRNA and actually make polypeptides.

A ribosome consists of two subunits, each made up of proteins and a kind of RNA called **ribosomal RNA (rRNA).** In Figure 10.12A, you can see the actual shapes and relative sizes of the ribosomal subunits. You can also see where mRNA, tRNA, and the growing polypeptide are located during translation.

The ribosomes of prokaryotes and eukaryotes are very similar in function, but those of eukaryotes are slightly larger and are different in composition. The differences are medically significant. Certain antibiotic drugs can inactivate prokaryotic ribosomes while leaving eukaryotic ribosomes unaffected. These drugs, such as tetracycline and streptomycin, are used to combat bacterial infections.

The simplified drawings in Figures 10.12B and 10.12C indicate how tRNA anticodons and mRNA codons fit together on ribosomes. As Figure 10.12B shows, each ribosome has a binding site for mRNA and two binding sites for tRNA. Figure 10.12C shows tRNA molecules occupying these two sites. The subunits of the ribosome act like a vise, holding the tRNA and mRNA molecules close together, allowing the amino acids carried by the tRNA molecules to be connected into a polypeptide chain. In the next two modules, we examine the steps of translation in detail.

? How does a ribosome function in protein synthesis?

A ribosome holds mRNA and tRNAs together and connects amino acids from the tRNAs to the growing polypeptide chain.

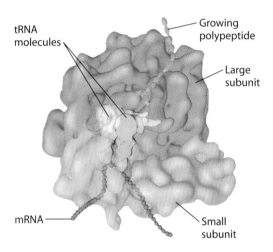

Figure 10.12A The true shape of a functioning ribosome

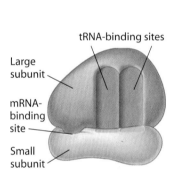

Figure 10.12B Binding sites of a ribosome

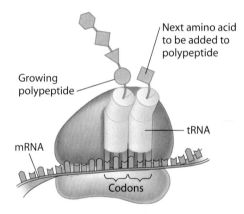

Figure 10.12C A ribosome with occupied binding sites

10.13 An initiation codon marks the start of an mRNA message

Translation can be divided into the same three phases as transcription: initiation, elongation, and termination. The process of polypeptide initiation brings together the mRNA, a tRNA bearing the first amino acid, and the two subunits of a ribosome.

As indicated in Figure 10.13A, an mRNA molecule transcribed from DNA is longer than the genetic message it carries. A sequence of nucleotides (light pink) at either end of the molecule is not part of the message but helps the mRNA bind to the ribosome. The role of the initiation process is to establish exactly where translation will begin, ensuring that the mRNA codons are translated into the correct sequence of amino acids.

Initiation occurs in two steps (Figure 10.13B, top of next page). ❶ An mRNA molecule binds to a small ribosomal subunit. A special initiator tRNA binds to the specific codon, called the **start codon,** where translation is to begin on the mRNA

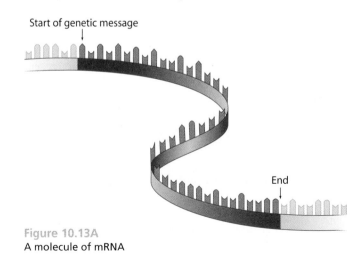

Figure 10.13A
A molecule of mRNA

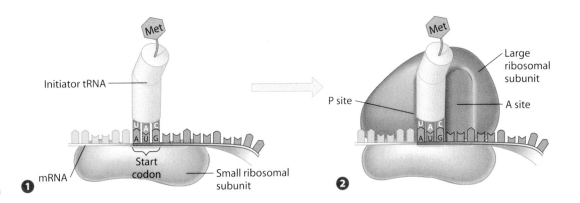

Figure 10.13B
The initiation of translation ❶

molecule. The initiator tRNA carries the amino acid methionine (Met); its anticodon, UAC, binds to the start codon, AUG. ❷ Next, a large ribosomal subunit binds to the small one, creating a functional ribosome. The initiator tRNA fits into one of the two tRNA-binding sites on the ribosome. This site, called the **P site,** will hold the growing polypeptide. The other tRNA-binding site, called the **A site,** is vacant and ready for the next amino-acid-bearing tRNA.

? What would happen if a genetic mutation changed a start codon to some other codon?

■ Any messenger RNA transcribed from the mutated gene would be nonfunctional because ribosomes could not initiate translation correctly.

10.14 Elongation adds amino acids to the polypeptide chain until a stop codon terminates translation

Once initiation is complete, amino acids are added one by one to the first amino acid. Each addition occurs in a three-step elongation process (Figure 10.14; the small green arrows indicate movement):

❶ **Codon recognition.** The anticodon of an incoming tRNA molecule, carrying its amino acid, pairs with the mRNA codon in the A site of the ribosome.

❷ **Peptide bond formation.** The polypeptide separates from the tRNA to which it was bound (the one in the P site) and attaches by a peptide bond to the amino acid carried by the tRNA in the A site. The ribosome catalyzes formation of the bond. Thus, one more amino acid is added to the chain.

❸ **Translocation.** The P site tRNA now leaves the ribosome, and the ribosome translocates (moves) the tRNA in the A site, with its attached polypeptide, to the P site. The codon and anticodon remain bonded, and the mRNA and tRNA move as a unit. This movement brings into the A site the next mRNA codon to be translated, and the process can start again with step 1.

Elongation continues until a **stop codon** reaches the ribosome's A site. Stop codons—UAA, UAG, and UGA—do not code for amino acids but instead act as signals to stop translation. This is the termination stage of translation. The completed polypeptide is released from the last tRNA and exits the ribosome, which then splits into its separate subunits.

Web Activity *Translation*

? What happens as a tRNA passes through the A and P binding sites on the ribosome?

■ In the A site, its amino acid receives the growing polypeptide from the tRNA that precedes it. In the P site, it gives up the polypeptide to the tRNA that follows it.

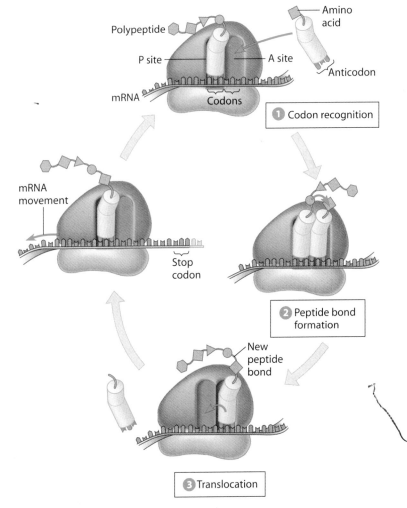

Figure 10.14 Polypeptide elongation

10.15 Review: The flow of genetic information in the cell is DNA → RNA → protein

Figure 10.15 summarizes the main stages in the flow of genetic information from DNA to RNA to protein. **①** In transcription (DNA → RNA), the RNA is synthesized on a DNA template. In eukaryotic cells, transcription occurs in the nucleus, and the messenger RNA must travel from the nucleus to the cytoplasm.

②–⑤ Translation (RNA → protein) can be divided into four steps, all of which occur in the cytoplasm. When the polypeptide is complete at the end of step 5, the two ribosomal subunits come apart, and the tRNA and mRNA are released (not shown in this figure). Translation is rapid; a single ribosome can make an average-sized polypeptide in less than a minute. Typically, an mRNA molecule is translated simultaneously by a number of ribosomes. Once the start codon emerges from the first ribosome, a second ribosome can attach to it; thus, several ribosomes may trail along on the same mRNA molecule.

Each polypeptide coils and folds, assuming a three-dimensional shape, its tertiary structure. Several polypeptides may come together, forming a protein with quaternary structure (see Module 3.14).

What is the overall significance of transcription and translation? These are the processes whereby genes control the structures and activities of cells, or, more broadly, the way the genotype produces the phenotype. The chain of command originates with the information in a gene, a specific linear sequence of nucleotides in DNA. The gene serves as a template, dictating transcription of a complementary sequence of nucleotides in mRNA. In turn, mRNA dictates the linear sequence in which amino acids appear in a specific polypeptide. Finally, the proteins that form from the polypeptides determine the appearance and the capabilities of the cell and organism.

? Which of the following molecules or structures does not participate directly in translation: ribosomes, transfer RNA, messenger RNA, DNA, ATP, enzymes?

■ DNA

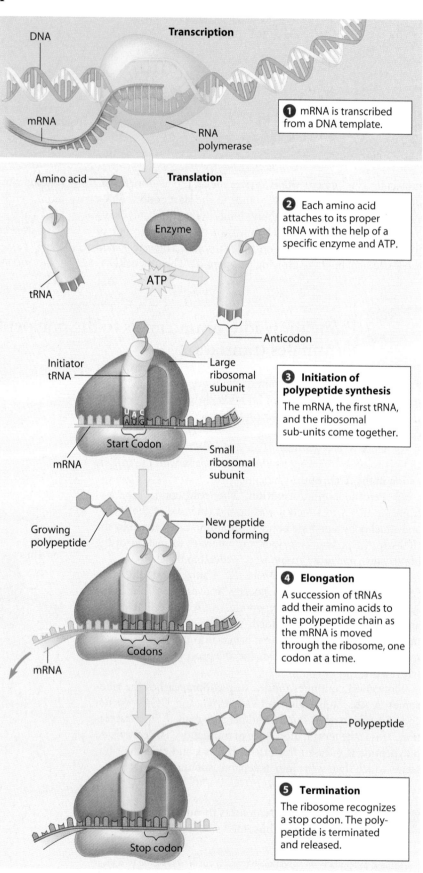

① mRNA is transcribed from a DNA template.

② Each amino acid attaches to its proper tRNA with the help of a specific enzyme and ATP.

③ **Initiation of polypeptide synthesis**
The mRNA, the first tRNA, and the ribosomal sub-units come together.

④ **Elongation**
A succession of tRNAs add their amino acids to the polypeptide chain as the mRNA is moved through the ribosome, one codon at a time.

⑤ **Termination**
The ribosome recognizes a stop codon. The polypeptide is terminated and released.

Figure 10.15 Summary of transcription and translation

Mutations can change the meaning of genes

Since discovering how genes are translated into proteins, scientists have been able to describe many inherited traits in molecular terms. For instance, when a child is born with sickle-cell disease (see Module 9.9), the condition can be traced back through a difference in a protein to one tiny change in a gene. In one of the two kinds of polypeptides in the hemoglobin protein, the sickle-cell child has a single different amino acid. This difference is caused by the change of a single nucleotide in the coding strand of DNA (Figure 10.16A). In the double helix, one base *pair* is changed.

We now know that the alternative alleles of many genes result from changes in single base pairs in DNA. Any change in the nucleotide sequence of DNA is called a **mutation.** Mutations can involve large regions of a chromosome or just a single nucleotide pair, as in sickle-cell disease. Here we consider how mutations involving only one or a few nucleotide pairs can affect gene translation.

Mutations within a gene can be divided into two general categories: base substitutions and base insertions or deletions (Figure 10.16B). A base substitution is the replacement of one nucleotide with another. For example, in the second row in Figure 10.16B, A replaces G in the fourth codon of the mRNA. What effect does a base substitution have? Because the genetic code is redundant, some substitution mutations have no effect at all. If a mutation causes an mRNA codon to change from GAA to GAG, for instance, no change in the protein product would result because GAA and GAG both code for the same amino acid (Glu). Other base substitutions may alter an amino acid but have little effect (perhaps because that amino acid is uninvolved in the protein's function). But as in the example in Figure 10.16A, base substitutions may cause changes in a protein that prevent it from functioning normally. Occasionally, a base substitution leads to an improved protein that enhances the success of the mutant organism and its descendants. Much more often, though, mutations are harmful. And if a base substitution changes an amino-acid codon to a stop codon, a shortened, probably nonfunctional, polypeptide will result.

Mutations involving the insertion or deletion of one or more nucleotides in a gene often have disastrous effects. Because mRNA is read as a series of nucleotide triplets (codons) during translation, adding or subtracting nucleotides may alter the **reading frame** (triplet grouping) of the message. All the nucleotides that are "downstream" of the insertion or deletion will be regrouped into different codons (Figure 10.16B, bottom). The result will most likely be a nonfunctional polypeptide.

The production of mutations, called **mutagenesis,** can occur in a number of ways. Errors that occur during DNA replication or recombination are called spontaneous mutations. Another source of mutation is a physical or chemical agent, called a **mutagen.** The most common physical mutagen in nature is high-energy radiation, such as X-rays and ultraviolet light. Chemical mutagens fall into several categories. One type, for example, consists of chemicals that are similar to normal DNA bases but that pair incorrectly.

Although mutations are often harmful, they are also extremely useful, both in nature and in the laboratory. It is because of mutations that there is such a rich diversity of genes in the living world, a diversity that makes evolution by natural selection possible. Mutations are also essential tools for geneticists. Whether naturally occurring (as in Mendel's peas) or created in the laboratory (Morgan used X-rays to make most of his fruit fly mutants), mutations create the different alleles needed for genetic research.

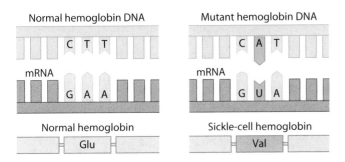

Figure 10.16A   The molecular basis of sickle-cell disease

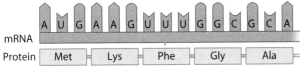

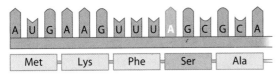

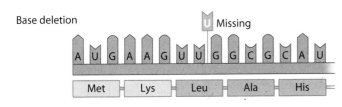

Figure 10.16B   Types of mutations and their effects

Web Process of Science *Connection: How Do You Diagnose a Genetic Disorder?*

? How could a single base substitution result in a shortened protein product?

■ A substitution that changed an amino acid codon into a stop codon would produce a prematurely terminated polypeptide.

10.17 Viral DNA may become part of the host chromosome

As we discussed in Module 10.1, bacteria and viruses served as models in experiments that uncovered the molecular details of heredity. Now let's take a closer look at viruses, focusing on the relationship between viral structure and the processes of nucleic acid replication, transcription, and translation. (Viruses, though not considered alive, are often grouped with microscopic organisms under the general term "microbes"; hence, this section is called "microbial genetics.")

In a sense, viruses are nothing more than "genes in a box": infectious particles consisting of nucleic acid enclosed in a protein coat called a **capsid** and, in some cases, a membrane envelope. Viruses are parasites that can reproduce only inside cells. In fact, the host cell provides most of the components necessary for replicating, transcribing, and translating the viral nucleic acid.

In Figure 10.1C, we described the reproductive cycle of phage T2. This sort of cycle is called a **lytic cycle** because it results in the lysis (breaking open) of the host cell and the release of the viruses that were produced within the cell. Some phages can also reproduce by an alternative route called the lysogenic cycle. During a **lysogenic cycle,** viral DNA replication occurs without destroying the host cell.

In Figure 10.17, you see the two kinds of cycles for a phage called lambda that infects *E. coli*. Both cycles begin when the phage DNA ❶ enters the bacterium and ❷ forms a circle. The DNA then embarks on one of the two pathways. In the lytic cycle (left), ❸ lambda's DNA immediately turns the cell into a

virus-producing factory, and ❹ the cell soon lyses and releases its viral products. In the lysogenic cycle, however, ❺ the DNA is inserted by genetic recombination into the bacterial chromosome. Once inserted, the phage DNA is referred to as a **prophage,** and most of its genes are inactive. ❻ Every time the *E. coli* cell prepares to divide, it replicates the phage DNA along with its own and passes the copies on to daughter cells. A single infected cell can quickly give rise to a large population of bacteria carrying the virus in prophage form. The lysogenic cycle enables viruses to propagate without killing the host cells on which they depend. The prophages may remain in the bacterial cells indefinitely. ❼ Occasionally, however, an environmental signal such as radiation or a certain chemical triggers a switchover from the lysogenic cycle to the lytic cycle.

Sometimes the few prophage genes active in a lysogenic bacterium can cause medical problems. For example, the bacteria that cause diphtheria, botulism, and scarlet fever would be harmless to humans if it were not for the prophage genes they carry. Certain of these genes direct the bacteria to produce the toxins responsible for making people ill. In the next module, we will explore animal viruses.

Web Activity *Phage Lysogenic and Lytic Cycles*

? Describe one way a virus can perpetuate its genes without destroying its host cell. What is this type of reproductive cycle called?

■ Some viruses can insert their DNA into a chromosome of the host cell, which replicates the viral genes when it replicates its own DNA prior to cell division. This is called the lysogenic cycle.

Figure 10.17 Two types of phage reproductive cycles

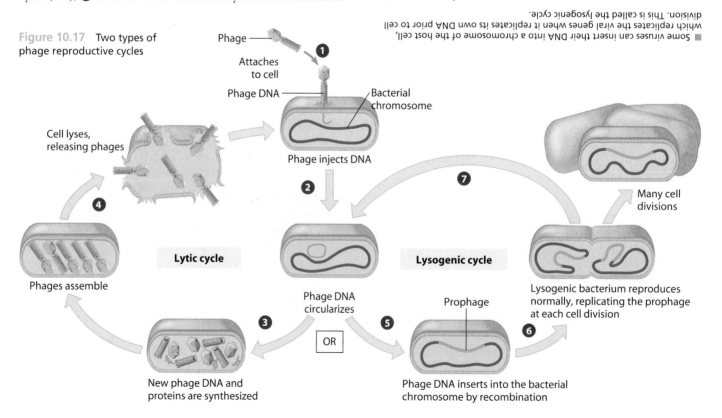

10.18 Many viruses cause disease in animals and plants

Viruses can cause disease in both animals and plants. A typical animal virus has a membranous outer envelope and projecting spikes of glycoprotein (protein with attached sugars). The envelope helps the virus enter and leave the host cell. Many animal viruses have RNA rather than DNA as their genetic material. Examples of RNA viruses include those that cause the common cold, measles, and mumps, as well as ones that cause more serious human diseases, such as AIDS and polio. Examples of diseases caused by DNA viruses are hepatitis, chicken pox, and herpes infections.

Figure 10.18 shows the reproductive cycle of a typical enveloped RNA virus (the mumps virus). When the virus contacts a host cell, the glycoprotein spikes attach to receptor proteins on the cell's plasma membrane. The envelope fuses with the cell's membrane, allowing the protein-coated RNA to ❶ enter the cytoplasm. ❷ Enzymes then remove the protein coat. ❸ An enzyme that entered the cell as part of the virus uses the virus's RNA genome as a template for making complementary strands of RNA (pink strand). The new strands have two functions: ❹ They serve as mRNA for the synthesis of new viral proteins, and ❺ they serve as templates for synthesizing new viral genome RNA. ❻ The new coat proteins assemble around the new viral RNA. ❼ Finally, the viruses leave the cell by cloaking themselves in plasma membrane. Thus, the virus obtains its envelope from the host cell, leaving the cell without necessarily lysing it.

Not all animal viruses reproduce in the cytoplasm. For example, herpesviruses, which you read about in the chapter introduction, are enveloped DNA viruses that reproduce in the host cell's nucleus; they acquire their envelopes from the cell's nuclear membranes. While inside the nuclei of certain nerve cells, herpesvirus DNA may remain latent, without destroying these cells. From time to time, physical stress, such as a cold or sunburn, or emotional stress may stimulate the herpesvirus DNA to begin production of the virus, which then infects cells at the body's surface and brings about cold sores or genital sores.

The amount of damage a virus causes our body depends partly on how quickly our immune system responds to fight the infection and partly on the ability of the infected tissue to repair itself. We usually recover completely from colds because our respiratory tract tissue can efficiently replace damaged cells by mitosis. In contrast, the poliovirus attacks nerve cells, which do not divide. The damage to such cells, unfortunately, is permanent. In such cases, we try to prevent the disease with vaccines (see Module 24.4).

Viruses that infect plants can stunt plant growth and diminish crop yields. Most plant viruses discovered to date are RNA viruses. To infect a plant, a virus must first get past the plant's outer protective layer of cells. Once a virus enters a plant cell and begins reproducing, it can spread throughout the entire plant through plasmodesmata, the cytoplasmic connections that penetrate the walls between adjacent plant cells (see Figure 4.22).

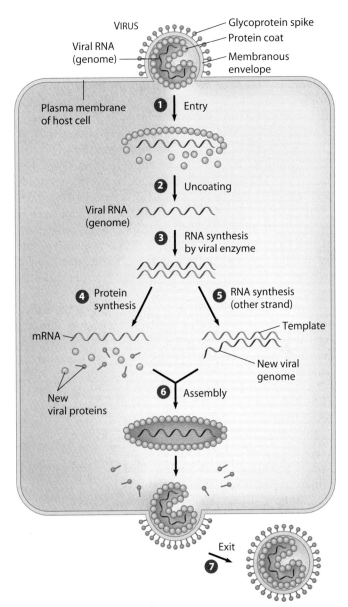

Figure 10.18 The reproductive cycle of an enveloped virus

Plant viruses may spread to other plants by insects, herbivores, humans, or farming tools. As with animal viruses, there are no cures for most viral diseases of plants. Agricultural scientists focus instead on preventing infections and on breeding resistant varieties of crop plants.

Web Activity *Simplified Reproductive Cycle of a DNA Virus*

❓ **Explain how some viruses replicate without having DNA.**

■ The genetic material of these viruses is RNA, which is replicated inside the host cell by special enzymes encoded by the virus. The viral genome (or its complement) serves as mRNA for the synthesis of viral proteins.

10.19 Emerging viruses threaten human health

Viruses that appear suddenly or are new to medical scientists are called **emerging viruses.** HIV, the AIDS virus, is a classic example: This virus appeared in San Francisco, Los Angeles, and New York in the early 1980s, seemingly out of nowhere. The deadly Ebola virus, recognized initially in 1976 in central Africa, is one of several emerging viruses that cause hemorrhagic fever, an often fatal syndrome characterized by fever, vomiting, massive bleeding, and circulatory system collapse. A number of other dangerous new viruses cause encephalitis, inflammation of the brain. One example is the West Nile virus, which appeared for the first time in North America in 1999 and has since spread to all 48 contiguous U.S. states. Severe acute respiratory syndrome (SARS) first appeared in China in 2002. Within eight months, about 8,000 people were infected, of whom some 10% died. Researchers quickly identified the infectious agent as a previously unknown, single-stranded RNA coronavirus, so named for its halo-like "corona" of spikes.

From where and how do such viruses burst on the human scene, giving rise to rare or previously unknown diseases? Three processes contribute to the emergence of viral diseases: mutations, contact between species, and spread from isolated populations.

The mutation of existing viruses is a major source of new viral diseases. RNA viruses tend to have unusually high rates of mutation because errors in replicating their RNA genomes are not subject to the kind of proofreading mechanisms that help reduce errors in DNA replication (see Module 10.5). Some mutations

enable existing viruses to evolve into new strains (genetic varieties) that can cause disease in individuals who have developed immunity to the ancestral virus. Flu epidemics, for instance, are caused by new influenza virus strains that are genetically different enough from earlier strains that people have little immunity to them.

New viral diseases often arise from the spread of existing viruses from one host species to another. Scientists estimate that about three-quarters of new human diseases have originated in other animals. For example, bats have been identified as the likely natural reservoir of the SARS virus: Large numbers of bats carry the virus and are able to transmit it to humans or to other animals that can infect humans.

In 1997, at least 18 people in Hong Kong were infected with a strain of flu virus previously seen only in birds. A mass culling of all of Hong Kong's 1.5 million domestic birds appeared to stop that outbreak. Beginning in 2002, however, new cases of human infection by this bird strain began to crop up around southeast Asia. As of 2007, the disease caused by this virus, now called "avian flu," has killed more than 180 people, and more than 100 million birds have either died from the disease or been killed to prevent the spread of infection (Figure 10.19). If this virus evolves so that it can spread easily from person to person, the potential for a major human outbreak is significant. Indeed, evidence is strong that the flu pandemic of 1918–1919, which killed about 40 million people, originated in birds.

The spread of a viral disease from a small, isolated population can lead to widespread epidemics. For instance, AIDS went unnamed and virtually unnoticed for decades before it began to spread around the world. In this case, technological and social factors, including affordable international travel, blood transfusions, sexual promiscuity, and the abuse of intravenous drugs, allowed a previously rare human disease to become a global scourge. It is likely that when we do find the means to control HIV and other emerging viruses, molecular biologists will be responsible for the discovery.

Colorized TEM 160,000×

Web Process of Science *Connection: Why Do AIDS Rates Differ Across the United States?*

? Why doesn't a bout of flu give us immunity to flu in subsequent years?

■ Influenza viruses evolve rapidly by frequent mutation; thus, the strains that later infect us will most likely be different from the ones to which we've developed immunity.

Figure 10.19 Ducks in Vietnam being checked for infection by the Avian flu virus (inset)

10.20 The AIDS virus makes DNA on an RNA template

The devastating disease AIDS is caused by a type of RNA virus with some special twists. In outward appearance, the AIDS virus (HIV) resembles the flu or mumps virus. As illustrated in Figure 10.20A, HIV has a membranous envelope and glycoprotein spikes. These components enable HIV to enter and leave a host cell much the way the mumps virus does (see Figure 10.18B). Notice, however, that HIV contains two identical copies of its RNA instead of one. HIV also has a different mode of reproduction. HIV carries molecules of an enzyme called **reverse transcriptase,** which catalyzes reverse transcription, the synthesis of DNA on an RNA template. This unusual phenomenon, which is opposite the usual DNA → RNA flow of genetic information, characterizes **retroviruses** ("retro" means "backward").

Figure 10.20B illustrates what happens after HIV RNA is uncoated in the cytoplasm of a host cell. ❶ The reverse transcriptase (⊙) uses the RNA as a template to make a DNA strand and then ❷ adds a second, complementary DNA strand. ❸ The resulting double-stranded DNA then enters the cell's nucleus and inserts itself into the chromosomal DNA, becoming a *provirus* (analogous to a prophage). Occasionally, the

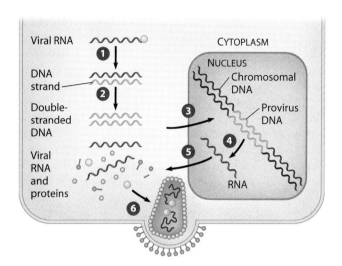

Figure 10.20B   The behavior of HIV nucleic acid in a host cell

provirus is ❹ transcribed into RNA and ❺ translated into viral proteins. ❻ New viruses assembled from these components leave the cell and can then infect other cells. This is the standard reproductive cycle for retroviruses.

AIDS stands for acquired immunodeficiency syndrome and **HIV** for human immunodeficiency virus; these terms describe the main effect of the virus on the body. HIV infects and eventually kills white blood cells that are important in immunity. We discuss AIDS in more detail when we take up the immune system in Chapter 24.

Web Activity Retrovirus (HIV) Reproductive Cycle

Web Process of Science Connection: What Causes Infections in AIDS Patients?

? Why is HIV classified as a retrovirus?

■ Because it synthesizes DNA from its RNA genome. This is the reverse ("retro") of the usual DNA → RNA information flow.

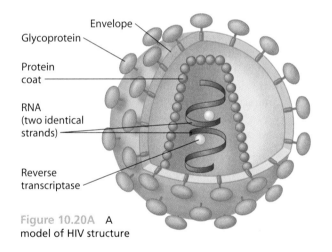

Envelope
Glycoprotein
Protein coat
RNA (two identical strands)
Reverse transcriptase

Figure 10.20A   A model of HIV structure

10.21 Viroids and prions are formidable pathogens in plants and animals

Viruses may be small and simple, but they dwarf another class of pathogens: viroids. **Viroids** are small circular RNA molecules that infect plants. Viroids do not encode proteins but can replicate in host plant cells, apparently using cellular enzymes. These small RNA molecules seem to cause errors in the regulatory systems that control plant growth. The typical signs of viroid diseases are abnormal development and stunted growth.

An important lesson to learn from viroids is that a single molecule can be an infectious agent. Viroids are nucleic acid, whose ability to be replicated is well known. Even more surprising is the evidence for infectious proteins, called **prions,** which appear to cause a number of degenerative brain diseases in various animal species, including scrapie in sheep, mad cow

disease, and Creutzfeldt-Jakob disease in humans. Although much remains to be learned, a prion is thought to be a misfolded form of a protein normally present in brain cells. When the prion enters a cell containing the normal form of protein, the prion somehow converts the normal protein molecules to the misfolded prion versions. To date, there is no known cure for prion diseases, and the only hope for developing effective treatments lies in understanding and preventing the process of infection.

? What makes prions different from all other known infectious agents?

■ Prions are proteins and have no nucleic acid.

10.22 Bacteria can transfer DNA in three ways

By studying viral reproduction, researchers also learn about the mechanisms that regulate DNA replication and gene expression in cells. Bacteria are equally valuable as microbial models in genetics research, but for different reasons. As prokaryotic cells, bacteria allow researchers to investigate molecular genetics in the simplest living organisms.

Most of a bacterium's DNA is found in a single chromosome, a closed loop of DNA with associated proteins. In the diagrams here, we show the chromosome much smaller than it actually is relative to the cell. A bacterial chromosome is hundreds of times longer than its cell; it fits inside the cell because it is tightly folded.

Bacterial cells divide by replication of the bacterial chromosome and then by binary fission (see Module 8.3). Because binary fission is an asexual process involving only a single parent, the bacteria in a colony are genetically identical to the parental cell. But this does not mean that bacteria lack ways to produce new combinations of genes. In fact, in the bacterial world, there are three mechanisms by which genes can move from one cell to another: transformation, transduction, and conjugation.

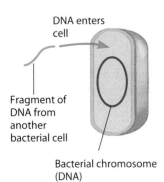

Figure 10.22A Transformation

Figure 10.22A illustrates **transformation,** the uptake of foreign DNA from the surrounding environment. In Frederick Griffith's "transforming factor" experiment (see Module 10.1), a harmless strain of bacteria took up pieces of DNA left from the dead cells of a disease-causing strain. The DNA from the pathogenic bacteria carried a gene that made the cells resistant to an animal's defenses, and when the previously harmless bacteria acquired this gene, it could cause pneumonia in infected animals.

Bacteriophages provide the second means of bringing together genes of different bacteria. The transfer of bacterial genes by a phage is called **transduction.** In this process, a fragment of DNA belonging to a phage's previous host cell is packaged within the phage's coat instead of the phage's DNA. When the phage infects a new bacterial cell, the DNA stowaway from the former host cell is injected into the new host (Figure 10.22B).

Figure 10.22C is an illustration of what happens at the DNA level when two bacterial cells mate. This physical union

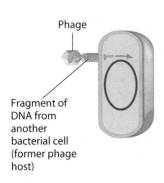

Figure 10.22B Transduction

of cells and the DNA transfer between them is called **conjugation** (from the Latin *conjugatus,* united). Notice that the "male" donor cell has hollow appendages called sex pili, one of which is attached to the "female" recipient cell. The outside layers of the cells have fused, and a cytoplasmic bridge has formed between them. Through this mating bridge, donor cell DNA (bright blue in the figure) passes to the recipient cell. The donor cell replicates its DNA as it transfers it, so the cell doesn't end up lacking any genes. The DNA replication is a special type that allows one copy to peel off and transfer into the recipient cell.

Once new DNA gets into a bacterial cell, by whatever mechanism, part of it may then integrate into the recipient's chromosome. As Figure 10.22D indicates, integration occurs by crossing over between the two DNA molecules, a process similar to crossing over between eukaryotic chromosomes (see Module 8.18). Here we see that two crossovers result in a piece of the donated DNA replacing part of the recipient cell's original DNA. The leftover pieces of DNA are broken down, leaving the recipient bacterium with a recombinant chromosome.

As we'll see in the next module, the transfer of genetic material between bacteria has important medical consequences.

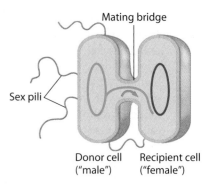

Figure 10.22C Conjugation

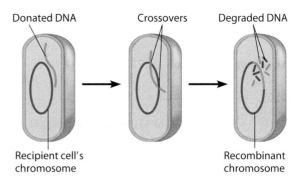

Figure 10.22D Integration of donated DNA into the recipient cell's chromosome

? The three modes of gene transfer between bacteria are _____, which is transfer via a virus; _____, which is the uptake of DNA from the surrounding environment; and _____, which is bacterial "sex."

transduction . . . transformation . . . conjugation

Bacterial plasmids can serve as carriers for gene transfer

The ability of a donor *E. coli* cell to carry out conjugation is usually due to a specific piece of DNA called the **F factor** (F for fertility). The F factor carries genes for making sex pili and other requirements for conjugation; it also contains an origin of replication, where DNA replication can start.

Let's see how the F factor behaves during conjugation. In Figure 10.23A, the F factor (light blue) is integrated into the male bacterium's chromosome. When this cell conjugates with a recipient cell, the male chromosome starts replicating at the F factor's origin of replication, indicated by the blue dot on the DNA. The growing copy of the DNA peels off the chromosome and heads into the recipient cell. Thus, part of the F factor serves as the leading end of the transferred DNA, but right behind it are genes from the donor's original chromosome. The rest of the F factor stays in the donor cell. Once inside the recipient cell, the transferred donor genes can recombine with the corresponding part of the recipient chromosome by crossing over. If crossing over occurs, the recipient cell may be genetically changed, but it usually remains female because the two cells break apart before the rest of the F factor transfers.

Alternatively, as Figure 10.23B shows, an F factor can exist as a **plasmid,** a small, circular DNA molecule separate from the bacterial chromosome. Every plasmid has an origin of replication, required for its replication within the cell. Some plasmids, including the F-factor plasmid, can bring about conjugation and move to another cell. When the male cell in Figure 10.23B mates with a recipient cell, the F factor replicates and at the same time transfers one whole copy of itself, in linear rather than circular form, to the recipient cell. The transferred plasmid re-forms a circle in the recipient cell, and the cell becomes male.

E. coli and other bacteria have many different kinds of plasmids. You can see several from one cell in Figure 10.23C, along with part of the bacterial chromosome, which extends in loops from the ruptured cell. Some plasmids carry genes that can affect the survival of the cell. Plasmids of one class, called **R plasmids,** pose serious problems for human medicine. Transferable R plasmids carry genes for enzymes that destroy antibiotics such as penicillin and tetracycline. Bacteria containing R plasmids are resistant (hence the designation R) to antibiotics that would otherwise kill them. The widespread use of antibiotics in medicine and agriculture has tended to kill off bacteria that lack R plasmids, whereas those with R plasmids have multiplied. As a result, an increasing number of bacteria that cause human diseases, such as food poisoning and gonorrhea, are becoming resistant to antibiotics (see Module 13.15).

We will return to the topic of plasmids in Chapter 12. But first, we continue our study of molecular genetics in Chapter 11, where we explore what is known about how genes themselves are controlled.

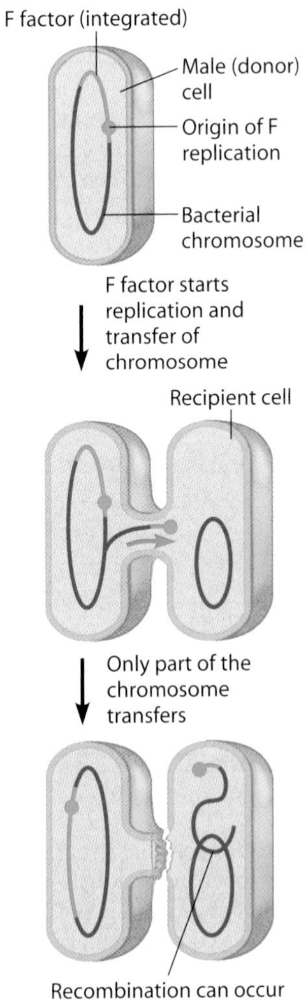

F factor (integrated)
Male (donor) cell
Origin of F replication
Bacterial chromosome

F factor starts replication and transfer of chromosome

Recipient cell

Only part of the chromosome transfers

Recombination can occur

Figure 10.23A Transfer of chromosomal DNA by an integrated F factor

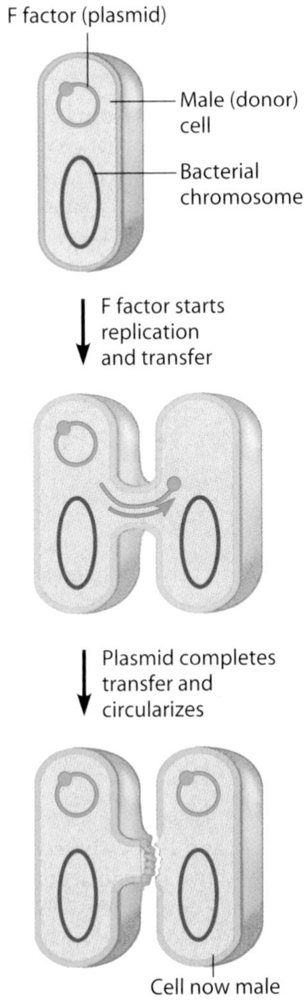

F factor (plasmid)
Male (donor) cell
Bacterial chromosome

F factor starts replication and transfer

Plasmid completes transfer and circularizes

Cell now male

Figure 10.23B Transfer of an F-factor plasmid

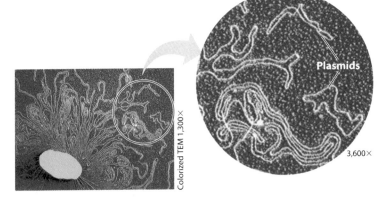

Colorized TEM 1,300×

Plasmids

3,600×

Figure 10.23C Plasmids and part of a bacterial chromosome released from a ruptured *E. coli* cell

? In Chapter 12, you will learn that plasmids are useful tools for genetic engineering. Can you guess why?

Web Process of Science *How Can Antibiotic-Resistant Plasmids Transform E. coli?*

■ Because plasmids can carry foreign genes and are duplicated and passed along by bacterial cells

Chapter Review

Reviewing the Concepts

The Structure of the Genetic Material (Introduction–10.3)

DNA as the genetic material.   Viruses provided some of the earliest evidence that genes are made of DNA. One key experiment showed that certain bacterial viruses (phages) reprogram host cells to produce more phages by injecting their DNA (**Introduction–10.1**).

DNA is a nucleic acid, made of long chains of nucleotide monomers (**10.2**):

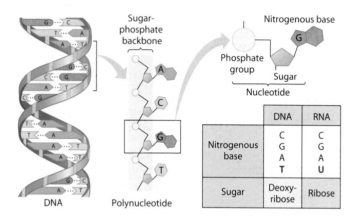

Watson and Crick worked out the three-dimensional structure of DNA: two polynucleotide strands wrapped around each other in a double helix. Hydrogen bonds between bases hold the strands together. Each base pairs with a complementary partner: A with T and G with C (**10.3**).

DNA Replication (10.4–10.5)

DNA replication starts with the separation of DNA strands. Enzymes then use each strand as a template to assemble new nucleotides into a complementary strand. Using the enzyme DNA polymerase, the cell synthesizes one daughter strand as a continuous piece, the other as a series of short pieces, which are then connected by the enzyme DNA ligase (**10.4–10.5**).

The Flow of Genetic Information from DNA to RNA to Protein (10.6–10.16)

From genotype to phenotype.   The information constituting an organism's genotype is carried in the sequence of its DNA bases. A particular gene—a linear sequence of many nucleotides—specifies a polypeptide. The DNA of the gene is transcribed into RNA, which is translated into the polypeptide (**10.6**). The "words" of the DNA "language" are triplets of bases called codons. The codons in a gene specify the amino acid sequence of a polypeptide. Nearly all organisms use exactly the same genetic code (**10.7–10.8**).

Transcription.   In the nucleus, the DNA helix unzips, and RNA nucleotides line up and hydrogen-bond along one strand of the DNA, following the base-pairing rules. As the single-stranded messenger RNA (mRNA) peels away from the gene, the DNA strands rejoin (**10.9**). Eukaryotic RNA is processed before leaving the nucleus as mRNA. Noncoding segments called introns are spliced out, and a cap and tail are added to the ends (**10.10**).

Translation takes place in the cytoplasm. A ribosome attaches to the mRNA and translates its message into a specific polypeptide, aided by transfer RNAs (tRNAs) that act as interpreters. Each tRNA is a folded molecule bearing a base triplet called an anticodon on one end; a specific amino acid is added to the other end (**10.11**). The mRNA moves a codon at a time relative to the ribosome, and a tRNA with a complementary anticodon pairs with each codon, adding its amino acid to the peptide chain (**10.12–10.14**):

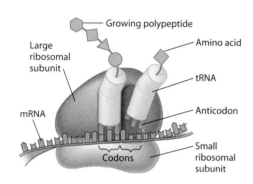

The sequence of codons in DNA, via the sequence of codons in mRNA, spells out the primary structure of a polypeptide (**10.15**).

Mutations are changes in the DNA base sequence, caused by errors in DNA replication or recombination, or by mutagens. Substituting, inserting, or deleting nucleotides alters a gene, with varying effects on the organism (**10.16**).

Microbial Genetics (10.17–10.23)

Viruses can be regarded as genes packaged in protein. When phage DNA enters a lytic cycle inside a bacterium, it is replicated, transcribed, and translated; the new viral DNA and protein molecules then assemble into new phages, which burst from the host cell. In the lysogenic cycle, phage DNA inserts into the host chromosome and is passed on to generations of daughter cells. Much later, it may initiate phage production (**10.17**). Many viruses cause disease when they invade animal or plant cells. Many, such as flu viruses and most plant viruses, have RNA, rather than DNA, as their genetic material. Some animal viruses steal a bit of host cell membrane as a protective envelope. Some viruses can remain latent in the host's body for long periods (**10.18**). Emerging viral diseases pose a threat to human health (**10.19**). HIV, the AIDS virus, is a retrovirus. Inside a cell it uses its RNA as a template for making DNA, which then inserts into a host chromosome (**10.20**). Viroids are RNA molecules that can infect plants. Prions are infectious proteins that can cause brain diseases in animals (**10.21**).

Bacteria can transfer genes from cell to cell by one of three processes: transformation, transduction, or conjugation (**10.22**). Plasmids, small circular DNA molecules separate from the bacterial chromosome, can serve as carriers for the transfer of genes (**10.23**).

Connecting the Concepts

1. Check your understanding of the flow of genetic information through a cell by filling in the blanks.

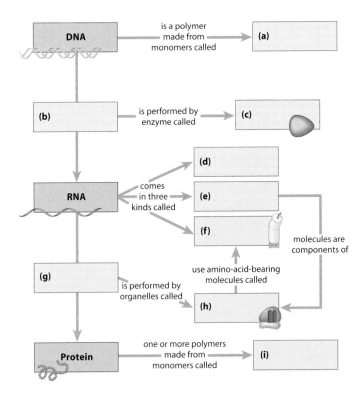

Testing Your Knowledge

Multiple Choice

2. Scientists have discovered how to put together a bacteriophage with the protein coat of phage T2 and the DNA of phage lambda. If this composite phage were allowed to infect a bacterium, the phages produced in the host cell would have _____. (*Explain your answer.*)
   a. the protein of T2 and the DNA of lambda
   b. the protein of lambda and the DNA of T2
   c. a mixture of the DNA and proteins of both phages
   d. the protein and DNA of T2
   e. the protein and DNA of lambda

3. A geneticist found that a particular mutation had no effect on the polypeptide coded by a gene. This mutation probably involved
   a. deletion of one nucleotide.
   b. alteration of the start codon.
   c. insertion of one nucleotide.
   d. deletion of the entire gene.
   e. substitution of one nucleotide.

4. Which of the following correctly ranks the structures in order of size, from largest to smallest?
   a. gene-chromosome-nucleotide-codon
   b. chromosome-gene-codon-nucleotide
   c. nucleotide-chromosome-gene-codon
   d. chromosome-nucleotide-gene-codon
   e. gene-chromosome-codon-nucleotide

5. The nucleotide sequence of a DNA codon is GTA. A messenger RNA molecule with a complementary codon is transcribed from the DNA. In the process of protein synthesis, a transfer RNA pairs with the mRNA codon. What is the nucleotide sequence of the tRNA anticodon?
   a. CAT   d. CAU
   b. CUT   e. GT
   c. GUA

Describing, Comparing, and Explaining

6. Describe the process of DNA replication: the ingredients needed, the steps in the process, and the final product.

7. Describe the process by which the information in a gene is transcribed and translated into a protein. Correctly use these words in your description: tRNA, amino acid, start codon, transcription, RNA splicing, exons, introns, mRNA, gene, codon, RNA polymerase, ribosome, translation, anticodon, peptide bond, stop codon.

Applying the Concepts

8. A cell containing a single chromosome is placed in a medium containing radioactive phosphate so that any new DNA strands formed by DNA replication will be radioactive. The cell replicates its DNA and divides. Then the daughter cells (still in the radioactive medium) replicate their DNA and divide, and a total of four cells are present. Sketch the DNA molecules in all four cells, showing a normal (nonradioactive) DNA strand as a solid line and a radioactive DNA strand as a dashed line.

9. The base sequence of the gene coding for a short polypeptide is CTACGCTAGGCGATTGACT. What would be the base sequence of the mRNA transcribed from this gene? Using the genetic code in Figure 10.8A, give the amino acid sequence of the polypeptide translated from this mRNA. (*Hint:* What is the start codon?)

10. Researchers on the Human Genome Project have determined the nucleotide sequences of human genes and in many cases identified the proteins encoded by the genes. Knowledge of the nucleotide sequences of genes might be used to develop lifesaving medicines or treatments for genetic defects. In the United States, both government agencies and biotechnology companies have applied for patents on their discoveries of genes. In Britain, the courts have ruled that a naturally occurring gene cannot be patented. Do you think individuals and companies should be able to patent genes and gene products? Before answering, consider the following: What are the purposes of a patent? How might the discoverer of a gene benefit from a patent? How might the public benefit? What might be some positive and negative results of patenting genes?

Answers to all questions can be found in Appendix 4.

For Practice Quizzes, BioFlix, MP3 Tutors, and Activities, go to www.mybiology.com.

Cloning to the Rescue?

All the animals shown on these pages share two unusual features. First, they are all members of endangered species. Second, they are all clones. In this usage, a **clone** is an individual created by asexual reproduction (that is, reproduction that does not involve fusion of sperm and egg) and thus genetically identical to a single parent. First demonstrated in the 1950s with frogs, animal cloning became much more commonplace after 1997, when Scottish researchers announced the first successful cloning of a mammal: the celebrated Dolly the sheep, cloned using a mammary cell from an adult ewe. In the years since Dolly's landmark birth, researchers have cloned a variety of other mammals, including mules, horses, cats, cows, goats, pigs, monkeys, mice, and, most recently, dogs.

Some researchers have concentrated their efforts on cloning members of endangered species. Among the rare animals that have been cloned are a gaur (an Asian ox) and a mouflon (a small European sheep). At the Audubon Center for Research of Endangered Species in New Orleans, cloners have focused their efforts on endangered felines, such as the African wildcat (*Felis silvestris*) shown at right. And in 2006, South Korean researchers announced the successful cloning of two gray wolves, a species that nearly became extinct in the United States in the early 20th century. One of these wolves is shown at left.

One remarkable case was the 2003 cloning of a banteng (a Javanese bovine whose numbers have dwindled to just a few in the wild) using frozen skin cells from a zoo-raised banteng that had died 23 years previously. Nuclei from the donor were inserted into denucleated eggs from ordinary cows. The resulting embryos were implanted into surrogate Angus cows, leading to the birth of a healthy baby banteng (above). This case proves the possibility of producing new animals even when a female of the donor species is unavailable. Scientists may someday be able to use similar cross-species methods to clone an animal from a recently extinct species.

The use of cloning to slow or reverse extinction and repopulate endangered species holds tremendous promise. However, conservationists object that cloning may trivialize the tragedy of extinction and detract from efforts to preserve natural habitats. They correctly point out that cloning, like other forms of asexual reproduction, does not increase genetic diversity and is therefore not as beneficial to endangered species as natural reproduction.

Researchers are also concerned that cloned animals are less healthy than those arising from a fertilized egg. In 2003, at age 6, Dolly was euthanized after suffering complications from a lung disease normally seen only in much older sheep. Dolly's premature death, as well as her arthritic condition, led to speculation that her cells were somehow "older" than those of normal sheep. Many other cloned animals exhibit defects. Cloned mice, for instance, are prone to obesity, pneumonia, liver failure, and premature death. Scientists believe that even cloned animals that appear normal are likely to have subtle defects. Recent work traces these problems to physical changes within the chromosomes of the cloned organism that derail the normal regulation of genes.

The ability to clone an animal using a transplanted nucleus demonstrates an important point: The nuclei of adult body cells contain a complete genome capable of directing the production of all the cell types in an organism. The development of a multicellular organism, with many different kinds of cells, thus depends on the turning on and off of genes—the control of gene expression. The health problems associated with cloned animals underscore the fact that proper gene regulation is important to the well-being of all organisms.

This chapter focuses first on how genes are controlled, beginning with simple prokaryotic systems and then describing gene regulation in eukaryotes. Next we look at the methods and applications of animal cloning. Finally, we discuss cancer, a disease that results from defects in the regulation of cell division. ■ ■ ■

Proteins interacting with DNA turn prokaryotic genes on or off in response to environmental changes

Gene regulation—the turning on and off of genes—can help organisms respond to environmental changes. But what do we actually mean when we say that genes are turned on or off? As we discussed in Chapter 10, genes determine the nucleotide sequences of specific RNA molecules; if this RNA is mRNA, it in turn determines the sequences of amino acids in protein molecules. Thus, a gene that is turned on is being transcribed into RNA, and that message is being translated into specific protein molecules. The overall process by which genetic information flows from genes to proteins—that is, from genotype to phenotype—is called **gene expression.** The control of gene expression makes it possible for cells to produce specific kinds of proteins when and where they are needed. The turning on and off of transcription is the main way that gene expression is regulated in all organisms.

Our earliest understanding of gene control came from studies of the bacterium *Escherichia coli* (Figure 11.1A). *E. coli* can change its metabolic activities in response to changes in its environment. For example, when a certain nutrient is plentiful, *E. coli* does not squander valuable resources to make it from scratch. Bacterial cells that can conserve resources and energy have an advantage over cells that are unable to do so. Thus, natural selection has favored bacteria that express only the genes whose products are needed by the cell. Let's look at how the regulation of gene transcription helps *E. coli* accomplish such control.

The lac Operon Picture an *E. coli* cell living in your intestine. It is dependent on your dietary whims for its nutrients. If

you eat a sweet roll for breakfast, the bacterium will soon be bathed in sugars and broken-down fats. Later on, if you have a glass of fat-free milk and a salad for lunch, *E. coli*'s environment will change drastically.

Let's focus on your glass of milk for a moment. One of the main nutrients in milk is the disaccharide sugar lactose (see Module 3.5). When lactose is plentiful in the intestine, *E. coli* makes the enzymes necessary to absorb the sugar and use it as an energy source. Conversely, when lactose is not plentiful, *E. coli* does not waste its energy producing these enzymes.

Remember that enzymes are proteins; their production is an outcome of gene expression. *E. coli* can make lactose-utilization enzymes because it has genes that code for these enzymes. Figure 11.1B (on the facing page) presents a model (first proposed in 1961 by French biologists François Jacob and Jacques Monod) to explain how an *E. coli* cell can turn genes coding for lactose-utilization enzymes off or on, depending on whether lactose is available.

E. coli uses three enzymes to take up and start metabolizing lactose, and the genes coding for these three enzymes are regulated as a single unit. The DNA at the top of Figure 11.1B represents a small segment of the bacterium's chromosome. Notice that the three genes that code for the lactose-utilization enzymes (light blue) are next to each other in the DNA.

Adjacent to the group of lactose enzyme genes are two *control sequences*, short sections of DNA that help control the enzyme genes. One stretch of nucleotides is a **promoter** (green), a site where the transcription enzyme, RNA polymerase, attaches and initiates transcription—in this case, transcription of all three lactose enzyme genes (as depicted in the bottom panel of Figure 11.1B). Between the promoter and the enzyme genes, a DNA segment called an **operator** (yellow) acts as a switch. The operator determines whether RNA polymerase can attach to the promoter and start transcribing the genes.

Such a cluster of genes with related functions, along with a promoter and an operator, is called an **operon;** with rare exceptions, operons exist only in prokaryotes. The key advantage to the grouping of related genes into operons is that a single "on-off switch" can control the whole cluster. The operon discussed here is called the *lac* operon, short for lactose operon. When an *E. coli* encounters lactose, all the enzymes needed for its use are made at once because the operon's genes are all controlled by a single switch, the operator. But what determines whether the operator switch is on or off?

The top panel of Figure 11.1B shows the *lac* operon in "off" mode, its status when there is no lactose in the cell's environment. Transcription is turned off by a molecule called a **repressor** ( ), a protein that functions by binding to the operator and physically blocking the attachment of RNA polymerase ( ) to the promoter. On the left side of the figure, you can see where the repressor comes from. A gene called a

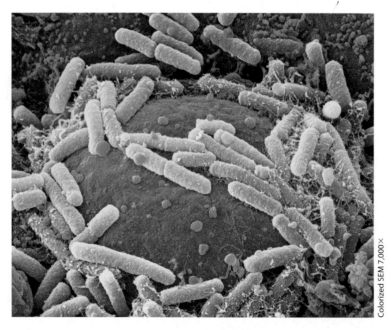

Colorized SEM 7,000×

Figure 11.1A Cells of *E. coli* bacteria

regulatory gene (dark blue), located outside the operon, codes for the repressor. The regulatory gene is expressed continually, so the cell always has a small supply of repressor molecules.

How can an operon be turned on if its repressor is always present? As the bottom panel of Figure 11.1B indicates, lactose (○) interferes with the attachment of the *lac* repressor to the operator by binding to the repressor and changing its shape. With its new shape (◉), the repressor cannot bind to the operator, and the operator switch remains on. RNA polymerase can now bind to the promoter (since it is no longer being blocked) and from there transcribe the genes of the operon. The resulting mRNA carries coding sequences for all three enzymes (purple) needed for lactose use. The cell can translate the message in this single mRNA into three separate polypeptides because the mRNA has multiple codons signaling the start and stop of translation.

The newly produced mRNA and protein molecules will remain intact for only a short time before cellular enzymes break them down. When the synthesis of mRNA and protein stops because lactose is no longer present, the molecules quickly disappear.

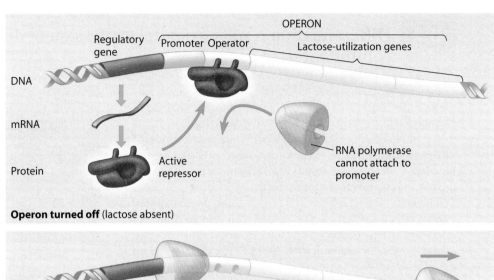

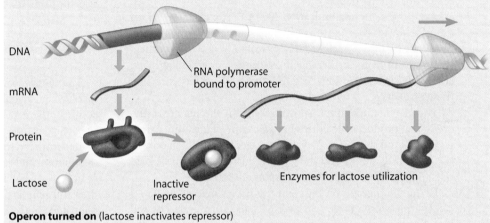

Figure 11.1B The *lac* operon

Other Kinds of Operons The *lac* operon is only one type of operon in bacteria. Other types also have a promoter, an operator, and several adjacent genes, but they differ in the way the operator switch is controlled. Figure 11.1C shows two types of repressor-controlled operons. The *lac* operon's repressor is active when alone and inactive when bound to lactose. A second type of operon, represented here by the *trp* operon, is controlled by a repressor that is *inactive* alone. To be active, this type of repressor must combine with a specific small molecule. In our example, the small molecule is tryptophan, an amino acid essential for protein synthesis. *E. coli* can make tryptophan from scratch, using enzymes encoded in the *trp* operon. But it will stop making tryptophan and simply absorb it in prefabricated form from the surroundings whenever possible. When *E. coli* is swimming in tryptophan, which occurs in large amounts in such foods as milk and poultry, tryptophan binds to the repressor of the *trp* operon. This activates the *trp* repressor, enabling it to switch off the operon. Thus, this type of operon allows bacteria to stop making certain essential molecules when the molecules are already present in the environment, saving materials and energy for the cells.

Another type of operon control involves **activators,** proteins that turn operons *on* by binding to DNA. These proteins act by making it easier for RNA polymerase to bind to the promoter, rather than by blocking RNA polymerase, as repressors do. Activators help control a wide variety of operons.

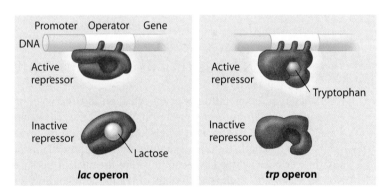

Figure 11.1C Two types of repressor-controlled operons

Armed with a variety of operons and regulated by repressors and activators, *E. coli* and other prokaryotes can thrive in frequently changing environments. Next we'll examine how more complex eukaryotes achieve gene regulation.

🎧 **MP3 Tutor** *Control of Gene Expression*

Web Activity *The* lac *operon in* E. coli

❓ A certain mutation in *E. coli* impairs the ability of the *lac* repressor to bind to the *lac* operator. How would this affect the cell?

■ The cell would wastefully produce the enzymes for lactose metabolism continuously, even when lactose is not present.

11.2 Differentiation results from the expression of different combinations of genes

Compared with bacteria, eukaryotic organisms, especially multicellular ones, face elaborate gene regulation challenges. During the repeated cell divisions that lead from a zygote to a multicellular adult, individual cells must undergo **differentiation**—that is, become specialized in structure and function. Differentiation results from selective gene expression, the turning on and off of specific genes.

The light micrographs in Figure 11.2 show several types of human cells. Each sample was stained with dyes to bring out important cellular details, and all three micrographs are printed at about the same magnification (900×). The micrograph on the left shows a short segment of a muscle cell, which is a long fiber with multiple nuclei (dark horizontal rods). The vertical stripes result from the arrangement of the proteins that bring about muscle contraction. As you'd expect, the genes encoding these proteins are active in muscle cells.

The center micrograph shows cells from the pancreas, the organ that produces the hormones glucagon and insulin, which regulate blood sugar. The gene for glucagon is turned on only in the alpha cells (pink) and the gene for insulin only in the beta cells (light purple). The dark blue spots are cell nuclei.

In the micrograph on the right, we see a single white blood cell (purple with a dark, two-part nucleus) surrounded by red blood cells (pink). In immature red blood cells, the genes for the oxygen-carrying protein hemoglobin are turned on full blast. Later in differentiation, mammalian red blood cells lose their nuclei and other organelles. Mature red blood cells (shown in the micrograph) are packed full with hemoglobin.

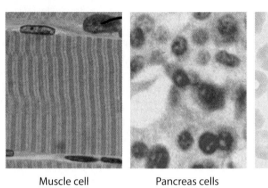

| Muscle cell | Pancreas cells | Blood cells |

Figure 11.2 Some types of human cells (all shown at 900×)

The genes for "housekeeping" enzymes, such as those involved in glycolysis, are active in all metabolizing cells. In contrast, the genes for specialized proteins are turned on only in particular types of cells.

In summary, the particular genes that are active in each type of differentiated cell are the source of its particular function and structure. As you'll see later, such differentiated cells still retain their full genetic potential. But first, let's examine how genes are regulated in eukaryotes.

? If a nerve cell and a skin cell in your body have the same genes, how can the cells be so different?

Each cell type must be expressing certain genes that are present in, but not expressed in, the other cell types.

11.3 DNA packing in eukaryotic chromosomes helps regulate gene expression

Let's begin our exploration of gene regulation in eukaryotes by looking at the chromosomes, where almost all of the cell's genes are located. The DNA in just a single human chromosome averages 4 cm in length, thousands of times more than the diameter of the nucleus. All this DNA can fit within the nucleus because of an elaborate, multilevel system of coiling and folding, or *packing*, of the DNA in each chromosome. A crucial aspect of DNA packing is the association of the DNA with small proteins called **histones.** In fact, histone proteins account for about half the mass of eukaryotic chromosomes. (Prokaryotes have analogous proteins, but lack the degree of DNA packing in eukaryotes.)

Figure 11.3 (top of next page) shows a model for the main levels of DNA packing. At the left, notice that the double-helical molecule of DNA has a diameter of 2 nm, which is not altered by packing. At the first level of packing, histones attach to the DNA. In electron micrographs, the DNA-histone complex has the appearance of beads on a string. Each "bead," called a **nucleosome,** consists of DNA wound around a protein core of eight histone molecules. Short stretches of DNA, called linkers, join consecutive nucleosomes (the "string" between the "beads").

At the next level of packing, the beaded string is wrapped into a tight helical fiber. Then this fiber coils further into a thick supercoil with a diameter of about 300 nm. Looping and folding can further compact the DNA, as you can see in the metaphase chromosome at the right of the figure. Viewed as a whole, Figure 11.3 gives a sense of how successive levels of coiling and folding enable a huge amount of DNA to fit into a cell nucleus.

DNA packing tends to prevent gene expression by preventing transcription proteins from contacting the DNA. Cells seem to use higher levels of packing for long-term inactivation of genes. Highly compacted chromatin, which is found not only in metaphase chromosomes but also in varying regions of interphase chromosomes, is generally not expressed at all. One intriguing instance of this phenomenon is described in the next module.

? How does dense packing of DNA in chromosomes prevent gene expression?

RNA polymerase and other proteins required for transcription do not have access to the DNA in tightly packed regions of a chromosome.

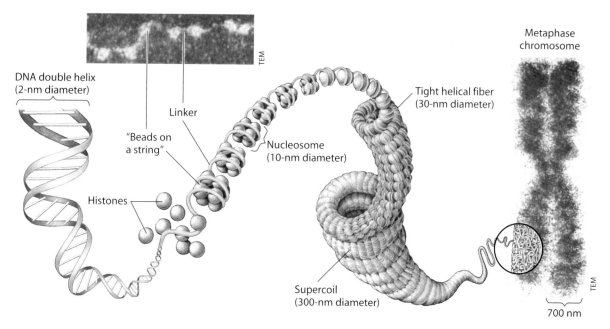

Figure 11.3 DNA packing in a eukaryotic chromosome

11.4 In female mammals, one X chromosome is inactive in each somatic cell

Female mammals, including humans, inherit two X chromosomes. So why don't females make twice as much of the proteins encoded by genes on the X chromosome compared to the amount in males? It turns out that in female mammals, one X chromosome in each somatic cell exists in a highly compacted and almost entirely inactive form. This **X chromosome inactivation** is initiated early in embryonic development, when one of the two X chromosomes in each cell is inactivated at random. As a result, the cells of females and males have the same effective dose (one copy) of these genes. The inactive X in each cell of a female condenses into a compact object called a **Barr body.**

Which X chromosome is inactivated is a matter of chance in each embryonic cell, but once an X chromosome is inactivated, all descendant cells have the same copy turned off. Consequently, females consist of a mosaic of two types of cells: those with the active X derived from the father and those with the active X derived from the mother. Thus, if a female is heterozygous for a gene on the X chromosome (a sex-linked gene; see Module 9.21), about half her cells will express one allele, while the others will express the alternate allele.

A striking example of this phenomenon is the tortoiseshell cat, which has orange and black patches of fur (Figure 11.4). The relevant fur-color gene is on the X chromosome, and the

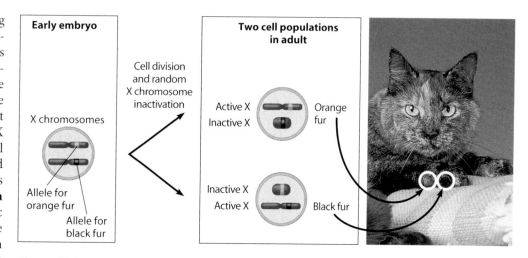

Figure 11.4 Tortoiseshell pattern on a cat, a result of X chromosome inactivation

tortoiseshell phenotype requires the presence of two different alleles, one for orange fur and one for black fur. Normally, only females can have both alleles because only they have two X chromosomes. If a female is heterozygous for the tortoiseshell gene, she will have the tortoiseshell phenotype. Orange patches are formed by populations of cells in which the X chromosome with the orange allele is active; black patches have cells in which the X chromosome with the nonorange allele is active.

? Why are tortoiseshell cats usually female?

■ In general, only females have two X chromosomes (but see Module 8.22).

11.5 Complex assemblies of proteins control eukaryotic transcription

The packing and unpacking of chromosomal DNA provide a coarse adjustment for eukaryotic gene expression by making a region of DNA either more or less available for transcription. The fine-tuning begins with the initiation of RNA synthesis—transcription. In both prokaryotes and eukaryotes, the initiation of transcription (whether transcription starts or not) is the most important stage for regulating gene expression.

Like prokaryotes (see Module 11.1), eukaryotes employ regulatory proteins that bind to DNA and turn the transcription of genes on and off. The eukaryotic control mechanisms involve proteins that, like prokaryotic repressors and activators, bind to specific segments of DNA. However, eukaryotic cells have more regulatory proteins and more control sequences in their DNA. The current model for the initiation of eukaryotic transcription features an intricate array of regulatory proteins that interact with DNA and with one another to turn genes on or off.

In contrast to the clustered genes of bacterial operons, each eukaryotic gene usually has its own promoter and other control sequences. Moreover, activator proteins seem to be more important in eukaryotes than repressors. In multicellular eukaryotes, the "default" state for most genes seems to be "off." A typical animal or plant cell needs to turn on (transcribe) only a small percentage of its genes, those required for the cell's specialized structure and function. Housekeeping genes, those continually active in virtually all cells for routine activities such as glycolysis, may be in an "on" state by default.

In order to function, eukaryotic RNA polymerase requires the assistance of proteins called **transcription factors.** In the model depicted in Figure 11.5, the first step in initiating gene transcription is the binding of activator proteins (green) to DNA sequences called **enhancers** (yellow). In contrast to the operators of prokaryotic operons, enhancers are usually far away from the gene they help regulate. The binding of activators to enhancers leads to bending of the DNA. Once the DNA is bent, the bound activators interact with other transcription factor proteins (purple), which then bind as a complex at the gene's promoter. This large assembly of proteins facilitates the correct attachment of RNA polymerase to the promoter and the initiation of transcription. As shown in the figure, several enhancers and activators may be involved. Not shown are *repressor* proteins called **silencers** that may bind to DNA sequences and *inhibit* the start of transcription.

If eukaryotic genomes only rarely have operons, how does the eukaryotic cell deal with genes of related function that

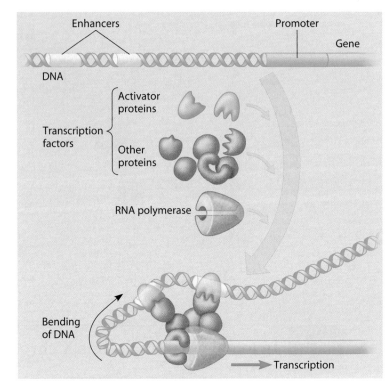

Figure 11.5   A model for the turning on of a eukaryotic gene

need to be turned on or off at the same time? Genes coding for the enzymes of a metabolic pathway, for example, are often scattered over different chromosomes. Coordinated gene expression in eukaryotes seems to depend on the association of a specific enhancer (or collection of enhancers) with every gene of a particular metabolic pathway. Copies of the transcription factors that recognize these DNA sequences bind to them all at once, promoting simultaneous transcription of the genes, no matter where they are in the genome.

Web Process of Science *How Do You Design a Gene Expression System?*

? What must occur before RNA polymerase can bind to a promoter and transcribe a specific eukaryotic gene?

■ Enhancers must bind to transcription factors to facilitate the attachment of RNA polymerase to the promoter.

11.6 Eukaryotic RNA may be spliced in more than one way

Once transcription of a eukaryotic RNA molecule is completed, the noncoding segments called introns are removed by splicing (see Module 10.10). RNA splicing provides several possible ways for regulating gene expression. Some scientists think that the splicing process itself may help control the flow of mRNA from nucleus to cytoplasm because until splicing is completed, the

RNA is attached to the molecules of the splicing machinery and cannot pass through the nuclear pores. Moreover, in some cases, the cell can carry out splicing in more than one way, generating different mRNA molecules from the same RNA transcript. Notice in Figure 11.6, for example, that one mRNA molecule ends up with the green exon and the other with the brown exon.

(The light blue and pink segments are the introns.) With this sort of **alternative RNA splicing,** an organism can get more than one type of polypeptide from a single gene.

One interesting example of two-way splicing is found in the fruit fly. In this animal, the differences between males and females are largely due to different patterns of RNA splicing. And as you will learn in Chapter 12, recent results from the Human Genome Project suggest that alternative splicing is very common in humans. Included among the many instances already known is one gene whose transcript can be spliced to encode *seven* alternative versions of a certain protein involved in cellular contraction. Each of the seven is made in a different type of cell.

> ? How does alternative RNA splicing enable a single gene to encode more than one kind of polypeptide?

◼ Each kind of polypeptide is encoded by an mRNA molecule containing a different combination of exons.

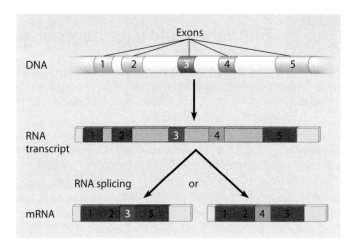

Figure 11.6 Production of two different mRNAs from the same gene

11.7 Small RNAs play multiple roles in controlling gene expression

Recall that only 1.5% of the human genome—and a similarly small percentage of the genomes of many other multicellular eukaryotes—codes for proteins. Another very small fraction of DNA consists of genes for ribosomal RNA and transfer RNA. Until recently, most of the remaining DNA was considered to be "noncoding," meaning that it neither coded for proteins nor was transcribed into functional RNA of the few known types. In other words, it was thought not to contain meaningful genetic information. However, a flood of recent data has contradicted this view. Biologists currently think that a significant amount of the genome may be transcribed into non-protein-coding RNAs, including a variety of small RNAs. While many questions about the functions of these RNAs remain unanswered, researchers are uncovering more evidence of their biological roles every day.

In 1993, researchers discovered small single-stranded RNA molecules, called microRNAs (miRNAs), that can bind to complementary sequences on mRNA molecules (Figure 11.7). Each miRNA, typically about 20 nucleotides long, ❶ associates with a large protein complex. The complex can ❷ bind to any mRNA molecule with the complementary sequence. Then the miRNA-protein complex either ❸ degrades the target mRNA or ❹ blocks its translation. It has been estimated that miRNAs may regulate the expression of up to one-third of all human genes, a striking figure given that miRNAs were unknown a mere 20 years ago.

Researchers can take advantage of the miRNA mechanism to artificially control gene expression. For example, injecting miRNAs into a cell can turn off expression of a gene with a sequence that matches the miRNA. This procedure is called **RNA interference (RNAi).** As you learned in Chapter 10, some viruses that infect cells have double-stranded RNA genomes. The RNAi pathway may have evolved as a natural defense against infection by such viruses.

Biologists are excited about these recent discoveries, which hint at a large, diverse population of RNA molecules in the cell

that play crucial roles in regulating gene expression—and have gone largely unnoticed until now. Clearly, we must revise the long-standing view that because they code for proteins, mRNAs are the most important RNAs in terms of cellular function.

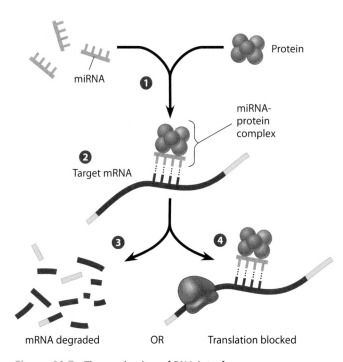

Figure 11.7 The mechanism of RNA interference

> ? If a gene has the sequence AATTCGCG, what would be the sequence of an miRNA that turns off the gene?

◼ The gene will be transcribed as the mRNA sequence UUAAGCGC; an miRNA of sequence AAUUCGCG would bind to and disable this mRNA.

Translation and later stages of gene expression are also subject to regulation

After a eukaryotic mRNA is fully processed and transported to the cytoplasm, there are additional opportunities for regulation. These include mRNA breakdown, initiation of translation, protein activation, and protein breakdown.

Breakdown of mRNA   Molecules of mRNA do not remain intact forever. Enzymes in the cytoplasm eventually break them down, and the timing of this event is an important factor regulating the amounts of various proteins that are produced in the cell. Long-lived mRNAs can be translated into many more protein molecules than short-lived ones. Prokaryotic mRNAs have very short lifetimes; they are degraded by enzymes within a few minutes after their synthesis. This is one reason bacteria can change their proteins relatively quickly in response to environmental changes. In contrast, the mRNA of eukaryotes can have lifetimes of hours or even weeks.

A striking example of long-lived mRNA is found in vertebrate red blood cells, which manufacture large quantities of the protein hemoglobin. In most species of vertebrates, the mRNAs for hemoglobin are unusually stable. They probably last as long as the red blood cells that contain them—about a month or a bit longer in reptiles, amphibians, and fishes—and are translated again and again. Mammals are an exception, as you learned in Module 11.2. When their red blood cells mature, they lose their ribosomes (along with their other organelles) and thus cease to make new hemoglobin. However, mammalian hemoglobin itself lasts about as long as the red blood cells last, around four months.

Initiation of Translation   The process of translating an mRNA into a polypeptide also offers opportunities for regulation. Among the molecules involved in translation are a great many proteins that control the start of polypeptide synthesis. Red blood cells, for instance, have an inhibitory protein that prevents translation of hemoglobin mRNA unless the cell has a supply of heme, the iron-containing chemical group essential for hemoglobin function. (It is the iron atom of the heme group to which oxygen molecules actually attach.)

Protein Activation   After translation is complete, polypeptides may require alteration to become functional. Post-translational control mechanisms in eukaryotes often involve the cleavage (cutting) of a polypeptide to yield a smaller final product that is the active protein, able to carry out a specific function in the organism. In Figure 11.8 we see the example of the hormone insulin, which is a protein. Insulin is synthesized in the cells of the pancreas as one long polypeptide that has no hormonal activity. After translation is completed, the polypeptide folds up, and covalent bonds form between the sulfur (S) atoms of sulfur-containing amino acids (see Figure 3.12C). (Two H atoms are lost as each S—S bond forms.) The result is that parts of the polypeptide are linked together in a specific way. Finally, a large center portion is cut away, leaving two shorter chains held together by the sulfur linkages. This combination of two shorter polypeptides is the form of insulin that functions as a hormone.

Protein Breakdown   The final control mechanism operating after translation is the selective breakdown of proteins. Though mammalian hemoglobin may last as long as the red blood cell housing it, the lifetimes of many other proteins are closely regulated. Some of the proteins that trigger metabolic changes in cells are broken down within a few minutes or hours. This regulation allows a cell to adjust the kinds and amounts of its proteins in response to changes in its environment. It also enables the cell to maintain its proteins in prime working order. Indeed, when proteins are damaged, they are usually broken down right away and replaced by new ones that function properly.

Web Activity *Gene Regulation in Eukaryotes*

? Review Figure 11.8. If the enzyme responsible for cleaving inactive insulin is deactivated, what effect will this have on the form and function of insulin?

■ The final molecule will have a different shape and therefore not be able to perform the proper function of insulin.

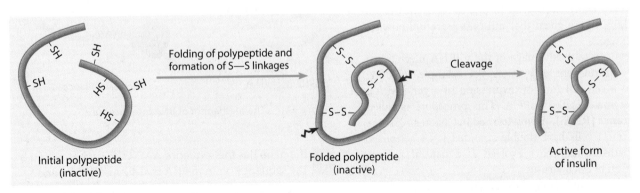

Figure 11.8   Protein activation: The role of polypeptide cleavage in producing the active insulin protein

Initial polypeptide (inactive)

Folding of polypeptide and formation of S—S linkages

Folded polypeptide (inactive)

Cleavage

Active form of insulin

11.9 Review: Multiple mechanisms regulate gene expression in eukaryotes

Figure 11.9 provides a review of eukaryotic gene expression and highlights the multiple control points where the process can be turned on or off, speeded up, or slowed down. Picture the series of pipes that carry water from your local water supply, perhaps a reservoir, to a faucet in your home. At various points, valves control the flow of water. We use this model in the figure to illustrate the flow of genetic information from a chromosome—a reservoir of genetic information—to an active protein that has been synthesized in the cell's cytoplasm. The multiple mechanisms that control gene expression are analogous to the control valves in water pipes. In the figure, each gene expression "valve" is indicated by a control knob. Note that these knobs represent *possible* control points; for most proteins, only a few control points are probably important. As we have seen, the most important control point, in both eukaryotes and prokaryotes, is usually the start of transcription. In the diagram, the large yellow knob represents the mechanisms that regulate the start of transcription.

After transcription, RNA processing in the nucleus adds nucleotides to the ends of the RNA (cap and tail) and splices out introns. As we discussed in Module 11.7, a growing body of evidence suggests the importance of control at this stage. Once mRNA reaches the cytoplasm, additional stages subject to regulation include mRNA translation and eventual breakdown, possible alteration of the polypeptide to give the active protein, and the eventual breakdown of the protein.

Despite its numerous steps, Figure 11.9 oversimplifies the control of gene expression. What it does not show is the web of control that connects different genes, often through their products. We have seen examples in both prokaryotes and eukaryotes of the actions of gene products (usually proteins) on other genes or on other gene products within the same cell. The genes of operons in *E. coli,* for instance, are controlled by repressor or activator proteins encoded by regulatory genes on the same DNA molecule. In eukaryotes, many genes are controlled by proteins encoded by regulatory genes on different chromosomes.

In eukaryotes, cellular differentiation results from the selective turning on and off of genes. In the next module, we will examine the stage in the life cycle of a multicellular eukaryote when cellular differentiation by selective gene expression is most vital: the development of a multicellular embryo from a unicellular zygote.

Web Activity Review: Gene Regulation in Eukaryotes

? Of the nine regulatory "valves" in Figure 11.9, which five can also operate in a prokaryotic cell?

■ (1) Control of transcription; (2) control of mRNA breakdown; (3) control of translation; (4) control of protein activation; and (5) control of protein breakdown.

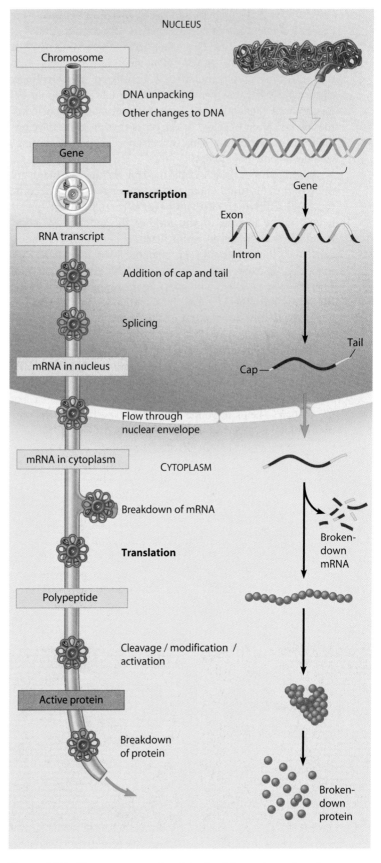

Figure 11.9 The gene expression "pipeline" in a eukaryotic cell

11.10 Cascades of gene expression direct the development of an animal

Some of the first glimpses into the relationship between gene expression and embryonic development came from studies of mutants of the fruit fly *Drosophila melanogaster* (see Module 9.18). Figure 11.10A, showing front views of the heads of two fruit flies, includes a mutant that developed in a strikingly abnormal way: It has two legs where its antennae should be! Research on this and other developmental mutants has led to the identification of many of the genes that program development in the normal fly. This genetic approach has revolutionized developmental biology.

Among the earliest events in fruit fly development are ones that determine which end of the egg cell will become the head and which end will become the tail. These events occur in the ovaries of the mother fly and involve communication between an unfertilized egg cell and cells adjacent to it in its follicle (egg chamber). As Figure 11.10B indicates, ❶ early genes in the egg produce proteins and mRNAs that establish which end of the embryo will become the head and which the tail. In a similar way, other egg cell genes direct the positioning of the top-to-bottom and side-to-side axes.

After the egg is fertilized and laid, repeated mitoses transform the zygote into an embryo. ❷ Translation within the early embryo produces further cascades of proteins that diffuse through the cell layers. Cell signaling—now among the cells of the embryo—helps drive the process. ❸ The result is the subdivision of the embryo's body into segments.

Now finer details of the fly can take shape. Protein products of some of the axis-specifying genes and segment-forming genes activate yet another set of genes. These genes, called homeotic genes, determine what body parts will develop from each segment. A **homeotic gene** is a master control gene that regulates batteries of other genes that actually determine the anatomy of parts of the body. For example, one set of homeotic genes in fruit flies instructs cells in the segments of the head and thorax (midbody) to form antennae and legs, respectively. Elsewhere, these homeotic genes remain turned off, while others are turned on. (To learn more about the evolutionary

significance of homeotic genes, see Module 27.14) ❹ The eventual outcome is an adult fly. Notice that the adult's body segments correspond to those of the embryo in step 3. It was mutation of a homeotic gene that was responsible for the abnormal fly in Figure 11.9A.

Cascades of gene expression, with the protein products of one set of genes activating another set of genes, and so on, are a common theme in development. Next, we'll look at how new DNA technologies can help elucidate gene expression in any type of cell.

Web Activity *Development of Head-Tail Polarity*

Web Process of Science *How Can the "Head" Gene Be Regulated to Alter Development?*

? What determines which end of a developing fruit fly will become the head?

■ A specific kind of mRNA localizes at the end of the unfertilized egg that will become the head.

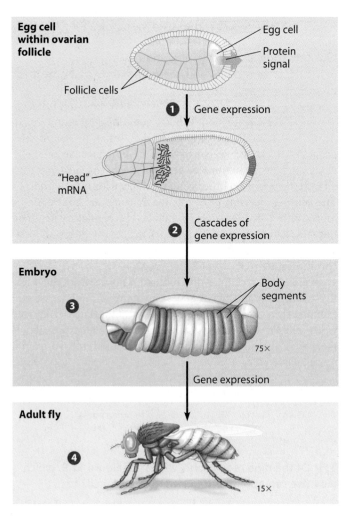

Figure 11.10B Key steps in the early development of head-tail polarity in a fruit fly

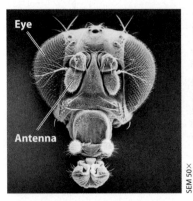

Head of a normal fruit fly   Head of a developmental mutant

Figure 11.10A   A mutant fruit fly with legs coming out of its head, compared with a normal fruit fly

11.11 DNA microarrays test for the transcription of many genes at once

A major goal of biologists is to learn how genes act together within a functioning organism. Now that a number of whole genomes have been sequenced, it is possible to study the expression of large groups of genes. Researchers can use gene sequences as probes to investigate which genes are transcribed in different situations, such as in different tissues or at different stages of development. They also look for groups of genes that are expressed in a coordinated manner, with the aim of identifying networks of gene expression across an entire genome.

Genome-wide expression studies are made possible by the use of DNA microarrays. A **DNA microarray** is a glass slide with thousands of different kinds of single-stranded DNA fragments fixed to it in a tightly spaced array, or grid. (A DNA microarray is also called a *DNA chip* or *gene chip* by analogy to a computer chip.) Each fixed DNA fragment is obtained from a particular gene; a single microarray thus carries DNA from thousands of genes, perhaps even all the genes of an organism.

Figure 11.11 outlines how microarrays are used. ❶ A researcher isolates all of the mRNA transcribed from genes in a particular type of cell. ❷ This collection of mRNAs is mixed with reverse transcriptase (see Module 10.20) to produce a mixture of DNA fragments. These fragments are called cDNAs (complementary DNAs) because each one is complementary to one of the mRNAs. The cDNAs are produced in the presence of nucleotides that have been modified to fluoresce (glow). The fluorescent cDNA collection thus represents all of the genes being transcribed in the cell under investigation. ❸ A small amount of the fluorescently labeled cDNA mixture is added to each of the fixed DNA fragments in the microarray. If a molecule in the cDNA mixture is complementary to a DNA fragment

at a particular location on the grid, the cDNA molecule binds to it, becoming fixed there. ❹ After nonbinding cDNA is rinsed away, the remaining cDNA produces a detectable glow in the microarray. The pattern of glowing spots enables the researcher to determine which genes are being transcribed in the starting cells.

DNA microarrays are a potential boon to medical research. For example, a 2002 study showed that DNA microarray data can classify different types of leukemia into specific subtypes based on the activity of 17 genes. This information can be used to predict which of several available regimens of chemotherapy is likely to be most effective. Further research suggests that many cancers have a variety of subtypes with different gene expressions that can be identified with DNA microarrays. Indeed, some oncologists predict that DNA microarrays will usher in a new era where medical treatment is customized to each patient.

DNA microarrays can also reveal general profiles of gene expression over the lifetime of an organism. In one example of a global expression study using this technique, researchers performed DNA microarray experiments on more than 90% of the genes of the nematode *C. elegans* during every stage of its life cycle. The results showed that expression of nearly 60% of the genes changed dramatically during development. This study supported the model held by most developmental biologists that embryonic development involves a complex and elaborate program of gene expression, rather than simple expression of a small number of important genes.

❓ What is learned from a DNA microarray?

■ What genes are active (transcribed) in a particular sample of cells

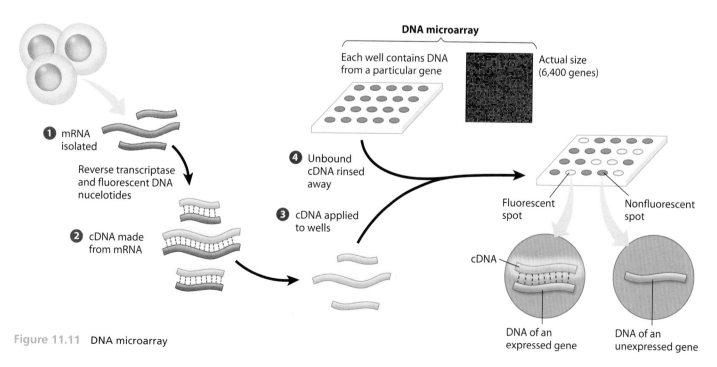

DNA microarray

Each well contains DNA from a particular gene

Actual size (6,400 genes)

❶ mRNA isolated

Reverse transcriptase and fluorescent DNA nucelotides

❷ cDNA made from mRNA

❹ Unbound cDNA rinsed away

❸ cDNA applied to wells

Fluorescent spot

Nonfluorescent spot

cDNA

DNA of an expressed gene

DNA of an unexpressed gene

Figure 11.11   DNA microarray

11.12 Signal transduction pathways convert messages received at the cell surface to responses within the cell

Cell-to-cell signaling, with proteins or other kinds of molecules carrying messages from signaling cells to receiving (target) cells, is a key mechanism in the coordination of cellular activities. In most cases, a signal molecule acts by binding to a receptor protein in the plasma membrane of the target cell and initiating a **signal transduction pathway** in the cell. A signal transduction pathway is a series of molecular changes that converts a signal on a target cell's surface to a specific response inside the cell.

Figure 11.12 shows the main elements of a signal transduction pathway in which the target cell's response is the transcription of a gene. ❶ The signaling cell secretes a signal molecule. ❷ This molecule binds to a receptor protein embedded in the target cell's plasma membrane. ❸ The binding activates the first in a series of relay proteins within the target cell. Each relay molecule activates another. ❹ The last relay molecule in the series activates a transcription factor that ❺ triggers transcription of a specific gene. ❻ Translation of the mRNA produces a protein.

Signal transduction pathways are crucial to many cellular functions. Throughout your study of biology, you'll see their importance again and again. We encountered them when we studied the cell cycle control system in Module 8.9; we'll revisit them when we discuss cancer later in this chapter; and we'll see how they are used to control hormone function in Chapters 26 and 33.

Web Activity *Signal Transduction Pathway*

? How can a signal molecule from one cell alter gene expression in a target cell without even entering the target cell?

By binding to a receptor protein in the membrane of the target cell and triggering a signal transduction pathway that activates transcription factors

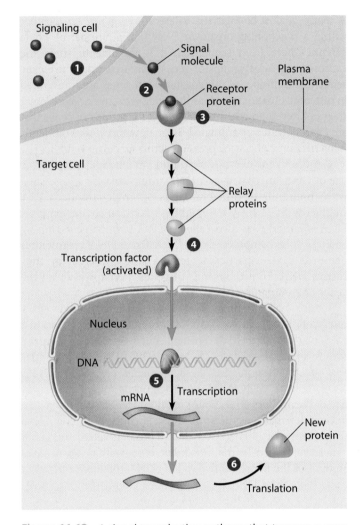

Figure 11.12 A signal transduction pathway that turns on a gene

11.13 Cell-signaling systems appeared early in the evolution of life

As explained in Module 11.12, one cell can communicate with another by secreting molecules that bind to surface proteins on a target cell. How ancient and widespread are such signaling systems among Earth's organisms? To answer these questions we can look at communication among microorganisms, for modern microbes are a window on the role of cell signaling in the evolution of life on Earth.

One topic of cell "conversation" is sex—at least for the yeast *Saccharomyces cerevisiae,* which people have used for millenia to make bread, wine, and beer. Researchers have learned that cells of this yeast identify their mates by chemical signaling. There are two sexes, or mating types, called **a** and **α** (Figure 11.13, top of facing page). Cells of mating type **a** secrete a chemical signal called **a** factor, which can bind to

specific receptor proteins on nearby **α** cells. At the same time, **α** cells secrete **α** factor, which binds to receptors on **a** cells. Without actually entering the target cells, the two mating factors cause the cells to grow toward each other and bring about other cellular changes. The result is the fusion, or mating, of two cells of opposite type. The new **a/α** cell contains all the genes of both original cells, a combination of genetic resources that provides advantages to the cell's descendants, which arise by subsequent cell divisions.

The yeast mating pathway is just one of many signal transduction pathways that have been extensively studied in both yeast and animal cells. Amazingly, the molecular details of signal transduction in yeast and mammals are strikingly similar, even though the last common ancestor of

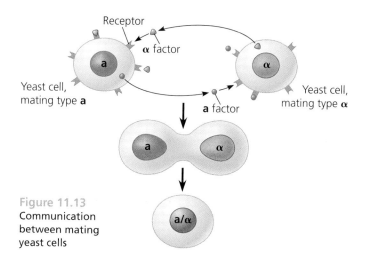

Receptor
α factor
Yeast cell, mating type **a**
a factor
Yeast cell, mating type **α**

Figure 11.13 Communication between mating yeast cells

these two groups of organisms lived over a billion years ago. These similarities—and others more recently uncovered between signaling systems in bacteria and plants—suggest that early versions of the cell-signaling mechanisms used today evolved well before the first multicellular creatures appeared on Earth. Scientists think that signaling mechanisms evolved first in ancient prokaryotes and single-celled eukaryotes and then became adapted for new uses in their multicellular descendants.

? In what sense is the joining of yeast mating types "sex"?

■ The process results in the creation of a diploid cell that is a genetic blend of two parental haploid cells.

Cloning of Plants and Animals

11.14 Plant cloning shows that differentiated cells may retain all of their genetic potential

One of the most important "take home lessons" from this chapter is that differentiated cells express only a small percentage of their genes. So then how do we know that all the genes are still present? And if all the genes are still there, do differentiated cells retain the potential to express them?

One way to approach these questions is to see if a differentiated cell can generate a whole new organism. In plants, this ability is common, as was first demonstrated during the 1950s by F. C. Steward and his students at Cornell University. As shown in Figure 11.14, they found that when they transferred cells from a carrot to a culture medium, a single cell could begin dividing and eventually grow into an adult plant, a genetic replica of the parent plant. The fact that a mature plant cell can dedifferentiate (reverse its differentiation) and then give rise to all the different kinds of specialized cells of a new plant shows that differentiation does not necessarily involve irreversible changes in the plant's DNA.

Plant cloning is now used extensively in agriculture. For some plants, such as orchids, cloning is the only commercially practical means of reproducing plants. In other cases, cloning has been used to reproduce a plant with valuable characteristics, such as the ability to resist a plant pathogen. You yourself may be a plant cloner if you have ever grown a new plant from a cutting.

But is this sort of cloning possible in animals? An indication that differentiation need not impair an animal cell's genetic potential is the natural process of **regeneration,** the regrowth of lost body parts. When a salamander loses a leg, for example, certain cells in the leg stump dedifferentiate, divide, and then

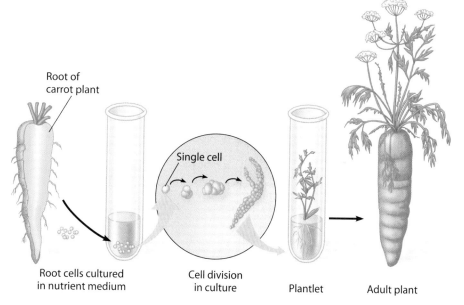

Root of carrot plant

Single cell

Root cells cultured in nutrient medium

Cell division in culture

Plantlet

Adult plant

Figure 11.14 Growth of a carrot plant from a differentiated root cell

redifferentiate, giving rise to a new leg. Many animals can regenerate lost parts, especially among the invertebrates, and in a few relatively simple animals, isolated differentiated cells can dedifferentiate and then develop into an organism. Further evidence for the complete genetic potential of animal cells comes from cloning experiments, our next topic.

? How does the cloning of plants from differentiated cells support the view that differentiation is based on the control of gene expression rather than on irreversible changes in the genome?

■ Cloning shows that all the genes of a fully differentiated plant are still present, but some may be turned off.

11.15 Nuclear transplantation can be used to clone animals

Animal cloning, including the various cases described in the chapter introduction, is achieved through a procedure called **nuclear transplantation** (Figure 11.15). First performed in the 1950s using cells from frog embryos, nuclear transplantation involves replacing the nucleus of an egg cell or a zygote with the nucleus of an adult somatic cell. The recipient cell may then begin to divide. About 5 days later, repeated cell divisions form a blastocyst, a ball of cells. At this point, the blastocyst may be used for different purposes, as indicated by the branching in Figure 11.15.

If the animal to be cloned is a mammal, further development requires implanting the blastocyst into the uterus of a surrogate mother (Figure 11.15, upper branch). The resulting animal will be genetically identical to the donor of the nucleus—a "clone" of the donor. This type of cloning, which results in birth of a new individual, is called **reproductive cloning.** Scottish researcher Ian Wilmut and his colleagues used reproductive cloning to produce the celebrated sheep Dolly. A fully differentiated mammary cell of an adult ewe was the source of the nucleus.

In a different cloning procedure (Figure 11.15, lower branch), **embryonic stem (ES) cells** are harvested from the blastocyst. In nature, embryonic stem cells give rise to all the different kinds of specialized cells of the body. In the laboratory, embryonic stem cells are easily grown in culture, where, given the right conditions, they can perpetuate themselves indefinitely. When the major aim is to produce embryonic stem cells for therapeutic treatments, the process is called **therapeutic cloning.** In the next two modules, we will discuss applications of reproductive and therapeutic cloning.

? What are the intended products of reproductive cloning and therapeutic cloning?

∎ Reproductive cloning is used to produce new individuals. Therapeutic cloning is used to harvest embryonic stem cells.

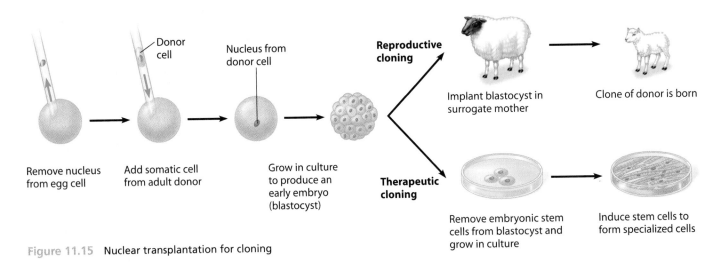

Figure 11.15   Nuclear transplantation for cloning

Remove nucleus from egg cell

Add somatic cell from adult donor

Grow in culture to produce an early embryo (blastocyst)

Reproductive cloning

Implant blastocyst in surrogate mother

Clone of donor is born

Therapeutic cloning

Remove embryonic stem cells from blastocyst and grow in culture

Induce stem cells to form specialized cells

Donor cell

Nucleus from donor cell

Connection

11.16 Reproductive cloning has valuable applications, but human reproductive cloning raises ethical issues

Since Dolly's birth in 1997, researchers have cloned many other mammals, including mice, cats, horses, cows, mules, pigs, and dogs. We have already learned much from such experiments. For example, cloned animals of the same species do not always look or behave identically. In a herd of cows cloned from the same line of cultured cells, certain cows are dominant and others are more submissive. Another example is the first cloned cat, named CC for Carbon Copy (Figure 11.16). She has a calico coat, like her single female parent, but the color and pattern are different due to random X chromosome inactivation (see Module 11.4). Besides appearance, CC and her lone parent behave differently; CC is playful while her mother is more reserved. You probably know that identical human twins, which are naturally occurring "clones," are always slightly different. Clearly, environmental influences and random phenomena can play a significant role during development.

Reproductive cloning has potential for many practical applications. On an experimental basis, agricultural scientists are cloning farm animals with specific sets of desirable traits in the hope of creating high-yielding, genetically identical herds. The pharmaceutical industry is experimentally cloning mammals for the production of potentially valuable drugs. Figure 12.6 shows cloned sheep that secrete a human blood protein into their milk, one that could prove useful in treating cystic fibrosis. Other researchers have produced piglet clones that lack one of two copies of a gene for a protein that can cause immune system rejection in humans. Such pigs may one day provide organs for transplant into humans. Some wildlife biologists hope that reproductive cloning can be used to restock the populations of endangered animals (although others caution against such an approach, as discussed in the chapter introduction).

The successful cloning of various mammals has heightened speculation that humans could be cloned. Critics point out that there are many obstacles—both practical and ethical—to human cloning. Practically, animal cloning is extremely difficult and inefficient. Only a small percentage of cloned embryos develop normally. Ethically, creating human embryos for such purposes raises many troubling questions. An ethical consensus on the status of the human blastocyst is unlikely any time soon. Meanwhile, the research and the debate continue.

? If you cloned your dog, would you expect the mother and baby to look and act just alike?

■ No. While cloning produces genetically identical dogs, appearance and behavior are affected by environment.

Figure 11.16   CC, the world's first cloned cat (right), and her lone parent (left)

Connection

11.17   Therapeutic cloning can produce stem cells with great medical potential

Therapeutic cloning produces embryonic stem cells, cells that in the early animal embryo differentiate to give rise to all the cell types in the body. When grown in laboratory culture, embryonic stem cells can divide indefinitely (like cancer cells; see Module 8.10). Furthermore, the right conditions can induce changes in gene expression that cause differentiation into a variety of cell types (Figure 11.17).

The adult body also has stem cells, which serve to replace nonreproducing specialized cells as needed. In contrast to embryonic stem cells, **adult stem cells** are able to give rise to many but not all cell types in the organism. For example, bone marrow contains several types of stem cells, including one that can generate all the different kinds of blood cells. Although adult animals have only tiny numbers of stem cells, scientists are learning to identify and isolate these cells from various tissues and, in some cases, to grow them in culture.

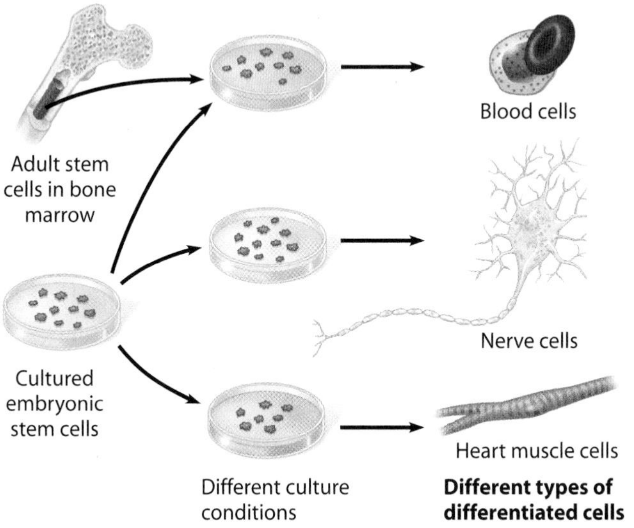

The ultimate aim of therapeutic cloning is to supply cells for the repair of damaged or diseased organs: for example, insulin-producing pancreatic cells for people with diabetes or certain kinds of brain cells for people with Parkinson's disease or Huntington's disease. Adult stem cells from donor bone marrow have long been used as a source of immune system cells in patients whose own immune systems are destroyed by genetic disorders or radiation treatments for cancer. More recently, clinical trials using bone marrow stem cells have shown slight success in promoting regeneration of heart tissue in patients whose hearts have been damaged by heart attacks.

The developmental potential of adult stem cells is limited to certain tissues, so embryonic stem cells are considered more promising than adult stem cells for medical applications, at least for now. However, the derivation of embryonic stem cells from donated human embryos raises ethical and political issues. In late 2007, two research groups discovered a procedure for using viruses to turn human skin cells into ES-like cells with the potential to develop into many kinds of body cells. The procedure does not involve the destruction of embryos. In the future, a donor nucleus from a patient with a particular disease could allow production of embryonic stem cells for treatment that match the patient and are thus not rejected by his or her immune system.

While many people believe that reproductive cloning of humans is unethical, opinions vary more widely about the morality of therapeutic cloning using embryonic stem cells. As with reproductive cloning, the research and the debate continue.

? In nature, how do embryonic stem cells differ from adult stem cells?

■ Embryonic cells give rise to all the different kinds of cells in the body. Adult stem cells generate only a few related types of cells.

Blood cells

Adult stem cells in bone marrow

Nerve cells

Cultured embryonic stem cells

Heart muscle cells

Different culture conditions

Different types of differentiated cells

Figure 11.17   Differentiation of stem cells in culture

11.18 Cancer results from mutations in genes that control cell division

In Chapter 8, you learned that cancerous cells have escaped from the control mechanisms that normally limit their growth. Scientists have learned that such escape is often due to changes in gene expression.

The abnormal behavior of cancer cells was observed years before anything was known about the cell cycle, its control, or the role genes play in making cells cancerous. One of the earliest clues to the cancer puzzle was the discovery, in 1911, of a virus that causes cancer in chickens. Recall that viruses are simply molecules of DNA or RNA coated with protein and in some cases a membranous envelope. Viruses that cause cancer can become permanent residents in host cells by inserting their nucleic acid into the DNA of host chromosomes. Researchers have identified several examples of viruses that harbor cancer-causing genes. When inserted into a host cell, these genes can make the cell cancerous. Such a gene, which can cause cancer when present in a single copy in the cell, is called an **oncogene** (from the Greek *onkos,* tumor).

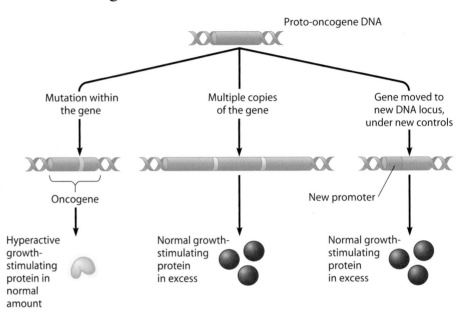

Figure 11.18A Alternative ways to make oncogenes from a proto-oncogene (all leading to excessive cell growth)

Proto-Oncogenes In 1976, American molecular biologists J. Michael Bishop, Harold Varmus, and their colleagues made a startling discovery. They found that a virus that causes cancer in chickens contains an oncogene that is an altered version of a gene found in normal chicken cells. Subsequent research has shown that the chromosomes of many animals, including humans, contain genes that can be converted to oncogenes. A normal gene that has the potential to become an oncogene is called a **proto-oncogene.** Thus, a cell can acquire an oncogene either from a virus or from the mutation of one of its own genes.

The work by Bishop and Varmus focused cancer research on proto-oncogenes. Searching for the normal role of these genes, researchers found that many proto-oncogenes code for growth factors—proteins that stimulate cell division—or for other proteins that somehow affect growth factor function or some other aspect of the cell cycle. When all these proteins are functioning normally, in the right amounts at the right times, they help control cell division and cellular differentiation.

How might a proto-oncogene—a gene that has an essential function in normal cells—become an oncogene, a cancer-causing gene? In general, an oncogene arises from a genetic change that leads to an increase either in the amount of the proto-oncogene's protein product or in the rate of activity of each protein molecule. Figure 11.18A illustrates three kinds of changes in somatic (non-gamete-forming) cell DNA that can produce active oncogenes. Let's assume that the starting proto-oncogene codes for a protein that stimulates cell division. On the left in the figure, a mutation (green) in the proto-oncogene itself creates an oncogene that codes for a hyperactive protein, one whose stimulating effect is stronger than normal. In the center, an error in DNA replication or recombination generates multiple copies of the gene, which are all transcribed and translated; the result is an excess of the normal stimulatory protein. On the right, the proto-oncogene has been moved from its normal location in the cell's DNA to another location. At its new site, the gene is under the control of a different promoter, one that causes it to be transcribed more often than normal; and the normal protein is again made in excess. So in all three cases, normal gene expression is changed, and the cell is stimulated to divide excessively.

Tumor-Suppressor Genes In addition to genes whose products normally *promote* cell division, cells contain genes whose normal products *inhibit* cell division. Such genes are called **tumor-suppressor genes** because the proteins they encode help prevent uncontrolled cell growth. Any mutation that decreases the normal activity of a tumor-suppressor protein may contribute to the onset of cancer, in effect stimulating growth through the absence of suppression (Figure 11.18B, top of the facing page). Recently, scientists have discovered a new class of tumor-suppressor genes that function in the repair of damaged DNA. When these genes are mutated, other cancer-causing mutations are more likely to accumulate.

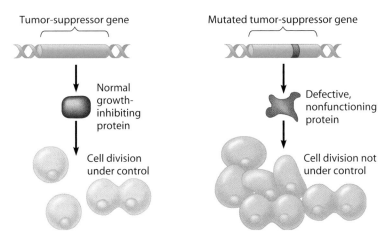

Figure 11.18B The effect of a mutation in a tumor-suppressor gene

11.19 Multiple genetic changes underlie the development of cancer

Over 150,000 Americans will be stricken by cancer of the colon (large intestine) or rectum this year, perhaps including some of your own relatives or friends. One of the best-understood types of human cancer, colon cancer illustrates an important principle about how cancer develops: *More than one somatic mutation is needed to produce a full-fledged cancer cell.* As in many cancers, the development of malignant colon cancer is gradual. (See Module 8.10 to review cancer terms.)

The fact that more than one somatic mutation is generally needed to produce a full-fledged cancer cell may help explain why the incidence of cancer increases greatly with age. If cancer results from an accumulation of mutations and if mutations occur throughout life, then the longer we live, the more likely we are to develop cancer.

Figure 11.19A illustrates this idea using colorectal cancer as an example. ❶ Colorectal cancer begins when an oncogene is activated, causing unusually frequent division of apparently normal cells in the colon lining. ❷ Later, additional DNA mutations (such as the inactivation of a tumor-suppressor gene) causes the growth of a small benign tumor (polyp) in the colon wall. ❸ Still more mutations eventually lead to formation of a malignant tumor (a carcinoma). The requirement for several mutations—the actual number is usually four or more—explains why cancers can take a long time to develop.

The development of a malignant tumor is paralleled by a gradual accumulation of mutations that convert proto-oncogenes to oncogenes and knock out tumor-suppressor genes. Multiple changes must occur at the DNA level for a cell to become fully cancerous. Such changes usually include the appearance of at least one active oncogene and the mutation or loss of several tumor-suppressor genes. In Figure 11.19B, colors distinguish the normal cell (tan) from cells (red) with one or more mutations leading to increased cell division and cancer. Once a cancer-promoting mutation occurs (red band on chromosome), it is passed to all the descendants of the cell carrying it. Cancer researchers hope to learn about mutations that cause cancer through the Cancer Genome Project, a 10-year multi-million dollar effort to map all human cancer-causing genes.

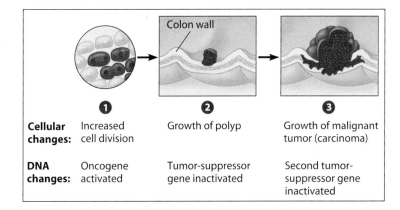

Figure 11.19A Stepwise development of a typical colon cancer

	❶	❷	❸
Cellular changes:	Increased cell division	Growth of polyp	Growth of malignant tumor (carcinoma)
DNA changes:	Oncogene activated	Tumor-suppressor gene inactivated	Second tumor-suppressor gene inactivated

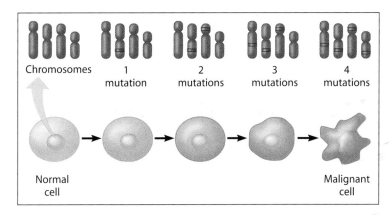

Figure 11.19B Accumulation of mutations in the development of a cancer cell

11.20 Faulty proteins can interfere with normal signal transduction pathways

To understand how oncogenes and defective tumor-suppressor genes can contribute to uncontrolled cell growth, we need to look more closely at the normal functions of proto-oncogenes and tumor-suppressor genes. Genes in both categories often code for proteins involved in signal transduction pathways leading to gene expression (see Module 11.12).

The figures below (excluding, for the moment, the white boxes) illustrate two types of signal transduction pathways leading to the synthesis of proteins that influence the cell cycle. In Figure 11.20A, the pathway leads to the stimulation of cell division. The initial signal is a growth factor (●), and the target cell's ultimate response is the production of a protein that stimulates the cell to divide. By contrast, Figure 11.20B shows an inhibitory pathway, in which a growth-*inhibiting* factor (▽) causes the target cell to make a protein that inhibits cell division. In both cases, the newly made proteins function by interacting with components of the cell cycle control system (see Module 8.9). The figures here do not show these interactions.

Now let's see what can happen when the target cell undergoes a cancer-causing mutation. The white box in Figure 11.20A shows the protein product of an oncogene resulting from mutation of a proto-oncogene called *ras*. The normal product of *ras* is a relay protein. Ordinarily, a stimulatory pathway like this will not operate unless the growth factor is available. However, an oncogene protein that is a hyperactive version of a protein in the pathway may trigger the pathway even in the absence of a growth factor. In this example, the oncogene protein is a hyperactive version of the *ras* relay protein that issues signals on its own. In fact, abnormal versions or amounts of any of the pathway's components—from the growth factor itself to the transcription factor—could have the same final effect: overstimulation of cell division.

The white box in Figure 11.20B indicates how a mutant tumor-suppressor protein can affect cell division. In this case, the mutation affects a gene called *p53*, which codes for the transcription factor. This mutation leads to the production of a faulty transcription factor, one that the signal transduction pathway cannot activate. As a result, the gene for the inhibitory protein at the bottom of the figure remains turned off, and excessive cell division may occur.

Mutations of the *ras* and *p53* genes have been implicated in many kinds of cancer. In fact, mutations in *ras* occur in about 30% of human cancers, and mutations in *p53* occur in more than 50%. As we see next, carcinogens are responsible for causing many mutations that lead to cancer.

? Contrast the action of an oncogene with that of a cancer-causing mutation in a tumor-suppressor gene.

■ An oncogene produces an abnormal protein that stimulates cell division via a signal transduction pathway; a mutant tumor-suppressor gene produces a defective protein unable to function in a pathway that normally inhibits cell division.

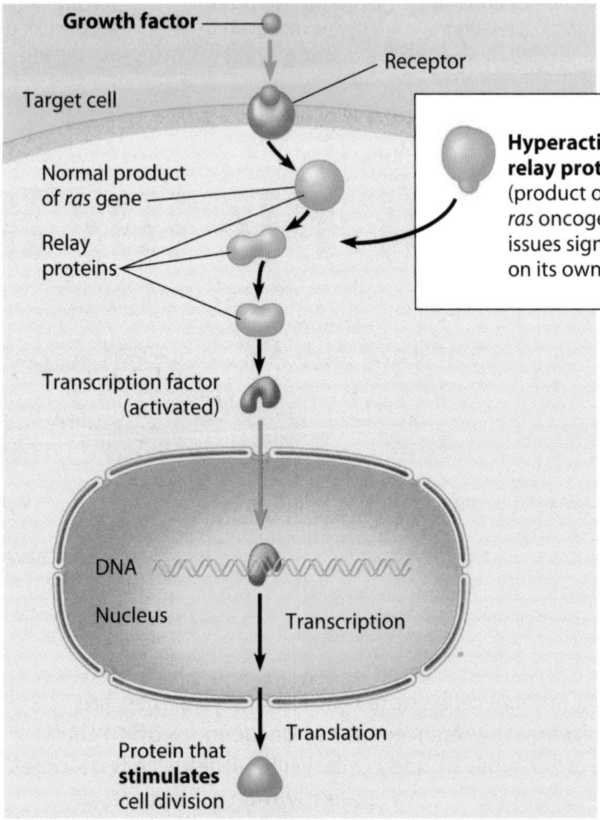

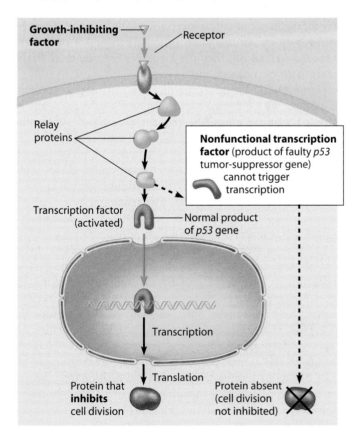

Figure 11.20A   A stimulatory signal transduction pathway and the effect of an oncogene protein

Figure 11.20B   An inhibitory signal transduction pathway and the effect of a faulty tumor-suppressor protein

11.21 Lifestyle choices can reduce the risk of cancer

Cancer is the second-leading cause of death in the United States, exceeded only by heart disease. Death rates due to certain forms of cancer (including stomach, cervical, and uterine cancers) have decreased in recent years, but the overall cancer death rate is still on the rise, currently increasing at about 1% per decade. Table 11.21 lists the most common cancers in the United States and associated risk factors for each.

The fact that multiple genetic changes are required to produce a cancer cell helps explain the observation that cancers can run in families. An individual inheriting an oncogene or a mutant allele of a tumor-suppressor gene is one step closer to accumulating the necessary mutations for cancer to develop than an individual without any such mutations.

But the majority of cancers cannot be associated with a mutation that is passed from parent to offspring; they arise from new mutations caused by environmental factors. Cancer-causing agents, factors that alter DNA and make cells cancerous, are called **carcinogens.** Most mutagens (substances that promote mutations) are carcinogens. Two of the most potent carcinogens (and mutagens) are X-rays and ultraviolet radiation in sunlight. X-rays are a significant cause of leukemia and brain cancer. Exposure to UV radiation from the sun is known to cause skin cancer, including a deadly type called melanoma.

The one substance known to cause more cases and types of cancer than any other single agent is tobacco. More people die of lung cancer (over 160,000 Americans in 2007) than any other form of cancer. Most tobacco-related cancers come from actually smoking, but the passive inhalation of secondhand smoke is also a risk. As Table 11.21 indicates, tobacco use, sometimes in combination with alcohol consumption, causes a number of other types of cancer in addition to lung cancer. In nearly all cases, cigarettes are the main culprit, but smokeless tobacco products (snuff and chewing tobacco) are linked to cancer of the mouth and throat.

How do carcinogens cause cancer? In many cases, the genetic changes that cause cancer result from decades of exposure to the mutagenic effects of carcinogens. Carcinogens can also produce their effect by promoting cell division. Generally, the higher the rate of cell division, the greater the chance for mutations resulting from errors in DNA replication or recombination. Some carcinogens seem to have both effects. For instance, the hormones that cause breast and uterine cancers promote cell division and may also cause genetic changes that lead to cancer. In other cases, several different agents, such as viruses and one or more carcinogens, may together produce cancer.

Avoiding carcinogens is not the whole story, for there is growing evidence that some food choices significantly reduce cancer risk. For instance, eating 20–30 grams (g) of plant fiber daily (about twice the amount the average American consumes) and at the same time reducing animal fat intake may help prevent colon cancer. There is also evidence that other substances in fruits and vegetables, including vitamins C and E and certain compounds related to vitamin A, may offer protection against a variety of cancers. Cabbage and its relatives, such as broccoli and

TABLE 11.21	CANCER IN THE UNITED STATES	
Cancer	Risk Factors	Estimated Number of Cases in 2007
Prostate	African heritage; possibly dietary fat	218,900
Lung	Tobacco smoke	213,400
Breast	Estrogen	180,500
Colon, rectum	High dietary fat; smoking; alcohol	153,800
Lymphomas	Viruses (for some types)	71,400
Urinary bladder	Cigarette smoke	67,200
Melanoma of skin	Ultraviolet light	59,900
Kidney	Cigarette smoke	51,200
Leukemias	X-rays; benzene; virus (for one type)	44,200
Uterus	Estrogen	39,000
Pancreas	Tobacco smoke; obesity	37,200
Mouth and throat	Tobacco in various forms; alcohol	34,400
Ovary	Obesity; many ovulation cycles	22,400
Stomach	Table salt; cigarette smoke	21,300
Liver	Alcohol; hepatitis viruses	19,200
Brain and nerve	Trauma; X-rays	20,500
Cervix	Sexually transmitted viruses; tobacco	11,200
All others		179,400

cauliflower (see Figure 13.2), are thought to be especially rich in substances that help prevent cancer, although the identities of these substances are not yet established. Determining how diet influences cancer has become a major research goal.

The battle against cancer is being waged on many fronts, and there is reason for optimism in the progress being made. It is especially encouraging that we can help reduce our risk of acquiring and increase our chance of surviving some of the most common forms of cancer by the choices we make in our daily life. Not smoking, exercising adequately, avoiding overexposure to the sun, and eating a high-fiber, low-fat diet can all help prevent cancer. Furthermore, seven types of cancer can be easily detected: skin and oral (via physical exam), breast (via self-exams and mammograms for higher-risk women), prostate (via rectal exam), cervical (via Pap smear), testicular (via self-exam), and colon (via colonoscopy) cancers. Regular visits to the doctor can help identify tumors early, thereby significantly increasing the possibility of successful treatment.

Web Activity *Connection: Causes of Cancer*

? Which type of cancer is the most frequently diagnosed in the United States? Which causes the most deaths?

▪ Prostate; lung

Chapter Review

Reviewing the Concepts

A clone is an individual created by asexual reproduction and thus genetically identical to a single parent (**Introduction**).

Control of Gene Expression (11.1–11.13)

In prokaryotes, genes for related enzymes are often controlled together in units called operons. Regulatory proteins bind to control sequences in the DNA and turn operons on or off in response to environmental changes (**11.1**):

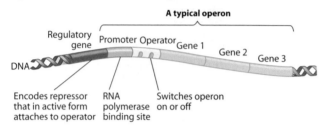

A typical operon

Regulatory gene • Promoter • Operator • Gene 1 • Gene 2 • Gene 3 • DNA

Encodes repressor that in active form attaches to operator | RNA polymerase binding site | Switches operon on or off

In multicellular eukaryotes, cells become specialized, or differentiated, as a zygote develops into a mature organism. Different types of cells make different proteins because different combinations of genes are active in each type (**11.2**). A chromosome contains DNA wound around clusters of histone proteins, forming a string of bead-like nucleosomes. This beaded fiber is further wound and folded. DNA packing tends to block gene expression, presumably by preventing access of transcription proteins to the DNA (**11.3**). An extreme example of DNA packing in interphase cells is X chromosome inactivation in the cells of female mammals (**11.4**). A variety of regulatory proteins interact with DNA and with each other to turn the transcription of eukaryotic genes on or off (**11.5**). After transcription, alternative RNA splicing may generate two or more types of mRNA from the same transcript (**11.6**). Small RNAs help regulate gene expression in a variety of ways (**11.7**). The lifetime of an mRNA molecule helps determine how much protein is made, as do protein factors involved in translation. A protein may need to be activated in some way, and eventually the cell will break it down (**11.8**). Figure 11.9 reviews the multiple stages of eukaryotic gene expression, each stage offering an opportunity for regulation (**11.9**).

Cascades of gene expression control the development of an animal from a fertilized egg (**11.10**). A DNA microarray can test for the transcription of many genes at once (**11.11**).

Signal transduction pathways convert molecular messages to cell responses (**11.12**). Similarities among organisms suggest that signal transduction pathways evolved early in the history of life (**11.13**).

Cloning of Plants and Animals (11.14–11.17)

Differentiated cells retain a complete set of genes (**11.14**).

Nuclear transplantation is used to clone animals (**11.15**):

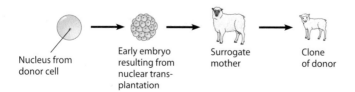

Nucleus from donor cell → Early embryo resulting from nuclear transplantation → Surrogate mother → Clone of donor

Reproductive cloning of nonhuman mammals has practical applications (**11.16**). The goal of therapeutic cloning is to produce embryonic stem cells. Like embryonic stem cells, adult stem cells can both perpetuate themselves in culture and give rise to differentiated cells. Unlike embryonic stem cells, however, adult stem cells normally give rise to only a limited range of cell types (**11.17**):

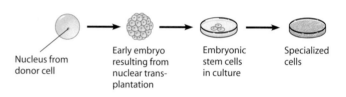

Nucleus from donor cell → Early embryo resulting from nuclear transplantation → Embryonic stem cells in culture → Specialized cells

The Genetic Basis of Cancer (11.18–11.21)

Cancer cells, which divide uncontrollably, result from mutations in genes whose protein products affect the cell cycle. A mutation can change a proto-oncogene (a normal gene that promotes cell division) into an oncogene, which causes cells to divide excessively. Mutations that inactivate tumor-suppressor genes have similar effects (**11.18**). Cancers result from a *series* of genetic changes in a cell lineage (**11.19**). Many proto-oncogenes and tumor-suppressor genes code for proteins active in signal transduction pathways regulating cell division (**11.20**). Reducing exposure to carcinogens (which induce cancer-causing mutations) and making other lifestyle choices can help reduce cancer risk (**11.21**).

Connecting the Concepts

1. Complete the following concept map to test your knowledge of gene regulation.

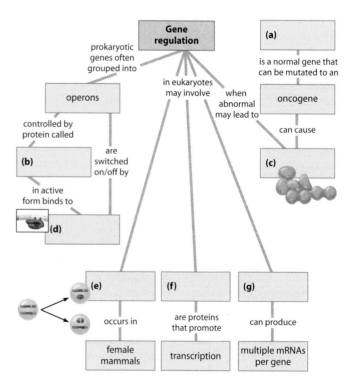

Testing Your Knowledge

Multiple Choice

2. The control of gene expression is more complex in multi-cellular eukaryotes than in prokaryotes because_____.
   (*Explain your answer.*)
   a. eukaryotic cells are much smaller
   b. in a multicellular eukaryote, different cells are specialized for different functions
   c. prokaryotes are restricted to stable environments
   d. eukaryotes have fewer genes, so each gene must do several jobs
   e. eukaryotic genes code for proteins

3. Your bone cells, muscle cells, and skin cells look different because
   a. each cell contains different kinds of genes.
   b. they are present in different organs.
   c. different genes are active in each kind of cell.
   d. they contain different numbers of genes.
   e. each cell has different mutations.

4. Which of the following methods of gene regulation do eukaryotes and prokaryotes have in common?
   a. elaborate packing of DNA in chromosomes
   b. activator and repressor proteins, which attach to DNA
   c. the addition of a cap and tail to mRNA after transcription
   d. *lac* and *trp* operons
   e. the removal of noncoding portions of RNA

5. A eukaryotic gene was inserted into the DNA of a bacterium. The bacterium then transcribed this gene into mRNA and translated the mRNA into protein. The protein produced was useless; it contained many more amino acids than the protein made by the eukaryotic cell. Why?
   a. The mRNA was not spliced, as it is in eukaryotes.
   b. Eukaryotes and prokaryotes use different genetic codes.
   c. Repressor proteins interfered with transcription and translation.
   d. The lifetime of the bacterial mRNA was too short.
   e. Ribosomes were not able to bind to the mRNA.

6. A homeotic gene does which of the following?
   a. It serves as the ultimate control for prokaryotic operons.
   b. It regulates the expression of groups of other genes during development.
   c. It represses the histone proteins that package eukaryotic DNA.
   d. It helps splice mRNA after transcription.
   e. It inactivates one of the X chromosomes in a female mammal.

7. All your cells contain proto-oncogenes, which can change into cancer-causing genes. Why do cells possess such potential time bombs?
   a. Viruses infect cells with proto-oncogenes.
   b. Proto-oncogenes are genetic "junk" with no known function.
   c. Proto-oncogenes are unavoidable environmental carcinogens.
   d. Cells produce proto-oncogenes as a by-product of mitosis.
   e. Proto-oncogenes normally control cell division.

8. Which of the following is a valid difference between embryonic stem cells and the stem cells found in adult tissues?
   a. In laboratory culture, only adult stem cells are immortal.
   b. In nature, only embryonic stem cells give rise to all the different types of cells in the organism.
   c. Only adult stem cells can differentiate in culture.
   d. Embryonic stem cells are generally more difficult to grow in culture than adult stem cells.
   e. Only embryonic stem cells are found in every tissue of the adult body.

Describing, Comparing, and Explaining

9. A mutation in a single gene may cause a major change in the body of a fruit fly, such as an extra pair of legs or wings. Yet it probably takes the combined action of hundreds or thousands of genes to produce a wing or leg. How can a change in just one gene cause such a big change in the body?

Applying the Concepts

10. Study the illustrations of the *lac* operon in Module 11.1. Normally, the genes are turned off when lactose is not present. Lactose activates the genes, which code for enzymes that enable the cell to use lactose. Mutations can alter the function of this operon; in fact, it was the effects of various mutations that enabled Jacob and Monod to figure out how the operon works. Predict how the following mutations would affect the function of the operon in the presence and absence of lactose:
    a. Mutation of regulatory gene; repressor will not bind to lactose.
    b. Mutation of operator; repressor will not bind to operator.
    c. Mutation of regulatory gene; repressor will not bind to operator.
    d. Mutation of promoter; RNA polymerase will not attach to promoter.

11. A chemical called dioxin is produced as a by-product of some chemical manufacturing processes. Trace amounts of this substance were present in Agent Orange, a defoliant sprayed on vegetation during the Vietnam War. There has been a continuing controversy over its effects on soldiers exposed to it during the war. Animal tests have suggested that dioxin can be lethal and can cause birth defects, cancer, liver and thymus damage, and immune system suppression. But its effects on humans are unclear, and even animal tests are inconclusive; a hamster is not affected by a dose that can kill a guinea pig. Researchers have discovered that dioxin enters a cell and binds to a protein that in turn attaches to the cell's DNA. How might this mechanism help explain the variety of dioxin's effects on different body systems and in different animals? How might you determine whether a particular individual became ill as a result of exposure to dioxin?

Answers to all questions can be found in Appendix 4.

For Practice Quizzes, BioFlix, MP3 Tutors, and Activities, go to www.mybiology.com.

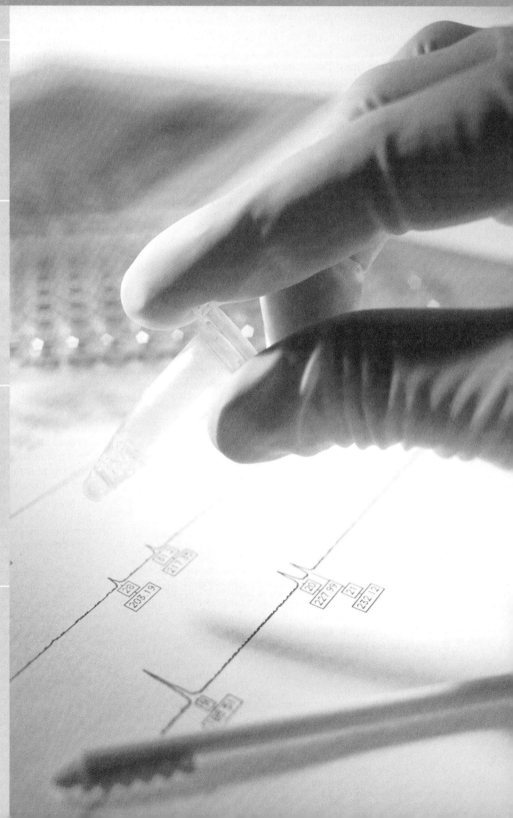

chapter 12

DNA Technology and Genomics

DNA and Crime Scene Investigations

On November 22, 1983, the sleepy English village of Narborough awoke to news of a horrific crime: A 15-year-old girl had been raped and murdered on a country lane. Despite an extensive investigation and the presence of semen on the victim's body and clothes, the trail of evidence ran cold and the crime went unsolved.

Three years later, the horror resurfaced when another 15-year-old girl was raped and murdered less than a mile away from the first crime scene. When tests indicated that semen from the new crime scene could match the 1983 sample, police began to search for a double murderer. After another extensive investigation, a local man was arrested and charged with both crimes. Under considerable pressure from police, the suspect confessed to the second murder, but denied committing the first.

In an attempt to pin both murders on the suspect, investigators turned to Alec Jeffreys, a biology professor at nearby Leicester University, who had recently developed the first DNA identification system. Because the DNA sequence of every person is unique (except for identical twins), DNA analysis can be used to determine with near certainty whether two DNA samples are from the same individual. Jeffreys compared DNA from the 1983 and 1986 semen samples. The DNA analysis confirmed that the same person had committed both crimes. However, when Jeffreys analyzed the suspect's DNA, it did not match either crime scene sample, proving the suspect's innocence. He was released, becoming the first person ever to be exonerated by DNA evidence.

The police were back at square one. In an attempt to collect more evidence, they asked every young male from the surrounding area to donate blood for DNA testing. Although nearly 5,000 men were sampled, none matched the evidence from the crime scenes. The case finally broke when a pub-goer described how a local named Colin Pitchfork had bullied him into submitting

blood on Pitchfork's behalf. The police arrested Pitchfork (above) and took a sample of his blood. Indeed, his DNA matched the samples from the two crime scenes. Colin Pitchfork pleaded guilty to both crimes, closing the first murder case ever to be solved by DNA evidence.

The Narborough murders were the first of a great many criminal investigations that have relied on DNA evidence. **DNA technology**—methods for studying and manipulating genetic material—has rapidly revolutionized the field of forensics, the scientific analysis of evidence for legal investigations. Since its introduction, DNA analysis has become a standard law enforcement tool and has provided crucial evidence (of both innocence and guilt) in many famous cases.

As we will see in this chapter, DNA technology has practical applications beyond its use in forensic science. Applications of DNA technology include the use of gene cloning in the production of medical and industrial products, the development of genetically modified organisms for agriculture, and even the investigation of genealogical questions. Equally important, DNA technology is invaluable in many areas of biological research, such as cancer and evolution. As we discuss these various applications, we'll consider the specific techniques involved, how they are applied, and some of the social, legal, and ethical issues that are raised by the new technologies. ■ ■ ■

12.1 Genes can be cloned in recombinant plasmids

Although it may seem like a modern field, **biotechnology,** the manipulation of organisms or their components to make useful products, actually dates back to the dawn of civilization. Consider such ancient practices as the use of microbes to make wine and cheese and the selective breeding of animals (livestock, dogs, and so forth). But when people use the term *biotechnology* today, they are usually referring to DNA technology, modern laboratory techniques that involve the manipulation of DNA. Using these techniques scientists can, for instance, modify specific genes and move them between organisms as different as bacteria, plants, and animals.

The field of DNA technology grew out of discoveries made about 60 years ago by American geneticists Joshua Lederberg and Edward Tatum. They performed a series of experiments with *E. coli* that demonstrated that two individual bacteria can combine genes—a phenomenon that was previously thought to be limited to sexually reproducing eukaryotic organisms. With this work, they pioneered bacterial genetics, a field that within 20 years made *E. coli* the most thoroughly studied and understood organism at the molecular level.

In the 1970s, the uses of biotechnology exploded with the invention of methods for making recombinant DNA in the laboratory. **Recombinant DNA** is formed when scientists combine nucleotide sequences (pieces of DNA) from two different sources—often different species—to form a single DNA molecule. Today, recombinant DNA technology is widely used to alter the genes of many types of cells for practical purposes. Scientists have genetically engineered bacteria to mass-produce a variety of useful chemicals, from cancer drugs to pesticides. Scientists have also transferred genes from bacteria into plants and from humans into farm animals (see Table 12.6).

To manipulate genes in the laboratory, biologists often use bacterial **plasmids,** which are small, circular DNA molecules that replicate separately from the much larger bacterial chromosome (see Module 10.23). Because plasmids can carry virtually any gene and are passed on from one generation of bacteria to the next, they are key tools for **gene cloning,** the production of multiple identical copies of a gene-carrying piece of DNA. Gene-cloning methods are central to **genetic engineering,** the branch of biotechnology that involves the direct manipulation of genes for practical purposes.

Consider a typical genetic engineering challenge: A molecular biologist at a pharmaceutical company has identified a gene that codes for a valuable product: a hypothetical substance called protein V. The biologist wants to manufacture large amounts of this protein. Figure 12.1 (on the facing page) illustrates how the techniques of gene cloning are used to make many copies of the gene that codes for protein V.

To begin, the biologist isolates two kinds of DNA: ❶ a bacterial plasmid that will serve as the **vector** (gene carrier) and ❷ the DNA containing gene *V*. In this example, the plasmid comes from the bacterium *E. coli*. The DNA containing gene *V* could come from a variety of sources such as a different bacterium, a plant, a nonhuman animal, or even human tissue cells growing in laboratory culture.

The researcher treats both the plasmid and the gene *V* source DNA with an enzyme that cuts each DNA. ❸ An enzyme is chosen that cleaves the plasmid in only one place. ❹ The other DNA, which is usually much longer in sequence, may be cut into many fragments, one of which carries gene *V*. The figure shows the processing of just one DNA fragment and one plasmid, but actually millions of plasmids and DNA fragments (most of which do *not* contain gene *V*) are treated simultaneously. The cuts leave single-stranded ends, as we'll explain in Module 12.2.

❺ The cut DNA from both sources (plasmid and target) are mixed. The single-stranded ends of the plasmid base-pair with the complementary ends of the target DNA fragment. ❻ The enzyme **DNA ligase** joins the two DNA molecules by covalent bonds. This enzyme, which the cell normally uses in DNA replication (see Module 10.5), is a "DNA pasting" enzyme that catalyzes the formation of covalent bonds between adjacent nucleotides, joining the strands. The result is a recombinant DNA plasmid containing gene *V* (as well as many other recombinant DNA plasmids carrying other genes not shown here).

❼ The recombinant plasmid containing the targeted gene is mixed with a culture of bacteria. Under the right conditions, a bacterium takes up the plasmid DNA by transformation (see Module 10.22). ❽ This recombinant bacterium then reproduces to form a **clone** of cells (a group of identical cells descended from a single ancestral cell), each carrying a copy of gene *V*. This step is the actual gene cloning. In our example, the biologist will eventually grow a cell clone large enough to produce protein V in marketable quantities.

As shown along the right side of the figure, gene cloning can be used to produce a variety of desirable products. In the examples in the upper box, copies of the gene itself are the immediate product, to be used in further genetic engineering projects. In the examples in the lower box, the protein product of the cloned gene is harvested and used.

In the next four modules, we will discuss the methods outlined in Figure 12.1. You may find it useful to turn back to this summary figure as each technique is discussed.

🎧 MP3 Tutor *DNA Technology*

Web Activity *Connection: Applications of DNA Technology*

Web Process of Science *How Can Antibiotic-Resistant Plasmids Transform* E. coli?

❓ Why does the rapid reproduction of bacteria make them a good choice for cloning a foreign gene?

■ A foreign gene located within plasmid DNA inside a bacterium is replicated each time the cell divides, resulting in rapid accumulation of many copies of the gene.

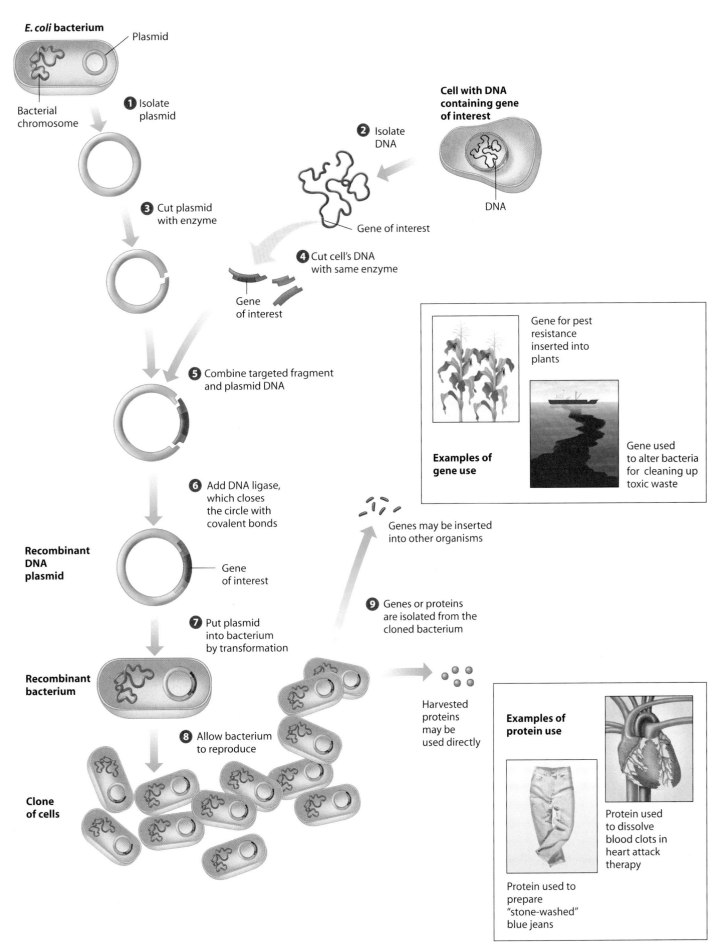

Figure 12.1 An overview of gene cloning

12.2 Enzymes are used to "cut and paste" DNA

In the gene-cloning procedure outlined in Figure 12.1, a piece of DNA containing the gene of interest must be cut out of a chromosome and "pasted" into a bacterial plasmid. The cutting tools used are bacterial enzymes called **restriction enzymes.** In nature, these enzymes protect bacterial cells against intruding DNA from other organisms or viruses. They work by chopping up the foreign DNA, a process that *restricts* the ability of the invader to infect the bacterium. (The bacterial cell's own DNA is protected from restriction enzymes through chemical modification by other enzymes.)

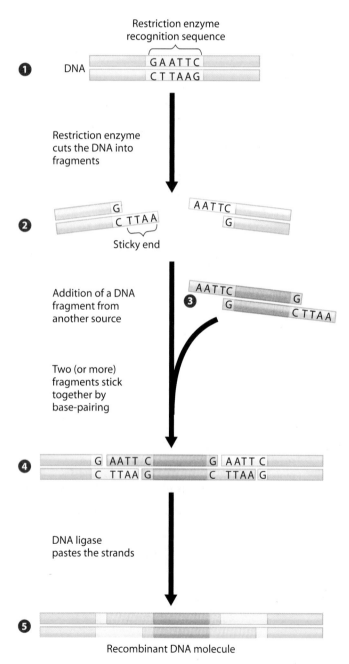

Figure 12.2 Creating recombinant DNA using a restriction enzyme and DNA ligase

Biologists have identified hundreds of different restriction enzymes. Each restriction enzyme is specific, recognizing a particular short DNA sequence (usually four to eight nucleotides long). For example, a restriction enzyme called *Eco*RI (found naturally in *E. coli*) only recognizes the DNA sequence GAATTC, whereas the enzyme called *Bam*HI recognizes GGATCC. The DNA sequence recognized by a particular restriction enzyme is called a **restriction site.** Once a restriction site is recognized, the restriction enzyme cuts both strands of the DNA at specific points within the sequence.

In Figure 12.2, ❶ we start with a piece of DNA containing one recognition sequence for the restriction enzyme *Eco*RI. In this case, the restriction enzyme cuts the DNA strands between the bases A and G within the sequence, producing pieces of DNA called **restriction fragments.** ❷ The staggered cuts yield two double-stranded DNA fragments with single-stranded ends, called "sticky ends." Sticky ends are the key to joining DNA restriction fragments originating from different sources. These short extensions can form hydrogen-bonded base pairs with complementary single-stranded stretches of DNA.

❸ A piece of DNA (gray) from another source is now added. Notice that the gray DNA has single-stranded ends identical in base sequence to the sticky ends on the blue DNA. The gray, "foreign" DNA has ends with this particular base sequence because it was cut from a larger molecule by the same restriction enzyme used to cut the blue DNA. ❹ The complementary ends on the blue and gray fragments allow them to stick together by base-pairing. (The hydrogen bonds are not shown.) The union between the blue and gray DNA fragments shown in step 4 of Figure 12.2 is then made permanent by the "pasting" enzyme DNA ligase. ❺ The final outcome is a stable molecule of recombinant DNA.

The ability to cut DNA with restriction enzymes and then paste it back together with DNA ligase is the key to the gene-cloning procedure outlined in Figure 12.1. This particular cloning procedure, which uses a mixture of fragments from the entire genome of an organism, is called a "shotgun" approach. Thousands of different recombinant plasmids are produced, and a clone of each is made. The complete set of plasmid clones, each carrying copies of a particular segment from the initial genome, is a type of library. The next three modules discuss such libraries in more detail.

Web Activity *Restriction Enzymes*

Web Activity *Cloning a Gene in Bacteria*

? What are "sticky ends"?

■ Single-stranded regions whose unpaired bases can hydrogen-bond to the complementary sticky ends of other fragments created by the same restriction enzyme

12.3 Cloned genes can be stored in genomic libraries

Each bacterial clone from the procedure in Figure 12.1 consists of identical cells with recombinant plasmids carrying one particular fragment of target DNA. The entire collection of all the cloned DNA fragments from a genome is called a **genomic library.** On the left side of Figure 12.3, the red, yellow, and green DNA segments represent three of the thousands of different library "books" that are "shelved" in plasmids inside bacterial cells. A typical cloned DNA fragment is big enough to carry one or a few genes, and together the fragments include the entire genome of the organism from which the DNA was derived.

In addition to bacterial plasmids, phages can also serve as vectors (DNA carriers) when cloning genes (Figure 12.3, right). When a phage is used, the DNA fragments are inserted into phage DNA molecules. The recombinant phage DNA can then be introduced into a bacterial cell through the normal infection process (see Figure 10.22B). Inside the cell, phage DNA is replicated, producing new phage particles, each carrying the foreign DNA. Another type of vector commonly used in library construction is a bacterial artificial chromosome (BAC). BACs are essentially large plasmids containing only the genes necessary to ensure replication. The primary advantage of BACs is that they can carry much larger pieces of foriegn DNA than plasmids or phages. In the next module, we look at another source of DNA for cloning: eukaryotic mRNA.

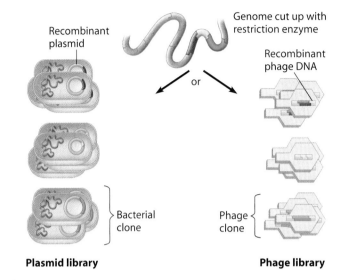

Figure 12.3 Genomic libraries

? In what sense does a genomic library have multiple copies of each "book"?

■ Each "book"—a piece of DNA from the genome that was the source of the library—is present in every recombinant bacterium or phage in a clone.

12.4 Reverse transcriptase can help make genes for cloning

Rather than starting with an entire eukaryotic genome, a researcher can focus on the genes *expressed* in a particular kind of cell by using its mRNA as the starting material for cloning. As shown in Figure 12.4, ❶ the chosen cells transcribe their genes and ❷ process the transcripts to produce mRNA. ❸ The researcher isolates the mRNA and makes single-stranded DNA transcripts from it using the enzyme **reverse transcriptase** (yellow in the figure), which is obtained from retroviruses (see Module 10.20). ❹ Another enzyme is added to break down the mRNA, and ❺ DNA polymerase is used to synthesize a second DNA strand.

The DNA that results from such a procedure, called **complementary DNA (cDNA),** represents only the subset of genes that had been transcribed into mRNA in the starting cells. Among other purposes, a cDNA library is useful for studying the genes responsible for the specialized functions of a particular cell type, such as brain or liver cells. And because cDNAs lack introns, they are shorter than the full versions of the genes, and therefore easier to work with.

In the next module, you will learn how to find one particular "book" (piece of DNA) from among the thousands that are stored in a genomic library.

? Why is a cDNA gene made using reverse transcriptase often shorter than the natural form of the gene?

■ Because cDNAs are made from spliced mRNAs, which lack introns

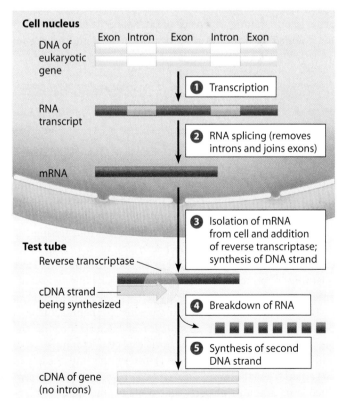

Figure 12.4 Making an intron-lacking gene from eukaryotic mRNA

12.5 Nucleic acid probes identify clones carrying specific genes

Often the most difficult task in gene cloning is finding the right "shelf" in a genomic library—that is, identifying clones containing a desired gene from among all those created (for example, pulling out just the bacteria from Figure 12.3 that have the red gene). If bacterial clones containing a specific gene actually translate the gene into protein, they can be identified by testing for the protein product. This is not always the case, however. Fortunately, researchers can also test directly for the gene itself.

Methods for detecting genes directly depend on base pairing between the gene and a complementary sequence on another nucleic acid molecule, either DNA or RNA. When at least part of the nucleotide sequence of a gene is already known or can be guessed, this information can be used to advantage. Taking a simplified example, if we know that a hypothetical gene contains the sequence TAGGCT, a biochemist can synthesize a short single strand of DNA with the complementary sequence (ATCCGA) and label it with a radioactive isotope or fluorescent tag. This labeled, complementary molecule is called a **nucleic acid probe** because it is used to find a specific gene or other nucleotide sequence within a mass of DNA. (In practice, probe molecules are usually considerably longer than six nucleotides.)

Figure 12.5 shows how a probe works. The DNA sample to be tested is treated with heat or alkali to separate the DNA strands. When the DNA probe is added to these strands, it tags the correct molecule—finds the correct shelf in the library—by hydrogen-bonding to the complementary sequence in the gene of interest. Such a probe can be simultaneously applied to many bacterial colonies to screen all of them at once for a desired gene. In one technique, a piece of filter paper is pressed against the colonies growing on a solid growth medium, picking up

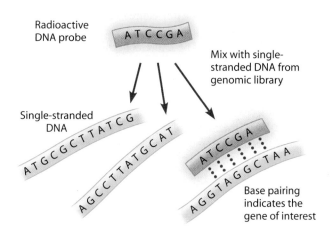

Figure 12.5 How a DNA probe tags a gene by base pairing

cells from each colony, treated to break the cells and separate the DNA strands, and then soaked in probe solution. Any bacterial colonies carrying the gene of interest will be tagged on the filter paper, marking them for easy identification. Once the researcher identifies a colony carrying the desired gene, the cells can be grown further and the gene of interest (and/or its protein product) isolated in large amounts.

? How does a probe consisting of radioactive DNA or RNA enable a researcher to find the bacterial clones carrying a particular gene?

■ The probe molecules bind to and label DNA only in the cells containing the gene of interest, which contains a complementary DNA sequence.

Genetically Modified Organisms

12.6 Recombinant cells and organisms can mass-produce gene products

Recombinant cells and organisms constructed by DNA technology are used to manufacture many useful products, chiefly proteins (Table 12.6, on the facing page). Most of these products are made by cells grown in culture. By transferring the gene for a desired protein into a bacterium, yeast, or other kind of cell that is easy to grow, a genetic engineer can produce large quantities of proteins that are present naturally in only minute amounts.

Bacteria are often the best organisms for manufacturing a protein product. Major advantages of bacteria include the plasmids and phages available for use as gene-cloning vectors and the fact that bacteria can be grown rapidly and cheaply in large tanks. Furthermore, bacteria can be engineered to produce large amounts of particular proteins and in some cases to secrete the proteins directly into their growth medium,

simplifying the task of collecting and purifying the products. As Table 12.6 shows, a number of proteins of importance in human medicine and agriculture are being produced in the bacterium *Escherichia coli*.

Despite the advantages of bacteria, it is sometimes desirable or necessary to use eukaryotic cells to produce a protein product. Often, the first-choice eukaryotic organism for protein production is the yeast used in making bread and beer, *Saccharomyces cerevisiae*. As bakers and brewers have recognized for centuries, yeast cells are easy to grow. And like *E. coli*, yeast cells can take up foreign DNA and integrate it into their genomes. Yeast cells also have plasmids that can be used as gene vectors, and yeast is often better than bacteria at synthesizing and secreting eukaryotic proteins. *S. cerevisiae* is currently used to produce a number of proteins. In certain cases, the same product (for example, interferons used in cancer research) can be made in either yeast or bacteria. In other cases, such as the hepatitis B vaccine, yeast alone is used.

The cells of choice for making some gene products come from mammals. For example, recombinant mammalian cells growing in laboratory cultures are currently used to produce human erythropoietin (EPO), a hormone that stimulates the production of red blood cells. EPO is used as a treatment for anemia; the drug is also abused by some athletes who seek the advantage of artificially high levels of oxygen-carrying red blood cells (called "blood doping").

Genes for products produced in mammalian cells are often cloned in bacteria as a preliminary step. For example, the gene for Factor VIII, a protein that affects blood clotting, is cloned in a bacterial plasmid before transfer to mammalian cells for large-scale production. Many proteins that mammalian cells normally secrete are glycoproteins, proteins with chains of sugars attached. Because only mammalian cells can attach the sugars correctly, mammalian cells must be used for making these products.

Recently, pharmaceutical researchers have been exploring the mass production of gene products by whole animals or plants rather than cultured cells. For example, using recombinant DNA technology, genetic engineers can add a gene for a desired human protein to the genome of a mammal in such a way that the gene's product is secreted in the animal's milk.

Figure 12.6 shows a herd of recombinant sheep that carry a gene for a human blood protein. The human protein can be harvested from the sheep's milk and is being tested as a treatment for cystic fibrosis. Because recombinant animals are difficult to produce, researchers may create a single one and then

Figure 12.6 "Pharm" animals that produce a human protein

clone it. The resulting herd of genetically identical recombinant animals, all carrying a recombinant human gene, could then serve as a grazing pharmaceutical factory—"pharm" animals. We continue an exploration of the medical applications of DNA technology in the next module.

? Why can't glycoproteins be mass-produced by engineered bacteria or yeast cells?

■ Because bacteria and yeast cells cannot correctly attach the sugar groups to the protein of glycoproteins

TABLE 12-6	SOME PROTEIN PRODUCTS OF RECOMBINANT DNA TECHNOLOGY	
Product	Made In	Use
Human insulin	E. coli	Treatment for diabetes
Human growth hormone (HGH)	E. coli	Treatment for growth defects
Epidermal growth factor (EGF)	E. coli	Treatment for burns, ulcers
Interleukin-2 (IL-2)	E. coli	Possible treatment for cancer
Bovine growth hormone (BGH)	E. coli	Improving weight gain in cattle
Cellulase	E. coli	Breaking down cellulose for animal feeds
Taxol	E. coli	Treatment for ovarian cancer
Interferons (alpha and gamma)	S. cerevisiae; E. coli	Possible treatment for cancer and viral infections
Hepatitis B vaccine	S. cerevisiae	Prevention of viral hepatitis
Erythropoietin (EPO)	Mammalian cells	Treatment for anemia
Factor VIII	Mammalian cells	Treatment for hemophilia
Tissue plasminogen activator (TPA)	Mammalian cells	Treatment for heart attacks and some strokes

12.7 DNA technology has changed the pharmaceutical industry and medicine

DNA technology, and gene cloning in particular, is widely used to produce medicines and to diagnose diseases.

Therapeutic Hormones Consider the first two products in Table 12.6 on the previous page, human insulin and human growth hormone (HGH). About 2 million people with diabetes in the United States depend on insulin treatment. Before 1982, the main sources of this hormone were slaughtered pigs and cattle. Insulin extracted from these animals is chemically similar, but not identical, to human insulin, and it causes harmful side effects in some people. Genetic engineering has largely solved this problem by developing bacteria that synthesize and secrete the human form of insulin. In 1982, Humulin (Figure 12.7A)—human insulin produced by bacteria—became the first recombinant DNA drug approved by the Food and Drug Administration.

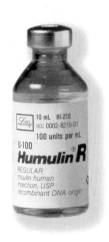

Figure 12.7A
Human insulin produced by bacteria

Treatment with human growth hormone is a boon to children born with a form of dwarfism caused by inadequate amounts of HGH. Because growth hormones from other animals are not effective in humans, HGH is urgently needed. In 1985, molecular biologists made an artificial gene for HGH by joining a human DNA fragment to a chemically synthesized piece of DNA; using this gene, they were able to produce HGH in *Escherichia coli*. Before this genetically engineered hormone became available, children with an HGH deficiency had to rely on scarce supplies from human cadavers or else face dwarfism.

Another important pharmaceutical product produced by genetic engineering is tissue plasminogen activator (TPA). If administered soon after a heart attack, this protein helps dissolve blood clots and reduces the risk of subsequent heart attacks.

Diagnosis and Treatment of Disease DNA technology is being used increasingly in disease diagnosis. Among the genes for human diseases that have been identified are those for sickle-cell disease, hemophilia, cystic fibrosis, and Huntington's disease. Affected individuals with such diseases often can be identified before the onset of symptoms, even before birth. It is also possible to identify symptomless carriers of potentially harmful recessive alleles. Additionally, DNA technology can pinpoint infections. For example, DNA analysis can help track down and identify elusive viruses such as HIV, the virus that causes AIDS.

Vaccines DNA technology is also helping medical researchers develop vaccines. A **vaccine** is a harmless variant (mutant) or derivative of a pathogen (usually a bacterium or virus) that is used to stimulate the immune system to mount a defense against that pathogen (see Module 24.4). For many

Figure 12.7B Equipment used in the production of a vaccine against hepatitis B

viral diseases, prevention by vaccination is the only medical way to prevent illness.

Genetic engineering can be used in several ways to make vaccines. One approach is to use genetically engineered cells (or organisms) to produce large amounts of a protein molecule that is found on the pathogen's outside surface. This method has been used to make the vaccine against hepatitis B virus. Hepatitis is a disabling and sometimes fatal liver disease, and the hepatitis B virus may also cause liver cancer. Figure 12.7B shows a tank for growing yeast cells that have been engineered to carry hepatitis B genes; the protein produced by these genes will make up part of the vaccine.

Another way to use DNA technology in vaccine development is to make a harmless artificial mutant of the pathogen by altering one or more of its genes. When a harmless mutant is used as a vaccine, it multiplies in the body and may trigger a strong immune response. Artificial-mutant vaccines may cause fewer side effects than those that have traditionally been made from natural mutants.

Yet another scheme for making vaccines employs a virus related to the one that causes smallpox. Smallpox was once a dreaded human disease, but it was eradicated worldwide in the 1970s by widespread vaccination with a harmless variant of the smallpox virus. Using this harmless virus, genetic engineers could replace some of the genes encoding proteins that induce immunity to smallpox with genes that induce immunity to other diseases. In fact, the virus could be engineered to carry genes needed to vaccinate against several diseases simultaneously. In the future, one inoculation may prevent a dozen diseases.

? Human growth hormone and insulin produced by DNA technology are used in the treatment of _____ and _____, respectively.

dwarfism . . . diabetes ■

12.8 Genetically modified organisms are transforming agriculture

Since ancient times, people have selectively bred agricultural crops to make them more useful. Today, DNA technology is quickly replacing traditional breeding programs as scientists work to improve the productivity of agriculturally important plants and animals. Genetic engineers have produced many varities of **genetically modified (GM) organisms,** ones that have acquired one or more genes by artificial means. If the newly acquired gene is from another species, the recombinant organism is called a **transgenic organism.**

To make genetically modified plants, researchers can often manipulate the DNA of a single somatic cell and then grow an entire plant from it in culture. Already in commercial use are a number of crop plants carrying new genes for desirable traits, such as delayed ripening and resistance to spoilage and disease.

The most common vector used to introduce new genes into plant cells is a plasmid from the soil bacterium *Agrobacterium tumefaciens* called the **Ti plasmid** (Figure 12.8A). ❶ With the help of a restriction enzyme and DNA ligase, the gene for the desired trait (red) is inserted into a modified version of the plasmid. ❷ Then the recombinant plasmid is put into a plant cell, where the DNA carrying the new gene integrates into a plant chromosome. ❸ Finally, the recombinant cell is cultured and grown into a plant.

In the United States today, roughly half the corn crop and over three-quarters of the soybean and cotton crops are genetically modified. Many of these GM plants have received bacterial genes that make the plants resistant to herbicides or pests. This modification increases profit for farmers by reducing the need for tillage and chemical insecticides. In India, the insertion of a salinity-resistance gene has enabled new varieties of rice to grow in water three times as salty as seawater.

Genetic engineering also has great potential for improving the nutritional value of crop plants. "Golden rice," a transgenic variety with a few daffodil genes, produces yellow grains containing beta-carotene, which our body uses to make vitamin A (Figure 12.8B). This rice could help prevent vitamin A deficiency—and resulting blindness—among the half of the world's people who depend on rice as their staple food.

Agricultural researchers are also making transgenic animals. To do this, scientists remove egg cells from a female and fertilize them *in vitro*. They then inject a previously cloned gene directly into the nuclei of the fertilized eggs. Some of the cells integrate the foreign DNA into their genomes. The engineered embryos are then surgically implanted in a surrogate mother. If an embryo develops successfully, the result is an animal containing a gene from a third "parent," which may even be of another species.

The goals in creating a transgenic animal are often the same as the goals of traditional breeding—for instance, to make a sheep with better quality wool or a cow that will mature in less time. In 2006, researchers succeeded in transferring a fat metabolism gene from a roundworm into a pig. Meat from the resulting swine had levels of healthy omega-3 fatty acids (which are believed to reduce the risk of heart disease) four to five times higher than meat from normal pigs. Unlike transgenic plants, however, transgenic animals are currently used only to produce potentially useful proteins; at present, no transgenic animals are sold as food.

Transgenic animals also have been engineered to be pharmaceutical "factories" that produce otherwise rare biological substances for medical use (see also Module 12.6). Recently, researchers have engineered transgenic chickens that express large amounts of desired foreign protein in their eggs. This success suggests that transgenic chickens may emerge as relatively inexpensive pharmaceutical factories in the near future. But, as we'll discuss next, genetically modified organisms can pose potential hazards as well as rewards.

Figure 12.8B   A mix of "golden rice" and standard rice

Web Activity *Connection: DNA Technology and Golden Rice*

? What is the function of the Ti plasmid in the creation of transgenic plants?

■ It is used as the vector for introducing foreign genes into a plant cell.

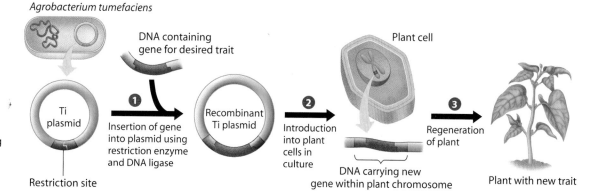

Agrobacterium tumefaciens

DNA containing gene for desired trait

Plant cell

Figure 12.8A   Using the Ti plasmid as a vector for genetically engineering plants

Ti plasmid

Restriction site

❶ Insertion of gene into plasmid using restriction enzyme and DNA ligase

Recombinant Ti plasmid

❷ Introduction into plant cells in culture

DNA carrying new gene within plant chromosome

❸ Regeneration of plant

Plant with new trait

12.9 Genetically modified organisms raise concerns about human and environmental health

As soon as scientists realized the power of DNA technology, they began to worry about potential dangers. Early concerns focused on the possibility that recombinant DNA technology might create new pathogens. What might happen, for instance, if cancer cell genes were transferred into bacteria or viruses? To guard against such rogue microbes, scientists developed a set of guidelines that were adopted as formal government regulations in the United States and some other countries. One safety measure is a set of strict laboratory procedures designed to protect researchers from infection by engineered microbes and to prevent the microbes from accidentally leaving the laboratory (Figure 12.9A). In addition, strains of microorganisms to be used in recombinant DNA experiments are genetically crippled to ensure that they cannot survive outside the laboratory. Finally, certain obviously dangerous experiments have been banned.

Today, most public concern about possible hazards centers not on recombinant microbes but on genetically modified (GM) organisms used for food. Advocates of a cautious approach fear that some crops carrying genes from other species might be hazardous to human health or the environment.

One specific concern is that genetic engineering could transfer allergens, which are molecules to which some people are allergic (see Module 24.17), to plants people eat. Although there is some evidence that this could happen, advocates for GM plants claim that transgenic proteins could be tested for their ability to cause allergic reactions.

Nevertheless, because of health concerns, activists continue to lobby for the clear labeling of all foods containing products of GM organisms. Early in 2000, negotiators from 130 countries (including the United States) agreed on a Biosafety Protocol that requires all exporters to identify GM organisms present in bulk food shipments and allows importing

countries to decide whether they pose environmental or health risks. Some biotechnology advocates, however, point out that similar demands were not made when "transgenic" crop plants produced by traditional breeding techniques were put on the market. One example of such a plant is triticale, which was created decades ago by combining the genomes of wheat and rye—two plants that do not interbreed in nature. Triticale is grown worldwide.

Advocates of a cautious approach toward GM crops also fear that transgenic plants might pass their new genes to close relatives in nearby wild areas (Figure 12.9B). We know that lawn and crop grasses, for example, commonly exchange genes with wild relatives via pollen transfer. If crop plants carrying genes for resistance to herbicides, diseases, or insect pests pollinated wild ones, the offspring might become "superweeds" that would be very difficult to control. However, researchers may be able to prevent the escape of such plant genes in various ways—for example, by engineering plants so that they cannot hybridize.

Today, governments and regulatory agencies throughout the world are grappling with how to facilitate the use of biotechnology in agriculture, industry, and medicine while ensuring that new products and procedures are safe. In the United States, all projects are evaluated for potential risks by regulatory agencies such as the Food and Drug Administration, Environmental Protection Agency, National Institutes of Health, and Department of Agriculture. These agencies are under increasing pressure from some consumer groups.

In the case of GM plants and certain other applications of DNA technology, zero risk is probably unattainable. Scientists and the public need to weigh the possible benefits versus risks on a case-by-case basis. The best scenario would be for us to proceed with caution, basing our decisions on sound scientific information rather than on either irrational fear or blind optimism.

Figure 12.9B Genetically engineered crop plants growing near their wild relatives

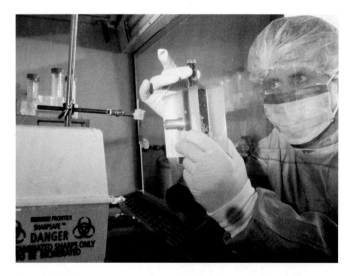

Figure 12.9A A maximum-security laboratory at the Pasteur Institute in Paris

? What is one of the concerns about engineering crop plants by adding genes for herbicide resistance?

■ The possibility that the genes could escape, via cross-pollination, to weeds that are closely related to the crop species

12.10   Gene therapy may someday help treat a variety of diseases

In this chapter, we have so far discussed transgenic viruses, bacteria, yeast, plants, and animals. What about transgenic humans? Why would, and should, anyone want to insert genes into a living person?

One reason to tamper with the human genome is the potential for treating a variety of diseases by **gene therapy**—alteration of an afflicted individual's genes. In people with disorders traceable to a single defective gene, it should theoretically be possible to replace or supplement the defective gene with a normal allele. The new allele could be inserted into somatic cells of the tissue affected by the disorder.

For gene therapy to be permanent, the normal allele would have to be transferred to cells that multiply throughout a person's life. Bone marrow cells, which include the stem cells that give rise to all the cells of the blood and immune system, are prime candidates (see Modules 11.17 and 23.17). Figure 12.10 outlines one possible procedure for gene therapy in an individual whose bone marrow cells do not produce a vital protein product because of a defective gene. ❶ The normal gene is cloned and then inserted into the nucleic acid of a retrovirus vector that has been rendered harmless. ❷ Bone marrow cells are taken from the patient and infected with the virus. ❸ The virus inserts its nucleic acid, including the human gene, into the cells' DNA (see Module 10.22). ❹ The engineered cells are then injected back into the patient. If the procedure succeeds, the cells will multiply throughout the patient's life and produce the missing protein. The patient will be cured.

A procedure like the one shown in Figure 12.10 was used in the first human gene therapy trial, which began in 1990. This trial sought to treat severe combined immunodeficiency disease (SCID), a disorder in which the patient lacks a funtional immune system (see Module 24.16). But the clinical results from this trial did not confirm the effectiveness of the treatment. In another trial beginning in 2000, ten young children with SCID were treated by the same procedure. Nine of these patients showed significant improvement after two years, providing the first definitive success of gene therapy. However, three of the patients subsequently developed leukemia, a cancer of the blood cells, and one died. Researchers discovered that in two of the cases, the virus used to carry the normal allele into bone marrow cells had inserted DNA near a gene involved in proliferation and development of blood cells. This insertion somehow caused the leukemia. Thus, although the concept of gene therapy remains promising, scientifically strong evidence of its effectiveness has yet to appear. Active research into human gene therapy continues with new, tougher safety guidelines.

The use of gene therapy raises several technical questions. For example, how can researchers build in gene control mechanisms to ensure that cells with the transferred gene make appropriate amounts of the gene product at the right time and in the right parts of the body? And how can they be sure that the gene's insertion does not harm the cell's normal function?

In addition to technical challenges, gene therapy raises difficult ethical questions. Some critics suggest that tampering with human genes in any way will inevitably lead to the practice of eugenics, the deliberate effort to control the genetic makeup of human populations. Other observers see no fundamental difference between the transplantation of genes into somatic cells and the transplantation of organs.

The implications of genetically manipulating germ cells or zygotes (already accomplished in lab animals) are more problematic. This possibility raises the most difficult ethical questions of all: Should we try to eliminate genetic defects in our children and their descendants? Should we interfere with evolution in this way? From a biological perspective, the elimination of unwanted alleles from the gene pool could backfire. Genetic variety is a necessary ingredient for the survival of a species as environmental conditions change with time. Genes that are damaging under some conditions may be advantageous under others (one example is the sickle-cell allele; see Module 9.13). Are we willing to risk making genetic changes that could be detrimental to our species in the future? We may have to face this question soon.

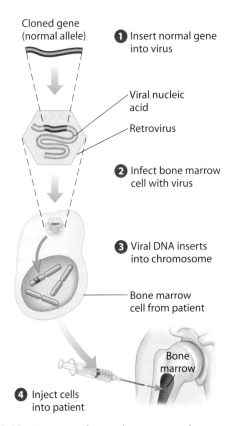

Figure 12.10   One type of gene therapy procedure

?   What characteristic of retroviruses makes them candidate vectors for gene therapy?

■ They integrate DNA into the DNA of host cells.

12.11 The analysis of genetic markers can produce a DNA profile

As discussed in the chapter introduction, DNA technology has rapidly transformed the field of **forensics,** the scientific analysis of evidence for crime scene investigations and other legal proceedings. The most important application of biology to forensics is **DNA profiling,** the analysis of DNA fragments to determine whether they come from a particular individual. (The term "DNA profiling" is preferred over "DNA fingerprinting" by forensics scientists, who want to emphasize the genetic, inheritable basis for this procedure.)

How do you prove that two samples of DNA come from the same person? You could compare the entire genomes found in the two samples. But such an approach would be extremely impractical, requiring a lot of time and money. Instead, scientists compare genetic markers, sequences in the genome that vary from person to person. Like a gene (which is also a type of genetic marker), a noncoding genetic marker is more likely to be a match between relatives than between unrelated individuals.

Figure 12.11 summarizes the basic steps in creating a DNA profile. ❶ First, DNA samples are collected from the crime scene, suspects, victims, or stored evidence. ❷ Next, the selected markers from each DNA sample are amplified (copied many times), producing an adequate supply for testing. ❸ Finally, the amplified DNA markers are compared. In the next four modules, we'll explore the methods behind these steps.

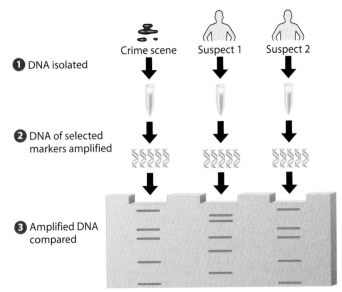

Figure 12.11   An overview of DNA fingerprinting

? Why does DNA profiling rely on comparing specific genetic markers rather than the entire genome?

■ It would be too costly and time consuming to compare whole genomes.

12.12 The PCR method is used to amplify DNA sequences

Cloning DNA in host cells is often the best method for preparing large quantities of DNA from a particular gene (see Module 12.1). However, when the source of DNA is scanty or impure, the **polymerase chain reaction (PCR)** is a much better method. In this technique, a specific segment of a DNA molecule can be targeted and quickly amplified in a test tube. Starting with a single DNA segment, automated PCR can generate billions of copies in just a few hours.

In the PCR procedure (Figure 12.12), a three-step cycle brings about a chain reaction that doubles the population of identical DNA molecules each round. The key to amplifying one particular segment of DNA (and no others) are **primers,** short (usually 15–20 nucleotides long), chemically synthesized single-stranded DNA molecules with sequences that are complementary to sequences at each end of the target sequence. One primer is complimentary to one strand at one end of the target sequence; the second primer is complementary to the other strand at the other end of the sequence. The primers thus bind to sequences that flank the target sequence, marking the start and end points for the segment of DNA being amplified. ❶ In the first step of each PCR cycle, the reaction mixture is heated to separate the strands of the DNA double helices. ❷ Next, the strands are cooled. As they cool, primer molecules hydrogen-bond to their target sequences on the DNA. ❸ In the third step,

a heat-stable DNA polymerase builds new DNA strands by extending the primers in the 5′→3′ direction. These three steps are repeated over and over, doubling the amount of DNA after each 3-step cycle. PCR became automated after the discovery of an unusual heat-stable DNA polymerase, first isolated from a bacterium living in hot springs, that could withstand the heat at the start of each cycle; standard DNA polymerases would denature (unfold) during the heating step of each cycle.

Just as impressive as the speed of PCR is its sensitivity. Only minute amounts of DNA need to be present in the starting material, and this DNA can even be in a partially degraded state. The key to the high sensitivity is the primers. Because DNA polymerase extends the primers, only the desired segments of DNA are duplicated. Other DNA will not be bound by primers and thus not copied by DNA polymerase.

Devised in 1985, PCR has had a major impact on biological research and biotechnology. PCR has been used to amplify DNA from a wide variety of sources: fragments of ancient DNA from a 40,000-year-old frozen woolly mammoth; DNA from fingerprints or from tiny amounts of blood, tissue, or semen found at crime scenes; DNA from single embryonic cells for rapid prenatal diagnosis of genetic disorders; and DNA of viral genes from cells infected with viruses that are difficult to detect, such as HIV.

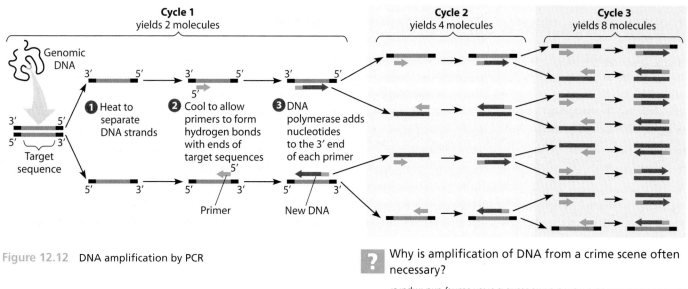

Cycle 1 yields 2 molecules
Cycle 2 yields 4 molecules
Cycle 3 yields 8 molecules

Genomic DNA

Target sequence

❶ Heat to separate DNA strands

❷ Cool to allow primers to form hydrogen bonds with ends of target sequences

❸ DNA polymerase adds nucleotides to the 3′ end of each primer

Primer

New DNA

Figure 12.12 DNA amplification by PCR

Why is amplification of DNA from a crime scene often necessary?

■ The DNA retrievable from a crime scene is often scanty and impure.

12.13 Gel electrophoresis sorts DNA molecules by size

Many approaches for studying DNA molecules make use of **gel electrophoresis.** This technique uses a gel (a thin slab of jellylike material) as a molecular sieve to separate macromolecules—usually proteins or nucleic acids—on the basis of size, electrical charge, or other physical properties.

Figure 12.13 outlines how gel electrophoresis can be used to separate mixtures of DNA fragments obtained from three different sources. A sample of each mixture is placed in a well (hole) at one end of a flat, rectangular gel. A negatively charged electrode from a power supply is attached near the end of the gel containing the DNA, and a positive electrode is attached near the other end. Because phosphate (PO_4^-) groups give DNA a negative charge (see Module 10.2), the DNA molecules all travel through the gel toward the positive pole. However, the longer DNA fragments are held back by a thicket of polymer fibers within the gel, so they move more slowly than the shorter fragments. Over time, shorter molecules move further through the gel than longer fragments. Gel electrophoresis thus separates DNA molecules by length, with shorter molecules migrating toward the bottom faster than longer molecules.

When the current is turned off, a series of bands is left in each "lane" of the gel. Each band is a collection of DNA molecules of the same length. The bands can be made visible by staining, by exposure onto photographic film (if the DNA is radioactively labeled), or by measuring fluorescence (if the DNA is labeled with a fluorescent dye).

Web Activity *Gel Electrophoresis of DNA*

Web Process of Science *How Can Gel Electrophoresis Be Used to Analyze DNA?*

(a) What causes DNA molecules to move toward the positive pole during electrophoresis? (b) Why do large molecules move more slowly than smaller ones?

■ (a) The negatively charged phosphate groups of the DNA are attracted to the positive pole. (b) The gel restricts their movement more.

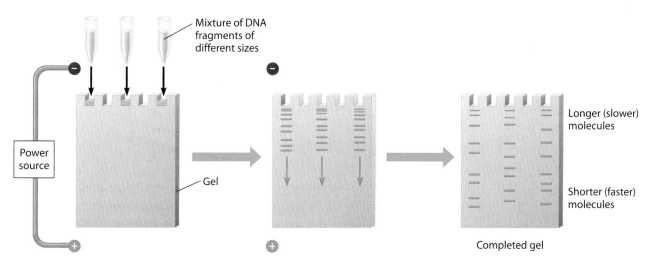

Mixture of DNA fragments of different sizes

Power source

Gel

Longer (slower) molecules

Shorter (faster) molecules

Completed gel

Figure 12.13 Gel electrophoresis of DNA

If you take another look at the overview of DNA profiling in Figure 12.11, you will see that we have learned about DNA amplification by PCR (step 2) and gel electrophoresis (step 3). Now let's put them together to see how a DNA profile is made.

To create a DNA profile, a forensic scientist must compare genetic markers from two or more DNA samples. The genetic markers most often used in DNA profiling are inherited variations in the lengths of repetitive DNA segments. **Repetitive DNA** consists of nucleotide sequences that are present in multiple copies in the genome; much of the DNA that lies between genes in humans is of this type. Some regions of repetitive DNA vary considerably from one individual to the next.

For DNA profiling, the relevant type of repetitive DNA consists of short sequences repeated many times in a row (tandemly); such sequences are called **short tandem repeats (STRs).** For example, one person might have the sequence AGAT repeated 12 times in a row at one place in the genome, the sequence GATA repeated 45 times in a row at a second place, and so on. Another person is likely to have the same sequences at the same places but with different numbers of repeats.

STR analysis is a method of DNA profiling that compares the lengths of STR sequences at specific sites in the genome. Most commonly, STR analysis compares the number of repeats of specific four-nucleotide DNA sequences at 13 sites scattered throughout the genome. Each of these repeat sites, which typically contain from 3 to 50 four-nucleotide repeats in a row, vary widely from person to person. In fact, some of the short tandem repeats used in the standard procedure can be found in up to 80 different variations in the human population.

Consider the two samples of DNA shown in Figure 12.14A. Imagine that the top DNA was obtained at a crime scene and the bottom DNA from a suspect's blood. The two segments have the same number of repeats at the first site: 7 repeats of the four-nucleotide DNA sequence AGAT (shown in orange). Notice, however, that they differ in the number of repeats at the second site: 8 repeats of GATA (shown in purple) in the crime scene DNA, compared to 13 repeats in the suspect's DNA. To create a DNA profile, a scientist uses PCR to specifically amplify the regions of DNA that include these STR sites. This can be done by using primers matching nucleotide sequences known to flank the STR sites. The resulting DNA molecules are then compared by gel electrophoresis.

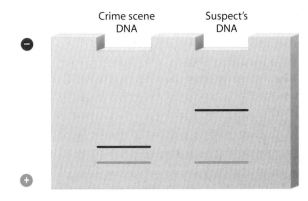

Figure 12.14B DNA profiles generated from the STRs in Figure 12.14A

Figure 12.14B shows a gel that could have resulted from the STR fragments in Figure 12.14A. The differences in the locations of the bands reflect the different lengths of the DNA fragments. (A gel from an actual DNA profile would typically contain more than just two bands in each lane.) This gel would provide evidence that the crime scene DNA did not come from the suspect. Notice that electrophoresis allows us to see similarities as well as differences between mixtures of DNA molecules. Thus, data from DNA profiling can provide evidence of either innocence or guilt.

Although other methods have been used in the past, STR analysis of 13 predetermined STR sites is the current standard for DNA profiling in forensic and legal systems. Once determined, the number of repeats at each of these sites can be entered into the Combined DNA Index System (CODIS) database, administered by the Federal Bureau of Investigation. In the next module, we'll examine several real-world examples of how this technology has been used.

Web Activity Analyzing DNA Fragments Using Gel Electrophoresis

? (a) What are STRs? (b) What is STR analysis?

■ (a) Regions of the genome that contain varying numbers of in-a-row repeats of a short nucleotide sequence (b) A technique for determining whether two DNA samples have identical STRs

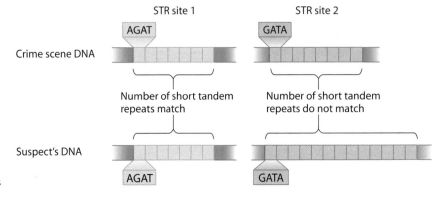

Figure 12.14A
Two representative STR sites from crime scene DNA samples

12.15 DNA profiling has provided evidence in many forensic investigations

When a violent crime is committed, body fluids or small pieces of tissue may be left at the crime scene or on the clothes of the victim or assailant. If rape has occurred, semen may be recovered from the victim's body. DNA profiling can match such samples to the person they came from with a high degree of certainty because the DNA sequence of every person is unique (except for identical twins). And with PCR amplification of DNA, a tissue sample as small as 20 cells can be sufficient for testing.

Since its introduction in 1986, DNA profiling has become a standard tool of criminalistics and has provided crucial evidence in many famous cases. In the O. J. Simpson murder trial, DNA analysis proved that blood in Simpson's car belonged to the victims and that blood at the crime scene belonged to Simpson. (The jury in this case did not find the DNA evidence alone to be sufficient to convict the suspect, and Simpson was found not guilty.) During the investigation that led up to his impeachment, President Bill Clinton repeatedly denied that he had sexual relations with Monica Lewinsky—until DNA profiling proved that his semen was on her dress.

Figure 12.15A STR analysis proved that convicted murderer Earl Washington was innocent, freeing him after 17 years in prison.

Of course, DNA evidence can prove innocence as well as guilt. The Innocence Project, associated with the Yeshiva University School of Law in New York City, has used DNA technology and legal work to exonerate over 200 convicted criminals since 1989, including 15 on death row (Figure 12.15A). In more than a third of these cases, DNA evidence pointed toward the true suspects. These facts have prompted several states to impose or consider moratoriums on the death penalty.

The use of DNA profiling extends beyond crimes. For instance, a comparison of the DNA of a child and the purported father can conclusively settle a question of paternity. Sometimes paternity is of historical interest: DNA profiling proved that Thomas Jefferson or one of his close male relatives fathered at least one child with his young slave, Sally Hemings. Going back much farther, one of the strangest cases of DNA profiling is that of Cheddar Man, a 9,000-year-old skeleton found in a cave near Cheddar, England (Figure 12.15B). DNA was extracted from his tooth and analyzed. The DNA profile showed that Cheddar Man was a direct ancestor (through approximately 300 generations) of a present-day schoolteacher who lived only a half mile from the cave!

DNA profiling can also identify victims of mass casualties. The largest such effort in history occurred after the World Trade Center attack on September 11, 2001. Forensic scientists under the coordination of the Office of the Chief Medical Examiner of New York City worked for years to identify over 20,000 samples of victims' remains. Tissue samples from the disaster site were matched to known DNA profiles (when available) or to DNA-carrying items provided by families, such as toothbrushes. When no sample of a victim's DNA was available, blood samples from close relatives were used to confirm identity through near matches. Over half of the identified victims at the World Trade Center site were identified solely by DNA evidence, providing closure to many grieving families.

Figure 12.15B Cheddar Man and his modern-day descendant

Just how reliable is DNA profiling? When the standard CODIS set of 13 STR sites (see Module 12.14) is used, the probability of finding the same DNA profile in randomly selected, unrelated individuals is less than one in 10 billion. Put another way, a standard DNA profile can provide a statistical match of a particular DNA sample to just *one living human*. For this reason, DNA analyses are now accepted as compelling evidence by legal experts and scientists alike. In fact, DNA analysis on stored forensic samples has provided the evidence needed to solve many "cold cases" in recent years. DNA analysis has also been used to probe the origin of nonhuman materials. In 1998, the U.S. Fish and Wildlife Service began testing the DNA in caviar to determine if the fish eggs originated from the species claimed on the label. By conclusively proving the origin of contraband animal products, DNA profiling could help protect endangered species. In another example, a 2005 study determined that DNA extracted from a 27,000-year-old Siberian mammoth was 98.6% identical to DNA from modern African elephants.

Although DNA profiling has provided definitive evidence in many investigations, the method is far from foolproof. Problems can arise from insufficient data, human error, or flawed evidence. While the science behind DNA profiling is irrefutable, the human element remains a possible confounding factor.

Web Activity *Connection: DNA Fingerprinting*

? In what way is DNA profiling valuable for determining innocence as well as guilt?

■ A DNA profile can prove with near certainty that a sample of DNA does or does not come from a particular individual. DNA profiling therefore can provide evidence in support of guilt or innocence.

12.16 RFLPs can be used to detect differences in DNA sequences

Recall that a genetic marker is a DNA sequence that varies in a population. Like different alleles of a gene, the DNA sequence at a specific place on a chromosome may exhibit small nucleotide differences, or polymorphisms (from the Greek for "many forms"). Geneticists have catalogued many single base-pair variations in the genome. Such a variation found in at least 1% of the population is called a **single nucleotide polymorphism** (**SNP**, pronounced "snip"). SNPs occur on average about once in 100 to 300 base pairs in the human genome in either the coding sequence of a gene or in a noncoding sequence. Methods have been developed that can detect a particular SNP.

SNPs may alter a restriction site (the sequence recognized by a restriction enzyme). Such alterations change the lengths of the restriction fragments formed by that enzyme when it cuts the DNA. This type of sequence change is called a **restriction fragment length polymorphism** (**RFLP**, pronounced "rif-lip"). Thus, RFLPs serve as genetic markers for particular loci in the genome. RFLPs have many uses. For example, disease-causing alleles can be diagnosed with reasonable accuracy if a closely linked SNP or RFLP marker has been found. Alleles for Huntington's disease and a number of other genetic diseases were first detected by means of RFLPs in this indirect way.

Restriction fragment analysis involves two of the methods you have learned about: DNA fragments produced by restriction enzymes (see Module 12.2) are sorted by gel electrophoresis (see Module 12.13). The number of restriction fragments and their sizes reflect the specific sequence of nucleotides in the starting DNA.

At the top of Figure 12.16, we see corresponding segments of DNA from two DNA samples prepared from human tissue. Notice that the DNA sequences differ by a single base pair (highlighted in gold). In this case, the restriction enzyme cuts the DNA between two cytosine (C) bases in the sequence CCGG and in its complement, GGCC. Because DNA from the first sample has two recognition sequences for the restriction enzyme, it is cleaved in two places, yielding three restriction fragments (labeled *w, x,* and *y*). DNA from the second sample, however, has only one recognition sequence and yields only two restriction fragments (*z* and *y*). Notice that the lengths of restriction fragments, as well as their numbers, differ depending on the exact sequence of bases in the DNA.

To detect the differences between the collections of restriction fragments, we need to separate the restriction fragments in the two mixtures and compare their lengths. This process, called RFLP analysis, is accomplished through gel electrophoresis. As shown in the bottom of the figure, the three kinds of restriction fragments from sample 1 separate into three bands in the gel, while those from sample 2 separate into only two bands. Notice that the shortest fragment from sample 1 (*y*) produces a band at the same location as the identical short fragment from the sample 2. So you can see that electrophoresis allows us to see similarities as well as differences between mixtures of restriction fragments—and

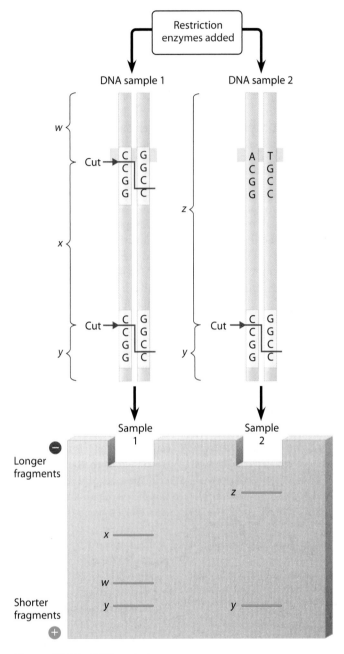

Figure 12.16 RFLP analysis

similarities as well as differences between the base sequences in DNA from two individuals. The restriction fragment analysis in Figure 12.16 clearly shows that the two DNA samples differ in sequence.

? Explain the meaning of the statement "A population has restriction fragment length polymorphisms."

■ There are differences (polymorphisms) in the lengths of restriction fragments produced when restriction enzymes cut comparable DNA samples from different individuals.

12.17   Genomics is the scientific study of whole genomes

By the 1980s, biologists were using RFLPs to help map important genes in humans and some other organisms. It didn't take long for biologists to think on a larger scale. In 1995, a team of scientists announced that they had determined the nucleotide sequence of the entire genome of *Haemophilus influenzae,* a bacterium that can cause several human diseases, including pneumonia and meningitis. **Genomics,** the science of studying a complete set of genes (a genome) and their interactions, was born.

Since 1995, researchers have used the tools and techniques of DNA technology to develop more and more detailed maps of the genomes of a number of species. The first targets of genomics research were bacteria, which have relatively little DNA. The genome of *H. influenzae,* for example, contains only 1.8 million nucleotides and 1,709 genes. But soon the attention of genomics researchers turned toward more complex organisms with much larger genomes. As of 2007, the genomes of over 600 species have been published, and over 1,000 more are in progress. Table 12.17 lists a few of the completed genomes. The vast majority of genomes under study are from prokaryotes, including *Escherichia coli* and a few hundred other bacteria (some of medical importance), and a few dozen archaea. There are 50 or so eukaryotic species that have been sequenced, including vertebrates, invertebrates, fungi, and plants.

Baker's yeast (*Saccharomyces cerevisiae*) was the first eukaryote to have its full sequence determined, and the roundworm *Caenorhabditis elegans* was the first multicellular organism. Other sequenced animals include the fruit fly (*Drosophila melanogaster*) and the lab mouse (*Mus musculus*), both model organisms for genetics. Plants, such as *Arabidopsis thaliana* (another important research organism) and rice (*Oryza sativa,* one of the world's most economically important crops), have also been completed. Other recently completed eukaryotic genomes include the honeybee, the dog, the chicken, and the sea urchin.

In 2005, researchers completed the genome sequence for our closest living relative on the evolutionary tree of life, the chimpanzee (*Pan troglodytes*). Comparisons with human DNA revealed that we share 96% of our genome with our closest animal relative. As you will see in Module 12.21, genomic scientists are currently finding and studying the important differences, shedding scientific light on the age-old question of what makes us human.

Why map so many genomes? Not only are all genomes of interest in their own right, but comparative analysis provides invaluable insights into the evolutionary relationships among organisms. Also, the maps of a variety of genomes help scientists interpret the human genome. For example, when scientists find a nucleotide sequence in the human genome similar to a yeast gene whose function is known, they have a valuable clue to the function of the human sequence. Indeed, several yeast protein-coding genes are so similar to certain human disease-causing genes that researchers have figured out the functions of the disease genes by studying their normal yeast counterparts. Many genes of disparate organisms are turning out to be astonishingly similar, to the point that one researcher has joked that he now views fruit flies as "little people with wings."

?   Why is it useful to sequence nonhuman genomes?

■ Besides their value in understanding evolution, comparative analysis of nonhuman genes helps scientists interpret human data.

TABLE 12.17	SOME IMPORTANT COMPLETED GENOMES		
Organism	Year Completed	Size of Genome (in Base Pairs)	Approximate Number of Genes
Haemophilus influenzae (bacterium)	1995	1.8 million	1,700
Saccharomyces cerevisiae (yeast)	1996	13 million	6,200
Escherichia coli (bacterium)	1997	4.6 million	4,400
Caenorhabditis elegans (nematode)	1998	97 million	19,000
Drosophila melanogaster (fruit fly)	2000	180 million	13,700
Arabidopsis thaliana (mustard plant)	2000	118 million	25,500
Mus musculus (mouse)	2001	2.6 billion	22,000
Oryza sativa (rice)	2002	430 million	60,000
Homo sapiens (humans)	2003	3.2 billion	21,000
Rattus norvegius (lab rat)	2004	2.8 billion	25,000
Pan troglodytes (chimpanzee)	2005	3.1 billion	22,000
Macaca mulatta (macaque)	2007	2.9 billion	22,000

12.18 The Human Genome Project revealed that most of the human genome does not consist of genes

The most ambitious genomics project to date has been the **Human Genome Project (HGP)**. The goals of the HGP were to determine the nucleotide sequence of all DNA in the human genome and to identify the location and sequence of every gene. The HGP began in 1990 as an effort by a consortium of 20 government-funded research centers in six countries. Several years into the project, private companies, chiefly Celera Genomics in the United States, joined the effort. At the completion of the final draft of the sequence in 2004, over 99% of the genome had been determined to 99.999% accuracy. As of 2007, there remain a few hundred gaps of unknown sequences within the human genome that will require special methods to figure out. The DNA sequences determined by the HGP have been deposited in a database, called Genbank, that is publically available via the Internet.

The chromosomes in the human genome (22 autosomes plus the X and Y sex chromosomes) contain approximately 3.2 billion nucleotide pairs of DNA. To try to get a sense of this quantity of DNA, imagine that its nucleotide sequence is printed in letters (A, T, C, and G) like the letters in this book. At this size, the sequence would fill a stack of books 18 stories high! The biggest surprise from the HGP is the small number of human genes. The current estimate is about 21,000 genes—very close to the number found in the nematode worm. How, then, do we account for human complexity? Part of the answer may lie in alternative RNA splicing (see Module 11.6); scientists think that a typical human gene probably specifies several polypeptides.

In humans, like most complex eukaryotes, only a small amount of our total DNA (about 1.5%) is contained in genes that code for proteins, tRNAs, or rRNAs (Figure 12.18). Most multicellular eukaryotes have a huge amount of noncoding DNA—about 98.5% of human DNA is of this type. Some noncoding DNA is made up of gene control sequences such as promoters and enhancers (see Chapter 11). The remaining noncoding DNA has been dubbed "junk DNA," a tongue-in-cheek way of saying that scientists don't fully understand its functions.

Much of the DNA between genes consists of repetitive DNA, nucleotide sequences present in many copies in the genome. Some of this noncoding DNA is used to create DNA profiles by STR analysis, as discussed in Module 12.14. Stretches of DNA with thousands of such repetitions are also prominent at the centromeres and ends of chromosomes—called **telomeres**—suggesting that this DNA plays a role in chromosome structure.

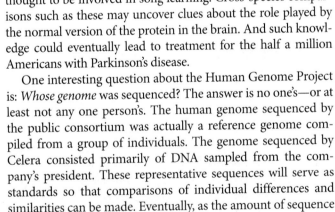

Figure 12.18 Composition of the human genome

In the second main type of repetitive DNA, each repeated unit is hundreds of nucleotides long, and the copies are scattered around the genome. Most of these sequences seem to be associated with **transposable elements** ("jumping genes"), DNA segments that can move or be copied from one location to another in a chromosome and even between chromosomes. Researchers believe that transposable elements, through their copy-and-paste mechanism, are responsible for the proliferation of dispersed repetitive DNA in the human genome.

The potential benefits of having a complete map of the human genome are enormous. For instance, hundreds of disease-associated genes have already been identified. One example is the gene that is mutated in an inherited type of Parkinson's disease, a debilitating brain disorder that causes tremors of increasing severity. Until recently, Parkinson's disease was thought to have only an environmental basis; there was no evidence of a hereditary component. But data from the Human Genome Project mapped a very small number of cases of Parkinson's disease to a specific gene. Interestingly, an altered version of the protein encoded by this gene has also been tied to Alzheimer's disease, suggesting a link between these two brain disorders. Moreover, the same gene is also found in rats, where it plays a role in the sense of smell, and in zebra finches, where it is thought to be involved in song learning. Cross-species comparisons such as these may uncover clues about the role played by the normal version of the protein in the brain. And such knowledge could eventually lead to treatment for the half a million Americans with Parkinson's disease.

One interesting question about the Human Genome Project is: *Whose genome* was sequenced? The answer is no one's—or at least not any one person's. The human genome sequenced by the public consortium was actually a reference genome compiled from a group of individuals. The genome sequenced by Celera consisted primarily of DNA sampled from the company's president. These representative sequences will serve as standards so that comparisons of individual differences and similarities can be made. Eventually, as the amount of sequence data multiplies, the small differences that account for individual variation within our species will come to light.

? The human genome consists of about _____ nucleotides and _____ genes spread over _____ different chromosomes (provide three numbers).

3.2 billion . . . 21,000 . . . 24 (22 autosomes plus two sex chromosomes)

12.19 The whole-genome shotgun method of sequencing a genome can provide a wealth of data quickly

Sequencing an entire genome is a complex task that requires careful work. The Human Genome Project proceeded through three stages that provided progressively more detailed views of the human genome. First, geneticists combined pedigree analyses of large families to map over 5,000 genetic markers (mostly RFLPs) spaced throughout all of the chromosomes. The resulting low-resolution *linkage map* provided a framework for mapping other markers and for arranging later, more detailed maps of particular regions. Next, researchers determined the number of base pairs between the markers in the linkage map. These data helped them construct a *physical map* of the human genome. Finally came the most arduous part of the project: determining the nucleotide sequences of the set of DNA fragments that had been mapped. Advances in automated DNA sequencing were crucial to this endeavor.

This three-stage approach is logical and thorough. However, in 1992, molecular biologist J. Craig Venter proposed an alternative strategy called the **whole-genome shotgun method** and set up the company Celera Genomics to implement it. His idea was essentially to skip the genetic and physical mapping stages and start directly with the sequencing step. In the whole genome shotgun method, an entire genome is chopped by restriction enzymes into fragments that are cloned and sequenced in just one stage (Figure 12.19). High-performance computers running specialized mapping software can assemble the millions of overlapping short sequences into a single continuous sequence.

Today, the whole-genome shotgun approach is the method of choice for genomic researchers because it is fast and relatively inexpensive. However, recent research has revealed some limitations of this method, suggesting that a hybrid approach that combines whole genome shotgunning with physical or genetic maps may prove to be the most useful method in the long run.

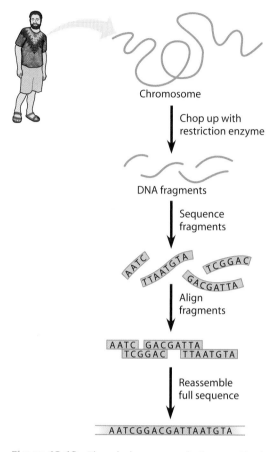

Figure 12.19 The whole-genome shotgun method

? What is the primary advantage of the whole-genome shotgun method?

■ It is faster and cheaper than the three-stage method of genome sequencing.

12.20 Proteomics is the scientific study of the full set of proteins encoded by a genome

The successes in the field of genomics have encouraged scientists to attempt similar systematic study of the full protein sets (proteomes) encoded by genomes, an approach called **proteomics.** The number of proteins in humans far exceeds the number of genes (about 100,000 proteins versus about 21,000 genes). And since proteins, not genes, actually carry out most of the activities of the cell, scientists must study when and where proteins are produced in an organism and how they interact in order to understand the functioning of cells and organisms. Assembling and analyzing proteomes pose many experimental challenges, but ongoing advances are providing the tools to meet those challenges.

Genomics and proteomics are enabling biologists to approach the study of life from an increasingly holistic perspective. Biologists are now in a position to compile catalogs of genes and proteins—that is, a listing of all the "parts" that contribute to the operation of cells, tissues, and organisms. With such catalogs in hand, researchers are shifting their attention from the individual parts to how they function together in biological systems.

? If every protein is encoded by a gene, how can humans have many more proteins than genes? *Hint:* See Module 11.6.

■ The mRNA from one gene may be spliced several different ways to produce several different proteins.

12.21 Genomes hold clues to the evolutionary divergence of humans and chimps

Comparisons of genome sequences from different species allow geneticists to evaluate the evolutionary relationships between those species. The more similar in sequence the same gene is in two species, the more closely related those species are in their evolutionary history. Comparing genes of closely related species sheds light on recent evolutionary events, whereas comparing those of distantly related species helps us understand more ancient evolutionary history. Indeed, comparisons of the completed genome sequences of bacteria, archaea, and eukarya strongly support the theory that these are the three fundamental domains of life.

The small number of gene differences between closely related species makes it easier to correlate phenotypic differences between the species with particular genetic differences. The completion of the chimpanzee genome in 2005 allows us to compare human, chimp, and other sequenced mammalian genomes, such as those of the mouse or rat. Identifying the genes shared by all of these species, but not by non-mammals, may give clues about what it takes to make a mammal, while the genes shared by chimps and humans, but absent from the rodents, may tell us something about primates (Figure 12.21).

An analysis of the overall composition of the human and chimp genomes reveals that they differ by 1.2% in single-base substitutions. Researchers were surprised when they found a further 2.7% difference due to insertions or deletions of larger regions in the genome of one or the other species, with many of the insertions being duplications or other repetitive DNA. In fact, a third of the human duplications are not present in the chimp genome, and some of these duplications contain regions associated with human diseases. All of these observations provide clues to the forces that might have swept the two genomes along different paths, but we don't have a complete picture yet.

What about specific genes and types of genes that differ between humans and chimps? Using evolutionary analyses, biologists have identified a number of genes that have evolved or are evolving faster in humans than in either mouse or chimp. Among them are genes involved in defense against malaria and tuberculosis and at least one gene regulating brain size.

One gene that shows evidence of rapid change in the human lineage is called *FOXP2*. Several lines of evidence suggest that *FOXP2* functions in speech and vocalization, characteristics that obviously distinguish the two species. For example, humans with mutations in this gene

Figure 12.21 What genomic information makes us human or chimp?

generally have severe speech and language impairment. Additionally, there is evidence that the *FOXP2* gene is expressed in the brains of zebra finches and canaries at the time when these songbirds are learning their songs, showing that the role of this gene in communication may be more widespread than in just mammals. These results support the idea that the differences between the *FOXP2* gene in humans and chimps may play a role in the ability of humans, but not chimps, to communicate by speech. The functional effect of the specific amino acid differences between the human and chimp FOXP2 proteins remains to be determined and is of great interest to scientists.

Other research efforts are under way to extend genomic studies to many more microbial species, additional primates, and neglected species from diverse branches of the tree of life. These studies will advance our understanding of all aspects of biology, including health and ecology as well as evolution.

 How can cross-species comparisons of the nucleotide sequences of a gene provide insight into evolution?

■ Similarities in gene sequences correlate with evolutionary relatedness; greater genetic similarities reflect a more recent shared ancestry.

Chapter Review

Reviewing the Concepts

DNA technology, methods for studying and manipulating genetic material, has revolutionized the field of forensics **(Introduction).**

Gene Cloning (12.1–12.5)

Gene cloning is one application of biotechnology, the manipulation of organisms or their components to make useful products. Researchers can insert desired genes into plasmids, creating recombinant DNA, and insert those plasmids into bacteria. If the recombinant bacteria multiply into a clone, the foreign genes are also copied and the desired product (more copies of the gene or its protein product) can be harvested **(12.1).** The tools used to make recombinant DNA include restriction enzymes, which cut DNA at specific sequences (called restriction sites) forming restriction fragments, and DNA ligase, which "pastes" DNA fragments together **(12.2).**

Genomic libraries, sets of DNA fragments containing all of an organism's genes, can be constructed and stored in cloned bacterial plasmids, phages, or bacterial artificial chromosomes (BACs) **(12.3):**

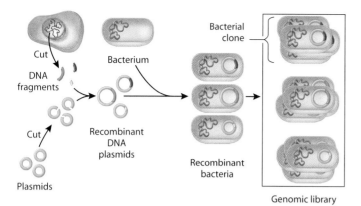

Reverse transcriptase can be used to make cDNA libraries containing only the genes that are transcribed by a particular type of cell (**12.4**). A nucleic acid probe, a short, single-stranded molecule of labeled DNA or RNA, can tag a desired gene in a library (**12.5**).

Genetically Modified Organisms (12.6–12.10)

Applications of gene cloning include the mass production of gene products for medical and other uses. Different organisms, including bacteria, yeast, and mammals, can be used for this purpose (**12.6**). Researchers use gene cloning to produce hormones, diagnose and treat diseases, and produce vaccines (**12.7**).

Recombinant DNA technology can be used to produce new genetic varieties of plants and animals; a number of important crop plants are genetically modified (**12.8**). Genetic engineering involves risks, such as ecological damage from GM crops (**12.9**). Gene therapy may one day be used to treat both genetic diseases and nongenetic disorders; progress is slow, however (**12.10**).

DNA Profiling (12.11–12.16)

DNA profiling, the analysis of DNA fragments, can conclusively determine whether two samples of DNA come from the same individual (**12.11**). The polymerase chain reaction (PCR) can be used to clone a small sample of DNA quickly, producing enough copies for analysis. The use of specific primers that flank the desired sequence ensures that only a particular subset of the DNA sample will be copied (**12.12**). Gel electrophoresis can sort DNA molecules by size (**12.13**):

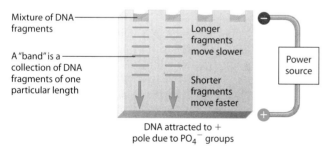

Short tandem repeats (STRs) are stretches of DNA within the genome that contain short nucleotide sequences repeated many times in a row. DNA profiling by STR analysis involves amplifying a standard set of 13 STRs and visualizing them on a gel (**12.14**). DNA profiling has many practical applications, including helping to solve crimes, establishing paternity, and identifying crime victims (**12.15**). Restriction fragment length polymorphisms (RFLPs) reflect differences in the sequences of DNA samples. After being cut by restriction enzymes, the fragments are run through a gel (**12.16**).

Genomics (12.17–12.21)

Genomics, the study of whole genomes, has advanced rapidly in recent years with the sequencing of many prokaryotic and eukaryotic genomes. Besides being interesting themselves, nonhuman genomes can be compared with the human genome (**12.17**). Data from the Human Genome Project (HGP) revealed that the haploid human genome contains about 21,000 genes and a huge amount of noncoding DNA. Much of the noncoding DNA consists of repetitive nucleotide sequences and transposable elements that can move about within the genome (**12.18**). The HGP used genetic and physical mapping of chromosomes followed by DNA sequencing. Modern genomic analysis often uses the faster whole-genome shotgun method (**12.19**). Proteomics is the study of organisms' full sets of proteins (**12.20**). Genomic data provides insight into recent and ancient evolution (**12.21**).

Connecting the Concepts

1. Imagine you have found a small quantity of DNA. Fill in the following diagram, which outlines a series of DNA technology experiments you could perform to study this DNA.

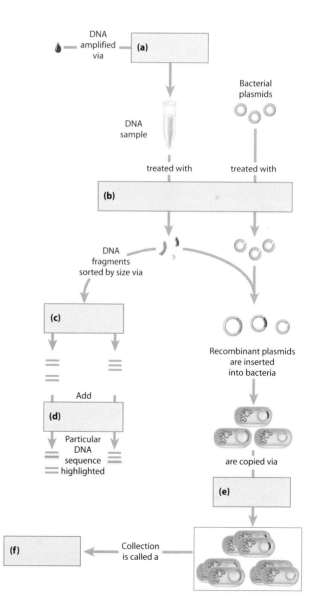

Testing Your Knowledge

Multiple Choice

2. Which of the following would be considered a transgenic organism?
   a. a bacterium that has received genes via conjugation
   b. a human given a corrected human blood-clotting gene
   c. a fern grown in cell culture from a single fern root cell
   d. a rat with rabbit hemoglobin genes
   e. a human treated with insulin produced by bacteria

3. When a typical restriction enzyme cuts a DNA molecule, the cuts are uneven so that the DNA fragments have single-stranded ends. These ends are useful in recombinant DNA work because
   a. they enable a cell to recognize fragments produced by the enzyme.
   b. they serve as starting points for DNA replication.
   c. the fragments will bond to other fragments with complementary ends.
   d. they enable researchers to use the fragments as molecular probes.
   e. only single-stranded DNA segments can code for proteins.

4. The DNA profiles used as evidence in a murder trial look something like supermarket bar codes. The pattern of bars in a DNA profile shows
   a. the order of bases in a particular gene.
   b. the presence of various-sized fragments from chopped-up DNA.
   c. the presence of dominant or recessive alleles for particular traits.
   d. the order of genes along particular chromosomes.
   e. the exact location of a specific gene in a genomic library.

5. A biologist isolated a gene from a human cell, attached it to a plasmid, and inserted the plasmid into a bacterium. The bacterium made a new protein, but it was nothing like the protein normally produced in a human cell. Why? (*Explain your answer.*)
   a. The bacterium had undergone transformation.
   b. The gene did not have sticky ends.
   c. The gene contained introns.
   d. The gene did not come from a genomic library.
   e. The biologist should have cloned the gene first.

6. A paleontologist has recovered a tiny bit of organic material from the 400-year-old preserved skin of an extinct dodo. She would like to compare DNA from the sample with DNA from living birds. Which of the following would be most useful for increasing the amount of DNA available for testing?
   a. restriction fragment analysis
   b. polymerase chain reaction
   c. molecular probe analysis
   d. electrophoresis
   e. Ti plasmid technology

7. How many genes are there in a human sperm cell?
   a. 23
   b. 46
   c. 5,000–10,000
   d. about 25,000
   e. about 3 billion

Describing, Comparing, and Explaining

8. Explain how you might engineer *E. coli* to produce human growth hormone (HGH) using the following: *E. coli* containing plasmids, DNA carrying a gene for HGH, DNA ligase, a restriction enzyme, equipment for manipulating and growing bacteria, a method for extracting and purifying the hormone, an appropriate DNA probe. (Assume that the human HGH gene lacks introns.)

9. Recombinant DNA techniques are used to custom-build bacteria for two main purposes: to obtain multiple copies of certain genes and to obtain useful proteins produced by certain genes. Give an example of each of these applications in medicine and agriculture.

Applying the Concepts

10. A biochemist hopes to find a gene in human liver cells that codes for an important blood-clotting protein. She knows that the nucleotide sequence of a small part of the gene is CTGGACTGACA. Briefly explain how to obtain the desired gene.

11. What is left for genetic researchers to do now that the Human Genome Project has determined the complete nucleotide sequences of nearly all of the human chromosomes? Explain.

12. Today, it is fairly easy to make transgenic plants and animals. What are some important safety and ethical issues raised by this use of recombinant DNA technology? What are some of the possible dangers of introducing genetically engineered organisms into the environment? What are some reasons for and against leaving decisions in these areas to scientists? To business owners and executives? What are some reasons for and against more public involvement? How might these decisions affect you? How do you think these decisions should be made?

13. In the not-too-distant future, gene therapy may be an option for the treatment and cure of some inherited disorders. What do you think are the most serious ethical issues that must be dealt with before human gene therapy is used on a large scale? Why do you think these issues are important?

14. The possibility of extensive genetic testing raises questions about how personal genetic information should be used. For example, should employers or potential employers have access to such information? Why or why not? Should the information be available to insurance companies? Why or why not? Is there any reason for the government to keep genetic files? Is there any obligation to warn relatives who might share a defective gene? Might some people avoid being tested for fear of being labeled genetic outcasts? Or might they be compelled to be tested against their wishes? Can you think of other reasons to proceed with caution?

Answers to all questions can be found in Appendix 4.

For Practice Quizzes, BioFlix, MP3 Tutors, and Activities, go to www.mybiology.com.

Concepts of Evolution

chapter

13

How Populations Evolve

Clown, Fool, or Simply Well Adapted?

Is the blue-footed booby really as awkward or foolish as its name implies? With its bright blue feet and trusting demeanor, boobies were certainly noticed by early travelers to the Galápagos Islands, a chain of relatively young volcanic islands located about 900 km off the Pacific coast of South America. Spanish sailors may have called them "clown" (bobo in Spanish). British seamen may have called them "booby" (slang for "stupid"), as the birds were so approachable that they were easily killed. Either way, their comical feet are hard to miss.

Charles Darwin, whose observations in the Galápagos contributed greatly to his theory of evolution, no doubt encountered this friendly bird. And like the other species he observed, boobies have physical features that suit them to their environment. For example, their large, webbed feet make great flippers, propelling the birds through the water at high speeds. In the clear Galápagos waters, boobies gracefully "fly" beneath the surface. Their feet are thus a huge advantage when hunting fish. On land, however, those same feet are awkward, making for klutzy walking and for an even clumsier flight takeoff; their wings often touch the ground before they become airborne.

Boobies have other traits that serve them well in their seafaring life. The booby's body and bill are streamlined, like a torpedo, minimizing friction when it dives from heights of up to 24 meters into the shallow water below. To pull themselves out of this high-speed dive once they hit the water, boobies use their large tail. Specialized glands help boobies manage salt intake. A gland in the bird's eye socket accumulates excess salt from body fluids. This salty liquid then trickles into the nasal cavity. The bird expels this liquid by shaking its head from side to side, a characteristic clown-like movement.

The booby's big webbed feet, streamlined shape, large tail, and specialized salt-excreting glands are all good examples of evolutionary **adaptations,** inherited traits that enhance an organism's ability to survive and reproduce in a particular environment. As useful as adaptations may be, however, they often represent a trade-off between different needs. The booby's webbed feet are a case in point: extremely functional in water, but clumsy on dry land. But because boobies spend a great deal of time in the water, and especially because they find most of their food there, this adaptation represents a net advantage.

In this chapter, we examine how such adaptations evolve. We thus begin our study of **evolution,** the core theme of biology. We explore Darwin's theory of evolution—that Earth's many species are descendants of ancestral organisms that were different from those living today. In Chapter 14 we trace Darwin's ideas on the origin of new species. And in Chapter 15 we present the milestones in the long history of life on Earth and consider the mechanisms involved in such large evolutionary changes. It is these changes in species over time that have transformed life on Earth from its earliest forms to the vast diversity that we see today. Let's begin this grand journey with Darwin's sea voyage. ▪ ▪ ▪

13.1 A sea voyage helped Darwin frame his theory of evolution

If you visit the Galápagos Islands today, you will see many of the same sights that fascinated Darwin over a century ago: blue-footed boobies waddling around; lumbering, long-necked giant tortoises (whose Spanish name is *galápagos*) (Figure 13.1A); and iguanas basking on dark lava rocks (Figure 13.1B). The marine iguanas feed on algae in the ocean. Their partially webbed feet and flattened tail suit their aquatic life. You might also observe some of the finches Darwin collected—closely related small birds that represent more than a dozen different species. (For now, we can define *species* as a group of organisms with similar anatomical features.) The most striking differences among these birds are the shapes of their beaks, which are adapted for crushing seeds, feeding on cactus flowers, or snagging insects. Some finches find their meals on tortoises, who raise themselves up on their legs and stretch out their necks to allow the birds to hop aboard and rid them of ticks and other parasites.

One of Darwin's lasting contributions is the scientific explanation for the striking ways in which organisms, such as these diverse inhabitants of the Galápagos Islands, are suited for life in their environment. Let's trace the path Darwin took to his theory of evolution.

Darwin's Cultural and Scientific Context Some early Greek philosophers suggested that life might change gradually over time. But the Greek philosopher Aristotle, whose views had an enormous impact on Western culture, generally viewed species as perfect and permanent. Judeo-Christian culture fortified this idea with a literal interpretation of the Book of Genesis, which holds that species were individually designed by a divine creator. The two ideas dominating the intellectual and cultural climate of the Western world for centuries were that Earth was only about 6,000 years old and that all living species were static in their form.

In the century prior to Darwin, only a few scientists questioned the belief that species are fixed, unchanging, and perfect.

In the mid-1700s, the study of **fossils** (the imprints or remains of organisms that lived in the past) revealed a succession of fossil forms in layers of sedimentary rock that differed from current life forms. In the early 1800s, French naturalist Jean Baptiste Lamarck suggested that the best explanation for the relationship of fossils to current organisms is that life evolves. Today, we remember Lamarck mainly for his erroneous view of *how* species evolve. He proposed that by using or not using its body parts, an individual may change its traits and then pass those changes on to its offspring. Lamarck's idea is known as the inheritance of acquired characteristics. He suggested, for instance, that the ancestors of the giraffe had lengthened their necks by stretching higher and higher into the trees to reach leaves. Our understanding of genetics refutes Lamarck's acquired characteristic mechanism. But the important fact remains that by strongly advocating evolution and by proposing that species evolve as a result of interactions with their environment, Lamarck helped set the stage for Darwin.

Darwin's Sea Voyage Charles Darwin was born in 1809. Even as a boy, Darwin had a consuming interest in nature. When not reading nature books, he was fishing, hunting, and collecting insects. His father, an eminent physician, could see no future for his son as a naturalist and sent him to medical school. But Darwin found medicine boring and surgery before the days of anesthesia horrifying. He enrolled at Cambridge University, intending to become a clergyman. At that time, many scholars of science belonged to the clergy.

Soon after graduation, Darwin's botany professor and mentor recommended him to the captain of the HMS *Beagle*, who was preparing a voyage to chart poorly known stretches of the South American coast. In December 1831, at the age of 22, Darwin began the round-the-world voyage (Figure 13.1C) that profoundly influenced his thinking and eventually the thinking of the world.

Figure 13.1A A giant tortoise, one of the unique inhabitants of the Galápagos Islands

Figure 13.1B A marine iguana basking on lava rocks

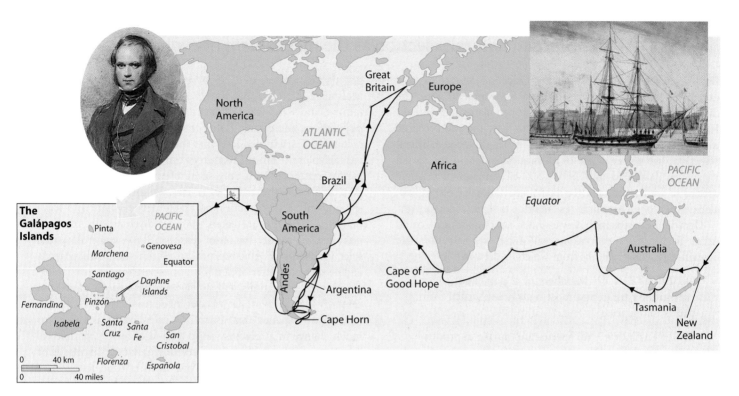

Figure 13.1C   The voyage of the *Beagle* (1831–1836). The insets show a young Charles Darwin and his ship.

While the ship's crew surveyed the coast, Darwin spent most of his time onshore, collecting thousands of plants and animals, as well as fossils. He noted the unique adaptations of these organisms in places as different as the Brazilian jungle, the grasslands of Argentina, and the desolate and frigid lands at the southern tip of South America. Darwin asked himself why fossils found on the South American continent were more similar to present-day South American species than to fossils found on other continents. During his visit to the Galápagos Islands, Darwin observed that these islands had many unique organisms, most of which were similar to, but different from, the plants and animals of the nearest mainland. Even the individual islands had species different from those on other islands.

While on the voyage, Darwin was strongly influenced by Charles Lyell's newly published *Principles of Geology*. Having read Lyell's book and witnessed an earthquake that raised part of the coastline of Chile almost a meter, Darwin realized that natural forces gradually changed Earth's surface and that these forces are still operating in modern times. Thus, the growth of mountains as a result of earthquakes could account for the presence of the marine snail fossils that he had collected on mountaintops in the Andes.

By the time Darwin returned to Great Britain five years after the Beagle first set sail, his experiences and reading had led him to seriously doubt that Earth and all its living organisms had been specially created only a few thousand years earlier. Darwin had come to realize that Earth was very old and constantly changing. Once in Great Britain, Darwin began to analyze his collections and to discuss them with colleagues. He continued to read, correspond with other scientists, and to maintain extensive journals of his observations, studies, and thoughts.

Darwin's Writings   By the early 1840s, Darwin had composed a long essay describing the major features of his theory of evolution. He realized that his ideas would cause a social furor, however, and he delayed publishing his essay. Even as he procrastinated, Darwin continued to compile evidence in support of his hypothesis. In the mid-1850s, Alfred Wallace, a British naturalist doing fieldwork in Indonesia, conceived a theory almost identical to Darwin's. Wallace asked Darwin to evaluate the manuscript he had written about his theory to see if it merited publication. In 1858, two of Darwin's colleagues presented Wallace's paper and excerpts of Darwin's earlier essay together to the scientific community. With the publication in 1859 of his book, *On the Origin of Species by Means of Natural Selection,* Darwin presented the world with a strong, logical explanation for evolution.

Darwin presented evidence that present-day species arose from a succession of ancestors. He called this evolutionary history of life "descent with modification." As the descendants of a remote ancestor spread into various habitats over millions and millions of years, they accumulated diverse modifications, or adaptations, that fit them to specific ways of life. Darwin's proposed mechanism for this descent with modification is natural selection, the topic of our next module.

Web Activity *Darwin and the Galápagos Islands*

Web Activity *The Voyage of the* Beagle: *Darwin's Trip Around the World*

? What was Darwin's phrase for evolution? What does it mean?

■ Descent with modification. An ancestral species could diversify into many descendant species by the accumulation of adaptations to various environments.

Darwin devoted much of *On the Origin of Species* to how organisms become adapted to their environment. First, he discussed familiar examples of domesticated plants and animals. Humans have modified other species by selecting and breeding individuals that possess desired traits—a process called **artificial selection.** We see evidence of artificial selection in the varied vegetables illustrated in Figure 13.2, all varieties of a single species of wild mustard. Crop plants and animals bred as livestock or pets often bear little resemblance to their wild ancestors.

Darwin then explained how a similar selection process could occur in nature, a process he called **natural selection.** He described two observations from which he drew two inferences:

OBSERVATION #1: Members of a population often vary in their traits, and most traits are inherited from parent to offspring.

OBSERVATION #2: All species are capable of producing more offspring than the environment can support.

INFERENCE #1: Individuals whose inherited traits give them a higher probability of surviving and reproducing in a given environment tend to leave more offspring than other individuals.

INFERENCE #2: This unequal production of offspring will cause favorable traits to accumulate in a population over generations.

Darwin recognized the connection between the capacity of organisms to produce excessive numbers of offspring and the process of natural selection. He had read an essay on human population written in 1798 by economist Thomas Malthus. Malthus contended that much of human suffering—disease, famine, and war—was the inescapable consequence of human populations increasing faster than food supplies and other resources. Darwin deduced that the production of more individuals than the limited resources can support leads to a struggle for existence, with only some offspring surviving in each generation. Of the many eggs laid, young born, and seeds spread, only a tiny fraction complete development and leave offspring. The rest are eaten, starved, diseased, unmated, or unable to reproduce for other reasons. The essence of natural selection is this unequal reproduction. Individuals whose traits enable them to better obtain food or escape predators or tolerate physical conditions will survive and reproduce more successfully than individuals without those traits.

Darwin reasoned that if artificial selection brings about so much change in a relatively short period of time, then natural selection could modify species considerably over hundreds or thousands of generations. Over vast spans of time, many traits that adapt a population to its environment will accumulate. If the environment changes, however, or if individuals move to a new environment, natural selection will select for adaptations to these new conditions. Major environmental changes, however, often lead to **extinction,** a frequent event in Earth's history resulting in the irrevocable loss of a species.

It is important to emphasize three key but subtle points about evolution by natural selection. The first point is that although natural selection occurs through interactions between individual organisms and their environment, individuals do not evolve. Evolution refers to generation-to-generation changes in populations.

A second key point is that natural selection can amplify or diminish only heritable traits. Certainly an organism may become modified through its own interactions with the environment during its lifetime, and those acquired characteristics may help the organism survive there. But unless coded for in the genes of an organism's gametes, such acquired characteristics cannot be passed on to offspring.

Third, evolution is not goal directed; it does not lead to perfectly adapted organisms. Natural selection is the result of environmental factors that vary from place to place and from time to time. A trait that is favorable in one situation may be useless—or even detrimental—in different circumstances. And adaptations are often compromises: A blue-footed booby's feet may be clumsy on land but efficient in water.

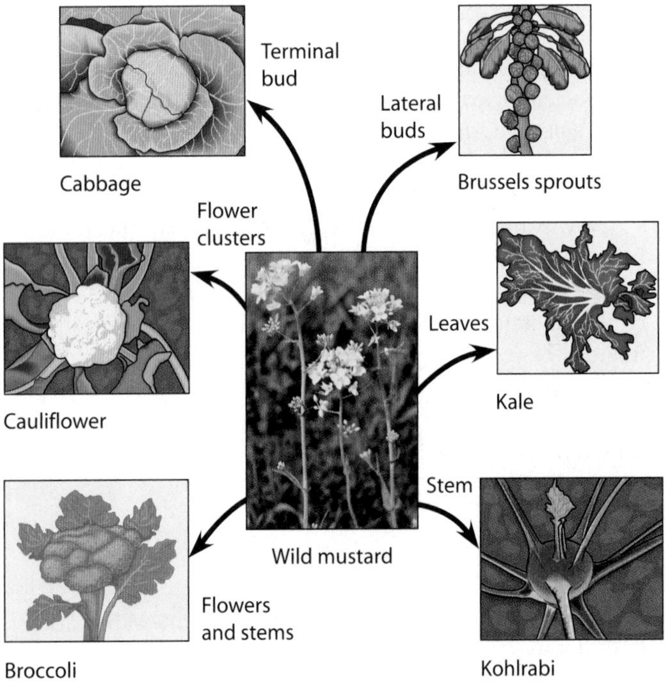

🎧 **MP3 Tutor** *Natural Selection*

Web Process of Science *How Do Environmental Changes Affect a Population?*

❓ Compare artificial selection and natural selection.

Figure 13.2 Artificial selection: different vegetables produced by selecting variations in different parts of the plant

■ In artificial selection, humans choose the desirable traits and breed only organisms with those traits. In natural selection, the environment does the choosing: individuals with traits best suited to the environment survive and reproduce most successfully, passing those adaptive traits to offspring.

Scientists can observe natural selection in action

The blue-footed boobies described in the chapter introduction exhibit traits that are evolutionary adaptations to their ocean-based life. The exquisite camouflage adaptations shown in Figure 13.3A by insects that evolved in different environments are also examples of the results of natural selection. But do we have examples of natural selection in action?

Indeed, biologists have documented evolutionary change in thousands of scientific studies. A classic example involves Peter and Rosemary Grant's work with finches in the Galápagos Islands over a period of 20 years (see Module 14.9). As part of their research, they measured changes in beak size in a population of a ground finch species. These birds eat mostly small seeds. In dry years, when all seeds are in short supply, birds must eat more large seeds. Birds with larger, stronger beaks have a feeding advantage and greater reproductive success, and the average beak depth for the population increases. During wet years, smaller beaks are more efficient for eating the now abundant small seeds, and the average beak depth decreases.

An unsettling example of natural selection in action is the evolution of pesticide resistance in hundreds of insect species. Pesticides are poisons used to kill insect pests in farmlands, swamps, backyards, and homes. Whenever a new type of pesticide is used to control agricultural pests, the story is similar (Figure 13.3B): A relatively small amount of poison dusted onto a crop may kill 99% of the insects, but subsequent sprayings are less and less effective. The few survivors of the first pesticide wave are insects that are genetically resistant, carrying an allele (alternative form of a gene, colored red in the figure) that somehow enables them to resist the chemical attack. So the poison kills most members of the population, leaving the resistant individuals to reproduce and pass the alleles for pesticide resistance to their offspring. The proportion of pesticide-resistant individuals increases in each generation. Like the finches, the insect population has adapted to environmental change through natural selection.

These two examples of evolutionary adaptation highlight three key points about natural selection. First, natural selection is more of an editing process than a creative mechanism. A pesticide does not create alleles that allow insects to resist it. Rather the presence of the pesticide leads to natural selection for insects already in the population who have those alleles. Second, natural selection is contingent on time and place: It favors those characteristics in a varying population that fit the current, local environment. For instance, mutations that endow houseflies with resistance to the pesticide DDT also reduce their growth rate. Before DDT was introduced, such mutations were a handicap to the flies that had them. But once DDT was part of the environment, the mutant alleles were advantageous, and natural selection increased their frequency in fly populations. Finally, these examples show that significant evolutionary change can occur in a short time.

> **?** In what sense is natural selection more of an editing process than a creative process?

■ Natural selection cannot create beneficial traits on demand, but "edits" variation in a population by selecting for those traits that work best in the current environment.

Figure 13.3A Camouflage as an example of evolutionary adaptation. Look carefully for the mantids in these pictures.

A flower mantid in Malaysia

A leaf mantid in Costa Rica

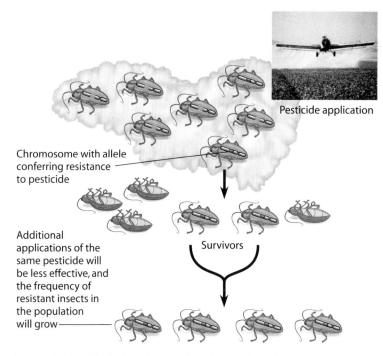

Pesticide application

Chromosome with allele conferring resistance to pesticide

Additional applications of the same pesticide will be less effective, and the frequency of resistant insects in the population will grow

Survivors

Figure 13.3B Evolution of pesticide resistance in an insect population

Darwin developed his theory of descent with modification mainly with evidence from the geographic distribution of species, examples of artificial selection, comparative anatomy (comparing the body structures of different species), and the fossil record. Fossils document differences between past organisms and present organisms and the fact that many species have become extinct.

The photographs on this page illustrate a number of fossils, each of which formed in a somewhat different way. The organic substances of a dead organism usually decay rapidly, but hard parts of an animal that are rich in minerals, such as the bones and teeth of dinosaurs and the shells of clams and snails, may remain as fossils. The fossilized skull in Figure 13.4A is from one of our early relatives, *Homo erectus,* who lived some 1.5 million years ago in Africa.

Many of the fossils that **paleontologists** (scientists who study fossils) find in their digs are not the actual remnants of organisms at all, but are replicas of past organisms. Such fossils result when a dead organism captured in sediment decays and leaves an empty mold that is filled by minerals dissolved in water. The casts that form when the minerals harden are replicas of the organism, as seen in the 375-million-year-old casts of ammonites (which were shelled marine organisms) shown in Figure 13.4B.

Trace fossils are footprints, burrows, and other remnants of an ancient organism's behavior. The boy in Figure 13.4C is standing in a 150-million-year-old dinosaur track in Colorado.

Some fossils actually retain organic material. The leaf in Figure 13.4D is about 40 million years old. It is a thin film pressed in rock, still greenish with remnants of its chlorophyll and well enough preserved that biologists can analyze its molecular and cellular structure. In rare instances, an entire organism, including its soft parts, is fossilized. That can happen only when the individual is buried in a medium that prevents bacteria and fungi from decomposing the corpse. The insect in Figure 13.4E got stuck in the resin of a tree about 35 million years ago. The resin hardened into amber (fossilized resin), preserving the insect. Other media can preserve whole organisms. Explorers have discovered mammoths, bison, and even prehistoric humans frozen in ice or preserved in acid bogs. Such rare discoveries make the news, as did the 1991 discovery of the "Ice Man" in Figure 13.4F, who had died 5,000 years ago. However, biologists rely mainly on the fossils formed in sedimentary rock to reconstruct the history of life.

The **fossil record**—the sequence in which fossils appear within layers of sedimentary rocks—provides some of the strongest evidence of evolution. Sedimentary rocks form from layers of sand and mud that settle to the bottom of seas, lakes,

Figure 13.4 A–F A gallery of fossils

A Skull of *Homo erectus*

B Ammonite casts

C Dinosaur tracks

D Fossilized organic matter of a leaf

E Insect in amber

F "Ice Man"

Figure 13.4G   Strata of sedimentary rock at the Grand Canyon

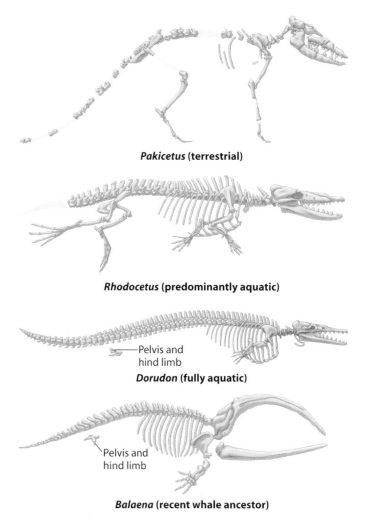

Pakicetus (terrestrial)

Rhodocetus (predominantly aquatic)

Pelvis and
hind limb

Dorudon (fully aquatic)

Pelvis and
hind limb

Balaena (recent whale ancestor)

Figure 13.4H   The transition to life in the sea: fossils of four species that link living whales with their land-dwelling ancestors

and swamps. Over millions of years, deposits pile up and compress the older sediments below into rock. When organisms die, they settle along with the sediments and may be preserved as fossils. The rate of sedimentation and the types of particles that settle vary over time. As a result, the rock forms in **strata,** or layers. Younger strata are on top of older ones; thus, the relative ages of fossils can be determined by the layer in which they are found. Figure 13.4G shows strata of sedimentary rock at the Grand Canyon. The Colorado River has cut through over 2,000 m of rock, exposing sedimentary layers that can be read like huge pages from the book of life. Scan the canyon wall from rim to floor, and you look back through hundreds of millions of years. Each layer entombs fossils that represent some of the organisms from that period of Earth's history. Of course, the fossil record is incomplete: Not all organisms live in areas that favor fossilization, many rocks are distorted by geologic processes, and not all fossils that have been preserved will be found.

The fossil record reveals the historical sequence in which organisms evolved. The oldest known fossils, dating from about 3.5 billion years ago, are prokaryotes. Molecular and cellular evidence also indicates that prokaryotes were the ancestors of all life. As you will learn in Chapter 15, over a billion years passed before fossils of single-celled eukaryotes formed and approximately a billion more before multicellular eukaryotes did. Fossils in younger layers of rock reveal the sequential evolution of various groups of eukaryotic organisms.

Darwin lamented the lack of fossils showing how existing groups of organisms gave rise to new groups. In the 150 years since publication of *The Origin of Species,* however, many fossil discoveries illustrate the transition of existing groups to new groups. For example, a series of fossils traces the gradual modification of jaws and teeth in the evolution of mammals from a reptilian ancestor.

Another series of fossils traces the evolution of whales from four-legged land mammals. Whales living today have forelegs in the form of flippers but lack hind legs, although they do have small hind-leg and foot bones that do not extend from the body. In the past few decades, a series of remarkable fossils of extinct mammals have been discovered in Pakistan, Egypt, and North America that document the transition from life on land to life in the sea (Figure 13.4H). Additional fossils show that *Pakicetus* and *Rodhocetus* had a type of ankle bone that is otherwise unique to the group of land mammals that includes pigs, hippos, cows, camels, and deer. The ankle bone similarity strongly suggests that whales (and dolphins and porpoises) are most closely related to this group of land mammals.

In the next module, we look at other sources of evidence for evolution.

? What types of animals do you think would be most represented in the fossil record? Explain your answer.

■ Animals with hard parts, such as shells or bones that readily fossilize, and those that lived in areas where sedimentary rock may form are most likely to be represented in the fossil record.

13.5 A mass of other evidence reinforces the evolutionary view of life

Darwin's careful documentation convinced many of the scientists of his day that organisms do indeed evolve. Let's continue to explore Darwin's evidence as well as some new types of data that provide overwhelming support for evolution.

Biogeography   It was the geographic distribution of species, known as **biogeography,** that first suggested to Darwin that organisms evolve from ancestral species. Darwin noted that Galápagos animals resembled species of the South American mainland more than they resembled animals on islands that were similar but much more distant. The logical explanation was that the Galápagos species evolved from animals that migrated from South America. These immigrants eventually gave rise to new species as they became adapted to their new environments. There are many other examples from biogeography that make sense only in the historical context of evolution. The unique collection of marsupials (pouched mammals) in Australia, such as kangaroos and koalas, evolved in isolation after geologic activities separated that island continent from the landmasses on which placental mammals diversified.

Comparative Anatomy   Also providing support for evolution and cited extensively by Darwin is comparative anatomy. Anatomical similarities between many species give signs of common descent. Similarity in characteristics that results from common ancestry is known as **homology.** As Figure 13.5A shows, the same skeletal elements make up the forelimbs of humans, cats, whales, and bats. The functions of these forelimbs differ. A whale's flipper does not do the same job as a bat's wing, so if these structures had been uniquely engineered, we would expect that their basic designs would be very different. The logical explanation is, in fact, that the arms, forelegs, flippers, and wings of these different mammals are variations on the anatomical structures of an ancestral organism, structures that over millions of years have become adapted to different functions.

Biologists call such anatomical similarities in different organisms **homologous structures**—features that often have different functions but are structurally similar because of common ancestry.

Comparative anatomy illustrates that evolution is a remodeling process in which ancestral structures that originally functioned in one capacity become modified as they take on new functions—the kind of process that Darwin called descent with modification.

Comparing early stages of development in different animal species reveals additional homologies not visible in adult organisms. For example, at some point in their development, all vertebrate embryos have a tail posterior to the anus, as well as structures called pharyngeal (throat) pouches. These homologous pharyngeal pouches ultimately develop into structures with very different functions, such as gills in fishes and parts of the ears and throat in humans. Note the pharyngeal pouches and tails of the bird embryo (left) and the human embryo (right) in Figure 13.5B, on facing page.

Some of the most interesting homologous structures are **vestigial organs,** structures that are of marginal or perhaps no importance to the organism. Vestigial organs are remnants of structures that served important functions in the organism's ancestors. The small hind-leg and foot bones of modern whales discussed in Module 13.4 are examples of vestigial organs.

Molecular Biology   Anatomical homology is not helpful in linking very distantly related organisms such as plants and animals and microorganisms. But in recent decades, advances in molecular biology have enabled biologists to read a molecular history of evolution in the DNA sequences of organisms. As we saw in Chapter 10, the hereditary background of an organism is documented in its DNA and in the proteins encoded there. Siblings have greater similarity in their DNA and proteins than do unrelated individuals of the same species. And if two species have homologous genes with sequences that match closely, biologists conclude that these sequences must have been inherited from a relatively recent common ancestor. In contrast, the greater the number of sequence differences between species, the more distant is their last common ancestor. Molecular comparisons between diverse organisms have allowed biologists to develop hypotheses about the evolutionary divergence of major branches on the tree of life.

Darwin's boldest hypothesis is that all life-forms are related. Molecular biology provides strong evidence for this claim: All forms of life use the same genetic language of DNA and

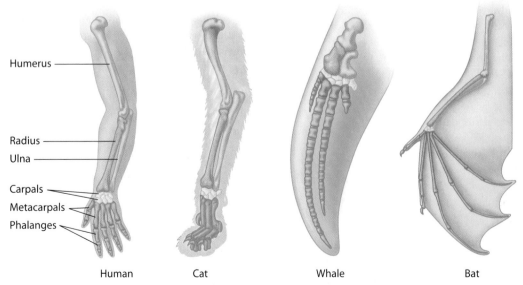

Humerus

Radius
Ulna

Carpals
Metacarpals
Phalanges

Human   Cat   Whale   Bat

Figure 13.5A   Homologous structures: vertebrate forelimbs

RNA, and the genetic code (how RNA triplets are translated into amino acids) is essentially universal. Thus it is likely that all species descended from one or a few common ancestors that used this code. But molecular homologies go beyond a shared code. For example, organisms as dissimilar as humans and bacteria share genes that have been inherited from a very distant common ancestor. Once again, homologies help identify common ancestry.

Web Activity *Reconstructing Forelimbs*

? What is homology? How does the concept of homology relate to molecular biology?

■ Homology is similarity in different species due to evolution from a common ancestor. Similarities in genes or other DNA sequences or proteins reflect the evolutionary relationship that is the basis of homology.

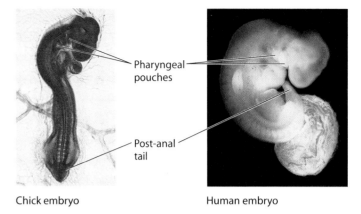

Pharyngeal pouches

Post-anal tail

Chick embryo   Human embryo

Figure 13.5B Homologous structures in vertebrate embryos

13.6 Homologies indicate patterns of descent that can be shown on an evolutionary tree

Darwin was the first to view the history of life as a tree, with multiple branchings from a common ancestral trunk to the descendant species at the tips of the twigs. Biologists represent these patterns of descent with an **evolutionary tree,** although today they usually turn the trees sideways as in Figure 13.6.

Homologous structures, both anatomical and molecular, can be used to determine the branching sequence of such a tree. Some homologous characteristics, such as the genetic code, are shared by all species because they date to the deep ancestral past. In contrast, characteristics that evolved more recently are shared only within smaller groups of organisms. For example, all tetrapods (from the Greek *tetra,* four, and *pod,* foot) possess the same basic limb bone structure illustrated in Figure 13.5A, but their ancestors do not.

Figure 13.6 is an evolutionary tree of tetrapods (amphibians, mammals, and reptiles, including birds) and their closest living relatives, the lungfishes. In this diagram, each branch point represents the common ancestor of all species that descended from it. For example, lungfishes and all tetrapods descended from ancestor ❶. Three homologies are shown by the purple hatch marks on the tree—tetrapod limbs, the amnion (a protective embryonic membrane), and feathers. Tetrapod limbs were present in common ancestor ❷ and hence are found in all of its descendants (the tetrapods). The amnion was present only in ancestor ❸ and thus is shared only by mammals and reptiles. Feathers were present only in ancestor ❻, and hence are found only in birds.

Evolutionary trees are hypotheses reflecting our current understanding of patterns of evolutionary descent. Some trees are more speculative because less data may be available. Others, are based on strong combinations of fossil, anatomical, and DNA sequence data.

? Refer to the evolutionary tree in Figure 13.6. Are crocodiles more closely related to lizards or birds?

■ Look for the most recent common ancestor of these groups. Crocodiles are more closely related to birds because they share a more recent common ancestor with birds (ancestor ❺) than with lizards (ancestor ❹).

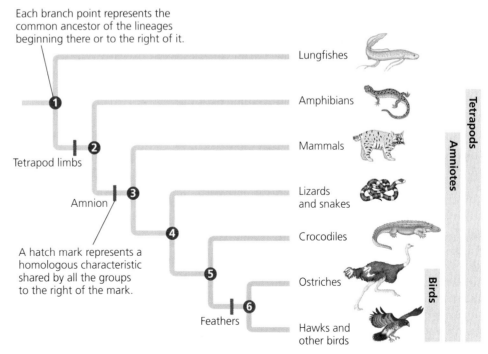

Each branch point represents the common ancestor of the lineages beginning there or to the right of it.

Tetrapod limbs

Amnion

A hatch mark represents a homologous characteristic shared by all the groups to the right of the mark.

Feathers

Lungfishes

Amphibians

Mammals

Lizards and snakes

Crocodiles

Ostriches

Hawks and other birds

Tetrapods

Amniotes

Birds

Figure 13.6 Evolutionary tree for tetrapods and their closest living relatives, the lungfishes. The purple hatch marks indicate the origin of three important homologies.

13.7 Populations are the units of evolution

One common misconception about evolution is that individual organisms evolve during their lifetimes. It is true that natural selection acts on individuals: Each organism's combination of traits affects its survival and reproductive success. But the evolutionary impact of natural selection is only apparent in the changes in a population of organisms over time.

A **population** is a group of individuals of the same species living in the same place at the same time. We can measure evolution as a change in the prevalence of certain heritable traits in a population over a span of generations. The increasing proportion of resistant insects in areas sprayed with pesticide is one example (see Module 13.3). Natural selection favored insects with alleles for pesticide resistance; these insects left more offspring, and the population changed, or evolved.

Let's examine some key features of populations. Different populations of the same species may be isolated from one another, with little interbreeding and thus little exchange of genes between them. Such isolation is common for populations confined to widely separated islands or in different lakes. However, populations are not usually so isolated, and they rarely have sharp boundaries. Figure 13.7 is a nighttime satellite photograph showing the lights of population centers in North America. We know that these populations are not really isolated; people move around, and there are suburban and rural communities between cities. Nevertheless, people are most likely to choose mates locally. As a result, for humans and other species, individuals in one population are generally more closely related to one another than to members of other populations.

In studying evolution at the population level, biologists focus on what is called the **gene pool,** the total collection of genes in a population at any one time. The gene pool consists of all alleles in all the individuals making up a population. For many genes, there are two or more alleles in the gene pool. For example, in an insect population, there may be two alleles relating to pesticide

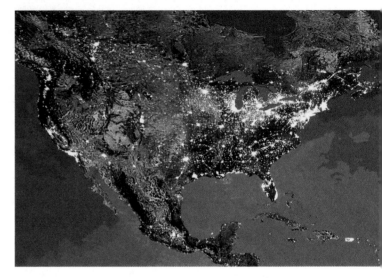

Figure 13.7 Human population centers in North America

breakdown, one that codes for an enzyme that breaks down a certain pesticide and one for a version of the enzyme that does not. In fields sprayed with pesticide, the allele for the enzyme conferring resistance will increase in frequency and the other allele will decrease in frequency. When the relative frequencies of alleles in a population change like this over a number of generations, evolution is occurring on its smallest scale. Such a change in a gene pool is often called **microevolution.**

Next let's look at the genetic variation that makes microevolution possible.

? Why can't an individual evolve?

■ Evolution is only apparent as the changes in the genetic makeup of a population over time. An individual's genes don't change during its lifetime.

13.8 Mutation and sexual reproduction produce genetic variation, making evolution possible

In *The Origin of Species,* Darwin provided evidence that life on Earth has evolved over time, and he proposed that natural selection, in favoring some heritable traits over others, was the primary mechanism for that change. But, he could not explain the cause of variation among the individuals making up a population, nor could he account for how those variations passed from parents to offspring. By breeding peas in his abbey garden, Mendel discovered the very hereditary processes required for natural selection (see Chapter 9). Although Darwin and Mendel were contemporaries, the significance of Mendel's work was not recognized during his or Darwin's lifetime.

Genetic Variation We have no trouble recognizing our friends in a crowd. Each person has a unique genome, reflected in individual variations in appearance and other traits. Indeed, individual variation occurs in all species. In addition to anatomical differences, most populations have a great deal of variation that can be observed at the molecular level.

Of course, not all variation in a population is heritable. The phenotype results from a combination of the genotype, which is inherited, and many environmental influences. For instance, a strength-training program can build up your muscle mass, though you would not pass this environmentally produced physique on to your offspring. Only the genetic component of variation is relevant to natural selection.

Many of the variable traits of the individuals in a population result from the combined effect of several genes. As we saw in Module 9.14, polygenic inheritance produces traits that vary more or less continuously—in human height, for instance, from very short individuals to very tall ones. By contrast, other features, such as human ABO blood groups (see Module 9.12), are determined by a single gene locus, with different alleles producing distinct phenotypes. But where do these alleles come from?

Mutation   New alleles originate by **mutation,** a change in the nucleotide sequence of DNA. Thus, mutation is the ultimate source of the genetic variation that serves as raw material for evolution. In multicellular organisms, however, most mutations occur in body cells and are lost when the individual dies. Only mutations in cells that produce gametes can be passed to offspring and affect a population's genetic variability.

A mutation that substitutes one nucleotide for another will be harmless if it does not affect the function of the protein the DNA encodes. However, if it does affect the protein's function, the mutation will probably be harmful. An organism is a refined product of thousands of generations of past selection, and a random change in its DNA is not likely to improve its genome any more than randomly changing some words on a page is likely to improve a story.

On rare occasions, however, a mutant allele may actually improve the adaptation of an individual to its environment and enhance its reproductive success. This kind of effect is more likely when the environment is changing in such a way that mutations that were once disadvantageous are favorable under the new conditions. The evolution of DDT-resistant houseflies (see Module 13.3) illustrates this point.

Chromosomal mutations that delete, disrupt, or rearrange many gene loci at once are almost certain to be harmful (see Figures 8.24A and 8.24B). But duplication of part of a chromosome is an important source of genetic variation. If a repeated segment of a chromosome can persist over the generations, it may provide a bigger genome with extra genes that may eventually take on new functions by further mutation. For example, the remote ancestors of mammals carried a single gene for an olfactory receptor, which is involved in detecting odors. This gene has been duplicated repeatedly. As a result, mice have about 1,300 different olfactory receptor genes, and humans have about 1,000. It is likely that such dramatic increases in the number of olfactory receptor genes helped early mammals by enabling them to distinguish among many different smells.

In prokaryotes, mutations can quickly generate genetic variation. Bacteria multiply so rapidly that a beneficial mutation can increase its frequency in descendant populations in a matter of hours or days. Because bacteria are haploid, with a single gene for each inherited character, a new allele can have an effect immediately.

Mutation rates in animals and plants average about one in every 100,000 genes per generation. For these organisms, low mutation rates, long time spans between generations, and diploid genomes prevent most mutations from significantly affecting genetic variation from one generation to the next.

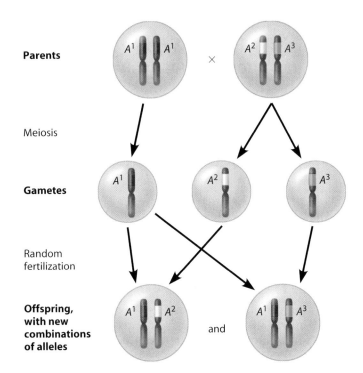

Figure 13.8   New allele combinations from sexual reproduction

Sexual Reproduction   In organisms that reproduce sexually, most of the genetic variation in a population results from the unique combination of alleles that each individual inherits. (Of course, the origin of that allele variation is past mutation.)

As we saw in Modules 8.16 and 8.18, fresh assortments of existing alleles arise every generation from three random components of sexual reproduction: independent orientation of homologous chromosomes at metaphase I of meiosis, random fertilization, and crossing over. During prophase I of meiosis, pairs of homologous chromosomes, one set inherited from each parent, trade some of their genes by crossing over, and then each pair separates into gametes independently of other chromosome pairs. Gametes from one individual vary extensively in their genetic makeup, and each zygote made by a mating pair has a unique assortment of alleles resulting from the random union of sperm and egg.

Figure 13.8 shows a simple example of the production of new allele combinations in sexual reproduction. We see the results of a mating of parents with the genotypes A^1A^1 and A^2A^3, where A^1, A^2, and A^3 are different alleles for a gene. Even without considering crossing over or the independent assortment of chromosomes, the offspring (A^1A^2 and A^1A^3) have different combinations of alleles than were present in their parents.

Genetic variation is necessary for a population to evolve, but variation alone does not guarantee that microevolution will occur. In the next module, we'll explore how to test whether evolution is occurring in a population.

Web Activity *Genetic Variation from Sexual Recombination*

? What is the ultimate source of genetic variation? What is the source of most genetic variation in a population that reproduces sexually?

■ Mutation. Unique combinations of alleles resulting from sexual reproduction

13.9 The Hardy-Weinberg equation can be used to test whether a population is evolving

To understand how microevolution works, let us first examine a simple population whose gene pool is not changing. Consider an imaginary, nonevolving population of blue-footed boobies with individuals that differ in foot webbing (Figure 13.9A). Let's assume that foot webbing is controlled by a single gene and that the allele for nonwebbed feet (W) is completely dominant to the allele for webbed feet (w). The term *dominant* (see Module 9.3) may seem to suggest that over many generations of sexual reproduction, the W allele will somehow come to "dominate," becoming more and more common at the expense of the recessive w allele. In fact, this is not what happens. The shuffling of alleles that accompanies sexual reproduction does not alter the genetic makeup of the population. In other words, sexual reproduction alone does not lead to evolution. No matter how many times alleles are segregated into different gametes by meiosis and united in different combinations by fertilization, the frequency of each allele in the gene pool will remain constant unless other factors are operating. This principle is known as **Hardy-Weinberg equilibrium,** named for the two scientists who derived it independently in 1908.

To test Hardy-Weinberg equilibrium, let's look at two generations of our booby population. Figure 13.9B shows the gene pool of the original population. We have a total of 500 boobies; of these, 320 birds have the genotype WW (nonwebbed feet), 160 have the heterozygous genotype, Ww (also nonwebbed feet, since W is dominant), and 20 have the genotype ww (webbed feet). The proportions of the three possible genotypes (the genotype frequencies) are shown in the middle of Figure 13.9B: 0.64 for WW ($\frac{320}{500}$), 0.32 for Ww ($\frac{160}{500}$), and 0.04 for ww ($\frac{20}{500}$).

From genotype frequencies, we can calculate the frequency of each allele in the population. Each booby carries two alleles for foot type, so this population of 500 has 1,000 alleles for foot type. To find the number of W alleles, we add the number carried by the WW boobies, $2 \times 320 = 640$, to the number carried by the Ww boobies, 160. The total number of W alleles is thus 800. The frequency of the W allele, which we will call p, is $\frac{800}{1,000}$, or 0.8. We can calculate the frequency of the w allele in a similar way; this frequency, called q, is 0.2. (The letters p and q

are often used to represent allele frequencies. Note that $p + q = 1$. The combined frequencies of all alleles must be 100% of the alleles for that gene in the population. If there are only two alleles and we know the frequency of one, we can calculate the frequency of the other.)

What happens when the boobies of this parent population form gametes? At the end of meiosis, each gamete has one allele for foot type, either W or w. The frequency of the two alleles in the gametes will be the same as it is in the gene pool of the parental population, 0.8 for W and 0.2 for w.

Figure 13.9C shows a Punnett square that uses these gamete allele frequencies and the rule of multiplication (see Module 9.7) to calculate the frequencies of the three possible genotypes in the next generation. The probability of producing a WW individual (by combining two W alleles from the pool of gametes) is $p \times p = p^2$, or $0.8 \times 0.8 = 0.64$. Thus, the frequency of WW boobies in the next generation would be 0.64. Likewise, the frequency of ww individuals would be $q^2 = 0.04$. For heterozygous individuals, Ww, the genotype can form in two ways, depending on whether the sperm or egg supplies the dominant allele. In other words, the frequency of Ww would be $2pq = 2 \times 0.8 \times 0.2 = 0.32$. Notice that the three possible genotypes have the same frequency in the next generation as they did in the parent generation.

Finally, what about the frequencies of the alleles in this next generation? Since the genotype frequencies are the same as in the parent population, the allele frequencies p and q are also the same. In fact, we could follow the frequencies through many generations, and the results would continue to be the same. Thus, the gene pool of this population is in a state of equilibrium—Hardy-Weinberg equilibrium.

Now let's write a general formula for calculating the frequencies of genotypes in a gene pool from the frequencies of alleles. In our imaginary blue-footed booby population, the frequency of the W allele (p) is 0.8, and the frequency of the w allele (q) is 0.2. (Again note that $p + q = 1$.)

Figure 13.9A
Imaginary blue-footed boobies, with and without foot webbing

Webbing   No webbing

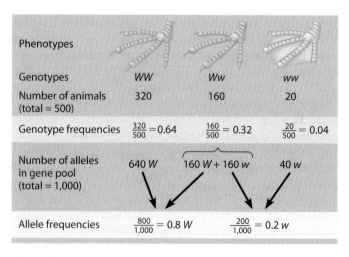

Phenotypes			
Genotypes	WW	Ww	ww
Number of animals (total = 500)	320	160	20
Genotype frequencies	$\frac{320}{500}$=0.64	$\frac{160}{500}$ = 0.32	$\frac{20}{500}$ = 0.04
Number of alleles in gene pool (total = 1,000)	640 W	160 W + 160 w	40 w
Allele frequencies	$\frac{800}{1,000}$ = 0.8 W		$\frac{200}{1,000}$ = 0.2 w

Figure 13.9B   Gene pool of original population of boobies

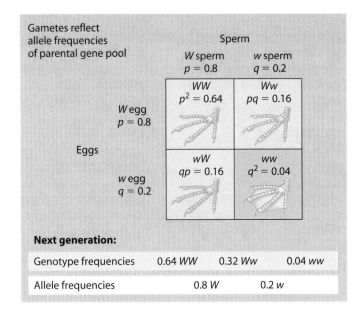

Gametes reflect allele frequencies of parental gene pool	Sperm	
	W sperm $p = 0.8$	w sperm $q = 0.2$
W egg $p = 0.8$	WW $p^2 = 0.64$	Ww $pq = 0.16$
w egg $q = 0.2$	wW $qp = 0.16$	ww $q^2 = 0.04$

Next generation:

Genotype frequencies	0.64 WW	0.32 Ww	0.04 ww
Allele frequencies		0.8 W	0.2 w

Figure 13.9C   Gene pool of next generation of boobies

Notice in Figures 13.9B and 13.9C that the frequencies of all possible genotypes in the populations also add up to 1 (that is, 0.64 + 0.32 + 0.04 = 1). We can represent these relationships symbolically with the Hardy-Weinberg equation:

$$\underset{\substack{\text{Frequency} \\ \text{of homozygous} \\ \text{dominants}}}{p^2} + \underset{\substack{\text{Frequency} \\ \text{of heterozygotes}}}{2pq} + \underset{\substack{\text{Frequency} \\ \text{of homozygous} \\ \text{recessives}}}{q^2} = 1$$

If a population is in Hardy-Weinberg equilibrium and its members continue to mate randomly, allele and genotype frequencies will remain constant generation after generation. Hardy-Weinberg equilibrium tells us that something other than the reshuffling processes of sexual reproduction is required to change allele frequencies in a population (result in microevolu-

tion). One way to find out what factors *can* change a gene pool is to identify the conditions that must be met if genetic equilibrium is to be maintained. For a population to be in Hardy-Weinberg equilibrium, it must satisfy five main conditions:

1. Very large population
2. No gene flow between populations
3. No mutations
4. Random mating
5. No natural selection

These five conditions are rarely met, and thus, we do not really expect a natural population to be in Hardy-Weinberg equilibrium. Let's expand on those conditions: (1) The smaller the population, the more likely it is that allele frequencies will fluctuate by chance from one generation to the next. (2) When individuals move into or out of populations, they add or remove alleles. Thus, gene flow can alter the gene pool. (3) By altering alleles or deleting or duplicating entire genes, mutations modify the gene pool. (4) If individuals mate preferentially, such as with close relatives (inbreeding), random mixing of gametes does not occur, and genotype frequencies change. Allele frequencies, however, usually stay the same. (5) The differential survival and reproductive success of individuals (natural selection) can alter allele frequencies.

The Hardy-Weinberg equation is often used to test whether evolution is occurring in a population. The equation also has many human medical applications, as we see next.

Web Process of Science *How Can Frequency of Alleles Be Calculated?*

? Which is least likely to alter allele and genotype frequencies in a few generations of a large, sexually reproducing population: gene flow, mutation, or natural selection?

■ Because mutations are rare, their effect on allele and genotype frequencies from one generation to the next is likely to be small.

13.10   The Hardy-Weinberg equation is useful in public health science

Public health scientists use the Hardy-Weinberg equation to estimate how many people carry alleles for certain inherited diseases. Consider the case of phenylketonuria (PKU), which is an inherited inability to break down the amino acid phenylalanine. PKU occurs in about one out of 10,000 babies born in the United States and, if untreated, results in severe mental retardation. Newborn babies are now routinely tested for PKU. For those who test positively for PKU, symptoms can be prevented or lessened if individuals follow a diet that strictly regulates the intake of phenylalanine. Packaged foods with ingredients such as aspartame, a common artificial sweetener that contains phenylalanine, must list warnings for people with PKU.

PKU is due to a recessive allele, so the frequency of individuals in the U.S. population born with PKU corresponds to the q^2 term in the Hardy-Weinberg equation. Given one PKU

occurrence per 10,000 births, $q^2 = 0.0001$. Therefore, the frequency of the recessive allele for PKU in the population, q, equals the square root of 0.0001, or 0.01. And the frequency of the dominant allele, p, equals $1 - q$, or 0.99. The frequency of carriers, heterozygous people who do not have PKU but may pass the PKU allele on to offspring, is $2pq$, which equals $2 \times 0.99 \times 0.01$, or 0.0198. Thus, the equation tells us that about 2% (actually 1.98%) of the U.S. population carries the PKU allele. Estimating the frequency of a harmful allele is part of any public health program dealing with genetic diseases.

? Which term in the Hardy-Weinberg equation—p^2, $2pq$, or q^2—corresponds to the frequency of individuals who have no alleles for the disease PKU?

■ The frequency of individuals with no PKU alleles is p^2.

13.11 Natural selection, genetic drift, and gene flow can alter allele frequencies in a population

Deviations from the five conditions for Hardy-Weinberg equilibrium can cause changes in gene pools (microevolution). Although new genes and new alleles originate by mutation, these random and rare events probably change allele frequencies little from one generation of sexually reproducing organisms to the next. Nonrandom mating can affect the frequencies of homozygous and heterozygous genotypes but by itself usually does not affect allele frequencies. The three main causes of evolutionary change are natural selection, genetic drift, and gene flow.

Natural Selection   The condition for Hardy-Weinberg equilibrium that there be no natural selection—that all individuals in a population be equal in their ability to reproduce—is probably never met in nature. Populations consist of varied individuals, and some variants leave more offspring than others. In our imaginary blue-footed booby population, birds with webbed feet (genotype *ww*) might survive better and produce more offspring because they are more efficient at swimming and catching food than birds without webbed feet (genotype *Ww* or *WW*). Genetic equilibrium would be disturbed as the frequency of the *w* allele increased in the gene pool.

Genetic Drift   Flip a coin a thousand times, and a result of 700 heads and 300 tails would make you very suspicious about that coin. But flip a coin ten times, and an outcome of seven heads and three tails would seem within reason. The smaller the sample, the greater the chance of deviation from an idealized result—in the case of the flipped coin, an equal number of heads and tails. Let's apply that logic to a population's gene pool. The frequencies of alleles in a gene pool will be more stable from one generation to the next when a population is large. If a population of organisms is small, its allele frequencies may vary unpredictably from generation to generation.

Genetic drift is a change in the gene pool of a population due to chance. The smaller the population, the more impact genetic drift is likely to have. In many cases, an allele may be lost from a population by such chance fluctuations. Thus, genetic drift tends to reduce genetic variation through such losses. Two situations in which genetic drift can have a significant impact on a population are the bottleneck effect and the founder effect.

Earthquakes, floods, or fires may kill large numbers of individuals, leaving a small surviving population that is unlikely to have the same genetic makeup as the original population. Such a drastic reduction in population size and change in allele frequencies is called a **bottleneck effect,** analogous to shaking just a few marbles through a bottleneck. As illustrated in Figure 13.11A, certain alleles (purple marbles) may be present at higher frequency in the surviving population than in the original population, others (green marbles) may be present at lower frequency, and some (orange marbles) may not be present at all. After a bottleneck, genetic drift may continue for many generations until the population is again large enough for fluctuations due to chance to have less of an impact. Even if a population that has passed through a bottleneck ultimately recovers its size, it may have low levels of genetic variation—a legacy of the genetic drift that occurred when the population was small.

One reason it is important to understand the bottleneck effect is that human actions may create severe bottlenecks for other species, such as the endangered Florida panther and the African cheetah (see Figure 13.12). The greater prairie chicken (Figure 13.11B) is another example. Millions of these birds once lived on the prairies of Illinois. As these prairies were converted to farmland and other uses during the 19th and 20th centuries, the number of greater prairie chickens plummeted. By 1993, only two Illinois populations remained, with a total of fewer than 50 birds. Less than 50% of the eggs of these birds hatched. Researchers compared the DNA of the 1993 population with DNA extracted from museum specimens dating back to the 1930s. They surveyed six gene loci and found that the modern birds had lost 30% of the alleles that were present in the museum specimens. Thus, genetic drift as a result of the bottleneck reduced the genetic variation of the population and may have increased the frequency of harmful alleles, leading to the low egg-hatching rate.

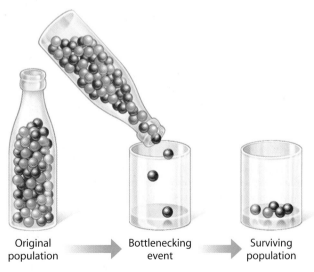

Original population → Bottlenecking event → Surviving population

Figure 13.11A   The bottleneck effect

Figure 13.11B
Greater prairie chicken (*Tympanuchus cupido*). Illinois populations have reduced genetic variation as a result of a bottleneck.

Genetic drift is also likely when a few individuals colonize an isolated island or other new habitat. The smaller the group, the less likely the genetic makeup of the colonists will represent the gene pool of the larger population they left. Such differences in the gene pool of a small colony compared with the original population is called the **founder effect.**

The founder effect explains the relatively high frequency of certain inherited disorders among some human populations established by small numbers of colonists. In 1814, 15 people founded a British colony on Tristan da Cunha, a group of small islands in the middle of the Atlantic Ocean. Apparently, one of the colonists carried a recessive allele for retinitis pigmentosa, a progressive form of blindness. Of the 240 descendants who still lived on the islands in the 1960s, four had retinitis pigmentosa, and at least nine others were known to be heterozygous carriers of the allele. The frequency of this allele is ten times higher on Tristan da Cunha than in the population from which the founders came.

Gene Flow Allele frequencies in a population can also change as a result of **gene flow,** where a population may gain or lose alleles when fertile individuals move into or out of a population or when gametes (such as plant pollen) are transferred between populations. Gene flow tends to reduce differences between populations. For example, humans today move more freely about the world than in the past, and gene flow has become an important agent of evolutionary change in previously isolated human populations.

Let's return to the Illinois greater prairie chickens and see how gene flow improved their fate. Between 1992 and 1996 researchers added a total of 271 birds from neighboring states to the Illinois populations. This strategy succeeded. New alleles entered the population, and the egg-hatching rate improved to over 90%.

Web Activity *Causes of Microevolution*

? How might gene flow between populations living in different habitats actually interfere with each population's adaptation to its local environment?

■ The introduction of alleles that may not be beneficial in a particular habitat prevents the population living there from becoming fully adapted to its local conditions.

13.12 Natural selection is the only mechanism that consistently leads to adaptive evolution

Genetic drift, gene flow, and even mutation can cause microevolution. But these are chance events, and only blind luck could result in their improving a population's fit to its environment. Evolution by natural selection, on the other hand, is a blend of chance and "sorting": chance in the random collection of genetic variation packaged in gametes and combined in offspring and sorting in that some alleles are favored over others. Because of this sorting effect, only natural selection consistently leads to adaptive evolution—evolution that results in a better fit between organisms and their environment.

The adaptations of organisms include many striking examples. Remember the adaptations of blue-footed boobies to their marine environment and the amazing camouflage of mantids shown in Figure 13.3A. Or consider the remarkable speed of the cheetah (Figure 13.12). The cheetah can accelerate from 0 to 40 mph in three strides and to full speed of 70 mph in seconds. Its entire body—skeleton, muscles, joints, heart, and lungs—is built for speed.

Such adaptations for speed are the result of natural selection. By consistently favoring some alleles over others, natural selection improves the match between organisms and their environment. However, an organism's environment may change over time. As a result, what constitutes a "good match" between an organism and its environment is a moving target, making adaptive evolution a continuous, dynamic process.

Let's take a closer look at natural selection. The commonly used phrases "struggle for existence" and "survival of the fittest" are misleading if we take them to mean direct competitive contests between individuals. There *are* animal species in which individuals lock horns or otherwise do combat to determine mating privilege. But reproductive success is generally more subtle and passive. In a varying population of moths, certain individuals may produce more offspring than others because their wing colors hide them from predators better. Plants in a wildflower population may differ in reproductive success because some attract more pollinators, owing to slight variations in flower color, shape, or fragrance. These examples point to a biological definition of **fitness:** the contribution an individual makes to the gene pool of the next generation relative to the contributions of other individuals. Thus, the fittest individuals in the context of evolution are those that produce the largest number of viable, fertile offspring and thus pass on the most genes to the next generation.

? Explain how the phrase "survival of the fittest" differs from the biological definition of fitness.

■ Survival alone does not guarantee reproductive success. An organism's fitness is determined by the number of fertile offspring it leaves and thus its contribution to the gene pool of the next generation.

Figure 13.12 The flexible spine of a cheetah stretches out in the middle of its 7 m (23 ft) stride.

Natural selection can alter variation in a population in three ways

Evolutionary fitness is related to genes, but it is an organism's phenotype—its physical traits, metabolism, and behavior—that is directly exposed to the environment. Let's see how natural selection can affect the distribution of phenotypes using an imaginary deer mouse population. The bell-shaped curve in the top graph of Figure 13.13 depicts the frequencies of individuals that could result from a polygenic inheritance pattern for variation in fur color. In this initial population, fur color varies along a continuum from very light (only a few individuals) through various intermediate shades (many individuals) to very dark (few individuals). The other graphs show three ways in which natural selection could alter the phenotypic variation in the deer mouse population. The blue downward arrows symbolize the pressure of natural selection working against certain phenotypes.

Stabilizing selection favors intermediate phenotypes. It typically occurs in relatively stable environments, where conditions tend to reduce phenotypic variation. In the mouse population depicted in the graph on the bottom left, stabilizing selection has eliminated the extremely light and dark individuals, and the population has a greater number of intermediate phenotypes, which may be best suited to a stable environment. Stabilizing selection probably prevails most of the time in most populations. For example, this type of selection keeps the majority of human birth weights in the range of 3–4 kg (6.5–9 pounds). For babies a lot smaller or larger than this size, infant mortality may be greater.

Directional selection shifts the overall makeup of the population by acting against individuals at *one* of the phenotypic extremes. For the mouse population in the bottom center graph, the trend is toward darker fur color, as might occur if the landscape consists of dark rocks, where darker fur would more readily camouflage the animal. Directional selection is most common during periods of environmental change or when members of a species migrate to some new habitat with different environmental conditions. The changes we described in populations of insects exposed to pesticides is an example of directional selection.

Disruptive selection typically occurs when environmental conditions are varied in a way that favors individuals at *both* extremes of a phenotypic range. For the mice in the graph on the bottom right, individuals with light and dark fur have increased numbers. Perhaps the mice colonized a patchy habitat where a background of light soil was studded with areas of dark rocks. Disruptive selection can lead to two or more contrasting phenotypes in a population. For example, in a population of African seedcracker finches, large-billed birds, who specialize in cracking hard seeds, and small-billed birds, who feed mainly on soft seeds, survive better than birds with intermediate-sized bills, who are fairly inefficient at cracking both types of seeds.

Next we consider a special case of selection, one that leads to phenotypic differences between males and females.

? Of the three modes of natural selection, which is most common?

■ Stabilizing selection

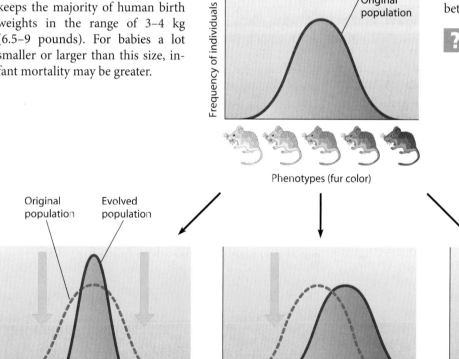

Figure 13.13 Three possible effects of natural selection on a phenotypic character

13.14 Sexual selection may lead to phenotypic differences between males and females

The males and females of an animal species obviously have different reproductive organs. But they may also have other noticeable differences, not directly associated with reproduction or survival, called secondary sexual characteristics. This distinction in appearance is called **sexual dimorphism.** It is often manifested in a size difference, but can also be evident in the form of male adornment, such as manes on lions, antlers on deer, or colorful plumage in birds (Figure 13.14A). Males are usually the showier sex, at least among vertebrates.

Darwin was the first to explore the implications of **sexual selection,** a form of natural selection in which individuals with certain characteristics are more likely than other individuals to obtain mates. Sexual selection can result in sexual dimorphism.

In some species, secondary sex structures may be used to compete with members of the same sex for mates (Figure 13.14B). This type of sexual selection is called intrasexual selection (within the same sex, most often the males). Contests may involve physical combat, but are more often ritualized displays (see Chapter 35). Intrasexual selection is common in species where the winning individual garners a harem of mates.

In a more common type of sexual selection, called intersexual selection (between sexes) or mate choice, individuals of one sex (usually females) are choosy in selecting their mates. Males with the largest or most colorful adornments are often the most attractive to females. The extraordinary feathers of a peacock's tail are an example of this sort of "choose me" statement (Figure 13.14A). What intrigued Darwin is that some of these mate-attracting features do not seem to be otherwise adaptive and may in fact pose some risks. For example, showy plumage may make male birds more visible to predators. But if such secondary sexual characteristics help a male gain a mate, then they will be reinforced over the generations for the most Darwinian of reasons—because they enhance reproductive success. Every time a female chooses a mate based on a certain appearance or behavior, she perpetuates the alleles that caused her to make that choice and allows a male with that particular phenotype to perpetuate his alleles.

What is the advantage to females of being choosy? One hypothesis is that females prefer male traits that are correlated with "good genes." In several bird species, research has shown that traits preferred by females, such as bright beaks or long tails, are related to overall male health. The "good genes" hypothesis was also tested in gray tree frogs. Female frogs prefer to mate with males that give long mating calls (Figure 13.14C). Researchers collected eggs from wild gray tree frogs. One half of each female's eggs was fertilized with sperm from long-calling males, the other with sperm from short-calling males. The offspring of long-calling male frogs grew bigger, faster, and survived better than their half-siblings fathered by short-calling males. The duration of a male's mating call was shown to be indicative of the male's overall genetic quality, supporting the hypothesis that female mate choice can be based on a trait that indicates whether the male has "good genes."

Figure 13.14A   Extreme sexual dimorphism (peacock at right; peahen at left)

Figure 13.14B   A contest for access to mates

Figure 13.14C   A male gray tree frog calling for mates

Next, we return to the concept of directional selection with the evolution of antibiotic resistance in bacteria.

? Males with the most elaborate ornamentation may garner the most mates. How might choosing such a mate be advantageous to a female?

■ An elaborate display may signal good health and therefore good genes, which would be provided to the female's offspring.

13.15 The evolution of antibiotic resistance in bacteria is a serious public health concern

Antibiotics are drugs that kill infectious microorganisms. Many antibiotics are naturally occurring chemicals derived from soil-dwelling fungi or bacteria. For these fungi microbes, antibiotic production is an adaptation that helps them destroy bacteria that compete with them for space and nutrients. Penicillin, originally isolated from a fungus, has been widely prescribed since the 1940s. A revolution in human health rapidly followed its introduction, rendering many previously fatal diseases easily curable (such as strep throat and surgical infections). During the 1950s, some doctors even predicted the end of human infectious disease.

Why hasn't that optimistic prediction come true? It did not take into account the force of evolution. In the same way that pesticides select for resistant insects (see Module 13.3), antibiotics select for resistant bacteria. Penicillin works by preventing the formation of bacterial cell walls. Resistant bacteria contain a gene that codes for a protein that breaks down penicillin. Many antibiotics bind to bacterial ribosomes and block protein synthesis. But existing or new mutations may alter the sites on the ribosomes where the antibiotic binds, making the bacterium and its offspring resistant to that antibiotic. Again we see the chance and sorting aspects of natural selection (see Module 13.12): the chance genetic variations in bacteria and the sorting effect of antibiotics as nonresistant bacteria are killed and resistant strains are left to flourish.

In what ways do we contribute to the problem of antibiotic resistance? Livestock producers add antibiotics to animal feed as a growth promoter and to prevent illness. These practices may select for bacteria that are resistant to standard antibiotics.

Doctors contribute to the problem by overprescribing antibiotics—for example, to patients with viral infections, which do not respond to antibiotic treatment. And patients misuse prescribed antibiotics, for example, by prematurely stopping the medication because they feel better. This allows mutant bacteria that may be killed more slowly by the drug to survive and multiply. Subsequent mutations in such bacteria may lead to full-blown antibiotic resistance.

Difficulty in treating common human infections is a serious public health concern. Penicillin, effective against many bacterial infections in the 1940s, is virtually useless today in its original form. New drugs have since been developed, but they continue to be rendered ineffective as resistant bacteria evolve. Natural selection for antibiotic resistance is particularly strong in hospitals, where antibiotic use is extensive. Nearly 100,000 people die each year in the United States from infections they contract in the hospital, often because the bacteria that cause hospital-acquired infections are resistant to antibiotics. The medical community and pharmaceutical companies are engaged in an ongoing race against the powerful force of bacterial evolution.

Web Process of Science *What Are the Patterns of Antibiotic Resistance?*

? Explain why the following statement is incorrect: "Antibiotics have created resistant bacteria."

■ The use of antibiotics has increased the frequency of alleles for resistance that were already naturally present in bacterial populations.

13.16 Diploidy and balancing selection preserve genetic variation

Natural selection acting on variations within a population adapts that population to its environment. But what prevents natural selection from eliminating the variation as it selects against unfavorable genotypes? Why aren't less adaptive alleles eliminated as the best alleles are passed to the next generation? The tendency for natural selection to reduce variation in a population is countered by mechanisms that maintain variation.

Most eukaryotes are diploid. Having two sets of chromosomes helps to prevent populations from becoming genetically uniform. The effects of recessive alleles are not often displayed in diploid organisms. A recessive allele is subject to natural selection only when it influences the phenotype, and that occurs only in individuals who are homozygous for that allele. In a heterozygote, a recessive allele is, in effect, hidden, or protected, from natural selection. The hiding of recessive alleles in the presence of dominant ones can allow a large number of recessive alleles to remain in a gene pool. Although these alleles may not be favored under present conditions, some could prove advantageous when the environment changes.

Genetic variability may also be preserved by natural selection, the very same force that generally reduces it. **Balancing selection** occurs when natural selection maintains stable frequencies of two or more phenotypic forms in a population.

In the type of balancing selection known as **heterozygote advantage,** heterozygous individuals have greater reproductive success than homozygotes, with the result that two or more alleles for a character are maintained in the population. An example of heterozygote advantage is the protection from malaria conferred by the sickle-cell allele (see Module 9.13). The frequency of the sickle-cell allele in Africa is generally highest in areas where malaria is a major cause of death. The environment in those areas favors heterozygotes, who are protected from the most severe effects of malaria. Less favored are individuals homozygous for the normal hemoglobin allele, who are susceptible to malaria, and individuals homozygous for the sickle-cell allele, who develop sickle-cell disease.

Frequency-dependent selection is a type of balancing selection that maintains two different phenotypic forms in a population. In this case, selection acts against either phenotypic form if it becomes too common in the population. An example of frequency-dependent selection is the scale-eating fish in Lake Tanganyika, Africa, which attack other fish from behind, darting in to remove a few scales from the side of its prey. As you can see in Figure 13.16, some of these fish are "left-mouthed" and others are "right-mouthed." Because their mouth twists to the left, left-mouthed fish always attack their prey's right side—to see why, twist your lower jaw and lips to the left and imagine which side of a fish you could take a bite from. Similarly, right-mouthed fish attack from the left. Prey fish guard more effectively against attack from whichever phenotype is most common in the lake. Thus, fish with the less common phenotype have a feeding advantage that enhances survival and reproductive success. As you can see on the graph, the frequency of left-mouthed fish oscillates over time, as frequency-dependent selection keeps each phenotype close to 50%.

Much of the DNA variation in populations probably has little or no impact on reproductive success. In humans, many of the nucleotide differences found in noncoding sequences of DNA (see Module 10.10) appear to confer no selective advantage or disadvantage. Such differences may be considered **neutral variation.** Even mutational changes that alter proteins can be neutral if they have little effect on protein function. Over time, the frequencies of alleles that are not affected by natural selection will vary by the chance effects of genetic drift. But even if only a fraction of the variation in a gene pool affects reproductive success, that is still an enormous resource of raw material for natural selection and the adaptive evolution it causes.

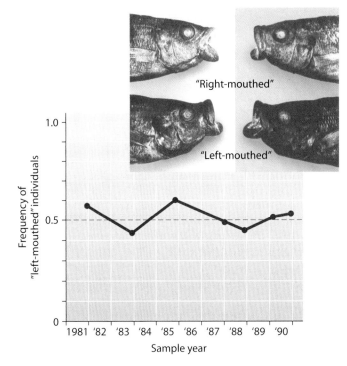

Figure 13.16 Frequency-dependent selection for left-mouthed and right-mouthed scale-eating fish

? Why would natural selection tend to reduce genetic variation more in populations of haploid organisms than in populations of diploid organisms?

■ All alleles in a haploid organism are phenotypically expressed and are hence screened by natural selection.

13.17 Natural selection cannot fashion perfect organisms

Though natural selection leads to adaptation, there are several reasons why nature abounds with organisms that seem to be less than ideally "engineered" for their lifestyles.

1. *Selection can act only on existing variations.* Natural selection favors only the fittest variations from the phenotypes that are available, which may not be the ideal traits. New, advantageous alleles do not arise on demand.

2. *Evolution is limited by historical constraints.* Each species has a legacy of descent with modification from a long line of ancestral forms. Evolution does not scrap ancestral anatomy and build each new complex structure from scratch; it co-opts existing structures and adapts them to new situations. If a terrestrial four-legged animal were to become adapted to an environment in which flight was advantageous, it might be best to grow an extra pair of limbs to serve as wings. However, evolution operates on the traits an organism already has. Thus, as birds and bats evolved from Earth-bound ancestors, their existing forelimbs took on new functions for flight.

3. *Adaptations are often compromises.* Each organism must do many different things. A blue-footed booby uses its webbed feet to swim as it dives into the ocean for prey, but these same feet make for clumsy travel on land.

4. *Chance, natural selection, and the environment interact.* Chance probably affects the genetic structure of populations to a greater extent than was once believed. For instance, when a storm blows insects hundreds of miles over an ocean to an island, the wind does not necessarily transport the specimens that are best suited to the new environment. And not all alleles fixed by genetic drift in the gene pool of a small population are better suited to the environment than alleles that are lost.

With all these constraints, we cannot expect evolution to craft perfect organisms. Natural selection operates on a "better than" basis. We can see evidence for evolution in the imperfections of the organisms it produces.

? Humans owe much of their physical versatility and athleticism to their flexible limbs and joints. But we are prone to sprains, torn ligaments, and dislocations. Why?

■ Adaptations are compromises: Structural reinforcement has been compromised as agility was selected for.

Reviewing the Concepts

Darwin's Theory of Evolution (Introduction–13.6)

Adaptations are inherited characteristics that enhance an organism's ability to survive and reproduce in a particular environment. According to Darwin's theory of evolution, living species are descended from earlier life-forms that differed from present day organisms (**Introduction**).

Descent with modification is what Darwin called his theory, which explains that all of life is connected by common ancestry and that descendants have accumulated adaptations to changing environments over vast spans of time. Darwin's theory differed greatly from the long-held notion of a young Earth inhabited by unchanging species (**13.1**).

Natural selection (13.2):

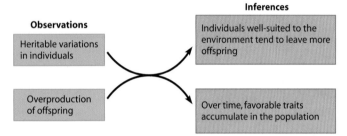

Evidence of evolution. Natural selection has been observed in populations of birds, insects, and many other organisms (**13.3**). The fossil record reveals the historical sequence in which organisms have evolved and many fossils linking ancestral species with those living today (**13.4**). Further evidence for evolution comes from biogeography, comparative anatomy, and molecular biology. Homologous structures and DNA sequences provide evidence of evolutionary relationships (**13.5**).

Evolutionary trees are diagrams based on homologies that reflect the evolutionary relationships among groups of organisms (**13.6**).

The Evolution of Populations (13.7–13.10)

Populations evolve. Microevolution is a change in the relative frequencies of alleles in a population's gene pool (**13.7**).

Genetic variation is generated by mutation and sexual reproduction. Mutations are important in the adaptation of prokaryotes to the environment. In diploid organisms, the shuffling of alleles in sexual reproduction provides most variation (**13.8**).

The Hardy-Weinberg equilibrium states that the allele and genotype frequencies of a population will remain constant if the population is large, mating is random, and there is no mutation, gene flow, or natural selection. The Hardy-Weinberg equation calculates allele and genotype frequencies in a population (**13.9**). Public health scientists use the Hardy-Weinberg equation to estimate frequencies of disease-causing alleles (**13.10**).

Allele frequencies	$p + q = 1$
Genotype frequencies	$p^2 + 2pq + q^2 = 1$

Dominant homozygotes — Heterozygotes — Recessive homozygotes

Mechanisms of Microevolution (13.11–13.17)

Allele frequencies change due to natural selection (reproductive success of individuals best suited to the environment), genetic drift (change in a gene pool due to chance), and gene flow (movement of individuals or gametes between populations). The bottleneck effect and founder effect lead to genetic drift (**13.11**).

Fitness is the relative contribution an individual makes to the gene pool of the next generation. As a result of natural selection, favorable traits increase in a population (**13.12**).

Outcomes of natural selection (13.13):

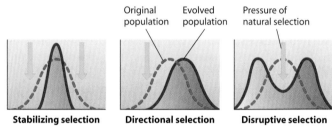

| Stabilizing selection | Directional selection | Disruptive selection |

Sexual selection leads to the evolution of secondary sex characteristics, which can give individuals an advantage in mating (**13.14**).

Antibiotic resistance in bacteria illustrates directional selection resulting from excessive or incorrect use of antibiotics (**13.15**).

Preservation of variation. Diploidy preserves variation by "hiding" recessive alleles. Balancing selection may result from heterozygote advantage or frequency-dependent selection (**13.16**).

Perfect organisms? Natural selection can act only on available variation; anatomical structures result from modified ancestral forms; adaptations are often compromises; and chance, natural selection, and the environment interact (**13.17**).

Connecting the Concepts

1. Summarize the key points of Darwin's theory of descent with modification, including his proposed mechanism of evolution.

2. Complete this map describing potential causes of evolutionary change within populations.

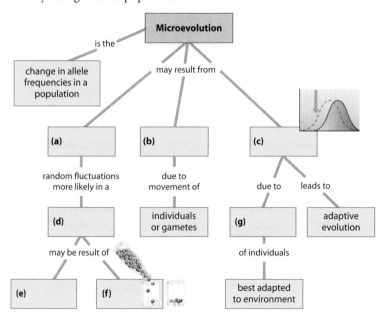

Testing Your Knowledge

Multiple Choice

3. Which of the following did not influence Darwin as he synthesized the theory of evolution by natural selection?
   a. examples of artificial selection that produce large and relatively rapid changes in domesticated species
   b. Lyell's *Principles of Geology* on gradual geologic changes
   c. comparisons of fossils with living organisms
   d. the biogeographic distribution of organisms such as the unique species on the Galápagos Islands
   e. Mendel's paper describing the "laws of inheritance"

4. Natural selection is sometimes described as "survival of the fittest." Which of the following best measures an organism's fitness?
   a. how many fertile offspring it produces
   b. its mutation rate
   c. how strong it is when pitted against others of its species
   d. its ability to withstand environmental extremes
   e. how much food it is able to make or obtain

5. Mutations are rarely the cause of evolution in populations of plants and animals because
   a. they are often harmful and do not get passed on.
   b. they do not directly produce most of the genetic variation present in a diploid population.
   c. they occur very rarely.
   d. they are only passed on when they occur in cells that lead to gametes.
   e. of all of the above.

6. In an area of erratic rainfall, a biologist found that grass plants with alleles for curled leaves reproduced better in dry years, and plants with alleles for flat leaves reproduced better in wet years. This situation would tend to (*Explain your answer.*)
   a. cause genetic drift in the grass population.
   b. preserve genetic variation in the grass population.
   c. lead to stabilizing selection in the grass population.
   d. lead to uniformity in the grass population.
   e. cause gene flow in the grass population.

7. Which of the following pairs of structures is *least* likely to represent homology?
   a. the hemoglobin of a human and that of a baboon
   b. the mitochondria of a plant and that of an animal
   c. the wings of a bird and those of an insect
   d. the tail of a cat and that of an alligator
   e. the foreleg of a pig and the flipper of a whale

8. If an allele is recessive and lethal in homozygotes before they reproduce,
   a. the allele is present in the population at a frequency of 0.001.
   b. the allele will be removed from the population by natural selection in approximately 1,000 years.
   c. the allele will most likely remain in the population at a low frequency because it cannot be selected against in heterozygotes.
   d. the fitness of the homozygous recessive genotype is 0.
   e. both c and d are correct.

9. Darwin's claim that all life is descended from a common ancestor is best supported with evidence from
   a. the fossil record.   d. comparative anatomy.
   b. molecular biology.   e. comparative embryology.
   c. evolutionary trees.

10. In a population with two alleles, B and b, the allele frequency of b is 0.4. B is dominant to b. What is the frequency of individuals with the dominant phenotype if the population is in Hardy-Weinberg equilibrium?
    a. 0.16   c. 0.48   e. You cannot tell from this
    b. 0.36   d. 0.84   information.

11. Within a few weeks of treatment with the drug 3TC, a patient's HIV population consists entirely of 3TC-resistant viruses. How can this result best be explained?
    a. HIV can change its surface proteins and resist vaccines.
    b. The patient must have become reinfected with a resistant virus.
    c. A few drug-resistant viruses were present at the start of treatment, and natural selection increased their frequency.
    d. The drug caused the HIV genes to change.
    e. HIV began making drug-resistant versions of its enzymes in response to the drug.

Describing, Comparing, and Explaining

12. Write a paragraph briefly describing the kinds of evidence for evolution.

13. Sickle-cell disease is caused by a recessive allele. Roughly one out of every 400 African-Americans (0.25%) is afflicted with sickle-cell disease. Use the Hardy-Weinberg equation to calculate the percentage of African-Americans who are carriers of the sickle-cell allele. (*Hint:* $q^2 = 0.0025$.)

14. It seems logical that natural selection would work toward genetic uniformity; the genotypes that are most fit produce the most offspring, increasing the frequency of adaptive alleles and eliminating less adaptive alleles. Yet there remains a great deal of genetic variation within populations. Describe some of the factors that contribute to this variation.

Applying the Concepts

15. A population of snails is preyed upon by birds that break the snails open on rocks, eat the soft bodies, and leave the shells. The snails occur in both striped and unstriped forms. In one area, researchers counted both live snails and broken shells. Their data are summarized below:

	Striped	Unstriped	Total	Percent Striped
Living	264	296	560	47.1
Broken	486	377	863	56.3

Which snail form seems to be better adapted to this environment? Why? Predict how the frequencies of striped and unstriped snails might change in the future.

16. Advocates of "scientific creationism" and "intelligent design" lobby school districts for such things as a ban on teaching evolution, equal time in science classes to teach alternative versions of the origin and history of life, or disclaimers in textbooks stating that evolution is "just a theory." They argue that it is only fair to let students evaluate both evolution and the idea that all species were created by God as the Bible relates or that, because organisms are so complex and well adapted, they must have been created by an intelligent designer. Do you think that alternative views of evolution should be taught in science courses? Why or why not?

Answers to all questions can be found in Appendix 4.
For Practice Quizzes, BioFlix, MP3 Tutors, and Activities, go to www.mybiology.com.

The Rise and Fall of Cichlids

They are so colorful and so diverse. There are algae, insect, and fish eaters, leaf biters, muck feeders, snail crushers, even scale scrapers who steal scales off the flanks of other fish. They are all colors of the rainbow, making them favorites of the aquarium trade (see above). Cichlids are found in many tropical lakes and rivers of Central America, Asia, and Africa. Until recently, there were more than 500 species of these small fish in Lake Victoria, the largest of the Great Lakes of East Africa (see map on the facing page). Where did they all come from? And why are they disappearing?

Scientists call the Lake Victoria cichlids—this physically, ecologically, and behaviorally diverse assemblage of closely related genera and species—a superflock. Their species number more than all the fish species in all the lakes and rivers of Europe combined. And if you add the more than 1,000 species in Lakes Victoria, Tanganyika, and Malawi, you have the biggest diversification of species of any animal group in one geographic region. And the species are young. Studies show that the cichlids of Lake Victoria diversified about 100,000 years ago. (As a comparative time reference, bony fish have been evolving on Earth for about 400 million years.)

How did so many cichlid species evolve so fast? There are several hypotheses. The jaw structure and teeth of cichlids vary widely, and these highly adaptable characteristics may have allowed ancestral species to specialize on different food sources, such as algae or snails or other fishes. Or perhaps the bright color patterns of cichlids may have led to species diversity as females selected mates of specific coloration. The separation of sub-populations by feeding preferences or by female mate choice could have reduced or elim-

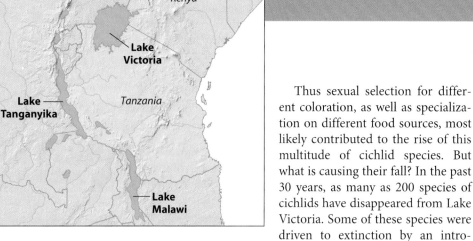

Africa

Uganda

Kenya

Lake Victoria

Lake Tanganyika

Tanzania

Lake Malawi

inated breeding between the populations. The now isolated gene pools of these subpopulations could have changed as a result of different mutations, natural selection, and genetic drift. (Remember from Chapter 13 that these are all causes of microevolution.) Eventually, the isolated groups may have lost the ability to interbreed. Biologists often define a **species** as a group of organisms whose members can breed and produce fertile offspring, but who do not produce fertile offspring with members of other groups.

Is there evidence that the different colors of cichlids keep the species from interbreeding? In Lake Victoria there are many pairs of similar species that differ in color but little else. As shown below, males of *Pundamilia nyererei* have a bright red back and dorsal fin (see below left). Males of *Pundamilia pundamilia* have a bright blue dorsal fin (see below right). Females of both species are smaller and less colorful. Various mate choice experiments have shown that females prefer brightly colored— and rightly colored—males. Remember the "good genes" hypothesis from Module 13.14? As one might predict, the more brightly colored males have fewer parasites and are in better overall condition.

A 2006 study found differences in the visual sensitivities of *P. nyererei* and *P. pundamilia* females. *P. nyererei* females, who prefer to mate with red males, are more sensitive to red light; *P. pundamilia* females are more sensitive to blue light. These two species inhabit different depths and experience different light conditions. For the deeper dwelling *P. nyererei,* the increased sensitivity to red light makes red males more conspicuous and males of other colors less noticeable.

Thus sexual selection for different coloration, as well as specialization on different food sources, most likely contributed to the rise of this multitude of cichlid species. But what is causing their fall? In the past 30 years, as many as 200 species of cichlids have disappeared from Lake Victoria. Some of these species were driven to extinction by an introduced predator, the Nile perch (see Chapter 36 introduction). But many species not eaten by Nile perch have also disappeared. For decades, the waters of Lake Victoria have become increasingly murky due to pollution, deforestation, and a huge growth in algal populations (linked to the loss of algae-eating cichlids). Researchers have found that female cichlids from turbid water are less choosy in regard to male coloration, and males from these areas are less brightly colored. Thus, the mate choice barriers to breeding that separate closely related species are crumbling. Many viable hybrid offspring are produced by interbreeding between two species, and the once isolated gene pools of the parent species are mixing—two species are fusing into one. Thus, the great speciation event in Lake Victoria may be coming to an end as the loss of cichlid species continues.

Speciation—the emergence of a new species—is the bridge between microevolution (the changes over time in allele frequencies in a population) and macroevolution (the broader pattern of evolutionary change over long periods of time). We examined microevolution in Chapter 13. And we'll explore macroevolution in Chapter 15. In this chapter, we consider the origin of species—the source of these diverse cichlids and of all the biological diversity on Earth. ■ ■ ■

Males: *Pundamilia nyererei* (left) and *Pundamilia pundamilia* (right)

14.1 The origin of species is the source of biological diversity

Even though Darwin titled his seminal work *On the Origin of Species,* he had relatively little to say about how this process occurred. Most of Darwin's theory of evolution focused on natural selection and the gradual adaptation of a population to its environment. We call this process microevolution—changes within a population over generations. But if microevolution were *all* that happened, then Earth would be inhabited only by a highly adapted version of the first form of life.

Figure 14.1 shows one species evolving into two. Each time speciation occurs, the diversity of life increases. Over the course of 3.5 billion years, an ancestral life-form first gave rise to two or more different types of cells, which then branched to form new lineages, which branched again, until we arrive at the millions of species that have existed and now exist on Earth.

A new species often closely resembles its parent species. Occasionally, however, a new species has properties novel enough to define a major branch in the tree of life, such as the legs of land vertebrates or the flowers of flowering plants. Macroevolution traces such large-scale changes in the history of life.

In this chapter, we explore the mechanisms of speciation. Let's begin by defining what we mean by species.

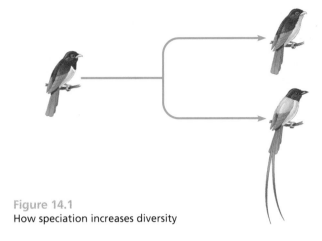

Figure 14.1
How speciation increases diversity

🎧 MP3 Tutor *Speciation*

❓ What is the difference between microevolution and macroevolution?

Microevolution deals with changes in the gene pool of a single population. Macroevolution considers broad patterns of evolutionary change over long periods of time and includes the origin of new groups.

Concepts of Species

14.2 There are several ways to define a species

The word *species* is from the Latin for "kind" or "appearance," and indeed, even young children learn to distinguish between kinds of plants or animals—between dogs and cats, for instance—from differences in their appearance. Although the basic idea of species as distinct life-forms seems intuitive, devising a more formal definition is not so easy.

Taxonomy is the branch of biology that names and classifies species and groups them into broader categories. In the 18th century, Swedish physician and botanist Carolus Linnaeus introduced the two-part, or binomial, system of naming organisms. (We describe this system in more detail in Chapter 15.) For our own species, the binomial designation is *Homo sapiens* (Latin for "wise human being"). Linnaeus named over 11,000 species based on each one's physical characteristics.

In some cases, the physical differences between two species are not obvious. The two skunks in Figure 14.2A look much the same—sometimes even experts have trouble telling them apart. The top skunk is an eastern spotted skunk (*Spilogale putoris*), and the bottom one is a western spotted skunk (*Spilogale gracilis*). The eastern species breeds in the late winter and the western species breeds in fall. Thus, even though their geographic ranges overlap in the Great Plains, each type of skunk breeds only with individuals of its own species.

Whereas the individuals of these spotted skunk species seem to exhibit fairly limited variation in physical appearance, certain other species—our own, for example—seem extremely

Figure 14.2A Similarity between the eastern spotted skunk (top) and western spotted skunk (bottom)

Figure 14.2B Diversity within one species

varied. If we did not know that humans all belong to one species, *Homo sapiens,* the physical diversity within our species (partly illustrated in Figure 14.2B) might lead us to guess that there are several human species. We see that populations of the same species, as well as individuals of the same population, may be very similar, or they may vary greatly in appearance.

The Biological Species Concept   How do biologists define a species? And what keeps one species distinct from others? The primary definition of species used in this book, called the **biological species concept,** was described in 1942 by biologist Ernst Mayr. It defines a species as a group of populations whose members have the potential to interbreed in nature and produce fertile offspring (offspring who themselves can reproduce). A businesswoman in Manhattan may be unlikely to meet a dairy farmer in Mongolia, but if the two should happen to meet and mate, they could have viable babies that develop into fertile adults.

Members of different species do not usually mate with each other. And, if members of one species do mate with members of another species, the offspring (or those of later generations) will probably not be fertile. In effect, **reproductive isolation** prevents genetic exchange (gene flow) and maintains the gap between species.

But what about the two species of cichlids described in the chapter introduction that mate indiscriminately in murky water and produce viable offspring? Determining a clear-cut species identification on the basis of reproductive isolation is sometimes more complex than it may seem.

There are other instances in which applying the biological species concept is problematic. For example, there is no way to determine whether organisms that are now fossils were once able to interbreed. Also, this criterion is useless for organisms such as prokaryotes that are asexual in their reproduction.

Some asexual organisms can exchange genes (for instance, the transfer of DNA between even distantly related bacteria, described in Chapter 10), but this is not part of their reproductive process.

Even with most living, sexually reproducing species, we lack sufficient information about interbreeding to use reproductive isolation as the sole criterion for distinguishing species. Because of the limitations of the biological species concept, alternative species concepts are useful in certain situations.

Other Definitions of Species   Biologists have developed several other ways to define species. For most organisms—sexual, asexual, and fossils alike—classification is based mainly on observable and measurable physical traits such as shape, size, and other features of morphology (form). This **morphological species concept** has been used to identify most of the 1.8 million species that have been named to date. The advantage of this concept is that it can be applied to asexual organisms and fossils and does not require information on possible interbreeding. The disadvantage of this method, however, is that it relies on subjective criteria, and researchers may disagree on which structural features distinguish a species.

Another species definition, the **ecological species concept,** identifies species in terms of their ecological niches, focusing on unique adaptations to particular roles in a biological community. (We will examine the concept of ecological niche in more detail in Chapter 37.) For example, two species of cichlids in Lake Victoria may be similar in appearance but distinguishable based on what they eat or the depth of water in which they are usually found.

Finally, the **phylogenetic species concept** defines a species as the smallest group of individuals that shares a common ancestor and that forms one branch on the tree of life. Biologists trace the phylogenetic history of a species by comparing its characteristics, such as morphology or DNA sequences, with those of other organisms. Such analysis can distinguish groups that are different enough to be considered separate species. Of course, agreeing on the amount of difference required to distinguish separate species remains a problem.

Each species definition is useful, depending on the situation and the questions being asked. The biological species concept, however, helps focus on how these discrete groups of organisms arise and are maintained by reproductive isolation. Because reproductive isolation is an essential factor in the evolution of many species, we look at it more closely next.

? Which species concept would be most useful for identifying species in the field? Explain.

■ The morphological species concept, because it is based only on appearance. Information on reproductive isolation, ecological role, or evolutionary history are not required.

14.3 Reproductive barriers keep species separate

Clearly, a fly will not mate with a frog or a fern. But what prevents species that are closely related from interbreeding? Although being separated by geographic barriers may prevent similar species from interbreeding, geography is not an intrinsic part of an organism. It takes a **reproductive barrier**—a biological feature of the organism itself—to prevent individuals of closely related species from interbreeding when their ranges overlap. As shown in Table 14.3, the various types of reproductive barriers that isolate the gene pools of species can be categorized as either prezygotic or postzygotic, depending on whether they function before or after zygotes (fertilized eggs) form.

Prezygotic Barriers   **Prezygotic barriers** prevent mating or fertilization between species. There are five main types of prezygotic barriers. The first, called *temporal isolation,* occurs when species breed at different times—during different seasons, at different times of the day, or even in different years. Remember that the eastern and western spotted skunks (Figure 14.2A) breed at different times of the year. Many plants also exhibit seasonal differences in reproduction. Two species of pine trees inhabit some of the same areas of central California, but the Monterey pine (*Pinus radiata*) releases pollen in February, and the Bishop's pine (*P. muricata*) does so in April. Some flowering plants are temporally isolated because their flowers open at different times of the day, so pollen cannot be transferred from one to another.

In a second type of prezygotic barrier, *habitat isolation,* two species live in the same general area but not in the same kinds of places. Two species of garter snake in the genus *Thamnophis* are found in western North America, but one lives mainly in water and the other on land (Figure 14.3A). Habitat isolation also affects parasites that are confined to certain plant or animal host species. Two species of parasites living in different hosts will not have the opportunity to interbreed.

In *behavioral isolation,* a third type of prezygotic barrier, there is little or no sexual attraction between females and males of different species. Special signals that attract mates and elaborate mating behaviors that are unique to a species are probably

TABLE 14.3   REPRODUCTIVE BARRIERS BETWEEN SPECIES

Prezygotic Barriers: Prevent Mating or Fertilization

Temporal isolation:	Mating or flowering occurs at different seasons or times of day.
Habitat isolation:	Populations live in different habitats and do not meet.
Behavioral isolation:	There is little or no sexual attraction between different species.
Mechanical isolation:	Structural differences in genitalia or flowers prevent copulation or pollen transfer.
Gametic isolation:	Male and/or female gametes die before uniting or fail to unite.

Postzygotic Barriers: Prevent the Development of Fertile Adults

Reduced hybrid viability:	Hybrids fail to develop or to reach sexual maturity.
Reduced hybrid fertility:	Hybrids fail to produce functional gametes.
Hybrid breakdown:	Offspring of hybrids are weak or infertile.

the most important reproductive barriers between closely related animals. For example, male fireflies of various species signal to females of their kind by blinking their lights in particular rhythms. Females respond only to signals of their own species, flashing back and attracting the males. The preference of female cichlids for males of a particular color, as described in the chapter introduction, also functions as a behavioral isolation, keeping species separate.

Figure 14.3B shows a form of behavioral isolation called a courtship ritual. Many species will not mate until the male and

Figure 14.3A   **Habitat isolation:** Two closely related species of garter snake reproductively isolated because one lives on land (left) and the other lives mainly in water (right)

Figure 14.3B   **Behavioral Isolation:** Courtship ritual in blue-footed boobies as a behavioral barrier between species

Figure 14.3C **Mechanical isolation:** Because these snails' shells spiral in opposite directions, their genital openings (indicated by arrows) cannot be aligned and mating cannot occur.

Figure 14.3D **Gametic isolation:** Gametes of these red and purple urchins are unable to fuse because proteins on the surfaces of the eggs and sperm cannot bind to one another.

female have performed an elaborate ritual that is unlike that of any other species. These blue-footed boobies are involved in a courtship dance in which the male points his beak, tail, and wing tips to the sky. Part of the "script" calls for the male to high-step, a dance that advertises his bright blue feet.

Mechanical isolation, a fourth type of prezygotic barrier, occurs when female and male sex organs are not compatible; for instance, the male copulatory organs of many insect species have a unique and complex structure that fits the female parts of only one species. Figure 14.3C shows how the spiraling of the shells of two species of snails in different directions results in a mechanical barrier to mating.

In the plant kingdom, mechanical barriers contribute to reproductive isolation of flowering plants. Many species have flower structures that are adapted to specific insect or animal pollinators that transfer pollen only between plants of the same species. In some cases, the beak of a particular species of hummingbird is just the right length for the flower tube of the one plant species it pollinates, ensuring that pollen is transferred only between plants of that species.

Gametic isolation is a fifth type of prezygotic barrier. A male and a female from two different species may copulate, but the gametes do not unite to form a zygote. In many mammals, for example, the sperm cannot survive in the reproductive tract of a female of a different species. Gametic isolation is very important when fertilization is external. Male and female sea urchins (Figure 14.3D) of many different species release eggs and sperm into the sea, but fertilization occurs only if species-specific molecules on the surfaces of egg and sperm attach to each other. A similar mechanism of molecular recognition enables a flower to discriminate between pollen of its own species and pollen of a different species.

Postzygotic Barriers In contrast to prezygotic barriers, **postzygotic barriers** operate after hybrid zygotes are formed. (Hybrids result from the union of gametes of two different species.) In some cases, there is *reduced hybrid viability;* that is, most hybrid offspring do not survive. For example, certain salamanders of the genus *Ensatina* that live in the same habitats may occasionally hybridize, but most of the hybrids do not complete development, and those that do are very frail.

Another type of postzygotic barrier is *reduced hybrid fertility,* in which hybrid offspring of two different species reach maturity and are vigorous but sterile, and therefore unable to bring about gene flow between the parent species. A mule, for example, is the robust offspring of a female horse and a male donkey (Figure 14.3E). The horse and donkey remain separate species because a mule virtually never interbreeds with a horse or a donkey. Therefore, the gene pools of the horse and donkey remain isolated.

In a third type of postzygotic barrier, called *hybrid breakdown,* the first-generation of hybrid offspring are viable and fertile, but when these hybrids mate with one another or with either parent species, the offspring are feeble or sterile. For example, different species of cotton plants can produce fertile hybrids, but the offspring of the hybrids do not survive.

In summary, reproductive barriers form the boundaries around many closely related species. Next we examine situations that make reproductive isolation and speciation possible.

? Two closely related cichlids live in Lake Victoria, but one feeds on detritus along the shoreline and the other is a bottom feeder in deep water. This is an example of _____ isolation, which is a _____ reproductive barrier.

■ habitat . . . prezygotic

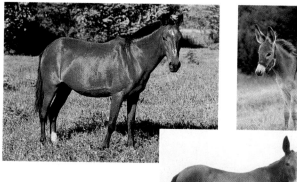

Figure 14.3E **Reduced hybrid fertility:** A horse (above) and a donkey (above right) may produce a hybrid, sterile offspring, a mule (right).

14.4 In allopatric speciation, geographic isolation leads to speciation

A key event in the origin of many species is the separation of a population—with its gene pool—from other populations of the same species. With its gene pool isolated, the splinter population can follow its own evolutionary course. Changes in its allele frequencies caused by natural selection, genetic drift, and mutations will not be diluted by alleles entering from other populations (gene flow). In the formation of many species living today, the initial block to gene flow seems to have been a geographic barrier that isolated a population. This mode of speciation is called **allopatric speciation** (from the Greek *allos,* other, and *patra,* fatherland). Populations separated by a geographic barrier are known as allopatric populations.

Several geologic processes can fragment a population into two or more isolated populations. A mountain range may emerge and gradually split a population of organisms that can inhabit only lowlands. A large lake may subside until there are several smaller lakes, isolating certain fish populations. A land bridge such as the Isthmus of Panama may form and separate the marine life on either side.

How large must a geographic barrier be to keep allopatric populations apart? The answer depends on the ability of the organisms to move about. Birds, mountain lions, and coyotes can easily cross mountain ranges, rivers, and canyons. The windblown pollen of pine trees is also not hindered by such barriers, and the seeds of many plants may be carried back and forth by animals. In contrast, small rodents may find a deep canyon or a wide river a formidable barrier. The Grand Canyon and Colorado River (Figure 14.4) separate two species of antelope squirrels. Harris's

antelope squirrel (*Ammospermophilus harrisi*) inhabits the south rim. Just a few miles away on the north rim is the closely related white-tailed antelope squirrel (*Ammospermophilus leucurus*).

The likelihood of allopatric speciation increases when a population is both small and isolated. A small population is less likely to have a gene pool that reflects its parent population (the founder effect, see Module 13.11). And a small population in a new habitat is more likely to have its gene pool changed substantially by factors such as genetic drift and natural selection. For example, in less than 2 million years, the few animals and plants from the South American mainland that colonized the Galápagos Islands gave rise to all the new species now found there.

Geographic isolation creates opportunities for speciation, but it does not necessarily lead to new species. Indeed, even when gene pool changes result in the adaptation of an isolated population to a local environment, speciation may or may not occur. Speciation occurs only when the gene pool undergoes changes that establish *reproductive* barriers between the isolated population and its parent population.

But speciation can also occur within populations that are not separated geographically, as we see next.

? A new species will not arise just because a population becomes geographically isolated. For _____ speciation to occur, changes in the gene pool must produce _____.

allopatric ⋯ reproductive isolation

Figure 14.4 Allopatric speciation of geographically isolated antelope squirrels

South

North

In sympatric speciation, speciation takes place without geographic isolation

Not all species arise as a result of geographic isolation. In **sympatric speciation** (from the Greek *syn*, together, and *patra*, fatherland), a new species arises within the same geographic area as a parent species. How can reproductive isolation develop when members of sympatric populations remain in contact with each other? Sympatric speciation may occur when mating and the resulting gene flow between populations are reduced by factors such as polyploidy, habitat differentiation, and sexual selection.

Many plant species have originated from accidents during cell division that resulted in extra sets of chromosomes. New species formed in this way are **polyploid,** meaning that their cells have more than two complete sets of chromosomes. Figure 14.5A shows one way in which a tetraploid plant (4n, with four sets of chromosomes) can arise from a parent species that is diploid. ❶ A failure of cell division after chromosome duplication could double a cell's chromosomes. ❷ If this 4n cell gives rise to a tetraploid branch, flowers produced on this branch would produce diploid gametes. ❸ If self-fertilization occurs (as it commonly does in plants), the resulting tetraploid zygotes may develop into plants that produce fertile tetraploid offspring by self-pollination or by mating with other tetraploids.

Tetraploids cannot, however, produce fertile offspring by mating with the parent plants. The fusion of a diploid (2n) gamete from the tetraploid plant and a haploid (n) gamete from the diploid parent would produce triploid (3n) offspring. Triploid individuals are sterile; they cannot produce normal gametes because the odd number of chromosomes cannot form homologous pairs and separate normally during meiosis. Thus, the formation of a tetraploid (4n) plant is an instantaneous speciation event: A new species, reproductively isolated from its parent species, is produced in just one generation.

Most polyploid species, however, arise from hybridization of two different species. Figure 14.5B illustrates one way in which this can happen. ❶ When haploid gametes from two different species combine, the resulting hybrid is normally sterile because its chromosomes cannot pair during meiosis. ❷ However, the hybrid may reproduce asexually (as many plants can do). ❸ Subsequent errors in cell division may produce chromosome duplications that result in a fertile polyploid species. The new species has a chromosome number equal to the sum of the diploid chromosome numbers of the parent species. Again, this new species is reproductively isolated from both parent species.

Does polyploid speciation occur in animals? It appears to happen occasionally. One example is the gray tree frog (see Figure 13.14C), which is thought to have originated in this way.

Sympatric speciation in animals is more likely to happen through habitat differentiation and sexual selection. You read in the chapter introduction that the speciation of cichlids in

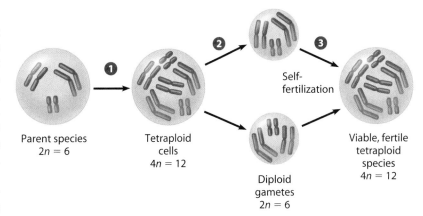

Figure 14.5A Sympatric speciation by polyploidy within a single species

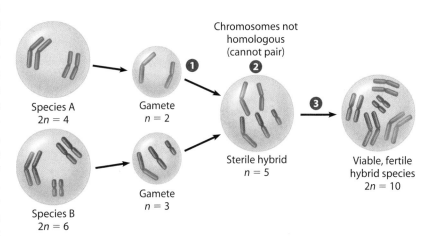

Figure 14.5B Sympatric speciation producing a hybrid polyploid from two different species

Lake Victoria may have occurred as subgroups of the original population evolved adaptations for exploiting different food sources. If those sources were in different habitats, mating between the two specializing populations would become rare, isolating their gene pools as they became adapted to different resources.

The type of sexual selection in which females choose mates based on coloration appears also to have contributed to reproductively isolating populations of cichlids and keeping the gene pools of newly forming species separate.

We'll return to the evolution of reproductive barriers in Module 14.7, but next we explore some common examples of speciation by polyploidy in plants.

Web Process of Science How Do New Species Arise by Genetic Isolation?

? Return to the reproductive barriers in Table 14.3 and choose the barrier that isolates a viable, fertile polyploid plant from its parental species.

■ Reduced hybrid fertility

14.6 Most plant species trace their origin to polyploid speciation

Plant biologists estimate that 80% of living plant species are descendants of ancestors that formed by polyploid speciation. Hybridization between two species accounts for most of these species, perhaps because of the adaptive advantage of the diverse genes a hybrid inherits from different parental species.

Many of the plants we grow for food are polyploids, including oats, potatoes, bananas, peanuts, barley, plums, apples, sugarcane, coffee, and wheat. Cotton, also a polyploid, remains the source of one of the world's most popular clothing fibers. Cotton cloth is made from the long white plumes that extend from the seeds of the plant.

Wheat, the most widely cultivated plant in the world, occurs as 20 different species of *Triticum*. We know that humans were cultivating wheat at least 11,000 years ago because wheat grains of *Triticum monococcum* ($2n = 14$) have been found in the remains of Middle Eastern farming villages from that time. This species has small seed heads and is not highly productive, but some varieties are still grown in the Middle East.

Our most important wheat species today is bread wheat (*Triticum aestivum*, Figure 14.6A), a polyploid with 42 chromosomes. Figure 14.6B illustrates how this species may have evolved; the uppercase letters represent not genes but *sets of chromosomes* that have been traced through the lineage.

❶ The process began with hybridization between two wheats, one the cultivated species *T. monococcum* (AA), the other one of several wild species that probably grew as weeds at the edges of fields (BB). Chromosome sets A and B of the two species would not have been able to pair at meiosis, making the AB hybrid sterile. ❷ However, an error in cell division and self-fertilization produced a new species (AABB) with 28 chromosomes. Today, we know this species as emmer wheat (*T. turgidum*), varieties of which are grown widely in Eurasia and western North America. It is used mainly for making macaroni and other noodle products because its proteins hold their shape better than bread-wheat proteins.

The final steps in the evolution of bread wheat are believed to have occurred in early farming villages on the shores of European lakes over 8,000 years ago. ❸ The cultivated emmer wheat, with its 28 chromosomes, hybridized spontaneously with the closely related wild species *T. tauschii* (DD), which has 14 chromosomes. The hybrid (ABD, with 21 chromosomes) was sterile, ❹ but a cell division error in this hybrid and self-fertilization doubled the chromosome number to 42. The result was bread wheat, with two each of the three ancestral sets of chromosomes (AABBDD).

Today, plant geneticists generate new polyploids in the laboratory by using chemicals that induce meiotic and mitotic errors. They can produce new hybrids with special qualities, such as a hybrid combining the high yield of wheat with the hardiness of rye.

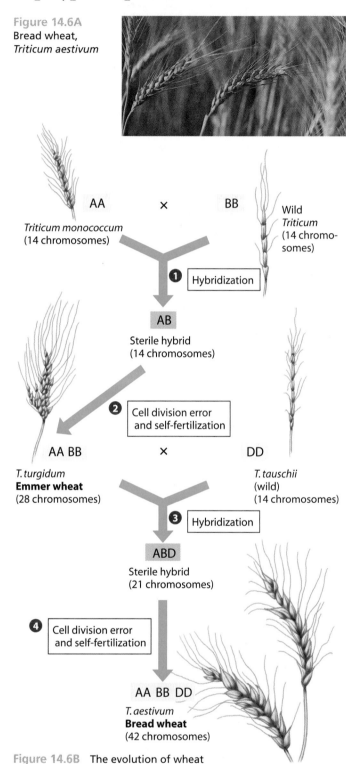

Figure 14.6A Bread wheat, *Triticum aestivum*

AA × BB

Triticum monococcum (14 chromosomes) — Wild *Triticum* (14 chromosomes)

❶ Hybridization

AB — Sterile hybrid (14 chromosomes)

❷ Cell division error and self-fertilization

AA BB × DD

T. turgidum **Emmer wheat** (28 chromosomes) — *T. tauschii* (wild) (14 chromosomes)

❸ Hybridization

ABD — Sterile hybrid (21 chromosomes)

❹ Cell division error and self-fertilization

AA BB DD — *T. aestivum* **Bread wheat** (42 chromosomes)

Figure 14.6B The evolution of wheat

? Why are errors in mitosis or in meiosis a necessary part of speciation by hybridization between two species?

■ If a hybrid has a single copy of the sets of chromosomes from two species, its chromosomes cannot form homologous pairs and separate normally during meiosis to produce gametes. Chromosomes must somehow duplicate so that there is a diploid number of each set and normal gametes can form.

Reproductive barriers may evolve as populations diverge

We have seen that speciation can occur when populations either are geographically isolated (allopatric) or live in the same area (sympatric). Both types require the restriction of gene flow between populations and the emergence of one or more reproductive barriers. How might such barriers evolve?

Researchers attempt to document the evolution of reproductive isolation with laboratory experiments. While at Yale University, Diane Dodd tested this hypothesis: Reproductive barriers can evolve as a by-product of changes in populations as they adapt to different environments. She raised fruit flies on different food sources. Some populations were fed starch, others were fed maltose (malt sugar). After many generations, populations raised on starch digested starch more efficiently, and those raised on maltose digested maltose more efficiently.

Dodd then combined flies from various populations in mating experiments. Figure 14.7A shows some of her results. When flies from "starch populations" were mixed with flies from "maltose populations," the flies mated more frequently with like partners (left grid), even when the like partners came from different populations. In one of the control tests (right grid), flies taken from different populations adapted to starch were about as likely to mate with each other as with flies from their own populations; similar results were obtained for maltose-adapted control groups. The mating preference shown in the experimental group is an example of a prezygotic barrier. The reproductive barrier was not absolute—some mating between "maltose flies" and "starch flies" did occur—but reproductive isolation was under way as these allopatric populations became adapted to different environments.

What types of genetic changes are involved in the evolution of reproductive barriers? In some cases, the reproductive isolation appears to be due to a change in a single gene. In the Japanese land snails shown in Figure 14.3C, for example, one gene controls the direction in which the shells spiral. When

Figure 14.7B A gene for flower color affects pollinator choice. (Left: typical monkey flowers, *M lewisii* top, *M cardinalis* bottom; right: flowers with color allele transfer from the other species.)

their shells spiral in opposite directions, the orientation of the snails' genitalia prevents mating.

A reproductive barrier that involves pollinator choice has been linked to a gene that influences flower color in two closely related species of monkey flower. Bumblebees prefer the pink-flowered *Mimulus lewisii,* and hummingbirds prefer the red-flowered *Mimulus cardinalis.* When scientists experimentally exchanged the alleles for flower color between these two species, the color of their flowers changed (Figure 14.7B). In a field experiment, the now light-orange *M. lewisii* received many more visits from hummingbirds than did the normal pink-flowered *M. lewisii,* and the now pink-purple *M. cardinalis* flowers received many more visits from bumblebees than usual. Thus, a change in a single gene can influence pollinator preference, which provides a reproductive barrier between species.

In other organisms, experiments have shown that the speciation process may be linked to larger numbers of genes and gene interactions. Researchers studying cichlid speciation (see chapter introduction) are beginning to identify and compare genes for jaw morphology, coloration, and color vision to understand the genetic changes involved in speciation.

? Females of the Galápagos finch *Geospiza difficilis* respond to the songs of males from the same island, but ignore songs of males of their species from other islands. How would you interpret these findings?

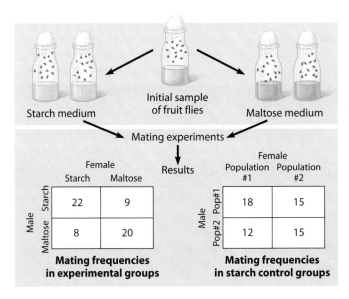

Mating experiments

Results

| | Female | | | Female | |
	Starch	Maltose		Population #1	Population #2
Male Starch	22	9	Male Pop#1	18	15
Maltose	8	20	Pop#2	12	15

Mating frequencies in experimental groups   **Mating frequencies in starch control groups**

Figure 14.7A Evolution of reproductive barriers in lab populations of fruit flies adapted to different food sources

Behavioral barriers to reproduction have begun to develop in these allopatric (geographically separated) ground finch populations.

14.8 Hybrid zones provide opportunities to study reproductive isolation

What happens when separated populations of two closely related species come back into contact with one another? Will reproductive barriers be strong enough to keep the species separate? Or will the two species interbreed and become one? Biologists attempt to answer such questions by studying **hybrid zones,** regions in which members of different species meet and mate, producing at least some hybrid offspring.

Figure 14.8A illustrates the formation of a hybrid zone, starting with the ancestral species. ❶ Four populations are connected by gene flow. ❷ A barrier to gene flow separates one population. ❸ This population diverges from the other three, and speciation occurs. ❹ Gene flow is reestablished in the hybrid zone. Let's consider several possible outcomes for the hybrid zone over time.

 Reinforcement When hybrids are less fit than members of both parent species, we might expect natural selection to strengthen, or *reinforce,* reproductive barriers, thus reducing the formation of unfit hybrids. And we would predict that barriers between species should be stronger where the species overlap.

As an example, consider the closely related pied flycatcher and collared flycatcher illustrated in Figure 14.8B. When populations of these two species do not overlap, males closely resemble each other, with similar black and white coloration (see left side of Figure 14.8B). However, in sympatric populations, male collared flycatchers are still black but with enlarged patches of white, whereas male pied flycatchers are a dull brown (see right side of Figure 14.8B). The photographs in Figure 14.8B show two pied flycatchers, the one on the left from a population that has no overlap with collared flycatchers and the one on the right from a sympatric population. In mate choice experiments, female flycatchers never selected mates from the other species when presented with males from a sympatric population. But females frequently made mistakes when presented with males from allopatric populations (which look similar).

 Fusion What happens when the reproductive barriers between two species are not strong and species contact one another in a

Allopatric populations   Sympatric populations

Male collared flycatcher

Male pied flycatcher

Pied flycatcher from allopatric population

Pied flycatcher from sympatric population

Figure 14.8B   Reinforcement of reproductive barriers

hybrid zone? So much gene flow may occur that the speciation process reverses, causing the two hybridizing species to fuse into one. Many similar cichlid species in Lake Victoria are reproductively isolated by female mate choice based on male coloration (see chapter introduction). But the murky water caused by pollution has reduced the ability of females to distinguish males of their own species from males of closely related species. In some cases, two separate species appear to be fusing into a single hybrid species. This breakdown in a behavioral reproductive barrier is contributing to the rapid disappearance of cichlid species in Lake Victoria.

 Stability One might predict that either reinforcement of reproductive barriers or fusion of gene pools would be the likely outcome in a hybrid zone. However, many hybrid zones turn out to be fairly stable in the sense that hybrids continue to be produced. Although these hybrids may allow for some gene flow between populations, each species maintains its own integrity. We examine one such example in the next module.

? Why might hybrid zones be called "natural laboratories" in which to study speciation?

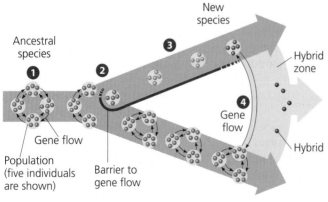

Figure 14.8A   Formation of a hybrid zone

■ By studying the fate of hybrids over time, scientists can directly observe factors that cause (or fail to cause) reproductive isolation.

14.9 Peter and Rosemary Grant study the evolution of Darwin's finches

Some hypotheses wait a long time to be tested. Such was the case with Darwin's 150-year-old hypothesis that the beaks of the diverse Galápagos finch species had adapted to different food sources through natural selection. Then came the classic research of Peter and Rosemary Grant (see Module 13.3).

How did the Grants (Figure 14.9A) come to work with Darwin's finches? They were looking for a pristine, undisturbed place to study variation within populations. In 1973, Peter banded about 60 medium ground finches on Daphne Major, an isolated, uninhabited island in the Galápagos (Figure 14.9B). Returning eight months later with Rosemary and their young daughters, he found all but two of the banded birds. With such an opportunity to work with a small, isolated population, the Grants decided to study these birds for 3 years. One evolutionary question led to another, and for 30 years they have spent up to three months a year on Daphne Major, an island about the size of a football stadium. Each year, the Grants and their students have captured, marked, measured, and studied every finch on the island. Here is how Peter Grant describes their research site:

> There is no beach on Daphne. There are just steep rocks. To land on the island, you have to find some little platform that the waves have cut out of the rock and then climb on from the boat when there are no waves. Then you climb up until you reach a slope where you can actually stand up and walk. And you have to get supplies up there too—something on the order of 30 5-gallon water jugs, cans of food, . . . plus a stove and cylinder of gas for cooking, as well as other camping supplies.

What were some of the questions that kept the Grants returning to this rocky island for so many years? In addition to their study of beak size and natural selection, the Grants focused on another evolutionary mystery: What keeps two finch species distinct despite their ability to interbreed? They found that the occasional interbreeding between the medium ground finch and the cactus finch (see Figure 14.10A) happens when a male learns to sing the song of the other species. Nestlings whose father dies or does not sing much may learn a neighbor's song, even if the neighbor is a different finch species. Thus, a medium ground finch female might breed with a cactus finch because he sings her song.

To find out whether these interspecies couples would create a new hybrid species, the Grants followed the survival of the offspring. They found that the hybrids have intermediate bill sizes and thus can only survive during wet years when there are plenty of soft, small seeds. During dry years, the hybrids can't crack the larger, harder seeds that the medium ground finches can eat and can't compete with cactus finches for cactus seeds. As Rosemary Grant explains:

> There is this occasional hybridization through a breakdown of a learned cultural trait, the song. And so you get this balance between an input of genes and then selection, during drought years, keeping the populations on divergent trajectories in spite of the episodes of hybridization.

Figure 14.9A   Rosemary and Peter Grant with mist nets for catching birds

Figure 14.9B
The rugged
terrain of
Daphne Major

In other words, when hybrids breed with members of the parent species, they introduce new genes on which natural selection can act. But the severe selection during drought years (when the populations of both finch species are greatly reduced and the lack of soft, small seeds causes the hybrid finches to die off) keeps the medium ground finch and the cactus finch on separate evolutionary paths.

Peter Grant conjectures about hybrid finches:

> Perhaps hybrids occasionally disperse . . . to another island that has neither parent species. The hybrids could start a new population with a range of genetic variation different from the parent species. . . . I see no reason why hybridization hasn't been important right from the beginning, from the first divergence of the ancestral finch stock that reached the islands. We don't have the early stages, but that's the big challenge of evolutionary biology—trying to infer from modern clues what happened in the past.

Another challenge of evolutionary biology, at least as practiced by the Grants, is to enjoy field research, even when it means camping on the rocks. In the next module, we continue to explore the divergence of Darwin's finches.

? Despite the rocks, what were the advantages of Daphne Major as a research site?

■ The resident finch populations were small and isolated, and individual birds and their offspring could be followed over several years.

14.10 Adaptive radiation may occur when new opportunities arise

The Galápagos island chain has a total of 14 species of closely related finches. They are often called Darwin's finches because Darwin collected them during his around-the-world voyage on the *Beagle* (see Module 13.1.) These birds have many similarities but differ in their feeding habits and their beak type, which is correlated with what they eat. The three examples shown in Figure 14.10A illustrate the range of beak shape and size, each adapted for a specific diet.

The evolution of many diverse species from a common ancestor is called **adaptive radiation.** The adaptations of these species allow them to fill new habitats or roles in their communities. Adaptive radiation typically occurs when a few organisms colonize new, unexploited areas or when environmental changes cause numerous extinctions, opening up a variety of opportunities for the survivors. For example, fossil evidence indicates that mammals underwent a dramatic adaptive radiation after the extinction of the dinosaurs 65 million years ago.

Isolated island chains with physically diverse habitats are often the sites of explosive adaptive radiations. Colonizers may undergo multiple allopatric and sympatric speciation events, producing species that are found nowhere else on Earth. The Galápagos Archipelago, located west of Ecuador, is one of the world's great showcases of adaptive radiation. Each island was born naked from underwater volcanoes and was gradually clothed by plants, animals, and microorganisms derived from strays that rode the ocean currents and winds from other islands and the South American mainland.

How might Darwin's finch species have evolved from a small population of ancestral birds that colonized one of the islands? Completely isolated on the island, the founder population may have significantly changed as it adapted to its new environment and become a new species. Later, a few individuals of this species may have migrated to a neighboring island, where, under different conditions, this new founder population also accumulated enough changes to become a new species. Some of these birds may have recolonized the first island and coexisted there with the original ancestral species if reproductive barriers kept the species distinct. Multiple colonizations and speciations on the many separate islands of the Galápagos probably followed.

Today, each of the Galápagos Islands has several species of finches, with as many as ten on some islands. The effects of the adaptive radiation of Darwin's finches are evident in their many types of beaks and in the birds' different habitats—some live in trees and others spend most of their time on the ground. Reproductive isolation due to species-specific songs helps keep the species separate.

But some of these islands are also hybrid zones. The Grants (see Module 14.9) have found evidence of hybridization and the exchange of genes between species. In stable hybrid zones, the added genetic variation may facilitate the rapid changes in beak size during fluctuations in the food supply.

Adaptative radiations are linked to new opportunities—no competitors, new habitats and food sources, and in some cases, the evolution of new structures. Recall the rapid adaptive

Figure 14.10A Examples of differences in beak shape and size in Galápagos finches, each adapted for a specific diet

Cactus-seed-eater (cactus finch)

Tool-using insect-eater (woodpecker finch)

Seed-eater (medium ground finch)

radiation of cichlids discussed in the chapter introduction. One of the factors leading to their diversification was their unique mouth (Figure 14.10B). Cichlids have an extra set of jaws and teeth. In different species, these jaws became adapted to processing different foods. Some are specialized for crushing snails or shredding prey or mashing algae. As a result, cichlids evolved as efficient eaters of a variety of foods.

Figure 14.10B Cichlids have a unique mouth and jaw structure

Web Activity *Exploring Speciation on Islands*

? What three types of opportunities might set the stage for adaptive radiation?

■ Colonization of a new, varied habitat; lack of competition and open niches following a mass extinction; evolution of a new morphological feature

14.11 Speciation can occur rapidly or slowly

Biologists continue to make field observations and devise experiments to study evolution in progress. However, much of the evidence for evolution comes from the fossil record, the chronicle of extinct organisms engraved in layers of rock over millions of years of geologic time. What does this record say about the process of speciation?

Many fossil species appear suddenly in a layer of rock and persist essentially unchanged through several layers (strata) until disappearing as suddenly as they appeared. Paleontologists Niles Eldredge and the late Stephen Jay Gould coined the term **punctuated equilibria** to describe these long periods of little change, or equilibrium, punctuated by abrupt episodes of speciation. Figure 14.11A illustrates the evolution of two lineages of hypothetical butterflies in a punctuated pattern. Notice that Figure 14.11A shows no transitional stages in the lineages. The butterfly species change little, if at all, once they appear.

Other fossil species appear to have diverged gradually over long periods of time. Figure 14.11B illustrates this gradual pattern of evolution. Differences gradually evolve in populations as they become adapted to their local environments; and new species (represented by the two butterflies at the far right) evolve gradually from the ancestral population.

What do punctuated and gradual patterns tell us about how long it takes new species to form? Suppose that a species survived for 5 million years, but that most of the changes in its body features occurred during the first 50,000 years of its existence, just 1% of the overall history of the species. Time periods this short often cannot be distinguished in fossil strata. Thus, based on its fossils, the species would seem to have appeared suddenly and continued with little change before becoming extinct. Even though such a species may have originated more slowly than its fossils suggest, a punctuated pattern indicates that speciation occurred relatively rapidly. For species whose fossils show much more gradual change, we cannot tell exactly when a new biological species formed. It is likely that speciation in such groups occurred relatively slowly, perhaps taking millions of years.

Is it likely that most species evolve abruptly and then remain essentially unchanged for most of their existence? Rapid speciation certainly occurs in some cases. As we saw earlier, abrupt speciation can occur by polyploidy in plants and even in a few animals. The diversification of cichlids in only about 100,000 years suggests that new species can form rapidly once divergence begins. It appears that genetic drift and natural selection can significantly alter the gene pool of a small population isolated in a challenging new environment in a few hundred to a few thousand generations.

Figure 14.11A   Punctuated equilibrium model

Figure 14.11B   Gradualism model

But what is the total length of time between speciation events—between when a new species forms and when its populations diverge enough to produce a new species? In one survey of 84 groups of plants and animals, this time ranged from 4,000 years to 40 million years. Overall, the time between speciation events averaged 6.5 million years and rarely took less than 500,000 years. What are the implications of such long time frames? They tell us that it has taken vast spans of time for life on Earth to evolve and that it takes a long time for life to recover from mass extinctions, as have occurred in the past and may be occurring now.

As you've seen, speciation may begin with differences seemingly as small as the color on a cichlid's back. However, as species diverge and speciate again and again, these differences accumulate and become more pronounced, eventually leading to the formation of new groups of organisms that differ greatly from their ancestors. Furthermore, as one group produces many new species, another group may lose species to extinction. The cumulative effects of multiple speciation and extinction events have helped to shape the dramatic changes documented in the fossil record. Such macroevolutionary changes are the subject of our next chapter.

Web Activity *Mechanisms of Microevolution*

? How does the punctuated equilibrium model account for the relative rarity of transitional fossils linking newer species to older ones?

■ If speciation takes place in a relatively short time compared with the overall time the species exists, the transition of one species to another may be difficult to find in the fossil record.

Chapter Review

Reviewing the Concepts

Speciation. The origin of new species connects microevolution and macroevolution and has led to Earth's incredible biological diversity (**Introduction–14.1**).

Concepts of Species (14.2–14.3)

Defining species. Linnaeus used physical characteristics to distinguish species. His binomial system is the basis of taxonomy, the naming and classification of life's forms.

The biological species concept defines a species as a group of populations whose members can interbreed and produce fertile offspring with each other but not with members of other species. Most organisms are classified based on observable phenotypic traits—the morphological species concept. The ecological species concept defines a species by its ecological niche. According to the phylogenetic species concept, a species is the smallest group that shares a common ancestor and forms one branch on the tree of life (**14.2**).

Reproductive barriers serve to isolate a species' gene pool and prevent interbreeding (**14.3**).

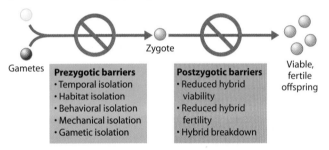

Gametes

Zygote

Viable, fertile offspring

Prezygotic barriers	**Postzygotic barriers**
• Temporal isolation	• Reduced hybrid viability
• Habitat isolation	
• Behavioral isolation	• Reduced hybrid fertility
• Mechanical isolation	
• Gametic isolation	• Hybrid breakdown

Mechanisms of Speciation (14.4–14.11)

Allopatric speciation. Geographically separated from other populations, a small population may become genetically unique as its gene pool is changed by natural selection, mutation, or genetic drift (**14.4**).

Sympatric speciation occurs when a new species originates while remaining in a geographically overlapping area with the parent species. Many plant species have evolved by polyploidy (duplications of the chromosome number due to errors in cell division). Many plants, including food plants such as bread wheat, are the result of hybridization and polyploidy. Habitat differentiation and sexual selection (usually involving mate choice) can also result in sympatric speciation (**14.5–14.6**).

Evolution of reproductive barriers. A laboratory study has documented the beginning of reproductive isolation as fruit fly populations adapted to a new food source. Researchers have identified the specific genes involved in some cases of speciation (**14.7**).

Hybrid zones are regions in which members of different species overlap and produce at least some hybrid offspring. Over time, reinforcement may strengthen prezygotic barriers to reproduction, or fusion may reverse the speciation process as reproductive barriers weaken and extensive gene flow occurs. In stable hybrid zones, a limited number of hybrid offspring continue to be produced (**14.8**). Peter and Rosemary Grant have documented hybridization of some Galápagos finch species (**14.9**).

Adaptive radiation can occur when populations are provided with expanded opportunities following mass extinctions, the colonization of a diverse new environment, or the evolution of new structures (**14.10**).

Speciation rates. The punctuated equilibrium model draws on the fossil record, where many species change most as they arise from an ancestral species and then change relatively little for the rest of their existence. But some species have evolved by the gradual accumulation of changes. The time interval between speciation events varies considerably, from a few thousand years to tens of millions of years (**14.11**).

Connecting the Concepts

1. Name the two types of speciation represented by this diagram. For each type, describe how reproductive barriers may develop between the new species.

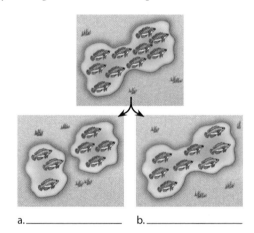

a. _____ b. _____

2. Fill in the blanks in the following concept map.

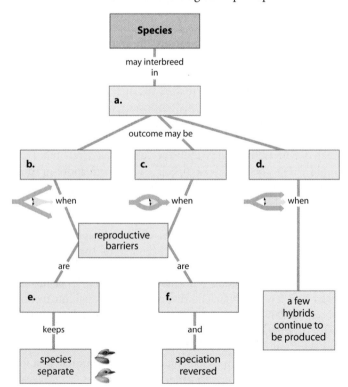

Testing Your Knowledge

Multiple Choice

3. Which concept of species would be most useful to a field biologist identifying new species in a tropical forest?
   a. biological
   b. ecological
   c. morphological
   d. phylogenetic
   e. a and b combined would prove most valuable.

4. The *largest* unit within which gene flow can readily occur is a
   a. population.
   b. species.
   c. genus.
   d. hybrid.
   e. taxon.

5. Bird guides once listed the myrtle warbler and Audubon's warbler as distinct species that lived side by side in parts of their ranges. However, recent books show them as eastern and western forms of a single species, the yellow-rumped warbler. Most likely, it has been found that the two kinds of warblers
   a. live in similar habitats and eat similar foods.
   b. interbreed often in nature, and the offspring are viable and fertile.
   c. are almost identical in appearance.
   d. have many genes in common.
   e. sing similar songs.

6. Which of the following is an example of a postzygotic reproductive barrier?
   a. One *Ceanothus* shrub lives on acid soil, another on basic soil.
   b. Mallard and pintail ducks mate at different times of year.
   c. Two species of leopard frogs have different mating calls.
   d. Hybrid offspring of two species of jimsonweeds always die before reproducing.
   e. Pollen of one kind of tobacco cannot fertilize another kind.

7. Biologists have found more than 500 species of fruit flies on the various Hawaiian Islands, all apparently descended from a single ancestor species. This example illustrates
   a. polyploidy.
   b. temporal isolation.
   c. adaptive radiation.
   d. sympatric speciation.
   e. postzygotic barriers.

8. A new plant species C, which formed from hybridization of species A ($2n = 16$) with species B ($2n = 12$), would probably produce gametes with a chromosome number of
   a. 12.
   b. 14.
   c. 16.
   d. 28.
   e. 56.

9. A horse ($2n = 64$) and a donkey ($2n = 62$) can mate and produce a mule. How many chromosomes would there be in a mule's cells?
   a. 31
   b. 62
   c. 63
   d. 126
   e. 252

10. What prevents horses and donkeys from hybridizing to form a new species?
    a. limited hybrid fertility
    b. limited hybrid viability
    c. hybrid breakdown
    d. gametic isolation
    e. prezygotic barrier

11. Which of the following most likely contributed to the rapid speciation of the hundreds of cichlid species in Lake Victoria?
    a. the introduction of the Nile perch
    b. the increasingly murky waters resulting from pollution
    c. female mate choice based on male coloration
    d. a unique mouth and jaw structure that was adaptable to diverse food resources.
    e. Both c and d probably contributed to cichlid speciation.

12. Which of the following factors would *not* contribute to allopatric speciation?
    a. A population becomes geographically isolated from the parent population.
    b. The separated population is small, and genetic drift occurs.
    c. The isolated population is exposed to different selection pressures than the parent population.
    d. Different mutations begin to distinguish the gene pools of the separated populations.
    e. Gene flow between the two populations continues to occur.

Describing, Comparing, and Explaining

13. Explain how each of the following makes it difficult to clearly define a species: variation within a species, geographically isolated populations, asexual species, fossil organisms.

14. Explain why allopatric speciation would be less likely on an island close to a mainland than on a more isolated island.

15. What does the term *punctuated equilibria* describe?

16. Differentiate between microevolution and speciation.

Applying the Concepts

17. Cultivated American cotton plants have a total of 52 chromosomes ($2n = 52$). In each cell, there are 13 pairs of large chromosomes and 13 pairs of smaller chromosomes. Old World cotton plants have 26 chromosomes ($2n = 26$), all large. Wild American cotton plants have 26 chromosomes, all small. Propose a testable hypothesis to explain how cultivated American cotton probably originated.

18. Explain how the murky waters of Lake Victoria may be contributing to the "fall" of cichlid species (see chapter introduction). How might these polluted waters affect the formation of new species? How might the loss of cichlid species be slowed?

19. The red wolf, *Canis rufus,* once widespread in the southeastern and southcentral United States, nearly became extinct in the late 1970s. Saved by a captive breeding program under the authority of the Endangered Species Act (ESA), it has been reintroduced in areas such as the Great Smoky Mountains National Park. Recent genetic evidence indicates that the red wolf may not be a separate species, but a hybrid of the coyote, *Canis latrans,* and the gray wolf, *Canis lupus.* Though the original intent of the ESA was to protect all endangered groups—whether species, subspecies, or hybrids—the costs may be prohibitive. What criteria should be applied if we must decide which organisms to protect? Are there reasons to preserve hybrids, subspecies, or local populations when the species as a whole is not at risk?

Answers to all questions can be found in Appendix 4.

For Practice Quizzes, BioFlix, MP3 Tutors, and Activities, go to www.mybiology.com.

On the Wings of Eagles, Bats, and Pterosaurs

Have you ever wished you could fly? When you were a child, did you ever want to join Wendy, Michael, and John as they swooped around their bedroom with Peter Pan? Unfortunately (or perhaps fortunately), we remain earthbound because our ancestors were busy using their arms for other functions that would not easily permit them to become adapted as wings.

But in some species, limbs did become adapted for flight. Forelimbs have been remodeled into wings three separate times in tetrapods (vertebrates with four limbs). Pterosaurs (left) were the first tetrapods in the air. These flying reptiles evolved about 225 million years ago, persisted for about 140 million years, and then went extinct along with most of the dinosaurs. Birds evolved from feathered dinosaurs about 150 million years ago and are still flying. The fossil record of the evolution of bats, the third group of flying vertebrates, is not very complete. The hypothesis is that bats diverged from nocturnal, insectivorous, tree-dwelling mammals some 60 million years ago. They, too, are still flying today.

Different times, different ancestors, same ability—powered flight. Are the wings of pterosaurs, birds, and bats the same or different? The requirements of powered flight include wings as broad airfoils that can move up and down, forward and backward, and fold against the body when not in use. It appears that in vertebrates, natural selection favored three different designs to meet these requirements.

Pterosaurs ruled the skies in the Jurassic and Cretaceous periods. Some were as small as songbirds; one was the size of a small airplane, with a wingspan of 12 m (40 feet). As you can see in the artist's reconstruction of a pterosaur to the left, the wing was supported along much of its length by one greatly elongated finger. The skin membrane making up the wing was reinforced with thickened fibers.

Ranging from hovering hummingbirds to soaring eagles, most birds are adept flyers. The bird wing is composed of feathers, which are outgrowths of the skin along the whole length of the forelimb. Much of the wingspan is provided by the feathers. The supporting structure is an elongated forearm and modified wrist and hand bones. Finger bones are reduced and fused.

Bat wings can be described as "hand wings," made of a membrane supported by the arm and four elongated fingers. The membrane extends to the hind limbs and is attached at the heel. These small mammals are agile flyers, capable of swooping after insects and navigating around obstacles (including humans) using echolocation.

As different as the architectures of the wings of pterosaurs, birds, and bats are, however, all three designs represent the remodeling of the same ancestral tetrapod limb. These examples of Darwin's "descent with modification" illustrate again that evolution is an editing process, which can gradually adapt existing structures to new functions. In this case, natural selection led to three different "solutions" to the challenge of powered flight.

Indeed, these three kinds of flyers lead to a multitude of evolutionary questions. What were their ancestors like? Were they bipedal, fast-running animals who leaped into the air to capture prey with their arms? Were they tree dwellers whose extended forelimbs helped them glide from limb to limb? How has body form changed as the ability to fly became an adaptation on which natural selection could act? And how did the amazing diversity of pterosaurs, birds, and bats evolve? These questions and more are addressed by biologists who study fossils, genes, development, and phylogeny (the evolutionary history of groups).

In this chapter we consider **macroevolution,** the major changes (such as the evolution of flight in three different vertebrate lineages) recorded in the history of life over vast tracts of time. We will trace this history and consider some of the major mechanisms of macroevolution. And we will explore how scientists organize the amazing diversity of life by attempting to discover the evolutionary relationships among living and extinct groups, tracing backwards to the first living organisms on Earth.

To approach these wide-ranging topics, we begin with the most basic of questions: How did life arise on planet Earth? ■ ■ ■

293

15.1 Conditions on early Earth made the origin of life possible

Earth is one of eight planets orbiting the sun, which is one of billions of stars in the Milky Way. The Milky Way, in turn, is one of billions of galaxies in the universe. Gazing at stars gives us a look back in time. The star closest to our sun is 4 light years—or 40 trillion kilometers—away; in other words, the light we see today is the light the star emitted four years ago. Some stars you can see in the night sky are so far away that their light is just reaching us, though they burned out millions of years ago.

The universe has not always been so spread out. Physicists have evidence that before the universe existed in its present form, all matter was concentrated in one mass. The mass seems to have blown apart with a "big bang" sometime between 10 and 20 billion years ago and to have been expanding ever since.

Scientific evidence indicates that Earth and the other planets of our solar system formed about 4.6 billion years ago from a vast swirling cloud of dust. Most of the dust condensed in the center to form the sun, but some matter was left orbiting the sun. As dust particles collided and stuck together, larger rocks formed. Additional collisions created even larger bodies, whose gravity attracted more matter, eventually forming the planets.

Conditions on Early Earth   Immense heat would have been generated by the impact of meteorites and compaction by gravity, and young planet Earth probably began as a molten mass. The mass then sorted into layers of varying densities, with the least dense material on the surface, solidifying into a thin crust.

As the bombardment of early Earth slowed, conditions on the planet were extremely different from those of today. The first atmosphere was probably thick with water vapor, along with various compounds released by volcanic eruptions, including nitrogen and its oxides, carbon dioxide, methane, ammonia, hydrogen, and hydrogen sulfide. As Earth slowly cooled, the water vapor condensed into oceans, and much of the hydrogen quickly escaped into space. Not only was the atmosphere of young Earth very different from the one we know today, but lightning, volcanic activity, and ultraviolet radiation were much more intense. It was in such an environment that life began.

When Did Life Begin?   The earliest evidence of life on Earth comes from fossils that are about 3.5 billion years old. Figure 15.1 shows one of these fossils, a layered rock called a **stromatolite** (from the Greek *stroma*, bed, and *lithos*, rock). These fossils resemble similar layered mats formed by present-day photosynthetic prokaryotes (see the introduction to Chapter 16). As they grow, the prokaryotes bind thin films of sediment together, then migrate to the surface and start the next layer.

Photosynthesis is not a simple process, so significant time must have elapsed before life could have become as complex as the organisms that formed the ancient stromatolites. It is unlikely that these photosynthetic prokaryotes were the first forms of life. The evidence that these prokaryotes lived 3.5 billion years

Figure 15.1   A cross-section of a fossilized stromatolite

ago is strong support for the hypothesis that life in a simpler form arose much earlier, perhaps as early as 3.9 billion years ago.

How Did Life Arise?   From the time of the ancient Greeks until well into the 19th century, it was commonly believed that life regularly arises from nonliving matter. Many people believed, for instance, that flies came from rotting meat and fish from ocean mud. Experiments performed in the 1600s showed that relatively large organisms, such as insects, cannot arise spontaneously from nonliving matter. And in 1862, experiments by the French scientist Louis Pasteur confirmed that all life arises only by the reproduction of preexisting life.

Pasteur ended the argument over spontaneous generation of present-day organisms, but he did not address the question of how life arose in the first place. To attempt to answer that question, for which there is no fossil evidence available, scientists develop hypotheses and test their predictions.

Observations and experiments have led scientists to hypothesize that chemical and physical processes on early Earth could have produced very simple cells through a sequence of four main stages:

1. The abiotic (nonliving) synthesis of small organic molecules, such as amino acids and nucleotides
2. The joining of these small molecules into macromolecules, including proteins and nucleic acids
3. The packaging of these molecules into "protobionts," droplets with membranes that maintain an internal chemistry different from that of their surroundings
4. The origin of self-replicating molecules that eventually made inheritance possible

In the next two modules, we examine some of the evidence for each of these four stages.

? Why do 3.5-billion-year-old stromatolites suggest that life originated *before* 3.5 billion years ago?

■ If photosynthetic prokaryotes existed by 3.5 billion years ago, a simpler, nonphotosynthetic type of cell probably originated before that time.

15.2 Stanley Miller's experiments showed that the abiotic synthesis of organic molecules is possible

In 1953, when Stanley Miller was a 23-year-old graduate student in the laboratory of Harold Urey at the University of Chicago, he performed some experiments that would attract global attention. Miller was the first to show that amino acids and other organic molecules could be formed under conditions believed to simulate those of early Earth (Figure 15.2).

Miller's experiments were a test of a hypothesis about the origin of life developed in the 1920s by Russian chemist A. I. Oparin and British scientist J. B. S. Haldane. Oparin and Haldane independently proposed that conditions on early Earth could have generated organic molecules. They reasoned that present-day conditions on Earth do not allow the spontaneous synthesis of organic compounds simply because the atmosphere is rich in O_2. As a strong oxidizing agent, O_2 tends to disrupt chemical bonds. However, before the early prokaryotes added O_2 to the air, Earth may have had a reducing (electron-adding) atmosphere. The energy for this abiotic synthesis of organic compounds could have come from lightning and intense UV radiation. Haldane suggested that the early oceans were a solution of organic molecules, a "primitive soup" from which life arose. As Miller, now a professor at the University of California, San Diego, told us several years ago,

> Oparin proposed that the primitive atmosphere contained the gases methane, ammonia, hydrogen, and water, and that chemical reactions in that primitive atmosphere produced the first organic molecules. That hypothesis had a good deal of appeal, but without the experiments, it was talked about but not very well accepted.

The apparatus shown with Dr. Miller in Figure 15.2 is similar to the one he used to test the prediction. A flask of warmed water represented the primeval sea. The water was heated so that some vaporized and moved into a second, higher flask. The "atmosphere" consisted of water vapor, H_2, CH_4, and NH_3—the gases that scientists in the 1950s thought prevailed in the ancient world. Electrodes discharged sparks into the flask to mimic lightning. A condenser cooled the atmosphere, raining water and any dissolved compounds back down into the miniature sea. As material circulated through the apparatus, the solution in the flask slowly changed color. As Dr. Miller described it,

> The first time I did the experiment, it turned red. Very dramatic! And then, after it turned red, it got more yellow and then brown as the sparking went on.

After the experiment proceeded for a week, Miller found a variety of organic compounds in the solution, including some of the amino acids that make up the proteins of organisms:

> The surprise was that we . . . got mainly organic compounds of biological significance. And the amino acids were formed, not in trace quantities, but abundantly! The experiment went beyond our wildest hopes.

Miller's early experiments stimulated a great deal of interest and research on the prebiotic (before-life) origin of organic

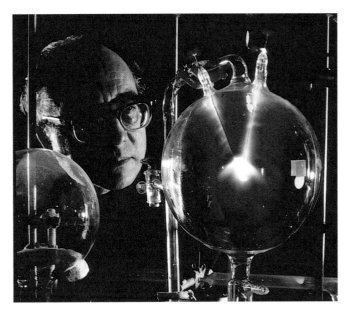

Figure 15.2 Stanley Miller re-creating his 1953 experiment

compounds. Similar experiments using various atmospheric conditions have also produced mixtures of organic compounds.

Scientists now think that the composition of the atmosphere of early Earth was somewhat different from what Miller assumed in his historic first experiment. There is growing evidence that the early atmosphere was made up primarily of N_2 and CO_2, and so far, Miller-Urey-type experiments using such atmospheres have not produced organic molecules. Still, it is possible that small "pockets" of the early atmosphere—perhaps near volcanic openings—were similar to those used by Miller.

Alternatively, submerged volcanoes and deep-sea hydrothermal vents—gaps in the Earth's crust where hot water and minerals gush into deep oceans—may have provided the initial chemical resources for life. Such environments are among the most extreme in which life exists today, and some researchers favor the hypothesis that life may have begun in similar regions on early Earth.

Miller-Urey-type experiments demonstrate that the abiotic synthesis of organic molecules is possible. Support for this idea also comes from analyses of the chemical composition of meteorites. Fragments of a 4.5-billion-year-old meteorite collected in 1969 contain more than 80 amino acids. Remarkably, the proportions of these amino acids are similar to those produced in the Miller-Urey experiments.

Web Process of Science How Might Conditions on Early Earth Have Created Life?

? Which of the four stages in the hypothetical scenario of the origin of simple cells was Stanley Miller testing with his experiments?

■ Stage 1: Conditions on early Earth favored synthesis of organic molecules from inorganic ingredients.

The formation of polymers, membranes, and self-replicating molecules represent stages in the origin of the first cells

The abiotic synthesis of small organic molecules is a first step in the origin of life. But could macromolecules, such as proteins and nucleic acids, have formed on early Earth?

Abiotic Synthesis of Macromolecules Polymers are synthesized by dehydration reactions that add monomers to a growing chain (see Module 3.3). In the cell, specific enzymes catalyze these reactions. But scientists have produced polymers in the laboratory without enzymes—for example, by dripping dilute solutions of organic monomers onto hot sand, clay, or rock. The heat vaporizes the water and concentrates the monomers on the underlying substance. Some of the monomers then spontaneously bond together in chains, forming polymers. On early Earth, waves may have splashed dilute solutions of organic monomers onto fresh lava or other hot rocks and then rinsed polypeptides and other polymers back into the sea.

Formation of Protobionts A key step in the origin of life would have been the isolation of a collection of abiotically created molecules within a membrane, an entity called a **protobiont.** Within a confined space, certain combinations of molecules could be concentrated and interact more efficiently. The internal environment of a protobiont could differ from its surroundings.

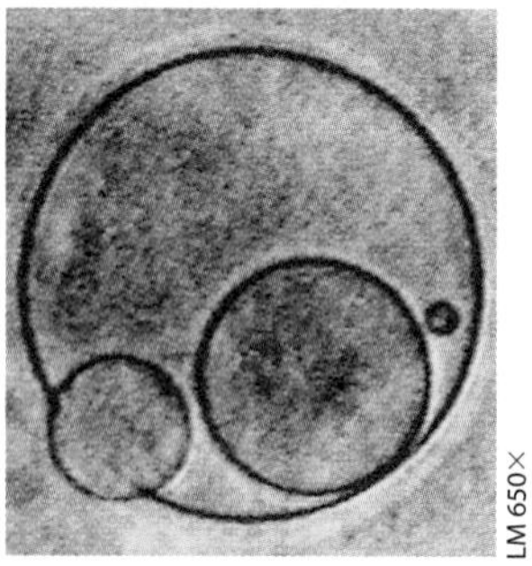

Laboratory experiments demonstrate that protobionts could have formed spontaneously from abiotically produced organic compounds. For example, small membrane-bounded droplets can form when lipids are added to water (Figure 15.3A). The hydrophobic molecules in the mixture organize into a layer at the surface of the droplet, much like the lipid bilayer of a plasma membrane. These spheres are not alive, but they display some of the properties of living cells. They have a selectively permeable membrane surface and swell or shrink osmotically when placed in solutions of different solute concentrations. If enzymes are included in these droplets, some spheres can carry out simple metabolic reactions. In a similar fashion, protobionts may have formed in the waters of a young Earth.

Figure 15.3A
Microscopic spheres with membranes made of lipids

LM 650×

Self-Replicating RNA Today's cells store their genetic information as DNA, transcribe the information into RNA, and then translate RNA messages into specific enzymes and other proteins. As we have seen in earlier chapters, this DNA → RNA → protein assembly system is extremely intricate. Most likely, it emerged gradually through a series of refinements to much simpler processes. What were the first genes like?

One hypothesis is that the first genes were short strands of RNA that replicated themselves without the assistance of proteins. Laboratory experiments support this idea. Short RNA molecules can assemble spontaneously from nucleotide monomers in the absence of enzymes. Furthermore, when RNA is added to a solution containing a supply of RNA monomers, new RNA molecules complementary to parts of the starting RNA sometimes assemble. So we can imagine a scenario on early Earth like the one in Figure 15.3B: ❶ RNA monomers—nucleotides—spontaneously join, forming the first small "genes." ❷ Then an RNA chain complementary to one of these genes assembles. If the new chain, in turn, serves as a template for another round of RNA assembly, the result is a replica of the original gene.

This RNA replication process might have been aided by RNA molecules that acted as catalysts. Scientists have discovered that some RNAs, which they call **ribozymes,** can carry out a number of enzyme-like functions.

Scientists use the term "RNA world" for the hypothetical period in the evolution of life when RNA served as both rudimentary genes and catalytic molecules.

It is easy to imagine that certain of these cell-like entities on early Earth might have contained self-replicating RNA molecules. Once that happened, natural selection would have begun to shape the properties of these protobionts. Those that grew and replicated more efficiently than others would have increased in number, passing their abilities on to subsequent generations. In other words, protobionts could have come closer to evolving in a Darwinian sense.

? Why would the formation of protobionts represent a key step in the evolution of life?

the next generation.
evolved and passed their successful properties to
selection once self-replicating molecular systems
assisting chemical reactions, and allow for natural
compartments could concentrate organic molecules,
■ Segregating mixtures of molecules within

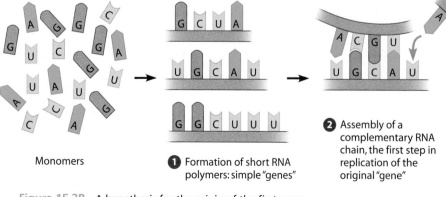

Monomers

❶ Formation of short RNA polymers: simple "genes"

❷ Assembly of a complementary RNA chain, the first step in replication of the original "gene"

Figure 15.3B A hypothesis for the origin of the first genes

15.4 The origins of single-celled and multicelled organisms and the colonization of land are key events in life's history

Figure 15.4 uses the analogy of a clock ticking down from the origin of Earth 4.6 billion years ago to the present to show some major events in the history of life. Earth's history is divided into three eons of geologic time. The first two—the Archaean and the Proterozoic—together lasted approximately 4 billion years. The Phanerozoic eon, roughly the last half billion years, is divided into three eras: the Paleozoic, Mesozoic, and Cenozoic. (We discuss the geologic record in more detail in Module 15.6.)

Origin of Prokaryotes As we discussed, the earliest evidence of life comes from fossil stromatolites (see Figure 15.1). Early prokaryotes were Earth's sole inhabitants from at least 3.5 billion years ago to about 2 billion years ago. During this time, prokaryotes transformed the biosphere. For instance, atmospheric oxygen (indicated by the salmon-colored band in Figure 15.4) began to appear 2.7 billion years ago as a result of prokaryotic photosynthesis. By 2.2 billion years ago, atmospheric O_2 began to increase rapidly. This "oxygen revolution" had an enormous impact on life: Many prokaryotes probably became extinct, while other species survived in anaerobic habitats. The evolution of cellular respiration, which uses O_2 in harvesting energy from organic molecules, allowed other prokaryotes to flourish.

Origin of Single-celled Eukaryotes The oldest widely accepted fossils of eukaryotes are about 2.1 billion years old (the dark gray band). You will learn in Module 16.12 about how the more complex eukaryotic cell originated when small prokaryotic cells capable of aerobic respiration or photosynthesis began living in larger cells. After the first eukaryotes appeared, a great range of unicellular forms evolved, giving rise to the diversity of single-celled eukaryotes that continue to flourish today.

Origin of Multicellular Eukaryotes Another wave of diversification followed: the origin of multicellular forms whose descendants include a variety of algae, plants, fungi, and animals. Molecular comparisons suggest that the common ancestor of multicellular eukaryotes lived 1.5 billion years ago (the light gray band). The oldest known fossils of multicellular eukaryotes are of relatively small algae that lived about 1.2 billion years ago.

Larger and more diverse multicellular eukaryotes do not appear in the fossil record until about 600 million years ago. These fossils were of soft-bodied animals (the dark blue band). A great increase in the diversity of animal forms occurred 535–525 million years ago, in a period known as the Cambrian explosion.

Colonization of Land The colonization of land was another milestone in the history of life (the light green band in Figure 15.4). There is fossil evidence that photosynthetic prokaryotes coated damp terrestrial surfaces well over a billion years ago. However, larger forms of life such as fungi, plants, and animals did not begin to colonize land until about 500 million years ago.

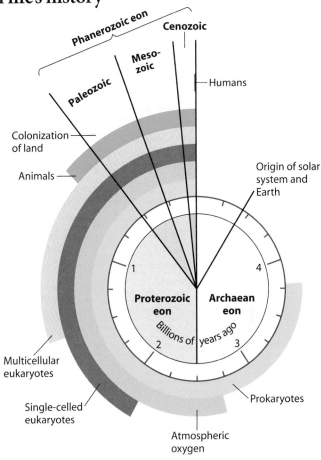

Figure 15.4 A clock analogy for some key events in the history of Earth and its life

Plants colonized land in the company of fungi. Even today, the roots of most plants are associated with fungi that aid in absorption and receive nutrients in return. Such mutually beneficial associations are evident in some of the oldest plant fossils.

The most widespread and diverse land animals are arthropods (particularly insects and spiders) and tetrapods. Tetrapods include humans, but we are late arrivals on the scene—the human lineage diverged from other hominoids (apes) around 6 to 7 million years ago, and our species originated about 195,000 years ago. If the clock of Earth's history were rescaled to represent an hour, humans appeared less than 0.2 seconds ago.

In the next two modules, we see how scientists have determined when these key episodes in Earth's history have occurred in geologic time.

Web Activity *The History of Life*

? For how long did life on Earth consist solely of single-celled organisms?

■ More than 2 billion years: The oldest known fossils of multicellular organisms date from about 1.2 billion years ago.

15.5 The actual ages of rocks and fossils mark geologic time

Geologists use several techniques to determine the ages of rocks and the fossils they contain. The method most often used, called **radiometric dating,** is based on the measurement of certain radioactive isotopes (see Module 2.5). Fossils contain isotopes of elements that accumulated when the organisms were alive. For example, the carbon in a living organism includes both the most common isotope, carbon-12, and a less common radioactive isotope, carbon-14, in the same ratio as is present in the atmosphere. Once an organism dies, it stops accumulating carbon, and its carbon-14 starts slowly to decay to another element. The rate of decay is expressed as a half-life, the time required for 50% of the isotope in a sample to decay. Each radioactive isotope has a characteristic half-life. With a half-life of 5,730 years, half the carbon-14 in a specimen decays in about 5,730 years, half the remaining carbon-14 decays in the next 5,730 years, and so on (Figure 15.5). Knowing both the half-life of a radioactive isotope and the ratio of radioactive to stable isotope in a fossil enables us to tell the age of the fossil.

Carbon-14 is useful for dating relatively young fossils (up to about 75,000 years old). Radioactive isotopes with longer half-lives are used to date older fossils.

There are indirect ways to estimate the age of much older fossils. For example, potassium-40, with a half-life of 1.3 billion years, can be used to date volcanic rocks hundreds of millions of years old. A fossil's age can be inferred from the ages of the rock layers above and below the strata in which the fossil is found.

By dating rocks and fossils, scientists have established a geologic record of Earth's history, the topic of our next module.

? Estimate the age of a fossil found in a sedimentary rock layer between two layers of volcanic rock that are determined to be 530 and 520 million years old.

■ We can infer that the organism lived approximately 525 million years ago.

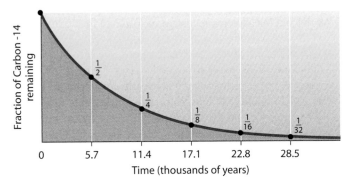

Figure 15.5 Radiometric dating

15.6 The fossil record documents the history of life

The fossil record, the sequence in which fossils appear in rock strata, is an archive of evolutionary history (see Module 13.4). Based on this sequence and the ages of rocks and fossils, geologists have established a **geologic record,** as shown in Table 15.6 on the facing page. As you saw in Figure 15.4, Earth's history is divided into three eons, the Archaean, Proterozoic, and Phanerozoic. The timeline on the left-hand side of Table 15.6 shows the lengths of these eons. Note that the Phanerozoic eon, which is roughly only the last half-billion years, is expanded in the table to show the key events in the evolution of multicellular eukaryotic life. This eon is divided into three eras: the Paleozoic, Mesozoic, and Cenozoic. Most eras are subdivided into periods. The boundaries between eras are marked by mass extinctions, when many forms of life disappeared from the fossil record and were replaced by species that diversified from the survivors. Lesser extinctions often mark the boundaries of the periods that make up an era.

Rocks from the Archaean and Proterozoic eons have undergone extensive change over time, and much of their fossil content is no longer visible. Nonetheless, paleontologists have pieced together ancient events in life's history. The oldest known fossils, dating from 3.5 billion years ago, are of prokaryotes; the oldest known eukaryotes are from 2.1 billion years ago. Strata from the Ediacaran period (635–542 million years ago) bear diverse fossils of algae and soft-bodied animals.

Dating from about 542 million years ago, rocks of the Paleozoic ("ancient animal") era contain fossils of lineages that gave rise to present-day organisms, as well as many lineages that have become extinct. During the early Paleozoic, virtually all life was aquatic, but by about 400 million years ago, plants and animals were well established on land.

Following the Paleozoic era was the Mesozoic ("middle animal") era, also known as the age of reptiles because of its abundance of reptilian fossils, including those of dinosaurs. The Mesozoic also saw the beginnings of mammals and flowering plants (angiosperms). By the end of the Mesozoic, dinosaurs had become extinct, except for one lineage—the birds.

An explosive period of evolution of mammals, birds, and angiosperms began at the dawn of the Cenozoic ("recent animal") era, about 65 million years ago. Because much more is known about the Cenozoic than about earlier eras, our table subdivides the two Cenozoic periods into finer intervals called epochs.

The chapters of Unit 4 describe the diversity of life forms that have evolved on Earth. In the next section, we examine some of the processes that have produced the distinct changes seen in the geologic record.

Web Activity *A Scrolling Geologic Record*

? What were the dominant animals during the Carboniferous period? When were gymnosperms the dominant plants? (*Hint:* Look at Table 15.6.)

■ Amphibians. Gymnosperms were dominant during the Triassic and Jurassic periods (251–145.5 million years ago).

TABLE 15.6		THE GEOLOGIC RECORD			
Relative Duration of Eons	**Era**	**Period**	**Epoch**	**Age (Millions of Years Ago)**	**Some Important Events in the History of Life**
Phanerozoic	Cenozoic	Neogene	Holocene		Historical time
				0.01	
			Pleistocene		Ice ages; humans appear
				1.8	
			Pliocene		Origin of genus *Homo*
				5.3	
			Miocene		Continued radiation of mammals and angiosperms; apelike ancestors of humans appear
				23	
		Paleogene	Oligocene		Origins of many primate groups, including apes
				33.9	
			Eocene		Angiosperm dominance increases; continued radiation of most present-day mammalian orders
				55.8	
			Paleocene		Major radiation of mammals, birds, and pollinating insects
				65.5	
Proterozoic	Mesozoic	Cretaceous			Flowering plants (angiosperms) appear; many groups of organisms, including most dinosaurs, become extinct at end of period (Cretaceous extinctions)
				145.5	
		Jurassic			Gymnosperms continue as dominant plants; dinosaurs abundant and diverse
				199.6	
		Triassic			Cone-bearing plants (gymnosperms) dominate landscape; origin and radiation of dinosaurs; origin of mammals
				251	
	Paleozoic	Permian			Radiation of reptiles; origin of most present-day groups of insects; extinction of many marine and terrestrial organisms at end of period
				299	
		Carboniferous			Extensive forests of vascular plants; first seed plants; origin of reptiles; amphibians dominant
				359.2	
		Devonian			Diversification of bony fishes; first tetrapods and insects
				416	
		Silurian			Diversification of early vascular plants
				443.7	
		Ordovician			Marine algae abundant; colonization of land by fungi, plants, and animals
				488.3	
		Cambrian			Sudden increase in diversity of many animal phyla (Cambrian explosion)
				542	
Archaean		Ediacaran			Diverse algae and soft-bodied invertebrate animals
				635	
				2,100	Oldest fossils of eukaryotic cells
				2,500	
				2,700	Concentration of atmospheric oxygen begins to increase
				3,500	Oldest fossils of cells (prokaryotes)
				3,800	Oldest known rocks on Earth's surface
				Approx. 4,600	Origin of Earth

15.7 Continental drift has played a major role in macroevolution

In 1912, German meteorologist Alfred Wegener proposed the hypothesis of **continental drift.** Wegener postulated that all land on Earth was once one great mass, which broke up into continents that drifted like rafts to their present positions. Wegener suggested that the shapes of our modern continents, like pieces of a jigsaw puzzle, reflect their former positions in the original supercontinent. Like many ideas generated before their time, Wegener's were not taken seriously for decades. Until the 1960s, the continents were thought to have always been fixed in their present positions.

But indeed, if photographs of Earth were taken from space every 10,000 years and spliced together to make a movie, it would show a different story. Wegener was right! The seemingly "rock solid" continents we live on move over time. Since the origin of eukaryotes roughly 1.5 billion years ago, there have been three occasions (1.1 billion, 600 million, and 250 million years ago) in which all the landmasses of Earth came together to form a supercontinent, then later broke apart. Each time the landmasses split, they yielded a different configuration of continents. Looking into the future, geologists estimate that the continents will come together again and form a new supercontinent roughly 250 million years from now.

The continents and seafloors form a thin outer layer of planet Earth, called the crust, which covers a mass of hot material called the mantle. This crust is divided into giant, irregularly shaped plates (outlined in black in Figure 15.7A). Because the mantle circulates constantly, the continental plates move about slowly but incessantly—they essentially float—on the underlying mantle. Many important geologic processes often occur at plate boundaries. In some cases, the plates are moving away from one another. North America and Europe, for example, are presently drifting apart at a rate of about 2 cm per year. In other cases, two plates are sliding past one another, forming regions where earthquakes are common. In still other cases, two plates are colliding. Massive upheavals may occur, forming mountains along the plate boundaries. The red dots in Figure 15.7A indicate zones of violent geologic activity.

Throughout geologic time, continental movements have reshaped the physical features of the planet and altered the habitats in which organisms live. Figure 15.7B on the facing page shows two chapters in the continuing saga of continental drift that have been especially significant in their influence on life.

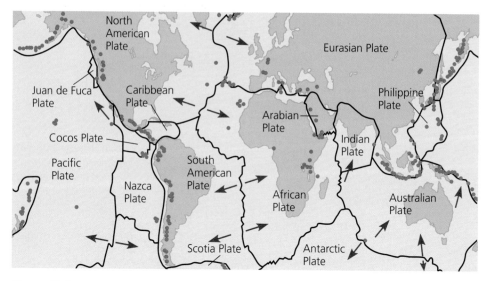

Figure 15.7A   Earth's continental plates. Red arrows indicate direction of movement, red dots are zones of violent geologic activity.

Formation of Pangaea   About 250 million years ago, near the end of the Paleozoic era, plate movements brought all the previously separated landmasses together into a supercontinent named **Pangaea,** meaning "all land" (bottom of Figure 15.7B) Imagine some of the possible effects on life as massive continents joined. Species that had been evolving in isolation came together and competed. When the landmasses fused, ocean basins became deeper, lowering the sea level and draining the shallow coastal seas. Then, as now, most marine species inhabited shallow waters, and many of these organisms probably died out as their habitats shrank. The formation of Pangaea also would have altered terrestrial environments. The interior of the vast continent was cold and dry, probably an even more severe environment than that of central Asia today. Overall, the fossil record indicates that the formation of Pangaea reshaped biological diversity, causing great numbers of extinctions. These, in turn, provided new opportunities for organisms that survived the changes.

Breakup of Pangaea   The second dramatic chapter in continental drift began about 180 million years ago, during the Mesozoic era, when Pangaea started to break apart again, causing a geographic isolation of colossal proportions. As the continents drifted apart, each became a separate evolutionary arena—a huge island on which organisms evolved in isolation from their previous neighbors. At first, Pangaea split into northern and southern landmasses, which we call Laurasia and Gondwana, respectively. This split was complete by about 135 million years ago, as shown in Figure 15.7B. By the end of the Mesozoic era (and the Cretaceous period), some 65 million years ago, the modern continents were beginning to take shape.

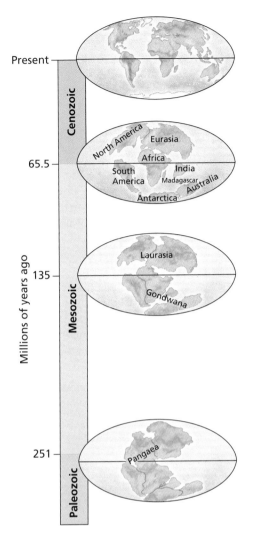

Figure 15.7B Continental drift during the Phanerozoic eon

Then, around 55 million years ago, India started colliding with Eurasia, and the slow, steady crunching of the Indian and Eurasian plates formed the Himalayas, the tallest and youngest of Earth's mountain ranges.

The pattern of continental mergings and separations solves many puzzles, including Australia's great diversity of marsupials (pouched mammals). Marsupials probably originated in what is now Asia and North America and reached Australia via South America and Antarctica while the continents were still joined. The subsequent breakup of continents set Australia "afloat" like a great ark of marsupials. The few early eutherians (placental mammals) that lived there became extinct, while on other continents, most marsupials became extinct. Isolated on Australia, marsupials evolved and diversified, filling ecological roles analogous to those filled by eutherians on other continents.

Continental drift also explains the distribution of a group of ancient vertebrates called lungfishes (Figure 15.7C). Today, there are six species of lungfishes in the world, four in Africa and one each in Australia and South America (yellow striped areas in Figure 15.7D). What is the evolutionary history of these animals? As the orange triangles in Figure 15.7D indicate, fossil lungfishes have been found on all continents except Antarctica. This widespread fossil record indicates that lungfishes evolved when Pangaea was intact.

In the next module, we consider some of the perils associated with the movements of Earth's crustal plates.

? Paleontologists have discovered fossils of the same species of Permian reptiles in West Africa and Brazil, regions that are separated by 3,000 km of ocean. How could continental drift explain such finds?

■ West Africa and Brazil were connected during the early Mesozoic era, and these reptiles must have ranged across both areas.

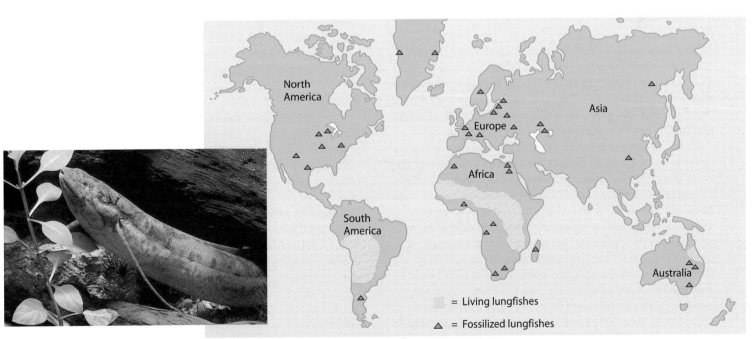

Figure 15.7C
An African lungfish

Figure 15.7D Lungfish distribution, a result of continental drift

= Living lungfishes

△ = Fossilized lungfishes

15.8 The effects of continental drift may imperil human life

Not only do moving crustal plates cause continents to collide, pile up, and build mountain ranges, they also produce volcanoes and earthquakes. The boundaries of plates are hotspots of such geologic activity. California's frequent earthquakes result from

Figure 15.8 The San Andreas Fault (an aerial view north of Los Angeles), a boundary between two crustal plates

movement along the infamous San Andreas Fault, part of the border where the Pacific and North American plates grind together and gradually slide past each other (Figure 15.8). Two major earthquakes have occurred in the region in the past century. The San Francisco earthquake and resulting fire of April 18, 1906, took about 700 lives and caused millions of dollars worth of damage. The 1989 Loma Prieta earthquake caused 62 deaths and damage topped $6 billion dollars.

Undersea earthquakes can cause giant waves, such as the devastating 2004 tsunamis that resulted when a large area of a fault in the Indian Ocean ruptured near the meeting point of the Indian, Eurasian, and Australian plates.

Erupting volcanoes, emitting hot, molten rock from beneath Earth's crust, can cause tremendous devastation, as when Mt. Vesuvius in southern Italy erupted in 79 AD, burying Pompeii in a layer of ash. But sometimes volcanoes imperil more than just local life, as we see in the next module.

? While volcanoes usually destroy life, how might undersea volcanoes create new opportunities for life?

■ By creating new landmasses on which life can evolve, such as the Galápagos and Hawaiian Islands

15.9 Mass extinctions destroy large numbers of species

Extinction is inevitable in a changing world. A species may become extinct because its habitat has been destroyed, because of unfavorable climatic changes, or because of changes in its biological community, such as the evolution of new predators or competitors. Extinctions occur all the time, but extinction rates have not been steady.

Mass Extinctions The fossil record chronicles a number of occasions when global environmental changes were so rapid and disruptive that a majority of species were swept away in a relatively short amount of time. Five mass extinctions have occurred over the past 500 million years. In each of these events, 50% or more of Earth's species became extinct.

Of all the mass extinctions, the ones marking the ends of the Permian and Cretaceous periods have received the most attention. The Permian mass extinction, which occurred about 251 million years ago and defines the boundary between the Paleozoic and Mesozoic eras, claimed about 96% of marine animal species and took a tremendous toll on terrestrial life as well. At the end of the Cretaceous period, about 65 million years ago, the world again lost an enormous number of species—more than half of all marine species and many lineages of terrestrial plants and animals. At that point, dinosaurs had dominated the land and pterosaurs had ruled the air for some 150 million years. Then, in less than 10 million years—a brief period in geologic time—all the dinosaurs were gone, leaving behind only the descendants of one lineage, the birds.

Causes of Mass Extinctions The Permian mass extinction occurred at a time of enormous volcanic eruptions in what is now Siberia, constituting the most extreme volcanic activity in the past half-billion years. Besides spewing lava and ash into the atmosphere, the eruptions may have produced enough carbon dioxide to warm the global climate. Reduced temperature differences between the equator and the poles would have slowed the mixing of ocean water, which in turn would have reduced the amount of oxygen available to marine organisms. This oxygen deficit in the oceans may have played a large role in the Permian extinction.

One clue to a possible cause of the Cretaceous mass extinction is a thin layer of clay enriched in iridium that separates sediments from the Mesozoic and Cenozoic eras. Iridium is an element very rare on Earth, but common in meteorites and other extraterrestrial objects that occasionally fall to Earth. During the molten phase of Earth's formation, most of the iridium traveled with iron to the core of the planet. Some is released by volcanoes if their lava comes from a deep source. But the rocks of the Cretaceous boundary layer have upwards of a million times more iridium than normal Earth levels. Many paleontologists conclude that the iridium layer is the result of fallout from a huge cloud of dust that billowed into the atmosphere when an asteroid or large comet hit Earth. The cloud would have blocked light and severely disturbed the global climate for months.

Is there evidence of such an asteroid? A large crater, the 65-million-year-old Chicxulub crater, has been found in the

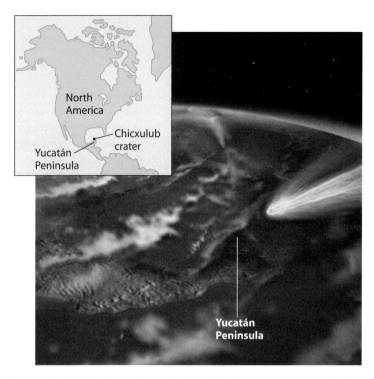

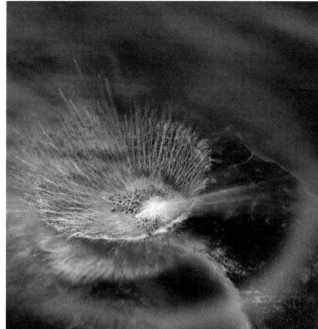

Figure 15.9 The impact hypothesis for the Cretaceous mass extinction

Caribbean Sea near the Yucatán Peninsula of Mexico (Figure 15.9). About 180 km in diameter, the crater is the right size to have been caused by an object with a diameter of 10 km. The horseshoe shape of the crater and the pattern of debris in sedimentary rocks indicate that an asteroid or comet struck at a low angle from the southeast. The artist's interpretation in Figure 15.9 represents the impact and its immediate effect—a cloud of hot vapor and debris that could have killed most of the plants and animals in North America within minutes.

Some researchers propose that climatic changes due to continental drift could have caused the Cretaceous extinctions, whether an asteroid collided with Earth or not. Still others point to evidence that a spike in volcanic activity took place in what is now India at the time of the Cretaceous mass extinction. The various hypotheses are not mutually exclusive, and researchers continue to debate the extent to which each contributed to the extinctions.

Consequences of Mass Extinctions Whatever their causes, mass extinctions affect biological diversity profoundly. By removing large numbers of species, a mass extinction can decimate a thriving and complex ecological community. Mass extinctions are random events that act on species indiscriminately. They can permanently remove species with highly advantageous features and change the course of evolution forever. Consider what would have happened if our early primate ancestors living 65 million years ago had died out in the Cretaceous mass extinction!

How long does it take the diversity of life to recover after a mass extinction? Recall from Module 14.11 that the time between speciation events averages about 6.5 million years. The fossil record shows that it typically takes 5 to 10 million years for species numbers to return to previous levels. In some cases, it has taken much longer: It took about 100 million years for the number of marine families to recover after the Permian mass extinction.

Is a Sixth Mass Extinction Underway? As we explore in Chapter 38, human actions that result in habitat destruction and climate change are modifying the global environment to such an extent that many species are currently threatened with extinction. In the past 400 years—a very short time on a geologic scale, more than a thousand species are known to have become extinct. Scientists estimate that this rate is 100 to 1,000 times the normal rate seen in the fossil record. Does this represent the beginning of a mass extinction?

This question is difficult to answer, partly because it is hard to document both the total number of species on Earth and the number of extinctions that are occurring. It is clear that losses have not reached the level of the other "big five" extinctions. Monitoring, however, does show that many species are declining at an alarming rate, suggesting that a sixth (human-caused) mass extinction could occur within the next few centuries or millennia. And, as seen with prior mass extinctions, it may take millions of years for life on Earth to recover.

But the fossil record also shows a creative side to the destruction. Mass extinctions can pave the way for adaptive radiations in which new groups rise to prominence, as we see next.

Web Activity *Mechanisms of Macroevolution*

? The Permian and Cretaceous mass extinctions mark the ends of the _____ and _____ eras, respectively. (*Hint:* Refer back to Table 15.6.)

■ Paleozoic . . . Mesozoic

15.10 Adaptive radiations have increased the diversity of life

As described in Module 14.10, adaptive radiations are periods of evolutionary change in which many new species form whose adaptations allow them to fill new habitats or community roles. Adaptive radiations followed each mass extinction when survivors became adapted to the many vacant ecological niches.

For example, fossil evidence indicates that mammals underwent a dramatic adaptive radiation after the extinction of terrestrial dinosaurs 65 million years ago (Figure 15.10). Although mammals originated 180 million years ago, fossils older than 65 million years indicate they were mostly small and not very diverse. Early mammals may have been eaten or outcompeted by the larger and more diverse dinosaurs. With the disappearance of the dinosaurs (except for the bird lineage), mammals expanded greatly in both diversity and size, filling the ecological roles once occupied by dinosaurs.

The history of life has also been altered by radiations that followed the evolution of new adaptations, such as the wings of pterosaurs, birds, and bats (see chapter introduction). Major new adaptations facilitated the colonization of land by plants, insects, and tetrapods. The radiation of land plants, for example, was associated with key features such as stems that supported the plant against gravity and a waxy coat that protected leaves from water loss. Finally, note that organisms that arise in an adaptive radiation can serve as a new source of food for still other organisms. In this way, the diversification of land plants stimulated a series of adaptive radiations in insects that ate or pollinated plants—helping to make insects the most diverse group of animals on Earth today.

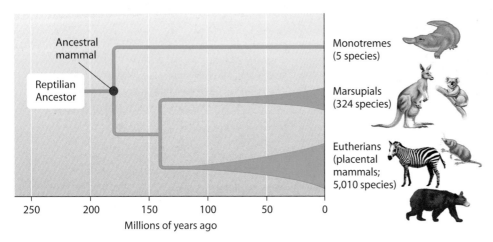

Figure 15.10 Adaptive radiation of mammals (width of line reflects numbers of species). Where on this time line did the dinosaurs become extinct?

? In addition to the new resources of plants, what other factors likely promoted the adaptive radiation of insects?

■ Many unfilled niches on land and the evolution of a waterproof exoskeleton and the origin of flight

15.11 Genes that control development play a major role in evolution

The fossil record can tell us *what* the great changes in the history of life have been and *when* they occurred. Continental drift, mass extinction, and adaptive radiation provide a big-picture view of *how* those changes came about. But now we are increasingly able to understand the genetic mechanisms that form the basis for the changes seen in the fossil record.

Scientists working at the interface of evolutionary biology and developmental biology—the research field abbreviated "**evo-devo**"—are studying how slight genetic changes can become magnified into major morphological differences between species. Genes that program development control the rate, timing, and spatial pattern of changes in an organism's form as it develops from a zygote into an adult. A great many of these developmental genes appear to have been conserved throughout evolutionary history: The same or very similar genes are involved in the development of form across multiple lineages. Changes in the number, nucleotide sequence, and regulation of these genes have led to the huge diversity in body forms.

Changes in Rate and Timing Many striking evolutionary transformations are the result of a change in the rate or timing of developmental events. Figure 15.11A shows a photograph of an axolotl, a salamander that illustrates a phenomenon called **paedomorphosis** (from the Greek *paedos*, child, and *morphosis*, shaping), the retention in the adult of features that were juvenile in an ancestral species. Most salamander species have aquatic larvae (with gills) that undergo metamorphosis in becoming terrestrial adults (with lungs). The axolotl is a salamander that grows to full size and reproduces without losing its external gills.

Slight changes in the relative growth of different body parts can change an adult form substantially. As the skulls in Figure 15.11B on the next page

Figure 15.11A An axolotl, a paedomorphic salamander

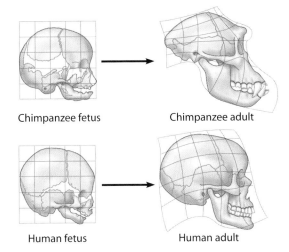

Chimpanzee fetus   Chimpanzee adult

Human fetus   Human adult

Figure 15.11B Chimpanzee and human skulls compared

show, humans and chimpanzees are much more alike as fetuses than they are as adults. In the fetuses of both species, skulls are rounded and jaws are small, making the face flat and rounded. As development proceeds, accelerated growth in the jaw produces the characteristic elongated skull, sloping forehead, and massive jaws of an adult chimpanzee. In contrast, the more even rate of growth of facial and skull bones produces an adult human skull with more rounded contours.

Our large skull is one of our most distinctive features. Our large, complex brain is another. Compared to the slow growth of a chimpanzee brain after birth, our brain continues to grow at the same rapid rate of a fetus for the first year of life, which can be interpreted as the prolonging of a juvenile process.

Changes in Spatial Pattern

Homeotic genes, the master control genes described in Chapter 11, determine such basic features as where a pair of wings or legs will develop on a fruit fly. Changes in such genes or in where such genes are expressed can have a profound impact on body form. Consider, for example, the evolution of terrestrial vertebrates from fishes. The location within a developing limb where certain homeotic genes are expressed is initially the same in both fish and tetrapods. A second region of expression in the developing tetrapod limb, however, produces the extra skeletal elements that develop into foot bones. Thus changes in the expression of these genes appear to have led to the evolution of walking legs from the paired fins of fishes.

New Genes and Changes in Genes

Duplications of developmental genes are very likely to have facilitated the origin of new morphological forms. For example, a fruit fly (an invertebrate) has a single cluster of certain homeotic genes that direct the development of major body parts. A mouse (a vertebrate) has four clusters of these genes that occur in the same linear order on chromosomes and direct the development of the same body regions as the fly genes (see Figure 27.14B). Two duplications of these gene clusters appear to have occurred in the evolution of vertebrates from invertebrate animals. Mutations in these duplicated genes may then have led to the origin of novel vertebrate characteristics. For instance, some genes may have taken on new roles, such as directing the development of a backbone, jaws, and limbs.

Changes in Gene Regulation

Researchers are finding that changes in the form of organisms often are caused by mutations that affect the regulation of developmental genes. As we just discussed, such a change in gene expression was shown to correlate with the evolution of tetrapod limbs from fish fins.

Additional evidence for this type of change in gene regulation is seen in studies of the threespine stickleback fish. In western Canada, these fish live in the open ocean and also in lakes that formed when the coastline receded in the past 12,000 years. Ocean populations have bony plates that make up a kind of body armor and a large set of pelvic spines that help to deter predatory fish. The body armor and pelvic spines are reduced or absent in threespine sticklebacks that live in lakes. The loss of the pelvic spine appears to have been driven by natural selection, because freshwater predators such as dragonfly larvae capture juvenile sticklebacks by grasping onto the spines. Figure 15.11C shows specimens of an ocean and a lake stickleback that have been stained to show their bony plates and spines.

Researchers have identified a key developmental gene (known as *Pitx1*) that influences the development of these spines. Was the reduction of spines in lake populations due to changes in the *Pitx1* gene or to changes in how the gene is expressed? It turns out that the gene is identical in the two populations, and it is expressed in the mouth region and other tissues of embryos from both populations. Studies have shown, however, that while the *Pitx1* gene is expressed in the developing pelvic region of ocean sticklebacks, it is not turned on in this region in lake sticklebacks. This example shows how morphological change can be caused by altering the expression of a developmental gene in some parts of the body but not others.

Web Activity *Paedomorphosis: Morphing Chimps and Humans*

? Research shows that many differences in body forms are caused by changes in gene regulation and not changes in the nucleotide sequence of the developmental gene itself. Why might this be the case?

■ A change in sequence may affect a gene's function wherever that gene is expressed—with potentially harmful effects. Changes in the regulation of gene expression can be limited to specific areas in a developing embryo.

Missing pelvic spine

Figure 15.11C Stickleback fish from ocean (top) and lake (bottom) stained to show bony plates and spines. (Arrow indicates the absence of the pelvic spine in the lake fish.)

15.12 Evolutionary novelties may arise in several ways

The Darwinian theory of gradual change can account for the evolution of intricate structures such as eyes or of new body structures such as wings. In most cases, complex structures have evolved in increments from simpler versions having the same basic function—a process of refinement. In others, we can trace the origin of evolutionary novelties to the gradual adaptation of existing structures to new functions.

As an example of the process of gradual refinement, consider the amazing camera-like eyes of vertebrates and squids. Although the eyes of vertebrates evolved independently of those of squids, both evolved from a simple ancestral patch of photoreceptor cells through a series of incremental modifications that benefited their owners at each stage. Indeed, there appears to have been a single evolutionary origin of light-sensitive cells, and all animals with eyes share the same master genes that regulate eye development.

Figure 15.12 illustrates the range of complexity in the structure of eyes among molluscs living today. Simple patches of light-sensitive cells enable limpets to distinguish light from dark, and they cling more tightly to their rock when a shadow falls on them—a behavioral adaptation that reduces the risk of being eaten. Other molluscs have eye cups that have no lenses or other means of focusing images but can indicate light direction. In those molluscs that do have complex eyes, the organs probably evolved in small steps of adaptation. Examples of such small steps may be seen in Figure 15.12.

Although eyes have retained their basic function of vision throughout their evolutionary history, evolutionary novelty can also arise when structures that originally played one role gradually acquire a different one. Structures that evolve in one context but become co-opted for another function are sometimes called *exaptations*. This term suggests that a structure can become adapted to alternative functions; it does not mean that a structure somehow evolves in anticipation of future use. Natural selection cannot predict the future; it can only improve a structure in the context of its current use. Novel features can arise gradually via a series of intermediate stages, each of which has some function in the organism's current context.

Consider the evolution of birds from a dinosaur ancestor. Feathers could not have evolved as an adaptation for upcoming flights. Their first utility may have been for insulation. It is possible that longer, winglike forelimbs and feathers, which increased the surface area of these forelimbs, were co-opted for flight after functioning in some other capacity, such as mating displays, thermoregulation, and camouflage (functions that feathers still serve today). The first flights may have been only extended hops to pursue prey or escape from a predator. Once flight itself became an advantage, natural selection would have remodeled feathers and wings to fit their additional function.

The flippers of penguins are another example of the modification of existing structures for different functions. Penguins cannot fly, but their modified wings are powerful oars that make them strong, fast, underwater swimmers.

Web Activity *Mechanisms of Macroevolution*

? Explain why the concept of exaptation does not imply that a structure evolves *in anticipation* of some future environmental change.

■ Although a structure is co-opted for new or additional functions in a new environment, it existed because it worked as an adaptation in the old environment.

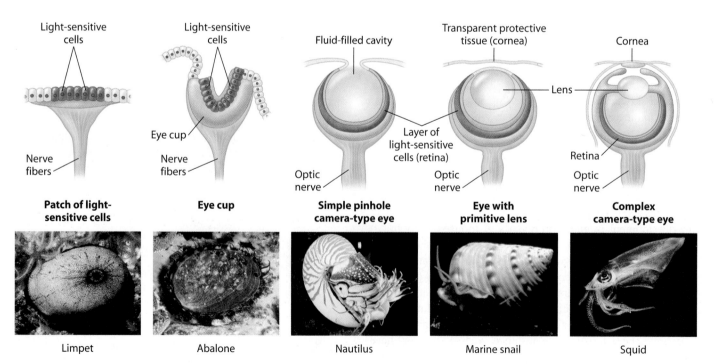

Figure 15.12 A range of eye complexity among molluscs

Evolutionary trends do not mean that evolution is goal directed

The fossil record seems to reveal trends in the evolution of many species. For example, some lineages show a trend toward larger or smaller body size. The modern horse is a descendant of an ancestor about the size of a large dog that lived some 40 million years ago. Named *Hyracotherium*, this ancestor had four toes on its front feet, three toes on its hind feet, and teeth adapted to browsing on shrubs and trees. In contrast, modern horses (*Equus*) are larger, have only one toe on each foot, and have teeth modified for grazing on grasses.

As you can see in Figure 15.13, the fossil record of horses includes many species that descended from *Hyracotherium*. The yellow-highlighted names track a sequence of fossil horses that were intermediate in form between *Hyracotherium* and *Equus*. If these were the only fossils known, they could create the illusion of a single trend in an unbranched lineage, progressing toward larger size, reduced number of toes, and teeth modified for grazing. However, if we consider *all* fossil horses known today, this apparent trend vanishes. Clearly, there was no "trend" toward grazing. Notice that only those lineages derived from *Parahippus* include grazers; the other lineages derived from *Miohippus* consisted of multi-toed browsers. These groups spanned 35 million years before they all became extinct. The genus *Equus* is the only surviving twig of an evolutionary tree that is so branched that it is more like a bush.

What accounts for the continuance of some evolutionary trends and the cessation of others? One model of long-term trends considers species to be analogous to individuals: Speciation is their birth, extinction their death, and new species that diverge from them are their offspring. According to this model of *species selection*, unequal generation of new species and unequal survival of species play a role in macroevolution similar to the role of unequal reproduction in microevolution (see Module 13.6). In other words, the species that generate the greatest number of new species determine the direction of major evolutionary trends.

Evolutionary trends can also result directly from natural selection. For example, when horse ancestors invaded the grasslands that spread during the Middle Cenozoic, there was strong selection for grazers that could escape predators by running faster. This trend would not have occurred without open grasslands.

Whatever the cause of an evolutionary trend, it is important to recognize that a trend does not imply that evolution is goal directed. Evolution is the result of interactions between organisms and their current environments. If conditions change, an apparent evolutionary trend may cease or even reverse itself.

Thus far we have broadly traced the evolutionary history of life and looked at some of the processes of macroevolution. In the concluding section, we explore how biologists arrange life's amazing diversity into an evolutionary tree of life.

? A trend in the evolution of mammals was toward larger brain size relative to body size. Use the species selection model to explain how such a trend could occur.

■ This could occur if, on average, those species with larger brains persisted longer before extinction and gave rise to more "offspring" species than did species with smaller brains.

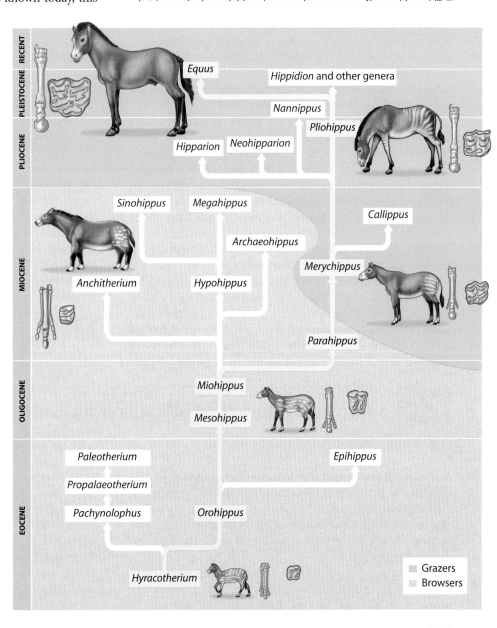

Figure 15.13 The branched evolution of horses

Equus

Hippidion and other genera

Nannippus

Pliohippus

Hipparion   Neohipparion

Sinohippus   Megahippus

Callippus

Archaeohippus

Merychippus

Anchitherium   Hypohippus

Parahippus

Miohippus

Mesohippus

Paleotherium

Epihippus

Propalaeotherium

Pachynolophus   Orohippus

Hyracotherium

■ Grazers
■ Browsers

RECENT
PLEISTOCENE
PLIOCENE
MIOCENE
OLIGOCENE
EOCENE

15.14 Phylogenies are based on homologies in fossils and living organisms

The evolutionary history of a species or group of species is called **phylogeny** (from the Greek *phylon*, tribe, and *genesis*, origin). The fossil record provides a substantial chronicle of evolutionary change that can help trace the phylogeny of many groups. It is, however, an incomplete record, as many of Earth's species probably never left fossils; many fossils that formed were probably destroyed by later geologic processes; and only a fraction of existing fossils have likely been discovered. Even with its limitations, however, the fossil record is a remarkably detailed account of biological change over the vast scale of geologic time.

In addition to evidence from the fossil record, phylogeny can also be inferred from morphological and molecular homologies among living organisms. As we discussed in Module 13.6, homologies are similarities due to shared ancestry. Homologous structures may look and function differently in different species, but they exhibit fundamental similarities because they evolved from the same structure in a common ancestor. For instance, the whale limb is adapted for steering in water; the bat wing is adapted for flight. Nonetheless, the bones that support these two structures, which were present in their common mammalian ancestor, are basically similar (see Figure 13.6A).

Generally, organisms that share similar morphologies are likely to be closely related. The search for homologies is not without pitfalls, however, for not all likenesses are inherited from a common ancestor. In a process called **convergent evolution,** species from different evolutionary branches may come to resemble one another if they live in similar environments and natural selection favors similar adaptations. In such cases, body structures and even whole organisms may resemble each other.

Similarity due to convergent evolution is called **analogy.** For example, the two mole-like animals shown in Figure 15.14 are very similar in external appearance. They both have enlarged front paws, small eyes, and a pad of protective thickened skin on the nose. However, the Australian "mole" (top) is a marsupial, meaning that its young complete their embryonic development in a pouch outside the mother's body. The North American mole (bottom) is a eutherian, which means its young complete

Figure 15.14 Convergent evolution of burrowing adaptations in Australian "mole" (top) and North American mole (bottom)

development in the mother's uterus. Genetic and fossil evidence indicate that the last common ancestor of these two animals lived 170 million years ago. And in fact, this ancestor and most of its descendants were not mole-like. Analogous traits evolved independently in these two mole lineages as they adapted to similar lifestyles.

In addition to molecular similarities and fossil evidence, another clue to distinguishing between homology and analogy is to consider the complexity of the structure being compared. For instance, the skulls of a human and a chimpanzee (see Figure 15.11B) consist of many bones fused together, and the composition of these skulls match almost perfectly, bone for bone. It is highly improbable that such complex structures have separate origins. More likely, the genes involved in the development of both skulls were inherited from a common ancestor, and these complex structures are homologous.

? Human forearms and a bat's wings are _____.
A bat's wings and a bee's wings are _____.

██ homologous · · · analogous

15.15 Systematics connects classification with evolutionary history

Systematics is a discipline of biology that focuses on classifying organisms and determining their evolutionary relationships. The system of naming and classifying species was introduced by Linnaeus in the 18th century. Although Linnaeus's system was not based on evolutionary relationship, many of its features remain useful in systematics. Two of these are the binomial designation of species and hierarchical classification.

Taxonomists (biologists who identify, name, and classify species) assign Latin scientific names to each species. Common names, such as eagle, bat, and pterosaur, may work well in

everyday communication, but they can be ambiguous because there are many species of each of these kinds of organisms. And some common names are downright misleading. Consider these three "fishes": jellyfish (a cnidarian), crayfish (a crustacean), and silverfish (an insect).

As we have seen, Linnaeus's system assigns to each species a two-part name, or **binomial.** The first part of a binomial is the **genus** (plural, *genera*) to which a species belongs. The second part identifies one **species** within that genus. The two parts must be used together to name a species. For example, the scientific name for the domestic cat is *Felis catus*. Notice

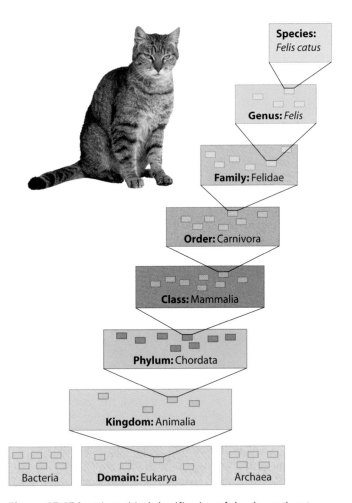

Figure 15.15A Hierarchical classification of the domestic cat

Classifying species into higher taxa, however, is ultimately arbitrary. Higher classification levels are generally defined by various morphological characteristics chosen by taxonomists rather than by quantitative measurements that could apply to the same taxon level across all lineages. Because of such difficulties with determining higher taxa, some biologists propose that classification be based entirely on evolutionary relationships, using a "phylocode," as it is called, which only names groups that include a common ancestor and all its descendants. While phylocode would change the way taxa are defined, the names of most groups would remain the same, just without "ranks," such as family, order, or class, attached to them.

Ever since Darwin, systematics has had a goal beyond simple organization: to have classification reflect evolutionary relationships. Biologists traditionally use **phylogenetic trees** to depict hypotheses about the evolutionary history of species. These branching diagrams reflect the hierarchical classification of groups nested within more inclusive groups. Figure 15.15B below illustrates the connection between classification and phylogeny. This tree shows the classification of some of the taxa in the order Carnivora and the probable evolutionary relationships among these groups. Note that the most inclusive taxon is at the left, and each branch point represents the divergence of two lineages from a common ancestor.

We explore how phylogenetic trees are constructed in the next module.

? How much of the classification in Figure 15.15A do we share with the domestic cat?

■ We are classified the same down to the class level: Both cats and humans are mammals. We do not belong to the same order.

that the first letter of the genus name is capitalized and that the binomial is italicized and latinized.

In addition to naming species, Linnaeus also grouped species into a hierarchy of categories. Beyond the grouping of species within genera (as indicated by the binomial), the Linnaean system extends to progressively broader categories of classification. It places similar genera in the same **family,** puts families into **orders,** orders into **classes,** classes into **phyla** (singular, *phylum*), phyla into **kingdoms,** and, more recently, kingdoms into **domains.**

Figure 15.15A illustrates the progressively more comprehensive classification of the domestic cat. The genus *Felis*, which includes the domestic cat (*Felis catus*) and several closely related small wild cats, is grouped in the cat family, Felidae, along with the genus *Panthera*. (*Panthera* includes the tiger, leopard, jaguar, and African lion.) Family Felidae belongs to the order Carnivora, which also includes the family Canidae (for example, the wolf and coyote) and several other families. Order Carnivora is grouped with many other orders in the class Mammalia, the mammals. Class Mammalia is one of several classes belonging to the phylum Chordata in the kingdom Animalia, which is one of several kingdoms in the domain Eukarya. Each taxonomic unit at any level—family Felidae or class Mammalia, for instance—is called a **taxon** (plural, *taxa*).

Grouping organisms into more inclusive categories seems to come naturally to humans—it is a way to structure our world.

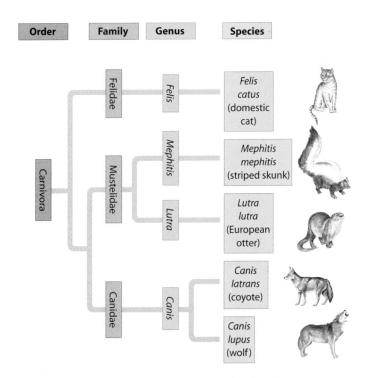

Figure 15.15B The relationship between classification and phylogeny

Biologists seek to reconstruct the evolutionary history of groups of organisms. To do this, they sort homologous features (which reflect evolutionary relationship) from analogous features. They then infer phylogeny from these homologous characters.

Cladistics The most widely used method in systematics is called **cladistics**. Evolutionary descent is the primary criterion used to group organisms into **clades** (from the Greek *clados*, branch). A clade is a group of species that includes an ancestral species and all its descendants. Such an inclusive group of ancestor and descendants, be it a genus, family, or some higher taxon, is said to be **monophyletic** (meaning "single tribe"). Identifying clades makes it possible to construct classification schemes that reflect the branching pattern of evolution.

Cladistics is based on the Darwinian concept that evolution proceeds when a new heritable trait develops in an organism and is passed on to its descendants. Groups of organisms that share such a new, or derived, trait are more closely related to each other than to groups that have only ancestral traits. New traits are called **shared derived characters,** and the original traits present in ancestral groups are called **shared ancestral characters.** Shared derived characters distinguish clades and thus the branch points in the tree of life.

For example, all mammals have backbones, but the presence of a backbone does not distinguish mammals from other vertebrates. The backbone predates the branching of the mammalian clade from other vertebrates. Thus, we say that for mammals, the backbone is a shared ancestral character that originated in an ancestor of the taxon. In contrast, hair, a character shared by all mammals but not found in their ancestors, is a shared derived character, an evolutionary novelty unique to mammals.

Inferring Phylogenies Using Shared Derived Characters
The simplified example in Figure 15.16A illustrates that the sequence in which shared derived characters appear can be used to construct a phylogenetic tree. The figure compares four taxa (all vertebrate animal species) according to the presence or absence in these taxa of a set of three traits, or characters. The color coding highlights how these three characters are shared among the four groups.

An important part of cladistics is a comparison between a so-called ingroup and an outgroup. The **ingroup** (in this example, the three mammals) is the group of taxa that is actually being analyzed. The **outgroup** (in this example, reptiles) is a species or group of species that is known to have diverged before the lineage that contains the groups we are studying.

In our example, the iguana (representing reptiles, the outgroup) and the mammals (collectively the ingroup) are all related in that they are vertebrates. By comparing members of the ingroup to each other and to the outgroup, we can determine which characters are the evolutionary innovations that define the sequence of branch points in the phylogeny of the ingroup.

The character table in Figure 15.16A indicates with a "0" that the character is not present in a group; a "1" indicates that the character is present. For example, all the mammals in the ingroup have hair and mammary glands. This character was present in the ancestral mammal, but not in the outgroup. Now consider the next character in the table—gestation, the carrying of offspring in the uterus within the female parent. The outgroup does not exhibit gestation. Instead, iguanas and most other reptiles lay eggs with a shell. One of the mammals, the duck-billed platypus, also lays eggs with a shell; and from this we might infer that the duck-billed platypus represents an early branch point in the mammalian clade. In fact, this hypothesis is strongly supported by structural, fossil, and molecular evidence.

Proceeding in this manner, we can translate the data in our table of characters into a phylogenetic tree. As we saw in Module 13.6, such a tree is constructed from a series of two-way branch points. Each branch (also called a node) represents

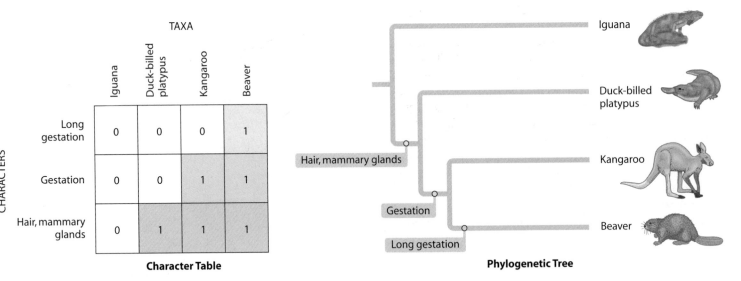

Figure 15.16A Constructing a phylogenetic tree using cladistics

the divergence of two groups from a common ancestor, with the emergence of a lineage possessing a new set of derived traits. The sequence of branching represents the order in which new traits evolved and the historical chronology of when groups last shared a common ancestor. For instance, hair and mammary glands evolved in the lineage that led to all the mammals; gestation and a long period of gestation were derived later in the course of mammalian evolution. The most recent common ancestor to all mammals lived longer ago than did the last ancestor shared by kangaroos and beavers.

Parsimony   Useful in many areas of science, **parsimony** is the quest for the simplest explanation for observed phenomena. Parsimony in systematics means that the simplest hypotheses that are consistent with the data are likely to be the correct ones.

Systematists use the principle of parsimony to construct phylogenetic trees that represent the smallest number of evolutionary changes. For instance, parsimony leads to the hypothesis that a beaver is more closely related to a kangaroo than to a platypus because the beaver and the kangaroo both have gestation. It is possible that gestation evolved twice, once in the kangaroo lineage and independently in the beaver lineage, but this explanation is more complicated and less likely. Typical cladistic analyses involve much more complex data sets than we presented in Figure 15.16A and are usually handled by computer programs designed to construct parsimonious trees.

Phylogenetic Trees as Hypotheses   Systematists use many kinds of evidence, such as structural and developmental features, molecular data, and behavioral traits, to reconstruct evolutionary histories. However, even the best tree represents only the most likely hypothesis based on available evidence. As new data accumulate, hypotheses are revised and new trees drawn.

An example of a redrawn tree is shown in Figure 15.16B. In traditional vertebrate taxonomy, crocodiles, snakes, lizards, and other reptiles were classified in the class Reptilia, while birds were placed in the separate class Aves. However, such a reptilian clade is not monophyletic—in other words, it does not include an ancestral species and all of its descendants, one group of which includes the birds. Studies of shared derived characters and the fossil record result in the tree shown in Figure 15.16B.

Thinking of phylogenetic trees as hypotheses allows us to use them to make and test predictions. For example, if our phylogeny is correct, then features shared by two groups of closely related organisms should be present in their common ancestor. Using this reasoning, consider the novel predictions that can be made about dinosaurs. As seen in the tree in Figure 15.16B, the closest *living* relatives of birds are crocodiles. Birds and crocodiles share numerous features: They have four-chambered hearts, they "sing" to defend territories and attract mates (although a crocodile "song" is more like a bellow), and they build nests. Both birds and crocodiles care for and warm their eggs by brooding. Birds brood by sitting on their eggs, whereas crocodiles cover their eggs with their neck. Reasoning that any feature shared by birds and crocodiles is likely to have been present in their common ancestor (denoted by the red circle in Figure 15.16B) and all of its descendants, biologists predict that

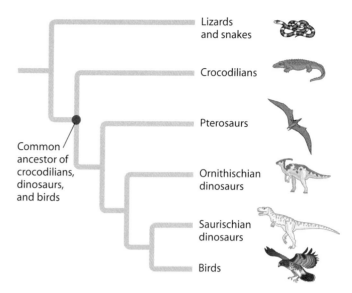

Figure 15.16B   A phylogenetic tree of reptiles

dinosaurs had four-chambered hearts, sang, built nests, and exhibited brooding.

Internal organs such as hearts rarely fossilize, and it is, of course, difficult to determine whether dinosaurs sang. However, fossilized dinosaur nests have been found. Figure 15.16C shows a fossil of an *Oviraptor* dinosaur thought to have died in a sandstorm while incubating or protecting its eggs. The prediction that dinosaurs built nests and exhibited brooding has been supported by additional fossil discoveries that show other species of dinosaurs caring for their eggs.

The more we know about an organism and its relatives, the more accurately we can portray its phylogeny. In the next module, we consider how molecular biology is providing valuable data for tracing evolutionary history.

?   To distinguish a particular clade of mammals within the larger clade that corresponds to class Mammalia, why is hair not a useful characteristic?

▀ Hair is a shared ancestral character common to all mammals and cannot be helpful in distinguishing different mammalian subgroups.

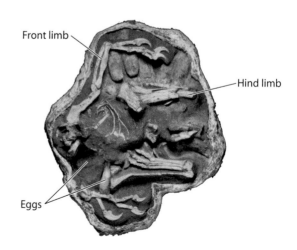

Figure 15.16C   Fossil remains of *Oviraptor* and eggs. The orientation of the bones, which surround the eggs, suggests that the dinosaur died while incubating or protecting its eggs.

An organism's evolutionary history is documented in its genome

The more recently two species have branched from a common ancestor, the more similar their DNA sequences should be. The longer two species have been on separate evolutionary paths, the more their DNA is expected to have diverged.

Molecular Systematics Comparing nucleic acids or other molecules to infer relatedness, a method called **molecular systematics** is a valuable approach for tracing evolutionary histories. Scientists have sequenced more than 100 billion bases of nucleic acid data from thousands of species. This enormous collection of data has fueled a boom in the study of phylogeny and clarified many evolutionary relationships. The phylogenetic tree for the family Ursidae (bears) and the family Procyonidae (raccoons) shown in Figure 15.17 was constructed from comparisons of DNA and blood proteins. The molecular evidence indicates that the giant panda is more closely related to bears than to raccoons and that the lesser panda is a member of the raccoon family. Notice that this phylogenetic tree can include a time-line because the fossil record has provided evidence for the timing of some of the branchings of ancestral groups.

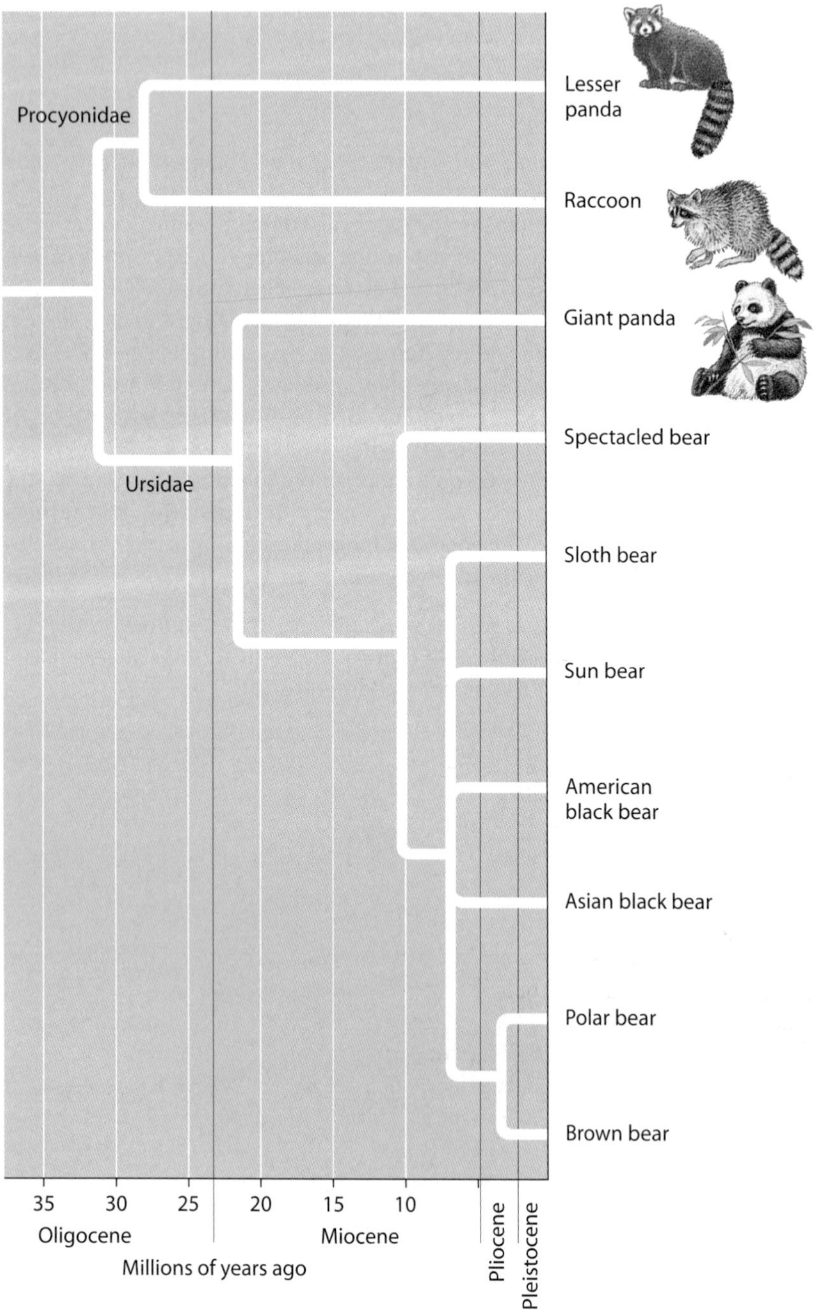

Bears and raccoons are closely related mammals, but systematists can also use DNA analyses to assess relationships between groups of organisms that are so phylogenetically distant that structural similarities are absent. It is also possible to trace phylogenies among groups of present-day prokaryotes and other microorganisms for which we have no fossil record at all. Molecular biology has helped to extend systematics to the extremes of evolutionary relationships far above and below the species level, ranging from major branches of the tree of life to its finest twigs.

The ability of molecular trees to encompass both short and long periods of time is based on the observation that different genes evolve at different rates. The DNA coding for ribosomal RNA (rRNA) changes relatively slowly, so comparisons of DNA sequences in these genes are useful for investigating relationships between taxa that diverged hundreds of millions of years ago. Studies of rRNA sequences, for example, have shown that fungi are more closely related to humans than to green plants—something that morphological comparisons alone certainly could not determine.

In contrast, the DNA in mitochondria (mtDNA) evolves relatively rapidly and can be used to investigate more recent evolutionary events. For example, researchers have used mtDNA sequences to study the relationships between different Native American groups. Their studies support earlier evidence that the Pima of Arizona, the Maya of Mexico, and the Yanomami of Venezuela are closely related, probably descending from the first wave of immigrants to cross the Bering Land Bridge from Asia to the Americas about 13,000 years ago.

Genome Evolution Now that we can compare entire genomes, including our own, some interesting facts have emerged. As you may have heard, the genomes of humans and chimpanzees are amazingly similar. An even more remarkable fact is that homologous genes (similar genes that species share because of descent from a common ancestor) are widespread and can extend over

Figure 15.17 A phylogenetic tree based on molecular data

huge evolutionary distances. While the genes of humans and mice are certainly not identical, 99% of them are detectably homologous. And 50% of human genes are homologous with those of yeast. This remarkable commonality demonstrates that all living organisms share many biochemical and developmental pathways and provides overwhelming support for Darwin's theory of "descent with modification."

Gene duplication has played a particularly important role in evolution because it increases the number of genes in the genome, providing additional opportunities for further evolutionary changes (see Module 15.11). Molecular techniques now allow scientists to trace the evolutionary history of such duplications—in which lineage they occurred and how the multiple copies of genes have diverged from each other over time. As you learned in Module 13.8, humans and mice each have huge families of more than 1,000 genes for smell receptors, allowing them to detect many different odors. These genes appear to have arisen as duplications of an ancestral olfactory gene.

Another interesting fact evident from genome comparisons is that the number of genes has not increased at the same rate as the complexity of organisms. Humans have only about four times as many genes as yeasts. Yeasts are simple, single-celled eukaryotes; humans have a complex brain and a large, multicellular body that contains more than 200 different types of tissues. Evidence is emerging that many human genes are more versatile than those of yeast, but explaining the mechanisms of such versatility remains an exciting scientific challenge.

Web Process of Science How Is Phylogeny Determined Using Protein Comparisons?

? What types of molecules should be compared to help determine whether fungi are more closely related to plants or to animals?

■ Because these organisms diverged so long ago, molecules that change or evolve very slowly, such as ribosomal RNA (or the DNA that codes for rRNA)

15.18 Molecular clocks help track evolutionary time

Some regions of genomes appear to accumulate changes at constant rates. Comparisons of certain homologous DNA sequences for taxa known to have diverged during a certain time period have shown that the number of nucleotide substitutions is proportional to the time that has elapsed since the lineages branched. For example, homologous genes of bats and dolphins are much more alike than are homologous genes of sharks and tuna. This is consistent with the fossil evidence that sharks and tuna have been on separate evolutionary paths much longer than have bats and dolphins. In this case, molecular divergence has kept better track of time than have changes in morphology.

For a gene shown to have a reliable average rate of change, a **molecular clock** can be calibrated in actual time by graphing the number of nucleotide differences against the dates of evolutionary branch points known from the fossil record. The graph line can then be used to estimate the dates of other evolutionary episodes not documented in the fossil record.

Figure 15.18 shows how a molecular clock has been used to date the origin of HIV infection in humans. HIV, the virus that causes AIDS, is descended from viruses that infect chimpanzees and other primates. When did HIV jump to humans? The most widespread strain in humans is HIV-1 M. To pinpoint the earliest infection of this strain, researchers compared samples of the virus from various times during the epidemic, including one partial viral sequence from 1959. The samples showed that the virus has evolved in a clocklike fashion since 1959. Extrapolating backward from this molecular clock suggests that HIV-1 M first spread to humans during the 1930s.

Some biologists are skeptical about the accuracy of molecular clocks because the rate of molecular change may vary at different times, in different genes, and in different groups. The judicial use of molecular clocks, however, may provide approximate markers of elapsed time. An abundant fossil record extends back only about 550 million years, and molecular

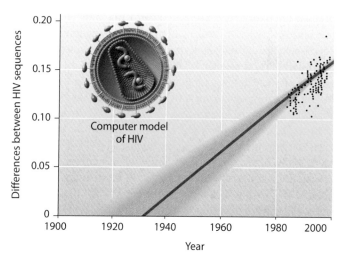

Figure 15.18 Dating the origin of HIV-1 M with a molecular clock. The data points in the upper-right corner represent different HIV samples taken at known times.

clocks have been used to date evolutionary divergences that occurred a billion or more years ago. But the estimates assume that the clocks have been constant for all that time. Thus, such estimates are highly uncertain.

Evolutionary theory holds that all of life has a common ancestor. Molecular systematics is helping to link all living organisms into a comprehensive tree of life, as we see next.

? What is a molecular clock? What assumption underlies the use of such a clock?

■ A molecular clock estimates the actual time of evolutionary events based on the number of genetic changes. It is based on the assumption that some regions of genomes evolve at constant rates.

15.19 Constructing the tree of life is a work in progress

As you have learned, phylogenetic trees are hypotheses about evolutionary history. Like all hypotheses, they are revised, or in some cases rejected, in accordance with new evidence. Molecular systematics and cladistics are remodeling phylogenetic trees and challenging some traditional classifications.

Over the years, many schemes have been proposed for classifying organisms into kingdoms. Historically, a two-kingdom system divided all organisms into plants and animals. But it was beset with problems. Where do bacteria fit? Or photosynthetic unicellular organisms that move? And what about fungi?

By the late 1960s, many biologists recognized five kingdoms: Monera (prokaryotes), Protista (a diverse kingdom consisting mostly of unicellular eukaryotes), Plantae, Fungi, and Animalia. This system was one attempt to classify the diversity of life into a scheme that was useful and reflective of evolutionary history.

However, molecular studies highlighted serious flaws in the five-kingdom system. Biologists have instead adopted a **three-domain system,** which recognizes three basic groups: two domains of prokaryotes, Bacteria and Archaea, and one domain of eukaryotes, called Eukarya. Kingdoms Fungi, Plantae, and Animalia are still recognized, but kingdoms Monera and Protista are obsolete because they are not monophyletic.

Molecular and cellular evidence indicates that the two lineages of prokaryotes (bacteria and archaea) diverged very early in the evolutionary history of life. Molecular evidence also suggests that archaea are more closely related to eukaryotes than to bacteria. Figure 15.19A is an evolutionary tree that is based largely on rRNA genes. As you just learned, rRNA genes have evolved so slowly that homologies between distantly related organsisms can still be detected. This tree shows that the ❶ first major split in the history of life was the divergence of the bacteria from the other two domains, followed by the divergence of domains Archaea and Eukarya.

Comparisons of complete genomes from the three domains, however, show that, especially during the early history of life, there have been substantial interchanges of genes between organisms in the different domains. These took place through **horizontal gene transfer,** a process in which genes are transferred from one genome to another through mechanisms such as plasmid exchange, viral infection (see Module 10.22), and even fusion of different organisms.

Figure 15.19A shows two major episodes of horizontal gene transfer: ❷ Gene transfer between a mitochondrial ancestor and the ancestor of eukaryotes and ❸ gene transfer between a chloroplast ancestor and the ancestor of green plants. Module 16.12 discusses these transfers in more detail.

Some scientists have argued that horizontal gene transfers were so common that the early history of life should be represented as a tangled network of connected branches. Others have suggested that the early history of life is best represented by a ring, not a tree (Figure 15.19B). Based on an analysis of hundreds of genes, researchers have hypothesized that eukaryotes arose through the fusion of an early bacterium and an early archaean. As a result, eukaryotes are simultaneously

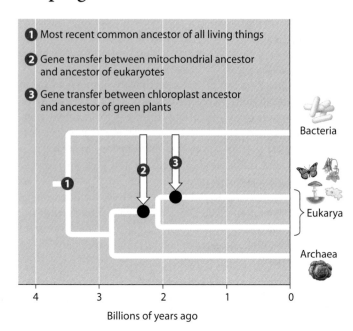

❶ Most recent common ancestor of all living things

❷ Gene transfer between mitochondrial ancestor and ancestor of eukaryotes

❸ Gene transfer between chloroplast ancestor and ancestor of green plants

Billions of years ago

Figure 15.19A A phylogenetic tree depicting the origin of the three domains of life

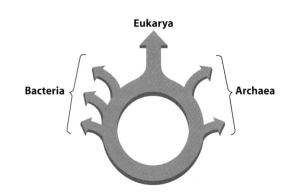

Figure 15.19B Is the tree of life really a ring?

most closely related to bacteria and archaea—an evolutionary relationship that can best be shown in a ring of life. As new data and new methods for analyzing that data emerge, constructing a comprehensive tree of life will continue to challenge systematists.

In the next unit, we examine the enormous diversity of organisms that have populated Earth since life first arose more than 3.5 billion years ago.

Web Activity *Classification Schemes*

❓ Why might the evolutionary history of the earliest organisms be best represented by a ring of life?

◾ There appear to have been multiple horizontal gene transfers among these earliest organisms before the three domains of life eventually emerged from the ring to give rise to Earth's tremendous diversity of life.

Chapter Review

Reviewing the Concepts

Macroevolution traces the pattern of evolutionary change over large time scales (**Introduction**).

Early Earth and the Origin of Life (15.1–15.3)

Early Earth, which formed some 4.6 billion years ago, had an atmosphere very different from that of today. Fossilized prokaryotes date back 3.5 billion years (**15.1**).

Stages in the origin of life. Organic molecules may have formed in the conditions on early Earth (**15.2**). These molecules may have polymerized on hot rocks. The first genes may have been RNA molecules that catalyzed their own replication. Protobionts containing self-replicating molecules could be acted on by natural selection (**15.3**).

Major Events in the History of Life (15.4–15.6)

Key events in life's history (15.4):

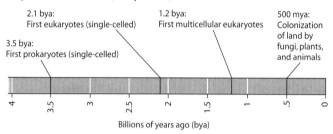

2.1 bya:
First eukaryotes (single-celled)

1.2 bya:
First multicellular eukaryotes

500 mya:
Colonization of land by fungi, plants, and animals

3.5 bya:
First prokaryotes (single-celled)

Billions of years ago (bya)

The fossil record documents the rise and fall of different groups over time. Radiometric dating can gauge the actual ages of fossils and the rocks in which they are found. In the geologic record, major transitions in life forms caused by extinctions separate eras (**15.5–15.6**).

Mechanisms of Macroevolution (15.7–15.13)

Continental drift is the slow movement of Earth's crustal plates. The formation of Pangaea altered habitats and triggered extinctions. The separation of the continents affected the distribution and diversification of organisms (**15.7**). Volcanoes and earthquakes result from the movements of Earth's plates (**15.8**).

Mass extinctions occurred at the end of the Permian and Cretaceous periods. The Cretaceaous extinction, which included most dinosaurs, may have been caused by an asteroid (**15.9**).

Adaptive radiations may increase diversity following mass extinctions, colonization of new habitats, or new adaptations (**15.10**).

Genes controlling development. "Evo-devo" combines evolutionary and developmental biology. New forms can evolve by changes in the number, sequences, or regulation of developmental genes (**15.11**).

Evolutionary novelties. Complex structures may evolve in stages from simpler versions with the same basic function. Other novel structures result from gradual adaptation of structures to new functions (**15.12**).

Evolutionary trends may be a result of species selection or natural selection in changing environments (**15.13**).

Phylogeny and the Tree of Life (15.14–15.19)

Phylogeny, the evolutionary history of a group, is based on identifying homologous structures and molecular sequences that provide evidence of common ancestry. Analogous similarities result from convergent evolution in similar environments or roles (**15.14**).

Systematics focuses on classification and determining phylogeny. Taxonomists assign each species a binomial—a genus and species name. Genera are grouped into progressively larger categories: family, order, class, phylum, kingdom, and domain. A phylogenetic tree is a hypothesis of evolutionary relationships (**15.15**).

Cladistics uses shared derived characters to define monophyletic clades. Shared ancestral characters are common to ancestral groups. The simplest (most parsimonious) hypothesis creates the most likely phylogenetic tree. Trees can be used to make predictions about traits found in common ancestors (**15.16**).

Molecular systematics develops phylogenetic hypotheses based on molecular comparisons. Some regions of DNA change at a rate consistent enough to serve as molecular clocks to date evolutionary events. Homologous genes are found in many diverse species (**15.17–15.18**).

A tree of all life suggests that domain Bacteria was the first to diverge, followed by the divergence of domains Archaea and Eukarya. Horizontal gene transfers among early organisms suggest that the early history of life may be best represented by a ring (**15.19**).

Connecting the Concepts

1. Using the figure below, describe the stages that may have led to the origin of life.

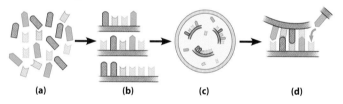

(a)   (b)   (c)   (d)

2. Fill in this concept map, which summarizes some of the key ideas about systematics.

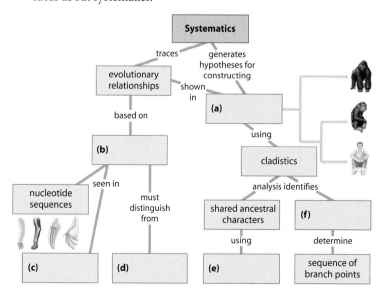

Testing Your Knowledge

Multiple Choice

3. You set your time machine for 3 billion years ago and push the start button. When the dust clears, you look out the window. Which of the following describes what you would probably see?
   a. plants and animals very different from those alive today
   b. a cloud of gas and dust in space
   c. green scum in the water
   d. land and water sterile and devoid of life
   e. an endless expanse of red-hot molten rock

4. Ancient photosynthetic prokaryotes, found in fossil stromatolites, were very important in the history of life because they
   a. were probably the first living things to exist on Earth.
   b. produced the oxygen in the atmosphere.
   c. are the oldest known archaea.
   d. were the first multicellular organisms.
   e. showed that life could evolve around deep-sea vents.

5. The animals and plants of India are very different from the species in nearby Southeast Asia. Why might this be true?
   a. They have become separated by convergent evolution.
   b. The climates of the two regions are different.
   c. India is in the process of separating from the rest of Asia.
   d. Life in India was wiped out by ancient volcanic eruptions.
   e. India was a separate continent until about 55 mya.

6. Adaptive radiations may be promoted as a direct consequence of all of the following *except* one. Which one?
   a. mass extinctions that result in vacant ecological niches
   b. colonization of an isolated region with few competitors
   c. the formation of a land bridge between two previously isolated and already inhabited continents
   d. a novel adaptation
   e. adaptive radiation that produces new food sources

7. A swim bladder is a gas-filled sac that helps fish maintain buoyancy. The evolution of the swim bladder from lungs of an ancestral fish is an example of
   a. an evolutionary trend.
   b. paedomorphosis.
   c. changes in homeotic gene expression.
   d. the gradual refinement of a structure with the same function.
   e. exaptation.

8. If you were using cladistics to build a phylogenetic tree of cats, which would be the best choice for an outgroup?
   a. wolf
   b. leopard
   c. domestic cat
   d. turtle
   e. lion

9. Which of the following could provide the best data for determining the phylogeny of very closely related species?
   a. the fossil record
   b. a comparison of embryological development
   c. an analysis of their morphological differences and similarities
   d. a comparison of nucleotide sequences in homologous genes and mitochondrial DNA
   e. a comparison of their ribosomal DNA sequences

10. Major divisions in the geologic record are marked by
    a. radioactive dating.
    b. distinct changes in the types of fossilized life.
    c. continental drift.
    d. regular time intervals measured in millions of years.
    e. the appearance, in order, of prokaryotes, eukaryotes, protists, plants, fungi, and animals.

Describing, Comparing, and Explaining

11. Distinguish between microevolution and macroevolution.

12. Which are more likely to be closely related: two species with similar appearance but divergent gene sequences, or two species with different appearances but nearly identical genes? Explain.

13. How can the Darwinian concept of descent with modification explain the evolution of such complex structures as an eye?

14. Explain why changes in the regulation of developmental genes may have played such a large role in the evolution of new forms.

15. What types of molecular comparisons are used to determine the very early branching of the tree of life? Explain.

Applying the Concepts

16. Measurements indicate that a fossilized skull you unearthed has a carbon-14/carbon-12 ratio about one-sixteenth that of the skulls of present-day animals. What is the approximate age of the fossil? (The half-life of carbon-14 is 5,730 years.)

17. A paleontologist compares fossils from three dinosaurs and *Archaeopteryx*, the earliest known bird. The following table shows the distribution of characters for each species, where 1 means that the trait is present and 0 means it is not. The outgroup (not shown in the table) had none of the traits. Arrange these species on the phylogenetic tree below and indicate the derived character that defines each branch point.

Trait	Velociraptor	Coelophysis	Archaeopteryx	Allosaurus
Hollow bones	1	1	1	1
Three-fingered hand	1	0	1	1
Half-moon-shaped wristbone	1	0	1	0
Reversed first toe	0	0	1	0

Outgroup

18. Experts estimate that human activities cause the extinction of hundreds of species every year. The normal rate of extinction is thought to average a few species per year. As we continue to alter the environment, especially by destroying tropical rain forests and altering Earth's climate, the resulting wave of extinctions may rival previous mass extinctions. Considering that life has endured five mass extinctions before, should we be concerned that we may cause a sixth? Are these current extinctions different? Why or why not? What might be the consequences for the surviving species, including ourselves?

Answers to all questions can be found in Appendix 4.

For Practice Quizzes, BioFlix, MP3 Tutors, and Activities, go to www.mybiology.com.

The Evolution of Biological Diversity

The Origin and Evolution of Microbial Life: *Prokaryotes and Protists*

How Ancient Bacteria Changed the World

Imagine visiting Earth some 3 billion years ago. It might have looked as illustrated above, a planet bristling with volcanoes spewing gases into the atmosphere and molten rock onto the surface. Life is already present—the "stepping stones" that dominate the shoreline, which are called stromatolites, have been built by thick mats of prokaryotes. (You learned about stromatolites in Module 15.1.) Greenish lawns of prokaryotes cover virtually every wet, sunlit surface on the planet. Animals that might graze the fields of prokaryotes have not yet evolved, and huge expanses of stromatolites are rising as prokaryotes multiply unchecked. Prokaryotes will continue to have the planet to themselves for perhaps another 800 million years. It's an alien world compared to the one we know now, and it's no place for an oxygen-breathing human. If we were to return half a billion years later, however, we would find a more hospitable atmosphere. By that time, photosynthetic prokaryotes had become so numerous that the gaseous oxygen (O_2) they

Facing Page: An artist's conception of a common scene on Earth 3 billion years ago

Below: Modern stromatolites at Shark Bay, Australia

Right: Layers in fossilized stromatolite

produced was changing Earth's anaerobic atmosphere to an aerobic atmosphere. This change set the stage for the evolution of all other aerobic life.

Few traces of that ancient landscape remain, but Shark's Bay in western Australia offers a rare glimpse of the prokaryotic life that changed Earth's atmosphere. Living stromatolites still stand in the warm, shallow waters, giving scientists a chance to see how the structures are formed. The photosynthetic prokaryotes live on the surface of the stromatolite, where they are exposed to sunlight. The sticky coating of these organisms concentrates sand grains and other fine particles from the seawater. While a sediment layer accumulates, the prokaryotes keep migrating to the surface and growing over it. A layered mat builds up as this process is repeated over and over. In this way, large mounds form over thousands of years, hardening as their older sediment layers solidify and become rocklike structures composed of many layers of prokaryotes and sediment. During their long reign as Earth's dominant organisms, ancient prokaryotes built stromatolites by a similar process. The distinctive pattern of layers can be seen in fossils such as the one shown in the photo above.

The direct descendants of the photosynthetic prokaryotes that dominated Earth from 3 billion years ago to about 1 billion years ago are the cyanobacteria and other photosynthetic bacteria of today. They remain abundant in freshwater lakes and ponds and in shallow oceans, but they only form thick mats or mounds in environments that are inhospitable to most other life. Thus, great expanses of stromatolites are a thing of the past. Nevertheless, the grand, global legacy of ancient photosynthetic prokaryotes—the aerobic atmosphere—remains with us today. And the aerobic atmosphere is but one illustration of the profound effects that microscopic organisms can have on the environment.

In this unit of chapters, we examine a small sample of the remarkable variety of species—now numbered in the millions—that arose from those early prokaryotes over the course of time. We focus most of our attention on the current diversity of life, but we also retrace evolutionary journeys that changed the face of the Earth, such as the colonization of land by plants and vertebrates and the emergence of *Homo sapiens*. We begin our exploration of life in this chapter with the prokaryotes, the simplest organisms, and the protists, an assortment of organisms so diverse that researchers are just beginning to decipher their evolutionary relationships. ■ ■ ■

16.1 Prokaryotes are diverse and widespread

Stromatolites reach an impressive size because the prokaryotes that construct them live in colonies consisting of trillions of individuals. Individual prokaryotes are extremely small. You can get an idea of the size of most prokaryotes from Figure 16.1, a colorized scanning electron micrograph of the point of a pin (purple) covered with numerous bacteria (orange). Most prokaryotic cells have diameters in the range of 1–5 μm, much smaller than most eukaryotic cells (typically 10–100 μm).

Despite their small size, prokaryotes have an immense impact on our world. They are found wherever there is life, including in and on the bodies of multicellular organisms. The collective biological mass (biomass) of prokaryotes is at least ten times that of all eukaryotes! More prokaryotes inhabit a handful of fertile soil than the total number of people who have ever lived. Prokaryotes also thrive in habitats too cold, too hot, too salty, too acidic, or too alkaline for any eukaryote. And scientists are just beginning to investigate the extensive prokaryotic diversity in the oceans.

Although prokaryotes are a constant presence in our environment, we hear most about the relatively few species that cause illnesses. Bacteria cause about half of all human diseases, including potentially fatal diseases such as plague and cholera. Strep throat and some ear infections, certain types of food poisoning, and the sexually transmitted diseases chlamydia, syphilis, and gonorrhea are also bacterial diseases. We focus on bacterial **pathogens,** disease-causing agents, in Module 16.8. In Module 16.9, you'll learn about the use of bacterial pathogens as biological weapons.

Far more common than harmful bacteria are those that are benign or beneficial. For example, each of us harbors several hundred species of bacteria in and on our bodies. Many of the bacteria that live on our skin perform helpful housekeeping functions such as decomposing dead skin cells. Some of our intestinal residents supply essential vitamins and enable us to obtain nutrition from food molecules that we can't otherwise digest. Bacteria also guard the body against pathogenic intruders. Prokaryotes in the soil help to decompose dead organisms and other organic waste material, returning vital chemical elements to the environment. If prokaryotic decomposers were to disappear, the chemical cycles that sustain life would halt, and all forms of eukaryotic life would also be doomed. In contrast, prokaryotic life would undoubtedly persist in the absence of eukaryotes, as it once did for billions of years. As you'll learn in Module 16.10, prokaryotes also serve vital roles in a wide array of human enterprises.

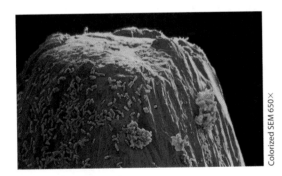

Figure 16.1
Bacteria on the point of a pin

Colorized SEM 650×

MP3 Tutor *Microbial Life*

? The number of bacterial cells that live in and on our bodies is greater than the number of cells that comprise the body. Why aren't we aware of these trillions of cells?

■ *We can't see our own eukaryotic cells individually, and bacterial cells are much smaller than that. Also, our "companion" microbes are adapted for coexisting with us.*

16.2 Bacteria and archaea are the two main branches of prokaryotic evolution

Prokaryotes have a cellular organization fundamentally different from that of eukaryotes, as we saw in Modules 4.3 and 4.4. Whereas eukaryotic cells have a membrane-enclosed nucleus and numerous other membrane-enclosed organelles, prokaryotic cells lack these structural features.

As we discussed in earlier chapters, however, there are two very different kinds of prokaryotes. Researchers first began comparing nucleotide sequences of prokaryotic genes in the 1970s, by studying a type of ribosomal RNA (rRNA) that is found in all prokaryotes and eukaryotes as a marker for evolutionary relationships. They concluded that many prokaryotes once classified as bacteria are actually more closely related to eukaryotes and belong in a domain of their own. Thus, prokaryotes are now classified in the domains **Bacteria** and **Archaea.** The name Archaea comes from the Greek *archaios* ("ancient").

In recent years, researchers have focused on DNA and have completely sequenced many bacterial and archaeal genomes (see Module 12.17). When compared with each other and with the genomes of eukaryotes, these genome sequences strongly support the three-domain view of life (see Module 15.19). Some genes of archaea are similar to bacterial genes and others to eukaryotic genes; still others seem unique to archaea.

A current hypothesis is that present-day archaea and eukaryotes evolved from a common ancestor (see Figure 15.19A). But as described in Module 15.19, tracing their evolution is complicated by gene transfer between prokaryotic lineages. Over hundreds of millions of years, prokaryotes have acquired genes from distantly related species, and they continue to do so today. As a result, significant portions of the genomes of many prokaryotes are mosaics of genes imported from other species.

Table 16.2 summarizes some of the main differences among the three domains. In addition to the rRNA sequences, several other differences involve the cellular machinery for gene expression. These include differences in RNA polymerases (the enzymes that catalyze the synthesis of RNA), in the presence of introns within genes, and in sensitivity to certain antibiotics that inhibit protein synthesis. Subtle differences between bacterial and archaeal ribosomes—in both rRNA and proteins—undoubtedly account for the insensitivity of archaea to these antibiotics.

Other differences between bacteria and archaea show up in their cell walls and membranes. Nearly all prokaryotes have a cell wall outside their plasma membrane. As in plants, the wall maintains cell shape and provides physical protection. Bacterial cell walls contain a unique material called **peptidoglycan,** a polymer of sugars cross-linked by short polypeptides. No archaea have true peptidoglycan. Furthermore, the lipids forming the backbone of plasma membranes differ between the two domains.

For most of the features listed in the table, archaea are actually more like eukaryotes than like bacteria. In fact, archaea have at least as much in common with eukaryotes as they do with bacteria. The main point here is that there are two very different kinds of prokaryotic organisms. We will discuss the diversity within these two groups after a look at some more general features of prokaryotes.

TABLE 16.2	DIFFERENCES AMONG THE DOMAINS BACTERIA, ARCHAEA, AND EUKARYA		
Characteristic	Bacteria	Archaea	Eukarya
rRNA sequences	Some unique to bacteria	Some unique to archaea; some match eukaryotic sequences	Some unique to eukarya; some match archaea sequences
RNA polymerase	One kind; relatively small and simple	Several kinds; complex	Several kinds; complex
Introns	Rare	In some genes	Present
Response to antibiotics streptomycin and chloramphenicol	Growth inhibited	Growth not inhibited	Growth not inhibited
Peptidoglycan in cell wall	Present	Absent	Absent
Histones associated with DNA	Absent	Present in some species	Present

? As different as bacteria and archaea are, both groups are characterized by _____ cells, which lack nuclei and other membrane-enclosed organelles.

■ prokaryotic

16.3 Prokaryotes come in a variety of shapes

Determining cell shape by microscopic examination is an important step in identifying prokaryotes. The micrographs below show three of the most common prokaryotic cell shapes. Spherical prokaryotic cells are called **cocci** (from the Greek word for "berries"). Cocci (singular, *coccus*) that occur in chains, like the ones in Figure 16.3A, are called streptococci (from the Greek *streptos,* twisted). The bacterium that causes strep throat in humans is a streptococcus. Other cocci occur in clusters; they are called staphylococci (from the Greek *staphyle,* cluster of grapes).

Figure 16.3B shows rod-shaped prokaryotes, which are called **bacilli** (singular, *bacillus*). Most bacilli occur singly, but the cells of some species occur in pairs (diplobacilli) and in chains (streptobacilli). The species shown here, which is common in fertile soil, exists as solitary cells.

A third prokaryotic cell shape is curved or spiral. Some bacteria in this category resemble commas and are called *vibrios.* Other bacteria and archaea have a helical shape, like a corkscrew. Helical prokaryotes that are relatively short and rigid are called *spirilla;* those with longer, more flexible cells are called *spirochetes* (Figure 16.3C). The bacterium that causes syphilis, for example, is a spirochete. Spirochetes include some giants by prokaryotic standards—cells 0.5 mm long (though very thin).

? How could a microscope help you distinguish the cocci that cause "staph" infections from those that cause "strep" throat?

■ It would show clusters of cells for staphylococcus and chains of cells for streptococcus.

Figure 16.3A   Cocci

Colorized SEM 12,000×

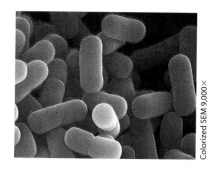

Figure 16.3B   Bacilli

Colorized SEM 9,000×

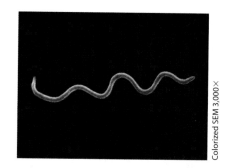

Figure 16.3C   Spirochete

Colorized SEM 3,000×

Various structural features contribute to the success of prokaryotes

Most prokaryotes are unicellular, but they are able to carry out all of the functions of life within that single cell. In addition, prokaryotes thrive in a great variety of habitats. In this module, we discuss some of the structural and functional adaptations that contribute to the diversity and success of prokaryotes.

External Structures   One of the most important features of nearly all prokaryotic cells is their cell wall. The wall maintains cell shape, provides physical protection, and prevents the cell from bursting in a hypotonic environment (see Module 5.5). In a hypertonic environment, most prokaryotes lose water and shrink away from their wall. Severe water loss inhibits reproduction, which explains why salt can be used to preserve certain foods, such as pork and fish.

As mentioned in Module 16.2, the cell walls of archaea differ from those of bacteria. The cell walls of bacteria fall into two general types, which scientists can identify with a technique called the **gram stain** (Figure 16.4A). Gram-positive bacteria have simpler walls with a relatively thick layer of peptidoglycan. The walls of gram-negative bacteria stain differently. They have less peptidoglycan and are more complex, with an outer membrane that contains lipids bonded to carbohydrates.

Gram staining is a valuable identification tool in medicine. Among disease-causing bacteria, gram-negative species are generally more threatening than gram-positive species. The lipid molecules of the outer membrane of these bacteria are often toxic. The membrane also protects them against the body's defenses and impedes the entry of antibiotic drugs. The effectiveness of many antibiotics is due to their inhibition of cross-linking in the peptidoglycan cell wall. Such drugs can cripple many types of bacteria without adversely affecting human cells, which do not contain peptidoglycan.

The cell wall of many prokaryotes is covered by a capsule, a sticky layer of polysaccharides or protein. The capsule enables prokaryotes to adhere to their substrate or to other individuals in a colony. Capsules can also shield pathogenic prokaryotes from attacks by their host's immune system. The capsule surrounding the *Streptococcus* bacterium shown in Figure 16.4B enables it to attach to cells that line the human respiratory tract—in this image, a tonsil cell.

Some prokaryotes stick to their substrate or to one another by means of hairlike appendages called **pili** (singular, **pilus**). Pili show up clearly in the highly magnified micrograph in Figure 16.4C. Pili help prokaryotes stick to each other and to surfaces, such as rocks in flowing streams or the lining of human intestines. Specialized pili, called sex pili, link prokaryotes during conjugation, a process in which one cell transfers DNA to another (see Module 10.22).

Motility   Many bacteria and archaea are equipped with flagella, which enable them to move about. In response to chemical or physical signals in their environment, prokaryotes can move toward nutrients or other members of their species or away from a toxic substance. Flagella may be scattered over the entire cell surface or concentrated at one or both ends of the cell. Entirely different in structure from the flagellum of eukaryotic cells (described in Module 4.18), the prokaryotic flagellum is a naked protein structure that lacks microtubules. It is attached to the cell surface by a system of rotating rings anchored in the plasma membrane and cell wall. The rings give the flagellum a propeller-like rotary movement, as shown in Figure 16.4D, at the top of the next page. The flagellated organism in the photo is the bacterium *Proteus,* an especially fast swimmer.

Reproduction and Adaptation   Certainly a large part of the success of prokaryotes is their potential to reproduce quickly in a favorable environment. Dividing by binary fission, a single cell becomes 2 cells, which then become 4, 8, 16, and so on. While most prokaryotes produce a new generation within 1–3 hours, some species can produce a new generation in only 20 minutes under optimal conditions. If reproduction continued unchecked at this rate, a single prokaryote could give rise to a colony outweighing Earth in only three days! In reality, of course, prokaryotic reproduction is limited, as the cells eventually exhaust their nutrient supply, poison themselves with metabolic wastes, or are consumed by other organisms. Prokaryotes in nature also face competition from other microorganisms, many of which produce antibiotic chemicals that slow prokaryotic reproduction.

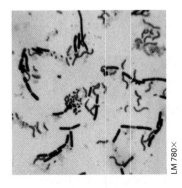

LM 780×

Figure 16.4A Gram positive (purple) and gram negative (pink) bacteria

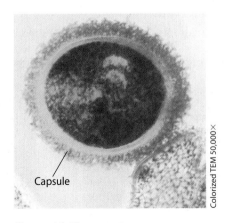

Capsule

Colorized TEM 50,000×

Figure 16.4B Capsule

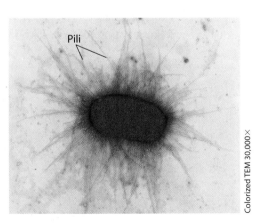

Pili

Colorized TEM 30,000×

Figure 16.4C Pili

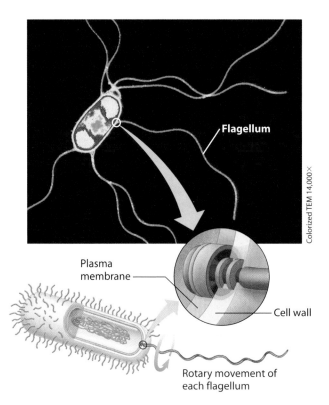

Figure 16.4D Prokaryotic flagella

Plasma membrane

Flagellum

Cell wall

Rotary movement of each flagellum

Colorized TEM 14,000×

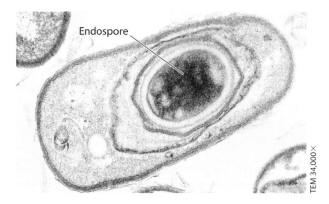

Endospore

TEM 34,000×

Figure 16.4E An endospore within a cell of the anthrax bacterium

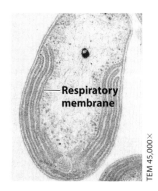

Respiratory membrane

TEM 45,000×

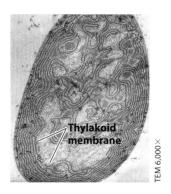

Thylakoid membrane

TEM 6,000×

Figure 16.4F Infoldings of plasma membrane specialized for respiration in an aerobic prokaryote (left) and photosynthesis in a cyanobacterium (right)

Some prokaryotes are able to withstand harsh conditions. Certain bacteria, for example, can form specialized resistant cells when food is depleted or the environment has otherwise become inhospitable. Figure 16.4E shows an example of such an organism, *Bacillus anthracis*, the bacterium that produces a disease called anthrax in cattle, sheep, and humans. There are actually two cells here, one inside the other. The outer cell, which will later disintegrate, produced the specialized inner cell, called an **endospore.** The endospore, which has a thick, protective coat, dehydrates and becomes dormant. It can survive all sorts of trauma, including extreme heat or cold. When the endospore receives cues that the environmental conditions have improved, it absorbs water and resumes growth.

Some endospores can remain dormant for centuries. Not even boiling water kills most of these resistant cells, making it difficult to get rid of spores in a contaminated area. An island off the coast of Scotland that was used for anthrax testing in 1942 was finally declared safe 48 years later, after tons of formaldehyde were applied and huge amounts of top soil were removed. To sterilize laboratory equipment, microbiologists use an autoclave, a pressure cooker that kills endospores by heating to a temperature of 121°C (250°F) with high-pressure steam. The food-canning industry uses similar methods to kill endospores of dangerous bacteria such as *Clostridium botulinum*, the source of the potentially fatal disease botulism.

Internal Organization As seen in Module 4.3, the cells of prokaryotes are simpler than those of eukaryotes in both their internal structure and organization of their genome. However, some prokaryotic cells do have specialized membranes that perform metabolic functions. The left side of Figure 16.4F

shows an aerobic prokaryote with infoldings of the plasma membrane that are specialized to function in cellular respiration. The right side of the figure shows a cyanobacterium with thylakoid membranes that function in photosynthesis.

The genome of a prokaryote has on average only about one-thousandth as much DNA as a eukaryotic genome. The prokaryotic DNA usually forms a circular chromosome. Smaller rings of DNA, called plasmids, carry genes that may provide resistance to antibiotics, direct the metabolism of rarely encountered nutrients, or have other "contingency" functions. The ability of many prokaryotes to transfer genes within and even between species contributes to the growing problem of antibiotic resistance in disease-causing bacteria (see Module 13.15).

Prokaryotic ribosomes are slightly smaller than eukaryotic ribosomes and differ in their protein and RNA content. These differences are great enough that certain antibiotics bind to prokaryotic ribosomes and block protein synthesis but do not affect eukaryotic ribosomes. As a result, we can use these antibiotics to kill bacteria without harming ourselves.

Another feature that contributes to the success of prokaryotes is the diversity of ways in which they obtain their nourishment, which we consider in the next module.

Web Activity *Prokaryotic Cell Structure and Function*

? Why do microbiologists autoclave lab instruments and glassware, rather than simply boiling them?

■ To kill bacterial endospores, which can survive boiling water

16.5 Prokaryotes obtain nourishment in a variety of ways

When classifying diverse organisms, biologists often use the phrase "mode of nutrition" to describe how an organism obtains two main resources: energy and carbon. As a group, prokaryotes exhibit much more nutritional diversity than eukaryotes.

Source of Energy As shown in Figure 16.5A, two sources of energy are used by prokaryotes. Like plants, prokaryotic *phototrophs* capture energy from sunlight. Prokaryotes called *chemotrophs* harness the energy stored in either organic molecules or inorganic chemicals such as hydrogen sulfide (H_2S), elemental sulfur (S), iron (Fe)-containing compounds, or ammonia (NH_3).

Source of Carbon Organisms that make their own organic compounds from inorganic sources are autotrophic (see Module 7.1). **Autotrophs,** including plants and some prokaryotes and protists, obtain their carbon atoms from carbon dioxide (CO_2). Most prokaryotes, as well as animals, fungi, and some protists, are **heterotrophs,** meaning they obtain their carbon atoms from organic compounds.

Mode of Nutrition The terms used to describe how a prokaryote obtains energy and carbon are combined to describe its mode of nutrition. **Photoautotrophs** harness sunlight for energy and use CO_2 for carbon. Cyanobacteria, such as the example shown in Figure 7.1D, are photoautotrophs. **Photoheterotrophs** obtain energy from sunlight but get their carbon atoms from organic sources. A group of bacteria called purple nonsulfur bacteria, which are found in anaerobic aquatic sediments, are examples of this nutritional type. **Chemoautotrophs** harvest energy from inorganic chemicals and use carbon from CO_2 to make organic molecules. This group includes sulfur bacteria that inhabit deep-sea vents (see the introduction to Chapter 34) and soil bacteria that are important to recycling nitrogen (see Module 37.21). **Chemoheterotrophs,** which acquire both energy and carbon from organic molecules, are by far the largest and most diverse group of prokaryotes. Almost any organic molecule is food for some species of chemoheterotrophic prokaryote.

Biofilms Although prokaryotes may exist as single cells, many species live in surface-coating colonies called **biofilms.** A biofilm may include just one species, or it may consist of several different species. It forms when cells in a colony secrete signaling molecules that recruit nearby cells. Once the colony becomes sufficiently large, the cells produce proteins that stick them to the substrate and to each other. Channels in the biofilm allow nutrients to reach cells in the interior and wastes to leave. Biofilms are common among bacteria that cause disease in humans. For instance, a biofilm known as dental plaque (Figure 16.5B) can cause tooth decay. Ear infections and urinary tract infections may be caused by biofilms. Biofilms that form on catheters or replacement joints or heart valves can cause serious medical problems. On the other hand, biofilms are used to remove contaminants from water in sewage treatment plants (see Module 16.10).

Biofilms are common in all types of aquatic habitats—for example, on rocks in streams and rivers and on the surface of stagnant water. In complex biofilm communities found in the mud of the ocean floor, sulfate-consuming bacteria and methane-consuming archaea coexist in ball-shaped clumps. The bacteria appear to use the archaea's organic waste products and, in turn, produce compounds that facilitate methane consumption by the archaea. This partnership has global ramifications. Each year, these archaea consume an estimated 300 billion kilograms of methane, a major greenhouse gas (see Module 38.5).

SEM 2,000×

Figure 16.5B Dental plaque, a biofilm that forms on teeth

Web Process of Science *What Are the Modes of Nutrition in Prokaryotes?*

? What term describes the mode of nutrition of stromatolite-building prokaryotes discussed in the chapter introduction?

Photoautotroph ∎

	Energy source	
	Light	Chemical
Carbon source — CO_2	Photoautotrophs	Chemoautotrophs
Organic compounds	Photoheterotrophs	Chemoheterotrophs

Figure 16.5A Sources of energy and carbon in prokaryotic modes of nutrition

Archaea thrive in extreme environments—and in other habitats

Archaea are abundant in many habitats, including places where few other organisms can survive. The archaeal inhabitants of extreme environments have unusual proteins and other molecular adaptations that enable them to metabolize and reproduce effectively. Scientists are only beginning to learn about these adaptations.

A group of archaea called the **extreme halophiles** ("salt lovers") thrive in very salty places, such as the Great Salt Lake in Utah, the Dead Sea, and seawater-evaporating ponds used to produce salt. Figure 16.6A shows an aerial view of ponds of this sort next to San Francisco Bay. The colors of the ponds result from the dense growth of the archaea that thrive when the salinity of the water reaches 15–20%. (Before evaporation, seawater has a salt concentration of about 3%.) The purplish color of the ponds near the top of the photo is due to an archaean called *Halobacterium halobium*. A unique photosynthesizer, *H. halobium* lacks chlorophyll; it has a purple molecule called bacteriorhodopsin that traps solar energy.

Another group of archaea, the **extreme thermophiles** ("heat lovers"), thrive in very hot water; some even live near deep-ocean vents where temperatures are above 100°C, the boiling point of water at sea level! One such habitat is the Nevada geyser shown in Figure 16.6B. Other thermophiles thrive in acid. Many hot, acidic pools in Yellowstone National Park harbor such archaea, which give the pools a vivid greenish color. One of these organisms, *Sulfolobus*, can obtain energy by oxidizing sulfur or a compound of sulfur and iron; the mechanisms involved may be similar to those used billions of years ago by the first cells.

A third group of archaea, the **methanogens,** live in anaerobic environments and give off methane as a waste product. Many thrive in anaerobic mud at the bottom of lakes and swamps. You may have seen methane, also called marsh gas, bubbling up from a swamp. Great numbers of methanogens also inhabit the digestive tracts of animals. In humans, intestinal gas is largely the result of their metabolism. More importantly, methanogens aid digestion in cattle, deer, and other animals that depend heavily on cellulose for their nutrition. Unfortunately, these animals belch out large volumes of methane, contributing to the global release of this major greenhouse gas.

Accustomed to thinking of archaea as "extremophiles," scientists have been surprised to discover their abundance in more moderate environments, especially in the oceans. Archaea live at all depths, making up a substantial fraction of the prokaryotes in waters below 150 m and equaling the number of bacteria below 1,000 m. Archaea are thus one of the most abundant cell types in Earth's largest habitat.

Because bacteria have been the subject of most prokaryotic research throughout the history of microbiology, much more is known about them than about archaea. Now that the evolutionary and ecological importance of archaea has come into focus, we can expect research on this domain to turn up many more surprises about the history of life and the roles of microbes in ecosystems.

Figure 16.6A "Salt-loving" archaea, extreme halophiles, growing in seawater-evaporating ponds near San Francisco Bay

Figure 16.6B Orange and yellow colonies of "heat-loving" archaea, extreme thermophiles, growing in a Nevada geyser

? Some archaea are referred to as "extremophiles." Why?

■ Because they can thrive in extreme environments too hot, too salty, or too acidic for other organisms

16.7 Bacteria include a diverse assemblage of prokaryotes

In this module we sample some of the diversity of bacteria. Until the late 20th century, systematists based prokaryotic taxonomy on phenotypic criteria such as shape, motility, nutritional mode, and Gram staining. These criteria are still valuable in some contexts, such as the rapid identification of bacteria cultured from a patient's blood. But they do not reveal relationships. Although still a work in progress, molecular systematics is starting to unravel bacterial phylogeny.

Domain Bacteria is currently divided into nine groups, five of which are considered subgroups of a clade of gram-negative bacteria called **proteobacteria.** Members of the subgroup alpha proteobacteria include *Rhizobium* species, which live in nodules in the roots of legumes, where they fix atmospheric N₂ (see Module 32.13), and *Agrobacterium* species, which produce tumors in plants and are used by genetic engineers to carry foreign DNA into the genomes of crop plants (see Module 12.8).

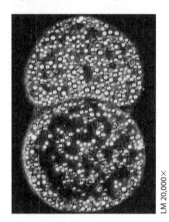

The photosynthetic members of the subgroup gamma proteobacteria include sulfur bacteria that oxidize H₂S. The small greenish globules you see in Figure 16.7A are sulfur wastes. Other gamma proteobacteria include many species that inhabit animal intestines, such as *Salmonella*, one cause of food poisoning; *Vibrio cholerae*, which causes cholera; and *Escherichia coli*, a common resident of the intestines of humans and other mammals and a favorite research organism.

Figure 16.7A *Thiomargarita namibiensis,* showing globules of sulfur wastes

The delta proteobacteria include the slime-secreting myxobacteria, which form elaborate colonies. When the soil dries out or food is scarce, the cells congregate into a fruiting body that releases resistant spores. *Bdellovibrio* bacteria are delta proteobacteria that attack other bacteria. They charge their prey at up to 100 μm/sec (comparable to a human running 600 km/hr) and bore into the prey by spinning at 100 revolutions per second (Figure 16.7B).

The **chlamydias,** which live inside eukaryotic host cells, form a second bacterial group. *Chlamydia trachomatis* is a common cause of blindness in developing countries and also causes nongonococcal urethritis, the most common sexually transmitted disease in the United States.

Spirochetes are a group of helical bacteria (see Figure 16.3C) that spiral through their environment by means of rotating, internal filaments. Some spirochetes are notorious pathogens: *Treponema pallidum* causes syphilis, and *Borrelia burgdorferi* causes Lyme disease.

The group of **gram-positive bacteria** rivals the proteobacteria in diversity. One subgroup, the actinomycetes (from the Greek *mykes*, fungus, for which these bacteria were once mistaken), form colonies of branched chains of cells. Actinomycetes are very common in the soil, where they decompose organic matter. Soil-dwelling species in the genus *Streptomyces* (Figure 16.7C) are cultured by pharmaceutical companies as a source of many antibiotics, including streptomycin. Gram-positive bacteria also include many solitary species, such as *Bacillus anthracis* (see Figure 16.4E). *Staphylococcus* and *Streptococcus* are also gram-positive bacteria. Mycoplasmas are gram-positive bacteria that lack cell walls and are the tiniest of all known cells, with diameters as small as 0.1 μm, only about five times as large as a ribosome.

The **cyanobacteria** are the only prokaryotes with plant-like, oxygen-generating photosynthesis. Ancient cyanobacteria generated the stromatolites discussed in the chapter introduction and made the atmosphere aerobic. Today, both solitary and colonial cyanobacteria provide an enormous amount of food for freshwater and marine ecosystems. Figure 16.7D shows the cyanobacterium *Anabaena* which has specialized cells that convert nitrogen gas into a form usable by other organisms.

Web Activity *Diversity of Prokaryotes*

? Which group of bacteria uses H₂S as an electron source in its photosynthetic production of organic molecules?

■ Sulfur bacteria of the subgroup gamma proteobacteria

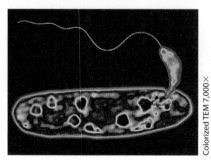

Figure 16.7B *Bdellovibrio bacteriovorus* (flagellated cell) attacking a larger bacterium

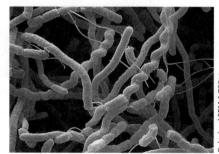

Figure 16.7C *Streptomyces,* the source of many antibiotics

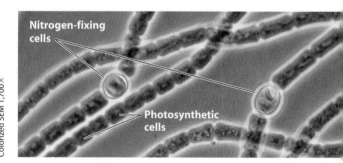

Figure 16.7D *Anabaena,* a filamentous cyanobacterium

16.8 Some bacteria cause disease

All organisms, humans included, are almost constantly exposed to pathogenic bacteria. Most often, our body's defenses prevent pathogens from affecting us (see Chapter 24). Occasionally, however, a pathogen establishes itself in the body and causes illness. Even some of the bacteria that are normal residents of the human body can make us ill when our defenses have been weakened by poor nutrition or by a viral infection.

Most bacteria that cause illness do so by producing a poison—either an exotoxin or an endotoxin. **Exotoxins** are proteins secreted by bacterial cells and include some of the most powerful poisons known. For example, the exotoxin of *Clostridium tetani* produces muscle spasms, including "lockjaw," which prevents the mouth from opening. *Staphylococcus aureus*, shown in Figure 16.8A, produces several exotoxins. Although *S. aureus* is commonly found on the skin and in the nasal passages, if it enters the body through a wound, it can cause serious diseases. One of the exotoxins that *S. aureus* produces can cause layers of skin to slough off; another causes the potentially deadly toxic shock syndrome. Food may also be contaminated with *S. aureus* exotoxins, which are so potent that less than a microgram (3.5 × 10^{-8} ounce) causes vomiting and diarrhea.

Occasionally, bacterial species that are generally harmless acquire an exotoxin-producing gene through transfer from another species (see Module 10.22). *Escherichia coli*, a normally benign resident of our intestines, is an example. An exotoxin-producing strain of *E. coli* designated O157:H7 has caused a number of outbreaks of bloody diarrhea and hundreds of deaths. Although many cases are associated with contaminated ground beef, outbreaks have also been traced to fresh produce such as spinach and lettuce, as well as to other sources. The recently sequenced genome of *E. coli* O157:H7 shows that more than 20% of its genes are not found in the genome of a harmless *E. coli* strain.

Endotoxins are components of the outer membrane of gram-negative bacteria that are released when the cell dies or is digested by a defensive cell. All endotoxins induce the same general symptoms: fever, aches, and sometimes a dangerous drop in blood pressure (septic shock). An endotoxin of *Neisseria meningitidis* causes bacterial meningitis, a disease that can kill a healthy person in a matter of days, or even hours. Other examples of endotoxin-producing bacteria include the species of *Salmonella* that cause food poisoning and typhoid fever.

During the last century, improvements in sanitation in the developed world greatly reduced the incidence of bacterial diseases. The installation of water treatment and sewage systems continues to be a public health priority throughout the world. Medical advances, especially the development of antibiotics, have also helped to control bacterial diseases. However, many pathogenic bacteria have evolved resistance to widely used antibiotics, becoming newly dangerous (see Module 13.15).

Education is another powerful weapon for combating disease. For example, sexually active adults who understand the importance of condom use are more likely to protect themselves against sexually transmitted diseases such as nongonococcal urethritis (caused by a type of chlamydia), syphilis, and gonorrhea (beta proteobacteria). Another case in point is Lyme disease, currently the most widespread pest-carried disease in the United States. The disease is caused by *Borrelia burgdorferi*, a spirochete carried by ticks that live on deer and field mice. As shown in Figure 16.8B, the disease usually starts as a red rash shaped like a bull's-eye around a tick bite. Antibiotics can cure the disease if administered within about a month after exposure. If untreated, Lyme disease can cause debilitating arthritis, heart disease, and nervous disorders. The best prevention is public education about avoiding tick bites and the importance of seeking treatment if a rash develops.

While billions of dollars and countless hours of research have been devoted to disease prevention and cures, people also have developed methods of using pathogens to deliberately cause harm. We look at biological weapons in the next module.

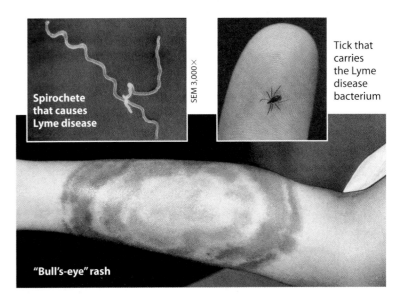

Tick that carries the Lyme disease bacterium

Spirochete that causes Lyme disease

SEM 3,000×

"Bull's-eye" rash

Figure 16.8B Lyme disease, a bacterial disease transmitted by ticks

? Contrast exotoxins with endotoxins.

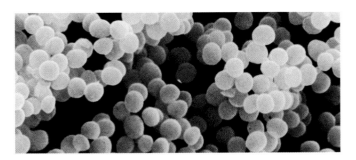

Figure 16.8A *Staphylococcus aureus*, an exotoxin producer

■ Exotoxins are proteins secreted by pathogenic bacteria; endotoxins are components of the outer membranes of pathogenic bacteria.

16.9   Bacteria can be used as biological weapons

In October 2001, endospores of *Bacillus anthracis*, the bacterium that causes anthrax, were found in envelopes mailed to members of the news media and the U.S. Senate (Figure 16.9A). Eighteen people developed cases of anthrax, and five died. Unfortunately, while these attacks were shocking, they were not unique. There is a long and ugly history of using biological agents as weapons, especially viruses and bacteria. *B. anthracis*, *Yersinia pestis* (the bacterium that causes plague), and the exotoxin of *Clostridium botulinum* are among the biological agents that are considered the highest-priority threats today.

B. anthracis forms hardy endospores (see Figure 16.4D) that are commonly found in the soil of agricultural regions, where large grazing animals can become infected. People who work in agriculture, leather tanning, or wool processing occasionally catch the cutaneous (skin) form of anthrax from infected animals. Cutaneous anthrax produces skin lesions and is relatively easy to treat. Inhalation anthrax, which affects the lungs, is the deadly form of the disease. "Weaponizing" anthrax involves manufacturing a preparation of endospores that disperses easily in the air, where the endospores will be inhaled by the target population. Endospores that enter the lungs germinate and the bacteria multiply, producing an exotoxin that eventually accumulates to lethal levels in the blood. *B. anthracis* can be controlled by certain antibiotics, including Ciprofloxacin, which was in huge demand during the 2001 crisis. But antibiotics only kill the bacteria—they can't eliminate the toxin already in the body. As a result, inhalation anthrax has a very high death rate. Researchers have recently made a great deal of progress toward developing antitoxins.

Plague conjures images of medieval epidemics, when people died by the thousands and corpses piled up in the streets. But plague is a modern disease as well. Bacteria that cause plague, *Y. pestis*, are carried by rodents in many parts of the world and are transmitted by fleas. People contract the disease when they are bitten by infected fleas. If diagnosed in time, plague can be treated by antibiotics. If the disease progresses, however, symptoms become strikingly gruesome. Egg-size

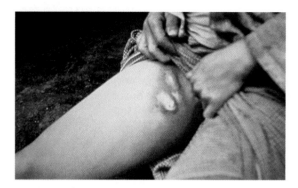

Figure 16.9B   Swellings (buboes) characteristic of the bubonic form of plague

swellings grow in the armpits, groin, and neck (Figure 16.9B), and black blotches appear where the bacteria proliferate in clots under the skin. If the bacteria spill into the bloodstream, death from the bacterial toxins is certain. As a weapon, though, pneumonic plague is the worst case scenario. In the pneumonic form of plague, *Y. pestis* infect cells in the lungs, becoming airborne as the disintegrating lung tissue is coughed out in droplets. As a result, pneumonic plague is easily transmitted from person to person and can become epidemic.

Unlike other biological agents, the weapon form of *C. botulinum* is the exotoxin it produces, botulinum, rather than the living microbes. Botulinum is the deadliest poison on earth. Thirty grams of pure toxin, a bit more than an ounce, could kill every person in the United States. Botulinum blocks transmission of the nerve signals that cause muscle contraction, resulting in paralysis of the muscles required for breathing. As a weapon, it would most likely be broadcast as an aerosol. An effective antitoxin is available, but existing methods of preparing it yield only small quantities. The U.S. government recently contracted with a biotechnology firm to develop mass production methods.

The United States has a long-standing bioweapons program. During World War II, the government established a research laboratory at Camp Detrick, Maryland, that is still in operation today. Fearing biological attack by the Soviets during the Cold War, the United States worked on developing its own arsenal of bioweapons as well as on defensive strategies. In 1969, however, the government abandoned its pursuit of offensive biological weapons and concentrated solely on defending against possible biological attacks. In 1975, the United States signed the Biological Weapons Convention, pledging never to develop or store biological weapons. Eventually, 103 nations joined the ban, although not all signatories have honored it.

Figure 16.9A   Cleaning up after an anthrax attack in October 2001

?   **Why is *Bacillus anthracis* an effective bioweapon?**

■ It is easy to obtain from the soil and forms potentially deadly endospores that resist destruction and can be easily dispersed.

16.10 Prokaryotes help recycle chemicals and clean up the environment

Despite the attention they demand, pathogenic prokaryotes are in the minority. Far more common are prokaryotes that are vital to our well-being. All life depends on the cycling of chemical elements between organisms and the nonliving parts of our environment. As you'll learn in Chapter 37, prokaryotes are indispensable components of the chemical cycle that makes nitrogen available to plants and, in turn, to animals. Prokaryotes also participate in other chemical cycles through decomposition of organic wastes and dead organisms to inorganic chemicals.

The varied metabolic talents of prokaryotes also enable them to help us solve environmental problems through **bioremediation,** the use of organisms to remove pollutants from soil, air, or water. Prokaryotic decomposers are the mainstays of our sewage treatment facilities. Raw sewage is first passed through a series of screens and shredders, and solid matter settles out from the liquid waste. This solid matter, called sludge, is then gradually added to a culture of anaerobic prokaryotes, including both bacteria and archaea. The microbes decompose the organic matter in the sludge to material that can be used as landfill or fertilizer.

Liquid wastes are treated separately from the sludge. In Figure 16.10A, you can see a trickling filter system, one type of mechanism for treating liquid wastes. The long horizontal pipes rotate slowly, spraying liquid wastes through the air onto a thick bed of rocks, the filter. Biofilms of aerobic bacteria and fungi growing on the rocks remove much of the organic material dissolved in the waste. Outflow from the rock bed is sterilized and then released, usually into a river or ocean.

In Figure 16.10B, workers are spraying fertilizers on an oil-polluted beach in Alaska. The fertilizers stimulate the growth of "oil-eating" bacteria that occur naturally in the soil, in some cases speeding the natural breakdown process fivefold. Researchers are trying to genetically engineer bacteria to degrade oil even more efficiently.

Figure 16.10B   Treatment of an oil spill in Alaska

Bacteria may also help us clean up old mining sites. The water that drains from mines is highly acidic and is also laced with poisons—often compounds of arsenic, copper, zinc, and the heavy metals lead, mercury, and cadmium. Contamination of our soils and groundwater by these toxic substances poses a widespread threat, and cleaning up the mess can be extremely expensive. Prokaryotes may be able to help. Bacteria called *Thiobacillus* thrive in the acidic waters that drain from mines, accumulating metals from the mine waters. Unfortunately, *Thiobacillus* bacteria also add sulfuric acid to the water. If this problem is solved, perhaps through genetic engineering, *Thiobacillus* and other prokaryotes may help us overcome some serious environmental problems.

Some mining companies use microbes to extract valuable metals from low-grade ores. Bacteria assist in extracting over 30 billion kilograms of copper from copper sulfides each year. Using prokaryotes that can extract gold from ore, one factory in the African nation of Ghana processes 1 million kilograms of gold concentrate a day.

 What is bioremediation?

■ The use of organisms to clean up pollution

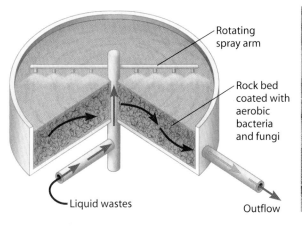

Rotating spray arm

Rock bed coated with aerobic bacteria and fungi

Liquid wastes

Outflow

Figure 16.10A   The trickling filter system at a sewage treatment plant

16.11 Protists are an extremely diverse assortment of eukaryotes

The photograph in Figure 16.11A—a drop of pond water viewed with the light microscope—illustrates a variety of **protists,** a diverse collection of mostly unicellular eukaryotes. Biologists used to classify all protists in a kingdom called Protista, but now it is thought that these organisms constitute multiple kingdoms within domain Eukarya. While our knowledge of the evolutionary relationships among these diverse groups remains incomplete, *protist* is still useful as a convenient term to refer to eukaryotes that are not plants, animals, or fungi.

Protists obtain their nutrition in a variety of ways. Some protists are autotrophic, producing their food by photosynthesis; these are called **algae** (another useful term that is not taxonomically meaningful). Others, informally called **protozoans** (from the Greek *protos*, first, and *zoion*, animal), are heterotrophic, eating bacteria and other protists. Some heterotrophic protists are fungus-like and obtain organic molecules by absorption. And some protists can be either heterotrophic or autotrophic, depending on availability of light and nutrients.

Protist habitats are also diverse. Like the sample shown in Figure 16.11A, most protists are aquatic, and they are found almost anywhere there is water, including moist terrestrial habitats, such as damp soil and leaf litter. Others inhabit the bodies of various host organisms. For example, Figure 16.11B shows some of the protists that live in the intestinal tract of termites. These protists are *endosymbionts.* **Symbiosis** is a close association between organisms of two or more species, and *endo*symbiosis refers to one species, called the endosymbiont, living *within* another. Termite endosymbionts digest the tough cellulose in the wood eaten by their host. Some of these protists even have endosymbionts of their own—prokaryotes that metabolize the cellulose.

As eukaryotes, protists are more complicated than any prokaryotes. Their cells have a membrane-enclosed nucleus (containing multiple chromosomes) and other organelles characteristic of eukaryotic cells. The flagella and cilia of protistan cells have a 9 + 2 pattern of microtubules, another typical eukaryotic trait (see Module 4.18).

Because most protists are unicellular, they are justifiably considered to be the simplest eukaryotes. However, the cells of many protists are among the most elaborate in the world. This level of cellular complexity is not really surprising, for each unicellular protist is a complete eukaryotic organism analogous to an entire animal or plant.

During the past 15 years, molecular and cellular studies have shaken the foundations of protistan taxonomy as much as they have that of the prokaryotes. The more intuitive groupings of protozoans and algae and fungus-like protists are phylogenetically meaningless, because the nutritional modes used to categorize them are spread across many different lineages. It is now clear that there are multiple clades of protists, with some lineages more closely related to plants, fungi, or animals than they are to other protists. Several protistan lineages appear to have arisen through secondary endosymbiosis, as you will learn in the next module.

? What is a general definition for protist?

■ A eukaryote that is not an animal, fungus, or plant

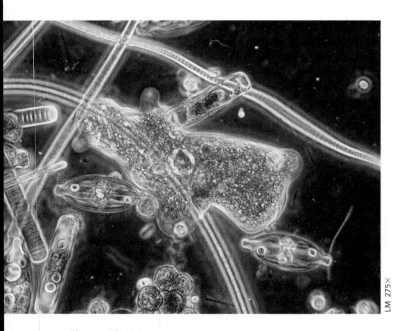

Figure 16.11A   Protists in pond water

LM 275×

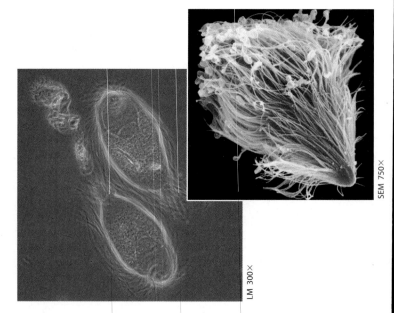

SEM 750×

LM 300×

Figure 16.11B   Protists in termite gut covered by thousands of flagella, viewed with light microscope (left) and scanning electron microscope (right).

16.12 Secondary endosymbiosis is the key to much of protist diversity

As Module 16.11 indicates (and a quick look at Figure 16.13 on the next page will confirm), protists are a bewilderingly diverse group. What is the origin of this enormous diversity? To explain the current hypothesis, let's first review the endosymbiosis hypothesis for the origin of eukaryotic cells (see Module 4.16). According to that hypothesis, known as primary endosymbiosis, eukaryotic cells evolved when prokaryotes established residence within other, larger prokaryotes. The evidence for this hypothesis includes the structural and molecular similarities between prokaryotic cells and present-day mitochondria and chloroplasts. Mitochondria and chloroplasts even replicate their own DNA and reproduce by a process similar to that of prokaryotes.

Scientists think that heterotrophic eukaryotes evolved first; these had mitochondria but not chloroplasts. Autotrophic eukaryotes are thought to have arisen later from a lineage of heterotrophic eukaryotes descended from an individual that engulfed an autotrophic cyanobacterium (left side of Figure 16.12). If the cyanobacterium continued to function within its host cell, its photosynthesis would have provided a steady source of food for the heterotrophic host and thus given it a significant selective advantage. And because the cyanobacterium had its own DNA, it could reproduce to make multiple copies of itself within the host cell. In addition, cyanobacteria could be passed on when the host reproduced. Over time, the descendants of the original cyanobacterium evolved into chloroplasts. This chloroplast-bearing lineage of eukaryotes later diversified into green algae and red algae (see Module 16.20).

On subsequent occasions during eukaryotic evolution, green algae and red algae themselves became endosymbionts following ingestion by heterotrophic eukaryotes. Heterotrophic host cells enclosed the algal cells in food vacuoles, but the algae—or parts of them—survived and became cellular organelles. The presence of the endosymbionts, which also had the ability to replicate themselves, gave their hosts a selective advantage. This process, in which an autotrophic eukaryotic protist became endosymbiotic in a heterotrophic eukaryotic protist, is called **secondary endosymbiosis.**

Secondary endosymbiosis appears to be a major key to protist diversity. In green algae, shown across the top of Figure 16.12, secondary endosymbiosis gave rise to the Euglenozoans (see Module 16.15). Some members of this group are primarily autotrophic, but can become heterotrophic when sunlight is limited. Secondary endosymbiosis of red algae (bottom of Figure 16.12) led to nutritional diversity in dinoflagellates (see Module 16.16) and stramenopiles (see Module 16.17). Some species in these two groups are exclusively autotrophic; other species are exclusively heterotrophic. The dinoflagellates also include species that can obtain their food by either method. In apicomplexans (see Module 16.16), which are all parasitic, an organelle acquired by secondary endosymbiosis is vital to the protist's survival. Its function, however, is not known for certain.

? Distinguish between primary and secondary endosymbiosis.

■ In primary endosymbiosis, the endosymbiont is a prokaryote. In secondary endosymbiosis, the endosymbiont is a eukaryote.

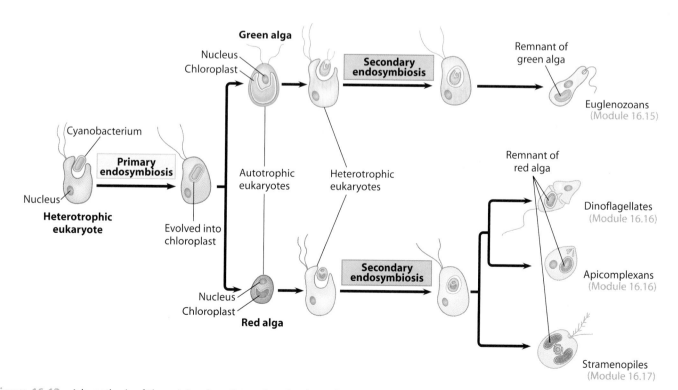

Figure 16.12   A hypothesis of the origin of protistan diversity through endosymbiosis (mitochondria not shown)

16.13 A tentative phylogeny of eukaryotes includes multiple clades of protists

The classification of protists, like all scientific inquiries, remains a work in progress. Figure 16.13 illustrates an abbreviated version of one hypothesis of the phylogeny of kingdoms Fungi, Animalia, and Plantae and the groups that were traditionally placed in kingdom Protista. Don't be overwhelmed by the complexity of the scheme. It simply shows that protists are an extremely diverse group that are difficult to categorize. Indeed, both in terms of number of species and number of living individuals, most eukaryotes are protists. Parts of Figure 16.13 are in dispute (dotted lines), but we present the phylogenetic tree as a means of providing a framework for the protistan groups that we will survey in the next several modules. Certainly, the names, boundaries, and placement of these clades will continue to change as the genomes of more protists are sequenced and compared.

? Why are protists especially important to biologists investigating the evolution of eukaryotic life?

■ Because the first eukaryotes were protists, and protists were ancestral to all other eukaryotes, including plants, fungi, animals, and modern protists.

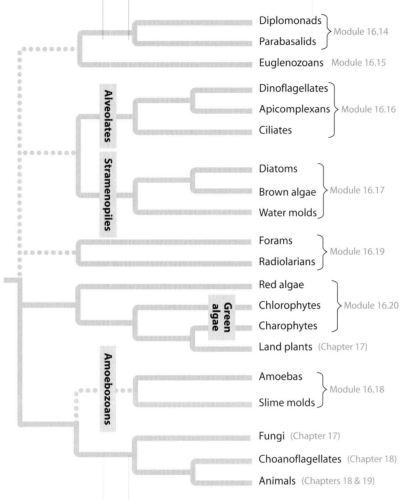

Figure 16.13 A tentative phylogeny of eukaryotes, with indication of the module or chapter in which each group is covered

- Diplomonads } Module 16.14
- Parabasalids
- Euglenozoans Module 16.15
- Dinoflagellates }
- Apicomplexans } Module 16.16
- Ciliates
- Diatoms }
- Brown algae } Module 16.17
- Water molds
- Forams }
- Radiolarians } Module 16.19
- Red algae }
- Chlorophytes } Module 16.20
- Charophytes
- Land plants (Chapter 17)
- Amoebas }
- Slime molds } Module 16.18
- Fungi (Chapter 17)
- Choanoflagellates (Chapter 18)
- Animals (Chapters 18 & 19)

Alveolates, Stramenopiles, Green algae, Amoebozoans

16.14 Diplomonads and parabasalids have modified mitochondria

The **diplomonads** are heterotrophic protists that may represent the most ancient surviving lineage of eukaryotes. They have modified mitochondria with no DNA or electron transport chains. Most diplomonads are anaerobic.

An infamous example of a diplomonad is *Giardia intestinalis*, a common waterborne parasite that causes severe diarrhea. **Parasites** derive their nutrition from living hosts, which are harmed by the interaction. People most often pick up *Giardia* by drinking water contaminated with feces containing the parasite. For example, a swimmer in a lake or river might accidentally ingest water, or a hiker might drink contaminated water from a seemingly pristine stream. (Boiling the water first will kill *Giardia*.)

Parabasalids are heterotrophic protists that have modified mitochondria that generate some energy anaerobically. The termite endosymbionts shown in Module 16.11 are parabasalids. A common sexually transmitted parasite, *Trichomonas vaginalis* (Figure 16.14), also belongs to this group. An estimated

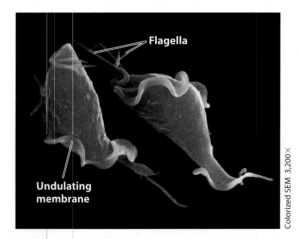

Flagella

Undulating membrane

Colorized SEM 3,200×

Figure 16.14 A parabasalid: *Trichomonas vaginalis*

5 million new infections by this parasite occur each year. *T. vaginalis* travels through the reproductive tract by moving its flagella and undulating part of its membrane. In women, the protists feed on white blood cells and bacteria living on the cells lining the vagina. *T. vaginalis* also infects the cells lining the male urethra, but limited availability of food results in very small population sizes. Consequently, males typically have no symptoms of infection, although they can repeatedly infect their female part- ners. The only treatment available is a drug called metronidazole. Disturbingly, drug resistance seems to be evolving in *T. vaginalis*, especially on college campuses.

? **What characteristics do diplomonads and parabasalids have in common?**

■ Both have modified mitochondria and generate energy mostly anaerobically.

16.15 Euglenozoans have flagella with a unique internal structure

The **euglenozoans** are a diverse clade of protists whose common feature is the presence of a crystalline rod of unknown function inside their flagella. Other characteristics give little clue that the organisms in this group are related. Euglenozoans include heterotrophs, photosynthetic autotrophs, and pathogenic parasites. In Figure 16.15A, the squiggly "worms" are cells of *Trypanosoma;* the red cells are human red blood cells. This trypanosome causes sleeping sickness, a potentially fatal disease spread by the African tsetse fly. *Euglena*, a common inhabitant of pond water, is a member of a different euglenozoan lineage (Figure 16.15B).

? **How do the nutritional modes of diplomonads and euglenozoans differ?**

■ Most diplomonads are anaerobic heterotrophs; euglenozoans exhibit all nutritional modes.

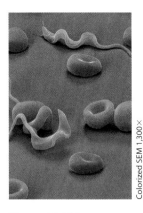

Figure 16.15A
A euglenozoan: *Trypanosoma* (with blood cells)

Colorized SEM 1,300×

Figure 16.15B
A euglenozoan: *Euglena*

Colorized SEM 1,300×

16.16 Alveolates have sacs beneath the plasma membrane

Dinoflagellates, apicomplexans, and ciliates form another clade of protists, the **alveolates,** whose identity is emerging from molecular systematics. Alveolates are characterized by membrane-enclosed sacs just beneath the plasma membrane. Researchers hypothesize that these sacs, called alveoli, help stabilize the cell surface or regulate water and ion content.

Dinoflagellates (Figure 16.16A) are very common components of marine and freshwater phytoplankton (microscopic autotrophs). Some reside within coral animals, providing much of the food for coral reef communities. Other dinoflagellates are heterotrophic. Dinoflagellate blooms—population explosions—sometimes cause warm coastal waters to turn pinkish orange, a phenomenon known as red tide. Toxins produced by some red-tide dinoflagellates have killed large numbers of fish. Humans who consume molluscs that have accumulated the toxins by feeding on dinoflagellates may be affected as well.

Ciliates are a large, varied group of alveolates, named for their use of cilia to move and feed. Nearly all ciliates are free-living, including the common freshwater protist *Paramecium* (see Figure 4.12B). Ciliates are unique in having two types of nuclei: a single, large macronucleus, which controls everyday activities, and multiple tiny micronuclei, which function in sexual reproduction. You can see the macronucleus, which resembles a string of beads running the length of the cell, in the trumpet-shaped ciliate *Stentor* (Figure 16.16B).

All **apicomplexans** are parasites of animals, and some cause serious human diseases. *Plasmodium*, which causes malaria, is one example of an apicomplexan.

? **Where might autotrophic dinoflagellates be found?**

■ Some are floating phytoplankton; others are symbionts within coral.

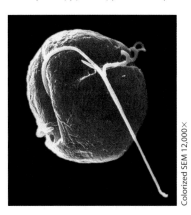

Figure 16.16A *Gymnodinium,* a dinoflagellate that causes red tides

Colorized SEM 12,000×

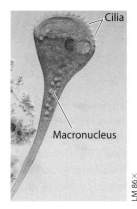

Figure 16.16B *Stentor*, a ciliate

Cilia

Macronucleus

LM 86×

16.17 Stramenopiles have "hairy" and smooth flagella

Stramenopiles (from the Latin *stramen,* straw, and *pilos,* hair) are protists named for their "hairy" flagellum, which has numerous hairlike projections. This flagellum is usually paired with a "smooth" flagellum. The clade is diverse, including both heterotrophs and autotrophs.

Water molds are fungus-like stramenopiles that typically decompose dead plants and animals in freshwater habitats (Figure 16.17A). Although they were long classified as fungi, molecular comparisons have revealed that water molds are more closely related to protists than to fungi. Parasitic water molds sometimes grow on the skin or gills of fish. Downy mildews, which are plant parasites, are another example of water molds. Late blight of potatoes, a disease caused by a downy mildew, led to a devastating famine in Ireland in the mid-1800s.

Diatoms are unicellular stramenopiles with a unique, glassy cell wall containing silica. They belong to the informal category called algae, which include all autotrophic protists (see Module 16.11). The cell wall of a diatom consists of two halves that fit together like the bottom and lid of a shoe box (Figure 16.17B). Both freshwater and marine environments are rich in diatoms, and the organic molecules these microscopic algae produce are a key source of food in all aquatic environments. Diatoms store their food reserves in the form of oil droplets. In addition to being a rich source of energy, oil makes the diatoms buoyant, which keeps them floating near the surface in the sunlight. Massive accumulations of fossilized diatoms make up thick sediments known as diatomaceous earth, which is mined for use as a filtering medium and as a grinding and polishing agent.

Brown algae are large, complex stramenopiles. Like diatoms, they are autotrophic. Brown algae owe their characteristic brown color to some of the pigments in their chloroplasts. All are multicellular, and most are marine. Brown algae include many of the species commonly called seaweeds. We use the word *seaweeds* here to refer to marine algae that have large multicellular bodies. (Some red and green algae are also referred to as seaweeds.) However, seaweeds lack the true stems, leaves, and roots found in most plants. Figure 16.17C shows an underwater bed of brown algae called **kelp** off the coast of California. Anchored to the seafloor by rootlike structures, kelp may grow to heights of 100 m. Fish, sea lions, sea otters, and gray whales regularly use these kelp "forests" as their feeding grounds.

Figure 16.17A   Water mold (the white threads decomposing an insect)

Figure 16.17B   Diatom, a unicellular alga

Colorized SEM 2,400×

Figure 16.17C   Brown algae: a kelp "forest"

? Which stramenopiles are not autotrophic?

● Water molds

16.18 Amoebozoans have lobe-shaped pseudopodia

Amoebas move and feed by means of **pseudopodia** (singular, *pseudopodium*), which are temporary extensions of the cell. Molecular systematics now indicates that amoebas are dispersed across many taxonomic groups. Members of the clade **amoebozoans,** including many species of free-living amoebas, some parasitic amoebas, and the slime molds, have lobe-shaped pseudopodia.

The amoeba in Figure 16.18A is ingesting a small protist. Its pseudopodia arch around the prey and will enclose it in a food vacuole (see Module 4.11). Free-living amoebas creep over rocks, sticks, or mud at the bottom of a pond or ocean. One species of parasitic amoeba causes amebic dysentery, a potentially fatal disease.

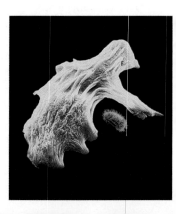

Figure 16.18A   An amoeba ingesting a smaller protist

The yellow, branching growth on the dead log in Figure 16.18B is an amoebozoan called a **plasmodial slime mold.** These protists are common where there is moist, decaying organic matter and are often brightly pigmented. Although it is large and has many extensions, the organism in Figure 16.18B is not multicellular. Rather, it is a **plasmodium,** a single, multinucleate mass of cytoplasm undivided by plasma membranes. (Don't confuse this word with the apicomplexan *Plasmodium*, which causes malaria.) Because most of the nuclei go through mitosis at the same time, plasmodial slime molds are used to study molecular details of the cell cycle.

The plasmodium extends pseudopodia through soil and rotting logs, engulfing food by phagocytosis as it grows. Within the fine channels of the plasmodium, cytoplasm streams first one way and then the other in pulsing flows that probably help distribute nutrients and oxygen. When food and water are in short supply, the plasmodium stops growing and differentiates into reproductive structures (shown in the inset) that produce spores. When conditions become favorable, the spores release haploid cells that fuse to form a zygote, and the life cycle continues.

Cellular slime molds are also common on rotting logs and decaying organic matter. Most of the time, these organisms exist as solitary amoeboid cells. When food is scarce, the amoeboid cells swarm together, forming a slug-like aggregate that wanders around for a short time. Some of the cells then dry up and form a stalk supporting an asexual reproductive structure in which yet other cells develop into spores. The cellular slime mold *Dictyostelium* is a useful model for researchers studying the genetic mechanisms and chemical changes underlying cellular differentiation (Figure 16.18C).

Figure 16.18C   A group of amoeboid cells (left) and reproductive structure of a cellular slime mold, *Dictyostelium*

? Contrast the plasmodium of a plasmodial slime mold with the slug-like stage of a cellular slime mold.

Figure 16.18B   A plasmodial slime mold: *Physarum*

■ A plasmodium is not multicellular, but is one cytoplasmic mass with many nuclei; the slug-like stage of a cellular slime mold consists of many cells.

16.19   Foraminiferans and radiolarians have threadlike pseudopodia

Like amoebozoans, **foraminiferans** (forams) and radiolarians are protists that move and feed by means of pseudopodia. However, they can be distinguished from amoebozoans morphologically by their threadlike (rather than lobe-shaped) pseudopodia (Figure 16.19A). Molecular studies have confirmed that forams and radiolarians are only distantly related to amoebozoans.

Forams are found both in the ocean and in fresh water. They have porous shells, called tests, composed of organic material hardened by calcium carbonate. The pseudopodia, which function in feeding and locomotion, extend through small pores in the test (inset of Figure 16.19A). Ninety percent of forams that have been identified are fossils. The fossilized tests, which are a component of sedimentary rock, are excellent markers for correlating the ages of rocks in different parts of the world.

Like forams, **radiolarians** produce a mineralized support structure, in this case an internal skeleton made of silica (Figure 16.19B). The cell is also surrounded by a test composed of organic material. Most species of radiolarians are marine. When they die, their hard parts, like those of forams, settle to the bottom of the ocean and become part of the sediments. In some areas, radiolarians are so abundant that sediments, known as radiolarian ooze, are hundreds of meters thick.

? What minerals compose foram tests and radiolarian skeletons?

■ Forams, calcium carbonate; radiolarians, silica

Figure 16.19A   A foraminiferan (inset SEM shows a foram test)

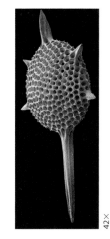

Figure 16.19B A radiolarian skeleton

As you learned in Module 16.12, autotrophic eukaryotes are thought to have arisen by primary endosymbiosis of a cyanobacterium that evolved into chloroplasts. The descendants of this ancient protist evolved into red algae and green algae.

The warm coastal waters of the tropics are home to the majority of species of **red algae.** Their red color comes from an accessory pigment that masks the green of chlorophyll. Red algae are typically soft-bodied, but some have cell walls encrusted with hard, chalky deposits. Encrusted species, such as the one in Figure 16.20A, are common on coral reefs, and their hard parts are important in reef building. Other red algae are commercially important. Carrageenan, a gel that is used to stabilize many products, including yogurt, chocolate milk, and pudding, is derived from species of red algae. Sheets of a red alga known as nori are used to wrap sushi. Agar, a polysaccharide used to grow bacteria, also comes from red algae.

Green algae, which are named for their grass-green chloroplasts, are split into two groups, the chlorophytes and the charophytes (see the phylogenetic tree in Figure 16.13). As you'll learn in Chapter 17, charophytes are the closest living relatives of land plants. The micrographs in Figure 16.20B show two types of green algae. *Chlamydomonas* is a unicellular alga common in freshwater lakes and ponds. It is propelled through the water by two flagella. (Such cells are said to be biflagellated.) *Volvox* is a colonial green alga. Each *Volvox* colony is a hollow ball composed of hundreds or thousands of biflagellated cells. The large colonies shown here will eventually release the small daughter colonies within them. Some marine green algae are large and complex enough to qualify as seaweeds.

Most green algae have complex life cycles. The life cycle of *Ulva*, or sea lettuce (Figure 16.20C), follows a pattern called **alternation of generations,** in which a multicellular diploid (2n) form alternates with a multicellular haploid (n) form. Alternation of generations occurs in a number of multicellular algae and in all plants. Notice in the figure that multicellular haploid forms are called **gametophytes.** The gametophyte generation alternates with a diploid generation that features a multicellular diploid form called a **sporophyte.** In *Ulva*, the gametophyte and sporophyte organisms are identical in appearance; both look like the one in the photograph, although they differ in chromosome number. The haploid gametophyte produces gametes by mitosis, and fusion of the gametes begins the sporophyte generation. In turn, cells in the sporophyte undergo meiosis and produce haploid, flagellated spores. The life cycle is completed when a spore settles to the bottom of the ocean and develops into a gametophyte.

Web Process of Science *What Kinds of Protists Are Found in Various Habitats?*

? What were the probable origins of the chloroplasts of green, red, and brown algae?

■ A cyanobacterial endosymbiont in both green and red algae; a red algal endosymbiont in brown algae

Figure 16.20A   A red alga: an encrusted type, on a coral reef

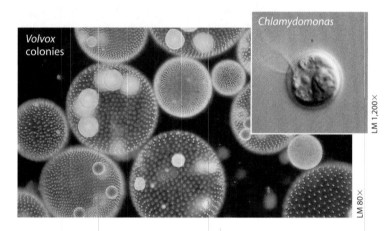

Volvox colonies

Chlamydomonas

LM 1,200×

LM 80×

Figure 16.20B   Green algae, colonial and unicellular (inset)

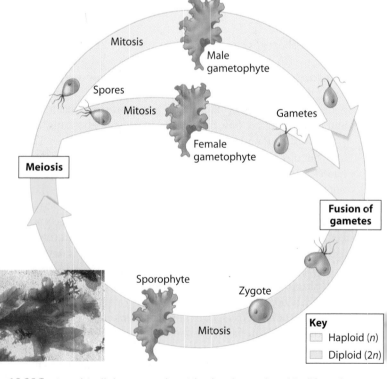

Mitosis

Male gametophyte

Spores

Mitosis

Female gametophyte

Gametes

Meiosis

Fusion of gametes

Sporophyte

Zygote

Mitosis

Key
☐ Haploid (n)
☐ Diploid (2n)

Figure 16.20C   A multicellular green alga: *Ulva* (sea lettuce) and its life cycle

16.21 Multicellularity evolved several times in eukaryotes

Increased complexity often makes more variations possible. Thus, the origin of the eukaryotic cell led to an evolutionary radiation of new forms of life. As you saw in this chapter, unicellular protists, which are structurally complex eukaryotic cells, are much more diverse in form than the simpler prokaryotes. The evolution of multicellular bodies broke through another threshold in structural organization.

Multicellular organisms—seaweeds, plants, animals, and most fungi—are fundamentally different from unicellular ones. In a unicellular organism, all of life's activities occur within a single cell. In contrast, a multicellular organism has various specialized cells that perform different functions and are dependent on each other. For example, some cells give the organism its shape, while others make or procure food, transport materials, enable movement, or reproduce.

The most widely held view is that the organisms linking multicellular organisms to their unicellular ancestors were probably unicellular protists that lived as colonies, federations of independent cells sticking loosely together. Figure 16.21 suggests how a unicellular protist with flagellated cells may have formed colonies that eventually gave rise to multicellular organisms. ❶ An ancestral colony may have formed, as colonial protists do today, when a cell divided and its offspring remained attached to one another. ❷ Next, the cells in the colony may have become somewhat specialized and interdependent, with different cell types becoming more and more efficient at performing specific, limited tasks. ❸ Later on, additional specialization among the cells in the colony may have led to distinctions between sex cells (gametes) and nonreproductive cells (somatic cells).

We see specialization and cooperation among cells today in several colonial protists, such as the green alga *Volvox* in Figure 16.20B. *Volvox* produces gametes, which depend on somatic cells while developing. Cells in truly multicellular organisms, as we know them today, are specialized for many more nonreproductive functions, including feeding, waste disposal, gas exchange, and protection, to name a few. Evolution of the division of labor to this extent involved many additional steps in somatic cell specialization.

As you saw in Figure 16.13, at least three major lineages from the ancestral eukaryote led to multicellular forms: the stramenopile lineage led to brown algae, a currently unnamed lineage led to red algae, green algae, and land plants, and another lineage led to fungi and animals. In Chapter 18, you will meet the choanoflagellates, the protists that are the closest relatives of animals.

The oldest known fossils of multicellular eukaryotes are small multicellular algae that date from 1.2 billion years ago. For the next 600 million years or so, the fossil record remains somewhat scanty, but by about 600 million years ago, a variety of multicellular algae had evolved, along with some soft-bodied animals resembling corals, jellies, and worms.

A period of mass extinctions separated the Precambrian from the Paleozoic era, but multicellular life again flourished soon thereafter. By about 500 million years ago, diverse animals, fungi, and multicellular algae populated aquatic environments. All life was still aquatic.

Around 500 million years ago, the move onto land began, probably as certain green algae living along the edges of lakes gave rise to primitive plants. In the next chapter, we trace the long evolutionary movement of plants onto land and their diversification there. After that, we pick up the threads of animal evolution in Chapter 18.

? In what way do multicellular organisms differ fundamentally from unicellular ones?

◼ In unicellular organisms, all the functions of life are carried out within a single cell. Multicellular organisms have specialized cells that perform different functions.

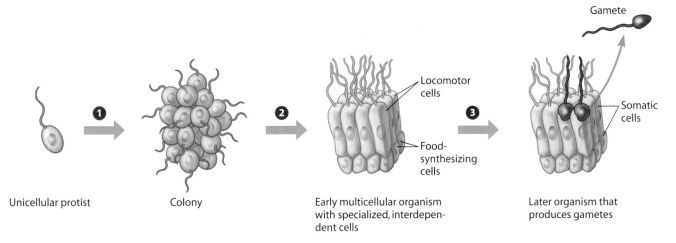

Figure 16.21   A model for the evolution of a multicellular organism from a unicellular protist

Unicellular protist

Colony

Locomotor cells

Food-synthesizing cells

Early multicellular organism with specialized, interdependent cells

Gamete

Somatic cells

Later organism that produces gametes

Reviewing the Concepts

Prokaryotes (16.1–16.10)

Prokaryotes are the oldest life-forms and remain the most numerous and widespread organisms (**16.1**).

Domains Bacteria and Archaea are distinguished on the basis of nucleotide sequences and other molecular and cellular features (**16.2**). Prokaryotes may be shaped as spheres (cocci), rods (bacilli), and curves or spirals (**16.3**).

Structural features that help prokaryotes thrive virtually everywhere are protective cell walls, sticky capsules, pili that cling to surfaces, flagella that provide motility, and resistant endospores. Rapid growth contributes to success (**16.4**).

Nutritional modes of prokaryotes are diverse (**16.5**):

Nutritional Mode	Energy Source	Carbon Source
Photoautotroph	Sunlight	CO₂
Chemoautotroph	Inorganic chemicals	CO₂
Photoheterotroph	Sunlight	Organic compounds
Chemoheterotroph	Organic compounds	Organic compounds

Archaea are common in extreme environments, such as salt lakes, acidic hot springs, and deep-sea hydrothermal vents. They are also a major life-form in the ocean (**16.6**).

Bacteria are currently organized into several subgroups of proteobacteria, chlamydias, spirochetes, gram-positive bacteria, and cyanobacteria. Cyanobacteria photosynthesize in a plant-like way (**16.7**).

Pathogenic bacteria cause disease by producing exotoxins or endotoxins (**16.8**). Bacteria such as the species that cause anthrax and plague, and bacterial toxins such as botulinum, can be used as biological weapons (**16.9**).

Bioremediation is the use of organisms to clean up pollution. Prokaryotes are decomposers in sewage treatment and can clean up oil spills and toxic mine wastes. Prokaryotes are vital to Earth's chemical cycles (**16.10**).

Protists (16.11–16.21)

Protists are a diverse collection of mostly unicellular eukaryotes that are found in a variety of aquatic or moist habitats. Molecular systematics is exploring protistan phylogeny (**16.11**).

Endosymbiosis of prokaryotic cells resulted in the evolution of eukaryotic cells containing mitochondria. Eukaryotic autotrophs arose by primary endosymbiosis of a heterotroph and cyanobacterium, leading to green algae and red algae. Secondary endosymbiosis of green algae by heterotrophic eukaryotes gave rise to the Euglenozoans. Secondary endosymbiosis of red algae by heterotrophic eukaryotes produced dinoflagellates, apicomplexans, and stramenopiles. Thus, protist diversity is largely the result of secondary endosymbiosis followed by the divergence of lineages (**16.12**).

Eukaryotic phylogeny, based on molecular systematics, has been proposed, but much of it is highly tentative. These organisms are very difficult to classify (**16.13**).

Diplomonads and parabasalids are mostly anaerobic protists that have modified mitochondria. The parasitic *Giardia* is a diplomonad. Termite endosymbionts and *Trichomonas vaginalis* are parabasalids (**16.14**).

Euglenozoans are a diverse clade of flagellated protists that includes trypanosomes (parasites that cause sleeping sickness) and *Euglena* (**16.15**).

Alveolates have membrane sacs under the plasma membrane. Dinoflagellates are unicellular algae; apicomplexans are parasites, such as *Plasmodium,* which causes malaria; ciliates use cilia to move and feed (**16.16**).

Stramenopiles are named for their "hairy" flagella. This clade includes fungus-like water molds, photosynthetic, unicellular diatoms, and brown algae—large, complex seaweeds. Kelp forms important marine habitats (**16.17**).

Amoebozoans include amoebas with lobe-shaped pseudopodia and slime molds. A plasmodial slime mold is a multinucleate plasmodium that forms reproductive structures under adverse conditions. Cellular slime molds have unicellular and multicellular stages (**16.18**).

Foraminiferans and radiolarians have threadlike pseudopodia. Forams have a calcium carbonate test and radiolarians have a skeleton composed of silica. The hard parts of both groups are a significant component of sediments wherever they occur (**16.19**).

Red and green algae are in the lineage that includes plants. Red algae contribute to coral reefs. Green algae may be unicellular, colonial, or multicellular. The life cycles of many algae involve the alternation of haploid gametophyte and diploid sporophyte generations (**16.20**).

Multicellularity evolved in several different lineages, probably by specialization of the cells of colonial protists. Multicellular life arose over a billion years ago (**16.21**).

Connecting the Concepts

1. Explain how each of the following characteristics contributes to the success of prokaryotes: cell wall, capsule, motility, rapid reproduction, endospores.

2. For each organism shown below, name the group it belongs to, and explain how the photo represents the group's characteristics.

(a)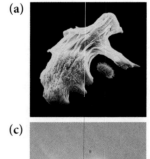

(b)

(c)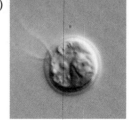

(d)

Testing Your Knowledge

Multiple Choice

3. You set your time machine for 3 billion years ago and push the start button. When the dust clears, you look out the window. Which of the following describes what you would probably see?
   a. plants and animals very different from those alive today
   b. a cloud of gas and dust in space
   c. green scum in the water
   d. land and water sterile and devoid of life
   e. an endless expanse of red-hot molten rock

4. In terms of nutrition, autotrophs are to heterotrophs as
   a. kelp are to diatoms.
   b. archaea are to bacteria.
   c. slime molds are to algae.
   d. algae are to slime molds.
   e. pathogenic bacteria are to harmless bacteria.

5. The bacteria that cause tetanus can be killed only by prolonged heating at temperatures considerably above boiling. This suggests that tetanus bacteria
   a. have cell walls containing peptidoglycan.
   b. protect themselves by secreting antibiotics.
   c. secrete endotoxins.
   d. are autotrophic.
   e. produce endospores.

6. Glycolysis is the only metabolic pathway common to nearly all organisms. To scientists, this suggests that it
   a. evolved many times during the history of life.
   b. was first seen in early eukaryotes.
   c. appeared early in the history of life.
   d. must be very complex.
   e. appeared rather recently in the evolution of life.

7. A new organism has been discovered. Tests have revealed that it is unicellular, autotrophic, and has a cell wall that contains peptidoglycan. Based on this evidence, it should be classified as a(n)
   a. algae.
   b. archean.
   c. stramenopile.
   d. bacterium.
   e. apicomplexan.

8. Which pair of protists has support structures composed of silica?
   a. dinoflagellates and diatoms
   b. diatoms and radiolarians
   c. radiolarians and forams
   d. forams and diplomonads
   e. diplomonads and diatoms

9. Of the following groups, which is thought to have most recently shared a common ancestor with animals?
   a. red algae
   b. cellular slime molds
   c. brown algae
   d. dinoflagellates
   e. green algae

10. Which of the following groups is incorrectly paired with an example of that group?
    a. euglenozoans—trypanosome causing sleeping sickness
    b. alveolates—apicomplexan such as *Plasmodium*
    c. stramenopiles—brown algae
    d. amoebozoans—water mold
    e. diplomonads—*Giardia*

11. Which of the following prokaryotes is not pathogenic?
    a. *Chlamydia*
    b. *Rhizobium*
    c. *Streptococcus*
    d. *Salmonella*
    e. *Bacillus anthracis*

Describing, Comparing, and Explaining

12. How can antibiotics kill bacteria without harming the cells in our bodies?

13. Explain how protist diversity may have arisen through secondary endosymbiosis.

14. *Chlamydomonas* is a unicellular green alga. How does it differ from a photosynthetic bacterium, which is also single-celled? How does it differ from a protozoan, such as an amoeba? How does it differ from larger green algae, such as sea lettuce (*Ulva*)?

15. What characteristic distinguishes true multicellularity from colonies of cells?

Applying the Concepts

16. Imagine you are on a team designing a moon base that will be self-contained and self-sustaining. Once supplied with building materials, equipment, and organisms from Earth, the base will be expected to function indefinitely. One of the team members has suggested that everything sent to the base be sterilized so that no bacteria of any kind are present. Do you think this is a good idea? Predict some of the consequences of eliminating all bacteria from an environment.

17. The buildup of CO_2 in the atmosphere resulting from the burning of fossil fuels is regarded as a major contributor to global warming (see Module 7.13). Diatoms and other microscopic algae in the oceans counter this buildup by using large quantities of atmospheric CO_2 in photosynthesis, which requires small quantities of iron. Experts suspect that a shortage of iron may limit algal growth in the oceans. Some scientists have suggested that one way to reduce CO_2 buildup might be to fertilize the oceans with iron. The iron would stimulate algal growth and thus the removal of more CO_2 from the air. A single supertanker of iron dust, spread over a wide enough area, might reduce the atmospheric CO_2 level significantly. Do you think this approach would be worth a try? Why or why not?

Answers to all questions can be found in Appendix 4.

For Practice Quizzes, BioFlix, MP3 Tutors, and Activities, go to www.mybiology.com.

LM 5×

Plants and Fungi—A Beneficial Partnership

We tend to take our orange juice for granted, but it is no small feat for citrus growers to produce it at a reasonable cost. Orange groves are found in Florida, Texas, and California. An enormous investment, trees like the ones pictured to the left take three to seven years to start producing fruit and require a rich supply of fertilizer. They are also vulnerable to freezing and to a long list of pathogenic bacteria (see Module 16.8), insects, and especially fungi.

Fungi are not always harmful to plants, however. There is another kind of association between fungi and plants in which the fungus plays the role of vital benefactor. You can see an example of this relationship in the photograph to the right, which shows two pine seedlings. A dense network of white strands of fungus ensheathes the roots of the seedling on the right. The microscopic image above shows a closer view of this intimate relationship. Together, the root and the fungus form an intimate, mutually beneficial association called a **mycorrhiza** (meaning "fungus root"; plural, *mycorrhizae*). For its part, a mycorrhizal fungus absorbs phosphorus and other essential minerals as well as water from the soil and makes them available to the plant. The sugars produced by the plant through photosynthesis nourish the fungus. Notice in the photo of the pines that the seedling on the left, which has no mycorrhizal fungi, appears stunted in comparison to the seedling that has mycorrhizae.

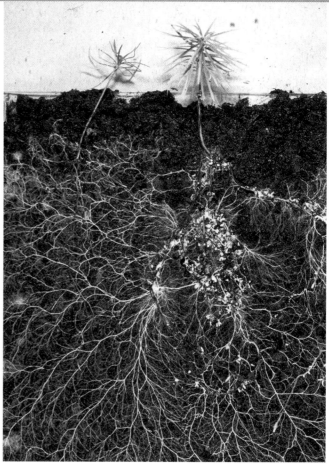

Citrus trees can also have mycorrhizae, and the fungi may offer a way to cut the high economic and environmental costs of producing citrus fruits. Mycorrhizae can make a tree more resistant to disease, thereby reducing the need for pesticides to kill disease-causing organisms. Also, by enhancing a tree's uptake of nutrients, mycorrhizae can reduce, or even eliminate, the need for fertilizers. Unfortunately, the conditions in a typical citrus grove undermine the growth of mycorrhizae. Many citrus growers use fungus-killing chemicals (fungicides) to control fungi that cause disease, and the fungicides poison the mycorrhizal fungi as well. As a result, the grower loses the benefits of mycorrhizae and must apply expensive fertilizers. The environment also suffers because fungicides harm many kinds of organisms, and excess fertilizers pollute streams, lakes, and groundwater.

Some growers have tried to mitigate these problems by adding mycorrhizal fungi back into the soil of their groves, alternating between destroying and replenishing the fungi. Organic citrus growers, who avoid fungicides, often supply mycorrhizal fungi to nursery plants or to the soil of established groves, happy to find a natural boost for their trees.

Cultivated citrus groves treated with fungicides are unnatural in lacking mycorrhizae. In nature, nearly all plants have mycorrhizae. In fact, mycorrhizae appear in fossils of some of the oldest known plants, suggesting that this beneficial relationship between plants and fungi may have played an important role in plant evolution, as the first plants adapted to land.

The colonization of land by plants was a major event in the history of life. Plants transformed the landscape, creating new environmental opportunities for prokaryotes and protists and making it possible for herbivorous animals and their predators to evolve on land. This chapter continues our account of the evolution of life's diversity, focusing on plants and fungi. ■ ■ ■

17.1 Plants have adaptations for life on land

The algal ancestors of plants may have carpeted moist fringes of lakes or coastal salt marshes over 500 million years ago. These shallow-water habitats were subject to occasional drying, and natural selection would have favored algae that could survive periodic droughts. Some species accumulated adaptations that enabled them to live permanently above the water line. The modern-day green alga *Coleochaete* (Figure 17.1A), which grows at the edges of lakes as disklike, multicellular colonies, may resemble one of these early plant ancestors. *Coleochaete* and the more elaborate pond alga *Chara*, shown in Figure 17.1B, belong to a lineage of green algae called the **charophytes.** Plants and present-day charophytes probably evolved from a common ancestor. Morphological, biochemical, and genetic evidence identify the charophytes as the closest living relatives of land plants.

Both plants and algae are multicellular eukaryotes that make organic molecules by photosynthesis. However, a set of derived characters distinguishes land plants as a clade (a monophyletic group), setting them apart from their closest algal relatives: alternation of haploid and diploid generations, walled spores produced in sporangia, male and female gametangia, and multicellular, dependent embryos. You will learn more about some of these traits as we explore plant adaptations for life on land.

Adaptations making life on dry land possible had accumulated by about 475 million years ago, the age of the oldest known plant fossils. The evolutionary novelties of these first land plants opened the new frontier of a terrestrial habitat. Early plant life would have thrived in the new environment. Bright sunlight was virtually limitless on land; the atmosphere had an abundance of carbon dioxide (CO_2); and at first there were relatively few pathogens and plant-eating animals.

Figure 17.1C on the facing page compares how plants such as moss, ferns, and pines differ from multicellular algae like *Chara*. Many algae are anchored by a holdfast, but generally they have no rigid tissues and are supported by the surrounding water. The whole algal body obtains CO_2 and minerals directly from the water. Almost all of the organism receives light and can perform photosynthesis. For reproduction, flagellated sperm swim to fertilize an egg. The offspring are dispersed by water as well. Because terrestrial organisms are surrounded by air rather than water, they face a number of challenges, including maintaining moisture inside their cells, obtaining resources from both soil and air, supporting the body in a nonbuoyant medium, and reproducing and dispersing offspring without water. Thus the water-to-land transition required fundamental changes in algal structure and life cycle.

Maintaining Moisture   The aboveground parts of most land plants are covered by a waxy cuticle that prevents water loss. Gas exchange cannot occur directly through the cuticle, but CO_2 and O_2 diffuse across the leaf surfaces through the tiny pores called

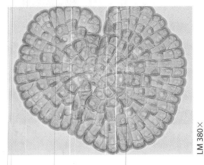

Figure 17.1A   *Coleochaete*, a simple charophyte (above)

Figure 17.1B   *Chara*, an elaborate charophyte

stomata (see Module 7.2). Two surrounding cells regulate each stoma's opening and closing. Stomata are usually open during sunlight hours, allowing gas exchange, and closed at other times, preventing water loss by evaporation. You'll learn more about the role of stomata in water regulation in Chapter 32.

Obtaining Resources from Two Locations   A typical plant must obtain chemicals from both soil and air, two very different media. Water and mineral nutrients are mainly found in the soil; light and CO_2 are available aboveground. Most plants have discrete organs—roots, stems, and leaves—that help meet this resource challenge.

Plant roots provide anchorage and absorb water and mineral nutrients from the soil. In most plants, as noted in the chapter introduction, mycorrhizae greatly enhance this absorption. Aboveground, a plant's stems bear leaves, which obtain CO_2 from the air and light from the sun, enabling them to perform photosynthesis. Growth-producing regions of cell division, called **apical meristems,** are found near the tips of stems and roots. The elongation and branching of a plant's stems and roots maximize exposure to the resources in the soil and air.

A plant must be able to connect its subterranean and aerial parts, conducting water and minerals upward from its roots to its leaves and distributing sugars produced in the leaves throughout its body. Most plants, including ferns, pines, and flowering plants, have **vascular tissue,** a network of thick-walled cells joined into narrow tubes that extend throughout the plant body (traced in red in Figure 17.1C). The photograph of part of an aspen leaf in Figure 17.1D shows the leaf's network of veins, which are fine branches of the vascular tissue. There are two types of vascular tissue. **Xylem** includes dead cells that form microscopic pipes conveying water and minerals up from the roots. **Phloem,** which consists entirely of living cells, distributes sugars throughout the plant. In contrast to plants with elaborate vascular tissues, mosses

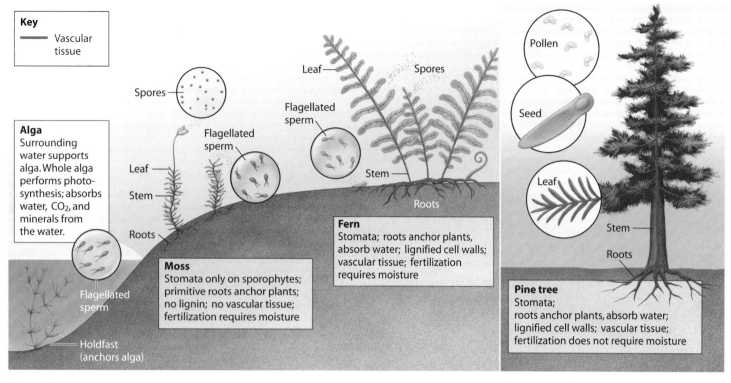

Figure 17.1C Comparing the terrestrial adaptations of moss, fern, and pine with *Chara*, a multicellular green alga

Key
— Vascular tissue

Alga
Surrounding water supports alga. Whole alga performs photosynthesis; absorbs water, CO_2, and minerals from the water.

Spores

Flagellated sperm

Holdfast (anchors alga)

Spores

Leaf

Stem

Roots

Flagellated sperm

Moss
Stomata only on sporophytes; primitive roots anchor plants; no lignin; no vascular tissue; fertilization requires moisture

Leaf

Spores

Flagellated sperm

Stem

Roots

Fern
Stomata; roots anchor plants, absorb water; lignified cell walls; vascular tissue; fertilization requires moisture

Pollen

Seed

Leaf

Stem

Roots

Pine tree
Stomata; roots anchor plants, absorb water; lignified cell walls; vascular tissue; fertilization does not require moisture

lack a complex transport system (although some mosses do have simple vascular tissue). With limited means for distributing water and minerals from soil to the leaves, the height of nonvascular plants is severely restricted.

Supporting the Plant Body Because air provides much less support than water, plants must be able to hold themselves up against the pull of gravity. The cell walls of some plant tissues, including xylem, are thickened and reinforced by a chemical called **lignin.** The absence of lignified cell walls in mosses and other plants that lack vascular tissue is another limitation on their height.

Reproduction and Dispersal Reproduction on land presents complex challenges. For *Chara* and other algae, the surrounding water ensures that gametes and offspring stay moist. Plants, however, must keep their gametes and developing embryos from drying out in the air. Like the earliest land plants, mosses and ferns still produce gametes in male and female **gametangia** (singular, *gametangium*), structures that consist of protective jackets of cells surrounding the gamete-producing cells. The egg remains in the female gametangium and is fertilized there by sperm that swim through a film of water. As a result, mosses and ferns can only reproduce in a moist environment. Pines and flowering plants have **pollen grains,** structures that contain the sperm-producing cells. Pollen

grains are conveyed close to the egg by wind or animals. Moisture is not required for bringing sperm and egg together.

In all plants, the fertilized egg (zygote) develops into an embryo while attached to and nourished by the parent plant. This multicellular, dependent embryo is the basis for designating plants as **embryophytes** (*phyte* means plant), distinguishing them from algae.

The life cycles of all plants involve the alternation of a haploid generation, which produces eggs and sperm, and a diploid generation, which produces spores within protective structures called **sporangia** (singular, *sporangium*). A **spore** is a cell that can develop into a new organism without fusing with another cell. The earliest land plants relied on tough-walled spores for dispersal, a trait retained by mosses and ferns today. Pines and flowering plants have seeds for launching their offspring. Seeds are elaborate embryo-containing structures that are well protected from the elements and are dispersed by wind or animals (see Modules 17.7–17.10). To emphasize this important reproductive difference, plants that disperse their offspring as spores are often referred to as seedless plants. Now let's take a closer look at the evolution of the major plant groups.

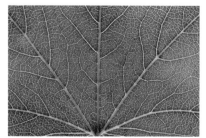

Figure 17.1D The network of veins in a leaf

Web Activity *Terrestrial Adaptations of Plants*

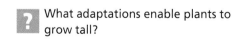

? What adaptations enable plants to grow tall?

■ Vascular tissues to transport materials from belowground to aboveground parts and vice versa; lignified cell walls to provide structural support

17.2 Plant diversity reflects the evolutionary history of the plant kingdom

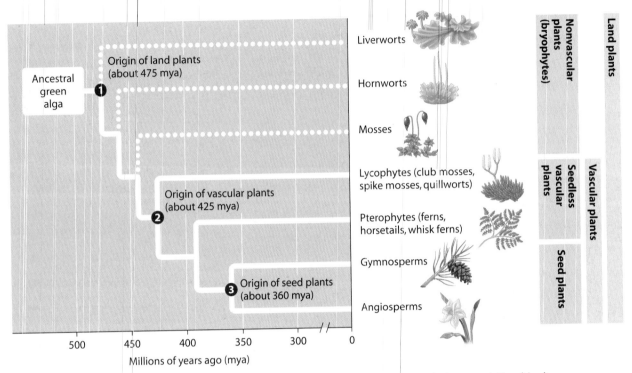

Figure 17.2A   Some highlights of plant evolution (Dotted lines indicate uncertain evolutionary relationships.)

Figure 17.2A ❶ ❷ ❸ highlights some of the major events in the history of the plant kingdom and presents a widely held view of the relationships between surviving lineages of plants.

After plants originated from an algal ancestor approximately 475 million years ago, early diversification gave rise to seedless, nonvascular plants, including mosses, liverworts, and hornworts (Figure 17.2B). These plants, which are informally called **bryophytes,** resemble other plants in having apical meristems and embryos that are retained on the parent plant, but they lack true roots and leaves. Without lignified cell walls, bryophytes with an upright growth habit lack support. Thus, a mat of moss actually consists of many plants growing in a tight pack, holding one another up. The mat is spongy and retains water. Other bryophytes grow flat against the ground, also in dense mats that facilitate fertilization by flagellated sperm swimming through a film of water left by rain or dew.

The **vascular plants** originated about 425 million years ago. Their lignin-hardened vascular tissues provide strong support, enabling stems to stand upright and grow tall on land. Two clades of vascular plants are informally called **seedless vascular plants** (Figure 17.2C, facing page): the lycophytes (such as club mosses) and the widespread pterophytes (ferns and their relatives). A fern has well-developed roots and rigid stems. Ferns are common in temperate forests, but they are most diverse in the tropics. In some tropical species, called tree ferns, upright stems can grow several meters tall. Like bryophytes, however, ferns and club mosses require moist conditions for fertilization, and they disperse their offspring as spores that are carried by air currents.

As indicated in Figure 17.2A, the first vascular plants with seeds evolved about 360 million years ago. Today, the seed plant lineage accounts for over 90% of the approximately 290,000 species of living plants. Seeds and pollen are key adaptations that improved the ability of plants to diversify in terrestrial habitats. A **seed** consists of an embryo packaged with a food supply within a protective covering. This survival packet facilitates wide dispersal of plant embryos (see Module 17.10). Pollen, as you learned in

Figure 17.2B   Bryophytes: moss (left), liverwort (top right), hornwort (bottom right)

Figure 17.2C Seedless vascular plants: fern (above) and club moss (right)

Figure 17.2D Gymnosperms: (clockwise, from top) blue spruce (a conifer), *Ephedra*, ginkgo, *Cycad*

Module 17.1, brings sperm-producing cells into contact with egg-producing parts. A water layer is not required for fertilization to occur. And unlike flagellated sperm, which can swim a few centimeters at most, pollen can travel great distances.

Among the earliest seed plants were the **gymnosperms** (from the Greek *gymnos*, naked, and *sperma*, seed). Seeds of gymnosperms are said to be "naked" because they are not produced in specialized chambers. The largest clade of gymnosperms are the conifers, consisting mainly of cone-bearing trees, such as pine, spruce, and fir. (The term *conifer* means "cone-bearing.") Some examples of gymnosperms that are less common are the ornamental ginkgo tree, desert shrubs in the genus *Ephedra*, and the palmlike cycads (Figure 17.2D). Gymnosperms flourished alongside the dinosaurs in the Mesozoic Era.

The most recent major episode in plant evolution was the appearance of flowering plants, or **angiosperms** (from the Greek *angion*, container, and *sperma*, seed), at least 140 million years ago. Flowers are complex reproductive structures that develop seeds within protective chambers. The great majority of living plants—some 250,000 species—are angiosperms and include a wide variety of plants, such as grasses, flowering shrubs, and flowering trees (Figure 17.2E).

In summary, four key adaptations for life on land distinguish the main lineages of the plant kingdom. (1) Dependent embryos are present in all plants. (2) Lignified vascular tissues mark a lineage that gave rise to most living plants. (3) Seeds are found in a lineage that includes all living gymnosperms and angiosperms and that dominates the plant kingdom today. (4) Flowers mark the angiosperm lineage. As we will see in the next several modules, the life cycles of living plants reveal additional details about plant evolution.

⊙ **MP3 Tutor** *Evolution of Plants*

Web Activity *Highlights of Plant Phylogeny*

Figure 17.2E Angiosperms: barley grass (top) and jacaranda trees (bottom)

? Identify which of the following traits is shared by all plants: flowers, seeds, retained embryo, vascular tissue.

■ Embryo retained on parent plant

17.3 Haploid and diploid generations alternate in plant life cycles

Plants have life cycles that are very different from ours. In Module 8.13, you learned that humans are diploid individuals. That is, each of us has two sets of chromosomes, one from each parent. The only haploid stages in the human life cycle are sperm and eggs. Plants have an **alternation of generations:** the diploid and haploid stages are distinct, multicellular bodies. The haploid generation of a plant produces gametes and is called the **gametophyte.** The diploid generation produces spores and is called the **sporophyte.** In a plant's life cycle, these two generations alternate in producing each other.

The life cycles of all plants follow the pattern shown in Figure 17.3. Be sure that you understand this diagram; then review it after studying each life cycle to see how the pattern applies. Haploid gametophyte plants produce gametes (sperm and eggs) by mitosis. Fertilization results in a diploid zygote. The zygote divides by mitosis and develops into the multicellular diploid sporophyte plant. The sporophyte produces haploid spores by meiosis. A spore then develops by mitosis into a multicellular haploid gametophyte.

Although some algae exhibit alternation of generations, the closest algal relatives of plants, the charophytes, do not. Thus, this life cycle appears to have evolved independently in plants; it is a derived character of land plants.

The next module highlights the life cycle of mosses. In mosses, as in all nonvascular plants, the gametophyte is the larger, more obvious stage of the life cycle.

MP3 Tutor *Alternation of Generations*

? Gametophyte is to _____ as _____ is to diploid.

haploid · · · sporophyte ■

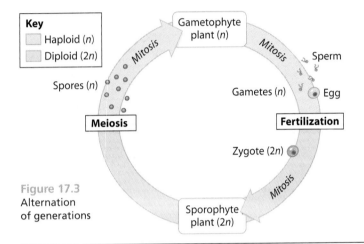

Figure 17.3
Alternation of generations

17.4 Mosses have a dominant gametophyte (See text at top of facing page.)

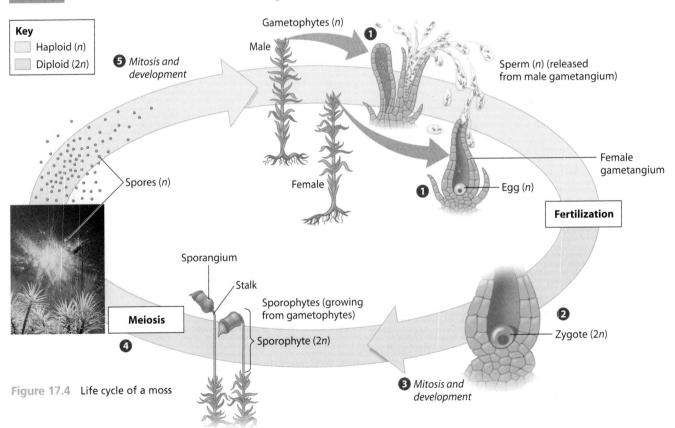

Figure 17.4 Life cycle of a moss

In a moss, the green, cushiony growth we see consists of gameto-phytes. Follow the moss life cycle in Figure 17.4 on the previous page. ❶ Gametes develop in male and female gametangia, usually on separate plants. The flagellated sperm swim through a film of water to the egg in the female gametangium. After fertilization, ❷ the zygote remains in the gametangium. ❸ There it divides by mitosis, developing into a sporophyte embryo and then a mature sporophyte, which remains attached to the gametophyte. ❹ Meiosis occurs in the sporangia at the tips of the sporophyte stalks. As you can see in the photograph, after meiosis, haploid spores are released from the sporangium. ❺ The spores undergo mitosis and develop into gametophyte plants.

Web Activity Moss Life Cycle

? How do moss sperm travel from male gametangia to female gametangia, where fertilization of eggs occurs?

■ The flagellated sperm swim through a film of water.

17.5 Ferns, like most plants, have a dominant sporophyte

All we usually see of a fern is the sporophyte. But let's start the fern life cycle with the gametophyte. ❶ Fern gametophytes often have a distinctive heart-like shape (shown at the top of Figure 17.5), but they are quite small (about 0.5 cm across) and inconspicuous. Like mosses, ferns have flagellated sperm that require moisture to reach an egg. Although eggs and sperm are usually produced in separate locations on the same gametophyte, a variety of mechanisms promote cross-fertilization between gametophytes. ❷ The zygote remains on the gametophyte, where ❸ it develops into an independent sporophyte. ❹ The yellow dots in the photograph are clusters of sporangia, in which cells undergo meiosis, producing haploid spores. The spores are released and ❺ develop into gametophytes by mitosis.

Today, about 95% of all plants, including all seed plants, have a dominant sporophyte in their life cycle. But before we resume this story, let's look back at a time in plant history, before seed plants rose to dominance, when ferns and other seedless vascular plants covered much of the land surface.

Web Activity Fern Life Cycle

Web Process of Science What Are the Different Stages of a Fern Life Cycle?

? How is it possible for the fern gametophyte to produce haploid gametes without meiosis?

■ All the gametophyte's cells are haploid, so there is no need to reduce chromosome number by meiosis to produce haploid gametes.

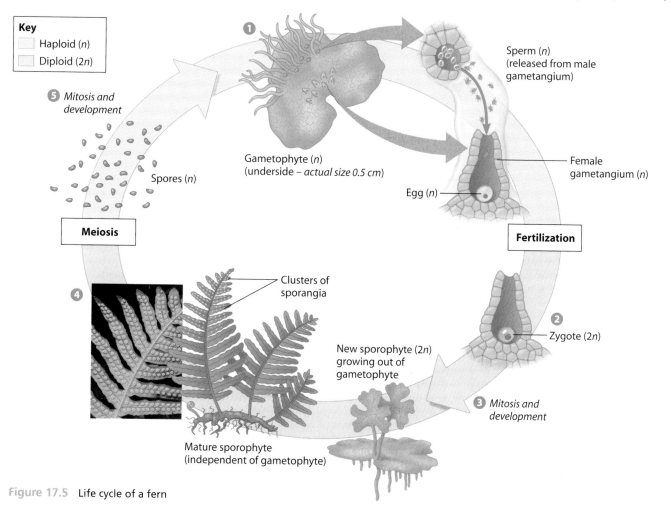

Key
- ▢ Haploid (n)
- ▢ Diploid (2n)

❺ Mitosis and development

Spores (n)

Meiosis

❹

Clusters of sporangia

Mature sporophyte (independent of gametophyte)

Gametophyte (n) (underside – *actual size 0.5 cm*)

Sperm (n) (released from male gametangium)

Female gametangium (n)

Egg (n)

Fertilization

❷ Zygote (2n)

New sporophyte (2n) growing out of gametophyte

❸ Mitosis and development

Figure 17.5 Life cycle of a fern

17.6 Seedless vascular plants dominated vast "coal forests"

During the Carboniferous period (about 360–299 million years ago), the two clades of seedless vascular plants, lycophytes (introduced in Module 17.2 as club mosses) and ferns and their relatives, grew in vast forests in the low-lying wetlands of what is now Eurasia and North America. At that time, these continents were close to the equator and had warm, humid climates that supported broad expanses of lush vegetation.

Figure 17.6 shows a reconstruction of one of these forests based on fossil evidence. Tree ferns are visible in the foreground. Most of the large trees are lycophytes, giants that grew to diameters of more than 2 m and heights that soared to 40 m. (For a sense of scale, dragonflies such as the one shown in the foreground had wing spans of up to 1 m.) As you'll learn in Chapter 19, vertebrates were adapting to terrestrial habitats in parallel with plants; amphibians and early reptiles lived among these trees.

Photosynthesis in these immense swamp forests fixed large amounts of carbon from CO_2 into organic molecules, dramatically reducing CO_2 levels in the atmosphere. Because atmospheric CO_2 traps heat, this change caused global cooling. Photosynthesis generated great quantities of organic matter. As the plants died, they fell into stagnant swamps and did not decay completely. Their remains formed thick organic deposits called peat. Later, seawater covered the swamps, marine sediments covered the peat, and pressure and heat gradually converted the peat to coal, black sedimentary rock made up of fossilized plant material. Coal deposits from the Carboniferous period are the most extensive ever formed. (The name Carboniferous comes from the Latin *carbo*, coal, and *fer*, bearing.) Coal, oil, and natural gas are called **fossil fuels**—fuels formed from the remains of ancient organisms. (Oil and natural gas were formed from marine organisms.) Since the Industrial Revolution, coal has been a crucial source of energy for human society. However, burning these fossil fuels releases CO_2 and other greenhouse gases into the atmosphere, which are now causing a warming climate (see Modules 38.5 and 38.6).

As temperatures dropped during the late Carboniferous period, glaciers formed. The global climate turned drier, and the vast swamps and forests began to disappear. The climate change provided an opportunity for the early seed plants, which grew along with the seedless plants in the Carboniferous swamps. With their wind-dispersed pollen and protective seeds, seed plants could complete their life cycles on dry land.

Figure 17.6 A reconstruction of an extinct forest dominated by seedless plants

? How did the tropical swamp forests contribute to global cooling in the Carboniferous period?

■ Photosynthesis by the abundant plant life reduced atmospheric CO_2, a gas that traps heat.

17.7 A pine tree is a sporophyte with gametophytes in its cones

In seed plants, a specialized structure within the sporophyte houses all reproductive stages, including spores, eggs, sperm, zygotes, and embryos. In gymnosperms such as pines and other conifers, this structure is called a cone. Viewed in longitudinal section (Figure 17.7, facing page), the cone resembles a short stem bearing thick leaves. The resemblance is not surprising—cones are shoots that were modified by natural selection to serve a different function. Each "leaf," or scale, of the cone contains

sporangia that produce spores by meiosis. Unlike seedless plants, however, the spores are not released. Rather, spores give rise to gametophytes within the shelter of the cone. The gametophytes later produce gametes, which unite to form a new sporophyte.

A pine tree bears two types of cones, which produce spores that develop into male and female gametophytes. The smaller ones, called pollen cones (❶ in Figure 17.7), will produce the male gametophytes. Pollen cones contain many sporangia, each of which makes numerous haploid spores by meiosis. Male gametophytes, or pollen grains, develop from the spores. Mature pollen cones release millions of microscopic pollen grains in great clouds. You may have seen yellowish conifer pollen covering cars or floating on puddles after a spring rain. Pollen grains, which are carried by the wind, house cells that will develop into sperm if by chance they land on a cone that contains a female gametophyte.

❷ An ovulate cone, which will produce the female gametophyte, is larger than a pollen cone. Each of its stiff scales bears a pair of ovules. (Only one is shown in the diagram.) Each ovule contains a sporangium surrounded by a protective covering called the integument. ❸ **Pollination** occurs when a pollen grain lands on an ovulate scale and enters an ovule. After pollination, the scales grow together, sealing the cone until the seeds are mature. Meiosis then occurs in a spore "mother cell" within the ovule. Over the course of many months, ❹ one surviving haploid spore cell develops into the female gametophyte, which makes eggs. ❺ Meanwhile, a tiny tube grows out of each pollen grain, digests its way through the ovule, and eventually releases sperm near an egg. Fertilization typically occurs more than a year after pollination.

❻ Usually, all eggs in an ovule are fertilized, but ordinarily only one zygote develops fully into a sporophyte embryo. The ovule matures into a seed, which contains the embryo's food supply (the remains of the female gametophyte) and has a tough **seed coat** (the ovule's integument). In a typical pine, seeds are shed about two years after pollination. ❼ The seed is dispersed by wind, and when conditions are favorable, it germinates, and its embryo grows into a pine seedling.

In summary, all the reproductive stages of conifers are housed in cones borne on sporophytes. The ovule is a key adaptation—a protective device for all the female stages in the life cycle, as well as the site of pollination, fertilization, and embryonic development. The ovule becomes the seed, an important terrestrial adaptation and a major factor in the success of the conifers and flowering plants.

Next we consider flowering plants, the most diverse and geographically widespread of all plants. Angiosperms dominate most landscapes today, and it is their flowers that account for their unparalleled success.

Web Activity *Pine Life Cycle*

? What key adaptations have contributed to the overwhelming success of seed plants?

■ Pollen transfers sperm to eggs without the need for water. Seeds protect, nourish, and help disperse plant embryos. The reduced gametophytes of seed plants are protected on the sporophyte plant.

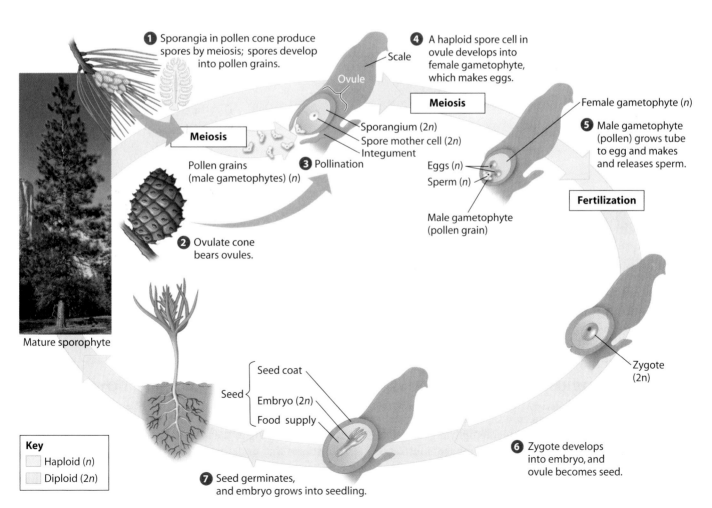

Figure 17.7 Life cycle of a pine tree

17.8 The flower is the centerpiece of angiosperm reproduction

No organisms make a showier display of their sex life than angiosperms (Figure 17.8A). From roses to cherry blossoms, flowers are the sites of pollination and fertilization. Like pine cones, flowers house separate male and female sporangia and gametophytes, and the mechanisms of sexual reproduction, including pollination and fertilization, are similar. And like cones, flowers are also short stems bearing modified leaves. However, as you can see in Figure 17.8B, the modifications are quite different from the scales of a pine cone. Each floral structure is highly specialized for a different function, and the structures are attached in a circle to a receptacle at the base of the flower. The outer layer of the circle consists of the **sepals**, which are usually green. They enclose the flower before it opens. When the sepals are peeled away, the next layer is the **petals**, which are conspicuous and attract animal pollinators.

As we'll explore further in Module 17.12, showy petals are a major reason for the overwhelming success of angiosperms.

Plucking off a flower's petals reveals the filaments of the **stamens**. The **anther**, a sac at the top of each filament, contains male sporangia and will eventually release pollen. At the center of the flower is the **carpel**, the female reproductive structure. It includes the **ovary**, a unique angiosperm adaptation that encloses the ovules. As in pines, each ovule contains a sporangium that will produce a female gametophyte and eventually become a seed. The ovary matures into a fruit, which aids in seed dispersal, as we'll discuss shortly. In the next module, you will learn how the alternation of generations life cycle proceeds in angiosperms.

? Where are the male and female sporangia located in flowering plants?

■ Anthers contain the male sporangia; ovules contain the female sporangia.

Figure 17.8A   Some examples of floral diversity: lupine (left, a cluster of flowers), orchid (center), water lily (right)

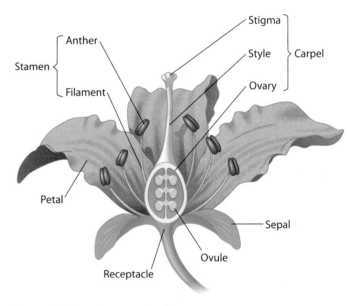

Figure 17.8B   The parts of a flower

17.9 The angiosperm plant is a sporophyte with gametophytes in its flowers

In angiosperms, as in gymnosperms, the sporophyte generation is dominant and produces the gametophyte generation within its body. Figure 17.9, on the facing page, illustrates the life cycle of a flowering plant and highlights features that have been especially important in angiosperm evolution. (We will discuss these features, as well as double fertilization in angiosperms, in more detail in Modules 31.9–31.13.)

Starting at the "Meiosis" box at the top of Figure 17.9, ❶ meiosis occurring in the anthers of the flower produces haploid spores that undergo mitosis and form the male gametophytes, or pollen grains. ❷ Meiosis in the ovule produces a haploid spore that undergoes mitosis and forms the few cells of the female gametophyte, one of which becomes an egg. ❸ Pollination occurs when a pollen grain, carried by the wind or an animal, lands on the stigma. As in gymnosperms, a tube grows from the pollen grain to the ovule, and a sperm fertilizes the egg, ❹ forming a zygote. Also as in gymnosperms, ❺ a seed develops from each ovule. Each seed consists of an embryo (a new sporophyte) surrounded by a food supply and a seed coat derived from the integuments. While the seeds develop, ❻ the ovary's wall thickens, forming the fruit that encloses the seeds. When conditions are favorable, ❼ a seed germinates. As the embryo begins to grow, it uses the food supply from the seed until it can begin to photosynthesize. Eventually it develops into a mature sporophyte, completing the life cycle.

The evolution of flowers that attract animals, which carry pollen more reliably than the wind, was a key adaptation of angiosperms. The success of angiosperms was also enhanced by their ability to reproduce rapidly. Fertilization in angiosperms usually occurs about 12 hours after pollination, making it possible for the plant to produce seeds in only a few days or weeks. As we mentioned in Module 17.7, a typical pine takes years to produce seeds. Rapid seed production is especially advantageous in harsh environments such as deserts, where growing seasons are extremely short.

Another feature contributing to the success of angiosperms is the development of fruits, which protect and help disperse the seeds, as we see in the next module.

Web Activity *Angiosperm Life Cycle*

? What is the difference between pollination and fertilization?

■ Pollination is the transfer of pollen by wind or animals from stamens to the tips of carpels. Fertilization is the union of egg and sperm, which are released from the pollen tube after the tube grows and makes contact with an ovule.

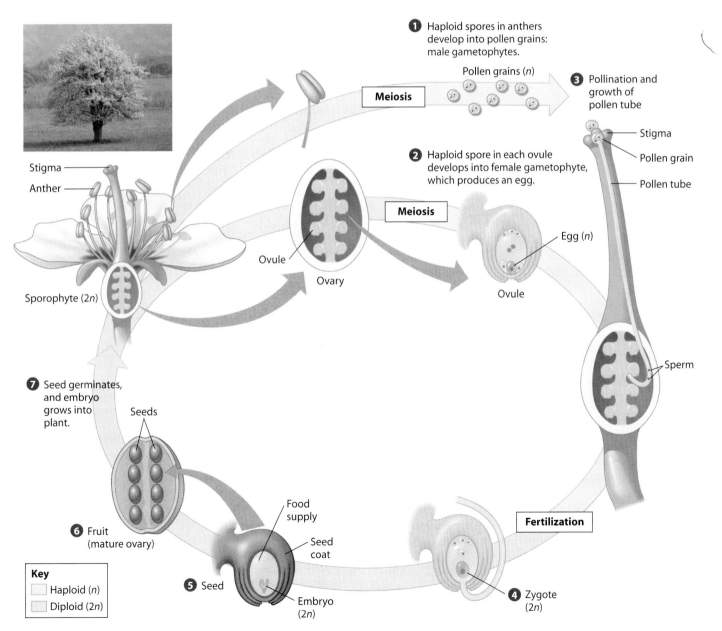

Figure 17.9 Life cycle of an angiosperm

❶ Haploid spores in anthers develop into pollen grains: male gametophytes.

Pollen grains (*n*)

Meiosis

❸ Pollination and growth of pollen tube

Stigma

Pollen grain

Pollen tube

❷ Haploid spore in each ovule develops into female gametophyte, which produces an egg.

Meiosis

Egg (*n*)

Ovule

Ovule

Ovary

Sperm

Stigma

Anther

Sporophyte (2*n*)

❼ Seed germinates, and embryo grows into plant.

Seeds

❻ Fruit (mature ovary)

Food supply

Seed coat

Fertilization

❺ Seed

Embryo (2*n*)

❹ Zygote (2*n*)

Key

☐ Haploid (*n*)

☐ Diploid (2*n*)

17.10 The structure of a fruit reflects its function in seed dispersal

A **fruit,** the ripened ovary of a flower, is an adaptation that helps disperse seeds. Some angiosperms depend on wind for seed dispersal. For example, the fruit of a maple tree (Figure 17.10A) spins like a propeller, carrying a seed away from the parent tree on wind currents. Hook-like modifications of the outer layer of the fruit or the seed coat allow some angiosperms to hitch a ride on animals. The fruits of the cocklebur plant (Figure 17.10B), for example, may be carried for miles before they open and release their seeds.

Many angiosperms produce fleshy, edible fruits that are attractive to animals as food. While the seeds are developing, these fruits are green and effectively camouflaged against green foliage. When ripe, the fruit turns a bright color such as red or yellow, advertising its presence to birds and mammals. When the catbird in Figure 17.10C eats a berry, it digests the fleshy part of the fruit, but most of the tough seeds pass unharmed through its digestive tract. The bird may then deposit the seeds, along with a supply of natural fertilizer, some distance from where it ate the fruit.

The dispersal of seeds in fruits is one of the main reasons that angiosperms are so successful. Humans have also made extensive use of fruits and seeds, as we see next.

? What is a fruit?

■ A ripened ovary of a flower, which contains, protects, and aids in the dispersal of seeds

Figure 17.10A Maple fruit can be dispersed by the wind

Figure 17.10B Cockleburs fruits may be carried by animal fur

Figure 17.10C Seeds within edible fruits are often dispersed in animal feces

Connection

17.11 Angiosperms sustain us—and add spice to our diets

We depend on the fruits and seeds of angiosperms for much of our food. Corn, rice, wheat, and the other grains, the main food sources for most of the world's people and their domesticated animals, are dry fruits. Many food crops are fleshy fruits, including apples, cherries, oranges, tomatoes, squash, and cucumbers. (In botanical terms, a fruit is an angiosperm structure containing seeds, so some vegetables are also fruits.) While most people can easily recognize grains and fleshy fruits as plant products, fewer realize that spices such as nutmeg, cinnamon, cumin, cloves, ginger, and licorice come from angiosperms. Figure 17.11 shows the source of a condiment found on most American dinner tables: black pepper. The pepper fruits are harvested before ripening, then dried and ground into powder or sold whole as "peppercorns." In medieval Europe, peppercorns were so valuable that they were used as currency. Rent and taxes could be paid in pep-percorns; as a form of wealth, peppercorns were included in dowries and left in wills. The search for a sea route to obtain pepper and other precious spices from India and southeast Asia led to the Age of Exploration and had a lasting impact on European history.

? Suppose you found a cluster of pepper berries like the ones in Figure 17.11. How would you know that they are fruits?

Figure 17.11 Berries (fruits) on *Piper nigrum*

■ Each berry has seeds inside it.

17.12 Pollination by animals has influenced angiosperm evolution

Most of us associate flowers with colorful petals and sweet fragrances, but not all flowers have these accessories. Figure 17.12A, for example, shows flowers of a red maple, which have many anthers but no petals (carpels are borne on separate flowers). Compare those flowers to the large, vibrantly colored columbine in Figure 17.12B. Such an elaborate flower costs the columbine an enormous amount of energy to produce, but the investment pays off when a pollinator, attracted by the flower's color or scent, carries the plant's pollen to another flower of the same species. Red maple, on the other hand, devotes substantial energy to making massive amounts of pollen for release into the wind, a far less certain method of pollination. Both species have adaptations to achieve pollination, which is necessary for reproductive success, but they allocate their resources differently.

Botanists estimate that about 90% of angiosperms employ animals to transfer their pollen. Birds, bats, and many different species of insects, notably bees, butterflies, moths, and beetles, serve as pollinators. These animals visit flowers in search of a meal, which the flowers provide in the form of nectar, a high-energy fluid. For pollinators, the colorful petals and alluring odors are signposts that mark food resources.

The cues that flowers offer are keyed to the sense of sight and smell of certain types of pollinators. For example, birds are attracted by bright red and orange flowers but not to particular scents, while most beetles are drawn to fruity odors but are indifferent to color. The petals of bee-pollinated flowers may be marked with guides in contrasting colors that lead to the nectar. In some flowers, the nectar guides are pigments that reflect ultraviolet light, a part of the electromagnetic spectrum that is invisible to us and most other animals, but readily apparent to bees. Flowers that are pollinated by night-flying animals such as bats and moths typically have large, light-colored, highly scented flowers that can easily be found at night. Some flowers even produce an enticing imitation of the smell of rotting flesh, thereby attracting pollinators such as carrion flies and beetles.

Many flowers have additional adaptations that improve pollen transfer, and thus reproductive success, once a pollinator arrives. The location of the nectar, for example, may manipulate the visitor's position in a way that maximizes pollen pickup and deposition. In Figure 17.12C, the pollen-bearing stamens of a scotch broom flower arch over the bee as it harvests nectar. Some of the pollen the bee picks up here will rub off onto the stigmas of other flowers it visits. In the columbine, as well as in many other flowers, the nectar can only be reached by pollinators with long tongues, a group that includes butterflies, moths, birds, and some bees.

Pollination is only effective if the pollen transfer occurs between members of the same species, but relatively few pollinators visit one species of flower exclusively. Biologists hypothesize that it takes time, through trial and error, for a pollinator to learn to extract nectar from a flower. Insects, for instance, can only remember one extraction technique at a time. Thus, pollinators are most successful at obtaining food if they visit another flower with the same cues immediately after mastering a technique for nectar extraction.

Although floral characteristics are adaptations that attract pollinators, they are a source of enjoyment to people, as well. Humans use flowering plants, including many species of trees, for a variety of purposes. But in the next module, we consider how plant diversity is threatened by some human activities.

? What type of pollinator do you think would be attracted to the columbine in Figure 17.12B?

■ Long-tongued birds, because the flower is red and has a long, tubular corolla.

Figure 17.12A Wind-pollinated flowers of red maple

Figure 17.12B Showy columbine flower

Figure 17.12C A bee picking up pollen from a scotch broom flower as it feeds on nectar

17.13 Plant diversity is an irreplaceable resource

The human population, with its demand for space and natural resources, is extinguishing plant species and other vital species at an unprecedented rate. The threat to biological diversity is especially visible in the world's forests. From the stands of conifers in North America to swampy rain forest groves in the tropics, clear-cutting, burning, and other human activities have relentlessly reduced the planet's forest cover. As trees fall, the plants and wildlife that live alongside them also perish. What is lost is irreplaceable—entire ecosystems that provide medicinal plants, food, timber, and clean water and air. More than 25% of prescription drugs are extracted from plants, and researchers have investigated fewer than 5,000 of the world's approximately 290,000 known plant species as sources of medicine. Table 17.13 lists only a few of the unique medicinal compounds derived from plants.

People have been pushing into forestlands for thousands of years, but in the last century, scientists say, the rate of global forest reduction has reached alarming levels. More than 50,000 square miles of forest—an area about the size of Lousiana—are cleared every year. Much of Europe's original forests are gone. The forests of North America, which once dominated the landscape, have shrunk by almost 40% in the last two centuries to make room for people and meet the demand for lumber and paper (Figure 17.13). Not only have many of the animals that depend on these ecosystems—bears, wolves, birds of prey—disappeared, but hardy species of trees have also been replaced by species that grow quickly but are less resilient. Timber farms on land that once sustained natural forests have been harvested and replanted with species that are economically profitable. These replanted areas have little of the biological diversity of the original forests.

TABLE 17.13	A SAMPLING OF MEDICINES DERIVED FROM PLANTS	
Compound	**Example of Source**	**Example of Use**
Atropine	Belladonna plant	Pupil dilator in eye exams
Digitalin	Foxglove	Heart medication
Menthol	Eucalyptus tree	Ingredient in cough medicines
Morphine	Opium poppy	Pain reliever
Quinine	Cinchona tree	Malaria preventive
Taxol	Pacific yew	Ovarian cancer drug
Tubocurarine	Curare tree	Muscle relaxant during surgery
Vinblastine	Periwinkle	Leukemia drug

About 20% of tropical forests were destroyed in the last third of the 20th century. About half of all the world's forests are found in the tropics, and they contain a large portion of the world's plant and animal species. The diversity of life in these forests is astonishing, and its rapid loss has escalated into a crisis.

Scientists are now rallying to stem the loss of biological diversity and to offer less destructive ways for humans to work with forests. One ambitious effort, the All Species Foundation, is seeking to catalog every species on Earth within the next 25 years, from the smallest insect to the largest forest tree. The United Nations is also working to conserve the vast majority of plant species and to develop better ways of managing forests and plants by the year 2010. The goal of such efforts is to encourage management practices that use forests as resources without damaging them.

There is little doubt that people will continue to cut forests. But the search is on for ways of harvesting forests while still preserving tree cover and protecting plants and wildlife, allowing them to be studied and used for medicines, food, and tourism. If we begin to see forests and other ecosystems as living treasures that regenerate slowly, we may learn to work with them in ways that preserve their biological diversity for the future.

In the remainder of this chapter, we look at another essential component of forests and many other ecosystems—the fungi.

Web Activity *Connection: Madagascar and the Biodiversity Crisis*

? Why does the loss of forests lead to reduced biological diversity?

Figure 17.13 Clear-cutting in Alaska

■ Many different species depend on forest ecosystems. When forests are cleared, animal species such as bears, wolves, and birds are threatened along with trees and other plants.

17.14 Fungi absorb food after digesting it outside their bodies

Members of the kingdom **Fungi** have body structures and modes of reproduction unlike those of any other organism. Although they are heterotrophs like animals, fungi don't eat (ingest) their food, but rather they acquire their nutrients by **absorption.** Fungi secrete powerful enzymes that digest their food outside their bodies and then absorb the small nutrient molecules into their cells.

Discussing fungi in the same chapter as plants may seem to indicate that these two kingdoms are close relatives. Actually, fungi are more closely related to animals than to plants. But the success of plants on land and the great diversity of fungi are interconnected. Mycorrhizae, the associations between plant roots and fungi that we discussed in the chapter introduction, helped make the colonization of land possible for both plants and fungi. Fungi cannot make their own food and must obtain their organic molecules from other organisms. Before plants colonized land and began stocking soil with organic molecules, fungi may have thrived only in aquatic environments. And without fungi as mycorrhizal partners and as decomposers that return vital nutrients to the soil, plants could not have survived on land.

Fungi are found virtually everywhere, both in soil and in water, and are essential decomposers in most ecosystems. The orange umbrellas in Figure 17.14A are the reproductive structures of a fungus that is breaking down a dead log. Not all fungi are beneficial, however. Some are **parasites,** obtaining their nutrients at the expense of living plants or animals.

A typical fungus consists of threadlike filaments that are called **hyphae** (singular, *hypha*). Hyphae branch repeatedly, forming a feeding network known as a **mycelium** (plural, *mycelia*), illustrated in Figure 17.14B. The photograph in Figure 17.14C shows the mycelium of a fungus growing on decaying leaves. The mushrooms in Figure 17.14A, solid as they seem, are actually made of tightly packed hyphae. Mushrooms arise from a mycelium that is usually hidden from view. While the mycelium is obtaining food from organic material in the soil or a rotting log, the function of the mushroom is reproduction. It must be aboveground in order to disperse its spores on air currents.

Fungal hyphae are surrounded by a cell wall. Unlike plants, which have cellulose cell walls, most fungi have cell walls made of chitin, a strong, flexible nitrogen-containing polysaccharide, identical to the chitin found in the external skeletons of insects. In most fungi, the hyphae consist of chains of cells separated by cross-walls that have pores large enough to allow ribosomes, mitochondria, and even nuclei to flow from cell to cell. Some fungi lack cross-walls entirely and have many nuclei within a single mass of cytoplasm.

Fungi cannot run or fly in search of food. But their mycelium makes up for the lack of mobility by being able to grow at a phenomenal rate, branching throughout a food source and extending its hyphae into new territory. Because its

Figure 17.14A Fungi decomposing a log

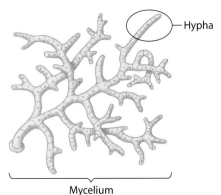

Figure 17.14B A mycelium, made of numerous hyphae

Figure 17.14C White, threadlike mycelium of a fungus growing on decaying leaves

hyphae grow longer without getting thicker, the fungus develops a huge surface area from which it can secrete digestive enzymes and through which it can absorb food. A mycelium can add as much as a kilometer of hyphae each day. The mycelium of one giant fungus in Oregon spreads through 890 hectares (2,200 acres) of forest (equivalent to over 1,600 football fields). Scientists estimate that this fungus has been growing for 2,600 years, expanding outward in search of food. Next we consider how fungi reproduce.

Web Process of Science *How Does the Fungus* Pilobolus *Succeed as a Decomposer?*

? Contrast how fungi digest and absorb their food with your own digestion.

■ A fungus digests its food externally by secreting enzymes onto the food and then absorbing the small nutrients that result from digestion. In contrast, humans and most other animals eat relatively large pieces of food and digest the food within their bodies.

17.15 Fungi produce spores in both asexual and sexual life cycles

Many fungal species can reproduce either sexually or asexually (Figure 17.15). Reproduction typically involves the release of vast numbers of haploid spores, which are transported easily over great distances by wind or water. Fungal spores have been found more than 160 km above Earth. A spore that lands in a moist place with food germinates and produces a new fungus.

In the sexual reproduction of many fungi, two haploid mycelia of different mating types release sexual signaling molecules, grow toward each other, and fuse. But this cytoplasmic fusion is often not followed immediately by the fusion of "parental" nuclei. Thus, many fungi have what is called a **heterokaryotic stage** (from the Greek, meaning "different nuclei"), in which cells contain two genetically distinct haploid nuclei. Hours, days, or even centuries may pass before the parental nuclei fuse, forming the usually short-lived diploid phase. Zygotes undergo meiosis inside specialized reproductive structures, from which haploid spores are dispersed.

Many fungi, including those commonly called molds and yeasts, typically reproduce asexually. The term **mold** refers to any rapidly growing fungus that reproduces asexually by producing spores, often at the tips of specialized hyphae. These familiar furry carpets often appear on aging fruit, bread, and other foods. The term **yeast** refers to any single-celled fungus that reproduces asexually by cell division or by budding—pinching off small "buds" from a parent cell. Yeasts inhabit liquid or moist habitats, such as plant sap and animal tissues.

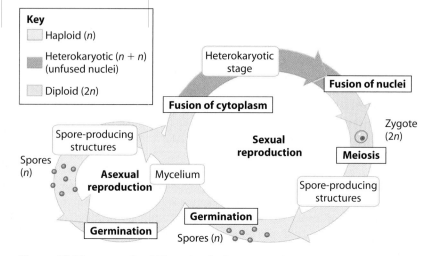

Figure 17.15   Generalized life cycle of a fungus

We will examine fungal reproduction more closely in Module 17.17, after we look at the classification of fungi.

? What is the heterokaryotic stage of a fungus?

■ The stage in which each cell has two different nuclei (from two different parents), with the nuclei not yet fused

17.16 Fungi are classified into five groups

Biologists who study fungi have described over 100,000 species, and there may be as many as 1.5 million. Sexual reproductive structures are often used to classify species. Fungi that have no known sexual stage, including many molds and yeasts, are informally known as **imperfect fungi.** Increasingly, scientists use molecular data to try to re-create the phylogeny of fungi.

All but one group of fungi lack flagella. Phylogenetic systematics, however, suggests that fungi evolved from a flagellated ancestor shared with animals. Based on molecular clock analysis (see Module 15.18), scientists estimate that the ancestors of animals and fungi diverged into separate lineages 1.5 billion years ago. The oldest undisputed fossils of fungi, however, are only about 460 million years old, perhaps because the ancestors of terrestrial fungi were microscopic and fossilized poorly.

Figure 17.16A shows a current hypothesis of fungal phylogeny. The thin lines leading to the chytrids indicate that this group is probably not monophyletic. Though there are still areas of uncertainty, most biologists recognize five groups of fungi.

The **chytrids,** the only fungi with flagellated spores, are thought to represent the earliest lineage of fungi. They are com-

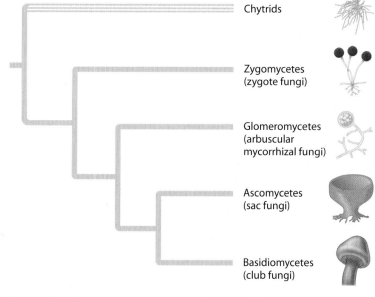

Chytrids

Zygomycetes
(zygote fungi)

Glomeromycetes
(arbuscular
mycorrhizal fungi)

Ascomycetes
(sac fungi)

Basidiomycetes
(club fungi)

Figure 17.16A   A proposed phylogenetic tree of fungi

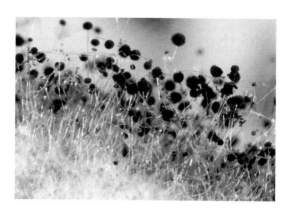

Figure 17.16B Zygomycete: *Rhizopus stolonifer,* the black bread mold

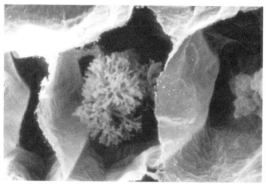

SEM 6,500×

Figure 17.16C Glomeromycete: arbuscule in a root cell

mon in lakes, ponds, and soil. Some species are decomposers; others parasitize protists, plants, or animals. Some researchers have linked the widespread decline of amphibian species to a highly infectious fungal disease caused by a species of chytrid. For example, populations of frogs in mountainous regions of Central America and Australia have suffered massive mortality from this emerging disease.

The **zygomycetes,** or **zygote fungi,** are characterized by their resistant zygosporangium, in which haploid spores form by meiosis. This diverse group includes fast-growing molds, such as black bread mold (Figure 17.16B) and molds that rot produce such as peaches, strawberries, and sweet potatoes. Some zygote fungi are parasites on animals.

The **glomeromycetes** (from Latin *glomer-,* ball) form a distinct type of mycorrhiza in which hyphae that invade plant roots branch into tiny treelike structures known as arbuscules (Figure 17.16C). About 90% of all plants have such symbiotic partnerships with glomeromycetes, which deliver phosphate and other minerals to plants while receiving organic nutrients in exchange.

The **ascomycetes,** or **sac fungi,** are named for saclike structures called asci (from Greek *asco-,* pouch) that produce spores in sexual reproduction. They live in a variety of marine, freshwater, and terrestrial habitats and range in size from unicellular yeasts to elaborate morels and cup fungi (Figure 17.16D). Ascomycetes include some of the most devastating plant pathogens. Other species of ascomycetes live with green algae or cyanobacteria in symbiotic associations called lichens, which we discuss in Module 17.19.

The **basidiomycetes,** or **club fungi,** are probably the most familiar fungi—the mushrooms, the puffballs, and the shelf fungi (Figure 17.16E). They are named for their club-shaped, spore-producing structure, called a basidium (meaning "little pedestal" in Latin, plural, *basidia;* see Figure 17.17B). Many species excel at breaking down the lignin found in wood and thus play key roles as decomposers. For example, shelf fungi often break down the wood of weak or damaged trees and continue to decompose the wood after the tree dies. The basidiomycetes also include two groups of particularly destructive plant parasites, the rusts and smuts, which we discuss in Module 17.18.

In the next module, we explore the life cycles of some representative fungi.

Figure 17.16D Ascomycetes: edible morels (left) and cup fungus (right)

Figure 17.16E Basidiomycetes (club fungi): mushrooms (left), puffball (center), and shelf fungi (right)

? What is one reason that chytrids are thought to have diverged earliest in fungal evolution?

■ Chytrids are the only fungi that have flagellated spores, a characteristic of the ancestor of fungi.

17.17 Fungal groups differ in their life cycles and reproductive structures

The life cycle of black bread mold (see Figure 17.16B) is typical of zygomycetes. As hyphae expand through its food, the fungus reproduces asexually, forming spores in sporangia at the tips of upright hyphae. When the food is depleted, the fungus reproduces sexually. As shown in Figure 17.17A, mycelia of different mating types ❶ join and produce ❷ a cell containing nuclei from two parents. This young zygosporangium develops into a thick-walled structure ❸ that can tolerate dry or harsh environments. When conditions are favorable, the parental nuclei fuse, and the diploid nucleus undergoes meiosis, ❹ forming haploid spores.

Like the zygomycetes, ascomycetes, such as those shown in Figure 17.16D reproduce asexually when conditions are suitable. In many species, sexual reproduction takes place in the fall, with the haploid spores maturing in the early spring. The genetic diversity of these sexually produced spores increases the likelihood that at least one genotype will successfully establish itself in the environment encountered in the new season. The surviving individuals then reproduce asexually for several generations through the spring and summer before undergoing sexual reproduction.

Now let's follow the life cycle of a mushroom, a basidiomycete, starting at the center bottom of Figure 17.17B. ❶ The heterokaryotic stage begins when hyphae of two different mating types fuse, forming ❷ a heterokaryotic mycelium, which grows and produces the mushroom. In the club-shaped cells called basidia, which line the gills of a mushroom, haploid nuclei fuse, forming ❸ diploid nuclei. Each diploid nucleus undergoes meiosis, and ❹ haploid spores are formed. A mushroom can release as many as a billion spores. If spores land on

moist matter that can serve as food, ❺ they germinate and grow into haploid mycelia.

Much of the success of fungi is due to their reproductive capacity, both in the asexual production of spores and in sexual reproduction, which increases genetic variation and facilitates adaptation to harsh or changing conditions.

Web Activity *Fungal Reproduction and Nutrition*

Web Activity *Fungal Life Cycles*

? Under what conditions is asexual reproduction advantageous? Under what conditions is sexual reproduction advantageous?

■ In asexual reproduction, a successful genotype can be propagated to take advantage of consistent, favorable conditions. Sexual reproduction reshuffles alleles, providing numerous genotypes that will be "tested" by the environment. This is an advantage when environmental conditions are changing.

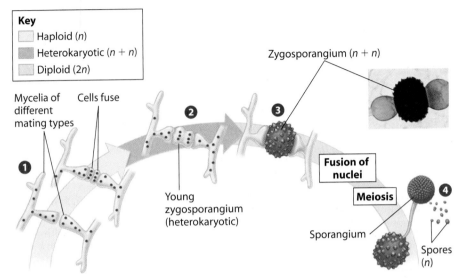

Key
- ☐ Haploid (*n*)
- ☐ Heterokaryotic (*n* + *n*)
- ☐ Diploid (2*n*)

Mycelia of different mating types
Cells fuse
Zygosporangium (*n* + *n*)
Young zygosporangium (heterokaryotic)
Fusion of nuclei
Meiosis
Sporangium
Spores (*n*)

Figure 17.17A   Sexual reproduction of a zygote fungus

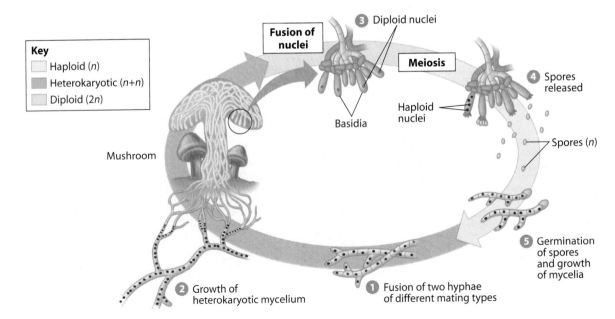

Figure 17.17B
Life cycle of a mushroom

Key
- ☐ Haploid (*n*)
- ☐ Heterokaryotic (*n*+*n*)
- ☐ Diploid (2*n*)

Mushroom
Fusion of nuclei
❸ Diploid nuclei
Meiosis
❹ Spores released
Basidia
Haploid nuclei
Spores (*n*)
❺ Germination of spores and growth of mycelia
❷ Growth of heterokaryotic mycelium
❶ Fusion of two hyphae of different mating types

17.18 Parasitic fungi harm plants and animals

Of the 100,000 known species of fungi, about 30% make their living as parasites or pathogens, mostly in or on plants. In some cases, fungi have literally changed the landscape. In 1926, a fungus that causes Dutch elm disease was accidentally introduced into the United States on logs sent from Europe to make furniture. (The name refers to The Netherlands, where the disease was first identified—the fungus originated in Asia.) Over the course of several decades, the fungi destroyed around 70% of elm trees across the eastern United States. English elms (a different species), such as those in Figure 17.18A, fared even worse. They were completely annihilated. Recently, scientists studying the DNA of English elms have found evidence that they were all genetically identical, derived by asexual reproduction from a single ancestor brought by the Romans 2,000 years ago. As a result, they were all equally susceptible to the ravages of Dutch elm disease.

Fungi are a serious problem as agricultural pests. Crop fields typically contain genetically identical individuals of a single species planted close together—ideal conditions for the spread of disease. About 80% of plant diseases are caused by fungi, causing tremendous economic losses each year. Between 10% and 50% of the world's fruit harvest is lost each year to fungal attack. A variety of fungi, including smuts and rusts, are common on grain crops. The ear of corn shown in Figure 17.18B is infected with a club fungus called corn smut. The grayish growths, known as galls, are made up of heterokaryotic hyphae that invade a developing corn kernel and eventually displace it. When a gall matures, it breaks open and releases thousands of blackish spores. In parts of Central America, the smutted ears are cooked and eaten as a delicacy, but generally corn smut is regarded as a scourge.

Some of the fungi that attack food crops are toxic to humans. The seed heads of many kinds of grain, including rye, wheat, and oats, are sometimes infected with fungal growths called ergots, the dark structures on the seed head of rye shown in Figure 17.18C. Consumption of flour made from ergot-infested grain can cause gangrene, nervous spasms, burning sensations, hallucinations, temporary insanity, and death. Several toxins have been isolated from ergots. One, called lysergic acid, is the raw material from which the hallucinogenic drug LSD is made. Certain others are medically useful in small doses. For instance, an ergot compound is useful in treating high blood pressure and in stopping maternal bleeding after childbirth.

Animals are much less susceptible to parasitic fungi than are plants. Only about 50 species of fungi are known to be parasitic in humans and other animals. In humans, fungi cause infections ranging from annoyances such as athlete's foot to deadly lung diseases. The general term for a fungal infection is **mycosis.** Skin mycoses include the disease called ringworm, so named because it appears as circular red areas on the skin. The ringworm fungus can infect virtually any skin surface. Most commonly, it attacks the feet, causing the intense itching and sometimes blistering known as athlete's foot. Systemic mycoses are fungal infections that spread throughout the body, usually from spores that are inhaled. These can be very serious diseases. Coccidioidomycosis is a systemic mycosis that produces tuberculosis-like symptoms in the lungs. It is so deadly that it is now considered a potential biological weapon.

The yeast that causes vaginal infections (*Candida albicans*) is an example of an opportunistic pathogen—a normal inhabitant of the body that causes problems only when some change in the body's microbiology, chemistry, or immunology allows the yeast to grow unchecked. Many other opportunistic mycoses have increased in recent decades, in part because of AIDS, which compromises the immune system.

In the next two modules, we look at close associations between fungi and other organisms.

? What is a mycosis? What is an opportunistic pathogen?

Figure 17.18B Corn smut

Figure 17.18C
Ergots on rye

Ergots

Figure 17.18A   Stately English elms in Australia, unaffected by Dutch elm disease

■ A mycosis is a fungal infection. An opportunistic pathogen is a normal inhabitant of the body that grows out of control when there is a change in the body's microbiology, chemistry, or immunology.

17.19 Lichens consist of fungi living in close association with photosynthetic organisms

The rock in Figure 17.19A below is covered with a living crust of **lichens,** associations of millions of green algae or cyanobacteria held in a mass of fungal hyphae. As Figure 17.19B shows, the fungal hyphae are wrapped tightly around their partners. The merger is so complete that lichens are actually named as species, as though they were single organisms. In many lichens that have been studied, each partner provides something the other could not obtain on its own. The fungus receives food from its photosynthetic partner. The fungal mycelium, in turn, provides a suitable habitat for the alga, helping it to absorb and retain water and minerals. In other species of lichen, it is not clear whether the relationship benefits the alga or is only advantageous to the fungus.

Lichens are rugged and able to live where there is little or no soil. As a result, they are important pioneers on new land. Lichens grow into tiny rock crevices, where the acids they secrete help to break down the rock to soil, paving the way for future plant growth. Some lichens can tolerate severe cold, and

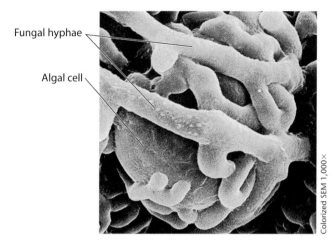

Fungal hyphae

Algal cell

Colorized SEM 1,000×

Figure 17.19B The close relationship between fungal and algal partners in a lichen

carpets of them cover the arctic tundra. Caribou feed on lichens in their winter feeding grounds in Alaska.

Lichens can also withstand severe drought. They are opportunists, growing in spurts when the conditions are favorable. When it rains, a lichen quickly absorbs water and photosynthesizes at a rapid rate. In dry air, it dehydrates and photosynthesis may stop, but the lichen remains alive more or less indefinitely. Some lichens are thousands of years old, rivaling the oldest plants and fungi as the oldest organisms on Earth.

As tough as lichens are, many do not withstand air pollution. Because they get most of their minerals from the air, in the form of dust or compounds dissolved in raindrops, lichens are very sensitive to airborne pollutants such as sulfur dioxide. The death of lichens is often a sign that air quality in an area is deteriorating.

? What benefits do algae in lichens receive from their fungal partners?

Figure 17.19A Several of the 200 to 300 species of lichen that live in Antarctica

■ A suitable habitat for growth; absorption and retention of water and minerals

17.20 Some fungi have mutually beneficial relationships with ants

As decomposers, many species of fungi have powerful enzymes that break down plant material. Some animals, including several species of ants and termites, take advantage of the cellulose-digesting enzymes of fungi by raising them in "farms." For example, Central American leaf-cutting ants construct huge underground nests with chambers in which they cultivate fungal gardens. The ants scour tropical forests in search of leaves. A colony of ants, often well over a million individuals, can strip a large tree of its foliage in a single night, carrying it back to the nest (Figure 17.20). In the nest, fungi feed on the leaves, using their enzymes to break down cellulose, which ants cannot digest. The ants harvest the swollen tips of hyphae as their food. Some members of the colony

Figure 17.20 Leaf-cutting ants carrying leaves to feed fungi

"weed" undesirable fungi out of the fungal gardens. When a queen ant establishes a new colony, she takes fungal hyphae along in a pouch in her mouth.

Such farmer insects and their fungal "crops" have been evolving together for well over 50 million years. Some of these fungi have become so dependent on their caretakers that they can no longer survive without the ants.

17.21 Fungi have enormous ecological benefits and practical uses

As you have read, fungi have been major players in terrestrial communities ever since they moved onto land in the company of plants. As symbiotic partners in mycorrhizae (see the chapter introduction), fungi supply essential nutrients to plants and are enormously important in natural ecosystems and agriculture.

Fungi, along with prokaryotes, are essential decomposers in ecosystems, breaking down organic matter and restocking the environment with vital nutrients essential for plant growth. The air is so full of fungal spores that as soon as a leaf falls or an insect dies, it is covered with spores and is soon infiltrated by fungal hyphae. If fungi and prokaryotes in a forest suddenly stopped decomposing, leaves, logs, feces, and dead animals would pile up on the forest floor. Plants and the animals they feed would starve because elements taken from the soil would not be replenished through decomposition.

Almost any organic (carbon-containing) substance can be consumed by fungi. During World War II, the moist tropical heat of Southeast Asia and islands in the Pacific Ocean provided ideal conditions for fungal decomposition of wood and organic fibers such as canvas and cotton. Packing crates, military uniforms, and tents quickly disintegrated, causing supply problems for the military forces. But decomposition of synthetic substances is beneficial in some situations. Some fungi can break down toxic pollutants, including the pesticide DDT and certain chemicals that cause cancer. Scientists are also investigating the possibility of using fungi that can digest petroleum products to clean up oil spills and other chemical messes.

Fungi have a number of practical uses for humans. Most of us have eaten mushrooms, although we may not have realized that we were ingesting reproductive structures of subterranean fungi. The distinctive flavors of certain cheeses, including

Roquefort and blue cheese (Figure 17.21A), come from fungi used to ripen them. Highly prized by gourmets are truffles, which are produced by certain mycorrhizal fungi associated with tree roots. Humans have used yeasts to produce alcoholic beverages and cause bread to rise for thousands of years.

Fungi are medically valuable as well. Like the bacteria called actinomycetes (see Module 16.7), some fungi produce antibiotics that we use to treat bacterial diseases. In fact, the first antibiotic discovered was penicillin, which is made by the common mold called *Penicillium*. In Figure 17.21B, the clear area between the mold and the bacterial growth is where the antibiotic produced by *Penicillium* has inhibited the growth of the bacteria (*Staphylococcus aureus*).

Fungi also figure prominently in research in molecular biology and in biotechnology. Researchers often use yeasts to study the molecular genetics of eukaryotes because they are easy to culture and manipulate. Yeasts have been genetically modified to produce human proteins. And researchers are studying fungi that may be used to convert plant biomass to ethanol for fuel.

Fungi are the third group of eukaryotes we have surveyed so far. (Protists and plants were the first two groups of eukaryotes we discussed.) Strong evidence suggests that they evolved from protistan ancestors that also gave rise to the fourth and most diverse group of eukaryotes, the animals, which we study next.

Figure 17.21A Blue cheese

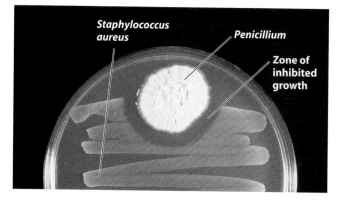

Staphylococcus aureus

Penicillium

Zone of inhibited growth

Figure 17.21B A culture of *Penicillium* and bacteria

Reviewing the Concepts

Mycorrhizae, mutually beneficial associations of plant roots and fungi, are common and may have helped the first plants adapt to land (**Introduction**).

Plant Evolution and Diversity (17.1–17.2)

Green algal ancestors. Molecular, physical, and biochemical evidence indicate that the green algae called charophytes are the closest living relatives of plants (**17.1**).

Terrestrial adaptations of plants. Plants are multicellular photosynthetic eukaryotes, and most have the following adaptations for living on land (**17.1**):

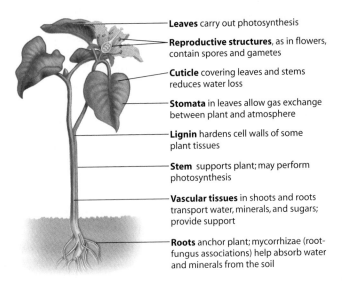

Leaves carry out photosynthesis

Reproductive structures, as in flowers, contain spores and gametes

Cuticle covering leaves and stems reduces water loss

Stomata in leaves allow gas exchange between plant and atmosphere

Lignin hardens cell walls of some plant tissues

Stem supports plant; may perform photosynthesis

Vascular tissues in shoots and roots transport water, minerals, and sugars; provide support

Roots anchor plant; mycorrhizae (root-fungus associations) help absorb water and minerals from the soil

Plant diversity. Nonvascular plants (bryophytes) include the mosses, hornworts, and liverworts. Vascular plants have supportive conducting tissues. Ferns are seedless vascular plants with flagellated sperm. Seed plants have sperm–transporting pollen grains and protect embryos in seeds. Gymnosperms, such as pines, produce seeds in cones. The seeds of angiosperms develop within protective ovaries (**17.2**).

Alternation of Generations and Plant Life Cycles (17.3–17.13)

Alternation of generations. The haploid gametophyte produces eggs and sperm by mitosis. The zygote develops into the diploid sporophyte, in which meiosis produces haploid spores. Spores grow into gametophytes (**17.3**).

Moss life cycle. A mat of moss is mostly gametophytes, which produce eggs and swimming sperm. The zygote develops on the gametophyte into the smaller sporophyte (**17.4**).

Fern life cycle. The sporophyte is the dominant generation. Sperm, produced by the gametophye, swim to the egg (**17.5**). Ferns and other seedless plants once dominated ancient forests; their remains formed coal (**17.6**).

Conifer life cycle. A pine tree is a sporophyte; tiny gametophytes grow in its cones. A sperm from a pollen grain fertilizes an egg in the female gametophyte. The zygote develops into a sporophyte embryo, and the ovule becomes a seed, with stored food and a protective coat (**17.7**).

Angiosperms have independent sporophytes with tiny gametophytes protected in flowers, which usually consist of sepals, petals, stamens (produce pollen), and carpels (produce ovules) (**17.8–17.9**). Ovules become seeds, and ovaries become fruits. Angiosperms provide most of our food (**17.10–17.11**).

Plant diversity has been influenced by interactions with animals and is an irreplaceable resource (**17.12–17.13**).

Fungi (17.14–17.21)

Fungal nutrition and body structure. Fungi are heterotrophic eukaryotes that digest their food externally and absorb the resulting nutrients. A fungus usually consists of a mass of threadlike hyphae, called a mycelium (**17.14**).

Fungal reproduction usually includes the production of spores. In some fungi, fusion of haploid hyphae produces a heterokaryotic stage containing nuclei from two parents. After the nuclei fuse, meiosis produces haploid spores (**17.15**).

Fungal phylogeny. Fungi evolved from an aquatic, flagellated ancestor. Fungal groups include chytrids, zygomycetes, glomeromycetes, ascomycetes, and basidiomycetes (**17.16**).

Fungal life cycles often include both asexual and sexual stages. Fungal groups have characteristic reproductive structures (**17.17**).

Parasitic fungi cause 80% of plant diseases and some serious human mycoses (**17.18**).

Mutually beneficial associations. Lichens consist of algae or cyanobacteria within a fungal network. Some animals benefit from the digestive abilities of fungi (**17.19–17.20**).

Ecological and other benefits. Fungi are essential decomposers and provide food and antibiotics (**17.21**).

Connecting the Concepts

1. In this abbreviated diagram, identify the four major plant groups and the key terrestrial adaptation associated with each of the three major branch points.

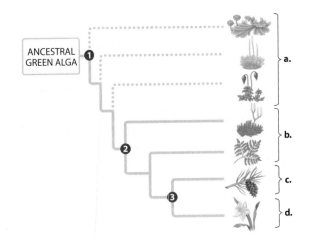

2. Identify the cloud seen in each photograph. Describe the life cycle events associated with each cloud.

A. Pine tree, a gymnosperm

B. Puffball, a club fungus

Testing Your Knowledge

Multiple Choice

3. Angiosperms are different from all other plants because only they have
   a. a vascular system.
   b. flowers.
   c. a life cycle that involves alternation of generations.
   d. seeds.
   e. a dominant sporophyte phase.

4. Which of the following produce eggs and sperm? (*Explain your answer.*)
   a. the sexual reproductive structures of a fungus
   b. fern sporophytes
   c. moss gametophytes
   d. the anthers of a flower
   e. moss sporangia

5. The eggs of seed plants are fertilized within ovules, and the ovules then develop into
   a. seeds.
   b. spores.
   c. gametophytes.
   d. fruit.
   e. sporophytes.

6. The diploid sporophyte stage is dominant in the life cycles of all of the following except
   a. a pine tree.
   b. a dandelion.
   c. a rose bush.
   d. a fern.
   e. a moss.

7. Under a microscope, a piece of a mushroom would look most like
   a. jelly.
   b. a tangle of string.
   c. grains of sugar or salt.
   d. a piece of glass.
   e. foam.

8. Which of the following is an opportunistic pathogen that can cause a mycosis?
   a. HIV, the AIDS virus
   b. the fungus that produces ergots on rye, which can cause serious symptoms if milled into flour
   c. smuts, serious pathogens of grain crops
   d. coccidioidomycosis, a deadly systemic infection
   e. *Candida albicans*, which causes vaginal yeast infections

9. Which of the following terms includes all the others?
   a. angiosperm
   b. gymnosperm
   c. vascular plant
   d. fern
   e. seed plant

10. Which of the following is a plant with flagellated sperm and a sporophyte-dominated life cycle?
    a. chytrid
    b. moss
    c. charophyte
    d. fern
    e. liverwort

Describing, Comparing, and Explaining

11. Compare a seed plant with an alga in terms of adaptations for life on land versus life in the water.

12. How do animals help flowering plants reproduce? How do the animals benefit?

13. Why are fungi and plants classified in different kingdoms?

Applying the Concepts

14. Many fungi produce antibiotics, such as penicillin, which are valuable in medicine. But of what value might the antibiotics be to the fungi? Similarly, fungi often produce compounds with unpleasant tastes and odors as they digest their food. What might be the value of these chemicals to the fungi? How might production of antibiotics and odors have evolved?

15. In April 1986, an accident at a nuclear power plant in Chernobyl, Ukraine, scattered radioactive fallout for hundreds of miles. In assessing the biological effects of the radiation, researchers found mosses to be especially valuable as organisms for monitoring the damage. As mentioned in Module 10.16, radiation damages organisms by causing mutations. Explain why it is faster to observe the genetic effects of radiation on mosses than on plants from other groups. Imagine that you are conducting tests shortly after a nuclear accident. Using potted moss plants as your experimental organisms, design an experiment to test the hypothesis that the frequency of mutations decreases with the organism's distance from the source of radiation.

16. Much of the conifer forest in the U.S. Pacific Northwest has been clear-cut; less than 10% of the original ancient forest, dominated by giant firs and hemlocks, remains. There is no law protecting endangered habitats, so to protect the northern spotted owl, which lives only in old-growth conifers, conservationists sued to stop logging under the Endangered Species Act. The lawsuits halted logging in many national forest areas. Lumber companies buy trees from national forests, loggers work there, and the economies of many small communities depend on logging. The reduction in timber supply has driven up the cost of lumber. Imagine that it is your task to deal with this situation. What are the opposing issues? What would you suggest to resolve this conflict, and how would you defend your policy?

Answers to all questions can be found in Appendix 4.

For Practice Quizzes, MP3 Tutors, and Activities, go to www.mybiology.com.

chapter 18

The Evolution of Invertebrate Diversity

Fatal or Fake?

The shallow rock pools of Sydney Harbor, Australia, offer a fascinating glimpse of invertebrate diversity. The tentacles of colorful sea anemones sway lazily in the water. A pencil urchin, wedged into a crevice, waits out the daylight hours for feeding time to roll around. An elephant snail grazes briskly along the underside of a rock, its jet-black, sluglike body capped by an incongruously small white shell. A shrimp darts out from behind a rock but quickly retreats to the safety of its hiding place. The large blue sea star draped over a rock would be spotted easily by a casual observer, but it takes a sharp eye to see its tiny cousin, camouflaged against the background and surrounded by its even tinier offspring. It would also take a keen eye to see the pale yellow-brown octopus the size of a golf ball that is loitering inconspicuously on the bottom of the pool. But this octopus changes color in a flash when threatened. A sudden fierce glow of sapphire-blue circles proclaims its identity—a blue-ringed octopus, one of the deadliest creatures in the ocean.

Blue-ringed octopus is a common name for members of the genus *Hapalochlaena*, which live in shallow reefs and tide pools from Japan to Australia. It is a daytime hunter, moving among the rocks of the tide pool in search of its favorite prey, shrimp and small crabs. Because its body contains no hard parts—no bones, no shell—it squeezes easily through narrow spaces, its tentacles snaking around rocks and exploring crevices. Its behavior is typical of octopuses, but its method of killing its prey is not. The blue-ringed octopus injects poison with a sharp bite of its beak-like mouth or by simply pumping the poison into the prey's vicinity. The poison it uses for killing prey, however, is different from the more potent substance it uses for self-defense—tetradotoxin, or TTX, a toxin 10,000 times more lethal than cyanide.

Predators, including humans, have little to fear from most octopuses, which tend to rely on nonaggressive defense mechanisms such as camouflage. Octopuses are generally reclusive, and most can expel an ink-like fluid that confuses a would-be predator, enabling the octopus to escape. The fascinating mimic octopus (*Thaumoctopus mimicus*), however, employs a completely different method of self-defense. An inhabitant of the Indian

Facing Page/Top: Blue-ringed octopus, about 7 cm from tentacle tip to tentacle tip

Facing Page/Bottom: Mimic octopus in natural colors and posture (may be as large as 60 cm)

Ocean just off the coast of Indonesia, the mimic octopus rests out in the open, on the seafloor. When threatened, it transforms itself to resemble a dangerous creature such as the highly venomous banded sea snake. You can see the octopus's imitation of the sea snake above. While two tentacles carry out the deception, the other six tentacles are tucked into a burrow in the sediment. The striking light/dark bands are part of the disguise. Like most other cephalopods (the group that includes octopuses and squids), the mimic octopus can change its color pattern, which is usually a dull brown.

The deadly sea snake is not the only imitation this cephalopod can do. Skimming the seafloor with its tentacles swept back and held together, the mimic octopus passes as a toxic flatfish. With tentacles extended at various angles, its posture and behavior resemble a lionfish brandishing its poisonous spines. Its repertoire also includes a sea anemone and a jellyfish, both of which can deliver a painful sting. Although the antipredator defenses of many animals include deceptive behaviors or markings (see Module 18.12 for some insect examples), the mimic octopus is the only animal known to play multiple roles.

Many species of cephalopod employ an impressive color palette for communication and camouflage and display a remarkable assortment of behaviors. With their shifting colors and fluid bodies, these are among the most intriguing members of the kingdom Animalia. But you can decide for yourself. This chapter is a brief tour of the vast diversity found in the animal kingdom—diversity that has been evolving for perhaps a billion years. We will encounter a spectacular variety of forms ranging from corals to cockroaches to chordates. We will look at 9 major phyla of the roughly 35 phyla in the kingdom. And we will see that identifying, classifying, and arranging this diversity remain a work in progress. But first let's define what an animal is! ■ ■ ■

18.1 What is an animal?

Animals are multicellular, heterotrophic eukaryotes that (with a few exceptions) obtain nutrients by ingestion. Now that's a mouthful—and speaking of mouthfuls, Figure 18.1A shows a rock python just beginning to ingest a gazelle. **Ingestion** means eating food. This mode of nutrition contrasts animals with fungi, which absorb nutrients after digesting food outside their body. Animals digest food within their body after ingesting other organisms, dead or alive, whole or by the piece.

Animals also have other distinctive features. Animal cells lack the cell walls that provide strong support in the bodies of plants and fungi. Animal cells are held together by extracellular structural proteins and by unique types of intercellular junctions (see Module 4.21). And most animals have muscle cells for movement and nerve cells for conducting impulses.

Other unique features are seen in animal life cycles. Most animals are diploid and reproduce sexually; eggs and sperm are the only haploid cells, as shown in the life cycle of a sea star in Figure 18.1B. ❶ Male and female adult animals make haploid gametes by meiosis, and ❷ an egg and a sperm fuse, producing a zygote. ❸ The zygote divides by mitosis, forming ❹ an early embryonic stage called a **blastula,** which is usually a hollow ball of cells. ❺ In the sea star and most other animals, one side of the blastula folds inward, forming a stage called a **gastrula.** ❻ The internal sac formed by gastrulation becomes the digestive tract, lined by a cell layer called the **endoderm.** The embryo also has an **ectoderm,** an outer cell layer that gives rise to the outer covering of the animal and, in some phyla, to the central nervous system. Most animals have a third embryonic layer, known as the **mesoderm,** that forms the muscles and most internal organs.

After the gastrula stage, many animals develop directly into adults. Others, such as the sea star, develop into one or more larval stages first. ❼ A **larva** is an immature individual that

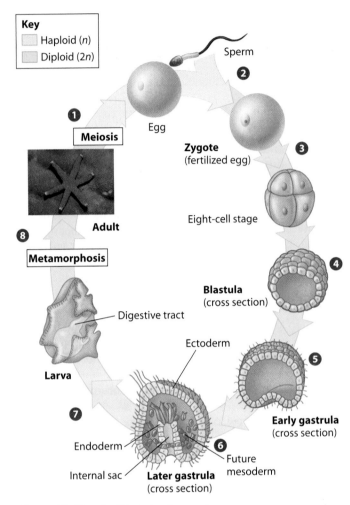

Key
- Haploid (*n*)
- Diploid (2*n*)

Sperm ❷

❶ Egg

Meiosis

Zygote (fertilized egg) ❸

Eight-cell stage

Adult

Blastula (cross section) ❹

Metamorphosis ❽

Digestive tract

Ectoderm

Larva

Early gastrula (cross section) ❺

❼

Endoderm

Future mesoderm

Internal sac

Later gastrula (cross section) ❻

Figure 18.1B The life cycle of a sea star

looks different from the adult animal. The larva undergoes a major change of body form, called **metamorphosis,** ❽ in becoming an adult capable of reproducing sexually.

This transformation of a zygote into an adult animal is controlled by clusters of homeotic genes called *Hox* genes (see Modules 11.10 and 15.11). The study of these master control genes has helped scientists investigate the phylogenetic relationships among the highly diverse animal forms we are about to survey. And, as you will learn when you meet biologist Sean Carroll in Module 18.16, the discovery of *Hox* genes has also provided key insights into the mechanisms of evolution.

? List the distinguishing characteristics of animals.

■ Multicellular, eukaryotic heterotrophs that ingest their food; no cell walls; unique cell junctions; nerve and muscle cells; sexual reproduction and life cycles with unique embryonic stages; unique developmental genes

Figure 18.1A Ingestion, the animal way of life

The ancestor of animals was probably a colonial, flagellated protist

Biologists have long speculated about the origins of animals. Some molecular clock calculations (see Module 15.18) estimate that the common ancestor of living animals lived about a billion years ago. This ancestor may have resembled modern choanoflagellates, colonial protists that are the closest living relatives of animals.

Figure 18.2A shows one hypothesis for how a colonial flagellated protist may have evolved into a simple animal with specialized cells arranged in two layers. ❶ The earliest colonial aggregates may have been only a few cells. ❷ Some larger colonies may have formed hollow spheres. ❸ Eventually, cells in the colony may have become specialized for functions such as reproduction, locomotion, and feeding. ❹ A simple multicellular organism with cell layers might have evolved as cells on one side of the colony folded inward. ❺ Eventually, a gastrula-like "proto-animal" may have evolved.

There are no fossils of this early evolutionary event. The oldest known animal fossils date from the late Precambrian time, about 575 million years ago. These fossils represent several different soft-bodied forms that were too complex to have been the first animals; these must have arisen earlier.

Beginning at the dawn of the Cambrian period, about 542 million years ago, the fossil record marks a dramatic increase in animal diversity. So many animal body plans and new phyla appear in the fossils from such an evolutionarily short time span (about 15 million years) that biologists call this episode the Cambrian explosion. Many Cambrian animals may seem bizarre (Figure 18.2B), but most biologists classify the Cambrian fossils as ancient representatives or at least relatives of animal phyla living today.

What ignited the Cambrian explosion? One hypothesis emphasizes ecological causes: The evolution of hard body coverings led to increasingly complex predator-prey relationships and diverse adaptations for feeding, motility, and protection. A second hypothesis focuses on geologic changes: Perhaps atmospheric oxygen had finally reached a high enough concentration to support the metabolism of more active, mobile animals. Whatever the ecological impetus might have been, it is highly probable that the genetic framework for complex

Figure 18.2B A drawing based on fossils from the early Cambrian period

bodies was already in place—the *Hox* complex of regulatory genes. As you learned in Module 15.11, much of the diversity in body form among the animal phyla is associated with variations in where and when *Hox* genes are expressed within developing embryos. The role of these master control genes in the evolution of animal diversity will be discussed further in Module 18.16.

Most animals are **invertebrates,** so called because they lack a vertebral column (backbone): Of the 35 or so animal phyla (systematists disagree on the precise number), the animals in all but one are invertebrates. You'll learn about invertebrates in this chapter and survey the vertebrates in Chapter 19. But first we look at some of the anatomical features biologists use to classify this vast animal diversity.

? **What is the main difference between a colonial organism and an organism that is truly multicellular?**

■ The cells of a multicellular organism are more extensively specialized and interdependent than are the cells of a colonial organism.

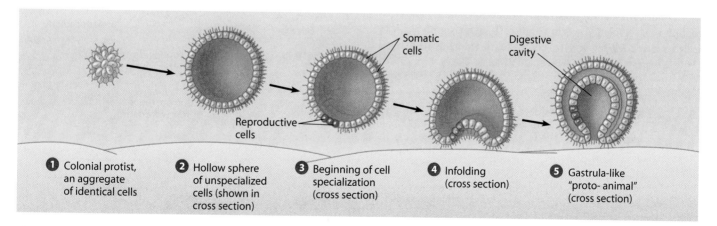

Somatic cells

Digestive cavity

Reproductive cells

❶ Colonial protist, an aggregate of identical cells

❷ Hollow sphere of unspecialized cells (shown in cross section)

❸ Beginning of cell specialization (cross section)

❹ Infolding (cross section)

❺ Gastrula-like "proto-animal" (cross section)

Figure 18.2A A hypothesis for the evolution of animals from a colonial flagellated protist (the arrows symbolize evolutionary time)

Animals can be characterized by basic features of their "body plan"

One way that biologists categorize the diversity of animals is by certain general features of body structure, which together describe what is referred to as an animal's "body plan." Distinctions between body plans are used to help infer the phylogenetic relationships between animal groups.

Symmetry is one feature of the body plan. Some animals have **radial symmetry:** As illustrated by the sea anemone on the left in Figure 18.3A, the body parts radiate from the center. Any imaginary slice through the central axis divides a radially symmetrical animal into mirror images. Thus, the animal has a top and a bottom, but not right and left sides. As shown by the lobster in the figure, an animal with **bilateral symmetry** has mirror-image right and left sides; a distinct head, or **anterior,** end; a tail, or **posterior,** end; a back, or **dorsal,** surface; and a bottom, or **ventral,** surface.

The symmetry of an animal fits its lifestyle. A radial animal is typically sedentary or passively drifting, meeting its environment equally on all sides. In bilaterally symmetrical animals, the brain, sense organs, and mouth are usually located in the head. This arrangement facilitates mobility. As the animal travels headfirst through the environment, its eyes and other sense organs contact the environment first.

Body plans also vary in the organization of tissues. True tissues are collections of specialized cells, usually isolated from other tissues by membrane layers, that perform specific functions (an example is the nervous tissue of your brain and spinal cord). Sponges lack true tissues, but in other animals, the cell layers formed in the process of gastrulation (see Figure 18.1B) give rise to true tissues and to organs. Some animals have only ectoderm and endoderm; most animals also have mesoderm, making a body with three tissue layers.

Animals with three tissue layers may be characterized by the presence or absence of a **body cavity.** This fluid-filled space between the digestive tract and outer body wall cushions the internal organs and enables them to grow and move independently of the body wall. In soft-bodied animals, a noncompressible fluid in the body cavity forms a **hydrostatic skeleton** that provides a rigid structure against which muscles contract, moving the animal.

In the figures to the right, colors indicate the tissue layers: ectoderm (blue), mesoderm (red), and endoderm (yellow). The cross section of a segmented worm (Figure 18.3B) has a body cavity called a **true coelom,** which is completely lined by tissue derived

from mesoderm. A roundworm (Figure 18.3C) has a body cavity called a **pseudocoelom** (from the Greek *pseudes,* false, and *koilos,* hollow). A pseudocoelom is not completely lined by tissue derived from mesoderm. Despite the name, pseudocoeloms function just like coeloms. A flatworm (Figure 18.3D) reveals a body that is solid except for the cavity of the digestive sac.

Animals with three tissue layers can be separated into two groups based on details of their embryonic development, such as the fate of the opening formed during gastrulation that leads to the developing digestive tract. In **protostomes** (from the Greek *protos,* first, and *stoma,* mouth), this opening becomes the mouth; in **deuterostomes** (from the Greek *deutero,* second), this opening becomes the anus, and the mouth forms from a second opening. Other differences between protostomes and deuterostomes include the pattern of early cell divisions and the way the coelom forms.

Next we see how these general features of body plan are used to infer relationships between animal groups.

? List four features that can describe an animal's body plan.

■ Symmetry, number of tissue layers, body cavity type, embryonic development

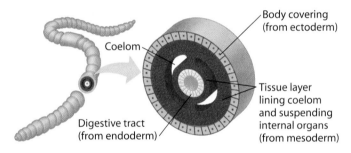

Figure 18.3B   True coelom (a segmented worm)

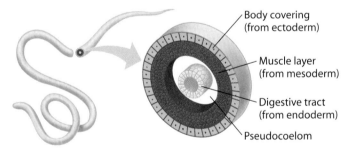

Figure 18.3C   Pseudocoelom (a roundworm)

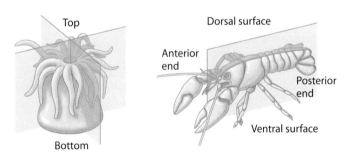

Figure 18.3A   Radial (left) and bilateral (right) symmetry

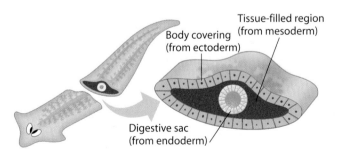

Figure 18.3D   No body cavity (a flatworm)

18.4 The body plans of animals can be used to build phylogenetic trees

Because animals diversified so rapidly on the scale of geologic time, it is difficult to sort out the evolutionary relationships between the various phyla using only the fossil record. Traditionally, biologists have proposed hypotheses about animal phylogeny based on morphological studies of living animals as well as fossils, often using the characteristics of body plan and embryonic development described in the preceding module.

Figure 18.4 presents a morphology-based phylogenetic tree of the major phyla of the animal kingdom. At the far left is the ancestral colonial protist. The tree has a series of branch points. The first branch point splits the sponges from the clade of **eumetazoans** ("true animals"), the animals with true tissues. The next branch point separates the animals with radial symmetry from those with bilateral symmetry. Most animal phyla have bilateral symmetry and thus belong to the clade called **bilaterians.** This morphology-based tree then divides the bilaterians into two clades based on embryology: deuterostomes and protostomes.

All phylogenetic trees are hypotheses for the key events in the evolutionary history that led to the animal phyla now living on Earth. Increasingly, researchers are adding molecular comparisons to their data sets for identifying clades. In Module 18.15, we will see that such data are leading to new hypotheses for grouping animal phyla.

? Humans are members of the phylum Chordata. Which phylum shared a common ancestor with chordates most recently?

■ Echinodermata

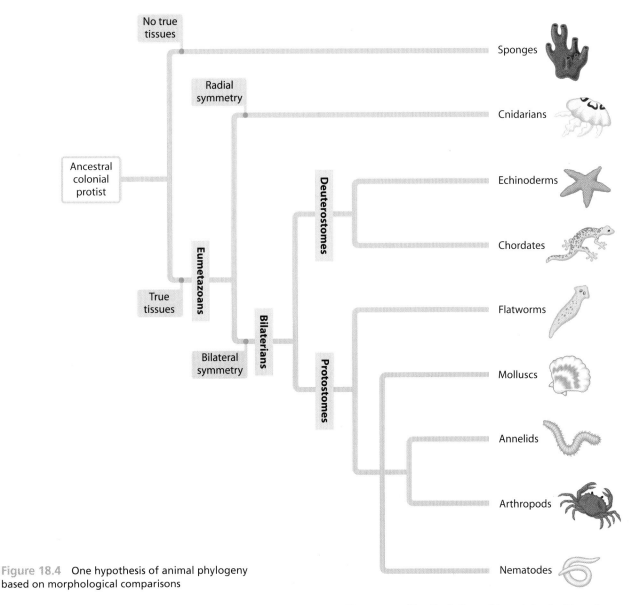

Figure 18.4 One hypothesis of animal phylogeny based on morphological comparisons

18.5 Sponges have a relatively simple, porous body

Sponges (phylum Porifera) are stationary animals that are so sedentary that the ancient Greeks believed them to be plants. The majority of species are marine, although some are found in fresh water. Figure 18.5A shows two individuals of the genus *Scypha,* a small sponge only about 1–3 cm high. The purple tube sponge in Figure 18.5B, on the other hand, can reach heights of 1.5 m. Some sponges, such as *Scypha,* resemble simple sacs. Others, such as the azure vase sponge in Figure 18.5C, have folded body walls and irregular shapes. Some sponges are radially symmetrical, but most lack body symmetry.

A simple sponge resembles a thick-walled sac perforated with holes. (*Porifera* means "pore-bearer" in Latin.) Water is drawn through the pores into a central cavity, then flows out through a larger opening (Figure 18.5D). More complex sponges have branching water canals.

The body of a sponge consists of two layers of cells separated by a gelatinous region. The inner layer of flagellated cells called **choanocytes** (purple in Figure 18.5D) help to sweep water through the sponge's body. Wandering through the middle body region are **amoebocytes** (blue), which produce supportive skeletal fibers (yellow) composed of a flexible protein called spongin and mineralized particles called spicules. Sharp spicules may also protect the large opening, as in Figure 18.5D. Most sponges have both types of skeletal components, but bath sponges only contain spongin.

Sponges are examples of **suspension feeders** (also known as filter feeders), animals that collect food particles from water passed through some type of food-trapping equipment. Sponges feed by collecting food particles from water that streams through their porous bodies. To obtain enough food to grow by 100 g (about 3 ounces), a sponge must filter roughly 1,000 kg (about 275 gallons) of seawater. Choanocytes trap food particles in mucus on the membranous collars that

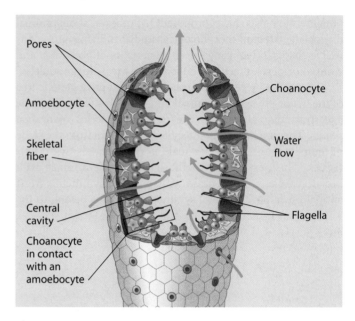

Figure 18.5D Structure of a simple sponge

surround the base of their flagella and then engulf the food by phagocytosis (see Module 5.9). Amoebocytes pick up food packaged in food vacuoles from choanocytes, digest it, and carry the nutrients to other cells.

Adult sponges are **sessile,** meaning they are anchored in place; therefore, they cannot escape from predators. Researchers have found that sponges produce defensive compounds such as toxins and antibiotics that deter pathogens, parasites, and predators. Some of these compounds may prove useful to humans as new drugs.

Sponges are the simplest of all animals. They have no nerves or muscles, though their individual cells can sense and react to changes in the environment. Since the cell layers are loose federations of cells, they are not considered true tissues. Biologists hypothesize that sponge lineages arose very early from the multicellular organisms that gave rise to the animal kingdom. The choanocytes of sponges and the cells of living choanoflagellates are similar, supporting the molecular evidence that animals evolved from a flagellated protist ancestor.

? Why is it thought that sponges represent the earliest branch of the animal kingdom?

Figure 18.5A
Scypha

Figure 18.5B   A purple tube sponge

Figure 18.5C   An azure vase sponge

■ Sponges lack true tissues, and their choanocytes resemble certain flagellated protists.

18.6 Cnidarians are radial animals with tentacles and stinging cells

All animals except sponges have true tissues and belong to the clade Eumetazoa ("true animals"). Among eumetazoans, one of the oldest groups is phylum Cnidaria, which includes the hydras, jellies (also called "jellyfish"), sea anemones, and corals. **Cnidarians** are characterized by radial symmetry and only two tissue layers. The simple body of most cnidarians has an outer epidermis and an inner cell layer that lines the digestive cavity. A jelly-filled middle region may have scattered amoeboid cells. Contractile tissues and nerves occur in their simplest forms in cnidarians.

Cnidarians exhibit two kinds of radially symmetrical body forms. Hydras, common in freshwater ponds and lakes, have a cylindrical body with tentacles projecting from one end. This body form is a **polyp** (Figure 18.6A). The other type of cnidarian body is the **medusa,** exemplified by the marine jelly in Figure 18.6B. While polyps are mostly stationary, medusae move freely about in the water. They are shaped like an umbrella with a fringe of tentacles around the lower edge. Some jellies have tentacles over 100 m long dangling from an umbrella up to 2 m in diameter.

Some cnidarians pass sequentially through both a polyp stage and a medusa stage in their life cycle. Others exist only as medusae; still others, such as hydras and sea anemones (Figure 18.6C), exist only as polyps.

Cnidarians are carnivores that use their tentacles to capture small animals and protists and to push the prey into their mouths. In a polyp, the mouth is on the top of the body, at the hub of the radiating tentacles (see Figure 21.3A). In a medusa, the mouth is in the center of the undersurface. The mouth leads into a digestive compartment called a **gastrovascular cavity** (from the Greek *gaster,* belly, and Latin *vas,* vessel). The mouth is the only opening in the body, so undigested food and other wastes exit through it. The gastrovascular cavity also circulates fluid that services internal cells (hence the "vascular" in gastrovascular; see Module 23.1). Acting as a hydrostatic skeleton, fluid in the cavity provides body support and helps give a cnidarian its shape, much like water in a balloon. When the animal closes its mouth, the volume of the cavity is fixed. Then contraction of selected cells changes the shape of the animal and can produce movement.

Phylum Cnidaria (from the Greek *cnide,* nettle) is named for its unique stinging cells, called **cnidocytes,** that function in defense and in capturing prey. Each cnidocyte contains a fine thread coiled within a capsule (Figure 18.6D). When it is discharged, the thread can sting or entangle prey. Some large marine cnidarians use their stinging threads to catch fish. A group of cnidarians called cubozoans have highly toxic cnidocytes. The sea wasp, a cubozoan found off the coast of northern Australia, is the deadliest organism on Earth: One animal may produce enough poison to kill as many as 60 people.

Coral animals are cnidarians that secrete a hard external skeleton. Each generation builds on top of the skeletons of previous generations, constructing characteristic shapes of "rocks" we call coral. Reef-building corals depend on sugars produced by symbiotic algae to supply them with enough energy to maintain the reef structure in the face of erosion and reef-boring animals.

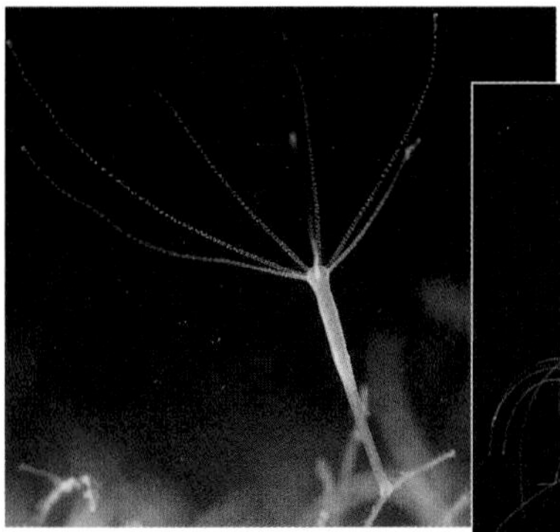

? What are three functions of a cnidarian's gastrovascular cavity?

■ (1) Digestion, (2) circulation, and (3) physical support and movement

Figure 18.6A Polyp body form: a hydra (about 2–25 mm high)

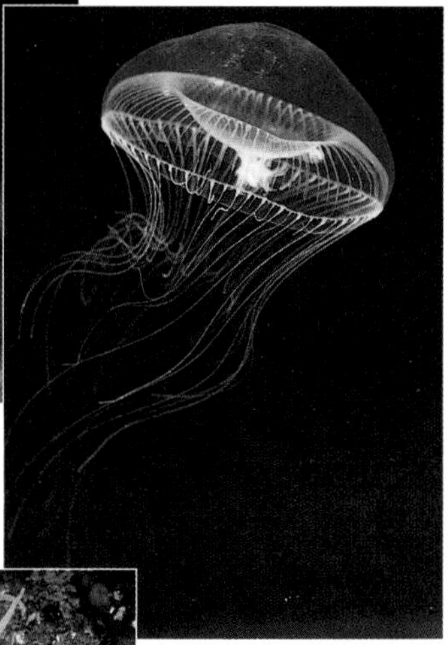

Figure 18.6B Medusa body form: a marine jelly called a sea nettle (about 5 cm in diameter)

Figure 18.6C Sea anemones, such as this *Anthopleura* (about 6 cm in diameter), exist only as polyps

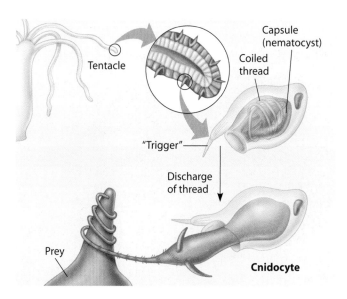

Figure 18.6D Cnidocyte action

18.7 Flatworms are the simplest bilateral animals

The vast majority of animal species belong to the clade Bilateria, which consists of animals with bilateral symmetry and three embryonic tissue layers. **Flatworms,** phylum Platyhelminthes (from the Greek *platys,* flat, and *helmis,* worm), are the simplest of the bilaterians. These thin, often ribbonlike animals range in length from about 1 mm to 20 m and live in marine, freshwater, and damp terrestrial habitats. In addition to free-living forms, there are many parasitic species. Like cnidarians, most flatworms have a gastrovascular cavity with only one opening. The fine branches of the gastrovascular cavity distribute food throughout the animal.

There are three major groups of flatworms. Worms called planarians represent the group **free-living flatworms** (Figure 18.7A). A planarian has a head with a pair of light-sensitive eyespots and a flap at each side that detects chemicals. Dense clusters of nerve cells form a simple brain, and a pair of nerve cords connect with small nerves that branch throughout the body.

The gastrovascular cavity of a planarian is highly branched. When the animal feeds, it sucks food in through a mouth at the tip of a muscular tube that projects from the mid-ventral surface of the body (as shown in the figure). Planarians live on the undersurfaces of rocks in freshwater ponds and streams. Using cilia on their ventral surface, they crawl about in search of food. They also have muscles that enable them to twist and turn.

A second group of flatworms, the **flukes,** live as parasites in other animals. Many flukes have suckers that attach to their host and a tough protective covering. Reproductive organs occupy nearly the entire interior of these worms.

Many flukes have complex life cycles with an intermediate host in which larvae develop. The larvae then infect the final host in which they live as adults. For example, blood flukes that parasitize humans spend part of their life cycle in snails. These flukes cause a long-lasting disease called schistosomiasis that affects 200 million people around the world, causing severe abdominal pain, anemia, and dysentery.

Tapeworms are another parasitic group of flatworms. Adult tapeworms inhabit the digestive tracts of vertebrates, including humans. In contrast with planarians and flukes, most tapeworms have a very long, ribbonlike body with repeated units. As Figure 18.7B shows, the anterior end, called the scolex, is armed with hooks and suckers that grasp the host. Notice that there is no mouth. Bathed in the partially digested food in the intestines of their hosts, they simply absorb nutrients across their body surface and have no digestive tract. Thus, tapeworms are an exception to our definition of animals in Module 18.1—other animals ingest nutrients. Behind the scolex is a long ribbon of repeated units filled with both male and female reproductive structures. The units at the posterior end, which are full of ripe eggs, break off and pass out of the host's body in feces.

Like parasitic flukes, tapeworms have a complex life cycle, usually involving more than one host. Most species benefit from the predator-prey relationships of their hosts. A prey species—a sheep or a rabbit, for example—may become infected by eating grass contaminated with tapeworm eggs. Larval tapeworms develop in these hosts, and a predator—a coyote or a dog, for instance—becomes infected when it eats an infected prey animal. The adult tapeworms develop in the predator's intestines.

Humans can be infected with tapeworms by eating undercooked beef, pork, or fish infected with tapeworm larvae. The larvae are microscopic, but the adults of some species can reach lengths of 2 m in the human intestine.

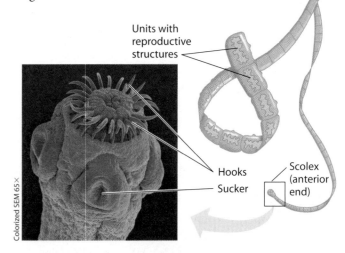

Figure 18.7B A tapeworm, a parasitic flatworm

? Flatworms and cnidarians differ in symmetry, with flatworms being _____ and cnidarians being _____, but the animals of both phyla have a _____.

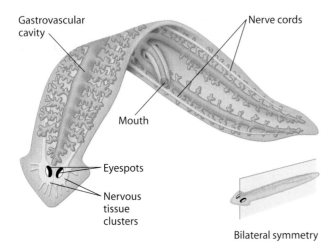

Figure 18.7A A free-living flatworm, the planarian (most are about 5–10 mm long)

■ bilateral ⋯ radial ⋯ gastrovascular cavity

18.8 Nematodes have a pseudocoelom and a complete digestive tract

Nematodes, also called roundworms, make up the phylum Nematoda. As bilaterians, these animals have bilateral symmetry and a three-tissue layer construction. In contrast with flatworms, roundworms have a fluid-filled body cavity (a pseudocoelom, not completely lined with mesoderm) and a digestive tract with two openings.

Nematodes are cylindrical with a blunt head and tapered tail. They range in size from less than 1 mm to more than a meter. Several layers of tough, nonliving material called a **cuticle** cover the body and prevent the nematode from drying out. In parasitic species, the cuticle protects the nematode from the host's digestive system. When the worm grows, it periodically sheds its cuticle (molts) and secretes a new, larger one. What looks like a corduroy coat on the nematode in Figure 18.8A is its cuticle.

You can also see the mouth at the tip of the blunt anterior end of the nematode in Figure 18.8A. Nematodes have a **complete digestive tract,** extending as a tube from the mouth to the anus near the tip of the tail. Food travels only one way through the system and is processed as it moves along. In animals with a complete digestive tract, the anterior regions of the tract churn and mix food with enzymes, while the posterior regions absorb nutrients and then dispose of wastes. This division of labor allows each part of the digestive tract to be specialized for its particular function.

Fluid in the pseudocoelom of a nematode distributes nutrients absorbed from the digestive tract throughout the body. The pseudocoelom also functions as a hydroskeleton, and contraction of longitudinal muscles produces the characteristic thrashing motion of nematodes.

Although only about 25,000 species of nematodes have been named, estimates of the total number of species range as high as 500,000. Free-living nematodes live virtually everywhere there is rotting organic matter, and their numbers are huge. Ninety thousand individuals were found in a single rotting apple lying on the ground; an acre of topsoil contains billions of nematodes. Nematodes are important decomposers in soil and on the bottom of lakes and oceans. Some are predators, eating other microscopic animals.

Little is known about most free-living nematodes. A notable exception is the soil-dwelling species *Caenorhabditis elegans,* an important research organism. A *C. elegans* adult consists of only about 1,000 cells—in contrast with the human body, which consists of some 60 trillion cells. By following every cell division in the developing embryo, biologists have been able to trace the lineage of every cell in the adult worm. The genome of *C. elegans* has been sequenced, and the ongoing research contributes to our understanding of

Figure 18.8B Dog heart infested by heartworm, a parasitic nematode

how genes control animal development, the functioning of nervous systems, and even some of the mechanisms involved in aging.

Other nematodes thrive as parasites in the moist tissues of plants and in the body fluids and tissues of animals. (The largest known nematodes are parasites of whales and measure more than 7 m long.) Many species are serious agricultural pests that attack the roots of plants or parasitize animals. The dog heartworm (Figure 18.8B), a common parasite, is deadly to dogs, and can also infect other pets such as cats and ferrets. It is spread by mosquitoes, which pick up heartworm eggs in the blood of an infected host and transmit them when sucking the blood of another animal. Although dog heartworms were once found primarily in the southeastern United States, they are now common throughout the contiguous United States. Regular doses of a preventative medication can protect dogs from heartworm.

Humans are host to at least 50 species of roundworms, including a number of disease-causing organisms. *Trichinella spiralis* causes a disease called trichinosis in a wide variety of mammals, including humans. People usually acquire the worms by eating undercooked pork or wild game containing the juvenile worms. Cooking meat until it is no longer pink kills the worms. Hookworms, which grapple onto the intestinal wall and suck blood, infect millions of people worldwide. Although hookworms are small (about 10 mm long), a heavy infestation can cause severe anemia.

We might expect that an animal group as numerous and widespread as nematodes would include a great diversity of body form. In fact, the opposite is true. Most species look very much alike. In sharp contrast, animals in the phylum Mollusca, which we examine next, exhibit enormous diversity in body form.

? What is the advantage of a complete digestive tract?

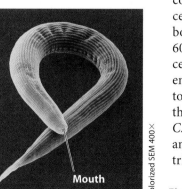

Mouth

Colorized SEM 400×

Figure 18.8A A free-living nematode

■ Different parts of the digestive tract can be specialized for different functions.

Diverse molluscs are variations on a common body plan

Snails, slugs, oysters, clams, octopuses, and squids are just a few of the great variety of animals known as **molluscs** (phylum Mollusca). Molluscs are soft-bodied animals (from the Latin *molluscus,* soft), but most are protected by a hard shell.

It may seem that animals as different as octopuses and clams could not belong in the same phylum, but these and other molluscs have inherited several common features from their ancestors. Figure 18.9A illustrates the basic body plan of a mollusc, consisting of three main parts: a muscular **foot** (gray in the drawing), which functions in locomotion; a **visceral mass** (orange) containing most of the internal organs; and a **mantle** (purple), a fold of tissue that drapes over the visceral mass and secretes a shell in molluscs such as clams and snails. In many molluscs, the mantle extends beyond the visceral mass, producing a water-filled chamber called the mantle cavity, which houses the gills (left side in Figure 18.9A).

Figure 18.9A shows yet another body feature found in many molluscs—a unique rasping organ called a **radula,** which is used to scrape up food. In a snail, for example, the radula extends from the mouth and slides back and forth like a backhoe, scraping and scooping algae off rocks.

Most molluscs have separate sexes, with reproductive organs located in the visceral mass. The life cycle of many marine molluscs includes a ciliated larva called a trochophore (Figure 18.9B). Trochophore larvae are a significant characteristic in determining phylogenetic relationships among the invertebrate phyla (see Module 18.15).

In contrast with flatworms, which have no body cavity, and nematodes, which have a pseudocoelom, molluscs have a true coelom (brown in Figure 18.9A). Also unlike flatworms and nematodes, molluscs have a **circulatory system**—an organ system that pumps blood and distributes nutrients and oxygen throughout the body.

These basic body features have evolved in markedly different ways in different groups of molluscs. The three most diverse groups (classes) are the gastropods (including snails and slugs), bivalves (including clams, scallops, and oysters), and cephalopods (including squids and octopuses).

Gastropods The largest group of molluscs are called the **gastropods** (from the Greek *gaster,* belly, and *pous,* pl. *podos,* foot), found in fresh water, salt water, and terrestrial environments. In fact, they include the only molluscs that live on land.

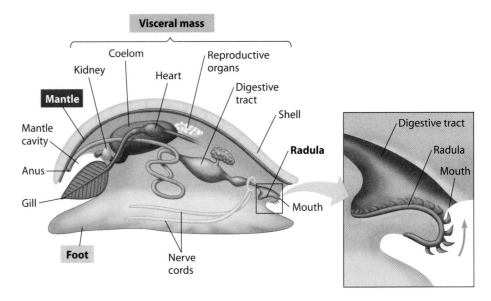

Figure 18.9A   The general body plan of a mollusc

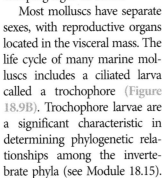

Figure 18.9B   Trochophore larva

Most gastropods are protected by a single, spiraled shell into which the animal can retreat when threatened. Many gastropods have a distinct head with eyes at the tips of tentacles, like the land snail in Figure 18.9C. Terrestrial snails lack the gills typical of aquatic molluscs; instead, the lining of the mantle cavity functions as a lung, exchanging gases with the air.

Most gastropods are marine, and shell collectors delight in their variety. Slugs, however, are unusual molluscs in that they have lost their mantle and shell during their evolution. The long, colorful projections on the sea slug shown in Figure 18.9D (about 5 cm long) function as gills.

Figure 18.9C   A terrestrial gastropod: a land snail

Figure 18.9D   A marine gastropod: a sea slug

Figure 18.9E A bivalve: a scallop

Figure 18.9F A cephalopod with an internal shell: a squid

Bivalves The **bivalves** (from the Latin *bi*, double, and *valva*, leaf of a folding door) include numerous species of clams, oysters, mussels, and scallops. They have shells divided into two halves that are hinged together. Most bivalves are suspension feeders. The mantle cavity contains gills that are used for feeding as well as gas exchange. The mucus-coated gills trap fine food particles suspended in the water, and cilia sweep the particles to the mouth. Most bivalves are sedentary, living in sand or mud. They may use their muscular foot for digging and anchoring. Mussels are sessile, secreting strong threads that attach them to rocks, docks, and boats. The scallop in Figure 18.9E (about 10 cm in diameter) can skitter along the seafloor by flapping its shell, rather like the mechanical false teeth sold in novelty shops. Notice the many bluish eyes peering out between the two halves of its hinged shell. The eyes are set into the fringed edges of the animal's mantle.

Cephalopods The **cephalopods** (from the Greek *kephale*, head, and *pous*, foot) differ from gastropods and bivalves in being adapted to the lifestyle of fast, agile predators. The chambered nautilus is a descendant of ancient groups with external shells, but in other cephalopods, the shell is small and internal (as in squids) or missing altogether (as in octopuses). If you have a pet bird, you may have hung the internal shell of another cephalopod, the cuttlefish, in its cage. Such "cuttlebones" are commonly given to caged birds as a source of calcium. Cephalopods use beak-like jaws and a radula to crush or rip prey apart. The mouth is at the base of the foot, which is drawn out into several long tentacles for catching and holding prey.

The squid in Figure 18.9F (about 20 cm long) ranks with fishes as a fast, streamlined predator. It darts about by drawing water into its mantle cavity and then forcing a jet of water out through a muscular siphon. It steers by pointing the siphon in different directions. Octopuses, such as the ones described in the chapter introduction and the one shown in Figure 18.9G (about 30 cm long), live on the seafloor, where they creep about in search of crabs and other food.

All cephalopods have large brains and sophisticated sense organs that contribute to their success as mobile predators. Cephalopod eyes are among the most complex sense organs in

Figure 18.9G A cephalopod without a shell: an octopus

the animal kingdom. Each eye contains a lens that focuses light and a retina on which clear images form. Octopuses are considered among the most intelligent invertebrates and have shown remarkable learning abilities in laboratory experiments.

The so-called *colossal squid,* which lives in the ocean depths near Antarctica, is the largest of all invertebrates. Specimens are rare, but in 2007, a male colossal squid measuring 10 m and weighing 450 kg was hauled in by a fishing boat. Females are generally even larger, and scientists estimate that colossal squid average around 13 m in overall length. The *giant squid,* which rivals the colossal in length, is thought to average about 10 m. Most information about these impressive animals has been gleaned from dead specimens that washed ashore. In 2004, however, Japanese zoologists succeeded in capturing the first video footage of a giant squid in its natural habitat.

? As representatives of classes of molluscs, a garden snail is an example of a _____; a clam is an example of a _____; and an octopus is an example of a _____.

■ gastropod · · · bivalve · · · cephalopod

18.10 Annelids are segmented worms

A segmented body resembling a series of fused rings is the hallmark of phylum Annelida (from the Latin *anellus*, ring). **Segmentation,** the subdivision of the body along its length into a series of repeated parts (segments), played a central role in the evolution of many complex animals. A segmented body allows for greater flexibility and mobility, and it probably evolved as an adaptation facilitating movement. An earthworm, a typical **annelid,** uses its flexible, segmented body to crawl and burrow rapidly into the soil.

Annelids range in length from less than 1 mm to 3 m, the length of some giant Australian earthworms. They are found in damp soil, in the sea, and in most freshwater habitats. Some aquatic annelids swim in pursuit of food, but most are bottom-dwelling scavengers that burrow in sand and mud. There are three main groups of annelids: earthworms and their relatives, polychaetes, and leeches.

Earthworms and Their Relatives Figure 18.10A illustrates the segmented anatomy of an earthworm. Internally, the coelom is partitioned by membrane walls (only a few are fully shown here). Many of the internal body structures are repeated within each segment. The nervous system (yellow) includes a simple brain and a ventral nerve cord with a cluster of nerve cells in each segment. Excretory organs (green), which dispose of fluid wastes, are also repeated in each segment (only a few are shown in this diagram). The digestive tract, however, is not segmented; it passes through the segment walls from the mouth to the anus.

Many invertebrates, including most molluscs and all arthropods (which you will meet next), have what is called an **open circulatory system,** in which blood is pumped through vessels that open into body cavities where organs are bathed directly in blood. Annelids and vertebrates, in contrast, have a **closed circulatory system,** in which blood remains enclosed in vessels as it distributes nutrients and oxygen throughout the body. As you can see in the diagram at the lower left, the main vessels of the earthworm circulatory system—a dorsal blood vessel and a ventral blood vessel—are connected by segmental vessels. The pumping organ, or "heart," is simply an enlarged region of the dorsal blood vessel plus five pairs of segmental vessels near the anterior end.

Each segment is surrounded by longitudinal and circular muscles. Earthworms move by coordinating the contraction of these two sets of muscles (see Figure 30.1D). These muscles work against the coelomic fluid in each segment, which acts as a hydrostatic skeleton. Each segment also has four pairs of stiff bristles that provide traction for burrowing.

Earthworms are hermaphrodites; that is, they have both male and female reproductive structures. But they mate and cross-fertilize by exchanging sperm. A specialized organ, visible as the thickened region of the worm in Figure 18.10A, secretes a cocoon made of mucus. The cocoon slides along the worm, picking up the eggs and the received sperm. The cocoon then slips off the worm into the soil, where the embryos develop.

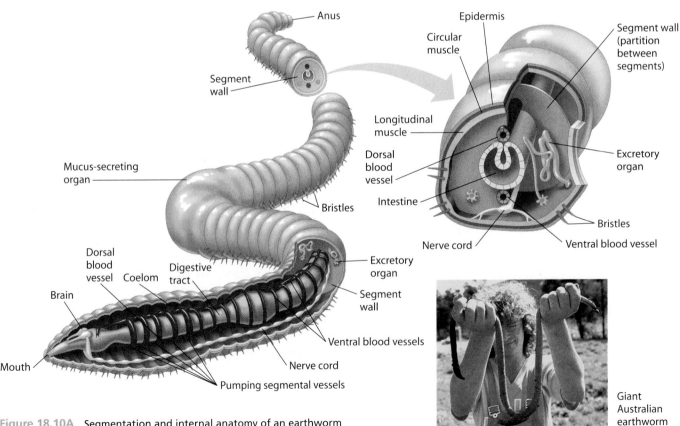

Figure 18.10A Segmentation and internal anatomy of an earthworm

Giant Australian earthworm

Earthworms eat their way through the soil, extracting nutrients as soil passes through their digestive tube. Undigested material, mixed with mucus secreted into the digestive tract, is eliminated as castings (feces) through the anus. Farmers value earthworms because the animals aerate the soil and their castings improve the soil's texture. Darwin estimated that a single acre of British farmland had about 50,000 earthworms, producing 18 tons of castings per year.

Polychaetes The **polychaetes** (from the Greek *polys*, many, and *chaeta*, hair) form the largest group of annelids. The polychaete in Figure 18.10B swims in the open ocean by moving the paddle-like appendages on each segment. In polychaetes that live on the seafloor, stiff bristles (called *chaetae*) on the appendages help the worm wriggle about in search of small invertebrates to eat. In many polychaetes, the appendages are richly supplied with blood vessels and are either associated with the gills, or modified as gills themselves.

Most polychaetes are marine. Many live in tubes and extend feathery appendages coated with mucus that trap suspended food particles. Tube-dwellers usually build their tubes by mixing mucus with bits of sand and broken shells. Some species of tube-dwellers are colonial, such as the group shown in Figure 18.10C. The circlet of feathery appendages seen at the mouth of each tube extends from the head of the worm inside.

Leeches The third main group of annelids is the **leeches,** which are notorious for their bloodsucking habits. However, most species are free-living carnivores that eat small invertebrates such as snails and insects. The majority of leeches inhabit fresh water, but there are also marine species and a few terrestrial species that inhabit moist vegetation in the tropics. They range in length from 1 to 30 cm.

Some bloodsucking leeches use razor-like jaws to slit the skin of an animal. The host is usually oblivious to this attack because the leech secretes an anesthetic as well as an anticoagulant into the wound. The leech then sucks as much blood as it can hold, often more than ten times its own weight. After this gorging, a leech can last for months without another meal.

Until the 1920s, physicians used leeches for bloodletting. For centuries, illness was thought to result from an imbalance in the body's fluids, and the practice of bloodletting was originally conceived to restore the natural balance. Later, physicians viewed bloodletting as a kind of spring cleaning for the body to remove any toxins or "bad blood" that had accumulated. Leeches are still occasionally applied to remove blood from bruised tissues (Figure 18.10D) and to help relieve swelling in fingers or toes that have been sewn back on after accidents. Blood tends to accumulate in a reattached finger or toe until small veins have a chance to grow back into it.

The anticoagulant produced by leeches has also proven to be medically useful. It is used to dissolve blood clots that form during surgery or as a result of heart disease. Because it is difficult to obtain this chemical from natural sources, it is now being produced through genetic engineering.

The segments of an annelid are all very similar. In the next group we explore, the arthropods, body segments and their appendages have become specialized, serving a variety of functions.

Figure 18.10B   A free-swimming marine polychaete

Figure 18.10C   A tube-building polychaete

Figure 18.10D   A medicinal leech applied to drain blood from a patient

? Tapeworms and leeches are both parasites. What are the key differences between these two?

■ Whereas both are composed of repeated segments, the segments of a tapeworm are filled mostly with reproductive organs and are shed from the posterior end of the animal. Tapeworms are flatworms with no body cavity and, in their parasitic lifestyle, not even a gastrovascular cavity. Leeches have a true coelom and a complete digestive tract.

18.11 Arthropods are segmented animals with jointed appendages and an exoskeleton

Over a million species of **arthropods**—including crayfish, lobsters, crabs, barnacles, spiders, ticks, and insects—have been identified. It is estimated that the arthropod population of the world numbers about a billion billion (10^{18}) individuals! In terms of species diversity, geographic distribution, and sheer numbers, phylum Arthropoda must be regarded as the most successful animal phylum.

The diversity and success of arthropods are largely related to their segmentation, their hard exoskeleton, and their jointed appendages, for which the phylum is named (from the Greek *arthron*, joint, and *pous*, pl. *podos*, foot). As indicated in the drawing of a lobster in Figure 18.11A, the appendages are variously adapted for sensory reception, defense, feeding, walking, and swimming. The arthropod body, including the appendages, is covered by an **exoskeleton,** an external skeleton that protects the animal and provides points of attachment for the muscles that move the appendages. This exoskeleton is a cuticle, a nonliving covering that in arthropods is hardened by layers of protein and chitin, a polysaccharide. The exoskeleton is thick around the head, where its main function is to house and protect the brain. It is paper-thin and flexible in other locations, such as the joints of the legs. As it grows, an arthropod must periodically shed its old exoskeleton and secrete a larger one, a complex process called **molting.**

In contrast with annelids, which have similar segments throughout their body, the body of most arthropods is formed of several distinct groups of segments that fuse during development: the head, thorax, and abdomen. (In some arthropods, including the lobster, the exoskeleton of the head and thorax is partly fused, forming a body region called the cephalothorax.) Each of the segment groups is specialized for a different function. In a lobster, the head bears sensory antennae, eyes, and jointed mouthparts on the ventral side. The thorax bears a pair of defensive appendages (the pincers) and four pairs of legs for walking. The abdomen has swimming appendages.

Like molluscs, arthropods have an open circulatory system in which a tube-like heart pumps blood through short arteries into spaces surrounding the organs. A variety of gas exchange organs have evolved. Most aquatic species have gills. Terrestrial insects have internal air sacs that branch throughout the body (see Module 22.4).

Fossils and molecular evidence suggest that living arthropods represent four major lineages that diverged early in the evolution of arthropods. The figures in this module illustrate representatives of three of these lineages. The fourth, the insects, will be discussed in Module 18.12.

Chelicerates Figure 18.11B shows a number of **horseshoe crabs,** members of a species that has survived with little change for hundreds of millions of years. This "living fossil" is the only surviving member of a group of marine chelicerates that were abundant in the sea some 300 million years ago. One member of this group, the water scorpion, could grow up

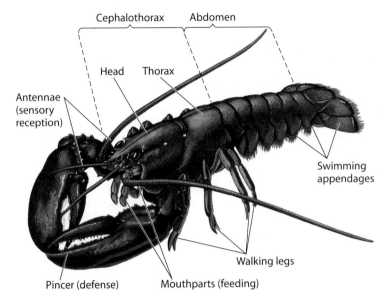

Figure 18.11A The structure of a lobster, an arthropod

to 3 m long. Horseshoe crabs are common on the Atlantic and Gulf coasts of the United States.

Living **chelicerates** (from the Greek *chele*, claw and *keras*, horn) also include the scorpions, spiders, ticks, and mites, collectively called **arachnids.** Most arachnids live on land. Scorpions (Figure 18.11C, on facing page) are nocturnal hunters. Their ancestors were among the first terrestrial carnivores, preying on other arthropods that fed on early land plants. Scorpions have a large pair of pincers for defense and capturing prey. The tip of the tail bears a poisonous stinger. Scorpions eat mainly insects and spiders and attack people only when prodded or stepped on. Only a few species are dangerous to humans, but the sting is painful nonetheless.

Spiders, a diverse group of arachnids, are usually active during the daytime, hunting insects or trapping them in webs of silk that they spin from specialized glands on their abdomen

Figure 18.11B Horseshoe crabs (up to about 30 cm wide)

A scorpion (about 8 cm long)

A black widow spider (about 1 cm wide)

A dust mite (about 420 μm long)

Colorized SEM 900×

Figure 18.11C   Arachnids

(see Figure 18.11C, center). Mites make up another large group of arachnids. On the right in Figure 18.11C is a micrograph of a dust mite, a ubiquitous scavenger in our homes. Thousands of these microscopic animals can thrive in a few square centimeters of carpet or in one of the dust balls that form under a bed. Dust mites do not carry infectious diseases, but many people are allergic to them.

Millipedes and Centipedes   The animals in this lineage have similar segments over most of their body and superficially resemble annelids; however, their jointed legs identify them as arthropods. **Millipedes** (Figure 18.11D) are wormlike terrestrial creatures that eat decaying plant matter. They have two pairs of short legs per body segment. **Centipedes** (Figure 18.11E) are terrestrial carnivores with a pair of poison claws used in defense and to paralyze prey such as cockroaches and flies. Each of their body segments bears a single pair of long legs.

Crustaceans   The **crustaceans** are nearly all aquatic. Lobsters and crayfish are in this group, along with numerous crabs, shrimps, and barnacles. Barnacles (Figure 18.11F) are marine crustaceans with a cuticle that is hardened into a shell containing calcium carbonate, which may explain why they were once classified as molluscs. Their jointed appendages project from their shell to strain food from the water. Most barnacles anchor themselves to rocks, boat hulls, pilings, or even whales. The adhesive they produce is as strong as any glue ever invented. Other crustaceans are small copepods and krill, which serve as food sources for many fishes and whales.

We turn next to the fourth lineage of arthropods, the insects, whose numbers dwarf all other groups combined.

? List the characteristics that arthropods have in common.

■ Segmentation, exoskeleton, specialized jointed appendages

Figure 18.11D   A millipede (about 7 cm long)

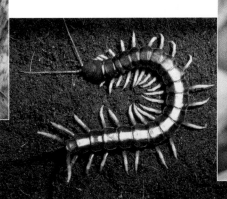

Figure 18.11E   Giant Peruvian centipede (can reach 30 cm in length)

Figure 18.11F   Crustaceans: goose barnacles (about 2 cm high)

18.12 Insects are the most successful group of animals

The evolutionary success of insects is unrivaled by any other group of animals. More than a million species of insects have been identified, comprising over 70% of all animal species. Entomologists (scientists who study insects) think that fewer than half the total number of insect species have been identified, and some believe there could be as many as 30 million. Insects are distributed worldwide and have a remarkable ability to survive challenging terrestrial environments. Although they have also flourished in freshwater habitats, insects are rare in the seas, where crustaceans are the dominant arthropods.

What characteristics account for the extraordinary success of insects? The features that they share with other arthropods—body segmentation, an exoskeleton, and jointed appendages—have contributed to their success. Other key features include flight, a waterproof coating on the cuticle, and a complex life cycle. In addition, many insects have short generation times and large numbers of offspring. For example, *Culex pipiens*, the most widely distributed species of mosquito, has a generation time of roughly 10 days, and a single female can lay many hundreds of eggs over the course of her lifetime. Thus, natural selection acts rapidly, and alleles that offer a reproductive advantage can quickly be established in the population.

Life Cycles One factor in the success of insects is a life cycle that includes **metamorphosis,** during which the animal takes on different body forms as it develops from larva to adult. Only the adult insect is sexually mature and has wings. More than 80% of insect species, including beetles, flies, bees, and moths and butterflies, undergo what is called **complete metamorphosis.** The larval stage (such as caterpillars, which are the larvae of moths and butterflies, and maggots, which are fly larvae) is specialized for eating and growing. A larva typically goes through several molts as it grows, then exists as an encased, nonfeeding pupa while its

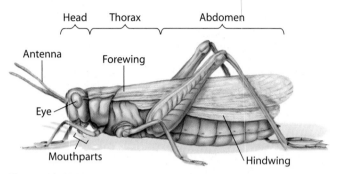

Figure 18.12B Insect anatomy, as seen in a grasshopper

body rebuilds from clusters of embryonic cells that have been held in reserve. The insect then emerges as an adult that is specialized for reproduction and dispersal. Adults and larvae eat different foods, permitting the species to make use of a wider range of resources and avoiding intergenerational competition. Figure 18.12A shows the caterpillar, pupa, and adult of the regal moth (*Citheronia regalis*). The caterpillar is known as the hickory horned devil for the type of leaves it prefers and its appearance.

Other insect species undergo **incomplete metamorphosis,** in which the transition from larva to adult is achieved through multiple molts, but without forming a pupa. In some species, including grasshoppers and cockroaches, the juvenile forms resemble the adults. In others, the body forms and lifestyles are very different. The larvae of dragonflies, for example, are aquatic, but the adults live on land.

Modular Body Plan Like the grasshopper in Figure 18.12B, most adult insects have a three-part body, consisting of a head, a thorax, and an abdomen. The parts are formed by the fusion of embryonic segments during development. Early in development, the embryonic segments are identical to each other. However, as discussed in Module 11.10, they soon diverge as different genes are expressed in different segments, giving rise to the three distinct body parts and to a variety of appendages, including antennae, mouthparts, legs, and wings. The insect body plan is essentially modular—each embryonic segment is a separate building block that develops independently of the other segments. As a result, a mutation in a homeotic gene can change the structure of one segment or its appendages without affecting any of the others. In the evolution of the grasshopper, for example, genetic changes in one thoracic segment produced the specialized jumping legs but did not affect the other two leg-producing segments. Wings, antennae, and mouthparts have all evolved in a similar fashion, by the specialization of independent segments through changes in the timing and location of homeotic

Figure 18.12A Complete metamorphosis of regal moth. Caterpillar (hickory horned devil, above left), is about 13 cm long; pupa, about 7 cm long (above right); adult moth, up to 15 cm in wingspan (right)

gene expression (see Module 15.11). Much of the extraordinary diversification of insects resulted from modifications of the appendages that adapted them for specialized functions.

The head typically bears a pair of sensory antennae, a pair of eyes, and several pairs of mouthparts. The mouthparts are adapted for particular kinds of eating—for example, for chewing plant material (in grasshoppers); for biting and tearing prey (praying mantis); for lapping up fluids (houseflies); and for piercing into and sucking the fluids of plants (aphids) or animals (mosquitoes). When flowering plants appeared, adaptations for nectar feeding were advantageous (see Module 17.12). As a result of this variety in mouthparts, insects have adaptations that exploit almost every conceivable food source.

Most adult insects have three pairs of legs, which may be adapted for walking, jumping, grasping prey, digging into the soil or even paddling on water. Insects are the only invertebrates that can fly; most adult insects have one or two pairs of wings. (Some insects such as fleas are wingless.) Flight, which is an effective means of dispersal and escape from predators, was a major factor in the success of insects. And because the wings are extensions of the cuticle, insects have acquired the ability to fly without sacrificing any legs. By contrast, the wings of birds and bats are modified limbs. With a single pair of walking legs, those animals are generally clumsy when on the ground.

Protective Color Patterns In many groups of insects, adaptations of body structures have been coupled with protective coloration. Many different animals, including insects, have camouflage, color patterns that blend into the background. But insects also have elaborate disguises that include modifications to their antennae, legs, wings, and bodies. For instance, there are insects that resemble twigs, leaves, and bird droppings (Figure 18.12C). Some even do a passable imitation of vertebrates. The "snake" in Figure 18.12D is actually a hawk moth caterpillar. The colors of its dorsal side are an effective camouflage. When disturbed, however, it flips over to reveal the snake-like eyes of its ventral side, even puffing out its thorax to enhance the deception. Eyespots that resemble vertebrate eyes are common in several groups of moths and butterflies. Figure 18.12E shows a member of a genus known informally as owl butterflies. A flash of these large eyes startles would-be predators. In other species, the eyespot deflects the predator's attack away from vital body parts.

How could evolution have produced these complex color patterns? It turns out that the genetic mechanism by which eyespots evolve is very similar to the mechanism by which specialized appendages evolve. Butterfly wings have a modular construction similar to that of embryonic body segments. Each section can change independently of the others and can therefore have a unique pattern. And like the specialization of appendages, eyespots result from different patterns of homeotic gene expression during development.

The study of insects has given evolutionary developmental biologists valuable insight into the genetic mechanisms that have generated the amazing diversity of life. You'll meet one of the leading researchers in this field in Module 18.16. But first, we complete our survey of the invertebrate phyla.

Figure 18.12C   Remarkable resemblances. A stick insect (upper left); leaf mimic (right); caterpillar resembling bird dropping (lower left)

Figure 18.12D Hawk moth caterpillar

Figure 18.12E   Owl butterfly (above) and long-eared owl (right)

Web Process of Science *How Are Insect Species Identified?*

? Contrast incomplete and complete metamorphosis.

■ In complete metamorphosis, there is a pupal stage; in incomplete metamorphosis, there is not.

18.13 Echinoderms have spiny skin, an endoskeleton, and a water vascular system for movement

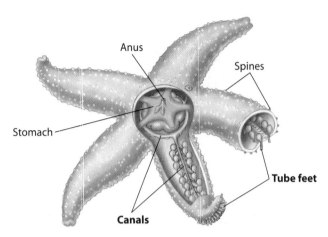

Figure 18.13A The water vascular system (canals and tube feet) of a sea star

Echinoderms, such as sea stars, sand dollars, and sea urchins, are slow-moving or sessile marine animals. Most are radially symmetrical as adults. Both the external and the internal parts of a sea star, for instance, radiate from the center like spokes of a wheel. The bilateral larval stage of echinoderms, however, tells us that echinoderms are not closely related to cnidarians or other animals that never show bilateral symmetry.

The phylum name Echinodermata (from the Greek *echin,* spiny, and *derma,* skin) refers to the prickly bumps or spines of a sea star or sea urchin. These are extensions of the hard calcium-containing plates that form the **endoskeleton,** or internal skeleton, under the thin skin of the animal.

Unique to echinoderms is the **water vascular system,** a network of water-filled canals that branch into extensions called tube feet. Tube feet function in locomotion, feeding, and gas exchange (Figure 18.13A). A sea star pulls itself slowly over the seafloor using its suction-cup-like tube feet. Its mouth is centrally located on its undersurface. When a sea star encounters an oyster or clam, its favorite food, it grips the mollusc with its tube feet and positions its mouth next to the narrow opening between the two valves of the shell. The sea star then turns its stomach inside out, pushing it through its mouth and into the opening of the mollusc's shell. The sea star's stomach proceeds to digest the soft parts of the mollusc inside the mollusc's shell (Figure 18.13B).

Sea stars and some other echinoderms are capable of regeneration. Arms that are lost are readily regrown.

In contrast with sea stars, sea urchins are spherical and have no arms. They do have five rows of tube feet that project through tiny holes in the animal's globe-like case. If you look carefully in Figure 18.13C, you can see the long, threadlike tube feet projecting among the spines of the sea urchin. Sea urchins move by pulling with their tube feet. They also have muscles that pivot their spines, which can aid in locomotion. Unlike the carnivorous sea stars, most sea urchins eat algae.

Other echinoderm classes include brittle stars, which move by thrashing their long, flexible arms; sea lilies, which live attached to the substrate by a stalk; and sea cucumbers, elongated animals that resemble their name more than they resemble other echinoderms.

Though echinoderms have many unique features, we see evidence of their relation to other animals in their embryonic development. As we discussed in Modules 18.3 and 18.4, differences in patterns of development have led biologists to identify echinoderms and chordates (which include vertebrates) as a clade of bilateral animals called deuterostomes. Thus, echinoderms are more closely related to our phylum, the chordates, than to the protostome animals, such as molluscs, annelids, and arthropods. We examine the chordates next.

Web Activity *Characteristics of Invertebrates*

? Contrast the skeleton of an echinoderm with that of an arthropod.

■ An echinoderm has an endoskeleton; an arthropod has an exoskeleton.

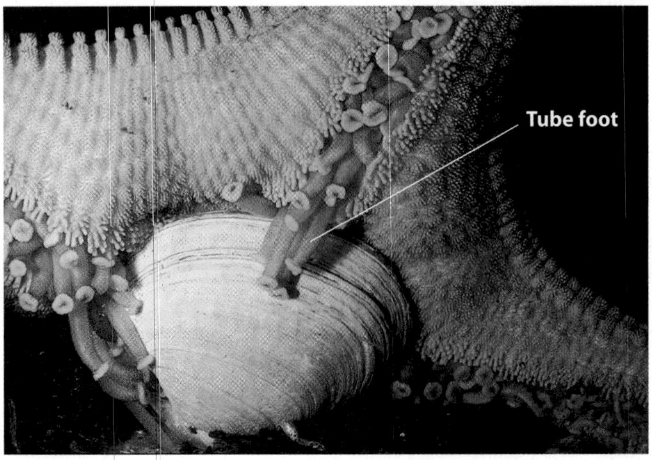

Figure 18.13B A sea star feeding on a clam

Figure 18.13C A sea urchin (about 12 cm in diameter)

18.14 Our own phylum, Chordata, is distinguished by four features

You may be surprised to find the phylum that includes humans in a chapter on invertebrate diversity. However, vertebrates evolved from invertebrate ancestors and continue to share the distinctive features that identify members of the phylum Chordata. The embryos, and often the adults, of chordates possess: (1) a **dorsal, hollow nerve cord;** (2) a **notochord,** a flexible, supportive, longitudinal rod located between the digestive tract and the nerve cord; (3) **pharyngeal slits** located in the pharynx, the region just behind the mouth; and (4) a muscular **post-anal tail** (a tail posterior to the anus). You can see these four features in the diagrams in Figures 18.14A and 18.14B. The two chordates shown, a tunicate and a lancelet, are called invertebrate chordates because they do not have a backbone.

Adult **tunicates** are stationary and look more like small sacs than anything we usually think of as a chordate (see Figure 18.14A). Tunicates often adhere to rocks and boats, and they are common on coral reefs. The adult has no trace of a notochord, nerve cord, or tail, but it does have prominent pharyngeal slits that function in feeding. Very different from the adult, the tunicate larva is a swimming, tadpole-like organism that exhibits all four chordate trademarks.

Tunicates are suspension feeders. Seawater enters the adult animal through an opening at the top, passes through the pharyngeal slits into a large cavity in the animal, and exits back into the ocean via an excurrent siphon on the side of the body (see the photo in Figure 18.14A). Food particles are trapped in a net made of mucus and then transported to the intestine, where they are digested. Because they shoot a jet of water through their excurrent siphon when threatened, tunicates are often called sea squirts.

Lancelets, another group of marine invertebrate chordates, also feed on suspended particles. Lancelets are small, bladelike chordates that live in marine sands (see Figure 18.14B). When feeding, a lancelet wriggles backward into the sand with its head sticking out. As in tunicates, a net of mucus secreted across the pharyngeal slits traps food particles. Water flowing through the slits exits via an opening in front of the anus.

Lancelets clearly illustrate the four chordate features. They also have segmental muscles that flex their body from side to side, producing slow swimming movements. These serial muscles are evidence of the lancelet's segmentation. Although not unique to chordates, body segmentation is another chordate characteristic.

What is the relationship between the invertebrate chordates and the vertebrates? The tunicates likely represent the deepest branch of the chordate lineage. Molecular evidence indicates that the lancelets are the closest living nonvertebrate relatives of vertebrates. Research has shown that the same genes that organize the major regions of the vertebrate brain are expressed in the same pattern at the anterior end of the lancelet nerve cord.

The invertebrate chordates have helped us identify the four chordate hallmarks. We'll look at the vertebrate members of this phylum in the next chapter. Next, we review the invertebrates by revisiting the phylogenetic tree presented in Module 18.4.

Web Activity *Characteristics of Chordates*

? What four features do we share with invertebrate chordates, such as lancelets?

■ (1) Dorsal, hollow nerve cord; (2) notochord; (3) pharyngeal slits; (4) post-anal tail

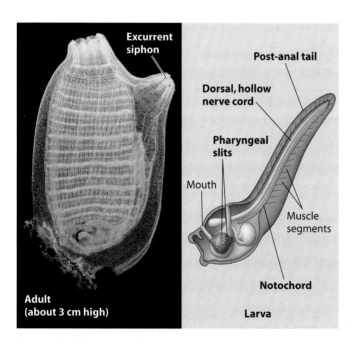

Figure 18.14A A tunicate

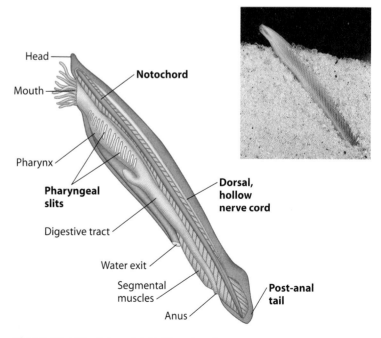

Figure 18.14B A lancelet (5–15 cm long)

18.15 An animal phylogenetic tree is a work in progress

We have looked at examples of an immense diversity of animals in this chapter—representatives of about one-third of the known animal phyla. The diversity of the animal kingdom is spectacular, but it can also be overwhelming to a student. One way to help organize all this diversity is to arrange these groups on a phylogenetic tree, as you saw in Figure 18.4. Such trees summarize the proposed evolutionary threads that tie all this diversity together. Biologists continue to sort out the evolutionary history of the animals using evidence from the fossil record, morphology, embryology, and molecular data. Thus, phylogenetic trees are hypotheses that will continue to be tested and refined.

Figure 18.4 presented a traditional, morphology-based set of hypotheses about the relationships between the nine phyla we surveyed in this chapter. Figure 18.15 presents a slightly revised tree based on molecular comparisons. Let's see how the hypotheses based on certain sets of molecular data agree with and differ from the morphology-based tree.

At the base (far left) of both trees is the ancestral colonial protist that was the probable ancestor of animals. Both trees agree on the early divergence of phylum Porifera (the sponges), with their lack of true tissues and body symmetry. However, because molecular data indicate that sponges are not a clade, the revised tree shows multiple lines for that branch.

The rest of the animal kingdom represents the clade of eumetazoans, animals with true tissues. The eumetazoans split into two distinct lineages that differ in body symmetry and the number of cell layers formed in gastrulation. The hydras, jellies, sea anemones, and corals of phylum Cnidaria are radially symmetrical and have two cell layers. The other lineage consists of bilateral animals, the bilaterians. The two trees agree on these early branchings.

The traditional phylogenetic tree in Figure 18.4 divides the bilaterians into deuterostomes and protostomes based on patterns of embryological development. The molecular-based tree also recognizes the deuterostomes, which include the echinoderms and chordates, as a clade.

How do the trees differ? As you can see in the branches highlighted in light blue, the molecular data distinguish two clades within the protostomes: lophotrochozoans and ecdysozoans. The lophotrochozoans, while grouped based on molecular similarities, are named for the feeding apparatus (called a lophophore) of some phyla in the group (which we did not discuss) and for the trochophore larva found in molluscs and annelids (see Figure 18.9B). This group includes the flatworms, molluscs, annelids, and many other phyla that we did not survey. The ecdysozoans include the nematodes and arthropods. Both of these phyla have exoskeletons that must be shed for the animal to grow. This molting process is called ecdysis, the basis for the name ecdysozoan.

Both the tree here and the one in Figure 18.4 represent hypotheses for the key events in the evolutionary history that led to the animal phyla now living on Earth. Like other

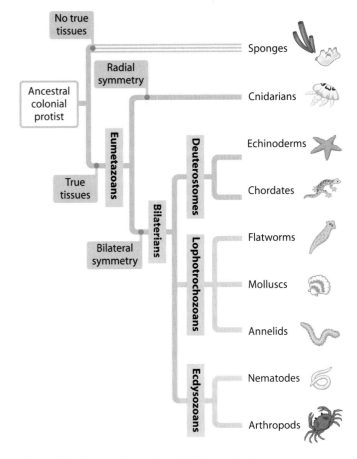

Figure 18.15   A molecular-based phylogenetic tree

phylogenetic trees, even this latest one serves to stimulate research and discussion, and it is subject to revision as new information is acquired. As zoologists continue to revise some branch points, however, the trees' overall message remains the same: The animal kingdom's great diversity arose through the process of evolution, and all animals exhibit features reminiscent of their evolutionary history. In the next module, researcher Sean Carroll discusses one of the common threads of evolutionary history, the generation of diversity from shared ancestral genes.

Web Activity *Animal Phylogenetic Tree*

Web Process of Science *How Do Molecular Data Fit Traditional Phylogenies?*

? Compare the placement of annelids, arthropods, and molluscs in Figures 18.4 and 18.15. How do they differ?

■ In the morphology-based tree, annelids and arthropods are shown to be more closely related to each other than to molluscs, largely based on their segmented bodies. In the molecular-based tree, arthropods are separated from both annelids and molluscs and are placed in the ecdysozoan clade.

18.16   Sean Carroll talks about the evolution of animal diversity

... from so simple a beginning endless forms most beautiful and most wonderful have been, and are being evolved.

—Charles Darwin, *The Origin of Species*

This chapter, a brief tour of invertebrate life, introduced you to some of the immense variety of animal bodies—flower-like sea anemones, echinoderms in the shapes of stars and cucumbers, and arthropods with countless configurations of appendages and color patterns, to name just a few. Data from molecular biology are providing new insights into the evolutionary relationships among these animal groups (see Module 18.15). But how can we explain the evolution of such strikingly different forms from a common protistan ancestor? Dr. Sean Carroll (Figure 18.16) of the Howard Hughes Medical Institute and the University of Wisconsin-Madison is a pioneer in the field of evolutionary developmental biology ("evo devo"). His book *Endless Forms Most Beautiful: The New Science of Evo Devo*, details the new understanding of the evolution of diversity that emerged when exciting new discoveries in the genetics and evolution of development were integrated with existing lines of evidence from the fossil record, comparative anatomy, and molecular biology. In a recent interview, Dr. Carroll described how surprising revelations forced scientists to think differently about the evolution of animal diversity.

> The conventional wisdom was that humans were very complex and really—in terms of genetic material—fundamentally different from things like fruit flies, butterflies, or worms. Maybe we'd have something more in common with a mouse or an ape, but not necessarily with the rest of the animal kingdom. But the huge discovery of evo devo was that the genes that build the bodies and body parts and organs of fruit flies are shared with us and with virtually every other animal in the kingdom. And that was just stunning.

This discovery forced a profound shift in thinking about the genetic instructions that generate complexity. Scientists had assumed that the construction of complex organs such as eyes and hearts required more instructions than simple organs. Consequently, they expected the results of the Human Genome Project (see Module 12.18) to reflect the complexity of the human body, estimating that we have about 100,000 genes. As you may recall, geneticists were taken aback to learn how few genes humans have (roughly 21,000), especially in comparison to other animals (about 14,000 in fruit flies).

> There has been an idea around biology for a very long time that to get new things you would need new genes. New genes for new structures, or new genes for new capabilities, new genes for new types of animals ... It doesn't look like that at all. It looks like a lot of things come about because very old genes learn new tricks. ... The genes involved in building our eyes and our hearts and our bones, these genes have been around for 500 million years and have been building all the rest of the members of the animal kingdom.

These ancient genes are the master control genes called homeotic genes, the body-building genes you learned about earlier. Researchers were amazed to discover that the same genes

Figure 18.16   Sean Carroll, talking with Jane Reece

control the events that lead to the different body forms of different animals. How is that possible? Dr. Carroll explains.

> It turns out that these body-building genes have very sophisticated and extensive sets of regulatory instructions [control sequences in the DNA that regulate gene expression; see Module 11.5] that surround them or are embedded within them.

The key genetic differences are not in the segments of DNA that are transcribed and translated into proteins, but in the segments of DNA that control when and where genes are transcribed and translated into proteins, or expressed (see Module 11.5). Given different sets of instructions, the same gene can direct the formation of a fly eye or a human eye. The same principle applies to gene expression throughout the development of an individual.

> A gene might act in two dozen places in the body in the course of development. And its action in that individual body part, of that individual moment of development, is controlled independently from its action in other places.

The great variety of specialized appendages and color patterns in insects, as you learned in Module 18.12, is an example of this. It is now clear that significant phenotypic changes have resulted from mutations in the segments of DNA that control gene expression, and that these changes in the regulatory instructions have played the leading role in the evolution of "endless forms."

The work of Dr. Carroll and other "evo devo" researchers is deepening our understanding of how Earth's enormous wealth of biological diversity evolved. With the next chapter, we conclude our exploration of diversity with the evolution of our own group, the vertebrates.

?   What does Sean Carroll mean by the "regulatory instructions" of a homeotic gene?

■ The DNA sequences that control gene expression

Chapter Review

Reviewing the Concepts

Animal Evolution and Diversity (18.1–18.4)

Animals ingest their food. Animal development may include a blastula, gastrula, and larval stage (**18.1**).

Ancestor. Animals evolved from a colonial protist. Animal diversity exploded during the Cambrian period (**18.2**).

Animal body plans may vary in symmetry (radial or bilateral), body cavity (none, pseudocoelom, or true coelom), and embryonic development (protostomes or deuterostomes). These differences may be used to infer phylogenetic relationships (**18.3–18.4**).

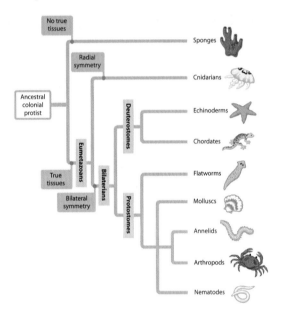

Invertebrate Diversity (18.5–18.14)

Sponges (phylum Porifera) are the simplest animals and have no true tissues. Their flagellated choanocytes filter food from water passing through the porous sponge body (**18.5**).

Cnidarians (phylum Cnidaria) have true tissues and radial symmetry. Their two body forms are polyps (such as hydra) and medusae (jellies). They have cnidocytes on tentacles that sting prey and a gastrovascular cavity (**18.6**).

Flatworms (phylum Platyhelminthes) are bilateral animals with no body cavity. A planarian has a gastrovascular cavity and a simple nervous system. Flukes and tapeworms are parasitic flatworms with complex life cycles (**18.7**).

Nematodes (phylum Nematoda) have a pseudocoelom and a complete digestive tract and are covered by a protective cuticle. Many nematodes (roundworms) are free-living decomposers; others are plant or animal parasites (**18.8**).

Molluscs (phylum Mollusca) include gastropods (such as snails and slugs), bivalves (such as clams), and cephalopods (such as squids). All have a muscular foot and a mantle, which may secrete a shell and which encloses the visceral mass. Many molluscs feed with a rasping radula (**18.9**).

Annelids (phylum Annelida) are segmented worms and include earthworms, polychaetes, and leeches (**18.10**).

Arthropods (phylum Arthropoda) are segmented animals with exoskeletons and jointed appendages. The four lineages are chelicerates (arachnids such as spiders), the aquatic crustaceans (lobsters and crabs), the lineage of millipedes and centipedes, and the terrestrial insects (**18.11**).

Insects have a three-part body (head, thorax, and abdomen) and three pairs of legs; most have wings. These most successful arthropods are grouped in about 26 orders. Their development often includes metamorphosis. Specialized appendages and protective color patterns, which frequently result from changes in the timing and location of homeotic gene expression, have played a major role in this group's success (**18.12**).

Echinoderms (phylum Echinodermata), such as sea stars, have spiny skins, an endoskeleton, and a water vascular system with tube feet that aids in respiration and locomotion. They are radially symmetrical as adults (**18.13**).

Chordates (phylum Chordata) have a dorsal, hollow nerve cord, a stiff notochord, pharyngeal slits, and a muscular post-anal tail. The simplest chordates are lancelets and tunicates, marine invertebrates that use their pharyngeal slits for suspension feeding (**18.14**).

Animal Phylogeny and Diversity Revisited (18.15–18.16)

Molecular-based phylogenetic trees distinguish two protostome clades: the lophotrochozoans and the ecdysozoans (**18.15**).

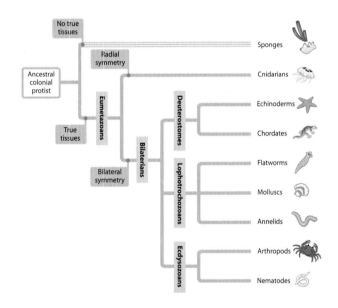

Research in evolutionary developmental biology has led to a deeper understanding of the evolution of organismal diversity (**18.16**).

Connecting the Concepts

1. The table below lists the common names of the nine animal phyla surveyed in this chapter. For each phylum, identify the key characteristics and some representatives.

Phylum	Characteristics	Representatives
Sponges		
Cnidarians		
Flatworms		
Nematodes		
Molluscs		
Annelids		
Arthropods		
Echinoderms		
Chordates		

Testing Your Knowledge

Multiple Choice

2. Bilateral symmetry in animals is best correlated with
   a. an ability to see equally in all directions.
   b. the presence of a skeleton.
   c. motility and active predation and escape.
   d. development of a true coelom.
   e. adaptation to terrestrial environments.

3. Jon found an organism in a pond, and he thinks it's a freshwater sponge. His friend Liz thinks it looks more like an aquatic fungus. How can they decide whether it is an animal or a fungus?
   a. See if it can swim.
   b. Figure out whether it is autotrophic or heterotrophic.
   c. See if it is a eukaryote or a prokaryote.
   d. Look for cell walls under a microscope.
   e. Determine whether it is unicellular or multicellular.

4. Which of the following groupings includes the largest number of species? (*Explain your answer.*)
   a. invertebrates
   b. chordates
   c. arthropods
   d. insects
   e. vertebrates

5. Which of the following animal groups does not have tissues derived from mesoderm?
   a. annelids
   b. amphibians
   c. echinoderms
   d. cnidarians
   e. flatworms

6. Molecular comparisons place nematodes and arthropods in clade Ecdysozoa. What characteristic do they share that is the basis for the name Ecdysozoa?
   a. a complete digestive tract
   b. body segmentation
   c. molting of an exoskeleton
   d. bilateral symmetry
   e. a true coelom

Matching

7. Include the vertebrates
8. Medusa and polyp body forms
9. The simplest bilateral animals
10. The most primitive animal group
11. Earthworms, polychaetes, and leeches
12. Largest phylum of all
13. Closest relatives of chordates
14. Body cavity is a pseudocoelom
15. Have a muscular foot and a mantle

   a. annelids
   b. nematodes
   c. sponges
   d. arthropods
   e. flatworms
   f. cnidarians
   g. molluscs
   h. echinoderms
   i. chordates

Describing, Comparing, and Explaining

16. Compare the structure of a planarian (a flatworm) and an earthworm with regard to the following: digestive tract, body cavity, and segmentation.

17. Name two phyla of animals that are radially symmetrical and two that are bilaterally symmetrical. How do the overall lifestyles of radial and bilateral animals differ?

18. One of the key characteristics of arthropods is their jointed appendages. Describe four functions of these appendages in four different arthropods.

19. Compare the phylogenetic tree from Module 18.4 with the one from Module 18.15 (see previous page for both phylogenetic trees). What are the similarities and differences? Why did scientists revise the tree?

Applying the Concepts

20. A marine biologist has dredged up an unknown animal from the seafloor. Describe some of the characteristics that could be used to determine the animal phylum to which the creature should be assigned.

21. Coral reefs harbor a greater diversity of animals than any other environment in the sea. Australia's Great Barrier Reef has been protected as a marine reserve and is a mecca for scientists and nature enthusiasts. Elsewhere, such as Indonesia and the Philippines, coral reefs are in danger. Many reefs have been depleted of fish, and runoff from the shore has covered coral heads with sediment. Nearly all the changes in the reefs can be traced back to human activities. What kinds of activities do you think might be contributing to the decline of the reefs? What are some reasons to be concerned about this decline? Do you think the situation is likely to improve or worsen in the future? Why? What might the local people do to halt the decline? Should the developed countries help? Why or why not?

Answers to all questions can be found in Appendix 4.

For Practice Quizzes, BioFlix, Mp3 Tutors, and Activities, go to www.mybiology.com.

The Evolution of Vertebrate Diversity

What Am I?

Of some 1.7 million species of organisms known to scientists, 1.3 million are animals. This large number of identified animal species not only indicates how successful animals are as a group, but also reflects our keen interest in the other members of our kingdom. Humans have a long history of studying, appreciating, and using animals. But classifying a new animal isn't always easy. Imagine you were the first European zoologist to encounter the aquatic animal at the far right (facing page) in its native Australia. What would you make of it? It has a bill and webbed feet similar to a duck's, but the rest of its furry body looks very much like that of a muskrat or other aquatic rodent. To make the case even more confusing, this animal lays eggs. So how would you classify it? Is it a bird or a mammal? The decision is easier once you study it a little more closely. This animal, called a duck-billed platypus, has mammary glands that produce milk for its young. That trait, along with its hair, is enough to place it in the mammalian class of animals.

Scientists investigating the platypus bill, which does look similar to that of a duck, found that it is not composed of hard, inert materials like a bird's bill but rather is covered by soft skin filled with sensitive nerve endings. While the duck and the platypus both use their bills to dig for food in muddy waters, the platypus's bill serves an additional purpose as a sensory organ to help it locate food and avoid obstacles underwater. When a platypus dives, it closes its eyes, so it relies heavily on its bill to "see" its surroundings. Indeed, biologists have found that a large portion of the platypus brain is devoted to processing sensory information from its bill. Part of the interest in studying animals is exploring their many fascinating adaptations. The incredible diversity of animal life arose through hundreds of millions of years of evolution as natural selection shaped animal adaptations to Earth's diverse and changing environments.

The duck-billed platypus belongs to a small group of egg-laying mammals, called monotremes, that are found only in Australia and New Guinea. Unlike the rest of the world, Australia has relatively few placental mammals (mammals that bear fully developed live young). Most

A Tasmanian tiger, 1928

Australian mammals are marsupials, such as the kangaroos pictured below, whose young complete their development in the mother's pouch.

Why do marsupials represent the majority of mammalian species in Australia, but are rare in other parts of the world? Marsupials are an ancient group of mammals that used to be common on other continents. But it appears that they did not compete well with placental mammals, and most became extinct. However, when Australia broke off from Pangaea over 60 million years ago (see Module 15.7), its isolated marsupials could flourish, filling the roles, or niches, that placental mammals fill on other continents. For example, the now-extinct Tasmanian tiger (or thylacine), shown above, was a marsupial that once filled the large-predator niche. The quoll, a small marsupial catlike animal that you will encounter in the last module of this chapter, fills a small-predator niche. And the kangaroo, which grazes on grass and other plants, fills the niche occupied by horses or antelopes on other continents. These are just a few examples of convergent evolution (see Module 15.14), in which unrelated species have adaptations that enable them to fill the same ecological niche in different parts of the world. In some cases, the species look different—nothing else looks quite like a kangaroo or a platypus. But in other cases, similar burrowing or gliding or even long-snouted ant-eating forms have evolved in unrelated animal species.

In this chapter, we continue our tour of the animal kingdom by exploring our own group, the vertebrates. We end the chapter, and our unit on the diversity of life, with a look at our predecessors—the primates who first walked on two legs, evolved a large, sophisticated brain, and eventually dominated the Earth. ■ ■ ■

19.1 Derived characters define the major clades of chordates

Using a combination of anatomical, molecular, and fossil evidence, biologists have developed hypotheses for the evolution of chordate groups. Figure 19.1 illustrates a current view of the major clades of chordates and lists some of the derived characters that define the clades. You can see that the tunicates are thought to be the first group to branch from the chordate lineage. Unlike tunicates, all other chordates have a brain, albeit a small one in the lancelets (only a swollen tip of the nerve cord).

The next transition was the development of a head that consists of a brain at the anterior end of the dorsal nerve cord, eyes and other sensory organs, and a skull. These innovations opened up a completely new way of feeding for chordates: active predation. All chordates with a head are called **craniates** (from the word *cranium*, meaning "skull").

The origin of a backbone came next. The **vertebrates** are distinguished by a more extensive skull and a backbone, or **vertebral column,** composed of a series of bones called **vertebrae** (singular, *vertebra*). These skeletal elements enclose the main parts of the nervous system. The skull forms a case for the brain, and the vertebrae enclose the nerve cord. The verte-brate skeleton is an endoskeleton, made of either flexible cartilage or a combination of hard bone and cartilage. Bone and cartilage are mostly nonliving material. But because there are living cells that secrete the nonliving material, the endoskeleton can grow with the animal.

The next major transition was the origin of jaws, which opened up new feeding opportunities. The evolution of lungs or lung derivatives, followed by muscular lobed fins with skeletal support, opened the possibility of life on land. **Tetrapods,** jawed vertebrates with two pairs of limbs, were the first vertebrates on land. The evolution of **amniotes,** tetrapods with a terrestrially adapted egg, completed the transition to land.

In the next several modules, we survey the vertebrates, from the jawless lampreys and hagfishes to the fishes to the tetrapods to the amniotes.

? List the hierarchy of clades to which mammals belong.

■ Chordates, craniates, vertebrates, jawed vertebrates, tetrapods, amniotes

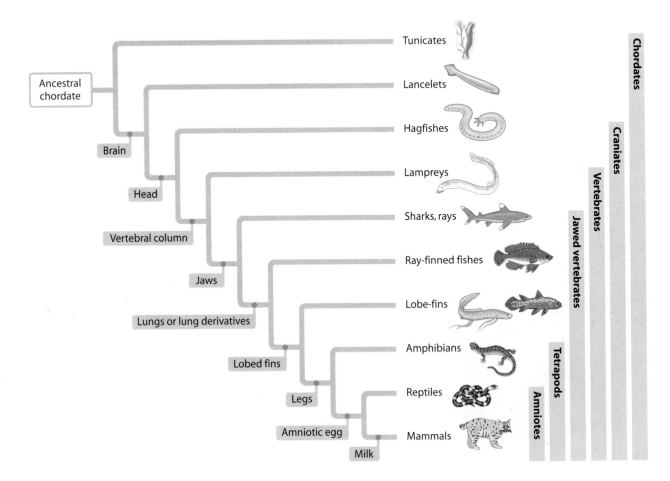

Figure 19.1 A phylogenetic tree of chordates showing key derived traits

The most primitive surviving vertebrates are hagfishes and lampreys. Both groups are craniates (chordates with heads), but neither hagfish nor lampreys have jaws. In hagfishes (Figure 19.2A), the notochord (a strong, flexible rod that runs most of the length of the body) is the body's main support in the adult. The notochord also persists in the adult lamprey (Figure 19.2B), but rudimentary vertebral structures are also present. Consequently, lampreys are considered vertebrates.

Present-day hagfishes (roughly 40 species) scavenge dead or dying vertebrates on the cold, dark seafloor. Although nearly blind, they have excellent senses of smell and touch. They feed by entering the animal through an existing opening or by creating a hole using sharp, toothlike structures on the tongue that grasp and tear flesh. For leverage, the hagfish may tie its tail in a knot, then slide the knot forward to tighten it against the prey's body. The knot trick is also part of its antipredator behavior. When threatened, a hagfish exudes an enormous amount of slime from special glands on the sides of its body (Figure 19.2A inset). The slime may make the hagfish difficult to grasp, or it may repel the predator. After the danger has passed, the hagfish ties itself into a knot and slides the knot forward, peeling off the layer of slime.

Fisherman who use nets have long been familiar with hagfishes. With their keen chemical senses, hagfishes are quick to detect bait and entrapped fish. Many fishermen have hauled in a net filled with feasting hagfishes, unsalable fish, and bucketsful of slime. But hagfish have gained economic importance recently. Asian fisheries have been harvesting them for decades. Both the meat and the skin, which is used to make faux-leather "eel-skin" belts, purses, and boots, are valuable commodities. As Asian fishing grounds have been depleted, the industry has moved to North America. Some populations of hagfish have been eradicated along the West Coast, and fisheries are now looking to the East Coast and South America for fresh stocks.

Lampreys represent the oldest living lineage of vertebrates. Lamprey larvae resemble lancelets. They are suspension feeders that live in freshwater streams, where they spend much of their time buried in sediment. Most lampreys migrate to the sea or lakes as they mature into adults.

Most species of lamprey are parasites, and just seeing the mouth of a sea lamprey suggests what it can do. The lamprey attaches itself to the side of a fish, uses its rasping tongue to penetrate the skin, and feeds on its victim's blood and tissues. After invading the Great Lakes via canals, these voracious vertebrates multiplied rapidly, decimating fish populations as they spread. Since the 1960s, streams that flow into the lakes have been treated with a chemical that reduces lamprey numbers, and fish populations have been recovering.

? Why are hagfishes described as craniates rather than vertebrates?

■ They do not have vertebrae; instead, their body is supported by a notochord.

Figure 19.2A Hagfish and slime (inset)

Slime glands

Figure 19.2B A sea lamprey, with its rasping mouth (inset)

Jawed vertebrates with gills and paired fins include sharks, ray-finned fishes, and lobe-finned fishes

Jawed vertebrates appeared in the fossil record in the mid-Ordovician period, about 470 million years ago, and steadily became more diverse. Their success probably relates to their paired fins and tail, which allowed them to swim after prey, as well as to their jaws, which enabled them to catch and eat a wide variety of prey instead of feeding as mud-suckers or suspension feeders. Sharks, fishes, amphibians, reptiles (including birds), and mammals—the vast majority of living vertebrates—have jaws supported by two skeletal parts held together by a hinge. Where did these hinged jaws come from? According to one hypothesis, they evolved by modification of skeletal supports of the anterior pharyngeal (gill) slits. The first part of Figure 19.3A shows the skeletal rods supporting the gill slits in a hypothetical ancestor. The main function of these gill slits was trapping suspended food particles. The other two parts of the figure show changes that may have occurred as jaws evolved. By following the red and green structures, you can see how two pairs of skeletal rods near the mouth now have become the jaws and their supports. The remaining gill slits, no longer required for suspension feeding, remained as sites of gas exchange.

Three lineages of jawed vertebrates with gills and paired fins are commonly called fishes. The sharks and rays of the class Chondrichthyes, which means "cartilage fish," have changed little in over 300 million years. As shown in Figure 19.1, lungs or lung derivatives are the key derived character of the clade that includes the ray-finned fishes and the lobe-fins. Muscular fins supported by stout bones further characterize the lobe-fins.

Chondrichthyans Sharks and rays, the **chondrichthyans,** have a flexible skeleton made of cartilage. The largest sharks are suspension feeders that eat small, floating plankton. Most sharks, however, are adept predators—fast swimmers with a streamlined body and powerful jaws (Figure 19.3B). A shark

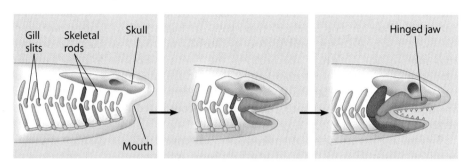

Figure 19.3A A hypothesis for the origin of vertebrate jaws

has sharp vision and a keen sense of smell. On its head it has electrosensors, organs that can detect minute electrical fields produced by muscle contractions in nearby animals. Sharks and most other aquatic vertebrates have a **lateral line system,** a row of sensory organs running along each side that are sensitive to changes in water pressure and can detect minor vibrations caused by animals swimming nearby.

While the bodies of sharks are streamlined for swimming in the open ocean, rays are adapted for life on the bottom. Their bodies are dorsoventrally flattened, with the eyes on the top of the head. Once settled, they flip sand over their bodies with their broad pectoral fins and lie half-buried for much of the day. The tails of stingrays bear sharp spines with venom glands at the base. Where stingrays are common, swimmers and divers must take care not to step on or swim too closely over a concealed ray. The sting is painful, and in rare cases, fatal. Steve Irwin, a wildlife expert and television personality, died when a ten-inch stingray barb pierced his heart while he was filming on The Great Barrier Reef in Australia.

The largest rays swim through the open ocean filtering plankton (Figure 19.3C). Some of these fishes are truly gigantic, measuring up to 6 m across the fins. The fin extensions in front of the mouth, which led to the common name devilfish, helps funnel in water for suspension feeding.

Figure 19.3B A sand bar shark, a chondrichthyan. Notice the gill openings on the side of the head.

Figure 19.3C A manta ray, a chondrichthyan

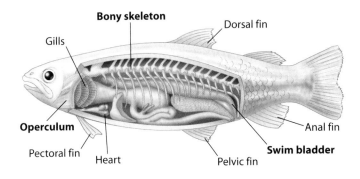

Figure 19.3D Anatomical features of a ray-finned fish

Figure 19.3E Variety of ray-finned fishes. Balloon fish (right); flounder (below, right); seahorse (below)

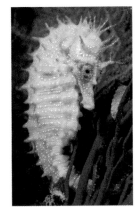

Ray-finned Fishes In **ray-finned fishes,** which include the familiar tuna, trout, and goldfish, the skeleton is made of bone—cartilage reinforced with a hard matrix of calcium phosphate. Their fins are supported by thin, flexible skeletal rays. Most have flattened scales covering their skin and secrete a coating of mucus that reduces drag during swimming. Figure 19.3D highlights key features of a ray-finned fish such as the rainbow trout shown in the photograph. On each side of the head, a protective flap called an **operculum** covers a chamber housing the gills. Movement of the operculum allows the fish to breathe without swimming. (By contrast, sharks must generally swim to pass water over their gills.) Ray-finned fishes also have a lung derivative that helps keep them buoyant—the **swim bladder,** a gas-filled sac. Swim bladders evolved from balloon-like lungs, which the ancestral fishes may have used in shallow water to supplement their gas exchange by gills.

Ray-finned fishes, which emerged during the Devonian period along with the lobe-fins, include the greatest number of species of any vertebrate group, more than 27,000, and more species are discovered all the time. They have adapted to virtually every aquatic habitat on Earth. From the basic structural adaptations that gave them great manueverability, speed, and feeding efficiency, various groups have evolved specialized body forms, fins, and feeding adaptations. (Recall the cichlids from the introduction to Chapter 14, for example.) Figure 19.3E shows a sample of the variety. The balloon fish doesn't always look like a spiky beachball. It raises its spines and inflates its body to deter predators. The flounder's flattened body is nearly invisible on the seabed. Pigment cells in its skin match the background for excellent camouflage. Notice that both eyes are on the top of its head. The larvae of flounders and other flatfish have eyes on both sides of the head. During development, one eye migrates to join the other on the side that will become the top. The small fins of the seahorse help it

manuever in dense vegetation, and the long tail is used for grasping onto a support. Seahorses have an unusual method of reproduction. The female deposits eggs in the male's abdominal brood pouch. His sperm fertilize the eggs, which develop inside the pouch.

Lobe-finned Fishes The key derived character of the **lobe-fins** is a series of rod-shaped bones in their muscular pectoral and pelvic fins. During the Devonian, they lived along coastal wetlands and may have used their lobed fins to "walk" underwater. Today, three lineages of lobe-fins survive: The coelacanth is a deep-sea dweller once thought extinct. The lungfishes are represented by a few Southern Hemisphere genera (Figure 19.3F) that generally inhabit stagnant waters and gulp air into lungs connected to the pharynx. And the third lineage, the tetrapods, adapted to life on land during the mid-Devonian and gave rise to terrestrial vertebrates, as we see next.

Figure 19.3F A lobe-finned lungfish, about a meter long

? From what structure might the swim bladder of ray-finned fishes have evolved?

■ Simple lungs of an ancestral species

19.4 New fossil discoveries are filling in the gaps of tetrapod evolution

During the late Devonian period, a line of lobe-finned fishes gave rise to **tetrapods** (meaning "four feet" in Greek), which today are defined as jawed vertebrates that have limbs and feet that can support their weight on land. Adaptation to life on land was a key event in vertebrate history; all subsequent groups of vertebrates—amphibians, mammals, and reptiles (including birds)—are descendants of these early land-dwellers.

In Chapter 17, we examined how the dramatic differences between aquatic and terrestrial environments shaped plant bodies and life cycles. Like plants, vertebrates faced obstacles on land in regard to gas exchange, water conservation, structural support, and reproduction. But vertebrates had other challenges, as well. Sensory organs that worked in water had to be adapted or replaced by structures that received stimuli transmitted through air. And, crucially, a new means of locomotion was required.

Lobe-finned fishes were long considered the most likely immediate ancestors of tetrapods. Their fleshy paired fins contain bones that appear to be homologous to tetrapod limb bones, and some of the modern lobe-fins have lungs that extract oxygen from the air. Alfred Romer, a renowned paleontologist, hypothesized that these features enabled lobe-fins to survive by moving from one pool of water to another as aquatic habitats shrank during periods of drought. With Romer's gift for vivid imagery, it was easy to imagine the fish dragging themselves short distances across the Devonian landscape, those with the best locomotory skills surviving such journeys to reproduce. In this way, according to the hypothesis, vertebrates gradually became fully adapted to a terrestrial existence. But fossil evidence of the transition was scarce.

For decades, the most informative fossils available were *Eusthenopteron*, a 385-million-year-old specimen that was clearly a fish, and *Ichthyostega*, which lived 365 million years ago and had advanced tetrapod features (Figure 19.4A). The ray-finned tail and flipper-like hind limbs of *Ichthyostega* indicated that it spent considerable time in the water, but its well-developed front limbs with small, finger-like bones and powerful shoulders showed that it was capable of locomotion on land. Unlike the shoulder bones of lobe-finned fishes, which are connected directly to the skull, *Ichthyostega* had a neck, a feature advantageous for terrestrial life. *Eusthenopteron* and *Ichthyostega* represented two widely separated points in the transition from fins to limbs. But what happened in between?

Recent fossil discoveries have begun to fill in the gap. Lobe-fin fossils that are more similar to tetrapods than to *Eusthenopteron* have been discovered, including a 380-million-year-old fish called *Panderichthyes* (Figure 19.4A). With its long snout, flattened body shape, and eyes on top of its head, *Panderichthyes* looked a bit like a crocodile. It had lungs as well as gills and an opening that allowed water to enter through the top of the skull, a possible indication of a shallow-water habitat. Its paired fins had fishlike rays, but the

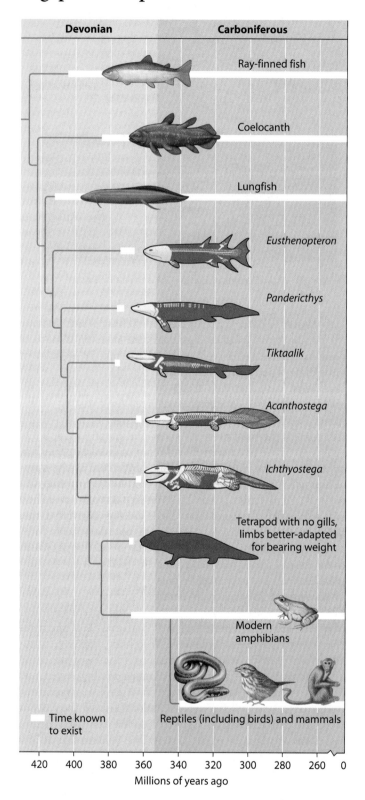

Figure 19.4A Some of the transitional forms in tetrapod evolution

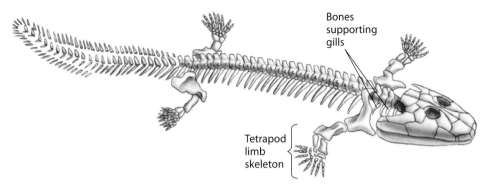

Figure 19.4B *Acanthostega*, a Devonian transitional form

dorsal and anal fins had been lost, and the tail fin was much smaller than *Eusthenopteron*. Certain features of its skull were more like those of a tetrapod. Although it had no neck, the shapes of the bones connecting forelimb to skull were intermediate between a fish and a tetrapod. *Pandericthys* could have been capable of leveraging its fins against the bottom as it propelled itself through shallow water. It was a fish, but a tetrapod-like fish.

On the other hand, *Acanthostega* (Figure 19.4B) was more of a fishlike tetrapod, and it turned scientists' ideas about tetrapod evolution upside down. Like *Ichthyostega*, *Acanthostega* had a neck, structural modifications that strengthened its backbone and skull, and four limbs with fingers and toes. But its limbs could not have supported the animal on land, nor could its ribs have prevented its lungs from collapsing out of water. The startling conclusion was that the first tetrapods were not fish with lungs that had gradually evolved legs as they dragged themselves from pool to pool in search of water. Instead, they were fish with necks and four limbs that raised their heads above water and could breathe oxygen from the air. Figure 19.4C shows an artist's rendering, with fanciful colors.

In 2006, a team of scientists added another important link to the chain of evidence. Using information from the dates and habitats of previous specimens, they predicted that transitional forms might be found in rock formed from the sediments of shallow river environments during a particular time period in the late Devonian. They found a suitable area to search in Greenland, which was located near the equator during that time. There, they discovered several remarkable fossils of an animal they named Tiktaalik ("large shallow water fish" in the language of the Nunavut Inuit tribe in Arctic Canada, where it was discovered). The specimens were exquisitely preserved; even its fishlike scales are clearly visible. Tiktaalik straddled the border between *Pandericthyes* and *Acanthostega* (Figure 19.4A). Its paddle-like forelimbs were part fin, part feet. The fin

rays had not been replaced by fingers and toes. The joints would have served to prop the animal up, but not enable it to walk. It had well-developed gills like a fish, but a tetrapod-like neck. It was a perfectly intermediate form.

With these images of early tetrapods gleaned from the fossil record in mind, let's look at the environmental conditions that drove their evolution. Plants had colonized the land 100 million years earlier, followed by arthropods. By the time Tiktaalik appeared, shallow water would have been a complex environment, with fallen trees and other debris from land plants, along with rooted aquatic plants, providing food and shelter for a variety of organisms. Even a meter-long predator like Tiktaalik would have found plenty to eat. But warm, stagnant water is low in oxygen. The ability to supplement its oxygen intake by air breathing—by lifting the head out of the water—would have been an advantage.

Once tetrapods had adaptations that enabled them to leave the water for extended periods of time, they diversified rapidly. Food and oxygen were plentiful in the Carboniferous swamp forests (see Module 17.6). From one of the many lines of tetrapods that settled ashore, modern amphibians evolved.

? How did *Acanthostega* change scientists' concept of tetrapod evolution?

■ It showed that the first tetrapods were more fishlike than previously thought. They did not spend time on land.

Figure 19.4C The first tetrapod, a four-limbed fish that lived in shallow water and could breathe air

19.5 Amphibians are tetrapods—vertebrates with two pairs of limbs

The early amphibians probably feasted on insects and other invertebrates in the lush forests of the Carboniferous period (see Module 17.6). As a result, amphibians became so widespread and diverse that the Carboniferous period is sometimes called the age of amphibians.

Amphibians include salamanders, frogs, and caecilians. Some present-day salamanders are entirely aquatic, but those that live on land walk with a side-to-side bending of the body that may resemble the swagger of early terrestrial tetrapods (Figure 19.5A). Frogs are more specialized for moving on land, using their powerful hind legs to hop along the terrain. Caecilians (Figure 19.5B) are nearly blind and are legless, looking like earthworms. However, they evolved from a legged ancestor. Most species burrow in moist soil in tropical areas.

In Greek, the word *amphibios* means "living a double life," a reference to the metamorphosis of many frogs. A frog spends much of its time on land, but it lays its eggs in water. As you can see in Figure 19.5C, the eggs are encapsulated in a jelly-like material. Consequently, they must be surrounded by moisture to prevent them from drying out. The larval stage, called a tadpole, is a legless, aquatic algae-eater with gills, a lateral line system resembling that of fishes, and a long, finned tail. In changing into a frog, the tadpole undergoes a radical metamorphosis (Figure 19.5D). When a young frog crawls onto shore and continues life as a terrestrial insect-eater, it has four legs and air-breathing lungs instead of gills. Not all amphibians live such a double life, however. Some species are strictly terrestrial, and others are exclusively aquatic. *Toad* is a term generally used to refer to frogs that have rough skin and live entirely in terrestrial habitats.

Most amphibians are found in damp habitats, where their moist skin supplements their lungs for gas exchange. Amphibian skin usually has poison glands that may play a role in defense. Poison arrow frogs have particularly deadly poisons, and their vivid coloration may warn away potential predators (Figure 19.5E). Frogs are usually quiet creatures, but during the breeding season, many species fill the air with their mating calls.

Figure 19.5A A spotted salamander

Figure 19.5B A caecilian

For the past 25 years, zoologists have been documenting a rapid and alarming decline in amphibian populations throughout the world. The causes may be multiple, including habitat degradation and the spread of a pathogenic fungus (see Module 17.16). The environmental assaults also include acid rain (see Module 2.16), which is especially damaging to amphibians because of their dependence on wet places for completing their life cycles.

Amphibians were the first vertebrates to colonize the land, but the distribution of most species is limited by their vulnerability to dehydration. The next clade of vertebrates we discuss, the amniotes, are able to complete their life cycles entirely on land.

? In what ways are amphibians not completely adapted for terrestrial life?

■ Their eggs are not well-protected against dehydration; many species have an aquatic larval form; their skin is not waterproof and must remain moist to permit gas exchange.

Figure 19.5C Frog eggs. A tadpole is developing in the center of each ball of jelly.

Figure 19.5D Tadpole undergoing metamorphosis

Figure 19.5E An adult poison arrow frog

Reptiles are amniotes—tetrapods with a terrestrially adapted egg

Figure 19.6A   A bull snake laying eggs

Reptiles (including birds) and mammals are **amniotes.** The major derived character of this clade is the **amniotic egg,** inside of which the embryo develops within a protective, fluid-filled sac called the amnion (see Chapter 27). The evolution of the amniotic egg, which functions as a "self-contained pond," enabled reptiles to complete their life cycles on land. Figure 19.6A shows a bull snake laying its eggs on dry land. As we saw in Module 17.2, the seed played a similar role in the evolution of plants.

The clade of amniotes called **reptiles** includes lizards, snakes, turtles, crocodilians, and birds, along with a number of extinct groups such as most of the dinosaurs. Lizards are the most numerous and diverse reptiles other than birds. Snakes, which are closely related to lizards, may have become limbless as their ancestors adapted to a burrowing lifestyle. Turtles have changed little since they evolved, although their ancestral lineage is still uncertain. Crocodiles and alligators (crocodilians) are the largest living reptiles, although some turtles are heavier. Crocodilians spend most of the time in water, breathing through upturned nostrils.

In addition to an amniotic egg protected in a waterproof shell, reptiles have several other adaptations for terrestrial living not found in amphibians. Reptilian skin, covered with scales waterproofed with the tough protein keratin, keeps the body from drying out. Reptiles cannot breathe through their dry skin and obtain most of their oxygen with their lungs, using their rib cage to help ventilate their lungs.

Nonbird reptiles are sometimes said to be "cold-blooded" because they do not use their metabolism to produce body heat. Nonetheless, reptiles do have ways to regulate their temperature. The bearded dragon of the Australian outback (Figure 19.6B) commonly warms up in the morning by sitting on warm rocks and basking in the sun. If the lizard gets too hot, it seeks shade. Because nonbird reptiles absorb the external heat rather than generating much of their own, they are said to be **ectothermic** (from the Greek *ektos,* outside, and *therme,* heat), a term that is more appropriate than the term *cold-blooded.*

Figure 19.6B
A bearded dragon

Like the amphibians from which they evolved, reptiles were once much more prominent than they are today. Following the decline of amphibians, reptilian lineages expanded rapidly, creating a dynasty that lasted 200 million years. Dinosaurs, the most diverse group, included the largest animals ever to inhabit land. Some were "gentle giants" like the large dinosaur in Figure 19.6C, lumbering about while browsing on vegetation. Others, like the 3-m-long *Deinonychus* (Greek for "terrible claw"), were voracious carnivores that ran on two legs. Unlike living reptiles (excluding birds), *Deinonychus* and other small dinosaurs may have been **endothermic,** using heat generated by metabolism to maintain a warm, steady body temperature. Paleontologists have also discovered signs of parental care in some dinosaurs.

Most dinosaurs died out during the period of mass extinctions about 65 million years ago (see Module 15.9). Descendants of one dinosaur lineage, however, survive today as the reptilian group we know as birds.

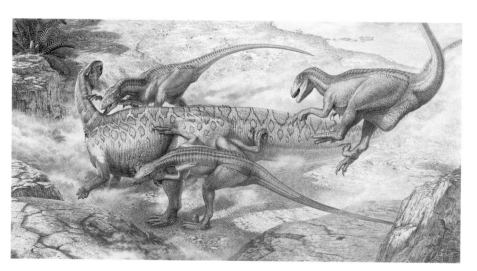

Figure 19.6C   Artist's reconstruction of a pack of *Deinonychus* attacking a tenontosaurus

? What is an amniotic egg?

■ An egg with a shell, housing an embryo in a fluid-filled sac, the amnion

19.7 Birds are feathered reptiles with adaptations for flight

Strong fossil evidence indicates that **birds** evolved from a lineage of small, two-legged dinosaurs called theropods. Chinese paleontologists have recently unearthed several fossils appearing to be feathered theropods. Such findings imply that feathers evolved long before powered flight. Among the possible functions of these early feathers were insulation and courtship displays.

Figure 19.7A is an artist's reconstruction based on a 150-million-year-old fossil of the oldest known, most primitive bird, called *Archaeopteryx* (from the Greek *archaios*, ancient, and *pteryx*, wing). Like living birds, it had feathered wings, but otherwise it was more like a small bipedal dinosaur of its era, with its teeth, wing claws, and tail with many vertebrae. Fossils of a number of other birds from the Cretaceous period show a gradual loss of certain features of dinosaurs, such as teeth, as well as the acquisition of traits shared by all birds today, including a short tail. Living birds seem to have evolved from a lineage of birds that survived the Cretaceous mass extinctions.

Nearly every part of the body of most birds reflects adaptations that enhance flight. Many features help reduce weight for flight: Present-day birds lack teeth; their tail is supported by only a few small vertebrae; their feathers have hollow shafts; and their bones have a honeycombed structure, making them strong but light. For example, the huge seagoing frigate bird (Figure 19.7B) has a wingspan of more than 2 m, but its whole skeleton weighs only about 113 g (4 oz).

Flight feathers shape bird wings into airfoils, providing lift and maneuverability in the air (see Figure 30.1E). Providing power for flight are large breast (flight) muscles, which are anchored to a keel-like breastbone. Most of what we call white meat on a turkey or chicken is the flight muscles.

Figure 19.7B   A soaring frigate bird, with a wingspan of 2 m

Flying requires a great amount of energy, and present-day birds have a high rate of metabolism and are endothermic. Insulating feathers help to maintain their warm body temperature. Supporting their high metabolic rate, birds have a highly efficient circulatory system, and their lungs are even more efficient at extracting oxygen from the air than are the lungs of mammals.

Flying safely requires acute senses. Birds have excellent vision, perhaps the best of all vertebrates. They have relatively large brains and display very complex behaviors, particularly during breeding season. The amniotic eggs of birds are covered with a hard shell, and in many species, male and female birds take turns incubating the eggs and then feeding the young. Some birds migrate great distances each year to different feeding or breeding grounds.

With wings driven by powerful flight muscles, present-day birds are masterful flyers. Some species, such as the frigate bird, have wings adapted to soaring on air currents, and they flap their wings only occasionally. Others, such as hummingbirds, excel at maneuvering but must flap almost continuously to stay aloft. Flight ability is typical of birds, but there are a few flightless species, including penguins and the ostrich and emu of Australia (Figure 19.7C).

Web Process of Science   *How Does Bone Structure Shed Light on the Origin of Birds?*

  List some adaptations of birds that enhance flight.

■ Reduced weight, endothermic with high metabolism, efficient respiratory and circulatory systems, feathered wings shaped like airfoils, good eyesight

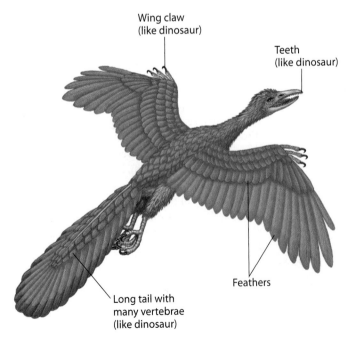

Wing claw
(like dinosaur)

Teeth
(like dinosaur)

Feathers

Long tail with
many vertebrae
(like dinosaur)

Figure 19.7A   *Archaeopteryx*, an extinct bird

Figure 19.7C
An emu, a
flightless bird

Mammals are amniotes that have hair and produce milk

There are two major lineages of amniotes: one that led to the reptiles (including birds) and one that produced the mammals. The first true **mammals** arose about 200 million years ago and were probably small, nocturnal insect-eaters. Of the three main groups of living mammals, monotremes are the oldest lineage. Marsupials diverged from eutherians (placental mammals) about 180 million years ago. During the Mesozoic era, mammals remained about the size of today's shrews, which are very small insectivores. After the extinction of large dinosaurs at the end of the Cretaceous period, however, mammals underwent an adaptive radiation, giving rise to large terrestrial herbivores and predators, as well as bats and aquatic porpoises and whales. The blue whale, an endangered species, grows to lengths of nearly 30 m and is the largest animal that has ever existed.

Two features—hair and mammary glands that produce milk to nourish young—are mammalian hallmarks. The main function of hair is insulation that helps maintain a warm body temperature. Like birds, mammals are endothermic. Efficient respiratory and circulatory systems (including a four-chambered heart) support a high rate of metabolism. A sheet of muscle called the diaphragm helps ventilate the lungs (see Module 22.5). Differentiation of teeth adapted for eating many kinds of foods is another mammalian trait.

Mammals generally have a larger brain than other vertebrates of the same size. The relatively long period of parental care provides time for offspring to learn from their parents.

Echidnas (spiny anteaters) and the duck-billed platypus, described in the chapter introduction, are the only existing **monotremes,** the egg-laying mammals. The female platypus usually lays two eggs and incubates them in a nest. After hatching, the young lick up milk secreted onto the mother's fur (Figure 19.8A).

Most mammals are born rather than hatched. During gestation, the embryos are nurtured inside the mother's body. The lining of the uterus and some membranes from the embryo (which are homologous to those of the amniotic egg of reptiles) form a **placenta,** a structure in which nutrients from the mother's blood diffuse into the embryo's blood.

Marsupials have a brief gestation and give birth to tiny, embryonic offspring that complete development while attached to the mother's nipples. The nursing young are usually housed in an external pouch, called the marsupium (Figure 19.8B). Nearly all marsupials live in Australia, New Zealand, and North and South America. As we discussed in the chapter introduction, Australia has been a marsupial sanctuary for much of the past 60 million years. Marsupials have diversified there, filling terrestrial habitats that on other continents are occupied by eutherians.

Eutherians are mammals that bear fully developed live young. They are commonly called **placental mammals** because their placentas are more complex than those of marsupials, and the young complete their embryonic development in the mother's uterus attached to the placenta. The large silvery membrane still clinging to the newborn zebra in Figure 19.8C is the amniotic sac, which held the developing embryo in a bath of protective amniotic fluid.

Elephants, rodents, rabbits, dogs, cows, whales, and bats are all examples of eutherians. Humans, also eutherians, belong to the order Primates, along with monkeys and apes. We begin our study of human evolution with the next module.

Web Activity *Characteristics of Chordates*

? What are two hallmarks of mammals?

■ Hair and mammary glands

Figure 19.8B   Marsupials: a gray kangaroo with her young in her pouch

Figure 19.8A   Monotremes: a duck-billed platypus with newly hatched young

Figure 19.8C   Eutherians: a zebra with newborn

19.9 The human story begins with our primate heritage

The mammalian order Primates includes the lemurs, tarsiers, monkeys, and apes. Humans are members of the ape group. The earliest primates were probably small arboreal (tree-dwelling) mammals that arose before 65 million years ago, when dinosaurs still dominated the planet. Most living primates are still arboreal, and the primate body has a number of features that were shaped, through natural selection, by the demands of living in trees. Although humans never lived in trees, the human body retains many of the traits that evolved in our arboreal ancestors.

The squirrel-sized slender loris in Figure 19.9A illustrates a number of primate features. It has limber shoulder and hip joints, enabling it to climb and brachiate (swing from one branch to another). The five digits of its grasping feet and hands are highly mobile; the separation of its big toe from the other toes and its flexible thumb give the lemur the ability to hang onto branches and manipulate food. The great sensitivity of the hands and feet to touch also aids in manipulation. Lorises have a short snout and eyes set close together on the front of the face. The position of the eyes makes their two fields of vision overlap, enhancing depth perception, an important trait for maneuvering in trees. We humans share all these basic primate traits with the slender loris except for the widely spaced big toe.

As shown in the phylogenetic tree in Figure 19.9B, the slender loris belongs to one of three main groups of living primates. The lorises and pottos of tropical Africa and southern Asia are placed in one group along with the lemurs. Ranging from the pygmy mouse lemur, which weighs 25 g (1 oz), to the sifakas (Figure 19.9C, facing page), which may weigh as much as 8 kg (17.6 lbs), lemurs are a diverse group. They are found only in Madagascar, a Texas-sized island in the Indian Ocean about 420 km off the eastern coast of Africa. Severed from the African continent by plate movements, the island has been an isolated hotbed of

Figure 19.9A   A slender loris

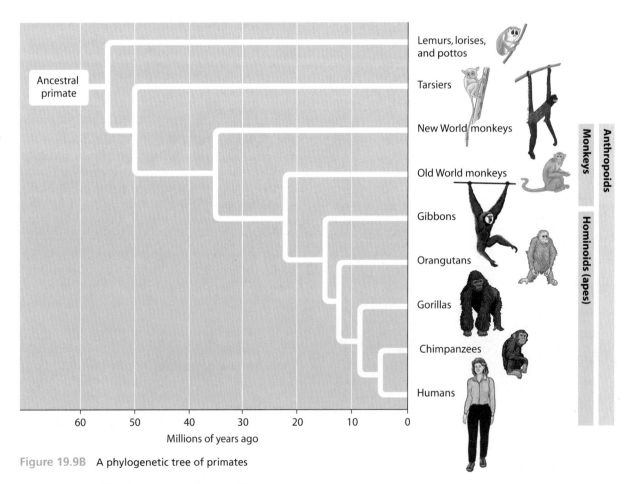

Figure 19.9B   A phylogenetic tree of primates

speciation for well over 100 million years. But of about 50 species of lemurs originally present, 18 have become extinct since humans first colonized Madagascar about 2,000 years ago. Most lemurs are agile climbers and leapers that spend nearly all their time in trees. Thus, they are threatened by the destruction of their forest homes.

The tarsiers form a second group of primates. Limited to Southeast Asia, these small, nocturnal tree-dwellers have flat faces with large eyes (Figure 19.9D). Fossil evidence indicates that tarsiers are more closely related to anthropoids, the third group of primates, than to the loris–potto group.

The **anthropoids** (from the Greek *anthropos,* man, and *eidos,* form) include monkeys and apes. The ape group, the **hominoids,** includes humans. Anthropoids generally have a larger brain relative to body size and rely more on eyesight and less on sense of smell than other mammals. Anthropoids have a fully **opposable thumb;** that is, they can touch the tip of all four fingers with their thumb. In monkeys and most apes, the opposable thumb functions in a grasping "power grip," but in humans, a distinctive bone structure at the base of the thumb allows it to be used for more precise manipulation.

The fossil record indicates that anthropoids began diverging from other primates about 50 million years ago. Notice that monkeys do not constitute a monophyletic group. The first monkeys probably evolved in the Old World (Africa and Asia) and may have reached South America by rafting on logs from Africa. The monkeys of the Old World and New World (the Americas) have been evolving separately for over 30 million years.

New World monkeys, found in Central and South America, are all arboreal. Their nostrils are wide open and far apart, and many, such as the woolly spider monkey, an inhabitant of rain forests in eastern Brazil (Figure 19.9E, left), have a long tail that is prehensile—specialized for grasping tree limbs. The squirrel-sized golden lion tamarin (Figure 19.9E, right) is a New World monkey that inhabits lowland rain forests of eastern Brazil. Most of its habitat having been replaced by housing developments, this species was reduced to fewer than 200 individuals in the wild in the 1970s. With an intense international conservation effort to save the species from extinction, its numbers have rebounded to about 1,200.

In contrast, Old World monkeys lack a prehensile tail, and their nostrils open downward. They include macaques (Figure 19.9F), mandrills, baboons, and rhesus monkeys. Many species are arboreal, but some, such as the baboons of the African savanna, are ground-dwelling. Both New and Old World monkeys are active during the day and usually live in bands held together by social behavior.

Monkeys differ from most apes in having forelimbs that are about equal in length to their hind limbs. Old World monkeys and apes (hominoids) diverged about 20–25 million years ago. The human lineage probably diverged from an ancestor shared with chimpanzees somewhere between 5 and 7 million years ago. We look at the apes next.

Figure 19.9C A sifaka, a type of lemur, with its young

Figure 19.9D A tarsier, member of a distinct primate group

? To which mammalian order do we belong? What are the three main groups of this order?

■ Primates: Lorises, pottos, and lemurs; tarsiers; and anthropoids

Figure 19.9E New World monkeys: The woolly spider monkey (left; note prehensile tail), and the golden lion tamarin (right; note nostrils that open to the side)

Figure 19.9F Old World monkey: a macaque with its young

19.10 Hominoids include humans and four other groups of apes

Hominoids (apes) include the gibbons, orangutans, gorillas, chimpanzees (and bonobos), and humans. Apart from the human lineage, the apes have a smaller geographic range than the monkeys; they evolved and diversified only in Africa and Southeast Asia and are confined to tropical regions (mainly rain forests). Apes lack a tail and have relatively long arms and short legs. They are chiefly vegetarians, although chimpanzees also eat insects and some vertebrates, such as young antelopes, pigs, and monkeys. Humans also are omnivorous. Apes have larger brains relative to body size than other primates, and their behavior is consequently more flexible. Gorillas, chimpanzees, and humans have a high degree of social organization.

Nine species of gibbons, all found in Southeast Asia, are the only entirely arboreal apes (Figure 19.10A). Gibbons are the smallest, lightest, and most acrobatic of the apes. They are also the only nonhuman apes that are monogamous, with mated pairs remaining together for life.

The orangutan is a shy, solitary species that lives in the rain forests of Sumatra and Borneo. The largest living arboreal mammal, it moves rather slowly through the trees, supporting its stocky body with all four limbs (Figure 19.10B). Orangutans may occasionally venture onto the forest floor.

The gorilla (Figure 19.10C) is the largest ape: Some males are almost 2 m tall and weigh about 200 kg (440 lb). Found only in African rain forests, gorillas usually live in groups of up to about 20 individuals. They spend nearly all their time on the ground. Gorillas can stand upright on their hind legs. But when they walk on all fours, their knuckles contact the ground.

Figure 19.10A A gibbon

Figure 19.10B An orangutan

Figure 19.10C A gorilla and offspring

Figure 19.10D A chimpanzee

Like the gorilla, the chimpanzee and a closely related species called the bonobo are knuckle walkers. These apes spend as much as a quarter of their time on the ground. Both species inhabit tropical Africa. Chimpanzees have been studied extensively, and many aspects of their behavior resemble human behavior. For example, chimpanzees make and use simple tools. The individual in Figure 19.10D is using a blade of grass to "fish" for termites. Chimpanzees also raid other social groups of their own species, exhibiting behavior formerly thought to be uniquely human. Researchers have demonstrated repeatedly that chimpanzees can learn human sign language. However, we do not yet know what role symbolic communication plays in the behavior of wild chimpanzees.

One of our most entrenched beliefs is that humans are the only thinking, self-aware beings. The behavior of chimpanzees in front of mirrors, however, challenges this belief. When first introduced to a mirror, a chimpanzee responds the way most other animals do—as if it were seeing another individual of its species. After several days, though, a chimp will begin using a mirror in ways that indicate it has a concept of self. It will inspect its face and other parts of its body that it cannot see without the mirror. It will also make faces at the mirror, using expressions different from those used in communicating with others.

Molecular evidence indicates that chimpanzees and gorillas are more closely related to humans than they are to other apes. Humans and chimpanzees are especially closely related; their genomes are 99% identical. Primate researchers are acutely aware of the special significance of the great apes to us. In the words of chimpanzee authority Jane Goodall, "The most important spin-off of the chimp research is probably the humbling effect it has on us who do the research. We are not, after all, the only aware, reasoning beings on this planet."

Web Activity *Primate Diversity*

? Which primate groups are classified as anthropoids, and which groups are classified as hominoids?

Anthropoids include the monkeys and hominoids. Hominoids include the gibbons, orangutans, gorillas, chimpanzees, and humans.

19.11 The hominid branch of the primate tree is only a few million years old

Humans and chimpanzees diverged from a common ancestor, probably between 5 and 7 million years ago (see Figure 19.9B). **Paleoanthropology,** the study of human origins and evolution, focuses on this tiny slice of biological history. If we compress the history of life to a year, the human branch has only existed for less than 18 hours. Paleoanthropologists have unearthed fossils of approximately 20 species of extinct **hominids,** species that are more closely related to humans than to chimpanzees and are therefore on the human branch of the evolutionary tree. (*Hominin* is a synonym of *hominid.* But remember that the term *hominoid* refers to all apes, including humans, and that *anthropoid* is an even broader term, since it includes monkeys.)

Although thousands of hominid fossils have been discovered, paleoanthropologists are still vigorously debating their classification and phylogenetic relationships. Therefore Figure 19.11 presents some of the known hominids in a time line rather than in a tree diagram like Figure 19.9. The vertical bars indicate the approximate time period when each species existed, as judged from the fossil record. The oldest possible hominid yet discovered, *Sahelanthropus tchadensis,* lived from about 7 to 6 million years ago, around the time when the lines that led to humans and chimpanzees diverged. Some experts think there is too little evidence to judge whether *Sahelanthropus, Orrorin,* and

Ardipithecus are the earliest hominids, or part of a different branch. The fossil record suggests that hominid diversity increased dramatically in the period between 4 and 2 million years ago. Many of the groups from that time are collectively called **australopiths.** Australopiths got their name from the 1924 discovery in South Africa of *Australopithecus africanus* ("southern ape of Africa"), which lived from 3 to 2.4 million years ago.

A common misconception is to think of human evolution as a parade of hominids leading directly from an ancestral hominoid to *Homo sapiens.* If human evolution is a parade, it is a very disorderly one, with many groups breaking away from the march to wander down dead-end alleyways. At times several hominid species coexisted, but all except one—our own—ended in extinction. In the next module, we take a closer look at one of the ancient members of our family tree.

🎧 MP3 Tutor *Human Evolution*

? Based on the fossil evidence represented in Figure 19.11, how many hominid species existed 1.7 million years ago?

■ *Five: P. boisei, P. robustus, H. habilis, H. ergaster, H. erectus*

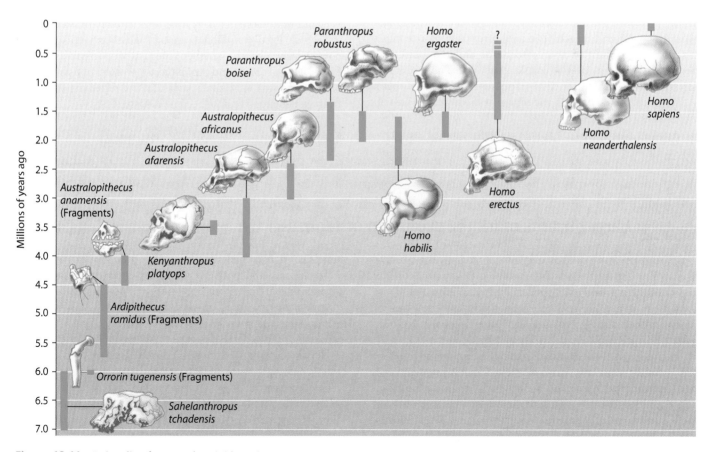

Figure 19.11 A time line for some hominid species

19.12 Upright posture evolved long before an enlarged brain

In the early 20th century, paleoanthropologists hypothesized that increased brain size was the initial change that separated hominids from apes. Bipedalism (upright walking) and other adaptations came later, as hominid intelligence led to changes in food-gathering methods, parental care, and social interactions. But fossil evidence was scarce, and until the discovery of a knee joint in 1973, the hypothesis was virtually untested. The knee joint, found in the Afar region of Ethiopia, was from a bipedal hominid—and it was more than 3 million years old. The following year, the same researchers found a 3.24-million-year-old female skeleton in the same area. As the team celebrated their discovery at their desert campsite, a tape of the Beatles song "Lucy in the Sky with Diamonds" played repeatedly, and they nicknamed the skeleton Lucy. Lucy was petite, only about 1 m tall, with a brain the size of a softball. Lucy and similar fossils (hundreds have since been discovered) were assigned to the species *Australopithecus afarensis*. Fossils show that *A. afarensis* walked on two legs and existed as a species for at least 1 million years.

In 1978, another team of paleoanthropologists discovered unique evidence of bipedalism in ancient hominids. While working in what is now Tanzania in East Africa, they found a 3.6-million-year-old layer of hardened volcanic ash crisscrossed with tracks of hyenas, giraffes, and several extinct species of mammals—including upright-walking hominids (Figure 19.12). After the ash had settled, it was dampened by rain. The feet of two hominids, one large and one small, walking close together, made impressions in the ash as if it were wet sand on a beach. The ash, composed of a cement-like material, solidified soon afterwards and was buried by more ash from a later volcanic eruption. The hominids strolling across that ancient landscape may have been *A. afarensis*, which lived in the region at the time, but we will never know for certain.

Another clue to upright stance is the location of the opening in the base of the skull through which the spinal cord exits. In early hominids (and in humans), this opening is located

Ancient footprints

Figure 19.12 Evidence of bipedalism in early hominids

underneath the skull, allowing the head to be held directly over the body. In chimpanzees and other species that are primarily quadipeds, the spinal cord exits farther toward the rear of the skull, at an angle.

Paleoanthropologists are now certain that bipedalism is a very old trait, dating back at least to *Australopithecus anamensis*, which lived 4 million years ago. Fossil evidence has not conclusively shown whether the earlier hominids, *Sahelanthropus, Orrorin,* and *Ardipithecus,* walked upright. But in any case, it was only much later that the other major human trait—an enlarged brain—became evident.

? How can paleoanthropologists conclude that a species was bipedal based on only a fossil skull?

■ By the location of the opening where the spinal cord exits the skull

19.13 Larger brains mark the evolution of *Homo*

By measuring fossil skulls, paleoanthropologists can estimate the size of the brain, which, relative to body size, roughly indicates the animal's intelligence. The brain volume of *Homo sapiens*, at an average 1,350 cc, is approximately triple that of australopithecines. As evolutionary biology Stephen J. Gould put it, "Mankind stood up first, and got smart later."

Enlargement of the hominid brain is first evident in fossils of *Homo habilis* from East Africa dating to about 2.4 million years ago. Its name, which means "handy man," refers to the sharp stone tools sometimes found with these fossils. *H. habilis* lived at the same time as *Paranthropus* (see Figure 19.11), husky hominids with large molars and enormous jaws and chewing muscle, adaptations for eating tough plant material. In con-

trast, *H. habilis* had smaller teeth and shorter jaws than *Paranthropus*, and probably ate meat acquired by scavenging. At 500 to 800 cc, its brain volume was nearly double that of *Australopithecus*. Based on new findings from the human and chimpanzee genomes, some scientists speculate that a mutation weakening the jaw muscles gave the brain more space to grow.

Fossils dating to 1.9 to 1.6 million years ago, considered by many paleoanthropologists to be those of a distinct species, *Homo ergaster*, mark not only a further increase in brain size, ranging from 850 to 1,100 cc, but a new stage in hominid evolution. *H. ergaster* has been associated with more sophisticated stone tools. Its long, slender legs with hip joints were well adapted for long-distance walking. The narrow pelvis placed an upper limit on fetal head circumference (and thus brain development at birth) as it does today, an indication that *H. ergaster* babies may have needed a long period of parental care.

Fossils of *H. ergaster* were originally considered early members of another species, *Homo erectus,* and they are certainly closely related. In *H. erectus* ("upright man"), brain volume had increased to 1,000 cc. These were the first hominids to extend their range beyond Africa. The oldest known fossils of hominids outside Africa, discovered in 2000 in the former Soviet Republic of Georgia, date back 1.8 million years. *H. erectus* eventually migrated as far as Indonesia. What was the fate of these larger-brained hominids who traveled across Asia? Although most fossil evidence indicates that *H. erectus* became extinct at some point about 200,000 years ago, recent discoveries raised the possibility that a population in Java (Indonesia) survived until 25,000 years ago.

More shocking, however, was the 2003 discovery of the skeleton of a tiny hominid on Flores, an island east of Java. Its discoverers think it is a new species of *Homo* that evolved from a population of *Homo erectus* into a dwarf form on the island and lived there as recently as 13,000 years ago. Although it was given the name *Homo floresiensis,* fierce controversy has erupted over its status as a separate species. While many scientists now agree that *H. floresiensis* is a previously undiscovered species, others think it is simply *Homo sapiens* that had a genetic disorder. The final verdict awaits the discovery of more fossils.

? Place the following hominids in order of increasing brain volume: *Australopithecus, H. erectus, H. ergaster, H. habilis, H. sapiens*

■ *Australopithecus, H. habilis, H. erectus, H. ergaster, H. sapiens*

19.14 Neanderthals and our human ancestors diverged half a million years ago

Of all ancient people, none have captured the popular imagination like Neanderthals. Since the discovery of their fossilized remains in the Neander Valley in Germany 150 years ago (Figure 19.14), Neanderthal reconstructions have created an image of this "caveman" as a heavy-browed, stocky brute partially clothed in animal skins. Novels and screenplays have imagined their daily lives and even speculated that remnant populations still exist in remote regions. Above all, people have wondered how Neanderthals are related to us. Are they the ancestors of Europeans? Close cousins? Or part of a different branch of evolution altogether?

Scientists have long puzzled over certain similarities between Neanderthals and modern humans. Neanderthals were muscular and robust, with a brain similar to ours in size but slightly different in shape. They had large noses, as well as heavy brows and cheekbones, and made hunting tools from stone and wood. Neanderthals lived alongside Cro-Magnons, a somewhat different-looking group who were the direct ancestors of present-day Europeans. Neanderthals and Cro-Magnons likely competed for territory, food, and other resources. For decades, researchers debated whether the two groups were biologically separate or interbreeding.

Genetic analysis is answering some questions about the Neanderthals and could answer many more in the future. In 1997, researchers isolated mitochondrial DNA from Neanderthal bones. Analysis showed that the DNA is different from that of all living humans. A greater similarity to living Europeans would be expected if Neanderthals had indeed contributed genes to Europeans. Instead, it seems that there was little or no genetic contribution by Neanderthals to present-day humans. Similar results were later obtained from several more specimens. All of modern Neanderthals studied form their own distinct genetic group, while modern Europeans are more closely related to Africans and Asians. This genetic evidence supports previously known fossil evidence that Neanderthals arose from a distinct species that arrived in Europe long before the ancestors of modern-day humans and that Neanderthals became extinct without contributing to the gene pool of modern humans. Researchers estimate that the last common ancestor of humans and Neanderthals lived around 500,000 years ago.

A new project to sequence Neanderthal chromosomal DNA may uncover more information about the Neanderthals than we could ever learn from bones and artifacts. A research team has extracted nuclear DNA from fossilized bones and so far has managed to identify a million nucleotides from it. The Neanderthal genome could potentially reveal details about their physical appearance, their physiology, and their brain structure.

In the next module, we look at the origin and worldwide spread of our own species, *Homo sapiens.*

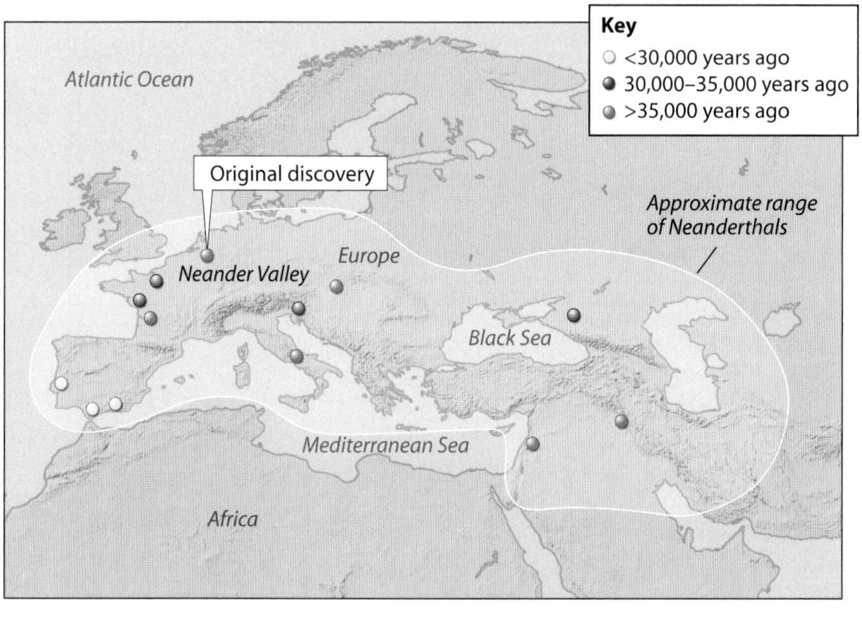

? What evidence supports the conclusion that Neanderthals were not the ancestors of Europeans?

Figure 19.14 Sites where Neanderthal fossils have been found

Key
○ <30,000 years ago
● 30,000–35,000 years ago
◉ >35,000 years ago

Atlantic Ocean

Original discovery

Europe

Neander Valley

Black Sea

Approximate range of Neanderthals

Mediterranean Sea

Africa

■ Both fossil evidence and analysis of mitochondrial DNA

19.15 From origins in Africa, *Homo sapiens* spread around the world

Evidence from fossils and DNA studies is coming together to support a compelling hypothesis about how our own species, *Homo sapiens,* emerged and spread around the world.

The ancestors of humans originated in Africa. Older species (perhaps *H. ergaster* or *H. erectus*) gave rise to newer species and ultimately *H. sapiens.* The oldest known fossils of our own species have been discovered in Ethiopia and include fossils that are 160,000 and 195,000 years old. These early humans lacked the heavy brow ridges of *H. erectus* and Neanderthals (*H. neanderthalensis*) and were more slender.

Molecular evidence about the origin of humans supports the conclusions drawn from fossils. In addition to showing that living humans are more closely related to one another than to Neanderthals, DNA studies indicate that Europeans and Asians share a more recent common ancestor and that many African lineages represent earlier branches on the human tree. These findings strongly suggest that all living humans have ancestors that originated as *H. sapiens* in Africa.

This conclusion is further supported by analyses of mitochondrial DNA, which is maternally inherited, and Y chromosomes, which are transmitted from fathers to sons. Such studies suggest that all living humans inherited their mitochondrial DNA from a common ancestral woman who lived approximately 160,000–200,000 years ago. Mutations on the Y chromosomes can serve as markers for tracing the ancestry and relationships among males alive today. By comparing the Y chromosomes of males from various geographic regions, researchers were able to infer divergence from a common African ancestor.

Evidence suggests that our species emerged from Africa in one or more waves, spreading first into Asia 50–60,000 years ago, then to Europe, Southeast Asia, and Australia (Figure 19.15). The date of the first arrival of humans in the New World is hotly debated.

The rapid expansion of *H. sapiens* may have been spurred by the evolution of human cognition as our species evolved in Africa. Although Neanderthals and other hominids were able to produce sophisticated tools, they showed little creativity and not much capability for symbolic thought, as far as we can tell. In contrast, researchers are beginning to find evidence of increasingly sophisticated thought as *H. sapiens* evolved. In 2002, researchers reported the discovery in South Africa of 77,000-year-old art, and by 36,000 years ago, humans were producing spectacular cave paintings.

The capacity for speech, perhaps the most uniquely human characteristic, must also have been a factor in the evolutionary success of *H. sapiens.* As you will learn in the next module, researchers are beginning to understand how speech evolved.

Web Activity *Human Evolution*

? What types of evidence indicate that *Homo sapiens* originated in Africa?

■ Fossils and analyses of mitochondrial DNA and Y chromosomes

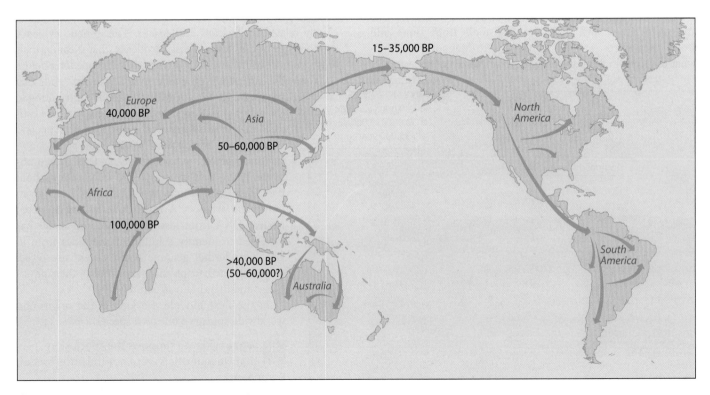

Figure 19.15 Migration of modern *Homo sapiens*

19.16 A genetic difference helped humans start speaking

Human language is unique. Other primates vocalize sounds and can use them to communicate. Chimpanzees, for example, utter cries that warn others of impending danger. But many nonprimate animals also share this form of simple communication. In contrast, humans are not only able to vocalize sounds, they also conduct complex communication, creating intricate societies that function through the use of shared language. Furthermore, humans use language in abstract ways—referring to objects that are not present, talking about the future, and generalizing.

Researchers studying a family with a rare developmental disorder that causes severe speaking difficulties linked a specific gene to language ability. The gene they identified, called *FOXP2*, has broad power to shape brain development. Rather than controlling just one function, *FOXP2* codes for a transcription factor (see Module 11.5) that controls the expression of many genes. In the family with the speech disorder, affected individuals carry a mutated version of the gene that interferes with the development of brain areas linked to speech.

That finding prompted comparisons to other species. Other animals, including crocodiles, birds, mice, and monkeys, also have a *FOXP2* gene. The animals whose gene is most similar to humans are song-learning birds, such as canaries and parakeets. Like humans, these birds learn their vocal behavior from others of their species. But the human version of *FOXP2* is unique, perhaps explaining our unique linguistic abilities. Further research has determined that the human form of *FOXP2* likely arose within the last 100,000 years, a time period that matches the emergence of *Homo sapiens*.

FOXP2 is not the only gene that allows humans to speak; other genes shape other parts of the brain and anatomy critical for talking. But researchers think that *FOXP2* is a key to the linguistic abilities that enable the expression of human thought and make it possible for older members of the species to pass large amounts of knowledge on to younger members.

? **FOXP2 shapes the development of which organ?**

■ The brain

Evolution Connection

19.17 Human skin color reflects adaptations to varying amounts of sunlight

In today's multicultural society, skin color is one of the most striking differences among individuals. For centuries, people assumed that these differences reflected more fundamental genetic distinctions, but modern genetic analysis has soundly disproven those assumptions. Is there an evolutionary explanation for skin color differences? To develop hypotheses, scientists began with the observation that human skin color varies geographically. People indigenous to tropical regions have darker skin pigmentation than people from more northerly latitudes.

Skin color results from a pigment called melanin that is produced by specialized skin cells. We all have melanin-producing cells, but the cells are less active in people who have light-colored skin. In addition to absorbing visible light, and therefore appearing dark-colored, melanin absorbs ultraviolet wavelengths. We know that ultraviolet radiation (UV) causes mutations (see Module 11.21), but it has other effects in skin, as well.

UV radiation helps catalyze the synthesis of vitamin D in the skin. This vitamin is essential for proper bone development, so it is especially important for pregnant women and small children to receive adequate amounts. By blocking UV radiation, melanin prevents vitamin D synthesis. Dark-skinned humans evolving in equatorial Africa received sufficient UV radiation to make vitamin D, but northern latitudes receive less sunlight. The loss of skin pigmentation in humans that migrated north from Africa probably helped their skin receive adequate UV radiation to produce enough vitamin D.

Why did humans evolve dark skin in the first place? UV radiation degrades folate (folic acid), a vitamin that is vital for fetal development and spermatogenesis. Researchers hypothesize that dark skin was selected for because melanin protects folate from the intense tropical sunlight. The evolution of differing skin pigmentations likely provided a balance between folate protection and vitamin D production (Table 19.17).

Because skin color was the product of natural selection, similar environments produced similar degrees of pigmentation. Widely separated populations may have the same adaptation, regardless of how closely they are related. As a result, skin color is not a useful characteristic for judging phylogenetic relationships.

TABLE 19.17	CORRELATION OF UV RADIATION WITH RISK OF VITAMIN D AND FOLATE DEFICIENCIES		
Latitude	UV Radiation	Risk of Vitamin D Deficiency	Risk of Folate Deficiency
Tropical latitudes 0–23.5° N and S	High	Low	High
Northern latitudes 23.5–90° N and S	Low	High	Low

? **Why didn't folate degradation select against lightly pigmented people in northern latitudes?**

■ UV radiation is less intense in northern latitudes than in the tropics, so it did not have an adverse effect on folate levels.

19.18 Humans threaten animal diversity by introducing non-native species

The animal kingdom's amazing diversity arose through hundreds of millions of years of evolution. But this diversity can be threatened by the actions of a latecomer on the scene—humans—as when a small, catlike creature meets an unstoppable toad imported from overseas. That catlike creature is the quoll (Figure 19.18A), one of the Australian marsupials discussed in the chapter introduction. Like the majority of the more than 1 million species of plants and animals found in Australia, quolls are endemic to that geographically isolated continent; that is, they are found only there. And when such local species encounter invasive non-native species, the results are often disastrous.

Quolls are predators, hunting smaller animals, including Australia's many types of frogs. Their diet didn't present a problem until 1935, when sugarcane growers in northern Australia decided to import a boxful of foreign toads to fight beetles that were damaging sugar crops. The non-native amphibians from South America, known as cane toads (shown in Figure 19.18B), didn't do much to stop the beetles. Instead, the 102 toads in that box quickly turned into one of Australia's biggest wildlife disasters. The invaders bred and spread quickly, with female cane toads able to lay as many as 20,000 eggs in a single season. Cane toads now inhabit vast stretches of northern and eastern Australia.

A number of the cane toad's behaviors and characteristics spell doom for native inhabitants. Cane toad tadpoles develop more quickly than those of native Australian frogs, beating them in the search for food. Adult cane toads have voracious appetites, eating everything from dog food to mice and devouring many native Australian insects and small animals along the way. They can grow to weigh more than 4 pounds! And, more importantly, they are poisonous to almost all their predators, including quolls. Quolls that eat cane toads die. Together with habitat loss and other environmental pressures, the introduction of cane toads to Australia has driven native quolls into serious decline.

This scenario of invasive non-native species harming native species has been repeated over and over in Australia, as a rush of human-introduced species in the last 200 years has threatened to turn the landscape upside down. In the mid-1800s, a would-be hunter brought over a few dozen rabbits, creating a population explosion of more than 200 million rabbits that have devoured native vegetation and turned grasslands into dusty deserts that offer little food for native animals. Non-native foxes, also brought in by sport hunters, have devoured populations of bilbies (Figure 19.18C), wallabies, and other small marsupials not adapted to confront such predators. Cats have also created havoc among the native animals.

Australia now spends millions of dollars each year combating non-native plants and animals, trying everything from mass hunting programs to biological control efforts. One effort involves moving species to safer territory. In 2003, 65 quolls were relocated from toad-infested lands in northern Australia to uninhabited offshore islands beyond the cane toad's reach. The transplanted quolls have thrived, and biologists hope that the island shelters will protect at least some members of the species while the cane toad problem receives further study. The damage from invasive species in Australia has become a continent-wide example of how precarious diversity can be: Millions of years worth of evolutionary changes can be threatened by just a few years of human-caused changes to native habitats.

The introduction of non-native species is just one of the environmental problems that humans brought with them as they spread around the globe and that are now threatening our own existence. We revisit the effects of non-native species in Chapter 37 and consider other human-related threats to the environment in Chapter 38.

? Why are native species often threatened by non-native species?

■ Non-native species may compete with or prey on (or poison, in the case of cane toads) native species that did not evolve with the newcomers and thus have not been able to adapt to their presence.

Figure 19.18A A quoll, an endangered species in Australia that preys on cane toads

Figure 19.18B The poisonous cane toad, an invasive species in Australia

Figure 19.18C Bilbies, Australian marsupials preyed on by non-native foxes

Chapter Review

Reviewing the Concepts

Vertebrate Evolution and Diversity (19.1–19.8)

A chordate phylogenetic tree is based on a sequence of derived characters. Most chordates are vertebrates, with a head and a backbone made of vertebrae (**19.1**).

Lampreys and hagfishes lack hinged jaws. Hagfishes also lack a vertebral column (**19.2**).

Jawed fishes, like most vertebrates, have hinged jaws, which may have evolved from skeletal supports of the gill slits. Jawed fishes include chondrichthyans (sharks and rays with skeletons made of cartilage), ray-finned fishes (with operculi that move water over the gills and a buoyant swim bladder), and lobe-finned fishes (with muscular fins supported by bones) (**19.3**).

Tetrapods evolved from lobe-finned fish that adapted to the low oxygen of shallow-water habitats by propping up on muscular forelimbs and breathing air (**19.4**).

Amphibians, represented by frogs, salamanders, and caecilians, were among the first terrestrial tetrapods, but their embryos and larvae still develop in water (**19.5**).

Reptiles are amniotes. The shelled amniotic egg and waterproof scales of reptiles (snakes, lizards, turtles, crocodilians, extinct dinosaurs, and birds) are terrestrial adaptations. Living reptiles other than birds are ectothermic (**19.6**).

Birds are reptiles that have wings, feathers, an endothermic metabolism, and many other adaptations related to flight (**19.7**).

Mammals are endothermic amniotes with hair, which insulates their bodies, and mammary glands, which produce milk. Monotremes lay eggs. The embryos of marsupials and eutherians are nurtured by the placenta within the uterus. Marsupial offspring complete development attached to the mother's nipples, usually in a pouch. Eutherians (placental mammals) complete development before birth (**19.8**).

Primate Diversity (19.9–19.10)

Primates had evolved as small arboreal mammals by 65 million years ago. Primate characters include limber joints, grasping hands and feet with flexible digits, a short snout, and forward-pointing eyes that enhance depth perception. The three groups of living primates are the lorises, pottos, and lemurs; the tarsiers; and the anthropoids (monkeys and apes) (**19.9**).

Hominoids include gibbons, orangutans, gorillas, chimpanzees, and humans. Hominoids have larger brains than other primates and lack tails. Humans share 99% of our genome with chimpanzees, our closest living relatives (**19.10**).

Hominid Evolution (19.11–19.18)

Hominids, which include species on the human evolutionary branch, and chimpanzees probably diverged from a common ancestor between 5 and 7 million years ago. Major milestones in hominid evolution included the appearance of bipedalism and a larger brain. The oldest known hominid, *Sahelanthropus tchadensis,* dates from 7 million years ago. Paleontologists have discovered fossils of approximately 20 species of extinct hominids. *Homo sapiens* is the only hominid that has not become extinct (**19.11**).

Bipedalism had evolved by 4 million years ago and preceded the evolution of the enlarged brain characteristic of later hominids. Fossils demonstrate that *Australopithecus* species walked upright, but the evidence on earlier hominids is inconclusive (**19.12**).

The genus *Homo* includes hominids with larger brains and evidence of tool use. *Homo ergaster* had a larger brain than *H. habilis*. *H. erectus*, with a larger brain than *H. ergaster,* was the first hominid to spread out of Africa (**19.13**).

Neanderthals were a species of *Homo* that lived in Europe as recently as 30,000 years ago. However, they were not closely related to modern humans. Genetic analysis has shown that the last common ancestor of Neanderthals and *Homo sapiens* lived roughly 500,000 years ago (**19.14**).

Homo sapiens evolved in Africa and later migrated to Asia, then to Europe, and later to North and South America. Evidence from fossils and DNA studies have enabled scientists to trace early human history (**19.15**).

Language is a uniquely human trait that permits the creation of human cultures. Linguistic ability has been linked to the human version of the *FOXP2* gene (**19.16**).

Human skin color variations probably resulted from natural selection balancing dark skin's ability to block UV radiation, which destroys folate, with the need for UV radiation to synthesize vitamin D (**19.17**).

Invasive species, transported by humans beyond their native habitats, are a serious threat to biological diversity (**19.18**).

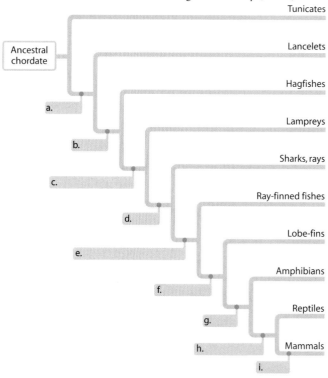

Connecting the Concepts

1. In the chordate phylogenetic tree on the previous page, fill in the key derived character that defines each clade.

2. List some of the derived characters of these groups.

Group	Derived characters
a. Primates	
b. Hominoids	
c. Hominids	

Testing Your Knowledge

Multiple Choice

3. A lamprey, a shark, a lizard, and a rabbit share all the following characteristics except
   a. pharyngeal slits in the embryo or adult.
   b. vertebrae.
   c. hinged jaws.
   d. a dorsal, hollow nerve cord.
   e. a post-anal tail.

4. Why were the Tiktaalik fossils an exciting discovery for scientists studying tetrapod evolution?
   a. They are the earliest frog-like animal discovered to date.
   b. They show that tetrapods successfully colonized land much earlier than previously thought.
   c. They have a roughly equal combination of fishlike and tetrapod-like characteristics.
   d. They demonstrate conclusively that limbs evolved as lobe-fins dragged themselves from pond to pond during droughts.
   e. They reveal new information about the breeding behavior of early tetrapods.

5. In Australia, marsupials fill the niches that placental mammals fill in other parts of the world because
   a. they are better adapted and have outcompeted placental mammals (eutherians).
   b. they originated in Australia.
   c. they evolved from monotremes that migrated to Australia about 50 million years ago.
   d. human-caused environmental changes have favored the success of marsupials.
   e. after Pangaea broke up, they diversified in physical isolation from placental mammals.

6. Fossils suggest that the first major trait distinguishing hominids from other primates was
   a. a larger brain.
   b. erect posture.
   c. forward-facing eyes with depth perception.
   d. grasping hands.
   e. toolmaking.

7. Which of the following correctly lists possible ancestors of humans from the oldest to the most recent?
   a. *Homo erectus, Australopithecus, Homo habilis*
   b. *Australopithecus, Homo habilis, Homo erectus*
   c. *Australopithecus, Homo erectus, Homo habilis*
   d. *Homo ergaster, Homo erectus, Homo neanderthalensis*
   e. *Homo habilis, Homo erectus, Australopithecus*

8. Which of these is not a member of the anthropoids?
   a. chimpanzee   d. human
   b. ape   e. New World monkey
   c. tarsier

9. Studies of DNA support which of the following?
   a. Members of the group called australopiths were the first to migrate from Africa.
   b. Neanderthals are more closely related to humans in Europe than to humans in Africa.
   c. *Homo sapiens* originated in Africa.
   d. *Sahelanthropus* was the earliest hominid.
   e. Chimpanzees are more similar to gorillas and orangutans than to humans.

10. The earliest members of the genus *Homo*
    a. had a large brain compared to other hominids.
    b. probably hunted dinosaurs.
    c. lived in Europe.
    d. lived about 4 million years ago.
    e. were the first hominids to be bipedal.

11. To which of the following human traits has the *FOXP2* gene been linked?
    a. opposable thumb enabling a precision grip
    b. brain development associated with language
    c. an extended period of parental care
    d. sophisticated tool manufacture and use
    e. the development of dark skin pigmentation

Describing, Comparing, and Explaining

12. Compare the adaptations of amphibians and reptiles for terrestrial life.

13. Birds and mammals are both endothermic, and both have four-chambered hearts. Most reptiles are ectothermic and have three-chambered hearts. Why don't biologists group birds with mammals? Why do most biologists now consider birds to be reptiles?

14. What adaptations inherited from our primate ancestors enable humans to make and use tools?

15. Contrast hominoids with hominids.

16. Summarize the hypotheses that explain variation in human skin color as adaptations to variation in UV radiation.

Applying the Concepts

17. A good scientific hypothesis is based on existing evidence and leads to testable predictions. What hypothesis did the paleontologists who discovered Tiktaalik test? What evidence did they use to predict where they would find fossils of transitional forms?

18. Explain some of the reasons why humans have been able to expand in number and distribution to a greater extent than most other animals.

19. Anthropologists are interested in locating areas in Africa where fossils 4–8 million years old might be found. Why?

Answers to all questions can be found in Appendix 4.

For Practice Quizzes, BioFlix, MP3 Tutors, and Activities, go to www.mybiology.com.

Animals: Form and Function

chapter

20 Unifying Concepts of Animal Structure and Function

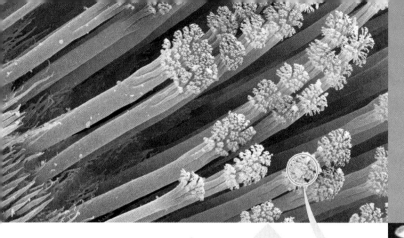

Left: Rows of setae on a gecko's foot
Below: Spatulae coming from a single seta

Climbing the Walls

Spiderman is a familiar character well known for his ability to climb walls, but few vertebrates have similar talents. One exception is the gecko, a small lizard commonly found in the tropics. Although perhaps not skilled at fighting crime or rescuing those in danger, geckos have no trouble walking up walls and even across ceilings. How do they do it? Several hypotheses, including either a sticky adhesive or suction cups on their toes, have turned out to be wrong. Instead, the explanation relates to hairs, called setae, on the gecko's toes. These hairs are made of keratin, the same protein found in our hair.

The micrographs at top of the page reveal the microscopic structure of setae. They are arranged in rows, and each seta ends in many split ends called spatulae, which have flattened tips. It took a multidisciplinary team of biologists and engineers to work out how the setae stick to surfaces with enough strength to support the animal's weight. In a study using the Tokay gecko (*Gekko gecko*), engineers designed an apparatus to measure the force of attraction between individual setae and the surface they touched—a difficult task because of the microscopic size of setae. This force of attraction turned out to be ten times greater than had been predicted.

But what is causing this attraction? The researchers attribute it to attractions between molecules at the tips of the spatulae and molecules making up the surface. Even uncharged molecules have regions that temporarily carry charges, and a region of positive charge on one molecule will be attracted to a region

of negative charge on another. (These attractions, called van der Waals interactions, also help hold individual protein and nucleic acid molecules in the characteristic shapes you saw in Chapter 3.) Each instance of attraction is fleeting and very weak, but there are so many setae—about half a million on each toe, each ending in hundreds of spatulae—that the combined strength of these forces becomes significant. In fact, a single seta could hold up an ant!

The gecko's remarkable ability to walk on walls is thus a function that emerges from special structural adaptations of its body, adaptations that extend to the microscopic level (shown at the top of the page). Other structural features of the gecko's body correlate with their functions, from the scales (also made of the protein keratin) that protect its body from drying out to the arrangement of the muscles and bones that move its feet as it walks up walls.

The correlation of structure and function is an important overarching theme of biology that is evident at all levels of life's hierarchy. We recognized this theme repeatedly in Unit I: The Life of the Cell. It also helps us understand animals. The chapters in this unit explore animal form and function in the context of the various problems animals must solve: how to nourish themselves, obtain oxygen from their environment and distribute it throughout their body, excrete wastes, sense and respond to the environment, move, and reproduce. The adaptations that represent the various solutions to these problems have been fashioned by natural selection, fitting structure to function by selecting, over many generations, for what works best within a particular population in its environment.

This chapter opens the unit with an overview of animal structure and function. ■ ■ ■

20.1 Structure fits function at all levels of organization in the animal body

When discussing structure and function, biologists distinguish anatomy from physiology. **Anatomy** is the study of the form of an organism's structures; **physiology** is the study of the functions of those structures. A biologist interested in anatomy, for instance, might focus on the arrangement of muscles and bones in a gecko's legs or on the shape and number of setae on its toes that allow it to climb walls (see chapter introduction). A physiologist might study the functioning of the muscles in the gecko's legs or the production of setae by epidermal cells. Despite their different approaches, both scientists are working toward a better understanding of the connection between structure and function—such as how structural adaptations give the gecko its remarkable ability to walk on walls.

Structure in the living world is organized in hierarchical levels. We followed the progression from molecules to cells in Unit I. Now let's trace the hierarchy from cells to organisms. (In Unit VII, we pick up the trail again, moving from organisms to ecosystems.) As we discussed in Chapter 1, emergent properties arise at each successive level of the hierarchy through the structural and functional organization of component parts.

Figure 20.1 illustrates structural hierarchy in a pelican. Part A shows a single muscle cell in the bird's heart. This cell's main function is to contract, and the stripes in the cell indicate the precise alignment of strands of proteins that perform that function. Each muscle cell is also branched, providing for multiple connections to other cells that ensure coordinated contractions of all the muscle cells in the heart.

Together these heart cells make up a tissue (Part B), the second structural level. A **tissue** is an integrated group of similar cells that perform a common function. The cells of a tissue are specialized, and their structure enables them to perform a specific task—in this instance, coordinated contraction.

Part C, the heart itself, illustrates the organ level of the hierarchy. An **organ** is made up of two or more types of tissues that together perform a specific task. In addition to muscle tissue, the heart includes nervous tissue and connective tissue.

Part D shows the circulatory system, the organ system of which the heart is a part. An **organ system** consists of multiple organs that together perform a vital body function. The other parts of the circulatory system are the blood and blood vessels: arteries, veins, and capillaries.

In Part E, the pelican itself forms the final level of this hierarchy. An organism contains a number of organ systems, each specialized for certain functions and all functioning together as an integrated, coordinated unit. For example, the pelican's circulatory system cannot function without oxygen supplied by the respiratory system and nutrients supplied by the digestive system. And it takes the coordination of several other organ systems to enable this bird to fly.

The ability to fly or to climb walls emerges from the specific arrangement of specialized structures. As we see throughout our study of the anatomy and physiology of animals, form fits

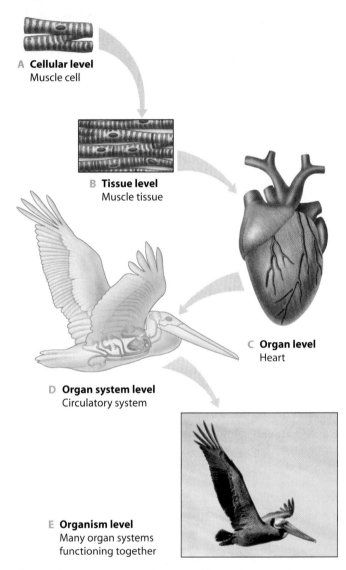

A **Cellular level**
Muscle cell

B **Tissue level**
Muscle tissue

C **Organ level**
Heart

D **Organ system level**
Circulatory system

E **Organism level**
Many organ systems functioning together

Figure 20.1 An example of structural hierarchy in a pelican

function at each level of the structural hierarchy. In several modules to come, we focus on the tissue level of this biological hierarchy. But first let's explore how evolution shapes an animal's body form.

🎧 MP3 Tutor *Animal Structure*

Web Activity *Correlating Structure and Function of Cells*

Web Activity *The Levels of Life Card Game*

? Explain how the ability to pump blood is an emergent property of a heart, which is at the organ level of the biological hierarchy.

■ The specific structural organization and integration of the individual muscle, connective, and nervous tissues of a heart enable the function of pumping blood.

20.2 An animal's form reflects natural selection

An animal's size and shape are fundamental aspects of form that significantly affect the way an animal interacts with its environment. Although biologists often refer to "body plan" or "design" in discussing animal form, they are not implying a process of conscious invention. The body plan of an animal is the result of millions of years of evolution. Natural selection fits form to function by selecting the variations that best meet the challenges of an animal's environment.

How might physical laws affect the evolution of an animal's form? Consider some of the properties of water that limit the shapes that are possible for fast swimmers. Water is about a thousand times denser than air. Thus, any bump on the body surface that causes drag slows a swimmer much more than it would a runner or a flyer. What shape, then, is best suited for rapid movement through the dense medium of water?

Tuna and other fast fishes can swim at speeds up to 80 km/hr (about 50 mph). Sharks, penguins, dolphins, and seals are also fast swimmers. As illustrated by the examples in Figure 20.2, such animals share a streamlined, tapered body form. The similar shape found in speedy fish, sharks, and aquatic birds and mammals is an example of convergent evolution (see Module 15.14). Natural selection often shapes similar adaptations when diverse organisms face the same environmental challenge, such as the resistance of water to fast swimming.

Physical laws also influence body form with regard to maximum size. As body dimensions increase, thicker skeletons are required to support the body. In addition, as bodies increase in size, the muscles required for movement become an ever-larger proportion of the total body mass. At some point, mobility becomes limited. By considering the fraction of body mass in leg muscles and the effective force such muscles generate, scientists can estimate maximum running speed for a wide range of body plans. Such calculations have indicated that *Tyrannosaurus rex*, a dinosaur that was more than 6 m tall, could probably only lumber from place to place, a far cry from the thunderous sprint depicted in the movie *Jurassic Park*.

Shark

Penguin

Seal

Figure 20.2 Convergent evolution of body shape in fast swimmers

? Explain how a seal, a penguin, and a shark illustrate convergent evolution.

■ All three have a streamlined, tapered shape, the result of natural selection adapting each to fast swimming in its dense, aquatic environment.

20.3 Tissues are groups of cells with a common structure and function

In almost all animals, most of the cells of the body are organized into tissues. The term *tissue* is from a Latin word meaning "weave," and some tissues resemble woven cloth in that they consist of a meshwork of nonliving fibers and other extracellular substances surrounding living cells. Other tissues are held together by a sticky glue that coats the cells or by special junctions between adjacent plasma membranes (see Module 4.21). The structure of tissues relates to their specific functions.

The specialization of complex body parts such as organs and organ systems is largely based on varied combinations of a lim-

ited set of cells and tissue types. For example, lungs and blood vessels have distinct functions but are lined by tissues that are of the same basic type.

An animal has four major categories of tissues: epithelial tissue, connective tissue, muscle tissue, and nervous tissue. We examine each of these separately in the next four modules.

? How is a tissue different from a cell and an organ?

■ Tissues are collections of similar cells that perform a common function. Several different tissue types usually produce the structure of an organ.

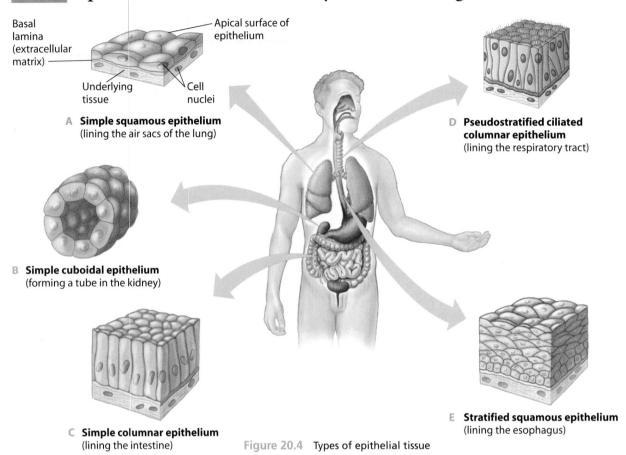

A Simple squamous epithelium
(lining the air sacs of the lung)

Basal lamina (extracellular matrix)

Apical surface of epithelium

Underlying tissue

Cell nuclei

B Simple cuboidal epithelium
(forming a tube in the kidney)

C Simple columnar epithelium
(lining the intestine)

D Pseudostratified ciliated columnar epithelium
(lining the respiratory tract)

E Stratified squamous epithelium
(lining the esophagus)

Figure 20.4 Types of epithelial tissue

Epithelial tissue, also called **epithelium,** occurs as sheets of tightly packed cells that cover body surfaces and line internal organs and cavities. The basal surface of an epithelium is attached to a *basal lamina,* a dense mat of extracellular matrix consisting of fibrous proteins and sticky polysaccharides that separates the epithelium from the underlying tissues. The free or apical surface faces the outside of an organ or the inside of a tube or passageway. Together, the tightly knit cells and basal lamina form a protective barrier and, in some cases, a surface for exchange with the fluid or air on the other side.

Epithelial tissues are named according to the number of cell layers they have and according to the shape of the cells on their apical surface. A *simple epithelium* has a single layer of cells, whereas a *stratified epithelium* has multiple layers. A *pseudostratified epithelium* is single-layered but appears stratified because the cells vary in length. The shape of the cells may be *squamous* (flat like floor tiles), *cuboidal* (like dice), or *columnar* (like bricks on end). Figure 20.4 shows examples of different types of epithelia. In each case, the pink color identifies the cells of the epithelium itself.

The structure of each type of epithelium fits its function. Simple squamous epithelium (Part A) is thin and suitable for exchanging materials by diffusion. We find it lining our capillaries (smallest blood vessels) and the air sacs of our lungs.

Both cuboidal and columnar epithelia have cells with a relatively large amount of cytoplasm, facilitating their role of secretion or absorption of materials. Part B shows a simple cuboidal epithelium forming a tube in the kidney. Such epithelia are also found in glands such as the thyroid and salivary glands. A simple columnar epithelium (Part C) lines the intestines. It secretes digestive juices and absorbs nutrients.

The pseudostratified ciliated columnar epithelium in Part D forms a mucous membrane that lines the respiratory tract. It secretes a slimy solution called mucus that lubricates the surface and keeps it moist. The mucous membrane of our air tubes helps keep our lungs clean by trapping dust, pollen, and other particles in its secretions. The beating of cilia on this mucous membrane then sweeps the mucus-trapped materials upward and out of the breathing passageways.

The many layers of the stratified squamous epithelium in Part E make it well suited for lining surfaces subject to abrasion, such as the outer skin and the esophagus, which can be abraded by rough food. Stratified squamous epithelium regenerates rapidly by division of the cells near the basal lamina. New cells move toward the apical surface as older cells slough off.

Web Activity *Epithelial Tissue*

? Epithelial tissues are named according to the _____ of cells on their apical surface and the number of cell _____.

■ shape . . . layers

Connective tissue binds and supports other tissues

In contrast to epithelium, **connective tissue** consists of a sparse population of cells scattered through an extracellular matrix. The cells produce and secrete the matrix, which usually consists of a web of fibers embedded in a liquid, jelly, or solid. Connective tissues may be grouped into six major types. Figure 20.5 shows micrographs of each type and illustrates where each would be found in an arm, for example.

The most common connective tissue in the human body is called **loose connective tissue** (Part A) because its matrix is a loose weave of fibers. Many of the fibers consist of the strong, ropelike protein collagen. Other fibers are more elastic, making the tissue resilient as well as strong. Loose connective tissue serves mainly as a binding and packing material, holding other tissues and organs in place. In the figure, we show the loose connective tissue that lies directly under the skin, where it helps bind the skin to underlying muscles.

Fibrous connective tissue (Part B) has a matrix of densely packed parallel bundles of collagen fibers, an arrangement that maximizes its nonelastic strength. Fibrous connective tissue forms tendons, which attach muscles to bone, and ligaments, which join bones together.

Adipose tissue (Part C) stores fat in large, closely packed adipose cells held in a sparse matrix of fibers. This tissue pads and insulates the body and stores energy. Each adipose cell contains a large fat droplet that swells when fat is stored and shrinks when fat is used as fuel.

The matrix of **cartilage** (Part D), a connective tissue that forms a strong but flexible skeletal material, consists of an abundance of collagen fibers embedded in a rubbery material. Cartilage commonly surrounds the ends of bones, where it forms a shock-absorbing surface. It also supports the nose and ears and forms the cushioning disks between our vertebrae.

Bone has a matrix of collagen fibers embedded in a hard mineral substance made of calcium, magnesium, and phosphate. The combination of fibers and minerals makes bone strong without being brittle. The microscopic structure of compact regions of bones, as shown in Part E, contains repeating circular units of matrix, each with a central canal containing blood vessels and nerves. Like other tissues, bone contains living cells and can therefore grow with the animal.

Blood (Part F) functions differently from other connective tissues. Its extensive extracellular matrix is a liquid called plasma that consists of water, salts, and dissolved proteins. Red blood cells, white blood cells, and platelets are suspended in the plasma. Blood functions mainly in transporting substances from one part of the body to another and in immunity.

Web Activity *Connective Tissue*

? Why does blood qualify as a type of connective tissue?

■ Because it consists of a relatively sparse population of cells surrounded by a noncellular matrix, which in this case is plasma

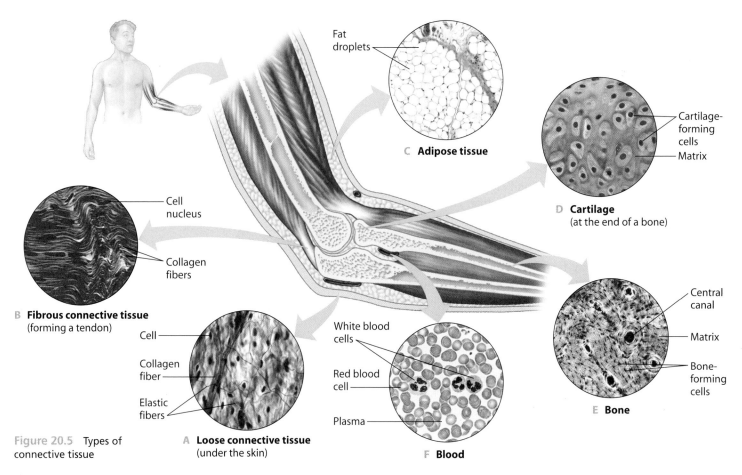

Figure 20.5 Types of connective tissue

C **Adipose tissue**

D **Cartilage** (at the end of a bone)
Cartilage-forming cells
Matrix
Fat droplets

Cell nucleus
Collagen fibers

B **Fibrous connective tissue** (forming a tendon)

Cell
Collagen fiber
Elastic fibers

A **Loose connective tissue** (under the skin)

White blood cells
Red blood cell
Plasma

F **Blood**

Central canal
Matrix
Bone-forming cells

E **Bone**

20.6 Muscle tissue functions in movement

Muscle tissue consists of bundles of long cells called muscle fibers and is the most abundant tissue in most animals. Within the cytoplasm of muscle fibers are large numbers of contractile proteins arranged in parallel. Geckos, birds, humans, and all other vertebrates have three types of muscle tissue. Figure 20.6 shows micrographs of these three types.

Skeletal muscle (Part A) is attached to bones by tendons and is responsible for voluntary movements of the body. The arrangement of the contractile units along the length of muscle cells gives them a striped or striated appearance, as you can see in the micrograph below.

Cardiac muscle (Part B) forms the contractile tissue of the heart. It is striated like skeletal muscle, but its cells are branched, interconnecting at specialized junctions that rapidly relay the signal to contract from cell to cell during the heartbeat.

Smooth muscle (Part C) gets its name from its lack of striations. Smooth muscle is found in the walls of the digestive tract, urinary bladder, arteries, and other internal organs. It is responsible for involuntary body activities, such as the movement of food through the intestines. The cells (fibers) are shaped like spindles. They contract more slowly than skeletal muscles, but they can sustain contractions for a longer period of time.

Web Activity *Muscle Tissue*

? The muscles responsible for a gecko climbing a wall are _____ muscles.

■ skeletal

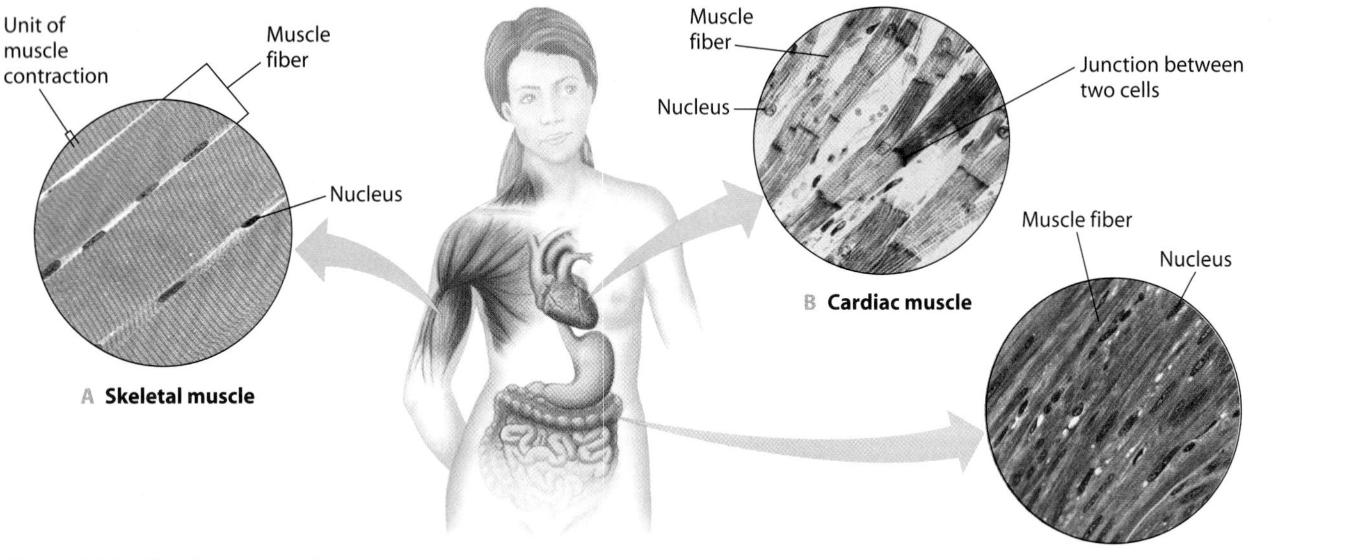

Unit of muscle contraction

Muscle fiber

Nucleus

A Skeletal muscle

Muscle fiber

Nucleus

Junction between two cells

B Cardiac muscle

Muscle fiber

Nucleus

C Smooth muscle

Figure 20.6 The three types of muscle

20.7 Nervous tissue forms a communication network

Nervous tissue senses stimuli and rapidly transmits information from one part of an animal to another. The structural and functional unit of nervous tissue is the nerve cell, or **neuron**, which is uniquely specialized to conduct electrical nerve impulses. The micrograph in Figure 20.7 shows a neuron consisting of a cell body (containing the cell's nucleus) and a number of slender extensions. One type of extension, called a dendrite, conveys signals from its tip toward the rest of the neuron; another type, the axon, transmits signals toward another neuron or to a muscle cell.

Nervous tissue is not made up entirely of neurons. It actually contains many more supporting cells than neurons. Some of these cells surround and insulate axons, promoting faster transmission of signals. Others nourish the neurons. The small dark nuclei of supporting cells are visible in the micrograph.

Web Activity *Overview of Animal Tissue*

Web Activity *Nervous Tissue*

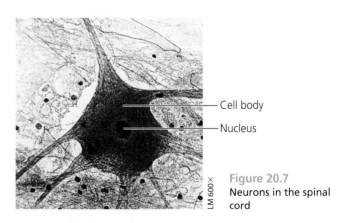

Cell body

Nucleus

LM 600×

Figure 20.7
Neurons in the spinal cord

? How does the long length of some axons (such as those that extend from your lower spine to your toes) relate to the function of a neuron?

■ It allows for the transmission of a nerve signal over a long distance directly to specific muscle cells.

20.8 Organs are made up of tissues

In all but the simplest animals, tissues are arranged into organs that perform specific functions. The heart, for example, while mostly muscle, also has epithelial, connective, and nervous tissues. Epithelial tissue lining the heart chambers prevents leakage and provides a smooth surface over which blood can flow. Connective tissue makes the heart elastic and strengthens its walls. Neurons regulate the contractions of cardiac muscles.

In some organs, tissues are organized in layers, as you can see in the diagram of the small intestine in Figure 20.8. The lumen, or interior space, of the small intestine is lined by a columnar epithelium that secretes digestive juices and absorbs nutrients. Notice the fingerlike projections that increase the surface area of this lining. Surrounding this layer (and extending into the projections) is connective tissue, which contains blood and lymph vessels. The two layers of smooth muscle (oriented in different directions) move food through the intestine. The smooth muscle, in turn, is surrounded by another layer of connective tissue and epithelial tissue.

An organ represents a higher level of structure than the tissues composing it, and it performs functions that none of its component tissues can carry out alone. These functions emerge from the coordinated interactions of tissues.

? Explain why a disease that damages connective tissue can impair most of the body's organs.

■ Connective tissue is a component of most organs.

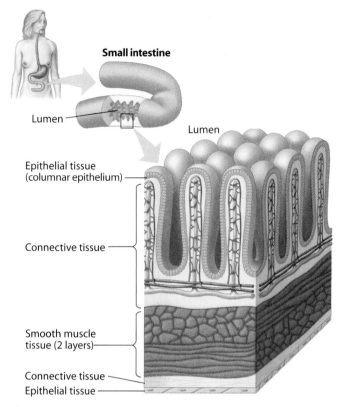

Figure 20.8   Tissue layers of the small intestine wall

20.9 Bioengineers are learning to produce tissues and organs for transplants

Severe burns, injuries, diseases, and birth defects often leave individuals with disfiguring or life-threatening conditions. Scientists are increasingly turning to bioengineering in their search for ways to repair or replace damaged tissues and organs.

One of the most successful tissue-engineering advances has come in the form of artificial skin, a type of human-engineered tissue designed for everyone from burn victims to diabetic patients with skin ulcers. The tissue is grown from human fibroblasts, tissue-generating cells often harvested from newborn foreskin tissue. These cells are applied along a tiny protein scaffolding, where they multiply and produce a three-dimensional skin substitute containing active living cells.

In 2006, researchers reported the first successful transplantation and long-term functioning of laboratory-grown bladders (Figure 20.9). This initial study involved only seven individuals, but it showed that organ engineering is possible. These organs were grown from the patients' own cells, thus removing the risk of rejection. Organs grown from a patient's cells may someday reduce the shortage of organs available for transplants.

In another remarkable advance in tissue engineering, some imaginative researchers filled recycled inkjet print cartridges with a cell solution instead of ink and, using desktop printers,

Figure 20.9   A laboratory-grown bladder

actually "printed" cells in layers on a biodegradable gel. The gel can be removed, leaving a three-dimensional cellular structure. Sheets of artificial tissues or even pieces of organs may someday be produced by such printing techniques. Researchers may also use such cellular systems for testing drugs or basic research into cell growth and interactions.

? Why is it beneficial to grow replacement tissues or organs from a patient's own cells?

■ The patient's body would not reject the tissue as foreign.

Chapter 20 *Unifying Concepts of Animal Structure and Function*   **419**

Organ systems work together to perform life's functions

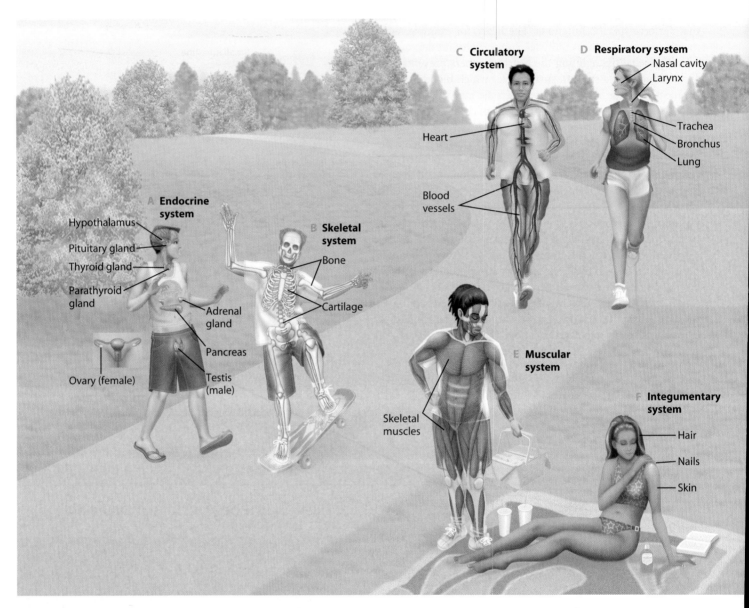

Figure 20.10 Human organ systems and their component parts

Just as it takes several different tissues to build an organ, it takes the integration of several organs into organ systems to perform the body's functions. Figure 20.10 illustrates the components of the 12 organ systems found in vertebrate animals. As you read through the brief descriptions of each system, remember that all of the organ systems are interdependent and work together to create a functional organism.

A The **endocrine system** secretes chemicals, called hormones, that regulate body activities such as digestion, metabolism, growth, reproduction, heart rate, and water balance.

B The **skeletal system** supports the body, protects certain internal organs such as the brain and lungs, and provides the framework for muscles to produce movement.

C The **circulatory system** delivers nutrients and oxygen (O_2) to body cells and carries carbon dioxide (CO_2) to the lungs and metabolic wastes to the excretory organs, the kidneys.

D The **respiratory system** exchanges gases with the environment, supplies the blood with O_2, and disposes of CO_2.

E The **muscular system** produces movement, maintains posture, and produces heat.

F The **integumentary system** protects against mechanical injury, infection, excessive heat or cold, and drying out.

G The **lymphatic system** returns excess body fluid to the circulatory system and functions as part of the immune system.

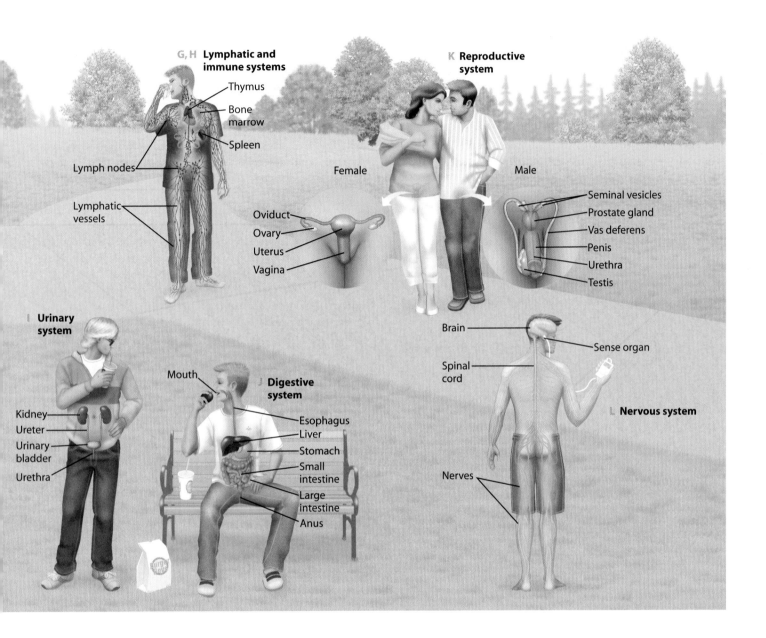

G, H Lymphatic and immune systems
- Thymus
- Bone marrow
- Spleen
- Lymph nodes
- Lymphatic vessels

K Reproductive system

Female
- Oviduct
- Ovary
- Uterus
- Vagina

Male
- Seminal vesicles
- Prostate gland
- Vas deferens
- Penis
- Urethra
- Testis

I Urinary system
- Kidney
- Ureter
- Urinary bladder
- Urethra

J Digestive system
- Mouth
- Esophagus
- Liver
- Stomach
- Small intestine
- Large intestine
- Anus

L Nervous system
- Brain
- Sense organ
- Spinal cord
- Nerves

H The **immune system** defends the body against infections and cancer.

I The **urinary system** (or **execretory system**) removes nitrogen-containing waste products from the blood and regulates the chemical makeup, pH, and water balance of the blood.

J The **digestive system** ingests and breaks down food, absorbs nutrients, and eliminates undigested material.

K The **reproductive system** produces gametes and sex hormones. The female system provides organs to support a developing embryo and glands for producing milk.

L The **nervous system** coordinates body activities by detecting stimuli, integrating information, and directing the body's responses.

These organ systems illustrate two important themes of biology: the correlation of structure with function and emergent properties. For example, the structural arrangement of bones and muscles enables the function of movement. The ability to perform life's functions emerges from the intricate organization and coordination of all the body's organ systems. Indeed, the whole is definitely greater than the sum of its parts.

? Which two organ systems are most directly involved in regulating all other systems?

■ The nervous system and the endocrine system

New imaging technology reveals the inner body

Among the most useful developments in medical technology are techniques that allow physicians to "see" the organ systems we have just surveyed without resorting to surgery. We mentioned one of these techniques—ultrasound—in Module 9.10. Here we look at some others.

X-Rays X-rays, discovered in 1895, were the first means of producing a photographic image of internal organs. X-rays are a type of high-energy radiation (see Module 7.6) that passes readily through soft tissues. The features that show up most distinctly on X-rays are shadows of hard structures, such as bones and dense tumors that block the rays. Conventional X-rays are used routinely to check for broken bones and tooth cavities.

CT A newer X-ray technology is called computed tomography (CT). This computer-assisted technique produces images of a series of thin cross sections through the body. The patient is often given a special liquid to improve the contrast of the images and is then slowly moved through a doughnut-shaped CT machine as the X-ray source circles around the body, illuminating successive sections from many angles. The CT scanner's computer then produces high-resolution video images of the cross sections, which can be studied individually or combined into various three-dimensional views.

CT scans are excellent diagnostic tools. They can detect small differences between normal and abnormal tissues in many organs, but are especially useful for evaluating problems that affect the abdomen and brain—areas where conventional X-ray procedures are of little help. The CT scan on the left in Figure 20.11 reveals details of internal organs.

Another useful diagnostic technique is one that uses ultra-fast CT scanners to show the actual movements and changes in volumes of body organs, such as the heart beating and blood flowing through vessels. Physicians use this technique to identify heart defects and constricted or blocked blood vessels and to monitor the status of coronary bypass grafts.

MRI A completely different technique, magnetic resonance imaging (MRI) takes advantage of the behavior of the hydrogen atoms in water molecules. MRI uses powerful magnets to align the hydrogen nuclei, then knocks the nuclei out of alignment with a brief pulse of radio waves. In response, the hydrogen atoms give out faint radio signals of their own, which are picked up by the MRI scanner and translated by computer into an image. Since water is a major component of all of our soft tissues, MRI visualizes them well. For example, MRI allows physicians to see delicate nerve fibers in the spinal cord.

Three-dimensional CT and MRI scans are often used before surgery to map out the surgical procedure or design artificial implants for reconstructive surgery. MRI scans are also used during surgery to guide delicate procedures.

PET Positron-emission tomography (PET) is an imaging technology that differs from both CT and MRI in its ability to yield information about metabolic processes at specific locations in the body. In preparation for a PET scan, the patient is injected with a biological molecule—glucose, for example—labeled with a radioactive isotope (see Module 2.5). Used only in small quantities, the isotope is not dangerous. Metabolically active cells take up more of the labeled glucose than less active cells. A PET scanner (see Figure 2.5A) pinpoints metabolic "hot spots" by highlighting the sites of most intense radiation. Cancerous tissue that is actively growing is very noticeable, as seen in the PET scan in the center of Figure 20.11 as dark vertebrae.

PET is proving most valuable for measuring the metabolic activity of various parts of the brain. This technique is providing insights into brain activity in people affected by illnesses such as schizophrenia, epilepsy, and Alzheimer's disease and in stroke patients. Equally exciting is the use of PET to learn about the healthy brain.

A recent advance is a combination CT-PET scanner. This machine performs both types of scans at the same time. Combining the images benefits from the better anatomical representation of a CT scan and the metabolic information of a PET scan. The image on the right of Figure 20.11 is a CT-PET scan revealing the location of bone cancer.

? Why are the imaging techniques described in this module regarded as relatively noninvasive in contrast to such diagnostic methods as exploratory surgery or biopsy?

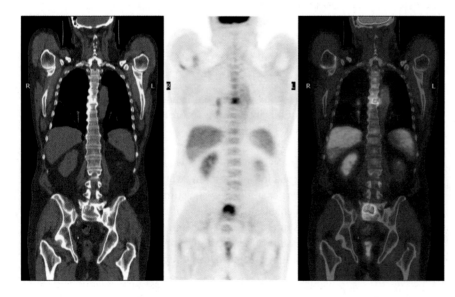

Figure 20.11 Scans showing bone cancer in two locations in the vertebral column. CT scan (left), PET scan (center), and combined CT-PET scan (right)

■ Although they may involve injections, these imaging techniques require no penetration of the body.

The integumentary system protects the body

Most of the organ systems introduced in Module 20.10 are examined in more detail in the chapters of this unit. Here we take a brief look at the integumentary system, which consists of the skin, hair, and nails.

Structure of Skin As shown in Figure 20.12, the skin consists of two layers: the epidermis and the dermis. The outer epidermis is a stratified squamous epithelium (many layers of flat cells, see Module 20.4). Rapid cell division near the base of the epidermis serves to replenish the skin cells that are constantly abraded from the body surface. As these new cells are pushed upward in the epidermis by the addition of new cells below them, they fill with the fibrous protein keratin and release a waterproofing glycolipid, and eventually die. These dead, tightly joined cells remain at the surface of the skin for up to two weeks. This continuous process means that we get a brand new epidermis every few weeks.

The dermis is the inner layer of the skin. It consists of a fairly dense connective tissue with many resilient elastic fibers and strong collagen fibers. The dermis contains hair follicles, oil and sweat glands, muscles, nerves, sensory receptors (shown in Figure 29.3A), and blood vessels. Beneath the skin lies the hypodermis, a layer of adipose tissue. (The hypodermis is the site of many of the injections you receive with a hypodermic needle.)

Functions of Skin The structure of the skin corresponds to its functions. The keratin-filled, tightly joined cells at the surface of the epidermis provide a waterproof covering that protects the body from dehydration and prevents penetration by microbes. And as close encounters with the environment abrade the epidermis, replacement cells move up.

The sensory receptors in the skin provide important environmental information to the brain: Is this surface too hot or cold or sharp? Your touch receptors help you manipulate food and tools and feel your way in the dark. The profusion of small blood vessels and the 2.5 million sweat glands in the dermis also facilitate the important function of temperature regulation, as Figure 20.15 illustrates.

A metabolic function of the skin is the synthesis of vitamin D, which is required for absorbing calcium. Adequate sunlight is needed for this synthesis. But sunlight also has damaging effects on the skin. DNA changes caused by the UV radiation in sunlight can lead to skin cancers.

Hair and Nails In mammals, hair is an important component of the integumentary system. Hair is a flexible shaft of

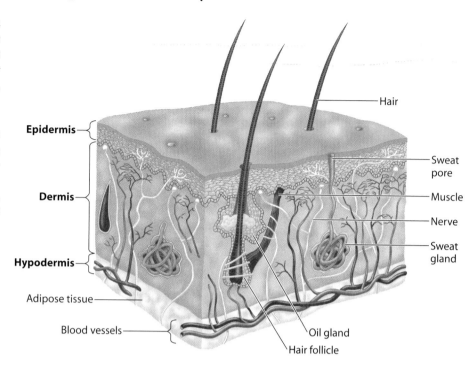

Figure 20.12 A section of skin, the major organ of the integumentary system

flattened, keratin-filled dead cells, which are produced by a hair follicle. As shown in Figure 20.12, oil glands are associated with hair follicles. The oil gland's secretions lubricate the hair, condition the surrounding skin, and inhibit the growth of bacteria. Look again at the hair follicle in the figure and you will see that it is wrapped in nerve endings. Hair also plays important sensory functions, as its slightest movement is relayed to the nervous system. (You can get a sense of this sensitivity by lightly touching the hair on your head.)

Hair insulates the bodies of most mammals—although only our heads have such insulation. Land mammals react to cold by raising their fur, which traps a layer of air and increases the insulating power of the fur. Goose bumps are a vestige of this hair-raising inherited from our furry ancestors.

Fingernails and toenails are the final component of our integumentary system. These protective coverings are also composed of keratin. Fingernails facilitate fine manipulation and scratching (and are useful for chewing, when nervous). In other mammals, the digits may end in claws or hooves.

The integumentary system encloses and protects an animal from its environment. But animals cannot be isolated systems; they must exchange gases, take in food, and eliminate wastes. In the next module, we explore some of the structural adaptations that provide for these functions.

? Look at the hair follicle in Figure 20.12. Name three structures that contribute to the functions of hair.

■ The nerve endings sense when the hair is moved; the muscle raises the hair (producing goose bumps in humans but warming other mammals); and the oil gland produces lubricating and antibacterial secretions.

20.13 Structural adaptations enhance exchange between animals and their environment

Animals must exchange materials with their environment to survive, and this exchange must extend to the level of each individual cell. Oxygen and nutrients must enter the cell, and carbon dioxide and other metabolic wastes must exit. Because a cell must be bathed in aqueous fluid for its plasma membrane to remain intact, only molecules dissolved in water can move across the plasma membrane.

A freshwater hydra has a body wall only two cell layers thick (see Figure 21.3A). The outside layer is in contact with its water environment; the inner layer is bathed by fluid in its saclike body cavity. This internal fluid circulates in and out of the hydra's mouth. Thus, every body cell exchanges materials directly with an aqueous environment.

Another common body plan that maximizes exposure to the surroundings is a flat, thin shape. For instance, a parasitic tapeworm (see Module 18.7) may be several meters long, but because it is very thin, most of its cells are bathed in the intestinal fluid of its host—the source of its nutrients.

The saclike body of a hydra or the paper-thin one of the tapeworm works well for animals with a simple body structure. However, most animals are composed of compact masses of cells and have an outer surface that is relatively small compared with the animal's overall volume. As an extreme example, the ratio of outer surface area to volume (see Module 4.2) of a whale is hundreds of thousands of times smaller than that of a small animal like a hydra. Still, every cell in the whale's body must be bathed in fluid, have access to oxygen and nutrients, and be able to dispose of its wastes. How is all this accomplished?

In whales and most other animals, the evolutionary adaptation that provides for sufficient exchange with the environment is in the form of extensively branched or folded surfaces. In almost all cases, these surfaces lie within the body, protected by the integumentary system from dehydration or damage and allowing for a streamlined body shape. In humans, the digestive, respiratory, and circulatory systems each rely on exchange surfaces within the body that create an area more than 25 times that of the skin! Indeed, if we lined up all the tiny capillaries across which exchange between the blood and body cells occurs, they would circle the globe.

Figure 20.13A is a schematic model illustrating four of the organ systems of a compact, complex animal. Each system has a large, specialized internal exchange surface. We have placed the circulatory system in the middle because of its central role in transporting substances between the other three systems. The blue arrows indicate exchange of materials between the circulatory system and the other systems.

Actually, direct exchange does not occur between the blood and the cells of tissues and organs. Body cells are bathed in a solution called **interstitial fluid** (see the circular enlargement in Figure 20.13A). Materials are exchanged

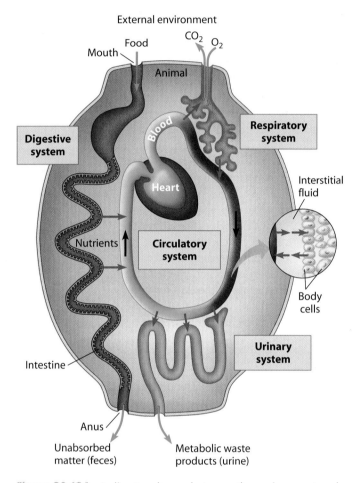

Figure 20.13A Indirect exchange between the environment and the cells of a complex animal

between the blood and the interstitial fluid and between the interstitial fluid and the body cells. In other words, to get from the blood to body cells, or vice versa, materials must pass through interstitial fluid.

The digestive system, especially the intestine, has an expanded surface area resulting from folds and projections of its inner lining (see Figure 20.8). Nutrients from digested food are absorbed from the lumen into the cells of this large surface area. They then pass into the interstitial fluid and from there into capillaries, tiny branched blood vessels that form an exchange network with the digestive surfaces. This system of exchange from the cells lining the intestine to the blood is so effective that enough nutrients move into the circulatory system to support the rest of the cells in the body.

The extensive, epithelium-lined tubes of the urinary system are equally effective at increasing the surface area for exchange. Enmeshed in capillaries, excretory tubes extract metabolic wastes that the blood brings from throughout the body. The wastes move out of the blood into the excretory tubes and pass out of the body in urine.

The respiratory system also has an enormous internal surface area. Your lungs are not shaped like big balloons, as they are sometimes depicted, but rather like millions of tiny balloons at the tips of finely branched air tubes. Figure 20.13B is a resin model of the human lungs. Blood vessels that convey blood from the heart to the lungs divide into tiny blood vessels that radiate throughout the lungs, shown in the model as the red branches. The white branches represent tiny air tubes that end in multilobed sacs lined with thin squamous epithelium and surrounded by capillaries. Oxygen readily moves from the air in the lungs across this epithelium and into the blood in the capillaries. The blood returns to the heart and is then pumped throughout the body to supply all cells with oxygen.

Both Figures 20.13A and 20.13B highlight two basic concepts in animal biology: First, any animal with a complex body—one with most of its cells not in direct contact with its external environment—must have internal structures that provide sufficient surface area to service those cells. Second, the organ systems of the body are interdependent; it takes their coordinated actions to produce a functional organism.

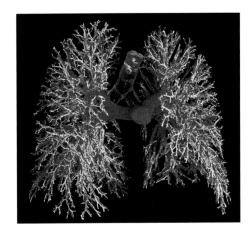

Figure 20.13B
A resin model of the finely branched air tubes and blood vessels of the human lungs

? How does the structure of the lungs, small intestine, and kidneys relate to the function of exchange with the environment?

■ These organs all have a huge number of sacs, projections, or tubes that greatly increase the area across which exchange of materials can occur.

20.14 Animals regulate their internal environment

More than a century ago, French physiologist Claude Bernard recognized that *two* environments are important to an animal: the external environment surrounding the animal and the internal environment, where its cells actually live. The internal environment of a vertebrate is the interstitial fluid that fills the spaces around the cells. Many animals maintain relatively constant conditions in their internal environment. Our own bodies maintain the salt and water balance of our internal fluids and also keep the fluids at about 37°C (98.6°F). A bird, such as the ptarmigan, also maintains salt and water balance and temperature (about 40°C), even in winter (Figure 20.14A). The bird uses energy from its food to generate body heat, and it has a thick, insulating coat of down feathers. A gecko does not generate its own body heat, but it can maintain a fairly constant body temperature by basking in the sun or resting in the shade. And it does regulate the salt and water balance of its internal fluids.

Today, Bernard's concept of the constant internal environment is included in the broader principle of **homeostasis,** which means "a steady state." Figure 20.14B to the right illustrates this principle. Conditions may fluctuate widely in the animal's external environment, but homeostatic mechanisms regulate internal conditions, resulting in much smaller fluctuations in the animal's internal environment. For example, birds and mammals have a control system that keeps

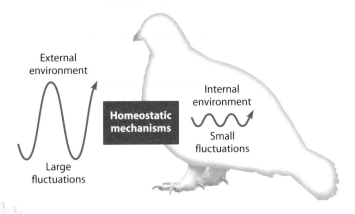

Figure 20.14B A model of homeostasis

body temperature within a narrow range, despite wide fluctuations in the temperature of the external environment.

The internal environment of an animal always fluctuates slightly. Homeostasis is a dynamic state, an interplay between outside forces that tend to change the internal environment and internal control mechanisms that oppose such changes. An animal's homeostatic control systems maintain internal conditions within a range where life's metabolic processes can occur.

? Look back at Figure 20.13A. What are some ways in which the circulatory system contributes to homeostasis?

■ By its exchanges with the digestive, respiratory, and excretory systems, the blood helps maintain the proper balance of materials in the interstitial fluid surrounding body cells.

Figure 20.14A A white-tailed ptarmigan in its snowy habitat

20.15 Homeostasis depends on negative feedback

Most of the control mechanisms of homeostasis are based on **negative feedback,** in which a change in a variable triggers mechanisms that reverse that change. To identify the components of a negative-feedback system, consider the simple example of the regulation of room temperature. You set the thermostat at a comfortable temperature—call this its *set point*. When a sensor in the thermostat detects that the temperature has dropped below this set point, the thermostat turns on the furnace. The response (heat) reverses the drop in temperature. Then, when the temperature rises to the set point, the thermostat turns the furnace off. Physiologists would call the sensor a *receptor* that is triggered by a *stimulus* (room temperature below the set point) and the furnace an *effector*, which produces a *response* (heat). The thermostat represents a *control center*, which processes information from the receptor and directs the response by the effector.

Many of the control centers that maintain homeostasis in animals are located in the brain. For example, your "thermostat" operates by negative feedback to switch on and off mechanisms that maintain body temperature around 37°C. As shown in the upper part of Figure 20.15, when the thermostat senses a rise in temperature above the set point, it activates cooling mechanisms, such as the dilation of blood vessels in the skin and sweating. Once body temperature returns to normal, the thermostat shuts off these cooling mechanisms. When body temperature falls below the set point (lower part of the figure), the thermostat activates warming mechanisms, such as constriction of blood vessels to reduce heat loss and shivering to generate heat. Again, a return to normal temperature shuts off these mechanisms.

As we examine the body's organ systems in detail in the chapters of this unit, we will see many examples of homeostatic control and negative feedback, as well as constant reminders of the relationship between structure and function.

Web Activity *Regulation: Negative and Positive Feedback*

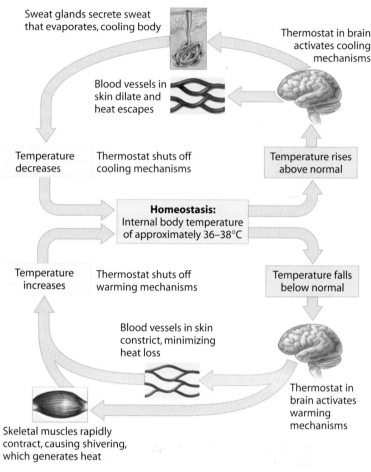

Figure 20.15   Control of body temperature

? Explain how homeostatic control is achieved by negative-feedback mechanisms.

■ Negative feedback maintains an internal balance by triggering mechanisms that reverse the movement of variables away from a set point.

Chapter Review

Reviewing the Concepts

The Hierarchy of Structural Organization in an Animal (Introduction–20.12)

Correlation between structure and function is a fundamental biological concept. Anatomy is the study of structure; physiology studies how structures function. The structural hierarchy of the body goes from cell to tissue, organ, organ system, and organism (**Introduction–20.1**).

An animal's form (size and shape) is influenced by the physical constraints of its environment (**20.2**).

Tissues are groups of many similar cells that perform a specific function (**20.3**).

	Epithelial (20.4)	Connective (20.5)	Muscle (20.6)	Nervous (20.7)
Tissue				
Structure	Sheets of closely packed cells	Sparse cells in extracellular matrix	Long cells (fibers) with contractile proteins	Neurons with branching extensions
Function	Protection, exchange, secretion	Binding and support of other tissues	Movement of body parts	Transmission of nerve signals

Organs are made of several tissues that collectively perform specific functions **(20.8)**. Tissue engineering holds promise for basic research and for medical applications **(20.9)**.

Organ systems consist of several coordinated organs. The integumentary system covers the body. Skeletal and muscular systems support and move it. The digestive and respiratory systems obtain food and oxygen, and the circulatory system transports these materials. The urinary system disposes of wastes, and the immune and lymphatic systems protect the body from infection. The nervous and endocrine systems control and coordinate body functions. The reproductive system produces offspring **(20.10)**. Technologies, such as CT, MRI, and PET, are used in medical diagnosis and research **(20.11)**.

The integumentary system consists of the skin, hair, and nails. The structure of the epidermis and dermis are correlated with the skin's functions, which include protection, sensation, and temperature regulation **(20.12)**.

Exchanges with the External Environment (20.13–20.15)

Surface areas for exchange. Complex animals have specialized structures that increase surface area. Exchange of materials between blood and body cells takes place through the interstitial fluid **(20.13)**.

Homeostasis. Animals regulate their internal environment to achieve homeostasis, an internal steady state **(20.14)**. Control systems detect change and direct responses. Negative-feedback mechanisms keep internal variables fairly constant, with small fluctuations around set points **(20.15)**.

Connecting the Concepts

1. There are several key concepts introduced in this chapter: Structure correlates with function; an animal body has a hierarchy of organization with emergent properties at each level; and complex bodies have structural adaptations to increase surface area for exchange. Label the tissue layers shown in this section of the small intestine and describe how this diagram illustrates these concepts.

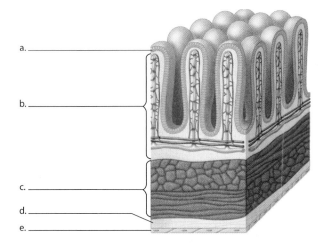

a. _____
b. _____
c. _____
d. _____
e. _____

Testing Your Knowledge

Multiple Choice

2. Which of the following pairs of body systems primarily regulates the activities of the other systems?
   a. circulatory and muscular systems
   b. nervous and endocrine systems
   c. lymphatic and integumentary systems
   d. endocrine and lymphatic systems
   e. integumentary and nervous systems

3. The cells in the human body are in contact with an internal environment consisting of
   a. blood.
   b. connective tissue.
   c. interstitial fluid.
   d. matrix.
   e. mucous membranes.

4. Negative-feedback mechanisms are
   a. most often involved in maintaining homeostasis.
   b. activated only when a variable rises above a set point.
   c. analogous to a furnace that produces heat.
   d. found only in birds and mammals.
   e. all of the above.

5. Which of the following best illustrates homeostasis? (*Explain your answer.*)
   a. Most adult humans are between 5 and 6 feet tall.
   b. The lungs and intestines have large surface areas.
   c. All the cells of the body are about the same size.
   d. When blood salt concentration goes up, the kidney expels more salt.
   e. When oxygen in the blood decreases, you feel dizzy.

Matching (*Terms in the right-hand column may be used more than once.*)

6. Closely packed cells covering a surface
7. Neurons
8. Adipose tissue, blood, and cartilage
9. May be simple or stratified
10. Scattered cells embedded in matrix
11. Senses stimuli and transmits signals
12. Cells are called fibers
13. Cells may be squamous, cuboidal, or columnar
14. Skeletal, cardiac, or smooth

a. connective tissue
b. muscle tissue
c. nervous tissue
d. epithelial tissue

Describing, Comparing, and Explaining

15. Briefly explain how the structure of each of these tissues is well suited to its function: stratified squamous epithelium in the skin, neurons in the brain, simple squamous epithelium lining the lung, bone in the skull.

16. Describe ways in which the bodies of animals are structured for exchanging materials with the environment. Why do some small animals not need such features?

Applying the Concepts

17. Some companies promote whole-body CT scans as health screens that consumers may purchase. Many parents request CT scans when they take an injured child to the ER. There is growing concern that widespread and repeated use of this technology is exposing adults and children to potentially high levels of radiation for procedures that may not be medically necessary. How can consumers evaluate the health risks versus benefits of these procedures? What is the government's role in regulating these commercial ventures?

Answers to all questions can be found in Appendix 4.

For Practice Quizzes, BioFlix, MP3 Tutors, and Activities, go to www.mybiology.com.

chapter

21 Nutrition and Digestion

Getting Their Fill of Krill

Whales are the largest animals in the world. Few other species, living or extinct, even approach their great size. Humpback whales are medium-sized members of the whale clan. They can be 16 m long and weigh up to 65,000 kg (72 tons), about as much as 70 midsized cars.

It takes an enormous amount of food to support a 72-ton animal. Humpback whales eat small fishes and crustaceans called krill, shown above. The painting on the left shows a remarkable technique humpbacks often use to corral food organisms before gulping them in. Beginning about 20 m below the ocean surface, a humpback swims slowly in an upward spiral, blowing air bubbles as it goes. The rising bubbles form a cylindrical screen, or "bubble net." Krill and fish inside the bubble net swim away from the bubbles and become concentrated in the center of the cylinder. The whale then surges up through the center of the net with its mouth open, harvesting the catch in one giant gulp.

Humpback whales strain their food from seawater. Instead of teeth, these giants have an array of comblike plates called baleen on each side of their upper jaw (visible as the white fringe in the open mouth of the whale in the photograph to the right). To start feeding, a humpback whale opens its mouth, expands its throat, and takes a huge gulp of seawater. When its mouth closes, the baleen acts as a sieve: Water is forced back out through spaces in the baleen, trapping a mass of food in the mouth. The food is then swallowed whole, passing into the stomach, where digestion begins. The humpback's stomach can hold about half a ton of food at a time, and in a typical day, the animal's digestive system will process as much as 2 tons of krill and fish.

The humpback and most other large whales are endangered species, having been hunted almost to extinction for meat and whale oil by the 1960s. Today, most nations honor an international ban on whaling, and some species are showing signs of recovery. Humpbacks still roam the Atlantic and Pacific oceans. They feed in polar regions during summer months and migrate to warmer oceans to breed when temperatures begin to fall. The photograph on this page was taken during summer in Glacier Bay, Alaska. Food is so abundant there that humpbacks harvest much

more energy than they burn each day. Much of the excess is stored as a thick layer of fat, or blubber, just under their skin. After a summer of feasting, humpback whales leave Glacier Bay and head south to breeding and calving grounds off the Hawaiian Islands, some 6,000 km away. Living off body fat, they eat little, if at all, until they return to Alaskan waters eight months later, to begin the process again. In the next four months, a humpback whale eats and digests more than 200 tons of food and stores enough fat to keep its 72-ton body active throughout its migration period—a remarkable feat and a fitting introduction to this chapter on animal nutrition and digestion. ■ ■ ■

21.1 Animals ingest their food in a variety of ways

All animals eat other organisms—dead or alive, whole or by the piece. In general, animals fall into one of three dietary categories. **Herbivores** (from the Latin *herba*, green crop, and *vorus*, devouring), such as cattle, gorillas, sea urchins, and snails, eat mainly autotrophs (plants and algae). **Carnivores** (from the Latin *carne*, flesh), such as lions, hawks, spiders, and whales, mostly eat other animals. Animals that ingest *both* plants and animals are called **omnivores** (from the Latin *omnis*, all). Omnivores include humans, who evolved as hunters, gatherers, and scavengers, as well as crows, cockroaches, and raccoons.

In what way do animals obtain and ingest their food? **Suspension feeders** extract food particles suspended in the surrounding water. For example, the humpback whale described in the chapter introduction uses its baleen to sift krill and small fish from the water. Clams and oysters are also suspension feeders. A film of mucus on their gills traps tiny morsels suspended in the water, and cilia on the gills sweep the food along to the mouth. Figure 21.1A shows the feathery tentacles of a suspension-feeding tube worm.

Substrate feeders live in or on their food source and eat their way through it. Figure 21.1B below shows a leaf miner caterpillar, the larva of a moth. The dark spots are a trail of feces that the caterpillar leaves in its wake. Earthworms are also substrate feeders. They eat their way through the soil, digesting partially decayed organic material as they go. In doing so, they help aerate and fertilize the soil.

Fluid feeders obtain food by sucking nutrient-rich fluids from a living host, either a plant or an animal. Aphids, for example, tap into the sugary sap in plants. Bloodsuckers, such as mosquitoes and ticks, pierce animals with needlelike mouthparts. The female mosquito in Figure 21.1C below has just filled her abdomen with a meal of human blood. (Only female mosquitoes suck blood; males live on plant nectar.) In contrast to such parasitic fluid feeders, which harm their hosts, some fluid feeders actually benefit their hosts. For example, hummingbirds and bees move pollen between flowers as they fluid-feed on nectar.

Rather than filtering food from water, eating their way through a substrate, or sucking fluids, most animals are **bulk feeders,** meaning they ingest large pieces of food. Figure 21.1D shows a grey heron preparing to swallow its prey whole. A bulk feeder may use such utensils as tentacles, pincers, claws, poisonous fangs, or jaws and teeth to kill its prey, to tear off pieces of meat or vegetation, or to take mouthfuls of animal or plant products.

Whatever the type of food or feeding mechanism, the processing of food involves four stages, as we see next.

? Blue whales, the largest animals ever to live, feed on krill. Name their diet category and type of feeding.

carnivore and suspension feeder

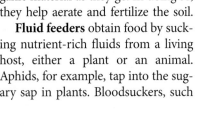
Figure 21.1A A suspension feeder: a tube worm filtering food through its tentacles

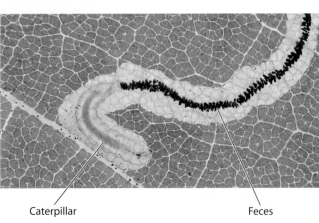
Caterpillar   Feces
Figure 21.1B A substrate feeder: a caterpillar eating its way through the soft green tissues inside an oak leaf

Figure 21.1C A fluid feeder: a mosquito sucking blood

Figure 21.1D A bulk feeder: a grey heron preparing to swallow a fish headfirst

21.2 Overview: Food processing occurs in four stages

So far we have discussed what animals eat and how they ingest it. As shown in Figure 21.2A, ❶ **ingestion,** the act of eating, is only the first of four main stages of food processing. ❷ The second stage, **digestion,** is the breaking down of food into molecules small enough for the body to absorb. Digestion typically occurs in two phases. First, food may be mechanically broken into smaller pieces. In animals with teeth, the process of chewing or tearing breaks large chunks of food into smaller ones. The second phase of digestion is the chemical breakdown process called hydrolysis. Catalyzed by specific enzymes, hydrolysis breaks chemical bonds in food molecules by adding water to them (see Module 3.3).

Most of the organic matter in food consists of proteins, fats, and carbohydrates—all large molecules. Animals cannot use these materials directly for two reasons. First, these molecules are too large to pass through plasma membranes and enter the cells. Second, an animal needs small components to make the molecules of its own body. Most of the molecules in food, for instance the proteins in the cat's food (shown in the figure below), are different from those that make up an animal's body.

All organisms use the same components to make their macromolecules. For instance, cats, humans, and trees all make their proteins from the same 20 kinds of amino acids. Digestion in an animal breaks the polymers in food into monomers. As shown in Figure 21.2B, proteins are split into amino acids, polysaccharides and disaccharides are split into monosaccharides, and nucleic acids are split into nucleotides. Fats are not polymers, but they are split into their components, glycerol and fatty acids. The animal can then use these small molecules to make the specific large molecules it needs (see Module 6.16).

The last two stages of food processing occur after digestion. ❸ In the third stage, **absorption,** the cells lining the digestive tract take up (absorb) the products of digestion—small molecules such as amino acids and simple sugars. From the digestive tract, these nutrients travel in the blood to body cells, where they are joined together to make the macromolecules of the cells or broken down further to provide energy. In an animal

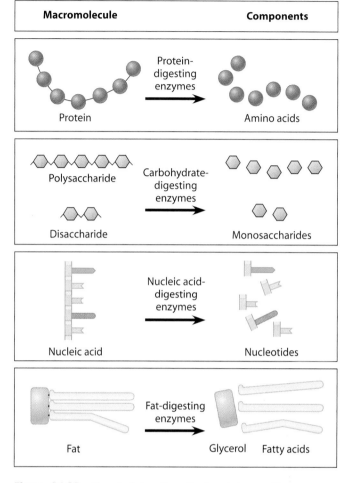

Figure 21.2B Chemical digestion: the breakdown of large organic molecules to their components

that eats much more than its body immediately uses, many of the nutrient molecules are converted to fat for storage. ❹ In the fourth and last stage of food processing, **elimination,** undigested material passes out of the digestive tract.

How can an animal digest food without digesting its own cells and tissues? After all, digestive enzymes hydrolyze the same biological materials (such as proteins, carbohydrates, and fats) that animals are made of—and it is obviously important to avoid digesting oneself! Most animal species process food in specialized compartments—an evolutionary adaptation that avoids the risk of self-digestion. We discuss digestive compartments in the next module.

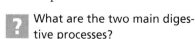 What are the two main digestive processes?

■ Mechanical breakdown and chemical breakdown (enzymatic hydrolysis)

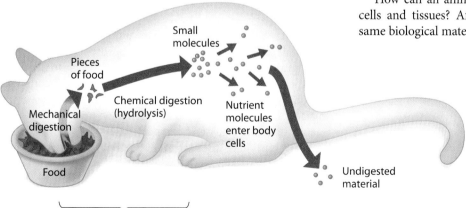

❶ Ingestion   ❷ Digestion   ❸ Absorption   ❹ Elimination

Figure 21.2A The four main stages of food processing

Digestion occurs in specialized compartments

Food vacuoles are the simplest digestive compartments. A cell engulfs food by phagocytosis (see Figure 5.9), and the newly formed food vacuole fuses with a lysosome containing hydrolytic enzymes (see Module 4.11). Sponges (see Module 18.5) digest their food entirely in food vacuoles. In contrast, most animals have an internal compartment in which digestion occurs outside of cells, enabling an animal to devour much larger food than could fit in a food vacuole.

As we saw in Chapter 18, cnidarians and flatworms have a **gastrovascular cavity,** a digestive compartment with a single opening, the **mouth.** Figure 21.3A below shows a hydra digesting a small crustacean called *Daphnia* that it has captured with its tentacles and stuffed into its mouth. ❶ Gland cells lining the gastrovascular cavity secrete digestive enzymes that ❷ break down the soft tissues of the prey. ❸ Other cells engulf small food particles, which ❹ are broken down in food vacuoles. Undigested materials are expelled through the mouth.

Most animals have an **alimentary canal,** a digestive tract with two openings, a mouth and an anus. Because food moves in one direction, specialized regions of the tube can carry out digestion and absorption of nutrients in sequence.

Food entering the mouth usually passes into a **pharynx,** or throat. Depending on the species, the **esophagus** may channel food to a crop, gizzard, or stomach. A **crop** is a pouch-like organ in which food is softened and stored. **Stomachs** and **gizzards** may also store food temporarily, but they are more muscular and they churn and grind the food. Chemical digestion and nutrient absorption occur mainly in the **intestine.** Undigested materials are expelled through the **anus.**

Figure 21.3B to the right illustrates three examples of alimentary canals. The digestive tract of an earthworm includes a muscular pharynx that sucks food in through the mouth. Food passes through the esophagus and is stored in the crop. The muscular gizzard, which contains small bits of sand and gravel, pulverizes the food. Digestion and absorption occur in the

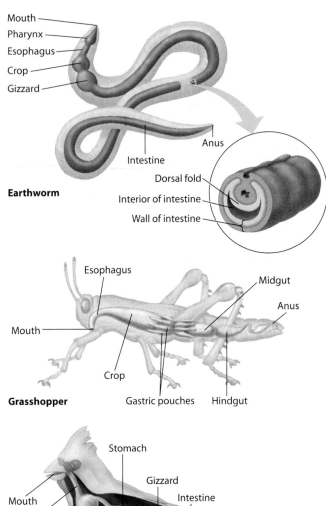

Earthworm

Grasshopper

Bird

Figure 21.3B   Three examples of alimentary canals

intestine. As the enlargement shows, a dorsal fold of the intestinal wall increases the surface area for absorption.

A grasshopper also has a crop where food is stored. Most chemical digestion in a grasshopper occurs in the midgut region, where projections called gastric pouches increase the surface area for nutrient absorption. The hindgut functions mainly to absorb water and compact wastes.

Many birds have three separate chambers: a crop, a stomach, and a gravel-filled gizzard, in which food is pulverized. Chemical digestion and absorption occur in the intestine.

? What is an advantage of an alimentary canal, compared to a gastrovascular cavity?

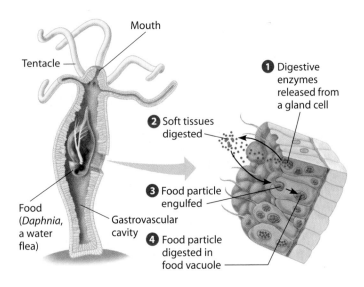

Figure 21.3A   Digestion in the gastrovascular cavity of a hydra

Specialized regions can carry out digestion and absorption sequentially. ■

21.4 The human digestive system consists of an alimentary canal and accessory glands

As an introduction to our own digestive system, Figure 21.4 illustrates the human alimentary canal and the accessory glands associated with it. The schematic diagram below, left, gives you an overview of the sequence of the organs and the locations of the associated glands. These glands—the salivary glands, pancreas, and liver—are labeled in blue on the figure. They secrete digestive juices that enter the alimentary canal through ducts. Secretions from the liver are stored in the gallbladder before they are released into the intestine.

Food enters the mouth, is chewed in the oral cavity, and then pushed by the tongue into the pharynx. Once food is swallowed, muscles propel it through the alimentary canal by **peristalsis,** alternating waves of contraction and relaxation of the smooth muscles lining the canal (see Module 21.6). It is peristalsis that enables us to process and digest food even while lying down. In only 5–10 seconds, food passes from the pharynx down the esophagus and into the stomach.

As shown in the blow-up diagram below, right, muscular ringlike valves, called **sphincters,** regulate the passage of food into and out of the stomach. The sphincter controlling the passage out of the stomach works like a drawstring to close it off, keeping food there for about 2–6 hours, long enough for stomach acids and enzymes to begin digestion. The final steps of digestion and nutrient absorption occur in the small intestine over a period of 5–6 hours. Undigested material moves slowly through the large intestine (taking 12–24 hours), and feces are stored in the rectum and then expelled through the anus.

In the next several modules, we follow a snack—an apple and some crackers and cheese—through the alimentary canal to see in more detail what happens to the food in each of the processing stations along the way.

🎧 MP3 Tutor *The Human Digestive System*

? By what process does food move from the pharynx to the stomach of an astronaut in the weightless environment of a space station?

■ peristalsis

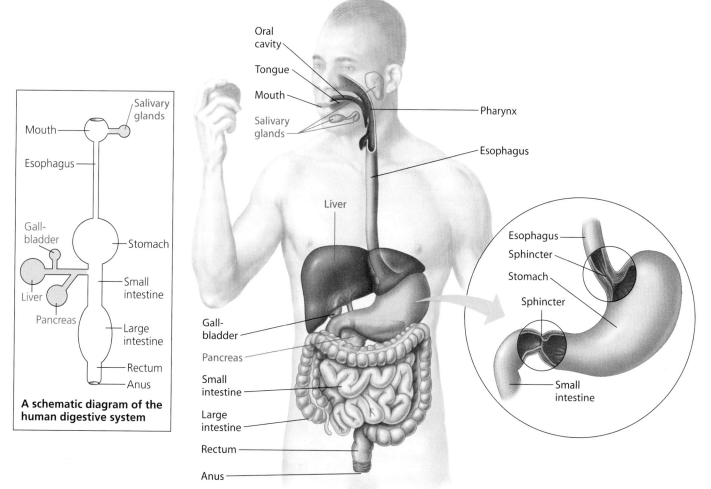

A schematic diagram of the human digestive system

Figure 21.4 The human digestive system

21.5 Digestion begins in the oral cavity

As you anticipate your apple, cheese, and crackers, your salivary glands may start delivering **saliva** through ducts to the **oral cavity** even before you take a bite. This is a response to the sight or smell (or even thought) of food. The presence of food in the oral cavity continues to stimulate salivation. In a typical day, your **salivary glands** secrete more than a liter of saliva. You can see the duct opening and salivary glands in Figure 21.5. (The third pair of glands is visible in Figure 21.4.)

Saliva contains several substances important in food processing. A slippery glycoprotein (carbohydrate-protein complex)

protects the soft lining of the mouth and lubricates food for easier swallowing. Buffers neutralize food acids, helping prevent tooth decay. Antibacterial agents kill many of the bacteria that enter the mouth with food. Saliva also contains the digestive enzyme amylase, which begins hydrolyzing the starch in your cracker.

Mechanical and chemical digestion begin in the oral cavity. Chewing cuts, smashes, and grinds food, making it easier to swallow and exposing more food surface to digestive enzymes. As Figure 21.5 shows, you have four kinds of teeth. Starting at the front on one side of the upper or lower jaw, there are two blade-like incisors. These you use for biting into your apple. Behind the incisors is a single pointed canine tooth. (Canine teeth are much bigger in carnivores, which use them to kill and rip apart prey.) Next come two premolars and three molars, which grind and crush your apple and crackers. (The third molar, a "wisdom" tooth, does not appear in all people, and in some people it pushes against the other teeth and must be removed.)

Also prominent in the oral cavity is the tongue, a muscular organ covered with taste buds. Besides enabling you to taste your meal, the tongue manipulates food and helps shape it into a ball called a **bolus.** When swallowing, the tongue pushes the bolus to the back of the oral cavity and into the pharynx.

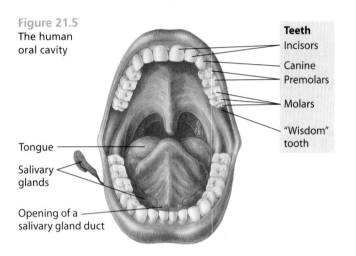

Figure 21.5
The human oral cavity

Teeth
Incisors
Canine
Premolars
Molars
"Wisdom" tooth

Tongue
Salivary glands
Opening of a salivary gland duct

? Chewing functions in _____ digestion. What does the amylase in saliva do?

■ mechanical . . . begins the chemical digestion of starch

21.6 After swallowing, peristalsis moves food through the esophagus to the stomach

Openings into both the esophagus and the trachea (windpipe) are in the pharynx, the region we call our throat. Most of the time, as shown on the left in Figure 21.6A, the esophageal opening is closed off by a sphincter (blue arrows). Air enters the larynx, the voice box containing the vocal cords, and flows through the trachea to the lungs (black arrows).

This situation changes when you start to swallow. The tongue pushes the bolus of food into the pharynx, triggering the swallowing reflex (center of Figure 21.6A); the esophageal sphincter relaxes and allows the bolus to enter the esophagus

(green arrow). At the same time, the larynx moves upward and tips the epiglottis (a flap of cartilage and fibrous connective tissue) down over the opening to the larynx. In this position, the epiglottis prevents food from passing into the trachea. You can see this motion in the bobbing of your larynx (also called your Adam's apple) during swallowing.

After the bolus enters the esophagus, the larynx moves back downward, the epiglottis tips up again, and the breathing passage reopens (right side of the figure). The esophageal sphincter contracts above the bolus.

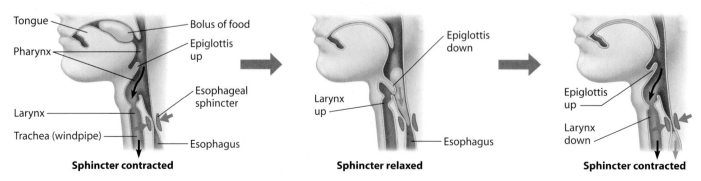

Tongue
Pharynx
Larynx
Trachea (windpipe)
Bolus of food
Epiglottis up
Esophageal sphincter
Esophagus
Sphincter contracted

Epiglottis down
Larynx up
Esophagus
Sphincter relaxed

Epiglottis up
Larynx down
Sphincter contracted

Figure 21.6A The human swallowing reflex

The esophagus is a muscular tube that conveys food boluses from the pharynx to the stomach. The muscles at the very top of the esophagus are under voluntary control; thus, the act of swallowing begins voluntarily. But then involuntary waves of contraction by smooth muscles in the rest of the esophagus take over. Figure 21.6B shows how waves of muscle contraction—peristalsis—squeeze a bolus toward the stomach. As food is swallowed, muscles above the bolus contract (blue arrows), pushing the bolus downward. At the same time, muscles around the bolus relax, allowing the passageway to open. Muscle contractions continue in waves until the bolus enters the stomach.

As with other parts of the digestive system, the structure of the esophagus fits its function. It has tough yet elastic connective tissues that allow it to stretch to accommodate a bolus, layers of circular and longitudinal smooth muscles for peristalsis, and a stratified epithelial lining that replenishes cells abraded off during swallowing. The length of the esophagus varies by species. For example, fishes have no lungs to bypass and have a very short esophagus. And it will come as no surprise that giraffes have a very long esophagus.

Usually our breathing and swallowing are carefully coordinated. Next, we consider what happens when they are not.

? What prevents food from going down the wrong tube?

■ The epiglottis tips over the opening to the trachea when swallowing.

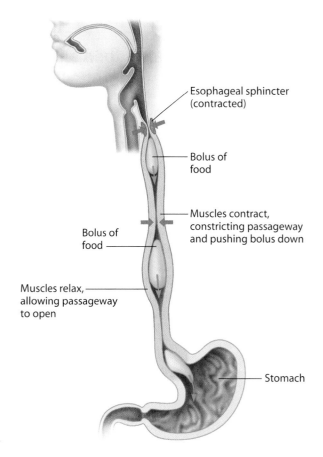

Figure 21.6B Peristalsis moving a food bolus down the esophagus

Connection

21.7 The Heimlich maneuver can save lives

Sometimes our swallowing mechanism goes awry. A person may eat too quickly or fail to chew food thoroughly. Or an infant may swallow an object too big to pass through the esophagus. Such mishaps can lead to a blocked pharynx or trachea. Air cannot flow into the trachea, causing the person to choke. If breathing is not restored within minutes, brain damage or death may result.

To save someone who is choking, it is essential to dislodge any foreign objects in the throat and get air flowing. This quick assistance often comes through the use of the Heimlich maneuver. The procedure, invented by Dr. Henry Heimlich in the 1970s, allows people with little medical training to step in and aid a choking victim.

The maneuver is often performed on someone who is seated or standing up. Stand behind the victim and place your arms around the victim's waist. Make a fist with one hand, and place it against the victim's upper abdomen, well below the rib cage. Then place the other hand over the fist and press into the victim's abdomen with a quick upward thrust. When done correctly, the diaphragm is forcibly elevated, pushing air into the trachea. Repeat this procedure until the object is forced out of the victim's airway (Figure 21.7). The maneuver can be performed on drowning victims to clear the lungs of water before beginning CPR. And individuals can use their own fists or the back of a chair to force air upward and dislodge an object from their trachea.

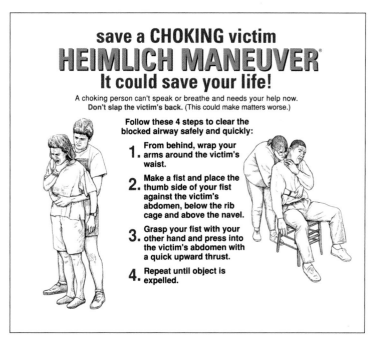

Figure 21.7 The Heimlich maneuver for helping choking victims

? If food is stuck in the pharynx, what effect could it have on nearby structures? (*Hint:* See Figure 21.6A.)

■ The epiglottis may be tipped down, blocking the opening to the trachea.

21.8 The stomach stores food and breaks it down with acid and enzymes

Having a stomach is the main reason we do not need to eat constantly. With its accordion-like folds and highly elastic wall, our stomach can stretch to accommodate about 2 liters of food and drink, usually enough to satisfy our needs for many hours.

Some chemical digestion occurs in the stomach. The stomach secretes **gastric juice,** which is made up of mucus, enzymes, and strong acid. The pH of gastric juice is about 2 (see Module 2.15). One function of the acid is to break apart the cells in food. The acid also kills most bacteria and other microbes that are swallowed with food.

The interior surface of the stomach wall is highly folded, and as the diagram in Figure 21.8 shows, it is dotted with pits leading into tubular gastric glands. The gastric glands have three types of cells that secrete different components of the gastric juice. Mucous cells (shown here in dark pink) secrete mucus, which lubricates and protects the cells lining the stomach. Parietal cells (yellow) secrete hydrogen ions and chloride ions, which combine in the lumen (cavity) of the stomach to form hydrochloric acid (HCl). Chief cells (light pink) secrete pepsinogen, an inactive form of the enzyme pepsin.

The diagram on the far right of the figure indicates how pepsinogen, HCl, and pepsin interact. ❶ Pepsinogen and HCl are secreted into the lumen of the stomach. ❷ Next, the HCl converts pepsinogen to pepsin. ❸ Pepsin then activates more pepsinogen, starting a chain reaction. This series of events is an example of positive feedback, in which the end product of a process promotes the formation of more end product.

Pepsin begins the chemical digestion of proteins—those in your cheese snack, for instance. It splits the polypeptide chains of the proteins into smaller polypeptides. This action primes the proteins for further digestion, which will occur in the small intestine.

What prevents gastric juice from digesting away the stomach lining? Secreting pepsin in the inactive form of pepsinogen helps protect the cells of the gastric glands, and mucus helps protect the stomach lining from both pepsin and acid. Regardless, the epithelium is constantly eroded. Enough new cells are generated by mitosis to replace the stomach lining completely about every three days.

Another protection for the stomach is that the cells in our gastric glands do not secrete acidic gastric juice constantly. Their activity is regulated by a combination of nerve signals and hormones. When you see, smell, or taste food, a signal from your brain to your stomach stimulates your gastric glands to secrete gastric juice. Once you have food in your stomach, substances in the food stimulate cells in the stomach wall to release the hormone **gastrin** into the circulatory system. Gastrin circulates in the bloodstream, returning to the stomach wall (green dotted line in the top section of Figure 21.8). When it arrives there, it stimulates additional secretion of gastric juice. As much as 3 L of gastric juice may be secreted a day. A negative-feedback mechanism like the one we described in Module 20.14 inhibits the secretion of gastric juice when the stomach contents become too acidic. The acid inhibits the release of gastrin, and with less gastrin in the blood, the gastric glands secrete less gastric juice. Other hormones control the release of digestive enzymes in the small intestine.

About every 20 seconds, the stomach contents are mixed by the churning action of muscles in the stomach wall. You may feel hunger pangs when your stomach has been empty for hours and strongly contracts. (Other sensations of

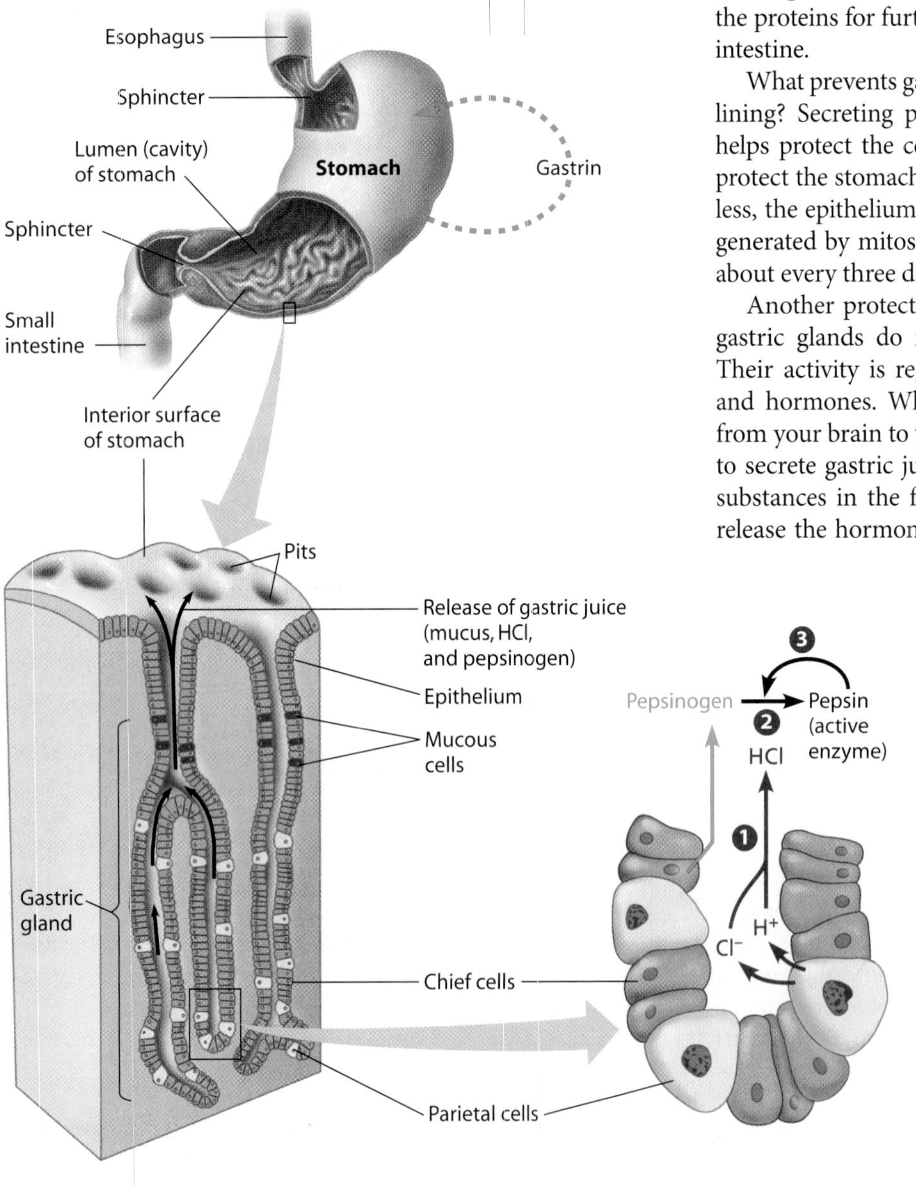

Figure 21.8   The stomach and its production of gastric juice

hunger result from appetite-controlling hormones that we will discuss in Module 21.22.) As a result of mixing and enzyme action, what begins in the stomach as a recently swallowed apple, cracker, and cheese snack becomes an acidic, nutrient-rich broth known as **chyme.**

The sphincter between the stomach and the small intestine helps regulate the passage of chyme from the stomach into the small intestine. With the chyme leaving the stomach only a squirt at a time, the stomach takes about 2–6 hours to empty after a meal.

We'll continue with the digestion of your snack in Module 21.10. But first, we'll consider some digestive problems.

? If you add pepsinogen to a test tube containing protein dissolved in distilled water, not much protein will be digested. What inorganic chemical could you add to the tube to accelerate protein digestion? What effect will it have?

■ Hydrochloric acid or some other acid will convert inactive pepsinogen to active pepsin, which will begin the digestion of the protein.

Connection

21.9   Digestive ailments include acid reflux and gastric ulcers

A stomachful of digestive juice laced with strong acid breaks apart the cells in our food, kills bacteria, and begins the digestion of proteins. At the same time, these chemicals, acidic enough to dissolve iron nails, can be harmful. The opening between the esophagus and the stomach is usually closed until a bolus arrives. Occasionally, however, there is acid reflux. This backflow of chyme into the lower end of the esophagus causes the feeling we call heartburn (which should more accurately be called esophagus-burn).

Some people suffer acid reflux frequently and severely enough to harm the lining of the esophagus, a condition called GERD (gastroesophageal reflux disease). GERD can often be treated with lifestyle changes. Doctors usually recommend that patients stop smoking, avoid alcohol, lose weight, eat small meals, refrain from lying down for 2–3 hours after eating, and sleep with the head of the bed raised. Medications to treat GERD include antacids, which reduce stomach acidity, and drugs called H2 blockers, such as Pepcid AC or Zantac, which impede acid production. Proton pump inhibitors, such as Prilosec, have a different mechanism of action and are very effective at stopping acid production. When lifestyle changes and medications fail to alleviate the symptoms, surgery to strengthen the lower esophageal sphincter may be an option.

Can all that acid also cause problems in the stomach? A gel-like coat of mucus normally protects the stomach wall from the corrosive effect of digestive juice. When it fails, open sores called gastric ulcers can develop in the stomach wall. The symptoms are usually a gnawing pain in the upper abdomen, which may occur a few hours after eating.

Gastric ulcers were formerly thought to result from the production of too much acid or too little mucus. For years, the blame was put on factors that may cause these effects, such as aspirin, ibuprofen, smoking, alcohol, coffee, and stress. However, a spiral-shaped bacterium called *Helicobacter pylori* has now been identified as the primary culprit (Figure 21.9).

The low pH of the stomach kills most microbes, but not the acid-tolerant *H. pylori*. This bacterium burrows beneath the mucus and releases harmful chemicals. Growth of *H. pylori* seems to result in a localized loss of protective mucus and damage to the cells lining the stomach. Numerous white blood cells move into the stomach wall to fight the infection, and their

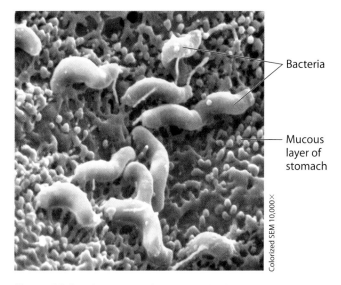

Colorized SEM 10,000×

Figure 21.9   Ulcer-causing bacteria, *Helicobacter pylori*

presence is associated with mild inflammation of the stomach, called gastritis. Gastric ulcers develop when pepsin and hydrochloric acid destroy cells faster than the cells can regenerate. Eventually, the stomach wall may erode to the point that it actually has a hole in it. This hole can lead to a life-threatening infection within the abdomen or internal bleeding. *H. pylori* is found in 70–90% of ulcer and gastritis sufferers. Some studies also link *H. pylori* to certain kinds of stomach cancer.

Gastric ulcers usually respond to a combination of antibiotics and bismuth (the active ingredient in Pepto-Bismol), which eliminates the bacteria and promotes healing. Drugs that reduce stomach acidity may also help.

When digesting food leaves the stomach, it is accompanied by gastric juices, and so the first section of the small intestine—the duodenum—is also susceptible to ulcers, as is the esophagus in cases of severe GERD.

? In contrast to most microbes, the species that causes ulcers thrives in an environment with a very low _____.

■ pH

21.10 The small intestine is the major organ of chemical digestion and nutrient absorption

Returning to our journey through the digestive tract, what is the status of your snack as it passes out of the stomach into the small intestine? The food has been mechanically reduced to smaller pieces and mixed with digestive juices; it now resembles a thick soup. Chemically, the digestion of starch in the cracker began in the mouth, and the breakdown of protein in the cheese began in the stomach. The rest of the digestion of the large molecules in the snack occurs in the **small intestine.** The nutrients that result from this digestion are absorbed into the blood from the small intestine. With a length of more than 6 m, the small intestine is the longest organ of the alimentary canal. (Its name is based not on its length but on its diameter, which is only about 2.5 cm; the large intestine is much shorter but has twice the diameter.)

Sources of Digestive Enzymes and Bile Two large organs, the pancreas and the liver, contribute to digestion in the small intestine (Figure 21.10A). The **pancreas** produces pancreatic juice, a mixture of digestive enzymes and an alkaline solution rich in bicarbonate. The bicarbonate acts as a buffer to neutralize the acidity of chyme as it enters the small intestine. As you will learn in Chapter 26, the pancreas also produces hormones that regulate blood-glucose levels.

In addition to its many other functions, the **liver** produces bile. **Bile** contains bile salts that emulsify fats, making them more susceptible to attack by digestive enzymes. The **gallbladder** stores bile until it is needed in the small intestine.

As you can see in Figure 21.10A, the intestinal wall itself also produces digestive enzymes. Some of these enzymes are secreted into the lumen of the small intestine; others are bound to the surface of epithelial cells. The first 25 cm or so of the small intestine is called the **duodenum.** This is where chyme squirted from the stomach mixes with bile from the gallbladder,

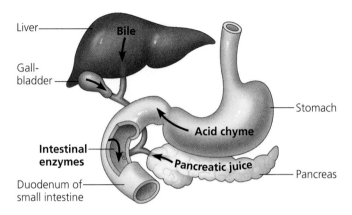

Figure 21.10A The small intestine and related digestive organs

pancreatic juice from the pancreas, and digestive enzymes from gland cells in the intestinal wall.

Digestion in the Small Intestine Table 21.10 summarizes the processes of enzymatic digestion that occur in the small intestine. All four types of large molecules (carbohydrates, proteins, nucleic acids, and fats) are digested in the small intestine. As we discuss the digestion of each, the table will help you keep track of the enzymes involved (shown in red).

The digestion of carbohydrates that began in the oral cavity is completed in the small intestine. An enzyme called pancreatic amylase hydrolyzes starch (a polysaccharide) into the disaccharide maltose. The enzyme maltase then splits maltose into the monosaccharide glucose. Maltase is one of a family of enzymes, each specific for the hydrolysis of a different disaccharide. For example, sucrase hydrolyzes table sugar (sucrose), and lactase digests lactose. Lactose is common in milk and cheese. As we saw in the introduction to Chapter 3, children generally have

TABLE 21.10	ENZYMATIC DIGESTION IN THE SMALL INTESTINE		
Carbohydrates			
Starch	→ Pancreatic amylase →	Maltose (and other disaccharides)	→ Maltase, sucrase, lactase, etc. → Monosaccharides
Proteins			
Polypeptides	→ Trypsin, chymotrypsin →	Smaller polypeptides	→ Aminopeptidase, carboxypeptidase, dipeptidase → Amino acids
Nucleic acids			
DNA and RNA	→ Nucleases →	Nucleotides	→ Other enzymes → Nitrogenous bases, sugars, and phosphates
Fats			
Fat globules	→ Bile salts →	Fat droplets (emulsified)	→ Lipase → Fatty acids and glycerol

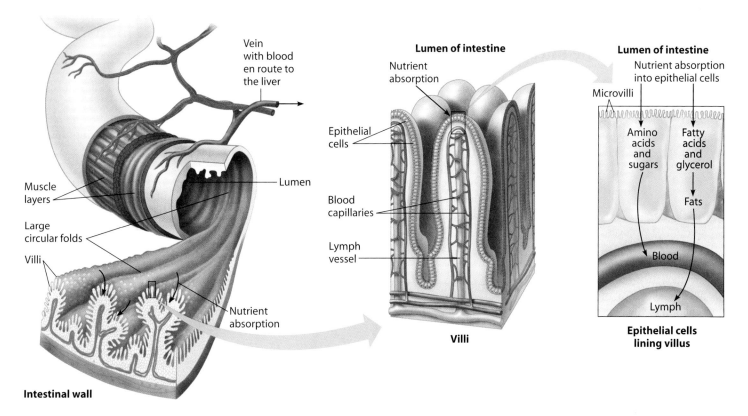

Intestinal wall

Vein with blood en route to the liver

Muscle layers

Large circular folds

Villi

Lumen

Nutrient absorption

Lumen of intestine

Nutrient absorption

Epithelial cells

Blood capillaries

Lymph vessel

Villi

Lumen of intestine

Nutrient absorption into epithelial cells

Microvilli

Amino acids and sugars

Fatty acids and glycerol

Fats

Blood

Lymph

Epithelial cells lining villus

Figure 21.10B   Structure of the small intestine

much more lactase than adults, and as a result, many adults have difficulty digesting milk products.

The small intestine also completes the digestion of proteins that was begun in the stomach. The pancreas and the duodenum secrete hydrolytic enzymes that completely dismantle polypeptides into amino acids. The enzymes trypsin and chymotrypsin break polypeptides into smaller polypeptides. Two other enzymes, aminopeptidase and carboxypeptidase, split off one amino acid at a time, working from opposite ends of a polypeptide. Another enzyme, dipeptidase, hydrolyzes fragments only two or three amino acids long. Working together, this enzyme team digests proteins much faster than any single enzyme could.

Yet another team of enzymes, the nucleases, hydrolyzes nucleic acids. Nucleases from the pancreas split DNA and RNA (which are present in the cells of food items) into their component nucleotides. The nucleotides are then broken down into nitrogenous bases, sugars, and phosphates by other enzymes.

In contrast to starch and proteins, fats remain undigested until they reach the duodenum. Hydrolysis of fats is a special problem because fats are insoluble in water. How is this problem solved? First, bile salts in bile cause fat globules to be physically broken up into smaller fat droplets, a process called emulsification. When there are many small droplets, a larger surface area of fat is exposed to lipase, a pancreatic enzyme that breaks fat molecules down into fatty acids and glycerol.

By the time peristalsis has moved the mixture of chyme and digestive juices through the duodenum, chemical digestion of your snack is just about complete. The main function of the rest of the small intestine is the absorption of nutrients and water.

Absorption in the Small Intestine   Structurally, the small intestine is well suited for its task of absorbing nutrients. Its lining has a huge surface area—roughly 300 m^2, about the size of a tennis court. As Figure 21.10B shows, this extensive surface area results from several kinds of folds and projections. Around the inner wall of the small intestine are large circular folds with numerous small, fingerlike projections called **villi** (singular, villus). Each of the epithelial cells lining a villus has many tiny surface projections, called **microvilli.** The microvilli extend into the lumen of the intestine and greatly increase the surface area across which nutrients are absorbed.

Some nutrients are absorbed by simple diffusion; other nutrients are pumped against concentration gradients into the epithelial cells. Notice that a small lymph vessel (yellow) and a network of capillaries (red, purple, and blue) penetrate the core of each villus. After fatty acids and glycerol are absorbed by an epithelial cell, these building blocks are recombined into fats, which are then transported into a lymph vessel. Other absorbed nutrients, such as amino acids and sugars, pass out of the intestinal epithelium and then across the thin walls of the capillaries into the blood.

Where does this nutrient-laden blood go? To the liver, where we also head in the next module.

Web Activity *Digestive System Function*

Web Process of Science *What Role Does Amylase Play in Digestion?*

? Amylase is to _____ as _____ is to DNA.

▪ starch . . . nuclease

21.11 One of the liver's many functions is processing nutrient-laden blood from the intestines

The liver has a strategic location in the body—between the intestines and the heart. As indicated in Figure 21.11, capillaries from the small and large intestines converge into veins that lead into the **hepatic portal vein.** This large vessel transports nutrients absorbed by the intestines directly to the liver. The liver thus gets first access to nutrients absorbed from a meal. One of its main functions is to remove excess glucose from the blood and convert it to glycogen (a polysaccharide), which is stored in liver cells. In balancing the amount of glycogen it stores with the amount of glucose it releases to the blood, the liver plays a key role in regulating body metabolism. You will read about the hormonal control of this important function in Chapter 26.

The liver also converts many of the nutrients it receives into new substances. Liver cells synthesize proteins essential to many of the body's functions. Among these are various plasma proteins important in blood clotting and in maintaining the osmotic balance of the blood, as well as lipoproteins that transport fats and cholesterol to body cells.

The liver has a chance to modify and detoxify substances absorbed by the digestive tract before the blood carries these materials to the heart for distribution to the rest of the body. It converts toxins such as alcohol and other drugs into inactive products that can be excreted in urine. (These products are what are looked for in urine tests for various drugs.) As they detoxify alcohol or process some over-the-counter and prescription drugs, however, liver cells can be damaged. The combination of alcohol and some drugs is particularly harmful to the liver.

The liver produces bile, which aids in the digestion of fats (see Module 21.10). And as you learn in Chapter 25, it also processes nitrogen wastes from the breakdown of proteins for disposal in urine.

Now, let's return to the digestive tract.

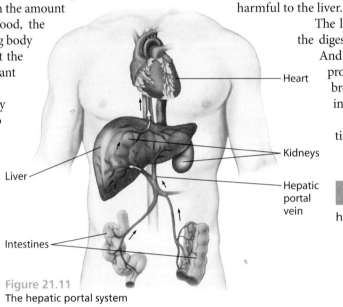

Heart

Kidneys

Liver

Hepatic portal vein

Intestines

Figure 21.11
The hepatic portal system

? What two functions of the liver relate to the hepatic portal vein?

■ As the hepatic portal vein delivers blood directly from the intestines, the liver can process and regulate the absorbed nutrients and remove toxic substances.

21.12 The large intestine reclaims water and compacts the feces

The **large intestine,** or **colon,** is about 1.5 m long and 5 cm in diameter. As the enlargement in Figure 21.12 shows, it joins the small intestine at a T-shaped junction, where a sphincter controls the passage of unabsorbed food material out of the small intestine. One arm of the T is a blind pouch called the **cecum.** The **appendix,** a small, fingerlike extension of the cecum, contains a mass of white blood cells that make a minor contribution to immunity. Despite this role, the appendix itself is prone to infection (appendicitis). If this occurs, the appendix can be surgically removed without weakening the immune system.

One major function of the colon is to absorb water from the alimentary canal. Altogether, about 7 L of fluid enters the lumen of the digestive tract each day as the solvent of the various digestive juices. About 90% of this water is absorbed back into the blood and tissue fluids by the small intestine and colon. As the water is absorbed, the remains of the digested food become more solid as they move along the colon by peristalsis. These waste products, the **feces,** consist mainly of indigestible plant fibers (cellulose from your apple, for instance) and prokaryotes that normally live in the colon. Some of our colon bacteria, such

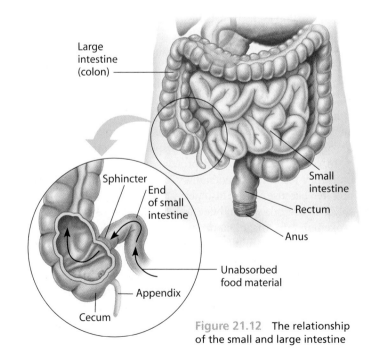

Large intestine (colon)

Sphincter
End of small intestine

Small intestine

Rectum

Anus

Unabsorbed food material

Appendix

Cecum

Figure 21.12 The relationship of the small and large intestine

as *E. coli,* produce important vitamins, including biotin, folic acid, several other B vitamins, and vitamin K. These vitamins are absorbed into the bloodstream from the colon.

Feces are stored in the final portion of the colon, the **rectum,** until they can be eliminated. Strong contractions of the colon create the urge to defecate. Two rectal sphincters, one voluntary and the other involuntary, regulate the opening of the anus.

If the lining of the colon is irritated—by a viral or bacterial infection, for instance—the colon is less effective in reclaiming water, and diarrhea may result. The opposite problem, constipation, occurs when peristalsis moves the feces along too slowly; the colon reabsorbs too much water, and the feces become too compacted. Constipation often results from a diet that does not include enough plant fiber.

? Explain why treatment with antibiotics for an extended period may cause a vitamin K deficiency.

■ The antibiotics may kill the bacteria that synthesize vitamin K in the colon.

21.13 Evolutionary adaptations of vertebrate digestive systems often relate to diet

We have used our own alimentary canal to illustrate the basic plan of vertebrate digestive systems. However, different groups exhibit variations on this plan. Natural selection has favored adaptations that fit the structure of an animal's digestive system to the function of digesting the kind of food the animal eats.

Large, expandable stomachs are common adaptations in carnivores, which may go a long time between meals and must eat as much as they can when they do catch prey. A 200-kg African lion can consume 40 kg (almost 90 lb) of meat in one meal!

The length of an animal's digestive tract is often correlated with diet. In general, herbivores and omnivores have longer alimentary canals, relative to their body size, than carnivores. Vegetation is more difficult to digest than meat because it contains cell walls. A longer canal provides more time for digestion and more surface area for the absorption of nutrients.

Most herbivorous animals also have special chambers that house great numbers of microbes—bacteria and protists. The animals themselves lack the enzymes needed to digest the cellulose in plants. The microbes break down cellulose to simple sugars, which the animals then absorb directly or obtain by digesting the microbes.

Many herbivorous mammals—horses, elephants, rabbits and some rodents, for example—house cellulose-digesting microbes in the colon and in a large cecum, the pouch where the small and large intestines connect. Some of the nutrients produced by the microbes are absorbed in the cecum and colon. Most of the nutrients, however, are lost in the feces because they do not go through the small intestine, the main site of nutrient absorption. Rabbits and some rodents obtain these nutrients by eating some of their feces, thus passing the food through the alimentary canal a second time. The feces from the second round of digestion, the familiar rabbit "pellets," are more compact and are not reingested. Many desert rodents conserve water by eating their first round of feces.

Figure 21.13 compares the digestive tract of a carnivore, the coyote, with that of an herbivore, the koala. The koala is an Australian marsupial (see Module 19.8). These two mammals are about the same size, but the koala's intestine is much longer and includes the longest cecum (about 2 m) of any animal of its size. Bacteria in the cecum digest plant material, making it possible for the koala to get almost all its food and water from the leaves of eucalyptus trees.

The most elaborate adaptations for an herbivorous diet have evolved in the mammals called **ruminants,** which include cattle, sheep, and deer. The stomach of a ruminant has four chambers containing symbiotic microbes. A cow periodically regurgitates food from the first two chambers and "chews its cud," exposing more plant fibers to its microbes for digestion. The cud is then swallowed and moves to the last stomach chambers, where digestion is completed. A cow actually obtains many of its nutrients by digesting the microbes along with the nutrients they produce. The microbes reproduce so rapidly that their numbers remain stable despite this constant loss.

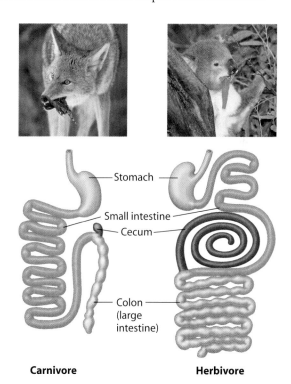

Carnivore   **Herbivore**

Figure 21.13   The alimentary canal in a carnivore (coyote) and an herbivore (koala)

? What are two advantages of a longer alimentary canal in herbivores?

■ It provides increased time for processing of difficult-to-digest plant material and increased surface area for absorption of nutrients.

21.14 Overview: A healthy diet satisfies three needs

All animals—whether herbivores like koalas, carnivores like coyotes, or omnivores like humans—have the same basic nutritional needs. All animals must obtain (1) fuel to power all body activities; (2) organic molecules to build the animal's own molecules; and (3) essential nutrients, or substances the animal cannot make for itself from raw materials but must obtain from food, with "no assembly required."

We have seen that digestion breaks the large molecules in food into their components. Cells can then oxidize these smaller molecules for energy or assemble them into their own complex molecules—the proteins, carbohydrates, lipids, and nucleic acids needed to build and maintain cell structure and function.

Eating too little food, too much food, or the wrong mixture of foods can all endanger an animal's health. Starting with the need for fuel and paying particular attention to humans, we discuss basic nutritional needs in the rest of this chapter.

? What are the three needs that a healthy diet fills?

■ Fuel, organic building materials, and essential nutrients

21.15 Chemical energy powers the body

Reading a book, walking to class, eating and digesting food, and every other activity your body performs require fuel in the form of chemical energy. Cellular metabolism produces the body's energy currency, ATP, by oxidizing organic molecules digested from food (see Module 6.12). Usually, cells use carbohydrates and fats as fuel sources. But if these are in short supply, cells will use proteins as an energy source. Fats are especially rich in energy: The oxidation of a gram of fat liberates more than twice the energy liberated from a gram of carbohydrate or protein. The energy content of food is measured in **kilocalories** (1 **kcal** = 1,000 calories). The calories listed on food labels or referred to elsewhere regarding nutrition are actually kilocalories and are written as *Calories* (with a capital C).

The rate of energy consumption by the body is called **metabolic rate.** It is the sum of all the energy-requiring biochemical reactions over a given time interval. Cellular metabolism must continuously drive several processes for an animal to remain alive. These include cell maintenance, breathing, the beating of the heart, and, in birds and mammals, the maintenance of body temperature. The number of kilocalories (Calories) a resting animal requires to fuel these essential processes for a given time is called the **basal metabolic rate (BMR).** The BMR for humans averages 1,300–1,500 kcal per day for adult females and about 1,600–1,800 kcal per day for adult males. This is about equivalent to the energy requirement of a 75-watt light bulb. But this is only a basal (base) rate—the amount of energy we "burn" lying motionless. Any activity, even working quietly at your desk, consumes kilocalories in addition to the BMR. The more strenuous the activity, the greater the energy demand. Table 21.15 gives you an idea of the amount of activity it takes for a 68-kg (150-lb) person to use up the kilocalories contained in several common foods.

What happens when we take in more kilocalories than we use? Rather than discarding the extra energy, our cells store it in various forms. Our liver and muscles store energy in the form of glycogen, a polymer of glucose molecules. Most of us store enough glycogen to supply about a day's worth of basal metabolism. Our cells also store excess energy as fat. This happens even if our diet contains little fat because the liver converts excess carbohydrates and proteins to fat. The average human's energy needs can be fueled by the oxidation of only 0.3 kg of fat per day. Most healthy people have enough stored fat to sustain them through several weeks of starvation. We discuss fat storage and its consequences in Module 21.21. But first we consider the essential nutrients that must be supplied in the diet.

? What is the difference between metabolic rate and basal metabolic rate?

■ Metabolic rate is the total energy used for all activities in a unit of time; BMR is the minimum number of kilocalories that a resting animal needs to maintain life's basic processes.

TABLE 21.15	EXERCISE REQUIRED TO "BURN" THE CALORIES (KCAL) IN COMMON FOODS		
	Jogging	**Swimming**	**Walking**
Speed of exercise	9 min/mi	30 min/mi	20 min/mi
kcal "burned"/hour	775	408	245
Cheeseburger (quarter-pound) 417 kcal	32 min	1 hr, 1 min	1 hr, 42 min
Pepperoni pizza (1 large slice) 280 kcal	22 min	42 min	1 hr, 8 min
Soft drink (12 oz) 152 kcal	12 min	22 min	37 min
Whole wheat bread (1 slice) 65 kcal	5 min	10 min	16 min

These data are for a person weighing 68 kg (150 lb).

21.16 An animal's diet must supply essential nutrients

Besides providing fuel and organic raw materials, an animal's diet must also supply **essential nutrients.** These are materials that must be obtained in preassembled form because the animal's cells cannot make them from *any* raw material. There are four classes of essential nutrients: essential fatty acids, essential amino acids, vitamins, and minerals. We will discuss vitamins and essential minerals in Module 21.18.

Undernourishment is a condition resulting from a diet that is chronically deficient in calories. **Malnourishment** results from the long-term absence from the diet of one or more essential nutrients. Because a diet of a single staple such as rice or corn can often provide sufficient calories, undernourishment is generally common only where drought, war, or some other crisis has severely disrupted the food supply. Another cause of undernourishment is anorexia nervosa, an eating disorder that leads individuals, most often female, to starve themselves compulsively. In human populations, malnutrition is much more common than undernutrition, and it is even possible for an overnourished (obese) individual to be malnourished.

Our cells make fats and other lipids by combining fatty acids with other molecules, such as glycerol (see Module 3.8). We can make most of the fatty acids we need. Those we cannot make, called **essential fatty acids,** we must obtain in our diet. One essential fatty acid, linoleic acid, is especially important because it is needed to make some of the phospholipids of cell membranes. Most diets furnish ample amounts of essential fatty acids, and deficiencies are rare.

Adult humans cannot make eight of the 20 kinds of amino acids needed to synthesize proteins. These eight, known as the **essential amino acids,** must be obtained from the diet. (Infants also require a ninth, histidine.) Because the body cannot store excess amino acids, a deficiency of a single essential amino acid limits the use of other amino acids, impairs protein synthesis, and can lead to protein deficiency, a serious type of malnutrition. This is the most common type of malnutrition among humans. The victims are usually children, who often have impaired physical and sometimes mental development if they survive infancy.

The simplest way to get all the essential amino acids is to eat meat and animal by-products such as eggs, milk, and cheese. The proteins in these products are said to be "complete," meaning they provide all the essential amino acids in the proportions needed by the body. In contrast, most plant proteins are "incomplete," or deficient in one or more essential amino acids. In the next module, we discuss how vegetarians can obtain all the essential amino acids in their diet.

? What is the difference between undernourishment and malnourishment? Which of these conditions is more common in human populations and why?

■ Undernourishment results from a diet deficient in calories; malnourishment results from a diet deficient in essential nutrients. A diet of a single staple usually can supply sufficient calories but may be deficient in essential vitamins or amino acids.

21.17 Vegetarians must be sure to obtain all eight essential amino acids

Vegetarian diets may range from avoiding meat to the vegan diet of avoiding meat and meat by-products such as eggs, milk, and cheese. Is it possible for a person to be a vegetarian and obtain adequate nutrition? The answer is yes, but vegetarians have to know how to get all the essential nutrients.

People may become vegetarians by choice or, more commonly in developing countries, because they simply cannot afford animal protein. Animal protein is more expensive to produce, and usually to buy, than plant protein, and most of the human population is primarily vegetarian. Nutritional problems can result when people have to rely on a single type of plant food—just corn, rice, or potatoes, for instance. On such a limited diet, people are likely to become protein-deficient because they lack certain essential amino acids.

The key to being a healthy vegetarian is to eat a variety of plant foods that together supply sufficient quantities of all the essential amino acids. Simply by eating a combination of beans and corn, for example, vegetarians can get all the essential amino acids (Figure 21.17). Most societies have, by trial and error, developed balanced diets that prevent protein deficiency. The Mexican staple diet of corn tortillas and beans is one example.

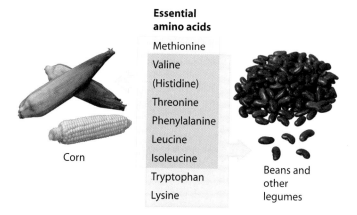

Essential amino acids

Methionine
Valine
(Histidine)
Threonine
Phenylalanine
Leucine
Isoleucine
Tryptophan
Lysine

Corn

Beans and other legumes

Figure 21.17 Essential amino acids from a vegetarian diet

? Look carefully at Figure 21.17. A diet consisting strictly of corn would probably result in a deficiency of which essential amino acids?

■ tryptophan and lysine

21.18 A healthy diet includes 13 vitamins and many essential minerals

A **vitamin** is an organic nutrient that we must obtain from our diet, but is required in minute amounts. For example, one tablespoon of vitamin B_{12} can provide the daily requirement for nearly a million people. Table 21.18A lists 13 essential vitamins and their major dietary sources. Though needed in only tiny amounts, vitamins are absolutely necessary, as you can see from the functions and symptoms of deficiencies listed in the table. Extreme excesses, however, can also be dangerous, as indicated in green type in the table.

Water-soluble vitamins include the B complex and vitamin C. Many B vitamins function as coenzymes; they have catalytic functions and are used over and over in metabolic reactions (see Module 5.15). Vitamin C is required to produce connective tissue. Fat-soluble vitamins include vitamins A, D, E, and K. In general, excess water-soluble vitamins will be eliminated in urine. Excessive amounts of fat-soluble vitamins, however, build up in body fat. Thus, overdoses may have toxic effects.

Minerals are simple inorganic nutrients, usually required in small amounts. We must acquire the essential minerals listed in Table 21.18B (next page) from our diet; some of the major dietary sources are listed. The table also lists the functions in the body and symptoms of deficiency for each mineral.

TABLE 21.18A VITAMIN REQUIREMENTS OF HUMANS

Vitamin	Major Dietary Sources	Functions in the Body	Symptoms of Deficiency / Symptoms of Extreme Excess
Water-Soluble Vitamins			
Vitamin B_1 (thiamine)	Pork, legumes, peanuts, whole grains	Coenzyme used in removing CO_2 from organic compounds	Beriberi (nerve disorders, emaciation, anemia)
Vitamin B_2 (riboflavin)	Dairy products, meats, enriched grains, vegetables	Component of coenzyme FAD	Skin lesions such as cracks at corners of mouth
Niacin (B_3)	Nuts, meats, grains	Component of coenzymes NAD^+ and $NADP^+$	Skin and gastrointestinal lesions, nervous disorders Liver damage
Vitamin B_6 (pyridoxine)	Meats, vegetables, whole grains	Coenzyme used in amino acid metabolism	Irritability, convulsions, muscular twitching, anemia Unstable gait, numb feet, poor coordination
Pantothenic acid (B_5)	Most foods: meats, dairy products, whole grains, etc.	Component of coenzyme A	Fatigue, numbness, tingling of hands and feet
Folic acid (folacin) (B_9)	Green vegetables, oranges, nuts, legumes, whole grains	Coenzyme in nucleic acid and amino acid metabolism; neural tube development in embryo	Anemia, gastrointestinal problems May mask deficiency of vitamin B_{12}
Vitamin B_{12}	Meats, eggs, dairy products	Coenzyme in nucleic acid metabolism; maturation of red blood cells	Anemia, nervous system disorders
Biotin	Legumes, other vegetables, meats	Coenzyme in synthesis of fat, glycogen, and amino acids	Scaly skin inflammation, neuromuscular disorders
Vitamin C (ascorbic acid)	Fruits and vegetables, especially citrus fruits, broccoli, cabbage, tomatoes, green peppers	Used in collagen synthesis (e.g., for bone, cartilage, gums); antioxidant; aids in detoxification; improves iron absorption	Scurvy (degeneration of skin, teeth, blood vessels), weakness, delayed wound healing, impaired immunity Gastrointestinal upset
Fat-Soluble Vitamins			
Vitamin A (retinol)	Dark green and orange vegetables and fruits, dairy products	Component of visual pigments; maintenance of epithelial tissues; antioxidant; helps prevent damage to cell membranes	Vision problems; dry, scaly skin Headache, irritability, vomiting, hair loss, blurred vision, liver and bone damage
Vitamin D	Dairy products, egg yolk (also made in human skin in presence of sunlight)	Aids in absorption and use of calcium and phosphorus; promotes bone growth	Rickets (bone deformities) in children; bone softening in adults Brain, cardiovascular, and kidney damage
Vitamin E (tocopherol)	Vegetable oils, nuts, seeds	Antioxidant; helps prevent damage to cell membranes	None well documented; possibly anemia
Vitamin K	Green vegetables, tea (also made by colon bacteria)	Important in blood clotting	Defective blood clotting Liver damage and anemia

Along with other vertebrates, we humans require relatively large amounts of calcium and phosphorus to construct and maintain our skeleton. Too little calcium can result in the degenerative bone disease osteoporosis. Calcium is also necessary for the normal functioning of nerves and muscles, and phosphorus is an ingredient of ATP and nucleic acids. Iron is a component of hemoglobin, the oxygen-carrying protein of red blood cells, and of several electron carrier molecules that function in cellular respiration. Vertebrates need iodine to make the hormone thyroxine, which regulates metabolic rate. Many minerals are components of various enzymes.

Sodium, potassium, and chlorine are important in nerve function and help maintain the osmotic balance of cells.

Most people ingest far more salt (sodium chloride) than they need. The average U.S. citizen eats enough salt to provide about 20 times the required amount of sodium. Ingesting too much sodium may be associated with high blood pressure.

? Which of the vitamins and minerals listed in these tables are involved with the formation or maintenance of bones and teeth?

■ Vitamin C, vitamin D, calcium, phosphorus, and fluorine

TABLE 21.18B	MINERAL REQUIREMENTS OF HUMANS		
Mineral	**Dietary Sources**	**Functions in the Body**	**Symptoms of Deficiency***
Calcium (Ca)	Dairy products, dark green vegetables, legumes	Bone and tooth formation, blood clotting, nerve and muscle function	Stunted growth, possibly loss of bone mass
Phosphorus (P)	Dairy products, meats, grains	Bone and tooth formation, acid-base balance, nucleotide synthesis	Weakness, loss of minerals from bone, calcium loss
Sulfur (S)	Proteins from many sources	Component of certain amino acids	Symptoms of protein deficiency
Potassium (K)	Meats, dairy products, many fruits and vegetables, grains	Acid-base balance, water balance, nerve function	Muscular weakness, paralysis, nausea, heart failure
Chlorine (Cl)	Table salt	Acid-base balance, water balance, nerve function, formation of gastric juice	Muscle cramps, reduced appetite
Sodium (Na)	Table salt	Acid-base balance, water balance, nerve function	Muscle cramps, reduced appetite
Magnesium (Mg)	Whole grains, green leafy vegetables	Component of certain enzymes	Nervous system disturbances
Iron (Fe)	Meats, eggs, legumes, whole grains, green leafy vegetables	Component of hemoglobin, of certain enzymes, and of electron carriers in energy metabolism	Iron-deficiency anemia, weakness, impaired immunity
Fluorine (F)	Fluoridated drinking water, tea, seafood	Maintenance of tooth (and probably bone) structure	Higher frequency of tooth decay
Zinc (Zn)	Meats, seafood, grains	Component of certain digestive enzymes and other proteins	Growth failure, scaly skin inflammation, reproductive failure, impaired immunity
Copper (Cu)	Seafood, nuts, legumes, organ meats	Component of enzymes in iron metabolism, electron transport, melanin synthesis	Anemia, bone and cardiovascular changes
Manganese (Mn)	Nuts, grains, vegetables, fruits, tea	Component of certain enzymes	Abnormal bone and cartilage
Iodine (I)	Seafood, dairy products, iodized salt	Component of thyroid hormones	Goiter (enlarged thyroid)
Cobalt (Co)	Meats, dairy products	Component of vitamin B-12	None, except as B_{12} deficiency
Selenium (Se)	Seafood, meats, whole grains	Component of enzymes; functions in association with vitamin E	Muscle pain, possible heart muscle deterioration
Chromium (Cr)	Brewer's yeast, liver, seafood, meats, some vegetables	Involved in glucose and energy metabolism	Impaired glucose metabolism
Molybdenum (Mo)	Legumes, grains, some vegetables	Component of certain enzymes	Disorder in excretion of nitrogen-containing compounds

*All of these minerals can be harmful when consumed in excess.

21.19 Do you need to take vitamin and mineral supplements?

A varied diet usually includes enough vitamins and minerals and is considered the best source of these nutrition mainstays. Such diets meet the **Recommended Dietary Allowances (RDAs)**, minimum amounts of nutrients that are needed each day, as determined by a national scientific panel.

The subject of vitamin dosage, however, can cause heated scientific and popular debate. Some people argue that RDAs are set too low for some vitamins, and some of these people believe, probably mistakenly, that *massive* doses of vitamins confer health benefits. Research is far from complete, and debate continues, especially over optimal doses of vitamins C and E. Evidence indicates, however, that excessive amounts of some vitamins and minerals, such as vitamin A and iron, can definitely be harmful.

The USDA makes specific recommendations for certain population groups, such as additional B_{12} for people over age 50, folic acid for pregnant women, and extra vitamin D for people with dark skin or who are exposed to insufficient sunlight. However, unless recommended by a doctor, people should generally avoid megavitamins—supplements that far exceed daily recommended doses.

? Why is an excess of fat-soluble vitamins more dangerous than an excess of water-soluble vitamins?

■ Excess fat-soluble vitamins can accumulate in the body to toxic levels; excess water-soluble vitamins are excreted in the urine.

21.20 What do food labels tell us?

Have you ever found yourself sitting at the breakfast table reading the label on a cereal box? What does it all mean? The Food and Drug Administration (FDA) requires that various types of information be given on packaged-food labels, as shown in Figure 21.20. You'll find the ingredients listed in order from the greatest amount (by weight) to the least. You may want to note how often high fructose corn syrup is listed (see Module 3.6).

Several kinds of "nutrition facts" are found on food labels. First, a serving size of the food is defined according to standards set by the FDA. The energy content in Calories is listed per serving. Selected nutrients are listed as amounts per serving and as percentages of a daily value based on a 2,000-kcal diet. For example, the 1.5 g of fat in a slice of this bread provides 2% of the daily fat allowance for a person needing 2,000 kcal per day.

Food labels emphasize nutrients believed to be associated with disease risks (fats, cholesterol, and sodium) and with a healthy diet (such as dietary fiber, protein, and certain vitamins and minerals). From the data shown, you can tell that each serving of this bread contains 19 g of total carbohydrate. Dietary fiber consists of indigestible complex carbohydrates, mainly cellulose. Subtracting 3 g of dietary fiber and 3 g of sugars (simple carbohydrates) from the 19 g of total carbohydrate tells you that each serving of this bread contains 13 g of digestible complex carbohydrate. This is chiefly starch.

The FDA labeling regulations change from time to time; for example, beginning in 2006, manufacturers were required to list *trans* fat levels (see Module 3.8).

To help consumers compare nutrient amounts in a food with their total daily needs, food labels also provide some general nutritional information. For example, the lower part of the label recommends less than 20 g of saturated fat and at least 25 g of dietary fiber for those with a 2,000-kcal daily diet.

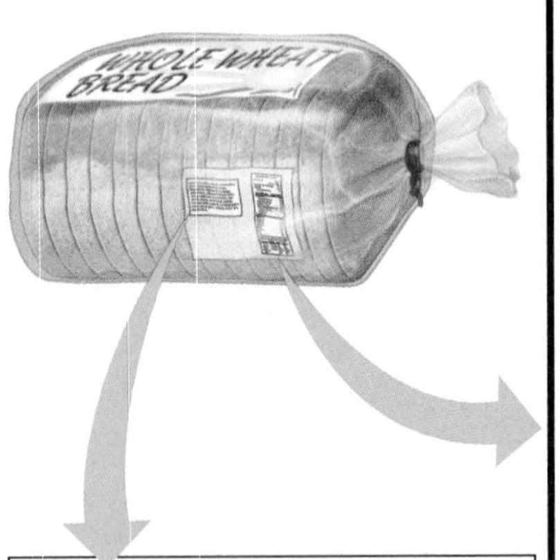

? What percentage of the daily requirements for the fat-soluble vitamins is provided by a slice of the bread in Figure 21.20? (*Hint:* Review Table 21.18A.)

■ 0%

Ingredients: whole wheat flour, water, high fructose corn syrup, wheat gluten, soybean or canola oil, molasses, yeast, salt, cultured whey, vinegar, soy flour, calcium sulfate (source of calcium).

Nutrition Facts

Serving Size 1 slice (43g)
Servings Per Container 16

Amount Per Serving

Calories 100   Calories from Fat 10

	% Daily Value*
Total Fat 1.5g	**2%**
Saturated Fat 0g	**0%**
Trans Fat 0g	**0%**
Cholesterol 0mg	**0%**
Sodium 190mg	**8%**
Total Carbohydrate 19g	**6%**
Dietary Fiber 3g	**12%**
Sugars 3g	
Protein 4g	

Vitamin A 0%	•	Vitamin C 0%
Calcium 2%	•	Iron 4%
Thiamine 6%	•	Riboflavin 2%
Niacin 6%	•	Folic Acid 0%

* Percent Daily Values are based on a 2,000 calorie diet. Your daily values may be higher or lower depending on your calorie needs:

		Calories:	2,000	2,500
Total Fat	Less than		65g	80g
Sat. Fat	Less than		20g	25g
Cholesterol	Less than		300mg	300mg
Sodium	Less than		2,400mg	2,400mg
Total Carbohydrate			300g	375g
Dietary Fiber			25g	30g

Calories per gram:
Fat 9 • Carbohydrate 4 • Protein 4

Figure 21.20 Whole wheat bread labels

21.21   The human health problem of obesity may reflect our evolutionary past

Overnourishment, consuming more food energy than the body needs for normal metabolism, causes obesity, the excessive accumulation of fat. The World Health Organization now recognizes obesity as a major global health problem. The increased availability of fattening foods and large portions in many countries combined with more sedentary lifestyles puts excess weight on bodies. In the United States, the percentage of obese (very overweight) people has doubled to 30% in the past two decades, and another 35% are overweight. Weight problems often begin at a young age; 15% of children and adolescents in the United States are overweight.

Obesity contributes to a number of health problems, including the most common type of diabetes, cancer of the colon and breasts, and cardiovascular disease. Obesity is estimated to be a factor in 300,000 deaths per year in the United States.

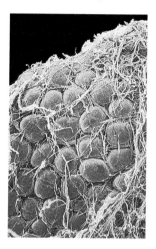

Figure 21.21A   Fat cells from the abdomen of a human

The obesity epidemic has stimulated an increase in scientific research on the causes and possible treatments for weight-control problems. Inheritance is one factor in obesity, which helps explain why certain people have to struggle so hard to control their weight while others can eat and eat without gaining weight. Researchers have identified dozens of the genes that code for weight-regulating hormones. As researchers continue to study the signaling pathways that regulate both long-term and short-term appetite and the body's storage of fat, there is reason to be somewhat optimistic that obese people who have inherited defects in these weight-controlling mechanisms may someday be treated with a new generation of drugs. But so far, the diversity of defects in these complex systems has made it difficult to develop drugs that are effective and free from serious side effects.

The complexity of weight control in humans is evident from studies of the hormone leptin, one of the key long-term appetite regulators in mammals. Leptin is produced by adipose (fat) cells (Figure 21.21A). As adipose tissue increases, leptin levels in the blood rise, which normally cues the brain to suppress appetite. This is one of the feedback mechanisms that usually keep people from becoming obese in spite of access to an abundance of food. Conversely, loss of body fat decreases leptin levels, signaling the brain to increase appetite. Mice that inherit a defect in the gene for leptin become very obese (Figure 21.21B). Researchers found that they could treat these obese mice by injecting leptin.

The discovery of the leptin-deficiency mutation in mice made headlines and initially generated excitement because humans also have a leptin gene. And indeed, obese children who have inherited a mutant form of the leptin gene do lose weight after leptin treatments. However, relatively few obese people have such defects in leptin production. In fact, most obese humans have an abnormally high level of leptin, which, after all, is produced by adipose tissue. For some reason, the brain's satiety center does not respond to the high leptin levels in many obese people. One hypothesis is that in humans, perhaps in contrast to many other mammals, the function of the leptin system is to prevent weight loss, not protect against weight gain. Thus, the decline in leptin when body fat is lost stimulates appetite; but the high levels of leptin produced by large stores of body fat do not function to depress appetite. This physiological nuance may be a consequence of our evolutionary history.

Most of us crave foods that are fatty: fries, chips, burgers, cheese, and ice cream. Though fat hoarding can be a health liability today, it may actually have been advantageous in our evolutionary past. Only in the past few centuries have large numbers of people had access to a reliable supply of high-calorie food. Our ancestors on the African savanna were hunter-gatherers who probably survived mainly on seeds and other plant products, a diet only occasionally supplemented by hunting game or scavenging meat from animals killed by other predators. In such a feast-and-famine existence, natural selection may have favored those individuals with a physiology that induced them to gorge on rich, fatty foods on those rare occasions when such treats were available. Individuals with genes promoting the storage of fat during feasts may have been more likely than their thinner friends to survive famines.

So perhaps our modern taste for fats and sugars reflects the selective advantage it conveyed in our evolutionary history. Although we know it is unhealthful, many of us find it difficult to overcome the ancient survival behavior of stockpiling for the next famine.

?   What are two roles of the hormone leptin? Which of these roles does leptin not play in obese humans?

Figure 21.21B   A mouse with a defect in a gene for leptin, an appetite-suppressing hormone (left); a normal mouse (right)

■ A drop in leptin due to a loss of adipose tissue stimulates appetite; a high level of leptin, produced by increased body fat, depresses appetite. The second mechanism does not seem to function in some humans.

21.22 What are the health risks and benefits of weight loss plans?

Is it our evolutionary past or our sedentary lifestyle? Is it super-sized fast food or the addition of high fructose corn syrup to processed foods? Whatever is fueling the rise in numbers of people who are overweight or obese is also igniting our interest in ways to shed body fat. According to some estimates, the U.S. market for weight loss products and services, worth about $60 million in 1999, expanded to $49 billion in 2006. But has this huge increase in expenditures bought us thinner, healthier bodies? Not yet.

More than 1,000 diet books are now available on the shelves. Diet commercials, magazine testimonials, and Internet advertisements bombard us. How can we know which diet plan is fastest, safest, most long-lasting? Weight-loss plans fall into several broad categories. Table 21.22 presents some popular examples of these categories. New diets continue to appear, such as the Mediterranean diet high in olive oil, fresh fruits, vegetables, and grains.

In recent years, many popular weight loss schemes have focused on reduced intake of carbohydrates. Backers of these diets say that with fewer carbohydrates, the body must burn stored fat instead. Such diets dramatically restrict carbohydrate consumption, some allowing as little as 20 g of carbohydrates a day—less than 10% of the current RDA. People following "low-carb" diets often drop sugar, bread, fruits, and potatoes from their diets, swapping in cheese, nuts, and meat instead.

Some people have lost weight quickly on a low-carbohydrate diet, and because of these success stories and the fatty foods the diet allows, the approach has surged in popularity. Americans spend as much as $15 billion a year on "low-carb" diet aids and foods. Although some studies have found these diets to be effective, others have found that they offer only short-lived benefits. Much of the initial weight reduction comes through water loss. The fatty foods encouraged in such diets may contribute to heart disease and kidney problems. Reductions in fruits and vegetables cut a person's intake of vitamins, minerals, and fiber. As a result, few doctors recommend low-carbohydrate diets as a healthy way to long-term weight loss.

Low-carb diets unseated low-fat diets, an earlier dieting trend with its own flood of low-fat (but often high-sugar) processed foods. Low-fat regimens dramatically reduce consumption of dairy products, meat, nuts, and oils. Low-fat diets often lack adequate amounts of fatty acids and protein and may make it difficult for the body to absorb fat-soluble vitamins.

Gastric bypass surgery, which surgically reduces the size of the stomach and the length of the small intestine, is an increasingly popular weight-loss solution for very obese individuals. This surgery limits food intake capacity and nutrient absorption. Although it has documented risks, studies show that gastric bypass surgery reduces some obesity-related health risks.

Scientific studies of weight-loss diets indicate that sustainability is the major shortcoming of all diets. There appears to be no silver bullet for losing weight and keeping it off without lifestyle changes. These changes involve a combination of increased exercise and a restricted but balanced diet that provides at least 1,200 kcal per day and adequate amounts of all essential nutrients. Such a combination can trim the body gradually and keep extra fat off without harmful side effects.

? In what sense is maintaining a stable body weight a matter of caloric bookkeeping?

■ When we burn as many calories a day through BMR and activities as we take in with our food, a stable body weight will result.

TABLE 21.22	TYPES OF WEIGHT LOSS PLANS
Diet Type and Examples	**Brief Description**
Low-carb diets	
Atkins diet	High-protein, high-fat, extremely low carbohydrate—restricting sugar, bread, pasta, milk, fruits, and vegetables.
Zone diet	40% of calories from carbs, 30% from protein, and 30% from fats. Encourages high-fiber fruits, vegetables, beans, and grains.
Low-fat diets	
Pritikin Plan	High carbohydrate. Fats less than 10% of daily fat intake; focus on vegetables, fruits, and high-fiber grains
Ornish diet	Vegetarian diet low in salt; 10% of calories from fat; focus on vegetables and grains.
Glycemic-index diets	
South Beach diet	Focus on carbohydrates with a low glycemic-index (GI) ranking to lower blood sugar levels. (Glycemic index [GI] ranks carbohydrates based on their effect on blood glucose levels in the first two hours after consumption.)
Formula diets	
Slim-Fast	Based on packaged products; nutritionally sound, low-calorie shakes/bars
Group-approach diets	
Weight-Watchers, eDiets.com	Group meetings (or on-line chats) that provide diet plans, exercise plans, and group support

21.23 Diet can influence cardiovascular disease and cancer

Food influences far more than our size and appearance. Diet also plays an important role in a person's risk of developing serious illnesses, including cardiovascular disease and cancer. Some risk factors associated with cardiovascular disease, such as family history, are unavoidable, but others, such as smoking and lack of exercise, we can influence through our behavior. Diet is another behavioral factor that may affect cardiovascular health. For instance, a diet rich in saturated fats is linked to high blood cholesterol levels, which in turn are linked to cardiovascular disease.

Cholesterol travels through the body in blood lipoproteins, which are particles made up of thousands of molecules of cholesterol and other lipids and one or more protein molecules. High blood levels of a family of lipoproteins called **low-density lipoproteins (LDLs)** generally correlate with a tendency to develop blocked blood vessels, high blood pressure, and consequent heart attacks. In contrast to LDLs, cholesterol carriers called **high-density lipoproteins (HDLs)** may decrease the risk of vessel blockage, perhaps because some HDLs convey blood cholesterol to the liver, where it is broken down. Some research indicates that reducing LDLs while maintaining or increasing HDLs lowers the risk of cardiovascular disease. Exercise tends to increase HDL levels; smoking lowers them.

A diet high in saturated fats tends to increase LDL levels. Saturated fats (see Module 3.8) are found in eggs, butter, and most meats. Saturated fats are also found in artificially saturated ("hydrogenated") vegetable oils. The hydrogenation process, which solidifies vegetable oils, also produces a type of fatty acid called *trans* fats. *Trans* fats tend not only to increase LDL levels but also to lower HDL levels, a two-pronged attack on cardiovascular health. By contrast, eating mainly unsaturated fats, such as fish oil and most liquid vegetable oils (including corn, soybean, and olive oils), tends to lower LDL levels and raise HDL levels. These oils are also important sources of vitamin E, whose antioxidant effect may help prevent blood vessel blockage.

As discussed in Module 11.20, diet also seems to be involved in some forms of cancer. Some research suggests a link between diets heavy in fats or carbohydrates and the incidence of breast cancer. The incidence of colon cancer and prostate cancer may

Figure 21.23   Foods that may contribute to good health

be linked to a diet rich in saturated fat or red meat. Other foods may help fight cancer. For example, some fruits and vegetables (Figure 21.23) are rich in antioxidants, chemicals that help protect cells from damaging molecules known as free radicals. Antioxidants may help prevent cancer, although this link is still debated by scientists.

Despite progress researchers have made in studying nutrition and health, it is often difficult to design controlled experiments that establish the link between the two. Experiments that may damage participants' health are clearly unethical. Some studies rely on self-reported food intake, and the accuracy of participants' memories may influence the outcome. Scientists often rely on epidemiology—studies that correlate certain health characteristics with groups that have particular diets or lifestyles. For example, French people eat high-fat diets and drink wine, yet have lower rates of obesity and heart disease than do Americans. When researchers notice apparent contradictions like these, they attempt to control for other variables and isolate the factors responsible for such observations, such as the fact that the French eat smaller portions; more unprocessed, fresh foods; and snack infrequently.

Even with large, controlled intervention trials, results may be contradictory or inconclusive. For instance, an 8-year study of almost 49,000 postmenopausal women found that low-fat diets failed to reduce the risk of breast and colon cancer and did not affect the incidence of cardiovascular disease. LDL and cholesterol levels did decrease slightly in the low-fat group, however.

The relationship between foods and health is complex, and much remains to be learned. The American Cancer Society (ACS) suggests that following the dietary guidelines listed in Table 21.23, in combination with physical activity, can help lower cancer risk. The ACS's main recommendation is to "eat a variety of healthful foods, with an emphasis on plant sources."

TABLE 21.23	DIETARY GUIDELINES FOR REDUCING CANCER RISK

Maintain a healthy weight throughout life.

Eat five or more servings of a variety of fruits and vegetables daily.

Choose whole grain rice, bread, pasta, and cereals.

Limit consumption of processsed and red meats, especially those high in fat. Prepare meats by baking, broiling, or poaching rather than by frying or charbroiling.

If you drink alcoholic beverages, limit yourself to a maximum of one or two drinks a day (a drink = 12 oz beer, 5 oz wine, 1.5 oz 80% distilled spirits).

? If you are trying to minimize the damaging effects of blood cholesterol on your cardiovascular system, your goal is to _____ your LDLs and _____ your HDLs.

■ decrease . . . increase

Chapter Review

Reviewing the Concepts

Obtaining and Processing Food (Introduction–21.3)

Animal feeding mechanisms include suspension, substrate, fluid, and bulk feeding. Animals may be herbivores, carnivores, or omnivores **(Introduction–21.1)**.

Food processing includes four stages: ingestion, digestion, absorption, and elimination **(21.2)**.

Digestive compartments may be food vacuoles (sponges), a gastrovascular cavity (cnidarians and flatworms), or, in most animals, an alimentary canal running from mouth to anus with specialized regions **(21.3)**.

Human Digestive System (21.4–21.13)

The human digestive system consists of an alimentary canal and digestive glands. The rhythmic muscle contractions of peristalsis squeeze food along the alimentary canal. Sphincters regulate passage of food into and out of the stomach **(21.4)**.

Oral cavity.   The teeth break up food, saliva moistens it, and salivary enzymes begin the hydrolysis of starch. The tongue pushes the bolus of food into the pharynx **(21.5)**.

Pharynx and esophagus.   The swallowing reflex moves food from the pharynx into the esophagus, while keeping it out of the trachea. Peristalsis in the esophagus moves food into the stomach **(21.6)**. The Heimlich maneuver can dislodge food from the pharynx or trachea during choking **(21.7)**.

The stomach stores food and mixes it with acidic gastric juice. Pepsin in gastric juice begins the hydrolysis of protein **(21.8)**. Acid reflux produces heartburn and GERD. Bacterial infections in the stomach are associated with ulcers **(21.9)**.

The small intestine is the site of most digestion and absorption. Alkaline pancreatic juice neutralizes the acid chyme, and its

enzymes digest food molecules. Bile, made in the liver and stored in the gallbladder, emulsifies fat for attack by enzymes. Enzymes from cells of the intestine complete the digestion of many nutrients. Folds of the intestinal lining and tiny, fingerlike villi (with microscopic microvilli) increase the absorptive surface. Nutrients pass across the epithelium and into the blood, which flows to the liver, where nutrients are processed **(21.10)**.

The liver receives nutrient-rich blood from the small intestine through the hepatic portal vein. It processes nutrients and regulates their levels in the blood. The liver also produces bile, detoxifies alcohol and drugs, and synthesizes blood proteins **(21.11)**.

The large intestine, or colon, reabsorbs water from undigested material. Some bacteria in the colon produce vitamins. Feces are stored in the rectum before elimination **(21.12)**.

Dietary adaptations of herbivores include longer alimentary canals and cellulose-digesting microbes housed in special chambers. Ruminants such as cows process food with the aid of microbes in four stomach chambers **(21.13)**.

Nutrition (21.14–21.23)

A healthy diet provides fuel for activities, raw materials for biosynthesis, and essential nutrients **(21.14)**.

Metabolic rate, the rate of energy consumption, includes the basal metabolic rate (BMR), the energy a resting animal requires for essential body processes, and any additional energy used for other activities. Excess energy is stored as glycogen or fat **(21.15)**.

Essential nutrients are those that an animal must obtain from its diet. The eight essential amino acids can be obtained from animal protein or from the proper combination of plant foods **(21.16–21.17)**.

Vitamins and minerals are essential in the human diet. Most vitamins function as coenzymes. Minerals are inorganic nutrients that play a variety of roles **(21.18)**. Supplements ensure a sufficient quantity of vitamins and nutrients; megadoses may be dangerous **(21.19)**. Food labels provide important nutritional information **(21.20)**.

Obesity is a serious health problem, caused by lack of exercise and an abundance of fattening foods, and may partly stem from an evolutionary advantage of fat hoarding **(21.21)**.

Weight loss plans that restrict important nutrients may help individuals lose weight but may have health risks. A healthy diet may reduce the risk of cardiovascular disease and cancer **(21.22–21.23)**.

Connecting the Concepts

1. Label the parts of the human digestive system below and indicate the functions of these organs and glands.

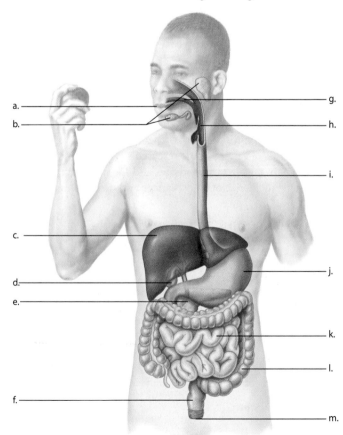

a.

b.

c.

d.

e.

f.

g.

h.

i.

j.

k.

l.

m.

2. Complete the following map summarizing the nutritional needs of animals that are met by a healthy diet.

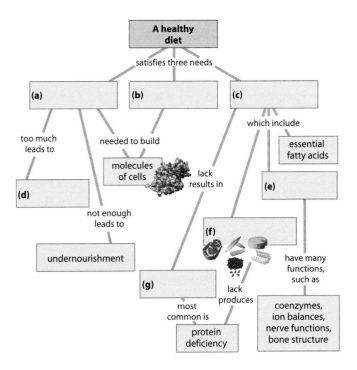

Testing Your Knowledge

Multiple Choice

3. Earthworms, which are substrate feeders,
   a. feed mostly on mineral substrates.
   b. filter small organisms from the soil.
   c. are bulk feeders.
   d. are herbivores that eat autotrophs.
   e. eat their way through the soil, feeding on partially decayed organic matter.

4. The energy content of fats
   a. is released by bile salts.
   b. may be lost unless an herbivore eats some of its feces.
   c. is more than two times that of carbohydrates or proteins.
   d. can reverse the effects of malnutrition.
   e. Both c and d are correct.

5. Which of the following statements is false?
   a. The average human has enough stored fat to supply calories for several weeks.
   b. An increase in leptin levels leads to an increase in appetite and weight gain.
   c. The interconversion of glucose and glycogen takes place in the liver.
   d. After glycogen stores are filled, excessive calories are stored as fat, regardless of their original food source.
   e. Carbohydrates and fats are preferentially used as fuel before proteins are used.

6. Which of the following is mismatched with its function?
   a. most B vitamins—coenzymes
   b. vitamin E—antioxidant
   c. vitamin K—blood clotting
   d. iron—component of thyroid hormones
   e. phosphorus—bone formation, nucleotide synthesis

7. Why do many vegetarians combine different protein sources or eat some eggs or milk products?
   a. to make sure they obtain sufficient calories
   b. to provide sufficient vitamins
   c. to make sure they ingest all essential fatty acids
   d. to make their diet more interesting
   e. to provide all essential amino acids at the same time

Describing, Comparing, and Explaining

8. A peanut butter and jelly sandwich contains carbohydrates, proteins, and fats. Describe what happens to the sandwich when you eat it. Discuss ingestion, digestion, absorption, and elimination.

Applying the Concepts

9. How might our craving for fatty foods, which is helping to fuel the obesity crisis, have evolved through natural selection?

10. Use this Nutrition Facts label to answer these questions:
    a. What percentage of the total Calories in this product is from fat?
    b. Is this product a good source of vitamin A and calcium? Explain.
    c. Each gram of fat supplies 9 Calories. Based on the grams of saturated fat and its % Daily Value, calculate the upper limit of saturated fat (in grams and Calories) that an individual on a 2,000-Calorie/day diet should consume.

Nutrition Facts		
Serving Size 1/2 Cup (83g)		
Servings Per Container 8		
Amount Per Serving		
Calories 190 Calories from Fat 110		
		% Daily Value*
Total Fat 12g		18%
Saturated Fat 8g		40%
Trans Fat 0g		0%
Cholesterol 45mg		15%
Sodium 75mg		3%
Total Carbohydrate 18g		6%
Dietary Fiber 0g		0%
Sugars 17g		
Protein 3g		
Vitamin A 10%	•	Vitamin C 8%
Calcium 10%	•	Iron 0%

*Percent Daily Values (DV) are based on a 2,000 calorie diet.

11. One common piece of dieting advice is to replace energy-dense food with nutrient-dense food. What does this mean?

12. The media report numerous claims and counterclaims about the benefits and dangers of certain foods, dietary supplements, and diets. Have you modified your eating habits on the basis of nutritional information disseminated by the media? Why or why not? How should one evaluate whether such nutritional claims are valid?

13. It is estimated that 10% of Americans don't get enough to eat on a regular basis. Worldwide, at least 850 million people go to bed hungry most nights, and millions of people have starved to death in recent decades. In some cases, war, poor crop yields, and disease epidemics strip people of food. Many also blame global food distribution systems, saying it is not inadequate food production but unequal food distribution that causes food shortages. What responsibility do nations have for feeding their citizens? For feeding the people of other countries? What do you think you can do to lessen world hunger?

Answers to all questions can be found in Appendix 4.

For practice Quizzes, BioFlix, MP3 Tutors, and Activities, go to www.mybiology.com.

chapter

22 Gas Exchange

Surviving in Thin Air

The high mountains of the Himalayas have claimed the lives of even the world's top mountain climbers; the journey into thin air can weaken their muscles, cloud their minds, and sometimes fill their lungs with fluid. The air at the height of the world's highest peak, 9,700-m Mount Everest, is so low in oxygen (O_2) that most people would pass out instantly if exposed to it.

But if you were ever to make it to the top of Mount Everest, you might see birds flying by. Twice a year, flocks of geese migrate over the Himalayas, traveling between winter quarters in India and summer breeding grounds in Russia. These geese, along with other species of migratory birds, can travel easily at heights that would leave most people drowsy, lethargic, or dead.

How do geese and ducks manage to fly at such heights? One factor is the efficiency of their lungs, which can draw far more oxygen from the air than our own lungs can. These birds also have blood containing hemoglobin that has a very high affinity for oxygen, picking it up in the lungs and carrying it to tissues throughout the body. Their circulatory system has a large number of capillaries (tiny blood vessels) that carry oxygen-rich blood to their flight muscles, and the muscles themselves pack a protein called myoglobin that stores a ready supply of oxygen. These adaptations allow the high-flying birds you see in the photo above to travel even where the air is very thin.

Humans can try to adapt to higher elevations, but their success is less certain. Most people function well only below 3,300 m and are helpless at higher elevations without an oxygen mask. There are permanent villages at extremely high elevations in the Himalayas and the Andes, but the people living there have adapted in ways that allow them to function with relatively little oxygen, including large lungs, a large heart, and blood that carries additional hemoglobin and red blood cells. Even then, thin air can take its toll. Local inhabitants such as the Nepalese Sherpas have a reputation for their strength as porters and guides in the high Himalayas. But many die doing such work, their bodies succumbing to altitude-related illnesses under the burdens of long travel and heavy loads.

Altitude-caused disorders include everything from mild headaches, dizziness, and nausea to life-threatening fluid buildup in the lungs and swelling of the brain. Avoiding these disorders requires careful conditioning for high altitudes, and the higher one goes, the longer the adjustment takes. As you move from sea level up into the mountains, your body starts adjusting immediately. Your heart pumps faster, and some blood vessels may

increase in diameter if you stay in the mountains more than a few days. Within weeks, the rate and depth of your breathing increase to bring more air into your lungs. At the same time, your body may develop more capillaries, and your red blood cell count may go up, allowing your blood to carry more oxygen. After long-term training, some Everest climbers have been able to survive for a short time at the top of the world's highest peak without oxygen masks.

Such gradual adaptation can be used in other contexts. For example, runners and cyclists may use this type of training, moving to high altitudes to build the strength of their lungs and to increase the oxygen in their blood, and then returning to sea level to blow past competitors who trained at lower elevations.

Our study of cellular respiration in Chapter 6 showed why animals require oxygen. Without O_2, the metabolic machinery that releases energy from food molecules shuts down. It is the continuous supply of O_2 to body cells that makes the difference between life and death in the thin air of the Himalayas.

The process of **gas exchange,** often called respiration, is the interchange of O_2 and the waste product CO_2 between an animal and its environment. In this chapter, we will explore the structures and functions of the respiratory systems of animals, which supply the oxygen essential for life. ■ ■ ■

22.1 Overview: Gas exchange in an animal with lungs involves breathing, transport of gases, and exchange of gases with tissue cells

Gas exchange makes it possible for animals to put to work the food molecules the digestive system provides. Figure 22.1 presents an overview of the three phases of gas exchange in an animal with lungs. ❶ Breathing is the first phase of the gas exchange process. When an animal breathes, a large, moist internal surface is exposed to air. O_2 diffuses across the cells lining the lungs and into surrounding blood vessels. At the same time, CO_2 diffuses out of the blood and into the lungs. As the animal exhales, CO_2 is removed from the body.

❷ A second phase of gas exchange is the transport of gases by the circulatory system. The O_2 that diffused into the blood attaches to hemoglobin in red blood cells. The red vessels in the figure are transporting O_2-rich blood from the lungs to capillaries in the body's tissues. CO_2 is also transported in blood from the tissues back to the lungs, shown here by the blue vessels.

❸ In the third phase of gas exchange, body cells take up O_2 from the blood and release CO_2 to the blood. As we learned in Module 6.5, O_2 functions in cellular respiration in the mitochondria as the final electron acceptor in the stepwise breakdown of fuel molecules. H_2O and CO_2 are waste products, and ATP is produced to power cellular work. (Gas exchange is sometimes called *respiration;* do not confuse the exchange of gases with the process of cellular respiration.)

Cellular respiration requires a continuous supply of O_2 and the disposal of CO_2. Gas exchange involves the respiratory system and the circulatory system in servicing the cells of the body.

? Humans cannot survive for more than a few minutes without O_2. Why?

■ Cells require a steady supply of ATP in order to function. Cellular respiration requires O_2 to produce this ATP. Without ATP, cells and the organism die.

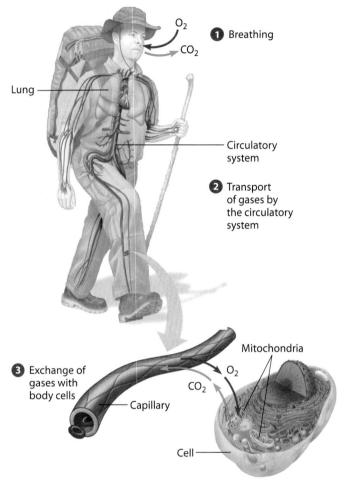

Figure 22.1 The three phases of gas exchange

22.2 Animals exchange O_2 and CO_2 across moist body surfaces

The part of an animal where gases are exchanged with the environment is called the *respiratory surface.* Respiratory surfaces are made up of living cells whose plasma membranes must be wet to function properly. Thus, the respiratory surfaces of terrestrial as well as aquatic animals must be moist.

Gas exchange takes place by diffusion. The surface area of the respiratory surface must be extensive enough to take up sufficient O_2 for every cell in the body and to dispose of waste CO_2. Usually, a single layer of cells covers or lines the entire respiratory surface. This thin and moist layer allows O_2 to diffuse rapidly into the circulatory system or directly into body tissues and allows CO_2 to diffuse out.

The four figures on the facing page illustrate, in simplified form, four types of respiratory organs, structures where gas exchange with the external environment occurs. In each case, the circle represents a cross section of the animal's body through the respiratory surface. The yellow areas represent the respiratory surfaces; the green outer circles represent body surfaces with little or no role in respiration. The boxed enlargements show gas exchange occurring across the respiratory surface.

Some animals use their entire outer skin as a gas exchange organ. The earthworm in Figure 22.2A is an example. The cross-sectional diagram shows its whole body surface as yellow; there are no specialized gas exchange surfaces. Oxygen diffuses into a dense net of thin-walled capillaries lying just beneath the skin. Earthworms and other "skin breathers" must live in damp places or in water because their whole body surface has to stay moist. Animals that breathe only through their skin and lack

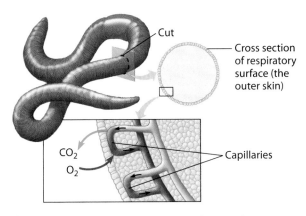

Figure 22.2A The entire outer skin of an earthworm serves as its respiratory surface.

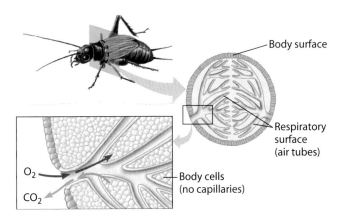

Figure 22.2C The tracheal system of an insect consists of tubes that extend throughout the body.

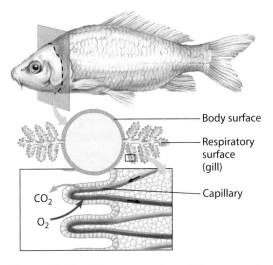

Figure 22.2B Gills are extensions of the body surface that function in gas exchange with the surrounding water.

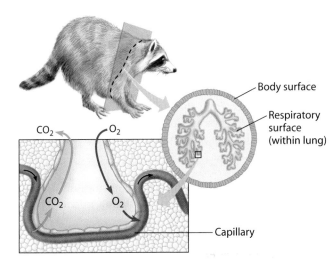

Figure 22.2D Lungs are internal thin-walled sacs.

specialized gas exchange organs are generally small, and many are long and thin or flattened. Small size or flatness provides a high ratio of respiratory surface to body volume, allowing for sufficient gas exchange for the entire body.

In most animals, the skin surface is not extensive enough to exchange gases for the whole body. Consequently, certain parts of the body have become adapted as highly branched respiratory surfaces with large surface areas. Such gas exchange organs include gills, tracheal systems, and lungs.

Gills have evolved in most aquatic animals. **Gills** are extensions, or outfoldings, of the body surface specialized for gas exchange. Many marine worms have flap-like gills that extend from each body segment. The gills of clams and crayfish are clustered in one body location. A fish (Figure 22.2B) has a set of feather-like gills on each side of its head. As indicated in the enlargement, O_2 diffuses across the gill surfaces into capillaries, and CO_2 diffuses in the opposite direction, out of the capillaries and into the external environment. Because the respiratory surfaces of aquatic animals extend into the surrounding water, keeping the surface moist is not a problem.

In most terrestrial animals, the respiratory surfaces are folded into the body rather than projecting from it. The infolded surfaces open to the air only through narrow tubes, an arrangement that helps retain the moisture that is essential for the cells of the respiratory surfaces to function.

The **tracheal system** of insects is an extensive system of branching internal tubes (Figure 22.2C) with the respiratory surface found at their tips. As we will see in Module 22.4, the smallest branches exchange gases directly with body cells. Thus, gas exchange in insects requires no assistance from the circulatory system.

Most terrestrial vertebrates have **lungs** (Figure 22.2D), which are internal sacs lined with moist epithelium. As the diagram indicates, the inner surfaces of the lungs branch extensively, forming a large respiratory surface. Gases are carried between the lungs and the body cells by the circulatory system.

We examine gills, tracheae, and lungs more closely in the next several modules.

? How does the structure of the respiratory surface of a gill or lung fit its function?

■ The respiratory surfaces of gills and lungs are moist and thin so that gases can easily diffuse across them and into or out of their closely associated capillaries. They are highly branched or subdivided, providing a large surface area for exchange.

Gills are adapted for gas exchange in aquatic environments

Oceans, lakes, and other bodies of water contain O_2 in the form of dissolved gas. The gills of fishes and many invertebrate animals, including lobsters and clams, tap this source of O_2.

An advantage of exchanging gases in water is that there is no problem keeping the respiratory surface wet. On the other hand, the amount of available oxygen (dissolved O_2) in water is only about 3–5% of what it is in the air, and the warmer and saltier the water, the less O_2 it holds. Thus, gills—especially those of large, active animals in warm oceans—must be very efficient to obtain enough oxygen from water. The total surface area of the gills is often much greater than that of the rest of the body surface.

Structure of Fish Gills The drawings in Figure 22.3 show the architecture of fish gills, which are among the most efficient gas exchange organs in the aquatic world. There are four supporting gill arches on each side of the body. Two rows of gill filaments project from each gill arch. Each filament bears many platelike structures called lamellae (singular, *lamella*), which are the actual respiratory surfaces. A lamella is full of tiny capillaries that are separated from the outside by only one or a few layers of cells. Capillaries are so narrow that red blood cells must pass through them in single file. As a result, every red blood cell comes in close contact with oxygen dissolved in the surrounding water.

What you can't see in the drawings are the movements that ventilate the gills. We use the term **ventilation** to refer to any

mechanism that increases the flow of the surrounding water or air over the respiratory surface (gills, tracheae, or lungs). Increasing this flow ensures a fresh supply of O_2 and the removal of CO_2. The blue arrows in the drawings represent the one-way flow of water into the mouth, across the gills, and out the side of the fish's body. Swimming fish simply open their mouths and let water flow over the gills. Fish also pump water across the gills by the coordinated opening and closing of the mouth and operculum, the stiff flap that covers and protects the gills. Because water is dense and contains so little oxygen, most fish must expend considerable energy in ventilating their gills.

Countercurrent Exchange The arrangement of capillaries in a fish gill enhances gas exchange. Blood flows opposite the movement of water past the gills. This makes it possible to transfer oxygen to the blood by an efficient process called countercurrent exchange. **Countercurrent exchange** is the transfer of a substance from a fluid moving in one direction to another fluid moving in the opposite direction. The name reflects the fact that the two fluids are moving *counter* to each other. Let's see how this arrangement enhances the exchange of gases in a fish gill.

In the enlargement on the top right of Figure 22.3, notice that the direction of water flow over the surface of a lamella (blue arrows) is opposite that of the blood flow within the lamella (black arrows). The diagram in the lower right

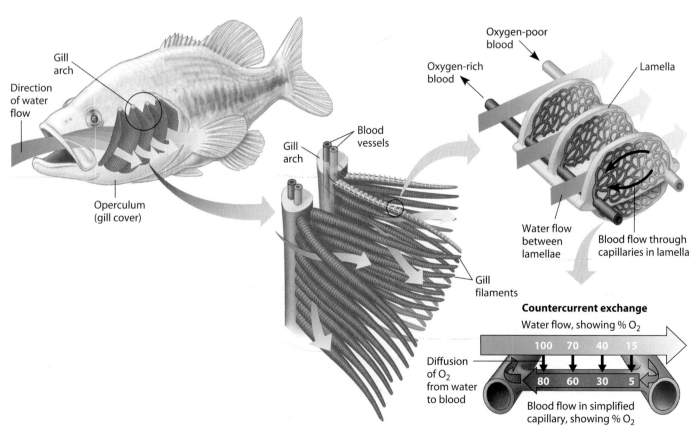

Figure 22.3 The structure of fish gills

illustrates the countercurrent exchange of O_2 from water to blood. The percentages shown indicate the changing amount of O_2 dissolved in each fluid. Notice that as blood flows through a lamella and picks up more and more O_2, it comes in contact with water that is closer to beginning its passage over the gills and thus has more O_2 available. As a result, a concentration gradient is maintained that favors the diffusion of O_2 from the water to the blood along the entire length of the capillary.

This countercurrent exchange mechanism is so efficient that fish gills can remove more than 80% of the oxygen dissolved in the water flowing through them. (In comparison, humans extract only 25% of the oxygen in the air we breathe.) The basic mechanism of countercurrent exchange is also important in temperature regulation, as you will see in Chapter 25.

However, gills are unsuitable for an animal living on land. An expansive surface of wet membrane extending out from the body and exposed to air would lose too much water to evaporation. Therefore, most terrestrial animals house their respiratory surfaces within the body, opening to the atmosphere through narrow tubes, as we see next.

? What would be the maximum percentage of the water's O_2 a gill could extract if its blood flowed in the same direction as the water instead of counter to it? (This is a challenging one! It may help to sketch it out.)

■ 50%. As O_2 diffused from the water into the blood as they flowed in the same direction, the concentration gradient would become less and less steep until there was the same amount of O_2 dissolved in both, and O_2 could no longer diffuse from water to blood.

22.4 The tracheal system of insects provides direct exchange between the air and body cells

There are two big advantages to exchanging gases by breathing air: Air contains a much higher concentration of O_2, and air is much lighter and easier to move than water. Thus, a terrestrial animal expends much less energy than an aquatic animal ventilating its respiratory surface. The main problem facing any air-breathing animal is the loss of water to the air by evaporation.

The tracheal system of insects, with respiratory surfaces at the tips of tiny branching tubes inside the body, greatly reduces evaporative water loss. Figure 22.4A illustrates the tracheal system in a grasshopper. The largest tubes, called tracheae, open to the outside and are reinforced by rings of chitin, as shown in the blowup on the bottom right of the figure. (An insect's tough exoskeleton is also made of chitin.) Enlarged portions of tracheae form air sacs near organs that require a large supply of O_2.

The micrograph in Figure 22.4A shows how these tubes branch repeatedly. The smallest branches, called tracheoles, extend to nearly every cell in the insect's body. The tiny tips of the tracheoles are closed and contain fluid (dark blue in the drawing). Gas is exchanged with body cells by diffusion across the moist epithelium that lines these tips. The structure of a tracheal system matches its function of delivering O_2 directly to body cells. Thus, the circulatory system of insects is not involved in transporting oxygen.

For a small insect, diffusion through the tracheae brings in enough O_2 and removes enough CO_2 to support cellular respiration. Larger insects may ventilate their tracheal systems with rhythmic body movements that compress and expand the air tubes like bellows. An insect in flight (Figure 22.4B) has a very high metabolic rate and consumes 10 to 200 times more O_2 than it does at rest. In many insects, alternating contraction and relaxation of the flight muscles rapidly pumps air through the tracheal system.

? In what basic way does the process of gas exchange in insects differ from that in both fish and humans?

■ The circulatory system of insects is not involved in transporting gases to and from the body cells.

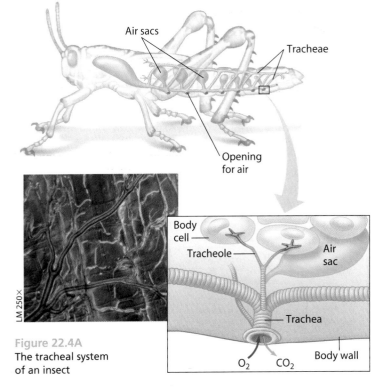

Air sacs

Tracheae

Opening for air

Body cell

Tracheole

Air sac

Trachea

O_2 CO_2 Body wall

LM 250×

Figure 22.4A
The tracheal system of an insect

Figure 22.4B
A grasshopper in flight

22.5 The evolution of lungs facilitated the movement of tetrapods onto land

The colonization of land by vertebrates was one of the pivotal milestones in the history of life. The evolution of legs from fins may be the most obvious change in body design, but the refinement of lung breathing was just as important. And while skeletal changes were undoubtedly required in the transition from fins to legs, the evolution of lungs for breathing on land also required skeletal changes. Interestingly, current fossil evidence supports the hypothesis that the beginning changes in the front fins and shoulder girdle of tetrapod ancestors may actually have been breathing adaptations that enabled the kind of push up needed in shallow water for a fish to gulp air.

Paleontologists have uncovered numerous transitional forms in tetrapod evolution (see Module 19.4). It now seems clear that tetrapods first evolved in shallow water from what some researchers jokingly call "fishapods." These ancient forms had both gills and lungs. The adaptations for air-breathing evident in their fossils include a stronger and elongated snout and a muscular neck that enabled the animal to lift the head clear of water and into the unsupportive air. Strengthening of the lower jaw may have facilitated the pumping motion presumed to be used by early air-breathing tetrapods and still employed by frogs to inflate their lungs. The recently discovered 375 million year old fossil of *Tiktaalik* (Figure 22.5) illustrates some of these air-breathing adaptations.

The first tetrapods on land diverged into three major lineages: amphibians, reptiles (including birds), and mammals. Most amphibians have small lungs and rely heavily on the diffusion of gases across body surfaces. Reptiles (including all

Figure 22.5 A cast of a fossil of *Tiktaalik*. Note the elongated snout and strong shoulder support.

birds) and mammals rely on lungs for gas exchange. In general, the size and complexity of lungs are correlated with an animal's metabolic rate (and thus oxygen need). For example, the lungs of birds and mammals, whose high body temperatures are maintained by a high metabolic rate, have a greater area of exchange surface than the lungs of similar-sized amphibians and nonbird reptiles, who have a much lower metabolic rate.

We explore the mammalian respiratory system next.

? How might adaptations for breathing air be linked to the evolution of tetrapod limbs?

■ Fossil evidence indicates that changes in the shoulder girdle and limb bones may have helped early tetrapod ancestors lift their heads above water to gulp air.

22.6 In the human respiratory system, branching tubes convey air to lungs located in the chest cavity

Mammalian lungs are located in the chest or thoracic cavity and protected by the supportive rib cage. The thoracic cavity is separated from the abdominal cavity by a sheet of muscle called the **diaphragm.** We will see how the diaphragm helps ventilate our lungs in Module 22.8.

Figure 22.6A on the facing page shows the human respiratory system (along with the esophagus and heart, for orientation). Air enters our respiratory system through the nostrils. It is filtered by hairs and warmed, humidified, and sampled for odors as it flows through a maze of spaces in the nasal cavity. We can also draw in air through the mouth, but mouth breathing does not allow the air to be processed by the nasal cavity.

From the nasal cavity or mouth, air passes to the **pharynx,** where the paths for air and food cross. When food is swallowed, the upper part of the respiratory tract moves upward and tips the epiglottis over the opening of the windpipe (see Figure 21.6A). The rest of the time, the air passage in the pharynx is open for breathing.

From the pharynx, air is inhaled into the **larynx** (voice box). When we exhale, the outgoing air rushes by a pair of **vocal cords** in the larynx, and we can produce sounds by voluntarily tensing muscles in the voice box, stretching the cords and making them vibrate. We produce high-pitched sounds when our vocal cords are tightly stretched and vibrating very fast. When the cords are less tense, they vibrate slowly and produce low-pitched sounds.

From the larynx, inhaled air passes toward the lungs through the **trachea,** or windpipe. Rings of cartilage (shown in the figure in blue) reinforce the walls of the larynx and trachea, keeping this part of the airway open. The trachea and major branches of the respiratory system are lined by a moist epithelium covered by cilia and a thin film of mucus. The cilia and mucus are the system's cleaning elements. The mucus traps dust, pollen, and other contaminants, and the beating cilia move this mucus upward to the pharynx, where it is usually swallowed.

The trachea forks into two **bronchi** (singular, *bronchus*), one leading to each lung. Within the lung, the bronchus branches

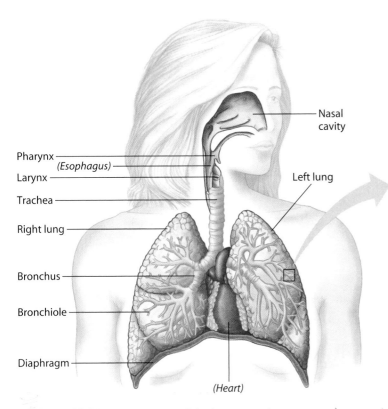

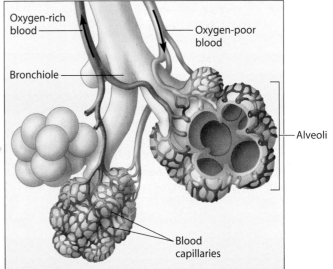

Figure 22.6A The anatomy of the human respiratory system (left) and details of the structure of alveoli (right)

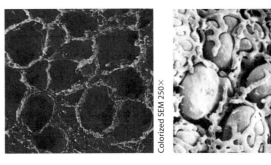

Figure 22.6B Scanning electron micrographs of the air spaces in alveoli (left) and the capillaries that envelop the alveoli (right)

repeatedly into finer and finer tubes called **bronchioles.** Bronchitis is a condition in which these small tubes become inflamed and constricted, making breathing difficult.

As the enlargement on the right of Figure 22.6A shows, the bronchioles dead-end in grapelike clusters of air sacs called **alveoli** (singular, *alveolus*). Each of our lungs contains millions of these tiny sacs. Together they have a surface area of about 100 m², 50 times that of the skin. On the left side of Figure 22.6B is a micrograph showing a cutaway view of the air spaces within the alveoli. The inner surface of each alveolus is lined with a thin layer of epithelial cells. The O_2 in inhaled air dissolves in a film of moisture on the epithelial cells. It then diffuses across the epithelium and into the dense web of blood capillaries that surrounds each alveolus (shown on the right side in Figure 22.6B). CO_2 diffuses the opposite way—from the capillaries, across the epithelium of the alveolus, into the air space of the alveolus, and finally out in the exhaled air.

Respiratory Problems Alveoli are so small that specialized secretions called **surfactants** are required to keep them from sticking shut due to the surface tension of their moist surface. A lack of lung surfactant is a major problem for babies born very prematurely. Surfactants typically appear in the lungs after 33 weeks of embryonic development. (Birth normally occurs at 38 weeks.) Among infants born before week 28, half suffer serious respiratory distress. Artificial surfactants may be administered through a breathing tube to treat such preterm infants.

Alveoli are highly susceptible to contaminants. Defensive blood cells called macrophages patrol them and engulf foreign particles. However, if too much particulate matter reaches the

alveoli, the delicate lining of these small sacs becomes damaged and the efficiency of gas exchange drops. Studies have shown a significant association between exposure to fine particles and premature death. Air pollution and tobacco smoke are sources of these lung-damaging particles.

Exposure to such pollutants can cause continual irritation and inflammation of the lungs and lead to chronic obstructive pulmonary diseases (COPD). These diseases include emphysema and chronic bronchitis. In emphysema, the delicate walls of alveoli become permanently damaged and the lungs lose the elasticity that helps expel air during exhalation. With COPD, both lung ventilation and gas exchange are severely impaired. COPD is a major cause of death and disability in the United States. Patients experience labored breathing, coughing, frequent lung infections, and most develop respiratory failure, which can be fatal.

🎧 MP3 Tutor *Human Respiration*

Web Activity *The Human Respiratory System*

? How does the structure of alveoli match their function?

■ Alveoli have a thin, moist epithelium across which dissolved O_2 and CO_2 can easily diffuse into or from the surrounding capillaries. The collective surface area of all the alveoli is huge.

22.7 Smoking is a serious assault on the respiratory system

One of the worst sources of lung-damaging air pollutants is tobacco smoke, which is mainly microscopic particles of carbon coated with toxic chemicals. A single drag on a cigarette exposes a person to more than 4,000 chemicals.

Remember that cilia on cells lining the respiratory tract sweep contaminant-laden mucus up and out of the airways. Tobacco smoke irritates these cells, inhibiting or destroying their cilia. Frequent coughing—common in heavy smokers—is the respiratory system's attempt to clear the mucus no longer moved by the cilia. Smoke's toxins also kill the macrophages that reside in the respiratory tract and engulf fine particles and microorganisms. Thus, smoking disables the normal cleansing and protective mechanisms of the respiratory system.

Smoking is a leading cause of emphysema (see Module 22.6), which causes breathlessness and constant fatigue as the body is forced to spend more and more energy just breathing.

Some of the toxins in tobacco smoke cause lung cancer. The photographs in Figure 22.7 show a cutaway view of a pair of healthy human lungs (left) and the lungs of a cancer victim (right), whose lungs are black from the long-term buildup of smoke particles. Smokers account for 90% of all lung cancer cases. Most victims die within one year of diagnosis. Smokers also have a greater risk of developing other cancers.

The second highest number of smoking-related deaths come from cardiovascular disease. Smokers have a higher rate of heart attacks and stroke. Smoking raises blood pressure and increases harmful cholesterol levels in the blood.

Every year in the United States, smoking kills about 440,000 people, more than all the deaths caused by accidents, alcohol

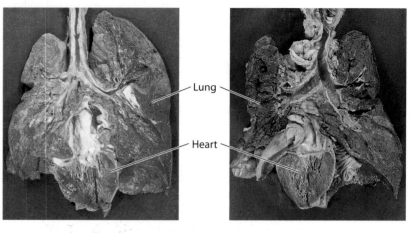

Figure 22.7 Healthy lungs (left) and cancerous lungs (right)

Lung

Heart

and drug abuse, HIV, and murders combined. On average, adults who smoke die 13 to 14 years earlier than nonsmokers. Studies show that nonsmokers exposed to secondary cigarette smoke are also at risk. Young children are particularly susceptible, with increased risk of asthma, bronchitis, and pneumonia.

About 15 years after quitting, a former smoker's risk of lung cancer and heart disease is similar to that of people who have never smoked. No lifestyle choice has a more positive impact on the long-term health of oneself and those with whom one lives than not smoking.

? What causes "smoker's cough"?

■ Smoke damages cilia, inhibiting their ability to sweep mucus and trapped particles from the respiratory tract. The body tries to compensate by coughing.

22.8 Negative pressure breathing ventilates our lungs

Breathing is the alternate inhalation and exhalation of air. This ventilation of our lungs maintains high O_2 and low CO_2 concentrations at the respiratory surface.

Figure 22.8 shows the changes that occur in our rib cage, chest cavity, and lungs during breathing. During inhalation, the rib cage expands as muscles between the ribs contract. At the same time, the diaphragm contracts, moving downward and expanding the chest cavity. The volume of the lungs increases with the expanding chest cavity during inhalation, which lowers the air pressure in the alveoli to less than atmospheric pressure. Flowing from a region of higher pressure to one of lower pressure, air rushes through the nostrils and down the breathing tubes to the alveoli. This type of ventilation is called **negative pressure breathing.**

The diagram on the right in Figure 22.8 shows exhalation. The rib muscles and diaphragm both relax, decreasing the volume of the rib cage and chest cavity, which increases the air pressure inside the lungs, forcing air out. Notice

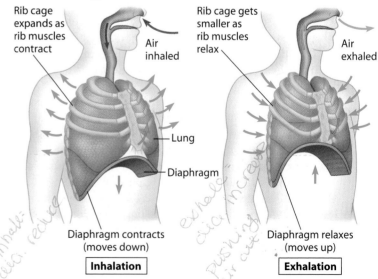

Rib cage expands as rib muscles contract

Air inhaled

Lung

Diaphragm

Diaphragm contracts (moves down)

Inhalation

Rib cage gets smaller as rib muscles relax

Air exhaled

Diaphragm relaxes (moves up)

Exhalation

Figure 22.8 Negative pressure breathing draws air into the lungs.

that the diaphragm curves upward into the chest cavity when relaxed.

Each year, a human adult may take between 4 million and 10 million breaths. The volume of air in each breath is about 500 mL when we breathe quietly. The maximum volume of air that we can inhale and exhale during forced breathing is called **vital capacity.** It averages about 3.4 L and 4.8 L for college-age females and males, respectively. (Women tend to have smaller rib cages and lungs.) The lungs actually hold more air than the vital capacity. Because the alveoli do not completely collapse, a residual volume of "dead" air remains in the lungs even after we blow out as much air as we can. As lungs lose resilience (springiness) with age or as the result of disease, such as emphysema, less air exits on exhalation and residual volume increases at the expense of vital capacity.

Because the lungs do not completely empty, each inhalation mixes fresh air with oxygen-depleted air. We extract only about 25% of the O_2 in the air we inhale. As we mentioned in the chapter introduction, the gas exchange system of birds is much more efficient. Unlike the in-and-out flow of air in the human alveoli, the air in the lungs of birds moves in one direction through tiny passageways where gas exchange occurs. Because of this one-way flow of air, oxygen-depleted air does not remain in the lungs after exhalation. Thus, birds can extract more oxygen from a volume of inhaled air than we can.

? Explain how negative pressure breathing ventilates our lungs.

■ The lungs expand along with the expanding chest cavity during inhalation. This increase in volume makes the pressure in the lungs lower than the outside air pressure, which draws air into the lungs.

[handwritten: 2009E13760 of CO₂ in air/year]

22.9 Breathing is automatically controlled

Although we can voluntarily hold our breath or breathe faster and deeper, most of the time our breathing is under automatic control. **Breathing control centers** (shown as yellow circles in Figure 22.9) are located in parts of the brain called the pons and medulla oblongata. The control center in the pons smooths out the basic rhythm of breathing set by the medulla. ❶ Nerves from the medulla's control center signal the diaphragm and rib muscles to contract, making us inhale. When we are at rest, these nerve signals result in about 10 to 14 inhalations per minute. Between inhalations, the muscles relax, and we exhale.

❷ The control center regulates breathing rate in response to changes in the CO_2 level of the blood. When we exercise vigorously, for instance, our metabolism speeds up and our body cells generate more CO_2 as a waste product. The CO_2 goes into the blood, where it reacts with water to form carbonic acid. The acid slightly lowers the pH of the blood and the fluid bathing the brain (cerebrospinal fluid). When the medulla senses this pH drop, its breathing control center increases the breathing rate and depth. As a result, more CO_2 is eliminated in the exhaled air, and the pH returns to normal.

The O_2 concentration in the blood usually has little effect on the breathing control centers. Because the same process that consumes O_2, cellular respiration, also produces CO_2, a rise in CO_2 (drop in pH) is generally a good indication of a drop in blood oxygen. Thus, by responding to lowered pH, the breathing control center also controls blood oxygen level.

❸ Secondary control over breathing is exerted by sensors in the aorta and carotid arteries that monitor concentrations of O_2 as well as CO_2. When the O_2 level in the blood is severely depressed, these sensors signal the control center via nerves to increase the rate and depth of breathing. This response may occur, for example, at high altitudes, where the air is so thin that we cannot get enough O_2 by breathing normally.

The breathing control center responds to a variety of nervous and chemical signals to keep the breathing rate and depth in tune with the changing metabolic needs of the body. Breathing rate must also be coordinated with the activity of the circulatory

system, which supplies blood to the alveolar capillaries. We examine the role of the circulatory system in gas exchange more closely in the next module.

? How is the increased need for O_2 during exercise accommodated by the breathing control centers?

■ During exercise, cells release more CO_2 to the blood, which forms carbonic acid, lowering the pH of the blood. The breathing centers sense the decrease in pH and send impulses to increase breathing, thus supplying more O_2.

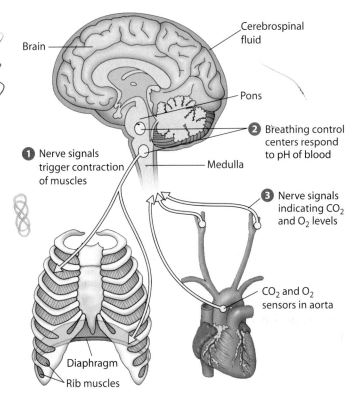

Figure 22.9 Control centers that regulate breathing respond to the pH of blood and nervous stimulation from sensors that detect CO_2 and O_2 levels.

22.10 Blood transports respiratory gases

How does O_2 get from our lungs to all the other tissues in our body, and how does CO_2 travel from the tissues to the alveoli? To answer these questions, we must jump ahead a bit to the subject of Chapter 23 and look at the basic organization of the human circulatory system.

Figure 22.10 is a schematic diagram showing the main components of our circulatory system and their role in gas exchange. Let's start with the heart, in the middle of the diagram. One side of the heart handles oxygen-poor blood (colored blue). The other side handles oxygen-rich blood (red). As indicated in the lower left of the diagram, oxygen-poor blood returns to the heart from capillaries in body tissues. The heart pumps this blood to the alveolar capillaries in the lungs. Gases are exchanged between air in the alveolar spaces and blood in the capillaries (top of diagram). Blood leaves the alveolar capillaries, having lost CO_2 and gained O_2. This oxygen-rich blood returns to the heart and is pumped out to body tissues.

The exchange of gases between capillaries and the cells around them occurs by the diffusion of gases down gradients of pressure. A mixture of gases, such as air, exerts pressure. (You see evidence of gas pressure whenever you open a can of soda, releasing the pressure of the CO_2 it contains.) Each kind of gas in a mixture accounts for a portion, called the **partial pressure**, of the mixture's total pressure. Molecules of each kind of gas will diffuse down a gradient of their own partial pressure independently of the other gases. At the bottom of the figure, for instance, O_2 moves from oxygen-rich blood, through the interstitial fluid, and into tissue cells because it diffuses from a region of higher partial pressure to a region of lower partial pressure. The tissue cells maintain this gradient as they consume O_2 in cellular respiration. The CO_2 produced as a waste product of cellular respiration diffuses down its own partial-pressure gradient out of the cells and into the capillaries. Diffusion also accounts for gas exchange in the alveoli.

? What is the physical process underlying gas exchange?

■ Diffusion of each gas down its partial-pressure gradient.

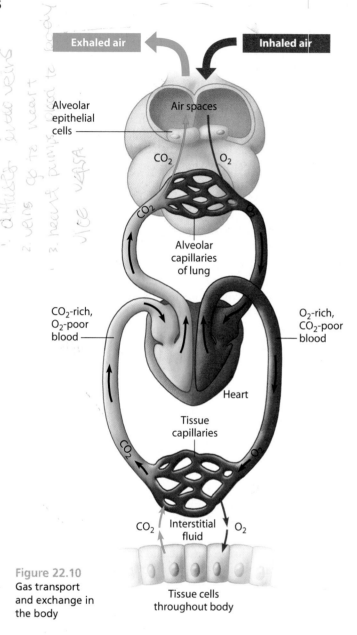

Figure 22.10
Gas transport and exchange in the body

22.11 Hemoglobin carries O_2, helps transport CO_2, and buffers the blood

Oxygen is not very soluble in water, and most animals transport O_2 bound to proteins called respiratory pigments. Most of these molecules have distinctive colors, hence the name *pigment*. Many molluscs and arthropods use a blue, copper-containing pigment. Almost all vertebrates and many invertebrates use **hemoglobin,** an iron-containing pigment that turns red when bound with O_2.

Each of our red blood cells is packed with about 250 million molecules of hemoglobin. A hemoglobin molecule consists of four polypeptide chains of two different types, depicted with

two shades of purple in Figure 22.11 on the next page. Attached to each polypeptide is a chemical group called a heme (colored green in the figure), at the center of which is an iron atom (black). Each iron atom can carry one O_2 molecule. Thus, every hemoglobin molecule can carry up to four oxygen molecules. Hemoglobin loads up with O_2 in the lungs and transports it to the body's tissues. There, hemoglobin unloads some or all of its cargo, depending on the O_2 needs of the cells. (The partial pressure of O_2 in the tissue reflects how much O_2 the cells are using.)

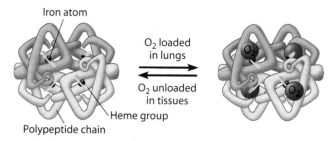

Iron atom

O₂ loaded
in lungs

O₂ unloaded
in tissues

Heme group

Polypeptide chain

Figure 22.11 Hemoglobin loading and unloading of O₂

Hemoglobin is a multipurpose molecule. It also helps transport CO_2 and assists in buffering the blood. When CO_2 leaves a tissue cell, it diffuses through the interstitial fluid, across a capillary wall, and into the blood plasma. Most of the CO_2 enters red blood cells, where some of it combines with hemoglobin. The rest reacts with water, forming carbonic acid (H_2CO_3), which then breaks apart into a hydrogen ion (H^+) and a bicarbonate ion (HCO_3^-). Hemoglobin binds most of the H^+, minimizing the change in blood pH. (As discussed in Module 22.9, the pH drop during exercise is the stimulus to increase breathing rate.)

The bicarbonate ions diffuse into the plasma, where they are carried to the lungs. This reversible reaction is shown here:

$$CO_2 + H_2O \rightleftharpoons H_2CO_3 \rightleftharpoons H^+ + HCO_3^-$$

Carbon dioxide   Water   Carbonic acid   Hydrogen ions   Bicarbonate

As blood flows through capillaries in the lungs, the reaction is reversed. Bicarbonate combines with H^+ to form carbonic acid; carbonic acid is converted to CO_2 and water, and CO_2 diffuses from the blood to the alveoli and out of the body in exhaled air.

We have seen how O_2 and CO_2 are transported between the lungs and body tissue cells via the bloodstream. In the next module, we consider a special case of gas exchange between two circulatory systems.

Web Activity *Transport of Respiratory Gases*

? O_2 in the blood is transported bound to _____ within _____ _____ cells, and CO_2 is mainly transported as _____ ions within the plasma.

■ hemoglobin . . . red blood . . . bicarbonate

Connection

22.12 The human fetus exchanges gases with the mother's bloodstream

Figure 22.12 shows a human fetus inside the mother's uterus. The fetus literally swims in a protective watery bath, the amniotic fluid. Its lungs are full of fluid and are nonfunctional. How does the fetus exchange gases with the outside world? The answer lies in the function of the placenta, a composite organ that includes tissues from both the fetus and the mother. A large net of capillaries fans out into the placenta from blood vessels in the umbilical cord of the fetus. These fetal capillaries exchange gases with the maternal blood that circulates in the placenta, and the maternal circulatory system carries the gases to and from the mother's lungs. Aiding O_2 uptake by the fetus is fetal hemoglobin, a special type that attracts O_2 more strongly than does adult hemoglobin.

The reason that smoking is considered a health risk during pregnancy is because it reduces, perhaps by as much as 25%, the supply of oxygen reaching the placenta.

What happens when a baby is born? Suddenly placental gas exchange ceases, and the baby's lungs must begin to work. Carbon dioxide acts as a signal. As soon as CO_2 stops diffusing from the fetus into the placenta, a CO_2 rise in fetal blood causes blood pH to fall, stimulating the breathing control centers in the infant's brain, and the newborn takes its first breath.

A human birth and the radical changes in gas exchange mechanisms that accompany it are extraordinary events. Resulting from millions of years of evolutionary adaptation, these events are on a par with the remarkable flying ability of the geese we discussed in the chapter introduction. For a goose to breathe the thin air and fly great distances high above Earth, or for a human baby to switch almost instantly from living in water and exchanging gases with maternal blood to breathing air directly, requires truly remarkable adaptations in the organism's respiratory system. Also required are adaptations of the circulatory system, which, as we have seen, supports the respiratory system in its gas exchange function. We turn to the circulatory system in Chapter 23.

? How does fetal hemoglobin enhance oxygen transfer from mother to fetus across the placenta?

■ Fetal hemoglobin has a greater affinity for O_2 than does adult hemoglobin, which helps "pull" the O_2 from maternal blood to fetal blood.

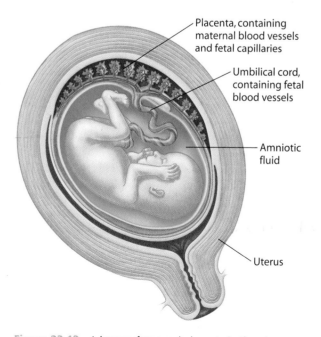

Placenta, containing maternal blood vessels and fetal capillaries

Umbilical cord, containing fetal blood vessels

Amniotic fluid

Uterus

Figure 22.12 A human fetus and placenta in the uterus

Reviewing the Concepts
Mechanisms of Gas Exchange (Introduction–22.9)

Gas exchange, the interchange of O_2 and CO_2 between an organism and its environment, provides O_2 for cellular respiration and removes its waste product, CO_2. Gas exchange often involves breathing, transport of gases, and exchange of gases with body cells (**Introduction–22.1**).

Respiratory surfaces must be thin and moist for diffusion of O_2 and CO_2 to occur. Some animals use their entire skin as a gas exchange organ. In most animals, specialized body parts—such as gills, tracheal systems, or lungs—provide large respiratory surfaces for gas exchange (**22.2**).

Gills are extensions of the body that absorb O_2 dissolved in water. In a fish, gas exchange is enhanced by ventilation and the countercurrent flow of water and blood (**22.3**).

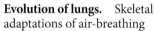

Lamella

Water flow

Blood flow

Tracheal systems in insects transport O_2 directly to body cells through a network of finely branched tubes (**22.4**).

Evolution of lungs. Skeletal adaptations of air-breathing fish enabled tetrapods to move onto land. The size and complexity of lungs correlate with metabolic rate and thus oxygen needs (**22.5**).

Human respiratory system. Air inhaled through the nostrils passes through the pharynx and larynx into the trachea, bronchi, and bronchioles to the alveoli, where gas exchange occurs. Mucus and cilia in the respiratory passages protect the lungs, but smoking can destroy these protections. Smoking causes lung cancer, heart disease, and emphysema (**22.6–22.7**).

Breathing is the alternation of inhalation and exhalation. The contraction of rib muscles and diaphragm expands the chest cavity and reduces air pressure in the alveoli (negative pressure breathing). Vital capacity is the maximum volume of air that can be inhaled and exhaled, but the lungs still hold a residual volume. Air flows in one direction through the more efficient lungs of birds (**22.8**).

Breathing control centers in the brain keep breathing in tune with body needs, sensing and responding to the CO_2 level in the blood. A drop in blood pH triggers an increase in the rate and depth of breathing (**22.9**).

Transport of Gases in the Human Body (22.10–22.12)

Circulation. The heart pumps oxygen-poor blood to the lungs, where it picks up O_2 and drops off CO_2. Then the heart pumps the oxygen-rich blood to body cells, where it drops off O_2 and picks up CO_2. Gases diffuse down partial-pressure gradients in lungs and body tissues (**22.10**).

Hemoglobin in red blood cells transports oxygen, helps buffer the blood, and carries some CO_2. Most CO_2 is transported as bicarbonate ions in the plasma (**22.11**). A human fetus exchanges gases with maternal blood in the placenta. Fetal hemoglobin enhances oxygen transfer from maternal blood. At birth, rising CO_2 in fetal blood stimulates the breathing control centers to initiate breathing (**22.12**).

Connecting the Concepts

1. Complete this map to review some of the concepts of gas exchange

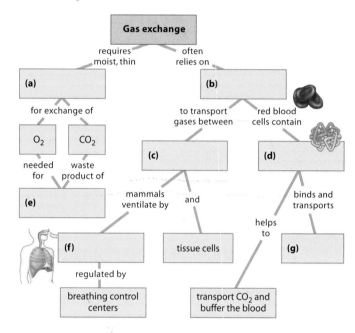

2. Label the parts of the human respiratory system.

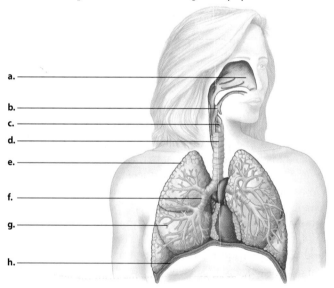

a.
b.
c.
d.
e.
f.
g.
h.

Testing Your Knowledge
Multiple Choice

3. When you hold your breath, which of the following first leads to the urge to breathe?
   a. falling CO_2
   b. falling O_2
   c. rising CO_2
   d. rising pH of the blood
   e. both c and d

4. Countercurrent gas exchange in the gills of a fish
   a. speeds up the flow of water through the gills.
   b. maintains a gradient that enhances diffusion.
   c. enables the fish to obtain oxygen without swimming.
   d. means that blood and water flow at different rates.
   e. allows O_2 to diffuse against its partial-pressure gradient.

5. When you inhale, the diaphragm
   a. relaxes and moves upward.
   b. relaxes and moves downward.
   c. contracts and moves upward.
   d. contracts and moves downward.
   e. is not involved in the breathing movements.

6. In which of the following organisms does oxygen diffuse directly across a respiratory surface to cells, without being carried by the blood?
   a. a grasshopper   d. a sparrow
   b. a whale   e. a mouse
   c. an earthworm

7. What is the function of the cilia in the trachea and bronchi?
   a. to sweep air into and out of the lungs
   b. to increase the surface area for gas exchange
   c. to vibrate when air is exhaled to produce sounds
   d. to dislodge food that may have slipped past the epiglottis
   e. to sweep mucus with trapped particles up and out of the respiratory tract

8. What do the alveoli of mammalian lungs, the gill filaments of fish, and the tracheal tubes of insects have in common?
   a. use of a circulatory system to transport gases
   b. respiratory surfaces that are infoldings of the body wall
   c. countercurrent exchange
   d. a large, moist surface area for gas exchange
   e. all of the above

9. Which of the following is the best explanation for why birds can fly over the Himalayas while most humans require oxygen masks to climb these mountains?
   a. Birds are much smaller and require less oxygen.
   b. Birds use positive pressure breathing, whereas humans use negative pressure breathing.
   c. With their one-way flow of air and efficient ventilation, the lungs of birds extract more O_2 from the air.
   d. The circulatory system of birds is much more efficient at delivering oxygen to tissues than is that of humans.
   e. Humans are endotherms and thus require more oxygen than do birds, which are ectotherms.

Describing, Comparing, and Explaining

10. What are two advantages of breathing air, compared with obtaining dissolved oxygen from water? What is a comparative disadvantage of breathing air?

11. Trace the path of an oxygen molecule from the air to a muscle cell in your arm, naming all the structures involved along the way.

12. Carbon monoxide (CO) is a colorless, odorless gas found in furnace and automobile engine exhaust and cigarette smoke. CO binds to hemoglobin 210 times more tightly than does O_2. (You also learned in Chapter 6 that CO binds with an electron transport protein and disrupts cellular respiration.) Explain why CO is such a deadly gas.

Applying the Concepts

13. Partial pressure reflects the relative amount of gas in a mixture and is measured in millimeters of mercury (mm Hg). Llamas are native to the Andes Mountains in South America. The partial pressure of O_2 (abbreviated P_{O_2}) in the atmosphere where llamas live is about half of the P_{O_2} at sea level. As a result, the P_{O_2} in the lungs of llamas is about 50 mm Hg, whereas it is about 100 mm Hg in human lungs at sea level.

    A dissociation curve for hemoglobin shows the percentage of saturation (the amount of O_2 bound to hemoglobin) at increasing values of P_{O_2}. As you see in the graph below, the dissociation curves for llama and human hemoglobin differ. Compare these two curves and explain how the hemoglobin of llamas is an adaptation to living where the air is "thin."

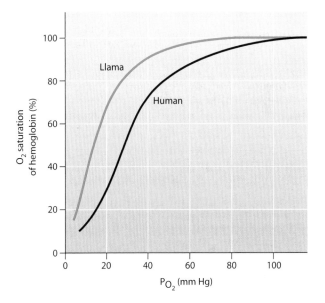

14. One of the many mutant opponents that the movie monster Godzilla contends with is Mothra, a giant moth-like creature with a wingspan of 7 to 8 m. Science fiction creatures like these can be critiqued on the grounds of biomechanical and physiological principles. Focusing on the principles of gas exchange that you learned about in this chapter, what problems would Mothra face? Why do you think truly giant insects are improbable?

15. Hundreds of studies have linked smoking with cardiovascular and lung disease. According to most health authorities, smoking is the leading cause of preventable, premature death in the United States. Antismoking and health groups have proposed that cigarette advertising in all media be banned entirely. What are some arguments in favor of a total ban on cigarette advertising? What are arguments in opposition? Do you favor or oppose such a ban? Defend your position.

Answers to all questions can be found in Appendix 4.

For Practice Quizzes, BioFlix, MP3 Tutors, and Activities, go to www.mybiology.com.

Mechanisms of Internal Transport

How Does Gravity Affect Blood Circulation?

Few animals seem less alike than the ones you see in this chapter introduction. The giraffes pictured on these pages live in central Africa. They are herbivores, using their long necks to browse trees. On the bottom right of this page is a photo of a corn snake, found throughout much of the United States. This nonpoisonous predator eats mostly rats and mice, but it can also climb trees and dine on bird eggs.

Despite their differences, giraffes and snakes have many features in common. As land vertebrates, they both have a backbone, lungs, and a circulatory system. They also have something in common with *all* animals that live on land: Every part of their body is subject to the persistent, unwavering force of gravity.

Gravity does not greatly affect aquatic animals because their body is supported by water, but it has profound effects on terrestrial species. Our own body shows signs of constant, long-term exposure to gravity, such as the tendency of the skin on our face to sag with age. Our circulatory system is strongly affected by gravity, which tends to pull blood downward into the lower parts of the body. The pull of gravity is a problem for us because we stand upright on two legs; the giraffe is similarly affected because of its long neck and legs; the corn snake faces the problem of gravity when it climbs a tree looking for bird eggs.

What are the solutions to these problems of gravity? Mammals, including humans and giraffes, have a very strong heart that keeps blood circulating despite gravity's pull. When we are standing, our heart must pump blood against gravity from our heart to our brain. The challenge is even greater for a giraffe. A standing giraffe requires a great deal more pressure to pump blood the 2.5 m from its heart to its head. Indeed, the blood pressure of a giraffe is about twice that of a human. But when a giraffe bends down to drink, as shown above, the pull of gravity almost doubles the blood pressure in the arteries leading to its head. Special check valves, saclike sinuses, and other mechanisms protect the giraffe's brain from this potentially dangerous spike in blood pressure.

How does blood travel uphill in the veins of a giraffe's long legs, or in our own legs, to return to the heart? Skeletal muscle contractions help blood move along its way. As we walk or run, our leg muscles squeeze the veins and force the blood upward toward the heart. Our veins also have valves that allow the blood to flow in only one direction, preventing it from flowing back down the legs.

How does gravity affect a corn snake when it is climbing a tree? Because its heart is located close to its head, the snake's brain receives enough blood even when it is vertical. Also, blood vessels in its tail constrict when it climbs, helping to maintain blood flow to the head. After a climb, a corn snake wriggles vigorously. This motion contracts muscles all over its body, squeezing veins and increasing circulation.

Like these land vertebrates, all organisms must exchange materials with their environment and distribute materials within their body. Most animals have a system of internal transport—a **circulatory system**—that transports oxygen and carbon dioxide, distributes nutrients to body cells, and conveys the waste products of metabolism to specific sites for disposal. Several types of transport systems have evolved that facilitate this exchange between the environment and the cells of an animal. This chapter focuses on the human circulatory system, but we will also survey some of the other solutions to exchange and transport that natural selection has favored. Let's begin our study of the evolution, structure, and function of circulatory systems. ■ ■ ■

23.1 Circulatory systems facilitate exchange with all body tissues

To sustain life, an animal must gain nutrients, exchange gases, and shed waste products, and these needs ultimately extend to every cell in the body. As we saw in Figure 20.13A, a circulatory system is necessary in any animal whose body is too large or too complex for such exchange to occur by diffusion alone. Diffusion is inadequate for transporting materials over distances greater than a few cell widths—far less than the distance oxygen must travel between our lungs and brain or the distance nutrients must go between our small intestine and the muscles in our arms and legs. An internal transport system must bring resources close enough to cells for diffusion to occur.

Several types of internal transport have evolved in animals. For example, in cnidarians and most flatworms, a central gastrovascular cavity serves both in digestion and in distribution of substances throughout the body. As we saw in Module 21.3, the body wall of a hydra is only two or three cells thick, so all the cells can exchange materials directly with the water surrounding the animal or with the water in its gastrovascular cavity. Digestion occurs in the gastrovascular cavity and in the cells lining it. Only these cells have direct access to nutrients, but nutrients have only a short distance to diffuse to cells of the outer layer.

A gastrovascular cavity is not adequate for animals with thick, multiple layers of cells. Such animals require a true circulatory system, which consists of a muscular pump (**heart**), a circulatory fluid (**blood**), and a set of tubes or vessels to carry the blood.

Two basic types of circulatory systems have evolved in animals. Many invertebrates, including most molluscs and all arthropods, have an **open circulatory system.** The system is called "open" because fluid is pumped through open-ended vessels and flows out among the cells; there is no distinction between blood and interstitial fluid. In an insect, such as the grasshopper (Figure 23.1A), pumping of the tubular heart drives "blood" into the head and the rest of the body (black arrows). Body movements help circulate the fluid as exchange occurs with body cells. When the heart relaxes, fluid returns to it through several pores. Each pore has a valve that closes when the heart contracts, preventing backflow of the circulating fluid. As we saw in Module 22.4, respiratory gases are conveyed to and from the insect's body cells by the tracheal system (not shown here), not by the circulatory system.

Earthworms, squids, octopuses, and vertebrates all have **closed circulatory systems.** The vertebrate circulatory system is often called a **cardiovascular system** (from the Greek *kardia,* heart, and Latin *vas,* vessel). The blood is confined to vessels, which keep it distinct from the interstitial fluid. There are three kinds of vessels: **Arteries** carry blood away from the heart to body organs and tissues; **veins** return blood to the heart; and **capillaries** convey blood between arteries and veins within each tissue.

The cardiovascular system of a fish (Figure 23.1B) illustrates key features of a closed circulatory system. The heart of a fish has two main chambers. The **atrium** (plural, *atria*) receives

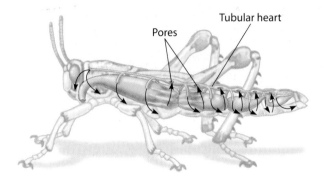

Figure 23.1A   The open circulatory system (vessels in gold) in a grasshopper

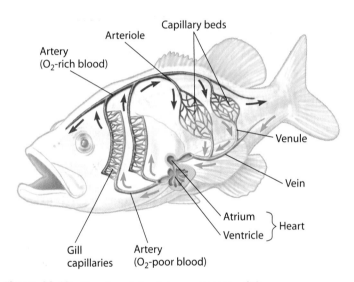

Figure 23.1B   The closed circulatory system in a fish

blood from the veins, and the **ventricle** pumps blood to the gills via large arteries. As in all figures depicting closed circulatory systems in this chapter, red represents oxygen-rich blood and blue represents oxygen-poor blood. After passing through the gill capillaries, oxygen-rich blood flows into large arteries that carry it to all other parts of the body. The large arteries branch into **arterioles,** small vessels that give rise to capillaries. Networks of capillaries called **capillary beds** infiltrate every organ and tissue in the body. The thin walls of the capillaries allow chemical exchange between the blood and the interstitial fluid. The capillaries converge into **venules,** which in turn converge into veins that return blood to the heart.

? How do the vessels and "blood" of an open circulatory system differ from those of a closed circulatory system?

■ The vessels in an open circulatory system do not form a complete circuit from the heart, through the body, and back to the heart. The "blood" is just interstitial fluid that is circulated through the body.

23.2 Vertebrate cardiovascular systems reflect evolution

The colonization of land by vertebrates was a major episode in the history of life. As aquatic vertebrates became adapted for terrestrial life, nearly all of their organ systems underwent major changes. One of these was the change from gill breathing to lung breathing (see Module 22.5), and this switch was accompanied by important changes in the cardiovascular system.

As shown in Figure 23.1B and diagrammed in Figure 23.2A, a fish has a single circuit of blood flow and two heart chambers. Blood pumped from the ventricle travels first to the gill capillaries. Blood pressure drops considerably while blood passes through the gill capillaries for reasons we will explain shortly, but the blood is helped on its way by the animal's swimming movements. An artery carries the oxygen-rich blood to systemic (body) capillaries in the tissues and organs of the body, from which the blood then returns to the atrium of the heart.

A single circuit would not supply enough pressure to move blood through the capillaries of the lungs and then to the systemic capillaries of a terrestrial vertebrate. The evolutionary adaptation that resulted in a more vigorous flow of blood to body organs is called **double circulation,** in which blood is pumped a second time after it loses pressure in the capillaries of the lungs. The **pulmonary circuit** carries blood between the heart and gas exchange tissues in the lungs, and the **systemic circuit** carries blood between the heart and the rest of the body.

You can see an example of these two circuits in Figure 23.2B. (Notice that the right side of the animal's heart is on the left in the diagram. It is customary to draw the system as though in a body facing you from the page.) Frogs and other amphibians have a three-chambered heart. The right atrium receives blood returning from the systemic capillaries. The ventricle pumps blood to capillary beds in the lungs and skin. Because gas exchange occurs both in the lungs and across the thin, moist skin, this is called a *pulmocutaneous circuit.* Oxygen-rich blood returns to the left atrium. Although blood from the left and right atria mixes in the single ventricle, most of the oxygen-poor blood is diverted to the pulmocutaneous circuit and most of the oxygen-rich blood goes to the systemic circuit.

In the three-chambered heart of turtles, snakes, and lizards, the ventricle is partially divided, and less mixing of blood occurs. The ventricle is completely divided in crocodilians.

In all birds and mammals, the heart has four chambers: two atria and two ventricles (Figure 23.2C). The right side of the heart handles only oxygen-poor blood, while the left side receives and pumps only oxygen-rich blood. The evolution of a powerful four-chambered heart was an essential adaptation to support the high metabolic rate characteristic of birds and mammals, which are endothermic. Endotherms use about ten times as much energy as equal-sized ectotherms (see Module 19.6). Therefore, their circulatory system needs to deliver about ten times as much fuel and oxygen to body tissues. This requirement is met by a large and powerful heart that is able to pump a large volume of blood and by separate systemic and pulmonary circulations. Birds and mammals descended from different reptilian ancestors, and their four-chambered hearts evolved independently—an example of convergent evolution in which natural selection favors the same adaptation in response to similar environmental challenges.

> ? What is the difference between the single circulation of fish and the double circulation of land vertebrates?
>
> ■ In fish, blood travels from gill capillaries to systemic capillaries before returning to the heart. In land vertebrates, blood returns to the heart and is pumped a second time between the pulmonary and systemic circuits.

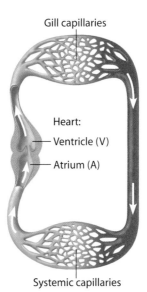

Figure 23.2A The single circulation and two-chambered heart of a fish

Gill capillaries

Heart:
— Ventricle (V)
— Atrium (A)

Systemic capillaries

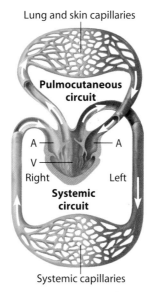

Figure 23.2B The double circulation and three-chambered heart of an amphibian

Lung and skin capillaries

Pulmocutaneous circuit

A — A
V — V
Right   Left

Systemic circuit

Systemic capillaries

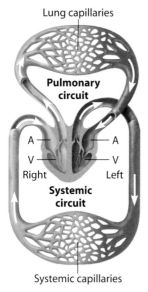

Figure 23.2C The double circulation and four-chambered heart of a bird or mammal

Lung capillaries

Pulmonary circuit

A — A
V — V
Right   Left

Systemic circuit

Systemic capillaries

23.3 The human cardiovascular system illustrates the double circulation of mammals

Let's follow the flow of blood through the human circulatory system. Let's start at the right ventricle in Figure 23.3A, below, and follow the pulmonary (lung) circuit first. ❶ The right ventricle pumps oxygen-poor blood to the two lungs via ❷ the **pulmonary arteries.** As the blood flows through ❸ capillaries in the lungs, it takes up oxygen and unloads carbon dioxide. Oxygen-rich blood flows back through ❹ the **pulmonary veins** to ❺ the left atrium. Next, the oxygen-rich blood flows from the left atrium into ❻ the left ventricle.

The left ventricle pumps blood to the systemic circuit. As Figure 23.3A shows, oxygen-rich blood leaves the left ventricle through ❼ the **aorta.** The aorta is our largest blood vessel, with a diameter of roughly 2.5 cm, about the same diameter as a quarter. The first branches from the aorta are the coronary arteries (not shown), which supply blood to the heart muscle itself. Next are branches leading to ❽ the head, chest, and arms. The aorta descends into the abdomen, supplying oxygen-rich blood to this region and the legs. For simplicity, Figure 23.3A does not show the individual organs, but within each organ, arteries lead to arterioles that branch to capillaries. The capillaries rejoin as venules, which lead to veins. ❾ Oxygen-poor blood from the upper body is channeled into a large vein called the

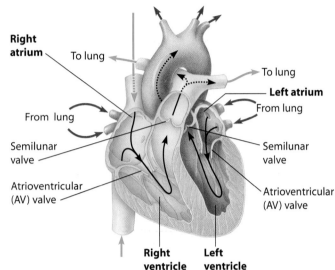

Figure 23.3B Blood flow through the human heart

superior vena cava, and the **inferior vena cava,** another large vein, returns blood from the lower body. The two venae cavae empty into ❿ the right atrium. As the blood flows from the right atrium into the right ventricle, we complete our journey.

Remember that the path of any single blood cell is always heart to lung capillaries to heart to body tissue capillaries and back to heart. In one systemic circuit, a blood cell may travel to the brain; in the next (after a pulmonary circuit), it may travel to the legs. It never travels from the brain to the legs without first returning to the heart and being pumped to the lungs to be recharged with oxygen.

Figure 23.3B shows the path of blood through the heart. About the size of a clenched fist, the human heart is enclosed in a sac just under the breastbone. The heart is formed mostly of cardiac muscle tissue. Its thin-walled atria collect blood returning to the heart. The thicker-walled ventricles pump blood to the lungs and to all other body tissues. Notice that the left ventricle walls are thicker, a reflection of how much farther it pumps blood in the body. Flap-like valves between the atria and ventricles and at the openings to the pulmonary artery and the aorta regulate the direction of blood flow. We look at these valves and the functioning of the heart in the next module.

🎧 MP3 Tutor *The Human Circulatory System*

Web Activity *Mammalian Cardiovascular System Structure*

Web Activity *Path of Blood Flow in Mammals*

❓ Why does blood in the pulmonary veins have more O_2 than blood in the venae cavae, which are also veins?

■ Pulmonary veins carry blood from the lungs, where it picked up O_2, to the heart. The venae cavae carry blood returning to the heart after delivering O_2 to body tissues.

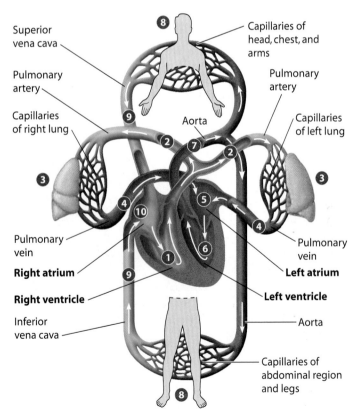

Figure 23.3A Blood flow through the double circulation of the human cardiovascular system

The heart contracts and relaxes rhythmically

The four-chambered heart is the hub of the circulatory system. It separately but simultaneously pumps oxygen-poor blood to the lungs and oxygen-rich blood to the body. Its pumping occurs as a rhythmic cycle of contraction and relaxation. When the heart contracts, it pumps blood; when it relaxes, blood fills its chambers. The sequence of pumping and filling is called the **cardiac cycle.**

Figure 23.4 shows a cardiac cycle that takes about 0.8 second, corresponding to a heart rate of 72 beats per minute. When the entire heart is relaxed, in the phase called ❶ **diastole,** blood flows into all four of its chambers. Blood enters the right atrium from the venae cavae and the left atrium from the pulmonary veins (see Figure 23.3A). The valves between the atria and the ventricles (atrioventricular, or AV, valves) are open. Diastole lasts about 0.4 second, during which the ventricles nearly fill with blood.

The contraction phase of the cardiac cycle is called **systole.** ❷ Systole begins with a very brief (0.1-second) contraction of the atria that completely fills the ventricles with blood (atrial systole). ❸ Then the ventricles contract for about 0.3 second (ventricular systole). The force of their contraction closes the AV valves, opens the semilunar valves located at the exit from each ventricle, and pumps blood into the large arteries. Blood flows into the atria during the second part of systole, as the green arrows in step 3 indicate.

Because it pumps blood to all body organs through the systemic circuit, the left ventricle contracts with greater force than the right ventricle. Both ventricles, however, pump the same volume of blood when they contract. The volume of blood that each ventricle pumps per minute is called **cardiac output.** This volume is equal to the amount of blood pumped by a ventricle each time it contracts (about 70 mL per beat for the average person) times the **heart rate** (number of beats per minute). An average person at rest might have a heart rate of 72 beats per minute. At this rate, cardiac output would be 70 mL/beat × 72 beats/min = 5,040 mL/min, or about 5 L/min, roughly equivalent to the total volume of blood in the human body. Thus, a drop of blood can travel the entire circuit in just 1 minute.

Heart rate and cardiac output vary, depending on age, fitness, and other factors. Both increase, for instance, during heavy exercise, in which cardiac output can increase fivefold, enabling the circulatory system to provide the additional oxygen needed by hardworking muscles. A well-trained athlete's heart may strengthen and enlarge with a resulting increase in the volume of blood a ventricle pumps. Thus, a resting heart rate of an athlete may be as low as 40 beats/min and still produce a normal cardiac output of about 5 L/min. During competition, a trained athlete's cardiac output may increase sevenfold.

Notice again the changes in the heart valves during a cardiac cycle, depicted in Figure 23.4. Made of flaps of connective tissue, these valves open when pushed from one side and close when pushed from the other. The powerful contraction of the ventricles forces blood against the AV valves, closing them and

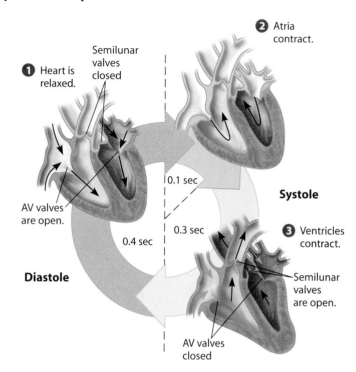

Figure 23.4 A cardiac cycle in a human with a heart rate of about 72 beats a minute

keeping blood from flowing back into the atria. The semilunar valves are pushed open when the ventricles contract. When the ventricles relax, blood in the arteries starts to flow back toward the heart, causing the flaps of the semilunar valves to close and preventing blood from flowing back into the ventricles.

The heart sounds we can hear with a stethoscope—"lub-dup, lub-dup, lub-dup"—are caused by the closing of these heart valves. The "lub" sound comes from the recoil of blood against the closed AV valves. The "dup" is produced by the recoil of blood against the closed semilunar valves.

A trained ear can detect the hissing sound of a **heart murmur,** which may indicate a defect in one or more of the heart valves. A murmur occurs when a stream of blood squirts backward through a valve. Some people are born with murmurs, while others have their valves damaged by infection (from rheumatic fever, for instance). Most valve defects do not reduce the efficiency of blood flow enough to warrant surgery. Those that do can be corrected by replacing the damaged valves with synthetic ones or with valves taken from an organ donor (human or other animal, usually a pig).

The next module explores the control of the cardiac cycle.

> **?** During a cardiac cycle of 0.8 second, the atria are generally relaxed for _____ second.

■ 0.7

23.5 The pacemaker sets the tempo of the heartbeat

In vertebrates, the cardiac cycle originates in the heart itself. Cardiac muscle cells "have their own beat," meaning they contract and relax repeatedly without any signal from the nervous system. One can even see these rhythmic contractions in tissue that has been removed from a heart and placed in a dish in the laboratory! Because each of these cells has its own intrinsic contraction rhythm, how are their contractions coordinated in the intact heart? The answer lies in a specialized region of cardiac muscle called the **pacemaker, or SA (sinoatrial) node,** which sets the rate at which all the muscle cells of the heart contract.

The pacemaker is situated in the upper wall of the right atrium (Figure 23.5A), below. ❶ The pacemaker generates electrical signals (black arrows) much like those produced by nerve cells. ❷ Cardiac muscle cells are electrically connected by specialized junctions between cells (see Module 20.6). Thus, signals (yellow color) spread quickly through both atria, making them contract in unison. The signals also pass to a relay point called the **AV (atrioventricular) node,** located in the wall between the right atrium and right ventricle. Here the signals are delayed about 0.1 second. The delay ensures that the atria contract and empty before the ventricles contract. ❸ Specialized cardiac muscle fibers (orange) then relay the signals to the apex of the heart and ❹ up through the walls of the ventricles, triggering the strong contractions that drive the blood out of the heart.

The electrical signals in the heart generate electrical changes in the skin, which can be detected by electrodes and recorded as an electrocardiogram (ECG or EKG). The yellow color in the graphs under the hearts indicates the part of an ECG that matches the electrical event shown in yellow in the heart. In step 4, the portion of the ECG to the right of the yellow "spike" is the electrical activity of the ventricles, becoming readied to conduct the next round of contraction signals.

Careful reading of an ECG can provide a wealth of data about the health of the heart. Arrhythmias are abnormal heart rhythms that an ECG can reveal. During a heart attack, the pacemaker is often unable to maintain a normal rhythm. Electrical shocks applied to the chest by a defibrillator may reset the pacemaker and restore proper cardiac function. The availability of automatic external defibrillators (AEDs) has

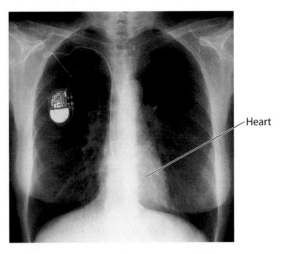

Figure 23.5B An artificial pacemaker implanted in the chest

saved thousands of lives. Unlike hospital defibrillators, AEDs are designed to be used by laypeople and are placed in public places (such as airports and shopping malls) where they are quickly accessible.

In certain kinds of heart disease, the heart's self-pacing system fails to maintain a normal heart rhythm. The remedy is an artificial pacemaker, a tiny electronic device surgically implanted near the heart (Figure 23.5B). Artificial pacemakers emit electrical signals that trigger normal heartbeats.

Two sets of nerves with opposite effects can direct the pacemaker to speed up or slow down, depending on physiological and emotional cues. Heart rate is also influenced by hormones, such as epinephrine, the "fight-or-flight" hormone released at times of stress. An increased heart rate provides more blood to muscles that may be needed to "fight or fly."

? A slight decrease in blood pH causes the pacemaker to speed up. What is the function of this control mechanism? (*Hint:* See Module 22.9.)

■ More CO_2 in the blood, as would occur with increased exercise, causes pH to drop. The increased heart rate enhances delivery of CO_2-rich blood to the lungs for removal of the CO_2. (Breathing rate is also speeded by this mechanism.) The accelerated heart rate also speeds delivery of O_2-rich blood to body tissues.

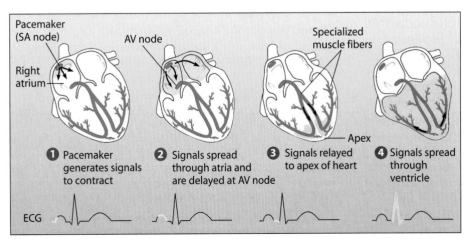

❶ Pacemaker generates signals to contract

❷ Signals spread through atria and are delayed at AV node

❸ Signals relayed to apex of heart

❹ Signals spread through ventricle

ECG

Figure 23.5A The sequence of electrical events in a heartbeat (top) and the corresponding electrocardiogram (bottom). (Yellow color represents electrical signals.)

23.6   What is a heart attack?

Like all of our cells, heart muscle cells require nutrients and oxygen-rich blood to survive. Indeed, their needs are high, as our heart contracts more than 100,000 times per day. When blood exits the heart via the aorta, several coronary arteries (shown in red in Figure 23.6A) immediately branch off to feed the heart muscle. If one or more of these blood vessels become blocked, heart muscle cells will quickly die (gray area in Figure 23.6A). A **heart attack,** also called a myocardial infarction, is the damage or death of cardiac muscle tissue as a result of such blockage. Approximately one-third of heart attack victims die almost immediately. For those who survive, the ability of the damaged heart to pump blood may be seriously impaired.

More than 40% of all deaths in the United States are caused by **cardiovascular disease**—disorders of the heart and blood vessels. A **stroke** is the death of brain tissue resulting from blockage of arteries in the head. The suddenness of a heart attack or stroke belies the fact that the arteries of most victims became impaired gradually by a chronic cardiovascular disease known as **atherosclerosis** (from the Greek *athero,* paste, and *sclerosis,* hardness). During the course of this disease, growths called plaques develop in the inner walls of arteries, narrowing the passages through which blood can flow (Figure 23.6B).

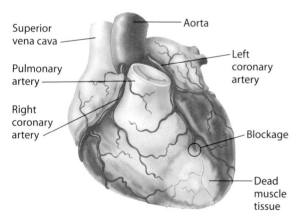

Figure 23.6A   Blockage of a coronary artery, resulting in a heart attack

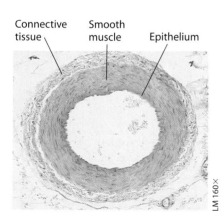

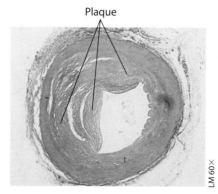

Figure 23.6B   Atherosclerosis: a normal artery (left) and an artery partially closed by plaque (right)

The artery wall becomes thickened and infiltrated with cholesterol and fibrous connective tissue. A blood clot is more likely to become trapped in a vessel that has been narrowed by plaques. Furthermore, plaques are common sites of blood clot formation.

There are treatments available for cardiovascular disease, and more continue to be developed. Heart attack victims are treated with clot-dissolving drugs, which stop many heart attacks and help prevent damage. Diagnostic tests ranging from cholesterol and blood pressure measurements to sophisticated imaging techniques such as CT and MRI help identify those at risk. Drugs can lower cholesterol and blood pressure, a risk factor we discuss in Module 23.9.

Recently, researchers have found that inflammation plays a central role in atherosclerosis and blood clot formation, and this has led to new ways of treating and diagnosing cardiovascular disease. A substance known as C-reactive protein (CRP) is produced by the liver during episodes of acute inflammation, and significant levels of CRP in the blood appear to be a predictor of cardiovascular disease. Aspirin, which blocks the body's inflammatory response, is given in treatment during a cardiac episode and has been found to help prevent the recurrence of heart attacks and stroke.

Angioplasty (inserting a tiny catheter with a balloon that is inflated to compress plaques and widen clogged arteries) and stents (small wire mesh tubes that prop arteries open) are common treatments for atherosclerosis. In bypass surgery, a more drastic remedy, blood vessels removed from a patient's legs are sewn into the heart to detour blood around clogged arteries. In extreme cases, a heart transplant may be necessary. With the severe shortage of donor hearts, various artificial pumping devices are under development.

Fortunately, the U.S. death rate from cardiovascular disease has been cut in half over the past 50 years. Health education, early diagnosis, and reduction of risk factors, particularly smoking, have contributed to this decline.

To some extent, the tendency to develop cardiovascular disease appears to be inherited. There are three everyday behaviors, however, that significantly impact the risk. Smoking doubles the risk of heart attack and harms the circulatory system in other ways. Exercise can cut the risk of heart disease in half. Eating a heart-healthy diet, low in cholesterol and *trans* and saturated fats, can reduce the risk of atherosclerosis (see Module 21.23).

?   Why is it important to be able to identify a person's risk of cardiovascular disease?

■ Although lifestyle choices (exercise, not smoking, healthy diet) are beneficial for everyone, they are particularly important for people with an increased risk of cardiovascular disease. Also, there are many drugs that may help reduce the risk of heart attack or stroke and the permanent damage or death that might ensue.

23.7 The structure of blood vessels fits their functions

Now that we have explored the structure and function of the heart, let's look at the amazingly extensive series of vessels that transport blood throughout the body. Indeed, the total length of blood vessels in an average adult human is twice Earth's circumference at the equator.

Functions of Blood Vessels The blood vessels of the circulatory system must have an intimate connection with all the body's tissues. This fact helps explain their remarkable total length. The micrograph in Figure 23.7A shows a capillary that supplies oxygenated, nutrient-rich blood to smooth muscle cells. Notice that red blood cells pass single file through the capillary, coming close enough to the surrounding tissue that O_2 can diffuse out of them into the muscle cells.

In Figure 23.7B, the downward arrows show the route that molecules take in diffusing from blood into tissue cells. As we discussed in Module 20.13, materials are not exchanged directly between blood and body cells. Each cell is immersed in interstitial fluid. Molecules such as O_2 (red dots in the figure) and nutrients (green dots) diffuse out of a capillary into the interstitial fluid and then from the interstitial fluid into a tissue cell.

In addition to transporting O_2 and nutrients, blood vessels convey metabolic wastes to waste disposal organs: CO_2 to the lungs and a variety of other metabolic wastes to the kidneys. The upward arrows in Figure 23.7B represent the diffusion of waste molecules (gray dots) out of a tissue cell, through the interstitial fluid, and into the capillary.

The circulatory system plays several key roles in maintaining a constant internal environment (homeostasis). By exchanging molecules with the interstitial fluid, it helps control the makeup of the environment in which the tissue cells live. And as we will see in later chapters, the circulatory system is involved in body defense, temperature regulation, and hormone distribution.

Structure of Blood Vessels Figure 23.7C illustrates the structures of the different kinds of blood vessels and how the vessels are connected. Look first at the capillaries (center). Appropriate to its function of exchange of materials, a capillary has a very thin wall formed of a single layer of epithelial cells, which is wrapped in a thin basal lamina (see Module 20.4). The inner surface of the capillary is smooth and keeps the blood cells from being abraded as they tumble along.

Arteries, arterioles, veins, and venules have thicker walls than those of capillaries. Their walls have the same epithelial layer, but they are reinforced by two other tissue layers. An outer layer of connective tissue with elastic fibers enables the vessels to stretch and recoil. The middle layer consists mainly of smooth muscle. Both these layers are thicker and sturdier in arteries, providing the strength and elasticity to accommodate the rapid flow and high pressure of blood pumped by the heart. Arteries are also able to regulate blood flow by constricting or relaxing their smooth muscle layer. The thinner-walled veins convey blood back to the heart at low velocity and pressure. Within large veins, flaps of tissue act as one-way valves, which permit blood to flow only toward the heart.

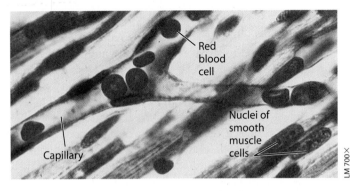

Figure 23.7A   A capillary in smooth muscle tissue

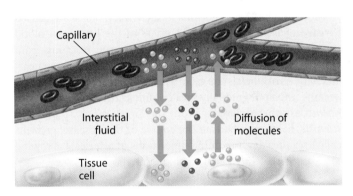

Figure 23.7B   Diffusion between blood and tissue cells

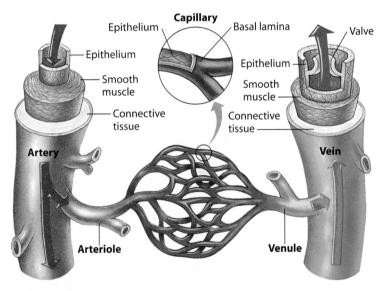

Figure 23.7C   Structural relationships of blood vessels

In the next module, we continue to explore how the structure of arteries and veins supports their role in blood transport.

? How does the structure of a capillary relate to its function?

■ The small diameter and thin walls of capillaries facilitate the exchange of substances between blood and interstitial fluid.

23.8 Blood pressure and velocity reflect the structure and arrangement of blood vessels

Blood pressure is the force that blood exerts against the walls of our blood vessels. Created by the pumping of the heart, blood pressure drives the flow of blood from the heart through arteries and arterioles to capillary beds.

When the ventricles contract, blood is forced into the arteries faster than it can flow into the arterioles. This stretches the elastic walls of the arteries. You can feel this effect of blood pressure when you measure your heart rate by taking your pulse. The **pulse** is the rhythmic stretching of the arteries. You can see this surge in pressure (expressed in millimeters of mercury, mm Hg) as the pressure peaks in the top graph of Figure 23.8A. The pressure caused by ventricular contraction is called systolic pressure. The elastic arteries snap back during diastole, maintaining pressure on the blood and a continuous flow of blood into arterioles and capillaries. The dips in pressure in the top graph represent diastolic pressure.

The center of Figure 23.8A diagrams the relative sizes and numbers of blood vessels as blood flows from the aorta through arteries to capillaries, and back through veins. Blood pressure is highest in the aorta and arteries and declines abruptly as the blood enters the arterioles. The pressure drop results mainly from the resistance to blood flow caused by friction between the blood and the large surface area it contacts in the walls of the numerous tiny arterioles.

Blood pressure depends on the volume of blood pumped into the aorta and also on the resistance to blood flow imposed by the narrow openings of the arterioles. These openings are controlled by smooth muscles. When the muscles relax, the arterioles dilate, and blood flows through them more readily, causing a fall in blood pressure. Physical and emotional stress can raise blood pressure by triggering nervous and hormonal signals that constrict these blood vessels. Regulatory mechanisms coordinate cardiac output and changes in the arteriole resistance to maintain adequate blood pressure as demands on the circulatory system change.

The blood's velocity (rate of flow, expressed in centimeters per second, cm/sec) is illustrated in the bottom graph of Figure 23.8A. As the figure shows, velocity declines rapidly in the arterioles, drops to almost zero in the capillaries, and then speeds up in the veins. What accounts for these changes? As larger arteries divide into smaller and more numerous arterioles, the total combined cross-sectional area of the many smaller vessels is much greater than the diameter of the artery that feeds into them. If there were only one small arteriole per artery, the blood would actually flow faster through the arteriole, the way water does when you add a narrow nozzle to a garden hose. However, there are many arterioles per artery, so the effect is like taking the nozzle off the hose: As you increase the diameter of a pipe, the flow rate goes down.

The cross-sectional area is greatest in the capillaries, and, as you can see, the velocity of blood is slowest through them. The overall result of this decline in velocity and pressure is a steady, leisurely flow of blood in the capillaries, enhancing the exchange of substances between blood and interstitial fluid.

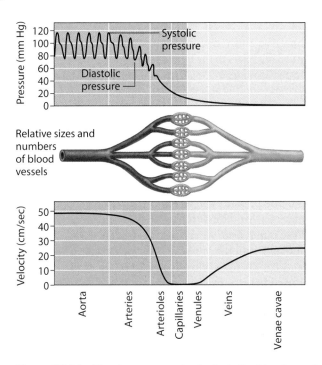

Figure 23.8A Blood pressure and velocity in the blood vessels

By the time blood reaches the veins, its pressure has dropped to near zero. The blood has encountered so much resistance as it passes through the millions of tiny arterioles and capillaries that the force from the pumping heart no longer propels it. How, then, does blood return to the heart, especially when it must travel up the legs against gravity? As we discussed in the chapter introduction, the veins of mammals such as humans and giraffes are sandwiched between skeletal muscles (Figure 23.8B). Consequently, whenever the body moves, the muscles pinch the veins and squeeze blood along toward the heart. The veins have valves that allow the blood to flow only toward the heart. Breathing also helps return blood to the heart. When we inhale, the change in pressure within our chest cavity causes the large veins near our heart to expand and fill.

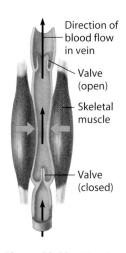

Figure 23.8B Blood flow in a vein

Web Activity *Mammalian Cardiovascular System Function*

? If blood pressure in the veins drops to zero, how is it that blood velocity increases as blood flows from venules to veins?

■ The combined diameter of the vessels through which the blood flows decreases. The velocity of the flow increases, just as water flows faster when a nozzle narrows the opening of a hose.

23.9   Measuring blood pressure can reveal cardiovascular problems

Blood exerts a force on the inside walls of blood vessels throughout the entire circulatory system, but the term *blood pressure* usually refers to the force pushing against arterial walls. As indicated in Figure 23.9, ❶ a typical blood pressure for a healthy young adult is about 120/70. The units are millimeters of mercury (mm Hg), indicating how tall a column of mercury the pressure could support. The first number is the systolic pressure; the second number is the diastolic pressure (see Module 23.8).

Blood pressure is an important indicator of cardiovascular health, and abnormal readings can indicate serious problems. Luckily, blood pressure can be easily measured using a sphygmomanometer, or blood pressure cuff. ❷ Once wrapped around the upper arm, where large arteries are accessible, the cuff is inflated until the pressure is strong enough to close the artery and cut off blood flow to the lower arm. ❸ A stethoscope is used to listen for sounds of blood flow below the cuff, and systolic blood pressure is the first measurement taken as the cuff is gradually deflated. The first sound of blood spurting through the constricted artery indicates that the blood pressure is stronger than the pressure exerted by the cuff. The pressure at this point is the systolic pressure. ❹ The sound of blood flowing unevenly through the artery continues until the pressure of the cuff falls below the pressure of the artery during diastole. Blood now flows continuously through the artery, and the sound of blood flow ceases. The reading on the pressure gauge at this point is the diastolic pressure.

Optimal blood pressure for adults is below 120 mm Hg for systolic pressure and below 80 mm Hg for diastolic pressure.

Lower values are generally considered better, except in rare cases where low blood pressure may indicate a serious underlying condition (such as endocrine disorders, malnutrition, or internal bleeding). Blood pressure higher than the normal range, however, may indicate a serious cardiovascular disorder.

High blood pressure, or **hypertension,** is persistent systolic blood pressure higher than 140 mm Hg and/or diastolic blood pressure higher than 90 mm Hg. Hypertension affects almost one-third of the adult population in the United States. It is sometimes called a "silent killer" because high blood pressure often displays no outward symptoms for years, but may be leading to severe health problems.

High blood pressure harms the cardiovascular system in several ways. Elevated pressure requires the heart to work harder to pump blood throughout the body, and over time the left ventricle may enlarge as a result. When the coronary blood supply does not keep up with the demands of this increase in muscle mass, the heart muscle weakens. In addition, the increased force on arterial walls throughout the body causes tiny ruptures that promote plaque formation, aggravating atherosclerosis (see Module 23.6) and increasing the risk of blood clot formation. Prolonged hypertension is the major cause of heart attack, heart disease, stroke, and also kidney failure, as renal arteries and areterioles may be damaged by high pressure.

In the vast majority of patients, the exact cause of hypertension cannot be firmly established. Some predispositions to hypertension cannot be avoided. For example, males have a greater risk of high blood pressure up to age 55; females have a greater risk over age 75. African Americans are more prone

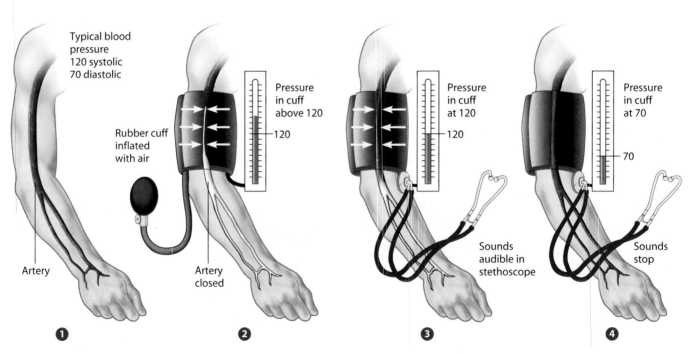

Figure 23.9   Measuring blood pressure

to hypertension than Caucasians. Blood pressure generally increases with age. Heredity also plays a role, since children of parents with hypertension are twice as likely to develop the condition.

However, no matter how many unavoidable predispositions a person may have, there are lifestyle changes that can prevent or control hypertension in just about everybody: eating a heart-healthy diet, not smoking, avoiding excess alcohol (more than two drinks per day), exercising regularly (30 minutes of moderate activity on most days), and maintaining a healthy weight. Many people associate salt with high blood pressure, but it is a contributing factor only in a small percentage of people. If

lifestyle changes don't lower blood pressure, there are several effective antihypertensive medications.

Web Process of Science *How Is Cardiovascular Fitness Measured?*

[?] Listening with a stethoscope below a sphygmomanometer cuff, you hear sounds that begin at 140 mm Hg and cease at 95 mm Hg. What are the systolic and diastolic blood pressures for this person? What is this person's blood pressure and does it indicate a health risk?

◾ Systolic = 140; diastolic = 95; blood pressure = 140/95; Yes, this person has hypertension and may be at risk for cardiovascular disease and kidney failure

23.10 Smooth muscle controls the distribution of blood

You learned in Module 23.8 that smooth muscles in arteriole walls can influence blood pressure by changing the resistance to blood flow out of the arteries and into arterioles. The smooth muscles in the arteriole walls also regulate the distribution of

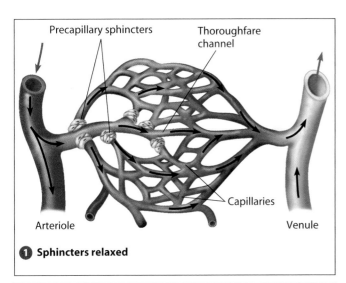

Precapillary sphincters   Thoroughfare channel

Capillaries

Arteriole   Venule

❶ **Sphincters relaxed**

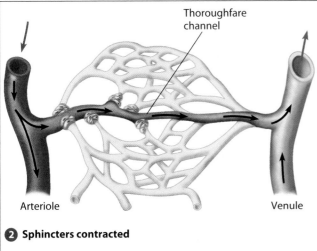

Thoroughfare channel

Arteriole   Venule

❷ **Sphincters contracted**

Figure 23.10   The control of capillary blood flow by precapillary sphincters

blood to the capillaries of the various organs. At any given time, only about 5–10% of the body's capillaries have blood flowing through them. However, each tissue has many capillaries, so every part of the body is supplied with blood at all times. Capillaries in a few organs, such as the brain, heart, kidneys, and liver, usually carry a full load of blood, but in many other sites, such as the muscles or intestines, blood supply varies as blood is diverted from one place to another, depending on need.

In addition to the smooth muscles that can constrict or dilate an arteriole leading into a capillary bed, a second mechanism, illustrated in Figure 23.10, regulates the distribution of blood. Notice that in both parts of this figure there is a capillary called a thoroughfare channel, through which blood streams directly from arteriole to venule. This channel is always open. Passage of blood into the branching capillaries is regulated by rings of smooth muscle. These are called precapillary sphincters because they are located at the entrance to capillary beds. As you can see in the figure, ❶ blood flows through a capillary bed when its precapillary sphincters are relaxed. ❷ It bypasses the capillary bed when the sphincters are contracted. After a meal, for instance, precapillary sphincters in the wall of the digestive tract let a large quantity of blood pass through the capillary beds. During strenuous exercise, many of the capillaries in the digestive tract are closed off, and blood is supplied more generously to skeletal muscles. As another example, blood flow to the skin is regulated to help control body temperature.

Nerve impulses, hormones, and chemicals produced locally influence the contraction of the smooth muscles that regulate the flow of blood to capillary beds. For example, release of histamine at a wound site causes smooth muscle relaxation, increasing blood flow and the supply of infection-fighting white blood cells.

Next we consider how substances are exchanged when blood flows through a capillary.

[?] What two mechanisms control the distribution of blood to the capillary beds of the body?

◾ Constriction of an arteriole, so that less blood reaches a capillary bed, and contraction of precapillary sphincters, so that blood flows through thoroughfare channels only, not capillary beds

23.11 Capillaries allow the transfer of substances through their walls

Capillaries are the only blood vessels with walls thin enough for substances to cross between the blood and the interstitial fluid that bathes the body cells. This transfer of materials is the most important function of the circulatory system, so let's take a closer look at it.

Figure 23.11A includes a TEM of a cross section of a capillary that serves skeletal muscle cells, along with a labeled drawing that will help you to interpret the micrograph. The capillary wall consists of adjoining epithelial cells that enclose a lumen, or space, which is just large enough for red blood cells to tumble through in single file. The nucleus you see here belongs to one of the two cells making up this portion of the capillary wall. (The other cell's nucleus does not appear in this particular cross section.) The blue area shown in the drawing around the capillary is a space containing interstitial fluid.

The exchange of substances between the blood and the interstitial fluid occurs in several ways. Some substances, such as O_2 and CO_2, simply diffuse through the epithelial cells of the capillary wall. Some larger molecules may be carried across an epithelial cell in vesicles that form by endocytosis on one side of the cell and then release their contents by exocytosis on the other side (see Module 5.9).

In addition, the capillary wall is leaky; there are narrow clefts between the epithelial cells making up the wall (see Figure 23.11A). Water and small solutes, such as sugars and salts, move freely through these clefts. Blood cells and dissolved proteins remain inside the capillary because they are too large to pass through these passageways. Much of the exchange between blood and interstitial fluid is the result of the pressure-driven flow of fluid (consisting of water and dissolved solutes) through these clefts.

The diagram in Figure 23.11B shows part of a capillary with blood flowing from its arterial end (near an arteriole) to its venous end (near a venule). What are the active forces that drive fluid into or out of the capillary? One of these forces is blood pressure, which tends to push fluid outward. The other is osmotic pressure, a force that tends to draw fluid into the capillary because the blood has a higher concentration of solutes than the interstitial fluid. Proteins dissolved in the blood account for much of this high solute concentration. (To review the principles of osmosis, see Module 5.4.)

The direction of fluid movement into or out of the capillary at any point depends on the difference between blood pressure and osmotic pressure. At the upstream (arterial) end of the capillary, the blood pressure exceeds the osmotic pressure, and there is a net movement of fluid out of the capillary. At the downstream (venous) end of the capillary, the situation is reversed. As blood moves through the tiny capillaries, the blood

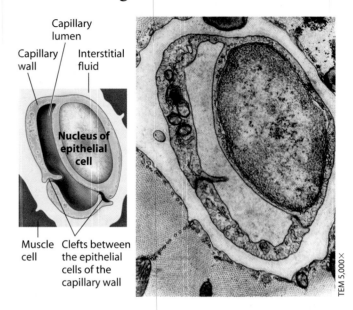

Figure 23.11A A capillary in cross section

pressure drops so much that the osmotic pressure outweighs it, and fluid reenters the capillary.

Most of the fluid that leaves the blood at the arterial end of a capillary bed reenters the capillaries at the venous end. The remaining fluid is returned to the blood by the vessels of the lymphatic system, which we discuss in Module 24.3.

Now that we have examined the structure and function of the heart and blood vessels, we turn our focus in the next several modules to the composition of blood itself.

> **?** Explain how edema, the accumulation of fluid in body tissues, can result from a severe protein deficiency in the diet that decreases the concentration of blood plasma proteins.

■ Decreased blood protein concentration reduces the osmotic gradient across the capillary, thus reducing the amount of fluid that moves back into the capillary.

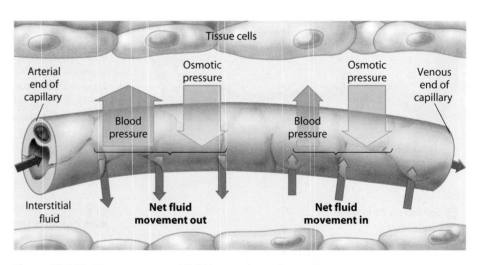

Figure 23.11B The movement of fluid into and out of a capillary

23.12 Blood consists of red and white blood cells suspended in plasma

Blood consists of several types of cells suspended in a liquid called **plasma.** When a blood sample is taken, the cells can be separated from the plasma by spinning the sample in a centrifuge (a chemical must be added to prevent the blood from clotting). The cellular elements (cells and cell fragments), which make up about 45% of the volume of blood, settle to the bottom of the centrifuge tube, underneath the transparent, straw-colored plasma (Figure 23.12).

Plasma is about 90% water. Among its many solutes are inorganic salts in the form of dissolved ions. These ions (also called electrolytes) have several functions, such as keeping the pH of blood at about 7.4, regulating the ions in interstitial fluid needed for nerve and muscle function, and maintaining the osmotic balance between blood and interstitial fluid.

Plasma includes proteins, which also help maintain osmotic balance. Some proteins act as buffers. Others have specific functions. For example, fibrinogen functions in blood clotting and immunoglobulins are important in body defense (immunity).

Plasma also contains a wide variety of substances in transit from one part of the body to another, such as nutrients, waste products, O_2, CO_2, and hormones.

Two classes of cells are suspended in blood plasma: these are **red blood cells** and **white blood cells.** A third cellular element, **platelets,** are cell fragments involved in clotting.

Red blood cells are also called **erythrocytes.** There are about 25 trillion of these cells in the average person's 5 L of blood. The structure of a red blood cell suits its main function, which is to carry oxygen. Human red blood cells are small biconcave disks, thinner in the center than at the sides. Their small size and shape create a large surface area across which oxygen can diffuse. Red blood cells lack a nucleus, allowing more room to pack in hemoglobin. Each tiny red blood cell contains about 250 million molecules of hemoglobin and, thus, can transport about a billion oxygen molecules.

There are five major types of white blood cells, or **leukocytes,** as pictured in Figure 23.12: monocytes, neutrophils, basophils, eosinophils, and lymphocytes. Their collective function is to fight infections and cancer. For example, monocytes and neutrophils are **phagocytes,** which engulf and digest bacteria and debris from our own dead cells. White blood cells actually spend much of their time moving through interstitial fluid, where most of the battles against infection are waged. There are also great numbers of white cells in the lymphatic system. You will learn more about the functions of leukocytes in body defense in the next chapter.

? For every white blood cell in normal human blood, there are about _____ to _____ red blood cells (see figure).

■ 500 . . . 1,000

Plasma (55%)	
Constituent	**Major functions**
Water	Solvent for carrying other substances
Ions (blood electrolytes)	Osmotic balance, pH buffering, and maintaining ion concentration of interstitial fluid
Sodium	
Potassium	
Calcium	
Magnesium	
Chloride	
Bicarbonate	
Plasma proteins	Osmotic balance and pH buffering
Fibrinogen	Clotting
Immunoglobulins (antibodies)	Defense
Substances transported by blood	
Nutrients (e.g., glucose, fatty acids, vitamins)	
Waste products of metabolism	
Respiratory gases (O_2 and CO_2)	
Hormones	

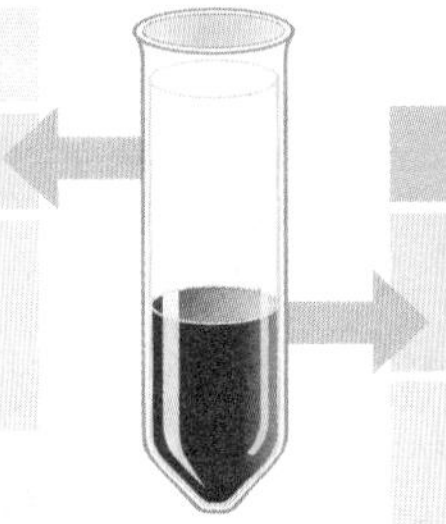

Centrifuged blood sample

Cellular elements (45%)		
Cell type	**Number** per μL (mm^3) of blood	**Functions**
Erythrocytes (red blood cells)	5–6 million	Transport of oxygen (and carbon dioxide)
Leukocytes (white blood cells)	5,000–10,000	Defense and immunity
Basophil		Lymphocyte
Eosinophil		
Neutrophil		Monocyte
Platelets	250,000–400,000	Blood clotting

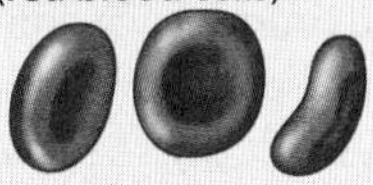

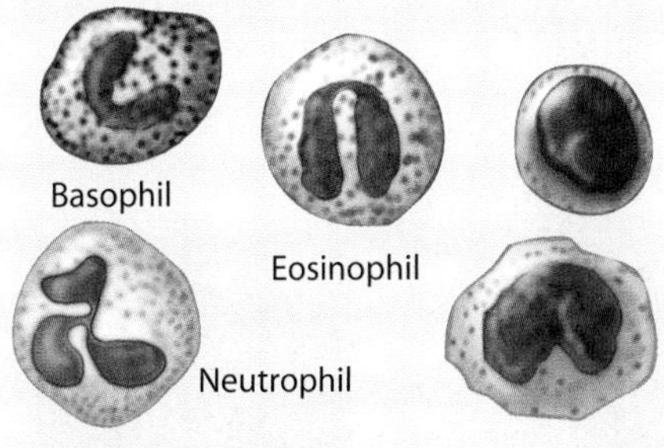

Figure 23.12 The composition of blood

23.13 Too few or too many red blood cells can be unhealthy

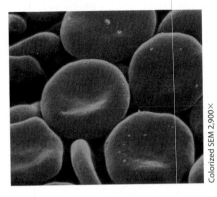

Figure 23.13 Human red blood cells

Colorized SEM 2,900×

Adequate numbers of red blood cells (Figure 23.13) are essential for healthy body function. After circulating for three or four months, red blood cells are broken down and their molecules recycled. Much of the iron removed from the hemoglobin is returned to the bone marrow, where new red blood cells are formed at the amazing rate of 2 million per second.

An abnormally low amount of hemoglobin or a low number of red blood cells is a condition called **anemia**. An anemic person feels constantly tired because body cells do not get enough oxygen. Anemia can result from a variety of factors, including excessive blood loss, vitamin or mineral deficiencies, and certain cancers. Iron deficiency is the most common cause. Women are more likely to develop iron deficiency than men because of blood loss during menstruation. Pregnant women generally benefit from iron supplements to support the developing fetus and placenta.

The production of red blood cells in the bone marrow is controlled by a negative-feedback mechanism that is sensitive to the amount of oxygen reaching the tissues via the blood. If the tissues are not receiving enough oxygen, the kidneys produce a hormone called **erythropoietin (EPO)** that stimulates the bone marrow to produce more red blood cells. Patients on kidney dialysis often have very low red blood cell counts because their kidneys do not produce enough erythropoietin. Genetically engineered EPO has significantly helped these patients, as well as cancer and AIDS patients, who often suffer from anemia.

One of the physiological adaptations of individuals who live at high altitudes, where oxygen levels are low, is the production of more red blood cells. Many athletes train at high altitudes to benefit from this effect. But other athletes take more drastic and illegal measures to increase the oxygen-carrying capacity of their blood and improve their performance. Injecting synthetic EPO can increase normal red blood cell volume from 45% to as much as 65%. Other athletes seek an unfair advantage by blood doping—withdrawing and storing their red blood cells and then reinjecting them before a competition. Athletic commissions test for cheaters by measuring the percentage of red blood cells in the blood volume or by testing for EPO-related drugs. In recent years, a number of well-known runners and cyclists have tested positive for these drugs and have forfeited both their records and their right to compete in the future.

But there can be even more serious consequences. In some athletes, a combination of dehydration from a long race and blood already thickened by an increased number of red blood cells has led to severe medical problems, such as clotting, stroke, heart failure, and even death. Indeed, EPO-related drugs have been blamed for the deaths of dozens of athletes.

? Why might increasing the number of red blood cells result in greater endurance and speed?

■ The additional red blood cells increase the oxygen-carrying capacity of blood and thus the oxygen supply to working muscles.

23.14 Blood clots plug leaks when blood vessels are injured

We all get cuts and scrapes from time to time, yet we don't bleed to death because our blood contains self-sealing materials that are activated when blood vessels are injured. These sealants are platelets and the plasma protein **fibrinogen**.

The immediate response to an injury is constriction of the damaged blood vessel, reducing blood loss and allowing time for repairs to begin. Figure 23.14A shows the stages of the clotting process. ❶ When the epithelium (shown as tan) lining a blood vessel is damaged, connective tissue in the vessel wall is exposed to blood. Platelets rapidly adhere to the exposed tissue and release chemicals that make nearby platelets sticky. ❷ Soon a cluster of sticky platelets forms a plug that provides fast protection against additional blood loss. Clotting factors released from the clumped platelets interact with clotting factors in the plasma, setting off a chain of reactions that culminates in the formation of a reinforced patch, called a scab when it's on the skin. In this complex process, which involves more than a dozen different clotting factors, an enzyme is activated that converts fibrinogen to a threadlike protein called **fibrin**. ❸ Threads of fibrin (white) trap blood cells and more platelets.

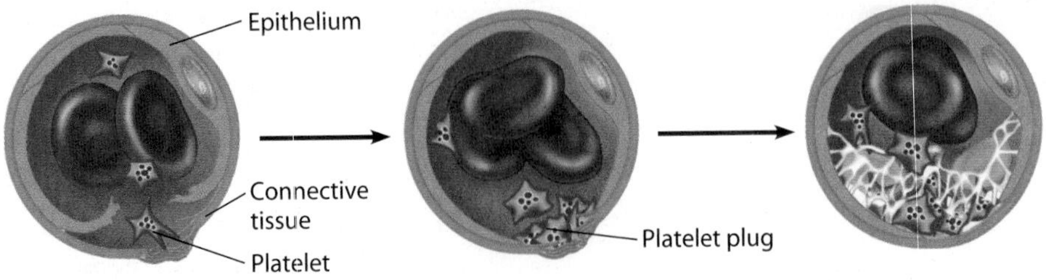

Figure 23.14B on the next page is a micrograph of a fibrin clot. Within an

❶ Platelets adhere to exposed connective tissue

❷ Platelet plug forms

❸ Fibrin clot traps blood cells

Epithelium

Connective tissue

Platelet

Platelet plug

Figure 23.14A The blood-clotting process

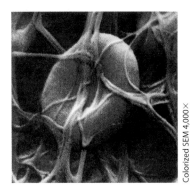

Colorized SEM 4,000×

Figure 23.14B A fibrin clot

hour after a fibrin clot forms, the platelets contract, pulling the torn edges closer together and reducing the size of the area in need of repair. Chemicals released by platelets also stimulate cell division in smooth muscle and connective tissue, initiating the healing process.

The clotting mechanism is so important that any defect in it can be life-threatening. In the inherited disease hemophilia, excessive, sometimes fatal bleeding occurs from even minor cuts and bruises. Anticlotting factors in the blood normally prevent spontaneous clotting in the absence of injury. Sometimes, however, clots form within a vessel and block the flow of blood. Such a clot, called a thrombus, can be dangerous if it blocks a key blood vessel of the heart or brain (see Module 23.6). Aspirin, heparin, and warfarin (Coumadin) are all anticoagulant drugs that work by different mechanisms to prevent undesirable clotting in patients at risk for heart attack or stroke.

? What is the role of platelets in blood clot formation?

■ Platelets adhere to exposed connective tissue and release various chemicals that help a platelet plug form, activate the pathway leading to a fibrin clot, and promote healing.

23.15 Stem cells offer a potential cure for blood cell diseases

The red marrow of bones such as the ribs, vertebrae, breastbone, and pelvis all contain a spongy tissue in which unspecialized cells called multipotent **stem cells** differentiate into blood cells. Multipotent stem cells are so named because they have the ability to form multiple types of cells, in this case all the blood cells and platelets. As shown in Figure 23.15, multipotent stem cells give rise to two different types of stem cells: Lymphoid stem cells give rise to lymphocytes, which function in the immune system (discussed in Chapter 24), and myeloid stem cells produce erythrocytes, other white blood cells, and platelets. After forming in the early embryo, these stem cells continually produce all the blood cells needed throughout life.

Leukemia is cancer of the white blood cells, or leukocytes. Leukocytes protect the body against infections and cancer cells. However, sometimes leukocytes become cancerous themselves. Because cancerous cells grow uncontrollably, a person with leukemia has an unusually high number of leukocytes, most of which do not function normally. The overabundant white blood cells crowd out red blood cells and platelets, causing severe anemia and impaired clotting.

Leukemia is usually fatal unless treated, and not all cases respond to the standard cancer treatments—radiation and chemotherapy. An alternative treatment is transplanting healthy bone marrow tissue from a suitable donor, often a sibling, into a patient whose own cancerous marrow has been purposely destroyed. Such a patient requires lifelong treatment with drugs that suppress the tendency of some of the transplanted marrow cells to "reject" the cells of the recipient. To avoid the rejection problem, patients may be treated with their own bone marrow: Marrow from the patient is harvested, processed to remove as many cancerous cells as possible, and then reinjected. Injection of as few as 30 stem cells can repopulate the blood and immune system.

Stem cells gathered from umbilical cord blood potentially may be used for stem cell transplants. To date, however, most attempts at cord blood therapy have not been successful.

Stem cell research holds great promise, and leukemia is just one of several blood diseases that may be treated by bone marrow

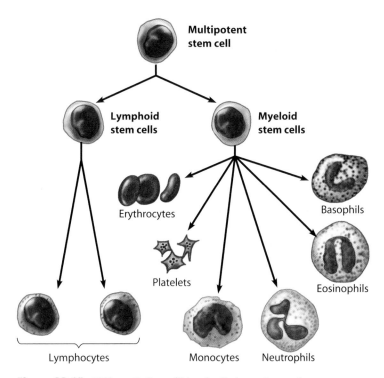

Figure 23.15 Differentiation of blood cells from stem cells

stem cells. Recently, researchers have been able to isolate bone marrow stem cells and grow them in the laboratory. In a few cases, they have induced these stem cells to differentiate into more than just blood cells. Thus, these adult stem cells may eventually provide cells for human tissue and organ transplants. (See Module 11.12 for more on adult and embryonic stem cells.)

In Chapter 24, we explore the diverse roles of white blood cells in the immune system.

? Why would a leukemia patient's bone marrow need to be destroyed and then replaced with a transplant?

■ Destruction of the bone marrow would be necessary to kill the cancerous leukocytes. The patient would need replacement multipotent stem cells to continue making both red and white blood cells.

Chapter Review

Reviewing the Concepts

Mechanisms of Internal Transport (Introduction–23.2)

A circulatory system transports O_2 and nutrients to cells and takes away CO_2 and other wastes (**Introduction**).

Types of internal transport. Gastrovascular cavities function in both digestion and transport. In the open circulatory systems of arthropods and many molluscs, a heart pumps blood through open-ended vessels to bathe tissue cells directly. In closed circulatory systems, a heart pumps blood, which travels through arteries to capillaries to veins and back to the heart (**23.1**).

Cardiovascular systems of vertebrates. The two-chambered heart of a fish pumps blood in a single circuit from gill capillaries to systemic capillaries and back to the heart. Land vertebrates have double circulation with separate pulmonary and systemic circuits. Amphibians and reptiles (except crocodilians and birds) have three-chambered hearts; birds and mammals have four-chambered hearts (**23.2**).

The Human Cardiovascular System (23.3–23.11)

The mammalian heart has two thin-walled atria that pump blood into the ventricles and two thick-walled ventricles. The right side of the heart receives and pumps oxygen-poor blood; the left side receives oxygen-rich blood from the lungs and pumps it to all other body organs (**23.3**).

Cardiac cycle. During diastole (relaxation), blood flows from the veins into the heart chambers; during systole, contractions of the atria push blood into the ventricles, and then stronger contractions of the ventricles propel blood into the large arteries. Cardiac output is the amount of blood per minute pumped by a ventricle. Heart valves prevent the backflow of blood (**23.4**).

The pacemaker (SA node) generates electrical signals that trigger contraction of the atria. The AV node relays these signals to the ventricles. An electrocardiogram records the electrical changes (**23.5**). A heart attack is damage to cardiac muscle, usually resulting from a blocked coronary artery (**23.6**).

Blood vessel structure correlates with function. Capillaries are the sites of exchange between blood and interstitial fluid (**23.7**).

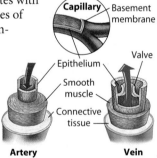

Blood pressure, the force blood exerts on vessel walls, depends on cardiac output and the resistance of vessels. Pressure is highest in the arteries and lowest in the veins. Blood velocity is slowest in the capillaries. Muscle contractions and one-way valves keep blood moving through veins to the heart (**23.8**). Blood pressure is measured as systolic and diastolic pressures. Hypertension is a serious cardiovascular problem (**23.9**).

Capillary exchange. Constriction of arterioles and precapillary sphincters controls blood flow through capillary beds (**23.10**). Blood pressure forces fluid out of the capillary at the arterial end, and osmotic pressure draws fluid in at the venous end (**23.11**).

Structure and Function of Blood (23.12–23.15)

Blood consists of cells in a fluid plasma, which contains various inorganic ions, proteins, nutrients, wastes, gases, and hormones. Red blood cells (erythrocytes) transport O_2 bound to hemoglobin. White blood cells (leukocytes) function both inside and outside the circulatory system to fight infections and cancer (**23.12**). The hormone erythropoietin regulates red blood cell production. Some athletes artificially increase their red blood cell number, a dangerous practice (**23.13**).

Blood clotting. When a blood vessel is damaged, platelets help trigger the conversion of fibrinogen to fibrin, forming a clot that plugs the leak (**23.14**).

Stem cells divide in bone marrow to produce all types of blood cells. Stem cells are used to treat some blood disorders (**23.15**).

Connecting the Concepts

1. Use this diagram to review the flow of blood through a human cardiovascular system. Label the indicated parts, highlight the vessels that carry oxygen-rich blood, and then trace the flow of blood by numbering the circles from 1 to 10, starting with 1 in the right ventricle. (When two locations are equivalent in the pathway, such as right or left lung capillaries, capillaries of top or lower portion of the body, assign them both the same number.)

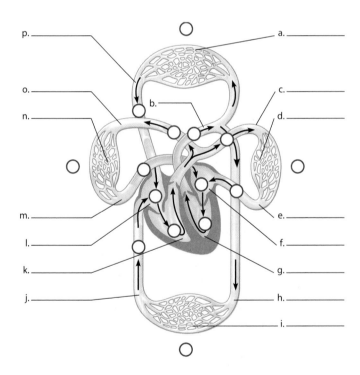

Testing Your Knowledge

Multiple Choice

2. Blood pressure is highest in _____, and blood moves most slowly in _____.
   a. veins; capillaries
   b. arteries; capillaries
   c. veins; arteries
   d. capillaries; arteries
   e. arteries; veins

3. When the doctor listened to Janet's heart, he heard "lub-hisss, lub-hiss" instead of the normal "lub-dup" sounds. The hiss is most likely due to _____. (*Explain your answer.*)
   a. a clogged coronary artery
   b. a defective atrioventricular (AV) valve
   c. a damaged pacemaker (SA node)
   d. a defective semilunar valve
   e. high blood pressure

4. Which of the following is the biggest difference between your cardiovascular system and that of a fish?
   a. In a fish, blood is oxygenated by passing through a capillary bed.
   b. Your heart has two chambers; a fish heart has four.
   c. Your circulation has two circuits; fish circulation has one.
   d. Your heart chambers are called atria and ventricles.
   e. Yours is a closed system; the fish's is an open system.

5. Paul's blood pressure is 150/90. The 150 indicates _____, and the 90 indicates _____.
   a. pressure in the left ventricle; pressure in the right ventricle
   b. arterial pressure; heart rate
   c. pressure during ventricular contraction; pressure during heart relaxation
   d. systemic circuit pressure; pulmonary circuit pressure
   e. pressure in the arteries; pressure in the veins

6. Which of the following *initiates* the process of blood clotting?
   a. damage to the lining of a blood vessel
   b. exposure of blood to the air
   c. conversion of fibrinogen to fibrin
   d. attraction of leukocytes to a site of infection
   e. conversion of fibrin to fibrinogen

7. Blood flows more slowly in the arterioles than in the artery that supplies them because the arterioles
   a. must provide opportunity for exchange with the interstitial fluid.
   b. have thoroughfare channels to venules that are often closed off, slowing the flow of blood.
   c. have sphincters that restrict flow to capillary beds.
   d. are narrower than the artery.
   e. collectively have a larger cross-sectional area than does the artery.

8. Which of the following is *not* a true statement about open and closed circulatory systems?
   a. Both systems have some sort of a heart that pumps blood through the body.
   b. A fish has an open circulatory system, land vertebrates have closed circulatory systems.
   c. The blood and interstitial fluid are separate in a closed system but are indistinguishable in an open system.
   d. The open circulatory system of an insect does not transport O_2 to body cells; closed circulatory systems do transport O_2.
   e. Some of the circulation of blood in both systems results from body movements.

9. If blood was supplied to all of the body's capillaries at the same time,
   a. blood pressure would fall dramatically.
   b. resistance to blood flow would increase.
   c. blood would move too rapidly through the capillaries.
   d. the amount of blood returning to the heart would increase.
   e. the increased gas exchange in the lungs and in the supply of O_2 to muscles would allow for strenuous exercise.

Describing, Comparing, and Explaining

10. Trace the path of blood starting in a pulmonary vein, through the heart, and around the body, returning to the pulmonary vein. Name, in order, the heart chambers and types of vessels through which the blood passes.

11. Explain how the structure of capillaries relates to their function of exchanging substances with the surrounding interstitial fluid. Describe how that exchange occurs.

12. Here is a blood sample that has been spun in a centrifuge. List, as completely as you can, the components you would find in the straw-colored fluid at the top of this tube and in the dense red portion at the bottom.

a. _____
b. _____

Applying the Concepts

13. Some babies are born with a small hole in the wall between the left and right ventricles. How might this affect the oxygen content of the blood pumped out of the heart into the systemic circuit?

14. Juan has a disease in which damaged kidneys allow some of his normal plasma proteins to be removed from the blood. How might this condition affect the osmotic pressure of blood in capillaries, compared to that of the surrounding interstitial fluid? One of the symptoms of this kidney malfunction is an accumulation of excess interstitial fluid, which causes Juan's arms and legs to swell. Can you explain why this occurs?

15. Recently, a 19-year-old woman received a bone marrow transplant from her 1-year-old sister. The woman was suffering from a deadly form of leukemia and was almost certain to die without a transplant. The parents had decided to have another child in a final attempt to provide their daughter with a matching donor. Although the ethics of the parents' decision were criticized, doctors report that this situation is not uncommon. In your opinion, is it acceptable to have a child in order to provide an organ or tissue donation? Why or why not?

16. As we discussed in the chapter introduction, gravity affects blood circulation. Physiologists speculate about cardiovascular adaptations in dinosaurs—some of which had necks almost 10 m long, which would have required a systolic pressure of nearly 760 mm Hg to pump blood to the brain when the head was fully raised. Some analyses suggest that dinosaurs' hearts were not powerful enough to generate such pressures, leading to the speculation that long-necked dinosaurs fed close to the ground rather than raising their heads to feed on high foliage. Scientists also debate whether dinosaurs had a "reptile-like" or "bird-like" heart. Most modern reptiles have a three-chambered heart with just one ventricle. Birds, which evolved from a lineage of dinosaurs, have a four-chambered heart. Some scientists believe that the circulatory needs of these long-necked dinosaurs provide evidence that dinosaurs must have had a four-chambered heart. Why might they conclude this?

Answers to all questions can be found in Appendix 4.

For Practice Quizzes, BioFlix, MP3 Tutors, and Activities, go to www.mybiology.com.

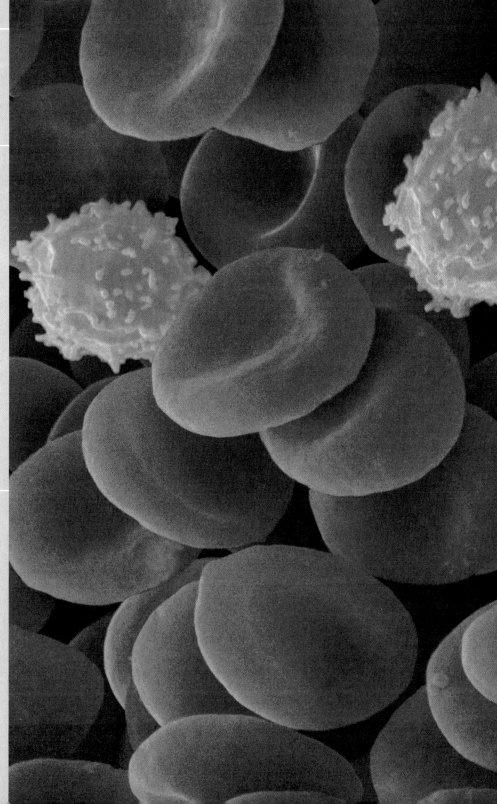

chapter

24

The Immune System

The Kissing Disease?!?

Ewwww, the kissing disease! It's true: Mononucleosis ("mono") can indeed be passed from person to person by kissing. What causes it? The answer to this question will take us on a tour of this chapter's main topic: the human body's defenses.

Mononucleosis results from infection by the Epstein-Barr virus (EBV), one of the most common human viruses. This virus is so easily spread that by the time of adulthood 95% of us are infected. If infection occurs during childhood, EBV usually causes very mild symptoms that often go unnoticed. But in about half of newly infected adolescents or young adults, EBV causes mono.

There are three primary symptoms of mono: fever, sore throat, and swollen lymph glands around the neck and armpits. Occasionally, the spleen, an important immune system organ, also swells. Mono is almost never life-threatening, except in people who have severely impaired immune systems, such as those with AIDS or people receiving immune-suppressing drugs after a transplant or during chemotherapy.

There is no vaccine to prevent mono, nor is there an effective treatment. Antibiotics are not effective against viral infections, so the only option is to treat the symptoms by resting, drinking lots of fluids, and taking over-the-counter pain relievers. Occasionally, doctors will prescribe corticosteroids to help reduce throat inflammation. Because EBV weakens the immune system, it leaves a person susceptible to opportunistic bacterial infections—ones normally defeated by a healthy immune system—of the throat, sinuses, or tonsils; these opportunistic infections can be treated with antibiotics. Symptoms of mono almost always fade on their own after a few weeks, with no long-term effects.

How does the Epstein-Barr virus cause mono? EBV usually enters the body via contact with another person's saliva through kissing or sharing food, glasses, dishes, or utensils. EBV first enters epithelial cells that line the oral cavity. From there, the virus multiplies and enters the bloodstream. In the blood, EBV selectively infects B cells, one type of the body's defensive white blood cells (visible at the top of the photo on the facing page). Many B cells die, and the resulting weakened immune system causes the symptoms of the disease. Once in a person's body, EBV remains dormant in cells of the throat and blood forever. Periodically, the virus reactivates. Such recurrences do not cause the symptoms of

mono to reappear, but the virus can enter the saliva and then be spread to other people. Many healthy people permanently carry the virus and intermittently spread it for life.

Luckily, you can only get mono once because the body acquires long-lasting immunity after a first exposure. This acquired immunity prevents symptoms from developing upon future exposures. As we will see in this chapter, our body's acquired immune system recognizes and fights back against specific **pathogens,** infectious agents that can cause disease, such as the Epstein-Barr virus. Once a pathogen is recognized, the acquired immune response rallies large numbers of dedicated immune cells that patrol the body, searching out and destroying the invader. In a healthy individual, acquired immunity backs up several forms of innate immunity. Some proteins of innate immunity punch holes in bacterial cell membranes; others block viruses from entering body cells. By combining acquired and innate defenses into a coordinated **immune system,** animals avoid or limit many infections.

In this chapter, you will learn about the body's varied defenses. Following a discussion of innate immunity in the first three modules, we concentrate on the mechanisms of acquired immunity. At the end of the chapter, we examine some of the problems—from seasonal nuisances to fatal illnesses—that may arise when the intricate interplay of our body's defenses goes awry. ■ ■ ■

24.1 Both invertebrates and vertebrates have innate defenses against infection

All animals have some form of **innate immunity,** a first line of defense against potential invaders (Figure 24.1A). Innate immune responses are defenses that act the same whether or not an invader has been previously encountered.

Invertebrate Innate Immunity Invertebrates rely solely on innate immunity. For example, insects have an exoskeleton, which is a tough, dry barrier that keeps out bacteria and viruses. Pathogens that breach these external defenses confront a set of internal defenses. A low pH and the secretion of lysozyme, an enzyme that digests the cell walls of many bacteria, protect the insect digestive system. Circulating insect immune cells are capable of **phagocytosis,** engulfing and destroying foreign substances by forming a vacuole that fuses with a lysosome (see Figure 5.9). The insect innate immune system also includes recognition proteins that bind to molecules found only on the outside of bacteria, fungi, and other pathogens. Recognition of the invading microbes triggers the production of antimicrobial peptides that bring about the destruction of the invaders.

Vertebrate Innate Immunity In vertebrates, innate immunity coexists with the more recently evolved system of acquired immunity (discussed later in this chapter). In mammals, innate defenses include skin and mucous membranes (see Module 20.4) that protect organ systems open to the external environment such as the digestive, respiratory, reproductive, and urinary systems. Nostril hairs filter incoming air, and mucus in the respiratory tract traps most microbes and dirt that get past the nasal filter. Cilia on cells lining the respiratory tract then sweep the mucus and any trapped microbes upward and out.

Microbes that breach a mammal's external barriers, such as those that enter through a cut in the skin, are confronted by innate defensive cells. These are all classified as white blood cells (see Module 23.12), although they are found in interstitial fluid as well as blood vessels. Most, such as abundant **neutrophils,** are phagocytes. **Macrophages** ("big eaters") are large phagocytic cells that wander through the interstitial fluid, "eating" any bacteria and virus-infected cells they encounter. **Natural killer (NK) cells,** another type of white blood cell, are not phagocytes. They attack cancer cells and virus-infected cells by releasing chemicals that promote programmed cell death (see Module 27.13).

Other innate defenses of vertebrates include proteins that either attack microbes directly or impede their reproduction. **Interferons** are proteins produced by virus-infected cells that help other cells resist viruses. Figure 24.1B shows how the interferon mechanism limits the cell-to-cell spread of viruses. ❶ The virus infects a cell, which causes ❷ interferon genes in the cell's nucleus to be turned on. ❸ The cell makes interferon. The infected cell then dies, but ❹ its interferon molecules may diffuse to neighboring healthy cells, ❺ stimulating them to produce other proteins that inhibit viral reproduction.

Innate immunity (24.1-3) Response is the same whether or not pathogen has been previously encountered		Acquired immunity (24.4-15) Found only in vertebrates; previous exposure to pathogen enhances immune response
External barriers	**Internal defenses**	
• Skin/exoskeleton • Secretions • Mucous membranes	• Phagocytic cells • NK cells • Defensive proteins • Inflammatory response (24.2)	• Antibodies (24.8-10) • Lymphocytes (24.11-14)
	The lymphatic system (24.3)	

Figure 24.1A An overview of animal immune systems

Additional innate immunity in vertebrates is provided by the **complement system,** a group of about 30 different kinds of proteins that circulate in an inactive form in the blood. These proteins can act together (in complement) with other defense mechanisms. Substances on the surfaces of many microbes trigger a cascade of steps that activate the complement system, leading to the lysis (bursting) of the invaders. Certain complement proteins also help trigger the inflammatory response, another vertebrate innate defense and the subject of the next module.

? Which innate defenses (a) actually help prevent infection and (b) come into play only after infection has occurred?

● (a) Skin/exoskeleton, secretions, mucous membranes; (b) Phagocytic cells (macrophages, neutrophils), NK cells, defensive proteins (interferons, complement system), inflammatory response

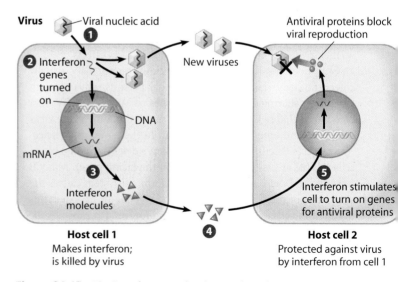

Figure 24.1B The interferon mechanism against viruses

24.2 The inflammatory response mobilizes innate defenses

The **inflammatory response** is a major component of our innate immunity. Any damage to tissue, whether caused by microorganisms or by physical injury—even just a scratch or insect bite—triggers this response. You can see signs of the inflammatory response if you fail to properly treat a cut. The area becomes red, warm, and swollen. This reaction is inflammation, which literally means "setting on fire." You read about an example of this in the chapter opener: one of the primary symptoms of mono is inflammation of the cells lining the throat.

Figure 24.2 shows the chain of events that make up the inflammatory response in a case where a pin has broken the skin, allowing infection by bacteria. ❶ The damaged cells soon release chemical alarm signals such as **histamine.** ❷ The chemicals spark the mobilization of various defenses. Histamine, for instance, induces neighboring blood vessels to dilate and become leakier. Blood flow to the damaged area increases, and blood plasma passes out of the leaky vessels into the interstitial fluid of the affected tissues. Other chemicals (some that are part of the complement system) attract phagocytes to the area. Squeezing between the cells of the blood vessel wall, these phagocytic white blood cells (yellow in the figure) migrate out of the blood into the tissue spaces. The local increase in blood flow, fluid, and cells produces the redness, heat, and swelling characteristic of inflammation.

The major results of the inflammatory response are to disinfect and clean injured tissues. ❸ The white blood cells mustered into the area engulf bacteria and the remains of any body cells killed by them or by the physical injury. Many of the white cells die in the process, and their remains are also engulfed and digested. The pus that often accumulates at the site of an infection consists mainly of dead white cells and fluid that has leaked from the capillaries during the inflammatory response.

The inflammatory response also helps prevent the spread of infection to surrounding tissues. Clotting proteins (see Module 23.14) present in blood plasma pass into the interstitial fluid during inflammation. Along with platelets, these substances form local clots that help seal off the infected region and allow healing to begin.

The inflammatory response may be localized, as we have just described, or widespread (systemic). Sometimes microorganisms such as bacteria or protozoans get into the blood or release toxins that are carried throughout the body in the bloodstream. The body may react with several inflammatory weapons. For instance, the number of white blood cells circulating in the blood may increase; an elevated "white cell count" is one way to diagnose mono. In more serious cases, severe infections such as meningitis or appendicitis may cause the number of circulating defensive cells to increase severalfold within just a few hours. Another response to systemic infection is fever, an abnormally high body temperature. Toxins themselves may trigger the fever, or macrophages may release compounds that cause the body's thermostat to be set at a higher temperature. A very high fever is dangerous, but a moderate one may stimulate phagocytosis and hasten tissue repair. (*How* a moderate fever aids immune defenses is still a subject of research and debate within the medical community.)

Sometimes bacterial infections bring about an overwhelming systemic inflammatory response leading to a condition called septic shock. Characterized by very high fever and low blood pressure, septic shock is a common cause of death in hospital critical care units. Clearly, while local inflammation is an essential step toward healing, widespread inflammation can be devastating.

> **?** Why is the inflammatory response considered a form of *innate* immunity?
>
> ■ Because the response is the same regardless of whether the invader has been previously encountered

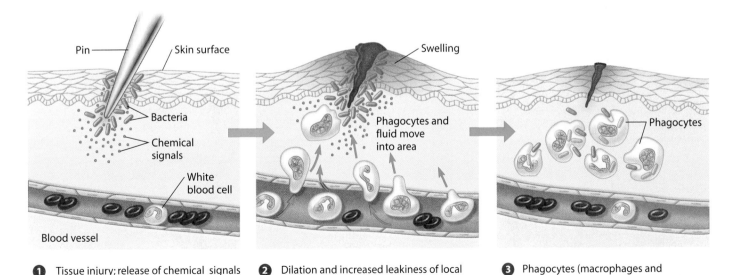

❶ Tissue injury; release of chemical signals such as histamine

❷ Dilation and increased leakiness of local blood vessels; migration of phagocytes to the area

❸ Phagocytes (macrophages and neutrophils) consume bacteria and cell debris; tissue heals

Figure 24.2 The inflammatory response

24.3 The lymphatic system becomes a crucial battleground during infection

The **lymphatic system,** which is involved in both innate and acquired immunity, consists of a branching network of vessels, numerous lymph nodes (rounded organs packed with macrophages and white blood cells called lymphocytes), the tonsils and adenoids, the appendix, and the spleen (Figure 24.3). It also includes the bone marrow and the thymus (green labels in the figure), which are the sites where white blood cells develop. The lymphatic vessels carry a fluid called **lymph,** which is similar to interstitial fluid but contains less oxygen and fewer nutrients. The lymphatic system has two main functions: to return tissue fluid to the circulatory system and to fight infection.

As we noted in Module 23.11, a small amount of the fluid that enters the tissue spaces from the blood in a capillary bed does not reenter the blood capillaries. Instead, this fluid is returned to the blood via lymphatic vessels. The enlargement in Figure 24.3 (bottom right) shows a branched lymphatic vessel in the process of taking up fluid from tissue spaces in the skin. As shown here, fluid enters the lymphatic system by diffusing into tiny, dead-end lymphatic capillaries that are intermingled among the blood capillaries.

Lymph drains from the lymphatic capillaries into larger and larger lymphatic vessels. It reenters the circulatory system via two large lymphatic vessels, the thoracic duct and the right lymphatic duct, which fuse with veins in the chest. As the close-up indicates, the lymphatic vessels resemble veins in having valves that prevent the backflow of fluid toward the capillaries (see Module 23.7C). Also, like veins, lymphatic vessels depend mainly on the movement of skeletal muscles to squeeze their fluid along. The black arrows in the close-ups indicate the flow of lymph.

As lymph circulates through the lymphatic organs (such as the lymph node shown in Figure 24.3, top right), it carries microbes, parts of microbes, and/or their toxins picked up from infection sites anywhere in the body. Once inside lymphatic organs (pink labels in the figure), macrophages that reside there permanently may engulf the invaders as part of the innate immune response. Additionally, lymphocytes may be activated to mount an aquired immune response against specific invaders. When your body is fighting an infection such as mononucleosis, the organs of the lymphatic system become a major battleground. Lymph nodes fill with huge numbers of defensive cells, causing the tender "swollen glands" in your neck and armpits that your doctor looks for as a sign of mono and other infections.

? What are the two main functions of the lymphatic system?

■ To return fluid from interstitial spaces to the circulatory system and to combat infection

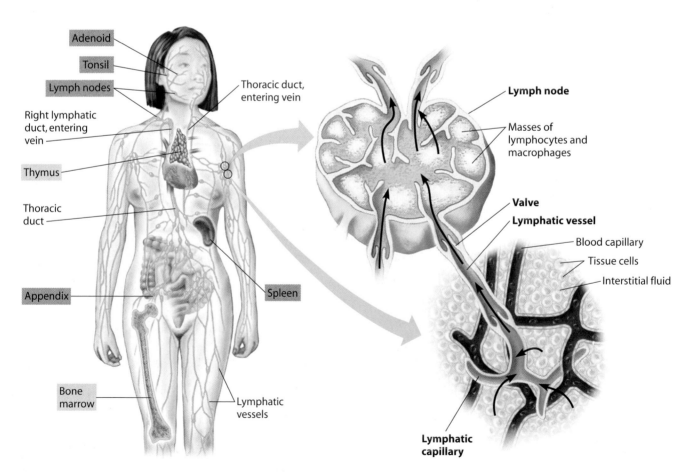

Figure 24.3   The human lymphatic system

24.4 The acquired immune response counters specific invaders

When the innate immune response fails to ward off a pathogen, the acquired immune response provides a second line of defense. **Acquired immunity,** found only in vertebrates, is a set of defenses that are activated only *after* exposure to pathogens. For example, you have acquired immunity against the Epstein-Barr virus (EBV) only if you have been exposed to it. Thus, unlike innate immunity, acquired immunity differs from individual to invididual (depending on what pathogens they have been previously exposed to). Once activated, the acquired immune response provides a strong defense against pathogens that is highly specific—that is, it acts against one infectious agent but not another. Moreover, acquired immunity can amplify certain innate responses, such as inflammation and the complement system.

Any foreign molecule that elicits an acquired immune response is called an **antigen.** Antigens may be molecules that protrude from pathogens or other particles, such as viruses, bacteria, mold spores, pollen, house dust, or the cell surfaces of transplanted organs. Antigens may also be substances released into the extracellular fluid, such as toxins secreted by bacteria. When the immune system detects an antigen, it responds with an increase in the number of cells that either attack the invader directly or produce immune proteins called antibodies. An **antibody** is a protein found in blood plasma that attaches to one particular kind of antigen and helps counter its effects. (The word *antigen* is a contraction of "*anti*body-*gen*erating," a reference to the fact that the foreign agent provokes an acquired immune response.) The defensive cells and antibodies produced against a particular antigen are usually specific to that antigen; they are ineffective against any other foreign substance.

Acquired immunity has a remarkable "memory." It can "remember" antigens it has encountered before and react against them more promptly and vigorously on subsequent exposures. For example, if a person is infected by EBV and contracts mono, the immune system "remembers" certain molecules on the virus. Should the virus enter the body again, the acquired immune response mounts a quick and decisive attack that usually destroys the virus before symptoms appear. Thus, the acquired immune response, unlike innate defenses, is adaptive; exposure to a foreign agent enhances future responses to that same agent.

Acquired immunity is usually obtained by natural exposure to antigens (that is, by being infected), but it can also be achieved by **vaccination** (also known as immunization). In this procedure, the immune system is confronted with a **vaccine** composed of a harmless variant or part of a disease-causing microbe, such as an inactivated bacterial toxin, a dead or weakened microbe, or a piece of a microbe. The vaccine stimulates the immune system to mount defenses against this harmless antigen, defenses that will also be effective against the actual pathogen because it has similar antigens. Once we have been successfully vaccinated, our immune system will respond quickly if it is exposed to the actual microbe.

In the United States, widespread vaccination of children has virtually eliminated some viral diseases, such as polio, mumps, and measles. Researchers are trying to develop a vaccine for AIDS, but so far success has been elusive. One of the major success stories of modern vaccination involves smallpox, a potentially fatal viral infection that affected over 50 million people per year worldwide in the 1950s. A subsequent vaccination effort was so effective that there have been no cases of smallpox since 1977. Since 2001, however, the U.S. government has stockpiled hundreds of millions of doses of smallpox vaccine and has begun to vaccinate high-risk health-care and military workers in case the smallpox virus is used in a bioterrorist attack (Figure 24.4).

Whether antigens enter the body naturally (if you catch the flu) or artificially (if you get a flu shot), the resulting immunity is called **active immunity** because the person's own immune system actively produces antibodies. It is also possible to acquire **passive immunity** by receiving premade antibodies. For example, a fetus obtains antibodies from its mother's bloodstream, a baby receives antibodies in breast milk that protect the digestive tract (although these antibodies are broken down there and do not enter the bloodstream), and travelers sometimes get a shot containing antibodies to pathogens they are likely to encounter. In yet another example, the effects of a poisonous snakebite may be counteracted by injecting the victim with antivenin, which contains antibodies extracted from animals previously immunized against the venom. Passive immunity is temporary because the recipient's immune system is not stimulated by antigens. Immunity lasts only as long as the antibodies do; after a few weeks or months, these proteins break down and are recycled by the body.

Figure 24.4 A soldier receiving a smallpox vaccination

? Why is protection resulting from a vaccination considered *active* immunity rather than *passive* immunity?

■ Because the body itself produces the immunity by mounting an immune response and generating antibodies, even though the stimulus consists of artificially introduced antigens

24.5 Lymphocytes mount a dual defense

Lymphocytes, white blood cells that spend most of their time in the tissues and organs of the lymphatic system, are responsible for the acquired immune response. Like all blood cells, lymphocytes originate from stem cells in the bone marrow (see Module 23.15). As shown in Figure 24.5A, some immature lymphocytes continue developing in the bone marrow; these become specialized as B lymphocytes, or **B cells.** Other immature lymphocytes are carried by the blood from the bone marrow to the thymus, a gland in the upper chest. There the lymphocytes become specialized as T lymphocytes, or **T cells.** Both B cells and T cells eventually make their way via the blood to the lymph nodes, spleen, and other lymphatic organs. (You may recall that the Epstein-Barr virus can remain dormant inside B cells for a long time;

thus, microscopic examination for unusually shaped "a typical lymphocytes" is one diagnostic test for mono).

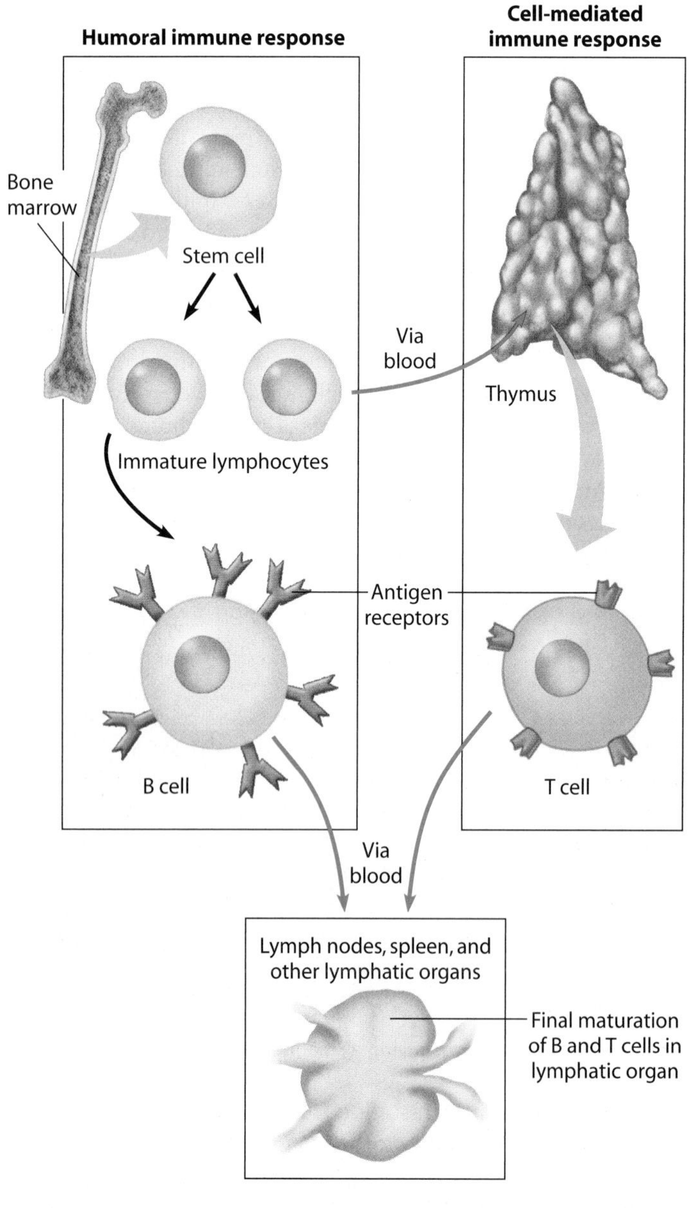

The B cells and T cells of the acquired immune response mount a dual defense. The first type of defense is called the **humoral immune response.** The humoral immune response involves the secretion of antibodies by B cells into the blood and lymph. (The humoral response is so named because blood and lymph were long ago called body "humors".) The humoral system defends primarily against bacteria and viruses present in body fluids. The humoral immune response can be passively transferred by injecting blood plasma (containing antibodies) from an immune individual into a non-immune individual. As we will see in Module 24.9, antibodies mark invaders by binding to them. The resulting antigen-antibody complexes are easily recognized for destruction and disposal by phagocytic cells.

The second type of acquired immunity, produced by T cells (Figure 24.5B), is called the **cell-mediated immune response.** As its name implies, this defensive system results from the action of defensive *cells* (rather than the defensive proteins of the humoral response). Certain T cells attack body cells that are infected with bacteria or viruses. Other T cells function indirectly by promoting phagocytosis by other white blood cells and by stimulating B cells to produce antibodies. Thus, T cells play a part in both the cell-mediated and humoral immune responses.

When a T cell develops in the thymus or a B cell develops in bone marrow, certain genes in the cell are turned on. This causes the cell to synthesize molecules of a specific protein, which are then incorporated into the plasma membrane. As indicated in Figure 24.5A, these protein molecules stick out from the cell's surface. The molecules are **antigen receptors,** capable of binding one specific type of antigen. Each T or B cell has about 100,000 antigen receptors, and all the receptors on a single cell are identical—they all recognize the same antigen. In the case of a B cell, the receptors are almost identical to the particular antibody that the B cell will secrete. Once a B cell or T cell has its surface proteins in place, it can recognize a specific antigen and mount an immune response against it. One cell may recognize an antigen on the mumps virus, for instance, while another detects a particular antigen on a tetanus-causing bacterium.

We see in Figure 24.5A that after the B cells and T cells have developed their antigen receptors, these lymphocytes leave the bone marrow and thymus and move via the bloodstream to the lymph nodes, spleen, and other parts of the lymphatic system. In these organs, many B and T cells take up residence and encounter infectious agents that have penetrated the body's outer defenses. Because lymphatic capillaries extend into virtually all the body's tissues, bacteria or viruses infecting nearly any part of the body eventually enter the lymph and are carried to the lymphatic organs. As we will describe in Module 24.7, when a B or T cell within a lymphatic organ first confronts the specific antigen it is programmed to recognize, it differentiates further and becomes a fully mature component of the immune system.

Figure 24.5A The development of B cells and T cells

An enormous diversity of B cells and T cells develops in each individual. Researchers estimate that each of us has millions of different kinds—enough to recognize and bind virtually every possible antigen. A small population of each kind of lymphocyte lies in wait in our body, genetically programmed to recognize and respond to a specific antigen. Only a tiny fraction of the immune system's lymphocytes will ever be used, but they are all available if needed. It is as if the immune system maintains a huge standing army of soldiers, each made to recognize one particular kind of invader. The majority of soldiers never encounter their target and remain idle. But when an invader does appear, chances are good that some lymphocytes will be able to recognize it, bind to it, and call in reinforcements. We'll take a closer look at the different types of T cells in Modules 24.11 and 24.12.

🎧 MP3 Tutor *The Human Immune System*

❓ Contrast the targets of the humoral immune response with those of the cell-mediated immune response

■ The humoral immune response works against pathogens in the body fluids; the cell-mediated immune response attacks infected cells.

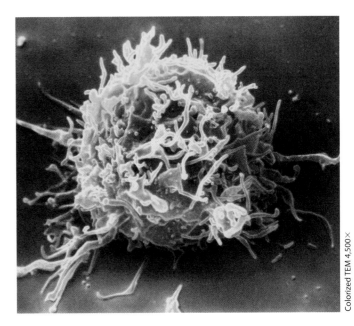

Figure 24.5B   A T cell

Colorized TEM 4,500×

24.6 Antigens have specific regions where antibodies bind to them

As molecules that elicit the acquired immune response, antigens usually do not belong to the host animal. Most antigens are proteins or large polysaccharides on the surfaces of viruses or foreign cells. Common examples are protein-coat molecules of viruses (such as the Epstein-Barr virus that causes mono), parts of the capsules and cell walls of bacteria, and macromolecules on the surface cells of other kinds of organisms, such as protozoans and parasitic worms. (Sometimes a particular microbe is called an antigen, but this usage is misleading because the microbe will almost always have several kinds of antigenic molecules.) Other sources of antigenic molecules include blood cells or tissue cells from other individuals (of the same species or a different species). Antigenic molecules are also found dissolved in body fluids; foreign molecules of this type include bacterial toxins and bee venom.

As shown in Figure 24.6, an antibody usually recognizes and binds to a small surface-exposed region of an antigen, called an **antigenic determinant** (also known as an epitope). An antigen-binding site, a specific region on the antibody molecule, recognizes an antigenic determinant by the fact that the binding site and antigenic determinant have complementary shapes, like an enzyme and substrate or a lock and key. An antigen usually has several different determinants (there are three in the diagram here), so different antibodies (two, in this case) can bind to the same antigen. A single antigen molecule may thus stimulate the immune system to make several distinct antibodies against it. Notice that each antibody molecule has two identical antigen-binding sites. We'll return to antibody structure in Module 24.8. But first, let's see how the body produces large quantities of antibodies and defensive cells in response to specific infections.

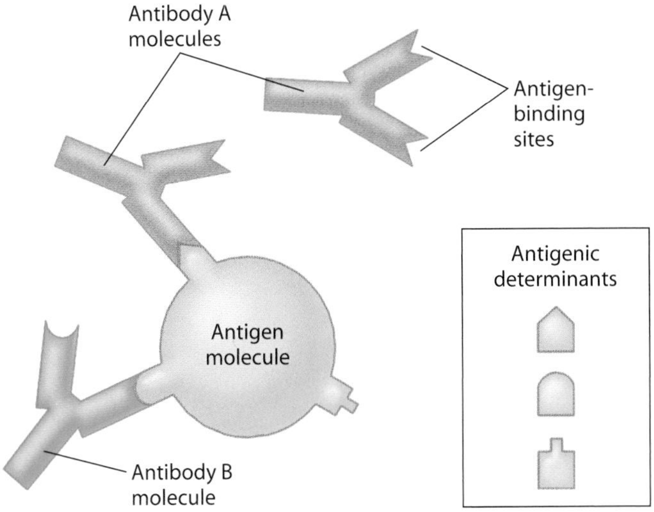

Antibody A molecules

Antigen-binding sites

Antigenic determinants

Antigen molecule

Antibody B molecule

Figure 24.6   The binding of antibodies to antigenic determinants

❓ Why is it inaccurate to refer to a pathogen, such as a virus, as an antigen?

■ It is inaccurate because antigens are not whole pathogens; they are molecules, which may be chemical components of a pathogen's surface. One pathogen may have several antigens.

Clonal selection musters defensive forces against specific antigens

The immune system's ability to defend against an enormous variety of antigens depends on a process known as **clonal selection.** Once inside the body, an antigen encounters a diverse pool of B and T lymphocytes. However, one particular antigen interacts only with the tiny fraction of lymphocytes bearing receptors specific to that antigen. Once activated by the antigen, these few "selected" cells proliferate, forming a clone (genetically identical population) of thousands of cells all specific for the stimulating antigen. This antigen-driven cloning of lymphocytes—clonal selection—is a vital step in the acquired immune response against infection.

The Steps of Clonal Selection Figure 24.7A indicates how clonal selection of B cells works in the humoral immune response. (A similar mechanism activates clonal selection for T cells in the cell-mediated immune response.) ❶ The row of three cells at the top of the figure represents a vast repertoire of B cells

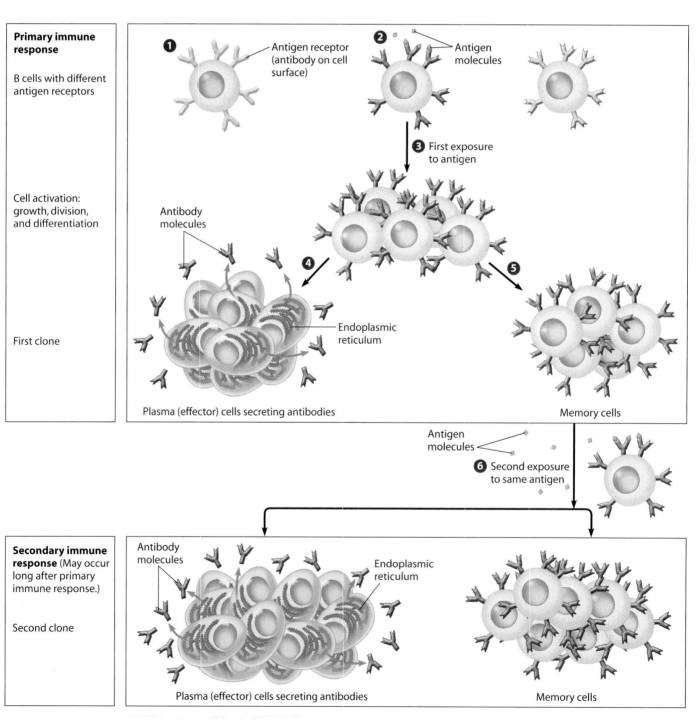

Figure 24.7A Clonal selection of B cells in the primary and secondary immune responses

in a lymph node. Notice that each lymphocyte has its own specific type of antigen receptor embedded in its surface. The cells' receptors are in place before they ever encounter an antigen.

The first time an antigen enters the body and is swept into a lymph node, ❷ antigenic determinants on its surface bind to the few B cells that happen to have complementary receptors. Other lymphocytes, without the appropriate binding sites, are not affected. Primed by the interaction with the antigen, ❸ the selected cell is activated: It grows, divides, and differentiates into two genetically identical yet physically distinct types of cells. Both newly produced types of cells are specialized for defending against the antigen that triggered the response. ❹ The first group of newly cloned cells are **effector cells,** which combat the antigen. Because the example in the figure involves B cells, the effector cells produced are **plasma cells.** Each plasma cell secretes antibody molecules, all of the same type. Each plasma cell makes as many as 2,000 copies of its antibody per second. (Plasma cells thus require large amounts of endoplasmic reticulum, a characteristic of cells actively synthesizing and secreting proteins.) The secreted antibodies circulate in the blood and lymphatic fluid, contributing to the humoral immune response. Although highly effective at combating infection, each effector cell lasts only 4 or 5 days before dying off.

❺ The second group of cells produced by the activated B cells are a smaller number of **memory cells,** which differ from effector cells in both appearance and function. In contrast to short-lived effector cells, memory cells may last for decades. They remain in the lymph nodes, poised to be activated by a second exposure to the antigen. In fact, in some cases, memory cells confer lifetime immunity, as they may after vaccination against such childhood diseases as mumps and measles and for those who contract mono. Steps 1–5 show the initial phase of acquired immunity, called the **primary immune response.** This phase occurs when lymphocytes are exposed to an antigen for the first time.

❻ When memory cells produced during the primary response are activated by a second exposure to the same antigen, they initiate the **secondary immune response.** This response is faster and stronger than the first. Another round of clonal selection ensues. The selected memory cells multiply quickly, producing a large second clone of lymphocytes that mount the secondary response. Like the first clone, the second clone includes effector cells that produce antibodies and memory cells capable of responding to future exposures to the antigen. In our example here with B cells, the secondary response produces very high levels of antibodies that, though they are short-lived, are often more effective against the antigen than those produced during the primary response.

The concept of clonal selection is so fundamental to understanding acquired immunity that it is worth restating: Each antigen, by binding to specific receptors, selectively activates a tiny fraction of lymphocytes; these few selected cells then give rise to a clone of many cells, all specific for and dedicated to eliminating the antigen that started the response. Thus, we see that the versatility of the acquired immune response depends on a great diversity of preexisting lymphocytes with different antigen receptors.

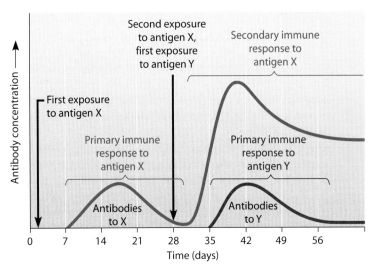

Figure 24.7B The two phases of the acquired immune response

Primary vs. Secondary Immune Response Now that we have seen how clonal selection works, let's take a look at the two phases of the acquired immune response in an individual. The blue curve in Figure 24.7B illustrates the difference between the two phases, triggered by two exposures to the same antigen. On the far left of the graph, you can see that the primary response does not start right away; it usually takes several days for the lymphocytes to become activated by an antigen (called X here) and form clones of effector cells. When the effector cell clone forms, antibodies start showing up in the blood, as the graph shows. During this delay, a stricken individual may become ill. The antibody level reaches its peak 2–3 weeks after initial exposure. As the antibody levels in the blood and lymph rise, the symptoms of the illness typically diminish and disappear. The primary response subsides as the effector cells die out.

The second exposure to antigen X (day 28 in the graph) triggers the secondary immune response. Notice that this secondary response starts faster than the primary response (typically in 2 to 7 days, versus 10 to 17 days). As mentioned, the secondary response is also of greater magnitude (produces higher levels of antibodies) and is more prolonged.

The red curve in Figure 24.7B illustrates the specificity of the immune response. If the body is exposed to a different antigen (Y), even after it has already responded to antigen X, it responds with another primary response, this one directed against antigen Y. The response to Y is not enhanced by the response to X; that is, acquired immunity is specific.

Although we have focused on the humoral immune response (produced by B cells) in this module, clonal selection, effector cells, and memory cells are features of the cell-mediated immune response (produced by T cells) as well. In the next several modules, we discuss the humoral immune response further. After that, we focus on how the cell-mediated arm of the immune system helps defend the body against pathogens.

? What is the immunological basis for referring to certain diseases, such as mumps, as *childhood* diseases?

■ One bout with the pathogen, which most often occurs during childhood, is usually enough to confer immunity for the rest of that individual's life.

24.8 Antibodies are the weapons of the humoral immune response

B cells are the "frontline warriors" of the humoral immune response. Plasma cells—the effector cells produced during clonal selection of B cells (as shown in Figure 24.7A)—make and secrete antibodies, the proteins that serve as molecular weapons of defense.

We have been using Y-shaped symbols to represent antibodies, and their shape actually does resemble a Y, as the antibody molecule in Figure 24.8A illustrates. Figure 24.8B is a simplified diagram explaining antibody structure. Each antibody molecule is made up of four polypeptide chains, two identical "heavy" chains and two identical "light" chains. In both figures, the parts colored in shades of pink represent the fairly long, heavy chains of amino acids that give the molecule its Y shape. Bonds (black lines in Figure 24.8B) at the fork of the Y hold these chains together. The two green regions in each figure are shorter chains of amino acids, the light chains. Each of the light chains is bonded to one of the heavy chains. As Figure 24.8A indicates, the bonded chains actually intertwine.

An antibody molecule has two related functions in the humoral immune response: to recognize and bind to a certain antigen, and to assist in neutralizing the antigen it recognizes. The structure of an antibody allows it to perform these functions. Notice in Figure 24.8B that each of the four chains of the molecule has a C (constant) region and a V (variable) region. At the tip of each arm of the Y, a pair of V regions forms an **antigen-binding site,** a region of the molecule responsible for the antibody's recognition-and-binding function. A huge variety in the three-dimensional shapes of the binding sites of different antibody molecules arises from a similarly large variety in the amino acid sequences in the V regions; hence the term *variable*. This structural variety accounts for the diversity of lymphocytes and gives the humoral immune system the ability to react to virtually any kind of antigen.

The tail of the antibody molecule, formed by the constant regions of the heavy chains, helps mediate the disposal of the bound antigen. Antibodies with different kinds of heavy-chain C regions are grouped into different classes. Humans and other mammals have five major classes of antibodies, called IgA, IgD, IgE, IgG, and IgM (where Ig stands for immunoglobulin, another name for antibody). Each of the five classes differs in where it's found in the body and how it works. However, all five classes of antibodies perform the same basic function: to mark invaders for elimination. We take a closer look at this process next.

? How is the specificity of an antibody molecule for an antigen analogous to an enzyme's specificity for its substrate?

■ Both antibodies and enzymes are proteins with binding sites of specific shape that recognize and bind to other molecules (antigens for antibodies, substrates for enzymes) with complementary shapes.

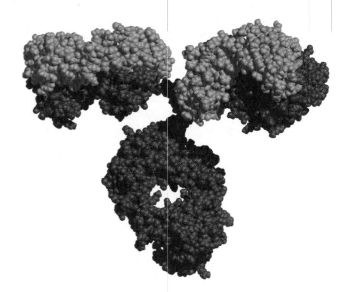

Figure 24.8A   A computer graphic of an antibody molecule

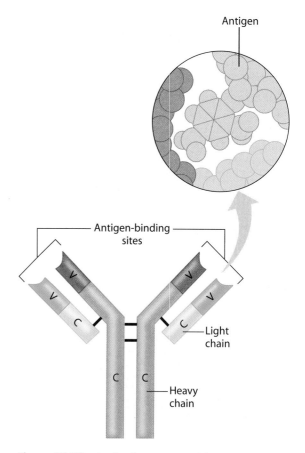

Figure 24.8B   Antibody structure with an antigen-binding site (inset)

The main role of antibodies in eliminating invading microbes or molecules is to mark the invaders. An antibody marks an antigen by combining with it to form an antigen-antibody complex. Weak chemical bonds between antigen molecules and the antigen-binding sites on antibody molecules hold the complex together.

As Figure 24.9 illustrates, it is the binding of antibodies to antigens that actually triggers mechanisms to neutralize or destroy an invader. Such a mechanism is called an effector mechanism. Several effector mechanisms are depicted in the figure. In viral neutralization, antibodies bind to certain surface proteins on a virus, thereby blocking its ability to infect a host cell. Similarly, antibodies may bind to surface molecules on bacterial cells. The bound antibodies enhance macrophage destruction of the bacteria.

Another effector mechanism is the agglutination (clumping together) of viruses, bacteria, or foreign eukaryotic cells. Because each antibody molecule has at least two binding sites, antibodies can hold a clump of invading cells together. Agglutination makes the cells easy for phagocytes to capture.

A third effector mechanism, precipitation, is similar to agglutination, except that the antibody molecules link *dissolved* antigen molecules together. This makes the antigen molecules precipitate out of solution as solids. Precipitation, like the other effector mechanisms discussed so far, enhances engulfment by phagocytes.

One of the most important effector mechanisms in the humoral immune response is the activation of the complement system (see Module 24.1) by antigen-antibody complexes. Activated complement system proteins (right side of figure) can attach to a foreign cell and poke holes in its plasma membrane, causing cell lysis (rupture).

Taken as a whole, this figure illustrates a fundamental concept of acquired immunity: All effector mechanisms involve a *specific* recognition-and-attack phase followed by a *nonspecific* destruction phase. Thus, the antibodies of the humoral immune response, which identify and bind to foreign invaders, work with innate defenses, such as phagocytes and complement, to form a complete defense system.

? How does acquired humoral immunity interact with the body's innate defense system?

■ Antibodies mark specific antigens for destruction, but it is usually the complement system, phagocytes, or other components of innate defense that destroy the antigens.

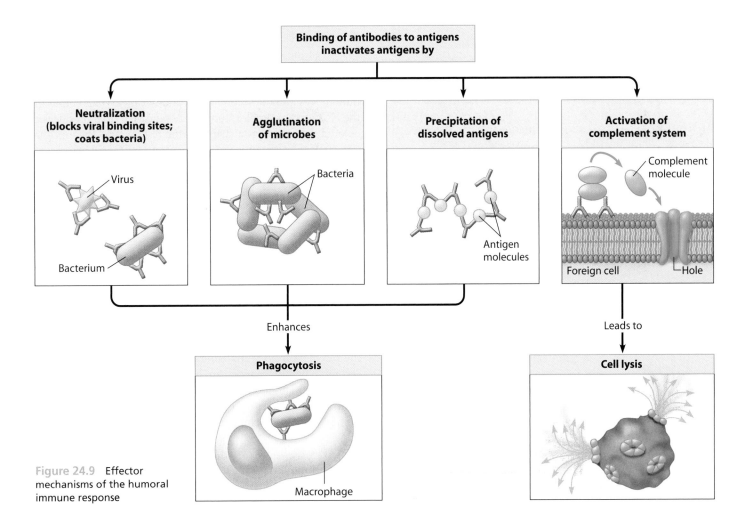

Binding of antibodies to antigens inactivates antigens by

Neutralization (blocks viral binding sites; coats bacteria)
Virus
Bacterium

Agglutination of microbes
Bacteria

Precipitation of dissolved antigens
Antigen molecules

Activation of complement system
Complement molecule
Foreign cell
Hole

Enhances

Leads to

Phagocytosis
Macrophage

Cell lysis

Figure 24.9 Effector mechanisms of the humoral immune response

24.10 Monoclonal antibodies are powerful tools in the lab and clinic

Because of their ability to tag specific molecules or cells, antibodies are widely used in laboratory research, clinical diagnosis, and the treatment of disease. In the original procedure for preparing antibodies, a small sample of antigen is injected into a rabbit or mouse. In response to the antigen, the animal produces antibodies, which can be collected directly from its blood. However, because the antigen usually has many different antigenic determinants, the result is polyclonal: a mixture of different antibodies produced by different clones of cells.

In the late 1970s, a technique was developed for making **monoclonal antibodies (mAb)**. The term *monoclonal* means that all the cells producing the antibodies are descendants of a single cell; thus, they all produce identical antibody molecules that are specific for the same antigenic determinant. Monoclonal antibodies are harvested from cell cultures rather than from live animals.

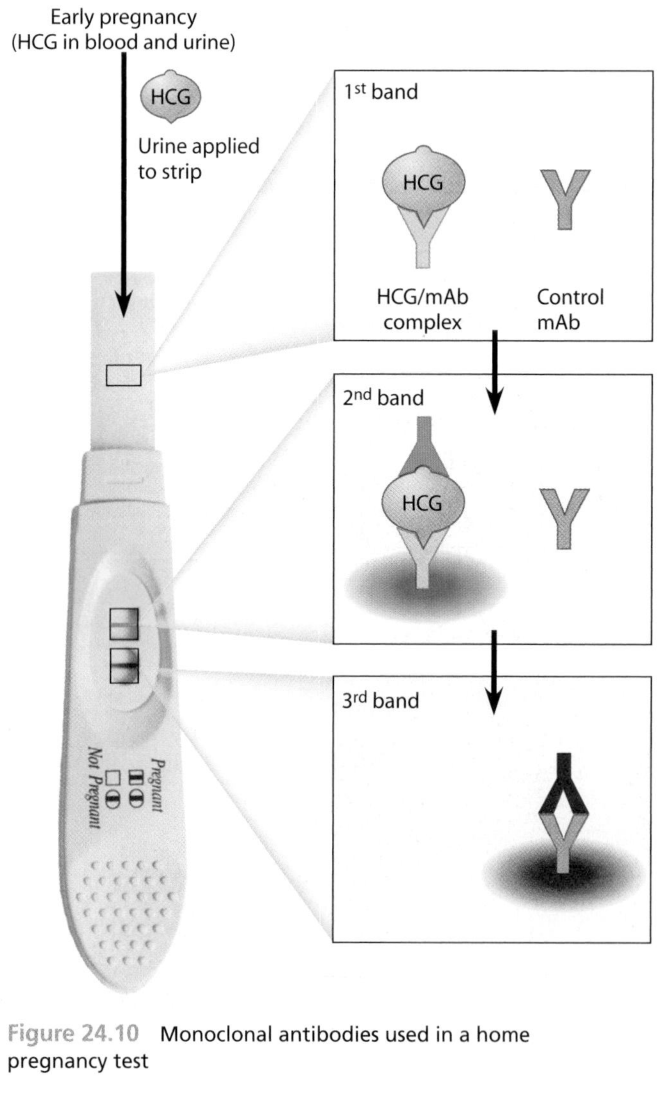

The trick to making monoclonal antibodies is the fusion of two cells to form a hybrid cell with a combination of desirable properties. First, an animal is injected with an antigen that will stimulate immune system B cells to make the desired antibody. At the same time, cancerous tumor cells, which can multiply indefinitely, are grown in a culture. The scientist then fuses a tumor cell with a normal antibody-producing B cell from the animal. The hybrid cell makes antibody molecules specific for a single antigenic determinant and is able to multiply indefinitely in a laboratory dish. Thus, large amounts of identical antibody molecules can be prepared.

As mentioned earlier, monoclonal antibodies are particularly useful for diagnosis. For example, monoclonal antibodies can be prepared that bind to a bacterium that causes a sexually transmissible disease or to the Epstein-Barr virus, which causes mono. If the antibodies have been labeled for easy detection (by a dye, for instance), they will reveal the presence of the bacterium or virus.

One common use of monoclonal antibodies (mAb's) is in home pregnancy tests. The most popular type of test uses mAb's to detect a hormone called human chorionic gonadotrophin (HCG; see Module 27.16) present in the urine of pregnant women. When urine is applied to the testing strip, it moves through a series of bands that contain different mAb's (Figure 24.10). In the first band, any HCG that is present in the urine binds to an mAb (orange) specific for HCG. This first band also contains a second mAb (purple) that serves as a control. As the fluid moves into the second band, another mAb (blue) binds to the HCG/mAb complex, forming a "sandwich" that produces a colored dye visible as a stripe on the testing strip (blue in the figure). If the woman is not pregnant, no HCG will be present in the urine, and no dye-producing complex will be formed. When the fluid moves into the third band, yet another type of monoclonal antibody (red) binds to the control mAb, producing a different colored stripe (red in the figure) that shows that the test strip is working.

Monoclonal antibodies also have great promise for use in the treatment of certain diseases, including cancer. For example, Herceptin, a genetically engineered monoclonal antibody, is used to treat a common form of aggressive breast cancer. The Herceptin antibody molecules act by binding to growth factor receptors that are present in excess on the cancer cells. The drug helps slow the progress of cancer by preventing the receptors from stimulating the cells to grow. Furthermore, certain types of cancer can be treated with tumor-specific monoclonal antibodies bound to toxin molecules. The toxin-linked antibodies carry out a precise search-and-destroy mission, selectively attaching to and killing tumor cells.

? How do home pregnancy tests based on monoclonal antibodies work?

Figure 24.10 Monoclonal antibodies used in a home pregnancy test

■ The monoclonal antibodies produce a color change in the test solution when they react with a hormone present in the female's urine during early pregnancy.

Helper T cells stimulate the humoral and cell-mediated immune responses

The antibody-producing B cells of the humoral immune response make up one army of the acquired immune response network. The humoral defense system identifies and helps destroy invaders that are in our blood, lymph, or interstitial fluid—in other words, outside our body cells. But many invaders, including all viruses, enter cells and reproduce there. It is the cell-mediated immune response produced by T cells that battles pathogens that have already entered body cells.

Whereas B cells respond to free antigens present in body fluids, T cells respond only to antigens present on the surfaces of the body's own cells. There are two main kinds of T cells. **Cytotoxic T cells** attack body cells that are infected with pathogens; we'll discuss these T cells in Module 24.12. **Helper T cells** play a role in many aspects of immunity. They help activate cytotoxic T cells and macrophages and even help stimulate B cells to produce antibodies. Other types of T cells include memory T cells.

Helper T cells interact with other white blood cells—including macrophages, B cells, and other types of immune cells—that function as **antigen-presenting cells.** All of the cell-mediated immune response and much of the humoral immune response depend on the precise interaction of antigen-presenting cells and helper T cells. This interaction activates the helper T cells, which can then go on to activate other cells of the immune system.

As its name implies, an antigen-presenting cell *presents* a foreign antigen to a helper T cell. Consider a typical antigen-presenting cell, a macrophage. As shown in Figure 24.11, ❶ the macrophage ingests a microbe (or other foreign particle) and breaks it into fragments—foreign antigens (orange). Then molecules of a special protein (green) belonging to the macrophage, which we will call a **self protein** (because it belongs to the body), ❷ bind the foreign antigens—**nonself molecules**—and ❸ display them on the cell's surface. (Each of us has a unique set of self proteins, which serve as identity markers for our body cells.) ❹ Helper T cells recognize and bind to the *combination* of a self protein and a foreign antigen—called a self-nonself complex—displayed on an antigen-presenting cell (the macrophage). This double-recognition system is like the system banks use for safe-deposit boxes: Opening your box requires the banker's key along with your specific key.

The ability of a helper T cell to recognize a unique self-nonself complex on an antigen-presenting cell depends on the receptors (purple) embedded in the T cell's plasma membrane. As indicated in the figure, a T cell receptor actually has two binding sites: one for antigen and one for self protein. The two binding sites enable a T cell receptor to recognize the overall shape of a self-nonself complex on an antigen-presenting cell. The immune response is highly specific because the receptors on each helper T cell can bind only one kind of self-nonself complex on an antigen-presenting cell.

The binding of a T cell receptor to a self-nonself complex activates the helper T cell. Several other kinds of signals can enhance this activation. For example, certain proteins secreted by the antigen-presenting cell, such as interleukin-1 (green arrow), diffuse to the helper T cell and stimulate it.

Activated helper T cells promote the immune response in several ways, with a major mechanism being the secretion of additional stimulatory proteins. One such protein, interleukin-2 (blue arrows), has three major effects. ❺ First, it makes the helper T cell itself grow and divide, producing both memory cells and additional active helper T cells. This positive-feedback loop amplifies the cell-mediated defenses against the antigen at hand. Second, interleukin-2 ❻ helps activate B cells, thus stimulating the humoral immune response. And third, ❼ it stimulates the activity of cytotoxic T cells, our next topic.

? How can one helper T cell stimulate both humoral and active immunity?

■ By releasing stimulatory proteins that activate both cytotoxic T cells and B cells

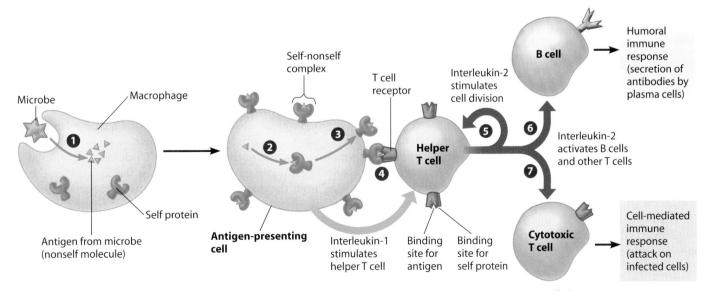

Figure 24.11   The activation of a helper T cell and its roles in immunity

24.12 Cytotoxic T cells destroy infected body cells

As you have just learned, two types of T cells participate in the cell-mediated immune response: helper T cells and cytotoxic T cells. Helper T cells activate cytotoxic T cells, the only T cells that actually kill infected cells.

Once activated, cytotoxic T cells identify infected cells in the same way that helper T cells identify antigen-presenting cells. An infected cell has foreign antigens—molecules belonging to the viruses or bacteria infecting it—attached to self proteins on its surface (Figure 24.12). Like a helper T cell, a cytotoxic T cell carries receptors that can bind with a self-nonself complex on the infected cell.

The self-nonself complex on an infected body cell is like a red flag to cytotoxic T cells that have matching receptors. As shown in the figure, ❶ the cytotoxic T cell binds to the infected

cell. The binding activates the T cell, which then synthesizes several new proteins that act on the bound cell, including one called perforin (🐾). ❷ Perforin is discharged and attaches to the infected cell's membrane, making holes in it. T cell enzymes (⋰) then enter the infected cell and promote its death by apoptosis. ❸ The infected cell dies and is destroyed.

Web Activity *Immune Responses*

> ? Compare and contrast the T cell receptor with the antigen receptor on the surface of a B cell.

■ Both receptors bind to a specific antigen, but the T cell receptor only recognizes that antigen when it is presented along with a "self" marker on the surface of one of the body's own cells.

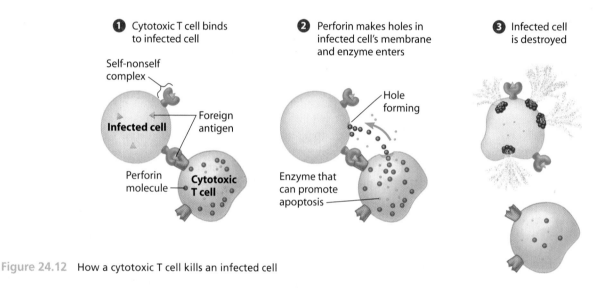

① Cytotoxic T cell binds to infected cell

Self-nonself complex
Infected cell
Foreign antigen
Perforin molecule
Cytotoxic T cell

② Perforin makes holes in infected cell's membrane and enzyme enters

Hole forming
Enzyme that can promote apoptosis

③ Infected cell is destroyed

Figure 24.12 How a cytotoxic T cell kills an infected cell

Connection

24.13 HIV destroys helper T cells, compromising the body's defenses

AIDS (acquired immunodeficiency syndrome) results from infection by **HIV, the human immunodeficiency virus.** Between 1981 and 2007, AIDS killed more than 27 million people. In 2007, 2.5 million people were newly infected with HIV, and over 2 million died, including over 300,000 children under 15. The vast majority of HIV infections and AIDS deaths occur in the nonindustrialized nations of southern Asia and sub-Saharan Africa. In some African nations, almost 40% of adults are HIV-positive. The virus is transmitted via body fluids carrying infected cells. Most often, HIV enters the body through small cuts or sores during sexual contact or by needles contaminated with infected blood.

Although HIV can infect a variety of cells (including brain cells and macrophages), it most often attacks helper T cells (Figure 24.13). As HIV depletes the body of helper T cells, both the cell-mediated and humoral immune responses are severely impaired. This drastically compromises the body's ability to fight infections.

How does HIV destroy helper T cells? Once HIV is in the bloodstream, proteins on the surface of the virus can bind to proteins on the surface of a helper T cell. Attached to the T cell, HIV may enter and begin to reproduce. Inside the host helper T cell, the RNA genome of HIV is reverse-transcribed, and the newly produced DNA is integrated into the T cell's genome (see Module 10.20). This viral genome can now direct the production of new viruses from inside the T cell, generating up to a thousand or more per day. Eventually, the host helper T cell dies from the damaging effects of virus reproduction and/or from virus-triggered apoptosis (programmed cell death).

After the new HIV is released into the bloodstream, HIV circulates, infecting and killing other helper T cells. As the number of T cells decreases, the body's ability to fight even the mildest infection is hampered, and AIDS eventually develops. It may take ten years or more for full-blown AIDS symptoms to appear after the initial HIV infection.

death sentence

Immune system impairment makes AIDS patients susceptible to **opportunistic infections** and cancers that can be successfully rebuffed by a person with a healthy immune system. For example, infection by a ubiquitous fungus called *Pneumocystis carinii* rarely occurs among healthy individuals. In a person with AIDS, however, infection by *P. carinii* can cause severe pneumonia and death. Likewise, Kaposi's sarcoma, a very rare skin cancer, used to be seen exclusively among the elderly or patients receiving chemotherapy treatments. It is now most frequently seen among AIDS patients. In fact, it was an unusual cluster of Kaposi's sarcoma and pneumocystis pneumonia cases in 1981 that first brought AIDS to the attention of the medical community. As mentioned in the opening essay, HIV infection can also turn a normally mild case of mono into a life-threatening opportunistic infection.

AIDS is incurable, although certain drugs can slow HIV reproduction and the progress of AIDS. One area of success has been in the development of drugs that can drastically lower transmission rates of HIV from mother to child. Unfortunately, these drugs are expensive and not available to all who need them. Combinations of drugs have proven effective at slowing the progress of AIDS, but the multidrug regimens are complicated, expensive, and may have debilitating side effects.

Until there is a vaccine or a cure, the best way to stop AIDS is to prevent the spread of HIV. This can only happen by educating people about how the virus is transmitted—through unprotected sex, through direct contact with blood, or from mother to baby. Safe sex behaviors, such as reducing promiscuity and using condoms, can save many lives. Although using a latex condom during sex does not completely eliminate the risk of transmitting HIV (or other sexually transmitted viruses), it does significantly reduce it. Practicing

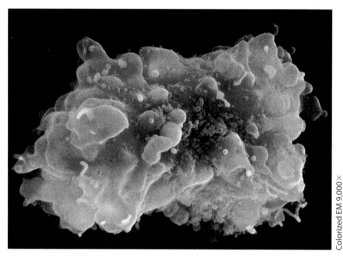

Colorized EM 9,000×

Figure 24.13 A human helper T cell (red) under attack by HIV (blue dots)

safe sex and avoiding intravenous drug use could save millions of lives.

Web Activity *HIV Reproductive Cycle*

Web Process of Science *Connection: What Causes Infections in AIDS Patients?*

Web Process of Science *Connection: Why Do AIDS Rates Differ Across the U.S.?*

? What is the function of the enzyme HIV reverse transcriptase?

■ HIV reverse transcriptase catalyzes the synthesis of a DNA version of HIV's RNA genome.

24.14 The rapid evolution of HIV complicates AIDS treatment

As HIV reproduces, mutational changes occur that can generate new strains of the virus. In fact, HIV has one of the fastest rates of mutation of any pathogen ever studied. This startling fact has led researchers to view the evolution of drug-resistant HIV strains as the number one obstacle to eradicating AIDS.

At one time, there was great hope that a "cocktail" of three anti-AIDS drugs, each of which attacks a different part of the HIV life cycle, could completely eliminate the virus in an infected person. It was hoped that virus strains resistant to one drug would be defeated by another. Such hope greatly underestimated the ability of HIV to evolve. Although people with access to such drugs do survive much longer and have a greatly improved quality of life, some HIV strains are resistent even to multidrug regimens. Thus, the virus is usually not totally eliminated from a patient's immune system; current

HIV treatment is thus only a temporary fix and not a cure for AIDS.

Disturbingly, drug-resistant HIV strains are now being found in newly infected patients. This demonstrates that HIV readily adapts through natural selection to a changing environment—one in which drug treatments are widely available. In other words, the presence of anti-AIDS drugs in the environment has created a selection pressure that favors the spread of drug-resistant strains. Thus, the battle continues, with medical science on one side and the constantly evolving HIV on the other.

? Why is it difficult to develop an AIDS vaccine?

■ Because HIV evolves rapidly

24.15　The immune system depends on our molecular fingerprints

As we have seen, the ability of lymphocytes to recognize the body's own molecules—that is, to distinguish *self* from *non-self*—enables our acquired immune response to battle foreign invaders without harming healthy body cells. Self proteins on cell surfaces are the key to this ability. Each person's cells have a particular collection of self proteins that provide the molecular "fingerprints" recognized by the immune system.

Each of us has two sets of self proteins on the surfaces of our cells. Class I proteins occur on almost all nucleated cells in the body. Class II proteins are found on only a few types of cells, including B cells, activated T cells, and macrophages. The particular collections of proteins in the two sets are specific to the individual in whom they are found, marking the body cells as "off-limits" to the immune system. As a result, our lymphocytes do not attack these molecules. How do lymphocytes "know" to avoid self cells? As lymphocytes mature, those with receptors that bind the body's own molecules are destroyed or deactivated, leaving only lymphocytes that react to foreign molecules.

The immune system not only distinguishes body cells from microbes, but also can tell your cells from those of other people. Genes at multiple chromosomal loci code for the main self proteins. The group of self-protein genes is called the **major histocompatibility complex**, or **MHC**.

Because there are hundreds of alleles in the human population for each MHC gene, it is virtually impossible for any two people (except identical twins) to have completely matching sets of self proteins.

The immune system's ability to recognize foreign antigens does not always work in our favor. For example, when a person receives an organ transplant or tissue graft, the person's immune system recognizes the MHC markers on the donor's cells as foreign and attacks them. To minimize rejection, doctors look for a donor with self proteins matching the recipient's as closely as possible. The best match is to transplant the patient's own tissue, as when a burn victim receives skin grafts removed from other parts of his or her body. Otherwise, identical twins provide the closest match, followed by nonidentical siblings. Sometimes doctors use drugs to suppress the immune response against the transplant. Unfortunately, these drugs may also reduce the ability to fight infections and cancer. A few drugs, however, such as cyclosporine, can suppress cell-mediated responses without crippling humoral immunity.

? In what sense is a cell's set of MHC surface markers analogous to a fingerprint?

■ The set of MHC ("self") markers is unique to each individual.

Disorders of the Immune System

24.16　Malfunction or failure of the immune system causes disease

Overall, our immune system is highly effective, protecting us against a vast array of potentially harmful invaders. When the immune system doesn't function properly, serious disease can result.

Autoimmune diseases result when the immune system goes awry and turns against some of the body's own molecules. In *systemic lupus erythematosus (lupus)*, for example, B cells make antibodies against a wide range of self molecules, such as histones and DNA released by the normal breakdown of body cells. Lupus is characterized by skin rashes, fever, arthritis, and kidney malfunction. *Rheumatoid arthritis* is another antibody-mediated autoimmune disease; it leads to damage and painful inflammation of the cartilage and bone of joints. In *insulin-dependent diabetes mellitus* (see Module 26.8), the insulin-producing cells of the pancreas are the targets of cytotoxic T cells. In *multiple sclerosis (MS)*, T cells react against the myelin sheath that surrounds many neurons (see Figure 28.2). People with MS experience a number of serious neurological abnormalities. Recent research suggests that Crohn's disease, a chronic inflammation of the digestive tract, may be caused by an autoimmune reaction against normal flora (bacteria) that inhabit the intestinal tract.

Most medicines currently available for treating autoimmune diseases either suppress immunity in general or are limited to the alleviation of specific symptoms. However, as research

scientists learn more about these diseases and about the normal operation of the immune system, they hope to develop more effective therapies.

In contrast to autoimmune diseases are a variety of defects called **immunodeficiency diseases.** Immunodeficient people lack one or more of the components of the immune system. This makes them susceptible to frequent and recurrent infections. In the rare congenital disease *severe combined immunodeficiency (SCID)*, both T cells and B cells are absent or inactive. People with SCID are extremely sensitive to even minor infections. Until recently, their only hope for survival was to live behind protective barriers (providing inspiration for "bubble boy" stories in the popular media) or to receive a successful bone marrow transplant that would continue to supply functional lymphocytes. Since the early 1990s, medical researchers have been testing a gene therapy for this disease, with some success (see Module 12.10).

Immunodeficiency is not always an inborn condition; it may be acquired later in life. For instance, *Hodgkin's disease,* a type of cancer that affects the lymphocytes, can depress the immune system. Radiation therapy and the drug treatments used against many cancers can have the same effect. Another well-known acquired immunodeficiency is AIDS (see Module 24.13).

There is growing evidence that physical and emotional stress can harm immunity. Hormones secreted by the adrenal glands during stress affect the numbers of white blood cells and may suppress the immune system in other ways. The association between emotional stress and immune function also involves the nervous system. Some neurotransmitters secreted when we are relaxed and happy may enhance immunity. In one study, college students were examined just after a vacation and again during final exams. Their immune systems were impaired in various ways during exam week; for example, interferon levels were lower. These and other observations indicate that general health and state of mind affect immunity.

? What is a probable side effect of autoimmune disease treatments that suppress the immune system?

■ Lowered resistance to infections

24.1 Allergies are overreactions to certain environmental antigens

Allergies are hypersensitive (exaggerated) responses to antigens in our surroundings. Antigens that cause allergies are called **allergens.** Protein molecules on pollen grains, on the feces of tiny mites that live in house dust, and in animal dander (shed skin cells) are common allergens. Many people who are allergic to cats and dogs are actually allergic to proteins in the animal's saliva that get deposited on the fur when the animal licks itself. Allergic reactions typically occur very rapidly and in response to tiny amounts of an allergen. A person allergic to cat or dog saliva, for instance, may react to a few molecules of the allergen in a matter of minutes. Allergic reactions can occur in many parts of the body, including the nasal passages, bronchi, and skin. Symptoms may include sneezing, runny nose, coughing, wheezing, and itching.

The symptoms of an allergy result from a two-stage reaction sequence outlined in Figure 24.17. The first stage, called sensitization, occurs when a person is first exposed to an allergen—pollen, for example. ❶ After an allergen enters the bloodstream, it binds to B cells with complementary receptors. ❷ The B cells then proliferate through clonal selection and secrete large amounts of antibodies to this allergen. ❸ Some of these antibodies attach to receptor proteins on the surfaces of **mast cells,** body cells that produce histamine and other chemicals that trigger the inflammatory response (see Module 24.2).

The second stage of an allergic response begins when the person is exposed to the same allergen later. The allergen enters the body and ❹ binds to the antibodies attached to mast cells. ❺ This causes the mast cells to release histamine, which triggers the allergic symptoms. As in inflammation, histamine causes blood vessels to dilate and leak fluid and so causes nasal irritation, itchy skin, and tears. **Antihistamines** are drugs that interfere with histamine's action and give temporary relief from an allergy.

Allergies range from seasonal nuisances to severe, life-threatening responses. **Anaphylactic shock** is an especially dangerous type of allergic reaction. Some people are extremely sensitive to certain allergens, such as bee venom, penicillin, or allergens in peanuts or shellfish. Any contact with these allergens makes their mast cells release inflammatory chemicals very suddenly. As a result, their blood vessels dilate abruptly, causing a rapid, potentially fatal drop in blood pressure. Fortunately, anaphylactic shock can be counteracted with injections of the hormone epinephrine.

? How do antihistamines relieve allergy symptoms?

■ By blocking the action of histamine, which produces the symptoms of the inflammatory response

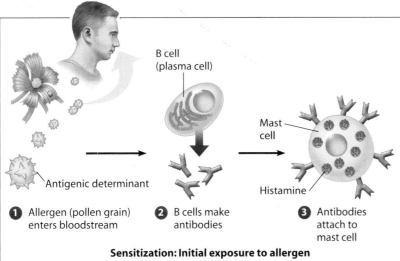

① Allergen (pollen grain) enters bloodstream
② B cells make antibodies
③ Antibodies attach to mast cell

B cell (plasma cell)
Mast cell
Antigenic determinant
Histamine

Sensitization: Initial exposure to allergen

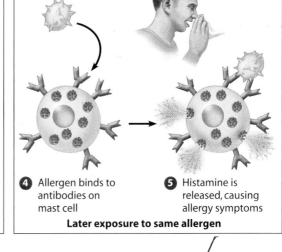

④ Allergen binds to antibodies on mast cell
⑤ Histamine is released, causing allergy symptoms

Later exposure to same allergen

Figure 24.17 The two stages of an allergic reaction

Chapter Review

Reviewing the Concepts

Innate Defenses Against Infection (24.1–24.3)

Innate defenses against infection are found in all animals and include the skin, mucous membranes, phagocytic cells, and antimicrobial proteins (**24.1**). Tissue damage triggers the inflammatory response, which can disinfect tissues and limit further infection (**24.2**). The lymphatic system is a network of vessels and organs. The vessels collect fluid from body tissues and return it as lymph to the blood. Lymph organs, such as the spleen and lymph nodes, are packed with white blood cells that fight infections (**24.3**).

Acquired Immunity (24.4–24.15)

Acquired immunity counters specific invaders by responding to antigens. Infection or vaccination triggers active immunity, which allows the immune system to "remember" an invader. We can also temporarily acquire passive immunity (**24.4**). Two kinds of lymphocytes carry out the acquired immune response:

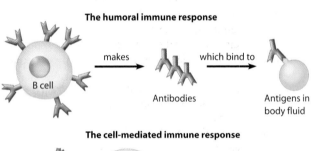

The humoral immune response

B cell — makes → Antibodies — which bind to → Antigens in body fluid

The cell-mediated immune response

T cell — Infected body cell — Self-nonself complex

Millions of kinds of B cells and T cells, each with different membrane receptors, wait in the lymphatic system, where they may respond to invaders (**24.5**). Antigen molecules have specific regions, called antigenic determinants, to which antibodies bind (**24.6**).

Clonal selection. When an antigen enters the body, it activates only a small subset of lymphocytes, those with complementary receptors. The selected cells multiply into clones of short-lived effector cells specialized for defending against the antigen that triggered the response and into memory cells, which confer long-term immunity. Activated by subsequent exposure to the antigen, memory cells mount a rapid and massive secondary immune response (**24.7**).

The humoral immune response. Antibodies are secreted by plasma (effector) B cells into the blood and lymph. An antibody has antigen-binding sites specific to the antigenic determinants that elicited its secretion (**24.8**). Antibodies promote antigen elimination through several mechanisms (**24.9**). Monoclonal antibodies are useful in research, diagnosis, and treatment of certain cancers (**24.10**).

The cell-mediated immune response. Helper T cells and cytotoxic T cells are primarily responsible for the cell-mediated immune response, and helper T cells also stimulate the humoral response. In the cell-mediated immune response, an antigen-presenting cell displays a foreign antigen (a nonself molecule) and one of the body's own self proteins to a helper T cell. The helper T cell's receptors recognize the self-nonself complexes, and the interaction activates the helper T cell. In turn, the helper T cell can activate cytotoxic T cells and B cells (**24.11**). Cytotoxic T cells bind to infected body cells and destroy them (**24.12**). The AIDS virus attacks helper T cells, opening the way for opportunistic infections (**24.13**). The rapid evolution of HIV complicates AIDS treatment (**24.14**). The immune system normally reacts only against nonself substances, not against self. Transplanted organs may be rejected because these cells lack the unique "fingerprint" of the recipient's self proteins (**24.15**).

Disorders of the Immune System (24.16–24.17)

When immunity malfunctions. In autoimmune diseases, the immune system turns against the body's own molecules. In immunodeficiency diseases, immune components are lacking and frequent infections occur (**24.16**). Allergies are abnormal sensitivities to antigens (allergens) in the surroundings (**24.17**).

Connecting the Concepts

1. Complete this concept map to summarize the key concepts concerning the body's defenses.

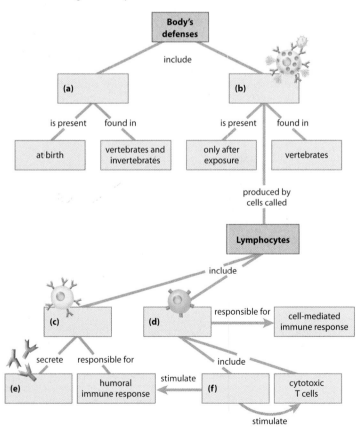

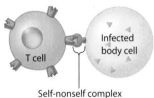

Testing Your Knowledge

Multiple Choice

2. Foreign molecules that elicit an immune response are called
   a. pathogens.
   b. antibodies.
   c. lymphocytes.
   d. histamines.
   e. antigens.

3. Which of the following is *not* part of the vertebrate innate defense system?
   a. natural killer cells
   b. antibodies
   c. interferons
   d. complement system
   e. inflammation

4. Which of the following best describes the difference in the way B cells and cytotoxic T cells deal with invaders?
   a. B cells confer active immunity; T cells confer passive immunity.
   b. B cells send out antibodies to attack; certain T cells can do the attacking themselves.
   c. T cells handle the primary immune response; B cells handle the secondary response.
   d. B cells are responsible for the cell-mediated immune response; T cells are responsible for the humoral immune response.
   e. B cells attack the first time the invader is present; T cells attack subsequent times.

5. The antigen-binding sites of an antibody molecule are formed from the molecule's variable regions. Why are these regions called variable?
   a. They can change their shapes on command to fit different antigens.
   b. They change their shapes when they bind to an antigen.
   c. Their specific shapes are unimportant.
   d. They have different shapes on antibodies to different antigens.
   e. Their sizes vary considerably from one antibody to another.

6. Cytotoxic T cells are able to recognize infected body cells because
   a. the infection changes the surfaces of infected cells.
   b. B cells help them.
   c. the infected cells produce antigens.
   d. infected cells release antibodies into the blood.
   e. helper T cells destroy them first.

Matching

7. Attacks infected body cells
8. Carries out the humoral immune response
9. Causes allergy symptoms
10. Phagocytic white blood cell
11. General name for a B or T cell
12. Carries out the secondary immune response
13. Cell most commonly attacked by HIV

a. lymphocyte
b. cytotoxic T cell
c. helper T cell
d. mast cell
e. macrophage
f. B cell
g. memory cell

Describing, Comparing, and Explaining

14. Describe (a) how AIDS is transmitted and (b) how immune system cells in an infected person are affected by HIV. Why is AIDS particularly deadly compared to other viral diseases? What are the most effective means of preventing HIV transmission?

15. What is inflammation? How does it protect the body? Why is inflammation considered part of the innate immune response?

16. Your roommate is rushed to the hospital after suffering a severe allergic reaction to a bee sting. After she is treated and released, she asks you (the local biology expert!) to explain what happened. She says, "I don't understand how this could have happened. I've been stung by bees before and didn't have a reaction." Suggest a hypothesis to explain what has happened to cause her severe allergic reaction and why she did not have the reaction after previous bee stings.

Applying the Concepts

17. Organ donation saves many lives each year. Even though some transplanted organs are derived from living donors, the majority come from patients who die but still have healthy organs that can be of value to a transplant recipient. Potential organ donors can fill out an organ donation card to specify their wishes. If the donor is in critical condition and dying, the donor's family is usually consulted to discuss the donation process. Generally, the next of kin must approve before donation can occur, regardless of whether the patient has completed an organ donation card. In some cases, the donor's wishes are overridden by a family member. Do you think that family members should be able to deny the stated intentions of the potential donor? Why or why not? Have you signed up to be an organ donor? Why or why not?

18. There is great concern about the rate of HIV infection among teenagers. Schools in some large cities have instituted programs to make condoms available to students, along with counseling about safer sex. These plans have divided school boards and communities. Some citizens and church groups are opposed to giving condoms to students on the grounds that it might appear to encourage sexual activity. By contrast, many school and public health officials view the situation as a health issue rather than a moral issue. The heart of the controversy seems to be whether the schools should take such a direct role in this part of student life. What are the reasons for and against distribution of condoms? What do you think the school's role should be?

Answers to all questions can be found in Appendix 4.

For Practice Quizzes, BioFlix, MP3 Tutors, and Activities, go to www.mybiology.com.

Control of Body Temperature and Water Balance

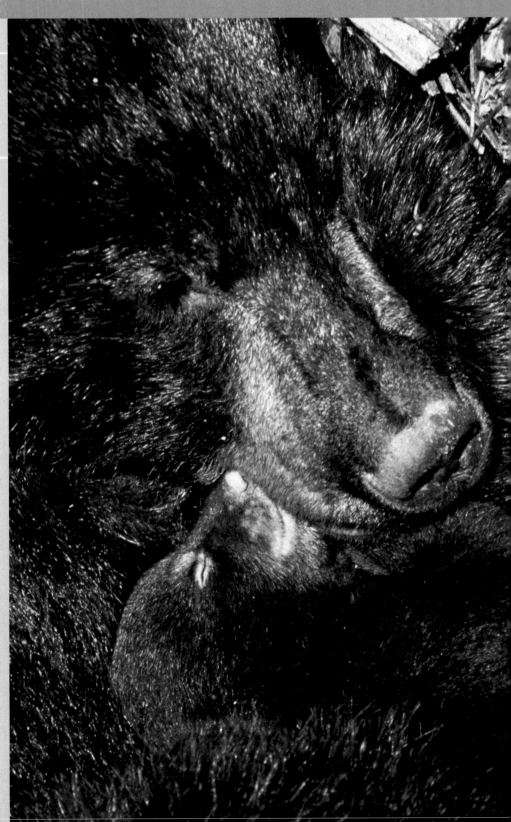

Chilling Out

When you think of hibernation, you may imagine a bear curled up in its den snoring the winter away. While bears do sleep a lot during winter, they do not actually hibernate. The low-activity state of a bear during winter is called dormancy. When dormant, a bear's body temperature is depressed from its normal value of 37°C to about 31–34°C. This is a lesser drop in temperature than true hibernators, whose body temperature may cool by 30°C. Also unlike a true hibernator, a dormant bear is easily awakened. Field biologists who have visited bears in their winter dens have been surprised—and distressed—to find how easily bears are roused and how annoyed they are at being awakened.

In contrast to a bear's relatively shallow sleep, other animals enter torpor, a state of reduced activity in which metabolic rate and body temperature decrease signficantly. Hibernation is a long-lasting torpor in cold weather. Unlike bears, many types of animals—including bats (above), squirrels and other rodents (such as the dormice at right), marsupials, and, most famously, the groundhog—do undergo true hibernation, entering a state of suspended animation from which they are not easily aroused. A hibernating animal's body temperature falls, breathing and heart rates decrease, and metabolism slows greatly. All of these physiological changes are evolutionary adaptations that help conserve energy as the animal taps into stored body fat during a long fast.

Most hibernating animals do so during winter, when food is scarce. Some animals do the opposite, entering torpor during the summer, protected from ambient heat and dry conditions. Long-lasting summertime torpor is called estivation. Animals that estivate include frogs, tortoises, lungfishes, and even one primate, the Madagascan fat-tailed dwarf lemur.

Other animals can lower their body temperature for just a few hours before resuming normal activity. For example, small birds, like chickadees and hummingbirds, feed during the day and enter torpor on cold nights; most bats feed at night and go into torpor during the day.

What changes occur in the body of an animal in a low-activity state? If you were to examine a dormant bear, you would find several physiological changes that aid homeostasis, the maintenance of nearly constant internal conditions despite fluctuations in the external environment (a concept first introduced in Module 20.14). First, the bear must thermoregulate. **Thermoregulation** is

the maintenance of internal temperature within narrow limits. Thermoregulation is aided by several adaptations. For example, a bear's stored body fat and dense fur provide superb insulation, its habit of curling up in its den helps keep heat loss to a minimum, and reduced blood flow to the bear's extremities decreases heat loss and maintains higher temperatures in its head and torso.

Besides regulating body temperature, dormant bears face several other challenges. While dormant, bears do not eat or drink, expel solid waste, or urinate. Their bodies compensate for this through **osmoregulation,** the control of the gain and loss of water and solutes, and the control of **excretion,** the disposal of nitrogen-containing wastes. A bear also has adaptations that aid in these processes. For example, dormant bears can metabolize nitrogen-containing wastes that accumulate in the bloodstream, converting them to harmless forms. This adaptation removes the need to urinate.

In this chapter, we explore the three homeostatic mechanisms mentioned in this chapter introduction: thermoregulation, osmoregulation, and excretion. We will see that most animals, like the bear, can survive fluctuations in the external environment because homeostatic control mechanisms such as these reduce large environmental fluctuations to small changes within a narrow range in the internal environment. ■ ■ ■

25.1 An animal's regulation of body temperature helps maintain homeostasis

The first homeostatic mechanism that we will examine is thermoregulation, the process by which animals maintain an internal temperature within a tolerable range. Thermoregulation is critical to survival because most of life's processes are sensitive to changes in body temperature. Each species of animal has an optimal temperature range. Thermoregulation helps keep body temperature within that range, enabling cells to function optimally even when the external temperature fluctuates.

Internal metabolism and the external environment provide the sources of heat for thermoregulation. Most birds and mammals are **endotherms,** meaning that they are warmed mostly by heat generated by their own metabolism. A few other reptiles, some fishes, and many insect species are also mainly

endothermic. In contrast, most amphibians, lizards, many fishes, and most invertebrates are **ectotherms,** meaning that they gain most of their heat from external sources.

Keep in mind, though, that endothermic and ectothermic modes of thermoregulation are not mutually exclusive. For example, a bird is mainly endothermic, but it may warm itself in the sun on a cold morning much as an ectothermic lizard does. In the following two modules, we'll examine some of the ways that animals regulate their body temperatures.

? A lizard basking on a warm rock is an example of an _____. Why?

ectotherm; Lizards absorb most of their body heat from their surroundings.

25.2 Heat is gained or lost in four ways

An animal can exchange heat with the environment by four physical processes. *Conduction* is the transfer of thermal motion (heat) between molecules of objects that are in direct contact, as when an animal is physically touching an object in its environment. Heat is always transferred from an object of higher temperature to one of lower temperature. In Figure 25.2, heat conducted from the warm rock is elevating the lizard's body temperature (red arrows).

Convection is the transfer of heat by the movement of air or liquid past a surface. In the figure, a breeze removes heat from the lizard's tail (orange arrows) by convection.

Radiation is the emission of electromagnetic waves. Radiation can transfer heat between objects that are not in direct contact, as when an animal absorbs heat radiating from the sun (yellow arrows). The lizard also radiates some of its own heat into the external environment.

Evaporation is the loss of heat from the surface of a liquid that is losing some of its molecules as a gas. A lizard loses heat as moisture evaporates from its nostrils (blue arrow).

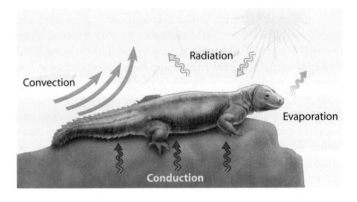

Figure 25.2   Mechanisms of heat exchange

? If you are sweating on a hot day and turn a fan on yourself, what two mechanisms contribute to your cooling?

Evaporation (sweating) and convection (fan moving air)

25.3 Thermoregulation involves adaptations that balance heat gain and loss

Different animals are adapted to different environmental temperatures. Within their optimal temperature range, endotherms and many ectotherms maintain a fairly constant internal temperature as the external temperature fluctuates. Five general categories of adaptations help animals thermoregulate.

Metabolic Heat Production   In cold weather, hormonal changes tend to boost the metabolic rate of birds and mammals, increasing their heat production. Simply moving around more or shivering also produces heat as a metabolic by-product

of the contraction of skeletal muscles. Honeybees survive cold winters by clustering together and shivering in their hive. The metabolic activity of all the bees together generates enough heat to keep the cluster alive.

Insulation   A major thermoregulatory adaptation in mammals and birds is insulation—hair (often called fur), feathers, or fat layers. Most land mammals and birds react to cold by raising their fur or feathers, which traps a thicker layer of air next to the warm skin, improving the insulation. (In humans, our muscles raise our hair in the cold, causing goose bumps, a vestige from our furry ancestors.) Aquatic mammals (such as seals) and aquatic birds (such as penguins) are insulated by a thick layer of fat called blubber.

Circulatory Adaptations   Heat loss can be altered by changing the amount of blood flowing to the skin. In a bird or mammal (and some ectotherms), nerves signal surface blood vessels to constrict or dilate, depending on the external temperature (see Module 20.15). When the vessels are constricted, less blood flows from the warm body core to the body surface, reducing the rate of heat loss. Conversely, dilation of surface blood vessels increases the rate of heat loss. Large, thin ears capable of radiating a lot of heat have evolved in elephants, helping to cool their large bodies (Figure 25.3A). Flapping increases heat dissipation by convection. In humans, dilation is what causes our faces to become flushed after rigorous exercise or a visit to a sauna.

Figure 25.3B illustrates a circulatory adaptation found in many birds and mammals. In **countercurrent heat exchange,** warm and cold blood flow in opposite (countercurrent) directions in two adjacent blood vessels. Warm blood (red) from the body core cools as it flows down the goose's leg or the dolphin's flipper. But the arteries carrying the warm blood are in close contact with veins (blue) conveying cool blood back toward the body core. As shown by the black arrows, heat passes from the warmer blood to the cooler blood along the whole length of these side-by-side vessels. By the time returning blood leaves the leg or the flipper, it is almost as warm as the body core. Thus, heat loss is minimal, even when the animal is standing on ice or swimming in frigid water.

Some endothermic fishes and sharks also have countercurrent exchange mechanisms. All fishes and sharks lose heat as blood passes through the gills. In large, powerful swimmers such as bluefin tuna, swordfish, and the great white shark, cold blood returning from the gills is transported in large vessels lying just under the skin. Small branches of these vessels deliver oxygenated blood to the deep muscles. Each branch runs side by side with a vessel carrying warm blood outward from the inner body. The resulting countercurrent heat exchange retains heat around the main swimming muscles and enables the vigorous, sustained activity of these endothermic animals.

Evaporative Cooling   Many animals live in places where thermoregulation requires cooling as well as warming. Some animals have adaptations that greatly increase evaporative cooling, such as panting, sweating, or even spreading saliva on body surfaces. Honeybees cool their hive during hot weather by transporting water into it and fanning with their wings, promoting heat loss by evaporation and convection. Humans sweat and other animals (such as dogs) lose heat as moisture evaporates from their nostrils and mouth during panting.

Behavioral Responses   Both endotherms and ectotherms control body temperature through behavioral responses. Some birds and butterflies migrate seasonally to more suitable climates. Other animals, such as desert lizards, bask in the sun when it is cold and find cool, damp areas or burrows when it is hot. Many animals bathe, which brings immediate cooling by

Figure 25.3A   Heat dissipation via the large ears of an elephant

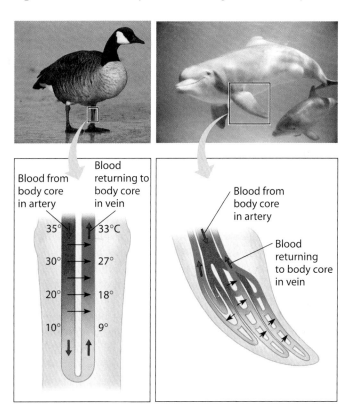

Figure 25.3B   Countercurrent heat exchange

convection and continues to cool for some time by evaporation. We humans dress for warmth.

While some animals can tolerate fluctuations in body temperature, few can withstand changes in the balance of water and solutes in body fluids. We consider osmoregulation next.

Web Process of Science *How Does Temperature Affect Metabolic Rate in* Daphnia?

? Compare countercurrent heat exchange with the countercurrent exchange of oxygen in fish gills. (See Module 22.3.)

■ In both cases, countercurrent exchange enhances transfer all along the length of a blood vessel—transfer of heat from one vessel to another in the case of a heat exchanger and transfer of oxygen between water and vessels in the case of gills.

| 25.4 | Animals balance the gain and loss of water and solutes through osmoregulation |

Osmoregulation, the second homeostatic control mechanism that we will discuss, involves balancing the uptake and loss of water and solutes, such as salt (NaCl) and other ions. Animal cells cannot survive a *net* water gain or loss: They swell and burst if there is a net uptake of water; they shrivel and die if there is a substantial net loss of water. Osmoregulation is based largely on regulating solutes because water follows the movement of solutes by osmosis. (Recall that osmosis is a net movement of water across a selectively permeable membrane from a solution with lower solute concentration to a solution with higher solute concentration—see Module 5.4.)

Some sea-dwelling animals—such as most marine invertebrates—have body fluids with a solute concentration equal to that of seawater. Called **osmoconformers,** such animals do not undergo a net gain or loss of water. However, the concentration of certain ions in their body fluids is different from that of seawater. For example, for cell membranes to function properly, the concentration of potassium ions (K^+) must be higher within cells than outside them. Because it takes energy to actively transport ions into or out of cells, even an osmoconformer expends some energy to maintain its ion concentrations.

Freshwater animals, land animals, and marine vertebrates have body fluids whose solute concentration differs from that of their environment. Therefore, they must regulate water loss or gain. Such animals are called **osmoregulators.**

The freshwater fish in Figure 25.4A has a much higher solute concentration in its internal fluids than that of fresh water. The fish constantly gains water by osmosis through its body surface, especially through its gills. It also loses salt by diffusion to its more dilute environment. How does a freshwater fish maintain water and solute balance? It does not drink water, its food helps supply some ions, and its gills actively take up salt. The fish's kidneys produce large amounts of **urine,** the waste material produced by its urinary system. By excreting dilute urine, the fish disposes of excess water and conserves solutes.

The saltwater fish in Figure 25.4B has the opposite osmoregulatory problem. Because its internal fluids are lower in total solutes than seawater, a saltwater fish loses water by osmosis across its body surfaces. It also gains salt both by diffusion and from the food it eats. The fish balances the water loss by drinking large amounts of seawater, and it pumps excess salt out through its gills. It also saves water by producing only small amounts of urine, in which it disposes of some excess ions.

What about land animals? They are osmoregulators and cannot directly exchange water with the environment by osmosis. Most terrestrial animals obtain water through food and drink while constantly losing water from moist respiratory surfaces, in urine and feces, and by evaporation across the skin.

The paramount osmoregulatory problem for a land animal is losing water and becoming dehydrated. For animals that have colonized land with great success, such as arthropods and

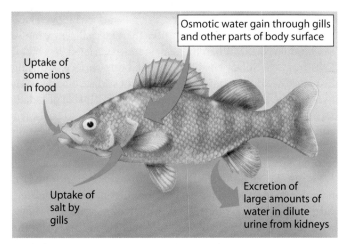

Figure 25.4A Osmoregulation in a perch, a freshwater fish

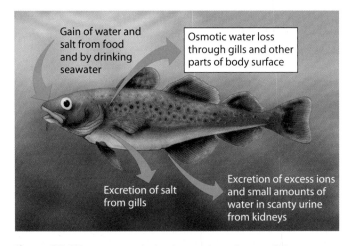

Figure 25.4B Osmoregulation in a cod, a saltwater fish

vertebrates, adaptations that prevent or reduce dehydration provide important evolutionary advantages. Insects have tough exoskeletons impregnated with waterproof wax. Most terrestrial vertebrates, including humans, have an outer skin formed of multiple layers of dead, water-resistant cells. Also essential to survival on land are adaptations that protect fertilized eggs and developing embryos from drying out. Many insects lay their eggs in moist areas, and the eggs of many species are surrounded by a tough, watertight shell. Likewise, the embryos of reptiles (including birds) and mammals develop in a water-filled amniotic sac surrounded by protective membranes. And, as we will see in Module 25.8, the kidney plays a major role in conserving water.

? Why are no freshwater animals osmoconformers?

■ Osmoconformers have solute concentrations equal to that of the environment. The body fluid of a freshwater osmoconformer would be too dilute to support life's processes.

25.5   A variety of ways to dispose of nitrogenous wastes have evolved in animals

Waste disposal is as important to homeostasis as osmoregulation. And because most wastes must be removed from the body dissolved in water, the type and quantity of an animal's wastes may have a large impact on its water balance. Metabolism produces a number of toxic by-products, such as the nitrogenous (nitrogen-containing) wastes that result from the breakdown of proteins and nucleic acids. An animal disposes of these metabolic wastes through the process of excretion, the third and final homeostatic control mechanism we will discuss.

The form of an animal's nitrogenous wastes depends on its evolutionary history and its habitat. As Figure 25.5 indicates, most aquatic animals dispose of their nitrogenous wastes as **ammonia.** Among the most toxic of all metabolic by-products, ammonia (NH_3) is formed when amino groups ($—NH_2$) are removed from amino acids and nucleic acids. Ammonia is too toxic to be stored in the body, but it is highly soluble in water and diffuses rapidly across cell membranes. If an animal is surrounded by water, ammonia readily diffuses out of its cells and body. Small, soft-bodied invertebrates, such as planarians (flatworms), excrete ammonia across their whole body surface. Fishes excrete it mainly across the gills.

Ammonia excretion does not work well for land animals. Ammonia does not diffuse readily into the air. Because it is so toxic, it must be transported and excreted in large volumes of very dilute solutions, and most terrestrial animals simply do not have access to that much water. Land animals convert ammonia into less toxic compounds, either urea or uric acid, that can be safely transported and stored in the body and released periodically by the urinary system. The disadvantage of excreting urea or uric acid is that the animal must use energy to produce these compounds from ammonia.

As shown in the figure, mammals, most amphibians, sharks, and some bony fishes excrete **urea.** Urea is produced in the vertebrate liver by a metabolic cycle that combines ammonia with carbon dioxide. The circulatory system transports urea to the kidneys. Urea is highly soluble in water. It is also some 100,000 times less toxic than ammonia, so it can be held in a concentrated solution in the body and disposed of with relatively little water loss. Some animals can switch between excreting ammonia and urea, depending on environmental conditions. Certain toads, for example, excrete ammonia (thus saving energy) when in water, but they excrete mainly urea (reducing excretory water loss) when on land.

Urea can be stored in a concentrated solution, but it still takes water to dispose of it. By contrast, land animals that excrete **uric acid** (birds and many other reptiles, insects, land snails, and a few amphibians living in deserts) avoid the water loss problem almost completely. As you can see in the figure, uric acid is a considerably more complex molecule than either urea or ammonia. Like urea, uric acid is relatively nontoxic. But unlike either ammonia or urea, uric acid is largely insoluble in water. In most cases, it is excreted as a semisolid paste. (The white material in bird droppings is mostly uric acid.) An animal must expend more energy to excrete uric acid than urea, but the higher energy cost is balanced by the great savings in body water.

An animal's type of reproduction also influences whether it excretes urea or uric acid. Urea can diffuse out of a shell-less amphibian egg or be carried away from a mammalian embryo in the mother's blood. However, the shelled eggs produced by birds and other reptiles are not permeable to liquids. In these animals, natural selection apparently favored the use of uric acid, which can be stored in the egg as a harmless solid left behind when the animal hatches.

In the next four modules, we'll focus on one specific example of a system that has evolved for osmoregulation and excretion: the human urinary system.

> **?** Aquatic turtles excrete both urea and ammonia; land turtles excrete mainly uric acid. What could account for this difference?
>
> ■ Although uric acid as a waste product evolved in terrestrial reptiles with their shelled eggs, natural selection favored the energy savings of ammonia and urea for aquatic turtles.

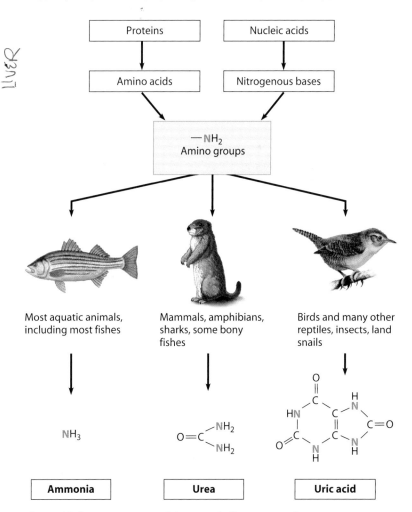

Figure 25.5   Nitrogen-containing metabolic waste products

The urinary system plays several major roles in homeostasis

Survival in any environment requires a precise balance between the competing demands of waste disposal and an animal's need for water. The urinary system plays a central role in homeostasis, forming and excreting urine while regulating the amount of water and ions in the body fluids.

In humans, the main processing centers of the urinary system are the two kidneys. Each is a compact organ, about the size of your fist, nearly filled with about 80 km of fine tubes (tubules) and an intricate network of blood capillaries. The human body contains only about 5 L of blood, but since this blood circulates repeatedly, about 1,100–2,000 L pass through the capillaries in our kidneys every day. From this enormous traffic of blood, our kidneys daily extract about 180 L of fluid, called **filtrate,** consisting of water, urea, and a number of

valuable solutes, including glucose, amino acids, ions, and vitamins. If we excreted all the filtrate as urine, we would lose vital nutrients and dehydrate rapidly. But our kidneys refine the filtrate, concentrating the urea and returning most of the water and solutes to the blood. In a typical day, we excrete only about 1.5 L of urine.

Figure 25.6 illustrates the "plumbing" plan and the blood supply of the human urinary system. Starting with the whole system in Part A, blood to be filtered enters each kidney via a renal artery, shown in red; blood that has been filtered leaves the kidney in the renal vein, shown in blue. Urine leaves each kidney through a duct called a **ureter** and passes into the **urinary bladder.** Periodically, the bladder empties during urination. Urine leaves the body through a tube called the

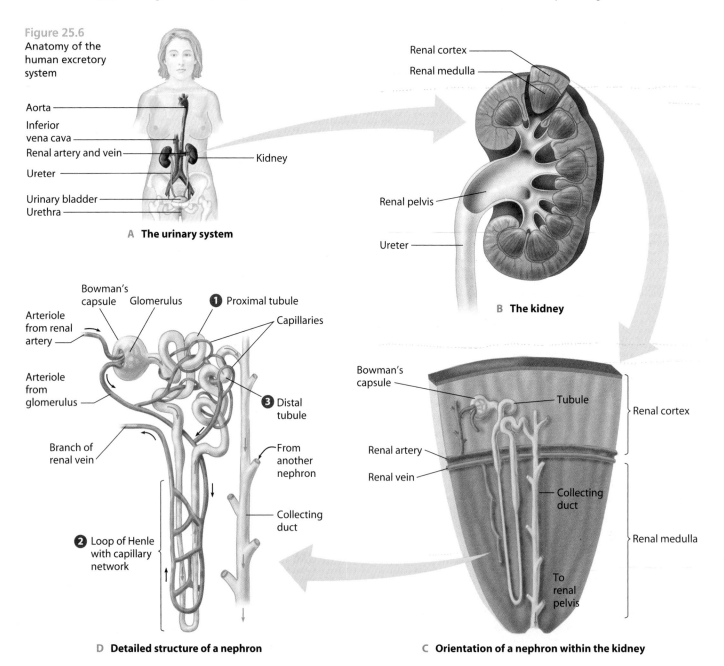

Figure 25.6
Anatomy of the human excretory system

Aorta
Inferior vena cava
Renal artery and vein
Ureter
Urinary bladder
Urethra
Kidney

A The urinary system

Renal cortex
Renal medulla
Renal pelvis
Ureter

B The kidney

Bowman's capsule   Glomerulus   **1** Proximal tubule
Arteriole from renal artery
Capillaries
Arteriole from glomerulus
3 Distal tubule
Branch of renal vein
From another nephron
2 Loop of Henle with capillary network
Collecting duct

D Detailed structure of a nephron

Bowman's capsule
Tubule
Renal cortex
Renal artery
Renal vein
Collecting duct
Renal medulla
To renal pelvis

C Orientation of a nephron within the kidney

urethra, which empties near the female vagina or through the male penis.

As shown in Part B, the kidney has two main regions, the **renal cortex** (outer layer) and the **renal medulla** (inner region). From the medulla, urine flows into a chamber called the renal pelvis, and from there into the ureter.

Each kidney contains about a million tiny functional units called **nephrons,** one of which is shown in Part C. A nephron consists of a tubule and its associated blood vessels. Performing the kidney's functions in miniature, the nephron extracts a tiny amount of filtrate from the blood and then refines the filtrate into a much smaller quantity of urine. Each nephron starts and ends in the kidney's cortex; some extend into the medulla, as in Part C. The receiving end of the nephron is a cup-shaped swelling, called **Bowman's capsule.** At the other end of the nephron is the **collecting duct,** which carries urine to the renal pelvis.

Part D shows a nephron in more detail, along with its blood vessels. Bowman's capsule envelops a ball of capillaries called the **glomerulus** (plural, *glomeruli*). The glomerulus and Bowman's capsule make up the blood-filtering unit of the nephron. Here, blood pressure forces water and solutes from the blood in the glomerular capillaries across the wall of Bowman's capsule and into the nephron tubule. This process creates the filtrate, leaving blood cells and large molecules such as plasma proteins behind in the capillaries.

The rest of the nephron refines the filtrate. The tubule has three sections: ❶ the **proximal tubule** (in the cortex); ❷ the **loop of Henle,** a hairpin loop carrying filtrate toward—and in some cases, into—the medulla and then back toward the cor-

tex; and ❸ the **distal tubule** (called distal because it is the most distant from Bowman's capsule). The distal tubule drains into a collecting duct, which receives filtrate from many nephrons. From the kidney's many collecting ducts, the processed filtrate, or urine, passes into the renal pelvis and then into the ureter.

The intricate association between blood vessels and tubules is the key to nephron function. As shown in Part D, the nephron has two distinct networks of capillaries. One network is the glomerulus, a finely divided portion of an arteriole that branches from the renal artery. Leaving the glomerulus, the arteriole re-forms and carries blood to the second capillary network, which surrounds the proximal and distal tubules. This second network functions with the tubule in refining the filtrate. Some of the vessels in this network parallel the loop of Henle, with blood flowing downward in one vessel and back up through another. Leaving the nephron, the capillaries converge to form a venule leading toward the renal vein.

With the structure of a nephron in mind, we focus next on what actually happens as our urinary system filters blood, refines the filtrate, and excretes urine.

🎧 **MP3 Tutor** *Kidney Function*

Web Activity *Structure of the Human Excretory System*

[?] Place these parts of a nephron in the order in which filtrate moves through them: proximal tubule, Bowman's capsule, collecting duct, distal tubule, loop of Henle.

■ Bowman's capsule, proximal tubule, loop of Henle, distal tubule, collecting duct

---

25.7 Overview: The key processes of the urinary system are filtration, reabsorption, secretion, and excretion

Our urinary system produces and disposes of urine in four major processes, shown in Figure 25.7. First of all, during **filtration,** water and virtually all other molecules small enough to be forced through the capillary wall enter the nephron tubule from the glomerulus.

Two processes then refine the filtrate. In **reabsorption,** water and valuable solutes, including glucose, salt, other ions, and amino acids, are returned to the blood from the filtrate. In **secretion,** substances in the blood are transported into the filtrate. When there is an excess of H^+ in the blood, for example, these ions are secreted into the filtrate, thus keeping the blood from becoming acidic. Secretion also eliminates certain drugs

and other toxic substances from the blood. In both reabsorption and secretion, water and solutes move between the tubule and capillaries by passing through the interstitial fluid (see Module 23.7).

Finally, in **excretion,** urine—the product of filtration, reabsorption, and secretion—passes from the kidneys to the outside via the ureters, urinary bladder, and urethra.

[?] Urine differs in composition from the fluid that enters a nephron tubule by filtration because of the processes of _____ and _____.

■ reabsorption . . . secretion

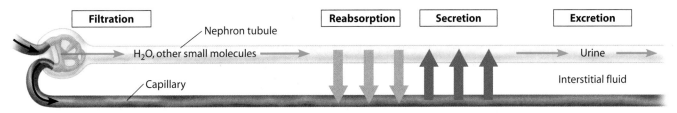

Figure 25.7 Major processes of the urinary system

25.8 Blood filtrate is refined to urine through reabsorption and secretion

Let's take a closer look at how a single nephron and collecting duct in the kidney produce urine from a blood filtrate that initially consists of both wanted substances (a large amount of water and a number of valuable solutes) and unwanted substances (waste molecules).

The broad arrows in Figure 25.8 indicate where reabsorption and secretion occur along the nephron tubule. The pink arrows pointing out of the tubule represent reabsorption, which may occur by active transport, passive diffusion, or osmosis. The blue arrows pointing into the tubule represent secretion. For simplicity, this figure omits the capillary network that surrounds the tubule.

The colored area of the figure represents the interstitial fluid, through which solutes and water move between the tubule and capillaries. The intensity of the color corresponds to the concentration of solutes in the interstitial fluid: The concentration is lowest in the cortex of the kidney and becomes progressively higher toward the inner medulla. We will see that it is by maintaining this solute gradient that the kidney can extract and save most of the water from the filtrate. All along the tubule, wherever you see water passing out of the filtrate into the interstitial fluid, the water is moving by osmosis. It does so because the solute concentration of the interstitial fluid exceeds that of the filtrate.

Let us first discuss the activities of the proximal and distal tubules. ❶ The proximal tubule actively transports nutrients such as glucose and amino acids from the filtrate into the interstitial fluid to be reabsorbed into the capillaries. NaCl (salt) is reabsorbed from both the proximal and distal tubules. As NaCl

is transported from the filtrate to the interstitial fluid, water follows by osmosis. Secretion of excess H^+ and reabsorption of HCO_3^- also occur at the proximal and distal tubules, helping to regulate the blood's pH. Potassium concentration in the blood is regulated by secretion of excess K^+ into the distal tubule. Drugs and poisons that were processed in the liver are secreted into the proximal tubule.

The loop of Henle and the collecting duct have one major function: water reabsorption. ❷ The long loop of Henle in the figure carries the filtrate deep into the medulla and then back to the cortex. The presence of NaCl and some urea in the interstitial fluid maintains the high concentration gradient in the medulla, which in turn increases water reabsorption by osmosis. Water leaves the tubule because the interstitial fluid in the medulla has a higher solute concentration than the filtrate. As soon as the water passes into the interstitial fluid, it moves into nearby blood capillaries and is carried away. This prompt removal is essential because the water would otherwise dilute the interstitial fluid surrounding the loop and destroy the concentration gradient necessary for water reabsorption.

Just after the filtrate rounds the hairpin turn in the loop of Henle, water reabsorption stops because the tubule there is impermeable to water. As the filtrate moves back toward the cortex, NaCl leaves the filtrate, first passively and then actively as the cells of the tubule pump NaCl into the interstitial fluid. It is primarily this movement of salt that creates the solute gradient in the interstitial fluid of the medulla.

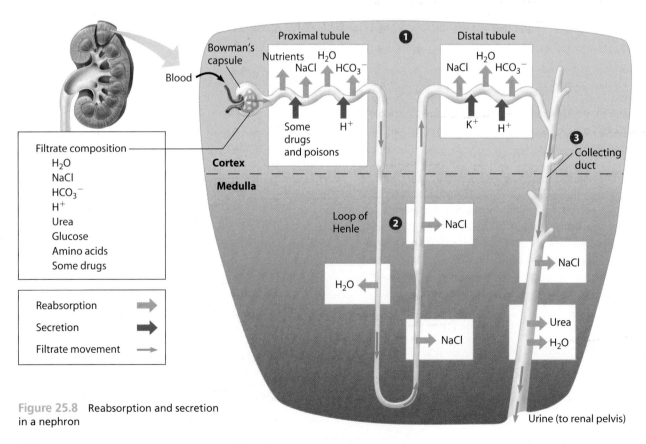

Figure 25.8 Reabsorption and secretion in a nephron

❸ Final refining of the filtrate occurs in the collecting duct. By actively reabsorbing NaCl, the collecting duct is important in determining how much salt is excreted in the urine. In the medulla, the collecting duct becomes permeable to urea and some leaks out, adding to the high concentration gradient in the interstitial fluid. As the filtrate moves through the collecting duct, more water is reabsorbed before the final product, urine, passes into the renal pelvis. In sum, the nephron returns much of the water that filters into it from the blood. This water conservation is one of the major functions of the kidneys.

Our kidneys also maintain a precise balance between water and solutes in our body fluids. When the solute concentration rises above a set point, a control center in the brain increases the blood level of **antidiuretic hormone (ADH),** which signals the nephrons to step up water reabsorption. When the solute concentration is diluted below the set point, as when we drink a lot of water, blood levels of ADH drop and water reabsorption is reduced, resulting in an increased discharge of dilute urine. (Increased urination is called diuresis, and it is because ADH opposes this state that it is called *anti*diuretic hormone.) Alcohol inhibits the release of ADH and can cause excessive urinary

water loss and dehydration, which may account for some of the symptoms of a hangover.

Still other hormones are involved in the kidney's regulation of blood volume and pressure. By adjusting both the flow of blood to the nephrons and the nephrons' reabsorption of Na^+ and water, these hormones alter the volume (and thus the pressure) of blood throughout the body.

We see that our kidneys' regulatory functions are controlled by an elaborate system of checks and balances. The coordination of all the body's regulatory systems by hormones is the subject of Chapter 26.

Web Activity *Nephron Function*

Web Activity *Control of Water Reabsorption*

Web Process of Science *What Affects Urine Production?*

? Some of the drugs classified as diuretics make the epithelium of the collecting duct less permeable to water. How would this affect kidney function?

■ The collecting ducts would reabsorb less water, and thus the diuretic would increase water loss in the urine.

Connection

25.9 Kidney dialysis can be a lifesaver

A person can survive with only one functioning kidney, but if both kidneys fail, the buildup of toxic wastes and the lack of regulation of blood pressure, pH, and ion concentrations will lead to certain death if untreated. Over 60% of kidney disease cases are caused by hypertension and diabetes, but the prolonged use of pain relievers, alcohol, and other drugs and medicines are also possible causes.

Knowing how the nephron works helps us understand how some of its functions can be performed artificially when the kidneys are damaged. Figure 25.9 illustrates a type of artificial kidney, called a dialysis machine. **Dialysis** means "separation" in Greek. Like the nephrons of the kidney, the machine sorts small molecules of the blood, keeping some and discarding others. The patient's blood is pumped from an artery through a series of tubes made of a selectively permeable membrane. The tubes are immersed in a dialyzing solution much like the interstitial fluid that bathes the nephrons. As the blood circulates through the tubing, urea and excess ions diffuse out. Needed substances, such as bicarbonate ions, diffuse from the dialyzing solution into the blood. The machine continually discards the used dialyzing solution as wastes build up.

Although dialysis is life sustaining, it is costly and time consuming (three times a week for 4–6 hours at a time). It also requires severe dietary and lifestyle restrictions. Many individuals benefit from a kidney transplant, from either a living donor (often a relative) or an organ donor who has just died. The waiting list for kidney transplants, unfortunately, is quite long.

? How would the composition of dialyzing solution compare with that of the patient's blood plasma?

■ Dialyzing solution has a solute concentration similar to interstitial fluid. The solution contains no urea, which allows urea from the patient's blood to diffuse into it.

Figure 25.9 Kidney dialysis

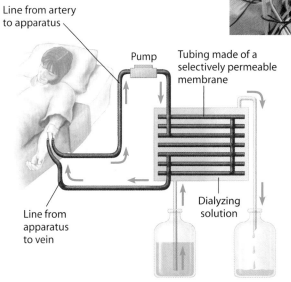

Line from artery to apparatus

Pump

Tubing made of a selectively permeable membrane

Line from apparatus to vein

Dialyzing solution

Fresh dialyzing solution

Used dialyzing solution (with urea and excess ions)

Chapter Review

Reviewing the Concepts

Thermoregulation (Introduction–25.3)

Thermoregulation maintains body temperature within a tolerable range. Osmoregulation balances the gain and loss of water and solutes. Excretion disposes of nitrogenous wastes (**Introduction**). **Endotherms** derive body heat mainly from their metabolism; **ectotherms** absorb heat from their surroundings (**25.1**). **Heat exchange** with the environment may occur by conduction, convection, radiation, and evaporation (**25.2**).

Adaptations for thermoregulation include increased metabolic heat production; insulation such as fur, feathers, and fat layers; circulatory adaptations such as adjusting blood flow to the skin or countercurrent heat exchange; evaporative cooling by sweating or panting; and behavioral responses such as moving to the sun or shade, migrating, or bathing (**25.3**).

Osmoregulation and Excretion (25.4–25.9)

Osmoregulation. Osmoconformers, such as many marine invertebrates, have the same internal solute concentration as seawater. Osmoregulators control their solute concentrations:

	Gain water	Lose water	Salt
Freshwater Fish	Osmosis	Excretion	Pump in
Saltwater Fish	Drinking	Osmosis	Excrete, pump out
Land Animal	Drinking, eating	Evaporation, urinary system	

The kidneys, waterproof skin, and reproductive and behavioral adaptations conserve water (**25.4**).

Nitrogenous wastes are toxic breakdown products of protein. Different adaptions for disposing of nitrogenous wastes have evolved in animals. Ammonia (NH_3) is poisonous but soluble and is easily disposed of by aquatic animals. Urea is less toxic and easier to store. Some land animals save water by excreting dry uric acid. Urea and uric acid take energy to produce (**25.5**).

Urea

The urinary system expels wastes and regulates water and ion balance. Nephrons, the functional units of kidneys, extract a filtrate from the blood and refine it to urine. Urine leaves the kidneys via the ureters, is stored in the urinary bladder, and is expelled through the urethra (**25.6**). The key processes of urine formation are filtration (blood pressure forces water and many small solutes into the nephron); reabsorption (valuable solutes are reclaimed from the filtrate); secretion (excess H^+ and toxins are added to the filtrate); and excretion of urine (**25.7**).

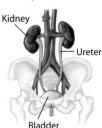

Filtrate to urine. Nutrients, salt, and water are reabsorbed from the proximal and distal tubules. Secretion of H^+ and reabsorption of HCO_3^- help regulate pH. High NaCl concentration in the medulla promotes reabsorption of water. Antidiuretic hormone (ADH) regulates the amount of water the kidneys excrete (**25.8**). Compensating for kidney failure, a dialysis machine removes wastes from the blood and maintains its solute concentration (**25.9**).

Connecting the Concepts

1. Complete this map, which presents the three main topics of this chapter.

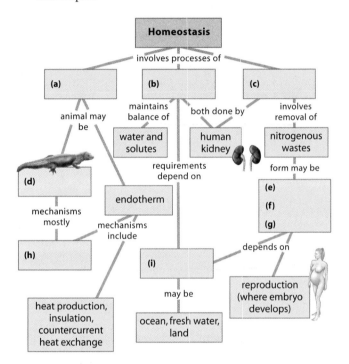

2. In this schematic of urine production in a nephron, label the four processes involved and list some of the substances that are moved in each process.

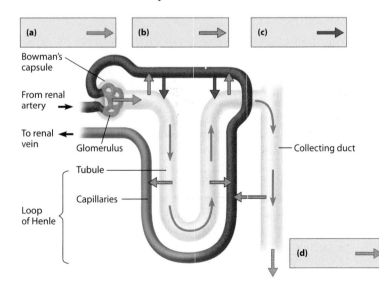

Testing Your Knowledge

Multiple Choice

3. The main difference between endotherms and ectotherms is
   a. how they conserve water.
   b. whether they are warm or cold.
   c. the source of most of their body heat.
   d. whether they live in a warm or cold environment.
   e. whether they maintain a fairly stable body temperature.

4. In each nephron of the kidney, the glomerulus and Bowman's capsule
   a. filter the blood and capture the filtrate.
   b. reabsorb water into the blood.
   c. break down harmful toxins and poisons.
   d. reabsorb ions and nutrients.
   e. refine and concentrate the urine for excretion.

5. As filtrate passes through the loop of Henle, salt is removed and concentrated in the interstitial fluid of the medulla. This high concentration enables nephrons to
   a. excrete the maximum amount of salt.
   b. neutralize toxins that might be found in the kidney.
   c. control the pH of the interstitial fluid.
   d. excrete a large amount of water.
   e. reabsorb water very efficiently.

6. Birds and insects excrete uric acid, whereas mammals and most amphibians excrete mainly urea. What is the chief advantage of uric acid over urea as a waste product?
   a. Uric acid is more soluble in water.
   b. Uric acid is a much simpler molecule.
   c. It takes less energy to make uric acid.
   d. Less water is required to excrete uric acid.
   e. More solutes are removed excreting uric acid.

7. Which process in the nephron is least selective?
   a. secretion
   b. reabsorption
   c. filtration
   d. active transport of salt
   e. passive diffusion of salt

8. You place your hand on a black car hood in bright summer sunshine. The car hood was heated by the process of _____ and your hand was warmed by the process of _____.
   a. conduction
   b. convection
   c. radiation
   d. evaporation

9. A freshwater fish would be expected to
   a. pump salt out through its gills.
   b. produce copious quantities of dilute urine.
   c. diffuse urea out through the gills.
   d. have scales and a covering of mucus that reduce water loss to the environment.
   e. do all of the above.

10. Which of the following is not an adaptation for reducing the rate of heat loss to the environment?
    a. feathers or fur
    b. reducing blood flow to surface blood vessels
    c. contraction of flight muscles before a moth takes off
    d. countercurrent heat exchange
    e. thick layer of fat

Matching

Match each of the following components of blood with what happens to it as the blood is processed by the kidney.

11. Water
12. Glucose
13. Plasma protein
14. Toxins or drugs
15. Red blood cell
16. Urea

a. passes into filtrate; almost all excreted in urine
b. remains in blood
c. passes into filtrate; mostly reabsorbed
d. secreted and excreted

Describing, Comparing, and Explaining

17. Compare the problems of water and salt regulation a salmon faces when it is swimming in the ocean and when it migrates into fresh water to spawn.

18. Can ectotherms have stable body temperatures? Explain.

Applying the Concepts

19. Assuming equal size, which of these organisms would produce the greatest amount of nitrogenous wastes? Explain.
    a. An endotherm or an ectotherm?
    b. A carnivore or an herbivore (assume both are endotherms)?

20. You are studying a large tropical reptile that has a high and relatively stable body temperature. How would you determine whether this animal is an endotherm or an ectotherm?

21. Riding by a lake in midwinter, you notice a small flock of geese standing on the ice. Imagine what it would be like for you to stand there with no boots or warm pants. Propose a hypothesis to explain why the birds' legs do not freeze. You may assume you have equipment for measuring temperatures in the birds' legs. What results would you expect if your hypothesis is correct?

22. The kidneys remove many drugs from the blood, and these substances show up in the urine. Some employers require a urine drug test at the time of hiring and/or at intervals during the term of employment. Why do some employers feel that drug testing is necessary? Do you think that passing a drug test is a valid criterion for employment? If so, for what types of jobs? Would you take a drug test to get or keep a job? Why or why not?

23. Kidneys were the first organs to be successfully transplanted. A donor can live a normal life with a single kidney, making it possible for individuals to donate a kidney to an ailing relative or even an unrelated individual. In some countries, poor people sell kidneys to transplant recipients through organ brokers. What are some of the ethical issues associated with organ commerce?

Answers to all questions can be found in Appendix 4.

For Practice Quizzes, BioFlix, MP3 Tutors, and Activities, go to www.mybiology.com.

chapter

26

Hormones and the Endocrine System

Gender Benders

Can you guess the common link in these bizarre occurrences from around the world?

- The alligator population of Florida's Lake Apopka plummeted after male juvenile alligators became "demasculinized" with abnormally small penises.

- One thousand miles north, most of the male smallmouth and largemouth bass in the Potomac River near Washington, DC, were found to be "intersex," having both male and female characteristics.

- A few thousand miles further north, researchers noted a significant decrease in the size and functioning of both male and female genitalia among arctic polar bears.

What's the link? Evidence suggests that all of these cases, in farflung locations and involving a wide variety of organisms, were caused by endocrine disruptors, environmental pollutants that interfere with the action of hormones. Hormones are chemicals that help regulate the body's functions such as energy use, metabolism, and growth. One class of hormones is the estrogens, sex hormones that help regulate the reproductive system. Each of the abnormalities described above has been linked to a group of endocrine disruptors called estrogen mimics. Estrogen mimics are chemical contaminants that disrupt the normal function of estrogens, often by binding to the same cellular protein receptors that estrogens bind to. Because the pollutants "look" like estrogens in a chemical sense, the physiology of an animal exposed to estrogen mimics changes as if their bodies had produced large amounts of these hormones.

How could estrogen disruptors get into the bodies of wild animals? It turns out that there are many sources of estrogen mimics, both natural (human estrogens from processed sewage and animal estrogens from farm manure) and synthetic (from chemicals found in pesticides, insecticides, health and beauty aids, pharmaceuticals, manufacturing chemicals, and flame retardants). In each case described above, the specific source of estrogen mimics was pinpointed. A heavy rain caused a wastewater pond at a chemical plant to overflow into the alligator habitat. The Potomac River water that was home to the bass tested positive for several known endocrine mimics, although no single one has been pinpointed as the primary cause. The polar bear diet contains large amounts of seal blubber, which tends to accumulate fat-soluble estrogen mimics.

Because estrogens act similarly in many types of animals, health officials fear that estrogen mimics may affect humans as well. The sites on the Potomac rich in intersex fish also provide drinking water to two states and the District of Columbia (although the water is extensively purified before consumption). And aquatic animals often serve as "canaries in a coal mine," providing early evidence of impending environmental problems. But so far, no clear link has been established between environmental pollutants and disruption of estrogen function in humans.

The concern about endocrine disruptors highlights how important chemical communication is to human health. The overarching role of hormones is to coordinate activities in different parts of the body, enabling the organ systems to function cooperatively. This chapter focuses on hormones and other kinds of chemical signals. However, our general theme is homeostasis, with an emphasis on how chemical signals maintain an animal body's dynamic steady state. We begin on the next page with a look at the organs that secrete hormones and the cells that respond to them. ■ ■ ■

26.1 Chemical signals coordinate body functions

Animals rely on many kinds of chemical signals to regulate their body activities. The estrogens are one kind of signal, a hormone (from the Greek *hormon,* to excite). An animal **hormone** is a chemical signal that is carried by the circulatory system (usually in the blood) and that communicates regulatory messages throughout the body. Hormones are made and secreted mainly by organs called **endocrine glands.** Collectively, all of an animal's hormone-secreting cells constitute its **endocrine system,** one of two bodily systems for communication and chemical regulation.

The other system of internal communication and regulation is the nervous system, the subject of Chapter 28. Unlike the endocrine system, which sends chemical signals through the bloodstream, the nervous system transmits electrical signals via nerve cells. These rapid messages control split-second responses. The flick of a frog's tongue catching a fly and the jerk of your hand away from a flame result from high-speed nerve signals. The endocrine system coordinates slower but longer-lasting responses. In some cases, the endocrine system takes hours or even days to act, partly because of the time it takes for hormones to be made and transported to all their target organs and partly because the cellular response may take time.

Hormones reach all cells of the body, and the endocrine system is thus especially important in controlling whole-body activities. For example, hormones coordinate responses to stimuli such as stress, dehydration, or low blood glucose levels. Hormones also regulate long-term developmental processes, such as growth, maturation, and reproduction.

Figure 26.1A sketches the activity of a hormone-secreting cell. Secretory vesicles in the endocrine cell are full of molecules of the hormone ( ). The endocrine cell secretes the molecules directly into the circulatory system. From there, hormones may travel to all parts of the body, but only certain types of cells, called **target cells,** are equipped to respond. A single hormone molecule may dramatically alter a target cell's metabolism by turning on or off the production of a number of enzymes. A tiny amount of a hormone can govern the activities of enormous numbers of target cells in a variety of organs. (In the next module, we'll look at *how* hormones trigger responses in their target cells.) A hormone is ignored by other types of cells (nontarget cells).

Chemical signals play a major role in coordinating the functioning of all animals. Hormones are the body's long-distance chemical regulators and convey information via the bloodstream to target cells throughout the body. Other chemical signals—**local regulators**—are secreted into the interstitial fluid and affect only nearby target cells. Still other chemical signals, called pheromones, carry messages between different individuals of a species, as in mate attraction.

While it is convenient to distinguish between the endocrine and nervous systems, in reality the lines between these two regulatory systems are blurred. In particular, certain specialized nerve cells called **neurosecretory cells** perform functions of both systems (Figure 26.1B). Like all nerve cells, neurosecretory cells conduct nerve signals, but they also make and secrete hormones into the blood.

A few chemicals serve both as hormones in the endocrine system and as chemical signals in the nervous system. Epinephrine (adrenaline), for example, functions in vertebrates as the "fight-or-flight" hormone (so called because it prepares the body for sudden action) and as a neurotransmitter. A neurotransmitter is a chemical that carries information from one nerve cell to another or from a nerve cell to another kind of cell that will react. When a nerve signal reaches the end of a nerve cell, it triggers the secretion of neurotransmitter molecules (Figure 26.1C). Unlike most hormones, however, neurotransmitters usually do not travel in the bloodstream. We discuss neurotransmitters further in Chapter 28. In the rest of this chapter, we'll explore the endocrine system and its hormones.

🎧 MP3 Tutor *Homeostasis and the Endocrine System*

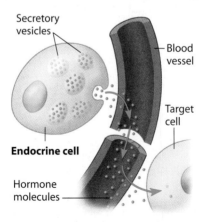

Figure 26.1A Hormone from an endocrine cell

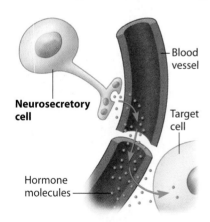

Figure 26.1B Hormone from a neurosecretory cell

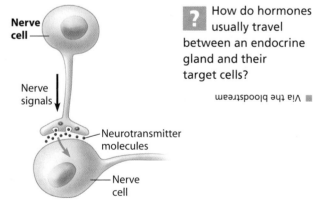

Figure 26.1C Neurotransmitter from a nerve cell

? How do hormones usually travel between an endocrine gland and their target cells?

■ Via the bloodstream

Hormones affect target cells by two main signaling mechanisms

Two major classes of molecules function as hormones in vertebrates (see Table 26.3). Amino-acid-derived hormones include proteins, peptides (small polypeptides containing 3–30 amino acids), and amines (molecules derived from amino acids). They are all hydrophilic (water-soluble). The second class of hormones, the steroids, are hydrophobic (lipid-soluble).

Regardless of their chemical structure, however, signaling by any of these molecules involves three key events: reception, signal transduction, and response. *Reception* of the signal occurs when a hormone binds to a specific receptor protein on or in the target cell. Each signal molecule has a specific shape that can be recognized by its target cell receptors. The binding of a signal molecule to a receptor protein triggers events within the target cell—*signal transduction*—that convert the signal from one form to another. The result is a *response,* a change in the cell's behavior. Cells that lack receptors for a particular chemical signal do not respond to that signal.

While both water- and lipid-soluble hormones carry out the three key steps outlined above, they do so by different mechanisms. We'll now take a closer look at how each type of hormone elicits cellular responses.

The receptors for most water-soluble hormones are embedded in the plasma membrane of target cells and project outward from the cell surface (Figure 26.2A). ❶ A hormone molecule (○) binds to the receptor protein, activating it. ❷ This initiates a signal transduction pathway, a series of changes in cellular proteins (relay molecules) that converts an extracellular chemical signal to a specific intracellular response. ❸ The final relay molecule (○) activates a protein (▲) that carries out the cell's response, either in the cytoplasm or in the nucleus. One hormone may trigger a variety of responses in target cells because the cells may contain different receptors for that hormone, diverse signal transduction pathways, or several proteins that can carry out the response.

While water-soluble hormones bind to receptors in the plasma membrane, lipid-soluble hormones bind to receptors *inside* the cell. **Steroid hormones,** such as the sex hormones testosterone and estrogen, are lipids made from cholesterol (see Module 3.9). Steroid hormones are small, nonpolar molecules that can diffuse through the phospholipid membranes of cells. As shown in Figure 26.2B, ❶ a steroid hormone (▽) enters a cell in this way. If the cell is a target cell, the hormone ❷ binds to a receptor protein in the cytoplasm or nucleus. Rather than triggering a signal transduction pathway, the hormone-receptor complex itself usually carries out the transduction of the hormonal signal: The complex acts as a transcription factor—a gene activator (see Module 11.5). ❸ The hormone-receptor complex attaches to specific sites on the cell's DNA in the nucleus. (These sites are enhancers; see Module 11.6.) ❹ The binding of the hormone-receptor complex to DNA stimulates transcription of certain genes into RNA, which is translated into new proteins. All steroid hormones act by turning genes on or off.

Because a hormone can bind to a variety of receptors in various kinds of target cells, different kinds of cells can respond differently to the same hormone. The main effect of epinephrine on heart muscle cells, for example, is cellular contraction, which speeds up the heartbeat; its main effect on liver and muscle cells, however, is glycogen breakdown, providing glucose (an energy source) to body cells.

? What are two major differences between the mechanisms of action of steroid and nonsteroid hormones?

■ (1) Steroid hormones bind to receptors inside the cell; most other hormones bind to plasma membrane receptors. (2) Steroid hormones always affect gene expression; other hormones have this or other effects.

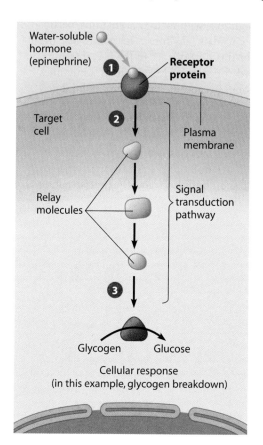

Figure 26.2A   A hormone that binds a plasma-membrane receptor

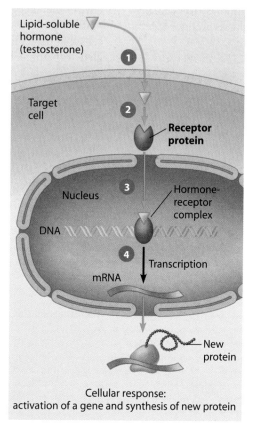

Figure 26.2B   A hormone that binds an intracellular receptor

26.3 Overview: The vertebrate endocrine system consists of more than a dozen major glands

Of the many glands of the vertebrate endocrine system, some, such as the thyroid and the pituitary gland, are endocrine specialists; their sole or main function is secreting hormones into the blood. Several other glands have both endocrine and nonendocrine functions. The pancreas, for example, secretes hormones that influence the level of glucose in the blood and also secretes digestive enzymes into the intestine (see Module 21.10). Still other organs, such as the stomach, are primarily nonendocrine but have some cells that secrete hormones.

Figure 26.3 shows the locations of the major human endocrine glands. Table 26.3 summarizes the actions of the main hormones they produce, as well as how the glands themselves are regulated. The table provides an overview of the human endocrine system, and you may wish to refer to it as we focus on the individual glands and their hormones in later modules. (Keep in mind, however, that this chapter covers only the major endocrine glands and hormones; there are other hormone-secreting structures—the heart, liver, and stomach, for example—and other hormones that will not be discussed in this chapter.)

Notice the distribution of the chemical classes of hormone (proteins, peptides, amines, and steroids) in Table 26.3. Only the sex organs and the cortex of the adrenal gland produce steroid hormones, the main type of hormone that actually enters target cells. Most of the endocrine glands produce water-soluble hormones, which generally bind to plasma membrane receptors and act via signal transduction.

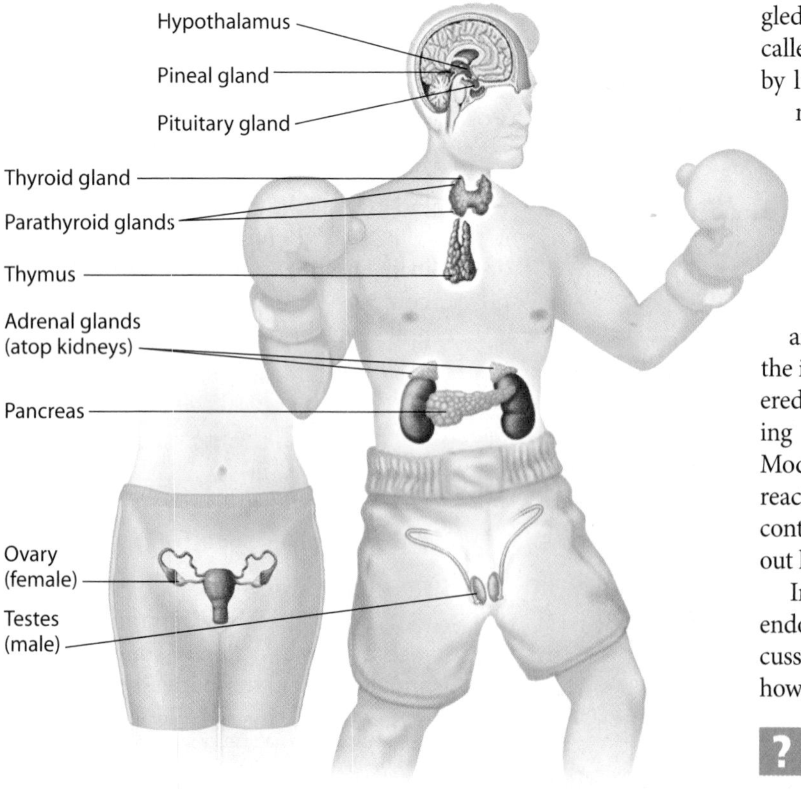

Hormones have a wide range of targets. Some, like the sex hormones, which promote male and female characteristics, affect most of the tissues of the body. Other hormones, such as glucagon from the pancreas, have only a few kinds of target cells (liver and fat cells for glucagon). Some hormones have other endocrine glands as their targets. For example, the pituitary gland produces thyroid-stimulating hormone, which promotes activity of the thyroid gland.

The close association between the endocrine system and the nervous system is apparent in both Figure 26.3 and Table 26.3. For example, the hypothalamus, which is part of the brain, secretes many hormones that regulate other endocrine glands, especially the pituitary. We'll explore structural and functional connections between the endocrine system and the nervous system further in Module 26.4.

Endocrine glands that we will not discuss at length in this chapter include the pineal gland and the thymus. Much remains to be learned about these organs. The **pineal gland** is a pea-sized mass of tissue near the center of the brain. The pineal gland synthesizes and secretes melatonin, a hormone that links environmental light conditions with biological rhythms, particularly the sleep-wake circadian rhythms. (In mammals, the cells that detect light for this purpose are in the eye, intermingled with the cells used for vision.) Melatonin is sometimes called "the dark hormone" because its production is inhibited by light. In most animals, melatonin production peaks in the middle of the night and then gradually falls. Some people ingest melatonin supplements as sleep aids, but the effectiveness of this treatment has not been established by scientists. In fact, we do not yet know exactly what effects melatonin has on body cells and the precise nature of how sleep-wake cycles are controlled.

The **thymus gland** lies under the breastbone in humans and is quite large during childhood. Not until the 1960s was the important role of the thymus in the immune system discovered. Thymus cells secrete several important hormones, including a peptide that stimulates the development of T cells (see Module 24.5). Beginning at puberty, when the immune system reaches its maturity, the thymus shrinks drastically. However, it continues to secrete its T-cell–stimulating hormones throughout life.

In the rest of this chapter, we'll explore several of the endocrine glands listed in Figure 26.3 and Table 26.3. The discussion will focus on the hormones produced by each organ and how they help to maintain homeostasis within the human body.

Hypothalamus
Pineal gland
Pituitary gland
Thyroid gland
Parathyroid glands
Thymus
Adrenal glands (atop kidneys)
Pancreas
Ovary (female)
Testes (male)

Figure 26.3 The major endocrine glands in humans

? Of the glands listed in Table 26.3, which ones secrete lipid-soluble hormones?

■ The testes, ovaries, and adrenal cortex

TABLE 26.3   MAJOR HUMAN ENDOCRINE GLANDS AND SOME OF THEIR HORMONES

Gland (module)	Hormone	Chemical Class	Representative Actions	Regulated by
Hypothalamus (26.4)	Hormones released by the posterior pituitary and hormones that regulate the anterior pituitary (see below)			
Pituitary gland (26.4) Posterior lobe (releases hormones made by hypothalamus)	Oxytocin	Peptide	Stimulates contraction of uterus and mammary gland cells	Nervous system
	Antidiuretic hormone (ADH)	Peptide	Promotes retention of water by kidneys	Water/salt balance
Anterior lobe	Growth hormone (GH)	Protein	Stimulates growth (especially bones) and metabolic functions	Hypothalamic hormones
	Prolactin (PRL)	Protein	Stimulates milk production	Hypothalamic hormones
	Follicle-stimulating hormone (FSH)	Protein	Stimulates production of ova and sperm	Hypothalamic hormones
	Luteinizing hormone (LH)	Protein	Stimulates ovaries and testes	Hypothalamic hormones
	Thyroid-stimulating hormone (TSH)	Protein	Stimulates thyroid gland	Thyroxine in blood; hypothalamic hormones
	Adrenocorticotropic hormone (ACTH)	Protein	Stimulates adrenal cortex to secrete glucocorticoids	Glucocorticoids; hypothalamic hormones
Pineal gland (26.3)	Melatonin	Amine	Involved in rhythmic activities (daily and seasonal)	Light/dark cycles
Thyroid gland (26.5–6)	Thyroxine (T_4) and triiodothyronine (T_3)	Amine	Stimulate and maintain metabolic processes	TSH
	Calcitonin	Peptide	Lowers blood calcium level	Calcium in blood
Parathyroid glands (26.5–6)	Parathyroid hormone (PTH)	Peptide	Raises blood calcium level	Calcium in blood
Thymus (26.3)	Thymosin	Peptide	Stimulates T cell development	Not known
Adrenal gland (26.9) Adrenal medulla	Epinephrine and norepinephrine	Amine	Increase blood glucose; increase metabolic activities; constrict certain blood vessels	Nervous system
Adrenal cortex	Glucocorticoids	Steroid	Increase blood glucose	ACTH
	Mineralocorticoids	Steroid	Promote reabsorption of Na^+ and excretion of K^+ in kidneys	K^+ in blood
Pancreas (26.7–8)	Insulin	Protein	Lowers blood glucose	Glucose in blood
	Glucagon	Protein	Raises blood glucose	Glucose in blood
Testes (26.10)	Androgens	Steroid	Support sperm formation; promote development and maintenance of male secondary sex characteristics	FSH and LH
Ovaries (26.10)	Estrogens	Steroid	Stimulate uterine lining growth; promote development and maintenance of female secondary sex characteristics	FSH and LH
	Progesterone	Steroid	Promotes uterine lining growth	FSH and LH

The hypothalamus, which is closely tied to the pituitary, connects the nervous and endocrine systems

The distinction between the endocrine system and the nervous system often blurs, especially when we consider the hypothalamus and its intricate association with the pituitary gland. As part of the brain (Figure 26.4A), the **hypothalamus** receives information from nerves about the internal condition of the body and about the external environment. It then responds to this information by sending out appropriate nervous or endocrine signals. Its hormonal signals directly control the pituitary gland, which in turn secretes hormones that influence numerous body functions. The hypothalamus thus exerts master control over the endocrine system by using the pituitary to relay directives to other glands.

As Figure 26.4A shows, the **pituitary gland** consists of two distinct parts: a posterior lobe and an anterior lobe, both situated in a pocket of skull bone at the base of the hypothalamus. The **posterior pituitary** is composed of nervous tissue and is actually an extension of the hypothalamus. It stores and secretes two hormones that are made in the hypothalamus. In contrast, the **anterior pituitary** is composed of endocrine cells that synthesize and secrete numerous hormones directly into the blood. Several of these hormones control the activity of other endocrine glands. The hypothalamus exerts control over the anterior pituitary by secreting two kinds of hormones—releasing hormones and inhibiting hormones—into short blood vessels that connect the two organs. **Releasing hormones** stimulate the anterior pituitary to secrete hormones, while **inhibiting hormones** induce the anterior pituitary to stop secreting hormones.

Figures 26.4B and 26.4C, on the facing page, emphasize the structural and functional connections between the hypothalamus and the pituitary. As Figure 26.4B indicates, a set of neurosecretory cells extends from the hypothalamus into the posterior pituitary. These cells synthesize the peptide hormones oxytocin and antidiuretic hormone (ADH). These hormones (⌄) are channeled along the neurosecretory cells into the posterior pituitary. When released into the blood from the posterior pituitary, oxytocin causes uterine muscles to contract during childbirth and mammary glands to eject milk during nursing. ADH helps cells of the kidney tubules reabsorb water, thus decreasing urine volume when the body needs to retain water (see Module 25.11). When the body has too much water, the hypothalamus responds to negative feedback, slowing the release of ADH from the posterior pituitary.

Figure 26.4C shows a second set of neurosecretory cells in the hypothalamus. These cells secrete releasing and inhibiting hormones (••) that control the anterior pituitary. A system of small blood vessels carries these hormones from the hypothalamus to the anterior pituitary. In response to hypothalamic releasing hormones, the anterior pituitary synthesizes and releases many different peptide and protein hormones (⌄), which influence a broad range of body activities. **Thyroid-stimulating hormone (TSH),** adrenocorticotropic hormone (ACTH), follicle-stimulating hormone (FSH), and luteinizing hormone (LH) all activate other endocrine glands. Feedback mechanisms control the secretion of these hormones by the anterior pituitary. Another anterior pituitary hormone, **prolactin (PRL),** produces very different effects in different species (see Module 26.11).

Of all the pituitary secretions, none has a broader effect than the protein called **growth hormone (GH).** GH promotes protein synthesis and the use of body fat for energy metabolism in a wide variety of target cells. In young mammals, GH promotes the development and enlargement of all parts of the body. Abnormal production of GH can result in several human disorders. Too much GH during childhood, usually due to a pituitary tumor, can lead to gigantism. Excessive production of GH in adulthood, a condition known as acromegaly, stimulates bony growth in the face, hands, and feet. In contrast, too little GH in childhood can lead to pituitary dwarfism. Administering growth hormone to children with GH deficiency can successfully prevent dwarfism. Once extracted only in minute quantities from the pituitary glands of cadavers, human GH is now produced in large amounts by bacteria modified to carry the human GH gene. Unfortunately, its increased availability has allowed some athletes to abuse human GH to bulk up their muscles. Such abuse is extremely dangerous and can lead to disfigurement, heart failure, and multiple cancers.

The **endorphins,** hormones produced by the anterior pituitary as well as the brain, are the body's natural painkillers. These chemical signals bind to receptors in the brain and dull the perception of pain. The effect on the nervous system is similar to that of the drug morphine, earning endorphins the nickname "natural opiates" (although it would be more accurate to call opiates "artificial endorphins"!). Some researchers speculate that the so-called "runner's high" results

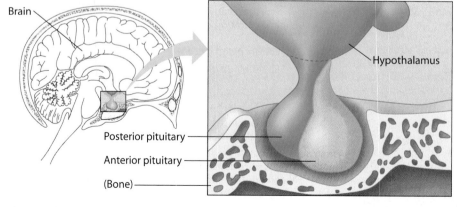

Brain
Hypothalamus
Posterior pituitary
Anterior pituitary
(Bone)

Figure 26.4A   Location of the hypothalamus and pituitary

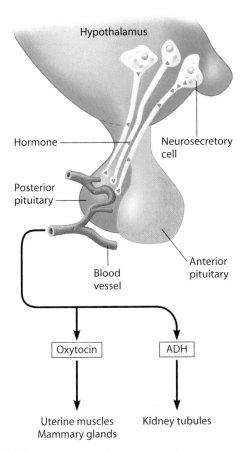

Figure 26.4B Hormones of the posterior pituitary

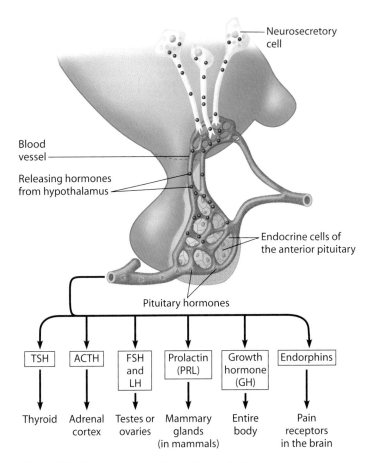

Figure 26.4C Hormones of the anterior pituitary

partly from the release of endorphins when stress and pain in the body reach critical levels. Some research also suggests that endorphins may be released during deep meditation, by acupuncture treatments, or even by eating very spicy foods.

Figure 26.4D shows how the hypothalamus operates through the anterior pituitary to direct the activity of another endocrine organ, the thyroid gland. The hypothalamus secretes a releasing hormone known as **TRH (TSH-releasing hormone).** In turn, TRH stimulates the anterior pituitary to produce thyroid-stimulating hormone (TSH). Under the influence of TSH, the thyroid secretes the hormone thyroxine into the blood. Thyroxine is converted into another hormone that increases the metabolic rate of most body cells, warming the body as a result.

Precise regulation of the TRH-TSH-thyroxine system keeps the hormones at levels that maintain homeostasis. The hypothalamus takes some cues from the environment; for instance, cold temperatures tend to increase its secretion of TRH. In addition, as the red arrows in Figure 26.4D indicate, negative-feedback mechanisms control the secretion of thyroxine. When thyroxine increases in the blood, it acts on the hypothalamus and anterior pituitary, inhibiting TRH and TSH secretion and consequently thyroxine synthesis.

The anterior and posterior pituitary, directed by the hypothalamus, stimulate a number of other endocrine glands with specific functions. In the next seven modules, we'll explore several of these endocrine glands, the hormones they produce, and their effects on the vertebrate body.

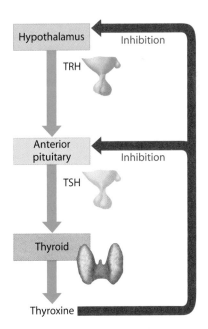

Figure 26.4D Control of thyroxine secretion

? Alcohol inhibits secretion of ADH by the anterior pituitary. Predict how this action of alcohol would affect urination.

■ Alcohol increases the volume of urine produced.

26.5 The thyroid regulates development and metabolism

Your **thyroid gland** is located just under your larynx (voice box). Thyroid hormones affect virtually all the tissues of the vertebrate body.

The thyroid produces two very similar amine hormones, both of which contain the element iodine. One of these, **thyroxine,** is often called T_4 because it contains four iodine atoms; the other, **triiodothyronine,** is called T_3 because it contains three iodine atoms. In target cells, most T_4 is converted to T_3.

T_3 and T_4 have essentially the same effects on their target cells. One of their crucial roles is in development and maturation. In a bullfrog, for example, they trigger the profound reorganization of body tissues that occurs as a tadpole—a strictly aquatic organism—transforms into an adult frog, which may spend much of its time on land. Thyroid hormones are equally important in mammals, especially in bone and nerve cell development. In humans, a thyroid deficiency known as cretinism results in retarded skeletal growth and poor mental development.

The thyroid gland has several important homeostatic functions. For example, T_3 and T_4 help maintain normal blood pressure, heart rate, muscle tone, digestion, and reproductive function. Throughout the body, these hormones tend to increase the rate of oxygen consumption and cellular metabolism. Too much or too little of these hormones in the blood can result in serious metabolic disorders. An excess of T_3 and T_4 in the blood (*hyper*thyroidism) can make a person overheat, sweat profusely, become irritable, develop high blood pressure, and lose weight. The most common form of hyperthyroidism is Graves' disease; protuding eyes caused by fluid accumulation behind the eyeballs are a typical symptom (Figure 26.5A). Conversely, insufficient amounts of T_3 and T_4 (*hypo*thyroidism) can cause weight gain, lethargy, and intolerance to cold.

Hypothyroidism can result from a defective thyroid gland or from dietary disorders. For example, severe iodine deficiency during

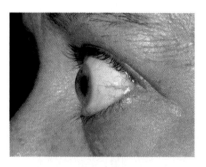

Figure 26.5A Graves' disease, a form of hyperthyroidism

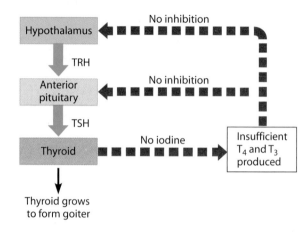

Figure 26.5B How iodine deficiency causes goiter

childhood can cause cretinism. And in adults, insufficient iodine in the diet can cause **goiter,** an enlargement of the thyroid (see Figure 2.2A). In such cases, the thyroid gland cannot synthesize adequate amounts of its T_3 and T_4 hormones. The lack of T_3 and T_4 interrupts the feedback loops that control thyroid activity (Figure 26.5B). The blood never carries enough of the T_3 and T_4 hormones to shut off the secretion of TRH (TSH-releasing hormone) or TSH. The thyroid enlarges because TSH continues to stimulate it.

Fortunately, both hypo- and hyperthyroidism can be successfully treated. For example, many cases of goiter can be improved simply by adding iodine to the diet. Seawater is a rich source of iodine, and goiter rarely occurs in people living near the seacoast, where the soil is iodine-rich and a lot of seafood is consumed. Goiter is less common today than in the past thanks to the incorporation of iodine into table salt, but it still affects thousands of people in developing nations.

Web Process of Science *How Do Thyroxine and TSH Affect Metabolism?*

 How does thyroxine switch off its own production?

By negative feedback: It inhibits the secretion of TRH from the hypothalamus and TSH from the pituitary.

26.6 Hormones from the thyroid and parathyroids maintain calcium homeostasis

An appropriate level of calcium in the blood and interstitial fluid is essential for many body functions. Without calcium, nerve signals cannot be transmitted from cell to cell, muscles cannot function properly, blood cannot clot, and cells cannot transport molecules across their membranes. The thyroid and parathyroid glands function in the homeostasis of calcium ions

(Ca^{2+}), keeping the concentration of the ions within a narrow range (about 10 mg of Ca^{2+} per 100 mL of blood).

There are four **parathyroid glands,** all embedded in the surface of the thyroid. Two peptide hormones, **calcitonin** from the thyroid gland and **parathyroid hormone (PTH),** secreted by the parathyroids, regulate blood calcium levels. Calcitonin and

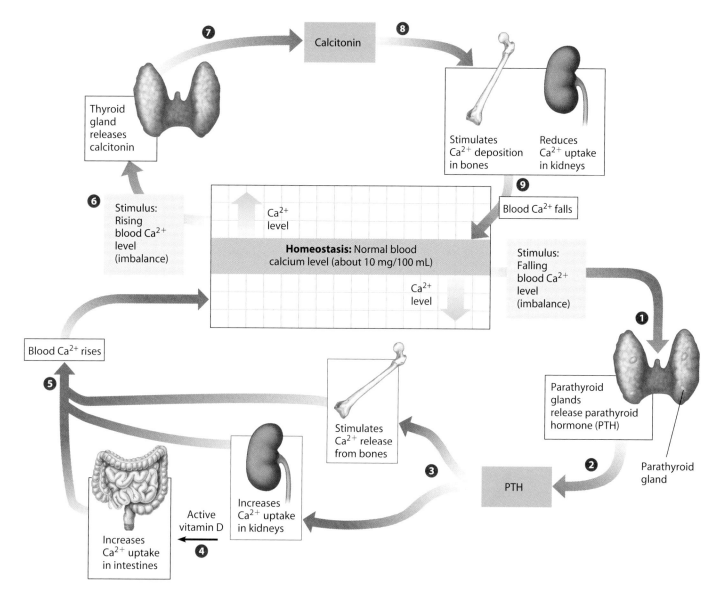

Figure 26.6 Calcium homeostasis

PTH are said to be **antagonistic hormones** because they have opposite effects: Calcitonin lowers the calcium level in the blood, whereas PTH raises it. As Figure 26.6 indicates, these two antagonistic hormones operate by means of feedback systems that keep the calcium level near the homeostatic set point of 10 mg/100 mL.

❶ When the blood Ca^{2+} level drops below 10 mg/100 mL, ❷ the parathyroids release PTH into the blood. ❸ PTH stimulates the release of calcium ions from bones and increases Ca^{2+} reabsorption by the kidneys. The kidneys also play an indirect role in calcium homeostasis, which involves vitamin D. We obtain this vitamin in inactive form from food and also from chemical reactions in our skin when it is exposed to sunlight. Transported in the blood, inactive vitamin D undergoes sequential steps of activation in the liver and kidneys. The active form of vitamin D, secreted by the kidneys, acts as a hormone. ❹ It stimulates the intestines to increase uptake of Ca^{2+} from food. ❺ The combined effects of PTH on the bones, kidneys, and intestines results in a higher Ca^{2+} level in the blood.

The top part of the diagram shows how calcitonin from the thyroid gland reverses the effects of PTH. ❻ A rise in blood Ca^{2+} above the set point induces the thyroid gland to ❼ secrete calcitonin. ❽ Calcitonin, in turn, has two main effects: It causes more Ca^{2+} to be deposited in the bones, and it makes the kidneys reabsorb less Ca^{2+} as they form urine. ❾ The result is a lower Ca^{2+} level in the blood.

In summary, a sensitive balancing system maintains calcium homeostasis. The system depends on feedback control by two antagonistic hormones. Failure of the system can have far-reaching effects in the body. For example, a shortage of PTH causes the blood calcium level to drop dramatically, leading to convulsive contractions of the skeletal muscles. This condition, known as tetany, can be fatal.

? In the control of calcium ion levels in the blood by calcitonin and PTH, what are the two main target organs of the hormones?

Bones and kidneys ■

The **pancreas** produces two hormones that play a large role in managing the body's energy supplies. Clusters of endocrine cells, called islets of Langerhans, are scattered throughout the pancreas. Each islet has a population of beta cells, which produce the hormone **insulin,** and a population of alpha cells, which produce another hormone, **glucagon.** Insulin and glucagon—both protein hormones—are secreted directly into the blood.

As shown in **Figure 26.7,** insulin and glucagon are antagonistic hormones that regulate the concentration of glucose in the blood. The two hormones counter each other in a feedback circuit that precisely manages the amount of circulating glucose available to use as cellular fuel versus the amount of glucose stored as the polymer glycogen in body cells. By negative feedback, the concentration of glucose in the blood determines the relative amounts of insulin and glucagon secreted by the islet cells.

In the top half of the diagram, you see what happens when the glucose concentration of the blood rises above the set point of about 90 mg of glucose per 100 mL of blood, as it does shortly after we eat a carbohydrate-rich meal. ❶ The rising blood glucose level ❷ stimulates the beta cells in the pancreas to secrete more insulin. ❸ The insulin stimulates nearly all body cells to take up glucose from the blood, thereby decreasing the blood glucose level. Liver cells (and skeletal muscle cells) take up much of the glucose and use it to form glycogen, which they store. Insulin also stimulates cells to metabolize the glucose for immediate energy use, for the storage of energy in fats, or for the synthesis of proteins. ❹ When the blood glucose level falls to the set point, the beta cells lose their stimulus to secrete insulin.

Following the bottom half of the diagram, you see what happens when ❺ the blood glucose level starts to dip below the set point, as it may between meals or during strenuous exercise. ❻ The pancreatic alpha cells respond by secreting more glucagon. ❼ Glucagon is a fuel mobilizer, signaling liver cells to break glycogen down into glucose, convert amino acids and

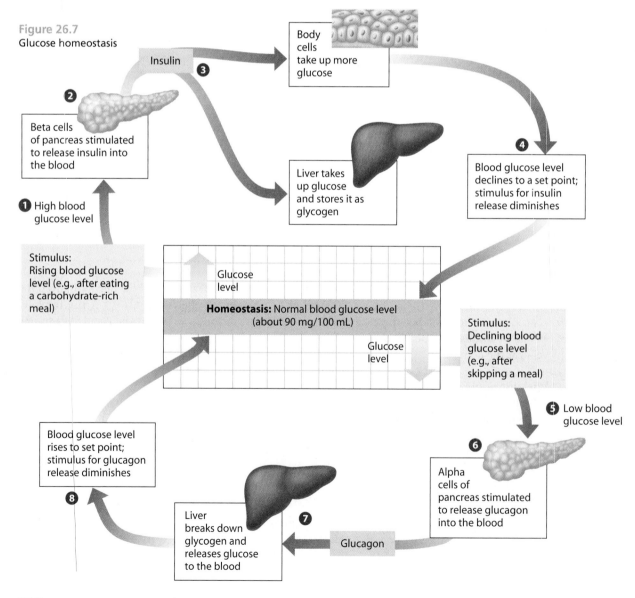

Figure 26.7
Glucose homeostasis

Insulin ❸

Body cells take up more glucose

❷

Beta cells of pancreas stimulated to release insulin into the blood

❶ High blood glucose level

Liver takes up glucose and stores it as glycogen

❹

Blood glucose level declines to a set point; stimulus for insulin release diminishes

Stimulus: Rising blood glucose level (e.g., after eating a carbohydrate-rich meal)

Glucose level

Homeostasis: Normal blood glucose level (about 90 mg/100 mL)

Glucose level

Stimulus: Declining blood glucose level (e.g., after skipping a meal)

❺ Low blood glucose level

Blood glucose level rises to set point; stimulus for glucagon release diminishes

❽

❻

Alpha cells of pancreas stimulated to release glucagon into the blood

Liver breaks down glycogen and releases glucose to the blood

❼

Glucagon

fat-derived glycerol to glucose, and release the glucose into the blood. ❽ Then, when the blood glucose level returns to the set point, the alpha cells slow their secretion of glucagon.

In the next module, we see what can happen when this delicately balanced system breaks down.

? How is the insulin-glucagon relationship similar to the calcitonin-PTH relationship?

■ In both cases, the two hormones are antagonists that help maintain homeostasis by counteracting one another's effects. Their actions keep the blood concentration of a key chemical (glucose for insulin-glucagon; calcium ions for calcitonin-PTH) near the set point.

26.8 Diabetes is a common endocrine disorder

Diabetes mellitus is a serious hormonal disease in which body cells are unable to absorb glucose from the blood. It affects about 21 million Americans—7% of the total population—and an estimated 6 million of them are undiagnosed and do not even know they are ill. Diabetes develops when there is not enough insulin in the blood or when body cells do not respond normally to blood insulin. In either case, the cells cannot obtain enough glucose from the blood, and thus, starved for fuel, they are forced to burn the body's supply of fats and proteins. Meanwhile, since the digestive system continues to absorb glucose from the diet, the glucose concentration in the blood can become extremely high—so high, in fact, that measurable amounts of glucose are excreted in the urine. (Normally, the kidneys leave no glucose in the urine.)

Deciphering the exact causes of diabetes remains elusive, but both genetic and environmental factors appear to be important. There are treatments for diabetes mellitus—insulin supplements and/or special diets—but no cure. Untreated diabetes can cause dehydration, blindness, cardiovascular and kidney disease, and nerve damage. Every year, almost 300,000 Americans die from the disease or its complications.

There are two common types of diabetes mellitus. Type 1 (insulin-dependent) diabetes is an autoimmune disease, in which white blood cells of the body's own immune system attack and destroy the pancreatic beta cells. As a result, the pancreas does not produce enough insulin, and glucose builds up in the blood. Type 1 diabetes generally develops during childhood. Treatment consists of injections of human insulin—produced by genetically engineered bacteria—several times daily.

Type 2 (non-insulin-dependent) diabetes is characterized either by a deficiency of insulin or, more commonly, by reduced responsiveness of target cells due to changes in insulin receptors. Type 2 diabetes is almost always associated with being overweight and underactive, although whether obesity causes diabetes (and, if so, how) remains unknown. This form of diabetes generally appears after age 40, but even young people who are overweight and inactive can develop the disease. In the U.S., more than 90% of diabetics are type 2. Many of them can manage their blood glucose with regular exercise and a healthy diet high in soluble fiber and low in fat and sodium; some require medications.

A third type of diabetes affects about 4% of pregnant women in the U.S. Called gestational diabetes, it can affect any pregnant woman, even one who has never shown symptoms of diabetes before. Left untreated, gestational diabetes can lead to dangerously large babies. If diagnosed, changes in diet and/or insulin injections can prevent problems in most women.

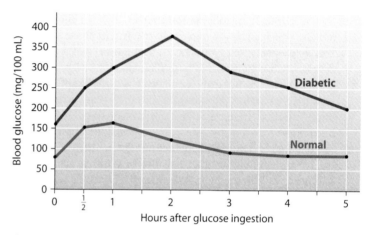

Figure 26.8 Results of glucose tolerance tests

How is diabetes detected? Early signs include a lack of energy, a craving for sweets, frequent urination, and persistent thirst. The diagnostic test for diabetes is a glucose tolerance test: The person swallows a sugar solution and then has blood drawn at prescribed time intervals. Each blood sample is tested for glucose. If the blood glucose level exceeds 200 mg/100 mL two hours after eating, the person tested has diabetes. In Figure 26.8, you can compare the glucose tolerance of a person with diabetes with that of a healthy individual. Note that a healthy body can maintain a nearly constant concentration of blood glucose; the body of a diabetic, in contrast, experiences a broad range of blood glucose concentrations.

Diabetes is not the only disease that can result from problems with insulin. Some people have hyperactive beta cells that put too much insulin into the blood when sugar is eaten. As a result, their blood glucose level can drop well below normal. This condition, called **hypoglycemia,** usually occurs 2–4 hours after a meal and may be accompanied by hunger, weakness, sweating, and nervousness. In severe cases, when the brain receives inadequate amounts of glucose, a person may develop convulsions, become unconscious, and even die. Hypoglycemia is uncommon, and most forms of it can be controlled by reducing sugar intake and eating smaller, more frequent meals.

? Three hours after glucose ingestion, the person with diabetes whose test is shown in the graph above has a blood glucose concentration about ____ times that of the normal individual (provide a number).

■ 3

The adrenal glands mobilize responses to stress

The endocrine system includes two **adrenal glands** sitting atop the kidneys. As you can see in Figure 26.9 (inset, top left), each adrenal gland is actually made up of two glands fused together: a central portion called the **adrenal medulla** and an outer portion called the **adrenal cortex.** Though the cells they contain and the hormones they produce are different, both the medulla and the cortex secrete hormones that enable the body to respond to stress.

The adrenal medulla produces the "fight-or-flight" hormones, which ensure a rapid, short-term response to stress. You've probably felt your heart beat faster and your skin develop goose bumps when sensing danger or approaching a stressful situation, such as entering a final exam. Positive emotions—extreme pleasure, for instance—can produce the same effects. These reactions are triggered by two amine hormones secreted by the adrenal medulla, **epinephrine** (adrenaline) and **norepinephrine** (noradrenaline).

Stressful stimuli, whether negative or positive, activate certain nerve cells in the hypothalamus. ❶ These cells send nerve signals via the spinal cord to the adrenal medulla, ❷ stimulating it to secrete epinephrine and norepinephrine into the blood.

Epinephrine and norepinephrine have somewhat different effects on tissues, but both contribute to the short-term stress response. Both hormones stimulate liver cells to release glucose, thus making more fuel available for cellular work. They also prepare the body for action by raising the blood pressure, breathing rate, and metabolic rate. In addition, epinephrine and norepinephrine change blood-flow patterns, making some organs more active and others less so. For example, epinephrine dilates blood vessels in the brain and skeletal muscles, thus increasing alertness and the muscles' ability to react to stress. At the same time, epinephrine and norepinephrine constrict blood vessels elsewhere, thereby reducing activities that are not immediately involved in the stress response, such as digestion. The short-term stress response occurs and subsides rapidly.

In contrast to hormones from the adrenal medulla, those secreted by the adrenal cortex can provide a slower, longer-lasting response to stress. The adrenal cortex responds to endocrine signals—chemical signals in the blood—rather than to nerve cell signals. As shown in Figure 26.9, ❸ the hypothalamus secretes a releasing hormone that ❹ stimulates target cells in the anterior pituitary to secrete the hormone

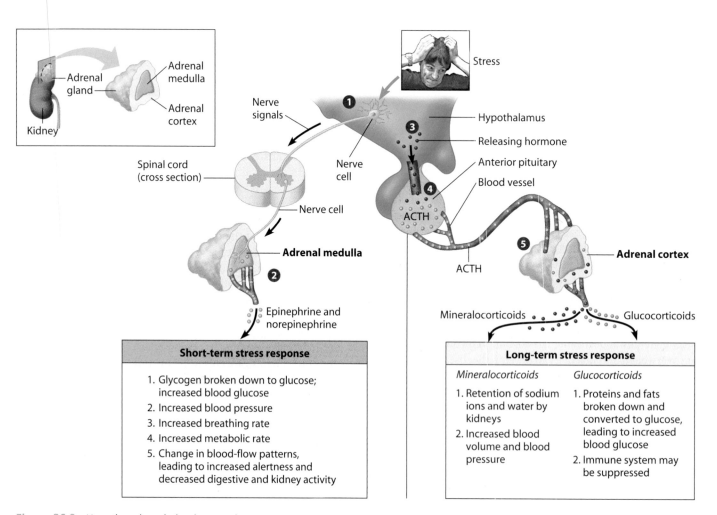

Figure 26.9   How the adrenal glands control our responses to stress

adrenocorticotropic hormone (ACTH). ❺ In turn, ACTH stimulates cells of the adrenal cortex to synthesize and secrete a family of steroid hormones called **corticosteroids.** The two main types in humans are the mineralocorticoids and the glucocorticoids. Both help maintain homeostasis when the body experiences long-term stress.

Mineralocorticoids act mainly on salt and water balance. One of these hormones (aldosterone) stimulates the kidneys to reabsorb sodium ions and water, with the overall effect of increasing the volume of the blood and raising blood pressure as a response to prolonged stress.

Glucocorticoids function mainly in mobilizing cellular fuel, thus reinforcing the effects of glucagon. Glucocorticoids promote the synthesis of glucose from noncarbohydrates such as proteins and fats. When the body cells consume more glucose than the liver can provide from glycogen stores, glucocorticoids stimulate the breakdown of muscle proteins, making amino acids available for conversion to glucose by the liver. This makes more glucose available in the blood as cellular fuel in response to stress.

Very high levels of glucocorticoids can suppress the body's defense system, including the inflammatory response that occurs at infection sites (see Module 24.2). For this reason, physicians may use glucocorticoids to treat excessive inflammation. The glucocorticoid cortisone, for example, was once regarded as a miracle drug for treating serious inflammatory conditions such as arthritis. Cortisone and other glucocorticoids can relieve swelling and pain from inflammation; but by suppressing immunity, they can also make a person highly susceptible to infection.

Physicians often prescribe oral glucocorticoids to relieve pain from athletic injuries. However, glucocorticoids are potentially very dangerous; prolonged use can depress the activity of the adrenal glands and cause side effects such as a weakened immune system, easy bruising, weak bones, weight gain, muscle breakdown, and increased risk of diabetes. It is safer, but still potentially dangerous, to inject a glucocorticoid at the site of an injury. With this treatment, the pain usually subsides, but its underlying cause remains. Masking the pain covers up the pain's message—that tissue is damaged. If an athlete exercises an injured site before the tissue has recovered, the added stress can cause more serious damage.

? How would a deficiency of receptors in the hypothalamus for adrenal steroids affect blood levels of those hormones ? (*Hint:* Apply to adrenal steroids what you learned in Module 26.5 about the regulation of thyroxine.)

■ This deficiency would cause abnormally high levels of adrenal steroids.

26.10 The gonads secrete sex hormones

The sex hormones are steroid hormones that affect growth and development and also regulate reproductive cycles and sexual behavior. The **gonads,** or sex glands (ovaries in the female and testes in the male), secrete sex hormones in addition to producing gametes.

The gonads of mammals produce three major categories of sex hormones: estrogens, progestins, and androgens. Both females and males have all three types, but in different proportions. Females have a high ratio of estrogens to androgens. **Estrogens** maintain the female reproductive system and promote the development of such female features as smaller body size, higher-pitched voice, breasts, and wider hips. In mammals, **progestins,** such as progesterone, are primarily involved in preparing and maintaining the uterus to support an embryo.

In general, **androgens** stimulate the development and maintenance of the male reproductive system. Males have a high ratio of androgens to estrogens, with their main androgen being **testosterone.** In humans, androgens produced by male embryos during the seventh week of development stimulate the embryo to develop into a male rather than a female. During puberty, high concentrations of androgens trigger the development of male characteristics, such as a lower-pitched voice, facial hair, and large skeletal muscles. The muscle-building action of androgens and other anabolic steroids has enticed some athletes to abuse them—at great risk to their health (see Module 3.10). As discussed in the chapter opener, environmental pollutants that mimic estrogens can disrupt the estrogen/androgen ratio, resulting in "demasculanization" of affected males.

In other animals, androgens can have somewhat different effects. In elephant seals, male androgens promote development of large bodies weighing 2 tons and more, an inflatable enlargement of the nasal cavity, a thick hide that can withstand bloody conflicts, and aggressive behavior toward other males. The two males in Figure 26.10 are fighting. One will establish dominance over the other and the right to mate with many females.

The synthesis of sex hormones by the gonads is regulated by the hypothalamus and anterior pituitary. In response to a releasing factor from the hypothalamus, the anterior pituitary secretes follicle-stimulating hormone (FSH) and luteinizing hormone (LH). These stimulate the ovaries or testes to synthesize and secrete the sex hormones, among other effects. We examine the complex effects of these hormones when we focus on human reproduction in the next chapter.

Figure 26.10 Male elephant seals in combat

Web Activity *Human Endocrine Glands and Hormones*

? Name the human male and female gonads and the primary hormone(s) that each produce(s).

■ Ovaries (estrogens and progestins) and testes (androgens)

26.11 A single hormone can perform a variety of functions in different animals

Hormones play important roles in all vertebrates, and some of the same hormones can be found in vertebrates that are only distantly related. Interestingly, the same hormone can have different actions in different animals—a strong indication that hormonal regulation was an early evolutionary adaptation.

The peptide hormone prolactin (PRL), produced and secreted by the anterior pituitary under the direction of the hypothalamus, is a good example. Prolactin produces diverse effects in different vertebrate species. In humans, PRL performs several important functions related to childbirth. During late

Figure 26.11 Suckling promotes prolactin production.

pregnancy, PRL stimulates mammary glands to grow and produce milk. (A brief surge in prolactin levels just before menstruation causes breast swelling and tenderness in some women.) Suckling by a newborn stimulates further release of PRL, which in turn increases the milk supply (Figure 26.11). High prolactin levels during nursing prevent the ovaries from releasing eggs, decreasing the chances of a new pregnancy occurring during the time of breast feeding. This may be an evolutionary adaptation that ensures adequate care is given to newborn babies.

PRL plays a wide variety of other rolls unrelated to childbirth. In some nonhuman mammals, PRL stimulates nestbuilding. In birds, PRL regulates fat metabolism and reproduction. In amphibians, it stimulates movement toward water in preparation for breeding and affects metamorphosis. In fish that migrate between salt and fresh water (salmon, for example), PRL helps regulate salt and water balance in the gills and kidneys.

Such diverse effects suggest that prolactin is an ancient hormone whose functions diversified through evolution. Over millions of years, the prolactin molecule stayed the same, but its role changed dramatically—a good example of how evolution can both preserve unity (the molecule itself) and promote diversity (the roles it plays).

? PRL promotes the production of milk. Newborn suckling promotes PRL production. This is an example of _____ feedback.

■ positive

Chapter Review

Reviewing the Concepts

The Nature of Chemical Regulation (26.1–26.2)

Hormones are chemical signals, usually carried in the blood, that cause specific changes in target cells. Endocrine glands and neurosecretory cells secrete hormones. All hormone-secreting cells make up the endocrine system, which works with the nervous system in regulating body activities (**26.1**). Hormones trigger changes in target cells by two general mechanisms (**26.2**):

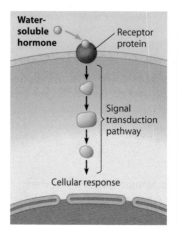
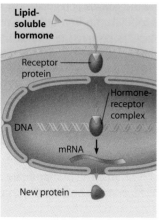

The Vertebrate Endocrine System (26.3–26.4)

The vertebrate endocrine system consists of more than a dozen glands secreting more than 50 hormones. Some glands are specialized for hormone secretion only; some also do other jobs. Some hormones have a very narrow range of targets and effects; others have numerous effects on many kinds of target cells (**26.3**).

The hypothalamus and pituitary.   Releasing and inhibiting hormones from the hypothalamus control the secretion of several other hormones (**26.4**).

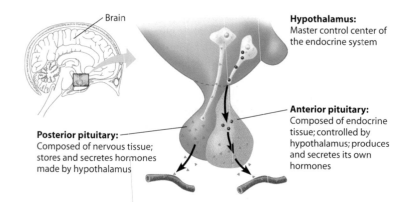

Hormones and Homeostasis (26.5–26.11)

Vertebrate hormones. Two amine hormones from the thyroid gland, T_4 and T_3, regulate an animal's development and metabolism. Negative feedback maintains homeostatic levels of T_4 and T_3 in the blood. Thyroid imbalance can cause disease (**26.5**). Blood calcium level is regulated by a tightly balanced antagonism between calcitonin from the thyroid and parathyroid hormone from the parathyroid glands (**26.6**). The pancreas secretes two hormones, insulin and glucagon, that control blood glucose levels. Insulin signals cells to use and store glucose. Glucagon causes cells to release stored glucose into the blood (**26.7**). Diabetes mellitus results from a lack of insulin or a failure of cells to respond to it (**26.8**). Hormones from the adrenal glands help maintain homeostasis when the body is stressed. Nerve signals from the hypothalamus stimulate the adrenal medulla to secrete epinephrine and norepinephrine, which quickly trigger the fight-or-flight response. ACTH from the pituitary causes the adrenal cortex to secrete glucocorticoids and mineralocorticoids, which boost blood pressure and energy in response to long-term stress (**26.9**). Estrogens, progestins, and androgens are steroid sex hormones produced by the gonads in response to signals from the hypothalamus and pituitary. Estrogens and progestins stimulate the development of female characteristics and maintain the female reproductive system. Androgens, such as testosterone, trigger the development of male characteristics (**26.10**). Over evolutionary history, a single hormone can assume diverse roles in different organisms (**26.11**).

Connecting the Concepts

Match each hormone (top) with the gland where it is produced (center) and its effect on target cells (bottom).

1. thyroxine	4. insulin	7. PTH
2. epinephrine	5. melatonin	8. ADH
3. androgens	6. FSH	

 Pineal gland

 Testes

 Parathyroid gland

Adrenal medulla

 Hypothalamus

Pancreas

 Anterior pituitary

Thyroid gland

- a. lowers blood glucose
- b. stimulates ovaries
- c. triggers fight-or-flight
- d. promotes male traits
- e. regulates metabolism
- f. related to daily rhythm
- g. raises blood calcium level
- h. boosts water retention

Testing Your Knowledge

Multiple Choice

9. Which of the following controls the activity of all the others?
   a. thyroid gland
   b. pituitary gland
   c. adrenal cortex
   d. hypothalamus
   e. ovaries

10. The pancreas increases its output of insulin in response to
    a. an increase in body temperature.
    b. changing cycles of light and dark.
    c. a decrease in blood glucose.
    d. a hormone secreted by the anterior pituitary.
    e. an increase in blood glucose.

11. Which of the following hormones have antagonistic (opposing) effects? Choose all that apply.
    a. parathyroid hormone and calcitonin
    b. glucagon and thyroxine
    c. growth hormone and epinephrine
    d. ACTH and cortisone
    e. epinephrine and norepinephrine

12. The body is able to maintain a relatively constant level of thyroxine in the blood because
    a. thyroxine stimulates the pituitary to secrete thyroid-stimulating hormone (TSH).
    b. thyroxine inhibits the secretion of TSH-releasing hormone (TRH) from the hypothalamus.
    c. TRH inhibits the secretion of thyroxine by the thyroid gland.
    d. thyroxine stimulates the hypothalamus to secrete TRH.
    e. thyroxine stimulates the pituitary to secrete TRH.

13. Which of the following hormones has the broadest range of targets?
    a. ADH   c. TSH   e. ACTH
    b. oxytocin   d. epinephrine

Describing, Comparing, and Explaining

14. Explain how the hypothalamus controls body functions through its action on the pituitary gland. How does control of the anterior and posterior pituitary differ?

15. Explain how the same hormone might have different effects on two different target cells and no effect on a third type of cell.

Applying the Concepts

16. A strain of transgenic mice remains healthy as long as you feed them regularly and do not let them exercise. After they eat, their blood glucose level rises slightly and then declines to a homeostatic level. However, if these mice fast or exercise at all, their blood glucose drops dangerously. Which hypothesis best explains their problem? (*Explain your choice.*)
    a. The mice have insulin-dependent diabetes.
    b. The mice lack insulin receptors on their cells.
    c. The mice lack glucagon receptors on their cells.
    d. The mice cannot synthesize glycogen from glucose.

17. How could a hormonal imbalance result in a person who is genetically male but physically female?

Answers to all questions can be found in Appendix 4.

For Practice Quizzes, BioFlix, MP3 Tutors, and Activities, go to www.mybiology.com.

27 Reproduction and Embryonic Development

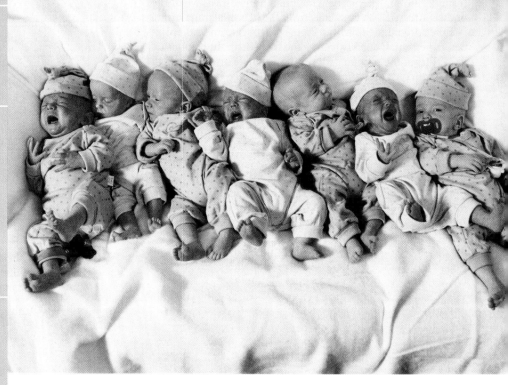

Baby Bonanza

If you walk through a busy park on a sunny day, chances are you'll be able to spot proud parents pushing a double-wide stroller. That's because in 2003, over 128,000 twins were born in the United States, more than in any previous year. Adding to that, over 7,500 "supertwins"—three or more children born at one time—also entered the world. Indeed, multiple birth rates have been on the rise for decades. Between 1980 and 2003, the rate of twin births in the United States rose by 87%. The rate of supertwin births rose even more dramatically during this period: almost 600%. Multiple pregnancies, multiple births, and hence multiple strollers have suddenly become much more common.

What is the cause of this remarkable baby bonanza? One answer: the increased use of fertility drugs. **Infertility,** the inability to bear children after one year of trying, affects about one in seven American couples. Many have sought medical assistance to conceive children. Although infertility is frequently traced to problems in the male reproductive system, such problems are rarely treated with medications. There are, however, several drugs that can effectively treat female infertility, which is often caused by a failure to ovulate. Ovulation, the release of a mature egg during the monthly ovarian cycle, is essential for becoming pregnant and is controlled by several

hormones. By altering the levels of one or more of these hormones in the female, fertility drugs have allowed thousands of infertile couples to have babies.

Fertility drugs have been prescribed in the United States for over 30 years and are relatively safe and effective. However, they are sometimes too effective: They can promote the release of multiple eggs. Each one of these eggs may then fuse with a sperm, resulting in multiple embryos. In fact, over 10% of women taking fertility drugs become pregnant with more than one embryo—sometimes quite a few more. Indeed, fertility drugs are responsible for some extraordinary multiple births, such as the "Iowa septuplets" (shown at left)—four boys and three girls, born in 1997—the world's first surviving septuplets.

Despite the successful births now enjoyed by previously infertile couples, multiple births are risky. Newborns from multiple births are premature more often than babies from single births.

They also have lower birth weights; are less likely to survive (the mortality rate is 5 times higher among twins and 12 times higher among supertwins); and are more likely to suffer life-long disabilities if they do survive. The Iowa septuplets, for example, were born more than two months premature and each weighed less than half of the national average birth weight. Thus, while modern medicine offers infertile couples new options, these options carry risks that require careful consideration.

Fertility drugs are one example of how reproductive technologies can alter the normal reproductive cycle. We investigate more of these reproductive options at the end of this chapter. Following a brief introduction to the diverse ways that animals reproduce, we'll focus on the reproductive system of our own species. In the second half of the chapter, we discuss the processes of fertilization and embryonic development in vertebrates and then focus on human embryonic development and birth. ■ ■ ■

27.1 Asexual reproduction results in the generation of genetically identical offspring

Individuals have a finite life span. A population transcends the limit of finite life spans only by **reproduction,** the creation of new individuals from existing ones. Animals reproduce in a great variety of ways, but there are two principal modes: asexual and sexual.

Asexual reproduction (reproduction without sex) is the creation of genetically identical offspring by a lone parent. Because asexual reproduction proceeds without **fertilization,** the fusion of egg and sperm, the resulting offspring are genetic copies of the one parent.

Several types of asexual reproduction are found among animals. Many invertebrates reproduce asexually by **budding,** splitting off new individuals from existing ones (see Figure 8.11C). The sea anemone in the center of Figure 27.1 is undergoing **fission,** the separation of a parent into two or more individuals of about equal size. Asexual reproduction can also result from the process of **fragmentation,** the breaking of the parent body into several pieces, some or all of which develop into complete adults. For reproduction to occur, fragmentation must be accompanied by **regeneration,** the regrowth of lost body parts. In sea stars of the genus *Linckia,* for example, a whole new individual can develop from a broken-off arm plus a bit of the central body. Thus, a single animal with five arms, if broken apart, could potentially give rise to five offspring via asexual reproduction in a matter of weeks. (Besides the natural means of asexual reproduction discussed here, many species have been the target of artificial asexual reproduction—see the discussion of cloning in Chapter 11).

In nature, asexual reproduction has several potential advantages. For one, it allows animals that do not move from place to place or that live in isolation to produce offspring without finding mates. Another advantage is that it enables an animal to produce many offspring quickly; no time or energy is lost in gamete production or fertilization. Asexual reproduction perpetuates a

Figure 27.1 Asexual reproduction of an aggregating sea anemone (*Anthopleura elegantissima*) by fission

particular genotype precisely and rapidly. Therefore, it can be an effective way for animals that are genetically well suited to a particular environment to quickly expand their populations and exploit available resources.

A potential disadvantage of asexual reproduction is that it produces genetically uniform populations. Genetically similar individuals may thrive in one particular environment, but if the environment changes and becomes less favorable, all individuals may be affected equally, and the entire population may die out.

? Would asexual reproduction be advantageous in a rapidly changing environment? Why or why not?

∎ Because asexual reproduction produces genetically identical offspring, it would not be advantageous in such an environment.

27.2 Sexual reproduction results in the generation of genetically unique offspring

Sexual reproduction is the creation of offspring by the fusion of two haploid (n) sex cells, or **gametes,** to form a diploid ($2n$) **zygote** (fertilized egg). (Recall from Chapter 8 that n refers to the haploid number of chromosomes and $2n$ refers to the diploid number; for humans, $n = 23$ and $2n = 46$.) The male gamete, the **sperm,** is a relatively small cell that moves by means of a flagellum. The female gamete, the **egg,** is a much larger cell that is not self-propelled. The zygote—and the new individual it develops into—contains a unique combination of genes carried from the parents via the egg and sperm.

Unlike asexual reproduction, sexual reproduction increases genetic variability among offspring. As we discussed in Modules

8.16 and 8.18, meiosis and random fertilization can generate enormous genetic variation. The variability produced by the reshuffling of genes in sexual reproduction may provide greater adaptability to changing environments. In theory, when an environment changes suddenly or drastically, there is a better chance that some of the offspring will survive and reproduce if they aren't all genetically very similar.

Some animals can reproduce both asexually and sexually, benefiting from both modes. In Figure 27.2A at the top of the facing page, you can see two sea anemones of the same species; the one on the left is reproducing asexually (via fission) while the one on the right is releasing eggs. Many other marine

Figure 27.2A   Asexual (left) and sexual (right) reproduction in the starlet sea anenome (*Nematostella vectensis*)

Figure 27.2C   Frogs in an embrace that triggers the release of eggs and sperm (the sperm are too small to be seen)

invertebrates can also reproduce by either mode. Why would such dual reproductive capabilities be advantageous to an organism? In several well-studied cases, it is known that certain animals reproduce asexually when there is ample food and when water temperatures are favorable for rapid growth and development. Asexual reproduction usually continues until cold temperatures signal the approach of winter or until the food supply dwindles or the habitat starts to dry up. At that point, the animals switch to a sexual reproduction mode, resulting in a generation of genetically varied individuals with better potential to adapt to the changing conditions.

Although sexual reproduction has advantages, it presents a problem for nonmobile animals and for those that live solitary lives: how to find a mate. One solution that has evolved is **hermaphroditism,** in which each individual has both female and male reproductive systems. (The term comes from the Greek myth in which Hermaphroditus, son of the gods Hermes and Aphrodite, fused with a woman to form a single individual of both sexes.) Although some hermaphrodites, such as

tapeworms, can fertilize their own eggs, most must mate with another member of the same species. When hermaphrodites mate (for example, the two earthworms in Figure 27.2B), each animal serves as both male and female, donating and receiving sperm. For hermaphrodites, every individual encountered is a potential mate, and mating can result in twice as many offspring than if only one individual's eggs were fertilized.

The mechanics of fertilization play an important part in sexual reproduction. Many aquatic invertebrates and most fishes and amphibians exhibit **external fertilization:** The parents discharge their gametes into the water, where fertilization then occurs, often without the male and female even making physical contact. Timing is crucial because the eggs must be ripe for fertilization when sperm contact them. For many species—certain clams that live in freshwater rivers and lakes, for instance—environmental cues such as temperature and day length cause a whole population to release gametes all at once. Males or females may also emit a chemical signal as they release their gametes. The signal triggers gamete release in members of the opposite sex. Most fishes and amphibians with external fertilization have specific courtship rituals that trigger simultaneous gamete release in the same vicinity by the female and male. An example of such a mating ritual is the clasping of a female frog by a male (Figure 27.2C).

In contrast to external fertilization, **internal fertilization** occurs when sperm are deposited in or close to the female reproductive tract, and gametes unite within the tract. Nearly all terrestrial animals exhibit internal fertilization, which is an adaptation that enables sperm to reach an egg in a dry environment. Internal fertilization usually requires **copulation,** or sexual intercourse. It also requires complex reproductive systems, including organs for gamete storage and transport and organs that facilitate intercourse. For examples of these complex structures, we turn next to the human female and male.

? In terms of genetic makeup, what is the most important difference between the outcome of sexual reproduction and that of asexual reproduction?

■ The offspring of sexual reproduction are genetically diverse, whereas the offspring of asexual reproduction are genetically identical.

Figure 27.2B   Hermaphroditic earthworms mating

27.3 Reproductive anatomy of the human female

Although we tend to focus on the anatomical differences between the human male and female reproductive systems, there are also some important similarities. Both sexes have a pair of gonads, the organs that produce gametes. And both sexes have ducts to deliver gametes as well as structures to facilitate copulation. In this and the next module, we'll examine the anatomical features of the human reproductive system, beginning with female anatomy.

A woman's **ovaries** are each about an inch long, with a bumpy surface (Figure 27.3A). The bumps are **follicles,** each consisting of one or more layers of follicle cells that surround, nourish, and protect a single developing egg cell. In addition to producing egg cells, the ovaries produce hormones, as we saw in Chapter 26. Specifically, the follicle cells produce the female sex hormone estrogen. (In this chapter, we use the word *estrogen* to refer collectively to several closely related chemicals that affect the body similarly.)

Most or all of the 400,000 follicles a woman will ever have are thought to be formed before her birth, but only several hundred will release egg cells during her reproductive years. Starting at puberty and continuing until menopause, one follicle (or rarely two or more) matures and releases its egg cell about every 28 days. An egg cell is ejected from the follicle in a process called **ovulation,** shown in Figure 27.3B.

After ovulation, the follicular tissue that had been surrounding the egg that was just ejected grows within the ovary to form a solid mass called the **corpus luteum;** you can see one in the ovary on the left in Figure 27.3A. The corpus luteum secretes additional estrogen as well as progesterone, a hormone that helps maintain the uterine lining during pregnancy. If the

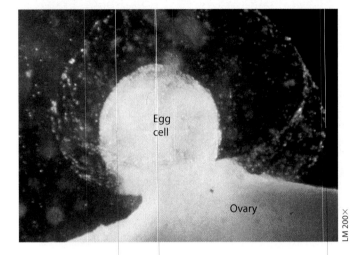

Figure 27.3B   Ovulation

released egg is not fertilized, the corpus luteum degenerates, and a new follicle matures during the next cycle. We discuss ovulation and female hormonal cycles further in later modules.

Notice in Figure 27.3A that each ovary lies next to the opening of an **oviduct,** also called a fallopian tube. The oviduct opening resembles a funnel fringed with fingerlike projections. The projections touch the surface of the ovary, but the ovary is actually separated from the opening of the oviduct by a tiny space. When ovulation occurs, the egg cell passes across

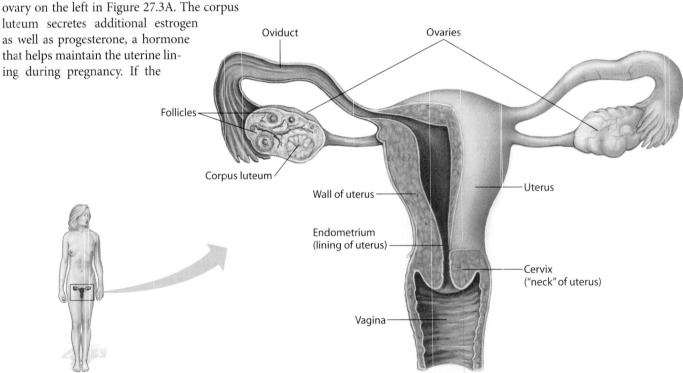

Figure 27.3A   Front view of female reproductive anatomy (upper portion)

the space and into the oviduct, where cilia sweep it toward the uterus. Fertilization usually occurs in the upper third of the oviduct. The resulting zygote starts to divide, thus becoming an embryo, as it moves along within the oviduct.

The **uterus,** also known as the womb, is the actual site of pregnancy. The uterus is only about 3 inches long in a woman who has never been pregnant, but during pregnancy it expands considerably to accommodate a baby. The uterus has a thick muscular wall, and its inner lining, the **endometrium,** is richly supplied with blood vessels. The embryo implants into the endometrium, and development is completed there. The term **embryo** is used for the stage in development from the first division of the *zygote* until body structures begin to appear, about the ninth week in humans. From the ninth week until birth, a developing human is called a **fetus.**

The uterus is the *normal* site of pregnancy. However, in about 1% of pregnancies, the embryo implants somewhere else, resulting in an **ectopic pregnancy.** Most ectopic pregnancies occur in the oviduct and are called tubal pregnancies. An ectopic pregnancy is a serious medical emergency that requires surgical intervention; otherwise, it can rupture surrounding tissues, causing severe bleeding and even death of the mother.

The narrow neck at the bottom of the uterus is the **cervix,** which opens into the vagina. It is recommended that a woman have a yearly Pap test in which cells are removed from around the cervix and examined under a microscope for signs of cervical cancer. Regular Pap smears greatly increase the chances of detecting cervical cancer early and therefore treating it successfully. The cervix opens to the **vagina,** a thin-walled, but strong, muscular chamber that serves as the birth canal through which the baby is born. The vagina is also the reposi-

tory for sperm during copulation. Glands near the vaginal opening secrete mucus during sexual arousal, lubricating the vagina and facilitating intercourse.

You can see more features of female reproductive anatomy in Figure 27.3C, a side view. Notice that the vagina opens to the outside just behind the opening of the urethra, the tube through which urine is excreted. A pair of slender skin folds, the **labia minora,** border the openings, and a pair of thick, fatty ridges, the **labia majora,** protect the vaginal opening. Until sexual intercourse or vigorous physical activity ruptures it, a thin piece of tissue called the hymen partly covers the vaginal opening.

Several female reproductive structures are important in sexual arousal, and stimulation of them can produce highly pleasurable sensations. The vagina, labia minora, and a small erectile organ called the **clitoris** all engorge with blood and enlarge during sexual activity. The sole function of the clitoris is sexual arousal. It consists of a short shaft supporting a rounded **glans,** or head, covered by a small hood of skin called the **prepuce.** In Figure 27.3C, blue highlights the spongy erectile tissue within the clitoris that fills with blood during arousal. The clitoris, especially the glans, has an enormous number of nerve endings and is very sensitive to touch. As we'll see in the next module, the reproductive organ of the human male contains several similar structures.

? Where does fertilization occur? In which organ does the fetus develop?

■ The oviduct; the uterus

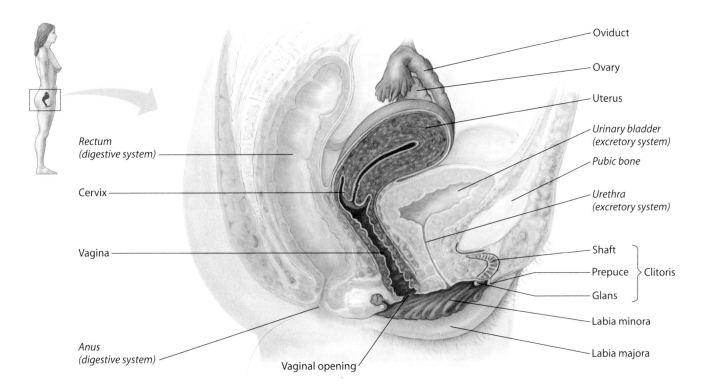

Figure 27.3C Side view of female reproductive anatomy (with nonreproductive structures in italic)

Reproductive anatomy of the human male

Figures 27.4A and 27.4B present two views of the male reproductive system. The male gonads, or **testes** (singular, *testis*), are each housed outside the abdominal cavity in a sac called the **scrotum.** Sperm do not develop optimally at human core body temperature; the scrotum keeps the sperm-forming cells cool enough to function normally.

Now let's track the path of sperm from one of the testes out of the male's body. From each testis, sperm pass into a coiled tube called the **epididymis,** which stores the sperm while they continue to develop. Sperm leave the epididymis during **ejaculation,** the expulsion of sperm-containing fluid from the penis. At that time, muscular contractions propel the sperm from the epididymis through another duct called the **vas deferens.** The vas deferens passes upward into the abdomen and loops around the urinary bladder. Next to the bladder, the vas deferens joins a short duct from a gland, the seminal vesicle. The two ducts unite to form a short **ejaculatory duct,** which joins its counterpart conveying sperm from the other testis. Each ejaculatory duct empties into the urethra, which conveys both urine and sperm out through the penis, although not at the same time. Thus, unlike the female, the male has a direct connection between the reproductive and urinary systems.

In addition to the testes and ducts, the reproductive system of human males contains three sets of glands: the seminal vesicles, the prostate gland, and the bulbourethral glands. The two **seminal vesicles** secrete a thick fluid that contains fructose, which provides most of the energy used by the sperm as they propel themselves through the female reproductive tract.

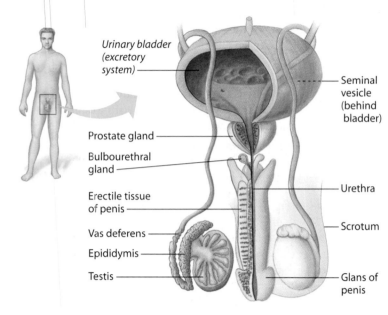

Figure 27.4A Front view of male reproductive anatomy

The **prostate gland** secretes a thin fluid that further nourishes the sperm. The two **bulbourethral glands** secrete a clear, alkaline mucus.

Together, the sperm and the glandular secretions make up **semen,** the fluid ejaculated from the penis during **orgasm,** a series of rhythmic, involuntary contractions of the reproductive structures. About 2–5 mL (1 teaspoonful) of semen are discharged

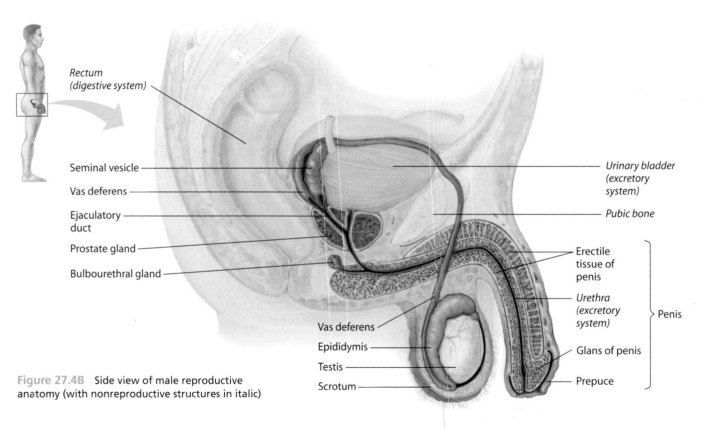

Figure 27.4B Side view of male reproductive anatomy (with nonreproductive structures in italic)

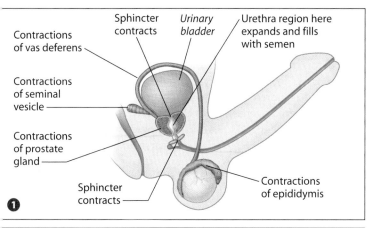

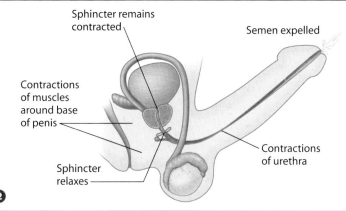

Figure 27.4C The two stages of ejaculation

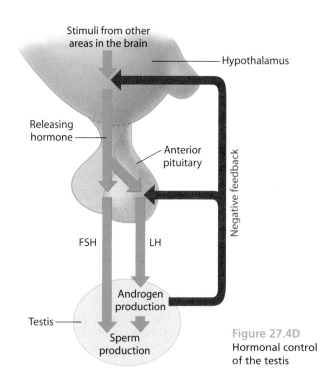

Figure 27.4D
Hormonal control of the testis

during a typical ejaculation. About 95% of the fluid consists of glandular secretions. The other 5% is made up of a few hundred million sperm, only one of which may eventually fertilize an egg. The alkalinity of the semen balances the acidity of any traces of urine in the urethra and neutralizes the acidic environment of the vagina, protecting the sperm and increasing their motility.

The human **penis** consists mainly of erectile tissue (shown in blue in Figures 27.4A and 27.4B) that can fill with blood to cause an erection during sexual arousal. Erection is essential for insertion of the penis into the vagina. (See Module 28.8 for a discussion of a signal molecule important for erection and how drugs such as Viagra can affect its action.) Like the clitoris, the penis consists of a shaft that supports the glans, or head. The glans is richly supplied with nerve endings and is highly sensitive to stimulation. As in the female, a fold of skin called the prepuce, or foreskin, covers the glans. Circumcision, the surgical removal of the prepuce, arose from religious traditions. The majority of American males are circumcised, although the practice remains somewhat controversial because of a slight (<1%) chance of complication from the surgery and questions about its benefit. Recent studies have suggested that circumcision does have a medical advantage: the procedure significantly reduces a man's chance of contracting and passing on sexually transmitted diseases, including AIDS.

Figure 27.4C illustrates the process of ejaculation and summarizes the production of semen and its expulsion. Ejaculation occurs in two stages. ❶ At the peak of sexual arousal, muscles in the epididymis, seminal vesicles, prostate gland, and vas

deferens contract. These contractions force secretions from the glands into the urethra and propel sperm from the epididymis. At the same time, a sphincter muscle at the base of the bladder contracts, preventing urine from leaking into the urethra from the bladder. Another sphincter also contracts, closing off the entrance of the urethra into the penis. The section of the urethra between the two sphincters fills with semen and expands. In the second stage of ejaculation, the expulsion stage, ❷ the sphincter at the base of the penis relaxes, admitting semen into the penis. Simultaneously, a series of strong muscle contractions around the base of the penis and along the urethra expels the semen from the body.

Figure 27.4D shows how hormones control sperm production by the testes. Influenced by signals from other parts of the brain, the hypothalamus secretes a releasing hormone that regulates release of follicle-stimulating hormone (FSH) and luteinizing hormone (LH) by the anterior pituitary (see Module 26.4). FSH increases sperm production by the testes, while LH promotes the secretion of androgens, mainly testosterone. Androgens stimulate sperm production. In addition, androgens carried in the blood help maintain homeostasis by a negative-feedback mechanism (red arrows), inhibiting secretion of both the releasing hormone and LH. Under the control of this chemical regulating system, the testes produce hundreds of millions of sperm every day, from puberty well into old age. Next we'll see how sperm and eggs are made.

Web Activity *Reproductive System of the Human Male*

Web Process of Science *Connection: What Might Obstruct the Male Urethra?*

? Arrange the following organs in the correct sequence for the travel of sperm: epididymis, testis, urethra, vas deferens.

■ Testis, epididymis, vas deferens, urethra

27.5 The formation of sperm and egg requires meiosis

Both sperm and egg are haploid (*n*) cells that develop by meiosis from diploid (*2n*) cells in the gonads. Recall that the diploid chromosome number in humans is 46; that is, $2n = 46$. Before we turn to the formation of gametes, **gametogenesis,** you may want to review Modules 8.12–8.14 as background for our discussion.

Figure 27.5A outlines **spermatogenesis,** the formation of sperm cells. Sperm develop in the testes in coiled tubes called the **seminiferous tubules.** Diploid cells that begin the process are located near the outer wall of the tubules (at the top of the enlarged wedge of tissue in Figure 27.5A). These cells multiply constantly by mitosis, and each day about 3 million of them differentiate into **primary spermatocytes,** the cells that undergo meiosis. Meiosis I of a primary spermatocyte produces two **secondary spermatocytes,** each with the haploid number of chromosomes ($n = 23$). The chromosomes are still in their duplicated state, each consisting of two attached sister chromatids. Meiosis II then forms four cells, each with the haploid number of single chromosomes. A sperm cell develops by differentiation of each of these haploid cells and is gradually

pushed toward the center of the seminiferous tubule. From there it passes into the epididymis, where it matures, becomes motile, and is stored until ejaculation. In human males, this process of spermatogenesis takes about ten weeks.

The left side of Figure 27.5B shows **oogenesis,** the development of **ova** (mature egg cells; singular *ovum*). Most of the process occurs in the ovary. Oogenesis actually begins prior to birth, when a diploid cell in each developing follicle begins

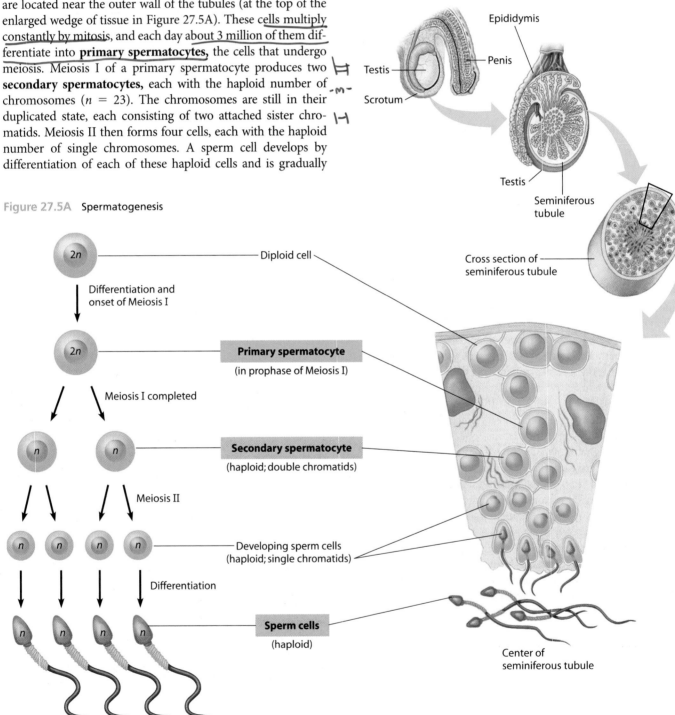

Figure 27.5A Spermatogenesis

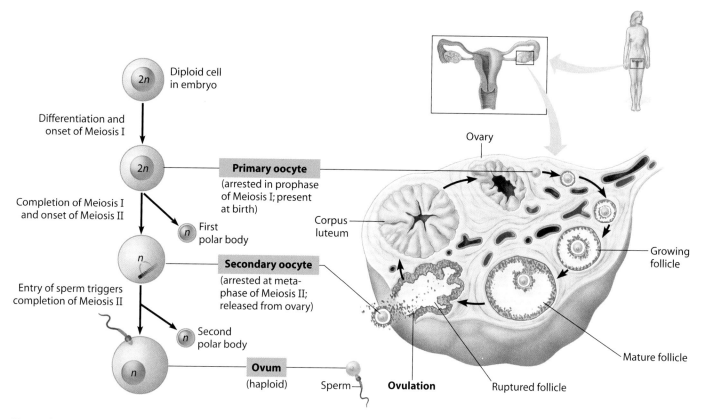

Figure 27.5B Oogenesis and the development of an ovarian follicle

meiosis. At birth, each follicle contains a dormant **primary oocyte,** a diploid cell that is resting in prophase of meiosis I. A primary oocyte can be hormonally triggered to develop further. After puberty, about every 28 days, FSH (follicle-stimulating hormone) from the pituitary stimulates one of the dormant follicles to develop. The follicle enlarges, and the primary oocyte completes meiosis I and begins meiosis II. Meiosis then halts again at metaphase II. In the female, the division of the cytoplasm in meiosis I is unequal, with a single **secondary oocyte** receiving almost all of it. The smaller of the two daughter cells, called the first polar body, receives almost no cytoplasm.

A secondary oocyte is what is actually released by the ovary during ovulation. It enters the oviduct, and if a sperm cell penetrates it, the secondary oocyte completes meiosis II. Meiosis II yields a second polar body and the actual ovum. The haploid nucleus of the ovum can then fuse with the haploid nucleus of the sperm cell, producing a zygote.

Although not shown in Figure 27.5B, the first polar body may also undergo meiosis II, forming two cells. These and the second polar body receive virtually no cytoplasm and quickly degenerate. Polar body formation leaves the ovum with nearly all the cytoplasm and thus the bulk of the nutrients contained in the original diploid cell.

The right side of Figure 27.5B is a cutaway view of an ovary. The series of follicles here represents the changes one follicle undergoes over time; the arrows indicate the sequence. An actual ovary would have thousands of dormant follicles, each containing a primary oocyte. Usually, only one follicle has a dividing oocyte at any one time, and as it develops, that follicle

stays in one place in the ovary. Meiosis I occurs as the follicle matures. About the time the secondary oocyte forms, the pituitary hormone LH (luteinizing hormone) triggers ovulation, the rupture of the follicle and expulsion of the secondary oocyte. The ruptured follicle then develops into a corpus luteum. Unless fertilization occurs, the corpus luteum degenerates before another follicle starts to develop.

Oogenesis and spermatogenesis are alike in that they both produce haploid gametes. However, these two processes differ in three important ways. First, only one ovum results from each diploid cell that undergoes meiosis. The other products of oogenesis, the polar bodies, degenerate. By contrast, in spermatogenesis, all four products of meiosis develop into mature gametes. Second, although the cells from which sperm develop continue to divide by mitosis throughout the male's life, this is thought not to be the case for the comparable cells in the human female. (Research in 2004 found ovarian stem cells in adult mice, but the implications of this research for human reproduction are not yet clear.) Third, oogenesis has long "resting" periods, whereas spermatogenesis produces mature sperm in an uninterrupted sequence.

🎧 MP3 Tutor *Spermatogenesis and Oogenesis*

❓ Which process in the development of sperm and ova is responsible for the genetic variation among gametes? (*Hint:* Review Module 8.17.)

■ The random alignment of homologous chromosomes during meiosis, specifically meiosis I

Hormones synchronize cyclic changes in the ovary and uterus

Oogenesis is one part of a female mammal's reproductive cycle, a recurring sequence of events that produces gametes, makes them available for fertilization, and prepares the body for pregnancy. The reproductive cycle is actually one integrated cycle involving cycles in two different reproductive organs: the ovaries and the uterus. In discussing oogenesis in the last module, we described the **ovarian cycle,** cyclic events that occur about every 28 days in the human ovary. Hormonal messages synchronize the ovarian cycle with related events in the uterus that are part of the **menstrual cycle.** The hormone story is complex and involves intricate feedback mechanisms. Table 27.6 lists the major hormones and their roles. You may find it useful to refer to this table while reading this module. Figure 27.6, on the facing page, shows how the events of the ovarian cycle (Part C) and menstrual cycle (Part E) are synchronized through the actions of multiple hormones (shown in Parts A, B, and D). Notice the time scale at the bottom of Part E; it also applies to Parts B–D. Follow Figure 27.5 carefully as you read the descriptions of the events in the ovarian and menstrual cycles.

An Overview of the Ovarian and Menstrual Cycles Let's begin with the structural events of the ovarian and menstrual cycles. For simplicity, we have divided the ovarian cycle (Part C of the figure) into two phases separated by ovulation: the pre-ovulatory phase, when a follicle is growing and a secondary oocyte is developing, and the post-ovulatory phase, after the follicle has become a corpus luteum.

Events in the menstrual cycle (Part E) are synchronized with the ovarian cycle. By convention, the first day of a woman's "period" is designated day 1 of the menstrual cycle. Uterine bleeding, called **menstruation,** usually persists for 3–5 days. Notice that this corresponds to the beginning of the pre-ovulatory phase of the ovarian cycle. During menstruation, the endometrium (inner lining of the uterus) breaks down and leaves the body through the vagina. The menstrual discharge consists of blood, small clusters of endometrial cells, and mucus. After menstruation, the endometrium regrows. It continues to thicken through the time of ovulation, reaching a maximum at about 20–25 days. If an embryo has not implanted in the uterine lining by this time, menstruation begins again, marking the start of the next ovarian and menstrual cycles.

Now let's consider the hormones that regulate the ovarian and menstrual cycles. The ebb and flow of the five hormones listed in Table 27.6 synchronize events in the ovarian cycle (the growth of the follicle and ovulation) with events in the menstrual cycle (preparation of the uterine lining for possible implantation of an embryo). A releasing hormone from the hypothalamus in the brain regulates secretion of the two pituitary hormones FSH and LH. The blood levels of FSH, LH, and two other hormones—estrogen and progesterone—coincide with specific events in the ovarian and menstrual cycles.

Hormonal Events Before Ovulation Focusing on Part A of Figure 27.6, we see that the releasing hormone from

TABLE 27.6	HORMONES OF THE OVARIAN AND MENSTRUAL CYCLES	
Hormone	**Secreted by**	**Major Roles**
Releasing hormone	Hypothalamus	Regulates secretion of LH and FSH by pituitary
Follicle-stimulating hormone (FSH)	Pituitary	Stimulates growth of ovarian follicle
Leuteinizing hormone (LH)	Pituitary	Stimulates growth of ovarian follicle and production of secondary oocyte; promotes ovulation; promotes development of corpus luteum and secretion of hormones
Estrogen	Ovarian follicle	Low levels inhibit pituitary; high levels stimulate hypothalamus; promotes endometrium
Estrogen and progesterone	Corpus luteum	Maintain endometrium; high levels inhibit hypothalamus and pituitary; sharp drops promote menstruation

the hypothalamus stimulates the anterior pituitary ❶ to increase its output of FSH and LH. True to its name, ❷ FSH stimulates the growth of an ovarian follicle, in effect starting the ovarian cycle. In turn, the follicle secretes estrogen. Early in the pre-ovulatory phase, the follicle is small (Part C) and secretes relatively little estrogen (Part D). As the follicle grows, ❸ it secretes more and more estrogen, and the rising but still relatively low level of estrogen exerts negative feedback on the pituitary. This keeps the blood levels of FSH and LH low for most of the pre-ovulatory phase (Part B). As the time of ovulation approaches, hormone levels change drastically, with estrogen reaching a critical peak (Part D) just before ovulation. This high level of estrogen exerts positive feedback on the hypothalamus (green arrow in Part A), which then ❹ makes the pituitary secrete bursts of FSH and LH. By comparing Parts B and D of the figure, you can see that the peaks in FSH and LH occur just after the estrogen peak. It may help to place a piece of paper over the figure and slide it slowly to the right. As you uncover the figure, you will see the follicle getting bigger and the estrogen level rising to its peak, followed almost immediately by the LH and FSH peaks. ❺ Then, just to the right of the peaks, comes the dashed line representing ovulation.

Hormonal Events at Ovulation and After LH stimulates the completion of meiosis, transforming the primary oocyte in the follicle into a secondary oocyte. It also signals enzymes to rupture the follicle, allowing ovulation to occur, and triggers the development of the corpus luteum from the ruptured follicle (hence its name, luteinizing hormone). LH also promotes the secretion of progesterone and estrogen by the corpus luteum. In Part D of the figure, you can see the progesterone peak and the second (lower and wider) estrogen peak after ovulation.

High levels of estrogen and progesterone in the blood following ovulation have a strong influence on both the ovary and uterus. The combination of the two hormones exerts negative feedback on the hypothalamus and pituitary, producing ❻ falling FSH and LH levels. The drops in FSH and LH prevent follicles from developing and ovulation from occurring during the post-ovulatory phase. Also, the LH drop is followed by the gradual degeneration of the corpus luteum. Near the end of the post-ovulatory phase, unless an embryo has implanted in the uterus, the corpus luteum stops secreting estrogen and progesterone. ❽ As blood levels of these hormones decline, the hypothalamus once again can stimulate the pituitary to secrete more FSH and LH, and a new cycle begins.

Control of the Menstrual Cycle Hormonal regulation of the menstrual cycle is simpler than that of the ovarian cycle. The menstrual cycle (Part E) is directly controlled by estrogen and progesterone alone. You can see the effects of these hormones by comparing Parts D and E of the figure. Starting around day 5 of the cycle, the endometrium thickens in response to the rising levels of estrogen and, later, progesterone. ❼ When the levels of these hormones drop, the endometrium begins to slough off. Menstrual bleeding begins soon thereafter, on day 1 of a new cycle.

■   ■   ■

We have now described what happens in the human ovary and uterus in the absence of fertilization. As we'll see later, the ovarian and menstrual cycles are put on hold if fertilization and pregnancy occur. Early in pregnancy, the developing embryo, implanted in the endometrium, releases a hormone (human chorionic gonadotropin, or HCG). This hormone acts like LH in that it maintains the corpus luteum, which continues to secrete progesterone and estrogen, keeping the endometrium intact. We'll return to the events of pregnancy in Modules 27.15 and 27.16.

🎧 MP3 Tutor *The Female Reproductive System*

Web Activity *Reproductive System of the Human Female*

? Which hormonal change triggers the onset of menstruation?

■ The drop in the levels of estrogen and progesterone. These changes are caused by negative feedback of these hormones on the hypothalamus and pituitary after ovulation.

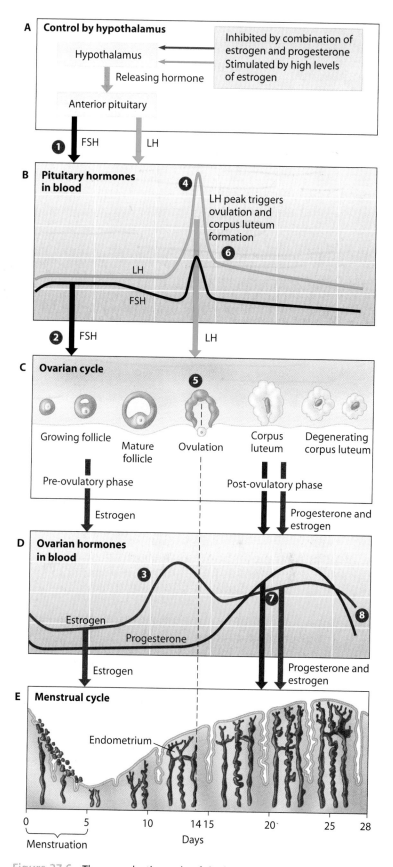

Figure 27.6 The reproductive cycle of the human female

27.7 Sexual activity can transmit disease

Sexually transmitted diseases (STDs) are contagious diseases spread by sexual contact. Table 27.7 lists the most common STDs, organized by the type of infectious agent. Notice that bacteria, viruses, protists, and fungi can all cause STDs.

Bacterial STDs are generally curable, but treatment must be given early, before any permanent damage is done. The most common bacterial STD (with nearly a million new cases reported annually in the United States and over 90 million worldwide) is **chlamydia**. Chlamydia poses a public health challenge since it is frequently "silent"—that is, it often produces no visible symptoms. Health officials estimate that for every reported case of chlamydia two other cases go unreported. The primary symptom of chlamydia is a burning sensation during urination and/or genital discharge, but half of infected men and three-quarters of infected women do not notice any symptoms. Long-term complications are rare among men, but up to 40% of infected women develop pelvic inflammatory disease (PID). The inflammation associated with PID may block the oviducts or scar the uterus, causing infertility.

Luckily, treatment for chlamydia is easy: a single dose of an antibiotic usually completely cures the disease. But early screening is required to catch the disease before any scarring occurs. Sexually active women under 25 are encouraged to be screened for chlamydia and other STDs annually.

In contrast to bacterial STDs, viral STDs are not curable. They can be controlled by medications, but symptoms and the ability to infect others remain a possibility through a person's lifetime. One in five American adolescents and adults are infected with genital herpes. **Genital herpes** is usually caused by the herpes simplex virus type 2 (HSV-2), a slight variant of the virus that causes oral cold sores (see the introduction to Chapter 10 and Module 10.18). Symptoms first appear about a week after exposure. Blisters form on the external genitalia. After a few days, the blisters change to scabs that fall off. Most outbreaks heal within a few weeks without leaving a scar. But the virus is not gone: It lies dormant within nearby nerve cells. Months or years later, the virus can reemerge, causing fresh sores that allow the virus to be spread to sexual partners. Abstinence during outbreaks, the use of condoms, and the use of antiviral medications that minimize symptoms can reduce the possibility of new infections. But there is no cure for genital herpes; infection lasts a lifetime.

AIDS, caused by HIV (see Modules 10.20 and 24.13), poses one of the greatest health challenges in the world today, particularly among the nonindustrialized nations of Africa and Asia. But even within the United States, there are 40,000 new infections each year, one-third of which result from heterosexual contact.

Many STDs can cause long-term problems or even death if left untreated. Anyone who is sexually active should have regular medical exams, be tested for STDs, and seek immediate help if any suspicious symptoms appear—even if they are mild. STDs are most prevalent among teenagers and young adults; nearly two-thirds of infections occur among people under 25. The best way to avoid the spread of STDs is, of course, abstinence. Alternatively, latex condoms provide the best protection for "safe sex."

TABLE 27.7	STDS COMMON IN THE UNITED STATES		
Disease	**Microbial Agent**	**Major Symptom and Effects**	**Treatment**
Bacterial			
Chlamydia	*Chlamydia trachomatis*	Genital discharge, itching, and/or painful urination; often no symptoms in women; pelvic inflammatory disease (PID)	Antibiotics
Gonorrhea	*Neisseria gonorrhoeae*	Genital discharge; painful urination; sometimes no symptoms in women; PID	Antibiotics
Syphilis	*Treponema pallidum*	Ulcer (chancre) on genitalia in early stages; spreads throughout body and can be fatal if not treated	Antibiotics can cure in early stages
Viral			
Genital herpes	Herpes simplex virus type 2, occasionally type 1	Recurring symptoms: small blisters on genitalia, painful urination, skin inflammation; linked to cervical cancer, miscarriage, birth defects	Valacyclovir can prevent recurrences
Genital warts	Papillomaviruses	Painless growths on genitalia; some of the viruses linked to cancer	Removal by freezing
AIDS and HIV infection	HIV	See Module 24.13	Combination of drugs
Protozoan			
Trichomoniasis	*Trichomonas vaginalis*	Vaginal irritation, itching, and discharge; usually no symptoms in men	Antiprotozoal drugs
Fungal			
Candidiasis (yeast infections)	*Candida albicans*	Similar to symptoms of trichomoniasis; frequently acquired nonsexually	Antifungal drugs

? How are bacterial STDs different from viral STDs in terms of their long-term prognosis?

■ Bacterial STDs can be cured; viral STDs can be controlled but not cured.

27.8 Contraception can prevent unwanted pregnancy

Contraception is the deliberate prevention of pregnancy. Table 27.8 lists common methods of contraception, with their failure rates when used correctly and when used typically. Note that these two rates are often quite different, emphasizing the importance of learning to use contraception correctly. It is also important to note that the "safe sex" provided by condoms can prevent both unwanted pregnancy and sexually transmitted diseases (STDs); other contraceptive methods do not prevent STDs.

Complete abstinence (avoiding intercourse) is the only totally effective method of birth control, but other methods are effective to varying degrees. Sterilization, surgery that prevents sperm from reaching an egg, is very reliable. A woman may have a **tubal ligation** ("having her tubes tied"), in which a doctor removes a short section from each oviduct (and may tie, or ligate, the remaining ends). A man may undergo a **vasectomy,** in which a doctor cuts a section out of each vas deferens to prevent sperm from reaching the urethra. Both forms of sterilization are relatively safe and free from side effects. Sterilization procedures are generally considered permanent. However, surgical reversals of tubal ligations or vasectomies are becoming increasingly successful, although these major surgeries carry some risk.

The effectiveness of other methods of contraception depends on how they are used. Temporary abstinence, also called the **rhythm method** or **natural family planning,** depends on refraining from intercourse during the days around ovulation, when fertilization is most likely. In theory, the time of ovulation can be determined by monitoring changes in body temperature and the composition of cervical mucus, but careful monitoring and record keeping is required. Additionally, the length of the reproductive cycle can vary from month to month, and sperm can survive for 3–5 days within the female reproductive tract, making natural family planning among the most unreliable methods of contraception in actual practice. Withdrawal of the penis from the vagina before ejaculation is also ineffective because sperm may be released before climax.

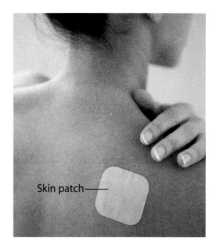

Skin patch

Figure 27.8
A contraceptive skin patch

If used correctly, barrier methods can be quite effective at physically preventing the union of sperm and egg. Condoms are sheaths, usually made of latex, that fit over the penis or within the vagina. A diaphragm is a dome-shaped rubber cap that covers the cervix; a cervical cap is similar but thimble-shaped and smaller. Both require a doctor's visit for proper fitting. Barrier devices (including condoms) are more effective when used in combination with **spermicides,** sperm-killing foam or jelly; spermicides used alone are unreliable.

Some of the most effective methods of contraception work by preventing the release of gametes. **Oral contraceptives,** or **birth control pills,** come in several different forms that contain synthetic estrogen and/or progesterone (or a synthetic progesterone-like hormone called progestin). In addition to pills, various combinations of these hormones are also available as an injection (Depo-Provera), a ring inserted into the vagina, or a skin patch (Figure 27.8). Steady intake of these hormones simulates their constant levels during pregnancy. In response, the hypothalamus fails to send the signals that start development of an ovarian follicle. Ovulation ceases, preventing pregnancy.

Certain drugs can prevent fertilization or implantation even after intercourse has occurred. Combination birth control pills can be prescribed in high doses for emergency contraception, also called **morning-after pills (MAPs).** If taken within three days after unprotected intercourse, MAPs are about 75% effective at preventing pregnancy. Such treatments should only be used in emergencies because they have significant side effects.

If pregnancy has already occurred, the drug mifepristone, or RU486, can induce an abortion, the termination of a pregnancy in progress. If taken within the first seven weeks, RU486 blocks progesterone receptors in the uterus, thus preventing progesterone from maintaining pregnancy. Mifepristone requires a doctor's prescription and several visits to a medical facility. → MISCARRIAGE

TABLE 27.8	**CONTRACEPTIVE METHODS**	
	Pregnancies/ 100 Women/Year*	
Method	Used Perfectly	Typically
Birth control pill (combination)	0.1	5
Vasectomy	0.1	0.15
Tubal ligation	0.2	0.5
Rhythm method	1–9	20
Withdrawal	4	19
Condom (male)	3	14
Diaphragm and spermicide	6	20
Spermicide alone	6	26

*Without contraception, about 85 pregnancies would occur.

? Based on typical usage, what are the most reliable contraceptive methods for women and men?

Tubal ligation for women and vasectomy for men

27.9 Fertilization results in a zygote and triggers embryonic development

The last six modules have focused on the anatomy and physiology of the human reproductive system. In the next six modules, we will examine the results of reproduction: the formation and development of an embryo. To understand these concepts, we will examine the processes of development in some well-studied animal models (sea urchin and frog). But the concepts presented in this section apply to most vertebrates. We will return to the human story in the final section of this chapter.

Embryonic development begins with fertilization, the union of a sperm and an egg to form a diploid zygote. Fertilization combines haploid sets of chromosomes from two individuals and also activates the egg by triggering metabolic changes that start embryonic development.

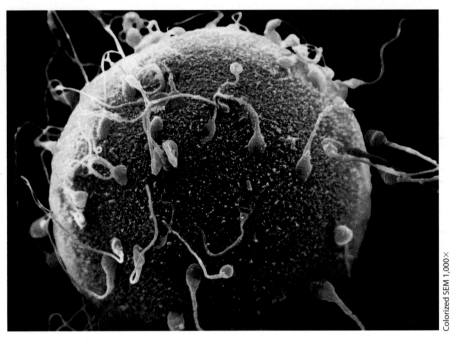

Colorized SEM 1,000×

Figure 27.9A   A human egg cell surrounded by sperm

The Properties of Sperm Cells   Figure 27.9A is a micrograph of an unfertilized human egg surrounded by sperm. Of all these sperm, only one will enter and fertilize the egg. All the other sperm—the ones shown here and millions more that were ejaculated with them—will die. The one sperm that penetrates the egg adds its unique set of genes to those of the egg and contributes to the next generation.

Figure 27.9B illustrates the structure of a mature human sperm cell, a clear case of form fitting function. The sperm's streamlined shape is an adaptation for swimming through fluids in the vagina, uterus, and oviduct of the female. It's thick head contains a haploid nucleus and is tipped with a vesicle, the **acrosome,** which lies just inside the plasma membrane. The acrosome contains enzymes that help the sperm penetrate the egg. The neck and middle piece of the sperm contain a long, spiral mitochondrion. The sperm absorbs high-energy nutrients, especially the sugar fructose, from the semen. Thus fueled, its mitochondrion provides ATP for movement of the tail, which is actually a flagellum. By the time a sperm has reached the egg, it has consumed much of the energy available to it. But a successful sperm will have enough energy left to penetrate the egg and deposit its nucleus in the egg's cytoplasm.

The Process of Fertilization   Figure 27.9C on the facing page illustrates the sequence of events in fertilization. This diagram is based on fertilization in sea urchins (phylum Echinodermata—see Module 18.13), on which a great deal of research has been done. Similar processes occur in other

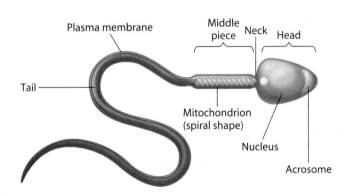

Figure 27.9B   The structure of a human sperm cell

animals, including humans. The diagram traces one sperm through the successive activities of fertilization. Notice that to reach the egg nucleus, the sperm nucleus must pass through three barriers: the egg's jelly coat (yellow), a middle region of glycoproteins called the vitelline layer (pink), and the egg cell's plasma membrane.

Let's follow the steps shown in the figure. ❶ As a sperm approaches an egg, it first must squeeze through cells left-over from the follicle. ❷ Next, the sperm contacts the jelly coat of the egg; as it does so, the acrosome in the sperm head releases a cloud of enzyme molecules that digest a cavity into the jelly. When the sperm head reaches the vitelline layer,

❸ species-specific protein molecules on its surface bind with specific receptor proteins on the vitelline layer. The binding between these proteins ensures that sperm of other species cannot fertilize the egg. This specificity is especially important when fertilization is external, because the sperm of other species may be present in the water. After the specific binding occurs, the sperm proceeds through the vitelline layer, and ❹ the sperm's plasma membrane fuses with that of the egg. Fusion of the two membranes ❺ makes it possible for the sperm nucleus to enter the egg.

Fusion of the sperm and egg plasma membranes triggers a number of important changes in the egg. Two such changes prevent other sperm from entering the egg. About 1 second after the membranes fuse, the entire egg plasma membrane becomes impenetrable to other sperm cells. Shortly thereafter, ❻ the vitelline layer hardens and separates from the plasma membrane. The space quickly fills with water, and the vitelline layer becomes the so-called fertilization envelope, another barrier impenetrable to sperm. If these events did not occur and an egg were fertilized by more than one sperm, the resulting zygote nucleus would contain too many chromosomes, and the zygote could not develop normally.

Membrane fusion also triggers a burst of metabolic activity in the egg. In preparation for the enormous growth and development that will follow fertilization, the egg gears up from near dormancy, increasing cellular respiration and protein synthesis. Next, ❼ the egg and sperm nuclei fuse, producing the diploid nucleus of the zygote.

Note that the sperm provided chromosomes to the zygote, but little else. The zygote's cytoplasm and various organelles were all provided by the mother through the egg. In the next module, we begin to trace the development of the zygote into a new animal.

? What is the function of the fertilization envelope?

■ It helps prevent the entry of more than one sperm into the egg.

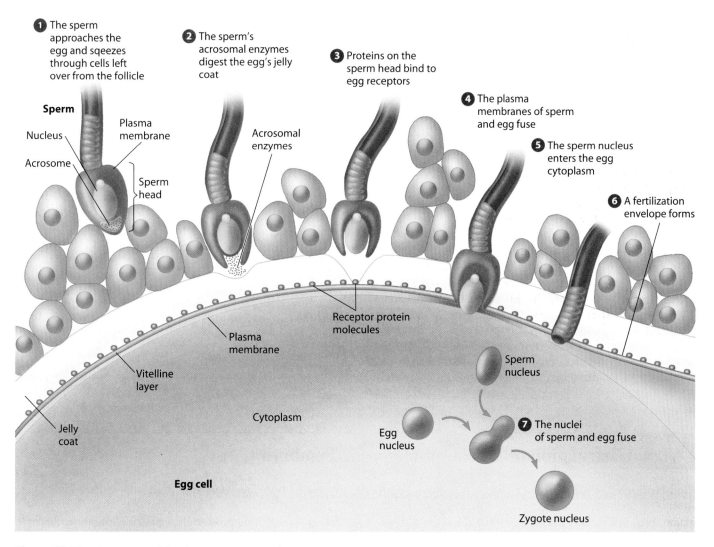

Figure 27.9C The process of fertilization in a sea urchin

27.10 Cleavage produces a ball of cells from the zygote

An animal consists of many thousands, millions, even trillions of cells organized into complex tissues and organs. The transformation from a zygote to this multicellular state is truly phenomenal. Order and precision are required at every step, and both are clearly displayed in the first two major phases of embryonic development: cleavage and gastrulation.

Cleavage is a rapid succession of cell divisions that produces a ball of cells—a multicellular embryo—from the zygote. Nutrients stored in the egg nourish the dividing cells. DNA replication, mitosis, and cytokinesis occur rapidly, but gene transcription virtually shuts down and few new proteins are synthesized. The embryo does not enlarge significantly; cleavage simply partitions the cytoplasm of one large cell, the zygote, into many smaller cells, each with its own nucleus.

Figure 27.10 illustrates cleavage in a sea urchin. (Similar processes occur in humans.) As the first three steps show, the number of cells doubles with each cleavage division. In a sea urchin, a doubling occurs about every 20 minutes, and the whole cleavage process takes about 3 hours to produce a solid ball of cells. Notice that each cell in the ball is much smaller than the zygote. As cleavage continues, a fluid-filled cavity called the **blastocoel** forms in the center of the embryo. At the completion of cleavage, there is a large cavity surrounded by one or more layers of cells. This hollow ball of cells is called the **blastula.**

Cleavage makes two important contributions to early development. It creates a multicellular embryo, the blastula, from a single-celled zygote. Cleavage is also an organizing process, partitioning the multicellular embryo into developmental regions. As we discussed in Module 11.10, the cytoplasm of the zygote contains a variety of chemicals that control gene expression during early development. During cleavage, regulatory chemicals become localized in particular groups of cells, where they later activate the genes that direct the formation of specific parts of the animal. Gastrulation, the next phase of development, further refines the embryo's cellular organization.

Rarely, and apparently at random, a cell in the early embryo may separate and "reset" to act as a zygote; the result is the development of identical (monozygotic) twins. (Nonidentical, or dizygotic, twins result from a completely

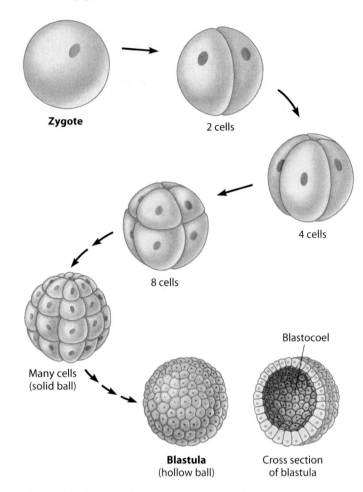

Figure 27.10 Cleavage in a sea urchin

different mechanism: Two separate eggs fuse with two separate sperm to produce two genetically unique zygotes that develop in simultaneously.)

🎧 **MP3 Tutor** *Embryonic Development*

? How does the reduction of cell size during cleavage increase oxygen supply to the cells' mitochondria? (*Hint:* Review Module 4.2.)

■ Smaller cells have a greater plasma membrane surface area relative to cellular volume, and this facilitates diffusion of oxygen from the environment to the cell's cytoplasm.

27.11 Gastrulation produces a three-layered embryo

After cleavage, the rate of cell division slows dramatically. Groups of cells then undergo **gastrulation,** the second major phase of embryonic development. During gastrulation, cells take up new locations that will allow later formation of all the organs and tissues. As gastrulation proceeds, the embryo is organized into a three-layer stage called a **gastrula.**

The three layers produced in gastrulation are embryonic tissues called **ectoderm, endoderm,** and **mesoderm.** The ectoderm forms the outer layer (skin) of the gastrula. The endoderm forms an embryonic digestive tract. And the mesoderm partly fills the space between the ectoderm and endoderm. Eventually, these three cell layers develop into all the parts of the adult animal. For instance, our nervous system and the outer layer (epidermis) of our skin come from ectoderm; the innermost lining of our digestive tract arises from endoderm; and most other organs and tissues, such as the kidney, heart, muscles, and the inner layer of our skin (dermis), develop from mesoderm. Table 27.11 lists the major organs and tissues that arise in most vertebrates from each of the three main embryonic tissue layers.

The mechanics of gastrulation vary somewhat depending on the species. We have chosen the frog, a vertebrate that has long been a favorite of researchers, to demonstrate how gastrulation produces three cell layers. The top of Figure 27.11 shows the frog blastula, formed by cleavage (as discussed in the previous module). The blastula is a partially hollow ball of unequally sized cells. The cells toward one end, called the animal pole, are smaller than those near the opposite end, the vegetal pole. The three colors on the blastula in the figure indicate regions of cells that will give rise to the primary cell layers in the gastrula at the bottom of the figure: ectoderm (blue), endoderm (yellow), and mesoderm (red). (Notice that each layer may be more than one cell thick.)

During gastrulation (shown in the center of Figure 27.11), cells migrate to new positions that will form the three layers. Gastrulation begins when a small groove, called the blastopore, appears on one side of the blastula. It is in the blastopore that cells of the future endoderm (yellow) move inward from the surface and fold over to produce a simple digestive cavity. Meanwhile, the cells that will form ectoderm (blue) spread downward over more of the surface of the embryo, and the cells that will

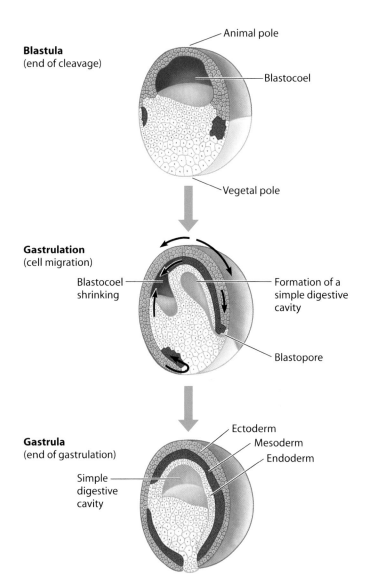

Figure 27.11   Development of the frog gastrula

form mesoderm (red) begin to spread into a thin layer inside the embryo, forming a middle layer between the other two.

As shown at the bottom of the figure, gastrulation is completed when cell migration has resulted in a three-layered embryo. Ectoderm covers most of the surface. Mesoderm forms a layer between the ectoderm and the endoderm.

Although gastrulation differs in detail from one animal group to another, the process is driven by the same general mechanisms in all species. The timing of these events also varies with the species. In many frogs, for example, cleavage and gastrulation together take about 15–20 hours.

Web Activity *Sea Urchin Development Video*

? The first two phases of embryonic development are _____ which forms the blastula followed by _____ which forms the _____.

cleavage · · · gastrulation · · · gastrula

TABLE 27.11	DERIVATIVES OF THE THREE EMBRYONIC TISSUE LAYERS
Embryonic Layer	**Organs and Tissues in the Adult**
Ectoderm	Epidermis of skin; epithelial lining of mouth and rectum; sense receptors in epidermis; cornea and lens of eye; nervous system
Endoderm	Epithelial lining of digestive tract (except mouth and rectum); epithelial lining of respiratory system; liver; pancreas; thyroid; parathyroids; thymus; lining of urethra, urinary bladder, and reproductive system
Mesoderm	Skeletal system; muscular system; circulatory system; excretory system; reproductive system (except gamete-forming cells); dermis of skin; lining of body cavity

27.12 Organs start to form after gastrulation

In organizing the embryo into three layers, gastrulation sets the stage for the shaping of an animal. Once the ectoderm, endoderm, and mesoderm form, cells in each layer begin to differentiate into tissues and embryonic organs. The cutaway drawing in Figure 27.12A shows the developmental structures that appear in a frog embryo a few hours after the completion of gastrulation. The orientation drawing at the upper left of the figure indicates a corresponding cut through an adult frog.

We see two structures in the embryo in Figure 27.12A that were not present at the gastrula stage described in the last module. An organ called the notochord has developed in the mesoderm, and a structure that will become the hollow nerve cord is beginning to form in the ectoderm (in the region colored green). Recall that the notochord and dorsal, hollow nerve cord are hallmarks of the chordates (see Module 18.14).

The notochord is visible in the mesoderm in the drawing in Figure 27.12A. Made of a substance similar to cartilage, the **notochord** extends for most of the embryo's length and provides support for other developing tissues. Later in development, the notochord will function as a core around which mesodermal cells gather and form the backbone.

You can also see in Figure 27.12A the beginnings of the frog's hollow nerve cord, formed from a portion of the ectoderm. The area shown in green in the cutaway drawing is a thickened region of ectoderm called the neural plate. From it arises a pair of pronounced ectodermal ridges, called neural folds, visible in both the drawing and the micrograph below it. If you now look at the series of diagrams in Figure 27.12B, you will see what happens as the neural folds and neural plate develop further. The neural plate rolls up and forms the neural tube, which then sinks beneath the surface of the embryo and is covered by an outer layer of ectoderm. If you look carefully at the figure, you'll see that cells of the ectoderm fold inward by changing shape, first elongating and then becoming wedge-shaped. The result is a tube of ectoderm—the **neural tube**—which is destined to become the brain and spinal cord.

Figure 27.12C shows a later frog embryo (about 12 hours older than the one in Figure 27.12A), in which the neural tube has formed. Notice in the drawing that the neural tube lies directly above the notochord. The relative positions of the neural tube, notochord, and digestive cavity give us a preview of the basic body plan of a frog. The spinal cord will lie within

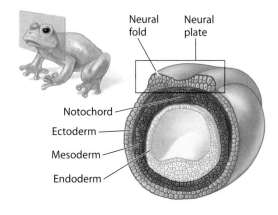

Neural fold   Neural plate

Notochord
Ectoderm
Mesoderm
Endoderm

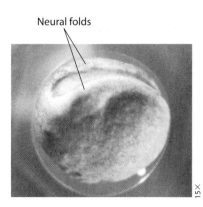

Neural folds

15×

Figure 27.12A The beginning of organ development in a frog: the notochord, neural folds, and neural plate

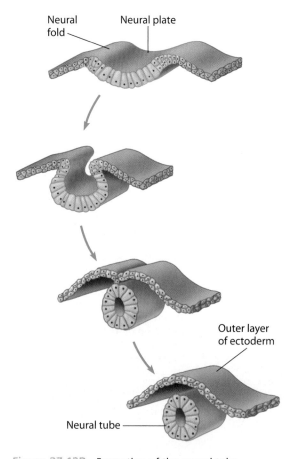

Neural fold   Neural plate

Outer layer of ectoderm

Neural tube

Figure 27.12B Formation of the neural tube

extensions of the dorsal (upper) surface of the backbone (which will replace the notochord), and the digestive tract will be ventral to (beneath) the backbone. We see this same arrangement of organs in all vertebrates.

The importance of these processes is underscored by birth defects that result from improper signaling. Spina bifida is a condition that results from the failure of the tube of ectoderm cells to close properly and form the spine during the first month of fetal development. Infants born with spina bifida often have permanent nerve damage that results in paralysis of the lower limbs.

Figure 27.12C shows several other fundamental changes. In the micrograph, which is a side view, you can see that the embryo is more elongated than the one in Figure 27.12A. You can also see the beginnings of an eye and a tail (called the tail bud). Part of the ectoderm has been removed to reveal a series of internal ridges called somites. The somites are blocks of mesoderm that will give rise to segmental structures (constructed of repeating units), such as the vertebrae and associated muscles of the backbone. In the cross-sectional drawing, notice that the mesoderm next to the somites is developing a hollow space—the body cavity, or **coelom.** Segmented body parts and a coelom are basic features of chordates (see Module 18.14).

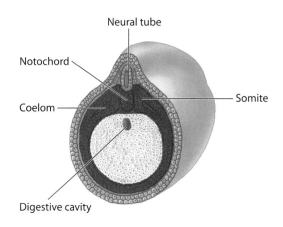

Neural tube

Notochord

Coelom

Somite

Digestive cavity

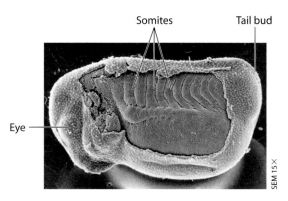

Somites Tail bud

Eye

SEM 15×

Figure 27.12C   An embryo with completed neural tube, somites, and coelom

Figure 27.12D   A tadpole

In this and the previous two modules, we have observed the sequence of changes that occur as an animal begins to take shape. To summarize, the key phases in embryonic development are cleavage (which creates a multicellular blastula from a zygote), gastrulation (which organizes the embryo into a gastrula with three discrete layers), and organ formation (which generates embryonic organs from the three embryonic tissue layers). These same three phases occur in nearly all animals.

If we followed a frog's development beyond the stage represented in Figure 27.12C, within a few hours we would be able to monitor muscular responses and a heartbeat and see a set of gills with blood circulating in them. A long tail fin would grow from the tail bud. The timing of the later stages in frog development varies enormously, but in many species, by 5–8 days after development begins, we would see all the body tissues and organs of a tadpole emerge from cells of the ectoderm, mesoderm, and endoderm. Eventually, the structures of the tadpole (Figure 27.12D) would transform into the tissues and organs of an adult frog.

Watching embryos develop helps us appreciate the enormous changes that occur as one tiny cell, the zygote, gives rise to a highly structured, many-celled animal. Your own body, for instance, is a complex organization of some 60 trillion cells, all of which arose from a zygote smaller than the period at the end of this sentence. Discovering how this incredibly intricate arrangement is achieved is one of biology's greatest challenges. Through research that combines the experimental manipulation of embryos with molecular biology and genetics, developmental biologists have begun to work out the mechanisms that underlie development. We examine several of these mechanisms in the next module.

Web Activity *Frog Development Video*

? What is the embryonic basis for the dorsal, hollow nerve cord that is common to all members of our phylum?

■ The nerve cord, which becomes the brain and spinal cord, develops from a dorsal ectodermal plate that folds to form an interior tube.

27.13 Multiple processes give form to the developing animal

The development of an animal embryo depends on several kinds of cellular processes. For example, in the last module you learned how changes in cell shape help form the neural tube (see Figure 27.12B).

Most developmental processes depend on signals passed between neighboring cells and cell layers, telling embryonic cells precisely what to do when. The mechanism by which one group of cells influences the development of an adjacent group of cells is called **induction.** Induction may be mediated by diffusible signals or, if the cells are in direct contact, by cell-surface interactions. Induction plays a major role in the early development of virtually all tissues and organs. Its effect is to switch on a set of genes whose expression makes the receiving cells differentiate into a specific tissue. Many inductions involve a sequence of inductive steps from different surrounding tissues that progressively determine the fate of cells. In the eye, for example, lens formation involves precisely timed inductive signals from ectodermal, mesodermal, and endodermal cells. In the developing animal, a sequence of inductive signals leads to increasingly greater specialization of cells as organs begin to take shape.

Cell migration is also essential in development. For example, during gastrulation, cells "crawl" within the embryo by extending and contracting cellular protrusions, similar to the pseudopodia of amoeboid cells. Migrating cells may follow inductive chemical trails secreted by cells near their specific destination. Once a migrating cell reaches its destination, surface proteins enable it to recognize similar cells. The cells join together and secrete glycoproteins that glue them in place.

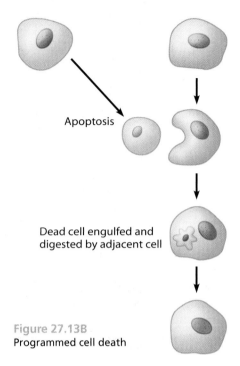

Figure 27.13B
Programmed cell death

Finally, they differentiate, taking on the characteristics of a particular tissue.

Another developmental process is **programmed cell death,** or **apoptosis,** the timely and tidy suicide of cells. Animals have genes coding for proteins that kill the cell that produces them. In humans, the timely death of specific cells in developing hands and feet creates the spaces between fingers and toes (Figure 27.13A). In Figure 27.13B, the cell on the left shrinks and dies because a suicide gene has been turned on. Meanwhile, signals from the dying cell make an adjacent cell phagocytic. This cell engulfs and digests the dead cell, keeping the embryo free of harmful debris.

Web Process of Science What Determines Cell Differentiation in the Sea Urchin?

? How do signal transduction pathways function in induction? (You may wish to review Module 11.12.)

■ They mediate between the chemical signal received by the cell and the resulting changes in gene expression and other responses by the cell.

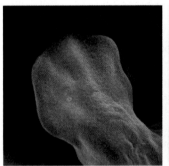 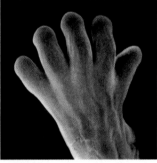

Figure 27.13A Apoptosis in a developing human hand

27.14 Pattern formation during embryonic development is controlled by ancient genes

So far, we have discussed the formation of individual organs. What directs the formation of larger scale structures? The shaping of an animal's major parts involves **pattern formation,** the emergence of a body form with specialized organs and tissues in the right places. Research indicates that master control genes (see Modules 11.10 and 15.11) respond to

chemical signals that tell a cell where it is relative to other cells in the embryo. These positional signals determine which master control genes will be expressed and, consequently, which body parts will form. Research has shown that such control genes arose early in the evolution of animals and so play similar roles across diverse animal groups. The field of

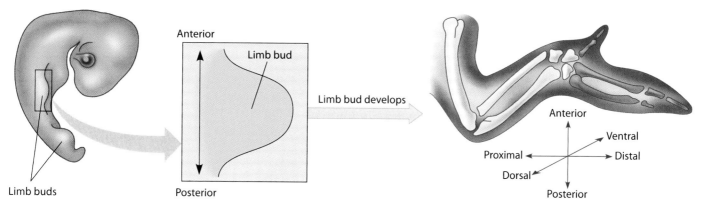

Figure 27.14A   The normal development of a wing

biology that studies the evolution of developmental processes is called "evo-devo."

Vertebrate limbs, such as bird wings, begin as embryonic structures called limb buds (Figure 27.14A). Each component of a chick wing, such as a specific bone or muscle, develops with a precise location and orientation relative to three axes: the proximal-distal axis (the "shoulder-to-fingertip" axis), the anterior-posterior axis (the "thumb-to-little finger" axis), and the dorsal-ventral axis (the "knuckle-to-palm" axis). The embryonic cells within a limb bud respond to positional information indicating location along these three axes. Only with this information will the cell's genes direct the synthesis of the proteins needed for normal differentiation in that cell's specific location.

Among the most exciting biological discoveries in recent years is that a class of similar genes—**homeotic genes**—help direct embryonic pattern formation in a wide variety of organisms. Researchers studying homeotic genes in fruit flies found a common structural feature: Every homeotic gene they looked at contained a common sequence of 180 nucleotides. Very similar sequences have since been found in virtually every eukaryotic organism examined so far, including yeasts, plants, and humans—and even some prokaryotes. These nucleotide sequences are called **homeoboxes,** and each is translated into a segment (60 amino acids long) of the protein product of the homeotic gene. The homeobox polypeptide segment binds to specific sequences in DNA, enabling homeotic proteins that contain it to turn groups of genes on or off during development.

Figure 27.14B highlights some striking similarities in the chromosomal locations and the developmental roles of some homeobox-containing homeotic genes in two quite different animals. The figure shows portions of chromosomes that carry homeotic genes in the fruit fly and the mouse. The colored boxes represent homeotic genes that are very similar in flies and mice. Notice that the order of genes on the fly chromosome is the same as on the four mouse chromosomes and that the gene order on the chromosomes corresponds to analogous body regions in both animals. These similarities suggest that the original version of these homeotic genes arose very early in the history of life and that the genes have remained remarkably unchanged for eons of animal evolution. By their presence in such diverse creatures, homeotic genes illustrate one of the central themes of biology: unity in diversity due to shared evolutionary history.

A major goal of developmental research is to learn how the one-dimensional information encoded in the nucleotide sequence of a zygote's DNA directs the development of the three-dimensional form of an animal. Pattern formation requires cells to receive and interpret environmental cues that vary from one location to another. These cues, acting together along three axes, tell cells where they are in the three-dimensional realm of a developing organ. In the next two modules, we'll see the results of this process as we watch an individual of our own species take shape.

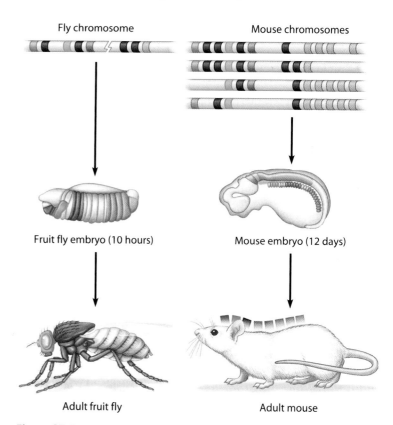

Figure 27.14B   Comparison of fruit fly and mouse homeotic genes

? How is pattern formation already apparent at the gastrula stage? (*Hint:* Review Figure 27.11.)

■ The major axes of the animal—anterior/posterior, dorsal/ventral, and proximal/distal—are already set at the gastrula stage.

27.15 The embryo and placenta take shape during the first month of pregnancy

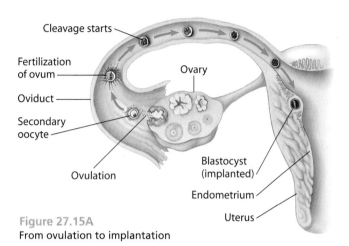

Figure 27.15A
From ovulation to implantation

Labels: Cleavage starts, Fertilization of ovum, Oviduct, Secondary oocyte, Ovulation, Ovary, Blastocyst (implanted), Endometrium, Uterus

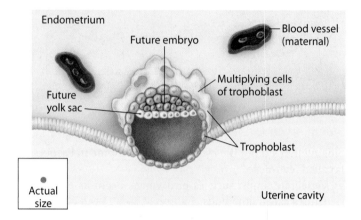

Figure 27.15C   Implantation under way (about 7 days)

Labels: Endometrium, Future embryo, Blood vessel (maternal), Multiplying cells of trophoblast, Future yolk sac, Trophoblast, Actual size, Uterine cavity

Pregnancy, or **gestation,** is the carrying of developing young within the female reproductive tract. It begins at the fertilization of the egg by a sperm and continues until birth. Duration of pregnancy varies considerably among animal species; gestation in mice lasts about 21 days, while elephants carry their young for 600 days. Human pregnancy averages 266 days (38 weeks) from fertilization (also called **conception** in humans), or 40 weeks from the start of the last menstrual cycle.

An Overview of Developmental Events   The figures in this module illustrate, in cross section, the changes that occur during the first month of human development. The insets at the lower left of Figures 27.15C–27.15F show the embryo's actual size at each stage.

Fertilization occurs in the oviduct (Figure 27.15A). Cleavage starts about 24 hours after fertilization and continues as the embryo moves down the oviduct toward the uterus. By the sixth or seventh day after fertilization, the embryo has reached the uterus, and cleavage has produced about 100 cells. The embryo is now a hollow sphere of cells called a **blastocyst** (the mammalian equivalent of the sea urchin blastula we saw in Figure 27.10).

The human blastocyst (Figure 27.15B) has a fluid-filled cavity, an inner cell mass that will actually form the baby, and an outer layer of cells called the **trophoblast.** The trophoblast secretes enzymes that enable the blastocyst to implant in the endometrium, the uterine lining (gray in all the figures).

The blastocyst starts to implant in the uterus about a week after conception. In Figure 27.15C, you can see extensions of the trophoblast spreading into the endometrium; these extensions consist of multiplying cells. The

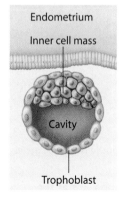

Figure 27.15B
Blastocyst (6 days after conception)

Labels: Endometrium, Inner cell mass, Cavity, Trophoblast

trophoblast cells eventually form part of the **placenta,** the organ that provides nourishment and oxygen to the embryo and helps dispose of its metabolic wastes. As we'll see, the placenta consists of both embryonic and maternal tissues.

In Figure 27.15C, the cells colored purple and yellow are derived from the inner cell mass. Most of the purple cells will give rise to the embryo. The yellow cells, some purple cells, and some trophoblast cells will give rise to four structures called the **extraembryonic membranes,** which develop as attachments to the embryo and help support it. You can see three of these membranes—the amnion (from purple cells), the yolk sac (from yellow cells), and the chorion (partly from the trophoblast)—starting to take shape in Figure 27.15D. A later stage (Figure 27.15E) shows the fourth extraembryonic membrane, the allantois, developing as an extension of the yolk sac.

Gastrulation, the stage shown in Figure 27.15D, is under way by 9 days after conception. There is already evidence of the three embryonic layers—ectoderm (blue), endoderm (yellow), and mesoderm (red). The embryo itself (not including the membranes) develops from the three inner cell layers shown in Figure 27.15E. The ectoderm layer will form the outer part of the embryo's skin. As indicated in the drawing, the ectoderm layer is continuous with the amnion. Similarly, the embryo's digestive tract will develop from the endoderm layer, which is continuous with the yolk sac. The bulk of most other organs will develop from the central layer of mesoderm.

Roles of the Extraembryonic Membranes   Figure 27.15F shows the embryo about a month after fertilization, along with its life-support system, made up largely of the four extraembryonic membranes. By this time, the **amnion** has grown to enclose the embryo. The amniotic cavity is filled with fluid, which protects the embryo. The amnion usually breaks just before childbirth, and the amniotic fluid leaves the mother's body through her vagina ("her water burst").

In humans and most other mammals, the **yolk sac** contains no yolk, but is given the same name as the homologous structure in other vertebrates. In a bird egg, the yolk sac contains a large mass of yolk. Isolated within a shelled egg outside the mother's body, a developing bird obtains nourishment from the yolk rather than from a placenta. In mammals, the yolk sac, which remains small, has other important functions: It produces the embryo's first blood cells and its first germ cells, the cells that will give rise to the gamete-forming cells in the gonads.

The **allantois** also remains small in mammals. It forms part of the umbilical cord—the lifeline between the embryo and the placenta. It also forms part of the embryo's urinary bladder. In birds and other reptiles, the allantois expands around the embryo and functions in waste disposal.

The outermost extraembryonic membrane, the **chorion,** completely surrounds the embryo and other extraembryonic membranes. The chorion becomes part of the placenta, where it functions in gas exchange. Cells in the chorion secrete a hormone called **human chorionic gonadotropin (HCG),** which maintains production of estrogen and progesterone by the corpus luteum of the ovary during the first few months of pregnancy. Without these hormones, menstruation would occur, and the embryo would abort spontaneously. Levels of HCG in maternal blood are so high that some is excreted in the urine, where it can be detected by pregnancy tests (see Module 24.10).

The Placenta Going back to Figure 27.15D, notice the knobby outgrowths on the outside of the chorion. In Figure 27.15E, these outgrowths, now called **chorionic villi,** are larger and contain mesoderm. In Figure 27.15F, the mesoderm cells have formed into embryonic blood vessels in the chorionic villi. By this stage, the placenta is fully developed. Starting with the chorion and extending outward, the placenta is a composite organ consisting of chorionic villi closely associated with the blood vessels of the mother's endometrium. The villi are actually bathed in tiny pools of maternal blood (purple in the figure). The mother's blood and the embryo's blood are not in direct contact. Instead, the chorionic villi absorb nutrients and oxygen from the mother's blood and pass these substances to the embryo via the chorionic blood vessels colored red. The chorionic vessels shown in blue carry wastes away from the embryo. The wastes diffuse into the mother's bloodstream and are excreted by her kidneys.

The placenta is a vital organ with both embryonic and maternal parts that mediates exchange of nutrients, gases, and the products of excretion between the embryo and the mother. However, the placenta cannot always protect the embryo from substances circulating in the mother's blood. A number of viruses—the German measles virus and HIV, for example—can cross the placenta. German measles can cause serious birth defects; HIV-infected babies usually die of AIDS within a few years without treatment. Most drugs, both prescription and not, also cross the placenta, and many can harm the developing embryo. Alcohol and the chemicals in tobacco smoke, for instance, raise the risk of miscarriage and birth defects. Alcohol can cause a set of birth defects called fetal alcohol syndrome, which includes mental retardation.

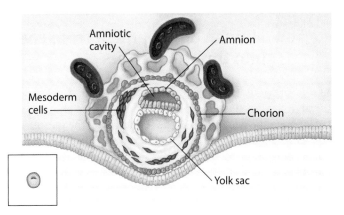

Figure 27.15D Embryonic layers and extraembryonic membranes starting to form (9 days)

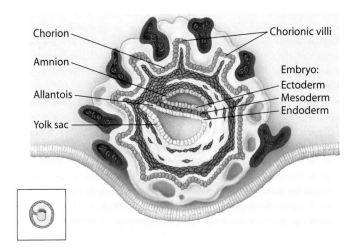

Figure 27.15E Three-layered embryo and four extraembryonic membranes (16 days)

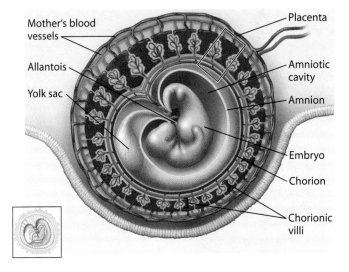

Figure 27.15F Placenta formed (31 days)

? Why does testing for HCG in a woman's urine or blood work as an early test of pregnancy?

■ Because this hormone is secreted by the chorion of an embryo

Human development from conception to birth is divided into three trimesters

In the previous module, we followed human development through the first 4 weeks. In this module, we use photographs to illustrate the rest of human development in the uterus. For convenience, we divide the period of human development from conception to birth into three **trimesters** of about 3 months each.

First Trimester The first trimester is the time of most radical change for both mother and embryo. Figure 27.16A shows a human embryo about 5 weeks after fertilization. In that brief time, this highly organized multicellular embryo has developed from a single cell. Not shown here are the extraembryonic membranes that surround the embryo or most of the umbilical cord that attaches it to the placenta. This embryo is about 7 mm (0.28 in.) long and has a number of features in common with the somite stage of a frog embryo (see Figure 27.12C). The embryo has a notochord and a coelom, both formed from mesoderm. Its brain and spinal cord have begun to take shape from a tube of ectoderm, as in the frog. The human embryo also has four stumpy limb buds, a short tail, and elements of gill pouches. The gill pouches appear during embryonic development in all chordates; in land vertebrates, they eventually develop into parts of the throat and middle ear. Overall, a month-old human embryo is similar to other vertebrates at the somite stage of development.

Figure 27.16B shows a developing human, now called a fetus, about 9 weeks after fertilization. The large pinkish structure on the left is the placenta, attached to the fetus by the umbilical cord. The clear sac around the fetus is the amnion. By this time, the fetus is decidedly human, rather than generally vertebrate. It is about 5.5 cm (2.2 in.) long and has all of its organs and major body parts, including a disproportionately large head. The somites have developed into the segmental muscles and the bones of the back and ribs. The limb buds have become tiny arms and legs with fingers and toes.

Beginning at about 9 weeks, the fetus can move its arms and legs, turn its head, frown, and make sucking motions with its lips. By the end of the first trimester, the fetus looks like a miniature human being, although its head is still oversized for

Timeline of Human Fetal Development

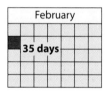

January
0 | Conception

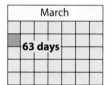

February
35 days

March
63 days

April
98 days

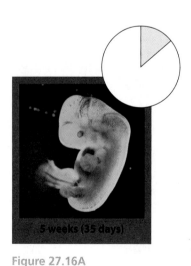

5 weeks (35 days)

Figure 27.16A

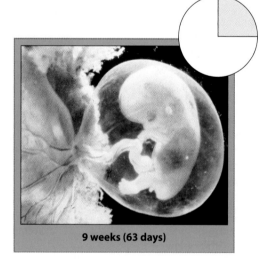

9 weeks (63 days)

Figure 27.16B

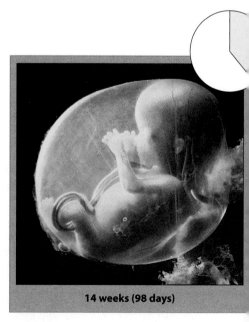

14 weeks (98 days)

Figure 27.16C

the rest of the body. The sex of the fetus is usually evident by this time, and its heartbeat can be detected with a stethoscope.

Second Trimester

The main developmental changes during the second and third trimesters include an increase in size and general refinement of the human features—nothing as dramatic as the changes of the first trimester. Figure 27.16C shows a fetus at 14 weeks, 2 weeks into the second trimester. The fetus is now about 6 cm (2.4 in.) long. During the second trimester, the placenta takes over the task of maintaining itself by secreting progesterone, rather than receiving it from the corpus luteum. At the same time, the placenta stops secreting HCG, and the corpus luteum, no longer needed to maintain pregnancy, degenerates.

At 20 weeks (Figure 27.16D), well into the second trimester, the fetus is about 19 cm (7.6 in.) long, weighs about half a kilogram (1 lb), and has the face of an infant, complete with eyebrows and eyelashes. Its arms, legs, fingers, and toes have lengthened. It also has fingernails and toenails and is covered with fine hair. Fetuses of this age are usually quite active. The mother's abdomen has become markedly enlarged, and she may often feel her baby move. Because of the limited space in the uterus, the fetus flexes forward into the so-called fetal position. By the end of the second trimester, the fetus's eyes are open and its teeth are forming.

Third Trimester

The third trimester (28 weeks to birth) is a time of rapid growth as the fetus gains the strength it will need to survive outside the protective environment of the uterus. Babies born prematurely—as early as 24 weeks—may survive, but they require special medical care after birth. During the third trimester, the fetus's circulatory system and respiratory system undergo changes that will allow the switch to air breathing (see Module 22.12). The fetus gains the ability to maintain its own temperature, and its bones begin to harden and its muscles thicken. It also loses much of its fine body hair, except on its head. The fetus becomes less active as it fills the space in the uterus. As the fetus grows and the uterus expands around it, the mother's abdominal organs become compressed and displaced, leading to frequent urination, digestive blockages, and strain in the back muscles. At birth (Figure 27.16E), babies average about 50 cm (20 in.) in length and weigh 3–4 kg (6–8 lb).

> **?** Certain drugs cause their most serious damage to an embryo very early in pregnancy, often before the mother even realizes she is pregnant. Why?
>
> ■ Because organ systems, such as the circulatory and nervous systems, begin to develop early in the first trimester

May	June	July	August	September	October
140 days					280 days

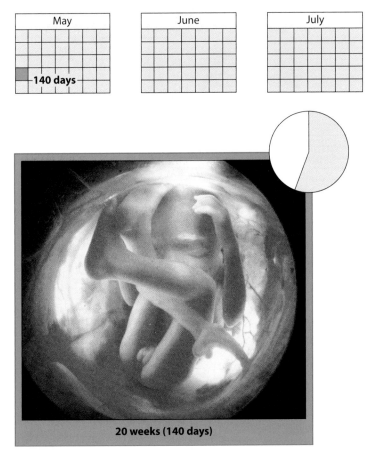

20 weeks (140 days)

Figure 27.16D

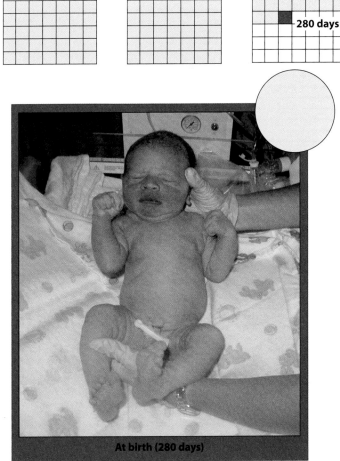

At birth (280 days)

Figure 27.16E

27.17 Childbirth is hormonally induced and occurs in three stages

The series of events that expel the infant from the uterus is called **labor.** Several hormones play key roles (Figure 27.17A). One hormone, estrogen, reaches its highest level in the mother's blood during the last weeks of pregnancy. An important effect of this estrogen is to trigger the formation of numerous oxytocin receptors on the uterus. Cells of the fetus produce the hormone oxytocin, and late in pregnancy, the mother's pituitary gland secretes it in increasing amounts. Oxytocin stimulates the smooth muscles in the wall of the uterus, producing the series of increasingly strong, rhythmic contractions characteristic of labor. It also stimulates the placenta to make prostaglandins, local tissue regulators that also stimulate uterine muscle cells to contract even more.

The induction of labor involves **positive feedback,** a type of control in which a change triggers mechanisms that amplify that change. In this case, the hormones oxytocin and prostaglandins cause uterine contractions that in turn stimulate the release of more and more oxytocin and prostaglandins. The result is a steady increase in intensity, climaxing in forceful muscle contractions that propel a baby from the uterus (womb).

Figure 27.17B shows the three stages of labor. As the process begins, the cervix (neck of the uterus) gradually opens, or dilates. ❶ The first stage, dilation, is the time from the onset of labor until the cervix reaches its full dilation of about 10 cm. Dilation is the longest stage of labor, lasting 6–12 hours or even considerably longer.

❷ The period from full dilation of the cervix to delivery of the infant is called the expulsion stage. Strong uterine contractions, lasting about 1 minute each, occur every 2–3 minutes, and the mother feels an increasing urge to push or bear down with her abdominal muscles. Within a period of 20 minutes to

an hour or so, the infant is forced down and out of the uterus and vagina. An attending physician or midwife (or nervous father!) clamps and cuts the umbilical cord after the baby is expelled. ❸ The final stage is the delivery of the placenta, usually within 15 minutes after the birth of the baby.

Hormones continue to be important after the baby and placenta are delivered. Decreasing levels of progesterone and estrogen allow the uterus to start returning to its prepregnancy

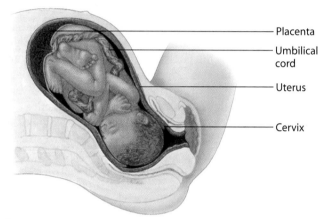

❶ Dilation of the cervix

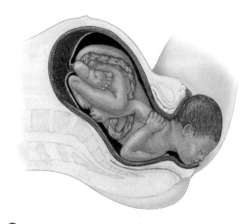

❷ Expulsion: delivery of the infant

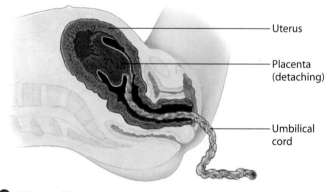

❸ Delivery of the placenta

Figure 27.17A The hormonal induction of labor

Figure 27.17B The three stages of labor

state. In response to suckling by the newborn, as well as falling levels of progesterone after birth, the pituitary secretes prolactin and oxytocin. These hormones promote milk production and release (called lactation) by the mammary glands. At first, a yellowish protein- and antibody-rich fluid called colostrum is secreted. After 2–3 days, normal milk production begins.

? The onset of labor is marked by dilation of the _____.

■ cervix

Connection

27.18 Reproductive technology increases our reproductive options

About 15% of couples who want children are infertile; that is, they cannot conceive a child after 12 months of unprotected intercourse. In about half of cases, such infertility can be traced to problems with the man. His testes may not produce enough sperm (a "low sperm count"), or those that are produced may be defective. Underproduction of sperm is frequently caused by the man's scrotum being too warm, so a switch of underwear from briefs (which hold the scrotum close to the body) to boxers may help. In other cases, infertility is caused by **impotence,** also called erectile dysfunction, the inability to maintain an erection. Female infertility can result from a lack of eggs, a failure to ovulate, or blocked oviducts (often caused by scarring due to sexually transmitted diseases). Other women are able to conceive, but cannot support a growing embryo in the uterus.

Reproductive technologies can help many cases of infertility. Drug therapies (including Viagra) and penile implants can be used to treat impotence. If a man produces no functioning sperm, the couple may elect to use another man's sperm that has been donated to a sperm bank.

If a woman has normal eggs that are not being released properly, hormone injections can induce ovulation. As discussed in the chapter opener, such treatments frequently result in multiple pregnancies. If a woman has no eggs of her own, they, too, can be obtained from a donor for fertilization and implantation into the uterus. While sperm can be collected without any danger to the donor, collection of eggs involves surgery and therefore pain and risk for the donating woman.

If a woman produces eggs but is unable to support a growing fetus, she and her partner may hire a surrogate mother. In such cases, the couple enters into a legal contract with a woman who agrees to be implanted with the couple's embryo and carry it to birth. However, a number of states have laws restricting surrogate motherhood owing to the serious ethical and legal problems that can arise.

Many infertile couples turn to fertilization procedures called **assisted reproductive technologies (ART).** In ART procedures, eggs (secondary oocytes) are surgically removed from a woman's ovaries after hormonal stimulation, fertilized, and returned to the woman's body. Eggs, sperm, and embryos from such procedures can be frozen for later pregnancy attempts.

With **in vitro fertilization (IVF),** the most common ART procedure, a woman's eggs are mixed with sperm in culture dishes (in vitro means "in glass") and incubated for several days

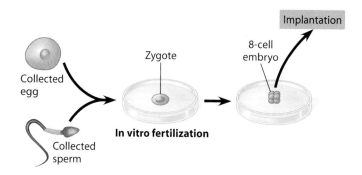

Figure 27.18 The process of in vitro fertilization

to allow fertilized eggs to start developing (Figure 27.18). When they have developed into embryos of at least eight cells each, the embryos are carefully inserted into the woman's uterus.

If mature sperm are defective, of low number, or even absent, fertility is often restored by a technique termed intracytoplasmic sperm injection (ICSI). In this form of IVF, the head of a sperm is drawn up into a needle and injected directly into an egg to achieve fertilization.

To date, the majority of evidence indicates that abnormalities arising as a consequence of an IVF procedure are quite rare, although some research has shown small but significant risks of lower birth weights and higher rates of birth defects. Despite such risks and the high cost (typically $10,000 per attempt, whether it succeeds or not), IVF techniques are now performed at medical centers throughout the world and result in the birth of thousands of babies each year.

? Explain how, through IVF, up to three different people can be involved in the birth of a child.

■ One woman (1) may become pregnant with an embryo created using the sperm of a man (2) and the ovum of a second woman (3).

■   ■   ■

In this chapter, we have watched a single-celled product of sexual reproduction, the zygote, become transformed into a new organism, complete with all organ systems. One of the first of those organ systems to develop is the nervous system. In the next chapter, we see how the nervous system functions together with the endocrine system to regulate virtually all body activities.

Chapter Review

Reviewing the Concepts

Asexual and Sexual Reproduction (27.1–27.2)

Asexual reproduction results in the production of genetically identical offspring from a lone parent. Asexual reproduction can proceed by budding, fission, or fragmentation/regeneration and enables an individual to produce many offspring rapidly **(27.1)**. **Sexual reproduction** involves the fusion of gametes from two parents, resulting in genetic variation among offspring. This may enhance reproductive success in changing environments **(27.2)**.

Human Reproduction (27.3–27.8)

The human reproductive system consists of a pair of ovaries (in females) or testes (in males), ducts that carry gametes, and structures for copulation. A woman's ovaries contain follicles that nurture eggs and produce sex hormones. Oviducts convey eggs to the uterus, where a fertilized egg develops. The uterus opens into the vagina, which receives the penis during intercourse, and forms the birth canal **(27.3)**. A man's testes produce sperm, which are expelled through ducts during ejaculation. Several glands contribute to the formation of fluid that nourishes and protects sperm. This fluid and the sperm constitute semen **(27.4)**.

Spermatogenesis and oogenesis produce sperm and eggs, respectively. Primary spermatocytes are made continuously in the testes; these diploid cells undergo meiosis to form four haploid sperm. Each month, one primary oocyte matures to form a secondary oocyte, which, if fertilized, completes meiosis and becomes a haploid ovum **(27.5)**:

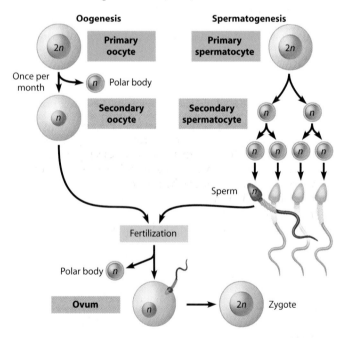

The female reproductive cycle.   Hormones synchronize cyclic changes in the ovaries and uterus. Approximately every 28 days, the hypothalamus signals the anterior pituitary to secrete FSH and LH, which trigger the growth of a follicle and ovulation, the release of an egg. The follicle becomes the corpus luteum, which secretes both estrogen and progesterone. These two hormones stimulate the endometrium (the uterine lining) to thicken,

preparing the uterus for implantation. They also inhibit the hypothalamus, reducing FSH and LH secretion. If the egg is not fertilized, the drop in LH shuts down the corpus luteum and its hormones. This triggers menstruation, the breakdown of the endometrium. The hypothalamus and pituitary then stimulate another follicle, starting a new cycle. If fertilization occurs, a hormone from the embryo maintains the uterine lining and prevents menstruation **(27.6)**.

Reproductive health.   Sexual intercourse carries the risk of exposure to sexually transmitted diseases. Bacterial diseases can often be cured, but viral diseases can only be controlled **(27.7)**. Several forms of contraception can prevent pregnancy, with varying degrees of success **(27.8)**.

Principles of Embryonic Development (27.9–27.14)

Fertilization and early embryonic development.   In fertilization, a sperm releases enzymes that pierce the egg's coat. Sperm surface proteins bind to egg receptor proteins, sperm and egg plasma membranes fuse, and the two nuclei unite. Changes in the egg membrane prevent entry of additional sperm, and the fertilized egg (zygote) develops into an embryo **(27.9)**. Cleavage is a rapid series of cell divisions that results in a blastula, a ball of cells **(27.10)**. In gastrulation, cells migrate and form a rudimentary digestive cavity. The resulting gastrula has three layers of cells **(27.11)**:

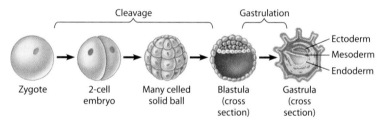

After gastrulation, the three embryonic tissue layers give rise to specific organ systems **(27.12)**. Tissues and organs take shape in a developing embryo as a result of cell shape changes, cell migration, and programmed cell death. In a process called induction, adjacent cells and cell layers influence each other's differentiation via chemical signals **(27.13)**. Pattern formation, the emergence of the parts of a structure in their correct relative positions, involves the response of genes to spatial variations of chemicals in the embryo. Homeotic genes contain nucleotide sequences, called homeoboxes, that appeared early in the evolutionary history of animals **(27.14)**.

Human Development (27.15–27.18)

Human development begins with fertilization in the oviduct. Cleavage produces a blastocyst, whose inner cell mass becomes the embryo. The blastocyst's outer layer, the trophoblast, implants in the uterine wall. Gastrulation occurs, and organs develop from the three embryonic layers. Meanwhile, the four extraembryonic membranes develop: the amnion, the chorion, the yolk sac, and the allantois. The embryo floats in the fluid-filled amniotic cavity, while the chorion and embryonic mesoderm form the embryo's part of the placenta. The placenta's chorionic villi absorb food and oxygen from the mother's blood **(27.15)**. Human embryonic development is divided into three trimesters of about 3 months each. The most rapid changes occur during the first trimester. By 9 weeks, the embryo is called a fetus. The second and third trimesters are times of growth and preparation for birth **(27.16)**. Hormonal changes induce birth. Estrogen makes the uterus more sensitive to oxytocin, which acts with prostaglandins to initiate labor. The cervix dilates, the baby is expelled by strong muscular contractions, and the placenta follows **(27.17)**.

Reproductive technologies.   New techniques can provide help to many infertile couples **(27.18)**.

Connecting the Concepts

1. This graph plots the rise and fall of pituitary and ovarian hormones during the human ovarian cycle. Identify each hormone (A–D) and the reproductive events with which each one is associated (P–S). For A–D, choose from: estrogen, LH, FSH, and progesterone. For P–S, choose from: ovulation, growth of follicle, menstruation, and development of corpus luteum. How would the right-hand side of this graph be altered if pregnancy occurred? What other hormone is responsible for triggering this change?

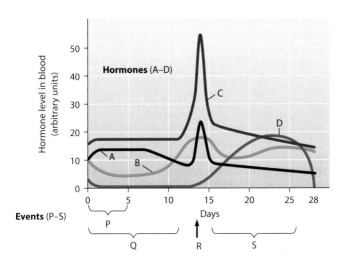

Testing Your Knowledge

2. After a sperm penetrates an egg, the fertilization envelope
   a. secretes important hormones.
   b. enables the fertilized egg to implant in the uterus.
   c. prevents more than one sperm from entering the egg.
   d. attracts additional sperm to the egg.
   e. activates the egg for embryonic development.

3. In an experiment, a researcher colored a bit of tissue on the outside of a frog gastrula with an orange fluorescent dye. The embryo developed normally. When the tadpole was placed under an ultraviolet light, which of the following glowed bright orange? (*Explain your answer.*)
   a. the heart   c. the brain   e. the liver
   b. the pancreas   d. the stomach

4. How does a zygote differ from an ovum?
   a. A zygote has more chromosomes.
   b. A zygote is smaller.
   c. A zygote consists of more than one cell.
   d. A zygote is much larger.
   e. A zygote divides by meiosis.

5. A woman had several miscarriages. Her doctor suspected that a hormonal insufficiency was causing the lining of the uterus to break down, as it does during menstruation, terminating her pregnancies. Treatment with which of the following might help her remain pregnant?
   a. oxytocin
   b. follicle-stimulating hormone
   c. testosterone
   d. luteinizing hormone
   e. prolactin

Matching

6. Turns into the corpus luteum   a. vas deferens
7. Female gonad   b. prostate gland
8. Site of spermatogenesis   c. endometrium
9. Site of fertilization in humans   d. testis
10. Human gestation occurs here   e. follicle
11. Sperm duct   f. uterus
12. Secretes seminal fluid   g. ovary
13. Lining of uterus   h. oviduct

Describing, Comparing, and Explaining

14. Compare sperm formation with egg formation. In what ways are the processes similar? In what ways are they different?

15. The embryos of reptiles (including birds) and mammals have systems of extraembryonic membranes. What are the functions of these membranes, and how do fish and frog embryos survive without them?

16. In an embryo, nerve cells grow out from the spinal cord and form connections with the muscles they will eventually control. What mechanisms described in this chapter might explain how these cells "know" where to go and which cells to connect with?

Applying the Concepts

17. As a frog embryo develops, the neural tube forms from ectoderm along what will be the frog's back, directly above the notochord. To study this process, a researcher carefully extracted a bit of notochord tissue and inserted it underneath the ectoderm where the belly of the frog would normally develop. What can the researcher hope to learn from this experiment? Predict the possible outcomes. What experimental control would you suggest?

18. Should parents undergoing in vitro fertilization have the right to choose which embryos to implant based on genetic criteria, such as the presence or absence of disease-causing genes? How about based on the sex of the embryo? How could you distinguish acceptable criteria from unacceptable ones? Do you think such options should be legislated? How would you formulate the law?

Answers to all questions can be found in Appendix 4.

For Activity Quizzes, BioFlix, MP3 Tutors, and Activities, go to www.mybiology.com.

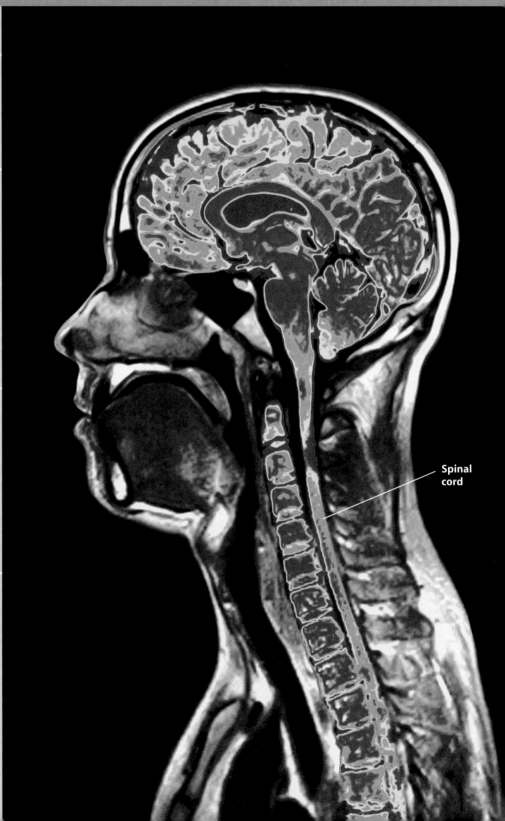

chapter

28 Nervous Systems

Spinal cord

Can an Injured Spinal Cord Be Fixed?

Protected inside the bony vertebrae of the spine is an inch-thick gelatinous bundle of nervous tissue called the spinal cord. Shown in the image on the facing page, the spinal cord acts as the central communication conduit between the brain and the rest of the body. Millions of nerve fibers carry motor information from the brain to the muscles, while other fibers bring sensory information (about touch, pain, and body position, for example) from the body to the brain. The spinal cord acts like a transcontinental telephone cable jam-packed with wires, each of which carries messages to or from the central hub and an outlying area.

But what happens if that cable is cut? Signals cannot get through, communication is lost, and the cable must be repaired or replaced. In humans, the spinal cord is rarely severed because the vertebrae provide rigid protection. However, a traumatic blow to the spinal column and subsequent bleeding, swelling, and scarring can crush the delicate nerve bundles and prevent signals from passing. The result may be a debilitating injury. Such trauma along the back can cause paraplegia—paralysis of the lower half of the body. Trauma higher up in the neck can cause quadriplegia—paralysis from the neck down, which may necessitate permanent breathing assistance from an artificial respirator. Such injuries are usually permanent because the spinal cord, unlike other body tissues, cannot repair itself.

The late actor Christopher Reeve (best known for playing Superman in several movies) suffered a spinal cord injury during a 1995 equestrian competition. He was thrown from his horse and landed headfirst. Two vertebrae in his neck were fractured, crushing the spinal cord at the base of his skull and causing quadriplegia. Reeve died of complications related to his injury in 2004.

Over 10,000 Americans suffer spinal cord injuries each year. The most common causes are car crashes, violence (usually from gunshots), falls, and sports. Because the majority of spinal cord injuries happen to people younger than 30, the subsequent disabilities often last for decades at great monetary and emotional cost.

Historically, spinal cord injuries have been considered untreatable. Recently, however, there has been some minor progress: In 1988, researchers discovered that administering a powerful steroid drug within hours of a spinal cord injury limits its severity. But reversing spinal cord damage is a formidable challenge.

Some researchers are coaxing damaged nerve cells to regenerate by administering growth factor proteins (see Module 8.9) or transplanting the cells that produce growth factors to the site of the injury. Other researchers are attempting to block proteins that inhibit growth. Still others, believing that damaged nerve cells cannot be fixed, are trying to find ways to replace them with either mature nerve cells from elsewhere in the body or fetal tissue. A recent version of this approach is the attempt to use embryonic stem cells—progenitor cells capable of developing into all other cell types (see Module 11.15)—or partially differentiated neural stem cells to grow or repair nerve connections. Several recent studies involving combinations of these proposed therapies have shown promise in rats, even partially restoring motor function below the injury. While none of these strategies may ever "cure" a damaged spinal cord, they may still offer benefits of great value, such as regained control of the bladder, bowels, respiration, or a limb. Thanks to the efforts of The Christopher Reeve Foundation and other individuals and organizations in raising public awareness, spinal damage is now the subject of much research. The years ahead hold great promise for improving the prognosis after spinal cord injuries.

In this chapter, we explore the structure, function, and evolution of nervous systems. We focus in particular on the vertebrate nervous system and the structure and function of the human brain. We'll begin with an introduction to the central nervous system (the brain and spinal cord) and peripheral nervous system (the nerves that carry information to and from the rest of the body). ■ ■ ■

28.1 Nervous systems receive sensory input, interpret it, and send out appropriate commands

Nervous systems are the most intricately organized data-processing systems on Earth. Your brain, for instance, contains an estimated 100 billion neurons (nerve cells), which are specialized for carrying signals from one location in the body to another. A **neuron** consists of a cell body, containing the nucleus and cell organelles, and long, thin extensions called neuron fibers that convey signals. Each neuron may communicate with thousands of others, forming networks that enable us to learn, remember, perceive our surroundings, and move.

With few exceptions, nervous systems have two main anatomical divisions. The first anatomical division, called the **central nervous system (CNS),** consists of the brain and, in vertebrates, the spinal cord. The other division of the nervous system, the **peripheral nervous system (PNS),** is made up mostly of communication lines called nerves that carry signals into and out of the CNS. A **nerve** is a cable-like bundle of neuron extensions tightly wrapped in connective tissue. In addition to nerves, the PNS also has **ganglia** (singular, *ganglion*), clusters of neuron cell bodies.

A nervous system has three interconnected functions (Figure 28.1A). **Sensory input** is the conduction of signals from sensory receptors, such as light-detecting cells of the eye, to integration centers. **Integration** is the interpretation of the sensory signals and the formulation of appropriate responses. **Motor output** is the conduction of signals from the integration centers to **effector cells,** such as muscle cells or gland cells, which perform the body's responses. The integration of sensory input and motor output is not usually rigid and linear, but involves the continuous background activity symbolized by the circular arrow in Figure 28.1A.

The relationship between neurons and nervous system structure and function is easiest to see in the relatively simple circuits that produce **reflexes,** or automatic responses to stimuli (Figure 28.1B). The small colored balls in the figure represent neuron cell bodies; the thin lines represent neuron fibers. Three functional types of neurons correspond to a nervous system's three main functions: **Sensory neurons** convey signals (information) from sensory receptors into the CNS. **Interneurons** are located entirely within the CNS. They integrate data and then relay appropriate signals to other interneurons or to motor neurons. **Motor neurons** convey signals from the CNS to effector cells. (For simplicity, this figure shows only one neuron of each functional type, but virtually any body activity actually involves many of each.)

When the knee is tapped, ❶ a sensory receptor detects a stretch in the tendon, and ❷ a sensory neuron conveys this information into the CNS (spinal cord). In the CNS, the information goes to ❸ a motor neuron and to ❹ one or more interneurons. One set of muscles (quadriceps) responds to motor signals conveyed by a motor neuron by contracting, jerking the lower leg forward. At the same time, another motor neuron, responding to signals from the interneuron, inhibits the flexor muscles, making them relax and not resist the action of the quadriceps.

As you can see from this example, the nervous system depends on the ability of neurons to receive and convey signals. Next, we'll examine how that is done.

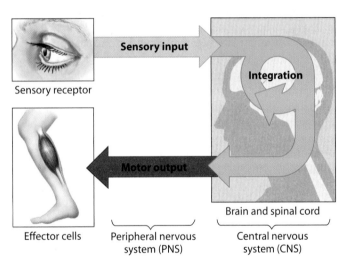

Figure 28.1A Organization of a nervous system

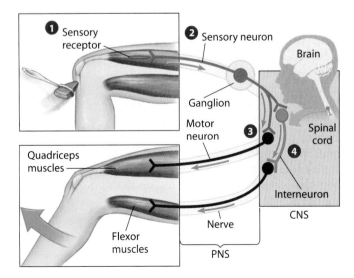

Figure 28.1B The knee-jerk reflex

> **?** Arrange the following neurons into the correct sequence for information flow during the knee-jerk reflex: interneuron, sensory neuron, motor neuron.

Sensory neuron → interneuron → motor neuron ■

28.2 Neurons are the functional units of nervous systems

The ability of neurons to receive and transmit information depends on their structure. Figure 28.2 depicts a motor neuron, like those that carry command signals from your spinal cord to your skeletal muscles. The inset shows an SEM of a neuron.

Most of a neuron's organelles, including its nucleus, are located in the **cell body.** Arising from the cell body are two types of extensions: numerous dendrites and a single axon. **Dendrites** (from the Greek *dendron*, tree) are highly branched extensions that receive signals from other neurons and convey this information toward the cell body. Dendrites are often short. In contrast, the **axon** is typically a much longer extension that transmits signals to other cells, which may be other neurons or effector cells. Some axons, such as the ones that reach from your spinal cord to muscle cells in your feet, can be over a meter long.

Neurons make up only part of a nervous system. To function normally, neurons require supporting cells called **glia.** Depending on the type, glia may nourish neurons, insulate the axons of neurons, or help maintain homeostasis of the extracellular fluid surrounding neurons. Glia may actually outnumber neurons by as many as 50 to 1.

Figure 28.2 shows one kind of glial cell called a Schwann cell, which is found in the PNS. (Analogous cells are found in the CNS.) In many vertebrates, axons that convey signals rapidly are enclosed along most of their length by a thick insulating material. This insulating material, called the **myelin sheath,** resembles a chain of oblong beads. Each bead is actually a Schwann cell, and the myelin sheath is essentially a chain of Schwann cells, each wrapped many times around the axon. The spaces between Schwann cells are called **nodes of Ranvier,** and they are the only

points along the axon that are "leaky," requiring nerve signals to be regenerated. Everywhere else, the myelin sheath insulates the axon, preserving the signal. Thus, when a nerve signal travels along a myelinated axon, it appears to jump from point to point as it is rejuvenated at each node. The resulting signal is much faster than one that needs to be regenerated constantly along the length of the axon. In the human nervous system, signals can travel along a myelinated axon about 150 m/sec (over 330 mi/hr!), which means that a command from your brain can make your fingers move in just a few milliseconds. Without myelin sheaths, the signals could go only about 5 m/sec.

The debilitating disease multiple sclerosis (MS) demonstrates the importance of myelin sheaths. MS leads to a gradual destruction of myelin sheaths by the individual's own immune system. The result is a progressive loss of signal conduction, muscle control, and brain function.

Notice in Figure 28.2 that the axon ends in a cluster of branches. A typical axon has hundreds or thousands of these branches, each with a **synaptic terminal** at the very end. The site of communication between a synaptic terminal and another cell is called a **synapse.** As we will see later, this is where information is passed between neurons. With the basic structure of a neuron in mind, let's take a closer look at the signals that neurons convey.

🎧 **MP3 Tutor** *Neurons and Electric Potentials*

Web Activity *Neuron Structure*

❓ **What is the function of the myelin sheath?**

■ It speeds up conduction of signals along axons.

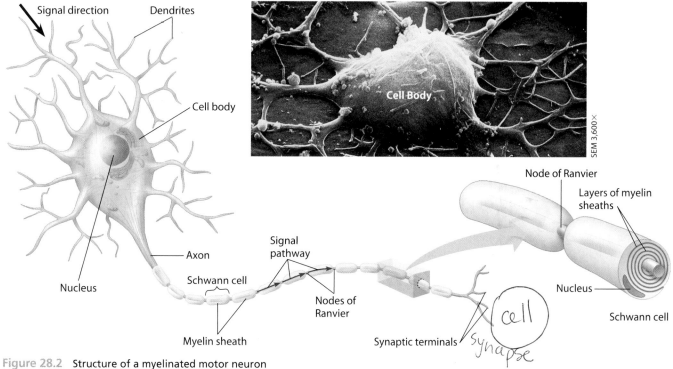

Figure 28.2 Structure of a myelinated motor neuron

Signal direction · Dendrites · Cell Body · Cell body · Node of Ranvier · Layers of myelin sheaths · Signal pathway · Axon · Schwann cell · Nodes of Ranvier · Nucleus · Nucleus · Myelin sheath · Synaptic terminals · Schwann cell · SEM 3,600×

28.3 A neuron maintains a membrane potential across its membrane

To understand nerve signals, we must first study a resting neuron, one that is not transmitting a signal. A resting neuron has potential energy, energy that can be put to work sending signals from one part of the body to another. This potential energy, called the **membrane potential,** exists as an electrical charge difference across the neuron's plasma membrane: The cytoplasm just inside the membrane is negative in charge, and the fluid just outside the membrane is positive. Since opposite charges tend to move toward each other, a membrane stores energy by holding opposite charges apart, like a battery. The strength (voltage) of a neuron's stored energy can be measured

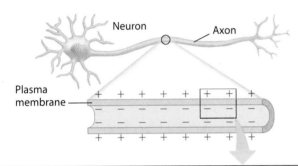

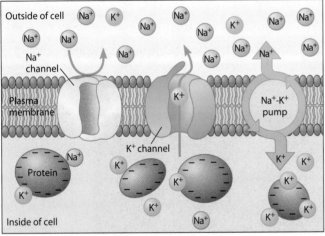

Figure 28.3   How the resting potential is generated

with microelectrodes connected to a voltmeter. The voltage across the plasma membrane of a resting neuron is called the **resting potential.** A neuron's resting potential is about −70 millivolts (mV)—about 5% of the voltage in a flashlight battery. (The minus sign indicates that the inside of the cell is negative relative to the outside.)

The resting potential exists because of differences in ionic composition of the fluids inside and outside the cell. The plasma membrane surrounding the neuron has channels and pumps made of proteins that regulate the passage of inorganic ions. One result is that a resting membrane allows much more potassium (K^+) than sodium (Na^+) to diffuse across the membrane. Notice in Figure 28.3 that Na^+ (Na⁺) is more concentrated outside the cell than inside; the Na^+ channels allow very little to diffuse in. But, K^+ (K⁺) which is more concentrated inside, can freely flow out through the K^+ channels. As the positively charged potassium ions diffuse out, they leave behind an excess of negative charge. Also helping maintain the resting potential are membrane proteins called **sodium-potassium (Na^+-K^+) pumps.** These pumps actively transport Na^+ out of the cell and K^+ in, thereby helping keep the concentration of Na^+ low in the cell and K^+ high.

This ionic gradient (high K^+/low Na^+ concentrations inside coupled with low K^+/high Na^+ concentrations outside) produces an electrical potential difference, or voltage, across the membrane—the resting potential. Notice an important point: *The membrane potential can change from its resting value if the membrane's permeability to particular ions changes.* As we will see in the next module, this is the basis of nearly all electrical signals in the nervous system.

? If a neuron's membrane suddenly becomes more permeable to sodium ions, there is a rapid net movement of Na^+ into the cell. What are the two forces that drive the ions inward?

■ The greater concentration of Na^+ outside the cell than inside and the membrane potential (negatively charged inside versus outside) favor the inward diffusion of Na^+.

28.4 A nerve signal begins as a change in the membrane potential

Turning on a flashlight uses the potential energy stored in a battery to create light. In a similar way, stimulating a neuron's plasma membrane can trigger the release and use of the membrane's potential energy to generate a nerve signal. A **stimulus** is any factor that causes a nerve signal to be generated. Examples of stimuli include light, sound, a tap on the knee, or a chemical signal from another neuron.

The discovery of giant axons in squids (up to 1 mm in diameter) gave researchers their first chance to study how

stimuli trigger signals in a living neuron. From microelectrode studies with squid neurons, British biologists A. L. Hodgkin and A. F. Huxley worked out the details of nerve signal transmission in the 1940s. Their findings, summarized in Figure 28.4, apply to neurons in all animals.

Measuring electrical changes in neuron membranes was the first step in discovering how nerve signals are generated. The multicolored line on the graph in the middle of the figure traces the electrical changes that make up the

action potential, a nerve signal that carries information along an axon. The graph records electrical events over time (in milliseconds) at a particular place on the membrane where a stimulus is applied. ❶ The graph starts out at the membrane's resting potential (−70 mV). ❷ The stimulus is applied. If it is strong enough, the voltage rises to what is called the **threshold** (−50 mV, in this case). The difference between the threshold and the resting potential is the minimum change in the membrane's voltage that must occur to generate the action potential. ❸ Once the threshold is reached, the action potential is triggered. The membrane polarity reverses abruptly, with the interior of the cell becoming positive with respect to the outside. ❹ The membrane then rapidly repolarizes as the voltage drops back down, ❺ undershoots the resting potential, and ❶ finally returns to it.

What actually causes the electrical changes of the action potential? The rapid flip-flop of the membrane potential results from the rapid movements of ions across the membrane at Na⁺ and K⁺ channels (called *voltage-gated channels* because they have special gates that open and close, depending on changes in membrane potential). The diagrams surrounding the graph show the ion movements. Starting at lower left, ❶ the resting membrane is positively charged on the outside, and the cytoplasm just inside the membrane is negatively charged. ❷ A stimulus triggers the opening of a few Na⁺ channels in the membrane, and a tiny amount of Na⁺ enters the axon. This makes the inside surface of the membrane slightly less negative. If the stimulus is strong enough, a sufficient number of Na⁺ channels open to raise the voltage to the threshold. ❸ Once the threshold is reached, more Na⁺ channels open, allowing even more Na⁺ to diffuse into the cell. As more and more Na⁺ moves in, the voltage soars to its peak. ❹ The peak voltage triggers closing and inactivation of the Na⁺ channels. Meanwhile, the K⁺ channels open, allowing K⁺ to diffuse out rapidly. These changes produce the downswing on the graph. ❺ A very brief undershoot of the resting potential results because the K⁺ channels close slowly. ❶ The membrane then returns to its resting potential. In a typical mammalian neuron, this entire process is very brief—only about 1–2 milliseconds in duration.

(Bio Flix) **BioFlix** *How Neurons Work*

Web Process of Science *What Triggers Nerve Impulses?*

? Is the generation of an action potential an example of positive or negative feedback?

■ The opening of Na⁺ gates caused by stimulation of the neuron changes the membrane potential, and this change causes more of the voltage-gated Na⁺ channels to open. This is an example of a positive feedback mechanism.

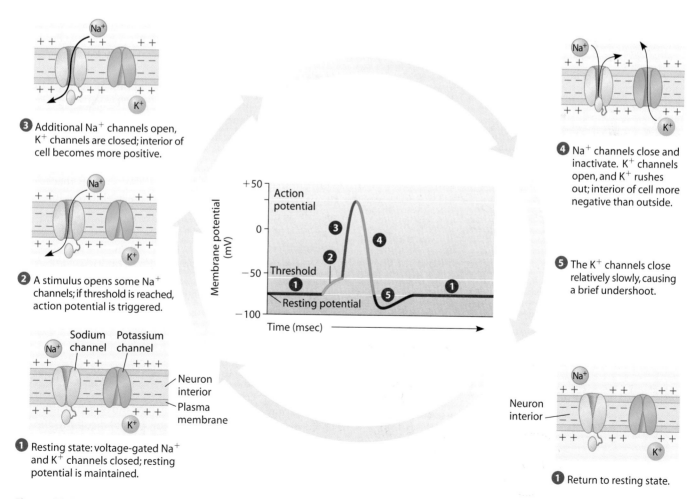

❸ Additional Na⁺ channels open, K⁺ channels are closed; interior of cell becomes more positive.

❷ A stimulus opens some Na⁺ channels; if threshold is reached, action potential is triggered.

❶ Resting state: voltage-gated Na⁺ and K⁺ channels closed; resting potential is maintained.

❹ Na⁺ channels close and inactivate. K⁺ channels open, and K⁺ rushes out; interior of cell more negative than outside.

❺ The K⁺ channels close relatively slowly, causing a brief undershoot.

❶ Return to resting state.

Figure 28.4 The action potential

The action potential propagates itself along the neuron

For an action potential to function as a long-distance signal, it must travel along the axon from the cell body to the synaptic terminals. It does so by regenerating itself along the axon. You can follow this process in Figure 28.5. A nerve signal starts out as an action potential generated in the axon near the cell body of the neuron. The effect of this action potential is like tipping the first of a row of standing dominoes: The first domino does not travel along the row, but its fall is relayed to the end of the row, one domino at a time.

The three steps in Figure 28.5 show the changes that occur in an axon segment as a nerve signal passes from left to right. Let's first focus on the axon region on the far left. ❶ When this region of the axon (blue) has its Na^+ channels open, Na^+ rushes inward ($\curvearrowright$), and an action potential is generated. This corresponds to the upswing of the curve (step 2) in Figure 28.4. ❷ Soon, the K^+ channels in that same region open, allowing K^+ to diffuse out of the axon ($\curvearrowleft$); at this time, its Na^+ channels are closed and inactivated, and we would see the downswing of the action potential. ❸ A short time later, we would see no signs of an action potential at this

(far-left) spot because the axon membrane here has returned to its resting potential.

Now let's see how these events lead to the "domino effect" of a nerve signal. In step 1 of the figure, the blue arrows pointing sideways within the axon indicate local spreading of the electrical changes caused by the inflowing Na^+ associated with the first action potential. These changes trigger the opening of Na^+ channels in the membrane just to the right of the action potential. As a result, a second action potential is generated, as indicated by the blue region in step 2. In the same way, a third action potential is generated in step 3, and each action potential generates another all the way down the axon.

So why are action potentials propagated in only one direction along the axon (left to right in the figure)? As the blue arrows indicate, local electrical changes do spread in both directions in the axon. However, these changes cannot open Na^+ channels and generate an action potential when the Na^+ channels are inactivated (step 4 in Figure 28.4). Thus, an action potential cannot be generated in the regions where K^+ is leaving the axon (green in the figure) and Na^+ channels are still inactivated. Consequently, the inward flow of Na^+ that depolarizes the axon membrane ahead of the action potential cannot produce another action potential behind it. Once an action potential starts, it moves only in the one direction, toward the synaptic terminals.

So we see that a nerve signal propagates itself in one direction by the electrical changes it produces in the neuron membrane. If you rap your finger on a desk, for instance, the contact is a stimulus that triggers action potentials in the tips of sensory neurons in your skin. The action potentials propagate along the axons, carrying the information (that your finger has hit a hard object) into your central nervous system.

Action potentials are *all-or-none* events; that is, they are the same no matter how strong or weak the stimulus that triggers them. How, then, do action potentials relay different intensities of information to your central nervous system? It is the *frequency* of action potentials that changes with the intensity of stimuli. If you rap your finger hard against the desk, your CNS receives many more action potentials per millisecond than after a soft tap.

In the past three modules, we've examined how nerve signals are conducted along a single neuron. In the next four modules, we focus on how signals pass from one neuron to another.

Web Activity *Nerve Signals: Action Potentials*

? During an action potential, ions move across the neuron membrane in a direction perpendicular to the direction of the impulse along the neuron. What is it that actually travels along the neuron as the signal?

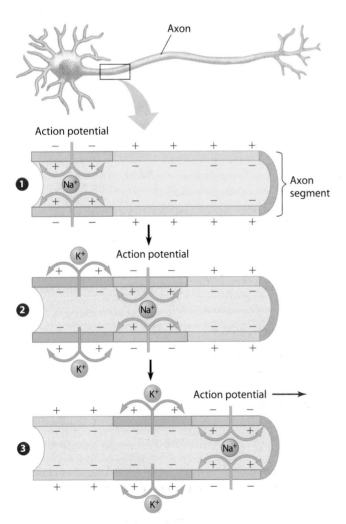

Axon

Action potential

Na⁺

Axon segment

Action potential

K⁺

Na⁺

K⁺

Action potential ➝

K⁺

Na⁺

K⁺

Figure 28.5 Propagation of the action potential along an axon

■ The signal is the wavelike change in membrane potential, the self-perpetuated action potential that regenerates sequentially at points farther and farther away from the site of stimulation.

28.6 Neurons communicate at synapses

Synapses—the regions of communication between a synaptic terminal and another cell—come in two varieties: electrical and chemical. In an electrical synapse, electrical current passes directly from one neuron to the next. The receiving neuron is stimulated quickly and at the same frequency of action potentials as the sending neuron. Lobsters and many fishes can flip their tails with lightning speed because the neurons that carry signals for these movements communicate by fast electrical synapses. In the human body, electrical synapses are found in the heart and digestive tract, where nerve signals maintain steady, rhythmic muscle contractions.

In contrast, when an action potential reaches the end of an axon with a chemical synapse, it stops there. At chemical synapses, action potentials in the form of electrical signal are *not* transmitted from cell to cell. However, *information* is transmitted.

Unlike electrical synapses, chemical synapses have a narrow gap, called the **synaptic cleft,** separating a synaptic terminal of the sending (presynaptic) neuron from the receiving (postsynaptic) neuron. The cleft prevents the action potential in the sending neuron from spreading directly to the receiving neuron. Instead, the action potential (an electrical signal) is first converted to a chemical signal consisting of molecules of **neurotransmitter** (see Module 26.1). The chemical signal may then generate an action potential in the receiving cell.

Now let's follow the events that occur at a chemical synapse in Figure 28.6. Molecules of the neurotransmitter (⦿) are contained in **synaptic vesicles** in the sending neuron's synaptic terminals. ❶ An action potential arrives at the synaptic terminal. ❷ The action potential triggers chemical changes that cause some synaptic vesicles to fuse with the plasma membrane of the sending cell. ❸ The fused vesicles release their neurotransmitter molecules by exocytosis into the synaptic cleft, and the neurotransmitter diffuses across the cleft.

What happens next varies among different types of chemical synapses. ❹ In one common type, the released neurotransmitter binds to receptor molecules on the receiving cell's plasma membrane. ❺ The binding of neurotransmitter to receptor opens *chemically gated ion channels* in the receiving cell's membrane. With the channels open, ions can diffuse into the receiving cell and trigger new action potentials. ❻ The neurotransmitter is broken down by an enzyme or is transported back into the signaling cell, and the ion channels close. Step 6 ensures that the neurotransmitter's effect is brief and precise.

You can review what we have covered so far in this chapter by thinking about what is happening right now in your own nervous system. Action potentials carrying information about the words on this page are streaming along sensory

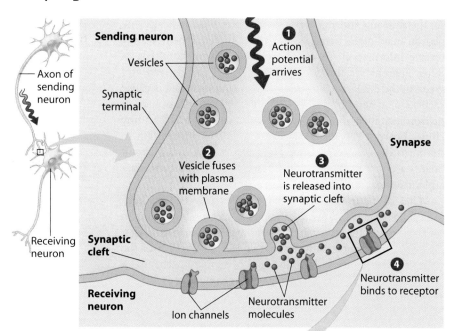

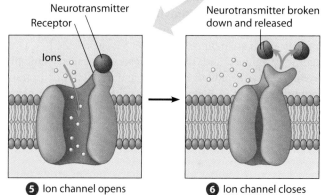

Figure 28.6 Neuron communication

neurons from your eyes to your brain. Arriving at synapses with receiving cells (interneurons in the brain), the action potentials are triggering the release of neurotransmitters at the ends of the sensory neurons. The neurotransmitters are diffusing across synaptic clefts and triggering changes in some of your interneurons—changes that lead to integration of the signals and ultimately to what the signals actually mean (in this case, the meaning of words and sentences). Next, motor neurons in your brain will send out action potentials to muscle cells in your fingers, telling them to contract in just the right way to turn the page.

Web Activity *Neuron Communication*

? How does a synapse ensure that signals pass only in one direction, from a sending neuron to a receiving cell?

■ The signal can go only one way at any one synapse because only the sending neuron releases neurotransmitter, and only the receiving cell has receptors for the neurotransmitter.

28.7 Chemical synapses make complex information processing possible

As the drawing and micrograph in Figure 28.7 indicate, one neuron may interact with many others. In fact, a neuron may receive information via neurotransmitters from hundreds of other neurons via thousands of synaptic terminals (red and green in the drawing). The inputs can be highly varied because each sending neuron may secrete a different quantity or kind of neurotransmitter. The membrane of a neuron resembles a tiny circuit board, receiving and processing bits of information in the form of neurotransmitter molecules. These living circuit boards account for the nervous system's ability to process data and formulate appropriate responses to stimuli.

What do neurotransmitters actually do to receiving neurons? The binding of a neurotransmitter to a receptor may open ion channels in the receiving cell's plasma membrane or trigger a signal transduction pathway that does so. The effect of the neurotransmitter depends on the kind of membrane channel it opens. Neurotransmitters that open Na^+ channels, for instance, may trigger action potentials in the receiving cell. Such neurotransmitters and the synapses where they are released are referred to as excitatory (green in the drawing). In contrast, many neurotransmitters open membrane channels for ions that *decrease* the tendency to develop action potentials in the receiving cell—such as channels that admit Cl^- or release K^+. These neurotransmitters and their synapses are called inhibitory (red). The effects of both excitatory and inhibitory neurotransmitters can vary in magnitude. In general, the more neurotransmitter molecules that bind to receptors on the receiving cell and the closer the synapse is to the base of the receiving cell's axon, the stronger the effect.

A receiving neuron's membrane may receive signals—both excitatory and inhibitory—from many different sending neurons. If the excitatory signals are collectively strong enough to raise the membrane potential to threshold, an action potential will be generated in the receiving cell. That neuron then passes signals along its axon to other cells at a rate that represents a summation of all the information it has received. (Signal frequency is key because action potentials are all-or-none events.) Each new receiving cell, in turn, processes this information along with all its other inputs.

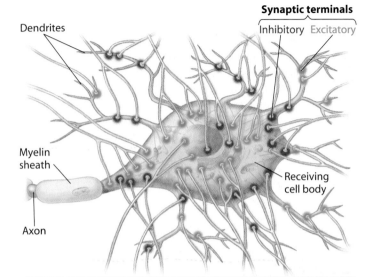

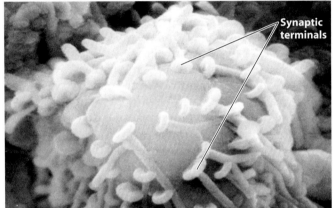

Figure 28.7   A neuron's multiple synaptic inputs

? Contrast how excitatory and inhibitory synapses change a receiving cell's membrane potential relative to triggering an action potential.

■ Neurotransmitters from an excitatory synapse open ion channels that move the receiving cell's membrane potential closer to threshold; neurotransmitters from an inhibitory synapse open ion channels that move the cell's membrane potential farther from threshold.

28.8 A variety of small molecules function as neurotransmitters

As we've discussed, the propagation of nerve signals across chemical synapses depends on neurotransmitters. A variety of small molecules serve this function.

Many neurotransmitters are small, nitrogen-containing organic molecules. One, called **acetylcholine,** is important in the brain and at synapses between motor neurons and muscle cells. Depending on the kind of receptors on receiving cells, acetylcholine may be excitatory or inhibitory. For instance, acetylcholine makes our skeletal muscles contract but slows the rate of contraction of cardiac muscles. Botulinum toxin (Botox), produced by certain bacteria that cause botulism food

poisoning, inhibits the release of acetylcholine. Botox injections disable the synapses that control particular facial muscles, thereby eliminating wrinkles around the eyes or mouth.

Biogenic amines are neurotransmitters derived from amino acids. Some examples of biogenic amines are epinephrine, norepinephrine, serotonin, and dopamine, the first three of which also function as hormones (see Chapter 26). Biogenic amines are important neurotransmitters in the central nervous system. Serotonin and dopamine affect sleep, mood, attention, and learning. Imbalances of biogenic amines are associated with various disorders. For example, the degenerative illness Parkinson's disease is associated with a lack of dopamine in the brain. Reduced levels of norepinephrine and serotonin seem to be linked with some types of depression. Some psychoactive drugs, including LSD and mescaline, apparently produce their hallucinatory effects by binding to serotonin and dopamine receptors in the brain.

Four other neurotransmitters—aspartate, glutamate, glycine, and GABA (gamma aminobutyric acid)—are actually amino acids. All are known to be important in the central nervous system. Aspartate and glutamate act primarily at excitatory synapses, while glycine and GABA act at inhibitory synapses.

Many neuropeptides, relatively short chains of amino acids, also serve as neurotransmitters. One, called substance P, is an excitatory neurotransmitter that mediates our perception of pain. The endorphins are peptides that function as both neurotransmitters and hormones, decreasing our perception of pain during times of physical or emotional stress.

Neurons also use some dissolved gases, notably nitric oxide (NO), as chemical signals. During sexual arousal in human males, certain neurons release NO into the erectile tissue of the penis, and the NO triggers an erection. (The erectile dysfunction drug Viagra promotes this effect of NO.) Neurons produce NO molecules on demand, rather than storing them in synaptic vesicles. The dissolved gas diffuses into neighboring cells, produces a change, and is quickly broken down.

? **What determines whether a neuron is affected by a specific neurotransmitter?**

■ To be affected by a particular neurotransmitter, a neuron must have specific receptors for that neurotransmitter.

Connection

28.9 Many drugs act at chemical synapses

Many psychoactive drugs, even common ones such as caffeine, nicotine, and alcohol, affect the action of neurotransmitters in the brain's billions of synapses (Figure 28.9). Caffeine, found in coffee, tea, chocolate, and many soft drinks, keeps us awake by countering the effects of inhibitory neurotransmitters. Nicotine acts as a stimulant by binding to and activating acetylcholine receptors. Alcohol is a strong depressant. Its precise effect on the nervous system is not yet known, but it seems to increase the inhibitory effects of the neurotransmitter GABA.

Many prescription drugs used to treat psychological disorders alter the effects of neurotransmitters (see Module 28.21). The most popular class of antidepressant medication works by affecting the action of serotonin. Called selective serotonin reuptake inhibitors (SSRIs), these medications block the removal of serotonin from synapses, increasing the amount of time this mood-altering neurotransmitter is available to receiving cells. Tranquilizers such as diazepam (Valium) and alprazolam (Xanax) activate the receptors for the neurotransmitter GABA, increasing its effect at inhibitory synapses. In other cases, a drug may bind to and block a receptor, reducing a neurotransmitter's effect. For instance, some antipsychotic drugs used to treat schizophrenia block receptors for the neurotransmitter dopamine. Some drugs used to treat attention deficit hyperactivity disorder (ADHD), such as methylphenidate (Ritalin), are chemically similar to dopamine and norepinephrine. ADHD medications are believed to block the reuptake of these neurotransmitters, but their precise actions are poorly understood.

What about illegal drugs? Stimulants such as amphetamines and cocaine increase the release and availability of norepinephrine and dopamine at synapses. Abuse of these drugs can produce symptoms resembling schizophrenia. The active ingredient in marijuana (tetrahydrocannabinol, or THC) binds to brain receptors normally used by other neurotransmitters that seem to play a role in pain, depression, appetite, memory, and fertility. Opiates—morphine, codeine, and heroin—bind to endorphin receptors, reducing pain and producing euphoria.

Not surprisingly, opiates can be used medicinally for pain relief. However, abuse of opiates may permanently change the brain's chemical synapses and reduce normal synthesis of neurotransmitters. As explained in Module 28.15, these drugs are also highly addictive.

The drugs discussed here are used for a variety of purposes, both medicinal and recreational. While they have the ability to increase alertness and sense of well-being or to reduce physical and emotional pain, they also have the potential to disrupt the brain's finely tuned neural pathways, altering the chemical balances that are the product of millions of years of evolution.

Figure 28.9 Alcohol, nicotine, and caffeine—drugs that alter the effects of neurotransmitters

? **When people say that "alcohol lowers a person's inhibitions," it is a behavioral description. At the neurological level, it is probably more accurate to say that "alcohol raises inhibitions." Why?**

■ Alcohol probably depresses the brain by enhancing the inhibitory effects of GABA.

28.10 The evolution of animal nervous systems reflects changes in body symmetry

One of the themes underlying all of biology is how evolution can account for life's simultaneous unity and diversity. To this point in the chapter, we have concentrated on the cellular mechanisms that are fundamental to the nervous system. There is remarkable uniformity throughout the animal kingdom in the way nerve cells function—a strong indication that the basic architecture of the neuron was an early evolutionary adaptation. But during subsequent animal evolution, great diversity emerged in the organization of nervous systems as a whole.

Animals on the earliest branches of the animal evolutionary tree (see Figure 18.4)—sponges, for instance—have no cells that are specialized for generating and transmitting nerve signals. In fact, they lack a nervous system altogether. The first modern phylum to evolve a nervous system was the hydras (cnidarians). Hydras have a **nerve net** (Figure 28.10A), a diffuse, weblike system of neurons extending throughout the body. Neurons of the nerve net control the contractions of the digestive cavity, movement of the tentacles, and other functions. However, a hydra has no head or brain, and its nerve net has no central or peripheral divisions.

Hydras and other cnidarians and adult echinoderms are radially symmetrical, and so are their nervous systems. Radially symmetrical nervous systems tend to be uncentralized, but this does not necessarily imply that they are structurally or functionally simple. Sea stars and many other echinoderms have a nerve net in each arm connected by radial nerves to a central nerve ring. This organization is better suited than a diffuse nerve net for controlling complex motion, such as the coordinated movements of hundreds of tube feet that allow a sea star to move in one direction.

The appearance of bilateral symmetry marks a key branch point in the evolution of animals and their nervous systems. Bilaterally symmetrical animals tend to move through the environment such that the head—usually equipped with sense organs and a brain—first encounters new stimuli. Flatworms are the simplest animals to have evolved two hallmarks of bilateral

symmetry: **cephalization,** an evolutionary trend toward concentration of the nervous system at the head end, and **centralization,** the presence of a central nervous system (CNS) distinct from a peripheral nervous system (PNS). For example, the planarian worm in Figure 28.10B has a small brain composed of ganglia (clusters of nerve cell bodies) and two parallel **nerve cords** that control the animal's movements. These elements constitute the simplest clearly defined central nervous system in the animal kingdom. Nerves that connect the CNS with the rest of the body make up the peripheral nervous system.

In subsequent animal phyla, the CNS tends to be more complex and contain many more neurons than in flatworms. For instance, the brains of leeches (Figure 28.10C) evolved to contain a greater concentration of neurons than those of flatworms, and leech ventral nerve cords contain segmentally arranged ganglia. The insect shown in Figure 28.10D has a brain composed of several fused ganglia, and its ventral nerve cord also has a ganglion in each body segment. Each of these ganglia directs the activity of muscles in its segment of the body.

Molluscs serve as a good illustration of how natural selection leads to correlation of the structure of a nervous system with an animal's interaction with the environment. Sessile or slow-moving molluscs such as clams have little or no cephalization and relatively simple sense organs. In contrast, squids and octopuses have quite sophisticated nervous systems. As shown in Figure 28.10E, the relatively large brain of a squid, accompanied by complex eyes and rapid signaling along giant axons, correlates well with the active predatory life of these animals.

In the next several modules, we explore the complex nervous systems that evolved in our own subphylum, the vertebrates.

> **?** Why is it advantageous for the brain of most bilateral animals to be located at the head end?

■ Cephalization places the brain close to major sense organs, which are concentrated on the end of the animal that leads the way as the animal moves through its environment.

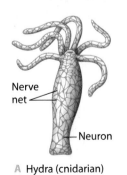

Nerve net

Neuron

A Hydra (cnidarian)

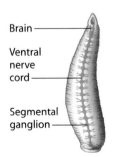

Eyespot

Brain

Nerve cord

Transverse nerve

B Flatworm (planarian)

Brain

Ventral nerve cord

Segmental ganglion

C Leech (annelid)

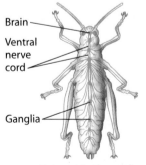

Brain

Ventral nerve cord

Ganglia

D Insect (arthropod)

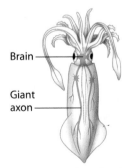

Brain

Giant axon

E Squid (mollusc)

Figure 28.10 Invertebrate nervous systems (not to scale)

Vertebrate nervous systems are diverse in both structure and level of sophistication. For instance, the brains of dolphins and humans are much more complex structurally than the brains of frogs or fishes; they are also much more powerful integrators. However, all vertebrate nervous systems have some fundamental similarities. All have distinct central and peripheral elements and are highly centralized and cephalized. In all vertebrates, the brain and spinal cord make up the CNS, while the PNS comprises the rest of the nervous system (Figure 28.11A). The **spinal cord,** which runs lengthwise inside the vertebral column, or spine, conveys information to and from the brain and integrates simple responses to certain kinds of stimuli (such as the knee-jerk reflex). The master control center, the **brain,** includes homeostatic centers that keep the body functioning smoothly; sensory centers that integrate data from the sense organs; and (in humans, at least) centers of emotion and intellect. The brain also sends out motor commands to muscles.

A vast network of blood vessels services the CNS. Brain capillaries are more selective than those elsewhere in the body.

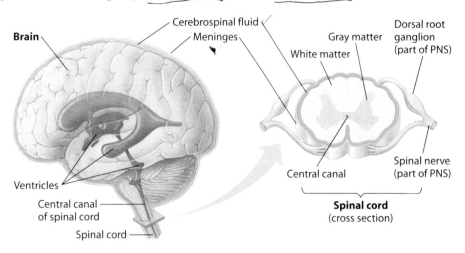

Figure 28.11B   Fluid-filled spaces of the vertebrate CNS

They allow essential nutrients and oxygen to pass freely into the brain, but keep out many other chemicals, such as metabolic wastes from other parts of the body. This selective mechanism, called the **blood-brain barrier,** maintains a stable chemical environment for the brain.

Both the brain and spinal cord contain fluid-filled spaces (Figure 28.11B). Fluid-filled spaces called **ventricles** in the brain are continuous with the narrow **central canal** of the spinal cord. Both of these cavities are filled with **cerebrospinal fluid,** which is formed in the brain by the filtration of blood. Circulating slowly through the central canal and ventricles (and then draining back into veins), the cerebrospinal fluid cushions the CNS and assists in the supply of nutrients and hormones and the removal of wastes. Also protecting the brain and spinal cord are layers of connective tissue, called **meninges.** In mammals, cerebrospinal fluid circulates between two of the meninges, providing an additional protective cushion for the CNS. As shown in the spinal cord diagram on the right side of Figure 28.11B, the CNS has two distinct areas. **White matter** is mainly composed of axons (with their whitish myelin sheaths); **gray matter** consists mainly of nerve cell bodies and dendrites.

The ganglia and nerves of the vertebrate PNS are a vast communication network. **Cranial nerves** originate in the brain and terminate mostly in structures of the head and upper body (your eyes, nose, and ears, for instance). **Spinal nerves** originate in the spinal cord and extend to parts of the body below the head. All spinal nerves and most cranial nerves contain both sensory neurons and motor neurons. Thousands of incoming and outgoing signals pass each other within the same nerves all the time.

In the next module, we'll take a closer look at the vertebrate PNS.

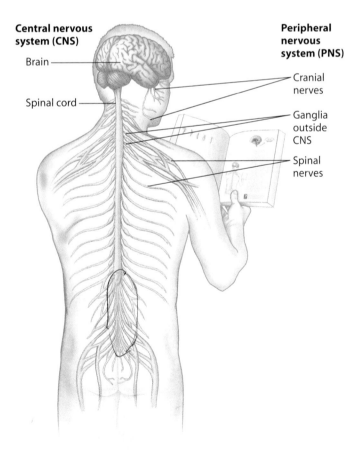

Central nervous system (CNS)
Brain
Spinal cord

Peripheral nervous system (PNS)
Cranial nerves
Ganglia outside CNS
Spinal nerves

? A vertebrate's central nervous system consists of the _____ and the _____.

28.12 The peripheral nervous system of vertebrates is a functional hierarchy

The PNS can be divided into two functional components: the somatic nervous system and the autonomic nervous system (Figure 28.12). The **somatic nervous system** carries signals to and from skeletal muscles, mainly in response to *external* stimuli. When you touch a hot stove, for instance, these neurons carry commands to your arm to pull away. The somatic nervous system is often considered voluntary because many of its actions (which we take up in Chapter 30) are under conscious control. But much skeletal muscle activity is actually controlled by reflexes mediated by the spinal cord or brainstem.

The **autonomic nervous system** regulates the *internal* environment by controlling smooth and cardiac muscles and the organs of the digestive, cardiovascular, excretory, and endocrine systems. This control is generally involuntary.

As you can see in Figure 28.12, the autonomic nervous system is composed of three divisions: sympathetic, parasympathetic, and enteric. In the next module, we discuss the varied functions of the autonomic nervous system in more detail.

? When you write the answer to this question, the muscles in your hand will be controlled by neurons of the _____ nervous system. When you eat lunch, the muscles in your digestive system will be controlled by neurons of the _____ nervous system.

■ somatic . . . autonomic

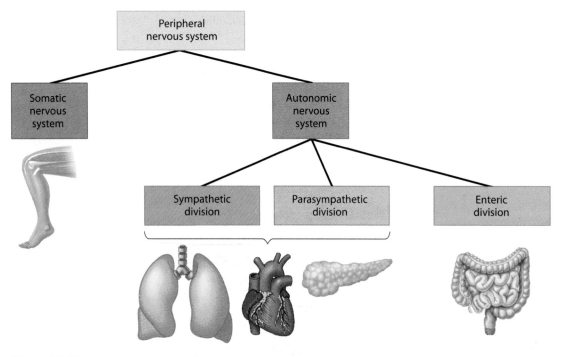

Figure 28.12 Functional divisions of the vertebrate PNS

28.13 Opposing actions of sympathetic and parasympathetic neurons regulate the internal environment

Your autonomic nervous system contains two sets of neurons that have largely antogonistic (opposite) effects on most body organs. The **parasympathetic division** primes the body for activities that gain and conserve energy for the body ("rest and digest"). A sample of the effects of parasympathetic signals appears on the left in Figure 28.13. These include stimulating the digestive organs, such as the salivary glands, stomach, and pancreas; decreasing the heart rate; and increasing glycogen production.

Neurons of the **sympathetic division** tend to have the opposite effect, preparing the body for intense, energy-consuming activities, such as fighting, fleeing, or competing in a strenuous

game (the "fight-or-flight" response). You see some of the effects of sympathetic signals on the right side of the figure. The digestive organs are inhibited, the bronchi dilate so that more air can pass through them, the heart rate increases, the liver releases the energy compound glucose into the blood, and the adrenal glands secrete the hormones epinephrine and norepinephrine.

Fight-or-flight and relaxation are opposite extremes. Your body usually operates at intermediate levels, with most of your organs receiving both sympathetic and parasympathetic signals. The opposing signals adjust an organ's activity to a suitable level. In regulating some bodily functions, however, the two divisions complement rather than antagonize each other. For example, in regulating reproduction, erection is promoted by the parasympathetic division while ejaculation is promoted by the sympathetic division.

The **enteric division** (not shown in Figure 28.13) of the autonomic nervous system consists of networks of neurons in the digestive tract, pancreas, and gallbladder. Within these organs, neurons of the enteric division control secretion as well as activity of the smooth muscles that produce peristalsis. Although the enteric division can function independently, it is normally regulated by the sympathetic and parasympathetic divisions.

Sympathetic and parasympathetic neurons emerge from different regions of the CNS, and they use different neurotransmitters. As the green dots in the figure indicate, neurons of the parasympathetic system emerge from the brain and the lower part of the spinal cord. Most parasympathetic neurons produce their effects by releasing the neurotransmitter acetylcholine at synapses within target organs. In contrast, neurons of the sympathetic system emerge from the middle regions of the spinal cord (red dots). Most sympathetic neurons release the neurotransmitter norepinephrine at target organs.

While it is convenient to divide the PNS into somatic and autonomic components, it is important to realize that these two divisions cooperate to maintain homeostasis. In response to a drop in body temperature, for example, the brain signals the autonomic nervous system to constrict surface blood vessels, which reduces heat loss. At the same time, the brain also signals the somatic nervous system to cause shivering, which increases heat production.

In the next several modules, we take a closer look at the highest level of the nervous system's structural hierarchy: the brain.

? How would a drug that inhibits the parasympathetic nervous system affect a person's pulse?

■ The pulse, or heart rate, would probably increase.

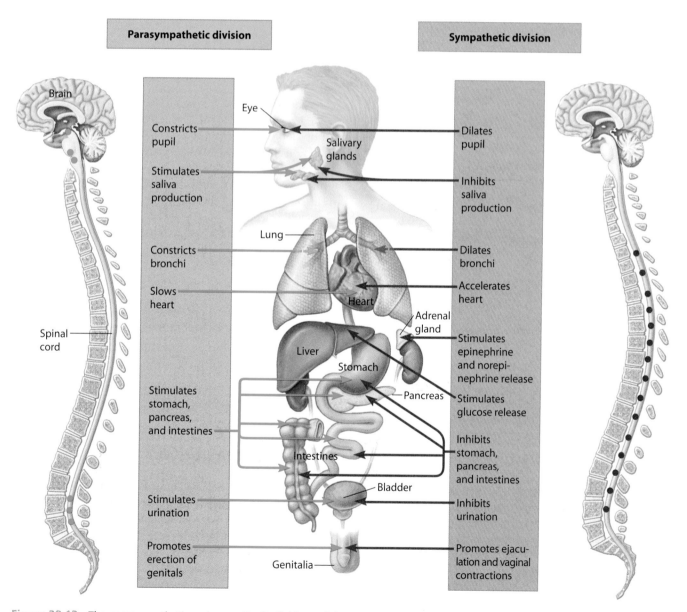

Figure 28.13 The parasympathetic and sympathetic divisions of the autonomic nervous system

28.14 The vertebrate brain develops from three anterior bulges of the neural tube

Before we focus on the human brain in the next section, we close our overview of vertebrate nervous systems by examining its embryonic development from the dorsal hollow nerve cord, one of the four distinguishing features of chordates (see Module 18.14). During early embryonic development in all vertebrates, three bilaterally symmetrical bulges—the **forebrain, midbrain, and hindbrain**—appear at the anterior end of the neural tube (Figure 28.14, left). In the course of vertebrate evolution, the forebrain and hindbrain gradually became subdivided—both structurally and functionally—into regions that assume specific responsibilities.

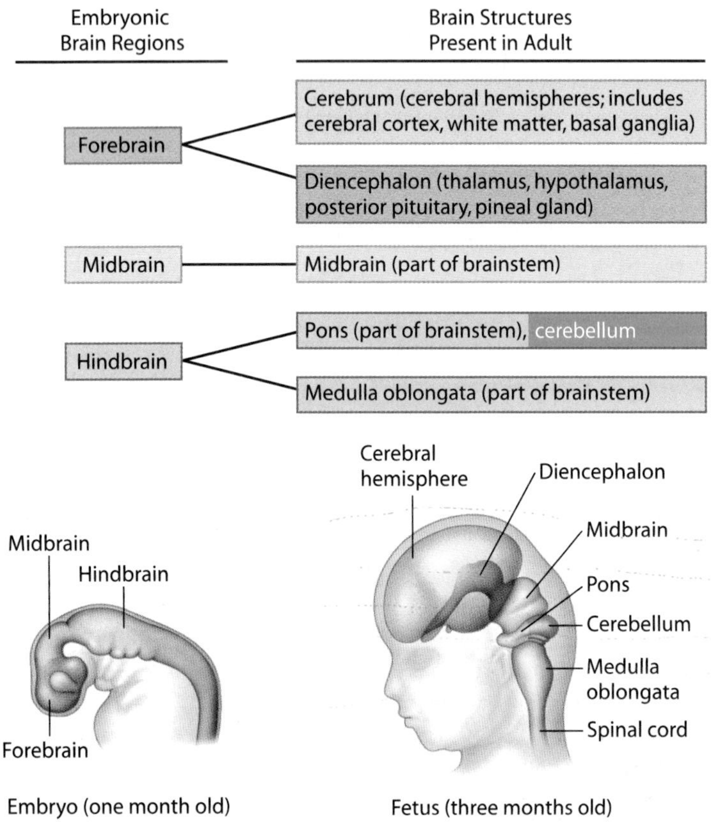

Another trend in brain evolution was the increasing integrative power of the forebrain. Evolution of the most complex vertebrate behavior paralleled the evolution of the **cerebrum,** an outgrowth of the forebrain that is the most sophisticated center of homeostatic control and integration. The cerebrum of birds and mammals is much larger relative to other parts of the brain than the cerebrum of fishes, amphibians, and other reptiles. And the sophisticated behavior of birds and mammals is directly correlated with their large cerebrum.

During the embryonic development of the human brain, the most profound changes occur in the region of the forebrain. Rapid, expansive growth of this region during the second and third months creates the two halves of the cerebrum, called the left and right cerebral hemispheres, which extend over and around much of the rest of the brain (Figure 28.14, right). By the sixth month of development, foldings increase the surface area of the outer cerebral hemispheres. This extensively convoluted outer region is called the cerebral cortex. (See Figure 28.15A on the facing page.)

Porpoises, whales, and primates also have large and complex cerebral cortices. Porpoises communicate using a large repertory of sounds. They also have the ability to locate objects, such as prey, using sound echoes (see the chapter introduction in Chapter 29). Much of a porpoise's cerebral cortex may be devoted to processing information about its sound-oriented world. In contrast, the brain of humans (one species of primate) is strongly oriented toward visual perceptions. Humans have the largest brain surface area, relative to body size, of all animals. Next we take a look at the main components of the human brain.

Embryonic Brain Regions	Brain Structures Present in Adult
Forebrain	Cerebrum (cerebral hemispheres; includes cerebral cortex, white matter, basal ganglia)
	Diencephalon (thalamus, hypothalamus, posterior pituitary, pineal gland)
Midbrain	Midbrain (part of brainstem)
Hindbrain	Pons (part of brainstem), cerebellum
	Medulla oblongata (part of brainstem)

Figure 28.14 Embryonic development of the human brain

? Which region of the brain has changed the most during the course of vertebrate evolution?

■ The cerebrum, in particular the cerebral cortex

The Human Brain

28.15 The structure of a living supercomputer: The human brain

Composed of perhaps as many as 100 billion intricately organized neurons, with a much larger number of supporting cells, the human brain is more powerful than the most sophisticated computer. Figure 28.15A (top of next page) shows the three ancestral brain regions in their fully-developed adult form. Table 28.15 summarizes the anatomy and physiology of each.

Looking first at the lower brain centers, two sections of the hindbrain (blue), the **medulla oblongata** and **pons,** and the midbrain (purple) make up a functional unit called the **brainstem.** Consisting of a stalk with cap-like swellings at the anterior end of the spinal cord, the brainstem is, evolutionarily speaking, one of the older parts of the vertebrate brain. The brainstem coordinates and filters the conduction of information from sensory and motor neurons to the higher brain regions. It also regulates sleep and arousal and helps coordinate body movements, such as walking. Table 28.15 lists some of the individual functions of the medulla oblongata, pons, and midbrain.

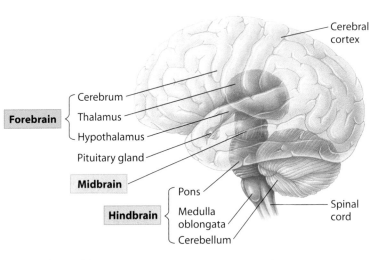

Figure 28.15A   The main parts of the human brain

TABLE 28.15   MAJOR STRUCTURES OF THE HUMAN BRAIN

Brain Structure	Major Functions
Brainstem	Conducts data to and from other brain centers; helps maintain homeostasis; coordinates body movement
Medulla oblongata	Controls breathing, circulation, swallowing, digestion
Pons	Controls breathing
Midbrain	Receives and integrates auditory data; coordinates visual reflexes; sends sensory data to higher brain centers
Cerebellum	Coordinates body movement; plays role in learning and in remembering motor responses
Thalamus	Serves as input center for sensory data going to the cerebrum, output center for motor responses leaving the cerebrum; sorts data
Hypothalamus	Functions as homeostatic control center; controls pituitary gland; serves as biological clock
Cerebrum	Performs sophisticated integration; plays major role in memory, learning, speech, emotions; formulates complex behavioral responses

Another part of the hindbrain, the **cerebellum** (light blue), is a planning center for body movements. It also plays a role in learning, decision making, and remembering motor responses. The cerebellum receives sensory information about the position of joints and the length of muscles, as well as information from the auditory and visual systems. It also receives input concerning motor commands issued by the cerebrum. The cerebellum uses this information to coordinate movement and balance. Hand-eye coordination is an example of such control by the cerebellum. If the cerebellum is damaged, the eyes can follow a moving object, but they will not stop at the same place as the object.

The most sophisticated integrating centers are those derived from the forebrain (orange and gold)—the thalamus, the hypothalamus, and the cerebrum. The **thalamus** contains most of the cell bodies of neurons that relay information to the cerebral cortex. The thalamus first sorts data into categories (all the touch signals from a hand, for instance). It also suppresses some signals and enhances others. The thalamus then sends information on to the appropriate higher brain centers for further interpretation and integration.

In Module 26.4, we saw that the hypothalamus controls the pituitary gland and the secretion of many hormones. The hypothalamus also regulates body temperature, blood pressure, hunger, thirst, the sex drive, and fight-or-flight responses, and it helps us experience emotions such as rage and pleasure. A "pleasure center" in the hypothalamus could also be called an addiction center, for it is strongly affected by certain addictive drugs, such as amphetamines and cocaine. As described in Module 28.9, these drugs increase the effects of norepinephrine and dopamine at synapses in the pleasure center, producing a short-term high, often followed by depression. Cocaine addiction may involve chemical changes in the pleasure center and elsewhere in the hypothalamus.

A pair of hypothalmic structures called the suprachiasmatic nuclei function as an internal timekeeper, our **biological clock.** Receiving visual input from the eyes (light/dark cycles, in particular), the clock maintains our **circadian rhythms**—patterns that are repeated daily—such as the sleep/wake cycle. Research with many different species has shown that without environmental cues, biological clocks keep time in a free-running way. For example, when humans are placed in artificial settings that lack environmental cues, our biological clocks and circadian rhythms maintain a cycle of approximately 24 hours 11 minutes with very little variation among individuals.

The cerebrum, which is the largest and most complex part of our brain, consists of right and left **cerebral hemispheres** (Figure 28.15B), each of which is responsible for the opposite side of the body. A thick band of nerve fibers called the **corpus callosum** facilitates communication between the cerebral hemispheres, enabling them to process information together. Under the corpus callosum, groups of neurons called the **basal nuclei** are important in motor coordination. If they are damaged, a person may be immobilized. Degeneration of the basal nuclei occurs in Parkinson's disease (see Module 28.21). The most extensive portion of our cerebrum, the cerebral cortex, is the subject of the next module.

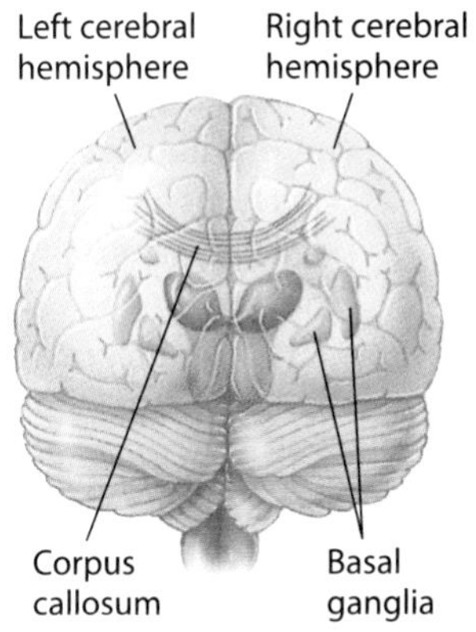

Left cerebral hemisphere   Right cerebral hemisphere

Corpus callosum   Basal ganglia

Figure 28.15B
A rear view of the brain

🎧 MP3 Tutor *The Human Brain*

❓ Choosing from the structures in Table 28.15, identify the brain part most important in solving an algebra problem.

■ Cerebrum

28.16 The cerebral cortex is a mosaic of specialized, interactive regions

Although less than 5 mm thick, the highly folded **cerebral cortex** accounts for about 80% of the total human brain mass. It contains some 10 billion neurons and hundreds of billions of synapses. Its intricate neural circuitry produces our most distinctive human traits: reasoning and mathematical abilities, language skills, imagination, artistic talent, and personality traits. Assembling information it receives from our eyes, ears, nose, taste buds, and touch sensors, the cortex also creates our sensory perceptions—what we are actually aware of when we see, hear, smell, taste, or touch. The cerebral cortex also regulates our voluntary movements.

Like the rest of the cerebrum, the cerebral cortex is divided into right and left sides connected by the corpus callosum. Each side of the cerebral cortex has four lobes. Each lobe (represented by a different color in Figure 28.16) is named for a nearby bone of the skull. Researchers have identified a number of functional areas within each lobe.

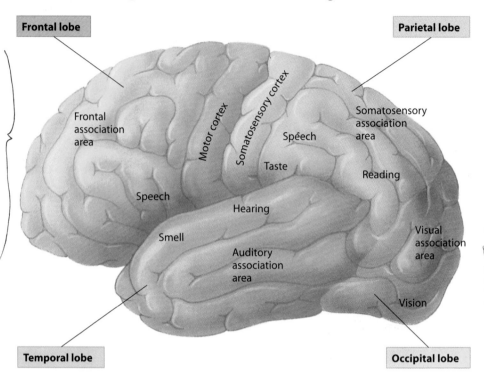

Figure 28.16 Functional areas of the left cerebral hemisphere

Two areas of known function form the boundary between the frontal and parietal lobes. One, called the motor cortex, functions mainly in sending commands to skeletal muscles, signaling appropriate responses to sensory stimuli. Next to the motor cortex, the somatosensory cortex receives and partially integrates signals from touch, pain, pressure, and temperature receptors throughout the body. The cerebral cortex also has centers that receive and begin processing sensory information concerned with vision, hearing, taste, and smell. Each of these centers, as well as the somatosensory cortex, cooperates with an adjacent area, called an association area. Imaging techniques, such as PET scanning (described in Module 20.11), are beginning to show how a complicated interchange of signals among receiving centers and association areas produces our sensory perceptions.

Making up most of our cerebral cortex, the association areas are the sites of higher mental activities—roughly, what we call thinking. In humans, a large association area in the frontal lobe uses varied inputs from many other areas of the brain to evaluate consequences, make considered judgments, and plan for the future.

Language results from extremely complex interactions among several association areas. For instance, the parietal lobe of the cortex has association areas used for reading and speech. These areas obtain visual information (the appearance of words on a page) from the vision centers. Then, if the words are to be spoken aloud, they arrange the information into speech patterns and tell another speech center, in the frontal lobe, how

to make the motor cortex move the tongue, lips, and other muscles to form words. When we hear words, the parietal areas perform similar functions using information from auditory centers of the cortex.

You may have heard people comment that they are "left-brained" or "right-brained." In a phenomenon known as **lateralization,** areas in the two hemispheres become specialized for different functions during infant and child brain development. In most people, the left hemisphere becomes adept at language, logic, and mathematical operations, as well as detailed skeletal motor control and processing of fine visual and auditory details. The right hemisphere is stronger at spatial relations, pattern and face recognition, and nonverbal thinking. (In about 10% of us, these roles of the left and right hemispheres are reversed or the hemispheres are less specialized.)

How have researchers identified the functions of different parts of the brain? We explore this question in the next two modules.

? A stroke that causes loss of speech and numbness of the right side of the body has probably damaged brain tissue in the _____ lobe of the _____ hemisphere.

■ parietal . . . left

28.17   Injuries and brain operations provide insight into brain function

The physiology of the human brain is exquisitely complex, making it one of the most difficult structures to study in all of biology. No animal model or computer simulation can accurately predict its complicated functions. New techniques, such as PET scans and fMRIs (see Modules 20.11 and 28.18), are allowing researchers to associate specific parts of the brain with various activities. Much of what has been learned about the brain, however, has come from rare individuals whose brains were altered through injury, illness, or surgery. By studying such "broken brains," researchers have gained insight into how healthy brains operate.

The first well-publicized case of this type involved a man named Phineas Gage. In 1848, while working as a railroad construction foreman, Gage accidentally exploded a dynamite charge that propelled a 3-foot-long spike through his head. The 13-pound steel rod entered his left cheek and traveled upward behind his left eye and out the top of his skull, landing several yards away. Incredibly, Gage walked away from the accident and appeared to have an intact intellect. However, his associates soon noticed drastic changes in his personality, with new propensities toward meanness, vulgarity, irresponsibility, and an inability to control his behavior.

At the time, Gage's doctor was able to note these changes, but understanding of the brain was insufficient to explain them. Luckily, the doctor preserved Gage's skull and the spike, allowing a group of researchers in 1994 to produce a computer model of the injury (Figure 28.17A). The modern analysis offered an explanation for Gage's bizarre behavior: The rod had pierced both frontal lobes of his brain. People with these sorts of injuries often exhibit irrational decision making and diffi-culty processing emotions. As you will learn in Module 28.20, the frontal lobes are part of the limbic system, a group of brain structures involved with emotions.

Beginning with the work of several neurosurgeons in the 1950s, many of the functional areas of the cerebral cortex have been identified during brain surgery. The cortex lacks cells that detect pain; thus, after anesthetizing the scalp, a neurosurgeon can operate on the cerebrum with the patient awake. Parts of the cortex can be stimulated with a harmless electrical current. Stimulation of specific areas can cause someone to experience different sensations or recall memories. Researchers can obtain information about the effects simply by questioning the conscious patient.

Neurophysiologists have also gained insight into the interrelatedness of the brain's two hemispheres. As discussed in Module 28.16, association areas in the left and right sides become specialized for different functions. Much of what we know about this lateralization stems from the work of Roger Sperry with patients whose corpus callosum (communicating fibers between the two hemispheres; see Module 28.15) had been surgically cut to treat severe epileptic seizures. In a series of ingenious experiments, Sperry demonstrated that his patients were unable to verbalize sensory information that was received by only the right hemisphere.

One of the most radical surgical alterations of the brain is a hemispherectomy (Figure 28.17B)—the removal of most of one half of the brain, excluding deep structures such as the thalamus, brainstem, and basal nuclei. This procedure is performed to alleviate severe seizure disorders that originate from one of the hemispheres as a result of illness, abnormal development, or stroke. Incredibly, with just half a brain, hemispherectomy patients recover quickly, often leaving the hospital within a few weeks. Their intellectual capacities are undiminished, although the side of the body opposite the surgery remains partially paralyzed. Higher brain functions that previously originated from the missing half of the brain begin to be controlled by the opposite side. The younger the patient is, ideally less than 5 or 6, the faster and more complete the recovery. Development after hemispherectomy is a striking example of the remarkable plasticity of the brain.

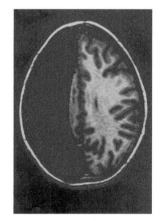

Figure 28.17B   X-ray of hemispherectomy patient after surgery

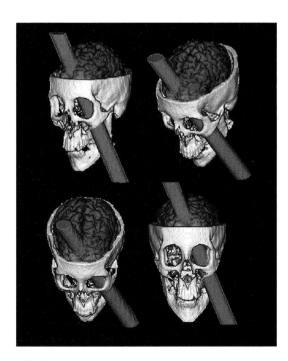

Figure 28.17A   Computer model of Phineas Gage's injury

? How are researchers able to investigate brain function during brain surgery?

■ The cortex lacks pain receptors. Regions of the brain can be stimulated during surgery, and the conscious patient can report sensations or memories.

28.18 fMRI scans can provide insight into brain structure and function

Functional magnetic resonance imaging (fMRI) is a scanning and imaging technology that can "light up" metabolic processes as they occur within living tissue. Because the procedure can be performed on a conscious patient, fMRI scans can provide significant insights into brain structure and function.

As first described in Module 20.11, MRI uses powerful magnets to align and then locate atoms within living tissue. During fMRI, a subject lies with his or her head in the hole of a large, doughnut-shaped magnet. When the brain is scanned with electromagnetic waves, changes in blood oxygen usage at sites of neuronal activity generate a signal that can be recorded. A computer then uses the data to construct a three-dimensional map of the subject's brain activity.

Figure 28.18 shows the areas of the brain that light up in an fMRI as a patient becomes aware of the intention to move a muscle. Metabolic hot spots in the brain are shaded red, revealing the regions of the brain that are most active during this process. Such studies confirm hypotheses based on older technologies about the roles of specific brain areas in movement and intention. Researchers have applied similar techniques to correlate specific brain regions with a wide range of behaviors, both simple and complex. Such visualization methods are at the forefront of neuroscience, one of biology's most fascinating and rapidly developing subdisciplines.

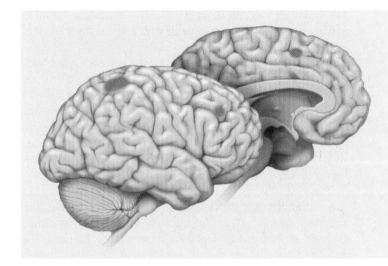

Figure 28.18 Regions of the brain highlighted during an fMRI scan of a person who is intending to move.

? **What does an fMRI scan actually measure?**

■ Changes in the use of oxygen by living tissues

28.19 Several parts of the brain regulate sleep and arousal

As anyone who has drifted off to sleep during a lecture (or while reading a book) knows, attentiveness and mental alertness can change rapidly. *Arousal* is a state of awareness of the outside world. Its counterpart is *sleep,* a state when external stimuli are received but not consciously perceived.

Several centers for controlling arousal and sleep are found in the brainstem. One such center is the reticular formation, a diffuse network of neurons that extends through the core of the brainstem. Acting as a sensory filter, the reticular formation receives data from sensory receptors and selects which information reaches the cerebral cortex. The more information the cortex receives, the more alert and aware a person is, although the brain often ignores certain familiar and repetitive stimuli—the feel of your clothes against your skin, for example—while actively processing other inputs. Sleep and wakefulness are also regulated by specific parts of the brainstem: The pons and medulla contain centers that cause sleep when stimulated, and the midbrain has a center that causes arousal. Serotonin may be one of the neurotransmitters of the sleep-producing centers. Drinking milk before bedtime may induce sleep because milk contains large amounts of tryptophan, the amino acid from which serotonin is synthesized.

Although we know very little about its function, sleep is essential for survival. Furthermore, sleep is an active state, at least for the brain. By placing electrodes at multiple sites on the scalp, researchers can record patterns of electrical activity called brain waves in an electroencephalogram, or EEG. These recordings reveal that brain wave frequency changes as the brain progresses through the distinct stages of sleep. During REM (rapid eye movement) sleep, the brain waves are rapid and less regular, like those of the awake state. We have most of our dreams during REM sleep, which typically occurs about six times a night for periods of 5 to 50 minutes each.

Understanding the function of sleeping and dreaming remains a compelling research problem. One hypothesis is that both sleep and dreams are involved in the consolidation of learning and memory, and experiments show that regions of the brain activated during a learning task can become active again during sleep.

? **What prevents the cerebral cortex from being overwhelmed by all the sensory stimuli arriving from sensory receptors?**

■ The reticular formation filters out unimportant stimuli.

The limbic system is involved in emotions, memory, and learning

Mapping the parts of the brain involved in human emotions, learning, and memory and studying the interactions between these parts are among the great challenges in biology today. Much of human emotion, learning, and memory depends on our **limbic system.** This functional unit includes parts of the thalamus and hypothalamus and two partial rings around them formed by portions of the cerebral cortex (Figure 28.20). Two cerebral structures, the amygdala and the hippocampus, play key roles in memory, learning, and emotions.

The limbic system is central to such behaviors as nurturing infants and bonding emotionally to other people. Primary emotions that produce laughing and crying are mediated by the limbic system, and it also attaches emotional "feelings" to basic survival mechanisms of the brainstem, such as feeding, aggression, and sexuality. The intimate relationship between our feelings and our thoughts results from interactions between the limbic system and the prefrontal cortex, which is involved in complex learning, reasoning, and personality.

Memory, which is essential for learning, is the ability to store and retrieve information derived from experience. The **amygdala** is central in recognizing the emotional content of facial expressions and laying down emotional memories. Sensory data converge in the amygdala, which seems to act as a memory filter, somehow labeling information to be remembered by tying it to an event or emotion of the moment. The **hippocampus** is involved in both the formation of memories and their recall. Portions of the frontal lobes are involved in associating primary emotions with different situations.

We sense our limbic system's role in both emotion and memory when certain odors bring back "scent memories." Have you ever had a particular smell suddenly make you nostalgic for something that happened when you were a child? As indicated in Figure 28.20, signals from your nose enter your brain through the olfactory bulb, which connects with the limbic system. Thus, a specific scent can immediately trigger emotional reactions and memories.

Short-term memory, as the name implies, lasts only a short time—usually only a few minutes. It is short-term memory that allows you to dial a phone number just after looking it up. You may, however, be able to recall the number weeks after you originally looked it up, or even longer. This is because you have stored the number in **long-term memory.** The transfer of information from short-term to long-term memory is enhanced by rehearsal, positive or negative emotional states mediated by the amygdala, and the association of new data with data previously learned and stored in long-term memory. For example, it's easier to learn a new card game if you already have "card sense" from playing other games.

Factual memories, involving names, faces, words, and places, are different from skill or procedural memories. Skill

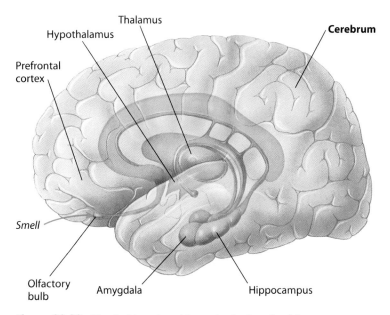

Figure 28.20 The limbic system (shown in shades of gold)

memories usually involve motor activities that are learned by repetition without consciously remembering specific information. You perform skills, such as tying your shoes, riding a bicycle, or hitting a baseball, without consciously recalling the individual steps required to do these tasks correctly. Once a skill memory is learned, it is difficult to unlearn. For example, a person who has played tennis with a self-taught, awkward backhand has a tougher time learning the correct form than a beginner just learning the game. Bad habits, as we all know, are hard to break.

Information processing by the brain generally seems to involve a complex interplay of several integrating centers. By experimenting with animals, studying amnesia (memory loss) in humans, and using brain-imaging techniques, scientists have begun to map some of the major brain pathways involved in memory. Their proposed pathway involves the hippocampus and amygdala, which receive sensory information from the cortex and convey it to other parts of the limbic system and to the prefrontal cortex. The memory storage is completed when signals return to the area in the cortex where the sensory perception originated.

In the final module of this chapter, we'll discuss four major disorders of the nervous system, including their symptoms and treatments.

? Which three factors help transfer information from short-term to long-term memory?

■ Rehearsal, emotional associations, and connection with previously learned data

Changes in brain physiology can produce neurological disorders

Neurological disorders (diseases of the nervous system) take an enormous toll on society. Examples of neurological disorders include schizophrenia, depression, Alzheimer's disease, and Parkinson's disease. While these conditions are not yet curable, there are a number of treatments available.

Schizophrenia About 1% of the world's population suffers from **schizophrenia,** a severe mental disturbance characterized by psychotic episodes in which patients lose the ability to distinguish reality. The symptoms of schizophrenia typically include hallucinations (most often "voices" telling patients that they are worthless and evil); delusions (generally paranoid); blunted emotions; distractibility; lack of initiative; and difficulty with verbal expression. Contrary to commonly held belief, schizophrenics do not necessarily exhibit a "split personality." There seem to be several different forms of schizophrenia, and it is unclear whether they represent different disorders or variations of the same underlying disease.

The causes of schizophrenia are unknown, although the disease has a strong genetic component. Studies of identical twins show that if one twin has schizophrenia, there is a 50% chance that the other twin will have it, too. Since identical twins share identical genes, this indicates that schizophrenia has an equally strong environmental component, the nature of which has not been identified.

There are several treatments for schizophrenia that can usually alleviate the major symptoms. Such treatments focus on brain pathways that use dopamine as a neurotransmitter. Unfortunately, these treatments often have substantial negative side effects, such as motor deficits that resemble those of Parkinson's disease, uncontrolled facial writhing movements, and a risk of a serious blood disorder. Identification of the genetic mutations responsible for schizophrenia may yield new insights about the causes of the disease, which may in turn lead to effective therapies with fewer drawbacks.

Depression Depression is a disorder characterized by depressed mood, as well as abnormalities in sleep, appetite, and energy level. Two broad forms of depressive illness have been identified: major depression and bipolar disorder.

People with **major depression** may experience persistent sadness, loss of interest in pleasurable activities, changes in body weight and sleeping patterns, loss of energy, and suicidal thoughts. While all of us experience feelings of sadness from time to time, major depression is extreme and more persistent, leaving a person unable to live a normal life. One of the most common forms of mental illness, major depression affects about one in every seven individuals at some point in their adult life. Women are affected twice as often as men. If left untreated, symptoms may become more frequent and severe over time.

Bipolar disorder, or manic-depressive disorder, involves extreme mood swings and affects about 1% of the population.

The manic phase is characterized by high self-esteem, increased energy and flow of ideas, overtalkativeness, inappropriate risk taking, promiscuity, and reckless spending. In its milder forms, this phase is sometimes associated with great creativity, and some well-known artists, musicians, and literary figures (including Keats, Tolstoy, and Hemingway) have had periods of intense output during their manic phases. The depressive phase is marked by lowered ability to feel pleasure, loss of motivation, sleep disturbances, and feelings of worthlessness. These symptoms can be so severe that some individuals attempt suicide.

Both bipolar disorder and major depression have a genetic component. As in schizophrenia, there is also a strong environmental influence; stress, especially severe stress in childhood, may be an important factor.

In recent years, researchers have begun to learn how brain physiology is involved in depression. The colored positron emission tomography scans (PET scans; see Module 20.11) in Figure 28.21A compare the brains of a depressed patient (top) and healthy patient (bottom). The red and yellow

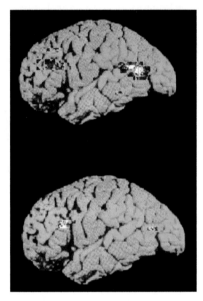

Figure 28.21A Brain activity in a depressed person (top) and healthy person (bottom)

colors indicate areas of low brain activity. Note that the PET scan from the depressed patient shows decreased activity in certain areas of the brain.

Several treatments for depression are available and can be quite effective. Many depressed people have an imbalance of neurotransmitters, particularly serotonin. Some medications are intended to correct such imbalances. The most commonly prescribed class of antidepressant drugs are selective serotonin reuptake inhibitors (SSRIs). As the name implies, these medications increase the amount of time that serotonin is available to stimulate certain neurons in the brain, which appears to relieve some symptoms. In 1987, fluoxetine (best known as Prozac) became the first SSRI approved to treat depression. Other frequently prescribed SSRIs include paroxetine (Paxil) and sertraline (Zoloft). As shown in Figure 28.21B, the number of prescriptions for SSRIs in the United States tripled over a 10-year period. SSRIs now represent one of the largest classes of medication in the United States, in terms of both dollars spent and number of prescriptions.

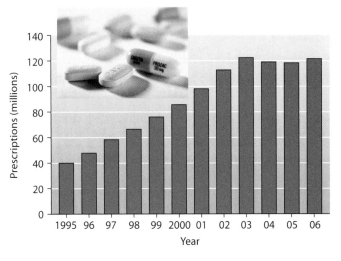

Figure 28.21B   SSRI (inset) prescriptions in the United States

Figure 28.21C   Actor Michael J. Fox (right) and boxer Muhammad Ali, both of whom suffer from Parkinson's disease, testifying before the Senate about funding for the disorder

Alzheimer's Disease   A form of mental deterioration, or dementia, **Alzheimer's disease (AD)** is characterized by confusion, memory loss, and a variety of other symptoms. Its incidence is usually age related, rising from about 10% at age 65 to about 35% at age 85. Thus, by helping humans live longer, modern medicine is increasing the proportion of AD patients in the population. The disease is progressive; patients gradually become less able to function and eventually need to be dressed, bathed, and fed by others. There are also personality changes, almost always for the worse. Patients often lose their ability to recognize people, even family members, and may treat them with suspicion and hostility.

At present, a firm diagnosis of AD is difficult to make while the patient is alive because it is one of several forms of dementia. However, AD results in a characteristic pathology of the brain: Neurons die in huge areas of the brain, and brain tissue often shrinks. (See Figure 2.5B to view a PET scan of the brain of an Alzheimer's patient.) The shrinkage is visible with brain imaging but is not enough to positively identify AD. What is diagnostic is the postmortem finding of two features—neurofibrillary tangles and senile plaques—in the remaining brain tissue. Neurofibrillary tangles are bundles of a protein called *tau*, a protein that normally helps regulate the movement of nutrients along microtubules within neurons. Senile plaques are aggregates of beta-amyloid, an insoluble peptide containing 40 to 42 amino acids that is cleaved from a membrane protein normally present in neurons. The plaques appear to trigger the death of surrounding neurons, but whether amyloid plaques cause Alzheimer's or are a symptom of it remains unclear.

Parkinson's Disease   Approximately 1 million people in the United States suffer from **Parkinson's disease** (Figure 28.21C), a motor disorder characterized by difficulty in initiating movements, slowness of movement, and rigidity. Patients often have a masked facial expression, muscle tremors, poor balance, a flexed posture, and a shuffling gait. Like Alzheimer's disease,

Parkinson's is progressive, and the risk increases with age. The incidence of Parkinson's disease is about 1% at age 65 and about 5% at age 85.

The symptoms of Parkinson's disease result from the death of neurons in the basal nuclei. These neurons normally release dopamine from their synaptic terminals. The disease itself appears to be caused by a combination of environmental and genetic factors. Evidence for a genetic role includes the fact that some families with an increased incidence of Parkinson's disease carry a mutated form of the gene for a protein important in normal brain function.

At present, there is no cure for Parkinson's disease, although various treatments can help control the symptoms. Treatments include drugs such as L-dopa, a precursor of the neurotransmitter dopamine, and surgery. One potential cure is to develop embryonic stem cells into dopamine-secreting neurons in the laboratory and implant them in patients' brains. In laboratory experiments, transplantation of such cells into rats with a Parkinson's-like condition can lead to a recovery of motor control. Whether this kind of regenerative medicine will also work in humans is one of many important questions on the frontier of modern brain research.

Unraveling the biological basis of neurological disorders remains one of the most challenging tasks of modern biology. Recent advances in locating genes that correlate with CNS disorders and creating animal models for these diseases have provided new insights, but many aspects of our nervous system remain mysterious. In the next chapter, we examine another aspect of nervous systems—how sense organs gather information about the environment.

? What do the initials SSRI stand for? Relate the name to its mechanism of action.

■ SSRIs (selective serotonin reuptake inhibitors) are antidepressant drugs that specifically (*selectively*) prevent (*inhibit*) the reabsorption (*reuptake*) of the neurotransmitter *serotonin in the brain*.

Chapter Review

Reviewing the Concepts

Nervous System Structure and Function (Introduction–28.2)

The **nervous system** obtains and processes sensory information and sends commands to effector cells (such as muscles) that carry out appropriate responses:

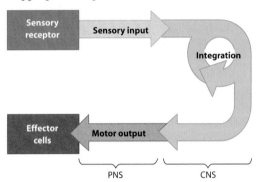

Sensory neurons conduct signals from sensory receptors to the central nervous system (CNS), which consists of the brain and, in vertebrates, the spinal cord. Interneurons in the CNS integrate information and send it to motor neurons. Motor neurons, in turn, convey signals to effector cells. Located outside the CNS, the peripheral nervous system (PNS) consists of nerves (bundles of fibers of sensory and motor neurons) and ganglia (clusters of cell bodies of the neurons) **(28.1).** The functional units of the nervous system are neurons, cells specialized for carrying signals **(28.2):**

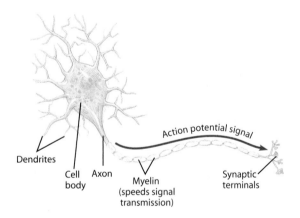

Nerve Signals and Their Transmission (28.3–28.9)

The conduction of nerve signals. At rest, a neuron's plasma membrane has an electrical voltage called the resting potential. The resting potential is caused by the membrane's ability to maintain a positive charge on its outer surface opposing a negative charge on its inner (cytoplasmic) surface **(28.3).** A stimulus alters the permeability of a portion of the membrane, allowing ions to pass through and changing the membrane's voltage. A nerve signal, called an action potential, is a change in the membrane voltage from the resting potential to a maximum level and back to the resting potential **(28.4).** Action potentials are self-propagated in a one-way chain reaction along a neuron. An action potential is an all-or-none event. The frequency of action potentials (but not their strength) changes with the strength of the stimulus **(28.5).**

Synapses. The transmission of signals between neurons or between neurons and effector cells occurs at junctions called synapses. Electrical signals pass between cells at electrical synapses. At chemical synapses, the sending (presynaptic) cell secretes a chemical signal, a neurotransmitter, which crosses the synaptic cleft and binds to a specific receptor on the surface of the receiving (postsynaptic) cell **(28.6).** Some neurotransmitters excite the receiving cell; others inhibit the receiving cell's activity by decreasing its ability to develop action potentials. A cell may receive differing signals from many neurons; the summation of excitation and inhibition determines whether or not it will transmit a nerve signal **(28.7).** Many small, nitrogen-containing molecules serve as neurotransmitters **(28.8).** Many psychoactive drugs act at synapses and affect neurotransmitter action **(28.9).**

An Overview of Animal Nervous Systems (28.10–28.14)

Animal nervous systems. Radially symmetrical animals have a nervous system arranged in a weblike system of neurons called a nerve net. Among bilaterally symmetrical animals, nervous systems evolved to exhibit cephalization, the concentration of the nervous system in the head end, and centralization, the presence of a central nervous system **(28.10).** Vertebrate nervous systems are highly centralized and cephalized. The brain and spinal cord contain fluid-filled spaces. Cranial and spinal nerves make up the peripheral nervous system **(28.11).**

Organization of the vertebrate nervous system (28.12–28.13):

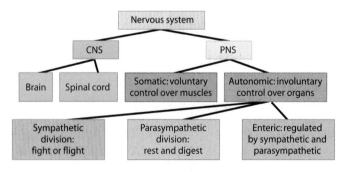

The vertebrate brain evolved by the enlargement and subdivision of three anterior bulges of the neural tube: the hindbrain, midbrain, and forebrain. The size and complexity of the cerebrum in birds and mammals correlates with their sophisticated behavior **(28.14).**

The Human Brain (28.15–28.21)

Anatomy of the human brain. The midbrain and subdivisions of the hindbrain, together with the thalamus and hypothalamus of the forebrain, function mainly in conducting information to and from higher brain centers. They regulate homeostatic functions, keep track of body position, and sort sensory information. The forebrain's cerebrum is the largest and most complex part of the brain. Most of the cerebrum's integrative power resides in the cerebral cortex of the two cerebral hemispheres **(28.15).** Specialized integrative regions of the cerebral cortex include the somatosensory cortex and centers for vision, hearing, taste, and smell. The motor cortex directs responses. Association areas, concerned with higher mental activities such as reasoning and language, make up most of the cerebrum. The right and left cerebral hemispheres tend to specialize in different mental tasks (28.16).

Higher brain functions. Brain injuries and surgeries have been used to study brain function (28.17). fMRI scans can provide insight into brain structure and function (28.18). Sleep and arousal involve activity by the hypothalamus, medulla oblongata, pons, and neurons of the reticular formation (28.19). The limbic system, a functional group of integrating centers in the cerebral cortex, thalamus, and hypothalamus, is involved in emotions, memory, and learning (28.20). Many neurological disorders can be linked to changes in brain physiology. Treatment can often improve patients' lives, but cures are elusive (28.21).

Connecting the Concepts

1. Test your understanding of the nervous system by matching the following labels with their corresponding letters: CNS, effector cells, interneuron, motor neuron, PNS, sensory neuron, sensory receptor, spinal cord, synapse.

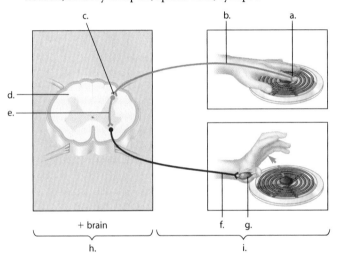

Testing Your Knowledge

Multiple Choice

2. Joe accidentally touched a hot pan. His arm jerked back, and an instant later, he felt a burning pain. How would you explain that his arm moved before he felt the pain?
   a. His limbic system blocked the pain momentarily, but the important pain signals eventually got through.
   b. His response was a spinal cord reflex that occurred before the pain signals got to the brain.
   c. It took a while for his brain to search long-term memory and figure out what was going on.
   d. Motor neurons are myelinated; sensory neurons are not. The signals traveled faster to his muscles.
   e. This scenario is not actually possible. The brain must register pain before a person can react.

3. Which of the following mediates sleep and arousal?
   a. the reticular formation, along with the hypothalamus and thalamus
   b. the limbic system, which includes the amygdala and hippocampus
   c. the left hemisphere of the cerebral cortex
   d. the midbrain and cerebellum
   e. the parasympathetic and sympathetic divisions of the nervous system

4. Anesthetics block pain by blocking the transmission of nerve signals. Which of these three chemicals might work as anesthetics? (*Circle all that apply and explain your selections.*)
   a. a chemical that prevents the opening of sodium channels in membranes
   b. a chemical that inhibits the enzymes that degrade neurotransmitters
   c. a chemical that blocks neurotransmitter receptors

Describing, Comparing, and Explaining

5. As you hold this book, nerve signals are generated in nerve endings in your fingertips and sent to your brain. Once a touch has caused an action potential at one end of a neuron, what causes the nerve signal to move from that point along the length of the neuron to the other end? What is the nerve signal, exactly? Why can't it go backward? How is the nerve signal transmitted from one neuron to the next across a synapse? Write a short paragraph that answers these questions.

Applying the Concepts

6. Using microelectrodes, a researcher recorded nerve signals in four neurons in the brain of a snail. The neurons are called A, B, C, and D in the table below. A, B, and C all can transmit signals to D. In three experiments, the animal was stimulated in different ways. The number of nerve signals transmitted per second by each of the cells is recorded in the table. Write a short paragraph explaining the different results of the three experiments.

	Signals/sec			
	A	**B**	**C**	**D**
Experiment #1	50	0	40	30
Experiment #2	50	0	60	45
Experiment #3	50	30	60	0

7. Brain injuries tend to be severe because most neurons, once damaged, will not regenerate. The use of embryonic stem cells has been proposed as a potential treatment for many neurological diseases. As mentioned in the text, neurons developed from such stem cells in laboratory culture might be implanted in the brain of a person with Alzheimer's or Parkinson's disease. These neurons might be able to replace those damaged by the disease. Do you favor or oppose research along those lines? Explain your answer.

8. Alcohol's depressant effects on the nervous system cloud judgment and slow reflexes. Alcohol consumption is a factor in most fatal traffic accidents in the United States. What are some other impacts of alcohol abuse on society? What are some of the responses of people and society to alcohol abuse? Do you think this is primarily an individual or societal problem? Do you think our responses to alcohol abuse are appropriate and proportional to the seriousness of the problem?

Answers to all questions can be found in Appendix 4.

For Practice Quizzes, BioFlix, MP3 Tutors, and Activities, go to www.mybiology.com.

Superhuman Senses

In this chapter we will focus on four human senses: hearing, vision, taste, and smell. But before we talk about ourselves, let's consider three senses found among animals but not humans: echolocation, electroreception, and magnetoreception.

Echolocation is the ability to locate objects by detecting the echoes of emitted sound waves. Many bat species, for example, produce high-pitched sounds (usually ultrasonic—that is, at frequencies above the range of human hearing) in their larynx and emit them through their mouths and noses, which are shaped in a way that focuses the sound waves. They receive the echoes of the emitted sounds via large ears. Their brains can process the time delay and spatial arrangement of the echoes to determine the size, shape, location, speed, and direction of objects in their environment. Bats emit about 15 pulses of sound per second while scanning for prey and up to 100 pulses per second while actively capturing insects. In another example of echolocation, porpoises (and a few other species of marine mammals) produce ultrasonic clicking sounds in their nasal passages. The sounds bounce off a skull bone and through an oil-filled structure in their forehead that focuses the sound. A porpoise receives echoes via a narrow window of bone behind the jaw.

Other animals detect prey by electroreception, the ability to sense electric fields. Electroreception is common among fish and is found in some amphibians and the monotremes (egg-laying mammals such as the duck-billed platypus). Many species of fish generate weak electric fields from organs in their tails. These fish can then detect distortions in the electric field caused by objects in their environment via sensors in the skin. This short-range sensing is important to fish in low-visibility environments, such as nocturnal hunters or fish that live in murky water.

The electroreception ability of some sharks rival the most accurate voltmeter for sensitivity to electric fields. A great hammerhead shark (*Sphyma mokarran*) can detect prey hiding on or under the ocean floor (such as flounder or sting rays) by sweeping its wide head back and forth, like a beachcomber scanning the shore with a metal detector. Cells in the shark's head can sense the minute electrical activity of muscles or nerves in nearby prey. Other organisms use electroreception for navigation, courtship, and social communication.

In contrast to echolocation and electroreception, magnetoreception—the ability to detect magnetic fields—is common throughout the animal kingdom. Migratory birds, fish (such as salmon and trout), turtles, amphibians, and even bees are believed to rely on magnetoreception to navigate relative to Earth's magnetic field. The physiological basis of magnetoreception has been widely studied but remains elusive. Some magnetoreceptive animals have neurons that contain the naturally occurring mineral magnetite (Fe_3O_4) arranged in a straight line. Such cells have been found in the abdomens of bees and in the skulls of a variety of vertebrates. Scientists speculate that these cells function somewhat like a compass needle. In fact, magnetite was once used by sailors as a primitive compass. But not all magnetoreceptive animals have magnetite-containing cells, and some animals without this sense (including humans) do have such cells.

Although the mechanisms underlying magnetoreception are poorly understood, the anatomical and physiological basis of several other senses are well known. This chapter focuses on the sensory structures of those senses and how they gather information. To begin, we examine the distinction between information gathering and information processing. Next, we will look at the general principles of sensory reception and then apply them to the human senses of hearing, vision, taste, and smell. ■ ■ ■

29.1 Sensory inputs become sensations and perceptions in the brain

Sensory receptor cells are tuned to the conditions of the external world and the internal organs. They detect stimuli, such as chemicals, light, tension in a muscle, sounds, electricity, cold, heat, and touch, and send information to the central nervous system. The sensory cells in a hammerhead shark's head, for instance, detect changes in the electric field surrounding its body and send reports about these changes to the brain. The reports take the form of action potentials, the same signals used throughout the nervous system (see Module 28.4). The sensory receptor's job is completed when it triggers action potentials that go to the central nervous system.

An action potential triggered in response to the electric field of a prey animal is the same as an action potential triggered in response to the sight of that animal. That is, all senses (smell, sight, detection of electric fields, etc.) trigger the same *type* of action potential. The ability to distinguish the particular type of stimulus, such as smell or sight, depends on the part of the brain that receives the action potential. Once the brain is aware of **sensations,** the action potentials it receives from sense receptors, the brain interprets them to produce a perception. **Perceptions,** such as colors, smells, sounds, and tastes, are constructed by the brain as it processes sensations and integrates them with other information, forming a meaningful interpretation or conscious understanding of sensory data.

Figure 29.1 may further demonstrate the difference between sensation and perception for you. What do you see when you first look at the figure? If you see only some black splotches on a blue background, you have developed a sensation—an awareness of this sensory information. What if we say that the figure shows a person riding a horse? With this clue, the brain forms a perception; it converts the sensation into a meaningful image. Your brain integrated the new information with some of its stored data.

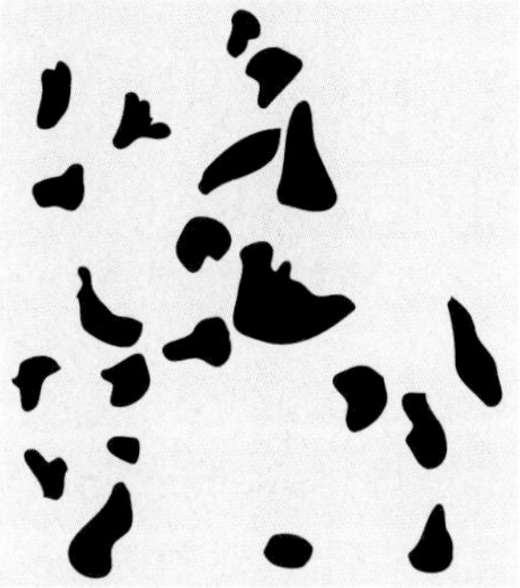

Figure 29.1 Black splotches or a person riding on a horse?

What does the brain actually do with sensory information in creating a perception? As we discussed in Chapter 28, researchers using brain-imaging techniques are beginning to find out. Perceptions appear to result from communication among neurons arranged in extremely complex circuits involving multiple areas of the cerebrum. These areas include sensory and association areas and the limbic system, which may access stored memories.

Perceptions are the product of a continuum of information processing, beginning with the detection of stimuli by sensory receptors. In this chapter, we focus on sensory receptors and how they function.

? Which would tend to vary more among humans faced with a specific environmental stimulus: sensation or perception?

■ Perception

29.2 Sensory receptors convert stimulus energy to action potentials

Sensory organs, such as your eyes or your taste buds, contain **sensory receptors,** specialized cells or neurons that detect stimuli. All stimuli represent forms of energy. The sensory receptors in your eyes detect light energy; those in your taste buds detect chemicals, such as salt or sugar.

What exactly do we mean when we say that a sensory receptor detects a stimulus—a photon of light or a molecule of sugar, for instance? Stimulus detection means that the receptor cell converts one type of signal (the stimulus) to an electrical signal. This conversion, called **sensory transduction,** produces a change in the membrane potential (the potential energy stored by the membrane; see Module 28.3) of the receptor cell. This change in the membrane potential is a result of the opening or closing of ion channels in the sensory receptor's plasma membrane.

Figure 29.2A (top of next page) shows sensory transduction occurring when sensory receptor cells in a taste bud detect sugar molecules. ❶ The sugar molecules (⁘) first enter the taste bud, where ❷ they bind to specific protein molecules on a taste receptor membrane. (Artificial sweeteners work by binding to these same taste receptors; see Module 3.6.) The binding initiates a signal transduction pathway (see Module 11.12) that causes ❸ some ion channels in the membrane to close and other ion channels to open. Changes in the flow of ions create a graded change in membrane potential called a **receptor potential.** In contrast to action potentials, which are all-or-none phenomena, receptor potentials vary; the stronger the stimulus, the greater the receptor potential.

Once a stimulus is converted to a receptor potential, the receptor potential usually results in signals entering the central nervous system. In our taste bud example, ❹ each receptor cell forms a synapse with a sensory neuron. This is a chemical synapse like the one between neurons described in Module 28.6. In many cases, a receptor cell constantly secretes neurotransmitter (⁘) into this synapse at a set rate, triggering a steady stream of action potentials in the sensory neuron. ❺ The graph on the left shows the rate at which the sensory neuron sends action potentials when the taste receptor is not detecting any sugar.

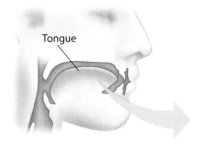

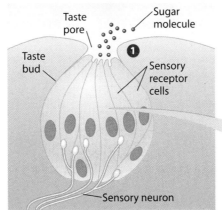

Figure 29.2A Sensory transduction at a taste bud

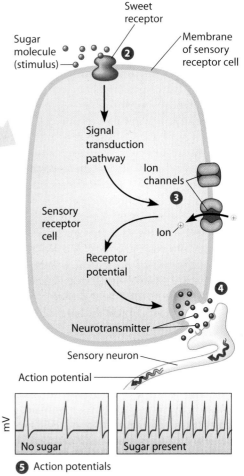

The graph on the right shows what happens when there are enough sugar molecules to trigger a strong receptor potential, causing the receptor cell to release more neurotransmitter than usual. This additional neurotransmitter increases the rate of action potential generation in the sensory neuron. It is the change in the rate of action potentials that signals the brain that the sensory receptor detects a stimulus.

Thus we see that sensory receptors transduce (convert) stimuli into receptor potentials, which trigger action potentials that enter the central nervous system for processing. Since action potentials are the same no matter where or how they are produced, how do they communicate a sweet taste instead of a salty one? In Figure 29.2B, the taste bud on the left has sensory receptors that respond to sugar, and the taste bud on the right responds to salt. The sensory neurons from the sugar-detecting taste bud synapse with different interneurons in the brain than those contacted by neurons from the salt-detecting taste bud. The brain distinguishes stimulus types (in this case, sugar from salt) by the patterns in which interneurons are stimulated.

The graphs in Figure 29.2B also indicate how action potentials communicate information about the *intensity* of stimuli (for example, very sweet or less sweet). In each case, the left part of the graph represents the rate at which the sensory neurons in the taste bud transmit action potentials when the taste receptors are not stimulated. The right side of each graph shows that the rate of transmission depends on the intensity of the stimulus. The stronger the stimulus, the more neurotransmitter released by the receptor cell and the more frequently the sensory neuron transmits action potentials to the brain. The brain interprets the intensity of the stimulus from the rate at which it receives action potentials.

There is an important qualification to what we have just said about stimulus intensity. Have you ever noticed how an odor that is strong at first seems to fade with time, even when you know the substance is still there? The same effect helps you adjust to a hot or cold shower and enables you to wear clothes without being constantly aware of them. The effect is called **sensory adaptation,** the tendency of some sensory receptors to become less sensitive when they are stimulated repeatedly. When receptors become less sensitive, they trigger fewer action potentials, and the brain may lose its awareness of stimuli as a result. Sensory adaptation keeps the body from reacting to normal background stimuli. Without it, our nervous system would become overloaded with useless information.

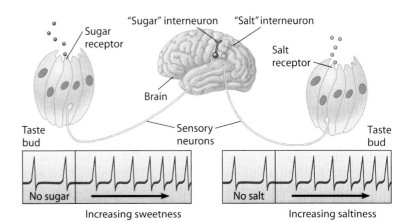

Figure 29.2B How action potentials transmit different taste sensations

This overview of sensory transduction, transmission, and adaptation explains how sensory receptors work in general. Now let's look at the receptors themselves.

🎧 MP3 Tutor *Sensory Receptors*

❓ What is meant by sensory transduction?

■ The conversion of a stimulus signal to an electrical signal (a receptor potential) by a sensory receptor cell

29.3 Specialized sensory receptors detect five categories of stimuli

Based on the type of signals to which they respond, we can group sensory receptors into five general categories: pain receptors, thermoreceptors, mechanoreceptors, chemoreceptors, and electromagnetic receptors.

Figure 29.3A, showing a section of human skin, reveals why the surface of our body is sensitive to such a variety of stimuli. Our skin contains pain receptors, thermoreceptors (sensors for both heat and cold), and mechanoreceptors (sensors for touch and pressure). Each of these receptors is a modified dendrite of a sensory neuron (see Module 28.2). The neuron both transduces stimuli and sends action potentials to the central nervous system. In other words, each receptor serves as both a receptor cell and a sensory neuron. Most of the dendrites in the dermis (the underlying region of the skin) are wrapped in one or more layers of connective tissue (purple areas in the figure); however, the pain and touch receptors in the epidermis (outer skin layer) and the touch receptors around the base of hairs are naked dendrites.

Pain Receptors   Probably all animals have **pain receptors** (labeled in red in the figure), although we cannot say what non-human perceptions of pain are like. Pain often indicates danger and usually makes an animal withdraw to safety. All parts of the human body except the brain have pain receptors. Pain can make us aware of injury or disease. Pain receptors may respond to excess heat or pressure or to chemicals released from damaged or inflamed tissues. Histamines and acids are some of the chemicals that trigger pain.

Thermoreceptors   **Thermoreceptors** (blue in the figure) in the skin detect either heat or cold. Other temperature sensors located deep in the body monitor the temperature of the blood. The hypothalamus is the body's major thermostat. Receiving action potentials from both surface and deep sensors, the hypothalamus keeps a mammal's body temperature within a narrow range (see Module 20.15).

Mechanoreceptors   Different types of **mechanoreceptors** (green in the figure) are stimulated by various forms of mechanical energy, such as touch and pressure, stretching, motion, and sound. All these forces produce their effects by bending or stretching the plasma membrane of a receptor cell. When the membrane changes shape, it becomes more permeable to sodium or potassium ions, and the mechanical energy of the stimulus is transduced into a receptor potential.

At the top of Figure 29.3A are two types of mechanoreceptors that detect light touch. Both types transduce very slight inputs of mechanical energy into action potentials. A third type of pressure sensor, lying deeper in the skin, is stimulated by strong pressure. A fourth type of mechanoreceptor, the touch receptor around the base of the hair, detects hair movements. Touch receptors at the base of the stout whiskers on a cat are extremely sensitive and enable the animal to detect close objects by touch in the dark. Another type of mechanoreceptor (not shown) is found in our skeletal muscles. Sensitive to changes in muscle length,

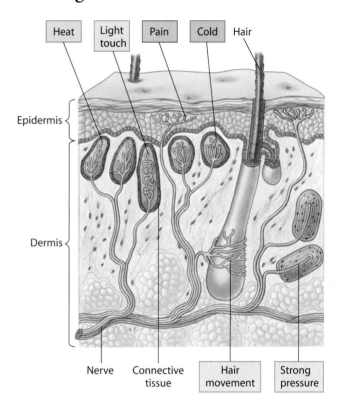

Figure 29.3A   Sensory receptors in the human skin

stretch receptors monitor the position of body parts (see Figure 28.1B).

A variety of mechanoreceptors collectively called **hair cells** detect sound waves and other forms of movement in water. The "hairs" on these sensors are either specialized types of cilia or cellular projections called microvilli. The sensory hairs project from the surface of a receptor cell into either the external environment, such as the water surrounding a fish, or an internal fluid-filled compartment, such as our inner ear. Figure 29.3B indicates how hair cells work, ❶ starting with a receptor cell at rest. ❷ When fluid movement bends the hairs in one direction, the hairs stretch the cell membrane, increasing its permeability to certain ions. This makes the hair cell secrete more neurotransmitter molecules and increases the rate of action potential production by a sensory neuron. ❸ When the hairs bend in the opposite direction, ion permeability decreases, the hair cell releases fewer neurotransmitter molecules, and the rate of action potential generation decreases. We'll see later that hair cells are involved in both hearing and balance.

Chemoreceptors   **Chemoreceptors** include the sensory receptors in our nose and taste buds, which are attuned to chemicals in the external environment, as well as some receptors that detect chemicals in the body's internal environment. Internal chemoreceptors include sensors in some of our arteries that can detect changes in the amount of O_2 in the blood. Osmoreceptors in the brain detect changes in the total solute

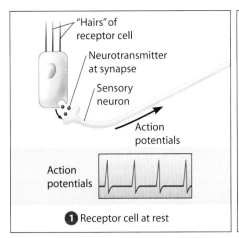
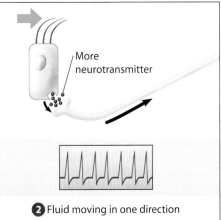
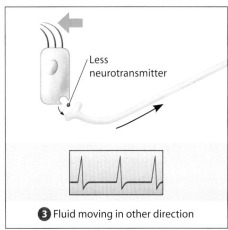

1 Receptor cell at rest

2 Fluid moving in one direction

3 Fluid moving in other direction

Figure 29.3B Mechanoreception by a hair cell

concentration of the blood and stimulate thirst when blood osmolarity increases (see Module 25.8).

One of the most sensitive chemoreceptors is found on the antennae of the male silkworm moth *Bombyx mori* (Figure 29.3C). The antennae are covered with thousands of sensory hairs (visible in the micrograph). The hairs have chemoreceptors that detect a sex pheromone released by the female.

Electromagnetic Receptors Energy of various wavelengths, occurring as electricity, magnetism, or light, may be detected by **electromagnetic receptors.** As mentioned in the chapter introduction, certain fishes discharge electrical currents into the water and use electroreceptors to detect nearby prey. A platypus has electroreceptors on its bill that can detect electrical fields generated by the muscles of prey, such as crustaceans, frogs, and small fishes.

Many animals appear to use Earth's magnetic field to orient themselves as they migrate. In addition to the animals men-tioned in the chapter introduction, certain protists and prokary-otes can orient with respect to Earth's magnetic field.

Photoreceptors, including eyes, are probably the most common type of electromagnetic receptor. Photoreceptors detect the electromagnetic energy we call light, which may be in the visible or ultraviolet part of the electromagnetic spectrum (see Module 7.6). The rattlesnake in Figure 29.3D has prominent eyes that detect visible light. Below its eyes are two infrared receptors. These specialized electromagnetic receptors detect the body heat of its preferred prey, small mammals and birds. The recep-tors can detect the infrared radiation emitted by a mouse a meter away.

A variety of light detectors have evolved in the animal king-dom. Despite their differences, however, all photoreceptors contain similar pigment molecules that absorb light, and evi-dence indicates that all photoreceptors may be homologous (having a common ancestry). We'll consider some of the differ-ent types of eyes found in animals in Module 28.7. But first, let's consider the senses of hearing and balance.

? For each of the following senses in humans, identify the type of receptor: seeing, tasting, hearing, smelling.

■ Photoreceptor; chemoreceptor; mechanoreceptor; chemoreceptor

Figure 29.3C Chemoreceptors on insect antennae

SEM 80×

Eye

Infrared receptor

Figure 29.3D Electromagnetic receptors in a snake

29.4 The ear converts air pressure waves to action potentials that are perceived as sound

The human ear is really two separate organs, one for hearing and the other for maintaining balance. We look at the structure and function of our hearing organ in this module and then turn to our sense of balance in Module 29.5. Both organs operate on the same basic principle, the stimulation of long microvilli-like projections on hair cells (mechanoreceptors) in fluid-filled canals.

The ear is complex, and it helps to learn its basic structure before studying how it functions (Figure 29.4A). The ear is composed of three regions: the outer ear, the middle ear, and the inner ear. The **outer ear** consists of the flap-like **pinna**—the fleshy structure we commonly refer to as our "ear"—and the **auditory canal.** The pinna and the auditory canal collect sound waves and channel them to the **eardrum,** a sheet of tissue that separates the outer ear from the **middle ear** (Figure 29.4B). Both the outer ear and middle ear are common sites of childhood infections (called swimmer's ear and otitis media, respectively).

When sound pressure waves strike the eardrum, it vibrates and passes the vibrations to three small bones: the hammer (more formally, the malleus), anvil (incus), and stirrup (stapes). The stirrup is connected to the **oval window,** a membrane-covered hole in the skull bone, through which vibrations pass into the inner ear. The middle ear also opens into the **Eustachian tube,** which connects with the pharynx (back of the throat) and equalizes pressure between the middle ear and the atmosphere. This tube is what enables you to move air in or out to equalize pressure on either side of the eardrum ("pop" your ears) when changing altitude rapidly or when diving underwater.

The **inner ear** consists of fluid-filled channels in the bones of the skull. Sound vibrations or movements of the head set the fluid in motion. One of the channels, the **cochlea** (Latin for "snail"), is a long, coiled tube. The cross-sectional view of the cochlea in Figure 29.4C, on the bottom of the next page, shows that inside it are three fluid-filled canals. Our hearing organ, the **organ of Corti,** is located within the middle canal. The organ of Corti consists of an array of hair cells embedded in a **basilar membrane** (the floor of the middle canal). The hair cells are the sensory receptors of the ear. As you can see in the enlargement, a shelflike projection called the tectorial membrane extends from the wall of the middle canal. Notice that the tips of the hair cells are in contact with the tectorial membrane. Sensory neurons synapse with the base of the hair cells and carry action potentials to the brain via the auditory nerve.

Now let's see how the parts of the ear function in hearing. Sound waves are pressure waves in the air that are collected by the pinna and auditory canal of the outer ear. Represented in Figure 29.4D, these pressure waves make your eardrum vibrate with the same frequency as the sound. The frequency, measured in hertz (Hz), is the number of vibrations per second (1 Hz is equal to one vibration per second).

From the eardrum, the vibrations are amplified as they are transferred through the hammer, anvil, and stirrup in the middle ear to the oval window. Vibrations of the oval window then produce pressure waves in the fluid within the cochlea. The vibrations first pass from the oval window into the fluid in the upper canal of the cochlea. Pressure waves travel through the upper canal to the tip of the cochlea, at the coil's center. The pressure

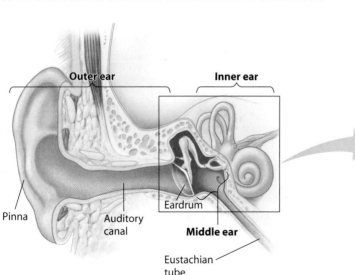

Figure 29.4A   An overview of the human ear

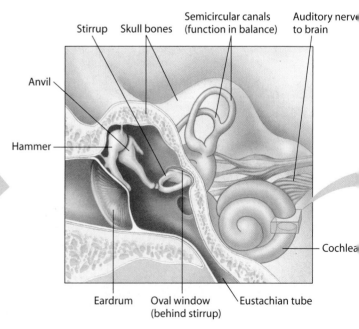

Figure 29.4B   The middle ear and the inner ear

waves then enter the lower canal and gradually fade away.

As a pressure wave passes through the upper canal of the cochlea, it pushes downward on the middle canal, making the basilar membrane vibrate. Vibration of the basilar membrane makes the hairlike projections on the hair cells alternately brush against and draw away from the overlying tectorial membrane. When a hair cell's projections are bent, ion channels in its plasma membrane open, and positive ions enter the cell. As a result, the hair cell develops a receptor potential and releases more neurotransmitter molecules at its synapse with a sensory neuron. In turn, the sensory neuron sends more action potentials to the brain through the auditory nerve.

Volume and Pitch The brain senses a sound as an increase in the frequency of action potentials it obtains from the auditory nerve. But how is the quality of the sound determined? The higher the volume (loudness) of sound, the higher the amplitude of the pressure wave it generates. In the ear, a higher amplitude of pressure waves produces more vigorous vibrations of fluid in the cochlea, more pronounced bending of the hair cells, and thus, more action potentials generated in the sensory neurons. The loudness of sound is measured in decibels (dB). The decibel scale for human hearing ranges from 0 to 120 dB, the loudest we can hear without intolerable pain.

The pitch of a sound depends on the frequency of the sound waves. High-pitched sounds, such as high notes sung by a soprano, generate high-frequency waves. Low-pitched sounds, like low notes sung by a bass, generate low-frequency waves. How does the cochlea distinguish sounds of different pitch? The key is that the basilar membrane is not uniform along its length. The end near the oval window is relatively narrow and stiff, while the other end, near the tip of the cochlea, is wider and more flexible. Each region of the basilar membrane is most sensitive to a particular frequency of vibration, and the region vibrating most vigorously at any instant sends the most action

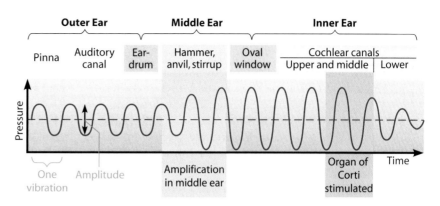

Figure 29.4D The route of sound wave vibrations through the ear

potentials to auditory centers in the brain. The brain interprets the information and gives us a perception of pitch. Young people with healthy ears can hear pitches in the range of 20–20,000 Hz. Dogs can hear sounds as high as 40,000 Hz, and bats can emit and hear clicking sounds as high-pitched as 100,000 Hz.

Deafness, the loss of hearing, can be caused by the inability to conduct sounds, resulting from middle-ear infections, a ruptured eardrum, or stiffening of the middle-ear bones (a common age-related problem). Deafness can also result from damage to sensory receptors or neurons. Few parts of our anatomy are more delicate than the organ of Corti. Frequent or prolonged exposure to sounds over 90 dB can damage or destroy hair cells (which are never replaced). For example, you may have noticed that a few hours in a very loud environment leaves you with ringing or buzzing sounds in your ears, a condition called tinnitus. In the United States, employees exposed to occupational noise above 90 dB must wear ear protection. Amplified rock music often reaches 120 dB. Ear plugs can provide protection to both rock musicians and their fans.

? How does the ear convert sound waves in the air to pressure waves of the fluid in the cochlea?

■ Sound waves in air cause the eardrum to vibrate. The small bones attached to the inside of the eardrum transmit the movement to the oval window on the wall of the inner ear. Vibrations of the oval window set in motion the fluid in the inner ear, which includes the cochlea.

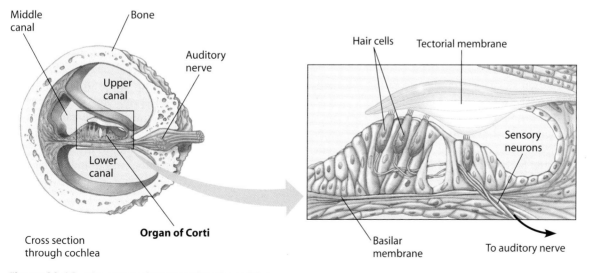

Figure 29.4C The organ of Corti, within the cochlea

29.5 The inner ear houses our organs of balance

Several organs in the inner ear detect body position and movement. These fluid-filled structures lie next to the cochlea (Figure 29.5) and include three semicircular canals and two chambers, the utricle and the saccule. All the equilibrium structures operate on the same principle, by the bending of hairs on hair cells.

The three **semicircular canals** detect changes in the head's rate of rotation or angular movement. As shown in the figure, the canals are arranged in three perpendicular planes and can therefore detect movement in all directions. A swelling at the base of each semicircular canal contains a cluster of hair cells with their hairs projecting into a gelatinous mass called a cupula (shown in the enlargements). When you rotate your head in any direction, the thick, sticky fluid in the canals moves more slowly than your head. Consequently, the fluid presses against the cupula, bending the hairs. The faster you rotate your head, the greater the pressure and the higher the frequency of action potentials sent to the brain. If you rotate your head at a constant speed, the fluid in the canals begins moving with the head, and the pressure on the cupula is reduced. But if you stop suddenly, the fluid continues to move and again stimulates the hair cells, which may make you feel dizzy.

Clusters of hair cells in the utricle and saccule detect the position of the head with respect to gravity. The hairs of these cells project into a gelatinous material containing many small calcium carbonate particles. When the position of the head changes, this heavy material bends the hairs in a different direction, causing an increase or decrease in the rate at which action potentials are sent to the brain. The brain determines the new position of the head by interpreting the altered flow of action potentials.

The equilibrium receptors provide data the brain needs to determine the position and movement of the head. Using this information, the brain develops and sends out commands that make the skeletal muscles balance the body.

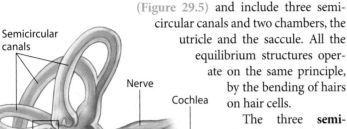

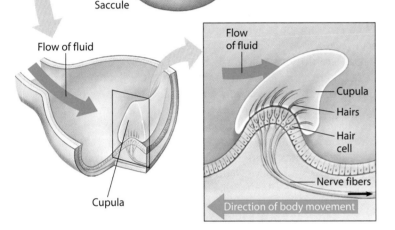

Figure 29.5 Equilibrium structures in the inner ear

? **What type of receptor cell is common to our senses of hearing and equilibrium?**

■ Hair cells, which are mechanoreceptors

Connection

29.6 What causes motion sickness?

Boating, flying, or even riding in a car can make us dizzy and nauseated, a condition called motion sickness. Some people start feeling ill just from thinking about getting on a boat or plane. Many others get sick only during storms at sea or during turbulence in flight. Motion sickness is thought to result from the brain's receiving signals from equilibrium receptors in the inner ear that conflict with visual signals from the eyes. When a susceptible person is inside a moving ship, for instance, signals from the equilibrium receptors in the inner ear indicate, correctly, that the body is moving (in relation to the environment outside the ship). In conflict with these signals, the eyes may tell the brain that the body is in a stationary environment, the cabin. Somehow the conflicting signals make the person feel ill. Symptoms may be relieved by closing the eyes, limiting head movements, or focusing on a stable horizon. Many sufferers of motion sickness take a sedative such as Dramamine or Bonine to relieve their symptoms.

Long-lasting, drug-containing skin patches prevent motion sickness by inhibiting input from the equilibrium sensors. Ginger tablets and pressure-point wristbands may also help.

Motion sickness can be a severe problem for astronauts, and the National Aeronautics and Space Administration (NASA) conducts research on the problem. One of NASA's most interesting findings is that some people can learn to consciously control body functions, such as the vomiting reflex. Astronauts receive intensive training in how to exert "mind over body" when zero gravity starts to induce motion sickness.

? **Explain how someone could suffer motion sickness when watching a film shot from the front of a roller coaster.**

■ There would be conflicting information between vision ("I'm moving") and the equilibrium sense ("I'm sitting still in my theater seat").

29.7 Several types of eyes have evolved among animals

Most invertebrates have some kind of light-detecting organ. One of the simplest is the **eye cup,** the visual sensory organs of flatworms (phylum Platyhelminthes), specifically within the genus *Dugesia,* free-living nonparasitic flatworms known as planarians (Figure 29.7A). Planarian eye cups provide information about light intensity and direction but not enough data for the brain to form an image. The eye cup contains photoreceptor cells that are partially shielded by darkly pigmented cells. Light can enter the eye cup only where there are no pigmented cells, and the openings of the two eye cups face opposite directions. The brain compares the rate of nerve impulses coming from the two eye cups, and the animal turns until the sensations are equal and minimal. The result is that the animal moves directly away from the light source and reaches a dark hiding place.

Two major types of image-forming eyes have evolved in invertebrates. Compound eyes are found in insects and crustaceans. A **compound eye** consists of many tiny light-detecting units called ommatidia. You can see some of the thousands of ommatidia (the tiny dots) making up the two compound eyes of a fly in Figure 29.7B. Each ommatidium has its own light-focusing lens and several photoreceptor cells. Every ommatidium picks up light from a tiny portion of the field of view. The animal's brain then forms a mosaic visual image by assembling the data from all the ommatidia.

Compound eyes are extremely acute motion detectors, an important advantage for flying insects and other small animals that are often threatened by predators. The compound eyes of most insects also provide excellent color vision. Some species, such as honeybees, can see ultraviolet light (invisible to humans), which helps them locate certain nectar-bearing flowers.

The second type of image-forming eye to have evolved in invertebrates is the **single-lens eye.** The single-lens eye also evolved independently in vertebrates (such as humans). The single-lens eye works on a principle similar to that of a camera. Figure 29.7C shows a human eye. (The single-lens eye found in squids and other invertebrates is similar, but differs in some of the details.) The human eye has a small opening at the center of the eye, the **pupil,** through which light enters. Analogous to a camera's shutter, an adjustable doughnut-shaped **iris** changes the diameter of the pupil to let in more or less light. After going through the pupil, light passes through a single disklike **lens.** The lens focuses light onto the **retina,** which consists of many photoreceptor cells.

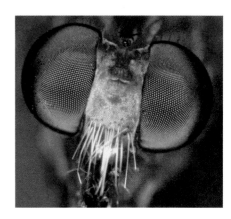

Figure 29.7B   The two compound eyes of a fly

Photoreceptor cells of the retina transduce light energy, and action potentials pass via sensory neurons in the optic nerve to the visual centers of the brain. Photoreceptor cells are highly concentrated at the retina's center of focus, called the **fovea.** There are no photoreceptor cells in the blind spot, the part of the retina where the optic nerve passes through the back of the eye. We cannot detect light that is focused on the blind spot, but having two eyes with overlapping fields of view enables us to perceive uninterrupted images. In the next module, we'll examine the eye's image-forming structures in more detail.

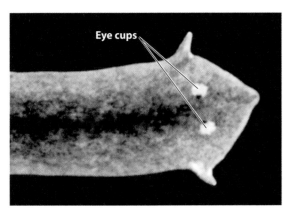

Figure 29.7A   The eye cups of a planarian

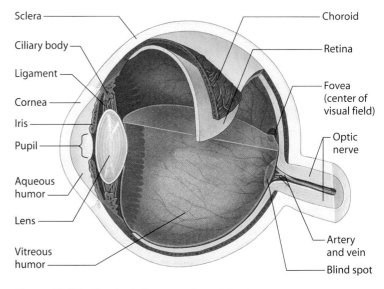

Figure 29.7C   The single-lens eye of a vertebrate

? What key optical feature is found in the eyes of both insects and squids but is not present in planarians?

■ Lenses, which focus light onto photoreceptor cells

Humans have single-lens eyes that focus by changing position or shape

The outer surface of the human eyeball is a tough, whitish layer of connective tissue called the **sclera** (see Figure 29.7C). At the front of the eye, the sclera becomes the transparent **cornea,** which lets light into the eye and also helps focus light. The sclera surrounds a pigmented layer called the **choroid.** The anterior choroid forms the iris, which gives the eye its color. After going through the pupil, the opening at the center of the iris, light passes through the lens, which is held in position by ligaments, and is focused onto the retina, a layer just inside the choroid.

Two chambers make up the bulk of the eye. The large chamber behind the lens is filled with a jellylike substance called **vitreous humor.** The much smaller chamber in front of the lens contains the thinner **aqueous humor.** The humors help maintain the shape of the eyeball. In addition, the aqueous humor circulates through its chamber. This fluid, secreted by the ciliary body, supplies nutrients and oxygen to the lens, iris, and cornea and carries off wastes. Blockage of the ducts that drain this fluid can cause glaucoma, increased pressure inside the eye that may lead to blindness. If diagnosed early, glaucoma can be treated with medications that increase the circulation of aqueous humor.

A thin mucous membrane (see Module 20.4) helps keep the outside of the eye moist. This membrane, called the conjunctiva, lines the inner surface of the eyelids and folds back over the white of the eye (but not the cornea). An infection or allergic reaction may cause inflammation of the conjuctiva, a condition called conjunctivitis or "pink eye." Bacterial conjuctivitis usually clears up after application of antibiotic eyedrops. Viral conjunctivitis usually clears up on its own, although it is very contagious, especially among young children.

A gland above the eye secretes a dilute salt solution that is spread across the eyeball by blinking and drains into ducts that lead into the nasal cavities. This fluid cleanses and moistens the eye surface. Excess secretion in response to eye irritation or strong emotions causes tears to spill over the eyelid and fill the nasal cavities, producing sniffles. Only humans shed emotional tears, which may play a role in reducing stress.

A lens focuses light onto a retina by bending light rays. Among animals, focusing can occur in two ways. The lens may be rigid, as in squids and many fishes, and focusing occurs as muscles move it back or forth, as you might focus on an object using a magnifying glass. Or, as in the mammalian eye, focusing is accomplished by changing the shape of the lens. The thicker the lens, the more sharply it bends light.

The shape of the lens is controlled by the ciliary muscles, which are attached to the choroid and the ligaments that suspend the lens (Figure 29.8). When the eye focuses on a nearby object, these muscles contract, pulling the choroid toward the lens, which reduces the tension on the ligaments. As these ligaments slacken, the elastic lens becomes thicker and rounder, as shown in the top diagram of Figure 29.8. This change, called **accommodation,** allows the diverging light rays from a close object to be bent and focused.

Light from distant objects approaches in parallel rays that require less bending for proper focusing on the retina. When the eye focuses on a distant object, the ciliary muscles relax, and the choroid moves away from the lens. This puts tension on the ligaments and flattens the elastic lens, as shown in the bottom diagram. Or, at least, that is how the human eye is *supposed* to work. In the next module, we'll consider ways that this mechanism might fail and ways to correct such failures.

Web Activity *Structure and Function of the Eye*

? Arrange the following eye parts into the correct sequence encountered by photons of light traveling into the eye: pupil, retina, cornea, lens, vitreous humor, aqueous humor.

■ Cornea → aqueous humor → pupil → lens → vitreous humor → retina

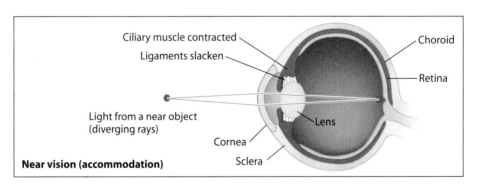

Near vision (accommodation)

Ciliary muscle contracted
Ligaments slacken
Choroid
Retina
Light from a near object (diverging rays)
Lens
Cornea
Sclera

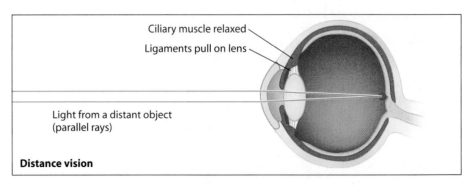

Distance vision

Ciliary muscle relaxed
Ligaments pull on lens
Light from a distant object (parallel rays)

Figure 29.8 How lenses focus light

29.9 Artificial lenses or surgery can correct focusing problems

When you have your vision tested, you are asked to read letters on a special chart. The chart measures your **visual acuity,** the ability of your eyes to distinguish fine detail. The examiner asks you to read a line of letters sized for legibility at a distance of 20 feet, using one eye at a time. If you can do this, you have so-called normal (20/20) acuity in each eye. This means that from a distance of 20 feet, each of your eyes can read the chart's line of letters designated for 20 feet.

Suppose you find out that your visual acuity is 20/10. This is better than normal; you can read letters from a distance of 20 feet that a person with 20/20 vision can only read at 10 feet. On the other hand, someone with 20/50 acuity has worse than normal vision. He or she must stand at a distance of 20 feet to read what a person with normal acuity can read at 50 feet.

Three of the most common visual problems are nearsightedness, farsightedness, and astigmatism. All three are focusing problems, easily corrected with artificial lenses. Nearsighted people cannot focus well on distant objects, although they can see well at short distances (the condition is named for the type of vision that is *unimpaired*). A nearsighted eyeball (Figure 29.9A) is longer than normal. The lens cannot flatten enough to compensate, and it focuses distant objects in front of the retina, instead of on it. As shown by the drawing on the right in Figure 29.9A, **nearsightedness** (also known as myopia) is corrected by glasses or contact lenses that are thinner in the middle than at the outside edge. The corrective lenses make the light rays from distant objects diverge as they enter the eye. The focal point formed by the lens in the eye then falls directly on the retina.

Farsightedness (also known as hyperopia) is the opposite of nearsightedness. It occurs when the eyeball is shorter than normal, causing the lens to focus images behind the retina (Figure 29.9B). Farsighted people see distant objects normally but cannot focus on close objects. Corrective lenses that are thicker in the middle than at the outside edge compensate for farsightedness by making light rays from nearby objects converge slightly before they enter the eye. Another type of farsightedness, called presbyopia, develops with age. Beginning around the mid-40s, the lens of the eye becomes less elastic. As a result, the lens gradually loses its ability to focus on nearby objects, and reading without glasses becomes difficult.

Astigmatism is blurred vision caused by a misshapen lens or cornea. Any such distortion makes light rays converge unevenly and not focus at any one point on the retina. Lenses that correct astigmatism are asymmetrical in a way that compensates for the asymmetry in the eye.

Surgical procedures are an option for treating vision disorders. In radial keratotomy (RK), a knife is used to cut slits in the cornea to change its shape. In laser-assisted in situ keratomileusis (LASIK), a laser is used to reshape the cornea and change its focusing ability. More than 1 million LASIK procedures are performed each year to correct a variety of vision problems.

? A person with 20/100 vision in both eyes must stand at _____ feet to read what someone with normal vision can read at _____ feet.

■ 20 . . . 100

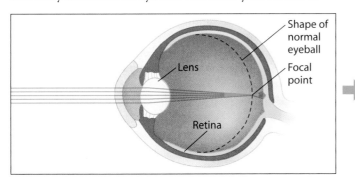

Figure 29.9A   A nearsighted eye (eyeball too long)

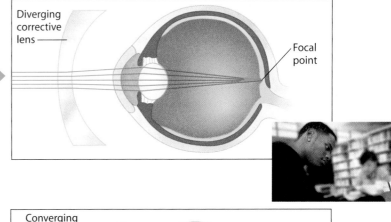

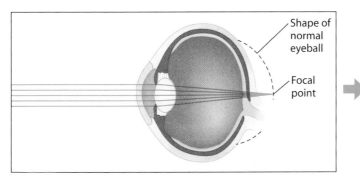

Figure 29.9B   A farsighted eye (eyeball too short)

29.10 The human retina contains two types of photoreceptors: rods and cones

Built into the human retina are about 125 million rod cells and 6 million cone cells, two types of photoreceptors named for their shapes (Figure 29.10A). **Cones** are stimulated by bright light and can distinguish color, but they contribute little to night vision. **Rods** are extremely sensitive to light and enable us to see in dim light at night, though only in shades of gray. The relative numbers of rods and cones correlate with whether an animal is most active during the day or night.

In humans, rods are found in greatest density at the outer edges of the retina and are completely absent from the fovea, the retina's center of focus. If you face directly toward a dim star in the night sky, the star is hard to see. Looking just to the side, however, makes your lens focus the starlight onto the parts of the retina with the most rods, and you can see the star. By contrast, you achieve your sharpest day vision by looking straight at the object of interest. This is because cones are densest (about 150,000 per square millimeter) in the fovea. Some birds, such as hawks, have more than a million cones per square millimeter, which enables them to spot small prey from high in the air.

How do rods and cones detect light? As Figure 29.10A shows, each rod and cone includes an array of membranous disks containing light-absorbing visual pigments. Rods contain a visual pigment called **rhodopsin,** which can absorb dim light. (Rhodopsin is derived from vitamin A, which is why vitamin A deficiency can cause "night blindness.") Cones contain visual pigments called **photopsins,** which absorb bright, colored light. We have three types of cones, each containing a different type of photopsin. These cells are called blue cones, green cones, and red cones, referring to the colors absorbed best by their photopsin. We can perceive a great number of colors because the light from each particular color triggers a unique pattern of stimulation among the three types of cones. Color blindness, more common in males than females because it is usually inherited as a sex-linked trait (see Module 9.22), results from a deficiency in one or more types of cones. The most common type is red-green color blindness, in which red and green are seen as the same color—either red or green—depending on which type of cone is deficient.

Figure 29.10B shows the pathway of light into the eye and through the cell layers of the retina. Notice that the rods and cones have their tips embedded in the back of the retina (pink cells). Light must pass through several relatively transparent layers of neurons before reaching the pigments in the rods and cones. Like all sensory receptors, rods and cones are stimulus transducers. When rhodopsin and photopsin absorb light, they change chemically, and the change alters the permeability of the cell's membrane. The resulting receptor potential triggers a change in the release of neurotransmitter, which initiates a complex integration process in the retina. As shown in Figure 29.10B, visual information transduced by the rods and cones passes from the photoreceptor cells through the network of neurons (black arrows). Notice the numerous synapses between the photoreceptor cells and the neurons and among the neurons themselves. Integration in this maze of synapses helps sharpen images and increases the contrast between light and dark areas. Action potentials carry the partly integrated information into the brain via the optic nerve. Three-dimensional perceptions (what we actually see) result when visual input coming from the two eyes is integrated further in several processing centers of the cerebral cortex.

? Explain why our night vision is mostly in shades of gray rather than in color.

■ Rods are more sensitive than cones to light, and thus the low light intensity at night stimulates far more rods than cones.

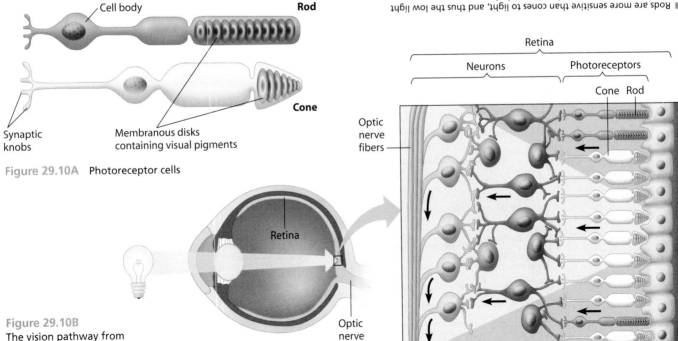

Figure 29.10A Photoreceptor cells

Figure 29.10B
The vision pathway from light source to optic nerve

29.11 Taste and odor receptors detect chemicals present in solution or air

Our senses of smell and taste depend on receptor cells that detect chemicals in the environment. Chemoreceptors in our taste buds detect molecules in solution; chemoreceptors in our nose detect airborne molecules.

The olfactory (smell) receptors are sensory neurons that line the upper portion of the nasal cavity and send impulses along their axons directly to the olfactory bulb of the brain (Figure 29.11). Notice the cilia extending from the tips of these chemoreceptors into the mucus that coats the nasal cavity. When an odorous substance (∴) diffuses into this region, it binds to specific receptor proteins on the cilia. The binding triggers a membrane depolarization and generates action potentials. Integration of the signals in the brain results in an odor perception. Humans can distinguish thousands of different odors.

Many animals rely heavily on their sense of smell for survival. Most mammals have a much more discriminating sense of smell than humans. Odors often provide more information than visual images about food, the presence of mates, or danger. In contrast, humans often pay more attention to sights and sounds than to smells. ("Seeing is believing" is thus a very humancentric adage!)

Our sense of taste depends on taste receptors organized into taste buds on the tongue. In addition to the four familiar taste perceptions—sweet, sour, salty, and bitter—a fifth, called *umami* (Japanese for "savory") is elicited by glutamate, an amino acid. Umami describes the savory flavor common in meats, cheeses, and other protein-rich foods. Umami is also what we sense from monosodium glutamate (MSG), often used as a flavor enhancer. Each type of taste receptor is most responsive to a particular type of substance. For example,

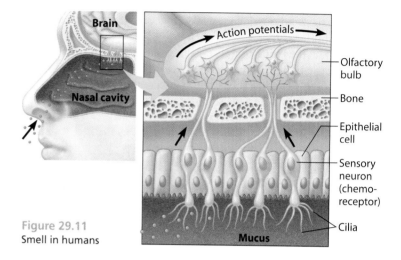

Figure 29.11
Smell in humans

Module 29.2 describes the functioning of a receptor stimulated by sugar. The brain integrates a variety of inputs from different receptors to create the broad pallet of flavors you perceive.

Although the receptors and brain pathways for taste and smell are independent, the two senses do interact. Indeed, much of what we call taste is really smell, as you have undoubtedly noticed when a stuffy head cold dulls your perception of taste.

? What is the key structural difference between taste receptors and olfactory receptors? (*Hint:* Review Module 29.2.)

■ Taste receptors release neurotransmitter, triggering action potentials carried to the brain by sensory neurons. Olfactory receptors are modified sensory neurons.

29.12 "Supertasters" have a heightened sense of taste

If you don't like your vegetables (and I mean *really* don't like your vegetables), there may be a genetic basis behind your aversion. About 25% of us are "supertasters" with three times the sensitivity to bitter tastes than other people. Supertasters tend to perceive bitter tastes in many more foods than do normal tasters. Hence, supertasters typically have more food dislikes, often avoiding such foods as coffee, alcoholic beverages, vegetables such as spinach, cabbage, and broccoli, and other common foods.

Are you a supertaster? There are two tests: hypersensitivity to a bitter-tasting chemical called propylthiouracil and more than normal numbers of fungiform papillae, the structures on your tongue that house tastebuds. Try sticking your tongue out, looking at it in a mirror, and comparing the number of taste buds you can see on your tongue to a friend's tongue.

There are consequences to being a supertaster beyond matters of food preference. Perhaps because they tend to avoid somewhat bitter but healthy compounds in vegetables and other foods, supertasters have a higher risk of obesity, colon cancer, and other serious health problems. Thus, taste-related food choices can have health implications.

Clues to the genetic basis of the supertaster trait have recently been discovered. This is a good example of how the genetic basis of seemingly random and complex behaviors, such as taste preference, may be elucidated as the human genome is deciphered.

? Why might being a supertaster be harmful to your health?

■ Supertasters tend to avoid slightly bitter but healthful vegetables, leading to higher risk of developing some health problems.

Review: The central nervous system couples stimulus with response

In this chapter and the previous one, we focused on information gathering and processing. Sensory receptors provide an animal's nervous system with vital data that enable the animal to avoid danger, find food and mates, and maintain homeostasis—in short, to survive.

A bat catching an insect helps us summarize the sequence of information flow in an animal. A bat hears its ultrasonic cries echoing off an insect. Within milliseconds, receptor cells in the bat's ears transduce the sound, and action potentials representing the echo of the insect enter the brain. Before the insect can fly out of reach, a vast network of neurons in the bat's brain, with millions of synapses, integrates the information and sends out command signals, again in the form of action potentials. The commands go out via motor neurons to muscles, and the bat swoops and grabs its prey.

In the next chapter, we see how muscles carry out the commands they receive from the nervous system.

? What three general types of neurons are involved when a bat catches an insect? (*Hint:* Review Module 28.1.)

Sensory neurons, interneurons, and motor neurons ■

Chapter Review

Reviewing the Concepts

Sensory Reception (Introduction–29.3)

Animal senses gather information that guides predation, migration, and other behaviors (**Introduction**). Perception is the brain's integration of sensations (**29.1**).

Sensory receptors are specialized cells or neurons that detect stimuli. Sensory transduction converts stimulus energy into receptor potentials, which trigger action potentials that are transmitted to the brain. Action potential frequency reflects stimulus strength:

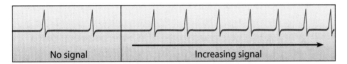

| No signal | Increasing signal |

Repeated stimuli may lead to adaptation, a decrease in sensitivity (**29.2**).

Types of sensory receptors. Pain receptors sense dangerous stimuli. Thermoreceptors detect heat or cold. Mechanoreceptors respond to mechanical energy (such as touch, pressure, and sound), and chemoreceptors to chemicals. Electromagnetic receptors respond to electricity, magnetism, and light (sensed by photoreceptors) (**29.3**).

Hearing and Balance (29.4–29.6)

The human ear channels sound waves through the outer ear to the eardrum to a chain of bones in the middle ear to the fluid in the coiled cochlea in the inner ear:

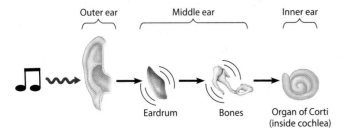

Outer ear   Middle ear   Inner ear

Eardrum   Bones   Organ of Corti (inside cochlea)

Pressure waves in the fluid bend hair cells of the organ of Corti against a membrane, triggering nerve signals to the brain.

Louder sounds generate more action potentials; pitches stimulate different regions of the organ of Corti (**29.4**).

The organs of balance, the semicircular canals and utricle and saccule located in the inner ear, sense body position and movement (**29.5**). Conflicting signals from the inner ear and eyes may cause motion sickness (**29.6**).

Vision (29.7–29.10)

Animal eyes range from simple eye cups that sense light intensity and direction to the many-lensed compound eyes of insects to the single-lens eyes of squids and vertebrates, including humans (**29.7**).

Vertebrate eyes are single-lens eyes. In the human eye, the cornea and flexible lens focus light on photoreceptor cells in the retina:

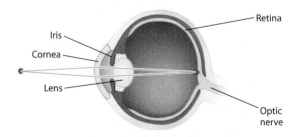

Retina
Iris
Cornea
Lens
Optic nerve

Focusing involves changing the shape of the lens (**29.8**). Nearsightedness and farsightedness result when the focal point is not on the retina. Corrective lenses bend the light rays to compensate (**29.9**).

Rods and cones allow us to see shades of gray in dim light and to see color in bright light, respectively (**29.10**).

Taste and Smell (29.11–29.13)

Taste and smell depend on chemoreceptors that bind specific molecules. Taste receptors, located in taste buds on the tongue, produce five taste sensations. Olfactory (smell) sensory neurons line the nasal cavity. Various odors and tastes result from the integration of input from many receptors (**29.11**). Supertasters have a heightened sensitivity to bitter tastes (**29.12**).

Stimulus and response. The nervous system receives sensory information, integrates it, and commands appropriate muscle responses (**29.13**).

Connecting the Concepts

1. Complete this map summarizing sensory receptors.

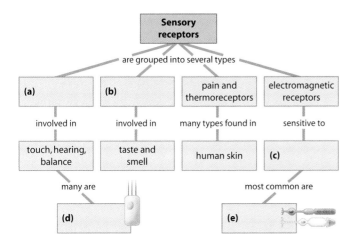

Testing Your Knowledge

Multiple Choice

2. Eighty-year-old Mr. Johnson was becoming slightly deaf. To test his hearing, his doctor held a vibrating tuning fork tightly against the back of Mr. Johnson's skull. This sent vibrations through the bones of the skull, setting the fluid in the cochlea in motion. Mr. Johnson could hear the tuning fork this way, but not when it was held away from the skull a few inches from his ear. The problem was probably in the (Explain your answer.)
   a. auditory center in Mr. Johnson's brain.
   b. auditory nerve leading to the brain.
   c. hair cells in the cochlea.
   d. bones of the middle ear.
   e. fluid of the cochlea.

3. Which of the following correctly traces the path of light into your eye?
   a. lens, cornea, pupil, retina
   b. cornea, pupil, lens, retina
   c. cornea, lens, pupil, retina
   d. lens, pupil, cornea, retina
   e. pupil, cornea, lens, retina

4. If you look away from this book and focus your eyes on a distant object, the eye muscles _____ and the lenses _____ to focus images on the retinas.
   a. relax . . . flatten
   b. relax . . . become more rounded
   c. contract . . . flatten
   d. contract . . . become more rounded
   e. contract . . . relax

5. Which of the following are *not* present in human skin?
   a. thermoreceptors
   b. electromagnetic receptors
   c. touch receptors
   d. pressure receptors
   e. pain receptors

6. Jim had his eyes tested and found that he has 20/40 vision. This means that
   a. the muscles in his iris accommodate too slowly.
   b. he is farsighted.
   c. the vision in his left eye is normal, but his right eye is defective.
   d. he can see at 40 feet what a person with normal vision can see at 20 feet.
   e. he can see at 20 feet what a person with normal vision can see at 40 feet.

7. What do the receptor cells on the skin of a fish and the cochlea of your ear have in common?
   a. They use hair cells to sense sound or pressure waves.
   b. They are organs of equilibrium.
   c. They use electromagnetic receptors to sense pressure waves in fluid.
   d. They use granules that signal a change in position and stimulate their receptor cells.
   e. They are homologous structures that share a common evolutionary origin with all organs of hearing.

Describing, Comparing, and Explaining

8. How does your brain determine the volume and pitch of sounds?

9. As you read these words, the lenses of your eyes project patterns of light representing the letters onto your retinas. There the photoreceptors respond to the patterns of light and dark and transmit nerve signals to the brain. The brain then interprets the words. In this example, which processes relate to sensation and which to perception?

10. For what purposes do animals use their senses of taste and smell?

Applying the Concepts

11. Sensory organs tend to come in pairs. We have two eyes and two ears. Similarly, a planarian worm has two eye cups, a rattlesnake has two infrared receptors, and a butterfly has two antennae. Propose a testable hypothesis that could explain the advantage of having two ears or eyes instead of one.

12. Sea turtles bury their eggs on the beach above the high-tide line. When the baby turtles hatch, they dig their way to the surface of the sand and quickly head straight for the water. How do you think the turtles know which way to go? Outline an experiment to test your hypothesis.

13. Have you ever felt your ears ringing after listening to loud music from a stereo or at a concert? Can this music be loud enough to permanently impair your hearing? Do you think people are aware of the possible danger of prolonged exposure to loud music? Should anything be done to warn or protect them? If you think so, what action would you suggest? What effect might warnings have?

Answers to all questions can be found in Appendix 4.

For Activity Quizzes, BioFlix, MP3 Tutors, and Activities, go to www.mybiology.com.

30

How Animals Move

Man Versus Horse

It started with a late night argument over a few beers, a dispute about whether a man could outrun a horse. Both sides acknowledged that we are pitifully slow sprinters compared to other animals. The cheetah, reputedly the fastest land animal, can run 400 m in 16 seconds, and at 19 seconds, the American quarter horse is not far behind. The world record for humans is 43.18 seconds. That's not even fast enough to outrun a grizzly bear, which can cover 400 m in 30 seconds. But which species wins in an endurance contest? We have several features in our favor. As we run, our long legs lengthen their stride to cover more ground. When each foot strikes the ground, tendons—the stretchy connective tissue that attaches muscle to bone—absorb energy, releasing it to add an extra bounce as the foot pushes off again. We also possess excellent thermoregulatory mechanisms to shed the enormous amount of heat generated by aerobic respiration as it powers muscle movement. Would humans prevail over horses on a longer course? Overhearing the debate, the innkeeper decided to put the question to the test. Thus began the annual Man vs. Horse Marathon, a challenging 22-mile run by men, women, and mounted horses over the rugged terrain near Llanwrtyd Wells, Wales.

Let's take stock of our competition. For their body size, horses have strikingly long, slender legs. You'll see why as you compare the skeletons below and facing page. The femur (blue) is the thighbone, and the tibia (shinbone, purple) and fibula (much reduced in the horse) support the lower leg. But where we have a number of small, short ankle bones separating the lower leg from the foot, the horse has a long bone called the cannon (yellow), a highly modified version of a bone that is part of the foot of most other tetrapods. The forelimbs are similarly constructed; each has a cannon bone that corresponds to one of the bones in our hands. The horse's hoof is a modified middle toe. A human in a stance equivalent to the horse would be poised on his fingertips and tiptoes.

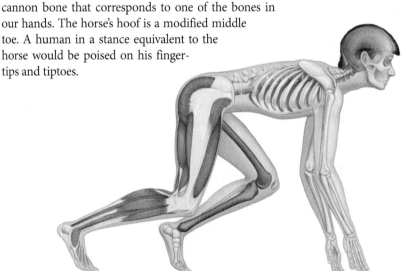

By extending the horse's legs, the cannon bones lengthen the animal's stride, one of the factors in running speed. Speed also depends on the number of strides the animal takes per minute. The advantage of longer limbs is generally offset by the added weight of the muscle required to move them—muscle is very heavy. But the cannon bone, like our hands and feet, carries only lightweight tendons. In fact, quite a few muscles that humans possess are absent in horses. For example, we have muscles that stabilize the mobile ball-and-socket joints in our shoulders and hips. (As you'll learn in Module 30.6, this type of joint enables the limb to rotate freely.) Other joints in our bodies, for example, the knees, work like hinges, bending in a single plane. Hinge joints do not require stabilizing muscles. In horses, the shoulder and hip joints also swing back and forth like hinges. Although this arrangement eliminates the need for some muscles, it also restricts the horse's leg motion to a single plane, comparable to our knee joints. In many ways, the horse's adaptations for speed have come at the expense of flexibility.

Endurance races are not won on the strength of anatomy alone. To power muscles, a marathoner must have a dependable supply of ATP, the energy source used by cells. Because most of the ATP will be generated by aerobic respiration (see Module 6.6), the ability to deliver enough oxygen to muscle cells is critical. As you learned in Module 23.13, some athletes have artificially (and illicitly) increased their concentration of red blood cells by transfusing themselves with extra cells or by using a hormone to stimulate overproduction of red blood cells. Horses have a natural method of blood doping. The spleen holds extra red blood cells in reserve, squeezing them into the bloodstream during strenuous exercise for a boost in oxygen-carrying capacity. The horse's heart is strong enough to handle the increase in blood viscosity, unlike the human heart, which may fail under the strain of pumping thicker blood.

So who wins the Man vs. Horse Marathon? In the nearly 30 years that the race has been held, only one human has taken first place. After all, 6 million years of natural selection and 6,000 years of selection by humans have endowed horses with an anatomy and physiology superbly adapted for a single purpose—running. Humans have evolved a more versatile body. In addition to our adaptations for running, we can grasp and throw objects, swim, crawl, jump, swing, and turn somersaults. However, as you'll learn in Module 30.12, our ability to increase our strength and endurance through athletic training may help even the odds. ■ ■ ■

30.1 Animals have evolved diverse means of locomotion

Animal movement is extremely diverse. Many animals stay in one place and move only certain parts of their body. For example, an adult sponge's only movements are the opening and closing of cellular pores on its surface and the beating of flagella, drawing suspended food particles in through the pores. Most animals, however, spend much of their time and energy moving about in search of food or mates or escaping from danger. Active travel from place to place, also called **locomotion,** requires that an animal expend energy to overcome two forces that tend to keep it stationary: friction and gravity. The relative importance of these two forces varies, depending on the environment.

Figure 30.1B Kangaroos hopping

Swimming Gravity is not much of a problem for a swimming animal, because water supports much or all of the animal's weight. On the other hand, overcoming friction is more difficult for a swimmer, because water is dense and offers considerable resistance to a body moving through it.

Figure 30.1C Dog running at full speed

Animals swim in diverse ways. Many insects, for example, swim the way we do, using their legs as oars to push against the water. Squids, scallops, and some jellies are jet-propelled, taking in water and squirting it out in bursts. Fishes swim by moving their body and tail from side to side

Figure 30.1A A fish swimming

(Figure 30.1A). Whales and other aquatic mammals move by undulating their body and tail from top to bottom. A sleek, streamlined shape, like that of seals, porpoises, penguins, and many fishes, is an adaptation that aids rapid swimming.

Locomotion on Land The problems of locomotion on land are more or less the opposite of those in the water. Air offers very little resistance to an animal moving through it, at least at moderate speeds. However, air provides little support for an animal's body, and a land animal must be able to support itself and overcome the force of gravity. When a land animal walks, runs, or hops, its leg muscles expend energy both to propel it and to keep it from falling down. To move on land, powerful muscles and strong skeletal support are more important than a streamlined shape.

Hopping Kangaroos travel mainly by hopping (Figure 30.1B). Large muscles in their hind legs generate a lot of power. Tendons (which connect muscle to bone) in the legs also momentarily store energy when the kangaroo lands—somewhat like the spring on a pogo stick. The higher the jump, the tighter the spring coils when a pogo stick lands and the greater

tension in the tendons when a kangaroo lands. In both cases, the stored energy is available for the next jump. For the kangaroo, the tension in its legs is a cost-free energy boost that reduces the total amount of energy the animal expends to travel. The pogo stick analogy applies to many land animals. The legs of an insect, a horse, or a human, for instance, retain some spring during walking or running, although less than those of a hopping kangaroo. At rest, the kangaroo sits upright with its tail and both hind feet touching the ground. This position stabilizes the animal's body and costs little energy to maintain.

Walking and Running A walking animal moves each leg in turn, overcoming friction between the foot and the ground with each step. In order to maintain balance, a four-legged animal usually keeps three feet on the ground at all times when walking slowly. Bipedal (two-footed) animals, such as birds and humans, are less stable on land and keep part of at least one foot on the ground when walking. A running four-legged animal may move two or three legs with each stride. At some gaits, all of its feet may be off the ground simultaneously (Figure 30.1C). At running speeds, momentum, more than foot contact, stabilizes the body's position, just as a moving bicycle stays upright.

Crawling A crawling animal, such as a snake or an earthworm, faces very different problems. Because much of the animal's body is in contact with the ground, friction offers considerable resistance to movement. Many snakes crawl rapidly by undulating the entire body from side to side. Aided by large, movable scales on their underside, a snake's body pushes against the ground, driving the animal forward. Boa constrictors and pythons creep forward in a straight line, driven by muscles that lift belly scales off the ground, tilt them forward, and then push them backward against the ground.

Earthworms crawl by peristalsis, a type of movement produced by rhythmic waves of muscle contractions passing from head to tail. (In Module 21.4, we saw how peristalsis squeezes food through our digestive tract.) To move by peristalsis, an animal needs a set of muscles that elongates the body and another set that shortens it. Also required are a way to anchor its body to the ground and a hydrostatic skeleton, which we will discuss more in Module 30.2. As illustrated in Figure 30.1D, the contraction of circular muscles, which encircle the circumference of the body, constricts and elongates some regions of the fluid-filled body segments of a crawling earthworm. At the same time, longitudinal muscles that run the length of the body shorten and thicken other regions. In position ❶, segments at the head and tail end of the worm are short and thick (longitudinal muscles contracted) and anchored to the ground by bristles. Just behind the head, a group of segments is thin and elongated (circular muscles contracted), with bristles held away from the ground. In position ❷, the head has moved forward because circular muscles in the head segments have contracted. Segments just behind the head and near the tail are now thick and anchored, thus preventing the head from slipping backward. In position ❸, the head segments are thick again and anchored to the ground in their new position, well ahead of their starting point. The rear segments of the worm now release their hold on the ground and are pulled forward.

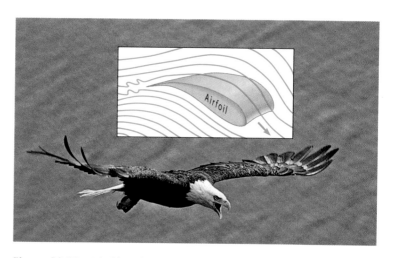

Figure 30.1E   A bald eagle in flight

Flying   Many phyla of animals include species that crawl, walk, or run, and almost all phyla include swimmers. But flying has evolved in only a few animal groups: insects, reptiles (including birds), and, among the mammals, bats. A group of large flying reptiles died out millions of years ago, leaving birds and bats as the only flying vertebrates.

For an animal to become airborne, its wings must develop enough lift to completely overcome the pull of gravity. The key to flight is the shape of wings. All types of wings, including those of airplanes, are airfoils—structures whose shape alters air currents in a way that creates lift. As Figure 30.1E shows, an airfoil has a leading edge that is thicker than the trailing edge. It also has an upper surface that is somewhat convex and a lower surface that is flattened or concave. This shape makes the air passing over the wing travel farther than the air passing under the wing. As a result, air molecules are spaced farther apart above the wing than below it, and the air pressure underneath the wing is greater. This pressure difference provides the "lift" for flight.

Birds can reach great speeds and cover enormous distances. Swifts, which can fly 170 km/hr (105 mph), are the fastest. The bird that migrates the farthest is the arctic tern, which flies round-trip between the North and South Poles each year.

Basis of Movement   All types of animal movement have certain underlying similarities. At the cellular level, every form of movement is based on one of two basic contractile systems: microtubules or microfilaments. Both of these systems consume energy moving protein strands against one another. The movements of cilia and flagella result from the bending of microtubules, as we discussed in Module 4.17. Microfilaments play a major role in amoeboid movement and are the contractile elements of muscle cells. Later in this chapter, we will look at the contraction of muscles and how muscle contraction translates into movement when the muscles work against a firm skeleton. First, let's look at skeletons.

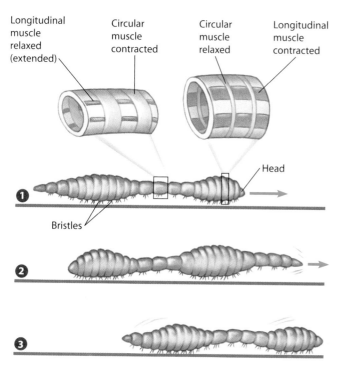

Longitudinal muscle relaxed (extended)   Circular muscle contracted   Circular muscle relaxed   Longitudinal muscle contracted

Head

Bristles

❶

❷

❸

Figure 30.1D   An earthworm crawling by peristalsis

?   Contrast swimming with walking in terms of the forces an animal must overcome to move.

■ Friction resists an animal moving through water, but gravity has little effect because of the animal's buoyancy; air poses little resistance to an animal walking on land, but the animal must support itself against the force of gravity.

30.2 Skeletons function in support, movement, and protection

A skeleton has many functions. An animal could not move without its skeleton, and most land animals would sag from their own weight if they had no skeleton to support them. Even an animal in water would be a formless mass without a skeletal framework to maintain its shape. Skeletons also may protect an animal's soft parts. For example, the vertebrate skull protects the brain, and the ribs form a cage around the heart and lungs.

There are three main types of skeletons: hydrostatic skeletons, exoskeletons, and endoskeletons. All three types have multiple functions.

Hydrostatic Skeletons A **hydrostatic skeleton** consists of fluid held under pressure in a closed body compartment. This is very different from the more familiar skeletons made of hard materials. Nonetheless, a hydrostatic skeleton helps protect other body parts, cushioning them from shocks. It also gives the body shape and provides support for muscle action.

Earthworms have a fluid-filled internal body cavity, or coelom (see Module 18.3). As a segmented animal, the earthworm has its coelom divided into separate compartments. The fluid in these segments functions as a hydrostatic skeleton, and the action of circular and longitudinal muscles working against the hydrostatic skeleton produces the peristaltic movement described in Module 30.1.

Cnidarians such as hydras and jellies (see Module 18.6) also have a hydrostatic skeleton. A hydra (Figure 30.2A), for example, has contractile cells in its body wall that enable it to alter its body shape by exerting pressure on the water-filled gastrovascular cavity. When a hydra closes its mouth and the contractile cells encircling its gastrovascular cavity contract, the body elongates, just as the earthworm elongates when its circular muscles contract (Figure 30.2A, left). The squeezing action also extends the tentacles. A hydra often sits in this position for hours, waiting for prey such as small worms or crustaceans that it can snare with its tentacles. If the hydra is disturbed, its mouth

opens, allowing water to flow out. At the same time, contractile cells arranged longitudinally in the body wall contract, causing the body to shorten (Figure 30.2A, right).

Hydrostatic skeletons work well for many aquatic animals and for terrestrial animals that crawl or burrow by peristalsis. Most animals with hydrostatic skeletons are soft and flexible. In addition to extending its body and tentacles, for example, a hydra can expand its body around ingested prey that are larger than the gastrovascular cavity. An earthworm can burrow through soil because it is flexible and has a hydrostatic skeleton. Similarly, having an expandable body and a hydrostatic skeleton enables tube-dwelling animals such as feather duster worms (see Figure 18.10C) to extend out of their tubes for feeding and gas exchange and then quickly squeeze back into the tube when threatened. However, a hydrostatic skeleton cannot support the forms of terrestrial locomotion in which an animal's body is held off the ground, such as walking.

Exoskeletons A variety of aquatic and terrestrial animals have a rigid external skeleton, or **exoskeleton**. Recall from Module 18.11 that the exoskeleton is a characteristic of the phylum Arthropoda, a group that includes insects, spiders, and crustaceans such as crabs. The arthropod exoskeleton is a tough covering composed of layers of protein and the polysaccharide chitin. The muscles are attached to knobs and plates on the inner surfaces of the exoskeleton. At the joints of legs, the exoskeleton is thin and flexible to allow movement. If you have eaten crab legs, you cracked the exoskeleton to extract the muscle within.

Because the exoskeleton is composed of nonliving material, it does not grow with the animal. It must be shed (molted) and replaced by a larger exoskeleton at intervals to allow for the animal's growth (Figure 30.2B). Depending on the species,

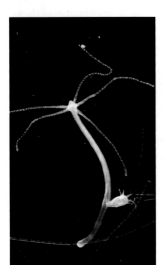
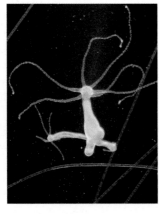

Figure 30.2A
The hydrostatic skeleton of a hydra in two states

Figure 30.2B A tailless whip scorpion that has shed its exoskeleton (top)

Figure 30.2C The exoskeleton of a mollusc: a cowrie (a marine snail)

Figure 30.2D A sea urchin (above) and its endoskeleton (right)

most insects molt from four to eight times before reaching adult size. A few insect species and certain other arthropods, such as lobsters and crabs, molt at intervals throughout life.

An arthropod is never without an exoskeleton of some sort. For instance, a newly molted crab has a new, soft, elastic exoskeleton, which formed under the old one. Soon after molting, the crab expands its body by gulping air or water. Its new exoskeleton then hardens in the expanded position, and the animal has room for further growth. Until its exoskeleton hardens, an arthropod is very susceptible to predation; besides being weakly armored, it is usually less mobile, because the soft exoskeleton cannot support the full action of its muscles. The soft-shelled crabs found seasonally on many menus are newly molted crabs.

The shells of molluscs such as clams, snails, and cowries (Figure 30.2C) are also exoskeletons, but unlike the chitinous arthropod exoskeleton, mollusc shells are made of a mineral, calcium carbonate. The mantle, a sheetlike extension of the animal's body wall, secretes the shell (see Module 18.9). As a mollusc grows, it does not molt; rather, it enlarges the diameter of its shell by adding to its outer edge.

Endoskeletons An **endoskeleton** consists of hard or leathery supporting elements situated among the soft tissues of an animal. Sponges, for example, are reinforced by a framework of tough protein fibers or by mineral-containing particles (see Module 18.5). Usually microscopic and sharp-pointed, the particles consist of inorganic material such as calcium salts or silica. Sea stars, sea urchins, and most other echinoderms have an endoskeleton of hard plates beneath their skin (see Module 18.13). In living sea urchins, about all you see are movable spines, which are attached to the endoskeleton by muscles (Figure 30.2D, left). A dead

urchin with its spines removed reveals the plates that form a rigid skeletal case (right photo).

Vertebrates have endoskeletons consisting of cartilage or a combination of cartilage and bone (see Module 20.5). One major lineage of vertebrates, the sharks, have skeletons of cartilage reinforced with calcium. Figure 30.2E shows the more common condition for vertebrates. Bone makes up most of a frog's skeleton, as it does in bony fishes and land vertebrates. The frog skeleton and the skeletons of most other vertebrates also include some cartilage (blue in the figure), mainly in areas where flexibility is needed. Next, let's take a closer look at endoskeletons.

> **?** What are the advantages and disadvantages of an exoskeleton as compared to an endoskeleton?

■ An exoskeleton may offer greater protection to body parts, but must usually be molted for the animal to grow.

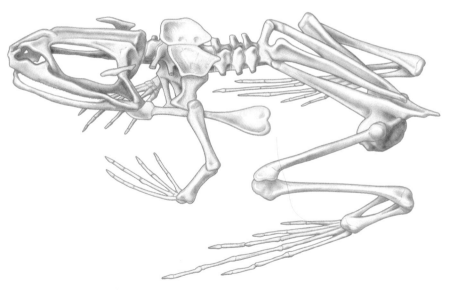

Figure 30.2E Bone (tan) and cartilage (blue) in the endoskeleton of a vertebrate: a frog

30.3 Vertebrate skeletons are variations on an ancient theme

As you learned in Module 19.4, the vertebrate skeletal system provided the structural support and means of locomotion that enabled tetrapods to colonize land. Subsequent evolution produced diverse groups of animals: amphibians, reptiles (including birds), and mammals. Each of those groups has diverse body forms whose skeletons are constructed from modified versions of the same parts. Even the skeletons of whales and dolphins, mammals that evolved from land-dwelling ancestors, are variations on the same theme.

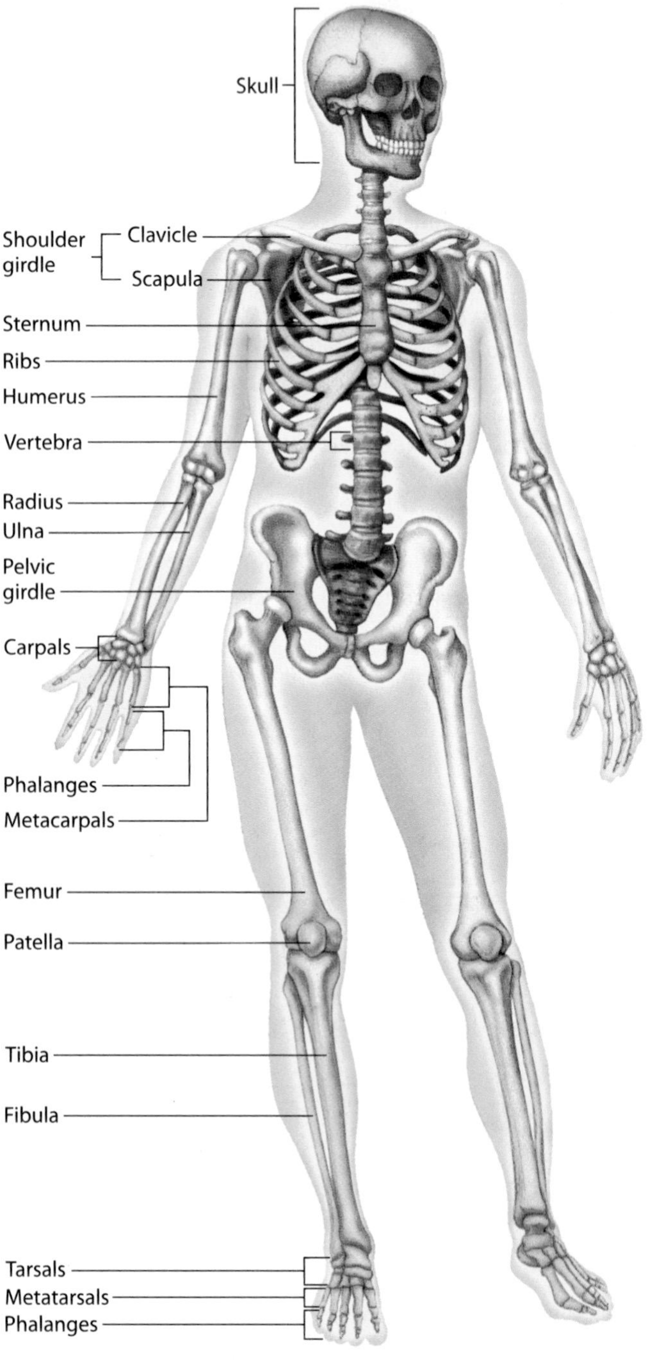

All vertebrates have an **axial skeleton** (orange in Figure 30.3A) supporting the axis, or trunk, of the body. The axial skeleton consists of the skull, enclosing and protecting the brain; the vertebral column (backbone), enclosing the spinal cord; and, in most vertebrates, a rib cage around the lungs and heart.

The backbone, the definitive characteristic of vertebrates, consists of a series of individual bones, the vertebrae, joined by pads of tough cartilage known as discs. The number of vertebrae varies among species. Pythons have 400, while a young human has a total of 33 before fusion. All vertebrae have the same basic structure, including a cylindrical block of bone that supports the body's weight, an arch that forms a protective tunnel for the spinal cord, and bony extensions that provide surface area for the attachment of muscles. Slight variations in structure reflect the position of each vertebra in the backbone. Anatomists group them into five types, shown in Figure 30.3B: cervical (neck), which support the head; thoracic (chest), which form joints with the ribs; lumbar (lower back); sacral, which fuse into a single bone; and coccygeal, which are reduced in humans and later partially fused into the coccyx or "tailbone."

Most vertebrates also have an **appendicular skeleton** (tan in Figure 30.3A), which is made up of the bones of the appendages and the bones that anchor the appendages to the axial skeleton. In a land vertebrate, the pectoral (shoulder) girdle and the pelvic girdle provide a base of support for the bones of the forelimbs and hind limbs. Modified versions of the same bones are found in all vertebrate limbs, whether they are arms, legs, fins, or wings (see Figure 13.5A). This variety of limbs equips vertebrates for every form of locomotion mentioned in Module 30.1—swimming, walking and running, hopping, crawling, and flying.

A few groups of vertebrates, including snakes, lost their limbs during their evolution. How did this happen? The identity of vertebrae is established during embryonic development by the pattern of master control (homeotic) genes expressed in the somites. (Recall from Module 27.12 that somites are the blocks of embryonic tissue that give rise to the vertebral column). Two of the homeotic genes that direct the differentiation of vertebrae are *Hoxc6* and *Hoxc8*. These genes are associated with the development of thoracic vertebrae, which support the ribs. Figure 30.3C shows the range of vertebrae formed by somites that expressed *Hoxc6* (red), *Hoxc8* (blue), or both (purple) in a chicken and a python.

Figure 30.3A   The human skeleton

Labels for Figure 30.3A:
- Skull
- Shoulder girdle — Clavicle, Scapula
- Sternum
- Ribs
- Humerus
- Vertebra
- Radius
- Ulna
- Pelvic girdle
- Carpals
- Phalanges
- Metacarpals
- Femur
- Patella
- Tibia
- Fibula
- Tarsals
- Metatarsals
- Phalanges

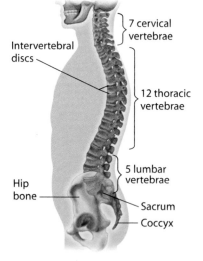

Figure 30.3B   Human backbone showing the groups of vertebrae

Labels:
- Intervertebral discs
- 7 cervical vertebrae
- 12 thoracic vertebrae
- 5 lumbar vertebrae
- Hip bone
- Sacrum
- Coccyx

In the python, both *Hoxc6* and *Hoxc8* are expressed in all somites for nearly the entire length of the vertebral column. As a result, the first rib-bearing thoracic vertebra is located immediately posterior to the head. Pythons have no cervical vertebrae. Chickens, on the other hand, have several cervical vertebrae, ending at the point where *Hoxc6* expression—and thoracic vertebrae—begin.

At some point during the evolution of snakes, mutation in the DNA segments that control the expression of *Hoxc6* and *Hoxc8* (see Modules 11.10 and 18.16) changed cervical vertebrae to thoracic. In all vertebrates, the forelimbs originate at the boundary between cervical and thoracic vertebrae. This position does not exist in snakes; as a result, forelimbs are not formed.

The loss of hindlimbs in snakes involved a different group of genes. Most snakes have lost their hindlimbs completely. Pythons, like whales (see Module 13.4), have retained tiny vestigial hindlimbs.

Web Activity *The Human Skeleton*

Figure 30.3C Expression of two *Hox* genes in python (left) and chicken (right)

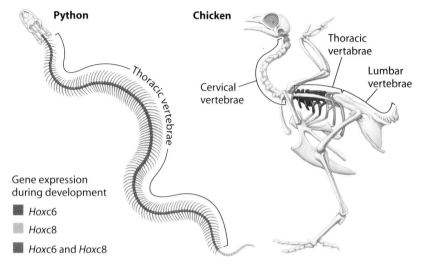

? What type of vertebra do chickens and humans both have that snakes do not have? How does this difference affect the appendicular skeleton of snakes?

■ Chickens and humans have cervical vertebrae; snakes do not (we did not compare vertebrae posterior to thoracic vertebrae). Because forelimbs form at the boundary between cervical and thoracic vertebrae, snakes lack forelimbs.

30.4 Bones are complex living organs

The expression "dry as a bone" should not be taken literally. Bones are actually complex organs consisting of several kinds of moist, living tissues. Figure 30.4 shows a human humerus (upper arm bone). A sheet of fibrous connective tissue, shown in pink (most visible in the enlargement on the lower right), covers most of the outside surface. This tissue helps form new bone in the event of a fracture. A thin sheet of cartilage (blue) forms a cushion-like surface for joints, protecting the ends of bones as they move against one another. The bone itself contains living cells that secrete a surrounding material, or matrix. Bone matrix consists of flexible fibers of the protein collagen with crystals of a mineral made of calcium and phosphate bonded to them (see Figure 20.5E). The collagen keeps the bone flexible and nonbrittle, while the hard mineral matrix resists compression.

The shaft of this long bone is made of compact bone, so named because it has a dense structure. Notice that the compact bone surrounds a central cavity. The central cavity contains **yellow bone marrow,** which is mostly stored fat brought into the bone by the blood. The ends, or heads, of the bone have an outer layer of compact bone and an inner layer of spongy bone, so named because it is honeycombed with small cavities. The cavities contain **red bone marrow** (not shown in the figure), a specialized tissue that produces our blood cells (see Module 23.13).

Like all living tissues, bone cells undergo metabolism. Blood vessels that extend through channels in the bone transport nutrients and regulatory hormones to its cells and remove waste materials. Nerves running parallel to the blood vessels help regulate the traffic of materials between the bone and the blood.

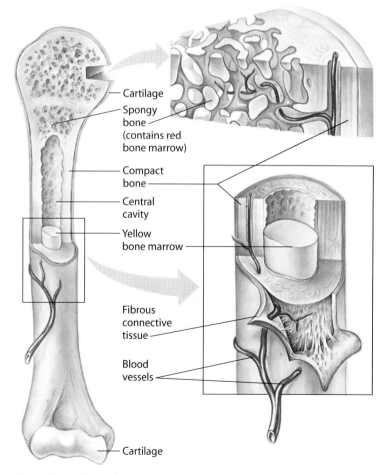

Figure 30.4 The structure of an arm bone

? What causes the colors of yellow and red bone marrow?

■ Stored fat and developing red blood cells, respectively

30.5 Healthy bones resist stress and heal from injuries

Bones are constantly subjected to stress as we go about our daily lives; exercise or physical labor cause additional stress. Bone fatigue can lead to so-called stress fractures, hair-line cracks in the bone, just as the accumulation of small amounts of stress on metals can cause a break. For example, when you bend a paper clip repeatedly, the metal fatigues and finally snaps. Unlike metal, however, bone is composed of living, dynamic tissue. Cells continually remove old bone matrix and replace it with new materials. Stress fractures only occur if this repair process cannot keep up with the amount of stress placed on a bone. A human athlete usually becomes aware of the problem and can allow time for healing, but in a racehorse, stress damage might go unnoticed until the fatigued bone breaks suddenly during a race.

A bone may also break when subjected to an external force that exceeds its resiliency. The average American will break two bones during his or her lifetime, most commonly the forearm or, for people over 75, the hip. Usually, this type of fracture occurs from a sudden impact such as a fall or car accident. Wearing protective gear, such as seat belts, helmets, or padding, can protect bones from high-energy trauma.

Treatment for a broken bone involves two steps: putting the bone back into its natural shape and then immobilizing it until the body's normal bone-building cells can repair the break. A splint or cast is used to protect the injured area, prevent movement, and promote healing. In severe cases, a fracture can be repaired surgically by inserting plates, rods, and/or screws that hold the broken pieces together (Figure 30.5A). In certain cases, however, severely injured or diseased bone is beyond repair and must be replaced. Broken hip joints, for example, can be replaced with artificial ones made of titanium or cobalt alloys. Researchers have recently developed new methods of bone replacement, including grafts (from the patient or from a cadaver) and synthetic polymers.

The risk of bone fracture increases if bones are porous and weak. Figure 30.5B contrasts healthy bone tissue (left) and bone diseased by osteoporosis (right). **Osteoporosis** is characterized by low bone mass and structural deterioration of bone tissue. This weakness emerges from an imbalance in the process of bone maintenance—the destruction of bone material exceeds the rate of replacement. Because the natural mechanism of bone maintenance responds to bone usage, weight-bearing exercise such as walking or running strengthens bones. On the other hand, disuse causes bones to become thinner. Strong bones also require an adequate intake of dietary calcium and enough vitamin D, which are both essential to bone replacement (see Modules 21.18 and 26.6).

Until recently, osteoporosis was mostly considered a problem for women after menopause because estrogen contributes to normal bone maintenance. But while osteoporosis remains a serious health problem for older women, it is also becoming a concern for men and younger people. Doctors have noted a dramatic increase in bone fractures in children and teenagers in recent years. Many scientists believe this is the result of exercising less and getting less calcium in their diets and less vitamin D. Prevention of osteoporosis in later years begins with exercise and sufficient calcium and vitamin D while bones are still increasing in density (up until about age 30).

Other lifestyle habits, such as smoking, may also contribute to osteoporosis. There is a strong genetic component as well—young women whose mothers or grandmothers suffer from osteoporosis should be especially concerned with maintaining good bone health. Treatments for osteoporosis include calcium and vitamin D supplements and drugs that slow bone loss.

? How do exercise and adequate calcium intake help prevent osteoporosis?

■ Bone tissue responds to the stress of exercise by stepping up the repair process, which builds greater bone density. Calcium is the primary component of the rigid mineral matrix of bone.

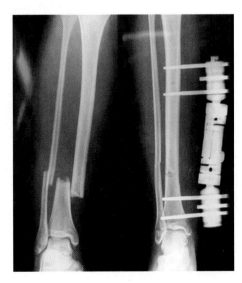

Figure 30.5A   X-rays of a broken leg (left) and the same leg after the bones were set with a plate and screws (right)

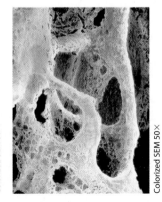

Colorized SEM 50×   Colorized SEM 50×

Figure 30.5B   Healthy spongy bone tissue (left) and bone diseased by osteoporosis (right)

30.6 Joints permit different types of movement

Much of the versatility of the vertebrate skeleton comes from its joints. Strong fibrous connective tissues called **ligaments** hold together the bones of movable joints. **Ball-and-socket joints**, such as are found where the humerus joins the pectoral girdle (Figure 30.6, left), enable us to rotate our arms and legs and move them in several planes. A ball-and-socket joint also joins the femur to the pelvic girdle. **Hinge joints** permit movement in a single plane, just as the hinge on a door enables it to open and close. Our elbows (shown in Figure 30.6, center) and knees are hinge joints. Hinge joints are especially vulnerable to injury in sports that demand quick turns, which can twist the joint

sideways. A **pivot joint** enables us to rotate the forearm at the elbow (Figure 30.6, right). Hinge and pivot joints in our wrists and hands enable us to make precise manipulations. As you'll learn in the next module, muscles supply the force to move the bones of each joint.

? Where we have ball-and-socket joints, horses have hinge joints. How does this affect the movements they can perform?

■ Hinge joints restrict the movement of their legs to a single plane, making horses less flexible than humans.

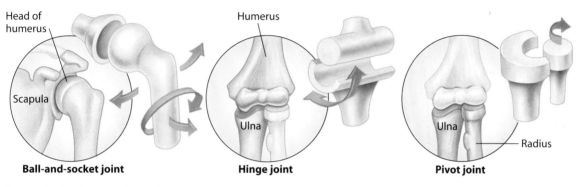

Ball-and-socket joint   **Hinge joint**   **Pivot joint**

Figure 30.6 Three kinds of joints

Muscle Contraction and Movement

30.7 The skeleton and muscles interact in movement

Figure 30.7 shows how an animal's muscles interact with its bones, which act as levers, to produce movement. Muscles are connected to bones by **tendons** (see Module 20.5). For instance, one end of the biceps muscle shown in the figure is attached by tendons to bones of the shoulder; the other end is attached across the hinge joint of the elbow—which acts as the fulcrum—to one of the bones in the forearm.

The action of a muscle is always to contract, or shorten. A muscle's relaxation to an extended position is a passive process. The ability to move an arm in opposite directions requires that muscles be attached to the arm bones in antagonistic pairs. As shown in the figure, contraction of the biceps muscle raises the forearm. The triceps muscle is the biceps's antagonist. The upper end of the triceps attaches to the shoulder, while its lower end attaches to the elbow. At the right, you see that contraction of the triceps lowers the forearm, extending the biceps in the process.

All animals—very small ones like ants and giant ones like elephants—have antagonistic pairs of muscles that apply opposite forces to move parts of their skeleton. Next, we see how a muscle's structure explains its ability to contract.

🎧 MP3 Tutor *Muscle*

? When exercising to strengthen muscles, why is it important to impose resistance while both flexing and extending the limbs?

■ This exercises both muscles of antagonistic pairs, which only do work when they are contracting.

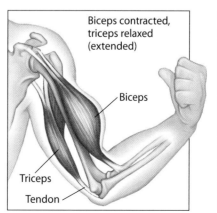
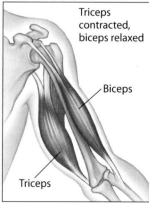

Biceps contracted, triceps relaxed (extended)

Biceps

Triceps

Tendon

Triceps contracted, biceps relaxed

Biceps

Triceps

Figure 30.7 Antagonistic action of muscles to pull bones up or down in the human arm

30.8 Each muscle cell has its own contractile apparatus

The skeletal muscle system is a beautiful illustration of the relationship between structure and function. Each muscle in the body is made up of a hierarchy of smaller and smaller parallel strands, from the muscle itself down to the contractile protein molecules that produce body movements.

Figure 30.8 shows the levels of organization of skeletal muscle. As indicated at the top of the figure, a muscle consists of many bundles of **muscle fibers**—roughly a million in a typical human calf muscle—oriented parallel to each other. Each muscle fiber is a single long, cylindrical cell that has many nuclei. Most of its volume is occupied by about a thousand **myofibrils,** discrete bundles of proteins that include the contractile proteins **actin** and **myosin.** Skeletal muscle is also called striated (striped) muscle because the arrangement of the proteins creates a repeating pattern of stripes along the length of a myofibril that is visible under a light microscope. Beneath the drawing of a myofibril in Figure 30.8 is an electron micrograph that shows one unit of the pattern, which is called a **sarcomere.** Structurally, a sarcomere is the region between two dark, narrow lines, called Z lines, in the myofibril. Each myofibril consists of a long series of sarcomeres, around 15,000 in a typical calf muscle. Functionally, the sarcomere is the contractile apparatus in a myofibril—the muscle fiber's fundamental unit of action.

The diagram of the sarcomere at the bottom of Figure 30.8 explains the features visible in the micrograph. The pattern of horizontal stripes is the result of the alternating bands of **thin filaments,** composed primarily of actin molecules, and **thick filaments,** which are made up of myosin molecules. The Z lines consist of proteins that connect adjacent thin filaments. The light band surrounding each Z line contains only thin filaments. The dark band centered in the sarcomere is the location of the thick filaments. The actin molecules in the thin filaments are globular proteins arrayed in long strands. In addition to actin, thin filaments include proteins called troponin and tropomyosin that play a key role in regulating muscle contraction.

In the next module, we examine the structure of a sarcomere in greater detail and see how its components function in muscle contraction.

Web Activity *Skeletal Muscle Structure*

? The two most abundant proteins of a myofibril are _____ and _____.

actin · · · myosin

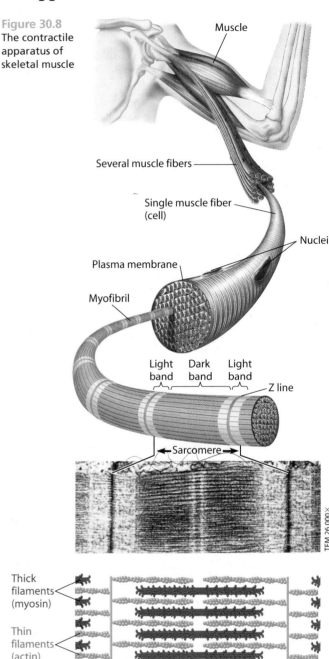

Figure 30.8
The contractile apparatus of skeletal muscle

Muscle

Several muscle fibers

Single muscle fiber (cell)

Nuclei

Plasma membrane

Myofibril

Light band Dark band Light band

Z line

Sarcomere

TEM 26,000×

Thick filaments (myosin)

Thin filaments (actin)

Z line

Z line

Sarcomere

30.9 A muscle contracts when thin filaments slide across thick filaments

Let's now relate the structure of a sarcomere to its function. According to the sliding-filament model of muscle contraction, a sarcomere contracts (shortens) when its thin filaments slide across its thick filaments. Figure 30.9A, on the next page, is a simplified diagram that shows a sarcomere in a relaxed muscle,

in a contracting one, and in a fully contracted one. Notice in the contracting sarcomere that the Z lines and the thin filaments (blue) have moved closer together. When the muscle is fully contracted, the thin filaments overlap in the middle of the sarcomere. Contraction shortens the sarcomere but does not

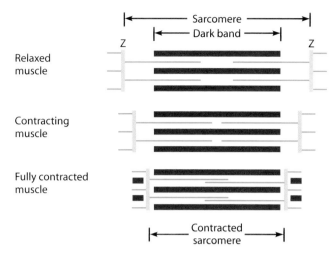

Figure 30.9A The sliding-filament model of muscle contraction

change the lengths of the thick and thin filaments. A whole muscle can shorten about 35% of its resting length when all of its sarcomeres contract.

What moves the thin filaments toward the middle of the sarcomere? Briefly, a part of each myosin molecule called the "head" pivots back and forth in a limited arc. As it swings toward the thin filament, it binds with an actin molecule, dragging the thin filament through the remainder of its arc. After releasing the actin molecule, it returns to it starting position and repeats the same motion with a different actin molecule. The combined actions of hundreds of myosin heads on each thick filament ratchets the thin filament toward the center of the sarcomere.

The details of the key events are shown in Figure 30.9B. The myosin head binds a molecule of ATP, as shown at ❶. When a molecule of ATP is bound, the myosin head is in a low-energy position. ❷ Myosin hydrolyzes the ATP to ADP and phosphate Ⓟ, releasing energy that extends the myosin head towards an actin molecule. The myosin head and actin molecule have complementary binding sites, much as an enzyme has a binding site that interacts with its substrate (see Module 5.15). ❸ The myosin head extends toward the thin filament, and the binding sites latch together, forming a cross-bridge. ❹ ADP and Ⓟ are released, and the myosin head pivots back to its low-energy configuration. This action, called the power stroke, pulls the thin filament toward the center of the sarcomere.

The cross-bridge remains intact until another ATP molecule binds to the myosin head, and the whole process repeats. On the next power stroke, the myosin head attaches to an actin molecule ahead of the previous one on the thin filament (closer to the Z line). This sequence—detach, extend, attach, pull—occurs again and again in a contracting muscle. Though we show only one myosin head in the figure, a typical thick filament has about 350 heads, each of which can bind and unbind to a thin filament about five times per second. Preventing the filaments from backsliding during contraction, some myosin heads hold the thin filaments in position, while others are reaching for new binding sites. As long as sufficient ATP is present, the process continues until the muscle is fully contracted or until the signal to contract stops.

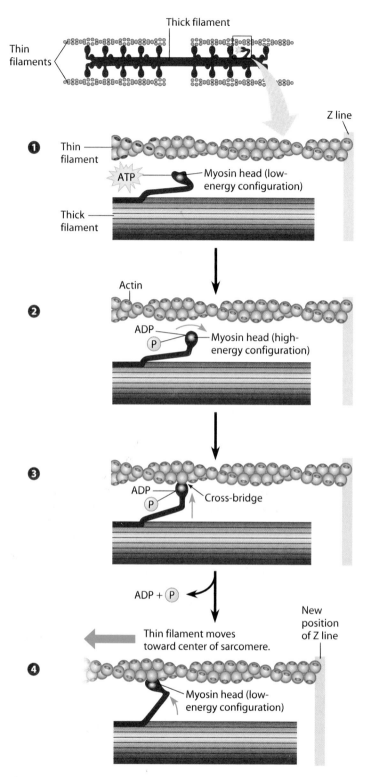

Figure 30.9B The mechanism of filament sliding.

Web Activity *Muscle Contraction*

? Which region of a sarcomere becomes shorter during contraction of a muscle: the dark band, the light bands, or both dark and light bands?

■ The light bands shorten and even disappear as the thin filaments slide (are pulled) toward the center of the sarcomere.

30.10 Motor neurons stimulate muscle contraction

The sarcomeres of a muscle fiber do not contract on their own. Signals from the central nervous system, conveyed by motor neurons (see Module 28.1), are required to initiate and sustain muscle contraction. When a motor neuron sends out an action potential, its synaptic terminals release the neurotransmitter acetylcholine, which diffuses across the synapse to the plasma membrane of the muscle fiber (Figure 30.10A).

The plasma membrane of muscle fibers is unusual in two ways. Like the plasma membrane of neurons, it is electrically excitable—it can propagate action potentials. Also, the plasma membrane extends deep into the interior of the muscle fiber via infoldings called transverse (T) tubules. As a result, when a motor neuron triggers an action potential in a muscle fiber, it spreads throughout the entire volume of the cell, rather than only along the surface. The T tubules are in close contact with the endoplasmic reticulum (ER; blue in Figure 30.10A), a network of interconnected tubules within the muscle fiber. The action potential causes channels in the ER to open, releasing calcium ions (Ca^{2+}) into the cytoplasmic fluid. Now let's see why Ca^{2+} is necessary for muscle contraction.

When a muscle fiber is in a resting state, the binding sites for myosin on the actin molecules are blocked by the regulatory proteins tropomyosin and troponin. As shown in Figure 30.10B, two strands of tropomyosin wrap around the thin filament, blocking access to the binding sites. The muscle fiber cannot contract while these sites are blocked. When Ca^{2+} binds to troponin, it moves tropomyosin away from the myosin-binding sites, allowing contraction to occur. As long as the cytoplasmic fluid is flooded with Ca^{2+}, contraction continues. When motor neurons stop sending action potentials to the muscle fibers, the ER pumps Ca^{2+} back out of the cytoplasmic fluid, binding sites

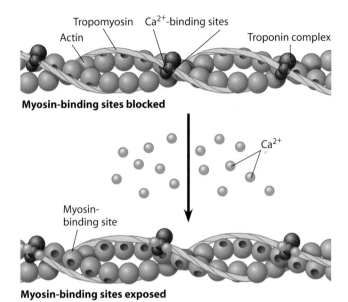

Myosin-binding sites blocked

Myosin-binding sites exposed

Figure 30.10B Thin filament, showing the interactions among actin, regulatory proteins, and Ca^{2+}

on the actin molecules are again blocked, the sarcomeres stop contracting, and the muscle relaxes.

As you learned in Module 30.8, a large muscle such as the calf muscle is composed of roughly a million muscle fibers. However, only about 500 motor neurons run to the calf muscle. Each motor neuron has axons that branch out to synapse with many muscle fibers distributed throughout the muscle. Thus an action potential from a single motor neuron in the calf causes the simultaneous contraction of roughly 2,000 muscle fibers.

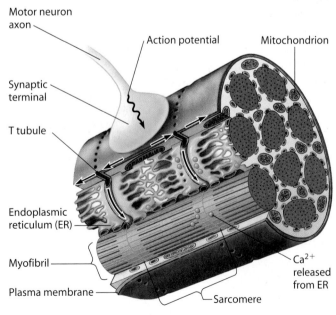

Figure 30.10A How a motor neuron stimulates muscle contraction

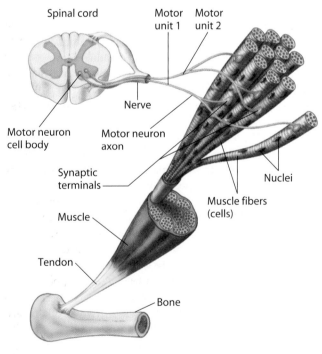

Figure 30.10C Motor units

A motor neuron and all the muscle fibers it controls is called a **motor unit.** Figure 30.10C, on the previous page, shows two motor units; one controls two muscle fibers (Motor unit 1 in the figure) and the other (Motor unit 2) controls three.

The organization of individual neurons and muscle cells into motor units is the key to the action of whole muscles. We know that we can vary the amount of force our muscles develop; an arm wrestler, for example, may change the amount of force developed by the biceps and triceps several times in the course of a match. The ability to do so depends mainly on the nature of motor units. More forceful contractions result when additional motor units are activated. Thus, depending on how many motor units your brain commands to contract, you can apply a small amount of force to lift a fork or considerably more to lift, say, this textbook. In muscles requiring precise control, such as those controlling eye movements, a motor neuron may control only a single muscle fiber.

Web Process of Science *How Do Electrical Stimuli Affect Muscle Contraction?*

? How does the endoplasmic reticulum help regulate muscle contraction?

■ By reversibly storing and releasing Ca^{2+}, the ER regulates the cytoplasmic concentration of this ion, which is required in the cytoplasmic fluid for the binding of myosin to actin.

Connection

30.11 Aerobic respiration supplies most of the energy for exercise

Many people exercise to get in better physical condition, to stay in shape, or to lose a few pounds before swimsuit season. Activities such as jogging, swimming, cycling, or other aerobic workouts are typically chosen for fitness routines (Figure 30.11). Aerobic exercise is an effective method of maintaining or losing weight—the goal is to burn at least as many calories as you consume. The number of calories burned varies with the intensity and duration of the exercise, as you saw in the examples listed in Tables 6.4 and 21.15. Notice that the examples in those tables use a standard body weight of 150 pounds. The more a body weighs, the more energy is required to move it. Thus, for an equivalent activity such as running a mile, a 150-pound person burns more calories than a 100-pound person. Most of the energy expended during exercise is used for muscle movement, specifically, to break the cross-bridges formed during contraction. We refer to this energy in terms of ATP rather than calories.

Muscles have a very small amount of ATP on hand, enough for about 4–6 seconds of muscle contractions. ATP can also be obtained using a high-energy molecule called phosphocreatine (PCr) that is stored in the muscles. The enzymatic transfer of a phosphate group from PCr to ADP makes ATP almost instantaneously. Together, ATP and PCr can provide enough energy for a 10- to 15-second burst of activity.

The bulk of the ATP for aerobic exercise comes from the oxygen-requiring process of aerobic respiration, which derives ATP by the breakdown of the energy-rich sugar glucose (see Module 6.3). Muscles are richly supplied with blood vessels that bring O_2 and glucose and carry away CO_2, the by-product of aerobic respiration. Breathing and heart rate increase during exercise to facilitate the exchange of gases. As you learned in the introduction to Chapter 6, muscles can also get oxygen from the hemoglobin-like molecule myoglobin.

If the demand for ATP outstrips the oxygen supply, muscle fibers can carry out the anaerobic process called *lactic acid fermentation* (see Module 6.13), which also uses glucose as a starting molecule. Fermentation supplies only a fraction of the ATP obtainable through aerobic respiration, but the process works 2.5 times as fast.

Glucose for ATP production is available from the bloodstream, and muscle tissue stores glycogen (see Module 3.7), which is broken down to provide more glucose. The liver contains glycogen, too, and can release glucose into the bloodstream. The body can also mobilize fats as a source of fuel, but the process of harvesting ATP from fatty-acid breakdown is slower than from glucose breakdown. With increasing exercise intensity, only glucose can be metabolized rapidly enough to keep pace with a high demand for ATP. Casual athletes are in no danger of running out of metabolic fuel during exercise, though glycogen stores are typically more than adequate.

Figure 30.11 Running is one form of aerobic exercise

When exercise begins, it takes a few minutes for the aerobic "machinery" to start producing enough ATP to meet the increased demand. After exercise, muscles must repay what is commonly called the oxygen debt that was incurred during the first few minutes of exercise. Rapid breathing and heart rate continue as energy harvested by aerobic respiration is used to replenish the supplies of ATP and PCr and restore myoglobin to its oxygenated state. Oxygen is also used to metabolize the lactic acid that was produced by fermentation.

In the next module, we consider some of the effects that exercise has on muscles.

? Compare the substances required, the ATP output, and the speed of aerobic respiration to lactic acid fermentation in muscles.

■ Aerobic respiration requires glucose and O_2; fermentation requires glucose, but not O_2. Aerobic respiration produces many more ATP than fermentation, but fermentation produces ATP faster.

30.12 Muscle fiber characteristics affect athletic performance

To understand how exercise affects muscles, we need to revisit the differences between the slow ("slow-twitch") and fast ("fast-twitch") muscle fibers described in the introduction to Chapter 6. The characteristics of slow and fast fibers are summarized in Table 30.12, along with the characteristics of intermediate fibers, which are also abundant in human muscle. Most of the features associated with fiber type reflect the pathway(s) the fiber uses to generate ATP from energy-rich molecules: aerobic respiration and fermentation, which is anaerobic.

Each muscle has a mixture of fast, slow, and intermediate fiber types, broadly correlated to the type of work it does. Muscles that maintain the body's posture, for example, are constantly active; they have a high proportion of slow, fatigue-resistant fibers. In other muscles, such as those used for walking and running, the proportion of each fiber type varies widely among individuals. Figure 30.12 compares the percentage of the three fiber types that are typical of elite athletes and average adults.

The differences among muscle fibers can be traced to the myosin that makes up the thick filaments. Fast, slow, and intermediate fibers contain different forms of myosin that hydrolyze ATP (see Module 30.9) at different speeds. The faster that ATP is consumed during muscle contraction, the faster the metabolic process needed to supply the ATP.

The fastest fibers cycle through cross-bridges rapidly, producing forceful contractions that power brief, explosive movements such as those that occur with the hitting of a baseball, hoisting of a heavy weight, or bursting out of the blocks for a sprint. For forceful contractions, the fibers use ATP at a breakneck pace. Therefore, the oxygen–independent process of fermentation is the primary mechanism for producing ATP. The combination of fast fibers with more fatigue-resistant intermediate fibers appears to be ideal for sprinters.

Hydrolysis of ATP occurs much more slowly in the myosin found in slow fibers. The slower pace at which cross-bridges are made and broken results in more sustained but less forceful contractions. Athletes who have a predominance of slow fibers

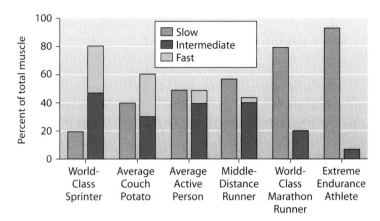

Figure 30.12   Percent of slow, intermediate, and fast muscle fibers in quadriceps (thigh) muscles of different athletes

excel at endurance events such as running, cycling, or swimming long distances. The slow fibers that give middle-distance runners aerobic endurance are complemented by intermediate fibers that sustain a higher power output over a longer period.

The comparison of active and inactive nonathletes in Figure 30.12 indicates that exercise can change the composition of muscle fibers to some extent. Indeed, muscles have a remarkable ability to adapt to how they are used. Muscle fibers respond to a variety of signals generated within the cell during exercise, including mechanical stretching, change in calcium concentration, and low oxygen levels. Hormones may also trigger cellular changes. The specific responses of muscle fibers correspond to the stresses placed on them. For example, weight-lifting stimulates muscle fibers to produce additional myofibrils. The thick filaments of the new myofibrils are composed of the fast form of myosin, increasing the force that the muscle can generate. Muscles adapt to aerobic exercise by synthesizing more myoglobin, increasing the number of mitochondria, and adding capillaries—changes that result in increased endurance and greater resistance to fatigue.

As human athletes increase their strength and endurance through training, they continue to set new standards in track events. On the other hand, for horses, millions of years of evolution and thousands of years of artificial selection have brought the performance of some breeds, notably thoroughbreds, near the limits of their potential. Despite new training methods based on scientific knowledge of the mechanics and physiology of running, the winning times in thoroughbred races such as the Kentucky Derby have remained virtually unchanged for many decades. As we continue to improve our understanding of how animals move, a human winner of the Man vs. Horse Marathon may not be such a rare occurrence.

? In Figure 30.12, how does becoming active affect the percentage of muscle fibers that use aerobic respiration?

■ The percentages of slow and intermediate fibers, which both use aerobic respiration, increase.

TABLE 30.12	CHARACTERISTICS OF MUSCLE FIBERS		
Characteristic	Slow Fibers	Intermediate Fibers	Fast Fibers
Speed of contraction	Slow	Fast	Fast
Rate of fatigue	Slow to fatigue	Intermediate	Fatigue rapidly
Primary pathway for making ATP	Aerobic respiration	Aerobic (some fermentation)	Anaerobic (fermentation)
Myoglobin content	High	High	Low
Mitochondria and capillaries	Many	Many	Few

Reviewing the Concepts
Movement and Locomotion (Introduction–30.1)

Movement is one of the most distinctive features of animals (**Introduction**).

Locomotion, active travel from place to place, requires that an animal use energy to overcome friction and gravity. Animals that swim are supported by water but are slowed by friction. Animals that walk, hop, or run on land are less affected by friction, but must support themselves against gravity. Burrowing or crawling animals must overcome friction. They may move by side-to-side undulation or by peristalsis. The wings of birds, bats, and flying insects are airfoils, which generate lift (**30.1**).

Skeletal Support (30.2–30.6)

Skeletons function in body support, movement as muscles pull against them, and protection of internal organs. Worms and cnidarians have hydrostatic skeletons—fluid held under pressure in closed body compartments. Exoskeletons are hard external cases, such as the chitinous, jointed skeletons of arthropods. The vertebrate endoskeleton is composed of cartilage and bone (**30.2**).

Vertebrate skeletons consist of an axial skeleton (skull, vertebrae, and ribs) and an appendicular skeleton (shoulder girdle, upper limbs, pelvic girdle, and lower limbs). There are many variations on this basic body plan, which may have evolved through changes in gene regulation. (**30.3**).

Bones are living organs. Cartilage at the ends of bones cushions the joints. Bone cells, serviced by blood vessels and nerves, live in a matrix of flexible protein fibers and hard calcium salts. Long bones have a fat-storing central cavity and spongy bone at their ends. Spongy bone contains red marrow, where blood cells are made (**30.4**). Bone cells continue to replace and repair bone throughout life. (**30.5**).

Osteoporosis, a bone disease characterized by weak, porous bones, occurs when bone destruction exceeds replacement (**30.5**).

Movable joints provide flexibility. Different types of joints allow different movements (**30.6**).

Muscle Contraction and Movement (30.7–30.12)

Muscles and bones interact to produce movement. Antagonistic pairs of muscles produce opposite movements. Muscles perform work only when contracting (**30.7**).

Muscle fibers, or cells, consist of bundles of myofibrils, which contain bundles of overlapping thick (myosin) and thin (actin) protein filaments. Sarcomeres, repeating groups of thick and thin filaments, are the contractile units (**30.8**).

The sliding-filament model explains muscle contraction. The myosin heads of the thick filaments bind ATP and extend to high-energy states. The heads then attach to binding sites on the actin molecules and pull the thin filaments toward the center of the sarcomere (**30.9**).

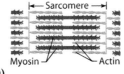

Motor neurons carry action potentials that initiate muscle contraction. A neuron and the muscle fibers it controls constitute a motor unit. Acetylcholine released at a synaptic terminal triggers an action potential that passes along T tubules into the center of the muscle cell. Calcium ions released from the endoplasmic reticulum initiate muscle contraction by moving regulatory proteins away from the myosin-binding sites on actin (**30.10**).

Exercise. Aerobic respiration, which requires a constant supply of glucose and oxygen, provides most of the ATP used for muscle movement in aerobic exercise. The anaerobic process of fermentation can provide ATP faster than aerobic respiration, but is less efficient (**30.11**). The classification of muscle fibers as slow, intermediate, and fast is based on the pathway(s) they use to generate ATP. Most muscles have a combination of fiber types, which can be affected by exercise. Exercise also affects other aspects of muscle composition (**30.12**).

Connecting the Concepts

1. Complete this concept map on animal movement.

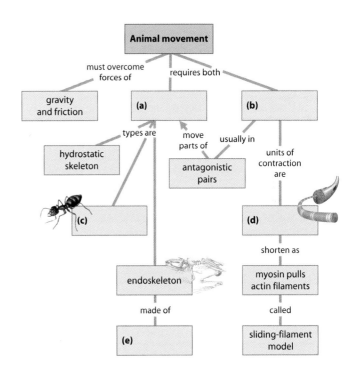

Testing Your Knowledge
Multiple Choice

2. A human's internal organs are protected mainly by the
   a. hydrostatic skeleton.   d. exoskeleton.
   b. motor unit.   e. appendicular skeleton.
   c. axial skeleton.

3. Arm muscles and leg muscles are arranged in antagonistic pairs. How does this affect their functioning?
   a. It provides a backup if one of the muscles is injured.
   b. One muscle of the pair pushes while the other pulls.
   c. A single motor neuron can control both of them.
   d. It allows the muscles to produce opposing movements.
   e. It doubles the strength of contraction.

4. Gravity would have the least effect on the movement of which of the following? (*Explain your answer.*)
   a. a salmon   d. a sparrow
   b. a human   e. a grasshopper
   c. a snake

5. Which of the following bones in the human arm would correspond to the femur in the leg?
   a. radius   d. metacarpal
   b. tibia   e. ulna
   c. humerus

6. Which of the following animals is correctly matched with its type of skeleton?
   a. fly—endoskeleton
   b. earthworm—exoskeleton
   c. dog—exoskeleton
   d. lobster—exoskeleton
   e. bee—hydrostatic skeleton

7. When a horse is running fast, its body position is stabilized by
   a. side-to-side undulation.
   b. energy stored in tendons.
   c. the lift generated by its movement through the air.
   d. foot contact with the ground.
   e. its momentum.

8. What is the role of calcium in muscle contraction?
   a. It moves the regulatory protein, exposing actin binding sites to the myosin heads.
   b. It provides energy for contraction.
   c. It blocks contraction when the muscle relaxes.
   d. It is the neurotransmitter released by a motor neuron, and it initiates an action potential in a muscle fiber.
   e. It forms the heads of the myosin molecules in the thick filaments inside a muscle fiber.

9. Muscle A and muscle B are the same size, but muscle A is capable of more precise control than muscle B. Which of the following is likely to be true of muscle A? (*Explain your answer.*)
   a. It is controlled by more neurons than muscle B.
   b. It contains fewer motor units than muscle B.
   c. It is controlled by fewer neurons than muscle B.
   d. It has larger sarcomeres than muscle B.
   e. Each of its motor units consists of more cells than the motor units of muscle B.

10. Which of the following statements about skeletons is true?
    a. Hydrostatic skeletons are soft and do not protect body parts.
    b. Chitin is a major component of vertebrate skeletons.
    c. Evolution of bipedalism involved little change in the axial skeleton.
    d. Most cnidarians must shed their skeleton periodically in order to grow.
    e. Vertebrate bones contain living cells.

Describing, Comparing, and Explaining

11. In terms of both numbers of species and numbers of individuals, insects are the most successful land animals. Write a paragraph explaining how their exoskeletons help them live on land. Are there any disadvantages to having an exoskeleton?

12. A hawk swoops down, seizes a mouse in its talons, and flies back to its perch. In a few sentences, explain how its wings enable it to overcome the downward pull of gravity as it flies upward.

13. The greatest concentration of thoroughbred horse farms in the world is in the bluegrass region of Kentucky. The grass growing in the limestone-based soil of this area is especially rich in calcium. How does this grass affect the development of championship horses?

14. Describe how you bend your arm, starting with action potentials and ending with the contraction of a muscle. How does a strong contraction differ from a weak one?

15. Using examples, explain this statement: "Vertebrate skeletons are variations on a theme."

Applying the Concepts

16. Drugs are often used to relax muscles during surgery. Which of the following chemicals do you think would make the best muscle relaxant, and why? Chemical A: Blocks acetylcholine receptors on muscle cells. Chemical B: Floods the cytoplasm of muscle cells with calcium ions.

17. An earthworm's body consists of a number of fluid-filled compartments, each with its own set of longitudinal and circular muscles. In a different kind of worm, the roundworm, a single fluid-filled cavity occupies the body, and there are only longitudinal muscles that run its entire length. Predict how the movement of a roundworm would differ from the movement of an earthworm.

18. When a person dies, muscles become rigid and fixed in position—a condition known as rigor mortis, which often figures importantly in mystery novels. Rigor mortis occurs because muscle cells are no longer supplied with ATP (when breathing stops, ATP synthesis ceases). Calcium also flows freely into dying cells. The rigor eventually disappears because the biological molecules break down. Explain, in terms of the mechanism of contraction described in Modules 30.9 and 30.10, why the presence of calcium and the lack of ATP would cause muscles to become rigid, rather than limp, after death.

19. A goal of the Americans with Disabilities Act is to allow people with physical limitations to fully participate in and contribute to society. Perhaps you have a disability or know someone with a disability. Imagine that a neuromuscular disease or injury makes it impossible for you to walk. Think about your activities during the last 24 hours. How would your life be different if you had to get around in a wheelchair? What kinds of barriers or obstacles would you encounter? What kinds of changes would have to be made in your activities and surroundings to accommodate your change in mobility?

Answers to all questions can be found in Appendix 4.

For Practice Quizzes, BioFlix, MP3 Tutors, and Activities, go to www.mybiology.com.

Plants: Form and Function

31

Plant Structure, Reproduction, and Development

Extreme Tree Climbing

Think there's nothing new under the sun? Think again! Over a span of eight weeks during the summer of 2006, two amateur "tree hunters" discovered the three tallest trees in the world. All are coast redwoods (*Sequoia sempervirens*) over 370 feet in height. The trees were found within some of the densest forests on Earth, in Humboldt County in northern California. To get to the extremely remote ravines where the trees live, the hunters traversed 60-degree slopes covered with jumbles of 10-foot-diameter logs, strewn randomly across the landscape like gigantic pickup sticks. The tallest redwood, named Hyperion (after a Greek Titan), is 379.1 feet tall. Estimated to be over 2,000 years old, which is actually young for a redwood, Hyperion is still growing at a rate of 2 to 5 inches per year.

The only way to accurately measure a tree is to climb it—a truly extreme version of a common childhood pastime! Imagine yourself standing at the trunk of Hyperion. Its base stretches 15 feet to either side of you. You look up. The trunk rises 200 feet straight up before the first limbs appear. You strap on a helmet, a tree-climbing safety harness, and soft-soled boots (to avoid damaging the bark). Using a crossbow, you shoot a climbing line over the first branch. You then pull yourself up the tree, foot by foot, using mechanical ascenders, which are rope-climbing devices used by rock climbers. The photo at left shows a coast redwood; if you look carefully, you can see a climber ascending the trunk.

As you climb upward, you discover a lost world high above the forest floor. Dotted with fruit-bearing berry bushes that grow in the crooks of branches, the redwood canopy supports its own unique ecosystem. Many species—such as not-yet-named golden-brown ants, pink earthworms, and salamanders—appear to spend their entire lives hundreds of feet in the air. The unique species of this redwood canopy are found nowhere else on Earth.

But these largely unexplored ecosystems are rapidly disappearing. In the 1840s, there were estimated to be 2 million acres of virgin old-growth forest in northern California. By the 1960s, after decades of virtually uncontrolled logging, only about 15% of the old-growth redwood tracts remained standing. Today, about 4% of the redwood's original habitat is left untouched. Almost all the remaining giant redwoods live in patches, totaling 90,000 acres, within a few small parks that lie in a 500-mile stretch from far southern Oregon through northern California. Nearly all of the tallest trees reside in the 83-square-mile Humboldt Redwoods State Park. The precise locations of these trees are kept secret, known to fewer than 20 park workers and botanists.

The coast redwood is one of about 600 living species of cone-bearing plants known as conifers. Conifers are gymnosperms, one of two groups of seed plants. Gymnosperms bear seeds in cones, whereas plants of the other group, the angiosperms (flowering plants), produce seeds enclosed in fruits.

Because angiosperms—the flowering plants—make up over 90% of the plant kingdom, we concentrate on them in this unit. (The gymnosperms and seedless plants such as mosses and ferns are covered in Chapter 17.) By studying angiosperm structure in relation to growth and reproduction, we can gain insight into the structure-function relationship of all plants and, by extension, all biological systems. ■ ■ ■

31.1 People have manipulated plant genetics since prehistoric times

To tell the story of human society, you have to talk about plants. From the dawn of civilization to the frontiers of genetic engineering, human progress has always depended upon expanding our use of plants—for food, fuel, clothing, and countless other trappings of modern life.

To illustrate, let's consider humankind's relationship with a single crop: wheat. Today, wheat accounts for about 20% of all calories consumed worldwide. In the United States, it is one of the most valuable cash crops; American farmers export 30 million tons of wheat each year. And we do more than just eat wheat. For example, researchers at the USDA developed a new material made from wheat that can be used to make biodegradable packaging materials, and a cold-protecting wheat protein added to ice cream helps prevent freezer burn.

The roots of today's prosperous relationship between humans and wheat can be traced back to the earliest human settlements. Throughout much of prehistory, humans were nomadic hunter-gatherers, moving and foraging with the changing seasons. People gathered and ate seeds from wild grasses and cereals, but didn't plant them. About 10,000 years ago, a major shift occurred: People in several parts of the world began to domesticate wild plants (see Module 14.6). This allowed for the production of surplus food and the formation of year-round farming villages. This, in turn, led to the establishment of cities and the emergence of modern civilizations.

When and where did this crucial shift take place? The origin of wheat cultivation has been traced to a region of the Middle East dubbed the "Fertile Crescent," near the upper reaches of the Tigris and Euphrates rivers (Figure 31.1). Archeological excavations of sites older than 10,000 years have found remnants of wild wheat. In sites younger than that, however, cultivated wheat also appears and, over the course of more than a millenium, displaces wild varieties in the archeological record.

Once begun, cultivation rapidly replaced hunting and gathering in the Middle East. These habits quickly spread. By 7,000 years ago, bread wheat was cultivated in Egypt, India, China, and northern Europe. Farming became the dominant way of life across Europe by 6,000 years ago. For most of the subsequent millenia, little changed in the cultivation of wheat. Individual farmers tinkered to produce slightly superior varieties, but these were small, incremental improvements.

A major leap forward occurred in the latter half of the 20th century. Faced with the rapidly expanding human population and a dwindling availability of farm land, an international effort was undertaken to improve wheat and other staple crops. The methods were straightforward: controlled cross-pollinations between varieties with desirable traits, selection of the few plants possessing the best combinations of characteristics, cross-pollination of these plants, and so on. Through such repeated selective breeding experiments, today's wheat varieties were created. The improved traits of modern wheat include a short stature with double the grain per acre of full-sized varieties; the ability to grow in a wide variety of climates and soil conditions (such as dry or acidic soils); short generation times; and resistance to over 30 pests and pathogens. As a result of the so-called "green revolution," wheat yield (tons of grain per acre planted) more than doubled from 1940 to 1980 while the cost of wheat production was cut in half.

Wheat continues to be the target of considerable research today, primarily through the methods of genetic engineering and genomics (see Chapter 12). One goal is to enhance the nutritional value of wheat. Many wild varieties of wheat are much more nutritious than domesticated varieties, with 10–15% higher concentrations of protein, iron, and zinc. In 2006, researchers identified and cloned a gene responsible for this difference. (Domestic wheat contains the gene, but at least one copy of it is inactive.) The hope is to transfer the active version of this gene into domesticated varieties, increasing their nutritional content.

Another group of researchers developed genetically modified (GM) wheat with triple the usual amount of amylose, a component of dietary fiber. They achieved this by turning off a gene that normally breaks down amylose in wheat. High-fiber strains may be a health boon since eating more fiber has been linked to reduced colorectal cancer, heart disease, diabetes, and obesity. Other genetic engineers are attempting to identify genes that confer resistance to a new strain of wheat rust, a virulent fungal pathogen that threatens the world's wheat crops.

As we continue to refine modern DNA technology methods, we will continue to deepen our understanding of wheat. Keeping in mind this staple crop as one example, we will explore the structure of angiosperms in the next five modules.

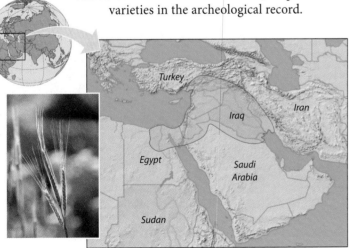

Figure 31.1 The fertile crescent region (shaded green), original site of the cultivation of wild wheat (inset)

? When and where was wheat first domesticated?

■ About 10,000 years ago in the fertile crescent

Angiosperms have dominated the land for over 100 million years, and there are about 250,000 known species of flowering plants living today. Most of our foods come from a few hundred domesticated species of flowering plants. Among these foods are roots, such as beets and carrots; the fruits of trees and vines, such as apples, nuts, berries, and squashes; the fruits and seeds of legumes, such as peas and beans; and grains, the fruits of grasses such as wheat, rice, and corn (maize).

On the basis of several structural features, botanists traditionally placed most angiosperms into two groups, called monocots and dicots (Figure 31.2). The names *monocot* and *dicot* refer to the first leaves that appear on the plant embryo. These embryonic leaves are called seed leaves, or **cotyledons.** A **monocot** embryo has one seed leaf; a **dicot** embryo has two seed leaves. The great majority of dicots, called the **eudicots** ("true" dicots), are evolutionarily related, having diverged from a common ancestor about 125 million years ago; a few smaller groups have evolved the dicot-type anatomy independently. In this chapter, we will focus on monocots and eudicots.

Monocots are a large group of related plants that include the orchids, bamboos, palms, and lilies, as well as the grains and other grasses. You can see the single cotyledon inside the seed on the top left in Figure 31.2. The leaves, stems, flowers, and roots of monocots are also distinctive. Most monocots have leaves with parallel veins. Monocot stems have vascular tissues (tissues that transport water and nutrients) arranged in a complex array of bundles. The flowers of most monocots have their petals and other parts in multiples of three. The roots of monocots form a fibrous system—a mat of threads—that spreads out below the soil surface. With most of their roots in the top few centimeters of soil, monocots, especially grasses, make excellent ground cover that reduces erosion. The roots of a rye grass plant, for instance, have more than 20 times the surface area of its stems and leaves.

Most angiosperms are eudicots, including most shrubs and trees (except for the conifers, which are gymnosperms), as well as the majority of our ornamental plants and many of our food crops. You can see the two cotyledons of a typical eudicot in the seed on the lower left in Figure 31.2. Eudicot leaves have a multibranched network of veins, and eudicot stems have vascular bundles arranged in a ring. The eudicot flower usually has petals and other parts in multiples of four or five. The large, vertical root of a eudicot, known as a taproot, goes deep into the soil, as you know if you've ever tried to pull up a dandelion.

As we saw in the preceding unit on animals, a close look at a structure often reveals its function. Conversely, function provides insight into the "logic" of a structure. In the modules that follow, we'll take a detailed look at the correlation between plant structure and function.

? The terms *monocot* and *dicot* refer to the number of _____ in the seed.

■ cotyledons (seed leaves)

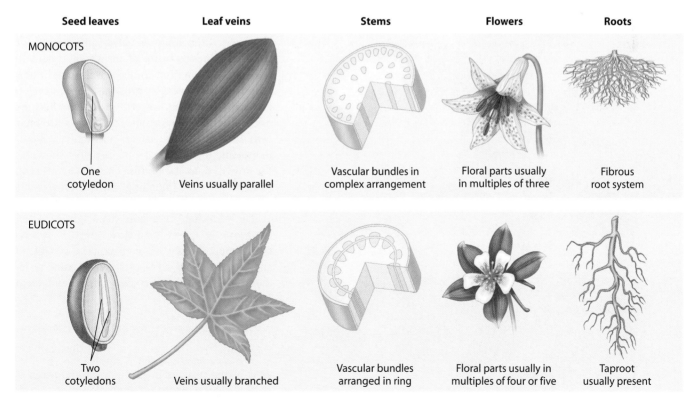

Seed leaves	**Leaf veins**	**Stems**	**Flowers**	**Roots**

MONOCOTS

One cotyledon — Veins usually parallel — Vascular bundles in complex arrangement — Floral parts usually in multiples of three — Fibrous root system

EUDICOTS

Two cotyledons — Veins usually branched — Vascular bundles arranged in ring — Floral parts usually in multiples of four or five — Taproot usually present

Figure 31.2   A comparison of monocots and eudicots

A typical plant body contains three basic organs: roots, stems, and leaves

Plants, like most animals, have organs composed of different tissues, which in turn are composed of cells of different types. In this and the next module, we'll focus on plant organs. An **organ** consists of several types of tissues that together carry out particular functions. We will then work our way down the structural hierarchy and examine plant tissues (see Module 31.5) and cells (see Module 31.6).

The basic structure of plants reflects their evolutionary history as land-dwelling organisms. Most plants must draw resources from two very different environments: They must absorb water and minerals from the soil, simultaneously obtaining CO_2 and light from aboveground. The subterranean roots and aerial shoots (stems and leaves) of a typical land plant, such as the generalized flowering plant shown in Figure 31.3, perform these vital functions. Neither roots nor shoots can survive without the other. Lacking chloroplasts and living in the dark, most roots would starve without sugar and other organic nutrients transported from the photosynthetic leaves of the shoot system. Conversely, stems and leaves depend on the water and minerals absorbed by roots.

A plant's **root system** anchors it in the soil, absorbs and transports minerals and water, and stores food. The fibrous root system of a monocot consists of a mat of generally thin roots spread out shallowly in the soil. In contrast, eudicots have one main vertical taproot with many small secondary lateral roots growing outward. Near the root tips in both eudicots and monocots, a vast number of tiny projections called **root hairs** enormously increase the root surface area for absorption of water and minerals. As shown on the far right of the figure, each root hair is an extension of an epidermal cell (a cell in the outer layer of the root).

The **shoot system** of a plant is made up of stems, leaves, and adaptations for reproduction—flowers, in angiosperms. (We'll return to the angiosperm flower in Module 31.9.) As indicated in Figure 31.3, the **stems** are the parts of the plant that are generally above the ground and that support the leaves and flowers. In the case of a tree, the stems are the trunk and all the branches, including the smallest twigs. A stem has **nodes,** the points at which leaves are attached, and **internodes,** the portions of the stem between nodes. The **leaves** are the main photosynthetic organs in most plants, although green stems also perform photosynthesis. Most leaves consist of a flattened blade and a stalk, or petiole, which joins the leaf to a node of the stem.

The two types of buds in the figure are undeveloped shoots. When a plant stem is growing in length, the **terminal bud,** at the apex (tip) of the stem, has developing leaves and a compact series of nodes and internodes. The **axillary buds,** one in each of the angles formed by a leaf and the stem, are usually dormant. In many plants, the terminal bud produces hormones that inhibit growth of the axillary buds, a phenomenon called **apical dominance.** By concentrating resources on growing taller, apical dominance is an evolutionary adaptation that increases the plant's exposure to light. This is especially important where vegetation is dense. However, branching is also important for increasing the exposure of the shoot system to the environment, and under certain conditions, the axillary buds begin growing. Some develop into shoots bearing flowers, and others become nonreproductive branches complete with their own terminal buds, leaves, and axillary buds. Removing the terminal bud usually stimulates the growth of axillary buds. This is why pruning fruit trees and "pinching back" houseplants makes them bushier.

The drawing in Figure 31.3 gives us an overview of plant structure, but it by no means represents the enormous diversity of angiosperms. Next, let's look briefly at some variations on the basic themes of root and stem structure.

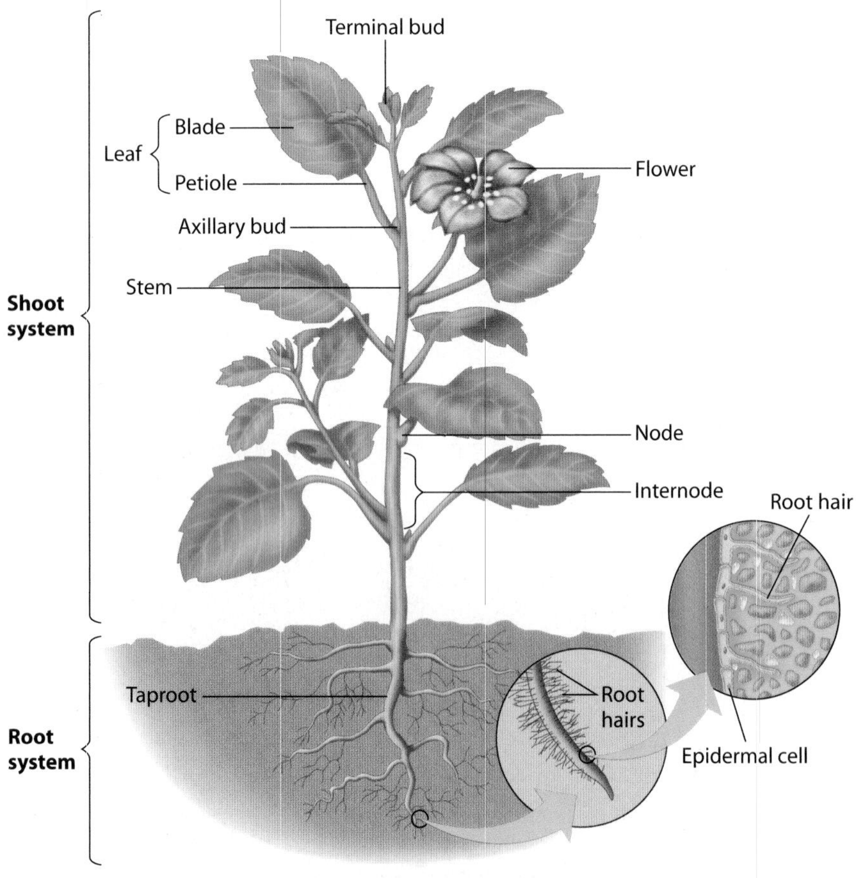

Web Process of Science *How Are Trees Identified by Their Leaves?*

? Name the two organ systems and three basic organs found in all plants.

Figure 31.3 The body plan of a flowering plant (a eudicot)

■ Root system and shoot system; roots, stems, leaves

31.4 Many plants have modified roots, stems, and leaves

The three basic plant organs—roots, stems, and leaves—come in a variety of sizes and shapes, and they are adapted for a variety of functions. Many are adapted for storage. Carrots, turnips, sugar beets, and sweet potatoes, for instance, all have unusually large taproots that store food in the form of carbohydrates such as starch (Figure 31.4A). The plants consume the stored sugars during flowering and fruit production. For this reason, root crops are harvested before flowering.

Figure 31.4B shows three examples of modified stems. The strawberry plant has a horizontal stem called a stolon (or "runner") that grows along the ground surface. Stolons enable a plant to reproduce asexually, as plantlets form at nodes along their length. You've seen a different stem modification if you have ever dug up an iris plant or cooked with fresh ginger; the large, brownish, rootlike structures are actually **rhizomes,** horizontal stems that grow just below or along the soil surface. The rhizomes store food and, having buds, they can also spread and form new plants. The potato plant has rhizomes that end in enlarged structures specialized for storage called **tubers** (the

Figure 31.4A The modified root of a sugar beet plant

potatoes we eat). Potato "eyes" are clusters of axillary buds on the tubers that mark the nodes. Plant bulbs are underground shoots containing swollen leaves that store food. As you peel an onion, you are removing layers of leaves attached to a short stem.

Plant leaves, too, are highly varied. Grasses and many other monocots, for instance, have long leaves without petioles. Some eudicots, such as celery, have enormous petioles—the stalks we eat—which contain a lot of water and stored food. The upper photograph in Figure 31.4C shows a modified leaf called a **tendril,** with its tips coiled around a stem. Tendrils help plants climb. (Some tendrils, as in grapevines, are modified stems.) The spines of the barrel cactus (lower photo) are modified leaves that protect the plant from being eaten by animals. The main part of the cactus is the stem, which is adapted for photosynthesis and water storage.

So far we have examined plants as we see them with the unaided eye. Next, we move further down the structural hierarchy to explore plant tissues.

? If a potato "sprouts" from an eye, Are the resulting appendages part of the root system or shoot system?

■ The potato tuber is a modified stem, so the eyes are part of the shoot system.

Strawberry plant

Potato plant

Ginger plant

Stolon (runner)

Taproot

Rhizome

Tuber

Rhizome

Figure 31.4B Three kinds of modified stems

Figure 31.4C Modified leaves: the tendrils of a pea plant (top) and cactus spines (bottom)

31.5 Three tissue systems make up the plant body

A **tissue** is a group of cells with a common structure, function, or both. A **tissue system** consists of one or more tissues organized into a functional unit within a plant. Each plant organ—root, stem, or leaf—is made up of three tissue systems: the dermal, vascular, and ground tissue systems. Each tissue system is continuous throughout the entire plant body, but the systems are arranged differently in leaves, stems, and roots (Figure 31.5). We examine the tissue systems of young roots and shoots in this module. Later, we will see that the tissue systems are somewhat different in older roots and stems.

The **dermal tissue system** (brown in the figure) forms an outer protective covering. Like our own skin, it acts as the first line of defense against physical damage and infectious organisms. In many plants, the dermal tissue system consists of a single layer of tightly packed cells called the **epidermis.** The epidermis of leaves and most stems has a waxy coating called the **cuticle,** which helps prevent water loss. The second tissue system is the **vascular tissue system** (purple). It is made up of xylem and phloem tissues and provides support and long-distance transport. Tissues that are neither dermal nor vascular make up the **ground tissue system** (yellow). It accounts for most of the bulk of a young plant, filling the spaces between the epidermis and vascular tissue system. Ground tissue internal to the vascular tissue is called **pith,** and ground tissue external to the vascular tissue is called **cortex.** The ground tissue system has diverse functions, including photosynthesis, storage, and support.

The close-up views on the right side of Figure 31.5 show how these three tissues systems are organized in typical plant roots, stems, and leaves. The micrograph at the bottom shows in cross section the three tissue systems in a young eudicot root. Water and minerals absorbed from the soil must enter through the epidermis. In the center of the root, the vascular tissue system forms a **vascular cylinder,** with xylem cells radiating from the center like spokes of a wheel and phloem cells filling in the wedges between the spokes. The ground tissue system of the root, the region between the vascular cylinder and epidermis, consists entirely of cortex. The cortex cells store food as starch and take up minerals that have entered the root through the epidermis. The innermost layer of the cortex is the **endodermis,** a cylinder one cell thick. The endodermis is a selective barrier, determining which substances pass between the rest of the cortex and the vascular tissue (see Module 32.2).

As the center of Figure 31.5 indicates, the young stem of a eudicot looks quite different from that of a monocot. Both stems have their vascular tissue system arranged in numerous **vascular bundles.** However, the monocot stem has vascular bundles scattered throughout the ground tissue,

whereas the eudicot stem has vascular bundles arranged in a ring. Unlike roots, the ground tissue of eudicot stems consists of both a cortex region and a pith region. The cortex fills the space between the vascular ring and the epidermis. The pith fills the center of the stem and is often important in food storage. The ground tissue of a monocot stem is not divided into cortex and pith.

The top of Figure 31.5 illustrates the arrangement of the three tissue systems in a typical eudicot leaf. The epidermis is interrupted by pores called **stomata** (singular, *stoma*), which allow CO_2 exchange between the surrounding air and the photosynthetic cells inside the leaf. Each stoma is flanked by two **guard cells,** which regulate the size of the stoma.

The ground tissue system of a leaf, called the **mesophyll,** is sandwiched between the upper and lower epidermis. Mesophyll consists mainly of photosynthetic parenchyma cells. The green structures in the diagram are their chloroplasts. In this eudicot leaf, notice that cells in the lower mesophyll are loosely arranged, with a labyrinth of air spaces through which CO_2 and O_2 circulate. The air spaces are particularly large in the vicinity of stomata, where gas exchange with the outside air occurs. In many monocot leaves and in some eudicot leaves, the mesophyll is not arranged in distinct upper and lower layers.

In both monocots and eudicots, the leaf's vascular tissue system is made up of a network of veins. As you can see in Figure 31.5, each **vein** is a vascular bundle composed of xylem and phloem surrounded by a sheath of parenchyma cells. The veins' xylem and phloem, continuous with the vascular bundles of the stem, are in close contact with the leaf's photosynthetic tissues. This ensures that those tissues are supplied with water and mineral nutrients from the soil and that sugars made in the leaves are transported throughout the plant. The vascular structure also functions as a skeleton that reinforces the shape of the leaf.

In the last three modules, we have examined plant structure at the level of organs and tissues. In the next module, we will complete our descent into the structural hierarchy of plants by examining cells.

Web Activity *Root, Stem, and Leaf Sections*

Web Process of Science *What Are the Functions of Monocot Tissues?*

? For each of the following structures in your body, name the most analogous plant tissue system: circulatory system, skin, adipose tissue (body fat).

▪ Vascular tissue system, dermal tissue system, ground tissue system

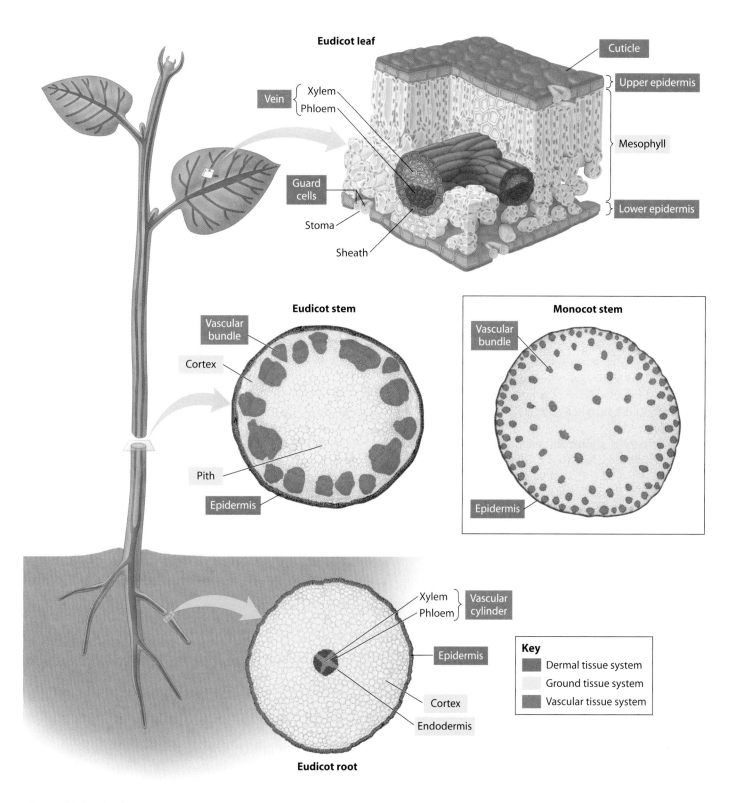

Eudicot leaf

Cuticle

Upper epidermis

Vein { Xylem / Phloem }

Mesophyll

Guard cells

Stoma

Sheath

Lower epidermis

Eudicot stem

Vascular bundle

Cortex

Pith

Epidermis

Monocot stem

Vascular bundle

Epidermis

Eudicot root

Xylem / Phloem } Vascular cylinder

Epidermis

Cortex

Endodermis

Key
Dermal tissue system
Ground tissue system
Vascular tissue system

Figure 31.5 The three tissue systems

In addition to features shared with other eukaryotic cells (see Module 4.4), most plant cells have three unique structures (Figure 31.6A): chloroplasts, the sites of photosynthesis; a central vacuole containing fluid that helps maintain cell turgor (firmness); and a cell wall made from the structural carbohydrate cellulose surrounding the plasma membrane.

The enlargement on the right in Figure 31.6A highlights the adjoining cell walls of two cells. Many plant cells, especially those that provide structural support, have a two-part cell wall; a primary cell wall is laid down first, and then a more rigid secondary cell wall is secreted between the plasma membrane and the primary wall. The primary walls of adjacent cells in plant tissues are held together by a sticky layer called the middle lamella. Pits, where the cell wall is relatively thin, allow migration of water between adjacent cells. Plasmodesmata are channels of communication and circulation between adjacent plant cells.

The structure of a plant cell and the nature of its wall often correlate with the cell's main functions. As you consider the five major types of plant cells shown in Figures 31.6B–31.6F, notice the structural adaptations that make their specific functions possible.

Parenchyma cells (Figure 31.6B) are the most abundant type of cell in most plants. They remain alive when mature and usually have only primary cell walls, which are thin and flexible. Parenchyma cells perform most of the metabolic functions of a plant, such as photosynthesis, aerobic respiration, and food storage. Most parenchyma cells can divide and differentiate into other types of plant cells under certain conditions, such as during the repair of an injury. In the laboratory, it is even possible to regenerate an entire plant from a single parenchyma cell (as we will see in Module 31.14).

Collenchyma cells (Figure 31.6C) resemble parenchyma cells in lacking secondary walls, but they have unevenly thickened primary walls. Their main function is to provide flexible support in parts of the plant that are still growing; young stems and petioles often have collenchyma cells just below their surface (the "string" of a celery stalk, for example). These living cells elongate as stems and leaves grow.

Sclerenchyma cells have thick secondary cell walls (yellow in Figure 31.6D, top of next page) usually strengthened with lignin, which is the main chemical component of wood. Mature sclerenchyma cells cannot elongate, and they occur in regions of the plant that have stopped growing in length. When mature, most sclerenchyma cells are dead, their cell walls forming a rigid "skeleton" that supports the plant.

Figure 31.6D shows two types of sclerenchyma cells. One, called a **fiber,** is long and slender and is usually arranged in bundles. Some plant tissues with abundant fiber cells are commercially important; hemp fibers, for example, are used

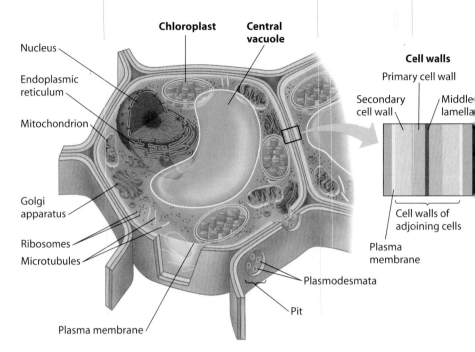

Figure 31.6A   The structure of a plant cell

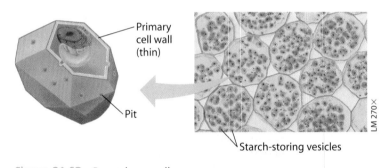

Figure 31.6B   Parenchyma cell

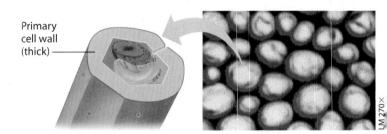

Figure 31.6C   Collenchyma cell

to make rope and clothing. **Sclereids,** which are shorter than fiber cells, have thick, irregular, and very hard secondary walls. Sclereids impart the hardness to nutshells and seed coats and the gritty texture to the soft tissue of a pear.

Angiosperms have two types of **water-conducting cells:** tracheids and vessel elements. Both have rigid, lignin-containing secondary cell walls. As Figure 31.6E on the next page shows, **tracheids** are long, thin cells with tapered ends.

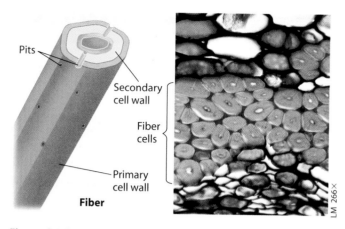

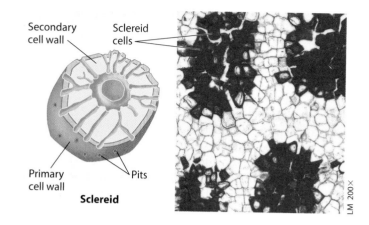

Figure 31.6D Sclerenchyma cells: fiber (left) and sclereid (right)

Vessel elements are wider, shorter, and less tapered. Chains of tracheids or vessel elements with overlapping ends form a system of tubes that conveys water from the roots to the stems and leaves as part of xylem tissue (discussed below). The tubes are hollow because both tracheids and vessel elements are dead when mature, with only their cell walls remaining. Water passes through pits in the walls of tracheids and vessel elements and through openings in the end walls of vessel elements. Because of their thick, rigid walls, these cells also function in support.

Food-conducting cells, known as **sieve-tube members,** are also arranged end to end, forming tubes as part of phloem tissue (Figure 31.6F). Unlike water-conducting cells, however, sieve-tube members remain alive at maturity, though they lose most organelles, including the nucleus and ribosomes. This reduction in cell contents enables nutrients to pass more easily through the cell. The end walls between sieve-tube members, called **sieve plates,** have pores that allow fluid to flow from cell to cell along the sieve tube. Alongside each sieve-tube member is at least one **companion cell,** which is connected to the sieve-tube member by numerous plasmodesmata. One companion cell may serve multiple sieve-tube members by producing and transporting proteins to all of them.

Now that we have reached the lowest level in the structural hierarchy of plants—cells—let's review by moving back up. Cells of plants are grouped into tissues with characteristic functions. For example, **xylem** tissue contains water-conducting cells that convey water and dissolved minerals upward from the roots, while **phloem** tissue contains sieve-tube members that transport sugars from leaves or storage tissues to other parts of the plant. In addition to conducting cells, phloem and xylem contain sclerenchyma cells, which provide support, and parenchyma cells, which store various materials. Xylem and phloem tissue are organized into the vascular tissue system, which provides structural support and long-term transportation throughout the plant body. The vascular tissue system, along with the dermal and ground tissue systems, makes up each plant organ, such as a leaf or a root. This review completes our survey of basic plant anatomy. Next, we examine how plants grow.

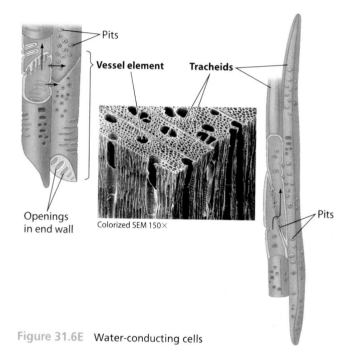

Figure 31.6E Water-conducting cells

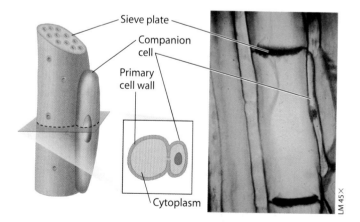

Figure 31.6F Food-conducting cell (sieve-tube member)

? Which of the following cell types has the potential to give rise to all others in the list: collenchyma, sclereid, parenchyma, vessel element, companion cell?

BioFlix *Tour of a Plant Cell*

■ Parenchyma

31.7 Primary growth lengthens roots and shoots

So far, we have looked at the structure of plant tissues and cells in mature organs. We will now examine how such organization arises through plant growth.

Most species of plants continue to grow as long as they live, a condition known as **indeterminate growth.** Most animals, in contrast, are characterized by **determinate growth;** that is, they cease growing after reaching a certain size. These differences underlie a broader distinction between plants and animals. Most animals *move* through their environment. Plants, in contrast, *grow* through their environment. Indeterminate growth can thus enable plants to increase their exposure to sunlight, air, and soil throughout life.

Indeterminate growth does not mean that plants are immortal; they do, of course, die. Flowering plants are categorized as annuals, biennials, or perennials, based on the length of their life cycle (the time from germination through flowering and seed production to death). **Annuals** complete their life cycle in a single year or less. Our most important food crops (including grains and legumes) are annuals, as are a great number of wildflowers. **Biennials** complete their life cycle in two years; flowering usually occurs during the second year. Beets and carrots are biennials, but we usually harvest them in their first year and so miss seeing their flowers.

Perennials are plants that live and reproduce for many years. They include trees, shrubs, and some grasses.

Some plants are among the oldest organisms alive. The coast redwoods we discussed in the chapter introduction are estimated to be 2,000–3,000 years old, and another conifer, the bristlecone pine, can live thousands of years longer than that (see Module 31.15). Some buffalo grass of the North American plains is believed to have been growing for 10,000 years from seeds that sprouted at the end of the last ice age. When a plant dies, it is not usually from old age, but from an infection or some environmental trauma, such as fire or severe drought.

Growth in plants is made possible by tissues called meristems. A **meristem** consists of cells that divide frequently, generating additional cells. Some products of this division remain in the meristem and produce still more cells, while others differentiate and are incorporated into tissues and organs of the growing plant. Meristems at the tips of roots and in the buds of shoots are called **apical meristems** (Figure 31.7A). Cell division in the apical meristems produces the new cells that enable a plant to grow in length. This process is known as **primary growth.** Primary growth enables roots to push through the soil and allows shoots to increase exposure to light and CO_2. Although apical meristems lengthen both roots and shoots, there are important differences in the mechanisms of primary growth in each system. We will examine them separately, starting at the bottom.

Figure 31.7B (on the facing page) illustrates primary growth in a longitudinal section through a growing onion root. The root tip is covered by a thimble-like **root cap** that protects the delicate, actively dividing cells of the apical meristem. Growth in length occurs just behind the root tip, where three zones of cells at successive stages of primary growth are located. Moving away from the root tip, they are called the zone of cell division, the zone of elongation, and the zone of maturation. The three zones of cells grade together, with no sharp boundaries between them.

The zone of cell division includes the root apical meristem and cells that derive from it. New root cells are produced in this region, including the cells of the root cap. In the zone of elongation, root cells elongate, sometimes to more than ten times their original length. It is cell elongation that pushes the root tip farther into the soil. The cells lengthen, rather than expand equally in all directions, because of the circular arrangement of cellulose fibers in parallel bands in their cell walls. The enlargement diagrams at the left of the root indicate how this works. The cells elongate by taking up water, and as they do, the cellulose fibers (shown in red) separate, somewhat like an expanding accordion. The cells cannot expand greatly in width because the cellulose fibers do not stretch much.

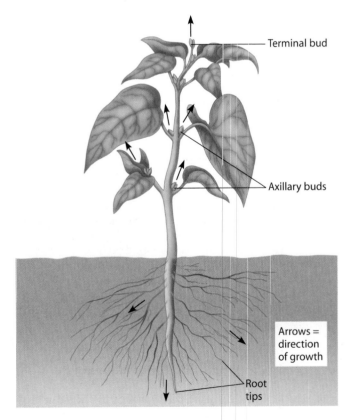

Figure 31.7A Locations of apical meristems, which are responsible for primary growth

Terminal bud

Axillary buds

Arrows = direction of growth

Root tips

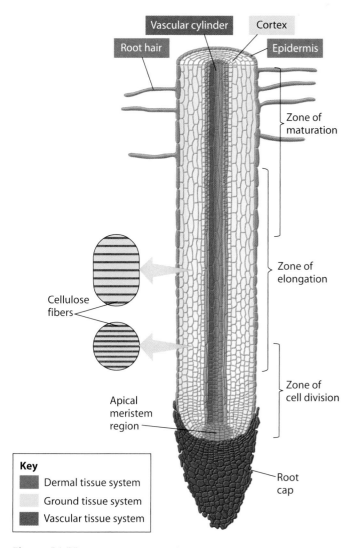

Figure 31.7B   Primary growth of a root

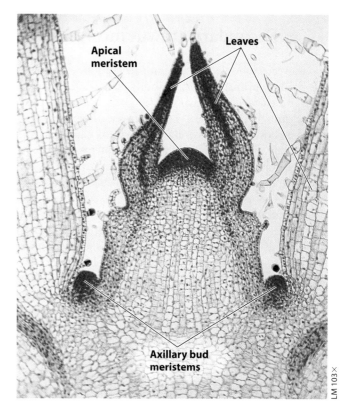

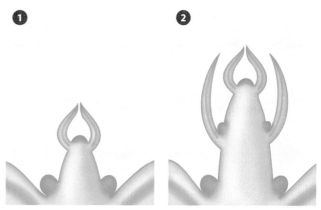

Figure 31.7C   Primary growth of a shoot

The three tissue systems of a mature plant (dermal, ground, and vascular) complete their development in the zone of maturation. Cells of the vascular cylinder differentiate into **primary xylem** and **primary phloem.** Differentiation of cells (specialization of their structure and function) results from differential gene expression (see Module 11.2). Cells in the vascular cylinder, for instance, develop into primary xylem or phloem cells because certain genes are turned on and are expressed as specific proteins, whereas other genes in these cells are turned off. Genetic control of differentiation is one of the most active areas of plant research.

The micrograph in Figure 31.7C shows a section through the end of a growing shoot that was cut lengthwise from its tip to just below its uppermost pair of axillary buds. You can see the apical meristem, which is a dome-shaped mass of dividing cells at the tip of the terminal bud. Elongation occurs just below this meristem, and the elongating cells push the apical meristem upward, instead of downward as in the root. As the apical meristem advances upward, some of its cells remain behind, and these become new axillary bud meristems at the base of the leaves.

The drawings in Figure 31.7C show two stages in the growth of a shoot. Stage ❶ is just like the micrograph. At the later stage shown in sketch ❷, the apical meristem has been pushed upward by elongating cells underneath.

Primary growth accounts for a plant's lengthwise growth. The stems and roots of many plants increase in thickness too, and in the next module, we see how this usually happens.

? A plant grows taller due to cell division within the _____ at the tips of the shoots. Such lengthening is called _____.

apical meristem . . . primary growth

Secondary growth increases the girth of woody plants

In the previous module, we saw how the apical meristems of plants produce primary growth. Woody plants (such as trees, shrubs, and vines) continue to grow in girth, thickening after primary growth has ceased. This increase in thickness of stems and roots, called **secondary growth,** is caused by the activity of dividing cells called **lateral meristems.** These dividing cells are arranged into two cylinders, known as the vascular cambium and the cork cambium, that extend along the length of roots and stems.

The **vascular cambium** is a cylinder of meristem cells one cell thick between the primary xylem and primary phloem (Figure 31.8A, left). This region of the stem is just beginning secondary growth. Except for the vascular cambium, the stem at this stage of growth is virtually the same as a young stem undergoing primary growth (compare this figure with the eudicot stem in Figure 31.5). Secondary growth adds layers of vascular tissue on either side of the vascular cambium, as indicated by the green arrows.

The drawings at the center and the right show the results of secondary growth. In the center drawing, the vascular cambium has given rise to two new tissues: **secondary xylem** to its interior and **secondary phloem** to its exterior. Each year, the vascular cambium gives rise to layers of secondary xylem and secondary phloem that are larger in circumference than the previous layer (see the drawing at the right). In this way, the vascular cambium thickens roots and stems.

Secondary xylem makes up the **wood** of a tree, shrub, or vine. Over the years, a woody stem gets thicker and thicker as its vascular cambium produces layer upon layer of secondary xylem. The cells of the secondary xylem have thick walls rich in lignin, giving wood its characteristic hardness and strength.

Annual growth rings, such as those in the locust tree in Figure 31.8B on the facing page, result from the layering of secondary xylem. The layers are visible as rings because of uneven activity of the vascular cambium during the year. In woody plants that live in temperate regions, such as most of the United States, the vascular cambium becomes dormant each year during winter, and secondary growth is interrupted. When secondary growth resumes in the spring, a cylinder of early wood forms. Made up of the first new xylem cells to develop, early wood cells are usually larger in diameter and thinner-walled than those produced later in summer. The boundary between the large cells of early wood and the smaller cells of the late wood produced during the previous growing season is usually a distinct ring visible in cross sections of tree trunks and roots. Therefore, a tree's age can be estimated by counting its annual rings. Dendrochronology (*dendros* is Greek for "tree") is the science of analyzing tree ring growth patterns. The rings may have varying thicknesses, reflecting climate conditions and therefore the amount of seasonal growth in a given year. In fact, growth ring patterns in older trees is one source of evidence for recent global climate change.

Now let's return to Figure 31.8A and see what happens to the parts of the stem that are *external* to the vascular cambium. Unlike xylem, the external tissues do not accumulate over the years. Instead, they are sloughed off at about the same rate they are produced.

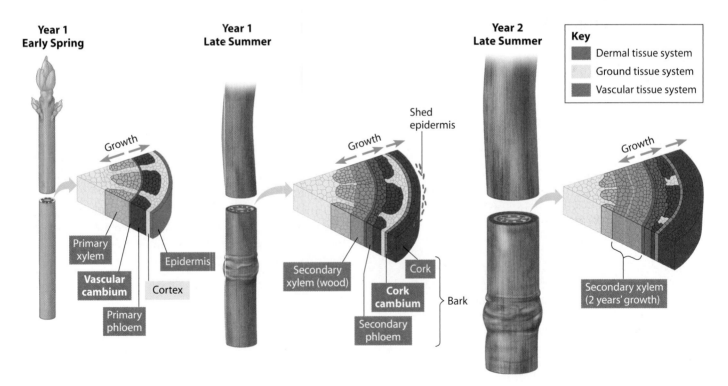

Figure 31.8A Secondary growth of a woody eudicot stem

Notice at the left of the figure that the epidermis and cortex, both the result of primary growth, make up the young stem's external covering. When secondary growth begins (center drawing), the epidermis is sloughed off and replaced with a new outer layer called **cork** (brown). Mature cork cells are dead and have thick, waxy walls that protect the underlying tissues of the stem from water loss, physical damage, and pathogens. Cork is produced by meristematic tissue called the **cork cambium** (light brown), which first forms from parenchyma cells in the cortex. As the stem thickens and the secondary xylem expands, the original cork and cork cambium are pushed outward and fall off, as is evident in the cracked, peeling bark of many tree trunks. A new cork cambium forms to the inside. When no cortex is left, it forms from parenchyma cells in the phloem.

Everything external to the vascular cambium is called **bark.** As indicated in the diagram at the right in Figure 31.8A, its main components are the secondary phloem, the cork cambium, and the cork. The youngest secondary phloem (next to the vascular cambium) functions in sugar transport. The older secondary phloem dies, as does the cork cambium you see here. Pushed outward, these tissues and cork produced by the cork cambium help protect the stem until they, too, are sloughed off as part of the bark. Keeping pace with secondary growth, cork cambium keeps regenerating from the secondary phloem and keeps producing a steady supply of cork.

The log on the left in Figure 31.8B, from a locust tree, shows the results of several decades of secondary growth. The bulk of a trunk like this is dead tissue. The living tissues in it are the vascular cambium, the youngest secondary phloem, the cork cambium, and cells in the wood rays, which you can see radiating from the center of the log in the drawing on the right. The **wood rays** consist of parenchyma cells that transport water and nutrients, store starch and other organic nutrients, and aid in wound repair. The **heartwood,** in the center of the trunk,

consists of older layers of secondary xylem. These cells no longer transport water and minerals (xylem sap); they are clogged with resins and other metabolic by-products that make the heartwood resistant to rotting. The lighter-colored **sapwood** consists of younger secondary xylem that does conduct xylem sap.

Thousands of useful products are made from wood—from construction lumber to fine furniture, musical instruments, paper, insulation, and a long list of chemicals, including turpentine, alcohols, artificial vanilla flavoring, and preservatives. Among the qualities that make wood so useful are a unique combination of strength, hardness, lightness, high insulating properties, durability, and workability. In many cases, there is simply no good substitute for wood. A wooden oboe, for instance, produces far richer sounds than a plastic one. Fence posts made of locust tree wood actually last much longer in the ground than metal ones. Ball bearings are sometimes made of a very hard wood called lignum vitae. Unlike metal bearings, they require no lubrication because a natural oil completely penetrates the wood.

In a sense, wood is analogous to the hard endoskeletons of many land animals. It is an evolutionary adaptation that enables a shrub or tree to remain upright and keep growing year after year on land—sometimes to attain enormous heights, as we saw in the chapter's introduction. In the next few modules, we examine some additional adaptations that enable plants to live on land—those that facilitate reproduction.

Web Activity *Primary and Secondary Growth*

? (a) What type of plant tissue makes up wood?
   (b) What is bark?

■ (a) Secondary xylem; (b) All tissues exterior to the vascular cambium—secondary phloem, cork cambium, and cork

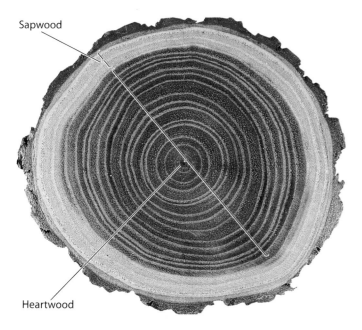

Sapwood

Heartwood

Figure 31.8B Anatomy of a locust log

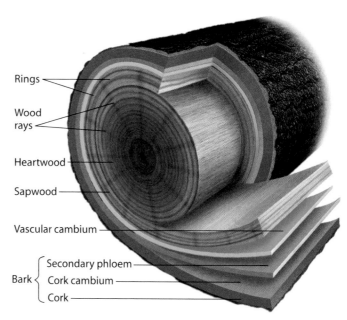

Rings

Wood rays

Heartwood

Sapwood

Vascular cambium

Bark { Secondary phloem
        Cork cambium
        Cork

Chapter 31 *Plant Structure, Reproduction, and Development* **633**

31.9 The flower is the organ of sexual reproduction in angiosperms

It has been said that an oak tree is merely an acorn's way of making more acorns. Indeed, evolutionary fitness for any organism is measured only by its ability to replace itself with healthy, fertile offspring. Thus, from an evolutionary viewpoint, all the structures and functions of a plant can be interpreted as mechanisms contributing to reproduction. In the remaining modules, we explore the reproductive biology of angiosperms, beginning here with a brief overview. (This would be a good time to review Modules 17.8–9, where this information was first presented.)

Flowers, the reproductive shoots of angiosperms, typically contain four types of modified leaves called floral organs: sepals, petals, stamens, and carpels (Figure 31.9A). **Sepals,** which enclose and protect the flower bud, are usually green and more leaflike than the other floral organs (picture the green at the base of a rosebud). The **petals** are often colorful and advertise the flower to pollinators. Stamens and carpels are the reproductive organs, containing sperm and eggs, respectively.

A **stamen** consists of a stalk (filament) tipped by an anther. Within the **anther** are sacs in which meiosis occurs and in which pollen is produced. Pollen grains house the cells that develop into sperm.

A **carpel** has a long slender neck (style) with a sticky stigma at its tip. The **stigma** is the landing platform for pollen. The base of the carpel is the **ovary,** within which are **ovules,** each containing a developing egg and supporting cells. The term **pistil** is sometimes used to refer to a single carpel or a group of fused carpels.

Figure 31.9B shows the life cycle of a generalized angiosperm. Fertilization occurs in the ovule, which then develops into a seed containing the embryo. Meanwhile, the ovary develops into a fruit, which protects the seed and aids in dispersing it. Completing the life cycle, the seed **germinates** (begins to grow) in a suitable habitat; the embryo develops into a seedling; and the seedling grows into a mature plant.

In the next four modules, we examine key stages in the angiosperm sexual life cycle in more detail. We will see that there are a number of variations in the basic themes presented here.

🎧 MP3 Tutor *From Flower to Fruit*

? Pollen develops within the _____ of _____. Ovules develop within the _____ of _____.

■ anthers · · · stamens · · · ovaries · · · carpels

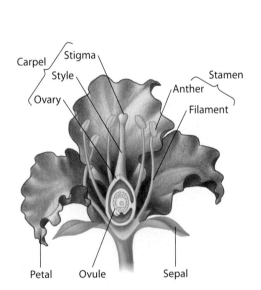

Figure 31.9A   The structure of a flower

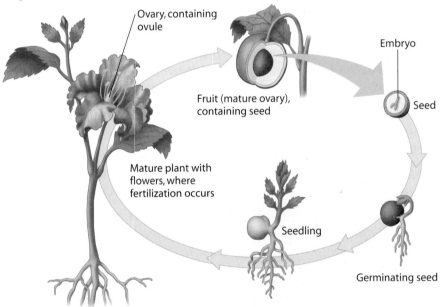

Figure 31.9B   Life cycle of a generalized angiosperm

31.10 The development of pollen and ovules culminates in fertilization

Recall from Chapter 17 that the life cycles of all plants include alternation of haploid (*n*) and diploid (2*n*) generations. The roots, stems, leaves, and most of the reproductive structures of a maple tree, a rose plant, or a grass—in fact, all angiosperms—are diploid. The diploid plant body is called the **sporophyte.** A sporophyte produces special structures, the anthers and ovules, in which cells undergo meiosis. Haploid spores result. Each of these then divides mitotically and becomes a multicellular

gametophyte, the plant's haploid generation. The gametophyte produces gametes by mitosis. At fertilization, gametes from the male and female gametophytes unite, producing a diploid zygote. The life cycle is completed when the zygote divides by mitosis and develops into a new sporophyte. Without a microscope, all we can see of the angiosperm life cycle is the sporophytes and occasionally the pollen (male gametophytes) produced by them. In this module, we take a microscopic look at the gametophytes of a flowering plant.

We begin with the development of the male gametophyte, the pollen grain. The cells that develop into pollen grains are found within a flower's anthers (top left in Figure 31.10). Each cell first undergoes meiosis, forming four haploid cells called spores. Each spore then divides mitotically, forming two haploid cells, called the tube cell and the generative cell. A thick wall forms around these cells, and the resulting pollen grain is ready for release from the anther.

Moving to the top right of the figure, we can follow the development of the flower parts that form the female gametophyte and eventually the egg. In most species, the ovary of a flower contains several ovules, but only one is shown here. An ovule contains a central cell (gold) surrounded by a protective covering of smaller cells (yellow). The central cell enlarges and undergoes meiosis, producing four haploid spores. Three of the spores usually degenerate, but the surviving one enlarges and divides mitotically, producing a multicellular structure called the **embryo sac** (gold area). Housed in several layers of protective cells (yellow) produced by the sporophyte plant, the embryo sac is the female gametophyte. The sac contains a large central cell with two haploid nuclei. One of its other cells is the haploid egg, ready to be fertilized.

The first step leading to fertilization is **pollination** (at the center of the figure), the transfer of pollen from anther to stigma. Most angiosperms are dependent on insects, birds, or other animals to transfer their pollen. But the pollen of some plants—such as grasses and many trees—is windborne, as anyone bothered by seasonal pollen allergies knows.

After pollination, the pollen grain germinates on the stigma. Its tube cell gives rise to the pollen tube, which grows downward into the ovary. Meanwhile, the generative cell divides mitotically, forming two sperm. When the pollen tube reaches the base of the ovule, it enters the ovary and discharges its two sperm near the embryo sac. One sperm fertilizes the egg, forming the zygote. The other contributes its haploid nucleus to the large diploid central cell of the embryo sac. This cell, now with a triploid ($3n$) nucleus, will give rise to a food-storing tissue called **endosperm.**

The union of two sperm with two different nuclei of the embryo sac is called **double fertilization,** and the resulting production of endosperm is unique to angiosperms. Endosperm will develop only in ovules containing a fertilized egg, thereby preventing angiosperms from squandering nutrients.

Web Activity Angiosperm Life Cycle

? Fertilization unites the sperm cell, which develops within the male gametophyte (or _____ _____), with the egg cell, which develops within the female gametophyte (or _____ _____).

▪ pollen grain . . . embryo sac

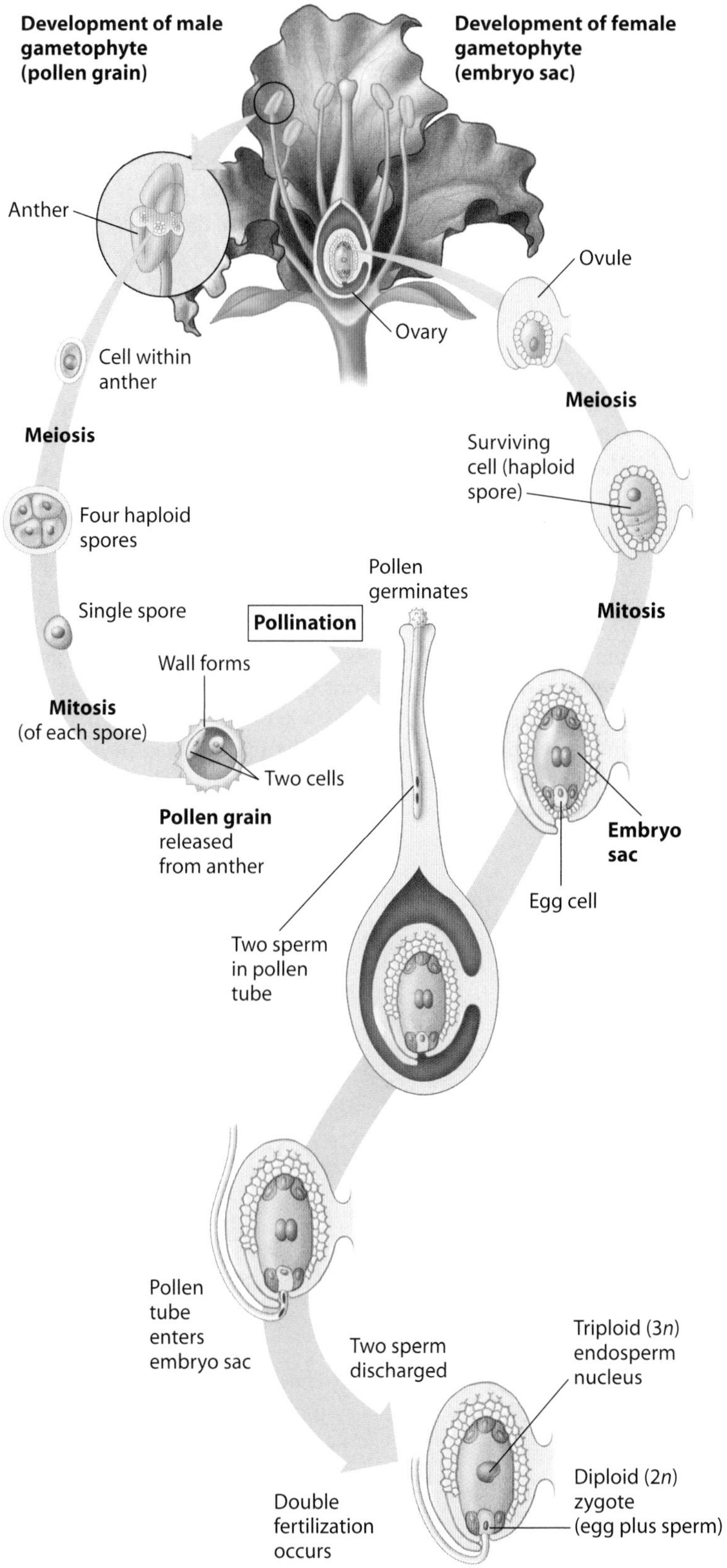

Figure 31.10 Gametophyte development and fertilization in an angiosperm

31.11 The ovule develops into a seed

After fertilization, the ovule, containing the triploid central cell and the zygote, begins developing into a seed. The triploid cell divides and develops into the nutrient-rich tissue called endosperm. The endosperm nourishes the embryo until it becomes a self-supporting seedling.

As shown in Figure 31.11A, embryonic development begins when the zygote divides into two cells. Repeated division of one of the cells then produces a ball of cells that becomes the embryo. The other cell divides to form a thread of cells that pushes the embryo into the endosperm. The bulges you see on the embryo are the developing cotyledons. You can tell that the plant in this drawing is a dicot since it has two cotyledons.

The result of embryonic development in the ovule is a mature seed, which you see on the bottom right of Figure 31.11A. Near the end of its maturation, the seed loses most of its water and forms a hard, resistant **seed coat** (brown). The embryo, surrounded by its endosperm food supply (gold), becomes dormant; it will not develop further until the seed germinates. **Seed dormancy,** a condition in which growth and development are suspended temporarily, is an important evolutionary adaptation. It allows time for a plant to disperse its seeds and increases the chance that a new generation of plants will begin growing only when environmental conditions, such as temperature and moisture, favor survival.

The dormant embryo contains a miniature root and shoot, each equipped with an apical meristem. After the seed germinates, the apical meristems will sustain primary growth as long as the plant lives. Also present in the embryo are the three tissue cylinders that will form the epidermis, cortex, and primary vascular tissues.

Figure 31.11B contrasts the internal structures of eudicot and monocot seeds. In the eudicot (a bean), the embryo is an elongated structure with two fleshy cotyledons (tan). The embryonic root develops just below the point at which the cotyledons are attached to the rest of the embryo. The embryonic shoot, tipped by a pair of miniature embryonic leaves, develops just above the point of attachment. The bean seed contains no endosperm because its cotyledons absorb the endosperm nutrients as the seed forms. The nutrients start passing from the cotyledons to the embryo when it germinates.

A kernel of corn, an example of a monocot, is actually a fruit containing one seed. Everything you see in the drawing is the seed, except the kernel's outermost covering. The covering is the dried tissue of the fruit, tightly bonded to the seed coat. Different from the bean, the corn seed contains a large endosperm and a single cotyledon. The cotyledon absorbs the endosperm's nutrients during germination. Also unlike the bean, the embryonic root and shoot in corn each have a protective sheath.

? What is the role of the endosperm in a seed?

■ The endosperm provides nutrients to the developing embryo.

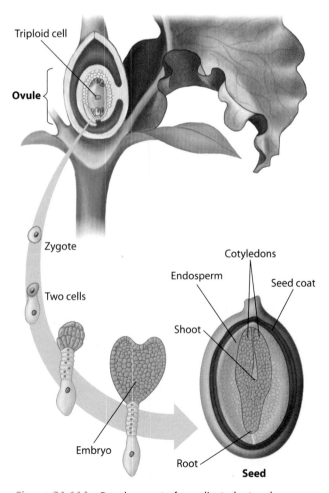

Figure 31.11A   Development of a eudicot plant embryo

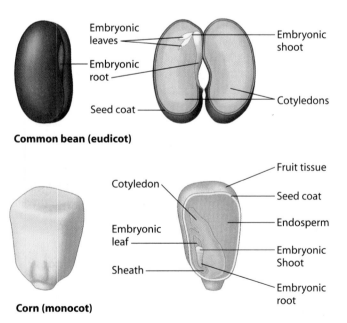

Common bean (eudicot)

Corn (monocot)

Figure 31.11B   Seed structure

31.12 The ovary develops into a fruit

In the previous two modules, we followed the angiosperm life cycle from the flower on the sporophyte plant through the transformation of an ovule into a seed. Simultaneous with seed development, fertilization triggers hormonal changes that cause a flower's ovary to mature into a **fruit**, a specialized vessel that houses and protects seeds and helps disperse them from the parent plant. A pea pod is a fruit, as is a peach, orange, tomato, cherry, or corn kernel.

The photographs in Figure 31.12A illustrate the changes in a pea plant leading to pod formation. ❶ Soon after pollination, ❷ the flower drops its petals, and the ovary starts to grow. The ovary expands tremendously, and its wall thickens, ❸ forming the pod, or fruit.

Figure 31.12B matches the parts of a pea flower with what they become in the pod. The wall of the ovary becomes the pod. The ovules, within the ovary, develop into the seeds. The small, threadlike structure at the end of the pod is what remains of the upper part of the flower's carpel. The sepals of the flower often stay attached to the base of the green pod. Peas are usually harvested at this stage of fruit development. If the pods are allowed to develop further, they become dry and brownish and will split open, releasing the seeds.

A fruit typically consists of a mature ovary, although it can include other flower parts as well. As seeds develop from ovules after fertilization, the wall of the ovary thickens. A pea pod is an example of a fruit, with seeds (mature ovules, the peas) encased in the ripened ovary (the pod). Fruits protect dormant seeds and aid in their dispersal.

Mature fruits can be either fleshy or dry. Oranges, plums, and grapes are examples of fleshy fruits, in which the wall of the ovary become soft during ripening. Dry fruits include beans, nuts, and grains. The dry, wind-dispersed fruits of grasses, harvested while on the plant, are major staple foods for humans. The cereal grains of wheat, rice, maize, and other grasses, though easily mistaken for seeds, are each actually a fruit with a

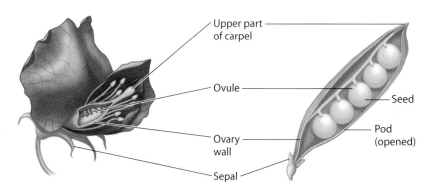

Figure 31.12B The correspondence between flower and fruit in the pea plant

Labels: Upper part of carpel · Ovule · Ovary wall · Sepal · Seed · Pod (opened)

dry outer covering (the former wall of the ovary) that adheres to the seed coat of the seed within.

As discussed in Module 17.10, various adaptations of fruits help disperse seeds (see Figures 17.10A–C). The seeds of some flowering plants, such as dandelions and maples, are contained within fruits that function like kites or propellers, adaptations that enhance dispersal by wind. Some fruits, such as coconuts, are adapted to dispersal by water. And many angiosperms rely on animals to carry seeds. Some of these plants have fruits modified as burrs that cling to animal fur (or the clothes of humans). Other angiosperms produce edible fruits, which are usually nutritious, sweet tasting, and vividly colored, advertising their ripeness. When an animal eats the fruit, it digests the fruit's fleshy part, but the tough seeds usually pass unharmed through the animal's digestive tract. Animals may deposit the seeds, along with a supply of fertilizer, kilometers from where the fruit was eaten.

Web Activity *Seed and Fruit Development*

? Seed is to _____ as _____ is to ovary.

■ ovule . . . fruit

Figure 31.12A Development of a fruit, a pea pod

31.13 Seed germination continues the life cycle

The germination of a seed is often used to symbolize the beginning of life, but as we have seen, the seed already contains a miniature plant, complete with embryonic root and shoot. Thus, at germination, the plant does not begin life but rather resumes the growth and development that was temporarily suspended during seed dormancy.

Germination usually begins when the seed takes up water. The hydrated seed expands, rupturing its coat. The inflow of water triggers metabolic changes in the embryo that make it start growing again. Enzymes begin digesting stored nutrients in the endosperm or cotyledons, and these nutrients are transported to the growing regions of the embryo.

The figures below trace germination in a eudicot (a garden bean) and a monocot (corn). In Figure 31.13A, notice that the embryonic root of a bean emerges first and grows downward from the germinating seed. Next, the embryonic shoot emerges, and a hook forms near its tip. The hook protects the delicate shoot tip by holding it downward, rather than pushing it up through the abrasive soil. As the shoot breaks through the soil surface, its tip is lifted gently out of the soil as exposure to light stimulates the hook to straighten. The first foliage leaves then expand from the shoot tip and begin making food by photosynthesis. The cotyledons emerge from the soil and become leaflike photosynthetic structures. In many other plants, such as peas, the cotyledons remain behind in the soil and decompose.

Corn and other monocots use a different mechanism for breaking ground at germination (Figure 31.13B). A protective sheath surrounding the shoot pushes upward and breaks through the soil. The shoot tip then grows up through the tunnel provided by the sheath. The corn cotyledon remains in the soil and decomposes.

In the wild, only a small fraction of fragile seedlings endure long enough to reproduce. Production of enormous numbers of seeds compensates for the odds against individual survival. Asexual reproduction, generally simpler and less hazardous for offspring than sexual reproduction, is an alternative means of plant propagation, as we see next.

Web Process of Science *What Tells Desert Seeds When to Germinate?*

? Which meristems provide additional cells for early growth of a seedling after germination?

■ The apical meristems of the shoot and root

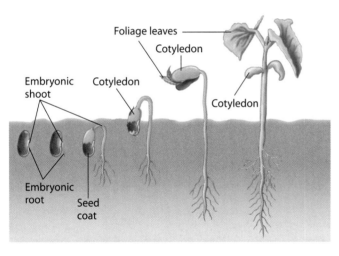

Figure 31.13A Pea germination (a eudicot)

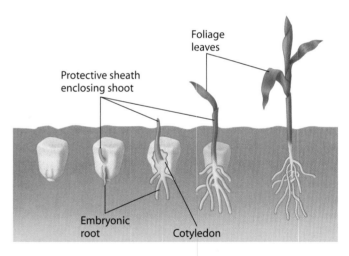

Figure 31.13B Corn germination (a monocot)

31.14 Asexual reproduction produces plant clones

Imagine chopping off your finger and watching it grow and develop into an exact copy of you. This would be an example of asexual reproduction (also called vegetative propagation in plants), in which offspring are derived from a single parent. The resulting asexually produced offspring, often called a **clone,** is genetically identical to its lone parent.

Asexual reproduction in angiosperms and other plants is an extension of their capacity to grow throughout life. A plant's meristematic tissues can sustain growth indefinitely. In addition, parenchyma cells throughout a plant can divide and differentiate into the various types of cells.

The photographs on the facing page show four examples of plant cloning. In nature, asexual reproduction in plants often involves **fragmentation,** the separation of a parent plant into parts that develop into whole plants. The garlic bulb (Figure 31.14A) is actually an underground stem that functions in storage. A single large bulb fragments into several parts, called cloves. Each clove can give rise to a separate plant, as indicated by the green shoots emerging from some of them. The white sheaths are leaves attached to the stem.

Each of the small trees you see in Figure 31.14B is a sprout from the roots of a coast redwood tree (discussed in the chapter introduction). Eventually, one or more of these root sprouts may take the place of its parent in the forest.

The ring of plants in Figure 31.14C is a clone of creosote bushes growing in the Mojave Desert in southern California. In this case, the word *clone* refers to a group of genetically identical organisms. All these bushes came from generations of asexual reproduction by roots. Making the oldest trees seem youthful, this clone apparently began with a single plant that germinated from a seed about 12,000 years ago. The original plant probably occupied the center of the ring.

Figure 31.14D shows a patch of dune grass in Massachusetts. Most grasses can propagate asexually by sprouting shoots and roots from runners. A small patch of grass can spread until it covers an acre or more of surface.

Many plants can reproduce both sexually and asexually. What advantages can asexual reproduction offer? For one thing, a parent plant well suited to its environment can clone many copies of itself, all of which would be equally well suited to current conditions. Also, the offspring may not be as fragile as seedlings produced by sexual reproduction. Both asexual reproduction and sexual reproduction have played important roles in the evolutionary adaptation of plant populations to their natural environments.

The ability of plants to reproduce asexually provides many opportunities for growers to produce large numbers of plants with minimal effort and expense. For example, most of our fruit trees and houseplants are asexually propagated from cuttings. Several other plants are propagated from root sprouts (for example, raspberries) or pieces of underground stems (such as potatoes).

Plants can also be propagated by test-tube methods. The laboratory tube in Figure 31.14E contains a geranium plantlet that was grown from a few meristem cells cut from a mature plant and cultured in a growth medium. Using this method, a single plant can be cloned into thousands of copies. Orchids and certain pine trees used for mass plantings are commonly propagated this way.

As discussed in Module 12.8, plant cell culture methods also enable researchers to grow plants from genetically engineered plant cells. Foreign genes are incorporated into a single parenchyma cell, and the cell is then cultured so that it multiplies and develops into a new plant. The resulting genetically modified (GM) plant may then be able to grow and reproduce normally. The commercial adoption by farmers of GM crops has been one of the most rapid cases of technology transfer in the history of agriculture, but as discussed in Module 12.9, it is not without controversy.

Figure 31.14E
Test-tube cloning

Aside from the issues raised by GM plants, modern agriculture faces some potentially serious problems. Nearly all of today's crop plants have very little genetic variability. Furthermore, we grow most crops in **monocultures,** large areas of land with a single plant variety. Given these conditions, plant scientists fear that a small number of diseases could devastate large crop areas. In response, plant breeders are working to maintain "gene banks," storage sites for seeds of many different plant varieties that can be used to breed new hybrids.

? Which mode of reproduction (sexual or asexual) would generally be more advantageous in a location where the composition of the soil is constantly changing? Why?

■ Sexual, because it generates genetic variation among the offspring, which enhances the potential for adaptation to a changing environment

Figure 31.14A Cloves of a garlic bulb

Figure 31.14B Sprouts from the roots of coast redwood trees

Figure 31.14C A ring of creosote bushes

Figure 31.14D Dune grass

31.15   Evolutionary adaptations allow some trees to live very long lives

We will end this chapter as it began: with a discussion of some extreme trees. In the chapter introduction, we discussed trees that are the *tallest* organisms on Earth. Here, we will discuss trees that are the *oldest*. The tree shown in Figure 31.15 is a 4,600-year-old bristlecone pine (*Pinus longaeva*). Bristlecone pines are found only in six western U.S. states (including the White Mountain region of California, pictured here). Another bristlecone pine, named Methuselah, is believed to be Earth's oldest living organism. Like the tallest trees, its location is kept secret to protect it.

Many other species of plants have life spans well beyond the longest-lived animals. A long life increases evolutionary fitness by increasing the number of reproductive opportunities. What evolutionary adaptations help trees to live so long? Adult trees, like most adult plants, retain meristem tissue (see Modules 31.7–8). Meristems allow for continued growth and repair throughout life. A tree thus has a remarkable ability to replace organs that have been lost or damaged by trauma or disease throughout its life. As we will see in the next chapter, the vascular (circulatory) system of a tree is decentralized, allowing part of a tree to survive damage and regrow. And, as we will see in Chapter 33, plants have a well-adapted hormonal control system that coordinates all these behaviors, and they produce defensive compounds that help protect them.

Figure 31.15   A bristlecone pine tree growing in California

? Why is the presence of meristems essential for the long life of many plants?

▪ Plant meristem can give rise to multiple types of cells throughout life.

Chapter Review

Reviewing the Concepts

Angiosperms, or flowering plants, are the most familiar and diverse group of plants **(Introduction).**

Plant Structure and Function (31.1–31.6)

Humans share a long and prosperous history with plants **(31.1).**

Plant structure.   The two main types of angiosperms, monocots and eudicots, differ in the number of seed leaves and the structure of roots, stems, leaves, and flowers **(31.2).** The structure of a flowering plant allows it to draw resources from both soil and air **(31.3):**

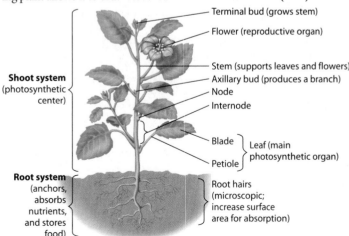

Terminal bud (grows stem)

Flower (reproductive organ)

Stem (supports leaves and flowers)
Axillary bud (produces a branch)
Node
Internode

Shoot system (photosynthetic center)

Blade ⎱ Leaf (main
Petiole ⎰ photosynthetic organ)

Root system (anchors, absorbs nutrients, and stores food)

Root hairs (microscopic; increase surface area for absorption)

Roots and stems may function in storing food, asexual reproduction, and protection **(31.4).**

Tissues and cells.   The five major types of plant cells are parenchyma, collenchyma, sclerenchyma (including fiber and sclereid cells), water-conducting cells (including tracheids and vessel elements), and food-conducting cells (sieve-tube members). Two kinds of vascular tissue are xylem, which conveys water and dissolved minerals, and phloem, which transports sugars **(31.5).** Roots, stems, and leaves are made up of three tissue systems: the dermal, vascular, and ground tissue systems. Dermal tissue covers and protects the plant. The vascular tissue system contains xylem and phloem. The ground tissue system consists of parenchyma cells and supportive collenchyma and sclerenchyma cells **(31.6).**

Plant Growth (31.7–31.8)

Meristems, areas of unspecialized, dividing cells, are where plant growth originates. Apical meristems at the tips of roots and in the terminal buds and axillary buds of shoots initiate primary (lengthwise) growth by producing new cells. A root or shoot lengthens further as the cells elongate and differentiate **(31.7).** An increase in a plant's girth, called secondary growth, arises from cell division in a cylindrical meristem called the vascular cambium. The vascular cambium thickens a stem by adding layers of secondary xylem, or wood, next to its inner surface. Outside the vascular cambium, the bark consists mainly of secondary phloem, cork cambium, and protective cork cells that are produced by the cork cambium (**31.8**).

Reproduction of Flowering Plants (31.9–31.15)

The angiosperm flower consists of sepals, petals, stamens, and carpels. Pollen grains develop in anthers, at the tips of stamens. The tip of the carpel, the stigma, receives pollen grains. The ovary, at the base of the carpel, houses the egg-producing structure, the ovule (**31.9**).

The angiosperm life cycle. Haploid spores are formed within ovules and anthers. The spores in the anthers give rise to male gametophytes—pollen grains—which produce sperm. A spore in an ovule produces the embryo sac, the female gametophyte. Each embryo sac contains an egg cell. Pollination is the arrival of pollen grains onto a stigma. A pollen tube grows into the ovule, and sperm pass through it and fertilize both the egg and a second cell. This process is called double fertilization (**31.10**). After fertilization, the ovule becomes a seed, and the fertilized egg within it divides and becomes an embryo. The other fertilized cell develops into the endosperm, which stores food for the embryo (**31.11**). The ovary develops into a fruit, which helps protect and disperse the seeds (**31.12**). A seed starts to germinate when it takes up water and expands. The embryo resumes growth and absorbs nutrients from the endosperm. An embryonic root emerges, and a shoot pushes upward and expands its leaves (**31.13**).

Asexual reproduction can be achieved via bulbs, sprouts, or runners. Propagating plants asexually from cuttings or bits of tissue can increase agricultural productivity but can also reduce genetic diversity (**31.14**). Asexual reproduction, as well as other adaptations such as meristem tissue, allow some trees to live very long lives (**31.15**).

Connecting the Concepts

1. Create a concept map or a detailed sketch that shows the relationships between the following parts of an angiosperm body: root system, root hairs, shoot system, leaf, petiole, blade, stem, node, internode, flower, stamen, carpel, sepal, petal, stigma, style, filament, ovary, ovule.

Testing Your Knowledge

Multiple Choice

2. Which of the following is closest to the center of a woody stem? (*Explain your answer.*)
   a. vascular cambium
   b. primary phloem
   c. secondary phloem
   d. primary xylem
   e. secondary xylem

3. A pea pod is formed from _____. A pea inside the pod is formed from _____.
   a. an ovule . . . a carpel
   b. an ovary . . . an ovule
   c. an ovary . . . a pollen grain
   d. an anther . . . an ovule
   e. endosperm . . . an ovary

4. While walking in the woods, you encounter an unfamiliar nonwoody flowering plant. If you want to know whether it is a monocot or eudicot, it would *not* help to look at the
   a. number of seed leaves, or cotyledons, present in its seeds.
   b. shape of its root system.
   c. number of petals in its flowers.
   d. arrangement of vascular bundles in its stem.
   e. size of the plant.

5. In angiosperms, each pollen grain produces two sperm. What do these sperm do?
   a. Each one fertilizes a separate egg cell.
   b. One fertilizes an egg, and the other fertilizes the fruit.
   c. One fertilizes an egg, and the other is kept in reserve.
   d. Both fertilize a single egg cell.
   e. One fertilizes an egg, and the other fertilizes a cell that develops into stored food.

Matching

6. Attracts pollinator
7. Develops into seed
8. Protects flower before it opens
9. Produces sperm
10. Produces pollen
11. Houses ovules

   a. pollen grain
   b. ovule
   c. anther
   d. ovary
   e. sepal
   f. petal

Describing, Comparing, and Explaining

12. The scent of apple blossoms and the buzzing of bees fill an orchard on a warm spring day. Describe the processes by which the pollen carried from flower to flower by the bees results in the apple you might pick in the fall.

13. Name three kinds of asexual reproduction. Explain two advantages of asexual reproduction over sexual reproduction. What is the primary drawback?

14. What part of a plant are you eating when you consume each of the following: tomato, celery stalk, peanut, strawberry, lettuce, artichoke, beet?

Applying the Concepts

15. Plant scientists are searching Peru, Mexico, and the Middle East for the wild ancestors of potatoes, corn, and wheat. Why is this search important?

16. Tropical forests contain a wealth of plants that are potential new sources of food, as well as sources of medicine and other useful products. The developing nations of the tropics cannot develop these resources themselves. Under pressure from growing populations and debt, they are cutting their forests for lumber and farmland, and many species are disappearing. Developed countries are pressuring the tropical countries to protect the forests before even more species are lost. Many people in the developing nations see little incentive to preserve the forests only to have corporations from industrialized countries profit from new products obtained from the forests. Is there a way to preserve the tropical forests so that both the developed and developing nations will benefit from their abundance?

Answers to all questions can be found in Appendix 4.

For Practice Quizzes, BioFlix, MP3 Tutors, and Activities, go to www.mybiology.com.

Planting Hope
in the Wake of Katrina

In the early morning of August 29, 2005, Hurricane Katrina—one of the strongest storms ever recorded—slammed into the Gulf coast of Louisiana and Mississippi. By the time it blew past, Katrina was the costliest natural disaster in U.S. history and the deadliest hurricane in 75 years, responsible for over 1,800 deaths.

The city of New Orleans was particularly hard hit. After several levees breached during the storm surge, about 80% of the city was flooded. Some areas remained underwater for over six weeks. The flood waters were a noxious stew containing raw sewage, heavy metals, toxic chemicals, and petroleum. The effect on the local environment was devastating, particularly to the topsoil in much of the city.

Cleanup efforts will take years. But in addition to the bulldozers and wrecking crews, help is coming from a natural source: plants. The use of plants to help clean up polluted soil and groundwater is called phytoremediation. Phytoremediation is one type of bioremediation, the use of living organisms (often prokaryotes and protists) to detoxify polluted sites.

Plants can efficiently move minerals and other compounds from the soil into the roots and eventually through the body of the plant. Most often, the compounds absorbed are nutrients that help the plant grow. But many species of plants are metal accumulators, capable of absorbing large amounts of lead, zinc, and other heavy metals. And some plant species are hyperaccumulators, able to absorb metals up to concentrations that would be deadly to typical plants. Such plants are ideally suited for detoxifying the soil because the heavy metals removed from the soil end up concentrated in the plants' own cells.

In New Orleans, a grass roots relief organization called Common Ground is using phytoremediation to help restore soil around devastated homes. Volunteers have planted tens of thousands of sunflowers at dozens of home sites in some of New Orleans' hardest hit wards. Their goal is to use the sunflowers (and other metal-accumulating plants, such as Indian mustard) to remove heavy metals, especially lead, from the soil. Once fully grown, the phytoremediation crops are harvested and hauled away to a hazardous-waste landfill. Because cleanup crews have to remove only a few cubic meters of plant material rather than thousands of cubic meters of contaminated soil, phytoremediation is a cost-effective and efficient way to clean the soil. Phytoremediation can also detoxify an area without dramatically disturbing the landscape. And sunflowers provide an uplifting message to the community, bringing beauty and life back into an area weary with grief and loss.

Common Ground's project is based on previously successful phytoremediation methods. For example, sunflowers and Indian mustard have been used to reduce lead contamination near an automobile factory in Detroit by 43%, down to acceptable levels. And sunflowers have helped clean up one of the most dangerous types of toxic substances: radioactive metals. In a contaminated pond near the destroyed nuclear power plant in Chernobyl (see Module 2.5), sunflower plants were set adrift on foam rafts. Within days, the concentrations of two radioactive metals, strontium-90 and cesium-137, reached levels in the sunflower roots several thousand times higher than in the water.

Despite its promise, phytoremediation is not without problems. Besides having to dispose of the plants, some researchers are concerned that some toxins absorbed by plants can evaporate from leaves and contaminate the air. They also worry about animals eating toxin-laden plants. However, they are encouraged by studies indicating that at least some animals avoid plants containing high concentrations of toxins. (This work suggests why toxin accumulation in plants may have evolved in the first place—as protection from hungry animals.) The biggest limita-

tion of phytoremediation is its slowness. While soil can be hauled away in a matter of days, albeit at high cost, plants can take months to grow. Furthermore, remediating crops must often be planted for multiple seasons to reduce soil toxicity to acceptable levels.

In using plants to clean up toxic wastes, we are benefiting from millions of years of plant evolution. Unable to move about in search of food, plants have evolved amazing abilities to pull water and nutrients out of the soil and air. While lead and other toxins are not normally considered plant nutrients, it's obvious that some plant species are adapted to survive, and even thrive, in their presence. In this chapter, we'll see how plants obtain essential nutrients and how they transport them throughout their roots, stems, and leaves. ■ ■ ■

32.1 Plants acquire their nutrients from soil and air

Watch a plant grow from a tiny seed, and you can't help wondering where all the mass comes from. Aristotle thought that soil provided all the substance for plant growth. The 17th-century physician Jan Baptista van Helmont performed an experiment to test this hypothesis. He planted a willow seedling in a pot containing 91 kg of soil. After five years, the willow had grown into a tree weighing 76.8 kg, but only 0.06 kg of soil had disappeared from the pot. Van Helmont concluded that the willow had grown mainly from added water. A century later, an English physiologist named Stephen Hales postulated that plants are nourished mostly by air.

As it turns out, there is some truth in all these early hypotheses about plant nutrition; air, water, and soil all contribute to plant growth (Figure 32.1A). A plant's leaves absorb carbon dioxide (CO_2) from the air; in fact, about 96% of a plant's dry weight is organic (carbon-containing) material built mainly from CO_2. The figure also points out that a plant gets water (H_2O), minerals, and some oxygen (O_2) from the soil.

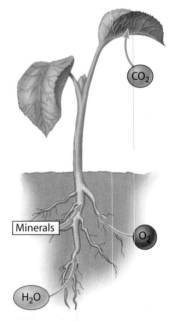

Figure 32.1A The uptake of nutrients by a plant

What happens to the materials a plant takes up from the air and soil? The sugars a plant makes by photosynthesis are composed of the elements carbon, oxygen, and hydrogen. In Chapter 7, we saw that the carbon and oxygen used in photosynthesis come from atmospheric CO_2 and that the hydrogen comes from water molecules. Plant cells use the sugars made by photosynthesis in constructing all the other organic materials they need, primarily carbohydrates. The giant trunks of the redwood trees in Figure 32.1B, for instance, consist mainly of sugar derivatives, such as the cellulose of cell walls.

Plants use cellular respiration to break down some of the sugars they make, obtaining energy from them in a process that consumes O_2. A plant's leaves take up some O_2 from the air, but we do not show this in Figure 32.1A because plants are actually net producers of O_2, giving off more of this gas than they use. When water is split during photosynthesis, O_2 gas is produced and released through the leaves. The O_2 being taken up from the soil by the plant's roots in Figure 32.1A is actually atmospheric O_2 that has diffused into the soil; it is used in cellular respiration in the roots themselves.

What does a plant do with the minerals it absorbs from the soil? A partial answer is provided by looking at three elements that plant roots take up as inorganic ions: nitrogen, magnesium, and phosphorus. Nitrogen is a component of all nucleic acids, all proteins, ATP, and many plant hormones and coenzymes. Nitrogen and magnesium are both components of chlorophyll, the plant's key light-absorbing molecule. Phosphorus is a major component of nucleic acids, phospholipids, and ATP.

A plant's ability to move water from its roots to its leaves and its ability to deliver sugars to specific areas of its body are staggering feats of evolutionary engineering. Figure 32.1B highlights the distance between the bottom of a tree and its leaves; the roots of a redwood can be over 100 m (330 feet) below the topmost leaves! In the next four modules, we follow the movement of water, dissolved mineral nutrients, and sugar throughout the plant body.

? Plants require inorganic nutrients from the air and soil. What inorganic nutrient is obtained in the greatest quantities from each source?

■ Carbon dioxide from the air and water from the soil

Figure 32.1B Redwood trees, giant products of photosynthesis

32.2 The plasma membranes of root cells control solute uptake

With its surface area enormously expanded by thousands of root hairs (Figure 32.2A), a plant root has a remarkable ability to extract water and other materials from soil. Recall from Module 31.3 that root hairs are extensions of epidermal cells that cover the root. Root hairs form a huge surface area; for example, the root hairs of a single sunflower plant, if laid end to end, could stretch for miles. All this surface area in contact with nutrient-containing soil allows a plant to absorb the water and minerals it needs for growth.

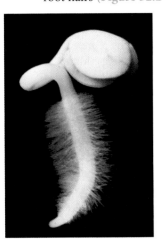

Figure 32.2A
Root hairs of radish seedling

All substances that enter a plant root are in solution (that is, dissolved in water). For water and solutes to be transported from the soil throughout the plant, they must move through the epidermis and cortex of the root and then into the water-conducting xylem tissue in the central cylinder of the root. Any route the water and solutes take from the soil to the xylem requires that they pass through some of the plasma membranes of the root cells. Because plasma membranes are selectively permeable, only certain solutes can enter the xylem. This selectivity helps regulate the mineral composition of a plant's vascular system.

You can see two possible routes to the xylem in the bottom part of Figure 32.2B. The blue arrows indicate an *intracellular* route. Water and selected solutes cross the cell wall and plasma membrane of an epidermal cell (usually at a root hair). The cells within the root are all interconnected by plasmodesmata (channels through the walls of adjacent cells); there is a continuum of living cytoplasm among the root cells. Therefore, once inside the epidermal cell, the solution can move inward from cell to cell without crossing any other plasma membranes, diffusing through the interconnected cytoplasm all the way into the root's endodermis. An endodermal cell then discharges the solution into the xylem (purple).

The red arrows indicate an alternative route. This route is *extracellular*; the solution moves inward within the hydrophilic walls and extracellular spaces of the root cells but does not enter the cytoplasm of the epidermis or cortex cells. The solution crosses no plasma membranes, and there is no selection of solutes until they reach the endodermis.

Here, a continuous waxy barrier called the **Casparian strip** stops water and solutes from entering the xylem via cell walls. The Casparian strip forces water and ions that travel the extracellular (red) route to cross a plasma membrane into an endodermal cell. Ion selection occurs at this membrane instead of in the epidermis, and once the selected solutes and water are in the endodermal cell, they can be discharged into the xylem.

Actually, water and solutes rarely follow just the two kinds of routes in Figure 32.2B. In a real plant, they may take any combination of these routes, and they may pass through numerous plasma membranes and cell walls en route to the xylem. Because of the Casparian strip, however, there are no nonselective routes; the water and solutes must cross a selectively permeable plasma membrane at some point. Next, we see how water and minerals move upward within the xylem from roots to shoots.

? What is the function of the Casparian strip?

■ It regulates the passage of minerals (inorganic ions) into the xylem by blocking access via cell walls and requiring all minerals to cross a selectively permeable plasma membrane.

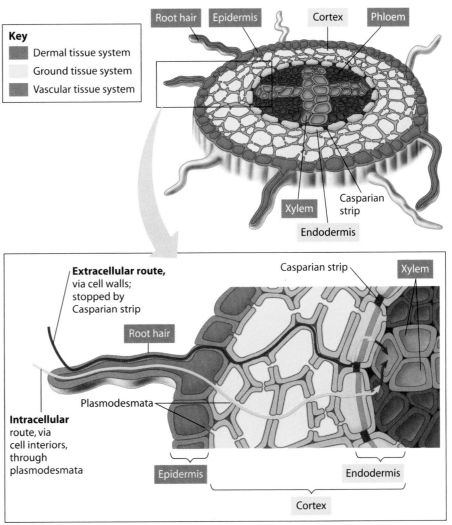

Key
- ■ Dermal tissue system
- ■ Ground tissue system
- ■ Vascular tissue system

Root hair · Epidermis · Cortex · Phloem

Xylem · Casparian strip · Endodermis

Extracellular route, via cell walls; stopped by Casparian strip

Root hair

Casparian strip · Xylem

Plasmodesmata

Intracellular route, via cell interiors, through plasmodesmata

Epidermis · Endodermis

Cortex

Figure 32.2B   Routes of water and solutes from soil to root xylem

32.3 Transpiration pulls water up xylem vessels

As a plant grows upward toward sunlight, it needs to extract an increasing supply of water and dissolved mineral ions from the soil. To thrive, a plant must be able to transport these resources from its roots to the rest of the plant.

We saw in Figure 31.6E that xylem tissue of angiosperms includes two types of conducting cells: tracheids and vessel elements. When mature, both types of cells are dead, consisting only of cell walls, and both are in the form of very thin tubes that are arranged end to end. Because the cells have openings in their ends, a solution of water and inorganic nutrients, called **xylem sap,** can flow through these tubes. Xylem sap flows all the way up from the plant's roots through the shoot system to the tips of the leaves.

What force moves xylem sap up against the downward pull of gravity? Is it pushed upward or pulled upward? Plant biologists have found that the roots of some plants do exert a slight upward push on xylem sap. The root cells actively pump inorganic ions into the xylem, and the root's endodermis holds the ions there.

As ions accumulate in the xylem, water tends to enter by osmosis, pushing xylem sap upward ahead of it. This force, called **root pressure,** can push xylem sap up a few meters.

For the most part, however, xylem sap is not pushed from below by root pressure but pulled upward by the leaves. Plant biologists have determined that the pulling force is **transpiration,** which is the loss of water from the leaves and other aerial parts of a plant.

Figure 32.3 illustrates transpiration and its effect on water movement in a tree. ❶ Water molecules (blue arrows) leave the leaf through a stoma, a microscopic pore on the surface of a leaf (see also Figures 31.5 and 32.4). When the stoma is open, water diffuses out of the leaves because the surrounding air is usually drier than the inside of the leaf. Under certain atmospheric conditions, this exiting moisture may condense to form dew.

Transpiration can pull xylem sap up the tree because of two special properties of water: cohesion and adhesion. Both of these properties arise from the polarity of water molecules (see Modules 2.9–2.11). **Cohesion** is the sticking together of molecules of the same kind. ❷ In the case of water, hydrogen bonds make the H_2O molecules stick to one another. ❸ The cohering water molecules in the xylem tubes form continuous strings, extending all the way from the leaves down to the roots. In contrast, **adhesion** is the sticking together of molecules of different kinds. ❹ Water molecules tend to adhere via hydrogen bonds to hydrophilic cellulose molecules in the walls of xylem cells.

What effect does transpiration have on a string of water molecules that tend to adhere to the walls of xylem tubes? Before a water molecule can leave the leaf, it must break off from the end of the string. In effect, it is pulled off by a steep diffusion gradient between the moist interior of the leaf and the drier surrounding air. Cohesion resists the pulling force of the diffusion gradient, but it is not strong enough to overcome it. The molecule breaks off, and the opposing forces of cohesion and transpiration put tension on the remainder of the string of water molecules. As long as transpiration continues, the string is kept tense and is pulled upward as one molecule exits the leaf and the one right behind it is tugged up into its place. Adhesion of the water molecules to the walls of the xylem cells

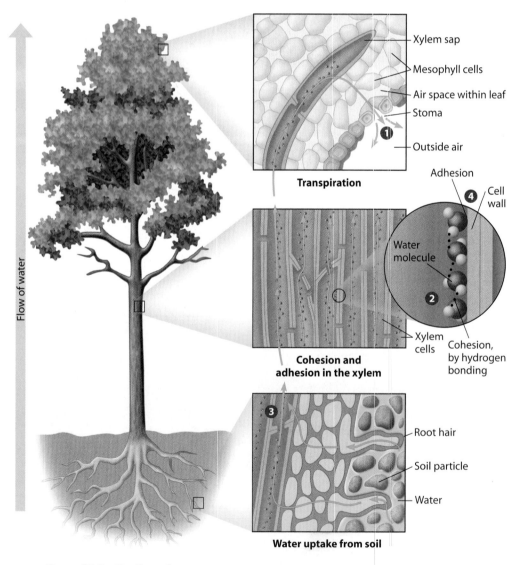

Flow of water

Xylem sap
Mesophyll cells
Air space within leaf
Stoma
Outside air

Transpiration

Adhesion
❹ Cell wall
Water molecule
Xylem cells
Cohesion, by hydrogen bonding

Cohesion and adhesion in the xylem

Root hair
Soil particle
Water

Water uptake from soil

Figure 32.3 The flow of water up a tree

assists the upward movement of the xylem sap by counteracting the downward pull of gravity.

Plant biologists call this explanation for the ascent of xylem sap the **transpiration-cohesion-tension mechanism.** We can summarize it as follows: Transpiration exerts a pull that is relayed downward along a string of water molecules held together by cohesion and helped upward by adhesion. Transpiration is an efficient means of moving large volumes of water upward from roots to shoots. A single corn plant, for example, transpires 125 L of water in a growing season. And yet, all this transport of xylem sap requires no energy expenditure by the plant. Physical properties—cohesion, adhesion, and evaporation—move water and dissolved minerals from a plant's roots to its shoots.

🎧 MP3 Tutor *Transposition*

Web Process of Science *How Is the Rate of Transpiration Calculated?*

? Describe the role of cohesion and adhesion in the ascent of xylem sap.

■ Cohesion maintains a continuous string of water during transpiration; adhesion helps to support xylem sap against the downward pull of gravity.

32.4 Guard cells control transpiration

Adaptations that increase photosynthesis—such as large leaf surface areas—have the serious drawback of increasing water loss by transpiration. Viewed this way, a plant's tremendous requirement for water is part of the cost of making food by photosynthesis. As long as water moves up from the soil fast enough to replace the water that is lost, transpiration presents no problem. But if the soil dries out and transpiration exceeds the delivery of water to the leaves, the leaves will wilt. Unless the soil and leaves are rehydrated, the plant will eventually die.

The leaf stomata, which can open and close, are adaptations that help plants regulate their water content and adjust to changing environmental conditions. As shown in Figure 32.4, a pair of guard cells flank each stoma. The guard cells control the opening of a stoma by changing shape, widening or narrowing the gap between the two cells.

What actually causes guard cells to change shape and thereby open or close stomata? Figure 32.4 illustrates the principle. A stoma opens (left) when its guard cells gain potassium ions (K^+, red dots) and water (blue arrows) from neighboring cells (shown in light gray). The cells actively take up K^+, and water then enters by osmosis. (For a review of osmosis, see Module 5.4.) When the vacuoles in the guard cells gain water, the cells become more turgid and bowed. The cell wall of a guard cell is not uniformly thick, as you can see in the drawing, and the cellulose molecules are oriented in such a way that the cell buckles away from its companion guard cell when it is turgid. The result is an increase in the size of the gap (stoma) between the two cells. Conversely, when the guard cells lose K^+, they also lose water by osmosis and become flaccid and less bowed, closing the space between them (right).

In general, guard cells keep the stomata open during the day and closed at night. During the day, CO_2 can enter the leaf from the atmosphere and thus keep photosynthesis going when sunlight is available. At night, when there is no light for photosynthesis and therefore no need to take up CO_2, the closed stomata save water.

At least three cues contribute to stomatal opening at dawn. One is sunlight, which stimulates guard cells to accumulate K^+ and become turgid. A low level of CO_2 in the leaf can have the same effect. A third cue is an internal timing mechanism—a biological clock—found in the guard cells; even if you keep a plant in a dark closet, stomata will continue their daily rhythm of opening and closing. (We'll return to biological clocks in plants in Module 33.10.)

Even during the day, the guard cells may close the stomata if the plant is losing water too fast. This response reduces further water loss and may prevent wilting, but it also slows down CO_2 uptake and photosynthesis—one reason that droughts reduce crop yields. In summary, guard cells arbitrate the photosynthesis-transpiration compromise on a moment-to-moment basis by integrating a variety of stimuli.

Web Activity *Transpiration*

? Some leaf molds, fungi that parasitize plants, secrete a chemical that causes guard cells to accumulate K^+. How does this adaptation help the mold infect the plant?

■ Accumulation of K^+ by guard cells results in osmotic water uptake, and the turgid condition of the cells keeps the stomata open. The mold can then grow into the leaf interior via the stomata.

Figure 32.4   How guard cells control stomata

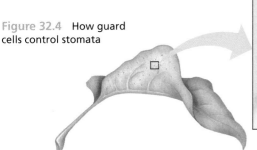

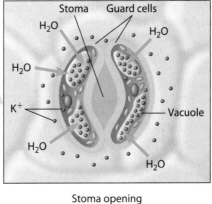

Stoma opening

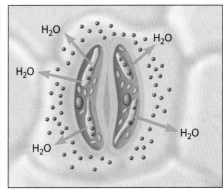

Stoma closing

32.5 Phloem transports sugars

A plant has two separate transport systems: xylem (the topic of Module 32.3) and phloem. Xylem transports xylem sap (water and dissolved minerals), while the main function of phloem is to transport the products of photosynthesis. In angiosperms, phloem contains food-conducting cells called sieve-tube members arranged end to end as tubes (see Figure 31.6F). The micrograph in Figure 32.5A shows two sieve-tube members and the sieve plate between them. Through the perforations in sieve plates, a sugary solution called **phloem sap** moves freely from one cell to the next (the cytoplasms of these living cells are continuous). Phloem sap may contain inorganic ions, amino acids, and hormones, but its main solute is usually the disaccharide sucrose. When highly concentrated by boiling, the sucrose-rich phloem sap produced by certain varieties of North American maple trees is better known as maple syrup.

In contrast to xylem sap, which only flows upward from the roots, phloem sap moves throughout the plant in various directions. However, sieve tubes always carry sugars from a sugar source to a sugar sink. A **sugar source** is a plant organ that is a net producer of sugar, by photosynthesis or by breakdown of starch. Leaves are the primary sugar sources in most mature plants. A **sugar sink** is an organ that is a net consumer or storer of sugar. Growing roots, buds, stems, and fruits are sugar sinks. A storage organ, such as a tuber or a bulb, may be a source or

sink, depending on the season. When stockpiling carbohydrates in the summer, it is a sugar sink. After breaking dormancy in the spring, it is a source as its starch is broken down to sugar, which is carried to the growing tips of the plant. Thus, each food-conducting tube in phloem tissue has a source end and a sink end, but these may change with the season or the developmental stage of the plant.

What causes phloem sap to flow from a sugar source to a sugar sink? Flow rates may be as high as 1 m/hr, which is much too fast to be accounted for by diffusion. (By diffusion alone, phloem sap travels less than 1 meter per year.) Plant biologists have tested a number of hypotheses for phloem sap movement. A hypothesis called the **pressure flow mechanism** is now widely accepted for angiosperms. Figure 32.5B, on the facing page, illustrates how this works, using a beet plant as an example. The pink dots in the phloem tube represent sugar molecules; notice their concentration gradient from top to bottom. The blue color represents a parallel gradient of water (hydrostatic) pressure in the phloem sap.

At the sugar source (leaves, in this example), ❶ sugar is loaded into a phloem tube by active transport. Sugar loading at the source end raises the solute concentration inside the phloem tube. ❷ The high solute concentration draws water into the tube by osmosis. The inward flow of water raises the water pressure at the source end of the tube.

At the sugar sink (the beet root, in this case), both sugar and water leave the phloem tube. ❸ As sugar departs from the phloem, ❹ water follows by osmosis. The exit of sugar lowers the sugar concentration in the sink end; the exit of water lowers the hydrostatic pressure in the tube. The building of water pressure at the source end and the reduction of that pressure at the sink end cause water to flow from source to sink—down a gradient of hydrostatic pressure. Since the sugar is dissolved in the water and the sieve plates allow free movement of solutes as well as water, the sugar is carried along from source to sink at the same rate as the water. As indicated on the right side of Figure 32.5B, xylem tubes recycle the water back from sink to source.

The pressure flow mechanism explains why phloem sap always flows from a sugar source to a sugar sink, regardless of their locations in the plant. However, the mechanism is somewhat difficult to test because most experimental procedures disrupt the structure and function of the phloem tubes. Some of the most interesting studies have taken advantage of natural phloem probes: insects called aphids, which feed on phloem sap.

The three photographs in Figure 32.5C show how plant biologists have used aphids to study phloem sap. On the left, an aphid feeds by inserting its needlelike mouthpart, called a stylet, into the phloem of a tree branch. The aphid is releasing from its anus a drop of so-called honeydew—actually, a tiny amount of phloem sap lacking some solutes that the insect's digestive tract has removed for food. The micrograph in the center shows an aphid's stylet inserted into one of the plant's food-conducting cells. The pressure within the phloem force-feeds the aphid, swelling it to several times its original size.

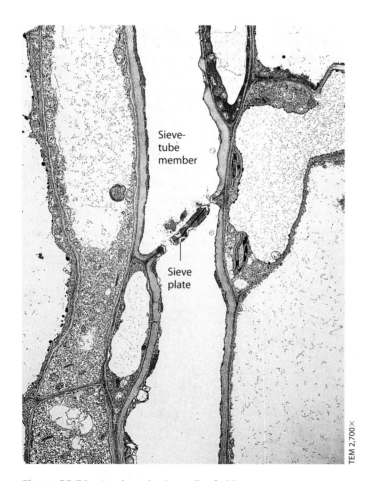

Sieve-tube member

Sieve plate

TEM 2,700×

Figure 32.5A   Food-conducting cells of phloem

While the aphid is feeding, it can be anesthetized and severed from its stylet. The stylet then serves the researcher as a miniature tap that drips phloem sap for hours. The photograph on the right shows a droplet of phloem sap on the cut end of a stylet. Studies using this technique support the pressure flow model: The closer the stylet is to a sugar source, the faster the sap flows out and the greater its sugar concentration. This is what we would expect if pressure is generated at the source end of the phloem tube by the active pumping of sugar into the tube.

Web Activity *Transport in Phloem*

? Contrast the forces that move phloem sap with the forces that move xylem sap.

■ Pressure is generated at the source end of a sieve tube by the loading of sugar and the resulting osmotic flow of water into the phloem. This pressure pushes phloem sap from the source end to the sink end of the tube. In contrast, transpiration generates a pulling force that drives the ascent of xylem sap.

We now have a broad picture of how a plant transports materials from one part of its body to another. Water and inorganic ions enter from the soil and are distributed by xylem. The xylem sap is pulled upward by transpiration. Carbon dioxide enters the plant through leaf stomata and is converted into sugars in the leaves. A second transport system, phloem, distributes the sugars. Pressure flow drives the phloem sap from leaves and storage sites to other parts of the plant, where the sugars are used or stored.

In Chapter 7, we discussed how plants convert raw materials into organic molecules by photosynthesis. We have yet to say much about the kinds of inorganic nutrients a plant needs and what it does with them. This is the subject of plant nutrition, which we discuss in the next section.

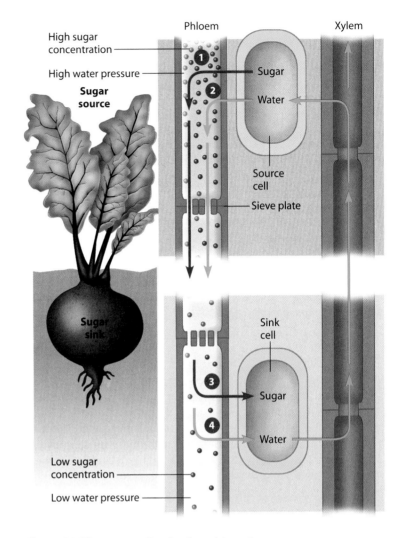

Figure 32.5B Pressure flow in plant phloem from a sugar source to a sugar sink (and the return of water to the source via xylem)

Aphid feeding on a small branch

Honeydew droplet

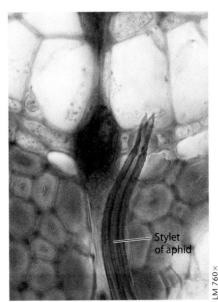

Stylet of aphid

LM 760×

Severed stylet dripping phloem sap

Figure 32.5C Tapping phloem sap with the help of an aphid

32.6 Plant health depends on a complete diet of essential inorganic nutrients

In contrast to animals, which require a complex diet of organic (carbon-containing) foods, plants survive and grow solely on inorganic substances (that is, plants are autotrophs; see Module 7.1). The ability of plants to assimilate CO_2 from the air, extract water and inorganic ions from the soil, and synthesize organic compounds is essential not only to the survival of plants but also to the survival of humans and other animals.

A chemical element is considered an **essential element** if a plant must obtain it to complete its life cycle—that is, to grow from a seed and produce another generation of seeds. A method called hydroponic culture can be used to determine which chemical elements are essential nutrients. As shown in Figure 32.6, plants are grown without soil by bathing the roots in mineral solutions. Air is bubbled into the water to give the roots oxygen for cellular respiration. By omitting a particular element, such as potassium, from the medium, a researcher can test whether that element is essential to the plant.

If the element left out of the solution is an essential nutrient, then the incomplete medium will make the plant abnormal in appearance compared to control plants grown on a complete nutrient medium. The most common symptoms of a nutrient deficiency are stunted growth and discolored leaves. Hydroponic culture studies have helped identify 17 essential elements needed by all plants. Most research has involved crop plants and houseplants; little is known about the nutritional needs of uncultivated plants.

Of the 17 essential elements, nine are called **macronutrients** because plants require relatively large amounts of them. Six of the nine macronutrients—carbon, oxygen, hydrogen, nitrogen, sulfur, and phosphorus—are the major ingredients of organic compounds forming the structure of a plant. These six elements make up almost 98% of a plant's dry weight. The other three macronutrients—potassium, calcium, and magnesium—make up another 1.5%.

How does a plant use calcium, potassium, and magnesium? Calcium has several functions. For example, it is important in the formation of cell walls, and it combines with certain proteins to form a glue that holds plant cells together in tissues. Calcium also helps maintain the structure of cell membranes and helps regulate their selective permeability. Potassium is crucial as a cofactor required for the activity of several enzymes. (Recall from Module 5.15 that a cofactor is an atom or molecule that cooperates with an enzyme in catalyzing a reaction.) Potassium is also the main solute for osmotic regulation in plants; we saw in Module 32.4 how potassium ion movements regulate the opening and closing of stomata. Magnesium is a component of chlorophyll and thus essential for photosynthesis. Magnesium is also a cofactor for several enzymes.

Elements that plants need in very small amounts are called **micronutrients.** The eight known micronutrients are chlorine, iron, manganese, boron, zinc, copper, nickel, and molybdenum. Micronutrients function in plants mainly as cofactors. Iron, for example, is a component of cytochromes, proteins in the electron transport chains of chloroplasts and mitochondria. Because micronutrients function mainly in catalysis (and are therefore used over and over), plants need only minute quantities of these elements. The requirement for molybdenum, for example, is so modest that there is only one atom of this rare element for every 60 million atoms of hydrogen in dried plant material. Yet a deficiency of molybdenum or any other micronutrient can weaken or kill a plant.

? You conduct an experiment like the one in Figure 32.6 to test whether a certain plant species requires a particular chemical element as a micronutrient. Why is it important that the glassware be completely clean?

Complete solution containing all minerals (control) Solution lacking potassium (experimental)

Figure 32.6 A hydroponic culture experiment

■ Because micronutrients are required in only minuscule amounts, even the smallest amount of dirt in the experimental flask may contain enough of the element you are testing to allow normal growth and invalidate your results.

32.7   Fertilizers can help prevent nutrient deficiencies

The quality of soil—especially the availability of nutrients—determines the health of a growing plant and the quality of our own nutrition. Figure 32.7A shows two experimental corn crops. The plants on the left are growing in soil rich in nitrogen. The small, lighter-colored plants on the right are growing in nitrogen-deficient soil. Even if the nitrogen-deficient plants produce grain, the crop will have a lower nutritional value, and its nutrient deficiences will then be passed on to livestock or human consumers.

Fortunately, the symptoms of nutrient deficiency are often distinctive enough for a grower to diagnose its cause. Many growers make visual diagnoses and then check their conclusions by having soil and plant samples chemically analyzed at a state or local laboratory.

Nitrogen shortage is the most common nutritional problem for plants. Soils are usually not deficient in nitrogen, but they are often deficient in the nitrogen compounds that plants can use: dissolved nitrate ions (NO_3^-) and ammonium ions (NH_4^+). Stunted growth and yellow-green leaves, starting at the tips of older leaves, are signs of nitrogen deficiency (Figure 32.7B). Other common nutrient shortfalls in plants include phosophorus and potassium deficiencies.

Once a diagnosis of a nutrient deficiency is made, treating the problem is usually simple. **Fertilizers** are compounds given to plants to promote growth. Fertilizers come in two basic types. Organic fertilizers are composed of chemically complex organic matter; inorganic fertilizers contain simple, inorganic minerals.

One source of organic fertilizer commonly used by home gardeners is **compost,** a soil-like mixture of decomposed organic matter. Many gardeners maintain a free-standing compost pile or an enclosed compost bin to which they add leaves, grass clippings, yard waste, and kitchen scraps (but not most animal products, such as meat, fat, bone, or animal droppings). Over time, the vegetable matter breaks down due to the action of naturally occurring microbes, fungi, and animals (Figure 32.7C). Occasional turning and watering will speed the composting process, producing homegrown fertilizer in several months. The compost can then be applied to outdoor gardens or indoor pots.

Inorganic fertilizers (also called mineral fertilizers) can contain naturally occuring inorganic compounds (such as mined limestone or phosphate rock) or synthethic inorganic compounds (such as ammonium nitrate). Inorganic fertilizers come in a wide variety of formulations, but most emphasize their "N-P-K ratio," the amounts of the three nutrients most often deficient in depleted soils: nitrogen (N), phosphorus (P), and potassium (K). For example, a 100-pound bag of 5–6–7 fertilizer contains 5 pounds of nitrogen (often as ammonium or nitrate), 6 pounds of phosphorus (as phosphoric acid), and 7 pounds of potassium (as the mineral potash), plus 82 pounds of filler. Many crops benefit from an all-purpose 5–5–5 formula, but some plants thrive only under special fertilizer formulations.

Figure 32.7A   The effect of nitrogen availability on corn growth: corn grown in nitrogen-rich soil (left) and nitrogen-poor soil (right)

Figure 32.7B   Nitrogen deficiency in a tomato leaf

Figure 32.7C   Steam produced by the metabolic activity of organisms within a compost pile

? (a) What is the most common nutrient deficiency in plants? (b) What are the signs?

■ Nitrogen deficiency; stunted growth and yellowing leaves

32.8 Fertile soil supports plant growth

Along with climate, the major factor determining whether a plant can grow well in a particular location is the quality of the soil. Fertile soil can support abundant plant growth by providing conditions that enable plant roots to absorb water and dissolved nutrients.

Different layers of soil are visible in a road cut or deep hole, such as the cross section shown in Figure 32.8A. You can see three distinct soil layers, called horizons, in the cut. The A horizon, or **topsoil,** is subject to extensive weathering (freezing, drying, and erosion, for example). Topsoil is a mixture of rock particles of various sizes, living organisms, and **humus,** the remains of partially decayed organic material. The rock particles provide a large surface area that retains water and minerals while also forming air spaces containing oxygen that can diffuse into plant roots. Fertile topsoil is home to an astonishing number and variety of bacteria, algae and other protists, fungi, and small animals such as earthworms, roundworms, and burrowing insects. Along with plant roots, these organisms loosen and aerate the soil and contribute organic matter to the soil as they live and die. Nearly all plants depend on bacteria and fungi in the soil to break down organic matter into inorganic molecules that roots can absorb. Besides providing nutrients, humus also tends to retain water while keeping the topsoil porous enough for good aeration of the plant roots. Topsoil is rich in organic materials and is therefore most important for plant growth. Plant roots branch out in the A horizon and usually extend into the next layer, the B horizon.

The soil's B horizon contains many fewer organisms and much less organic matter than the topsoil and is less subject to weathering. Fine clay particles and nutrients dissolved in soil water drain down from the topsoil and often accumulate in the B horizon. Below the B horizon, the C horizon is composed mainly of partially broken-down rock that serves as the "parent" material for the upper layers of soil.

Figure 32.8B illustrates the intimate association between a plant's root hairs, soil water, and the tiny particles of topsoil. The root hairs are in direct contact with the water that surrounds the

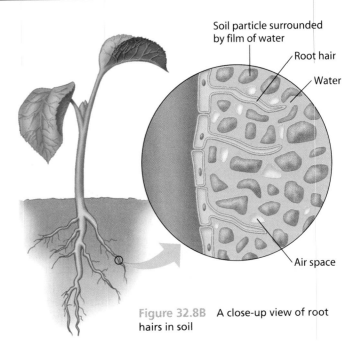

Soil particle surrounded by film of water
Root hair
Water
Air space

Figure 32.8B   A close-up view of root hairs in soil

particles. The soil water is not pure but a solution containing dissolved inorganic ions. Oxygen diffuses into the water from small air spaces in the soil. Roots absorb this soil solution.

Cation exchange is a mechanism by which root hairs take up certain positively charged ions (cations). Inorganic cations—such as calcium (Ca^{2+}), magnesium (Mg^{2+}), and potassium (K^+)—adhere by electrical attraction to the negatively charged surfaces of soil particles. This adherence helps prevent these positively charged nutrients from draining away during heavy rain or irrigation. In cation exchange (Figure 32.8C), root hairs release hydrogen ions (H^+) into the soil solution. The hydrogen ions displace cations on the clay particle surfaces, and root hairs can then absorb them.

In contrast to cations, negative ions (anions)—such as nitrate (NO_3^-)—are usually not bound tightly by soil particles. Unbound ions are readily available to plants, but they tend to drain out of the soil quickly. If they do, the soil may become deficient in nitrogen.

It may take centuries for a soil to become fertile through the breakdown of rock and the accumulation of organic material. The loss of soil fertility is one of our most pressing environmental problems, as we discuss next.

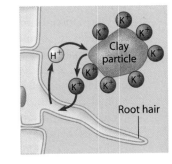

Figure 32.8C   Cation exchange

Web Activity *Absorption of Nutrients from Soil*

Web Process of Science *Connection: How Does Acid Precipitation Affect Mineral Deficiency?*

? How do roots actively increase the availability of mineral nutrients that are cations?

Figure 32.8A   Three soil horizons visible beneath grass

By secreting hydrogen ions, which displace cations from soil particles

32.9 Soil conservation is essential to human life

Our survival as a species depends on soil, and yet erosion and chemical pollution threaten this vital resource throughout the world. As the human population continues to grow and more land is cultivated, farming practices that conserve soil fertility will become essential to our survival. Three critical goals of soil conservation are proper irrigation, prevention of erosion, and prudent fertilization.

Irrigation can turn a desert into a garden, but farming in dry regions is a huge drain on water resources. Additionally, irrigation can gradually make the soil salty. The whitish deposits on the soil in the photograph in Figure 32.9A are salts that were dissolved in irrigation water flooded onto a field. Left behind when the excess water evaporated, the deposits will eventually make the soil too salty for crop plants to tolerate. Instead of flooding fields, modern irrigation often employs perforated pipes that drip water slowly into the soil close to plant roots. This drip irrigation uses less water, allows the plants to absorb most of the water, and reduces water loss from evaporation and drainage.

Preventing erosion—the blowing or washing away of soil—is one of the most important challenges of modern agriculture. Thousands of acres of farmland are lost to water and wind erosion each year in the United States alone. Precautions such as planting rows of trees as windbreaks, terracing hillside crops, and cultivating in a contour pattern can prevent loss of topsoil. The crops in Figure 32.9B are planted in rows that go around, rather than up and down, the hill in the field. This contour tillage helps slow the runoff of water and topsoil after heavy rains. Crops such as alfalfa and wheat provide good ground cover and protect the soil better than other crops that are usually planted in more widely spaced rows.

Prehistoric farmers may have started fertilizing their fields after noticing that grass grew faster and greener where animals had defecated. In developed nations today, most farmers use inorganic commercially produced fertilizers containing minerals that are either mined or prepared by industrial processes. These fertilizers are usually enriched in nitrogen, phosphorus, and potassium, the macronutrients most commonly deficient in farm and garden soils.

Figure 32.9B   Planting to prevent soil erosion in a hilly area

Manure, fish meal, and compost (decaying plant matter) are common fertilizers that contain decomposing organic material. Before the nutrients in these substances can be used by plants, the organic material must be broken down by bacteria and fungi to inorganic nutrients that roots can absorb. Whether from natural fertilizer or a chemical factory, the minerals a plant extracts from the soil are in the same form. The difference is that naturally derived fertilizers release nutrients gradually, whereas minerals in inorganic commercial fertilizers are available immediately. However, because they are soluble in water, minerals from inorganic fertilizers may not be retained in the soil for long. Problems arise when fields are overfertilized with inorganic products and excess nutrients are not taken up by plants. The excess minerals are often leached from the soil by rainwater or irrigation. This mineral runoff may pollute groundwater, streams, and lakes.

Agricultural researchers are developing ways to maintain crop yields while reducing the use of costly fertilizer. One approach is to genetically engineer "smart" plants that inform the grower when a nutrient deficiency is imminent but *before* damage has occurred. One such plant contains a reporter gene that leads to the production of a blue pigment in leaf cells when the phosphorus content of plant tissues declines. When leaves of these smart plants develop a blue tinge, the farmer knows it is time to add phosphorus-containing fertilizer.

? Why do fertilizers containing organic materials generally contaminate water resources less than inorganic fertilizers?

Figure 32.9A   Flood irrigation

■ Fertilizers containing organic materials release mineral nutrients gradually as they decompose, so there is less likelihood of the minerals leaching into the groundwater or running off into streams and lakes.

32.10 Organic farmers follow principles of sustainable agriculture

If you find tomatoes at a local farmers' market labeled "organic" and also find tomatoes at a giant grocery store chain marked the same way, can you be sure they were both *grown* the same way? What does an "organic" label mean? To use the term *organic* or to bear the "USDA Organic" seal, food must be grown and processed according to strict guidelines established and regulated by the U.S. Department of Agriculture (USDA). The goal of these guidelines is **sustainable agriculture,** a system embracing a variety of farming methods that are conservation-minded, environmentally safe, and profitable. Organic farming is meant to sustain biological diversity, maintain soil quality (as by crop rotation), manage pests with no or few synthetic pesticides (by, for example, providing habitat for predators of pests), avoid genetically modified organisms, and use few or no synthetic fertilizers. Yearly inspections ensure proper organic farming practices, accurate record keeping, and a buffer of land between organic farms and neighboring conventional farms.

Farmers in the United States have dedicated over 2 million acres to organic farming (Figure 32.10). Some is managed by small family operations, others by large businesses. The U.S. organic farming industry is growing at a rate of 20% per year, making it one of the fastest-growing segments of agriculture.

The ultimate aim of many organic farmers is to restore as much to the soil as is drawn from it, creating fields that are bountiful and self-sustaining. Many have chosen organic farming to both protect the environment and answer the growing demand for more naturally produced foods. The benefits of organic farming are clear: fewer synthetic chemicals in the environment and less risk of exposing farmworkers and wildlife to potential toxins. And since organic fruits and vegetables are often picked when ripe and sold locally, rather than treated with preservatives and shipped long distances, they can be fresher and better tasting than conventional produce.

But while organic farming is spreading, it hasn't replaced conventional agriculture: Only about 0.3% of U.S. cropland is

Figure 32.10 An organic farmer harvesting sweet corn

certified organic, and only about 2% of the U.S. food supply is grown using organic methods. Furthermore, an organic label is no guarantee of safety or extra health benefits. Scientists disagree, for example, about the nutritional differences, if any, between organic and conventional produce.

Still, organic farmers continue to improve their practices. Some are trying new growing methods that promote greater biological diversity among the plants and wildlife in their fields. Others are looking for better natural fertilizers that increase crop yields. The future of farming, they say, lies in working toward two goals simultaneously: feeding the world's people and promoting a healthy environment.

? If you buy some "organic" apples, does that tell you anything about how they were grown?

■ An organic label indicates that the grower has been certified to be following standards meant to promote long-term agricultural sustainability.

32.11 Agricultural research is improving the yields and nutritional values of crops

Eight hundred million people suffer from malnutrition. Every day, 40,000 people, including 20,000 children, die of hunger. Advocates of plant biotechnology believe that the genetic engineering of crop plants is the key to overcoming the pressing problem of world hunger.

The most limited resource for food production is land. The size of the human population is steadily increasing while, at the same time, the amount of farmland is decreasing. Thus, improving crop yields is a major goal of plant biotechnology.

The commercial adoption by farmers of genetically modified (GM) crops has been one of the most rapid advances in the history of agriculture. These crops include transgenic varieties of cotton and corn that contain genes from the bacterium

Bacillus thuringiensis. These transgenes encode for a protein (*Bt* toxin) that effectively controls a number of serious insect pests. The use of such plant varieties greatly reduces the need for spraying crops with chemical insecticides. Although *Bt* toxin is harmless to vertebrates, including humans, its use is controversial due, in part, to concerns about its effects on helpful insects.

Considerable progress also has been made in the development of transgenic plants of cotton, corn, soybeans, sugar beets, and wheat that are tolerant to a number of herbicides. The cultivation of these plants may reduce production costs and enable farmers to "weed" crops with herbicides that do not damage the transgenic crop plants. This can reduce tillage, which can cause erosion of soil. Researchers are also engineering plants with

Figure 32.11 Plant researchers with high-protein rice

increase quantities of vitamin A, is under development (see Module 12.18). Plant breeding has also resulted in new varieties of corn, wheat, and rice that are enriched in protein (Figure 32.11). Such modified crops may be particularly important because protein deficiency is the most common cause of malnutrition around the world. However, many of these "super" varieties have an extraordinary demand for nitrogen, usually supplied in the form of commercial fertilizer. Unfortunately, these fertilizers are expensive to produce. Thus, the countries that most need high-protein crops are usually the ones least able to afford to grow them.

Another strategy that could potentially increase protein yields in crops is to increase the activity of organisms that form symbiotic relationships with plants, the focus of the next section.

Web Activity *Connection: Genetic Engineering of Golden Rice*

? Why is research on the protein content of crop plants so important to human health worldwide?

■ Because the most common form of malnutrition is protein deficiency, and most people in the world get most of their protein from plants

enhanced resistance to disease. In one case, a transgenic papaya resistant to a ring spot virus was introduced into Hawaii, thereby saving its papaya industry.

The nutritional quality of plants is also being improved. Golden Rice, a transgenic variety with a few daffodil genes that

Plant Nutrition and Symbiosis

32.12 Most plants depend on bacteria to supply nitrogen

As discussed in Module 32.7, nitrogen deficiency is the most common nutritional problem in plants. The atmosphere is nearly 80% nitrogen. But atmospheric nitrogen is gaseous N_2, a form that plants cannot use. For plants to absorb nitrogen, it must first be converted to ammonium (NH_4^+) or nitrate (NO_3^-).

In contrast to other minerals, the (NH_4^+) and (NO_3^-) in soil are not derived from the breakdown of rock. Instead, ammonium and nitrate in the soil are produced from atmospheric N_2 or from organic matter by bacteria. As shown in Figure 32.12, certain soil bacteria, called nitrogen-fixing bacteria, convert atmospheric N_2 to ammonia (NH_3), a metabolic process called **nitrogen fixation.** In soil, ammonia picks up another H^+ to form an ammonium ion (NH_4^+). A second group of bacteria,

called ammonifying bacteria, adds to the soil's supply of ammonium by decomposing organic matter (humus).

Plant roots can absorb nitrogen as ammonium. However, plants acquire their nitrogen mainly in the form of nitrate (NO_3^-), which is produced in the soil by a third group of soil bacteria called nitrifying bacteria. After nitrate is absorbed by roots, plant enzymes convert the nitrate back into ammonium, which is then incorporated into amino acids. Amino acids are used to make proteins and other organic molecules.

? What is the danger in applying a bactericide to the soil around plants?

■ The bactericide might kill soil bacteria that make nitrogen available to plants, causing nitrogen deficiency.

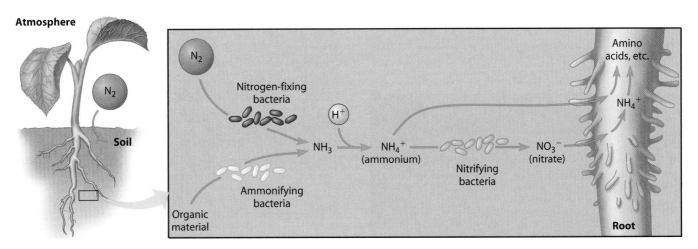

Figure 32.12 The roles of bacteria in supplying nitrogen to plants

32.13 Mutually beneficial relationships have evolved between plants and their symbionts

Reliance on the soil for nutrients that may be in short supply makes it imperative that the roots of plants have a large absorptive surface area. As we have seen, root hairs add a great deal of surface to plant roots. Most plants gain even more absorptive surface by teaming up with fungi.

The micrograph in Figure 32.13A shows a root of a eucalyptus tree. The root is covered with a twisted mat of fungal filaments. Together, the roots and the fungi form a mutually beneficial structure called a **mycorrhiza** ("fungus root"). The fungi benefit from a steady supply of sugar supplied by the host plant. In return, the fungi increase the surface area for water uptake and selectively absorb phosphate and other minerals from the soil and supply them to the plant. The fungi of mycorrhizae also secrete growth factors that stimulate roots to grow and branch, as well as antibiotics that may protect the plant from pathogens in the soil.

The plant-fungus symbiosis of mycorrhizae might have been one of the evolutionary adaptations that made it possible for plants to colonize land. Indeed, fossilized roots from some of the earliest plants include mycorrhizae. When terrestrial ecosystems were young, the soil was probably not very rich in nutrients. The fungi of mycorrhizae, which are more efficient at absorbing minerals than the roots, would have helped nourish the pioneering plants. Even today, the plants that first become established on nutrient-poor soils, such as abandoned farmland or eroded hillsides, are usually heavily colonized with mycorrhizae.

However, roots can be transformed into mycorrhizae only if they are exposed to the appropriate species of fungus. For example, if seeds are collected in one environment and planted in foreign soil, the plants may show signs of malnutrition resulting from the absence of the plants' natural mycorrhizal partners. Farmers may avoid this problem by inoculating seeds with spores of an appropriate mycorrhizal fungi.

Plants also form symbiotic relationships with other organisms besides fungus. Some plant species maintain close association with nitrogen-fixing bacteria. Such symbiotic relationships provide a built-in source of ammonium. For example, the roots of plants in the legume family—including peas, beans, peanuts,

alfalfa, and many other plants that produce their seeds in pods—have swellings called **nodules** (Figure 32.13B). Within these nodules, plant cells have been "infected" by nitrogen-fixing bacteria of the genus *Rhizobium* ("root living"). *Rhizobium* bacteria reside in cytoplasmic vesicles formed by the root cell.

Figure 32.13C shows a cross section of one such cell; notice the vesicles full

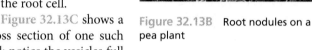

Figure 32.13B   Root nodules on a pea plant

of bacteria. Each legume is associated with a particular strain of *Rhizobium*. Other nitrogen-fixing bacteria (actinomycetes) are found in the root nodules of some plants that are not legumes, such as alders.

The relationship between a plant and its nitrogen-fixing bacteria is mutually beneficial. The plant provides the bacteria with carbohydrates and other organic compounds. The bacteria have enzymes that catalyze the conversion of atmospheric N_2 to ammonium ions (NH_4^+), a form readily used by the plant. When conditions are favorable, root nodule bacteria fix so much nitrogen that the nodules secrete excess NH_4^+, which increases the fertility of the soil. This is one reason farmers practice crop rotation, one year planting a nonlegume, such as corn, and the next year planting a legume, such as alfalfa. The legume crop may be plowed under so that it will decompose as "green manure," reducing the need for manufactured fertilizer.

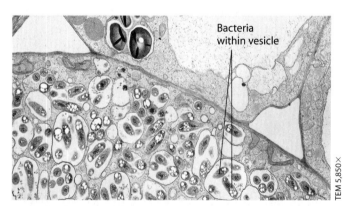

Bacteria within vesicle

Figure 32.13C   Bacteria within a root nodule cell

?   How do the nitrogen-fixing bacteria of root nodules benefit from their symbiotic relationship with plants?

■ The bacteria, which do not photosynthesize, depend on the host plant for a supply of certain organic compounds produced by the photosynthesizing plant.

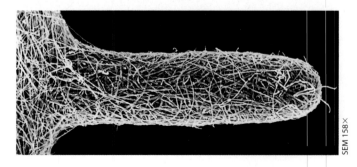

Figure 32.13A   A mycorrhiza on a eucalyptus root

The plant kingdom includes epiphytes, parasites, and carnivores

Almost all plant species have mutualistic symbiotic associations with soil fungi, bacteria, or both. Though rarer, there are also plant species with nutritional adaptations that use other organisms in nonmutalistic ways. For example, an epiphyte is a plant that grows on another plant, usually anchored to branches or trunks of living trees. Examples of epiphytes include staghorn ferns and orchids, like the one shown in Figure 32.14A. Epiphytes absorb water and minerals from rain.

Unlike epiphytes, parasitic plants absorb sugars and minerals from their living hosts. Figure 32.14B shows a parasitic plant called dodder (the yellow-orange threads wound around the green plant). Dodder cannot photosynthesize; it obtains organic molecules from other plant species, using specialized roots that tap into the host's vascular tissue.

Figure 32.14C shows part of an oak tree parasitized by mistletoe, the plant traditionally tacked above doorways during the Christmas season. All the leaves you see here are mistletoe; the oak has lost its leaves for winter. Mistletoe is photosynthetic, but it supplements its diet by siphoning sap from the vascular tissue of the host tree.

Certain plants are carnivores, obtaining some of their nutrients, especially nitrogen and minerals, by killing and digesting insects and other small animals. Carnivorous plants grow in habitats where soils are poor in nitrogen and other minerals (such as acid bogs). Organic matter decays so slowly in acidic soils that there is little inorganic nitrogen available for plant roots to take up. Though they are photosynthetic, the sundew and Venus' flytrap (Figures 32.14D and 32.14E) thrive by obtaining their nitrogen from insects.

Few species illustrate the correlation of structure and function better than carnivorous plants. The sundew plant (Figure 32.14D) has modified leaves, each bearing many club-shaped hairs. A sticky, sugary secretion at the tips of the hairs attracts insects and traps them. The presence of an insect triggers the hairs to bend and the leaf to enfold its prey. The hairs then secrete digestive enzymes, and the plant absorbs nutrients released as the insect is digested.

The Venus' flytrap (Figure 32.14E) has hinged leaves that close around small insects, usually ants and grasshoppers. As insects walk on the insides of these leaves, they touch sensory hairs that trigger closure of the trap. The leaf then secretes digestive enzymes and absorbs nutrients from the prey.

Using insects as a source of nitrogen is a nutritional adaptation that enables carnivorous plants to thrive in soils where most other plants cannot. Fortunately for animals, such predator-prey turnabouts are rare!

? Carnivorous plants are most common in locales where the soil is deficient in _____ and _____.

■ nitrogen . . . minerals

Figure 32.14A Orchids, a type of epiphyte, growing on the trunk of a tree

Figure 32.14B Dodder growing on a pickleweed

Figure 32.14C Mistletoe growing on an oak

Figure 32.14D A sundew plant trapping a damselfly

Figure 32.14E A Venus' flytrap digesting a fly

Chapter Review

Reviewing the Concepts

The Uptake and Transport of Plant Nutrients (Introduction–32.5)

Water and nutrient uptake. In phytoremediation, certain plant species that absorb toxic substances are used to help clean up polluted soil and groundwater (**Introduction**). As a plant grows, its roots absorb water, minerals (inorganic ions), and some O_2 from the soil. Its leaves take carbon dioxide from the air (**32.1**). Root hairs greatly increase a root's absorptive surface. Water and solutes can move through the root's epidermis and cortex by going either through cells or between them. However, all water and solutes must pass through the selectively permeable plasma membranes of cells of the endodermis to enter the xylem (water-conducting tissue) for transport upward (**32.2**).

Transport of water, minerals, and sugar. Transpiration can move xylem sap, consisting of water and dissolved inorganic nutrients, to the top of the tallest tree (**32.3–32.4**):

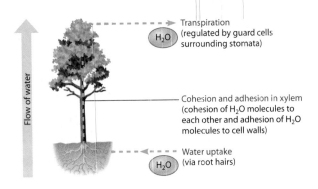

Phloem transports food molecules made by photosynthesis by a pressure flow mechanism. At a sugar source, sugar is loaded into a phloem tube. The sugar raises the solute concentration in the tube, and water follows, raising the pressure in the tube. As sugar is removed at a sugar sink, water follows. The increase in pressure at the sugar source and decrease at the sugar sink causes phloem sap to flow from source to sink (**32.5**).

Plant Nutrients and the Soil (32.6–32.11)

Plant nutrition. A plant must obtain usable sources of the chemical elements—"nutrients"—it requires from its surroundings. Macronutrients, such as carbon and nitrogen, are needed in large amounts, mostly to build organic molecules. Micronutrients, including iron and zinc, act mainly as cofactors of enzymes (**32.6**). Nutrient deficiencies can often be recognized and fixed by using appropriate fertilizers (**32.7**).

Fertile soil contains a mixture of small rock and clay particles that hold water and ions and also allow O_2 to diffuse into plant roots. Humus (decaying organic material) provides nutrients and supports the growth of organisms that enhance soil fertility. Anions (negatively charged ions), such as nitrate (NO_3^-), are readily available to plants because they are not bound to soil particles. However, anions tend to drain out of soil rapidly. Cations (positively charged ions), such as K^+, adhere to soil particles. In cation exchange, root hairs release H^+ ions, which displace cations from soil particles; the root hairs then absorb the free cations (**32.8**). Water-conserving irrigation, erosion

control, and the prudent use of herbicides and fertilizers are aspects of good soil management (**32.9**).

Nutrition and agriculture. Organic farmers are certified to follow ecologically sound practices (**32.10**). Through traditional plant breeding and DNA technologies, researchers are developing new varieties of crop plants with improved yields and nutritional value (**32.11**).

Plant Nutrition and Symbiosis (32.12–32.14)

Relationships with other organisms help plants obtain nutrients. Bacteria in the soil convert atmospheric N_2 to forms that can be used by plants (**32.12**):

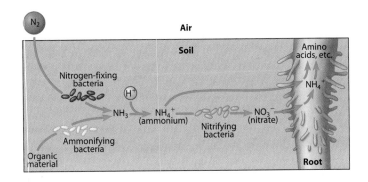

Many plants form mycorrhizae, mutually beneficial associations with fungi. A network of fungal threads increases a plant's absorption of nutrients and water, and the fungus receives some nutrients from the plant. Legumes and certain other plants have nodules in their roots that house nitrogen-fixing bacteria (**32.13**). Epiphytes are plants that grow on other plants. Parasitic plants siphon sap from host plants. Carnivorous plants can obtain nitrogen by digesting insects (**32.14**).

Connecting the Concepts

1. Fill in the blanks in this concept map to help you tie together key concepts concerning transport in plants.

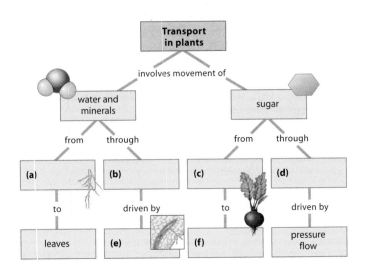

Testing Your Knowledge

Multiple Choice

2. Plants require the smallest amount of which of the following nutrients?
   - a. oxygen
   - b. phosphorus
   - c. carbon
   - d. iron
   - e. hydrogen

3. Which of the following activities of soil bacteria does *not* contribute to creating usable nitrogen supplies for plant use?
   - a. the fixation of atmospheric nitrogen
   - b. the conversion of ammonium ions to nitrate ions
   - c. the decomposition of dead animals
   - d. the assembly of amino acids into proteins
   - e. the generation of ammonium from proteins in dead leaves

4. By trapping insects, carnivorous plants obtain _____, which they need _____. (*Pick the best answer.*)
   - a. water . . . because they live in dry soil
   - b. nitrogen . . . to make sugar
   - c. phosphorus . . . to make protein
   - d. sugars . . . because they can't make enough by photosynthesis
   - e. nitrogen . . . to make protein

5. A major long-term problem resulting from flood irrigation is the
   - a. drowning of crop plants
   - b. accumulation of salts in the soil
   - c. erosion of fine soil particles
   - d. encroachment of water-consuming weeds
   - e. excessive cooling of the soil

Describing, Comparing, and Explaining

6. Explain how guard cells limit water loss from a plant on a hot, dry day. How can this be harmful to the plant?

Applying the Concepts

7. Acid rain contains an excess of hydrogen ions (H^+). One effect of acid rain is to deplete the soil of plant nutrients such as calcium (Ca^{2+}), potassium (K^+), and magnesium (Mg^{2+}). Offer a hypothesis to explain why acid rain washes these nutrients from the soil. How might you test your hypothesis?

8. In some situations, the application of nitrogen fertilizer to crops has to be increased each year because the fertilizer decreases the rate of nitrogen fixation in the soil. Propose a hypothesis to explain this phenomenon. Describe a test for your hypothesis. What results would you expect from your test?

9. Transpiration is fastest when humidity is low and temperature is high, but in some plants it seems to increase in response to light as well. During one 12-hour period when cloud cover and light intensity varied frequently, a scientist studying a certain crop plant recorded the data in the table below. (The transpiration rates are grams of water per square meter of leaf area per hour.)

Time (hr)	Temperature (°C)	Humidity (%)	Light (% of full sun)	Transpiration Rate (g/m²/hr)
8 A.M.	14	88	22	57
9	14	82	27	72
10	21	86	58	83
11	26	78	35	125
12 P.M.	27	78	88	161
1	33	65	75	199
2	31	61	50	186
3	30	70	24	107
4	29	69	50	137
5	22	75	45	87
6	18	80	24	78
7	13	91	8	45

Do these data support the hypothesis that the plants transpire more when the light is more intense? If so, is the effect independent of temperature and humidity? Explain your answer. (*Hint:* Look for overall trends in each column, and then compare pairs of data within each column and between columns.)

10. Agriculture is by far the biggest user of water in arid western states, including Colorado, Arizona, and California. The populations of these states are growing, and there is an ongoing conflict between cities and farm regions over water. To ensure water supplies for urban growth, cities are purchasing water rights from farmers. This is often the least expensive way for a city to obtain more water, and some farmers can make more money selling water than growing crops. Discuss the possible consequences of this trend. Is this the best way to allocate water for all concerned? Why or why not?

Answers to all questions can be found in Appendix 4.

For Practice Quizzes, BioFlix, MP3 Tutors, and Activities, go to www.mybiology.com.

What Are the Health Benefits of Soy?

Americans are discovering soy. In fact, soy product sales have almost quadrupled in the last decade. In addition to the long-popular soy sauce (made from fermented soybeans), many supermarkets carry other soy foods such as soy milk (ground soybeans mixed with water and flavorings), tofu (cooked pureed soybeans formed into cakes), soy flour (which lacks gluten and so must be mixed with wheat flour for baking), and miso (fermented soybean paste used for seasoning).

Soy offers a number of dietary benefits, so the increase in soy consumption is a positive trend. Soy protein is one of the few plant proteins that contain all the essential amino acids, making it a healthy meat substitute. In addition, 25 g of soy protein per day (for example, 2.5 glasses of soy milk or 8 ounces of tofu) has been shown to reduce the risk of heart disease. Soy is rich in antioxidants and fiber, low in fat, and has been shown to lower levels of LDL ("bad cholesterol") and triglycerides while maintaining HDL ("good cholesterol"). Not bad for a plant that until recently in the United States was mostly used for animal feed!

While the nutritional benefits of soybeans are not surprising, you may not be aware that soybeans also contain non-nutritive phytochemicals (literally, "plant chemicals") that may have significant metabolic effects on the human body. As you will learn in this chapter, plants, like humans, use hormones as chemical signals that control growth and development. It is only natural, then, that when we eat plants we consume plant hormones.

Phytoestrogens, a class of plant hormones, are found in soy. Their chemical structure is similar to the class of human female sex hormones collectively called estrogen, which allows them to bind to estrogen receptors on human cells. One type of phytoestrogen, the isoflavones, help regulate the soy plant's growth and appear to exert a weak hormonal effect on the human body as well. In menopausal women (whose ovaries greatly curtail estrogen production), isoflavones may help reduce the negative effects of lower estrogen production, such as hot flashes and the risk of osteoporosis. So some women choose dietary supplements with isoflavones instead of hormone replacement therapy (HRT), which often contains estrogen isolated from the urine of pregnant mares.

However, the health benefits of isoflavones are still being investigated. The strongest evidence comes from epidemiological studies (studies of the incidence and distribution of health-related problems in various populations). For example, women in China, who generally consume high levels of soy, have lower incidence of hot flashes and fewer hip fractures compared with Chinese women living in the West, who consume less soy. Controlled clinical trials, however, have shown only small effects.

While most health professionals agree that soy is a good addition to any diet, all the benefits and risks of isolated isoflavones have not been established. The metabolism of estrogen and the related phytoestrogens involves risks and benefits that are related to the amounts consumed. For example, while moderate levels of estrogen relieve menopausal symptoms, ward off heart disease, and sustain bone mass, high levels appear to increase the risk of breast and ovarian cancers. It is also hard for women to know if they are receiving beneficial quantities of isoflavones because amounts may vary widely among soy products. Soy flour, for example, contains over 15 times more isoflavones per weight than soy milk. In addition, dietary supplements are not subject to strict federal regulation, making it harder for women to assess their risk.

While scientists continue to explore the potential health risks and benefits of isoflavones and other plant hormones, we do know that hormones are crucial to the life of plants. In this chapter, we explore the diverse roles of plant hormones: how they affect plant movement, growth, flowering, fruit development, and even defense. ■ ■ ■

33.1 Experiments on how plants turn toward light led to the discovery of a plant hormone

The shoot of a houseplant on a windowsill grows toward light (Figure 33.1A). If you rotate the plant, it soon reorients its growth until the leaves again face the window. The growth of a shoot in response to light is called **phototropism** (from the Greek *photos*, light, and *tropos*, turn). Phototropism is an adaptive response, directing growing seedlings and the shoots of mature plants toward the sunlight that drives photosynthesis.

Microscopic observations of plants growing toward light indicate the cellular mechanism that underlies phototropism. Figure 33.1B shows a grass seedling curving toward light coming from one side. As the enlargement shows, cells on the darker side of the seedling are larger—actually, they have elongated faster—than those on the brighter side. The different cellular growth rates made the shoot bend toward the light. If a seedling is illumi-

Figure 33.1A A houseplant growing toward light

nated uniformly from all sides or if it is kept in the dark, the cells all elongate at a similar rate. In these situations, the seedling grows straight upward.

What causes plant cells in a shoot to grow at different rates? We saw in Chapter 26 that hormones help coordinate internal activities, such as growth rates, in animals. We might predict, therefore, that plants also have hormones that regulate growth. Actually, the idea that plants have hormones emerged from a series of classic experiments conducted by two scientists with a very famous name.

Showing That Light Is Detected by the Shoot Tip   In the late 19th century, Charles Darwin and his son Francis conducted some of the earliest experiments on phototropism. They observed that grass seedlings could bend toward light only if the tips of their shoots were present. The first five grass plants in Figure 33.1C (on the facing page) summarize the Darwins' findings. ❶ When they removed the tip of a grass shoot, the shoot did not curve toward the light. ❷ The shoot also remained straight when the Darwins placed an opaque cap on its tip. ❸ However, the shoot curved normally when they placed a transparent cap on its tip or ❹ an opaque shield around its base. The Darwins concluded that the tip of the shoot was responsible for sensing light. They also recognized that the growth response, the bending of the shoot, occurs below the tip. Therefore, they speculated that some signal was transmitted downward from the tip to the growth region of the shoot.

A few decades later, Danish botanist Peter Boysen-Jensen further tested the chemical signal idea of the Darwins. In one group of seedlings, Boysen-Jensen ❺ inserted a block of gelatin between the tip and the lower part of the shoot. The gelatin block prevented cellular contact but allowed chemicals to diffuse through. The seedlings with gelatin blocks behaved normally, bending toward light. In a second set of seedlings, ❻ Boysen-Jensen inserted a thin piece of the mineral mica under the shoot tip. Mica is an impermeable barrier, and the seedlings with mica had no phototropic response. These experiments supported the hypothesis that the signal for phototropism is a mobile chemical.

Isolating the Chemical Signal   In 1926, Frits Went, a Dutch graduate student, modified Boysen-Jensen's techniques and extracted the chemical messenger for phototropism in grasses. As shown in Figure 33.1D, Went first removed the tips of grass seedlings and placed them on blocks of agar, a gelatin-like material. He reasoned that the chemical messenger (pink in the figure) from the shoot tips should diffuse into the agar and that the blocks should then be able to substitute for the shoot tips. Went tested the effects of the agar blocks on tipless seedlings, which he kept in the dark to eliminate the effect of light. ❶ First, he centered the treated agar blocks on the cut tips of a batch of seedlings. These plants grew straight upward. They also grew faster than the decapitated control seedlings, which

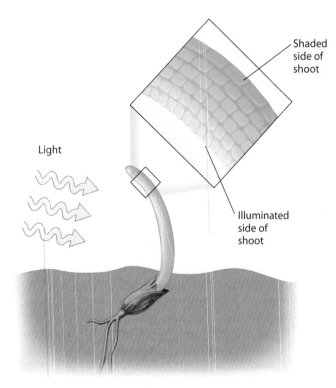

Figure 33.1B Phototropism in a grass seedling

Light

Shaded side of shoot

Illuminated side of shoot

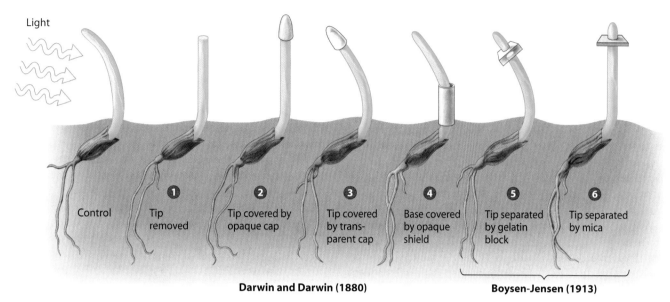

Light

① Control ② Tip removed ③ Tip covered by opaque cap ④ Tip covered by transparent cap ⑤ Base covered by opaque shield ⑥ Tip separated by gelatin block ⑦ Tip separated by mica

Darwin and Darwin (1880)   **Boysen-Jensen (1913)**

Figure 33.1C   Early experiments on phototropism: detection of light by shoot tips and evidence for a chemical signal

hardly grew at all. Went concluded that the agar had absorbed the chemical messenger produced in the shoot tip and that the chemical had passed into the shoot and stimulated it to grow. ❷ He then placed agar blocks off center on another batch of tipless seedlings. These plants bent away from the side with the chemical-laden agar block, as though growing toward light. ❸ Control seedlings with blank agar blocks (whether offset or not) grew no more than the first control. Went concluded that the agar block contained a chemical produced in the shoot tip, that this chemical stimulated growth as it passed down the shoot, and that a shoot curved toward light because of a higher concentration of the growth-promoting chemical on the darker side. For this chemical messenger, or hormone, Went chose the name *auxin*. In the 1930s, biochemists determined the chemical structure of Went's auxin.

The classical hypothesis for what causes grass shoots to grow toward light, based on these early experiments, is that an uneven distribution of auxin moving down from the shoot tip causes cells on the darker side to elongate faster than cells on the brighter side. Studies with plants other than grass shoots, however, do not always support this hypothesis. For example, there is no evidence that light from one side causes an uneven distribution of auxin in the stems of sunflowers, radishes, and other eudicots. There is, however, a greater concentration of substances that may act as growth inhibitors on the lighted side of a stem. Still, auxin's role in the phototropism of grass shoots opened up the field of research on plant hormones.

? How do the experiments illustrated in Figures 33.1C and 33.1D provide evidence that phototropism depends on a chemical signal—that is, a hormone?

■ Light is detected by the shoot tip, but the bending response occurs farther down the shoot. The fact that the signal can pass through a barrier that prevents cell contact but allows chemicals to pass suggests that the signal is a chemical.

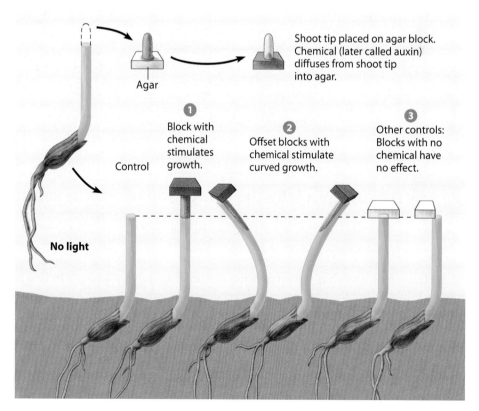

Agar

Shoot tip placed on agar block. Chemical (later called auxin) diffuses from shoot tip into agar.

No light

Control

① Block with chemical stimulates growth.

② Offset blocks with chemical stimulate curved growth.

③ Other controls: Blocks with no chemical have no effect.

Figure 33.1D   Went's experiments: isolation of the chemical signal

33.2 Five major types of hormones regulate plant growth and development

Found in all plants and animals, a **hormone** is a chemical signal produced in one part of the body and transported to other parts, where it triggers responses in target cells. In this way, hormones help coordinate many activities in an organism. The binding of a hormone to a specific receptor triggers a signal transduction pathway that amplifies the hormone's signal (see Module 11.12). Therefore, a minute quantity of a hormone may have a profound effect on target cells and tissues. Virtually every aspect of plant growth and development is under hormonal control to some degree.

Plant biologists have identified five major types of plant hormones, which are previewed in Table 33.2. (It is important to note that some of the "hormones" listed in the table actually represent a group of related hormones; such cases will be mentioned in the text.) Notice that all five types of hormones exert control by influencing growth (stimulating or inhibiting cell division and elongation).

As Table 33.2 indicates, each hormone has multiple effects, depending on its site of action, its concentration, and the devel-opmental stage of the plant. In most situations, no single hormone acts alone. Instead, it is usually the balance of several plant hormones—their relative concentrations—that controls the growth and development of a plant. These interactions will become apparent during this chapter's survey of hormone functions.

Not listed in Table 33.2 are the brassinosteroids (see Module 33.13). The brassinosteroids are chemicals similar to the steroids cholesterol and the animal sex hormones. Their effects are very similar to those of auxin. Additionally, many of the plant defense molecules we will discuss in Module 33.14 are probably plant hormones as well. The next five modules focus on the five major types of plant hormones.

Web Process of Science *What Plant Hormones Affect Organ Formation?*

? Hormones elicit cellular responses by binding to extra-cellular receptors and triggering _____.

■ signal transduction pathways

TABLE 33.2 MAJOR TYPES OF PLANT HORMONES

Hormone (Module)	Major Functions	Where Produced or Found in Plant
Auxins (33.3)	Stimulate stem elongation; affect root growth, differentiation, branching, development of fruit, apical dominance, phototropism and gravitropism (response to gravity)	Meristems of apical buds; young leaves; embryos within seeds
Cytokinins (33.4)	Affect root growth and differentiation; stimulate cell division and growth; stimulate germination; delay aging	Made in roots and transported to other organs
Gibberellins (33.5)	Promote seed germination, bud development, stem elongation, and leaf growth; stimulate flowering and fruit development; affect root growth and differentiation	Meristems of apical buds and roots; young leaves; embryos
Abscisic acid (ABA) (33.6)	Inhibits growth; closes stomata during water stress; helps maintain dormancy	Leaves, stems, roots, green fruits
Ethylene (33.7)	Promotes fruit ripening; opposes some auxin effects; promotes or inhibits growth and development of roots, leaves, and flowers, depending on species	Ripening fruits, nodes of stems, aging leaves and flowers

33.3 Auxin stimulates the elongation of cells in young shoots

The term **auxin** is used for any chemical substance that promotes seedling elongation. The most common natural auxin occurring in plants is indoleacetic acid, or IAA, but several other compounds, including some synthetic ones, have auxin activity. When we use the term *auxin* in this text, however, we are referring specifically to IAA.

Figure 33.3A (at the top of the next page) shows the effect of auxin on growing pea plants. All the seedlings in the photograph were grown under controlled conditions for the same length of time. The only difference is that the taller seedlings, on the right, were treated with auxin.

The apical meristem at the tip of a shoot (see Module 31.7) is a major site of auxin synthesis. As auxin moves downward, it stimulates growth of the stem by making cells elongate. As the blue curve in Figure 33.3B shows, auxin promotes cell elongation in stems only within a certain concentration range. Above a certain level (0.9 g of auxin per liter of solution, in this case), it usually inhibits cell elongation in stems, probably by inducing the production of ethylene, a hormone that generally counters the effects of auxin.

The red curve on the graph shows the effect of auxin on root growth. An auxin concentration too low to stimulate shoot cells will cause root cells to elongate. On the other hand, an auxin concentration high enough to make stem cells elongate is in the concentration range that inhibits root cell elongation. These effects of auxin on cell elongation reinforce two

Figure 33.3A   The effect of auxin (IAA) on pea plants (right)

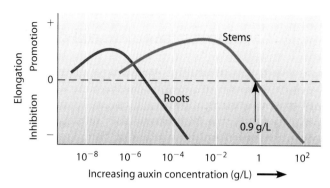

Figure 33.3B   The effect of auxin concentration on cell elongation

points: (1) the same chemical messenger may have different effects at different concentrations in one target cell, and (2) a given concentration of the hormone may have different effects on different target cells. In fact, some herbicides (including the infamous Vietnam War defoliant Agent Orange) contain synthetic auxins at high concentrations that kill plants via hormonal overdose.

How does auxin make plant cells elongate? One hypothesis is that auxin initiates elongation by weakening cell walls. As shown in Figure 33.3C, ❶ auxin may stimulate certain proteins in a plant cell's plasma membrane to pump hydrogen ions into the cell wall, lowering the pH within the wall. The low pH ❷ activates enzymes that separate cross-linking molecules from larger cellulose molecules in the wall (see Figure 3.7). ❸ The cell then swells with water and elongates because its weakened wall no longer resists the cell's tendency to take up water osmotically. After this initial elongation caused by the uptake of water, the cell sustains the growth by

synthesizing more cell wall material and cytoplasm. These processes are also stimulated by auxin.

Auxin produces a number of other effects in addition to stimulating cell elongation and causing stems and roots to lengthen. Auxin induces cell division in the vascular cambium, thus promoting growth in stem diameter (see Module 31.8). Furthermore, auxin is produced by developing seeds and promotes the growth of fruit. Under greenhouse conditions, a lack of insect pollinators can lead to poorly developed tomato fruits. To address this problem, synthetic auxins sprayed on greenhouse-grown tomato vines induce fruit development without a need for pollination, making it possible to grow seedless tomatoes. The use of synthetic auxins as herbicides is described in Module 33.8.

? Suppose you had a tiny pH electrode that could measure the pH of a plant cell's wall. How could you use it to test the hypothesis presented in Figure 33.3C for how auxin stimulates cell elongation?

■ The hypothesis predicts that addition of auxin to the cell should stimulate H^+ pumps and lower the pH of the wall (make it more acidic). You could test this prediction with your pH electrode by measuring the wall pH in the presence of (experimental group) or absence of (control group) auxin.

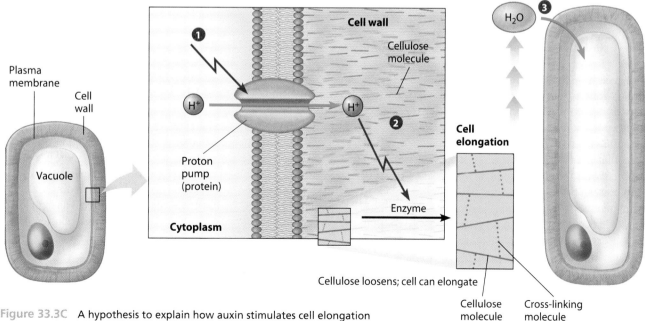

Figure 33.3C   A hypothesis to explain how auxin stimulates cell elongation

33.4 Cytokinins stimulate cell division

Cytokinins are growth regulators that promote cytokinesis, or cell division. A number of cytokinins have been extracted from plants—the most common is zeatin, so named because it was first discovered in corn (*Zea mays*). Also, several synthetic cytokinins have been artificially produced. Natural cytokinins are produced in actively growing tissues, particularly in roots, embryos, and fruits. Cytokinins made in the roots reach target tissues in stems by moving upward in xylem sap.

Plant biologists have found that cytokinins enhance the division, growth, and development of plant cells grown in culture. Cytokinins also retard the aging of flowers and leaves. Thus, florists use cytokinin sprays to keep cut flowers fresh.

Cytokinins and auxin interact in the control of apical dominance, the ability of the terminal bud to suppress the growth of axillary buds (see Module 31.3). The photographs in Figure 33.4 show the results of a simple experiment that demonstrates this effect. Both basil plants pictured are the same age. The one on the left has an intact terminal bud; the one on the right had its terminal bud removed several weeks earlier. In the plant on the left, apical dominance resulted in lengthwise growth but inhibited growth of the axillary buds (the buds that produce side branches). As a result, the shoot grew in height but did not branch out to the sides very much. In the plant on the right, the lack of a terminal bud resulted in the activation of the axillary buds, making the plant grow more branches and become bushy. Some Christmas tree growers use cytokinins in this way to produce attractive branching.

The leading hypothesis to explain the hormonal regulation of apical dominance—known as the direct inhibition hypothesis—proposes that auxin and cytokinin act antagonistically in regulating axillary bud growth. According to this view, auxin transported down the shoot from the terminal bud directly inhibits axillary buds from growing, causing a shoot to lengthen at the expense of lateral branching. Meanwhile, cytokinins entering the shoot system from roots counter the action of auxin by signaling axillary buds to begin growing. Thus, the ratio of auxin and cytokinin is viewed as the critical factor in controlling axillary bud inhibition. Many observations and experiments are consistent with the direct inhibition hypothesis. However, there are also some data that contradict it. The direct inhibition hypothesis, therefore, does not account for all experimental findings: It is likely that plant biologists have not uncovered all the pieces of this puzzle.

Terminal bud

No terminal bud

Figure 33.4 Apical dominance in a basil plant resulting from the action of auxin and cytokinins

? According to the direct inhibition hypothesis, the status of axillary buds—dormant or growing—depends on the relative concentrations of _____ moving down from the shoot tip and _____ moving up from the roots.

■ auxin . . . cytokinins

33.5 Gibberellins affect stem elongation and have numerous other effects

Farmers in Asia have long noticed that some rice seedlings in their paddies grew so tall and spindly that they toppled over before they could produce grain. In the 1920s, Japanese scientists discovered that a fungus of the genus *Gibberella* caused this "foolish seedling disease." By the 1930s, Japanese scientists had determined that the fungus produced hyperelongation of rice stems by secreting a chemical, which was given the name **gibberellin.** Researchers later discovered that gibberellin exists naturally in plants, where it is a growth regulator. Foolish seedling disease occurs when rice plants infected with the *Gibberella* fungus get an overdose of gibberellin.

More than 100 different gibberellins have been identified in plants. Roots and young leaves are major sites of gibberellin production. One of the main effects of gibberellins is to stimulate cell elongation and cell division in stems and leaves. This action generally enhances that of auxin and can be demonstrated when certain dwarf plant varieties (such as the dwarf pea plant variety studied by Mendel; see Module 9.2) grow to normal height after treatment with gibberellins (Figure 33.5A). Also in combination with auxin, gibberellins can influence fruit development, and gibberellin-auxin sprays can make apples, currants, and eggplants develop without fertilization. The most important commercial application of gibberellins is in the production of the Thompson variety of seedless grapes. Gibberellins make the grapes grow larger and farther apart in a cluster. The left cluster of grapes in Figure 33.5B is untreated; the right cluster shows the effect of gibberellin treatment.

Gibberellins are also important in seed germination in many plants. Many seeds that require special environmental conditions to germinate, such as exposure to light or cold temperatures, will germinate when sprayed with gibberellins. In nature, gibberellins in seeds are probably the link between

Figure 33.5A Reversing dwarfism in pea plants with gibberellin

Figure 33.5B An effect of gibberellin treatment on grapes (right)

environmental cues and the metabolic processes that renew growth of the embryo. For example, when water becomes available to a grass seed, it causes the embryo in the seed to release gibberellins, which promote germination by mobilizing nutrients stored in the seed. In some plants, gibberellins seem to interact antagonistically with another hormone, abscisic acid, which we discuss next.

? A gibberellin deficiency probably caused the dwarf variety of pea plants studied by Gregor Mendel. Given the role of gibberellin as a growth hormone, why does it make sense that the dwarf pea plant allele is recessive?

■ Because minute amounts of hormone are usually enough to elicit a strong response, a heterozygous plant would still produce enough gibberellin to result in normal growth.

33.6 Abscisic acid inhibits many plant processes

In the 1960s, one research group studying bud dormancy and another team investigating leaf abscission (the dropping of autumn leaves) isolated the same compound, **abscisic acid (ABA).** Ironically, ABA is no longer thought to play a primary role in either bud dormancy or leaf abscission (for which it was named), but it is a plant hormone of great importance in other functions. In contrast to the growth-stimulating hormones we have studied so far—auxin, cytokinins, and gibberellins—ABA *slows* growth.

One of the times in the life of a plant when it is advantageous to suspend growth is the onset of seed dormancy. Seed dormancy has great survival value because it ensures that the seed will germinate only when there are favorable conditions of light, temperature, and moisture. What prevents a seed

dispersed in autumn from germinating immediately only to be killed by winter? For that matter, what prevents seeds from germinating in the dark, moist interior of the fruit? ABA is the answer. Levels of ABA may increase 100-fold during seed maturation. Many types of dormant seeds will only germinate when ABA is removed or inactivated in some way. For example, some seeds require prolonged exposure to cold to trigger ABA inactivation. In these plants, the breakdown of ABA in the winter is required for seed germination in the spring.

The seeds of some desert plants remain dormant in parched soil until a downpour washes ABA out of the seeds, allowing them to germinate. For example, the evening primroses and purple sand verbena in Figure 33.6 grew from seeds that germinated just after a hard rain.

As we saw in the previous module, gibberellins promote seed germination. For many plants, the ratio of ABA to gibberellins determines whether the seed will remain dormant or germinate.

In addition to its role in dormancy, ABA is the primary internal signal that enables plants to withstand drought. When a plant begins to wilt, ABA accumulates in its leaves and causes stomata to close (see Module 32.4). This reduces transpiration and prevents further water loss. In some cases, water shortage can stress the root system before the shoot system. ABA transported from roots to leaves may function as an "early warning system."

Figure 33.6 The Mojave Desert in California blooming after a rain

? Which two hormones regulate seed dormancy and germination? What are their opposing effects?

■ Abscisic acid maintains seed dormancy. Gibberellins promote germination.

Early in the 20th century, oranges and grapefruits were ripened for market in sheds equipped with kerosene stoves. Fruit growers thought it was the heat that ripened the fruit, but when they tried newer, cleaner-burning stoves, the fruit did not ripen fast enough. Plant biologists learned later that ripening in the sheds was actually due to **ethylene**, a gaseous by-product of kerosene combustion. We now know that plants produce their own ethylene, which functions as a hormone that triggers a variety of aging responses, including fruit ripening and programmed cell death. Ethylene is also produced in response to stresses such as drought, flooding, mechanical pressure, injury, and infection.

Fruit Ripening A burst of ethylene production in a fruit triggers its ripening. Because ethylene is a gas, the signal to ripen spreads from fruit to fruit: One bad apple really does spoil the lot! The ripening process includes the enzymatic breakdown of cell walls, which softens the fruit, and the conversion of starches and acids to sugars, which makes the fruit sweet. The production of new scents and colors attracts animals, which eat the fruits and disperse the seeds.

You can make some fruits ripen faster if you store them in a bag so that the ethylene gas can accumulate. Figure 33.7A shows the results of a fruit-ripening demonstration. Three unripe bananas were stored for the same time period in plastic bags: ❶ with an ethylene-releasing orange, ❷ with a beaker of an ethylene-releasing chemical, and ❸ alone. As you can see, the more ethylene present, the riper (darker) the banana. On a commercial scale, many kinds of fruit—tomatoes, for instance—are often picked green and then ripened in huge storage bins into which ethylene gas is piped—a modern variation on the old storage shed.

In other cases, growers take measures to *retard* the ripening action of natural ethylene. Stored apples are often flushed with CO_2, which inhibits ethylene synthesis. Also, gas is circulated around the apples to prevent ethylene from accumulating. In this way, apples picked in autumn can be stored for sale the following summer.

The Falling of Leaves Like fruit ripening, the changes that occur in deciduous trees each autumn—color changes, drying, and the loss of leaves—are also aging processes. Leaves lose their green color because chlorophyll is broken down during autumn. Fall colors result from a combination of new red pigments made in autumn and the exposure of yellow and orange pigments that were already present in the leaf but masked by dark green chlorophyll. Autumn leaf drop is an adaptation that helps keep the tree from drying out in winter. Without its leaves, a tree loses less water by evaporation when its roots cannot take up water from the frozen ground. Before leaves fall, many essential elements are salvaged from them and stored in the stem.

When an autumn leaf falls, the base of the leaf stalk separates from the stem. The separation region is called the abscission layer. As indicated in Figure 33.7B, the abscission layer consists of a narrow band of cells with thin walls that are further weakened when enzymes digest the cell walls. The leaf drops off when its weight, often helped by wind, splits the abscission layer apart. Notice the layer of cells next to the abscission layer. Even before the leaf falls, these cells form a leaf scar on the stem. Dead cells covering the scar help protect the plant from infectious organisms.

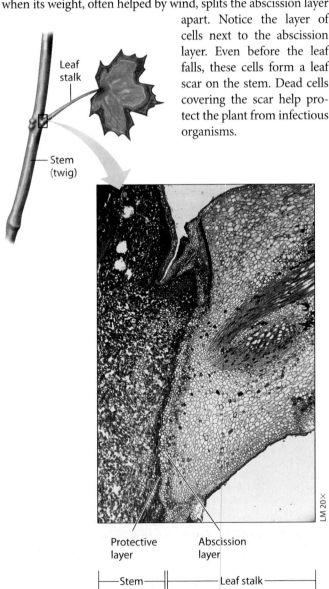

Figure 33.7B Abscission layer at the base of a leaf

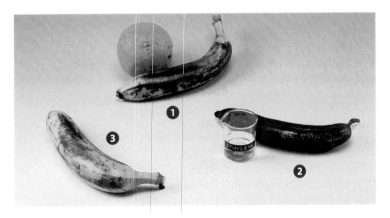

Figure 33.7A The effect of ethylene on the ripening of bananas

Leaf drop is triggered by environmental stimuli, including the shortening days of autumn and, to a lesser extent, cooler temperatures. These stimuli apparently cause a change in the balance of ethylene and auxin. The auxin prevents abscission and helps maintain the leaf's metabolism, but as a leaf ages, it produces less auxin. Meanwhile, cells begin producing ethylene, which stimulates formation of the abscission layer. The ethylene primes the abscission layer to split by promoting the synthesis of enzymes that digest cell walls in the layer.

We have now completed our survey of the five major types of plant hormones. Before moving on to the topic of plant behavior, let's look at some agricultural uses of these chemical regulators.

Web Activity *Leaf Abscission*

? Botanists sometimes refer to ethylene as the "senescence hormone." What is the basis for this term?

■ Many of ethylene's functions, including fruit ripening and leaf abscission, are associated with aging-like changes in cells.

33.8 Plant hormones have many agricultural uses

Although a lot remains to be learned about plant hormones, much of what we do know has a direct application to agriculture. As already mentioned, the control of fruit ripening and the production of seedless fruits are two of several major uses of these chemicals. Plant hormones also enable farmers to control when plants will drop their fruit. For instance, synthetic auxins are often used to prevent orange and grapefruit trees from dropping their fruit before they can be picked. Figure 33.8 shows a fruit grower spraying a grove with plant hormones. The quantity of auxin must be carefully monitored because too much of the hormone may stimulate the plant to release more ethylene, making the fruit ripen and drop off sooner. Indeed, large doses of auxins are sometimes used to *promote* premature fruit drop. For example, auxin may be sprayed on apple and olive trees to thin the developing fruits; the remaining fruits will grow larger. Ethylene is used to thin peaches and prunes, and it is sometimes sprayed on berries, grapes, and cherries to loosen the fruit so it can be picked by machines.

In combination with auxin, gibberellins are used to produce seedless fruits, as mentioned in Module 33.5. Sprayed on other kinds of plants, at an earlier stage, gibberellins can have the opposite effect: the *promotion* of seed production. A large dose of gibberellins will induce many biennial plants, such as carrots, beets, and cabbage, to flower and produce seeds during their first year of growth. Ordinarily, biennials do not produce seeds until their second year of growth.

Research on plant hormones has had other spin-offs. One of the most widely used herbicides, or weed killers, is 2,4-D, a synthetic auxin that disrupts the normal balance of hormones that regulate plant growth. Because dicots are more sensitive than monocots to this herbicide, 2,4-D can be used to selectively remove dandelions and other broadleaf dicot weeds from a lawn or grainfield. By applying herbicides to cropland, a farmer can reduce the amount of tillage required to control weeds, thus reducing soil erosion, fuel consumption, and labor costs.

Modern agriculture relies heavily on the use of synthetic chemicals. Without chemically synthesized herbicides to control weeds and synthetic plant hormones to help grow and preserve fruits, less food would be produced, and food prices could increase considerably. At the same time, there is growing concern that the heavy use of artificial chemicals in food

Figure 33.8 Using auxins to prevent early fruit drop

production may pose environmental and health hazards. A chemical called dioxin, for example, is a by-product of 2,4-D synthesis. Though 2,4-D itself does not appear to be toxic to mammals, dioxin causes birth defects, liver disease, and leukemia in laboratory animals. Therefore, dioxin is a serious hazard when it leaks into the environment. Also, many consumers are concerned that foods produced using artificial hormones may not be as tasty or nutritious as those raised naturally. At present, however, organic foods are relatively expensive to produce. As we discussed in Module 32.10, these issues involve both economics and ethics: Should we continue to produce cheap, plentiful food using artificial chemicals and tolerate the potential problems, or should we put more of our agricultural effort into farming without these potentially harmful substances, recognizing that foods may be less plentiful and more expensive as a result?

? (a) What is the main commercial incentive for agribusiness to treat its products with plant hormones? (b) What behavior of consumers helps drive this use of hormone sprays in agriculture?

■ (a) The need to compete in the market by increasing production and lowering cost; (b) Shopping for the lowest price on produce

33.9 Tropisms orient plant growth toward or away from environmental stimuli

Having surveyed the hormones that carry signals within a plant, we now shift our focus to the responses of plants to physical stimuli from the environment. A plant cannot migrate to water or a sunny spot, and a seed cannot maneuver itself into an upright position if it lands upside down in the soil. Because of their immobility, plants must respond to environmental stimuli through developmental and physiological mechanisms. **Tropisms** are directed growth responses that cause parts of a plant to grow toward (a positive tropism) or away from (a negative tropism) a stimulus. In Module 33.1 we discussed positive phototropism, the growth of a plant shoot toward light. Two other types of tropisms are gravitropism, a response to gravity, and thigmotropism, a response to touch.

Response to Light As we saw in Module 33.1, the mechanism for phototropism is a greater rate of cell elongation on the darker side of a stem. In grass seedlings, the signal linking the light stimulus to the cell elongation response is auxin. Researchers have shown that illuminating a grass shoot from one side causes auxin to migrate across the tip from the bright side to the dark side. The shoot tips contain a protein pigment that detects the light and somehow passes the "message" to molecules that affect auxin transport. (We discuss protein light receptors in Module 33.12.)

Response to Gravity A plant's growth response to gravity, **gravitropism,** is illustrated by the corn seedlings in Figure 33.9A. These seedlings were both germinated in the dark. The one on the left was left untouched; notice that its shoot grew straight up and its root straight down. The seedling on the right was germinated in the same way, but two days later it was turned on its side so that the shoot and root were horizontal. By the time the photo was taken, the shoot had turned back upward, exhibiting a negative response to gravity, and the root had turned down, exhibiting positive gravitropism.

One hypothesis for how plants tell up from down is that gravity pulls special organelles containing dense starch grains to the low points of cells. The uneven distribution of organelles may in turn signal the cells to redistribute auxin. This effect has been documented in roots. A higher auxin concentration on the lower side of a root inhibits cell elongation (see the red line in Figure 33.3B). As cells on the upper side continue to elongate, the root curves downward. This tropism continues until the root is growing straight down. Gravitropism is an important adaptation, making the shoot of a germinating plant grow upward into the light and the root grow into the soil, no matter how the seed lands or is planted.

Response to Touch Directional growth in response to touch, **thigmotropism,** is illustrated by the tendril of a pea plant coiling around a support (see Figure 31.4C, top). Tendrils (actually modified leaves) grow straight until they touch an object. Contact then stimulates the cells to grow at different rates on opposite sides of the tendril (slower in the contact area), making the tendril coil around the support. Most climbing plants have tendrils that respond by coiling and grasping when they touch rigid objects. Thigmotropism enables these plants to use such objects for support while growing toward sunlight.

Some plants show remarkable displays of thigmotropism. When touched, the "sensitive plant" (*Mimosa pudica*) rapidly folds its leaflets together (Figure 33.9B). This response, which takes only a second or two, results from a rapid loss of turgor by cells within the joints of the leaf. It takes about 10 minutes for the cells to regain their turgor and restore the "unstimulated" form of the leaf.

Tropisms all have one function in common: They help plant growth stay in tune with the environment. In the next module, we see that plants also have a way of keeping time with their environment.

? Why are tropisms called "growth responses"?

■ Because the movement of a plant organ toward or away from an environmental stimulus takes place by growing. An organ bends when cells on one side grow faster than cells on the other side, and it elongates in one direction when cells grow evenly.

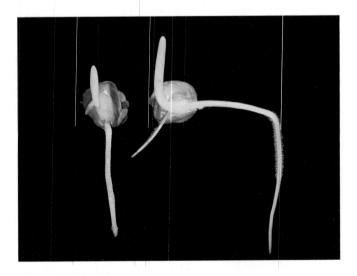

Figure 33.9A Gravitropism

Figure 33.9B The "sensitive plant" *Mimosa pudica*

Plants have internal clocks

Your pulse rate, blood pressure, body temperature, rate of cell division, blood cell count, alertness, urine composition, metabolic rate, sex drive, and responsiveness to medications all fluctuate rhythmically with the time of day. Plants also display rhythmic behavior; examples include the opening and closing of stomata (see Module 32.4) and the "sleep movements" of many species that fold their leaves or flowers in the evening and unfold them in the morning.

Innate Biological Rhythms   An innate biological cycle of about 24 hours is called a **circadian rhythm** (from the Latin *circa,* about, and *dies,* day). A circadian rhythm persists even when an organism is sheltered from environmental cues. A bean plant, for example, exhibits sleep movements at about the same intervals even if kept in constant light or darkness. Thus, circadian rhythms occur with or without external stimuli such as sunrise and sunset. Research on many organisms indicates that circadian rhythms are controlled by internal timekeepers called **biological clocks.**

Although a biological clock continues to mark time in the absence of environmental cues, to remain tuned to a period of *exactly* 24 hours, it requires daily signals from the environment. This is because innate circadian rhythms generally differ somewhat from a 24-hour period. Consider bean plants, for instance. As shown in Figure 33.10, the leaves of a bean plant are normally horizontal at noon and folded downward at midnight. When the plant is held in darkness, however, its sleep movements change to a cycle of about 26 hours.

The light/dark cycle of day and night provides the cues that usually keep biological clocks precisely synchronized with the outside world. But a biological clock cannot immediately adjust to a sudden major change in the light/dark cycle. We observe this problem ourselves when we cross several time zones in an airplane: When we reach our destination, we have "jet lag"; our internal clock is not synchronized with the clock on the wall. Moving a plant across several time zones produces a similar lag. In the case of either the plant or the human traveler, resetting the clock usually takes several days.

The Nature of Biological Clocks   Just what is a biological clock? Researchers are actively investigating this question. In humans and other mammals, the clock is located within a cluster of nerve cells in the hypothalamus of the brain (see Module 28.15). But for most other organisms, including plants, we know little about where the clocks are located or what kinds of cells are involved. A leading hypothesis is that biological timekeeping in plants may depend on the synthesis of a protein that regulates its own production through

Noon   Midnight

Figure 33.10   Sleep movements of a bean plant

feedback control. Once the protein accumulates to a sufficient concentration, it turns off its own gene. When the concentration of the protein falls, transcription restarts. The result would be a cycling of the protein's concentration over a roughly 24-hour period—a clock! Some research indicates that such a molecular mechanism may be common to all eukaryotes. However, much research remains to be done in this area.

Unlike most metabolic processes, biological clocks and the circadian rhythms they control are affected little by temperature. Somehow, a biological clock compensates for temperature shifts. This adjustment is essential, for a clock that speeds up or slows down with the rise and fall of outside temperature would be an unreliable timepiece.

In attempting to answer questions about biological clocks, it is essential to distinguish between the clock and the processes it controls. You could think of the sleep movements of leaves as the "hands" of a biological clock, but they are not the essence of the clockwork itself. You can restrain the leaves of a bean plant for several hours so that they cannot move. But on release, they will rush to the position appropriate for the time of day. Thus, we can interfere with an organism's rhythmic activity, but its biological clock goes right on ticking. In the next module, we'll learn more about the interface between a plant's internal biological clock and the external environment.

? It has been hypothesized that biological clocks in plants are controlled by the synthesis of a protein that, once it accumulates to a high enough concentration, shuts off its own gene. What kind of feedback does such a mechanism represent?

■ Negative feedback, where the result of a process (in this case, a protein) shuts off that process (the production of the protein by the gene)

A biological clock not only times a plant's everyday activities, but may also influence seasonal events that are important in a plant's life cycle. Flowering, seed germination, and the onset and ending of dormancy are all stages in plant development that usually occur at specific times of the year. The environmental stimulus plants most often use to detect the time of year is called **photoperiod,** the relative lengths of day and night.

Plants whose flowering is triggered by photoperiod fall into two groups. One group, the **short-day plants,** generally flower in late summer, fall, or winter, when light periods shorten. Chrysanthemums and poinsettias are examples of short-day plants. In contrast, **long-day plants,** such as spinach, lettuce, iris, and many cereal grains, usually flower in late spring or early summer, when light periods lengthen. Spinach, for instance, flowers only when daylight lasts at least 14 hours. (Some plants, such as dandelions, are day-neutral; their flowering is unaffected by photoperiod.)

In the 1940s, researchers discovered that flowering and other responses to photoperiod are actually controlled by *night* length, not day length. In fact, the so-called short-day plants are actually long-night plants, and the so-called long-day plants are actually short-night plants. However, the day-length terms are embedded firmly in the literature of plant biology.

Figure 33.11 illustrates the evidence for the night-length effect and also shows the difference between the flowering response of a short-day plant and a long-day plant. The left side of the figure represents short-day plants. Notice that a short-day plant ❶ will not flower ❷ until it is exposed to a *continuous* dark period exceeding a critical length (about 14 hours, in this case). The continuity of darkness is important. ❸ The short-day plant will not blossom if the nighttime part of the

photoperiod is interrupted by even a brief flash of light. (There is no effect if the daytime portion of the photoperiod is broken by a brief exposure to darkness.)

Florists apply this information about short-day plants to bring us flowers out of season. Chrysanthemums, for instance, are short-day plants that normally bloom in the autumn, but their blooming can be stalled until Mother's Day in May by punctuating each long night with a flash of light, thus turning one long night into two short nights.

The right side of the figure demonstrates the effect of night length on a long-day plant. In this case, ❹ flowering occurs when the night length is *shorter* than a critical length (less than 10 hours, in this example). ❺ A dark interval that is too long will not produce flowers. Additionally, ❻ flowering can be induced in a long-day plant by a flash of light during the night.

Notice that every species of plant has a critical night length, but how that critical night length affects flowering varies with the type of plant. In short-day plants, the critical night length is the *minimum* number of hours of darkness required for flowering; less darkness prevents flowering. By contrast, in long-day plants, this critical night length is the *maximum* number of hours of darkness required for flowering; less darkness promotes flowering.

? A particular short-day plant won't flower in the spring. We try to induce flowering by using a short dark interruption to split the long-light period of spring into two short-light periods. What result do you predict?

■ The plants still won't flower because it is actually night length, not day length, that counts in the photoperiodic control of flowering.

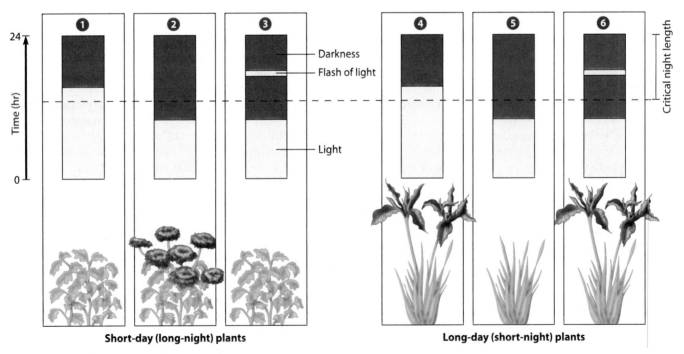

Short-day (long-night) plants **Long-day (short-night) plants**

Figure 33.11 Photoperiodic control of flowering

33.12 Phytochrome is a light detector that may help set the biological clock

The discovery that photoperiod (specifically night length) determines the seasonal responses of plants poses another question: How does a plant actually measure photoperiod? Much remains to be learned about this, but photoreceptive pigments called phytochromes are part of the answer. **Phytochromes** are proteins with a light-absorbing component. Because the light absorbed is at the red end of the spectrum (while light in the blue/green range is reflected), the molecules appear blue or bluish green (see Module 7.6).

Phytochromes were discovered during studies on how different wavelengths of light affect seed germination. Red light, with a wavelength of 660 nanometers (nm), was found to be most effective at increasing germination. Light of a longer wavelength, called far-red light (730 nm, near the edge of human visibility), both inhibited germination and reversed the effect of red light.

How do phytochromes respond differently to different wavelengths of light? The key to this ability is that a phytochrome molecule changes back and forth between two forms that differ slightly in structure. One form absorbs red light and the other absorbs far-red light. The two forms of a phytochrome are designated P_r (red absorbing) and P_{fr} (far-red absorbing). As diagrammed in Figure 33.12A, when the P_r form absorbs red light (660 nm), it is quickly converted to P_{fr}, and when P_{fr} absorbs far-red light (730 nm), it is slowly converted back to P_r.

The P_r form of phytochrome slowly accumulates in the continuous darkness that follows sunset. Each night, new phytochrome is synthesized in the P_r form, and P_{fr} is broken down by enzymes more readily than P_r. Also, P_{fr} in some plant species slowly reverts to P_r in the dark. After sunrise, the red wavelengths of sunlight cause much of the phytochrome to be rapidly converted from the P_r form to P_{fr}. It is this sudden increase in P_{fr} each day at dawn that resets a plant's biological clock. Interactions between phytochrome and the biological clock enable plants to measure the passage of night and day. In doing so, the clock monitors photoperiod and cues appropriate seasonal physiological responses, such as seed germination, flowering, and the beginning and ending of bud dormancy.

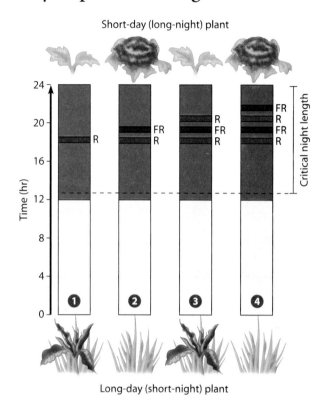

Figure 33.12B The reversible effects of red and far-red light

The consequences of this phytochrome switch are shown in Figure 33.12B. Bar ❶ shows the results we saw in the previous module for both short-day and long-day plants that receive a flash of light during their critical dark period. The letter R on the light flash stands for red light. The other three bars show how flashes of far-red (FR) light affect flowering. Bar ❷ reveals that the effect of a flash of red light that interrupts a period of darkness can be reversed by a subsequent flash of far-red light: Both types of plants behave as though there is no interruption in the night length. Bars ❸ and ❹ indicate that no matter how many flashes of light a plant receives, only the wavelength of the *last* flash affects the plant's measurement of night length.

Plants also have a group of blue-light photoreceptors that control such light-sensitive plant responses as phototropism and the opening of stomata at daybreak. Light is an especially important environmental factor in the lives of plants, and diverse receptors and signaling pathways have evolved that mediate a plant's responses to light.

Web Activity *Flowering Lab*

❓ How do phytochrome molecules help the plant recognize dawn each day?

■ Phytochrome molecules are mainly in the P_r form during the night. The sudden conversion of P_r to P_{fr}, due to the absorption of the red wavelengths of sunlight, signals dawn.

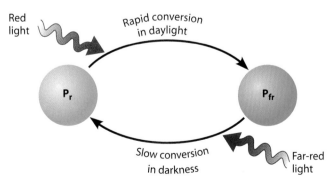

Figure 33.12A Interconversion of the two forms of phytochrome

33.13 Joanne Chory studies the effects of light and hormones in the model plant *Arabidopsis*

Figure 33.13   Joanne Chory

Joanne Chory (Figure 33.13) is a professor of biology and Howard Hughes Medical Institute Investigator at the Salk Institute for Biological Studies in La Jolla, California. Dr. Chory's research has revealed key steps in the signal transduction pathways by which light regulates the development of plants. As is often the case in biology, Dr. Chory's success has depended in part on selecting an appropriate organism as a research model—in this case, the mustard plant *Arabidopsis,* a model organism for the study of plants in the laboratory. In a recent interview, Dr. Chory commented on how the study of this tiny mustard plant has had agricultural applications:

One example is what we've learned from *Arabidopsis* mutants about how the hormone ethylene functions in fruit ripening. The same genes responsible for the ethylene pathway in *Arabidopsis* are found in such fruits as tomatoes, and understanding how these genes work enables us to control the ripening process. Another application is that identifying genes in *Arabidopsis* can help breeders of crop plants. . . . For instance, sorghum would not normally grow in Texas, but breeders have selected for a mutation affecting a photoreceptor in the plant that we know, based on *Arabidopsis* research, would allow sorghum to complete its life cycle in Texas fields.

Plants use various photoreceptors to detect light. These photoreceptors then trigger complex signal transduction pathways that mediate a plant's many important responses to light. Dr. Chory's research with *Arabidopsis* mutants that are affected abnormally by light has led to her discovery of the role of steroid hormones in this process and the identification of a plant steroid hormone receptor. What is she learning about such hormones?

Plant steroids, which are called brassinosteroids, do a lot of the same kinds of things as sex steroids do in humans. The more steroid a plant has, the bigger and tougher and more robust it is. Steroids also regulate sexual reproduction in plants. I think it's interesting how a certain group of molecules began functioning in diverse organisms as signaling molecules. Many of the enzymes a plant uses to make its steroids are also found in animals that make their own types of steroids. So some of the genes for these enzymes have probably been conserved since plants and animals diverged from a common ancestor over a billion years ago.

Dr. Chory's research relates to the signal transduction pathways you learned about in Chapters 11 and 26, to the importance of light in regulating the lives of plants, which you read about in this chapter, and to the central theme of biology—evolution.

? **Why is *Arabidopsis* called the "laboratory mouse" of plant biology?**

■ This plant serves as a model system in which researchers can study many complex processes.

33.14 Defenses against herbivores and infectious microbes have evolved in plants

Plants are at the base of most food webs and are therefore subject to attack by a wide range of **herbivores** (plant-eating animals). There are also a number of pathogens that have the potential to damage or kill plants. A wide variety of defensive systems that counter these threats have evolved in plants.

Defenses Against Herbivores   Plants counter herbivores with both physical defenses, such as thorns, and chemical defenses, such as distasteful or toxic compounds. For example, some plants produce an unusual amino acid called canavanine. Canavanine resembles arginine, one of the 20 amino acids normally used to make proteins. If an insect eats a plant containing canavanine, the molecule is incorporated into the insect's proteins in place of arginine. Because canavanine is different enough from arginine to change the shape and hence the function of proteins, the insect is harmed.

Some plants even recruit predatory animals that help defend the plants against certain herbivores. Such helpful predators include wasps that kill caterpillars feeding on plants. The recruitment process is outlined in Figure 33.14A. ❶ When a caterpillar bites into the plant, the physical damage to the plant and a chemical in the caterpillar's saliva together trigger ❷ a signal transduction pathway within the plant cells. The pathway leads to a specific cellular response: ❸ the synthesis and release

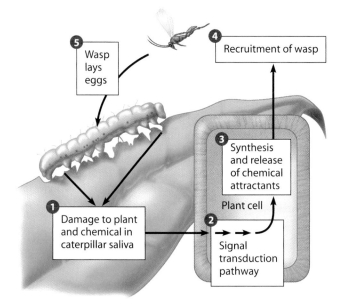

Figure 33.14A Recruitment of a wasp in response to an herbivore

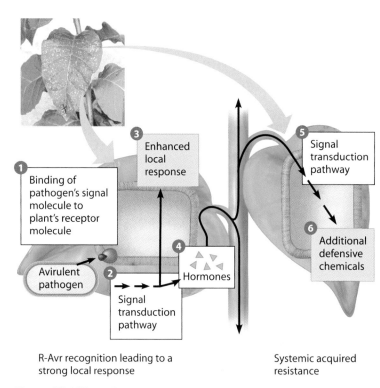

R-Avr recognition leading to a strong local response

Systemic acquired resistance

Figure 33.14B Defense responses against an avirulent pathogen

of volatile (gaseous) chemicals that ❹ attract ("recruit") the wasp. ❺ The wasp injects its eggs into the caterpillar. When the eggs hatch, the wasp larvae eat their way out of the caterpillar, killing it.

Defenses Against Pathogens

A plant, like an animal, is subject to infection by pathogenic microbes: viruses, bacteria, and fungi. And as in animals, defense systems that hinder infection and counter pathogens that do manage to infect have also evolved in plants.

A plant's first line of defense against infection is the physical barrier of the plant's "skin," the epidermis (see Chapter 31). However, microbes can cross this barrier through wounds or through natural openings such as stomata. Once infected, the plant uses chemicals as a second line of defense. Plant cells damaged by the infection release microbe-killing molecules and, in addition, chemicals that signal nearby cells to mount a similar chemical defense. Infection also stimulates chemical changes in the plant cell walls, which toughen the walls and thus slow the spread of the microbe within the plant.

The plant's chemical defense system is enhanced by the plant's inherited ability to recognize certain pathogens. A kind of "compromise" has coevolved between plants and most of their pathogens: The pathogen gains enough access to its host to perpetuate itself without severely harming the plant. The plant is said to be resistant to that pathogen, and the pathogen is said to be avirulent to the plant.

This resistance to destruction by a specific pathogen is based on the ability of the plant and the microbe to make a complementary pair of molecules. A plant has many *R* genes (for *resistance*), and each pathogen has a set of *Avr* genes (for *avirulence*). Researchers hypothesize that an *R* gene encodes a receptor protein on the plant's cells and that the complementary *Avr* gene leads to the production of some "signal" molecule of the pathogen that binds specifically to that receptor. Recognition

of pathogen-derived molecules by R proteins triggers a signal transduction pathway that leads to a defense response in the infected plant tissue.

Figure 33.14B shows this interaction and subsequent events in the plant. ❶ The binding of the pathogen's signal molecule (magenta) to the plant's receptor (purple) triggers ❷ a signal transduction pathway, which leads to ❸ a defense response that is much stronger than would occur without the R-Avr matchup. The cells at the site of infection mount a vigorous chemical defense, tightly seal off the area, and then kill themselves. The spots on one of the leaves in the photo in Figure 33.14B result from this sort of local response. As sick as such a leaf appears, it will survive.

The defense response at the site of infection helps protect the rest of the plant in yet another way. Among the signal molecules produced there are ❹ hormones that sound an alarm throughout the plant. ❺ At destinations distant from the original site, these hormones trigger signal transduction pathways leading to ❻ the production of additional defensive chemicals. This defense response, called **systemic acquired resistance,** provides protection against a diversity of pathogens for days.

Researchers suspect that one of the alarm hormones is salicylic acid, a compound whose pain-relieving effects led early cultures to use the salicylic acid-rich bark of willows (*Salix*) as a medicine. Aspirin is a chemical derivative of this compound. With the discovery of systemic acquired resistance, biologists have learned one function of salicylic acid in plants.

? What is released at a site of infection that triggers the development of general resistance to pathogens elsewhere in the plant?

■ A hormone

33.15 Plant biochemist Eloy Rodriguez studies how animals use defensive chemicals made by plants

Figure 33.15
Eloy Rodriguez in his research laboratory

Dr. Eloy Rodriguez (Figure 33.15), James A. Perkins Professor of Environmental Biology at Cornell University, is one of the world's leading experts on defensive chemicals produced by plants. In a recent interview, he described the significance of these chemicals this way:

> A plant's ability to survive is really due to chemistry. A large array of organisms eat plants, and a plant can't just get up and run. Natural selection favors those plants with the right kinds of chemical compounds that ward off fungi, bacteria, viruses, insects, and large herbivores.

Dr. Rodriguez spends much of his time in his laboratory, but he is also a field biologist. His studies in tropical rain forests have had far-reaching influence:

> One project that I got involved with has now developed into a discipline. It's called zoopharmacognosy, the study of how animals possibly medicate themselves with plants. I got involved with Richard Wrangham, a primatologist who was studying primates in the Kibale Forest of Uganda. In one observation, researchers followed a particular chimp for several days, and for two whole days this animal concentrated on one plant species. It wouldn't eat the whole plant. It would take off the leaves, crack the stalk, and then suck out the juice.
>
>   Richard pointed out to me that sometimes chimps seem to select young leaves from certain plant species. These animals get up in the morning, make a beeline toward these plants, and take a certain amount of the young leaves. We calculated that they were more or less getting a set dosage of the drug or drugs in the leaf.

Some of the compounds that Rodriguez and other researchers have discovered by following apes and other animals have potential use in human medicine:

> Some of the compounds we've identified kill parasitic worms, and some may be useful against tumors. . . . A lot of plant collection is done randomly—you go out and collect a bunch of bark, for example. It is a very tedious way to go about getting drugs. It's nice when you have animals basically telling you, "Here, try these leaves."

Additionally, as Dr. Rodriguez relates, modern scientists also benefit from observing native peoples:

> Indigenous people have been extremely successful [at using medicinal plants]. Anthropologists have documented that they extract plant materials, grind them up, filter them, and treat them with burned leaves (which act as a base)—almost the same process that I use. They get out materials that are relatively pure for use as medicine. A lot of the drug companies started out thanks to those folks.

Though his work focuses on the medicinal potential of rain forest chemicals, Rodriguez is quick to emphasize that the importance of rain forests goes beyond human concerns:

> The more we search the rain forest, the more in awe of it we are. We see how important it is. In the tropical rain forest, we are talking about the ultimate diversity in plants, animals, fungi—life! The arguments that have been made for preserving the forests are excellent. We are talking about the health of the planet.

Rodriguez's words lead us to our next and final unit, Ecology. In these chapters, we will explore connections between organisms and see how organisms interact with their environments.

? What do researchers hope to learn by observing the feeding behavior of sick chimpanzees?

■ The selective feeding of the ill animals on certain parts of specific plants might lead researchers to medicines that could help humans, too.

Chapter Review

Reviewing the Concepts
Plant Hormones (Introduction–33.8)

Plant hormones, such as isoflavones from soy, may provide human health benefits **(Introduction)**. Hormones coordinate the activities of plant cells and tissues. Experiments carried out by Darwin and others showed that the tip of a grass seedling detects light and transmits a signal down to the growing region of the shoot. This signal is a hormone named auxin. Auxin promotes faster cell elongation on the shaded site of the root **(33.1)**. By triggering signal transduction pathways, very small amounts of plant hormones regulate plant growth and development **(33.2)**.

Major types of hormones. Plants produce auxin (IAA) in the apical meristems at the tips of shoots. At different concentrations, auxin stimulates or inhibits the elongation of shoots and roots. It may act by weakening cell walls, allowing them to stretch when cells take up water. Auxin also stimulates the development of vascular tissues and cell division in the vascular cambium, promoting growth in stem diameter **(33.3)**. Cytokinins, produced by growing roots, embryos, and fruits, are hormones that promote cell division. Cytokinins from roots may balance the effects of auxin from apical meristems, causing lower buds to develop into branches **(33.4)**. Gibberellins stimulate the elongation of stems and leaves and the development of fruit. Gibberellins released from embryos function in some of the early events of seed germination **(33.5)**. Abscisic acid (ABA) inhibits the germination of seeds. The ratio of ABA to gibberellins often determines whether a seed will remain dormant or germinate. Seeds of many plants remain dormant until their ABA is inactivated or washed away. ABA also acts as a "stress hormone," causing stomata to close when a plant is dehydrated **(33.6)**. As fruit cells age, they give off ethylene gas, which hastens ripening. A changing ratio of auxin to ethylene, triggered mainly by shorter days, probably causes autumn color changes and the loss of leaves from deciduous trees **(33.7)**. Plant hormones have a variety of agricultural uses. Farmers use auxin to delay or promote fruit drop. Auxin and gibberellins are used to produce seedless fruits. A synthetic auxin called 2,4-D is used to kill weeds. There are questions about the safety of using such chemicals **(33.8)**.

Growth Responses and Biological Rhythms in Plants (33.9–33.13)

Tropisms.   Plants sense and respond to environmental changes in a variety of ways. Tropisms are growth responses that change the shape of a plant or make it grow toward or away from a stimulus:

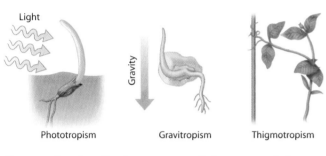

Phototropism   Gravitropism   Thigmotropism

Phototropism, bending in response to light, may result from auxin moving from the light side to the dark side of a stem. A response to gravity, or gravitropism, may be caused by the settling of special organelles on the low sides of shoots and roots, which may trigger a change in the distribution of hormones. Thigmotropism, a response to touch, is responsible for the coiling of tendrils and vines around objects **(33.9)**.

Biological clocks and circadian rhythms.   An internal biological clock controls sleep movements and other daily cycles in plants. These cycles, called circadian rhythms, persist with periods of about 24 hours even in the absence of environmental cues, but such cues are needed to keep them synchronized with day and night **(33.10)**. Plants mark the seasons by measuring photoperiod, the relative lengths of night and day. The timing of flowering is one of the seasonal responses to photoperiod **(33.11)**:

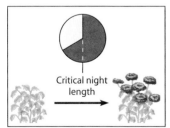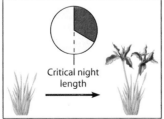

Short-day (long-night) plants   Long-day (short-night) plants

Light-absorbing proteins called phytochromes may help plants set their biological clock and monitor photoperiod **(33.12)**. A small, wild mustard called *Arabidopsis* is a popular model organism for plant molecular biologists **(33.13)**.

Plant Defenses (33.14–33.15)

Defensive chemicals.   Plants use chemicals to defend themselves against both herbivores and pathogens. So-called avirulent plant pathogens interact with host plants in a specific way that stimulates both local and systemic defenses in the plant. Local defenses include microbe-killing chemicals and sealing off the infected area. Hormones trigger generalized defense responses in other organs (systemic acquired resistance) **(33.14)**. Some animals may medicate themselves by eating plants containing certain defensive chemicals **(33.15)**.

Connecting the Concepts

1. Test your knowledge of the five major classes of plant hormones (auxins, cytokinins, gibberellins, abscisic acid, ethylene) by matching one hormone to each lettered box. (Note that some hormones will match up to more than one box.)

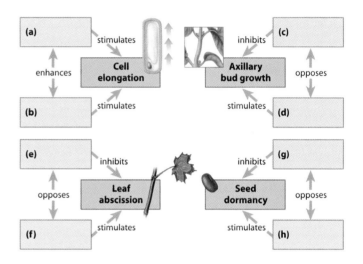

Testing Your Knowledge
Multiple Choice

2. During winter or periods of drought, which one of the following plant hormones inhibits growth and seed germination?
   a. ethylene
   b. abscisic acid
   c. gibberellin
   d. auxin
   e. cytokinin

3. A certain short-day plant flowers only when days are less than 12 hours long. Which of the following would cause it to flower? (*Explain your answer.*)
   a. a 9-hour night and 15-hour day with 1 minute of darkness after 7 hours
   b. an 8-hour day and 16-hour night with a flash of white light after 8 hours
   c. a 13-hour night and 11-hour day with 1 minute of darkness after 6 hours
   d. a 12-hour day and 12-hour night with a flash of red light after 6 hours

4. Auxin causes a shoot to bend toward light by
   a. causing cells to shrink on the dark side of the shoot.
   b. stimulating growth on the dark side of the shoot.
   c. causing cells to shrink on the lighted side of the shoot.
   d. stimulating growth on the lighted side of the shoot.
   e. inhibiting growth on the dark side of the shoot.

5. In the autumn, the amount of _____ increases and _____ decreases in fruit and leaf stalks, causing a plant to drop fruit and leaves.
a. ethylene . . . auxin
b. gibberellin . . . abscisic acid
c. cytokinin . . . abscisic acid
d. auxin . . . ethylene
e. gibberellin . . . auxin

6. Plant hormones act by affecting the activities of
a. genes.
b. membranes.
c. enzymes.
d. genes, membranes, and enzymes.
e. genes and enzymes.

7. Buds and sprouts often form on tree stumps. Which of the following hormones would you expect to stimulate their formation?
a. auxin
b. cytokinin
c. abscisic acid
d. ethylene
e. gibberellin

8. A plant's defense response at the site of initial infection by a pathogen will be especially strong if
a. the pathogen is virulent.
b. the plant makes a receptor protein that recognizes a signal molecule from the microbe.
c. the pathogen is a fungus.
d. the plant has an *Avr* gene that is the right match for one of the microbe's *R* genes.
e. the right combination of hormones travel from the infection site to other parts of the plant.

Matching

9. Bending of a shoot toward light
10. Growth response to touch
11. A cycle with a period of about 24 hours
12. Pigment that helps control flowering
13. Relative lengths of night and day
14. Growth response to gravity
15. Folding of plant leaves at night

a. phytochrome
b. photoperiod
c. sleep movement
d. circadian rhythm
e. thigmotropism
f. phototropism
g. gravitropism

Describing, Comparing, and Explaining

16. If apples are to be stored for long periods, it is best to keep them in a place with good air circulation. Explain why.

17. Write a short paragraph explaining why a houseplant becomes bushier if you pinch off its terminal buds.

Applying the Concepts

18. Jon just started a new job as night watchman at a plant nursery. His boss told him to stay out of a room where chrysanthemums (which are short-day plants) were about to flower. Around midnight, looking for the restroom, Jon accidentally opened the door to the chrysanthemum room and turned on the lights for a moment. How might this affect the chrysanthemums? What could Jon do to correct his mistake?

19. A plant biologist observed a peculiar pattern when a tropical shrub was attacked by caterpillars. The biologist noticed that after a caterpillar ate a leaf, it would skip over nearby leaves and attack a leaf some distance away. The researcher found that when a leaf was eaten, nearby leaves started making a chemical that deterred the caterpillars. Simply removing a leaf did not trigger the same change nearby. The biologist suspected that a damaged leaf sent out a chemical that signaled other leaves. How could this hypothesis be tested?

20. In the 1950s, scientists discovered that many plants emit a gaseous compound called isoprene into the atmosphere. Isoprene is a precursor of the cytokinin hormones, but why plants emit it remains an open question. Researchers have found that plants emit isoprene when they are photosynthesizing, and the amount emitted increases as the temperature of a plant's leaves increases. Plants also synthesize more isoprene when they cannot obtain enough water. Darkness, substances that inhibit photosynthesis, and pure nitrogen gas (N_2) slow or stop isoprene emission. One hypothesis is that isoprene emissions help prevent damage to plants caused by high temperatures. Chlorophyll fluorescence has been used as a measure of irreversible leaf damage; the more fluorescence, the greater the damage. The graph below displays the results of some experimental tests of this hypothesis.

The red line on the graph shows the effect of increasing temperature on leaves in air with no isoprene. The green line shows the effect on leaves from the same species of plant in air containing isoprene. Do these results support the hypothesis? Explain your answer. The leaves were exposed to N_2 while the recordings were made. Why do you suppose the experimenters did this?

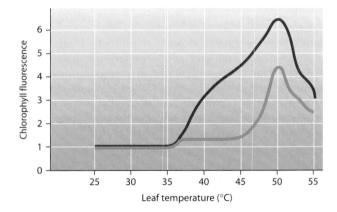

21. Imagine the following scenario: A plant scientist has developed a synthetic chemical that mimics the effects of a plant hormone. The chemical can be sprayed on apples before harvest to prevent flaking of the natural wax that is formed on the skin. This makes the apples shinier and gives them a deeper red color. What kinds of questions do you think should be answered before farmers start using this chemical on apples? How might the scientist go about finding answers to these questions?

Answers to all questions can be found in Appendix 4.

For Practice Quizzes, BioFlix, MP3 Tutors, and Activities, go to www.mybiology.com.

Ecology

Life from Top to Bottom

Describing his unique perspective from Earth orbit, Apollo astronaut Rusty Schweickart wrote: "On that small blue-and-white planet below is everything that means anything to you. National boundaries and human artifacts no longer seem real. Only the biosphere, whole and home of life." The **biosphere,** which extends from the atmosphere several kilometers above the Earth to the depths of the oceans, is all of the Earth that is inhabited by life. It encompasses a multitude of wildly diverse environments, some hospitable to life and others so hostile that the presence of any organism is surprising.

At the top of the world, the rocky slopes of Mount Everest, in the eastern Himalayas, seems an unlikely place for living things to flourish. But in spite of the frigid air, thin soil, and brief growing season, meter-high stalks of *Rheum nobile* ("noble rhubarb") stand like pale candles against the bleak landscape (see photo on facing page). In springtime, the alpine meadows nearby are filled with the colorful blooms of dozens of species, including Himalayan blue poppies, whose flowers are so stunning that gardeners throughout the world aspire to meet the plants' exacting requirements. And where there are plants to eat, animals thrive, too. Nimble-footed blue sheep and other hoofed animals graze on the hardy herbaceous plants of the meadow. The graceful snow leopard (see photo on preceding page) preys on the sheep and other herbivorous mammals.

Tube worms (Riftia pachyptila) near hydrothermal vent

Kiwa hirsuta (Yeti crab)

Over millions of years, natural selection by this harsh environment of the Himalayas has resulted in adaptations that equip the resident plants and animals to cope with the rigorous conditions. For example, the overlapping, cream-colored leaves on the noble rhubarb are an adaptation that moderates the environment of the plant's reproductive structures. Like miniature greenhouses, the leaves trap heat that warms the tiny flowers beneath. The leaves also contain sunscreen-like chemicals that block the strong ultraviolet radiation that is a danger to life at high altitudes. The adaptations of snow leopards include an exceptionally long, dense coat and large fur-covered paws that act as snowshoes. Large nasal cavities warm the air the snow leopard inhales. Its ears, small in comparison to those of its relatives in warm climates, reduce heat loss by minimizing surface area. The thickly furred tail, almost as long as its body, curls cozily around the entire animal at naptime.

While the top of the world may seem like an improbable place to find life, the bottom of the world—the darkest depths of the ocean—seems like an impossible environment. For the alpine meadows of the Himalayas, and for most other life in the biosphere, the sun is the main source of energy. In 1977, scientists exploring the seafloor at a depth of 2,500 m (roughly 1.5 miles), far beyond the reach of sunlight, were astonished to discover organisms whose energy source comes from the interior of the planet. In the unique world of hydrothermal vents, sites near the adjoining edges of giant plates of Earth's crust, molten rock and hot gases surge upward from Earth's interior. Towering chimneys, perhaps as much as 30 m high, emit scalding water and hot gases such as hydrogen sulfide (H_2S). Yet this environment, too, is home to organisms capable of overcoming its challenges.

Life at hydrothermal vents depends on chemoautotrophic sulfur bacteria, most of which obtain energy by oxidizing hydrogen sulfide. Many of the animals in these communities, including the tube worms shown in the photo at the top of the page, harbor sulfur bacteria within their bodies. The animals absorb sulfur compounds from the water; the bacteria use the compounds as an energy source to convert CO_2 from seawater into organic food molecules that, in turn, provide nutrition for their hosts. These animals are food for others in the thriving hydrothermal vent community, which also includes sea anemones, giant clams, shrimps, crabs, and a few fishes. In 2005, a group of researchers studying the animals around hydrothermal vents near Easter Island, in the Pacific Ocean, were intrigued by the sight of unusually large (15-cm long) crabs. The crew scooped up a specimen (photo above) to investigate further. After genetic analysis, the scientists concluded that the crab was so unusual that it belonged to an entirely new family. They called the crab *Kiwa* (after a Polynesian goddess of crustaceans) *hirsuta* (from the Latin *hirsute,* hairy), and gave it the common name "Yeti crab." Other studies showed that the Yeti crab's bristly hairs are coated with colonies of filamentous bacteria. Some scientists speculate that the Yeti crab gathers the bacteria to eat, but for now, the bacteria's presence is just one of many puzzles about the bizarre-looking crabs. Each new species that is discovered brings new mysteries, and thousands more await in the vast realm of the seafloor.

From the roof of the world to the deepest ocean, the biosphere is a treasure trove of life that is still being explored. **Ecology** (from the Greek *oikos,* home), the scientific study of the interactions of organisms with their environments, offers a context for new discoveries, a means of understanding familiar forms of life, and insight into our own role in the biosphere. ■ ■ ■

Rheum nobile (noble rhubarb)

34.1 Ecologists study how organisms interact with their environment at several levels

Ecologists describe the distribution and abundance of organisms—where they live and how many live there. In order to explain these observations, ecologists investigate the interactions between organisms and their environment. Because the environment is complex, organisms can potentially be affected by many different variables. Ecologists group these variables into two major types, biotic factors and abiotic factors. **Biotic factors,** which include all of the organisms in the area, are the living component of the environment. **Abiotic factors** are the environment's nonliving component, the physical and chemical factors such as temperature, forms of energy available, water, and nutrients. An organism's **habitat,** the specific environment it lives in, includes the biotic and abiotic factors present in its surroundings.

As you might expect, field research—both discovery and hypothesis-based science—is fundamental to ecology. But ecologists also test hypotheses using laboratory experiments, where conditions can be simplified and controlled. Some ecologists take a theoretical approach, devising mathematical and computer models that enable them to simulate large-scale experiments that are impossible to conduct in the field.

Ecologists study environmental interactions at several levels. At the **organism** level, they may examine how one kind of organism meets the challenges and opportunities of its environment through its physiology or behavior. An ecologist working at this level might study, for instance, adaptations of the Himalayan blue poppy (*Meconopsis betonicifolia,* Figure 34.1A) to the freezing temperatures and short days of its abiotic environment.

Another level of study in ecology is the **population,** a group of individuals of the same species living in a particular geographic area. The Himalayan blue poppies living in a particular alpine meadow would constitute a population (Figure 34.1B). An ecologist studying blue poppies might investigate factors that affect the size of the population, such as the availability of chemical nutrients or seed dispersal.

A third level, the **community** (Figure 34.1C), is an assemblage of all the populations of organisms living close enough together for potential interaction—all of the biotic factors in the

environment. All the organisms in a particular alpine meadow would constitute a community. An ecologist working at this level might focus on interspecies interactions, such as the competition between poppies and other plants for soil nutrients or the effect of plant-eaters on poppies.

The fourth level of ecological study, the **ecosystem** (Figure 34.1D), includes both the biotic and abiotic components of the environment. Some critical questions at the ecosystem level concern how chemicals cycle and how energy flows between organisms and their surroundings. For an alpine meadow, one ecosystem-level question would be, How rapidly does the decomposition of decaying plants release inorganic molecules?

Some ecologists take a wider perspective by studying **landscapes,** which are arrays of ecosystems. Landscapes are usually visible from the air as distinctive patches. For example, the Himalayan alpine meadows are part of a mountain landscape that also includes conifer and broadleaf forests. A landscape perspective emphasizes the lack of clearly defined ecosystem boundaries; energy, materials, and organisms may be exchanged by ecosystems within a landscape.

🎧 MP3 Tutor *Ecological Hierarchy*

? List the biotic and abiotic factors in the hydrothermal vent ecosystems described in the opening essay.

■ The biotic factors include chemosynthetic bacteria, sea anemones, giant clams, shrimp, crabs, fish, and tube worms. Abiotic factors include hydrogen sulfide and temperatures ranging from boiling to extremely cold.

Figure 34.1C Community

Figure 34.1A
Organism

Figure 34.1B Population

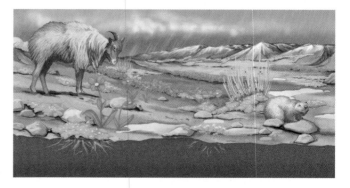

Figure 34.1D Ecosystem

34.2 The science of ecology provides insight into environmental problems

The "control of nature" is a phrase conceived in arrogance, born of the Neanderthal age of biology and philosophy, when it was supposed that nature exists for the convenience of man.

— Rachel Carson, *Silent Spring*

An airplane trip almost anywhere in the United States reveals the extent to which humans have changed the landscape (Figure 34.2A). But the cities, farms, and highways visible from the air are only the tip of the vast and largely unseen iceberg of human impact on the environment. Overgrazing, deforestation, and overcultivation of land have caused massive soil erosion. Industrial and agricultural practices have spread chemical pollutants throughout the biosphere. Manmade products have contaminated the groundwater, water that has penetrated the soil and may eventually resurface in the drinking supply. In some areas, groundwater is being depleted by the ever-expanding human population. Hundreds of species of organisms are in danger of joining the growing list of those that we have already driven to extinction. Most worrisome of all, human activities are producing potentially catastrophic changes in the global climate.

Although most of our environmental problems have been many decades in the making, few people were aware of them until the 1960s. Technology had been advancing rapidly, and humankind seemed poised on the brink of freedom from several age-old bonds. New chemical fertilizers and pesticides, for example, showed great promise for increasing agricultural productivity. Massive applications of fertilizer and extensive aerial spraying of pesticides, such as DDT, brought astonishing increases in crop yields. Pesticides were also used to combat disease-carrying insects, including mosquitoes (which transmit malaria), body lice (typhus), and fleas (plague).

But two events raised questions about the long-term effects of widespread DDT use. The first event was the evolution of pesticide resistance in insects, rendering this widely used pesticide less effective (see Module 13.3). The second event was the publication of a book called *Silent Spring* by Rachel Carson (Figure 34.2B). As DDT lost its killing power and the chemical companies introduced ever more potent pesticides, Carson sought to alert the general public to the dangers of global pesticide use. Although Carson and her work were harshly criticized by the agriculture and pesticide industries, scientists had already begun to document the harmful effects of DDT on predatory birds. Moreover, traces of the pesticide were turning up worldwide, thousands of miles from where it had been applied. DDT even showed up in human milk.

Figure 34.2B Rachel Carson

Public awareness of the problems caused by pesticides quickly developed into a general concern about a host of environmental issues. The 1970s brought a series of legislative acts aimed at curbing pollution and cleaning up the environment. Despite these efforts, serious issues clearly remain.

The science of ecology can provide the understanding needed to resolve environmental problems. But these problems cannot be solved by ecologists alone, because they require making decisions based on values and ethics. On a personal level, each of us makes daily choices that affect our ecological impact (see Module 36.11). Legislators and corporations, motivated by environmentally aware voters and consumers, must address questions that have wider implications. "How should land use be regulated? Should we try to save all species, or just certain ones? What alternatives to environmentally destructive practices can be developed? How can we balance environmental impact with economic needs?"

We will examine some of our environmental problems in Chapter 38. But analyzing environmental issues and planning for better practices begin with an understanding of the basic concepts of ecology, which we start to explore now.

Web Activity *Connection: DDT and the Environment*

? Why can't ecologists alone solve environmental problems?

■ The science of ecology can inform the decision-making process, but solving environmental problems involves making ethical, economic, and political judgments that are outside the realm of science.

Figure 34.2A Aerial view of a landscape changed by humans

Physical and chemical factors influence life in the biosphere

You have learned that life thrives in a wide variety of habitats, from the mountaintops to the seafloor. To be successful, the organisms that live in each place must be adapted for the abiotic factors present in those environments.

Energy Source   All organisms require a source of energy to live. Solar energy from sunlight, captured during the process of photosynthesis, powers most ecosystems. Lack of sunlight is seldom the most important factor limiting plant growth for terrestrial ecosystems (though shading by trees does create intense competition for light among plants growing on forest floors). In many aquatic environments, however, light is not uniformly available. Microorganisms, as well as the water itself, absorb light and prevent it from penetrating beyond certain depths. As a result, most photosynthesis occurs near the water's surface.

In dark environments such as caves or hydrothermal vents, bacteria that extract energy from inorganic chemicals power ecosystems. As we mentioned in the opening essay, sulfur bacteria perform this function in hydrothermal vent communities, where many of the animals either feed directly on the sulfur bacteria or derive nutrition from bacteria living inside their

Figure 34.3A   Respiratory surface of giant tube worm

bodies. For example, tube worms (Figure 34.3A) have no mouth or digestive tract. The red tip extending from the white casing is a respiratory surface that acquires oxygen and sulfide from the water. Bacteria living in a specialized organ in the worm's body use the sulfide to produce energy—a lot of energy. These worms can grow to be over 2 m long.

Temperature   Temperature is an important abiotic factor because of its effect on metabolism (see Module 5.15). Few organisms can maintain a sufficiently active metabolism at temperatures close to 0°C, and temperatures above 45°C destroy the enzymes of most organisms. However, extraordinary adaptations enable some species to live outside this temperature range. For example, bacteria living in hot springs have enzymes that function optimally at extremely high temperatures. Mammals and birds, such as the snowy owl in Figure 34.3B, can remain considerably warmer than their surroundings and can be active at a fairly wide range of temperatures. Amphibians and reptiles, which gain most of their warmth by absorbing heat from their surroundings, have a more limited distribution.

Water   Water is essential to all life. Thus, for terrestrial organisms, drying out in the air is a major danger. As we discussed

in Modules 17.1 and 19.6, watertight coverings were key adaptations for plants and vertebrates to be successful on land. Aquatic organisms are surrounded by water, but solute concentration is a problem. Freshwater organisms live in a hypotonic medium, while the environment of marine organisms is hypertonic. As we saw in Module 25.4, animals maintain fluid balance by a variety of mechanisms.

Figure 34.3B   Snowy owl

Nutrients   The distribution and abundance of photosynthetic organisms, including plants, algae, and photosynthetic bacteria, depends on the availability of inorganic nutrients such as nitrogen and phosphorus. Plants obtain these nutrients from the soil. Soil structure, pH, and nutrient content often play major roles in determining the distribution of plants. In many aquatic ecosystems, low levels of nitrogen and phosphorus limit the growth of algae and photosynthetic bacteria.

Other Aquatic Factors   Several abiotic factors are important in aquatic, but not terrestrial, ecosystems. While terrestrial organisms have a plentiful supply of O_2 from the air, aquatic organisms must depend on oxygen dissolved in water. This is a critical factor for many species of fish. Trout, for example, require high levels of dissolved oxygen. Cold, fast-moving water has a higher oxygen content than warm or stagnant water. Salinity, current, and tides may also play a role in aquatic ecosystems.

Other Terrestrial Factors   On land, wind is often an important abiotic factor. Local wind damage may create openings in forests that provide opportunities for colonization. Wind also increases an organism's rate of water loss by evaporation. The resulting increase in evaporative cooling (see Module 25.3) can be advantageous on a hot summer day, but it can cause dangerous wind chill in the winter. In some ecosystems, fire occurs frequently enough that many plants have adapted to this disturbance.

Next, we examine the interaction between one animal species and the abiotic and biotic factors of its environment.

Web Process of Science *How Do Abiotic Factors Affect Distribution of Organisms?*

? Why are birds and mammals, but not amphibians and reptiles other than birds, found in the Himalayan alpine meadows?

■ As ectotherms (see Module 25.1), reptiles and amphibians do not have adaptations that enable them to withstand the cold temperatures of the alpine habitat.

34.4 Organisms are adapted to abiotic and biotic factors by natural selection

One of the fundamental goals of ecology is to explain the distribution of organisms. The presence of a species in a particular place has two possible explanations: The species may have evolved from ancestors living in that location, or it may have dispersed to that location and been able to survive once it arrived. The pronghorn "antelope," pictured in Figure 34.4, is the descendant of ancestors that inhabited the open plains and shrub deserts of North America millions of years ago. The animal is found nowhere else and is only distantly related to the numerous species of antelope in Africa. What selective factors in the abiotic and biotic environments of its ancestors produced the adaptations we see in the pronghorn that roams the same region today?

The pronghorn's habitat is arid, windswept, and subject to extreme temperature fluctuations both daily and seasonally. Individuals able to survive and reproduce under these conditions left offspring that carried their alleles forward into subsequent generations. We can identify several adaptations in present-day pronghorns that contribute to their success in their abiotic environment. As you may recall from Module 25.3, a mammal's fur is often instrumental in thermoregulation. The pronghorn has a thick coat made of hollow hairs that trap air, insulating the animal in cold weather. Thus, if you drive through Wyoming or parts of Colorado in the winter, you will see herds of these animals foraging in the open when temperatures are well below 0°C. In hot weather, the pronghorn can raise patches of this stiff hair to release body heat. Water is rarely a problem for a pronghorn because it obtains a great deal of moisture from the vegetation it eats.

The biotic environment, which includes what the animal eats and any predators that threaten it, is also a factor in determining which members of a population survive and reproduce. The pronghorn's main foods are forbs (small broadleaf plants), grasses, and woody shrubs. Over time, characteristics that enabled the ancestors of the pronghorn to exploit these food sources more efficiently became established through natural selection. As a result, the teeth of a pronghorn are specialized for biting and chewing tough plant material. Like cows, the pronghorn's stomach contains cellulose-digesting bacteria. As the pronghorn eats plants, the bacteria digest the cellulose, and the animal obtains most of its nutrients from the bacteria. Selective pressure of wolves, coyotes, and cougars led to the pronghorn's impressive speed and endurance. Capable of sprinting about 95 km/hr (60 mph) on flat ground, it is one of the fastest mammals. An adult pronghorn can also keep up a pace of about 65 km/hr for at least 30 minutes—a definite benefit when being chased by long-distance runners such as wolves. Like many large herbivores that live in open grasslands, the pronghorn also derives protection from living in herds. When one pronghorn starts to run, its white rump patch seems to alert other herd members to danger. Other adaptations that help the pronghorn foil predators include its tan and white coat, which provides camouflage, and its keen eyes, which can detect movement at great distances.

Figure 34.4 Pronghorns (*Antilocapra americana*)

The pronghorn is a highly successful herbivorous running mammal of open country. If its environment changed significantly, the adaptations that contribute to the pronghorn's current success might not be as advantageous. For example, if an increase in rainfall turned the open plains into woodlands, where predators would be more easily hidden by vegetation and could stalk their prey at close range, the pronghorn's adaptations for escaping predators might not be as effective. Thus, in adapting populations to local environmental conditions, natural selection may limit the distribution of organisms. The absence of the pronghorn outside North America, however, does not necessarily imply that the species could not survive elsewhere; it may only mean that it never dispersed beyond this region. In the next module, we see how global climatic patterns determine temperature and rainfall, the major abiotic factors that influence the distribution of organisms. We examine the biotic components of the environment more closely in other chapters in this unit.

Web Activity Adaptations to Biotic and Abiotic Factors

? What is the role of the environment in adaptive evolution?

■ The individuals whose phenotypes are best suited to the environment (including both abiotic and biotic factors) will pass their alleles to the next generation. But individuals with other phenotypes may not. For example, if the biotic environment includes wolves, a pronghorn that is not able to run as long as the rest of the herd will probably not survive to reproduce.

Regional climate influences the distribution of terrestrial communities

When we ask what determines whether a particular organism or community of organisms lives in a certain area, the climate of the region—especially temperature and rainfall—is often a crucial part of the answer. Earth's global climate patterns are largely determined by the input of radiant energy from the sun and the planet's movement in space.

Figure 34.5A shows that because of its curvature, Earth receives an uneven distribution of solar energy. The sun's rays strike equatorial areas most directly (perpendicularly). Away from the equator, the rays strike Earth's surface at oblique angles. As a result, the same amount of solar energy is spread over a larger area. Thus, any particular area of land or ocean near the equator absorbs more heat than comparable areas in the more northern or southern latitudes.

The seasons of the year result from the permanent tilt of the planet on its axis as it orbits the sun. As Figure 34.5B shows, the globe's position relative to the sun changes through the year. The Northern Hemisphere, for instance, is tipped most toward the sun in June, creating the long days of summer in that hemisphere; at the same time, days are short and it is winter in the Southern Hemisphere. Conversely, the Southern Hemisphere is tipped farthest toward the sun in December, creating summer there and causing winter in the Northern Hemisphere. The **tropics,** the region surrounding the equator between latitudes 23.5° north (the Tropic of Cancer) and 23.5° south (the Tropic of Capricorn), experience the greatest annual input and least seasonal variation in solar radiation.

Figure 34.5C (facing page) shows some of the effects of the intense solar radiation near the equator on global patterns of rainfall and winds. Arrows indicate air movements. High temperatures in the tropics evaporate water from Earth's surface. Heated by the direct rays of the sun, moist air at the equator rises, creating an area of calm or of very light winds known as the **doldrums.** As warm equatorial air rises, it cools and releases much of its water content, creating the abundant pre-cipitation typical of most tropical regions. High temperatures throughout the year and ample rainfall largely explain why rain forests are concentrated near the equator.

After losing their moisture over equatorial zones, high-altitude air masses spread away from the equator until they cool and descend again at latitudes of about 30° north and south. This descending dry air absorbs moisture from the land. Thus, many of the world's great deserts—the Sahara in North Africa and the Arabian on the Arabian Peninsula, for example—are centered at these latitudes. As the dry air descends, some of it spreads back toward the equator. This movement creates the cooling **trade winds,** which dominate the tropics. As the air moves back toward the equator, it warms and picks up moisture until it ascends again.

The latitudes between the tropics and the Arctic Circle in the north and the Antarctic Circle in the south are called **temperate zones.** Generally, these regions have seasonal variations in climate and more moderate temperatures than the tropics or the polar zones. Notice in Figure 34.5C that some of the descending air heads into the latitudes above 30°. At first these air masses pick up moisture, but they tend to drop it as they cool at higher latitudes. This is why the north and south temperate zones, especially latitudes around 60°, tend to be moist. Broad expanses of coniferous forest dominate the landscape at these fairly wet but cool latitudes.

Figure 34.5D (facing page) shows the major global air movements, called the **prevailing winds.** Prevailing winds (pink arrows) result from the combined effects of the rising and falling of air masses (blue and brown arrows) and Earth's rotation (gray arrows). Because Earth is spherical, its surface moves faster at the equator (where its diameter is greatest) than at other latitudes. In the tropics, Earth's rapidly moving surface deflects vertically circulating air, making the trade winds blow from east to west. In temperate zones, the slower-moving surface produces the **westerlies,** winds that blow from west to east.

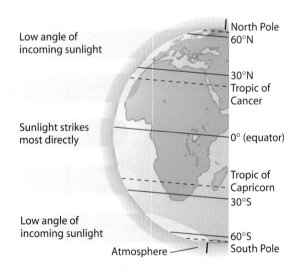

Figure 34.5A How solar radiation varies with latitude

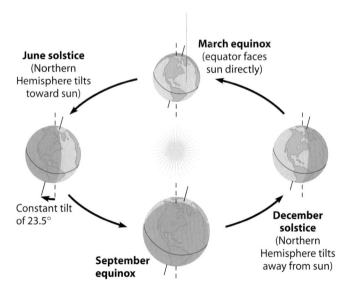

Figure 34.5B How Earth's tilt causes the seasons

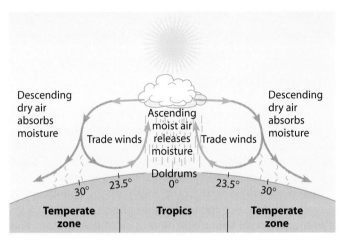

Figure 34.5C How uneven heating causes rain and winds

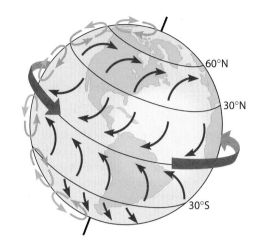

Figure 34.5D Prevailing wind patterns

A combination of the prevailing winds, the planet's rotation, unequal heating of surface waters, and the locations and shapes of the continents creates **ocean currents,** river-like flow patterns in the oceans (Figure 34.5E). Ocean currents have a profound effect on regional climates. For instance, the Gulf Stream circulates warm water northward from the Gulf of Mexico and makes the climate on the west coast of Great Britain warmer during winter than the coast of New England, which is actually farther south but is cooled by a current flowing south from the coast of Greenland.

Landforms can also affect local climate. Air temperature declines by about 6°C with every 1,000-m increase in elevation, an effect you've probably experienced if you've hiked up a mountain. Figure 34.5F illustrates the effect of mountains on rainfall. This drawing represents major landforms across the state of California, but mountain ranges cause similar effects elsewhere. California is a temperate area in which the prevailing winds are westerlies. As moist air moves in off the Pacific Ocean and encounters the westernmost mountains (the Coast Range), it flows upward, cools at higher altitudes, and drops a large amount of water. The world's tallest trees, the coastal redwoods, thrive here. Farther inland, precipitation increases again as the air moves up and over higher mountains (the Sierra Nevada). Some of the world's deepest snow packs occur here. On the eastern side of the Sierra, there is little precipitation, and the dry descending air also absorbs moisture. This effect, called a rain shadow, is responsible for the desert that covers much of central Nevada.

Climate and other abiotic factors of the environment control the global distribution of organisms. The influence of these abiotic factors results in **biomes,** major types of ecological associations that occupy broad geographic regions of land or water. Terrestrial biomes are determined primarily by temperature and rainfall—similar assemblages of plant and animal types are found in areas that have similar climates. Aquatic biomes are defined by different abiotic factors; the primary distinction is based on salinity. Marine biomes, which include oceans, intertidal zones, coral reefs, and estuaries, generally have salt concentrations around 3%, while freshwater biomes (lakes, streams and rivers, and wetlands) typically have a salt concentration of less than 1%. We describe several aquatic biomes in the next two modules.

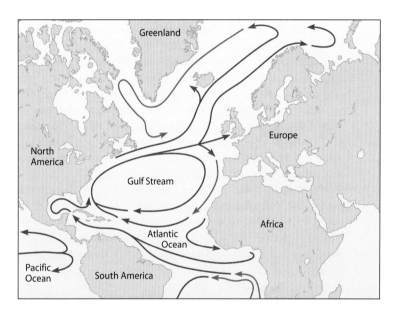

Figure 34.5E Atlantic Ocean currents (Red arrows indicate warming currents and blue arrows indicate cooling currents.)

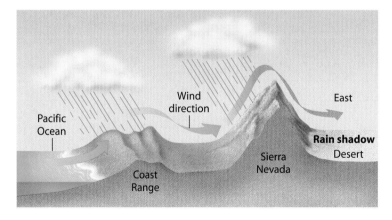

Figure 34.5F How mountains affect rainfall (California)

? What causes summer in the Northern Hemisphere?

■ Because of the fixed angle of Earth's axis relative to the orbital plane around the sun, the Northern Hemisphere is tilted toward the sun during the portion of the annual orbit that corresponds to the summer months.

34.6 Sunlight and substrate are key factors in the distribution of marine organisms

Gazing out over a vast ocean, you might think that it is the most uniform environment on Earth. But marine ecosystems can be as different as night and day. The deepest ocean, where hydrothermal vents are located, is perpetually dark. In contrast, the vivid coral reefs are utterly dependent on sunlight. Habitats near shore are different from those in mid-ocean, and the seafloor hosts different communities from the open waters.

The **pelagic realm** of the oceans includes all open water, and the seafloor is known as the **benthic realm** (Figure 34.6A). The depth of light penetration, a maximum of 200 m, marks the **photic zone.** In shallow areas such as the submerged parts of continents, called **continental shelves,** the photic zone includes both the pelagic and benthic realms. In these sunlit regions, photosynthesis by **phytoplankton** (microscopic algae and cyanobacteria) and multicellular algae provides energy for a diverse community of animals. Sponges, burrowing worms, clams, sea anemones, crabs, and echinoderms inhabit the benthic realm of the photic zone. **Zooplankton** (small, drifting animals), fish, marine mammals, and many other types of animals are abundant in the pelagic photic zone.

Coral reefs, a visually spectacular and biologically diverse biome, are scattered around the globe in the photic zone of warm tropical waters above continental shelves, as shown in Figure 34.6B on the facing page. The reef is built up slowly by successive generations of coral animals—a diverse group of

cnidarians that secrete a hard external skeleton—and by multicellular algae encrusted with limestone. Unicellular mutualistic algae live within the corals, providing the coral with food. Coral reefs support a huge variety of invertebrates and fishes.

Below the photic zone of the ocean lies the **aphotic zone.** Although there is not enough light for photosynthesis between 200 and 1,000 m, some light does reach these depths. This dimly lit world, sometimes called the twilight zone, is dominated by a fascinating variety of small fishes and crustaceans. Food sinking from the photic zone provides some sustenance for these animals. In addition, many of them migrate to the surface at night to feed. Some fishes in the twilight zone have enlarged eyes, enabling them to see in the very dim light, and luminescent organs that attract mates and prey.

Below 1,000 m, the ocean is completely and permanently dark. Adaptation to this environment has produced bizarre-looking creatures, such as the angler fish shown in Figure 34.6C, facing page. The scarcity of food probably explains the strangely outsized mouths of the angler and other fishes that inhabit this region of the ocean, a feature that allows them to grab any available prey, large or small. Inwardly angled teeth ensure that once caught, prey do not escape. The angler fish improves its chances of encountering prey by dangling a lure lit by bioluminescent bacteria. Most benthic organisms here are *deposit feeders*, animals that consume detritus on the substrate. Crustaceans, polychaete

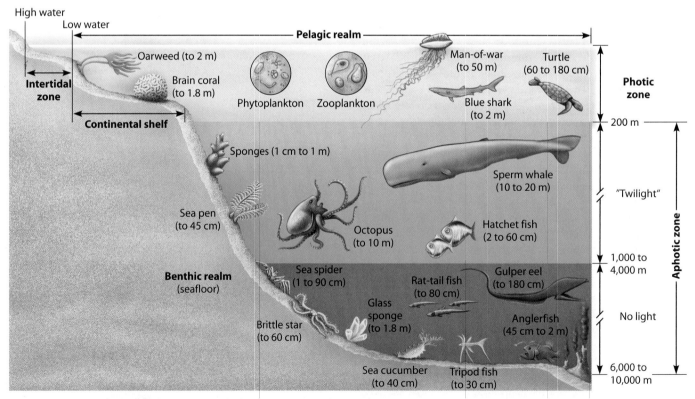

Figure 34.6A   Ocean life (zone depths and organisms not drawn to scale)

worms, sea anemones, and echinoderms such as sea cucumbers, sea stars, and sea urchins are common. Because of the scarcity of food, however, the density of animals is low—except at hydrothermal vents, where chemoautotrophic bacteria are abundant.

The marine environment also includes distinctive biomes where the ocean interfaces with land or with fresh water. In the **intertidal zone,** where the ocean meets land, the shore is pounded by waves during high tide and exposed to the sun and drying winds during low tide. The rocky intertidal zone is home to many sedentary organisms, such as algae, barnacles, and mussels, which attach to rocks and are thus prevented from being washed away. On sandy beaches, suspension-feeding worms, clams, and predatory crustaceans bury themselves in the ground.

Figure 34.6D shows an **estuary,** a biome that occurs where a freshwater stream or river merges with the ocean. The saltiness of estuaries ranges from nearly that of fresh water to that of the ocean. With their waters enriched by nutrients from the river, estuaries are among the most productive biomes on Earth. Oysters, crabs, and many fishes live in estuaries or reproduce in them. Estuaries are also crucial nesting and feeding areas for waterfowl.

Figure 34.6B   A coral reef with its immense variety of invertebrates and fishes

Figure 34.6C   Angler fish

Wetlands constitute a biome that is transitional between an aquatic ecosystem—either marine or freshwater—and a terrestrial one. Covered with water either permanently or periodically, wetlands support the growth of aquatic plants. Mudflats and salt marshes are extensive coastal wetlands that often border estuaries.

For centuries, people viewed the ocean as a limitless resource, harvesting its bounty and using it as a dumping ground for wastes. The impact of these practices is now being felt in many ways, large and small. From worldwide declines in commercial fish species to dying coral reefs to beaches closed by pollution, danger signs abound. Because of their proximity to land, estuaries and wetlands are especially vulnerable. Many have been completely replaced by development on landfill. Other threats include nutrient pollution, contamination by pathogens or toxic chemicals, alteration of freshwater inflow, and introduction of non-native species. Coral reefs have suffered from many of the same problems. In addition, overfishing has upset the species balance in some reef communities and greatly reduced diversity, and the widespread deaths of reef-building corals in some regions have been attributed to global warming.

Freshwater biomes share many characteristics with marine biomes and experience some of the same threats. We introduce freshwater biomes in the next module.

? _____, small, photosynthetic organisms inhabiting the _____ zone of the pelagic biome, provide most of the food for oceanic life.

■ Phytoplankton . . . photic

Figure 34.6D   An estuary in Georgia

Current, sunlight, and nutrients are important abiotic factors in freshwater biomes

Freshwater biomes include lakes and ponds (standing water), rivers and streams (running water), and a variety of wetlands. Because these biomes are embedded in terrestrial landscapes, their characteristics are intimately connected with the soils and organisms of the ecosystems that surround them.

Light has a significant impact on freshwater biomes, just as it does in oceans. In all but the smallest lakes and ponds, there are usually distinct photic and aphotic zones. Phytoplankton grow in the photic zone, and rooted plants often inhabit shallow waters. Large populations of microorganisms in the benthic community decompose dead organisms that sink to the bottom. Respiration by microbes removes oxygen from water near the bottom, and in some lakes, benthic areas are unsuitable

Figure 34.7A A stream in the Great Smoky Mountains, Tennessee

for any organisms except anaerobic microbes.

Temperature may also have a profound effect on freshwater communities. During the summer, lakes often have a distinct upper layer of water that has been warmed by the sun and does not mix with underlying, cooler water. Fish often spend much of their time in the deep, cool waters of a lake unless oxygen levels there have become depleted by decomposers.

Nitrogen and phosphorus are the nutrients that usually limit the amount of phytoplankton growth in a lake or pond. When there are temperature layers in a lake, for instance, nutrients released by decomposers can become trapped near the bottom, out of reach of the phytoplankton. During the summer months, this may limit the growth of algae (and thus photosynthesis) in the photic zone. As winter approaches, the surface water becomes denser as it cools and tends to mix with the deeper water, allowing nutrients to return to the surface, where phytoplankton can again use them. Seasonal mixing also restores oxygen to the depths.

Today, many lakes and ponds are affected by large inputs of nitrogen and phosphorus from sewage and runoff from fertilized lawns and agricultural fields. These nutrients often produce blooms, population explosions of algae. Heavy algal growth reduces light penetration into the water, and when the algae die and decompose, a pond or lake can suffer serious oxygen depletion.

Rivers and streams generally support communities of organisms quite different from those of lakes and ponds. A river or a stream changes greatly between its source (perhaps a spring or snowmelt) and the point at which it empties into a lake or the ocean. Near the source, the water is usually cold, low in nutrients, and clear (Figure 34.7A). The channel is often narrow, with a swift current that does not allow much silt to accumulate on the bottom. The current also inhibits the growth of phytoplankton; most of the organisms found here are supported by the photosynthesis of algae attached to rocks or by organic material (such as leaves) carried into the stream from the surrounding land. The most abundant benthic animals are usually arthropods, such as small crustaceans and insect larvae, that have physical and behavioral adaptations that enable them to resist being swept away. Trout are often the predominant fishes, locating their food, including insects, mainly by sight in the clear water.

Downstream, a river or stream generally widens and slows. The water is usually warmer and may be murkier because of sediments and phytoplankton suspended in it. Worms and insects that burrow into mud are often abundant, as are waterfowl, frogs, and catfish and other fishes that find food more by scent and taste than by sight.

Freshwater wetlands range from marshes, as shown in Figure 34.7B, to swamps and bogs. They may form in shallow basins or along the banks of rivers or lakes. Wetlands are among the richest of biomes in terms of species diversity. They provide water storage areas that reduce flooding and improve water quality by filtering pollutants. The recognition of their ecological and economic value has led to governmental and private efforts to protect and restore wetlands.

Web Activity *Aquatic Biomes*

? Why does sewage cause algal blooms in lakes?

■ The sewage adds nutrients, such as nitrates and phosphates, that stimulate growth of algae.

Figure 34.7B The Okefenokee National Wetland Reserve, Georgia

34.8   Terrestrial biomes reflect regional variations in climate

Terrestrial ecosystems are grouped into eight major types of biomes, which are distinguished primarily by their predominant vegetation. By providing food, shelter, nesting sites, and much of the organic material for decomposers, plants build the foundation for the communities of animals, fungi, and microorganisms that are characteristic of each biome. The geographic distribution of plants, and thus of terrestrial biomes, largely depends on climate, with temperature and rainfall often the key factors determining the kind of biome that exists in a particular region.

Figure 34.8 shows the locations of the biomes. If the climate in two geographically separate areas is similar, the same type of biome may occur in both places; notice on the map that each kind of biome occurs on at least two continents. Each biome is characterized by a type of biological community, rather than an assemblage of particular species. For example, the species living in the deserts of the American Southwest and in the Sahara Desert of Africa are different, but all are adapted to desert conditions. Widely separated biomes may look alike because of convergent evolution, the appearance of similar traits in independently evolved species living in similar environments (see Module 15.14).

Within each biome there is local variation, giving the vegetation a patchy, rather than uniform, appearance. For example, in northern coniferous forests, snowfall may break branches and small trees, causing openings where broadleaf trees such as aspen and birch can grow. Local storms and fires also create openings in many biomes.

Today, global warming is generating intense interest in the effect of climate on vegetation patterns. Using powerful new tools such as satellite imagery, scientists are documenting latitudinal shifts in biome borders, snow and ice coverage, and changes in length of the growing season. At the same time, many natural biomes have been fragmented and altered by human activity. As we discuss in Chapter 38, a high rate of biome alteration by humans is correlated with an unusually high rate of species loss throughout the globe.

We now begin a survey of the major terrestrial biomes. To help you locate the biomes, we include with each module an orientation map that is color-coded to match Figure 34.8.

? Test your knowledge of world geography: Which biome is most closely associated with a "Mediterranean climate"?

■ Chaparral

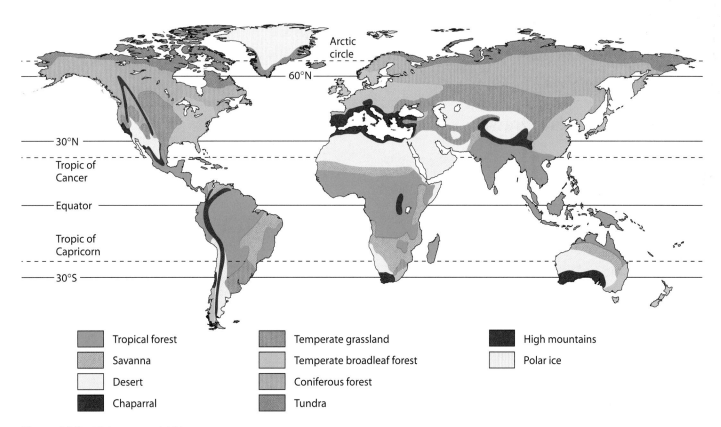

Figure 34.8   Major terrestrial biomes

Tropical forest

Savanna

Desert

Chaparral

Temperate grassland

Temperate broadleaf forest

Coniferous forest

Tundra

High mountains

Polar ice

34.9 Tropical forests cluster near the equator

Tropical forests occur in equatorial areas where the temperature is warm and days are 11–12 hours long year-round. Rainfall in these areas is quite variable, and this variability, rather than temperature or day length, generally determines the vegetation that grows in a particular tropical forest. In areas where rainfall is scarce or there is a prolonged dry season, tropical dry forests predominate. The plants found there are a mixture of thorny shrubs and deciduous trees and succulents. Tropical rain forests are found in very humid equatorial areas where rainfall is abundant (200–400 cm per year).

The tropical rain forest, such as the lush area shown in Figure 34.9, is among the most complex of all biomes, harboring enormous numbers of different species. Up to 300 species of trees can be found in a single

hectare (2.5 acres). The forest structure consists of distinct layers that provide many different habitats: emergent trees growing above a closed upper canopy, one or two layers of lower trees, a shrub understory, and a sparse ground layer of herbaceous plants. Because of the closed canopy, little sunlight reaches the forest floor. Many trees are covered by woody vines growing toward the light. Other plants, including bromeliads and orchids, gain access to sunlight by growing on the branches or trunks of tall trees. Many of the animals also dwell in trees, where food is abundant. Monkeys, birds, insects, snakes, bats, and frogs find food and shelter many meters above the ground.

The soils of tropical rain forests are typically poor. High temperatures and rainfall lead to rapid decomposition, and the nutrients released are quickly taken up by the luxuriant vegetation or leached away by the frequent rains.

Human impact on the world's tropical rain forests is currently a source of great concern. It is a common practice to clear the forest for lumber or simply burn it, farm the land for a few years, and then abandon it. Mining has also devastated large tracts of rain forest. Once stripped, the tropical rain forest recovers very slowly because the soil is so nutrient-poor. We discuss the potential consequences of destroying the tropical forests in Chapter 38, including the impact on world climate.

> **?** Why are the soils in most tropical rain forests so poor in nutrients that they can only support farming for a few years after the forest is cleared?

■ The tropical conditions favor rapid decomposition of organic litter in the forest soil, and most of the ecosystem's nutrients are tied up in the vegetation that is cleared away before farming.

Figure 34.9 Tropical rain forest

34.10 Savannas are grasslands with scattered trees

Figure 34.10, a photograph taken in Kenya, shows a typical **savanna,** a biome dominated by grasses and scattered trees. The temperature is warm year-round. Rainfall averages 30–50 cm per year, with dramatic seasonal variation. Poor soils and lack of moisture inhibit the establishment of most trees. Grazing animals and frequent fires, caused by lightning or human activity, further inhibit invasion by trees. Grasses survive burning because the growing points of their shoots are below ground. Savanna plants have also been selected for their ability to survive prolonged periods of drought.

Grasses and forbs (small broadleaf plants) grow rapidly during the rainy season, providing a good food source for many animal species. Large grazing mammals must migrate to greener

Figure 34.10 Savanna

pastures and scattered watering holes during seasonal drought. The dominant herbivores in savannas are actually insects, especially ants and termites. Also common are many burrowing animals, including mice, moles, gophers, snakes, ground squirrels, worms, and numerous arthropods.

Many of the world's large herbivores and their predators inhabit savannas. African savannas are home to giraffes, zebras, and many species of antelope, as well as to lions and cheetahs. Several species of kangaroo are the dominant large herbivores of Australian savannas.

? How do fires help to maintain savannas as grassland ecosystems?

■ By repeatedly preventing the spread of trees and other woody plants

34.11 Deserts are defined by their dryness

Deserts are the driest of all terrestrial biomes, characterized by low and unpredictable rainfall (less than 30 cm per year). The large deserts in central Australia and northern Africa have average annual rainfalls of less than 2 cm, and in the Atacama Desert in Chile, the driest place on Earth, there is often no rain at all for decades at a time. But not all desert air is dry. Coastal sections of the Atacama and of the Namib Desert in Africa are often shrouded in fog, although the ground remains extremely dry.

As we discussed in Module 34.5, large tracts of desert occur in two regions of descending dry air centered around the 30° north and 30° south latitudes. At higher latitudes, large deserts may occur in the rain shadows of mountains (Figure 34.5F); these include much of central Asia east of the Caucasus Mountains and Washington and Oregon east of the Cascade Mountains. The Mojave Desert, shown in Figure 34.11, is in the rain shadow of the Sierra Nevada, along with much of the rest of southern California and Nevada.

Figure 34.11 Desert

Some deserts are very hot, with daytime soil surface temperatures above 60°C (140°F) and large daily temperature fluctuations. Other deserts, such as those west of the Rocky Mountains, are relatively cold. Air temperatures in cold deserts may fall below −30°C. Antarctica, not shown on our map, is largely desert.

The cycles of growth and reproduction in the desert are keyed to rainfall. The driest deserts have no perennial vegetation at all, but less arid regions have scattered deep-rooted shrubs, often interspersed with water-storing succulents such as cacti. The leaves of some plants, including the Joshua tree shown in Figure 34.11, have a waxy coating to prevent water loss. Desert plants typically produce great numbers of seeds, which may remain dormant until a heavy rain triggers germination. After periods of rainfall (often in late winter), deserts display spectacular blooms of annual plants.

Like desert plants, desert animals are adapted to drought and extreme temperatures. Many live in burrows and are active only during the cooler nights, and most have special adaptations that conserve water. Seedeaters such as ants, many birds, and rodents are common in deserts. Lizards, snakes, and hawks eat the seedeaters.

The process of **desertification,** the conversion of semiarid regions to desert, is a significant environmental problem. In northern Africa, for example, a burgeoning human population, overgrazing, and dry land farming are converting large areas of savanna to desert.

? Why isn't "cold desert" an oxymoron?

■ Because deserts are defined by relatively little precipitation and dry soil, not by temperature

34.12 Spiny shrubs dominate the chaparral

Chaparral (the Spanish word for "place of evergreen scrub oaks") is characterized by dense, spiny shrubs with tough, evergreen leaves. The climate that supports chaparral vegetation results mainly from cool ocean currents circulating offshore, which produce mild, rainy winters and hot, dry summers. As a result, this biome is limited to small coastal areas, including California, where the photograph in Figure 34.12 was taken. The largest region of chaparral surrounds the Mediterranean Sea; "Mediterranean" is another name for this biome. In addition to the perennial shrubs that dominate chaparral, annual plants are also commonly seen, especially during the wet winter and spring months. Animals characteristic of the chaparral are browsers such as deer, fruit-eating birds, and seed-eating rodents, as well as lizards and snakes.

Chaparral vegetation is adapted to periodic fires, most often caused by lightning. Many plants contain flammable chemicals and burn fiercely, especially where dead brush has accumulated. After a fire, shrubs use food reserves stored in the surviving roots to support rapid shoot regeneration. Some chaparral plant species produce seeds that will germinate only after a hot fire. The ashes of burned vegetation fertilize the soil with mineral nutrients, promoting regrowth of the plant community. Houses do not fare as well, and firestorms that race through the densely populated canyons of Southern California can be devastating.

? What is one way that homeowners in chaparral areas can protect their neighborhoods from fire?

Figure 34.12 Chaparral

■ Keep the area clear of dead brush, which is highly flammable.

34.13 Temperate grasslands include the North American prairie

Temperate grasslands have some of the characteristics of tropical savannas, but they are mostly treeless, except along rivers or streams, and are found in regions of relatively cold winter temperatures. Rainfall, averaging between 25 and 75 cm per year, with periodic severe droughts, is too low to support forest growth. Fires and grazing by large mammals also inhibit growth of woody plants but do not harm the below-ground grass shoots.

Large grazing mammals, such as the bison and pronghorn of North America and the wild horses and sheep of the Asian steppes, are characteristic of grasslands. Without trees, many birds nest on the ground, and some small mammals dig burrows to escape predators. Enriched by glacial deposits and mulch

Figure 34.13 Temperate grassland

from decaying plant material, the soil of grasslands supports a great diversity of microorganisms and small animals, including annelids and arthropods.

The amount of annual rainfall influences the height of grassland vegetation. The photograph in Figure 34.13 (previous page) was taken on the relatively dry, short-grass prairie of South Dakota. Tallgrass prairie occurs in wetter areas, such as eastern Kansas. Little remains of North American prairies today. Most of the region is intensively farmed, and it is one of the most productive agricultural regions in the world.

? How do humans now use most of the North American land that was once temperate grasslands?

■ For farming

34.14 Broadleaf trees dominate temperate forests

Temperate broadleaf forests grow throughout midlatitude regions, where there is sufficient moisture to support the growth of large trees. These regions include most of the eastern United States, most of central Europe, and parts of eastern Asia and Australia. In the Northern Hemisphere, deciduous trees (trees that drop their leaves seasonally) characterize temperate broadleaf forests. Some of the dominant trees include species of oak, hickory, birch, beech, and maple. The mix of tree species varies widely, depending on such factors as the climate at different latitudes, topography, and local soil conditions. Figure 34.14 features a photograph taken in the autumn in Great Smoky Mountains National Park in North Carolina, where maples, oaks, hickories, and sweetgum contribute to the spectacular display of color.

Temperatures in temperate broadleaf forests range from very cold in the winter to hot in the summer ($-30°C$ to $30°C$). Annual precipitation is relatively high—75 to 150 cm—and usually evenly distributed throughout the year. These forests typically have a growing season of 5 to 6 months and a distinct annual rhythm, in which the trees drop leaves and become dormant in late autumn, then produce new leaves each spring. The loss of leaves in winter prevents evaporation of water from the tree at a time when freezing reduces the available water.

Temperate broadleaf forests are more open than tropical rain forests and are not as tall or as diverse. However, their soils are rich in inorganic and organic nutrients. Rates of decomposition are lower in temperate forests than in the tropics, and the thick layer of leaf litter accumulates on forest floors and conserves many of the biome's nutrients. Numerous invertebrates live in the soil and leaf litter. Some vertebrates, such as mice, shrews, and ground squirrels, bur-

Figure 34.14 Temperate broadleaf forest

row for shelter and food, while others, including many species of birds, live in the trees. Predators include bobcats, foxes, black bears, and mountain lions.

Virtually all the original broadleaf forests in North America have been destroyed by logging and by the clearing of land for agriculture and urban development. In contrast to drier ecosystems, these forests tend to recover after disturbance, and today we see broadleaf trees dominating undeveloped areas over much of their former range.

? How does the soil of a temperate broadleaf forest differ from that of a tropical rain forest?

■ The soil in temperate broadleaf forests is rich in inorganic and organic nutrients, while the soil in tropical rain forests is low in nutrients.

34.15 Coniferous forests are often dominated by a few species of trees

Cone-bearing evergreen trees, such as spruce, pine, fir, and hemlock, dominate **coniferous forests.** The northern coniferous forest, or taiga, is the largest terrestrial biome on Earth, stretching in a broad band across North America and Asia south of the Arctic Circle. Taiga (from the Russian word for "mountain") is also found at cool, high elevations in more temperate latitudes, as in much of the mountainous region of western North America. Figure 34.15 shows part of a coniferous forest in Oregon.

The taiga is characterized by long, cold winters and short, wet summers, which are sometimes warm. The soil is thin and acidic, and the slow decomposition of conifer needles makes few nutrients available. There may be considerable precipitation, mostly in the form of snow. The snow usually falls before the coldest temperatures occur, and it insulates the soil, keeping it from freezing to such depths that it would never thaw out during the short summers. Animals of the taiga include moose, elk, hares, bears, wolves, grouse, and migratory birds.

The temperate rain forests of coastal North America (from Alaska to Oregon) are also coniferous forests. Warm, moist air from the Pacific Ocean supports this unique biome, which, like most coniferous forests, is dominated by a few tree species, such as hemlock, Douglas fir, and redwood. These forests are heavily logged, and the old-growth stands of trees may soon disappear.

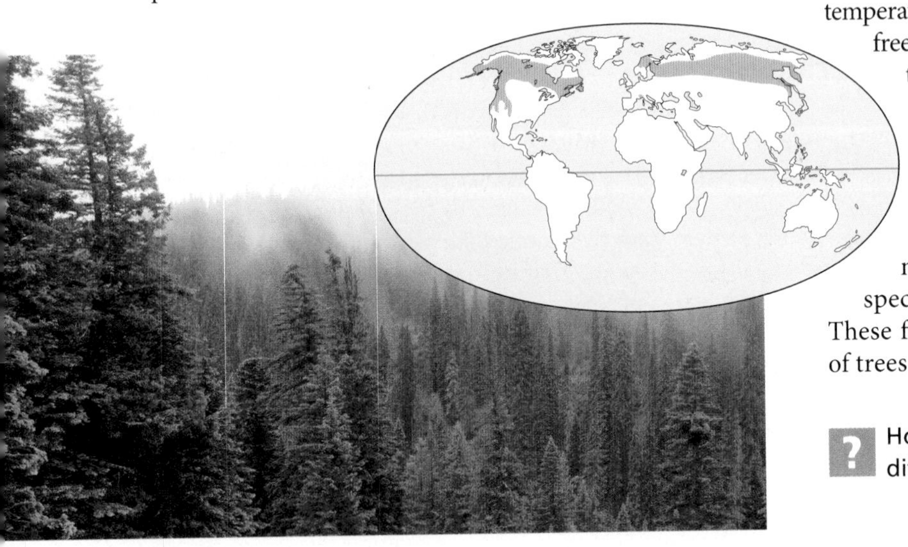

? How does the soil of the northern coniferous forests differ from that of a broadleaf forest?

■ The soil is thinner, nutrient-poor, and acidic because conifer needles decompose slowly in the low temperatures.

Figure 34.15 Coniferous forest

34.16 Long, bitter-cold winters characterize the tundra

At the northernmost limits of plant growth and at high altitudes is the **tundra** (from the Russian word for "marshy plain"). The arctic tundra encircles the North Pole, extending southward to the coniferous forests. Alpine tundras, including the alpine meadows described in the opening essay, are found above the treeline (the point above which trees cannot grow) on high mountains, even in the tropics.

Figure 34.16 shows the arctic tundra in central Alaska in the autumn. The climate here is often extremely cold, with little light for much of the autumn and winter. During the brief, warm summers, when there is nearly constant daylight, plants grow quickly and flower in a rapid burst.

The arctic tundra is characterized by **permafrost,** continuously frozen subsoil. Only the upper part of tundra soil thaws in the summer. The arctic tundra may receive as little precipitation as some deserts. But poor drainage, due to the permafrost, and slow evaporation keep the soil continually saturated. The permafrost also prevents the roots of plants from penetrating very far into the soil, which is one factor that explains the absence of trees. Extremely cold winter air temperatures and high winds also contribute to the exclusion of trees. Vegetation in the tundra includes dwarf shrubs, grasses, mosses, and lichens.

Animals of the tundra withstand the cold by having good insulation that retains heat. Large herbivores include musk oxen and caribou. The principal smaller animals are rodents called lemmings and a few predators, such as the arctic fox and snowy owl. Many animals are migratory, using the tundra as a summer breeding ground. During the brief warm season, the marshy ground supports the aquatic larvae of insects, providing food for migratory waterfowl, and clouds of mosquitoes often fill the tundra air.

Web Activity *Terrestrial Biomes*

? What three abiotic factors account for the rarity of trees in arctic tundra?

■ Long, very cold winters (short growing season), high winds, and permafrost

Figure 34.16 Tundra

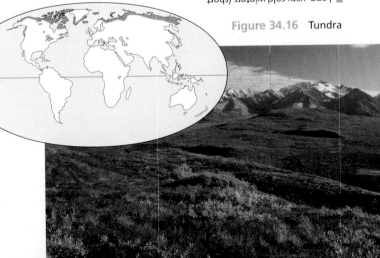

34.17 The global water cycle connects aquatic and terrestrial biomes

When Rusty Schweickart (the astronaut quoted in the chapter opening essay) viewed Earth from space, he saw the biosphere as it truly is, an undivided whole. Ecological subdivisions such as biomes are not self-contained units. Rather, all parts of the biosphere are linked by the global water cycle, illustrated in Figure 34.17, and by nutrient cycles, which you will learn about in Chapter 37. Consequently, events in one biome may reverberate throughout the biosphere.

Recall from Module 34.5 that water and air move in global patterns driven by solar energy. Precipitation and evaporation, as well as transpiration from plants (see Module 32.3), continuously move water between the land, oceans, and the atmosphere. Over the oceans (left side of Figure 34.17), evaporation exceeds precipitation. The result is a net movement of water vapor to clouds that are carried by winds from the oceans across the land. On land (right side of the figure), precipitation exceeds evaporation and transpiration. Excess precipitation forms systems of surface water (such as lakes and rivers) and groundwater, all of which flow back to the sea, completing the water cycle.

Just as the water draining from your shower carries dead skin cells from your body along with the day's grime, the water washing over and through the ground carries traces of the land and its history. For example, water flowing from land to sea carries with it silt (fine soil particles) and chemicals such as fertilizers and pesticides. The accumulation of silt, aggravated by the development of coastal areas, has muddied the waters of some coral reefs, dimming the light available to the photosynthetic algae that power the reef community. Chemicals in surface water may travel hundreds of miles by stream and river to the ocean, where currents then carry them even farther from their point of origin. For instance, in the 1960s, researchers began finding traces of DDT in marine mammals in the Arctic, far from any places DDT had been used (see Module 34.2). Airborne pollutants such as nitrogen oxides and sulfur oxides, which combine with water to form acid precipitation (see Module 2.16), are also distributed by the water cycle.

Human activity also affects the global water cycle itself in a number of important ways. One of the main sources of atmospheric water is transpiration from the dense vegetation making up tropical rain forests. The destruction of these forests changes the amount of water vapor in the air. This, in turn, will likely alter local, and perhaps global, weather patterns. Pumping large amounts of groundwater to the surface for irrigation affects the water cycle, too. This practice can increase the rate of evaporation over land and may deplete groundwater supplies.

? What is the main way that living organisms contribute to the water cycle?

▨ Plants move water from the ground to the atmosphere via transpiration.

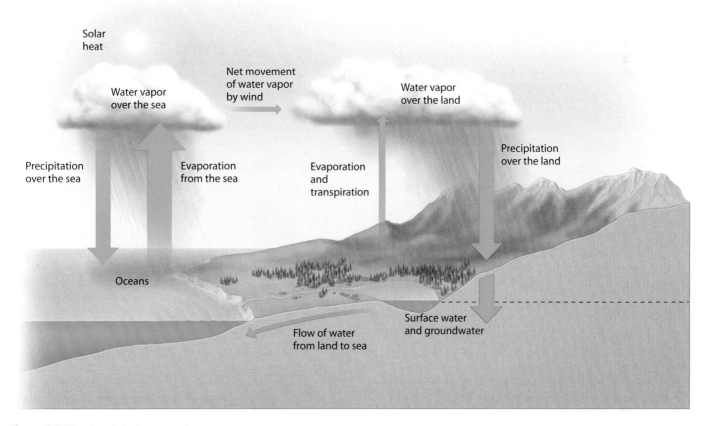

Figure 34.17 The global water cycle

Reviewing the Concepts

The Biosphere (Introduction–34.5)

The biosphere is the global ecosystem. Each habitat has a unique community of species (**Introduction–34.1**).

Ecology. Ecologists study the interactions of organisms with their environments at the organismal, population, community, and ecosystem levels. Ecosystem interactions involve living (biotic) factors and nonliving (abiotic) physical and chemical factors. Knowledge of ecology is essential for recognizing and solving environmental problems (**Introduction–34.2**).

Abiotic and biotic factors. Energy sources, temperature, water, and nutrients are among the abiotic factors determining the biosphere's structure and dynamics (**34.3**). The process of natural selection results in adaptations to both abiotic factors (such as temperature and water) and biotic factors (such as predation and competition) (**34.4**).

Climate largely determines the distribution of terrestrial communities. Most climatic variations are due to the uneven heating of Earth's surface as it orbits the sun, setting up patterns of precipitation and prevailing winds. Ocean currents influence coastal climate. Landforms such as mountains affect rainfall (**34.5**).

Aquatic Biomes (34.6–34.7)

Marine biomes. Marine biomes include a variety of habitats to which organisms are adapted, including the pelagic and benthic realms, which are further distinguished by the availability of light. Coral reefs are found in warm, shallow waters above continental shelves. Other marine biomes are estuaries, wetlands, and the intertidal zone (**34.6**).

Freshwater biomes. Sunlight, temperature, and the availability of nutrients and oxygen shape lake and pond communities. Abiotic factors change from the source of a river to its mouth, and communities vary accordingly. Wetlands include marshes and swamps (**34.7**).

Terrestrial Biomes (34.8–34.17)

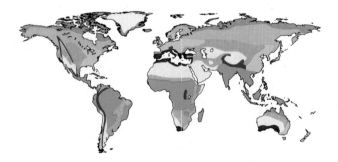

Eight terrestrial biomes. Temperature and rainfall mainly determine the terrestrial biomes (**34.8**). Tropical forests occur along the equator. Tropical rain forests are the most diverse ecosystem. Savannas are grasslands with scattered trees. Deserts are the driest biomes. The chaparral is a shrubland with cool, rainy winters and dry, hot summers. Temperate grasslands are found where winters are cold. Forests of broadleaf trees grow in some temperate areas. The northern coniferous forest, or taiga, is found where there are short summers and long, snowy winters. Arctic tundra is a treeless biome characterized by extreme cold, wind, and permafrost. Alpine tundra occurs above the treeline on high mountains (**34.9–34.16**). The global water cycle links terrestrial and aquatic biomes (**34.17**).

Connecting the Concepts

1. You have seen that Earth's terrestrial biomes reflect regional variations in climate. But what determines these climatic variations? Interpret the following diagrams in reference to global patterns of temperature, rainfall, and winds.
   a. Solar radiation and latitude:

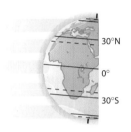

   b. Earth's orbit around the sun:

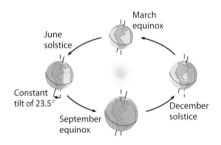

   c. Global patterns of air circulation and rainfall:

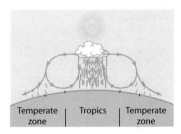

2. Aquatic biomes differ in levels of light, nutrients, oxygen, and water movement. These abiotic factors influence the productivity and diversity of freshwater ecosystems.
   a. Productivity, roughly defined as photosynthetic output, is high in estuaries, coral reefs, and shallow ponds. Describe the abiotic factors that contribute to high productivity in these ecosystems.
   b. How does extra input of nitrogen and phosphorus (for instance, by fertilizer runoff) affect the productivity of lakes and ponds? Is this nutrient input beneficial for the ecosystem? Explain.

Testing Your Knowledge

Matching

3. The most complex and diverse biome
4. Ground permanently frozen
5. Broadleaf trees such as hickory and birch
6. Mediterranean climate
7. Spruce, fir, pine, and hemlock trees
8. Home of ants, antelopes, and lions
9. North American plains

a. chaparral
b. savanna
c. taiga
d. temperate forest
e. temperate grassland
f. tropical rain forest
g. arctic tundra

Multiple Choice

10. Changes in the seasons are caused by
    a. the tilt of Earth's axis toward or away from the sun.
    b. annual cycles of temperature and rainfall.
    c. variation in the distance between Earth and the sun.
    d. an annual cycle in the sun's energy output.
    e. the periodic buildup of heat energy at the equator.

11. What makes the Gobi Desert of Asia a desert?
    a. The growing season there is very short.
    b. Its vegetation is sparse.
    c. It is hot.
    d. Temperatures vary little from summer to winter.
    e. It is dry.

12. Which of the following sea creatures might be described as a pelagic animal of the aphotic zone?
    a. a coral reef fish
    b. a giant clam near a deep-sea hydrothermal vent
    c. an intertidal snail
    d. a deep-sea squid
    e. a harbor seal

13. Why do the tropics and the windward side of mountains receive more rainfall than areas around latitudes 30° north and south and the leeward side of mountains?
    a. Rising warm, moist air cools and drops its moisture as rain.
    b. Descending air condenses, creating clouds and rain.
    c. The tropics and the windward side of mountains are closer to the ocean.
    d. There is more solar radiation in the tropics and on the windward side of mountains.
    e. Earth's rotation creates seasonal differences in rainfall.

14. Phytoplankton are the major photosynthesizers in
    a. swamps.
    b. streams.
    c. the ocean photic zone.
    d. the intertidal zone.
    e. coral reefs.

15. An ecologist monitoring the number of great apes in a wildlife refuge over a 5-year period is studying ecology at which level?
    a. organism
    b. population
    c. community
    d. ecosystem
    e. biosphere

16. Many plant species have adaptations for dealing with periodic fires. Such fires are typical of a
    a. chaparral.
    b. savanna.
    c. temperate grassland.
    d. temperate broadleaf forest.
    e. a, b, or c

Describing, Comparing, and Explaining

17. Tropical rain forests are the most diverse biomes. What factors contribute to this diversity?

18. What biome do you live in? Describe your climate and the factors that have produced that climate. What plants and animals are typical of this biome? If you live in an urban or agricultural area, how have human interventions changed the natural biome?

Applying the Concepts

19. In the climograph below, biomes are plotted by their range of annual mean temperature and annual mean precipitation. Identify the following biomes: arctic tundra, coniferous forest, desert, grassland, temperate forest, and tropical forest. Explain why there are areas in which biomes overlap on this graph.

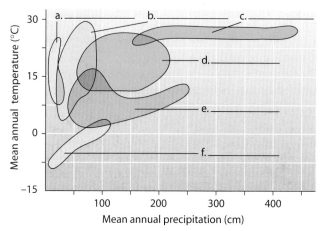

20. The North American pronghorn looks and acts like the antelopes of Africa. But the pronghorn is the only survivor of a family of mammals restricted to North America. Propose a hypothesis to explain how these widely separated animals came to be so much alike.

21. Near Lawrence, Kansas, there was a rare patch of the original North American temperate grassland that had never been plowed. It was home to numerous native grasses and annual plants, including two endangered plants. Environmental activists thought that the area should be set aside as a nature reserve. In 1990, the owner of the land plowed it, stating that there are no federal laws protecting endangered plants on private grasslands. What issues and values are in conflict in this situation? How could this story have had a more satisfactory ending for all concerned?

Answers to all questions can be found in Appendix 4.

For Practice Quizzes, BioFlix, MP3 Tutors, and Activities, go to www.mybiology.com.

35 Behavioral Adaptations to the Environment

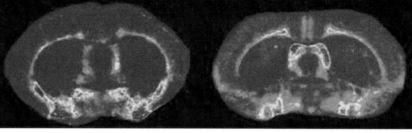

Facing Page: Prairie voles (*Microtus ochrogaster*)

Left: Meadow vole (*Microtus pennsylvanicus*)

Below: Density of hormone receptors (in red) in brain of prairie vole (left) and montane vole (right)

Of Mice and Monogamy

When composer Cole Porter wrote the famous lines, "Birds do it, bees do it/Even educated fleas do it," he was referring to one of the most fundamental activities in the animal world—mating. As evolutionary theory predicts, many animals, including humans, expend enormous amounts of energy attracting and persuading members of the opposite sex to mate with them. But these relationships are often fleeting. **Promiscuous** animals mate with multiple partners and form no lasting bonds. The mating pattern of other animals is **polygamous,** a type of relationship in which an individual of one sex mates with several of the other. **Monogamous** animals form a bond with a single partner, and both partners care for their offspring.

Some people associate a monogamous mating pattern with a "happily ever after" scenario of a faithful partner and a two-parent family, but this vision of monogamy is extremely rare among animals. For example, although most birds are considered monogamous, in many species, the relationship only lasts until the offspring have flown the nest, and sexual fidelity is not part of the deal. Only a tiny minority of mammals—fewer than 5%—are monogamous in any sense. Of those, one species stands out for its lifelong monogamy. It is not humans, though we can certainly point to many remarkable examples of long-lasting pair bonds between *Homo sapiens.* Rather, it is the small, mouse-like creatures known as the prairie vole (*Microtus ochrogaster*), whose monogamous relationships are renowned in the scientific community. After they mate, prairie voles associate closely and exclusively with each other, engaging in mutual grooming and staying in close physical contact. The male becomes intensely aggressive toward any strange vole, male or female. The relationship endures throughout life. If one partner dies, the other is unlikely to bond with another mate.

Scientists have learned a great deal about the biological mechanisms underlying the formation of this extraordinary bond. A prairie vole relationship begins when a female sniffs the scent of a potential male partner. The smell causes her to become sexually receptive, which leads to repeated matings over the next 24 hours or more. During this honeymoon period, biochemical responses in the brains of both partners cement a lasting bond between them.

The act of mating stimulates the release of dopamine, a neurotransmitter (see Module 28.6) that plays a prominent role in motivation and reward. At the same time, the posterior pituitary (see Module 26.4) releases hormones that activate membrane receptors in a region of the brain that is associated with creating social memories—recognizing other individuals. (Recall from Modules 26.2 and 28.6 that receptors are protein molecules located in or on the target cells of a chemical messenger such as a hormone or neurotransmitter.) Social memories are essential to behaviors such as caring for offspring, showing aggression toward strangers, and forming social attachments. As a result of their extended bout of sexual activity, prairie voles form a strong memory of their sexual partner, whom they recognize by scent and associate with the dopamine reward.

There are many species of voles (which are sometimes called *field mice),* and they display a spectrum of mating behaviors. Prairie voles are one of two vole species known to be monogamous. Meadow voles (*Microtus pennsylvanicus*) and montane voles (*M. montanus*) are promiscuous. Researchers have discovered a crucial difference in the sensitivity of prairie and meadow voles to the hormones that are released during mating. While the brains of prairie voles have dense clusters of hormone receptors in the area associated with social memory, meadow voles have very few receptors in the same region. Consequently, although meadow voles experience the same biochemical reward during the act of mating, they do not form a memory that connects the reward to the current partner's scent. If the pair should happen to meet again, they will not recognize each other. It turns out that the difference in the density of hormone receptors in vole brains results from a small difference in a segment of DNA that controls the expression of the receptor gene. Thus, the difference between a lifelong bond and a one night stand may hinge on a snippet of DNA.

Behavior encompasses a wide range of activities. At its most basic level, a behavior is an action carried out by muscles or glands under the control of the nervous system in response to an environmental cue. Collectively, behavior is the sum of an animal's responses to internal and external environmental cues. Investigations of vole mating patterns illustrate the multiple levels at which scientists study the mechanisms of behavior. **Behavioral ecologists** study behavior in an evolutionary context. In this chapter, you will learn about several categories of behavior. We will also explore the roles of genetics and the environment in determining behavior. ■ ■ ■

35.1 Behavioral ecologists ask both proximate and ultimate questions

We have defined ecology as the study of interactions between organisms and their environment. Behavior is a significant portion of an animal's immediate response to its environment. Although we commonly think of behavior in terms of observable actions that result from muscle movements, for instance, a courtship dance or an aggressive posture, other activities are also considered behaviors. Chemical communication such as secreting a chemical that attracts mates or marks a territory is a form of behavior. Learning is also considered a behavioral process. Behavioral ecologists draw on the knowledge of a variety of disciplines to describe the details of animal behaviors and investigate how they develop, evolve, and contribute to the animal's survival and reproductive success.

The questions investigated by behavioral ecologists fall into two broad categories. Proximate questions concern the immediate reason for a behavior—how it is triggered by **stimuli** (environmental cues that cause a response), what physiological or anatomical mechanisms play a role, and what underlying genetic factors are at work. For example, researchers studying the mating pattern of prairie voles might ask "How do voles choose their mates?" or "How does the act of mating cause voles to form lifelong bonds with their partners?" or "How are genetic factors involved in monogamous behavior?" Generally, proximate questions help us understand *how* a behavior occurs. **Proximate causes** are the answers to such questions about the immediate mechanism for a behavior.

Ultimate questions address *why* a particular behavior occurs. As a component of the animal's phenotype, behaviors are adaptations that have been shaped by natural selection. The answers to ultimate questions lie in the adaptive value of the behavior. For example, researchers think that at some point in their evolutionary history, the ancestors of prairie voles were nonmonogamous, like the overwhelming majority of mammals. Why did natural selection favor the switch to monogamy?

In order to explore this ultimate question, researchers test hypotheses on the adaptive value of monogamous behavior. For example, perhaps male prairie voles that form lasting bonds have greater reproductive success because they prevent other males from getting close to their mates. An experiment to determine whether a female vole would have sexual intercourse with other males if she were not constantly accompanied by her mate would shed light on one possible adaptive value of monogamous behavior by males. **Ultimate causes** are the evolutionary explanations for behavior, or the answers to ultimate questions.

The foundation for the scientific study of behavior was established in the mid-20th century with the work of Karl von Frisch and Konrad Lorenz, of Austria, and Niko Tinbergen, of the Netherlands. These researchers shared a Nobel Prize in 1973 for their discoveries in animal behavior. Von Frisch, who pioneered the use of experimental methods in behavior, studied honeybee behavior in detail. You'll learn about some of his work in Module 35.13. Konrad Lorenz, often regarded as the founder of behavioral biology, emphasized the importance of studying and comparing the behavior of various animals in response to different stimuli. Tinbergen worked closely with Lorenz, concentrating on experimental studies of inborn behavior and on simple forms of learning. In the next module, we look at a type of behavior that demonstrates the complementary nature of proximate and ultimate questions.

? When you touch a hot plate, your arm automatically recoils. What might be the proximate and ultimate causes of this behavior?

■ The proximate cause is a simple reflex, a neural pathway linking stimulation of receptors in your finger to motor response by muscles of your arm and hand; the ultimate cause is the natural selection for a behavior that minimizes damage to the body, thereby contributing to survival and reproductive success.

35.2 Fixed action patterns are innate behaviors

One important proximate question is how a behavior develops during an animal's life span. Lorenz and Tinbergen were among the first to demonstrate the importance of **innate behavior,** behavior that is under strong genetic control and is performed in virtually the same way by all individuals of a species. Many of Lorenz's and Tinbergen's studies were concerned with behavioral sequences called **fixed action patterns (FAPs).** A FAP is an unchangeable series of actions triggered by a specific stimulus. Once initiated, the sequence is performed in its entirety, regardless of any changes in circumstances. Consider a coffee vending machine as an analogy. The purchaser feeds money into the machine and presses a button. Having received this stimulus,

the machine performs a series of actions: drops a cup into place; releases a specific volume of coffee; adds cream; adds sugar. Once the stimulus—in this case, the money—triggers the mechanism, it carries out its complete program. Likewise, FAPs are behavioral routines that are completed in full.

Figure 35.2A (top of facing page) illustrates one of the FAPs that Lorenz and Tinbergen studied in detail. The bird is the graylag goose, a common European species that nests in shallow depressions on the ground. If the goose happens to bump one of her eggs out of the nest, she always retrieves it in the same manner. As shown in the figure, she stands up, extends her neck, uses a side-to-side head motion to nudge the egg back with her beak,

Figure 35.2A A graylag goose retrieving an egg—a FAP

and then sits down on the nest again. If the egg slips away (or is pulled away by an experimenter) while the goose is retrieving it, she continues as though the egg were still there. Only after she sits back down on her eggs does she seem to notice that the egg is still outside the nest. Then she begins another retrieval sequence. If the egg is again pulled away, the goose again completes the retrieval motion without the egg. A goose would even perform the sequence when Lorenz and Tinbergen placed a foreign object, such as a small toy or a ball, near her nest. The goose performs a series of actions regardless of the absence of an egg, just as a coffee vending machine does when the cup dispenser is empty—the coffee, cream, and sugar are poured anyway.

In its simplest form, a fixed action pattern is an innate response to a certain stimulus. For the graylag goose, the stimulus for egg retrieval is the presence of an egg (or other object) near the nest. These relatively simple, innate behaviors seem to occur in all animals. When a baby bird senses that an adult bird is near, it responds with a FAP: It begs for food by raising its head, opening its mouth, and cheeping. In turn, the parent responds with another FAP: It stuffs food in the gaping mouth. Humans perform FAPs, too. Infants grasp strongly with their hands in response to a touch stimulus on the hand. They smile in response to a face or even something that vaguely resembles a face, such as two dark spots in a circle.

Although a single FAP is typically a simple behavior, complex behaviors can result from several FAPs performed sequentially. Many vertebrates engage in courtship rituals that consist of chains of FAPs, as you'll learn in Module 35.14. The completion of a single FAP by one partner cues the other partner to begin its next FAP. There is some flexibility in these patterns. For example, a segment of the pattern might be repeated if the partner does not readily respond.

What might be the ultimate causes of FAPs? Automatically performing certain behaviors may maximize fitness to the point that genes that result in variants of that behavior do not persist in the population. For example, there are some things that a young animal has to get right on the first try if it is to stay alive. Consider kittiwakes, gulls that nest on cliff ledges. Unlike other gull species, kittiwakes show an innate aversion to cliff edges; they turn away from the edge.

Chicks in earlier generations that did not show this edge-aversion response would not have lived to pass the genes for their risk-taking behavior on to the next generation.

Fixed action patterns for reproductive behaviors are also under strong selection pressure. One example is the behavior of a pair of king penguins, which each take a turn incubating their egg while their mate feeds (Figure 35.2B). Standing face-to-face, the pair must execute a delicate series of maneuvers to pass the egg from the tops of one penguin's feet to the tops of its partner's feet, where the egg will incubate in a snug fold of the abdominal skin. (You may have seen emperor penguins engage in this behavior in the film *The March of the Penguins*.) If either partner makes a mistake, the egg may roll onto the ice, where it can freeze in seconds, eliminating the pair's only chance of successful reproduction for the year.

Innate behaviors are under strong genetic control, but the animal's performance of most innate behaviors also improves with experience. And despite the genetic component, input from the environment—an object or sensory stimulus, for example—is required to trigger the behavior. In the next module, we look at the interaction of genes and environment in producing a behavior.

? How would you explain FAPs in the context of proximate and ultimate causes?

■ The proximate cause of a FAP is often a simple environmental cue. The ultimate cause is that natural selection favors behaviors that enable animals to perform tasks essential to survival without any previous experience.

Figure 35.2B A pair of king penguins transferring their egg

Behavior is the result of both genetic and environmental factors

Evolutionary explanations for behavior assume that it has a genetic basis. Many scientific studies have corroborated that specific behaviors do, indeed, have a genetic component. Until recently, though, the heritability of a trait could only be estimated by traditional methods such as constructing pedigrees and performing breeding experiments. With the tools of molecular genetics, scientists have begun to investigate the roles of specific genes in behavior.

Groundbreaking experiments with fruit flies have led to the discovery of genes that govern learning, memory, internal clocks, and courtship and mating behaviors (Figure 35.3A). For example, male fruit flies that possess a mutation in a master gene known as *fruitless* (*fru* for short) attempt to court other male flies. In normal males without the mutation, *fru* encodes a protein that switches on a suite of genes responsible for male courtship behavior. However, neither male *fru* mutants nor wild-type females produce the normal male protein; females make a different protein. When researchers used genetic engineering to create female flies that made the male protein, the females behaved like normal males, vigorously courting other females.

Some of the genes governing fruit fly behavior have counterparts in mice, and even in humans. The genetic underpinnings of courtship and mating behaviors in mammals are probably quite different from those of flies, but research on genes implicated in learning and memory has yielded promising results. By studying fruit fly mutants with colorful names like *dunce, amnesiac,* and *rutabaga,* scientists have identified key components of memory storage, and have even used genetic engineering to create fruit flies that have exceptionally good memories. Similar genes and their protein products have been identified in mice and humans, sparking a flood of research on memory-enhancing drugs. Such drugs could improve the quality of life for people suffering from Alzheimer's and other neurological diseases that impair memory.

The mating behavior of prairie voles, described in the chapter introduction, offers another example of a behavioral trait linked to a specific genetic factor. We mentioned that monogamous prairie voles and promiscuous meadow voles differ in a small segment of DNA that controls the abundance of hormone receptors in part of the brain. If the brains of meadow voles had the same density of receptors as the brains of prairie voles, would meadow voles also be monogamous? To answer this question, researchers inserted that segment of DNA into meadow voles. Along with matching the prairie vole pattern of receptors, the behavior of these meadow voles did indeed become significantly more monogamous. Thus, it appears that a single genetic factor may account for a considerable portion of this social behavior in voles.

As we discussed in Chapter 9, however, phenotype depends on both genes and the environment. Many environmental factors, including diet and social interactions, can modify how genetic instructions are carried out. In some animals, even an individual's sex can be determined by the environment. For example, sex determination in some reptiles depends on the temperature of the egg during embryonic development.

Let's look at a study that illustrates the influence of environment on behavior (Figure 35.3B on facing page). Some female Norway rats (*Rattus norvegicus*) spend a great deal of time licking and grooming their offspring (called pups), while others interact little with their pups. As adults, pups that receive less maternal attention tend to be more sensitive to stimuli that trigger the "fight or flight" stress response (see Module 26.9) and thus are more fearful and anxious in new situations. Female pups from these litters become low-interaction mothers themselves. On the other hand, pups whose mothers spent more time licking and grooming them are more relaxed in stressful situations as adults, and the female pups become high-interaction mothers.

To investigate whether these responses to stress are entirely determined by genetics or are influenced by the interactions with their mothers, researchers performed "cross-fostering" experiments. They placed pups born to high-interaction mothers in nests with low-interaction mothers, and pups born to low-interaction mothers in the nests of high-interaction mothers. The cross-fostered pups responded to stress more like their foster mothers than like their biological mothers. The results showed that the pups' environment, in this case, maternal behavior, was the determining factor in their anxiety level.

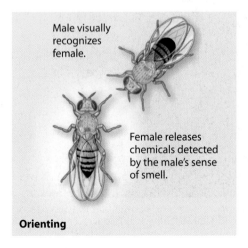

Male visually recognizes female.

Female releases chemicals detected by the male's sense of smell.

Orienting

Male taps female's abdomen with a foreleg.

Tapping

Male extends and vibrates wing, producing a courtship song.

"Singing"

Figure 35.3A Male fruit fly performing its elaborate courtship behavior

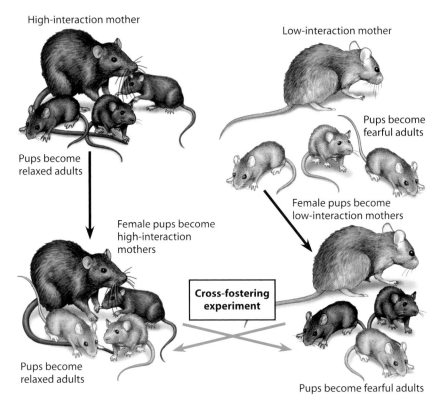

High-interaction mother

Low-interaction mother

Pups become relaxed adults

Pups become fearful adults

Female pups become high-interaction mothers

Female pups become low-interaction mothers

Cross-fostering experiment

Pups become relaxed adults

Pups become fearful adults

Figure 35.3B Cross-fostering experiment with rats

Remarkably, these experiments also demonstrated that behavioral changes can be passed to future generations, not through genes, but through the social environment. When the female rats that had been fostered gave birth, they showed the same degree of interaction with their pups as their foster mothers had shown with them. In further studies, researchers learned that interaction with the mother changes the pattern of gene expression in the pups, thus affecting the development of parts of the neuroendocrine system that regulate the fight or flight response.

These experiments and others provide evidence that behavior is the product of both genetic *and* environmental factors. Indeed, the interaction of genes and the environment appears to determine most animal behaviors. One of the most powerful ways that the environment can influence behavior is through learning, the topic we consider next.

? Without doing any experiments, how might you distinguish between a behavior that is mostly controlled by genes and one that is mostly determined by the environment?

◾ In many cases, genetically controlled behavior would not differ much between populations, regardless of the environment, while the environmentally controlled behavior would differ widely across populations located in different environments.

Learning

35.4 Learning establishes specific links between experience and behavior

Learning is modification of behavior as a result of specific experiences. Learning enables animals to change their behaviors in response to changing environmental conditions. As Table 35.4 indicates, there are various forms of learning, ranging from a simple behavioral change in response to a single stimulus, to complex problem solving that uses entirely new behaviors.

One of the simplest forms of learning is **habituation,** in which an animal learns not to respond to a repeated stimulus that conveys little or no information. There are many examples of habituation in both invertebrate and vertebrate animals. The cnidarian *Hydra* (see Module 18.6), for example, contracts when disturbed by a slight touch; it stops responding, however, if disturbed repeatedly by such a stimulus. Similarly, a scarecrow stimulus will usually make birds avoid a tree with ripe fruit for a few days. But the birds soon become habituated to the scarecrow and may even land on it on their way to the fruit tree. Once habituated to a particular stimulus, an animal still senses the stimulus—its sensory organs detect it—but the animal has learned not to respond to it.

In terms of ultimate causation, habituation may increase fitness by allowing an animal's nervous system to focus on stimuli that signal food, mates, or real danger and not waste time or energy on a vast number of other stimuli that are irrelevant to survival and reproduction.

Proximate and ultimate causes of behavior are also evident in imprinting, the type of learning we introduce in the next module.

TABLE 35.4	TYPES OF LEARNING
Learning Type	**Defining Characteristic**
Habituation	Loss of response to a stimulus after repeated exposure
Imprinting	Learning that is irreversible and limited to a sensitive time period in an animal's life
Spatial learning	Use of landmarks to learn the spatial structure of the environment
Cognitive mapping	An internal representation of the spatial relationships among objects in the environment
Associative learning	Behavioral change based on linking a stimulus or behavior with a reward or punishment; includes trial-and-error learning
Social learning	Learning by observing and mimicking others
Problem solving	Inventive behavior that arises in response to a new situation

? In a natural setting, squirrels flee at the approach of a human. On a college campus, squirrels appear to be unconcerned as students walk by them. What kind of learning accounts for the behavior of the squirrels on campus?

◾ Habituation

35.5 Imprinting requires both innate behavior and experience

Learning often interacts closely with innate behavior. Some of the most interesting examples of such an interaction involve the phenomenon known as imprinting. **Imprinting** is learning that is limited to a specific time period in an animal's life and that is generally irreversible. The limited phase in an animal's development when it can learn certain behaviors is called the **sensitive period.**

In classic experiments done in the 1930s, Konrad Lorenz used the graylag goose to demonstrate imprinting. When incubator-hatched goslings spent their first few hours with Lorenz, rather than with their mother, they steadfastly followed Lorenz (Figure 35.5A) and showed no recognition of their mother or other adults of their species. Even as adults, the birds continued to prefer the company of Lorenz and other humans to that of geese.

Lorenz demonstrated that the most important imprinting stimulus for graylag geese was movement of an object (normally the parent bird) away from the hatchlings. The effect of movement was increased if the moving object emitted some sound. The sound did not have to be that of a goose, however; Lorenz found that a box with a ticking clock in it was readily and permanently accepted as a "mother."

Just as a young bird requires imprinting to know its parents, the adults must also imprint to recognize their young. For a day or two after their own chicks hatch, adult herring gulls will accept and even defend a strange chick introduced into their nesting territory. However, once imprinted on their offspring, adults will kill any strange chicks that arrive later.

Not all examples of imprinting involve parent-offspring bonding. Newly hatched salmon, for instance, do not receive parental care but seem to imprint on the complex mixture of odors unique to their stream. This imprinting enables adult salmon to find their way back to their home stream to spawn after spending a year or more at sea.

For many kinds of birds, imprinting plays a role in song development. For example, researchers studying song development in white-crowned sparrows (Figure 35.5B) found that male birds memorize the song of their species during a sensitive period (the first 50 days of life). They do not sing during this phase, but several months later they begin to practice this song, eventually learning to reproduce it correctly. The birds do not need to hear the adult song during their practice phase; isolated males raised in soundproof chambers learned to sing normally as long as they had heard a recorded song of their species during the sensitive period. In contrast, isolated males that did not hear their species' song until they were more than 50 days old sang an abnormal song. Researchers also discovered a purely genetic component of white-crowned sparrow song development: Isolated males exposed to recorded songs of other species during the sensitive period did not adopt these foreign songs. When they later learned to sing, these birds sang an abnormal song similar to that of the isolated males of their own species that had heard no recorded bird songs.

The ability of parents and offspring to keep track of each other and the ability of male songbirds to attract mates are examples of behaviors that have direct and immediate effects on survival and reproduction. Imprinting provides a way for such behavior to become more or less fixed in an animal's nervous system.

? Explain why we say that imprinting has both innate and learned components.

■ Its innate component is the tendency to imprint on a stimulus during a sensitive period. The imprinting itself is a form of learning.

Figure 35.5A   Konrad Lorenz with geese imprinted on him

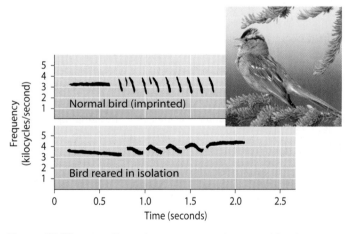

Figure 35.5B   The effect of imprinting on the songs of male white-crowned sparrows

35.6 Imprinting poses problems and opportunities for conservation programs

In attempting to save species that are at the edge of extinction, biologists sometimes try to increase their numbers in captivity. Generally, the strategy of a captive breeding program is to provide a safe environment for infants and juveniles, the stages at which many animals are most vulnerable to predation and other risks. In some programs, adult animals are caught and kept in conditions that are conducive to breeding. Offspring are usually raised by the parents and may be kept for breeding or released back to the wild. In other programs, parents are absent, as when eggs are removed from a nest. Artificial incubation is often successful, but without parents available as models for imprinting, the offspring may not learn appropriate behaviors. The effort to save the whooping crane is one example of a program that has successfully used surrogate parents, though they are a bit unusual.

The whooping crane (*Grus americana*) is a migratory waterfowl that reaches a height of about 1.5 m and has a white body with black-tipped wings that spread out over 2 m. Its name comes from its distinctive call. Once common in North American skies during their north-south migrations, whooping cranes were almost killed off by habitat loss and hunting. By the 1940s, only 16 wild birds returned from their summer breeding ground in Canada to one of their wintering areas on the Texas coast. A young whooping crane doesn't reach breeding maturity until it is 4 years old, preventing new generations of birds from reproducing quickly. And although whooping cranes often lay two eggs, parents usually successfully rear only one chick. Protections for whooping cranes were established, and in 1967, U.S. wildlife officials launched long-term recovery and captive breeding efforts.

At first, biologists used sandhill cranes as surrogate parents for whooping crane chicks. All went well until the whooping cranes reached maturity. Having imprinted on the sandhill cranes, they showed no interest in breeding with their own kind. Biologists realized that they needed another approach. With plans to set up a separate breeding colony of whooping cranes in Wisconsin, they turned to Operation Migration to help get the birds to a winter nesting site in Florida. This bird advocacy group had developed ways to hatch geese and sandhill cranes, teach the birds to recognize a small, lightweight plane as a parent figure, and train them to follow the plane along migratory routes.

In 2001, Operation Migration applied its techniques to whooping cranes. Incubating crane eggs were serenaded with recorded sounds of the plane's engine. When the chicks emerged from their shells, the first thing they saw was a hand puppet, shaped and painted in the form of an adult whooping crane (Figure 35.6A). As the chicks grew, the same type of puppet guided them through exercise and training. But now the puppet was attached to a plane that rolled along the ground, coaxing the chicks to follow it. To make sure the birds bonded with the puppet and plane and not humans, pilots and other members of Operation Migration were cloaked in hooded suits.

Eventually, the birds started following the plane on short flights (Figure 35.6B).

October 2001 brought the real test. Would the young whooping cranes follow the plane from their protected grounds in Wisconsin along a migratory route to Florida? The trip took 48 days but ultimately proved successful. Each flight day, the young cranes lined up eagerly behind the plane, wings raised, ready to follow their "parent" to the next stop. And the next spring, five of the eight young cranes retraced the route to Wisconsin on their own. Operation Migration has since taught several more generations of whooping cranes to migrate, boosting the species' chances of survival. At the end of 2006, there were about 300 adult and 75 young whooping cranes in the wild.

? **What features of whooping cranes have made their recovery difficult?**

■ They do not breed until 4 years of age and usually only raise one chick a breeding season. As migratory birds, they require resources and protection in two habitats.

Figure 35.6A Whooping crane chick interacting with puppet "parent"

Figure 35.6B Whooping cranes following a surrogate parent

35.7 Animal movement may be a simple response to stimuli or require spatial learning

Moving in a directed way enables animals to avoid predators, migrate to a more favorable environment, obtain food, and find mates and nest sites. The simplest kinds of movement do not involve learning. A random movement in response to a stimulus is called a **kinesis** (plural, *kineses;* from the Greek word for

Figure 35.7A Sow bug (also known as roly-poly or wood louse)

"movement"). A kinesis may be merely starting or stopping, changing speed, or turning more or less frequently. Sow bugs, (Figure 35.7A), which are the only terrestrial crustaceans, are not as well protected from drying out as their insect cousins. Consequently, they typically live in moist habitats, such as the underside of a rock or log. In a dry area, sow bugs exhibit kinesis, becoming more active and moving about randomly. The more they move, the greater their chance of finding a moist area. Once they are in a more favorable environment, their decreased activity tends to keep them there.

In contrast to kineses, a **taxis** (plural, *taxes;* from the Greek *tasso,* put in order) is a response directed toward (positive taxis) or away from (negative taxis) a stimulus. For example, many stream fish, such as trout, exhibit positive taxis in the current; they automatically swim or orient in an upstream direction (Figure 35.7B). This orientation keeps them from being swept away by the current and keeps them facing in the direction food is likely to come from.

In **spatial learning,** animals establish memories of landmarks in their environment that indicate the locations of food, nest sites, prospective mates, and potential hazards. Figure 35.7C illustrates a classic experiment Tinbergen performed to investigate spatial learning in an insect called the digger wasp. The female digger wasp builds its nest in a small burrow in the ground. She will often excavate four or five separate nests and fly to each one daily, cleaning them and bringing food to the larvae in the nests. To test his hypothesis that the digger wasp uses landmarks to keep track of her nests, Tinbergen ❶ placed a circle of pinecones around a nest opening and waited for the mother wasp to return and tend the nest. After the wasp flew away, Tinbergen ❷ moved the pinecones a few feet to one side of the nest opening.

The next time the wasp returned, she flew to the center of the pinecone circle instead of to the actual nest opening.

This experiment indicated that the wasp did use landmarks and that she could learn new ones to keep track of her nests. Next, to test whether the wasp was responding to the pinecones themselves or to their circular arrangement, Tinbergen ❸ arranged the pinecones in a triangle around the nest and made a circle of small stones to one side of the nest opening. This time, the wasp flew to the stones, indicating that her cue was the *arrangement* of the landmarks rather than their appearance.

While many animals find their way by learning the particular set of landmarks in their area, others appear to use more sophisticated navigation mechanisms, as we see next.

Web Process of Science *How Can Sow Bug Responses to Environments Be Tested?*

? Planarians (see Figure 18.7A) move directly away from light into dark places. What type of movement is this?

■ Negative taxis (directed movement away from light)

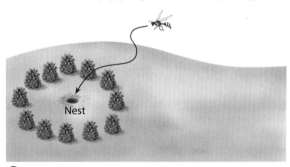

❶

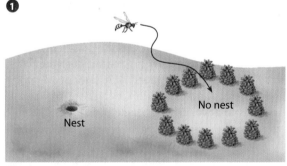

❷

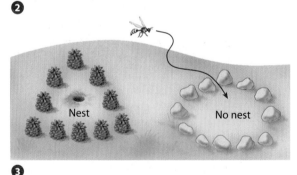

❸

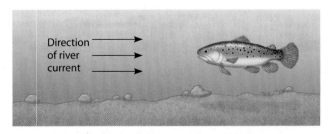

Figure 35.7B Trout exhibiting positive taxis to current

Figure 35.7C Nest-locating behavior of the digger wasp

35.8 Movements of animals may depend on internal maps

An animal can move around its environment using landmarks alone. Honeybees, for instance, might learn ten or so landmarks and locate their hive and flowers in relation to those features. A more powerful mechanism is a **cognitive map,** an internal representation, or code, of the spatial relationships among objects in an animal's surroundings.

It is actually very difficult to distinguish experimentally between an animal that is simply using landmarks and one that is using a true cognitive map. The best evidence for cognitive maps comes from research on the family of birds that includes jays, crows, and nutcrackers. Many of these birds store food in caches. A single bird may store nuts in thousands of caches that may be widely dispersed. The bird not only finds each cache, but also keeps track of food quality, bypassing caches in which the food was relatively perishable and would have decayed. It would seem that these birds use cognitive maps to memorize the locations of their food stores.

The most extensive studies of cognitive maps have involved animals that exhibit **migration,** the regular back-and-forth movement of animals between two geographic areas. Migration enables many species, such as the whooping cranes discussed in Module 35.6, to access food resources throughout the year and to breed or winter in areas that favor survival.

Another long-distance traveler is the gray whale. During the summer, these giant mammals feast on small, bottom-dwelling invertebrates that abound in northern oceans. In the autumn, they leave their feeding grounds north of Alaska and begin the long trip south along the North American coastline to winter in the warm lagoons of Baja California (Mexico). Females give birth there before migrating back north with their young. The yearly round-trip, some 20,000 km, is the longest made by any mammal.

Researchers have found that migrating animals stay on course by using a variety of cues. Gray whales seem to use the coastline to pilot their way north and south. Whale watchers sometimes see gray whales stick their heads straight up out of the water, perhaps to obtain a visual fix on land. Gray whales may also use the topography of the ocean floor, cues from the temperature and chemistry of the water, and magnetic sensing to guide their journey.

Many birds migrate at night, navigating by the stars the way ancient human sailors did. Navigating by the sun or stars requires an internal timing device to compensate for the continuous daily movement of celestial objects. Consider what would happen if you started walking one day, orienting yourself by keeping the sun on your left. In the morning, you would be heading south, but by evening you would be heading back north, having made a circle and gotten nowhere. A calibration mechanism must also allow for the apparent change in position of celestial objects as the animal moves over its migration route.

At least one night-migrating bird, the indigo bunting, seems to avoid the need for a timing mechanism by fixing on the North Star, the one bright star in northern skies that appears almost stationary. Figure 35.8 illustrates an experimental setup that was used to study the bunting's navigational mechanism.

During the migratory season, wild and laboratory-reared birds were placed in funnel-like cages in a planetarium. Each funnel had an ink pad at its base and was lined with blotting paper. When a bird stepped on the ink pad and then tried to fly in a certain direction, it tracked ink on the paper. The researchers found that wild buntings tracked ink in the direction of the North Star, as did those raised in the lab and introduced to the northern sky in a planetarium. Furthermore, researchers found that buntings raised under a planetarium sky with a different fixed-location star oriented to that star. Apparently, buntings learn to construct a star map and fix on a stationary star.

Some animals appear to migrate using only innate responses to environmental cues. For example, each fall, a new generation of monarch butterflies flies about 4,000 km over a route they've never flown before to specific wintering sites. Studies of other animals, including some songbirds, show the interaction of genes and experience in migration. Research efforts continue to reveal the complex mechanisms by which animals traverse Earth.

? Why is a timekeeping mechanism essential for stellar navigation?

■ Because the positions of the stars change with time of night and season

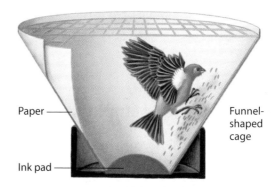

Paper —
Funnel-shaped cage
Ink pad —

Figure 35.8 Indigo bunting in experimental funnel (top); researcher sets up funnels in planetarium (bottom)

35.9 Animals may learn to associate a stimulus or behavior with a response

Associative learning is the ability to associate one environmental feature with another. In one type of associative learning, an animal learns to link a particular stimulus to a particular outcome. If you keep a pet, you have probably observed this type of associative learning firsthand. A dog or cat will learn to associate a particular sound, word, or gesture (stimulus) with a specific punishment or reward (outcome). For example, the sound of a can being opened may bring a cat running for food.

In the other type of associative learning, called **trial-and-error learning,** an animal learns to associate one of its own behaviors with a positive or negative effect. The animal then tends to repeat the response if it is rewarded or avoid the response if it is harmed. For example, predators quickly learn to associate certain kinds of prey with painful experiences. A porcupine's sharp quills and ability to roll into a quill-covered ball are strong deterrents against many predators. Coyotes, mountain lions, and domestic dogs often learn the hard way to avoid attacking porcupines nose-first (Figure 35.9).

Memory is the key to all associative learning. The mating behavior of voles, described in the chapter introduction, provides insight into the mechanisms involved. Researchers hypothesize that partner recognition is the crucial difference between monogamous and promiscuous behavior in voles. During mating, three neurochemical events occur simultaneously: release of dopamine activates neural reward circuits; olfactory signals identify the partner; and hormones are released in the brain. Because the brains of monogamous voles contain dense clusters of hormone receptors, the partners forge memories associating each others' scents with reward. With fewer hormone receptors, promiscuous voles do not form the same type of memories during their sexual encounters.

? How might the fact that many bad-tasting or stinging insect species have similar color patterns benefit both the insects and the animals that may prey on them?

■ Potential predators associate insects displaying that coloration with a negative effect, and consequently, all these insects are less likely to be preyed upon.

Figure 35.9 Trial-and-error learning by a coyote

35.10 Social learning employs observation and imitation of others

Another form of learning is **social learning**—learning by observing the behavior of others. Many predators, including cats, coyotes, and wolves, seem to learn some of their basic hunting tactics by observing and imitating their mothers.

Studies of the alarm calls of vervet monkeys in Amboseli National Park in Kenya provide an interesting example of how performance of a behavior can improve through social learning. Vervet monkeys (*Cercopithecus aethiops*) are about the size of a domestic cat. They give distinct alarm calls when they see leopards, eagles, or snakes, all of which prey on vervets. When a vervet sees a leopard, it gives a loud barking sound; when it sees an eagle, it gives a short two-syllable cough; and the snake alarm call is a "chutter." Upon hearing a particular alarm call, other vervets in the group behave in an appropriate way: They run up a tree on hearing the alarm for a leopard (vervets are nimbler than leopards in trees); look up on hearing the alarm for an eagle; and look down on hearing the alarm for a snake (Figure 35.10).

Infant vervet monkeys give alarm calls, but in a relatively undiscriminating way. For example, they give the "eagle" alarm

Figure 35.10 On seeing a python (foreground), a vervet monkey gives a distinct "snake" alarm call (inset).

on seeing any bird, including harmless birds such as bee-eaters. With age, the monkeys improve their accuracy. In fact, adult vervet monkeys give the eagle alarm only on seeing an eagle belonging to either of the two species that eat vervets. Infants probably learn how to give the right call by observing other members of the group and receiving social confirmation. For instance, if the infant gives the call on the right occasion—an eagle alarm when there is an eagle overhead—another member of the group will also give the eagle call. But if the infant gives the call when a bee-eater flies by, the adults in the group are silent. Thus, vervet monkeys have an initial, unlearned tendency

to give calls on seeing potentially threatening objects in the environment. Learning fine-tunes the calls so that by adulthood, vervets give calls only in response to genuine danger and are prepared to fine-tune the alarm calls of the next generation. However, neither vervets nor any other species comes close to matching the social learning and cultural transmission that occurs among humans, a topic we'll explore later in the chapter.

? What type of learning in humans is exemplified by identification with a role model?

■ Social learning (observation and imitation)

35.11 Problem-solving behavior relies on cognition

A broad definition of **cognition** is the process carried out by an animal's nervous system to perceive, store, integrate, and use information gathered by the senses. One area of research in the study of animal cognition is how an animal's brain represents physical objects in the environment. For instance, some researchers have discovered that many animals, including insects, are capable of categorizing objects in their environment according to concepts such as "same" and "different." One research team has trained honeybees to match colors and black-and-white patterns. Other researchers have developed innovative experiments, in the tradition of Lorenz and Tinbergen, for demonstrating pattern recognition in birds called nuthatches. These studies suggest that nuthatches apply simple geometric rules to locate their many seed caches.

Some animals have complex cognitive abilities that include **problem solving**—the process of applying past experience to overcome obstacles in novel situations. Problem-solving behavior is highly developed in some mammals, especially dolphins and primates. If a chimpanzee is placed in a room with a banana hung high above its head and several boxes on the floor, the chimp will "size up" the situation and then stack the boxes in order to reach the food. One way many animals learn to solve problems is by observing the behavior of other individuals. For example, young chimpanzees can learn from watching their elders how to crack oil palm nuts by using two stones as a hammer and anvil (Figure 35.11A).

Problem-solving behavior has also been observed in some bird species. For example, researchers placed ravens in situations in which they had to obtain food hanging from a string. Interestingly, the researchers observed a great deal of variation in the ravens' solutions. The raven in Figure 35.11B used one foot to pull up the string incrementally and the other foot to secure the string so the food didn't drop. An excellent test of human cognition and problem-solving behavior is in the construction of experiments that allow us to explore the cognition and problem-solving behavior of other animals!

? Besides problem solving, what other type of learning is illustrated by Figure 35.11A?

■ Social learning (observation)

Figure 35.11A   A chimpanzee solving a problem

Figure 35.11B
A raven solving
a problem

35.12 Behavioral ecologists use cost-benefit analysis in studying foraging

Because adequate nutrition is essential to an animal's survival and reproductive success, we should expect natural selection to refine behaviors that enhance the efficiency of feeding. Food-obtaining behavior, or **foraging,** includes not only eating, but also any mechanism an animal uses to search for, recognize, and capture food items.

Animals forage in a great many ways. Some animals are "generalists," whereas others are "specialists." Crows, for instance, are extreme generalists; they will eat just about anything that is readily available—plant or animal, alive or dead. In sharp contrast, the koala of Australia, an extreme feeding specialist, eats only the leaves of a few species of eucalyptus trees. As a result, it is restricted to certain areas and is extremely vulnerable to habitat loss. Most animals are somewhere in between crows and koalas in the range of their diet. The pronghorn (Module 34.4), for example, eats a variety of plants, including forbs, grasses, and woody shrubs.

Often, even a generalist will concentrate on a particular item of food when it is readily available. The mechanism that enables an animal to find particular foods efficiently is called a **search image.** If the favored food item becomes scarce, the animal may develop a search image for a different food item. (Humans often use search images; for example, when you look for something on a kitchen shelf, you probably scan rapidly to find a package of a certain size and color rather than reading all the labels.)

Whenever an animal has food choices, there are trade-offs involved in the selection. The amount of energy required to locate, capture, subdue, and prepare prey for consumption may vary considerably among the items available. Some behavioral ecologists use an approach known as cost-benefit analysis, comparing the positive and negative aspects of the alternative choices, to evaluate the efficiency of foraging behaviors. According to the predictions of **optimal foraging theory,** an animal's feeding behavior should provide maximal energy gain with minimal energy expense and minimal risk of being eaten while foraging. A researcher tested part of this theory by studying insectivorous birds called wagtails (Figure 35.12A).

Figure 35.12A
Wagtail

Figure 35.12B
Dung fly

In England, wagtails are commonly seen in cow pastures foraging for dung flies (Figure 35.12B). The researcher collected data on the time required for a wagtail to catch and consume dung flies of different sizes (Figure 35.12C). He then calculated the number of calories the bird gained per second of "handling" time for the different sizes of flies. The smallest flies were easily handled, but yielded few calories, while the caloric value of large flies was offset by the energy required to catch and consume them. An optimal forager would be expected to choose medium-sized flies most often. The researcher tested this prediction by observing wagtails as they foraged for dung flies in a cow pasture. As Figure 35.12D shows, he found that wagtails did select medium-size flies most often, even though they were not the most abundant size class.

Predation is one of the most significant potential costs of foraging. Studies have shown that foraging in groups, as done by herds of antelopes, flocks of birds, or schools of fish, reduces the individual's risk of predation. And for some predators, such as wolves and spotted hyenas, hunting in groups improves their success. Thus, group behavior may increase foraging efficiency by both reducing the costs and increasing the benefits of foraging.

? Early humans were hunter-gatherers, but evidence suggests that they obtained more nutrition from gathering than hunting. How does this finding relate to optimal foraging theory?

■ Meat is very nutritious, but hunting also poses relatively high costs in effort and risk compared to the gathering of plant products and dead animals.

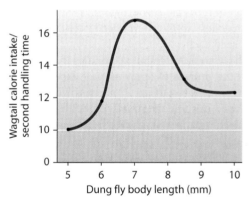

Figure 35.12C Relationship between calories gained by the wagtail and size of its prey (the dung fly)

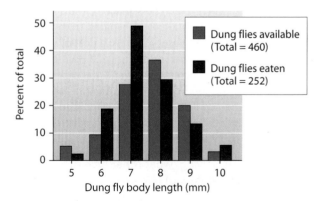

Figure 35.12D Prey sizes selected by wagtails compared with prey sizes available

Communication is an essential element of interactions between animals

Interactions between animals depend on some form of signaling between the participating individuals. In behavioral ecology, a **signal** is a stimulus transmitted by one animal to another animal. The sending of, reception of, and response to signals constitute animal **communication,** an essential element of interactions between individuals. In general, the more complex the social organization of a species, the more complex the signaling required to sustain it.

What determines the type of signal animals use to communicate? Most terrestrial mammals are nocturnal (active at night), which makes visual displays relatively ineffective. Many nocturnal mammals use odor and auditory (sound) signals, which work well in the dark. Birds, by contrast, are mostly diurnal (active in daytime) and use visual and auditory signals. Humans are also diurnal and likewise use mainly visual and auditory signals. Therefore, we can detect the bright colors and songs birds use to communicate. If we had the well-developed olfactory abilities of most mammals and could detect their rich world of odor cues, mammal-sniffing might be as popular with us as bird-watching.

What types of signals are effective in aquatic environments? A common visual signal used by territorial fishes is to erect their fins, which is generally enough to drive off intruders. Electrical signals produced by certain fishes communicate hierarchy or status. And chemical alarm substances released from an injured fish may signal nearby fish to form a tightly packed school, often near the bottom, where they are safer from attack.

Animals often use more than one type of signal simultaneously. Figure 35.13A shows a ring-tailed lemur, a tree-dwelling primate of Madagascar that lives in social groups averaging 15 individuals. Lemurs use visual displays, scent, and vocalizations to communicate with other members of the group. The animal shown here is communicating aggression with its prominent tail. Prior to this display, it smeared its tail with odorous secretions from glands on its forelegs. By waving its scented tail over its head, the lemur transmits both visual and chemical signals.

One of the most amazing examples of animal communication is found in honeybees, which have a complex social organization characterized by division of labor. Adult worker bees leave the hive to forage, bringing back pollen and nectar for the nest. When a forager locates a patch of flowers, she regurgitates some nectar that the others taste and smell, then communicates the location of the food source by performing a "dance."

Beginning with studies by Karl von Frisch, researchers have conducted numerous experiments to decipher the meanings of different dances. A pattern of movements called the waggle dance communicates the distance and the direction of food (Figure 35.13B). The dancer runs a half circle, then turns and runs in a straight line back to her starting point, buzzing her wings and waggling her abdomen as she goes. She then runs a half circle in the other direction, followed by another waggling run to the starting point. The length of the straight run and the number of waggles indicate the distance to the food source. The angle of the straight run relative to the vertical surface of the hive (30° in Figure 35.13B) is the same as the horizontal angle of the food in relation to the sun. Once the other workers have learned the location of the food source, they leave the hive to forage there.

Web Activity *Honeybee Waggle Dance*

? What types of signals do bees use?

■ Visual and chemical

Figure 35.13A
A lemur
communicating
aggression

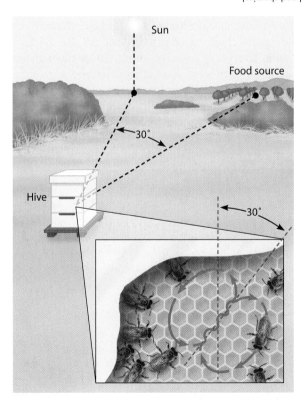

Figure 35.13B Returning honeybee forager performing the waggle dance inside the hive

35.14 Mating behavior often includes elaborate courtship rituals

Animals of many species tend to view members of their own species as competitors to be driven away. Even animals that forage and travel in groups often maintain a distance from their companions. Thus, careful communication is an essential prerequisite for mating. In many species, prospective mates must perform an elaborate courtship ritual, which confirms that individuals are of the same species, of the opposite sex, physically primed for mating, and not threats to each other. Courtship rituals are common among vertebrates, as well as some groups of invertebrates such as insects and cephalopods. The male cuttlefish in Figure 35.14A, for instance, is using part of his enormous repertoire of color changes to signal the female.

Figure 35.14B shows the courtship behavior of the common loon, which breeds on secluded lakes in the northern United States and Canada. The courting male and female swim side by side, performing a series of displays, ❶ frequently turning their heads away from each other. (In contrast, a male loon defending his territory charges at an intruder with his beak pointed straight ahead.) ❷ The birds then dip their beaks in the water and submerge their heads. ❸ The male invites the female onto land by turning his head backward with his beak down. There, they copulate. Each movement in this complex behavior is a FAP (see Module 35.2). When a FAP is executed successfully by one partner, it triggers the next FAP in the other partner. Thus the entire routine is a chain of FAPs that must be performed flawlessly if mating is to occur.

In some species, courtship is a group activity in which members of one or both sexes choose mates from a group of candidates. (See Module 13.14 to review mate choice and sexual selection.) For example, consider sage grouse, chicken-like birds that inhabit high sagebrush plateaus in the western United States. Every day in early spring, 50 or more males congregate in an open area, where they strut about, erecting their tail feathers in a bright, fanlike display (Figure 35.14C). A booming sound produced in the male's inflated air sac accompanies the show. Dominant males usually defend a prime territory near the center of the area. Females arrive several

Figure 35.14B Courtship of the common loon

Figure 35.14A Male cuttlefish displaying colors for female

Figure 35.14C Courtship display by a male sage grouse

weeks after the males. After watching the males perform, a female selects one, and the pair copulates. Usually, all the females choose dominant males, so only about 10% of the males actually mate. Is there an advantage to this group courtship display? In choosing a dominant male, a female sage grouse may be giving her offspring, and thus her own genes, the best chance for survival. Research on several species of animals has shown a connection between a male's physical characteristics and the quality of his genes (see Module 13.14).

? What categories of signals do the squid, loon, and sage grouse use to communicate with potential mates?

■ All three use visual signals. The sage grouse also uses an auditory signal.

35.15 Mating behaviors and parental care enhance reproductive success

Natural selection favors mating behaviors that enhance reproductive success. As you learned in the chapter introduction, animal mating systems fall into three major categories: promiscuous (no strong pair-bonds or lasting relationships between males and females), monogamous (a bond between one male and one female, with shared parental care), or polygamous (an individual of one sex mating with several of the other). Polygamous relationships most often involve a single male and many females, although in some species this is reversed, and a single female mates with several males.

The needs of the young are an important factor in the evolution of mating systems. Most newly hatched birds, for example, cannot care for themselves and require a large, continuous food supply that a single parent may not be able to provide. This may explain why most birds are monogamous: A male may leave more viable offspring by helping a single mate than by going off to seek more mates. (In many species of birds, however, one or both members of the pair also seeks other sexual partners.) Birds whose young can feed and care for themselves almost immediately after hatching, such as pheasants and quail, have less need for parents to stay together. In these species, polygamy is common, most likely because males can maximize their reproductive success by seeking other mates.

In the case of mammals, the lactating female is often the only food source for the young, and males usually play no role in caring for their offspring. In species such as lions, where males protect the females and young, a male or small group of males typically guard many females at once in a harem. The prairie voles described in the chapter introduction are highly unusual mammals in regard to both monogamy and parental care.

Certainty of paternity may also be a factor in the evolution of mating behavior and parental care. By helping their offspring survive to maturity, males help ensure the perpetuation of their own genes. Young born or eggs laid definitely contain the mother's genes. But even in the case of a normally monogamous relationship, the young may have been fathered by a male other than the female's usual mate. The certainty of paternity is relatively low in most species with internal fertilization because the acts of mating and birth (or mating and egg laying) are separated over time. This could help explain why species in which males are the sole parental caregiver are rare in birds and mammals. However, the males of many species with internal fertilization do engage in behaviors that appear to increase their certainty of paternity, such as guarding females from other males.

Certainty of paternity is much higher when egg laying and mating occur together, as happens when fertilization is external. This connection may explain why parental care in aquatic invertebrates, fishes, and amphibians, when care occurs at all, is at least as likely to be by males as by females. Figure 35.15 shows a male jawfish exhibiting paternal care of eggs. Jawfish, which are found in tropical marine habitats, hold the eggs they have fertilized in their mouths, keeping them aerated (by spitting them out and sucking them back in) and protected from predators until they hatch. Seahorses (see Module 19.3) have the most extreme method of ensuring paternity. Females lay their eggs in a brood pouch in the male's abdomen, where they are fertilized by his sperm. When the eggs hatch a few weeks later, the pouch opens and the male pushes them out with pumping movements.

It is important to realize that when behavioral ecologists use the phrase *certainty of paternity,* they do not mean that animals are aware of paternity when they behave a certain way. Parental behaviors associated with certainty of paternity exist because they have been reinforced over generations by natural selection. Individuals with genes for such behaviors reproduced more successfully and passed those genes on to the next generation.

? A male lion is polygamous but not promiscuous. Explain.

■ A male lion mates with several females and is thus polygamous. But he forms long-lasting bonds with these females and thus is not considered promiscuous.

Figure 35.15   A male jawfish with his mouth full of eggs

35.16 Chemical pollutants can cause abnormal behavior

Appropriate behavior is the cornerstone of success in the animal world. So, something is amiss when fish are lackadaisical about territorial defense, salamanders ignore mating cues, birds exhibit sloppy nest-building techniques, and mice take inexplicable risks. Scientists have linked observations of these abnormal behaviors, as well as many others, to endocrine-disrupting chemicals. As you learned in the introduction to Chapter 26, endocrine disruptors are a diverse group of substances that affect the vertebrate endocrine system by mimicking a hormone or by enhancing or inhibiting hormone activity. Endocrine disruptors enter ecosystems from a variety of sources, including discharge from paper and lumber mills and factory wastes such as dioxin (a by-product of many industrial processes) and PCBs (organic compounds used in electrical equipment until 1977). Agriculture is another major source of pollutants—DDT and other pesticides are endocrine disruptors. Traces of birth control pills and other hormones are commonly found in waste water from sewage treatment plants. Endocrine disruptors are especially worrisome pollutants because they persist in the environment for decades and become concentrated through food chain interactions (see Module 38.4).

Hundreds of studies have demonstrated the effects of endocrine disruptors on vertebrate reproduction and development. Like hormones, endocrine disruptors also affect behavior.

For example, some male fish attract females during the breeding season by defending territories. Males have high levels of androgens (male hormones) during this time. Researchers showed that the intensity of nest-guarding behavior in male sticklebacks (a type of fish) dropped after they were exposed to pollutants that mimic the female hormone estrogen. Male sticklebacks' performance of courtship rituals was also impaired.

Another series of studies showed that the anatomy of female mosquitofish was masculinized by endocrine disruptors. Female mosquitofish that were exposed to pollutants discharged from a paper mill (Figure 35.16A) developed the fin modification that males use to transfer sperm to females (Figure 35.16B). The masculinized females also behaved like males, waving the fin back and forth in front of females in the typical courtship behavior. Female fish living downstream from the paper mill were masculinized at contaminated sites, while female fish living in uncontaminated water near the same mill were normal.

Although the effects of endocrine disruptors on reproductive behavior have received the most attention, endocrine disruptors also affect other kinds of behavior by acting on thyroid hormones and neurological functions. For example, spatial learning ability was impaired in young monkeys exposed to PCBs.

Could endocrine disruptors in drinking water or food affect humans, too? Answers are not yet clear, but the Safe Drinking Water Act and the Food Quality Protection Act of 1996 directed the Environmental Protection Agency (EPA) to establish a screening and testing program for endocrine disruptors. As of 2007, the EPA was still working on the program's development.

? How would ineffective courtship behavior affect the fitness of male fish?

■ A male whose courtship display is perceived by females as inferior will not be successful in attracting mates and will thus be less likely to produce offspring.

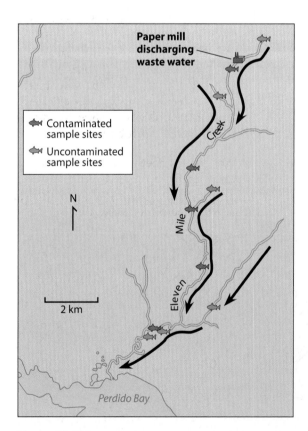

Figure 35.16A   Map of Eleven Mile Creek in Escambia County, Florida, an area used to study the effect of endocrine disruptors on mosquitofish

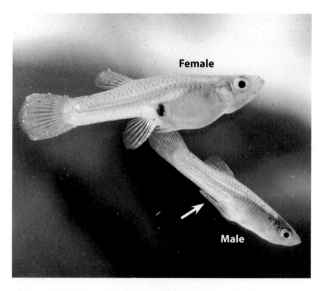

Figure 35.16B   Normal female mosquitofish (top); male showing modified fin used in courtship and sperm transfer

35.17 Sociobiology places social behavior in an evolutionary context

Biologists define **social behavior** as any kind of interaction between two or more animals, usually of the same species. The courtship behaviors of loons and sage grouse are examples of social behavior. Other social behaviors observed in animals are aggression and cooperation.

Many animals migrate and feed in large groups (flocks, packs, herds, or schools). The pronghorns you learned about in Module 34.4, for example, derive protection from feeding in herds. Many watchful eyes increase the chance that a predator will be spotted before it can strike. When alarmed, a pronghorn flares out the white hairs on its rump, sending a danger signal to other members of its herd. Predators, too, may benefit from traveling in a group. Wolves usually hunt in a pack consisting of

a tightly knit group of family members. Hunting in packs enables them to kill large animals, such as moose or elk, that would be unattainable by an individual wolf.

The discipline of **sociobiology** applies evolutionary theory to the study and interpretation of social behavior—the study of how social behaviors are adaptive and how they could have evolved by natural selection. We discuss several aspects of social behavior in the next several modules.

? Why is communication essential to social behavior?

■ Group members must be able to transfer information, for example, to signal danger.

35.18 Territorial behavior parcels out space and resources

Many animals exhibit territorial behavior. A **territory** is an area, usually fixed in location, which individuals defend and from which other members of the same species are usually excluded. The size of the territory varies with the species, the function of the territory, and the resources available. Territories are typically used for feeding, mating, rearing young, or combinations of these activities.

Figure 35.18A shows a nesting colony of gannets in Quebec, Canada. Space is at a premium, and the birds defend territories just large enough for their nests by calling out and pecking at other birds. As you can see, each gannet is literally only a peck away from its closest neighbors. Such small nesting territories are characteristic of many colonial seabirds. In contrast, most cats, including jaguars, leopards, cheetahs, and even domestic cats, defend much larger territories, which they use for foraging as well as breeding.

Individuals that have established a territory usually proclaim their territorial rights continually; this is the function of most bird songs, the noisy bellowing of sea lions, and the chattering

of squirrels. Scent markers are frequently used to signal a territory's boundaries. The male cheetah in Figure 35.18B, a resident of Africa's Serengeti National Park, is spraying urine on a tree. The odor will serve as a chemical "No Trespassing" sign. Other males that approach the area will sniff the marked tree and recognize that the urine is not their own. Usually, the intruder will avoid the marked territory and a potentially deadly confrontation with its proprietor.

Not all species are territorial. However, for those that are, the territory can provide exclusive access to food supplies, breeding areas, and places to raise young. Familiarity with a specific area may help individuals avoid predators or forage more efficiently. In a territorial species, such benefits increase fitness and outweigh the energy costs of defending a territory.

? Why is the territory of a gannet so much smaller than the territory of a cheetah?

■ The gannet uses its territory only for raising young, not for foraging. Cheetahs use their territories for foraging as well as for breeding.

Figure 35.18A Gannet territories

Figure 35.18B A cheetah marking its territory

35.19 Agonistic behavior often resolves confrontations between competitors

In many species, conflicts that arise over limited resources, such as food, mates, or territories, are settled by **agonistic behavior** (from the Greek *agon,* struggle), including threats, rituals, and sometimes combat that determine which competitor gains access to the resource. An agonistic encounter may involve a test of strength, such as when male moose lower their heads, lock antlers, and push against each other. More commonly, animals engage in exaggerated posturing and other symbolic displays that make the individual look large or aggressive. For example, the colorful fish *Betta splendens* (also known as Siamese fighting fish) spread their fins dramatically when they encounter a rival, thus appearing much bigger than they actually are. Eventually, one individual stops threatening and becomes submissive, exhibiting some type of appeasement display—in effect, surrendering.

Because violent combat may injure the victor as well as the vanquished in a way that reduces reproductive fitness, we would predict that natural selection would favor ritualized contests. And, in fact, this is what usually happens in nature. The rattlesnakes pictured in Figure 35.19, for example, are rival males wrestling over access to a mate. If they bit each other, both would die from the toxin in their fangs, but they are engaged in a pushing, rather than a biting, match. One snake usually tires before the other, and the winner pins the loser's head to the ground. In a way, the snakes are like two people who settle an argument by arm wrestling instead of resorting to fists or guns. In a typical case, the agonistic ritual inhibits further aggressive activity. Once two individuals have settled a dispute by agonistic behavior, future encounters between them usually involve less conflict, with the original loser giving way to the original victor. Often the victor of an agonistic ritual gains first or exclusive access to mates, and so this form of social behavior can directly affect an individual's evolutionary fitness.

Figure 35.19 Ritual wrestling by rattlesnakes

? Why is "fighting to the death" an unusual form of agonistic behavior among animals?

Because ritualized posturing or nonlethal combat can usually produce a winner without injuries that would lower reproductive fitness for the winner and eliminate it altogether for the loser

35.20 Dominance hierarchies are maintained by agonistic behavior

Many animals live in social groups maintained by agonistic behavior. Chickens are an example. If several hens unfamiliar to one another are put together, they respond by chasing and pecking each other. Eventually, they establish a clear "pecking order." The alpha, or top-ranked, hen in the pecking order (the one on the left in Figure 35.20) is dominant; she is not pecked by any other hens and can usually drive off all the others by threats rather than actual pecking. The alpha hen also has first access to resources such as food, water, and roosting sites. The beta, or second-ranked, hen similarly subdues all others except the alpha, and so on down the line to the omega, or lowest, animal.

Pecking order in chickens is an example of a **dominance hierarchy,** a ranking of individuals based on social interactions. Once a hierarchy is established, each animal's status in the group is fixed, often for several months or even years. Dominance hierarchies are common, especially in vertebrate populations. In a wolf pack, for example, there is a dominance hierarchy among the females, and this hierarchy may control the pack's size. When food is abundant, the alpha female mates and also allows others to do so. When food is scarce, she usually monopolizes males for herself and keeps other females from mating. In other species, such as savanna baboons and red deer, the dominant male monopolizes fertile females. As a result, his reproductive success is much greater than that of lower-ranking males.

Next we hear from a scientist who has conducted long-term studies of dominance hierarchies in nature.

? Dog trainers advise dog owners to be sure the dog understands that the owner is the alpha. How might this facilitate obedience?

Like wolves, dogs are pack animals. They are more likely to respond to commands from a higher-ranking individual.

Figure 35.20 Chickens exhibiting pecking order

35.21 Behavioral biologist Jane Goodall discusses dominance hierarchies and reconciliation behavior in chimpanzees

Chimpanzees are humans' closest relatives. Dr. Jane Goodall (Figure 35.21A), one of the world's best-known biologists, has studied these remarkable primates in their natural habitat in East Africa since the early 1960s. She has described her discoveries in her many books, her appearances on National Geographic Society television specials, and her frequent lecture tours. In all of her writings and interviews, Dr. Goodall promotes a better understanding of animal behavior, especially that of primates. For more than 30 years, the Jane Goodall Institute has promoted issues related to the conservation of primates and their habitats. Dr. Goodall also works tirelessly to encourage better living conditions for animals in medical research labs and zoos.

In an interview, Dr. Goodall described dominance hierarchies:

Some male chimpanzees devote much time and effort to improving or maintaining their position in the hierarchy. For the most part, the male uses the impressive charging display, during which he races across the ground, hurls rocks, drags branches, leaps up and shakes the vegetation. In other words, he makes himself look larger and more dangerous than he may actually be. In this way he can often intimidate a rival without having to risk an actual fight, which could be dangerous for him as well as for his rival. The more frequent, the more vigorous, and the more imaginative his charging display, the more likely it is that he will attain a high social position.

And what about female chimpanzees?

Females have a hierarchy too. . . . The reproductive advantage to the high-ranking female is clear. She can better appropriate choice food items and thus make her milk richer. In addition, her offspring are likely to become high-ranked since she will support them. In the supportive family group situation, all have a better chance of survival.

Chimpanzees live in fairly permanent social groups, and there are benefits to maintaining friendly relations within the group. Following a conflict, some kind of reconciliation behavior usually occurs. For example, a chimpanzee that has threatened another member of its group may invite reconciliation by a hand gesture (Figure 35.21B), leading to a bout of friendly grooming. In Dr. Goodall's words:

Social grooming is the single most important social activity in the chimp community. It improves bad relationships and maintains good ones. A few brief grooming movements serve to reassure, to appease a higher-ranking individual, or to calm a subordinate. A mother pacifies her child by embracing and then grooming him or her. Adult males enjoy particularly long grooming sessions—this is important. Males do sometimes compete quite vigorously for dominance rank, and their relationship may then become tense. Yet it is crucial that they be able to cooperate in order to jointly protect the territory of their community.

Social primates seem to spend substantial time in reconciliation and pacification-type behavior.

Studies of chimpanzee behavior can make us more aware of what we have in common with other species. Jane Goodall's years of chimpanzee research have convinced her that chimpanzees are truly conscious beings. As she explains:

Science has been very quick to recognize the incredible similarity in [the anatomy] of the chimpanzee brain and human brain . . . and all the other amazing physiological similarities. So it stands to reason that you would find similarities in the emotions . . . and in certain kinds of behavior and intellect.

? Why is the study of chimpanzee behavior relevant to understanding the origins of certain human behaviors?

■ Because chimpanzees and humans share a common ancestor

Figure 35.21A Jane Goodall with Goblin, an alpha male

Figure 35.21B Reconciliation in chimpanzees

35.22 Altruistic acts can often be explained by the concept of inclusive fitness

Many social behaviors are selfish. Behavior that maximizes an individual's survival and reproductive success is favored by selection, regardless of how much the behavior may harm others. For example, superior foraging ability by one individual may leave less food for others. **Altruism,** on the other hand, is defined as behavior that reduces an individual's fitness while increasing the fitness of others in the population. How can what appears to be altruistic, or selfless, behavior be explained by natural selection?

Consider the Belding's ground squirrel, which lives in regions of the western United States and is vulnerable to predators such as coyotes and hawks. Upon seeing a predator approach, a squirrel often gives a high-pitched alarm call (Figure 35.22A), which alerts nearby squirrels, who then retreat to their burrows. Field observations have confirmed that the conspicuous alarm call identifies the caller's location and increases the risk of being killed.

Altruistic behavior is often evident in animals that live in cooperative colonies. For example, workers in a honeybee hive are sterile females who labor all their lives on behalf of the queen. When a worker stings an intruder in defense of the hive, the worker usually dies.

The animals in Figure 35.22B are highly social rodents called naked mole rats. Almost hairless and nearly blind, they live in colonies in underground chambers and tunnels in southern and northeastern Africa. With a social structure resembling that of honeybees, each colony has only one reproducing female, called the queen. The queen mates with one to three males, called kings. The rest of the colony consists of nonreproductive females and males who forage for roots and care for the queen, her young, and the kings. While trying to protect the queen or kings from a snake that invades the colony, a nonreproductive naked mole rat may sacrifice its own life.

You might wonder how altruistic behavior can evolve if it reduces the reproductive success of self-sacrificing individuals. It is easy to see how selfless behavior might be selected for when it involves parents and offspring. When parents sacrifice their own well-being to ensure the survival of their young, they are maximizing the survival of their own genes. But reproducing is only one way to pass along genes; helping a close relative reproduce is another. Siblings, like parents and offspring, have half their genes in common, and an individual shares one-fourth of its genes with offspring of a sibling. The concept of **inclusive fitness** describes an individual's success at perpetuating its genes by producing its

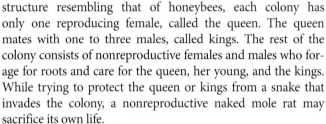

Figure 35.22A Belding's ground squirrel giving an alarm call

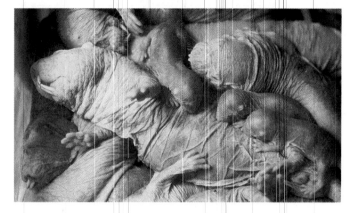

Figure 35.22B The queen of a naked mole rat colony nursing offspring while surrounded by other individuals of the colony

own offspring and by helping close relatives, who likely share many of those genes, to produce offspring.

Altruism increases inclusive fitness when it maximizes the reproduction of close relatives. The natural selection favoring altruistic behavior that benefits relatives is called **kin selection.** Thus, the genes for altruism may be propagated if individuals that benefit from altruistic acts are themselves carrying those genes.

If kin selection explains altruism, then the examples of unselfish behavior we observe should involve close relatives. This is in fact the case. Most alarm calls are given by female Belding's squirrels, whose close relatives live nearby. Researchers have also found that all the individuals in a naked mole rat colony are closely related. The nonreproductive members are the queen's descendants or siblings; by enhancing a queen's chances of reproducing, they increase the chances that some genes identical to their own will be propagated. Likewise, bees in a hive all share genes with the queen. Their work (or even death) in support of the queen helps ensure that many of their genes will survive.

Kin selection does not explain all types of altruism. Chimpanzees sometimes save the lives of nonrelatives, and female dolphins without young often help unrelated mothers care for their young. In these cases, there can be no immediate enhancement of the altruists' fitness. However, the beneficiary may reciprocate—that is, "return the favor" someday. Thus, we can explain altruism toward nonrelatives as **reciprocal altruism:** an altruistic act that may be repaid at a later time by the beneficiary. Reciprocal altruism appears to be fairly rare, limited largely to species with social groups stable enough that individuals have many chances to exchange aid. Reciprocal altruism is often used to explain altruism in humans. In the last module of this chapter, we take a look at human behavior.

 What is the ultimate cause for altruism between kin?

■ Natural selection reinforces such altruistic behavior through the reproductive success of closely related individuals that have many genes in common with the altruist, including genes for altruism.

Human behavior is the result of both genetic and environmental factors

Variations in behavioral traits such as personality, temperament, talents, and intellectual abilities make each person a unique individual. What are the roles of nature (genes) and nurture (environment) in shaping these behaviors? Let's look at how scientists distinguish between genetic and environmental influences on behavioral variations in humans.

Twins provide a natural laboratory for investigating the origins of complex behavioral traits. In general, researchers attempt to estimate the heritability of a trait, or how much of the observed variation can be explained by inheritance. Twin studies compare identical twins (Figure 35.23A), who have the same DNA sequence and are raised in the same environment, with fraternal twins (Figure 35.23B), who share an environment but only half of their DNA sequence. Some twin studies compare identical twins who were raised in the same household to twins who were separated at birth, a design that allows researchers to study the interactions of different environments with the same genotype. However, separated twins are very rare.

Results from twin studies consistently show that for complex behavioral traits such as general intelligence and personality characteristics, genetic differences account for roughly half the variation among individuals. The remainder of the variation can be attributed mostly to each individual's unique environment.

Determining that a trait is heritable does not mean that scientists have identified a gene "for" the trait. Genes do not dictate behavior. Instead, genes cause tendencies to react to the environment in a certain way. For example, the hormone receptor gene in prairie voles does not produce a protein that causes them to be monogamous. Rather, the receptor protein produced by the gene connects the dopamine reward experienced during mating with the scent of its partner, and the prairie vole responds to the presence of that scent with behaviors that keep it in close contact with its mate.

Let's look at genes related to social bonding as an example of how genetics could play a role in human behavioral variations. Like the brains of other mammals, human brains produce the same hormones and hormone receptors that are implicated in social recognition and bonding in voles. Humans also have considerable individual variation in the key segment of DNA that determines receptor density. As a result, human brains vary in their sensitivity to hormones involved in bonding. Researchers are not investigating whether these differences are related to human mating behavior. Rather, the studies on voles are helping scientists understand how the human brain processes social information. For example, it has been hypothesized that variation in this receptor gene may play a role in some types of autism, a disorder characterized by difficulty forming social attachments.

The mechanisms and underlying genetics of behavior are proximate causes (see Module 35.1). Scientists are also exploring the ultimate causes of human behavior. Sociobiology, the area of research introduced in Module 35.17, centers on the idea that social behavior evolves, like anatomical traits, as an expression of genes that have been perpetuated by natural selection. When applied to humans, this idea might seem to imply that life is predetermined. But it's unlikely that most human behavior is programmed by our genome. Unlike other animals, human offspring have an extraordinarily long period of development after birth. Children interact with a rich social environment, consisting of parents and other family members, peers, teachers, and society in general. The abilities to learn, to innovate, to advance technologically, and to participate in complex social networks have been key elements in the phenomenal success of the human species. It is much more likely that natural selection favored mechanisms that enabled humans to operate on the fly, that is, to use experience and feedback from the environment to adjust their behavior according to the circumstances.

Figure 35.23A Identical twins Ronde Barber, cornerback for the Tampa Bay Buccaneers, and Tiki Barber, sportscaster and former professional football player

Figure 35.23B Fraternal twins Mary-Kate and Ashley Olsen, who are actors and fashion designers

? A researcher conducted a study on "pseudo-twins," unrelated children of the same age who were raised in the same household. Results showed no correlation between the IQs of pseudo-twins. What do these results indicate about the influence of genes and environment on IQ?

■ The results indicate a strong genetic component to IQ. The children shared an environment but none of their DNA sequences. If environment were the main influence, you would expect the IQ scores to be similar.

Reviewing the Concepts

The Scientific Study of Behavior (35.1–35.3)

Behavioral ecology is the study of behavior in an evolutionary context, considering both proximate (immediate) and ultimate (evolutionary) causes of an animal's actions. Natural selection preserves behaviors that enhance fitness (**35.1**).

Innate behavior is performed the same way by all members of a species. A fixed action pattern (FAP) is an unchangeable series of actions triggered by a specific stimulus. FAPs ensure that activities essential to survival are performed correctly without practice (**35.2**).

Determinants of behavior. Behavior usually involves both genetic and environmental influences (**35.3**).

Learning (35.4–35.11)

Learning is a change in behavior resulting from experience. Habituation is learning to ignore a repeated, unimportant stimulus (**35.4**).

Imprinting is irreversible learning limited to a sensitive period in the animal's life. Captive breeding programs for endangered species must provide proper imprinting models (**35.5–35.6**).

Spatial learning involves using landmarks to move through the environment. Kineses and taxes are simple movements in response to a stimulus (**35.7**).

Cognitive maps are internal representations of spatial relationships of objects in the surroundings. Migratory animals may move between areas using the sun, stars, landmarks, or other cues (**35.8**).

Associative learning. Many animals can learn by associating external stimuli or their own behavior with positive or negative effects (**35.9**).

Social learning involves changes in behavior that result from observation and imitation of others (**35.10**).

Cognition is the process of perceiving, storing, integrating, and using information. Some animals exhibit problem-solving behavior, which involves complex cognitive processes (**35.11**).

Survival and Reproductive Success (35.12–35.16)

Foraging includes identifying, obtaining, and eating food. Optimal foraging theory predicts that feeding behavior will maximize energy gain and minimize energy expenditure and risk (**35.12**).

Signaling in the form of sounds, scents, displays, or touches provides communication needed for interactions between members of the same species (**35.13**).

Courtship rituals advertise the species, sex, and physical condition of potential mates (**35.14**).

Mating systems may be promiscuous, monogamous, or polygamous. The needs of offspring and certainty of paternity help explain differences in mating systems and parental care by males (**35.15**).

Endocrine disruptors in the environment may cause abnormal behavior as well as reproductive abnormalities (**35.16**).

Social Behavior and Sociobiology (35.17–35.23)

Sociobiology is the study of social behavior, the interactions of two or more animals, in the context of evolution (**35.17**).

Territorial behavior allocates space and resources. Animals exhibiting this behavior defend their territories (**35.18**).

Agonistic behavior, including threats, rituals, and sometimes combat, settles disputes over resources (**35.19**).

Dominance hierarchies partition resources among members of a social group (**35.20**). Chimpanzees exhibit dominance hierarchies and reconciliation behaviors (**35.21**).

Altruism can usually be explained by the concepts of inclusive fitness and kin selection: An animal can propagate its own genes by helping relatives reproduce. In reciprocal altruism, individuals do favors that may later be repaid (**35.22**).

Human behavior has a genetic basis but is strongly influenced by learning (**35.23**).

Connecting the Concepts

1. Complete this map, which reviews behavioral ecology.

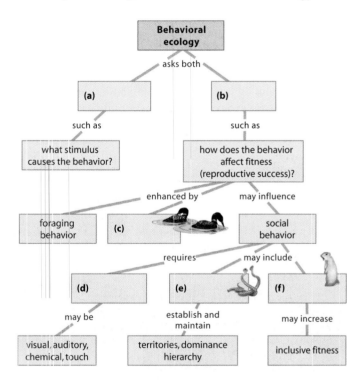

2. Create your own concept map to organize your understanding of the genetic and environmental components of animal behavior and their relationship to learning. Include examples where possible.

Testing Your Knowledge

Multiple Choice

3. Although many chimpanzee populations live in environments containing oil-palm nuts, members of only a few populations use stones to crack open the nuts. The most likely explanation for this behavioral difference between populations is that
   a. members of different populations differ in manual dexterity.
   b. members of different populations have different nutritional requirements.
   c. members of different populations differ in learning ability.
   d. the cultural tradition of using stones to crack nuts has arisen in only some populations.
   e. the behavioral difference is caused by genetic differences between populations.

4. Pheasants do not feed their chicks. Immediately after hatching, a pheasant chick starts pecking at seeds and insects on the ground. How might a behavioral ecologist explain the ultimate cause of this behavior?
   a. Pecking is a fixed action pattern.
   b. Pheasants learned to peck, and their offspring inherited this behavior.
   c. Pheasants that pecked survived and reproduced best.
   d. Pecking is a result of imprinting during a sensitive period.
   e. Pecking is an example of habituation.

5. A blue jay that aids its parents in raising its siblings is increasing its
   a. reproductive success.
   b. status in a dominance hierarchy.
   c. altruistic behavior.
   d. inclusive fitness.
   e. certainty of paternity.

6. Ants carry dead ants out of the anthill and dump them on a "trash pile." If a live ant is painted with a chemical from dead ants, other ants repeatedly carry it, kicking and struggling, to the trash pile, until the substance wears off. Which of the following best explains this behavior?
   a. The chemical triggers a fixed action pattern.
   b. The ants have become imprinted on the chemical.
   c. The ants continue the behavior until they become habituated.
   d. The ants can learn only by trial and error.
   e. The chemical triggers a negative taxis.

Describing, Comparing, and Explaining

7. Almost all the behaviors of a housefly are innate. What are some advantages and disadvantages to the fly of innate behaviors compared with behaviors that are mainly learned?

8. In Module 35.3, you learned that Norway rat offspring whose mothers don't interact with them much grow up to be fearful and anxious in new situations. Suggest a possible ultimate cause for this link between maternal behavior and stress response of offspring. (*Hint:* Under what circumstances might high reactivity to stress be more adaptive than being relaxed?)

9. A chorus of frogs fills the air on a spring evening. The frog calls are courtship signals. What are the functions of courtship behaviors? How might a behavioral ecologist explain the proximate cause of this behavior? The ultimate cause?

Applying the Concepts

10. Crows break the shells of certain molluscs before eating them by dropping them onto rocks. Hypothesizing that crows drop the molluscs from a height that gives the most food for the least effort (optimal foraging), a researcher dropped shells from different heights and counted the drops it took to break them.

Height of drop (m)	Average number of drops required to break shell	Total flight height (number of drops × height per drop)
2	55	110
3	13	39
5	6	30
7	5	35
15	4	60

   a. The researcher measured the average drop height for crows and found it was 5.23 m. Does this support the researcher's hypothesis? Explain.
   b. Describe an experiment to determine whether this feeding behavior of crows is learned or innate.

11. Scientists studying scrub jays found that it is common for "helpers" to assist mated pairs of birds in raising their young. The helpers lack territories and mates of their own. Instead, they help the territory owners gather food for their offspring. Propose a hypothesis to explain what advantage there might be for the helpers to engage in this behavior instead of seeking their own territories and mates. How would you test your hypothesis? If your hypothesis is correct, what kind of results would you expect your tests to yield?

12. Researchers are very interested in studying identical twins who were raised apart. Among other things, they hope to answer questions about the roles of inheritance and upbringing in human behavior. So far, data suggest that identical twins raised apart are much more alike than researchers would have predicted based on their different environments. They have similar personalities, mannerisms, habits, and interests. Why do identical twins make such good subjects for this kind of research? What do the results suggest to you? What are the potential pitfalls of this research? What abuses might occur in the use of these data if the studies are not evaluated critically?

13. Do animals think and feel the same kinds of things we do? These questions bear on animal rights, a subject much in the news. Many important biological discoveries have come from experiments performed on animals, yet some animal rights activists believe that all animal experimentation is cruel and should be stopped. They have harassed researchers, vandalized laboratories, and set animals free. Why are animals used in experiments? Are there uses of animals that should be discontinued? What kinds of guidelines should researchers follow in using animals in experiments, and who should establish and enforce the guidelines?

Answers to all questions can be found in Appendix 4.

For Practice Quizzes, BioFlix, MP3 tutors, and Activities, go to www.mybiology.com.

36 Population Ecology

A Tale of Two Fishes

In the early part of the twentieth century, people drew a bountiful harvest from Lake Victoria, one of the Great Lakes of East Africa (see map on page 277, and see facing page for a photo of wooden fishing boats on the lake). The rich biodiversity of the lake provided a variety of fishes that were eagerly sought for food. A member of the cichlid family called ngege (*Oreochromis esculentus*), was one of the most important species. Typically about 30 cm long and weighing just a couple of kilograms, ngege were prized for their excellent flavor and tender flesh. As you may recall from the opening essay in Chapter 14, the diversity of fish species in Lake Victoria also attracted the attention of ecologists, who discovered that the lake was the site of a spectacular burst of adaptive radiation of the cichlid family. (Recall from Module 14.10 that adaptive radiation is the evolution of many diverse species from a common ancestor.) Cichlids account for the

vast majority of Lake Victoria's fish fauna.

As the century progressed, new types of fishing gear replaced traditional methods, enabling larger catches. The abundance of food fueled human population growth, which in turn intensified fishing. When populations of the most favored species began to dwindle, British colonial administrators conceived a scheme to boost Lake Victoria's economic value. They released Nile perch, a voracious predator, into the lake. The small cichlids, so elusive to fishermen, would be easy prey for the Nile perch. Thus the cichlids would, in effect, be repackaged in the ample flesh of the easily caught giant Nile perch, which grow to 2 m in length and weigh over 100 kg (left).

The plan, implemented in the 1950s and 1960s, was initially thought to be a failure. The population of Nile perch remained small. The local people found its oily flesh unsuitable for sun-drying, their traditional method of preservation (above).

By the 1980s, however, the tide had turned. The Nile perch population skyrocketed. Simultaneously, populations of the cichlids that composed the bulk of the Nile perch's diet crashed. Approximately 200 cichlid species either vanished or approached the brink of extinction, including ngege.

As Nile perch became more plentiful, commercial fisheries moved in to exploit them, hauling out more than 100,000 tons of fish each year. Shoreline factories sprang up to process and package the fish for export to markets in Europe and Asia. The human population around the lake mushroomed with hundreds of thousands of people who depended, either directly or indirectly, on the Nile perch industry for their livelihoods. Local people no longer dined on the lake's bounty, however. Their access to Nile perch was limited to the stripped carcasses and heads left over from processing plants. The fish took a further ecological toll, because the scraps are preserved by smoking over wood fires. Before long, the shores of Lake Victoria were deforested to provide wood for fires and for other purposes such as cooking and building.

By the 1990s, the intense fishing pressure again led to diminishing returns. The huge Nile perch that were commonplace in the 1980s were replaced by smaller, immature specimens, leaving the fishing industry of Lake Victoria with a questionable future. Scientists are now using population data gathered over several decades to develop recommendations for sustainable fisheries. On the other hand, overfishing has been good news for the cichlids. In the wake of the declining Nile perch population, some of the remaining cichlid species are bouncing back. Unfortunately, the delicious ngege is not among them.

Meanwhile, the human population around Lake Victoria continues to increase, as it does worldwide. Like any population, it can't grow indefinitely. But when will it stop, and what will stop it?

Population ecology, the subject of this chapter, is concerned with changes in population size and the factors that regulate populations over time. As ecologists learn more about the structure and dynamics of natural populations, we become better equipped to develop sustainable food sources, assess the impact of human activities, and balance human needs with the conservation of biodiversity and resources. ■ ■ ■

36.1 Population ecology is the study of how and why populations change

Ecologists usually define a **population** as a group of individuals of a single species that occupy the same general area. These individuals rely on the same resources, are influenced by the same environmental factors, and are likely to interact and breed with one another. A population can be described by the number and distribution of individuals. When a researcher chooses a population to study, he or she defines it by boundaries appropriate to the species being studied and to the purposes of the investigation. For example, a population ecologist interested in a species of cichlid in Lake Victoria might study a population in Mwanza Gulf, a narrow inlet at the southern end of the lake.

Population ecologists also examine population dynamics, the interactions between biotic and abiotic factors that cause variation in population sizes. One important aspect of population dynamics—and a major topic for this chapter—is population growth. A cichlid population in Mwanza Gulf increases through births and the immigration of fish from elsewhere in the lake. Deaths and the emigration of individuals out of Mwanza Gulf decrease the population. Population ecologists might investigate how various environmental factors, such as availability of food, predation by Nile perch, or degradation of water quality by shoreline towns and factories, affect the density, distribution, or dynamics of the population.

Population ecology plays a key role in applied research. For example, population ecology provides critical information for identifying and saving endangered species such as the Lake Victoria cichlids. Population ecology is being used to develop sustainable fisheries throughout the world, including Lake Victoria, and to manage wildlife populations. Studying the population ecology of pests and pathogens provides insight into controlling their spread. Population ecologists also study human population growth, one of the most critical environmental issues of our time.

? What is the relationship between a population and a species?

■ A population is a localized group of individuals of a species.

36.2 Density and dispersion patterns are important population variables

Two important aspects of population structure are population density and dispersion pattern. **Population density** is the number of individuals of a species per unit area or volume—the number of oak trees per square kilometer (km^2) in a forest, for instance, or the number of earthworms per cubic meter (m^3) in forest soil. Because it is impractical or impossible to count all individuals in a population in most cases, ecologists use a variety of sampling techniques to estimate population densities. For example, they might base an estimate of the density of alligators in the Florida Everglades on a count of individuals in a few sample plots of 1 km^2 each. The larger the number and size of sample plots, the more accurate the estimates. In some cases, population densities are estimated not by counts of organisms but by indirect indicators, such as number of bird nests or rodent burrows.

Within a population's geographic range, local densities may vary greatly. The **dispersion pattern** of a population refers to the way individuals are spaced within their area. A **clumped dispersion pattern,** in which individuals are grouped in patches, is the most common in nature. Clumping often results from an unequal distribution of resources in the environment. For instance, plants or fungi may be clumped in areas where soil conditions and other factors favor germination and growth. Clumping of animals is often associated with uneven food distribution or with mating or other social behavior. For example, fish are often clumped in schools (Figure 36.2A), which may reduce predation risks and increase feeding efficiency.

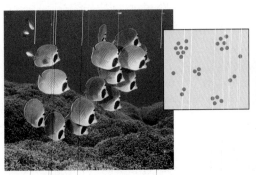

Figure 36.2A Clumped dispersion of schooling fish

Figure 36.2B Uniform dispersion of nesting king penguins

Figure 36.2C Random dispersion of dandelions

A **uniform dispersion pattern** (an even one) often results from interactions between the individuals of a population. For instance, some plants secrete chemicals that inhibit the germination and growth of nearby plants that could compete for resources. Animals may exhibit uniform dispersion as a result of territorial behavior. Figure 36.2B (previous page) shows uniform dispersion in king penguins.

In a **random dispersion pattern,** individuals in a population are spaced in an unpredictable way, actually without a pattern. Plants, such as dandelions (Figure 36.2C, previous page), that grow from windblown seeds might be randomly dispersed. However, varying habitat conditions and social interactions make random dispersion rare.

Estimates of population density and dispersion patterns enable researchers to monitor changes in a population and to compare and contrast the growth and stability of populations in different areas. The next module describes another tool that ecologists use to study populations.

Web Activity *Techniques for Estimating Population Density and Size*

? What dispersion pattern would you predict in a forest population of termites, which live in damp, rotting wood?

■ Clumped (in fallen logs or dead trees)

36.3   Life tables track survivorship in populations

Life tables track survivorship, the chance of an individual in a given population surviving to various ages. Starting with a population of 100,000 people, Table 36.3 shows the number who are expected to be alive at the beginning of each age interval, based on death rates in 2003. For example, 93,585 out of 100,000 people are expected to live to age 50. The chance of surviving to age 60, shown in the last column of the table, is 0.938. The chance of surviving to age 90, however, is only 0.405. The life insurance industry uses life tables to predict how long, on average, a person will live. Population ecologists have adopted this technique and constructed life tables for various plant and animal species. By identifying the most vulnerable stage of the life cycle, life table data help conservationists develop effective measures for maintaining a viable population.

Life tables can be used to construct **survivorship curves,** which plot survivorship as the proportion of individuals from an initial population that are alive at each age (Figure 36.3). By using a percentage scale instead of actual ages on the *x*-axis, we can compare species with widely varying life spans on the same graph. The curve for the human population shows that most people survive to the older age intervals, as we saw in the life table. Ecologists refer to the shape of this curve as Type I survivorship. Species that exhibit a Type I curve—humans and many other large mammals—usually produce few offspring but give them good care, increasing the likelihood that they will survive to maturity.

In contrast, a Type III curve indicates low survivorship for the very young, followed by a period when survivorship is high for those few individuals who live to a certain age. Species with this type of survivorship curve usually produce very large numbers of offspring but provide little or no care for them. Nile perch, for example, can produce 16 million eggs at a time, but most offspring die as larvae from predation or other causes. Many invertebrates, such as oysters, also have Type III survivorship curves.

A Type II curve is intermediate, with survivorship constant over the life span. That is, individuals are no more vulnerable at one stage of the life cycle than another. This type of survivorship has been observed in some invertebrates, lizards, and rodents.

Web Activity *Investigating Survivorship Curves*

TABLE 36.3	LIFE TABLE FOR THE U.S. POPULATION IN 2003		
Age Interval	Number Living at Start of Age Interval (*N*)	Number Dying During Interval (*D*)	Chance of Surviving Interval (1−D/N)
0–10	100,000	884	0.991
10–20	99,116	423	0.996
20–30	98,693	941	0.990
30–40	97,752	1,308	0.987
40–50	96,444	2,859	0.970
50–60	93,585	5,825	0.938
60–70	87,760	12,225	0.861
70–80	75,535	22,794	0.698
80–90	52,741	31,401	0.405
90+	21,340	21,340	0.000

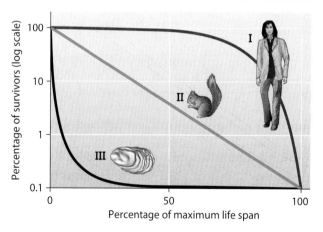

Figure 36.3   Three types of survivorship curves

? How does the chance of survival change with age in organisms with a Type III survivorship curve?

■ Chance of survival is initially low but increases after an individual reaches maturity.

36.4 Idealized models predict patterns of population growth

Population size fluctuates as new individuals are born or immigrate into an area and others die or emigrate. Some populations, for example, trees in a mature forest, are relatively constant over time. Other populations change rapidly, even explosively. Consider a single bacterium that divides every 20 minutes. There would be two bacteria after 20 minutes, four after 40 minutes, eight after 60 minutes, and so on. In just 12 hours, the population would approach 70 billion cells. If reproduction continued for a day and a half—a mere 36 hours—there would be enough bacteria to form a layer a foot deep over the entire Earth. Using idealized models, population ecologists can predict how the size of a particular population will change over time under different conditions.

The Exponential Growth Model The rate of population increase under ideal conditions, called exponential growth, is calculated using the simple equation $G = rN$. The G stands for the growth rate of the population (the number of new individuals added per time interval); N is the population size (the number of individuals in the population at a particular time); and r stands for the **per capita rate of increase** (the average contribution of each individual to population growth; per capita means "per person").

How do we estimate the per capita rate of increase? Population growth reflects the number of births minus the number of deaths (the model assumes that immigration and emigration are equal). Suppose a population of rabbits has 100 individuals, and there are 50 births and 20 deaths in one month. The net increase is 30 rabbits. The per capita increase in the population, or r, is 30/100, or 0.3.

In a population growing in an ideal environment with unlimited space and resources, r is the maximum capacity of members of that population to reproduce. Thus, the value of r depends on the kind of organism. For example, rabbits have a higher r than elephants, and bacteria have a higher r than rabbits.

When a population is expanding without limits, r remains constant and the rate of population growth depends on the number of individuals already in the population (N). In Table 36.4A, a population begins with 20 rabbits. The growth rate (G) for this population, using $r = 0.3$, is shown in the right-hand column. Notice that the larger the population size, the more new individuals are added during each time interval.

TABLE 36.4A	EXPONENTIAL GROWTH OF RABBITS, $r = 0.3$	
Time (months)	N	G = rN
0	20	6
1	26	8
2	34	10
3	44	13
4	57	17
5	74	22
6	96	29
7	125	38
8	163	49
9	212	64
10	276	83
11	359	108
12	467	140

Graphing these data in Figure 36.4A produces a J-shaped curve, which is typical of exponential growth. The lower part of the J, where the slope of the line is almost flat, results from the relatively slow growth when N is small. As the population increases, the slope becomes steeper. The **exponential growth model** gives an idealized picture of unregulated population growth. There is no restriction on the abilities of the organisms to live, grow, and reproduce. Even elephants, the slowest breeders on the planet, would increase exponentially if enough resources were available. Although elephants typically produce only six young in a 100-year life span, Charles Darwin estimated that it would take only 750 years for a single pair to give rise to a population of 19 million. Obviously, no population—neither bacteria nor rabbits nor elephants—can grow exponentially indefinitely.

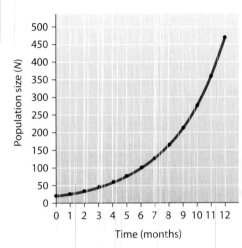

Figure 36.4A Exponential growth of rabbits

Limiting Factors and the Logistic Growth Model In nature, a population that is introduced to a new environment or is rebounding from a catastrophic decline in numbers may grow exponentially for a while, but eventually, one or more environmental factors will limit its growth. Population size then stops increasing or may even crash. Environmental factors that restrict population growth are called **limiting factors.**

You can see the effect of population-limiting factors in the graph in Figure 36.4B (see top of facing page), which illustrates the growth of a population of fur seals on St. Paul Island, off the coast of Alaska. (For simplicity, only the mated bulls were counted. Each has a harem of a number of females, as shown in the photograph.) Before 1925, the seal population on the island remained low because of uncontrolled hunting, although it changed from year to year. After hunting was controlled, the population increased rapidly until about 1935, when it began to level off and started fluctuating around a population size of about 10,000 bull seals. At this point, a number of limiting factors, including some hunting and the amount of space suitable for breeding, restricted population growth.

The fur seal growth curve resembles the **logistic growth model,** a description of idealized population growth that is slowed by limiting factors as the population size increases. Figure 36.4C (at the top of facing page) compares the logistic growth model (red) with the exponential growth model (blue). As you can see, the logistic curve is J-shaped at first, but gradually levels off to resemble an S shape.

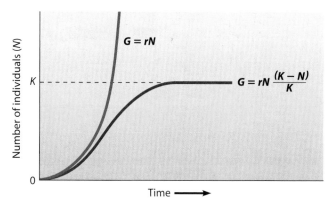

Figure 36.4B Growth of a population of fur seals

To model logistic growth, the formula for exponential growth, *rN,* is multiplied by an expression that describes the effect of limiting factors on an increasing population size:

$$G = rN\frac{(K - N)}{K}$$

This equation is actually simpler than it may appear. The only new symbol in the equation is *K,* which stands for carrying capacity. **Carrying capacity** is the maximum population size that a particular environment can sustain ("carry"). For the fur seal population on St. Paul Island, for instance, *K* is about 10,000 mated males. The value of *K* varies, depending on the species and the resources available in the habitat. *K* might be considerably less than 10,000 for a fur seal population on a smaller island with fewer breeding sites. Even in one location, *K* is not a fixed number. Organisms interact with other organisms in their communities, including predators, parasites, and food sources, that may affect *K.* Changes in abiotic factors may also increase or decrease carrying capacity. In any case, the concept of carrying capacity expresses an essential fact of nature: Resources are finite.

Table 36.4B demonstrates how the expression $(K - N)/K$ in the logistic growth model produces the S-shaped curve. At the outset, *N* (the population size) is very small compared to *K* (the carrying capacity). Thus, $(K - N)/K$ nearly equals K/K, or 1, and population growth (*G*) is close to *rN*—that is, exponential growth. As the population increases and *N* gets closer to carrying capacity, $(K - N)/K$ becomes an increasingly smaller fraction. The growth rate slows as *rN* is multiplied by that fraction. At carrying capacity, the population is as large as it can theoretically get in its environment; at this point, $N = K$, and $(K - N)/K = 0$. The population growth rate (*G*) becomes zero.

What does the logistic growth model suggest to us about real populations in nature? The model predicts that a population's growth rate will be small when the population size is *either* small or large, and highest when the population is at an intermediate level relative to the carrying capacity. At a low population level, resources are abundant, and the population is able to grow nearly exponentially. At this point, however, the increase is small because *N* is small. In contrast, at a high population level, limiting factors strongly oppose the population's potential to increase. There might be less food available per individual or fewer breeding territories, nest sites, or shelters. These limiting factors cause the birth rate to decrease, the death rate to increase, or both. Eventually, the population stabilizes at the carrying capacity (*K*), when the birth rate equals the death rate.

It is important to realize that the logistic growth model presents a mathematical ideal that is a useful starting point for studying population growth and for constructing more complex models. Like any good starting hypothesis, the logistic model has stimulated research, leading to a better understanding of the factors affecting population growth. We take a closer look at some of these factors next.

? In logistic growth, at what population size is the population increasing most rapidly? Explain why.

■ When *N* is $\frac{1}{2}$ *K.* At this population size, there are more reproducing individuals than at lower population sizes and still lots of space or other resources available for growth.

TABLE 36.4B	EFFECT OF *K* ON GROWTH RATE AS *N* APPROACHES *K,* $K = 1,000, r = 0.1$		
N	*rN*	(*K*−*N*)/*K*	*G* = *rN*(*K*−*N*)/*K*
10	1	0.99	0.99
100	10	0.9	9.00
400	40	0.6	24.00
500	50	0.5	25.00
600	60	0.4	24.00
700	70	0.3	21.00
900	95	0.05	0.25
1,000	100	0.00	0.00

36.5 Multiple factors may limit population growth

The logistic growth model predicts that population growth slows and eventually ceases as population density increases. That is, at higher population densities, the birth rate decreases, the death rate increases, or both. What are the possible causes of **density-dependent** rates—declining birth rates and rising death rates in response to increasing population density?

Several factors appear to regulate growth in natural populations. The most obvious is competition among members of a growing population for limited resources. As a limited food supply is divided among more and more individuals, birth rates may decline. Field studies of songbirds have demonstrated this effect. Figure 36.5A shows one such study of a song sparrow population on a small island in British Columbia. As the density of females increases, the clutch size (number of eggs laid) decreases. A shortage of food appears to cause this decrease. In an experiment in which females were given extra food when population densities were high, they did not show this decrease in clutch size.

A limited resource may be something other than food or nutrients. In many vertebrates that defend a territory, the availability of space may limit reproduction. For instance, the number of nesting sites on rocky islands may limit the population size of oceanic birds such as gannets.

Population density also influences the health and thus the survival of organisms. Plants grown under crowded conditions tend to be smaller and less likely to survive. And those that do survive produce fewer flowers, fruits, and seeds. Gardeners who understand this density-dependent result thin their seedlings to produce the best possible yield. Animals, too, experience increased mortality at high population densities. These deaths may be a result of increased disease transmission under crowded conditions or the accumulation of toxic waste products. Predation may also be an important cause of density-dependent mortality. A predator may concentrate on and capture more of a particular kind of prey as that prey becomes abundant, thus limiting further growth of the population.

For some animal species, physiological factors may regulate population size. White-footed mice in a small field enclosure will multiply from a few to a colony of 30 to 40 individuals, but reproduction then declines until the population ceases to grow, even when additional food and shelter are provided. High population densities in mice appear to induce a stress syndrome in which hormonal changes can delay sexual maturation, cause reproductive organs to shrink, and depress the immune system. In this case, high densities cause both an increase in mortality and a decrease in birth rate. Similar effects of crowding have been observed in wild populations of other rodents.

In these examples of population regulation, we have seen how increased density causes population growth rate to decline by reducing birth rate and/or increasing death rate. In many natural populations, however, abiotic factors such as weather may limit or reduce population size well before other limiting factors become important. If we look at the growth

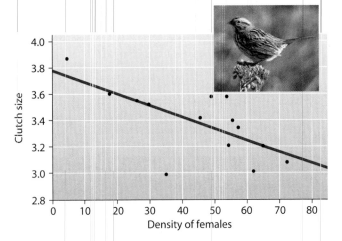

Figure 36.5A Decrease in song sparrow clutch size as population density increases

curve of such a population, we see something like exponential growth followed by a rapid decline, rather than a leveling off. Figure 36.5B shows this effect for a population of aphids, insects that feed on the phloem sap of plants. These and many other insects often show virtually exponential growth in the spring and then rapid die-offs when the weather turns hot and dry in the summer. A few individuals may remain, and these may allow population growth to resume again if favorable conditions return. In some populations of insects— many mosquitoes and grasshoppers, for instance—adults die off entirely, leaving only eggs, which initiate population growth the following year. In addition to seasonal changes in the weather, environmental factors, such as fire, floods, storms, and habitat disruption by human activity, can affect a population's size regardless of its density.

Over the long term, most populations are probably regulated by a mixture of factors. Some populations remain fairly stable in size and are presumably close to a carrying

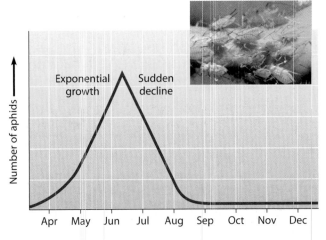

Figure 36.5B The effect of an abiotic factor (climate) on aphid population size

capacity that is determined by biotic factors such as competition or predation. Most populations for which we have long-term data, however, show fluctuations in numbers. For example, though density has a population-regulating influence on clutch size in a song sparrow population (see Figure 36.5A), over a 25-year period this natural population shows alternating spurts of growth and drastic decline (in periods of severe winter weather) (Figure 36.5C). Thus, the dynamics of many populations result from a complex interaction of both density-dependent birth and death rates and abiotic factors such as climate and disturbances.

? List some of the factors that may reduce birth rate or increase death rate as population density increases.

■ Food and nutrient limitations, insufficient territories, increase in disease and predation, accumulation of toxins

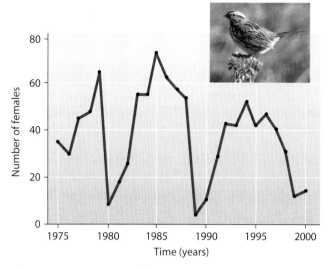

Figure 36.5C Fluctuations in a song sparrow population, with periodic catastrophic reductions due to severe winter weather

---

36.6 Some populations have "boom-and-bust" cycles

Some populations of insects, birds, and mammals undergo dramatic fluctuations in density with remarkable regularity. "Booms" characterized by rapid exponential growth are followed by "busts," during which the population falls back to a minimal level. A striking example is the boom-and-bust growth cycles of lemming populations that occur every three to four years. (Lemmings are small rodents that live in the tundra.) Some researchers hypothesize that natural changes in the lemmings' food supply may be the underlying cause. Another hypothesis, as discussed in Module 36.5, is that stress from crowding during the "boom" may reduce reproduction, causing a "bust."

Figure 36.6 illustrates another well-known example—the cycles of snowshoe hare and lynx. The lynx is one of the main predators of the snowshoe hare in the far northern forests of Canada and Alaska. About every ten years, both hare and lynx populations show a rapid increase followed by a sharp decline.

What causes these boom-and-bust cycles? Since ups and downs in the two populations seem to almost match each other on the graph, does this mean that changes in one directly affect the other? For the hare cycles, there are three main hypotheses. First, cycles may be caused by winter food shortages that result from overgrazing. Second, cycles may be due to predator-prey interactions. Many predators other than lynx, such as coyotes, foxes, and great-horned owls, eat hares, and the combination of predators might overexploit their prey. Third, cycles could be affected by a combination of food resource limitation and excessive predation. Recent experimental studies performed in the field support the hypothesis that the ten-year cycles of the snowshoe hare are largely driven by excessive predation, but are also influenced by fluctuations in the hare's food supplies.

For the lynx and many other predators that depend heavily on a single species of prey, the availability of prey can influence population size. Thus, the ten-year cycles in the lynx population probably do result at least in part from the ten-year cycles in the hare population. And when prey become scarce, predators often turn on one another, accelerating the collapse of predator populations. Prey populations, released from predator pressure, can then climb again. Long-term studies are the key to unraveling the complex causes of such population cycles.

Now that we have looked at patterns of population growth, we turn our attention to the differences in reproductive patterns of populations and how they are shaped by natural selection.

? In one experiment, increasing food supply to hares increased their population density, but the population continued to show cyclic collapses. What might you conclude from these results?

■ Hare population cycles are not primarily caused by food shortage.

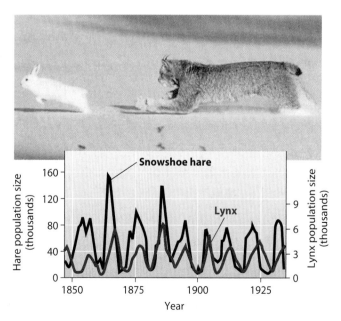

Figure 36.6 Population cycles of the snowshoe hare and the lynx

36.7 Evolution shapes life histories

The traits that affect an organism's schedule of reproduction and death make up its **life history.** Some key life history traits are the age of first reproduction, the frequency of reproduction, the number of offspring, and the amount of parental care given. Natural selection cannot optimize all of these traits simultaneously because an organism has limited time, energy, and nutrients. For example, an organism that gives birth to a large number of offspring will not be able to provide a great deal of parental care. Consequently, the combination of life history traits in a population represents tradeoffs that balance the demands of reproduction and survival. Because selective pressures vary, life histories are very diverse. Nevertheless, ecologists have observed some patterns that are useful for understanding how life history characteristics have been shaped by natural selection.

One life history pattern is typified by small-bodied, short-lived animals (for example, insects and small rodents) that develop and reach sexual maturity rapidly, have a large number of offspring, and offer little or no parental care. A similar pattern is seen in small, nonwoody plants such as dandelions that produce thousands of tiny seeds. Ecologists hypothesize that selection for this set of life history traits occurs in environments where resources are abundant, permitting exponential growth. It is called *r*-selection because *r* (the per capita rate of increase) is maximized (see Module 36.4). Habitats that favor *r*-selected species may experience unpredictable disturbances such as fire, flood, hurricanes, droughts, or cold weather that create new opportunities by suddenly reducing the population to low levels. Human activity is a major cause of disturbance, so *r*-selected plants and animals commonly colonize road cuts, freshly cleared fields and woodlots, and poorly maintained lawns.

In contrast, large-bodied, long-lived animals (such as bears and elephants) develop slowly and produce few, but well-cared-for, offspring. Plants with comparable life history traits include trees; for example, coconut palms produce relatively few seeds that are well-stocked with nutrient-rich material—the plant's version of parental care. Ecologists hypothesize that selection for this set of life history traits occurs in environments where the population size is near carrying capacity (*K*), so it is called **K-selection.** Population growth in these situations is limited by density-dependent factors (see Module 36.5). Because competition for resources is keen, *K*-selected organisms gain an advantage by allocating energy to their own survival and to the survival of their descendants. Environments that favor *K*-selected organisms typically have a stable climate.

The hypothesis of *r*- and *K*-selection has been criticized as an oversimplification, and most organisms fall somewhere between the extremes. However, this hypothesis has stimulated a vigorous subfield of ecological research on the evolution of life histories.

As with any adaptation, permanent changes in the environment may turn an advantage into a disadvantage. For example, most of the Lake Victoria cichlids have few offspring at a time, brood the eggs in their mouths, and defend their young until they are self-sufficient. When the Nile perch was introduced to their environment, cichlids lacked the reproductive capacity to make up for the population's losses to predation.

A long-term project has provided direct evidence that life history traits can be shaped by natural selection. For years, researchers have been studying guppy populations living in small, relatively isolated pools. As shown in Figure 36.7, certain guppy populations live in pools with predators called killifish, which eat mainly small, immature guppies. Other guppy populations live where larger fish, called pike-cichlids, eat mostly mature, large-bodied guppies. Guppies in populations exposed to these pike-cichlids tend to be smaller, mature earlier, and produce more offspring at a time than those in areas with killifish. Thus, guppy populations differ in certain life history traits,

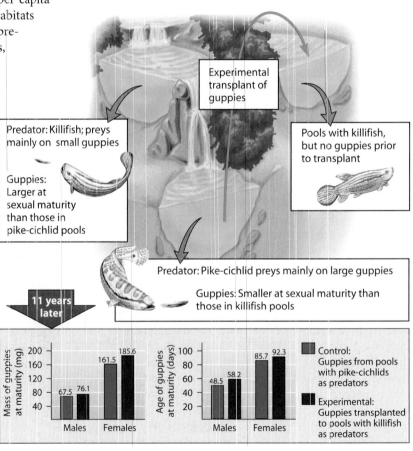

Figure 36.7 Effect of predation on life history traits of guppies

depending on the kind of predator in their environment. For these differences to be the result of natural selection, the traits should be heritable. And indeed, guppies from both populations raised in the laboratory without predators retained their life history differences.

To test whether the feeding preferences of different predators caused these differences in life histories by natural selection, researchers introduced guppies from a pike-cichlid habitat into a guppy-free pool inhabited by killifish. The scientists tracked the weight and age at sexual maturity in the experimental guppy populations for 11 years, comparing these guppies with control guppies that remained in the pike-cichlid pools. The average weight and age at sexual maturity of the transplanted populations increased significantly as compared with the control populations. These studies demonstrate not only that life history traits are heritable and shaped by natural selection, but also that questions about evolution can be tested by field experiments.

As we have seen, population ecology involves theoretical model building as well as observations and experiments in the field. Next we look at how the principles of population ecology can be applied in conservation and management.

> **?** Refer back to Module 36.3. Which type of survivorship curve would you expect to find in a population experiencing *r*-selection? *K*-selection?

■ Type III for a population experiencing *r*-selection; Type I for *K*-selection

36.8 Principles of population ecology have practical applications

We often attempt to manage natural resources—to increase populations we wish to harvest or save from extinction or to decrease populations we consider pests. Principles of population ecology can help guide us toward such resource management goals.

Wildlife managers, fishery biologists, and foresters try to practice **sustainable resource management:** harvesting crops without damaging the resource. According to the concept of **maximum sustained yield,** harvesting should be done at a level that produces a consistent yield without forcing a population into decline. A population growing according to the logistic growth model increases the fastest when its density is at an intermediate level relative to its carrying capacity (see Module 36.4). One approach has been to harvest populations down to this level to ensure high growth rates.

Human economic and political pressures, however, often outweigh ecological concerns, and the scientific information available is frequently inadequate. For example, in the collapse of the northern cod fishery, shown in Figure 36.8, estimates of cod stocks were too high, and the practice of discarding young cod (not of legal size) at sea caused a higher mortality rate than had been predicted. Following the collapse of many other fish and whale populations, resource managers are trying to minimize the risk of resource collapse by setting minimum population sizes or imposing protected, harvest-free areas. For species that are in decline or facing extinction, resource managers may try to provide additional habitat or improve the quality of existing habitat to raise the carrying capacity, *K*, and thus increase population growth.

Reducing the size of a population is also a challenging task. Simply killing many individuals will not usually decrease the size of pest populations. Many insect and weed species have life history traits that are *r*-selected (see Module 36.7), adapted to produce rapid population growth. Also, most pesticides kill both the pest and their natural predators. Because prey species often have a higher reproductive rate than predators, pest populations rapidly rebound before their predators can.

Integrated pest management (IPM) uses a combination of biological, chemical, and culturing methods to control agricultural pests. IPM relies on knowledge of the population ecology of the pest and its associated predators and parasites, as well as crop growth dynamics.

As you've learned, there are many factors that influence a population's size. To effectively manage any population, we must identify those variables, account for the unpredictability of the environment, consider interactions with other species, and weigh the economic and political issues. These same issues apply to the growth of the human population, which we explore next.

> **?** Explain why managers often try to maintain populations of fish and game species at about half their carrying capacity.

■ To protect wildlife from overharvest yet maintain lower population levels so that growth rate is high and mortality from resource limitation is reduced

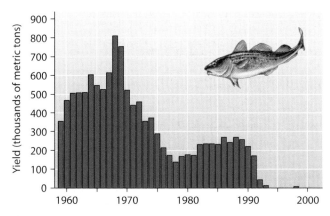

Figure 36.8 Collapse of northern cod fishery off Newfoundland

36.9   The human population continues to increase, but the growth rate is slowing

In the second or two it took to read this sentence, 21 babies were born somewhere in the world and nine people died. The statistics may have changed a bit by the time you read this, but births will still far outnumber deaths. An imbalance between births and deaths is the cause of population growth (or decline), and as the red curve in Figure 36.9A shows, the human population is expected to continue increasing for at least the next several decades. The bar graph in Figure 36.9A tells a different part of the story. The number of people added to the population each year has been declining since the 1980s. How do we explain these patterns of human population growth?

Let's begin with the rise in population from 480 million people in 1500 to the current population of more than 6.5 billion. In our simplest model (see Module 36.4), population growth depends on r (per capita rate of increase) and N (population size). Because the value of r was assumed to be constant in a given environment, the growth rate in the examples we used in Module 36.4 depended wholly on the population size. Throughout most of human history, the same was true of humans. Although parents had many children, mortality was also high, so r (birth rate − death rate) was only slightly higher than 0. As a result, population growth was very slow. (If we extended the x-axis of Figure 36.9A back in time to year 1, when the population was roughly 300 million, the line would be almost flat for 1,500 years.) The 1 billion mark was not reached until the early 19th century.

As economic development in Europe and the United States led to advances in nutrition and sanitation, and later, medical care, humans took control of their population's r. At first, the death rate decreased, while the birth rate remained the same. The net rate of increase rose, and population growth began to pick up steam as the 20th century began. By mid-century, improvements in nutrition, sanitation, and health care had spread to the developing world, spurring growth at a breakneck pace as birth rates far outstripped death rates.

As the world population skyrocketed from 2 billion in 1927 to 3 billion just 33 years later, some scientists became alarmed. They feared that Earth's carrying capacity would be reached, and density-dependent factors (see Module 36.5) would maintain that population size through human suffering and death. But the overall growth rate peaked in 1962. In the more developed nations, advanced medical care continued to improve survivorship, but effective contraceptives held down the birth rate. As a result, the overall growth rate of the world's population began a downward trend.

The world population is undergoing a change known as a **demographic transition,** a shift from zero population growth in which birth rates and death rates are high but roughly equal, to zero population growth characterized by low birth and death rates. Figure 36.9B shows the demographic transition of Mexico, which is projected to approach zero population growth with low birth and death rates in the next few decades. Notice that the death rate dropped sharply from 1925 to 1975 (the spike corresponds to the worldwide flu epidemic of 1918–1919), while the birth rate remained high until the 1960s. This is a typical pattern for demographic transitions.

Because economic development has occurred at different times in different regions, worldwide demographic transition is a mosaic of the changes occurring in different countries. The most developed nations have completed or are nearing completion of their demographic transitions. In these countries collectively, r was 0.7 in 2006 (see Table 36.9 on facing page). In the developing world, death rates have dropped, but high birth rates persist. As a result, these populations are growing rapidly. Of the more than 77 million people added to the world each year, more than 73 million are in developing nations.

Reduced family size is the key to the demographic transition. As women's status and education increase, they delay reproduction and choose to have fewer children. This phenomenon has been observed in both developed and developing

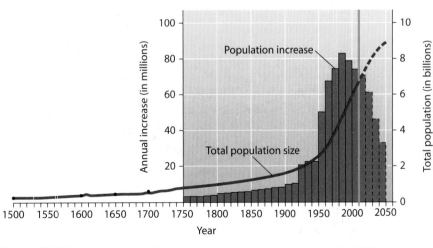

Figure 36.9A   Five centuries of human population growth, projected to 2050

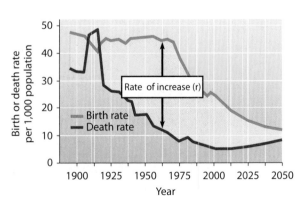

Figure 36.9B   Demographic transition in Mexico

TABLE 36.9	POPULATION CHANGES IN 2006		
Population	Birth Rate per 1,000	Death Rate per 1,000	Per Capita Rate of Increase (r)
World	20.3	8.5	11.8
More developed nations	11.1	10.4	0.7
Less developed nations	22.4	8.0	14.4

countries, anywhere that the lives of women have improved. Family sizes are smaller when women begin having children in their early 30s instead of in their teens, and the time span between generations is longer. Given access to affordable contraceptive methods, women generally practice birth control, and many countries now subsidize family planning services and have official population policies. In many other countries, however, issues of family planning remain socially and politically charged, with heated disagreement over how much support should be provided for family planning.

A demographic tool called an **age structure** is helpful for predicting a population's future growth. The age structure of a population is the number of individuals in different age-groups. Figure 36.9C shows the age structures of Mexico's population in 1980 and 2005 and its projected age structure in 2030. In these diagrams, green represents the portion of the population in their prereproductive years (0–14), purple indicates the part of the population in prime reproductive years (15–44), and dark blue is the proportion in postreproductive years (45 and older). Within each of these broader groups, each horizontal bar represents the population in a 5-year age-group. The area to the left of each vertical center line represents the number of

males in each age-group; females are represented on the right side of the line.

An age structure with a broad base, such as Mexico's in 1980, reflects a population that has a high proportion of children and a high birth rate. On average, each woman is substantially exceeding the replacement rate of two children per couple. As Figure 36.9B shows, the birth rate and the rate of increase have dropped 25 years later, but the population continues to be affected by its earlier expansion. This situation, which results from the increased proportion of women of childbearing age in the population, is known as **population momentum.** Girls 0–14 in the 1980 age structure (outlined in yellow) are in their reproductive years in 2005, and girls who are 0–14 in 2005 (outlined in orange) will carry the legacy of rapid growth forward to 2030. Putting the brakes on a rapidly expanding population is like stopping a freight train—the end result takes place long after the decision to do it was made. Even when fertility is reduced to replacement rate, the total population will continue to increase for several decades. The percentage of individuals under the age of 15 gives a rough idea of future growth. In the developing countries, about 30% of the population is in this age group. In contrast, roughly 17% of the population of developed nations is under the age of 15. Population momentum also explains why the population size in Figure 36.9A continues to increase even though fewer people are added to the population each year.

In the next module, we examine the age structure of the United States.

Web Activity *Human Population Growth*

Web Activity *Analyzing Age-Structure Diagrams*

> ? During the demographic transition from high birth and death rates to low birth and death rates, countries usually undergo rapid population growth. Explain why.

■ The death rate declines before the birth rate declines, creating a period when births greatly outnumber deaths. This also sets up population momentum.

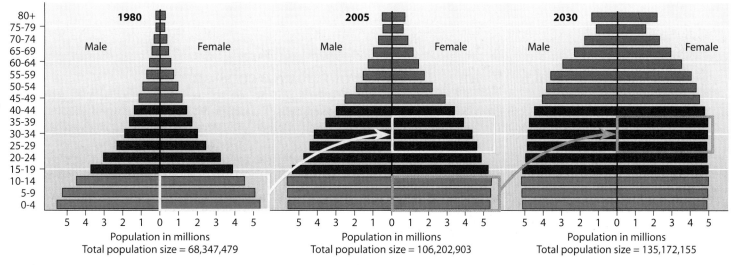

Figure 36.9C Population momentum in Mexico

36.10   Age structures reveal social and economic trends

Age structure diagrams not only reveal a population's growth trends; they also indicate social conditions. For instance, an expanding population has an increasing need for schools, employment, and infrastructure. A large elderly population requires that extensive resources be allotted to health care. Let's look at trends in the age structure of the United States from 1980 to 2030 (Figure 36.10).

The large lower bulge in the 1980 age structure corresponds to the "baby boom" that lasted for about two decades after World War II ended in 1945. The large number of children swelled school enrollments, prompting construction of new schools and creating a demand for teachers. On the other hand, graduates who were born near the end of the boom faced stiff competition for jobs. Because they make up such a large segment of the population, boomers have had an enormous influence on social and economic trends. They also produced a

boomlet of their own—in the 2005 age structure, notice the bump from 1981–1995.

Where are the baby boomers now? In a few years, the leading edge will reach retirement age, placing pressure on programs such as Medicare and Social Security. In 2005, 59% of the population was between 20 and 64, the ages most likely to be in the workforce, and 12% were over 65. In 2030, the percentages are projected to be 54 and 19. In part, the increase in the elderly population is because people are living longer. The percentage of the population over 80, which was 2% in 1980, is projected to rise to 8%—close to 34 million people—in 2050.

? Point out an example of population momentum in Figure 36.10.

■ The 1981–1995 "boomlet" is a consequence of rapid reproduction in 1946–1965, as girls born during the baby boom entered their reproductive years.

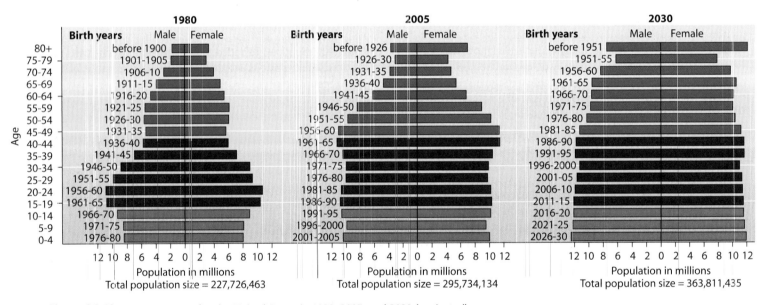

Figure 36.10   Age structures for the United States in 1980, 2005, and 2030 (projected)

36.11   An ecological footprint is a measure of resource consumption

How large a population of humans can Earth hold? In Module 36.9 we saw that the world's population is growing exponentially, though at a slower rate than it did in the last century. The rate of increase, as well as population momentum, predict that the populations of most developing nations will continue to increase for the foreseeable future. The U.S. Census Bureau projects a global population of 8 billion within the next 20 years, and 9.5 billion by the mid-21st century. But these numbers are only part of the story. Trillions of bacteria can live in a petri dish *if* they have sufficient resources. Do we have sufficient resources to sustain 8 or 9 billion people? To accommodate all the people

expected to live on our planet by 2025, the world will have to double food production. Already, agricultural lands are under pressure. Overgrazing by the world's growing herds of livestock is turning vast areas of grassland into desert. Water use has risen sixfold over the past 70 years, causing rivers to run dry, water for irrigation to be depleted, and levels of groundwater to drop. And because so much open space will be needed to support the expanding human population, many thousands of other species are expected to become extinct.

The concept of an **ecological footprint** is one approach to understanding resource availability and usage. An ecological

Figure 36.11A Families in India (above) and the United States (right) display their possessions

footprint is an estimate of the amount of land required to provide the raw materials an individual or a nation consumes, including food, fuel, water, housing, and waste disposal. When the total area of ecologically productive land on Earth is divided by the global population, we each have a share of about 2 hectares (1 hectare, or ha, = 2.47 acres). Reserving some land for parks and conservation reduces this to 1.7 ha per person. In 2003, the average ecological footprint for the world's population was 2.23 ha—we have already overshot the planet's capacity to sustain us.

The United States has a bigger ecological footprint (8.4 ha per person) than its own land and resources can support (6.2 ha per person)—it has a large ecological deficit. Looking at Figure 36.11A, it is not difficult to understand why. Compared to a family in rural India, Americans have an abundance of possessions. We also consume a disproportionate amount of food and fuel. By this measure, the ecological impact of affluent nations such as the United States is potentially as damaging as the unrestrained population growth in the developing world. So the problem is not just overpopulation, but overconsumption. Figure 36.11B is a graphic representation of the disparity in consumption. Regions depicted in purple have the largest ecological footprints. Tan areas have the lowest per capita consumption—the smallest ecological footprint. The world's richest countries, with 20% of the global population, use 86% of the world's resources, leaving just 14% of global resources—energy, food, water, and other essentials—for the other 80% of the world's population to share. Indeed, the poorest 20% of the popula-

tion accounts for just 1.3% of resource consumption. The Lake Victoria fisheries described in the chapter introduction are a typical example. Many people in the region are malnourished, starving for protein. After the fillets are sliced off the Nile perch, they are exported to Europe, Israel, and Japan. The local population, no longer able to afford the bountiful harvest from Lake Victoria, purchase the heads and stripped-down carcasses. Some researchers estimate that providing everyone with the same standard of living as the United States would require the resources of three more planet Earths.

? What is your ecological footprint? Do a web search to find a site that calculates personal resource consumption.

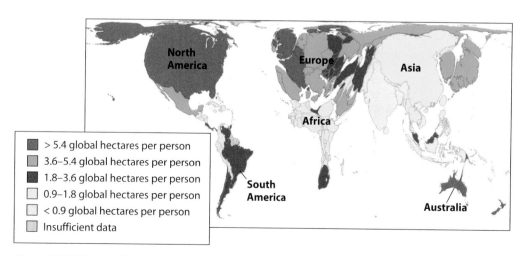

> 5.4 global hectares per person
3.6–5.4 global hectares per person
1.8–3.6 global hectares per person
0.9–1.8 global hectares per person
< 0.9 global hectares per person
Insufficient data

Figure 36.11B World map with area corresponding to ecological footprint

Reviewing the Concepts

Population Structure and Dynamics (36.1–36.8)

Population ecology is concerned with changes in population size and the factors that regulate populations over time. A population consists of members of a species living in the same place at the same time **(Introduction–36.1)**.

Population characteristics. Population density is the number of individuals in a given area or volume. Environmental and social factors influence the spacing of individuals in various dispersion patterns: clumped (most common), uniform, or random

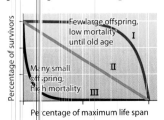

(36.2). Life tables and survivorship curves predict an individual's statistical chance of dying or surviving during each interval in its life. The three types of survivorship curves reflect species' differences in reproduction and mortality **(36.3).**

Exponential growth is the accelerating increase that occurs when growth is unlimited. The equation $G = rN$ describes this J-shaped growth curve; G = the population growth rate, r = an organism's inherent capacity to reproduce, and N = the population size **(36.4).**

Logistic growth is the model that represents the slowing of population growth as a result of limiting factors and the leveling off at carrying capacity, which is the number of individuals the environment can support. The equation $G = rN(K − N)/K$ describes a logistic growth curve, where K = carrying capacity and the term $(K − N)/K$ accounts for the leveling off of the curve **(36.4).**

Limiting factors. As a population's density increases, factors such as limited food supply and increased disease or predation may increase the death rate, decrease the birth rate, or both. Abiotic factors such as severe weather may limit many natural populations. Most populations are probably regulated by a mixture of factors, and fluctuations in numbers are common **(36.5).** Some populations undergo regular boom-and-bust cycles of growth and decline **(36.6).**

Diversity of life histories. Natural selection shapes a species' life history, the series of events from birth through reproduction to death. Populations with so-called r-selected life history traits produce many offspring and grow rapidly in unpredictable environments. Populations with K-selected traits raise few offspring and maintain relatively stable populations. Most species fall between these extremes **(36.7).** Principles of population ecology are useful in managing natural resources **(36.8).**

The Human Population (36.9–36.11)

Human population growth. The human population grew rapidly during the 20th century and currently stands at more than 6.5 billion. Demographic transition, the shift from high birth and death rates to low birth and death rates, has lowered the rate of growth in developed countries. In the developing nations, deaths rates have dropped, but birth rates are still high. The age structure of a population—the proportion of individuals in different age-groups—affects its future growth.

Population momentum is the continued growth that occurs despite reduced fertility and as a result of girls in the 0–14 age group of a previously expanding population reaching their childbearing years **(36.9).** Age structures for the United States indicate social and economic trends **(36.10).**

Earth's carrying capacity. An ecological footprint estimates the amount of land required by each person or country to produce all the resources it consumes and to absorb all its wastes. The carrying capacity of the world may already be smaller than the population's ecological footprint. There is a huge disparity between resource consumption in more developed and less developed nations **(36.11).**

Connecting the Concepts

1. Use this graph of the idealized exponential and logistic growth curves to complete the following.
   a. Label the axes and curves on the graph.
   b. Give the formula that describes the blue curve.
   c. What does the dotted line represent?
   d. For each curve, indicate and explain where population growth is the most rapid.
   e. Which of these curves best represents global human population growth?

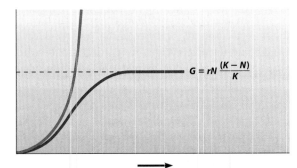

$$G = rN \frac{(K - N)}{K}$$

2. The graph below shows the demographic transition for a hypothetical country. Many developed countries that have achieved a stable population size have undergone a transition similar to this. Answer the following questions concerning this graph.
   a. What does the blue line represent? The red line?
   b. This diagram has been divided into four sections. Describe what is happening in each section.
   c. In which section(s) is the population size stable?
   d. In which section is the population growth rate the highest?

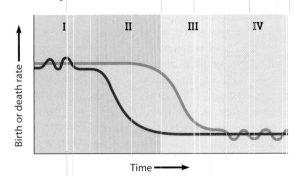

Testing Your Knowledge

Multiple Choice

3. To figure out the human population density of your community, you would need to know the number of people living there and
   a. the land area in which they live.
   b. the birth rate of the population.
   c. whether population growth is logistic or exponential.
   d. the dispersion pattern of the population.
   e. the carrying capacity.

4. The term $(K - N)/K$
   a. is the carrying capacity for a population.
   b. is greatest when K is very large.
   c. is zero when population size equals carrying capacity.
   d. increases in value as N approaches K.
   e. accounts for the overshoot of carrying capacity.

5. Which of the following represents a demographic transition?
   a. A population switches from exponential to logistic growth.
   b. A population reaches zero population growth when the birth rate drops to zero.
   c. There are equal numbers of individuals in all age-groups.
   d. A population exhibits boom-and-bust cycles.
   e. A population switches from high birth and death rates to low birth and death rates.

6. With regard to its rate of growth, a population that is growing logistically
   a. grows fastest when density is lowest.
   b. has a high intrinsic rate of increase.
   c. grows fastest at an intermediate population density.
   d. grows fastest as it approaches carrying capacity.
   e. is always slowed by abiotic factors.

7. Skyrocketing growth of the human population appears to be mainly a result of
   a. migration to thinly settled regions of the globe.
   b. a drop in death rate due to sanitation and health care.
   c. better nutrition boosting the birth rate.
   d. the concentration of humans in cities.
   e. social changes that make it desirable to have more children.

8. According to the ecological footprint study produced in 2005,
   a. the carrying capacity of the world is 10 billion.
   b. the carrying capacity of the world would increase if all people ate more meat.
   c. the current demand on global resources by industrialized countries is less than the resources available in those countries.
   d. the United States has a larger ecological footprint than its own resources can provide.
   e. nations with the largest ecological footprints have the fastest population growth rates.

Describing, Comparing, and Explaining

9. What are some factors that might have a density-dependent limiting effect on population growth?

10. What is survivorship? What does a survivorship curve show? Explain what the three survivorship curves tell us about humans, squirrels, and oysters.

11. Describe the factors that might produce the following three types of dispersion patterns in populations.

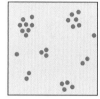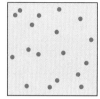

Applying the Concepts

12. A sampling technique commonly used to estimate the sizes of wildlife populations is the mark-recapture method. The researcher traps animals in the study area and marks them with tags, bands, or spots of dye. The marked animals are released, and traps are set a second time. This second capture will yield both marked and unmarked individuals. The proportion of marked (recaptured) animals to unmarked animals in the second trapping is assumed to be equivalent to the proportion of the animals marked in the first catch to the total population. For example, suppose a researcher captures, tags, and releases 50 voles. Two weeks later, 100 voles are captured, of which 10 are marked. What percent of the total population is estimated to have been marked? Using this information, how could you estimate the total population size?

13. Considering the learning behavior of animals, what might be some limits to the accuracy of the mark-recapture method?

14. The mountain gorilla, spotted owl, giant panda, snow leopard, and grizzly bear are all endangered by human encroachment on their environments. Another thing these animals have in common is that they all have K-selected life history traits. Why might they be more easily endangered than animals with r-selected life history traits? What general type of survivorship curve would you expect these species to exhibit? Explain your answer.

15. How does the age structure of the U.S. population explain the current surplus in the Social Security fund? If the system is not changed, why will the surplus give way to a deficit sometime in the next few decades?

16. Many people regard the rapid population growth of developing countries as our most serious environmental problem. Others think that the growth of developed countries, though slower, is actually a greater threat to the environment. What kinds of environmental problems result from population growth in (a) developing countries and (b) developed countries? Which do you think is the greater threat? Why?

Answers to all questions can be found in Appendix 4.

For Practice Quizzes, BioFlix, MP3 Tutors, and Activities, go to www.mybiology.com.

chapter 37

Communities and Ecosystems

Hungry Hippos

Tsavo National Park in Kenya is famous for many things: the grisly tale of man-eating lions that terrorized railroad workers in 1898; picturesque herds of elephants, giraffes, and buffalo; and breathtaking scenery that includes the spectacular rise of Mount Kilamanjaro, which is a few miles from the park's western border. Not far from Tsavo are the Chyulu Hills, the youngest volcanic range in Africa, where recent lava flows are visible in some areas. Rainfall on Chyulu does not linger at the surface. It quickly enters the porous volcanic rock and then slowly percolates underground before emerging, 25 years later, some 30 miles away to create a unique wetland called Mzima Springs.

The cool, clear water of Mzima Springs is hippo heaven (facing page bottom photo). The skin of hippopotamuses (*Hippopotamus amphibious*) is almost hairless, providing little protection from the tropical sun (see photo above and top image on the facing page). Hippos have no sweat glands for evaporative cooling. (They do, however, exude a reddish skin-conditioning fluid through their pores, giving rise to the myth that they sweat blood.) To avoid the heat, hippos typically leave the water only at night, when they forage. Feeding hippos cut wide swaths through the countryside, traveling more than a mile from the springs in search of the grasses that are the mainstay of their diet. By eating steadily for as long as five hours, a typical hippo manages to consume over 100 pounds of plants each night.

At more than 3,000 kg (3.3 tons), a hippopotamus moves awkwardly on land. In the buoyant water of Mzima Springs, however, hippos appear light-footed as they gallop along the bottom underwater. In their wake, a group of fish called barbels converge eagerly on the prodigious piles of waste the hippos excrete into the water—each hippo produces roughly 40 pounds of dung per day. Even after passing through the hippo's digestive tract, plenty of nutritional value remains in the waste, and the barbels make use of it. Shovel-nosed gobis, another species of fish, also appear, their round mouths sucking in dung like vacuum cleaners. Snails join the feast too, and the barbels gladly add them to the menu. The fish in turn become meals for crocodiles and long-tailed cormorants, which dive into the spring from perches overlooking the water, spearing fish with their bills.

The abundant hippo waste (yellow on the substrate in the bottom photo on the facing page) is also a plentiful source of food and energy for a thriving microbial community. The actions of these microorganisms on the complex organic molecules in the dung release chemical nutrients that support plant growth in and around the spring. Water pears and fig trees grow on the banks, their roots extending into the water, from which they extract vital minerals. The shoreline vegetation supports a variety of animal life, including many species of birds and monkeys.

The variety of life that is nourished by hippo waste is astonishing, but we must give credit where it is due. Hippos are not solely responsible for the bounty of dung. Recall from Module 21.13 that the presence of cellulose makes the digestion of plant material very difficult. Like other herbivores, hippopotamuses rely on an internal community of microorganisms to metabolize ingested plant material. The animals then absorb the metabolites and utilize some of the microbes themselves as food. In hippos, as in ruminants (for example, cows), this microbial community inhabits the stomach. But hippopotamus stomachs are not as elaborate as ruminant stomachs, and consequently, the digestive systems of hippos are not as efficient. Plenty of unabsorbed nutrients remain in hippo dung to fuel the food chain.

Dung is not the only contribution hippos make to the diversity of life in Mzima. In some areas of the springs, a variety of fish species dine on the bounty found on hippos' external surfaces. Carp move easily from scraping algae from rocks to scraping hippo skin, which bears a banquet of dead skin and parasites in addition to algae. Carp, like living toothbrushes, even scrub the inside of hippo mouths. But carp mouths are too large to clean the smaller crevices of the hippo, so those areas are left for others to exploit. A species of cichlid specializes in cleaning the tail bristles, while the ubiquitous barbels probe between the toes.

The diversity of life at Mzima stems from its hippopotamuses. Outside the boundaries of the national park, springs similar to Mzima are devoid of hippos, possibly as a result of farmers protecting their livelihoods. Hippo dung deposited along the feeding trails may enrich the soil, but the fertilizer is little compensation for fields cropped as close as newly mown lawns or trampled underfoot as the hippos make their way to tastier fare. In any case, springs that lack hippopotamuses—and their nutritious dung—are nearly barren of other forms of life as well.

Whether communities and ecosystems are as large as Mzima Springs or small enough to be contained in a hippopotamus's stomach, they have certain features in common. In this chapter, you will learn about the structure and dynamics of ecosystems and communities and how human activities can disrupt the balance. ■ ■ ■

37.1 A community includes all the organisms inhabiting a particular area

In the previous chapter, we saw that a population is a group of interacting individuals of a particular species. We now move one step up the hierarchy of nature to the level of the **community.** A biological community is an assemblage of all the populations of organisms living close enough together for potential interaction. Ecologists define the boundaries of the community according to the research questions they want to investigate. For example, an ecologist interested in wetland communities might study the organisms that live in and around Mzima Springs, while a microbial ecologist might investigate the inhabitants of the hippopotamus's stomach.

A community can be described by its species composition. Community ecologists seek to understand how abiotic factors and interactions between populations affect the composition and distribution of communities. For example, a community ecologist might compare the stomach microbes of a hippopotamus in a zoo with those of a hippopotamus in the wild. Community ecologists also investigate community dynamics,

the variability or stability in the species composition of a community caused by biotic and abiotic factors. For example, a community ecologist might study changes in species composition of a spring community where the hippopotamus population is declining.

Community ecology is necessary for the conservation of endangered species and management of wildlife, game, and fisheries resources. It is vital for controlling diseases, such as malaria, bird flu, and Lyme disease, that are carried by animals. Community ecology also has applications in agriculture, where humans attempt to control the species composition of their communities.

Web Process of Science How Are Impacts on Community Diversity Measured?

? **What is the relationship between a community and a population?**

■ A community is a group of populations that interact with each other.

37.2 Interspecific interactions are fundamental to community structure

In Chapters 35 and 36 we discussed interactions between members of the same species. Organisms also have **interspecific interactions**—relationships with individuals of other species in the community—that greatly affect population structure and dynamics. In Table 37.2, interspecific interactions are classified according to the effect on the populations concerned, which may be helpful ($+$) or harmful ($-$).

Recall from Module 36.5 that members of a population may compete for limited resources such as food or space. **Interspecific competition** occurs when populations of two different species compete for the same limited resource. For example, desert plants compete for water, while plants in a tropical rain forest compete for light. Squirrels and black bears are among the animals that feed on acorns in a temperate deciduous forest in autumn. When acorn production is low, the nut is a limited resource for which squirrels and bears compete. In general, the effect of interspecific competition is negative for both populations ($-/-$). However, it may be far more harmful for one population than the other. Interspecific competition is responsible for some of the disastrous effects of introducing non-native species into a community, a topic we will explore further in Module 37.13.

In **mutualism,** both populations benefit ($+/+$). Plants and mycorrhizae (see Chapter 17 introduction) and the hippopotamus and the microbes that inhabit its stomach are examples of mutualism between symbiotic species. Mutualism can also occur between species that are not symbiotic. For example, flowers and their pollinators are mutualists (see Module 17.12) and so are hippos and the fish that clean them.

Predation refers to an interaction in which one species (the predator) kills and eats another (the prey). **Herbivory** is con-

sumption of plant parts or algae by an animal. Both of these are interspecific interactions that clearly benefit one population and harm the other ($+/-$). In the same vein, both plants and animals may be victimized by parasites (see Module 16.14) or pathogens (see Module 16.1). Thus, parasite–host and pathogen–host interactions are also $+/-$.

In the next several modules, you will learn more about these interspecific interactions and how they affect communities. We will also look at some examples of how interspecific interactions can act as powerful agents of natural selection.

Web Activity Interspecific Interactions

? **Barbels, shovel-nosed gobis, and snails all feed on hippopotamus dung. Is this an example of interspecific competition?**

■ Probably not. Hippo dung is plentiful, so it is not likely to become a limited resource.

TABLE 37.2	INTERSPECIFIC INTERACTIONS		
Interspecific Interaction	Effect on Species 1	Effect on Species 2	Example
Competition	−	−	Squirrels/black bears
Mutualism	+	+	Hippo/microbes in hippo stomach
Predation	+	−	Crocodile/fish
Herbivory	+	−	Hippo/grasses
Parasites and pathogens	+	−	Heartworm/dog; *Salmonella*/humans

37.3 Competition may occur when a shared resource is limited

Each species in a community has an **ecological niche,** defined as the sum of its use of the biotic and abiotic resources in its environment. For example, the ecological niche of a small bird called the Virginia's warbler (Figure 37.3A) includes its nest sites and nest-building materials, the insects it eats, and climatic conditions such as the amount of precipitation and the temperature and humidity that enable it to survive. In other words, the ecological niche encompasses everything the Virginia's warbler needs for its existence.

Interspecific competition occurs when the niches of two populations overlap and both populations need a resource that is in short supply. Ecologists can study the effects of competition by removing all the members of one species from a study site. For example, in central Arizona, the niche of the orange-crowned warbler (Figure 37.3B) overlaps in some respects with the niche of Virginia's warbler. When researchers removed either species, the remaining species was significantly more successful in raising their offspring. Thus, interspecific competition has a direct effect on reproductive fitness.

In general, competition lowers the carrying capacity (see Module 36.4) of competing populations because the resources used by one population are not available to the other population. In 1934, Russian ecologist G. F. Gause demonstrated the effects of interspecific competition with an elegant series of experiments using three closely related species of ciliates (see Module 16.16): *Paramecium caudatum, P. aurelia,* and *P. bursaria.* He established the carrying capacity for each species under the culturing condi-

Figure 37.3A (above) Virginia's warbler (*Vermivora virginiae*)

Figure 37.3B (below) Orange-crowned warbler (*Vermivora celata*)

tions used to grow them in the laboratory and then cultured combinations of species. In a mixed culture of *P. caudatum* and *P. bursaria,* population sizes stabilized at lower numbers than they achieved in the absence of a competing species. On the other hand, in a mixed culture of *P. caudatum* and *P. aurelia,* only *P. aurelia* survived. Gause concluded that the requirements of these two species were so similar that they could not coexist; *P. aurelia* outcompeted *P. caudatum* for essential resources.

? Which do you think has more severe effects, competition between members of the same species or competition between members of different species? Explain why.

■ Competition between members of the same species is more severe because members of the same species have exactly the same niche. Thus they compete for exactly the same resources.

37.4 Mutualism benefits both partners

Reef-building corals and photosynthetic dinoflagellates (unicellular algae, see Module 16.16) provide a good example of how $+/+$ mutualists benefit from their relationship. As you learned in Module 34.6, coral reefs are constructed by successive generations of colonial coral animals that secrete an external calcium carbonate ($CaCO_3$) skeleton. Deposition of the skeleton must outpace erosion and competition for space from fast-growing seaweeds. Corals could not build and sustain the massive reefs that provide the food, living space, and shelter to support the splendid diversity of the reef community without the millions of dinoflagellates that live in the cells of each coral polyp (Figure 37.4). The sugars that the dinoflagellates produce by photosynthesis provide at least half of the energy used by the coral animals. In return, the dinoflagellates gain a secure shelter that provides access to light. They also use the coral's waste products, including CO_2 and ammonia (NH_3), a valuable source of nitrogen.

? When corals are stressed by environmental conditions, they expel their dinoflagellates in a process called bleaching. How is widespread bleaching likely to affect coral reefs?

■ Without their dinoflagellate mutualists, corals do not have enough energy to maintain the reef structure. Bleached reefs will die.

Figure 37.4 Coral polyps

37.5 Predation leads to diverse adaptations in prey species

Predation benefits the predator but kills the prey. Because predation has such a negative impact on reproductive success, numerous adaptations for predator avoidance have evolved in prey populations through natural selection.

+/−

In Module 18.12, you learned how insect color patterns, including camouflage, provide protection against predators. Camouflage is also common in other animals. As Figure 37.5A shows, the gray tree frog (*Hyla arenicolor*), an inhabitant of the southwestern United States, becomes almost invisible on a gray tree trunk.

Other protective devices include mechanical defenses, such as the porcupine's sharp quills (see Figure 35.9) or the hard shells of clams and oysters. Chemical defenses are also widespread. Animals with effective chemical defenses usually have bright color patterns, often yellow, orange, or red in combination with black. Predators learn to associate these color patterns with undesirable consequences, such as noxious taste or painful sting, and avoid potential prey with similar markings. The vivid colors of the poison-arrow frog (Figure 37.5B), an inhabitant of rain forests in Costa Rica, warn of noxious chemicals in the frog's skin. In some parts of South America, human hunters in the rain forest tip their arrows with poisons from similar species of frogs to bring down large mammals.

? Explain why predation is a powerful factor in the adaptive evolution of prey species.

■ The prey that avoid being eaten will be most likely to survive and reproduce and pass alleles for antipredator adaptations on to their offspring.

Figure 37.5A   Camouflage: a gray tree frog on bark

Figure 37.5B   Chemical defenses: the poison-arrow frog

37.6 Herbivory leads to diverse adaptations in plants

Although herbivory is not usually fatal, a plant whose body parts have been eaten by an animal must expend energy to replace the loss. Consequently, numerous defenses against herbivores have evolved in plants. Spines and thorns are obvious antiherbivore devices, as anyone who has plucked a rose from a thorny rosebush or brushed against a spiky cactus knows. Chemical toxins are also very common in plants. Like the chemical defenses of animals, toxins in plants are distasteful, and herbivores learn to avoid them. Among such chemical weapons are the poison strychnine, produced by a tropical vine called *Strychnos toxifera;* morphine, from the opium poppy; nicotine, produced by the tobacco plant; mescaline, from peyote cactus; and tannins, from a variety of plant species. A variety of sulfur compounds, including those that give Brussels sprouts and cabbage their distinctive taste, are also toxic to herbivorous insects and mammals such as cattle. (The vegetables we eat are not toxic because the amount of chemicals in them were reduced by crop

+/−

breeders.) Some plants even produce chemicals that cause abnormal development in insects that eat them. Chemical companies have taken advantage of the poisonous properties of certain plants to produce the pesticides called pyrethrin and rotenone. Nicotine is also used as an insecticide.

Some herbivore-plant interactions illustrate the concept of **coevolution,** a series of reciprocal evolutionary adaptations in two species. Coevolution occurs when a change in one species acts as a new selective force on another species, and counteradaptation of the second species in turn affects the selection of individuals in the first species. Figure 37.6 (top right of next page) illustrates an example of coevolution of an herbivorous insect (the caterpillar of the butterfly *Heliconius,* top left) and a plant (the passionflower *Passiflora,* a tropical vine). *Passiflora* produces toxic chemicals that protect its leaves from most insects, but *Heliconius* caterpillars have digestive enzymes that break down the toxins. As a result, *Heliconius* gains access to a food source that few other insects can eat.

These poison-resistant caterpillars seem to be a strong selective force for *Passiflora* plants, some of which have evolved defenses against them. For instance, the leaves of some *Passiflora* species produce yellow sugar deposits that look like *Heliconius* eggs. You can see two eggs in the top right photograph of Figure 37.6 and two egg-like sugar deposits in the bottom photo. Female butterflies avoid laying their eggs on leaves that already have eggs, presumably ensuring that only a few caterpillars will hatch and feed on any one leaf. Because the butterfly often mistakes the yellow sugar deposits for eggs, *Passiflora* species with the yellow deposits are less likely to be eaten.

The story of *Passiflora* is even more complicated, however. The egg-like sugar deposits, as well as smaller ones scattered over the leaf, attract ants and wasps that prey on *Heliconius* eggs and larvae. Thus, adaptations that appear to be coevolutionary responses between just two species may in fact involve interactions among many species in a community.

? People find most bitter-tasting foods objectionable. Why do you suppose we have taste receptors for bitter-tasting chemicals?

■ Individuals having bitter taste receptors presumably survived better because they could identify potentially toxic food when they foraged.

Figure 37.6 Coevolution: *Heliconius* and the passionflower vine (*Passiflora*)

37.7 Parasites and pathogens can affect community composition

A parasite lives on or in a host from which it obtains nourishment. You may recall learning about several invertebrate parasites, including flukes and tapeworms in Module 18.7 and a variety of nematodes in Module 18.8, which live inside a host organism's body. External parasites include arthropods such as ticks, lice, mites, and mosquitoes (see Module 18.11), which attach to their victims temporarily to feed on blood or other body fluids. Plants are also attacked by parasites, including nematodes and aphids, tiny insects that tap into the phloem and suck plant sap (Figure 37.7). Pathogens are disease-causing bacteria, viruses, fungi, or protists that can be thought of as microscopic parasites.

Figure 37.7 Aphids parasitizing a plant

The potentially devastating effects of parasites and pathogens on cultivated plants, livestock, and humans are well-known, but ecologists know little about how these interactions affect natural communities. Non-native pathogens, whose impact is rapid and often dramatic, have provided some opportunities to investigate the effects of pathogens on communities. The investigation of Dutch elm disease (see Module 17.18) is one example. In another instance, ecologists have studied the consequences of an epidemic of chestnut blight that wiped out virtually all American chestnut trees during the first half of the 20th century; the disease is caused by a protist. The loss of chestnuts, massive canopy trees that dominated many forest communities in North America, had a significant impact on species composition and community structure. Trees such as oaks and hickories that had formerly competed with chestnuts became more prominent; overall, the diversity of tree species increased. Dead trees also furnished niches for other organisms such as insects, cavity-nesting birds, and eventually, decomposers. On the other hand, many other animals depend on living trees for their food and habitat. A fungus-like protist that causes a disease called sudden oak death is currently spreading on the West Coast. More than a million oaks have been lost so far, causing the decline of bird populations. Despite its common name, sudden oak death affects many other species as well, including the majestic redwood and Douglas fir trees and flowering shrubs such as rhododendron and camellia. Because the epidemic is just beginning, its full effect will not be known for some time.

? Use your knowledge of interspecific interactions to explain why tree diversity increased after all the chestnuts died.

■ Chestnuts had many of the same niche characteristics as other trees, but apparently chestnuts were superior competitors. After they died, the remaining species may have had fewer niche similarities, or they may have been more equal as competitors, allowing more species to coexist.

Every community has a trophic structure, a pattern of feeding relationships consisting of several different levels. The sequence of food transfer up the trophic levels is known as a **food chain.** In Figure 37.8, the trophic levels are arranged vertically, and the names of the levels appear in colored boxes. The arrows connecting the organisms point from the food to the consumer. This transfer of food moves chemical nutrients and energy from the producers up through the trophic levels in a community.

Figure 37.8 compares a terrestrial food chain and an aquatic food chain. Starting at the bottom, the trophic level that supports all others consists of autotrophs, which ecologists call **producers.** Photosynthetic producers use light energy to power the synthesis of organic compounds. Plants are the main producers on land. In water, the producers are mainly photosynthetic protists and cyanobacteria, collectively called phytoplankton. Multicellular algae and aquatic plants are also important producers in shallow waters. In a few communities, such as the hydrothermal vents you learned about in Chapter 34, the producers are chemosynthetic prokaryotes.

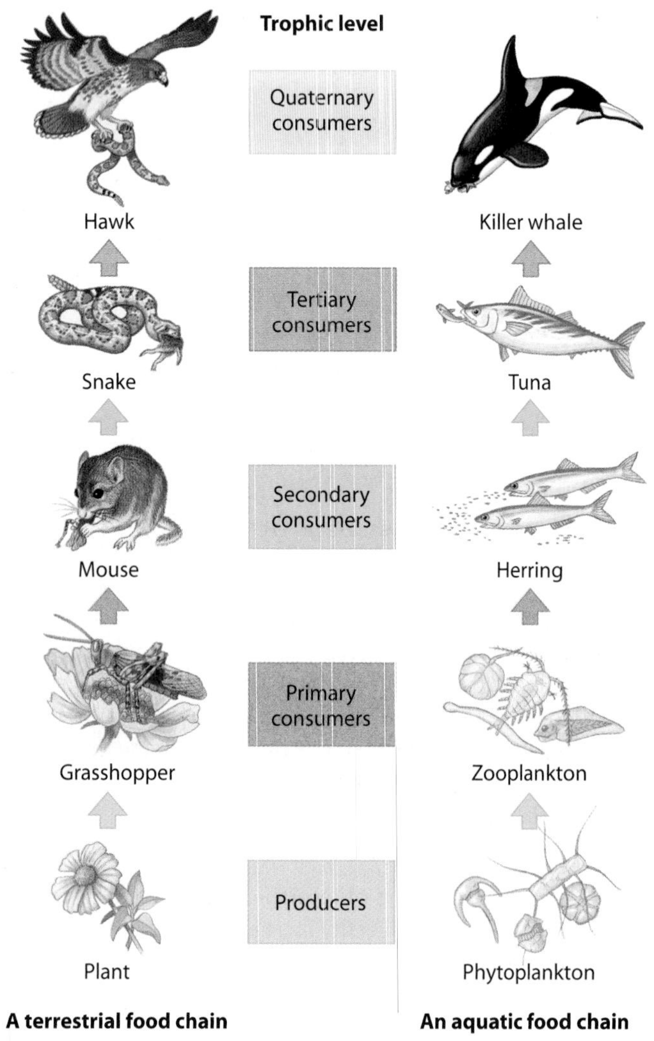

All organisms in trophic levels above the producers are heterotrophs, or consumers, and all consumers are directly or indirectly dependent on the output of producers. Herbivores, which eat plants, algae, or phytoplankton, are the **primary consumers.** Primary consumers on land include grasshoppers and many other insects, snails, and certain vertebrates, such as grazing mammals and birds that eat seeds and fruits. In aquatic environments, primary consumers include a variety of zooplankton (mainly protists and microscopic animals such as small shrimps) that eat phytoplankton.

Above the primary consumers, the trophic levels are made up of carnivores and insectivores, which eat the consumers from the level below. On land, **secondary consumers** include many small mammals, such as the mouse shown here eating an herbivorous insect, and a great variety of birds, frogs, and spiders, as well as lions and other large carnivores that eat grazers. In aquatic ecosystems, secondary consumers are mainly small fishes that eat zooplankton.

Higher trophic levels include **tertiary consumers,** such as snakes that eat mice and other secondary consumers. Most ecosystems have secondary and tertiary consumers. As the figure indicates, some also have a higher level, **quaternary consumers.** These include hawks in terrestrial ecosystems and killer whales in the marine environment.

Not shown in Figure 37.8 is another trophic level of consumers consisting of **detritivores** and **decomposers,** which derive their energy from **detritus,** the dead material produced at all the trophic levels. Detritus includes animal wastes, plant litter, and all sorts of dead organisms. Detritivores, which are often called scavengers, consume detritus. A great variety of animals, including earthworms, many rodents and insects, crows, and vultures, are detritivores. Detritivores in aquatic communities include catfish and crayfish. Decomposers are prokaryotes and fungi, which secrete enzymes that digest molecules in organic material and convert them into inorganic forms. Enormous numbers of microscopic fungi and prokaryotes in the soil and in the mud at the bottom of lakes and oceans break down most of the community's organic materials to inorganic compounds that plants or phytoplankton can use. The breakdown of organic materials to inorganic ones is called **decomposition.** In a sense, all organisms perform decomposition. In cellular metabolism, they all break down organic material and release inorganic products, such as carbon dioxide and ammonia, to the environment. But the decomposition by prokaryotes and fungi links all trophic levels. It is essential for all communities and, indeed, for the continuation of life on Earth.

Trophic level	
Hawk	Quaternary consumers	Killer whale
Snake	Tertiary consumers	Tuna
Mouse	Secondary consumers	Herring
Grasshopper	Primary consumers	Zooplankton
Plant	Producers	Phytoplankton
A terrestrial food chain		**An aquatic food chain**

Figure 37.8 Two food chains

? I'm eating a cheese pizza. At which trophic level(s) am I feeding?

■ Primary consumer (flour and tomato sauce) and secondary consumer (cheese, a product from cows, which are primary consumers)

37.9 Food chains interconnect, forming food webs

A more realistic view of the trophic structure of a community is a **food web,** a network of interconnecting food chains. Figure 37.9 shows a simplified example of a food web in a Sonoran desert community. As in the food chains of Figure 37.8, the arrows indicate the direction of nutrient transfer ("who eats whom") and are color-coded by trophic levels.

Notice that a consumer may eat more than one type of producer, and several species of primary consumers may feed on the same species of producer. Some animals weave into the web at more than one trophic level. The lizard and mantid are strictly secondary consumers, eating insects. The woodpecker on the left, however, is a primary consumer when it eats cactus seeds and a secondary consumer when it eats ants or grasshoppers. The hawk at the top of the web is a secondary, tertiary, or quaternary consumer, depending on its prey. Food webs, like food chains, do not typically show detritivores, which consume dead organic material from all trophic levels.

We have now looked at how populations in a community interact with each other. In the next modules, we consider factors that affect the community as a whole.

Web Activity *Food Webs*

? Even consumers at the highest level of a food web eventually become food for _____.

■ detritivores and decomposers

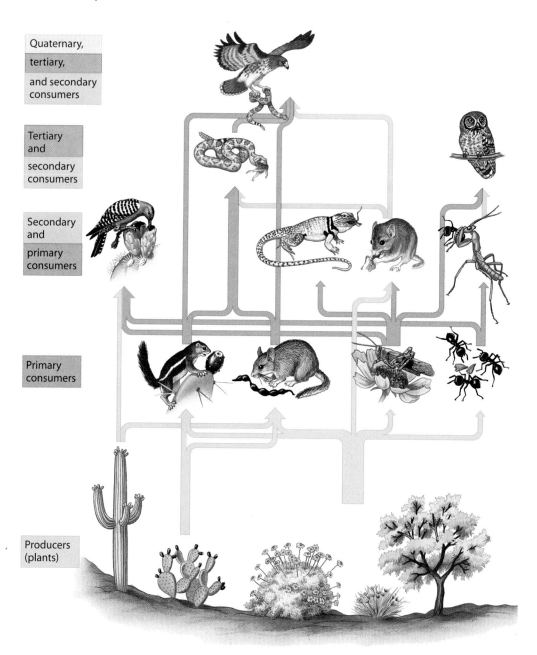

Figure 37.9 A food web

37.10 Species diversity includes relative abundance and species richness

Species diversity of a community is defined by two components: species richness, or the number of different species in a community, and relative abundance, the proportional representation of a species in a community. To understand why both components are important for describing species diversity, imagine walking through the woodlot shown in Figure 37.10A. On the path through woodlot A, you would pass by four different species of trees. Most of the trees you encounter, however, would be the same species. Now imagine walking on a path through woodlot B (Figure 37.10B). You would see the same four species of trees that you saw in woodlot A—the species richness of the two woodlots is the same. However, woodlot B would probably seem more diverse to you, because no single species predominates. As Table 37.10 shows, the relative abundance of one species in woodlot A is much higher than the relative abundances of the other three species. In woodlot B, all four species are equally abundant. As a result, species diversity is greater in woodlot B.

Plant species diversity in a community often has consequences for the species diversity of animals in the community. For example, suppose a species of caterpillar only eats the leaves of a tree that makes up just 5% of woodlot A. If the caterpillar is present at all, its population may be small and scattered. Birds that depend on those caterpillars to feed their young may be absent. But the caterpillars would easily be able to locate their food source in woodlot B, and their abundance would attract birds, as well. By providing a broader range of habitats, a diverse tree community promotes animal diversity.

Species diversity also has consequences for pathogens. Most pathogens infect a limited range of host species, or may even be restricted to a single host species. When many potential hosts are living close together, it is easy for a pathogen to spread from one to another. In woodlot A, for example, a pathogen that infects the most abundant tree would rapidly be transmitted through the entire forest. On the other hand, the more isolated trees in woodlot B are more likely to escape infection.

Low species diversity is characteristic of agricultural ecosystems. For efficiency, most crops and trees are planted in monocultures—a single species grown over a wide area. Monocultures are especially vulnerable to attack by pathogens and herbivorous insects. Also, plants grown in monoculture have been bred for certain desirable characteristics, so their genetic variation is typically low, too. As a result, a pathogen can potentially devastate an entire field, or more. Between 1845 and 1849, a pathogen wiped out a monoculture of genetically uniform potatoes throughout Ireland. As a result, thousands of people died of starvation, and thousands more left the country.

To combat potential losses, many farmers and forest managers rely heavily on chemical methods of controlling pests. Modern crop scientists have bred varieties of plants that are genetically resistant to common pathogens, but these varieties can suddenly become vulnerable, too. In 1970, pathogen evolution led to an epidemic of a disease called corn leaf blight that resulted in a billion dollars of crop damage in the United States. Some researchers are now investigating the use of more diverse agricultural ecosystems—polyculture—as an alternative to monoculture.

? Which would you expect to have higher species diversity, a well-maintained lawn or one that is poorly maintained? Explain.

■ A lawn that is poorly maintained would have higher species diversity. A well-maintained lawn should have low species diversity. While a lawn that is cared for may not be a perfect monoculture, any weeds that are present would have low relative abundance. The opposite is true if the lawn is not cared for.

Figure 37.10A Species composition of woodlot A

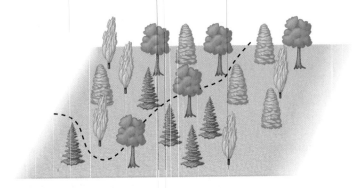

Figure 37.10B Species composition of woodlot B

TABLE 37.10	RELATIVE ABUNDANCE OF TREE SPECIES IN WOODLOTS A AND B	
Species	Relative Abundance in Woodlot A (%)	Relative Abundance in Woodlot B (%)
	80	25
	10	25
	5	25
	5	25

Keystone species have a disproportionate impact on diversity

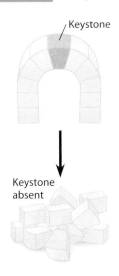

Keystone

Keystone
absent

Figure 37.11A Arch collapse with removal of keystone

In Module 37.10, you saw that the abundance of dominant species such as forest trees can have an impact on the diversity of other species in the community. But less abundant species may also exert control over community composition. A **keystone species** is a species whose impact on its community is much larger than its biomass or abundance indicate. The word "keystone" comes from the wedge-shaped stone at the top of an arch that locks the other pieces in place. If the keystone is removed, the arch collapses (Figure 37.11A). A keystone species occupies a niche that holds the rest of its community in place.

To investigate the role of a potential keystone species in a community, ecologists compare diversity when the species is present with diversity when it is absent. Experiments by Robert Paine in the 1960s were among the first to provide evidence of the keystone species effect. Paine manually removed a predator, a sea star of the genus *Pisaster* (Figure 37.11B), from experimental areas within the intertidal zone of the Washington coast. The result was that *Pisaster*'s main prey, a mussel of the genus *Mytilus,* outcompeted many of the other shoreline organisms (algae, barnacles, and snails, for instance) for the important resource of space on the rocks. The number of different organisms present in experimental areas dropped from more than 15 species to fewer than five species.

The keystone concept has practical application in efforts to restore or rehabilitate damaged ecosystems. In 1983, a disease swept through the coral reefs of the Caribbean, killing massive numbers of the long-spined sea urchin, *Diadema antillarum* (Figure 37.11C). In the following decade, the area of reef covered by living coral animals plummeted, along with overall species diversity. Fleshy seaweeds replaced the low turf of encrusted red algae that is vital to reef building (Figure 37.11D). The thick growth of seaweed also prevented light from reaching

Figure 37.11B A *Pisaster* sea star, a keystone species, eating a mussel

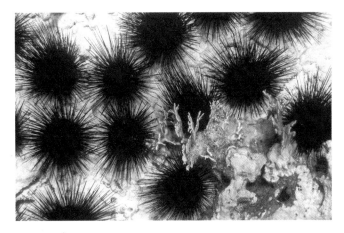

Figure 37.11C *Diadema* sea urchins grazing on reef

Figure 37.11D Reef overgrown by fleshy seaweeds

the symbiotic dinoflagellates that corals depend upon for food. These dramatic changes in the reef community revealed that *Diadema* is a keystone species whose herbivorous habits have two major effects. First, its grazing suppresses the seaweed populations. (Because *Diadema* normally shares this role with herbivorous fish, it is an especially important species on the many Caribbean reefs that have been overfished.) In addition, the urchins scrape patches of substrate clear of algae, providing platforms for coral larvae to settle. *Diadema* populations have been slow to rebound from the catastrophic die-off. However, recognition of this organism's key role in the community has prompted conservationists to artificially replenish urchin populations in some areas to help restore damaged reefs.

? Removing saguaro cacti from the Sonoran desert community (see Module 37.9) would have a drastic impact, yet saguaro is not considered a keystone species. Why not?

■ Saguaro is abundant and makes up a large part of the community. Keystone species have a large effect relative to their representation in the community, just as a keystone is a small but vital piece of the arch.

A traditional view of biological communities is that they are characterized by a more or less stable structure and composition of species. But in many communities, at least on a local scale, change seems to be more common than stability. A more recent model describes communities as constantly changing in response to disturbances.

Disturbances are events such as storms, fire, floods, droughts, overgrazing, or human activity that damage biological communities, remove organisms from them, and alter the availability of resources. The types of disturbances and their frequency and severity vary from community to community. By gathering data from specific communities over many years, ecologists are beginning to appreciate and understand the impact of disturbances.

We tend to think of disturbances in negative terms, but this is not always the case. Small-scale disturbances often have positive effects. For example, when a large tree falls in a windstorm, it disturbs the immediate surroundings, but it also creates new habitats. For instance, more light may now reach the forest floor, giving small seedlings the opportunity to grow; or the depression left by its roots may fill with water and be used as egg-laying sites by frogs, salamanders, and numerous insects. Small-scale disturbances may enhance environmental patchiness, which can contribute to species diversity in a community.

Communities change drastically following a severe disturbance that strips away vegetation and even soil. The disturbed area may be colonized by a variety of species, which are gradually replaced by a succession of other species, in a process called **ecological succession.**

When ecological succession begins in a virtually lifeless area with no soil, it is called **primary succession.** Examples of such areas are new volcanic islands or the rubble left by a retreating glacier. Often the only life-forms initially present are autotrophic bacteria. Lichens and mosses, which grow from windblown spores, are commonly the first large photosynthesizers to colonize the area. Soil develops gradually as rocks weather and organic matter accumulates from the decomposed remains of the early colonizers. Lichens and mosses are gradually overgrown by grasses and shrubs that sprout from seeds blown in from nearby areas or carried in by animals. Eventually, the area is colonized by plants that become the community's prevalent form of vegetation. Primary succession can take hundreds or thousands of years.

Secondary succession occurs where a disturbance has destroyed an existing community but left the soil intact. For example, secondary succession occurs as areas recover from fires or floods. Disturbances that lead to secondary succession are also caused by human activities. Even before colonial times, humans were clearing the temperate deciduous forests of eastern North America for agriculture and settlements. Some of this land was later abandoned as the soil was depleted of its chemical nutrients or the residents moved west to new territories. Whenever human intervention stops, secondary succession begins.

Numerous studies have documented the stages by which an abandoned farm field becomes a hardwood forest (Figure 37.12). A recently disturbed site provides an envi-

ronment that is favorable to *r*-selected species (see Module 36.7)—plants and animals that reach reproductive age rapidly, produce huge numbers of offspring, and provide little or no parental care. Interspecific competition is not a major factor during the very early stages of succession, which are dominated by weedy annual species such as crabgrass and ragweed. Within a few years, perennial grasses and small broadleaf plants cover the field. (An annual plant completes it life cycle in a single year. Perennial plants live for many years.) Softwood species, especially pines, begin to invade within five years, turning the area into a pine forest in roughly 10–15 years. But pine seedlings, which need high levels of light to grow, don't do well in the understory. The seedlings of many deciduous species are more shade-tolerant, and thus hardwoods such as oak and hickory begin to replace pine as competition becomes a significant force in determining the composition of the community. The final mixture of species depends on abiotic factors such as soil and topography. Because animals depend on plants for food and shelter, the animal community undergoes successional changes, too. The diversity of bird species, for example, increases dramatically as trees replace herbaceous plants.

Understanding the effects of disturbance in communities is especially important today; as we discuss in Chapter 38, humans are the most widespread and significant agents of disturbance. Disturbances may also create opportunities for undesirable plants and animals that have been transported by humans to new habitats.

Web Activity *Primary Succession*

? What is the main abiotic factor that distinguishes primary from secondary succession?

■ Absence (primary) versus presence (secondary) of soil at the onset of succession

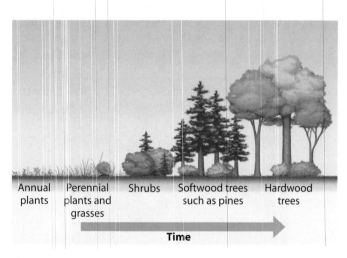

| Annual plants | Perennial plants and grasses | Shrubs | Softwood trees such as pines | Hardwood trees |

Time

Figure 37.12 Stages in secondary succession of abandoned farm field

37.13 Invasive species can devastate communities

As you may recall from Module 19.18, humans have transported plants and animals around the globe, sometimes intentionally and sometimes by accident. Many of these non-native species have established themselves firmly in their new locations. Furthermore, many have become **invasive species,** spreading far beyond the original point of introduction and causing environmental or economic damage by colonizing and dominating wherever they find a suitable habitat. In the United States alone, there are hundreds of invasive species, including plants, mammals, birds, fish, arthropods, and molluscs. Worldwide, there are thousands more. Invasive species are a leading cause of local extinctions, a topic we'll return to in Chapter 38. The economic costs of invasive species are enormous—an estimated $136 billion a year in the United States. Regardless of where you live, an invasive plant or animal is probably living nearby.

Not every organism that is introduced to a new habitat is successful, and not every species that is able to survive in its new habitat becomes invasive. There is no single explanation for why any non-native species turns into a destructive pest, but community ecology offers some insight. Interspecific interactions act as a system of checks and balances on the populations in a community. Every population is subject to multiple negative effects, whether from competitors, predators, herbivores, or pathogens, that curb its growth rate (see Module 36.4). For example, predators increase the death rate and decrease the number of individuals in a prey population. Competitors diminish the amount of resources available. Without biotic factors such as these to check population growth, a population will continue to expand until limited by abiotic factors.

In Module 36.4, we illustrated exponential population growth with rabbits, which are notoriously prolific breeders. In 1859, 12 pairs of European rabbits (*Oryctolagus cuniculus*) were released on a ranch in southern Australia by a European who wanted to hunt familiar game. The animals quickly became a nuisance. In 1865, 20,000 rabbits were killed on the ranch. By 1900, several hundred million rabbits were distributed over most of the continent (Figure 37.13A). The rabbit invasion was a catastrophe in several ways. Their activities destroyed farm and grazing land by eating vegetation down to, and sometimes including, the roots (Figure 37.13B). Especially in arid regions, the loss of plant cover led to soil erosion. In addition, rabbits dug extensive underground burrows that made grazing treacherous for cattle and sheep. They also competed directly with native herbivorous marsupials. After many fruitless attempts to control the rabbit population, in 1950, the Australian government turned to **biological control,** the intentional release of a natural enemy to attack a pest population. A virus lethal to rabbits was introduced into the environment. The rabbits and virus then underwent several coevolutionary cycles as the rabbits became more resistant to the disease and the virus became less lethal. The government managed to stave off a complete resurgence of the rabbit population by introducing new viral strains, but in 1995, they had to switch to a different pathogen to maintain control.

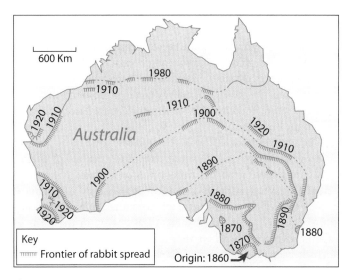

Figure 37.13A The spread of rabbits in Australia

Coevolution is just one potential pitfall of biological control. The imported enemy may not be as successful in the new environment as the target species. It may not disperse widely enough, or its population growth rate may not be high enough to overtake a rapidly expanding population. Caution is especially warranted because the control agent may turn out to be as invasive as its target. As you may recall from Module 19.18, the cane toad invasion of Australia began when it was imported to control an insect pest.

In the next modules, we broaden our scope to look at ecosystems, the highest level of ecological complexity.

? **What is the ecological basis for biological control of pests?**

■ By having a negative effect on population growth rate, a natural enemy keeps the target population in check.

Figure 37.13B A familiar sight in early 20th-century Australia

37.14 Ecosystem ecology emphasizes energy flow and chemical cycling

An **ecosystem** consists of all the organisms in a community as well as the abiotic environment with which the organisms interact. Ecosystem ecologists are especially interested in **energy flow,** the passage of energy *through* the components of the ecosystem, and **chemical cycling,** the transfer of materials *within* the ecosystem.

The terrarium in Figure 37.14 represents the most familiar type of ecosystem and illustrates the fundamentals of energy flow. Energy enters the terrarium in the form of sunlight (yellow arrows). Plants (producers) convert light energy to chemical energy through the process of photosynthesis. Animals (consumers) take in some of this chemical energy in the form of organic compounds when they eat the plants. Decomposers such as bacteria and fungi in the soil obtain chemical energy when they decompose the dead remains of plants and animals. Every use of chemical energy by organisms involves a loss of some energy to the surroundings in the form of heat (see Module 6.1). Because so much of the energy captured by photosynthesis is lost as heat, the ecosystem would run out of energy if it were not powered by a continuous inflow of energy from the sun. A few ecosystems, for example, hydrothermal vents (see Chapter 34's introduction), are powered by chemical energy obtained from inorganic compounds. Some others obtain their energy from an external source. For example, the energy source for the ecosystem in the hippo's stomach is the plant material that the hippo consumes (see this chapter's introduction).

In contrast to energy flow, chemical cycling (blue arrows in Figure 37.14) involves the transfer of materials *within* the ecosystem. While most ecosystems have a constant input of energy from sunlight, the supply of the chemical elements used to construct molecules is limited. Chemical elements such as carbon and nitrogen are cycled between the abiotic component of the ecosystem, including the air, water, and soil, and the biotic component of the ecosystem (the community). Plants acquire these elements in inorganic form from the air and soil and fix them into organic molecules. Animals, such as the snail in Figure 37.14, consume some of these organic molecules. When the plants and animals become detritus, decomposers

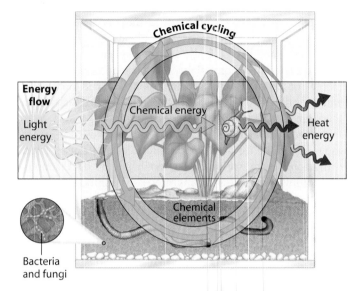

Figure 37.14 A terrarium ecosystem

return most of the elements to the soil and air in inorganic form. Some elements are also returned to the soil as the by-products of plant and animal metabolism.

In summary, both energy flow and chemical cycling involve the transfer of substances through the trophic levels of the ecosystem. However, energy flows through, and ultimately out of, ecosystems, whereas chemicals are recycled within ecosystems. We explore these fundamental ecosystem dynamics in the rest of the chapter.

Web Activity *Energy Flow and Chemical Cycling*

? Why is the transfer of energy in an ecosystem referred to as energy *flow,* not energy *cycling?*

■ Because energy passes through an ecosystem, entering as sunlight and leaving as heat. It is not recycled within the ecosystem.

37.15 Primary production sets the energy budget for ecosystems

Each day, Earth receives about 10^{19} kcal of solar energy, the energy equivalent of about 100 million atomic bombs. Most of this energy is absorbed, scattered, or reflected by the atmosphere or by Earth's surface. Of the visible light that reaches plants, algae, and cyanobacteria, only about 1% is converted to chemical energy by photosynthesis.

Ecologists call the amount, or mass, of living organic material in an ecosystem the **biomass.** The amount of solar energy converted to chemical energy (in organic compounds) by an

ecosystem's producers for a given area and during a given time period is called **primary production.** It can be expressed in units of energy or of mass. The primary production of the entire biosphere is roughly 165 billion tons of biomass per year.

Different ecosystems vary considerably in their primary production as well as in their contribution to the total production of the biosphere. Figure 37.15, at the top of the next page, contrasts the net primary production of a number of different ecosystems. (Net primary production refers to the amount of biomass produced minus the amount used by producers as fuel for their own cellular respiration.) Tropical rain forests are among the most productive terrestrial ecosystems and contribute a large portion of the planet's overall production of biomass. Coral reefs also have very high production, but their contribution to global production is small because they cover such a small area. Interestingly, even though the open ocean has very low production, it contributes the most to Earth's total net primary production because of its huge size—it covers 65% of Earth's surface area.

Web Process of Science How Do Temperature and Light Affect Primary Production?

? Deserts and semidesert scrub cover about the same amount of surface area as tropical forests but contribute less than 1% of Earth's net primary production, while rain forests contribute 22%. Explain this difference.

■ The primary production of tropical rain forests is over 20 times greater than that of deserts and semidesert scrub ecosystems.

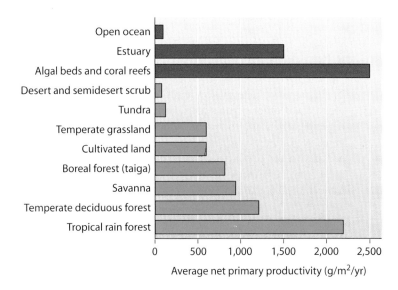

Figure 37.15 Net primary production of various ecosystems

37.16 Energy supply limits the length of food chains

When energy flows as organic matter through the trophic levels of an ecosystem, much of it is lost at each link in a food chain. Consider the transfer of organic matter from producers to primary consumers (herbivores), represented in Figure 37.16 by a grasshopper. In most ecosystems, herbivores manage to eat only a fraction of the plant material produced, and they can't digest all of what they do consume. For example, a grasshopper feeding on a flower might digest and absorb only about half the organic material it eats, passing the indigestible wastes as feces. Of the organic compounds it does absorb, the grasshopper typically uses two-thirds as fuel for cellular respiration. Only the chemical energy left over after respiration can be converted to grasshopper biomass. Only this biomass or amount of energy is available to the next trophic level.

Figure 37.16, called a *pyramid of production,* illustrates the cumulative loss of energy with each transfer in a food chain. Each tier of the pyramid represents all of the organisms in one trophic level, and the width of each tier indicates how much of the chemical energy of the tier below is actually incorporated into the organic matter of that trophic level. Note that producers convert only about 1% of the energy in the sunlight available to them to primary production. In this idealized pyramid, 10% of the energy available at each trophic level becomes incorporated into the next higher level. The efficiencies of energy transfer usually range from 5–20%. In other words, 80–95% of the energy at one trophic level never transfers to the next.

An important implication of this stepwise decline of energy in a trophic structure is that the amount of energy available to top-level consumers is small compared with that available to lower-level consumers. Only a tiny fraction of the energy stored by photosynthesis flows through a food chain

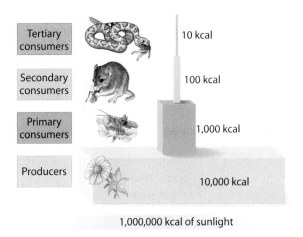

Figure 37.16 An idealized pyramid of production

all the way to a tertiary consumer. This explains why top-level consumers such as lions and hawks require so much geographic territory; it takes a lot of vegetation to support trophic levels so many steps removed from photosynthetic production. Pyramids of production help us understand why most food chains are limited to three to five levels; there is simply not enough energy at the very top of an ecological pyramid to support another trophic level.

○ *MP3 Tutor Energy Flow in Ecosystems*

? Approximately what proportion of the energy produced by photosynthesis makes it to the snake in Figure 37.16?

■ 1/1000 of the 10,000 kcal produced by photosynthesis $[(0.1 \times 0.1)(10,000 \text{ kcal})] = 10 \text{ kcal}$

37.17 A production pyramid explains why meat is a luxury for humans

The dynamics of energy flow apply to the human population as much as to other organisms. Like most other consumers, we depend entirely on production by plants for our food. As omnivores, we eat both plant material and meat. When we eat grain or fruit, we are primary consumers; when we eat beef or other meat from herbivores, we are secondary consumers. When we eat fish like trout and salmon (which eat insects and other small animals), we are tertiary or quaternary consumers.

The production pyramid on the left in Figure 37.17 indicates energy flow from primary producers to humans as vegetarians (herbivores). The energy in the producer trophic level comes from a corn crop. The pyramid on the right illustrates energy flow from the same corn crop, with humans as secondary consumers, eating beef. These two pyramids are generalized models, based on the rough estimate that about 10% of the energy available in a trophic level appears at the next higher trophic level. Thus, the pyramids indicate that the human population has about ten times more energy available to it when people eat corn than when they process the same amount of corn through another trophic level and eat corn-fed beef.

Eating meat of any kind is an expensive luxury, both economically and environmentally. Compared to growing plant crops for human consumption, producing meat usually requires that more land be cultivated, more water be used for irrigation, and more chemical fertilizers and pesticides be applied to croplands used for growing grain. In many countries, people cannot afford to buy much meat and are vegetarians by necessity. In others, religion also plays a role in the decision. In India, for example, about 80% of the population practice Hinduism, a religion that discourages meat-eating. India's meat consumption was roughly 5.2 kg (11.5 pounds) per person annually in 2002 (the most recent year for which statistics are available). In Mexico, where many people are too poor to eat meat daily, per capita consumption in 2002 was 58.6 kg (129 pounds) per year. That is a large amount compared to India, but less than half that of the United States, where the per capita rate was 124.8 kg (275 pounds) in 2002.

We turn next to the subject of chemical nutrients. Unlike energy, which is ultimately lost from an ecosystem, all chemical nutrients cycle within ecosystems.

Web Activity *Energy Pyramids*

? Why does demand for meat also tend to drive up prices of grains such as wheat and rice, fruits, and vegetables?

■ The potential supply of plants for direct consumption as food for humans is diminished by the use of agricultural land to grow feed for cattle, chickens, and other meat sources.

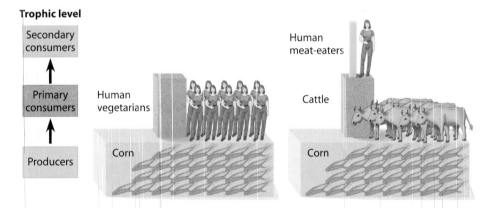

Trophic level

Figure 37.17 Food energy available to the human population at different trophic levels

37.18 Chemicals are cycled between organic matter and abiotic reservoirs

The sun (or in some cases Earth's interior) supplies ecosystems with a continual influx of energy, but aside from an occasional meteorite, there are no extraterrestrial sources of chemical elements. Life, therefore, depends on the recycling of chemicals. While an organism is alive, much of its chemical stock changes continuously, as nutrients are acquired and waste products are released. Atoms present in the complex molecules of an organism at the time of its death are returned to the environment by the action of decomposers, replenishing the pool of inorganic nutrients that plants and other producers use to build new organic matter.

Because chemical cycles in an ecosystem include both biotic and abiotic (geological and atmospheric) components,

they are called **biogeochemical cycles.** Figure 37.18, at the top of the next page, is a general scheme for the cycling of a nutrient within an ecosystem. Note that the cycle has an **abiotic reservoir,** where a chemical accumulates or is stockpiled outside of living organisms. The atmosphere, for example, is an abiotic reservoir for carbon. Phosphorus, on the other hand, is available only from the soil.

Let's trace the general biogeochemical cycle in Figure 37.18. ❶ Producers incorporate chemicals from the abiotic reservoir into organic compounds. ❷ Consumers feed on the producers, incorporating some of the chemicals into their own bodies. ❸ Both producers and consumers release some chemicals back to the environment in waste products (CO_2 and nitrogen wastes of animals). ❹ Decomposers play a central role by breaking down the complex organic molecules in detritus such as plant litter, animal wastes, and dead organisms. The products of this metabolism are inorganic compounds such as nitrates (NO_3^-), phosphates (PO_4^{3-}), and CO_2, which replenish the abiotic reservoir. Geologic processes such as erosion and the weathering of rock also contribute to the abiotic reservoirs. Producers use the inorganic molecules from abiotic reservoirs as raw materials for synthesizing new organic molecules (carbohydrates and proteins, for example), and the cycle continues.

Biogeochemical cycles can be local or global. The chemical phosphorus, for example, cycles almost entirely within local areas, at least over the short term. Soil is the main reservoir for nutrients in a local cycle. In contrast, for those chemicals that spend part of their time in gaseous form—carbon and nitrogen are examples—the cycling is essentially global. For instance, some of the carbon a plant acquires from the air may have been released into the atmosphere by the respiration of a plant or animal on another continent.

In the next three modules, we look at the cyclic movements of carbon, nitrogen, and phosphorus. As you study the cycles, look for the four basic steps we have cited, as well as the geologic processes that may move chemicals around and between ecosystems. In the diagrams, the main abiotic reservoirs are highlighted in white boxes.

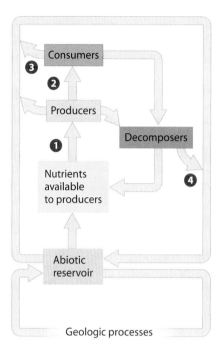

Figure 37.18   A general model of biogeochemical cycling of nutrients

? Which boxes in Figure 37.18 represent biotic components of an ecosystem?

■ Consumers, producers, and decomposers.

37.19 **The carbon cycle depends on photosynthesis and respiration**

Carbon, the major ingredient of all organic molecules, has an atmospheric reservoir and cycles globally. Other abiotic reservoirs of carbon include fossil fuels, dissolved carbon compounds in the oceans, and sedimentary rocks such as limestone ($CaCO_3$).

As shown in Figure 37.19, the reciprocal metabolic processes of photosynthesis and cellular respiration are mainly responsible for the cycling of carbon between the biotic and abiotic worlds. ❶ Photosynthesis removes CO_2 from the atmosphere and incorporates it into organic molecules, which are ❷ passed along the food chain by consumers. ❸ Cellular respiration returns CO_2 to the atmosphere. ❹ Decomposers break down the carbon compounds in detritus; that carbon, too, is eventually released as CO_2.

On a global scale, the return of CO_2 to the atmosphere by respiration closely balances its removal by photosynthesis. However, ❺ the increased burning of wood and fossil fuels (coal and petroleum) is raising the level of CO_2 in the atmosphere. As we will discuss in Module 38.5, this increase in CO_2 is leading to significant global warming.

Web Activity *The Carbon Cycle*

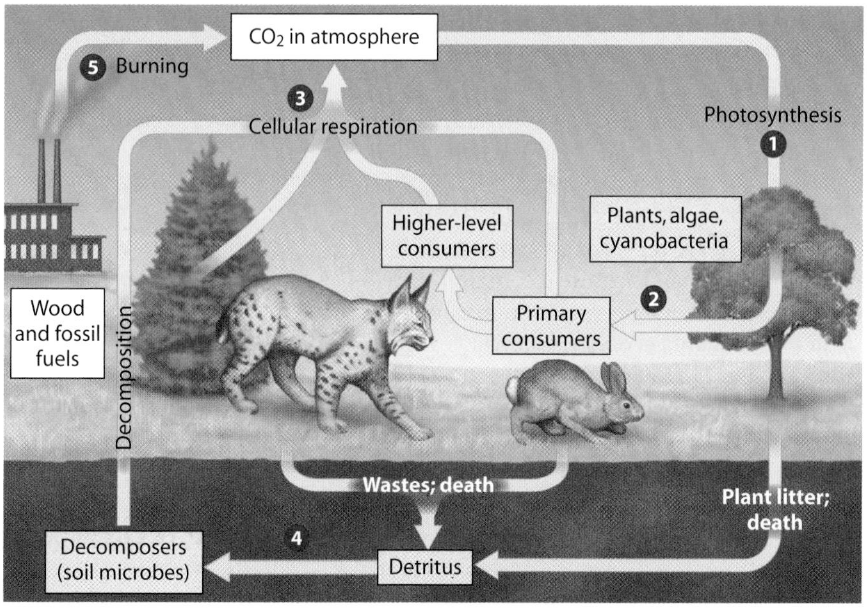

Figure 37.19   The carbon cycle

? What would happen to the carbon cycle if all the decomposers suddenly went on "strike" and stopped working?

■ Carbon would accumulate in organic mass, the atmospheric reservoir of carbon would decline, and plants would eventually be starved for CO_2.

37.20 The phosphorus cycle depends on the weathering of rock

Organisms require phosphorus as an ingredient of nucleic acids, phospholipids, and ATP and (in vertebrates) as a mineral component of bones and teeth. In contrast to the carbon cycle and the other major biogeochemical cycles, the phosphorus cycle does not have an atmospheric component. Rocks are the only source of phosphorus for terrestrial ecosystems; in fact, rocks that have high phosphorus content are mined for agricultural fertilizer.

At the center of Figure 37.20, ❶ the weathering (breakdown) of rock gradually adds inorganic phosphate (PO_4^{3-}) to the soil. ❷ Plants assimilate the dissolved phosphate ions in the soil and build them into organic compounds. ❸ Consumers obtain phosphorus in organic form from plants. ❹ Phosphates are returned to the soil by the action of decomposers on animal waste and the remains of dead plants and animals. ❺ Some phosphorus drains from terrestrial ecosystems into the sea, where it may settle and eventually become part of new rocks. This phosphorus will not cycle back into living organisms until ❻ geologic processes uplift the rocks and expose them to weathering.

Because weathering is generally a slow process, the amount of phosphates available to plants in natural ecosystems is often quite low and commonly a limiting factor. Farmers and gardeners may use crushed phosphate rock, bone meal (finely ground bones from slaughtered livestock), or guano to add phosphorus to the soil. Guano, the droppings of sea birds and bats, is mined from densely populated colonies or caves, where meters-deep deposits have accumulated.

The level of dissolved phosphates in aquatic ecosystems is typically low enough to be a limiting factor for algae and cyanobacteria. In many areas, however, excess phosphates from large amounts of fertilizer used in agriculture are a problem. Phosphates are also a common ingredient in pesticides. Other major sources of phosphates in aquatic ecosystems include outflow from sewage treatment facilities and runoff of

animal waste from livestock feedlots (where hundreds of animals are penned together). Phosphate pollution of lakes and rivers results in heavy growth of algae and cyanobacteria, making the water murky. Microbes consume a great deal of oxygen (O_2) as they decompose the extra biomass, a process that depletes the water of dissolved oxygen. These changes lead to reduced species diversity—and a much less appealing body of water.

? Over the short term, why does phosphorus cycling tend to be more localized than either carbon or nitrogen cycling?

■ Because phosphorus is cycled almost entirely within the soil rather than transferred over long distances via the atmosphere

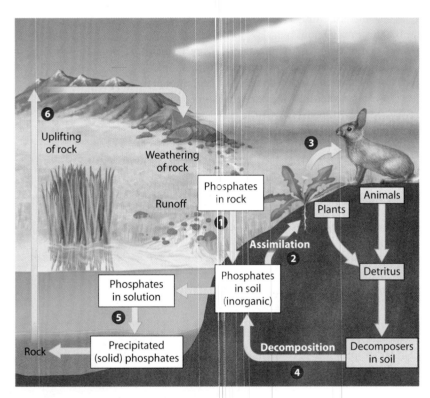

Figure 37.20 The phosphorus cycle

37.21 The nitrogen cycle depends on bacteria

As an ingredient of proteins and nucleic acids, nitrogen is essential to the structure and functioning of all organisms. In particular, it is a crucial and often limiting plant nutrient. Nitrogen has two abiotic reservoirs, the atmosphere and the soil. The atmospheric reservoir is huge; almost 80% of the atmosphere is nitrogen gas (N_2). However, plants cannot absorb nitrogen in the form of N_2. The process of **nitrogen fixation,** which is performed by some bacteria and cyanobacteria, con-

verts N_2 to compounds of nitrogen that can be used by plants. Without these organisms, the reservoir of usable soil nitrogen would be extremely limited.

Figure 37.21, on the facing page, illustrates the actions of two types of nitrogen-fixing bacteria. Starting at the far right in the figure, ❶ some bacteria live symbiotically in the roots of certain species of plants, supplying their hosts with a direct source of usable nitrogen. The largest group of plants with this mutualistic relationship is the legumes, a family that includes peanuts, soybeans, and alfalfa (see Module 32.13). A number of nonleguminous plants that live in nitrogen-poor soils have a similar relationship with bacteria or cyanobacteria. ❷ Free-living nitrogen-fixing bacteria and cyanobacteria in soil or water convert N_2 from air pockets in the soil to ammonium (NH_4^+).

❸ After nitrogen is "fixed," some of the NH_4^+ is taken up and used by plants. ❹ Nitrifying bacteria in the soil also convert some of the NH_4^+ to nitrate (NO_3^-), which is more readily ❺ absorbed by plants. Plants use the nitrogen they assimilate to synthesize molecules such as amino acids, which are then incorporated into proteins.

❻ When an herbivore (represented by the rabbit in Figure 37.21) eats a plant, it digests the proteins into amino acids, then uses the amino acids to build the proteins it needs. Higher order consumers gain nitrogen from their prey. Recall from Module 25.6 that nitrogen-containing waste products are formed during protein metabolism; consumers excrete some nitrogen as well as incorporating it into their body tissues. Mammals such as the rabbit excrete nitrogen as urea, a substance that is widely used as an agricultural fertilizer.

Organisms that are not consumed eventually die and become detritus, which is decomposed by bacteria and fungi. ❼ Decomposition releases NH_4^+ from organic compounds back into the soil, replenishing the soil reservoir of NH_4^+ and, with the help of nitrifying bacteria (step 4), NO_3^-. Under low-oxygen conditions, however, ❽ soil bacteria known as denitrifiers strip the oxygens from NO_3^-, releasing N_2 back into the atmosphere and depleting the soil reservoir of usable nitrogen.

Although not shown in the figure, some NH_4^+ and NO_3^- are made in the atmosphere by chemical reactions involving N_2 and ammonia gas (NH_3). The ions produced by these chemical reactions reach the soil in precipitation and dust, which are a crucial source of nitrogen for plants in some ecosystems.

Human activity has altered the nitrogen cycle balance in many areas. Sewage treatment facilities often empty large amounts of dissolved inorganic nitrogen compounds into rivers or streams. Agricultural sources of nitrogen cycle imbalance include feedlots and the large amounts of inorganic nitrogen fertilizers that are routinely applied to crops. Lawns and golf courses also receive sizable applications of nitrogen fertilizers. Crop and lawn plants take up some of the nitrogen compounds, and denitrifiers convert some into atmospheric N_2, but nitrate is not bound tightly by soil particles and is easily washed out of the soil by rain or irrigation (see Module 32.8). As a result, the introduction of chemical fertilizers overwhelms the soil's natural recycling capacity.

The excess nitrogen compounds from fertilizers frequently enter streams and lakes, where nitrogen may be a limiting factor. As in the case of phosphate pollution, nitrogen influx causes heavy growth of algae and cyanobacteria. Some nitrogen compounds also wind up in the groundwater, a serious problem in many agricultural areas. Nitrates in drinking water are converted to nitrites (NO_2^-), which can be toxic in the human digestive tract. In Module 32.10, we discussed some alternatives to the extensive use of agricultural fertilizers.

In the next two modules, we discuss some long-term ecological studies of nutrient cycling that demonstrate some of the effects of human alteration of nutrient cycles.

Web Activity *The Nitrogen Cycle*

Web Activity *Water Pollution from Nitrates*

? What are the abiotic reservoirs of nitrogen? In what form does nitrogen occur in each reservoir?

■ Atmosphere: N_2;
Soil: NH_4^+ and NO_3^-

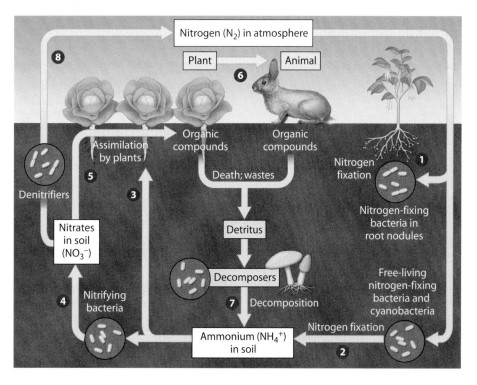

Figure 37.21 The nitrogen cycle

37.22 Ecosystem alteration can upset chemical cycling

The cycling of any chemical element in an ecosystem depends on the web of feeding relationships between plants, animals, and detritivores and on geologic processes such as erosion and weathering of rocks. Obtaining an accurate picture of chemical cycling requires long-term study, and a number of research groups have been conducting such studies for several decades. In one of the longest-running examples of long-term ecological research, a team has been monitoring nutrient cycling in a forest ecosystem since 1963. The study site, the Hubbard Brook Experimental Forest in the White Mountains of New Hampshire, is a deciduous forest with several valleys, each drained by a small creek that is a tributary of Hubbard Brook. Bedrock impenetrable to water is close to the surface of the soil, and each valley constitutes a watershed that can drain only through its creek.

The Hubbard Brook team set out to study water and nutrient dynamics under natural conditions and after severe human intrusion. The team first determined the amounts of water and key nutrients that normally move in and out of six of the area's watersheds. Figure 37.22A shows a small concrete dam with a V-shaped spillway built across the stream at the bottom of a watershed to monitor water and nutrient losses. When monitoring began, about 60% of the water that fell as rain and snow exited through the streams, and the remaining 40% was lost by transpiration from plants and evaporation from the soil. Preliminary data also indicated that the flow of nutrients into and out of the watersheds was nearly balanced and was relatively small compared with the quantity of nutrients being recycled within the forest itself.

Figure 37.22A A dam at the Hubbard Brook study site

In 1966, one of the valleys was completely logged and then sprayed with herbicides for three years to prevent regrowth of plants. All the original plant material was left in place to decompose. This severely altered watershed is the brown area in the center of Figure 37.22B. The inflow and outflow of water and minerals for this watershed were compared with a control (unaltered) watershed for three years. Water runoff from the altered system increased by 30–40%, apparently because there were no plants to absorb and transpire water from the soil. Net losses of nutrients were huge, as shown for nitrate in Figure 37.22C. Following deforestation, the nitrate concentration in the creek was 60 times greater in the altered watershed than in the control watershed. Not only was this vital nutrient drained from the ecosystem, but nitrate in the stream reached a level considered unsafe for drinking water.

After 40 years, data from Hubbard Brook point to some other long-term trends. For instance, since the 1950s, acid rain and snow have dissolved most of the calcium ions (Ca^{2+}) from the forest soil, and the streams have carried it away. By the 1990s, forest plants at Hubbard Brook had virtually stopped adding new growth, apparently because of a lack of Ca^{2+} in the soil. In 1998, ecologists began a massive experiment to test the hypothesis that depletion of calcium in the soil hinders plant growth. They monitored control and experimental watersheds in Hubbard Brook for two years and then added Ca^{2+} by helicopter. Initial results show significant increases in pH, nitrate concentrations, and soil respiration in the experimental watershed. Changes in tree growth may take longer to record.

In the next module, we discuss another ecologist's work, studying the effects of nutrients in freshwater ecosystems.

? How can clear-cutting a forest (removing all trees) damage the water quality of nearby lakes?

■ Without the growing trees to assimilate minerals from the soil, more of the minerals run off and end up polluting water resources.

Figure 37.22B Logged watershed in the Hubbard Brook Forest

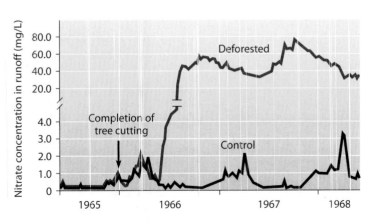

Figure 37.22C The loss of nitrate from a deforested watershed

37.23 David Schindler talks about the effects of nutrients on freshwater ecosystems

Figure 37.23A
David Schindler

The Hubbard Brook experiment shows that major change in a terrestrial ecosystem disrupts chemical cycling, moving nutrients to other areas, such as streams and lakes. How do nutrients from deforested lands and agricultural areas affect aquatic ecosystems? Nutrient levels may naturally increase over time, making a lake more productive. However, in a process known as *cultural eutrophication,* added nutrients cause algae and cyanobacteria to multiply rapidly, resulting in a "bloom." Heavy bacterial and algal growth greatly reduces oxygen levels at night, when the photosynthesizers respire. As these organisms die and accumulate at the bottom of the lake, decomposers can use up much of the oxygen dissolved in deep waters. As a result, the pond or lake may lose much of its species diversity. Human-caused eutrophication, for example, wiped out commercially important fishes in Lake Erie by the 1960s.

Dr. David Schindler (Figure 37.23A) is a professor of ecology at the University of Alberta, Canada. Before becoming a professor, he was a government scientist at the Experimental Lakes Project in northern Ontario. There he performed classic experiments that led to the banning of phosphates in detergents. He describes his research:

> When the project started in 1968, our mandate was to test water management issues at the level of whole-lake ecosystems. The main objective was the study of eutrophication, the overfertilization of lakes with mineral nutrients. The big issue at the time was whether phosphorus was the main culprit in eutrophication. Laboratory experiments implicated phosphorus, but my bosses at the time had a hard time convincing politicians and managers to invest millions or billions of dollars in phosphorus management schemes based solely on these small-scale experiments.

He explains why there was resistance to managing the input of phosphorus into freshwater ecosystems:

> One of the big sources of phosphorus in those days was phosphate detergents, and there was a big political lobby defending the use of these detergents. The companies that produced detergents were pointing the finger at carbon as the main problem.

Dr. Schindler tested whether it was carbon, phosphorus, or some other nutrient that was causing the eutrophication and algal blooms:

> In our best-known experiment, we divided a lake into two basins. We added just carbon and nitrogen to one basin. We included phosphorus along with carbon and nitrogen in the other basin. We got a tremendous algal bloom within weeks after adding phosphorus, but no change with just the carbon and nitrogen.

The results of that experiment, shown in Figure 37.23B, prompted government regulation of phosphate detergents, which helped solve the problems of eutrophication. Here, Dr. Schindler describes the most serious current threats to freshwater ecosystems:

> Acid precipitation is one. Warming of the climate is another. Changes in land use can affect lakes. For example, if you bulldoze a forest to pasture cows on the land, you increase the runoff of nutrients into the water four- or fivefold at least. And if you plow that land, plant crops, and add fertilizer, you increase the yield of nutrients to the water even more. And now we have these huge, intensive livestock operations—up to 30,000 cattle or 80,000 hogs. Very few of these big livestock operations have sewage treatment. The animals have a lot of the same intestinal microbes that humans have. We would no longer think of discharging raw sewage from a city into a river without treating it. But we do it all the time with intensive livestock operations. We're not handling our fresh water very well. And I think a big problem is that we've tended to look at the water issues, such as land use and acid precipitation, in isolation, when it's usually a combination of factors damaging the ecosystems.

Human disruption of both aquatic and terrestrial ecosystems is a global problem. In the next chapter, we look at conservation biology, efforts to conserve the diversity of the ecosystems on which humans depend.

? How does excessive addition of mineral nutrients to a lake eventually lead to the loss of most fish species?

■ The eutrophication initially causes population explosions of algae and cyanobacteria. Their respiration and that of the decomposers of all the organic refuse, consumes most of the lake's oxygen, which the fish require.

Figure 37.23B Experimental eutrophication of part of a lake (phosphorus added to basin at bottom of photo)

Chapter Review

Reviewing the Concepts

Community Structure and Dynamics (37.1–37.13)

A community includes all the organisms in a particular area. Community ecology is concerned with factors that influence the species composition and distribution of communities and with factors that affect community stability (**37.1**).

Interspecific interactions affect the structure and dynamics of populations within a community and can be categorized according to their effect on the interacting populations (**37.2**).

Interspecific competition $(-/-)$ occurs when two populations require the same limited resource (**37.3**). **Mutualism** $(+/+)$ benefits both partners (**37.4**). The strongly negative effects of **predation** $(+/-)$ and **herbivory** $(+/-)$ have selected for protective adaptations in prey and plants. Some herbivore-plant interactions illustrate coevolution or reciprocal evolutionary adaptations (**37.5, 37.6**). **Parasites** and pathogens $(+/-)$ can dramatically alter community composition (**37.7**).

Trophic structure of a community can be represented by a food chain—the stepwise flow of energy and nutrients from plants (producers), to herbivores (primary consumers), to carnivores (secondary and higher-level consumers). Decomposers break down the complex organic molecules that make up the bodies of organisms and return inorganic substances to the soil, air, and water (**37.8**). A food web is a network of interconnecting food chains (**37.9**).

Species diversity takes into account both the number of species in a community and the relative abundance of each species (**37.10**). **Keystone species** have low biomass or relative abundance, but have a significant impact on species diversity (**37.11**).

Disturbance, whether human caused or not, is characteristic of most communities. Ecological succession is a transition in species composition of a community. Primary succession is the gradual colonization of barren rocks. Secondary succession occurs after a disturbance has destroyed a community but left the soil intact (**37.12**).

Invasive species have been introduced to non-native habitats by human actions and have established themselves at the expense of native communities. Because they have no natural enemies, population growth is rapid (**37.13**).

Ecosystem Structure and Dynamics (37.14–37.23)

An ecosystem includes a community and the abiotic factors with which it interacts. Ecosystem ecologists study energy flow and chemical cycling (**37.14**).

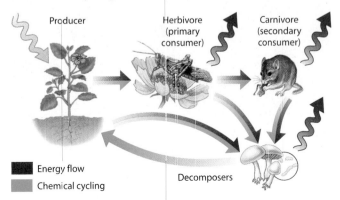

Producer · Herbivore (primary consumer) · Carnivore (secondary consumer) · Decomposers

■ Energy flow
■ Chemical cycling

Primary production is the rate at which producers convert sunlight to chemical energy in organic matter (biomass). A pyramid of production shows the flow of energy from producers to primary consumers and to higher trophic levels. Only about 10% of the energy stored at each trophic level is available to the next level. Thus, a field of corn can support many more human vegetarians than meat-eaters (**37.15–37.17**).

Biogeochemical cycles. Nutrients recycle between organic matter and abiotic reservoirs (**37.18**). Carbon is taken from the atmosphere by photosynthesis, used to make organic molecules, and returned to the atmosphere by cellular respiration (**37.19**). Phosphorus and other soil minerals are recycled locally; they are stored long term in rocks (**37.20**). Various bacteria in soil (and root nodules of some plants) convert gaseous N_2 to compounds that plants can use: ammonium (NH_4^+) and nitrate (NO_3^-). Decomposers recycle nitrogen to plants (**37.21**).

Ecosystem studies show that drastic alterations, such as the total removal of vegetation, can increase the runoff of water and loss of soil nutrients. Environmental changes caused by humans, such as acid rain, can unbalance nutrient cycling over the long term (**37.22**). Nutrient runoff from agricultural lands and large livestock operations may cause excessive algal growth. This cultural eutrophication reduces species diversity and harms water quality (**37.23**).

Connecting the Concepts

1. Fill in this concept map summarizing ecosystem dynamics.

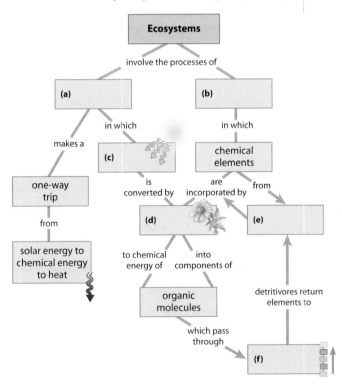

2. Fill in the blanks in the table below summarizing the interspecific interactions in a community.

Interspecific Interaction	Effect on species 1	Effect on species 2	Example
	+	−	
	−	−	
	+	−	
	+	−	
	+	+	

Testing Your Knowledge
Multiple Choice

3. Which of the following groups is absolutely essential to the functioning of an ecosystem?
   a. producers
   b. producers and herbivores
   c. producers, herbivores, and carnivores
   d. detritivores
   e. producers and decomposers

4. The open ocean and tropical rain forests contribute the most to Earth's net primary production because
   a. both have high rates of net primary production.
   b. both cover huge surface areas of Earth.
   c. nutrients cycle fastest in these two ecosystems.
   d. the ocean covers a huge surface area and the tropical rain forest has a high rate of production.
   e. Both a and b are correct.

5. Which of the following organisms is mismatched with its trophic level?
   a. algae—producer
   b. fungi—decomposer
   c. phytoplankton—primary consumer
   d. carnivorous fish larvae—secondary consumer
   e. eagle—tertiary or quaternary consumer

6. Which of the following best illustrates ecological succession?
   a. A mouse eats seeds, and an owl eats the mouse.
   b. Decomposition in soil releases nitrogen that plants can use.
   c. Grasses grow in a deserted field, followed by shrubs and then trees.
   d. Imported pheasants increase in numbers, while local quail disappear.
   e. Overgrazing causes a loss of nutrients from soil.

7. To ensure adequate nitrogen for a crop, a farmer would want to *decrease* _____ by soil bacteria.
   a. cellular respiration
   b. nitrification
   c. denitrification
   d. nitrogen fixation
   e. b and d

Describing, Comparing, and Explaining

8. Explain how seed dispersal by animals is an example of mutualism in some cases.

9. What is cultural eutrophication? What steps might be taken to slow this process?

10. Local conditions, such as heavy rainfall or the removal of plants, may limit the amount of nitrogen, phosphorus, or calcium available, but the amount of carbon available in an ecosystem is seldom a problem. Explain.

11. In Southeast Asia, there's an old saying: "There is only one tiger to a hill." In terms of energy flow in ecosystems, explain why big predatory animals such as tigers and sharks are relatively rare.

12. For which chemicals are biogeochemical cycles global? Explain.

13. What roles do bacteria play in the nitrogen cycle?

Applying the Concepts

14. An ecologist studying plants in the desert performed the following experiment. She staked out two identical plots, which included a few sagebrush plants and numerous small, annual wildflowers. She found the same five wildflower species in roughly equal numbers on both plots. She then enclosed one of the plots with a fence to keep out kangaroo rats, the most common grain-eaters of the area. After two years, to her surprise, four of the wildflower species were no longer present in the fenced plot, but one species had increased dramatically. The control plot had not changed. Using the principles of ecology, propose a hypothesis to explain her results. What additional evidence would support your hypothesis?

15. Sometime in 1986, near Detroit, a freighter pumped out water ballast containing larvae of European zebra mussels. The molluscs multiplied wildly, spreading through Lake Erie and entering Lake Ontario. In some places, they have become so numerous that they have blocked the intake pipes of power plants and water treatment plants, fouled boat hulls, and sunk buoys. What makes this kind of population explosion occur? What might happen to native organisms that suddenly must share the Great Lakes ecosystem with zebra mussels? How would you suggest trying to solve the mussel population problem?

Answers to all questions can be found in Appendix 4.

For Practice Quizzes, BioFlix, MP3 Tutors, and Activities, go to www.mybiology.com.

Saving the Tiger

Sniffing the air as it searches for prey, a massive orange and black cat prowls a forest in Myanmar, the Southeast Asian country formerly known as Burma. The tiger is free to hunt and roam in this dark valley, a lowland stretch of trees where people are still relatively few. But just because humans don't live near this forest doesn't mean the tiger goes unnoticed. As the big cat glides toward a large tree, a hidden camera senses its body heat and snaps a photo. And so another endangered tiger has been tracked—part of the effort to protect tigers in Myanmar and keep them from joining the legions of species sliding toward extinction around the world.

Tigers (*Panthera tigris*) once roamed across Asia from Turkey to the Russian Far East. A hundred years ago, scientists estimate, about 100,000 tigers could be found in the wild. Now that number has plummeted as low as 5,000. Three of the world's eight tiger subspecies have disappeared entirely.

But the big cats persist in 14 Asian countries today, and some of these nations, such as Myanmar, are making an effort to protect this majestic part of their natural heritage. For Myanmar, that has meant focusing on the Hukawng Valley, an isolated forest not very hospitable to people but an ideal habitat for tigers. The cats ruled here for thousands of years, feeding on deer and other wildlife without much interference from humans.

The modern era shook that serenity. Better roads brought loggers and miners who destroyed parts of the forest and ruined habitat for tigers and their prey. Human arrival also brought the tiger's greatest nemesis—hunting. While such killing may be illegal, a dead tiger still brings big money. Tiger skins, heads, and claws are prized as trophies, and the cats' bones and internal organs are considered potent ingredients for some traditional Asian medicines. In recent years, tiger bones have fetched as much as $200 per kilogram on the black market—a tempting price in a country as poor as Myanmar. Between people hoping to make a living off the forest and people hoping to make a living off tiger parts, the big cats' future looked grim.

In 2001, officials in Myanmar decided to protect tigers by turning 6,500 square kilometers of the Hukawng Valley into a reserve with strict prohibitions against mining, logging, and hunting. International wildlife biologists, working with the Myanmar Forest Department, helped determine how many tigers remained. They set up remote cameras to locate tigers, and the results weren't encouraging. Using photos to identify individual cats by their stripes, the scientists determined that only about 100 tigers remained.

Myanmar responded by giving the cats even more protection. In 2004, government officials tripled the size of the reserve, expanding its size to more than 20,000 square kilometers, nearly as large as the state of Vermont. Plans are under way to provide training and jobs to people living in the area, either making them part of the tiger protection efforts or giving them other alternatives to hunting tigers. Scientists estimate that if properly protected, the tiger population in the newly expanded reserve could grow about tenfold.

The effort to save Myanmar's tigers is part of a worldwide struggle to preserve **biodiversity,** the variety of living things. We are now in the midst of an alarming **biodiversity crisis,** a rapid decrease in Earth's great diversity of organisms. Biology, the science of life, is a critical part of modern attempts to reverse this trend and conserve biodiversity.

In this final chapter, it is fitting to discuss **conservation biology,** a goal-oriented science that seeks to counter the biodiversity crisis. Conservation biologists may focus on the preservation of a single species, such as the Myanmar tigers. Or they may reach more broadly, trying to protect many species at once by preserving habitats and ecosystems. Some biologists are even working to reverse habitat loss by restoring ecosystems that have been altered and polluted by human activities. In this chapter, we examine some of the major factors behind the biodiversity crisis. We then look at some of the strategies biologists are using to try to slow the rate of species loss. As we look at the fight to save our biological heritage, we will see that conservation biology touches all levels of ecology, from a single tiger to the forest it roams. ■ ■ ■

Hukawng Valley, Myanmar

38.1 Biodiversity is a vital resource that is being lost

In many ways, the plight of the tiger described in the chapter introduction illustrates the effects of modern human culture worldwide. Throughout the biosphere, human activities are altering trophic structure, energy flow, chemical cycling, and natural disturbances—ecosystem processes on which we and other species depend. By some estimates, we are doing more damage to the biosphere and pushing more species toward extinction than the changes that triggered the mass extinctions of dinosaurs about 65 million years ago. To date, scientists have described and formally named about 1.8 million species. Some biologists believe that the total number of species is about 10 million, but others estimate it to be as high as 200 million. Because we do not know the number of species currently in existence, we cannot determine the actual rate of species loss, but it may be as much as 1,000 times higher than at any time in the past 100,000 years. Several researchers estimate that at the current rate of destruction, over half of all currently living plant and animal species will be gone by the end of this century.

Why should we care about the loss of biodiversity? Perhaps the purest reason is what Harvard biologist E. O. Wilson calls *biophilia,* our sense of connection to nature and other forms of life. Many people also share a moral belief that other species have an inherent right to life. But in addition to ethical and aesthetic reasons for preserving biodiversity, there are practical ones as well. We depend on other species for food, clothing, shelter, oxygen, soil fertility—the list goes on and on. In the United States, 25% of all prescriptions dispensed from pharmacies contain substances derived from

plants. For instance, drugs called vincristine and vinblastine, which come from the rosy periwinkle (Figure 38.1), are effective against Hodgkin's disease and certain other forms of cancer. Rosy periwinkle is a flowering plant native to Madagascar, an island in the Indian Ocean east of Africa. Separated from the mainland for more than 150 million years, Madagascar is home to some 8,000 species of flowering plants, 80% of which occur nowhere else in the world. With an estimated 200,000 species of plants and animals, Madagascar is among the top five most biologically diverse countries in the world. Unfortunately, most of Madagascar's species are in serious trouble. In the 2,000 years that humans have lived on the island, Madagascar has lost 80% of its forests and about 50% of its native species.

When species are lost, so are their unique genes. The enormous genetic diversity of all the organisms on Earth has great potential benefit. Consider PCR (see Module 12.12), the DNA-replicating technology based on an enzyme extracted from thermophilic prokaryotes. Currently, biotechnology companies are searching for other commercially useful enzymes in the prokaryotes found in the numerous hot springs in Yellowstone National Park. Many researchers and biotechnology leaders are enthusiastic about the potential that such "bioprospecting" holds for future development of new medicines, industrial chemicals, and other products.

We must also be concerned about the ecological changes that underlie the biodiversity crisis because humans, too, depend on Earth's ecosystems for survival. To help policymakers appreciate the value of the biosphere's life-sustaining features, ecologists point out the "services" provided by ecosystems, such as the contribution of wetlands to reducing the severity of floods or the control of agricultural pests by their natural enemies. Other ecosystem services include purification of air and water, decomposition of wastes, pollination of crops, nutrient cycling, and protection from UV rays, to name just a few. Some scientists have attempted to assign an economic value to these benefits, arriving at an average annual value of ecosystem services at 33 trillion U.S. dollars. In contrast, the global gross national product is currently about $18 trillion. Although rough, these estimates make the important point that we cannot afford to take ecosystems for granted.

In the next several modules, we describe the extent of the biodiversity crisis and explore some of the factors responsible for it.

Web Activity *Connection: Madagascar and the Biodiversity Crisis*

? What are two reasons to be concerned about the impact of the biodiversity crisis on human welfare?

■ The environmental degradation threatening other species may also harm us. We are dependent on biodiversity, both directly through use of organisms and their products and indirectly through ecosystem services.

Figure 38.1 The rosy periwinkle (*Catharanthus roseus*), a source of anticancer drugs

Biodiversity includes genetic, species, and ecosystem diversity

The biodiversity crisis encompasses more than just the fate of individual species. Biodiversity has three levels: genetic diversity, species diversity, and ecosystem diversity. The *genetic diversity* within and between populations of a species is the raw material that makes microevolution and adaptation to the environment possible. If local populations are lost and the total number of individuals of a species declines, so, too, do the genetic resources for that species. As you learned in Module 13.11, a severe reduction in genetic variation threatens the survival of a species.

Much of the discussion about the biodiversity crisis in government and in the press centers on *species diversity*—the variety of species in an ecosystem or throughout the biosphere. The U.S. Endangered Species Act (ESA) defines an **endangered species** as one that is "in danger of extinction throughout all or a significant portion of its range." Also defined for protection by the ESA, **threatened species** are those that are likely to become endangered in the foreseeable future. Here are just a few examples that illustrate the extent of the problem:

About 12% of the 9,934 known bird species and 20% of the 5,416 known mammalian species in the world are threatened with extinction.

About 20% of the known freshwater fishes in the world have either become extinct during historical times or are seriously threatened.

About 32% of all known amphibian species are either near extinction or endangered.

Of the approximately 20,000 known plant species in the United States, 200 species have become extinct since dependable records have been kept, and 730 species are endangered or threatened.

Extinctions may be local; for example, a species of cichlid may be lost from Lake Victoria but survive in an adjacent lake.

Global extinction means that the species has been lost from all the ecosystems in which it lived.

Ecosystem diversity is the third component of biological diversity. Because of the network of community interactions among populations of different species within an ecosystem, the local extinction of one species, especially a keystone species, can have a negative impact on the overall species richness of the ecosystem (see Module 37.11). And each ecosystem has characteristic patterns of energy flow and chemical cycling that can affect the whole biosphere, a critical factor in understanding global warming, as we'll discuss later.

Figure 38.2A shows an all-too-common example of ecosystem damage and loss—the destruction of tropical forests to make room for and support an expanding human population. The coral reef shown in Figure 38.2B is another ecosystem known for its species richness and productivity. An estimated 20% of the world's coral reefs have been destroyed by human activities, and 24% are in imminent danger of collapse. Scientists predict that another 26% of coral reefs will succumb in the next few decades if they are not protected.

? Why is it too narrow to define the biodiversity crisis as a loss of species?

■ In addition to species loss, the biodiversity crisis includes the loss of genetic diversity within populations and species and the degradation of entire ecosystems.

Figure 38.2A Virgin rain forest in Brazil, with areas recently cleared for farming

Figure 38.2B A coral reef, one of the most diverse, and one of the most endangered, ecosystems

Habitat destruction, invasive species, and overexploitation are major threats to biodiversity

Human alteration of habitats poses the single greatest threat to biodiversity throughout the biosphere. Agriculture, urban development, forestry, mining, and environmental pollution have brought about massive destruction and fragmentation of habitats. Deforestation has been occurring at an alarming rate in tropical and coniferous forests.

The amount of human-altered land surface is approaching 50%, and we use over half of all accessible surface fresh water. Some of the most productive aquatic habitats in estuaries and intertidal wetlands are also prime locations for commercial and residential development. The loss of marine habitat is severe, especially in coastal areas and coral reefs. According to the International Union for the Conservation of Nature and Natural Resources, which compiles information on the conservation status of species worldwide, habitat destruction is implicated in 73% of the species in modern history that have become extinct, endangered, vulnerable, or rare.

Ranking second behind habitat loss as a cause of the biodiversity crisis are invasive species, which disrupt communities by competing with, preying on, or parasitizing native species. In Modules 19.18 and 37.13, we described a few of the hundreds of incidents in which species introduced by humans to non-native habitats caused havoc. The lack of interspecific interactions (see Module 37.2) that keep the newcomer populations in check is often a key factor in a non-native species becoming invasive. On the other side of the coin, newly arrived species are an unfamiliar biotic factor in the environment of native species. Natives are especially vulnerable when a new species poses an unprecedented threat. In the absence of an evolutionary history with predators, for example, animals may lack defense mechanisms, or even a fundamental recognition of danger, as illustrated by the invasion of Guam by the brown tree snake.

The Pacific island of Guam was home to 13 species of forest birds when the brown tree snake (Figure 38.3A) arrived as a stowaway on a cargo plane. With no competitors, predators, or parasites to hinder it, the snake proliferated rapidly on a diet of unwary birds. Four of the native species of birds became locally extinct, although they survive on nearby islands. Three species of birds that lived nowhere else but Guam are now globally extinct. As the populations of two other species of birds became perilously low, officials took the remaining individuals into protective custody; they now exist only in zoos. The brown tree snake also eliminated species of seabirds and lizards.

The third major threat to biodiversity is overexploitation of wildlife by harvesting at rates that exceed the ability of populations to rebound. Such overharvesting has threatened some rare trees that produce valuable wood. Animal species whose numbers have been drastically reduced by excessive commercial harvest, poaching, or sport hunting include tigers, whales, the American bison, Galápagos tortoises, and numerous fishes. In parts of Africa, Asia, and South America, wild animals are heavily hunted for food, and the African term *bushmeat* is now used to refer generally to such meat. As once-impenetrable forests are opened to exploitation, the commercial bushmeat trade has become one of the greatest threats to primates, including gorillas, chimpanzees, and many species of monkeys, as well as other mammals and birds (Figure 38.3B). No longer hunted only for local use, large quantities of bushmeat are sold at urban markets or exported worldwide, including to the United States. An expanding, often illegal world trade in wildlife products (such as rhinoceros horns, elephant tusks, and grizzly bear gallbladders) also threatens many species.

Aquatic species are suffering overexploitation, too, and are further threatened by pollution, habitat loss, and climate change. Many edible marine fish and seafood species are in a precarious state. North Atlantic cod fisheries have already collapsed, and catches of other commercially desirable species in that region are dwindling. Worldwide, fishing fleets are working farther offshore and harvesting fish from greater depths in order to obtain hauls comparable to those of previous decades.

Figure 38.3B Results of bushmeat hunt (baboon and antelope) in the Democratic Republic of Congo

Web Activity *Connection: Fire Ants: An Introduced Species*

? How might interspecific interactions prevent an introduced species from becoming invasive?

■ If an introduced species is subjected to a natural control such as a predator or pathogen in its new environment, its population will be kept in check, and it will probably not spread far from the site of introduction.

Figure 38.3A Brown tree snake (*Boiga irregularis*)

38.4 Pollution of the environment compounds our impact on other species

Another way in which the human population contributes to the biodiversity crisis is through the release of pollutants, which can have local, regional, and global effects. Some pollutants, such as oil spills, contaminate local areas. Recall from Module 34.17 that the global water cycle can transport pollutants—for instance, pesticides used on land—from terrestrial to aquatic ecosystems hundreds of miles away. Pollutants that are emitted into the atmosphere, such as sulfur and nitrogen oxides from the burning of fossil fuels, may be carried aloft for thousands of miles before descending to earth in the form of acid precipitation (see Module 2.16).

As you learned in Module 7.14, the **ozone layer** in the upper atmosphere protects Earth from the harmful ultraviolet rays in sunlight. Beginning in the mid-1970s, scientists realized that the ozone layer was gradually thinning. Evidence implicated chemicals called chlorofluorocarbons (CFCs), which were used as refrigerants, as propellants in aerosol cans, and in certain manufacturing processes, in ozone depletion. The consequences of ozone depletion for life on Earth may be quite severe, not only increasing skin cancers and cataracts among humans, but harming crops and natural communities, especially the phytoplankton that are responsible for a large proportion of Earth's primary production. In 1996, an international agreement was reached to phase out the production of CFCs. Recently, the rate of ozone depletion has begun to slow, but complete ozone recovery is probably decades away.

In Module 37.23, we discussed the effects of nutrient pollution, which "fertilizes" algae and cyanobacteria in aquatic ecosystems. In a far-reaching example of this problem, nitrogen runoff from Midwestern farm fields has been linked to an annual summer "dead zone" in the Gulf of Mexico. Vast algal blooms extend outward from where the Mississippi River deposits its nutrient-laden waters. As the algae die, decomposition of the huge quantities of biomass depletes the water of oxygen over an area that ranges from 13,000 to 21,000 square km (roughly 5,000 to 8,000 square miles). Oxygen starvation disrupts benthic communities, displacing fish and invertebrates that can move and killing the sessile organisms left behind. More than 150 recurring and permanent coastal dead zones have been documented in seas worldwide.

As we discussed in Module 34.2, chemical pesticides have helped us grow more food and fight insect-borne diseases, but they cause environmental problems, too. In addition to being transported to areas far from where they are applied, many of these toxins become concentrated as they pass through the food chain. This concentration, or **biological magnification,** occurs because the biomass at any given trophic level is produced from a much larger toxin-containing biomass ingested from the level below (see Module 37.16). Thus, top-level predators are usually the organisms most severely damaged by toxic compounds in the environment. In the Great Lakes food chain shown in Figure 38.4, the concentration of industrial chemicals called PCBs measured in the eggs of herring gulls, a top-level

consumer, was almost 5,000 times higher than that measured in phytoplankton. The concentration increased at each successive trophic level. Current research implicates many of the chlorinated hydrocarbons such as DDT and PCBs as endocrine disruptors in a large number of animal species (see Module 35.16).

Many of the synthetic chemicals produced by human activities cannot be degraded by microorganisms and consequently persist in the environment. In other cases, chemicals may be converted to more toxic products by reaction with other substances or by the metabolism of microorganisms. For example, mercury, a by-product of plastic production and coal-fired power generation, was once routinely expelled into rivers and the sea. Bacteria in the bottom mud converted the waste to methyl mercury, an extremely toxic soluble compound that then accumulated in the tissues of organisms, including humans who consumed fish from the contaminated waters.

Web Activity Connection: DDT and the Environment

? How does the use of nitrogen fertilizers lead to the development of coastal dead zones?

■ Excess nutrients produce algal blooms. As these algae die, vast numbers of bacteria use up the O_2 from the bottom water as they decompose the algae.

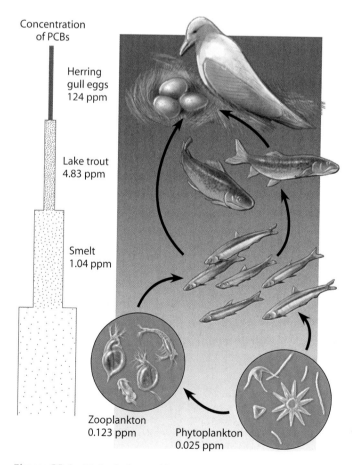

Figure 38.4 Biological magnification of PCBs in a food web

Concentration of PCBs

Herring gull eggs 124 ppm

Lake trout 4.83 ppm

Smelt 1.04 ppm

Zooplankton 0.123 ppm

Phytoplankton 0.025 ppm

38.5 Rapid warming is changing the global climate

The scientific debate about global warming is over. The vast majority of scientists now agree that rising concentrations of greenhouse gases (see Module 7.13) such as carbon dioxide (CO_2), methane (CH_4), and nitrous oxide (N_2O) in the atmosphere are changing global climate patterns. This was the overarching conclusion of the assessment report released by the Intergovernmental Panel on Climate Change (IPCC) in 2007. Thousands of scientists and policymakers from more than 100 countries participated in producing the report, which is based on data published in hundreds of scientific papers.

The signature effect of increasing greenhouse gases is the steady increase in the average global temperature, which has risen 0.8°C over the last 100 years, with 0.6°C of that increase occurring over the last three decades. Further increases of 2 to 4.5°C are likely by the end of the 21st century, depending on the rate of future greenhouse gas emissions. Ocean temperatures are also rising, in deeper layers as well as at the surface. But the temperature increases are not distributed evenly around the globe. Warming is greater over land than sea, and the largest increases are in the northernmost regions of the Northern Hemisphere. In Figure 38.5A, dark red areas indicate the greatest temperature increases. In northern Alaska and Canada, for example, the temperature has risen 1.4°C just since 1961. The only continent that has not yet shown overall warming (on average) is Antarctica. Some of the consequences of the global warming trend are already clear from rising temperatures, ununusual precipitation patterns, and melting ice.

Many of the world's glaciers are receding rapidly. Himalayan glaciers are retreating at an average rate of 15 m (50 ft) a year, while glaciers are projected to disappear entirely from Glacier National Park in northwest Montana by 2030. For example, almost all of the Grinell Glacier is now a meltwater lake. (Figure 38.5B).

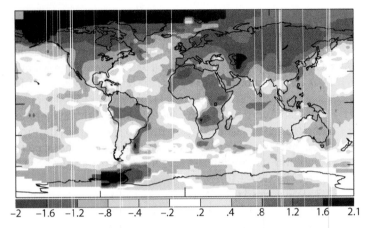

Figure 38.5A   2000–2005 temperature deviations relative to 1951–1980 (°C)

Permanent Arctic sea ice is shrinking; each summer brings increased melting and thinner ice. The massive ice sheets of Greenland and Antarctica are thinning and collapsing. If melting were to accelerate, rising sea levels would cause catastrophic flooding of coastal areas worldwide. The permafrost that characterizes the tundra biome is melting.

Warm temperatures are beginning earlier each year. Cold days and nights and frosts have become less frequent; hot days and nights have become more frequent. Deadly heat waves are increasing in frequency and duration.

Precipitation patterns are changing, bringing longer and more intense drought to some areas. In other regions, a greater proportion of the total precipitation is falling in torrential downpours that cause flooding. Hurricane intensity is increasing, fueled by higher sea surface temperatures.

Many of these changes will have a profound impact on biodiversity, as we explore in Modules 38.7 and 38.8. In the next module, we examine the cause of global warming.

🎧 MP3 Tutor *Global Warming*

❓ From the map in Figure 38.5A, which biomes are likely to be most affected by global warming, and why?

■ The high-latitude biomes of the Northern Hemisphere, tundra and taiga, will be most affected. Those biomes are experiencing the greatest temperature change. Also, the organisms that live there are adapted to cold weather and a short growing season, so their survival is on the line.

Figure 38.5B   Grinell Glacier in Glacier National Park, 1938 (left), 1981 (center), and 2005 (right)

38.6   Human activities are responsible for rising concentrations of greenhouse gases

Without its blanket of natural greenhouse gases such as CO_2 and water vapor to trap heat, the Earth would be too cold to support life. It is the increase in the thickness of the blanket that is making the Earth uncomfortably warm, and that increase is occurring rapidly. For 650,000 years, the atmospheric concentration of CO_2 did not exceed 300 parts per million (ppm); the pre-Industrial concentration was 280 ppm. Today, atmospheric CO_2 is approximately 385 ppm. The levels of methane (CH_4) and nitrous oxide (N_2O), which also trap heat in the atmosphere, have increased dramatically, too (Figure 38.6A). CO_2 and N_2O are released when fossil fuels—oil, coal, and natural gas—are burned. N_2O is also released when nitrogen fertilizers are used in agriculture. Livestock and landfills are among the factors responsible for increases of atmospheric CH_4. The consensus of scientists, reported by the Intergovernmental Panel on Climate Change, is that rising concentrations of greenhouse gases—and thus, global warming—are the result of human activities.

Let's take a closer look at CO_2, the dominant greenhouse gas. Recall from Module 37.19 that atmospheric CO_2 is a major reservoir for carbon. (CH_4 is also part of that reservoir.) CO_2 is removed from the atmosphere by the process of photosynthesis and stored in organic molecules such as carbohydrates (Figure 38.6B). Thus biomass, the organic molecules in an ecosystem, is a biotic carbon reservoir. The carbon-containing molecules in living organisms may be used in the process of cellular respiration, which releases carbon in the form of CO_2. Nonliving biomass may be decomposed by microorganisms or fungi that also release CO_2. Overall, uptake of CO_2 by photosynthesis roughly equals the release of CO_2 by cellular respiration. CO_2 is also exchanged between the atmosphere and the surface waters of the oceans.

Fossil fuels consist of biomass that was buried under sediments without being completely decomposed (see Module 17.6). The burning of fossil fuels and wood, which is also an organic material, can be thought of as a rapid form of decomposition. While cellular respiration releases energy from organic molecules slowly and harnesses it to make ATP, combustion liberates the energy rapidly as heat and light. In both processes, the carbon atoms that make up the organic fuel are released in CO_2.

The CO_2 flooding into the atmosphere from combustion of fossil fuels may be absorbed by photosynthetic organisms and incorporated into biomass. But deforestation has significantly decreased the number of CO_2 molecules that can be accommodated by this pathway. Alternatively, CO_2 could be absorbed into the ocean. For decades, the oceans have been absorbing considerably more CO_2 than they have released, and will continue to do so, but the excess CO_2 is beginning to affect ocean chemistry. When CO_2 dissolves in water, it becomes carbonic acid (see Module 22.11). Recently, measurable decreases in ocean pH have begun to raise concern among biologists. Organisms that construct shells or exoskeletons out of calcium carbonate ($CaCO_3$), including corals and many plankton, are most likely to be affected as decreasing pH reduces the concentration of the carbonate ions.

Greenhouse gas emissions are accelerating. From 2000 to 2005, global CO_2 emissions increased four times faster than in the preceding 10-year span. At this rate, further climate change is inevitable.

Web Activity *The Greenhouse Effect*

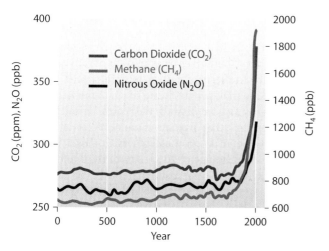

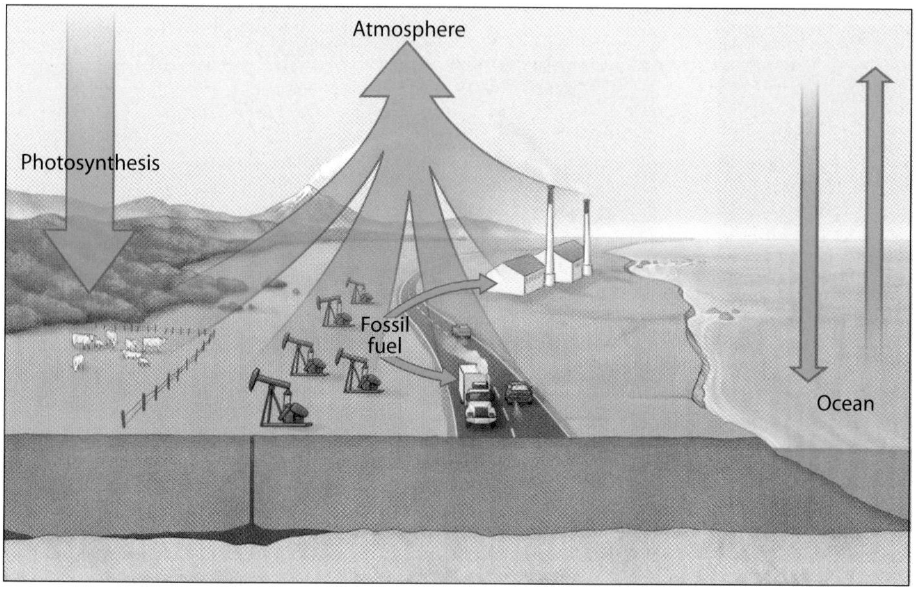

Figure 38.6A   Atmospheric concentrations of CO_2, N_2O (*y*-axis on left) and CH_4 (*y*-axis on right)

?   The amount of CO_2 you are responsible for releasing every year is called your *carbon footprint*. Search for an online calculator that estimates your carbon footprint. What are the primary sources of the CO_2 you generate?

Figure 38.6B   Carbon cycling

■ Transportation and home energy use are the two major categories used in the footprint.

Global climate change affects biomes, ecosystems, communities, and populations

The distribution of terrestrial biomes, which is primarily determined by temperature and rainfall (see Module 34.8), is changing as a consequence of global warming. Melting permafrost is shifting the boundary of the tundra as shrubs and conifers are able to stretch their ranges into the previously frozen ground. Prolonged droughts will increasingly extend the boundaries of deserts. Great expanses of the Amazonian tropical rain forest will gradually become savanna as increased temperatures dry out the soil.

The combined effects of climate change on components of forest ecosystems in western North America have spawned catastrophic wildfire seasons (Figure 38.7A). In these mountainous regions, spring snowmelt releases water into streams that sustain forest moisture levels over the summer dry season. With the earlier arrival of spring, snowmelt begins earlier and dwindles away before the dry season ends. As a result, the fire season has been getting longer since the 1980s. In addition, drought conditions have made trees more vulnerable to insect and pathogen attack; vast numbers of dead trees add fuel to the flames. Fires burn longer, and the number of acres burned has increased dramatically. As dry conditions persist and snowpacks diminish, the problem will worsen.

The earlier arrival of warm temperatures in the spring is disturbing ecological communities. In many plants and animals, spring events are triggered by warm temperatures. With temperatures rising earlier in the year, a variety of species, including birds and frogs, have begun their breeding season earlier. Satellite images show earlier greening of the landscape, and flowering occurs sooner. For other species, day length is the environmental cue that spring has arrived. Because global climate change affects temperature but not day length, interactions between species may become out of sync. For example, plants may bloom before pollinators have emerged, or eggs may hatch before a dependable food source is available. Because the magnitude of seasonal shifts increases from the tropics to the poles, migratory birds may also experience timing mismatches. For instance, birds arriving in the Arctic to breed may find that the period of peak food availability has already passed.

Warming oceans threaten tropical coral reef communities. When stressed by high temperatures, corals expel their symbiotic algae in a phenomenon called bleaching (see Module 37.4). Corals can recover if temperatures return to normal, but they cannot survive prolonged temperature increases. When corals die, the community is overrun by large algae, and species diversity plummets.

The distributions of populations and species are also changing in response to climate change. Recall from Module 34.4 that the distribution of a species may be determined by its adaptations for the abiotic conditions in its environment. With rising temperatures, the ranges of many species have already shifted toward the poles or to higher elevations. For example, researchers in Europe and the United States have reported that the ranges of more than two dozen species of butterflies have moved north by as much as 150 miles. Shifts in the ranges of many bird species have also been reported; the Inuit peoples living north of the Arctic Circle have sighted birds such as robins in the region for the first time.

However, species that live on mountaintops or in polar regions have nowhere to go. Researchers in Costa Rica have reported the disappearance of 20 species of frogs and toads as warmer Pacific Ocean temperatures reduce the dry-season mists in their mountain habitats. Polar bears (Figure 38.7B), which stalk their prey on ice and need to store up body fat for the warmer months, are showing signs of starvation as their hunting grounds melt away. The disappearance of sea ice in the Antarctic is blamed for recent decreases in populations of Emperor and Adelie penguins.

Figure 38.7A   Firefighter battles one of several wildfires that scorched half a million acres of Southern California in October, 2007

Figure 38.7B   Polar bear (*Ursus maritimus*) crossing pack ice in Canada

Global climate change has been a boon to some organisms, but so far the beneficiaries have been species that have a negative impact on humans. For example, in mountainous regions of Africa, Southeast Asia, and Central and South America, the range of mosquitoes that carry diseases such as malaria, yellow fever, and dengue are restricted to lower elevations by frost. With warming temperatures and fewer days of frost, these mosquitoes—and the diseases they carry—are appearing at higher elevations. Longer summers in western North America have enabled bark beetles to complete their life cycle in one year instead of two, promoting devastating beetle outbreaks that have destroyed

millions of acres of conifers. Undesirable plants such as poison ivy and kudzu have also benefited from rising temperatures.

Environmental change has always been a part of life; in fact, it is a key ingredient of evolutionary change. In the next module, we consider the evidence of evolutionary adaptation to global warming.

 How might timing mismatches caused by climate change affect an individual's reproductive fitness?

■ Any factor that reduces the number of offspring an organism produces may affect fitness. Examples: flowers emerging too soon or too late for pollinators; birds that arrive too late in the season to find food for offspring

Evolution Connection

38.8 Global climate change is an agent of natural selection

In the previous module, we described several ways in which organisms have responded to global climate change. For the most part, those examples can be attributed to **phenotypic plasticity,** the ability to change phenotype in response to local environmental conditions. Differences resulting from phenotypic plasticity are within the normal range of expression for an individual's genotype. Phenotypic plasticity allows organisms to cope with short-term environmental changes. On the other hand, phenotypic plasticity is itself a trait that has a genetic basis and can evolve. Researchers studying the effects of climate change on populations have detected microevolutionary changes in phenotypic plasticity.

A common bird in Europe, the great tit (Figure 38.8A), is the third link of a food chain that has been altered by climate change. As warm temperatures arrive earlier in the spring, tree

leaves emerge earlier and caterpillars, which use the swelling buds and unfolding leaves as their food source, hatch sooner. The reproductive success of great tits depends on having an ample supply of these nutritious caterpillars to feed their offspring. Like many other birds, great tits have some phenotypic plasticity in the timing of their breeding, which helps them synchronize their reproduction

Figure 38.8A Great tit (*Parus major*)

with the availability of caterpillars. The range and degree of plasticity vary among great tits, and this variation has a genetic basis. Researchers have found evidence of directional selection (see Module 13.13) favoring individuals that have the greatest phenotypic plasticity and lay their eggs earlier, when the abundance of food gives their offspring a better chance of survival.

In another example, scientists studied reproduction in a population of red squirrels (Figure 38.8B) in the Yukon Territory of Canada, where spring temperatures have increased by approximately 2°C in the last three decades. These researchers also found earlier breeding times in the spring. Over a period of 10 years, the date on which female squirrels gave birth advanced by 18 days, a change of about 6 days per generation. Using statistical analysis, the scientists determined that phenotypic

plasticity was responsible for most of the shift in breeding times. However, a small but significant portion of the change (roughly 15%) could be attributed to microevolution, directional selection for earlier breeding. The researchers hypothesize that red squirrels born earlier in the year are larger and more capable of gathering and storing food in the autumn and thus have a better chance of successful reproduction the following spring.

From the scant evidence available at this time, it appears that some populations, especially those with high genetic variability and short life spans, may avoid extinction by means of evolutionary adaptation. In addition to the studies on phenotypic plasticity in great tits and red squirrels, researchers have also documented microevolutionary changes in traits such as dispersal ability and timing of life cycle events in insect populations. However, evolutionary adaptation is unlikely to save long-lived species such as polar bears and penguins that are experiencing rapid habitat loss. The rate of climate change is incredibly fast compared to major climate shifts in evolutionary history (see Module 17.6), and if it continues on its present course, thousands of species—the Intergovernmental Panel on Climate Change (IPCC) estimates that as many as 30% of plants and animals—will go extinct.

 How does a short generation time hasten the process of evolutionary adaptation?

■ Each generation has the potential for testing new phenotypes in the environment, allowing natural selection to proceed rapidly.

Figure 38.8B Red squirrel (*Tamiasciurus hudsonicus*) eating the seeds from a spruce cone

38.9 Protecting endangered populations is one goal of conservation biology

In order to stem the loss of biodiversity, some conservation biologists focus on protecting populations of endangered or threatened species. Because such populations face different types of threats and may be in different stages of decline, biologists tailor their strategies to fit the specific circumstances.

One of the most harmful effects of habitat loss is population fragmentation, the splitting and consequent isolation of portions of populations. The aerial photograph in Figure 38.9A illustrates fragmentation of a coniferous forest ecosystem in the Mount Hood National Forest in northwestern Oregon. The forest was originally continuous. The open, snow-covered areas in the photo were logged, carving the forest into smaller, discontinuous patches with reduced populations. The northern spotted owl, which inhabits coniferous forests of the U.S. Pacific Northwest (Figure 38.9B), was a victim of this type of habitat alteration. Owl populations declined markedly when these forests were logged.

After factors such as habitat loss have taken their toll on population size, the reduced population may enter a downward spiral toward smaller and smaller population size, leading to extinction. The key factors driving this downward spiral are inbreeding and genetic drift. Both factors reduce genetic variation, which is a prerequisite for adaptive evolutionary responses to environmental change (see Module 13.9). To halt the extinction process, some conservation measures focus on maintaining sufficient habitat to support a minimum viable population size. When possible, individuals from other populations are imported to increase the genetic variation of small, isolated populations.

A more proactive conservation strategy attempts to stop declines in population sizes before they begin spiraling downward toward extinction. This approach requires that researchers carefully dissect the causes of a decline before trying corrective measures. Once corrective measures are taken, the population's recovery is monitored until the decline is stopped.

The case of the endangered red-cockaded woodpecker (Figure 38.9C) provides an example of the proactive approach to conservation. The red-cockaded woodpecker requires mature pine forests, preferably ones dominated by the longleaf pine. Most woodpeckers nest in dead trees, but the red-cockaded woodpecker drills its nest holes in mature, living pine trees. Originally found throughout the southeastern United States, the numbers of red-cockaded woodpeckers declined as a result of the destruction or fragmentation of suitable habitats by logging and agriculture. Low growth of plants among the mature pine trees is another critical factor for this bird. Historically, periodic fires swept through longleaf pine forests, keeping the undergrowth low. Fire suppression in recent times, however, has altered the forest composition. Breeding birds tend to abandon nests when vegetation among the pines is thick and higher than about 4.5 m. Apparently, the birds require a clear flight path between their home trees and the neighboring feeding grounds. The recent recovery of the red-cockaded woodpecker from near extinction to sustainable populations is largely due to recognizing and providing its key habitat requirements.

Because we will not be able to save every endangered species, we must determine which are most important for conserving biodiversity as a whole. For instance, identifying and protecting keystone species (see Module 37.11) may help preserve whole communities. And in many situations, conservation must look beyond individual species to communities.

? How did scientists engineer the recovery of the red-cockaded woodpecker?

■ Researchers identified and then provided the key habitat requirements of mature pine forests with low understory.

Figure 38.9A Fragmentation of a forest ecosystem in Mount Hood National Forest, western United States

Figure 38.9B The northern spotted owl (*Strix occidentalis caurina*)

Figure 38.9C The red-cockaded woodpecker (*Picoides borealis*)

Sustaining ecosystems and landscapes is a conservation priority

Most conservation efforts in the past have focused on saving individual species. But increasingly, conservation biology aims to sustain the biodiversity of entire ecosystems and landscapes. Ecologically, a landscape is a regional assemblage of interacting ecosystems, such as a forest, adjacent fields, wetlands, streams, and streamside habitats (see Module 34.1). **Landscape ecology** is the application of ecological principles to the study of the structure and dynamics of a collection of ecosystems. One goal of landscape ecology is to study human land-use patterns in the past, present, and foreseeable future and to make biodiversity conservation a priority.

Edges, or boundaries, between ecosystems are prominent features of landscapes. The photograph in Figure 38.10A shows a landscape area in Yellowstone National Park that includes grassland and forest. Human activities, such as logging and road building, often create edges that are more abrupt than those delineating natural landscapes.

Edges have their own sets of physical conditions and thus their own communities of organisms. Some organisms thrive in edges because they require resources of the two adjacent areas. For instance, whitetail deer browse on woody shrubs found in edge areas between woods and fields, and their populations often expand when forests are logged or interrupted with housing developments.

Communities where human activities have generated many edges often have less diversity and are dominated by a few species that are adapted to edges. In one example, populations of the brown-headed cowbird (Figure 38.10B), an edge-adapted species that lays its eggs in the nests of other birds, are currently expanding in many areas of North America. Cowbirds forage in open fields on insects disturbed by or attracted to cattle and other large herbivores; the cowbirds also need forests, where they can parasitize the nests of other birds. Increasing cowbird parasitism and loss of habitats are correlated with declining populations of several songbird species.

Where habitats have been severely fragmented, a **movement corridor,** a narrow strip or series of small clumps of high-quality habitat connecting otherwise isolated patches, can be a deciding factor in conserving biodiversity. Streamside habitats often serve as corridors, and government policy in some nations prohibits destruction of these areas. In areas of heavy human use, artificial corridors are sometimes constructed. In many areas, bridges or tunnels have reduced the number of animals killed as they try to cross highways (Figure 38.10C).

Corridors can also promote dispersal and reduce inbreeding in declining populations. Corridors are especially important to species that migrate between different habitats seasonally. In some European countries, amphibian tunnels have been constructed to help frogs, toads, and salamanders cross roads to access their breeding territories. On the other hand, a corridor can be harmful—as, for example, in the spread of diseases, especially among small subpopulations in closely situated habitat patches. The effects of movement corridors between habitats in a landscape are not yet well understood, and researchers continue to study them.

Figure 38.10A   A landscape in Yellowstone National Park with distinct edges

Figure 38.10B   A male brown-headed cowbird (*Molothrus ater*)

Figure 38.10C   Animal bridge in Banff National Park, Canada

? How can "living on the edge" be a good thing for some species, such as whitetail deer and cowbirds?

■ Such animals use a combination of resources from the two ecosystems on either side of the edge.

38.11 Protected areas are established to slow the loss of biodiversity

Conservation biologists are applying their understanding of population, community, ecosystem, and landscape dynamics in establishing parks, wilderness areas, and other legally protected nature reserves. Choosing locations for protection often focuses on **biodiversity hot spots.** These relatively small areas have a large number of endangered and threatened species and an exceptional concentration of **endemic species,** those that are found nowhere else. Together, the "hottest" of Earth's biodiversity hot spots, shown in Figure 38.11A, total less than 1.5% of Earth's land but are home to a third of all species of plants and vertebrates. For example, all lemurs are endemic to Madagascar, which is home to more than 50 species. In fact, almost all of the mammals, reptiles, amphibians, and plants that inhabit Madagascar are endemic. There are also hot spots in aquatic ecosystems, such as certain river systems and coral reefs.

Because endemic species are limited to specific areas, they are highly sensitive to habitat degradation. At the current rate of human development, some biologists estimate that loss of habitat will cause the extinction of about half of the species in terrestrial biodiversity hot spots in the next 10 to 15 years. Thus, biodiversity hot spots can also be hot spots of extinction. They rank high on the list of areas demanding strong global conservation efforts.

Concentrations of species provide an opportunity to protect many species in very limited areas. However, the "hot spot" designation tends to favor the most noticeable organisms, especially vertebrates and plants. Invertebrates and microorganisms are often overlooked. Furthermore, species endangerment is a truly global problem, and it is important that a focus on hot spots not detract from efforts to conserve habitats and species diversity in other areas.

Migratory species pose a special problem for conservationists. For example, monarch butterflies occupy much of the United States and Canada during the summer months, but migrate in the autumn to specific local sites in Mexico and California, where they congregate in huge numbers. Overwintering populations are particularly susceptible to habitat disturbances because they are concentrated in small areas. Thus, habitat preservation must extend across all of the sites that monarchs inhabit in order to protect them. The situation is similar for many species of migratory songbirds, waterfowl, marine mammals, and sea turtles.

Sea turtles, such as the loggerhead turtle (Figure 38.11B), are threatened both in their ocean feeding grounds and on land. Loggerheads take about 20 years to reach sexual maturity, and great numbers of juveniles and adults are drowned at sea when caught in fishing nets. The adults mate at sea, and the females migrate to specific sites on sandy beaches to lay their eggs. Buried in shallow depressions, the eggs are susceptible to predators, especially raccoons. And many egg-laying sites have become housing developments and beachside resorts. An ongoing international effort to conserve sea turtles focuses on protecting egg-laying sites and minimizing the death rate of adults and juveniles at sea.

Currently, governments have set aside about 7% of the world's land in various forms of reserves. One major conservation question is whether it is better to create one large reserve or a group of smaller ones. Far-ranging animals with low-density populations, such as tigers, require extensive habitats. As conservation biologists learn more about the requirements for achieving minimum viable population sizes for endangered species, it is becoming clear that most national parks and other reserves are far too small. Given political and economic realities, it is unlikely that many existing parks will be enlarged, and most new reserves will also be too small. In the next module, we look at one approach to this problem, called a zoned reserve.

? What is a biodiversity hot spot?

■ A relatively small area with a disproportionate number of endangered and threatened species, many of which are endemic

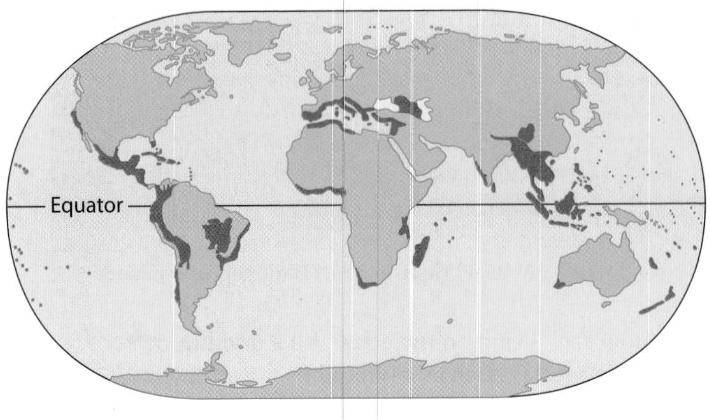

Figure 38.11A Earth's terrestrial biodiversity hot spots (purple)

Figure 38.11B An adult loggerhead turtle (*Caretta caretta*) swimming in the Caribbean Sea

Equator

38.12 Zoned reserves are an attempt to reverse ecosystem disruption

Today, few, if any, ecosystems remain unaltered by human activities. Accelerated eutrophication (see Module 37.23) reduces species diversity in lakes and rivers because many organisms cannot tolerate the rapid changes in water quality. Large tracts of forest in taiga biomes are still being clear-cut in the United States, Canada, and Siberia to satisfy demands for lumber and urban development. The current rate of conversion of tropical forests to farmland threatens the survival of thousands of species and contributes to global climate change.

In an attempt to slow the disruption of ecosystems, a number of countries are setting up what they call zoned reserves. A **zoned reserve** is an extensive region of land that includes one or more areas undisturbed by humans. The lands surrounding these areas continue to be used to support the human population, but they are protected from extensive alteration. As a result, they serve as a buffer zone, or shield, against further intrusion into the undisturbed areas. A primary goal of the zoned reserve approach is to develop a social and economic climate in the buffer zone that is compatible with the long-term viability of the protected area.

The small Central American nation of Costa Rica has become a world leader in establishing zoned reserves. In exchange for a reduction in its international debt, the Costa Rican government established eight zoned reserves, called "conservation areas," outlined in black in Figure 38.12A. The green areas on the map are designated national parklands, which remain relatively unchanged by human activity; the yellow buffer zones are privately owned areas where people live and work.

Zoned reserves contribute to **sustainable development,** the long-term prosperity of human societies and the ecosystems that support them. Costa Rica is making progress in managing its reserves so that the buffer zones provide a steady, lasting supply of forest products, water, and hydroelectric power and also support sustainable agriculture. Costa Rica looks to its zoned reserve system to maintain at least 80% of its native species and to make this rich resource accessible to tourists and to its own citizens (Figure 38.12B). An important goal is providing a stable economic base for people living there. Destructive practices that are not compatible with long-term ecosystem stability and from which there is often little local profit are gradually being discouraged. Such destructive practices include massive logging, large-scale single-crop agriculture, and extensive mining.

A recent analysis showed mixed results for Costa Rica's system of zoned reserves. The good news was that negligible deforestation occurred within and just beyond protected parkland boundaries. However, some deforestation has occurred in the buffer zones, with plantations of cash crops such as banana and palm replacing the natural vegetation. Conservationists fear that continuation of these practices will isolate protected areas, restricting gene flow and decreasing species and genetic diversity.

Establishment of the zoned reserve system in Costa Rica required leadership by the national government as well as partnerships between government, nongovernmental organizations, and private citizens. How has the focus on conservation affected the Costa Rican people? Living conditions have improved in Costa Rica over the past half century; the literacy rate is near 100%. Thus, we can infer that Costa Rica's conservation initiatives have not compromised human welfare. Nevertheless, maintaining a commitment to conservation in the face of a growing population will be a major challenge. Although Costa Rica is in the middle of a rapid demographic transition (see Module 36.9), the population is predicted to grow from its current 4 million until it levels off at 6 million by the middle of this century. As always, a growing human population increases the demand on resources.

? In zoned reserves, regulations prevent large-scale alterations of habitat in the buffer zones but do support sustainable development for the people living there. Why?

■ Large-scale disruptions could impact the nearby undisturbed areas, and preservation is a realistic goal only if it is compatible with an acceptable standard of living for the local people.

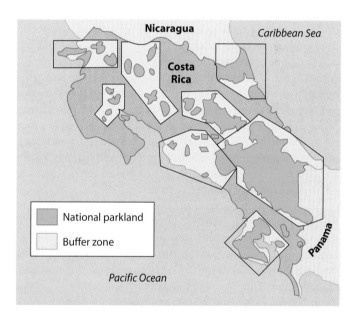

Figure 38.12A   Zoned reserves in Costa Rica

Figure 38.12B   Local schoolchildren marvel at the diversity of life in one of Costa Rica's reserves

38.13 The Yellowstone to Yukon Conservation Initiative seeks to preserve biodiversity by connecting protected areas

If many existing reserves are too small to sustain a large number of threatened species, how can biologists draw the land around reserves into conservation efforts? In North America, one ambitious biodiversity plan is creating innovative ways to give wild creatures more room. The plan builds on lessons from research on a howling predator that once roamed a vast stretch of the northern Rocky Mountains.

This predator was a single individual, a gray wolf dubbed Pluie that was captured by scientists in western Canada in 1991. The biologists fitted the five-year-old female with a radio tracking collar and released her—routine work in studies of threatened animals. But the scientists were stunned by what they learned from Pluie. Over the next two years, this wolf roamed over an area more than 100,000 km^2 (38,600 mi^2) in size, crossing between Canada and the United States and traveling between protected reserves and lands where she was fair game for killing. In 1995, her story came to a bloody end. While moving through lands outside the boundary of a nearby national park, Pluie, her mate, and one of her pups were shot (legally) by a hunter.

Biologists who had studied Pluie realized that the wolf's life captured all the promise—and all the pitfalls—of efforts to protect her. She had thrived for years within the sporadic shelter of parks and other protected territory. But such lands were never big enough to hold her. Like others of her species (*Canis lupus*), Pluie needed more room. Reserves could shield animals briefly, the scientists realized. True protection would have to include paths of safe passage between reserves.

This conclusion inspired the creation of the Yellowstone to Yukon Conservation Initiative (Y2Y), one of the world's most ambitious conservation biology efforts. The initiative aims at nothing less than preserving the web of life that has long defined the Rocky Mountains of Canada and the northern United States. This area is dotted with famous parks—Canada's Banff National Park and Wyoming's Yellowstone National Park among them—but scientists behind Y2Y now say that those areas alone cannot protect native species from human threats.

Y2Y seeks to knit together a string of parks and reserves, creating a vast 3,200-km wildlife corridor stretching down from Alaska across Canada to northern Wyoming (Figure 38.13A). The idea is not to create one giant park, but rather to connect parks with protected corridors where wildlife can travel safely.

Many of the signature species that live in this vast region, such as grizzly bears (Figure 38.13B, on facing page), lynx, moose, and elk, don't confine themselves to human boundaries. But few have as great a range as the wolf. If Y2Y can provide safe passage for gray wolves, it will have also created secure zones for other animals in the Rockies.

Gray wolves (Figure 38.13C, on facing page) once roamed all of North America. These carnivorous hunters live in packs that protect pups and search cooperatively for food. A pack may have a territory of about 130 km^2 or range much farther to find prey. The wolf's hunting prowess kept it the top predator of

North American ecosystems as long as the human population was small. Things changed when large numbers of people migrated from Europe and pushed into the continent.

Deeming wolves a dangerous predator and competitor that threatened people and livestock, settlers in the United States launched widespread campaigns to wipe out wolves. By the early 20th century, gray wolves were nearly extinct in the lower 48 states, with only a few hundred surviving in northern Minnesota. More managed to stay alive in the wilds of less populated western Canada and Alaska.

Scientists gradually realized that widespread damage rippled through habitats after wolves had been removed. Without a predator to control their numbers, populations of elk and deer grew unchecked. As these increasing numbers of herbivores

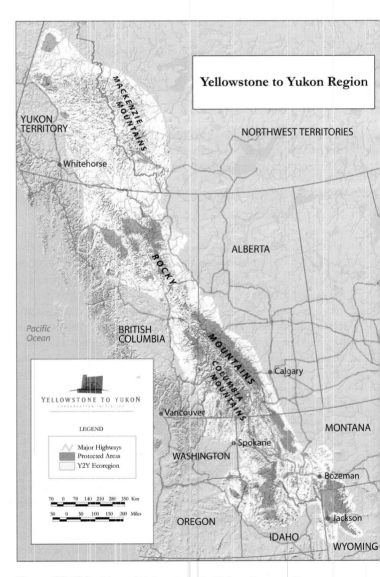

Figure 38.13A Map of Yellowstone to Yukon Conservation Initiative with protected areas shown in green

Figure 38.13B   Grizzly bear with cubs in Yellowstone Park

Figure 38.13C   Gray wolf

foraged for food, vegetation that sheltered smaller animals was damaged. Other animals, such as ravens and foxes, had once fed on the carcasses of wildlife killed by wolves and were now left without an important source of food. Wolves, biologists determined, were a keystone species—a species critical to the balance and maintenance of an ecosystem.

That understanding led to one of the most important and controversial conservation biology efforts in the Yellowstone to Yukon area. In 1991, the U.S. Fish and Wildlife Service launched a campaign to bring wolves back to Yellowstone National Park, a reserve that hadn't sheltered the animals in at least 50 years. After careful planning, which included compensation for ranchers who feared losing their cattle and sheep, about 60 wolves from Canada were released in the park in 1995 and 1996.

As wolf howls once again echoed through the Wyoming darkness, the wolf quickly became a hopeful symbol for Y2Y backers. Yellowstone's wolves formed new packs and raised pups. By 2004, scientists counted 12 wolf packs inside the park, totaling about 300 wolves. And the wolf's return has brought more than howling. Park officials noted significant environmental improvements as wolves once again roamed Yellowstone. As wolves killed elk, moose, and deer, streambeds and other lands near waterways started to shelter a greater variety of plants and animals. Fewer hoofed animals meant more grasses and taller trees. Those plants brought more birds, along with more water-dwelling beaver. In all, park biologists report that the wolf's return has affected at least 25 different species.

True to their nature, Yellowstone's wolves haven't followed human borders; six packs have been found just outside the park. Meanwhile, the migrations of Canadian wolves, along with smaller release programs, have brought the animals back to Idaho and Montana. In June 2004, a Yellowstone wolf was found hundreds of kilometers away in Colorado—a reminder that travels like Pluie's are common to wolves throughout the Y2Y region.

Such successes also bring risks—and reminders from scientists that wolves need safe corridors. The Colorado wolf

was discovered dead by the side of a highway, most likely the victim of a car. Its appearance sparked angry protests from ranchers in the state, who said that any new wolves that appear should either be shot or shipped back to Yellowstone. Meanwhile, wildlife advocates maintain that wolves should be allowed to migrate naturally.

That argument reflects a broader debate about how to treat wolves as they return to their old ranges. Federal laws have protected them in the northern United States, but as wolves thrive, federal officials are preparing to remove them from the Endangered Species list. They asked states where wolves are found to submit management plans, but delayed removing federal protection because some of the proposed plans seemed designed to kill off wolves all over again. As of late 2007, however, only Wyoming's plan had not been approved.

The biologists involved in the Yellowstone to Yukon Conservation Initiative are studying wildlife population dynamics on a landscape scale to help support regional conservation planning. The initiative is backing a range of research projects to determine the requirements for maintaining terrestrial and aquatic ecosystems in the Y2Y region. Their efforts to connect habitats include wildlife bridges, such as the one in Banff National Park (Figure 38.10C). This cross-border initiative is also providing broader lessons for global conservation. In Myanmar, for example, plans for the tiger reserve discussed in the chapter introduction include proposals to link the Myanmar reserve with others in neighboring countries.

In addition to creating reserves to protect species and their habitats from human disruptions, conservation efforts also attempt to restore ecosystems degraded by human activities. We look at the field of restoration ecology next.

> **?** In what ways are wolves considered a keystone species?

> ■ Wolves regulate the populations of their prey, preventing these herbivores from damaging vegetation and degrading habitat for other members of the community. Wolf kills also provide food for other animals.

38.14 The study of how to restore degraded habitats is a developing science

For centuries, humans have altered and degraded natural areas without considering the consequences. But as people have gradually come to realize the severity of some of the consequences of ecosystem alteration, they have sought ways to return degraded areas to their natural state. The new and expanding field of **restoration ecology** uses ecological principles to develop methods of achieving this goal.

One of the major strategies in restoration ecology is bioremediation, the use of living organisms to detoxify polluted ecosystems. For example, in Module 16.10, you learned that bacteria have been used to clean up oil spills and old mining sites. Bacteria are also employed to metabolize toxins in dump sites. As you read in the introduction to Chapter 32, plants have successfully extracted potentially toxic metals such as zinc, nickel, lead, and cadmium from contaminated soil. Researchers are working with trees and even lichens to clean up soil polluted with uranium.

Some restoration projects have the broader goal of returning ecosystems to their natural state, which may involve replanting vegetation, fencing out non-native animals, or removing dams that restrict water flow. Hundreds of restoration projects are currently underway in the United States. One of the most ambitious endeavors is the Kissimmee River project in south central Florida.

The Kissimmee River was once a meandering shallow river that wound its way through diverse wetlands from Lake Kissimmee southward into Lake Okeechobee (Figure 38.14A). Periodic flooding of the river covered a wide floodplain during about half of the year, creating wetlands that provided critical habitat for vast numbers of birds, fishes, and invertebrates. As often happens, however, people saw the floodplain as wasted land that could be developed if the flooding were controlled. Between 1962 and 1971, the U.S. Army Corps of Engineers converted the 166-km wandering river into a straight canal 9 m deep, 100 m wide, and 90 km long. This project drained approximately 31,000 acres of wetlands, with significant negative impacts on fish and wetland bird populations. Spawning and foraging habitats for fishes were eliminated, and important sport fishes, such as largemouth bass, were replaced by nongame species more tolerant of the lower O_2 concentration in the deeper canal. The populations of waterfowl declined by 92%, and the number of bald eagle nesting territories decreased by 70%. Without the marshes to help filter and reduce agricultural runoff, phosphorus and other excess nutrients were transported through Lake Okeechobee into the Everglades ecosystem to the south.

As these negative ecological effects began to be recognized, public pressure to restore the river grew. In 1992, Congress authorized the Kissimmee River Restoration Project, one of the largest landscape restoration projects and ecological experiments in the world. As Figure 38.14A shows, the plan involves removing water control structures such as dams, reservoirs, and channel modifications and filling in about

35 km of the canal. The first phase of the project was completed in 2004, with the second phase slated to begin in 2010. In Figure 38.14B, the natural curves of the river are a pleasing contrast to the artificial linearity of the backfilled canal. Birds and other wildlife have returned in unexpected

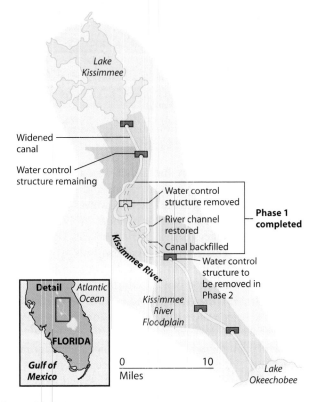

Figure 38.14A Kissimmee River restoration project

Figure 38.14B Restoring the natural water flow patterns of the Kissimmee River

numbers to the 11,000 acres of wetlands that have been restored. The marshes are filled with native vegetation, and game fishes again swim in the river channels. However, 2006 was the driest season on record in south central Florida. As the drought continued, the southward flow of the Kissimmee River stopped. Lake Okeechobee, which depends on the Kissimmee basin for more than half its water supply, hit record lows. The rapidly worsening water shortage in south Florida has renewed attention to the urgent need to complete an even more ambitious project, the restoration of the Everglades.

Web Process of Science *How Are Potential Restoration Sites Analyzed?*

? How will the Kissimmee River Restoration Project improve water quality in the Everglades ecosystem?

■ The wetlands filter agricultural runoff and prevent excess nutrients from entering the Everglades.

38.15 Sustainable development is an ultimate goal

In numbers, geographic range, and capacity to alter the biosphere, our species is clearly one of the most successful ones ever to inhabit planet Earth. Yet, we seem to have set ourselves and the rest of the biosphere on a precarious path into the future. Facing increasing degradation of ecosystems, fragmentation of habitats, and loss of biodiversity, how can we best manage Earth's resources?

We must understand the complex interconnections of the biosphere in order to make sensible decisions about how to conserve these networks. To this end, many nations, scientific societies, and private foundations have embraced the concept of sustainable development. The Ecological Society of America, the world's largest organization of ecologists, endorses a research agenda called the Sustainable Biosphere Initiative. The goal of this initiative is to acquire the basic ecological information necessary for the intelligent and responsible development, management, and conservation of Earth's resources. The research agenda includes ways to sustain the productivity of natural and artificial ecosystems and studies of the relationship between biological diversity, global climate change, and ecological processes.

Sustainable development does not depend only on continued research and application of ecological knowledge. It also requires us to connect the life sciences with the social sciences, economics, and humanities. Conservation and restoration of biodiversity is only one side of sustainable development; the other key facet is improving the human condition. Public education and the political commitment and cooperation of nations, especially the United States, are essential to the success of this endeavor.

The marvelous image of the leopard on this book's cover and in Figure 38.15 serves as a reminder of what we stand to lose if we fail to recognize and solve the ecological crises at hand. Flourishing in habitats as dissimilar as rainforests and deserts, the leopard owes its success to its flexible diet. It is the most widespread of Africa's big cats and continues to thrive in areas where lions and cheetahs have disappeared. Yet, the leopard's future is not guaranteed. Populations in North Africa are already vulnerable to local extinction. The relentless growth and expansion of the human population in sub-Saharan Africa continues to encroach on leopard habitats. Humans compete with leopards for prey animals and hunt leopards as livestock predators or a danger to humans.

An awareness of our unique ability to alter the biosphere and jeopardize the existence of other species, as well as our own,

Figure 38.15 The African leopard (*Panthera pardus*)

may help us choose a path toward a sustainable future. The goal is a world in which each generation inherits an adequate supply of natural and economic resources and a relatively stable environment. Although the current state of the biosphere is grim, the situation is far from hopeless. Now is the time to aggressively pursue more knowledge about life and to work toward long-term sustainability.

Biology is the scientific expression of the human desire to know nature. We are most likely to save what we appreciate, and we are most likely to appreciate what we understand. By learning about the processes and diversity of life, we also become more aware of ourselves and our place in the biosphere.

Web Activity *Conservation Biology Review*

? Why is a concern for the well-being of future generations essential for progress toward sustainable development?

■ Sustainable development is a long-term goal—longer than a human lifetime. Preoccupation with the here and now is an obstacle to sustainable development because it discourages behavior that benefits future generations.

Reviewing the Concepts

The Biodiversity Crisis (Introduction–38.8)

Conservation biology is a goal-driven science that seeks to counter the biodiversity crisis, the rapid decrease in the number of species on Earth (**Introduction**).

Biodiversity, while valuable for its own sake, also provides food, fiber, medicines, and ecosystem services (**38.1**). Biodiversity includes genetic diversity, within and between populations; species diversity; and ecosystem diversity (**38.2**). Human activities threaten diversity at all levels (**38.3**).

Threats to biodiversity

Habitat destruction

Invasive species

Overexploitation

Another threat is pollution, which harms the environment. Effects include acid rain, ozone depletion, and dead zones. Chemical pesticides may be concentrated by biological magnification (**38.4**).

Global warming is perhaps the greatest threat to biodiversity. Increased global temperature caused by rising concentrations of greenhouse gases is changing climatic patterns, with grave consequences. The increases in greenhouse gases are the result of burning fossil fuels. Many organisms and ecosystems are currently experiencing the effects of global warming. Phenotypic plasticity has minimized the impact on some species, and a few cases of microevolutionary changes have been observed. However, the rapidity of the environmental changes makes it unlikely that evolutionary processes will save many species from extinction (**38.5–38.8**).

Conservation Biology and Restoration Ecology (38.9–38.15)

Protecting populations is one goal of conservation biology. Habitat degradation often breaks populations into small fragments that are vulnerable to extinction. Conservationists attempt to identify and sustain the minimum viable population size and preserve genetic variation. In other cases, conservationists are able to identify and treat the causes of a population's decline. Preserving critical habitat may help endangered species recover (**38.9**).

Landscape structure and biodiversity. Conservation efforts are increasingly aimed at sustaining ecosystems and landscapes. Edges between ecosystems have distinct sets of features and species. The increased frequency and abruptness of edges caused by human activities can increase species loss. Movement corridors connecting isolated habitats may be helpful to fragmented populations (**38.10**).

Nature reserves. Biodiversity hot spots have large concentrations of endemic species (**38.11**). **Zoned reserves** are undisturbed wildlands surrounded by buffer zones of compatible economic development. Costa Rica has established many zoned reserves (**38.12**). The Yellowstone to Yukon Conservation

Initiative is an international research and conservation effort that seeks to connect reserves and protect species and ecosystems (**38.13**).

Restoration ecology uses ecological principles to return degraded areas to their natural state, a process that may include detoxifying polluted ecosystems, replanting native vegetation, and returning waterways to their natural course. Large-scale restoration projects attempt to restore damaged landscapes. The Kissimmee River Restoration Project is restoring river flow and wetlands, thus improving wildlife habitat (**38.14**).

Sustainable development seeks to improve the human condition while conserving biodiversity. It depends on increasing and applying ecological knowledge as well as valuing our linkages to the biosphere (**38.15**).

Connecting the Concepts

1. Complete the following map, which organizes some of the key concepts of conservation biology.

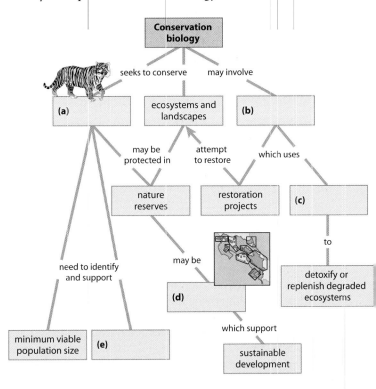

Testing Your Knowledge

Multiple Choice

2. Which of these statements best describes what conservation biologists mean by the "biodiversity crisis"?
   a. Introduced species, such as starlings and zebra mussels, have rapidly expanded their ranges.
   b. Harvests of marine fishes, such as cod and bluefin tuna, are declining.
   c. The current species extinction rate is as much as 1,000 times greater than at any time in the last 100,000 years.
   d. Many potential medicines are being lost as plant species become extinct.
   e. The number of hot spots worldwide is rapidly declining.

3. Which of the following poses the single greatest threat to biodiversity?
   a. introduced species
   b. overhunting
   c. movement corridors
   d. habitat loss
   e. global warming

4. Which of the following is characteristic of endemic species?
   a. They are often found in biodiversity hot spots.
   b. They are distributed widely in the biosphere.
   c. They require edges between ecosystems.
   d. Their trophic position makes them very susceptible to the effects of biological magnification.
   e. They are often keystone species whose presence helps to structure a community.

5. Ospreys and other top predators are most severely affected by pesticides such as DDT because they
   a. are especially sensitive to chemicals.
   b. have rapid reproductive rates.
   c. have very long life spans.
   d. store the pesticides in their tissues.
   e. consume prey in which pesticides are concentrated.

6. Movement corridors are
   a. the routes taken by migratory animals.
   b. strips or clumps of habitat that connect isolated fragments of habitat.
   c. landscapes that include several different ecosystems.
   d. edges or boundaries between ecosystems.
   e. buffer zones that promote the long-term viability of protected areas.

7. With limited resources, conservation biologists need to prioritize their efforts. Of the following choices, which should receive the greatest attention for the goal of conserving biodiversity?
   a. the northern spotted owl
   b. a commercially important species
   c. threatened and endangered vertebrate species
   d. a declining keystone species in a community
   e. all endangered species

8. Which of the following statements about protected areas is *not* correct?
   a. We now protect 25% of the land areas of the planet.
   b. National parks are only one type of protected area.
   c. Most reserves are smaller in size than the ranges of some of the species they are meant to protect.
   d. Management of protected areas must coordinate with the management of lands outside the protected zone.
   e. Biodiversity hot spots are important areas to protect.

Describing, Comparing, and Explaining

9. What are the three levels of biological diversity? Explain how human activities threaten each of these levels.

10. What is the so-called greenhouse effect? How is it important to life on Earth?

11. What are the causes and possible consequences of global warming? Why is international cooperation necessary if we are to solve this problem?

Applying the Concepts

12. Biologists in the United States are concerned that populations of many migratory songbirds are declining. Evidence suggests that some of these birds might be victims of pesticides. Most of the pesticides implicated in songbird mortality have not been used in the United States since the 1970s. Suggest a hypothesis to explain the current decline in songbird numbers.

13. You may have heard that human activities cause the extinction of one species every hour. Such estimates vary widely because we do not know how many species exist or how fast their habitats are being destroyed. You can make your own estimate of the rate of extinction. Start with the number of species that have been identified. To keep things simple, ignore extinction in the temperate latitudes and focus on the 80% of plants and animals that live in the tropical rain forest. Assume that destruction of the forest continues at a rate of 1% per year, so the forest will be gone in 100 years. Assume (optimistically) that half the rain forest species will survive in preserves, forest remnants, and zoos. How many species will disappear in the next century? How many species is that per year? Per day? Recent studies of the rain forest canopy have led some experts to predict that there may be as many as 30 million species on Earth. How does starting with this figure change your estimates?

14. The price of energy does not reflect its real costs. What kinds of hidden environmental costs are not reflected in the price of fossil fuels? How are these costs paid, and by whom? Do you think these costs could or should be figured into the price of oil? How might that be done?

15. Research your country's per capita carbon emissions. Compare your carbon footprint with the average for your country. (See the question at the end of Module 38.6.) How can individuals reduce the carbon emissions for which they are directly responsible? Make a list of actions that you are willing to take to reduce your carbon footprint.

16. Until recently, response to environmental problems has been fragmented—an antipollution law here, incentives for recycling there. Meanwhile, the problems of the gap between the rich and poor nations, diminishing resources, and pollution continue to grow. Now people and governments are starting to envision a sustainable society. The Worldwatch Institute, a respected environmental monitoring organization, estimates that we must reach sustainability by the year 2030 to avoid economic and environmental disaster. To get there, we must begin shaping a sustainable society during this decade. In what ways is our present system not sustainable? What might a sustainable society be like? Do you think a sustainable society is an achievable goal? Why or why not? What is the alternative? What might we do to work toward sustainability? What are the major roadblocks to achieving sustainability? How would your life be different in a sustainable society?

Answers to all questions can be found in Appendix 4.

For Practice Quizzes, BioFlix, MP3 Tutors, and Activities, go to www.mybiology.com.

Metric Conversion Table

Measurement	Unit and Abbreviation	Metric Equivalent	Approximate Metric-to-English Conversion Factor	Approximate English-to-Metric Conversion Factor
Length	1 kilometer (km)	= 1000 (10^3) meters	1 km = 0.6 mile	1 mile = 1.6 km
	1 meter (m)	= 100 (10^2) centimeters	1 m = 1.1 yards	1 yard = 0.9 m
		= 1000 millimeters	1 m = 3.3 feet	1 foot = 0.3 m
			1 m = 39.4 inches	
	1 centimeter (cm)	= 0.01 (10^{-2}) meter	1 cm = 0.4 inch	1 foot = 30.5 cm
				1 inch = 2.5 cm
	1 millimeter (mm)	= 0.001 (10^{-3}) meter	1 mm = 0.04 inch	
	1 micrometer (μm)	= 10^{-6} meter (10^{-3} mm)		
	1 nanometer (nm)	= 10^{-9} meter (10^{-3} μm)		
	1 angstrom (Å)	= 10^{-10} meter (10^{-4} μm)		
Area	1 hectare (ha)	= 10,000 square meters	1 ha = 2.5 acres	1 acre = 0.4 ha
	1 square meter (m^2)	= 10,000 square centimeters	1 m^2 = 1.2 square yards	1 square yard = 0.8 m^2
			1 m^2 = 10.8 square feet	1 square foot = 0.09 m^2
	1 square centimeter (cm^2)	= 100 square millimeters	1 cm^2 = 0.16 square inch	1 square inch = 6.5 cm^2
Mass	1 metric ton (t)	= 1000 kilograms	1 t = 1.1 tons	1 ton = 0.91 t
	1 kilogram (kg)	= 1000 grams	1 kg = 2.2 pounds	1 pound = 0.45 kg
	1 gram (g)	= 1000 milligrams	1 g = 0.04 ounce	1 ounce = 28.35 g
			1 g = 15.4 grains	
	1 milligram (mg)	= 10^{-3} gram	1 mg = 0.02 grain	
	1 microgram (mg)	= 10^{-6} gram		
Volume (Solids)	1 cubic meter (m^3)	= 1,000,000 cubic centimeters	1 m^3 = 1.3 cubic yards	1 cubic yard = 0.8 m^3
			1 m^3 = 35.3 cubic feet	1 cubic feet = 0.03 m^3
	1 cubic centimeter (cm^3 or cc)	= 10^{-6} cubic meter	1 cm^3 = 0.06 cubic inch	1 cubic inch = 16.4 cm^3
	1 cubic millimeter (mm^3)	= 10^{-9} cubic meter (10^{-3} cubic centimeter)		
Volume (Liquids and Gases)	1 kililiter (kL or kl)	= 1000 liters	1 kL = 264.2 gallons	1 gallon = 3.79 L
	1 liter (L)	= 1000 milliliters	1 L = 0.26 gallon	1 quart = 0.95 L
			1 L = 1.06 quarts	
	1 milliliter (mL or ml)	= 10^{-3} liter	1 mL = 0.03 fluid ounce	1 quart = 946 mL
		= 1 cubic centimeter	1 mL = approx. $\frac{1}{4}$ teaspoon	1 pint = 473 mL
			1 mL = approx. 15–16 drops	1 fluid ounce = 29.6 mL
				1 teaspoon = approx. 5 mL
Volume (Liquids and Gases)	1 microliter (ml or mL)	= 10^{-6} liter (10^{-3} milliliters)		
Time	1 second (s)	= $\frac{1}{60}$ minute		
	1 millisecond (ms)	= 10^{-3} second		
Temperature	Degrees Celsius (°C)		°F = $\frac{9}{5}$ °C − 32	°C = $\frac{5}{9}$(°F − 32)

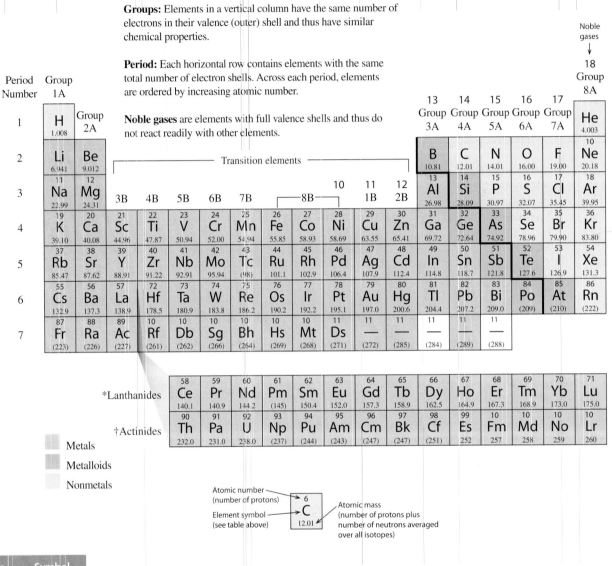

Groups: Elements in a vertical column have the same number of electrons in their valence (outer) shell and thus have similar chemical properties.

Period: Each horizontal row contains elements with the same total number of electron shells. Across each period, elements are ordered by increasing atomic number.

Noble gases are elements with full valence shells and thus do not react readily with other elements.

Transition elements

Metals
Metalloids
Nonmetals

*Lanthanides
†Actinides

Atomic number (number of protons) → 6
Element symbol (see table above) → C
12.01
Atomic mass (number of protons plus number of neutrons averaged over all isotopes)

Name	Symbol
Actinium	Ac
Aluminum	Al
Americium	Am
Antimony	Sb
Argon	Ar
Arsenic	As
Astatine	At
Barium	Ba
Berkelium	Bk
Beryllium	Be
Bismuth	Bi
Bohrium	Bh
Boron	B
Bromine	Br
Cadmium	Cd
Calcium	Ca
Californium	Cf
Carbon	C
Cerium	Ce
Cesium	Cs
Chlorine	Cl
Chromium	Cr
Cobalt	Co
Copper	Cu
Curium	Cm
Darmstadtium	Ds
Dubnium	Db
Dysprosium	Dy
Einsteinium	Es
Erbium	Er
Europium	Eu
Fermium	Fm
Fluorine	F
Francium	Fr
Gadolinium	Gd
Gallium	Ga
Germanium	Ge
Gold	Au
Hafnium	Hf
Hassium	Hs
Helium	He
Holmium	Ho
Hydrogen	H
Indium	In
Iodine	I
Iridium	Ir
Iron	Fe
Krypton	Kr
Lanthanum	La
Lawrencium	Lr
Lead	Pb
Lithium	Li
Lutetium	Lu
Magnesium	Mg
Manganese	Mn
Meitnerium	Mt
Mendelevium	Md
Mercury	Hg
Molybdenum	Mo
Neodymium	Nd
Neon	Ne
Neptunium	Np
Nickel	Ni
Niobium	Nb
Nitrogen	N
Nobelium	No
Osmium	Os
Oxygen	O
Palladium	Pd
Phosphorous	P
Platinum	Pt
Plutonium	Pu
Polonium	Po
Potassium	K
Praseodymium	Pr
Promethium	Pm
Protactinium	Pa
Radium	Ra
Radon	Rn
Rhenium	Re
Rhodium	Rh
Rubidium	Rb
Ruthenium	Ru
Rutherfordium	Rf
Samarium	Sm
Scandium	Sc
Seaborgium	Sg
Selenium	Se
Silicon	Si
Silver	Ag
Sodium	Na
Strontium	Sr
Sulfur	S
Tantalum	Ta
Technetium	Tc
Tellurium	Te
Terbium	Tb
Thallium	Ti
Thorium	Th
Thulium	Tm
Tin	Sn
Titanium	Ti
Tungsten	W
Uranium	U
Vanadium	V
Xenon	Xe
Ytterbium	Yb
Yttrium	Y
Zinc	Zn
Zirconium	Zr

The Amino Acids of Proteins

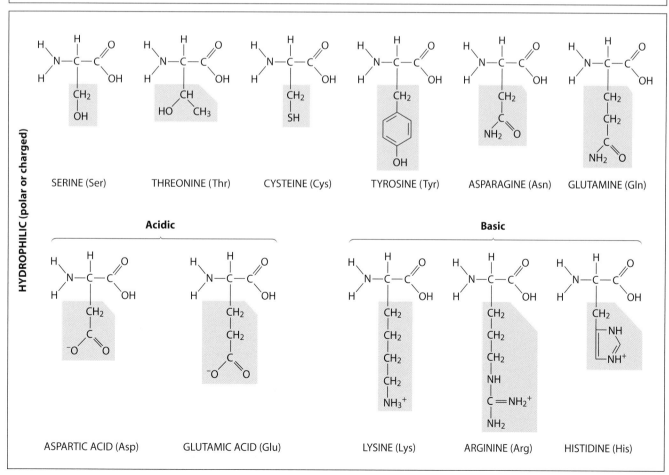

HYDROPHOBIC (Nonpolar)

GLYCINE (Gly) ALANINE (Ala) VALINE (Val) LEUCINE (Leu) ISOLEUCINE (Ile)

METHIONINE (Met) PHENYLALANINE (Phe) TRYPTOPHAN (Trp) PROLINE (Pro)

HYDROPHILIC (polar or charged)

SERINE (Ser) THREONINE (Thr) CYSTEINE (Cys) TYROSINE (Tyr) ASPARAGINE (Asn) GLUTAMINE (Gln)

Acidic

Basic

ASPARTIC ACID (Asp) GLUTAMIC ACID (Glu) LYSINE (Lys) ARGININE (Arg) HISTIDINE (His)

Chapter 1

1. The vertical scale of biology refers to the hierarchy of biological organization: from molecules to organelles, cells, tissues, organs, organ systems, organisms, populations, communities, ecosystems, and biosphere. At each level, emergent properties arise from the interaction and organization of component parts. The horizontal scale of biology refers to the incredible diversity of living organisms, past and present, including the 1.8 million species that have been named so far, which can be grouped into three domains—Bacteria, Archaea, and Eukarya—and divided among numerous kingdoms.

2. a. life; b. evolution; c. natural selection; d. unity of life; e. three domains (or numerous kingdoms; 1.8 million species)

3. d **4.** c **5.** e (You may have been tempted to choose b, the molecular level. However, protists may have chemical communication or interactions with other protists. No protists, however, have organs.) **6.** d **7.** c **8.** b **9.** d

10. Both energy and chemical nutrients are passed through an ecosystem from producers to consumers to decomposers. But energy enters an ecosystem as sunlight and leaves as heat. Chemical nutrients are recycled from the abiotic soil or atmosphere through plants, consumers, and decomposers and returned to the air, soil, and water.

11. Darwin described how natural selection operates in populations whose individuals have varied traits that are inherited. When natural selection favors the reproductive success of certain individuals in a population more than others, it changes the proportions of heritable variations in the population, gradually adapting a population to its environment.

12. In pursuit of answers to questions about nature, a scientist uses a logical thought process involving the key elements: observations about natural phenomena, questions derived from observations, hypotheses posed as tentative explanations of observations, logical predictions of the outcome of tests if the hypotheses are true, and actual tests of hypotheses. Scientific research is not a rigid method because a scientist must adapt these key components to the set of conditions particular to each study. Intuition, chance, and luck are also part of science.

13. Technology is the application of scientific knowledge. For example, the use of solar power to run a calculator or heat a home is an application of our knowledge, derived by the scientific process, of the nature of light as a type of energy and how light energy can be converted to other forms of energy. Another example is the use of pieces of DNA removed from bacteria to insert new genes into crop plants. This process, often called genetic engineering, stems from decades of scientific research on the structure and function of DNA from many kinds of organisms.

14. Natural selection screens heritable variations by favoring the reproductive success of some individuals over others. These individuals pass more genes to the next generation than individuals that are not favored. As a result, the genetic makeup of a population changes. The change results from a screening (editing) of individuals (and consequently their genes), not from the creation of new genes or new individuals.

15. a. Hypothesis: Giving rewards to mice will improve their learning. Prediction: If mice are rewarded with food, they will learn to run a maze faster.
    b. The control group was the mice that were not rewarded. Without them, it would be impossible to know if the mice that were rewarded decreased their time running the maze only because of practice.
    c. Both groups of mice should be about the same age. Both experiments should be run at the same time of day and under the same conditions.
    d. Yes, the data show that the rewarded mice began to run the maze faster by day 3, and improved their performance (ran faster than the control mice) each day thereafter.

16. The researcher needed to compare the number of attacks on artificial king snakes with attacks on artificial brown snakes. It may be that there were simply more predators in the coral snake areas or that the predators were hungrier than the predators in the other areas. The experiment needed a control and proper data analysis.

17. If these cell division control genes are involved in producing the larger tomato, they may have similar effects if transferred to other fruits or vegetables. Cancer is a result of uncontrolled cell division. One could see if there are similarities between the tomato genes and any human genes that could be related to human development or disease. The control of cell division is a fundamental process in growth, repair, and asexual reproduction—all important topics in biology.

18. Virtually any news report or magazine contains stories that are mainly about biology or at least have biological connections. How about biological connections in advertisements?

Chapter 2

1. a. protons; b. neutrons; c. electrons; d. different isotopes; e. covalent bonds; f. ionic bonds; g. polar covalent bonds; h. hydrogen bonding

2.

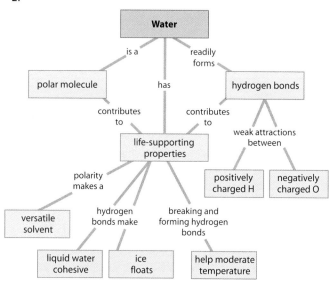

3. c 4. b 5. c 6. e 7. a (It needs to share 2 more electrons for a full outer shell of 8.) 8. a 9. c 10. d 11. F (Only salt and water are compounds; carbon is an element.) 12. F (The smallest particle of an element is an atom.) 13. T 14. T 15. F (Water molecules in ice are farther apart.) 16. T 17. F (Most acid precipitation results from burning fossil fuels.) 18. T

19. For a diagram, see Figure 2.10. Water molecules form hydrogen bonds because they are polar. Their slightly negative O atoms are attracted to the slightly positive H atoms of neighboring molecules. The unique properties of water that result from hydrogen bonding are cohesion, adhesion, surface tension, the ability to absorb and store large amounts of heat, a solid form (ice) that is less dense than liquid water, and solvent properties.

20. First: Because increasing the temperature of water (the average speed of its molecules) requires breaking hydrogen bonds, a process that uses heat, a large amount of heat can be added to water before the water's temperature starts to rise. Conversely, when the surrounding temperature falls, new hydrogen bonds form in water, with the release of heat that slows the cooling process. Second: When the body becomes overheated, water evaporating from its surface decreases the body's temperature (evaporative cooling) because the hotter water molecules leave.

21. A covalent bond forms when atoms complete their outer shells by sharing electrons. Atoms can also complete their outer shells by gaining or losing electrons. This leaves the atoms as ions, with + and − charges. The oppositely charged ions are attracted to each other, forming an ionic bond.

22. An acid is a compound that donates hydrogen ions (H^+) to a solution. A base is a compound that accepts hydrogen ions and removes them from solution. Acidity is described by the pH scale, which measures H^+ concentration on a scale of 0 (most acidic) to 14 (most basic).

23. Fluorine needs 1 electron for a full outer shell of 8, and if potassium loses 1 electron, its outer shell will have 8. Potassium will lose an electron (becoming a + ion), and fluorine will pick it up (becoming a − ion). The ions will form an ionic bond.

24. Give a mouse sugar or oxygen gas containing a radioactive isotope of oxygen. See whether the carbon dioxide it exhales is radioactive.

25. The elements in a row all have the same number of electron shells. In a column, all the elements have the same number of electrons in their outer shell.

26. Some issues and questions to consider: Which is less expensive, power from nuclear power plants or power from fossil fuel plants? Does the price of electricity reflect its actual cost, including environmental costs? Which would be more harmful: the environmental effects of acid rain and global warming from fossil fuel power plants or the effects of nuclear wastes and potential nuclear accidents? Do you favor development of nuclear energy or fossil fuel power plants? Which would you prefer to have near you? Do your answers to these last two questions differ? If so, why?

Chapter 3

1. Carbon forms four covalent bonds, either with other carbon atoms, producing chains or rings of various lengths and shapes, or with other atoms, such as characteristic chemical groups that confer specific properties on a molecule. This is the basis for the incredible diversity of organic compounds. Organisms can link a small number of monomers into different arrangements to produce a huge variety of polymers.

2. a. glucose; b. energy storage; c. cellulose; d. fats; e. cell membrane component; f. steroids; g. amino group; h. carboxyl group; i. R group; j. enzyme; k. structural protein; l. movement; m. hemoglobin; n. defense; o. phosphate group; p. nitrogenous base; q. ribose or deoxyribose; r. DNA; s. code for enzymes

3. d (The second kind of molecule is a polymer of the first.) 4. c 5. d 6. c 7. a 8. b 9. e 10. d

11. Fats (triglycerides)—store energy. Phospholipids—are major components of membranes. Steroids—one kind, cholesterol, is a component of cell membranes; other kinds function as hormones.

12. Weak bonds that stabilize the three-dimensional structure of a protein are disrupted, and the protein unfolds. Function depends on shape, so if the protein is the wrong shape, it won't function properly.

13. Proteins are made of 20 amino acids arranged in many different sequences into chains of many different lengths. Genes, defined stretches of DNA, dictate the amino acid sequences of proteins in the cell.

14. Proteins function as enzymes to catalyze chemical reactions. They also function in structure, contraction, transport,

defense, signaling, signal reception, and storage of amino acids (see Module 3.11).

15. The sequence of nucleotides in DNA is transcribed into a sequence of nucleotides in RNA, which determines the sequence of amino acids that will be used to build a polypeptide. Proteins mediate all the activities of a cell; thus, by coding for proteins, DNA controls the functions of a cell.

16. This is a hydrolysis reaction, which consumes water. It is essentially the reverse of the diagram in Figure 3.5, except note that fructose has a different shape than glucose.

17. Circle NH_2, an amino group; COOH, a carboxyl group; and OH, a hydroxyl group on the R group. This is an amino acid, a monomer of proteins. The OH group makes it a polar amino acid.

18. Some issues and questions to consider: How will you choose your test subjects? How many subjects should you have? Will you give them all vitamin C, or just some of them? What criteria will you use to divide the test subjects into groups? What is a control group? Should the subjects know whether they are getting vitamin C or not? Should the experimenters who are giving out the drug and measuring the severity of cold symptoms know which of the subjects are getting vitamin C? What is a double-blind study? If there is a difference between your groups, can you be sure it can be attributed to vitamin C dosing?

19. a. A: at about 37°C; B: at about 78°C.
    b. A: from humans (human body temperature is about 37°C; B: from thermophilic bacteria.
    c. Above 40°C, the human enzyme denatures and loses its shape and thus its function. The increased thermal energy disrupts the weak bonds that maintain secondary and tertiary structure in an enzyme.

20. Silicon has four electrons in its outer electron shell, as does carbon. One would predict that silicon could thus form complex molecules by binding with four partners. Neon has a filled outer shell and is nonreactive. Sulfur can only form two covalent bonds and thus would not have the versatility of carbon or silicon.

Chapter 4

1. a. nucleus; b. nucleolus; c. ribosomes; d. Golgi apparatus; e. plasma membrane; f. mitochondrion; g. cytoskeleton; h. peroxisome; i. centriole; j. lysosome; k. rough endoplasmic reticulum; l. smooth endoplasmic reticulum. For functions, see Table 4.23.

2. flagellum or cilia (in some cells), lysosome, centriole (involved with microtubule formation)

3. chloroplast, central vacuole, cell wall

4. d 5. c (Small cells have a greater ratio of surface area to volume.) 6. b 7. d 8. e 9. a 10. c 11. b 12. d 13. c 14. d

15. DNA as genetic material, ribosomes, plasma membrane

16. Tight junctions form leakproof sheets of cells. Anchoring junctions link cells to each other but allow materials to pass

between them; they form strong sheets of cells. Gap junctions are channels that allow flow from cell to cell.

17. Both process energy. A chloroplast converts light energy to chemical energy (sugar molecules). A mitochondrion converts chemical energy (food molecules) to another form of chemical energy (ATP).

18. Different conditions and conflicting processes can occur simultaneously within separate, membrane-enclosed compartments. Also, there is increased area for membrane-attached enzymes that carry out metabolic processes.

19. Cilia may propel a cell through its environment or sweep a fluid environment past the cell.

20. A protein inside the ER is packaged inside transport vesicles that bud off the ER and then join to the Golgi apparatus. A transport vesicle containing the finished protein product then buds off the Golgi and travels to and joins with the plasma membrane, expelling the protein from the cell.

21. Part true, part false. Both animal and plant cells have mitochondria.

22. The plasma membrane is a phospholipid bilayer with the hydrophilic heads facing the aqueous environment on both sides and the hydrophobic fatty acid tails mingling in the center of the membrane. Proteins are embedded in and attached to this membrane. Microfilaments form a three-dimensional network just inside the plasma membrane. The extracellular matrix outside the membrane is composed largely of glycoproteins, which may be attached to membrane proteins called integrins. Integrins can transmit information from the ECM to microfilaments on the other side of the membrane.

23. Cell 1: $S = 1{,}256\ \mu m^2$; $V = 4{,}187\ \mu m^3$; $S/V = 0.3$. Cell 2: $S = 5{,}024\ \mu m^2$; $V = 33{,}493\ \mu m^3$; $S/V = 0.15$. The smaller cell has a larger surface area relative to volume for absorbing food and oxygen and excreting waste. Small cells thus perform these activities more efficiently.

24. Individuals with PCD have nonfunctional cilia and flagella due to a lack of dynein motor proteins. This defect would also mean that the cilia involved in left-right pattern formation in the embryo would not be able to set up the fluid flow that initiates the normal arrangement of organs.

25. A single layer of phospholipids surrounding the oil droplet would have their hydrophobic fatty acid tails associated with the hydrophobic oil and their hydrophilic heads facing the aqueous environment of the cell outside the droplet.

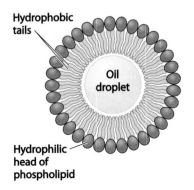

Hydrophobic tails

Oil droplet

Hydrophilic head of phospholipid

26. Some issues and questions to consider: Were the cells the patient's property, a gift, or just surplus? Was he asked to donate the cells? Was he informed about how the cells might be used? Is it important to ask permission or inform the patient in such a case? How much did the researchers modify the cells? What did they have to do to them to sell the product? Do the researchers and the university have a right to make money from patient's cells? Is the fact that they saved the patient's life a factor? Does the patient have the right to sell his cells? Would he have been able to sell the cells without the researchers' help?

Chapter 5

1. a. active transport; b. concentration gradient; c. small nonpolar molecules; d. facilitated diffusion; e. transport proteins

2. a. enzyme; b. active site of enzyme; c. substrate; d. substrate in active site; induced fit strains substrate bonds; e. substrate converted to products; f. product molecules released; enzyme is ready for next catalytic cycle

3. b 4. d 5. e 6. c (Only active transport can move solute against a concentration gradient.) 7. d 8. b

9. Aquaporins are water transport channels that allow for very rapid diffusion of water through a cell membrane. They are found in cells that have high water-transport needs, such as blood cells, kidney cells, and plant cells.

10. Energy is neither created nor destroyed but can be transferred and transformed. Plants transform the energy of sunlight into chemical energy stored in organic molecules. Almost all organisms rely on the products of photosynthesis for the source of their energy. In every energy transfer or transformation, disorder increases as some energy is lost to the random motion of heat.

11. The work of cells falls into three main categories: mechanical, chemical, and transport. ATP provides the energy for cellular work, usually by transferring a phosphate group to a protein (movement and transport) or a substrate (chemical).

12. Energy is stored in the chemical bonds of organic molecules. The barrier of E_A prevents these molecules from spontaneously breaking down and releasing that energy. When a substrate fits into an enzyme's active site with an induced fit, its bonds may be strained and thus easier to break, or the active site may orient two substrates in such a way as to facilitate the reaction.

13. Cell membranes are composed of a lipid bilayer with embedded proteins. The bilayer creates the hydrophobic boundary between cells and their surroundings (or between organelles and the cytoplasm). The proteins perform the many functions of membranes, such as enzyme action, transport, attachment, and signaling.

14. Inhibitors that are toxins or poisons irreversibly inhibit key cellular enzymes. Inhibitors that are designed as drugs are beneficial if they interfere with the enzymes of bacterial or viral invaders or cancer cells. Cells use feedback inhibition

of enzymes in metabolic pathways as important mechanisms that conserve resources.

15. Heating, pickling, and salting denature enzymes, changing their shapes so they do not fit substrates. Freezing decreases the kinetic energy of molecules, so they lack energy of activation, even in the presence of enzymes.

16. a. The more enzyme present, the faster the rate of reaction, because it is more likely that enzyme and substrate molecules will meet.

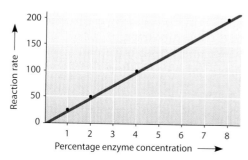

b. The more substrate present, the faster the reaction, for the same reason, but only up to a point. An enzyme molecule can work only so fast; once it is saturated (working at top speed), more substrate does not increase the rate.

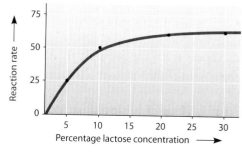

17. The black curve on the left would correspond to the stomach enzyme pepsin, which has a lower optimal pH as is found in the stomach; the red curve on the right would correspond to trypsin, which has a higher optimal pH. The curve for a lysosomal enzyme should have an optimal pH at 4.5.

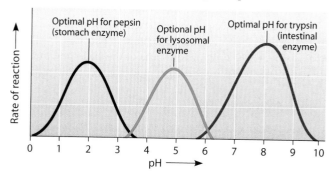

18. Some issues and questions to consider: Does a woman have a right to work in an unsafe environment, even if it may put her child at risk? What are the rights of the child? Does the company have the right to "protect" women who are not pregnant? Is the company trying to protect the mother and child or protect itself from a potential lawsuit? Who is responsible for protecting employees and their children? The employees? The company? The government?

Chapter 6

1. a. glycolysis; b. citric acid cycle; c. oxidative phosphorylation; d. oxygen; e. electron transport chain; f. CO_2 and H_2O

2. e 3. c (NAD^+ and FAD, which are recycled by electron transport, are limited.) 4. e 5. d 6. c 7. b (at the same time NADH is oxidized to NAD^+) 8. a

9. Glycolysis is considered the most ancient because it occurs in all living cells and doesn't require oxygen or membrane-enclosed organelles.

10. Oxygen picks up electrons from the oxidation of glucose at the end of the electron transport chain. Carbon dioxide results from the oxidation of glucose. It is released in the grooming of pyruvate and the citric acid cycle.

11. In lactic acid fermentation (in muscle cells), pyruvate is reduced by NADH to form lactate, and NAD^+ is recycled. In alcohol fermentation, pyruvate is broken down to CO_2 and ethanol as NADH is oxidized to NAD^+. Both types of fermentation allow glycolysis to continue to produce 2 ATP/glucose by recycling NAD^+.

12. As carbohydrates are broken down in glycolysis and the grooming of pyruvate, glycerol can be made from G3P and fatty acids from acetyl CoA. Amino groups, containing N atoms, must be supplied to various intermediates of glycolysis and the citric acid cycle to produce amino acids.

13. 100 kcal per day is 700 kcal per week. On the basis of Table 6.4, walking 3 mph would require $\frac{700}{245}$ = about 2.8 hr; swimming, 1.7 hr; running, 0.7 hr.

14. NAD^+ and FAD are not used up but are recycled. NAD^+ and FAD are recycled between the electron transport chain and glycolysis or the citric acid cycle. We need a small additional supply to replace those that are damaged.

15. a. No, this shows the blue color getting more intense. The reaction decolorizes the blue dye.

    b. No, this shows the dye being decolorized, but it also shows the three mixtures with different initial color intensities. The intensities should have started out the same, since all mixtures used the same concentration of dye.

    c. Correct. The mixtures all start out the same, and then the ones with more succinate (reactant) decolorize faster. The mixture with the highest concentration of succinate decolorizes the fastest.

16. The presence of ATP synthase enzymes in prokaryotic plasma membranes and the inner membrane of mitochondria provides support for the hypothesis of endosymbiosis—that mitochondria evolved from an engulfed prokaryote that used aerobic respiration (see Module 4.16).

17. Some issues and questions to consider: Is your customer aware of the danger? Do you have an obligation to protect the customer, even against her wishes? Does your employer have the right to dismiss you for informing the customer? For refusing to serve the customer? Could you or the restaurant later be held liable for injury to the fetus? Or is the mother responsible for willfully disregarding warnings about drinking?

Chapter 7

1. a. electron transport chain; b. ATP synthase; c. thylakoid space (higher H^+ concentration); d. stroma; e. ATP

2. In chemiosmosis in mitochondria: a. Electrons come from food molecules. b. Electrons get their energy from the chemical energy of bonds in organic molecules. c. Electrons are passed to oxygen, which picks up H^+ and forms water.

    In chemiosmosis in chloroplasts: a. Electrons come from splitting of water. b. Light energy excites the electrons. c. Electrons flow from water to the reaction center chlorophyll in photosystem II to the reaction center chlorophyll in photosystem I to NADPH.

    In both processes: d. Energy released by redox reactions in the electron transport chain is used to transport H^+ across a membrane. The flow of H^+ down its concentration gradient back through ATP synthase drives the phosphorylation of ADP to make ATP.

3. a. light energy; b. light reactions; c. Calvin cycle; d. O_2 released; e. electron transport chain; f. NADPH; g. ATP; h. 3-PGA is reduced

4. d 5. c 6. c 7. b 8. a 9. e 10. c (NADPH and ATP from the light reactions are used by the Calvin cycle.) 11. d 12. e

13. Light reactions: light and water are inputs; ATP, NADPH, and O_2, are outputs. Calvin cycle: CO_2, ATP, and NADPH are inputs; G3P, ADP, and $NADP^+$ are outputs.

14. Plants can break down the sugar for energy in cellular respiration or use the sugar as a raw material for making other organic molecules. Excess sugar is stored as starch.

15. Some issues and questions to consider: What are the risks that we take and costs we must pay if global warming continues? How certain do we have to be that warming is caused by human activities before we act? What can we do to reduce CO_2 emissions? What are the risks and costs if we reduce CO_2 emissions and greenhouse warming turns out not to be a real threat? Is it possible that the costs and sacrifices of reducing CO_2 emissions might actually improve our lifestyle?

16. Some issues and questions to consider: How much land would be required for large-scale conversion to biomass energy? Would different power plants have to be built? How do the benefits of energy plantations (habitat for wildlife, reduced erosion, diversification for farmers) compare with the benefits of fossil fuel harvesting? Is there a difference in net output of CO_2 between production and use of fossil fuels and production and use of biomass energy? Which one offers a more long-term solution to energy needs?

Chapter 8

1.	**Mitosis**	**Meiosis**
Number of chromosomal duplications	*1*	*1*
Number of cell divisions	*1*	*2*
Number of daughter cells produced	*2*	*4*
Number of chromosomes in daughter cells	*Diploid* (2n)	*Haploid* (n)
How chromosomes line up during metaphase	*Singly*	*In tetrads (metaphase I), then singly (metaphase II)*
Genetic relationship of daughter cells to parent cell	*Genetically identical*	*Genetically unique*
Functions performed in the human body	*Growth, development, repair*	*Production of gametes*

2. c 3. a 4. b 5. c 6. e (A diploid cell would have an even number of chromosomes; the odd number suggests that meiosis I has been completed. Sister chromatids are together only in prophase and metaphase of meiosis II.) 7. c 8. b 9. c 10. b 11. d

12. Various orientations of chromosomes at metaphase I of meiosis lead to different combinations of chromosomes in gametes. Crossing over during prophase I results in an exchange of chromosome segments and new combinations of genes. Random fertilization of eggs by sperm further increases possibilities for variation in offspring.

13. Interphase (for example, third column from left in micrograph, third cell from top): Growth; metabolic activity; DNA synthesis. Prophase (for example, second column, cell at bottom): Chromosomes shorten and thicken; mitotic spindle forms. Metaphase (for example, first column, middle cell): Chromosomes line up on a plane going through the cell's equator. Anaphase (for example, third column, second cell from top): Sister chromatids separate and move to the poles of the cell. Telophase (for example, fourth column, fourth complete cell from top): Daughter nuclei form around chromosomes; cytokinesis usually occurs.

14. In culture, normal cells usually divide only when they are in contact with a surface but not touching other cells on all sides (the cells usually grow to form only a single layer). The density-dependent inhibition of cell division apparently results from local depletion of substances called growth factors. Growth factors are proteins secreted by certain cells that stimulate other cells to divide; they act via signal transduction pathways to signal the cell cycle control system of the affected cell to proceed past its checkpoints. The cell cycle control systems of cancer cells do not function properly. Cancer cells generally do not require externally supplied growth factors to complete the cell cycle, and they divide indefinitely (in contrast to normal mammalian cells, which stop dividing after 20–50 generations)—two reasons why they are relatively easy to grow in the lab. Furthermore, cancer cells can often grow without contacting a solid surface, making it possible to culture them in suspension in a liquid medium.

15. A ring of microfilaments pinches an animal cell in two, a process called cleavage. In a plant cell, membranous vesicles form a disk called the cell plate at the midline of the parent cell, cell plate membranes fuse with the plasma membrane, and a cell wall grows in the space, separating the daughter cells.

16. See Figures 8.21A and 8.21B.

17. a. No. For this to happen, the chromosomes of the two gametes that fused would have to represent, together, a complete set of the donor's maternal chromosomes (the ones that originally came from the donor's mother) and a complete set of the donor's paternal chromosomes (from the donor's father). It is much more likely that the zygote would be missing one or more maternal chromosomes and would have an excess of paternal chromosomes, or vice versa.

b. Correct. Consider what would have to happen to produce a zygote genetically identical to the gamete donor: The zygote would have to have a complete set of the donor's maternal chromosomes and a complete set of the donor's paternal chromosomes. The first gamete in this union could contain any mixture of maternal and paternal chromosomes, but once that first gamete was "chosen," the second one would have to have one particular combination of chromosomes—the combination that supplies whatever the first gamete did not supply. So, for example, if the first three chromosomes of the first gamete were maternal, maternal, and paternal, the first three of the second gamete would have to be paternal, paternal, and maternal. The chance that all 23 chromosome pairs would be complementary in this way is only one in 22^3 (that is, one in 8,388,608). Because of independent assortment, it is much more likely that the zygote would have an unpredictable combination of chromosomes from the donor's father and mother.

c. No. First, the zygote could not be genetically identical to the gamete donor (see b). Second, the zygote could not be identical to either of the gamete donor's parents because the donor only has half the genetic material of each of his or her parents. For example, even if the zygote were formed by two gametes containing only paternal chromosomes, the combined set of chromosomes could not be identical to that of the donor's father because it would still be missing half of the father's chromosomes.

d. No. See answer c.

18. Some possible hypotheses: The replication of the DNA of the bacterial chromosome takes less time than the replication of the DNA in a eukaryotic cell. The time required for a growing bacterium to roughly double its cytoplasm is much less than for a eukaryotic cell. Bacteria

have a cell cycle control system much simpler than that of eukaryotes.

19. 1 cm³ = 1,000 mm³, so 5,000 mm³ of blood contain 5,000 × 1,000 × 5,000,000 = 25,000,000,000,000, or 2.5×10^{13}, red blood cells. The $\frac{1}{120}$ of the cells that are replaced each day = $2.5 \times 10^{13}/120 = 2.1 \times 10^{11}$ cells. There are 24 × 60 × 60 = 86,400 seconds in a day. Therefore, the number of cells replaced each second = $2.1 \times 10^{11}/86,400$ = about 2×10^{6}, or 2 million. Thus, about 2 million cell divisions must occur each second to replace red blood cells that are lost.

20. Each chromosome is on its own in mitosis; chromosome replication and the separation of sister chromatids occur independently for each horse or donkey chromosome. Therefore, mitotic divisions, starting with the zygote, are not impaired.

   In meiosis, however, homologous chromosomes must pair in prophase I. This process of synapsis cannot occur properly because horse and donkey chromosomes do not match in number or content.

21. Some issues and questions to consider: Could it be that less money is spent on prevention because effective prevention is so much cheaper? Or because prevention has been tried, and it does not work well? Are lifestyle changes the kind of measures that could benefit from a shift in resources? Is prevention an individual matter of avoiding exposure or a social matter of preventing exposure? How might the answer to this question shape prevention policy? If more money were devoted to prevention, how could it be used to encourage you or others to make lifestyle changes? Would prevention work better for younger or older people? Might older people, already exposed to cancer-causing agents, actually be harmed by a shift of resources to prevention?

Chapter 9

1. a. alleles; b. loci; c. homozygous; d. dominant; e. recessive; f. incomplete dominance

2. c 3. d 4. d (Neither parent is ruby-eyed, but some offspring are, so it is recessive. Different ratios among male and female offspring show that it is sex-linked.) 5. a 6. e

7. Genes on the single X chromosome in males are always expressed because there are no corresponding genes on the Y chromosome to mask them. A male needs only one recessive color blindness allele (from his mother) to show the trait; a female must inherit the allele from both parents, which is less likely.

8. See Figure 9.18A. The parental gametes are WS and ws. Recombinant gametes are Ws and wS, produced by crossing over.

9. Height appears to be a quantitative trait resulting from polygenic inheritance, like human skin color. See Module 9.14.

10. The trait of freckles is dominant, so Tim and Jan must both be heterozygous. There is a $\frac{3}{4}$ chance that they will produce a child with freckles and a $\frac{1}{4}$ chance that they will produce a child without freckles. The probability that the next two children will have freckles is $\frac{3}{4} \times \frac{3}{4} = \frac{9}{16}$.

11. As in problem 10, both Tim and Jan are heterozygous, and Mike is homozygous recessive. The probability of the next child having freckles is $\frac{3}{4}$. The probability of the next child having a straight hairline is $\frac{1}{4}$. The probability that the next child will have freckles and a straight hairline is $\frac{3}{4} \times \frac{1}{4} = \frac{3}{16}$.

12. The genotype of the black short-haired parent rabbit is BBSS. The genotype of the brown long-haired parent rabbit is bbss. The F1 rabbits will all be black and short-haired, BbSs. The F₂ rabbits will be $\frac{9}{16}$ black short-haired, $\frac{3}{16}$ black long-haired, $\frac{3}{16}$ brown short-haired, and $\frac{1}{16}$ brown long-haired.

13. If the genes are not linked, the proportions among the offspring will be 25% gray red, 25% gray purple, 25% black red, 25% black purple. The actual percentages show that the genes are linked. The recombination frequency is 6%.

14. The recombination frequencies are: black dumpy 36%, purple dumpy 41%, and black purple 6% (see problem 13). Since these recombination frequencies reflect distances between the genes, the sequence must be purple-black-dumpy (or dumpy-black-purple).

15. $\frac{1}{4}$ will be boys suffering from hemophilia, and $\frac{1}{4}$ will be female carriers. (The mother is a heterozygous carrier, and the father is normal.)

16. For a woman to be color-blind, she must inherit X chromosomes bearing the color blindness allele from both parents. Her father has only one X chromosome, which he passes on to all his daughters, so he must be color-blind. A male need only inherit the color blindness allele from a carrier mother; both his parents are usually phenotypically normal.

17. Start out by breeding the cat to get a population to work with. If the curl allele is recessive, two curl cats can have only curl kittens. If the allele is dominant, curl cats can have "normal" kittens. If the curl allele is sex-linked, ratios will differ in male and female offspring of some crosses. If the curl allele is autosomal, the same ratios will be seen among males and females. Once you have established that the curl allele is dominant and autosomal, you can determine if a particular curl cat is true-breeding (homozygous) by doing a testcross with a normal cat. If the curl cat is homozygous, all offspring of the testcross will be curl; if heterozygous, half of the offspring will be curl and half normal.

Chapter 10

1. a. nucleotides; b. transcription; c. RNA polymerase; d. mRNA; e. rRNA; f. tRNA; g. translation; h. ribosomes; i. amino acids

2. e (Only the phage DNA enters a host cell; lambda DNA determines both DNA and protein.) 3. e 4. b 5. c

6. Ingredients: original DNA, nucleotides, several enzymes and other proteins, including DNA polymerase and DNA ligase. Steps: Original DNA strands separate at a specific

site (origin of replication), nucleotides line up along each strand according to base-pairing rules, DNA polymerase links the nucleotides to form new strands. New nucleotides are added only to the 3′ end of a growing strand. One new strand is made in one continuous piece; the other new strand is made in a series of short pieces that are then joined by DNA ligase. Product: two identical DNA molecules, each with one old strand and one new strand.

7. A gene is the polynucleotide sequence with information for making one polypeptide. Each codon—a triplet of bases in DNA or RNA—codes for one amino acid. Transcription occurs when RNA polymerase produces RNA using one strand of DNA as a template. In prokaryotic cells, the RNA transcript may immediately serve as mRNA. In eukaryotic cells, the RNA is processed: A cap and tail are added, and RNA splicing removes introns and links exons together to form a continuous coding sequence. A ribosome is the site of translation, or polypeptide synthesis, and tRNA molecules serve as interpreters of the genetic code. Each tRNA molecule has an amino acid attached at one end and a three-base anticodon at the other end. Beginning at the start codon, mRNA moves relative to the ribosome a codon at a time. A tRNA with a complementary anticodon pairs with each codon, adding its amino acid to the polypeptide chain. The amino acids are linked by peptide bonds. Translation stops at a stop codon, and the finished polypeptide is released. The polypeptide folds to form a functional protein, sometimes in combination with other polypeptides.

8.

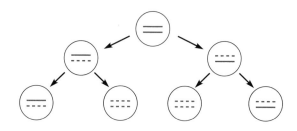

9. mRNA: GAUGCGAUCCGCUAACUGA. Amino acids: Met-Arg-Ser-Ala-Asn.

10. Some issues and questions to consider: Is it fair to issue a patent for a gene or gene product that occurs naturally in every human being? Or should a patent be issued only for something new that is invented rather than found? Suppose another scientist slightly modifies the gene or protein. How different does the gene or protein have to be to avoid patent infringement? Might patents encourage secrecy and interfere with the free flow of scientific information? What are the benefits to the holder of a patent? When research discoveries cannot be patented, what are the scientists' incentives for doing the research? What are the incentives for the institution or company that is providing financial support?

Chapter 11

1. a. proto-oncogene; b. repressor (or activator); c. cancer; d. operator; e. X inactivation; f. transcription factors; g. alternative RNA splicing

2. b (Different genes are active in different kinds of cells.) 3. c 4. b 5. a 6. b 7. e 8. b

9. A mutation in a single gene can influence the actions of many other genes if the mutated gene is a control gene, such as a homeotic gene. A single control gene may encode a protein that affects (activates or represses) the expression of a number of other genes. In addition, some of the affected genes may themselves be control genes that in turn affect other batteries of genes. Cascades of gene expression are common in embryonic development.

10. a. If the mutated repressor could still bind to the operator on the DNA, it would continuously repress the operon; enzymes for lactose utilization would not be made, whether or not lactose was present.
   b. The lac genes would continue to be transcribed and the enzymes made, whether or not lactose was present.
   c. Same predicted result as for b.
   d. RNA polymerase would not be able to transcribe the genes; no proteins would be made, whether or not lactose was present.

11. The protein to which dioxin binds in the cell is probably a transcription factor that regulates multiple genes (see Module 11.6). If the binding of dioxin influences the activity of this transcription factor (either activating or inactivating it), dioxin could thereby affect multiple genes and thus have a variety of effects on the body. The differing effects in different animals might be explained by differing genetic details in the different species. It would be extremely difficult to demonstrate conclusively that dioxin exposure was the cause of illness in a particular individual, even if dioxin had been shown to be present in the person's tissues. However, if you had detailed information about how dioxin affects patterns of gene expression in humans and were able to show dioxin-specific abnormal patterns in the patient (perhaps using DNA microarrays; see Module 11.11), you might be able to establish a strong link between dioxin and the illness.

Chapter 12

1. a. PCR; b. restriction enzymes; c. gel electrophoresis; d. nucleic acid probe; e. cloning; f. genomic library

2. d 3. c 4. b 5. c (Bacteria lack the RNA-splicing machinery needed to delete eukaryotic introns.) 6. b 7. d

8. Isolate plasmids from *E. coli*. Cut the plasmids and the human DNA containing the HGH gene with restriction enzyme to produce molecules with sticky ends. Join the plasmids and the fragments of human DNA with ligase. Allow *E. coli* to take up recombinant plasmids. Bacteria will then replicate plasmids and multiply, producing clones of bacterial cells. Identify a clone carrying and expressing the HGH gene using a nucleic

acid probe. Grow large amounts of the bacteria and extract and purify HGH from the culture.

9. Medicine: Genes can be used to engineer viruses for vaccines, to produce transgenic lab animals for AIDS research, or for research related to human gene therapy. Proteins can be hormones, enzymes, blood-clotting factor, or the starting material for making vaccines. Agriculture: Foreign genes can be inserted into plant cells or animal eggs to produce transgenic crop plants or farm animals or inserted into viruses for making animal vaccines. Animal growth hormones are examples of agriculturally useful proteins that can be made using recombinant DNA technology.

10. She could start with DNA isolated from liver cells (the entire genome) and carry out the procedure outlined in Module 12.1 to produce a collection of recombinant bacterial clones, each carrying a small piece of liver cell DNA. To find the clone with the desired gene, she could then make a probe of radioactive RNA with a nucleotide sequence complementary to part of the gene: GACCUGACUGU. This probe would bind to the gene, labeling it and identifying the clone that carries it. Alternatively, the biochemist could start with mRNA isolated from liver cells and use it as a template to make DNA (using reverse transcriptase). Cloning this DNA rather than the entire genome would yield a smaller library of genes to be screened—only those active in liver cells. Furthermore, the genes would lack introns, making the desired gene easier to manipulate after isolation.

11. Determining the nucleotide sequences is just the first step. Once researchers have written out the DNA "book," they will have to try to figure out what it means—what the nucleotide sequences code for and how they work.

12. Some issues and questions to consider: What are some of the unknowns in recombinant DNA experiments? Do we know enough to anticipate and deal with possible unforeseen and negative consequences? Do we want this kind of power over evolution? Who should make these decisions? If scientists doing the research were to make the decisions about guidelines, what factors might shape their judgment? What might shape the judgment of business executives in the decision-making process? Does the public have a right to a voice in the direction of scientific research? Does the public know enough about biology to get involved in this decision-making process? Who represents "the public," anyway?

13. Some issues and questions to consider: What kinds of impact will gene therapy have on the individuals who are treated? On society? Who will decide what patients and diseases will be treated? What costs will be involved, and who will pay them? How do we draw the line between treating disorders and "improving" the human species?

14. Some issues and questions to consider: Should genetic testing be mandatory or voluntary? Under what circumstances? Why might employers and insurance companies be interested in genetic data? Since genetic characteristics differ among ethnic groups and between the sexes, might such information be used to discriminate? Which of these questions do you think is most important? Which issues are likely to be the most serious in the future?

Chapter 13

1. The two major components of Darwin's theory of descent with modification are that all life has descended from a common ancestral form as a result of natural selection. Individuals in a population have hereditary variations. The overproduction of offspring in the face of limited resources leads to a struggle for existence. Individuals that are well-suited to their environment tend to leave more offspring than other individuals, leading to the gradual accumulation of adaptations to the local environment in the population.

2. a. genetic drift; b. gene flow; c. natural selection; d. small population; e. founder effect; f. bottleneck effect; g. unequal reproductive success

3. e 4. a 5. e 6. b (Erratic rainfall and unequal reproductive success would ensure that a mixture of both forms remained in the population.) 7. c 8. e 9. b (All of these provide evidence of evolution, but DNA and the universal genetic code are best able to connect all of life's diverse forms through common ancestry.) 10. d 11. c

12. Your paragraph should include such evidence as fossils and the fossil record, biogeography, comparative anatomy, comparative embryology, DNA and protein comparisons, artificial selection, and examples of natural selection.

13. If $q^2 = 0.0025$, then $q = 0.05$. Since $p + q = 1, p = 1 - q = 0.95$. The proportion of heterozygotes is $2pq = 2 \times 0.95 \times 0.05 = 0.095$. About 9.5% of African-Americans are carriers.

14. Genetic variation is retained in a population by diploidy and balanced selection. Recessive alleles are hidden from selection when in the heterozygote; thus, less adaptive or even harmful alleles are maintained in the gene pool and are available should environmental conditions change. Both heterozygote advantage and frequency-dependent selection tend to maintain alternate alleles in a population.

15. The unstriped snails appear to be better adapted. Striped snails make up 47% of the living population but 56% of the broken shells. Assuming that all the broken shells result from the meals of birds, we would predict that bird predation would reduce the frequency of striped snails and the frequency of unstriped individuals would increase.

16. Some issues and questions to consider: Who should decide curriculum, scientific experts in a field or members of the community? Are these alternative versions scientific ideas? Who judges what is scientific? If it is fairer to consider alternatives, should the door be open to all alternatives? Are constitutional issues (separation of church and state) involved here? Can a teacher be compelled to teach an idea he or she disagrees with? Should a student be required to learn an idea he or she thinks is wrong?

Chapter 14

1. a. Allopatric speciation: Reproductive barriers may evolve between these two geographically separated populations as a by-product of the genetic changes associated with each population's adaptation to its own environment or caused by genetic drift or mutation.
   b. Sympatric speciation: Some change, perhaps in resource use or female mate choice, may lead to a reproductive barrier that isolates the gene pools of these two populations, which are not separated geographically. If speciation occurred by polyploidy (common in plants but unusual in animals), then the new species is instantly isolated from the parent species.

2. a. hybrid zone
   b. reinforcement
   c. fusion
   d. stability
   e. strengthened
   f. weakened or eliminated

3. c   4. b   5. b   6. d   7. c   8. b   9. c   10. a   11. e   12. e

13. Different physical appearance may indicate different species or just differences within a species. Isolated populations may or may not be able to interbreed. Organisms that reproduce only asexually and fossil organisms do not have the potential to interbreed and produce fertile offspring; therefore, the biological species concept does not apply to them.

14. There is more chance for gene flow between populations on a mainland and nearby island. This interbreeding would make it more difficult for reproductive isolation to develop and separate the two populations.

15. *Punctuated equilibria* refers to a common pattern seen in the fossil record, in which most species diverge relatively quickly as they arise from an ancestral species and then remain fairly unchanged for the rest of their existence as a species.

16. Microevolution is the change in the gene pool of a population from one generation to the next. Speciation is the origin of a new species when a population becomes reproductively isolated from its parent species.

17. A broad hypothesis would be that cultivated American cotton arose from a sequence of hybridization, mistakes in cell division, and self-fertilization. We can divide this broad statement into at least three testable hypotheses. *Hypothesis 1:* The first step in the origin of cultivated American cotton was hybridization between a wild American cotton plant (with 13 pairs of small chromosomes) and an Old World cotton plant (with 13 pairs of large chromosomes). If this hypothesis is correct, we would predict that the hybrid offspring would have 13 small chromosomes and 13 large chromosomes. *Hypothesis 2:* The second step in the origin of cultivated American cotton was a failure of cell division in the hybrid offspring, such that all chromosomes are duplicated (now 26 small and 26 large). If this hypothesis is true, we would expect the resulting gametes to each have 13 large chromosomes and 13 small chromosomes.

Hypothesis 3: The third step in the origin of cultivated American cotton was self-fertilization of these gametes. If this hypothesis is true, we would expect the outcome of self-fertilization to be a hybrid plant with 52 chromosomes: 13 pairs of large ones and 13 pairs of small ones. This is the genetic makeup of cultivated American cotton.

18. By decreasing the ability of females to distinguish males of their own species, the polluted turbid waters have increased the frequency of mating between members of species that had been reproductively isolated from one another. As the number of hybrid fish increase, the parent species' gene pools may fuse, resulting in a loss of the two separate parent species and the formation of a new hybrid species. Future speciation events in Lake Victoria cichlids are now less likely to occur because females are less able to base mate choice on male breeding color. Reducing the pollution in the lake may help reverse this trend.

19. Some issues and questions to consider: The rationale behind protecting all endangered groups is the desire to preserve genetic diversity. Each species, subspecies, and hybrid group may represent a unique mix of genes. Studies of the degree of genetic distinctiveness of a subspecies or hybrid group may help decision makers if cost is an issue. If the species as a whole is not at risk, it seems appropriate to determine how distinctive the gene pool of a subspecies or hybrid group is before assigning it a lower priority for saving. Questions for society in general are: What is the value of any particular species and its genetically distinct subgroups? And how far are we willing (should we be willing) to go to preserve a genetically distinct group of organisms? How should the costs of preserving genetic diversity compare with the costs of other public projects?

Chapter 15

1. a. Abiotic synthesis of organic molecules using chemicals in atmosphere and lightning or other source of energy (or around deep-sea vents)
   b. Polymerization of monomers, perhaps on hot rocks
   c. Enclosure within a lipid membrane, which maintains distinct internal environment
   d. Beginnings of heredity as RNA molecules may have replicated themselves. Natural selection could act on protobionts that enclosed self-replicating RNA.

2. a. phylogenetic tree; b. homologies; c. morphology; d. analogies; e. outgroup; f. shared derived characters

3. c   4. b   5. e   6. c   7. e   8. a   9. d   10. b

11. Microevolution is the change in the gene pool of a population from one generation to the next. Macroevolution involves the pattern of evolutionary changes over large time spans and includes the origin of new groups and evolutionary novelties, as well as mass extinctions.

12. The latter are more likely to be closely related, since even small genetic changes can produce divergent

physical appearances. But if genes have diverged greatly, it implies that lineages have been separate for some time, and the similar appearances may be analogous, not homologous.

13. Complex structures can evolve by the gradual refinement of earlier versions of those structures, all of which served a useful function in the ancestor.

14. Where and when in a developing embryo that key developmental genes are expressed can greatly alter body geography and structures. Regulating gene expression allows these genes to continue to be expressed in some areas, turned off in other areas, and expressed at different times during development.

15. The ribosomal RNA genes, which code for the RNA parts of ribosomes, have evolved so slowly that homologies between even distantly related organisms can still be detected. Analysis of other homologous genes is also used.

16. 22,920 years old, a result of four half-life reductions

17.

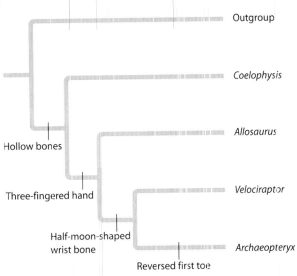

18. Some issues and questions to consider: Whereas previous mass extinctions have resulted from catastrophic events, such as asteroid collisions or volcanism, the current mass extinction is the result of human-caused environmental alteration. Mass extinctions can reduce complex ecological communities to much simpler ones. It can take millions of years for diversity to recover from a mass extinction. Do we have any ethical responsibility to preserve other species? By disrupting ecological communities throughout the world, a sixth mass extinction would have great consequences for all species alive today—including humans.

Chapter 16

1. Cell wall: maintains cell shape, provides physical protection, prevents cell from bursting in a hypotonic environment

   Capsule: enables cell to stick to substrate or to other individuals in a colony; shields pathogens from host's defensive cells

Motility: responds to chemical or physical signals in the environment that lead to nutrients or other members of their species and away from toxic substances

Rapid reproduction: colonizes favorable environment rapidly

Endospores: withstands harsh conditions

2. (a) Amoebozoan: has lobe-shaped pseudopodia

   (b) Ciliate: cilia visible on wide end of cell

   (c) Green alga: chloroplasts visible

   (d) Either a radiolarian or foraminiferan: threadlike pseudopodia

3. c 4. d (Algae are autotrophs; slime molds are heterotrophs.)
   5. e 6. c 7. d 8. b 9. b 10. d 11. b

12. Antibiotics target structures or processes that differ in bacteria and eukaryotic cells. Some antibiotics interfere with cross-linking in peptidoglycan; other antibiotics inhibit prokaryotic ribosomes.

13. Small, free-living prokaryotes were probably engulfed by a larger cell and took up residence inside. A symbiotic relationship developed between the host cell and engulfed cells, which became mitochondria. By a similar process, heterotrophic eukaryotic cells engulfed cyanobacteria, which became chloroplasts. Lineages of these autotrophic cells then diverged into several groups of protists.

14. *Chlamydomonas* is a eukaryotic cell, much more complex than a prokaryotic bacterium. It is autotrophic, while amoebas are heterotrophic. It is unicellular, unlike multicellular sea lettuce.

15. Multicellular organisms have a greater extent of cellular specialization and more interdependence of cells. New organisms are produced from a single cell, either an egg or an asexual spore.

16. Not a good idea; all life depends on bacteria. You could predict that eliminating all bacteria from an environment would result in a buildup of toxic wastes and dead organisms (both of which bacteria decompose), a shutdown of all chemical cycling, and the consequent death of all organisms.

17. Some issues and questions to consider: Could we determine beforehand whether the iron would really have the desired effect? How? Would the "fertilization" need to be repeated? Could it be a cure for the problem, or would it merely treat the symptoms? Might the iron treatment have side effects? What might they be?

Chapter 17

1. a. bryophytes; b. ferns and relatives; c. gymnosperms; d. angiosperms; 1. lignin-hardened vascular tissue; 2. seeds that protect and disperse embryos; 3. flowers that enclose gametophyte generation; fruits

2. A. This is a cloud of pollen being released from a pollen cone of a pine tree. The pine tree is the sporophyte generation. Sporangia in ovulate and pollen cones produce spores by meiosis. In ovulate cones, spores develop into the female gametophyte in the ovule. In pollen cones, spores produce

millions of the male gametophytes—the pollen grains. Pollination followed by fertilization produces a sporophyte embryo that is protected within the seed.

B. This is a cloud of haploid spores produced by a puffball fungus. Each spore may germinate to produce a haploid mycelium. Haploid hyphae of different mating types will fuse, creating a heterokaryotic mycelium, which is the feeding structure. Eventually, another puffball may develop, in which nuclei fuse in reproductive cells called basidia. Meiosis will produce spores that will be released, as shown in the photo.

3. b **4.** c (It is the only gametophyte among the possible answers.) **5.** a **6.** e **7.** b **8.** e **9.** c **10.** d

11. The alga is surrounded and supported by water, and it has no supporting tissues, vascular system, or special adaptations for obtaining or conserving water. Its whole body is photosynthetic, and its gametes and embryos are dispersed into the water. The seed plant has lignified vascular tissues that support it against gravity and carry food and water and organs that absorb water and minerals (roots), provide support (stems and roots), and photosynthesize (leaves and stems). It is covered by a waterproof cuticle and has stomata for gas exchange. Its sperm are carried by pollen grains, and embryos develop on the parent plant and are then protected and provided for by seeds.

12. Animals carry pollen from flower to flower and thus help fertilize eggs. They also disperse seeds by consuming fruit or carrying fruit that clings to their fur. In return, they get food (nectar, pollen, fruit).

13. Plants are autotrophs; they have chlorophyll and make their own food by photosynthesis. Fungi are heterotrophs that digest food externally and absorb nutrient molecules. There are also many structural differences; for example, the threadlike fungal mycelium is different from the plant body, and their cell walls are made of different substances. Plants evolved from green algae, fungi from a unicellular, flagellated, protistan ancestor. Evidence indicates that fungi are more closely related to animals than to plants.

14. Antibiotics probably kill off bacteria that compete with fungi for food. Similarly, bad tastes and odors deter animals that eat or compete with fungi. Those fungi that produce antibiotics and bad-smelling and bad-tasting chemicals would survive and reproduce more successfully than fungi unable to inhibit competitors. Animals that could recognize the smells and tastes also would survive and reproduce better than their competitors. Thus, natural selection would favor fungi that produce the chemicals and, to some extent, the competitors deterred by them.

15. Moss gametophytes, the dominant stage in the moss life cycle, are haploid plants. The diploid (sporophyte) generation is dominant in most other plants. Recessive mutations are not expressed in a diploid organism unless both homologous chromosomes carry the mutation. In haploid organisms, recessive mutations are apparent in the phenotype of the organism because haploid organisms have only one set of chromosomes. Some factors to consider in designing your experiment: What are the advantages/disadvantages of performing the experiment in the laboratory? In the field? What variables would be important to control? How many potted plants should you use? At what distances from the radiation source should you place them? What would serve as a control group for the experiment? What age of plants should you use?

16. Some issues and questions to consider: What are the other functions of forestland? How are other uses affected by logging? Must all the trees in an area be clear-cut? Should a particular area have multiple uses, or should different areas be used for different purposes? How much timber do we need? Could we conserve and recycle more? Are government-managed forests subsidizing private industry? Should we protect habitats as well as species? Does the rate of regrowth of trees match the rate of harvest? Will the ancient forests grow back? How do we balance conservation of the owl with the need for jobs and a productive economy?

Chapter 18

1. Sponges: sessile, saclike body with pores, suspension feeder; sponges

    Cnidarians: radial symmetry, gastrovascular cavity, cnidocytes, polyp or medusa body form; hydras, sea anemones, jellies, corals

    Flatworms: bilateral symmetry, gastrovascular cavity, no body cavity; free-living planarians, flukes, tapeworms

    Nematodes: pseudocoelom, covered with cuticle, complete digestive tract, ubiquitous, free-living and parasitic; roundworms, heartworms, hookworms, trichinosis worms

    Molluscs: muscular foot, mantle, visceral mass, circulatory system, many with shells, radula in some; snails and slugs, bivalves, cephalopods (squids and octopuses)

    Annelids: segmented worms, closed circulatory system, many organs repeated in each segment; earthworms, polychaetes, leeches

    Arthropods: exoskeleton, jointed appendages, segmentation, open circulatory system; chelicerates (spiders), crustaceans (lobsters, crabs), millipedes and centipedes, insects

    Echinoderms: radial symmetry as adult, water vascular system with tube feet, endoskeleton, spiny skin; sea stars, sea urchins

    Chordates: notochord; dorsal, hollow nerve cord; pharyngeal slits; post-anal tail; tunicates, lancelets, hagfish, and all the vertebrates (lampreys, sharks, ray-finned fishes, lobefins, amphibians, reptiles (including birds), mammals

2. c **3.** d **4.** a (The invertebrates include all animals except the vertebrates.) **5.** d **6.** c **7.** i **8.** f **9.** e **10.** c **11.** a **12.** d **13.** h **14.** b **15.** g

16. The gastrovascular cavity of a flatworm is an incomplete digestive tract; the worm takes in food and expels waste through the same opening. An earthworm has a complete digestive tract; food travels one way, and different areas

are specialized for different functions. The flatworm's body is solid and unsegmented. The earthworm has a coelom, allowing its internal organs to grow and move independently of its outer body wall. Fluid in the coelom cushions internal organs, acts as a skeleton, and aids circulation. Segmentation of the earthworm, including its coelom, allows for greater flexibility and mobility.

17. Cnidarians and most adult echinoderms are radially symmetrical, while most other animals, such as arthropods and chordates, are bilaterally symmetrical. Most radially symmetrical animals stay in one spot or float passively. Most bilateral animals are more active and move headfirst through their environment.

18. For example, the legs of a horseshoe crab are used for walking, while the antennae of a grasshopper have a sensory function. Some appendages on the abdomen of a lobster are used for swimming, while the scorpion catches prey with its pincers. (Note that the scorpion stinger and insect wings are not considered jointed appendages.)

19. Both trees agree on the early branching of eumetazoans into two groups based on body symmetry and the number of cell layers formed in gastrulation. Both trees recognize deuterostomes as a clade of bilaterians. In the morphological tree, protostomes are a second clade of bilaterians. The molecular tree distinguishes two clades within the protostomes, lophotrochozoans and ecdysozoans.

20. Important characteristics include symmetry, the presence and type of body cavity, segmentation, type of digestive tract, type of skeleton, and appendages.

21. Some issues and questions to consider: How does the decline of the reefs relate to agriculture? Deforestation? Overfishing? Rapid population growth? What value are the reefs to the local people? What is the value of reefs as a world biological resource? What might be the consequences if the reefs disappear? What is likely to make the situation worse? What is likely to improve the situation? In what ways might developed countries contribute to this problem? In what ways might developed countries be able to help the local people preserve the reefs? How might developed countries benefit from helping?

Chapter 19

1. a. brain; b. head; c. vertebral column; d. jaws; e. lungs or lung derivatives; f. lobed fins; g. legs; h. amniotic egg; i. milk

2. a. Primates: limber shoulder and hip joints, mobile digits of hands and feet, flexible thumb and big toe, short snout, and eyes set close together
   b. Hominoids: larger brains relative to body size than other primates, lack tails, more flexible behavior. (Along with other anthropoids, hominoids have an opposable thumb.)
   c. Hominids: bipedalism, larger brains, reduced sexual dimorphism, shorter jaws, flatter faces, language, symbolic thought, manufacture and use of complex tools, longer period of parental care. (Note that these traits appeared at various times in hominid evolution.)

3. c 4. c 5. e 6. b 7. b 8. c 9. c 10. a 11. b

12. Amphibians have four limbs adapted for locomotion on land, a skeletal structure that supports the body in a non-buoyant medium, and lungs. However, most amphibians are tied to water because they obtain some of their oxygen through thin, moist skin and they require water for fertilization and development. Reptiles are completely adapted to life on land. They have amniotic eggs that contain food and water for the developing embryo and a shell to protect it from dehydration. Reptiles are covered by waterproof scales that enable them to resist dehydration (more efficient lungs eliminate the need for gas exchange through the skin).

13. Fossil evidence supports the evolution of birds from a small, bipedal, feathered dinosaur, which was probably endothermic. The last common ancestor that birds and mammals shared was the ancestral amniote. The four-chambered hearts of birds and mammals must have evolved independently.

14. Several primate characteristics make it easy for us to make and use tools—mobile digits, opposable fingers and thumb, and great sensitivity of touch. Primates also have forward-facing eyes, which enhances depth perception and eye-hand coordination, and a relatively large brain.

15. Hominoids include all apes (gibbons, orangutans, gorillas, chimpanzees, and humans). Hominids include the species on the human evolutionary branch, more closely related to humans than to chimpanzees.

16. UV radiation is most intense in tropical regions and decreases farther north. Skin pigmentation is darkest in people indigenous to tropical regions and much lighter in northern latitudes. Scientists hypothesize that depigmentation was an adaptation to permit sufficient exposure to UV radiation, which catalyzes the production of vitamin D, a vitamin that permits the calcium absorption needed for both maternal and fetal bones. Dark pigmentation is hypothesized to protect against degradation of folate, a vitamin essential to normal embryonic development.

17. The paleontologists who discovered Tiktaalik hypothesized the existence of transitional forms between fishlike tetrapods such as *Pandericthyes* and tetrapod-like fish such as *Acanthostega*. From the available evidence, they knew the time periods when fishlike tetrapods and tetrapod-like fish lived. From the rocks in which the fossils had been found, they knew the geographic region and the type of habitat these creatures occupied. With this knowledge, they predicted the type of rock formation where transitional fossils might be found.

18. Our intelligence and culture—accumulated and transmitted knowledge, beliefs, arts, and products—have enabled us to overcome our physical limitations and alter the environment to fit our needs and desires.

19. Most anthropologists think that humans and chimpanzees diverged from a common ancestor 5–7 million years ago. Primate fossils 4–8 million years old might help us understand how the human lineage first evolved.

Chapter 20

1. a. epithelial tissue; b. connective tissue; c. smooth muscle tissue; d. connective tissue; e. epithelial tissue

   The structure of the specialized cells in each type of tissue fits their function. For example, columnar epithelial cells are specialized for absorption and secretion; the fibers and cells of connective tissue provide support and connect the tissues. The hierarchy from cell to tissue to organ is evident in this diagram. The functional properties of a tissue or organ emerge from the structural organization and coordination of its component parts. The many projections of the lining of the small intestine greatly increase the surface area for absorption of nutrients.

2. b 3. c 4. a 5. d (Expelling salt opposes the increase in blood salt concentration, thereby maintaining a constant internal environment.) 6. d 7. c 8. a 9. d 10. a 11. c 12. b 13. d 14. b

15. Stratified squamous epithelium consists of many cell layers. The outer cells are flattened, filled with the protein keratin, and dead, providing a protective, waterproof covering for the body. Neurons are cells with long branches that conduct signals to other cells, making multiple connections in the brain. Simple squamous epithelium is a single, thin layer of cells that allows for diffusion of gases across the lining of the lung. Bone cells are surrounded by a matrix that consists of fibers and mineral salts, forming a hard protective covering around the brain.

16. The surfaces of the intestine, excretory system, and lungs are highly folded and divided, increasing their surface area for exchange. These surfaces interface with many blood capillaries. Smaller creatures have a greater surface-to-volume ratio, and their cells are closer to the surface, enabling direct exchange between cells and the outside environment.

17. Some issues and questions to consider: Should a doctor's prescription be required for a whole-body CT scan? Should such scans be available only to those who can pay for them? Are CT scan machines calibrated so that they expose children or small adults to less radiation? Have there been research studies to test the effect of repeated exposures to radiation from CT scans? Whose responsibility is it to perform such studies and then publicize the results?

Chapter 21

1. a. mouth—ingests food; b. salivary glands—produce saliva; c. liver—produces bile and processes nutrient-laden blood from intestines; d. gall bladder—stores bile; e. pancreas—produces digestive enzymes and bicarbonate; f. rectum—storage of feces before elimination; g. oral cavity—where food is chewed, moistened, and formed into a bolus; h. pharynx—site of openings into esophagus and trachea; i. esophagus—transports bolus to stomach by peristalsis; j. stomach—food storage, mixes food with acid, begins digestion of proteins; k. small intestine—digestion and absorption; l. large intestine—absorption of water, compacts feces; m. anus—elimination of feces

2. a. fuel, chemical energy; b. raw materials, monomers; c. essential nutrients; d. overnourishment or obesity; e. vitamins and minerals; f. essential amino acids; g. malnourishment

3. e 4. c 5. b 6. d 7. e

8. You ingest the sandwich one bite at a time. In the oral cavity, chewing begins mechanical digestion, and salivary amylase action on starch begins chemical digestion. When you swallow, food passes through the pharynx and esophagus to the stomach. Mechanical and chemical digestion continue in the stomach, where HCl in gastric juice breaks apart food cells and pepsin begins protein digestion. In the small intestine, enzymes from the pancreas and intestinal wall break down starch, protein, and nucleic acids to monomers. Bile from the liver and gallbladder emulsifies fat droplets for attack by enzymes. Most nutrients are absorbed into the bloodstream through the walls of the small intestine. Fats travel through lymph. In the large intestine, water is absorbed from undigested material, and feces are produced and eliminated.

9. Our craving for fatty foods may have evolved from the feast-and-famine existence of our ancestors. Natural selection may have favored individuals who gorged on and stored high-energy molecules, as they were more likely to survive famines.

10. a. 58%
    b. Based on a 2,000-Calorie diet, this product would supply about 9.5% of daily Calories, and it supplies 10% of vitamin A and calcium. If all food consumed supplied a similar quantity, the daily requirement for these two nutrients would be met.
    c. The 8 g of saturated fat in this product represent 40% of the daily value. Thus, the daily value must be 20 g (8/0.4 = 20). This represents 180 Calories from saturated fat.

11. Sodas, chips, crackers, and cookies provide many calories (high energy) but few vitamins, minerals, proteins, or other nutrients. Unprocessed, fresh foods such as fruits and vegetables are considered nutrient-dense—they provide substantial amounts of vitamins, minerals, and other nutrients, and relatively few calories.

12. Some issues and questions to consider: What are the roles of family, school, advertising, media, and government in providing nutritional information? How might the available information be improved? What types of scientific studies form the foundation of various nutritional claims?

13. Some issues and questions to consider: In wealthy countries, what are the factors that make it difficult for some people to get enough food? In your community, what types of help exist to feed hungry people? Think of two recent food crises in other countries and what caused them. Did other countries or international organizations provide aid? Which ones, and how did they help? Did that aid address the underlying causes of malnutrition and starvation in the stricken area or only provide temporary relief? How might that aid be changed to offer more permanent solutions to food shortages?

Chapter 22

1. a. respiratory surface; b. circulatory system; c. lungs; d. hemoglobin; e. cellular respiration; f. negative pressure breathing; g. O_2

2. a. nasal cavity; b. pharynx; c. larynx; d. trachea; e. right lung; f. bronchus; g. bronchiole; h. diaphragm

3. c 4. b 5. d 6. a 7. e 8. d 9. c

10. Advantages of breathing air: It has a higher concentration of O_2 than water and is easier to move over the respiratory surface. Disadvantage of breathing air: It dries out the respiratory surface; living cells on the respiratory surface must remain moist.

11. Nasal cavity, pharynx, larynx, trachea, bronchus, bronchiole, alveolus, through wall of alveolus into blood vessel, blood plasma, into red blood cell, attaches to hemoglobin, carried by blood through heart, blood vessel in muscle, dropped off by hemoglobin, out of red blood cell, into blood plasma, through capillary wall, through interstitial fluid, and into muscle cell.

12. Both these effects of carbon monoxide interfere with cellular respiration and the production of ATP. By binding more tightly to hemoglobin, CO would decrease the amount of O_2 picked up in the lungs and delivered to body cells. Without sufficient O_2 to act as the final electron acceptor, cellular respiration would slow. And by blocking electron flow in the electron transport chain, cellular respiration and ATP production would cease. Without ATP, cellular work stops and cells and organisms die.

13. Llama hemoglobin has a higher affinity for O_2 than does human hemoglobin. The dissociation curve shows that its hemoglobin becomes saturated with O_2 at the lower PO_2 of the high altitudes to which llamas are adapted. At that PO_2, human hemoglobin is only 80% saturated.

14. Insects have a tracheal system for gas exchange. In order to provide O_2 to all the body cells in such a huge moth, the tracheal tubes would have to be wider (to provide enough ventilation across larger distances) and very extensive (to service large flight muscles and other tissues), thus presenting problems of water loss and increased weight. Both the tracheal system and the weight of the exoskeleton would limit the size of insects.

15. Some issues and questions to consider: Would a total ban on advertising decrease the number of cigarette smokers? If cigarettes are legal, can the right of cigarette manufacturers to advertise their product be restricted in this manner? Have similar bans on advertising of other legal, but potentially deadly products (such as alcohol) been tried? How have they worked? Do health concerns outweigh commercial concerns? If cigarettes are so bad, should they be declared illegal?

Chapter 23

1. a. capillaries of head, chest, and arms; b. aorta; c. pulmonary artery; d. capillaries of left lung; e. pulmonary vein; f. left atrium; g. left ventricle; h. aorta; i. capillaries of abdominal region and legs;

j. inferior vena cava; k. right ventricle; l. right atrium; m. pulmonary vein; n. capillaries of right lung; o. pulmonary artery; p. superior vena cava

See text Figure 23.3A for numbers and red vessels.

2. b 3. d (The second sound is the closing of the semilunar valves as the ventricles relax.) 4. c 5. c 6. a 7. e 8. b 9. a

10. Pulmonary vein, left atrium, left ventricle, aorta, artery, arteriole, body tissue capillary bed, venule, vein, vena cava, right atrium, right ventricle, pulmonary artery, capillary bed in lung, pulmonary vein

11. Capillaries are very numerous, producing a large surface area for exchange close to body cells. The capillary wall is only one epithelial cell thick. Clefts between epithelial cells allow fluid with small solutes to move out at the arteriole end and back in at the venous end of the capillary.

12. a. Plasma (the straw-colored fluid) would contain water, inorganic salts (ions such as sodium, potassium, calcium, magnesium, chloride, and bicarbonate), plasma proteins such as fibrinogen and immunoglobulins (antibodies), and substances transported by blood, such as nutrients (for example, glucose, amino acids, vitamins), waste products of metabolism, respiratory gases (O_2 and CO_2), and hormones. b. The red portion would contain erythrocytes (red blood cells), leukocytes (white blood cells—basophils, eosinophils, neutrophils, lymphocytes, and monocytes), and platelets.

13. Oxygen content is reduced as oxygen-poor blood returning to the right ventricle from the systemic circuit mixes with oxygen-rich blood of the left ventricle.

14. Proteins are important solutes in blood, accounting for much of the osmotic pressure that draws fluid back into the blood. If protein concentration is reduced, the inward pull of osmotic pressure will fail to balance the outward push of blood pressure. The net pressure at the arterial end of a capillary forces fluid out into the interstitial fluid. Net pressure at the venous end is not great enough to draw fluid back in.

15. Some issues and questions to consider: Is it ethical to have a child in order to save the life of another? Is it right to conceive a child as a means to an end—to produce a tissue or organ? Is this a less acceptable reason than most reasons parents have for bearing children? Do parents even need a reason for conceiving a child? Do parents have the right to make decisions like this for their young children? How will the donor (and recipient) feel about this when the donor is old enough to know what happened?

16. With a three-chambered heart, there is some mixing of oxygen-rich blood returning from the lungs with oxygen-poor blood returning from the systemic circulation. Thus, the blood of a dinosaur might not have supplied enough O_2 to support the higher metabolism and strong cardiac muscle contractions needed to generate such a high systolic blood pressure. Also, with a single ventricle pumping simultaneously to both pulmonary and systemic circuits, the blood pumped to the lungs would be at such a high pressure that it would damage the lungs.

Chapter 24

1. a. innate immunity; b. acquired immunity; c. B cells;
   d. T cells; e. antibodies; f. helper T cells

2. e 3. b 4. b 5. d 6. a 7. b 8. f 9. d 10. e 11. a 12. g 13. c

14. AIDS is mainly transmitted in blood and semen. It enters the
    body through slight wounds during sexual contact or via
    needles contaminated with infected blood. AIDS is deadly
    because it infects helper T cells, crippling both the humoral
    and cell-mediated immune responses and leaving the body
    vulnerable to other infections. The most effective way to
    avoid HIV transmission is to prevent contact with bodily
    fluids by practicing safe sex and avoiding intravenous drugs.

15. Inflammation is triggered by tissue injury. Damaged cells
    release histamine and other chemicals, which cause nearby
    blood vessels to dilate and become leakier. Blood plasma
    leaves vessels, and phagocytes are attracted to the site of
    injury. An increase in blood flow, fluid accumulation, and
    increased cell population cause redness, heat, and swelling.
    Inflammation disinfects and cleans the area and curtails the
    spread of infection from the injured area. Inflammation is
    considered part of the innate immune response because
    similar defenses are presented in response to any infection.

16. One hypothesis is that your roommate's previous bee stings
    caused her to become sensitized to the allergens in bee
    venom. During sensitization, antibodies to allergens attach
    to receptor proteins on mast cells. During this sensitization
    stage, she would not have experienced allergy symptoms.
    When she was exposed to the bee venom again at a later
    time, the bee venom allergens bound to the mast cells,
    which triggered her allergic reaction.

17. Some issues and questions to consider: There is no correct
    answer to this question. Possible directions include the idea
    that if the donor felt strongly about the process, then his
    or her wishes should be respected. The opposite direction
    would be that the next of kin should be able to approve
    or deny the procedure. Other considerations may be
    appropriate, including religious beliefs.

18. Some issues and questions to consider: How important is it to
    protect students from AIDS? Is this a function of schools? Do
    schools serve other such "noneducational" purposes? Should
    parents or citizens' and church groups—on either side of the
    issue—have a say in this, or is it a matter between the school
    and the student? Does the distribution of condoms condone
    or sanction sexual activity or promiscuity? Is a school legally
    liable if a school-issued condom fails to protect a student? Are
    there alternative measures, such as education, that might be as
    effective for slowing the spread of AIDS?

Chapter 25

1. a. thermoregulation; b. osmoregulation; c. excretion;
   d. ectotherm; e.–g. ammonia, urea, uric acid; h. behavioral
   responses; i. environment

2. a. filtration: water; NaCl, HCO_3^-, H^+, urea, glucose,
   amino acids, some drugs; b. reabsorption: nutrients,
   NaCl, water, HCO_3^-, urea; c. secretion: some drugs

and toxins, H^+, K^+; d. excretion: urine containing water,
urea, and excess ions

3. c 4. a 5. e 6. d 7. c 8. c, a 9. b 10. c 11. c 12. c

13. b 14. d 15. b 16. a

17. In salt water, the fish loses water by osmosis. It drinks salt
    water and disposes of salts through its gills. Its kidneys con-
    serve water and excrete excess ions. In fresh water, it gains
    water by osmosis. Its kidneys excrete a lot of dilute urine. Its
    gills take up salt, and some ions are ingested with food.

18. Yes. Ectotherms that live in very stable environments,
    such as tropical seas or deep oceans, have stable body
    temperatures. And terrestrial ectotherms can maintain
    relatively stable temperatures by behavioral means.

19. a. An endotherm would produce more nitrogenous wastes
    because it must eat more food to maintain its higher
    metabolic rate.
    b. A carnivore, because it eats more protein and thus
    produces more breakdown products of protein
    digestion—nitrogenous wastes.

20. You could take it back to the laboratory and measure its
    body temperature under different ambient temperatures.

21. A countercurrent heat exchange in the birds' legs reduces
    the loss of heat from the body. You would expect the
    temperature of blood flowing back to the body from the
    legs to be only slightly cooler than the blood flowing
    from the body to the legs.

22. Some issues and questions to consider: Could drug use
    endanger the safety of the employee or others? Is drug test-
    ing relevant to jobs where safety is not a factor? Is drug
    testing an invasion of privacy, interfering in the private life
    of an employee? Is an employer justified in banning drug
    use off the job if it does not affect safety or ability to do the
    job? Do the same criteria apply to employers requiring the
    test? Could an employer use a drug test to regulate other
    employee behavior that is legal, such as smoking?

23. Some issues and questions to consider: Should human
    organs be sold? If the donor is poor, he or she may also
    not be in the best of health. What are the health risks to
    the donor? Will the recipient be assured of buying a healthy
    organ? How much does the organ broker make from this
    transaction? Both parties in such a transaction are probably
    fairly desperate. Should regulations be in place when people
    may not be able to make reasoned decisions? Or do people
    have a right to sell parts of their bodies, just as they can now
    sell other possessions or their labor?

Chapter 26

1–8. Pineal gland: 5, f
     Testes: 3, d
     Parathyroid gland: 7, g
     Adrenal medulla: 2, c
     Hypothalamus: 8, h
     Pancreas: 4, a
     Anterior pituitary: 6, b
     Thyroid gland: 1, e

9. d **10.** e **11.** a **12.** b (Negative feedback: When thyroxine increases, it inhibits TSH, which reduces thyroxine secretion.) **13.** d

14. The hypothalamus secretes releasing hormones and inhibiting hormones, which are carried by the blood to the anterior pituitary. In response to these signals from the hypothalamus, the anterior pituitary increases or decreases its secretion of a variety of hormones that directly affect body activities or influence other glands. Neurosecretory cells that extend from the hypothalamus into the posterior pituitary secrete hormones that are stored in the posterior pituitary until they are released into the blood.

15. Only cells with the proper receptors will respond to a hormone. For a steroid hormone, the presence (or absence) and types of receptor proteins inside the cell determine the hormone's effect. For a nonsteroid hormone, the types of receptors on the cell's plasma membrane are key, and the proteins of the signal transduction pathway may have different effects inside different cells.

16. a. No. Blood sugar level goes too low. Diabetes would tend to make the blood sugar level go too high after a meal.
   b. No. Insulin is working, as seen by the homeostatic blood sugar response to feeding.
   c. Correct. Without glucagon, exercise and fasting lower blood sugar, the cells cannot mobilize any sugar reserves, and their blood sugar level drops. Insulin (which lowers blood sugar) has no effect.
   d. No. If this were true, blood sugar level would increase too much after a meal.

17. If cells within a male embryo do not secrete testosterone at the proper time during development, the embryo will develop into a female despite being genetically male.

Chapter 27

1. A. FSH; B. estrogen; C. LH; D. progesterone; P. menstruation; Q. growth of follicle; R. ovulation; S. development of corpus luteum.

If pregnancy occurs, the embryo produces human chorionic gonadotrophin (HCG), which maintains the corpus luteum, keeping levels of estrogen and progesterone high.

2. c **3.** c (The outer layer in a gastrula is the ectoderm; of the choices given, only the brain develops from ectoderm.) **4.** a **5.** d **6.** e **7.** g **8.** d **9.** h **10.** f **11.** a **12.** b **13.** c

14. Both produce haploid gametes. Spermatogenesis produces four small sperm; oogenesis produces one large ovum. In humans, the ovary contains all the primary oocytes at birth, while testes can keep making primary spermatocytes throughout life. Oogenesis is not complete until fertilization, but sperm mature without eggs.

15. The extraembryonic membranes provide a moist environment for the embryos of terrestrial vertebrates and enable the embryos to absorb food and oxygen and dispose of wastes. Such membranes are not needed when an embryo is surrounded by water, as are those of fishes and amphibians.

16. The nerve cells may follow chemical trails to the muscle cells and identify and attach to them by means of specific surface proteins.

17. The researcher might find out whether chemicals from the notochord stimulate the nearby ectoderm to become the neural tube, a process called induction. Transplanted notochord tissue might cause ectoderm anywhere in the embryo to become neural tissue. Control: Transplant non-notochord tissue under the ectoderm of the belly area.

18. Some issues and questions to consider: What characteristics might parents like to select for? If parents had the right to choose embryos based on these characteristics, what are some of the possible benefits? What are potential pitfalls? Could an imbalance of the population result?

Chapter 28

1. a. sensory receptor; b. sensory neuron; c. synapse; d. spinal cord; e. interneuron; f. motor neuron; g. effector cell; h. CNS; i. PNS

2. b **3.** a **4.** Both a and c would prevent action potentials from occurring; b could actually increase the generation of action potentials.

5. At the point where the action potential is triggered, sodium ions rush into the neuron. They diffuse laterally and cause sodium gates to open in the adjacent part of the membrane, triggering another action potential. The moving wave of action potentials, each triggering the next, is a moving nerve signal. Behind the action potential, sodium gates are temporarily inactivated, so the action potential can only go forward. At a synapse, the transmitting cell releases a chemical neurotransmitter, which binds to receptors on the receiving cell and may trigger a nerve signal in the receiving cell.

6. The results show the cumulative effect of all incoming signals on neuron D. Comparing experiments 1 and 2, we see that the more nerve signals D receives from C, the more it sends; C is excitatory. Because neuron A is not varied here, its action is unknown; it may be either excitatory or mildly inhibitory. Comparing experiments 2 and 3, we see that neuron B must release a strongly inhibitory neurotransmitter, because when B is transmitting, D stops.

7. Some issues and questions to consider: Some people might be against the use of embryonic stem cells for any disorder. This may be related to religious or moral beliefs. Potentially, some people may not be aware of the source of these stem cells and may be against their use because they believe that the cells come from elective abortions. Other people may agree with the use of stem cell research because of the potential to cure diseases that are currently fatal. Some people who may have a neutral opinion on the issue might be swayed by the thought of a loved one who might be helped by stem cell therapy.

8. Some issues and questions to consider: What is the role of alcohol in crime? What are its effects on families and in the workplace? In what ways is the individual responsible for alcohol abuse? The family? Society? Who is affected by alcohol abuse? How effective are treatment and punishment in curbing alcohol abuse? Who pays for alcohol abuse and consequent treatment or punishment? Is it possible to enjoy alcohol without abusing it?

Chapter 29

1. a. mechanoreceptors; b. chemoreceptors; c. electricity, magnetism, light; d. hair cells; e. photoreceptors

2. d (He could hear the tuning fork against his skull, so the cochlea, nerve, and brain are OK. Apparently, sounds are not being transmitted to the cochlea; therefore, the bones are the problem.) **3.** b **4.** a **5.** b **6.** e **7.** a

8. Louder sounds create pressure waves with greater amplitude, moving hair cells more and generating a greater frequency of action potentials. Different pitches affect different parts of the basilar membrane; different hair cells stimulate different sensory neurons that transmit action potentials to different parts of the brain.

9. Sensation is the detection of stimuli (light) by the photoreceptors of the retina and transmission of action potentials to the brain. Perception is the interpretation of these nerve signals—sorting out the patterns of light and dark and determining their meaning.

10. Taste is used to sample food and determine its quality. Smell is used for many functions—communication of territories (scent marking), navigation (salmon), mate location (moths), sensing danger (predators, fires), and finding food.

11. Some possible hypotheses: Paired sensory receptors enable an animal to determine the direction from which stimuli come. Paired receptors enable comparison of the intensity of stimuli on either side. Paired receptors enable comparison of slightly different images seen by the eyes or sounds heard by the ears (thus enabling the brain to perceive depth and distance).

12. Do the turtles hear the surf? Plug the ears of some turtles and not others. If turtles without earplugs head for the water and turtles with earplugs get lost, they probably hear the ocean. Or do they smell the water? Plug their nostrils and follow the same process.

13. Some issues and questions to consider: Assuming that the sound is loud enough to impair hearing, how long an exposure is necessary for this to occur? Does exposure have to occur all at once, or is damage cumulative? Who is responsible, concert promoters or listeners? Should there be regulations regarding sound exposure at concerts (as there are for job-related noise)? Are young people sufficiently mature and aware to heed such warnings?

Chapter 30

1. a. skeleton; b. muscles; c. exoskeleton; d. sarcomeres; e. bone and cartilage

2. c **3.** d **4.** a (Water supports aquatic animals, reducing the effects of gravity.) **5.** c **6.** d **7.** e **8.** d **9.** a (Each neuron controls a smaller number of muscle fibers.) **10.** e

11. Advantages of an insect exoskeleton include strength, good protection for the body, flexibility at joints, and protection from water loss. The major disadvantage is that the exoskeleton must be shed periodically as the insect grows, leaving the insect temporarily weak and vulnerable.

12. The bird's wings are airfoils, with convex upper surfaces and flat or concave lower surfaces. As the wings beat, air passing over them travels farther than air beneath. Air molecules above the wings are more spread out, lowering pressure. Higher pressure beneath the wings pushes them up.

13. Calcium is needed for healthy bone development. It strengthens bones and makes them less susceptible to stress fractures.

14. Action potentials from the brain travel down the spinal cord and along a motor neuron to the muscle. The neuron releases a neurotransmitter, which triggers action potentials in a muscle fiber membrane. These action potentials initiate the release of calcium ions from the ER of the cell. Calcium enables myosin heads of the thick filaments to bind with actin of the thin filaments. ATP provides energy for the movement of myosin heads, which causes the thick and thin filaments to slide along one another, shortening the muscle fiber. The shortening of muscle fibers pulls on bones, bending the arm. If more motor units are activated, the contraction is stronger.

15. The fundamental vertebrate body plan includes an axial skeleton (skull, backbone, and rib cage) and an appendicular skeleton (bones of the appendages). Species vary in the numbers of vertebrae and the numbers of different types of vertebrae they possess. For example, pythons have no cervical vertebrae. Mammals (almost) all have seven cervical vertebrae, but may have different numbers of other types. For example, human coccygeal vertebrae are small and fused together, but horses and other animals with long tails have many coccygeal vertebrae. Limb bones have been modified into a variety of appendages, such as wings, fins, and limbs. Snakes have no appendages.

16. Chemical A would work better, because acetylcholine triggers contraction. Blocking it would prevent contraction. Chemical B would actually increase contraction, because Ca^{2+} allows contraction to occur.

17. Circular muscles in the earthworm body wall decrease the diameter of each segment, squeezing internal fluid and lengthening the segment. Longitudinal muscles shorten and thicken each segment. Different parts of the earthworm can lengthen while others shorten, producing a crawling motion. The whole roundworm body moves at once because of a lack of segmentation. The body can only shorten or bend, not lengthen, because of a lack of circular muscles. Roundworms simply thrash from side to side.

18. Calcium ions move the regulatory protein tropomyosin out of the way so myosin heads can bind to actin, resulting in muscle contraction. ATP causes the myosin heads of the thick filaments to detach from the thin filaments (Figure 30.9B, step 1). If there is no ATP present, the myosin heads remain attached to the thin filaments, and the muscle fiber remains fixed in position.

19. Some issues and questions to consider: Are the places where you live, work, or attend class accessible to a person in a wheelchair? If you were in a wheelchair, would you have trouble with doors, stairs, drinking fountains, toilet facilities, and eating facilities? What kinds of transportation would be available to you, and how convenient would they be? What activities would you have to forgo? How might your disability alter your relationships with your friends and family? How well would you manage on your own?

Chapter 31

1. Here is one possible concept map:

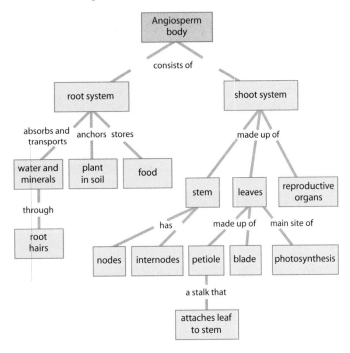

2. d (The vascular cambium forms to the outside of the primary xylem. The secondary xylem forms between primary xylem and the vascular cambium. The secondary phloem and primary phloem are outside the vascular cambium.)
   3. b **4.** e **5.** e **6.** f **7.** b **8.** e **9.** a **10.** c **11.** d

12. Bees deposit the pollen on the stigma of a carpel, and a pollen tube grows to the ovary at the base of the carpel. Sperm travel down the pollen tube and fertilize egg cells in ovules. The ovules grow into seeds, and the ovary grows into the flesh of the fruit (actually, in an apple, just the core). As the seeds mature, the fruit ripens and falls (or is picked).

13. Bulbs, root sprouts, and runners are all examples of asexual reproduction. Asexual reproduction is less wasteful and costly than sexual reproduction and less hazardous for young plants. The primary disadvantage of asexual reproduction is that it produces genetically identical offspring, decreasing genetic variability that can help a species survive times of environmental change.

14. Tomato: fruit (ripened ovary); celery stalk: leaf stalk (petiole); peanut: seed (ovule); strawberry: fruit (ripened ovary); lettuce: leaf blades; artichoke: terminal bud; beet: root

15. Modern methods of plant breeding and propagation have increased crop yields but have decreased genetic variability, so plants have become more vulnerable to epidemics. Primitive varieties of crop plants could contribute to gene banks and be used for breeding new strains.

16. Some issues and questions to consider: Why are the forests being cut when preserving them is more beneficial in the long run? Why can't the developing nations create new forest products themselves? Should developing and developed nations share in the profits from the forest? What incentive is there for a company to create a new product if it can't keep the profits? What about companies paying an exploration fee for exclusive rights to take plants from a certain area? What about a contract for the developing nation to get a percentage of the profits from new products? Are there reasons to preserve the forest other than profits and beneficial new products?

Chapter 32

1. a. roots; b. xylem; c. sugar source; d. phloem; e. transpiration; f. sugar sink

2. d **3.** d **4.** e **5.** b

6. If the plant starts to dry out, K^+ is pumped out of the guard cells. Water follows by osmosis, the guard cells become flaccid, and the stomata close. This prevents wilting, but it keeps leaves from taking in carbon dioxide, needed for photosynthesis.

7. Hypothesis: The hydrogen ions in acid precipitation displace positively charged nutrient ions from negatively charged clay particles. Test: In the laboratory, place equal amounts and types of soil in separate filters. The pore size of the filter must not allow any undissolved soil particles to pass through. Spray (to simulate rain) soil samples in the filters with solutions of different pH (for example, pH 5, 6, 7, 8, 9). Determine the concentration of nutrient ions in the solutions. (The only variable in the solutions should be the hydrogen ion concentration. Ideally, the solutions would contain no dissolved nutrient ions.) Collect fluid that drips through soil samples and filters. Determine the hydrogen ion concentration and the nutrient ion concentration in each sample of fluid. Prediction: If the hypothesis is correct, the fluid collected from the soil samples exposed to pH lower than 5.6 (acid rain) will contain the highest concentration of positively charged nutrient ions.

8. Hypothesis: When fixed nitrogen levels increase to a certain level in the soil, it slows the metabolism of (or otherwise harms or kills) the nitrogen-fixing bacteria that provide usable nitrogen to the crops. Test: Expose cultures of nitrogen-fixing bacteria (symbiotic and nonsymbiotic ones found in soil) to solutions of different concentrations of fixed nitrogen (that is, NO_3^- and NH_4^+). Determine the concentrations of nitrogen-fixing enzymes produced by the surviving bacteria in each sample. Prediction: If your hypothesis is correct, and the fixed nitrogen concentration is high enough to cause

harm in some of the samples, you would expect the enzyme concentration to be measurably lower in samples whose fixed nitrogen concentration is above the level that causes harm.

9. The hypothesis is supported if transpiration varies with light intensity when humidity and temperature are about the same. These conditions are seen at two places in the table; at hours 11 and 12, recordings for temperature and humidity are about the same, but light intensity increased markedly from 11 to 12, as did the transpiration rate. The recordings made at hours 3 and 4 show the same effects. Also, the recordings made at hours 1 and 2 generally support the hypothesis. Here, both temperature and humidity decreased, so you might expect the transpiration rate to stay about the same or perhaps increase because the temperature decrease is small; however, the transpiration rate dropped, as did the light intensity.

10. Some issues and questions to consider: How were the farmers assigned or sold "rights" to the water? How is the price established when a farmer buys or sells water rights? Is there enough water for everyone who "owns" it? What kinds of crops are these farmers growing? What will the water be used for in the city? Are there other users with no rights, such as wildlife? Is any effort being made to curb urban growth and conserve water? Should millions of people be living in what is essentially a desert? What are the reasons for farming desert land?

Chapter 33

1. a. auxin; b. gibberellin; c. auxin; d. cytokinin; e. auxin; f. ethylene; g. gibberellin; h. abscisic acid

2. b 3. c (A short-day plant requires a long night. The 13-hour night of answer c is the only uninterrupted night longer than 12 hours.) 4. b 5. a 6. d 7. b 8. b 9. f 10. e 11. d 12. a 13. b 14. g 15. c

16. Fruit produces ethylene gas, which triggers the ripening and aging of the fruit. Ventilation prevents a buildup of ethylene and delays its effects.

17. The terminal bud produces auxins, which counter the effects of cytokinins from the roots and inhibit the growth of axillary buds. If the terminal bud is removed, the cytokinins predominate, and lateral growth occurs at the axillary buds.

18. The red wavelengths in the room's lights quickly convert the phytochrome in the mums to the P_{fr} form, which inhibits flowering in a long-night plant. The mums will not flower unless Jon can rig up some far-red lights. Exposure to a burst of far-red light would convert the phytochrome to the P_r form, allowing flowering to occur.

19. The biologist could remove leaves at different stages of being eaten to see how long it takes for changes to occur in nearby leaves. The "hormone" could be captured in an agar block, as in the phototropism experiments in Module 33.1, and applied to an undamaged plant. Another experiment would be to block "hormone" movement out of a damaged leaf or into a nearby leaf.

20. The results illustrated in the graph generally support the hypothesis; measured by chlorophyll fluorescence, greater leaf damage occurred in air with no isoprene than in air with isoprene. The experimenters exposed leaves to N_2 to prevent the leaves themselves from emitting isoprene. By doing so, they were able to control the amount of isoprene in the air to which the leaves were exposed.

21. Some issues and questions to consider: Is the hormone safe for human consumption? What are its effects in the environment? Could its production produce impurities or wastes that might be harmful? What kinds of tests need to be done to demonstrate its safety? How much does it cost to make and use? Are the benefits worth the costs and risks? Is it worth using an artificial chemical on food simply to improve its appearance?

A scientist could seek answers by studying the stability of the hormone in a variety of laboratory simulations of natural conditions. The toxicity of the hormone, the materials used to produce it, and its breakdown products could be determined in laboratory tests.

Chapter 34

1. a. The shape of Earth results in uneven heating, such that the tropics are warm and polar regions are cold.
   b. The seasonal differences of winter and summer in temperate and polar regions are produced as Earth tips toward or away from the sun during its orbit around the sun.
   c. Intense solar radiation in the tropics evaporates moisture; warm air rises, cools, and drops its moisture as rain; air circulates, cools, and drops around 30°N and S, warming as it descends and evaporating moisture from land, which creates arid regions.

2. a. All three areas have plenty of sunlight and nutrients.
   b. The addition of nitrogen or phosphorus "fertilizes" the algae living in ponds and lakes, leading to explosive population growth. (These nutrients are typically in short supply in aquatic ecosystems.) The effects are harmful to the ecosystem. Algae cover the surface, reducing light penetration. When the algae die, bacterial decomposition of the large amount of biomass can deplete the oxygen available in the pond or lake, which may adversely affect the animal community.

3. f 4. g 5. d 6. a 7. c 8. b 9. e 10. a 11. e 12. d 13. a 14. c 15. b 16. e

17. Tropical rain forests have a warm, moist climate, with favorable growing conditions year-round. The diverse plant growth provides various habitats for other organisms.

18. After identifying your biome by looking at the map in Figure 34.8, review Module 34.5 on climate and the module that describes your biome (Modules 34.9–34.16).

19. a. desert; b. grassland; c. tropical forest; d. temperate forest; e. coniferous forest; f. arctic tundra. Areas of

overlap have to do with seasonal variations in temperature and precipitation.

20. Through convergent evolution, these unrelated animals adapted in similar ways to similar environments—temperate grasslands and savanna.

21. Some issues and questions to consider: What reasons are there for protecting areas like this? Is a public refuge or reserve the only way to protect species and habitats? Is there some way to protect organisms without confiscating land? Does protection take precedence over property rights? Would the prairie have been better off without the publicity? Should there be state or federal laws to protect endangered plants on private land? Should there be laws to protect endangered habitats, like this patch of prairie?

Chapter 35

1.

2. Your concept map should include: a. proximate questions; b. ultimate questions; c. mating behavior; d. communication/signals; e. agonistic behaviors; f. altruism

3. d 4. c 5. d 6. a

7. Main advantage: Flies do not live long. Innate behaviors can be performed the first time without learning, enabling flies to find food, mates, etc., without practice. Main disadvantage: Innate behaviors are rigid; flies cannot learn to adapt to specific situations.

8. In a stressful environment, for example, where predators are abundant, rats that behave cautiously are more likely to survive long enough to reproduce.

9. Courtship behaviors reduce aggression between potential mates and confirm their species, sex, and physical condition. Environmental changes such as rainfall, temperature, and day length probably lead frogs to start calling, so these would be the proximate causes. The ultimate cause relates to evolution. Fitness (reproductive success) is enhanced for frogs that engage in courtship behaviors.

10. a. Yes. The experimenter found that 5 m provided the most food for the least energy because the total flight height (number of drops × height per drop) was the lowest. Crows appear to be using an optimal foraging strategy.
    b. An experiment could measure the average drop height for juvenile and adult birds, or it could trace individual birds during a time span from juvenile to adult and see if their drop height changed.

11. One likely hypothesis is that the helper is closely related to one or both of the birds in the mated pair. Because closely related birds share relatively many genes, the helper bird is indirectly enhancing its own fitness by helping its relatives raise their young. (In other words, this behavior evolved by kin selection.) The easiest way to test the hypothesis would be to determine the relatedness of the birds by comparing tissue samples using molecular methods such as those described in Module 15.18. If birds are closely related, their DNA and proteins should be more similar than those of more distantly related or unrelated birds.

12. Identical twins are genetically the same, so any differences between them are due to environment. Thus, the study of identical twins enables researchers to sort out the effects of "nature" and "nurture" on human behavior. The data suggest that many aspects of human behavior are inborn. Some people find these studies disturbing because they seem to leave less room for free will and self-improvement than we would like. Results of such studies may be carelessly cited in support of a particular social agenda.

13. Animals are used in experiments for various reasons. A particular species of animal may have features that make it well suited to answer an important biological question. Squids, for example, have a giant nerve fiber that made possible the discovery of how all nerve cells function. Animal experiments play a major role in medical research. Many vaccines that protect humans against deadly diseases, as well as drugs that can cure diseases, have been developed using animal experiments. Animals also benefit, as vaccines and drugs are developed for combating their own pathogens. Some researchers point out that the number of animals used in research is a small fraction of those killed as strays by animal shelters and a minuscule fraction of those killed for human food. They also maintain that modern research facilities are models of responsible and considerate treatment of animals. Whether or not this is true, the possibility that at least some kinds of animals used in research suffer physical pain as a result, and perhaps mental anguish as well, raises serious ethical issues. Some questions to consider: What are some medical treatments or products that have undergone testing in animals? Have you benefited from any of them? Are there alternatives to using animals in experiments?

Would alternatives put humans at risk? Are all kinds of animal experiments equally valuable? In your opinion, what kinds of experiments are acceptable, and what kinds are unnecessary? What kinds of treatment are humane, and what kinds are inhumane?

Chapter 36

1. a. The x-axis is time; the y-axis is the number of individuals (N). The blue curve represents exponential growth; red is logistic growth.
   b. $G = rN$
   c. Carrying capacity of the environment (K).
   d. In exponential growth, population growth continues to increase as the population size increases. In logistic growth, the population grows fastest when the population is about $\frac{1}{2}$ the carrying capacity—when N is large enough so that rN produces a large increase, but the expression $(K - N)/K$ has not yet slowed growth as much as it will as N gets closer to K.
   e. Exponential growth curve, although the worldwide growth rate is slowing.

2. a. Birth rate is blue line; death rate is red line.
   b. I. Both birth and death rates are high. II. Birth rate remains high; death rate decreases, perhaps as a result of increased sanitation and health care. III. Birth rate declines, often coupled with increased opportunities for women and access to birth control; death rates are low. IV. Both birth and death rates are low.
   c. I and IV
   d. II, when death rate has fallen but birth rate remains high

3. a 4. c 5. e 6. c 7. b 8. d

9. Food and resource limitation, such as food or nesting sites; accumulation of toxic wastes; disease; increase in predation; stress responses, such as seen in some rodents

10. Survivorship is the fraction of individuals in a given age interval that survive to the next interval. It is a measure of the probability of surviving at any given age. A survivorship curve shows the fraction of individuals in a population surviving at each age interval during the life span. Oysters produce large numbers of offspring, most of which die young, with a few living a full life span. Few humans die young; most live out a full life span and die of old age. Squirrels have approximately constant mortality and about an equal chance of surviving at all ages.

11. Clumped is the most common dispersion pattern, usually associated with unevenly distributed resources or social grouping. Uniform dispersion may be related to territories or inhibitory interactions between plants. A random dispersion is least common and may occur when other factors do not influence the distribution of organisms.

12. 10% of the animals in the second capture were marked. Knowing that 50 animals were initially marked, and assuming that 10% of the whole population is marked, then the

population is estimated to be 500. The equation for the mark-recapture method is:

$$N = \frac{\text{Number marked in first catch} \times \text{total second catch}}{\text{recaptured marked individuals}}$$

13. An animal that has been trapped before may be wary of the traps or, having learned that traps contain food, may have deliberately returned for more. This would cause estimates to be either higher or lower, respectively, than actual population numbers.

14. Populations with K-selected life history traits tend to live in fairly stable environments held near carrying capacity by density-dependent limiting factors. They reproduce later and have fewer offspring than species with r-selected traits. Their lower reproductive rate makes it hard for them to recover from human-caused disruption of their habitat. We would expect species with K-selected life histories to have a Type I survivorship curve (see Module 36.3).

15. The largest population segment, the baby boomers, is currently in the workforce in their peak earning years, paying into the Social Security system. However, they are approaching the end of their contributing years, and the smaller working numbers following them will provide less money, driving the fund into a deficit.

16. Some issues and questions to consider: How does population growth in developing countries relate to food supply, pollution, and the use of natural resources such as fossil fuels? How are these things affected by population growth in developed countries? Which of these factors are most critical to our survival? Are they affected more by the growth of developing or developed countries? What will happen as developing countries become more developed? Will it be possible for everyone to live at the level of the developed world?

Chapter 37

1. a. energy flow; b. chemical cycling; c. solar energy;
   d. producers; e. abiotic reservoirs; f. trophic levels (food chain or web)

2.

Interspecific interaction	Effect on species 1	Effect on species 2	Example (other answers possible)
Predation	+	−	Crocodile/fish
Competition	−	−	Squirrel/black bear
Herbivory	+	−	Hippo/grasses
Parasites and pathogens	+	−	Heartworm/dog; Salmonella/human
Mutualism	+	+	Hippo/microbes in hippo stomach

(Many other examples possible)

3. e 4. d 5. c 6. c 7. c

8. Plants benefit by having their seeds distributed away from the parent. Animals benefit when the seed contains food, for example, in fleshy fruits.

9. Cultural eutrophication is excessive fertilization of bodies of water from agricultural or suburban runoff that results in blooms of cyanobacteria and algae. Respiration from these organisms and their decomposers depletes oxygen levels, leading to fish kills.

10. The abiotic reservoir of the first three nutrients is the soil. Carbon is available as carbon dioxide in the atmosphere.

11. These animals are secondary or tertiary consumers, at the top of the production pyramid. Stepwise energy loss means not much energy is left for them; thus, they are rare and require large territories in which to hunt.

12. Chemicals with a gaseous form in the atmosphere, such as carbon and nitrogen, have a global biogeochemical cycle.

13. Nitrogen fixation of atmospheric N_2 into ammonium; decomposition of detritus into ammonium; nitrification of ammonium into nitrate; denitrification (by denitrifiers) of nitrates into N_2

14. Hypothesis: The kangaroo rat is a keystone species in the desert. (Apparently, herbivory by the rats kept the one plant from outcompeting the others; removing the rats reduced plant diversity.) Additional supporting evidence: Observations of the rats preferentially eating dominant plants; finding that the dominant plant recovers from herbivore damage faster

15. Some issues and questions to consider: What relationships (predators, competitors, parasites) might exist in the mussels' native habitat that are altered in the Great Lakes? How might the mussels compete with Great Lakes organisms? Might the Great Lakes species adapt in some way? Might the mussels adapt? Could possible solutions present problems of their own?

Chapter 38

1. endangered and threatened species; b. restoration ecology; c. bioremediation and augmentation; d. zoned reserves; e. critical habitat needs

2. c 3. d 4. a 5. e 6. b 7. d 8. a

9. Genetic, species, and ecosystem diversity. As populations become smaller, genetic diversity is usually reduced. Genetic biodiversity is also threatened when local populations of a species become extinct owing to habitat destruction or other assaults or when entire species are lost. Many human activities have led to the extinction of species. The greatest threats include habitat loss, introduced species, and overexploitation. Species extinction may alter the structure of whole communities. Pollution and other widespread disruptions may lead to the loss of entire ecosystems.

10. Carbon dioxide and several other gases in the atmosphere absorb infrared radiation and thus slow the escape of heat from Earth. This is called the greenhouse effect. The greenhouse effect is beneficial to life on Earth; without CO_2 in the atmosphere, the temperature at the surface of Earth would be much colder and less hospitable for life.

11. Fossil fuel consumption, industry, and agriculture are increasing the quantity of greenhouse gases—such as CO_2, methane, and nitrous oxide—in the atmosphere. These gases are trapping more heat and raising atmospheric temperatures. Increases of 2–5°C are projected over the next century. Logging and the clearing of forests for farming contribute to global warming by reducing the uptake of CO_2 by plants (and adding CO_2 to the air when trees are burned). Global warming is having numerous effects already, including melting polar ice, permafrost, and glaciers, shifting patterns of precipitation, causing spring temperatures to arrive earlier, and reducing the number of cold days and nights. Future consequences include rising sea levels and the extinction of many plants and animals. Global warming is an international problem; air and climate do not recognize international boundaries. Greenhouse gases are primarily produced by industrialized nations. Cooperation and commitment to less use of fossil fuels and to reduce deforestation will be necessary if the problem of global warming is to be solved.

12. These birds might be affected by pesticides while in their wintering grounds in Central and South America, where such chemicals may still be in use. The birds are also affected by deforestation throughout their range.

13. About 1.8 million species have been named and described. Assume that 80% of all living things (not just plants and animals) live in tropical rain forests. This means that there are 1.44 million species there. If half the species survive, this means that 0.72 million species will be extinct in 100 years, or 7,400 per year. This means that 19+ species will disappear per day, or almost one per hour. If there are 30 million species on Earth, 24 million live in the tropics, and 12 million will disappear in the next century. This is 120,000 per year, 329 per day, or 14 per hour.

14. Some issues and questions to consider: How does the use of fossil fuels affect the environment? What about oil spills? Disruption of wildlife habitat for construction of oil fields and pipelines? Burning of fossil fuels and possible climate change and flooding from global warming? Pollution of lakes and destruction of property by acid precipitation? Health effects of polluted air on humans? How are we paying for these "side effects" of fossil fuel use? In taxes? In health insurance premiums? Do we pay a nonfinancial price in terms of poorer health and quality of life? Could oil companies be required to pick up the tab for environmental effects of fossil fuel use? Could these costs be covered by an oil tax? How would this change the price of oil? How would a change in the price of oil change our pattern of energy use, our lifestyle, and our environment?

15. Data on carbon emissions can be found at a number of websites, including the United Nations Statistics Division (http://unstats.un.org/unsd/default.htm) and the World Resources Institute (http://www.wri.org/climate/). The per capita rank of the United States and Canada are generally very high, along with other developed nations. Transportation and energy use are the major contributors to the carbon

footprint. Any actions you can take to reduce these will help. For example, if you have a car, you can try to minimize the number of miles you drive by consolidating errands into fewer trips and using an alternative means of transportation (public transportation, walking, biking, etc) whenever possible. To reduce energy consumption, be aware of the energy you use—turn off lights, disconnect electronics that draw power when on standby, do laundry in cold water, for example. Websites such as http://www.ucsusa.org/ (Union of Concerned Scientists) and http://climatecrisis.net offer simple suggestions such as changing to energy-efficient lightbulbs. A Web search will turn up plenty of sites.

16. Some issues and questions to consider: How do population growth, resource consumption, pollution, and reduction in biodiversity relate to sustainability? How do poverty, economic growth and development, and political issues relate to sustainability? Why might developed and developing nations take different views of a sustainable society? What would life be like in a sustainable society? Have any steps toward sustainability been taken in your community? What are the obstacles to sustainability in your community? What steps have you taken toward a sustainable lifestyle? How old will you be in 2030? What do you think life will be like then?

	BioFlix	MP3 Tutors	Web Activities	Web Process of Science	Discovery Channel Videos	GraphIt!/You Decide
Chapter 1 *Biology: Exploring Life*		The Process of Science (1.7)	The Levels of Life Card Game (1.1) Classification Schemes (1.5)	How Do Environmental Changes Affect a Population? (1.6) How Does Acid Precipitation Affect Trees? (1.8)	Antibiotic Resistance (1.10)	GraphIt!: An Introduction to Graphing (1.8)
Chapter 2 *The Chemical Basis of Life*		The Properties of Water (2.11)	Structure of the Atomic Nucleus (2.4) Electron Arrangement (2.6) Build an Atom (2.6) Ionic Bonds (2.7) Covalent Bonds (2.8) Nonpolar and Polar Molecules (2.9) Water's Polarity and Hydrogen Bonding (2.10) Cohesion of Water (2.11) Acids, Bases, and pH (2.15)	How Are Space Rocks Analyzed for Signs of Life? (2.1) How Does Acid Precipitation Affect Trees? (2.16)	Early Life (2.17)	
Chapter 3 *The Molecules of Cells*		Protein Structure and Function (3.13) DNA Structure (3.16)	Diversity of Carbon-Based Molecules (3.1) Functional Groups (3.2) Making and Breaking Polymers (3.3) Models of Glucose (3.4) Carbohydrates (3.7) Lipids (3.9) Protein Functions (3.11) Protein Structure (3.14) Nucleic Acid Structure (3.16)	Connection: What Factors Determine the Effectiveness of Drugs? (3.16)		You Decide: Low-Fat or Low-Carb Diets—Which is Healthier? (3.6)

(*Continued*)

APPENDIX 5

	BioFlix	MP3 Tutors	Web Activities	Web Process of Science	Discovery Channel Videos	GraphIt!/You Decide
Chapter 6 *How Cells Harvest Chemical Energy*	Cellular Respiration (6.6)	Cellular Respiration Part 1: Glycolysis (6.7) Cellular Respiration Part 2: Citric Acid and Electron Transport Chain (6.9)	Build a Chemical Cycling System (6.1) Overview of Cellular Respiration (6.6) Glycolysis (6.7) The Citric Acid (Krebs) Cycle (6.9) Electron Transport and Chemiosmosis (6.10) Fermentation (6.13)	How is the Rate of Cellular Respiration Measured? (6.12)	Tasty Bacteria (6.13)	
Chapter 7 *Photosynthesis: Using Light to Make Food*	Photosynthesis (7.2)	Photosynthesis (7.5)	The Sites of Photosynthesis (7.2) Overview of Photosynthesis (7.5) Light Energy and Pigments (7.6) The Light Reactions (7.9) The Calvin Cycle (7.10) Photosynthesis in Dry Climates (7.12)	How Does Paper Chromatography Separate Plant Pigments? (7.6) How is the Rate of Photosynthesis Measured? (7.11)	Space Plants (7.11) Early Life (7.13)	
Chapter 8 *The Cellular Basis of Reproduction and Inheritance*	Mitosis (8.5) Meiosis (8.14)	Mitosis (8.5) Meiosis (8.14) Mitosis-Meiosis Comparison (8.15)	The Cell Cycle (8.5) Mitosis and Cytokinesis Video (8.7) Asexual and Sexual Life Cycles (8.13) Origins of Genetic Variation (8.18)	How Much Time Do Cells Spend in Each Phase of Mitosis?(8.6) How is Crossing Over Measured in the Fungus *Sordaria*? (8.18)	Fighting Cancer (8.1) Cells (8.2)	
Chapter 9 *Patterns of Inheritance*		Chromosomal Basis of Inheritance (9.16)	Monohybrid Cross (9.3) Dihybrid Cross (9.6) Gregor's Garden (9.7) Incomplete Dominance (9.11) Linked Genes and Crossing Over (9.19) Sex-Linked Genes (9.21)	How is the Chi-Square Test Used in Genetic Analysis? (9.21)	Colored Cotton (9.2)	

Chapter 10 *Molecular Biology of the Gene*	Protein Synthesis (10.6)	DNA to RNA to Protein (10.6)	The Hershey-Chase Experiment (10.1) Phage T2 Reproductive Cycle (10.1) DNA and RNA Structure (10.3) DNA Double Helix (10.3) DNA Replication (10.5) Transcription (10.9) Translation (10.14) Phage Lysogenic and Lytic Cycles (10.17) Simplified Reproductive Cycle of a DNA Virus (10.18) Retrovirus (HIV Reproductive Cycle) (10.20)	What is the Correct Model for DNA Replication? (10.4) How Are Nutritional Mutations Identified? (10.6) How Do You Diagnose a Genetic Disorder? (10.16) Why Do AIDS Rates Differ Across the United States? (10.19) What Causes Infections in AIDS Patients? (10.20) How Can Antibiotic-Resistant Plasmids (10.23) Transform *E. coli*?	Emerging Diseases (10.19)	
Chapter 11 *How Genes Are Controlled*	Control of Gene Expression (11.9)		The *lac* Operon in *E. coli* (11.1) Gene Regulation in Eukaryotes (11.8) Review: Gene Regulation in Eukaryotes (11.9) Development of Head-Tail Polarity (11.10) Signal Transduction Pathway (11.12) Connection: Causes of Cancer (11.21)	How Do You Design a Gene Expression System? (11.5) How Can the "Head" Gene be Regulated to Alter Development? (11.10)	Cloning (11.16) Fighting Cancer (11.18)	You Decide: Do Cell Phones Cause Brain Cancer? (11.21)

(Continued)

	BioFlix	MP3 Tutors	Web Activities	Web Process of Science	Discovery Channel Videos	GraphIt!/You Decide
Chapter 12 *DNA Technology and Genomics*		DNA Technology (12.1)	Connection: Applications of DNA Technology (12.1) Cloning a Gene in Bacteria (12.2) Restriction Enzymes (12.2) Connection: DNA Technology and Golden Rice (12.8) Gel Electrophoresis of DNA (12.13) Analyzing DNA Fragments Using Gel Electrophoresis (12.14) Connection: DNA Fingerprinting (12.15)	How Can Antibiotic-Resistant Plasmids Transform *E. coli*? (12.1) How Can Gel Electrophoresis Be Used to Analyze DNA? (12.13)	Transgenics (12.8) DNA Forensics (12.11)	
Chapter 13 *How Populations Evolve*		Natural Selection (13.2)	Darwin and the Galápagos Islands (13.1) The Voyage of the *Beagle*: Darwin's Trip Around the World (13.1) Reconstructing Forelimbs (13.5) Genetic Variation from Sexual Recombination (13.8) Causes of Microevolution (13.11)	How Do Environmental Changes Affect a Population? (13.2) How Can Frequency of Alleles Be Calculated? (13.9) What Are the Patterns of Antibiotic Resistance? (13.15)	Charles Darwin (13.1) Antibiotic Resistance (13.15)	You Decide: What Can We Do About Antibiotic-Resistant Bacteria? (13.15)
Chapter 14 *The Origin of Species*		Speciation (14.1)	Polyploid Plants (14.6) Exploring Speciation on Islands (14.10)	How Do New Species Arise By Genetic Isolation? (14.5)		

Chapter 15 *Tracing Evolutionary History*		The History of Life (15.4) A Scrolling Geologic Record (15.6) Paedomorphosis: Morphing Chimps and Humans (15.11) Mechanisms of Macroevolution (15.12) Classification Schemes (15.19)	How Might Conditions on Early Earth Have Created Life? (15.2) How is Phylogeny Determined Using Protein Comparisons? (15.17)	Mass Extinction (15.9)
Chapter 16 *The Origin and Evolution of Microbial Life: Prokaryotes and Protists*	Microbial Life (16.1)	Prokaryotic Cell Structure and Function (16.4) Diversity of Prokaryotes (16.7)	What Are the Modes of Nutrition in Prokaryotes? (16.5) What Kinds of Protists Are Found in Various Habitats? (16.20)	Bacteria (16.1) Tasty Bacteria (16.1) Early Life (16.6) Antibiotic Resistance (16.8)
Chapter 17 *Plants, Fungi, and the Colonization of Land*	Evolution of Plants (17.2) Alternation of Generations (17.3)	Terrestrial Adaptations of Plants (17.1) Highlights of Plant Phylogeny (17.2) Moss Life Cycle (17.4) Fern Life Cycle (17.5) Pine Life Cycle (17.7) Angiosperm Life Cycle (17.9) Connection: Madagascar and the Biodiversity Crisis (17.13) Fungal Reproduction and Nutrition (17.17) Fungal Life Cycles (17.17)	What Are the Different Stages of a Fern Life Cycle? (17.5) How Does the Fungus *Pilobolus* Succeed as a Decomposer? (17.14)	Plant Pollination (17.12) Trees (17.14) Leafcutter Ants (17.20) Fungi (17.21)

(Continued)

	BioFlix	MP3 Tutors	Web Activities	Web Process of Science	Discovery Channel Videos	GraphIt!/You Decide
Chapter 18 *The Evolution of Invertebrate Diversity*			Characteristics of Invertebrates (18.13) Characteristics of Chordates (18.14) Animal Phylogenetic Tree (18.15)	How Are Insect Species Identified? (18.12) How Do Molecular Data Fit Traditional Phylogenies? (18.15)	Invertebrates (18.5) Leafcutter Ants (18.12)	
Chapter 19 *The Evolution of Vertebrate Diversity*		Human Evolution (19.11)	Characteristics of Chordates (19.8) Primate Diversity (19.10) Human Evolution (19.15)	How Does Bone Structure Shed Light on the Origin of Birds? (19.7)	Introduced Species (19.18)	
Chapter 20 *Unifying Concepts of Animal Structure and Function*		Animal Structure (20.1)	Correlating Structure and Function of Cells (20.1) The Levels of Life Card Game (20.1) Epithelial Tissue (20.4) Connective Tissue (20.5) Muscle Tissue (20.6) Nervous Tissue (20.7) Overview of Animal Tissue (20.7) Regulation: Negative and Positive Feedback (20.15)		An Introduction to the Human Body (20.1)	
Chapter 21 *Nutrition and Digestion*		The Human Digestive System (21.4)	Digestive System Function (21.10)	What Role Does Amylase Play in Digestion? (21.10)	Nutrition (21.21)	You Decide: Is Ephedra Safe and Effective? (21.22)
Chapter 22 *Gas Exchange*		Human Respiration (22.6)	The Human Respiratory System (22.6) Transport of Respiratory Gases (22.11)			You Decide: Is Second-Hand Smoke Dangerous? (22.7)

Chapter 23 *Circulation*	The Human Circulatory System (23.3)	Mammalian Cardiovascular System Structure (23.3) Path of Blood Flow in Mammals (23.3) Mammalian Cardiovascular System Function (23.8)	How Is Cardiovascular Fitness Measured? (23.9)	Blood (23.12)
Chapter 24 *The Immune System*	The Human Immune System (24.5)	Immune Responses (24.12) HIV Reproductive Cycle (24.13)	What Causes Infections in AIDS Patients? (24.13) Why Do AIDS Rates Differ Across the U.S.? (24.13)	Vaccines (24.4) Fighting Cancer (24.5)
Chapter 25 *Control of Body Temperature and Water Balance*	Kidney Function (25.6)	Structure of the Human Excretory System (25.6) Nephron Function (25.8) Control of Water Reabsorption (25.8)	How Does Temperature Affect Metabolic Rate in *Daphnia*? (25.3) What Affects Urine Production? (25.8)	
Chapter 26 *Hormones and the Endocrine System*	Homeostasis and the Endocrine System (26.1)	Overview of Cell Signaling (26.2) Nonsteroid Hormone Action (26.2) Steroid Hormone Action (26.2) Human Endocrine Glands and Hormones (26.10)	How Do Thyroxine and TSH Affect Metabolism? (26.5)	The Endocrine System (26.3)
Chapter 27 *Reproduction and Embryonic Development*	Spermatogenesis and Oogenesis (27.5) The Female Reproductive Cycle (27.6) Embryonic Development (27.10)	Reproductive System of the Human Male (27.4) Reproductive System of the Human Female (27.6) Sea Urchin Development Video (27.11) Frog Development Video (27.12)	Connection: What Might Obstruct the Male Urethra? (27.4) What Determines Cell Differentiation in the Sea Urchin? (27.13)	

(Continued)

	BioFlix	MP3 Tutors	Web Activities	Web Process of Science	Discovery Channel Videos	GraphIt!/You Decide
Chapter 28 *Nervous Systems*	How Neurons Work (28.4)	Neurons and Electric Potentials (28.2) The Human Brain (28.15)	Neuron Structure (28.2) Nerve Signals: Action Potentials (28.5) Neuron Communication (28.6)	What Triggers Nerve Impulses? (28.4)	Teen Brains (28.15)	
Chapter 29 *The Senses*		Sensory Receptors (29.2)	Structure and Function of the Eye (29.8)			
Chapter 30 *How Animals Move*		Muscle (30.7)	The Human Skeleton (30.3) Skeletal Muscle Structure (30.8) Muscle Contraction (30.9)	How Do Electrical Stimuli Affect Muscle Contraction? (30.10)	Muscles and Bones (30.7)	
Chapter 31 *Plant Structure, Reproduction, and Development*	Tour of a Plant Cell (31.6)	From Flower to Fruit (31.9)	Root, Stem, and Leaf Sections (31.5) Primary and Secondary Growth (31.8) Angiosperm Life Cycle (31.10) Seed and Fruit Development (31.12)	How Are Trees Identified by Their Leaves? (31.3) What Are the Functions of Monocot Tissues? (31.5) What Tells Desert Seeds When to Germinate? (31.13)	Colored Cotton (31.1) Plant Pollination (31.10)	
Chapter 32 *Plant Nutrition and Transport*		Transpiration (32.3)	Transpiration (32.4) Transport in Phloem (32.5) Absorption of Nutrients from Soil (32.8) Connection: Genetic Engineering of Golden Rice (32.11)	How is the Rate of Transpiration Calculated? (32.3) Connection: How Does Acid Precipitation Affect Mineral Deficiency? (32.8)		GraphIt!: Global Soil Degradation (32.9)
Chapter 33 *Control Systems in Plants*			Leaf Abscission (33.7) Flowering Lab (33.12)	What Plant Hormones Affect Organ Formation? (33.2)		

Chapter 34 *The Biosphere: An Introduction to Earth's Diverse Environments*	Ecological Hierarchy (34.1)	Connection: DDT and the Environment (34.2) Adaptations to Biotic and Abiotic Factors (34.4) Aquatic Biomes (34.7) Terrestrial Biomes (34.16)	How Do Abiotic Factors Affect Distribution of Organisms? (34.3)		
Chapter 35 *Behavioral Adaptations to the Environment*		Honeybee Waggle Dance (35.13)	How Can Sow Bug Responses to Environments Be Tested? (35.7)	Novelty Gene (35.23)	
Chapter 36 *Population Ecology*		Techniques for Estimating Population Density and Size (36.2) Investigating Survivorship Curves (36.3) Human Population Growth (36.9) Analyzing Age-Structure Diagrams (36.9)			GraphIt: Age Pyramids and Population Growth (36.9)
Chapter 37 *Communities and Ecosystems*	Energy Flow in Ecosystems (37.16)	Interspecific Interactions (37.2) Food Webs (37.9) Primary Succession (37.12) Energy Flow and Chemical Cycling (37.14) Energy Pyramids (37.17) The Carbon Cycle (37.19) The Nitrogen Cycle (37.21) Water Pollution from Nitrates (37.21)	How Are Impacts on Community Diversity Measured? (37.1) How Do Temperature and Light Affect Primary Production? (37.15)	Leafcutter Ants (37.4) Rain Forests (37.9) Trees (37.14) Space Plants (37.19)	GraphIt: Species Area Effect and Island Geography (37.10) GraphIt: Animal Food Production Efficiency and Food Policy (37.17)

(Continued)

	BioFlix	MP3 Tutors	Web Activities	Web Process of Science	Discovery Channel Videos	GraphIt!/You Decide
Chapter 38 *Conservation Biology*		Global Warming (38.5)	Connection: Madagascar and the Biodiversity Crisis (38.1) Connection: Fire Ants: An Invasive Species (38.3) Connection: DDT and the Environment (38.4) Connection: The Greenhouse Effect (38.6) Conservation Biology Review (38.15)	How Are Potential Restoration Sites Analyzed? (38.14)	Introduced Species (38.3)	GraphIt!: Forestation Change (38.3) GraphIt!: Global Fisheries and Overfishing (38.3) GraphIt!: Global Fresh Water Resources (38.3) GraphIt!: Atmospheric CO_2 and Temperature Changes (38.6) You Decide: Does Human Activity Cause Global Warming? (38.2) You Decide: Can We Prevent Species Extinction? (38.11)

Photo Credits

Front matter: xxii top Anup Shah/Photolibrary. **bottom** Mitch Ranger/Workbook Stock/Jupiter Images. **xxiii** Y. Kito/Image Quest 3-D. **xxiv** Getty Images. **xxv** Andoni Canela/agefotostock. **xxvi left** David Mack/SPL/Photo Researchers, Inc. **right** Lee Jae Won/Reuters. **xxvii** Wolfgang Kaehler/CORBIS. **xxviii** Joe Tucciarone/Photo Researchers, Inc. **xxix left** Peter Titmuss/Alamy. **right** Jez Tryner/Image Quest Marine. **xxx** Columbia/Marvel/Picture Desk, Inc./Kobal Collection. **xxxi left** Robert Holmes/CORBIS. **right** Andrew Syred/Photo Researchers, Inc. **xxxii** Michael Breuer/agefotostock. **xxxiii** ullstein-Nill/Peter Arnold Inc. **xxxiv** Mavis Yorks. **xxxv** Garry Weare/Lonely Planet Images. **xxxvii** Belinda Wright/Photolibrary.

Unit Openers: Unit 1 Dr. Torsten Wittmann/SPL/Photo Researchers, Inc. **Unit 2** Michael Clutson/SPL/Photo Researchers, Inc. **Unit 3** Louie Psihoyos/CORBIS. **Unit 4** Reinhard Dirscheri/Visuals Unlimited/Getty Images. **Unit 5** Pete Oxford/Nature Picture Library. **Unit 6** Purestock/Getty Images. **Unit 7** J & C Sohns/age fotostock.

Chapter 1: Chapter Opener 1 Peter Blackwell/npl/Minden Pictures. **Chapter Opener 2** Roger De La Harpe/Photolibrary. **Chapter Opener 3** Anup Shah/Photolibrary. **1.1 top** NASA/Goddard Space Flight Center. **1.1 center** Dan Guravich/CORBIS. **1.1 bottom** CORBIS. **1.2** Shin Yoshino/Minden Pictures. **1.3** (eukaryotic cell) Don W. Fawcett/Visuals Unlimited. **1.3** (prokaryotic cell) S. C. Holt/Biological Photo Service. **1.4B** (1) Image State/International Stock Photography Ltd. (2) Joe McDonald/CORBIS. (3) Frans Lanting/Minden Pictures. (4) Michael DeYoung/CORBIS. (5) Kim Taylor and Jane Burton/Dorling Kindersley. (6) Frans Lanting/Minden Pictures. (7) Fred Bavendam/Minden Pictures. **1.5A** National Museum of Natural History, Smithsonian Institution. **1.5B top left** Oliver Meckes/Nicole Ottawa/Photo Researchers. **1.5B top center** D. P. Wilson/Photo Researchers. **1.5B top right** Michael & Patricia Fogden/Minden Pictures. **1.5B bottom left** Ralph Robinson/Visuals Unlimited. **1.5B bottom center** Frank Young/Papilio/CORBIS. **1.5B bottom right** Michael and Patricia Fogden/CORBIS. **1.6A** Courtesy Department of Library Services, American Museum of Natural History. **1.6C left** Frans Lanting/Minden Pictures. **1.6C right** Fred Felleman/Stone/Getty Images. **1.8B** E. R. Degginger/Photo Researchers. **1.8C** Breck P. Kent. **1.8D left** David Pfennig. **1.8D right** David Pfennig. **Chapter Review** Shin Yoshino/Minden Pictures.

Chapter 2: Chapter Opener 1 Mark Moffett/Minden Pictures. **Chapter Opener 2** Martin Dohrn/BBC Natural History Unit. **Chapter Opener 3** Martin Dohrn/BBC Natural History Unit. **2.2A** Ivan Polunin/Bruce Coleman, Inc. **2.2B** Kristin Piljay. **2.3** Chip Clark. **2.5A** CC Studio/Photo Researchers, Inc. **2.5B** W. E. Klunk and C. A. Mathis, University of Pittsburgh. **2.11** Bernard Photo Productions/Animals Animals/Earth Scenes. **2.12** Jose Luis Pelaeq, Inc./CORBIS. **2.13B** BIOS Klein & Hubert/Peter Arnold, Inc. **2.16** P. J. Wagner/Photo Access/Taxi/Getty Images, Inc. **2.17** NASA.

Chapter 3: Chapter Opener 1 Mitch Ranger/Workbook Stock/Jupiter Images. **Chapter Opener 2** Peter Dean/Getty Images. **3.2 top** Stockbyte/Getty Images. **3.2 bottom** John Downer/Taxi/Getty Images. **3.4A** Scott Camazine/Photo Researchers. **3.6** Kristin Piljay. **3.7 top & bottom** Biophoto Associates/Photo Researchers. **3.7 center** Courtesy of Dr. L. M. Beidler. **3.8A** Tony Hamblin/Frank Lane Picture Agency/CORBIS. **3.10** Mike Neveux Photography. **3.11** Cancan Chu/Getty Images Sport. **3.14** Denny Allen/Getty Images. **3.15** Courtesy of the Archives, California Institute of Technology.

Chapter 4: Chapter Opener 1 James King-Holmes/Photo Researchers, Inc. **Chapter Opener 2** Tina Carvalho/Visuals Unlimited. **Chapter Opener 3** ISM/Phototake. **4.1A** Leica Microsystems Inc. **4.1B** Dennis Kunkel/Phototake. **4.1C** Dennis Kunkel/Phototake. **4.1D** Microworks Color/Phototake. **4.1E** M. I. Walker/Photo Researchers, Inc. **4.1F** Dr. Torsten Wittmann/Photo Researchers, Inc. **4.3** S. C. Holt, University of Texas Health Center/Biological Photo Service. **4.6** Courtesy of Richard Rodewald/Biological Photo Service. **4.7** D. W. Fawcett/Photo Researchers, Inc. **4.9A** R. Bolender, D. Fawcett/Photo Researchers. **4.10** Don Fawcett/Visuals Unlimited. **4.12A** Biophoto Associates/Photo Researchers, Inc. **4.12B** Roland Birke/Peter Arnold, Inc. **4.14** Nicolae Simionescu. **4.15** Courtesy of W. P. Wergin and E. H. Newcomb, University of Wisconsin/BPS **4.17 left** Dr. Frank Solomon. **4.17 center** Mark S. Ladinsky and J. Richard McIntosh, University of Colorado. **4.17 right** Dr. Mary Osborn. **4.18A** Science Photo Library/Photo Researchers, Inc. **4.18B** Biophoto Associates/Photo Researchers. **4.18C** W. L. Dentler/Biological Photo Service. **4.19** Bjorn Afzelius.

Chapter 5: Chapter Opener 1 Y. Kito/Image Quest 3-D. **Chapter Opener 2** Peter Herring/Image Quest 3-D. **Chapter Opener 3** Edith Widder/Ocean Research & Conservation Association. **5.2** Peter B. Armstrong. **5.7** Reuters. **5.9 top** Michael Abbey/Visuals Unlimited. **5.9 center** D. W. Fawcett/Science Source/Photo Researchers, Inc. **5.9 bottom** M. M. Perry and A. B. Gilbert, *J. Cell Sci.* (1979) 39: 257. Copyright 1979 The Company of Biologists Ltd. **5.10A** Dennis Curran/Photolibrary. **5.10B** Alain McGlaughlin, Pearson Benjamin Cummings. **5.10C** Scott Markewitz/Getty Images.

Chapter 6: Chapter Opener 1 Joe Sohm/Chromosohm/The Stock Connection. **Chapter Opener 2** Getty Images. **Chapter Opener 3** Duomo/CORBIS. **6.2** BananaStock/Jupiter Images. **6.13C** Charles O'Rear/CORBIS. **6.15** Simon Smith/Dorling Kindersley. **6.16** Gerry Ellis/Minden Pictures.

Chapter 7: Chapter Opener 1 James King-Holmes/Photo Researchers, Inc. **Chapter Opener 2** Martin Bond/Photo Researchers, Inc. **7.1A** Theresa DeSalis/Peter Arnold, Inc. **7.1B** EyeWire/Photodisc/Getty Images. **7.1C** Jeff Rotman/Nature Picture Library. **7.1D** Susan M. Barns, Ph.D. **7.2 top** Graham Kent. **7.2 bottom** Courtesy of W. P. Wergin and E. H. Newcomb, University of Wisconsin/BPS. **7.3A** Runk/Schoenberger/Grant Heilman Photography, Inc. **7.6B** Adamsmith/Taxi/Getty Images. **7.7A** Christine L. Case/Skyline College. **7.13A** Doug Martin/Photo Researchers, Inc. **7.14A** Stella Johnson. **7.14B** NASA.

Chapter 8: Chapter Opener 1 Greg Vaughn/Pacific Stock. **Chapter Opener 2** Andoni Canela/age fotostock. **Chapter Opener 3** National Tropical Botanical Garden. **8.1A** Biophoto Associates/Photo Researchers, Inc. **8.1B** Bob Thomas/Photographers Choice/Getty Images. **8.3B** Lee D. Simon/Photo Researchers, Inc. **8.4A** Andrew S. Bajer. **8.4B** Biophoto/Photo Researchers, Inc. **8.6** Conly L. Rieder, Ph.D. **8.7A** David M. Phillips/Visuals Unlimited. **8.7B** Eldon H. Newcomb. **8.11A left** Garcia/photocuisine/CORBIS. **8.11A center** Joseph F. Gennaro Jr./Photo Researchers, Inc. **8.11A right** Brian Capon. **8.11B** Biophoto/Science Source/Photo Researchers, Inc. **8.11C** Biophoto/Science Source/Photo Researchers, Inc. **8.17B left** F. Schussler/Getty Images. **8.17B right** Chris Collins/Bettmann/CORBIS. **8.18A** Cabisco/Visuals Unlimited. **8.19 left** Veronique Burger/Phanie Agency/Photo Researchers, Inc. **8.19 right** CNRI/SPL/Photo Researchers, Inc. **8.20A** CNRI/Science Photo Library/Photo Researchers, Inc. **8.20B** Lauren Shear/Photo Researchers, Inc. **Chapter Review** Carolina Biological Supply/Phototake.

Chapter 9: Chapter Opener 1 Jon Gordon/Getty Images. **Chapter Opener 2** Dorling Kindersley. **9.2A** Hulton Archive/Getty Images, Inc. **9.8A top left** CORBIS. **9.8A top right** Miep van Damm/Masterfile Stock Image Library. **9.8A center left** PhotoDisc/Getty Images. **9.8A center right** PhotoDisc/Getty Images. **9.9B** Michael Ciesielski Photography. **9.10A** Science Source/CNRI/SPL/Photo Researchers, Inc. **9.10B** Eric J. Simon. **9.10B inset** Stone/Getty Images. **9.13** Bill Longcore/Photo Researchers. **9.15** Eric J. Simon. **9.20A** Andrew Syred/Photo Researchers,

Inc. **9.21A** Cabisco/Visuals Unlimited. **9.22** FPG International/Taxi/Getty Images, Inc. **Chapter Review** Norma Jubinville.

Chapter 10: Chapter Opener 1 David Mack/SPL/Photo Researchers, Inc. **Chapter Opener 2** Dr. Linda Stannard, UCT/SPL/Photo Researchers, Inc. **Chapter Opener 3** Peter Hince/Image Bank/Getty Images Inc. **10.1A** Robley C. Williams, University of California Berkeley/Biological Photo Services. **10.3A top** Courtesy of the Library of Congress. **10.3A bottom** From the *Double Helix* by James D. Watson, Atheneum Press, N.Y., 1968, p. 215. Courtesy CSHL Archive. **10.3B** National Institute of Health. **10.6B** Arthur M. Siegelman/Visuals Unlimited. **10.12A** Joachim Frank. **10.19** KHAM/Reuters/CORBIS. **10.19 inset** NIBSC/Science Photo Library/Photo Researchers, Inc. **10.23C** Huntington Potter, University of South Florida College of Medicine.

Chapter 11: Chapter Opener 1 Lee Jae Won/Reuters. **Chapter Opener 2** Advanced Cell Technology, Inc. **Chapter Opener 3** Audubon Nature Institute. **11.1A** SPL/Photo Researchers, Inc. **11.3 left** A. L. Olins, Univ. of Tennessee/Biological Photo Service. **11.3 right** G. F. Bahr, Armed Forces Institute of Pathology. **11.4** Grant Heilman/Grant Heilman Photography, Inc. **11.10A** F. Rudolf Turner, Indiana University. **11.11** Reproduced with permission from R. F. Service, *Science* (1998) 282:396–399. Copyright 1998 American Association for the Advancement of Science. Incyte Pharmaceuticals, Inc., Palo Alto, Calif. **11.16** Pat Sullivan/Associated Press.

Chapter 12: Chapter Opener 1 Tek Image/Photo Researchers, Inc. **Chapter Opener 2** Peter Marlow/Magnum Photos, Inc. **Chapter Opener 3** Neville Chadwick Photography. **Chapter Opener 4** Peter Marlow/Magnum Photos, Inc. **12.6** PPL Therapeutics. **12.7A** SIU/Visuals Unlimited. **12.7B** Hank Morgan/Photo Researchers, Inc. **12.8B** Peter Berger. **12.9A** Philippe Plailly/Eurelios/SPL/Photo Researchers, Inc. **12.9B** Bill Beatty/Visuals Unlimited. **12.15A** Steve Helber/Associated Press **12.15B** South West News Service, Bristol, England. **12.21** Cyril Ruoso/JH Editorial/Minden Pictures.

Chapter 13: Chapter Opener 1 Wolfgang Kaehler/CORBIS. **Chapter Opener 2** Damschen/ARCO/Nature Picture Library. **Chapter Opener 3** Tom Brakefield/CORBIS. **13.1A** Gerry Ellis/Getty Images. **13.1B** Galen Rowell/CORBIS. **13.1C left inset** ARCHIV/Photo Researchers. **13.1C right inset** National Maritime Museum Picture Library, London, England. **13.2** Jack Wilburn/Animals Animals/Earth Scenes. **13.3A top** Edward S. Ross, California Academy of Sciences. **13.3A bottom** Michael & Patricia Fogden/CORBIS **13.3B** CORBIS. **13.4A** Colin Keates/Dorling Kindersley, Courtesy of the Natural History Museum, London. **13.4B** Chip Clark. **13.4C** Martin Lockley. **13.4D** Manfred Kage/Peter Arnold, Inc. **13.4E** Alfred Pasieka/SPL/Photo Researchers, Inc. **13.4F** Hanny Paul/Gamma Press USA, Inc. **13.4G** CORBIS. **13.5B left** Dwight R. Kuhn Photography. **13.5B right** Photo Lennart Nilsson/Albert Bonniers Forlag AB. **13.7** Earth Imaging/Getty Images, Inc. **13.11B** William Ervin/Photo Researchers, Inc. **13.12** Tom Brakefield/CORBIS. **13.14A** Frans Lanting/Minden Pictures. **13.14B** George D. Lepp/CORBIS. **13.14C** Barry Mansell/Nature Picture Library. **13.16** Dr. Michio Hori, Kyoto University.

Chapter 14: Chapter Opener 1 Don Northrup/IllinoisPhoto.com. **Chapter Opener 2** Ole Seehausen. **Chapter Opener 3** Ole Seehausen. **14.2A top** USDA/APHIS Animal and Plant Health Inspection Service. **14.2A bottom** W. J. Weber/Visuals Unlimited. **14.2B top left** Zefa/CORBIS. **14.2B top center** PhotoDisc/Getty Images, Inc. **14.2B top right** PhotoDisc/Getty Images, Inc. **14.2B bottom left** PhotoDisc/Getty Images, Inc. **14.2B bottom center** PhotoDisc/Getty Images, Inc. **14.2B bottom right** Masterfile. **14.3A left** Joe McDonald/CORBIS. **14.3A right** Joe McDonald/Bruce Coleman Inc. **14.3B** Wolfgang Kaehler/CORBIS. **14.3C** Ueshima R, Asami T. "Evolution: single-gene speciation by left-right reversal." *Nature.* 2003 Oct 16;425(6959):679; Fig. 1. **14.3D** William E. Ferguson. **14.3E top left** Ralph A. Reinhold/Animals Animals/Earth Scenes. **14.3E top right** EyeWire/Photodisc/Getty Images, Inc. **14.3E bottom** Grant Heilman/Grant Heilman Photography, Inc. **14.4 backdrop** CORBIS. **14.4 left inset** John Shaw/Bruce Coleman, Inc. **14.4 right inset** Michael Fogden/Bruce Coleman, Inc. **14.6A** Marge Lawson. **14.7B** H. D. Bradshaw and D. W. Schemske. "Allele substitution at a flower colour locus produces pollinator shift in monkeyflowers." *Nature.* 2003 Nov. 13;426(6963):176–78. **14.8B left** Melvin Grey/NHPA. **14.8B right** Juan

Mart'n Simon. **14.9A left** P. R. Grant. **14.9A right** B. R. Grant. **14.9B** Yann Arthus-Bertrand/CORBIS. **14.10A top** Kevin Schafer/CORBIS. **14.10A center** Tui de Roy/Bruce Coleman, Inc. **14.10A bottom** Morales/age fotostock. **14.10B** Don Northrup.

Chapter 15: Chapter Opener 1 Joe Tucciarone/Photo Researchers, Inc. **Chapter Opener 2** Theo Allofs/CORBIS. **Chapter Opener 3** BIOS Huguet Pierre/Peter Arnold, Inc. **15.1** Stanley Awramik/Biological Photo Service. **15.2** Roger Ressmeyer/CORBIS. **15.3A** Fred M. Menger and Kurt Gabrielson, Emory University. **15.7C** Zig Leszczynski/Animals Animals Earth Scenes. **15.8** Tom Bean/CORBIS. **15.11A** Stephen Dalton/Photo Researchers, Inc. **15.11C** William A. Cresko. **15.12 left** Hal Beral/V&W/Image Quest Marine. **15.12 2nd from left** Norbert Wu/Minden Pictures. **15.12 3rd from left** Flip Nicklin/Minden Pictures. **15.12 2nd from right** Jim Greenfield/Image Quest Marine. **15.12 right** Chris Newbert/Minden Pictures. **15.15A** PhotoDisc/Getty Images, Inc. **15.16C** Courtesy Dept. of Library Services, American Museum of Natural History.

Chapter 16: Chapter Opener 1 Peter Sawyer/NMNH Smithsonian Inst. **Chapter Opener 2** Francois Gohier/Photo Researchers, Inc. **Chapter Opener 3** Georgette Douwma/Nature Picture Library. **16.1** Dr. Tony Brain/David Parker/Science Photo Library/Photo Researchers, Inc. **16.3A** Dr. Dennis Kunkel/Visuals Unlimited. **16.3B** Dr. Dennis Kunkel/Visuals Unlimited. **16.3C** Stem Jems/Photo Researchers, Inc. **16.4A** Jack Bostrack/Visuals Unlimited. **16.4B** Dr. Immo Rantala/SPL/Photo Researchers, Inc. **16.4C** Fran Heyl Associates. **16.4D** Lee D. Simon/Science Source/Photo Researchers, Inc. **16.4E** H. S. Pankratz, T. C. Beaman/BPS/Biological Photo Service. **16.4F left** S. W. Watson. © Journal of Bacteriology, American Society of Microbiology. **16.4F right** N. J. Lang/Biological Photo Service. **16.5B** David Scharf/Photo Researchers, Inc. **16.6A** Helen E. Carr/Biological Photo Service. **16.6B** Jack Dykinga/Stone/Getty Images. **16.7A** Helde Schulz/Max Planck Institute. **16.7B** Alfred Pasieka/Peter Arnold, Inc. **16.7C** David Scharf/Photo Researchers, Inc. **16.7D** Susan M. Barns, Ph.D. **16.8A** Dr. Richard Kessel & Dr. Gene Shih/Visuals Unlimited. **16.8B left inset** David M. Phillips/Photo Researchers, Inc. **16.8B right inset** Scott Camazine/Photo Researchers, Inc. **16.8B bottom** Centers for Disease Control and Prevention. **16.9A** Alex Wong/Liaison/Getty Images, Inc. **16.9B** Centers for Disease Control and Prevention. **16.10A** Martin Bond/The National Audubon Society Collection/Photo Researchers, Inc. **16.10B** ExxonMobil Corporation. **16.11A** M. I. Walker/Photo Researchers. **16.11B left** Richard Kessel & Dr. Gene Shih/Visuals Unlimited. **16.11B right** Wim van Egmond/Visuals Unlimited. **16.14** Dr. David M. Phillips/Visuals Unlimited. **16.15A** Eye of Science/Science Source/Photo Researchers, Inc. **16.15B** Dr. Richard Kessel & Dr. Gene Shih/Visuals Unlimited. **16.16A** Dr. David M. Phillips/Visuals Unlimited. **16.16B** Carolina Biological/Visuals Unlimited. **16.17A** James Richardson/Visuals Unlimited. **16.17B** Eric Condliffe/Visuals Unlimited. **16.17C** W. Lewis Trusty/Animals Animals Earth Scenes. **16.18A** Dr. K. W. Jeon/Visuals Unlimited. **16.18B** George Barron. **16.18B inset** Ray Simons/Photo Researchers, Inc. **16.18C** Robert Kay. **16.19A** Manfred Kage/Peter Arnold, Inc. **16.19A inset** Visuals Unlimited. **16.19B** Steve Gschmeissner/Photo Researchers, Inc. **16.20A** Gary Robinson/Visuals Unlimited. **16.20B** Manfred Kage/Peter Arnold, Inc. **16.20B inset** David L. Kirk, Washington University. **16.20C** D. P. Wilson/Eric and David Hosking/Photo Researchers, Inc. **Chapter Review** (a) Dr. K. W. Jeon/Visuals Unlimited. (b) Carolina Biological/Visuals Unlimited. (c–d) Manfred Kage/Peter Arnold, Inc.

Chapter 17: Chapter Opener 1 Peter Titmuss/Alamy. **Chapter Opener 2** Dana Richter/Visuals Unlimited. **Chapter Opener 3** PlantWorks Limited, courtesy of www.plantworksuk.co.uk and www.friendlyfungi.co.uk. **17.1A** Linda Graham, University of Wisconsin-Madison. **17.1B** John D. Cummingham/Visuals Unlimited. **17.1D** CORBIS. **17.2B left** Tony Wharton/Frank Lane Picture Agency/CORBIS. **17.2B top right** Runk/Schoenberger/Grant Heilman Photography. **17.2B bottom right** Hidden Forest. **17.2C left** Colin Walton/Dorling Kindersley. **17.2C right** Barry Runk/Stan/Grant Heilman Photography. **17.2D bottom left** Grant Heilman Photography. **17.2D bottom right** Gerald and Buff Corsi/Visuals Unlimited. **17.2D top right** Ned Therrien/Visuals Unlimited. **17.2D top left** George Louin/Visuals Unlimited. **17.2E top** Terry W. Eggers/CORBIS. **17.2E bottom** Charlotte Thege/Das fotoarchiv/Peter Arnold Inc. **17.4**

Dwight Kuhn/Dwight Kuhn Photography. **17.5** Milton Rand/Tom Stack and Associates. **17.6** The Open University. **17.7** Andrew Brown/Ecoscene/CORBIS. **17.8A left** Craig Lovell/CORBIS. **17.8A right** Howard Rice/Dorling Kindersley Media Library. **17.8A center** Merri Cyr/Workbook Stock/Jupiter Images. **17.9** Shin Yoshino/Minden Pictures. **17.10A** C. P. George/Visuals Unlimited. **17.10B** Scott Camazine/Photo Researchers, Inc. **17.10B inset** Derek Hall/Dorling Kindersley. **17.10C** Cal Vornberger/age fotostock. **17.11** Bryan R. Brunner. **17.12A** Daniel Tigner, Canadian Forest Tree Essences 2007. **17.12B** Greg Vaughn/Photolibrary. **17.12C** D. Wilder/Wilder Nature Photography. **17.13** Tom Bean/DRK Photo. **17.14A** Jim Brandenburg/Minden Pictures. **17.14C** Kjell B. Sandved/Visuals Unlimited. **17.16B** Gregory G. Dimijian/Photo Researchers, Inc. **17.16C** M. F. Brown/Biological Photo Service. **17.16D left** David M. Dennis/Animals Animals/Earth Scenes. **17.16D right** Frank Young/Papilio/CORBIS. **17.16E left** Phil Dotson/Photo Researchers, Inc. **17.16E center** Adrian Davies/Bruce Coleman Inc. **17.16E right** Konrad Wothe/Minden Pictures. **17.17A** Ed Reschke/Peter Arnold. **17.18A** MT Media/LPI. **17.18B** Brad Mogen/Visuals Unlimited. **17.18C** David Cavagnaro/Visuals Unlimited. **17.19A** Wolfgang Kaehler/CORBIS. **17.19B** V. Ahmadijian/Visuals Unlimited. **17.20** Mark Moffett/Minden Pictures. **17.21A** Ron Fehling/Masterfile. **17.21B** Christine Case/Skyline College. **Chapter Review left** R. J. Erwin/Photo Researchers, Inc. **right** Adrian Davies/Bruce Coleman Inc.

Chapter 18: Chapter Opener 1 Gary Bell/CORBIS. **Chapter Opener 2** Norbert Wu/Getty Images. **Chapter Opener 3** Jez Tryner/Image Quest Marine. **Chapter Opener 4** Roger Steene/Image Quest Marine. **18.1A** Gunter Ziesler/Peter Arnold, Inc. **18.1B** Georgette Douwma/Nature Picture Library. **18.2B** J. Sibbick/The Natural History Museum, London. **18.5A** Charles R. Wyttenbach/Biological Photo Service. **18.5B** CORBIS. **18.5C** Andrew J. Martinez/Photo Researchers. **18.6A** Gwen Fidler/Comstock Images. **18.6B** Claudia Mills. **18.6C** Ken Lucas. **18.7B** Dennis Kunkel/Phototake. **18.8A** Reproduced with permission from A. Eizinger and R. Sommer, Max Planck Institut fur entwicklungsbiologie, Tubingen. Copyright 2000 American Association for the Advancement of Science. Cover 278(5337) 17 Oct 97. **18.8B** Reproduced by permission from Howard Shiang, D.V.M., *Journal of the American Veterinary Medical Association* 163:981, Oct. 1973. **18.9C** Photodisc/Getty Images, Inc. **18.9D** CORBIS. **18.9E** H. W. Pratt/Biological Photo Service. **18.9F** Mike Severns/Tom Stack & Associates, Inc. **18.9G** Charles R. Wyttenbach/Biological Photo Service. **18.10A** A.N.T./Photoshot/NHPA Limited. **18.10B** Peter Herring/Image Quest 3-D. **18.10C** Jurgen Freund/Nature Picture Library. **18.10D** Astrid & Hanns-Frieder Michler/SPL/Photo Researchers, Inc. **18.11A** Milton Tierney, Jr./Visuals Unlimited. **18.11C left** William Dow/CORBIS. **18.11C center** D. Suzio/Photo Researchers, Inc. **18.11C right** Oliver Meckes/Ottawa/Photo Reseachers, Inc. **18.11D** Wolfgang Kaehler/CORBIS. **18.11E** Mark Smith/Photo Researchers, Inc. **18.11F** Anthony Bannister/NHPA/Photo Researchers, Inc. **18.12A left** Phil Degginger/Alamy. **18.12A top right** Millard H. Sharp/Photo Researchers, Inc. **18.12A bottom right** Jack Dermid/Photolibrary. **18.12C top left** Michael & Patricia Fogden/Minden Pictures. **18.12C right** Art Wolfe. **18.12C bottom left** Ingo Arndt/Nature Picture Library. **18.12D** Stephen J. Krasemann/Photo Researchers, Inc. **18.12E right** Purestock/Getty Images. **18.12E left** Bruce Coleman Inc./Alamy. **18.13B** Gary Milburn/Tom Stack & Associates, Inc. **18.13C** Biophoto Associates/Photo Researchers, Inc. **18.14A** Robert Brons/Biological Photo Service. **18.14B** Runk/Schoenberger/Grant Heilman Photography. **18.16** Brent Nicastro.

Chapter 19: Chapter Opener 1 Digital Vision/Getty Images. **Chapter Opener 2** Australian Museum, Nature Focus. **Chapter Opener 3** Dave Watts/Tom Stack & Associates. **19.2A top** Brandon D. Cole/CORBIS. **19.2A bottom** Tom McHugh/Photo Researchers, Inc. **19.2B** Breck P. Kent/Animals Animals Earth Scenes. **19.2B inset** Gary Meszaros/Photo Researchers, Inc. **19.3B** George Grall/National Geographic Image Collection. **19.3C** Brian J. Skerry/National Geographic Image Collection. **19.3D** Tom and Pat Leeson/Photo Researchers, Inc. **19.3E top right** Stephen Frink/Digital Vision/Getty Images. **19.3E bottom right** Jeff Rotman/Getty Images. **19.3E left** Marevision/age fotostock America, Inc. **19.3F** Rudie Kuiter/OceanwideImages.com. **19.4C** Publiphoto/Photo Researchers, Inc. **19.5A** Dwight Kuhn/Dwight R. Kuhn Photography. **19.5B** A. Noellert/WILDLIFE/Peter Arnold, Inc. **19.5C** Stephen Dalton/Minden

Pictures. **19.5D** Hans Pfletschinger/Peter Arnold, Inc. **19.5E** Michael Fogden/Bruce Coleman, Inc. **19.6A** Robert and Linda Mitchell. **19.6B** GK Hart, Vikki Hart/Getty Images. **19.6C** The Natural History Museum, London. **19.7B** John Gerlach/Visuals Unlimited. **19.7C** Russell Mountford/Alamy Images. **19.8A** Jean Phillipe Varin/Jacana/Photo Researchers, Inc. **19.8B** Michael S. Yamashita/CORBIS. **19.8C** Mitch Reardon/Photo Researchers, Inc. **19.9A** E. H. Rao/Photo Researchers, Inc. **19.9C** Frans Lanting/Minden Pictures. **19.9D** Frans Lanting/Minden Pictures. **19.9E left** Kevin Schafer/Photo Researchers, Inc. **19.9E right** Kevin Schafer/CORBIS. **19.9F** Digital Vision/Getty Images. **19.10A** Steve Bloom/Alamy Images. **19.10B** Gerry Ellis/Digital Vision/Getty Images. **19.10C** Image Source/Alamy Images. **19.10D** Gerry Ellis/Digital Vision/Getty Images. **19.12** John Reader/SPL/Photo Researchers, Inc. **19.18A** Eric & David Hosking/CORBIS. **19.18B** Eric & David Hosking/CORBIS. **19.18C** Peter Mead/Tom Stack & Associates.

Chapter 20: Chapter Opener 1 Columbia/Marvel/Picture Desk, Inc./Kobal Collection. **Chapter Opener 2** Reuters/CORBIS. **Chapter Opener 3** Andrew Syred/Photo Researchers, Inc. **Chapter Opener 4** Kellar Autumn. **20.1** CORBIS. **20.2 top** CORBIS. **20.2 middle** Tui De Roy/Minden Pictures. **20.2 bottom** Masa Ushioda/Stephen Frink Collection/Alamy Image. **20.5A** Nina Zanetti. **20.5B** Science/Visuals Unlimited. **20.5C** Nina Zanetti. **20.5D** Chuck Brown/Photo Researchers, Inc. **20.5E** Nina Zanetti. **20.5F** Dr. Gopal Murti/SPL/Photo Researchers, Inc. **20.6A** Nina Zanetti. **20.6B** Manfred Kage/Peter Arnold, Inc. **20.6C** Gladden Willis, M.D./Visuals Unlimited. **20.7** Ed Reschke. **20.9** Brian Walker/AP Photo. **20.11** ISM/Phototake. **20.13B** Science Photo Library/Photo Researchers, Inc. **20.14A** CORBIS. **Chapter Review left** Nina Zanetti; **center** Gladden Willis, M. D./Visuals Unlimited; **right** Ed Reschke

Chapter 21: Chapter Opener 1 Richard Schlecht/National Geographic Image Collection. **Chapter Opener 2** David Tipling/Nature Picture Library. **Chapter Opener 3** Brandon Cole, www.brandoncole.com. **21.1A** Pete Atkinson/Taxi/Getty Images, Inc. **21.1B** Thomas Eisner, Cornell University. **21.1C** Martin Dohrn/npl/Minden Pictures. **21.1D** Jennette van Dyk/Getty Images. **21.7** The Heimlich Institute Foundation **21.9** Oliver Meckes/Ottawa/Photo Researchers, Inc. **21.13A left** Joe McDonald/CORBIS. **21.13A right** Eyewire Collection/Photodisc/Getty Images, Inc. **21.17** CORBIS. **21.21A** Susumu Nishinaga/SPL/Photo Researchers, Inc. **21.21B** Photo courtesy of The Jackson Laboratory, Bar Harbor, Maine. **21.23** Steve Allen/Brand X Pictures/Jupiter Images.

Chapter 22: Chapter Opener 1 Robert Holmes/CORBIS. **Chapter Opener 2** John Downer/Nature Picture Library. **Chapter Opener 3** Jupiter Images/Goodshoot/Alamy. **22.4A** Thomas Eisner, Cornell University. **22.4B** Stephen Dalton/Minden Pictures. **22.5** Ted Daeschler/AFP/Getty Images. **22.6B left** CNRI/SPL/Photo Researchers, Inc. **22.6B right** Dr. Richard Kessel & Dr. Randy Kardon/Visuals Unlimited. **22.7** Martin Rotker.

Chapter 23: Chapter Opener 1 Werner Bollmann/age fotostock. **Chapter Opener 2** Wolfgang Kaehler/CORBIS. **Chapter Opener 3** Chris Taylor, Cordaiy Photo Library Ltd/CORBIS. **23.5B** Photodisc/Getty Images, Inc. **23.6B left** Ed Reschke. **23.6B right** W. Ober/Visuals Unlimited. **23.7A** Photo Lennart Nilsson/Albert Bonniers Forlag AB. The Body Victorious, Dell Publishing Company. **23.11A** D. W. Fawcett/Photo Researchers, Inc. **23.13** David M. Phillips/Visuals Unlimited. **23.14B** Courtesy, The Gillette Company.

Chapter 24: Chapter Opener 1 Andrew Syred/Photo Researchers, Inc. **Chapter Opener 2** First Light/Getty Images. **24.4** David McNew/Getty Images. **24.5B** NIBSC/Photo Researchers. **24.10** Raimund Koch/Getty Images. **24.13** Lennart Nilsson/Albert Bonniers Forlag.

Chapter 25: Chapter Opener 1 Lynn Rogers/Peter Arnold, Inc. **Chapter Opener 2** BIOS Feve Frederic/Peter Arnold, Inc. **Chapter Opener 3** Terry Whittaker/Alamy. **25.3A** Mitsuaki Iwago/Minden Pictures. **25.3B left** Jeff Lepore/Photo Researchers, Inc. **25.3B right** O. Alamany & E. Vicens/CORBIS. **25.9** BSIP/Phototake NYC.

Chapter 26: Chapter Opener 1 Michael Breuer/age fotostock. **Chapter Opener 2** Robert Brook/Photo Researchers, Inc. **Chapter Opener 3** Patricio Robles Gil/Sierra Madre/Minden Pictures. **26.5A** NMSB/Custom Medical Stock Photo, Inc. **26.10** Jonathan Blair/CORBIS. **26.11** CORBIS.

Chapter 27: Chapter Opener 1 Brooks Kraft/Sygma. **Chapter Opener 2** Annie Griffiths Belt/CORBIS. **27.1** David Wrobel. **27.2A** Adam Reitzel, John R. Finnerty, Boston University **27.2B** Hans Pfletschinger/Peter Arnold, Inc. **27.2C** John Cancalosi/Nature Picture Library. **27.3B** C. Edelman/La Vilette/Photo Researchers, Inc. **27.8** Gusto/Photo Researchers, Inc. **27.9A** D. Phillips/Photo Researchers, Inc. **27.12A** Cabisco/Visuals Unlimited. **27.12C** Thomas Poole, SUNY Health Science Center. **27.12D** G.I. Bernard/Photo Researchers, Inc. **27.13A left** David Barlow. **27.13A right** David Barlow/BBC Photo Library. **27.16A** Lennart Nilsson/A Child is Born, Dell Publishing. **27.16B** Lennart Nilsson/A Child is Born, Dell Publishing. **27.16C** Lennart Nilsson/A Child is Born, Dell Publishing. **27.16D** Lennart Nilsson/Albert Bonniers Forlag AB. **27.16E** Eric J. Simon.

Chapter 28: Chapter Opener 1 Lester Lefkowitz/CORBIS. **Chapter Opener 2** AP Wide World Photos. **28.2** Manfred Kage/Peter Arnold, Inc. **28.7** Edwin R. Lewis, University of California at Berkeley. **28.9** Steve Gorton/Dorling Kindersley and PhotoDisc/Getty Images, Inc. **28.17A** From H. Damasio, et al., "The return of Phineas Cage: Clues about the brain from a famous patient." *Science,* 264:1102–1105, 1994. Department of Neurology and Image Analysis Facility, University of Iowa. **28.17B** Dana Boatman & John Freeman, Johns Hopkins School of Medicine. **28.21A** WDCN/Univ. College London/Photo Researchers, Inc. **28.21B** Jonathan Nourak/Stone/Getty Images. **28.21C** Douglas Graham/Roll Call/Sygma/CORBIS.

Chapter 29: Chapter Opener 1 Jim Watt/Pacific Stock. **Chapter Opener 2** F. Graner/WILDLIFE/Peter Arnold. **Chapter Opener 3** ullstein-Nill/Peter Arnold Inc. **29.3C** Dr. R.A. Steinbrecht. **29.3D** Joe McDonald/Animals Animals Earth Scenes. **29.7A** Tom Adams/Visuals Unlimited. **29.7B** Thomas Eisner, Cornell University. **29.9A** CORBIS. **29.9B** CORBIS.

Chapter 30: Chapter Opener Juniors Bildarchiv/age fotostock. **30.1A** David B. Fleetham/Visuals Unlimited. **30.1B** Dave Watts/NHPA/Photo Researchers, Inc. **30.1C** Arco Digital Images/Wegner/Jupiter Images. **30.1E** Stephen J. Krasemann/DRK Photo. **30.2A** Dwight Kuhn. **30.2B** Piotr Naskrecki/Minden Pictures. **30.2C** Carlos Villoch/Image Quest 3-D. **30.2D left** Darrell Gulin/CORBIS. **30.2D right** Kaj R. Svensson/Photo Researchers, Inc. **30.5A** Jochen Tack Alamy. **30.5B left** Professor P. Motta, Department of Anatomy, University lá Sapienzá Rome/Photo Researchers, Inc. **30.5B right** P. Motta/Photo Researchers, Inc. **30.8** Professor Clara Franzini-Armstrong. **30.11** Whit Richardson/Aurora Photos.

Chapter 31: Chapter Opener 1 Steve Sillett. **Chapter Opener 2** Michael Hutchinson/Nature Picture Library. **Chapter Opener 3** Henk Wallays/Photographer's Direct. **31.1** D. Cavagnaro/Visuals Unlimited. **31.4A** Eric Simon. **31.4B top** Dorling Kindersley. **31.4B right** Grant Heilman Photography, Inc. **31.4B bottom** Nigel Cattlin/Photo Researchers Inc. **31.4C top** Scott Camazine/Photo Researchers, Inc. **31.4C bottom** CORBIS. **31.6B** Dwight Kuhn/Dwight R. Kuhn Photography. **31.6C** Graham Kent. **31.6D left** Graham Kent. **31.6D right** Bruce Iverson, Photomicrography. **31.6E** R. Kessel-Shih/Visuals Unlimited. **31.6F** Visuals Unlimited. **31.7C** Ed Reschke. **31.8B** Runk/Schoenberger/Grant Heilman Photography. **31.12A** W. H. Hodge/Peter Arnold, Inc. **31.14A** Kevin Schafer Photography. **31.14B** Frank Balthis. **31.14C** Mountain Light Photography, Inc. **31.14D** Francois Gohier/Photo Researchers. **31.14E** Rosenfeld Images Ltd/Photo Researchers, Inc. **31.15** David Welling/Nature Picture Library.

Chapter 32: Chapter Opener 1 Mavis Yorks. **Chapter Opener 2** Kayte Deioma/Photographer's Direct. **Chapter Opener 3** Common Ground Collective. **32.1B** Chris Falkenstein/Photodisc/Getty Images. **32.2A** Brian Capon. **32.5A** Ray F. Evert. **32.5C** M. H. Zimmermann. **32.7A** Grant Heilman Photography. **32.7B** Nigel Cattlin/Holt Studios International Ltd. **32.7C** Paul Rapson/Photo Researchers, Inc. **32.8A** Agricultural Research Service/USDA. **32.9A** Runk/Schoenberger/Grant Heilman Photography. **32.9B** Kevin Horan/Stone/Getty Images. **32.10** Ralf-Finn Hestoft/CORBIS. **32.11** Louisiana State University Press. **32.13A** R. L. Peterson/Biological Photo Service. **32.13B** Breck P. Kent/Animals Animals Earth Scenes. **32.13C** E. H. Newcomb and S. R. Tandon/BPS. **32.14A** Clay

Perry/Photolibrary. **32.14B** Kevin Schafer/CORBIS. **32.14C** Jim Strawser/Grant Heilman Photography **32.14D** Fritz Polking/Frank Lane Picture Agency/CORBIS. **32.14E** Fabio Colombini/Animals Animals

Chapter 33: Chapter Opener 1 D. Hurst/Alamy Images. **Chapter Opener 2** CORBIS. **Chapter Opener 3** Michael Pohuski/Botanica/Jupiter Images. **33.1A** Dorling Kindersley. **33.3A** David Newman/Visuals Unlimited. **33.4** Chandoha Photography. **33.5A** Alan Crozier. **33.5B** Fred Jensen. **33.6** Charles Krebs/CORBIS. **33.7A** Runk/Schoenberg/Grant Heilman Photography. **33.7B** Ed Reschke. **33.8** Stockbyte/Getty Images. **33.9A** Michael Evans. **33.9B** Scott Camazine/Photo Researchers, Inc. **33.10** Malcolm B. Wilkins, University of Glasgow, Glasgow, Scotland, U.K. **33.13** James Aronovsky. **33.14B** Barbara Baker. **33.15** Mary De Chirico.

Chapter 34: Chapter Opener 1 Garry Weare/Lonely Planet Images. **Chapter Opener 2** Verena Tunnicliffe. **Chapter Opener 3** AP Photo/Ifremer, A. Fifis. **Chapter Opener 4** Tom Kornack. **34.2A** Rich Reid/NGS Image Collection. **34.2B** Alfred Eisenstaedt/Time Life Pictures/Getty Images. **34.3A** Peter Batson/Imagequest Marine. **34.3B** Richard Kolar/Animals Animals. **34.4** CORBIS. **34.6B** Digital Vision/Getty Images. **34.6C** SOC/Imagequest Marine. **34.6D** James Randklev/Image Bank/Getty Images. **34.7A** Jay Dickman/CORBIS. **34.7B** David Muench/CORBIS. **34.9** Frans Lanting/Minden Pictures. **34.10** Wolfgang Kaehler/CORBIS. **34.11** Dennis Frates/Workbook Stock/Jupiter Images. **34.12** John D. Cunningham/Visuals Unlimited. **34.13** Jake Rajs/Stone/Getty Images, Inc. **34.14** Kennan Ward/CORBIS. **34.15** Steve Terrill/CORBIS. **34.16** Darrell Gulin/CORBIS.

Chapter 35: Chapter Opener 1 Lowell L. Getz and Lisa Davis. **Chapter Opener 2** John R. MacGregor/Peter Arnold, Inc. **Chapter Opener 3** Zuoxin Wang, Ph.D., Florida State University. **35.2B** Theo Allofs/Danita Delimont. **35.5A** Thomas McAvoy/Time & Life Pictures/Getty Images. **35.5B** Frank Lane Picture Agency/CORBIS. **35.6A** Ruth Cole/Animals Animals. **35.6B** Star Banner, Doug Engle/AP Photo. **35.7A** George Grall/National Geographic Image Collection. **35.8** Jonathan Blair/Woodfin Camp. **35.9** Harry Engels/Animals Animals Earth Scenes. **35.10** Richard Wrangham, Harvard University. **35.10 inset** Alissa Crandall/CORBIS. **35.11A** Clive Bromhall/OSF/Animals Animals. **35.11B** Bernd Heinrich/University of Vermont. **35.12A** Jose B. Ruiz/Nature Picture Library. **35.12B** Michael Hutchinson/Nature Picture Library. **35.13A** J. P. Varin/Jacana/Photo Researchers, Inc. **35.14A** Georgette Douwma/ImageState/Jupiter Images. **35.14C** Carol Walker/Nature Picture Library. **35.15** Fred Bavendam/Minden Pictures. **35.16B** W. Mike Howell. **35.18A** Wolfgang Kaehler/CORBIS. **35.18B** C & M Denis-Huot/Peter Arnold, Inc. **35.19** Gordon Wiltsie/AlpenImage Ltd. **35.20** Renne Lynn. **35.21A** Ken Regan/Camera 5. **35.21B** Frans B. M. De Waal. **35.22A** Stephen Kraseman/Peter Arnold, Inc. **35.22B** Jennifer Jarvis. **35.23A** Ted S. Warren/AP Photo. **35.23B** Lisa O'Connor/ZUMA/CORBIS.

Chapter 36: Chapter Opener 1 Franz Pagot/Alamy. **Chapter Opener 2** Euan Denholm/Reuters/CORBIS. **Chapter Opener 3** Trygve Bolstad/Peter Arnold Inc. **36.2A** Fred Bavendam/Minden Pictures. **36.2B** Frans Lanting/Minden Pictures. **36.2C** Ross M. Horowitz/Iconica/Getty Images. **36.4B** Roy Corral/CORBIS. **36.5A** Tom Bean/CORBIS. **36.5B** B. Borrell Casals/Frank Lane Picture Agency/CORBIS. **36.5C** Tom Bean/CORBIS. **36.6** Alan Carey/Photo Researchers, Inc. **36.11A left** Peter Ginter. **36.11A right** Peter Menzel. **36.11B** SASI Group, University of Sheffield.

Chapter 37: Chapter Opener 1 Frans Lanting/Minden Pictures. **Chapter Opener 2** Doug Allan/Nature Picture Library. **Chapter Opener 3** Jonathan and Angela/Getty Images. **37.3A** Doug Backlund. **37.3B** Terry Sohl, South Dakota Birds. **37.4** Jurgen Freund/Nature Picture Library. **37.5A** Thomas Gula/Visuals Unlimited. **37.5B** Mark Moffett/Minden Pictures. **37.6** Lawrence E. Gilbert/Biological Photo Service. **37.7** Joel Sartore/Photolibrary. **37.11B** William E. Townsend/Photo Researchers, Inc. **37.11C** Aletta Yniguez, National Center for Coral Reef Research (NCORE). **37.11D** Reinhard Dirscherl/Alamy. **37.13D** Embassy of Australia. **37.22A** Hubbard Brook Research Foundation. **37.22B** USDA Forest Service. **37.23A** Richard Siemens. **37.23B** David Schindler.

Chapter 38: Chapter Opener 1 Belinda Wright/Photolibrary. **Chapter Opener 2** altrendo nature/Getty Images. **Chapter Opener 3** Steve Winter/National Geographic Image Collection. **38.1** Scott Camazine/Photo Researchers, Inc. **38.2A** Rickey Rogers/Reuters/CORBIS. **38.2B** Fred Bavendam/Minden Pictures. **38.3A** Michael Fodgen/Animals Animals/Earth Scenes. **38.3B** Michel Gunther/Peter Arnold, Inc. **38.5A** NASA Goddard Institute for Space Studies. **38.5B left** T. J. Hileman, Glacier National Park Archives. **38.5B center** Carl Key, USGS. **38.5B right** Blase Reardon, USGS. **38.7A** Justin Sullivan/Getty Images. **38.7B** Tom Murphy/WWI/Peter Arnold, Inc. **38.8A** William Osborn/Nature Picture Library. **38.8B** Paul McCormick/Getty Images. **38.9A** Gary Braasch/CORBIS. **38.9B** T. Davis/Photo Researchers, Inc. **38.9C** Tim Thompson/CORBIS. **38.10A** Yann Arthus-Bertrand/CORBIS. **38.10B** Calvin Larsen/Photo Researchers, Inc. **38.10C** Alan Sirulnikoff/SPL/Photo Researchers, Inc. **38.11B** Mark Conlin/Planet Earth Pictures. **38.12B** Frans Lanting/Minden Pictures. **38.13B** NPS Photo. **38.13C** Tim Davis/CORBIS. **38.14B** South Florida Water Management District (WPB). **38.15** Heinrich van den Berg/Getty Images.

Illustration and Text Credits

The following figures are adapted from Elaine N. Marieb, *Human Anatomy and Physiology,* 5th ed. Copyright © 2001 Pearson Education, Inc., publishing as Pearson Benjamin Cummings: **4.4A, 4.6, 4.9A,** and **EOC UN4.3.**

The following figures are adapted from Elaine N. Marieb, *Human Anatomy and Physiology,* 4th ed. Copyright © 1998 Pearson Education, Inc., publishing as Pearson Benjamin Cummings: **22.10, 22.12, 25.8, 27.3A, 27.3C, 27.4B, 27.17B, 29.9A, 29.9B, 30.4**

The following figures are adapted from Gerard J. Tortora, Berdell R. Funke, and Christine L. Case, *Microbiology: An Introduction,* 9th ed. Copyright © 2007 Pearson Education, Inc., publishing as Pearson Benjamin Cummings: **16.10A, 24.8B,** and **Table 27.7.**

The following figures are adapted from Lawrence G. Mitchell, John A. Mutchmor, and Warren D. Dolphin, *Zoology.* Copyright © 1988 Pearson Education, Inc., publishing as Pearson Benjamin Cummings: **18.1B, 18.14B, 21.10B, 25.5, 30.2E,** and **35.7C.**

Chapter 1: 1.8E Data in graph based on D. W. Pfennig et al. 2001. Frequency-dependent Batesian mimicry. *Nature* 410: 323.

Chapter 2: Chapter 2 introduction Adapted from an interview in Neil Campbell and Jane Reece, *Biology,* 6th ed. Copyright © 2002 Pearson Education, Inc., publishing as Pearson Benjamin Cummings.

Chapter 3: Figure 3.13A Adapted from D. W. Heinz, W. A. Baase, F. W. Dahlquist, B. W. Matthews, 1993. How amino-acid insertions are allowed in an alpha-helix of T4 lysozyme. *Nature* 361: 561. **Unnumbered figure, page 44** Illustration, Irving Geis. Image from Irving Geis Collection, Howard Hughes Medical Institute. Rights owned by HHMI. Reprinted by permission. Not to be reproduced without permission. **Module 3.15: Talking About Science** Adapted from an interview in Neil Campbell *Biology,* 1st Copyright © 1987 Pearson Education, Inc., publishing as Pearson Benjamin Cummings.

Chapter 5: Module 5.7: Talking About Science Adapted from an interview in Neil Campbell and Jane Reece, *Biology,* 7th ed. Copyright © 2005 Pearson Education, Inc., publishing as Pearson Benjamin Cummings.

Chapter 6: Table 6.4 Data from C. M. Taylor and G. M. McLeod, *Rose's Laboratory Handbook for Dietetics,* 5th ed. (New York: Macmillan, 1949), p. 18; J. V. G. A. Durnin and R. Passmore, *Energy and Protein Requirements in FAO/WHO Technical Report* No. 522, 1973; W. D. McArdle, F. I. Katch, and V. L. Katch, *Exercise Physiology* (Philadelphia, PA: Lea & Feibiger, 1981); R. Passmore and J. V. G. A. Durnin, *Physiological Reviews* vol. 35, pp. 801–840 (1955).

Chapter 7: 7.8B Adapted from Richard and David Walker. *Energy, Plants and Man,* fig. 4.1, p. 69. Sheffield: University of Sheffield. Courtesy of Richard Walker (http://www.oxygraphics.co.uk/. **Module 7.14: Talking About Science** Adapted from an interview in Neil Campbell, Jane Reece, and Lawrence Mitchell, *Biology,* 5th ed. Copyright © 1999 Pearson Education, Inc., publishing as Pearson Benjamin Cummings.

Chapter 9: 9.13 Adapted from *Introduction to Genetic Analysis,* 4th ed. by Suzuki, Griffiths, Miller, and Lewontin. Copyright © 1976, 1981, 1986, 1989, 1993, 1996 by W. H. Freeman and Company. Used with permission.

Chapter 11: 11.3 Adapted from C. K. Mathews and K. E. van Holde, *Biochemistry,* 2nd ed. Copyright © 1996 Pearson Education, Inc., publishing as Pearson Benjamin Cummings. **11.14** Adapted from W. H. Becker, *The World of the Cell,* p. 592. Copyright © 1986 Pearson Education, Inc., publishing as Pearson Benjamin Cummings. **Table 11.21** Data from *American Cancer Society. Cancer Facts & Figures 2007.* Atlanta: American Cancer Society, 2007.

Chapter 13: 13.14a Adapted from J. G. M. Thewissen et al. 2001. Skeletons of terrestrial cetaceans and the relationship of whales to artiodactyls. *Nature* 413:277–281, fig. 2a. **13.14b** Adapted from P. D. Gingerich et al. 2001. Origin of whales from early artiodactyls: Hands and feet of eocene protocetidae from Pakistan. *Science* 293: 2239–2242, fig. 3. **13.14c** and **d** Adapted from C. de Muizon. 2001. Walking with whales, *Nature* 413: 259–260, fig. 1.

Chapter 14: 14.7A Adapted from D. M. B. Dodd, *Evolution,* vol. 11, pp. 1308–1311. Reprinted by permission of the Society for the Study of Evolution. **Module 14.9: Talking About Science** Adapted from an interview in Neil Campbell and Jane Reece, *Biology,* 6th ed. Copyright © 2002 Pearson Education, Inc., publishing as Pearson Benjamin Cummings.

Chapter 15: Module 15.2: Talking About Science Adapted from an interview in Neil Campbell, *Biology,* 2nd ed. Copyright © 1990 Pearson Education, Inc. publishing as Pearson Benjamin Cummings. **15.7A** Map adapted from http://geology.er.usgs.gov/eastern/plates.html. **15.7D** Adapted from W. K. Purves and G. H. Orians, *Life: The Science of Biology,* 2nd ed., fig. 16.5c, p. 1180 (New York: Sinauer Associates, 1987.) Used with permission. **15.10** Adapted from Hickman, Roberts, and Larson. 1997, *Zoology,* 10/e, Wm. C. Brown, fig. 31.1. **15.12** Adapted from M. Strickberger. 1990. *Evolution.* Boston: Jones & Bartlett. **15.18** Adapted from B. Korber et al., June 9, 2000. Timing the ancestor of the HIV-1 pandemic strains. *Science* 288: 1789–1796, fig. 1b. **15.19A** Adapted from S. Blair Hedges. The origin and evolution of model organisms. *Nature Reviews Genetics* 3: 838–848, fig. 1, p. 840. **15.19B** Adapted from M. C. Rivera and J. A. Lake. 2004. The ring of life provides evidence for a genome fusion origin of eukaryotes. *Nature* 431: 152–155, fig. 3.

Chapter 16: CO16 Artist: Peter Sawyer © NMNH Smithsonian Institution. **16.12** From Archibald and Keeling, "Recycled Plastids," *Trends in Genetics,* Vol. 18, No. 11, 2002. Copyright © 2002, with permission from Elsevier.

Chapter 17: Table 17.13 Adapted from Randy Moore et al., *Botany,* 2nd ed. Dubuque, IA: Brown, 1998, Table 2.2, p. 37.

Chapter 18: Module 18.16: Talking About Science Adapted from an interview in Neil Campbell and Jane Reece, *Biology,* 8th ed. Copyright © 2008 Pearson Education, Inc., publishing as Pearson Benjamin Cummings.

Chapter 19: 19.4A Adapted from B. Holmes. Sept. 9, 2006. Meet your ancestor—the fish that crawled. *New Scientist,* 2568: 35–39. By permission of *New Scientist* magazine. **19.4B** From C. Zimmer. *At the Water's Edge.* Free Press, Simon & Schuster, p. 90. Copyright © 1999 by Carl Zimmer. Reprinted by permission of the author. **19.11** Drawn from photos of fossils: *O. tugenensis* photo in Michael Balter, Early hominid sows division, *ScienceNow,* Feb. 22, 2001, © 2001 American Association for the Advancement of Science. *A. ramidus kadabba* photo by Timothy White, 1999/Brill Atlanta. *A. anamensis, A. garhi,* and *H. neanderthalensis* adapted from *The Human Evolution Coloring Book. K. platyops* drawn from photo in Meave Leakey et al., New hominid genus from eastern Africa shows diverse middle Pliocene lineages, *Nature,* March 22, 2001, 410: 433. *P. boisei* drawn from a photo by David Bill. *H. ergaster* drawn from a photo at www.inhandmuseum.com. *S. tchadensis* drawn from a photo in Michel Brunet et al., A new hominid from the Upper Miocene of Chad, Central Africa, *Nature,* July 11, 2002, 418: 147, fig. 1b. **19.14** Adapted from D. Jones. Nov. 11, 2006. Blueprint for a Neanderthal, *New Scientist,* Issue

2577. By permission of *New Scientist* magazine. **19.15** Adapted from L. Luca Cavalli-Sforza and M. W. Feldman. 2003. The application of molecular genetic approaches to the study of human evolution. *Nature Genetics* 33: 266–275, fig. 3.

Chapter 21: 21.4 Adapted from R.A. Rhoads and R. G. Pflanzer. 1996. *Human Physiology*, 3/e., p. 666, fig. 22–1. Copyright 1996 Saunders. **21.17** Adapted from Murray W. Nabors, *Introduction to Botany*. Copyright © 2004 Pearson Education, Inc., publishing as Pearson Benjamin Cummings. **Table 21.18A** Data from RDA Subcommittee, *Recommended Dietary Allowances* (Washington, D. C.: National Academy Press, 1989); M. E. Shils and V. R. Young, *Modern Nutrition in Health and Disease* (Philadelphia, PA: Lea & Feibiger, 1988). **Table 21.18B** Data from (1) M. E. Shils, "Magnesium," in M. E. Shils and V. R. Young, eds. *Modern Nutrition in Health and Disease* (Philadelphia, PA: Lea & Feibiger, 1988); (2) V. F. Fairbanks and E. Beutler, "Iron" [same as 1]; (3) N. W. Solomons, "Zinc and Copper" [same as 1]; (4) RDA Subcommittee, *Recommended Dietary Allowances* (Washington, D. C.: National Academy Press, 1989); (5) E. J. Underwood, *Trace Elements in Human and Animal Nutrition* (New York: Academic Press, 1977).

Chapter 22: 22.11 Illustration, Irving Geis. Image from Irving Geis Collection. Howard Hughes Medical Institute. Rights owned by HHMI. Reprinted by permission. Not to be reproduced without permission.

Chapter 26: 26.8 Data from C. J. Bryne, D. F. Saxton et al., *Laboratory Tests: Implications for Nursing Care*, 2nd ed. Copyright © 1986 Pearson Education, Inc., publishing as Pearson Benjamin Cummings.

Chapter 27: 27.4C Adapted from Robert Crooks and Karla Baur, *Our Sexuality*, 5th ed. Copyright © 1993 Pearson Education, Inc., publishing as Pearson Benjamin Cummings. **27.14B** Adapted from an illustration by William McGinnis, UCSD. **Table 27.8** Data from R. Hatcher et al., *Contraceptive Technology: 1990–1992*, p. 134 (New York: Irvington, 1990).

Chapter 28: 28.6 Adapted from Becker, Kleinsmith, and Hardin, *The World of the Cell*, 6th ed. Copyright © 2006 Pearson Education, Inc., publishing as Pearson Benjamin Cummings.

Chapter 30: 30.12 Adapted from J. L. Andersen, et al. 2000. Muscles, Genes, and Athletic Performance. *Scientific American*, September, 2000, p. 49. Copyright 2000 Scientific American, Inc.

Chapter 33: Module 33.13: Talking About Science Adapted from an interview in Neil Campbell and Jane Reece, *Biology*, 6th ed. Copyright © 2002 Pearson Education, Inc., publishing as Pearson Benjamin Cummings. **33.14A** Adapted from Edward Farmer, "Plant Biology: New Fatty Acid-based Signals: A Lesson from the Plant World," *Science* 276: 912 (1997). Reprinted with permission from AAAS. **Module 33.15: Talking About Science** Adapted from an interview in Neil Campbell, *Biology*, 4th ed. Copyright © 1996 Pearson Education, Inc., publishing as Pearson Benjamin Cummings.

Chapter 34: Module 34.2 Text quotation from Rachel Carson, *Silent Spring* (New York: Houghton Mifflin, 1962). **34.8** Adapted from Heinrich Walter and Siegmar-Walter Brecle. 2003. *Walter's Vegetation of the Earth*, fig. 16, p. 36. Springer-Verlag, © 2003.

Chapter 35: 35.5B Courtesy of Masakazu Konishi. **35.8 (top)** Adapted from *Animal Navigation*, p. 109, by Talbot H. Waterman. Copyright © 1989 by Scientific American Books, Inc. Used with permission by W. H. Freeman and Company. **35.12C and D** Graphs from N. B. Davies. Prey selection and social behaviour in wagtails (Aves: Motacillidae). *Journal of Animal Ecology* 46: 37–57. Reprinted by permission of Blackwell Publishing Ltd. **35.14B** From Judith W. McIntyre, drawings by Anne Olson, *The Common Loon: Spirit of Northern Lakes*, p. 97 (University of Minnesota Press, 1988). Copyright © 1988 by University of Minnesota Press. Reprinted by permission. **35.16A** Map adapted from W. M. Howell et al. 1980. Abnormal expression of secondary sex characters in a population of mosquitofish, *Gambusia affinis holbrooki*, *Copeia* Vol. 1980, No. 4 (Dec. 5, 1980), pp. 676–681. Reprinted by permission of ASIH. **Module 35.21: Talking About Science** Adapted from an interview in Neil Campbell, *Biology*, 3rd ed. Copyright © 1993 Pearson Education, Inc., publishing as Pearson Benjamin Cummings.

Chapter 36: Table 36.3 Data from Center for Health Statistics, 2003 CDC. **36.5A** Data courtesy of J. N. M. Smith and P. Arcese. **36.5C** Data courtesy of P. Arcese and J. N. M. Smith. **36.8** Data from Fisheries and Oceans, Canada, 1999. **Table 36.9** Data from United Nations, the World at Six Billion, 2007. **36.9A** Data from United Nations, the World at Six Billion, 2007. **36.9B** Data from Population Reference Bureau, 2000 and U.S. Census Bureau International Data Base, 2003. **36.9C and 36.10** Data from U.S. Census Bureau, International Data Base. **36.11B** Adapted from information at phtbb.org/natural/footprint.

Chapter 37: 37.9 From Robert Leo Smith, *Ecology and Field Biology*, 4th ed. Copyright © 1990 by Robert Leo Smith. Reprinted by permission of Addison Wesley Longman, Inc. **37.13A** Source: www.dest.gov.au/archive/Science/pmsec/14meet/rcd1.html. **37.22C** From G. E. Likens et al., "Effects of Forest Cutting and Herbicide Treatment on Nutrient Budgets in the Hubbard Brook Watershed Ecosystem," *Ecological Monographs*, vol. 50, pp. 22–32 (1966). Copyright © 1966 by The Ecological Society of America. Reprinted by permission. **Module 37.23: Talking About Science** Adapted from an interview with Neil Campbell and Jane Reece, *Biology*, 6th ed. Copyright © 2002 Pearson Education, Inc., publishing as Benjamin Cummings.

Chapter 38: 38.5A From "Global Temperature Change," J. Hansen et al., *PNAS*, Sept. 26, 2006, Vol. 103, No. 39. Copyright 2006 National Academy of Sciences, U.S.A. Reprinted by permission. **38.6A** From Climate Change 2007: *The Physical Science Basis*, Intergovernmental Panel on Climate Change. Reprinted by permission of IPCC, c/o World Meteorological Organization. **38.11A** From N. Myers et al., "Biodiversity hotspots for conservation priorities," *Nature*, Vol. 403, p. 853. 2/24/2000. Copyright © 2000 Nature Publishing, Inc. Used with permission. **38.12A** Adapted from W. Purves and G. Orians, *Life: The Science of Biology*, 5th ed., fig. 55.23, p. 1239. Coypright © 1998 by Sinauer and W. H. Freeman & Company. Reprinted by permission. **38.13A** Reprinted by permission of Yellowstone to Yukon Conservation Initiative.

Glossary

A

abiotic factor (ā′-bī-ot′-ik) A nonliving component of an ecosystem, such as air, water, or temperature.

abiotic reservoir The part of an ecosystem where a chemical, such as carbon or nitrogen, accumulates or is stockpiled outside of living organisms.

ABO blood groups Genetically determined classes of human blood that are based on the presence or absence of carbohydrates A and B on the surface of red blood cells. The ABO blood group phenotypes, also called blood types, are A, B, AB, and O.

abscisic acid (ABA) (ab-sis′-ik) A plant hormone that inhibits cell division and promotes dormancy; interacts with gibberellins in regulating seed germination.

absorption The uptake of small nutrient molecules by an organism's own body; the third main stage of food processing, following digestion.

accommodation The automatic changes made by the eye as it focuses on nearby objects.

acetyl CoA (acetyl coenzyme A) The entry compound for the citric acid cycle in cellular respiration; formed from a fragment of pyruvate attached to a coenzyme.

acetylcholine (as′-uh-til-kō′-lēn) A nitrogen-containing neurotransmitter. Among other effects, it slows the heart rate and makes skeletal muscles contract.

achondroplasia (uh-kon′-druh-plā′-zhuh) A form of human dwarfism caused by a single dominant allele; the homozygous condition is lethal.

acid A substance that increases the hydrogen ion concentration in a solution.

acid precipitation Rain, snow, or fog with a pH below 5.6.

acquired immunity The kind of defense that is mediated by B lymphocytes (B cells) and T lymphocytes (T cells). It exhibits specificity, memory, and self-nonself recognition.

acrosome (ak′-ruh-som) A membrane-enclosed sac at the tip of a sperm; contains enzymes that help the sperm penetrate an egg.

actin A globular protein that links into chains, two of which twist helically around each other, forming microfilaments in muscle cells.

action potential A self-propagating change in the voltage across the plasma membrane of a neuron; a nerve signal.

activator A protein that switches on a gene or group of genes.

active immunity Immunity conferred by recovering from an infectious disease or by receiving a vaccine.

active site The part of an enzyme molecule where a substrate molecule attaches (by means of weak chemical bonds); typically, a pocket or groove on the enzyme's surface.

active transport The movement of a substance across a biological membrane against its concentration gradient, aided by specific transport proteins and requiring input of energy (often as ATP).

adaptation An inherited characteristic that enhances an organism's ability to survive and reproduce in a particular environment.

adaptive radiation Period of evolutionary change in which groups of organisms form many new species whose adaptations allow them to fill new or vacant ecological roles in their communities.

adenine (A) (ad′-uh-nēn) A double-ring nitrogenous base found in DNA and RNA.

adenosine triphosphate (ATP) Main energy source for cells.

adhesion The attraction between different kinds of molecules.

adipose tissue A type of connective tissue whose cells contain fat.

adrenal cortex (uh-drē′-nul) The outer portion of an adrenal gland, controlled by ACTH from the anterior pituitary; secretes hormones called glucocorticoids and mineralocorticoids.

adrenal gland One of a pair of endocrine glands, located atop each kidney in mammals, composed of an outer cortex and a central medulla.

adrenal medulla (uh-drē′-nul muh-dul′-uh) The central portion of an adrenal gland, controlled by nerve signals; secretes the fight-or-flight hormones epinephrine and norepinephrine.

adrenocorticotropic hormone (ACTH) (uh-drē′-nō-cōr′-ti-kō-trop′-ik) A protein hormone secreted by the anterior pituitary that stimulates the adrenal cortex to secrete corticosteroids.

adult stem cell A cell present in adult tissues that generates replacements for nondividing differentiated cells.

age structure The relative number of individuals of each age in a population.

agonistic behavior (a′-gō-nis′-tik) Confrontational behavior involving a contest waged by threats, displays, or actual combat, which settles disputes over limited resources, such as food or mates.

AIDS Acquired immunodeficiency syndrome; the late stages of HIV infection, characterized by a reduced number of T cells and the appearance of characteristic opportunistic infections.

alcohol fermentation The conversion of pyruvate from glycolysis to carbon dioxide and ethyl alcohol.

alga (al′-guh) (plural, **algae**) A protist that produces its food by photosynthesis.

alimentary canal (al′-uh-men′-tuh-rē) A digestive tract consisting of a tube running between a mouth and an anus.

allantois (al′-an-tō′-is) In animals, an extraembryonic membrane that develops from the yolk sac; helps dispose of the embryo's nitrogenous wastes and forms part of the umbilical cord in mammals.

allele (uh-lē′-ul) An alternative version of a gene.

allergen (al′-er-jen) An antigen that causes an allergy.

allergy A disorder of the immune system caused by an abnormal sensitivity to an antigen. Symptoms are triggered by histamines released from mast cells.

allopatric speciation The formation of new species in populations that are geographically isolated from one another.

alpha helix (al′-fuh hē′-liks) The spiral shape resulting from the coiling of a polypeptide in a protein's secondary structure.

alternation of generations A life cycle in which there is both a multicellular diploid form, the sporophyte, and a multicellular haploid form, the gametophyte; a characteristic of plants and multicellular green algae.

alternative RNA splicing A type of regulation at the RNA-processing level in which different mRNA molecules are produced from the same primary transcript, depending on which RNA segments are treated as exons and which as introns.

altruism (al′-trū-iz-um) Behavior that reduces an individual's fitness while increasing the fitness of another individual.

alveolates A clade of protists that includes dinoflagellates, apicomplexans, and ciliates.

alveolus (al-vē′-oh-lus) (plural, **alveoli**) One of millions of tiny dead-end sacs within the vertebrate lungs where gas exchange occurs.

Alzheimer's disease (AD) An age-related dementia (mental deterioration) characterized by confusion, memory loss, and other symptoms.

amine (uh-mēn′) An organic compound with one or more amino groups.

amino acid (uh-mēn′-ō) An organic molecule containing a carboxyl group and an amino group; serves as the monomer of proteins.

amino group In an organic molecule, a functional group consisting of a nitrogen atom bonded to two hydrogen atoms.

ammonia A small and very toxic nitrogenous waste produced by metabolism.

amniocentesis (am'-nē-ō-sen-tē'-sis) A technique for diagnosing genetic defects while a fetus is in the uterus. A sample of amniotic fluid, obtained by a needle inserted into the amnion, is analyzed for telltale chemicals and defective fetal cells.

amnion (am'-nē-on) In vertebrate animals, the extraembryonic membrane that encloses the fluid-filled amniotic sac containing the embryo.

amniote Member of a clade of tetrapods that have an amniotic egg containing specialized membranes that protect the embryo. Amniotes include mammals and birds and other reptiles.

amniotic egg (am'-nē-ot'-ik) A shelled egg in which an embryo develops within a fluid-filled amniotic sac and is nourished by yolk; produced by reptiles, birds, and egg-laying mammals, it enables them to complete their life cycles on dry land.

amoeba (uh-mē'-buh) Protist that moves and feeds by means of pseudopodia.

amoebocyte (uh-mē'-buh-sīt) An amoeba-like cell that moves by pseudopodia, found in most animals; depending on the species, may digest and distribute food, dispose of wastes, form skeletal fibers, fight infections, and change into other cell types.

amoebozoan A member of a clade of protists that includes amoebas and slime molds and is characterized by lobe-shaped pseudopodia.

amphibian Member of the tetrapod class Amphibia. Amphibians include frogs, toads, and salamanders.

amygdala (uh-mig'-duh-la) An integrative center of the cerebrum; functionally the part of the limbic system that seems to label information to be remembered.

anabolic steroid (an'-uh-bol'-ik stār'-oyd) A synthetic variant of the male hormone testosterone that mimics some of its effects.

analogy The similarity between two species that is due to convergent evolution rather than to descent from a common ancestor with the same trait.

anaphase The fourth stage of mitosis, beginning when sister chromatids separate from each other and ending when a complete set of daughter chromosomes arrives at each of the two poles of the cell.

anaphylactic shock (an'-uh-fi-lak'-tik) A potentially fatal allergic reaction caused by extreme sensitivity to an allergen; involves an abrupt dilation of blood vessels and a sharp drop in blood pressure.

anatomy The study of the structure of an organism.

anchorage dependence The requirement that to divide, a cell must be attached to a solid surface.

androgen (an'-drō-jen) A steroid sex hormone secreted by the gonads that promotes the development and maintenance of the male reproductive system and male body features.

anemia (uh-nē'-me-ah) A condition in which an abnormally low amount of hemoglobin or a low number of red blood cells results in the body cells receiving too little oxygen.

angiosperm (an'-jē-ō-sperm) A flowering plant, which forms seeds inside a protective chamber called an ovary.

annelid (uh-nel'-id) A segmented worm. Annelids include earthworms, polychaetes, and leeches.

annual A plant that completes its life cycle in a single year or growing season.

antagonistic hormones Two hormones that have opposite effects.

anterior Pertaining to the front, or head, of a bilaterally symmetrical animal.

anterior pituitary (puh-tū'-uh-tār-ē) An endocrine gland, adjacent to the hypothalamus and the posterior pituitary, that synthesizes several hormones, including some that control the activity of other endocrine glands.

anther A sac in which pollen grains develop, located at the tip of a flower's stamen.

anthropoid (an'-thruh-poyd) A member of a primate group made up of the apes (gibbons, orangutans, gorillas, chimpanzees, bonobos, and humans) and monkeys.

antibody (an'-tih-bod'-ē) A protein dissolved in blood plasma that attaches to a specific kind of antigen and helps counter its effects.

anticodon (an'-tī-kō'-don) On a tRNA molecule, a specific sequence of three nucleotides that is complementary to a codon triplet on mRNA.

antidiuretic hormone (ADH) (an'-tē-dī'-yū-ret'-ik) A hormone made by the hypothalamus and secreted by the posterior pituitary that promotes water retention by the kidneys.

antigen (an'-tuh-jen) A foreign (nonself) molecule that elicits an acquired immune response.

antigen receptor A transmembrane version of an antibody molecule that B cells and T cells use to recognize specific antigens; also called a membrane antibody.

antigen-binding site A region of the antibody molecule responsible for the antibody's recognition and binding function.

antigenic determinant A region on the surface of an antigen molecule to which an antibody binds.

antigen-presenting cell (APC) One of a family of white blood cells (for example, a macrophage) that ingests a foreign substance or a microbe and attaches antigenic portions of the ingested material to its own surface, thereby displaying the antigens to a helper T cell.

antihistamine (an'-tē-his'-tuh-mēn) A drug that interferes with the action of histamine, providing temporary relief from an allergic reaction.

anus The opening through which undigested materials are expelled.

aorta (ā-or'-tuh) An artery that conveys blood directly from the left ventricle of the heart to other arteries.

aphotic zone (ā-fō'-tik) The region of an aquatic ecosystem beneath the photic zone, where light does not penetrate enough for photosynthesis to take place.

apical dominance (ā'-pik-ul) In a plant, the hormonal inhibition of axillary buds by a terminal bud.

apical meristem (ā'-pik-ul mer'-uh-stem) A meristem at the tip of a plant root or in the terminal or axillary bud of a shoot.

apicomplexan (ap'-ē-kom-pleks'-un) A member of a protistan group of parasitic alveolates, some of which cause human diseases.

appendicular skeleton (ap'-en-dik'-yū-ler) Components of the skeletal system that support the fins of a fish or the arms and legs of a land vertebrate: cartilage and bones of the shoulder girdle, pelvic girdle, and forelimbs and hind limbs. *See also* axial skeleton.

appendix (uh-pen'-dix) A small, fingerlike extension of the vertebrate cecum; contains a mass of white blood cells that contribute to immunity.

aquaporin A transport protein in the plasma membrane of some plant or animal cells that facilitates the diffusion of water across the membrane (osmosis).

aqueous humor (ā'-kwē-us hyū'-mer) Plasma-like liquid in the space between the lens and the cornea in the vertebrate eye; helps maintain the shape of the eye, supplies nutrients and oxygen to its tissues, and disposes of its wastes.

aqueous solution (ā'-kwā-us) A solution in which water is the solvent.

arachnid A member of a major arthropod group (chelicerates) that includes spiders, scorpions, ticks, and mites.

Archaea (ar'-kē-uh) One of two prokaryotic domains of life, the other being Bacteria.

arteriole (ar-ter'-ē-ōl) A vessel that conveys blood between an artery and a capillary bed.

artery A vessel that carries blood away from the heart to other parts of the body.

arthropod (ar'-thrō-pod) A member of the most diverse phylum in the animal kingdom. Arthropods include the horseshoe crab, arachnids (for example, spiders, ticks, scorpions, and mites), crustaceans (for example, crayfish, lobsters, crabs, and barnacles), millipedes, centipedes, and insects. Arthropods are characterized by a chitinous exoskeleton, molting, jointed appendages, and a body formed of distinct groups of segments.

artificial selection The selective breeding of domesticated plants and animals to promote the occurrence of desirable traits.

ascomycete *See* sac fungus.

asexual reproduction The creation of offspring by a single parent, without the participation of sperm and egg.

assisted reproductive technology (ART) Procedure that involves surgically removing eggs from a woman's ovaries, fertilizing them, and then returning them to the woman's body. *See also* in vitro fertilization.

associative learning Learning that a particular stimulus or response is linked to a reward or punishment; includes classical conditioning and trial-and-error learning.

astigmatism (uh-stig′-muh-tizm) Blurred vision caused by a misshapen lens or cornea.

atherosclerosis (ath′-uh-rō′-skluh-rō′-sis) A cardiovascular disease in which fatty deposits called plaques develop on the inner walls of the arteries, narrowing their inner diameters.

atom The small unit of matter that retains the properties of an element.

atomic mass The approximate total mass of an atom; also called atomic weight. Given as a whole number, the atomic mass approximately equals the mass number.

atomic number The number of protons in each atom of a particular element.

ATP synthase A cluster of several membrane proteins that function in chemiosmosis with adjacent electron transport chains, using the energy of a hydrogen ion concentration gradient to make ATP.

atrium (ā′-trē-um) (plural, **atria**) A heart chamber that receives blood from the veins.

auditory canal Part of the vertebrate outer ear that channels sound waves from the pinna or outer body surface to the eardrum.

australopith (os-trā′-lō-pith) The first hominids; scavenger-gatherer-hunters who lived on African savannas between about 4.4 million years ago and 1.5 million years ago.

autoimmune disease An immunological disorder in which the immune system attacks the body's own molecules.

autonomic nervous system (ot′-ō-nom′-ik) The component of the vertebrate peripheral nervous system that regulates the internal environment; made up of sympathetic and parasympathetic subdivisions.

autosome A chromosome not directly involved in determining the sex of an organism; in mammals, for example, any chromosome other than X or Y.

autotroph (ot′-ō-trōf) An organism that makes its own food (often by photosynthesis), thereby sustaining itself without eating other organisms or their molecules. Plants, algae, and numerous bacteria are autotrophs.

auxin (ok′-sin) A plant hormone, indoleacetic acid or a related compound, whose chief effect is to promote seedling elongation.

AV (atrioventricular) node A region of specialized heart muscle tissue between the left and right atria where electrical impulses are delayed for about 0.1 second before spreading to both ventricles and causing them to contract.

axial skeleton (ak′-sē-ul) Components of the skeletal system that support the central trunk of the body: the skull, backbone, and rib cage in a vertebrate. *See also* appendicular skeleton.

axillary bud (ak′-sil-ār-ē) An embryonic shoot present in the angle formed by a leaf and stem.

axon (ak′-son) A neuron fiber that conducts signals to another neuron or to an effector cell.

B

B cell A type of lymphocyte that matures in the bone marrow and later produces antibodies; responsible for the humoral immune response. *See also* T cell.

bacillus (buh-sil′-us) (plural, **bacilli**) A rod-shaped prokaryotic cell.

Bacteria One of two prokaryotic domains of life, the other being Archaea.

bacteriophage (bak-tēr′-ē-ō-fāj) A virus that infects bacteria; also called a phage.

balancing selection Natural selection that maintains stable frequencies of two or more phenotypic forms in a population.

ball-and-socket joint A joint that allows rotation and movement in several planes. Examples in humans are the hip and shoulder joints.

bark All the tissues external to the vascular cambium in a plant that is growing in thickness. Bark is made up of secondary phloem, cork cambium, and cork.

Barr body A dense body formed from a deactivated X chromosome found in the nuclei of female mammalian cells.

basal body (ba′-sul) A eukaryotic cell organelle consisting of a 9 + 0 arrangement of microtubule triplets; may organize the microtubule assembly of a cilium or flagellum; structurally identical to a centriole.

basal metabolic rate (BMR) The number of kilocalories a resting animal requires to fuel its essential body processes for a given time.

basal nuclei (ba′-sul nū′-klē-ī) Clusters of nerve cell bodies located deep within the cerebrum that are important in motor coordination.

base A substance that decreases the hydrogen ion (H) concentration in a solution.

basidiomycete *See* club fungus.

basilar membrane The floor of the middle canal of the inner ear.

behavior Individually, an action carried out by the muscles or glands under control of the nervous system in response to a stimulus; collectively, the sum of an animal's responses to external and internal stimuli.

behavioral ecology The scientific field concerned with behavior in an evolutionary context.

benign tumor An abnormal mass of cells that remains at its original site in the body.

benthic realm A seafloor, or the bottom of a freshwater lake, pond, river, or stream.

biennial A plant that completes its life cycle in two years.

bilateral symmetry An arrangement of body parts such that an organism can be divided equally by a single cut passing longitudinally through it. A bilaterally symmetrical organism has mirror-image right and left sides.

bilaterian Member of the clade of animals Bilateria exhibiting bilateral symmetry.

bile A mixture of substances that is produced by the liver and stored in the gallbladder; it emulsifies fats and aids in their digestion.

binary fission A means of asexual reproduction in which a parent organism, often a single cell, divides into two individuals of about equal size.

binomial A two-part, latinized name of a species; for example, *Homo sapiens*.

biodiversity The variety of living things, encompassing genetic diversity, species diversity, and ecosystem diversity.

biodiversity crisis The current rapid decline in the variety of life on Earth, largely due to the effects of human culture.

biodiversity hot spot A small geographic area with an exceptional concentration of endangered and threatened species, especially endemic species (those found nowhere else).

biofilm A surface-coating colony of prokaryotes that engage in metabolic cooperation.

biogenic amine A neurotransmitter derived from an amino acid.

biogeochemical cycle Any of the various chemical circuits that involve both biotic and abiotic components of an ecosystem.

biogeography The study of past and present distribution of organisms.

biological clock An internal timekeeper that controls an organism's biological rhythms; marks time with or without environmental cues, but often requires signals from the environment to remain tuned to an appropriate period. *See also* circadian rhythm.

biological control The intentional release of a natural enemy to attack a pest population.

biological magnification The accumulation of persistent chemicals in the living tissues of consumers in food chains.

biological species concept Definition of a species as a population or groups of populations whose members have the potential to interbreed in nature and produce viable, fertile offspring, but do not produce viable, fertile offspring with members of other such populations.

biology The scientific study of life.

biomass The amount, or mass, of organic material in an ecosystem.

biome (bī′-ōm) Major types of ecological associations that occupy broad geographic regions of land or water and are characterized by organisms adapted to the particular environments.

bioremediation The use of living organisms to detoxify and restore polluted and degraded ecosystems.

biosphere The entire portion of Earth inhabited by life; the sum of all the planet's ecosystems.

biotechnology The use of living organisms (often microbes) to perform useful tasks; today, usually involves DNA technology.

biotic factor (bī-o′-tik) A living component of a biological community; an organism, or a factor pertaining to one or more organisms.

bipolar disorder Depressive mental illness characterized by swings of mood from high to low; also called manic-depressive disorder.

birth control pill A chemical contraceptive that inhibits ovulation, retards follicular development, or alters a woman's cervical mucus to prevent sperm from entering the uterus.

bivalve A member of a group of molluscs that includes clams, mussels, scallops, and oysters.

blastocoel (blas′-tuh-sēl) In a developing animal, a central, fluid-filled cavity in a blastula.

blastocyst (blas′-tō-sist) A mammalian embryo (equivalent to an amphibian blastula) made up of a hollow ball of cells that results from cleavage and that implants in the mother's endometrium.

blastula (blas′-tyū-luh) An embryonic stage that marks the end of cleavage during animal development; a hollow ball of cells in many species.

blood A type of connective tissue with a fluid matrix called plasma in which blood cells are suspended.

blood pressure The force that blood exerts against the walls of blood vessels.

blood-brain barrier A system of capillaries in the brain that restricts passage of most substances into the brain, thereby preventing large fluctuations in the brain's environment.

body cavity A fluid-containing space between the digestive tract and the body wall.

bolus A lubricated ball of chewed food.

bone A type of connective tissue consisting of living cells held in a rigid matrix of collagen fibers embedded in calcium salts.

bottleneck effect Genetic drift resulting from a drastic reduction in population size; typically, the surviving population is no longer genetically representative of the original population.

Bowman's capsule A cup-shaped swelling at the receiving end of a nephron in the vertebrate kidney; collects the filtrate from the blood.

brain The part of the central nervous system involved in regulating and controlling bodily activity and interpreting information from the senses transmitted through the nervous system.

brainstem A functional unit of the vertebrate brain, composed of the midbrain, the medulla oblongata, and the pons; serves mainly as a sensory filter, selecting which information reaches higher brain centers.

breathing Ventilation of the lungs through alternating inhalation and exhalation.

breathing control center A brain center that directs the activity of organs involved in breathing.

bronchiole (bron′-kē-ōl) A thin breathing tube that branches from a bronchus within a lung.

bronchus (bron′-kus) (plural, **bronchi**) One of a pair of breathing tubes that branch from the trachea into the lungs.

brown alga One of a group of marine, multicellular, autotrophic protists belonging to the stramenopile clade; the most common and largest type of seaweed. Brown algae include the kelps.

bryophyte (brī′-uh-fīt) One of a group of plants that lack xylem and phloem; a nonvascular plant. Bryophytes include mosses, liverworts, and hornworts.

budding A means of asexual reproduction whereby a new individual developed from an outgrowth of a parent splits off and lives independently.

buffer A chemical substance that resists changes in pH by accepting hydrogen ions from or donating hydrogen ions to solutions.

bulbourethral gland (bul′-bō-yū-rē′-thrul) One of a pair of glands near the base of the penis in the human male that secrete fluid that lubricates and neutralizes acids in the urethra during sexual arousal.

bulk feeder An animal that eats relatively large pieces of food.

C

C₃ plant A plant that uses the Calvin cycle for the initial steps that incorporate CO_2 into organic material, forming a three-carbon compound as the first stable intermediate.

C₄ plant A plant that prefaces the Calvin cycle with reactions that incorporate CO_2 into four-carbon compounds, the end product of which supplies CO_2 for the Calvin cycle.

calcitonin (kal′-sih-tōn′-in) A peptide hormone secreted by the thyroid gland that lowers the blood calcium level.

Calvin cycle The second of two stages of photosynthesis; a cyclic series of chemical reactions that occur in the stroma of a chloroplast, using the carbon in CO_2 and the ATP and NADPH produced by the light reactions to make the energy-rich sugar molecule G3P.

CAM plant A plant that uses an adaptation for photosynthesis in arid conditions in which carbon dioxide entering open stomata during the night is converted to organic acids, which release CO_2 for the Calvin cycle during the day, when stomata are closed.

capillary (kap′-il-er-ē) A microscopic blood vessel that conveys blood between an arteriole and a venule; enables the exchange of nutrients and dissolved gases between the blood and interstitial fluid.

capillary bed One of the networks of capillaries that infiltrate every organ and tissue in the body.

capsid The protein shell that encloses a viral genome. It may be rod-shaped, polyhedral, or more complex in shape.

carbohydrate (kar′-bō-hi′-drāt) Member of the class of biological molecules consisting of simple single-monomer sugars (monosaccharides), two-monomer sugars (disaccharides), and other multiunit sugars (polysaccharides).

carbon fixation The incorporation of carbon from atmospheric CO_2 into the carbon in organic compounds. During photosynthesis in a C₃ plant, carbon is fixed into a three-carbon sugar as it enters the Calvin cycle. In C₄ and CAM plants, carbon is fixed into a four-carbon sugar.

carbon skeleton The chain of carbon atoms that forms the structural backbone of an organic molecule.

carbonyl group (kar′-buh-nēl′) In an organic molecule, a functional group consisting of a carbon atom linked by a double bond to an oxygen atom.

carboxyl group (kar′-bok-sil) In an organic molecule, a functional group consisting of an oxygen atom double-bonded to a carbon atom that is also bonded to a hydroxyl group.

carboxylic acid An organic compound containing a carboxyl group.

carcinogen (kar-sin′-uh-jin) A cancer-causing agent, either high-energy radiation (such as X-rays or UV light) or a chemical.

carcinoma (kar′-sih-nō′-muh) Cancer that originates in the coverings of the body, such as skin or the lining of the intestinal tract.

cardiac cycle (kar′-dē-ak) The alternating contractions and relaxations of the heart.

cardiac muscle Striated muscle that forms the contractile tissue of the heart.

cardiac output The volume of blood pumped per minute by each ventricle of the heart.

cardiovascular disease (kar′-dē-ō-vas′-kyū-ler) Diseases of the heart and blood vessels.

cardiovascular system A closed circulatory system with a heart and branching network of arteries, capillaries, and veins.

carnivore An animal that mainly eats other animals.

carpel (kar′-pul) The female part of a flower, consisting of a stalk with an ovary at the base and a stigma, which traps pollen, at the tip.

carrier An individual who is heterozygous for a recessively inherited disorder and who therefore does not show symptoms of that disorder but who may pass on the recessive allele to offspring.

carrying capacity In a population, the number of individuals that an environment can sustain.

cartilage (kar′-ti-lij) A flexible connective tissue consisting of living cells and collagenous fibers embedded in a rubbery matrix.

Casparian strip (kas-par′-ē-un) A waxy barrier in the walls of endodermal cells in a plant root that prevents water and ions from entering the xylem without crossing one or more cell membranes.

cation exchange A process in which positively charged minerals are made available to a plant when hydrogen ions in the soil displace mineral ions from the clay particles.

cecum (sē′-kum) (plural, **ceca**) A blind outpocket at the beginning of the large intestine.

cell A basic unit of living matter separated from its environment by a plasma membrane; the fundamental structural unit of life.

cell body The part of a cell, such as a neuron, that houses the nucleus.

cell cycle An ordered sequence of events (including interphase and the mitotic phase) that extends from the time a eukaryotic cell is first formed from a dividing parent cell until its own division into two cells.

cell cycle control system A cyclically operating set of proteins that triggers and coordinates events in the eukaryotic cell cycle.

cell division The reproduction of a cell.

cell plate A double membrane across the midline of a dividing plant cell, between which the new cell wall forms during cytokinesis.

cell theory The theory that all living things are composed of cells and that all cells come from other cells.

cell wall A protective layer external to the plasma membrane in plant cells, bacteria, fungi, and some protists; protects the cell and helps maintain its shape.

cell-mediated immune response The type of specific immunity brought about by T cells; fights body cells infected with pathogens. *See also* humoral immune response.

cellular metabolism (muh-tab′-uh-lizm) The chemical activities of cells.

cellular respiration The aerobic harvesting of energy from food molecules; the energy-releasing chemical breakdown of food molecules, such as glucose, and the storage of potential energy in a form that cells can use to perform work; involves glycolysis, the citric acid cycle, and oxidative phosphorylation (the electron transport chain and chemiosmosis).

cellular respiration The aerobic harvesting of energy from food molecules; the breakdown of organic molecules, such as glucose, for the production of ATP; involves glycolysis, the citric acid cycle, and oxidative phosphorylation (the electron transport chain and chemiosmosis).

cellular slime mold A type of protist that has unicellular amoeboid cells and aggregated reproductive bodies in its life cycle; members of amoebozoan clade.

cellulose (sel′-yū-lōs) A large polysaccharide composed of many glucose monomers linked into cable-like fibrils that provide structural support in plant cell walls.

centipede A carnivorous terrestrial arthropod that has one pair of long legs for each of its numerous body segments, with the front pair modified as poison claws.

central canal The narrow cavity in the center of the spinal cord that is continuous with the fluid-filled ventricles of the brain.

central nervous system (CNS) The integration and command center of the nervous system; the brain and, in vertebrates, the spinal cord.

central vacuole (vak′-yū-ōl) A membrane-enclosed sac occupying most of the interior of a mature plant cell, having diverse roles in reproduction, growth, and development.

centralization The presence of a central nervous system (CNS) distinct from a peripheral nervous system.

centriole (sen′-trē-ōl) A structure in an animal cell composed of cylinders of microtubule triplets arranged in a 9 and 0 pattern. An animal usually has a centrosome with a pair of centrioles involved in cell division.

centromere (sen′-trō-mēr) The region of a duplicated chromosome where two sister chromatids are joined and where spindle microtubules attach during mitosis and meiosis. The centromere divides at the onset of anaphase during mitosis and anaphase II during meiosis.

centrosome (sen′-trō-sōm) Material in the cytoplasm of a eukaryotic cell that gives rise to microtubules; important in mitosis and meiosis; also called microtubule-organizing center.

cephalization (sef′-uh-luh-zā′-shun) The concentration of a nervous system at the anterior (head) end.

cephalopod A member of a group of molluscs that includes squids and octopuses.

cerebellum (sār-ruh-bel′-um) Part of the vertebrate hindbrain; mainly a planning center that interacts closely with the cerebrum in coordinating body movement.

cerebral cortex (suh-rē′-brul kor′-teks) A folded sheet of gray matter forming the surface of the cerebrum. In humans, it contains integrating centers for higher brain functions such as reasoning, speech, language, and imagination.

cerebral hemisphere The right or left half of the vertebrate cerebrum.

cerebrospinal fluid (suh-rē′-brō-spī′-nul) Blood-derived fluid that surrounds, protects against infection, nourishes, and cushions the brain and spinal cord.

cerebrum (suh-rē′-brum) The largest, most sophisticated, and most dominant part of the vertebrate forebrain, made up of right and left cerebral hemispheres.

cervix (ser′-viks) The neck of the uterus, which opens into the vagina.

chaparral (shap′-uh-ral′) A biome dominated by spiny evergreen shrubs adapted to periodic drought and fires; found where cold ocean currents circulate offshore, creating mild, rainy winters and long, hot, dry summers.

character A heritable feature that varies among individuals within a population, such as flower color in pea plants.

charophyte (kār′-uh-fīt) A member of the green algal group that are considered the closest relatives of land plants.

chelicerate A lineage of arthropods that includes horseshoe crabs, scorpions, ticks, and spiders.

chemical bond An attraction between two atoms resulting from a sharing of outer-shell electrons or the presence of opposite charges on the atoms. The bonded atoms gain complete outer electron shells.

chemical cycling The use and reuse of chemical elements such as carbon within an ecosystem.

chemical energy Energy available in molecules for release in a chemical reaction; a form of potential energy.

chemical reaction The making and breaking of chemical bonds, leading to changes in the composition of matter.

chemiosmosis (kem′-ē-oz-mō′-sis) Energy-coupling mechanics that uses the energy of hydrogen ion (H^+) gradients across membranes to phosphorylate ADP; powers most ATP synthesis in cells.

chemoautotroph An organism that obtains both energy and carbon from inorganic chemicals. A chemoautotroph makes its own organic compounds from CO_2 without using light energy.

chemoreceptor (kē′-mō-rē-sep′-ter) A sensor (sensory receptor) that detects chemical changes within the body or a specific kind of molecule in the external environment.

chiasma (kī-az′-muh) (plural, **chiasmata**) The microscopically visible site where crossing over has occurred between chromatids of homologous chromosomes during prophase I of meiosis.

chitin A structural polysaccharide found in many fungal cell walls and in the exoskeletons of arthropods.

chlamydia A group of bacteria that live inside eukaryotic host cells. Includes human pathogens that cause blindness and nongonococcal urethritis, a common sexually transmitted disease.

chlorophyll A green pigment located within the chloroplasts of plants, algae, and certain prokaryotes. Chlorophyll *a* can participate directly in the light reactions, which convert solar energy to chemical energy.

chloroplast (klō′-rō-plast) An organelle found in plants and photosynthetic protists that absorbs sunlight and uses it to drive the synthesis of organic molecules (sugars) from carbon dioxide and water.

choanocyte (kō-an′-uh-sīt) A flagellated feeding cell found in sponges. Also called a collar cell, it has a collar-like ring that traps food particles around the base of its flagellum.

cholesterol (k-les′-tuh-rol) A steroid that is an important component of animal cell membranes and that acts as a precursor molecule for the synthesis of other steroids such as hormones.

chondrichthyan Member of the class Chondrichthyes, vertebrates with skeletons made mostly of cartilage, such as sharks and rays.

Chordate (kor′-date) Member of the phylum Chordata, animals that at some point during their development have a dorsal hollow nerve cord, a notochord, pharyngeal slits, and a post-anal tail. Chordates include lancelets, tunicates, and vertebrates.

chorion (kō′r-ē-on) In animals, the outermost extraembryonic membrane, which becomes the mammalian embryo's part of the placenta.

chorionic villus (kōr′-ē-on′-ik vil′-us) An outgrowth of the chorion, containing embryonic blood vessels. As part of the placenta, chorionic villi absorb nutrients and oxygen from, and pass wastes into, the mother's bloodstream.

chorionic villus sampling (CVS) A technique for diagnosing genetic defects while the fetus is in an early development stage within the uterus. A small sample of the fetal portion of the placenta is removed and analyzed.

choroid (kōr′-oyd) A thin, pigmented layer in the vertebrate eye, surrounded by the sclera. The iris is part of the choroid.

chromatin (krō′-muh-tin) The complex of DNA and proteins that constitutes eukaryotic chromosomes; often used to refer to the diffuse, very extended form taken by chromosomes when a cell is not dividing.

chromosome (krō′-muh-sōm) A threadlike, gene-carrying structure found in the nucleus of a eukaryotic cell and most visible during mitosis and meiosis; also, the main gene-carrying structure of a prokaryotic cell. Chromosomes consist of chromatin, a combination of DNA and protein.

chromosome theory of inheritance A basic principle in biology stating that genes are located on chromosomes and that the behavior of chromosomes during meiosis accounts for inheritance patterns.

chyme (kīm) The mixture of partially digested food and digestive juices formed in the stomach.

chytrid Member of a group of fungi that are mostly aquatic and have flagellated spores. They probably represent the most primitive fungal lineage.

cilia A short cellular appendage specialized for locomotion formed from a core of nine outer doublet microtubules and two single microtubules covered by the cell's plasma membrane.

ciliate (sil′-ē-it) A type of protist that moves and feeds by means of cilia. Ciliates belong to the alveolate clade.

circadian rhythm (ser-kā′-dē-un) In an organism, a biological cycle of about 24 hours that is controlled by a biological clock, usually under the influence of environmental cues; a pattern of activity that is repeated daily. *See also* biological clock.

circulatory system The organ system that transports materials such as nutrients, O_2, and hormones to body cells and transports CO_2 and other wastes from body cells.

citric acid cycle The metabolic cycle fueled by acetyl CoA formed after glycolysis in cellular respiration. Chemical reactions in the citric acid cycle complete the metabolic breakdown of glucose molecules to carbon dioxide. The cycle occurs in the matrix of mitochondria and supplies most of the NADH molecules that carry energy to the electron transport chains. The second major stage of cellular respiration.

clade A group of species that includes an ancestral species and all its descendants.

cladistics (kluh-dis′-tiks) An approach to systematics in which common descent is the primary criterion used to classify organisms by placing them into groups called clades.

class In classification, the taxonomic category above order.

cleavage (klē′-vij) (1) Cytokinesis in animal cells and in some protists, characterized by pinching in of the plasma membrane. (2) In animal development, the succession of rapid cell divisions without cell growth that converts the animal zygote into a ball of cells.

cleavage furrow The first sign of cytokinesis during cell division in an animal cell; a shallow groove in the cell surface near the old metaphase plate.

clitoris An organ in the female that engorges with blood and becomes erect during sexual arousal.

clonal selection (klōn′-ul) The production of a lineage of genetically identical cells that recognize and attack the specific antigen that stimulated their proliferation. Clonal selection is the mechanism that underlies the immune system's specificity and memory of antigens.

clone As a verb, to produce genetically identical copies of a cell, organism, or DNA molecule. As a noun, the collection of cells, organisms, or molecules resulting from cloning; *also* (colloquially), a single organism that is genetically identical to another because it arose from the cloning of a somatic cell.

closed circulatory system A circulatory system in which blood is confined to vessels and is kept separate from the interstitial fluid.

club fungus Member of a group of fungi characterized by club-shaped, spore-producing structures called basidia.

clumped Describing a dispersion pattern in which individuals are aggregated in patches.

cnidarian (nī-dār′-ē-un) An animal characterized by cnidocytes, radial symmetry, a gastrovascular cavity, and a polyp and medusa body form. Cnidarians include the hydras, jellyfishes, sea anemones, corals, and related animals.

cnidocyte (nī′-duh-sīt) A specialized cell for which the phylum Cnidaria is named; consists of a capsule containing a fine coiled thread, which, when discharged, functions in defense and prey capture.

coccus (kok′-us) (plural, **cocci**) A spherical prokaryotic cell.

cochlea (kok′-lē-uh) A coiled tube in the inner ear of birds and mammals that contains the hearing organ, the organ of Corti.

codominant Inheritance pattern in which a heterozygote expresses the distinct trait of both alleles.

codon (kō′-don) A three-nucleotide sequence in mRNA that specifies a particular amino acid or polypeptide termination signal; the basic unit of the genetic code.

coelom (sē′-lom) A body cavity completely lined with mesoderm.

coenzyme An organic molecule serving as a cofactor. Most vitamins function as coenzymes in important metabolic reactions.

coevolution Evolutionary change in which adaptations in one species act as a selective force on a second species, inducing adaptations that in turn act as a selective force on the first species; mutual influence on the evolution of two different interacting species.

cofactor A nonprotein molecule or ion that is required for the proper functioning of an enzyme. *See also* coenzyme.

cognition The process carried out by an animal's nervous system, which includes perceiving, storing, integrating, and using information obtained by its sensory receptors.

cognitive map A representation within the nervous system of spatial relations among objects in an animal's environment.

cohesion (kō-hē′-zhun) The binding together of like molecules, often by hydrogen bonds.

collecting duct A tube in the vertebrate kidney that concentrates urine while conveying it to the renal pelvis.

collenchyma cell (kō-len′-kim-uh) In plants, a cell with a thick primary wall and no secondary wall, functioning mainly in supporting growing parts.

colon (kō'-lun) Large intestine; the tubular portion of the vertebrate alimentary tract between the small intestine and the anus; functions mainly in water absorption and the formation of feces.

communication Animal behavior including transmission of, reception of, and response to signals.

community An assemblage of all the organisms living together and potentially interacting in a particular area.

companion cell In a plant, a cell connected to a sieve-tube member whose nucleus and ribosomes provide proteins for the sieve-tube member.

competitive inhibitor A substance that reduces the activity of an enzyme by binding to the enzyme's active site in place of the substrate. A competitive inhibitor's structure mimics that of the enzyme's substrate.

complement system A family of innate defensive blood proteins that cooperate with other components of the vertebrate defense system to protect against microbes; can enhance phagocytosis, directly lyse pathogens, and amplify the inflammatory response.

complementary DNA (cDNA) A DNA molecule made *in vitro* using mRNA as a template and the enzyme reverse transcriptase. A cDNA molecule therefore corresponds to a gene but lacks the introns present in the DNA of the genome.

complete digestive tract A digestive tube with two openings, a mouth and an anus.

complete dominance A type of inheritance in which the phenotypes of the heterozygote and dominant homozygote are indistinguishable.

complete metamorphosis (met'-uh-mōr'-fuh-sis) A type of development in certain insects in which development from larva to adult is achieved by multiple molts that are followed by a pupal stage. While encased in its pupa, the body rebuilds from clusters of embryonic cells that have been held in reserve. The adult emerges from the pupa.

compost Decomposing organic material that can be used to add nutrients to soil.

compound A substance containing two or more elements in a fixed ratio. For example, table salt (NaCl) consists of one atom of the element sodium (Na) for every atom of chlorine (Cl).

compound eye The photoreceptor in many invertebrates; made up of many tiny light detectors, each of which detects light from a tiny portion of the field of view.

concentration gradient An increase or decrease in the density of a chemical substance in an area. Cells often maintain concentration gradients of ions across their membranes. When a gradient exists, substances tend to move from where they are more concentrated to where they are less concentrated.

conception The fertilization of the egg by a sperm cell in humans.

cone (1) In vertebrates, a photoreceptor cell in the retina, stimulated by bright light and enabling color vision. (2) In conifers, a reproductive structure bearing pollen or ovules.

coniferous forest A biome characterized by conifers, cone-bearing evergreen trees.

conjugation The union (mating) of two bacterial cells or protist cells and the transfer of DNA between the two cells.

connective tissue Tissue consisting of a sparse population of cells held in an abundant extracellular matrix, which they produce.

conservation biology A goal-oriented science that endeavors to sustain biological diversity.

consumer An organism that obtains its food by eating plants or by eating animals that have eaten plants.

continental drift A change in the position of continents resulting from the incessant slow movement of the plates of Earth's crust on the underlying molten mantle.

continental shelf The submerged part of a continent.

contraception The deliberate prevention of pregnancy.

controlled experiment A component of the process of science whereby a scientist carries out two parallel tests, an experimental test and a control test. The experimental test differs from the control by one factor, the variable.

convergent evolution Adaptive change resulting in nonhomologous (analogous) similarities among organisms.

Species from different evolutionary lineages come to resemble each as a result of living in very similar environments.

copulation Sexual intercourse, usually necessary for internal fertilization to occur.

cork cambium Meristematic tissue that produces cork cells during secondary growth of a plant.

cork The outermost protective layer of a plant's bark, produced by the cork cambium.

cornea (kor'-nē-uh) The transparent frontal portion of the sclera, which admits light into the vertebrate eye.

corpus callosum (kor'-pus kuh-lō'-sum) The thick band of nerve fibers that connect the right and left cerebral hemispheres in placental mammals, enabling the hemispheres to process information together.

corpus luteum (kor'-pus lū'-tē-um) A small body of endocrine tissue that develops from an ovarian follicle after ovulation; secretes progesterone and estrogen during pregnancy.

cortex In plants, the ground tissue system of a root, made up mostly of parenchyma cells, which store food and absorb minerals that have passed through the epidermis. *See also* adrenal cortex; cerebral cortex; renal cortex.

corticosteroid A family of hormones synthesized and secreted by the adrenal cortex, consisting of the mineralocorticoids and glucocorticoids.

cotyledon (kot'-uh-lē'-don) The first leaf that appears on an embryo of a flowering plant; a seed leaf. Monocot embryos have one cotyledon; dicot embryos have two.

countercurrent exchange The transfer of a substance or heat between two fluids flowing in opposite directions.

countercurrent heat exchange Parallel blood vessels that convey warm and cold blood in opposite directions, maximizing heat transfer to the cold blood.

covalent bond (ko-vā'-lent) An attraction between atoms that share one or more pairs of outer-shell electrons; symbolized by a single line between the atoms.

cranial nerve A nerve that leaves the brain and innervates an organ of the head or upper body.

craniate A chordate with a head.

crista (kris'-tuh) (plural, **cristae**) An infolding of the inner membrane of a mitochondrion in which is embedded the electron transport chain and the enzyme catalyzing the synthesis of ATP.

crop A pouchlike organ in a digestive tract where food is softened and may be stored temporarily.

cross A mating of two sexually reproducing individuals; often used to describe a genetics experiment involving a controlled mating.

cross-fertilization The fusion of sperm and egg derived from two different individuals.

crossing over The exchange of segments between chromatids of homologous chromosomes during synapsis in prophase I of meiosis; also, the exchange of segments between DNA molecules in prokaryotes.

crustacean A member of a major arthropod group that includes lobsters, crayfish, crabs, shrimps, and barnacles.

cuticle (kyū'-tuh-kul) (1) In animals, a tough, nonliving outer layer of the skin. (2) In plants, a waxy coating on the surface of stems and leaves that helps retain water.

cyanobacteria (sī-an'-ō-bak-tēr'-ē-uh) Photoautotrophic prokaryotes with plantlike, oxygen-generating photosynthesis.

cystic fibrosis (sis'-tik fī-brō'-sis) A genetic disease that occurs in people with two copies of a certain recessive allele; characterized by an excessive secretion of mucus and vulnerability to infection; fatal if untreated.

cytokinesis (sī'-tō-kuh-nē-sis) The division of the cytoplasm to form two separate daughter cells. Cytokinesis usually occurs during telophase of mitosis. Mitosis and cytokinesis make up the mitotic (M) phase of the cell cycle.

cytokinin (sī'-tō-kī'-nin) One of a family of plant hormones that promotes cell division, retards aging in flowers and fruits, and may interact antagonistically with auxins in regulating plant growth and development.

cytoplasm (sī-tō-plaz′-um) Everything inside a cell between the plasma membrane and the nucleus; consists of a semifluid medium and organelles.

cytosine (C) (sī′-tuh-sin) A single-ring nitrogenous base found in DNA and RNA.

cytoskeleton A network of protein fibers in the cytoplasm of a eukaryotic cell; includes microfilaments, intermediate filaments, and microtubules.

cytotoxic T cell (sī′-tō-tok′-sik) A type of lymphocyte that attacks body cells infected with pathogens.

decomposer Prokaryotes and fungi that secrete enzymes that digest nutrients from organic material and convert them into inorganic forms.

decomposition The breakdown of organic materials into inorganic ones.

dehydration reaction (dē-hī-drā′-shun) A chemical process in which two molecules become covalently bonded to each other with the removal of a water molecule. Also called condensation.

dehydrogenase (dē′-hī-droj′-uh-nās) An enzyme that catalyzes a chemical reaction during which one or more hydrogen atoms are removed from a molecule.

deletion The loss of one or more nucleotides from a gene by mutation; the loss of a fragment of a chromosome.

demographic transition A shift from zero population growth in which birth rates and death rates are high to zero population growth characterized instead by low birth and death rates.

denaturation (dē-nā′-chur-ā′-shun) A process in which a protein unravels, losing its specific structure and hence function; can be caused by changes in pH or salt concentration or by high temperature. Also refers to the separation of the two strands of the DNA double helix, caused by similar factors.

dendrite (den′-drīt) A neuron fiber that conveys signals from its tip inward, toward the rest of the neuron; in a motor neuron, one of several short, branched extensions that convey nerve signals toward the cell body.

density-dependent Referring to a decline in birth rates or a rise in death rates in response to an increase in the number of individuals living in a designated area.

density-dependent inhibition The arrest of cell division that occurs when cells grown in a laboratory dish touch one another.

deoxyribonucleic acid (DNA) (dē-ok′-sē-rī′-bō-nū-klā′-ik) A double-stranded helical nucleic acid molecule consisting of nucleotide monomers with deoxyribose sugar and the nitrogenous bases adenine (A), cytosine (C), guanine (G), and thymine (T). Capable of replicating, is an organism's genetic material. *See also* gene.

dermal tissue system The outer protective covering of plants.

desert A biome characterized by organisms adapted to sparse rainfall (less than 30 cm per year) and rapid evaporation.

desertification The conversion of semi-arid regions to desert.

determinate growth Termination of growth after reaching a certain size, as in most animals. *See also* indeterminate growth.

detritivore (duh-trī′-tuh-vor) An organism that consumes organic wastes and dead organisms.

detritus (duh-trī′-tus) Dead organic matter.

deuterostome (dū-ter′-ō-stōm) An animal with a coelom that forms from hollow outgrowths of the digestive tube of the early embryo. The deuterostomes include the echinoderms and the chordates.

diabetes mellitus (dī′-uh-bē′-tis me-li′-tis) A human hormonal disease in which body cells cannot absorb enough glucose from the blood and become energy starved; body fats and proteins are then consumed for their energy. Type 1 (insulin-dependent) diabetes results when the pancreas does not produce insulin; type 2 (non-insulin-dependent) diabetes results when body cells fail to respond to insulin.

dialysis (dī-al′-uh-sis) Separation and disposal of metabolic wastes from the blood by mechanical means; an artificial method of performing the functions of the kidneys.

diaphragm (dī′-uh-fram) The sheet of muscle separating the chest cavity from the abdominal cavity in mammals. Its contraction expands the chest cavity, and its relaxation reduces it.

diastole (dī-as′-tō-lē) The stage of the heart cycle in which the heart muscle is relaxed, allowing the chambers to fill with blood. *See also* systole.

diatom (dī′-uh-tom) A unicellular, autotrophic protist that belongs to the stramenopile clade. Diatoms possess a unique, glassy cell wall containing silica.

dicot (dī′-kot) A term traditionally used to refer to flowering plants that have two embryonic seed leaves, or cotyledons.

differentiation The specialization in the structure and function of cells that occurs during the development of an organism; results from selective activation and deactivation of the cells' genes.

diffusion The spontaneous tendency of a substance to move down its concentration gradient from where it is more concentrated to where it is less concentrated.

digestion The mechanical and chemical breakdown of food into molecules small enough for the body to absorb; the second main stage of food processing in animals.

digestive system The organ system involved in ingestion and digestion of food, absorption of nutrients, and elimination of wastes.

dihybrid cross (dī′-hī′-brid) An experimental mating of individuals differing at two genetic loci.

dinoflagellate (dī′-nō-flaj′-uh-let) A member of a group of protists belonging to the alveolate clade. Dinoflagellates are common components of marine and freshwater phytoplankton.

diploid cell In an organism that reproduces sexually, a cell containing two homologous sets of chromosomes, one set inherited from each parent; a $2n$ cell.

diplomonad A member of a group of heterotrophic protists that have modified mitochondria. The group diplomonads includes the waterborne parasite *Giardia*.

directional selection Natural selection in which individuals at one end of the phenotypic range survive and reproduce more successfully than do other individuals.

disaccharide (dī-sak′-uh-rīd) A sugar molecule consisting of two monosaccharides linked by a dehydration reaction.

dispersion pattern The manner in which individuals in a population are spaced within their area. Three types of dispersion patterns are clumped (individuals are aggregated in patches), uniform (individuals are evenly distributed), and random (unpredictable distribution).

disruptive selection Natural selection in which individuals on both extremes of a phenotypic range are favored over intermediate phenotypes.

distal tubule In the vertebrate kidney, the portion of a nephron that helps refine filtrate and empties it into a collecting duct.

disturbance In ecology, a force that changes a biological community and usually removes organisms from it.

DNA fingerprinting *See* DNA profiling.

DNA ligase (lī′-gās) An enzyme, essential for DNA replication, that catalyzes the covalent bonding of adjacent DNA strands; used in genetic engineering to paste a specific piece of DNA containing a gene of interest into a bacterial plasmid or other vector.

DNA microarray A glass slide carrying thousands of different kinds of single-stranded DNA fragments arranged in an array (grid). A DNA microarray is used to detect and measure the expression of thousands of genes at one time. Tiny amounts of a large number of single-stranded DNA fragments representing different genes are fixed to the glass slide. These fragments, ideally representing all the genes of an organism, are tested for hybridization with various samples of cDNA molecules.

DNA polymerase (puh-lim′-er-ās) An enzyme that assembles DNA nucleotides into polynucleotides using a preexisting strand of DNA as a template.

DNA profiling A procedure that analyzes DNA fragments to determine whether they come from a specific individual.

DNA technology Methods used to study and/or manipulate DNA, including recombinant DNA technology.

doldrums (dol'-drums) An area of calm or very light winds near the equator, caused by rising warm air.

domain A taxonomic category above the kingdom level. The three domains of life are Archaea, Bacteria, and Eukarya.

dominance hierarchy The ranking of individuals within a group, based on social interactions; usually maintained by agonistic behavior.

dominant allele The allele that determines the phenotype of a gene when the individual is heterozygous for that gene.

dorsal Pertaining to the back of a bilaterally symmetrical animal.

dorsal, hollow nerve cord One of the four hallmarks of chordates, a tube that forms on the dorsal side of the body, above the notochord.

double bond A type of covalent bond in which two atoms share two pairs of electrons; symbolized by a pair of lines between the bonded atoms.

double circulation Circulation with separate pulmonary and systemic circuits, in which blood passes through the heart after completing each circuit; ensures vigorous blood flow to all organs.

double fertilization In flowering plants, the formation of both a zygote and a cell with a triploid nucleus, which develops into the endosperm.

double helix The form of native DNA, referring to its two adjacent polynucleotide strands wound into a spiral shape.

Down syndrome A human genetic disorder resulting from the presence of an extra chromosome 21; characterized by heart and respiratory defects and varying degrees of mental retardation.

Duchenne muscular dystrophy (duh-shen' dis'-truh-fē) A human genetic disease caused by a sex-linked recessive allele; characterized by progressive weakening and a loss of muscle tissue.

duodenum (dū-ō-dē'-num) The first portion of the vertebrate small intestine after the stomach, where chyme from the stomach mixes with bile and digestive enzymes.

duplication Repetition of part of a chromosome resulting from fusion with a fragment from a homologous chromosome; can result from an error in meiosis or from mutagenesis.

E

eardrum A sheet of connective tissue separating the outer ear from the middle ear that vibrates when stimulated by sound waves and passes the waves to the middle ear.

echinoderm (uh-kī'-nō-derm) Member of a phylum of slow-moving or sessile marine animals characterized by a rough or spiny skin, a water vascular system, an endoskeleton, and radial symmetry in adults. Echinoderms include sea stars, sea urchins, and sand dollars.

ecological footprint A method of using multiple constraints, including food, fuel, water, housing, and waste deposits, to estimate the human carrying capacity of the Earth.

ecological niche (nich) The role of a species in its community; the sum total of a species use of the biotic and abiotic resources of its environment.

ecological species concept A definition of species in terms of ecological roles (niches).

ecological succession The process of biological community change resulting from disturbance; transition in the species composition of a biological community, often following a flood, fire, or volcanic eruption. *See also* primary succession; secondary succession.

ecology The scientific study of how organisms interact with their environments.

ecosystem (ē'-kō-sis-tem) All the organisms in a given area, along with the nonliving (abiotic) factors with which they interact; a biological community and its physical environment.

ectoderm (ek'-tō-derm) The outer layer of three embryonic cell layers in a gastrula; forms the skin of the gastrula and gives rise to the epidermis and nervous system in the adult.

ectopic pregnancy (ek-top'-ik) The implantation and development of an embryo outside the uterus.

ectotherm (ek'-tō-therm) An animal that warms itself mainly by absorbing heat from its surroundings.

ectothermic Referring to organisms that do not produce enough metabolic heat to have much effect on body temperature. *See also* ectotherm.

effector cell A cell capable of carrying out some action in response to a command from the nervous system.

egg The female gamete; when unfertilized, also called an ovum.

ejaculation (ih-jak'-yū-lā'-shun) Discharge of semen from the penis.

ejaculatory duct The short section of the ejaculatory route in mammals formed by the convergence of the vas deferens and a duct from the seminal vesicle. The ejaculatory duct transports sperm from the vas deferens to the urethra.

electromagnetic receptor A sensor (sensory receptor) that detects energy of different wavelengths, such as electricity, magnetism, and light.

electromagnetic spectrum The entire spectrum of radiation ranging in wavelength from less than a nanometer to more than a kilometer.

electron A subatomic particle with a single negative electrical charge. One or more electrons move around the nucleus of an atom.

electron microscope (EM) An instrument that focuses an electron beam through, or onto the surface of, a specimen. An electron microscope achieves a hundredfold greater resolution than a light microscope.

electron shell An energy level representing the distance of an electron from the nucleus of an atom.

electron transport chain A series of electron carrier molecules that shuttle electrons during the redox reactions that release energy used to make ATP; located in the inner membrane of mitochondria, the thylakoid membranes of chloroplasts, and the plasma membranes of prokaryotes.

electronegativity The attraction of a given atom for the electrons of a covalent bond.

element A substance that cannot be broken down to other substances by chemical means.

elimination The passing of undigested material out of the digestive compartment; the fourth and final stage of food processing in animals.

embryo (em'-brē-ō) A developing stage of a multicellular organism. In humans, the stage in the development of offspring from the first division of the zygote until body structures begin to appear, about the ninth week of gestation.

embryo sac The female gametophyte contained in the ovule of a flowering plant.

embryonic stem cell (ES cell) Cell in the early animal embryo that differentiates during development to give rise to all the different kinds of specialized cells in the body.

embryophyte Another name for land plants, recognizing that land plants share the common derived trait of multicellular, dependent embryos.

emergent properties New properties that emerge with each step upward in the hierarchy of life, owing to the arrangement and interactions of parts as complexity increases.

emerging virus A virus that has appeared suddenly or has recently come to the attention of medical scientists.

endangered species As defined in the U.S. Endangered Species Act, a species that is in danger of extinction throughout all or a significant portion of its range.

endemic species A species whose distribution is limited to a specific geographic area.

endergonic reaction (en'-der-gon'-ik) An energy-requiring chemical reaction, which yields products with more potential energy than the reactants. The amount of energy stored in the products equals the difference between the potential energy in the reactants and that in the products.

endocrine gland (en'-dō-krin) A ductless gland that synthesizes hormone molecules and secretes them directly into the bloodstream.

endocrine system The organ system consisting of ductless glands that secrete hormones and the molecular receptors on or in target cells that respond to the hormones; cooperates with the nervous system in regulating body functions and maintaining homeostasis.

endocytosis (en'-dō-sī-tō'-sis) Cellular uptake of molecules or particles via formation of new vesicles from the plasma membrane.

endoderm (en'-dō-derm) The innermost of three embryonic cell layers in a gastrula; forms the archenteron in the gastrula and gives rise to the innermost linings of the digestive tract and other hollow organs in the adult.

endodermis The innermost layer (a one-cell-thick cylinder) of the cortex of a plant root; forms a selective barrier determining which substances pass from the cortex into the vascular tissue.

endomembrane system A network of membranes inside and around a eukaryotic cell, related either through direct physical contact or by the transfer of membranous vesicles.

endometrium (en'-dō-mē'-trē-um) The inner lining of the uterus in mammals, richly supplied with blood vessels that provide the maternal part of the placenta and nourish the developing embryo.

endoplasmic reticulum (ER) An extensive membranous network in a eukaryotic cell, continuous with the outer nuclear membrane and composed of ribosome-studded (rough) and ribosome-free (smooth) regions. *See also* rough ER; smooth ER.

endorphin (en-dōr'-fin) A pain-inhibiting hormone produced by the brain and anterior pituitary; also serves as a neurotransmitter.

endoskeleton A hard skeleton located within the soft tissues of an animal; includes spicules of sponges, the hard plates of echinoderms, and the cartilage and bony skeletons of vertebrates.

endosperm In flowering plants, a nutrient-rich mass formed by the union of a sperm cell with two polar nuclei during double fertilization; provides nourishment to the developing embryo in the seed.

endospore A thick-coated, protective cell produced within a bacterial cell; endospore becomes dormant and is able to survive harsh environmental conditions.

endosymbiosis (en'-dō-sim'-bē-ō-sis) A process by which the mitochondria and chloroplasts of eukaryotic cells probably evolved from symbiotic associations between small prokaryotic cells living inside larger cells.

endotherm An animal that derives most of its body heat from its own metabolism.

endothermic Referring to animals that use heat generated by metabolism to maintain a warm, steady body temperature. *See also* endotherm.

endotoxin A poisonous component of the outer membrane of gram-negative bacteria that is released only when the bacteria die.

energy The capacity to perform work, or to rearrange matter.

energy coupling In cellular metabolism, the use of energy released from an exergonic reaction to drive an endergonic reaction.

energy flow The passage of energy through the components of an ecosystem.

energy of activation (E_A) The amount of energy that reactants must absorb before a chemical reaction will start.

enhancer A eukaryotic DNA sequence that helps stimulate the transcription of a gene at some distance from it. An enhancer functions by means of a transcription factor called an activator, which binds to it and then to the rest of the transcription apparatus. *See* silencer.

enteric division Part of the autonomic nervous system consisting of complex networks of neurons in the digestive tract, pancreas, and gallbladder.

entropy (en'-truh-pē) A measure of disorder. One form of disorder is heat, which is random molecular motion.

enzyme (en'-zīm) A protein (or RNA molecule) that serves as a biological catalyst, changing the rate of a chemical reaction without itself being changed into a different molecule in the process.

epidermis (ep'-uh-der'-mis) (1) In animals, the living layer or layers of cells forming the protective covering, or outer skin.

(2) In plants, the tissue system forming the protective outer covering of leaves, young stems, and young roots.

epididymis (ep'-uh-did'-uh-mus) A long coiled tube into which sperm pass from the testis and are stored until mature and ejaculated.

epinephrine (ep'-uh-nef'-rin) An amine hormone (also called adrenaline) secreted by the adrenal medulla that prepares body organs for action (fight or flight); also serves as a neurotransmitter.

epithelial tissue (ep'-uh-thē'-lē-ul) A sheet of tightly packed cells lining organs, body cavities, and external surfaces; also called epithelium.

epithelium (plural, epithelia) Epithelial tissue.

erythrocyte (eh-rith'-rō-sīt) A blood cell containing hemoglobin, which transports O_2. Also called red blood cell.

erythropoietin (EPO) A hormone that stimulates the production of erythrocytes. It is secreted by the kidney when tissues of the body do not receive enough oxygen.

esophagus (eh-sof'-uh-gus) The channel that conducts food by peristalsis, usually from the pharynx to the stomach.

essential amino acid An amino acid that an animal cannot synthesize itself and must obtain from food. Eight amino acids are essential for the human adult.

essential element In plants, a chemical element required for the plant to complete its life cycle (to grow from a seed and produce another generation of seeds).

essential fatty acid An unsaturated fatty acid that animals need but cannot make.

essential nutrient A substance that an organism must absorb in preassembled form because it cannot be synthesized from any other material. In humans, there are essential vitamins, minerals, amino acids, and fatty acids.

estrogen (es'-trō-jen) One of several chemically similar steroid hormones secreted by the gonads; maintains the female reproductive system and promotes the development of female body features.

estuary (es'-chū-ār-ē) The area where a freshwater stream or river merges with the ocean.

ethylene A gas that functions as a hormone in plants, triggering aging responses such as fruit ripening and leaf drop.

eudicot (yū-dē-kot) A group consisting of the vast majority of flowering plants that have two embryonic seed leaves, or cotyledons.

euglenozoan A member of a diverse clade of flagellated protists that includes trypanosomes and *Euglena*.

Eukarya The domain that includes all eukaryotic organisms.

eukaryotic cell (yū-kār-ē-ot'-ik) A type of cell that has a membrane-enclosed nucleus and other membrane-enclosed organelles. All organisms except bacteria and archaea are composed of eukaryotic cells.

eumetazoan Member of the clade of "true animals," the animals with true tissues (all animals except sponges).

eustachian tube (yū-stā'-shun) An air passage between the middle ear and throat of vertebrates that equalizes air pressure on either side of the eardrum.

eutherian (yū-thēr'-ē-un) Placental mammal; mammal whose young complete their embryonic development within the uterus, joined to the mother by the placenta.

"evo-devo" The research field that combines evolutionary biology with developmental biology.

evolution Descent with modification; the idea that living species are descendants of ancestral species that were different from present-day ones; also the genetic changes in a population over generations.

evolutionary tree A branching diagram that reflects a hypothesis about evolutionary relationships among groups of organisms.

excretion (ek-skrē'-shun) The disposal of nitrogen-containing metabolic wastes.

excretory system (ek'-skruh-tōr-ē) The organ system that disposes of nitrogen-containing waste products of cellular metabolism.

exergonic reaction (ek'-ser-gon'-ik) An energy-releasing chemical reaction in which the reactants contain more potential energy than the products. The reaction releases an amount of energy equal to the difference in potential energy between the reactants and the products.

exocytosis (ek′-sō-sī-tō′-sis) The movement of materials out of the cytoplasm of a cell by the fusion of vesicles with the plasma membrane.

exon (ek′-son) In eukaryotes, a coding portion of a gene. *See* intron.

exoskeleton A hard, external skeleton that protects an animal and provides points of attachment for muscles.

exotoxin A poisonous protein secreted by certain bacteria.

exponential growth model A mathematical description of idealized, unregulated population growth.

external fertilization The fusion of gametes that parents have discharged into the environment.

extinction The irrevocable loss of a species.

extracellular matrix (ECM) A substance in which the cells of an animal tissue are embedded; consists of protein and polysaccharides.

extraembryonic membranes Four membranes (the yolk sac, amnion, chorion, and allantois) that form a life-support system for the developing embryo of a reptile, bird, or mammal.

extreme halophile A microorganism that lives in a highly saline environment, such as the Great Salt Lake or the Dead Sea.

extreme thermophile A microorganism that thrives in a hot environment (often 60–80°C).

eye cup The simplest type of photoreceptor, a cluster of photo-receptor cells shaded by a cuplike cluster of pigmented cells; detects light intensity and direction.

F

F factor A piece of DNA that can exist as a bacterial plasmid. The F factor carries genes for making sex pili and other structures needed for conjugation, as well as a site where DNA replication can start. F stands for fertility.

F_1 generation The offspring of two parental (P generation) individuals; F_1 stands for first filial.

F_2 generation The offspring of the F_1 generation; F_2 stands for second filial.

facilitated diffusion The passage of a substance through a specific transport protein across a biological membrane down its concentration gradient.

facultative anaerobe (fak′-ul-tā′-tiv an′-uh-rōb) An organism that makes ATP by aerobic respiration if oxygen is present, but that switches to fermentation when oxygen is absent.

family In classification, the taxonomic category above genus.

farsightedness An inability to focus on close objects; occurs when the eyeball is shorter than normal and the focal point of the lens is behind the retina. Also called hyperopia.

fat A large lipid molecule made from an alcohol called glycerol and three fatty acids; a triglyceride. Most fats function as energy-storage molecules.

feces The wastes of the digestive tract.

feedback inhibition A method of metabolic control in which a product of a metabolic pathway acts as an inhibitor of an enzyme within that pathway.

fertilization The union of the nucleus of a sperm cell with the nucleus of an egg cell, producing a zygote.

fertilizer Compounds given to plants to promote their growth.

fetus (fē′-tus) A developing human from the ninth week of gestation until birth; has all the major structures of an adult.

fiber (1) In animals, an elongate, supportive thread in the matrix of connective tissue; an extension of a neuron; a muscle cell. (2) In plants, a long, slender sclerenchyma cell that usually occurs in a bundle.

fibrin (fī′-brin) The activated form of the blood-clotting protein fibrinogen, which aggregates into threads that form the fabric of a blood clot.

fibrinogen (fī-brin′-uh-jen) The plasma protein that is activated to form a clot when a blood vessel is injured.

fibrous connective tissue A dense tissue with large numbers of collagenous fibers organized into parallel bundles. This is the dominant tissue in tendons and ligaments.

filtrate Fluid extracted by the excretory system from the blood or body cavity. The excretory system produces urine from the filtrate after extracting valuable solutes from it and concentrating it.

filtration In the vertebrate kidney, the extraction of water and small solutes, including metabolic wastes, from the blood by the nephrons.

first law of thermodynamics The principle of conservation of energy. Energy can be transferred and transformed, but it cannot be created or destroyed.

fission A means of asexual reproduction whereby a parent separates into two or more genetically identical individuals of about equal size.

fitness The contribution an individual makes to the gene pool of the next generation relative to the contribution of other individuals in the population.

fixed action pattern (FAP) A genetically programmed, virtually unchangeable behavioral sequence performed in response to a certain stimulus.

flagellum (fluh-jel′-um) (plural, **flagella**) A long cellular appendage specialized for locomotion. The flagella of prokaryotes and eukaryotes differ in both structure and function. Like cilia, eukaryotic flagella have a 9 + 2 arrangement of microtubules covered by the cell's plasma membrane.

flatworm A member of the phylum Platyhelminthes.

fluid feeder An animal that lives by sucking nutrient-rich fluids from another living organism.

fluid mosaic A description of membrane structure, depicting a cellular membrane as a mosaic of diverse protein molecules embedded in a fluid bilayer made of phospholipid molecules.

fluke One of a group of parasitic flatworms.

follicle (fol′-uh-kul) A cluster of cells that surround, protect, and nourish a developing egg cell in the ovary; also secretes estrogen.

food chain A sequence of food transfers from producers through one to four levels of consumers in an ecosystem.

food web A network of interconnecting food chains.

food-conducting cell A specialized, living plant cell with thin primary walls; arranged end to end, such cells collectively form phloem tissue. Also called sieve-tube member.

foot In an invertebrate animal, a structure used for locomotion or attachment, such as the muscular organ extending from the ventral side of a mollusc.

foraging Behavior used in recognizing, searching for, capturing, and consuming food.

foraminiferan (foram) A protist that moves and feeds by means of threadlike pseudopodia and has porous shells composed of calcium carbonate.

forebrain One of three ancestral and embryonic regions of the vertebrate brain; develops into the thalamus, hypothalamus, and cerebrum.

forensics The scientific analysis of evidence for crime scene and other legal proceedings. Also referred to as forensic science.

fossil fuel An energy-containing deposit of organic material formed from the remains of ancient organisms.

fossil record The chronicle of evolution over millions of years of geologic time engraved in the order in which fossils appear in rock strata.

fossils A preserved remnant or impression of an organism that lived in the past.

founder effect Genetic drift that occurs when a few individuals become isolated from a larger population, with the result that the composition of the new population's gene pool is not reflective of that of the original population.

fovea (fō′-vē-uh) An eye's center of focus and the place on the retina where photoreceptors are highly concentrated.

fragmentation A means of asexual reproduction whereby a single parent breaks into parts that regenerate into whole new individuals.

free-living flatworm One of a group of nonparasitic flatworms.

frequency-dependent selection A decline in the reproductive success of individuals that have a phenotype that has become too common in a population.

fruit A ripened, thickened ovary of a flower, which protects developing seeds and aids in their dispersal.

functional group An assemblage of atoms commonly attached to the carbon skeletons of organic molecules and usually involved in chemical reactions.

Fungi (fun'-jē) The kingdom that contains the fungi.

fungus (plural, **fungi**) A heterotrophic eukaryote that digests its food externally and absorbs the resulting small nutrient molecules. Most fungi consist of a netlike mass of filaments called hyphae. Molds, mushrooms, and yeasts are examples of fungi.

G

gallbladder An organ that stores bile and releases it as needed into the small intestine.

gametangium (gam'-uh-tan'-jē-um) (plural, **gametangia**) A reproductive organ that houses and protects the gametes of a plant.

gamete (gam'-ēt) A sex cell; a haploid egg or sperm. The union of two gametes of opposite sex (fertilization) produces a zygote.

gametogenesis The creation of gametes within the gonads.

gametophyte (guh-mē'-tō-fīt) The multicellular haploid form in the life cycle of organisms undergoing alternation of generations; mitotically produces haploid gametes that unite and grow into the sporophyte generation.

ganglion (gang'-glē-un) (plural, **ganglia**) A cluster (functional group) of nerve cell bodies in a centralized nervous system.

gas exchange The exchange of O_2 and CO_2 between an organism and its environment. An aerobic organism takes up O_2 and gives off CO_2.

gastric juice The collection of fluids (mucus, enzymes, and acid) secreted by the stomach.

gastrin A digestive hormone that stimulates the secretion of gastric juice.

gastropod A member of the largest group of molluscs, including snails and slugs.

gastrovascular cavity A digestive compartment with a single opening, the mouth; may function in circulation, body support, waste disposal, and gas exchange, as well as digestion.

gastrula (gas'-trū-luh) The embryonic stage resulting from gastrulation in animal development. Most animals have a gastrula made up of three layers of cells: ectoderm, endoderm, and mesoderm.

gastrulation (gas'-trū-lā'-shun) The phase of embryonic development that transforms the blastula into a gastrula. Gastrulation adds more cells to the embryo and sorts the cells into distinct cell layers.

gel electrophoresis (jel' ē-lek'-trō-fōr-ē'-sis) A technique for separating and purifying macromolecules, either DNAs or proteins. A mixture of the macromolecules is placed on a gel between a positively charged electrode and a negatively charged one. Negative charges on the molecules are attracted to the positive electrode, and the molecules migrate toward that electrode. The molecules separate in the gel according to their rates of migration, which is mostly determined by their size.

gene A discrete unit of hereditary information consisting of a specific nucleotide sequence in DNA (or RNA, in some viruses). Most of the genes of a eukaryote are located in its chromosomal DNA; a few are carried by the DNA of mitochondria and chloroplasts.

gene cloning The production of multiple copies of a gene.

gene expression The process whereby genetic information flows from genes to proteins; the flow of genetic information from the genotype to the phenotype.

gene flow The transfer of alleles from one population to another, as a result of the movement of individuals or their gametes.

gene pool All the alleles for all the genes in a population.

gene therapy A treatment for a disease in which the patient's defective gene is supplemented or altered.

genetic code The set of rules that dictates the correspondence between RNA codons in an mRNA molecule and amino acids in protein.

genetic drift A change in the gene pool of a population due to chance; effects of genetic drift are most pronounced in small populations.

genetic engineering The direct manipulation of genes for practical purposes.

genetic recombination The production, by crossing over and/or independent assortment of chromosomes during meiosis, of offspring with allele combinations different from those in the parents. The term may also be used more specifically to mean the production by crossing over of eukaryotic or prokaryotic chromosomes with gene combinations different from those in the original chromosomes.

genetically modified (GM) organism An organism that has acquired one or more genes by artificial means. If the gene is from another species, the organism is also known as a transgenic organism.

genital herpes A sexually transmitted disease caused by the herpes simplex virus type 2.

genome (jē'-nōm) A complete (haploid) set of an organism's genes; an organism's genetic material.

genomic library (juh-nō'-mik) A set of DNA segments representing an organism's entire genome; each segment is usually carried by a plasmid or phage.

genomics The study of whole sets of genes and their interactions.

genotype (jē'-nō-tīp) The genetic makeup of an organism.

genus (jē'-nus) (plural, **genera**) In classification, the taxonomic category above species; the first part of a species' binomial; for example, *Homo*.

geologic record A time scale established by geologists that divides Earth's history into time periods, grouped into three eons—Archaean, Proterozoic, and Phanerozoic—and further subdivided into eras, periods, and epochs.

germinate To start developing or growing.

gestation (jes-tā'-shun) Pregnancy; the state of carrying developing young within the female reproductive tract.

gibberellin (jib'-uh-rel'-in) One of a family of plant hormones that triggers the germination of seeds and interacts with auxins in regulating growth and fruit development.

gill An extension of the body surface of an aquatic animal, specialized for gas exchange and/or suspension feeding.

gizzard A pouch-like organ in a digestive tract where food is mechanically ground.

glans The rounded, highly sensitive head of the clitoris in females and penis in males.

glia A supporting cell that is essential for the structural integrity and normal functioning of the nervous system.

global warming A slow but steady rise in Earth's surface temperature, caused by increasing concentrations of greenhouse gases (such as CO_2 and CH_4) in the atmosphere.

glomeromycete Member of a group of fungi characterized by a distinct branching form of mycorrhizae (symbiotic relationships with plant roots) called arbuscules.

glomerulus (glō-mer'-ū-lus) (plural, **glomeruli**) In the vertebrate kidney, the part of a nephron consisting of the capillaries that are surrounded by Bowman's capsule; together, a glomerulus and Bowman's capsule produce the filtrate from the blood.

glucagon (glū'-kuh-gon) A peptide hormone, secreted by the islets of Langerhaus in the pancreas, that raises the level of glucose in the blood.

glucocorticoid (glū'-kuh-kor'-tih-koyd) A corticosteroid hormone secreted by the adrenal cortex that increases the blood glucose level and helps maintain the body's response to long-term stress.

glycogen (glī'-kō-jen) An extensively branched polysaccharide of many glucose monomers; serves as an energy-storage molecule in liver and muscle cells; the animal equivalent of starch.

glycolysis (glī-kol'-uh-sis) The multistep chemical breakdown of a molecule of glucose into two molecules of pyruvate; the first stage of cellular respiration in all organisms; occurs in the cytoplasmic fluid.

glycoprotein (glī'-kō-prō'-tēn) A macromolecule consisting of one or more polypeptides linked to short chains of sugars.

goiter An enlargement of the thyroid gland resulting from a dietary iodine deficiency.

Golgi apparatus (gol'-jē) An organelle in eukaryotic cells consisting of stacks of membranous sacs that modify, store, and ship products of the endoplasmic reticulum.

gonad A sex organ in an animal; an ovary or testis.

gram stain Microbiology technique to identify the cell wall composition of bacteria. Results categorize bacteria as gram-positive or gram-negative.

gram-positive bacteria Diverse group of bacteria with a cell wall that is structurally less complex and contains more peptidoglycan than that of gram-negative bacteria. Gram-positive bacteria are usually less toxic than gram-negative bacteria.

granum (gran'-um) (plural, **grana**) A stack of hollow disks formed of thylakoid membrane in a chloroplast. Grana are the sites where light energy is trapped by chlorophyll and converted to chemical energy during the light reactions of photosynthesis.

gravitropism (grav'-uh-trō'-pizm) A plant's growth response to gravity.

gray matter Regions of dendrites and clusters of nerve cell bodies within the CNS.

green alga A member of a group of photosynthetic protists that includes chlorophytes and charophyceans, the closest living relatives of land plants. Green algae include unicellular, colonial, and multicellular species.

greenhouse effect The warming of the atmosphere caused by CO_2, CH_4, and other gases that absorb infrared radiation and slow its escape from Earth's surface.

ground tissue system A tissue of mostly parenchyma cells that makes up the bulk of a young plant and is continuous throughout its body. The ground tissue system fills the space between the epidermis and the vascular tissue system.

growth factor A protein secreted by certain body cells that stimulates other cells to divide.

growth hormone (GH) A protein hormone secreted by the anterior pituitary that promotes development and growth and stimulates metabolism.

guanine (G) (gwa'-nēn) A double-ring nitrogenous base found in DNA and RNA.

guard cell A specialized epidermal cell in plants that regulates the size of a stoma, allowing gas exchange between the surrounding air and the photosynthetic cells in the leaf.

gymnosperm (jim'-nō-sperm) A naked-seed plant. Its seed is said to be naked because it is not enclosed in an ovary.

H

habitat A place where an organism lives; an environment situation in which an organism lives.

habituation Learning not to respond to a repeated stimulus that conveys little or no information.

hair cell A type of mechanoreceptor that detects sound waves and other forms of movement in air or water.

haploid cell In the life cycle of an organism that reproduces sexually, a cell containing a single set of chromosomes; an *n* cell.

Hardy-Weinberg equilibrium The principle that the shuffling of genes that occurs during sexual reproduction by itself cannot change the overall genetic makeup of a population.

heart A muscular pump that propels a circulatory fluid (blood) through vessels to the body.

heart attack The damage or death of cardiac muscle cells and the resulting failure of the heart to deliver enough blood to the body.

heart murmur A hissing sound that most often results from blood squirting backward through a leaky valve in the heart.

heart rate The frequency of heart contraction.

heartwood In the center of trees, the darkened, older layers of secondary xylem made up of cells that no longer transport water and are clogged with resins. *See also* sapwood.

heat Thermal energy; the amount of energy associated with the movement of the atoms and molecules in a body of matter. Heat is energy in its most random form.

helper T cell A type of lymphocyte that helps activate other types of T cells and may help stimulate B cells to produce antibodies.

hemoglobin (hē'-mō-glō-bin) An iron-containing protein in red blood cells that reversibly binds O_2 and transports it to body tissues.

hemophilia (hē'-mō-fil'-ē-uh) A human genetic disease caused by a sex-linked recessive allele; characterized by excessive bleeding following injury.

hepatic portal vein A blood vessel that conveys nutrient-laden blood from capillaries surrounding the intestine directly to the liver.

herbivore An animal that eats only plants or algae. *See also* carnivore; omnivore.

herbivory Consumption of plant parts or algae by an animal.

hermaphroditism (her-maf'-rō-dī-tizm) A condition in which an individual has both female and male gonads and functions as both a male and female in sexual reproduction by producing both sperm and eggs.

heterokaryotic stage A fungal life cycle stage that contains two genetically different nuclei in the same cell.

heterotroph (het'-er-ō-trōf) An organism that cannot make its own organic food molecules and must obtain them by consuming other organisms or their organic products; a consumer or a decomposer in a food chain.

heterozygote advantage Greater reproductive success of heterozygous individuals compared to homozygotes; tends to preserve variation in gene pools.

heterozygous (het'-er-ō-zī'-gus) Having two different alleles for a given gene.

high-density lipoprotein (HDL) A cholesterol-carrying particle in the blood, made up of cholesterol and other lipids surrounded by a single layer of phospholipids in which proteins are embedded. An HDL particle carries less cholesterol than a related lipoprotein, LDL, and may be correlated with a decreased risk of blood vessel blockage.

hindbrain One of three ancestral and embryonic regions of the vertebrate brain; develops into the medulla oblongata, pons, and cerebellum.

hinge joint A joint that allows movement in only one plane. In humans, examples include the elbow and knee.

hippocampus (hip'-uh-kam'-pus) An integrative center of the cerebrum; functionally, the part of the limbic system that plays a central role in memory and learning.

histamine (his'-tuh-mēn) A chemical alarm signal released by injured cells of vertebrates that causes blood vessels to dilate during an inflammatory response.

histone (his'-tōn) A small protein molecule associated with DNA and important in DNA packing in the eukaryotic chromosome.

HIV Human immunodeficiency virus, the retrovirus that attacks the human immune system and causes AIDS.

homeobox (hō'-mē-ō-boks') A 180-nucleotide sequence within a homeotic gene and some other developmental genes.

homeostasis (hō'-mē-ō-stā'-sis) The steady state of body functioning; a state of equilibrium characterized by a dynamic interplay between outside forces that tend to change an organism's internal environment and the internal control mechanisms that oppose such changes.

homeotic gene (hō'-mē-ot'-ik) A master control gene that determines the identity of a body structure of a developing organism, presumably by controlling the developmental fate of groups of cells. (In plants, such genes are called organ identity genes.)

hominid (hah'-mi-nid) A species on the human branch of the evolutionary tree; a member of the family Hominidae, including *Homo sapiens* and our ancestors.

homologous chromosomes (hō-mol'-uh-gus) The two chromosomes that make up a matched pair in a diploid cell. Homologous chromosomes are of the same length, centromere position, and staining pattern and possess genes for the same characteristics at corresponding loci. One homologous chromosome is inherited from the organism's father, the other from the mother.

homologous structures Structures in different species that are similar because of common ancestry.

homology Similarity in characteristics resulting from a shared ancestry.

homozygous (hō′-mō-zī′-gus) Having two identical alleles for a given gene.

horizontal gene transfer The transfer of genes from one genome to another through mechanisms such as transposable elements, plasmid exchange, viral activity, and perhaps, fusions of different organisms.

hormone (1) In animals, a regulatory chemical that travels in the blood from its production site, usually an endocrine gland, to other sites, where target cells respond to the regulatory signal. (2) In plants, a chemical that is produced in one part of the plant that travels to another part, where it coordinates certain plant activities.

horseshoe crab A bottom-dwelling marine chelicerate, a member of the phylum Arthropoda.

human chorionic gonadotropin (HCG) (kōr′-ē-on′-ik gon′-uh-dō-trō′-pin) A hormone secreted by the chorion that maintains the corpus luteum of the ovary during the first three months of pregnancy.

Human Genome Project (HGP) An international collaborative effort to map and sequence the DNA of the entire human genome.

humoral immune response The type of specific immunity brought about by antibody-producing B cells; fights bacteria and viruses in body fluids. *See also* cell-mediated immune response.

humus (hyū′-mus) Decomposing organic material found in topsoil.

Huntington's disease A human genetic disease caused by a dominant allele; characterized by uncontrollable body movements and degeneration of the nervous system; usually fatal 10–20 years after the onset of symptoms.

hybrid The offspring of parents of two different species or of two different varieties of one species; the offspring of two parents that differ in one or more inherited traits; an individual that is heterozygous for one or more pairs of genes.

hybrid zone A geographic region in which members of different species meet and mate, producing at least some hybrid offspring.

hydrocarbon A chemical compound composed only of the elements carbon and hydrogen.

hydrogen bond A type of weak chemical bond formed when the partially positive hydrogen atom participating in a polar covalent bond in one molecule is attracted to the partially negative atom participating in a polar covalent bond in another molecule (or in another part of the same macromolecule).

hydrolysis (hī-drol′-uh-sis) A chemical process in which polymers are broken down by the chemical addition of water molecules to the bonds linking their monomers; an essential part of digestion.

hydrophilic (hī′-drō-fil′-ik) "Water-loving"; pertaining to polar, or charged, molecules (or parts of molecules) that are soluble in water.

hydrophobic (hī′-drō-fō′-bik) "Water-fearing"; pertaining to nonpolar molecules (or parts of molecules) that do not dissolve in water.

hydrostatic skeleton A skeletal system composed of fluid held under pressure in a closed body compartment; the main skeleton of most cnidarians, flatworms, nematodes, and annelids.

hydroxyl group (hī-drok′-sil) In an organic molecule, a functional group consisting of a hydrogen atom bonded to an oxygen atom.

hypertension Abnormally high blood pressure; a persistent blood pressure above 140/90.

hypertonic solution In comparing two solutions, the one with the greater concentration of solutes; cells in such a solution will lose water to their surroundings.

hypha (hī′-fuh) (plural, **hyphae**) One of many filaments making up the body of a fungus.

hypoglycemia (hī′-pō-glī-sē′-mē-uh) An abnormally low level of glucose in the blood that results when the pancreas secretes too much insulin into the blood.

hypothalamus (hī-pō-thal′-uh-mus) The master control center of the system, located in the ventral portion of the vertebrate forebrain. The hypothalamus functions in maintaining homeostasis, especially in coordinating the endocrine and nervous systems; secretes hormones of the posterior pituitary and releasing hormones that regulate the anterior pituitary.

hypothesis (hī-poth′-uh-sis) (plural, **hypotheses**) A tentative explanation a scientist proposes for a specific phenomenon that has been observed.

hypotonic solution In comparing two solutions, the one with the lower concentration of solutes; cells in such a solution will take up water from their surroundings.

immune system An animal body's system of defenses against agents that cause disease.

immunodeficiency disease An immunological disorder in which the immune system lacks one or more components, making the body susceptible to infectious agents that would ordinarily not be pathogenic.

imperfect fungus A fungus with no known sexual stage.

impotence The inability to maintain an erection; also called erectile dysfunction.

imprinting Learning that is limited to a specific critical period in an animal's life and that is generally irreversible.

in vitro fertilization (IVF) (vē′-tro) Uniting sperm and egg in a laboratory container, followed by the placement of a resulting early embryo in the mother's uterus.

inbreeding Mating between close relatives.

inclusive fitness An individual's success at perpetuating its genes by producing its own offspring and by helping close relatives to produce offspring.

incomplete dominance A type of inheritance in which the phenotype of a heterozygote (*Aa*) is intermediate between the phenotypes of the two types of homozygotes (*AA* and *aa*).

incomplete metamorphosis A type of development in certain insects in which development from larva to adult is achieved by multiple molts, but without forming a pupa.

indeterminate growth Growth that continues throughout life, as in most plants. *See also* determinate growth.

induced fit The change in shape of the active site of an enzyme, induced by entry of the substrate so that it binds more snugly to the substrate.

induction During embryonic development, the influence of one group of cells on another group of cells.

inferior vena cava (vē′-nuh kā′-vuh) A large vein that returns O₂-poor blood to the heart from the lower, or posterior, part of the body. *See also* superior vena cava.

infertility The inability to conceive after one year of regular, unprotected intercourse.

inflammatory response An innate body defense in vertebrates caused by a release of histamine and other chemical alarm signals that trigger increased blood flow, a local increase in white blood cells, and fluid leakage from the blood. The resulting inflammatory response includes redness, heat, and swelling in the affected tissues.

ingestion The act of eating; the first main stage of food processing in animals.

ingroup In a cladistic study of evolutionary relationships among taxa of organisms, the group of taxa that is actually being analyzed. *See also* outgroup.

inhibiting hormone A kind of hormone released from the hypothalamus that makes the anterior pituitary stop secreting hormone.

innate behavior Behavior that is under strong genetic control and is performed in virtually the same way by all members of a species.

innate immunity The kind of immunity that is present in an animal before exposure to pathogens and is effective from the time of birth. Innate immune defenses include barriers, phagocytic cells, antimicrobial proteins, the inflammatory response, and natural killer cells.

inner ear One of three main regions of the vertebrate ear; includes the cochlea, organ of Corti, and semicircular canals.

insulin A protein hormone, secreted by the islets of Langerhans in the pancreas, that lowers the level of glucose in the blood.

integration The interpretation of sensory signals within neural processing centers of the central nervous system.

integrins A transmembrane protein that interconnects the extracellular matrix and the cytoskeleton.

integumentary system (in-teg′-yū-men′-ter-ē) The organ system consisting of the skin and its derivatives, such as hair and nails in mammals; helps protect the body from drying out, mechanical injury, and infection.

interferon (in′-ter-fēr′-on) An innate defensive protein produced by virus-infected vertebrate cells and capable of helping other cells resist viruses.

intermediate One of the compounds that form between the initial reactant and the final product in a metabolic pathway, such as between glucose and pyruvate in glycolysis.

intermediate filament An intermediate-sized protein fiber that is one of the three main kinds of fibers making up the cytoskeleton of eukaryotic cells. Intermediate filaments are ropelike, made of fibrous proteins.

intermembrane space One of the two fluid-filled internal compartments of the mitochondrion. The intermembrane space is the narrow region between the inner and outer membranes.

internal fertilization Reproduction in which sperm are typically deposited in or near the female reproductive tract and fertilization occurs within the tract.

interneuron (in′-ter-nūr′-on) A nerve cell, entirely within the central nervous system, that integrates sensory signals and may relay command signals to motor neurons.

internode The portion of a plant stem between two nodes.

interphase The period in the eukaryotic cell cycle when the cell is not actually dividing. Interphase constitutes the majority of the time spent in the cell cycle. *See also* mitotic phase.

interspecific competition Competition between individuals or populations of two or more species requiring a limited resource.

interspecific interactions Relationships among individuals of different species in a community.

interstitial fluid (in′-ter-stish′-ul) An aqueous solution that surrounds body cells and through which materials pass back and forth between the blood and the body tissues.

intertidal zone (in′-ter-tīd′-ul) A shallow zone where the waters of an estuary or ocean meet land.

intestine The region of a digestive tract located between the gizzard or stomach and the anus and where chemical digestion and nutrient absorption usually occur.

intron (in′-tron) In eukaryotes, a nonexpressed (noncoding) portion of a gene that is excised from the RNA transcript. *See* exon.

invasive species Non-native species that spread beyond the original point of introduction and cause environmental or economic damage.

inversion A change in a chromosome resulting from reattachment of a chromosome fragment to the original chromosome, but in a reverse direction. Mutagens and errors during meiosis can cause inversions.

invertebrate An animal that lacks a backbone.

ion (ī-on) An atom that has gained or lost one or more electrons, thus acquiring a charge.

ionic bond (ī-on′-ik) A chemical bond resulting from the attraction between oppositely charged ions.

iris The colored part of the vertebrate eye, formed by the anterior portion of the choroid.

isomers (ī′-sō-mers) Organic compounds with the same molecular formula but different structures and, therefore, different properties.

isotonic solution (ī-sō-ton′-ik) A solution having the same solute concentration as nother solution, thus having no effect on passage of water in or out of the cell.

isotope (ī′-sō-tōp) A variant form of an atom. Isotopes of an element have the same number of protons but different numbers of neutrons.

K

karyotype (kār′-ē-ō-tīp) A display of micrographs of the metaphase chromosomes of a cell, arranged by size and centromere position.

keystone species A species that is not usually abundant in a community yet exerts strong control on community structure by the nature of its ecological role or niche.

kilocalorie (kcal) A quantity of heat equal to 1,000 calories. Used to measure the energy content of food, it is usually called a "Calorie."

kin selection The natural selection that favors altruistic behavior by enhancing reproductive success of relatives.

kinesis (kuh-nē′-sis) Random movement in response to a stimulus.

kinetic energy (kuh-net′-ik) The energy of motion; the energy of a mass of matter that is moving. Moving matter does work by imparting motion to other matter.

kingdom In classification, the broad taxonomic category above phylum.

***K*-selection** The concept that in certain (*K*-selected) populations, life history is centered around producing relatively few offspring that have a good chance of survival.

L

labia majora (lā′-bē-uh muh-jor′-uh) A pair of outer thickened folds of skin that protect the female genital region.

labia minora (lā′-bē-uh mi-nor′-uh) A pair of inner folds of skin, bordering and protecting the female genital region.

labor A series of strong, rhythmic contractions of the uterus that expel a baby out of the uterus and vagina during childbirth.

lactic acid fermentation The conversion of pyruvate to lactate with no release of carbon dioxide.

lancelet One of a group of invertebrate chordates.

landscape Several different ecosystems linked by exchanges of energy, materials, and organisms.

landscape ecology The application of ecological principles to the study of the structure and dynamics of a collection of ecosystems; the scientific study of the biodiversity of interacting ecosystems.

large intestine *See* colon.

larva (lar′-vuh) (plural, **larvae**) A free-living, sexually immature form in some animal life cycles that may differ from the adult in morphology, nutrition, and habitat.

larynx (lār′-inks) The voice box, containing the vocal cords.

lateral line system A row of sensory organs along each side of a fish's body. Sensitive to changes in water pressure, it enables a fish to detect minor vibrations in the water.

lateral meristem A meristem that thickens the roots and shoots of woody plants. The vascular cambium and cork cambium are lateral meristems.

lateralization The phenomenon in which the two hemispheres of the brain become specialized for different functions.

law of independent assortment A general rule in inheritance that when gametes form during meiosis, each pair of alleles for a particular characteristic segregate independently of other pairs; also known as Mendel's second law of inheritance.

law of segregation A general rule in inheritance that individuals have two alleles for each gene and that when gametes form by meiosis, the two alleles separate, each resulting gamete ending up with only one allele of each gene; also known as Mendel's first law of inheritance.

leaf The main site of photosynthesis in a plant; consists of a flattened blade and a stalk (petiole) that joins the leaf to the stem.

learning Modification of behavior as a result of specific experiences.

leech A member of one of the three large groups of annelids. *See* annelid.

lens The structure in an eye that focuses light rays onto the retina.

leukemia (lū-kī′-mē-ah) A type of cancer of the blood-forming tissues, characterized by an excessive production of white blood cells and an abnormally high number of them in the blood; cancer of the bone marrow cells that produce leukocytes.

leukocyte (lū′-kō-sīt) A blood cell that functions in defending the body against infections and cancer cells. Also called white blood cell.

lichen (lī′-ken) A close association between a fungus and an alga or between a fungus and a cyanobacterium, some of which are known to be beneficial to both partners.

life cycle The entire sequence of stages in the life of an organism, from the adults of one generation to the adults of the next.

life history The series of events from birth through reproduction to death.

life table A listing of survivals and deaths in a population in a particular time period and predictions of how long, on average, an individual of a given age will live.

ligament A type of fibrous connective tissue that joins bones together at joints.

light microscope (LM) An optical instrument with lenses that refract (bend) visible light to magnify images and project them into a viewer's eye or onto photographic film.

light reactions The first of two stages in photosynthesis; the steps in which solar energy is absorbed and converted to chemical energy in the form of ATP and NADPH. The light reactions power the sugar-producing Calvin cycle but produce no sugar themselves.

lignin A chemical that hardens the cell walls of plants.

limbic system (lim′-bik) A functional unit of several integrating and relay centers located deep in the human forebrain; interacts with the cerebral cortex in creating emotions and storing memories.

limiting factors Environmental factors that restrict population growth.

linked genes Genes located near each other on the same chromosome that tend to be inherited together.

lipid An organic compound consisting mainly of carbon and hydrogen atoms linked by nonpolar covalent bonds, making the compound mostly hydrophobic. Lipids include fats, phospholipids, and steroids and are insoluble in water.

liver The largest organ in the vertebrate body. The liver performs diverse functions, such as producing bile, preparing nitrogenous wastes for disposal, and detoxifying poisonous chemicals in the blood.

lobe-fin A bony fish with strong, muscular fins supported by bones.

local regulator A chemical messenger that is secreted into the interstitial fluid and causes changes in cells very near the point of secretion. Neurotransmitters are local regulators.

locomotion Active movement from place to place.

locus (plural, **loci**) The particular site where a gene is found on a chromosome. Homologous chromosomes have corresponding gene loci.

logistic growth model A mathematical description of idealized population growth that is restricted by limiting factors.

long-day plant A plant that flowers in late spring or early summer, when day length is long. Long-day plants actually flower in response to a short night (and so are sometimes called short-night plants).

long-term memory The ability to hold, associate, and recall information over one's life.

loop of Henle (hen′-lē) In the vertebrate kidney, the portion of a nephron that helps concentrate the filtrate while conveying it between a proximal tubule and a distal tubule.

loose connective tissue The most widespread connective tissue in the vertebrate body. It binds epithelia to underlying tissues and functions as packing material, holding organs in place.

low-density lipoprotein (LDL) A cholesterol-carrying particle in the blood, made up of cholesterol and other lipids surrounded by a single layer of phospholipids in which proteins are embedded. An LDL particle carries more cholesterol than a related lipoprotein, HDL, and high LDL levels in the blood correlate with a tendency to develop blocked blood vessels and heart disease.

lung An infolded respiratory surface of terrestrial vertebrates that connects to the atmosphere by narrow tubes.

lymph A colorless fluid, derived from interstitial fluid, that circulates in the lymphatic sytem.

lymphatic system (lim-fat′-ik) The vertebrate organ system through which lymph circulates; includes lymph vessels, lymph nodes, and the spleen. The lymphatic system helps remove toxins and pathogens from the blood and interstitial fluid and returns fluid and solutes from the interstitial fluid to the circulatory system.

lymphocyte (lim′-fuh-sīt) A type of white blood cell that is chiefly responsible for the acquired immune response; found mostly in the lymphatic system. *See* B cell; T cell.

lymphoma (lim-fō′-muh) Cancer of the tissues that form white blood cells.

lysogenic cycle (lī′-sō-jen′-ik) A type of bacteriophage replication cycle in which the viral genome is incorporated into the bacterial host chromosome as a prophage. New phages are not produced, and the host cell is not killed or lysed unless the viral genome leaves the host chromosome.

lysosome (lī-sō-sōm) A digestive organelle in eukaryotic cells; contains hydrolytic enzymes that digest the cell's food and wastes.

lytic cycle (lit′-ik) A type of viral replication cycle resulting in the release of new viruses by lysis (breaking open) of the host cell.

macroevolution Evolutionary change on a grand scale, encompassing the origin of new taxonomic groups, evolutionary trends, adaptive radiation, and mass extinction.

macromolecule A giant molecule in a living organism formed by the joining of smaller molecules: a protein, carbohydrate, or nucleic acid.

macronutrient A chemical substance that an organism must obtain in relatively large amounts. *See also* micronutrient.

macrophage (mak′-rō-fāj) A large, amoeboid, phagocytic white blood cell that functions in innate immunity by destroying microbes and in acquired immunity as an antigen-presenting cell.

major depression Depressive mental illness characterized by experiencing a low mood most of the time.

major histocompatibility complex (MHC) *See* self protein.

malignant tumor An abnormal tissue mass that can spread into neighboring tissue and to other parts of the body; a cancerous tumor.

malnourishment The long-term absence from the diet of one or more essential nutrients.

mammal Member of the class Mammalia, amniotes that possess mammary glands and hair.

mantle In a mollusc, the outgrowth of the body surface that drapes over the animal. The mantle produces the shell and forms the mantle cavity.

marsupial (mar-sū′-pē-ul) A pouched mammal, such as a kangaroo, opossum, or koala. Marsupials give birth to embryonic offspring that complete development while housed in a pouch and attached to nipples on the mother's abdomen.

mass number The sum of the number of protons and neutrons in an atom's nucleus.

mast cell In vertebrates, a type of circulating body cell that produces histamine, triggering the inflammatory response.

matter Anything that occupies space and has mass.

maximum sustained yield The level of harvest that produces a consistent yield without forcing a population into decline.

mechanoreceptor (mek′-uh-nō-ri-sep′-ter) A sensor (sensory receptor) that detects physical deformations in the environment associated with pressure, touch, stretch, motion, or sound.

medulla oblongata (meh-duh′-luh ob′-long-got′-uh) Part of the vertebrate hindbrain continuous with the spinal cord; passes data between the spinal cord and forebrain and controls autonomic, homeostatic functions, including breathing, heart rate, swallowing, and digestion.

medusa (med-ū′-suh) (plural, **medusae**) One of two types of cnidarian body forms; an umbrella-like body form.

meiosis (mī-ō′-sis) In a sexually reproducing organism, the division of a single diploid nucleus into four haploid daughter nuclei. Meiosis and cytokinesis produce haploid gametes from diploid cells in the reproductive organs of the parents.

membrane potential The charge difference between a cell's cytoplasm and extracellular fluid due to the differential distribution of ions.

memory The ability to store and retrieve information. *See also* long-term memory, short-term memory.

memory cell A clone of long-lived lymphocytes formed during the primary immune response; remains in a lymph node until activated by exposure to the same antigen that triggered its formation. When activated, a memory cell forms a large clone that mounts the secondary immune response.

meninges (muh-nin′-jēz) Layers of connective tissue that enwrap and protect the brain and spinal cord.

menstrual cycle (men′-strū-ul) The hormonally synchronized cyclic buildup and breakdown of the endometrium of some primates, including humans.

menstruation (men′-strū-ā′-shun) Uterine bleeding resulting from shedding of the endometrium during a menstrual cycle.

meristem (mār′-eh-stem) Plant tissue consisting of undifferentiated cells that divide and generate new cells and tissues.

mesoderm (mez′-ō-derm) The middle layer of the three embryonic cell layers in a gastrula; gives rise to muscles, bones, the dermis of the skin, and most other organs in the adult.

mesophyll (mes′-ō-fil) The green tissue in the interior of a leaf; a leaf's ground tissue system; the main site of photosynthesis.

messenger RNA (mRNA) The type of ribonucleic acid that encodes genetic information from DNA and conveys it to ribosomes, where the information is translated into amino acid sequences.

metabolic pathway A series of chemical reactions that either builds a complex molecule or breaks down a complex molecule into simpler compounds.

metabolic rate The total amount of energy an animal uses in a unit of time.

metabolism The totality of an organism's chemical reactions.

metamorphosis (met′-uh-mōr′-fuh-sis) The transformation of a larva into an adult. *See* complete metamorphosis; incomplete metamorphosis.

metaphase (met′-eh-fāz) The third stage of mitosis, during which all the cell's duplicated chromosomes are lined up at an imaginary plane equidistant between the poles of the mitotic spindle.

metastasis (muh-tas′-tuh-sis) The spread of cancer cells beyond their original site.

methanogens A type of Archaea that produces methane as a metabolic waste product.

methyl group In an organic molecule, a carbon bonded to three hydrogens.

microevolution A change in a population's gene pool over generations.

microfilament The thinnest of the three main kinds of protein fibers making up the cytoskeleton of a eukaryotic cell; a solid, helical rod composed of the globular protein actin.

micrograph A photograph taken through a microscope.

micronutrient An element that an organism needs in very small amounts and that functions as a component or cofactor of enzymes. *See also* macronutrient.

microtubule The thickest of the three main kinds of fibers making up the cytoskeleton of a eukaryotic cell; a straight, hollow tube made of globular proteins called tubulins. Microtubules form the basis of the structure and movement of cilia and flagella.

microvillus (plural, **microvilli**) A microscopic projection on the epithelial cells in the lumen of the small intestine that increase its surface area.

midbrain One of three ancestral and embryonic regions of the vertebrate brain; develops into sensory integrating and relay centers that send sensory information to the cerebrum.

middle ear One of three main regions of the vertebrate ear; a chamber containing three small bones (the hammer, anvil, and stirrup) that convey vibrations from the eardrum to the oval window.

migration The regular back-and-forth movement of animals between two geographic areas at particular times of the year.

millipede A terrestrial arthropod that has two pairs of short legs for each of its numerous body segments and that eats decaying plant matter.

mineral In nutrition, a simple inorganic nutrient that an organism requires for proper body functioning.

mineralocorticoid (min′-er-uh-lō-kort′-uh-koyd) A corticosteroid hormone secreted by the adrenal cortex that helps maintain salt and water homeostasis and may increase blood pressure in response to long-term stress.

mitochondrial matrix (mī′-tō-kon′-drē-ul mā′-triks) The fluid contained within the inner membrane of a mitochondrion.

mitochondrion (mī′-tō-kon′-drē-on) (plural, **mitochondria**) An organelle in eukaryotic cells where cellular respiration occurs. Enclosed by two concentric membranes, it is where most of the cell's ATP is made.

mitosis (mī′-tō-sis) The division of a single nucleus into two genetically identical daughter nuclei. Mitosis and cytokinesis make up the mitotic (M) phase of the cell cycle.

mitotic phase (M phase) The part of the cell cycle when the nucleus is divided (via mitosis), its chromosomes are distributed to the daughter nuclei, and the cytoplasm divided (via cytokinesis), producing two daughter cells.

mitotic spindle A football-shaped structure formed of microtubules and associated proteins that is involved in the movements of chromosomes during mitosis and meiosis.

mold A rapidly growing fungus that reproduces asexually by producing spores.

molecular biology The study of the molecular basis of genes and gene expression; molecular genetics.

molecular clock Evolutionary timing method based on the observation that at least some regions of genomes evolve at constant rates.

molecular systematics A scientific discipline that uses nucleic acids or other molecules in different species to infer evolutionary relationships.

molecule A group of two or more atoms held together by covalent bonds.

mollusc (mol′-lusk) A soft-bodied animal characterized by a muscular foot, mantle, mantle cavity, and radula; includes gastropods (snails and slugs), bivalves (clams, oysters, and scallops), and cephalopods (squids and octopuses).

molting The process of shedding an old exoskeleton or cuticle and secreting a new, larger one.

monoclonal antibody (mon′-ō-klōn′-ul) An antibody secreted by a clone of cells and, consequently, specific for the one antigen that triggered the development of the clone.

monocot (mon′-ō-kot) A flowering plant whose embryos have a single seed leaf, or cotyledon.

monoculture The cultivation of a single plant variety in a large land area.

monogamous Referring to a type of relationship in which one male mates with just one female, and both parents care for the children.

monohybrid cross An experimental mating of individuals differing at one genetic locus.

monomer (mon′-uh-mer) A chemical subunit that serves as a building block of a polymer.

monophyletic (mon′-ō-fī-let′-ik) Pertaining to a taxon derived from a single ancestral species that gave rise to no species in any other taxa.

monosaccharide (mon′-ō-sak′-uh-rīd) The simplest carbohydrate; a simple sugar with a molecular formula that is generally some multiple of CH_2O. Monosaccharides are the building blocks of disaccharides and polysaccharides.

monotreme (mon′-uh-trēm) An egg-laying mammal, such as the duck-billed platypus.

morning after pill (MAP) Birth control pill taken within three days of unprotected intercourse to prevent fertilization or implantation.

morphological species concept A definition of species in terms of measurable anatomical criteria.

motor neuron A nerve cell that conveys command signals from the central nervous system to effector cells, such as muscle cells or gland cells.

motor output The conduction of signals from a processing center in a central nervous system to effector cells.

motor unit A motor neuron and all the muscle fibers it controls.

mouth An opening through which food is taken into an animal's body.

movement corridor A series of small clumps or a narrow strip of quality habitat (usable by organisms) that connects otherwise isolated patches of quality habitat.

mRNA *See* messenger RNA.

muscle fiber Muscle cell.

muscle tissue Tissue consisting of long muscle cells that are capable of contracting when stimulated by nerve impulses; the most abundant tissue in a typical animal. *See* skeletal muscle; cardiac muscle; smooth muscle.

muscular system All the skeletal muscles in the body. (Cardiac muscle and smooth muscle are components of other organ systems.)

mutagen (myū′-tuh-jen) A chemical or physical agent that interacts with DNA and causes a mutation.

mutagenesis (myū′-tuh-jen′-uh-sis) The creation of a mutation.

mutation A change in the nucleotide sequence of an organism's DNA; mutation also can occur in the DNA or RNA of a virus; the ultimate source of genetic diversity.

mutualism An interspecific relationship in which both partners benefit.

mycelium (mī-sē′-lē-um) (plural, **mycelia**) The densely branched network of hyphae in a fungus.

mycorrhiza (mī′-kō-rī′-zuh) (plural, **mycorrhizae**) A close association of plant roots and fungi that is beneficial to both partners.

myelin sheath (mī′-uh-lin) A series of cells, each wound around, and thus insulating, the axon of a nerve cell in vertebrates. Each pair of cells in the sheath is separated by a space called a node of Ranvier.

myofibril (mī′-ō-fī′-bril) A contractile strand in a muscle cell (fiber) made up of many sarcomeres. Longitudinal bundles of myofibrils make up a muscle fiber.

myosin A type of protein filament that interacts with actin filaments to cause cell contraction.

N

NAD+ Nicotinamide adenine dinucleotide; a coenzyme that can accept electrons during the redox reactions of cellular metabolism. The plus sign indicates that the molecule is oxidized and ready to pick up hydrogens; the reduced, hydrogen- (electron-) carrying form is NADH.

natural family planning A form of contraception that relies on refraining from sexual intercourse when conception is most likely to occur; also called the rhythm method.

natural killer cell A cell type that provides an innate immune response by attacking cancer cells and infected body cells, especially those harboring viruses.

natural selection A process in which organisms with certain inherited characteristics are more likely to survive and reproduce than are organisms with other characteristics.

nearsightedness An inability to focus on distant objects; occurs when the eyeball is longer than normal and the lens focuses distant objects in front of the retina. Also called myopia.

negative feedback A common control mechanism in which a chemical reaction, metabolic pathway, or hormone-secreting gland is inhibited by the products of the reaction, pathway, or gland. As the concentration of the products builds up, the product molecules themselves inhibit the process that produced them.

negative pressure breathing A breathing system in which air is pulled into the lungs.

nematode (nem′-uh-tōde) A roundworm, characterized by a pseudocoelom, a cylindrical, wormlike body form, and a tough cuticle.

nephron The tubular excretory unit and associated blood vessels of the vertebrate kidney; extracts filtrate from the blood and refines it into urine.

nerve A cable-like bundle of neuron fibers (axons and dendrites) tightly wrapped in connective tissue.

nerve cord An elongated bundle of axons and dendrites, usually extending longitudinally from the brain or anterior ganglia. One or more nerve cords and the brain make up the central nervous system in many animals.

nerve net A weblike system of neurons, characteristic of radially symmetrical animals such as a hydra.

nervous system The organ system that forms a communication and coordination network between all parts of an animal's body.

nervous tissue Tissue made up of neurons and supportive cells.

neural tube (nyūr′-ul) An embryonic cylinder that develops from ectoderm after gastrulation; gives rise to the brain and spinal cord.

neuron (nyūr′-on) A nerve cell; the fundamental structural and functional unit of the nervous system, specialized for carrying signals from one location in the body to another.

neurosecretory cell A nerve cell that synthesizes hormones and secretes them into the blood and also conducts nerve signals.

neurotransmitter A chemical messenger that carries information from a transmitting neuron to a receiving cell, either another neuron or an effector cell.

neutral variation Genetic variation that does not appear to provide a selective advantage or disadvantage.

neutron An electrically neutral particle (a particle having no electrical charge), found in the nucleus of an atom.

neutrophil (nyū′-truh-fil) An innate, defensive, phagocytic white blood cell that can engulf bacteria and viruses in infected tissue; has a multilobed nucleus.

nitrogen fixation The conversion of atmospheric nitrogen (N_2) into nitrogen compounds (NH_4^+, NO_3^-) that plants can absorb and use.

node The point of attachment of a leaf on a stem.

node of Ranvier (ron′-vē-ā) An unmyelinated region on a myelinated axon of a nerve cell, where nerve signals are regenerated.

nodule A swelling on a plant root consisting of plant cells that contain nitrogen-fixing bacteria.

noncompetitive inhibitor A substance that impedes the activity of an enzyme without entering an active site. By binding elsewhere on the enzyme, a noncompetitive inhibitor changes the shape of the enzyme so that the active site no longer functions.

nondisjunction An accident of meiosis or mitosis in which a pair of homologous chromosomes or a pair of sister chromatids fail to separate at anaphase.

nonpolar covalent bond A covalent bond in which electrons are shared equally between two atoms of similar electronegativity.

nonself molecule A foreign antigen; a protein or other macromolecule that is not part of an organism's body. *See also* self protein.

norepinephrine (nor′-ep-uh-nef′-rin) An amine hormone (also called noradrenaline) secreted by the adrenal medulla that prepares body organs for action (fight or flight); also serves as a neurotransmitter.

notochord (nō′-tuh-kord) A flexible, cartilage-like, longitudinal rod located between the digestive tract and nerve cord in chordate animals; present only in embryos in many species.

nuclear envelope A double membrane, perforated with pores, that encloses the nucleus and separates it from the rest of the eukaryotic cell.

nuclear transplantation A technique in which the nucleus of one cell is placed into another cell that already has a nucleus or in which the nucleus has been previously destroyed.

nucleic acid (nū-klā′-ik) A polymer consisting of many nucleotide monomers; serves as a blueprint for proteins and, through the actions of proteins, for all cellular structures and activities. The two types of nucleic acids are DNA and RNA.

nucleic acid probe In DNA technology, a labeled single-stranded nucleic acid molecule used to find a specific gene or other nucleotide sequence within a mass of DNA. The probe hydrogen-bonds to the complementary sequence in the targeted DNA.

nucleoid (nū'-klē-oyd) A dense region of DNA in a prokaryotic cell.

nucleolus (nū-klē'-ō-lus) A structure within the nucleus of a eukaryotic cell where ribosomal RNA is made and assembled with proteins imported from the cytoplasm to make ribosomal subunits.

nucleosome (nū'-klē-ō-sōm) The bead-like unit of DNA packaging in a eukaryotic cell; consists of DNA wound around a protein core made up of eight histone molecules.

nucleotide (nū'-klē-ō-tīd) An organic monomer consisting of a five-carbon sugar covalently bonded to a nitrogenous base and a phosphate group. Nucleotides are the building blocks of nucleic acids.

nucleus (plural, **nuclei**) (1) An atom's central core, containing protons and neutrons. (2) The genetic control center of a eukaryotic cell.

obligate anaerobes An organism that only carries out fermentation; such organisms cannot use oxygen and also may be poisoned by it.

ocean current One of the riverlike flow patterns in the oceans.

omnivore An animal that eats animals as well as plants or algae.

oncogene (on'-kō-jēn) A cancer-causing gene; usually contributes to malignancy by abnormally enhancing the amount or activity of a growth factor made by the cell.

oogenesis (ō'-uh-jen'-uh-sis) The formation of ova (egg cells).

open circulatory system A circulatory system in which blood is pumped through open-ended vessels and bathes the tissues and organs directly. In an animal with an open circulatory system, blood and interstitial fluid are one and the same.

operator In prokaryotic DNA, a sequence of nucleotides near the start of an operon to which an active repressor can attach. The binding of a repressor prevents RNA polymerase from attaching to the promoter and transcribing the genes of the operon.

operculum (ō-per'-kyū-lum) (plural, **opercula**) A protective flap on each side of a fish's head that covers a chamber housing the gills.

operon (op'-er-on) A unit of genetic regulation common in prokaryotes; a cluster of genes with related functions, along with the promoter and operator that control their transcription.

opportunistic infections Infections that can be controlled by a normally functioning immune system but that cause illness in a person with an immunodeficiency.

opposable thumb An arrangement of the fingers such that the thumb can touch the ventral surface of the fingertips of all four fingers.

optimal foraging theory The basis for analyzing behavior as a compromise of feeding costs versus feeding benefits.

oral cavity The mouth of an animal.

oral contraceptive *See* birth control pills.

order In classification, the taxonomic category above family.

organ A structure consisting of several tissues adapted as a group to perform specific functions.

organ of Corti (kor'-tē) The hearing organ in birds and mammals, located within the cochlea.

organ system A group of organs that work together in performing vital body functions.

organelle (ōr-guh-nel') A membrane-enclosed structure with a specialized function within a cell.

organic compound A chemical compound containing the element carbon and usually synthesized by cells.

organism An individual living thing, such as a bacterium, fungus, protist, plant, or animal.

orgasm A series of rhythmic, involuntary contractions of the reproductive structures.

osmoconformer (oz'-mō-con-form'-er) An organism whose body fluids have a solute concentration equal to that of its surroundings. Osmoconformers do not have a net gain or loss of water by osmosis.

osmoregulation Method by which organisms regulate solute concentrations and balance the gain and loss of water.

osmoregulator An organism whose body fluids have a solute concentration different from that of its environment and that must use energy in controlling water loss or gain.

osmosis (oz-mō'-sis) The diffusion of water across a selectively permeable membrane.

osteoporosis (os'-tē-ō-puh-rō'-sis) A skeletal disorder characterized by thinning, porous, and easily broken bones; most common among women after menopause.

outer ear One of three main regions of the ear in reptiles, birds, and mammals; made up of the auditory canal and, in many birds and mammals, the pinna.

outgroup In a cladistic study of evolutionary relationships among taxa of organisms, a taxon or group of taxa known to have diverged before the lineage that contains the group of species being studied. *See also* ingroup.

oval window In the vertebrate ear, a membrane-covered gap in the skull bone, through which sound waves pass from the middle ear into the inner ear.

ovarian cycle (ō-vār'-ē-un) Hormonally synchronized cyclic events in the mammalian ovary, culminating in ovulation.

ovary (1) In animals, the female gonad, which produces egg cells and reproductive hormones. (2) In flowering plants, the basal portion of a carpel in which the egg-containing ovules develop.

overnourishment The consumption of more food energy than the body needs for normal metabolism.

oviduct (ō'-vuh-dukt) The tube that conveys egg cells away from an ovary; also called a fallopian tube.

ovulation (ah'-vyū-lā'-shun) The release of an egg cell from an ovarian follicle.

ovum (ō'-vum) (plural, **ova**) The female gametes; the haploid, unfertilized egg.

oxidation The loss of electrons from a substance involved in a redox reaction; always accompanies reduction.

oxidative phosphorylation (fos'-fōr-uh-lā'-shun) The production of ATP using energy derived from the redox reactions of an electron transport chain; the third major stage of cellular respiration.

ozone layer The layer of ozone (O_3) in the upper atmosphere that protects life on Earth from the harmful ultraviolet rays in sunlight.

P generation The parent individuals from which offspring are derived in studies of inheritance. (P stands for parental.)

P site One of two of a ribosome's binding sites for tRNA during translation. The P site holds the tRNA carrying the growing polypeptide chain. (P stands for peptidyl tRNA.)

pacemaker The SA (sinoatrial) node; a specialized region of cardiac muscle in the right atrium that maintains the heart's pumping rhythm (heartbeat) by setting the rate at which the heart contracts.

paedomorphosis (pē'-duh-mōr'-fuh-sis) The retention in an adult of juvenile features of its evolutionary ancestors.

pain receptor A sensor (sensory receptor) that detects pain.

paleoanthropology (pā'-lē-ō-an'-thruh-pol'-uh-jē) The study of human origins and evolution.

paleontologist (pā'-lē-on-tol'-uh-jist) A scientist who studies fossils.

pancreas (pan'-krē-us) A gland with dual functions: The digestive portion secretes digestive enzymes and an alkaline solution into the small intestine via a duct. The endocrine portion secretes the hormones insulin and glucagon into the blood.

Pangaea (pan-jē'-uh) The supercontinent consisting of all the major landmasses of Earth fused together. Continental drift formed Pangaea near the end of the Paleozoic era.

parabasilid A heterotrophic protist that has modified mitochondria that generates some energy anaerobically.

parasite Organism that derives its nutrition from a living host, which is harmed by the interaction.

parasympathetic division A set of neurons in the autonomic nervous system that generally promotes body activities that gain and conserve energy, such as digestion and reduced heart rate. *See also* sympathetic division.

parathyroid glands (par′-uh-thī′-royd) Four endocrine glands embedded in the surface of the thyroid gland that secrete parathyroid hormone.

parathyroid hormone (PTH) A peptide hormone secreted by the parathyroid glands that raises blood calcium level.

parenchyma cell (puh-ren′-kim-uh) In plants, a relatively unspecialized cell with a thin primary wall and no secondary wall; functions in photosynthesis, food storage, and aerobic respiration and may differentiate into other cell types.

Parkinson's disease A motor disorder caused by a progressive brain disease and characterized by difficulty in initiating movements, slowness of movement, and rigidity.

parsimony (par′-suh-mō′-nē) In scientific studies, the search for the least complex explanation for an observed phenomenon.

partial pressure A measure of the relative amount of gas in a mixture.

passive immunity Temporary immunity obtained by acquiring ready-made antibodies; lasts only a few weeks or months because the immune system has not been stimulated by antigens.

passive transport The diffusion of a substance across a biological membrane, without any input of energy.

pathogen An agent such as a virus, bacteria, or fungus, that causes disease.

pattern formation During embryonic development, the emergence of a spatial organization in which the tissues and organs of the organism are all in their correct places.

pedigree A family tree representing the occurrence of heritable traits in parents and offspring across a number of generations.

pelagic realm (puh-laj′-ik) The region of an ocean occupied by seawater.

penis The copulatory structure of male mammals.

peptide bond The covalent linkage between two amino acid units in a polypeptide; formed by a dehydration reaction.

peptidoglycan (pep′-tid-ō-glī′-kan) A polymer of complex sugars cross-linked by short polypeptides; a material unique to bacterial cell walls.

per capita rate of increase The average contribution of each individual in a population to population growth.

perception The brain's meaningful interpretation, or conscious understanding, of sensory information.

perennial (puh-ren′-ē-ul) A plant that lives for many years.

peripheral nervous system (PNS) The network of nerves and ganglia carrying signals into and out of the central nervous system.

peristalsis (per′-uh-stal′-sis) Rhythmic waves of contraction of smooth muscles. Peristalsis propels food through a digestive tract and also enables many animals, such as earthworms, to crawl.

permafrost Continuously frozen ground found in the tundra.

peroxisome An organelle containing enzymes that transfer hydrogen from various substrates to oxygen, producing and then degrading hydrogen peroxide.

petal A modified leaf of a flowering plant. Petals are the often colorful parts of a flower that advertise it to pollinators.

pH scale A measure of the relative acidity of a solution, ranging in value from 0 (most acidic) to 14 (most basic). The letters pH stand for potential hydrogen and refer to the concentration of hydrogen ions (H).

phage (fāj) See bacteriophage.

phagocyte (fag′-ō-sīt) A white blood cell (for example, a neutrophil or a monocyte) that engulfs bacteria, foreign proteins, and the remains of dead body cells.

phagocytosis (fag′-ō-sī-tō′-sis) Cellular "eating"; a type of endocytosis whereby a cell engulfs macromolecules, other cells, or particles into its cytoplasm.

pharyngeal slit (fa-rin′-jē-ul) A gill structure in the pharynx; found in chordate embryos and some adult chordates.

pharynx (far′-inks) The organ in a digestive tract that receives food from the oral cavity; in terrestrial vertebrates, the throat region where the air and food passages cross.

phenotype (fē′-nō-tīp) The expressed traits of an organism.

phenotypic plasticity An individual's ability to change phenotype in response to local environmental conditions.

phloem (flō′-um) The portion of a plant's vascular tissue system that conveys phloem sap throughout a plant. Phloem tissue is made up of sieve-tube members.

phloem sap The solution of sugars, other nutrients, and hormones conveyed throughout a plant via phloem tissue.

phosphate group (fos′-fāt) A functional group consisting of a phosphorus atom covalently bonded to four oxygen atoms.

phospholipid (fos′-fō-lip′-id) A lipid made up of glycerol joined to two fatty acids and a phosphate group, giving the molecule a nonpolar hydrophobic tail and a polar hydrophilic head. Phospholipids form bilayers that function as biological membranes.

phosphorylation (fos′-fōr-uh-lā′-shun) The transfer of a phosphate group, usually from ATP, to a molecule. Nearly all cellular work depends on ATP energizing other molecules by phosphorylation.

photic zone (fō′-tik) The region of an aquatic ecosystem into which light penetrates and where photosynthesis occurs.

photoautotroph An organism that obtains energy from sunlight and carbon from CO_2 by photosynthesis.

photoheterotroph An organism that obtains energy from sunlight and carbon from organic sources.

photon (fō′-ton) A fixed quantity of light energy. The shorter the wavelength of light, the greater the energy of a photon.

photoperiod The length of the day relative to the length of the night; an environmental stimulus that plants use to detect the time of year.

photophosphorylation (fō′-tō-fos′-fōr-uh-lā′-shun) The production of ATP by chemiosmosis during the light reactions of photosynthesis.

photopsin (fō-top′-sin) One of a family of visual pigments in the cones of the vertebrate eye that absorb bright, colored light.

photoreceptor A type of electromagnetic sensor (sensory receptor) that detects light.

photorespiration In a plant cell, the breakdown of a two-carbon compound produced by the Calvin cycle. The Calvin cycle produces the two-carbon compound, instead of its usual three-carbon product G3P, when leaf cells fix O_2, instead of CO_2. Photorespiration produces no sugar molecules or ATP.

photosynthesis (fō′-tō-sin′-thuh-sis) The process by which plants, autotrophic protists, and some bacteria use light energy to make sugars and other organic food molecules from carbon dioxide and water.

photosystem A light-capturing unit of a chloroplast's thylakoid membrane, consisting of a reaction center complex surrounded by numerous light-harvesting complexes.

phototropism (fō′-tō-trō′-pizm) The growth of a plant shoot toward (positive phototropism) or away from (negative phototropism) light.

phyla In classification, the taxonomic category above class.

phylogenetic species concept A definition of species as the smallest group of individuals that shares a common ancestor and forms one branch on the tree of life.

phylogenetic tree (fī′-lō-juh-net′-ik) A branching diagram that represents a hypothesis about the evolutionary history of a group of organisms.

phylogeny (fī-loj′-uh-nē) The evolutionary history of a species or group of related species.

physiology (fi′-zi-ol′-uh-ji) The study of the functions of an organism's structures.

phytochrome (fī′-tuh-krōm) A colored protein in plants that contains a special set of atoms that absorbs light.

phytoplankton (fī′-tō-plank′-ton) Algae and photosynthetic bacteria that drift passively in aquatic environments.

pilus (pī′-lus) (plural, **pili**) A short projection on the surface of a prokaryotic cell that helps the prokaryote attach to other surfaces. Specialized sex pili are used in conjugation to hold the mating cells together.

pineal gland (pin′-ē-ul) An outgrowth of the vertebrate brain that secretes the hormone melatonin, which coordinates daily and seasonal body activities such as the sleep/wake circadian rhythm with environmental light conditions.

pinna (pin'-uh) The flap-like part of the outer ear, projecting from the body surface of many birds and mammals; collects sound waves and channels them to the auditory canal.

pinocytosis (pē'-nō-sī-tō'-sis) Cellular "drinking"; a type of endocytosis in which the cell takes fluid and dissolved solutes into small membranous vesicles.

pistil Part of the reproductive organ of an angiosperm, a single carpel or a group of fused carpels.

pith Part of the ground tissue system of a dicot plant. Pith fills the center of a stem and may store food.

pituitary gland An endocrine gland at the base of the hypothalamus; consists of a posterior lobe, which stores and releases two hormones produced by the hypothalamus, and an anterior lobe, which produces and secretes many hormones that regulate diverse body functions.

pivot joint A joint that allows precise rotations in multiple planes. An example in humans is the wrist.

placenta (pluh-sen'-tuh) In most mammals, the organ that provides nutrients and oxygen to the embryo and helps dispose of its metabolic wastes; formed of the embryo's chorion and the mother's endometrial blood vessels.

placental mammal (pluh-sen'-tul) Mammal whose young complete their embryonic development in the uterus, nourished via the mother's blood vessels in the placenta; also called a eutherian.

plasma The liquid matrix of the blood in which the blood cells are suspended.

plasma cell An antibody-secreting B cell.

plasma membrane The membrane that sets a cell off from its surroundings and acts as a selective barrier to the passage of ions and molecules into and out of the cell; consists of a phospholipid bilayer in which are embedded molecules of protein and cholesterol.

plasmid A small ring of independently replicating DNA separate from the main chromosome(s). Plasmids are found in prokaryotes and yeast.

plasmodesma (plaz'-mō-dez'-muh) (plural, **plasmodesmata**) An open channel in a plant cell wall through which strands of cytoplasm connect from adjacent cells.

plasmodial slime mold (plaz-mō'-dē-ul) A type of protist that has amoeboid cells, flagellated cells, and an amoeboid plasmodial feeding stage in its life cycle.

plasmodium (1) A single mass of cytoplasm containing many nuclei. (2) The amoeboid feeding stage in the life cycle of a plasmodial slime mold.

platelet A pinched-off cytoplasmic fragment of a bone marrow cell; platelets circulate in the blood and are important in blood clotting.

pleated sheet The folded arrangement of a polypeptide in a protein's secondary structure.

pleiotropy (plī'-uh-trō-pē) The control of more than one phenotypic characteristic by a single gene.

polar covalent bond A covalent bond between atoms that differ in electronegativity. The shared electrons are pulled closer to the more electronegative atom, making it slightly negative and the other atom slightly positive.

polar molecule A molecule containing polar covalent bonds.

pollen See pollen grain.

pollen grain The structure that will produce the sperm in seed plants; the male gametophyte.

pollination In seed plants, the delivery, by wind or animals, of pollen from the male parts of a plant to the stigma of a carpel on the female.

polychaete (pol'-ē-kēt) A member of the largest group of annelids. See annelid.

polygamous Referring to a type of relationship in which an individual of one sex mates with several of the other.

polygenic inheritance (pol'-ē-jen'-ik) The additive effect of two or more gene loci on a single phenotypic characteristic.

polymer (pol'-uh-mer) A large molecule consisting of many identical or similar molecular units, called monomers, covalently joined together in a chain.

polymerase chain reaction (PCR) (puh-lim'-uh-rās) A technique used to obtain many copies of a DNA molecule or part of a DNA molecule. A small amount of DNA mixed with a heat-resistant DNA polymerase, DNA nucleotides, and a few other ingredients replicates repeatedly in a test tube.

polynucleotide (pol'-ē-nū'-klē-ō-tīd) A polymer made up of many nucleotides covalently bonded together.

polyp (pol'-ip) One of two types of cnidarian body forms; a columnar, hydra-like body.

polypeptide A polymer (chain) of amino acids linked by peptide bonds.

polyploidy An organism that has more than two complete sets of chromosomes as a result of an accident of cell division.

polysaccharide (pol'-ē-sak'-uh-rīd) A carbohydrate polymer consisting of hundreds to thousands of monosaccharides (sugars) linked by dehydration synthesis.

pons (pahnz) Part of the vertebrate hindbrain that functions with the medulla oblongata in passing data between the spinal cord and forebrain and in controlling autonomic, homeostatic functions.

population A group of individuals belonging to one species and living in the same geographic area.

population density The number of individuals of a species per unit area or volume.

population ecology The study of how members of a population interact with their environment, focusing on factors that influence population density and growth.

population momentum In a population in which $r = 0$, the continuation of population growth as girls in the prereproductive age group reach their reproductive years.

positive feedback A control mechanism in which the products of a process stimulate the process that produced them.

post-anal tail A tail posterior to the anus; found in chordate embryos and most adult chordates.

posterior Pertaining to the rear, or tail, of a bilaterally symmetrical animal.

posterior pituitary An extension of the hypothalamus composed of nervous tissue that secretes hormones made in the hypothalamus; a temporary storage site for hypothalamic hormones.

postzygotic barrier A reproductive barrier that prevents hybrid zygotes produced by two different species from developing into viable, fertile adults. Includes reduced hybrid viability, reduced hybrid fertility, and hybrid breakdown.

potential energy The energy that matter possesses because of its location or arrangement. Water behind a dam and chemical bonds possess potential energy.

predation An interaction between species in which one species, the predator, eats the other, the prey.

prepuce (prē'-pyūs) A fold of skin covering the head of the clitoris and penis.

pressure flow mechanism The method by which phloem sap is transported through a plant from a sugar source, where sugars are produced, to a sugar sink, where sugars are used.

prevailing winds Winds that result from the combined effects of Earth's rotation and the rising and falling of air masses.

prezygotic barrier A reproductive barrier that impedes mating between species or hinders fertilization if mating between two species is attempted. Includes temporal, habitat, behavioral, mechanical, and gametic isolation.

primary consumer In the trophic structure of an ecosystem, an organism that eats plants or algae.

primary growth Growth in the length of a plant root or shoot, produced by an apical meristem.

primary immune response The initial immune response to an antigen, which appears after a lag of several days.

primary oocyte (ō'-uh-sīt) A diploid cell, in prophase I of meiosis, that can be hormonally triggered to develop into an ovum.

primary phloem See phloem.

primary production The amount of solar energy converted to chemical energy (in organic compounds) by autotrophs in an ecosystem during a given time period.

primary spermatocyte (sper-mat'-eh-sīt') A diploid cell in the testis that undergoes meiosis I.

primary structure The first level of protein structure; the specific sequence of amino acids making up a polypeptide chain.

primary succession A type of ecological succession in which a biological community arises in an area without soil. *See also* secondary succession.

primary xylem *See* xylem.

primers Short, artificially created, single-stranded DNA molecules that bind to each end of a target sequence to drive a PCR procedure.

prion An infectious form of protein that may multiply by converting related proteins into more prions. Prions cause several related diseases in different animals, including scrapie in sheep and mad cow disease.

problem solving The activity of applying past experiences to overcome obstacles in novel situations.

producer An organism that makes organic food molecules from CO_2, H_2O, and other inorganic raw materials: a plant, alga, or autotrophic prokaryote.

product An ending material in a chemical reaction.

progestin (prō-jes'-tin) One of a family of steroid hormones, including progesterone, produced by the mammalian ovary. Progestins prepare the uterus for pregnancy.

programmed cell death The timely death (and disposal of the remains) of certain cells, triggered by certain genes; an essential process in normal development; *also called* apoptosis.

prokaryotic cell (prō-kār'-ē-ot'-ik) A type of cell lacking a membrane-enclosed nucleus and other membrane-enclosed organelles; found only in the domains Bacteria and Archaea.

prolactin (PRL) (prō-lak'-tin) A protein hormone secreted by the anterior pituitary that stimulates different responses in different animals.

prometaphase The second stage of mitosis, during which the nuclear envelope fragments and the spindle microtubules attach to the kinetochores of the sister chromatids.

promiscuous Referring to a type of relationship in which mating occurs with no strong pair-bonds or lasting relationships.

promoter A specific nucleotide sequence in DNA located at the start of a gene that is the binding site for RNA polymerase and the place where transcription begins.

prophage (prō'-fāj) Phage DNA that has inserted by genetic recombination into the DNA of a prokaryotic chromosome.

prophase The first stage of mitosis, during which the chromatin condenses to form structures (sister chromatids) visible with a light microscope and the mitotic spindle begins to form, but the nucleus is still intact.

prostate gland (pros'-tāt) A gland in human males that secretes an acid-neutralizing component of semen.

protein A functional biological molecule consisting of one or more polypeptides folded into a specific three-dimensional structure.

proteobacteria A diverse clade of gram-negative bacteria that includes five subgroups known as alpha, beta, gamma, delta, and epsilon.

proteomics The study of whole sets of proteins and their interactions.

protist A member of the Kingdom Protista. Most protists are unicellular, though some are colonial or multicellular.

protobiont An aggregate of abiotically produced molecules surrounded by a membrane or membrane-like structure.

proton A subatomic particle with a single positive electrical charge, found in the nucleus of an atom.

proto-oncogene (prō'-tō-on'-kō-jēn) A normal gene that can be converted to a cancer-causing gene.

protostome An animal with a coelom that develops from solid masses of cells that arise between the digestive tube and the body wall of the embryo. The protostomes include the molluscs, annelids, and arthropods.

protozoan (prō'-tō-zō'-un) (plural, **protozoa**) A protist that lives primarily by ingesting food; a heterotrophic, "animal-like" protist.

proximal tubule In the vertebrate kidney, the portion of a nephron immediately downstream from Bowman's capsule that conveys and helps refine filtrate.

proximate cause In animal behavior, a condition in an animal's internal or external environment that is the immediate reason or mechanism for a behavior.

pseudocoelom (sū'-dō-sē'-lōm) A body cavity that is in direct contact with the wall of the digestive tract.

pseudopodium (sū'-dō-pō'-dē-um) (plural, **pseudopodia**) A temporary extension of an amoeboid cell. Pseudopodia function in moving cells and engulfing food.

pulmonary artery A large blood vessel that conveys blood from the heart to a lung.

pulmonary circuit One of two main blood circuits in terrestrial vertebrates; conveys blood between the heart and the lungs. *See also* systemic circuit.

pulmonary vein A blood vessel that conveys blood from a lung to the heart.

pulse The rhythmic stretching of the arteries caused by the pressure of blood during contraction of ventricles in systole.

punctuated equilibria In the fossil record, long periods of apparent stasis in which a species undergoes little or no morphological change interrupted by relatively brief periods of sudden change.

Punnett square A diagram used in the study of inheritance to show the results of random fertilization.

pupil The opening in the iris that admits light into the interior of the vertebrate eye. Muscles in the iris regulate the pupil's size.

Q

quaternary consumer (kwot'-er-ner-ē) An organism that eats tertiary consumers.

quaternary structure The fourth level of protein structure; the shape resulting from the association of two or more polypeptide subunits.

R

R plasmid A bacterial plasmid that carries genes for enzymes that destroy particular antibiotics, thus making the bacterium resistant to the antibiotics.

radial symmetry An arrangement of the body parts of an organism like pieces of a pie around an imaginary central axis. Any slice passing longitudinally through a radially symmetrical organism's central axis divides it into mirror-image halves.

radioactive isotope An isotope whose nucleus decays spontaneously, giving off particles and energy.

radiolarian A protist that moves and feeds by means of threadlike pseudopodia and has a mineralized support structure composed of silica.

radiometric dating A method for determining the absolute ages of fossils and rocks, based on the half-life of radioactive isotopes.

radula (rad'-yū-luh) A toothed, rasping organ used to scrape up or shred food; found in many molluscs.

random Describing a dispersion pattern in which individuals are spaced in a patternless, unpredictable way.

ray-finned fish A bony fish having fins supported by thin, flexible skeletal rays.

reabsorption In the vertebrate kidney, the reclaiming of water and valuable solutes from the filtrate.

reactant A starting material in a chemical reaction.

reaction center complex In a photosystem in a chloroplast, the chlorophyll *a* molecules and the primary electron acceptor that trigger the light reactions of photosynthesis. The chlorophyll donates an electron excited by light energy to the primary electron acceptor, which passes an electron to an electron transport chain.

reading frame The way in which a cell's mRNA-translating machinery groups the mRNA nucleotides into codons.

receptor potential The electrical signal produced by sensory transduction.

receptor-mediated endocytosis (en'-dō-sī-tō'-sis) The movement of specific molecules into a cell by the inward budding of membranous vesicles. The vesicles contain proteins with receptor sites specific to the molecules being taken in.

recessive allele An allele that has no noticeable effect on the phenotype of a gene when the individual is heterozygous for that gene.

reciprocal altruism (al'-trū-izm) In animal behavior, a selfless act repaid at a later time by the beneficiary or by another member of the beneficiary's social system.

recombinant DNA A DNA molecule carrying genes derived from two or more sources.

recombination frequency With respect to two given genes, the number of recombinant progeny from a mating divided by the total number of progeny. Recombinant progeny carry combinations of alleles different from those in either of the parents as a result of independent assortment of chromosomes or crossing over.

Recommended Dietary Allowance (RDA) A recommendation for daily nutrient intake established by nutritionists.

rectum The terminal portion of the large intestine where the feces are stored until they are eliminated.

red alga A member of a group of marine, mostly multicellular, autotrophic protists, which includes the reef-building coralline algae.

red blood cell *See* erythrocyte.

red bone marrow A specialized tissue found in cavities in the ends of bones that produces blood cells.

red-green color blindness A category of common, sex-linked human disorders involving several genes on the X chromosome; characterized by a malfunction of light-sensitive cells in the eyes; affects mostly males but also homozygous females.

redox reaction Short for oxidation-reduction; a chemical reaction in which electrons are lost from one substance (oxidation) and added to another (reduction). Oxidation and reduction always occur together.

reduction The gain of electrons by a substance involved in a redox reaction; always accompanies oxidation.

reflex An automatic reaction to a stimulus, mediated by the spinal cord or lower brain.

regeneration The regrowth of body parts from pieces of an organism.

regulatory gene A gene that codes for a protein, such as a repressor, that controls the transcription of another gene or group of genes.

releasing hormone A kind of hormone released from the hypothalamus that makes the anterior pituitary secrete hormones.

renal cortex The outer portion of the vertebrate kidney.

renal medulla The inner portion of the vertebrate kidney, beneath the renal cortex.

repetitive DNA Nucleotide sequences that are present in many copies in the DNA of a genome. The repeated sequences may be long or short and may be located next to each other (tandomly) or dispersed in the DNA.

repressor A protein that blocks the transcription of a gene or operon.

reproduction The creation of new individuals from existing ones.

reproductive barrier A biological feature of a species that prevents it from interbreeding with other species even when populations of the two species live together.

reproductive cloning Using a somatic cell from a multicellular organism to make one or more genetically identical individuals.

reproductive isolation The existence of biological factors (barriers) that impede members of two species from producing viable, fertile hybrids.

reproductive system The body organ system responsible for reproduction.

reptile Member of the clade of amniotes that includes snakes, lizards, turtles, crocodilians, and birds, along with a number of extinct groups such as dinosaurs.

respiratory system The organ system that functions in exchanging gases with the environment. It supplies the blood with O_2 and disposes of CO_2.

resting potential The voltage across the plasma membrane of a resting neuron.

restoration ecology The use of ecological principles to develop ways to return degraded ecosystems to conditions as similar as possible to their natural, predegraded state.

restriction enzyme A bacterial enzyme that cuts up foreign DNA (at specific *restriction sites*), thus protecting bacteria against intruding DNA from phages and other organisms. Restriction enzymes are used in DNA technology to cut DNA molecules in reproducible ways.

restriction fragment length polymorphism (RFLP) (rif'-lips) The differences in homologous DNA sequences that are reflected in different lengths of restriction fragments produced when the DNA is cut up with restriction enzymes.

restriction fragments Molecules of DNA produced from a longer DNA molecule cut up by a restriction enzyme; used in genome mapping and other applications.

restriction site A specific sequence on a DNA strand that is recognized as a "cut site" by a restriction enzyme.

retina (ret'-uh-nuh) The light-sensitive layer in an eye, made up of photoreceptor cells and sensory neurons.

retrovirus An RNA virus that reproduces by means of a DNA molecule. It reverse-transcribes its RNA into DNA, inserts the DNA into a cellular chromosome, and then transcribes more copies of the RNA from the viral DNA. HIV and a number of cancer-causing viruses are retroviruses.

reverse transcriptase (tran-skrip'-tās) An enzyme used by retroviruses that catalyzes the synthesis of DNA on an RNA template.

rhizome (rī'-zōm) A horizontal stem that grows below the ground.

rhodopsin (ro-dop'-sin) A visual pigment that is located in the rods of the vertebrate eye and that absorbs dim light.

ribonucleic acid (RNA) (rī-bō-nū-klā'-ik) A type of nucleic acid consisting of nucleotide monomers with a ribose sugar and the nitrogenous bases adenine (A), cytosine (C), guanine (G), and uracil (U); usually single-stranded; functions in protein synthesis and as the genome of some viruses.

ribosomal RNA (rRNA) (rī'-buh-sōm'-ul) The type of ribonucleic acid that, together with proteins, makes up ribosomes; the most abundant type of RNA in most cells.

ribosome (rī'-buh-sōm) A cell structure consisting of RNA and protein organized into two subunits and functioning as the site of protein synthesis in the cytoplasm. The ribosomal subunits are constructed in the nucleolus.

ribozyme (rī'-bō-zīm) An enzyme-like RNA molecule that catalyzes chemical reactions.

RNA interference (RNAi) A biotechnology technique used to silence the expression of specific genes. Synthetic RNA molecules with sequences that correspond to particular genes trigger the breakdown of the gene's mRNA.

RNA polymerase (puh-lim'-uh-rās) An enzyme that links together the growing chain of RNA nucleotides during transcription, using a DNA strand as a template.

RNA splicing The removal of introns and joining of exons in eukaryotic RNA, forming an mRNA molecule with a continuous coding sequence; occurs before mRNA leaves the nucleus.

rod A photoreceptor cell in the vertebrate retina, enabling vision in dim light.

root cap A cone of cells at the tip of a plant root that protects the root's apical meristem.

root hair An outgrowth of an epidermal cell on a root, which increases the root's absorptive surface area.

root pressure The upward push of xylem sap in a vascular plant, caused by the active pumping of minerals into the xylem by root cells.

root system All of a plant's roots, which anchor it in the soil, absorb and transport minerals and water, and store food.

rough endoplasmic reticulum (reh-tik'-yuh-lum) A network of interconnected membranous sacs in a eukaryotic cell's cytoplasm. Rough ER membranes are studded with ribosomes that make membrane proteins and secretory proteins.

rRNA *See* ribosomal RNA.

r-selection The concept that in certain (*r*-selected) populations, a high reproductive rate is the chief determinant of life history.

rule of addition A rule stating that the probability that an event can occur in two or more alternative ways is the sum of the separate probabilities of the different ways.

rule of multiplication A rule stating that the probability of a compound event is the product of the separate probabilities of the independent events.

ruminant (rū′-min-ent) An animal, such as a cow or sheep, with a multichambered stomach housing microorganisms that can digest cellulose.

S

SA (sinoatrial) node (sī-nō′-ā′-trē-ul) The pacemaker of the heart, located in the wall of the right atrium, that sets the rate and timing at which all cardiac muscle cells contract. *See* pacemaker.

sac fungus Member of a group of fungi characterized by saclike structures called asci that produce spores in sexual reproduction.

saliva Salivary gland secretion that contains substances to lubricate food, buffers, antibacterial agents, and the digestive enzyme amylase.

salivary glands Glands associated with the oral cavity that secrete substances to lubricate food and begin the process of chemical digestion.

salt A compound resulting from the formation of ionic bonds; also called an ionic compound.

sapwood Light-colored, water-conducting secondary xylem in a tree. *See also* heartwood.

sarcoma (sar-kō′-muh) Cancer of the supportive tissues, such as bone, cartilage, and muscle.

sarcomere (sar′-kō-mēr) The fundamental unit of muscle contraction, composed of thin filaments of actin and thick filaments of myosin; in electron micrographs, the region between two narrow, dark lines, called Z lines, in a myofibril.

saturated Pertaining to fats and fatty acids whose hydrocarbon chains contain the maximum number of hydrogens and therefore have no double covalent bonds. Saturated fats and fatty acids solidify at room temperature.

savanna A biome dominated by grasses and scattered trees.

scanning electron microscope (SEM) A microscope that uses an electron beam to study the surface architecture of a cell or other specimen.

schizophrenia Severe mental disturbance characterized by psychotic episodes in which patients lose the ability to distinguish reality from hallucination.

sclera (sklār′-uh) A layer of connective tissue forming the outer surface of the vertebrate eye. The cornea is the frontal part of the sclera.

sclereid (sklār′-ē-id) In plants, a very hard, dead sclerenchyma cell found in nutshells and seed coats; a stone cell.

sclerenchyma cell (skluh-ren′-kē-muh) In plants, a supportive cell with rigid secondary walls hardened with lignin.

scrotum A pouch of skin outside the abdomen that houses a testis; functions in cooling sperm, thereby keeping them viable.

search image The mechanism that enables an animal to find a particular kind of food efficiently.

second law of thermodynamics The principle whereby every energy conversion reduces the order of the universe, increasing its entropy. Ordered forms of energy are at least partly converted to heat.

secondary consumer An organism that eats primary consumers.

secondary endosymbiosis A process by which protist diversity is hypothesized to have evolved from a symbiotic association that arose when an autotrophic eukaryotic protist was engulfed by a heterotrophic eukaryotic protist.

secondary growth An increase in a plant's girth, involving cell division in the vascular cambium and cork cambium.

secondary immune response The immune response elicited when an animal encounters the same antigen at some later time. The secondary immune response is more rapid, of greater magnitude, and of longer duration than the primary immune response.

secondary oocyte (ō′-uh-sīt′) A haploid cell resulting from meiosis I in oogenesis, which will become an ovum after meiosis II.

secondary phloem *See* phloem.

secondary spermatocyte (sper-mat′-uh-sīt′) A haploid cell resulting from meiosis I in spermatogenesis, which will become a sperm cell after meiosis II.

secondary structure The second level of protein structure; the regular local patterns of coils or folds of a polypeptide chain.

secondary succession A type of ecological succession that occurs where a disturbance has destroyed an existing biological community but left the soil intact. *See also* primary succession.

secondary xylem *See* xylem.

secretion (1) The discharge of molecules synthesized by a cell. (2) In the vertebrate kidney, the discharge of wastes from the blood into the filtrate from the nephron tubules.

seed A plant embryo packaged with a food supply within a protective covering.

seed coat A tough outer covering of a seed, formed from the outer coat (integuments) of an ovule. In a flowering plant, the seed coat encloses and protects the embryo and endosperm.

seed dormancy The temporary suspension of growth and development of a seed.

seedless vascular plants The informal collective name for lycophytes (club mosses and their relatives) and pterophytes (ferns and their relatives).

segmentation Subdivision along the length of an animal body into a series of repeated parts called segments.

selective permeability (per′-mē-uh-bil′-uh-tē) A property of biological membranes that allows some substances to cross more easily than others and blocks the passage of other substances altogether.

self protein A protein on the surface of an antigen-presenting cell that can hold a foreign antigen and display it to helper T cells. Each individual has a unique set of self proteins that serve as molecular markers for the body. Lymphocytes do not attack self proteins unless the proteins are displaying foreign antigens; therefore, self proteins mark normal body cells as off-limits to the immune system. The technical name for self proteins is *major histocompatibility complex (MHC) proteins. See also* nonself molecule.

self-fertilize The fusion of sperm and egg produced by the same individual organism.

semen (sē′-mun) The sperm-containing fluid that is ejaculated by the male during orgasm.

semicircular canals Fluid-filled channels in the inner ear that detect changes in the head's rate of rotation or angular movement.

semiconservative model Type of DNA replication in which the replicated double helix consists of one old strand, derived from the old molecule, and one newly made strand.

seminal vesicle (sem′-uh-nul ves′-uh-kul) A gland in males that secretes a fluid component of semen that lubricates and nourishes sperm.

seminiferous tubule (sem′-uh-nif′-uh-rus) A coiled sperm-producing tube in a testis.

sensation A feeling, or general awareness, of stimuli resulting from sensory information reaching the central nervous system.

sensitive period A limited phase in an individual animal's development when learning of particular behaviors can take place.

sensory adaptation The tendency of sensory neurons to become less sensitive when they are stimulated repeatedly.

sensory input The conduction of signals from sensory receptors to processing centers in the central nervous system.

sensory neuron A nerve cell that receives information from sensory receptors and conveys signals into the central nervous system.

sensory receptor A specialized cell or neuron that detects stimuli and sends information to the central nervous system.

sensory transduction The conversion of a stimulus signal to an electrical signal by a sensory receptor.

sepal (sē′-pul) A modified leaf of a flowering plant. A whorl of sepals encloses and protects the flower bud before it opens.

sessile An organism that is anchored to its substrate.

sex chromosome A chromosome that determines whether an individual is male or female.

sex-linked gene A gene located on a sex chromosome. In humans, the vast majority of sex-linked genes are located on the Y chromosome.

sexual dimorphism (dī-mōr´-fizm) Marked differences between the secondary sex characteristics of males and females.

sexual reproduction The creation of offspring by the fusion of two haploid sex cells (gametes), forming a diploid zygote.

sexual selection A form of natural selection in which individuals with certain inherited characteristics are more likely than other individuals to obtain mates.

sexually transmitted disease (STD) A contagious disease spread by sexual contact.

shared ancestral characters A character, shared by members of a particular clade, that originated in an ancestor that is not a member of that clade.

shared derived characters An evolutionary novelty that is unique to a particular clade.

shoot system All of a plant's stems, leaves, and reproductive structures.

short-day plant A plant that flowers in late summer, fall, or winter, when day length is short. Short-day plants actually flower in response to long nights (and so are sometimes called long-night plants).

short-term memory The ability to hold information, anticipations, or goals for a time and then release them if they become irrelevant.

sieve plate An end wall in a sieve-tube member that facilitates the flow of phloem sap.

sieve-tube member A food-conducting cell in a plant. Chains of sieve-tube members make up phloem tissue.

signal In behavioral ecology, a stimulus transmitted by one animal to another animal.

signal transduction pathway In cell biology, a series of molecular changes that converts a signal on a target cell's surface to a specific response inside the cell.

silencer A eukaryotic DNA sequence that functions to inhibit the start of gene transcription; may act analogously to an enhancer by binding a repressor.

single-lens eye The camera-like eye found in some jellies, polychaetes, spiders, many molluscs, and in vertebrates.

sister chromatid (krō´-muh-tid) One of the two identical parts of a duplicated chromosome in a eukaryotic cell.

skeletal muscle Striated muscle attached to the skeleton. The contraction of striated muscles produces voluntary movements of the body.

skeletal system The organ system that provides body support and protects body organs such as the brain, heart, and lungs.

small intestine The longest section of the alimentary canal. It is the principal site of the enzymatic hydrolysis of food macromolecules and the absorption of nutrients.

smooth endoplasmic reticulum A network of interconnected membranous tubules in a eukaryotic cell's cytoplasm. Smooth ER lacks ribosomes.

smooth muscle Muscle made up of cells without striations, found in the walls of organs such as the digestive tract, urinary bladder, and arteries.

SNP (single nucleotide polymorphism) A variation in DNA sequence found within the genomes of at least 1% of a population.

social behavior Any kind of interaction between two or more animals, usually of the same species.

social learning Modification of behavior through the observation of other individuals.

sociobiology The study of the evolutionary basis of social behavior.

sodium-potassium (Na-K) pump A membrane protein that transports sodium ions out of, and potassium ions into, a cell against their concentration gradients. The process is powered by ATP.

solute (sol´-yūt) A substance that is dissolved in a solution.

solution A liquid consisting of a homogeneous mixture of two or more substances, consisting of a dissolving agent, called the solvent, and a substance that is dissolved, called the solute.

solvent The dissolving agent of a solution. Water is the most versatile solvent known.

somatic cell (sō-mat´-ik) Any cell in a multicellular organism except a sperm or egg cell or a cell that develops into a sperm or egg.

somatic nervous system The component of the vertebrate peripheral nervous system that carries signals to and from skeletal muscles.

spatial learning Modification of behavior based on experience of the spatial structure of the environment.

speciation The evolution of a new species.

species A group whose members possess similar anatomical characteristics and have the ability to interbreed and produce viable, fertile offspring. *See* biological species concept.

species diversity The variety of species that make up a community; includes both species richness (the total number of different species) and the relative abundance of the different species in the community.

sperm A male gamete.

spermatogenesis (sper-mat´-ō-jen´-uh-sis) The formation of sperm cells.

spermicide A sperm-killing chemical (cream, jelly, or foam) that works with a barrier device as a method of contraception.

sphincter (sfink´-ter) A ringlike valve, consisting of modified muscles in a muscular tube, that regulates passage between some compartments of the alimentary canal.

spinal cord The dorsal hollow nerve cord in vertebrates, located within the vertebral column; with the brain, makes up the central nervous system.

spinal nerve In the vertebrate peripheral nervous system, a nerve that carries signals to or from the spinal cord.

spirochete A member of a group of helical bacteria that spiral through the environment by means of rotating, internal filaments.

sponge An aquatic animal characterized by a highly porous body.

sporangium (spuh-ranj´-ē-um´) (plural, **sporangia**) A structure in fungi and plants in which meiosis occurs and haploid spores develop.

spore (1) In plants and algae, a haploid cell that can develop into a multicellular individual without fusing with another cell. (2) In prokaryotes, protists, and fungi, any of a variety of thick-walled life cycle stages capable of surviving unfavorable environmental conditions.

sporophyte (spōr´-uh-fīt) The multicellular diploid form in the life cycle of organisms undergoing alternation of generations; results from a union of gametes and meiotically produces haploid spores that grow into the gametophyte generation.

stabilizing selection Natural selection that favors intermediate variants by acting against extreme phenotypes.

stamen (stā´-men) A pollen-producing male reproductive part of a flower, consisting of a filament and an anther.

starch A storage polysaccharide found in the roots of plants and certain other cells; a polymer of glucose.

start codon (kō´-don) On mRNA, the specific three-nucleotide sequence (AUG) to which an initiator tRNA molecule binds, starting translation of genetic information.

stem The part of a plant's shoot system that supports the leaves and reproductive structures.

stem cell An unspecialized cell that can divide to produce an identical daughter cell and a more specialized daughter cell, which undergoes differentiation.

steroid (ster´-oyd) A type of lipid whose carbon skeleton is in the form of four fused rings with various chemical groups attached; examples are cholesterol, testosterone, and estrogen.

steroid hormone A lipid made from cholesterol that acts as a regulatory chemical, activating the transcription of specific genes in target cells.

stigma (stig´-muh) (plural, **stigmata**) The sticky tip of a flower's carpel, which traps pollen grains.

stimulus (plural, **stimuli**) (1) In the context of a nervous system, a factor that triggers sensory transduction. (2) In behavioral biology, a factor that triggers a specific response.

stoma (stō'-muh) (plural, **stomata**) A pore surrounded by guard cells in the epidermis of a leaf. When stomata are open, CO_2 enters a leaf, and water and O_2 exit. A plant conserves water when its stomata are closed.

stomach A pouch-like organ in a digestive tract that grinds and churns food and may store it temporarily.

stop codon In mRNA, one of three triplets (UAG, UAA, UGA) that signal gene translation to stop.

STR analysis (short tandem repeat analysis) A method of DNA profiling that involves the comparison of the lengths of short tandem repeat (STR) sequences selected from specific sites within the genome.

stramenopile A member of a clade of protists that includes water mold, diatoms, and brown algae and is characterized by a "hairy" flagellum.

strata Rock layers formed when new layers of sediment cover older ones and compress them.

stretch receptor A type of mechanoreceptor sensitive to changes in muscle length; detects the position of body parts.

stroke The death of nervous tissue in the brain, usually resulting from rupture or blockage of arteries in the head.

stroma (strō'-muh) The fluid of the chloroplast surrounding the thylakoid membrane; involved in the synthesis of organic molecules from carbon dioxide and water; Sugars are made in the stroma by the enzymes of the Calvin cycle.

stromatolite (strō-mat'-uh-līt) Layered rocks that result from the activities of prokaryotes that bind thin films of sediment together.

STRs (short tandem repeats) Short DNA sequences that are repeated many times in a row in the genome.

substrate (1) A specific substance (reactant) on which an enzyme acts. Each enzyme recognizes only the specific substrate or substrates of the reaction it catalyzes. (2) A surface in or on which an organism lives.

substrate feeder An organism that lives in or on its food source, eating its way through the food.

substrate-level phosphorylation The formation of ATP by an enzyme directly transferring a phosphate group to ADP from an organic molecule (for example, one of the intermediates in glycolysis or the citric acid cycle).

sugar sink A plant organ that is a net consumer or storer of sugar. Growing roots, shoot tips, stems, and fruits are sugar sinks supplied by phloem.

sugar source A plant organ in which sugar is being produced by either photosynthesis or the breakdown of starch. Mature leaves are the primary sugar sources of plants.

sugar-phosphate backbone The alternating chain of sugar and phosphate to which the DNA and RNA nitrogenous bases are attached.

superior vena cava (vē'-nuh kā'-vuh) A large vein that returns O_2-poor blood to the heart from the upper body and head. See also inferior vena cava.

surface tension A measure of how difficult it is to stretch or break the surface of a liquid. Water has a high surface tension because of the hydrogen bonding of surface molecules.

surfactant A substance secreted by alveoli that decreases surface tension in the fluid that coats the alveoli.

survivorship curve A plot of the number of members of a cohort that are still alive at each age; one way to represent age-specific mortality.

suspension feeder An aquatic animal that sifts small food particles from the water.

sustainable agriculture Long-term productive farming methods that are environmentally safe.

sustainable development The long-term prosperity of human societies and the ecosystems that support them.

sustainable resource management Management of a natural resource so as not to damage the resource.

swim bladder A gas-filled internal sac that helps bony fishes maintain buoyancy.

symbiosis A close association between organisms of two or more species.

sympathetic division A set of neurons in the autonomic nervous system that generally prepares the body for energy-consuming activities, such as fleeing or fighting. See also parasympathetic division.

sympatric speciation The formation of new species in populations that live in the same geographic area.

synapse (sin'-aps) A junction between two neurons, or between a neuron and an effector cell. Electrical or chemical signals are relayed from one cell to another at a synapse.

synaptic cleft (sin-ap'-tik) In a chemical synapse, a narrow gap separating the synaptic terminal of a transmitting neuron from a receiving neuron or an effector cell.

synaptic terminal The tip of a transmitting neuron's axon, where signals are sent to another neuron or to an effector cell.

synaptic vesicle A membranous sac containing neurotransmitter molecules at the tip of the presynaptic axon.

systematics A scientific discipline focused on classifying organisms and determining their evolutionary relationships.

systemic acquired resistance A defensive response in plants infected with a pathogenic microbe; helps protect healthy tissue from the microbe.

systemic circuit One of two main blood circuits in terrestrial vertebrates; conveys blood between the heart and the rest of the body. See also pulmonary circuit.

systems biology An approach to studying biology that aims to model the dynamic behavior of whole biological systems.

systole (sis'-tō-lē) The contraction stage of the heart cycle, when the heart chambers actively pump blood. See also diastole.

T cell A type of lymphocyte that matures in the thymus and is responsible for the cell-mediated immune response; also involved in humoral immunity. See also B cell.

tapeworm A parasitic flatworm characterized by the absence of a digestive tract.

target cell A cell that responds to a regulatory signal, such as a hormone.

taxis (tak'-sis) (plural, **taxes**) Virtually automatic orientation toward or away from a stimulus.

taxon A named taxonomic unit at any given level of classification.

taxonomy The branch of biology that identifies, names, and classifies species.

technology The practical application of scientific knowledge.

telomere (tel'-uh-mēr) The repetitive DNA at each end of a eukaryotic chromosome.

telophase The fifth and final stage of mitosis, during which daughter nuclei form at the two poles of a cell. Telophase usually occurs together with cytokinesis.

temperate broadleaf forest A biome located throughout midlatitude regions where there is sufficient moisture to support the growth of large, broadleaf deciduous trees.

temperate grassland A grassland region maintained by seasonal drought, occasional fires, and grazing by large mammals.

temperate zones Latitudes between the tropics and the Arctic Circle in the north and the Antarctic Circle in the south; regions with milder climates than the tropics or polar regions.

temperature A measure of the intensity of heat in degrees, reflecting the average kinetic energy or speed of molecules.

tendon Fibrous connective tissue connecting a muscle to a bone.

tendril A modified leaf used by some plants to climb around a fixed structure.

terminal bud Embryonic tissue at the tip of a shoot, made up of developing leaves and a compact series of nodes and internodes.

terminator A special sequence of nucleotides in DNA that marks the end of a gene. It signals RNA polymerase to release the newly made RNA molecule and then to depart from the gene.

territory An area that one or more individuals defend and from which other members of the same species are usually excluded.

tertiary consumer (ter'-shē-ār-ē) An organism that eats secondary consumers.

tertiary structure The third level of protein structure; the overall, three-dimensional shape of a polypeptide due to interactions of the R groups of the amino acids making up the chain.

testcross The mating between an individual of unknown genotype for a particular characteristic and an individual that is homozygous recessive for that same characteristic.

testis (plural, testes) The male gonad in an animal; produces sperm and, in many species, reproductive hormones.

testosterone (tes-tos'-tuh-rōn) An androgen hormone that stimulates an embryo to develop into a male and promotes male body features.

tetrad A paired set of homologous chromosomes, each composed of two sister chromatids. Tetrads form during prophase I of meiosis.

tetrapod A vertebrate with two pairs of limbs. Tetrapods include mammals, amphibians, and birds and other reptiles.

thalamus (thal'-uh-mus) An integrating and relay center of the vertebrate forebrain; sorts and relays selected information to specific areas in the cerebral cortex.

theory A widely accepted explanatory idea that is broad in scope and supported by a large body of evidence.

therapeutic cloning The cloning of human cells by nuclear transplantation for therapeutic purposes, such as the generation of embryonic stem cells. *See* nuclear transplantation; reproductive cloning.

thermodynamics The study of energy transformation that occurs in a collection of matter. *See* first law of thermodynamics; second law of thermodynamics.

thermoreceptor A sensor (sensory receptor) that detects heat or cold.

thermoregulation The maintenance of internal temperature within a range that allows cells to function efficiently.

thick filament A filament composed of staggered arrays of myosin molecules; a component of myofibrils in muscle fibers.

thigmotropism (thig'-mō-trō'-pizm) Growth movement of a plant in response to touch.

thin filament The thinner of the two myofilaments, consisting of two strands of actin and two strands of regulatory protein coiled around each other.

threatened species As defined in the U.S. Endangered Species Act, a species that is likely to become endangered in the foreseeable future throughout all or a significant portion of its range.

three-domain system A system of taxonomic classification based on three basic groups: Bacteria, Archaea, and Eukarya.

threshold The minimum change in a membrane's voltage that must occur to generate a nerve signal (action potential).

thylakoid (thī'-luh-koyd) One of a number of disk-shaped membranous sacs inside a chloroplast. Thylakoid membranes contain chlorophyll and the enzymes of the light reactions of photosynthesis. A stack of thylakoids is called a granum.

thymine (T) (thī'-min) A single-ring nitrogenous base found in DNA.

thymus gland (thī'-mus) An endocrine gland in the neck region of mammals that is active in establishing the immune system; secretes several hormones that promote the development and differentiation of T cells.

thyroid gland (thī'-royd) An endocrine gland that secretes thyroxine (T_4), triiodothyronine (T_3), and calcitonin.

thyroid-stimulating hormone (TSH) A protein hormone secreted by the anterior pituitary that stimulates the thyroid gland to secrete its hormones.

thyroxine (T_4) (thī-rok'-sin) An amine hormone secreted by the thyroid gland that stimulates metabolism in virtually all body tissues.

Ti plasmid A bacterial plasmid that induces tumors in plant cells that the bacterium infects; often used as a vector to introduce new genes into plant cells. Ti stands for tumor-inducing.

tissue An integrated group of cells with a common function, structure, or both.

tissue system One or more tissues organized into a functional unit within a plant or animal.

tonicity The ability of a solution surrounding a cell to cause that cell to gain or lose water.

topsoil A mixture of particles derived from rock, living organisms, and humus.

trace element An element that is essential for life but required in extremely minute amounts.

trachea (trā'-kē-uh) (plural, **tracheae**) The windpipe; the portion of the respiratory tube that has C-shaped cartilagenous rings and passes from the larynx to two bronchi.

tracheal system A system of branched, air-filled tubes in insects that extend throughout the body and carry oxygen directly to cells.

tracheid (trā'-kē-id) A tapered, porous, water-conducting and supportive cell in plants. Chains of tracheids or vessel elements make up the water-conducting, supportive tubes in xylem.

trade winds The movement of air in the tropics (those regions that lie between 23.5° north latitude and 23.5° south latitude).

trait A variant of a character found within a population, such as purple flowers in pea plants.

transcription factor In the eukaryotic cell, a protein that functions in initiating or regulating transcription. Transcription factors bind to DNA or to other proteins that bind to DNA.

transcription The synthesis of RNA on a DNA template.

transduction (1) The transfer of bacterial genes from one bacterial cell to another by a phage. (2) *See* sensory transduction. (3) *See* signal transduction.

transfer RNA (tRNA) A type of ribonucleic acid that functions as an interpreter in translation. Each tRNA molecule has a specific anticodon, picks up a specific amino acid, and conveys the amino acid to the appropriate codon on mRNA.

transformation The incorporation of new genes into a cell from DNA that the cell takes up from the surrounding environment.

transgenic organism An organism that contains genes from another species.

translation The synthesis of a polypeptide using the genetic information encoded in an mRNA molecule. There is a change of "language" from nucleotides to amino acids.

translocation (1) During protein synthesis, the movement of a tRNA molecule carrying a growing polypeptide chain from the A site to the P site on a ribosome. (The mRNA travels with it.) (2) A change in a chromosome resulting from a chromosomal fragment attaching to a nonhomologous chromosome; can occur as a result of an error in meiosis or from mutagenesis.

transmission electron microscope (TEM) A microscope that uses an electron beam to study the internal structure of thinly sectioned specimens.

transpiration The evaporative loss of water from a plant.

transpiration-cohesion-tension mechanism A transport mechanism that drives the upward movement of water in plants: transpiration exerts a pull that is relayed downward along a string of molecules held together by cohesion and helped upward by adhesion.

transport vesicle A tiny membranous sac in a cell's cytoplasm carrying molecules produced by the cell. The vesicle buds from the endoplasmic reticulum or Golgi and eventually fuses with another membranous organelle or the plasma membrane, releasing its contents.

transposable element (tranz-pō'-zon) A transposable genetic element, or "jumping gene"; a segment of DNA that can move from one site to another within a cell and serve as an agent of genetic change.

TRH-releasing hormone A peptide hormone that triggers the release of TSH (thyroid-stimulating hormone), which in turn stimulates the thyroid gland.

trial-and-error learning Learning to associate a particular behavioral act with a positive or negative effect.

triiodothyronine (T₃) (trī'-ī-ō-dō-thī'-rō-nīn) An amine hormone secreted by the thyroid gland that stimulates metabolism in virtually all body tissues.

trimester In human development, one of three 3-month-long periods of pregnancy.

triplet code A set of three-nucleotide-long "words" that specify the amino acids for polypeptide chains. *See* genetic code.

trisomy 21 *See* Down syndrome.

tRNA *See* transfer RNA.

trophoblast (trōf'-ō-blast) In mammalian development, the outer portion of a blastocyst. Cells of the trophoblast secrete enzymes that enable the blastocyst to implant in the endometrium of the mother's uterus.

tropical forest A terrestrial biome characterized by high levels of precipitation and warm temperatures year-round.

tropics Latitudes between 23.5° north and south.

tropism (trō'-pizm) A growth response that makes a plant grow toward or away from a stimulus.

true-breeding Referring to organisms for which sexual reproduction produces offspring with inherited traits identical to those of the parents; the organisms are homozygous for the characteristics under consideration.

tubal ligation A means of sterilization in which a woman's two oviducts (fallopian tubes) are tied closed to prevent eggs from reaching the uterus; a segment of each oviduct is removed.

tuber An enlargement at the end of a rhizome in which food is stored.

tumor An abnormal mass of cells that forms within otherwise normal tissue.

tumor-suppressor gene A gene whose product inhibits cell division, thereby preventing uncontrolled cell growth.

tundra A biome at the northernmost limits of plant growth and at high altitudes, characterized by dwarf woody shrubs, grasses, mosses, and lichens.

tunicate One of a group of invertebrate chordates.

U

ultimate cause In animal behavior, the evolutionary reason for a behavior.

ultrasound imaging A technique for examining a fetus in the uterus. High-frequency sound waves echoing off the fetus are used to produce an image of the fetus.

undernourishment A condition that results from a diet that consistently supplies less chemical energy than the body requires.

uniform Describing a dispersion pattern in which individuals are evenly distributed.

unsaturated Pertaining to fats and fatty acids whose hydrocarbon chains lack the maximum number of hydrogen atoms and therefore have one or more double covalent bonds. Unsaturated fats and fatty acids do not solidify at room temperature.

uracil (U) (yū'-ruh-sil) A single-ring nitrogenous base found in RNA.

urea (yū-rē'-ah) A soluble form of nitrogenous waste excreted by mammals and most adult amphibians.

ureter (yū-rē'-ter or yū'-reh-ter) A duct that conveys urine from the kidney to the urinary bladder.

urethra (yū-rē'-thruh) A duct that conveys urine from the urinary bladder to the outside. In the male, the urethra also conveys semen out of the body during ejaculation.

uric acid (yū'-rik) An insoluble precipitate of nitrogenous waste excreted by land snails, insects, birds, and some reptiles.

urinary bladder The pouch where urine is stored prior to elimination.

urinary system The organ system that forms and excretes urine while regulating the amount of water and ions in the body fluids.

urine Concentrated filtrate produced by the kidneys and excreted by the bladder.

uterus (yū'-ter-us) In the reproductive system of a mammalian female, the organ where the development of young occurs; the womb.

V

vaccination (vak'-suh-nā'-shun) A procedure that presents the immune system with a harmless variant or derivative of a pathogen, thereby stimulating the immune system to mount a long-term defense against the pathogen.

vaccine (vak-sēn') A harmless variant or derivative of a pathogen used to stimulate a host organism's immune system to mount a long-term defense against the pathogen.

vacuole (vak'-ū-ōl) A membrane-enclosed sac that is part of the endomembrane system of a eukaryotic cell, having diverse functions.

vagina (vuh-jī'-nuh) Part of the female reproductive system between the uterus and the outside opening; the birth canal in mammals; also accommodates the male's penis and receives sperm during copulation.

vas deferens (vas def'-er-enz) (plural, **vasa deferentia**) Part of the male reproductive system that conveys sperm away from the testis; the sperm duct; in humans, the tube that conveys sperm between the epididymis and the common duct that leads to the urethra.

vascular bundle (vas'-kyū-ler) A strand of vascular tissues (both xylem and phloem) in a plant stem.

vascular cambium (vas'-kyū-ler kam'-bē-um) During secondary growth of a plant, the cylinder of meristematic cells, surrounding the xylem and pith, that produces secondary xylem and phloem.

vascular cylinder The central cylinder of vascular tissue in a plant root.

vascular plant A plant with xylem and phloem, including club mosses, ferns, gymnosperms, and angiosperms.

vascular tissue Plant tissue consisting of cells joined into tubes that transport water and nutrients throughout the plant body.

vascular tissue system A system formed by xylem and phloem throughout the plant, serving as a transport system for water and nutrients, respectively.

vasectomy (vuh-sek'-tuh-mē) Surgical removal of a section of the two sperm ducts (vasa deferentia) to prevent sperm from reaching the urethra; a means of sterilization in the male.

vector In molecular biology, a piece of DNA, usually a plasmid or a viral genome, that is used to move genes from one cell to another.

vein (1) In animals, a vessel that returns blood to the heart. (2) In plants, a vascular bundle in a leaf, composed of xylem and phloem.

ventilation A mechanism that provides contact between an animal's respiratory surface and the air or water to which it is exposed.

ventral Pertaining to the underside, or bottom, of a bilaterally symmetrical animal.

ventricle (ven'-truh-kul) (1) A heart chamber that pumps blood out of a heart. (2) A space in the vertebrate brain, filled with cerebrospinal fluid.

venule (ven'-yūl) A vessel that conveys blood between a capillary bed and a vein.

vertebra (ver'-tuh-bruh) (plural, **vertebrae**) One of a series of segmented skeletal units that enclose the nerve cord, making up the backbone of a vertebrate animal.

vertebral column backbone; composed of a series of segmented units called vertebrae.

vertebrate (ver'-tuh-brāt) A chordate animal with a backbone. Vertebrates include lampreys, chondricthyans, ray-finned fishes, lobe-fin fishes, amphibians, reptiles (including birds), and mammals.

vesicle A sac made of membrane in the cytoplasm of a eukaryotic cell.

vessel element A short, open-ended, water-conducting and supportive cell in plants. Chains of vessel elements or tracheids make up the water-conducting, supportive tubes in xylem.

vestigial organ A structure of marginal or no importance to an organism. Vestigial organs are historical remnants of structures that had important function in ancestors.

villus (vil'-us) (plural, **villi**) (1) A fingerlike projection of the inner surface of the small intestine. (2) A fingerlike projection of the chorion of the mammalian placenta. Large numbers of villi increase the surface areas of these organs.

viriod (vī′-royd) A plant pathogen composed of molecules of naked, circular RNA several hundred nucleotides long.

virus A microscopic particle capable of infecting cells of living organisms and inserting its genetic material. Viruses are generally not considered to be alive because they do not display all of the characteristics associated with life.

visceral mass (vis′-uh-rul) One of the three main parts of a mollusc, containing most of the internal organs.

visual acuity The ability of the eyes to distinguish fine detail.

vital capacity The maximum volume of air that a mammal can inhale and exhale with each breath.

vitamin An organic nutrient that an organism requires in very small quantities. Vitamins generally function as coenzymes.

vitreous humor (vit′-rē-us hyū′-mer) A jellylike substance filling the space behind the lens in the vertebrate eye; helps maintain the shape of the eye.

vocal cord One of a pair of bands of elastic tissues in the larynx. Air rushing past the tensed vocal cords makes them vibrate, producing sounds.

water mold A fungus-like protist in the stramenopile clade.

water vascular system In echinoderms, a radially arranged system of water-filled canals that branch into extensions called tube feet. The system provides movement and circulates water, facilitating gas exchange and waste disposal.

water-conducting cell A specialized, dead plant cell with lignin-containing secondary walls, arranged end to end, forming xylem tissue. *See also* tracheid; vessel element.

wavelength The distance between crests of adjacent waves, such as those of the electromagnetic spectrum.

westerlies Winds that blow from west to east.

wetland An ecosystem intermediate between an aquatic ecosystem and a terrestrial ecosystem. Wetland soil is saturated with water permanently or periodically.

white blood cell *See* leukocyte.

white matter Tracts of axons within the central nervous system.

whole-genome shotgun method A method for determining the DNA sequence of an entire genome. After a genome is cut into small fragments, each fragment is sequenced and then placed in the proper order.

wood ray A column of parenchyma cells that radiates from the center of a log and transports water to its outer living tissues.

wood Secondary xylem of a plant. *See also* heartwood; sapwood.

X chromosome inactivation In female mammals, the inactivation of one X chromosome in each somatic cell.

xylem (zī′-lum) The nonliving portion of a plant's vascular system that provides support and conveys xylem sap from the roots to the rest of the plant. Xylem is made up of vessel elements and/or tracheids, water-conducting cells. Primary xylem is derived from the procambium. Secondary xylem is derived from the vascular cambium in plants exhibiting secondary growth.

xylem sap The solution of inorganic nutrients conveyed in xylem tissue from a plant's roots to its shoots.

yeast A single-celled fungus that inhabits liquid or moist habitats and reproduces asexually by simple cell division or by the pinching of small buds off a parent cell.

yellow bone marrow A tissue found within the central cavities of long bones, consisting mostly of stored fat.

yolk sac An extraembryonic membrane that develops from endoderm; produces the embryo's first blood cells and germ cells and gives rise to the allantois.

zoned reserve An extensive region of land that includes one or more areas that are undisturbed by humans. The undisturbed areas are surrounded by lands that have been altered by human activity.

zooplankton (zō′-ō-plank′-tun) Animals that drift in aquatic environments.

zygomycete Member of a group of fungi characterized by a sturdy structure called a zygosporangium during sexual reproduction.

zygote (zī′-gōt) The fertilized egg, which is diploid, that results from the union of a sperm cell nucleus and an egg cell nucleus.

zygote fungus *See* zygomycete.

Index

Autotrophs, 108, 324
Auxins, 663, 664 *table*, 669
    antagonistic interactions of, 666, 669
    cell elongation in shoots regulated by, 664–65,
        665 *fig.*
AV (atrioventricular) node, 472
Avian flu, 202, 202 *fig.*
Axial skeleton, 608, 608 *fig.*
Axillary buds, 624
Axolotl, 304, 304 *fig.*
Axons, 418, 565, 565 *fig.*
    propagation of action potential along, 568, 568 *fig.*

B

Bacili, 321, 321 *fig.*
Bacillus anthracis, 323, 323 *fig.*, 326, 328
Bacillus coagulans, 55 *fig.*
Bacteria (domain), 6, 7 *fig.*, 314
    antibiotic resistance in, 205, 272, 323
    archaea compared to, 320–21, 321 *table*
    binary fission in, 127, 127 *fig.*
    bioterrorism using, 328
    cell structure and size in, 54
    chromosomes in, 127 *fig.*
    customization using plasmids, 232
    cyanobacteria, 319
    as denitrifiers, 757
    disease causing, 327, 328, 437, 437 *fig.*, 487
        See also Disease; Human disease
    diversity of, 326
    flagella and cilia, 66, 322
    at hydrothermal vents, 681
    mutation rates in, 265
    natural means of DNA transfer in, 204
    nitrogen cycle and, 756–57, 757 *fig.*
    nitrogen-fixing, 655, 656, 756–57, 757 *fig.*
    oil spill remediation with, 778
    plasmids of (*see* Plasmids)
    recycling and cleaning environment with, 329
    as tool for manipulating DNA, 232–38
Bacterial cell wall, 55
Bacteriophage(s), 182
    genomic library of, 235, 235 *fig.*
    Hershey and Chase experiments on,
        182–83, 183 *fig.*
    lambda, 200
    reproductive cycle of, 183 *fig.*, 200 *fig.*
Bacteriorhodopsin, 325
Balance, 592–94
Balancing selection, 272–73
Ball-and-socket joints, 611, 611 *fig.*
Bamboo, 103
Banff National Park, Canada, animal bridge in,
    773 *fig.*
Bark, 633, 633 *fig.*
Barnacles, 379, 379 *fig.*
Barr body, 213
Barrier methods (contraception), 545
Basal body, 66
Basal ganglia, 577, 577 *fig.*
Basal lamina, 416
Basal metabolic rate (BMR), 442
Base, 27
    nitrogenous, 46, 47 *fig.*
Basidiomycetes, 356 *fig.*, 357, 357 *fig.*
Basilar membrane, 592, 593, 593 *fig.*
Basophils, 479, 479 *fig.*
Bateson, William, 172
Bats, 293, 399, 505, 587, 587 *fig.*
B cells (B lymphocytes), 490–91, 490 *fig.*
Bdellovibrio bacteriophorus, 51, 326, 326 *fig.*
Beadle, George, 190
Bears
    global warming and polar, 770, 770 *fig.*
    systematics of raccoons and, 312, 312 *fig.*
    thermoregulation in, 505

Bees. *See also* Honeybees
Beetles, 408, 771
Behavior, 700–723
    abnormal, chemical pollutants causing, 716
    innate, 702–3
    learning, 705–11
    scientific study of, 702–5
    sociobiology and social, 717–21
    survival and reproductive success, 712–16
    thermoregulation and, 507
Behavioral ecology/ecologists, 701, 702
    animal movement, 709
    feeding behavior, 712
Behavioral isolation, 280, 280 *fig.*
Belding's ground squirrel, 720, 720 *fig.*
Benign tumors, 135
Benthic realm, 688
Benzene, 34 *fig.*
Bernard, Claude, 425
Beta-carotene, conversion to vitamin A, 239
Biennials, 630
Big bang theory, 294
Bilateral symmetry, 368, 368 *fig.*, 372, 572
Bilaterians, 369, 372, 373, 384
Bilbies, 408, 408 *fig.*
Bile, 438, 440
Binary fission, 127
Binomial naming system, 308–9
Biodiversity
    in communities, 749
    endangered species protection and, 763
    evolution of (*see* Evolution)
    human activities as threat to, 764
    human welfare and, 764
    in plants as irreplaceable resource, 354
    protected areas to slow loss of, 774
Biodiversity crisis, 764–71
    endangered species and, 429, 763
    global climate change and, 768–69
    habitat destruction, invasive species, and
        overexploitation as causes of, 766
    ozone layer depletion and, 120,
        120 *fig.*, 767
    reasons for concern about, 764
    technology and human overpopulation as
        factors in, 354
    (E. O.) Wilson on, 764
Biodiversity hotspots, 774, 774 *fig.*
Biofilms, 324, 324 *fig.*
Biogenic amines, 570–71
Biogeochemical cycles, 754–55, 755 *fig.*
Biogeography, evidence for evolution in, 262
Biological clocks, 577, 671
    phytochromes and, 673
    plant circadian rhythms, 671
    in plant guard cells, 647
    seasonal, in plants, 672
Biological control, 751
Biological magnification, 767, 767 *fig.*
Biological species concept, 279
Biological weapons, 328
Biological Weapons Convention, 328
Biology, 1, 2. *See also* Conservation biology;
        Ecology; Molecular biology
    at atomic and molecular level, 18
    everyday life and, 10, 12
    scope of, 2–3
Bioluminescence, 73
Biomass, 752–53, 769
Biomass energy, 107
Biomes, 687
    chaparral, 694
    coniferous forests, 696, 696 *fig.*
    deserts, 693
    estuary, 689
    freshwater, 690
    global climate change and, 770–71

global terrestrial, 691 *fig.*
global water cycle connecting, 697, 697 *fig.*
oceans, 688–89
savannas, 692–93, 692 *fig.*
temperate broadleaf forests, 695, 695 *fig.*
temperate grasslands, 694–95
tropical forest, 692
tundra, 696, 696 *fig.*
wetlands, 689
Biophilia, 764
Bioremediation, 329, 778
Biosafety Protocol, 240
Biosphere, 2, 2 *fig.*, 680–99
    abiotic and biotic factors in, 682, 685
    aquatic ecosystems of, 681, 688–90
    autotrophs and producers in, 108
    chemical cycling in, 754–58
    climate and distribution of communities
        in, 686–87
    definition of, 680
    ecology as study of, 681
    environmental problems, ecology and
        insights into, 683
        See also Environmental problems
    global water cycle and, 697, 697 *fig.*
    physical and chemical factors in, 684
    terrestrial ecosystems of, 691–97
Biotechnology, 232
    biodiversity and bioprospecting for, 764
    yeasts and, 361
Biotic components of ecosystems. *See also*
        Organisms
Biotic factors, 682, 752
    adaptation to, 685
Bipedalism, 404, 404 *fig.*
Bipolar disorder, 582
Birds, 398. *See also names of individual birds*
    alimentary canal of, 432 *fig.*
    blue-footed booby, 254–55, 266–67
    bones of, 306
    breathing in, 461
    cardiovascular system, 469, 469 *fig.*
    chemical pesticides and, 767
    courtship and mating behaviors of, 280,
        280 *fig.*, 714, 714 *fig.*
    digestion in, 432
    edge-adapted species of, 773
    evolution of, 293, 306
    excretion in, 509 *fig.*
    extinct, 398 *fig.*
    finches, 259, 285, 287, 288 *fig.*
    flight in, 605, 605 *fig.*
    foraging behavior of, 712, 712 *fig.*
    habituation in, 705
    in high altitudes, 453
    lungs of, 461
    migration and navigation in, 709
    phylogenetic tree of, 311, 311 *fig.*
    problem-solving behavior in, 711, 711 *fig.*
    ptarmigan, 425, 425 *fig.*
    respiration and gas exchange in, 461
    respiratory system of, 458
    songbird species, 706
    song sparrow populations, 730, 730 *fig.*, 731,
        731 *fig.*
    territorial behavior of, 717 *fig.*
    thermoregulation in, 506–7
    wing formation in, 553, 553 *fig.*
Bird song, imprinting and learning of, 706
Birth control pills, 545
Birth defects, 148
Birth rates, human population growth
    and, 735
Births, multiples, 532–33
Bishop, J. Michael, 224
Bivalves, 374, 375, 375 *fig.*
Blastocoel, 548, 548 *fig.*

Corpus luteum, 536
Cortex, plant tissue, 626
Corticosteroids, 529
Costa Rica
zoned reserves in, 775, 775 *fig.*
Cotton cloth, 284
Cotyledons, 623
Countercurrent exchange, 456–57
Countercurrent heat exchanger, 507, 507 *fig.*
Counter-illumination, 73
Courtship rituals, 280, 280 *fig.*, 704, 704 *fig.*, 714–15, 714 *fig.*
Covalent bonds, 23
polar, 24
Cowbirds as edge-adapted species, 773, 773 *fig.*
Coyotes, trial-and-error learning by, 710, 710 *fig.*
Cranial nerves, 573
Craniates, 390
Crawling as locomotion, 604–5
Cretaceous mass extinction, 302, 303, 303 *fig.*, 397
Crick, Francis, 46, 186–87, 186 *fig.*
Cri du chat syndrome, 148
Crime scene investigations, DNA technology and, 231
Cristae, 63
Crocodiles, 311
Cro-Magnon humans, 405
Crop (animal), 432
Crop plants
improving protein content of, 655
Crop rotation, 656
Cross (genetic), 155–59
dihybrid, 158–59
monohybrid, 156
Cross fertilization in plants, 154–55, 155 *fig.*
Crossing over, 142–43, 143 *fig.*
mapping genes using data from, 174, 174 *fig.*
new allele combinations produced by, 172–73
Crustaceans, 379, 379 *fig.*
CT (computed tomography), 422, 422 *fig.*
CT-PET scanner, 422, 422 *fig.*
Cuboidal epithelium, simple, 416, 416 *fig.*
Cultural eutrophication, 759
Cuticle, 342, 373, 378, 626
Cyanea kuhihewa, 125, 125 *fig.*
Cyanide, 99, 99 *fig.*
Cyanobacteria, 108, 319, 326
chemical cycling and, 329
chloroplasts and, 331
as photoautotrophs, 108 *fig.*
Cycads, 345
Cyclohexane, 34 *fig.*
Cystic fibrosis, 162, 222
Cytokinesis, 129, 131 *fig.*
in animal versus plant cells, 132
telophase I and, 139 *fig.*
Cytokinins, 664 *table*, 666
Cytoplasm, 55
Cytosine (C), 185, 185 *fig.*
Cytoskeleton, 56 *fig.*, 57 *fig.*, 65–66
Cytotoxic T cells, 497, 497 *fig.*
destruction of infected cell by, 498, 498 *fig.*

D

Daphnia, 432
Darwin, Charles R., 8, 8 *fig.*, 385
evolutionary theory of, 256–63
modern systematics and, 310
sea voyage and studies of, 256–57, 257 *fig.*
studies of phototropism and hormones by, 662, 663 *fig.*
Darwin, Francis, 662, 663 *fig.*
Daughter cells
in meiosis, 139
in mitosis, 132, 132 *fig.*

Daughter DNA, 188, 188 *fig.* See also DNA polymerases
DDT (pesticide), 683, 697, 767
Deafness, 162, 593
Death rates, human population growth and lower, 734
Decomposers, 3–4, 361, 746
Decomposition, 746
Deductive reasoning, 9
Deep-sea firefly squid (*Watasenia scintillans*), 73, 73 *fig.*
Defenses, animal, 744, 744 *fig.*
Defenses, plant, 674–76, 744–45
Defensive proteins, 42. *See also* Antibody(ies)
Deforestation, 354, 692, 758, 765, 766, 769
Dehydration reactions, 36, 36 *fig.*, 38 *fig.*, 39
Dehydrogenase, 92
Deletion, chromosomal, 148, 148 *fig.*
Delta proteobacteria, 326
Demographic transition, 734–35, 734 *table*
Denaturation, 43
Dendrites, 418, 565, 565 *fig.*
Density-dependent factors limiting population growth, 730
Density-dependent inhibition of cell division, 133, 133 *fig.*
Dental plaque, 324 *fig.*
Deoxyribonucleic acid, 46. *See also* DNA
Deposit feeders, 688
Depression, 582
Derived characters, shared, 310–11
Dermal tissue system, plants, 626
Dermis, 423, 423 *fig.*
Descent with modification, principle of, 257, 293
Desertification, 693
Deserts, 693, 693 *fig.*
Determinate growth, 630
Detritivores, 746
Detritus, 746, 757 *fig.*
Deuterostomes, 368, 369, 384
"Devil's garden", 17, 17 *fig.*
Devonian period, 393
Diabetes mellitus, 527
insulin-dependent, 500
Dialysis, kidney, 480, 513, 513 *fig.*
Diaphragm, respiratory, 458
Diastole, 471, 476
Diatoms, 334, 334 *fig.*
Dicots, 623. *See also* Eudicots
Dictyostelium (cellular slime mold), 335
Diet, animal, 430–32
adaptations of digestive system reflect, 441
as chemical energy, 442
needs satisfied by healthy, 442
Diet, human
cancer and, 449, 449 *table*
cardiovascular disease and, 473
meat in, 754
nutrition and, 442–49
obesity and, 447
soy in, 661
vitamin and mineral supplements in, 446
Differential interference-contrast microscopy, 53, 53 *fig.*
Differentiation. *See* Cellular differentiation
Diffusion, 75 *fig.*
between blood and tissue cells, 474 *fig.*
facilitated, 77
passive, 75
Digestion, 431, 431 *fig.* *See also* Digestive system
Digestive system, 421, 421 *fig.*
diet and adaptations of vertebrate, 441
exchanges with environment and role of, 424–25, 424 *fig.*
human, 433–41, 433 *fig.*
in nematodes, 373

nutrition and (*see* Nutrition)
pharynx and, 458
specialized compartments of animal, 432, 432 *fig.*
Dihybrid cross, 158–59
Dinitrophenol (DNP) as disruptor of cellular respiration, 99, 99 *fig.*
Dinoflagellates, 333, 333 *fig.*
corals and, 743, 743 *fig.*
Dinosaurs, 306, 397
tracks, 260 *fig.*
Dipeptides, 43
Diploid cells, 137
plant, 346, 349 *fig.*, 634–35
See also Sporophytes
Diploid organisms, balancing selection and, 272–73
Diplomonads, 332–33
Directional selection, 270, 270 *fig.*
Disaccharides, 38, 38 *fig.*
Discovery science, 9
Disease. *See also* names of specific diseases; Human disease
amoebic dysentery, 334
bacterial, 205, 272, 320, 322, 326, 327, 437, 437 *fig.*, 487
bioterrorism, 328
fungal, 359
immune system malfunctions and, 500–501
viral, 202–3
Dispersion patterns, population, 726–27
Disruptive selection, 270, 270 *fig.*
Distal tubule, of kidney nephron, 510 *fig.*, 511, 512, 512 *fig.*
Disturbances, in biological communities, 750
DNA (deoxyribonucleic acid), 46
bacterial plasmids and transfer of, 234–36
bacterial transfer of, in nature, 204
complementary, 235
double helix structure of, 186, 187 *fig.*
experiments demonstrating, as genetic material, 182–83, 183 *fig.*
gel electrophoresis of, 243, 246
gene regulation aided by packing of, in eukaryotes, 212, 213 *fig.*
genotype of, expressed as proteins, 190
human evolution and mitochondrial, 406
microarrays, 219, 219 *fig.*
molecular structure of, 5
Neanderthal, 405
nitrogenous bases of, 185 *fig.*
as nucleotide polymer, 184–85
polynucleotide, 184 *fig.*
repetitive, 244, 248
replication of, and multiple "bubbles" and daughter strands, 189, 189 *fig.*
replication of, dependent on base pairing, 188–89, 188 *fig.*
replication of, in cell cycle, 129
restriction enzymes used to create recombinant, 234, 234 *fig.*
transcription of, into RNA, 190, 193, 198
viruses, 201
DNA chip, 219, 219 *fig.*
DNA fingerprints, 242 *fig.*
DNA ligase, 189, 232
DNA polymerases, 189
DNA probes, 236
DNA profiling, 242–46
with STR analysis, 244, 244 *fig.*
DNA sequence(s)
amplifying, by PCR method, 242, 243 *fig.*
analysis of, in systematics, 312–13
analysis of, using restriction fragments, 246
enhancer, 214, 214 *fig.*
mutations in, 199
operator, 210, 211 *fig.*
promoter, 193, 210, 211 *fig.*
silencer, 214

INDEX

INDEX

INDEX

Intermediate filaments, 56 *fig.*, 57 *fig.*, 65, 65 *fig.*
Intermembrane space, 63
Internal environment, animal regulation of. *See also* Body temperature; Osmoregulation; Water balance
    excretion, 505, 508–13
    homeostatic functions of liver, 511
    sympathetic and parasympathetic neurons and, 574–75
    thermoregulation, 505
Internal fertilization, 535
Internal transport, mechanisms of animal, 468–69. *See also* Circulatory system
Interneurons, 564
Internodes, 624
Interphase
    meiosis, 138 *fig.*
    mitosis, 129, 129 *fig.*, 130, 130 *fig.*
Intersexual selection, 271
Interspecific interactions, 742, 742 *table*, 743
Interstitial fluid, 424
    exchanges between blood and, 478, 478 *fig.*
    urine production and, 512
Intertidal zones, 689
Intestines, 432
Intrasexual selection, 271
Introduced (exotic) species, 408, 725, 766
Introns, 194
    making gene with no, 235 *fig.*
Invasive species, 751, 766
Inversion, chromosomal, 148, 148 *fig.*
Invertebrates, 370–83
    annelids, 376–77
    arthropods, 378–79
    body plans of, 368, 369, 371, 371 *fig.*, 372, 374 *fig.*
    body segmentation in, 376–77, 376 *fig.*
    as chordates, 383
    cnidarians, 371
    diversity of, 365, 370–83
    echinoderms, 382
    evolution of, from protists, 367 *fig.*
    eyes of, 595
    flatworms, 372
    innate immunity in, 486
    insects, 380–81
    molluscs, 374–75
    nervous system of, 572 *fig.*
    roundworms, 373
    sponges, 370
In vitro fertilization (IVF), 559, 559 *fig.*
Iodine, as trace element, 18
Ion channels, chemically-gated, 569
Ionic bonds, 22–23
Iris (eye), 595, 595 *fig.*
Iron, as trace element, 18
Irrigation, 653, 653 *fig.*
Islets of Langerhans, 526
Isobutane, 34, 34 *fig.*
Isoflavones, 661
Isomers, 34
Isotonic solutions, 76, 77 *fig.*
Isotopes, 20
    carbon, 20 *table*
    radioactive, 20, 21

J

Jacob, François, 210
Jaws
    hinged, 391
    human evolution and, 404
Jefferson, Thomas, 245
Jeffreys, Alec, 231
Jellies (jellyfish), 371
Joints, three types of, 611 *fig.*

K

Kahn, Genghis, 177
Kangaroo, 389, 399 *fig.*
    locomotion in, 604, 604 *fig.*
Kaposi's sarcoma, 499
Karyotype, 144, 144 *fig.*, 145 *fig.*
Katrina, Hurricane, 643
Kelp, 108, 108 *fig.*, 334
Keystone species, 749, 749 *fig.*
Kidneys
    conversion of blood filtrate to urine by, 511
    dialysis of, 480, 513, 513 *fig.*
    nephron structure and orientation in, 510 *fig.*
    structure of, 510 *fig.*
    urinary system and, 510–11
Killer whales, 9, 9 *fig.*
Killifish, predation on guppy populations by, 732–33, 732 *fig.*
Kilocalories, 91, 442
    expense of, in select activities, 91 *table*
Kinesis, 708
Kinetic energy, 80, 80 *fig.*
Kingdoms (classification), 6, 309
Kin selection, altruism and, 720
Kissimmee River Restoration Project, 778–79, 778 *fig.*
Kittiwakes, fixed action patterns in, 703
Kiwa hirsuta, 681, 681 *fig.*
Klinefelter syndrome, 147, 147 *table*
Knee-jerk reflex, 564 *fig.*
Koalas, foraging behavior of, 712
Krebs, Hans, 96
Krebs cycle, cellular respiration and process of, 96–97. *See also* Citric acid cycle
Krill, 429
K-selection, 732

L

Labia majora, 537
Labia minora, 537
Labor stages in childbirth, 558–59, 558 *fig.*
lac operon, 210–11, 211 *fig.*
Lactase, 33, 42, 47
Lactic acid fermentation, 101, 101 *fig.*, 615
Lactose, metabolism of, in prokaryotes, 210–11
Lactose intolerance, 33, 47
Lakes and streams
    acid precipitation and, 28, 759
    eutrophication in, 759
    as freshwater ecosystem, 690
Lamarck, Jean Baptiste, 256
Lamellae, 456
Lampreys, 391, 391 *fig.*
Lancelets, 383, 383 *fig.*
Land
    animal locomotion on, 604–5
    climate and landforms, 687
    plant colonization of, 297
    vertebrates and life on, 396, 458
Landmarks, animal behavior using, 708
Landscape ecology
    edges and corridors affecting, 773
    habitat restoration, 778–79
Landscapes, 682
Language, evolution of human, 407
Large intestine, 440–41
Larva, 366
Larynx, 458
Laser-assisted in situ keratomileusis (LASIK), 597
Lateralization, brain, 578
Lateral line system, 392
Lateral meristems, 632
Laurasia, 300–301
Law of independent assortment, 158–59, 171. *See also* Mendelian inheritance, laws of

Law of segregation, 156–57, 171. *See also* Mendelian inheritance, laws of
Laws of thermodynamics, 81
Lead contamination, phytoremediation of, 643
Leaf (leaves), 624–25
    falling of, 667, 668–69, 668 *fig.*
    modified, 625, 625 *fig.*
    stomata on (*see* Stomata)
    tissue systems in, 627, 627 *fig.*
    veins of, 626
Learning, 705–11
    associative, 710
    cognition and problem-solving behavior in, 711
    habituation, 705
    imprinting, 706–7, 706 *fig.*
    limbic system role in, 581
    sleep and, 580
    social, 710–11
    trial-and-error, 710, 710 *fig.*
    types of, 705 *table*
Lederberg, Joshua, 232
Leeches, 376, 377, 377 *fig.*
    nervous system, 572, 572 *fig.*
Legal applications of DNA technology, 231, 242, 244, 245
Legumes, 443 *fig.*
    nitrogen-fixing bacteria on, 656
Lemba people, 177
Lemon ant (*Myrmelachista schumanni*), 17
Lemon ant tree (*Duroia hirsute*), 17
Lemurs, 400, 401, 713, 713 *fig.*
Lens (eye), 595, 595 *fig.*
    accommodation in, 596 *fig.*
    corrective, 597
Leopard (*Panthera pardus*), 1, 779, 779 *fig.*
Leptin, obesity and, 447
Leucine, 42, 42 *fig.*
Leukemia(s), 135, 481
Leukocytes, 479, 479 *fig. See also* White blood cells (leukocytes)
LH (luteinizing hormone), 522, 529, 542–43, 543 *fig.*
Lichens, 360, 778
Life. *See also* Organisms
    arranging, into kingdoms, 314
    cells as structural and functional units of, 4
    chemistry of (*see* Chemistry of life)
    common features, 4–5
    diversity of, 6–7
    evolution of, 8–9
        See also Evolution
    levels of organization in, 2–3, 2 *fig.*, 414
    prokaryote evolution and, 320–29
    scope of biology and, 2–4
Life cycle, 125
    angiosperm, 350–51, 351 *fig.*, 634 *fig.*, 635 *fig.*
    fern, 347 *fig.*
    fungi, 356, 356 *fig.*
    moss, 346 *fig.*, 347
    mushrooms, 358, 358 *fig.*
    pine, 348–49, 349 *fig.*
    sea star, 366 *fig.*
    zygote fungi, 358, 358 *fig.*
Life history of organisms, evolution and, 732–33
Life tables, population survivorship tracked in, 727, 727 *table*
Light
    distribution of marine organisms and, 688–89
    Earth's tilt as cause of variations in, 686 *fig.*
    phytochromes as detector of, 673, 673 *fig.*
    plant responses to, 662–63, 670, 674
    vision and, 598, 598 *fig.*
Light microscopes (LMs), 52, 52 *fig.*, 53
Light reactions of photosynthesis, 111, 111 *fig.*, 112–15
    ATP synthesis through chemiosmosis in, 115, 115 *fig.*
    mechanical analogy of, 114 *fig.*

Metabolism. *See* Cellular metabolism
Metal accumulators, plants as, 643
Metals, bioremediation of toxic, 778
Metamorphosis, 366, 380
Metaphase
    cell division II, of meiosis, 139 *fig.*
    cell division I of meiosis I, 138 *fig.*
    of mitosis, 130–31, 131 *fig.*
Metastasis, 135
Methane, 34, 34 *fig.*
    alternate ways to represent, 23 *table*
    structure of, 34 *fig.*
Methanogens, 325
Methyl group, 35, 35 *fig.*
Mexico
    demographic transition in, 734, 735 *fig.*
    population momentum in, 735, 735 *fig.*
Microarrays, DNA, 219, 219 *fig.*
Microevolution, 264, 277, 278
    gene pool and, 267
    Hardy-Weinberg equilibrium in gene pools
        and, 267
    mechanisms of, 268–73
    in phenotypic plasticity, 771
    potential causes of, 269
Microfilaments, 56 *fig.*, 57 *fig.*, 65, 65 *fig.*
Micrographs, 51, 52, 53 *fig.*
Micronutrients, plant, 650
MicroRNAs (miRNAs), 215
Microscopes, 52–53
Microtubule-organizing centers, 131
Microtubules, 56 *fig.*, 57 *fig.*, 65, 65 *fig.*, 131
    bending of, in cilia and flagella, 66
Microvilli, 439, 439 *fig.*
Midbrain, 576, 577 *fig.*
Middle ear, 592–93, 592 *fig.*
Migration
    electromagnetic receptors and, 587, 591
Migration, internal maps and animal, 709
Milk, mammalian production of, 399
Miller, Stanley, 295
Millipedes, 379, 379 *fig.*
Mimicry, 10–11, 11 *fig.*
Mineralocorticoids, 529
Minerals, essential, 444–45, 445 *table*
Mining wastes, bioremediation of, 778
Mitchell, Peter, 93
Mites, 378, 379, 379 *fig.*
Mitochondria, 56 *fig.*, 57 *fig.*, 63, 63 *fig.*
    chemiosmosis and ATP synthesis
        in, 98 *fig.*
    origins and evolution of, 64, 64 *fig.*
    poisons and function of, 99
Mitochondrial matrix, 63
Mitosis, 128–36
    interphase and, 129
    meiosis versus, 140, 140 *fig.*
    review of, 136
    stages of, 130–31, 130–31 *fig.*
Mitotic phase (M phase) of cell cycle,
    129, 130–31 *fig.*
Mitotic spindle, 130–31
Mold, 356
Molecular biology, 180–207
    DNA and RNA as nucleotide
        polymers, 184–85
    DNA as genetic material, 182–83
    DNA replication, 188–89
    DNA structure, 186–87
    evidence for evolution in, 262–63
    genetic information flow from DNA to RNA to
        protein, 190–99
    phylogenetic tree based on, 384 *fig.*
    as tool in systematics, 312
    of viruses, 200–205
    yeasts and research in, 361
Molecular clock, 313

Molecules, 2–3, 2 *fig.*
    alternative ways to represent, 23, 23 *table*
    covalent bonds and, 23
    polar and nonpolar, 24
Mole rats, altruism among, 720, 720 *fig.*
Molina, Mario, 120
Molluscs (Mollusca), 374–75
    exoskeleton of, 607, 607 *fig.*
    nervous system, 572, 572 *fig.*
    range of eye complexity among, 306, 306 *fig.*
Molting, 378
Monera (kingdom), 314
Monkey flower, 285
Monkeys, 400, 401, 401 *fig.*
    alarm calls of vervet, 710–11, 710 *fig.*
Monoclonal antibodies, 496, 496 *fig.*
Monocots
    embryo development in, 636, 636 *fig.*
    eudicots versus, 623 *fig.*
    seed germination in, 638, 638 *fig.*
    tissues in, 626, 627 *fig.*
Monocytes, 479, 479 *fig.*
Monod, Jacques, 210
Monogamous mating behavior, 701, 715
Monohybrid cross, 156
Monomers, 36
Mononucleosis, 485
Monophyletic taxa, 310
Monosaccharides, 37, 37 *fig.*
Monotremes, 388–89, 399, 399 *fig.*
Morgan, Thomas Hunt, 173, 174 *fig.*
Morning after pills (MAPs), 545
Morphine, 744
Morphological species concept, 279
Mosquito(es), 771
Mosquitofish, effect of endocrine disruptors
        on, 716, 716 *fig.*
Mosses, 344, 344 *fig.*
    life cycle of, 346 *fig.*, 347
Moths, 380
    chemoreceptors in, 591 *fig.*
Motion sickness, 594
Motor neurons, 564
    muscle contraction stimulated by, 614–15
    relationship of muscle fibers to, 614 *fig.*
    structure of, 565 *fig.*
Motor output, nervous system, 564
Motor units, 614 *fig.*
Mountains, effect of, on rainfall, 687, 687 *fig.*
Mount Hood National Forest, 772, 772 *fig.*
Mouth, 432
    human, 434, 434 *fig.*
Movement, 602–18
    animal locomotion and, 604–5
    cell, 65–66, 66 *fig.*, 69, 549 *fig.*, 552
    internal maps in animal, 709
    muscles and animal, 611–16
        See also Muscle
    skeletal support for animal, 606–11
    stimuli, landmarks, and animal, 708
Movement corridors, 773
Mucous cells (stomach), 436, 436 *fig.*
Mucous membrane, 416
Multicellular green algae, 336, 336 *fig.*
Multicellular organisms
    evolution of, from unicellular protists,
        337, 337 *fig.*
    seaweed as, 334
Multiple sclerosis (MS), 500, 565
Muscle
    antagonistic pairs of, 611, 611 *fig.*
    blood distribution controlled by
        smooth, 477
    contractile apparatus of, 612
    interaction of skeleton and, in movement, 611
    motor neurons and contraction of, 614
    types of, 418 *fig.*

Muscle cells
    aerobic function in slow muscle, 89
    anaerobic function in slow muscle, 89
    contractile apparatus in each, 612
Muscle contraction, 612–13
    motor neurons and, 614
    sliding filament model of, 612–13, 613 *fig.*
Muscle fibers, 612
    characteristics of, 616 *table*
    relationship between motor neurons
        and, 614 *fig.*
    slow versus fast, 89, 616
Muscle tissue, 418
    capillaries in, 474 *fig.*
    contraction of, 612–13
    motor neurons and, 614
    types of, 418 *fig.*
Muscular dystrophy, Duchenne, 177
Muscular system, 420, 420 *fig.*
Mushrooms, 357, 357 *fig.*, 361
    life cycle, 358, 358 *fig.*
Mus musculus (mouse), 247
Mustard plant (*Arabidopsis thaliana*), 247
Mutagenesis, 199
Mutagens, 199
Mutation(s), 199
    cancer caused by, 224–25, 225 *fig.*, 226
    emerging viruses due to, 202–3
    in gene expression in Drosophila, 218
    genetic variation due to, 199, 265
    by HIV virus, 499
    types of, and effects of, 199 *fig.*
Mutualism, 742. *See also* Symbiosis
    benefits to both partners, 743
    fungi and animals, 360, 360 *fig.*
    lichens and algae, 360, 360 *fig.*
    mycorrhiza and plants association as, 341
Myanmar tigers, conservation of, 763
Mycelium, 355, 355 *fig.*
Mycoplasmas, 326
Mycorrhiza, 341, 355, 656, 656 *fig.*
Mycosis, 359
Myelin sheath, 565, 565 *fig.*
Myofibrils, 612
Myosin filaments, muscle contraction and, 612,
        612 *fig.*, 613, 613 *fig.*, 616
Myxobacteria, 326

N

NAD^+ (nicotinamide adenine dinucleotide),
        cellular respiration and, 92, 100, 101
NADH, cellular respiration and role of, 92, 92 *fig.*,
        93, 93 *fig.*, 94, 96, 97 *fig.*
$NADP^+$, photosynthesis and, 111, 115
NADPH, photosynthesis and, 111, 114, 115, 116
Nails, 423
Naked mole rats, altruism among, 720, 720 *fig.*
Natural family planning, 545
Natural killer cells, 486
Natural selection, 8–9, 8 *fig.*
    adaptations to abiotic or biotic factors and, 685
    allele frequencies in population and, 268
    animal's form as reflection of, 415
    evolutionary adaptation resulting from, 259,
        259 *fig.*, 269
    general outcomes of, 270, 270 *fig.*
    genetic variation in populations affected by,
        272–73
    global climate change as agent of, 771
    as mechanism of evolution, 258
    perfection not produced by, 273
    population variation through, 270
    reproductive success and, 715
    scientific observation of, 259
Neanderthals, 405, 406
Nearsightedness, 597, 597 *fig.*

interaction of, with environment, 682
See also Ecology
life tables of survivorship in, 727, 727 table
natural selection and evolution of, 258
as units of evolution, 264
Population density, 726–27
factors limiting population growth dependent on, 730–31
Population dynamics
behavior and (see Social behavior)
Population ecology, 724–39. See also Population dynamics
human, 734–37
life histories and evolution in, 732–33
non-native species and, 725
practical applications of principles of, 733
structure of populations and, 726–33
Population growth
boom-and-bust cycles of, 731, 731 fig.
density-dependent factors limiting, 730–31
exponential growth model of, 728, 728 fig.
human, 734–37
logistic growth model and limiting factors in, 728–29
social and economic trends, age structures and, 736, 736 fig.
Population momentum, 735, 735 fig.
Porifera, 370
Porpoises, 399, 587
Post-anal tail, 383, 383 fig.
Posterior pituitary, 522
hormones of, 523 fig.
Posterior surface of animals, 368, 368 fig.
Postzygotic barriers, 280 table, 281
Potassium–40, for radiometric dating, 298
Potassium as plant nutrient, 650, 651
Potential energy, 80, 80 fig.
Pottos, 400
Prairies, 694–95, 694 fig.
Prairie voles, behavior in, 704
monogamy, 701
Precambrian period, 337
Precapillary sphincters, 477, 477 fig.
Precipitation
acid, 28
climate change and, 768
in deserts, 693
of dissolved antigens, 495, 495 fig.
mountains' effects on, 687, 687 fig.
uneven heating as cause of, 687 fig.
Predation
adaptations of prey/predators for, 744
coevolution and, 744
group foraging behaviors and, 712
as interspecific interaction, 742
life history shaping through evolution and, 732–33, 732 fig.
plant defenses against, 674–75, 675 fig., 744–45
population density and, 730
Predators
coevolution and prey interactions with, 744–45
group foraging behaviors of, 712
plant recruitment of predatory animals, 674–75, 675 fig.
population cycles and, 731
prey defenses against, 744
scientific method and study of prey and, 10–11
Pregnancy, human, 536–37
childbirth and, 530, 558–59
development of embryo and placenta, 554–55
prolactin and, 530
trimesters of, 556–57, 556–57 fig.
Pregnancy tests, home, 496
Prenatal testing, 164–65
Prepuce, 537, 539
Pressure-flow mechanism in plant phloem tissue, 648, 649 fig.

Prevailing winds, 686–87
Prey
coevolution and predator interactions with, 744–45
population cycles and, 731
predator defenses by, 744
scientific method and study of predators and, 10–11
Prezygotic barriers, 280–81, 280 table
Primary consumers, 746, 746 fig., 747, 747 fig.
Primary endosymbiosis, 331
Primary immune response, 492 fig., 493, 493 fig.
Primary oocytes, 541, 541 fig.
Primary phloem, 631
Primary production, 752–53, 753 fig.
Primary spermatocytes, 540
Primary structure of proteins, 44, 45 fig.
Primary succession, 750
Primary xylem, 631
Primates, 400–402
apes as closest relatives of humans, 402
hominid branch of, 403
hominid evolution and, 403–8
phylogenetic tree of, 400 fig.
reasoning in, 711
signaling behavior in, 713, 713 fig.
Primers, 242
Principles of Geology (Lyell), 257
Prions, 44, 203
PRL (prolactin), 522
Probability, Mendelian inheritance and rules of, 160
Probes, nucleic acid, 236, 236 fig.
Problem solving, 711
Producers, 3, 3 fig., 108, 746, 746 fig.
Production, pyramid of, 753 fig.
Products of chemical reactions, 29
Progesterone, ovarian and menstrual cycles and, 542–43, 543 fig.
Progestins, 529
Programmed cell death, 552, 552 fig.
Progressive retinal atrophy (PRA), 158
Prokaryotes, 6, 314, 320–29
archaea and bacteria as main branches in evolution of, 320–21
cyanobacteria, 319
as detritivores, 701, 746
diversity and extent of, 320
early origins of, 297
fossilized, 261
gene regulation in, 210–11
genetic variation in, 265
nutritional modes of, 324, 324 fig.
recycling and cleaning environment using, 329
shapes of, 321, 321 fig.
structural features of, 322–23
Prokaryotic cells, 4, 4 fig., 55 fig.
binary fission of, 127, 127 fig.
size and structure of, 55, 322–23
Prokaryotic flagella, 55, 55 fig., 322, 323 fig.
Prolactin (PRL), 522, 530
Prometaphase, mitosis, 130–31, 130 fig.
Promiscuous mating behavior, 701, 715
Promoter DNA sequences, 193, 210, 211 fig.
Pronghorn antelopes, 685, 685 fig.
Propagation, vegetative, 638
Propane, 34, 34 fig.
Prophages, 200
Prophase
cell division I, of meiosis I, 138 fig.
cell division II, of meiosis II, 139 fig.
mitosis, 130, 130 fig.
Propylthiouracil, 599
Prostaglandins, 558, 558 fig.
Prostate gland, 538
Protease inhibitors, 85

Protein(s), 36, 42–45. See also Amino acids
amino acids and formation of, 42–43
biosynthesis of, 103 fig.
classes of, 42
Class I and II, 500
comparison of, in systematics, 312
denaturation of, 43
digestion of, 438 table, 439
DNA technology and mass production of, 237 table
essential amino acids and, 443
as essential to life, 42
as fuel for cellular respiration, 102–3, 102 fig.
in membranes, 74
nucleic acids as polymer blueprints for, 46
peptide bonds in, 43 fig.
post-translational breakdown of, 216
post-translational controls on activation of, 216
receptors, 74
secretory, 60, 60 fig.
self and nonself, 497, 500
shape and function, 43
shape and function of, 43
signal transduction and role of, 220
structure of, 44
structures, 44–45
translation of RNA information into, 190
Protein hormones, 519
Proteobacteria, 326
Proteomics, 249
Proterozoic eon, 298
Proteus (bacterium), 322 fig.
Protista (kingdom), 314
Protists, 7 fig., 330–37, 330 fig.
alveolates, 333
amoebozoans, 334–35
classification of, 332, 332 fig.
diplomonads, 332–33
euglenozoans, 333, 333 fig.
evolution of animal kingdom from colonial, 367
foraminiferans, 335, 335 fig.
multicellular life originating in, 337
Paramecium, 62, 62 fig., 743
radiolarians, 335, 335 fig.
red algae and green algae, 336, 336 fig.
secondary endosymbiosis in, 331, 331 fig.
stramenopiles, 334
as unicellular eukaryotes, 330, 330 fig.
Protobionts, 296
Protons, 20
Proto-oncogenes, 224
ras, 226
Protostomes, 368, 369, 384
Protozoans, 6, 330
Provirus, 203
Proximal tubule, of kidney nephron, 510 fig., 511, 512, 512 fig.
Proximate causes, of behavior, 702
Pseudocoelom, 368, 368 fig.
Pseudopodia (pseudopodium), 334, 335
Pseudostratified epithelium, 416
Ptarmigan, 425, 425 fig.
Pterophytes, 344
Pterosaurs, 293
Public health science, Hardy-Weinberg equation useful in, 267
Pulmocutaneous circuit, 469
Pulmonary arteries, 470
Pulmonary circuit, 469
Pulmonary veins, 470
Pulse, 475
Punctuated equilibrium model of speciation, 289, 289 fig.
Punnett, Reginald, 172
Punnett square, 157
Pupil (eye), 595, 595 fig.

Purines, 185, 185 *fig.*
Pyramid of production
    human meat-eating as luxury explained by, 754
    idealized, 753, 753 *fig.*
Pyrimidines, 185, 185 *fig.*
Pyruvate
    cellular respiration and production of,
        93, 94, 95 *fig.*
    chemical grooming for citric acid cycle,
        96, 96 *fig.*
Python, *Hox* gene expression in, 609, 609 *fig.*

Q

Quaternary consumers in communities, 746,
    746 *fig.*, 747 *fig.*
Quaternary structure of proteins, 44, 45 *fig.*
Quolls, 389, 408, 408 *fig.*

R

Rabbits
    exponential growth model applied to,
        728, 728 *fig.*
    as non-native pest species, 408
Raccoons, systematics of bears and, 312, 312 *fig.*
Radial keratotomy (RK), 597
Radial symmetry, 368, 368 *fig.*, 371
Radiation
    heat gains or losses by, 506, 506 *fig.*
    as mutagen, 199
Radiation therapy, 135
Radioactive isotopes, 20, 21
Radiolarians, 335, 335 *fig.*
Radiometric dating, 298, 298 *fig.*
Radon, 21
Radula, mollusc, 374
Rainfall. *See* Precipitation
Rain forests. *See* Tropical rain forests
Random population-dispersion patterns,
    726 *fig.*, 727
Ras proto-oncogene, 226
Ray-finned fishes, 393, 393 *fig.*
Reabsorption function of excretory system,
    511, 511 *fig.*, 512–13, 512 *fig.*
Reactants in chemical reactions, 29
Reaction center of photosynthetic pigments, 113
Reading frame, 199
Receptor-mediated endocytosis, 79, 79 *fig.*
Receptor potential, 588–89
Receptors, protein, 42
Receptors for hormones, 519, 519 *fig.*
Recessive alleles, 156, 161
Recessive traits, genetic disorders as, 162, 162 *fig.*,
    163 *table*
Reciprocal altruism, 720
Recombinant cells and organisms, 236–37
Recombinant DNA, use of restriction enzymes
    and DNA ligase to create, 234, 234 *fig.*
Recombinant DNA technology, 232. *See also*
    DNA technology
Recombinant gametes, production of,
    143, 172, 173 *fig.*
Recombination frequency, 173
Recommended Daily Allowances (RDAs), 446
Reconciliation behavior, 719
Rectum, 440 *fig.*, 441
Red algae, 336, 336 *fig.*
Red blood cells (erythrocytes), 479, 479 *fig.*, 480 *fig.*
    healthy body function and levels of, 480
Red bone marrow, 609
Red-cockaded woodpecker (*Picoides borealis*),
    772, 772 *fig.*
Red-green color blindness, 177
Redox reactions, 92 *fig.*, 95 *fig.*, 97 *fig.*
    of cellular respiration, 92, 110 *fig.*
    of photosynthesis, 110–11, 116, 116 *fig.*

Red squirrel (*Tamiasciurus hudsonicus*),
    771, 771 *fig.*
Red-tide dinoflagellates, 333
Reduced hybrid fertility, 281
Reduced hybrid viability, 281
Reduction, 92, 116, 116 *fig.*
Redwoods (*Sequoia sempervirens*), 620–21
Reeve, Christopher, 563, 563 *fig.*
Reflexes, 564
    knee-jerk, 564 *fig.*
Regeneration, 221, 534
    by sea stars and echinoderms, 382
Regulation, as property of all organisms, 5
Regulatory genes, 211, 211 *fig.*
Regulatory systems, animal, 504–15
    chemical regulation, overview of, 518–19
    endocrine system as, 520–23
    homeostasis and (*see* Homeostasis)
    hormones and, 521 *table*
    osmoregulation and excretion as, 508–13
    sympathetic and parasympathetic nervous
        systems and, 574–75
    thermoregulation as, 505, 506–8
Regulatory systems, plant, 660–78
    defenses, 674–76
    growth responses, 670
    hormones, 662–69
    internal clocks, 671–74
Releasing hormones, 522
REM sleep, 580
Renal cortex, 510 *fig.*, 511
Renal medulla, 510 *fig.*, 511
Repetitive DNA, 244, 248
Repressor protein, 210–11
Reproduction, 532–61. *See also* Inheritance
    alterations of chromosome number and
        structure during, 144–48
    asexual, 126, 534
    cellular (cell division), 126–27
    connections between cell division and, 126–27
    endocrine disruptors effects on, 716
    in green algae, 336, 336 *fig.*
    human, 536–45
    meiosis and crossing over in, 136–43
    mitosis and eukaryotic cell cycle in, 128–36
    natural selection success in, 715
    plant, 634–40
    of prokaryotes, 322
    as property of all organisms, 5
    sexual, 125, 534–35
        See also Sexual reproduction
    of viruses, 201 *fig.*
Reproductive barriers, 280–81, 280 *table*
    evolution of, 285
    hybrid zones and, 286, 286 *fig.*
    speciation and, 282
Reproductive cloning, 222–23
    applications of, 222–23
Reproductive isolation, 279, 286, 286 *fig.*
Reproductive systems, 421, 421 *fig.*
Reptiles (Reptilia), 397
    cardiovascular system, 469
    dinosaurs, 397
    excretion in, 509 *fig.*
    gravity and circulatory system of, 467
    lizards, 397
    phylogenetic tree of, 311 *fig.*
    respiratory system of, 458
    snakes, 397, 397 *fig.*
Research
    applications of reproductive cloning
        to, 222–23
    radioactive isotopes used in, 21
Resolution, of microscopes, 52
Resources
    competition over shared, 743
    human overexploitation of, 766

Respiration, 452–65. *See also* Cellular respiration
    aerobic, 615
    in birds, 461
    breathing and, 90, 460–61
    gas exchange and, overview of, 454
    gas exchange through moist body surfaces as,
        454–55, 455 *fig.*
    gills in aquatic animals and, 456–57, 456 *fig.*
    in high altitudes, 453
    insect tracheal system and, 457, 457 *fig.*
    lungs in terrestrial animals and
        See also Lung(s)
    smoking effects on, 460
    transport of gases in body and, 462–63
Respiratory surfaces, 454–55, 455 *fig.*
Respiratory system, 420, 420 *fig.*, 454–55
    exchanges with external environment and role
        of, 424 *fig.*, 425
    in high altitudes, 453
    human, 458–59, 459 *fig.*
    smoking as assault on, 460
Resting potential, neuron membrane,
    566, 566 *fig.*
    nerve signal and changes in, 566–67
Restoration ecology, 772, 778–79
    Kissimmee River Project, 778–79, 778 *fig.*
Restriction enzymes, creating recombinant DNA
    with, 234
Restriction fragment length polymorphisms
    (RFLPs), 246, 246 *fig.*
Restriction fragments, 234, 246
    analyzing, to detect differences in DNA
        sequences, 246, 246 *fig.*
    detecting harmful alleles using, 246
    sticky ends and, 234
Restriction sites, 234
Reticular formation, 580
Retina, 595, 595 *fig.*
    photoreceptor cells (rods and cones)
        in, 598 *fig.*
Retinitis pigmentosa, 269
Retrovirus, 203, 241 *fig.*
Reverse transcriptase, 203
    gene cloning and role of, 235
Rheumatoid arthritis, 500
Rhizobium spp., 326, 656, 656 *fig.*
Rhizomes, 625, 625 *fig.*
Rhodopsin, 598
Rhythm method (contraception), 545
Ribonucleic acid (RNA), 46. *See also* RNA
Ribosomal RNA (rRNA), 196
    in bacteria vs. in archaea, 320
Ribosomes, 55, 55 *fig.*, 56 *fig.*, 57 *fig.*, 59 *fig.*
    as factories for polypeptides, 196
    protein synthesis by, 59
Ribozymes, 296
Rice (*oryza sativa*), 247
Ringworm, 359
Rituals
    agonistic behaviors and, 718
    courtship, 280, 280 *fig.*, 714–15, 714 *fig.*
Rivers. *See* Lakes and streams
RNA (ribonucleic acid), 47
    comparing DNA and, in systematics, 312
    messenger RNA (mRNA), 194
    as nucleotide polymer, 185 *fig.*
    polynucleotide, 184–85, 185 *fig.*
    roles in controlling gene expression, 215
    self-replicating, 296
    transcription of DNA into, 190, 193
    translation of, into protein, 190
    viruses, 201, 202–3
RNA interference, 215, 215 *fig.*
RNA polymerase, 193, 211
RNA splicing, 194
    alternative, 214–15
RNA world, 296

Rocks
    age of, and geologic time, 298
    phosphorus cycle and weathering of, 758
Rodhocetus, 261, 261 *fig.*
Rodriguez, Eloy, 676, 676 *fig.*
Rods, retina, 598, 598 *fig.*
Romer, Alfred, 394
Root(s)
    modified, 625, 625 *fig.*
    mycorrhiza on, 341, 355
    nutrient uptake by, 645, 645 *fig.*
    primary growth in, 630–31, 631 *fig.*
    tissue systems in, 626, 627 *fig.*
Root cap, 630, 631 *fig.*
Root hairs, 624, 624 *fig.*
    in soil, 652, 652 *fig.*
    solute uptake and, 645, 645 *fig.*
Root nodules, nitrogen-fixing bacteria on,
    656, 656 *fig.*
Root pressure, 646
Root system, 624, 624 *fig.*, 625
Rotenone as disruptor of cellular respiration,
    99, 99 *fig.*
Rough endoplasmic reticulum (rough ER), 56 *fig.*,
    57 *fig.*, 60–61, 60 *fig.*
Roundworms, 368, 368 *fig.*, 373
R plasmids, 205
r-selection, 732
Rubisco, 118
    Calvin cycle and, 116, 116 *fig.*
RuBP, Calvin cycle and regeneration of,
    116, 116 *fig.*
Rule of addition, 160
Rule of multiplication, 160
Runners, as modified stems, 625, 625 *fig.*
Runners, sprinter versus marathoner, 89
Running as locomotion, 603, 604
Rusts, 359

S

Saccharomyces cerevisiae (yeast), 220–21,
    221 *fig.*, 247
Saccule, 594, 594 *fig.*
Sac fungi, 357
Safe sex, 499, 544, 545
Sage grouse, courtship and mating behaviors of,
    714–15, 714 *fig.*
Sahelanthropus tchadensis, 403
Salamanders, 396, 396 *fig.*
    paedomorphic axolotl, 304, 304 *fig.*
Saliva, 434
Salivary glands, 434, 434 *fig.*
Salmon, sense perception and migration in, 706
Salmonella, of gamma proteobacteria group, 326
Salt
    ionic bonds as, 23, 23 *fig.*
San Andreas fault, 302, 302 *fig.*
Sapwood, 633, 633 *fig.*
Sarcomas, 135
Sarcomeres, 612–13
SARS (severe acute respiratory syndrome), 202
SA (sinoatrial) node, 472
Saturated fats, 40
Savannas, 692–93, 692 *fig.*
Scale-eating fish, 273, 273 *fig.*
Scanning electron microscopes (SEMs), 52–53
Scarlet king snake, 11 *fig.*
Schindler, David, 759, 759 *fig.*
Schistosomiasis, 372
Schizophrenia, 571, 582
Schweickart, Rusty, 680, 697
Science. *See also* Biology
    discovery, using inductive reasoning, 9
    hypothesis-based, 9
    politics and, 120
    theories and, 8

Sclera, 595 *fig.*, 596, 596 *fig.*
Sclereids, 628, 629 *fig.*
Sclerenchyma cells, 628, 629 *fig.*
Scolex, 372
Scorpions, 378, 379 *fig.*
Scrotum, 538
Scypha (sponge), 370, 370 *fig.*
Sea anemones, 371
    fission in, 534 *fig.*
Sea cucumbers, 382
Sea ice, melting of, 768
Seals, 415 *fig.*
    combat between male elephant, 529 *fig.*
    population-limiting factors on fur,
        728–29, 729 *fig.*
Search image, 712
Seasonal cycles, plant, 672
Season(s), Earth's tilt as cause of, 686 *fig.*
Sea star, 366, 382, 382 *fig.*
    endoskeleton of, 607
    as keystone species, 749, 749 *fig.*
    life cycle of, 366 *fig.*
Sea turtles, conservation efforts for, 774, 774 *fig.*
Sea urchins, 382, 382 *fig.*
    egg fertilization in, 546, 547 *fig.*
    endoskeleton of, 607, 607 *fig.*
    as keystone species, 749, 749 *fig.*
    zygote cleavage in, 548 *fig.*
Seaweeds, 334
Secondary consumers in communities, 746,
    746 *fig.*, 747, 747 *fig.*
Secondary endosymbiosis, 331, 331 *fig.*
Secondary growth in plants, 632–33, 632 *fig.*
Secondary immune response, 492 *fig.*, 493, 493 *fig.*
Secondary oocytes, 541, 541 *fig.*
Secondary phloem, 632–33
Secondary spermatocytes, 540
Secondary structure of proteins, 44, 45 *fig.*
Secondary succession, 750, 750 *fig.*
Secondary xylem, 632
Second law of thermodynamics, 81
Secretion function of urinary system, 511,
    511 *fig.*, 512, 512 *fig.*
Secretory proteins, 60, 60 *fig.*
Sedimentary rock, strata of, 261, 261 *fig.*
Seed(s), 344–45
    abscisic acid and dormancy of, 667
    development of, 636, 636 *fig.*, 637
    fruits and dispersal of, 352 *fig.*
    germination of, 638, 638 *fig.*, 666–67
    structure of, 636 *fig.*
Seed coat, 636
Seed dormancy, 636, 667
Seedless plants, 348
    carboniferous coal forests formed by, 348
Seedless vascular plants, 344–45, 345 *fig.*
Seed plants, 345, 345 *fig. See also* Angiosperms;
    Gymnosperms
Segmentation, animal body, 376–77, 376 *fig.*
Segregation, Mendel's law of, 156–57, 171
Selective breeding, 153
Selective permeability of plasma membrane,
    74, 76
Selective serotonin uptake inhibitors (SSRIs),
    571, 582
    prescriptions in U.S. by year, 582, 583 *fig.*
Self-fertilizing plants, 154–55
Self proteins, 497
    Class I and II, 500
Self-recognition and non-self recognition,
    immune system and, 497, 500
Semen, 538–39
Semicircular canals, 594, 594 *fig.*
Semiconservative model of DNA replication, 188
Seminal vesicles, 538
Seminiferous tubules, 540, 540 *fig.*
Sensations, 588

Senses, 586–601
    hearing and balance, 592–94
    movement and, 587
    sensory inputs into nervous system and, 588, 600
    sensory reception, 588–91
    taste and smell, 599
    vision, 595–98
Sensitive period, 706
Sensory adaptation, 589
Sensory input, central nervous system, brain, and,
    564, 588
Sensory neurons, 564
Sensory organ, platypus bill as, 388
Sensory receptors, 588–89, 590 *fig.*
    for balance, 594
    for electromagnetism, 591
    for light (vision), 595–98
    for pain, 590
    in skin, 423
    for sound (hearing), 592–93
    specialized, for five categories of stimuli, 590–91
    stimuli converted to action potentials by,
        588–89
    for taste and smell, 589 *fig.*, 599
    for temperature and touch, 590
Sensory transduction, 588, 589 *fig.*
Sepals, 350, 634
September 11 terrorist attacks, 245
Septic shock, 487
Septuplets, Iowa, 532–33
Serine, 42, 42 *fig.*
Serotonin, 580
Sessile organisms, 370
Setae, 413, 413 *fig.*
Set point, homeostasis and, 426
Severe acute respiratory syndrome (SARS), 202
Severe combined immunodeficiency disease
    (SCID), 500
    gene therapy for, 241
Sewage treatment, 329, 329 *fig.*
Sex, safe, 499, 544, 545
Sex chromosome(s), 137, 174–75
    abnormal numbers of, 146–47, 147 *table*
    sex determination due to, 174–75
    sex-linked genes and unique inheritance
        patterns, 176
    sex-linked genetic disorders, 177
Sex determination, 175, 175 *fig.*
Sex hormones, 67, 520, 529
    functional group differences in, 35 *fig.*
    steroids and, 41
Sex-linked genes, 176–77
Sex pili, 322
Sexual dimorphism, 271
Sexually transmitted diseases (STDs),
    544, 544 *table*
Sexual reproduction, 125, 126
    in animals, 534–35
    description of, 534
    in fungi, 356
    genetic variation from, 265, 265 *fig.*
    in humans, 536–45
    meiosis and, 540–41, 540 *fig.*, 541 *fig.*
    in plants, 346, 634–40
    prezygotic and postzygotic barriers to, 280–81,
        280 *table*
    sexual selection and sexual dimorphism in, 271
    zygote fungi, 358, 358 *fig.*
Sexual selection, 271
Shared ancestral characters, 310
Shared derived characters, 310–11
Sharks, 392, 392 *fig.*, 415 *fig.*, 587
Shoots
    auxin stimulation of cell elongation in, 664–65,
        665 *fig.*
    modified, 625, 625 *fig.*
    primary growth in, 630–31, 631 *fig.*

INDEX